ASTRONOMY AND ASTROPHYSICS ABSTRACTS

A Publication of the Astronomisches Rechen-Institut Heidelberg
Member of the Abstracting Board
of the International Council of Scientific Unions
Astronomy and Astrophysics Abstracts is Prepared
Under the Auspices of the International Astronomical Union

Volume 32
Literature 1982, Part 2

Edited by
S. Böhme W. Fricke H. Hefele I. Heinrich W. Hofmann
D. Krahn V. R. Matas L. D. Schmadel G. Zech

Springer-Verlag Berlin Heidelberg GmbH 1983

Astronomisches Rechen-Institut, Mönchhofstraße 12–14,
D-6900 Heidelberg, Fed. Rep. of Germany
Director: Professor Dr. Walter Fricke

Astronomy and Astrophysics Abstracts
Editors-in-Chief: Inge Heinrich, Dr. Lutz D. Schmadel

ISBN 978-3-662-12339-3 ISBN 978-3-662-12337-9 (eBook)
DOI 10.1007/978-3-662-12337-9

Originally published by Astronomisches Rechen-Institute Heidelberg in 1983
Softcover reprint of the hardcover 1st edition 1983 Library of Congress

Library of Congress Catalog Card Number 72-104650.

Preface

Astronomy and Astrophysics Abstracts, which has appeared in semi-annual volumes since 1969, is devoted to the recording, summarizing and indexing of astronomical publications throughout the world. It is prepared under the auspices of the International Astronomical Union (according to a resolution adopted at the 14th General Assembly in 1970).

Astronomy and Astrophysics Abstracts aims to present a comprehensive documentation of literature in all fields of astronomy and astrophysics. Every effort will be made to ensure that the average time interval between the date of receipt of the original literature and publication of the abstracts will not exceed eight months. This time interval is near to that achieved by monthly abstracting journals, compared to which our system of accumulating abstracts for about six months offers the advantage of greater convenience for the user.

Volume 32 contains literature published in 1982 and received before February 11, 1983; some older literature which was received late and which is not recorded in earlier volumes is also included.

We acknowledge with thanks contributions to this volume by Dr. J. Bouška, Prague, who surveyed journals and publications in Czech and supplied us with abstracts in English.

We express our warmest thanks again to Ms. Helga Ballmann, Ms. Mona El-Choura, Ms. Monika Kohl, Ms. Sylvia Matyssek and Ms. Angelika Meßmer for typing the text of this volume on IBM 72 Composers, for compiling the pages from abstract slips in a perfect form for offset reproduction, and for punching material for the author index and for the subject index, which finally were printed with a TN chain on a 1403 IBM high-speed printer. Finally, we have to thank Mr. Uwe Reichert and Mr. Roland Zanella who supported our task by careful proofreading.

Heidelberg, April 1983

Siegfried Böhme
Walter Fricke
Herbert Hefele
Inge Heinrich
Wilfried Hofmann
Dietlinde Krahn
Vladimir R. Matas
Lutz D. Schmadel
Gert Zech

Contents

Earth

Planetary System

Stars

Introduction

Astronomical bibliographies

Astronomy and Astrophysics Abstracts begins documentation and abstracting from the year 1969. For information on astronomical literature before this date consultation of one of the following bibliographies is suggested:
(1) J. J. de Lalande, Bibliographie Astronomique, Paris 1803 (this work covers the time from 480 B. C. to the year 1803, VIII + 966 pages).
(2) J. C. Houzeau, A. Lancaster, Bibliographie générale de l'astronomie, Volume I (in two parts), Bruxelles 1887, 1889, Volume II, Bruxelles 1882. The complete title of Volume II is "Bibliographie générale de l'astronomie ou catalogue méthodique des ouvrages, des mémoires et des observations astronomiques, publiés depuis l'origine de l'imprimerie jusqu'en 1880". A new edition of these volumes was prepared by D. W. Dewhirst in 1964.
(3) Bibliography of Astronomy, 1881 - 1898. The literature of this period was recorded on standard slips by the Observatoire Royal de Belgique. From the material (some 52,000 items) a microfilm version was produced by University Microfilms Limited, Tylers Green, High Wycombe, Buckinghamshire, England, in 1970.
(4) Astronomischer Jahresbericht, 1899 gegründet von Walter Wislicenus, herausgegeben vom Astronomischen Rechen-Institut in Heidelberg (formerly in Berlin), Verlag W. de Gruyter, Berlin. For the period from 1899 to 1968 sixty-eight volumes were published, each of which, in general, covers the literature of one year.
(5) Bulletin Signalétique – Section 120: Astronomie, Physique Spatiale, Géophysique. Published by Centre de Documentation du Centre National de la Recherche Scientifique, Paris. This publication is a continuation of "Bibliographie Mensuelle de l'Astronomie" founded in 1933 by the Société Astronomique de France. The publication is continued.
(6) Referativnyj Zhurnal. Founded in 1953 and published by Vsesoyuznyj Institut Nauchnoj i Tekhnicheskoj Informatsii, Akademiya Nauk, Moskva. The publication is continued.

Concept of Astronomy and Astrophysics Abstracts

This abstracting service aims to present a comprehensive documentation of the literature in all fields of astronomy and astrophysics and their border fields. It appears in semi-annual volumes. Two of these volumes cover the literature of one calendar year. The half-yearly period of issue is regarded as an optimal period for summarizing papers into subject categories and for the presentation of abstracts as quickly as possible after the publication of the original literature. The recording summarizing and indexing of astronomical publications of the year 1982 received from July 1982 to February 1983 are subjects of **Volume 32.** It also records a number of papers issued before 1982 but received within this period.
The main characteristics of the concept of Astronomy and Astrophysics Abstracts may be summarized as follows:
(1) The subdivision of astronomy and its border fields into subject categories is facilitated by the fact that the astronomical objects appear to be particularly well suited for the formation of categories. It may be assumed that such subdivisions can be maintained for a long period. Experience shows, however, that progress in research might imply minor changes in the classification scheme.
(2) Each paper has been classified into one of 108 numbered subject categories and given a serial number within the category. In this way each item is numbered by six figures: the first three indicate the number of the category, the following three the serial number within the category. Reference to an abstract in Volume 1 is indicated by "01" before the number of the category; for example: 01.074.028, denotes Volume 1, category 074, abstract 028.
A paper might be classified into more than one category. In this case, its abstract is placed only in one category, whereas in the other categories only cross references are given. These are listed at the end of each category.
(3) Authors' abstracts are used whenever possible. Popular articles are not abstracted.
(4) If possible, titles of papers and abstracts are given in English. A special reference is made to titles which we have not taken in the original language.

Transliteration scheme for the Russian alphabet

The transliteration of the Russian alphabet in use in Astronomy and Astrophysics Abstracts is presented here.

А	а	a	П	п	p
Б	б	b	Р	р	r
В	в	v	С	с	s
Г	г	g	Т	т	t
Д	д	d	У	у	u
Е	е	e	Ф	ф	f
Ё	ё	e	Х	х	kh
Ж	ж	zh	Ц	ц	ts
З	з	z	Ч	ч	ch
И	и	i	Ш	ш	sh
Й	й	j	Щ	щ	shch
К	к	k	Ы	ы	y
Л	л	l	Ь	ь	'
М	м	m	Э	э	eh
Н	н	n	Ю	ю	yu
О	о	o	Я	я	ya

This transliteration was recommended by the Abstracting Board of the International Council of Scientific Unions in 1969. It corresponds essentially to the transliteration proposed by the Academy of Sciences, Moscow, which is used by the Referativnyj Zhurnal. In this case the letters can be read and printed by usual data processing machines.
If the names of Russian authors in the literature are transliterated in a different scheme, we present the names as they are given in the references cited and in addition in brackets according to our transliteration table.

Sources of information

The majority of sources of information for this volume is given in section **001 Periodicals** and in section **008 Observa-**

tories, Institutes. Section 001 records 710 periodicals indicating full titles and publishers. It may be noted that the titles of the periodicals are given in the original languages, and that Russian titles have been transliterated applying the transliteration scheme given above. Section 008 records 91 periodicals; these are publication series of observatories and astronomical institutes. Titles of the periodicals have been given following the recommendations of the "International List of Periodical Title Word Abbreviations" and its additions (see also **Abbreviations**, p. 10). In most cases they permit recognition of the full title without recourse to the key in section 001.
If other abstracting journals have been consulted in order to examine the degree of completeness of our service, we cite these papers and give reference to the abstracting service.

Author index and subject index

The subject category and the serial number have been used as a reference both in the author index and the subject index. These references are more precise than page references. They offer considerable advantages in indexing by means of data processing machines, and are more convenient for the user.

The author index of this volume contains 10360 names. A complete reference comprises six figures, three for the subject category and three for the serial number within the category. In the case of more than one reference to abstracts in one category, the number of the category is given only once and not repeated in the immediately following references. The total number of papers (some do not give names of authors) recorded in this volume amounts to 8826.
We consider the subject index as an approximation to an optimal index covering all fields of astronomy and astrophysics and their border fields. The assigning of one or more key words to a paper is, undoubtedly, a difficult task. Some journals have started giving key words together with the titles of papers. These key words are chosen by the authors themselves. Starting with Volume 18, the subject index was enlarged to a certain extent in order to provide a thesaurus of astronomical and astrophysical terms. This is done not only for the users' convenience, but also with the intention to propose the use of special key words to authors and publishers.

While each volume is scheduled to contain an author index and a subject index, the magnetic tapes containing the index information will be used to produce separate index volumes (authors and subjects) at intervals of five years.

The sorting program for the author and subject indexes is based on the IBM SORT/MERGE Program. This program sorts blank before hyphen (–) and before letters. Apostrophes are ignored by a special routine. The computations and printing were carried out on an IBM 360/44.

The two most common and widely used classification systems in astronomy and astrophysics are given by Class 9 of the International Classification System for Physics, published by the International Council of Scientific Unions Abstracting Board (Second edition 1978. ICSU-AB, 17 Rue Mirabeau, 75017 Paris, France, ISSN 0305-9618), and the Astronomy and Astrophysics Abstracts classification. In order to facilitate literature searches, we introduce a concordance relation between these two very different systems. This solution is only a unilateral one. Starting from the fourth hierarchical level of the ICSU-AB system, the appropriate Astronomy and Astrophysics Abstracts chapter numbers are listed. This cannot imply an identical content of the respective chapters in both systems. In many cases there is only a rather partial concordance, and therefore the Astronomy and Astrophysics Abstracts numbers are enclosed in parentheses. Due to the fact that our service only aims to present a comprehensive documentation of the literature in all fields of astronomy and astrophysics, only a certain part of Class 9 of the ISCU-AB scheme is covered.

The users are requested to inform us on spelling errors within the author and subject indexes in order to assist us in eliminating mistakes in future cumulative indexes.

Concordance Relation

between the ICSU-AB International Classification System for Physics
and the *Astronomy and Astrophysics Abstracts* Classification Scheme

ICSU-AB International Classification System for Physics	Astronomy and Astrophysics Abstracts Classification Scheme
0 General	
01.10 Announcements, news, and organizational activities	
01.10.C	006 (010)
01.10.F	011
01.10.H	013 (010)
01.30 Physics literature and publications	
01.30.B	012 (014)
01.30.C	012
01.30.E	003
01.30.K	002 (003)
01.30.M	003
01.30.P	003
01.30.R	014
01.30.T	002
01.40 Education	014
01.50 Educational aids	014
01.60 Biographical, historical, and personal notes	004 (005, 006, 007)
01.65 History of science	004
01.90 Other topics of general interest	015

ICSU-AB International Classification System for Physics	Astronomy and Astrophysics Abstracts Classification Scheme
9 Geophysics, Astronomy, and Astrophysics	
91.10 Geodesy and gravity	
91.10.B	046
91.10.N	044 (045)
91.10.Q	081
91.25 Geomagnetism and paleomagnetism; geoelectricity	084
91.35 Earth's interior structure and properties	081
91.90 Other topics in solid Earth physics	081
92.60 Meteorology	082
92.65 Atmospheric optics	082
94.10 Physics of the neutral atmosphere	
94.10.B	082
94.10.D	082
94.10.F	082
94.10.G	082 (063)
94.10.H	082
94.10.L	082
94.10.N	106
94.10.Q	082
94.10.S	084
94.20 Physics of the ionosphere	
94.20.B	083
94.20.D	083
94.20.M	083 (084)
94.20.P	083 (084)
94.20.W	083 (062)
94.20.Y	083 (084)
94.30 Physics of the magnetosphere	
94.30.C	084
94.30.D	084

ICSU-AB International Classification System for Physics	Astronomy and Astrophysics Abstracts Classification Scheme
94.30.E	084
94.30.F	084 (062)
94.30.G	084 (062)
94.30.H	084
94.30.L	084
94.30.M	084
94.30.S	084
94.30.V	084 (074)
94.30.W	084 (078, 143)
94.40 Cosmic rays	
94.40.C	143
94.40.E	143 (078, 106)
94.40.H	078
94.40.K	143 (085)
94.40.L	143 (078)
94.40.V	105 (143)
94.60 Interplanetary space	
94.60.D	074
94.60.F	074
94.60.G	074 (062)
94.60.K	106
94.60.M	106
94.60.Q	074 (091, 094.0)
94.60.R	106 (062)
94.80 Aerospace facilities and techniques, space research	
94.80.P	053 (051)
94.80.R	054 (051)
94.80.W	032.5
95.10 Fundamental astronomy	
95.10.C	042 (043, 052)
95.10.E	042 (052)
95.10.G	041 (079, 095, 096)
95.10.J	041

ICSU-AB International Classification System for Physics	Astronomy and Astrophysics Abstracts Classification Scheme
95.30 Fundamental aspects of astrophysics	
95.30.C	061 (022)
95.30.E	022
95.30.G	022 (061)
95.30.J	063
95.30.L	062
95.30.Q	062
95.30.S	066.0 (162)
95.45 Observatories	008 (009)
95.55 Astronomical instruments	
95.55.B	032.0
95.55.C	032.0 (031.0)
95.55.E	032.0
95.55.J	033
95.55.L	032.5
95.65 Auxiliary and recording instruments	034
95.70 Other instrumentation and techniques (including clocks, frequency standards, etc.)	035 (031.5, 034, 036)
95.75 Techniques of observation and reduction	
95.75.D	031.5
95.75.F	031.5
95.75.H	031.5
95.75.K	031.5
95.75.M	031.5 (021)
95.75.P	021
95.80 Catalogues, atlases, etc.	002 (047)
96.10 General, solar nebula, and cosmogony	107 (091)
96.20 Moon	
96.20.B	094.0
96.20.D	094.0 (094.5)
96.20.J	094.0

ICSU-AB International Classification System for Physics	Astronomy and Astrophysics Abstracts Classification Scheme
96.30 Planets and satellites (excluding the moon)	
96.30.D	092
96.30.E	093
96.30.G	097
96.30.H	098
96.30.K	099
96.30.M	100
96.30.T	101
96.50 Other objects in the planetary system	
96.50.D	106
96.50.G	102 (103)
96.50.K	104
96.50.M	105
96.60 Solar physics	
96.60.C	080 (075)
96.60.F	071 (080)
96.60.K	080
96.60.M	071
96.60.N	073 (074)
96.60.P	074
96.60.Q	072
96.60.R	073 (076, 077)
96.60.S	073
96.60.V	078 (074)
97.10 Stellar characteristics and properties	
97.10.B	131
97.10.C	065 (061)
97.10.E	064 (063)
97.10.F	112 (064)
97.10.H	064 (112)
97.10.K	116 (065)
97.10.L	116 (065)
97.10.N	115
97.10.Q	115
97.10.R	115 (113, 114).
97.10.T	114
97.10.V	111
97.10.W	111

ICSU-AB International Classification System for Physics	Astronomy and Astrophysics Abstracts Classification Scheme
97.20 Normal stars (by class): general or individual	
97.20.D	121
97.20.R	126
97.30 Variable and peculiar stars (including novae)	
97.30.E	114
97.30.F	116
97.30.G	122 (123)
97.30.J	122 (123)
97.30.K	122 (123)
97.30.N	122 (123)
97.30.Q	124 (122, 123)
97.30.S	122 (123)
97.60 Late stages of stellar evolution (including black holes)	
97.60.B	125
97.60.G	141.5
97.60.J	066.5
97.60.L	066.0
97.60.S	066.0 (065)
97.80 Binary and multiple stars (including extrasolar planetary systems)	
97.80.D	118 (002)
97.80.F	120
97.80.H	119
97.80.J	142.0 (117)
97.80.K	118
97.80.M	118
98.10 Stellar dynamics	151
98.20 Stellar clusters and associations	
98.20.C	152
98.20.E	153
98.20.H	154

ICSU-AB International Classification System for Physics	Astronomy and Astrophysics Abstracts Classification Scheme
98.40 Interstellar matter and nebulae	
98.40.B	131
98.40.C	131
98.40.F	132
98.40.H	132 (134)
98.40.J	133 (112)
98.40.K	134 (131)
98.40.M	134
98.40.N	125
98.50 The Galaxy; extragalactic objects and systems	
98.50.C	158
98.50.E	151 (158)
98.50.H	158 (162)
98.50.K	160
98.50.L	155 (156, 157)
98.50.M	160 (159)
98.50.R	158
98.50.T	161
98.70 Other objects and background radiations of unknown origin or distances	
98.70.D	141.0
98.70.J	141.0
98.70.L	133
98.70.Q	142.0 (142.5)
98.70.S	143
98.70.V	066.0 (142.0, 142.5, 162)
98.80 Cosmology	
98.80.B	162
98.80.D	162 (066.0)
98.80.F	061 (162)

Abbreviations

Abbreviations used in *Astronomy and Astrophysics Abstracts* are primarily based on the 'International List of Periodical Title Word Abbreviations', prepared for the UNISIST/ICSU-AB Working Group on Bibliographic Descriptions (1970).

A.A.B.	Associazione Astrofili Bolognesi
Aarg.	Aargang
AAS	American Astronomical Society
AAVSO	American Association of Variable Star Observers
Abh.	Abhandlung–
Abstr.	Abstract–
Abt.	Abteilung
Acad.	Academi–, Academy
Accad.	Accademi–
Act.	Active, Activit–
Adm.	Administr–
Adv.	Advanc–
Aehron.	Aehronomi–
Aeron.	Aeronom–
Aeronaut.	Aeronauti–
Aerosp.	Aerospace
AG	Astronomische Gesellschaft
AIAA	American Institute of Aeronautics and Astronautics
AJB	Astronomischer Jahresbericht
Akad.	Akadem–
Ala.	Alabama
Alm.	Almanac–, Almanak–
Amat.	Amateur–
An.	Anais, Anale–, Anali–, Anals
Anal.	Analis–, Analit–, Analys–, Analyt–
Angew.	Angewandt–
Ann.	Annaes, Annal–
Annu.	Annu–
Anst.	Anstalt
Anu.	Anual–, Anuar–
Anz.	Anzeiger
Appl.	Applied
Arb.	Arbeit
Arch.	Archiv–
Årg.	Årgang
Ariz.	Arizona
Ark.	Arkiv–
Arkh.	Arkhiv–
Artif.	Artifici–
ASA	Astronomical Society of Australia
Asoc.	Asocia–
ASP	Astronomical Society of the Pacific
ASSA	Astronomical Society of Southern Africa
Assem.	Assembl–
Assoc.	Associ–
Assoz.	Assozi–
Astrofis.	Astrofisic–
Astrofiz.	Astrofizi–
Astrometr.	Astrometr–
Astron.	Astronom–
Astronaut.	Astronauti–, Astronauty–
Astrophys.	Astrophys–
ASV	Astronomical Society of Victoria
ASWA	Astronomical Society of Western Australia
At.	Atom–
Atmos.	Atmosf–, Atmosph–
BAA	British Astronomical Association
Bayer.	Bayerisch–
Beitr.	Beitrag, Beiträge
Beob.	Beobacht–
Ber.	Bericht–
Bibl.	Bibliot–
Bibliogr.	Bibliograf–, Bibliograph–
BIH	Bureau International de l'Heure (Paris)
Bimest.	Bimestr–
Bl.	Blatt, Blätter
Bol.	Boletin
Boll.	Bolletino
Bul.	Buleten–, Buletin–, Bulten
Bull.	Bulletin–, Bullettino
Bur.	Bureau–
Byul.	Byuleten–, Byuletin–
Byull.	Byulleten–
C.R.	Comptes Rendus
Cah.	Cahier–
Calif.	California
Cas.	Casopis
Cent.	Center–, Central, Centrale, Centrally, Centre
Cercet.	Cercetary
Chem.	Chemi–
Chim.	Chimi–
Chron.	Chronic–, Chronik, Chronique
Chronom.	Chronometr–
Cie.	Compagnie
Cienc.	Ciencia–
Cient.	Cientific–
Circ.	Circolar–, Circolo, Circolaire–, Circular–, Circulo
Cirk.	Cirkulaer–
Cl.	Clasa, Classe–
Co.	Companies, Company
Coll.	College
Collect.	Collect–
Colloq.	Colloqui–
Colo.	Colorado
Comet.	Cometary
Commentat.	Commentat–
Commun.	Communica–
Comput.	Computation, Computer–, Computing
Comun.	Comunica–
Conf.	Conferen–
Congr.	Congres–
Conn.	Connecticut
Contract.	Contract–
Contrib.	Contribu–
Cosm.	Cosmic–
Cosmochim.	Cosmochimi–
COSPAR	Committee on Space Research
Crystallogr.	Crystallograph–
CSIRO	Commonwealth Scientific and Industrial Research Organization
Cult.	Cultur–, Cultuur

Curr.	Current
D.C.	District of Columbia
DDR	Deutsche Demokratische Republik
Del.	Delaware
Dep.	Departament, Département, Department
Dev.	Development–, Développement–
Diss.	Disserta–
Div.	Divis–
Doc.	Document–
Dok.	Dokument–
Dokl.	Doklad–
Ehksp.	Ehksperiment–
Eidg.	Eidgenössisch–
Eksp.	Eksperiment–
Electron.	Electroni–
Eng.	Engineer–
Environ.	Environment–
Equip.	Equipement, Equipment
Ergeb.	Ergebnis–
ESA	European Space Agency
ESO	European Southern Observatory
ESRO	European Space Research Organization
Eval.	Evaluation–
Exp.	Experiment–
Extraterr.	Extraterrestr–
F. R. Germany	Federal Republic of Germany
Fac.	Facolt–, Faculd–, Facult–
Fak.	Fakult
Fasc.	Fascicul–
Fenn.	Fenni–
Fis.	Fisic–, Fisik–
Fiz.	Fizic–, Fizik–, Fizyk–
Fla.	Florida
Fluid.	Fluidi–
Fond.	Fondation–, Fondazione
Fortschr.	Fortschritt–
Fotogr.	Fotograf–
Found.	Foundation–
Freq.	Frequen–
Fundam.	Fundamenta–
Fys.	Fysik–, Fysisch, Fysisk–
Fyz.	Fyzik–
G.	Giornale
Ga.	Georgia
Gaz.	Gazeta, Gazette
Gazz.	Gazzetta
Gen.	General
Geochem.	Geochem–
Geochim.	Geochim–
Geod.	Geodaes–, Geodaet–, Geodes–, Geodet–, Geodez–
Geofis.	Geofis–
Geofiz.	Geofiz–
Geofys.	Geofys–
Geogr.	Geograf–, Geograph–
Geokhim.	Geokhim–
Geol.	Geolog–, Geolosk–
Geomagn.	Geomagneti–
Geophys.	Geophys–
Ges.	Gesellschaft
Gesch.	Geschichte
Gl.	Glavno–
Glas.	Glasnik
Gos.	Gosudarst–
Gov.	Government–
Grenzgeb.	Grenzgebiet–
GSFC	Goddard Space Flight Center
H. M.	Her Majesty's, His Majesty's
Handb.	Handbook, Handbuch
Her.	Herald–
Hist.	History
Hochsch.	Hochschule
Hoegsk.	Hoegskol–
HR-diagram	Hertzsprung-Russell diagram
Hydrogr.	Hydrograf–, Hydrograph–
IAF	International Astronautical Federation
IAU	International Astronomical Union
IBM	International Business Machines Corporation
ICSU	International Council of Scientific Unions
ICSU-AB	International Council of Scientific Unions–Abstracting Board
IEEE	Institute of Electrical and Electronics Engineers
Ill.	Illinois
Inc.	Incorporated
Ind.	Industr–
Inf.	Informat–, Informaz–, Informe–
Ing.	Ingenieur
INIS	International Nuclear Information System
INSPEC	International Information Services for the Physics and Engineering Communities
Inst.	Institut–, Instytut–
Instn.	Institution
Instrum.	Instrument–
Int.	Internationa–, Internazional–
Inter.	Intérieur–, Interior
Interplanet.	Interplanetary
Intez.	Intezet–
Invest.	Investiga–
Ionos.	Ionosfer–, Ionospher–
Iskusstv.	Iskusstvenn–
Issled.	Issledovan–
Ist.	Istitut–
Izd.	Izdatel–
Izv.	Izvesti–
J.	Joernaal–, Jornal–, Journal–
Jaarb.	Jaarboek–
Jahrb.	Jahrbuch, Jahrbücher
Jahresber.	Jahresbericht–
Jahresschr.	Jahresschrift
Jahrg.	Jahrgang
JPL	Jet Propulsion Laboratory
K.	Königlich–, Koninklijk–, Kunglig–
Kans.	Kansas
Kartogr.	Kartograf–
Kernforsch.	Kernforschung
Kernphys.	Kernphysik–
Khem.	Khemyi–
Khim.	Khimi–
Kim.	Kimija–, Kimya
Kl.	Klass–
Kolloq.	Kolloquium–
Komet.	Kometnyj
Komm.	Kommission–
Konf.	Konfer–
Kongr.	Kongress
Kosm.	Kosmich–
Kosmog.	Kosmogon–
Kozp.	Kozponti
KPNO	Kitt Peak National Observatory
Kut.	Kutato
Ky.	Kentucky
La.	Louisiana
Lab.	Laborato–
Lett.	Letter–, Lettra, Lettre
Libr.	Librair–, Librar–
Mag.	Magasin, Magazin–
Magn.	Magneti–, Magnitn–
Mass.	Massachusetts
Mat.	Matemaat–, Matemat–
Mater.	Material–
Math.	Mathemat–
Md.	Maryland

Meas.	Measur–
Mec.	Mecani–
Mech.	Mechani–
Medd.	Meddelande–, Meddelelse
Meded.	Mededeeling–, Mededeling–
Mekh.	Mekhani–
Mem.	Memento–, Memoir–, Memori–, Memory–, Memuary
Memo.	Memorand–
Mens.	Mensile, Mensual–, Mensuel–
Messtech.	Messtechni–
Meteorol.	Meteorolog–
Mich.	Michigan
Micromec.	Micromecaniq–
Miner.	Mineral, Minerale–, Minerali–
Mineral.	Mineralog–
Minn.	Minnesota
Miss.	Mississippi
MIT	Massachusetts Institute of Technology
Mitt.	Mitteilung–
Mo.	Missouri
Mod.	Modern–
Mol.	Molecul–, Molekul–
Mon.	Monat, Monatlich–, Month–
Monogr.	Monograph–
Mont.	Montana
MPI	Max-Planck-Institut
Mus.	Museum
N. C.	North Carolina
N. D.	North Dakota
N. H.	New Hampshire
N. J.	New Jersey
N. M.	New Mexico
N. Y.	New York
Nablyud.	Nablyudeni–
Nac.	Nacion–
Nachr.	Nachricht–
NASA	National Aeronautics and Space Administration
Nat.	Natur–
Natl.	National–
Naturforsch.	Naturforsch–
Naturwiss.	Naturwissenschaft–
Natuurkd.	Natuurkunde
Nauchn.	Nauchny–
Nauk.	Nauka, Naukite, Naukov–, Naukow–
Naut.	Nautic–
Nav.	Naval–
Navig.	Navigat–
Naz.	Nazion–
Nebr.	Nebraska
Nev.	Nevada
Newsl.	Newsletter–
Not.	Notationes, Notic–, Notise–, Notizi–
Nouv.	Nouveau–, Nouvell–
Nov.	Novoe
Nucl.	Nucléaire–, Nuclear–, Nucl–
Nukl.	Nukle–
Numer.	Numeri–
O-va	Obshchestva
O-vo	Obshchestvo
Obs.	Observ–
Obz.	Obzor–
Okla.	Oklahoma
Opt.	Optic–, Optik–, Optique
Oreg.	Oregon
Oss.	Osserva–
Pa.	Pennsylvania
Paleontol.	Paleontolog–
Pap.	Paper–, Papier
Part.	Particle
Perem.	Peremenn–
Period.	Periodi–
Petrol.	Petrolog–
Philos.	Philosoph–
Photogr.	Photograf–, Photograph–
Photogramm.	Photogrammetr–
Photom.	Photometr–
Phys.	Physic–, Physik–, Physique–, Physisch–
Pict.	Picture–
Planet.	Planetary
Pr.	Prac–
Prelim.	Prelimin–
Prepr.	Preprint
Prib.	Pribor–
Prikl.	Prikladnoj
Prir.	Prirodn–
Prirodoved.	Prirodoved–
Probl.	Problem–
Proc.	Proceedings
Prod.	Prodott–, Produc–, Produkt,
Prog.	Progres–
Propag.	Propagation
Prov.	Provinc–, Provints–, Provinz–
Pubbl.	Pubblicazion–
Publ.	Publicac–, Publicas–, Publicat–, Publikas–, Publikat–
Q.	Quarterly
Quant.	Quantit–
R.	Royal
R. I.	Rhode Island
Radiat.	Radiati–
Radioact.	Radioactiv–, Radioaktiv–
Radioisot.	Radioisotop–
Rap.	Raport–
Rapp.	Rapport–
RAS	Royal Astronomical Society
Rec.	Record–
Rech.	Recherche–
Ref.	Referat–, Reference–, Referieren
Relat.	Related, Relation–
Relativ.	Relativit–
Rend.	Rendicont–
Rep.	Report–
Repr.	Reprint–
Repub.	Republi–
Res.	Research–
Result.	Resultad–, Resultat–
Rev.	Review–, Revisio, Revista, Revue–
Rezul't.	Rezul'tat–
Ric.	Ricerca, Ricerche
Riv.	Rivist–
Rundsch.	Rundschau
S. C.	South Carolina
S. D.	South Dakota
SAF	Société Astronomique de France
SAI	Società Astronomica Italiana
Samml.	Sammlung–
SAO	Smithsonian Astrophysical Observatory
SAS	Société Astronomique de Suisse
Satell.	Satellite
Sb.	Sbornik–
Schr.	Schrift–
Schriftenr.	Schriftenreihe
Sci.	Scienc–, Scient–, Scienz–
Scr.	Scripta, Scritt–
Secc.	Seccion–
Sect.	Secti–
Sekc.	Sekci–, Sekcj–
Sekt.	Sektion–, Sektor–
Sekts.	Sektsi–
Sel.	Seleccion–, Select–, Selek–, Selezione
Selsk.	Selskab–, Selskap–
Semin.	Séminair–, Seminar–

Sep.	Separat–
Ser.	Seria–, Serie–, Seriya
Serv.	Servic–, Serviz–
Sess.	Sessi–
Signal.	Signalétique–
Simp.	Simpoz–
Sitzungsber.	Sitzungsbericht–
Skr.	Skrift–
Soc.	Sociedad–, Societ–
Sol.	Solar
Soln.	Solnechn–
Sonderdr.	Sonderdruck–
Soobshch.	Soobshchen–
South.	Southern
Spacecr.	Spacecraft
Spat.	Spatial–
Spec.	Special–
Spectrosc.	Spectroscop–
Spectrosk.	Spectroskop–
Spets.	Spetsial–
Spez.	Spezial–, Speziell–
SSR	Sovetskaya Sotsialisticheskaya Respublika
SSSR	Soyuz Sovetskikh Sotsialisticheskikh Respublik
St.	Saint–, Sankt–, Sant–
–St.	–Straße, Street
Stand.	Standard–, Standart–
Sternw.	Sternwarte–
Stiint.	Stiintific–
Stn.	Station, Stazione
Stud.	Studia, Studie–, Studii
Supl.	Suplement–, Supliment–
Suppl.	Supplement–
Surv.	Survey–
Symp.	Sympos–, Sympoz–
Syst.	System–
Sz.	Szemle
Teach.	Teacher–, Teaching
Tec.	Tecni–
Tech.	Techni–
Technol.	Technolog–
Tecnol.	Tecnolog–
Teh.	Tehnic–, Tehnika, Tehnisk–
Tehnol.	Tehnolog–, Tehnolosk–
Tek.	Tekni–
Tekh.	Tekhni–
Tekhnol.	Tekhnolog–
Teknol.	Teknolog–
Telesc.	Telescop–
Telev.	Television–
Tenn.	Tennessee
Teor.	Teoret–, Teori–
Terr.	Terrestr–
Test.	Testing
Tex.	Texas
TH	Technische Hochschule
Theor.	Theoret–, Theori–
Tidschr.	Tidschrift–

Tidskr.	Tidskrift–
Tidsskr.	Tidsskrift–
Top.	Topic–
Tr.	Trudy
Trans.	Transactions, Transazione
Tsentr.	Tsentral–
Tsirk.	Tsirkulyar–
TU	Technical University
Uch.	Uchen–
Uchebn.	Uchebn–
UK	United Kingdom
Umsch.	Umschau
UN	United Nations
Univ.	Universidad–, Universit–, Univerzitet–
US	United States
USA	United States of America
USSR	Union of Soviet Socialist Republics
Va.	Virginia
Var.	Various
Ver.	Verein–, Verenig–
Veränderl.	Veränderlich–
Verh.	Verhandl–
Vermess.	Vermessung–
Vermessungswes.	Vermessungswesen
Veröff.	Veröffentlich–
Vesn.	Vesnik
Vestn.	Vestnik
Vetensk.	Vetenskap–
Vidensk.	Videnskab–, Videnskap
Vierteljahresschr.	Vierteljahresschrift–
Vierteljahrsschr.	Vierteljahrsschrift–
VLB	Very Long Baseline
Volcanol.	Volçanolog–
Vopr.	Vopros–
Vortr.	Vorträge
Vses.	Vsesoyuzn–
Vt.	Vermont
Vyp.	Vypusk–
Vyssh.	Vyssh–
Vyzk.	Vyzkum–
W. Va.	West Virginia
Wash.	Washington
West.	Western
Wet.	Wetenschap–, Wetenskap–
Wis.	Wisconsin
Wiss.	Wissenschaft–
Wyo.	Wyoming
Yad.	Yadern–
Z.	Zeitschrift–
ZA	Zero Age
ZAED	Zentralstelle für Atomkernenergie-Dokumentation
Zap.	Zapisk–, Zapyisk–
Zaved.	Zaveden–
Zent.	Zentral
Zentralbl.	Zentralblatt
Zesz.	Zeszyt
Zh.	Zhurnal–
Zirk.	Zirkular

Periodicals, Proceedings, Books, Activities

001 Periodicals

A. A. O. Newsl.
Anglo-Australian Observatory Newsletter. Published by the Anglo-Australian Observatory, PO Box 296, Epping, NSW 2121, Australia.

AAS Photo-Bull.
AAS (American Astronomical Society) Photo-Bulletin. Published by the Working Group on Photographic Materials. Produced by Eastman Kodak Co., Rochester, N.Y.

AAVSO Bull.
American Association of Variable Star Observers Bulletin. 187 Concord Avenue, Cambridge, Mass., 02138, U.S.A.

Abh. Hamburger Sternw.
Abhandlungen aus der Hamburger Sternwarte. Hamburg-Bergedorf. ISSN 0374-1583.

Acad. R. Belgique, Bull. Cl. Sci.
Académie Royale de Belgique, Bulletin de la Classe des Sciences (Koninklijke Academie van België, Mededelingen van de Klasse der Wetenschappen). 5ᵉ Série, Palais des Académies, Bruxelles.

Acta Acust.
Acta Acustica. Science Press, Beijing, People's Republic of China. Subscription address: Guozi Shudian, P. O. Box 399, Beijing.

Acta Astron.
Acta Astronomica. An international quarterly journal. Publisher: Polska Akademia Nauk, Komitet Astronomii (Polish Academy of Sciences, Committee of Astronomy), Warszawa – Wrocław.

Acta Astron. Sinica
Acta Astronomica Sinica. Published by Purple Mountain Observatory, Academia Sinica, Nanking, China.

Acta Astronaut.
Acta Astronautica. Journal of the International Academy of Astronautics. Publisher: Pergamon Press Inc., Elmsford, New York, U.S.A.; Pergamon Press Ltd., Oxford, England.

Acta Astrophys. Sinica
Acta Astrophysica Sinica. Editorial Board "Acta Astrophysica Sinica", P. O. Box 399, Beijing, China.

Acta Cosmologica
Acta Cosmologica. Published by Obserwatorium Astronomiczne Universytetu Jagiellońskiego, Kraków, Poland.

Acta Fac. Rerum Nat. Univ. Comenianae Phys.
Acta Facultatis Rerum Naturalium Universitatis Comenianae, Physica. Ústredná knižnica PFUK, 886 11 Bratislava, Ul. 29 Augusta č.5, Czechoslovakia.

Acta Geod. Geophys.
Acta Geodaetica et Geophysica. Edited by the Institute of Geodesy and Geophysics, Academia Sinica. Published by Science Press, Beijing, Cao Yan Men Nei Street No. 137, China.

Acta Geod. Geophys. Montan.
Acta Geodaetica, Geophysica et Montanistica. Akademiai Kiado, 1054 Budapest, Alkotmany utca 21, Hungary.

Acta Geophys. Polonica
Acta Geophysica Polonica. ARS Polona-Ruch, 00-068 Warszawa, Krakowskie Przedmiescie 7, P.O. Box 1001, Poland.

Acta Geophys. Sinica
Acta Geophysica Sinica. Chinese Academy of Sciences, Department of Geophysical Research. Published by Science Press, Peking, People's Republic of China.

Acta Mech. Sinica
Acta Mechanica Sinica,Science Press, Peking, People's Republic of China. Subscription address: Guozi Shudian, P.O. Box 399, Peking.

Acta Phys. Acad. Sci. Hungaricae
Acta Physica Academiae Scientiarum Hungaricae. Kultura, Hungarian Trading Co., H-1389 Budapest 62, P.O. Box 149, Hungary.

Acta Phys. Austriaca
Acta Physica Austriaca. Springer-Verlag, A-1011 Wien, Molkerbastei 5, Postfach 367, Austria.

Acta Phys. Polonica B
Acta Physica Polonica B. ARS Polona-Ruch, Warszawa 1, P.O. Box 154, Poland.

Acta Phys. Sinica
Acta Physica Sinica. Chinese Academy of Sciences, Institute of Physics, Peking, People's Republic of China. [English translation in: Chinese J. Phys. (*USA*)].

Acta Phys. Slovaca
Acta Physica Slovaca. VEDA Publishing House of the Slovak Academy of Sciences, 895 30 Bratislava, Klemensova 27, Czechoslovakia.

Acta Polytech. Scandinavica, Appl. Phys. Ser.
Acta Polytechnica Scandinavica, Applied Physics Series. Finnish Academy of Technical Sciences, Kansakoulukatu 10 A, 00100 Helsinki 10, Finland.

Acta Polytech. III
Acta Polytechnica. Series III. Elektrotechnická fakulta ČVUT v Praze, Technická ul. 2, Praha 6-Dejvice, Czechoslovakia.

Acta Sci. Nat. Univ. Pekinensis
Acta Scientiarum Naturalium Universitatis Pekinensis. Peking, People's Republic of China.

Acta Sci. Nat. Univ. Sunyatseni
Acta Scientiarum Naturalium Universitatis Sunyatseni (Zhongshandaxue Xuebao). Canton Post Office, Canton, People's Republic of China.

Acta Tech. Acad. Sci. Hungaricae
Acta Technica Academiae Scientiarum Hungaricae. Akademiai Kiado, Budapest 1363, P. O. Box 24, Hungary.

Acta Tech. CSAV
Acta Technica Československá akademie věd. Academia, Publishing House of the Czechoslovak Academy of Sciences, Vodickova 40, 112 29 Praha 1, Czechoslovakia. John Benjamins N.V., Periodical Trade, Warmoesstraat 54, Amsterdam, Netherlands.

Acta Univ. Carolinae Math. Phys.
Acta Universitatis Carolinae, Mathematica et Physica. Administrace: Matematicko-fyzikální fakulta University Karlovy, Praha.

Adv. Phys.
Advances in Physics. Taylor & Francis Ltd., 10–14 Macklin Street London, WC2B 5NF, England.

Adv. Space Res.
Advances in Space Research. The official journal of the Committee on Space Research (COSPAR). Pergamon Press, Oxford – New York – Toronto – Sydney – Paris - Frankfurt. ISSN 0273-1177.

Aeronaut. Astronaut.
L'Aeronautique et l'Astronautique. Editions Air et Cosmos, 6 Rue Anatole de la Forge, 75017 Paris, France.

AIAA J.
AIAA Journal. A Publication of the American Institute of Aeronautics and Astronautics devoted to Aerospace Research and Development. Published by the American Institute of Aeronautics and Astronautics, New York, N.Y.

AIP Conf. Proc.
AIP Conference Proceedings. American Institute of Physics, 335 East 45th Street, New York, N.Y. 10017, USA.

Alta Freq.
Alta Frequenza. Ufficio Centrale AEI-CEI, Viale Monza 259, 20126 Milano, Italy.

American J. Phys.
American Journal of Physics. Published for the American Association of Physics Teachers by the American Institute of Physics, 335 East 45th Street, New York, N. Y. 10017, USA.

American Mineral.
American Mineralogist. Mineralogical Society of America, 2000 Florida Avenue, N. W., Washington, DC 20009, USA.

American Sci.
American Scientist. Society of Sigma XI, 345 Whitney Avenue, New Haven, CT 06510, USA.

An. Acad. Brasil. Cienc.
Anais da Academia Brasileira de Ciencias. Caixa Postal 229, ZC-00 Rio de Janeiro gb, Brazil.

An. Fis.
Anales de Física. Real Sociedad Española de Física y Química (Facultad de Ciencias), Ciudad Universitaria, Madrid-3, Spain.

Anal. Chem.
Analytical Chemistry. American Chemical Society, 1155 Sixteenth Street N. W. Washington, DC 20036, USA.

Ann. Acad. Sci. Fennicae, Ser. A. VI
Annales Academiae Scientiarum Fennicae, Series A VI (Physica). Snellmaninkato 9-11, 00170 Helsinki-17, Finnland.

Ann. Geofis.
Annali di Geofisica. Istituto Nazionale di Geofisica, Citta Universitaria, Via Ruggero Bonghi 11/B, 00184 Roma, Italy.

Ann. Géophys.
Annales de Géophysique. Service des Publications du CNRS, 15 Quai Anatole-France, 75700 Paris, France.

Ann. Inst. Henri Poincaré A
Annales de l'Institut Henri Poincaré, Section A (Physique Théorique). 11 Rue Pierre-Curie, Paris 5, France.

Ann. Israel Phys. Soc.
Annals of the Israel Physical Society. c/o Department of Physics, Bar-Ilan University, Ramat-Gan, Israel. Adam Hilger Ltd., Techno House, Recliffe Way, Bristol BS1 6NX England.

Ann. N.Y. Acad. Sci.
Annals of the New York Academy of Sciences. The New York Academy of Sciences, 2 East 63rd Street, New York, N.Y. 10021, USA.

Ann. Nucl. Energy
Annals of Nuclear Energy. Pergamon Press Ltd., Headington Hill Hall Oxford, OX3 0BW, England.

Ann. Physics
Annals of Physics. Academic Press Inc., 111 Fifth Avenue, New York, NY 10003, USA.

Ann. Physik
Annalen der Physik. 7. Folge. Publisher: Johann Ambrosius Barth, Salomonstr. 18B, Leipzig 701, German Democratic Republic.

Ann. Physique
Annales de Physique. Publisher: Masson et Cie., 120 Boulevard Saint-Germain, Paris 6, France.

Ann. Sci.
Annals of Science. Taylor & Francis Ltd., 10-14 Macklin Street, London, WC2B 5NF, England.

Ann. Shanghai Obs., Acad. Sinica
Annuals of Shanghai Observatory, Academia Sinica. Published by Shanghai Scientific and Technical Publishers, Shanghai, Rei Jing Er Street, No. 450.

Ann. Soc. Sci. Bruxelles I
Annales de la Société Scientifique de Bruxelles. Série I: Sciences Mathématiques, Astronomiques et Physiques. Rue de Bruxelles 61, B 5000 Namur, Belgium.

Ann. Télécommun.
Annales des Télécommunications. Centre National

d'Études des Télécommunications, 38 rue du Général Leclerc, 92 Issy-les-Moulineaux, France.

Ann. Tokyo Astron. Obs., Second Ser.
Annals of the Tokyo Astronomical Observatory, Second Series. University of Tokyo, Mitaka, Tokyo, Japan. ISSN 0082-4704.

Annu. Rep. Astron. Inst. Greece
Annual Reports of the Astronomical Institutes of Greece. Published by the Greek National Committee for Astronomy. Academy of Athens, Research Center for Astronomy and Applied Mathematics.

Annu. Rev. Astron. Astrophys.
Annual Review of Astronomy and Astrophysics. Annual Reviews Inc., 4139 El Camino Way, Palo Alto, Calif. 94306, USA. ISSN 0066-4146.

Annu. Rev. Earth Planet. Sci.
Annual Review of Earth and Planetary Sciences. Annual Reviews Inc., 4139 El Camino Way, Palo Alto, Calif. 94306, USA. ISSN 0084-6597.

Annu. Univ. Sofia Fac. Phys.
Annuaire de l'Université de Sofia Faculté de Physique, Sofiya, Bulgaria.

Antenna
L'Antenna. Via Monte Generoso 6/a, 20155 Milano, Italy.

Anz. Österreich. Akad. Wiss. Math.-Naturwiss. Kl.
Anzeiger. Österreichische Akademie der Wissenschaften. Mathematisch-Naturwissenschaftliche Klasse. Publisher: Springer-Verlag, Wien.

Appl. Opt.
Applied Optics. A monthly publication of the Optical Society of America. Published for the Optical Society of America by the American Institute of Physics, 335 East 45th Street, New York, NY 10017, USA.

Appl. Phys.
Applied Physics. Springer-Verlag, Heidelberger Platz 3, D-1000 Berlin 33, F. R. Germany.

Appl. Phys. Lett.
Applied Physics Letters. American Institute of Physics, 335 East 45th Street, New York, N.Y. 10017, USA.

Appl. Spectrosc.
Applied Spectroscopy. 428 East Preston Street, Baltimore, MD 21202, USA.

Appl. Spectrosc. Rev.
Applied Spectroscopy Reviews. Marcel Dekker Inc., 95 Madison Avenue, New York, NY 10016, USA.

Arch. Hist. Exact Sci.
Archive for History of Exact Sciences. Springer-Verlag, Berlin - Heidelberg - New York. ISSN 0003 - 9519.

Arch. Mech.
Archives of Mechanics (Archiwum Mechaniki Stosowanej). Polish Scientific Publishers, Swietokrzyska 21, Warszawa, Poland.

Arch. Sci.
Archives des Sciences, éditées par la Société de Physique et d'Histoire Naturelle de Genève. Publisher: Imprimerie Kundig, Genève. Subscription address: Librairie Payot, Genève.

Archaeoastronomy (*England*)
Archaeoastronomy. Supplement to Journal for the History of Astronomy. Published by Science History Publications Ltd, Halfpenny Furze, Mill Lane, Chalfont St Giles, Bucks, England, HP8 4NR. ISSN 0142–7253.

Archaeoastronomy (*U.S.A.*)
Archaeoastronomy. The Bulletin of the Center for Archaeoastronomy, Space Sciences Building, University of Maryland, College Park, MD 20742 (301) 454-4460. ISSN 0190–9940.

Archaeometry
Archaeometry. Cambridge University Press, P.O. Box 92, London, NW1 2DB, England.

Ark. Fys. Semin. Trondheim
Arkiv for det Fysiske Seminar i Trondheim. c/o Institutt for Teoretisk Fysikk, Universitetet i Trondheim, NTH, N-7034 Trondheim, Norway.

Ark. Mat.
Arkiv för Matematik. Published by Institut Mittag-Leffler, Auravägen 17, S-182 62 Djursholm, Sweden.

Artif. Satell.
Artificial Satellites. Publication of Polish Scientific Institutions. Polish Academy of Sciences, National Committee of Geophysics and Geodesy, National Committee for Space Research, Warsaw. Space Research Centre, Pałac Kultury i Nauki 2301, 00-901 Warszawa, Poland.

Astrofiz. Issled. Izv. Spets. Astrofiz. Obs.
Astrofizicheskie Issledovaniya. Izvestiya Spetsial'noj Astrofizicheskoj Observatorii. Akademiya Nauk SSSR. Publishers: Izdatel'stvo "Nauka", Leningradskoe Otdelenie, Leningrad. [An English translation is published in Bulletin of the Special Astrophysical Observatory – North Caucasus].

Astrofizika
Astrofizika. Izdatel'stvo Akademii Nauk Armyanskoj SSR, Erevan. [An English translation is published in "Astrophysics"].

Astrometr. Astrofiz.
Astrometriya i Astrofizika. Respublikanskij Mezhvedomstvennyj Sbornik. Akademiya Nauk Ukrainskoj SSR, Glavnaya Astronomicheskaya Observatoriya. Naukova Dumka, Kiev.

Astron. Astrophys.
Astronomy and Astrophysics. A European Journal. Published by Springer-Verlag, Berlin–Heidelberg–New York.

Astron. Astrophys., Suppl. Ser.
Astronomy and Astrophysics. Supplement Series. A European Journal. Published by les Editions de Physique, Orsay, France, on behalf of the Board of Directors of the European Southern Observatory. ISSN 0365-0138.

Astron. Circ.
Astronomical Circular. Compiled by the editor section of Acta Astronomica Sinica, Purple Mountain Observatory, Nanking, China. Edited by the Chinese Astronomical Society.

Astron. Data Cent. Bull.
Astronomical Data Center Bulletin. National Space Science Data Center/World Data Center A for Rockets and Satellites. National Aeronautics and Space Administration, Goddard Space Flight Center, Greenbelt, Maryland 20771, USA.

Astron. Her.
Astronomical Herald. Astronomical Society of Japan, Tokyo Astronomical Observatory, Oosawa Mitaka, Tokyo, Japan.

Astron. J.
The Astronomical Journal. Published for the American Astronomical Society by the American Institute of Physics, New York, N. Y. Editorial Office: Department of Astronomy, Columbia University, New York, N. Y.

Astron. Nachr.
Astronomische Nachrichten. Publisher: Akademie-Verlag, Berlin.

Astron. Pap.
Astronomical Papers prepared for the use of the American Ephemeris and Nautical Almanac. Published by the Nautical Almanac Office, U.S. Naval Observatory by direction of the Secretary of the Navy and under the authority of Congress. U.S. Government Printing Office, Washington, D.C.

Astron. Q.
The Astronomy Quarterly. Pachart Publishing House, P.O. Box 35549, Tucson, Ariz. 85740. ISSN 0364-9229.

Astron. Rep.
The Astronomical Reports. Polish Amateur Astronomical Society. Polskie Towarzystwo Miłośników Astronomii, Kraków, Poland.

Astron. Schule
Astronomie in der Schule. Zeitschrift für die Hand des Astronomielehrers. Herausgegeben vom Verlag Volk und Wissen, Berlin. Redaktion: Sternwarte Bautzen.

Astron. Soc. Western Australia, Circ.
The Astronomical Society of Western Australia, Circular.

Astron. Tidsskr.
Astronomisk Tidsskrift. Edited by Astronomisk Selskab, København; Norsk Astronomisk Selskap, Oslo; Svenska Astronomiska Sällskapet, Stockholm. Printed by John Griegs Boktrykkeri, Bergen.

Astron. Tsirk.
Astronomicheskij Tsirkulyar, izdavaemyj Byuro Astronomicheskikh Soobshchenij Akademii Nauk SSSR. Tipografiya Astrosoveta AN SSSR, Moskva.

Astron. Vestn.
Astronomicheskij Vestnik. Publishers: Izdatel'stvo "Nauka", Moskva. [An English translation is published in Solar System Research, USA].

Astron. Zh.
Astronomicheskij Zhurnal. Akademiya Nauk SSSR. Publishers: Izdatel'stvo "Nauka", Moskva. [An English translation is published in "Soviet Astronomy"].

Astronomia
Astronomia. Periodico trimestrale dell'Unione Astrofili Italiani.

Astronomie
L'Astronomie et Bulletin de la Société Astronomique de France. Société Astronomique de France, Paris.

Astronomy
Astronomy. AstroMedia Corp., 757 North Broadway, Suite 204, Milwaukee, WI 53202, USA.

Astrophys. J.
The Astrophysical Journal. Published for the American Astronomical Society by the University of Chicago Press, Chicago, Illinois.

Astrophys. J., Lett.
The Astrophysical Journal. Letters to the Editors. Published for the American Astronomical Society by the University of Chicago Press, Chicago, Illinois.

Astrophys. J., Suppl. Ser.
The Astrophysical Journal. Supplement Series. Published for the American Astronomical Society by the University of Chicago Press, Chicago, Illinois.

Astrophys. Lett.
Astrophysical Letters. Published by NASA–Goddard Space Flight Center. Gordon and Breach Science Publishers Ltd., New York–London–Paris.

Astrophys. Space Sci.
Astrophysics and Space Science. An International Journal of Cosmic Physics. Published by D. Reidel Publishing Company, Dordrecht, Holland.

Astrophysics
Astrophysics. A cover-to-cover translation of Astrofizika (USSR). Consultants Bureau, New York, N. Y.

Atmos. Environ.
Atmospheric Environment. Pergamon Press Ltd., Headington Hill Hall, Oxford, OX3 OBW, England.

Atomkernenerg. Kerntech.
Atomkernenergie Kerntechnik. Verlag Karl Thiemig AG, Pilgersheimerstrasse 38, Postfach 900740, D-8000 München 90, F.R. Germany.

Atti Accad. Naz. Lincei, Mem. Ser. Ottava
Atti della Accademia Nazionale dei Lincei. Serie Ottava. Memorie. Classe di Scienze fisiche, matematiche e naturali. Sezione I: Matematica, Meccanica, Astronomia, Geodesia e Geofisica. Published by Accademia Nazionale dei Lincei, Roma.

Atti Accad. Naz. Lincei, Rend. Ser. Ottava
Atti della Accademia Nazionale dei Lincei. Serie Ottava. Rendiconti. Classe di Scienze fisiche, matematiche e naturali. Published by Accademia Nazionale dei Lincei, Roma.

Atti Accad. Sci. Torino I
Atti della Accademia delle Scienze di Torino. I. Classe di Scienze Fisiche, Mathematiche e Naturali. Via Accademia delle Scienze 6, Via Maria Vittoria 3, Torino (208), Italy.

Atti Fond. Giorgio Ronchi
Atti della Fondazione Giorgio Ronchi. Largo Enrico Fermi 1, 50125 Arcetri-Firenze, Italy.

Australian Comput. J.
Australian Computer Journal. Australian Trade Publications, 28 Chippen Street, Chippendale, N.S.W. 2008, Australia.

Australian J. Phys.
Australian Journal of Physics. Published by the Commonwealth Scientific and Industrial Research Organization, 372 Albert Street, East Melbourne, Victoria 3002, Australia.

Australian J. Phys., Astrophys. Suppl.
Australian Journal of Physics, Astrophysical Supplement. Published by Commonwealth Scientific and Industrial Research Organization, 372 Albert Street, East Melbourne, Victoria 3002, Australia.

Autom. Strum.
Automazione e Strumentazione. Associazione Nazionale Italiana per l'Automazione, Via Le Premuda 2, 21029 Milano, Italy.

B.I.H., Paris, Circ.
Bureau International de l'Heure, B.I.H., Paris, Circulars. 61, Avenue de l'Observatoire, 75014-Paris.

BAV Rundbrief
BAV Rundbrief. Mitteilungsblatt der Berliner Arbeitsgemeinschaft für Veränderliche Sterne. Editor: BAV Berliner Arbeitsgemeinschaft für Veränderliche Sterne eV., Berlin.

BBSAG Bull.
Bedeckungsveränderlichen Beobachter der Schweizerischen Astronomischen Gesellschaft, [Swiss Astronomical Society's Eclipsing Variable Observers], Bulletin. To be obtained from R. Diethelm, Winterthur, Switzerland.

Biul. Obs. Astron. Uniw. M. Kopernika Toruniu
Biuletyn Obserwatorium Astronomicznego Uniwersytetu M. Kopernika w Toruniu.

Blick Weltall
Blick in das Weltall. Monatsprogramm und Mitteilungen für Sternfreunde. Archenhold-Sternwarte, Berlin-Treptow.

Bol. Acad. Cienc. Fis. Mat. Nat.
Boletin de la Academia de Ciencias Fisicas Matematicas y Naturales. Printed by Italgrafica, S.R. L. Republica de Venezuela.

Bol. Asoc. Argentina Astron.
Boletin de la Asociación Argentina de Astronomía, La Plata, Argentina. ISSN 0571-3285.

Bol. Astron.
Boletin Astronômico. Observatório do Capricórnio, Prefeitura Municipal de Campinas–SP, Brazil.

Bol. Astron. Obs. Madrid
Boletín Astronómico del Observatorio de Madrid. Instituto Geografico Nacional, General Ibáñez de Ibero, 3. Madrid 3. Spain.

Bol. Inst. Tonantzintla
Boletin del Instituto de Tonantzintla. Instituto Nacional de Astrofisica, Optica y Electronica, Apartados Postales Nos. 216 y 51, Puebla, Pue, Mexico.

Bol. Obs. Ebro
Boletín del Observatorio del Ebro, Tortosa. Printed by Cooperativa Gráfica Dertosense, Tortosa.

Boll. Geod. Sci. Affini
Bolletino di Geodesia e Scienze Affini. Pubblicazione dell'Istituto Geografico Militare, Firenze.

Boundary-Layer Meteorol.
Boundary-Layer Meteorology. D. Reidel Publishing Co., P.O. Box 17, Dordrecht, Netherlands.

British Astron. Assoc. Circ.
British Astronomical Association, Circular. Editorial Office: S. W. Milbourn, Brookhill Road, Copthorne Bank, Crawley, West Sussex, RH10 3QJ.

British J. Philos. Sci.
British Journal for the Philosophy of Science. Cambridge University Press, Bentley House, 200 Euston Road, London, NW1 2DB, England.

British J. Photogr.
British Journal of Photography. Henry Greenwood & Co., 24 Wellington Street, London, WC2E 7DH, England.

Bul. Inst. Politeh. 'Gheorghe Gheorghiu-Dej' Bucuresti.
Buletinul Institutului Politehnic 'Gheorghe Gheorghiu-Dej' Bucuresti.Calea Grivitei 132, Bucuresti, Rumania. Journal split into three series, Bul. Inst. Politeh. 'Gheorghe Gheorghiu-Dej' Bucuresti Ser. Chim. – Metal., Ser. Electroteh. and Ser. Mec. (Rumania).

Bulgarian J. Phys.
Bulgarian Journal of Physics. Bulgarian Academy of Sciences, Faculty of Physics, 5 Anton Ivanov Blvd., 1126 Sofia, Bulgaria.

Bull. Acad. Polonaise Sci., Ser. Sci. Tech.
Bulletin de l'Académie Polonaise des Sciences. Série des Sciences Techniques. 00-901 Warszawa, Palac Kultury i Nauki, P. O. Box 20, Poland.

Bull. AFOEV
Bulletin de l'Association Française des Observateurs d'Etoiles Variables. Rédaction et publication: E. Schweitzer, "La Moineaudière", 16, rue de Plobsheim, 67100 Strasbourg–Neudorf, France.

Bull. American Astron. Soc.
Bulletin of the American Astronomical Society. Published for the American Astronomical Society by the American Institute of Physics, 335 East 45th Street, New York, N.Y. 10017, USA.

Bull. Astron. Inst. Czechoslovakia
Bulletin of the Astronomical Institutes of Czechoslovakia. Published under the auspices of the Czechoslovak Academy of Sciences by Academia, Praha. Editor: Astronomical Institute of the Czechoslovak Academy of Sciences, Praha.

Bull. Astron., Obs. R. Belgique
Bulletin Astronomique, Observatoire Royal de Belgique. (Astronomisch Bulletin, Koninklijke Sterrenwacht van België).

Bull. Astron. Soc. India
Bulletin of the Astronomical Society of India. Published by S. K. Trehan (Editor) on behalf of the Astronomical Society of India, Osmania University, Hyderabad.

Bull. Crimean Astrophys. Obs.
Bulletin of the Crimean Astrophysical Observatory. A cover-to-cover translation of Izvestiya Krymskoj Astrofizicheskoj Observatorii. Published by Allerton Press, Inc.,150 Fifth Avenue, New York, N.Y. 10011, USA.

Bull. Géod.
Bulletin Géodésique. The Journal of the International Association of Geodesy. Publié par le Bureau Central de l'Association Internationale de Géodésie, 39 Rue Gay-Lussac, 75005 Paris, France.

Bull. Geophys.
Bulletin of Geophysics. National Central University, Chung-Li, Taiwan, Republic of China.

Bull. Inf. Cent. Données Stellaires
Bulletin d'Information du Centre de Données Stellaires. Compiled at Observatoire de Strasbourg, 11, rue de l'Université, 67000-Strasbourg, France.

Bull. Inst. Space Aeronaut. Sci. Univ. Tokyo A
Bulletin of the Institute of Space and Aeronautical Science, University of Tokyo A. Tokyo, Japan.

Bull. Inst. Space Aeronaut. Sci., Univ. Tokyo B
Bulletin of the Institute of Space and Aeronautical Science, University of Tokyo B. Tokyo, Japan.

Bull. Obs. Astron. Belgrade
Bulletin de l'Observatoire Astronomique de Belgrade. Editor: Observatoire Astronomique de Belgrade. Printed by Naucna delo, Belgrade.

Bull. Res. Inst. Sci. Meas. Tôhoku Univ.
Bulletin of the Research Institute for Scientific Measurements, Tôhoku University, Sendai, Japan.

Bull. Sci. Yougoslavie
Bulletin Scientifique. Conseil des Academies des Sciences et des Arts de la RSF de Yougoslavie. Section A: Sciences Naturelles, Techniques et Médicales. Rédaction et Administration: Opatička ul. 18/II, Zagreb, Yougoslavie.

Bull. Signal.
Bulletin Signalétique. Section 120: Astronomie, physique spatiale, geophysique. Centre Nationale de la Recherche Scientifique, Informascience, Centre de Documentation Scientifique et Technique, 26, rue Boyer, 75971 Paris. ISSN 0007-5337.

Bull. Soc. R. Sci. Liège
Bulletin de la Société Royale des Sciences de Liège. L'Université, 15 Avenue des Tilleurs, Liège, Belgium.

Bull. Spec. Astrophys. Obs. – North Caucasus
Bulletin of the Special Astrophysical Observatory – North Caucasus. A cover-to-cover translation of Astrofizicheskie Issledovaniya – Izvestiya Spetsial'noj Astrofizicheskoj Observatorii. Allerton Press, Inc., 150 Fifth Avenue, New York, N.Y. 10011, USA.

Bull. Tokyo Gakugei Univ., Ser. IV
Bulletin of Tokyo Gakugei University. Series IV (Mathematics and Natural Sciences) 4-1-1 Nukui-kita-machi, Koganei, Tokyo, Japan.

Bull. Yamagata Univ. (Nat. Sci.)
Bulletin of the Yamagata University (Natural Science). Yamagata, Japan.

Byull. Abastumanskaya Astrofiz. Obs.
Abastumanskaya Astrofizicheskaya Observatoriya, Gora Kanobili. Byulleten'. Akademiya Nauk Gruzinskoj SSR. Publishers: Izdatel'stvo "Metsniereba", Tbilisi.

Byull. Inst. Astrofiz.
Byulleten' Instituta Astrofiziki, Akademiya Nauk Tadzhikskoj SSR. Izdatel'stvo Donish, Dushanbe.

Byull. Inst. Teor. Astron.
Byulleten' Instituta Teoreticheskoj Astronomii. Izdatel'-stvo Nauka, Leningradskoe Otdelenie, Leningrad.

C. R. Acad. Sci. Paris
Comptes Rendus des Séances de l'Académie des Sciences. Série II: mécanique, physique, chimie, sciences de l'univers, sciences de la terre. Publiés par MM. les Secrétaire perpétuels. Gauthier-Villars, C.D.R., Centrale des Revues, B.P. No. 119, 93104 Montreuil Cedex, France. ISSN 0567-6541.

Canadian J. Earth Sci.
Canadian Journal of Earth Sciences. National Research Council of Canada, Ottawa KIA OR6, Canada.

Canadian J. Phys.
Canadian Journal of Physics. Published by the National Research Council of Canada, Ottawa. Printed in Canada by the University of Toronto Press, Toronto, Ont.

Carter Obs., Astron. Bull.
Carter Observatory, Astronomical Bulletin. Carter Observatory, P.O. Box 2909, Wellington 1, New Zealand.

Celestial Mech.
Celestial Mechanics. An International Journal of Space Dynamics. Publishers: D. Reidel Publishing Company, Dordrecht, Holland.

Cent. Astrophys. Prepr. Ser.
Center for Astrophysics, Preprint Series. Harvard College Observatory, Smithsonian Astrophysical Observatory. Center for Astrophysics, 60 Garden St., Cambridge, Mass. 02138.

Centaurus
Centaurus. International magazine of the history of mathematics, science, and technology. Munksgaard, Copenhagen.

Ceskoslovensky Cas. Fyz., Sekce A
Československý časopis pro fyziku. Sekce A. Academia Publishing House of the Czechoslovak Academy of Sciences, Vodičkova 40, 112 29 Praha 1, Czechoslovakia.

Chinese Astron. Astrophys.
Chinese Astronomy and Astrophysics. A selected translation of Acta Astron. Sinica and Acta Astrophys. Sinica. Pergamon Press, Oxford – New York – Toronto – Paris – Frankfurt – Sydney. ISSN 0146-6364.

Chinese J. Phys.
Chinese Journal of Physics. Physical Society of the Republic of China, Physics Department, National Taiwan University, Taipei, Taiwan, China.

Chinese Phys.
Chinese Physics. American Institute of Physics. 335 East 45th Street, New York, NY 10017, USA. Contains selected translations from current issues of major Chinese physics and astronomy journals.

Ciel
Le Ciel. Bulletin de la Société Astronomique de Liège. Éditeur responsable: Françoise Rameau, 19 - 21, rue des genêts, 4310 St. Nicolas.

Ciel Terre
Ciel et Terre. Bulletin de la Société Royale Belge d'Astronomie, de Météorologie et de Physique du Globe. Administration: Avenue Circulaire 3, B-1180 Bruxelles.

Circ. Czechoslovak Obs., Time and Latitude
Circular of the Czechoslovak Observatories, Time and Latitude. Czechoslovak Academy of Sciences, Astronomical Institute, Prague, Czechoslovakia.

Circ. Inf.
Circulaire d'Information. Union Astronomique Internationale. Commission des Etoiles Doubles. Address: Observatoire de Meudon, Meudon, France.

Circ. Stn. Astron. Int. Latitudine, Carloforte-Cagliari
Circolari della Stazione Astronomica Internazionale di Latitudine, Carloforte-Cagliari. Serie A printed by Tipo-Offset "3T", Cagliari. Serie B printed by Multi Copy, Milano.

Circ. Time and Latitude Serv.
Circular Time and Latitude Service. Polish Academy of Sciences, Astronomical Latitude Observatory, Borowiec, Poland.

Clim. Change
Climatic Change. D. Reidel Publishing Co., P.O. Box 17, 3300 AA Dordrecht, Netherlands.

CODATA Bull.
CODATA Bulletin. Committee on Data for Science and Technology of the International Council of Scientific Unions. CODATA Secretariat, 51 Boulevard de Montmorency, 75016 Paris, France. Pergamon Press, Oxford–New York–Toronto–Sydney–Paris–Frankfurt. ISSN 0366-757 X.

Coelum
Coelum. Periodico bimestrale per la Divulgazione dell' Astronomia. Editor: Osservatorio Astronomico Universitario di Bologna.

Comments Astrophys.
Comments on Astrophysics. A Journal of Critical Discussion of the Current Literature. Comments on Modern Physics: Part C. Publishers: Gordon and Breach, Science Publishers Ltd., 42 William IV Street, London WC2, England.

Comments At. Mol. Phys.
Comments on Atomic and Molecular Physics. Gordon & Breach Science Publishers Ltd., 41 and 42 William IV Street, London, WC2, England.

Comments Nucl. Part. Phys.
Comments on Nuclear and Particle Physics. Gordon & Breach Science Publishers Ltd., 41 and 42 William IV Street, London, WC2, England.

Comments Plasma Phys. Controlled Fusion
Comments on Plasma Physics and Controlled Fusion. Gordon & Breach Science Publishers Ltd., 41 and 42 William IV Street, London, WC2, England.

Commun. Fac. Sci. Univ. Ankara
Communications de la Faculté des Sciences de l'Université d'Ankara, Série A_2: Physique, Série A_3: Astronomie.

Commun. Konkoly Obs.
Communications from the Konkoly Observatory of the Hungarian Academy of Sciences, Budapest. HU ISSN 0324-2234.

Commun. Math. Phys.
Communications in Mathematical Physics, Springer-Verlag, Postfach 105280, 6900 Heidelberg 1, F.R. Germany.

Comput. Geosci.
Computers & Geosciences. Pergamon Press Ltd., Headington Hill Hall, Oxford OX3 0BW, England.

Comput. Phys. Commun.
Computer Physics Communications. North-Holland Publishing Co., P. O. Box 211 Amsterdam, Netherlands.

Comun. Obs. Astron. Univ. Coimbra
Comunicações do Observatório Astronomico da Universidade de Coimbra, Portugal.

Contemp. Phys.
Contemporary Physics.Taylor and Francis Ltd., 10 - 14 Macklin Street, London, WC2B 5NF, England.

Contrib. Astron. Obs. Skalnaté Pleso.
Contributions of the Astronomical Observatory Skalnaté Pleso. VEDA, vydavatel'stvo Slovenskej akadémie vied, Bratislava, Czechoslovakia.

Contrib. Atmos. Phys.
Contributions to Atmospheric Physics – Beiträge zur Physik der Atmosphäre. Publisher: Friedrich Vieweg & Sohn, Braunschweig.

Contrib. Bosscha Obs.
Contributions from the Bosscha Observatory. Bandung Institute of Technology, Department of Science, Lembang, Indonesia.

Contrib. Geophys. Inst. Slovak Acad. Sci.
Contributions of the Geophysical Institute of the Slovak Academy of Sciences. Dubravska 9, 842 28 Bratislava, Czechoslovakia.

Czechoslovak J. Phys. B
Czechoslovak Journal of Physics, Section B. Czechoslovak Academy of Science, Akademia, Vodičkova 40, 112 29 Praha 1, Czechoslovakia.

Data Rep. Hydrogr. Obs., Ser. Astron. Geod., Tokyo
Data Report of Hydrographic Observations. Series of Astronomy and Geodesy. Maritime Safety Agency, Hydrographic Department Tsukiji-5, Chuo-ku, Tokyo, 104 Japan.

Debrecen Heliophys. Obs. Hungarian Acad. Sci., Repr.
Debrecen Heliophysical Observatory of the Hungarian Academy of Sciences, Reprints.

Def. Sci. J.
Defence Science Journal. Metcalfe House, Delhi 110054, India.

Deutsche Geod. Komm. Bayerisch. Akad. Wiss.
Deutsche Geodätische Kommission bei der Bayerischen Akademie der Wissenschaften. Reihe A: Höhere Geodäsie; Reihe B: Angewandte Geodäsie; Reihe C: Dissertationen; Reihe D: Tafelwerke; Reihe E: Geschichte und Entwicklung der Geodäsie. Published by Verlag der Bayerischen Akademie der Wissenschaften, München.

Dimensions NBS
Dimensions NBS. U.S. Department of Commerce, Washington, DC 20234, USA.

Dokl. Akad. Nauk SSSR
Doklady Akademii Nauk SSSR. Seriya Matematika, Fizika. Publishers: Izdatel'stvo "Nauka", Moskva.

Dokl. Bolg. Akad. Nauk
Doklady Bolgarskoj Akademii Nauk. Sofiya, Bulgaria.

Dudley Obs. Rep.
Dudley Observatory Reports. Dudley Observatory, Schenectady, N.Y., USA.

Earth Planet. Sci. Lett.
Earth and Planetary Science Letters. A Letter Journal devoted to the Development in Time of the Earth and Planetary System. Publisher: North-Holland Publishing Company, Amsterdam, Netherlands.

Earth-Sci. Rev.
Earth-Science Reviews. Elsevier Scientific Publishing Co., P. O. Box 330, 1000 AH Amsterdam, Netherlands.

Electro-Opt. Syst. Des.
Electro-Optical Systems Design. Milton S. Kiver Publications Inc., 222 West Adams, Chicago, IL 60606, USA.

Electron. Lett.
Electronics Letters. Institution of Electrical Engineers, Savoy Place, London, WC2R 0BL, England.

Electronics
Electronics. McGraw-Hill Publishing Co., 1221 Avenue of the Americas, New York, N.Y. 10020, USA.

Elektor
Elektor. Elektor Publishers Ltd., Elektor House, 10 Longport, Canterbury CT1 1PE, England.

Elektroteh. Vestn.
Elektrotehniski Vestnik. Editorial address: YU-61001 Ljubljana, P. O. Box 92 - II, Trzaska 25, Yugoslavia.

Endeavour New Ser.
Endeavour New Series. Pergamon Press Ltd., Headington Hill Hall, Oxford, OX3 0BW, England.

EOS Trans. American Geophys. Union
EOS Transactions of the American Geophysical Union. 1707 L Street, N.W., Washington, DC 20036, USA.

ESA Bull.
ESA Bulletin. Editorial Office: ESA Scientific and Technical Publications Branch, ESTEC, Noordwijk, The Netherlands. ISSN 0376-4265.

ESA IUE Newsl.
ESA IUE Newsletter. Published by The ESA IUE Observatory, Villafranca Satellite Tracking Station, Apartado 54065, Madrid, Spain.

ESA J.
ESA Journal. Editorial Office: ESA Scientific and Technical Publications Branch, ESTEC, Noordwijk, The Netherlands. ISSN 0379-2285.

ESO Sci. Prepr.
European Southern Observatory, Scientific Preprints. Published by European Southern Observatory, Karl-Schwarzschild-Straße 2, D-8046 Garching bei München.

ESO Tech. Rep.
European Southern Observatory, Technical Report. Published by European Southern Observatory, Karl-Schwarzschild-Straße 2, D-8046 Garching bei München.

European J. Phys.
European Journal of Physics. Institute of Physics, 47 Belgrave Square, London SW1X 8QX.

Europhys. News
Europhysics News. European Physical Society, P. O. Box 69, CH-1213 Petit-Lancy 2, Switzerland.

Exp. Tech. Phys.
Experimentelle Technik der Physik, VEB Deutscher Verlag der Wissenschaften, Traubenstrasse 10, 108 Berlin 8, German Democratic Republic.

Feingerätetechnik
Feingerätetechnik. VEB Verlag Technik, Oranienburger Strasse 13/14, 1020 Berlin, DDR.

Feinwerktech. Messtech.
F & M. Feinwerktechnik und Messtechnik. Fusion of "Feinwerktechnik" and "Messtechnik" (formerly Zeitschrift für Instrumentenkunde) beginning with Jahrgang 82, No. 5 (1974). Publishers: Karl Hanser Verlag, Kolbergerstr. 22, D-8000 München 80. F.R.Germany.

Fis. Tecnol.
Fisica e Tecnologia. Societa Italiana di Fisica, Via Loderingo Degli Andalo 2, 40124 Bologna, Italy.

Fiz. Sz.
Fizikai Szemle. Kiadja a Lapkiado Vallalat, Budapest VII, Lenin korut 9–11, Hungary.

Fizika
Fizika. 'Mladost' Export-Import, Zagreb, Ilica 30, Yugoslavia.

Folia Fac. Sci. Nat. Univ. Purkynianae Brunensis Phys.
Folia Facultatis Scientiarum Naturalium Universitatis Purkynianae Brunensis, Physica. University J. E. Purkyne, 61137 Brno-Kotlarska 2, Czechoslovakia.

Fortschr. Phys.
Fortschritte der Physik. Akademie-Verlag. DDR-108 Berlin, Leipzigerstrasse 3 - 4, Germany.

Found. Phys.
Foundations of Physics. Plenum Publishing Co., 8 Scrubs Lane, Harlesden, London, NW10 6SE, England.

Fra Fys. Verden
Fra Fysikkens Verden. Fysisk Institutt, Universitetet i Trondheim, Norges Laererhogskole, 7000 Trondheim, Norway.

Fujitsu Sci. Tech. J.
Fujitsu Scientific and Technical Journal. Fujitsu Ltd., 1015 Kamikodanaka, Nakahara-ku, Kawasaki 211, Kanagawa, Japan.

Fundam. Cosmic Phys.
Fundamentals of Cosmic Physics. Gordon and Breach Science Publishers Ltd., New York–London–Paris.

Funkschau
Funkschau. Francis-Verlag, 8 München 37, Postfach 37 01 20, Karlstrasse 37, Germany.

Fys. Tidsskr.
Fysisk Tidsskrift. Subscription address: Jul. Gjellerups Boghandel, Solvgade 87, 1307 Kobenhavn, Denmark.

G. A.A.B.
Giornale dell'A.A.B. Notiziario trimestrale delle attività culturali e scientifiche della Associazione Astrofili Bolognesi, Bologna, Italy.

G. Astron.
Giornale di Astronomia, Pubblicazione della Società Astronomica Italiana. Printed by Tipolitografia Lodigraf S.p.A. Lodi (MI).

Gen. Relativ. Gravitation
General Relativity and Gravitation. Published under the auspices of the International Committee on General Relativity and Gravitation GRG. Publishing Office: Plenum Publishing Corporation, 233 Spring Street, New York, N. Y. 10013, USA.

Geochim. Cosmochim. Acta
Geochimica et Cosmochimica Acta. Journal of the Geochemical Society and the Meteoritical Society. Publishing House: Pergamon Press, Ltd., Oxford.

Geod. Geophys. Veröff., Reihe III
Geodätische und Geophysikalische Veröffentlichungen. Reihe III: Physik der festen Erde. Herausgegeben vom Nationalkomitee für Geodäsie und Geophysik bei der Akademie der Wissenschaften der Deutschen Demokratischen Republik.

Geod. Kartogr.
Geodezja i Kartografia. Komitet Geodezji Polskiej Akademii Nauk. Publisher: Państwowe Wydawnictwo Naukowe, Warszawa.

Geol. Rundsch.
Geologische Rundschau. Ferdinand Enke Verlag, Stuttgart.

Geomagn. Aehron.
Geomagnetizm i Aehronomiya. Akademiya Nauk SSSR. Izdatel'stvo "Nauka", Moskva [An English translation is published in "Geomagnetism and Aeronomy", American Geophysical Union, Washington, D.C.].

Geomagn. Ser. Earth Phys. Branch
Geomagnetic Series, Earth Physics Branch. Energy, Mines and Resources Canada, 1 Observatory Crescent, Ottawa K1A OE4, Canada.

Geophys. Astrophys. Fluid Dyn.
Geophysical and Astrophysical Fluid Dynamics. Gordon and Breach Science Publishers Ltd., 41/42 William IV Street, London, WC2, England.

Geophys. J. R. Astron. Soc.
The Geophysical Journal of the Royal Astronomical Society. Published for the Royal Astronomical Society by Blackwell Scientific Publications, Oxford–Edinburgh.

Geophys. Prospect.
Geophysical Prospecting. European Association of Exploration Geophysicists, P. O. Box 162, 2501 AN The Hague, Netherlands.

Geophys. Res. Lett.
Geophysical Research Letters. Published monthly by the American Geophysical Union, Washington, D.C., U.S.A.

Geophys. Surv.
Geophysical Surveys. D. Reidel Publishing Co., P.O. Box 17, Dordrecht, Netherlands.

Geophysics
Geophysics. Society of Exploration Geophysicists, P.O. Box 3098, Tulsa, OK 74101, USA.

GEOS
GEOS. Department of Energy, Mines and Resources, Ottawa, Canada.

GEOS Circ.
GEOS (Groupe: Etude et Observation Stellaire and Gruppo Europeo di Osservazione Stellare) Circulars, Series: RR (RR Lyrae type variables), SR (red variables), EB (eclipsing binaries), SA (small-amplitude variables). Published by A. Figer, GEOS, 12 rue Bezout, 75014 Paris, France.

Gerlands Beitr. Geophys.
Gerlands Beiträge zur Geophysik. Publisher: Akademische Verlagsgesellschaft Geest & Portig K.-G., Leipzig.

Glasnik Mat.
Glasnik Matematicki. Published by the Society of Mathematicians and Physicists of the S. R. of Croatia. Publisher: Drustvo Matematicara i Fizicara S. R. Hrvatske, Zagreb.

Heavens
The Heavens. The Oriental Astronomical Association, Ōtsu-shi, Shiga-ken, Japan. In Japanese.

Helvetica Phys. Acta
Helvetica Physica Acta. Schweizerische Physikalische Gesellschaft. Publisher: Birkhäuser Verlag, Elisabethenstrasse 19, CH-4000 Basel 10, Switzerland.

HHI Sol. Data
HHI Solar Data. Heinrich-Hertz-Institut, Solare Beobachtungsergebnisse. Akademie der Wissenschaften der DDR, Zentralinstitut für Solar-Terrestrische Physik (Heinrich-Hertz-Institut), DDR-1199 Berlin-Adlershof.

HHI-STP-Rep.
HHI Solar-Terrestrische Physik Reports. Heinrich-Hertz-Institut. Akademie der Wissenschaften der DDR, Zentralinstitut für Solar-Terrestrische Physik (Heinrich-Hertz-Institut), DDR-1199 Berlin-Adlershof.

Hvar Obs. Bull.
Hvar Observatory Bulletin. Faculty of Geodesy. 41000 Zagreb, Kačićeva 26, Yugoslavia. ISSN 0351-2651.

Hydraul. Pneum.
Hydraulics and Pneumatics. Industrial Publishing Co., 614 Superior Avenue West, Cleveland, OH 44113, USA.

I. A. P. P. P. Commun.
International Amateur-Professional Photoelectric Photometry, Communication. R. M. Genet, Fairborn Observatory, 1247 Folk Road, Fairborn, Ohio 45324, USA.

I.U.A.A. Bull.
I.U.A.A. Bulletin. International Union of Amateur Astronomers, Contributions. I.U.A.A. c/o Achille Leani, via Bertesi 15, 26100 Cremona, Italy.

IAU Circ.
International Astronomical Union, Circular. Central

Bureau for Astronomical Telegrams, Smithsonian Astrophysical Observatory, Cambridge, Mass.

IAU Inf. Bull.
International Astronomical Union Information Bulletin, Published by D. Reidel Publishing Company, P. O. Box 17, 3300 AA Dordrecht, Holland.

Icarus
Icarus. International Journal of Solar System Studies. Publisher: Academic Press, New York – London.

IEE J. Microwave Opt. Acoust.
IEE Journal on Microwave, Optics and Acoustics. Institution of Electrical Engineers, Publishing Department, P.O. Box 8, Southgate House, Stevenage, Herts. SG1 1HQ, England.

IEEE J. Solid-State Circuits
IEEE Journal of Solid-State Circuits. Institute of Electrical and Electronics Engineers, 345 East 47th Street, New York, NY 10017, USA.

IEEE Spectrum
IEEE Spectrum. Published monthly by the Institute of Electrical and Electronics Engineers, 345 East 47th Street New York, N.Y. 10017, USA.

IEEE Trans. Aerosp. Electron. Syst.
IEEE Transactions on Aerospace and Electronic Systems. Published by the Institute of Electrical and Electronics Engineers, 345 East 47th Street, New York, N.Y. 10017, USA.

IEEE Trans. Antennas Propag.
IEEE Transactions on Antennas and Propagation. Published by the Institute of Electrical and Electronics Engineers, 345 East 47th Street, New York, N.Y. 10017, USA.

IEEE Trans. Commun.
IEEE Transactions on Communications. Institute of Electrical and Electronics Engineers, 345 East 47th Street, New York, NY 10017, USA.

IEEE Trans. Consum. Electron.
IEEE Transactions on Consumer Electronics. Institute of Electrical and Electronics Engineers, 345 East 47th Street, New York, N. Y. 10017, USA.

IEEE Trans. Electromagn. Compat.
IEEE Transactions on Electromagnetic Compatibility. Institute of Electrical and Electronics Engineers, 345 East 47th Street, New York, NY 10017, USA.

IEEE Trans. Electron Devices
IEEE Transactions on Electron Devices. Published by the Institute of Electrical and Electronics Engineers, 345 East 47th Street, New York, N. Y. 10017, USA.

IEEE Trans. Geosci. Electron.
IEEE Transactions on Geoscience Electronics. Published by the Institute of Electrical and Electronics Engineers, 345 East 47th Street, New York, N.Y. 10017, USA.

IEEE Trans. Ind. Electron.
IEEE Transactions on Industrial Electronics. Institute of Electrical and Electronics Engineers, 345 East 47th Street, New York, NY 10017, USA.

IEEE Trans. Instrum. Meas.
IEEE Transactions on Instrumentation and Measurement. Published by the Institute of Electrical and Electronics Engineers, 345 East 47th Street, New York, N.Y. 10017, USA.

IEEE Trans. Magn.
IEEE Transactions on Magnetics. Institute of Electrical and Electronics Engineers, 345 East 47th Street, New York, N. Y. 10017, USA.

IEEE Trans. Microwave Theory Tech.
IEEE Transactions on Microwave Theory and Techniques. Published by the Institute of Electrical and Electronics Engineers, 345 East 47th Street, New York, N.Y. 10017, USA.

IEEE Trans. Nucl. Sci.
IEEE Transactions on Nuclear Science. Institute of Electrical and Electronics Engineers, 345 East 47th Street, New York, N.Y. 10017, USA.

IEEE Trans. Plasma Sci.
IEEE Transactions on Plasma Science. Institute of Electrical and Electronics Engineers, 345 East 47th Street, New York, NY 10017, USA.

Indian J. Hist. Sci.
Indian Journal of History of Science. Published and printed by Indian National Science Academy, Bahadur Shah Zafar Marg, New Delhi 110002, at Mudranika, 13-A Bepin Pal Road, Calcutta 700026.

Indian J. Phys. Part B
Indian Journal of Physics Part B. Indian Association for the Cultivation of Science, 2 & 3 Raja Subodh Chandra Mallik Road, Calcutta 700032, India.

Indian J. Pure Appl. Math.
Indian Journal of Pure and Applied Mathematics. National Institute of Sciences India, Bahadur Shah Zafar Marg, New Delhi 1, India.

Indian J. Pure Appl. Phys.
Indian Journal of Pure and Applied Physics. Council of Scientific and Industrial Research, Hillside Road, New Delhi 110012, India.

Indian J. Radio Space Phys.
Indian Journal of Radio & Space Physics. Council of Scientific & Industrial Research. Editorial address: Publications & Information Directorate, Hillside Road, New Delhi 110012, India.

Indian J. Theor. Phys.
Indian Journal of Theoretical Physics. Institute of Theoretical Physics, Bognan Kutir, 4 - 1 Mohan Bagan Lane, Calcutta 700004, India

Inf. Bull. Variable Stars
Commission 27 of the I.A.U. Information Bulletin on Variable Stars. Konkoly Observatory, Budapest.

Infrared Phys.
Infrared Physics. An International Research Journal. Publisher: Pergamon Press Ltd., Oxford, England.

Inst. Theor. Astrophys., Blindern–Oslo, Rep.
Institute of Theoretical Astrophysics, Blindern–Oslo, Report. Universitetsforlagets trykningssentral, Oslo.

Int. Comet Q.
The International Comet Quarterly, Physics Department, Appalachian State University, Boone, NC 28608.

Int. Halley Watch Newsl.
The International Halley Watch Newsletter. Jet Propulsion Laboratory, California Institute of Technology, MS T-1166, 4800 Oak Grove Drive, Pasadena, Calif. 91109, USA.

Int. J. Electron.
International Journal of Electronics. Taylor and Francis Ltd., 10–14 Macklin Street, London, WC2B 5BF, England.

Int. J. Eng. Sci.
International Journal of Engineering Science. Pergamon Press Ltd., Headington Hill Hall, Oxford, OX3 0BW, England.

Int. J. Gen. Syst.
International Journal of General Systems. Gordon & Breach Science Publishers Ltd., 42 William IV Street, London WC2N 4DE, England.

Int. J. Heat Mass Transfer
International Journal of Heat and Mass Transfer, Pergamon Press Ltd., Headington Hill Hall, Oxford, OX3 0BW, England.

Int. J. Infrared Millim. Waves
International Journal of Infrared and Millimeter waves. Plenum Publishing Corp., 227 West 17th Street, New York, NY 10011, USA.

Int. J. Mass Spectrom. Ion Phys.
International Journal of Mass Spectrometry and Ion Physics. Elsevier Scientific Publishing Co., P.O. Box 211 Amsterdam, Netherlands.

Int. J. Theor. Phys.
International Journal of Theoretical Physics. Plenum Publishing Co. Ltd., Davis House, 8 Scrubs Lane, London, NW10 6SE, England.

Interdisciplinary Sci. Rev.
Interdisciplinary Science Reviews. Heyden & Son Ltd., Spectrum House, Alderton Crescent, London NW4 3XX, England.

Irish Astron. J.
The Irish Astronomical Journal. A Quarterly Publication under the auspices of the Observatories of Armagh and Dunsink. Armagh Observatory, Northern Ireland.

ISIS
ISIS. An international review devoted to the history of science and its cultural influences. Publication and Editorial Office, Department of History and Sociology of Science, University of Pennsylvania, Philadelphia 19104.

Istanbul Üniv. Fen Fak. Mec., Ser. C
Istanbul Üniversitesi Fen Fakultesi Mecmuasi, Serie C (Astronomie, Physique, Chimie). Istanbul, Turkey.

Izv. Akad. Nauk Armyansk. SSR
Izvestiya Akademii Nauk Armyanskoj SSR. Fizika. Publisher: Izdatel'stvo AN Armyanskoj SSR, Erevan.

Izv. Astron. Ehngel'gardt. Obs.
Izvestiya Astronomicheskoj Ehngel'gardtovskoj Observatorii. Izdatel'stvo Kazanskogo Universiteta, Kazan.

Izv. Glav. Astron. Obs. Pulkovo
Izvestiya Glavnoj Astronomicheskoj Observatorii v Pulkove. Akademiya Nauk SSSR. Izdanie Glavnoj astronomicheskoj observatorii v Pulkove, Leningrad.

Izv. Krymskoj Astrofiz. Obs.
Izvestiya Krymskoj Astrofizicheskoj Observatorii. Akademiya Nauk SSSR. Publishers: Izdatel'stvo "Nauka", Moskva. [An English translation is published in Bulletin of the Crimean Astrophysical Observatory].

J. Acoust. Soc. America
Journal of the Acoustical Society of America. Published for the Acoustical Society of America by the American Institute of Physics, 335 East 45th Street, New York, NY 10017, USA

J. American Assoc. Variable Star Obs.
The Journal of the American Association of Variable Star Observers. Published by The American Association of Variable Star Observers, 187 Concord Avenue, Cambridge, Mass. 02138, USA.

J. Appl. Meteorol.
Journal of Applied Meteorology. American Meteorological Society, 45 Beacon Street, Boston, MA 02108, USA.

J. Appl. Photogr. Eng.
Journal of Applied Photographic Engineering. Society of Photographic Scientists and Engineers, Suite 204, 1330 Massachusetts Avenue, N.W. Washington, D.C. 20005, USA.

J. Appl. Phys.
Journal of Applied Physics. American Institute of Physics, 335 East 45th Street, New York, NY 10017, USA.

J. Astron. Soc. Egypt
Journal of the Astronomical Society of Egypt. Published by Astronomical Society of Egypt, Astronomy Department, Faculty of Sciences, Cairo University, Egypt.

J. Astron. Soc. Western Australia
The Journal of the Astronomical Society of Western Australia. Edited by the Astronomical Society of Western Australia, Perth, W. A.

J. Astronaut. Sci.
Journal of the Astronautical Sciences. American Astronautical Society, 6060 Duke Street, Alexandria, VA 22304, USA.

J. Astrophys. Astron.
Journal of Astrophysics and Astronomy. Published by Indian Academy of Sciences, Post Box No. 8005, Bangalore 560080, India.

J. Atmos. Sci.
Journal of the Atmospheric Sciences. American Meteorological Society, 45 Beacon Street, Boston, MA 02108, USA.

J. Atmos. Terr. Phys.
Journal of Atmospheric and Terrestrial Physics. Pergamon Press Ltd., Oxford, England.

J. British Astron. Assoc.
Journal of the British Astronomical Association. Burlington House, Piccadilly, London, W1V 0NL, England.

J. British Interplanet. Soc.
Journal of the British Interplanetary Society. British

Interplanetary Society, 12 Bessborough Gardens, London SW1V 2JJ, England.

J. Colloid Interface Sci.
Journal of Colloid and Interface Science. Academic Press Inc., 111 Fifth Avenue, New York, N.Y. 10003, USA.

J. Comput. Phys.
Journal of Computational Physics. Academic Press Inc., 111 Fifth Avenue, New York, NY 10003, USA.

J. Electron Spectrosc. Relat. Phenom.
Journal of Electron Spectroscopy and Related Phenomena. Elsevier Scientific Publishing Co., P.O. Box 330, 1000 AH Amsterdam, Netherlands.

J. Electrostat.
Journal of Electrostatics. Elsevier Scientific Publishing Co., P.O.Box 330, 1000 AH Amsterdam, Netherlands.

J. Fluid Mech.
Journal of Fluid Mechanics. Cambridge University Press Bentley House, 200 Euston Road, London, NW1 2DB, England.

J. Geomagn. Geoelectr.
Journal of Geomagnetism and Geoelectricity. Society of Terrestrial Magnetism and Electricity of Japan, Geophysical Institute, Tokyo University, Tokyo 113, Japan.

J. Geophys.
Journal of Geophysics / Zeitschrift für Geophysik. Springer Verlag, D-6900 Heidelberg 1, Postfach 105280, F. R. Germany.

J. Geophys. Res.
Journal of Geophysical Research. Published by American Geophysical Union, 1909K Street, N.W. Washington D.C. First section: Space physics; Second section: Physics and chemistry of the solid earth, planetology, geodesy; Third section: Oceans and atmospheres.

J. Guid. Control
Journal of Guidance and Control. American Institute of Aeronautics and Astronautics, 1290 Avenue of the Americas, New York, NY 10019, USA.

J. Hist. Astron.
Journal for the History of Astronomy. Published by Science History Publications Ltd., Halfpenny Furze, Mill Lane, Chalfont St Giles, Buckinghamshire, England.

J. Inorg. Nucl. Chem.
Journal of Inorganic and Nuclear Chemistry. Pergamon Press Ltd., Headington Hill Hall, Oxford OX3 0BW, England.

J. Inst. Electron. Telecommun. Eng.
Journal of the Institution of Electronics and Telecommunication Engineers. Institution of Electronics & Telecommunication Engineers, 2 Institutional Area, Lodi Road, New Delhi 110003, India.

J. Inst. Math. Appl.
Journal of the Institute of Mathematics and its Applications. Academic Press Inc. (London) Ltd., 24 - 28 Oval Road, London NW1 7DX, England.

J. Korean Astron. Soc.
The Journal of the Korean Astronomical Society. Published by the Korean Astronomical Society.

J. Magn. Magn. Mater.
Journal of Magnetism and Magnetic Materials. North-Holland Publishing Co., P.O. Box 211, Amsterdam, Netherlands.

J. Math. Phys.
Journal of Mathematical Physics. American Institute of Physics, 335 East 45th Street, New York, N.Y. 10017, USA.

J. Meteor Res.
The Journal of Meteor Research. Space News Publishing Co., P. O. Box 66521, Baton Rouge, LA 70896. ISSN 0277-6057.

J. Mol. Spectrosc.
Journal of Molecular Spectroscopy. Academic Press Inc., 111 Fifth Avenue, New York, NY 10003, USA.

J. Nanjing Univ.
Journal of Nanjing University. Nanjing Daxue Xuebao. (Natural Science Edition). Nanking University, Nanking, China.

J. Navig.
The Journal of Navigation. The Royal Institute of Navigation at the Royal Geographical Society, Kensington Gore, London, SW7 2AT. Scottish Academic Press Ltd., 33 Montgomery Street, Edinburgh EH7 5JX.

J. Opt. *(France)*
Journal of Optics. Masson Editeur, 120 Boulevard Saint-Germain, 75280 Paris Cedex 06, France.

J. Opt. *(India)*
Journal of Optics. Optical Society of India, Department of Applied Physics, University of Calcutta, 92 Acharya Prafulla Chandra Road, Calcutta-9, India.

J. Opt. Soc. America
Journal of the Optical Society of America. American Institute of Physics, 335 East 45th Street, New York, N.Y. 10017, USA.

J. Phys. A
Journal of Physics A, (Mathematical, Nuclear and General). Institute of Physics, 47 Belgrave Square, London, SW1X 8QX, England.

J. Phys. B
Journal of Physics B, (Atomic and Molecular Physics). Institute of Physics, 47 Belgrave Square, London, SW1X 8QX, England.

J. Phys. Chem. Ref. Data
Journal of Physical and Chemical Reference Data. American Chemical Society, 1155 Sixteenth Street, N.W., Washington, DC 20036, USA.

J. Phys. Colloq.
Journal de Physique Colloque. Société Française de Physique, 87 bis Avenue du Général Leclerc, 75014 Paris, France.

J. Phys. E
Journal of Physics E, (Scientific Instruments). Formerly: J. Sci. Instrum. (GB). Institute of Physics, 47 Belgrave Square, London, SW1X 8QX, England.

J. Phys. F
Journal of Physics F, (Metal Physics). Institute of Physics, 47 Belgrave Square, London, SW1X 8QX, England.

J. Phys. G
Journal of Physics G, (Nuclear Physics). Institute of Physics, 47 Belgrave Square, London, SW1X 8QX, England.

J. Phys. Soc. Japan
Journal of the Physical Society of Japan. Room 211, Kikai Shinko Building, Shiba Koen, Minato-ku, Tokyo 105, Japan.

J. Physique
Journal de Physique. Z. I. de Courtaboeuf, B. P. 112, 91402 Orsay, France.

J. Plasma Phys.
Journal of Plasma Physics. Cambridge University Press, Bentley House, 200 Euston Road, London, NW1 2DB, England.

J. Proc. R. Soc. New South Wales
Journal and Proceedings of the Royal Society of New South Wales. Science Centre, 35 Clarence Street, Sydney, N.S.W. 2000, Australia

J. Quant. Spectrosc. Radiat. Transfer
Journal of Quantitative Spectroscopy & Radiative Transfer. Pergamon Press Ltd., Headington Hill Hall, Oxford, OX3 OBW, England.

J. R. Astron. Soc. Canada
The Journal of the Royal Astronomical Society of Canada, devoted to the advancement of astronomy and allied sciences. The Royal Astronomical Society of Canada, 124 Merton Street, Toronto, Ontario, Canada.

J. Radioanal. Chem.
Journal of Radioanalytical Chemistry. Elsevier Sequoia S. A., P. O. Box 851, 1001 Lausanne 1, Switzerland. Kultura, H-1389 Budapest, 62, P. O. Box 149, Hungary.

J. Res. Natl. Bur. Stand. B
Journal of Research of the National Bureau of Standards. Section B (Mathematics and Mathematical Physics). US Government Printing Office, Division of Public Documents, Washington, DC 20402, USA.

J. Sci. Ind. Res.
Journal of Scientific and Industrial Research. Sales & Distribution Office, Publications & Information Directorate, Hillside Road, New Delhi 110012, India.

J. Sci. Res. Banaras Hindu Univ.
Journal of Scientific Research of the Banaras Hindu University. PO-Banaras Hindu University, India.

J. Soc. Instrum. Control Eng.
Journal of the Society of Instrument and Control Engineers. Hongo 1-35-28-303, Bunkyo-ku, Tokyo 113, Japan.

J. Spacecr. Rockets
Journal of Spacecraft and Rockets. American Institute of Aeronautics and Astronautics, 1290 Avenue of the Americas, New York, N. Y. 10019, USA.

J. Spectrosc. Soc. Japan
Journal of the Spectroscopical Society of Japan. (Bunkyo Kenkyu). 2-15-1, Nakai, Shinjuku-ku, Tokyo 161, Japan.

J. Stat. Phys.
Journal of Statistical Physics. Plenum Publishing Corp., 227 West 17th Street, New York, NY 10011, USA.

J. Toyo Univ., Gen. Educ., Nat. Sci.
Journal of the Toyo University, General Education, Natural Science. Published by The Toyo University, 28, Hakusan 5-chôme, Bunkyo-ku, Tokyo, Japan.

Japanese J. Appl. Phys.
Japanese Journal of Applied Physics. Publication Office, 2nd Toya Kaiji Building, 24-8 Shinbashi, Minato-ku, Tokyo 105, Japan.

Jenaer Rundsch.
Jenaer Rundschau. Publisher: VEB Verlag Technik, DDR-1020 Berlin, Oranienburger Str. 13 - 14, German Democratic Republic.

JETP Lett.
JETP Letters. A translation of JETP Pis'ma v Redaktsiyu of the Academy of Sciences in the USSR. American Institute of Physics, 335 East 45th Street, New York, NY 10017, USA.

Johns Hopkins APL Tech. Dig.
Johns Hopkins APL Technical Digest. The Johns Hopkins University, Applied Physics Laboratory, Johns Hopkins Road, Laurel, Md. 20707. ISSN 0270-5214.

Kexue Tongbao
Kexue Tongbao. Academia Sinica, Peking, People's Republic of China [English translation in: Kexue Tongbao (Scientia)(USA)].

Kodaikanal Obs. Bull.
Kodaikanal Observatory Bulletins, Series A, Indian Institute of Astrophysics, Bangalore, India.

Komet. Tsirk.
Kometnyj Tsirkulyar. Gruppa po Issledovaniyu Komet Astrosoveta i Mezhduvedomstvennyj Geofizicheskij Komitet Akademii Nauk SSSR. Kievskij Universitet im. T. G. Shevchenko.

Komety i Meteory
Komety i Meteory. Akademiya Nauk Tadzhikskoj SSR. Astronomicheskij Sovet Akademii Nauk SSSR. Publishers: Izdatel'stvo "Donish", Dushanbe.

Kosm. Issled.
Kosmicheskie Issledovaniya. Akademiya Nauk SSSR. Publishers: Izdatel'stvo "Nauka", Moskva [An English translation is published as "Cosmic Research", Consultants Bureau, New York, N. Y.].

Kozmos
Kozmos. Popular Astronomical Journal. Publisher: Slovenské ústredie amatérskej astronomie, Hurbanovo.

Lett. Math. Phys.
Letters in Mathematical Physics. D. Reidel Publishing Co., P.O. Box 17, Dordrecht, Netherlands.

Lett. Nuovo Cimento
Lettere al Nuovo Cimento. Società Italiana di Fisica. Editrice Compositori, viale XII Giugno 1, 40124 Bologna, Italy.

L'Universo
L'Universo. Rivista dell'Istituto Geografico Militare.

Direzione, Redazione e Amministrazione: Istituto Geografico Militare, Firenze.

Mada
Mada (Science). Published by The Weizmann Science Press of Israel, Jerusalem.

Magy. Geofiz.
Magyar Geofizika. Lapkiado Vallalat, 1073 Budapest, Lenin korut 9 - 11, Hungary

Mat.-Fys. Medd.
Matematisk-fysiske Meddelelser. Published by Det Kongelige Danske Videnskabernes Selskab. Sold by Munksgaard, Nørre Søgade 35, DK-1370 Copenhagen K., Denmark. ISSN 0023-3323.

Math. Intelligencer
The Mathematical Intelligencer. Springer-Verlag, Berlin–Heidelberg–New York, 175 Fifth Avenue, New York, NY 10010, USA.

Math. Proc. Cambridge Philos. Soc.
Mathematical Proceedings of the Cambridge Philosophical Society. Formerly: Proceedings of the Cambridge, Philosophical Society (Mathematical and Physical Sciences). Cambridge University Press, Bentley House, 200 Euston Road, London, NW1 2DB, England.

Meded. K. Acad. Wet. Lett. Schone Kunsten België
Mededelingen van de Koninklijke Academie voor Wetenschappen, Letteren en Schone Kunsten van België. Paleis der Academien, Hertogsstraat 1, B-1000 Bruxelles, Belgium.

Mem. Astron. Soc. India
Memoirs of the Astronomical Society of India. Edited and published by M.S. Vardya, Tata Institute of Fundamental Research, Bombay 400005 on behalf of the Astronomical Society of India, Osmania University, Hyderabad 500007.

Mem. Fac. Eng. Kyoto Univ.
Memoirs of the Faculty of Engineering, Kyoto University, Kyoto, Japan.

Mem. Fac. Eng. Osaka City Univ.
Memoirs of the Faculty of Engineering, Osaka City University. 459 Sugimoto-cho, Sumi Yoshi-kum, Osaka, Japan.

Mem. Fac. Sci. Kyoto Univ.
Memoirs of the Faculty of Science, Kyoto University. Series of Physics, Astrophysics, Geophysics, and Chemistry. Printed by Yamashiro Printing Publishing Co. Ltd., Kamigyo, Kyoto.

Mem. Japan Astron. Study Assoc.
Memoirs of the Japan Astronomical Study Association. c/o National Science Museum, Ueno Park, Taito-ku, Tokyo, Japan.

Mem. Soc. Astron. Italiana
Memorie della Società Astronomica Italiana. Presso Laboratorio di Astrofisica Spaziale, Castella Postale 67, 00044 Frascati, Italy.

Mercury
Mercury. The Journal of the Astronomical Society of the Pacific. Published by the Astronomical Society of the Pacific, 1290 24th Avenue, San Francisco, California 94122, USA. (415) 661 - 8660.

Messenger
The Messenger – El Mensajero. Published by European Southern Observatory, Karl-Schwarzschild-Straße 2, D-8046 Garching bei München.

Meteoritics
Meteoritics. The Journal of the Meteoritical Society. Published quarterly by The Meteoritical Society and Arizona State University Bureau of Publications. Editorial address: Center for Meteorite Studies, The Arizona State University, Tempe, Ariz. 85281, USA.

Meteoritika
Akademiya Nauk SSSR. Komitet po Meteoritam. Publishers: Izdatel'stvo "Nauka", Moskva.

Meteorol. Rundsch.
Meteorologische Rundschau. Springer-Verlag, D-1000 Berlin 33, Heidelberger Platz 3, Germany.

Metrologia
Metrologia. Springer-Verlag, Heidelberger Platz 3, D-1000 Berlin 33, F. R. Germany.

Microwave J.
Microwave Journal. To be obtained from 610 Washington Street, Dedham Plaza, Dedham, Massachusetts, U.S.A.

Minor Planet Bull.
The Minor Planet Bulletin. Bulletin of the Minor Planets Section of the Association of Lunar and Planetary Observers. Editorial Office: R. G. Hodgson, Dordt College, Sioux Center, Iowa, U.S.A.

Minor Planet Circ., (M. P. C.)
The Minor Planet Circulars/Minor Planets and Comets. Edited under the supervision of B. G. Marsden. Published by Minor Planet Center, Smithsonian Astrophysical Observatory, Cambridge, Mass. 02138, USA.

Mitt. Astron. Ges.
Mitteilungen der Astronomischen Gesellschaft, Hamburg. Available from Astron. Instit. Univ. Bochum, Postfach 10 21 48, D-4630 Bochum. ISSN 0172–5483.

Mitt. Inst. Theor. Geod. Univ. Bonn
Mitteilungen aus dem Institut für Theoretische Geodäsie der Universität Bonn, Nußallee 17, 5300 Bonn 1, F. R. Germany.

Mitt. Karl-Schwarzschild-Obs. Tautenburg
Mitteilungen des Karl-Schwarzschild-Observatoriums Tautenburg der Deutschen Akademie der Wissenschaften der DDR. Zentralinstitut für Astrophysik.

Mitt. Satell.-Beobachtungsstn. Zimmerwald
Mitteilungen der Satelliten-Beobachtungsstation Zimmerwald. Hausdruckerei Institut für Exakte Wissenschaften, Bern, Switzerland.

Mitt. Sternw. Sonneberg
Akademie der Wissenschaften der DDR, Zentralinstitut für Astrophysik, Sternwarte Sonneberg. Mitteilungen der Sternwarte zu Sonneberg.

Mitt. Veränderl. Sterne (MVS)
Mitteilungen über Veränderliche Sterne. Herausgegeben von der Sternwarte Sonneberg der Akademie der Wissenschaften der DDR, Sonneberg, German Democratic Republic.

Mod. Geol.
Modern Geology. Gordon & Breach Science Publishers Ltd., 41 and 42 William IV Street, London WC2, England.

Mon. Not. R. Astron. Soc.
Monthly Notices of the Royal Astronomical Society. Published for the Royal Astronomical Society by Blackwell Scientific Publications, Oxford – London – Edinburgh – Melbourne.

Mon. Notes Astron. Soc. South. Africa
Monthly Notes of the Astronomical Society of Southern Africa. Published by the Astronomical Society of Southern Africa, S. A. Astronomical Observatory, Cape Province, South Africa.

Mon. Notes Int. Polar Motion Serv.
Monthly Notes of the International Polar Motion Service. Published by the Central Bureau, International Latitude Observatory of Mizusawa, Mizusawa-shi, Iwate-ken, Japan.

Moon Planets
The Moon and the Planets. An International Journal of Comparative Planetology. Publisher: D. Reidel Publishing Company, Dordrecht, Holland – Boston, USA. Formerly: Moon.

Nablyud. Iskusstv. Nebesn. Tel
Nablyudeniya Iskusstvennykh Nebesnykh Tel. Published by Astronomicheskij Sovet Akademii Nauk SSSR, Moskva.

Nachr. Akad. Wiss. Göttingen II
Nachrichten der Akademie der Wissenschaften in Göttingen. II. Mathematisch-Physikalische Klasse. Vandenhoeck & Ruprecht, Göttingen.

Nachr. Karten-, Vermessungswesen
Nachrichten aus dem Karten- und Vermessungswesen. Editor: Institut für Angewandte Geodäsie (Abt. II des Deutschen Geodätischen Forschungsinstituts). Published by Verlag des Instituts für Angewandte Geodäsie, Frankfurt a. M.

NASA Conf. Publ.
NASA Conference Publication. National Aeronautics and Space Administration. Scientific and Technical Information Branch, Washington, D.C. For sale by the National Technical Information Service, Springfield, Virginia 22161.

NASA Contract. Rep.
NASA Contractor Report. National Aeronautics and Space Administration, Washington, D.C. For sale by the National Technical Information Service, Springfield, Virginia 22161.

NASA Ref. Publ.
NASA Reference Publication. National Aeronautics and Space Administration. Scientific and Technical Information Office. Washington, D.C. 20546. For sale by the National Technical Information Service, Springfield, Virginia 22161.

NASA Tech Briefs
NASA Tech Briefs. NASA Technology Utilization Program, Technology Transfer Division, P. O. Box 8757, Baltimore/Washington International Airport, MD 21240, USA.

NASA Tech. Memo.
NASA Technical Memorandum. National Aeronautics and Space Administration, Washington, D.C. For sale by the National Technical Information Service, Springfield, Virginia 22161.

NASA Tech. Note
NASA Technical Note. National Aeronautics and Space Administration, Washington, D.C. For sale by the National Technical Information Service, Springfield, Virginia 22161.

NASA Tech. Pap.
NASA Technical Paper. National Aeronautics and Space Administration, Washington, D.C. For sale by the National Technical Information Service, Springfield, Virginia 22161.

Natl. Acad. Sci. Lett.
National Academy Science Letters. National Academy of Sciences, India, 5-Lajpat Rai Road, Allahabad 21 10 02, India.

Natl. Electron. Rev.
National Electronics Review. National Electronics Council, Abell House, John Islip Street, London, SW1P 4LN, England.

Natl. Geogr.
National Geographic. Official Journal of the National Geographic Society, Washington, D.C. 17th and M Sts. N.W., Washington, D.C. 20036.

Nature
Nature. Editorial and Publishing Offices: Macmillan Journals Limited, 4 Little Essex Street, London WC2R 3LF, 711 National Press Building, Washington, D.C. 20045.

Naturwissenschaften
Die Naturwissenschaften. Publisher: Springer-Verlag, Berlin – Heidelberg – New York.

Nauchn. Inf.
Nauchnye Informatsii. Astronomicheskij Sovet Akademii Nauk SSSR, Moskva.

Naučna Misao
Naučna Misao. Društvo za Unapredivanje i Širenje Nauke, Zagreb, Baboničeva 54, Yugoslavia. Scientific Idea. Society for Promotion and Propagation of Science.

Navigation *(France)*
Navigation. Institut Française de Navigation, 3 avenue Octave-Greard, Paris 7, France.

Navigation *(USA)*
Navigation. Journal of the Institute of Navigation, Institute of Navigation, Suite 832, 815 15th Street, N. W., Washington, DC 20005, USA.

NBS Monogr.
National Bureau of Standards Monograph. U.S. Government Printing Office, Washington, D.C. 20402.

Nederlands Tijdschr. Natuurkd. A
Nederlands Tijdschrift voor Natuurkunde, Publisher: Martinus Nijhoff, Lange Voorhout 9, Den Haag, Netherlands.

New Phys.
New Physics. Korean Physical Society, Seoul, Korea.

New Scientist
New Scientist. New Science Publications, 128 Long Acre, London, WC2E 9QH, England.

New Zealand J. Sci.
New Zealand Journal of Science. Department of Scientific and Industrial Research, Private Bag, Wellington, New Zealand.

News Lett. Astron. Soc. N.Y.
News Letter of the Astronomical Society of New York. A. G. D. Philip (Editor). Astronomical Society of New York, Dudley Observatory, 69 Union Avenue, Schenectady, New York 12308.

Nova Acta Leopoldina
Nova Acta Leopoldina, Abhandlungen der Deutschen Akademie der Naturforscher Leopoldina. Deutsche Akademie der Naturforscher Leopoldina, Halle (Saale). ISSN 0369-5034.

NSSDC/WDC-A-R&S
National Space Science Data Center/World Data Center A for Rockets and Satellites. National Aeronautics and Space Administration, Goddard Space Flight Center, Greenbelt, Maryland 20771, USA.

Nucl. Instrum. Methods Phys. Res.
Nuclear Instruments and Methods in Physics Research. Formerly: Nucl. Instrum. Methods. North-Holland Publishing Co., P. O. Box 211, 1000 AE Amsterdam, Netherlands.

Nucl. Phys. A
Nuclear Physics, Volume A. North-Holland Publishing Co., P. O. Box 211, Amsterdam, Netherlands.

Nucl. Sci. Appl. Sect. A
Nuclear Science Applications, Section A. Harwood Academic Publishers GmbH, Poststrasse 22, CH-7000 Chur, Switzerland.

Nucl. Tracks Methods Instrum. Appl.
Nuclear Tracks, Methods, Instruments and Application. Formerly: Nucl. Track Detect. (GB) Pergamon Press Ltd., Headington Hill Hall, Oxford OX3 0BW, England.

Numer. Math.
Numerische Mathematik. Springer-Verlag, Berlin–Heidelberg–New York.

Nuovo Cimento A
Il Nuovo Cimento A. Società Italiana di Fisica, Editrice Compositori, viale XII Giugno 1, 40124 Bologna, Italy.

Nuovo Cimento B
Il Nuovo Cimento B. Società Italiana di Fisica, Editrice Compositori, viale XII Giugno 1, 40124 Bologna, Italy.

Nuovo Cimento C
Il Nuovo Cimento C. Società Italiana di Fisica, Editrice Compositori, viale XII Giugno 1, 40124 Bologna, Italy.

Obs. Artif. Earth Satell.
Observations of Artificial Satellites of the Earth (Nablyudeniya Iskusstvennykh Sputnikov Zemli). Magyar Tudományos Akadémia Csillagvizsgáló Intézete. Budapest.

Obs. Astron. Antares, Contrib. Cient.
Universidade Estadual de Feira de Santana, Observatório Astronômico Antares, Contribuição Cientifica. Feira de Santana, Brazil.

Obs. Astrophys. Lab., Univ. Helsinki.Rep.
Observatory and Astrophysics Laboratory, University of Helsinki. Report. Helsinki, Finland.

Obs. Lyon Prepr.
Observatoire de Lyon, Preprints, Observatoire de Lyon, 69230 St. Genis Laval, France.

Observatory
The Observatory. A Review of Astronomy. Publishers: The Editors of 'The Observatory', Royal Greenwich Observatory, Herstmonceux Castle, Hailsham, Sussex, England, BN27 1RP.

Occas. Rep. R. Obs. Edinburgh
Occasional Reports of the Royal Observatory, Edinburgh, Blackford Hill, Edinburgh EH9 3HJ, Scotland. ISSN 0309-099X.

Occultation Newsl.
Occultation Newsletter. Published by the International Occultation Timing Association (I.O.T.A.). 6N106 White Oak Lane, St. Charles, Ill. 60174, USA.

Österreich. Z. Vermessungswes. Photogramm.
Österreichische Zeitschrift für Vermessungswesen und Photogrammetrie. Editor and Publisher: Österreichischer Verein für Vermessungswesen und Photogrammetrie, Wien, Austria.

Onde Electr.
L'Onde Electrique. Masson et Cie, 120 bd. Saint-Germaine, 75280 Paris Cedex 06, France.

Opt. Acta
Optica Acta. Taylor and Francis Ltd., 10 - 14 Macklin Street, London,WC2B 5NF, England.

Opt. Commun.
Optics Communications. North-Holland Publishing Co., P.O. Box 211, Amsterdam, Netherlands.

Opt. Eng.
Optical Engineering. Society of Photo-Optical Instrumentation Engineers, 337 Tejon Place, Palos Verdes Estates, CA 90274, USA.

Opt. Lett.
Optics Letters. A publication of the Optical Society of America. American Institute of Physics, 335 East 45th Street, New York, N. Y. 10017, USA.

Opt. Pura Apl.
Optica Pura y Aplicada. Serrano 121, Madrid 6, Spain.

Optik
Optik. Zeitschrift für das gesamte Gebiet der Licht- und Elektronenoptik. Publishers: Wissenschaftliche Verlagsgesellschaft mbH, Postfach 40, D-7000 Stuttgart, F. R. Germany.

Origins of Life
Origins of Life (Formerly Space Life Sciences). An International Journal. Publisher: D. Reidel Publishing Company, Dordrecht, Holland.

Orion
Orion. Zeitschrift der Schweizerischen Astronomischen Gesellschaft (SAG). Revue de la Société Astronomique

de Suisse (SAS). Printed by A. Schudel & Co. AG, 4125 Riehen, Switzerland.

Orione
Orione. Rivista Trimestrale di Divulgazione Astronomica. Via Roma 6, 10025 Pino Torinese (Torino).

Oss. Astrofis. Catania, Pubbl.
Osservatorio Astrofisico di Catania, Pubblicazione. Printed by Scuola Salesiana del Libro, Catania.

Oss. Mem. Oss. Astrofis. Arcetri
Osservazioni e Memorie dell'Osservatorio Astrofisico di Arcetri. Università Degli Studi di Firenze, Firenze, Italy.

Oyo Buturi
Oyo Buturi. Japan Society of Applied Physics, Room No. 209-2, Kikai-Shinko Building, 21 Shiba-Koen Minato-ku, Tokyo, Japan.

Perem. Zvezdy
Peremennye Zvezdy, Byulleten', izdavaemyj Astronomicheskim Sovetom Akademii Nauk SSSR. Published by Astronomicheskij Sovet Akademii Nauk SSSR, Moskva.

Perem. Zvezdy, Prilozhenie
Peremennye Zvezdy, Prilozhenie (The Variable Stars, Supplement). Astronomicheskij Sovet Akademii Nauk SSSR, Moskva.

Perspect. Comput.
Perspectives in Computing. International Business Machines Corp., Armonk, NY 10504, USA.

Philips Tech. Rev.
Philips Technical Review. Research Laboratories N. V. Philips Gloeilampenfabrieken, Eindhoven, Netherlands.

Philos. Mag. A
Philosophical Magazine A. (Physics of Condensed Matter. Defects and Mechanical Properties) Taylor & Francis Ltd., 10 - 14 Macklin Street, London, WC2B 5NF, England.

Philos. Trans. R. Soc. London, Ser. A
Philosophical Transactions of the Royal Society of London. Series A, Mathematical and Physical Sciences. Carlton House Terrace, London, SW1Y 5AG, England.

Photogr. J. Sun
Photographic Journal of the Sun. Supplement to Monthly Bulletin, Solar Phenomena. Osservatorio Astronomico di Roma.

Photogr. Sci. Eng.
Photographic Science and Engineering. Society of Photographic Scientists and Engineers, Suite 204, 1330 Massachusetts Avenue N.W., Washington, DC 20005, USA.

Photogramm. Eng. Remote Sensing
Photogrammetric Engineering and Remote Sensing. Formerly: Photogramm. Eng. American Society of Photogrammetry, 105 North Virginia Avenue, Falls Church, VA 22046, USA.

Photonics Spectra
Photonics Spectra. The Magazine of Optical/Electro-Optical/Laser Technology. Published by the Optical Publishing Co., Inc., 59 Bartlett Ave., P. O. Box 1146, Pittsfield, Mass. 01201, USA.

Photorin
Photorin, Mitteilungen der Lichtenberg-Gesellschaft e.V.

Phys. Abstr.
Physics Abstracts. Science Abstracts, Series A. An INSPEC Publication, published by The Institution of Electrical Engineers in Association with the Institute of Electrical and Electronics Engineers Inc. Printed by Pindar & Son Ltd., Scarborough, N. Yorkshire, England.

Phys. Bl.
Physikalische Blätter. Physik-Verlag GmbH, Pappelallee 3, Postfach 1260/1280, D-6940 Weinheim, F. R. Germany.

Phys. Briefs
Physics Briefs. Physikalische Berichte. Edited by Deutsche Physikalische Gesellschaft and Fachinformationszentrum Energie, Physik, Mathematik in cooperation with American Institute of Physics. Published by Physik Verlag GmbH, Postfach 1260/1280, D-6940 Weinheim, F. R. Germany. ISSN 0170-7434.

Phys. Chem. Earth
Physics and Chemistry of the Earth. Pergamon Press Ltd., Headington Hill Hall, Oxford OX3 0BW, England.

Phys. Chem. Miner.
Physics and Chemistry of Minerals. Springer-Verlag, D-1000 Berlin 33, Heidelberger Platz 3, Germany.

Phys. Earth Planet. Inter.
Physics of the Earth and Planetary Interiors. A journal devoted to observational and experimental studies of the Earth and Planetary interiors and their theoretical interpretation by the physical sciences. Publisher: North-Holland Publishing Company, Amsterdam, Netherlands.

Phys. Energ. Fortis Phys. Nucl.
Physica Energiae Fortis et Physica Nuclearis. Science Press, Peking, People's Republic of China. Subscription address: Guozo Shudian, P. O. Box 399, Peking.

Phys. Fluids
The Physics of Fluids. Published by the American Institute of Physics, 335 East 45th Street, New York, NY 10017, USA.

Phys. Lett.
Physics Letters. Volumes A and B. Publisher: North-Holland Publishing Company, Amsterdam.

Phys. Rep.
Physics Reports. North-Holland Publishing Co., P. O. Box 211, Amsterdam, Netherlands.

Phys. Rev. A
Physical Review A, General Physics. Published for the American Physical Society by the American Institute of Physics, 335 East 45th Street, New York, N. Y. 10017, USA.

Phys. Rev. B
Physical Review B, Solid State. Published for the American Physical Society by the American Institute of Physics, 335 East 45th Street, New York, NY 10017, USA.

Phys. Rev. C
Physical Review C, Nuclear Physics. Published for the American Physical Society by the American Institute of Physics, 335 East 45th Street, New York, NY 10017, USA.

Phys. Rev. D
Physical Review D, Particles and Fields. Published for the American Physical Society by the American Institute of Physics, 335 East 45th Street, New York, NY 10017, USA.

Phys. Rev. Lett.
Physical Review Letters. Published for the American Physical Society by the American Institute of Physics, 335 East 45th Street, New York, NY 10017, USA.

Phys. Scr.
Physica Scripta. (Formerly Arkiv för Fysik). Published by the Royal Swedish Academy of Sciences, S-104 05 Stockholm 50, Sweden.

Phys. Teach.
Physics Teacher. American Institute of Physics, 335 East 45th Street, New York, NY 10017, USA.

Phys. Technol.
Physics in Technology. Institute of Physics, 47 Belgrave Square, London SW1X 8QX, England.

Phys. Today
Physics Today. Published by the American Institute of Physics, 335 East 45th Street, New York, NY 10017, USA.

Phys. unserer Zeit
Physik in unserer Zeit. Verlag Chemie GmbH, Pappelallee 3, Postfach 1260/1280, D-6940 Weinheim, F. R. Germany.

Physica B, C
Physica B & C. Subscription address: North-Holland Publishing Co., P.O. Box 211, Amsterdam, Netherlands.

Physis
Physis. Rivista Internazionale di Storia della Scienza.

Pis'ma Astron. Zh.
Pis'ma v Astronomicheskij Zhurnal. Akademiya Nauk SSSR. Publishers: Izdatel'stvo 'Nauka', Moskva. Translation in Soviet Astron. Lett.

Planet. Space Sci.
Planetary and Space Science. Pergamon Press Ltd., Headington Hill Hall, Oxford, OX3 0BW, England.

Pokroky Mat. Fyz. Astron.
Pokroky Matematiky, Fyziky a Astronomie. Academia, Praha, Czechoslovakia. ISSN 0032-2423.

Postępy Astron.
Postępy Astronomii. Czasopismo Poświecone Upowszechnianiu Wiedzy Astronomicznej. Polskie Towarzystwo Astronomiczne, Warszawa. Printed in Poland by Pánstwowe Wydawnictwo Naukowe, Lódź.

Postępy Fiz.
Postępy Fizyki. Polskie Towarzystwo Fizykczne, 00-681 Warszawa, ul. Hoza 69, Poland.

Pramāna
Pramāṇa. Indian Academy of Sciences, Bangalore 560006, India.

Pré-publ. Inst. Astrophys. Paris
Pré-publication Institut d'Astrophysique de Paris, 98bis, Boulevard Arago, 75014 Paris.

Priroda
Priroda. Publishers: Izdatel'stvo "Nauka", Moskva.

Probl. Kosm. Fiz.
Problemy Kosmichskoj Fiziki. Mezhvedomstvennyj Nauchnoj Sbornik. Izdatel'skoe Obedinenie Vishcha Shkola. Izdatel'stvo pri Kievskom Universitete, Kiev.

Proc. Astron. Soc. Australia
Proceedings of the Astronomical Society of Australia. Published for the Society by Sydney University Press, Sydney.

Proc. IEEE
Proceedings of the IEEE. Published by the Institute of Electrical and Electronics Engineers, 345 East 47th Street, New York, NY 10017, USA.

Proc. Indian Acad. Sci. Earth Planet. Sci.
Proceedings of the Indian Academy of Sciences, Earth and Planetary Sciences. Bangalore 560 006, India.

Proc. Indian Natl. Sci. Acad., Part A.
Proceedings of the Indian National Science Academy, Part A. Bahadur Shah Zafar Marg, New Delhi 1, India.

Proc. Int. Latitude Obs. Mizusawa
Proceedings of the International Latitude Observatory of Mizusawa. Published by the International Latitude Observatory of Mizusawa, Japan.

Proc. Japan Acad., Ser. B
Proceedings of the Japan Academy. Series B, Physical and Biological Sciences, Ueno Park, Tokyo 110, Japan.

Proc. K. Nederlandse Akad. Wet. B
Koninklijke Nederlandse Akademie van Wetenschappen. Proceedings. Series B, Physical Sciences. Publisher: North-Holland Publishing Company, Amsterdam, Netherlands.

Proc. Natl. Acad. Sci. India, Sect. A
Proceedings of the National Academy of Sciences of India. Section A (Physical Sciences). Lajpatrai Road, Allahabad-2, India

Proc. Natl. Acad. Sci. U.S.A.
Proceedings of the National Academy of Sciences of the United States of America. National Academy of Sciences, 2101 Constitution Avenue, Washington, DC 20418, USA.

Proc. R. Soc. London, Ser. A
Proceedings of the Royal Society of London. Series A: Mathematical and Physical Sciences. Published by the Royal Society, 6 Carlton House Terrace, London, SW1Y 5AG, England.

Proc. Res. Inst. Atmos. Nagoya Univ.
Proceedings of the Research Institute of Atmospherics Nagoya University. Nagoya University, 13 Honohara, 3 Chrome, Toyokawa 442, Japan.

Proc. SPIE Int. Soc. Opt. Eng.
Proceedings of the SPIE – The International Society for Optical Engineering. Formerly: Proc. Soc. Photo-Opt. Instrum. Eng. P. O. Box 10, Bellingham, WA 98227, USA.

Prog. Part.Nucl. Phys.
Progress in Particle and Nuclear Physics. Pergamon Press Ltd., Headington Hill Hall, Oxford OX3 0BW, England.

Prog. Theor. Phys.
Progress of Theoretical Physics. Published for the Research Institute for Fundamental Physics and the Physical Society of Japan. Publication Office: Progress of Theoretical Physics, Yukawa Hall, Kyoto University, 606 Kyoto, Japan.

Prog. Theor. Phys. Suppl.
Supplement of the Progress of Theoretical Physics. Published for the Research Institute for Fundamental Physics and The Physical Society of Japan. Publication Office: Progress of Theoretical Physics. Yukawa Hall, Kyoto University, 606 Kyoto, Japan.

PTB Mitt.
PTB Mitteilungen. Forschen + Prüfen. Fachorgan für Wirtschaft und Wissenschaft. Amts- und Mitteilungsblatt der Physikalisch-Technischen Bundesanstalt. Braunschweig–Berlin. Deutscher Eichverlag, Postfach 3367, D-3300 Braunschweig, F.R. Germany.

Publ. Astron. Obs. Sarajevo
Publications of the Astronomical Observatory of Sarajevo. Published by Astronomical Society, University of Sarajevo on behalf of the Union of societies of mathematicians, physicists and astronomers of Yugoslavia. ISSN 0351-4587.

Publ. Astron. Soc. Japan
Publications of the Astronomical Society of Japan. Published by the Astronomical Society of Japan. Office of the Society: Tokyo Astronomical Observatory, Mitaka, Tokyo. Agent: Maruzen Co. Ltd. (Export Department), Nihonbashi, Tokyo, Japan.

Publ. Astron. Soc. Pacific
Publications of the Astronomical Society of the Pacific. Published by the Astronomical Society of the Pacific, 1290 24th Avenue, San Francisco, California 94122, USA. (415) 661 - 8660.

Publ. Beijing Astron. Obs.
Publications of the Beijing Astronomical Observatory. Beijing Astronomical Observatory, Academia Sinica, China.

Publ. Bosscha Obs.
Publications of the Bosscha Observatory. Bandung Institute of Technology, Department of Science. Lembang, Indonesia.

Publ. Debrecen Heliophys. Obs.
Publications of Debrecen Heliophysical Observatory of the Hungarian Academy of Sciences.

Publ. Dep. Astron., Univ. Beograd
Publications of the Department of Astronomy, University of Beograd, Faculty of Sciences (Publications de la Chaire d'Astronomie, Université de Beograd, Faculté des Sciences) Beograd. YU ISSN 0350-3283.

Publ. Dep. Astron. Univ. Cape Town
Publications of the Department of Astronomy of the University of Cape Town.

Publ. Dep. Geod. Astron., Univ. Thessaloniki
Publications of the Department of Geodetic Astronomy, University of Thessaloniki.

Publ. Dominion Astrophys. Obs.
Publications of the Dominion Astrophysical Observatory, Victoria, B. C. National Research Council of Canada.

Publ. Eidg. Sternw. Zürich
Publikationen der Eidgenössischen Sternwarte Zürich. Schulthess Polygraphischer Verlag, Zürich.

Publ. Inst. Geophys., Polish Acad. Sci.
Publications of the Institute of Geophysics, Polish Academy of Sciences. Państwowe Wydawnictwo Naukowe, Warszawa-Łódź. ISBN 83-01-02210-8. ISSN 0138-0214.

Publ. Inst. R. Meteorol. Belgique A
Publications, Institut Royal Meteorologique de Belgique. Serie A. 3 Avenue Circulaire, Uccle-Bruxelles 1180, Belgium.

Publ. Int. Latitude Obs. Mizusawa
Publications of the International Latitude Observatory of Mizusawa. Published by the International Latitude Observatory of Mizusawa, Japan.

Publ. Korean Natl. Astron. Obs.
Publications of the Korean National Astronomical Observatory. Published by the Korean National Astronomical Observatory, Seoul, Korea.

Publ. Obs. Astron. Beograd
Publications de l'Observatoire Astronomique de Beograd. Editeur: Observatoire Astronomique de Belgrade, 11050 Beograd, Volgina 7, Yougoslavie.

Publ. R. Obs. Edinburgh
Publications of the Royal Observatory, Edinburgh. Published by The Royal Observatory, Edinburgh, Scotland.

Publ. Shaanxi Astron. Obs.
Publications of the Shaanxi Astronomical Observatory. Academia Sinica, P. O. Box 18, Lintong, near Xian, China.

Publ. United States Naval Obs.
Publications of the United States Naval Observatory. Department of the Navy, U.S. Naval Observatory, Washington. U.S. Government Printing Office, Washington, D.C.

Publ. Variable Star Sect.,R. Astron. Soc. New Zealand
Publications of the Variable Star Section, Royal Astronomical Society of New Zealand. Director: F. M. Bateson, Greerton, Tauranga, New Zealand.

Publ. Warner Swasey Obs.
Publications of the Warner and Swasey Observatory, Case Western Reserve University, Cleveland, Ohio 44106.

Q. Appl. Math.
Quarterly of Applied Mathematics. American Mathematical Society, P. O. Box 1571, Providence, RI 02901, USA.

Q. Bull. Sol. Act.
International Astronomical Union, Quarterly Bulletin on Solar Activity. Published by the Tokyo Astronomical Observatory. Beginning with No. 197; formerly Zürich.

Q. J. R. Astron. Soc.
Quarterly Journal of the Royal Astronomical Society. Burlington House, London, W1V ONL, England.

R Muscae
R Muscae. Revista de Estrellas Variables. Published by Departamento de Astronomía, Instituto Copérnico,

Buenos Aires, Argentina. Editorial address: Montevideo 724, 1° "4", (1019), Buenos Aires, Argentina.

R. Greenwich Obs. Bull.
Royal Greenwich Observatory Bulletins. Royal Greenwich Observatory, Herstmonceux Castle, Hailsham, East Sussex BN27 1RP, England.

R. Greenwich Obs., Time Latitude Serv.
Royal Greenwich Observatory, Time and Latitude Service Royal Greenwich Observatory, Herstmonceux Castle, Hailsham, East Sussex BN27 1RP, England.

R. Obs. Ann.
Royal Observatory Annals, Royal Greenwich Observatory, Herstmonceux Castle, Hailsham, East Sussex, BN 27 1RP, England.

Rech. Aerosp.
Recherche Aerospatiale. Office National d'Études et de Recherches Aerospatiales, 29 Avenue de la Division Leclerc, 92320-Chatillon, France.

Recherche
Recherche, 4 Place de l'Odéon, Paris 6, France.

Ref. Zh., 51. Astron.
Referativnyi Zhurnal, 51. Astronomiya. Vsesoyuznyj Institut Nauchnoj i Tekhnicheskoj Informatsii. Moskva.

Ref. Zh., 52. Geod. Aehrosemka
Referativnyj Zhurnal, 52. Geodeziya i Aehrosemka. Vsesoyuznyj Institut Nauchnoj i Tekhnicheskoj Informatsii. Moskva.

Ref. Zh., 62. Issled. Kosm. Prostranstva
Referativnyj Zhurnal, 62. Issledovanie Kosmicheskogo Prostranstva. Vsesoyuznyj Institut Nauchnoj i Tekhnicheskoj Informatsii. Moskva.

Rep. Finnish Geod. Inst.
Reports of the Finnish Geodetic Institute. Suomen Geodeettisen Laitoksen Tiedonantoja. Helsinki, Finland.

Rep. Math. Phys.
Reports on Mathematical Physics, Pergamon Press Ltd., Headington Hill Hall, Oxford OX3 OBW, England. Subscription address: ARS Polona-Ruch Foreign Trade Enterprise, Krakowskie Przedmiescie 7, 00-068 Warszawa.

Rep. Obs. Lund
Reports from the Observatory of Lund. Lunds Universitet, Institutionen för Astronomie, S-22224 Lund, Sweden. ISSN 0349-4217.

Rep. Prog. Phys.
Reports on Progress in Physics. Published by the Institute of Physics, 47 Belgrave Square, London, SW1X 8QX, England.

Rep. Ser. Dep. Phys. Sci., Univ. Turku
Report Series, Department of Physical Sciences, Institute of Astronomy, University of Turku, SF-20500 Turku 50, Finland.

Rep. Univ. Electro-Commun.
Reports of the University of Electro-Communications. University of Electro-Communications, 1-5-1 Chofugaoka, Chofu-shi, Tokyo, Japan.

Res. Lab. Electron. Onsala Space Obs.
Research Laboratory of Electronics and Onsala Space Observatory, Chalmers University of Technology, Gothenburg, Sweden. Research Report.

Rev. Astron.
Revista Astronomica. Organo de la Asociación Argentina Amigos de la Astronomia, Avenida Patricias Argentinas 550, Buenos Aires 5, Argentina.

Rev. Brasil. Fis.
Revista Brasileira de Fisica. Sociedade Brasileira de Fisica, Cx. Postal 20553, Sao Paolo SP, Brazil.

Rev. Geophys. Space Phys.
Reviews of Geophysics and Space Physics (formerly Reviews of Geophysics). Published by the American Geophysical Union, 1909 K Street, N.W., Washington, DC 20006, USA.

Rev. Hist. Sci.
Revue d'Histoire des Sciences. Revue trimestrielle publiée avec le concours du C.N.R.S. Centre International de Synthèse, Section d'Histoire des Sciences.

Rev. Mexicana Astron. Astrofis.
Revista Mexicana de Astronomia y Astrofisica. Dirección: Instituto de Astronomia, Universidad Nacional Autónoma de México, Apartado Postal 70-264, Mexico 20, D. F., Mexico.

Rev. Mexicana Fis.
Revista Mexicana de Fisica. Sociedad Mexicana de Fisica, Apartado Postal No. 20-364, Mexico 20, D. F. Mexico.

Rev. Mod. Phys.
Reviews of Modern Physics. Published for the American Physical Society by the American Institute of Physics, 335 East 35th Street, New York, NY 10017, USA.

Rev. Phys. Appl.
Revue de Physique Appliquée. Editions de Physique, Z. I. de Courtaboeuf, B. P. 112, Avenue du Hoggar, 91944 Les Ulis Cedex, France.

Rev. Quest. Sci.
Revue des Questions Scientifiques. Publiée par la Société Scientifiques de Bruxelles, 61, rue de Bruxelles, 5000 Namur, Belgique.

Rev. Radio Res. Lab.
Review of the Radio Research Laboratories. Ministry of Posts & Telecommunications, Nukui-Kitamachi, Konganei-shi, Tokyo 184, Japan.

Rev. Roumaine Phys.
Revue Roumaine de Physique. Academie Republicii Populare Romine, Boite Postale 134 - 135, Bucuresti, Rumania.

Rev. Roumaine Sci. Tech. Ser. Mec. Appl.
Revue Roumaine des Sciences Techniques, Série de Mécanique Appliquée. Str. Constantin Mille 15, Bucuresti, Rumania.

Rev. Sci. Instrum.
Review of Scientific Instruments. American Institute of Physics, 335 East 45th Street, New York, NY 10017, USA.

Rev. Telegr. Electron.
Revista Telegrafica Electronica. Editorial address: Avenida Martin Garcia 653, Buenos Aires, Argentina.

Rezul't. Nablyud. Iskusstv. Sputnikov Zemli
Rezul'taty Nablyudenij Iskusstvennykh Sputnikov Zemli. Published by Astronomicheskij Sovet Akademii Nauk SSSR, Ryazanskij Gosudarstvennyj Pedagogicheskij Institut, Ryazan'.

Ric. Astron.
Ricerche Astronomiche. Specola Vaticana, Città del Vaticano.

Ric. Spettrosc.
Ricerche Spettroscopiche. Specola Vaticana, Città del Vaticano.

Říše hvězd
Říše hvězd. Czech popular astronomical journal. Publisher: Panorama, Praha.

Riv. Nuovo Cimento
La Rivisty del Nuovo Cimento. Società Italiana di Fisica. Editrice Compositori, viale XII Giugno 1, Bologna 40124, Italy.

Schweizerische Tech. Z.
Schweizerische Technische Zeitschrift. Schweizerischer Technischer Verband, Weinbergstrasse 41, 8023 Zürich, Switzerland.

Sci. American
Scientific American. 415 Madison Avenue, New York, NY 10017, USA.

Sci. Atmos. Sinica
Scientia Atmospherica Sinica. Science Press, Peking. Subscription address: Guozi Shudian, P.O. Box 399, Peking, Peoples's Republic of China.

Sci. Dimension
Science Dimension. National Research Council of Canada Ottawa K1A 0R6, Canada.

Sci. Pap. Inst. Phys. Chem. Res.
Scientific Papers of the Institute of Physical and Chemical Research. Rikagaku Kenkyusho, Wako-shi, Saitama 351, Japan.

Sci. Prog.
Science Progress. Blackwell Scientific Publications, Oxford, England.

Sci. Rep. Tôhoku Univ., Ser. 1.
The Science Reports of the Tôhoku-University. First Series (Physics, Chemistry, Astronomy). Published by the Faculty of Science, Tôhoku University, Sendai, Japan. ISSN 0040-8778.

Sci. Rep. Tôhoku Univ., Ser. 8.
The Science Reports of the Tôhoku University. Eighth Series (Physics and Astronomy). Published by the Faculty of Science, Tôhoku University, Sendai, Japan. ISSN 0388-5607.

Sci. Rev.
Scienca Revuo. Prof. B. Popovic, Ognjena Price 80, Beograd, Yugoslavia.

Sci. Sinica, Ser. A
Scientia Sinica, Series A (Mathematical, Physical, Astronomical and Technical Sciences). Published by Science Press, Beijing, China. Distributed by Scientific and Technical Books Service Ltd., P.O. Box 197, London WC2N 4DE, England.

Science
Science. American Association for the Advancement of Science, 1515 Massachusetts Avenue, N. W., Washington, D. C. 20005, USA.

Scr. Fac. Sci. Nat. Univ. Purkynianae Brunensis Phys.
Scripta Facultatis Scientiarum Naturalium Universitatis Purkynianae Brunensis, Physica. University J. E. Purkyne, 61137 Brno-Kotlarská 2, Czechoslovakia.

Shaanxi Astron. Obs. Repr.
Shaanxi Astronomical Observatory Reprints. Academia Sinica, P. O. Box 18, Lintong, near Xian, China.

SIAM J. Appl. Math.
SIAM Journal on Applied Mathematics. Society for Industrial and Applied Mathematics, 33 South 17th Street, Philadelphia, PA 19103, USA.

Sitzungsber. Akad. Wiss. DDR
Sitzungsberichte der Akademie der Wissenschaften der DDR. Mathematik-Naturwissenschaften-Technik. Akademie-Verlag, Berlin.

Sitzungsber. Bayerische Akad. Wiss.
Bayerische Akademie der Wissenschaften. Mathematisch-Naturwissenschaftliche Klasse. Sitzungsberichte. Publisher: Verlag der Bayerischen Akademie der Wissenschaften, München.

Sitzungsber. Heidelberger Akad. Wiss.
Sitzungsberichte der Heidelberger Akademie der Wissenschaften. Mathematisch-Naturwissenschaftliche Klasse. Publisher: Springer-Verlag, Heidelberg.

Sitzungsber. Österreich. Akad. Wiss.
Sitzungsberichte. Österreichische Akademie der Wissenschaften. Mathematisch-Naturwissenschaftliche Klasse. Abteilung II: Mathematik, Astronomie, Meteorologie und Technik. Publisher: Springer-Verlag, Wien.

Sky Telesc.
Sky and Telescope. Published by Sky Publishing Corporation, 49-50-51 Bay State Road, Cambridge, Mass. 02138, USA.

Smithsonian Astrophys. Obs., Spec. Rep.
Smithsonian Astrophysical Observatory, Special Report. Available from the Publications Division, Distribution Section, Smithsonian Astrophysical Observatory, Cambridge, Mass. 02138.

Smithsonian Contrib. Astrophys.
Smithsonian Contributions to Astrophysics. Smithsonian Institution Astrophysical Observatory, Cambridge, Mass. Printed by Smithsonian Institution Press, City of Washington. For sale by the Superintendent of Documents, U.S. Government Printing Office, Washington, D.C.

SMPTE J.
SMPTE Journal. Society of Motion Picture and Television Engineers, 868 Scarsdale Avenue, Scarsdale, NY 10583, USA.

Sol. Energy
Solar Energy, Pergamon Press, Maxwell House, Fairview Park, Elmsford, NY 10523, USA.

Sol. Phenom.
Solar Phenomena. Osservatorio Astronomico di Roma.

Sol. Phys.
Solar Physics. A Journal for Solar Research and the Study of Solar Terrestrial Physics. Publisher: D. Reidel Publishing Company, Dordrecht, Holland.

Sol. Syst. Res.
Solar System Research. Translation of Astronomicheskij Vestnik (USSR). Consultants Bureau, 227 West 17th Street, New York, N.Y. 10011, USA.

Sol. Terr. Environ. Res. Japan
Solar Terrestrial Environmental Research in Japan. Institute of Space and Aeronautical Science, University of Tokyo, 4-6-1, Komaba, Meguro-ku, Tokyo 153, Japan.

Soln. Dannye, Byull.
Solnechnye Dannye. Byulleten'. (*Solar Data*). Publishers: Izdatel'stvo "Nauka", Leningradskoe Otdelenie, Leningrad.

Sonne
Sonne. Mitteilungsblatt der Amateursonnenbeobachter. Peter Völker, c/o Wilhelm-Foerster-Sternwarte, Munsterdamm 90, 1000 Berlin 41.

Soobshch. Byurakan. Obs.
Soobshcheniya Byurakanskoj Observatorii. Akademiya Nauk Armyanskoj SSR, Erevan.

Soobshch. Gos. Astron. Inst. Shternberg
Soobshcheniya Gosudarstvennogo Astronomicheskogo Instituta im. P. K. Shternberga. Publishers: Izdatel'stvo Moskovskogo Universiteta, Moskva.

Soobshch. Spets. Astrofiz. Obs.
Soobshcheniya Spetsial'noj Astrofizicheskoj Observatorii. Izdanie Spetsial'noj Astrofizicheskoj Observatorii AN SSSR.

South African Astron. Obs. Circ.
South African Astronomical Observatory, Circulars. S.A. Astronomical Observatory, Observatory, Cape.

South African J. Phys.
South African Journal of Physics (Suid-Afrikaanse Tydskrif vir Fisika). Bureau for Scientific Publications, P.O. Box 1758, Pretoria 0001, South Africa.

South. Stars
Southern Stars. The Journal of the Royal Astronomical Society of New Zealand (Inc.). Address of the Society: P.O. Box 3181, Wellington C1, New Zealand.

Soviet Astron.
Soviet Astronomy. A translation of Astronomicheskij Zhurnal (Astronomical Journal). Published by the American Institute of Physics, New York, N.Y.

Soviet Astron. Lett.
Soviet Astronomy Letters. A translation of "Pis'ma v Astronomicheskij Zhurnal". Published by the American Institute of Physics.

Space Educ.
Space Education. Supplement to Spaceflight. Published by the British Interplanetary Society, 27/29 South Lambeth Road, London SW8 1SZ, England.

Space Sci. Rev.
Space Science Reviews. Publishers: D. Reidel Publishing Company, Dordrecht, Holland.

Spaceflight
Spaceflight. Published by the British Interplanetary Society. Printed by Unwin Brothers Ltd., at the Gresham Press, Old Woking, England.

Spaceworld
Spaceworld. Palmer Publications Inc., Amherst, WI 54406, USA.

Speculations Sci. Technol.
Speculations in Science and Technology. Elsevier Sequoia, S.A., P. O. Box 851, 1001 Lausanne 1, Switzerland.

Sterne
Die Sterne. Zeitschrift für alle Gebiete der Himmelskunde. Johann Ambrosius Barth, Leipzig, German Democratic Republic.

Sterne Weltraum
Sterne und Weltraum. Astronomische Monatsschrift. Publisher: Verlag Sterne und Weltraum Dr. Vehrenberg, Düsseldorf, F. R. Germany.

Sternenbote
Sternenbote. Monatsschrift für Österreichs Amateurastronomen. Publisher: Astronomisches Büro. Hermann Mucke, Wien, Austria.

Stockholms Obs. Ann.
Stockholms Observatorium Annaler. Printed by Almquist & Wiksell, Stockholm, Sweden.

Stockholms Obs. Rep.
Stockholms Observatorium, Saltsjöbaden, Sweden, Report.

Strolling Astron.
The Strolling Astronomer. The Journal of The Association of Lunar and Planetary Observers, Publication Office: The Strolling Astronomer, Box 3 AZ, University Park, New Mexico, USA.

Stud. Astron. Sinica
Studia Astronomica Sinica. Published by the Purple Mountain Observatory, Academia Sinica, Nanking, People's Republic of China.

Stud. Cercet. Fiz.
Studii si Cercetari de Fizica. Academia Republicii Populare Romine. P. O. Box 134-5, Calca Victoriei 126, Bucuresti, Rumania.

Stud. Geophys. Geod.
Studia geophysica et geodaetica. Published for the Geophysical Institute of the Czechoslovak Academy of Sciences by Academia, Praha.

Stud. Hist. Philos. Sci.
Studies in History and Philosophy of Science. Pergamon Press Ltd., Headington Hill Hall, Oxford OX3 0BW, England.

Stud. Soc. Sci. Torunensis
Studia Societatis Scientiarum Torunensis, Toruń – Polonia. Sectio F (Astronomia).

Stud. Univ. Babeş-Bolyai
Studia Universitatis Babeş-Bolyai. Series Mathematica-Physica. Publishers: Intreprinderea Poligrafica, Cluj.

Surv. High Energy Phys.
Surveys in High Energy Physics. Harwood Academic Publishers GmbH, P. O. Box 786, Cooper Station, New York, NY 10003, USA.

Syst. Int.
Systems International. IPC Electrical-Electronic Press Ltd., Dorset House, Stamford Street, London SE1 9LU, England.

Tartu Astrofüüs. Obs. Publ.
W. Struve nimelise Tartu Astrofüüsika Observatooriumi, Publikatsioonid. Eesti NSV Teaduste Akadeemia, Tartu.

Tartu Astrofüüs. Obs. Teated
Tartu Astrofüüsika Observatoorium Teated. Eesti NSV Teaduste Akadeemia W. Struve nim. Tartu Astrofüüsika Observatoorium, Tartu.

Tech. Mitt. Krupp Forschungsber.
Technische Mitteilungen Krupp Forschungsberichte. Fachbücherei Krupp, Postfach 917, 43 Essen 1, Germany.

Telecommun. J.
Telecommunication Journal.(English Edition). International Telecommunications Union, Place des Nations, 1211 Genève 20, Switzerland.

Tellus
Tellus, a bi-monthly Journal of Geophysics. Svenska Geofysiska Foreningen, Arrhenius laboratoriet, Fack, S - 104 05 Stockholm, Sweden.

Time Freq. Serv. Bull.
Time and Frequency Services Bulletin. Shaanxi Astronomical Observatory, Chinese Academy of Sciences, Lintong, Xian, China.

Tôhoku Geophys. J. Sci. Rep. Tôhoku Univ. Fifth Ser.
Tôhoku Geophysical Journal, Science Reports of the Tôhoku University, Fifth Series. Formerly: Sci. Rep. Tôhoku Univ. Fifth Ser. Geophys. Faculty of Science, Tôhoku University, Sendai 980, Japan.

Tokyo Astron. Bull., Second Ser.
Tokyo Astronomical Observatory, Japan. Tokyo Astronomical Bulletin, Second Series.

Tokyo Astron. Obs. Rep.
University of Tokyo, Tokyo Astronomical Observatory, Japan. Report. ISSN 0374-4639.

Tokyo Astron. Obs., Time and Latitude Bull.
Tokyo Astronomical Observatory, Time and Latitude Bulletins. Mitaka, Tokyo, Japan.

Tr. Astrofiz. Inst. Alma-Ata
Trudy Astrofizicheskogo Instituta, Alma-Ata. Akademiya Nauk Kazakhskoj SSR. Publishers: Izdatel'stvo "Nauka" Kazakhskoj SSR, Alma-Ata.

Tr. Astron. Obs., ***Leningrad***
Uchenye Zapiski Gosudarstvennogo Universiteta im. A. A. Zhdanova, Seriya matematicheskikh nauk = Trudy Astronomicheskoj Observatorii. Izdatel'stvo Leningradskogo Universiteta, Leningrad.

Tr. Glav. Astron. Obs. Pulkovo
Trudy Glavnoj Astronomicheskoj Observatorii v Pulkove. Akademiya Nauk SSSR, Izdanie Glavnoj astronomicheskoj observatorii v Pulkove, Leningrad.

Tr. Inst. Teor. Astron., ***Leningrad***
Trudy Instituta Teoreticheskoj Astronomii. Akademiya Nauk SSSR. Publishers: Izdatel'stvo "Nauka", Leningrad.

Tr. Kazan. Gorod. Astron. Obs.
Trudy Kazanskoj Gorodskoj Astronomicheskoj Observatorii. Izdatel'stvo Kazanskogo Universiteta, Kazan.

Tr. Tashkent. Astron. Obs.
Trudy Tashkentskoj Astronomicheskoj Observatorii. Akademiya Nauk Uzbekskoj SSR. Publishers: Izdatel'stvo "FAN" Uzbekskoj SSR, Tashkent.

Trans. American Nucl. Soc.
Transactions of the American Nuclear Society. American Nuclear Society, 555 North Kensington Avenue, La Grange Park, IL 60525, USA.

Trans. Astron. Obs. Yale Univ.
Transactions of the Astronomical Observatory of Yale University. Published by the Observatory, New Haven.

Trans. IAU
Transactions of the International Astronomical Union. Published and distributed for the IAU (UAI) by D. Reidel Publishing Company, Dordrecht, Holland – Boston, U.S.A.

Transp. Theory Stat. Phys.
Transport Theory and Statistical Physics. Marcel Dekker Inc., 270 Madison Avenue, New York, NY 10016, USA.

Tsirk. Astron. Inst. Tashkent
Tsirkulyar Astronomicheskogo Instituta. Akademiya Nauk Uzbekskoj SSR. Izdatel'stvo "FAN" Uzbekskoj SSR, Tashkent.

Tsirk. Astron. Obs. L'vov
Tsirkulyar. Astronomicheskaya Observatoriya. L'vovskij Ordena Lenina Gosudarstvennyj Universitet imeni Ivana Franko. Publisher: Izdatel'stvo L'vovskogo Universiteta, L'vov.

UKIRT Rep.
United Kingdom Infrared Telescope Report. ISSN 0260-9983.

Umschau
Umschau in Wissenschaft und Technik. Umschau Verlag, Stuttgarter Str. 18 - 24, D-6000 Frankfurt/M., F. R. Germany.

United States Naval Obs., Circ.
United States Naval Observatory, Circular. U.S. Naval Observatory, Washington, D.C. 20390.

Univ. Chile, Dep. Astron., Publ.
Universidad de Chile, Facultad de Ciencias Fisicas y Matematicas, Departamento de Astronomía, Publicaciones. Observatorio Astronómico Nacional, Cerro Calán, Santiago de Chile.

Urania Barcelona
Urania. Revista de Astronomía y Ciencias Afines. Órgano de la Sociedad Astronómica de España y América, Barcelona; Unión Nacional de Astronomía y Ciencias Afines, Madrid, Spain.

Urania Kraków
Urania. Miesięcznik Polskiego Towarzystwa Miłośników Astronomii, Kraków. Publisher: Z. N. im Ossolińskich, Kraków, Poland.

Vac News
Vac News. Indian Vacuum Society, c/o Technical Physics Division, Bhabha Atomic Research Centre, Trombay, Bombay 400 085, India.

Vasiona
Vasiona. Revue d'Astronomie et d'Astronautique. Bulletin de la Société Astronomique "R. Bosković", Beograd.

Vatican Obs. Publ.
Vatican Observatory Publications, Specola Vaticana, Città del Vaticano.

Veröff. Astron. Rechen-Inst. Heidelberg
Veröffentlichungen des Astronomischen Rechen-Instituts Heidelberg. Verlag G. Braun, Karlsruhe, F.R. Germany.

Veröff. Bayer. Komm. Int. Erdmessung, Bayer. Akad. Wiss., Astron.-Geod. Arb.
Veröffentlichungen der Bayerischen Kommission für die Internationale Erdmessung der Bayerischen Akademie der Wissenschaften. Astronomisch-Geodätische Arbeiten. Published by Verlag der Bayerischen Akademie der Wissenschaften, München, F. R. Germany. ISSN 0340-7691, ISBN 3-7696-9782-0.

Veröff. Remeis-Sternw. Bamberg
Veröffentlichungen der Remeis-Sternwarte Bamberg, Astronomisches Institut der Universität Erlangen-Nürnberg.

Veröff. Sternw. Sonneberg
Akademie der Wissenschaften der DDR, Zentralinstitut für Astrophysik, Veröffentlichungen der Sternwarte in Sonneberg. Publisher: Akademie-Verlag, Berlin, German Democratic Republic.

Veröff. Zentralinst. Phys. Erde
Akademie der Wissenschaften der DDR, Forschungsbereich Geo- und Kosmowissenschaften. Veröffentlichungen des Zentralinstituts für Physik der Erde, Potsdam, German Democratic Republic.

Vesmír
Vesmír. Přírodovědecký časopis Čs. akademie věd. Publisher: Academia, Praha.

Vestn. Khar'kov. Univ.
Vestnik Khar'kovskogo Universiteta. Seriya Astronomicheskaya. Publishers: Izdatel'stvo Khar'kovskogo Universiteta, Khar'kov.

Vestn. Kiev. Univ.
Vestnik Kievskogo Universiteta. Seriya Astronomii. Publishers: Izdatel'stvo Kievskogo Universiteta, Kiev.

Vignana Bharathi
Vignana Bharathi. Department of Publications & Extension, Bangalore University, Bangalore 560056, India.

Vistas Astron.
Vistas in Astronomy. An international review journal. Pergamon Press, Oxford – New York – Braunschweig.

Weather
Weather. James Glaisher House, Grenville Place, Bracknell, Berks RG12 1BX, England.

Wiss. Z. Friedrich-Schiller-Univ. Jena
Wissenschaftliche Zeitschrift der Friedrich-Schiller-Universität. Jena. Mathematisch-Naturwissenschaftliche Reihe; Edited by the Rektor der Friedrich-Schiller-Universität Jena, Am Anger 24, Jena, German Democratic Republic.

Wiss. Z. Humboldt-Univ. Berlin
Wissenschaftliche Zeitschrift der Humboldt-Universität zu Berlin. Mathematisch-Naturwissenschaftliche Reihe. Edited by the Rektor der Humboldt-Universität Berlin, Unter den Linden 6, 108 Berlin, German Democratic Republic.

Wiss. Z. Tech. Univ. Dresden
Wissenschaftliche Zeitschrift der Technischen Universität Dresden. Mommsenstraße 13, Dresden A227, Germany.

Wuli
Wuli. Science Press, Peking, People's Republic of China. Subscription address: Guozi Shudian, P.O. Box 399, Peking.

Yamamoto Circ.
Yamamoto Circular. Published by the Yamamoto Observatory, Kamitanakami-Kiryutyo, Otu, Siga-Ken, [520-21] Japan.

Z. angew. Math. Mech.
Zeitschrift für angewandte Mathematik und Mechanik. Akademie-Verlag GmbH, 108 Berlin, Leipziger Strasse 3–4, German Democratic Republic.

Z. angew. Math. Phys.
Zeitschrift für angewandte Mathematik und Physik. Verlag Birkhauser, Postfach 4000, Basel 24, Switzerland.

Z. Naturforsch.
Zeitschrift für Naturforschung. Teil A. A Europhysics Journal. Physik–Physikalische Chemie–Kosmophysik. Verlag der Zeitschrift für Naturforschung, Tübingen. P. O. Box 2645, D–7400 Tübingen, F. R. Germany.

Z. Phys. A
Zeitschrift für Physik A. Atoms and Nuclei. Springer-Verlag, Berlin–Heidelberg–New York.

Z. Phys. B
Zeitschrift für Physik B. Condensed Matter and Quanta. Springer-Verlag, Berlin–Heidelberg–New York.

Z. Phys. C
Zeitschrift für Physik C. Particles and Fields. Springer-Verlag, P.O. Box 105280, D-6900 Heidelberg 1, Germany.

Z. Vermessungswes.
Zeitschrift für Vermessungswesen. Verlag Konrad Wittwer, 7000 Stuttgart 1, Nordbahnhofstrasse 16, Postfach 147, F. R. Germany.

Zeiss Inf.
Zeiss Information. Carl Zeiss, Oberkochen. F. R. Germany.

Zemlya Vselennaya
Zemlya i Vselennaya. Astronomiya, Geofizika. Issledovaniya Kosmicheskogo Prostranstva, Nauchno-Populyarnyj Zhurnal Akademii Nauk SSSR. Publishers: Izdatel'stvo "Nauka", Moskva.

Zenit
Populair wetenschappelijk maandblad over sterrenkunde/

weerkunde/ruimtevaart/ruimte-onderzoek/aanverwante wetenschappen en technieken. Bureau: Stichting De Koepel, Utrecht.

Zentralbl. Math. Grenzgeb. – Math. Abstr.
Zentralblatt für Mathematik und ihre Grenzgebiete – Mathematics Abstracts. Publisher: Springer-Verlag, Berlin–Heidelberg–New York.

Zvaigžņota Debess
Latvijas PSR Zinātņu Akadēmijas Radioastrofizikas Observatorijas Populārzinatnisks Gadalaiku Izdevums. Izdevnieciba "Zinātne", Riga.

Journals abstracted completely.

A selected number of periodicals listed in category 001 are central to the subject scope of *Astronomy and Astrophysics Abstracts.* Depending on their relevance, almost all papers of the journals listed below are abstracted in our service.

AAVSO Bull.
Acta Astron.
Acta Astron. Sinica
Acta Astrophys. Sinica
Acta Cosmologica
Ann. Shanghai Obs., Acad. Sinica
Ann. Tokyo Astron. Obs.
Annu. Rep. Astron. Inst. Greece
Annu. Rev. Astron. Astrophys.
Astrofiz. Issled. Izv. Spets. Astrofiz. Obs.
Astrofizika
Astrometr. Astrofiz.
Astron. Astrophys.
Astron. Astrophys., Suppl. Ser.
Astron. J.
Astron. Nachr.
Astron. Pap.
Astron. Q.
Astron. Tidsskr.
Astron. Tsirk.
Astron. Vestn.
Astron. Zh.
Astronomia
Astronomie
Astrophys. J.
Astrophys. J., Lett.
Astrophys. J., Suppl. Ser.
Astrophys. Lett.
Astrophys. Space Sci.
Australian J. Phys., Astrophys. Suppl.
BAV Rundbrief
BBSAG Bull.
Bol. Astron. Obs. Madrid
Bol. Inst. Tonantzintla
British Astron. Assoc. Circ.
Bull. AFOEV
Bull. American Astron. Soc.
Bull. Astron. Inst. Czechoslovakia
Bull. Astron., Obs. R. Belgique
Bull. Astron. Soc. India
Bull. Inf. Cent. Données Stellaires
Bull. Obs. Astron. Belgrade
Byull. Abastumanskaya Astrofiz. Obs.
Byull. Inst. Astrofiz.
Byull. Inst. Teor. Astron.
Carter Obs., Astron. Bull.
Celestial Mech.
Circ. Inf.
Circ. Stn. Astron. Int. Latitudine, Carloforte-Cagliari
Coelum
Comments Astrophys.
Comun. Obs. Astron. Univ. Coimbra
Dudley Obs. Rep.
Dunsink Obs. Publ.
ESO Sci. Prepr.
ESO Tech. Rep.
G. Astron.
IAU Circ.
Icarus
Inf. Bull. Variable Stars
Inst. Theor. Astrophys., Blindern–Oslo, Rep.
Irish Astron. J.
Izv. Astron. Ehngel'gardt. Obs.
Izv. Glav. Astron. Obs. Pulkovo
Izv. Krymskoj Astrofiz. Obs.
J. American Assoc. Variable Star Obs.
J. British Astron. Assoc.
J. Hist. Astron.
J. R. Astron. Soc. Canada
Komet. Tsirk.
Komety i Meteory
Mem. Soc. Astron. Italiana
Mercury
Messenger
Meteoritics
Meteoritika
Minor Planet Bull.
Minor Planet Circ.
Mitt. Astron. Ges.
Mitt. Sternw. Sonneberg
Mitt. Veränderl. Sterne (MVS)
Mon. Not. R. Astron. Soc.
Mon. Notes Astron. Soc. South Africa
Mon. Notes Int. Polar Motion Serv.
Moon Planets
Nablyud. Iskusstv. Nebesn. Tel
Nauchn. Inf.
News Lett. Astron. Soc. N.Y.
Obs. Astrophys. Lab., Univ. Helsinki. Rep.
Observatory
Occas. Rep. R. Obs. Edinburgh
Occultation Newsl.
Orion
Oss. Astrofis. Catania, Pubbl.
Oss. Mem. Oss. Astrofis. Arcetri
Perem. Zvezdy
Perem. Zvezdy, Prilozhenie
Pis'ma Astron. Zh.
Postępy Astron.
Probl. Kosm. Fiz.
Proc. Astron. Soc. Australia
Proc. Int. Latitude Obs. Mizusawa
Publ. Astron. Soc. Japan
Publ. Astron. Soc. Pacific
Publ. Dominion Astrophys. Obs.
Publ. Eidg. Sternw. Zürich
Publ. Int. Latitude Obs. Mizusawa
Publ. R. Obs. Edinburgh
Publ. United States Naval Obs.
Publ. Variable Star Sect. R. Astron. Soc. New Zealand
Q. Bull. Sol. Act.
Q. J. R. Astron. Soc.
R. Obs. Ann.
Rep. Obs. Lund
Rev. Mexicana Astron. Astrofis.
Rezul't. Nablyud. Iskusstv. Sputnikov Zemli
Sky Telesc.
Smithsonian Astrophys. Obs., Spec. Rep.
Smithsonian Contrib. Astrophys.
Sol. Phys.
Soln. Dannye, Byull.
Soobshch. Byurakan. Obs.
Soobshch. Gos. Astron. Inst. Shternberg
Soobshch. Spets. Astrofiz. Obs.
South African Astron. Obs. Circ.
South. Stars
Space Sci. Instrum.
Space Sci. Rev.
Sterne
Sterne Weltraum

Stockholms Obs. Ann.
Stockholms Obs. Rep.
Strolling Astron.
Tartu Astrofüüs. Obs. Publ.
Tartu Astrofüüs. Obs. Teated
Tokyo Astron. Bull.
Tokyo Astron. Obs. Rep.
Tr. Astrofiz. Inst. Alma-Ata
Tr. Astron. Obs., *Leningrad*
Tr. Glav. Astron. Obs. Pulkovo
Tr. Inst. Teor. Astron., *Leningrad*
Tr. Kazan. Gorod. Astron. Obs.
Tr. Tashkent. Astron. Obs.
Trans. IAU
Tsirk. Astron. Inst. Tashkent
Tsirk. Astron. Obs. L'vov
United States Naval Obs., Circ.
Urania Barcelona
Urania Kraków
Vatican Obs. Publ.
Veröff. Astron. Rechen-Inst. Heidelberg
Veröff. Sternw. Sonneberg
Vestn. Khar'kov. Univ.
Vestn. Kiev. Univ.
Vistas Astron.
Yamamoto Circ.

002 Bibliographical Publications, Catalogues, Atlases

002.001 **ANS ultraviolet photometry, catalogue of point sources.**
P. R. Wesselius, R. J. van Duinen, A. R. W. de Jonge, J. W. G. Aalders, W. Luinge, K. J. Wildeman.
Astron. Astrophys., Suppl. Ser., Vol. 49, 427 - 474 (1982).
All reliable observations on point sources with the ultraviolet photometer in ANS, operating between end 1974 and beginning 1976, are presented. Extended objects, non-detected objects, and objects at the edge of the field of view of the instrument have been omitted. The catalogue contains 3573 objects, mostly stars; the total data base has 4800 observed positions. The instrumental characteristics, operational procedures, and reduction methods relevant for the interpretation are described. The added cross-reference data (name, *UBV* photometry, MK spectral class and luminosity) have been checked for consistency.

002.002 **The ESO Quick Blue Survey and ESO *(B)* Atlas.**
R. M. West, H.-E. Schuster.
Astron. Astrophys., Suppl. Ser., Vol. 49, 577 - 589 (1982).
Plate data and other pertinent information are given for the ESO Quick Blue Survey and the resulting ESO *(B)* Atlas (1973 - 78, mean epoch 1976.0).

002.003 **Detailed bibliography on the surface photometry of galaxies.**
E. Davoust, W. D. Pence.
Astron. Astrophys., Suppl. Ser., Vol. 49, 631 - 661 (1982).
This bibliography documents all publications on the surface photometry of external galaxies using a concise code to describe methods of observation and reduction, types of photometric information displayed for each galaxy, limiting surface brightness, and general purpose of the paper. Information on about 650 galaxies is given from more than 300 references.

002.004 **Catalog of CO radial velocities toward galactic H II regions.**
L. Blitz, M. Fich, A. A. Stark.
Astrophys. J., Suppl. Ser., Vol. 49, 183 - 206 (1982).
This is a catalog of 242 molecular cloud complexes which are associated with optical H II regions. CO observations were made toward all but five of the H II regions in the Sharpless catalog and toward 62 additional suspected H II regions. Radial velocities are tabulated for each molecular cloud complex found to be associated with an H II region. The CO antenna temperature and line width are given for the most intense CO line seen toward each source. The catalog also summarizes previous CO observations as well as the optical distances to the stars exciting the H II regions.

002.005 **The Copernicus ultraviolet spectral atlas of Beta Orionis.**
J. B. Rogerson, Jr., W. L. Upson II.
Astrophys. J., Suppl. Ser., Vol. 49, 353 - 404 (1982).
An ultraviolet spectral atlas is presented for the B8 Ia star β Orionis, which has been scanned from 999 to 1561 Å by the Princeton spectrometer aboard the Copernicus satellite. From 999 to 1420 Å the observations have a nominal resolution of 0.05 Å. At the longer wavelengths the resolution is 0.1 Å. The atlas is presented in graphs. Lines identified in the spectrum are also listed.

002.006 **Archives of Dunsink and Markree Observatories.**
M. Hoskin.
J. Hist. Astron., Vol. 13, 146 - 152 (1982).

002.007 **Prospects for automated solution of the subject characterization problem in the bibliographic services.**
S. Schiminovich.
Automated data retrieval in astronomy, (see 012.002), p. 147 - 154 (1982).
Evidence is presented on the feasibility of an automated generation of classification schemes and the subsequent indexing of papers with the automatically generated subject headings. The output generated for Astronomy and Astrophysics compares favorably with the schemes used by the services and is not inconsistent with recently proposed classifications.

002.008 **The bibliographical card-catalogue of the I.A.P. Library.**
C. Chagnard-Carpuat, A. Lagorce, S. Laloë.
Automated data retrieval in astronomy, (see 012.002), p. 155 - 158 (1982).
This bibliographical service is available at the Library of the "Institut d'Astrophysique de Paris". It is a hand-operated system, consisting of a card-catalogue concerning all astronomical objects (except those of the solar system).

002.009 **Astronomy and Astrophysics Abstracts: a status report.**
L. D. Schmadel.
Automated data retrieval in astronomy, (see 012.002), p. 159 - 161 (1982).
The astronomical bibliography 'Astronomy and Astrophysics Abstracts' (AAA) is the direct successor of the well-known 'Astronomischer Jahresbericht' (AJB). AAA is now going into its 13th year of existence. It is the aim of this contribution to outline the present status and the future plans of the abstracting and indexing work at the Rechen-Institut.

002.010 **INSPEC activities in astronomy and astrophysics.**
M. J. Collins.
Automated data retrieval in astronomy, (see 012.002), p. 163 - 165 (1982).

002.011 **Stellar bibliography retrieving system in Japan.**
S. Nishimura.
Automated data retrieval in astronomy, (see 012.002), p. 167 - 169 (1982).
On-line retrieving systems of BSI in Japan are briefly described. The BSI reference file was adapted to be retrieved by some keywords other than star names. Then the retrieval system of BSI was constructed on a general retrieval language CIR. Some efforts to make the system efficient and to save the disk space are discussed.

002.012 **The Bibliographical Star Index.**
F. Ochsenbein.
Automated data retrieval in astronomy, (see 012.002), p. 171 - 173 (1982).
The Bibliographical Star Index (B.S.I.) tries to solve the problem of retrieving the observations or any kind of notes about a star. This file was initiated by Prof. R. Cayrel in 1964 and is going on through a collaboration between Paris-Meudon Observatories and the Stellar Data Center.

002.013 **Copyright and use of computers for access to protected works.**
E. Guerassimov *(Gerasimov)*.
Automated data retrieval in astronomy, (see 012.002), p. 177 - 184 (1982).

002.014 **Editorial policies.**
J. Lequeux.
Automated data retrieval in astronomy, (see 012.002), p. 187 - 191 (1982).

A short summary of the way the main astronomy journals are produced is given. The different ways of publishing a scientific journal are described, their advantages and drawbacks are discussed and some projections towards the future are made. The author finally deals with a few specific editorial topics: publication of large collections of data, designation of astronomical objects, indexing and retrieval of information.

002.015 **Aids to the retrieval and evaluation of astronomical data.**
G. A. Wilkins.
Automated data retrieval in astronomy, (see 012.002), p. 193 - 198 (1982).

The main purpose of this paper is to draw attention to some of the ways by which authors, editors and data centres may significantly increase the probability that published data will be found and used by those who require such data. The paper also contains some information that may be helpful to those who wish to retrieve astronomical data.

002.016 **A list of standard star designation.**
F. Ochsenbein, F. Spite.
Automated data retrieval in astronomy, (see 012.002), p. 199 - 202 (1982).

Some rules about star designation have been edicted by Commission 5 of the IAU: they are briefly recalled and discussed. It is proposed to publish in the CDS Bulletin a list of designations considered as "standard".

002.017 **Dictionnaire de la nomenclature des objets célestes.**
A. Fernandez, M. C. Lortet, F. Spite.
Automated data retrieval in astronomy, (see 012.002), p. 203 - 205 (1982).

The confusing situation of the nomenclature of the astronomical objects led the authors to build, as a first step, a dictionary of the designations presently used. The future publication of this work is supported by Commission 5 of IAU. A brief description of this dictionary, as well as a sample of it, is given.

002.018 **On the nomenclature of diffuse fragmentary objects.**
P. Polechova.
Automated data retrieval in astronomy, (see 012.002), p. 207 - 210 (1982).

For the purpose of cataloguing diffuse objects in a wide range of dimensions, two alternative proposals are presented.

002.019 **The Catalogue of Stellar Identifications.**
F. Ochsenbein, M. Bischoff.
Automated data retrieval in astronomy, (see 012.002), p. 211 - 213 (1982).

The evolution of the Catalogue of Stellar Identifications (C.S.I.) developed by the Stellar Data Center (C.D.S.) is summarized and recent developments of this file are presented.

002.020 **Data in astronomy.**
B. Hauck.
Automated data retrieval in astronomy, (see 012.002), p. 217 - 225 (1982).

The present situation regarding data for astronomy and astrophysics is dominated by two factors: large increase of data and better accessibility, mainly due to the existence of data centres. An attempt to centralize the information of plate vaults is also presented.

002.021 **Proposal for centralization of information concerning plate vaults.**
B. Hauck.
Automated data retrieval in astronomy, (see 012.002), p. 227 - 228 (1982).

002.022 **The UK Schmidt Telescope plate catalogue and problems associated with increasing numbers of plates and users.**
S. B. Tritton.
Automated data retrieval in astronomy, (see 012.002), p. 229 - 231 (1982).

The UK 1.2m Schmidt Telescope is situated at Siding Spring Observatory in New South Wales, Australia. The telescope has been operational since 1973 and has, to date, taken about 7000 plates. This paper discusses some of the problems associated with cataloguing these plates and keeping track of their locations.

002.023 **Retrieval of astronomical information from Padova-Asiago Observatory plates archives.**
L. Benacchio.
Automated data retrieval in astronomy, (see 012.002), p. 233 - 236 (1982).

This paper shortly describes a software procedure for the on line information retrieval from astronomical plates archives. The procedure is active in Padova Asiago Observatory and will be improved linking them with other retrieval and graphic procedures acting on astronomical catalogues.

002.024 **Data and data retrieval in space astronomy.**
J. I. Vette.
Automated data retrieval in astronomy, (see 012.002), p. 243 - 252 (1982).

To illustrate the range of storage and retrieval facilities and data products that are involved in space astronomy, the present situation on data from some instruments on the following missions is presented: Small Astronomy Satellite-B, Orbiting Astronomical Observatory 3, High Energy Astrophysics Observatory 2, International Ultraviolet Explorer, Viking Orbiter, and Solar Maximum Mission.

002.025 **The "Catalogue of Stellar Groups": a progress report.**
D. Egret, M. Jaschek.
Automated data retrieval in astronomy, (see 012.002), p. 253 - 256 (1982).

The catalog provides lists of stars for about fifty groups of determined spectral peculiarity. The adopted selection procedure, the connection to the C.D.S. data base and the future developments are described.

002.026 **Current progress with the double star catalogues.**
C. E. Worley.
Automated data retrieval in astronomy, (see 012.002), p. 257 - 258 (1982).

Current progress on the visual double star Index and Observation Catalogues is discussed. In addition, a new edition of the visual binary orbit catalogue is being prepared.

002.027 **Metacatalogue of galaxies and large scale structures in the Universe.**
M. Kalinkov.
Automated data retrieval in astronomy, (see 012.002), p. 259 - 260 (1982).

002.028 **Data retrieval in the Metacatalogue of galaxies.**
M. Kalinkov, I. Kuneva, K. Stavrev.
Automated data retrieval in astronomy, (see 012.002), p. 261 - 263 (1982).

To study the large scale structure of the Universe, the authors started several years ago the compilation of a new type of catalogue – a metacatologue of optical extragalactic objects, on magnetic tape. The Metacatalogue will contain information for galaxies, clusters of galaxies, counts of galaxies and bibliographical data. Until now some files for individual catalogues and lists have been created, which may be used also before their merging into the Metacatalogue.

002.029 **Le Centre de Données des Amas de Galaxies.**
A. Fernandez, G. Mathez, A. Mazure, L. Nottale, R. Rothenflug.
Automated data retrieval in astronomy, (see 012.002), p. 265 - 267 (1982).

A Cluster of Galaxies Data Center is in the process of being settled. Its aim will be to collect the data presently scattered in the literature, concerning clusters of galaxies, in all wavelength ranges. The authors describe its diverse entries and they discuss some specific problems and its possible future utilization.

002.030 **Centralization of extragalactic data.**
G. Paturel.
Automated data retrieval in astronomy, (see 012.002), p. 269 - 271 (1982).

The author presents a project of centralization of extragalactic data concerning normal galaxies.

002.031 **The compiled catalogue of galaxies in machine-readable form and its statistical investigation.**
N. G. Kogoshvili.
Automated data retrieval in astronomy, (see 012.002), p. 273 - 275 (1982).

The compilation of a machine-readable catalogue of relatively bright galaxies was undertaken in Abastumani Astrophysical Observatory. This catalogue is mainly based on the data from the MCG catalogue supplied with information from the UGC, RCBG, CGCG catalogues and from other sources. The Abastumani Catalogue comprises about thirty-three thousand galaxies most of them brighter than $15.^m1$ and located in the area north of $-33°$.

002.032 **Data for the compilation of the "Third Catalogue of Nearby Stars".**
W. Gliese.
Automated data retrieval in astronomy, (see 012.002), p. 281 - 283 (1982).

002.033 **Fundamental data for FK4/FK4Sup stars.**
H. Jahreiss.
Automated data retrieval in astronomy, (see 012.002), p. 285 - 287 (1982).

002.034 **Remarks about the cataloguing of open cluster data.**
G. Lyngå.
Automated data retrieval in astronomy, (see 012.002), p. 289 - 291 (1982).

002.035 **Catalogue bibliographique de vitesses radiales stellaires.**
M. Barbier, M. Petit.
Automated data retrieval in astronomy, (see 012.002), p. 293 - 295 (1982).

The continuation of the Abt-Biggs catalog of stellar radial velocities from 1970 is presented. A first part (6000 references) has been published on microfiches (CDS, 1980). Five thousand new references should be published in a near future.

002.036 **The new catalogue of optical H II regions.**
P. Polechova.
Automated data retrieval in astronomy, (see 012.002), p. 297 - 300 (1982).

The principles of the work on the new catalogue of the optical H II regions are described.

002.037 **Catalogue of planetary nebulae and their nuclei.**
A. Acker, J. Marcout.
Automated data retrieval in astronomy, (see 012.002), p. 301 - 303 (1982).

002.038 **Catalogue of declinations of the zenith zone obtained on the Moscow zenith telescope ZTL-180.**
L. P. Gribko.
Tr. Gos. Astron. Inst. Shternberga, Tom 51, 234 - 246 (1982). In Russian.

This article contains the catalogue of declinations of 426 zenith stars obtained on the Moscow zenith-telescope ZTL-180 (1970.5 - 1977.0). The declinations of the stars were computed by the method of "the place of zenith" with the use of the pole coordinates.

002.039 **New catalogue of stars suspected of variability containing data on 14810 variable stars which have not received definite nomenclature until 1980.**
P. N. Kholopov (Editor).
Nauka, Moskva. 359 pp. (1982). In Russian. – From Ref. zh., 51. Astron., 7.51.59 (1982).

002.040 **Catalogue of right ascensions of 586 FKSZ stars in the declination zone from +90° to –20°.**
Ya. E. Gordon, L. F. Gorel', E. V. Khrutskaya.
Glav. astron. obs. AN SSSR. Leningrad, 1982, 57 pp. In Russian. – Abstr. in Ref. zh., 51. Astron., 7.51.128 (1982).

002.041 **Catalogue of declinations of Pulkovo latitude stars.**
S. P. Pulyaev.
Glav. astron. obs. AN SSSR. Leningrad, 1982. 25 pp. In Russian. – Abstr. in Ref. zh., 51. Astron., 7.51.129 (1982).

002.042 **Bibliographical index of objects observed by IUE.**
J. M. Mead, A. Boggess.
Advances in ultraviolet astronomy, (see 012.006), p. 322 - 325 (1982).

The authors have searched six astronomical journals and identified 343 papers describing studies using data obtained with the International Ultraviolet Explorer satellite. They have recorded the names of the astronomical objects discussed. These objects have been compiled into a list of 2460 entries, along with each reference, and sorted by object name or catalogue number.

002.043 **The JPL millimeter, sub-millimeter, and microwave line catalog.**
H. Pickett.
Vibrational-rotational spectroscopy for planetary atmospheres, (see 012.007), p. 191 - 193 (1982).

The catalog includes molecules of interest in the interstellar medium, in planetary atmospheres and in the earth's atmosphere.

002.044 **The AFGL molecular line atlases.**
L. Rothman.
Vibrational-rotational spectroscopy for planetary atmospheres, (see 012.007), p. 195 - 208 (1982).

The author describes the current status of the AFGL line compilations. There are two compilations extant, the main atlas and the trace gas atlas. The main atlas includes the major absorbers in the earth's atmosphere.

002.045 **Providing a data base with results of primary reduction of information from the Interkosmos 13 satellite.**
M. Slivka.
Obrab. inf., poluch. po programme Interkosmos, Moskva, 1982 p. 106 - 111. In Russian. – Abstr. in Ref. zh., 62. Issled. kosm. prostranstva, 7.62.104 (1982).

002.046 **Un nouveau catalogue d'étoiles variables suspectes.**
E. Schweitzer.
Bull. AFOEV, No. 21, p. 108 - 109 (1982).

002.047 **The Parkes 2700 MHz survey: a resurvey of the ± 4° declination zone.**
A. E. Wright, D. L. Jauncey, J. G. Bolton, A. Savage.
Australian J. Phys., Vol. 35, 177 - 206 (1982).
The ± 4° declination zone of the Parkes 2700 MHz survey has been resurveyed with the Parkes 64 m telescope. The catalogue of 822 sources is complete to a limiting flux density of 0.25 Jy ($\equiv 0.25 \times 10^{-26}$ Wm^{-2} Hz^{-1}), substantially lower than the original (1971) survey. All but one of the sources have flux density measurements at 5000 MHz. Appreciable changes in flux density have been found for 20% of the sources in the original survey and 59 new optical identifications have been made.

002.048 **New optical identifications from the Parkes 2700 MHz resurvey of the ± 4° declination zone.**
A. Savage, J. G. Bolton, J. Trett.
Australian J. Phys., Vol. 35, 207 - 219 (1982).
Identifications are suggested for 59 radio sources from the resurvey of declination zone ± 4°, 27 with galaxies and 32 with stellar objects; 19 of the latter are confirmed as quasi-stellar objects (QSOs). The identifications were made from the Palomar Sky Survey prints plus supplementary plates taken with the Palomar and U.K. Schmidt telescopes and the Anglo-Australian 3.9 m telescope.

002.049 **Meteoritical Bulletin. Index 1957 - 1982, Issues 1 - 60.**
Meteoritics, Vol. 17, 99 - 115 (1982).

002.050 **Resources for the history of astronomy.**
C. J. McKown.
Astron. Q., Vol. 4, 65 - 74, 123 - 131 (1982).

002.051 **Les cent ans de la "Bibliographie générale de l'astronomie" de J. C. Houzeau et A. Lancaster.**
J. Sauval.
Ciel Terre, Vol. 98, 180 (1982).

002.052 **Bibliographie d'Henri Michel.**
J. De Graeve.
Suppl. Ciel Terre, Vol. 98, 6 - 24 (1982).

002.053 **Catalogue of the meteorite collection of the Leningrad mining museum on January 1, 1979.**
V. D. Kolomenskij, E. G. Gus'kova.
Meteoritika, vyp. 40, p. 134 - 158 (1982). In Russian.

002.054 **A catalog of gamma-ray bursts with earth crossing times.**
R. Klebesadel, W. D. Evans, J. G. Laros, I. B. Strong, T. L. Cline, U. Desai, B. J. Teegarden, C. Barat, K. Hurley, M. Niel, G. Vedrenne, I. V. Estulin *(I. V. Ehstulin),* A. V. Kuznetsov, V. M. Zenchenko.
Astrophys. J., Lett., Vol. 259, L51 - L56 (1982).
A catalog of 111 confirmed gamma-ray bursts detected between 1967 July and 1979 June is presented. Both localization information and Earth crossing times for the gamma-ray wave fronts are given. Data which have appeared in some previous catalogs have been revised; many previously unpublished events detected by an international network of dedicated deep space and near-Earth experiments since 1976 are also presented.

002.055 **The radio properties of a complete sample of bright galaxies.**
J. I. Harnett.
Australian J. Phys., Vol. 35, 321 - 350 (1982).
Results are given for the radio continuum properties of an optically complete sample of 294 bright galaxies. The radio luminosity functions for all galaxies and for spiral galaxies alone are derived and the radio emission for different galaxy types is investigated. Spectral indices of 73 galaxies which had been detected at other frequencies were derived.

002.056 **Catalogue of planetary nebula radial velocities.**
S. E. Schneider, Y. Terzian.
News Lett. Astron. Soc. N.Y., Vol. 2, No. 2, p. 7 (1982). Abstract.

002.057 **Results from Volume 3 of the Michigan Catalogue of two-dimensional spectral types for the HD stars (−40° to −26°) and future plans.**
N. Houk, T. H. Fuentes-Williams.
Bull. American Astron. Soc., Vol. 14, 615 (1982). – Abstract.

002.058 **A compilation catalog in the direction of the galactic center.**
L. G. Taff, S. A. Stansfield.
Bull. American Astron. Soc., Vol. 14, 655 (1982). – Abstract.

002.059 **Results and problems of VINITI's information activity in astronomy, geodesy and space research.**
I. S. Shcherbina-Samojlova.
Informatics in astronomy and geodesy, (see 003.008), p. 5 - 11 (1982). In Russian. – Abstr. in Ref. zh., 51. Astron., 9.51.12 (1982).

002.060 **Structure and dynamics of the documentary information flow in astronomy.**
T. I. Zapol'skaya, I. S. Shcherbina-Samojlova.
Informatics in astronomy and geodesy, (see 003.008), p. 12 - 20 (1982). In Russian. – Abstr. in Ref. zh., 51. Astron., 9.51.13 (1982).

002.061 **Coverage of the astronomical world literature and information importance of periodical publications.**
T. I. Zapol'skaya, I. S. Shcherbina-Samojlova.
Informatics in astronomy and geodesy, (see 003.008), p. 21 - 28 (1982). In Russian. – Abstr. in Ref. zh., 51. Astron. 9.51.14 (1982).

002.062 **Structure of the information flow of scientific and technical literature on geodesy.**
D. D. Dzyaman.
Informatics in astronomy and geodesy, (see 003.008), p. 28 - 36 (1982). In Russian. – Abstr. in Ref. zh., 51. Astron., 9.51.15 (1982).

002.063 **Structural characteristics of the documentary flow in geodesy.**
D. D. Dzyaman.
Informatics in astronomy and geodesy, (see 003.008), p. 36 - 43 (1982). In Russian. – Abstr. in Ref. zh., 51. Astron., 9.51.16 (1982).

002.064 **Lexical and terminological structure of astronomical texts.**
B. N. Panovkin.
Informatics in astronomy and geodesy, (see 003.008), p. 44 - 60 (1982). In Russian. – Abstr. in Ref. zh., 51. Astron., 9.51.17 (1982).

002.065 **Retrospective retrieval using the Referativnyi zhurnal "Astronomiya".**
N. P. Slovokhotova, I. S. Shcherbina-Samojlova.
Informatics in astronomy and geodesy, (see 003.008), p. 61 - 71 (1982). In Russian. – Abstr. in Ref. zh., 51. Astron., 9.51.18 (1982).

002.066 **UDC as an information retrieval language in the Referativnyj zhurnal "Astronomiya".**
I. S. Shcherbina-Samojlova, T. N. Mandryka.
Informatics in astronomy and geodesy, (see 003.008), p. 71 -

76 (1982). In Russian. – Abstr. in Ref. zh., 51. Astron., 9.51.19 (1982).

002.067 **Information services for specialists at an astronomical observatory.**
T. I. Zapol'skaya, I. S. Shcherbina-Samojlova, G. S. Shvedova.
Informatics in astronomy and geodesy, (see 003.008), p. 77 - 81 (1982). In Russian. – Abstr. in Ref. zh., 51. Astron., 9.51.20 (1982).

002.068 **Investigation of publications on a fundamental scientific problem (using literature on pulsars, as an example).**
V. T. Fedorov, S. F. Komissarova.
Informatics in astronomy and geodesy, (see 003.008), p. 81 - 97 (1982). In Russian. – Abstr. in Ref. zh., 51. Astron., 9.51.21 (1982).

002.069 **On the preparation of a systematic index of scientific and technical information concerning the exploration of the moon.**
L. D. Revina, I. S. Shcherbina-Samojlova.
Informatics in astronomy and geodesy, (see 003.008), p. 98 - 103 (1982). In Russian. – Abstr. in Ref. zh., 51. Astron., 9.51.22 (1982).

002.070 **Glossary of planetological terms.**
L. D. Revina.
Informatics in astronomy and geodesy, (see 003.008), p. 104 - 119 (1982). In Russian. – Abstr. in Ref. zh., 51. Astron., 9.51.23 (1982).

002.071 **Structure of the information flow as an indicator of senseful differentiation of scientific trends in astronomy.**
V. A. Minin.
Informatics in astronomy and geodesy, (see 003.008), p. 120 - 127 (1982). In Russian. – Abstr. in Ref. zh., 51. Astron., 9.51.24 (1982).

002.072 **On a new approach to the analysis of the structure of astronomical research.**
V. A. Minin.
Informatics in astronomy and geodesy, (see 003.008), p. 127 - 132 (1982). In Russian. – Abstr. in Ref. zh., 51. Astron., 9.51.25 (1982).

002.073 **Traducción al castellano de términos de la astronomía moderna.**
H. Alvarez, L. E. Campusano.
Rev. Mexicana Astron. Astrofis., Vol. 6, (see 012.017), 379 - 382 (1981).
A dictionary of astronomical terms is proposed for translation into Spanish.

002.074 **Bibliography of the moon and the planets.** Articles entered into the data base at the Lunar and Planetary Institute Library, January - April, 1981.
Z. Kopal, M. Moutsoulas, F. B. Waranius (Editors).
Moon Planets, Vol. 26, 399 - 453 (1982).

002.075 **The archiving, distribution and utilization of planetary science data: an old concept that is possible with new technology.**
A. L. Lane.
Bull. American Astron. Soc., Vol. 14, 715 - 716 (1982). Abstract.

002.076 **A new version of TRIAD (*Tucson Revised Index of Asteroid Data*).**
B. Zellner.
Bull. American Astron. Soc., Vol. 14, 720 (1982). – Abstract.

002.077 **Asteroid data files for mission analysis.**
D. F. Bender.
Bull. American Astron. Soc., Vol. 14, 728 (1982). – Abstract.

002.078 **The fourth meridian catalogue of Besançon Observatory.**
M. Crézé, B. Mazodier, J. Clairemidi, J. Colin, S. Considère, G. Hilaire, E. Oblak, J. P. Parisot, F. Puel, R. Andrez, M. Athanase, D. Chabod, L. Godet, P. Jeandenans, B. Mougin, A. Vincent, M. Vincent.
Astron. Astrophys., Suppl. Ser., Vol. 50, 147 - 171 (1982). In French.
The "Quatrième catalogue de l'Observatoire de Besançon" gives differential meridian positions for 670 F type stars in the declination zone 15° to 45°. The transits have been observed photoelectrically through a Høg's grid (Sauzeat, 1974). Circle readings are classical. FK4 stars have been used as reference stars. Positions are reduced to equinox 1950.0 without proper motions. Mean corrections (Catalogue, FK4) for 283 reference stars are given.

002.079 **Supplement to the astronomical terminology. I.**
G. Ivanišević.
Publ. Astron. Obs. Sarajevo, Vol. 1, No. 1, (see 012.043), p. 65 - 68 (1981). In Croatian.

002.080 **A catalog of infrared magnitudes and H I velocity widths for nearby galaxies.**
M. Aaronson, J. Huchra, J. R. Mould, R. B. Tully, J. R. Fisher, H. van Woerden, W. M. Goss, P. Chamaraux, U. Mebold, B. Siegman, G. Berriman, S. E. Persson.
Astrophys. J., Suppl. Ser., Vol. 50, 241 - 262 (1982).
Integrated infrared (1.6 μm) magnitudes and 21 cm velocity widths are presented for 308 nearby spiral galaxies. Positions, types, inclinations, diameters, H I velocities and flux density integrals, and distances are also listed. These data form the basis for a recent analysis of the velocity field in the Local supercluster.

002.081 **A catalogue of parameters for eclipsing binaries.**
H. Brancewicz.
Ejection and accretion of matter in binary systems, (see 012.051), p. 109 - 110 (1982). – See also Abstr. 29.002.053.
Main parameters for 1048 binaries have been collected and summarized in the catalogue. Part of data as absolute radii, absolute bolometric luminosities, temperature, sum of mass and mass ratio for some binary systems were derived by the authors.

002.082 **Documentation for the machine-readable version of the Catalog of Galactic O-Type Stars.**
W. H. Warren, Jr.
NSSDC/WDC-A-R&S, 82-15, 16 pp. (1982).

002.083 **Documentation for the machine-readable version of the Third Santiago-Pulkovo Fundamental Stars Catalogue (SPF 3).**
W. H. Warren, Jr.
NSSDC/WDC-A-R&S, 82-17, 9 pp. (1982).

002.084 **Documentation for the machine-readable version of the Bright Star Catalogue, 4th edition.**
W. H. Warren, Jr.
NSSDC/WDC-A-R&S, 82-18, 23 pp. (1982).

002.085 **Documentation for the machine-readable version of the Lowell Proper Motion Survey, Northern Hemisphere, The G Numbered Stars.**
W. H. Warren, Jr.
NSSDC/WDC-A-R&S, 82-19, 17 pp. (1982).

002.086 **Documentation for the machine-readable version of the General Catalogue of Trigonometric Stellar Parallaxes and Supplement.**
W. H. Warren, Jr.
NSSDC/WDC-A-R&S, 82-20, 12 pp. (1982).

002.087 **NSSDC Data Listing.**
NSSDC/WDC-A-R&S, 82 - 23, 68 pp. (1982).
The NSSDC Data Listing provides a convenient reference to space science and supportive data available from the National Space Science Data Center (NSSDC). The first part of this listing, Satellite Data, is in an abbreviated form compared to the data catalogs published by NSSDC. The second part, Supplementary Data, contains a listing of ground-based data, models, computer routines, and composite spacecraft data that are available from NSSDC.

002.088 **Documentation for the machine-readable version of the Catalogue of Nearby Stars, Edition 1969.**
W. H. Warren, Jr.
NSSDC/WDC-A-R&S, 82-26, 15 pp. (1982).

002.089 **Catalogue of the Small Magellanic Cloud star members.**
M. Azzopardi, J. Vigneau.
Astron. Astrophys., Suppl. Ser., Vol. 50, 291 - 376 (1982).
This catalogue contains 524 Small Magellanic Cloud (SMC) members located between 0^h51 and 1^h22 in right ascension for the epoch 1975.0. Equatorial coordinates, localization of stars with respect to the complexes of ionized hydrogen and information about possible close optical companions are given. Photoelectric photometric UBV data, spectral classifications and radial velocities published by various authors are listed as well in this catalogue. A master set and 17 identification astrographic charts including one panelled chart for isolated members are provided.

002.090 **New catalogue of suspected variable stars,** containing information on 14810 variable stars not named until 1980 (including 2475 objects for which the discovery of light variability is doubtful or erroneous).
B. V. Kukarkin, P. N. Kholopov, N. M. Artyukhina, V. P. Goranskij, N. A. Gorynya, E. A. Karitskaya, N. N. Kireeva, N. P. Kukarkina, N. E. Kurochkin, G. I. Medvedeva, N. B. Perova, G. A. Ponomareva, N. N. Samus', V. P. Fedorovich, M. S. Frolov, S. Yu. Shugarov.
Astronomicheskij Sovet Akademii Nauk SSSR, Gosudarstvennyj Astronomicheskij Institut im. P. K. Shternberga pri MGU. Izdatel'stvo Nauka. 360 pp. (1982). In Russian and English.
The present edition contains data on 14810 stars suspected of variability which did not get final designations by 1980. This is a new, completely revised catalogue of suspected variables, replacing former editions of similar catalogues (Catalogue of Suspected Variable Stars, 1951, and the Second Catalogue of Suspected Variable Stars, 1965).

002.091 **A catalogue of peculiar and interacting galaxies.**
H. C. Arp, B. F. Madore.
J. R. Astron. Soc. Canada, Vol. 76, 315 (1982).
Abstract.

002.092 **A glossary of space terms. (Recueil de terminologie spatiale).**
Y. Taillefer.
ESA SP-1011. European Space Agency, Scientific and Technical Publications Branch, ESTEC, Noordwijk, The Netherlands. 432 pp. Price FF 200.00 (1982). ISSN 0379-6566.
This glossary of scientific and technical terms in space research contains translations of these terms in English, French and partially in German. Descriptions and explanatory notes including some 200 figures are given.

002.093 **Publications and research papers of William A. Fowler.**
Essays in nuclear astrophysics, (see 003.013), p. 533 - 544 (1982).

002.094 **Storage of the TRIAD-file (Tucson Revised Index of Asteroid Data) at the Centre de Données Stellaires (CDS)?**
H. J. Schober.
Mitt. Astron. Ges., Nr. 57, p. 328 (1982). – Abstract.

002.095 **Bibliography on Be stars.**
A. Feinstein, A. M. Hubert-Delplace, H. Hubert, J. Jugaku, P. Koubsky.
Be Star Newsl., No. 2, p. 11 - 17 (1980), No. 3, p. 20 - 25, No. 4, p. 27 - 32 (1981), No. 5, p. 11 - 19, No. 6, p. 37 - 41 (1982).

002.096 **A catalogue, spectrum and magnitude data bank for Be, Bp and Bpe stars.**
A. A. Page.
Be Star Newsl., No. 4, p. 2 - 6 (1981).

002.097 **Second Catalogue d'Éphémérides des vitesses radiales relatives des composantes des étoiles doubles visuelles dont l'orbite est connue.**
J. Dommanget, O. Nys.
Obs. R. Belgique, Commun., Sér. B, No. 124, 280 pp. (1982).

002.098 **Astronomical data in machine readable form.**
United States Naval Obs. Circ., No. 164, 19 pp. (1981).

002.099 **Organizing a data base for solving planetodesic problems on the ES computer.**
T. I. Murunova.
Sb. nauchn. tr. TsNII geod., aehrosemki i kartogr., 1981, No. 228, p. 110 - 120. In Russian. – Abstr. in Ref. zh., 52. Geod. Aehrosemka, 11.52.218 (1982).

002.100 **A library of digitized spectra of stars classified in the MK system.**
H. Greidanus, J. Tinbergen.
Bull. Inf. Cent. Données Stellaires, No. 23, p. 15 - 17 (1982).

002.101 **The best and the worse in nomenclature: a proposed list of basic designations.**
A. Fernandez, M. C. Lortet, F. Spite.
Bull. Inf. Cent. Données Stellaires, No. 23, p. 69 - 71 (1982).

002.102 **Errors in identification of two stars, HD 227310 and HD 237089 in photometric listings.**
D. Hoffleit.
Bull. Inf. Cent. Données Stellaires, No. 23, p. 72 - 73 (1982).

002.103 **Errata in the "Catalogue of Star Clusters and Associations", Suppl. 1, Part A.**
J. Ruprecht.
Bull. Inf. Cent. Données Stellaires, No. 23, p. 74 (1982).
See Abstr. 31.002.111.

002.104 **On the coverage in the Cordoba zones of the Astrographic Catalogue. Comments on a note by Herald.**
G. Torres.
Bull. Inf. Cent. Données Stellaires, No. 23, p. 76 - 77 (1982).

002.105 **Some comments on the Spectral Classification Catalogue by W. Buscombe.**

E. K. Kharadze.
Bull. Inf. Cent. Données Stellaires, No. 23, p. 78 (1982).

002.106 **A computer catalog of the Kiso Schmidt plates.**
H. Maehara, T. Noguchi, T. Soyano.
Bull. Inf. Cent. Données Stellaires, No. 23, p. 79 - 80 (1982).

002.107 **Supplement to the Second Catalogue of Am Stars with Known Spectral Types (End 1981).**
B. Hauck.
Bull. Inf. Cent. Données Stellaires, No. 23, p. 85 - 88 (1982).

002.108 **A catalogue of photometric sequences.**
A. N. Argue, E. W. Miller, W. H. Warren.
Bull. Inf. Cent. Données Stellaires, No. 23, p. 90 - 91 (1982).

002.109 **A first glance at the Catalogue of the Central Stars of Planetary Nebulae.**
A. Acker.
Bull. Inf. Cent. Données Stellaires, No. 23, p. 92 - 95 (1982).

002.110 **Space velocity vectors of 2595 B and A stars.**
J. Palouš.
Bull. Inf. Cent. Données Stellaires, No. 23, p. 96 - 97 (1982).

002.111 **A new catalog of ephemerides of v_r for orbital visual binaries.**
J. Dommanget, O. Nys.
Bull. Inf. Cent. Données Stellaires, No. 23, p. 98 - 100 (1982).

002.112 **Visual double star data base.**
C. E. Worley.
Bull. Inf. Cent. Données Stellaires, No. 23, p. 101 (1982).

002.113 **IAU Commission 45. Working Group on Spectroscopic and Photometric Data. Catalogs recently published, to be published or in preparation, List XV.**
Bull. Inf. Cent. Données Stellaires, No. 23, p. 102 - 103 (1982).

002.114 **Star catalogs and files available at the Stellar Data Center.**
Bull. Inf. Cent. Données Stellaires, No. 23, p. 108 - 123 (1982).

002.115 **IUE ultraviolet spectral atlas.**
C. C. Wu, A. Boggess, R. C. Bohlin, A. V. Holm, F. H. Schiffer III, B. E. Turnrose.
Ultraviolet stellar classification, (see 012.062), p. 25 - 54 (1982).

The IUE observatory at the Goddard Space Flight Center is preparing an atlas of low dispersion UV spectra for a group of normal stars. The atlas will present calibrated plots, tables of fluxes at 5 Å intervals and photographic representation of the spectra. A list of stars observed before October 15, 1981 and a sample of spectral plots and flux tables are presented.

002.116 **The Vilspa IUE data bank.**
P. Benvenuti.
Ultraviolet stellar classification, (see 012.062), p. 81 - 83 (1982).

002.117 **Supplement No. 3: Catalogue of Photometric Sequences.**
A. N. Argue, E. W. Miller.
Publ. Astron. Soc. Pacific, Vol. 94, 748 (1982). – Abstract.

002.118 **On an analysis of the BBC key to the "B6 Astronomy" section.**
L. L. Kostritskij.
Sov.-bibl.-bibliogr. klassifik. Leningrad, 1982, No. 4, p. 6 - 23. In Russian. – Abstr. in Ref. zh., 51. Astron., 10.51.34 (1982).

002.119 **Comparative analysis of subject indexes to the abstract journals "Astronomy" and "Astronomy and Astrophysics Abstracts".**
T. N. Mandryka, N. P. Slovokhotova, I. S. Shcherbina-Samojlova.
Vsesoyuznyj Institut Nauchnoj i Tekhnicheskoj Informatsii (VINITI). Moskva, 1982. 12 pp. In Russian. – Abstr. in Ref. zh., 51. Astron., 11.51.31 (1982).

002.120 **The SAOC and the AGK3.**
L. G. Taff, S. A. Stansfield.
Astron. J., Vol. 87, 1884 - 1892 (1982).

An analysis of a comparison of the AGK3 with the SAOC is presented. Stars in common to both catalogs, with both matching BD numbers and a positional (proper motion) difference of at most 2$''$.5 (2$''$.5/cent), provide the raw data. There are 128998 (129 744) such stars north of declination −2°.5. The results presented are position-independent comparisons for $\Delta\alpha \cos\delta$, $\Delta\delta$, $\Delta\mu_\alpha \cos\delta$, and $\Delta\mu_\delta$, along with the time dependence of these quantities, a classical analysis of the variation of these differences over the Northern hemisphere, and a position-independent exploration of color index and magnitude effects. In addition, an analysis of the above quantities based on an expansion of the systematic differences in a complete set of orthonormal basis functions is discussed. These results are examined by statistical means and the significant expansion coefficients tabulated. Tables of the systematic differences are also included.

002.121 **Catalogue of absolute right ascensions of 1960 bright (FK4) and faint (FKSZ) stars of the southern sky.**
M. P. Varin, V. A. Varina, V. S. Gubanov, L. I. Medvedeva, A. A. Nemiro, D. D. Polozhentsev, T. A. Polozhentseva, G. M. Timashkova.
Tr. Glav. Astron. Obs. Pulkovo, Ser. 2, Tom 84, p. 4 - 60 (1981). In Russian.

The present catalogue was compiled from absolute observations made with the Pulkovo large transit instrument at the Cerro Calan Observatory in 1969 - 1973. Altogether 62887 observations were made. The mean error of one determination equals ±0^s.019. The existence of large $\Delta\alpha_\delta \cos\delta$ and $\Delta\alpha_\alpha \cos\delta$ errors of the FK4 is affirmed.

002.122 **Catalogue of absolute right ascensions of 2256 stars from the Backlund-Hough list from +32° to −90° declination observed at the Melbourne Observatory during 1928 - 1941, Me_{50}I (α).**
K. N. Tavastsherna.
Tr. Glav. Astron. Obs. Pulkovo, Ser. 2, Tom 84, p. 61 - 118 (1981). In Russian.

002.123 **Catalogue of absolute right ascensions of 144 latitude stars from the Kimura List observed at the Melbourne Observatory during 1933 - 1935, Me_{50}II (α).**
K. N. Tavastsherna, L. I. Yagudin.
Tr. Glav. Astron. Obs. Pulkovo, Ser. 2, Tom 84, p. 119 - 123 (1981). In Russian.

002.124 **Catalogue of right ascensions of 35 stars from +32° to −90° declination from the A. Kopff List observed at the Melbourne Observatory during 1928 - 1940, Me_{50}III (α).**
K. N. Tavastsherna, L. I. Yagudin.
Tr. Glav. Astron. Obs. Pulkovo, Ser. 2, Tom 84, p. 124 - 126 (1981). In Russian.

002.125 **A preliminary RA catalogue of PZT stars on the International Parallel.**
Z. M. Malkin.
Izv. Glav. Astron. Obs. Pulkovo, Astrometr. Astrofiz., No. 199, p. 41 - 42 (1982). In Russian.

A preliminary RA catalogue of PZT stars on the International parallel has been compiled using observations with the PZT at Kitab.

002.126 **Michigan Catalogue of Two-dimensional Spectral Types for the HD stars. Volume 3. Declinations −40°0 to −26°0.**
N. Houk.
Published by the Department of Astronomy, University of Michigan, Ann Arbor, Michigan, USA. 12 + 390 pp. (1982).

This is the third of several volumes in a program of systematic reclassification of the Henry Draper stars on the MK system and contains all HD stars between $\delta_{1900} = -40°0$ and $-26°0$. The stars were classified visually on objective-prism plates taken with the Michigan Curtis Schmidt telescope at Cerro Tololo Inter-American Observatory. The 4° + 6° prisms yield a dispersion at Hγ of 108 Å/mm, and the resolution is about 2 Å/mm, comparable to that of the original MK system. The spectra were taken on IIaO plates and widened to 0.8 mm, with 20^m, 4^m, and 1^m exposures being obtained. Of the 30,314 stars in the catalogue, 29,652 or 98% were classifiable. Previously published types are listed for the 2% not classifiable on the plates. Internal classification errors were obtained using the stars classified independently on more than one plate. Both the internal and external errors are similar to those usually obtained in the classification of slit spectra.

002.127 **Atlas of South Pole IR solar spectra.**
R. D. Blatherwick, F. J. Murcray, F. H. Murcray, A. Goldman, D. G. Murcray.
Appl. Opt., Vol. 21, 2658 - 2659 (1982).

002.128 **Deuxième complément à la bibliographie des publications ayant rapport avec l'astrolabe Danjon ou avec d'autres astrolabes de types nouveaux.**
With an introduction by A. Stoyko.
Published by Observatoire de Paris, Paris, France, 13 pp. (1982).

002.129 **Viking Lander Atlas of Mars.**
S. Liebes, Jr.
NASA Contract. Rep., NASA CR-3568, 10 + 290 pp. (1982).

This Atlas contains half-size reproductions of the complete set of true-scale systematic map products that have been generated for the two Mars Viking landing sites from stereo pairs of images radioed to Earth from the landers. The maps cover from the immediate foreground to the remote limits of ranging capability, several hundred meters from the landers. The maps are of two kinds: 1) elevation contour, and 2) vertical profile. The Atlas includes background and explanatory material important for understanding and utilizing the maps. The collection consists of nearly 200 individual sheets, spanning eleven different scales from 1:1 to 1:2000. The Atlas includes an extensive set of both single and stereo mosaic images into which the contour and vertical profile data have been inlayed.

002.130 **NASA Catalogue of Lunar Nomenclature.**
L. E. Andersson, E. A. Whitaker.
NASA Ref. Publ., NASA RP-1097, 190 pp. (1982).

002.131 **A bibliography of planetary geology principal investigators and their associates, 1981 - 1982.**
J. B. Plescia.
NASA Contract. Rep., NASA CR-3593, 94 pp. (1982).

002.132 **Comments on the gamma-ray burst catalog of Mazets et al. (1981a).**
J. G. Laros, W. D. Evans, E. E. Fenimore, R. W. Klebesadel.
Astrophys. Space Sci., Vol. 88, 243 - 246 (1982).

The authors have discovered a correlation between the Venera spacecraft locations and the gamma-ray burst positions reported in the KONUS catalog (Mazets et al., 1981a). The reason for the correlation is not clear, but it could be due to spatial selection effects and/or large localization errors for weak or soft bursts. Whatever the cause, it seems likely that this systematic bias might significantly affect catalog results pertaining to gamma burst locations, intensity distributions, and spectra. For example, it can explain why the KONUS galactic latitude distribution is peaked significantly south of the galactic equator. Apparent discrepancies between the KONUS and Los Alamos gamma burst data bases are noted.

002.133 **A reply to 'Comments on the gamma-ray burst catalog' of Laros et al.**
E. P. Mazets, S. V. Golenetskii (*Golenetskij*).
Astrophys. Space Sci., Vol. 88, 247 - 251 (1982).

Laros et al. (1982) reported on a discovery of a correlation between the Venera spacecraft positions and the gamma-burst locations specified in the Catalog (Mazets et al., 1981). This correlation may indeed seem puzzling. Therefore the authors explain its nature and show that it does not affect the final results.

002.134 **New photographic sky atlases.**
P. Mayer.
Říše hvězd, Vol. 63, 158 - 159, 165 - 168 (1982). In Czech.

002.135 **Perth 75. A catalogue of positions of 2589 FK4 and FK4S stars.**
I. Nikoloff, E. Høg, in collaboration with C. Ayers, I. Nikoloff, C. Spencer, M. Sultana, A. Verveer, P. J. Høyer.
Published by Perth Observatory, Bickley, Western Australia. 71 pp. (1982).

This catalogue lists the results of about 60000 observations obtained with the photoelectric meridian circle at the Perth Observatory in Bickley for 1156 stars of FK4 and 1433 stars of FK4 Supplement with $\delta < +38°$.

002.136 **Documentation for the machine-readable version of A Finding List of Stars of Spectral Type F2 and Earlier in a North Galactic Pole Region.**
W. H. Warren, Jr.
NSSDC/WDC-A-R&S, 82 - 27, 15 pp. (1982).

002.137 **Documentation for the machine-readable version of The Revised New General Catalogue of Nonstellar Astronomical Objects.**
W. H. Warren, Jr.
NSSDC/WDC-A-R&S, 82 - 29, 33 pp. (1982).

002.138 **Documentation for the machine-readable version of the Uppsala General Catalogue of Galaxies.**
W. H. Warren, Jr.
NSSDC/WDC-A-R&S, 82 - 30, 15 pp. (1982).

002.139 **Documentation for the machine-readable version of the Morphological Catalogue of Galaxies (MCG) of Vorontsov-Velyaminov et al. 1962–1968.**
W. H. Warren, Jr.
NSSDC/WDC-A-R&S, 82 - 31, 19 pp. (1982).

002.140 **Ionization energies of some molecules found in interstellar clouds calculated by a Green's function method.**
W. Von Niessen, L. S. Cederbaum, J. Schirmer, G. H. F. Diercksen, W. P. Kraemer.
J. Electron. Spectrosc. Relat. Phenom., Vol. 28, No. 1, p. 45 - 78 (1982). – Abstr. in Phys. Abstr., Vol. 86, Abstr. 5502 (1983).

002.141 **AIP 50th anniversary physics Vade Mecum.**
H. L. Anderson (Editor).

AIP, New York, USA. 330 pp. (1981). ISBN 0-88318-289-0. Review in Phys. Abstr., Vol. 85, Abstr. 58565 (1982).

002.142 **The dictionary of space technology.**
J. A. Angelo, Jr.
Facts on File, Inc., 460 Park Ave. S., New York, N. Y. 10016. 383 pp. Price $ 19.95 (1982). ISBN 0-87196-583-6. – Review in Sky Telesc., Vol. 64, 153 (1982).

002.143 **Astronomy and Astrophysics Abstracts.** Vol. 31. Literature 1982, Part 1.
S. Böhme, W. Fricke, H. Hefele, I. Heinrich, W. Hofmann, D. Krahn, V. R. Matas, L. D. Schmadel, G. Zech (Editors).
Published for Astronomisches Rechen-Institut by Springer-Verlag, Berlin–Heidelberg–New York. 10 + 776 pp. Price DM 148.00, US $ 66.00 [Subscription price DM 118.40, US $ 52.80] (1982). ISBN 3-540-12072-6, ISBN 0-387-12072-6 (USA).

002.144 **The Cambridge photographic atlas of the planets.**
G. Briggs, F. Taylor.
Cambridge University Press, Cambridge – London – New York – New Rochelle – Melbourne – Sydney. 255 pp. Price £ 12.50, DM 65.00 (1982). ISBN 0-521-23976-1. Reviews in Astron. Tidsskr., Årg. 15, 191 (1982); Sky Telesc., Vol. 64, 559 (1982).

Contents: Introduction. Mercury. Venus. Earth and Moon. Mars. The Jovian system. The Saturnian system. Epilogue.

002.145 **Nébuleuses et galaxies – atlas du ciel profond.**
S. Brunier.
Editions Dunod, Paris. 192 pp. Price sFr. 24.00 (1981). ISBN 2-04-012188-9. – Review in Orion, 40. Jahrg., 166; 1982 (*W. Maeder*).

002.146 **The history of modern astronomy and astrophysics: a selected, annotated bibliography.**
D. H. De Vorkin.
Garland Publishing Inc., New York. 434 pp. Price $ 65.00 (1982). ISBN 0-8240-9283-X. – Reviews in J. British Astron. Assoc., Vol. 93, 42 - 43; 1982 (*D. Trombino*); Sky Telesc., Vol. 64, 152 (1982).

002.147 **Water vapour line parameters from microwave to medium infrared (an atlas of $H_2{}^{16}O$, $H_2{}^{17}O$ and $H_2{}^{18}O$ line positions and intensities between 0 and 4350 cm^{-1})**.
J.-M. Flaud, C. Camy-Peyret, R. A. Toth.
Pergamon Press, Oxford. 259 pp. Price £ 31.25 (1981). Review in Observatory, Vol. 102, 246 - 247; 1982 (*L. T. Little*).

002.148 **Atlas of the solar system.**
D. A. Hardy.
World's Work Ltd., The Windmill Press, Kingswood–Tadworth, Surrey. 96 pp. Price £ 12.50 (1982). – Review in Astron. Tidsskr., Årg. 15, 191 (1982).

002.149 **Atlas of photometric standards of stellar fields.**
M. S. Kazanasmas, L. A. Zavershneva, L. F. Tomak.
Naukova dumka, Kiev. 168 pp. (1982). In Russian. – Abstr. in Ref. zh., 51. Astron., 11.51.39 (1982).

002.150 **A new list of 450 deep-sky objects for amateur telescopes.**
F. W. Klein.
Published by F. W. Klein, Hawaiian Volcano Observatory, Hawaii Volcanoes National Park, HI 96718. 31 pp. Price $ 6.75 (1981). – Reviews in Sky Telesc., Vol. 64, 559 (1982); Strolling Astron., Vol. 29, 164 - 165; 1982 (*D. Machholz*).

002.151 **S 59 spectrophotometric atlas of the mid-ultraviolet spectra of 220 early-type stars with a resolution of 1.8 Å.**
H. J. G. L. M. Lamers, R. Hoekstra, K. A. van der Hucht, T. M. Kamperman.
Report ESA SP-1038, European Space Agency, Paris, France. 5 + 209 pp. (1981). – Review in Phys. Abstr., Vol. 85, Abstr. 63278 (1982).

002.152 **Key definitions in astronomy.**
J. Mitton.
Littlefield, Adams, Totowa, N. J., 168 pp. Price $ 4.95 (1982). From Phys. Today, Vol. 35, No. 10, p. 86 (1982).

002.153 **A catalogue of European scientific instruments in the Department of Medieval and Later Antiquities of the British Museum.**
F. A. B. Ward.
Published for the Trustees of the British Museum by British Museum Publications Ltd., London. 152 pp, 61 plates. Price £ 50.00 (1981). – Review in J. Hist. Astron., Vol. 13, 129 - 132; 1982 (*D. de Solla Price*).

002.154 **Lista de objetos celestes visibles en telescopios pequenos.**
J. G. H. von Wilhelms.
Juan G. Hernandez von Wilhelms, c/o Gonzalez Vilchis y Cia., S. C., Reforma No. 243-7° piso, 06500 Mexico, D. F., Mexico. 198 pp. Price $ 22.00 (1981). – Review in Sky Telesc., Vol. 64, 152 (1982).

002.155 **Concise dictionary of scientific biography.**
Charles Scribner's Sons, New York. 10 + 773 pp. Price $ 100.00 (1981). – Review in J. Hist. Astron., Vol. 13, 223 - 224; 1982 (*M. Hoskin*).

002.156 **Glossary of terms used in cosmology.**
American Institute of Physics, New York. 23 pp. (1982). – Review in Phys. Abstr., Vol. 85, Abstr. 89484 (1982).

The Stellar Data Center.
See Abstr. 013.004.

The astronomical data base and retrieval system at NASA.
See Abstr. 013.005.

The Soviet Center of Astronomical Data.
See Abstr. 013.006.

Management of astronomical data at Kanazawa Data Center.
See Abstr. 013.007.

Activities of the Potsdam Data Center.
See Abstr. 013.008.

IUE data distribution.
See Abstr. 013.009.

Infrared Astronomical Data Base and Catalog of Infrared Observations.
See Abstr. 013.010.

Data on time and polar motion: immediate accessibility.
See Abstr. 013.011.

The National Space Science Data Center (NSSDC) and the World Data Center A for Rockets and Satellites (WDC–A–R&S): their role in X-ray astronomy data.
See Abstr. 013.012.

The stellar data file at Dearborn Observatory.
See Abstr. 013.013.

Euronet-Diane, premier reseau européen de transfert de l'information.
See Abstr. 013.014.

The U.K. StarLink computer network.
See Abstr. 013.015.

ASTRONET. The network for analysis and retrieval of astronomical data in Italy.
See Abstr. 013.016.

The International URSIGRAM and World Days Service.
See Abstr. 013.017.

A data analysis facility for the Faint Object Camera.
See Abstr. 013.018.

Responsibilities and practical limitations in the operation of an astronomical data center.
See Abstr. 013.019.

Numerical data documentation in physics in the Federal Republic of Germany.
See Abstr. 013.020.

New developments in data storage.
See Abstr. 021.003.

System software approaches to the analysis of multidimensional data structures.
See Abstr. 021.004.

A fast method to retrieve data from a large star catalogue file.
See Abstr. 021.005.

On systems of standards.
See Abstr. 021.006.

Digital image processing and data retrieval: the needs.
See Abstr. 021.007.

International standards for software structures in astronomy.
See Abstr. 021.008.

Archiving and retrieval of data from the International Ultraviolet Explorer (IUE) mission.
See Abstr. 021.009.

Towards effective data processing: portable software models.
See Abstr. 021.010.

A process for retrieval of data from a compiled star catalogue.
See Abstr. 021.011.

The system SPORA *(Software Package Oriented to Research in Astronomy).*
See Abstr. 021.012.

Reduction of observations of declinations of bright stars and cepheids.
See Abstr. 041.002.

A pool of faint stars applied to star catalogue formation.
See Abstr. 041.017.

A comparison of the AGK 3 star catalogue declinations with the results of latitude observations with zenith telescopes and photographic zenith tubes.
See Abstr. 041.024.

Effect of different sources of variation of latitude data on meridian circle catalogues.
See Abstr. 045.001.

Catalogue of solar activity for the year 1977.
See Abstr. 072.047.

Catalogue of solar activity for the year 1978.
See Abstr. 072.048.

Catalogue of minor planet identities. I. Identities with planets (1)-(2297).
See Abstr. 098.001.

Radial velocities of 617 stars belonging to four stellar fields of 4° × 4°.
See Abstr. 111.001.

Be star terminology.
See Abstr. 114.184.

Variable stars in the General Catalogue of Trigonometric Parallaxes.
See Abstr. 123.030.

Distribution of the stars in the *General Catalogue of Variable Stars.*
See Abstr. 123.033.

An atlas of UV spectra of supernovae.
See Abstr. 125.016.

Selected area surveys at 5 GHz – I. Catalogue for the 22^h, $-18°$ region.
See Abstr. 141.031.

Selected area surveys at 5 GHz – II. Optical identifications for the 22^h, $-18°$ region.
See Abstr. 141.032.

IAU numbers for some recently discovered clusters.
See Abstr. 153.035.

The Maryland-Green Bank Galactic 21-cm Line Survey.
See Abstr. 156.001.

The small-angle galaxy correlation function as a function of magnitude.
See Abstr. 158.128.

Errata

002.901 **Erratum: "Catalogue of stars with Ca II H and K emissions"** [Acta Astron., Vol. 30, 453 - 478 (1980)].
R. Głebocki, G. Musielak, A. Stawikowski.
Acta Astron., Vol. 32, 145 (1982). – See Abstr. 29.002.052.

002.902 **Erratum: "Errors or omissions in star-identifications in the General Catalogue of Trigonometric Stellar Parallaxes"** [Bull. Inf. Cent. Données Stellaires, No. 22, p. 112 - 117 (1982)].
D. Hoffleit.
Bull. Inf. Cent. Données Stellaires, No. 23, p. 75 (1982).
See Abstr. 31.002.029.

002.903 **Erratum: "New and revised catalogues available from the Astronomical Data Center (ADC) at NASA/Goddard Space Flight Center"** [Bull. Inf. Cent. Données Stellaires, No. 22, p. 127 - 130 (1982)].
Bull. Inf. Cent. Données Stellaires, No. 23, p. 75 (1982).
See Abstr. 31.002.032.

003 Books

003.001 **Galactic X-ray sources.** Based on the proceedings of the NATO Advanced Study Institute on galactic X-ray sources, held in Sounion, Greece.
P. W. Sanford, P. Laskarides, J. Salton (Editors), with a foreword by R. L. F. Boyd.
John Wiley & Sons, Chichester – New York – Brisbane – Toronto – Singapore, 17 + 450 pp. Price $ 45.00, DM 132.60 (1982). ISBN 0-471-27963-3. – The individual contributions are included in their corresponding subject categories – see abstracts 022.012, 063.009, 064.006, 064.007, 066.512, 114.016, 117.009 - 117.011, 122.014, 124.201, 125.008, 126.007, 142.010 - 142.030.

003.002 **The problem of motion in Einstein's gravitation theory.** Collection of papers.
Kaz. univ., Alma-Ata. 149 pp. (1981). In Russian. – From Ref. zh., 51. Astron., 7.51.61 (1982). – See abstracts 066.005 - 066.008, 066.010 - 066.014.

003.003 **Physics of planetary atmospheres.** Collection of scientific papers.
Naukova dumka, Kiev. 164 pp. (1981). In Russian. – From Ref. zh., 51. Astron., 7.51.54 (1982). – See abstracts 099.019 - 099.022, 100.017 - 100.019, 101.009, 101.010.

003.004 **K. E. Ziolkovskij and scientific-technical progress.**
Nauka, Moskva. 232 pp. (1982). In Russian. – From Ref. zh., 62. Issled. kosm. prostranstva, 8.62.62 (1982).
See abstracts 051.006, 051.007.

003.005 **Radiation detectors in terrestrial astronomy.**
G. S. Khromov. Edited by V. V. Prokof'eva.
Itogi nauki i tekhniki, Akademiya Nauk SSSR. Seriya Astronomiya, Tom 19. Vsesoyuznyj Institut Nauchnoj i Tekhnicheskoj Informatsii. Moskva. 88 pp. (1982). In Russian. ISSN 0202-0742.

003.006 **Handbuch der Physik. Band XLIX/6, Geophysik III, Teil VI.**
K. Rawer (Editor).
Springer Verlag, Berlin – Heidelberg – New York. 6 + 385 pp. Price DM 228.00 (1982). ISBN 3-540-07080-X, 0-387-07080-X. – The individual contributions are included in their corresponding subject categories – see abstracts 076.005, 082.033, 082.034, 083.007.

003.007 **Plasma physics and nuclear fusion research.**
R. D. Gill (Editor).
Academic Press, London, England. 20 + 688 pp. (1981). ISBN 0-12-283860-2. – Review in Phys. Abstr., Vol. 85, Abstr. 68980 (1982). – See Abstr. 062.044.

003.008 **Informatics in astronomy and geodesy.**
Eh. R. Mustel' (Editor).
Nauka, Moskva. 136 pp. (1982). In Russian. – From Ref. zh., 51. Astron., 9.51.31 (1982). – The individual contributions are included in their corresponding subject categories – see abstracts 002.059 - 002.072.

003.009 **Annual Review of Astronomy and Astrophysics. Volume 20.**
G. Burbidge, D. Layzer, J. G. Phillips (Editors).
Annual Reviews Inc., 4139 El Camino Way, Palo Alto, Calif. 94306, USA. 10 + 626 pp. Price £ 25.00 (1982). ISBN 0-8243-0920-0. – The individual contributions are included in their corresponding subject categories – see abstracts 033.023, 064.043, 065.045, 071.009, 077.017, 082.046, 084.049, 091.024, 091.025, 099.042, 131.162 - 131.164, 142.084, 151.041, 155.027, 155.028, 158.173, 160.043, 162.121.

003.010 **Telescopes for the 1980s.**
G. Burbidge, A. Hewitt (Editors).
Annual Reviews Monograph. Annual Reviews Inc., 4139 El Camino Way, Palo Alto, Calif. 94306, USA. 10 + 278 pp. Price $ 28.00 (1981). ISBN 0-8243-2902-3.
The individual contributions are included in their corresponding subject categories – see abstracts 032.014, 032.536, 032.537, 033.024.

003.011 **Solar active regions.** A monograph from Skylab Solar Workshop III, held in Boulder, Colorado, 6 - 8 July 1978, 30 October - 1 November 1978, 12 - 14 March 1979.
F. Q. Orrall (Editor).
Colorado Associated University Press, Boulder, Colorado 80309, USA. 10 + 350 pp. Price DM 85.60 (1981).
ISBN 0-87081-085-5. – The individual contributions are included in their corresponding subject categories – see abstracts 011.014, 013.066, 031.590, 072.030 - 072.032, 073.064, 073.065, 074.058 - 074.060. – Review in Space Sci. Rev., Vol. 32, 471 - 472; 1982 (B. Valniček).

003.012 **Revealing the Universe.** Prediction and proof in astronomy.
J. Cornell, A. P. Lightman (Editors).
MIT Press, Cambridge, Mass. and London, England. 16 + 246 pp. Price $ 17.50, £ 12.25 (1982). ISBN 0-262-03080-2.
Contents: The historical tension between astronomical theory and observation (*O. Gingerich*). Einstein's perceptions of space and time: Newton, Einstein and gravity (*K. Brecher*); Rockets, clocks and gravity (*R. F. C. Vessot*). The evolution of the solar system: The material ingredients and structure (*F. L. Whipple*); Computer simulations of the beginning (*F. Franklin*). The puzzle of the sun's hot corona: A matter of degrees (*R. W. Noyes*); Solar scenarios (*R. Rosner*). Cosmic powerhouses – quasars and black holes: The farthest and the brightest (*H. Tananbaum*); Too much from too little? (*A. P. Lightman*). The mystery of the X-ray burst sources (*W. H. G. Lewin*). The age and structure of the universe: A

tale of two theories (*W. H. Press*), The cosmic calendar (*J. Huchra*). Three unanswered questions in astronomy (*G. B. Field*).

003.013 **Essays in nuclear astrophysics.**
Presented to William A. Fowler on the occasion of his seventieth birthday.
C. A. Barnes, D. D. Clayton, D. N. Schramm (Editors).
Cambridge University Press, Cambridge – London – New York – New Rochelle – Melbourne – Sydney. 13 + 562 pp. Price $ 75.00 (1982). ISBN 0-521-24410-2. The individual contributions are included in their corresponding subject categories – see abstracts 002.093, 004.065, 005.022, 022.145, 061.033 - 061.043, 065.091, 080.042, 105.077, 105.078, 124.011, 125.099, 125.100, 162.206, 162.207.

003.014 **Advances in planetary geology.**
A. Woronow (Editor), with a foreword by J. M. Boyce.
NASA Tech. Memo., NASA TM-84894, 5 + 578 pp. (1982). The individual contributions (dissertations) are included in their corresponding subject categories – see abstracts 022.161 - 022.163, 081.043, 092.007, 094.523, 097.064, 098.082.

003.015 **Cosmo- and geochemistry.**
F. L. Boschke (Editor).
Topics in current chemistry, Vol. 99. Springer-Verlag, Berlin–Heidelberg–New York. 7 + 133 pp. Price DM 68.00 (1981). ISBN 3-540-10920-X. USA: ISBN 0-387-10920-X. – The individual contributions within the subject scope of Astronomy and Astrophysics Abstracts are included in their corresponding categories – see abstracts 102.096, 105.104, 131.329.

003.016 **Exploration of the Universe.**
G. O. Abell.
4th edition. Saunders, Philadelphia. 8 + 722 pp. Price $ 28.95 (1982). – From Science, Vol. 217, 1280 (1982).

003.017 **Galakser og kosmologi.**
J. Andersen, L. Hansen, B. Nordström.
Akademisk Forlag, Köpenhamn. 213 pp. Price D. kr. 91.50 (1981). – Review in Astron. Tidsskr., Årg. 15, 141 (1982).

003.018 **Exploring the earth and the cosmos.** The growth and future of human knowledge.
I. Asimov.
Crown, New York. 12 + 340 pp. Price $ 13.95 (1982). ISBN 0-517-54667-1. – Review in Sky Telesc., Vol. 64, 152 (1982).

003.019 **Venus, near neighbor of the sun.**
I. Asimov.
Lothrop, Lee & Shepard Books, 105 Madison Ave., New York, N. Y. 10016. 223 pp. Price $ 8.95 (1981). ISBN 0-688-41976-3. – Review in Sky Telesc., Vol. 64, 153 (1982).

003.020 **Archaeoastronomy in the New World.**
A. F. Aveni (Editor).
Cambridge University Press, Cambridge. 11 + 219 pp. Price £ 19.00 (1982). – From J. British Astron. Assoc., Vol. 92, 290 (1982).

003.021 **Souvenirs d'un astronome (1908 - 1977).**
R. Baillaud.
Imprimerie Carrère, Rodez. 632 pp. (1980). – Review in Astronomie, Vol. 96, 370; 1982 *(J. Pernet)*.

003.022 **Space instrument engineering.**
V. M. Balebanov (Editor).
Nauka, Moskva. 215 pp. (1982). In Russian. – Review in Ref. zh., 62. Issled. kosm. prostranstva, 11.62.69 (1982).

003.023 **Kepler.**
J. Banville.
Secker and Warburg, London. 182 pp. Price £ 5.95 (1981). Review in J. Hist. Astron., Vol. 13, 135 - 136; 1982 *(W. Donahue)*.

003.024 **Essays of comparative planetology.**
V. L. Barsukov (Editor).
Nauka, Moskva. 326 pp. (1981). In Russian. – Reviews in Ref. zh., 51. Astron., 7.51.53 (1982); Ref. zh., 52. Geod. Aehrosemka, 11.52.211 (1982).

003.025 **Atomic energy levels and Grotrian diagrams.** Vol. 3. Vanadium I – Chromium XXIV.
S. Bashkin, J. O. Stoner, Jr.
North-Holland Publishing Company, Amsterdam–Oxford–New York. 549 pp (1981). – Review in Astron. Nachr., Band 303, 348; 1982 *(G. Scholz)*.

003.026 **Interior of the earth: its structure, constitution and evolution.**
M. H. P. Bott.
Second edition. Edward Arnold, London, England. 416 pp. (1982). ISBN 0-7131-2842-9. – Review in Phys. Abstr., Vol. 85, Abstr. 84402 (1982).

003.027 **Astronomy in colour.**
P. Brown.
Blandford Press, Poole, Dorset (U. S. dist. Sterling, Publ. G. New York). 263 pp. Price $ 9.95 (1982). – From Phys Today, Vol. 35, No. 11, p. 79 (1982).

003.028 **Star and planet spotting.**
P. L. Brown.
Sterling Publ. Co. 155 pp. Price $ 6.95 (1981). – Review in Mercury, Vol. 11, 98 (1982).

003.029 **Die Sonnenuhr des Augustus.**
E. Buchner.
Nachdruck aus RM 1976 und 1980 und Nachtrag über die Ausgrabung 1980/1981. Verlag Philipp von Zabern, Mainz am Rhein. 112 pp. + 32 plates (1982). ISBN 3-8053-0430-7.

003.030 **Infrared and millimeter waves. Vol. 6. Systems and components.**
K. J. Button (Editor).
Academic Press, New York, USA, 15 + 416 pp (1982). ISBN 0-12-147706-1. – Review in Phys. Abstr., Vol. 85, Abstr. 112668 (1982).

003.031 **Planets of rock and ice: from Mercury to the moons of Saturn.**
C. R. Chapman.
Scribner's, New York. 18 + 222 pp. Price $ 13.95 (1982). ISBN 0-684-17484-7. – Review in Sky Telesc., Vol. 64, 448 (1982).

003.032 **The cosmic serpent. A catastrophist view of earth history.**
V. Clube, B. Napier.
Faber & Faber, London, Universe Books, New York. 299 pp. Price £ 12.50, $ 17.95 (1982). ISBN 0-571-11816-X ISBN 0-87663-379-3. – Reviews in J. British Astron. Assoc., Vol. 93, 40 - 41; 1982 *(H. Miles)*; Phys. Abstr., Vol. 85, Abstr. 101149 (1982); Sky Telesc., Vol. 64, 559 - 560 (1982).

003.033 **Was sind Pulsare?**
G. Dautcourt.
BSB B. G. Teubner Verlagsgesellschaft, Leipzig. 108 pp. Price

M 4.90 (DDR), M 9.80 (foreign countries) (1981). – Review in Astron. Schule, 19. Jahrg., 94; 1982 *(H. Bernhard)*.

003.034 **Other Worlds.** Space, superspace and the quantum universe.
P. Davies.
A Touchstone Book, reprint of the 1980 edition. Simon and Schuster, New York. 208 pp. Price $ 5.15 (1982). – From Science, Vol. 216, 1435 (1982).

003.035 **The accidental universe.**
P. C. W. Davies.
Cambridge University Press. Cambridge – London – New York – New Rochelle – Melbourne – Sydney. 12 + 139 pp. Price £ 10.00, $ 19.95 hbk., £ 4.95, $ 9.95, DM 19.80 pbk. (1982). ISBN 0-521-242126 hbk, ISBN 0-521-286921 pbk. – Reviews in Astron Tidsskr., Årg. 15, 191 (1982); Nature, Vol. 300, 133 - 134; 1982 *(M. Berry)*; Phys. Abstr., Vol. 85, Abstr. 112673 (1982); Spaceflight, Vol. 24, 474 (1982).
Contents: The fundamental ingredients of nature. Scales of structure. The delicate balance. Cosmic coincidences. The anthropic principle.

003.036 **Astrofysica.**
J. A. de Boer, J. Scheper, J. Schweers, P. A. C. van Vianen.
Malmberg, 's Hertogenbosch. 56 pp. Price *f* 11.90 (1982). Review in Zenit, 9. Jaarg., 431; 1982 *(H. Woutersen)*.

003.037 **Theoretical foundations of stellar statistics.**
H. Eelsalu.
Eesti NSV Teaduste Akadeemia, W. Struve nimeline Tartu Astrofüüsika Observatoorium. "Valgus", Tallinn. 138 pp. Price 1 Rbl. 10 Kop. (1982). – Contents: Part I: Basic statistical methods applicable to observational data systems. Part II: Elements of a dynamic descriptive theory. Part III: Theoretical principles of descriptive transformations. Part IV: Kinematic moment equations. Part V: Stellar-statistical research system.

003.038 **A meeting with the universe.**
B. M. French, S. P. Maran (editors).
National Aeronautics and Space Administration. 221 pp. Price $ 14.00 (1981). – Review in Sky Telesc., Vol. 64, 448 (1982).

003.039 **Quarks – Urstoff unserer Welt.**
H. Fritzsch.
Piper, München. 300 pp. Price DM 39.80 (1981). – Review in Sterne Weltraum, 21. Jahrg., 491; 1982 *(T. Reddmann)*.

003.040 **Real time control with the TRS-80.**
R. Genet.
Howard W. Sams & Co., Inc., 4300 West 62nd St., Indianapolis. Indiana 46268, USA. 166 pp. Price $ 14.95 (1982). ISBN 0-672-21831-3.
Contents: Introduction. Building a simple control system. Remote keypad and video monitor. What time is it? Counting. Voltage/frequency converters. Stepper motors. You can do it! Astronomical photoelectric photometry. The Mark I photometric system. The Mark II photometric system. The Mark III photometric system. The Mark III software.

003.041 **Software for photometric astronomy.**
S. Ghedini.
Willmann-Bell, Inc., P. O. Box 3125, Richmond, Va. 23235. 221 pp. Price $ 26.95 (1982). ISBN 0-943396-00-X. Review in Sky Telesc., Vol. 64, 559 (1982).

003.042 **Einführung in die Astronomie.**
R.-H. Giese.
Wissenschaftliche Buchgesellschaft, Darmstadt, F. R. Germany. 12 + 393 pp + 32 plates. Price DM 55.00 (1981). ISBN 3-534-06713-4.
Contents: Einleitung. Sphärische Astronomie und Himmelsmechanik. Das Planetensystem. Die Sonne. Physik der Fixsterne. Die Milchstrasse. Galaxien. Das Weltall.

003.043 **Introduction to planetary geology.**
B. P. Glass.
Cambridge University Press, Cambridge England. 10 + 469 pp. Price $ 29.95 (1982). ISBN 0-521-23579-0. – Reviews in Phys. Abstr., Vol. 86, Abstr. 10542 (1983); Sky Telesc., Vol. 64, 558 (1982).

003.044 **Physics of stellar evolution and cosmology.**
H. S. Goldberg, M. D. Scadron.
Gordon & Breach, New York. 14 + 390 pp (1981). ISBN 0-677-05540-4. – Review in Phys. Abstr., Vol. 86, Abstr. 10545 (1983).

003.045 **Méthodes de l'astrophysique – comment connaître et comprendre l'Univers.**
L. Gouguenheim.
Hachette – CNRS. 303 pp. (1981). – Review in Astronomie, Vol. 96, 424; 1982 *(J.-C. Vial)*.

003.046 **Observe the Herschel objects.**
L. Guzman, B. Guzman, P. Jones.
Astronomical League. 40 pp. (1980). – Review in Mercury, Vol. 11, 98 (1982).

003.047 **Calendar of the correspondence of Pierre Simon Laplace.**
R. Hahn.
Office of History of Science and Technology, University of California, Berkeley. 92 pp. Price $ 5.00 (1982). – Review in J. British Astron. Assoc., Vol. 93, 43; 1982 *(C. Ronan)*.

003.048 **Photoelectric photometry of variable stars.**
A practical guide for the smaller observatory.
D. S. Hall, R. M. Genet, with a foreword by F. B. Wood and a history of photometry by G. E. Kron and contributions by W. H. Allen, R. Berry, R. Boyd, R. Cadmus, Jr., D. L. DuPuy, J. A. Eaton, J. L. Hopkins, J. P. Oliver, G. Persha, G. C. G. C. Roberts, D. J. Sauer, A. J. Stokes, R. C. Wolpert.
Published by International Amateur-Professional Photoelectric Photometry, IAPPP, Fairborn, Ohio. 10 + 272 pp. (1982).
Contents: Introduction. A search for our roots. The observatory and telescope. Photometer head. Supporting electronics. Automation. Dual-channel systems. Photoelectric observatories. The earth's atmosphere. Standard photometric systems. Differential photometry. Observing techniques. Data reduction. Achieving maximum accuracy. Factors relevant in the choice of an observing program. How to select an observing program.

003.049 **The Alfvén wave.**
A. Hasegawa, C. Uberoi.
U. S. Department of Energy Technical Information Center, Oak Ridge, Tenn. (available as DE82001702 from the National Technical Information Service, Springfield, Va.). 10 + 128 pp. Price $ 11.50 (1982). – From Phys. Today, Vol. 35, No. 10, p. 87 (1982); Science, Vol. 218, 370 (1982).

003.050 **I bane rundt solen.**
Ø. Hauge.
2nd edition. Universitetsforlaget, Oslo–Bergen–Tromsö. 209 pp. Price N. kr. 130.00 (1982). – Review in Astron. Tidsskr., Årg. 15, 191 (1982).

003.051 **Aristarchus of Samos.**
T. Heath.
Dover Publications Inc., New York. 425 pp. Price $ 7.00 (1981). ISBN 0-486-24188-2. Constable Publishers, London. Price £ 5.25 (1982). – Reviews in Sky Telesc., Vol. 64, 246 (1982); Space Educ., Vol. 1, 190 (1982).

003.052 **International Directory of Amateur Astronomical Societies. – IDAAS.**
A. Heck, J. Manfroid (Editors).
Heck, ESA Satellite Tracking Station, Apartado 54065, Madrid, Spain. Manfroid, Institut d'Astrophysique, avenue de Cointe 5, B-4200 Cointe-Ougrée, Belgium. 4 + 308 pp. (1982).

003.053 **Archaeoastronomy in the old world.**
D. C. Heggie.
Cambridge University Press, Cambridge. 8 + 280 pp. Price £ 20.00 (1982). – From J. British Astron. Assoc., Vol. 92, 290 (1982).

003.054 **Megalithic science: ancient mathematics and astronomy in north-west Europe.**
D. C. Heggie.
Thames & Hudson, London. 256 pp. Price £ 12.00 (1981). Review in Archaeoastronomy, *(England)*, No. 4, p. S71 - S72; 1982 *(A. Burl)*.

003.055 **Landolt-Börnstein.** Numerical data and functional relationships in science and technology. New Series. Group VI: Astronomy, astrophysics and space research. Volume 2: Astronomy and astrophysics. Extension and supplement to Volume 1. Subvolume c: Interstellar matter – Galaxy – Universe.
K.-H. Hellwege (Editor-in-chief), K. Schaifers, H. H. Voigt (Editors).
Springer-Verlag, Berlin – Heidelberg – New York. 16 + 478 pp. Price DM 780.00, $ 325.30 (1982). ISBN 3-540-10977-3 (Berlin, Heidelberg, New York), ISBN 0-387-10977-3 (New York, Heidelberg, Berlin).

Contents: Compact objects (*M. Grewing, J. Trümper, H. H. Fink*). X-ray and γ-ray sources (*H. H. Fink, J. Trümper*). Phenomena of the generally distributed medium (*H. Scheffler*). Cool interstellar clouds (*H. Scheffler*). H II regions (*H. Scheffler*). Physics of interstellar dust (*H. Scheffler*). Physics of the interstellar gas (*J. Schmid-Burgk*). Cosmic rays (*M. Grewing*). Interstellar magnetic field (*J. Schmid-Burgk*). Positions, motions, parallaxes of stars (*W. Gliese*). The nearest stars (*W. Gliese*). Structure of the Galaxy (*H. Scheffler*). Kinematics and dynamics (*R. Wielen*). General information and integral properties of galaxies (*B. F. Madore, H. H. Voigt*). Internal structure and dynamics of galaxies (*H. H. Voigt, W. K. Huchtmeier*). Galaxies with special peculiarities; pairs, groups and clusters of galaxies (*P. Biermann*). Evolution of galaxies (*H. H. Voigt, P. Biermann*). Quasars and active galactic nuclei (*H. Netzer*). Extragalactic radio sources (*K. J. Fricke, A. Witzel*). Cosmology (*G. A. Tammann*). Comets, Appendix (*J. Rahe*). Comprehensive index (*L. D. Schmadel, G. Zech*).

003.056 **Astronomical photometry.**
A. A. Henden, R. H. Kaitchuck.
Van Nostrand Reinhold Company, New York – Cincinatti – Toronto – London – Melbourne. 12 + 392 pp. Price $ 30.00, DM 137.20 (1982). ISBN 0-442-23647-6. – Review in Sky Telesc., Vol. 64, 246 (1982).

Contents: An introduction to astronomical photometry. Photometric systems. Statistics. Data reduction. Observational calculations. Constructing the photometer head. Pulse-counting electronics. DC electronics. Practical observing techniques. Applications of photoelectric photometry.

003.057 **Astronomie.**
H. R. Henkel.
Verlag Harri Deutsch, Thun–Frankfurt/Main. 274 pp (1982). ISBN 3-87144-650-5. – Review in Sterne Weltraum, 21. Jahrg., 543; 1982 *(A. Kunert)*.

003.058 **Karl-Friedrich Zöllner.**
D. B. Herrmann.
Biographien hervorragender Naturwissenschaftler, Techniker und Mediziner (Band 57). BSB B. G. Teubner-Verlagsgesellschaft, Leipzig. 96 pp. Price M 4.80 (1982). – Review in Astron. Schule, 19. Jahrg., 119; 1982 *(H. Bernhard)*.

003.059 **Stellar astronomy: historical studies.**
M. Hoskin.
Science History Publications, Chalfont St Giles, England. 4 + 197 pp. Price £ 9.95, $ 19.95 (1982). ISBN 0-905193-04-0. Reviews in J. British Astron. Assoc., Vol. 92, 290 (1982); J. Hist. Astron., Vol. 13, 209 - 211; 1982 *(R. Berendzen)*; Sky Telesc., Vol. 64, 559 (1982).

003.060 **Facts and dogmas in cosmology and elsewhere.**
F. Hoyle.
Cambridge University Press, Cambridge. 16 pp. Price £ 1.75 (1982). – Review in J. British Astron. Assoc., Vol. 92, 288; 1982 *(C. Ronan)*.

003.061 **Evolution from space** (The OMNI Lecture).
F. Hoyle.
Enslow, Hillside, N. J. 80 pp. Price $ 8.95 (1982). – From Phys. Today, Vol. 35, No. 10, p. 86 (1982).

003.062 **The cosmogony of the solar system.**
F. Hoyle.
University College Cardiff Press. 192 pp. Price £ 7.95. Review in Irish Astron. J., Vol. 15, 157; 1981 *(H. Deasy)*.

003.063 **The planet Venus.**
G. Hunt, P. Moore.
Faber and Faber, London. 207 pp. Price £ 10.00 (1982). From J. British Astron. Assoc., Vol. 93, 44 (1982).

003.064 **Diffraction gratings.**
M. C. Hutley.
Techniques of Physics Series, Vol. 6. Academic Press, Inc., New York–London. 334 pp. Price $ 49.60 (1982). ISBN 0-12-362980-2.

Contents: Introduction. Theory. The use of gratings in spectroscopic instruments. The spectroscopic properties of gratings. Anomalies and the electromagnetic theory of grating efficiencies. Concave gratings. X-ray gratings. Nonspectroscopic uses of diffraction gratings.

003.065 **Physik der Sterne.**
S. A. Kaplan.
Kleine Naturwissenschaftliche Bibliothek, Band 45, BSB B. G. Teubner Verlagsgesellschaft, Leipzig. 254 pp. Price M 13.00 (1980). – Review in Jenaer Rundsch., 27. Jahrg., 146; 1982 *(J. Neumann)*.

003.066 **Planets and moons.**
W. J. Kaufmann III.
Translated from the English edition. Mir, Moskva. 215 pp. (1982). In Russian. Review in Ref. zh., 51. Astron., 7.51.56 (1982).

003.067 **Planetary vehicles.**
A. L. Kemurdzhian (Editor).
Mashinostroenie, Moskva. 319 pp. (1982). In Russian. Review in Ref. zh., 62. Issled. kosm. prostranstva, 11.62.68 (1982).

003.068 **The problem of infinity in modern science.**
G. Sh. Khutsishvili.
Metsniereba, Tbilisi. 151 pp. (1981). In Russian. – Review in Ref. zh., 51. Astron., 7.51.63 (1982).

003.069 **Early emission line stars.**
C. R. Kitchin.
Monographs on astronomical subjects: 8. Adam Hilger Ltd, Bristol. 11 + 182 pp. Price £ 16.00, $ 35.00 (1982). ISBN 0-85274-402-1. – Review in Astron. Tidsskr., Arg. 15, 143 (1982).

Contents: Extended atmospheres and the origin of emission lines. Wolf-Rayet stars. Morphology. Cataclysmic variables. Be stars. Other emission line systems.

003.070 **Observation and fact in astronomy.**
I. G. Kolchinskij.
Naukova dumka, Kiev. 104 pp. (1982). In Russian. – From Ref. zh., 51. Astron., 9.51.29 (1982).

003.071 **Iron shower.** – Concerning the Sikhote-Alin meteorite.
E. L. Krinov.
Nauka, Moskva (1981). In Russian. – Review in Zemlya Veselennaya, No. 5, p. 75 - 76; 1982 *(N. B. Divari)*.

003.072 **Design of scientific space instrumentation.**
T. I. Kurmanaliev.
Nauka, Moskva. 191 pp. (1982). In Russian. – Review in Ref. zh., 62. Issled. kosm. prostranstva, 10.62.58 (1982).

003.073 **Relief, tectonics and volcanism of Mars.**
Yu. Ya. Kuznetsov, V. E. Khanin (Editors).
Tr. VNII geol. zarubezhn. stran, 1982, No. 39, 106 pp. In Russian. – Review in Ref. zh., 51. Astron., 7.51.55 (1982).

003.074 **An introduction to tensor calculus, relativity and cosmology.**
D. F. Lawden.
3rd Edition. John Wiley & Sons, Chichester–New York–Brisbane–Toronto–Singapore. 205 pp. Price £ 6.75, inb. £ 14.75, $ 35.95 (1982). ISBN 0-471-10082-X. – Reviews in Astron. Tidsskr., Årg. 15, 143 (1982); Sky Telesc., Vol. 64, 153 (1982).

003.075 **Solar wind and magnetospheric investigations.**
A. E. Levitin, I. G. Simakov (Editors).
Inst. zem. magn., ionos., i rasprostr. radiovoln. AN SSSR. 191 pp. (1980). In Russian. – From Ref. zh., 51. Astron. 8.51.40 (1982).

003.076 **Wie groß ist die Welt?**
P. Lübke.
Verlag Quelle & Meyer, Heidelberg. 32 pp. Price DM 14.80 (1980). ISBN 3-491-010390. – Review in Sterne Weltraum, 21. Jahrg., 490; 1982 *(A. Kunert)*.

003.077 **Utveckling i universum – En astrofysikalisk kurs för gymnasiet.**
G. Lyngå, L. Lyngå.
2nd edition. Eget förlag, Lund/Stockholm. 42 pp. Price Sv. kr. 42.00 (1981). – Review in Astron. Tidsskr., Årg. 15, 137 - 138; 1982 *(H. Lindholm)*.

003.078 **Réalisez vouz-même votre télescope.**
M. Lyonnet du Moutier, J. Funel.
Editions Eyrolles, 61 Blvd. Saint Germain, 75240 Paris 05 France. 116 pp. Price FF 65.00, $ 9.50, BF 494.00 (1982). Reviews in Ciel Terre, Vol. 98, 316; 1982 *(A. Koeckelenbergh)*; Sky Telesc., Vol. 64, 345 (1982).

003.079 **BMS meteor observer's handbook. Vol. 1. Naked eye work.**
R. A. Mackenzie.
6th edition. British Meteor Society, Dover, England. 27 pp. (1982). – Review in Phys. Abstr., Vol. 86, Abstr. 16100 (1983).

003.080 **L'universo nel tempo.**
P. Maffei.
Ed. Mondadori, Milano. 391 pp. Price L. 20.000. (1982). Review in Orione, Vol. 3, 178; 1982 *(W. Ferreri)*.

003.081 **Observatories of the world.**
S. Marx, W. Pfau.
Blandford Press, Poole, Dorset. England. 197 pp. Price £ 8.95 (1982). – Reviews in J. British Astron. Assoc., Vol. 92, 285; 1982 *(P. Moore)*; Spaceflight, Vol. 24, 420 (1982).

003.082 **A perspective of physics, Vol. 3.**
H. Massey (Editor).
Gordon and Breach Science Publishers, New York–London–Paris. 38 + 316 pp. Price £ 16.70 (1980). – Review in Sterne, 58. Band, 260 - 261; 1982 *(H. Lambrecht)*.

003.083 **Basaltic volcanism on the terrestrial planets.**
T. R. McGetchin, R. O. Pepin, R. J. Phillips.
Pergamon, New York. 1286 pp. Price $ 60.00 (1982). – From Phys. Today, Vol. 35, No. 11, p. 79 (1982).

003.084 **Riddles of the Stone Age: rock carvings of ancient Europe.**
J. McMann.
Thames and Hudson, London. 161 pp. Price £ 8.95 (1980). Review in Archaeoastronomy *(England)*, No. 4, p. S72 (1982).

003.085 **Albert Einstein's special theory of relativity: emergence (1905) and early interpretation (1905 - 1911).**
A. I. Miller.
Addison-Wesley, Reading, Massachusetts. 28 + 466 pp. Price $ 42.50, $ 19.50 pbk. (1981). – Review in J. Hist. Astron., Vol. 13, 135; 1982 *(O. Gingerich)*.

003.086 **The unfolding universe.**
P. Moore.
Michael Joseph/George Rainbird, London; Crown Publishers, New York. 256 pp. Price £ 12.50, $ 17.95 (1982). ISBN 0-517-548364. – Reviews in J. British Astron. Assoc., Vol. 93, 39; 1982 *(C. J. Watkis)*; Sky Telesc., Vol. 64, 246 (1982); Space Educ., Vol. 1, 191 - 192 (1982).

003.087 **Die Jugendjahre von Bernhard Schmidt und sein Briefwechsel mit dem Potsdamer Observatorium.**
P. Müürsepp.
Akademie der Wissenschaften der Estnischen SSR. Institut für Physik. "Valgus", Tallinn. 116 pp. Price 75 Kop. (1982).

003.088 **The Kingfisher astronomy handbook.**
J. Muirden.
Kingfisher Books, London. 189 pp. Price £ 3.95 (1982). From J. British Astron. Assoc., Vol. 92, 290 (1982).

003.089 **Libro del nuevo cometa.**
J. Muñoz, introduction and appendices by V. N. Brotóns, English translation by E. Ladd.
Artes Gráficas Sóler, Valencia. Available from Science History Publications, New York. 124+96 pp. Facsimiles. Price $ 20.00 (1981). – Review in J. Hist. Astron., Vol. 13, 224 - 225; 1982 *(O. Gingerich)*.

003.090 **The lighter side of gravity.**
J. V. Narlikar.

W. H. Freeman. 194 pp. Price $ 17.95, £ 13.95 hbk., $ 9.95, £ 6.95 pbk. (1982). ISBN 0-7167-1343-8 hbk.; 0-7167-1344-6 pbk. – Review in Nature, Vol. 300, 563; 1982 *(H. Bondi).*

003.091 **Schwarze Löcher im All.**
I. D. Nowikow *(Novikov).*
BSB B. G. Teubner Verlagsgesellschaft, Leipzig. 96 pp. Price M 5.50 (1981). – Review in Astron. Schule, 19. Jahrg., 94; 1982 *(H. Bernhard).*

003.092 **The 1769 transit of Venus.**
D. B. Nunis, Jr. (Editor).
Natural History Museum of Los Angeles County, Book Shop, 900 Exposition Blvd., Los Angeles, Calif. 90007. 185 pp. Price $ 62.50 (1982). ISBN 0-938644-18-1. – Review in Sky Telesc., Vol. 64, 560 (1982).

003.093 **The cosmic code.** Quantum physics as the language of nature.
H. R. Pagels.
Simon and Schuster, New York. 370 pp. Price $ 17.50 (1982). From Science, Vol. 216, 1312 (1982).

003.094 **Herinneringen.**
A. Pannekoek, with introductions by B. A. Sijes and E. P. J. van den Heuvel.
Van Gennep, Amsterdam. 398 pp. Price *f* 58.50 (1982). ISBN 90-6012-350-6. – Review in Zenith, 9. Jahrg., 483; 1982 *(G. W. E. Beekman).*

003.095 **Les étoiles variables.**
M. Petit, with a preface by P. Maffei.
Masson, Paris–New York–Barcelone–Milan–Mexico–Rio de Janeiro. 252 pp. Price FF 125.00 (1982). ISBN 2-225-78925-8.
Contents: Les variables pulsantes. Les variables éruptives et cataclysmiques. Les binaires à éclipses. Des variables particulières ou insolites.

003.096 **Traité de l'astrolabe.**
J. Philopon.
Astrolabica 2, Librarie Alain Brieux, Paris. 205 pp. (1981). Review in J. Hist. Astron., Vol. 13, 145 (1982).

003.097 **Cosmic electrodynamics.**
J. H. Piddington.
Robert E. Krieger Publishing Co., P.O. Box 9542, Melbourne, Fla. 32901. 361 pp. Price $ 27.50 (1981). ISBN 0-88275-587-0. – Review in Sky Telesc., Vol. 64, 246 (1982).

003.098 **General astrometry.**
V. V. Podobed, V. V. Nesterov.
2nd revised and suppl. edition. Handbook for students of astronomy. Nauka, Moskva. 576 pp. (1982). In Russian. From Ref. zh., 51. Astron., 11.51.34 (1982).

003.099 **The open universe.**
K. R. Popper.
Rowman and Littlefield, Totowa, N. J. 185 pp. Price $ 23.50 (1982). – From Phys. Today, Vol. 35, No. 9, p. 85 (1982).

003.100 **Solar magneto-hydrodynamics.**
E. R. Priest.
Geophysics and astrophysics monographs, Vol. 21.
D. Reidel Publishing Company, Dordrecht, Holland – Boston, U.S.A. – London, England. 19 + 469 pp. Price Dfl. 235.00, $ 99.00 (1982). ISBN 90-277-1374-X.
Contents: A description of the sun. The basic equations of magnetohydrodynamics. Magnetohydrostatics. Waves. Shock waves. Heating of the upper atmosphere. Instability. Sunspots. Dynamo theory. Solar flares. Prominences. The solar wind.

003.101 **Physics of the polar ionosphere.**
O. M. Raspopov (Editor).
Nauka, Leningrad. 148 pp. (1982). In Russian. – Review in Ref. zh., 62. Issled. kosm. prostranstva, 7.62.65 (1982).

003.102 **Horizons: exploring the universe tonight.**
G. Reed.
Wadsworth, Inc. 240 pp. Price $ 7.95 (1981). – Review in Mercury, Vol. 11, 97 (1982).

003.103 **Astronomy for the inquiring mind.** The growth and use of theory in science.
E. M. Rogers.
Princeton University Press, Princeton, N. J. 10 + 164 pp. Price $ 7.95 (1982). ISBN 0-691-02370-0. – Review in Sky Telesc., Vol. 64, 442; 1982 *(G. Lovi).*

003.104 **Deep space.**
C. A. Ronan.
Macmillan, London – New York. 208 pp. Price £ 10.95 (1982). – From J. British Astron. Assoc., Vol. 93, 45 (1982).

003.105 **Mercury's perihelion from Le Verrier to Einstein.**
N. T. Roseveare.
Clarendon Press, Oxford; Oxford University Press, New York. 8 + 208 pp. Price £ 20.00, $ 49.00 (1982). – Review in Astron. Tidsskr., Årg. 15, 143 (1982).

003.106 **Die Sekunde der Kosmologen.**
K. Rudnicki.
Vittorio Klostermann, Frankfurt am Main, F. R. Germany. 9 + 82 pp. Price DM 12.80 (1982). ISBN 3-465-01526-6.
Contents: Was ist Kosmologie? Der Einfluß der Weltanschauung auf die Ergebnisse der kosmologischen Forschung. Das kosmologische Prinzip der Ur-Inder. Das kosmologische Prinzip der Antike. Das kopernikanische kosmologische Prinzip. Was ist das: ein kosmologisches Modell? Eigenschaften einiger Modelle. Die Sekunde der Kosmologen. Zeit- und Raumbestimmungen der wirklichen Himmelsobjekte und ihrer Erscheinungen. Was weiter?

003.107 **Astronomy and society in Britain during the period 4000 - 1500 B.C.**
C. L. N. Ruggles, A. W. R. Whittle (Editors).
British Archaeological Reports, British Series, 88; B. A. R., 122 Banbury Road, Oxford, England, 342 pp. Price £ 12.00 (1981). – Review in Archaeoastronomy, (*England*), No. 4, p. S68 - S71; 1982 (*D. C. Heggie*).

003.108 **Bau und Physik der Galaxis.**
H. Scheffler, H. Elsässer.
Bibliographisches Institut Mannheim–Wien–Zürich. 642 pp. Price DM 68.00 (1982). ISBN 3-411-01641-8.
Contents: Positionen, Bewegungen und Entfernungen der Sterne – Begriffe und Methoden. Aufbau und Kinematik des Sternsystems. Interstellare Phänomene. Physik der interstellaren Materie. Dynamik der Galaxis.

003.109 **Astronomische Musterversuche.**
W. Schlosser, T. Schmidt-Kaler.
Hirschgrabenverlag, Frankfurt/Main. 224 pp. Price DM 39.00 (1982). ISBN 3-454-24939-9.– Review in Sterne Weltraum, 21 Jahrg., 542; 1982 (*A. Kunert*).

003.110 **Planeten und Sternbilder im Wandel der Geschichte.**
F. Schmeidler.
Deutsches Museum, Munich, 24 pp. (1980). – Review in J. Hist. Astron., Vol. 13, 143 (1982).

003.111 **Relativität – Gruppen – Teilchen.** Spezielle Relativitätstheorie als Grundlage der Feld- und Teilchenphysik.
R. U. Sexl, H. K. Urbantke.
2nd revised edition. Springer-Verlag, Wien – New York. 9 + 328 pp. Price ÖS 496.00, DM 69.00 (1982). ISBN 3-211-81680-1 (Springer-Verlag Wien – New York), ISBN 0-387-81680-1 (Springer-Verlag New York – Wien).

003.112 **Problems of modern astrophysics.**
I. S. Shklovskij.
Glavnaya redaktsiya fiziko-matematicheskoj literatury, Nauka, Moskva. 224 pp. Price 45 Kop. (1982). In Russian.

003.113 **Observational astronomy for amateurs.**
J. B. Sidgwick.
Fourth edition prepared by J. Muirden. Enslow, Hillside, N. J. 10 + 350 pp. Price cloth $ 19.95; paper $ 5.95 (1982). From Science, Vol. 218, 189 (1982).

003.114 **In the world of aggregates of moons (satellites of planets).**
B. I. Silkin. Edited by E. L. Ruskol.
Nauka, Glavnaya redaktsiya fiziko-matematicheskoj literatury, Moskva. 208 pp. Price 40 Kop. (1982). In Russian. – Review in Priroda, 1982, No. 11, p. 124; Zemlya Vselennaya, 1982, No. 6, p. 71 - 72 (*V. I. Chikmachev*).

003.115 **The Cambridge encyclopedia of Earth sciences.**
D. G. Smith (Editor).
Crown and Cambridge University Press. 496 pp. Price $ 35.00 (1981). ISBN 0-517-54370-2. – Review in Sky Telesc., Vol. 64, 152 (1982).

003.116 **The expanding universe: astronomy's 'Great Debate' 1900 - 1931.**
R. W. Smith.
Cambridge University Press, Cambridge. 14 + 220 pp. Price £ 19.00, $ 29.50 (1982). ISBN 0-521-23212-0. – Reviews in J. Hist. Astron., Vol. 13, 209 - 211; 1982 (*R. Berendzen*); Nature, Vol. 298, 776; 1982 (*O. Gingerich*).

003.117 **Metal-deficient stars.**
V. Straižys.
Institut Fiziki AN Litovskoj SSR. Mokslas, Vil'nyus. 312 pp. (1982). In Russian.

003.118 **Astrophysics and space physics.**
R. A. Syunyaev (Editor).
Nauka, Moskva. 320 pp. (1982). In Russian. – Review in Ref. zh., 51. Astron., 9.51.33 (1982).

003.119 **The three voyages of Edmond Halley in the "Paramore" 1698 - 1701.**
N. J. W. Thrower.
Londra, Hakluyt Society. 392 pp. Price £ 20.00 (1981). Review in Coelum, Vol. 50, 227 - 229; 1982 (*S. Zavatti*).

003.120 **Große kosmische Systeme.**
H.-J. Treder, J.-P. Mücket.
Veröffentlichungen des Forschungsbereichs Geo- and Kosmoswissenschaften der Akademie der Wissenschaften der DDR, Heft 10. Akademie-Verlag, Berlin. 91 pp. Price M 15.00 (1981). – Review in Sterne, 58. Band, 315; 1982 (*K.-H. Lotze*).

003.121 **Thermo field dynamics and condensed states.**
H. Umezawa, H. Matsumoto, M. Tachiki.
North Holland Publishing Co., Amsterdam, Netherlands. 16 + 592 pp. (1982). ISBN 0-444-86361-3. – Review in Phys. Abstr., Vol. 86, Abstr. 10535 (1983).

003.122 **Glimtar från astronomins forskningsfält.**
A. Wallenquist.
Bokförlaget Prisma, Stockholm. 184 pp. Price Sv. kr. 135.00 (1982). – Review in Astron. Tidsskr., Arg. 15, 192 (1982).

003.123 **Ciel, passé-présent.**
G. Walusinski.
Collection Axes sciences, Éditions Études vivantes. 222 pp. (1981). – Review in Astronomie, Vol. 96, 369; 1982 (*G. Oudenot*).

003.124 **Mathematisches Hilfsbuch für Studierende und Freunde der Astronomie.**
W. Wepner.
Treugesell-Verlag Dr. Vehrenberg K. G., 4000 Düsseldorf 1, Schillerstraße 17, F. R. Germany. 5 + 279 pp. Price DM 29.80. ISBN 3-87974-911-9.
Contents: Rechenhilfsmittel. Aus der Mathematik. Zeit. Beobachtungsort und Instrumente. Sphärische Koordinatensysteme. Reduktion von Katalogörtern und Beobachtungen. Ephemeridenrechnung. Photographische Ortsbestimmung. Ortsbestimmungen auf Sonne und Planeten. Rechnen mit Magnitudines. Auswertung von Beobachtungen Veränderlicher. Bahnbestimmung.

003.125 **Webb Society deep-sky observer's handbook. Vol. 5: Clusters of galaxies.**
G. S. Whiston, edited by K. G. Jones.
Enslow Publishers, Hillside, N. J.; Lutterworth, Guildford, England. 241 pp. Price $ 16.95 (1982). ISBN 0-89490-066-8. Review in Sky Telesc., Vol. 64, 448 (1982).

003.126 **Theory and experiment in gravitational physics.**
C. M. Will.
Cambridge University Press, Cambridge – London New York – New Rochelle – Melbourne – Sydney. 10 + 342 pp. Price £ 37.50, $ 75.00, DM 184.30 (1981). ISBN 0-521-23237-6. – Reviews in Nature, Vol. 299, 284; 1982 (J. H. Taylor). – Phys. Abstr., Vol. 85, Abstr. 68679 (1982).
Contents: Introduction. The Einstein equivalence principle and the foundations of gravitation theory. Gravitation as a geometric phenomenon. The parametrized post-Newtonian formalism. Post-Newtonian limits of alternative metric theories of gravity. Equations of motion in the PPN formalism. The classical tests. Tests of the Strong Equivalence Principle. Other tests of post-Newtonian gravity. Gravitational radiation as a tool for testing relativistic gravity. Structure and motion of compact objects in alternative theories of gravity. The binary pulsar. Cosmological tests.

003.127 **Modern cosmology.**
P. J. Willcox.
P. J. Willcox, P. O. Box 39, Huntington, Ind. 46750. 96 pp. Price $ 4.50 (1981). – Review in Sky Telesc., Vol. 64, 560 (1982).

003.128 **Kalender und Chronologie.**
H. Zemanek.
2nd revised edition, R. Oldenbourg Verlag, München-Wien. 156 pp. Price DM 29.80 (1981). ISBN 3-486-23292-4.
Contents: Einige Grundbegriffe der Kalenderwissenschaft. Vom Tag zum Jahr. Der Computus – die Kirchenrechnung. Joseph Scaligers Chronologie. Der Julianische Tag (JD). Verschiedene Kalender. Die verbesserte Sekunde. Rechenhilfen für die Kalenderrechnung.

003.129 **Astronomy and astrophysics for the 1980's, Vol. 1.**
Report of the Astronomy Survey Committee, 1982.
National Academy Press. 2101 Constitution Ave., NW., Washington, D.C. 20418. 20 + 189 pp. Price $ 14.75 (1982). ISBN 0-309-03249-0. – Review in Sky Telesc., Vol. 64, 246 (1982).

003.130 **Inhomogeneities in the ionosphere.**
Collected papers.
Inst. kosmofiz. issled. i aehron. Yakut. fil. SO AN SSSR. Yakutsk. 175 pp. (1981). In Russian. – Review in Ref. zh.. 62. Issled. kosm. prostranstva, 7.62.63 (1982).

003.131 **Materialien zu den Ephemeriden von Johannes Kepler.**
V. Bialas, E. Papadimitriou.
Nova Kepleriana, Neue Folge Heft 7. Verlag der Bayerischen Akademie der Wissenschaften, Munich. 218 pp. (1980). Review in J. Hist. Astron. Vol. 13, 144 (1982).

003.132 **Planetary exploration.**
British National Committee on Space Research.
The Royal Society, 6 Carlton House Terrace, London SW1Y 5AG, England. 167 pp. Price £ 21.55 (1982). ISBN 0-85403-185-5. – Review in Sky Telesc., Vol. 64, 153 (1982).

003.133 **The 'astronomical' chapters of the Ethiopic Book of Enoch (72 to 82).**
Translation and commentary by O. Neugebauer, with additional notes on the Aramaic fragments by M. Black.
Det Kongelige Danske Videnskabernes Selskab Matematisk-fysiske Meddelelser 40:10, Munksgaard, Copenhagen. 42 pp. Dk 50.00 paperback (1981). – Review in J. Hist. Astron., Vol. 13, 144 (1982).

004 History of Astronomy

004.001 **The Eros parallax, 1930 - 31.**
R. d'E. Atkinson.
J. Hist. Astron., Vol. 13, 77 - 83 (1982).

004.002 **The Castlehill Observatory, Aberdeen.**
J. S. Reid.
J. Hist. Astron., Vol. 13, 84 - 96 (1982).

004.003 **The first drawing of a spiral nebula.**
M. Hoskin.
J. Hist. Astron., Vol. 13, 97 - 101 (1982).

004.004 **Une instrument en voie de disparition: l'équatorial coudé.**
T. Weimer.
J. Hist. Astron., Vol. 13, 110 - 118 (1982).

004.005 **A note on polar refractors.**
D. W. Dewhirst.
J. Hist. Astron., Vol. 13, 119 - 120 (1982).

004.006 **Some astronomical observations from thirteenth-century Egypt.**
D. A. King, O. Gingerich.
J. Hist. Astron., Vol. 13, 121 - 128 (1982).

004.007 **Geometry and astronomy in prehistoric Ohio.**
R. Hively, R. Horn.
Archaeoastronomy, *(England)*, No. 4, p. S1 - S20 (1982).

004.008 **A reassessment of the high precision megalithic lunar sightlines, 1: Backsights, indicators and the archaeological status of the sightlines.**
C. L. N. Ruggles.
Archaeoastronomy, *(England)*, No. 4, p. S21 - S40 (1982).

004.009 **Orientation astronomique des tombes mérovingiennes du cimetière de Soyria (Jura).**
J. P. Parisot, P. Petrequin.
Archaeoastronomy, *(England)*, No. 4, p. S41 - S48 (1982).

004.010 **A statistical analysis of some Palenque dates.**
G. Rosenfeldt.
Archaeoastronomy, *(England)*, No. 4, p. S49 - S67 (1982).

004.011 **Keeler's gap in Saturn's A ring.**
D. E. Osterbrock, D. P. Cruikshank.
Sky Telesc., Vol. 64, 123 - 126 (1982).

004.012 **Barnard's satellite of Jupiter.**
D. P. Cruikshank.
Sky Telesc., Vol. 64, 220 - 224 (1982).

004.013 **The lamps of Atlantis.**
A. E. Roy.
Mem. Soc. Astron. Italiana, Vol. 52, 613 - 625 (1981).
Paper presented in June 1980 on the Greek island of Samos on the occasion of the Aristarchus of Samos Symposium.

004.014 **A prehistoric solar observatory in Veneto, Italy.**
G. Romano.
Mem. Soc. Astron. Italiana, Vol. 52, 627 - 631 (1981).
The orientation of the prehistoric construction called "Le Motte di Godego" in the province of Treviso (Veneto) suggests that this monument is an ancient solar observatory.

004.015 **The beginnings of solar-terrestrial studies.**
D. W. Hughes.
Nature, Vol. 298, 518 (1982).

004.016 **Ibn Sina and Ptolemy's "Almagest".**
A. A. Akhmedov.
Abu Ali Ibn Sina i estestv. nauk. Mater. yubil. nauchn. ses., posvyashch. 1000-letiyu so dnya rozhd. Abu Ali Ibn Siny (Avitsenny), Bukhara, 24 - 26 sent., 1980. Tashkent, 1981, p. 142 - 147. In Russian. – Abstr. in Ref. zh., 51. Astron., 7.51.27 (1982).

004.017 **Mathematics and astronomy of Ibn Sina.**
B. A. Rozenfel'd.
Abu Ali Ibn Sina i estestv. nauk. Mater. yubil. nauchn. ses., posvyashch. 1000-letiyu so dnya rozhd. Abu Ali Ibn Siny (Avitsenny), Bukhara, 24 - 26 sent., 1980. Tashkent, 1981, p. 101 - 105. In Russian. – Abstr. in Ref. zh., 51. Astron., 7.51.28 (1982).

004.018 **Scientific instruments of Ibn Sina.**
Z. K. Sokolovskaya.
Abu Ali Ibn Sina i estestv. nauk. Mater. yubil. nauchn. ses., posvyashch. 1000-letiyu so dnya rozhd. Abu Ali Ibn Siny (Avitsenny), Bukhara, 24 - 26 sent., 1980. Tashkent, 1981, p. 135 - 142. In Russian. – Abstr. in Ref. zh., 51. Astron., 7.51.29 (1982).

004.019 **"In the beginning . . . " The Ussher chronology and other renaissance ideas dating the creation.**
R. L. Reese, S. M. Everett, E. D. Craun.
Archaeoastronomy *(U. S. A.)*, Vol. 5, 20 - 23 (1982).

004.020 **Ptolemy's search for a law of refraction: a case-study in the classical methodology of "Saving the appearances" and its limitations.**
A. M. Smith.
Arch. Hist. Exact Sci., Vol. 26, 221 - 240 (1982).

004.021 **Algebraic formulae for the Moon, Saturn and Jupiter in the Pancasiddhantika.**
G. Abraham.
Arch. Hist. Exact Sci., Vol. 26, 287 - 297 (1982).

004.022 **The Galileo affair.**
O. Gingerich.
Sci. American, Vol. 247, No. 2, p. 118 - 127 (1982).

004.023 **Die Entdeckung der kosmischen Mikrowellenrest-strahlung.**
A. A. Penzias.
Sterne, 58. Band, 206 - 210 (1982).

004.024 **"Earthquakes in the air": the seismological theory of John Flamsteed (1693).**
J. E. Kennedy, W. A. S. Sarjeant.
J. R. Astron. Soc. Canada, Vol. 76, 213 - 223 (1982).

004.025 **Early days of astronomy at Toronto – Part IV. The total solar eclipse of August 7, 1869.**
H. Sawyer Hogg.
J. R. Astron. Soc. Canada, Vol. 76, 235 - 244 (1982).

004.026 **Gli astrofili e la luna.**
M. Boetto.
Orione, Vol. 3, 108 - 113 (1982).

004.027 **Photoelectric lunar occultations. I. History, uses and limitations.**
G. L. Blow.
South. Stars, Vol. 29, 109 - 114 (1982).
A historical summary of attempts to observe lunar occultations by photoelectric means is presented, and the uses to which these observations can be put are detailed. Some of the limitations of the occultation technique are also discussed.

004.028 **Wandel in der Kenntnis des Polarlichtes.**
W. Schröder.
Sterne Weltraum, 21. Jahrg., 358 - 361 (1982).

004.029 **La comète de Halley et Mira Ceti dans le Talmud?**
P. Véron.
Astronomie, Vol. 96, 351 - 355 (1982).

004.030 **Der grosse Tragstein des "Table des Marchands".**
Seine Zahlen – ihre Entschlüsselung und Deutung als Sonnen- und Mondkalendarium.
H. Hindrichs.
Orion, 40. Jahrg., 146 - 152 (1982).

004.031 **Geburtstag einer Entdeckung – zum 200. Jahrestag der Auffindung des Planeten Uranus.**
D. B. Herrmann.
Jenaer Rundsch., 27. Jahrg., 141 - 142 (1982).

004.032 **Historical eclipses.**
F. R. Stephenson.
Sci. American, Vol. 247, No. 10, p. 154 - 163 (1982).
Reliable records of solar and lunar eclipses go back as far as 750 B.C. They bear on such questions as whether the sun is shrinking or the earth is not spinning as fast as it once did.

004.033 **Das Polarlicht im Wandel der Zeiten.**
W. Schröder.
Phys. unserer Zeit, 13. Jahrg., 155 - 159 (1982).

004.034 **Histoire et singularités de la solution de Schwarzschild (1915 - 1923).**
J. Eisenstaedt.
Arch. Hist. Exact Sci., Vol. 27, 157 - 198 (1982).

004.035 **Il sistema di misurazione del tempo nella civiltà azteca precolombiana.**
A. M. Gallingani.
Coelum, Vol. 50, 247 - 253 (1982).

004.036 **Sesquicentenary of Gauss's first measurement of the absolute value of magnetic intensity.**
S. R. C. Malin.
Philos. Trans. R. Soc. London, Ser. A, Vol. 306, (see 012.014), 5 - 8 (1982).

004.037 **Historical observations of the geomagnetic field.**
D. R. Barraclough.
Philos. Trans. R. Soc. London, Ser. A, Vol. 306, (see 012.014), 71 - 78 (1982).

004.038 **Condensed matter – genesis of a concept.**
W. Sullivan.
Ninth Texas Symposium on Relativistic Astrophysics, (see 012.020), p. 402 - 407 (1980).

004.039 **La comète de Halley a son retour de 1835.**
C. Bertaud, E. Stram.
Astronomie, Vol. 96, 455 - 469 (1982).

004.040 **En astronomisk studieresa i Europa på 1840-talet, II. Utdrag ur J. M. Agardhs resedagbok 1843 - 1845.**
C. Schalén, N. Hansson.
Astron. Tidsskr., Årg. 15, 49 - 64 (1982).

004.041 **Barnard and his observations of Io.**
R. F. Griffin.
Sky Telesc., Vol. 64, 428 - 429 (1982).

004.042 **Chandrasekhar vs. Eddington – an unanticipated confrontation.**
K. C. Wali.
Phys. Today, Vol. 35, No. 10, p. 33 - 40 (1982).
For many years astronomers did not accept the validity of a young scientist's application of the new physics because it was ridiculed by a preeminent astronomer.

004.043 **J. E. Keeler's discovery of a gap in Saturn's A-ring.**
D. E. Osterbrock, D. P. Cruikshank.
Bull. American Astron. Soc., Vol. 14, 715 (1982). – Abstract.

004.044 **Kant's thoughts on the ageing of the Earth.**
O. Reinhardt, D. R. Oldroyd.
Ann. Sci., Vol. 39, 349 - 369 (1982). – Abstr. in Phys. Abstr., Vol. 85, Abstr. 89515 (1982).

004.045 **Les priorités douteuses des découvertes astronomiques.**
B. M. Ševarlić.
Publ. Astron. Obs. Sarajevo, Vol. 1, No. 1, (see 012.043), p. 29 (1981). – Abstract.

004.046 **Notes in Bosnian chronicles about comets of 1664 and 1769.**
J. Mulaomerović.
Publ. Astron. Obs. Sarajevo, Vol. 1, No. 1, (see 012.043), p. 30 (1981). – Abstract.

004.047 **Survey of the development of astronomy in Bosnia and Hertzegovina.**
J. Mulaomerović.
Publ. Astron. Obs. Sarajevo, Vol. 1, No. 1, (see 012.043), p. 31 - 38 (1981).

004.048 **G. K. Gilbert's contribution to selenology.**
W. G. Hoyt.
J. Hist. Astron., Vol. 13, 155 - 167 (1982).

004.049 **The significance of the Copernican orbs.**
N. Jardine.
J. Hist. Astron., Vol. 13, 168 - 194 (1982).

004.050 **Problems raised by Ptolemy's lunar tables.**
S. J. Goldstein, Jr.
J. Hist. Astron., Vol. 13, 195 - 205 (1982).

004.051 **The first aplanatic object glass.**
J. A. Bennett.
J. Hist. Astron., Vol. 13, 206 - 208 (1982).

004.052 **De geboorte van de radio-astronomie.**
P. D. Barthel.
Zenit, 9. Jaarg., 506 - 509 (1982).

004.053 **400 years mathematics and astronomy at the Würzburg University: Alma Julia Herbipolensis 1582 - 1982.**
O. Volk.
Celestial Mech., Vol. 28, (see 012.049), 243 - 250 (1982). In German.

004.054 **On the Computus paschalis of "Cassiodorus".**
O. Neugebauer.
Centaurus, Vol. 25, 292 - 302 (1982).

004.055 **Astronomical computations for 1299 from the Cairo Geniza.**
B. R. Goldstein, D. Pingree.
Centaurus, Vol. 25, 303 - 318 (1982).

004.056 **Venus 1882: public, parallax, and HNR (*Henry Norris Russell*).**
D. H. DeVorkin.
Sky Telesc., Vol. 64, 524 - 526 (1982).

004.057 **Radio astronomy's golden anniversary.**
W. T. Sullivan III.
Sky Telesc., Vol. 64, 544 - 550 (1982).

004.058 **A bibliophile's quest for Copernicus.**
W. J. Broad.
Science, Vol. 218, 661 - 662, 664 (1982).

004.059 **Vor 50 Jahren – ein Außenseiter entdeckt Radiostrahlung aus der Milchstraße.**
W. Schröder.
Sterne Weltraum, 21. Jahrg., 448 - 449 (1982).

004.060 **Astronomy in Ireland from earliest times to the eighteenth century.**
S. M. P. McKenna-Lawlor.
Vistas Astron., Vol. 26, 1 - 13 (1982).

004.061 **Die Entdeckung des Planeten Pluto.**
A. Tarnutzer.
Orion, 40. Jahrg., 182 - 184 (1982).

004.062 **Der Astronom Johannes Kepler.**
H. Mauder.
Phys. unserer Zeit, 13. Jahrg., 174 - 182 (1982).

004.063 **Herschel, Uranus et les autres.**
S. Débarbat.
Astronomie, Vol. 96, 561 - 569 (1982).

004.064 **Medieval islamic astronomical terminology.**
I. Ahmed, S. Yousef, A. Aiad.
J. Astron. Soc. Egypt, Vol. 3, 13 - 30 (1981).

004.065 **Nucleosynthesis in galaxies.**
E. M. Burbidge, G. R. Burbidge.
Essays in nuclear astrophysics, (see 003.013), p. 11 - 21 (1982).

004.066 **The evolution of the survey for latitude, longitude and meridian in ancient China.**
H.-r. Xiang, Z.-x. Niu, G.-q. Li, Y.-j. Ding.
Publ. Shaanxi Astron. Obs., No. 1, p. 23 - 34 (1982). In Chinese.

004.067 **Longitude of local meridian used in the first half period of Meiji.**
S. Ito.
Tokyo Astron. Obs. Rep., Vol. 19, 503 - 512 (1981). In Japanese.

004.068 **The Large Southern Telescope: Cape or Melbourne?**
B. Warner.
Q. J. R. Astron. Soc., Vol. 23, 505 - 514 (1982).

004.069 **The correspondence of the Rev. B. W. S. Vallack.**
D. W. Dewhirst.
Q. J. R. Astron. Soc., Vol. 23, 552 - 555 (1982).

004.070 **Kepler's cosmological theories: their agreement with observation.**
J. V. Field.
Q. J. R. Astron. Soc., Vol. 23, 556 - 568 (1982).

004.071 **The ancient Greek theory of trepidation.**
G. E. Kurtik.
Tr. 23. Nauchn. konf. po istor. estestvozn. i tekh., Moskva, mart, 1980. Sekts. istor. fiz. i mekh. Moskva, 1980, p. 19 - 25. In Russian. – Abstr. in Ref. zh., 51. Astron., 11.51.3 (1982).

004.072 **Precession in the as-Sufi star catalogue.**
G. E. Kurtik.
Tr. 23. Nauchn. konf. po istor. estestvozn. i tekh., Moskva, mart, 1980, Sekts. istor. fiz. i mekh. Moskva, 1980, p. 26 - 30. In Russian. – Abstr. in Ref. zh., 51. Astron., 11.51.4 (1982).

004.073 **Theories of the origin of the solar system: some historical remarks.**
A. Brahic.
Formation of planetary systems, (see 012.064), p. 15 - 58 (1982).

The author gives some insights on the history of the cosmogonical theories. He describes the birth of the main classes of theories, namely the tidal theories or catastrophic theories of which Buffon is the father and the nebular theory with Kant and Laplace as fathers. The author discusses the problems raised by each of these theories and why the Laplacian solar nebula seems now the most plausible theory. And finally, he enumerates a number of new data and considers their possible importance for a better understanding of the origin of planets.

004.074 **The history of the Rossbank Observatory, Tasmania.**
A. Savours, A. McConnell.
Ann. Sci., Vol. 39, 527 - 564 (1982).

Rossbank functioned from 1840 to 1854 as one of a chain of British Colonial Observatories which combined with European and Asian observatories in the study of terrestrial magnetism. The history and operation of the Rossbank Observatory is related, its instruments described, and the results discussed.

004.075 **Astronomy and probability: Forbes versus Michell on the distribution of the stars.**
B. Gower.
Ann. Sci., Vol. 39, 145 - 160 (1982).

James Forbes' critical examination of the probabilistic reasoning, which led John Michell to infer a physical connection between optically double and multiple stars, is analysed. It is argued that despite the interpretations of its nineteenth-century defenders, Michell's reasoning has some force which does not depend upon questionable Bayesian principles. Attention is drawn to some of the ambiguities concerning the notion of randomness, and it is shown that these ambiguities render Forbes' objections less than conclusive.

004.076 **Pluto and a transplutonian planet as predicted by Venkatesha Ketakara.**
S. D. Sharma.
Indian J. Hist. Sci., Vol. 16, 118 - 129 (1981).

004.077 **The gnomon in early Indian astronomy.**
G. Abraham.
Indian J. Hist. Sci., Vol. 16, 215 - 218 (1981).

004.078 **Observations on the Arabic reception of the astrolabe.**
P. Kunitzsch.
Arch. Int. Hist. Sci., Vol. 31, 243 - 252 (1981).

004.079 **The basic problems of the Babylonian lunar theory.**
Y. Maeyama.
Arch. Int. Hist. Sci., Vol. 31, 253 - 372 (1981).

004.080 **Tycho Brahe e l'astronomia. Per una nuova valutazione dell'astronomo danese.**
M. Di Bono.
Physis, Anno 24, 157 - 195 (1982).

004.081 **Kepler, elliptical orbits, and celestial circularity: a study in the persistence of metaphysical commitment. Part I.**
J. B. Brackenridge.
Ann. Sci., Vol. 39, 117 - 143 (1982).

004.082 **Kepler, elliptical orbits, and celestial circularity: a study in the persistence of metaphysical commitment. Part II.**
J. B. Brackenridge.
Ann. Sci., Vol. 39, 265 - 295 (1982).

004.083 **History of infrared detection – Part II. The first thermographs.**
E. H. Putley.
Infrared Phys., Vol. 22, 189 - 191 (1982).

004.084 **Astrophysik contra Astronomie.** Das Zurückdrängen einer alten Disziplin durch die Begründung einer neuen.
F. Krafft.
Ber. Wiss.-Gesch., Vol. 4, 89 - 110 (1981).

004.085 **A study of the sun's shadow observations of Guo Shou-jing.**
M.-d. Chen.
Acta Astron. Sinica, Vol. 23, 299 - 305 (1982). In Chinese.

004.086 **The folk-24-hour-system of chronometry and the early records of astronomical phenomena before 600AD.**
L.-x. Wang.
Acta Astron. Sinica, Vol. 23, 409 - 416 (1982). In Chinese.

004.087 **A recent valuable acquisition by the library of the Dominion Astrophysical Observatory.**
A. H. Batten.
J. R. Astron. Soc. Canada, Vol. 76, 382 - 391 (1982).
A copy of the two-volume Description de l'Observatoire astronomique central de Poulkova by F. G. W. Struve (published in 1845) has recently been acquired by the library of the Dominion Astrophysical Observatory. In this article, the Description is itself described and some account is given of people associated with this particular copy.

Die Sonnenuhr des Augustus.
See Abstr. 003.029.

Megalithic science: ancient mathematics and astronomy in north-west Europe.
See Abstr. 003.054.

Stellar astronomy: historical studies.
See Abstr. 003.059.

Albert Einstein's special theory of relativity: emergence (1905) and early interpretation (1905 - 1911).
See Abstr. 003.085.

Libro del nuevo cometa.
See Abstr. 003.089.

The 1769 transit of Venus.
See Abstr. 003.092.

Astronomy and society in Britain during the period 4000 - 1500 B.C.
See Abstr. 003.107.

Planeten und Sternbilder im Wandel der Geschichte.
See Abstr. 003.110.

The 'astronomical' chapters of the Ethiopic Book of Enoch (72 to 82).
See Abstr. 003.133.

Alexander William Bickerton: New Zealand's colourful astronomer.
See Abstr. 005.005.

Arthur Stanley Eddington – a centennial tribute.
See Abstr. 005.023.

The historical supernovae.
See Abstr. 125.052.

005 Biography

005.001 **In memory of Heino Albo.**
Publ. Tartu Astrofiz. Obs., Tom 48, 3 - 9 (1981). In Russian.

005.002 **On the occasion of the 75th birthday of G. N. Duboshin, 1979 December 25.**
Tr. Gos. Astron. Inst. Shternberga, Tom 51, 3 - 6 (1982). In Russian.

005.003 **Ludwig Boltzmann in Graz.**
P. Urban.
Phys. Bl., Vol. 38, 259 - 264 (1982).

005.004 **F. J. M. Stratton, DSO, OBE, TD, FRS, 1881 - 1960.**
W. H. McCrea.
Q. J. R. Astron. Soc., Vol. 23, 358 - 362 (1982).

005.005 **Alexander William Bickerton: New Zealand's colourful astronomer.**
G. Gilmore.
South. Stars, Vol. 29, 87 - 108 (1982).

005.006 **Der streitbare Editor. Zum 150. Todestag Franz Xaver von Zachs.**
P. Brosche.
Sterne Weltraum, 21. Jahrg., 344 (1982).

005.007 **Quatrième centenaire de la naissance de Peiresc (1580 - 1637).**
G. Courtès.
Astronomie, Vol. 96, 383 - 387 (1982).

005.008 **On the occasion of the 125th birthday of K. E. Ziolkovskij. 5/17.IX.1857 - 19.IX.1935.**

S. P. Korolev.
Priroda, 1982, No. 9, p. 2 - 5. In Russian.

005.009 **Harold Urey and the discovery of deuterium.**
F. G. Brickwedde.
Phys. Today, Vol. 35, No. 9, p. 34 - 39 (1982).

005.010 **E. Halley** (1656 - 1742).
A. I. Eremeeva.
Zemlya Vselennaya, 1982, No. 4, p. 46 - 49. In Russian.

005.011 **Thomas Jefferson Jackson See and the University of Missouri.**
C. J. Peterson.
Bull. American Astron. Soc., Vol. 14, 622 (1982). – Abstract.

005.012 **K. E. Ziolkovskij**, on the occasion of the 125th anniversary of his birthday (1857 - 1935).
G. S. Vetrov.
Zemlya Vselennaya, 1982, No. 5, p. 48 - 52. In Russian.

005.013 **Consul of the Danzig municipal council.**
J. Hevelius (1611 - 1687).
A. Lunin.
Zemlya Vselennaya, 1982, No. 5, p. 53 - 56. In Russian.

005.014 **Luigi Lilio and the Gregorian reform of the calendar.**
G. Moyer.
Sky Telesc., Vol. 64, 418 - 419 (1982).

005.015 **Bruno H. Bürgel. Vom Arbeiter zum Astronomen.**
A. Kunert.
Sterne Weltraum, 21. Jahrg., 396 - 397 (1982).

005.016 **Biographical note: Henri Chrétien and his magnificent contribution to the design of telescopes.**
J. S. Hall.
Bull. American Astron. Soc., Vol. 14, 834 - 836 (1982).

005.017 **Franjo Patricius' astronomical views.**
Ž. Dadić.
Publ. Astron. Obs. Sarajevo, Vol. 1, No. 1, (see 012.043), p. 23 - 28 (1981).

005.018 **Ćiro Carić, the author of the first tables for astronomic navigation in Yugoslavia.**
B. Franušić.
Publ. Astron. Obs. Sarajevo, Vol. 1, No. 1, (see 012.043), p. 45 - 50 (1981).

005.019 **Astronomers about their profession.**
Interviews given by A. A. Mikhajlov,
V. A. Ambartsumyan, A. B. Severnyj, V. V. Sobolev,
V. A. Krat, D. Ya. Martynov.
Zemlya Vselennaya, 1982, No. 6, p. 45 - 50. In Russian.

005.020 **Arthur Stanley Eddington – Pionier der Sternphysik.**
D. B. Herrmann.
Astron. Schule, 19. Jahrg., 112 - 113 (1982).

005.021 **Begegnung mit dem Entdecker des Pluto. Clyde W. Tombaugh – Legende zu Lebzeiten.**
S. Plozza.
Orion, 40. Jahrg., 181 - 182 (1982).

005.022 **Two decades of collaboration with Willy Fowler.**
F. Hoyle.
Essays in nuclear astrophysics, (see 003.013), p. 1 - 9 (1982).

005.023 **Arthur Stanley Eddington – a centennial tribute.**
J. S. Tenn.
Mercury, Vol. 11, 178 - 181, 186 - 188 (1982).

005.024 **In memoriam – Arthur Stanley Eddington (1882 - 1944).**
Z. Kopal.
Astrophys. Space Sci., Vol. 88, 267 (1982).

005.025 **Arthur Stanley Eddington (1882 - 1944).**
V. Vaculík.
Kozmos, Vol. 13, 190 (1982). In Slovak.

005.026 **A. S. Eddington and the 20th century astrophysics.**
V. Vanýsek.
Pokroky Mat. Fyz. Astron., Vol. 27, 308 - 317 (1982). In Czech.

005.027 **Sir Arthur Stanley Eddington.**
Říše hvězd, Vol. 63, 252 (1982). In Czech.

Concise dictionary of scientific biography.
See Abstr. 002.155.

Souvenirs d'un astronome (1908 - 1977).
See Abstr. 003.021.

Kepler.
See Abstr. 003.023.

Calendar of the correspondence of Pierre Simon Laplace.
See Abstr. 003.047.

Aristarchus of Samos.
See Abstr. 003.051.

Karl-Friedrich Zöllner.
See Abstr. 003.058.

Die Jugendjahre von Bernhard Schmidt und sein Briefwechsel mit dem Potsdamer Observatorium.
See Abstr. 003.087.

Herinneringen.
See Abstr. 003.094.

The three voyages of Edmond Halley in the "Paramore" 1698 - 1701.
See Abstr. 003.119.

Der Astronom Johannes Kepler.
See Abstr. 004.062.

006 Personal Notes

Dr. Paul Ahnert zum 85. Geburtstag.
G. A. Richter.
Sterne, 58. Band, 267 - 269 (1982).

Glückwünsche und Beiträge zur Feier des 75. Geburtstags von Wilhelm Becker 3. Juli 1982.
Extraprint Astron. Inst. Univ. Basel, 67 pp. (1982).

B. J. Bok received the Klumpke-Roberts Award 1982.
H. C. Arp, A. Fraknoi.
Mercury, Vol. 11, 155 - 156 (1982).

Bart J. Bok received the Dorothea Klumpke-Roberts Award.
Phys. Today, Vol. 35, No. 11, p. 81 - 82 (1982).

E. M. Burbidge received the Bruce Medal of the Astronomical Society of the Pacific.
Observatory, Vol. 102, 215 (1982).

E. Margaret Burbidge received the Catherine Wolf Bruce Medal.
Phys. Today, Vol. 35, No. 11, p. 81 (1982).

E. M. Burbidge received the Bruce Medal 1982.
H. C. Arp, A. Fraknoi.
Mercury, Vol. 11, 154 - 155 (1982).

Laudatio auf Jean Delhaye.
G. A. Tammann.
Mitt. Astron. Ges., Nr. 57, p. 121 - 122 (1982).

Margherita Hack received the Premio Linceo per l'astronomia, geodesia, geofisica e applicazioni, per il 1980.
Atti Accad. Naz. Lincei, Rend. Adunanze Solenni, Vol. 8, 206 (1981).

On the occasion of the 70th birthday of B. Yu. Levin, 1982 October 26.
Zemlya Vselennaya, 1982, No. 5, p. 38. In Russian.

B. Mayer received the Amateur Achievement Award 1982.
H. C. Arp, A. Fraknoi.
Mercury, Vol. 11, 156 (1982).

Ben Mayer received the Amateur Achievement Award.
Phys. Today, Vol. 35, No. 11, p. 83 (1982).

Laudatio auf Bohdan Paczynski.
T. Schmidt-Kaler.
Mitt. Astron. Ges., Nr. 57, p. 25 - 26 (1982).

On the occasion of the 90th birthday of O. Yu. Schmidt, 1891 - 1956.
Izv. AN SSSR. Fiz. Zemli, 1982, No. 6, p. 3 - 4. In Russian.
Abstr. in Ref. zh., 51. Astron., 10.51.9 (1982).

V. P. Shcheglov received the F. A. Bredikhin prize 1980.
Zemlya Vselennaya, 1982, No. 6, p. 50. In Russian.

I. S. Shklovskij received the annual prize of "Zemlya i Vselennaya" for 1982.
Zemlya Vselennaya, 1982, No. 4, p. 33. In Russian.

H. Terpstra ontvangt Simon Stevin-kijker.
T. A. M. Vermeesch.
Zenit, 9. Jaarg., 409 (1982).

B. Twarog received the Trumpler Award 1982.
H. C. Arp, A. Fraknoi.
Mercury, Vol. 11, 155 (1982).

Bruce Twarog received the Robert J. Trumpler Price.
Phys. Today, Vol. 35, No. 11, p. 82 - 83 (1982).

Victor G. Szebehely, 60th birthday.
U. Kirchgraber, J. Waldvogel.
Celestial Mech., Vol. 28, (see 012.049), 7 (1982).

Peter van de Kamp received the Prix Janssen 1982.
Astronomie, Vol. 96, 526 (1982).

007 Obituaries

Giorgio Abetti, 1882 - 1982.
M. G. Fracastoro.
Coelum, Vol. 50, 289 - 291 (1982).

M. K. V. Bappu, 1927 - 1982.
Messenger, No. 29, p. 1 (1982).

M. K. Vainu Bappu 1927 August 10 - 1982 August 19.
R. M. West.
Messenger, No. 30, p. 1 - 2 (1982).

C. Bertaud died 29 October 1982.
Astronomie, Vol. 96, 571 (1982).

Valdemar Axel Firsoff, 1910 - 1982.
P. Moore.
Q. J. R. Astron. Soc., Vol. 23, 629 - 630 (1982).

Kirill Pavlovich Florensky, 27 December 1915 - 9 April 1982.
Z. Kopal.
Moon Planets, Vol. 27, 3 - 4 (1982).

Seymour L. Hess died January 15, 1982.
P. J. Gierasch.
Icarus, Vol. 49, 297 (1982).

Ü. Ibrus, died 1981, May 27.
Tartu Astrofüüs. Obs. Publ., Köide 49, 3 - 6 (1982). In Russian.

Harold Lester Johnson, 1921 - 1980 April 2.
Recent advances in observational astronomy, (see 012.028), p. IX (1981).

I. Kuusik, 1921 February 18 - 1981 October 20.
Tartu Astrofüüs. Obs. Publ., Köide 49, 7 - 13 (1982). In Russian.

Gunnar Malmquist, 2 February 1893 - 27 June 1982.
T. Elvius.
Astron. Tidsskr., Årg. 15, 176 - 177 (1982).

O. A. Mel'nikov, 1912, April 2 - 1982, May 12.
Astron. Zh., Tom 59, 1036 - 1037 (1982). In Russian. English translation in Soviet Astron., Vol. 26, No. 5.

Henri Michel, 17 June 1885 - 18 January 1981.
E. Hoge.
Suppl. Ciel Terre, Vol. 98, 3 - 5 (1982).

Thomas A. Mutch, 1931 - 1980.
R. E. Arvidson.
J. Geophys. Res., Vol. 87, (see 012.071), 9716 (1982).

Cecilia Payne-Gaposchkin, 1900 May 10 - 1979 December 7.
O. Gingerich.
Q. J. R. Astron. Soc., Vol. 23, 450 - 451 (1982).

J. G. Porter (1900 - 1981).
B. G. Marsden.
Int. Comet. Q., Vol. 4, 55 (1982).

J. P. M. Prentice, 1903 March 14 - 1981 October 6.
B. Lovell.
Q. J. R. Astron. Soc., Vol. 23, 452 - 460 (1982).

Peter Read, 1923 June 18 - 1981.
M. L. Barton.
South. Stars, Vol. 29, 206 - 207 (1982).

Slavko Rozgaj, 1895 April 20 - 1978 February 27.
G. Ivanišević.
Publ. Astron. Obs. Sarajevo, Vol. 1, No. 1, (see 012.043), p. 39 - 44 (1981).

J. Schilt, 1894 February 3 - 1982 January 9.
J. H. Oort.
Zenit, 9. Jaarg., 375 (1982).

Krzysztof Serkowski, 1930 December 8 - 1981 October 7.
T. Gehrels.
Astron. Q., Vol. 4, 76 - 81 (1982).

Richard Sommer, 1888 May 4 - 1982 April 13.
A. Kunert.
Sterne Weltraum, 21. Jahrg., 330 (1982).

Richard Sommer, 4. 5. 1888 - 13. 4. 1982.
A. Kunert.
Sterne, 58. Band, 301 (1982).

Merlė A. Tuve, 1901 - 1982.
P. H. Abelson.
Phys. Today, Vol. 35, No. 9, p. 90 - 93 (1982).

John Whelan, 1945 July 30 - 1981.
G. T. Bath.
Q. J. R. Astron. Soc., Vol. 23, 460 - 462 (1982).

008 Observatories, Institutes

Reports, communications and publications of observatories and astronomical institutes are recorded in this section; included are numbered series of reprints. Whenever possible, the numbers of the abstracts referring to the publications are given. Observatories and institutes are listed in alphabetical order of their towns. In some cases observatory publications do not give the name of the town; the following list which gives names and towns of some institutions may serve as an aid in such cases.

Aarne Karjalainen Observatory	**Oulu**, Finland
Algonquin Radio Observatory	**Lake Traverse**, Ontario, Canada
Allegheny Observatory	**Pittsburgh**, Pennsylvania, USA
Anglo-Australian Observatory	**Epping**, N. S. W., Australia
Archenhold-Sternwarte	**Berlin-Treptow**, German Democratic Republic
Argentine Radioastronomy Institute	**Pereyra**, Iraola, Argentina
Arizona State University	**Tempe**, Arizona, USA
Arthur J. Dyer Observatory	**Nashville**, Tennessee, USA
Astronomical Latitude Station, Polish Academy of Sciences	**Borowiec**, Poland
Astronomisches Rechen-Institut	**Heidelberg**, F. R. Germany
Bell Laboratories	**Murray Hill**, New Jersey, USA
Bell Telephone Laboratories	**Holmdel**, New Jersey, USA
Bosscha Observatory	**Lembang**, Indonesia
Boyden Observatory	**Bloemfontein**, South Africa
Bureau International de l'Heure	**Paris**, France
Cajigal Observatory	**Caracas**, Venezuela
California Institute of Technology	**Pasadena**, California, USA
Carter Observatory	**Wellington**, New Zealand
Cavendish Laboratory	**Cambridge**, England
Centre de Données Stellaires	**Strasbourg**, France
Centro Astronomico Hispano-Aleman (Max-Planck Institut für Astronomie Station)	**Calar Alto, Almeria**, Spain
Centro de Investigación de Astronomía (CIDA)	**Mérida**, Venezuela
Cerro Tololo Interamerican Observatory	**La Serena**, Chile
Československá akademie věd, Astronomický ústav	**Prague**, Czechoslovakia
Chamberlin Observatory, University of Denver	**Denver**, Colorado, USA
Columbia University, Department of Astronomy	**New York**, New York, USA
Commonwealth Observatory	**Canberra**, Australia
Cornell University, Center for Radiophysics and Space Research	**Ithaca**, New York, USA
Corralitos Observatory	**Las Cruces**, New Mexico, USA
Crawford Hill Laboratory	**Holmdel**, New Jersey, USA
CSIRO, Radiophysics Division	**Sydney**, Australia
David Dunlap Observatory, University of Toronto	**Richmond Hill**, Ontario, Canada
Dearborn Observatory	**Evanston**, Illinois, USA
Department of Astronomy and Observatory, University of California	**Los Angeles, California**, USA
Department of Astronomy Swarthmore College	**Swarthmore**, Pennsylvania, USA
Department of Astronomy, University of Texas	**Austin**, Texas, USA
Deutsches Hydrographisches Institut (DHI)	**Hamburg**, F. R. Germany
Dominion Astrophysical Observatory	**Victoria**, B. C., Canada
Dominion Observatory	**Ottawa**, Ontario, Canada
Dominion Radio Astrophysical Observatory	**Penticton**, B. C., Canada
Dudley Observatory	**Schenectady**, New York, USA
Dunsink Observatory	**Dublin**, Ireland
Dyer Observatory, Vanderbilt University	**Nashville**, Tennessee, USA
Ege University Observatory	**Izmir**, Turkey
Engelhardt Observatory	**Kazan**, USSR
Erwin W. Fick Observatory, Iowa State University	**Ames**, Iowa, USA
European Southern Observatory	**La Silla**, Chile
Fabra Observatory	**Barcelona**, Spain
Felix Aguilar Observatory	**San Juan**, Argentina
Fernbank Observatory	**Atlanta**, Georgia, USA
Five College Observatories	**Amherst**, Massachusetts, USA
Florida State University Radio Observatory	**Tallahassee**, Florida, USA
Flower and Cook Observatories, University of Pennsylvania	**Philadelphia**, Pennsylvania, USA
George R. Wallace Jr. Astrophysical Observatory	**Cambridge**, Massachusetts, USA
Georgetown Observatory	**Washington**, D.C., USA
Glavnaya Astronomicheskaya Observatoriya AN SSSR	**Pulkovo**, USSR
Goddard Space Flight Center	**Greenbelt**, Maryland, USA
Goethe Link Observatory, Indiana University	**Bloomington**, Indiana, USA
H. M. Nautical Almanac Office, Royal Greenwich Observatory	**Greenwich**, England
Hale Observatories	**Pasadena**, California, USA
Harvard Radio Astronomy Station	**Cambridge**, Massachusetts, USA
Haute Provence Observatory	**Saint Michel**, France
Haystack Observatory	**Westford**, Massachusetts,USA
Heinrich-Hertz Institut	**Berlin-Adlershof**, German Democratic Republic
Herzberg Institute of Astrophysics	**Victoria**, B. C., **Ottawa**, Canada
High Altitude Observatory, University of Colorado	**Boulder**, Colorado, USA
Hopkins Observatory	**Williamstown**, Massachusetts, USA
Horn d'Arturo Observatory	**Bologna**, Italy
Hvar Observatory	**Zagreb**, Yugoslavia
IBM Thomas J. Watson Research Center	**Yorktown Heights**, New York USA
Indian Institute of Astrophysics	**Bangalore**, India
Infrared Telescope Facility	**Honolulu**, Hawaii, USA
Institute for Astronomy, University of Hawaii	**Honolulu**, Hawaii, USA
Institute for Theoretical Astronomy	**Leningrad**, USSR
Institute of Astronomy and Space Science, University of British Columbia	**Vancouver**, B. C., Canada

Institute of Theoretical Astrophysics, Blindern	**Oslo**, Norway
Instituto Argentino de Radioastronomía	**Villa Elisa**, Provincia de Buenos Aires, Argentina
Instituto de Astronomía y Fisica del Espacio (IAFE)	**Buenos Aires**, Argentina
Instituto y Observatorio de Marina	**San Fernando (Cadiz)**, Spain
International Latitude Observatory	**Mizusawa**, Japan
IUE Observatory, European Space Agency	**Villafranca**, Madrid, Spain
Jet Propulsion Laboratory, California Institute of Technology	**Pasadena**, California
Joint Institute for Laboratory Astrophysics (JILA)	**Boulder**, Colorado, USA
Judson B. Coit Observatory	**Boston**, Massachusetts, USA
Kandilli Observatory	**Istanbul**, Turkey
Kansas University Observatory	**Lawrence**, Kansas, USA
Kapteyn Astronomical Laboratory	**Groningen**, Netherlands
Karl-Schwarzschild-Observatorium	**Tautenburg**, German Democratic Republic
Kenneth Mees Observatory	**Rochester**, New York, USA
Kiepenheuer-Institut für Sonnenphysik, formerly Fraunhofer-Institut	**Freiburg**, F. R. Germany
Kitt Peak National Observatory	**Tucson**, Arizona, USA
Kodaikanal Observatory	**Bangalore**, India
Korean National Astronomical Observatory	**Seoul**, Korea
Kwasan and Hida Observatories	**Kyoto**, Japan
Lamont-Hussey Observatory	**Bloemfontein**, South Africa
Landessternwarte Heidelberg-Königstuhl	**Heidelberg**, F. R. Germany
Las Campanas Observatory	**Pasadena**, California, USA
Lawrence Livermore Laboratory, University of California	**Livermore**, California, USA
Leander McCormick Observatory, University of Virginia	**Charlottesville**, Virginia, USA
Lee Observatory	**Beirut**, Lebanon
Leopold-Figl-Observatorium	**Vienna**, Austria
Leuschner Observatory	**Berkeley**, California, USA
Lick Observatory	**Santa Cruz**, (Mount Hamilton), California, USA
Lindheimer Astronomical Research Center	**Evanston**, Illinois, USA
Lockheed Palo Alto Research Laboratory	**Palo Alto**, California, USA
Lockheed Solar Observatory	**Saugus**, California, USA
Lohrmann-Observatorium der Technischen Universität Dresden	**Dresden**, German Democratic Republic
Louisiana State University Observatory	**Baton Rouge**, Louisiana, USA
Lowell Observatory	**Flagstaff**, Arizona, USA
Lunar and Planetary Laboratory, Catalina Station	**Tucson**, Arizona, USA
Max-Planck-Institut für Astronomie	**Heidelberg**, F. R. Germany
Max-Planck-Institut für Physik und Astrophysik	**Munich**, F. R. Germany
Max-Planck-Institut für Radioastronomie	**Bonn**, F. R. Germany
McDonald Observatory	**Fort Davis**, Texas, USA
McMath Hulbert Observatory	**Pontiac**, Michigan, USA
Michigan State University Observatory	**East Lansing**, Michigan, USA
Molonglo Radio Observatory, University of Sydney	**Sydney**, Australia
Monterey Institute for Research in Astronomy	**Carmel Valley**, California, USA
Mount Cuba Observatory	**Wilmington**, Delaware, USA
Mount John Observatory	**Lake Tekapo**, New Zealand
Mount Stromlo Observatory	**Canberra**, Australia
Mount Wilson Observatory	**Pasadena**, California, USA
Mt. Laguna Observatory	**San Diego**, California, USA
Mullard Radio Astronomy Observatory	**Cambridge**, England
Mullard Space Science Laboratory	**London**, England
Multiple Mirror Telescope Observatory	**Tucson**, Arizona, USA
Narrabri Observatory, University of Sydney	**Sydney**, Australia
National Bureau of Standards	**Washington**, D.C., USA
National Radio Astronomy Observatory	**Charlottesville**, Virginia, USA **Green Bank**, West Virginia, USA **Socorro**, New Mexico, USA **Tucson**, Arizona, USA
National Research Council of Canada	**Ottawa**, Ontario, Canada
New Mexico State University Observatory	**Las Cruces**, New Mexico, USA
Nicholas Copernicus Observatory	**Brno**, Czechoslovakia
Nizamiah & Rangapur Observatories	**Hyderabad**, India
Nuffield Radio Astronomy Laboratories, Jodrell Bank University of Manchester	**Manchester**, England
Oak Ridge Observatory (formerly Harvard College Observatory)	**Cambridge**, Massachusetts, USA
Observatoire Royal de Belgique	**Uccle**, Belgium
Observatories of the University of Western Ontario	**London**, Canada
Observatório Astronômico Antares	**Feira de Santana**, Brazil
Observatório Astronômico do Instituto de Física da Universidade Federal do Rio Grande do Sul	**Porto Alegre**, Rio Grande do Sul, Brazil
Observatorio de Cartuja	**Granada**, Spain
Observatorio del Ebro	**Tortosa**, Spain
Observatorio Nacional	**Rio de Janeiro**, Brazil
Observatorio Nacional de Física Cósmica	**San Miguel**, Argentina
Observatory, University of Michigan	**Ann Arbor**, Michigan, USA
Ohio State University Radio Observatory	**Columbus**, Ohio, USA
Ole Roemer-Observatoriet	**Aarhus**, Denmark
Onsala Space Observatory	**Göteborg**, Sweden
Owens Valley Radio Observatory	**Big Pine**, California, USA
Palomar Observatory	**Pasadena**, California, USA
Perkins Observatory, Ohio State and Wesleyan Universities	**Delaware**, Ohio, USA
Physical Research Laboratory	**Ahmedabad**, India
Purple Mountain Observatory	**Nanking**, China
Radcliffe Observatory	**Pretoria**, South Africa
Raman Research Institute	**Bangalore**, India
Rattlesnake Mountain Observatory	**Richland**, Washington, USA

Remeis-Sternwarte	**Bamberg**, F. R. Germany
Ritter Astrophysical Research Center of the University of Toledo	**Toledo**, Ohio, USA
Rosemary Hill Observatory	**Gainesville**, Florida, USA
Rothney Astrophysical Observatory	**Calgary**, Canada
Royal Aircraft Establishment, Geophysical Studies in Space Department	**Farnborough**, England
Royal Radar Establishment, Radio Astronomy Division	**Malvern**, England
Sacramento Peak Observatory	**Sunspot**, New Mexico, USA
Sagamore Hill Radio Observatory	**Hamilton**, Massachusetts, USA
San Fernando Observatory	**Northridge**, California, USA
Shaanxi Astronomical Observatory	**Lintong**, Sian, China
Siding Spring Observatory	**Siding Spring**, N. S. W., Australia
Smithsonian Astrophysical Observatory	**Cambridge**, Massachusetts, USA
Sonnenobservatorium Kanzelhöhe	**Graz**, Austria
South African Astronomical Observatory	**Cape Town**, South Africa
Specola Vaticana	**Castel Gandolfo**, Vatican
Specola di Padova	**Asiago**, Italy
Sproul Observatory	**Swarthmore**, Pennsylvania, USA
Sternberg Astronomical Institute	**Moscow**, USSR
Steward Observatory, University of Arizona	**Tucson**, Arizona, USA
W. Struve Tartu Astrophysical Observatory	**Tartu**, USSR
Tata Institute of Fundamental Research	**Bombay**, India
United States Naval Observatory	**Washington**, D. C., USA
University of Alabama	**University**, Alabama, USA
University of California	**Berkeley**, California, USA
University of Florida Observatories	**Bronson**, Florida, USA
University of Florida, Radio Observatory	**Old Town**, Florida, USA
University of Hawaii	**Honolulu**, Hawaii, USA
University of Illinois Observatory	**Urbana**, Illinois, USA
University of Kansas Observatory	**Lawrence**, Kansas, USA
University of Maryland	**College Park**, Maryland, USA
University of Michigan Observatories	**Ann Arbor**, Michigan, USA
University of Minnesota	**Minneapolis**, Minnesota, USA
University of South Florida Observatory	**Tampa**, Florida, USA
University of Texas, Department of Astronomy	**Austin**, Texas, USA
University of Washington, Astronomy Department	**Seattle**, Washington, USA
Uttar Pradesh State Observatory	**Nainital**, India
Van Vleck Observatory	**Middletown**, Connecticut, USA
Vatican Observatory	**Castel Gandolfo**, Vatican
Wallace Astrophysical Observatory	**Cambridge**, Massachusetts, USA
Warner and Swasey Observatory	**Cleveland**, Ohio, USA
Washburn Observatory University of Wisconsin	**Madison**, Wisconsin, USA
West Melton Observatory	**Christchurch**, New Zealand
Wilhelm-Foerster Sternwarte	**Berlin**, F. R. Germany
Yale University Observatory	**New Haven**, Connecticut, USA
Yerkes Observatory	**Williams Bay**, Wisconsin, USA
Zentralinstitut für Astrophysik, Sternwarte Babelsberg	**Potsdam-Babelsberg**, German Democratic Republic
Zentralinstitut für Astrophysik, Sternwarte in Sonneberg	**Sonneberg**, German Democratic Republic
Zentralinstitut für solar-terrestrische Physik	**Berlin-Adlershof**, German Democratic Republic

008.001 **Ahmedabad**

Physical Research Laboratory. Report for the period 1980 April 1 - 1981 March 31.
D. Lal.
Bull. Astron. Soc. India, Vol. 10, 53 - 57 (1982).

008.002 **Baltimore, Md.**

Space Telescope Science Institute, Baltimore, Maryland 21218. Report for the period 30 April 1981 - 30 September 1981.
Bull. American Astron. Soc., Vol. 14, 699 (1982).

Space Telescope Science Institute, Preprint Series, No. 1 (32.155.048), 2 (32.135.048).

The Johns Hopkins University, Department of Physics, Baltimore, Maryland 21218. – Report for the period 1 January 1979 - 31 August 1981.
Bull. American Astron. Soc., Vol. 14, 671 - 674 (1982).

008.003 **Bamberg**

Dr.-Remeis-Sternwarte Bamberg, Astronomisches Institut der Universität Erlangen-Nürnberg. – Report for 1981.
I. Bues.
Mitt. Astron. Ges., Nr. 56, p. 7 - 13 (1982).

008.004 **Bangalore**

Indian Institute of Astrophysics. – Report for the period 1980 April 1 - 1981 March 31.
M. K. V. Bappu.
Bull. Astron. Soc. India, Vol. 10, 165 - 179 (1982).

Raman Research Institute. – Report for the period 1980 April 1 - 1981 March 31.
V. Radhakrishnan.
Bull. Astron. Soc. India, Vol. 10, 180 - 186 (1982).

008.005 **Basel**

Astronomisches Institut der Universität Basel. Report for 1981.
G. A. Tammann.
Mitt. Astron. Ges., Nr. 56, p. 15 - 21 (1982).

008.006 **Berlin**

Institut für Astronomie und Astrophysik der Technischen Universität Berlin. – Report for 1981.
R. Wielen.
Mitt. Astron. Ges., Nr. 56, p. 23 - 30 (1982).

Institut für Astronomie und Astrophysik der Technischen Universität Berlin, Mitteilungen, No. 3 (32.008. 006).

Zentralinstitut für solar-terrestrische Physik (Heinrich-Hertz-Institut).
Akad. Wiss. DDR, Jahrb. 1980, p. 26 - 27, 325 - 328 (1982).

Heinrich-Hertz-Institut, Solare Beobachtungsergebnisse. Akademie der Wissenschaften der DDR, Zentralinstitut für Solar-Terrestrische Physik (Heinrich-Hertz-Institut), Berlin-Adlershof, HHI Solar Data, Vol. 33, 1982 January - April (32.072.054).

Heinrich-Hertz-Institut. Supplement Series of Solar Data. Akademie der Wissenschaften der DDR, Zentralinstitut für Solar-Terrestrische Physik (Heinrich-Hertz-Institut), Berlin-Adlershof. **HHI Suppl. Ser. Solar Data,** Vol. 5, Nos. 2 - 3 (32.072.055).

008.007 **Bloemfontein**

Boyden Observatory Publications, No. 1 (32.034. 072).

Boyden Observatory Reprint Nos. 50 (31.073.048), 51 (31.082.006), 52 (32.013.080), 53 (31.158.306).

008.008 **Bochum**

Astronomisches Institut der Ruhr-Universität Bochum. – Report for 1981.
J. Dachs.
Mitt. Astron. Ges., Nr. 56, p. 31 - 37 (1982).

Bereich Extraterrestrische Physik in der Abteilung XII der Ruhr-Universität Bochum. – Report for 1981.
R. H. Giese.
Mitt. Astron. Ges., Nr. 56, p. 37 - 38 (1982).

Institut für Theoretische Physik, Lehrstuhl IV der Ruhr-Universität Bochum. – Report for 1981.
K. Schindler.
Mitt. Astron. Ges., Nr. 56, p. 39 - 40 (1982).

008.009 **Bonn**

Astronomische Institute der Universität Bonn. Sternwarte mit Observatorium Hoher List; Radioastronomisches Institut mit Radioobservatorium Stockert; Institut für Astrophysik und Extraterrestrische Forschung; Sonderforschungsbereich Radioastronomie. – Reports for 1981.
H. Schmidt, U. Mebold, W. Priester.
Mitt. Astron. Ges., Nr. 56, p. 41 - 73 (1982).

Max-Planck-Institut für Radioastronomie. Report for 1981.
P. G. Mezger, R. Wielebinski.
Mitt. Astron. Ges., Nr. 56, p. 75 - 110 (1982).

Mitteilungen aus den Geodätischen Instituten der Rheinischen Friedrich-Wilhelms-Universität Bonn, Nr. 62 (32.094.032; 32.021.047), 63, 65.

008.010 **Bucharest**

Observations solaires. Rotations 1703 - 1716 (32.072.053).

008.011 **Budapest**

Publications of the Astronomy Department of the Eötvös University, No. 6 (32.012.056).

008.012 **Byurakan**

Byurakan Astrophysical Observatory, Armenia, USSR, Reprints, Nos. 305 (32.158.013), 306 (32.158.014), 307 (32.158.015), 308 (32.158.021), 309 (32.158.022), 310 (32.156.002), 311 (32.121.003), 312 (32.113.006).

008.013 **Cambridge, England**

Institute of Astronomy. Annual Report 1981-1982.
M. J. Rees.
Published by The Observatories, Madingley Road, Cambridge CB3 0HA, England. 39 pp. (1982).

Institute of Astronomy, University of Cambridge. Report.
M. J. Rees.
Q. J. R. Astron. Soc., Vol. 23, 399 - 428 (1982).

008.014 **Cambridge, Mass.**

Massachusetts Institute of Technology, Cambridge, Massachusetts 02139. – Report.
J. L. Elliot.
Bull. American Astron. Soc., Vol. 14, 678 - 696 (1982).

Smithsonian Astrophysical Observatory, Special Report, No. 392, Vol. II (32.012.065).

IAU Circulars, Nos. 3708 - 3758 (1982).

Minor Planet Circulars, (M. P. C.), Nos. 6989 - 7514 (1982).

008.015 **Cape Town**

South African Astronomical Observatory. Report for year ending 1981 December 31.
M. W. Feast.
Q. J. R. Astron. Soc., Vol. 23, 608 - 623 (1982).

008.016 **Castel Gandolfo**

Vatican Observatory Publications, Vol. 2, Nos. 1 (32.113.048), 2 (32.034.073).

008.017 **Catania**

Osservatorio Astrofisico di Catania, Pubblicazione, N. 166 (32.072.056).

008.018 **Dresden**

Technische Universität Dresden, Lohrmann-Observatorium, Zirkular, Nos. 99 - 100 (32.045.018).

008.019 **Dublin**

Dunsink Observatory in 1980.
P. A. Wayman.
Irish Astron. J., Vol. 15, 138 - 142 (1981).

008.020 **Durham**

Department of Physics, University of Durham. Report for the period 1980 October 1 to 1981 September 30.
A. W. Wolfendale.
Q. J. R. Astron. Soc., Vol. 23, 429 - 436 (1982).

008.021 **Edinburgh**

Occasional Reports of the Royal Observatory, Edinburgh, Nos. 8 (32.034.074), 9 (32.012.058).

008.022 **Epping**

Anglo-Australian Telescope 1981/82. Report of the Anglo-Australian Telescope Board 1 July 1981 to 30 June 1982.
Australian Government Publishing Service, Canberra, 8 + 44 pp. (1982).

Anglo-Australian Observatory, Preprint, Nos. 169 (32.013.083), 170 (32.132.042), 171 (32.156.012), 172 (32.156.013), 173 (32.117.200), 174 (32.158.287), 175 (32.158.288), 176 (32.121.038).

A. A. O. Newsletter, Nos. 22 - 23 (1982).

008.023 **Erlangen**

Physikalisches Institut, Angewandte Optik. Report for 1981.
G. Weigelt.
Mitt. Astron. Ges., Nr. 56, p. 111 - 112 (1982).

008.024 **Falmer**

Astronomy Centre, University of Sussex. Report for the year ending 1981 September 30.
R. J. Tayler.
Q. J. R. Astron. Soc., Vol. 23, 438 - 449 (1982).

008.025 **Farnborough**

Royal Aircraft Establishment, Farnborough Geophysical Studies in Space Department. Report for the year ending 1982 March 31.
Q. J. R. Astron. Soc., Vol. 23, 624 - 628 (1982).

008.026 **Frankfurt**

Institut für Theoretische Physik (Astrophysik). Report for 1981.
W. H. Kegel.
Mitt. Astron. Ges., Nr. 56, p. 113 (1982).

008.027 **Freiburg**

Kiepenheuer-Institut für Sonnenphysik. – Report for 1981.
E. H. Schröter.
Mitt. Astron. Ges., Nr. 56, p. 115 - 127 (1982).

008.028 **Ghent**

Rijksuniversiteit Gent, Sterrenkundig Observatorium, Mededelingen, Nos. 46 (29.042.008), 47 (31.117.026), 48 (32.042.072), 49 (31.117.035), 50 (31.117.036), 51 (31.042.114).

008.029 **Göteborg**

Research Laboratory of Electronics and Onsala Space Observatory, Chalmers University of Technology, Gothenburg, Sweden. Research Report, No. 144 (32.082.070).

008.030 **Göttingen**

Universitäts-Sternwarte Göttingen und Institut für Sonnenforschung Locarno-Orselina (Tessin). – Reports for 1981.
K. J. Fricke, H. H. Voigt.
Mitt. Astron. Ges., Nr. 56, p. 129 - 141 (1982).

008.031 **Graz**

Institut für Astronomie der Universität Graz; Institut für Astronomie (Universitäts-Sternwarte); Observatorium Lustbühel; Sonnenobservatorium Kanzelhöhe. – Reports for 1981.
H. Haupt.
Mitt. Astron. Ges., Nr. 56, p. 143 - 149 (1982).

008.032 **Greenwich**

Greenwich Time Report, Royal Greenwich Observatory, Time and Latitude Service, 1981 October - 1982 March (32.044.023).

Gemini. Newsletter of the Royal Greenwich Observatory, Nos. 2 - 4 (1982).

008.033 **Hamburg**

Hamburger Sternwarte. – Report for 1981.
A. Weigert.
Mitt. Astron. Ges., Nr. 56, p. 153 - 160 (1982).

Deutsches Hydrographisches Institut (DHI).
Report for 1981.
H. Enslin.
Mitt. Astron. Ges., Nr. 56, p. 151 (1982).

Deutsches Hydrographisches Institut. 36. Jahresbericht 1981.
G. Zickwolff.
Deutsches Hydrogr. Inst., Hamburg, 120 pp. (1982).

008.034 **Hannover**

Astronomische Station am Institut für Theoretische Geodäsie der Universität. – Report for 1981.
G. Seeber.
Mitt. Astron. Ges., Nr. 56, p. 161 - 163 (1982).

Universität Hannover, Astronomische Station, Veröffentlichung, No. 14 (32.115.019).

008.035 **Heidelberg**

Astronomisches Rechen-Institut. – Report for 1981.
W. Fricke.
Mitt. Astron. Ges., Nr. 56, p. 165 - 175 (1982).

Astronomisches Rechen-Institut, Heidelberg, Mitteilungen Serie A, Nos. 140 (32.002.032), 141 (32.002.033), 142 (31.098.098), 143 (32.041.019).

Astronomisches Rechen-Institut Heidelberg, Mitteilungen Serie B, Nos. 109 (31.043.015), 110 (31.041.038), 111 (31.111.026), 112 (31.041.047), 113 (32.141.014), 114 (32.098.004), 115 (32.098.001).

Astronomische Grundlagen für den Kalender 1984 (32.047.023).

Astronomy and Astrophysics Abstracts. Vol. 31. Literature 1982, Part 1 (32.002.143).

Institut für Theoretische Astrophysik. – Report for 1981.
B. Baschek, G. Traving.
Mitt. Astron. Ges., Nr. 56, p. 177 - 180 (1982).

Landessternwarte. – Report for 1981.
I. Appenzeller.
Mitt. Astron. Ges., Nr. 56, p. 191 - 200 (1982).

Max-Planck-Institut für Astronomie. Deutsch-Spanisches Astronomisches Zentrum. – Report for 1981.
H. Elsässer.
Mitt. Astron. Ges., Nr. 56, p. 201 - 229 (1982).

Max-Planck-Institut für Kernphysik. Abteilung Kosmophysik. – Report for 1981.
H. Fechtig, H. Völk.
Mitt. Astron. Ges., Nr. 56, p. 181 - 190 (1982).

008.036 **Hyderabad**

Contributions from Nizamiah and Japal-Rangapur Observatories, Nos. 5 (32.119.085), 6 (32.119.086), 7 (32.119.087), 8 (32.119.088), 9 (32.119.089), 12 (32.119.090), 13 (32.119.091), 14 (32.119.092).

Nizamiah and Rangapur Observatories and Department of Astronomy, Osmania University, Reprint Nos. 76 (25.151.020), 77 (26.151.003), 78 (22.121.019), 79 (26.119.046), 80 (27.119.016), 81 (27.151.024), 82 (26.151.071), 83 (26.008.037), 84 (27.119.070), 85 (28.104.037), 86 (28.034.048), 87 (28.083.046), 88 (29.119.024), 89 (29.077.056), 90 (30.119.029), 91 (29.119.064), 94 (29.118.016), 96 (30.117.039), 97 (30.117.078), 98 (30.119.053), 99 (30.151.074), 100 (30.063.053), 101 (30.119.069).

008.037 **Innsbruck**

Institut für Astronomie. – Report for 1981.
J. Pfleiderer.
Mitt. Astron. Ges., Nr. 56, p. 231 - 235 (1982).

008.038 **Istanbul**

Kandilli Observatory, Heliophysics Service, Istanbul. Publications, Second Series, Nos. 1 (32.072.057), 2 (32.073.089), 4 (32.072.058), 5 (32.072.059), 6 (32.022.148), 7 (32.114.173), 8 (32.114.174).

008.039 **Ithaca, N. Y.**

National Astronomy and Ionosphere Center, Cornell University, Ithaca, Astronomy Publications, Nos. A82-1 (31.158.073), A82-2 (31.141.511), A82-3 (31.141.204), A82-4 (31.158.323), A82-5 (31.158.324), A82-6 (31.141.167), A82-7, A82-8, A82-9 (31.132.033), A82-10 (31.132.044), A82-11 (27.160.032), A82-12 (30.160.030), A82-13 (26.159.003), A82-14 (31.158.002), A82-15 (31.122.002), A82-16 (31.103.801), A82-17 (31.132.022), A82-18 (32.100.003), A82-19 (31.131.073), A82-20 (31.158.005),

A82-21 (31.097.115), A82-22 (31.066.010), A82-23 (31.158.003), A82-24 (31.134.045).

008.040 **Kiel**

Institut für Theoretische Physik und Sternwarte der Universität. – Report for 1981.
K. Hunger.
Mitt. Astron. Ges., Nr. 56, p. 237 - 242 (1982).

Institut für Reine und Angewandte Kernphysik, Arbeitsgruppe Mathematische Physik. – Report for 1981.
K. O. Thielheim.
Mitt. Astron. Ges., Nr. 56, p. 243 - 245 (1982).

008.041 **Kyoto**

Contributions from the Kwasan and Hida Observatories, No. 250 (32.097.051).

008.042 **La Plata**

Observatorio Astronomico de la Universidad Nacional de La Plata, Serie Astronómica, Tomo 41 (32.151.071).

008.043 **La Serena**

Cerro Tololo Inter-American Observatory, La Serena, Chile. – Report for the period 1 October 1980 - 30 September 1981.
P. S. Osmer.
Bull. American Astron. Soc., Vol. 14, 781 - 795 (1982).

008.044 **La Silla**

ESO Scientific Preprint, Nos. 201 (32.142.132), 202 (32.142.133), 203 (32.141.166), 204 (32.141.167), 205 (32.158.289), 206 (32.112.048), 207 (32.158.290), 208 (32.131.283), 209 (32.114.175), 210 (32.155.049), 211 (32.162.208), 212 (32.158.291), 213 (32.132.043), 214 (32.121.039), 215 (32.158.292), 216 (32.061.044), 217 (32.142.134), 218 (32.103.245), 219 (32.031.628), 220 (32.121.040), 221 (32.125.102), 222 (32.113.049), 223 (32.158.293), 224 (32.160.064), 225 (32.141.550), 226 (32.160.065).

Twenty years ESO.
L. Woltjer.
Messenger, No. 30, p. 1 (1982).

The Users Committee – where the grassroots talk.
G. Lyngå.
Messenger, No. 29, p. 1 (1982).

The Messenger – El Mensajero, Nos. 29 - 30 (1982).

008.045 **Lake Tekapo**

Astronomy at the University of Canterbury Physics Department and Mt. John University Observatory. Report for 1981.
A. G. McLellan.
South. Stars, Vol. 29, 145 - 150 (1982).

008.046 **Lawrence, Kans.**

University of Kansas, Lawrence, Kansas 66045. Report.
R. B. Phillips.
Bull. American Astron. Soc., Vol. 14, 675 - 677 (1982).

008.047 **Leicester**

University of Leicester X-ray Astronomy Group, Department of Physics. Report for the period 1980 October to 1981 September.
K. A. Pounds.
Q. J. R. Astron. Soc., Vol. 23, 575 - 592 (1982).

008.048 **Leningrad**

Trudy Instituta Teoreticheskoj Astronomii, Akademiya Nauk SSSR, Vyp. 18 (32.103.901; 32.099.150; 32.103.363).

008.049 **Lintong**

Publications of the Shaanxi Astronomical Observatory, 1982, No. 1 (32.013.084; 32.031.627; 32.043.007; 32.043.008; 32.004.066; 32.044.025; 32.083.032; 32.035.031).

Time and Frequency Services Bulletin, 1982, Nos. 4 - 9 (32.044.024).

008.050 **Madrid**

Boletín Astronómico del Observatorio de Madrid, Vol. 10, Nos. 5 (32.072.061), 6 (32.034.075; 32.015.031; 32.022.149).

008.051 **Manila**

Manila Observatory, Solar Division, Solar maps and activity, 1982 March - September (32.072.060).

008.052 **Mizusawa**

Monthly Notes of the International Polar Motion Service, 1982, No. 5 - 10 (32.045.020).

Publications of the International Latitude Observatory of Mizusawa, Vol. 15, No. 1 (32.045.021).

008.053 **Moscow**

Trudy Gosudarstvennogo Astronomicheskogo Instituta im. P. K. Shternberga, Izdatel'stvo Gosudarstvennogo Universiteta, Tom 51 (32.005.002; 32.097.004; 32.094.509; 32.094.006; 32.041.002; 32.002.038).

008.054 **Münster**

Astronomisches Institut der Universität. Außenstation Schalkenmehren. Forschergruppe Erde-Mond-System (Planetologie). – Reports for 1981.
W. Seitter, D. Stöffler.
Mitt. Astron. Ges., Nr. 56, p. 291 - 298 (1982).

008.055 **Munich**

Institut für Astronomie und Astrophysik der Universität München. Universitäts-Sternwarte. – Reports for 1981.
P. Wellmann.
Mitt. Astron. Ges., Nr. 56, p. 283 - 289 (1982).

Mitteilungen aus dem Institut für Astronomische und Physikalische Geodäsie der Technischen Universität München, Nr. 156 (32.042.073).

Max-Planck-Institut für Physik und Astrophysik. Institut für Astrophysik und Institut für extraterrestrische Physik. – Reports for 1981.
R. Kippenhahn, J. Trümper.
Mitt. Astron. Ges., Nr. 56, p. 247 - 282 (1982).

Max-Planck-Institut für Physik und Astrophysik, Institut für extraterrestrische Physik. MPE Preprint 5 (32.034.076), 6 (32.131.284).

008.056 **Nanking**

Publications of Purple Mountain Observatory, Vol. 1, No. 1 (32.073.094; 32.073.095; 32.072.065; 32.073.096; 32.071.018; 32.021.050; 32.031.028; 32.098.073; 32.051.069; 32.034.108).

008.057 **Oxford**

Department of Astrophysics, University of Oxford. Report for the period 1980 August 1 to 1981 July 31.
D. E. Blackwell.
Q. J. R. Astron. Soc., Vol. 23, 388 - 398 (1982).

University of Oxford, Department of Astrophysics, Publications, Nos. 301 (28.132.015), 302 (28.122.162), 303 (29.066.062), 304 (28.099.120), 305, 306 (29.066.177), 307 (30.066.082), 308 (29.162.040), 309 (29.131.105), 310 (29.064.007), 311 (29.063.017), 312 (29.161.003), 313 (29.158.037), 314 (30.158.109), 315 (30.064.069), 316, 317, 318 (29.021.003), 319 (30.071.027), 321 (29.115.006), 320 (30.143.072), 322, 323 (29.154.001), 324 (29.061.015), 325 (30.117.068).

008.058 **Paris**

Rapport d'activité de l'Observatoire de Paris. 1er janvier - 31 décembre 1980.
Observatoire de Paris, 61, Avenue de l'Observatoire, 75014 Paris. 157 pp. (1981).

Observatoire de Paris, Section d'Astrophysique, à Meudon, **Cartes synoptiques,** Vol. VI, Fasc. 4, années 1980 - 1981 (32.072.063).

Pré-publication.Institut d'Astrophysique de Paris, Nos. 7 (32.073.091), 8 (32.141.168), 9 (32.073.092), 10 (32.124.414), 11 (32.062.095), 12 (32.132.044), 13 (32.061.045), 14 (32.074.073), 15 (32.106.043), 16 (32.124.012), 17 (32.124.505), 18 (32.074.074), 19 (32.158.298), 20 (32.074.075), 21 (32.121.041), 22 (32.073.093), 23 (32.061.046).

Bureau International de l'Heure, (B. I. H.), Circular, A (32.044.028), D188 - D194 (32.044.029).

008.059 **Potsdam**

Zentralinstitut für Astrophysik.
Akad. Wiss. DDR, Jahrb. 1980, p. 23 - 24, 309 - 317 (1982).

Zentralinstitut für Physik der Erde.
Akad. Wiss. DDR, Jahrb. 1980, p. 24 - 26, 317 - 325 (1982).

Mitteilungen des Zentralinstituts für Physik der Erde, Potsdam, Nos. 966 (30.045.008), 976 (30.045.010).

Veröffentlichungen des Zentralinstituts für Physik der Erde, Potsdam, Nr. 63 (32.012.053), 67 (32.045.022).

Zeit- und Breitenbestimmungen, Zeitsysteme, Präzisionszeitvergleiche. Jahrg. 1981, Nos. 3 - 4 (32.044.027).

008.060 **Pulkovo**

Izvestiya Glavnoj Astronomicheskoj Observatorii v Pulkove, Astrofizika i Astrometriya, Nos. 199 (32.111.018; 32.118.042; 32.099.151; 32.041.022; 32.041.023; 32.041.024; 32.002.125; 32.032.054; 32.082.086; 32.021.056; 32.032.055; 32.032.056; 32.032.057; 32.071.019; 32.080.060; 32.122.164; 32.114.220; 32.154.045; 32.031.029; 32.031.030; 32.031.031; 32.034.135; 32.031.032), 200 (32.071.020; 32.075.021; 32.072.108; 32.122.165; 32.158.340; 32.119.107; 32.082.087; 32.031.669; 32.031.670; 32.032.058; 32.032.059; 32.045.025; 32.045.026; 32.045.027; 32.032.060; 32.044.039; 32.031.671; 32.097.062; 32.032.061; 32.032.062; 32.032.063; 32.051.083; 32.032.064; 32.032.065; 32.031.033; 32.032.066).

Trudy Glavnoj Astronomicheskoj Observatorii v Pulkove, Akademiya Nauk SSSR, Ser. 2, Tom 84 (32.002.121; 32.002.122; 32.002.123; 32.002.124).

Glavnaya Astronomicheskaya Observatoriya, Pulkovo, Catalogue of solar activity for the year 1977 (32.072.047), 1978 (32.072.048).

008.061 **San Diego, Calif.**

Mount Laguna Observatory, San Diego, California 92182. – Report.

M. S. Snowden.
Bull. American Astron. Soc., Vol. 14, 697 - 698 (1982).

008.062 **Santa Cruz, Calif.**

Lick Observatory Bulletin, Nos. 801 (29.065.105), 812 (32.154.009), 854 (29.116.017), 866 (29.158.244), 871 (30.122.021), 876 (29.158.075), 877 (30.114.022), 881 (30.154.037), 882 (30.120.003), 883 (29.107.026), 884 (30.141.042), 885 (30.041.004), 886 (30.158.159), 887 (30.121.019), 888 (30.158.172), 889 (30.141.108), 890 (30.100.048), 891 (32.034.079), 892 (32.034.080), 893 (30.031.579), 894 (30.141.143), 895 (30.121.022), 896 (31.141.007), 897 (31.141.011), 898 (30.158.226), 899 (31.131.014), 900 (31.158.289), 901 (30.114.177), 902 (31.158.027), 903 (31.158.009), 904 (31.141.043), 905 (31.158.292), 906 (31.122.030), 907 (31.114.097), 908 (31.141.221), 909 (31.062.116), 910 (31.062.117).

Contributions from the Lick Observatory, Nos. 424 (29.120.027), 425 (30.101.016).

008.063 **Sarajevo**

Publications of the Astronomical Observatory of Sarajevo, Vol. 1, No. 1 (32.012.043).

008.064 **Schenectady, N. Y.**

Dudley Observatory Reports No. 16 (32.013.085).

008.065 **Sendai**

Annual report of the Astronomy Group for the period 1981 April - 1982 March.
Sci. Rep. Tôhoku Univ., Eighth Ser., Vol. 3, 160 - 164 (1982).

Sendai Astronomiaj Raportoj, Nos. 229 (29.131.161), 230 (29.131.163), 237 (30.151.059), 238 (31.158.100), 239 (32.122.134), 240 (32.112.049), 242 (31.114.048), 244 (30.112.048), 245 (32.131.290), 246 (32.122.135).

008.066 **Sonneberg**

Akademie der Wissenschaften der DDR, Zentralinstitut für Astrophysik. Veröffentlichungen der Sternwarte in Sonneberg, Band 9, Heft 5 (32.153.033; 32.153.034; 32.123.028).

008.067 **Stockholm**

Stockholms Observatorium, Saltsjöbaden, Sweden, Report. Nos. 20 (32.118.036), 21 (32.031.631).

008.068 **Strasbourg**

Centre de Données Stellaires. Annual report of the director for 1981.
C. Jaschek.
Bull. Inf. Cent. Données Stellaires, No. 23, p. 81 - 83 (1982).

Bulletin d'Information du Centre de Données Stellaires. No. 23 (1982).

Be Star Newsletter, Nos. 1 (32.114.178; 32.113.051; 32.113.052; 32.114.179; 32.114.180), 2 (32.113.053; 32.002.095), 3 (32.113.054; 32.114.181; 32.114.182; 32.114.183; 32.031.632), 4 (32.113.055; 32.002.096; 32.119.094; 32.119.095; 32.114.184), 5 (32.114.185; 32.114.186), 6 (32.011.022; 32.113.056; 32.114.187; 32.114.188; 32.114.189; 32.113.057; 32.115.021; 32.065.094; 32.153.032; 32.114.190).

008.069 **Swarthmore, Penn.**

Swarthmore Astronomical Reprints Nos. 1 (22.118.005), 2 (21.118.008), 3 (22.118.004), 4 (22.118.006), 5 (26.118.008), 6 (26.118.007), 7 (26.118.019), 8 (26.118.015), 9 (26.118.034), 10, 11 (28.118.030), 12 (29.118.032), 13 (30.118.003), 14 (30.118.007), 15 (30.118.018), 16 (31.118.010), 17 (32.012.002), 18 (31.118.040).

008.070 **Sydney**

Sydney Observatory. Report for the year ending 1981 December 31.
W. H. Robertson.
Q. J. R. Astron. Soc., Vol. 23, 437 (1982).

008.071 **Tartu**

W. Struve nimelise Tartu Astrofüüsika Observatooriumi Publikatsioonid, Köide (Tom) 48 (32.005.001, 032.063.006, 32.063.007, 32.022.011, 32.063.008, 32.114.011, 32.114.012, 32.114.013, 32.114.014, 32.123.002, 32.122.013, 32.119.009, 32.119.010, 32.114.015, 32.113.012, 32.155.006, 32.155.007, 32.151.007, 32.094.005, 32.061.003), 49 (32.007.000, 32.007.000, 32.022.152, 32.063.063, 32.135.050, 32.114.217, 32.114.218, 32.122.160, 32.119.102, 32.131.307, 32.111.017, 32.158.334, 32.104.020, 32.160.072, 32.160.073, 32.034.134).

W. Struve nimeline Tartu Astrofüüsika Observatoorium Teated, Nos. 66 (32.064.060; 32.022.143; 32.063.059), 67 (32.022.144), 68 (32.158.299).

008.072 **Thessaloniki**

Contributions from the Department of Geodetic Astronomy, University of Thessaloniki, Nos. 41 (31.122.248), 42 (31.122.121).

008.073 **Tokyo**

Annals of the Tokyo Astronomical Observatory, University of Tokyo, Second Series, Vol. 19, No. 1 (32.122.136; 32.124.302; 32.122.137; 32.032.036; 32.122.138; 32.031.021; 32.044.031).

Tokyo Astronomical Observatory, Reprints Nos. 625 (31.117.056), 626 (32.158.051), 627 (32.158.052), 628 (32.155.009), 629 (32.131.037), 630 (32.115.004), 631 (32.124.301), 632 (32.031.513), 633 (32.075.005), 634 (31.077.050), 635 (32.002.011), 636 (31.036.014), 637 (32.036.014), 638 (32.077.019), 639 (32.126.021), 640 (32.155.035), 641 (32.111.014), 642 (32.159.015), 643 (32.158.211), 644 (31.080.109), 645 (32.073.019), 647 (32.131.289), 648 (32.115.020), 649 (32.113.050).

University of Tokyo, Tokyo Astronomical Observatory, Report (No. 75), Vol. 19, No. 4 (32.031.022; 32.033.046; 32.113.058; 32.004.067; 32.034.081; 32.034.082; 32.047.031; 32.034.083; 32.032.037; 32.034.084; 32.034.085; 32.031.637; 32.072.064; 32.036.023; 32.032.038).

Tokyo Astronomical Observatory, Time and Latitude Bulletins, Vol. 56, Nos. 1 - 2 (32.044.030).

Tokyo Astronomical Bulletin, Second Series, Nos. 266 (32.119.096), 267 (32.100.092).

Contributions from the Department of Astronomy, University of Tokyo, Nos. 291 (30.151.023), 292 (30.065.019), 293 (30.065.020), 294 (30.131.078), 295 (30.119.026), 296 (30.065.098), 297 (30.071.038), 298 (30.075.003), 299 (31.158.099), 300 (31.080.017), 301 (31.158.097), 302 (31.077.018), 303 (31.120.009), 304 (31.122.048), 305 (32.065.068), 306 (29.062.107).

008.074 **Torun**

Biuletyn Obserwatorium Astronomicznego Uniwersytetu M. Kopernika w Toruniu, Nr. 64 (29.131.209; 29.114.128; 29.135.041; 29.002.052), 65 (30.064.075; 31.122.065; 31.114.069; 31.114.068; 31.153.011).

008.075 **Tucson, Ariz.**

Kitt Peak National Observatory, Tucson, Arizona 85726. – Report for the period 1 October 1980 - 30 September 1981.
G. Burbidge.
Bull. American Astron. Soc., Vol. 14, 796 - 829 (1982).

Preprints of the Steward Observatory, University of Arizona, Tucson, Nos. 400 (32.154.038), 401 (32.126.032), 402 (32.160.066), 403 (32.131.291), 404 (32.162.209), 405 (32.103.348), 406 (32.031.023), 407 (32.133.019), 408 (32.117.204), 409 (32.158.302), 410 (32.141.169), 411 (32.103.349), 412 (32.158.303), 413 (32.134.032), 414 (32.141.170), 415 (32.162.210), 416 (32.121.042), 417 (32.158.304), 418 (32.131.292), 419 (32.131.293), 420 (32.131.294), 421 (32.126.033), 422 (32.141.171), 423 (32.066.199), 424 (32.117.205), 425 (32.126.034), 426 (32.141.172), 427 (32.142.137), 428 (32.158.305), 429 (32.131.295), 430 (32.117.206).

008.076 **Tübingen**

Astronomisches Institut der Universität und Lehrstuhl für Theoretische Astrophysik. – Reports for 1981.
M. Grewing.
Mitt. Astron. Ges.,Nr. 56, p. 299 - 309 (1982).

008.077 **Turku**

Turku University Observatory Informo, Nos. 66 (32.158.306), 67 (32.151.077), 68 (32.162.211), 69 (31.160.079), 70 (32.032.035).

008.078 **Uccle**

Observatoire Royal de Belgique, Communications. Koninklijke Sterrenwacht van België, Mededelingen, Série A, Nos. 66 (30.114.157), 67 (31.098.022), 68 (31.071.016), 69 (31.064.084).

Observatoire Royal de Belgique, Communications. Koninklijke Sterrenwacht van België, Mededelingen Série B, Nos. 123 (31.118.014), 124 (32.002.097), 125 (31.013.001), 126 (31.122.059).

008.079 **Victoria**

Dominion Astrophysical Observatory, Victoria, British Columbia. Herzberg Institute of Astrophysics, National Research Council, Canada. Report for the year 1981 April 1 to 1982 March 31.
S. van den Bergh.
Q. J. R. Astron. Soc., Vol. 23, 593 - 607 (1982).

Publications of the Dominion Astrophysical Observatory, Victoria, B. C., Vol. 16, Nos. 2 (32.031.633), 3 (32.031.634), 4 (32.031.635), 5 (32.031.636), 6 (32.021.049).

008.080 **Vienna**

Institut für Astronomie der Universität Wien.
Report for 1981.
W. Tscharnuter.
Mitt. Astron. Ges., Nr. 56, p. 311 - 321 (1982).

008.081 **Vilnius**

Vilniaus Astronomijos Observatorijos Biuletenis (Bulletin of the Vilnius Astronomical Observatory), Nos. 59 (32.113.063), 60 (32.113.064; 32.113.065; 32.113.066; 32.013.103), 61 (32.131.306; 32.113.067; 32.121.049; 32.034.133; 32.021.054; 32.034.901).

008.082 **Warsaw**

Politechnika Warszawska, Obserwatorium Astronomiczno-Geodezyjne w Józefosławiu, (Warsaw Technical University, Astronomic-Geodetical Observatory at Józefosław), **Latitude Circular**, Nos. 80 - 83 (32.045.023).

008.083 **Washington, D. C.**

Astronomical Papers prepared for the use of the American Ephemeris and Nautical Almanac, Vol. 22, Part IV (32.097.052).

U. S. Naval Observatory, Washington, D.C. Time Service Publications. Series 4, Nos. 804 - 830; Series 7, Nos. 757 - 783; Series 11, No. 226; Series 14, Nos. 33 - 34 (32.044.032 - 32.044.035).

United States Naval Observatory, Washington, D.C., Circular, Nos. 164 (32.002.098), 165 (32.079.401).

008.084 **Wellington**

Annual report of the Carter Observatory Board for the year ending 1982 March 31.
I. W. G. Cochrane.
Carter Obs., Astron. Bull., No. 98, 8 pp. (1982).

Carter Observatory. Report for the year ended March 1982.
R. J. Dodd.
South. Stars, Vol. 29, 208 - 211 (1982).

Carter Observatory, Astronomical Bulletin, Nos. 97 (32.047.032), 98 (32.008.084).

Carter Observatory, Wellington, Reprint Series 2, No. 13.

008.085 **Würzburg**

Institut für Astronomie und Astrophysik, Lehrstuhl Astronomie. – Report for 1981.
F.-L. Deubner.
Mitt. Astron. Ges., Nr. 56, p. 323 - 326 (1982).

008.086 **Zurich**

Institut für Astronomie. – Report for 1981.
J. O. Stenflo.
Mitt. Astron. Ges., Nr. 56, p. 327 - 332 (1982).

009 Notes on Observatories, Planetaria, Exhibitions

009.001 **Wisconsin's teaching observatory.**
W. Campbell.
Sky Telesc., Vol. 64, 231 (1982).

009.002 **On the occasion of the 150th anniversary of the Moscow University Observatory (State Astronomical Sternberg Institute).**
Yu. P. Pskovskij.
Vopr. istor. estestvozn. i tekh., 1982, No. 1, p. 113 - 117. In Russian. – From Ref. zh., 51. Astron., 7.51.26 (1982).

009.003 **Die frühe Geschichte der Radiosternwarte in Jodrell Bank.**
B. Lovell.
Sterne, 58. Band, 197 - 205 (1982).

009.004 **Recent research at Siding Spring.**
J. British Astron. Assoc., Vol. 92, 236 (1982).

009.005 **Compte rendu de notre excursion du 15 mai 1982 au radiotélescope géant de Bonn (Effelsberg).**
Ciel Terre, Vol. 98, 132 - 136 (1982).

009.006 **Das Observatório do Capricórnio in Campinas, Brasilien.**
A. Tarnutzer.
Orion, 40. Jahrg., 120 - 122 (1982).

009.007 **15 Jahre astronomische Forschung mit dem 2 m - Spiegelteleskop am Observatorium Ondřejov (ČSSR).**
P. Koubský, V. Rajský, J. Zicha.
Jenaer Rundsch., 27. Jahrg., 110 - 115 (1982).

009.008 **Das Nationalobservatorium der Volksrepublik Bulgarien.**
B. Z. Kovačev.
Jenaer Rundsch., 27. Jahrg., 118 - 120 (1982).

009.009 **Raumflugplanetarium RFP-DP 2.**
H. G. Beck, L. Meier.
Jenaer Rundsch., 27. Jahrg., 123 - 126 (1982).

009.010 **Die Urania-Volkssternwarte in Jena.**
H. Wolf.
Jenaer Rundsch., 27. Jahrg., 133 - 134, 145 (1982).

009.011 **The Fulton Observatory.**
G. A. Bower.
I.A.P.P.P. Commun., No. 8, p. 31 (1982).

009.012 **The Scuppernong Observatory.**
T. R. Renner.
I.A.P.P.P. Commun., No. 8, p. 32 - 33 (1982).

009.013 **The Dance Hill Observatory.**
C. Cunningham, M. Kaitting.
I.A.P.P.P. Commun., No. 8, p. 37 - 38 (1982).

009.014 **Het Zeiss Planetarium Amsterdam.**
K. Velt, G. Schilling.
Zenit, 9. Jaarg., 424 - 430 (1982).

009.015 **The Astronomical Observatory of the Kiev University before and after the Second World War.**
A. F. Bogorodskij, N. A. Chernega.
Vestn. Kiev. Univ., Astron. Vyp. 24, p. 3 - 18 (1982). In Russian.

009.016 **150 years Sternberg State Astronomical Institute.**
Zemlya Vselennaya, 1982, No. 4, p. 14 - 15. In Russian.

009.017 **De astronomiske observatoriene på Kanariøyene.**
T. E. Aslesen, S. Barth, P. Jorås, F. W. Vikanes.
Astron. Tidsskr., Årg. 15, 65 - 69 (1982).

009.018 **Solar physics at the Main Astronomical Observatory of the Ukrainian Academy of Sciences.**

E. A. (*Eh. A.*) Gurtovenko.
Sol. Phys., Vol. 80, 317 - 320 (1982).
A brief review of the research highlights at the Main Astronomical Observatory since 1968 is given.

009.019 **L'Observatoire de Nice à travers son siècle.**
P. Couteau.
Astronomie, Vol. 96, 489 - 499 (1982).

009.020 **The Kitab international latitude station "Ulugh Begh".**
V. P. Shcheglov.
Izdatel'stvo FAN Uzbekskoj SSR, Tashkent. 20 pp. (1981). In Russian.

009.021 **Sic itur ad astra: a history of the Norman Lockyer Observatory.**
A. R. Hutchings.
J. British Astron. Assoc., Vol. 93, 25 - 28 (1982).

009.022 **Courrier de Hong Kong.**
M. Laffineur.
Astronomie, Vol. 96, 572 - 574 (1982).

009.023 **A review of the current status of Canadian planetariums.**
T. Clarke.
J. R. Astron. Soc. Canada, Vol. 76, 317 (1982). – Abstract.

009.024 **Cent ans d'astronomie au Pic-du-Midi.**
J. Rösch.
Société Astronomique de France et les Observatoires du Pic-du-Midi et de Toulouse. 44 pp. (1982).

009.025 **Infrared instrumentation on Calar Alto.**
D. Lemke.
Second ESO infrared workshop, (see 012.060), p. 17 - 28 (1982).

009.026 **Stockholms Observatorium i Saltsjöbaden.** Ett byggnadsverk av Axel Anderberg.
I.-B. Byström.
Astron. Tidsskr., Årg. 15, 152 - 163 (1982).

009.027 **Vom Schloßturm zur Waldhäuser Höhe.** Astronomie in Tübingen in alter und neuer Zeit.
K. Walter.
Tübinger Bl., 69. Jahrg., 65 - 70 (1982).

009.028 **De Bosscha-Sterrenwacht in Indonesie. II. Over het leven en werken op de Bosscha Sterrenwacht bij Lembang een halve eeuw geleden.**
Å. A. E. Wallenquist.
Moesson 20, No. 10, p. 6 - 9 (1982).

009.029 **Annual report for magnetic observatories – 1979.**
G. J. Van Beek, E. I. Loomer.
Geomagn. Ser. Earth Phys. Branch, No. 22, p. 1 - 148 (1982). In English and French. – Abstr. in Phys. Abstr., Vol. 86, Abstr. 15331 (1983).

Observatories of the world.
See Abstr. 003.081.

Dunsink Observatory, Chinese astronomy and the IAU.
See Abstr. 013.040.

010 Societies, Associations, Organizations

010.001 **American Astronomical Society (AAS)**

The 160th meeting of the American Astronomical Society, held 6 - 9 June 1982 at Troy, New York. Abstracts of papers presented.
Bull. American Astron. Soc., Vol. 14, 585 - 670 (1982).

The 14th annual meeting of the Division for Planetary Sciences of the American Astronomical Society, held 19 - 22 October 1982 at Boulder, Colorado. Abstracts of papers presented.
Bull. American Astron. Soc., Vol. 14, 701 - 769 (1982).

Minutes of the annual business meeting of the Division for Planetary Sciences, held 14 October 1981 at Pittsburgh, Pennsylvania.
L. L. Wilkening.
Bull. American Astron. Soc., Vol. 14, 773 - 775 (1982).

Late-paper abstracts from the 160th meeting of the American Astronomical Society, held 6 - 9 June 1982 at Troy, New York.
Bull. American Astron. Soc., Vol. 14, 776 - 780 (1982).

AAS Photo-Bulletin, No. 28 (1981); Nos. 29, 30 (1982).

Bulletin of the American Astronomical Society, Vol. 14, Nos. 2, 3 (1982).

010.002 **Association Française des Observateurs d'Etoiles Variables (A.F.O.E.V.)**

La vie de l'Association.
E. Schweitzer.
Bull. AFOEV, No. 21, p. 110; No. 22, p. 167 (1982).

Activité de l'A.F.O.E.V.
E. Schweitzer.
Astronomie, Vol. 96, 356 - 363 (1982).

Bulletin de l'Association Française des Observateurs d'Etoiles Variables, Nos. 21, 22 (1982).

010.003 **Association of Lunar and Planetary Observers (A.L.P.O.)**

The Strolling Astronomer. The Journal of the Association of Lunar and Planetary Observers, Vol. 29, Nos. 7 - 8 (1982).

010.004 **Astronomical Society of Australia (ASA)**

Meetings of the Society.
Proc. Astron. Soc. Australia, Vol. 4, 274 - 275 (1981).

Proceedings of the Astronomical Society of Australia, Vol. 4, No. 2 (1981); Vol. 4, No. 3 (1982).

010.005 **Astronomical Society of Egypt**

Journal of the Astronomical Society of Egypt, Vol. 3 (1981).

010.006 **Astronomical Society of India**

Abstracts of papers presented at the seventh meeting 1981 November 29 - December 1 at Roorkee.
Bull. Astron. Soc. India, Vol. 10, 26 - 52 (1982).

Bulletin of the Astronomical Society of India, Vol. 10, Nos. 1, 2 (1982).

010.007 **Astronomical Society of Japan**

Publications of the Astronomical Society of Japan, Vol. 34, Nos. 2 - 4 (1982).

010.008 **Astronomical Society of New York**

Abstracts of papers given at the Spring meeting of the Astronomical Society of New York, held April 16, 1982, at the University of Rochester, Rochester, New York.
News Lett. Astron. Soc. N.Y., Vol. 2, No. 2, p. 1 - 26 (1982).

News Letter of the Astronomical Society of New York, Vol. 2, No. 2 (1982).

010.009 **Astronomical Society of the Pacific (ASP)**

Abstracts of papers presented at the University of California San Diego in La Jolla, Meeting of the Astronomical Society of the Pacific 28 June - 1 July 1982.
Publ. Astron. Soc. Pacific, Vol. 94, 748 - 762 (1982).

Publications of the Astronomical Society of the Pacific, Vol. 94, Nos. 559 - 562 (1982).

Mercury. The Journal of the Astronomical Society of the Pacific, Vol. 11, Nos. 3 - 6 (1982).

010.010 **Astronomical Society of Southern Africa (ASSA)**

Centre reports, 1980 - 1981.
Cape Centre *(P. Booth)*; **Natal Centre** *(J. G. Barker)*; **Natal Midlands Centre** *(J. Watson)*; **O. F. S. Centre** *(B. D. Viljoen)*; **Pretoria Centre** *(P. D. Bennewith)*; **Salisbury Centre** *(C. B. Archer)*; **Transvaal Centre** *(E. van Zyl)*.
Mon. Notes Astron. Soc. South. Africa, Vol. 41, 28 - 38 (1982).

Notices.
Mon. Notes Astron. Soc. South. Africa, Vol. 40, 47 (1981); Vol. 41, 27 (1982).

Monthly Notes of the Astronomical Society of Southern Africa, Vol. 40, Nos. 10 - 12 (1981); Vol. 41, Nos. 5, 6 (1982).

010.011 **Astronomical Society of Western Australia (ASWA)**

Journal of the Astronomical Society of Western Australia, Session 1981/82, April - July 1982; Vol. 33, Nos. 1 - 4 (1982).

010.012 **Astronomische Gesellschaft (AG)**

Internationale Astronomische Tagung Innsbruck 1981: Übersichtsvorträge IV.
Mitt. Astron. Ges., Nr. 57, p. 9 - 118 (1982).

Bericht über die wissenschaftliche Tagung in Straßburg, 24. - 26. Februar 1982.
Mitt. Astron. Ges., Nr. 57, p. 119 - 331 (1982).

Die Versammlungen der Astronomischen Gesellschaft 1863 - 1981.
W. Seggewiß.
Mitt. Astron. Ges., Nr. 57, p. 135 - 142 (1982).

Mitteilungen der Astronomischen Gesellschaft, Nr. 56, 57 (1982).

010.013 **British Astronomical Association (BAA)**

Meetings and activities of the Association.
J. British Astron. Assoc., Vol. 92, 237 - 246, 270 - 274, 291 - 300 (1982).

Journal of the British Astronomical Association, Vol. 92, Nos. 5, 6; Vol. 93, No. 1 (1982).

010.014 **British Interplanetary Society (BIS)**

JBIS. Journal of the British Interplanetary Society, Vol. 35, Nos. 8 - 12 (1982).

Space Education. A publication of the British Interplanetary Society, Vol. 1, No. 4 (1982).

Spaceflight. A publication of the British Interplanetary Society, Vol. 24, Nos. 9 - 12 (1982).

010.015 **Canadian Astronomical Society**

Thirteenth meeting of the Canadian Astronomical Society held at the University of Toronto, June 2 - 4, 1982.
Abstracts of papers presented.
J. R. Astron. Soc. Canada, Vol. 76, 315 - 326 (1982).

010.016 **Chinese Astronomical Society**

Celebrating meeting of the 60th anniversary of the founding of the Chinese Astronomical Society and the 4th congress of the Chinese Astronomical Society.

S.-z. Zhang.
Acta Astron. Sinica, Vol. 23, 306 - 307 (1982). In Chinese.

The general meeting for celebrating the 60th anniversary of the Chinese Astronomical Society and the 4th representative congress of the Society were convoked at Lin Tong, Shensi Province, March 15 - 18, 1982.
Acta Astrophys. Sinica, Vol. 2, 247 - 249, 251 - 252 (1982). In Chinese.

010.017 **European Space Agency (ESA)**

ESA Bulletin, Nos. 31, 32 (1982).

ESA IUE Newsletter, Nos. 14, 15 (1982).

ESA Journal, Vol. 5, Nos. 1 - 4 (1981); Vol. 6, Nos. 1, 2 (1982).

European Space Agency, SCI, (82)2 - (82) 6 (1982).

010.018 **International Amateur-Professional Photoelectric Photometry (I.A.P.P.P.)**

I.A.P.P.P. Communication, No. 8 (1982).

010.019 **International Astronomical Union (IAU)**

Appended, in toto, is a letter relating to the planned reconstitution of a WG on supernovae.
V. Trimble.
Inf. Bull. Variable Stars, No. 2222, 2 pp. (1982).

Circulaire d'Information, Nos. 87, 88 (1982).

IAU Circulars, Nos. 3708 - 3758 (1982).

Commission 27 of the I. A. U. Information Bulletin on Variable Stars, Nos. 2160 - 2239, 2251 - 2256 (1982).

Minor Planet Circulars, (M. P. C.), Nos. 6989 - 7514 (1982).

010.020 **Meteoritical Society**

Meteoritics. The Journal of the Meteoritical Society, Vol. 17, Nos. 2, 3 (1982).

010.021 **Nantucket Maria Mitchell Association**

The Nantucket Maria Mitchell Association.
Eightieth annual report for the year ending December 31, 1981.
Edited by the Nantucket Maria Mitchell Association, Vestal Street, Nantucket, Mass., USA. 45 pp. (1982).

010.022 **Oriental Astronomical Association**

The Heavens, Nos. 685 - 690, Vol. 63, Nos. 6 - 11 (1982). In Japanese.

010.023 **Royal Astronomical Society (RAS)**

Recent X-ray observations and the EXOSAT mission.
Summaries of papers presented at the RAS specialist discussion held 1981 November 13, with an introduction by P. A. Charles.
Observatory, Vol. 102, 108 - 116 (1982).

Solar flares. Summaries of papers presented at the RAS specialist discussion held 1981 December 11, with opening remarks by J. L. Culhane.
Observatory, Vol. 102, 116 - 124 (1982).

Expansion speeds in radio sources. Summaries of papers presented at the RAS specialist discussion held 1982 February 12 in the Scientific Societies' Lecture Theatre, Savile Row.
Observatory, Vol. 102, 125 - 133 (1982).

The interstellar medium with particular reference to other galaxies. Summaries of papers presented at the RAS specialist discussion, held 1981 October 9.
Observatory, Vol. 102, 170 - 178 (1982).

Proceedings of the meeting of the Royal Astronomical Society, Dublin, 5 - 7 April 1982.
See Abstr. 012.061.

Meetings of the Society.
Observatory, Vol. 102, 97 - 133, 157 - 178 (1982).

Meetings and activities of the Society.
Q. J. R. Astron. Soc., Vol. 23, 463 - 484 (1982).

Geophysical Journal of the Royal Astronomical Society, Vol. 70, Nos. 2, 3; Vol. 71, Nos. 1 - 3 (1982).

Monthly Notices of the Royal Astronomical Society, Vol. 200, Nos. 2, 3; Vol. 201, Nos. 1 - 3 (1982).

The Quarterly Journal of the Royal Astronomical Society, Vol. 23, Nos. 3, 4 (1982).

010.024 **Royal Astronomical Society of Canada (RAS Canada)**

Two centenaries: the Royal Society of Canada and the last transit of Venus, 1882.
H. Sawyer Hogg.
J. R. Astron. Soc. Canada, Vol. 76, 362 - 370 (1982).

Meetings of the Society.
J. R. Astron. Soc. Canada, Vol. 76, 261 - 265, 315 - 326 (1982).

The Journal of the Royal Astronomical Society of Canada, Vol. 76, Nos. 4 - 6 (1982).

National Newsletter. Supplement to the Journal of the Royal Astronomical Society of Canada, Vol. 76, Nos. 4 - 6 (1982).

010.025 **Royal Astronomical Society of New Zealand (RAS New Zealand)**

The Royal Astronomical Society of New Zealand (Inc.) 59th annual report of council for the 15 months ended

1981 December 30.
South. Stars, Vol. 29, 189 - 205 (1982).

Publications of Variable Star Section, Royal Astronomical Society of New Zealand, No. 9 (C81) (1981).

Southern Stars. Journal of the Royal Astronomical Society of New Zealand, Vol. 29, Nos. 5 - 7 (1982).

010.026 **Schweizerische Astronomische Gesellschaft (SAG)**

Mitteilungen.
Orion, 40. Jahrg., 157 - 160, 189 - 192 (1982).

BBSAG Bulletin, Nos. 61 - 63 (1982).

Orion. Zeitschrift der Schweizerischen Astronomischen Gesellschaft. Revue de la Société Astronomique de Suisse, 40. Jahrg., Nr. 191 - 193 (1982).

010.027 **Società Astronomica Italiana (S.A.It.)**

Memorie della Società Astronomica Italiana, Vol. 52, No. 4 (1981).

010.028 **Société Astronomique de France (SAF)**

Séances, commissions, activités de la Société.
Astronomie, Vol. 96, 332, 575 - 580 (1982).

L'Astronomie et Bulletin de la Société Astronomique de France, Vol. 96, juillet - décembre, Suppl. No. 2 (1982)

010.029 **Société Astronomique de Liège**

Le Ciel, Vol. 44, 145 - 304, septembre - décembre (1982).

010.030 **Société Royale Belge d'Astronomie, de Météorologie et de Physique du Globe.**

Ciel et Terre. Bulletin de la Société Royale Belge d'Astronomie, de Météorologie et de Physique du Globe, Vol. 98, Nos. 3 + Suppl., 4 - 6 (1982).

010.031 **VAGO (Astronomical-Geodetical Society of the USSR)**

Plenum of the Central Council of VAGO in the capital of Belorussia. 1982, February 23 - 25.
V. A. Bronshtehn.
Zemlya Vselennaya, 1982, No. 4, p. 58 - 60. In Russian.

010.032 **Vereinigung der Sternfreunde e. V.**

Nachrichten der Vereinigung der Sternfreunde e. V.
Sterne Weltraum, 21. Jahrg., 328 - 331, 382 - 385, 430 - 432, 486 - 487, 537 - 539 (1982).

Sonne. Mitteilungsblatt der Amateursonnenbeobachter, Jahrg. 6, Nos. 23, 24 (1982).

011 Reports on Colloquia, Congresses, Meetings, Symposia, Expeditions

011.001 **The outer regions of galaxies.**
A. Lawrence.
Nature, Vol. 298, 421 - 422 (1982).
Report of a conference held at the Royal Greenwich Observatory.

011.002 **Remote operation of telescopes.**
M. Penston.
Nature, Vol. 298, 603 (1982).
Report on a workshop on 'Remote operation of telescopes' held at Herstmonceaux Castle.

011.003 **Expedition zur Sonnenfinsternis vom 31. Juli 1981 an den Amur.**
J. Rose, T. Brosowski, T. Marold, H. Sanke.
Jenaer Rundsch., 27. Jahrg., 138 - 140 (1982).

011.004 **The origin of cosmic rays.** A report on the International Cosmic Ray Conference, Paris, July 1981.
D. J. Mullan.
Irish Astron. J., Vol. 15, 9 - 24 (1981).

011.005 **Results of cooperation according to the "Intercosmos" program. VIIth conference of the Working Group of the Socialist Countries.**
L. A. Vedeshin.
Issled. Zemli iz kosm., 1982, No. 2, p. 136 - 139. In Russian.
Abstr. in Ref. zh., 62. Issled. kosm. prostrastva, 8.62.14 (1982).

011.006 **Plasma astrophysics at Santa Barbara.**
R. Rosner, E. Zweibel, V. Trimble.
Nature, Vol. 299, 579 - 580 (1982).
Report on the workshop 'Space and Astrophysical Plasmas' held at the University of California, Santa Barbara.

011.007 **Activity in red-dwarf stars.**
R. Rosner.
Nature, Vol. 299, 680 - 681 (1982).
Report on IAU Colloquium No. 71, 10 - 13 August 1982, Catania, Italy.

011.008 **Colloquium in Abastumani.** Tbilisi, 1982, March.
A. G. Totochava.
Zemlya Vselennaya, 1982, No. 4, p. 57. In Russian.

011.009 **Supernovae.**
K. Nomoto.
Comments Astrophys., Vol. 9, 301 - 309 (1982).
Report on the NATO Advanced Study Institute, June 29 - July 10, 1981, Cambridge, England.

011.010 **Mapping the Universe.**
V. Trimble.
Nature, Vol. 300, 13 - 14 (1982).
Report on the IAU Symposium 104 on 'Early Evolution of the Universe and its present structure', held 30 August - 2 September in Chania, Crete.

011.011 **Planetary rings explained and unexplained.**
R. A. Kerr.
Science, Vol. 218, 141 - 144 (1982).
Report on "Planetary rings", Colloquium No. 75 of the International Astronomical Union, held 30 August - 2 September 1982 at the National Center for Space Studies, Toulouse, France.

011.012 **Moussaka, retsina, and the IAU.**
L. J. Robinson.
Sky Telesc., Vol. 64, 406 - 409 (1982).

011.013 **Workshop on gravitation and relativistic astrophysics,** held at Ahmedabad 1982 January 18 - 20.
A. R. Prasanna.
Bull. Astron. Soc. India, Vol. 10, 58 - 59 (1982).

011.014 **Introduction to "Solar active regions. A monograph from Skylab Solar Workshop III.**
F. Q. Orrall.
Solar active regions, (see 003.011), p. 1 - 15 (1981).
Contents: Solar active regions. The third Skylab Solar Workshop. Achievments of the workshop – an overview.

011.015 **Recent (1981-82) evolution of galactic haloes.**
V. Trimble.
Nature, Vol. 300, 315 - 316 (1982).
Report on the meeting of the International Astronomical Union, held in Patras, Greece on 16 - 26 August 1982.

011.016 **IAU Commission 41 at Patras.**
O. Gingerich.
J. Hist. Astron., Vol. 13, 227 - 228 (1982).

011.017 **Astronomy on ice.**
J. Harvey, M. Pomerantz, T. Duvall, Jr.
Sky Telesc., Vol. 64, 520 - 523 (1982).

011.018 **Die 18. Generalversammlung der Internationalen Astronomischen Union in Patras.**
H. Haupt.
Sternenbote, 25. Jahrg., 187 - 191 (1982).

011.019 **Very high-energy gamma-ray astronomy.**
P. V. Ramana Murthy, T. Weekes.
Nature, Vol. 300, 483 - 484 (1982). – Report on the International Workshop on Very High-Energy Gamma Ray Astronomy, held at Ootacamund, India, 20 - 25 September 1982.

011.020 **Mysteries of the ringed planets.**
J. N. Cuzzi.
Nature, Vol. 300, 485 - 486 (1982). – Report on the IAU Colloquium No. 75, Planetary Rings, held at Toulouse, France, 30 August - 2 September 1982.

011.021 **Bericht über die Wissenschaftliche Tagung in Straßburg, 24.- 26. Februar 1982.**
W. Seggewiß.
Mitt. Astron. Ges., Nr. 57, p. 119 - 120 (1982).

011.022 **Working Group (WG) on Be stars. Report of the chairperson (1979 - 1982).**
M. Jaschek.
Be Star Newsl., No. 6, p. 2 - 4 (1982).

011.023 **On the meetings in Moscow's Scientists Club.**
Vestn. An SSSR, 1982, No. 6, p. 3 - 68. In Russian.
Abstr. in Ref. zh., 51. Astron., 10.51.17 (1982).

011.024 **XVIIth International conference on cosmic rays.**
Yu. N. Vavilov, Yu. D. Kotov.
Vestn. AN SSSR, 1982, No. 7, p. 91 - 97. In Russian. – Abstr. in Ref. zh., 62. Issled. kosm. prostranstva, 11.62.7 (1982).

011.025 **Meeting of the Royal Astronomical Society, Dublin, 5 - 7 April, 1982.**
P. A. Wayman.
Irish Astron. J., Vol. 15, 93 - 94 (1981).

011.026 **The 1982 workshop on cometary astronomy: a review.**
C. S. Morris.
Int. Comet Q., Vol. 4, 39 - 42 (1982).

011.027 **Dynamics of planets and of their satellites on the XXIVth COSPAR General Assembly.**
M. Burša.
Říše hvězd, Vol. 63, 221 - 227, 248 - 249 (1982). In Czech.

011.028 **XVIIIth General Assembly of the IAU.**
P. Andrle.
Říše hvězd, Vol. 63, 245 - 246 (1982). In Czech.

011.029 **Symposium SETI in Tallin (Dec. 1981).**
R. Pešek, I. Budil.
Vesmír, Vol. 61, 207 - 211 (1982). In Czech.

011.030 **The symposium on the history of Chinese astronomy was held in Lin Tong, Shensi Province, March 8 - 12, 1982.**
S. Bo.
Acta Astrophys. Sinica, Vol. 2, 249 - 250 (1982). In Chinese.

011.031 **Academia Sinica-Max Planck Society workshop on high energy astrophysics was held in Nanjing, April 9 - 17, 1982.**
Acta Astrophys. Sinica, Vol. 2, 340 - 343 (1982). In Chinese.

012 Proceedings of Colloquia, Congresses, Meetings, Symposia

012.001 **Proceedings of the conference on multi-ring basins: formation and evolution**, held at Houston, Texas, November 10 - 12, 1980. Compiled by the Lunar and Planetary Institute, Houston, Texas.
R. B. Merrill (Managing Editor), P. H. Schultz (Science Editor). Proceedings of Lunar and Planetary Science, Vol. 12, Part A. Geochim. Cosmochim. Acta, Suppl. 15. Pergamon Press, New York – Oxford – Toronto – Sydney – Frankfurt – Paris. 9 + 295 pp. Price $ 35.00, £ 17.50 (1981). ISBN 0-08-028045-5. – The individual contributions are included in their corresponding subject categories – see abstracts 022.008 - 022.010, 091.001, 091.002, 092.001, 094.002 - 094.004, 094.501 - 094.507, 097.001, 099.004, 105.001, 105.002.

012.002 **Automated data retrieval in astronomy.** Proceedings of the 64th Colloquium of the International Astronomical Union, held in Strasbourg, France, July 7 - 10, 1981.
C. Jaschek, W. D. Heintz (Editors), with an inaugural address and a report of discussions by W. D. Heintz and a final comment by B. Hauck.
Astrophysics and Space Science Library, Vol. 97.
D. Reidel Publishing Company, Dordrecht, Holland – Boston, U.S.A. – London, England. 20 + 324 pp. Price Dfl. 110.00, $ 48.00 (1982). ISBN 90-277-1435-5. – The individual contributions are included in their corresponding subject categories – see abstracts 002.007 - 002.037, 013.004 - 013.020, 021.003 - 021.012, 031.505, 031.506.

012.003 **Quasi-periodic climate change on Mars and Earth.** Workshop held at the NASA Ames Research Center, Moffett Field, Calif., February 24 - 26, 1981.
With an introduction by J. A. Cutts, J. B. Pollack.
Icarus, Vol. 50, Nos. 2/3, p. 127 - 472 (1982). – The individual contributions are included in their corresponding subject categories – see abstracts 031.512, 081.004 - 081.006, 082.002, 082.003, 091.006, 097.005 - 097.013.

012.004 **Noyaux actifs des galaxies.** Compte-Rendu de l'École de Goutelas, 6 - 10 avril 1981.
E. Schatzman (Editor), with concluding remarks by S. Collin-Souffrin.
Published by Société Française des Spécialistes d'Astronomie. SFSA, Supplément au Journal des Astronomes Français n° 13. 225 pp. – The individual contributions are included in their corresponding subject categories – see abstracts 062.013, 065.011, 141.036, 151.010, 151.011, 158.054 - 158.057.

012.005 **Shock waves in the solar corona and interplanetary space.** Proceedings of the STIP (Study of Travelling Interplanetary Phenomena) Workshop, held in Smolenice, Czechoslovakia, 15 - 19 June, 1980.
M. A. Shea, D. F. Smart, S. T. Wu, S. Pintér (Editors).
Space Sci. Rev., Vol. 32, Nos. 1/2, p. 7 - 271 (1982). – The individual contributions are included in their corresponding subject categories – see abstracts 062.014 - 062.016, 073.015, 073.016, 074.015, 074.016, 075.006, 078.003, 106.004 - 106.010.

012.006 **Advances in ultraviolet astronomy: Four years of IUE Research.** Proceedings of a symposium held at NASA Goddard Space Flight Center, Greenbelt, Maryland, March 30 - April 1, 1982.
Y. Kondo, J. M. Mead, R. D. Chapman (Editors).
NASA Conf. Publ., NASA CP-2238, 15 + 643 pp. (1982). The individual contributions are included in their corresponding subject categories – see abstracts 002.042, 013.021, 022.017, 031.514 - 031.517, 032.503, 051.002, 051.003, 064.014 - 064.016, 082.004, 091.008, 099.008 - 099.011, 101.006, 103.601, 112.008 - 112.014, 113.016, 114.022 - 114.040, 116.003, 117.022 - 117.033, 119.011 - 119.015, 120.011, 120.012, 121.005, 121.006, 122.026 - 122.035, 123.003, 124.001, 124.002, 124.401, 124.501, 125.015, 125.016, 131.040 - 131.051, 132.007, 134.002 - 134.004, 135.009 - 135.011, 141.037 - 141.039, 142.040, 153.007, 158.058 - 158.073, 159.006.

012.007 **Vibrational-rotational spectroscopy for planetary atmospheres, Volumes I, II.** Proceedings of a workshop held at Annapolis, Maryland, March 17 - 19, 1980.
M. J. Mumma, K. Fox, J. Hornstein (Editors), with a preface by R. E. Murphy, an introduction by M. J. Mumma, L. Wallace, K. Fox, V. Kunde and a summary of the Working Session by J. Hornstein, K. Fox, M. J. Mumma.
NASA Conf. Publ., NASA CP-2223, 12 + 773 pp. (1982). The individual contributions are included in their corresponding subject categories – see abstracts 002.043, 002.044, 013.022, 022.018 - 022.036, 031.518, 031.519, 063.015, 063.016, 093.016 - 093.019, 097.015, 099.012 - 099.016, 100.013, 101.007, 101.008, 102.008.

012.008 **Proceedings of the meeting held by the Astronomical Science Group of Ireland,** held at Dunsink Observatory on 19th December, 1980.
Irish Astron. J., Vol. 15, 25 - 56 (1981). – The individual contributions are included in their corresponding subject categories – see abstracts 085.006, 114.043, 131.054, 160.019, 162.037.

012.009 **Comets.** International Astronomical Union. Colloquium No. 61, held in Tucson, Arizona, March 11 - 14, 1981.
L. L. Wilkening (Editor) with the assistance of M. Shapley Matthews, with an appendix by B. G. Marsden and E. Roemer.
The University of Arizona Press, Tucson. 10 + 766 pp. Price £ 26.28 (1982). ISBN 0-8165-0769-4. – The individual contributions are included in their corresponding subject categories see abstracts 022.044, 022.045, 031.537, 102.012 - 102.034, 106.017, 107.008.

012.010 **Neutrino physics and astrophysics.** Proceedings of 'Neutrino 80', an international conference on neutrino physics and astrophysics, held in Erice, Sicily, June 23 - 28, 1980.
E. Fiorini (Editor).
Ettore Majorana international science series, physical sciences, Vol. 12. Plenum Press. New York – London. 11 + 421 pp. Price $ 49.50, £ 27.50 (1982). ISBN 0-306-40746-9. – See Abstr. 061.010.

012.011 **Magnetospheric phenomena of celestial bodies.** Proceedings of a symposium held in Tokyo, 22 - 23 February 1982.
Y. Uchida (Editor), with a summary and future prospects by T. Sato.
Tokyo Astronomical Observatory. 8 + 167 pp. (1982). The individual contributions are included in their corresponding subject categories – see abstracts 062.035 - 062.040, 064.036, 064.037, 066.522, 072.021, 073.037, 074.037, 074.038, 075.014, 084.023, 084.024, 085.010, 099.036, 106.018, 116.008, 117.048, 121.017, 141.084, 141.085, 142.070, 143.017, 151.032, 158.148.

012.012 **Relict radiation and its observations from the earth and space.** Transactions of the 7th (2nd Inter-

national) Cracow Summer School of Cosmology, held at Jabłonna near Warsaw, 16 - 26 July 1980.
With a foreword by K. Rudnicki.
Acta Cosmologica, Zesz. 11, 151 pp. (1982). – The individual contributions are included in their corresponding subject categories – see abstracts 066.024 - 066.027, 142.071, 162.050 - 162.055.

012.013 **The Space Telescope Observatory.** Special session of Commission 44, IAU 18th General Assembly, Patras, Greece, August, 1982.
D. N. B. Hall (Editor).
NASA Conf. Publ., NASA CP-2244, 5 + 134 pp. (1982). – The individual contributions are included in their corresponding subject categories – see abstracts 013.046 - 013.048, 032.512 - 032.517, 051.015.

012.014 **The Earth's core: its structure, evolution and magnetic field.** A discussion held 27 and 28 January 1982.
S. K. Runcorn, K. M. Creer, J. A. Jacobs (Editors).
Philos. Trans. R. Soc. London, Ser. A, Vol. 306, No. 1492, p. 1 - 289 (1982). – The individual contributions within the subject scope of Astronomy and Astrophysics Abstracts are included in their corresponding categories – see abstracts 004.036, 004.037, 044.006, 062.041, 062.042, 081.013 - 081.017, 084.025 - 084.033.

012.015 **11th International Conference on Solid State Nuclear Track Detectors,** held at Bristol, England, 7 - 12 September 1981.
Nucl. Tracks Methods Instrum. Appl., Vol. 5, No. 4 (1981). Review in Phys. Abstr., Vol. 85, Abstr. 58532 (1982). – See abstracts 032.522, -32.523, 034.027, 034.028, 034.031, 034.034, 034.035, 094.511, 105.046, 105.047, 143.023 - 143.025.

012.016 **Laser spectroscopy V.** Proceedings of the fifth international conference, held at Jasper, Alta., Canada, 29 June - 3 July 1981.
A. R. W. McKellar, T. Oka, B. P. Stoicheff (Editors).
Springer-Verlag, Berlin, Germany. 11 + 495 pp. (1981). ISBN 3-540-10914-5. – Review in Phys. Abstr., Vol. 85, Abstr. 63748 (1982). – See abstracts 022.056, 031.554, 066.040, 066.041.

012.017 **Segunda Reunión Regional Latinoamericana de Astronomía,** held at Mérida, Venezuela, January 19 - 23, 1981.
G. Bruzual A., F. Fuenmayor, M. Peimbert, J. F. Rayo, S. Torres-Peimbert (Editors), with introductory remarks by R. M. West.
Rev. Mexicana Astron. Astrofis., Vol. 6, Spec. Issue, 389 pp. (1981). – The individual contributions are included in their corresponding subject categories – see abstracts 002.073, 013.051, 013.052, 014.013, 021.025, 022.057, 031.009, 031.551, 033.014, 033.015, 041.008, 041.009, 042.024, 064.039, 065.034, 066.034, 073.042, 078.005, 078.006, 080.024, 102.038, 102.039, 103.002, 103.261, 106.021, 111.011, 111.012, 112.024 - 112.026, 113.023 - 113.025, 114.074 - 114.080, 117.053, 117.054, 118.017, 119.026 - 119.028, 121.018, 121.019, 122.051, 122.052, 125.035, 131.150 - 131.153, 132.020, 132.021, 134.015, 135.023 - 135.026, 141.091 - 141.095, 141.525, 155.022 - 155.024, 156.008, 158.155, 158.156, 159.009, 160.036, 160.037, 162.067.

012.018 **Twelfth Congress of the International Commission for Optics 1981,** held at Graz, Austria, 31 August - 5 September 1981.
Opt. Acta, Vol. 29, No. 4, (1982). – Review in Phys. Abstr., Vol. 85, Abstr. 68956 (1982). – See abstracts 031.010 - 031.012, 031.014, 031.015, 031.017, 031.556, 031.557, 032.008, 032.247.

012.019 **Shock waves in condensed matter.** Conference held at Menlo Park, Calif., USA, 23 - 25 June 1981.
AIP Conf. Proc., No. 78, Price $ 43.00 (1981). – Review in Phys. Abstr., Vol. 85, Abstr. 74143 (1982). – See abstracts 022.069, 022.070, 091.018.

012.020 **Ninth Texas Symposium on Relativistic Astrophysics.** Proceedings of a conference held in Munich, West Germany, December 14 - 19, 1978.
J. Ehlers, J. J. Perry, M. Walker (Editors), with an introduction by J. Ehlers.
Annals of the New York Academy of Sciences, Vol. 336. New York Academy of Sciences, New York. 5 + 599 pp. Price $ 100.00 (1980). ISBN 0-89766-045-5. – The individual contributions are included in their corresponding subject categories – see abstracts 004.038, 061.021, 064.040, 065.036 - 065.039, 066.051 - 066.057, 117.055, 125.036, 125.037, 141.096 - 141.100, 141.527 - 141.529, 142.074 - 142.079, 142.512, 142.513, 157.003 - 157.005, 158.159 - 158.162, 162.083 - 162.089.

012.021 **Tenth Texas Symposium on Relativistic Astrophysics.** Proceedings of a conference held in Baltimore, Maryland, December 15 - 19, 1980.
R. Ramaty, F. C. Jones (Editors).
Annals of the New York Academy of Sciences, Vol. 375. New York Academy of Sciences, New York. 5 + 467 pp. Price $ 100.00 cloth (1981). ISBN 0-89766-139-7 cloth; ISBN 0-89766-140-0 paper. – The individual contributions are included in their corresponding subject categories – see abstracts 066.058 - 066.064, 117.056, 125.038, 141.101, 141.534, 142.080 - 142.082, 142.516, 142.517, 143.026, 143.027, 155.025, 157.006, 157.007, 158.163, 160.038, 160.039, 162.090 - 162.098.

012.022 **Supernovae: a survey of current research.** Proceedings of the NATO Advanced Study Institute, held at Cambridge, U.K., June 29 - July 10, 1981.
M. J. Rees, R. J. Stoneham (Editors), with an introduction by V. L. Trimble.
NATO Advanced Study Institutes, Series C, Vol. 90.
D. Reidel Publishing Company, Dordrecht, Holland – Boston, U.S.A. – London, England. 23 + 590 pp. Price Dfl. 160.00, US $ 69.00 (1982). ISBN 90-277-1442-8. – The individual contributions are included in their corresponding subject categories – see abstracts 031.559 - 031.561, 065.040 - 065.042, 066.065, 107.014, 125.039 - 125.062, 141.535, 162.099, 162.100.

012.023 **1981 Nuclear Science Symposium and the 1981 Symposium on Nuclear Power Systems,** held at San Francisco, Calif., USA, 21 - 23 October 1981.
IEEE Trans. Nucl. Sci., Vol. NS-29, No. 1 (1982). – Review in Phys. Abstr., Vol. 85, Abstr. 58528 (1982). – See abstracts 032.524 - 032.530, 034.033.

012.024 **Evolution in the Universe.** Symposium held on the occasion of the inauguration of the ESO Headquarters building in Garching, F. R. Germany, 5 - 6 May 1981.
With a foreword by P. O. Lindblad, L. Woltjer.
European Southern Observatory, Garching bei München, Federal Republic of Germany. 122 pp. Price DM 20.00 (1982). ISBN 3-923524-12-9. – The individual contributions are included in their corresponding subject categories – see abstracts 013.053, 015.007, 066.068, 107.015, 162.103, 162.104.

012.025 **The need for coordinated ground-based observations of Halley's comet.** Proceedings of an ESO workshop,

held at Paris, France, 29 - 30 April 1982.
P. Véron, M. Festou, K. Kjär (Editors), with a welcoming address by J. Audouze, and an opening address by L. Woltjer. Published and distributed by European Southern Observatory, Karl-Schwarzschild-Str. 2, D-8046 Garching, FRG. 12 + 309 pp. (1982). ISBN 3-923524-14-5. – The individual contributions are included in their corresponding subject categories – see abstracts 013.054 - 031.057, 015.008, 031.574 - 031.578, 034.041, 036.008, 051.017, 051.018, 052.025, 102.044 - 102.052, 103.201, 103.506, 103.507, 103.621.

012.026 **Particles and fields 1981; testing the Standard Model.** Conference held at Santa Cruz, Calif., USA, 9 - 11 September 1981. AIP Conf. Proc., No. 81, Price $ 40.25 (1981). – Review in Phys. Abstr., Vol. 85, Abstr. 79265 (1982). – See abstracts 022.072, 162.111, 162.114.

012.027 **Astrodynamics 1981.** Proceedings of the AAS/AIAA Astrodynamics Conference held August 3 - 5, 1981, North Lake Tahoe, Nevada.
A. L. Friedlander, P. J. Cefola, B. Kaufman, W. Williamson, G. T. Tseng (Editors), with a preface by H. Jacobs.
Advances in the Astronautical Sciences, Vol. 46, Parts I, II. Published for the American Astronautical Society by Univelt, Inc., P. O. Box 28130, San Diego, California 92128. 18 + 1088 pp. (1982). Part I: ISBN 0-87703-159-2 (hard cover); ISBN 0-87703-160-6 (soft cover); Part II: ISBN 0-87703-161-4 (hard cover); ISBN 0-87703-162-2 (soft cover). The individual contributions within the subject scope of Astronomy and Astrophysics Abstracts are included in their corresponding subject categories – see abstracts 042.032 - 042.036, 051.022, 051.023, 052.027 - 052.035, 052.039, 053.004.

012.028 **Recent advances in observational astronomy.** Symposium held in Ensenada, Mexico, September 18 - 19, 1979.
H. L. Johnson, C. Allen (Editors), with welcoming remarks by A. Poveda.
Universidad Nacional Autónoma de México, Ciudad Universitaria, México 20, D. F. 15 + 165 pp. (1981). – The individual contributions are included in their corresponding subject categories – see abstracts 031.582, 031.583, 032.011 - 032.013, 032.535, 034.046 - 034.049, 101.013, 115.010, 117.058, 121.021, 134.017, 141.106, 158.174, 158.175.

012.029 **Third European IUE Conference.** Proceedings of the Third International Ultraviolet Explorer Conference, held at Madrid, Spain, 10 - 13 May 1982.
E. Rolfe, A. Heck, B. Battrick (Editors), with a foreword by P. Benvenuti.
ESA SP-176. European Space Agency, Scientific and Technical Publications Branch, ESTEC, Noordwijk, The Netherlands. 13 + 621 pp. Price FF 175.00 (1982). ISSN 0379-6566. The individual contributions are included in their corresponding subject categories – see abstracts 013.061, 031.586, 051.025 - 051.029, 064.046, 065.053, 065.054, 091.027, 099.048, 101.014, 102.055, 102.056, 112.028 - 112.031, 114.091 - 114.103, 115.012, 116.010, 117.065 - 117.077, 119.038 - 119.040, 120.023, 120.024, 121.022 - 121.027, 122.062 - 122.068, 124.005, 124.311, 124.411, 124.421, 124.503, 124.602, 125.070, 125.071, 126.016-126.018, 131.172 - 131.186, 135.028, 135.029, 141.124 - 141.126, 142.091 - 142.096, 154.019 - 154.021, 158.192 - 158.207, 159.014, 162.132.

012.030 **The scientific importance of submillimetre observations.** Proceedings of a workshop organized by ESAs Astronomy Division, Space Science Department at Noordwijkerhout, The Netherlands, 10 - 12 May, 1982.
T. de Graauw, T. D. Guyenne (Editors).
ESA SP-189. European Space Agency, Scientific and Technical Publications Branch, ESTEC, Noordwijk, The Netherlands. 5 + 232 pp. Price FF 140.00 (1982). ISSN 0379-6566. – The individual contributions are included in their corresponding subject categories – see abstracts 013.062, 013.063, 032.016, 032.540 - 032.542, 033.025 - 033.033, 034.051 - 034.054, 051.030, 072.026, 091.028, 131.187 - 131.202, 141.127, 156.009, 158.208, 158.209.

012.031 **Submillimetre wave astronomy.** Proceedings of a conference held at Queen Mary College, London, September 1981.
J. E. Beckman, J. P. Phillips (Editors).
Cambridge University Press, Cambridge – London – New York – New Rochelle – Melbourne – Sydney. 11 + 370 pp. Price DM 128.20 (1982). ISBN 0-521-24733-0. – The individual contributions are included in their corresponding subject categories – see abstracts 022.112, 032.544, 033.035, 034.056, 034.057, 112.033, 131.213 - 131.233, 158.220 - 158.222.

012.032 **Morphology and dynamics of galaxies.** Twelfth Advanced Course of the Swiss Society of Astronomy and Astrophysics, held in Saas-Fee, Switzerland, March 29 - April 3, 1982.
L. Martinet, M. Mayor (Editors).
Published and sold by Geneva Observatory, CH-1290 Sauverny, Switzerland. 14 + 420 pp. Price SF 40.00 (1982). – The individual contributions are included in their corresponding subject categories – see abstracts 151.048 - 151.050.

012.033 **Proceedings of the INS International Symposium on Nuclear Radiation Detectors,** held at Tokyo, Japan, 23 - 26 March 1981.
Nucl. Instrum. Methods Phys. Res., Vol. 196, No. 1 (1982). Review in Phys. Abstr., Vol. 85, Abstr. 89434 (1982). – See abstracts 032.548, 032.549.

012.034 **Gamma ray transients and related astrophysical phenomena.** Proceedings of a workshop held at La Jolla, Calif., August 5 - 8, 1981.
R. E. Lingenfelter, H. S. Hudson, D. M. Worrall (Editors).
AIP Conf. Proc., No. 77. American Institute of Physics, New York. 10 + 500 pp. Price $ 39.00 (1982).
ISBN 0-88318-176-2. – The individual contributions are included in their corresponding subject categories – see abstracts 032.551 - 032.555, 066.127, 066.128, 066.526 - 066.536, 073.066 - 073.068, 076.010 - 076.013, 122.076, 142.103, 142.104, 142.524 - 142.541.

012.035 **Los Alamos Conference on Optics '81,** held at Los Alamos, New Mex., USA, 7 - 10 April 1981.
Proc. SPIE Int. Soc. Opt. Eng., Vol. 288 (1981). – Review in Phys. Abstr., Vol. 85, Abstr. 94889 (1982). – See abstracts 032.020, 034.063, 082.059.

012.036 **Secondary ion mass spectrometry. SIMS III.** Proceedings of the third international conference, held at Budapest, Hungary, 30 August - 5 September 1981.
A. Benninghoven, J. Giber, J. Laszlo, M. Riedel, H. W. Werner (Editors).
Springer-Verlag, Berlin, Germany. 11 + 444 pp. (1982). ISBN 3-540-11372-X. – Review in Phys. Abstr., Vol. 85, Abstr. 94910 (1982). – See Abstr. 105.062.

012.037 **Interpretation of climate and photochemical models. Ozone and temperature measurements.** Workshop held at San Diego, Calif., USA, 9 - 11 March 1981.
AIP Conf. Proc., No. 82. (1981). – Review in Phys. Abstr., Vol. 85, Abstr. 106993 (1982). – See abstracts 031.595, 085.022.

012.038 **Proton-antiproton collider physics – 1981.** Workshop held at Madison, Wis., USA, 10 - 12 December 1981.
V. Barger, D. Cline, F. Halzen (Editors).
AIP Conf. Proc., No. 85. American Institute of Physics, New York. 676 pp. Price $ 42.00 (1982). ISBN 0-88318-184-3. Review in Phys. Abstr., Vol. 85, Abstr. 106992 (1982). – See abstracts 143.042, 143.045 - 143.050.

012.039 **Mathematical methods in hydrodynamics and integrability in dynamical systems.** Workshop held at La Jolla, Calif., USA, 7 - 9 December 1981.
M. Tabor, Y. M. Treve (Editors).
AIP Conf. Proc., No. 88. American Institute of Physics, New York. 352 pp. Price $ 34.00 (1982). ISBN 0-88318-187-8. – See Abstr. 151.056.

012.040 **Imaging spectroscopy.** Conference held at Los Angeles, Calif., USA, 10 - 11 February 1981.
Proc. SPIE Int. Soc. Opt. Eng., Vol. 268 (1981). – Review in Phys. Abstr., Vol. 85, Abstr. 106998 (1982). – See abstracts 032.562, 032.564, 051.045, 097.048.

012.041 **Image quality.** Conference held at San Diego, Calif., USA, 27 - 28 August 1981.
Proc. SPIE Int. Soc. Opt. Eng., Vol. 310 (1981). – Review in Phys. Abstr., Vol. 85, Abstr. 107000 (1982). – See Abstr. 031.016.

012.042 **Wolf-Rayet stars: observations, physics, evolution.** International Astronomical Union Symposium No. 99, held at Cozumel, Mexico, September 18 - 22, 1981.
C. W. H. de Loore, A. J. Willis (Editors), with a concluding summary by L. F. Smith.
D. Reidel Publishing Company, Dordrecht, Holland – Boston, USA – London, England. 20 + 618 pp. Price Dfl. 160.00, $ 69.50 cloth, Dfl. 80.00, $ 34.50 paper (1982).
ISBN 90-277-1469-X cloth, ISBN 90-277-1470-3 paper. The individual contributions are included in their corresponding subject categories – see abstracts 064.051 - 064.056, 065.071 - 065.077, 112.035 - 112.038, 113.032 - 113.036, 114.115 - 114.140, 115.013 - 115.015, 116.012, 116.013, 117.091 - 117.096, 119.044 - 119.046, 120.029 - 120.033, 132.032 - 132.034, 134.024 - 134.028, 142.108, 142.109, 153.022, 155.040, 155.041, 158.236, 158.237.

012.043 **Proceedings of the IV National Conference of Yugoslav Astronomers**, held at Sarajevo 1979, October 22 - 26.
With an opening address by B. Ševarlić.
Publ. Astron. Obs. Sarajevo, Vol. 1, No. 1, 294 pp. (1981). The individual contributions are included in their corresponding subject categories – see abstracts 002.079, 004.045 - 004.047, 005.017, 005.018, 013.068, 013.069, 014.018 - 014.020, 022.125, 031.598 - 031.602, 032.022 - 032.024, 041.012 - 041.016, 042.043, 042.044, 045.007, 045.008, 046.005, 063.053, 071.012, 073.070 - 073.072, 082.063, 098.038, 098.039, 103.222, 113.037, 116.014, 118.023, 132.035, 156.011.

012.044 **Proceedings of the International Symposium on Time and Frequency**, held at New Delhi, India, 10 - 12 February 1981.
J. Inst. Electron. Telecommun. Eng., Vol. 27, No. 10, Part 1 (1981). – Review in Phys. Abstr., Vol. 85, Abstr. 95306 (1982). – See abstracts 013.071, 035.009 - 035.030, 044.015.

012.045 **The nature of symbiotic stars.** Proceedings of IAU Colloquium No. 70, held at the Observatoire de Haute Provence, France, 26 - 28 August, 1981.
M. Friedjung, R. Viotti (Editors), with an opening address by C. Fehrenbach, an introduction by J. Sahade and A. A. Boyarchuk and concluding remarks by J. P. Swings.
Astrophysics and Space Science Library, Vol. 95.
D. Reidel Publishing Company, Dordrecht, Holland – Boston, U.S.A. – London, England. 20 + 310 pp. Price Dfl. 100.00, US $ 43.50 (1982). ISBN 90-277-1422-3. The individual contributions are included in their corresponding subject categories – see abstracts 114.147 - 114.155, 116.017, 117.100 - 117.129, 119.052, 119.053, 122.084, 122.085, 124.511, 135.038, 142.113, 142.114.

012.046 **Binary and multiple stars as tracers of stellar evolution.** Proceedings of the 69th Colloquium of the International Astronomical Union, held in Bamberg, F. R. G., August 31 - September 3, 1981.
Z. Kopal, J. Rahe (Editors), with welcome addresses by R. Grafberger, A. Bollée, N. Fiebiger, J. Rahe and an introduction and concluding remarks by Z. Kopal.
Astrophysics and Space Science Library Vol. 98. D. Reidel Publishing Company, Dordrecht, Holland – Boston, U. S. A – London, England. 30 + 503 pp. Price Dfl. 155.00, US $ 67.50 (1982). ISBN 90-277-1436-3. – The individual contributions are included in their corresponding subject categories – see abstracts 065.083, 066.180, 117.130 - 117.159, 118.025 - 118.029, 119.054 - 119.060, 120.036 - 120.039, 121.031, 122.087 - 122.091, 125.085, 126.025 - 126.027, 131.268, 142.115, 152.004.

012.047 **Progress in cosmology.** Proceedings of the Oxford International Symposium held in Christ Church, Oxford, U. K., September 14 - 18, 1981.
A. W. Wolfendale (Editor).
Astrophysics and space science library, Vol. 99. D. Reidel Publishing Company, Dordrecht, Holland – Boston, U. S. A – London, England. 8 + 360 pp. Price Dfl. 125.00, US $ 54.50 (1982). ISBN 90-277-1441-X. – The individual contributions are included in their corresponding subject categories – see abstracts 061.031, 066.181 - 066.184, 141.147, 142.116, 142.117, 142.542, 143.053, 143.054, 151.060 - 151.062, 158.251, 160.057, 162.183 - 162.190.

012.048 **The Venus environment.**
An international conference held in Palo Alto, California, November 1981. Some of the contributed papers.
J. A. Burns, J. Veverka (Editors).
Icarus, Vol. 51, No. 2, p. 167 - 459, Vol. 52, No. 2, p. 209 - 372 (1982). – The individual contributions are included in their corresponding subject categories – see abstracts 093.070 - 093.085, 093.108 - 093.118.

012.049 **Proceedings of the seventh Oberwolfach conference on mathematical methods in celestial mechanics.** Conference held in Oberwolfach, F. R. G., 24 - 28 August 1981.
R. L. Duncombe (Editor), with a preface by U. Kirchgraber, H. Rüssmann, V. Szebehely.
Celestial Mech., Vol. 28, Nos. 1/2, p. 1 - 250 (1982). – The individual contributions are included in their corresponding subject categories – see abstracts 004.053, 006.000, 021.034 - 021.037, 022.132, 042.046 - 042.059, 052.040, 094.029, 098.053, 098.054.

012.050 **The Giotto spacecraft gas and plasma environments.** Proceedings of a series of workshops held at ESTEC, Noordwijk, The Netherlands, in 1980/1981.
W. R. Burke (Editor), with a foreword by R. Grard.
ESA SP-187. ESA Scientific and Technical Publications Branch, ESTEC, Noordwijk, The Netherlands. 6 + 58 pp. (1982). – The individual contributions are included in their corresponding subject categories – see abstracts 022.133, 051.046 - 051.050.

012.051 **Ejection and accretion of matter in binary systems.** Multilateral Cooperation of the Academies of

Sciences of Socialist Countries "Physics and Evolution of Stars", Meeting of the 5th Subcommission "Double Stars", Tatranská Lomnica, Czechoslovakia, April 22 - 24, 1980.
J. Tremko (Editor).
VEDA, Publishing House of the Slovak Academy of Sciences, Bratislava, Czechoslovakia. 126 pp. Price Kčs 26.00 (1982). The individual contributions are included in their corresponding subject categories – see abstracts 002.081, 031.609, 064.058, 066.188, 117.161 - 117.168, 119.062 - 119.064.

012.052 **The Big Bang and element creation.** Proceedings of a Royal Society discussion meeting held on 11 - 12 March 1982.
D. Lynden-Bell (Editor).
Philos. Trans. R. Soc. London, Ser. A, Vol. 307, No. 1497, p. 1 - 148 (1982). – The individual contributions are included in their corresponding subject categories – see abstracts 066.189 - 066.191, 160.060, 161.003, 162.193 - 162.199.

012.053 **Geodesy and physics of the earth.** Proceedings of the 4th International Symposium, G. D. R. Karl-Marx-Stadt, May 12th - 17th, 1980.
H. Kautzleben (Editor).
Veröff. Zentralinst. Phys. Erde, Nr. 63 Teil I - III, 836 pp. (1981). – The individual contributions within the subject scope of Astronomy and Astrophysics Abstracts are included in their corresponding categories – see abstracts 031.611 - 031.615, 032.026, 041.018, 043.005, 044.016 - 044.019, 045.009 - 045.015, 046.006 - 046.008, 081.031 - 081.035.

012.054 **The solar system and its exploration.** Papers presented at the Alpbach Summer School held at Alpbach, Austria, 29 July - 7 August 1981, organised by ASSA, sponsored by ESA, CNES and DFVLR.
W. R. Burke (Editor), with a foreword by E. Mondré.
ESA SP-164, ESA Scientific and Technical Publications Branch, ESTEC, Noordwijk, The Netherlands. 17 + 255 pp. (1981). – The individual contributions are included in their corresponding subject categories – see abstracts 032.568 - 032.570, 051.051 - 051.058, 052.042, 074.069, 080.036, 080.037, 084.083 - 084.085, 091.041 - 091.045, 094.030, 099.109, 099.110, 100.084, 102.063, 102.064, 106.042, 107.031 - 107.036.

012.055 **The comet Halley dust and gas environment.** Proceedings of a joint NASA/ESA working group meeting, Heidelberg, 26 - 27 August 1981.
B. Battrick, E. Swallow (Editors), with a foreword by R. Reinhard.
ESA SP-174. European Space Agency, Scientific and Technical Publications Branch, ESTEC, Noordwijk, The Netherlands. 3 + 85 pp. (1981). ISSN 0379-6566. – The individual contributions are included in their corresponding subject categories – see abstracts 102.077 - 102.081, 103.518 - 103.521.

012.056 **Dynamical astronomy.** Proceedings of a workshop held at Budapest, Hungary, August 31 - September 4, 1981.
B. Balázs, V. Szebehely (Editors).
Published by the Eötvös University, Budapest, Hungary. 164. pp. (1982) = Publ. Astron. Dep. Eötvös Univ., No. 6. The individual contributions are included in their corresponding subject categories – see abstracts 013.082, 042.063 - 042.066, 052.044, 052.045, 081.038, 084.092, 091.048, 100.086, 117.197, 151.069, 151.070, 155.047.

012.057 **Extraterrestrials: where are they?** Symposium held at the Center of Adult Education of the University of Maryland, College Park, Maryland, November 2 - 3, 1979.
M. H. Hart, B. Zuckerman (Editors).
Pergamon Press, New York - Oxford - Toronto - Sydney - Paris - Frankfurt. 182 pp. Price $ 22.50 cloth, $ 9.50 paper (1982). ISBN 0-08-026342-9 cloth, ISBN 0-08-026341-0 paper. – The individual contributions are included in their corresponding subject categories – see abstracts 015.016 - 015.030, 051.067, 118.034.

012.058 **Proceedings of the workshop on interstellar comets,** held at Edinburgh, 13 - 15 April 1982.
S. V. M. Clube, B. McInnes (Editors), with introductory remarks by R. D. Cannon and a brief review by W. H. McCrea.
Occas. Rep. R. Obs., Edinburgh, No. 9, 65 pp. (1982).
The individual contributions are included in their corresponding subject categories – see abstracts 051.068, 102.082 - 102.089, 105.079, 105.080.

012.059 **Instrumentation for astronomy with large optical telescopes.** Proceedings of IAU Colloquium No. 67, held at Zelenchukskaya, USSR, 8 - 10 September 1981.
C. M. Humphries (Editor).
Astrophysics and Space Science Library, Vol. 92.
D. Reidel Publishing Company, Dordrecht, Holland – Boston, U.S.A. – London, England. 18 + 322 pp. Price Dfl. 110.00, US $ 48.50 (1982). ISBN 90-277-1388-X. – The individual contributions are included in their corresponding subject categories – see abstracts 031.024 - 031.027, 031.638 - 031.642, 032.039 - 032.047, 034.088 - 034.107.

012.060 **Second ESO infrared workshop,** held at Garching, 20 - 23 April 1982.
A. F. M. Moorwood, K. Kjär (Editors).
European Southern Observatory, Karl-Schwarzschild-Str. 2, D-8046 Garching bei München, F.R.G. 12 + 446 pp. Price DM 50.00 (1982). ISBN 3-923524-15-3. The individual contributions are included in their corresponding subject categories – see abstract 009.025, 013.086 - 013.094, 031.643 - 031.646, 032.048 - 032.050, 032.571 - 032.574, 034.109 - 034.131, 051.070 - 051.076, 082.082.

012.061 **Proceedings of the meeting of the Royal Astronomical Society, Dublin, 5 - 7 April 1982.**
Opening address by P. A. Wayman.
Irish Astron. J., Vol. 15, 95 - 137 (1981). – The individual contributions are included in their corresponding subject categories – see abstracts 032.576, 062.099, 066.224, 080.049 - 080.052, 114.193, 114.194, 122.146, 141.175, 142.138, 142.546 - 142.548, 143.061 - 143.063, 162.227 - 162.231.

012.062 **Ultraviolet stellar classification.** Proceedings of a workshop held at Vilspa, Spain, 26 - 28 October 1981.
A. Heck, B. Battrick (Editors), with a foreword by P. Benvenuti and A. Heck.
ESA SP-182. European Space Agency, Scientific and Technical Publications Branch, ESTEC, 2200 AG Noordwijk, The Netherlands. 5 + 135 pp. Price FF 90.00 (1982). – The individual contributions are included in their corresponding subject categories – see abstracts 002.115, 002.116, 013.095, 113.059, 114.195 - 114.201.

012.063 **The physics of sunspots.** Sacramento Peak Observatory conference, held at Sunspot, New Mexico, 14 - 17 July 1981.
L. E. Cram, J. H. Thomas (Editors).
Available from Sacramento Peak Observatory, Sunspot, New Mexico 88349. 10 + 495 pp. Price $ 20.00 (individuals), $ 30.00 (institutions) (1981). The individual contributions are included in their corresponding subject categories – see abstracts 062.101, 064.068, 072.074 - 072.102, 075.020, 080.054 - 080.057, 122.148.

012.064 **Formation of planetary systems.** CNES International Summer School held in Grasse, France, August 1 - 27 (1980).
A.Brahic (Editor), with a foreword by H. Curien.
Cepadues - Editions, 111, rue Nicolas-Vauquelin, 31100 Toulouse, France. 893 pp. Price FF 295.00 (1982). ISBN 2-85428-080-6. – The individual contributions are included in their corresponding subject categories – see abstracts 004.073, 015.036, 031.664, 061.050, 091.053, 091.054, 093.096, 097.057 - 097.059, 098.077, 099.114, 100.096, 102.094, 107.045 - 107.051, 131.304, 131.305.

012.065 **Second Cambridge workshop on cool stars, stellar systems, and the sun.** Vol. II. Proceedings of a workshop held at the Harvard-Smithsonian Center for Astrophysics, Cambridge, Ma. 02138, October 21 - 23, 1981.
M. S. Giampapa, L. Golub (Editors), with summaries by G. H. Herbig, R. K. Ulrich.
Smithsonian Astrophys. Obs., Spec. Rep. No. 392, Vol. II, 9 + 262 pp. (1982). – The individual contributions are included in their corresponding subject categories – see abstracts 013.102, 051.080 - 051.082, 064.072, 080.059, 114.211 - 114.216, 116.022, 116.023, 117.209 - 117.213, 119.098, 121.044 - 121.047, 122.152, 122.153, 142.147, 153.039.

012.066 **Satellites of Jupiter.** Proceedings of IAU Colloquium No. 57, held in Kailua-Kona, Hawaii, U.S.A., May 13 - 16, 1980.
D. Morrison, M. S. Matthews (Editors).
The University of Arizona Press, Arizona, U.S.A. 10 + 972 pp. Price £ 44.80 (1982). ISBN 0-8165-0762-7. – The individual contributions are included in their corresponding subject categories – see abstracts 099.125 - 099.149.

012.067 **The source region of the solar wind.** Invited review papers presented at the IX. Lindau Workshop, held in Lindau, F. R. Germany, November 1981.
W. K. H. Schmidt, H. Grünwaldt (Editors).
Space Sci., Rev., Vol. 33, Nos. 1/2, p. 1 - 275 (1982). – The individual contributions are included in their corresponding subject categories – see abstracts 031.665, 031.666, 073.110, 074.082 - 074.088, 076.015.

012.068 **Mécanismes de production d'énergie dans le milieu interstellaire.** Colloquium held at La Baume-Sainte-Marie (Aix-en-Provence), France, 10 - 12 June 1981.
J.-P. Sivan (Editor), with a preface by J.-L. Vidal.
Available from the Laboratoire d'Astronomie Spatiale du CNRS, Marseille, France. 284 pp. Price FF 50.00 (1981). The individual contributions are included in their corresponding subject categories – see abstracts 062.102, 064.073, 131.311 - 131.316, 132.045 - 132.050, 142.549, 143.066.

012.069 **Accreting neutron stars.**
Proceedings of a workshop held at the Max-Planck-Institut für extraterrestrische Physik, Garching, F. R. Germany, July 19 - 23, 1982.
W. Brinkmann, J. Trümper (Editors).
MPE Report 177. Max-Planck-Institut für Physik und Astrophysik. Institut für extraterrestrische Physik, Karl-Schwarzschild-Straße 1, D-8046 Garching, F. R. Germany. 13 + 337 pp. (1982). – The individual contributions are included in their corresponding subject categories – see abstracts 051.084, 051.085, 062.112, 063.064, 064.074 - 064.077, 066.543 - 066.560, 117.218, 117.219, 141.554, 142.149 - 142.170, 142.550 - 142.556.

012.070 **17th International Cosmic Ray Conference.**
Proceedings of a conference held at Paris, France, July 13 - 25, 1981, by the Commissariat à l'Energie Atomique and the International Union of Pure and Applied Physics. Published by Centre d'Études Nucléaires de Saclay, DPh/EP/Ap, 91191 Gif-sur-Yvette, Cedex, France. Available from D. Reidel Publishing Company, Dordrecht, Holland– Boston, U. S. A.–London, England. 14 volumes, Price Dfl. 730.00, $ 315.00 complete set (1981/82).
Vol. 1. XG session: X- and gamma rays (non solar).
19 + 257 pp. Price Dfl. 65.00, $ 28.50. ISBN 90-277-9058-5.
Contents: 1. X- and gamma-ray continuum emission related to specific stellar and compact objects. 2. X- and gamma-ray bursts. 3. Hard X- and gamma-ray lines. 4. Galactic X- and gamma rays. 5. Extragalactic X- and gamma rays.
Vol. 2. OG session: origin and galactic phenomena.
26 + 396 pp. Price Dfl 85.00, $ 36.95. ISBN 90-277-9059-6.
Contents: HEAO-3 symposium. 1. Cosmic ray composition (up to 1000 GeV/n). 2. Cosmic rays above 1000 GeV-observations. 3. Interstellar propagation. 4. Cosmic ray sources, acceleration. 5. Secular variations and cosmogenic nuclei. 6. Others.
Vol. 3 and Vol. 4. SH sessions: Sun and heliosphere.
29 + 513 pp. and 21 + 352 pp. resp. Price Dfl. 170.00, $ 74.00. ISBN 90-277-9060-7.
Contents: 1. Electromagnetic signature of energetic particles on the Sun. 2. Solar energetic particle event observations. 3. Solar energetic particle observations: composition. 4. Solar-cycle modulation of galactic cosmic rays and other particles – observations. 5. Interplanetary propagation and the heliospheric medium. 6. Acceleration of particles on the Sun and in the heliosphere. 7. Forbush decreases (single or recurrent), non-recurrent and irregular variations. 8. Periodic and recurrent variations. 9. Geomagnetic and atmospheric effects. 10. Others.
Vol. 5. HE session: high-energy interactions.
29 + 438 pp. Price Dfl. 95.00, $ 41.50. ISBN 90-277-9061-8.
Contents: 1. Accelerator results. 2. New particles: theory and observation. 3. Nuclear interactions. 4. Cascade studies: propagation in atmosphere. 5. Gamma-ray families. 6. High-energy interaction theory. 7. Proton lifetime and others.
Vol. 6. EA session: extensive air showers.
21 + 325 pp. Price Dfl. 75.00, $ 32.50. ISBN 90-277-9062-9.
Contents: 1. Structure of large air showers. 2. Optical and radio emission. 3. Structure of medium and small showers. 4. High-energy interactions. 5. Cascade studies. 6. Others.
Vol. 7. MN session: muons and neutrinos.
16 + 199 pp. Price Dfl. 55.00, $ 23.95. ISBN 90-277-9063-0.
Contents: 1. Measured muon intensities. 2. Muon spectra, charge ratio. 3. Muon bundles. 4. Muon interactions. 5. Neutrinos and neutrino interactions. 6. Underwater projects (DUMAND). 7. Others.
Vol. 8. T session: techniques.
17 + 210 pp. Price Dfl. 55.00, $ 23.95. ISBN 90-277-9064-1.
Contents: 1. X- and gamma-ray detectors. 2. Detectors of cosmic ray particles (above atmosphere). 3. Trajectory location detectors. 4. Ground level detectors. 5. Others.
Vol. 9. Late papers on X- and gamma rays (non-solar), origin and galactic phenomena, and techniques.
19 + 392 pp. Price Dfl. 75.00, $ 32.50. ISBN 90-277-9065-2.
Vol. 10. Late papers on Sun and heliosphere, and muons and neutrinos.
20 + 416 pp. Price Dfl. 75.00, $ 32.50. ISBN 90-277-9066-3.
Vol. 11. Late papers on high-energy interactions, extensive air showers, and errata.
25 + 441 pp. Price Dfl. 75.00, $ 32.50. ISBN 90-277-9067-4.
Vol. 12. Invited papers.
6 + 295 pp. Price Dfl. 75.00, $ 32.50. ISBN 90-277-9068-5.
Contents: High-energy cosmic rays (*P. Auger*). The HEAO-3 high-resolution gamma-ray spectrometer (*A. S. Jacobson*). Source abundances and propagation of relativistic cosmic rays up to Z = 30: HEAO-3 results (*L. Koch-Miramond*). Ultraheavy cosmic rays: HEAO-3 results (*M. H. Israel*). The nature of the interstellar medium (*C. F. McKee*). Gamma rays and astronomy (*J. A. Paul*).

X rays associated with high energy particles (*P. Gorenstein*). Supernovae, young remnants, and nucleosynthesis (*R. P. Kirshner*). Nucleosynthesis, mass loss and stellar evolution (*C. de Loore*). Acceleration of cosmic rays by shock fronts (*W. I. Axford*). The Solar Maximum Mission (*G. M. Simnett*). Planetary magnetospheres: the in-situ astrophysical laboratories (*S. M. Krimigis*). Status of the fireball concept in theory and experiment (*E. L. Feinberg*).
Vol. 13. Rapporteur papers.
9 + 399 pp. Price Dfl. 75.00, $ 32.50. ISBN 90-277-9069-6.
Contents: Gamma-ray bursts, gamma-ray spectroscopy, pulsars and very high-energy gamma rays (*K. Hurley*). High-energy astronomy in Paris (*G. F. Bignami*). The elemental and isotopic composition of galactic cosmic rays (*R. A. Mewaldt*). Observations of the energy spectrum, composition and anisotropy of primary cosmic rays above 1000 GeV (*A. M. Hillas*). Propagation of cosmic ray nuclei, electrons and antiprotons (*S. A. Stephens*), Cosmic ray sources: I. Clues to the origin of galactic cosmic rays (*M. Cassé*). Cosmic ray sources: II. Acceleration (*H. J. Völk*). Solar energetic particle composition and electromagnetic signature of solar flares (*B. Klecker*). Observations of solar particle propagation and solar modulation (*R. B. McKibben*). Solar particle propagation and modulation theory (*M. A. Lee*). The acceleration of particles in the Sun and in the heliosphere (*F. B. McDonald*). Periodic and recurrent variations of cosmic rays (*A. J. Somogyi*). The geomagnetic and magnetospheric effects on primary and secondary cosmic rays (*R. Gall*). Rapporteur talk on HE 1, 2, 6 and 7 (*K. Niu*). Hadron-nucleus, nucleus-nucleus interactions and electromagnetic cascades (*A. D. Erlykin*). Emulsion chamber experiments and related topics (*A. Tomaszewski*). Structure of large air showers, optical and radio emission, and others (*K. Kamata*). Structure of extensive air showers for primary energies of 10^4–10^8 GeV (*S. C. Tonwar*). Rapporteur paper on the muon sessions (*T. Kitamura*). Neutrinos (*P. V. Ramana Murthy*).
Vol. 14. General Index.
7 + 261 pp. ISBN 90-277-9070-7.

012.071 **Third International Colloquium on Mars**, held in Pasadena, Calif., August 30 - September 2, 1982.
R. Greeley (Editor), with a foreword by B. I. Edelson.
J. Geophys. Res., Vol. 87, No. B12, p. 9715 - 10305 (1982). The individual contributions are included in their corresponding subject categories – see abstracts 022.167 - 022.169, 097.065 - 097.105.

012.072 **Ultraviolet and vacuum ultraviolet systems.** Conference held at Washington, D. C., USA, 21 - 22 April 1981.
Proc. SPIE Int. Soc. Opt. Eng., Vol. 279 (1981). – Review in Phys. Abstr., Vol. 85, Abstr. 106999 (1982). – See abstracts 032.594, 032.595, 032.598, 034.146, 034.147, 034.153, 051.091, 054.003, 082.096.

012.073 **1981 international conference on Fourier transform infrared spectroscopy**, held at Columbia, S.C., USA, 8 - 12 June 1981.
Proc. SPIE Int. Soc. Opt. Eng., Vol. 289 (1981). – Review in Phys. Abstr., Vol. 85, Abstr. 112640 (1982). – See abstracts 022.186, 031.680, 031.681, 032.596, 034.148, 082.097, 082.098.

012.074 **Modern utilization of infrared technology VII.** Conference held at San Diego, Calif., USA, 27 - 28 August 1981.
Proc. SPIE Int. Soc. Opt. Eng., Vol. 304 (1981). – Review in Phys. Abstr., Vol. 85, Abstr. 112643 (1982). – See abstracts 032.073, 032.597, 034.150.

012.075 **Dirac symposium.** Conference held at New Orleans, La., USA, May 1981.
Int. J. Theor. Phys., Vol. 21, Nos. 10 - 11 (1982). – Review in Phys. Abstr., Vol. 85, Abstr. 112646 (1982). – See abstracts 022.174, 066.240.

012.076 **The galactic center.**
Proceedings of a workshop held at the California Institute of Technology, Pasadena, January 7 - 8, 1982.
G. R. Riegler, R. D. Blandford (Editors).
AIP Conf. Proc. No. 83. 11 + 216 pp. Price $ 30.25 (1982). ISBN 0-88318-182-7. – The individual contributions are included in their corresponding subject categories – see abstracts 051.092, 066.251, 066.252, 131.323 - 131.325, 133.022 - 133.025, 151.086, 155.066 - 155.074, 156.016 - 156.021, 157.014 - 157.020.

012.077 **Neutron-capture gamma-ray spectroscopy and related topics 1981.** Proceedings of the fourth international symposium, held at Grenoble, France, 7 - 11 September 1981.
T. von Egidy, F. Gonnenwein, B. Maier (Editors).
IOP, Bristol, England. 19 + 728 pp. (1982). ISBN 0-85498-153-5. – Review in Phys. Abstr., Vol. 85, Abstr. 107014 (1982). – See abstracts 061.059, 061.060.

012.078 **New methods for optical, quasi-optical, acoustic, and electromagnetic synthesis.** Conference held at San Diego, Calif., USA, 25 - 26 August 1981.
Proc. SPIE Int. Soc. Opt. Eng., Vol. 294 (1981). – Review in Phys. Abstr., Vol. 86, Abstr. 11 (1983). – See Abstr. 031.063.

012.079 **Contemporary methods of optical fabrication.** Conference held at San Diego, Calif., USA, 25 - 26 August 1981.
Proc. SPIE Int. Soc. Opt. Eng., Vol. 306 (1981). – Review in Phys. Abstr., Vol. 86, Abstr. 5479 (1983). – See Abstr. 032.599.

012.080 **Polarizers and applications.** Conference held at San Diego, Calif., USA, 27 - 28 August 1981.
Proc. SPIE Int. Soc. Opt. Eng., Vol. 307 (1981). – Review in Phys. Abstr., Vol. 86, Abstr. 5482 (1983). – See Abstr. 034.155.

012.081 **Proceedings of the Second Marcel Grossmann Meeting on General Relativity**, held at Trieste, Italy, 5 - 11 July 1979.
R. Ruffini (Editor).
North-Holland Publishing Company, Amsterdam, Netherlands. 2 Vol. 14 + 1268 pp. (1982). ISBN 0-444-86357-5. – Review in Phys. Abstr., Vol. 86, Abstr. 5496 (1983). – See abstracts 066.275 - 066.325, 117.228, 117.229, 160.077, 162.286 - 162.294.

012.082 **Optical systems engineering II.** Conference held at Los Angeles, Calif., USA, 26 - 27 January 1982.
Proc. SPIE Int. Soc. Opt. Eng., Vol. 330 (1982). – Review in Phys. Abstr., Vol. 86, Abstr. 10501 (1983). – See Abstr. 032.602.

012.083 **Grand unified theories and related topics.**
Proceedings of the 4th Kyoto Summer Institute, held at Kyoto, Japan, 29 June - 3 July 1981.
M. Konuma, T. Maskawa (Editors).
World Scientific Publishing Company, Singapore. 469 pp. (1981). – Review in Phys. Abstr., Vol. 86, Abstr. 10509 (1983). – See abstracts 022.193, 162.309, 162.310.

012.084 **Unified theories of elementary particles. Critical assessment and prospects.** Proceedings of the

Heisenberg Symposium, held at München, F. R. Germany, 16 - 21 July 1981.
P. Breitenlohner, H. P. Dürr (Editors).
Springer-Verlag, Berlin–Heidelberg–New York. 6 + 217 pp. (1982). ISBN 3-540-11560-9. – Review in Phys. Abstr., Vol. 86, Abstr. 10516 (1983). – See abstracts 066.267, 162.311.

012.085 **XXXII Congress of the International Astronautical Federation**, held at Rome, Italy, 7 - 12 September 1981.
Acta Astronaut., Vol. 9, Nos. 6 - 7 (1982). – Review in Phys. Abstr., Vol. 86, Abstr. 16065 (1983). – See abstracts 015.038, 052.053, 094.045, 099.161.

012.086 **High resolution soft X-ray optics.** Conference held at Brookhaven, N. Y., USA, 18 - 20 November 1981.
Proc. SPIE Int. Soc. Opt. Eng., Vol. 316 (1981). – Review in Phys. Abstr., Vol. 86, Abstr. 16070 (1983). – See abstracts 031.067, 031.068.

012.087 **Nonlinear problems: present and future.** Proceedings of the First Los Alamos Conference on Nonlinear Problems, held at Los Alamos, N. M., USA, 2 - 6 March 1981.
A. Bishop, D. Campbell, B. Nicolaenko (Editors).
North-Holland Publishing Company, Amsterdam, Netherlands. 11 + 483 pp. (1982). ISBN 0-444-86395. – Review in Phys. Abstr., Vol. 86, Abstr. 16085 (1983). – See Abstr. 151.088.

012.088 **Monopoles in quantum field theory.** Proceedings of the Monopole Meeting, held at Trieste, Italy, 11 - 15 December 1981.
N. S. Craigie, P. Goddard, W. Nahm (Editors).
World Sci. Publishing, Singapore. 21 + 440 pp. (1982).
Review in Phys. Abstr., Vol. 86, Abstr. 16090 (1983).
See abstracts 022.196, 061.065, 066.331, 162.312.

012.089 **Ethnoastronomy and archaeoastronomy in the American tropics.** Papers from a conference, March 1981.
A. F. Aveni, G. Urton (Editors).
Annals of the New York Academy of Sciences, Vol. 385. New York Academy of Sciences, New York. 10 + 366 pp. Price $ 75.00 (1982). – From Science, Vol. 218, 704 (1982).

012.090 **Space in the 1980's and beyond.** 17th European Space Symposium, held in London, England, June 4 - 6, 1980.
P. M. Bainum (Editor), with an opening address by E. Quistgaard.
Science and Technology Series, Vol. 53, A supplement to Advances in the Astronautical Sciences. Published for the American Astronautical Society by Univelt, Inc., P.O. Box 28130, San Diego, California 92128. 10 + 292 pp. Price $ 40.00 cloth, £ 30.00 paper. (1981). ISBN 0-87703-154-1 cloth, ISBN 0-87703-155-X paper.

Contents: Long-term space program for Europe. Space communications. Space transportation. Space applications. Space technology. Lunar and planetary exploration.

012.091 **Haute resolution spectrale en astrophysique: applications au Telescope Spatial.** Deuxième Colloque National du Conseil Français du Telescope Spatial, Orsay, March 10 - 12, 1981.
R. Cayrel, P. Felenbok (Editors).
Observatoire de Paris. 214 pp. (1982). – Review in Observatory, Vol. 102, 155; 1982 (*R. F. Griffin*).

012.092 **Earth radiation science seminars.** Lectures from a series of seminars held at Langley Research Center, Hampton, Virgina, June 1980 to October 1981.
J. B. Hall, Jr. (Compiler).
NASA Conf. Publ., NASA CP-2239, 5 + 173 pp. (1982).

012.093 **Nonlinear dynamics.** Proceedings of the international conference on nonlinear dynamics, December 1979.
R. H. G. Helleman (Editor).
Annals of the New York Academy of Sciences, Vol. 357, New York. 11 + 507 pp. (1980). – Review in Celestial Mech., Vol. 28, 347 - 348; 1982 (*R. Broucke*).

012.094 **Relativity and gravitation.** Proceedings of the Third Latin-American Symposium.
S. Hojman, M. Rosenbaum, M. Ryan (Editors).
Universidad Nacional Autónoma de México. 280 pp. Price $ 10.00 cloth, $ 5.00 paper (1982). – From Phys. Today, Vol. 35, No. 10, p. 86 (1982).

012.095 **Spacecraft materials in a space environment.** Proceedings of a conference held at Toulouse, France, 8 - 11 June 1982.
A. Rolfo, J. Dauphin, T. D. Guyenne (Editors).
ESA SP-178. European Space Agency, Scientific and Technical Publications Branch, ESTEC, Noordwijk, The Netherlands. 5 + 323 pp. (1982).

012.096 **Atmospheric transmission.** Conference held at Washington, DC, USA, 21 - 22 April 1981.
Proc. SPIE Int. Soc. Opt. Eng., Vol. 277 (1981). – Review in Phys. Abstr., Vol. 85, Abstr. 84370 (1982).

012.097 **Infrared astronomy – scientific/military thrusts & instrumentation.** Conference held at Washington, DC, USA, 21 - 22 April 1981.
Proc. SPIE Int. Soc. Opt. Eng., Vol. 280 (1981). – Review in Phys. Abstr., Vol. 85, Abstr. 84371 (1982).

012.098 **Solid state imagers for astronomy.** Conference held at Cambridge, MA, USA, 10 - 11 June 1981.
Proc. SPIE Int. Soc. Opt. Eng., Vol. 290 (1981). – Review in Phys. Abstr., Vol. 85, Abstr. 79270 (1982).

012.099 **1982 Spring Meeting of the American Geophysical Union.** Conference held at Philadelphia, PA, USA, 31 May - 4 June 1982.
EOS Trans. American Geophys. Union, Vol. 63, No. 18 (1982). – Review in Phys. Abstr., Vol. 85, Abstr. 89436 (1982).

012.100 **5th workshop on electromagnetic induction in the earth and moon.** Conference held at Istanbul, Turkey, August 1980.
Geophys. Surv., Vol. 4, No. 4 (1982). – Review in Phys. Abstr., Vol. 85, Abstr. 74148 (1982).

013 Reports on Astronomy in Various Countries and Particular Fields, International Cooperation

013.001 **Astronomy and astrophysics in the United States.**
P. H. Abelson.
Science, Vol. 217, 691 (1982).

013.002 **The wanderings of a 25-inch refractor.**
D. Dialetis, N. Matsopoulos, T. Prokakis.
Sky Telesc., Vol. 64, 136 - 137 (1982).

013.003 **Outlines for a three year plan in astronomy.**
Mem. Soc. Astron. Italiana, Vol. 52, 601 - 611 (1981).

013.004 **The Stellar Data Center.**
C. Jaschek.
Automated data retrieval in astronomy, (see 012.002), p. 3 - 7 (1982).

The paper presents an overview of the problems astronomical data centers are facing today.

013.005 **The astronomical data base and retrieval system at NASA.**
J. M. Mead, T. A. Nagy, R. S. Hill, W. H. Warren Jr.
Automated data retrieval in astronomy, (see 012.002), p. 9 - 13 (1982).

More than 250 machine-readable catalogs of stars and extended celestial objects are now available at the NASA/Goddard Space Flight Center (GSFC) as the result of over a decade of catalog acquisition, verification and documentation. Retrieval programs are described which permit the user to obtain from a remote terminal bibliographical listings for stars; to find all celestial objects from a given list that are within a defined angular separation from each object in another list; to plot celestial objects on overlays for sky survey plate areas; and to search selected catalogs for objects by criteria of position, identification number, magnitude or spectral type.

013.006 **The Soviet Center of Astronomical Data.**
O. B. Dluzhnevskaya.
Automated data retrieval in astronomy, (see 012.002), p. 15 - 20 (1982).

The main tasks of the Soviet Center are those of accumulating, keeping and computer-processing of astronomical data files with the purpose of their subsequent distribution in the U.S.S.R., and centralizing the data exchange with Strasbourg. At present 140 catalogues are included into the files. All the data are recorded on EC computer magnetic tapes.

013.007 **Management of astronomical data at Kanazawa Data Center.**
Y. Terashita.
Automated data retrieval in astronomy, (see 012.002), p. 21 - 25 (1982).

The methods of controlling and disseminating astronomical data at the Kanazawa Data Center (operated under the Computer Center of the Kanazawa Institute of Technology) are described. Discussions include the coded data descriptions used in conjunction with a file processing language. Problems concerning remote data access are also discussed, and the future plan presented.

013.008 **Activities of the Potsdam Data Center.**
E. Schilbach.
Automated data retrieval in astronomy, (see 012.002), p. 27 - 28 (1982).

The task of the Potsdam Data Center is essentially the analysis of astronomical observations and the elaboration of theoretical models for astronomical problems. In 1974, a cooperation was started between this institution and the Stellar Data Centre at Strasbourg.

013.009 **IUE data distribution.**
A. Heck.
Automated data retrieval in astronomy, (see 012.002), p. 29 - 31 (1982).

The IUE project is described, with special emphasis on the IUE data management and archive retrieval.

013.010 **Infrared Astronomical Data Base and Catalog of Infrared Observations.**
M. Schmitz, D. Y. Gezari, J. M. Mead.
Automated data retrieval in astronomy, (see 012.002), p. 33 - 35 (1982).

A computer data base of infrared astronomical observations has been established at NASA/Goddard Space Flight Center. It contains a summary of all infrared (1 μm – 1000 μm) observations of celestial sources outside the solar system, published in the major scientific journals since 1960, as well as the contents of infrared surveys and catalogs. A Catalog of Infrared Observations (CIO) has been developed from the data base in printed and magnetic tape versions.

013.011 **Data on time and polar motion: immediate accessibility.**
D. D. McCarthy, G. Westerhout.
Automated data retrieval in astronomy, (see 012.002), p. 37 - 44 (1982).

The U.S. Naval Observatory is now implementing a versatile automated data retrieval system in the area of Earth orientation information and for precise time and time interval applications. Data are available through a series of weekly, monthly, and annual reports, as well as through a digital communications link to Time Service machine-readable files.

013.012 **The National Space Science Data Center (NSSDC) and the World Data Center A for Rockets and Satellites (WDC–A–R&S): their role in X-ray astronomy data.**
M. C. Locke.
Automated data retrieval in astronomy, (see 012.002), p. 45 - 47 (1982).

The combined role of NSSDC and WDC–A–R&S in collecting, preserving and disseminating data from 10 years of X-ray observations is presented along with a brief history of the organizations.

013.013 **The stellar data file at Dearborn Observatory.**
W. Buscombe.
Automated data retrieval in astronomy, (see 012.002), p. 49 - 50 (1982).

013.014 **Euronet-Diane, premier réseau européen de transfert de l'information.**
Y. Salle.
Automated data retrieval in astronomy, (see 012.002), p. 53 - 61 (1982).

Euronet is one of the world's most recent networks created for data transmission. Its purpose is to promote the transfer of scientific and technical information within Europe, and to develop the European information industry by encouraging the creation and availability of data bases and data banks.

013.015 **The U.K. StarLink computer network.**
D. J. Carnochan.

Automated data retrieval in astronomy, (see 012.002), p. 63 - 65 (1982).

Six VAX–11/780 computers have been set up at different centres of astronomical research throughout the United Kingdom. These computers are solely for the use of astronomers and they are linked together using special telephone lines. They are proving to be a powerful tool, both in the reduction of astronomical images and in the subsequent data analysis.

013.016 **ASTRONET. The network for analysis and retrieval of astronomical data in Italy.**
G. Sedmak.
Automated data retrieval in astronomy, (see 012.002), p. 67 - 71 (1982).

ASTRONET, the network for analysis and retrieval of astronomical data planned in Italy, is described with regard to the organization of the network, the hardware, and the standard software environment. ASTRONET starts with four VAX centers operating within 1981, to be expanded to six within 1983. The network includes colour graphics and PDS plates digitization facilities. The software and graphics standards will be set as compatible as possible to major existing astronomical standards.

013.017 **The International URSIGRAM and World Days Service.**
P. A. Simon.
Automated data retrieval in astronomy, (see 012.002), p. 73 - 75 (1982).

Broad outlines of a permanent service involved in the prompt dissemination of solar and geophysical data are given.

013.018 **A data analysis facility for the Faint Object Camera.**
F. Macchetto.
Automated data retrieval in astronomy, (see 012.002), p. 127 - 129 (1982).

The Faint Object Camera is the ESA instrument to be flown in the Space Telescope. A modern facility is being established to provide the software required to analyse the FOC data. This paper describes the basic hardware and software characteristics of the ESA facility.

013.019 **Responsibilities and practical limitations in the operation of an astronomical data center.**
W. H. Warren Jr., J. M. Mead, T. A. Nagy.
Automated data retrieval in astronomy, (see 012.002), p. 131 - 133 (1982).

013.020 **Numerical data documentation in physics in the Federal Republic of Germany.**
G. Ebel.
Automated data retrieval in astronomy, (see 012.002), p. 305 - 308 (1982).

013.021 **The IUE Observatory – a status report.**
Advances in ultraviolet astronomy, (see 012.006), p. 629 - 631 (1982).

013.022 **A round-table discussion on prospects for observations of planets in the next decade (1980-1990).**
Vibrational-rotational spectroscopy for planetary atmospheres, (see 012.007), p. 13 - 26 (1982).

The discussion concerns: The Space Telescope (*J. Caldwell*); The future of ground-based near-IR observations (*U. Fink*); The next decade in space-based observations (*R. Hanel*); Next generation infrared spectroscopy of the planets (*M. Mumma*); The 10 μm region (*A. Tokunaga*); Future developments (*R. Beer*).

013.023 **Blick in Geburtsstätten der Sterne.** Ein 10 Meter Radioteleskop erschließt den Astronomen den Bereich von Sub-Millimeterwellen.
Umschau, 82. Jahrg., 444 (1982).

013.024 **Fünfzig Jahre Radioastronomie.**
J. Gürtler.
Sterne, 58. Band, 195 - 196 (1982).

013.025 **Millimeter-Radioastronomie: ein neues Fenster in den Weltraum.**
J. W. M. Baars, J. Schmid-Burgk.
Sterne, 58. Band, 211 - 220 (1982).

013.026 **Erforschung der Sonne mit Mitteln der Radioastronomie.**
A. Krüger.
Sterne, 58. Band, 221 - 231 (1982).

013.027 **Zu Stand und Entwicklung der Radiointerferometrie in der Astronomie.**
E. Preuss.
Sterne, 58. Band, 232 - 251 (1982).

013.028 **Perspectives on space astronomy.**
J. L. Greenstein.
Sky Telesc., Vol. 64, 317 - 319 (1982).

013.029 **NASA and the planetary imperative.**
J. A. Van Allen.
Sky Telesc., Vol. 64, 320 - 322 (1982).

013.030 **The international space arena.**
K. Gatland.
Sky Telesc., Vol. 64, 323 - 326 (1982).

013.031 **The new ocean of space.**
H. J. Schmitt.
Sky Telesc., Vol. 64, 327 - 329 (1982).

013.032 **L'universo rivisitato.** I principi, le domande e le risposte della radioastronomia contemporanea.
G. Tanzella-Nitti.
Orione, Vol. 3, 91 - 106 (1982).

013.033 **Ballon-Röntgenmessungen an Neutronensternen und Quasaren.**
R. Staubert, J. Trümper.
Sterne Weltraum, 21. Jahrg., 286 - 290 (1982).

013.034 **Begegnungen mit dem Reich der Mitte.**
R. Schwartz.
Sterne Weltraum, 21. Jahrg., 291 - 293 (1982).

013.035 **Gastaufenthalt in Shanghai.**
D. Graham.
Sterne Weltraum, 21. Jahrg., 294 (1982).

013.036 **The ups and downs of coordinated observations.**
C. Bertout.
Messenger, No. 29, p. 12 - 13 (1982).

013.037 **Les instruments de l'amateur en 1882 et en 1982.**
R. Verseau.
Astronomie, Vol. 96, 333 - 336 (1982).

013.038 **Extraterrestrische Astronomie – eine aktuelle Herausforderung.**
H. Stiller, D. Möhlmann.
Jenaer Rundsch., 27. Jahrg., 108 - 109 (1982).

013.039 **Amateurastronomie in der DDR.**
K. Lindner.
Jenaer Rundsch., 27. Jahrg., 128 - 130 (1982).

013.040 **Dunsink Observatory, Chinese astronomy and the IAU.**
P. A. Wayman.
Irish Astron. J., Vol. 15, 2 - 8 (1981).

013.041 **Space at JPL.**
W. I. McLaughlin.
Spaceflight, Vol. 24, 393 - 395 (1982).

013.042 **Planetary science in Europe.** Present State and outlook for the future. Workshop on Planetology sponsored by the Space Science Committee of the European Science Foundation held at Strasbourg, 15 - 18 September 1980.
M. Combes, H. Fechtig, F. Mariani, F. M. Neubauer, S. K. Runcorn, P. Signer, with a preface by J. Geiss.
Published by the European Science Foundation. 4 + 178 pp. (1982). ISBN 2-903148-24-4.
Contents: Present status of Europe in the planetary sciences. European strategy in planetary sciences. Planetary science – scope, techniques and current understanding.

013.043 **Photometry at the Pulpit Rock Observatory.**
R. Wasatonic.
I.A.P.P.P. Commun., No. 8, p. 34 - 36 (1982).

013.044 **A proposed photometric program for A. L. P. O. lunar observers.**
P. Hédervári.
Strolling Astron., Vol. 29, 147 - 150 (1982).

013.045 **Assessment of the U. S. Space Research Program.**
W. E. Howard III, P. P. Chandler.
Bull. American Astron. Soc., Vol. 14, 622 (1982). – Abstract.

013.046 **Science operations with Space Telescope.**
R. Giacconi.
The Space Telescope Observatory, (see 012.013), p. 1 - 15 (1982).

013.047 **European astronomy and the Space Telescope: the Space Telescope European Coordinating Facility.**
F. Macchetto.
The Space Telescope Observatory, (see 012.013), p. 16 - 19 (1982).
To allow European astronomers to make full use of the extraordinary capabilities of the Space Telescope, ESA has decided to establish the European Coordinating Facility. Its role is described in the paper.

013.048 **Reflections on the Space Telescope.**
M. S. Longair.
The Space Telescope Observatory, (see 012.013), p. 121 - 134 (1982).

013.049 **Review of the scientific usage of INTERCOSMOS satellite observations at the Astronomical Institute of the Czechoslovak Academy of Sciences.**
L. Sehnal.
Bull. Astron. Inst. Czechoslovakia, Vol. 33, 253 - 258, plate 1 (1982).
The first Soviet "Sputnik" brought into space a man-made body which could be used as an astronomical orbiting object to solve various scientific problems by orbital analysis. This work carried out at the Astronomical Institute of the Czechoslovak Academy of Sciences with special use of the INTERCOSMOS satellite observations is reviewed.

013.050 **An infrared look at the universe.**
J. H. Duxbury.
Perspect. Comput., Vol. 1, No. 1, p. 9 - 16 (1981). – Abstr. in Phys. Abstr., Vol. 85, Abstr. 63207 (1982).

013.051 **Astronomía en Venezuela.**
J. Stock.
Rev. Mexicana Astron. Astrofis., Vol. 6, (see 012.017), 13 - 16 (1981).

013.052 **Quince años de actividad del Astrolabio Danjon de Santiago.**
F. Noël.
Rev. Mexicana Astron. Astrofis., Vol. 6, (see 012.017), 135 (1981). – Abstract.

013.053 **Space sciences and geosciences: evolution of two interactive fields of knowledge.**
H. Curien.
Evolution in the Universe, (see 012.024), p. 13 - 30 (1982).
The author retraces the modern, or rather the recent, history of the constructive interaction between two fields of scientific research; two fields the aims of which are closely akin but the approach to which has long remained separate: space sciences and geosciences.

013.054 **Why should we observe Halley's comet?**
R. Lüst.
The need for coordinated ground-based observations of Halley's comet, (see 012.025), p. 7 - 18 (1982).

013.055 **A worldwide photographic network for wide-field observations of Halley's comet in 1985–1986.**
M. B. Niedner, Jr., J. Rahe, J. C. Brandt.
The need for coordinated ground-based observations of Halley's comet, (see 012.025), p. 227 - 242 (1982).

013.056 **The contribution of French astronomers to the orbit determination of Halley's comet.**
B. Morando.
The need for coordinated ground-based observations of Halley's comet, (see 012.025), p. 297 - 298 (1982).

013.057 **The International Halley Watch.**
J. Rahe, R. L. Newburn, Jr.
The need for coordinated ground-based observations of Halley's comet, (see 012.025), p. 301 - 309 (1982).

013.058 **Report presented by the European Space Agency to the 24th COSPAR meeting, Ottawa, Canada, May 1982.**
W. R. Burke (Editor).
ESA SP-1040. European Space Agency, Scientific and Technical Publications Branch, ESTEC, Noorwijk, The Netherlands. 287 pp. (1982).
Contents: Satellites in orbit or decayed: Cos-B, Geos-1 and 2, ISEE, IUE. Projects under development: First Spacelab payload, Exosat, International Solar Polar Mission, ESA's participation in the NASA Space Telescope programme, Hipparcos, Giotto. Projects under study. Bibliography.

013.059 **The Universe in X-rays.**
N. Henbest.
New Scientist, Vol. 93, 720 - 725 (1982). – Abstr. in Phys. Abstr., Vol. 85, Abstr. 79247 (1982).

013.060 **Status of positional astronomy in survey of India.**
K. L. Khosla, M. G. Arur, P. S. Bains, S. Singh.
Bull. Astron. Soc. India, Vol. 10, 52 (1982). – Abstract.

013.061 **Review of galactic ultraviolet astronomy.**
C. de Jager.
Third European IUE Conference, (see 012.029), p. 43 - 46 (1982).
A review is given of the main results presented during the Third European IUE Conference in May 1982 at Madrid.

013.062 **Infrared and submillimetre astronomical spectroscopy at Berkeley.**
R. Genzel.
The scientific importance of submillimetre observations, (see 012.030), p. 141 (1982). – Abstract.

013.063 **The scientific significance of submillimetre astronomy.**
M. S. Longair.
The scientific importance of submillimetre observations, (see 012.030), p. 219 - 223 (1982).
A review is given of the scientific significance of submillimetre observations for a wide range of astrophysical problems.

013.064 **Space Telescope course.**
T. Page.
Bull. American Astron. Soc., Vol. 14, 712 (1982). – Abstract.

013.065 **Status of the International Halley Watch.**
R. L. Newburn, Jr.
Bull. American Astron. Soc., Vol. 14, 754 (1982). – Abstract.

013.066 **Future observations of active regions.**
R. H. Levine.
Solar active regions, (see 003.011), p. 319 - 342 (1981).
This paper reviews some of the areas in which future observations of solar active regions are likely to prove fruitful. The viewpoint has been that of the participants in the Skylab Solar Workshop on active regions, which emphasized plasma physics and the organizing role of solar magnetic field structures.

013.067 **Astrogeodetic work on the geoid in Austria.**
K. Bretterbauer, K. Rinner.
Z. Vermessungswes., Vol. 107, 333 - 340 (1982). In German. Abstr. in Phys. Abstr., Vol. 85, Abstr. 105899 (1982).

013.068 **Photographing of bright meteors-fireballs.**
M. Muminović.
Publ. Astron. Obs. Sarajevo, Vol. 1, No. 1, (see 012.043), p. 231 - 232 (1981).

013.069 **Hvar Observatory report on the FBS/SERF (*Flare Build-up Study/Study of Energy Release in Flares*) trial period of the solar maximum year May 1979.**
V. Ruždjak.
Publ. Astron. Obs. Sarajevo, Vol. 1, No. 1, (see 012.043), p. 293 - 294 (1981).

013.070 **Astrophysical investigations.**
D. Ya. Martynov.
Zemlya Vselennaya, 1982, No. 4, p. 21 - 26. In Russian.

013.071 **Role of International Radio Consultative Committee in co-ordination of time and frequency activities.**
C. Stettler.
J. Inst. Electron. Telecommun. Eng., Vol. 27, No. 10, Part 1, (see 012.044), p. 379 - 382 (1981). – Abstr. in Phys. Abstr., Vol. 85, Abstr. 89665 (1982).

013.072 **Le 25e anniversaire de la troisième opération des longitudes mondiales. (1 juillet 1957 - 31 décembre 1959).**
A. Stoyko.
Astronomie, Vol. 96, 522 - 525 (1982).

013.073 **Development of astronomy in the individual Republics of the USSR.**
A. E. Mal'tsev.
Zemlya Vselennaya, 1982, No. 6, p. 17 - 24. In Russian.

013.074 **Soviet astronomy today: a personal view.**
L. J. Robinson.
Sky Telesc., Vol. 64, 518 - 519 (1982).

013.075 **Planetary science: up from the ashes?**
M. Mitchell Waldrop.
Science, Vol. 218, 665 - 666 (1982).

013.076 **Ergebnisse aus der Meteorforschung in der Tschechoslowakei.**
H. Nováková.
Sterne Weltraum, 21. Jahrg., 452 (1982).

013.077 **High-energy astrophysics.**
G. W. Clark.
Phys. Today, Vol. 35, No. 11, p. 26 - 33 (1982).
Astrophysical observations of high-energy photons and particles enable us to study states of matter under conditions far more extreme than will ever be found in an earth-bound laboratory.

013.078 **Contemporary planetary science.**
M. J. S. Belton, E. H. Levy.
Phys. Today, Vol. 35, No. 11, p. 54 - 59 (1982).
A vigorous program of studies involving spacecraft missions, astronomical observations and laboratory work will tell us much about the formation of the solar system and the conditions that led to life.

013.079 **The A.S.P.'s new development program.**
A. Fraknoi.
Mercury, Vol. 11, 148 - 150 (1982).

013.080 **Ground-based and space astronomy.**
A. H. Jarrett.
Mon. Notes Astron. Soc. South. Africa, Vol. 41, 39 - 41 (1982).

013.081 **Early planning for the occultation of 1 Vulpeculae by Pallas on 1983 May 29.**
D. W. Dunham.
Occultation Newsl., Vol. 3, 2 - 3 (1982).

013.082 **Selected research projects in celestial mechanics at the University of Texas at Austin.**
V. Szebehely.
Dynamical astronomy, (see 012.056), p. 7 - 11 (1982).
This paper reports on recent research on the inverse problem, on stability, on accuracy and on undeterminacy in celestial mechanics.

013.083 **The potential for far-infrared astronomy in Australia.**
J. W. V. Storey.
Anglo-Australian Obs., Prepr. No. 169, 5 pp. (1982). – Submitted to Proc. Astron. Soc. Australia.

013.084 **Time service in China during the last three years.**
Y.-r. Miao.
Publ. Shaanxi Astron. Obs., No. 1, p. 1 - 4 (1982). In Chinese.

013.085 **Skywatch Lectures 1981 - 1982.**
J. K. Ludwig, H. C. Pollock, R. A. Spenser (Editors) with contributions by E. Chaisson, S. Edwards, W. Herbst, T. L. Page.
Dudley Obs. Rep., No. 16, p. 1 - 41 (1982).

013.086 **Infrared astronomy at ESO.**
A. F. M. Moorwood.
Second ESO infrared workshop, (see 012.060), p. 3 - 15 (1982).
The author reviews the status and future plans for IR instrumentation at the 1m and 3.6m telescopes of ESO, gives

an overview of the areas of astronomical research being most actively pursued, shows some illustrative results and finally draws attention to possibilities for expanding the IR observing possibilities within ESO created by the acquisition of a new 2.2 m telescope, the 3.5 m NTT and hopefully at some stage a VLT.

013.087 **Astronomical capability of the U.K. Infrared Telescope.**
T. J. Lee.
Second ESO infrared workshop, (see 012.060), p. 57 - 70 (1982).

The author describes what is arguably the most powerful ground based infrared facility in the world and a European one at that. First he talks about the status of the facility, then about future developments and follows them by examples of astronomical data from the telescope and how they relate to the results from other kinds of IR observatories.

013.088 **Very Large Telescope (VLT) studies at ESO.**
J. P. Swings.
Second ESO infrared workshop, (see 012.060), p. 249 - 253 (1982).

013.089 **VLT versus space.**
A. F. M. Moorwood.
Second ESO infrared workshop, (see 012.060), p. 255 - 258 (1982).

013.090 **Infrared observations from the NASA airborne observatories.**
M. Harwit.
Second ESO infrared workshop, (see 012.060), p. 275 - 283 (1982).

013.091 **ASTROPLANE: a European airborne observatory for infrared astronomy.**
C. B. Cosmovici.
Second ESO infrared workshop, (see 012.060), p. 285 - 306 (1982).

013.092 **ASTROPLANE: the Airbus proposal.**
P. Léna.
Second ESO infrared workshop, (see 012.060), p. 307 - 314 (1982).

013.093 **ASTROPLANE: a working group of the European Science Foundation.**
P. Léna.
Second ESO infrared workshop, (see 012.060), p. 315 - 319 (1982).

013.094 **Achievements and promise of balloon IR astronomy.**
S. Drapatz.
Second ESO infrared workshop, (see 012.060), p. 321 - 340 (1982).

013.095 **IUE observation statistics.**
A. Heck.
Ultraviolet stellar classification, (see 012.062), p. 55 - 79 (1982).

This paper presents statistics performed on the merged log of the IUE observations collected at the two ground stations (GSFC and VILSPA) from April 1978 to August 1981 inclusive.

013.096 **Space astronomy: the next 30 years.**
M. M. Waldrop.
Science, Vol. 218, 1101 - 1103 (1982).

013.097 **Das „Radiofenster" hat unser Weltbild erweitert.**
R. Beck.
Umschau, 82. Jahrg., 774 - 775, 778 - 779 (1982).

013.098 **Kosmische Mineralogie und Petrologie.**
K. Heide.
Sterne, 58. Band, 323 - 325 (1982).

013.099 **Space investigations made in the USSR in 1981. Reports of COSPAR, 24th plenum in Ottawa, 1982.**
Nauka, Moskva. 49 pp. (1982). In Russian. – Abstr. in Ref. zh., 62. Issled. kosm.prostranstva, 10.62.52 (1982).

013.100 **Radar astronomy.**
V. A. Kotel'nikov.
Vestn. AN SSSR, 1982, No. 6, p. 51 - 61. In Russian. – Abstr. in Ref. zh., 51. Astron., 10.51.36 (1982).

013.101 **Decametric radio astronomy.**
S. Ya. Braude.
Vestn. AN SSSR, 1982, No. 6, p. 83 - 89. In Russian. – Abstr. in Ref. zh., 51. Astron., 10.51.37 (1982).

013.102 **An Air Force/NSF initiative to study the origins of solar activity.**
S. P. Worden, R. B. Dunn, M. A. Smith, J. B. Zirker.
Smithsonian Astrophys. Obs., Spec. Rep. 392, Vol. II, (see 012.065), p. 221 - 235 (1982).

An astronomical telescope facility dedicated wholly to the study of stellar variability is proposed. The facility consists of a 90-inch (2.3 meter) telescope, high-resolution spectrograph, precision photometer, chromospheric spectrometer, and solid-state detectors. The facility would be located at an existing observatory building at Cloudcraft, N. M., and operated by the Sacramento Peak Observatory for all qualified U.S. and foreign astronomers. Operation would begin in 1986.

013.103 **Chronicle of astronomy in Lithuania in 1978 - 1981.**
V. Straižys.
Bull. Vilnius Astron. Obs., No. 60, p. 67 - 70 (1982). In Russian.

013.104 **Soviet astronomy and 60 years of the USSR.**
Astron. Zh., Tom 59, 1041 - 1043 (1982). In Russian. English translation in Soviet Astron., Vol. 26, No. 6.

013.105 **Pulsar research as normal science.**
M. N. McMorris.
Physis, Anno 24, 265 - 279 (1982).

013.106 **Amateur astronomy in Finland.**
O. Obůrka.
Říše hvězd, Vol. 63, 159 - 160 (1982). In Czech.

013.107 **Astronomy in Belgium.**
O. Obůrka.
Říše hvězd, Vol. 63, 134 - 136 (1982). In Czech.

013.108 **Advances in astronomy in 1981.**
J. Grygar.
Říše hvězd, Vol. 63, 113 - 118, 136 - 141 (1982). In Czech.

013.109 **General results of the Shanghai-Effelsberg VLBI experiment.**
T.-s. Wan, Z.-h. Qian, L.-d. Wu, D.-m. Hua, W.-d. Wu, D. A. Graham, R. A. Porcos.
Acta Astron. Sinica, Vol. 23, 376 - 384 (1982). In Chinese.

The first VLBI experiment between Shanghai and Effelsberg was carried out at 1430 MHz and lasted 54 hours in November 1981. The Mark II recording system and hydrogen masers were used at both stations. Fourteen extragalactic radio sources covering a wide range of declination were observed.

The analysis showed that the Shanghai-Effelsberg baseline vector could be estimated to a formal precision of 2 - 3 metres and the position of sources 1739 + 52 and 1928 + 73 were estimated to 0".05 - 0".08 for each coordinate, with the effects of the ionosphere not being removed. The source 1928 + 73 was found to be double with a component separation of 0".0089.

Proposal for centralization of information concerning plate vaults.
See Abstr. 002.021.

Le Centre de Donnees des Amas de Galaxies.
See Abstr. 002.029.

The archiving, distribution and utilization of planetary science data: an old concept that is possible with new technology.
See Abstr. 002.075.

NASA's Kuiper Airborne Observatory.
See Abstr. 032.508.

The High Resolution Spectrograph for the Space Telescope.
See Abstr. 032.516.

AST *(Australian Synthesis Telescope)* parameters and the AST Astrophysics Sub-Committee.
See Abstr. 033.004.

International Halley Watch.
See Abstr. 103.523.

Erratum

013.901 **Erratum: 'Hard X-ray astronomy at IAS'** [Mem. Soc. Astron. Italiana, Vol. 52, 117 - 135 (1981)].
P. Ubertini.
Mem. Soc. Astron. Italiana, Vol. 52, 657 (1981). – See Abstr. 30.013.010.

014 Teaching in Astronomy

014.001 **Physikalische Experimente für den Astronomieunterricht.**
W. Hering.
Astron. Schule, 19. Jahrg., 75 - 78 (1982).

014.002 **Zur Anwendung der Mathematik im Astronomieunterricht.**
H. Bienioschek.
Astron. Schule, 19. Jahrg., 78 - 80 (1982).

014.003 **Methodische Hilfen zur Unterrichtseinheit "Das Planetensystem" (II).**
L. Gräfe, E.-M. Schober.
Astron. Schule, 19. Jahrg., 80 - 82 (1982).

014.004 **Die koorbitalen Saturn-Monde 1980 S1 und 1980 S3.**
G. Groschopf.
Sterne Weltraum, 21. Jahrg., 304 - 306 (1982).

014.005 **Zur Methodik des Astronomie-Unterrichts in der reformierten Oberstufe.**
O. Zimmermann, with a reply by T. Schmidt.
Sterne Weltraum, 21. Jahrg., 363 - 365 (1982).

014.006 **Einführung in die Astronomie an Mittelschulen.**
H. Kaiser-Mauer.
Orion, 40. Jahrg., 137 - 139, 163 - 166 (1982).

014.007 **Astronomische Bildung.**
H. Groth.
Jenaer Rundsch., 27. Jahrg., 126 - 127, 140 (1982).

014.008 **Beobachtungen mit dem Schulfernrohr 63/840 TELEMENTOR.**
H. J. Nitschmann.
Jenaer Rundsch., 27. Jahrg., 131 - 133 (1982).

014.009 **Selecting your first telescope.**
S. Harrington.
Mercury, Vol. 11, 106 - 116 (1982).

014.010 **A laboratory exercise on resolving power.**
H. Kruglak.
Mercury, Vol. 11, 133 - 135 (1982).

014.011 **Some means for demonstration and practical work on the theme "Cosmic bodies and their motion" in the VIIIth class.**
V. Tomov.
Fizika, NRB, Vol. 7, 29 - 34 (1982). In Bulgarian. – Abstr. in Ref. zh., 51. Astron., 8.51.59 (1982).

014.012 **The analemma – a lab exercise for introductory astronomy.**
R. H. Allen.
Bull. American Astron. Soc., Vol. 14, 622 (1982). – Abstract.

014.013 **Informe sobre cursos de introducción a la astronomía.**
J. Cova S.
Rev. Mexicana Astron. Astrofis., Vol. 6, (see 012.017), 375 (1981). – Abstract.

014.014 **General relativity primer.**
R. H. Price.
American J. Phys., Vol. 50, 300 - 329 (1982). – Abstr. in Phys. Abstr., Vol. 85, Abstr. 63774 (1982).

014.015 **Laboratory exercises in astronomy – the rotation of the sun.**
O. Gingerich, R. Tresch-Fienberg.
Sky Telesc., Vol. 64, 433 - 438 (1982).

014.016 **Astronomie in der Volkshochschule.**
H. Hornung.
Sterne Weltraum, 21. Jahrg., 418 - 419, 519 - 520 (1982).

014.017 **Nuclear spins in the Earth's magnetic field.**
P. T. Callaghan, M. Le Gros.
American J. Phys., Vol. 50, 709 - 713 (1982). – Abstr. in Phys. Abstr., Vol. 85, Abstr. 107042 (1982).

014.018 **Organizacija usmerenog obrazovanja u nas i mesto astronomije u njoj.**
S. N. Sadžakov.
Publ. Astron. Obs. Sarajevo, Vol. 1, No. 1, (see 012.043), p. 51 - 56 (1981).

014.019 **Astronomical topics in primary school.**
M. Šuveljak, V. Vujnović.
Publ. Astron. Obs. Sarajevo, Vol. 1, No. 1, (see 012.043), p. 57 - 62 (1981).

014.020 **Extra-curriculum astronomical activities of the young in Croatia.**
M. Berić.
Publ. Astron. Obs. Sarajevo, Vol. 1, No. 1, (see 012.043), p. 63 - 64 (1981).

014.021 **Einige grundsätzliche Überlegungen zum Astronomieunterricht.**
E. Hörber, W. Haupt.
Sterne Weltraum, 21. Jahrg., 466 - 468 (1982).

014.022 **Methodische Hilfen zur Unterrichtseinheit "Das Planetensystem" (III).**
L. Gräfe, E.-M. Schober.
Astron. Schule, 19. Jahrg., 106 - 108 (1982).

014.023 **Die kleinen Monde des Saturn.**
M. Reichstein.
Astron. Schule, 19. Jahrg., 110 - 112 (1982).

014.024 **Geschichte der Astronomie im Astronomieunterricht.**
H. Bernhard.
Astron. Schule, 19. Jahrg., 140 - 144 (1982).

014.025 **La formation continue en astronomie des professeurs de l'enseignement secondaire.**
B. Hauck.
Orion, 40. Jahrg., 196 - 198 (1982).

014.026 **Astronomy and creationism.**
D. Morrison.
Mercury, Vol. 11, 144 - 147 (1982).

014.027 **The DES/Astronomical Societies Working Party on Astronomical Education 1979–1981.**
D. McNally.
Q. J. R. Astron. Soc., Vol. 23, 569 - 574 (1982).

014.028 **Astronomy and education.**
R. Hubert.
Kozmos, Vol. 13, 181 (1982). In Slovak.

Astronomie.
See Abstr. 003.057.

Wie groß ist die Welt?
See Abstr. 003.076.

Utveckling i universum – En astrofysikalisk kurs för gymnasiet.
See Abstr. 003.077.

Astronomische Musterversuche.
See Abstr. 003.109.

Wisconsin's teaching observatory.
See Abstr. 009.001.

015 Miscellaneous Papers (Philosophical Aspects, Extraterrestrial Civilizations, etc.)

015.001 **On the orientation of the Kaaba.**
G. S. Hawkins, D. A. King.
J. Hist. Astron., Vol. 13, 102 - 109 (1982).

015.002 **Opinautics: a matter of much gravity.**
E. Sheldon.
Sky Telesc., Vol. 64, 118 - 122 (1982).

015.003 **The "space age" (25 years and counting).**
J. K. Beatty.
Sky Telesc., Vol. 64, 310 - 312 (1982).

015.004 **Biological crises and asteroid impact.**
L. B. Galbraith.
South. Stars, Vol. 29, 139 - 142 (1982).

015.005 **Astronomy, astrology, Ptolemy, and us.**
R. R. Newton.
Johns Hopkins APL Tech. Dig., Vol. 3, 77 - 80 (1982).

015.006 **Search for intelligent life in the universe.**
L. M. Gindilis.
Zemlya Vselennaya, 1982, No. 4, p. 52 - 57. In Russian.

015.007 **The early evolution of life on earth.**
M. Eigen.
Evolution in the Universe, (see 012.024), p. 51 - 52 (1982). Abstract.

015.008 **Is comet Halley really dangerous?**
P. Veron.
The need for coordinated ground-based observations of Halley's comet, (see 012.025), p. 5 - 6 (1982).

015.009 **Images of other worlds.**
A. Chaikin.
Sky Telesc., Vol. 64, 423 - 427 (1982).

015.010 **Frauen in der Astronomie.**
H. Korneck.
Sterne Weltraum, 21. Jahrg., 412 - 414 (1982).

015.011 **The moon illusion.**
P. Devadas.
Bull. Astron. Soc. India, Vol. 10, 16 - 17 (1982).

015.012 **The decline and fall of SETI.**
R. Sheaffer.
Spaceflight, Vol. 24, 457 - 458 (1982).

015.013 **Extraterrestrial civilizations and terrestrial problems.**
A. D. Ursul.
Zemlya Vselennaya, 1982, No. 6, p. 51 - 53. In Russian.

015.014 **Cosmology and religion.**
J. Casanovas.
Astron. Q., Vol. 4, 117 - 122 (1982).

015.015 **L'étoile de Noël et la naissance du Christ.**
G. Gabriel.
Le Ciel, Vol. 44, 273 - 275 (1982).

015.016 **An explanation for the absence of extraterrestrials on Earth.**
M. H. Hart.
Extraterrestrials: where are they? (see 012.057), p. 1 - 8 (1982). – See also Abstr. 13.015.017.

015.017 **Searches for electromagnetic signals from extraterrestrial beings.**
B. Zuckerman.
Extraterrestrials: where are they? (see 012.057), p. 9 - 19 (1982).

There have been more than a dozen searches for radio signals from alien civilizations. Targets have included nearby stars and external galaxies. The author briefly reviews a few of the most ambitious and sensitive projects. Within the next few years infrared telescopes in orbit around the Earth should greatly enhance the ability to detect very advanced civilizations (Dyson spheres). He briefly considers the relative merits of infrared and radio wavelengths for interstellar communication between two civilizations that possess advanced space technologies.

015.018 **An examination of claims that extraterrestrial visitors to earth are being observed.**
R. Sheaffer.
Extraterrestrials: where are they? (see 012.057), p. 20 - 28 (1982).

015.019 **The likelihood of interstellar colonization, and the absence of its evidence.**
S. von Hoerner.
Extraterrestrials: where are they? (see 012.057), p. 29 - 33 (1982).

015.020 **Preemption of the Galaxy by the first advanced civilization.**
R. Bracewell.
Extraterrestrials: where are they? (see 012.057), p. 34 - 40 (1982).

015.021 **Settlements in space, and interstellar travel.**
C. Singer.
Extraterrestrials: where are they? (see 012.057), p. 46 - 61 (1982).

An upper limit to the amount of effort required for interstellar travel by extraterrestrial intelligence (ETI) can be deduced by examining the options available to mankind with foreseeable technology. On-board nuclear propulsion should allow travel at up to 1 percent of the speed of light, and recent concepts in remote propulsion may allow much faster travel. The requirements for constructing a self-replicating manned ecosystem from debris in the solar system have recently been examined in great detail, and the propulsion system needed to use such a habitat as an interstellar vehicle is readily determined. However, biologically advanced ETI may have additional options available for interstellar travel, and some of these options have not yet been given sufficiently careful consideration.

015.022 **Terraforming.**
J. Oberg.
Extraterrestrials: where are they? (see 012.057), p. 62 - 65 (1982).

015.023 **Estimates of expansion time scales.**
E. M. Jones.
Extraterrestrials: where are they? (see 012.057), p. 66 - 76 (1982).

Monte Carlo simulations of the expansion of a spacefaring civilization show that descendants of that civilization should be found near virtually every useful star in the Galaxy in a time much less than the current age of the Galaxy. Only extreme assumptions about local population growth rates, emigration rates, or ship ranges can slow or halt an expansion. The apparent absence of extraterrestrials from the solar system suggests that no such civilization has arisen in the Galaxy.

015.024 **Colonies in the asteroid belt, or a missing term in the Drake equation.**
M. D. Papagiannis.
Extraterrestrials: where are they? (see 012.057), p. 77 - 86 (1982).

The asteroid belt, an excellent source of raw materials, seems to be the best place to search for space colonies, while the missing term of the Drake equation is probably related to the difficulty of maintaining a clement environment on a planet for several billion years so that life will have the necessary time to evolve from primitive microorganisms to advanced civilizations.

015.025 **Primordial organic chemistry.**
C. Ponnamperuma.
Extraterrestrials: where are they? (see 012.057), p. 87 - 99 (1982).

A.Oparin contended in 1924 that there was no fundamental difference between a living organism and lifeless matter and that the complex combinations, manifestations and properties so characteristic of life must have arisen in the process of the evolution of matter. In 1928, Haldane had similar ideas. He described the formation of a primordial broth by the action of ultraviolet light on the Earth's primitive atmosphere. The Oparin–Haldane hypothesis is the basis of the scientific study of the origin of life.

015.026 **Chance and the origin of life.**
E. Argyle.
Extraterrestrials: where are they? (see 012.057), p. 100 - 112 (1982).

Random chemical reactions in the Earth's primitive hydrosphere could have generated no more than 200 bits of information, whereas the first Darwinian organism must have encoded about a million bits, and therefore could not have arisen by chance. This information gap is bridged by separating reproduction from organism, and postulating a reproductive chemical community that would generate information by proto-Darwinian evolution. The information content of the initial community could have been as low as 160 bits, and its evolution might have led to the first Darwinian cell.

015.027 **Possible forms of life in environments very different from the Earth.**
R. Shapiro, G. Feinberg.
Extraterrestrials: where are they? (see 012.057), p. 113 - 121 (1982).

Speculations concerning extraterrestrial life forms have generally assumed that their basic physical processes will be similar to those employed by life on Earth. Such assumptions limit the possible locations suitable for life to planets similar to Earth. The authors have considered the requirements for life from a broader point of view and find them to be less restricting. What is required is a flow of free energy, a system of matter capable of interacting with the energy and using it to become ordered, and enough time to build up the complexity of order that is associated with life. These conditions may be met in a variety of environments, some very different from Earth.

015.028 **Cosmology and life in the Universe.**
J. R. Gott, III.
Extraterrestrials: where are they? (see 012.057), p. 122 - 134 (1982).

The detectability of civilizations at cosmological distances is investigated. For communications at cosmological distances frequency standards tied to the T = 2.7°K cosmic microwave background are proposed: 28 GHz ($h\nu = 1/2$ kT) and 56 GHz ($h\nu$ = kT). Modest searches from the ground at 28 GHz could be carried out with present technology. These would provide an important supplement to searches for life in our galaxy alone.

015.029 **Nucleosynthesis and galactic evolution: implications for the origin of life.**
V. Trimble.
Extraterrestrials: where are they? (see 012.057), p. 135 - 141 (1982).

The terrestrial example suggests that the evolution of intelligent life requires large amounts of both heavy elements (those beyond hydrogen and helium) and time. How much of each is available in various places depends on the chemical and dynamical evolution of our galaxy. Non-detection of extraterrestrials is not explicable in terms of stellar and galactic evolution.

015.030 **Atmospheric evolution, the Drake equation, and DNA: sparse life in an infinite universe.**
M. H. Hart.
Extraterrestrials: where are they? (see 012.057), p. 154 - 165 (1982).

Computer simulations were made of the evolution of a planetary atmosphere. The results indicate that only for a very small range of parameters will the surface temperature of the planet remain continuously in a range which permits H_2O to exist in the liquid phase. Using a "Drake equation" approach, it is calculated that N, the number of technological civilizations independently arising in a galaxy, is $\ll 1$. This agrees with and explains the common observation that the Earth has not been colonized by extraterrestrials. The astronomical evidence favors an open (and apparently infinite) universe. There are probably, therefore, an infinite number of advanced extraterrestrial civilizations; but all are extremely far from us.

015.031 **La observación astronómica desde la tierra y desde el espacio.**
J. Pensado.
Bol. Astron. Obs. Madrid, Vol. 10, No. 6, p. 17 - 27 (1981).

015.032 **Die astronomische Forschung im Lichte der quantitativen Entwicklung der astronomischen Fachliteratur und ihre Bewertung.**
T. Schmidt-Kaler.
Mitt. Astron. Ges., Nr. 57, p. 9 - 23 (1982).

015.033 **Sky poles.**
E. C. Krupp.
Publ. Astron. Soc. Pacific, Vol. 94, 755 (1982). – Abstract.

015.034 **Astrology, religion and science.**
T. G. Cowling.
Q. J. R. Astron. Soc., Vol. 23, 515 - 526 (1982).

015.035 **Der Stern der Weisen.**
K.-F. Hoffmann.
Umschau, 82. Jahrg., 780 - 782 (1982).

015.036 **The biological point of view: the effect of life on the planet.**
L. Margulis.
Formation of planetary systems, (see 012.064), p. 889 - 893 (1982).

015.037 **Ist das Gravitationsgesetz eine Hypothese?**
B. Thüring.
Philosophia Naturalis, Band 19, 471 - 483 (1982).

015.038 **A symbiotic approach to SETI observations: use of maps from the Westerbork Synthesis Radio Telescope.**
J. C. Tarter, F. P. Israel.
Acta Astronaut., Vol. 9, (see 012.085), 415 - 419 (1982).
Abstr. in Phys. Abstr., Vol. 86, Abstr. 20259 (1983).

The problem of infinity in modern science.
See Abstr. 003.068.

Die Sekunde der Kosmologen.
See Abstr. 003.106.

Planetary atmospheres and the search for life.
See Abstr. 091.017.

A brief survey of the solar system.
See Abstr. 091.054.

Probing the presently tenuous link between comets and the origin of life.
See Abstr. 102.097.

The anthropic principle.
See Abstr. 162.055.

Applied Mathematics, Physics

021 Mathematical Papers Related to Astronomy and Astrophysics, Computing, Data Processing

021.001 **On the possibility of using Laplace series for the gravitational potential on the surface of a planet. II.**
V. A. Antonov, K. V. Kholshevnikov.
Astron. Zh., Tom 59, 763 - 768 (1982). In Russian. English translation in Soviet Astron., Vol. 25, No. 4.

The authors consider potential V representations by means of a segment V_n of the expansion in surface spherical harmonics and by means of a segment $\tilde{V}_n$ of the expansion in space spherical harmonics.

021.002 **A Monte Carlo approach to scattering off surfaces of triaxial ellipsoids.**
W. T. Thompson, D. Van Blerkom.
Icarus, Vol. 49, 387 - 397 (1982) = Contrib. Five College Astron. Dep., No. 493.

The authors developed a Monte Carlo procedure to calculate the scattering properties of ellipsoidal bodies. Although originally intended for application to Saturn's ring A, the technique can also be applied to surfaces of asteroids modeled as triaxial ellipsoids. A simple coordinate transformation allows the Monte Carlo approach to be quick and efficient By modifying the well-known acceptance–rejection technique to reflect the dependence of surface albedo on the angle of incidence, one can apply this technique to any surface scattering function.

021.003 **New developments in data storage.**
P. J. Grosbøl.
Automated data retrieval in astronomy, (see 012.002), p. 79 - 83 (1982).

It is estimated that up to 100 Gbytes of primary data from digital detectors have to be stored each year. The amount of reduced data is at least one order of magnitude smaller. Although the storage density for magnetic recording can be made higher, only the optical technique can provide a substantial denser medium than the photographic emulsion. It seems likely that optical read-only devices will be developed for archival storage of data in this decade. Magnetic recording of data will still be preferred whenever changing of the information is important.

021.004 **System software approaches to the analysis of multidimensional data structures.**
R. Albrecht.
Automated data retrieval in astronomy, (see 012.002), p. 87 - 97 (1982).

In this review paper on software techniques, the concepts of "software system" and "data base interface" are defined in the context of astronomical data analysis requirements. Principles of software design and maintenance are discussed.

021.005 **A fast method to retrieve data from a large star catalogue file.**
A. Schuetz.
Automated data retrieval in astronomy, (see 012.002), p. 99 - 101 (1982).

A method optimized for the access from a large catalogue of stars lying in a given cap of the celestial sphere is described. The performances of the method applied to the SAO catalogue are presented.

021.006 **On systems of standards.**
H. Eichhorn.
Automated data retrieval in astronomy, (see 012.002), p. 103 - 108 (1982) = Contrib. Dep. Astron. Univ. Florida, No. 38.

Rigorous formulations closely following current practice, are proposed for the absolute and the relative estimation of certain target quantities. A system of standards is defined as a set of estimates that has certain properties. This is followed by a discussion of the meaning of "reducing estimates to a specified system" and certain procedures are suggested for making this process less ambiguous. It is finally shown that relative estimates can be employed for replacing a system which is based on absolute measurements alone by another, more accurate system.

021.007 **Digital image processing and data retrieval: the needs.**
A. Bijaoui.
Automated data retrieval in astronomy, (see 012.002), p. 109 - 110 (1982).

021.008 **International standards for software structures in astronomy.**
M. Pucillo, P. Santin, G. Sedmak.
Automated data retrieval in astronomy, (see 012.002), p. 111 - 113 (1982).

The increasing request for astronomical data processing facilities forces the definition of standards for astronomical application software. Separating the application module from the interfaces to the data processing system allows to identify the standard user interaction interface, the standard data storage interface, the standard dedicated subsystems interface and the standard on-line documentation interface. The authors tested a tentative integrated solution that fulfills these requirements. This solution includes colour graphics and is available to the astronomical community.

021.009 **Archiving and retrieval of data from the International Ultraviolet Explorer (IUE) mission.**
W. H. Warren Jr., D. F. Alderman.
Automated data retrieval in astronomy, (see 012.002), p. 115 - 117 (1982).

The data base of IUE spectra available for dissemination to the scientific community presently contains approximately 22,000 separate images. It was decided to store the data on high density (6250 bpi) specially formatted tapes and to retrieve and reconvert them to the originally supplied format upon request. The archival/retrieval system for the data and an automatic data entry system for creation and maintenance of an indexing data base are described.

021.010 **Towards effective data processing: portable software models.**

I. Mistrik.
Automated data retrieval in astronomy, (see 012.002), p. 119 - 121 (1982).

The approach taken is to develop flexible software models in order to test for applicability in the real-world environment. Major characteristics of such software models are: 1) portability; 2) modifiability; 3) functionality; 4) operationality.

021.011 **A process for retrieval of data from a compiled star catalogue.**
W. Stein.
Automated data retrieval in astronomy, (see 012.002), p. 123 - 125 (1982).

021.012 **The system SPORA *(Software Package Oriented to Research in Astronomy).***
S. Lavrov.
Automated data retrieval in astronomy, (see 012.002), p. 135 - 143 (1982).

021.013 **Application of mathematical modelling to the estimation of errors of photographic determinations of proper motions of stars.**
A. I. Yatsenko.
Glav. astron. obs. AN USSR. Kiev, 1982, 18 pp. In Russian. Abstr. in Ref. zh., 51. Astron., 7.51.127 (1982).

021.014 **Computation of the lowest eigenvalue arising from the problem of stellar stability by the method of Schwartz iteration.**
W. E. Haymes.
Astrophys. Space Sci., Vol. 84, 307 - 316 (1982).

The Schwartz iteration procedure is used to compute the lowest eigenvalue from the equation for radial oscillations of stars. It was found that the convergence of the iteration had an oscillating convergence. This made it unsuitable for the computation of the higher eigenvalues. A series of associated Laguerre functions is suggested as a method of obtaining better results.

021.015 **Calculs astronomiques pour amateurs.**
J. Meeus.
Astronomie, Vol. 96, 365 - 366, 474 - 477, 509 - 514 (1982).

021.016 **Geometric interpretation of a covariance matrix.**
A. Drożyner.
Artif. Satell., Vol. 16, No. 3, p. 31 - 40 (1981).

021.017 **Dynamic exchange with external storage in UPP-systems.**
V. I. Skripnichenko.
Algoritmy nebesnoj mekhaniki (Materialy matematicheskogo obespecheniya EhVM), No. 36. Institut Teoreticheskoj Astronomii Akademii Nauk SSSR. Price 10 Kop. 30 pp. Leningrad (1982). In Russian.

A system of subroutines intended for automatic distribution of Poisson series between the main memory and the magnetic drum storage is described.

021.018 **Period search techniques for microcomputers.**
E. P. Belserene.
I.A.P.P.P. Commun., No. 8, p. 19 - 25 (1982).

021.019 **On an approach to representation of systematic differences of catalogues.**
I. F. Makarenko, V. V. Tel'nyuk-Adamchuk.
Vestn. Kiev. Univ., Astron. Vyp. 24, p. 80 - 84 (1982). In Russian.

021.020 **Techniques for the estimation of red power spectra. I. Context and methodology.**
J. E. Deeter, P. E. Boynton.
Astrophys. J., Vol. 261, 337 - 350 (1982).

In pursuing the application of pulsar timing techniques to the investigation of possible torque noise and the subsequent diagnosis of the internal structure of neutron stars, the authors have formulated a general mathematical framework for the computation of power spectra of irregularly sampled red noise processes.

021.021 **A graphical method for studying and modeling gravitational lens systems.**
E. E. Falco, M. V. Gorenstein, I. I. Shapiro, N. L. Cohen.
Bull. American Astron. Soc., Vol. 14, 658 (1982). – Abstract.

021.022 **On the solution of the problem of choice of the optimum integration algorithm for interpretation of kinematic rotation equations.**
S. V. Evsyukov, Yu. A. Plaksij.
Khar'k. politekh. inst. Khar'kov, 1982. 18 pp. In Russian. Abstr. in Ref. zh., 51. Astron., 9.51.65; 62. Issled. kosm. prostranstva, 9.62.110 (1982).

021.023 **Application of two-factor dispersion analysis to estimation of the quality of experimental observations.**
M. S. Kolchina.
Krivorozh. gornorud. inst. Krivoj Rog, 1982. 9 pp. In Russian. Abstr. in Ref. zh., 51. Astron., 9.51.789 (1982).

021.024 **Modelling of light curves on a computer.**
V. M. Mikhajlets.
Uzhgorod. univ. Uzhgorod, 1982. 10 pp. In Russian. – Abstr. in Ref. zh., 51. Astron., 9.51.790 (1982).

021.025 **Programas para el cálculo de anuarios astronómicos.**
M. A. Cardona Garcés, F. Contreras Botto.
Rev. Mexicana Astron. Astrofis., Vol. 6, (see 012.017), 377 - 378 (1981).

The authors describe computer programs in different languages to prepare and print astronomical ephemerides.

021.026 **A method for calculation of electric and magnetic fields in the high-latitude ionosphere.**
D. S. Faermark.
Soln. veter i magnitosfer. issled. Moskva, 1980, p. 122 - 128. In Russian. – Abstr. in Ref. zh., 62. Issled. kosm. prostranstva, 9.62.257 (1982).

021.027 **Present status of the theory and analytic evaluation of non-resonant thermonuclear reaction rates.**
H. J. Haubold, R. W. John.
Astron. Nachr., Band 303, 161 - 187 (1982). In German.

Concerning the analytic evaluation of reaction rates the authors propose to use the integration theory of special functions and show for the special case of thermonuclear reactions in non-degenerate, nonrelativistic plasma that the non-resonant reaction rate can be evaluated in closed form by means of generalized hypergeometric functions. In connection with these closed-form results one can give differential and functional relations and series expansions. Also approximation considerations can be performed.

021.028 **Computer-assisted reduction of photographic two-colour diagrams. Application to field stars near IC 348.**
W. Pfau, H. Lehmann.
Astron. Nachr., Band 303, 193 - 199 (1982).

A computer programme is described which derives amounts of reddening and distances of stars from *UBV* photometry. It is based on analytical expressions for the normal line in the two-colour diagram and for the absolute magnitude as a function of the intrinsic colour. The programme was applied to photographic *UBV* data of two fields of stars in the vicinity of the galactic cluster IC 348. It is shown that the interstellar dust is concentrated there in a thin slab situated at about 200 pc to 300 pc distance and giving rise to amounts of reddening, $E(B-V)$, of about 0.4 mag and 0.8 mag for the two fields, respectively.

021.029 **On the statistical uncertainties associated with line profile fitting.**
D. A. Landman, R. Roussel-Dupré, G. Tanigawa.
Astrophys. J., Vol. 261, 732 - 735 (1982).

The authors study the statistical uncertainties to be ascribed to the three principal parameters (amplitude, center, and width) associated with the fitting of model line profiles to observational data. A general procedure applicable to any profile type of parametric form is described.

021.030 **RETICENT – a command language for spectrophotometric data reduction.**
C. J. Pritchet, S. Mochnacki, S. Yang.
Publ. Astron. Soc. Pacific, Vol. 94, 733 - 741 (1982).

The authors describe a FORTRAN-coded computer program ("RETICENT") that has been written for spectrophotometric data reduction. RETICENT operates as a command language, and possesses a powerful "front-end" command interpreter that permits the coding of programmable structures such as MACROs and loops. A wide range of scientific application commands are available. Versions of RETICENT exist for a number of computer/operating system configurations.

021.031 **Geometry of the bifurcations of the normalized reduced Henon-Heiles family.**
R. Cushman.
Proc. R. Soc. London, Ser. A, Vol. 382, 361 - 371 (1982).
Abstr. in Phys. Abstr., Vol. 85, Abstr. 94981 (1982).

021.032 **Simulation of particle problems in astrophysics.**
R. A. James.
Comput. Phys. Commun., Vol. 26, 423 - 431 (1982). – Abstr. in Phys. Abstr., Vol. 85, Abstr. 101061 (1982).

021.033 **BIRP – a way to search through image engineering data.**
L. K. Bolef, E. A. Guinness, R. E. Arvidson.
IEEE Trans. Geosci. Remote Sensing, Vol. GE-20, 378 - 383 (1982). – Abstr. in Phys. Abstr., Vol. 85, Abstr. 112139 (1982).

021.034 **The concept of integrability on Cantor sets for Hamiltonian systems.**
J. Pöschel.
Celestial Mech., Vol. 28, (see 012.049), 133 - 139 (1982).

Differentiable Hamiltonian systems close to non-degenerate, integrable Hamiltonian systems are shown to be integrable on a Cantor set in the sense that on some Cantor set, (1) the invariant KAM-tori form a smooth foliation, (2) there exist smooth, independent integrals in involution, and (3) there exists a complete solution of the Hamilton Jacobi equation. The complement of the Cantor set is shown to be small in measure.

021.035 **Melnikov's method and averaging.**
J. A. Sanders.
Celestial Mech., Vol. 28, (see 012.049), 171 - 181 (1982).

Recently, P. J. Holmes (1980) proposed the use of the Melnikov integral in order to analyse averaged systems. The Melnikov integral is the first term of an asymptotic expansion. Besides computing the integral, it is necessary to estimate the error term. Formal calculations seem to indicate that higher order terms in the asymptotic expansion are very small (at least in a model problem), according to Holmes (1980). While this is encouraging, it is far from being a proof. The aim of this article is to provide the asymptotics behind the Melnikov approach and to discuss where the estimates might be sharpened.

021.036 **A procedure to compute level lines.**
U. Kirchgraber, W. Hein.
Celestial Mech., Vol. 28, (see 012.049), 183 - 187 (1982).

Using the Stiefel-Baumgarte stabilization technique a method to compute level lines is proposed.

021.037 **The Lyapunov transformation of Hill's equation and its dynamic interpretation.**
F. Nahon.
Celestial Mech., Vol. 28, (see 012.049), 233 - 238 (1982).
In French.

The Lyapunov transformation transforms Hill's equation $\ddot{y} + F(t)y = 0, F(t+T) = F(t)$ into another one which belongs to the same class in Yakubovich's classification. Let (C) be a closed trajectory of a Lagrangian conservative system with two degrees of freedom. Due to the "Principle of Least action" (C) is the image of a geodesic (Γ) of a certain two-dimensional surface (Σ). The author shows that the two Hill's equations associated with (C) and (Γ) are related by a certain Lyapunov transformation.

021.038 **A relation between Giorgilli-Galgani's and Deprit's algorithms.**
M. Rapaport.
Celestial Mech., Vol. 28, 291 - 293 (1982).

If $\mathbf{T} = \Sigma \epsilon^i \mathbf{T}_i$ and $\mathbf{W} = \Sigma n \epsilon^{n-1} \mathbf{W}^{(n)}$ are respectively the generators of Giorgilli-Galgani's and Deprit's transformations, the author shows that the change of variables generated by $\mathbf{T}$ is the inverse to the one generated by $\mathbf{W}$, if $\mathbf{T}_i = \mathbf{W}^{(i)}$ for any i. The method used shows that the recurrence which defines the first algorithm can also be obtained with the second one.

021.039 **Limits of stability for an area-preserving polynomial mapping.**
J. H. Bartlett.
Celestial Mech., Vol. 28, 295 - 317 (1982).

The Henon–Heiles mapping $x' = x + a(y - y^3)$, $y' = y - a(x' - x'^3)$ has been studied, with the aim of finding where the unstable regions of the (x, y) plane are. When this mapping is put into the normal form, it is found to be a typical twist mapping. The criteria of Moser (1971) are used to obtain an upper limit to the size of a stable region around the origin.

021.040 **The TAGFER routine set for computing the ephemerides of the Sun, Mercury, Venus and Mars for the RATAN-600 radiotelescope.**
N. I. Glebova.
Algoritmy nebesn. mekh. (Mater. mat. obespecheniya EhVM), No. 38. Institut Teoreticheskoj Astronomii Akademii Nauk SSSR. Price 11 Kop. 36 pp. (1982). In Russian.

The complete set of routines has been described for computing the ephemerides of the Sun, Mercury, Venus and Mars which are used for observations on the RATAN-600 radiotelescope. The computation procedures are based on Newcomb's theories of planetary motions.

021.041 **A system for automation of ephemeris calculations MAMONT-1981.**
G. A. Krasinskij.
Algoritmy nebesn. mekh. (Mater. mat.obespecheniya EhVM), No. 39. Institut Teoreticheskoj Astronomii Akademii Nauk SSSR. Price 10 Kop. 34 pp. (1982). In Russian.

A computer system designed to facilitate calculations of astrometric ephemerides is described. Any request to the system is expressed in terms of a specialized language which is very like to the language of the astrometric terminology.

021.042 **Close-approach predictions for comets and asteroids.** C. Townsend, S. Hanssen.
J. British Astron. Assoc., Vol. 93, 7 - 10 (1982).
The authors, after evaluating several computational techniques, have programmed an elementary equation set for the electronic calculator and/or microcomputer investigation of undisturbed two-body elliptical orbit and parabolic orbit close-encounter situations involving major and minor members of the Solar System. This article presents the key theoretical precepts and practical applications of their generalized methodology.

021.043 **Programmes de calcul – le programme: ASTRO.** J. P. Leburton, D. Houbrechts.
Le Ciel, Vol. 44, 155 - 159 (1982).

021.044 **Programmes de calcul – le programme: LVCOSL.** J. P. Leburton, D. Houbrechts.
Le Ciel, Vol. 44, 247 - 266 (1982).

021.045 **Zur Berechnung der Sternzeit mit Taschenrechnern.** W. Quester.
BAV Rundbrief, 31. Jahrg., 67 (1982).

021.046 **Discrete Fourier Transforms of nonuniformly spaced data.**
P. R. Swan.
Astron. J., Vol. 87, 1608 - 1615 (1982).
Time series or spatial series of measurements taken with nonuniform spacings have failed to yield fully to analysis using the Discrete Fourier Transform. Two original methods are presented for deconvolving such transforms for signals containing significant noise. The results of numerical experiments with noisy data are presented in order to demonstrate the capabilities and limitations of the methods.

021.047 **Ein FORTRAN-Programm zur Berechnung translatorischer Mondephemeriden nach der Theorie von E. W. Brown.**
B.-D. Teichert.
Mitt. Geod. Inst. Rheinischen Friedrich-Wilhelms-Univ. Bonn, Nr. 62, p. 26 - 53 (1980).
The solution of many problems of physical geodesy requires the knowledge of accurate Lunar Ephemerides. To avoid the use of the "Astronomical Ephemeris", there has been written a FORTRAN-Program based on the Lunar Theory of E. W. Brown. This program being described in the appendix computes ecliptical polar coordinates (Longitude, Latitude, Parallaxe) of the center of mass of the moon at any epoch. Furthermore, the principles of Brown's way of the solution has been explained, especially the treatment of his "Main Problem", e.g. the restricted three body problem.

021.048 **Calculation of the Laplace coefficients and their derivatives in the case of $0.5 \leqslant \alpha \leqslant 0.9$.**
I. A. Gerasimov.
Astron. Tsirk., No. 1174, p. 5 - 6 (1981). In Russian.

021.049 **INTEP, an effective interpolation subroutine.** G. Hill.
Publ. Dominion Astrophys. Obs., Vol. 16, 67 - 70 (1982) = NRC No. 20446.
The computer code for a most useful interpolation subroutine is presented, based on a little-known algorithm by Tsipouras and Cormier (1973).

021.050 **A note to the computation of perturbations with tesseral harmonics.**
L.-d. Wu, C.-b. Wang, P.-z. Rong.
Publ. Purple Mountain Obs., Vol. 1, 39 - 46 (1982). In Chinese.
Recursive relations of normalized inclination functions and their derivatives are given. A numerical method to compute Hansen coefficients and their derivatives is suggested.

021.051 **Some possibilities of a machine analysis of space photos of planets.**
M. V. Ostrovskij.
Sb. nauchn. tr. TsNII geod., aehrosemki i kartogr., 1981, No. 228, p. 65 - 74. In Russian. – Abstr. in Ref. zh., 52. Geod. Aehrosemka, 11.52.228 (1982).

021.052 **Studies in astronomical time series analysis. II. Statistical aspects of spectral analysis of unevenly spaced data.**
J. D. Scargle.
Astrophys. J., Vol. 263, 835 - 853 (1982).
Detection of a periodic signal hidden in noise is frequently a goal in astronomical data analysis. This paper studies the reliability and efficiency of detection with the most commonly used technique, the periodogram, in the case where the observation times are unevenly spaced. A modification of the classical definition of the periodogram is necessary in order to retain the simple statistical behavior of the evenly spaced case. An analysis of the statistical significance of signal detections is presented, with examples.

021.053 **Computer analysis of light curves of eclipsing binaries. Part 3. Description of the program complex of the direct method for the Nairi K computer.**
M. I. Lavrov.
Kazan. univ. M., Kazan', 1982. 31 pp. In Russian. – Abstr. in Ref. zh., 51. Astron., 10.51 837 (1982).

021.054 **Program for arbitrary smoothing of experimental data.**
A. Bogdanovič.
Bull. Vilnius Astron. Obs., No. 61, p. 52 - 55 (1982). In Russian.
The paper describes a program, which permits to calculate a smoothed function with arbitrarily given accuracy on the basis of observational data.

021.055 **Toroidal solutions of the Gegenbauer equation.** E. A. Evangelidis.
Astrophys. Space Sci., Vol. 87, 117 - 120 (1982).
New solutions are found for the Laplacian and Grad-Shafranov operators in toroidal coordinates suitable for the study of gravitational and plasma problems.

021.056 **A comparison of two methods of smoothing.** L. D. Kostina, V. I. Sakharov.
Izv. Glav. Astron. Obs. Pulkovo, Astrometr. Astrofiz., No. 199, p. 53 - 55 (1982). In Russian.
Filters of the moving average and Whittaker's filters have been compared for the cases of weak, moderate and strong smoothing with parameters $n = 3, 5, 7$ and $\epsilon \approx 1, 0.1, 0.01$ respectively, where n is the number of the smoothed ordinates (method of smoothing by moving averages) and ϵ is the measure of smoothing (Whittaker's method). In all the cases the obtained results agree well with each other.

021.057 **Several notes on the transformation methods in non-linear systems.**
L. Liu.
Acta Astron. Sinica, Vol. 23, 255 - 263 (1982). In Chinese.
With a view of practice, a comparative study of several transformations for the non-linear system is given in the paper. Having generalized the von Zeipel approach the author proposes a new transformation for the non-Hamiltonian system which is simpler than the Lie transform.

021.058 **A new definition in the design of digital filters and the variational solution of the extended Whittaker problem.**

K.-y. Huang, X. Zhou.
Acta Astron. Sinica, Vol. 23, 280 - 286 (1982). In Chinese.

021.059 **A simple and accurate on-line data acquisition program.**
M. P. FitzGerald, I. Shelton.
J. R. Astron. Soc. Canada, Vol. 76, 337 - 361 (1982) = Contrib. Univ. Waterloo Obs. No. 85.
The authors present a simple method by which photometric data may be obtained to any given reasonable accuracy with a telescope equipped with a small on-line computer.

Retrieval of astronomical information from Padova-Asiago Observatory plates archives.
See Abstr. 002.023.

Organizing a data base for solving planetodesic problems on the ES computer.
See Abstr. 002.099.

Mathematisches Hilfsbuch für Studierende und Freunde der Astronomie.
See Abstr. 003.124.

A data analysis facility for the Faint Object Camera.
See Abstr. 013.018.

A program for fabricating Schmidt corrector plates.
See Abstr. 031.013.

Geneva photometric boxes. IV. A refined method for direct access.
See Abstr. 031.506.

Quantization errors in the digital implementation of 'clean'.
See Abstr. 031.507.

Reduktion von Positionsbestimmungen durch Koordinatentransformation.
See Abstr. 031.533.

Processing astronomical images from space.
See Abstr. 031.593.

La classification spectrale automatique en tant que problème de reconnaissance de forme.
See Abstr. 031.653.

Some applications of microcomputers in observatory automation.
See Abstr. 032.001.

On the small divisors. I. The convergence of the modified Zeipel's series and the resonant quasi-periodic solutions.
See Abstr. 042.018.

022 Physical Papers Related to Astronomy and Astrophysics

022.001 **Experiments and calculations on the extinction of starlight by iron or magnetite grains.**
J. Hecht, J. Nuth.
Astrophys. J., Vol. 258, 878 - 885 (1982).
Laboratory measurements and theoretical calculations were made for the extinction by Fe and Fe_3O_4 particles. The measurements, taken from 195 to 650 nm, were consistent with the calculations. This implies that Fe is not responsible for the 220 nm feature in the interstellar extinction curve. Calculations also show that Fe_3O_4 particles have extinction features in the near-UV similar to those recently reported in spectra which lack the strong 220 nm bump. The feature in the extinction curve of HD 44179 which begins near 160 nm is identified as due to H_2O ice.

022.002 **Laboratory band strengths of methane and their application to the atmospheres of Jupiter, Saturn, Uranus, Neptune, and Titan. II. The red region 6000–7600 Å.**
B. L. Lutz, T. Owen, R. D. Cess.
Astrophys. J., Vol. 258, 886 - 898 (1982).
The authors report room-temperature laboratory studies of the red bands of methane at 40 different pressure–path-length combinations. Detailed equivalent width measurements were used to determine curves of growth and integrated band strengths for the bands at 6190, 6825, 7050, and 7250 Å. New spectrophotometric observations of the outer planets and Titan are presented, and equivalent width measurements of the planetary methane bands are used to show that the laboratory curves of growth are temperature independent to first order and that scattering is dominant in the radiative transfer in all five objects.

022.003 **The far-infrared spectrum of the OH radical.**
J. M. Brown, J. E. Schubert, K. M. Evenson, H. E. Radford.
Astrophys. J., Vol. 258, 899 - 903 (1982).
The frequencies, wavelengths, and line strengths for transitions between the lowest spin-rotation levels of the OH molecule have been calculated from the recently reported laser magnetic resonance spectra.

022.004 **High-energy collision-induced dissociation of H_2 by H.**
N. C. Blais, D. G. Truhlar.
Astrophys. J., Lett., Vol. 258, L79 - L81 (1982).
Cross sections are presented for collision-induced dissociation of H_2 by H with initial collision conditions representative of those in molecular clouds behind high-speed shock fronts. The collision-induced dissociation cross section appears to have a classical dynamic threshold about 1 eV above the energetic threshold.

022.005 **Oscillator strengths for lines of neutral titanium.**
R. I. Kostyk.
Astron. Zh., Tom 59, 693 - 698 (1982). In Russian. English translation in Soviet Astron., Vol. 25, No. 4.
Oscillator strengths of 133 Ti I lines have been determined to an accuracy of 0.10 dex.

022.006 **Laboratory simulation of interstellar chemical evolution.**
K. Kuriki, Y. Ishikawa.
Astrophys. J., Vol. 259, 411 - 422 (1982).
The chemical evolution in an interstellar cloud was simulated by a plasma flow experiment in the laboratory. The essential features of the experimental results are: the ion-molecule reaction was playing an important role, spatial patterns of molecular density were constructed in the plasma flow, and the transition from chemical equilibrium to non-equilibrium state was found in the rapidly cooling flow. The similarity law is established to connect the laboratory and interstellar phenomena. The relation to the circumstellar flow is also discussed.

022.007 **The theoretical KLL+KLM Auger spectrum of the free Na atom.**
D. Petrini.
Astron. Astrophys., Vol. 111, 279 - 283 (1982).
The KLL and KLM_1 Auger rates for the free atomic Na are obtained from the resonant behaviour of the scattering matrix defined in the theory of the $Na^{2+}+e^-$ collision process, and calculated by the close coupling and distorted wave methods. Positions, widths and relative intensities are obtained simultaneously and compare well with experimental results. Theoretical results for the Auger decay of the $1s2s^2\,2p^5\,3s$ terms are also given.

022.008 **Multi-ring basin formation: possible clues from impact cratering calculations.**
P. H. Schultz, D. Orphal, B. Miller, W. F. Borden, S. A. Larson.
Multi-ring basins, (see 012.001), p. 181 - 195 (1981).

022.009 **Z-model analysis of impact cratering: an overview.**
M. G. Austin, J. M. Thomsen, S. F. Ruhl, D. L. Orphal, W. F. Borden, S. A. Larson, P. H. Schultz.
Multi-ring basins, (see 012.001), p. 197 - 205 (1981).
The authors have taken the Maxwell Z-model developed for near-surface explosion cratering and applied it to calculations of laboratory-scale impact into plasticene clay and to calculations of planetary-scale impact into gabbroic anorthosite, a lunar material.

022.010 **The excavation stage of basin formation: a qualitative model.**
S. K. Croft.
Multi-ring basins, (see 012.001), p. 207 - 225 (1981).
Crater formation may be divided into high-pressure, excavation, and modification stages. The first two have been investigated by both numerical simulation and observation. A model of the transient crater, which marks the end of the excavation stage and represents the initial boundary conditions of the modification stage, is constructed by combining the properties of cratering flow fields during the excavation stage and the implications of both gravity and strength scaling applied to individual craters at basin dimensions.

022.011 **Simple algorithms to compute absorption line profiles.**
A. Sapar, L. Sapar.
Publ. Tartu Astrofiz. Obs., Tom 48, 41 - 62 (1981).
Spectral line profiles and synthetic stellar spectra have been calculated by a number of astrophysicists. In the present paper the authors try to propose a method, which is relatively concise and relatively free of over-simplifications. The algorithms proposed are oriented on atomic spectral lines.

022.012 **The influence of collective effects on the deceleration of an ion in an accretion column.**
J. G. Kirk.
Galactic X-ray sources, (see 003.001), p. 253 (1982).
Abstract.

022.013 **On the variation of Stark line widths within a supermultiplet.**
M. S. Dimitrijević.
Astron. Astrophys., Vol. 112, 251 - 256 (1982).

In order to find out if similarities among Stark broadening parameters within a supermultiplet are apparent to such a degree that accurate interpolation of new data and critical evaluation of experimental results are possible, the exceptions to this rule in the available theoretical data have been analysed and reasons for such situations have been discussed. For the O II $3s-3p$ quartets, semi-classical calculations for Stark broadening parameters have been performed taking into account the emitter structure properly.

022.014 **New lifetime measurements for Nb I and Rh I and the solar photospheric abundances of Nb and Rh.**
M. Kwiatkowski, P. Zimmermann, E. Biémont, N. Grevesse.
Astron. Astrophys., Vol. 112, 337 - 340 (1982).
New lifetime measurements for 11 levels in Nb I and 13 levels in Rh I have been performed by time-resolved observation of the reemitted fluorescence after pulsed laser excitation. New f-values have been deduced and used in order to redetermine the photospheric abundances of niobium and rhodium. The result for Rh, $A_{Rh} = 1.12 \pm 0.12$, is in excellent agreement with the meteoritic value, but the result for Nb, $A_{Nb} = 2.10 \pm 0.10$, is about four times larger than the meteoritic abundance.

022.015 **The excitation of the triplet lines of O^{2+} in nebulae.**
A. Dalgarno, A. Sternberg.
Mon. Not. R. Astron. Soc., Vol. 200, 77P - 80P (1982).
The excitation of the triplet lines of O^{2+} between 280 and 380 nm in nebulae is attributed to a combination of the Bowen fluorescence mechanism pumped by the He^+ resonance line and charge transfer into excited states of O^{2+} in collisions of O^{3+} with neutral hydrogen atoms. The lines at 377.4 and 375.7 nm are due almost entirely to charge transfer population of the $(2p3p)^3D_1$ level. The close agreement obtained between the calculated and observed line intensities confirms the accuracy of the charge transfer rate coefficients.

022.016 **Probability of observation of cosmic OH sources with linear and circular polarization.**
D. Ishankuliev, A. Khanberdiev.
Izv. AN TSSR. Ser. fiz.-tekh., khim. i geol. nauk, 1981, No. 6, p. 100 - 102. In Russian. – Abstr. in Ref. zh., 51. Astron., 7.51.163 (1982).

022.017 **Oscillator strengths and collision strengths for some ions of oxygen and sulphur.**
Y. K. Ho, R. J. W. Henry.
Advances in ultraviolet astronomy, (see 012.006), p. 315 - 317 (1982).
Collision strengths for electron impact excitation of the O II, O III, S II and S III sor some transitions in the ultraviolet are calculated in a close-coupling approximation for an energy rate up to 10^6 K.

022.018 **Prediction of molecular infrared band and line intensities.**
W. B. Person.
Vibrational-rotational spectroscopy for planetary atmospheres, (see 012.007), p. 89 - 123 (1982).
The general problem of predicting infrared intensities is examined, with an introduction intended to define the problem to be considered and also the concepts involved. After some rather non-specific discussion of the problem of predicting line intensities, attention is focussed specifically on the problem of predicting vibrational strengths (or total band intensities). It is suggested that the most important application of this work may be to the prediction of intensities for free radicals, ions, and reactive intermediates so that the infrared spectrum can be used in the quantitative analysis for these interesting species.

022.019 **New theoretical and experimental methods for pressure-broadened linewidths and their interpretation.**
J. J. Gelfand.
Vibrational-rotational spectroscopy for planetary atmospheres, (see 012.007),p. 125 - 147 (1982).
The author attempts to review recent progress in the theory of collisional line broadening, particularly the impact of recent advances in collision dynamics calculations. He discusses some new approaches to the interpretation of experimentally measured linewidths and their impact on planetary atmosphere research. He mentions some new experimental techniques which may have some advantage in providing pressure broadening data at very low temperatures.

022.020 **Laboratory molecular spectroscopy.**
J. Margolis.
Vibrational-rotational spectroscopy for planetary atmospheres, (see 012.007), p. 149 - 169 (1982).

022.021 **Microwave rotational spectroscopy.**
H. Pickett.
Vibrational-rotational spectroscopy for planetary atmospheres, (see 012.007), p. 171 - 187 (1982).
The author concentrates on two topics. He urges to take advantage of the region upward of 1 millimeter. He describes applications of this region: how microwave, millimeter and submillimeter spectra (the so-called rotational spectra) can fruitfully interact with infrared spectral measurements.

022.022 **Infrared absorption by OH-containing compounds.**
J. J. Hillman.
Vibrational-rotational spectroscopy for planetary atmospheres, (see 012.007), p. 295 - 309 (1982).
A review of the current state of spectroscopy of the relevant OH compounds as it applies to the inner planets is presented. Critical problems are addressed and a projection of the future spectroscopic needs are made.

022.023 **Infrared absorption of sulfur-bearing compounds.**
A. Maki.
Vibrational-rotational spectroscopy for planetary atmospheres, (see 012.007), p. 311 - 330 (1982).
The sulfur-bearing molecules of most interest in studies of planetary atmospheres are the simple, stable hydrides and oxides, H_2S, H_2SO_4, SO_2, and SO_3, and the carbon-containing species OCS and CS_2. This paper deals with those molecules.

022.024 **H_2 and HD spectroscopic status.**
J. S. Margolis.
Vibrational-rotational spectroscopy for planetary atmospheres, (see 012.007), p. 431 - 437 (1982).

022.025 **Collision-induced sharp features in the infrared spectrum of HD.**
R. M. Herman.
Vibrational-rotational spectroscopy for planetary atmospheres, (see 012.007), p. 439 - 447 (1982).
The author discusses the collision induced spectra of HD and H_2, and their application to the study of planetary (and planetary satellite) atmospheres.

022.026 **Far infrared spectra of H_2 and mixtures of H_2-CH_4 and H_2-He.**
G. Birnbaum.
Vibrational-rotational spectroscopy for planetary atmospheres, (see 012.007), p. 449 - 471 (1982).
The purpose of this report is to present some recent laboratory measurements of the far infrared absorption of H_2 and of mixtures of H_2 with CH_4 and He, and to describe fits to the pure H_2 spectra with a semi-empirical line shape. Such

results are needed in analyzing the thermal emission from the atmospheres of the outer planets.

022.027 **A review of acetylene, ethylene and ethane molecular spectroscopy for planetary applications.**
W. C. Maguire,
Vibrational-rotational spectroscopy for planetary atmospheres, (see 012.007), p. 473 - 501 (1982).
The author discusses selected examples of spectroscopic work in acetylene, ethylene and ethane, which are of particular interest since the Voyager IRIS team is currently analyzing observations of Jupiter. Acetylene and ethane but not ethylene are seen in the Jovian spectrum. Two fundamental bands of the observed gases are used to determine the spatial distribution of these hydrocarbons on Jupiter and to illuminate the photochemistry of these species.

022.028 **Current studies of CH_4 from 2.5 to 7 μm.**
L. Brown.
Vibrational-rotational spectroscopy for planetary atmospheres, (see 012.007), p. 503 - 527 (1982).

022.029 **Line intensities from band and diode laser measurements: ν_4 of CH_4.**
K. Fox.
Vibrational-rotational spectroscopy for planetary atmospheres, (see 012.007), p. 529 - 548 (1982).
The author reviews the experimental data and theoretical work on the 7.7 micron band of methane. This band is particularly relevant in studies of the atmospheres of Jupiter, Saturn and the other outer planets.

022.030 **Random walks with the 6420- and 6825-Å features of methane: lost in the wilderness.**
B. L. Lutz.
Vibrational-rotational spectroscopy for planetary atmospheres, (see 012.007), p. 549 - 557 (1982).

022.031 **Current band model studies of CH_4 at wavelengths less than 2.5 microns.**
U. Fink.
Vibrational-rotational spectroscopy for planetary atmospheres, (see 012.007), p. 559 - 583 (1982).
Band model coefficients for CH_4 below 1 μm are useful in the analysis of the atmospheres of the major planets. Band model coefficients are useful for the region from 1–3 μm because of the necessity of matching the deep CH_4 absorptions displayed by the major planets.

022.032 **Long-pathlength absorption spectroscopy of CH_4.**
P. Silvaggio.
Vibrational-rotational spectroscopy for planetary atmospheres, (see 012.007), p. 585 - 598 (1982).

022.033 **Current studies of CH_3D: ye olde line drive.**
B. L. Lutz.
Vibrational-rotational spectroscopy for planetary atmospheres, (see 012.007), p. 599 - 610 (1982).

022.034 **Spectra of ammonia.**
K. Narahari Rao.
Vibrational-rotational spectroscopy for planetary atmospheres, (see 012.007), p. 611 - 634 (1982).

022.035 **Current studies of PH_3.**
A. Goldman.
Vibrational-rotational spectroscopy for planetary atmospheres, (see 012.007), p. 635 - 653 (1982).
Recent quantitative spectroscopic studies of the PH_3 fundamentals in the 10 and 4.5 μm regions resulted in line parameters applicable to spectral radiative models of planetary atmospheres.

022.036 **Organic chemistry in Titan's atmosphere.**
T. Scattergood.
Vibrational-rotational spectroscopy for planetary atmospheres, (see 012.007), p. 679 - 697 (1982).
The author discusses: 1. The chemistry of methane, which is the major known constituent of Titan's atmosphere. 2. The composition of dust that comprises the haze layer, and 3. Isotope fractionation in planetary atmospheres.

022.037 **Forbidden emission lines of Fe VII.**
H. Nussbaumer, P. J. Storey.
Astron. Astrophys., Vol. 113, 21 - 26 (1982).
Excitation of [Fe VII] lines, in conditions prevalent in gaseous nebulae is discussed. For this purpose collision strengths and transition probabilities have been calculated, for all transitions within the ground configuration $3d^2$.

022.038 **On the "just overlapping line approximation" for molecular absorption.**
E.-M. Zeidler-K. T., D. Koester.
Astron. Astrophys., Vol. 113, 173 - 175 (1982).
The authors present a formulation of the just overlapping line approximation for molecular absorption that makes use only of the basic assumption of the model (treating the rotational lines as a continuous absorption) without further simplifications. P, R, and Q branches are handled separately; the case of headless bands is also included. In the final section the authors give a simple criterion for the validity of this approximation.

022.039 **NH^+ – a candidate for comets and interstellar space.**
A. A. de Almeida, P. D. Singh.
Astron. Astrophys., Vol. 113, 199 - 204 (1982).
Photoionization, charge-transfer, radiative association, and ion-molecule reactions in gas phase may lead to a significant concentration of the NH^+ ion in interstellar molecular clouds and in comets.

022.040 **The vacuum ultraviolet spectra of HCN, C_2N_2 and CH_3CN.**
J. A. Nuth, S. Glicker.
J. Quant. Spectrosc. Radiat. Transfer, Vol. 28, 223 - 231 (1982).
Quantitative vacuum ultraviolet absorption spectra for HCN, C_2N_2, and CH_3CN have been obtained over the wavelength range 60 nm $\lesssim \lambda \lesssim$ 160 nm. Where comparison is possible, the measurements of the absorption coefficients for HCN and C_2N_2 are consistent with previous studies. Because of the superior resolution of this work (0.05 nm), vibrational assignments in the valence and Rydberg transitions of HCN have been extended while higher members of the Rydberg series in CH_3CN have been identified.

022.041 **Dielectronic recombination rate coefficients for Fe^{23+} ion.**
D. J. McLaughlin, Y. Hahn.
J. Quant. Spectrosc. Radiat. Transfer, Vol. 28, 343 - 353 (1982).

022.042 **Absolute oscillator strengths for lines of neutral cobalt between 2276 Å and 9357 Å and a redetermination of the solar cobalt abundance.**
B. L. Cardon, P. L. Smith, J. M. Scalo, L. Testerman, W. Whaling.
Astrophys. J., Vol. 260, 395 - 412 (1982).
Absolute oscillator strengths of neutral cobalt have been determined from hook measurements for 159 transitions and emission intensity measurements for 314 transitions between 2276 Å and 9357 Å. Absolute oscillator strengths for 362 transitions and 36 lifetimes were determined. Equivalent widths were obtained for nineteen solar cobalt lines with the McMath solar telescope at Kitt Peak National Observatory.

These widths were used to redetermine the solar cobalt abundance.

022.043 **Microwave spectra of molecules of astrophysical interest. XX. Methane.**
I. Ozier, M. C. L. Gerry, A. G. Robiette.
J. Phys. Chem. Ref. Data, Vol. 10, 1085 - 1095 (1981).
Abstr. in Phys. Abstr., Vol. 85, Abstr. 75280 (1982).

022.044 **The infrared spectral properties of frozen volatiles.**
U. Fink, G. T. Sill.
Comets, (see 012.009), p. 164 - 202 (1982).

Since Whipple's dirty snowball model of comet nuclei, it has been generally accepted that volatile ices help to explain cometary phenomena. The authors have been pursuing the infrared spectral properties of many substances that are potential candidates for frozen volatiles in the solar system. They present here a review of laboratory spectra in the range 1 to 20 μm of H_2O, CO_2, SO_2, CH_4, NH_3, H_2S, CO,NH_4HS and $NH_3 \cdot H_2O$. Both reflection spectra of thick frosts and transmission spectra of thin films are shown, and their main characteristics are described.

022.045 **Laboratory studies of photochemical and spectroscopic phenomena related to comets.**
W. M. Jackson.
Comets, (see 012.009), p. 480 - 495 (1982).

The relationship between laboratory observations and cometary phenomena is reviewed. It is pointed out how supersonic expansion of the gas evaporating from the nucleus can affect the photochemical lifetime and the energy partitioning among the fragments. The use of the CS radical as a remote sensing probe of conditions in the inner coma is also discussed.

022.046 **Fragmentation of ^{56}Fe nuclei on CH_2, carbon, and hydrogen targets: individual charge changing and total cross sections.**
W. R. Webber, D. A. Brautigam.
Astrophys. J., Vol. 260, 894 - 908 (1982).

The fragmentation of ^{56}Fe at 710, 950, and 1050 MeV per nucleon has been studied in CH_2 and C targets with the objective of obtaining cross sections in H appropriate to the cosmic-ray propagation problem. The fragments of various charge and mass exiting the target were measured with a cosmic-ray telescope capable of determining the elemental and isotopic compositions of the fragments. The elemental cross sections, $\sigma(Z)$, were measured for $Z = 10$–25, and the total charge-changing and mass-changing cross sections were determined.

022.047 **An experimental determination of the cross section for photodesorption.**
E. B. Bourdon, R. H. Prince, W. W. Duley.
Astrophys. J., Vol. 260, 909 - 913 (1982).

Measurement of the cross section for photodesorption of an absorbing molecule on a transparent substrate shows that this process is highly inefficient. Photodesorption yields under these conditions are some four orders of magnitude lower than previously assumed. Photodesorption of molecules from interstellar dust grains will be a highly inefficient process except in certain specialized situations.

022.048 **Acetylene bands observed in carbon stars: a laboratory study and an illustrative example of its application to IRC + 10216.**
C. P. Rinsland, A. Baldacci, K. N. Rao.
Astrophys. J., Suppl. Ser., Vol. 49, 487 - 513 (1982).

Acetylene (C_2H_2) has been identified as a constituent of the atmospheres of cool carbon stars based on Fourier transform stellar spectra in the near-infrared. A systematic laboratory study of the absorption spectrum of acetylene occurring in this spectral region has been made at a resolution of ~ 0.03 cm^{-1}, both at room temperature and at an elevated temperature of 160° C. Interpretation of the acetylene lines in the 3 μm region has led to the identification of 27 vibration-rotation bands which have been assigned to transitions between 41 different vibrational states of $^{12}C_2H_2$ and $^{12}C^{13}CH_2$. Absolute strengths for 101 selected unblended spectral lines were measured, and by comparing with the stellar data for IRC + 10216, an estimate has been made for the acetylene column density in the circumstellar shell of this carbon star.

022.049 **Scaled collision strengths for $\Delta n \geqslant 1$ transitions involving $n \leqslant 5$ in hydrogenic ions.**
R. E. H. Clark, D. H. Sampson, S. J. Goett.
Astrophys. J., Suppl. Ser., Vol. 49, 545 - 554 (1982).

Scaled hydrogenic ion collision strengths $Z^2 \Omega_H$ and $Z^2 \Omega_H^e$ are calculated using time-dependent perturbation theory, with the perturbation consisting of the entire electrostatic interaction between the electrons. These results, together with results given in earlier work, provide values for all transitions of the kind nl - $n'l'$, with $n < n' \leqslant 5$. From the data one can obtain approximate cross sections and collision rates for transitions in He-like and Li-like ions, as well as hydrogenic ions.

022.050 **Studies of irradiated ice mixtures – laboratory synthesized residue.**
M. H. Moore, B. Donn.
Bull. American Astron. Soc., Vol. 14, 605 (1982). – Abstract.

022.051 **The effect of Fe XXIII on Fe XXIV satellite line ratios.**
J. R. Lemen, K. J. H. Phillips, G. A. Doschek, R. D. Cowan.
Bull. American Astron. Soc., Vol. 14, 608 (1982). – Abstract.

022.052 **Simulation of Martian chaotic terrain and outflow channels.**
J. P. Manker, A. P. Johnson.
Icarus, Vol. 51, 121 - 132 (1982).

Outflow channeling and associated chaotic terrain were created under temperature and pressure conditions suggested for a diluvian period on Mars 3.5 to 0.5 b. y. ago. Pressures under which both features were formed ranged from 130 to 34 mbar at a constant ambient temperature of 266° K. Analogs of the collapse structures and channels evolved in a high-altitude/low-temperature chamber are found on the Martian surface. Similarities exist not only in their overall morphology but in the finer details of the megastructures themselves.

022.053 **The mass distribution of an accreting system of particles – an analytic approach.**
S. Simons.
Mon. Not. R. Astron. Soc., Vol. 201, 127 - 135 (1982).

It is shown that subject to a certain approximation, an analytic technique can be used to obtain the mass distribution function for the steady state of a system of coagulating particles together with a source. The results give the limiting value for large masses of the mass distribution index S as $(3/2) + \alpha$ (independent of the form of the source term) for a collision kernel which is a homogeneous function of the interacting masses of degree 2α. Reasons are given why observational results greater than this may be obtained.

022.054 **Ground state mass of ^{81}Kr and the solar neutrino problem.**
R. T. Kouzes, M. M. Lowry, C. L. Bennett.
Phys. Rev. C, Vol. 25, 1076 - 1078 (1982). – Abstr. in Phys. Abstr., Vol. 85, Abstr. 59227 (1982).

022.055 **The ^{3}He(α, γ)^{7}Be reaction and the solar neutrino problem.**
H. Krawinkel, H. W. Becker, L. Buchmann, J. Gorres, K. U. Kettner, W. E. Keiser, R. Santo, P. Schmalbrock, H. P. Trautvetter, A. Vlieks, C. Rolfs, J. W. Hammer,

R. E. Azuma, W. S. Rodney.
Z. Phys. A, Vol. 304, 307 - 332 (1982). – Abstr. in Phys. Abstr., Vol. 85, Abstr. 59380 (1982).

022.056 **Molecular spectroscopy using diode and combined diode-CO_2 lasers.**
G. Winnewisser, K. Yamada, R. Schieder, R. Guccione-Gush, H. P. Gush.
Laser spectroscopy V, (see 012.016), p. 268 - 271 (1981). Abstr. in Phys. Abstr., Vol. 85, Abstr. 63271 (1982).

022.057 **Electron impact excitation of positive ions of astrophysical interest. I. Theoretical method.**
C. Mendoza.
Rev. Mexicana Astron. Astrofis., Vol. 6, (see 012.017), 285 - 299 (1981).
The current theoretical methods developed for the calculation of electron impact excitation cross sections of positive ions are described, and illustrated with recent results of astrophysical interest.

022.058 **Time-dependent quadratic constants of motion, symmetries, and orbit equations for classical particle dynamical systems with time-dependent Kepler potentials.**
G. H. Katzin, J. Levine.
J. Math. Phys., Vol. 23, 552 - 563 (1982). – Abstr. in Phys. Abstr., Vol. 85, Abstr. 63816 (1982).

022.059 **Energy-loss mechanisms and the annihilation of confined monopoles.**
F. A. Bais, P. Langacker.
Nucl. Phys. B, Vol. B197, 520 - 532 (1982). – Abstr. in Phys. Abstr., Vol. 85, Abstr. 64247 (1982).

022.060 **On the theory of pair production by accelerated charges.**
N. D. Hari Dass.
Nucl. Phys. B, Vol. B197, 399 - 412 (1982). – Abstr. in Phys. Abstr., Vol. 85, Abstr. 64264 (1982).

022.061 **QCD and neutron stars.**
P. D. Morley.
Statistical mechanics of quarks and hadrons. Proceedings of an international symposium, Bielefeld, F. R. Germany, 24 - 31 August 1980. H. Satz (Editor). North-Holland Publishing Company, Amsterdam, Netherlands. p. 321 - 328 (1981). Abstr. in Phys. Abstr., Vol. 85, Abstr. 64281 (1982).

022.062 **Effective charges of ions and the stopping power of dense media.**
W. Brandt.
Nucl. Instrum. Methods Phys. Res., Vol. 194, No. 1 - 3, p. 13 - 19 (1982). – Abstr. in Phys. Abstr., Vol. 85, Abstr. 64817 (1982).

022.063 **The interplay of molecular spectroscopy and astronomy.**
G. Herzberg.
J. Spectrosc. Soc. Japan, Vol. 30, 407 - 421 (1981). – Abstr. in Phys. Abstr., Vol. 85, Abstr. 73834 (1982).

022.064 **Laboratory simulation of volatilisation from melts induced by micrometeoroid impacts.**
M. D. Nussinov *(Nusinov)*, V. V. Vysochkin, V. I. Feldman *(Fel'dman)*.
Moon Planets, Vol. 26, 279 - 304 (1982).
The laboratory simulation experiments on volatilization from the melts formed due to micrometeoroid impacts on the lunar surface were carried out. The simulation was performed using modulated laser pulses incident on rocks and minerals in vacuum; in so doing microcraters on the targets, glass particles, condensates were formed and gases solved in the bulk of the rock and mineral grains were released.

022.065 **On the relationships between impact crater diameters and projectile kinetic energy.**
J. W. Bond.
Moon Planets, Vol. 26, 317 - 321 (1982).
By reforming a series of expressions derived by Öpik (1969), the author has shown explicitly the dependence of impact crater diameter on projectile kinetic energy. Comparisons between this reformed version of Öpik's theory and the models of Gault (1973) and Oberbeck and Aoyagi (1972) have demonstrated good agreement between experiment and theory over seventeen orders of magnitude of projectile kinetic energy.

022.066 **On spinor equations of motion and their "possible integrals".**
P. E. Kustaanheimo, J. Tutein.
Astron. Nachr., Band 303, 221 - 225 (1982).
The motion of one point mass of the classical mechanics is treated by means of the relativistic spinor regularization (Kustaanheimo 1975). Most general spinor equations of motion and the differential equations for the "possible integrals" of these equations of motion are deduced. If the force is a superposition of a conservative central force and of another force perpendicular to radius vector and velocity then the theory yields scalar and spinor integrals that enable a parametric representation of the orbit by quadratures, as soon as one solution of a Riccati differential equation has been found.

022.067 **Massive magnetic monopoles: indirect and direct limits on their number density and flux.**
M. J. Longo.
Phys. Rev. D, Vol. 25, 2399 - 2405 (1982). – Abstr. in Phys. Abstr., Vol. 85, Abstr. 74634 (1982).

022.068 **Nucleus-nucleus collisions and interpretation of cosmic-ray cascades above 100 TeV.**
T. K. Gaisser, T. Stanev, P. Freier, C. J. Waddington.
Phys. Rev. D, Vol. 25, 2341 - 2350 (1982). – Abstr. in Phys. Abstr., Vol. 85, Abstr. 74718 (1982).

022.069 **TOODY-WONDY calculations of penetration events *(meteorite craters)*.**
D. L. Hicks, F. R. Norwood, T. G. Trucano.
AIP Conf. Proc., No. 78, (see 012.019), p. 544 - 547 (1981). Abstr. in Phys. Abstr., Vol. 85, Abstr. 78654 (1982).

022.070 **Shock wave apparatus for studying minerals at high pressure and impact phenomena on planetary surfaces.**
T. J. Ahrens, M. B. Boslough, W. G. Ginn, M. S. Vassiliou, M. A. Lange, J. P. Watt, K.-i. Kondo, R. F. Svendsen, S. M. Rigden, E. M. Stolper.
AIP Conf. Proc., No. 78, (see 012.019), p. 631 - 633 (1981). Abstr. in Phys. Abstr., Vol. 85, Abstr. 78852 (1982).

022.071 **Massive photinos and ultraviolet astronomy.**
D. W. Sciama.
Phys. Lett. B, Vol. 112B, 211 - 212 (1982). – Abstr. in Phys. Abstr., Vol. 85, Abstr. 79796 (1982).

022.072 **Naturality problems *(QCD)*.**
F. Wilczek.
AIP Conf. Proc., No. 81, (see 012.026), p. 590 - 599 (1981). Abstr. in Phys. Abstr., Vol. 85, Abstr. 79817 (1982).

022.073 **Cross sections and thermonuclear reaction rates for $^{42}Ca(p, \gamma)^{43}Sc$, $^{44}Ca(p, \gamma)^{45}Sc$, $^{44}Ca(p, n)^{44}Sc$ and $^{45}Sc(p, n)^{45}Ti$.**

L. W. Mitchell, M. R. Anderson, S. R. Kennett, D. G. Sargood.
Nucl. Phys. A, Vol. A380, 318 - 334 (1982). – Abstr. in Phys. Abstr., Vol. 85, Abstr. 80019 (1982).

022.074 **Vacuum stability and symmetry breaking in non-Minkowskian space-times.**
D. J. Toms.
Phys. Rev. D, Vol. 25, 2536 - 2547 (1982). – Abstr. in Phys. Abstr., Vol. 85, Abstr. 84803 (1982).

022.075 **Strings in SO(10).**
T. W. B. Kibble, G. Lazarides, Q. Shafi.
Phys. Lett. B, Vol. 113B, 237 - 239 (1982). – Abstr. in Phys. Abstr., Vol. 85, Abstr. 84822 (1982).

022.076 **Supranuclear densities in physics and astrophysics.**
D. F. Falla.
Phys. Bull., Vol. 33, 53 - 54 (1982). – Abstr. in Phys. Abstr., Vol. 85, Abstr. 84986 (1982).

022.077 **An extension of RPA theory to strongly-interacting systems.**
J. M. C. Chen, J. W. Clark, D. G. Sandler.
Z. Phys. A, Vol. 305, 223 - 229 (1982). – Abstr. in Phys. Abstr., Vol. 85, Abstr. 84987 (1982).

022.078 **Absolute cross section for $^7Li(d, p)^8Li$ and solar neutrino capture rates.**
B. W. Filippone, A. J. Elwyn, W. Ray, Jr., D. D. Koetke.
Phys. Rev. C, Vol. 25, 2174 - 2180 (1982). – Abstr. in Phys. Abstr., Vol. 85, Abstr. 85063 (1982).

022.079 **Calculation of the absorption coefficient of the 15-μm ν_2 band of CO_2 using the theory of overlapping spectral lines.**
C. Braun.
J. Mol. Spectrosc., Vol. 93, 1 - 15 (1982). – Abstr. in Phys. Abstr., Vol. 85, Abstr. 85657 (1982).

022.080 **Observations of structuring in the downstream region of a large spherical model in a laboratory simulated solar wind plasma.**
D. S. Intriligator, G. R. Steele.
J. Geophys. Res., Vol. 87, 6053 - 6059 (1982).

Laboratory experiments have been performed that show the effects of inserting a spherical conducting model, large in comparison with the Debye length, into a free streaming high-energy (1 kV) unmagnetized hydrogen plasma. These experiments are the first laboratory measurements at energies and compositions directly relevant to solar wind and astrophysical plasma phenomena.

022.081 **Molecular analysis of tholins produced under simulated Titan conditions.**
B. N. Khare, C. Sagan, S. Shrader, E. T. Arakawa.
Bull. American Astron. Soc., Vol. 14, 714 (1982). – Abstract.

022.082 **Three dimensional Thomas-Fermi-Dirac calculations for dense binary mixtures.**
J. J. MacFarlane, W. B. Hubbard.
Bull. American Astron. Soc., Vol. 14, 721 (1982). – Abstract.

022.083 **Anomalous reflectance decrease with small particles.**
C. M. Pieters.
Bull. American Astron. Soc., Vol. 14, 726 - 727 (1982). Abstract.

022.084 **Effects of mineral grain size and physical particle size on spectral reflectance of basalts.**
R. B. Singer, P. L. Blake.
Bull. American Astron. Soc., Vol. 14, 727 (1982). – Abstract.

022.085 **Effects of temperature on mafic mineral crystal-field absorptions, 0.4 to 2.6 μm.**
R. B. Singer, T. L. Roush.
Bull. American Astron. Soc., Vol. 14, 727 (1982). – Abstract.

022.086 **A non-linear mineral mixing model: anisotropic scattering.**
P. E. Johnson, M. O. Smith, S. Taylor-George, J. B. Adams.
Bull. American Astron. Soc., Vol. 14, 727 (1982). – Abstract.

022.087 **Laboratory tests of proposed scattering laws for particulate surfaces.**
J. Gradie, J. Veverka.
Bull. American Astron. Soc., Vol. 14, 729 (1982). – Abstract.

022.088 **Laboratory spectral measurements at low temperatures for better abundance determinations of ethane, ethylene, and propane in outer solar system atmospheres.**
L. P. Giver, P. Varanasi, F. P. J. Valero.
Bull. American Astron. Soc., Vol. 14, 731 (1982). – Abstract.

022.089 **Line strengths of methane for the 2.3 μm region.**
L. R. Brown.
Bull. American Astron. Soc., Vol. 14, 731 (1982). – Abstract.

022.090 **High-resolution spectroscopy in the 6825 Å region.**
C. de Bergh, J. Chauville, J. Brault, B. L. Lutz, T. Owen, R. Danehy, J. P. Maillard.
Bull. American Astron. Soc., Vol. 14, 733 (1982). – Abstract.

022.091 **Improvements in the line parameters of molecules involved in the radiative transfer of the giant planets: line parameters of CH_4 and NH_3.**
N. Husson, A. Chedin.
Bull. American Astron. Soc., Vol. 14, 733 (1982). – Abstract.

022.092 **Experimental line strengths for the ν_4 and ν_2 bands of methane in the 7 μm region.**
L. R. Brown.
Bull. American Astron. Soc., Vol. 14, 733 - 734 (1982). Abstract.

022.093 **Thermal infrared absorption measurements for methane and acetylene at planetary atmospheric temperatures.**
P. Varanasi, L. P. Giver, F. P. J. Valero.
Bull. American Astron. Soc., Vol. 14, 734 (1982). – Abstract.

022.094 **Implications of the new H_2 pressure broadening data for models of the outer planet atmospheres.**
W. D. Cochran, W. H. Smith.
Bull. American Astron. Soc., Vol. 14, 734 (1982). – Abstract.

022.095 **Laboratory measurements of the 4–0 and 5–0 bands of HD.**
J. T. Trauger, M. E. Mickelson.
Bull. American Astron. Soc., Vol. 14, 734 (1982). – Abstract.

022.096 **Io: laboratory studies of possible surface materials.**
J. Gradie, J. Veverka, P. Thomas, S. Ostro.
Bull. American Astron. Soc., Vol. 14, 735 (1982). – Abstract.

022.097 **Sulfur dioxide on Io: physical state and distribution.**
R. R. Howell, D. P. Cruikshank, F. P. Fanale, R. H. Brown.
Bull. American Astron. Soc., Vol. 14, 735 (1982). – Abstract.

022.098 **Jupiter's satellite Io: a surface with sulfur trioxide?**
W. D. Smythe, R. M. Nelson, D. B. Nash.
Bull. American Astron. Soc., Vol. 14, 735 - 736 (1982). Abstract.

022.099 **Low temperature studies of proton irradiated SO_2: implications for Io.**
M. H. Moore.
Bull. American Astron. Soc., Vol. 14, 739 (1982). – Abstract.

022.100 **Direct computational of atmospheric normal modes and instabilities.**
H. Houben, R. E. Young.
Bull. American Astron. Soc., Vol. 14, 742 (1982). – Abstract.

022.101 **Spectral properties of mixtures of ice and particulates: implications for remote sensing.**
R. N. Clark, P. G. Lucey.
Bull. American Astron. Soc., Vol. 14, 744 (1982). – Abstract.

022.102 **Ice poor regolith development/destruction on small icy rocky bodies.**
F. P. Fanale, J. R. Salvail.
Bull. American Astron. Soc., Vol. 14, 745 - 746 (1982). Abstract.

022.103 **Observational evidence of errors in the currently accepted transition probabilities for [S II] λλ 6716, 6731.**
J. S. Morgan.
Bull. American Astron. Soc., Vol. 14, 764 (1982). – Abstract.

022.104 **Excitation and dielectronic recombination of Fe XVII.**
J. C. Raymond, B. W. Smith, J. Mann.
Bull. American Astron. Soc., Vol. 14, 779 (1982). – Abstract.

022.105 **Laboratory detection of the C_3N and C_4H radicals.**
C. A. Gottlieb, E. W. Gottlieb, P. Thaddeus, H. Kawamura.
Bull. American Astron. Soc., Vol. 14, 779 (1982). – Abstract.

022.106 **Oscillator strengths for Y I and Y II and the solar abundance of yttrium.**
P. Hannaford, R. M. Lowe, N. Grevesse, E. Biémont, W. Whaling.
Astrophys. J., Vol. 261, 736 - 746 (1982).
Oscillator strengths have been determined from measurements of radiative lifetimes and branching ratios for 154 lines of Y I and 66 lines of Y II. These data are used, together with equivalent widths measured on the Jungfraujoch solar atlas, to perform a new determination of the solar abundance of yttrium.

022.107 **Some lifetimes and transition probabilities in Cu(I).**
A. Kono, S. Hattori.
J. Quant. Spectrosc. Radiat. Transfer, Vol. 28, 383 - 387 (1982).

022.108 **Stark widths and transition probabilities of some multiplets of singly ionized carbon.**
A. Goly, S. Weniger.
J. Quant. Spectrosc. Radiat. Transfer, Vol. 28, 389 - 391 (1982).
Photoelectric measurements of lines emitted by a helium-carbon d.c. arc plasma of $n_e = 1.9 \times 10^{16}$ cm^{-3} at T = 12,800 K have been performed. Stark widths of 6 C(II) lines and transition probabilities for components of 3 C(II) multiplets were determined and compared with other experimental and theoretical data.

022.109 **Measurements of intensities and self- and foreign-gas-broadened half-widths of spectral lines in the CO fundamental band.**
T. Nakazawa, M. Tanaka.
J. Quant. Spectrosc. Radiat. Transfer, Vol. 28, 409 - 416 (1982).
Accurate measurements of the intensities and the self-, nitrogen-, oxygen-, and carbon-dioxide-broadened half-widths of spectral lines in the carbon monoxide fundamental band have been made at 300 K. Data reduction was made by using an iterative procedure to correct for overlapping with neighboring lines and the non-linearity of the curve-of-growth. Results are derived by using the Lorentz and the Benedict line profiles.

022.110 **One-electron spectrum of Lu^{2+}: relativistic energies, transition probabilities and dipole polarizability.**
J. Migdałek.
J. Quant. Spectrosc. Radiat. Transfer, Vol. 28, 417 - 424 (1982).

022.111 **An analytical description of populations of highly excited levels.**
I. L. Beigman (*Bejgman*), I. M. Gaisinsky.
J. Quant. Spectrosc. Radiat. Transfer, Vol. 28, 441 - 454 (1982).
Analytical methods are developed to describe populations of highly excited levels. The Kramers approximation was used for radiative transitions and the Born approximation for transitions induced by electron collisions. An exact solution is derived for the limiting case of small densities. In the limiting case of large densities, an analytical solution is obtained with constants determined from a comparison with computer solutions of the balance equations. A general interpolation formula is presented, which includes the limiting cases and describes results for intermediate densities with 20–30%.

022.112 **Millimeter and submillimeter laboratory spectroscopy: recent results of astrophysical interest.**
F. C. De Lucia, E. Herbst.
Submillimetre wave astronomy, (see 012.031), p. 233 - 251 (1982).
The authors review their recent laboratory spectroscopic results on the transient neutral species CCH, NH_2, and NaH, the molecular ions HCO^+, N_2H^+, and CO^+, and the light asymmetric internal rotor HOOH. They show that the shorter millimeter and submillimeter spectral region has many advantages even in the laboratory for this type of work. The requirements for theoretical models capable of the accurate spectral prediction required for many astrophysical applications are also discussed.

022.113 **Measurement of relative oscillator strengths for Fe I: transitions from levels b^3F_{2-4} (2.61 eV–2.56 eV). Use of a multipass optical system.**
D. E. Blackwell, A. D. Petford, G. J. Simmons.
Mon. Not. R. Astron. Soc., Vol. 201, 595 - 602 (1982).
The design of a three-pass optical system for use with the Oxford spectroscopic furnace is described. Relative oscillator strengths of 16 lines of Fe I (2.56 eV $< \chi <$ 2.61 eV, $-0.9 > \log gf > -2.7$) measured by its use are presented.

022.114 **Precision measurement of relative oscillator strengths for Ti II – I. Transitions from levels $a^4F_{3/2-9/2}$ (0.00–0.05 eV), $b^4F_{5/2-9/2}$ (0.12–0.15 eV).**
D. E. Blackwell, S. L. R. Menon, A. D. Petford.
Mon. Not. R. Astron. Soc., Vol. 201, 603 - 610 (1982).
Relative oscillator strengths measured with an accuracy of 0.5 per cent using the Oxford furnace technique are given for 18 lines of Ti II having excitation energies of up to 0.15 eV. The relative values have been put on an absolute scale by use of lifetime measurements.

022.115 **Precision measurement of relative oscillator strengths for Ti I – II. Transitions from levels a^5F_{1-4} (0.81–0.84 eV), a^1D_2 (0.90 eV) and a^3P_{0-2} (1.05–1.07 eV).**
D. E. Blackwell, S. L. R. Menon, A. D. Petford, M. J. Shallis.
Mon. Not. R. Astron. Soc., Vol. 201, 611 - 617 (1982).
Relative oscillator strengths are given for 22 lines of Ti I in the range 0.81 eV $< \chi <$ 1.07 eV, 428 nm $< \lambda <$ 843 nm and $+0.50 > \log gf > -1.47$ measured in absorption using the Oxford furnace technique with an accuracy of 0.5 per cent. The relative values have been placed on an absolute scale to an

estimated accuracy of 12 per cent by comparison with 0 eV Ti I lines, relative measures of which have previously been placed on an absolute scale using lifetime and atomic beam measurements. Comparisons are made with the data of other experimenters and with theoretical values.

022.116 **The Heisenberg uncertainty relation and the resolving power in optics.**
M. Di Jorio.
Lett. Nuovo Cimento, Vol. 34, Ser. 2, 147 - 151 (1982).
Abstr. in Phys. Abstr., Vol. 85, Abstr. 89784 (1982).

022.117 **The $^{15}N(p, \alpha_0)^{12}C$ reaction at stellar energies.**
A. Redder, H. W. Becker, H. Lorenz-Wirzba, C. Rolfs, P. Schmalbrock, H. P. Trautvetter.
Z. Phys. A, Vol. 305, 325 - 333 (1982). – Abstr. in Phys. Abstr., Vol. 85, Abstr. 90119 (1982).

022.118 **Reaction products from a microwave discharge in N_2 and H_2S. The microwave spectrum of NS.**
F. J. Lovas, R. D. Suenram.
J. Mol. Spectrosc., Vol. 93, 416 - 422 (1982). – Abstr. in Phys. Abstr., Vol. 85, Abstr. 90524 (1982).

022.119 **Far infrared fine structure lines in the interstellar medium.**
D. M. Watson, J. W. V. Storey.
Int. J. Infrared Millim. Waves, Vol. 1, 609 - 629 (1980).
Abstr. in Phys. Abstr., Vol. 85, Abstr. 94621 (1982).

022.120 **Energy values and sum rules for hydrogenic atoms in magnetic fields of arbitrary strength using numerical wave functions: comparison with variational results.**
G. Wunner, W. Rösner, H. Ruder, H. Herold.
Astrophys. J., Vol. 262, 407 - 411 (1982).
Energy-weighted sum rules are evaluated using highly accurate numerical wave functions. The wave functions are expanded in terms of spherical harmonics for $B \lesssim 10^{10}$ gauss, and in terms of Landau states for $B \gtrsim 10^{10}$ gauss. The energy values obtained reproduce the best ones known in the literature to date, or even provide a higher digit accuracy.

022.121 **The computer and the Universe.**
J. A. Wheeler.
Int. J. Theor. Phys., Vol. 21, 557 - 572 (1982). – Abstr. in Phys. Abstr., Vol. 85, Abstr. 95060 (1982).

022.122 **Preparation of lunar sample and meteorite mock-ups: or how to make rocks.**
D. R. Edwards, J. H. Reeves, R. L. Brodzinski.
Nucl. Instrum. Methods Phys. Res., Vol. 196, 507 - 509 (1982).
Abstr. in Phys. Abstr., Vol. 85, Abstr. 100747 (1982).

022.123 **Evidence of production of H_2CN^+-N_2. Application to the atmosphere of Titan.**
C. V. Speller, M. Fitaire, A.-M. Pointu.
J. Phys. Lett., Vol. 43, L499 - 503 (1982). – Abstr. in Phys. Abstr., Vol. 85, Abstr. 100826 (1982).

022.124 **Specific entropy and the sign of the energy.**
J. D. Bekenstein.
Phys. Rev. D, Vol. 26, 950 - 953 (1982). – Abstr. in Phys. Abstr., Vol. 85, Abstr. 112320 (1982).

022.125 **Stark broadening of some O III lines.**
M. S. Dimitrijević.
Publ. Astron. Obs. Sarajevo, Vol. 1, No. 1, (see 012.043), p. 215 - 218 (1981).

022.126 **Formaldehyde formation in a H_2O/CO_2 ice mixture under irradiation by fast ions.**
V. Pirronello, W. L. Brown, L. J. Lanzerotti, K. J. Marcantonio, E. H. Simmons.
Astrophys. J., Vol. 262, 636 - 640 (1982).
The authors report the measurement of formaldehyde (H_2CO) production in an ice mixture of equal parts of water and carbon dioxide under irradiation by fast (1.5 MeV) helium ions at temperatures of ~9 K and above. They find the production rate of H_2CO to be about 3.7 per incident ion for the ice layers used in the experiment.

022.127 **Erosion of frozen sulfur dioxide by ion bombardment: applications to Io.**
C. L. Melcher, D. J. LePoire, B. H. Cooper, T. A. Tombrello.
Geophys. Res. Lett., Vol. 9, 1151 - 1154 (1982).
The erosion of frozen SO_2 due to bombardment by both light and heavy ions (He and F) was measured for bombarding energies of 0.08 to 1.3 MeV/amu. The number of SO_2 molecules ejected from the target per incident ion (i. e., the sputtering yield) was 50 for 1.5 MeV He ions and 7300 for 6 MeV F ions. Ion bombardment followed by heating produced an oxygen/sulfur residue which was much more stable against subsequent ion bombardment than the initial frozen SO_2. The erosion rate of SO_2 frost on Jupiter's moon Io depends strongly on the elemental composition and energy spectra of the magnetospheric ion flux which bombards the surface. The combined effects of ion bombardment and heating which produced residues on the target substrates may also occur on Io from magnetospheric ion bombardment and heating by volcanism. The experimental results compare favorably with a new model of the sputtering process which considers the energy loss of the incident ion to electronic excitation in the target.

022.128 **Transitions in highly ionized silicon.**
R. Glass.
Sol. Phys., Vol. 80, 321 - 331 (1982).
Configuration interaction wavefunctions are used to calculate energy levels, wavelengths, oscillator strengths, line strengths, transition probabilities, lifetimes and branching ratios for absorption transitions between the $2lnl'$ $[n = 2, 3; l' \geqslant l]$ states for Si XI. Comparison with recent beam-foil experimental data for the $2s2p\ ^1P^\circ$, $2p^2\ ^1D$ and $2p^2\ ^1S$ states is generally good, though in some cases cascade effects may be larger than had been expected.

022.129 **Electron densities from the O IV] λ1401 multiplet.**
H. Nussbaumer, P. J. Storey.
Astron. Astrophys., Vol. 115, 205 - 206 (1982).
Transition probabilities for the O IV] intercombination multiplet at 1401 Å are calculated. It is shown that this multiplet is useful for electron density determinations for $10^2 < N_e\ [\mathrm{cm}^{-3}] < 10^{12}$.

022.130 **Some H/D exchange reactions involved in the deuteration of interstellar molecules.**
D. Smith, N. G. Adams, E. Alge.
Astrophys. J., Vol. 263, 123 - 129 (1982).
Laboratory investigations of the reaction $D^+ + H_2 \rightleftharpoons H^+ + HD$, which is of key importance for interstellar deuterium chemistry, are reported. Reactions involving more complex molecules are also considered. From the data reaction rate coefficients appropriate for an interstellar environment are calculated. The synthesis of complex deuterated interstellar molecules is briefly discussed.

022.131 **On the systematics of line ratios along the helium isoelectronic sequence.**
A. K. Pradhan.
Astrophys. J., Vol. 263, 477 - 482 (1982).
Temperature and charge dependence of the line ratios involving the forbidden, intercombination, and the resonance lines is investigated for helium-like ions of C, O, Ne, Si, S, Ca,

and Fe. A few recent observations from laboratory and astrophysical plasmas are discussed.

022.132 **Periodic travelling waves in a non-integrable one-dimensional lattice.**
T. P. Valkering.
Celestial Mech., Vol. 28, (see 012.049), 119 - 131 (1982).
Existence of a one-parameter family of periodic solutions representing longitudinal travelling waves is established for a one-dimensional lattice of identical particles with nearest-neighbour interaction. The potential is not given in closed form but is specified by only a few global properties. The lattice is either infinite or consists of N particles on a circle with fixed circumference. Waves with low energy are sinusoidal and their properties are studied using bifurcation methods. Waves of high energy, however, are of solitary type, i. e. the excitation is strongly localized.

022.133 **Update of the impact ionization problem.**
K. Hornung.
The Giotto spacecraft gas and plasma environments, (see 012.050), p. 15 - 19 (1982).
Previous model calculations on the impact ionization of Fe particles have been updated to account for the new impact velocity of 68 km/sec as well as for possible variations of the shield material (aluminum and gold).

022.134 **Electron collision strengths for the far-infrared lines of O III.**
K. M. Aggarwal, K. L. Baluja, J. A. Tully.
Mon. Not. R. Astron. Soc., Vol. 201, 923 - 932 (1982).
New collision strengths for electron impact excitation of the $^3P_J \rightarrow {}^3P_{J'}$ fine structure transitions within the ground configuration of O^{++} are presented. The results exhibit a complex pattern of resonances and are given graphically for electron energies below 7.36 eV. The thermally averaged collision strengths have been evaluated and are tabulated as functions of the electron temperature in the range $0 \leq T_e \leq 2 \times 10^4$ K.

022.135 **Dielectronic satellite spectra for highly-charged helium-like ions – VII. Calcium spectra – theory and comparison with SMM observations.**
F. Bely-Dubau, J. Dubau, P. Faucher, A. H. Gabriel, M. Loulergue, L. Steenman-Clark, S. Volonté, E. Antonucci, C. G. Rapley.
Mon. Not. R. Astron. Soc., Vol. 201, 1155 - 1169 (1982).
The atomic theory, developed in earlier papers in this series for iron, has been applied to the spectra from calcium. This includes the production of satellite lines by dielectronic recombination and inner-shell excitation, as well as the production of helium-like spectra by excitation, radiative and dielectronic recombination and through cascades. Two methods are described for fitting observed spectra, to determine the plasma physical parameters. The results are applied to the observations from the soft X-ray Polychromator of the flare on 1980 May 21. Analysis of the time variations shows that the flare is close to steady-state ionization balance throughout.

022.136 **Vibration-rotation transition probabilities for the ground electronic $X^1 \Sigma^+$ state of HD.**
H. Abgrall, E. Roueff, Y. Viala.
Astron. Astrophys., Suppl. Ser., Vol. 50, 505 - 522 (1982).
Transition probabilities are calculated for all the possible spontaneous electric dipole and quadrupole rotation-vibration transitions of the ground electronic state of HD originating in vibrational levels up to $v = 17$ and rotational levels up to $J = 13$. Comparison with experiment is performed when possible.

022.137 **Franck-Condon factors for some abundant umbral molecules.**
G. C. Joshi, U. C. Joshi, L. M. Punetha, M. C. Pande.
Bull. Astron. Soc. India, Vol. 10, 159 - 164 (1982).
Some of the molecular transitions in UV in sunspots are pointed out and Franck-Condon factors for some of these have been calculated.

022.138 **Absorptions of the $\rho\sigma\tau$, 0.8 μm and a bands of the water vapor.**
M. Tanaka, T. Nakazawa, M. Fukabori.
J. Quant. Spectrosc. Radiat. Transfer, Vol. 28, 463 - 470 (1982).
Absorptions of the $\rho\sigma\tau$, 0.8 μm and a bands of water vapor were measured using a long pathlength, multiple-traversal absorption cell. The measured equivalent widths gradually exceed calculated values for the Lorentz line profile and the self-broadening coefficients of 5 with increasing absorber amounts. Possible causes of this excess absorption are discussed.

022.139 **Intensities, half-widths and shapes of spectral lines in the fundamental band of CO at low temperatures.**
T. Nakazawa, M. Tanaka.
J. Quant. Spectrosc. Radiat. Transfer, Vol. 28, 471 - 480 (1982).
Intensities and half-widths of spectral lines in the R-branch of the $C^{12}O^{16}$ fundamental band have been measured at 200, 150, 99, and 81 K. The broadeners used were CO, N_2, O_2, and CO_2. The validity of some representative theories relevant to these line parameters have been examined in the light of the present results. The authors also discuss the light of the present results. The authors also discuss the shape of CO lines at low temperatures.

022.140 **Relativistic calculation of spectra of 2-2 transitions in O- and F-like atomic ions.**
M. N. Driker, E. P. Ivanova, L. N. Ivanov, A. F. Shestakov.
J. Quant. Spectrosc. Radiat. Transfer, Vol. 28, 531 - 535 (1982).

022.141 **Fast ion bombardment of ices and its astrophysical implications.**
W. L. Brown, L. J. Lanzerotti, R. E. Johnson.
Science, Vol. 218, 525 - 531 (1982).
Ices such as water, carbon dioxide, and methane are now known to be pervasive constituents of the solar system and probably of the interstellar medium as well. Many of these ices and ice-covered surfaces are exposed to bombardment by the energetic ions of space. Laboratory experiments have been carried out to study the effects of such bombardment. Surprisingly efficient erosion of ice layers is associated with electronic excitation of the ices by the ions. These results are a challenge to an understanding of the physical processes involved and have implications for a number of astrophysical problems of current interest.

022.142 **The loss-cone driven electron-cyclotron maser.**
R. G. Hewitt, D. B. Melrose, K. G. Rönnmark.
Australian J. Phys., Vol. 35, 447 - 471 (1982).
Electron-cyclotron instabilities may be classified in two ways depending on whether the relativistic correction to the gyrofrequency is important (class S) or not (class N), and whether the instability mechanism is of a maser type (class M) or due to bunching (class B). Renewed interest in class SM has followed the Wu and Lee application of it to the interpretation of terrestrial kilometric radiation. The maser is assumed to be driven by a one-sided loss-cone distribution of electrons. This mechanism seems particularly favourable for the interpretation of certain planetary, solar and stellar radio emissions.

An application of the mechanism to the interpretation of the Jovian decametric radio emissions is outlined.

022.143 **Rotational broadening of some ultraviolet lines in spectra of B stars.**
M. Ruusalepp.
Tartu Astrofüüs. Obs. Teated, No. 66, p. 16 - 37 (1982).
The results of calculating rotationally broadened line profiles for resonance lines of Si III λ 1206 Å, Si IV λλ 1393, 1402 Å and C IV λλ 1548, 1550 Å are presented for B stars. Distortions of stellar surface from spherical symmetry, gravity darkening and limb darkening have been taken into account. The dependence of half-intensity widths on fractional angular velocities and on inclinations of the rotational axes is studied.

022.144 **Calculation of the emission-line spectrum of He I.**
M. Ilmas, T. Nugis.
Tartu Astrofüüs. Obs. Teated, No. 67, 60 pp. (1982).
The purpose of the present study is to calculate the theoretical He I emission-line spectra for a wide range of conditions which are expected to exist in the gaseous nebulae and in the mass outflow envelopes of hot stars.

022.145 **Improving the reliability of experimental atomic transition probabilities.**
W. Whaling.
Essays in nuclear astrophysics, (see 003.013), p. 65 - 76 (1982).
The author describes a procedure for combining absorption and emission measurements of the same transitions to determine relative transition probabilities with no assumption concerning level populations. The method is applicable to broad spectrum recording whereby many lines are recorded simultaneously. Measurements from different experiments and different laboratories are readily incorporated so that large sets of transitions can be put on the same relative scale.

022.146 **Coronal line intensities for ions with fine-structured ground states: Si X.**
H. P. Saha, E. Trefftz.
Astron. Astrophys., Vol. 116, 224 - 231 (1982).
With the data published by the authors (1982) populations of the 15 lowest levels of Si X have been determined. A simple formula for the balance between the two ground state levels allows a quick estimate of the relative importance of radiative and collisional excitation. It also gives the cascade contribution to the effective impact excitation rate. The line intensity ratios quoted deviate from those of Flower and Nussbaumer (1975) by about 5%. It is shown that compared to the ^2D lines the intercombination lines of boron-like ions are less suited for density diagnostics.

022.147 **Experimental Stark broadening data of Si II and Si III lines.**
H. J. Kusch, K. Schröder.
Astron. Astrophys., Vol. 116, 255 - 259 (1982).
Line profiles of 11 Si II-lines and 6 Si III-lines have been determined from pulsed-discharge measurements in a silicon-hydride (SiH_4) plasma. The line widths are dominantly due to electron impact; the electron density values were deduced from the profile of H_β and varied in the range $0.5-2 \times 10^{17}$ cm^{-3} at temperatures between 22,500 K and 25,000 K. The experimental line broadening parameters are compared with theoretical values obtained from the straight-line non-adiabatic approximation, from the semi-empirical approach of Griem, and with his hyperbolic trajectory calculations.

022.148 **A compilation of some oscillator strengths for selected Fe I transitions.**
L. Altaş, T. Ataç.
Kandilli Obs., Heliophys. Serv., Istanbul. Publ., Second Ser., No. 6, 9 pp. (1982).

022.149 **Algunos aspectos sobre la interpretacion de las observaciones radioastronómicas de la molécula OH.**
V. Bujarrabal.
Bol. Astron. Obs. Madrid, Vol. 10, No. 6, p. 29 - 40 (1981).

022.150 **Fourier spectroscopy of the $^{12}C^{13}C$ and $^{13}C_2$ Ballik-Ramsay system.**
C. Amiot, J. Verges.
Astrophys. J., Vol. 263, 993 - 998 (1982).
The first laboratory observation of the $^{12}C^{13}C$ and $^{13}C_2$ Ballik-Ramsay electronic systems is reported. The observed wavenumbers for the 0–0 band have been reduced to molecular parameters using an iterative, nonlinear, least squares method.

022.151 **Line positions and strengths in the H_2 quadrupole spectrum.**
S. L. Bragg, J. W. Brault, W. H. Smith.
Astrophys. J., Vol. 263, 999 - 1004 (1982).
The positions and strengths of a number of H_2 quadrupole transitions in the 1–0 through 4–0 bands have been determined. Pressure shifts have been measured for the 1–0 and 2–0 bands. Calibrated transition frequencies allow a recalculation of the ground-state rotational constants. The newly determined values are $B = 59.33451$ cm^{-1}, $D = 0.04561$ cm^{-1}, and $H = 4.56 \times 10^{-5}$ cm^{-1}.

022.152 **Temperature in relativistic thermodynamics.**
L. Kofman.
Tartu Astrofüüs. Obs. Publ., Köide 49, 14 - 27 (1982). In Russian.
The formulae of temperature transformation in equilibrium thermodynamics, kinetics and statistic mechanics are compared. It is shown that the formulae from Tolman's, Ott-Meller's, Chernikov's theories are in correspondence. The phenomenological definition of temperature is discussed.

022.153 **Release of volatiles from possible Martian analogs.**
R. K. Kotra, E. K. Gibson, M. A. Urbancic.
Icarus, Vol. 51, 593 - 605 (1982).
Viking data suggest the presence of volatile-rich materials in the Martian regolith. The thermal stabilities of mineral phases and their volatile release profiles were studied in detail in the laboratory. Thermal analysis, combined with mass spectrometry, was applied to the study of the behavior of carbonates, sulfates, hydrates, and clays.

022.154 **An interpretation of the near-ultraviolet absorption spectrum of SO_2: implications for Venus, Io, and laboratory measurements.**
M. J. S. Belton.
Icarus, Vol. 52, 149 - 165 (1982).
The character of the line spectrum of SO_2 between 2000 and 3200 Å is described, and a crude band model for its absorption characteristics longward of 2280 Å is formulated. The model predicts planetary spectra that are substantially different from those predicted using a continuum model (Beer's law) for SO_2 absorption. As a result, the SO_2 scale height and mixing ratio in the Venus stratosphere could be substantially different from the values that have been previously supposed. Also, it is found that the absence of the signature of SO_2 vapor absorption in Io's spectrum can be treated as a physically reasonable lower limit to the mean column abundance of SO_2 on the satellite as well as an upper limit.

022.155 **Magnetic quadrupole transitions in the beryllium isoelectronic sequence.**
R. Glass.
Astrophys. Space Sci., Vol. 87, 41 - 50 (1982).

Relativistic intermediate-coupled wavefunctions are used to evaluate magnetic quadrupole transitions between the $2s^2$, $2s2p$, and $2p^2$ states in the beryllium isoelectronic sequence. It is found that the contribution from the magnetic quadrupole transitions is negligible, even for moderately high Z and will not be an important process in any laboratory or astrophysical plasma.

022.156 **Refractive index of LiF from 105 to 200 nm.**
P. Laporte, J. L. Subtil.
J. Opt. Soc. America, Vol. 72, 1558 - 1559 (1982).
The refractive index of LiF has been measured from 200 nm down to its ultraviolet cutoff by a thin-prism method. The data have been fitted to a dispersion equation whose low-energy term is consistent with reflection results.

022.157 **Complex index of refraction of fused silica and sapphire from 0.3 to 50 μm.**
M. L. Lang, W. W. Wolfe.
J. Opt. Soc. America, Vol. 72, 1789 (1982). – Abstract.

022.158 **Infrared transmittance and reflectance of crystalline quartz and high- and low-water-content fused silica.**
J. B. Heaney, K. P. Stewart, G. Hass.
J. Opt. Soc. America, Vol. 72, 1789 - 1790 (1982).
Abstract.

022.159 **Methane line parameters for the 2.3-μm region.**
L. R. Brown, L. S. Rothman.
Appl. Opt., Vol. 21, 2425 - 2427 (1982).
A new compilation of methane line parameters for the 4136 - 4666-cm^{-1} region has been obtained using experimental positions and intensities combined with known quantum assignments from the literature. Positions and intensities measured directly from the observed spectrum by computer are reported with accuracies of ± 0.0005 cm^{-1} and 2%, respectively, for single unblended absorptions.

022.160 **Magnetic dipole infrared atmospheric oxygen bands.**
L. S. Rothman.
Appl. Opt., Vol. 21, 2428 - 2431 (1982).

022.161 **Possible applications of time domain reflectometry in planetary exploration missions.**
S. Heckendorn.
Advances in planetary geology, (see 003.014), p. 545 - 555 (1982). – Abstract.

022.162 **The effects of varying surface roughness on aerodynamic and aeolian processes.**
L. M. Reding.
Advances in planetary geology, (see 003.014), p. 559 - 564 (1982). – Abstract.

022.163 **Two methods for determining volumes for explosion and impact craters.**
A. N. Verdes.
Advances in planetary geology, (see 003.014), p. 568 - 573 (1982). – Abstract.

022.164 **The Schumann-Runge continuum of oxygen at wavelengths greater than 175 nm.**
H. P. F. Gies, S. T. Gibson, A. J. Blake, D. G. McCoy.
J. Geophys. Res., Vol. 87, 8307 - 8310 (1982).
A highly temperature dependent continuum corresponding to absorption by thermally excited molecules underlies the higher bands of the Schumann-Runge system of oxygen. This continuum has been experimentally determined at 18 wavelengths in the range 175 nm to 179 nm, and at temperatures of 90°K and 294°K. The continuum cross-section has been calculated at a range of temperatures for use in studies of the atmospheric absorption of solar radiation.

022.165 **O_2 absorption cross sections (187–225 nm) from stratospheric solar flux measurements.**
J. R. Herman, J. E. Mentall.
J. Geophys. Res., Vol. 87, 8967 - 8975 (1982).
The paper presents the calculation of the O_2 and O_3 absorption cross sections from an analysis of solar flux measurements obtained within the stratosphere between 20 and 40 km altitude. The ozone column content is determined from the ratios of the attenuated solar flux over the wavelength range from 280 to 300 nm. Using this result and the simultaneously measured pressure and temperature, the absorption cross sections are determined for O_2 from 187 to 230 nm and the relative absorption cross sections for O_3 from 210 to 310 nm.

022.166 **Synthetic spectra for auroral studies: the N_2 Vegard–Kaplan band system.**
V. Degen.
J. Geophys. Res., Vol. 87, 10541 - 10547 (1982).
Synthetic spectra of the N_2 Vegard–Kaplan system are given for general reference. The material contains the spectrum across the entire range of emission (1500–6500 Å) calculated for auroral conditions and three typical resolutions (3.5, 10, and 25 Å); also included are individual band profiles calculated for 300–5000°K rotational temperature and 0–5 Å spectral halfwidth.

022.167 **Spectral evidence for the mineralogy of high-albedo soils and dust on Mars.**
R. B. Singer.
J. Geophys. Res., Vol. 87, (see 012.071), 10159 - 10168 (1982).
Because of the lack of direct mineralogic data available for Mars, spectral remote sensing techniques and, in particular, earth-based reflectance spectroscopy remain the primary source of this information. Laboratory observations which further constrain the mineralogy and origin of the high-albedo soils and dust are presented.

022.168 **Spectral characteristics of the iron oxides with application to the Martian bright region mineralogy.**
D. M. Sherman, R. G. Burns, V. M. Burns.
J. Geophys. Res., Vol. 87, (see 012.071), 10169 - 10180 (1982).
In an effort to determine the nature and significance of the iron oxide mineralogy on Mars, reflectance spectra of eight polymorphs of FeOOH and Fe_2O_3 are presented together with additional data on these materials.

022.169 **Seasonal carbon dioxide exchange between the regolith and atmosphere of Mars: experimental and theoretical studies.**
F. P. Fanale, W. B. Banerdt, R. S. Saunders, L. A. Johansen, J. R. Salvail.
J. Geophys. Res., Vol. 87, (see 012.071), 10215 - 10225 (1982).
Measurements of the rate of penetration of a CO_2 'pressure wave' through basalt-clay soils have been conducted under conditions of P_{CO_2} (~6 mbar), ΔP_{CO_2} (~2 mbar), soil density (ρ ~1.3), and T(–40° to –70°C) which simulate the penetration of the cap-induced seasonal CO_2 pressure wave through the topmost regolith on Mars. Results suggest that existing theoretical models for diffusion of a gas through a porous, highly adsorbing medium may be reliably used to assess the importance of Martian seasonal regolith-atmosphere CO_2 exchange. Results also suggest that although Mars topsoils with a substantial component of fine-grained weathering products could harbor, within a few meters of the surface, a substantial fraction of the atmospheric CO_2 inventory, seasonal cap-atmosphere CO_2 exchange is still much more important than seasonal regolith-atmosphere exchange.

022.170 **Mid- and far-infrared extinction coefficients of hydrous silicate minerals.**
C. Koike, H. Hasegawa, T. Hattori.
Astrophys. Space Sci., Vol. 88, 89 - 98 (1982).

Extinction coefficients were measured for three kinds of hydrous silicate minerals, montmorillonite, chlorite and serpentine, from 7 to 140 μm. The infrared extinction coefficients of these minerals show (1) a few broad bands in the mid-infrared region and (2) a less steep wavelength-dependence in the far-infrared region, in contrast to those of high-temperature magnesium silicates. The changes of mid-infrared spectra by heating were measured. Change in chlorite spectrum is the most significant. Many fine features appear by heating and then they disappear. Extinction coefficients at very low temperatures were measured in the far-infrared region.

022.171 **Calculated line positions and intensities for the $\nu_1 + \nu_3$ and $\nu_1 + \nu_2 + \nu_3 - \nu_2$ bands of $^{14}N^{16}O_2$.**
A. Perrin, J.-M. Flaud, C. Camy-Peyret.
Infrared Phys., Vol. 22, 343 - 348 (1982).

022.172 **Ein Äther für Elektrizität, Magnetismus und Gravitation - Faradays gravo-elektrische Induktion und Riemanns Gravitationsäther als Programm der großen Unitarisierung der Physik.**
H.-J. Treder.
Published by Physikalische Gesellschaft der Deutschen Demokratischen Republik, 11 pp. (1982).

022.173 **Toward an atomic model of the planetary system.**
A. Buţă.
Rev. Roumaine Phys., Tome 27, 321 - 329 (1982).

It is shown that the orbital motion of the principal planets in the solar system could possibly exhibit quantum aspects, similar to those existing for stationary orbits in the Bohr model of the hydrogen atom. A simple model is proposed to describe this situation.

022.174 **Covariant chronogeometry and extreme distances. III. Macro-micro relations.**
I. E. Segal.
Int. J. Theor. Phys., Vol. 21, (see 012.075), 851 - 869 (1982). Abstr. in Phys. Abstr., Vol. 85, Abstr. 113180 (1982).

022.175 **Roles of a 45-plet of Higgs particles in SU(5) grand unified theory.**
H. Hayashi, A. Murayama, M. J. Hayashi.
Phys. Rev. D, Vol. 26, 1185 - 1187 (1982). – Abstr. in Phys. Abstr., Vol. 85, Abstr. 113191 (1982).

022.176 **Mass generation in finite temperature QED.**
K. B. Joseph.
Phys. Lett. B, Vol. 115B, 252 - 256 (1982). – Abstr. in Phys. Abstr., Vol. 85, Abstr. 113198 (1982).

022.177 **Neutron-capture resonances for ^{82}Se.**
J. C. Browne, B. L. Berman.
Phys. Rev. C, Vol. 26, 969 - 976 (1982). – Abstr. in Phys. Abstr., Vol. 85, Abstr. 113445 (1982).

022.178 **Nuclear reactions in Rb, Sr, Y, and Zr targets.**
S. Regnier, B. Lavielle, M. Siminoff, G. N. Simonoff.
Phys. Rev. C, Vol. 26, 931 - 943 (1982). – Abstr. in Phys. Abstr., Vol. 85, Abstr. 113473 (1982).

022.179 **Unusually long bonds in CCO^+ and COC^+.**
N. N. Haese, R. C. Woods.
Chem . Phys. Lett., Vol. 91, 190 - 192 (1982). – Abstr. in Phys. Abstr., Vol. 85, Abstr. 113862 (1982).

022.180 **Metal-silicate extraction technique for the analysis of geological and meteoritic samples.**
W. Rammensee, H. Palme.
J. Radioanal. Chem., Vol. 71, 401 - 418 (1982). – Abstr. in Phys. Abstr., Vol. 85, Abstr. 117492 (1982).

022.181 **Rotational-state dependence of the collisional pumping rate for astrophysical masers.**
I. Shimamura.
Prog. Theor. Phys., Vol. 68, 178 - 190 (1982). – Abstr. in Phys. Abstr., Vol. 85, Abstr. 117583 (1982).

022.182 **Transmission electron microscopy on meteoritic troilite.**
J. Topel-Schadt, W. F. Müller.
Phys. Chem. Miner., Vol. 8, 175 - 179 (1982). – Abstr. in Phys. Abstr., Vol. 85, Abstr. 117641 (1982).

022.183 **Comment on the cyclotron emission rates in superstrong magnetic fields.**
H. Herold, H. Ruder, G. Wunner.
Phys. Lett. A, Vol. 91A, 272 - 274 (1982). – Abstr. in Phys. Abstr., Vol. 86, Abstr. 578 (1983).

022.184 **The ^{4}He(^{12}C, γ)^{16}O reaction at stellar energies.**
K. U. Kettner, H. W. Becker, L. Buchmann, J. Gorres, H. Krawinkel, C. Rolfs, P. Schmalbrock, H. P. Trautvetter, A. Vlieks.
Z. Phys. A, Vol. 308, 73 - 84 (1982). – Abstr. in Phys. Abstr., Vol. 86, Abstr. 818 (1983).

022.185 **Photodissociation processes in the HCl molecule.**
E. F. van Dishoeck, M. C. van Hemert, A. Dalgarno.
J. Chem. Phys., Vol. 77, 3693 - 3702 (1982). – Abstr. in Phys. Abstr., Vol. 86, Abstr. 1440 (1983).

022.186 **Spectral measurements of high temperature $^{13}C^{16}O_2$ and $^{13}C^{16}O^{18}O$.**
M. P. Esplin, R. J. Huppi, G. A. Vanasse.
Proc. SPIE Int. Soc. Opt. Eng., Vol. 289, (see 012.073), 226 - 230 (1981). – Abstr. in Phys. Abstr., Vol. 86, Abstr. 6629 (1983).

022.187 **The application of INAA to microsamples of meteoritic metal.**
P. A. Baedecker.
Trans. American Nucl. Soc., Vol. 41, 186 (1982). – Abstr. in Phys. Abstr., Vol. 86, Abstr. 5119 (1983).

022.188 **High-precision instrumental neutron activation analysis of meteoritic, lunar, and terrestrial samples.**
J. N. Grossman, G. W. Kallemeyn, J. T. Wasson.
Trans. American Nucl. Soc., Vol. 41, 192 (1982). – Abstr. in Phys. Abstr., Vol. 86, Abstr. 9378 (1983).

022.189 **Left-right symmetric theories and vacuum domain walls and strings.**
A. E. Everett, A. Vilenkin.
Nucl. Phys. B, Part. Phys., Vol. B207, 43 - 53 (1982). – Abstr. in Phys. Abstr., Vol. 86, Abstr. 11084 (1983).

022.190 **Z_2 vortex strings in grand unified theories.**
D. Olive, N. Turok.
Phys. Lett. B, Vol. 117B, 193 - 196 (1982). – Abstr. in Phys. Abstr., Vol. 86, Abstr. 11091 (1983).

022.191 **On supercooling and Ginzburg fluctuations in the Coleman-Weinberg mode of the SU(5) grand unification model.**
M. D. Pollock, M. Calvani.
Phys. Lett. B, Vol. 117B, 392 - 394 (1982). – Abstr. in Phys. Abstr., Vol. 86, Abstr. 11096 (1983).

022.192 **Peccei-Quinn symmetries without domains.**
S. M. Barr, X. C. Gao, D. B. Reiss.
Phys. Rev. D, Vol. 26, 2176 - 2177 (1982). – Abstr. in Phys. Abstr., Vol. 86, Abstr. 11098 (1983).

022.193 **Tales of the GUT age.**
D. V. Nanopoulos.
Grand unified theories and related topics, (see 012.083), p. 5 - 63 (1981). – Abstr. in Phys. Abstr., Vol. 86, Abstr. 11102 (1983).

022.194 **Mobilities of CH_n^+ (n = 0 to 3) in He and of CH_3^+ in Ar.**
K. Peska, H. Stori, F. Egger, G. Sejkora, H. Ramler, M. Kriegel, W. Lindinger.
J. Chem. Phys., Vol. 77, 5253 - 5255 (1982). – Abstr. in Phys. Abstr., Vol. 86, Abstr. 12811 (1983).

022.195 **Microwave spectra of molecules of astrophysical interest. XXI. Ethanol (C_2H_5OH) and propionitrile (C_2H_5CN).**
F. J. Lovas.
J. Phys. Chem. Ref. Data, Vol. 11, 251 - 312 (1982). – Abstr. in Phys. Abstr., Vol. 86, Abstr. 16092 (1983).

022.196 **Review of the experimental status (past and future) of monopole searches.**
G. Giacomelli.
Monopoles in quantum field theory, (see 012.088), p. 377 - 405 (1982). – Abstr. in Phys. Abstr., Vol. 86, Abstr. 16560 (1983).

022.197 **Broken supersymmetries in high energy physics.**
S. Rajpoot, J. G. Taylor.
Phys. Lett. B, Vol. 118B, 85 - 90 (1982). – Abstr. in Phys. Abstr., Vol. 86, Abstr. 16577 (1983).

The JPL millimeter, sub-millimeter, and microwave line catalog.
See Abstr. 002.043.

The AFGL molecular line atlases.
See Abstr. 002.044.

Ionization energies of some molecules found in interstellar clouds calculated by a Green's function method.
See Abstr. 002.140.

Water vapour line parameters from microwave to medium infrared (an atlas of $H_2{}^{16}O$, $H_2{}^{17}O$ and $H_2{}^{18}O$ line positions and intensities between 0 and 4350 cm^{-1}).
See Abstr. 002.147.

Atomic energy levels and Grotrian diagrams. Vol. 3. Vanadium I – Chromium XXIV.
See Abstr. 003.025.

Relativität – Gruppen – Teilchen. Spezielle Relativitätstheorie als Grundlage der Feld- und Teilchenphysik.
See Abstr. 003.111.

Thermo field dynamics and condensed states.
See Abstr. 003.121.

Primordial organic chemistry.
See Abstr. 015.025.

Present status of the theory and analytic evaluation of non-resonant thermonuclear reaction rates.
See Abstr. 021.027.

The concept of integrability on Cantor sets for Hamiltonian systems.
See Abstr. 021.034.

The modification stage of basin formation: conditions of ring formation.
See Abstr. 091.002.

Laboratory studies of charged particle erosion of SO_2 ice and applications to the frosts of Io.
See Abstr. 099.027.

Active experiments in space, laboratory experiments and numerical simulation.
See Abstr. 106.037.

Identification of some diffuse interstellar features.
See Abstr. 131.322.

Determination of the intrinsic Q(3)/S(1) line intensity ratio of molecular hydrogen.
See Abstr. 135.004.

Possible effects of positronium negative ion on 0.511 MeV γ-ray line in astrophysical sources.
See Abstr. 142.519.

Particle physics aspects in cosmology.
See Abstr. 162.186.

Errata

022.901 **Erratum: 'Recombination coefficients for iron ions'** [Astrophys. J., Vol. 249, 399 - 401 (1981)].
D. T. Woods, J. M. Shull, C. L. Sarazin.
Astrophys. J., Suppl. Ser., Vol. 49, 351 (1982). – See Abstr. 30.022.093.

022.902 **Erratum: 'Ionization equilibrium of astrophysically abundant elements'** [Astrophys. J., Suppl. Ser., Vol. 48, 95 - 107 (1982)].
J. M. Shull, M. Van Steenberg.
Astrophys. J., Suppl. Ser., Vol. 49, 351 (1982). – See Abstr. 31.022.045.

022.903 **Erratum: 'New rotational and vibrational analyses of CN violet system'** [Astrophys. J., Suppl. Ser.,Vol. 41, 669 - 674 (1979)].
L. Schoonveld, S. Sundaram.
Astrophys. J., Suppl. Ser., Vol. 49, 351 (1982). – See Abstr. 27.022.022.

022.904 **Erratum: 'Collisional excitation of OH by H_2 in the interstellar medium'** [Mon. Not. R. Astron. Soc., Vol. 199, 457 - 463 (1982)].
D. P. Dewangan, D. R. Flower.
Mon. Not. R. Astron. Soc., Vol. 200, 767 (1982). – See Abstr. 31.022.101.

022.905 **Erratum: "Intermediate coupling collision strengths for $\Delta n = 0$ transitions produced by electron impact on highly charged ions. I. Theoretical development and application to $n = 2$ levels in helium-like ions"** [Astrophys. J., Suppl. Ser., Vol. 44, 169 - 191 (1980)].
D. H. Sampson, R. E. H. Clark.
Astrophys. J., Suppl. Ser., Vol. 49, 593 (1982). – See Abstr. 28.022.161.

022.906 **Erratum: "Intermediate coupling collision strengths for $\Delta n = 0$ transitions produced by electron impact on highly charged ions. II. Transitions between states of the $1s^2 2s^2$ and $1s^2 2p^2$ configurations and those of the $1s^2 2s2p$ configuration in beryllium-like ions"** [Astrophys. J., Suppl. Ser., Vol. 44, 193 - 213 (1980)].
D. H. Sampson, R. E. H. Clark, L. B. Golden.
Astrophys. J., Suppl. Ser., Vol. 49, 593 (1982). – See Abstr. 28.022.162.

022.907 **Erratum: "On resonant thermonuclear reaction rate integrals – closed-form evaluation and approximation considerations"** [Astron. Nachr., Band 300, 63 - 65 (1979)].
H. J. Haubold, R. W. John.
Astron. Nachr., Band 303, 217 (1982). – See Abstr. 25.022.042.

Astronomical Instruments and Techniques

031 Astronomical Optics, Methods of Observation and Reduction

Astronomical Optics

031.001 **The liquid-mirror telescope as a viable astronomical tool.**
E. F. Borra.
J. R. Astron. Soc. Canada, Vol. 76, 245 - 256 (1982).
The surface of a rotating liquid takes the shape of a paraboloid. This fact can be used to build very large transit telescopes having as primary mirror a reflecting film of mercury. The practical aspects of this type of telescope are discussed. It is shown that a great deal of research can be done with the instrument, notwithstanding the fact that it is strictly a transit telescope and can only observe the zenith.

031.002 **Progress on the 3.5 m "New Technology Telescope" (NTT).**
R. N. Wilson.
Messenger, No. 29, p. 24 - 27 (1982).

031.003 **Piezoelectric pitch drive for adjustment of a combined telescope mirror.**
V. I. Kryukov, P. V. Nikolaev, Yu. S. Privezentsev, V. A. Resovskij, A. K. Rodionov, V. V. Sychev.
Opt.-mekh. prom-st', 1982, No. 1, p. 59 - 60. In Russian. Abstr. in Ref. zh., 51. Astron., 8.51.876 (1982).

031.004 **Results of the project for the comparison of the performance of astronomical telescopes.**
J. Hilton, J. Bond.
J. British Astron. Assoc., Vol. 92, 264 - 266 (1982).

031.005 **A high-speed, flat-field photovisual optical system for amateur telescope makers.**
J. S. Hanssen.
J. British Astron. Assoc., Vol. 92, 267 - 269 (1982).

031.006 **Determination of technological allowance in optical systems.**
S. V. Pozdnov.
Red. zh. Izv. vuzov. i Geod. i aehrofotosemka. Moskva, 1982, 35 pp. In Russian. – Abstr. in Ref. zh., 51. Astron., 9.51.738 (1982).

031.007 **On the problem of profile control of large-size optical polishing details with non-spherical surfaces.**
Yu. K. Orlov, I. P. Araev.
Nauchn.-issled. i proizv. tsentr "Priroda". Moskva, 1981, 8 pp. In Russian. – Abstr. in Ref. zh., 51. Astron., 9.51.739 (1982).

031.008 **Map construction of normal mirror deviations by a modified shadow method.**
E. N. Bol'shakov, E. S. Zhivopistsev, A. M. Prokhorov, I. N. Sisakyan.
Fiz. inst. AN SSSR. Prepr., 1981, No. 241, 18 pp. In Russian. Abstr. in Ref. zh., 51. Astron., 9.51.740 (1982).

031.009 **Use of a laser in the collimation of the CIDA Schmidt camera.**
D. J. MacConnell, H.-J. Schenner.
Rev. Mexicana Astron. Astrofis., Vol. 6, (see 012.017), 373 - 374 (1981).
The authors describe the use of a low-power laser for the optical alignment of the classical Schmidt system. The method is simple and allows rapid decrease of the plateholder tilt to 30 arcsec ≅ 0.02 mm at the edge of the plate.

031.010 **Problems of modern optical coating for astronomy.**
D. Enard.
Opt. Acta, Vol. 29, (see 012.018), 345 - 353 (1982). – Abstr. in Phys. Abstr., Vol. 85, Abstr. 73838 (1982).

031.011 **A flat-field reflecting focal reducer.**
D.-Q. Su, L.-J. Wang.
Opt. Acta, Vol. 29, (see 012.018), 391 - 394 (1982). – Abstr. in Phys. Abstr., Vol. 85, Abstr. 73840 (1982).

031.012 **The Leoncito coudé conversion.**
W. A. Grundmann, E. H. Richardson.
Opt. Acta, Vol. 29, (see 012.018), 549 - 554 (1982). – Abstr. in Phys. Abstr., Vol. 85, Abstr. 73841 (1982).

031.013 **A program for fabricating Schmidt corrector plates.**
N. J. Babu.
Electro-Opt. Syst. Des., Vol. 13, No. 12, p. 36 - 37 (1981). Abstr. in Phys. Abstr., Vol. 85, Abstr. 75653 (1982).

031.014 **Temos 4. A French solution to the mosaic telescope problem.**
A. Baranne, G. Lemaitre.
Opt. Acta, Vol. 29, (see 012.018), 847 - 860 (1982). – Abstr. in Phys. Abstr., Vol. 85, Abstr. 94600 (1982).

031.015 **Image quality considerations in ESO telescope projects.**
R. N. Wilson.
Opt. Acta, Vol. 29, (see 012.018), 985 - 992 (1982). – Abstr. in Phys. Abstr., Vol. 85, Abstr. 94602 (1982).

031.016 **Effects of mirror surface ripple on image quality.**
W. B. Wetherell.
Proc. SPIE Int. Soc. Opt. Eng., Vol. 310, (see 012.041), 77 - 93 (1981). – Abstr. in Phys. Abstr., Vol. 85, Abstr. 108994 (1982).

031.017 **A configuration of the optical system for a 5 m telescope.**
D.-Q. Su, L.-Z. Shao, M. Liang.
Opt. Acta, Vol. 29, (see 012.018), 1237 - 1242 (1982). Abstr. in Phys. Abstr., Vol. 85, Abstr. 112330 (1982).

031.018 **Thin mirrors for large optical telescopes.**
J. Cheng, C. M. Humphries.
Vistas Astron., Vol. 26, 15 - 35 (1982).
The use of thin mirrors in large aperture, optical

telescopes offers substantial benefits without necessarily causing any loss in optical performance. This paper discusses the design requirements for supporting thin mirrors in a telescope, the consequences which result, and the limitations of using thin mirrors.

031.019 **Optica e instrumentos astronómicos.**
A. Di Baja.
Rev. Astron., Tomo 54, No. 222, p. 17 - 22 (1982).

031.020 **Astrograph objectives related to the Petzval lens with field-flattener.**
N. J. Rumsey.
South. Stars, Vol. 29, 151 - 159 (1982).

031.021 **Design and adjustment of polarization compensators for coudé optics of stellar telescopes.**
K. Nariai.
Ann. Tokyo Astron. Obs., Second Ser., Vol. 19, 55 - 85 (1982).

031.022 **Derivatives of Seidel aberration coefficients.**
Y. Yamashita.
Tokyo Astron. Obs. Rep., Vol. 19, 475 - 478 (1981). In Japanese.

031.023 **The liquid mirror telescope as a viable astronomical tool.**
E. F. Borra.
Prepr. Steward Obs., No. 406, 26 pp. (1982).

031.024 **Honeycomb mirrors of borosilicate glass; current results and plans for 7–8 m diameter.**
J. R. P. Angel, D. Arganbright, L. Harmonson, J. M. Hill, N. Woolf.
Instrumentation for astronomy with large optical telescopes, (see 012.059), p. 33 - 36 (1982).

031.025 **Improved coudé efficiency of existing telescopes.**
W. A. Grundmann, E. H. Richardson.
Instrumentation for astronomy with large optical telescopes, (see 012.059), p. 61 - 66 (1982).

The ordinary coudê train has a single, large, aluminum coated mirror at each reflection point. The transmission can be increased and the blur decreased by a train of small mirrors with high reflectance coatings. The high reflectance coatings have the added advantages that they can reduce polarization effects and flatten response curves.

031.026 **High resolution imagery and slitless spectroscopy at the prime focus of the CFH telescope.**
J. C. Fouéré, G. Lelièvre, J. P. Lemonier, G. J. Odgers, E. H. Richardson, D. S. Salmon.
Instrumentation for astronomy with large optical telescopes, (see 012.059), p. 143 - 152 (1982).

The basic characteristics of the Wide-Field Corrector are presented. This arrangement with only three optical elements locates the gratings at a maximum distance from the focal surface resulting in improved resolution over a one degree field. The C.F.H.T. grating-plus-lens, or "grens" has a brightness limit which is fainter than the Palomar Sky Survey.

031.027 **The catadioptric camera as used for astronomical spectroscopy with image tubes and similar detectors: a limited review.**
C. F. W. Harmer.
Instrumentation for astronomy with large optical telescopes, (see 012.059), p. 153 - 167 (1982).

This paper compares the two major types of catadioptric camera, viz. the Maksutov camera and the Schmidt camera, when used for astronomical spectroscopy in instruments dedicated to image tubes, diode arrays and similar types of detectors.

031.028 **A method to determine the focal distance and distortion coefficient for a camera with a flattened field.**
M.-x. Wu, X.-f. Yan.
Publ. Purple Mountain Obs., Vol. 1, 47 - 53 (1982). In Chinese.

031.029 **Instrumental polarization of a horizontal solar telescope.**
O. V. Nikonov, E. S. Nikonova.
Izv. Glav. Astron. Obs. Pulkovo, Astrometr. Astrofiz., No. 199, p. 103 - 111 (1982). In Russian.

It is shown that the type of polarization matrix of a flat mirror depends on the orientation of the coordinate systems associated with the incident and reflected ray of light. The matrix of the flat mirror was obtained in the right-handed coordinate system. A detailed analysis is made of deducing the polarization matrix of the horizontal solar telescope. Expressions for the angles entering the matrix are found. A numerical example is considered.

031.030 **Vector conversion with a coelostat and additional mirror of a solar telescope.**
Yu. A. Bubnov, N. N. Bubnova, A. V. Shumakher.
Izv. Glav. Astron. Obs. Pulkovo, Astrometr. Astrofiz., No. 199, p. 112 - 120 (1982). In Russian.

A conversion of three unit orthogonal vectors with a system of two flat mirrors is considered. A method is proposed for determining the position of each mirror if the unit vectors of the direction of the incident and reflected rays are known. The paper gives formulas enabling to determine the three vectors in the image space if the three vectors of the system are known in the object space for different modes of the work of the system. Conclusions on the character of conversions of three unit vectors are made on the basis of digital computing using the obtained formulas for inclined and altazimuth mountings of the additional mirror.

031.031 **A three-mirror anastigmat with a coincidence of the main and the third mirror.**
N. N. Mikhel'son.
Izv. Glav. Astron. Obs. Pulkovo, Astrometr. Astrofiz., No. 199, p. 121 - 124 (1982). In Russian.

031.032 **On the stability of epoxy mirrors.**
E. G. Khablo.
Izv. Glav. Astron. Obs. Pulkovo, Astrometr. Astrofiz., No. 199, p. 129 - 131 (1982). In Russian.

The paper reports on the deformation of an epoxy mirror of a diameter 203 cm and thickness 30 mm during one year and 12 years. A conclusion is made that epoxy mirrors should be used only as energy collectors.

031.033 **A three-mirror telescope-anastigmat with a flat field.**
N. N. Mikhel'son.
Izv. Glav. Astron. Obs. Pulkovo, Astrofiz. Astrometr., No. 200, p. 146 - 153 (1982). In Russian.

031.034 **On the diffraction at a segmented telescope mirror.**
A. V. Lenskij.
Astron. Zh., Tom 59, 1225 - 1228 (1982). In Russian. English translation in Soviet Astron., Vol. 26, No. 6.

It is shown that a segmented telescope mirror may be used as a reflection echelon.

031.035 **Image quality criteria in the presence of moderately large aberrations.**
D. Kessler, R. V. Shack.
J. Opt. Soc. America, Vol. 72, 1105 - 1106 (1982). Abstract.

031.036 **Strehl ratio for primary aberrations: some analytical results for circular and annular pupils.**
V. N. Mahajan.
J. Opt. Soc. America, Vol. 72, 1258 - 1266 (1982).
Imaging systems with circular and annular pupils aberrated by primary aberrations are considered. Both classical and balanced (Zernike) aberrations are discussed. Closed-form solutions are derived for the Strehl ratio, except in the case of coma, for which the integral form is used. Numerical results are obtained and compared with Maréchal's formula for small aberrations.

031.037 **Superachromatic Petzval objectives with glasses selected using Buchdahl's chromatic coordinate.**
R. I. Mercado, P. N. Robb.
J. Opt. Soc. America, Vol. 72, 1725 (1982). – Abstract.

031.038 **Extrapolated least-squares optimization for lens design.**
E. D. Huber.
J. Opt. Soc. America, Vol. 72, 1729 (1982). – Abstract.

031.039 **Design of a glancing-incidence X-ray telescope–microscope system.**
D. L. Shealy, R. B. Hoover, A. Kassim, S. Chao.
J. Opt. Soc. America, Vol. 72, 1730 (1982). – Abstract.

031.040 **Description of a multiconfigurational optical-design program.**
B. Tatian.
J. Opt. Soc. America, Vol. 72, 1730 (1982). – Abstract.

031.041 **Unobscured two-element telescope.**
D. R. Shafer.
J. Opt. Soc. America, Vol. 72, 1730 (1982). – Abstract.

031.042 **Alignment compensation of surface-figure errors in a Ritchey-Chrétien telescope.**
K.-L. Shu.
J. Opt. Soc. America, Vol. 72, 1739 (1982).

031.043 **Optical design studies for the University of Texas 7.6-m telescope.**
A. B. Meinel, M. P. Meinel, D.-Q. Su, Y.-N. Wang.
J. Opt. Soc. America, Vol. 72, 1739 (1982). – Abstract.

031.044 **Shapes of telescopes to come: through a spyglass darkly.**
N. J. Woolf.
J. Opt. Soc. America, Vol. 72, 1759 (1982). – Abstract.

031.045 **Characteristics of a telescope.**
R. H. Noble, F. Cobos, F. Diego, J. M. Sasian.
J. Opt. Soc. America, Vol. 72, 1766 (1982). – Abstract.

031.046 **Fabrication of borosilicate honeycomb mirror blanks.**
J. M. Hill, J. R. P. Angel.
J. Opt. Soc. America, Vol. 72, 1767 (1982). – Abstract.

031.047 **Thermal-expansion homogeneity of future telescope materials.**
S. F. Jacobs, D. M. Shough.
J. Opt. Soc. America, Vol. 72, 1789 (1982). – Abstract.

031.048 **Trends in optical design with desk-top computers.**
M. J. Kidger.
J. Opt. Soc. America, Vol. 72, 1795 (1982). – Abstract.

031.049 **Application of microcomputers in the development of optical systems.**
R. R. Chirra, T. H. Bleakney.
J. Opt. Soc. America, Vol. 72, 1795 (1982). – Abstract.

031.050 **Advances in desk-top computer technology for optical desgin.**
D. C. Sinclair.
J. Opt. Soc. America, Vol. 72, 1795 (1982). – Abstract.

031.051 **Thermal gradient effects in optical testing of large astronomical mirrors: comments.**
J. C. Fouere, L. J. Golden.
Appl. Opt., Vol. 21, 2301 - 2303 (1982).

031.052 **Holographic Twyman-Green interferometer.**
C. W. Chen, J. B. Breckinridge.
Appl. Opt., Vol. 21, 2563 - 2568 (1982).
A dichromated gelatin off-axis Fresnel zone plane (OFZP) was designed, fabricated, and used in a new type of interferometer for optical metrology. This single hologram optical element combines the functions of a beam splitter, beam diverger, and aberrated null lens. Data presented show the successful application for an interferometric test of an $f/6$, 200-mm diam parabolic mirror.

031.053 **Deformable primary mirror for a space telescope.**
J. G. R. Hansen, R. M. Richard, R. R. Shannon.
Appl. Opt., Vol. 21, 2620 - 2630 (1982).
A 4 m aperture deformable primary mirror has been designed with the mirror and its supports integrated into a single structure. The integrated active mirror's minimal weight makes it desirable for a space telescope as well as a terrestrial application. This paper describes the results of finite element analyses.

031.054 **Figure control for a fully segmented telescope mirror.**
T. S. Mast, J. E. Nelson.
Appl. Opt., Vol. 21, 2631 - 2641 (1982).
A system is described to control the figure of a large telescope primary mirror that is composed of many individual segments. The geometry considered, employing hexagonal mirrors, allows a simple and economical control system. The system is shown to be reliable and effective in continuously maintaining the figure to the precision required for optical astronomy.

031.055 **Thick meniscus field correctors.**
T. H. Jamieson.
Appl. Opt., Vol. 21, 2799 - 2803 (1982).

031.056 **All-spherical catadioptric telescope with small corrector lenses.**
R. D. Sigler.
Appl. Opt., Vol. 21, 2804 - 2808 (1982).
A well-corrected telescope with a large spherical primary mirror can be realized by using three or four small, widely spaced, corrector lenses. An interesting solution uses the lenses in a relay configuration which results in an erect image, accessible real exit pupil, and an internal field stop. Lateral color is corrected by adding a field lens at an intermediate focus. Polychromatic axial correction is excellent despite use of only one type of glass. The useful field of view is typically limited to 1 deg. Examples are presented.

031.057 **Holographic phase conjugation and Strehl ratio.**
D. Armitage.
Appl. Opt., Vol. 21, 3197 - 3199 (1982).
Deformable-mirror adaptive optic systems strive to correct the phase at every point of the aperture. Calculations exist which relate the system performance to the spatial-frequency response of the mirror. Adaptive systems such as holographic devices, which form the complete complex conjugate of the

impressed wave front, require a different treatment. The Strehl ratio for such a system is calculated in terms of the device optical-transfer function. The systems are compared in the limit of photon shot noise.

031.058 **Theoretical fringe profiles with crossed Babinet compensators in testing concave aspheric surfaces.**
A. K. Saxena, J. P. Lancelot.
Appl. Opt., Vol. 21, 4030 - 4032 (1982).
This paper presents the theory for the use of crossed Babinet compensators in testing concave aspheric surfaces. Theoretical fringe profiles for a sphere and for an aspheric surface with primary aberration are shown. Advantages of this method are discussed.

031.059 **Optical aberration functions: chromatic aberrations and derivatives with respect to refractive indices for symmetrical systems.**
T. B. Andersen.
Appl. Opt., Vol. 21, 4040– 4044 (1982).
Formulas are presented for the computation of derivatives of the optical aberration functions *S, T, V, W,* and *K* with respect to the relative refractive indices, leading to the chromatic aberrations, for symmetrical optical systems. The formulas are suitable for use in computer programs for automatic computation of optical aberration coefficients.

031.060 **Optical figuring process, Dupont 2.6 m telescope.**
A. B. Meinel, M. P. Meinel.
Appl. Opt., Vol. 21, 4198 - 4200 (1982).

031.061 **Blaze characteristics of echelle gratings.**
S. Engman, P. Lindblom.
Appl. Opt., Vol. 21, 4356 - 4362 (1982).
A new model for the blaze function of the two in-plane mountings of echelle gratings is described and extended to account for defects in the edges of the grooves. Expressions for the relative efficiency are given. Results are given from measurements of the blaze angle and the groove edge defects of three echelles.

031.062 **A technique for measuring transverse lens aberrations.**
R. Silvennoinen.
Publ. Univ. Joensuu, Ser. BI, No. 22, 9 pp. (1981). ISBN 951-696-325-0.
A design technique is presented for a measure of magnitude of primary transverse aberrations. The technique applies optical ray trace procedure which is also easy to simulate with the aid of a computer.

031.063 **Comparison of coherent and incoherent imaging in the location of point sources.**
W. L. Wolfe.
Proc. SPIE Int. Soc. Opt. Eng., Vol. 294, (see 012.078), 146 - 150 (1981). – Abstr. in Phys. Abstr., Vol. 86, Abstr. 10211 (1983).

031.064 **A design study of an aspherical plate corrector for the field of view of a Cassegrain system.**
M.-I. Yi.
Acta Astron. Sinica, Vol. 23, 398 - 408 (1982). In Chinese.
Taking a system with an aperture of 1000, a primary mirror focal ratio of 3 and a Cassegrain focal ratio of 8 for an example, the author studied and designed four kinds of aspherical plate correctors for a quasi R-C, R-C, parabolic and a spherical primary mirror Cassegrain system, respectively.

031.065 **Ray tracing for tilting non-rotationally-symmetric aspherical surfaces.**
S. Lu, Z. Zhang.
Acta Astrophys. Sinica, Vol. 2, 159 - 164 (1982). In Chinese.

031.066 **Ray tracing for a Schmidt corrector aspherical surface expressed with strict parametric equations.**
S. Lu, Z. Zhang.
Acta Astrophys. Sinica, Vol. 2, 233 - 239 (1982). In Chinese.

031.067 **X-ray and extreme ultraviolet imaging using layered synthetic microstructures.**
J. H. Underwood, T. W. Barbee, D. L. Shealy.
Proc. SPIE Int. Soc. Opt. Eng., Vol. 316, (see 012.086), 79 - 89 (1981). – Abstr. in Phys. Abstr., Vol. 86, Abstr. 16456 (1983).

031.068 **Grazing incidence relay optics.**
R. C. Chase, J. M. Davis, A. S. Krieger, J. H. Underwood.
Proc. SPIE Int. Soc. Opt. Eng., Vol. 316, (see 012.086), 74 - 78 (1981). – Abstr. in Phys. Abstr., Vol. 86, Abstr. 20242 (1983).

The Heisenberg uncertainty relation and the resolving power in optics.
See Abstr. 022.116.

The performance of ground based telescopes.
See Abstr. 032.046.

Optical designs of plane aspherized grating spectrographs.
See Abstr. 034.092.

Methods of Observation and Reduction

031.501 **On the combination of partially overlapping sets of data.**
B. C. Reed, M. P. FitzGerald.
Astron. Astrophys., Vol. 111, 81 - 85 (1982) = Contrib. Univ. Waterloo Obs. No. 84.
The authors present a technique to combine partially overlapping sets of data making use of all the mutual observations available. This is accomplished in a one-step process and produces a considerable increase in accuracy over previously employed techniques. The authors give two examples: (1) Combination of data from 9 different sources of photoelectric *V* observations of OB stars in Puppis, for which only zero-point corrections were found necessary; and (2) Combination of data in the open cluster NGC 2483 from 3 different sources (one fictitious), for which zero-point and first order corrections are required.

031.502 **A digital image processing method for automatic reduction of échelle spectrograms.**
V. Moreno, F. Llorente de Andrés, J. Jiménez.
Astron. Astrophys., Vol. 111, 260 - 265 (1982).
The authors have developed a software to correct and process échelle spectrograms. The geometric deformations of

these spectrograms are obtained and corrected without knowing them a priori. The algorithms have been developed in *PL/I* and can be implemented in a minicomputer with a capability of 128 K bytes. One spectrum is reduced in less than 6 min.

031.503 **A new method of determination of the pole motion in a uniform system.**
S. Takagi.
Astron. Astrophys., Vol. 112, 11 - 15 (1982).

A new method of determination of the pole motion was investigated to combine in a uniform system all the results of worldwide astrometric observations. An observational equation is constructed to be used for routine work with an electronic computer.

031.504 **A method for evaluating the epoch of minimum of an eclipsing variable: the sliding integrations.**
S. Ghedini.
Mem. Soc. Astron. Italiana, Vol. 52, 633 - 645 (1981).

This paper proposes a new method for evaluating the epoch of minimum of eclipsing variables, which may be used in alternative to the widely known procedure of Kwee and Van Woerden. It is based on sliding integrations of the light curve, which allows to partially filter noises. Some hypothetic light curves with different shapes and asymmetry degrees have been simulated in order to compare the results of the sliding integration method and the Kwee and Van Woerden method. The first shows a greater immunity against the observative errors and a smaller deviation from the theoretical case in presence of asymmetries.

031.505 **Large scale photometric surveys using archival plates.**
K. Janes.
Automated data retrieval in astronomy, (see 012.002), p. 237 - 239 (1982).

Modern microdensitometers and image processing software suggest the possibility that it may soon be feasible to utilize the enormous amount of photometric and positional data potentially available on existing photographic plates. This paper describes the possibility of large-scale surveys and outlines a method for densitometric calibration of plates from an analysis of the detailed structure of the star images on the plate.

031.506 **Geneva photometric boxes. IV. A refined method for direct access.**
B. Nicolet.
Automated data retrieval in astronomy, (see 012.002), p. 277 - 280 (1982).

In the Geneva photometry, a photometric box around a so-called central star is the set of stars having almost the same colours as this central star. It is useful to construct the boxes of a few hundred stars. A method allowing to obtain these boxes rapidly is described. It is based on so-called direct access square files. A FORTRAN routine on a UNIVAC 1160 computer is widely used at the Geneva Observatory. The principles described in this paper may be easily adapted to build square files and to write routines in evolved languages for most computers.

031.507 **Quantization errors in the digital implementation of 'clean'.**
C. F. Chen.
Proc. Astron. Soc. Australia, Vol. 4, 256 - 259 (1981).

This paper investigates finite wordlength effects in the digital implementation of 'clean'. The investigation reveals that quantization error prevents 'clean' from converging to the noise level and is able to explain cases of 'clean' failure and so helps one to avoid drawing any misleading conclusions from the restored image.

031.508 **Temperature effect in zero-point drift.**
Z. Kviz.
Proc. Astron. Soc. Australia, Vol. 4, 260 - 262 (1981).

031.509 **The maximum entropy principle in two-dimensional spectral analysis.**
J. V. Pendrel, D. E. Smylie.
Astron. Astrophys., Vol. 112, 181 - 189 (1982).

The problem of formulating a two-dimensional maximum entropy spectral density is examined. It is shown that the derivative of the two-dimensional entropy rate with respect to the autocorrelations are related to the spectral density.

031.510 **On the linearity of the SWP (*short wavelength prime*) camera of the International Ultraviolet Explorer (IUE): a correction algorithm.**
A. Holm, R. C. Bohlin, A. Cassatella, D. P. Ponz, F. H. Schiffer.
Astron. Astrophys., Vol. 112, 341 - 349 (1982).

An algorithm is presented for correcting IUE low resolution spectral images obtained with the SWP camera for some of the non-linearity effects reported by Bohlin et al. (1980). The results obtained by applying the method to a representative set of spectra of both point and extended sources are reported. In addition, a new evaluation of linearity and reproducibility of the SWP spectral data is provided, based on an improved intensity transfer function.

031.511 **The application of Bates' algorithm to binary stars.**
W. G. Bagnuolo, Jr.
Mon. Not. R. Astron. Soc., Vol. 200, 1113 - 1122 (1982).

Bates' algorithm has been analysed for binary stars by means of 'theory' and computer simulation, for both low and high intensity cases. Low intensity results indicate that the algorithm can resolve the 180° position ambiguity of 15th magnitude stars with a large telescope. High intensity results indicate that Bates' algorithm is probably better for determining relative intensities than the autocorrelation for $\Delta m \leqslant 1.0$ mag.

031.512 **Photoclinometric determination of the topography of the Martian north polar cap.**
A. D. Howard, K. R. Blasius, J. A. Cutts.
Icarus, Vol. 50, (see 012.003), 245 - 258 (1982).

Photoclinometry is useful for the determination of topography for areas which have a uniform albedo. The technique is applied to early spring Viking images of the Martian north polar cap, taken when the surface was covered by a nearly uniform frost cover. Unlike earlier approaches, the topographic profiling can be used for surfaces with any photometric function, but the strike of the planetary surface relative to the illumination angle must be specified along the profile. The resultant profiles are relatively insensitive to misestimation of the photometric function and slope orientation, but are quite sensitive to the assumed values of the reflectance of an equivalent level surface and the atmospheric opacity (if it is large).

031.513 **Evaluation of the flux, position, size, and shape of a compact source beyond the Rayleigh resolution limit.**
T. Kosugi.
Publ. Astron. Soc. Japan, Vol. 34, 281 - 297 (1982).

A new method to evaluate the flux, position, size, and shape of a compact source beyond the Rayleigh resolution limit is described.

031.514 **IUE data reduction: wavelength determinations and line identifications using a VAX/750 computer.**
J. P. Davidson, D. J. Bord.
Advances in ultraviolet astronomy, (see 012.006), p. 326 - 330 (1982).

A fully automated, interactive system for determining the wavelengths of features in extracted IUE spectra using a VAX/750 computer is described. Wavelengths are recorded from video displays of expanded plots of individual orders using a movable cursor, and then corrected for IUE wavelength scale errors. The estimated accuracy of an individual wavelength in the final tabulation is 0.050 Å. Such lists are ideally suited for line identification work using the method of wavelength coincidence statistics (WCS), and the authors present the results of WCS studies of the ultraviolet spectra of the chemically peculiar stars ι CrB and κ Cnc.

031.515 **Faint object studies with IUE.**
T. R. Gull.
Advances in ultraviolet astronomy, (see 012.006), p. 331 - 334 (1982).

031.516 **Spectral anomalies in low-dispersion SWP images.**
R. L. Hackney, K. R. H. Hackney, Y. Kondo.
Advances in ultraviolet astronomy, (see 012.006), p. 335 - 338 (1982).

Physical interpretation of IUE spectra obtained with the SWP camera may be significantly affected by artificial spectral features of several types. In low-dispersion large-aperture SWP exposure of sources with nominally featureless spectra, the authors identify: a spectral imprint which alters the shape of the continuum, several spurious "emission features" which recur at fixed locations on the camera target, and fixed-pattern noise which can result in illusory emission and absorption features. The anomalies appear in spectra extracted from line-by-line files generated by IUESIPS.

031.517 **The photometric performance of the IUE.**
A. V. Holm.
Advances in ultraviolet astronomy, (see 012.006), p. 339 - 342 (1982).

Analysis of low dispersion spectra shows that there are systematic photometric errors which are a function of exposure level. These errors are referred to as linearity errors. Spectra from both the Long Wavelength Redundant (LWR) camera and the Short Wavelength Prime (SWP) camera show linearity errors. Examples of these errors are shown. Their stability in time is examined. Some data concerning their spatial uniformity on the cameras' image area is discussed.

031.518 **Molecular spectroscopy and planetary exploration from space.** R. A. Hanel.
Vibrational-rotational spectroscopy for planetary atmospheres, (see 012.007), p. 29 - 46 (1982).

The author shows examples of the role which molecular spectroscopy has played in the interpretation of the thermal emission spectra of Earth, Mars, and Jupiter. He discusses the IRIS instrument and summarizes the theory of remote sensing in the thermal infrared.

031.519 **Spectroscopic measurements of Mars and Venus with heterodyne techniques.**
A. Betz.
Vibrational-rotational spectroscopy for planetary atmospheres, (see 012.007), p. 277 - 293 (1982).

The author illustrates the new types of planetary observations made possible with the high spectral resolution of heterodyne spectroscopy. The author's applications of the heterodyne technique to planetary spectroscopy have included measurements of the exact shapes of CO_2 absorption lines in Mars and Venus to determine the vertical pressure-temperature structure, accurate measurements of absolute wind velocities in both the mesosphere and stratosphere of Venus from the Doppler shifts of narrow CO_2 lines, and , more recently, searches for minor molecular species of interest in modeling the stratospheric photochemistry of Mars and Venus.

031.520 **You can discover a comet.**
W. A. Bradfield.
Space Educ., Vol. 1, 161 - 163 (1982).

031.521 **Visual observations of geostationary satellites.**
G. E. Taylor.
J. British Astron. Assoc., Vol. 92, 216 - 219 (1982).

031.522 **Photometry among the climatically underprivileged.**
J. D. Fernie.
J. R. Astron. Soc. Canada, Vol. 76, 224 - 234 (1982).

The question of whether or not accurate photometry can be done on a worthwhile number of nights in a mediocre climate is addressed. Using data accumulated at the David Dunlap Observatory over a number of years, it is found that absolute photometry of brighter stars can be done there with a standard deviation of $0\overset{m}{.}018$ per observation. It is concluded that photometry done at this site can be a valuable supplement to work done at a good site.

031.523 **Ervaringen met een protuberansenkijker.**
B. van Slooten.
Zenit, 9. Jaarg., 376 - 377 (1982).

031.524 **Astronomy with a microscope.**
R. J. Dodd, F. P. Andrews.
South. Stars, Vol. 29, 136 - 138 (1982).

031.525 **Radioastronomische Beobachtungen für den Amateur. Teil 1: Die Radiosonne.**
W. Schippke.
Sterne Weltraum, 21. Jahrg., 324 - 326, 328 (1982).

031.526 **Zur Technik der Beobachtung veränderlicher Sterne.**
R. Lukas.
Sterne Weltraum, 21. Jahrg., 368 - 370 (1982).

031.527 **La mesure des étoiles doubles par interférométrie "speckle".**
P. Baize.
Astronomie, Vol. 96, 389 - 404 (1982).

031.528 **Choice of a transformation polynomial for reduction of photographic plates taken by an SBG camera.**
S. Świerkowska.
Artif. Satell., Vol. 16, No. 1, p. 57 - 63 (1981).

031.529 **Project of improved drive of carriage with plate in SBG cameras.**
I. Domiński, R. Serafin.
Artif. Satell., Vol. 16, No. 1, p. 65 - 75 (1981).

031.530 **On the possibility of application of a photoelectric zenith tube placed on the moon or planets for investigating their nutations.**
J. Bem.
Artif. Satell., Vol. 16, No. 3, p. 41 - 52 (1981).

The principle of measurement of the transit time of a star over a slit of the photoelectric micrometer, suitable for exterrestrial automatic observations is described. On this basis the possibility of determination of the parameters of nutation is investigated.

031.531 **Methods, algorithms and processing programmes of extended cosmic radio source images.**
S. K. Panishko, M. I. Ryabov.
Odes. univ. Odessa, 1982, 33 pp. In Russian. – Abstr. in Ref. zh., 51. Astron., 8.51.908 (1982).

031.532 **Very long-range radio interferometry.**
L. I. Matveenko.
Priroda, 1982, No. 7, p. 56 - 67. In Russian.

031.533 **Reduktion von Positionsbestimmungen durch Koordinatentransformation.**
H. Meyerdierks.
Sonne, Jahrg. 6, 130 - 132 (1982).

031.534 **Bestimmung der Längenausdehnung von Sonnenfleckengruppen.**
P. Kroupa.
Sonne, Jahrg. 6, 133 - 135 (1982).

031.535 **Observing the sun in safety.**
J. C. D. Marsh.
J. British Astron. Assoc., Vol. 92, 257 - 259 (1982).

031.536 **Speckle interferometry in the visible and infrared.**
C. Dainty.
News Lett. Astron. Soc. N.Y., Vol. 2, No. 2, p. 5 - 6 (1982). Abstract.

031.537 **Radar detectability of comets.**
P. G. Kamoun, G. H. Pettengill, I. I. Shapiro.
Comets, (see 012.009), p. 288 - 296 (1982).

Earth-based radar observations of comets are restricted to the detection of the nucleus since detection of an echo from the dust or from the plasma near a comet seems infeasible. The successful detection of the nucleus can provide information on its size, spin vector, surface-scattering properties, and orbit.

031.538 **Digital stellar speedometry.**
D. W. Latham, R. P. Stefanik.
Bull. American Astron. Soc., Vol. 14, 612 (1982). – Abstract.

031.539 **Speckle image reconstruction of solar features.**
R. V. Stachnik, P. Nisenson, R. W. Noyes.
Bull. American Astron. Soc., Vol. 14, 624 (1982). – Abstract.

031.540 **Precision radial velocities with a hydrogen fluoride absorption cell.**
B. Campbell, A. Irwin, G. A. H. Walker.
Bull. American Astron. Soc., Vol. 14, 625 - 626 (1982). Abstract.

031.541 **Criteria for positive planetary detection using the radial velocity method.**
P. H. Smith.
Bull. American Astron. Soc., Vol. 14, 626 (1982). – Abstract.

031.542 **Solar observations pertaining to the search for extra-solar planetary systems.**
R. Howard, D. H. Bruning.
Bull. American Astron. Soc., Vol. 14, 626 (1982). – Abstract.

031.543 **A search for low mass unseen companions in nearby binary star systems.**
H. A. McAlister, W. I. Hartkopf, O. G. Franz.
Bull. American Astron. Soc., Vol. 14, 626 (1982). – Abstract.

031.544 **Techniques for spatially resolving stellar chromospheres.**
R. R. Radick.
Bull. American Astron. Soc., Vol. 14, 646 (1982). – Abstract.

031.545 **Atmospheric field-of-view constraints on high-resolution imaging techniques.**
W. H. Waller.
Bull. American Astron. Soc., Vol. 14, 646 (1982). – Abstract.

031.546 **CCD observations of faint planetary satellites.**
D. Pascu, P. K. Seidelmann, W. A. Baum, R. E. Schmidt.
Bull. American Astron. Soc., Vol. 14, 657 (1982). – Abstract.

031.547 **Real-time infrared imaging of stars using multiplex encoding techniques.**
J. P. Corteggiani, J. Gay, A. Journet, Y. Rabbia.
J. Opt. *(France)*, Vol. 13, No. 1, p. 9 - 11 (1982). – Abstr. in Phys. Abstr., Vol. 85, Abstr. 63262 (1982).

031.548 **Measurements of the spatial-temporal statistics of stellar speckle patterns at Mauna Kea, Hawaii.**
K. A. O'Donnell, B. J. Brames, J. C. Dainty.
Opt. Commun., Vol. 41, 79 - 82 (1982). – Abstr. in Phys. Abstr., Vol. 85, Abstr. 63267 (1982).

031.549 **Radioastronomy *(resolution by aperture synthesis).***
W. N. Brouw.
Underwater Acoustics and Signal Processing. Proceedings of The NATO Advanced Study Institute, Copenhagen, Denmark, 18 - 29 August 1980. L. Bjorno (Editor). D. Reidel Publishing Company, Dordrecht, Holland. p. 683 - 692 (1981). – Abstr. in Phys. Abstr., Vol. 85, Abstr. 63268 (1982).

031.550 **Analysis of the possibilities of bistatic radiolocation of planets using sources of cosmic radio emission.**
A. I. Kucheryavenkov, A. G. Pavel'ev.
Radiotekh. i ehlektron., Tom 27, 914 - 920 (1982). In Russian. – Abstr. in Ref. zh., 51. Astron., 9.51.145 (1982).

031.551 **A derivation of the stellar luminosity function in the direction of the south galactic pole by a statistical method.**
S. L. Kipp.
Rev. Mexicana Astron. Astrofis., Vol. 6, (see 012.017), 95 - 98 (1981).

A method is developed to calculate absolute magnitude probability distributions for stars from a proper motion survey. This method uses ellipsoidal velocity distributions to predict tangential velocity distributions. The tangential velocity distributions are transformed into absolute magnitude distributions. The absolute magnitude distributions for the stars in a proper motion survey may be summed to produce a luminosity function. This method was applied to stars in the region of the south galactic pole and the resulting luminosity function is statistically identical to Luyten's 1968 luminosity function.

031.552 **Measurement of a stellar angular diameter by a high-time-resolving photometric observation of a lunar occultation.**
W. Tanaka.
Astron. Her., Vol. 74, 346 - 349 (1981). In Japanese. Abstr. in Phys. Abstr., Vol. 85, Abstr. 68627 (1982).

031.553 **Radio astronomical interferometry and computerized tomography of X-rays.**
L. F. Rodriguez.
Rev. Mexicana Fis., Vol. 28, 249 - 264 (1982). In Spanish. Abstr. in Phys. Abstr., Vol. 85, Abstr. 68631 (1982).

031.554 **Infrared heterodyne spectroscopy of ammonia and ethylene in stars.**
A. L. Betz.
Laser spectroscopy V, (see 012.016), p. 81 - 88 (1981). Abstr. in Phys. Abstr., Vol. 85, Abstr. 68635 (1982).

031.555 **Spectroscopy in radio astronomy.**
M. Morimoto.

J. Spectrosc. Soc. Japan, Vol. 30, 379 - 386 (1981). In Japanese. – Abstr. in Phys. Abstr., Vol. 85, Abstr. 73854 (1982).

031.556 **Differential speckle interferometry.**
J. M. Beckers.
Opt. Acta, Vol. 29, (see 012.018), 361 - 362 (1982). – Abstr. in Phys. Abstr., Vol. 85, Abstr. 73856 (1982).

031.557 **The Steward Observatory speckle interferometry system.**
E. K. Hege, E. N. Hubbard, P. A. Strittmatter, W. J. Cocke.
Opt. Acta, Vol. 29, (see 012.018), 701 - 715 (1982). – Abstr. in Phys. Abstr., Vol. 85, Abstr. 73857 (1982).

031.558 **Infrared heterodyne spectroscopy *(Mars atmosphere).***
M. J. Mumma, T. Kostiuk, D. Buhl, G. Chin, D. Zipoy.
Opt. Eng., Vol. 21, 313 - 319 (1982). – Abstr. in Phys. Abstr., Vol. 85, Abstr. 73858 (1982).

031.559 **Automated supernova search.**
S. A. Colgate.
Supernovae: a survey of current research, (see 012.022), p. 319 - 323 (1982).
An automated supernova search is under development at the New Mexico Institute of Mining and Technology to search several thousand galaxies per night to pick up SN at 15th and 16th mag.

031.560 **The Berkeley automated supernova search.**
J. T. Kare, C. R. Pennypacker, R. A. Muller, T. S. Mast, F. S. Crawford, M. S. Burns.
Supernovae: a survey of current research, (see 012.022), p. 325 - 339 (1982).
The Berkeley automated supernova search employs a computer controlled 36-inch telescope and CCD detector to image 2500 galaxies per night. A dedicated minicomputer compares each galaxy image with stored reference data to identify supernovae in real time. The threshold for detection is $m_v = 18.8$. The search should be operational in mid-1982.

031.561 **Automated supernova search from photographic plates.**
M. G. M. Cawson, E. J. Kibblewhite.
Supernovae: a survey of current research, (see 012.022), p. 341 - 353 (1982).
A new automated technique for detecting supernovae on pairs of IIIaJ Schmidt plates is described and the initial results presented. Analysis of 2144 galaxies on 4 plates produced 3 supernovae indicating a supernova rate per galaxy of one every 160 ± 50 years.

031.562 **Digital image processing at the Institute of Radio Electronics of the Ukrainian Academy of Sciences.**
A. A. Babichev, A. D. Egorov, N. K. Zaets, E. V. Zdor, Yu. V. Kornienko, V. G. Parusimov, A. Ya. Usikov.
Vestn. Khar'kov. Univ., No. 223. Metody obrabotki astron. i kosm. izobrazhenij, Vyp. 16, p. 6 - 16 (1981). In Russian.

031.563 **Analysis of possibilities of coherent-optical methods of image processing.**
V. N. Dudinov, V. S. Tsvetkova.
Vestn. Khar'kov. Univ., No. 223. Metody obrabotki astron. i kosm. izobrazhenij, Vyp. 16, p. 16 - 25 (1981). In Russian.

031.564 **On function restoration by Fourier transformation modules.**
V. A. Petrov.
Vestn. Khar'kov. Univ., No. 223. Metody obrabotki astron. i kosm. izobrazhenij, Vyp. 16, p. 25 - 33 (1981). In Russian.

031.565 **The structure of astronomical images. Speckle interferometry.**
V. N. Dudinov, V. S. Tsvetkova, V. N. Erokhin.
Vestn. Khar'kov. Univ., No. 223. Metody obrabotki astron. i kosm. izobrazhenij, Vyp. 16, p. 33 - 46 (1981). In Russian.

031.566 **Measurement of angular diameters of stars. Possibilities of the 6-m telescope.**
S. G. Kuz'menkov, V. G. Vakulik.
Vestn. Khar'kov. Univ., No. 223. Metody obrabotki astron. i kosm. izobrazhenij, Vyp. 16, p. 47 - 55 (1981). In Russian.

031.567 **On the use of an aperture splitter for astronomical image production through an inhomogeneous atmosphere.**
V. N. Uvarov.
Vestn. Khar'kov. Univ., No. 223. Metody obrabotki astron. i kosm. izobrazhenij, Vyp. 16, p. 55 - 59 (1981). In Russian.

031.568 **Possibilities of speckle interferometry for observations of eclipsing variable systems (computer modelling results).**
V. A. Zakhozhaj.
Vestn. Khar'kov. Univ., No. 223. Metody obrabotki astron. i kosm. izobrazhenij, Vyp. 16, p. 60 - 63 (1981). In Russian.

031.569 **Experience of construction of a lunar colour map by digital image processing methods.**
D. G. Stankevich.
Vestn. Khar'kov. Univ., No. 223. Metody obrabotki astron. i kosm. izobrazhenij, Vyp. 16, p. 63 - 67 (1981). In Russian.

031.570 **Apparatus aids of digital processing of astronomical images.**
V. G. Parusimov, A. V. Polyanskij.
Vestn. Khar'kov. Univ., No. 223. Metody obrabotki astron. i kosm. izobrazhenij, Vyp. 16, p. 67 - 77 (1981). In Russian.

031.571 **Half-tone indicator in a digital image processing system.**
E. V. Zdor.
Vestn. Khar'kov. Univ., No. 223. Metody obrabotki astron. i kosm. izobrazhenij, Vyp. 16, p. 77 - 79 (1981). In Russian.

031.572 **Plotting system on the cathode-ray tube in an image processing system.**
S. A. Kanygin.
Vestn. Khar'kov. Univ., No. 223. Metody obrabotki astron. i kosm. izobrazhenij, Vyp. 16, p. 79 - 81 (1981). In Russian.

031.573 **Application of a computer controlled intensity registering procedure to the spectral region at 4430 Å.**
H. Lehmann.
Astron. Nachr., Band 303, 263 - 268 (1982).
A computer controlled measuring procedure is presented ascertaining intensity courses. The method was applied to eight weakly or not reddened early type stars in order to obtain data on atomic lines in the central part of the diffuse interstellar line at 4430 Å. The results show the general influence of the stellar portion on the central depth and equivalent width of the whole diffuse feature.

031.574 **The radio and radar observations of comets: a review.**
E. Gérard.
The need for coordinated ground-based observations of Halley's comet, (see 012.025), p. 69 - 79 (1982).

031.575 **Proposed radio observations of inverted OH in Halley's comet.**
R. P. Norris.

The need for coordinated ground-based observations of Halley's comet, (see 012.025), p. 81 - 82 (1982).

031.576 **Large-scale photographic observations, photometry, colorimetry and polarimetry of cometary tails.**
P. L. Lamy, S. Koutchmy.
The need for coordinated ground-based observations of Halley's comet, (see 012.025), p. 243 - 256 (1982).

031.577 **Astrometric observations and the motion of comet Halley.**
D. K. Yeomans.
The need for coordinated ground-based observations of Halley's comet, (see 012.025), p. 277 - 285 (1982).

031.578 **Determination of precise photographic positions with the automatic measuring machine IRIS.**
G. Hahn, C.-I. Lagerkvist.
The need for coordinated ground-based observations of Halley's comet, (see 012.025), p. 287 - 291 (1982).

031.579 **A pulsar experiment – our experiences in time keeping.**
P. V. R. Murthy.
J. Inst. Electron. Telecommun. Eng., Vol. 27, 549 - 553 (1981). – Abstr. in Phys. Abstr., Vol. 85, Abstr. 79540 (1982).

031.580 **The instrumental error in radar detection of meteor heights.**
S. Gottardi, C. Franceschi, G. Cevolani.
Autom. Strum., Vol. 30, No. 3, p. 70 - 78 (1982). In Italian. Abstr. in Phys. Abstr., Vol. 85, Abstr. 89153 (1982).

031.581 **Time accuracy requirements for VLBI experiments in radio astronomy.**
R. V. Bhonsle, S. S. Degaonkar, S. K. Alurkar.
J. Inst. Electron. Telecommun. Eng., Vol. 27, 601 - 605 (1981). – Abstr. in Phys. Abstr., Vol. 85, Abstr. 89167 (1982).

031.582 **High spatial resolution in the infrared.**
F. J. Low.
Recent advances in observational astronomy, (see 012.028), p. 89 - 93 (1981).

Two complementary techniques are now available in the infrared for extracting the highest angular resolution possible from large ground based telescopes. Even with apertures of two meters or less, when "seeing" conditions are good or excellent, a large number of objects may be resolved and their structures may be studied in great detail by varying both the position angle and the wavelength. Similar work at radio frequencies demonstrates that, given adequate data over many position angles and spatial frequencies, image reconstruction procedures exist which may be applied to yield detailed images of infrared sources now only partially resolved.

031.583 **Speckle interferometry in astronomy.**
J. C. Dainty.
Recent advances in observational astronomy, (see 012.028), p. 95 - 111 (1981).

In this paper the basic principles of speckle interferometry are reviewed and some topics of current interest are discussed. An extensive bibliography on all aspects of stellar interferometry is given.

031.584 **Maximum entropy in two dimensions – illustrative examples.**
R. Nityananda, R. Narayan.
Bull. Astron. Soc. India, Vol. 10, 30 (1982). – Abstract.

031.585 **Source counts at and beyond the confusion limit.**
A. Franceschini.
Astrophys. Space Sci., Vol. 86, 3 - 11 (1982).

The author discusses the integral observables – intensity of the background radiation due to unresolved sources and its cell-to-cell fluctuations – that constrain a source distribution function below the resolution limit. It is shown in detail how these quantities are affected by the instrumental transmission and found that these effects vary qualitatively with the depth of the survey. One implication centers on the particular importance that the fluctuation studies still have at the present stage of X-ray astronomy to solve the question concerning the origin of the ≥ 2 keV background. A second, more general, implication amounts to a reformulation of the confusion limit.

031.586 **Interstellar lines in high resolution IUE spectra: paper I. Groningen data reduction package and technical results.**
D. P. Gilra, T. H. Pwa, E. M. Arnal, J. de Vries.
Third European IUE Conference, (see 012.029), p. 385 - 390 (1982).

To process and analyse high resolution IUE data on a large number of interstellar lines in a large number of images for a large number of stars the authors have developed several computer programs. They work on 115 'expected lines' in the short wavelength range and 40 in the long wavelength range. Programs include extraction, processing, plotting, averaging, and profile fitting. A brief description of the data reduction package is given.

031.587 **Image processing applied to CCD images of Saturn's rings.**
T. J. Kreidl.
Bull. American Astron. Soc., Vol. 14, 747 - 748 (1982). Abstract.

031.588 **A critical comparison of the accuracy of proposed methods for small-angle and wide-angle astrometry.**
S. H. Knowles.
Bull. American Astron. Soc., Vol. 14, 779 (1982). – Abstract.

031.589 **Color equations for the United Kingdom Schmidt Telescope.**
M. Blair, G. Gilmore.
Publ. Astron. Soc. Pacific, Vol. 94, 742 - 743 (1982).

Color equations relating the natural photographic system of the United Kingdom 1.2-m Schmidt Telescope to the Cousins photoelectric system are derived from 36 plates in the *UJBVR* and *I* photographic passbands.

031.590 **Spectroscopic diagnostics of the active region: transition zone and corona.**
K. P. Dere, H. E. Mason.
Solar active regions, (see 003.011), p. 129 - 164 (1981).

In this review, the authors try to summarize most of the techniques currently used to analyze X-ray and ultraviolet spectroscopic data of solar features, with special reference to active regions. In particular, they focus on the limitations and accuracy of the various methods.

031.591 **Astro speckle interferometry. A simple modification resolving the phase problem.**
A. M. J. Huiser.
Optik, Vol. 61, 433 - 436 (1982). – Abstr. in Phys. Abstr., Vol. 85, Abstr. 94622 (1982).

031.592 **Photogrammetry of the Viking Lander imagery.**
S. S. C. Wu, F. J. Schafer.
Photogramm. Eng. Remote Sensing, Vol. 48, 803 - 816 (1982). Abstr. in Phys. Abstr., Vol. 85, Abstr. 94624 (1982).

031.593 **Processing astronomical images from space.**
B. R. Martin, E. Dunford.
Comput. Phys. Commun., Vol. 26, 441 - 446 (1982). – Abstr. in Phys. Abstr., Vol. 85, Abstr. 100752 (1982).

031.594 **A procedure to correct the images of astronomical objects for the distortions due to atmospheric turbulence.**
A. M. J. Huiser.
Opt. Commun., Vol. 42, 226 - 230 (1982). – Abstr. in Phys. Abstr., Vol. 85, Abstr. 100756 (1982).

031.595 **Methods of analysis of stratospheric ozone data.**
G. S. Watson.
AIP Conf. Proc., No. 82, (see 012.037), p. 201 - 221 (1981). Abstr. in Phys. Abstr., Vol. 85, Abstr. 112150 (1982).

031.596 **On the isoplanatic patch size in stellar speckle interferometry.**
F. Roddier, J. M. Gilli, J. Vernin.
J. Opt. *(France)*, Vol. 13, 63 - 70 (1982). – Abstr. in Phys. Abstr., Vol. 85, Abstr. 112343 (1982).

031.597 **Free sight of Venus by radar.**
K. Arbenz.
Schweizerische Tech. Z., Vol. 79, No. 13, p. 10 - 12 (1982). In German. – Abstr. in Phys. Abstr., Vol. 85, Abstr. 112346 (1982).

031.598 **Analysis of the difference in accuracy of the photoelectrical time registration on one level and five discrimination levels.**
N. Solarić.
Publ. Astron. Obs. Sarajevo, Vol. 1, No. 1, (see 012.043), p. 115 - 124 (1981).

Analysis of the second series of test observations on the photoelectrical device in Zagreb has shown that the accuracy of the registration of the stellar transit, using the discrimination on five amplitude levels, is 22% greater than that obtained with the discrimination on one level only. This increase of measuring accuracy brings a significant saving in time of about 40%, if a longitude of a place is to be determined with a certain accuracy.

031.599 **The application of the cross-correlation function at the photoelectric time registration of the stellar transit.**
N. Solarić.
Publ. Astron. Obs. Sarajevo, Vol. 1, No. 1, (see 012.043), p. 125 - 130 (1981).

031.600 **A method of simultaneous determination of latitude and azimuth by stars' observations near elongation.**
V. Malanović.
Publ. Astron. Obs. Sarajevo, Vol. 1, No. 1, (see 012.043), p. 157 - 162 (1981).

031.601 **Polarimetry in astronomy.**
J. Arsenijević.
Publ. Astron. Obs. Sarajevo, Vol. 1, No. 1, (see 012.043), p. 173 - 181 (1981). In Croatian.

031.602 **Narrow band Hα observations of the chromospheric activity of stars.**
V. S. Oskanian *(Oskanyan)*.
Publ. Astron. Obs. Sarajevo, Vol. 1, No. 1, (see 012.043), p. 182 (1981). – Abstract.

031.603 **Computer controlled high speed photometry presented by the lunar occultation of 57 Orionis. I.**
B. Stecklum.
Astron. Nachr., Band 303, 305 - 307 (1982) = Mitt. Univ.-Sternw. Jena Nr. 154. In German.

For high speed photometric observations the photoelectric system was modified. The supplement of the interpreting programming language permits measurements down to 1 ms integration time with an overall measuring time of at least 4 s. The application in the case of lunar occultations is described and an outlook of the data reduction treatment is given.

031.604 **Corrections to Watts' datum from photoelectric occultations.**
G. Rossello, C. Jordi.
Moon Planets, Vol. 27, 131 - 134 (1982).

Reduction and analysis of photoelectric occultations of stars by the Moon made between 1955–1978 permit to obtain corrections to the Watts' reference surface, using the recent values for the correction of the equinox.

031.605 **The width of echelle orders in IUE images as derived with the Astronomical Image Display and Analysis (AIDA) system in Tübingen.**
K. S. de Boer, P.-R. Preussner, M. Grewing.
Astron. Astrophys., Vol. 115, 128 - 132 (1982).

This paper describes the image display and analysis facilities developed at the Astronomical Institute in Tübingen. With this system the authors have determined the width of the echelle orders in International Ultraviolet Explorer geometrically and photometrically corrected images.

031.606 **A note on Garcia-Alegre et al.'s article, "A model for constructing artificial integrated spectral lines and their Fourier transform properties relevant to the search for differential rotation of stars".**
D. H. Bruning.
Astron. Astrophys., Vol. 115, 203 - 204 (1982). – See Abstr. 31.031.517.

The numerical results obtained by Garcia-Alegre et al. (1982) for the Fourier transform properties of the line broadening due to differential rotation were reported to be different from those of Bruning (1981) by as much as 30%. This discrepancy may be removed by noting that the results of Garcia-Alegre et al. apply explicitly to the case of no limb darkening while those of Bruning refer to an assumed linear limb darkening. When the correction to the transform amplitudes is made for limb darkening; the results of Garcia-Alegre et al. and Bruning agree totally.

031.607 **Measurements of solar transition zone velocities and line broadening using the Ultraviolet Spectrometer and Polarimeter on the Solar Maximum Mission.**
G. Simon, P. Mein, J. C. Vial, R. A. Shine, B. E. Woodgate.
Astron. Astrophys., Vol. 115, 367 - 372 (1982).

The UVSP instrument on SMM is able to observe solar regions at two wavelengths in the same line with a band-pass of 0.3 Å. Intensity and Doppler velocity maps are derived. The authors describe a method called Double Dopplergram Determination from which they can derive both the Doppler width and the velocity. To illustrate the method, a set of C IV 1548 Å observations is analysed according to this procedure. The mean C IV Doppler width measured (0.15 Å) is comparable to previous determinations. The authors find a relation between bright regions and down-flows. Large Doppler widths correspond to strong velocity gradients.

031.608 **Computer controlled high speed photometry presented by the lunar occultation of 57 Orionis. II.**
B. Stecklum.
Astron. Nachr., Band 303, 341 - 347 (1982) = Mitt. Univ.-Sternw. Jena Nr. 155. In German.

The needed formulae for the reduction of lunar occultations are presented and used for the treatment of the occultation data of 57 Orionis. A computing method for the determination of the object profile is given and the influence of noise is discussed. The results for the spectroscopic binary 57 Orionis cannot be explained by the parameters published.

031.609 **Methods of photoelectric observations of lunar occultations and their application for the study of close binaries.**
M. B. Bogdanov, Yu. A. Sklyarov.
Ejection and accretion of matter in binary systems, (see 012.051), p. 111 - 122 (1982).
A review of methods of analysis of photoelectric occultation observations enabling to search for close binaries, to measure angular diameters of stars and to investigate the brightness distribution across their disks is presented. The problem of brightness distribution restoration across the source and the influence of instrumental effects have been examined. The efficiency of the regularization method for the investigation of complex angular structure sources has been demonstrated on the example of 16 Tau close binary.

031.610 **A revised and extended calibration for the Spinrad-Taylor scanner system.**
B. J. Taylor.
Astrophys. J., Suppl. Ser., Vol. 50, 391 - 401 (1982).
New spectrophotometric data are presented for the standards of the Spinrad-Taylor scanner photometric system. The new data are in the 4040–6180 Å wavelength range and are used to construct a calibration of Spinrad-Taylor data to absolute units in this range. In addition, a previously published calibration for $\lambda > 6100$ Å is revised and extende through use of data in the literature. The calibrations and the Spinrad-Taylor data themselves are tested for systematic error.

031.611 **Beobachtungsmethoden der photographischen Astrometrie für Zwecke des fundamentalen Referenzsystems – Einfluß des Kassettentyps auf die Genauigkeit der Resultate.**
D. Böhme.
Geodesy and physics of the earth, (see 012.053), p. 45 (1981). – Abstract.

031.612 **Examination of some components of the astrolabe temperature field at Potsdam and their relations to the geodetic-astronomical observations.**
J. Dittrich.
Geodesy and physics of the earth, (see 012.053), p. 82 - 87 (1981).

031.613 **Über Anwendung von Ergebnissen der Laserentfernungsmessungen zum Monde bei Auflösung einiger Probleme der Astronomie und Geodynamik.**
V. K. Abalakin, V. N. Bojko, M. A. Fursenko, O. M. Gromowa *(Gromova)*, J. L. *(Yu. L.)* Kokurin, W. F. Lobanow *(V. F. Lobanov)*, L. I. Rumjanzewa *(Rumyantseva)*, A. A. Schirjajew *(Shiryaev)*, S. W. Sserowa *(S. V. Serova)*, A. N. Ssuchanowskij *(Sukhanovskij)*.
Geodesy and physics of the earth, (see 012.053), p. 200 - 209 (1981).
The paper deals with some problems related to applications of lunar laser ranging results to improvement of parameters of the theories of the Moon's orbital motion and rotation and to determination of the polar motion, the Earth's rotation variations and the continental drift effects.

031.614 **Laser satellite ranging using the automated SBG Camera.**
H. Fischer, R. Neubert, L. Grunwaldt.
Geodesy and physics of the earth, (see 012.053), p. 267 - 275 (1981).

031.615 **On the investigation of geodynamic parameters by means of laser measurements.**
H. Montag.
Geodesy and physics of the earth, (see 012.053), p. 416 - 423 (1981).
The determination of polar motions by the aid of dynamic methods of satellite geodesy is treated in more detail. It is shown that the use of laser measurements to artificial Earth's satellites is one of the most accurate methods of studying polar motions and other parameters.

031.616 **Practical use of iris diaphragm photometers.**
B. E. Schaefer.
AAS Photo-Bull., No. 29, p. 14 - 15 (1982).
This paper fills the need for a detailed description of the practical use of an iris diaphragm photometer. To do meaningful photographic photometry: (1) the comparison sequence should have more than six stars with a range of over two magnitudes; (2) an iris diaphragm photometer reading of the background near each star should be taken; (3) a quantitative error analysis should be made. Under the best of conditions the internal error can be $0\overset{m}{.}03$. A more typical internal error is $0\overset{m}{.}06$ to $0\overset{m}{.}10$.

031.617 **Un elenco di stelle doppie per saggiare le capacità dei telescopi.**
W. Ferreri.
Orione, Vol. 3, 155 - 160 (1982).

031.618 **Stellar photometry in a city area.**
A. W. J. Cousins.
Mon. Notes Astron. Soc. South. Africa, Vol. 41, 50 - 52 (1982).

031.619 **Einführung in die Beobachtung veränderlicher Sterne: Die Argelandersche Stufenschätzungsmethode.**
M. Fernandes.
BAV Rundbrief, 31. Jahrg., 62 - 66 (1982).

031.620 **Some hints for timing occultations.**
D. Büttner.
Occultation Newsl., No. 3, 33 - 35 (1982).

031.621 **Experiences with the eye-and-ear method.**
D. Büttner.
Occultation Newsl., Vol. 3, 36 (1982).

031.622 **An absolute calibration of the night-sky photometer of the UK Schmidt telescope.**
R. J. Smyth.
Observatory, Vol. 102, 225 - 227 (1982).
A new calibration based on comparative UBV photometry permits photometric zero points of about $\pm 0\overset{m}{.}1$ per square arcsecond accuracy to be derived for plates taken by the UK Schmidt at field centres North of $\delta \sim -65°$.

031.623 **Precision radial velocities with a hydrogen-fluoride absorption cell.**
B. Campbell, A. Irwin, G. A. H. Walker.
J. R. Astron. Soc. Canada, Vol. 76, 316 (1982).
Abstract.

031.624 **A search for quasars at very faint magnitudes.**
G. Edwards, E. Borra, E. Hardy.
J. R. Astron. Soc. Canada, Vol. 76, 319 (1982). – Abstract.

031.625 **The eight-color asteroid survey: standard stars.**
E. F. Tedesco, D. J. Tholen, B. Zellner.
Astron. J., Vol. 87, 1585 - 1592 (1982).
A new photometric system has been established in order to obtain compositionally diagnostic information on the spectral reflectance of faint asteroids and planetary satellites.

Eight broad-band filters include standard *UBV* and range in effective wavelength from 0.34 to 1.04 μm. Magnitudes and color indices, with mean uncertainty ± 0.006 mag are presented for 50 standard stars. The zero point of the system is established by setting the mean colors of four of these standards, considered to be solar analogs, to zero.

031.626 **Seeing-independent definitions of the solar limb position.**
T. M. Brown.
Astron. Astrophys., Vol. 116, 260 - 264 (1982).

One of the most-used definitions of the solar limb position, the finite Fourier transform definition (FFTD), has many convenient properties. The most important of these is a low sensitivity to changes in the width of the atmospheric point spread function. This paper describes a class of similar edge definitions that share the advantages of the FFTD, and that may be tailored to any desired limb darkening function or set of observing circumstances. For some limb darkening functions, edge definitions may be found that are substantially less sensitive to seeing than the FFTD.

031.627 **Sources of errors in observations with the equal altitude method and their effects.**
J.-y. Xu.
Publ. Shaanxi Astron. Obs., No. 1, p. 5 - 9 (1982). In Chinese.

031.628 **One-dimensional high resolution image reconstruction on Eta Carinae at 4.6 μm with speckle data.**
A. Chelli, C. Perrier, Y. G. Biraud.
ESO Sci. Prepr., No. 219, 24 pp. (1982). – Submitted to Astron. Astrophys.

031.629 **Étude préliminaire à l'établissement d'un catalogue fondamental par observation diurne des étoiles brillantes.**
D. Benest, B. Milet.
Mitt. Astron. Ges., Nr. 57, p. 273 - 276 (1982).

031.630 **Observations of stars and nebulae with the TV tube superisokon.**
A. V. Didenko, V. A. Kachmin.
Astron. Tsirk., No. 1181, p. 3 - 4 (1981). In Russian.

031.631 **Determination of accurate radial velocities of early-type stars by cross-correlating spectra scanned on the ESO-S-3000 measuring machine.**
N. Zentelis.
Stockholms Obs. Rep., No. 21, 30 pp. (1982).

The ESO-S-3000 scanning microdensitometer is shown to provide sufficient positional accuracy to be used for radial-velocity measurements. The repeatability of the machine is better than 0.1 μ. A sequence of computer programs has been developed to treat and cross-correlate the digitized spectrograms. The method has been tested on a number of 20 Å/mm Coudé spectrograms of reference stars of spectral types B2-A0. The results yielded radial velocities with an accuracy better than 1.4 km/s.

031.632 **Profils de la raie Hα par télévision analogique.**
Y. Andrillat, C. Fehrenbach.
Be Star Newsl., No. 3, p. 6 - 11 (1981).

031.633 **The reduction of spectra. I. VELMEAS, an interactive radial velocity program.**
G. Hill, D. Ramsden, W. A. Fisher, S. C. Morris.
Publ. Dominion Astrophys. Obs., Vol. 16, 11 - 26 (1982) = NRC No. 20441.

The interactive radial-velocity measuring program VELMEAS is described. Results for 61 Cygni A compared with the ARCTURUS measures of Fletcher show that there are no systematic differences in the velocity systems.

031.634 **The reduction of spectra. II. VLINE, an interactive line fitting spectrophotometric reduction program.**
G. Hill, W. A. Fisher, R. Poeckert.
Publ. Dominion Astrophys. Obs., Vol. 16, 27 - 41 (1982) = NRC No. 20443.

The interactive line-fitting program VLINE is described. VLINE enables the user to measure equivalent widths, line positions and line widths from spectrophotometric data. An intercomparison of results based on various measures of 61 Cygni A indicate that there are no systematic differences between the velocity systems of Fletcher, VELMEAS or VLINE.

031.635 **The reduction of spectra. III. REDUCE, an interactive spectrophotometric program.**
G. Hill, W. A. Fisher, R. Poeckert.
Publ. Dominion Astrophys. Obs., Vol. 16, 43 - 58 (1982) = NRC No. 20444.

The interactive spectrophotometric reduction program REDUCE is described. REDUCE processes PDS scans of photographic material (arc, stellar, clear plate and calibration) reducing these data to linearized stellar wavelength files which may then be converted to intensity files and rectified. Spectrometric data from Reticon or IPCS systems, for example, can also be linearized and rectified if need be. VELMEAS and VLINE are also incorporated within REDUCE so that measures of radial velocity, equivalent width, line position, line width and rotation may also be made.

031.636 **The reduction of spectra. IV. VCROSS, an interactive cross-correlation velocity program.**
G. Hill.
Publ. Dominion Astrophys. Obs., Vol. 16, 59 - 66 (1982) = NRC No. 20445.

The cross-correlation relative velocity program VCROSS is described. The program is extensively tested against velocity measures of 61 Cygni A and a variety of synthetic data.

031.637 **The calibration of photographic photometry of faint stars with a sub-beam prism.**
T. Noguchi, T. Soyano.
Tokyo Astron. Obs. Rep., Vol. 19, 567 - 577 (1981). In Japanese.

031.638 **Use of large telescopes in fast focal ratio mode – application to the 6m telescope (BTA).**
G. Courtes, J. Boulesteix, J. P. Sivan, H. Petit, I. Karachentsev.
Instrumentation for astronomy with large optical telescopes, (see 012.059), p. 67 - 71 (1982).

Large aperture telescopes usually have an unavoidably long focal length so that stellar image diameters are often much larger than the resolution of the detector. The detection of extended sources can take advantage of a faster focal ratio and both stellar and extended monochromatic sources can be detected with great sensitivity by using interference filters and a focal reducer, without much loss in resolution. Extragalactic results obtained with the 6 meter telescope at focal ratio f/1 are given as examples.

031.639 **Speckle interferometry with the BTA telescope.**
V. N. Dudinov, V. V. Konichek, S. G. Kuz'menkov, V. S. Tsvetkova, V. S. Rylov, L. V. Gyavgyanen, V. N. Erokhin.
Instrumentation for astronomy with large optical telescopes, (see 012.059), p. 191 - 198 (1982).

The description of a speckle camera used with the BTA 6-meter telescope is given. Observational data recorded on photographic film are processed with a Fourier optical spectrum analyzer. Results of binary star measurement are presented.

031.640 **Experiments in differential speckle interferometry.**
J. M. Beckers, E. K. Hege.
Instrumentation for astronomy with large optical telescopes, (see 012.059), p. 199 - 206 (1982).

The authors propose to develop a technique to achieve submillisecond of arc resolution on stellar objects. It uses the fact that the spectrum of the light emitted by such an object often changes across its surface either because of Doppler shifts, Zeeman splitting, abundance anomalies or changes in the stellar atmosphere. It is possible to obtain differences between the position of speckles in the stellar image when viewed in slightly different wavelengths. It is possible to determine these differences in the position of the speckles with an accuracy much higher than the speckle size itself using techniques which are already developed for binary star-speckle research.

031.641 **High angular resolution with rotation shearing interferometers: preliminary results and future potentials.**
F. Roddier, C. Roddier.
Instrumentation for astronomy with large optical telescopes, (see 012.059), p. 207 - 211 (1982).

031.642 **Measurements of photon statistics with nanosecond resolution.**
D. Dravins.
Instrumentation for astronomy with large optical telescopes, (see 012.059), p. 229 - 237 (1982).

031.643 **Infrared speckle interferometry; results, true image reconstruction and instrumental plans.**
A. Chelli, C. Perrier, J. M. Mariotti.
Second ESO infrared workshop, (see 012.060), p. 153 - 158 (1982).

031.644 **The struggle against sky noise in ground based and air-borne observations.**
A. Tarrius, R. Papoular, T. Lebertre.
Second ESO infrared workshop, (see 012.060), p. 159 - 168 (1982).

031.645 **Infrared raster scans and a self optimising μ-processor system for IR-photometry.**
R. Riedinger, K. Haussecker, D. Lemke.
Second ESO infrared workshop, (see 012.060), p. 241 - 246 (1982).

031.646 **Aperture synthesis in the infrared.**
P. Léna.
Second ESO infrared workshop, (see 012.060), p. 259 - 268 (1982).

The paper first summarizes the knowledge of the difficulties associated with the retrieval of high frequencies in an infrared image distorted by the earth atmosphere. The author then discusses the specific point of beam combination in an aperture-synthesis scheme and concludes with a survey of present and future instrumental possibilities.

031.647 **Analysis of the possibilities of a bistatic radar investigation of planets using sources of cosmic radio radiation.**
A. I. Kucheryavenkov, A. G. Pavel'ev.
Radiotekh. i ehlektron., Tom 27, 914 - 920 (1982). In Russian. – Abstr. in Ref. zh., 62. Issled. kosm. prostranstva, 10.62.382 (1982).

031.648 **Automated spectral classification – a survey.**
T. Schmidt-Kaler.
Bull. Inf. Cent. Données Stellaires, No. 23, p. 2 - 12 (1982).

031.649 **Automated spectral classification.**
M. J. Kurtz.
Bull. Inf. Cent. Données Stellaires, No. 23, p. 13 - 14 (1982).

031.650 **Two-dimensional automated spectral classification.**
H. Zekl.
Bull. Inf. Cent. Données Stellaires, No. 23, p. 18 (1982).

031.651 **Classification spectrale: mise au point d'un système d'analyse rapide pour l'exploitation des "boîtes photométriques" de l'Observatoire de Genève.**
G. Goy.
Bull. Inf. Cent. Données Stellaires, No. 23, p. 19 - 25 (1982).

031.652 **Classification automatique des spectres de prisme objectif de l'Observatoire de Haute-Provence.**
F. Simien.
Bull. Inf. Cent. Données Stellaires, No. 23, p. 26 - 32 (1982).

031.653 **La classification spectrale automatique en tant que problème de reconnaissance de forme.**
A. Bijaoui.
Bull. Inf. Cent. Données Stellaires, No. 23, p. 33 - 36 (1982).

Progress in digital image processing allows now to develop automated spectral classifications. This problem is analysed as a pattern recognition problem. The definition of the representation is given and the interpretation spaces and the relation between the structures derived from the data and the concept.

031.654 **Le programme de classification stellaire UV.**
D. Egret.
Bull. Inf. Cent. Données Stellaires, No. 23, p. 37 - 38 (1982).

031.655 **Further experiments in automatic quantitative spectral classification at Tartu.**
V. Malyuto, J. Pelt.
Bull. Inf. Cent. Données Stellaires, No. 23, p. 39 - 40 (1982).

031.656 **Taking into account the conditions of orthogonality in planetodesic reduction of photographs.**
S. G. Valeev, A. L. Bezenovskaya, M. G. Shamarin.
Mol. uchenye i spets. Kuzbassa v 10 pyatiletke. Kemerovo, 1981, p. 17 - 20. In Russian. – Abstr. in Ref. zh., 52. Geod. Aehrosemka, 11.52.221 (1982).

031.657 **Numbering of photos and condensation points of a block in the construction of reference nets on the moon and planets.**
E. G. Belen'kij.
Sb. nauchn. tr. TsNII geod., aehrosemki i kartogr., 1981, No. 228, p. 88 - 109. In Russian. – Abstr. in Ref. zh., 52. Geod. Aehrosemka, 11.52.225 (1982).

031.658 **Main dependences of photogrammetry in the reduction of space photos of planets.**
Yu. S. Tyuflin.
Sb. nauchn. tr. TsNII geod., aehrosemki i kartogr., 1981, No. 228, p. 3 - 38. In Russian. – Abstr. in Ref. zh., 52. Geod. Aehrosemka, 11.52.226 (1982).

031.659 **A complex of algorithms and programs for the preparation of foundations for photographic maps of planetary surfaces.**
S. A. Kadnichanskij.
Sb. nauchn. tr. TsNII geod., aehrosemki i kartogr., 1981, No. 228, p. 75 - 87. In Russian. – Abstr. in Ref. zh., 52. Geod. Aehrosemka, 11.52.227 (1982).

031.660 **Seeing and surface photometry.**
A. Porter.
Publ. Astron. Soc. Pacific, Vol. 94, 757 (1982). – Abstract.

031.661 **Review of magnitude sources for visual cometary photometry. II.**
C. S. Morris, D. W. E. Green.
Int. Comet Q., Vol. 4, 62 - 63 (1982).

The authors have evaluated several references which have wide circulation among visual cometary observers. In connection with this review of magnitude sources, they have incorporated a new Reference Key for the Tabulation of Comet Observations.

031.662 **Measuring the anisotropy effect of the velocity of light with a double-beam interferometer.**
G. N. Izmajlov.
Moskva, 1982, 6 pp. In Russian. – Abstr. in Ref. zh., 51. Astron., 10.51.818 (1982).

031.663 **On a possibility of using the Michelson interferometer for measuring the anisotropy of the velocity of light in the gravitational field of the earth.**
G. N. Izmajlov.
Mosk. aviats. inst. Moskva, 1982. 7 pp. In Russian. – Abstr. in Ref. zh., 51. Astron., 10.51.819 (1982).

031.664 **Detection of extra solar planets.**
A. Labeyrie.
Formation of planetary systems, (see 012.064), p. 883 - 888 (1982).

The detection of extra solar planets has long remained beyond observational capabilities. The current dramatic improvement of optical instruments on the ground and in space makes it likely that this field will see major advances in the coming years. Initial discoveries should encourage observers to develop this as a major new field utilizing special techniques.

031.665 **Coronal investigations with occulted spacecraft signals.**
M. K. Bird.
Space Sci. Rev., Vol. 33, (see 012.067), 99 - 126 (1982).

The radio telemetry links between Earth and a spacecraft near superior conjunction penetrate the corona at ranges well within the acceleration regime of the solar wind. Occultation experiments in the solar corona have been performed on many interplanetary missions. The changes in group and phase velocity of the radio signal are measured to determine the total electron content of the corona and its fluctuations. The broadening of the carrier signal may be used in combination with the electron content data to derive a solar wind velocity profile. The wave number spectrum of electron density fluctuations in the corona may be inferred from amplitude and phase scintillations of the received signal. Linearly polarized signals, which are rotated along the propagation path by the Faraday effect, can provide information on the coronal magnetic field and its variations.

031.666 **Measurements of the properties of solar wind plasma relevant to studies of its coronal sources.**
M. Neugebauer.
Space Sci. Rev., Vol. 33, (see 012.067), 127 - 150 (1982).

Interplanetary measurements of the speeds, densities, abundances, and charge states of solar wind ions are diagnostic of conditions in the source region of the solar wind. The absolute values of the mass, momentum, and energy fluxes in the solar wind are not known to an accuracy of 20%. The principal limitations on the absolute accuracies of observations of solar wind protons and alpha particles arise from uncertain instrument calibrations, from the methods used to reduce the data, and from sampling biases. Sampling biases are very important in studies of alpha particles. Instrumental resolution and measurement ambiguities are additional major problems for the observation of ions heavier than helium. Progress in overcoming some of these measurement inadequacies is reviewed.

031.667 **A new search technique for M and C stars.**
L. G. Palmer, R. F. Wing.
Astron. J., Vol. 87, 1739 - 1750 (1982).

A method is described for identifying cool stars of types M and C by means of direct photography through narrowband interference filters. The relative magnitudes in filters centered at 7780 and 8120 Å reveal the presence of absorption by TiO or CN, respectively, while a third filter at 10400 Å allows the definition of a color index. Photographs of selected globular cluster fields have been taken through the 2-in.-diam filters with an S-1 image tube. Results are presented for the central region of 47 Tucanae in which the authors have found 14 new M stars, including five in the range M3–M5; no carbon stars were found.

031.668 **An interactive procedure to solve the background subtraction problems in IUE high resolution spectra: application to the image SWP 4616.**
C. Morossi, M. Ramella, M. Pucillo, P. Santin.
Astrophys. Space Sci., Vol. 87, 279 - 286 (1982).

The standard BKG subtraction and blaze efficiency correction in IUE high resolution images are unsatisfactory. The authors have developed an interactive procedure to eliminate these well-known inadequacies. A background statistically constant and a constant point spread function perpendicular to the direction of the dispersion have been assumed. The validity of these hypotheses is controlled during the run of the procedure for each image. The authors present an application to the IUE Hi Res SWP 4616 image.

031.669 **A mathematical reduction of astrolabe observations.**
V. S. Gubanov.
Izv. Glav. Astron. Obs. Pulkovo, Astrofiz. Astrometr., No. 200, p. 68 - 76 (1982). In Russian.

A theory of reducing calculations is given which can be used as an algorithm of the automatic computing of current observations with the astrolabe.

031.670 **On joint and separate reduction of equations in astrometry.**
E. I. Krejnin, S. A. Tolchel'nikova-Murri.
Izv. Glav. Astron. Obs. Pulkovo, Astrofiz. Astrometr., No. 200, p. 77 - 79 (1982). In Russian.

The conditions for equivalence of joint and separate solutions of the systems of equations used for reduction of observations in positional astronomy are derived. Formulae are given for calculating weights (errors) of unknowns depending upon the parameters derived by least squares.

031.671 **A comparison of reductions obtained from galaxies at Pulkovo.**
N. V. Fatchikhin.
Izv. Glav. Astron. Obs. Pulkovo, Astrofiz. Astrometr., No. 200, p. 107 - 108 (1982). In Russian.

A comparison is made of reductions of relative to absolute proper motions of stars obtained from galaxies with other data. The accuracy of the measurements of galaxies proved to be by 2.4 times smaller than that of 15^m stars.

031.672 **Speckle interferometry with the Multiple Mirror Telescope.**
E. K. Hege, D. McCarthy, P. Strittmatter, S. Worden, F. J. Lowe.
J. Opt. Soc. America, Vol. 72, 1759 (1982). – Abstract.

031.673 **Infrared imaging by interferometry with a large deployable reflector.**
J. B. Breckinridge.
J. Opt. Soc. America, Vol. 72, 1760 (1982). – Abstract.

031.674 **Interferometric alignment of optical surfaces and rotational axes.**
J. Davis, W. J. Tango, R. J. Thompson.
Appl. Opt., Vol. 21, 2867 - 2868 (1982).

031.675 **Rapid alternate detection system of the Rothney Astrophysical Observatory.**
E. F. Milone, R. M. Robb, F. M. Babott, C. H. Hansen.
Appl. Opt., Vol. 21, 2992 - 2995 (1982) = Publ. Rothney Astrophys. Obs., No. 19.

The system uses a single telescope, photometer, and detector and consists of gated pulse-counting electronics synchronized to a chopped secondary mirror. Within certain constraints the system is insensitive to the effects of transparency, sky-brightness, and detector-sensitivity variations occurring on time scales longer than the system's duty cycles.

031.676 **Computer simulation of a method for object reconstruction from stellar speckle interferometry data.**
J. G. Walker.
Appl. Opt., Vol. 21, 3132 - 3137 (1982).

A method for reconstructing a diffraction-limited image from data consisting of many short-exposure turbulence degraded images is described. The results of a computer simulation of the method are presented. The method should prove useful for obtaining high-angular resolution images from large earth bound telescopes. The simulation indicates that the method may be applicable to astronomical objects as faint as ~11th magnitude.

031.677 **Measurement of the line-of-sight velocity of high-altitude barium clouds: a technique.**
S. B. Mende, S. E. Harris.
Appl. Opt., Vol. 21, 3348 - 3353 (1982).

031.678 **Two-dimensional inversion technique for satellite airglow data.**
C. G. Fesen, P. B. Hays.
Appl. Opt., Vol. 21, 3784 - 3791 (1982).

A technique is described which inverts satellite airglow data producing volume emission rates as functions of altitude and position. The inversion is applied to data obtained when the spacecraft spins in the orbital plane. The altitude and height resolutions are constrained by the geometry chosen to simplify the inversion. The limitations of the method and its implementation on data from the Visual Airglow Experiment onboard the Atmosphere Explorer satellite are discussed. Sample maps of brightness and volume emission rates are shown.

031.679 **Digital speckle interferometry of Juno, Amphitrite and Pluto's moon Charon.**
G. Baier, N. Hetterich, G. Weigelt.
Messenger, No. 30, p. 23 - 26 (1982).

031.680 **Fourier transform spectroscopy in the submillimeter region.**
B. Carli.
Proc. SPIE Int. Soc. Opt. Eng., Vol. 289, (see 012.073), 394 - 403 (1981). – Abstr. in Phys. Abstr., Vol. 86, Abstr. 5887 (1983).

031.681 **Fourier spectroscopy on planetary missions including Voyager.**
R. A. Hanel.
Proc. SPIE Int. Soc. Opt. Eng., Vol. 289, (see 012.073), 331 - 344 (1981). – Abstr. in Phys. Abstr., Vol. 86, Abstr. 10182 (1983).

031.682 **A discussion on the measuring method of radial velocities with a direct vision mosaic objective prism.**
S. Jiang, S. Lu.
Acta Astrophys. Sinica, Vol. 2, 327 - 333 (1982). In Chinese.

031.683 **Will the next supernova in our Galaxy be discovered by a radio telescope?**
W. T. Sullivan III.
Publ. Astron. Soc. Pacific, Vol. 94, 901 - 904 (1982).

Given that a large fraction of young supernovae which occur in our Galaxy are probably too much obscured by dust to be detected optically and that little suitable radio monitoring now takes place, the chances of an accidental radio detection are estimated. An analysis of the characteristics and programs of the world's radio telescopes indicates that wavelengths in the 2- to 20-cm range are much more likely than others to make such a discovery, typically within 1–2 yr of the appearance of a new source.

031.684 **A technique for measuring magnetic fields on solar-type stars.**
G. W. Marcy.
Publ. Astron. Soc. Pacific, Vol. 94, 989 - 996 (1982) = Lick Obs. Bull., No. 935.

A method is developed and tested by which magnetic-field strengths and area filling-factors of magnetic regions on late-type stars may be inferred from high-resolution, absorption-line profiles that have been broadened by the Zeeman effect. The technique involves fitting such profiles with a triplet of components, the shape of which is derived from profiles of lines of low Zeeman sensitivity.

031.685 **The equivalent-width calibration of the KPNO échelle spectrograph.**
C. Pilachowski, C. Sneden, J. F. Dominy, P. Cottrell, D. C. May.
Publ. Astron. Soc. Pacific, Vol. 94, 1029 - 1036 (1982).

Equivalent widths measured from spectrograms obtained with the echelle spectrograph at the Kitt Peak National Observatory are compared in detail with measurements obtained from conventional coudé spectrographs. Equivalent-width measurements of the bright standard star Arcturus are used to recalculate the composition of that star. The results of all tests performed indicate that equivalent-width measurements from the echelle spectrographs are in reasonable agreement with those obtained from other instruments, and can be used reliably to derive element abundances in stars.

Real time control with the TRS-80.
See Abstr. 003.040.

Photoelectric photometry of variable stars.
See Abstr. 003.048.

Astronomical photometry.
See Abstr. 003.056.

RETICENT – a command language for spectrophotometric data reduction.
See Abstr. 021.030.

Information efficiencies of telescopes and their instrumentation.
See Abstr. 032.047.

Interferometry with very great base lines.
See Abstr. 033.020.

An outline of VLBI.
See Abstr. 033.021.

Significance of radio astronomy in cosmological research.
See Abstr. 033.022.

The response of a radio-astronomy synthesis array to interfering signals.
See Abstr. 033.037.

A SIT (*Silicon Intensified Target*) vidicon for surface photometry.
See Abstr. 034.010.

Das Zweikoordinatenmeßgerät ASCORECORD 3 DP im Dienste des Europäischen Feuerkugel-Programms.
See Abstr. 034.011.

Charge-coupled devices in astronomy.
See Abstr. 034.013.

A microcomputer-controlled photometer.
See Abstr. 034.059.

Modernizing Michelson's stellar interferometer.
See Abstr. 034.063.

Narrow-band GaAsSb heterojunction detectors for coronal polarimetry.
See Abstr. 034.068.

A coronographic mode for large telescopes.
See Abstr. 034.088.

Transit telescope designs optimized for multiple object spectroscopy with fibers.
See Abstr. 034.090.

A photometric system to search for optical variability on time-scales of 3×10^{-7} to 300s: main results.
See Abstr. 034.096.

Two dimensional interferometric photon counting observations with the 6m telescope.
See Abstr. 034.098.

Spectroscopy with photon-counting Reticons.
See Abstr. 034.101.

Experiments with silicon array detectors in the RGO coudé spectrograph.
See Abstr. 034.102.

Two-dimensional television system: application for direct acquisition of images and speckle interferometry.
See Abstr. 034.103.

TV spectrum scanner of the 6-meter telescope.
See Abstr. 034.104.

Application of infrared arrays to speckle interferometry.
See Abstr. 034.127.

Photoelastic-modulator polarimeters in astronomy.
See Abstr. 034.155.

A photoelectric radial-velocity spectrometer on the 1.2-m telescope of the Dominion Astrophysical Observatory.
See Abstr. 034.158.

Kodak IIIa-J plates – astrometry at Goethe Link Observatory.
See Abstr. 036.017.

Joint use of cosmic and astrogeodetic methods for geodynamical investigations.
See Abstr. 045.012.

PZT reductions based on the method of least-squares.
See Abstr. 045.021.

Requirements for observations of geostationary satellites.
See Abstr. 046.008.

Radiative transfer and remote sensing.
See Abstr. 063.015.

Results of investigations of internal gravitational waves with an optical method.
See Abstr. 082.007.

Internal gravitational waves near the mesopause. III. Method for determining the height of the radiating layer.
See Abstr. 082.008.

The importance of atmospheric differential refraction in spectrophotometry.
See Abstr. 082.056.

Ray tracing in a refracting spherically symmetric atmosphere.
See Abstr. 082.089.

Atmospheric ozone determination by solar occultation using the UV spectrometer on the Solar Maximum Mission.
See Abstr. 082.091.

Reassessment of net radiation measurements in the atmosphere of Venus.
See Abstr. 093.112.

Application of the method of collocation when constructing selenodetic reference nets.
See Abstr. 094.037.

Photoelectric lunar occultations. II. Data reduction and analysis.
See Abstr. 096.001.

Minor planets discoveries at the GPO, ESO-La Silla. Dependences of stars for catalogue improvement and future perturbation studies.
See Abstr. 098.002.

Spatially resolved methane band photometry of Saturn. I. Absolute reflectivity and center-to-limb variations in the 6190-, 7250-, and 8900-Å bands.
See Abstr. 100.028.

The IUE spectra of ET And – their reduction and variation.
See Abstr. 114.103.

Detection of other planetary systems.
See Abstr. 118.013.

A search for spectroscopic binaries among the runaway O type stars.
See Abstr. 120.015.

MERLIN spectral line observations of W3OH.
See Abstr. 131.144.

Quantitative measures of slitless spectra of QSOs.
See Abstr. 141.056.

The estimation of galaxy angular correlation functions.
See Abstr. 160.061.

031.901 **Erratum: 'An outline of a computer program for two-dimensional spectral classification'** [Astron. Astrophys., Vol. 108, 380 - 386 (1982)].
H. Zekl.
Astron. Astrophys., Vol. 113, 178 (1982). – See Abstr. 31.031.542.

032 Astronomical Instruments, Space Instrumentation

Astronomical Instruments

032.001 **Some applications of microcomputers in observatory automation.**
R. K. Honeycutt, J. E. Kephart.
Publ. Astron. Soc. Pacific, Vol. 94, 605 - 609 (1982).
The authors present some of the techniques used to automate many of the observing tasks on the 0.91 meter telescope of the Goethe Link Observatory. A description of the method used to calculate the dome position for a telescope which is mounted asymmetrically is included. Details of a novel autoguider are given. This autoguider uses a digitized television image of the star field to enable the microcomputer to generate error signals from a centroid calculation.

032.002 **The digital encoding system created for the 18-inch telescope.**
B. Wade, P. Ihnat.
Proc. Astron. Soc. Australia, Vol. 4, 254 - 256 (1981).

032.003 **A new system to eliminate gear backlash in telescopes.**
J. F. R. van der Ven.
Messenger, No. 29, p. 23 (1982).

032.004 **Astronomische Beobachtungen mit der Automatischen Kamera für Astrogeodäsie der JENOPTIK JENA GmbH.**
D. A. Rozkovskij *(Rozkhovskij)*, D. I. Gorodeckij *(Gorodetskij)*.
Jenaer Rundsch., 27. Jahrg., 121 - 122 (1982).

032.005 **El instrumental de la astronomía optica.**
H. Elsässer.
Rev. Astron., Tomo 54, No. 221, p. 6 - 11 (1982).

032.006 **Setting of the polar axis of the 40 cm reflector.**
H. Kubota, N. Owaki, T. Mizuno.
Bull. Tokyo Gakugei Univ., Ser. IV, Vol. 32, 209 - 219 (1980). In Japanese. – Abstr. in Phys. Abstr., Vol. 85, Abstr. 63248 (1982).

032.007 **Automatic photoelectric meridian circle.**
H. Yasuda.
Astron. Her., Vol. 74, 290 - 296 (1981). In Japanese. – Abstr. in Phys. Abstr., Vol. 85, Abstr. 68614 (1982).

032.008 **Wide-field imagery at the CFH telescope prime focus.**
J. C. Fouere, G. Lelievre, J. P. Lemonnier, D. A. Salmon, G. J. Odgers, E. H. Richardson.
Opt. Acta, Vol. 29, (see 012.018), 383 - 390 (1982). – Abstr. in Phys. Abstr., Vol. 85, Abstr. 73839 (1982).

032.009 **Possible applications of large solar arrays in astronomy and astrophysics.**
S. Danaher, D. J. Fegan. N. A. Porter, T. C. Weekes, T. Cole.
Sol. Energy, Vol. 28, 335 - 343 (1982). – Abstr. in Phys. Abstr., Vol. 85, Abstr. 73843 (1982).

032.010 **Experiences with the U.S. Naval Observatory glass circles.**
T. J. Rafferty, B. L. Klock.
Astron. Astrophys., Vol. 114, 95 - 101 (1982).
The intent of the paper is to present the experiences with the glass circles acquired for the six-inch transit circle at the U.S. Naval Observatory. Analyses of the diameter corrections obtained with a number of glass circles are given. A method is introduced which permits the monitoring of the stability of the circles without taking an entire set of diameter corrections.

032.011 **The Mexican 2.12-m telescope for the National Astronomical Observatory at San Pedro Martir, Baja California, Mexico.**
J. de la Herrán.
Recent advances in observational astronomy, (see 012.028), p. 133 - 139 (1981).

032.012 **Arrays vs. single telescopes.**
H. L. Johnson.
Recent advances in observational astronomy, (see 012.028), p. 141 - 146 (1981).
The question of the relative efficiencies of telescope arrays versus an equivalent mirror-area very large telescope is re-examined and summarized.

032.013 **The University of California Ten-Meter Telescope Project.**
E. J. Wampler.
Recent advances in observational astronomy, (see 012.028), p. 147 - 151 (1981).

032.014 **The Multiple Mirror Telescope.**
J. M. Beckers, B. L. Ulich, R. R. Shannon, N. P. Carleton, J. C. Geary, D. W. Latham, J. Roger, P. Angel, W. F. Hoffmann, F. J. Low, R. J. Weymann, N. J. Woolf.
Telescopes for the 1980s, (see 003.010), p. 63 - 128 (1981).
Contents: Early multiple objective telescopes. Early history of the MMT project. R. J. Weymann: Personal recollections. Design and performance of the MMT. MMT instrumentation. Astronomical research with the MMT. Future multiple objective telescopes.

032.015 **Optische telescopen gaan met hun tijd mee.**
H. van de Stadt.
Zenit, 9. Jaarg., 452 - 457 (1982).

032.016 **Astroplane: a European research aircraft for IR and submillimetre astronomy (the DFVLR proposal).**
C. B. Cosmovici, E. Schmidt, U. Stanggassinger.
The scientific importance of submillimetre observations, (see

012.030), p. 69 - 79 (1982).

A proposal for an European Aircraft carrying a competitive infrared-submillimeter telescope on board is presented. The DFVLR (German Aerospace Research Center) proposes a Challenger 601 (Canadair) Aircraft as carrier for a 120 cm telescope. New concepts for a light-weight updated airborne astrolaboratory are outlined.

032.017 **A portable system for high-speed photometry of occultations.**
R. A. Nye, R. C. Oliver, R. L. Millis.
Bull. American Astron. Soc., Vol. 14, 711 - 712 (1982).
Abstract.

032.018 **An accurate derivation of the division corrections in a photoelectric meridian circle.**
M. Miyamoto, C. Kühne.
Astron. Astrophys., Suppl. Ser., Vol. 50, 173 - 186 (1982).

In an accurate determination of the division corrections for the graduated circle (with 3600 divisions) of a photoelectric and automized meridian circle, a method is presented to derive simultaneously unbiased microscope scales and the corrections. The principle is a randomization of the measuring order of all the divisions whereby it is possible to separate clearly the time variation of the scales from each Fourier component of the graduation error.

032.019 **Inventory of major operational and planned ground-based astronomical telescopes of the countries in the European Science Foundation (second edition, 1982).**
European Science Foundation.
Astron. Astrophys., Suppl. Ser., Vol. 50, 187 - 193 (1982).

This paper gives an inventory of: Optical and infrared telescopes (100 cm diameter or larger) – Diameter of optical and infrared telescopes against year of completion (diagram) – Schmidt telescopes (70 cm aperture or larger) – Optical and infrared telescopes (histogram) – Solar optical telescopes (30 cm diameter or larger) – Large radio telescopes – List of addresses – Major operational and planned radio telescopes (diagram).

032.020 **Design and performance of a new emission-line coronagraph.**
R. N. Smartt, R. B. Dunn, R. R. Fisher.
Proc. SPIE Int. Soc. Opt. Eng., Vol. 288, (see 012.035), 395 - 402 (1981). – Abstr. in Phys. Abstr., Vol. 85, Abstr. 106528 (1982).

032.021 **Telescope control.**
Elektor, Vol. 7, No. 11, p. 21 - 22 (1981). – Abstr. in Phys. Abstr., Vol. 85, Abstr. 106529 (1982).

032.022 **An analysis of the division errors of the Belgrade Large Meridian Instrument.**
V. Trajkovska.
Publ. Astron. Obs. Sarajevo, Vol. 1, No. 1, (see 012.043), p. 85 - 91 (1981).

032.023 **New method of measurement for the corrections determination of the totality of circle diameters.**
D. Bozhickovich.
Publ. Astron. Obs. Sarajevo, Vol. 1, No. 1, (see 012.043), p. 93 - 101 (1981).

032.024 **Transistorized servo-regulator of the Askania astrograph drive mechanism of the Belgrade Observatory.**
L. Varga, V. Protić-Benišek.
Publ. Astron. Obs. Sarajevo, Vol. 1, No. 1, (see 012.043), p. 153 - 155 (1981).

032.025 **De Schmidt-Cassegraintelescoop. Een testrapport.**
S. Wadman.
Zenit, 9. Jaarg., 513 - 519 (1982).

032.026 **Influence of some instrument errors on time observations with transit instruments and their determination.**
T. Darakchiev.
Geodesy and physics of the earth, (see 012.053), p. 55 - 71 (1981).

Two instrument errors are subject of the study – ellipticity and inequality of transit instrument horizontal axis bearing working sections.

032.027 **Ultraviolet, optical and infrared astronomy.**
E. J. Wampler.
Phys. Today, Vol. 35, No. 11, p. 44 - 51 (1982).

A 15-m telescope on the ground and a far-ultraviolet spectrograph and 10-m infrared telescope in orbit would complement projects already planned and give us totally new insight into the nature of the universe.

032.028 **How to automate a 35 mm all-sky camera for an automatic fireball network.**
H. Betlem, H. E. Mostert.
J. British Astron. Assoc., Vol. 93, 11 - 16 (1982).

032.029 **A new infrared telescope.**
F. Pacini.
Messenger, No. 30, p. 21 - 22 (1982).

032.030 **Der Weg zum Vielzweck-Teleskop.**
Sternenbote, 25. Jahrg., 127 - 139 (1982).

032.031 **Das Vielzweck-Teleskop MPT 300.**
Sternenbote, 25. Jahrg., 146 - 159 (1982).

032.032 **The largest optical telescopes.**
J. Pasachoff.
Mercury, Vol. 11, 142 - 143 (1982).

032.033 **Le télescope de 2 mètres du Pic-du-Midi: premières observations des planètes.**
A. Dollfus.
Astronomie, Vol. 96, 539 - 552 (1982).

032.034 **Zur Leistungsfähigkeit von transportablen Zenitkameras bei der Lotabweichungsbestimmung.**
H. Wissel.
Wiss. Arb. Fachrichtung Vermessungswes. Univ. Hannover, Nr. 107, 5 + 130 pp. (1982).

032.035 **A technical report of the photographic astrolabe of Turku Observatory.**
A. Niemi.
Rep. Finnish Geod. Inst., No. 82:4, 57 pp. (1982). ISBN 951-711-081-2.

032.036 **High accuracy measurement of the instrumental polarization of the solar coudé telescope at the Okayama Astrophysical Observatory.**
M. Makita, S. Hamana, H. Kawakami, K. Nishi.
Ann. Tokyo Astron. Obs., Second Ser., Vol. 19, 24 - 37 (1982).

032.037 **On the motion of the PZT carriage (the mechanical precision and elastic deformation).**
H. Ishizaki, K. Nakajima.
Tokyo Astron. Obs. Rep., Vol. 19, 537 - 546 (1981). In Japanese.

032.038 **Systematic errors and the observing accuracy of the Gautier meridian circle of Tokyo Astronomical Observatory.**
H. Hara, K. Hurukawa.
Tokyo Astron. Obs. Rep., Vol. 19, 596 - 606 (1981). In Japanese.

032.039 **The Zelenchuk 6m telescope (BTA) of the USSR Academy of Sciences.**
B. K. Ioannisiani, E. M. Neplokhov, I. M. Kopylov, V. S. Rylov, L. I. Snezhko.
Instrumentation for astronomy with large optical telescopes, (see 012.059), p. 3 - 10 (1982).

032.040 **Progress report on the performance of the Multiple Mirror Telescope.**
J. M. Beckers, B. L. Ulich.
Instrumentation for astronomy with large optical telescopes, (see 012.059), p. 11 - 19 (1982).

032.041 **The Canada-France-Hawaii telescope.**
P. Y. Bely, G. Lelièvre.
Instrumentation for astronomy with large optical telescopes, (see 012.059), p. 21 - 32 (1982).

The general concepts and technical details of the telescope and observatory are discussed. The focal plane instrumentation includes more than 25 instruments and detectors, and these are also discussed.

032.042 **Optical system for the University of Texas 7.6m telescope.**
R. G. Tull, T. G. Barnes, D. N. Dittmar, F. B. Ray, E. Pearson.
Instrumentation for astronomy with large optical telescopes, (see 012.059), p. 37 - 44 (1982).

The University of Texas will have a Ritchey-Chrétien configuration, f/13.5 with a 7.6m diameter f/2 primary of ultra-thin construction, on an alt-azimuth mount. The authors describe the current concept for the mounting, the building, mirror handling techniques, and the optical system and give preliminary results of the finite element flexure analysis of the mirror on its support system.

032.043 **A 6-metre polar-viewing telescope.**
E. B. Gajur.
Instrumentation for astronomy with large optical telescopes, (see 012.059), p. 45 - 47 (1982).

032.044 **Design of the 4.2 m Herschel telescope and instruments.**
P. Murdin.
Instrumentation for astronomy with large optical telescopes, (see 012.059), p. 49 - 59 (1982).

032.045 **Aerodynamic and thermal problems around and inside a dome and telescope.**
J. Rösch.
Instrumentation for astronomy with large optical telescopes, (see 012.059), p. 79 - 84 (1982).

Much effort has been spent at Pic-du-Midi, with increasingly large telescopes, to reduce the instrumental causes of image deterioration and take best advantage of local atmospheric conditions. With the new 2-meter reflector, specifically intended for high resolution imaging, the concept has been, ideally, to separate three air-masses: outside, inside the tube, and inside the dome.

032.046 **The performance of ground based telescopes.**
D. S. Brown.
Instrumentation for astronomy with large optical telescopes, (see 012.059), p. 85 - 92 (1982).

The paper identifies causes of degradation of image quality in astronomical telescopes, describes how the effects of these can be calculated using MTFs, considers the validity of multiplication of MTFs for this purpose and gives the results of measurement of external and dome seeing, system MTF and image quality obtained during the commissioning of UKIRT.

032.047 **Information efficiencies of telescopes and their instrumentation.**
V. S. Rylov.
Instrumentation for astronomy with large optical telescopes, (see 012.059), p. 93 - 98 (1982).

A criterion based upon information content allows one to obtain objective estimates of the efficiencies of astronomical observations for the whole system of light receiving and recording equipment, including the conditions of observation. The proposed method for calculating this information efficiency relies on the analysis of the output observational material and not on the characteristics of optical and electronic components of the instrumentation.

032.048 **The C.F.H. (*Canada-France-Hawaii*) telescope as an infrared telescope.**
J. P. Maillard, H. M. Dyck.
Second ESO infrared workshop, (see 012.060), p. 29 - 43 (1982).

032.049 **The TIRGO observatory.**
P. Salinari.
Second ESO infrared workshop, (see 012.060), p. 45 - 55 (1982).

032.050 **Use of the 3.6 m telescope for millimeter observations.**
A. R. Gillespie.
Second ESO infrared workshop, (see 012.060), p. 179 - 180 (1982).

032.051 **Modernization of the clock drive of the horizontal solar telescope of the Pulkovo Observatory.**
O. V. Nikonov, L. M. Zatsiorskij, A. P. Kulish, E. S. Nikonova.
Soln. Dannye 1982 Byull., No. 8, p. 83 - 87 (1982). In Russian.

The improvements of the clock drive of the horizontal solar telescope of the Pulkovo Observatory are described.

032.052 **Sonnenforschungsanlagen. I.**
H.-J. Teske, J. Reiche.
Sterne, 58. Band, 270 - 282 (1982).

032.053 **Sub-millimetre aircraft telescope for astronomical investigations.**
V. D. Gromov, I. E. Ivanov, Yu. A. Kovalev, V. B. Lazarev, I. A. Maslov, M. Z. Khokhlov.
Inst. kosm. issled. AN SSSR. Prepr., 1982, No. 695, 28 pp. In Russian. – Abstr. in Ref. zh., 51. Astron., 10.51.827 (1982).

032.054 **A study of refraction anomalies in horizontal tubes of a horizontal meridian circle using a photoelectric wedge eye-piece micrometer.**
T. R. Kir'yan, G. I. Pinigin, G. M. Timashkova.
Izv. Glav. Astron. Obs. Pulkovo, Astrometr. Astrofiz., No. 199, p. 43 - 47 (1982). In Russian.

032.055 **Some photometric characteristics of the fast two-meniscus astrograph AZT-16.**
L. A. Panaiotov.
Izv. Glav. Astron. Obs. Pulkovo, Astrometr. Astrofiz., No. 199, p. 56 - 63 (1982). In Russian.

Photometric errors of the two-meniscus astrograph AZT-16 have been determined. Corrections have been obtained in the B-band as a function of the distance to the center of the plate. A colour equation giving photometric corrections depending on the colour of the star is found. The effect of the form of the image of a star and curvature of the plate on the accuracy of photometric measurements is considered.

032.056 **On the improvement of the accuracy of measurements of the inclination of the horizontal axis of a transit instrument.**
L. I. Lobanov, G. M. Petrov.
Izv. Glav. Astron. Obs. Pulkovo, Astrometr. Astrofiz., No. 199, p. 64 - 65 (1982). In Russian.

032.057 **Limb division of astronomical instruments in Nikolaev.**
N. A. Il'kiv.
Izv. Glav. Astron. Obs. Pulkovo, Astrometr. Astrofiz., No. 199, p. 66 - 70 (1982). In Russian.

032.058 **An instrument for absolute determinations of declinations of equatorial stars using observations in the prime vertical.**
A. A. Nemiro.
Izv. Glav. Astron. Obs. Pulkovo, Astrofiz. Astrometr., No. 200, p. 80 - 83 (1982). In Russian.
A constructive scheme of an astrometric instrument is proposed that can be used for determinations of absolute declinations of equatorial stars from observations in the prime vertical.

032.059 **The system of the Pulkovo vertical circle in the zenith zone.**
S. P. Pulyaev.
Izv. Glav. Astron. Obs. Pulkovo, Astrofiz. Astrometr., No. 200, p. 84 - 88 (1982). In Russian.
It is shown as a result of observations of FK 4 stars in 1977 - 1980 using the photographic vertical circle that the instrument system was stable during the period of observations.

032.060 **A determination of the value of a revolution of the micrometer screw of the ZTF-135 at Pulkovo using observations of large arcs in declination.**
L. D. Kostina, N. R. Persiyaninova.
Izv. Glav. Astron. Obs. Pulkovo, Astrofiz. Astrometr., No. 200, p. 101 - 104 (1982). In Russian.

032.061 **A photoelectric system of limb readings for the Pulkovo horizontal meridian circle.**
R. I. Gumerov, V. B. Kapkov, G. I. Pinigin, O. E. Shornikov.
Izv. Glav. Astron. Obs. Pulkovo, Astrofiz. Astrometr., No. 200, p. 114 - 117 (1982). In Russian.

032.062 **Preliminary results of investigation of the pendulum mirror horizon of the Pulkovo horizontal meridian circle.**
L. A. Sukharev, G. M. Timashkova, G. I. Pinigin, T. R. Kir'yan.
Izv. Glav. Astron. Obs. Pulkovo, Astrofiz. Astrometr., No. 200, p. 118 - 121 (1982). In Russian.

032.063 **A complex of the photoguide of the Pulkovo horizontal solar telescope.**
O. V. Nikonov, A. P. Kulish, L. M. Zatsiorskij, Yu. S. Muzalevskij.
Izv. Glav. Astron. Obs. Pulkovo, Astrofiz. Astrometr., No. 200, p. 122 - 131 (1982). In Russian.

032.064 **What a very large multi-purpose solar telescope should be.**
V. N. Karpinskij.
Izv. Glav. Astron. Obs. Pulkovo, Astrofiz. Astrometr., No. 200, p. 135 - 139 (1982). In Russian.

032.065 **A 600 mm vacuum solar tower telescope based on the ACU-26 coelostat system.**
Kh. I. Abdusamatov.
Izv. Glav. Astron. Obs. Pulkovo, Astrofiz. Astrometr., No. 200, p. 140 - 145 (1982). In Russian.
A most efficient vacuum solar tower telescope of moderate size is proposed. An attempt is made to minimize the thermal and aerodynamic effects of its tower and dome. The optics of the telescope and the tower are sufficiently stable and provide regularly for solar photographs and spectra with a high spatial resolution (~0".3).

032.066 **A method of determination of the error of mounting and flexure of an equatorial telescope.**
N. N. Mikhel'son.
Izv. Glav. Astron. Obs. Pulkovo, Astrofiz. Astrometr., No. 200, p. 154 - 159 (1982). In Russian.

032.067 **Experiments with the Multiple Mirror Telescope.**
J. M. Beckers.
J. Opt. Soc. America, Vol. 72, 1759 (1982). – Abstract.

032.068 **Optical characteristics of the UNAM *(Universidad Nacional Autonoma de Mexico)* 2-m Ritchey-Chretien telescope.**
R. Noble, F. Cobos, F. Diego, J. Sasian.
Appl. Opt., Vol. 21, 3181 - 3183 (1982).

032.069 **Solar Diameter Monitor: an instrument to measure long-term changes.**
T. M. Brown, D. F. Elmore, L. Lacey, H. Hull.
Appl. Opt., Vol. 21, 3588 - 3596 (1982).
Analyses of historical data suggest that the solar diameter may vary with time with an amplitude of a few tenths of a second of arc. The telescope is an *f*/50 transit instrument with an aperture of 10 cm and is almost completely automated to avoid observer bias. Each day at solar noon, it measures the sun's horizontal diameter by timing the solar disk transit time and the vertical diameter by comparing the image size to that of a stable length standard. Preliminary estimates suggest that these observations will allow a test of the solar diameter's constancy at the 1 sec of arc/century level in an observing time of 3 - 5 years.

032.070 **New development of large telescopes in the world. I.**
M. Simizu, S. Isobe.
Astron. Her., Vol. 75, 77 - 80 (1982). In Japanese. – Abstr. in Phys. Abstr., Vol. 86, Abstr. 5111 (1983).

032.071 **New development of large telescopes in the world. II.**
M. Simizu, S. Isobe.
Astron. Her., Vol. 75, 100 - 105 (1982). In Japanese. – Abstr. in Phys. Abstr., Vol. 85, Abstr. 117587 (1982).

032.072 **On the domeless solar telescope in Hida Observatory.**
I. Kawaguchi.
Astron. Her., Vol. 75, 188 - 194 (1982). In Japanese. – Abstr. in Phys. Abstr., Vol. 85, Abstr. 117588 (1982).

032.073 **Simple infrared telescope with stray-light rejection.**
D. Shafer.
Proc. SPIE Int. Soc. Opt. Eng., Vol. 304, (see 012.074), 171 - 175 (1981). – Abstr. in Phys. Abstr., Vol. 85, Abstr. 117596 (1982).

032.074 **Aluminization of mirrors at the Okayama Astrophysical Observatory.**
E. Watanabe.
Astron. Her., Vol. 75, 113 - 116 (1982). In Japanese. – Abstr. in Phys. Abstr., Vol. 85, Abstr. 117605 (1982).

032.075 **Fabrication of a honeycomb mirror using borosilicate glass.**
T. Noguchi.
Astron. Her., Vol. 75, 234 - 237 (1982). In Japanese. – Abstr. in Phys. Abstr., Vol. 86, Abstr. 5116 (1983).

032.076 **A capacitive bubble level and its application to the transit instrument.**
L.-y. Xie.
Acta Astron. Sinica, Vol. 23, 385 - 389 (1982). In Chinese.

Remote operation of telescopes.
See Abstr. 011.002.

Astronomical capability of the U.K. Infrared Telescope.
See Abstr. 013.087.

An Air Force/NSF initiative to study the origins of solar activity.
See Abstr. 013.102.

The liquid-mirror telescope as a viable astronomical tool.
See Abstr. 031.001.

Progress on the 3.5 m "New Technology Telescope" (NTT).
See Abstr. 031.002.

The Leoncito coudé conversion.
See Abstr. 031.012.

Temos 4. A French solution to the mosaic telescope problem.
See Abstr. 031.014.

Image quality considerations in ESO telescope projects.
See Abstr. 031.015.

A configuration of the optical system for a 5 m telescope.
See Abstr. 031.017.

Instrumental polarization of a horizontal solar telescope.
See Abstr. 031.029.

Vector conversion with a coelostat and additional mirror of a solar telescope.
See Abstr. 031.030.

Optical figuring process, Dupont 2.6 m telescope.
See Abstr. 031.060.

Comparison of coherent and incoherent imaging in the location of point sources.
See Abstr. 031.063.

Use of large telescopes in fast focal ratio mode – application to the 6m telescope (BTA).
See Abstr. 031.638.

Future UKIRT instrumentation.
See Abstr. 034.118.

High resolution imaging from the ground.
See Abstr. 082.046.

Space Instrumentation

032.501 **An amateur space telescope.**
R. J. Sawyer.
Sky Telesc., Vol. 64, 127 - 129 (1982).

032.502 **Detectors in astronomy: X-ray astronomy.**
B. Taylor.
Nature, Vol. 298, 798 - 800 (1982).

032.503 **The IUE échelle blaze function.**
T. B. Ake.
Advances in ultraviolet astronomy, (see 012.006), p. 318 - 321 (1982).

An improved form for the échelle blaze function derived by Ake (1981) has been used to study the apparent variation of grating constant K with order m for the long and short wavelength spectrographs. For the LWR camera, early images indicate that K should vary linearly with order. In four years of use, changes in sensitivity across the camera has changed the observed blaze function, inducing a curvature in the K versus m relation. This curvature varies in time with variations in camera response. The SWP camera shows these effects to a lesser degree.

032.504 **The Space Telescope.**
J. N. Bahcall, L. Spitzer, Jr.
Sci. American, Vol. 247, No. 1, p. 38 - 49 (1982).

The largest astronomical telescope designed to operate beyond the interfering effects of the earth's atmosphere is scheduled to be transported into orbit by the U. S. space shuttle in 1985.

032.505 **Mass spectrometers for investigation of the atmospheres of the earth and planets.**
V. A. Pavlenko, M. A. Berezhkovskij, V. G. Klimovitskij, G. N. Levina, L. N. Ozerov, M. A. Pushkina, M. E. Slutskij.
Pribory dlya nauchn. issled. i avtomatiz. ehksp. Leningrad, 1982, p. 94 - 104. In Russian. – Abstr. in Ref. zh., 51. Astron., 8.51.891 (1982).

032.506 **Laboratory calibration of the FOTON-1 spectrometer determined for measurement of atmospheric emission in the UV region aboard the IKB-1300 satellite.**
M. Yu. Kozlovskij, V. G. Kurt, M. D. Nusinov, T. V. Perevodchikova, Yu. S. Pogodin, L. P. Smirnova, M. Gogoshev, S. Sargojchev.
Kosm. Issled., Tom 20, 620 - 630 (1982). In Russian.

032.507 **The European X-ray observatory Exosat – its mission and scientific instruments.**
B. G. Taylor, R. D. Andresen, A. Peacock, R. Zobl.
ESA Bull., No. 31, p. 20 - 33 (1982).

032.508 **NASA's Kuiper Airborne Observatory.**
L. C. Haughney.

Bull. American Astron. Soc., Vol. 14, 608 - 609 (1982). Abstract.

032.509 **High resolution near-infrared spectroscopy from the Kuiper Airborne Observatory.**
H. P. Larson, D. S. Davis.
Bull. American Astron. Soc., Vol. 14, 610 (1982). – Abstract.

032.510 **The Wisconsin Ultraviolet Photo-Polarimeter Experiment.**
K. H. Nordsieck, A. D. Code.
Bull. American Astron. Soc., Vol. 14, 657 (1982). – Abstract.

032.511 **The Hopkins Ultraviolet Telescope: a far UV spectrophotometer for the Space Shuttle.**
A. F. Davidsen, K. S. Long.
Bull. American Astron. Soc., Vol. 14, 657 - 658 (1982). Abstract.

032.512 **The Space Telescope Observatory.**
C. R. O'Dell.
The Space Telescope Observatory, (see 012.013), p. 20 - 27 (1982).
The history of the observatory is briefly described, as is the overall system design. The principal design features that are important to the scientific user are described. The present status of preparation for launch in the spring of 1985 is reported on.

032.513 **The Wide Field/Planetary Camera.**
J. A. Westphal.
The Space Telescope Observatory, (see 012.013), p. 28 - 39 (1982).
A wide suite of potential astronomical and solar system scientific studies using the Wide Field/Planetary Camera on Space Telescope are described. The expected performance of the camera as it approaches final assembly and testing is also detailed.

032.514 **The Faint Object Camera.**
F. Macchetto.
The Space Telescope Observatory, (see 012.013), p. 40 - 54 (1982).
The Faint Object Camera will fully exploit the spatial resolution capability of the Space Telescope on the very faintest detectable objects over a broad wavelength range. A full complement of filters, objective prisms and polarizers, a choice of coronographic masks, and a variety of scan formats extend the scientific versatility of the direct imaging mode. In addition, the Faint Object Camera provides the unique facility of long-slit spectroscopy to Space Telescope observers.

032.515 **Astronomical capabilities of the Faint Object Spectrograph on Space Telescope.**
R. J. Harms.
The Space Telescope Observatory, (see 012.013), p. 55 - 75 (1982).
Examples of scientific observing programs planned with the Faint Object Spectrograph on Space Telescope are presented. An overview of the spectrograph design and operation is presented. The expected astronomical performance of the instrument is described in some detail.

032.516 **The High Resolution Spectrograph for the Space Telescope.**
J. C. Brandt.
The Space Telescope Observatory, (see 012.013), p. 76 - 105 (1982).
The High Resolution Spectrograph in conjunction with the Space Telescope, will extend ultraviolet astronomical spectroscopy to higher spectral, spatial, and time resolutions than previously achieved, as well as to fainter and more distant celestial objects. Other significant advances inherent in the instrument are high photometric accuracy and efficient operation via exposure meter control and real-time rejection of bad data.

032.517 **The High Speed Photometer for the Space Telescope.**
R. C. Bless.
The Space Telescope Observatory, (see 012.013), p. 106 - 113 (1982).
The High Speed Photometer exploits the capabilities of the Space Telescope by making photometric measurements over the visual and ultraviolet at rates up to 10^5 Hz and by measuring very low amplitude variability (especially for hotter stars in the UV). A secondary purpose of the instrument is to measure linear polarization in the near UV.

032.518 **Infrared telescope on Spacelab 2.**
D. Koch, G. G. Fazio, W. A. Traub, G. H. Rieke, T. N. Gautier, W. F. Hoffmann, F. J. Low, W. Poteet, E. T. Young, E. W. Urban, L. Katz.
Opt. Eng., Vol. 21, 141 - 147 (1982). – Abstr. in Phys. Abstr., Vol. 85, Abstr. 63250 (1982).

032.519 **Accelerator calibration of the Low Energy Telescope of the International Solar Polar Mission COSPIN experiment.**
J. F. Le Borgne, J. Henrion, R. G. Marsden, T. R. Sanderson, K.-P. Wenzel.
Report ESA STM-224, European Space Agency, Paris, France. 6 + 21 pp. (1981). – Abstr. in Phys. Abstr., Vol. 85, Abstr. 63251 (1982).

032.520 **Light ion mass spectrometer for space-plasma investigations.**
D. L. Reasoner, C. R. Chappell, S. A. Fields, W. J. Lewter.
Rev. Sci. Instrum., Vol. 53, 441 - 448 (1982). – Abstr. in Phys. Abstr., Vol. 85, Abstr. 64159 (1982).

032.521 **Hard X-ray telescope on the satellite Hinotori.**
K. Oki.
Astron. Her., Vol. 74, 316 - 320 (1981). In Japanese. – Abstr. in Phys. Abstr., Vol. 85, Abstr. 68615 (1982).

032.522 **A measurement of cosmic-ray LET-spectra using a microprocessor supported microscope.**
J. Beer, W. Heinrich.
Nucl. Tracks Methods Instrum. Appl., Vol. 5, (see 012.015), 389 - 390 (1981). – Abstr. in Phys. Abstr., Vol. 85, Abstr. 68623 (1982).

032.523 **Isotope-measurement of cosmic ray nuclei from Ca to Fe by a plastic-electronic-experiment.**
W. Hunger, G. Siegmon, G. Sermund, R. Beaujean, W. Enge, W. R. Webber.
Nucl. Tracks Methods Instrum. Appl., Vol. 5, (see 012.015), 402 (1981). – Abstr. in Phys. Abstr., Vol. 85, Abstr. 68624 (1982).

032.524 **Low temperature characteristics of solid state detectors for energetic X-ray, ion and electron spectrometers *(spacecraft instrumentation)*.**
H. D. Voss, J. B. Reagan, W. L. Imhof, D. O. Murray, D. A. Simpson, D. P. Cauffman, J. C. Bakke.
IEEE Trans. Nucl. Sci., Vol. NS-29, (see 012.023), 164 - 168 (1982). – Abstr. in Phys. Abstr., Vol. 85, Abstr. 75108 (1982).

032.525 **An imaging gas scintillation proportional counter for the detection of subkiloelectron-volt X-rays.**
C. J. Hailey, W. H.-M. Ku, M. H. Vartanian.
IEEE Trans. Nucl. Sci., Vol. NS-29, (see 012.023), 138 - 141

(1982). – Abstr. in Phys. Abstr., Vol. 85, Abstr. 78950 (1982).

032.526 **A gas scintillation camera for X-ray astronomy.**
J. Davelaar, A. Peacock, B. G. Taylor.
IEEE Trans. Nucl. Sci., Vol. NS-29, (see 012.023), 142 - 145 (1982). – Abstr. in Phys. Abstr., Vol. 85, Abstr. 78951 (1982).

032.527 **A sealed high pressure xenon-carbon dioxide filled position sensitive proportional counter.**
H. D. Thomas, M. R. Sims, M. J. L. Turner.
IEEE Trans. Nucl. Sci., Vol. NS-29, (see 012.023), 148 - 150 (1982). – Abstr. in Phys. Abstr., Vol. 85, Abstr. 78952 (1982).

032.528 **A bismuth germanate-shielded mercuric iodide X-ray detector for space applications.**
J. V. Vallerga, G. R. Ricker, W. S. Schnepple, C. Ortale.
IEEE Trans. Nucl. Sci., Vol. NS-29, (see 012.023), 151 - 154 (1982). – Abstr. in Phys. Abstr., Vol. 85, Abstr. 78953 (1982).

032.529 **A coded aperture gamma ray telescope.**
M. L. McConnell, D. J. Forrest, E. L. Chupp, P. P. Dunphy.
IEEE Trans. Nucl. Sci., Vol. NS-29, (see 012.023), 155 - 159 (1982). – Abstr. in Phys. Abstr., Vol. 85, Abstr. 78954 (1982).

032.530 **Electronic system for an impact mass spectrometer with high dynamic range for the Giotto probe.**
A. Glasmachers, P. Dullenkopf, J. W. Klein, J. Melbert.
IEEE Trans. Nucl. Sci., Vol. NS-29, (see 012.023), 160 - 163 (1982). – Abstr. in Phys. Abstr., Vol. 85, Abstr. 78955 (1982).

032.531 **Camera window for ultrasoft X-rays from celestial sources.**
J. A. M. Bleeker, W. H. Diemer, A. P. Huben, H. Huizenga.
Philips Tech. Rev., Vol. 40, No. 1, p. 12 - 18 (1982). – Abstr. in Phys. Abstr., Vol. 85, Abstr. 84096 (1982).

032.532 **Development, design, and characteristics of an automatic pointing system for infrared telescopes for use on stratospheric balloons.**
C. Alarcon, M. Pupareli.
Rev. Telegr. Electron., Vol. 69, 724 - 728 (1981). In Spanish. Abstr. in Phys. Abstr., Vol. 85, Abstr. 84097 (1982).

032.533 **High-temperature electronics applications in space exploration.**
R. F. Jurgens.
IEEE Trans. Ind. Electron., Vol. IE-29, 107 - 111 (1982). Abstr. in Phys. Abstr., Vol. 85, Abstr. 89106 (1982).

032.534 **Visible/infrared imaging systems for spacecraft.**
H. Koshiishi.
J. Soc. Instrum. Control Eng., Vol. 21, 253 - 259 (1982). In Japanese. – Abstr. in Phys. Abstr., Vol. 85, Abstr. 89110 (1982).

032.535 **The Space Telescope: the first imaging orbital observatory.**
R. J. Harms.
Recent advances in observational astronomy, (see 012.028), p. 153 - 156 (1981).

032.536 **The space telescope.**
C. R. O'Dell.
Telescopes for the 1980s, (see 003.010), p. 129 - 193 (1981).
Contents: Constraints on ground-based optical telescopes. History of the project. Design considerations. The planned spacecraft system. The scientific instruments. The operations system. Selecting the science for the space telescope.

032.537 **The Einstein Observatory and future X-ray telescopes.**
R. Giacconi, P. Gorenstein, S. S. Murray, E. Schreier, F. Seward, H. Tananbaum, W. H. Tucker, L. Van Speybroeck.
Telescopes for the 1980s, (see 003.010), p. 195 - 278 (1981).
Contents: Design philosophy of the observatory. Mirrors. Focal plane instrumentation. Observatory operations. Scientific impact. X-ray observatories for the future.

032.538 **A 75-cm balloon-borne telescope for far-infrared observations.**
R. R. Daniel, S. K. Ghosh, K. V. K. Iyengar, T. N. Rengarajan, S. N. Tandon, R. P. Verma.
Bull. Astron. Soc. India, Vol. 10, 29 - 30 (1982). – Abstract.

032.539 **Space Telescope: eye on the universe.**
J. K. Davies.
Spaceflight, Vol. 24, 434 - 441 (1982).

032.540 **Infrared Space Observatory (ISO) – an ESA study of a cryogenically-cooled telescope in space for infrared astronomy.**
R. Emery.
The scientific importance of submillimetre observations, (see 012.030), p. 81 - 87 (1982).
An ESA study for an infrared observatory facility in space is described. It consists of a 60 cm diameter cooled telescope with a range of spectroscopic and photometric instrumentation, mounted on a 3 axis stabilized platform. It will be launched into a 12 hour elliptical orbit using Ariane, and is designed to have a lifetime of at least 1.5 years.

032.541 **Heterodyne instrumentation for space.**
T. de Graauw.
The scientific importance of submillimetre observations, (see 012.030), p. 97 - 98 (1982).
A development programme, sponsored by ESA and carried out in several European laboratories, on submillimeter heterodyne components and instrumentation for applications in space is nearing completion and is briefly discussed. A study to identify critical areas of a space heterodyne receiver, operating between 300 and 500 GHz, has been completed.

032.542 **A concept for an orbiting submillimetre-infrared observatory.**
P. N. Swanson, M. Kiya.
The scientific importance of submillimetre observations, (see 012.030), p. 99 - 105 (1982).
NASA is in the preliminary planning stages for a large submillimeter and far infrared orbiting astronomical observatory. The telescope will be between 10 and 30 meters in diameter, the upper limit being defined by technology and cost. The operating wavelength will be between 1 mm and 30 μm. The telescope will be a Cassegrain design of approximately f/1. The primary reflector will be made up of a number of close packed panels. Each panel will be actively controlled with three degrees of freedom so that the overall optical figure of the primary reflector can be maintained within a small fraction of the shortest operating wavelength.

032.543 **Development of an advanced atmospheric sounder for future planetary missions.**
D. J. McCleese.
Bull. American Astron. Soc., Vol. 14, 764 - 765 (1982). Abstract.

032.544 **Observational aspects of the millimetre-wave cosmic background radiation.**

D. H. Martin.
Submillimetre wave astronomy, (see 012.031), p. 339 - 350 (1982).

Contents: A new radiometer for CBR measurements at $\lambda < 2$ mm. The antenna system. Interferometric Dicke switch. Thermal control for calibration. The ^{3}He cooler and detectors.

032.545 **Computer controlled operation of ultraviolet spectrometer and polarimeter on Solar Maximum Mission Satellite.**
R. A. Rehse, M. E. C. Bruner, H. D. Tinsley.
J. Spacecr. Rockets, Vol. 19, 186 - 188 (1982). – Abstr. in Phys. Abstr., Vol. 85, Abstr. 94578 (1982).

032.546 **Space Telescope orbital viewing constraints.**
T. J. Sherrill.
J. Spacecr. Rockets, Vol. 19, 118 - 124 (1982). – Abstr. in Phys. Abstr., Vol. 85, Abstr. 94599 (1982).

032.547 **The solar beacon, an experiment in solar astrometry.**
J. M. Beckers.
Opt. Acta, Vol. 29, (see 012.018), 913 - 922 (1982). – Abstr. in Phys. Abstr., Vol. 85, Abstr. 94601 (1982).

032.548 **Gas scintillation proportional counters for Japanese astronomical satellites.**
H. Inoue, K. Koyama, T. Mae, M. Matsuoka, T. Ohashi, Y. Tanaka, I. Waki.
Nucl. Instrum. Methods Phys. Res., Vol. 196, (see 012.033), 69 - 72 (1982). – Abstr. in Phys. Abstr., Vol. 85, Abstr. 94609 (1982).

032.549 **Experimental techniques for ion composition measurements in space.**
B. Wilken, T. A. Fritz, W. Studemann.
Nucl. Instrum. Methods Phys. Res., Vol. 196, (see 012.033), 161 - 167 (1982). – Abstr. in Phys. Abstr., Vol. 85, Abstr. 94612 (1982).

032.550 **The Giotto dust protection system.**
R. Lainé, F. Felici.
ESA Bull., No. 32, p. 36 - 42 (1982).

Giotto's planned encounter with comet Halley in March 1986 will take the spacecraft into the dust cloud produced by the interaction between the comet and the Sun. The spacecraft will be protected from hypervelocity dust-particle impacts by a dual-sheet bumper shield.

032.551 **The Burst and Transient Source Experiment for the Gamma-Ray Observatory.**
G. J. Fishman, C. A. Meegan, T. A. Parnell, R. B. Wilson.
AIP Conf. Proc., No. 77, (see 012.034), p. 443 - 451 (1982).

The Burst and Transient Source Experiment (BATSE) is a highly sensitive, all-sky monitor for the Gamma-Ray Observatory (GRO). The eight scintillation detector modules of BATSE are positioned around the GRO spacecraft to provide a uniform view of the entire sky above the horizon. Details of the design and capabilities of the BATSE are presented.

032.552 **Modulated Multiple Slit Camera for improved localization of gamma-ray bursts.**
P. Gorenstein.
AIP Conf. Proc., No. 77, (see 012.034), p. 453 - 467 (1982).

The addition of a modulation collimator technique to a one-dimensional coded aperture imaging system known as the Multiple Slit Camera (MSC) can improve the location precision for gamma-ray bursts and slower hard X-ray transients to about a tenth of a square arcminute. The basic features of this technique are described.

032.553 **A third-generation small spectroscopy experiment for hard transient events.**
R. W. Klebesadel, W. D. Evans, J. G. Laros, G. H. Nakano, D. W. Datlowe, W. L. Imhof, H. S. Hudson.
AIP Conf. Proc., No. 77, (see 012.034), p. 469 - 475 (1982).

The authors describe an experiment for monitoring high-energy transient events (X-ray and γ-ray bursts from both solar and stellar sources).

032.554 **Proposed hard X-ray imaging and gamma-ray burst studies for XTE.**
J. E. Grindlay, S. S. Murray.
AIP Conf. Proc., No. 77, (see 012.034), p. 477 - 487 (1982).

A new hard X-ray detector concept is proposed for XTE which would enable very high sensitivity studies of persistent hard X-ray (15 - 300 keV) sources and gamma-ray bursts (50 - 550 keV).

032.555 **Ion chamber gamma burst detector.**
S. A. Colgate.
AIP Conf. Proc., No. 77, (see 012.034), p. 489 - 496 (1982).

The principles of a high-sensitivity detector for gamma-ray bursts are described. Estimates for the possible detection of radiation bursts from supernova explosions are given.

032.556 **Pyroelectric-vidicon-aided infrared spectrometer for balloon experiments.**
A. Matsuzaki, Y. Nakamura, T. Itoh.
Rev. Sci. Instrum., Vol. 53, 978 - 983 (1982). – Abstr. in Phys. Abstr., Vol. 85, Abstr. 95493 (1982).

032.557 **Television systems and slow-scan vidicons on NASA space probes.**
M. M. Mirabito.
SMPTE J., Vol. 91, 542 - 546 (1982). – Abstr. in Phys. Abstr., Vol. 85, Abstr. 100704 (1982).

032.558 **A position sensitive detector for a gamma-ray imaging telescope.**
J. N. Carter, P. Charalambous, A. J. Dean, J. B. Stephen, R. C. Butler, G. Di Cocco, E. Morelli, G. Spada, A. Spizzichino, L. Barbareschi, G. Boella. F. Perotti, G. Villa, M. Badiali, C. La Padula, F. Polcaro, P. Ubertini.
Nucl. Instrum. Methods Phys. Res., Vol. 196, 477 - 482 (1982). – Abstr. in Phys. Abstr., Vol. 85, Abstr. 100738 (1982).

032.559 **Nearer the stars and galaxies. Space Telescope – new impulse for astronomy.**
B. Schwarz.
Elektron. Prax., Vol. 17, No. 4, p. 36 - 45 (1982). In German. Abstr. in Phys. Abstr., Vol. 85, Abstr. 100740 (1982).

032.560 **Performance of the hard X-ray imaging spectrometer.**
H. F. van Beek, R. J. Fryer.
J. Spacecr. Rockets, Vol. 19, 232 - 235 (1982). – Abstr. in Phys. Abstr., Vol. 85, Abstr. 100743 (1982).

032.561 **A balloon-borne solid state cosmic X-ray detector.**
R. Proctor, W. Pietsch, C. Reppin.
Nucl. Instrum. Methods Phys. Res., Vol. 197, 477 - 483 (1982). – Abstr. in Phys. Abstr., Vol. 85, Abstr. 100748 (1982).

032.562 **High resolution imaging systems for spin-stabilized Probe spacecraft.**
G. E. Danielson, M. C. Malin, W. A. Delamere.
Proc. SPIE Int. Soc. Opt. Eng., Vol. 268, (see 012.040), 42 - 48 (1981). – Abstr. in Phys. Abstr., Vol. 85, Abstr. 112301 (1982).

032.563 **Calibration of a gamma ray astronomy telescope in the 5-50 MeV energy range.**

J. M. Lavigne, M. Niel, G. Vedrenne, C. Doulade, G. Giordano, B. Agrinier, E. Bonfand, J. Andrejol, J. C. Courtois, M. Gorisse, A. Lecomte, B. Parlier, B. Mougin.
Nucl. Instrum. Methods Phys. Res., Vol. 199, 585 - 596 (1982). – Abstr. in Phys. Abstr., Vol. 85, Abstr. 112325 (1982).

032.564 **A hard X-ray and soft gamma ray telescope spectrometer.**
L. I. Yin, J. I. Trombka, R. L. Schmadebeck, S. M. Seltzer, M. J. Bielefeld.
Proc. SPIE Int. Soc. Opt. Eng., Vol. 268, (see 012.040), 97 - 102 (1981). – Abstr. in Phys. Abstr., Vol. 85, Abstr. 112326 (1982).

032.565 **Coronal observations from the Solar Maximum Mission satellite.**
A. Csoeke-Poeckh, R. H. Lee, W. J. Wagner, L. House, E. Hildner, C. Sawyer.
J. Spacecr. Rockets, Vol. 19, 345 - 348 (1982). – Abstr. in Phys. Abstr., Vol. 85, Abstr. 112328 (1982).

032.566 **Design and performance of the Solar Maximum Mission hard X-ray burst spectrometer.**
L. G. Workman, J. L. Wolfgang Jr.
J. Spacecr. Rockets, Vol. 19, 375 - 376 (1982). – Abstr. in Phys. Abstr., Vol. 85, Abstr. 112329 (1982).

032.567 **A solar flare X-ray polarimeter for the Space Shuttle.**
J. R. Lemen, G. A. Chanan, J. P. Hughes, M. R. Laser, R. Novick, I. T. Rochwarger, M. Sackson, L. J. Tramiel.
Sol. Phys., Vol. 80, 333 - 349 (1982).
The authors describe a Space Shuttle-borne polarimeter which they have recently designed and built, and which, by virtue of its large effective area and demonstrably small instrumental polarization, should provide an unambiguous determination of the continuum X-ray polarization in a solar flare.

032.568 **Magnetometer experiments on space missions.**
F. M. Neubauer.
The solar system and its exploration, (see 012.054), p. 61 (1981). – Abstract.

032.569 **Application of CCD techniques in planetary exploration.**
F. E. Vescelus.
The solar system and its exploration, (see 012.054), p. 199 (1981). – Abstract.

032.570 **Infrared sounding of comet Halley from the flyby Venera Halley probes.**
J. F. Crifo.
The solar system and its exploration, (see 012.054), p. 229 - 244 (1981).
One of the unique capabilities provided by the Soviet flyby mission to comet Halley will be the sounding of the innermost regions of the comet at far infrared wavelengths (2.5–12 microns) performed by the French instrument "IKS". The instrument includes two spectroscopic channels with a resolution $\lambda/\Delta\lambda = 50$ for analysis of the chemical composition of the gas and dust phases and an image spatial frequency modulating channel aimed at measuring the nucleus size, temperature and emissivity.

032.571 **IR-camera for GIRL.**
P.-R. Preußner, R. Mühlheim, R. G. Jaacks, M. Grewing.
Second ESO infrared workshop, (see 012.060), p. 237 - 240 (1982).
An IR-camera is presented to be used in the GIRL telescope. The design principles of that camera, however, may also be applied to other IR-instruments having detector arrays for the middle- or far-infrared spectral region. The authors give a short description of the cooled part of the hardware and the basic data of the instrument.

032.572 **The scientific case and feasibility of a three metre balloon telescope.**
R. D. Joseph, G. S. Wright.
Second ESO infrared workshop, (see 012.060), p. 341 - 353 (1982).

032.573 **A balloon-borne cooled telescope for far IR astronomy.**
D. Rouan, J. P. Baluteau, A. Marten, J. L. Puget, G. Serra, G. Malini, R. Papoular, C. Ryter.
Second ESO infrared workshop, (see 012.060), p. 355 - 357 (1982).

032.574 **The ISO Michelson.**
J. J. Wijnbergen, J. W. G. Aalders, G. Ploeger, T. J. Helmerhorst.
Second ESO infrared workshop, (see 012.060), p. 437 - 446 (1982).

032.575 **KRT-10 space telescope.**
V. M. Arsent'ev, V. I. Berzhatyj, V. D. Blagov, A. S. Gvamichava, Yu. I. Danilov, G. A. Dolgopolov, M. B. Zakson, K. K. Katsari, V. A. Krasnov, B. T. Kuznetsov, Yu. P. Kuleshov, V. A. Lyakhov, N. P. Mel'nikov, I. M. Moshkunov, V. V. Ryumin, A. I. Savin, A. G. Sokolov.
Dokl. AN SSSR, Tom 264, 588 - 591 (1982). In Russian. Abstr. in Ref. zh., 62. Issled. kosm. prostranstva, 10.62.119 (1982).

032.576 **Solar particles, Comet Halley and EPONA.**
S. McKenna-Lawlor, A. Thompson, D. O'Sullivan, E. Kirsch, D. B. Melrose, K. P. Wenzel.
Irish Astron. J., Vol. 15, (see 012.061), 111 - 112 (1981). Abstract.

032.577 **Investigation of the characteristics of the scintillation calorimeter of a gamma-telescope in an electron-beam accelerator.**
V. V. Akimov, I. D. Blokhintsev, M. B. Dobriyan, L. F. Kalinkin, V. D. Kozlov, B. T. Karimov, Yu. I. Nagornykh, V. E. Nesterov, A. V. Plyasheshnikov, V. L. Prokhin, V. S. Shmelev.
Inst. kosm. issled. AN SSSR. Prepr., 1982, No. 682, 43 pp. In Russian. – Abstr. in Ref. zh., 51. Astron., 10.51.831 (1982).

032.578 **Multiple-spacecraft Michelson spatial interferometer.**
R. V. Stachnik, E. F. McCormack, P. Melroy, D. Arnold.
J. Opt. Soc. America, Vol. 72, 1759 - 1760 (1982). Abstract.

032.579 **Image reconstruction for the COSMIC telescope array.**
W. F. Davis, W. A. Traub.
J. Opt. Soc. America, Vol. 72, 1760 (1982). – Abstract.

032.580 **Rocket-borne instrument with a high-resolution microchannel plate detector for planetary UV spectroscopy.**
W. E. McClintock, C. A. Barth, R. E. Steele, G. M. Lawrence, J. G. Timothy.
Appl. Opt., Vol. 21, 3071 - 3079 (1982).
A telescope-spectrograph employing a photon-counting microchannel plate (MCP)–CODACON detector has been built, tested, and flown on a sounding rocket. The detector uses a

curved-channel MCP proximity focused onto a coded anode array of 1024 channels spaced 25.4 mm center to center. The instrument has obtained an ultraviolet (1500 - 1800 Å) spectrum of Jupiter with a spectral resolution of 8 Å, which is higher than that of any previously reported observation in this wavelength range.

032.581 **Extreme ultraviolet imaging spectrometer for thermospheric emissions.**
M. R. Torr, R. C. Vitz.
Appl. Opt., Vol. 21, 3080 - 3090 (1982).
One of the instruments that has been developed to fly on the Space Shuttle is the Imaging Spectrometric Observatory, an array of five imaging spectrometers to cover the 300 - 12,000 Å wavelength range. The authors describe the spectrometer designed to operate in the extreme ultraviolet (300 - 1200 Å).

032.582 **Intensified charge coupled devices for use as a spaceborne spectrographic image-plane detector system.**
M. R. Torr, J. Devlin.
Appl. Opt., Vol. 21, 3091 - 3108 (1982).

032.583 **Imaging characteristics of a conical primary, aspheric secondary X-ray telescope.**
D. G. Burkhard, D. L. Shealy, G. L. Strobel.
Appl. Opt., Vol. 21, 3713 - 3718 (1982).

032.584 **Stable and rugged étalon for the Dynamics Explorer Fabry-Pérot interferometer. 1: Design and construction.**
D. Rees, T. J. Fuller-Rowell, A. Lyons, T. L. Killeen, P. B. Hays.
Appl. Opt., Vol. 21, 3896 - 3902 (1982).

032.585 **Stable and rugged étalon for the Dynamics Explorer Fabry-Pérot interferometer. 2: Performance.**
T. L. Killeen, P. B. Hays, B. C. Kennedy, D. Rees.
Appl. Opt., Vol. 21, 3903 - 3912 (1982).

032.586 **Quantum efficiency of opaque Cs I photocathodes with channel electron multiplier arrays in the extreme and far ultraviolet.**
C. Martin, S. Bowyer.
Appl. Opt., Vol. 21, 4206 - 4207 (1982).

032.587 **Grazing incidence relay optics.**
R. C. Chase, A. S. Krieger, J. H. Underwood.
Appl. Opt., Vol. 21, 4446 - 4452 (1982).

032.588 **Digitale Signalaufbereitung im Photopolarimeter auf GIRL.**
J. Riedinger.
Diss. Naturwiss.-Math. Gesamtfak. Univ. Heidelberg, 218 pp. (1982).

032.589 **The Pinhole/Occulter Facility.** Executive summary.
J. R. Dabbs, E. A. Tandberg-Hanssen, H. S. Hudson.
NASA Tech. Pap., NASA TP-2089, 16 pp. (1982).
The Pinhole/Occulter Facility is a novel instrumentation concept that will provide a major improvement in sensitivity and resolution for both X-ray and coronagraphic observations of the nonthermal phenomena central to understanding solar activity. It will also make possible high-energy cosmic X-ray observations with unprecedented angular resolution (0.2 arc sec) and will provide hard X-ray information complementary to the soft X-ray data from AXAF.

032.590 **Preamplifier for slow-scanning Fourier transform spectroscopy.**
R. Hofmann, H. Rothermel, D. Zasche.
Infrared Phys., Vol. 22, 251 - 253 (1982).
For balloon-borne applications a low noise bolometer-preamplifier was developed to be used with a slow-scanning Michelson interferometer.

032.591 **Cooled low-noise trans-impedance amplifier for infrared detectors.**
R. G. Jaacks, R. Mühlheim, P.-R. Preußner.
Infrared Phys., Vol. 22, 309 (1982).

032.592 **Calibration of an infrared spectrometer and a far-infrared photometer for astronomical applications under low background conditions.**
W. Luinge.
Infrared Phys., Vol. 22, 357 - 365 (1982).
A slitless low resolution infrared spectrometer and a chopped far-infrared photometer will be focal plane instruments in the liquid helium cooled telescope of the infrared astronomical satellite IRAS. A description of the compact test set-up simulating the low background and signal levels for these instruments and the calibration results are given.

032.593 **The UOSAT magnetometer experiment.**
M. H. Acuna.
Radio Electron. Eng., Vol. 52, 431 - 436 (1982). – Abstr. in Phys. Abstr., Vol. 85, Abstr. 117499 (1982).

032.594 **Proton-induced noise in Space Telescope digicon.**
L. C. Smith, J. Becher, W. B. Fowler, K. Flemming.
Proc. SPIE Int. Soc. Opt. Eng., Vol. 279, (see 012.072), 156 - 165 (1981). – Abstr. in Phys. Abstr., Vol. 85, Abstr. 117590 (1982).

032.595 **Progress report on the high resolution spectrograph for the Space Telescope.**
J. C. Brandt.
Proc. SPIE Int. Soc. Opt. Eng., Vol. 279, (see 012.072), 183 - 194 (1981). – Abstr. in Phys. Abstr., Vol. 85, Abstr. 117593 (1982).

032.596 **Helium-cooled Michelson interferometer for far-infrared astronomy aboard German Infrared Laboratory (GIRL).**
S. Drapatz, R. Hofmann, R. Katterloher.
Proc. SPIE Int. Soc. Opt. Eng., Vol. 289, (see 012.073), 66 - 69 (1981). – Abstr. in Phys. Abstr., Vol. 85, Abstr. 117595 (1982).

032.597 **Bidirectional reflectance distribution function (BRDF) measurements of sunshield and baffle materials for the Infrared Astronomy Satellite (IRAS) telescope.**
S. M. Smith.
Proc. SPIE Int. Soc. Opt. Eng., Vol. 304, (see 012.074), 205 - 213 (1981). – Abstr. in Phys. Abstr., Vol. 85, Abstr. 117597 (1982).

032.598 **Photon-counting array detectors for space and ground-based studies at ultraviolet and vacuum ultraviolet (VUV) wavelengths.**
J. G. Timothy, R. L. Bybee.
Proc. SPIE Int. Soc. Opt. Eng., Vol. 279, (see 012.072), 129 - 139 (1981). – Abstr. in Phys. Abstr., Vol. 85, Abstr. 117608 (1982).

032.599 **Design and tolerance analysis of two null corrector designs for the Space Telescope fine guidance aspheric collimating mirror.**
I. Friedman, R. E. Casas.
Proc. SPIE Int. Soc. Opt. Eng., Vol. 306, (see 012.079), 176 - 182 (1981). – Abstr. in Phys. Abstr., Vol. 86, Abstr. 10208 (1983).

032.600 **Progress in 800 × 800 charge-coupled device (CCD) imager development and applications.**
D. D. Norris, J. R. Janesick, M. Blouke, P. May, D. McGrath.
Proc. SPIE Int. Soc. Opt. Eng., Vol. 311, 42 - 45 (1981).
Abstr. in Phys. Abstr., Vol. 86, Abstr. 10209 (1983).

032.601 **Shuttle contamination effects on ultraviolet coronagraphic observations.**
J. L. Kohl, H. Weiser.
Proc. SPIE Int. Soc. Opt. Eng., Vol. 287, 35 - 40 (1981).
Abstr. in Phys. Abstr., Vol. 86, Abstr. 10210 (1983).

032.602 **Design and fabrication of the NASA 2.4 meter Space Telescope.**
D. J. McCarthy, T. A. Facey.
Proc. SPIE Int. Soc. Opt. Eng., Vol. 330, (see 012.082), 139 - 143 (1982). – Abstr. in Phys. Abstr., Vol. 86, Abstr. 15842 (1983).

032.603 **Spectroscopic equipment for the Space Telescope.**
D. S. Leckrone.
Philos. Trans. R. Soc. London, Ser. A, Vol. 307, 549 - 561 (1982). – Abstr. in Phys. Abstr., Vol. 86, Abstr. 15848 (1983).

032.604 **Pulse shape discrimination technique in a NaI(TI)-CsI(Na) X-ray detector.**
J. P. Malkar, P. K. Kunte.
Nucl. Instrum. Methods Phys. Res., Vol. 202, 465 - 468 (1982). – Abstr. in Phys. Abstr., Vol. 86, Abstr. 17179 (1983).

ANS ultraviolet photometry, catalogue of point sources.
See Abstr. 002.001.

Space instrument engineering.
See Abstr. 003.022.

Design of scientific space instrumentation.
See Abstr. 003.072.

Solid state imagers for astronomy.
See Abstr. 012.098.

Science operations with Space Telescope.
See Abstr. 013.046.

Deformable primary mirror for a space telescope.
See Abstr. 031.053.

X-ray and extreme ultraviolet imaging using layered synthetic microstructures.
See Abstr. 031.067.

Molecular spectroscopy and planetary exploration from space.
See Abstr. 031.518.

Optische telescopen gaan met hun tijd mee.
See Abstr. 032.015.

Ultraviolet, optical and infrared astronomy.
See Abstr. 032.027.

Further developments of electrographic image detectors.
See Abstr. 034.153.

Magellan: high resolution spectroscopy at FUV and EUV wavelengths.
See Abstr. 051.003.

An astrometric earth orbiter for detection of other planetary systems.
See Abstr. 051.014.

Astrometric observations with the Space Telescope.
See Abstr. 051.015.

Hinotori (ASTRO-A).
See Abstr. 051.016.

The Giotto mission to Halley's comet.
See Abstr. 051.017.

Current NASA studies for a Far-Ultraviolet Spectrographic Explorer (FUSE).
See Abstr. 051.025.

Magellan – far and extreme ultraviolet spectrographic observatory.
See Abstr. 051.026.

Submillimetre astronomy from space platforms.
See Abstr. 051.030.

The IRAS project.
See Abstr. 051.041.

Spectral mapping of Jupiter and the Galilean satellites in the near infrared.
See Abstr. 051.045.

The European X-ray Observatory Satellite – Exosat.
See Abstr. 051.061.

Stellar X-ray astronomy with the Advanced X-ray Astrophysics Facility.
See Abstr. 051.080.

The Extreme Ultraviolet Explorer.
See Abstr. 051.091.

The International Ultraviolet Explorer.
See Abstr. 054.003.

On the use of the moon for the calibration of spaceborne radiometers.
See Abstr. 094.005.

Far infrared polarimetry of W51A and M 42.
See Abstr. 131.036.

High resolution soft X-ray spectra with the objective grating spectrometer aboard Einstein Observatory.
See Abstr. 142.029.

Study of diffuse cosmic and atmospheric gamma radiation using a spark chamber in the energy range 4 MeV–100 MeV.
See Abstr. 142.521.

Erratum

032.901 **Erratum: "A scheme for determining division corrections of a photoelectric meridian circle"** [Publ. Astron. Soc. Japan, Vol. 34, 117 - 140 (1982)].
M. Miyamoto, H. Ishii.
Publ. Astron. Soc. Japan, Vol. 34, 311 (1982). – See Abstr. 31.032.012.

033 Radio Telescopes and Equipment

033.001 **Telescope beam characteristics and temperature scale of the Maryland-Green Bank 21-cm Line Survey.**
G. Westerhout, G. L. Mader, R. H. Harten.
Astron. Astrophys., Suppl. Ser., Vol. 49, 137 - 141 (1982).
Measurements are presented of the antenna pattern of the 300-foot telescope of the NRAO, used for the Maryland-Green Bank Galactic 21-cm Line Survey. The method is given by which the brightness temperature scale of this survey has been determined.

033.002 **Radio astronomy with the Very Large Array.**
R. M. Hjellming, R. C. Bignell.
Science, Vol. 216, 1279 - 1285 (1982).
The construction of the Very Large Array of radio telescopes has been completed, and this new research instrument is now being used to make radio images of astronomical objects with a resolution comparable to or better than that of ground-based optical telescopes. The role of the Very Large Array in current and future research is discussed both in principle and in terms of a sample of observing projects.

033.003 **The Molonglo Observatory Synthesis Telescope.**
B. Y. Mills.
Proc. Astron. Soc. Australia, Vol. 4, 156 - 159 (1981).

033.004 **AST *(Australian Synthesis Telescope)* parameters and the AST Astrophysics Sub-Committee.**
J. B. Whiteoak.
Proc. Astron. Soc. Australia, Vol. 4, 160 - 163 (1981).

033.005 **Annular synthesis telescopes: a novel class of multiple beam telescopes operating at mm-wavelengths.**
N. Fourikis.
Proc. Astron. Soc. Australia, Vol. 4, 259 - 260 (1981).

033.006 **18-25 GHz low-noise line receiver for the Tidbinbilla 64-m antenna.**
D. L. Jauncey, M. J. Batty, G. J. Gay, C. R. Moore, P. J. Batelaan, R. Clauss, D. Dickinson, S. Gulkis, M. Klein, T. Kuiper, G. Morris, D. Neff, W. B. Ricketts, P. Swanson.
Proc. Astron. Soc. Australia, Vol. 4, 262 - 264 (1981).

033.007 **A harmonic analysis solution to the Parkes spectrometer baseline ripple.**
D. R. Williams.
Proc. Astron. Soc. Australia, Vol. 4, 264 - 266 (1981).

033.008 **The pre-history of the Parkes 64-m telescope.**
E. G. Bowen.
Proc. Astron. Soc. Australia, Vol. 4, 267 - 273 (1981).

033.009 **Broad-band acousto-optical spectrum analyzer for radio astronomy.**
N. A. Esepkina, G. M. Blaer, Z. V. Dravskikh, A. A. Zhukov, A. P. Lavrov, I. I. Saenko.
Zh. tekh. fiz., Tom 52, 540 - 543 (1982). In Russian.
Abstr. in Ref. zh., 51. Astron., 7.51.954 (1982).

033.010 **Determination of the deformation state of the carrier frame of the mirror antenna of the RT-70 radio telescope under the action of its own weight.**
U. V. Mamedov, K. I. Gakhramanov, Eh. Eh. Gasanov, M. B. Rasulov.
Izv. AN AzSSR. Ser. fiz.-tekh. i mat. nauk, 1981, No. 6, p. 116 - 123. In Russian. – Abstr. in Ref. zh., 51. Astron., 8.51.896 (1982).

033.011 **Investigation of the fine structure of radio sources.**
L. I. Matveenko.
Astrofiz. i kosm. fiz. Moskva, 1982, p. 88 - 134. In Russian.
Abstr. in Ref. zh., 51. Astron., 8.51.897 (1982).

033.012 **Submillimeter observations using the MMT as a phased array.**
R. L. Berson, N. R. Erickson, B. L. Ulich, C. J. Lada, P. F. Goldsmith, G. R. Huguenin.
Bull. American Astron. Soc., Vol. 14, 638 (1982). – Abstract.

033.013 **The Miyun 460 MHz Solar Radio Compound Interferometer.**
Kexue Tongbao, Vol. 26, 1137 - 1140 (1981). – Abstr. in Phys. Abstr., Vol. 85, Abstr. 63249 (1982).

033.014 **Calibrador programable y automático para radiofuentes en 45 MHz.**
J. Aparici, J. May, F. Salas, J. Ventura.
Rev. Mexicana Astron. Astrofis., Vol. 6, (see 012.017), 363 - 366 (1981).
The design, construction and operation of a standard calibrator is presented. The calibrator consists of saturated diodes controlled by an indirect feed-back system and a digital-to-analog converter. The advantages over similar designs are described.

033.015 **Sistema computacional para la adquisición automática de datos en línea con un radiotelescopio de 4 haces simultaneos.**
J. Aparici, J. May, F. Salas, J. Ventura.
Rev. Mexicana Astron. Astrofis., Vol. 6, (see 012.017), 367 - 372 (1981).
The 45 MHz radiotelescope at the Maipú Radio Astronomy Observatory allows the simultaneous observation from 4 different directions. In order to use this feature, an automatic data acquisition system was developed. A description of the system, its main characteristics and the originality of some solutions are given.

033.016 **Frequency response of a synthesis array: performance limitations and design tolerances.**
A. R. Thompson, L. R. D'Addario.
Radio Sci., Vol. 17, 357 - 369 (1982). – Abstr. in Phys. Abstr., Vol. 85, Abstr. 68618 (1982).

033.017 **The design of a 30-m millimetre-wave radio telescope.**
P. Brandt, H. Gatzlaff.
Tech. Mitt. Krupp Forschungsber., Vol. 39, 111 - 124 (1981). In German. – Abstr. in Phys. Abstr., Vol. 85, Abstr. 73845 (1982).

033.018 **Radio astronomy changes our view of the Universe.**
R. Schwartz, A. Witzel.
Funkschau, No. 6, p. 45 - 50 (1982). In German. – Abstr. in Phys. Abstr., Vol. 85, Abstr. 89148 (1982).

033.019 **Mapping the radio sky.**
H. P. Palmer.
Natl. Electron. Rev., Vol. 17, 48 - 54 (1982). – Abstr. in Phys. Abstr., Vol. 85, Abstr. 89149 (1982).

033.020 **Interferometry with very great base lines.**
A. Eckart.
Funkschau, No. 6, p. 55 - 58 (1982). In German. – Abstr. in Phys. Abstr., Vol. 85, Abstr. 89165 (1982).

033.021 **An outline of VLBI.**
N. Kawajiri.
J. Soc. Instrum. Control Eng., Vol. 21, 240 - 246 (1982). In Japanese. – Abstr. in Phys. Abstr., Vol. 85, Abstr. 89166 (1982).

033.022 **Significance of radio astronomy in cosmological research.**
H. Geschwinde, M. Pilz.
Funkschau, No. 6, p. 51 - 54 (1982). In German. – Abstr. in Phys. Abstr., Vol. 85, Abstr. 89421 (1982).

033.023 **Millimeter- and submillimeter-wave receivers.**
T. G. Phillips, D. P. Woody.
Annu. Rev. Astron. Astrophys., Vol. 20, (see 003.009), 285 - 321 (1982).
Classical diode mixers. Quantum mixers. The InSb bolometer mixer.

033.024 **The Very Large Array.**
D. S. Heeschen.
Telescopes for the 1980s, (see 003.010), p. 1 - 61 (1981).
Contents: Performance goals and conceptual design. System design. Construction. Performance and expectations.

033.025 **Status report on the UK-NL 15 metre telescope.**
R. Hills.
The scientific importance of submillimetre observations, (see 012.030), p. 51 - 52 (1982).
The key features of the 15-m telescope are its homologous structure, designed to minimise the effects of gravitational deformations, the use of an enclosure to reduce wind forces and thermal disturbances, the fabrication of the panels by replication from highly accurate moulds and the setting of the surface using a laser measuring machine. There is a high degree of confidence that the specified surface accuracy of 50 μm rms will be achieved under all observing conditions. Provided funds can be made available at the right rate, construction of the telescope and installation on Mauna Kea should be completed in 1985.

033.026 **The Cologne 3 m millimetre and submillimetre radiotelescope.**
G. Winnewisser, M. Bester, E. Pofahl, M. Olberg, G. Rau, W. Zensen, P. Wratil, B. Vowinkel.
The scientific importance of submillimetre observations, (see 012.030), p. 53 - 57 (1982).
A 3 m radiotelescope with a surface accuracy of better than 50 μm is presently being constructed. The telescope will be used mainly for molecular line work, and is expected to be usable to a shortest wavelength of about 0.5 mm. It is being equipped with a cryogenic Schottky-barrier mixer, operating initially in the 90 GHz band. A 256 channel filter spectrometer and high resolution acousto-optic spectrometer will be used for spectral analysis. The entire system is expeted to be in operation early 1983.

033.027 **Projected installation of a 2.5 m antenna on the Plateau de Bure for 1.3 mm observations.**
R. Lucas.
The scientific importance of submillimetre observations, (see 012.030), p. 59 - 60 (1982).
Installation of a 2.5 m antenna on the Plateau de Bure is planned, to be operated at 1.3 mm wavelength for observations of molecular clouds, mostly in the J = 2–1 line of CO and isotopes.

033.028 **A 10-m telescope for submillimetre astronomy.**
J. W. M. Baars, P. G. Mezger.
The scientific importance of submillimetre observations, (see 012.030), p. 61 - 62 (1982).
The authors have started the design of a telescope for the sub-mm wavelength region, which will exhibit full performance at 350 μm. The diameter of the reflector is $\approx$ 10 m, the surface accuracy is $\leqslant$ 15 μm and the pointing accuracy 1″. Extensive use of carbon fiber reinforced plastic enables a light and stiff construction without significant thermal deformations.

033.029 **A project for a millimetre wave interferometer near Matterhorn.**
G. Silvestro.
The scientific importance of submillimetre observations, (see 012.030), p. 63 - 64 (1982).
Many fundamental problems of galactic astronomy require high resolution in the mm-wave band, which could be obtained with an interferometric system. A project for an interferometer involving the Alpine stations at Testa Grigia and Gornergrat is described, which would enable very high resolution observations of compact millimeter sources. The system expected performance is estimated and objects of investigation are discussed.

033.030 **Design of acousto-optical spectrometers.**
H. van de Stadt, J. A. Lindenbergh, J. Claas.
The scientific importance of submillimetre observations, (see 012.030), p. 197 - 199 (1982).

033.031 **Superconducting mixers.**
E. Kollberg.
The scientific importance of submillimetre observations, (see 012.030), p. 201 - 202 (1982).
A short review with results on superconducting quasiparticle, Josephson and Super-Schottky mixers is presented.

033.032 **A 185-GHz Josephson point-contact low-noise mixer.**
G. ter Horst, H. Schaeffer, T. Poorter.
The scientific importance of submillimetre observations, (see 012.030), p. 203 - 205 (1982).
The performance of a Josephson mixer at 185 GHz is reported. At 185 GHz the best single side band (SSB) mixer temperature was 160 K with a SSB conversion loss of 4.4 dB. The associated total receiver temperature (SSB) is found to be 415 K. The instantaneous bandwidth is 20 GHz.

033.033 **Lower-noise Mott diodes for cryogenic mixers at submillimetre-wavelengths.**
N. J. Keen.
The scientific importance of submillimetre observations, (see 012.030), p. 207 - 211 (1982).
The main noise contributions in Mott-barrier mixer diode are considered and it is concluded that improvements in epitaxial doping profiles for Mott diodes should lead to marked reductions in cooled mixer noise temperatures.

033.034 **A stable acousto-optical spectrometer for millimeter radio astronomy.**
C. R. Masson.
Astron. Astrophys., Vol. 114, 270 - 274 (1982).
A stable, low cost acousto-optical spectrometer has been developed for millimeter wave observations at the Owens Valley Radio Observatory. This instrument has a bandwidth of 102 MHz, divided into 1024 channels, with an effective resolution of 160 kHz. The spectrometer is stable enough to permit total-power observations with on- and off-source integrations of several minutes. The many measures taken to achieve this stability are described and astronomical tests of the spectrometer are presented.

033.035 **The submillimeter receiver of the future.**
P. Encrenaz.
Submillimetre wave astronomy, (see 012.031), p. 315 - 322 (1982).

An investigation of the different types of receivers used today at millimeter and submillimeter wavelength is carried out. Forecasts for their behavior at 300 GHz and above for the next five years are presented.

033.036 **Solar radio observation at millimeter wavelength – description of instruments and preliminary results of observation.**
H. Kumagai, S. Isozaki.
Rev. Radio Res. Lab., Vol. 27, 17 - 37 (1981). In Japanese. Abstr. in Phys. Abstr., Vol. 85, Abstr. 94603 (1982).

033.037 **The response of a radio-astronomy synthesis array to interfering signals.**
A. R. Thompson.
IEEE Trans. Antennas Propag., Vol. AP-30, 450 - 456 (1982). Abstr. in Phys. Abstr., Vol. 85, Abstr. 94625 (1982).

033.038 **Millimeter wave radio telescopes: gain and pointing characteristics.**
B. L. Ulich.
Int. J. Infrared Millim. Waves, Vol. 2, 293 - 310 (1981). Abstr. in Phys. Abstr., Vol. 85, Abstr. 100741 (1982).

033.039 **Radiation cones from feed-support struts of symmetric paraboloidal antennas.**
W. V. T. Rusch, O. Sorensen, J. W. M. Baars.
IEEE Trans. Antennas Propag., Vol. AP-30, 786 - 790 (1982). Abstr. in Phys. Abstr., Vol. 85, Abstr. 106530 (1982).

033.040 **New large radio telescope in Japan.**
Astron. Her., Vol. 75, 165 - 179 (1982). In Japanese. Abstr. in Phys. Abstr., Vol. 85, Abstr. 112324 (1982).

033.041 **The computer-controlled solar radio spectrometer 'IKARUS'.**
M. R. Perrenoud.
Sol. Phys., Vol. 81, 197 - 203 (1982).

The radio spectrometer IKARUS is a fully computer-controlled instrument covering the frequency band 0.11 to 1 GHz in steps of 1 MHz. The novelty of the instrument is its ability to measure broadband calibrated spectra (flux and degree of polarization) in the very interesting region of the lower corona.

033.042 **The Clark Lake Teepee-Tee telescope.**
W. C. Erickson, M. J. Mahoney, K. Erb.
Astrophys. J., Suppl. Ser., Vol. 50, 403 - 419 (1982).

A new, fully steerable, decametric array for radio astronomy has been built at the Clark Lake Radio Observatory near Borrego Springs, California. This array is a "T" of 720 conical spiral antennas (teepee-shaped antennas), 3.0 km by 1.8 km. It is capable of operating between 15 and 125 MHz but has best sensitivity in the 25 to 75 MHz range. A 1024 channel digital correlator has been built and attached to the array. This permits the simultaneous measurement of the complex visibility functions on 512 interferometer baselines between various portions of the array. These visibility data yield a 32 X 32 resolution element picture of the area of sky under observation, with frequency-dependent angular resolution ranging from 20′ to 2.′7 and a sensitivity of 1 Jy per beam. The system is described, and some initial observations are presented.

033.043 **Facilities for US radioastronomy.**
P. Thaddeus.
Phys. Today, Vol. 35, No. 11, p. 36 - 42 (1982).

At the fiftieth anniversary of the birth of their science, radioastronomers are looking forward to a very-long-baseline interferometric array, a millimeter-wave telescope and a submillimeter-wave antenna.

033.044 **Holographic studies of the Algonquin 46m telescope.**
C. H. Costain, J. L. Yen, C. J. Wu, A. W. Woodsworth, R. S. Richards, N. W. Broten, J. M. MacLeod.
J. R. Astron. Soc. Canada, Vol. 76, 317 - 318 (1982). Abstract.

033.045 **Universal radio spectrograph for the 8 - 12 GHz region.**
Yu. D. Panfilov, A. G. Serkin, V. A. Tikhomirov.
Astron. Tsirk., No. 1177, p. 6 - 7 (1981). In Russian.

033.046 **Improvement of angular resolution of the 17 GHz solar interferometer.**
H. Sekiguchi, H. Nakajima, Y. Shiomi, M. Sawa, S. Kawashima, K. Kosugi, K. Hashimoto.
Tokyo Astron. Obs. Rep., Vol. 19, 479 - 487 (1981). In Japanese.

033.047 **The digital multi-channel radiospectrograph in Nançay.**
G. Dumas, C. Caroubalos, J.-L. Bougeret.
Sol. Phys., Vol. 81, 383 - 394 (1982).

The authors describe the new solar radio spectrograph which has been operated at the Nançay Radio Astronomy Station since December 1978 for the analog part (which uses photographic film data acquisition) and since July 1979 using digital magnetic recording. This instrument was designed and built by the Space Research Department of the Paris Observatory and covers the range 469–110 MHz.

033.048 **Elimination of guidance errors of the antenna of the RT-22 radiotelescope of the Crimean Astrophysical Observatory during observations at 1.35 cm wavelength.**
A. M. Romanov, V. I. Kostenko.
Inst. kosm. issled. AN SSSR. Prepr., 1981, No. 679, 21 pp. In Russian. – From Ref. zh., 51. Astron., 10.51.822 (1982).

033.049 **A mathematical model of a radioastronomical measuring complex.**
P. V. Beketov, M. M. Gantsevich, D. I. Gazarova, A. Kh. Khanberdiev.
Izv. AN TSSR. Ser. fiz.-tekh., khim. i geol. nauk, 1982, No. 2, p. 20 - 25. In Russian. – Abstr. in Ref. zh., 51. Astron., 10.51.836 (1982).

033.050 **A new design for 1-cm rutile traveling wave masers.**
J. Askne, S. Galt, E. Kollberg.
IEEE Trans. Microwave Theory Tech., Vol. MTT-30, 1252 - 1255 (1982). – Abstr. in Phys. Abstr., Vol. 85, Abstr. 114226 (1982).

033.051 **The Meteor Wind Radar System at Waltair.**
M. I. Ahmed, P. C. S. Devara, M. Premkumar, M. Srirama Rao, B. Ramachandra Rao.
J. Inst. Electron. Telecommun. Eng., Vol. 28, 111 - 116 (1982). – Abstr. in Phys. Abstr., Vol. 85, Abstr. 117491 (1982).

033.052 **A disk based observing system for a radio telescope.**
P. J. Turner.
Australian Comput. J., Vol. 14, No. 1, p. 32 - 34 (1982). Abstr. in Phys. Abstr., Vol. 85, Abstr. 117589 (1982).

033.053 **The new digital multichannel radiospectrograph at the Nancay radioastronomical station.**
G. Dumas.
Onde Electr., Vol. 62, 83 - 94 (1982). In French. – Abstr. in Phys. Abstr., Vol. 85, Abstr. 117598 (1982).

033.054 **Very Long Baseline Interferometry (VLBI) system for observation of Jupiter's decametric emissions.**
N. Batori, I. Tomizawa, T. Yoshino.
Rep. Univ. Electro-Commun., Vol. 32, 295 - 302 (1982). In Japanese. – Abstr. in Phys. Abstr., Vol. 85, Abstr. 117599 (1982).

033.055 **Study of the solar wind with the interplanetary scintillation at UHF.**
M. Kojima, Y. Ishida, K. Maruyama, T. Kakinuma.
Sol. Terr. Environ. Res. Japan, Vol. 6, 3 - 13 (1982). – Abstr. in Phys. Abstr., Vol. 86, Abstr. 5113 (1983).

033.056 **A phase-stabilized local-oscillator system for a synthesis radio telescope.**
T. L. Landecker, J. F. Vaneldik.
IEEE Trans. Instrum. Meas., Vol. IM-31, 185 - 192 (1982). Abstr. in Phys. Abstr., Vol. 86, Abstr. 10212 (1983).

033.057 **Absolute calibration of the solar radio flux density at 35 GHz.**
Acta Astron. Sinica, Vol. 23, 110 - 117 (1982). In Chinese.
By means of a standard gain-horn and two precise standard noise sources, the absolute measurement of the solar radio flux density at 35 GHz was carried out.

033.058 **Absolute calibration of the solar emission in the centimeter band and the daily calibration.**
Q. Fu, L. Zheng, X. Li, Y. Liu, S. Wang.
Acta Astrophys. Sinica, Vol. 2, 191 - 200 (1982). In Chinese.

033.059 **Hydraulic extends radio observatory's celestial 'ear'** ***(VLA site).***
R. J. Wolf, R. L. Swansbro.
Hydraul. Pneum., Vol. 35, No. 10, p. 137 - 141 (1982). Abstr. in Phys. Abstr., Vol. 86, Abstr. 20239 (1983).

033.060 **Jodrell under MERLIN's spell.**
N. Henbest.
New. Scientist, Vol. 96, 442 - 448 (1982). – Abstr. in Phys. Abstr., Vol. 86, Abstr. 20243 (1983).

Zu Stand und Entwicklung der Radiointerferometrie in der Astronomie.
See Abstr. 013.027.

Time accuracy requirements for VLBI experiments in radio astronomy.
See Abstr. 031.581.

Will the next supernova in our Galaxy be discovered by a radio telescope?
See Abstr. 031.683.

Inventory of major operational and planned ground-based astronomical telescopes of the countries represented in the European Science Foundation (second edition, 1982).
See Abstr. 032.019.

Use of the 3.6 m telescope for millimeter observations.
See Abstr. 032.050.

Band-pass filters for submillimetre astronomy.
See Abstr. 034.056.

High frequency techniques in heterodyne astronomy.
See Abstr. 034.057.

Die Arbeiten des Sonderforschungsbereiches 78 Satellitengeodäsie der Technischen Universität München im Jahre 1981.
See Abstr. 046.010.

Recent submillimetre wavelength studies of molecules in star formation regions.
See Abstr. 131.187.

High dynamic range mapping of strong radio sources, with application to 3C84.
See Abstr. 141.045.

Radio observations of the galactic center: future directions.
See Abstr. 156.021.

034 Auxiliary Instrumentation

034.001 **An efficient low- and moderate-resolution spectrograph for the Hale telescope.**
J. B. Oke, J. E. Gunn.
Publ. Astron. Soc. Pacific, Vol. 94, 586 - 594 (1982).
A new low-to-moderate resolution spectrograph has been designed and built for the Cassegrain focus of the Hale 5.08-meter telescope. To maximize efficiency, resolution, and wavelength coverage the light is divided into two spectral regions by a dichroic filter behind the entrance slit, after which there are two completely separate spectrographs. The blue spectrograph operates from 3200 Å to 5200 Å while the red one goes from 5200 Å to 10,000 Å. The red detector is an 800 × 800 TI CCD while the blue detector is a 320 × 512 RCA CCD or a spectrograph image pulse-counting system. The overall efficiency of the Cassegrain telescope, spectrographs, and CCD's combined, ranges from 5% to 13% between 3600 Å and 8200 Å. The spectrograph is usable from 3200 Å to 10,400 Å.

034.002 **Photometric calibration of astronomical photographic plates. II. The Anglo-Australian Telescope prime focus spot sensitometer.**
D. F. Malin, B. A. Peterson.
Publ. Astron. Soc. Pacific, Vol. 94, 595 - 599 (1982).
The calibration of photographic sensitometers is beset with problems which can defeat photoelectric or densitometric procedures. The authors describe a direct and simple photographic technique using self-consistent multiple exposures which has been used to calibrate the projection sensitometers installed at the prime focus of the Anglo-Australian Telescope and at the U. K. Schmidt Telescope Unit. The method can be applied to any sensitometer and all photographic passbands. Values of relative log *I* for the AAT sensitometer are given for the IIIa-J emulsion with a Schott GG385 filter.

034.003 **A portable water-vapor monitor for day or night use.**
R. Landau.
Publ. Astron. Soc. Pacific, Vol. 94, 600 - 604 (1982).
The design, construction, and calibration of a small, portable, battery-operated, spectral hygrometer is described. The instrument is sensitive enough to use either the sun or the moon as a source; it contains no moving parts and will produce a measurement in a few seconds. Its use in conjunction with millimeter and submillimeter observations is briefly discussed.

034.004 **A simultaneous infrared/optical polarimeter.**
J. Bailey, J. H. Hough.
Publ. Astron. Soc. Pacific, Vol. 94, 618 - 623 (1982).
A polarimeter capable of making simultaneous observations in the optical and near infrared is described. The instrument achieves very small ($\lesssim 0.03\%$) instrumental polarization coupled with high efficiency over a very wide wavelength range. The authors present *UBVRIJHK* polarization measurements of a number of standard stars with large interstellar polarizations, and use them to compile composite interstellar polarization curves.

034.005 **What can infrared photometry tell you?**
D. A. Allen.
Proc. Astron. Soc. Australia, Vol. 4, 149 - 151 (1981).

034.006 **Detectors in astronomy: optical astronomy.**
A. Boksenberg.
Nature, Vol. 298, 795 - 796 (1982).

034.007 **Detectors in astronomy: infrared astronomy.**
E. T. Young.
Nature, Vol. 298, 796 - 798 (1982).

034.008 **Device for automatic recording of a radar echo from meteors.**
V. Kaksa.
Obrab. inf., poluch. po programme Interkosmos, Moskva, 1982, p. 38 - 41. In Russian. – Abstr. in Ref. zh., 62. Issled. kosm. prostranstva, 7.62.85 (1982).

034.009 **The MSFC vector magnetograph.**
M. J. Hagyard, N. P. Cumings, E. A. West, J. E. Smith.
Sol. Phys., Vol. 80, 33 - 51 (1982).
The authors present a description of the NASA/Marshall Space Flight Center's solar vector magnetograph system, discuss its calibration and performance characteristics, and describe some of the specific scientific investigations which are uniquely applicable to this instrument.

034.010 **A SIT (*Silicon Intensified Target*) vidicon for surface photometry.**
E. H. Geyer, A. Hänel.
Messenger, No. 29, p. 5 - 8 (1982).

034.011 **Das Zweikoordinatenmeßgerät ASCORECORD 3 DP im Dienste des Europäischen Feuerkugel-Programms.**
Z. Ceplecha.
Jenaer Rundsch., 27. Jahrg., 115 - 117 (1982).

034.012 **On an investigation of basic characteristics of the photomultiplier-115.**
Yu. V. Khan.
Izv. AN TSSR. Ser. fiz.-tekh., khim. i geol. nauk, 1981, No. 4, p. 121 - 123. In Russian. – Abstr. in Ref. zh., 51. Astron., 8.51.886 (1982).

034.013 **Charge-coupled devices in astronomy.**
J. Kristian, M. Blouke.
Sci. American, Vol. 247, No. 10, p. 48 - 56 (1982).
Microelectronic technology has presented astronomers with a sensitive new radiation detector that is expected to improve the accuracy of many crucial observations.

034.014 **Photodiodes versus photomultipliers: a theoretical comparison.**
R. H. Kaitchuck.
I.A.P.P.P. Commun., No. 8, p. 3 - 12 (1982).

034.015 **Photoelectric photometers of the future.**
R. K. Honeycutt.
I.A.P.P.P. Commun., No. 8, p. 13 - 18 (1982).

034.016 **On the instrumental profile of the echelle spectrograph of the horizontal solar telescope of the Astronomical Observatory of the Kiev University.**
N. I. Lozitskaya, V. G. Lozitskij.
Vestn. Kiev. Univ., Astron. Vyp. 24, p. 41 - 44 (1982). In Russian.

034.017 **On sources of discrepancy between theoretical and laboratory calibration dependencies of a solar magnetograph.**
N. I. Lozitskaya, Yu. A. Chesnok, V. G. Lozitskij.
Vestn. Kiev. Univ., Astron. Vyp. 24, p. 45 - 50 (1982). In Russian.

034.018 **Asteroid occultation astrometry with a lensless Schmidt.**
W. S. Penhallow.

Bull. American Astron. Soc., Vol. 14, 621 - 622 (1982). Abstract.

034.019 **The LPL extrasolar-planet-detection spectrometer: a new configuration.**
R. S. McMillan, J. E. Frecker, P. H. Smith, E. H. Roland, W. J. Merline.
Bull. American Astron. Soc., Vol. 14, 625 (1982). – Abstract.

034.020 **The MIT-NRL two-color dual astrometric stellar interferometer.**
M. Shao, M. M. Colavita, P. M. Garnavich, D. H. Staelin.
Bull. American Astron. Soc., Vol. 14, 627 (1982). – Abstract.

034.021 **The Allegheny Observatory's Multichannel Astrometric Photometer – progress report and early results.**
J. W. Stein, G. D. Gatewood, C. D. Di Fatta, J. H. Kiewiet de Jonge.
Bull. American Astron. Soc., Vol. 14, 645 (1982). – Abstract.

034.022 **The Yale plate measurement system.**
W. F. van Altena, J.-F. Lee.
Bull. American Astron. Soc., Vol. 14, 645 (1982). – Abstract.

034.023 **A circuit for correcting fatigue effects in photomultipliers.**
W. A. Rosen.
Bull. American Astron. Soc., Vol. 14, 646 (1982). – Abstract.

034.024 **Inexpensive digital systems for astronomy.**
D. B. Caton, J. B. Rafert, G. Cone.
Bull. American Astron. Soc., Vol. 14, 657 (1982). – Abstract.

034.025 **On the use of microprocessors in astronomy.**
P. C. Chen.
Bull. American Astron. Soc., Vol. 14, 657 (1982). – Abstract.

034.026 **Digital controlled time-gate of the Ondřejov laser radar.**
Z. Neumann.
Bull. Astron. Inst. Czechoslovakia, Vol. 33, 319 - 320 (1982).
The new digital controlled time-gate of the Ondřejov laser radar began its operation. Its properties and advantages for laser radar operation are described. The block diagram is added.

034.027 **Characteristics of CR-39 for detection of cosmic ray particles.**
T. Hayashi, R. Hamasaki, T. Doke.
Nucl. Tracks Methods Instrum. Appl., Vol. 5, (see 012.015), 347 (1981). – Summary. Abstr. in Phys. Abstr., Vol. 85, Abstr. 63255 (1982).

034.028 **An improved CR-39 track detector for cosmic ray applications.**
A. Thompson, D. O'Sullivan, D. L. Henshaw, S. Amin, V. M. Clapham, P. H. Fowler, D. J. Webster.
Nucl. Tracks Methods Instrum. Appl., Vol. 5, (see 012.015), 349 (1981). – Summary. Abstr. in Phys. Abstr., Vol. 85, Abstr. 63256 (1982).

034.029 **Diode rules for detection of photoelectrons.**
V. D. Alekseev, V. G. Debur, V. P. Zabrodskaya, V. S. Rylov.
Pribory i tekh. ehksp. 1982, No. 1, p. 172 - 174. In Russian. Abstr. in Ref. zh., 51. Astron., 9.51.760 (1982).

034.030 **Compton polarimeter and possibility of its application to astrophysics.**
G. M. Gorodinskij, B. I. Ermolaev, E. M. Kruglov, V. V. Khmylko.
Fiz.-tekh. inst. AN SSSR. Prepr., 1982, No. 761, 25 pp. In Russian. – Abstr. in Ref. zh., 51. Astron., 9.51.763 (1982).

034.031 **A precision measuring microscope employing laser interferometry.**
J. H. Adams, Jr., L. P. Beahm, B. Stiller.
Nucl. Tracks Methods Instrum. Appl., Vol. 5, (see 012.015), 406 - 407 (1981). – Summary. Abstr. in Phys. Abstr., Vol. 85, Abstr. 64841 (1982).

034.032 **Cosmic ray trigger circuit for large solenoidal particle detectors.**
A. Seiden, P. Shaffer, J. Warner.
Nucl. Instrum. Methods Phys. Res., Vol. 193, 597 - 600 (1982). – Abstr. in Phys. Abstr., Vol. 85, Abstr. 68621 (1982).

034.033 **Design optimization of liquid scintillation cosmic ray veto detector with BBQ shifter.**
H. W. Kruse, S. S. Egdorf, D. F. Simmons.
IEEE Trans. Nucl. Sci., Vol. NS-29, (see 012.023), 351 - 353 (1982). – Abstr. in Phys. Abstr., Vol. 85, Abstr. 68622 (1982).

034.034 **A large area cosmic ray isotope experiment using Cherenkov range detector techniques.**
W. Heinrich, M. Simon, H. O. Tittel, J. F. Ormes, V. K. Balasubrahmanyan, R. E. Streitmatter.
Nucl. Tracks Methods Instrum. Appl., Vol. 5, (see 012.015), 402 - 403 (1981). – Summary. Abstr. in Phys. Abstr., Vol. 85, Abstr. 68625 (1982).

034.035 **A method for measuring the isotopic abundances of cosmic rays.**
J. H. Adams, Jr., L. Beahm, J. A. Maturi, B. Stiller, A. Thompson.
Nucl. Tracks Methods Instrum. Appl., Vol. 5, (see 012.015), 403 (1981). – Summary. Abstr. in Phys. Abstr., Vol. 85, Abstr. 68626 (1982).

034.036 **Novel photodiode array for astronomical spectroscopy.**
A. R. Hedge.
Thesis Univ. Durham, England (1979). – Abstr. in Phys. Abstr., Vol. 85, Abstr. 73847 (1982).

034.037 **Determination of the characteristics of a spectrometer of cosmic ray relativistic electrons with the method of modelling.**
G. A. Glukhov, Yu. P. Kratenko, G. V. Lupenko, A. F. Titenkov, S. A. Charyshnikov.
Dokl. AN UzSSR, 1982, No. 1, p. 23 - 24. In Russian. – Abstr. in Ref. zh., 62. Issled. kosm. prostranstva, 9.62.38 (1982).

034.038 **Differential measurement system for direct determination of the electron temperature.**
V. G. Markov, V. S. Genov, H. D. Bliznakov.
Dokl. BAN, Vol. 34, 1103 - 1105 (1981). – Abstr. in Ref. zh., 62. Issled. kosm. prostranstva, 9.62.39 (1982).

034.039 **Instrument for determination of the difference of potentials between a plasma and a body in ionospheric probe experiments.**
V. G. Markov, V. S. Genov, H. D. Bliznakov.
Dokl. BAN, Vol. 34, 1265 - 1268 (1981). – Abstr. in Ref. zh., 62. Issled. kosm. prostranstva, 9.62.40 (1982).

034.040 **Electrometric current-voltage transformer.**
M. N. Gousheva.
Dokl. BAN, Vol. 34, 1371 - 1372 (1981). – Abstr. in Ref. zh., 62. Issled. kosm. prostranstva, 9.62.41 (1982).

034.041 **Observations of Halley's comet with intensified photodiode arrays.**
G. Moreels, J. Clairemidi, J. P. Parisot.
The need for coordinated ground-based observations of Halley's comet, (see 012.025), p. 185 - 190 (1982).

034.042 **Evaluation of a 32 X 32 InSb CID array.**
D. Stefanovitch, F. Sibille.
Rev. Phys. Appl., Vol. 17, 365 - 369 (1982). – Abstr. in Phys. Abstr., Vol. 85, Abstr. 74502 (1982).

034.043 **Proposal of a laser radiochemical Ga detector of solar neutrinos.**
V. S. Letokhov, V. I. Mishin, N. Eshkobilov, A. T. Tursunov.
Opt. Commun., Vol. 41, 331 - 334 (1982). – Abstr. in Phys. Abstr., Vol. 85, Abstr. 78957 (1982).

034.044 **Solid-state imagers move into the observatory.**
R. C. Cunningham.
Electro-Opt. Syst. Des., Vol. 14, No. 2, p. 53 - 55 (1982). Abstr. in Phys. Abstr., Vol. 85, Abstr. 78963 (1982).

034.045 **Shanghai Observatory's hydrogen maser.**
Z.-C. Zhai.
J. Inst. Electron. Telecommun. Eng., Vol. 27, 510 - 513 (1981). – Abstr. in Phys. Abstr., Vol. 85, Abstr. 80925 (1982).

034.046 **Television spectrophotometry.**
C. Firmani, E. Ruiz.
Recent advances in observational astronomy, (see 012.028), p. 25 - 30 (1981).

Television detectors have been successfully applied in astronomy during the past decade. Among the bidimensional detectors, there are very versatile devices because of their high sensitivity, wide spectral range, wide dynamical range, linearity and ease of data handling. The purpose of this paper is to demonstrate the application of this technique to astronomical spectrophotometry.

034.047 **CCDs.**
R. Lynds.
Recent advances in observational astronomy, (see 012.028), p. 31 - 38 (1981).

A general description of CCD imagers and their operational characteristics is presented. Special attention is given to some of the pathological properties which can, if not avoided, significantly compromise the otherwise superb observational advantages offered by the devices.

034.048 **A Michelson spectrophotometer instead of a coudé spectrograph.**
H. L. Johnson.
Recent advances in observational astronomy, (see 012.028), p. 113 - 121 (1981).

The rapid-scanning Michelson spectrophotometer system is reviewed and its performance as a spectrometer is illustrated in detail. Its usefulness instead of a coudé spectrograph is demonstrated.

034.049 **Very wide band stellar spectrophotometry.**
H. L. Johnson, S. T. Ridgway.
Recent advances in observational astronomy, (see 012.028), p. 123 - 130 (1981).

Virtually no broad bandwidth coverage spectrophotometry is currently available for stars. This may be attributed to a lack of suitable instruments and to uncertainties about the absolute flux calibration. The authors describe one of several recent instrumental developments (in this case a broad-band Michelson-Fourier transform spectrometer) suitable for very wide spectral coverage.

034.050 **Microprocessor based photon-counting photomultiplier system for astronomical use.**
K. S. B. Manian, F. M. Pathan.
Bull. Astron. Soc. India, Vol. 10, 30 (1982). – Abstract.

034.051 **Progress in airborne infrared and submillimetre spectroscopy.**
M. Harwit.
The scientific importance of submillimetre observations, (see 012.030), p. 65 - 67 (1982).

Recent progress in spectral resolution and in instrumental sensitivity for far infrared and submillimeter observations on NASAs Kuiper Airborne Observatory and Learjet are reviewed.

034.052 **Far infrared cryogenic spectrometers.**
J. J. Wijnbergen.
The scientific importance of submillimetre observations, (see 012.030), p. 183 - 187 (1982).

The author explains the basic ideas behind the development of cryogenic spectrometers in the far infrared. A number of instruments constructed in the last decade is reported.

034.053 **Heterodyne spectrometer for 10 microns working in the Cassegrain focus of astronomical telescopes.**
H. Rothermel, H. U. Käufl.
The scientific importance of submillimetre observations, (see 012.030), p. 189 - 196 (1982).

A quasi-optical receiver is using a grating tunable CO_2 laser as local oscillator and a photovoltaic detector for frontend mixing. The sensitivity of this system is approaching the quantum noise limit. The bandwidth (2 GHz) is sufficient for high resolving spectroscopy at selected astronomical objects. For illustration a subdoppler emission of Mars measured with a resolution of 5 000 000 is shown, which was obtained with the 2.2 m mirror on Calar Alto in April 1982.

034.054 **Submillimetre receiver with laser local oscillators.**
G. V. Schultz, E. J. Durwen, H. P. Röser, R. Wattenbach.
The scientific importance of submillimetre observations, (see 012.030), p. 213 - 216 (1982).

A heterodyne detection system for the sub-mm wavelength using a laser as local oscillator and an open structure mixer is described. At Jungfraujoch the system had a noise temperature of 4300 K and was used for astronomical observations of objects with continuum spectra in a bandwidth of 80 MHz and for narrow lines using a 1024 channel acousto-optical spectrometer with a resolution of 200 kHz per channel. The first results are presented.

034.055 **Progress report on the Spacewatch Camera.**
T. Gehrels, R. McMillan, J. Frecker, E. Roland, C. Stoll, L. Doose, E. Shoemaker, S. Nozette, H. Boesgaard.
Bull. American Astron. Soc., Vol. 14, 728 (1982). – Abstract.

034.056 **Band-pass filters for submillimetre astronomy.**
S. T. Chase, R. D. Joseph.
Submillimetre wave astronomy, (see 012.031), p. 309 - 313 (1982).

Resonant arrays of simple crosses offer an attractive solution to the problem of filter design for far-infrared and submillimetre astronomy.

034.057 **High frequency techniques in heterodyne astronomy.**
T. de Graauw.
Submillimetre wave astronomy, (see 012.031), p. 323 - 338 (1982).

The paper describes the advances in several techniques for heterodyne detection at frequencies above 200 GHz. Local oscillators and mixers in use for astronomical observations and current developments in these areas are reviewed.

034.058 **TAURUS: a wide-field imaging Fabry-Perot spectrometer for astronomy.**
P. D. Atherton, K. Taylor, C. D. Pike, C. F. W. Harmer, N. M. Parker, R. N. Hook.
Mon. Not. R. Astron. Soc., Vol. 201, 661 - 696 (1982).
This paper describes the design, development and operation of an imaging Fabry-Perot interferometer (known as TAURUS). The system is primarily intended for mapping velocity fields of astronomical emission-line sources and possesses considerable advantages over conventional grating spectrographs when studying the dynamics of extended objects.

034.059 **A microcomputer-controlled photometer.**
R. Elston, M. Zeilik.
Publ. Astron. Soc. Pacific, Vol. 94, 729 - 732 (1982).
The authors have incorporated a low-cost microcomputer into an automated photometer using readily available hardware. The photometer control program, written mostly in BASIC, interprets commands designed for photometry and allows the easy construction of observing routines for particular objects. The resulting system provides the user with real-time data reduction and flexible operation at a small cost of time and effort. It permits the user to make intelligent decisions during an observation and so increases the productivity of an observing session; and it reduces the burden of postobservation data reduction.

034.060 **An antinucleus detector with unprecedented collecting power and resolution.**
S. P. Ahlen, P. B. Price, M. H. Salamon, G. Tarle.
Nucl. Instrum. Methods Phys. Res., Vol. 197, 485 - 506 (1982). – Abstr. in Phys. Abstr., Vol. 85, Abstr. 100749 (1982).

034.061 **Low-noise In-Sb photodetector preamp for the infrared.**
M. A. M. Barros.
IEEE J. Solid-State Circuits, Vol. SC-17, 761 - 768 (1982). Abstr. in Phys. Abstr., Vol. 85, Abstr. 112333 (1982).

034.062 **Cellulose triacetate track detectors for the measurement of abundances of solar cosmic ray heavy ions of helium to iron.**
N. Durgaprasad, J. Nevatia, S. Biswas.
Nucl. Instrum. Methods Phys. Res., Vol. 199, 573 - 583 (1982). – Abstr. in Phys. Abstr., Vol. 85, Abstr. 112335 (1982).

034.063 **Modernizing Michelson's stellar interferometer.**
R. Hanbury Brown.
Proc. SPIE Int. Soc. Opt. Eng., Vol. 288, (see 012.035), 545 - 550 (1981). – Abstr. in Phys. Abstr., Vol. 85, Abstr. 112342 (1982).

034.064 **On the possibility of an analogous microdensitometer.**
B. F. Osak, P. G. Papushev.
Soln. Dannye 1982 Byull., No. 5, p. 93 - 97 (1982). In Russian.

034.065 **Measurements of the magnetic field in solar prominences with a spectrally scanning magnetograph.**
G. M. Nikolsky (*Nikol'skij*), I. S. Kim, S. Koutchmy.
Sol. Phys., Vol. 81, 81 - 89 (1982).
The authors describe observations with a new magnetograph capable of recording the whole profile of emission lines in prominences. The instrument is calibrated using 100% circularly polarized light and an emission line produced in laboratory conditions in a simulated longitudinal magnetic field. The magnetograph was attached to the large coronagraph of Kislovodsk to give a series of measurements of the Hβ line of several quiescent and active prominences. The observed values of the longitudinal component of the magnetic field are between: −25 G + 13 G with a noise level at ± 2 G for a corresponding resolution of 8 arc sec.

034.066 **Ein Mikroprozessor-gesteuertes, integrierendes Photometer.**
S. Witzigmann, U. Hopp.
Sterne Weltraum, 21. Jahrg., 481 - 484 (1982).

034.067 **A 'jamjar' magnetometer.**
R. J. Livesey.
J. British Astron. Assoc., Vol. 93, 17 - 19 (1982).

034.068 **Narrow-band GaAsSb heterojunction detectors for coronal polarimetry.**
C. W. Querfeld.
AAS Photo-Bull., No. 29, p. 3 - 8 (1982).
GaAsSb heterojunction photodiodes with low noise (3.9 $fAHz^{-1/2}$) and high quantum efficiency (≈90%) have been developed for measurements of the intensity and linear polarization of the FeXIII 10747 Å emission line emitted by the solar corona. Characteristics of the heterojunction diodes and their low noise, capacitance neutralizing preamplifier are discussed. The detectors have made it possible to extend emission line observations to 1.7 $R_\odot$ routinely and to detect emission line polarization to 2.4 $R_\odot$ from the ground outside of eclipse.

034.069 **The CCD on La Silla.**
H. Pedersen, M. Cullum.
Messenger, No. 30, p. 26 - 28 (1982).

034.070 **Un système modulaire d'interferométrie de Fabry-Pérot pour l'imagerie et la spectrométrie.**
J.-R. Roy, R. Arsenault, R. Bisson, G. Joncas.
J. R. Astron. Soc. Canada, Vol. 76, 277 - 302 (1982).
The authors have built a portable and microcomputer controlled Fabry-Pérot (F-P) interferometer system to obtain radial velocity field maps and photoelectric line profiles of galactic and extragalactic H II regions. The interferometer is piezo scanned by using capacitance micrometry to detect deviations from parallelism or changes in the absolute spacing.

034.071 **Un spectrophotometre interferentiel de Fabry-Pérot controlé par micro-ordinateur.**
R. Arsenault, J.-R. Roy, R. Bisson.
J. R. Astron. Soc. Canada, Vol. 76, 315 - 316 (1982). Abstract.

034.072 **The Fabry-Perot interferometer in astronomy.**
A. H. Jarrett, J. Stapelberg.
Boyden Obs. Publ., No. 1, 18 pp. (1982).

034.073 **An achromatic polarimeter for spectral scanning across atomic lines and molecular bands.**
E. J. Benedetti.
Vatican Obs. Publ., Vol. 2, 15 - 19 (1982).
A new polarimeter which can be used both for wide spectral band observations from about 3000 to 8000 Å and for spectral scanning at resolutions of about 2 Å has been constructed. The design has been directed to reducing instrumental polarization to a minimum. The system is controlled by a microprocessor which is also used for on-line data reduction and is designed for compactness so that it may conveniently travel to various observatories.

034.074 **The UK 1.2-m Schmidt Telescope and the Southern Sky Survey. V. An absolute calibration of the Night Sky Photometer of the UK 1.2 m Schmidt Telescope.**
R. J. Smyth.
Occas. Rep. R. Obs., Edinburgh, No. 8, 26 pp. (1982).
A comparison of parallel UBV observations obtained

using the Night Sky Photometer (NSP) of the UK 1.2-m Schmidt Telescope and a photometric telescope at the same site has led to recalibrations of the U-, B- and V-band surface brightness measurements of the NSP. Absolute calibration to an accuracy of $\pm 0\overset{m}{.}1$ arcsec^{-2} is feasible for plates taken at field centres north of $\delta \cong -65°$, subject to the exclusion of bright field stars.

034.075 **Fotometria fotoeléctrica en el O. A. N. (*Observatorio Astronómico Nacional*): Instrumentación y transformación de las medidas al sistema estándar.**
J. F. Lahulla, J. Pensado.
Bol. Astron. Obs. Madrid, Vol. 10, No. 6, p. 3 - 16 (1981).
A description of a photon counting photometer attached to the 1.5 mts. reflector of the O. A. N. is presented. UBVRI measurements made in 29 nights show the extinction coefficients and the colour transformation.

034.076 **A middle-infrared heterodyne spectrometer to be used at the Cassegrain-focus of medium-size and large astronomical telescopes.**
H. U. Käufl, H. Rothermel, S. Drapatz.
Max-Planck-Inst. Phys. Astrophys., Inst. Extraterr. Phys., MPE Prepr. 5, 10 pp. (1982). – Contributed talk given at the Second ESO Infrared Workshop, Munich 20. - 23. 4. 82.

034.077 **Experiences with the fiber-linked spectrograph of the Landessternwarte.**
J. G. Schiffer, I. Appenzeller, R. Östreicher.
Mitt. Astron. Ges., Nr. 57, p. 323 - 328 (1982).

034.078 **An instrument for spectroscopic and photographic observations with the UM-92 image tube.**
V. Yu. Terebizh.
Astron. Tsirk., No. 1175, p. 1 - 2 (1981). In Russian.

034.079 **The Lick Observatory charge-coupled device (CCD) and controller.**
L. B. Robinson.
Proc. SPIE Int. Soc. Opt. Eng., Vol. 290, 124 - 129 (1981) = Lick Obs. Bull., No. 891.
A flexible microprocessor-based controller for charge coupled device two-dimensional detectors is described. The controller can operate under manual control or as a slave to a remote computer linked by coaxial cable. The system is discussed along with data taken at the telescope and in the laboratory.

034.080 **Silicon photodiode array detectors for high resolution spectroscopy at Lick Observatory.**
S. S. Vogt.
Proc. SPIE Int. Soc. Opt. Eng., Vol. 290, 70 - 80 (1981) = Lick Obs. Bull., No. 892.
This paper describes the design and performance of two unintensified Reticon photodiode array detectors for high resolution spectroscopy. Reticon arrays of 2 X 936 and 1 X 1872 formats were chosen for the initial installation. Detailed descriptions are given of cooling techniques, temperature control, and signal processing electronics which represent major improvements in performance and simplicity of operation over many previous Reticon systems.

034.081 **On the spectrograph collimator with a long slit.**
Y. Yamashita.
Tokyo Astron. Obs. Rep., Vol. 19, 513 - 516 (1981). In Japanese.

034.082 **Experiment with a stellar magnetograph.**
K. Nariai, S. Ishibashi, M. Shimizu, Y. Norimoto, K. Okida, Y. Yamashita.
Tokyo Astron. Obs. Rep., Vol. 19, 517 - 524 (1981). In Japanese.

034.083 **Improvement of a visual photometer for measuring coronal line intensities.**
H. Miyazaki, T. Okamoto, H. Fukushima.
Tokyo Astron. Obs. Rep., Vol. 19, 532 - 536 (1981). In Japanese.

034.084 **Photoelectric photometric system for the new 30 cm reflector at Mitaka, Tokyo Astronomical Observatory (II).**
Y. Nishino, H. Sato, K.-H. Sato.
Tokyo Astron. Obs. Rep., Vol. 19, 547 - 557 (1981). In Japanese.

034.085 **Test of characteristics of RCA image intensifiers for photographic photometry (I).**
T. Okada.
Tokyo Astron. Obs. Rep., Vol. 19, 558 - 566 (1981). In Japanese.

034.086 **Bi-channel lamp for standardization of spectrograms.**
Yu. N. Bondarenko, Yu. A. Medvedev.
Astron. Tsirk., No. 1202, p. 7 - 8 (1981). In Russian.

034.087 **On mutual calibration of photoelectric seeing monitors.**
Yu. V. Khan, B. B. Khodzhadurdyev.
Astron. Tsirk., No. 1203, p. 7 - 8 (1982). In Russian.

034.088 **A coronographic mode for large telescopes.**
G. Courtes, P. Lamy, N. Mauron, M. Saisse.
Instrumentation for astronomy with large optical telescopes, (see 012.059), p. 73 - 78 (1982).
The study and efficiency test of the optics of the Space Telescope Faint Object Camera carried out at the Laboratoire d'Astronomie Spatiale, led the authors to design a similar coronographic mode for large ground based telescopes. Adaptation of this design to the Mount Chiran 1 meter and E.S.O. 3.6 meter Ritchey-Chrétien telescopes has given new levels of performance in the detection of faint features in stellar and planetary observations.

034.089 **Multi-order formats to increase the versatility of Cassegrain spectrographs.**
D. L. Harmer.
Instrumentation for astronomy with large optical telescopes, (see 012.059), p. 109 - 116 (1982).
"Grism" and compound prism cross-dispersers have been proposed for the new Cassegrain spectrograph of the 2.5m telescope at La Palma. The RGO image-tube spectrograph has been fitted with a grism.

034.090 **Transit telescope designs optimized for multiple object spectroscopy with fibers.**
J. R. P. Angel.
Instrumentation for astronomy with large optical telescopes, (see 012.059), p. 117 - 122 (1982).

034.091 **An integral field spectrograph (IFS) for large telescopes.**
G. Courtes.
Instrumentation for astronomy with large optical telescopes, (see 012.059), p. 123 - 128 (1982).
A combination of an array of small lenses together with fibre optics can be used to produce a high transmission spectrograph which can record the spectra of many image elements simultaneously. Such a design may play an important role in ground based observations preparing for the NASA-ESA Space Telescope programmes.

034.092 **Optical designs of plane aspherized grating spectrographs.**
E. H. Richardson.

Instrumentation for astronomy with large optical telescopes, (see 012.059), p. 129 - 136 (1982).

Spectrographs using plane aspherized gratings have excellent resolution over a wide field and require one less element than a conventional Schmidt or Maksutov camera. However, they are restricted to configurations where the grating is approximately normal to the camera axis.

034.093 **Un spectrographe à réseau asphérique pour télescope f/4.**
G. Lemaître.
Instrumentation for astronomy with large optical telescopes, (see 012.059), p. 137 - 141 (1982).

Aspheric diffraction gratings have been produced by the elastic relaxation method. These gratings lead to more efficient spectrograph mounting arrangements since, for a wide field instrument of fast aperture ratio, a smaller number of optical surfaces are required.

034.094 **A spectrophotometer devoted to stellar seismology.**
E. Fossat, Y. Decanini, G. Grec.
Instrumentation for astronomy with large optical telescopes, (see 012.059), p. 169 - 174 (1982).

034.095 **The 200-channel spectrometer of the Crimean Astrophysical Observatory.**
A. N. Abramenko, Ju. S. Alexandrin *(Yu. S. Aleksandrin)*, V. V. Prokof'eva, V. N. Yakushin.
Instrumentation for astronomy with large optical telescopes, (see 012.059), p. 175 - 180 (1982).

A simple, low cost, television spectrometer has been constructed at the Crimean Astrophysical Observatory using an I-Isocon image tube for analog-digital recording of stellar spectra. High quantum efficiency (0.04) together with good accuracy (± 0.3%) and repeatability are obtained.

034.096 **A photometric system to search for optical variability on time-scales of 3 $\times 10^{-7}$ to 300s: main results.**
G. M. Beskin, S. I. Neizvestnyi *(Neizvestnyj)*, A. A. Pimonov, V. L. Plakhotnichenko, V. F. Shvartsman.
Instrumentation for astronomy with large optical telescopes, (see 012.059), p. 181 - 184 (1982).

The system comprises the prime focus photometer of the 6m telescope, a unit which registers the time of arrival at the telescope of each incoming photon, a CAMAC crate and an off-line computer.

034.097 **A high resolution Fourier transform spectrometer for the Cassegrain focus at the CFH telescope.**
J. P. Maillard, G. Michel.
Instrumentation for astronomy with large optical telescopes, (see 012.059), p. 213 - 222 (1982).

A high resolution interferometer (60cm maximum path difference) for use from the visible region to 5μm has been designed and manufactured at Meudon Observatory to be part of the instrumentation of the Canada-France-Hawaii Telescope. The considerations which led to the choice of the interferometer being operated at the Cassegrain focus, and the design features which allow this to be achieved, are outlined.

034.098 **Two dimensional interferometric photon counting observations with the 6m telescope.**
J. Boulesteix, Y. Georgelin, M. Marcelin, J. A. Fort.
Instrumentation for astronomy with large optical telescopes, (see 012.059), p. 223 - 227 (1982).

034.099 **Solid state imagers for astronomy.**
D. W. Latham.
Instrumentation for astronomy with large optical telescopes, (see 012.059), p. 245 - 249 (1982).

The main emphasis of this report is a comparison of the performance characteristics of the CCD detectors now in use.

034.100 **Recent progress with Multi-Anode Microchannel Array detector systems.**
J. G. Timothy.
Instrumentation for astronomy with large optical telescopes, (see 012.059), p. 251 - 258 (1982).

The Multi-Anode Microchannel Arrays (MAMAs) are a family of photoelectric, photon-counting array detectors that are being developed specifically for use in instruments on space-borne and ground-based telescopes. The author reviews briefly the construction and modes-of-operation of the MAMA detectors and describes the current development status of the different detector systems.

034.101 **Spectroscopy with photon-counting Reticons.**
D. W. Latham.
Instrumentation for astronomy with large optical telescopes, (see 012.059), p. 259 - 270 (1982).

034.102 **Experiments with silicon array detectors in the RGO coudé spectrograph.**
C. F. W. Harmer, D. L. Harmer.
Instrumentation for astronomy with large optical telescopes, (see 012.059), p. 271 - 278 (1982).

034.103 **Two-dimensional television system: application for direct acquisition of images and speckle interferometry.**
Yu. Yu. Balega, N. N. Somov, A. F. Fomenko.
Instrumentation for astronomy with large optical telescopes, (see 012.059), p. 279 - 281 (1982).

034.104 **TV spectrum scanner of the 6-meter telescope.**
T. A. Somova, N. N. Somov, S. V. Markelov. V. B. Nebelitskii *(Nebelitskij)*, O. I. Spiridonova, A. F. Fomenko.
Instrumentation for astronomy with large optical telescopes, (see 012.059), p. 283 - 290 (1982).

The television multichannel spectrophotometer of the BTA, used for accumulating and recording the spectra of faint astronomical objects, is described. The image detection system comprises a three-stage image tube (UM-92) optically coupled to a high sensitivity SIT TV tube. The system operates in a photon counting mode.

034.105 **Utilisation des récepteurs électronographiques avec les grands télescopes.**
G. Wlérick, M. Duchesne, B. Servan, F. Gex, J. M. Munier.
Instrumentation for astronomy with large optical telescopes, (see 012.059), p. 291 - 298 (1982).

034.106 **Application of the Electronographic Image Converter for spectroscopy at the 2.6m reflector of the Crimean Astrophysical Observatory.**
P. P. Petrov.
Instrumentation for astronomy with large optical telescopes, (see 012.059), p. 299 - 304 (1982).

An Electronographic Image Converter (EIC) of the "Spectracon" type is being used for routine spectroscopic observations at the 2.6m reflector of the Crimean Astrophysical Observatory. Some details of the EIC and film-loop applicator are given and the spectrophotometric accuracy and detective quantum efficiency are discussed.

034.107 **First results with an electronographic camera used at the prime focus of the C. F. H. telescope in Hawaii.**
J. P. Picat, J. Baudrand, R. Bellenger, M. Cailloux, A. Chevillot, F. Crifo, J. P. Dupin, P. Felenbok, J. Guérin, C. Jouan, J. P. Picat, I. Raulet, G. Rousset.

Instrumentation for astronomy with large optical telescopes, (see 012.059), p. 305 - 310 (1982).

The camera is briefly described. It is an electronic camera with a new technological design and a tight valve allowing the recovery of the nuclear plates without destroying the photocathode. Some preliminary results are given.

034.108 **Detectors used in UV astronomy.**
Y.-x. Xu.
Publ. Purple Mountain Obs., Vol. 1, 81 - 87 (1982). In Chinese.

034.109 **The infrared photometer of the 2 m telescope at Pic du Midi Observatory.**
N. Epchtein.
Second ESO infrared workshop, (see 012.060), p. 73 - 76 (1982).

034.110 **Infrared broad-band filters optimised for observations of Wolf-Rayet stars.**
C. Sterken, C. de Loore.
Second ESO infrared workshop, (see 012.060), p. 77 - 79 (1982).

034.111 **An F/20 millimeter photometer for TIRGO.**
N. Mandolesi, G. Morigi, G. Spada, M. Ferrari-Toniolo, O. Leonetti, P. Persi, L. Spinoglio, F. S. Delli Santi, F. Delpino, M. Landini, P. Salinari.
Second ESO infrared workshop, (see 012.060), p. 81 - 91 (1982).

A new instrument for the detection of the millimeter radiation, mounted at the 1.5 m, F/20 infrared telescope (TIRGO), is described. The photometer uses a composite GaGe bolometer and an uncooled optical system.

034.112 **The IR spectrophotometer for the Gornergrat telescope.**
P. Saraceno, R. Messi, D. Lorenzetti, R. Orfei.
Second ESO infrared workshop, (see 012.060), p. 93 - 100 (1982).

034.113 **Reflective field optics for IR spectrophotometers.**
J. Schulte in den Bäumen.
Second ESO infrared workshop, (see 012.060), p. 101 - 107 (1982).

034.114 **A cooled grating spectrometer for TIRGO.**
P. Salinari.
Second ESO infrared workshop, (see 012.060), p. 109 - 112 (1982).

034.115 **IRSPEC.**
A. F. M. Moorwood.
Second ESO infrared workshop, (see 012.060), p. 113 - 117 (1982).

IRSPEC is a cooled, grating array spectrometer for 1 - 5 μm spectroscopy at R $\sim 3 \times 10^3$ being built within ESO for use initially at the 3.6 m telescope and possibly later on at the 3.5 m NTT (New Technology Telescope). The performance and design features are summarized and the overall optical/mechanical concept is illustrated.

034.116 **IFU: a multi-detector IR spectrometer.**
J. Slingerland.
Second ESO infrared workshop, (see 012.060), p. 119 - 123 (1982).

034.117 **Cooled grating spectrometer.**
R. Wade.
Second ESO infrared workshop, (see 012.060), p. 125 - 129 (1982).

A cooled grating spectrometer for 1 - 5 μm has been constructed for UKIRT. The paper describes the design and gives preliminary performance figures obtained on the telescope.

034.118 **Future UKIRT instrumentation.**
R. Wade.
Second ESO infrared workshop, (see 012.060), p. 131 - 136 (1982).

The present status and future plans for UKIRT common-user instrumentation are outlined.

034.119 **Fabry-Pérot spectrometers for the ground-based infrared.**
T. R. Geballe.
Second ESO infrared workshop, (see 012.060), p. 137 - 141 (1982).

034.120 **A middle-infrared heterodyne spectrometer to be used at the Cassegrain-focus of medium-size and large astronomical telescopes.**
H. U. Käufl, H. Rothermel, S. Drapatz.
Second ESO infrared workshop, (see 012.060), p. 143 - 152 (1982).

The design of the spectrometer is presented: grid-tuned and dither-stabilized CO_2-laser, CdHgTe photovoltaic detector, laserbeam diagnostics, calibration source, TV-guiding system. Spectra taken with the 1 m telescope on the Gornergrat and the 2.2 m telescope on Calar Alto are presented.

034.121 **An improved chopping secondary design.**
A. Krabbe, K. Haussecker, D. Lemke, K. Proetel.
Second ESO infrared workshop, (see 012.060), p. 169 - 172 (1982).

034.122 **Photoconductor developments at the Battelle – Institut e.V., Frankfurt am Main.**
U. Brunsmann.
Second ESO infrared workshop, (see 012.060), p. 183 - 187 (1982).

034.123 **Research on far-infrared photoconductors.**
P. van der Wal.
Second ESO infrared workshop, (see 012.060), p. 189 - 190 (1982).

034.124 **Tests of low-background infrared detectors.**
J. Wolf, D. Lemke.
Second ESO infrared workshop, (see 012.060), p. 191 - 195 (1982).

034.125 **He^3-bolometers for mm- and sub-mm-photometry.**
E. Kreysa, E. E. Haller
Second ESO infrared workshop, (see 012.060), p. 197 - 204 (1982).

034.126 **Infrared detector arrays.**
M. W. Werner, C. R. McCreight.
Second ESO infrared workshop, (see 012.060), p. 205 - 206 (1982).

034.127 **Application of infrared arrays to speckle interferometry.**
F. Sibille, D. Stefanovitch, A. Chelli.
Second ESO infrared workshop, (see 012.060), p. 207 - 213 (1982).

A short review of two-dimensional arrays of infrared detectors is presented. Performances under low background conditions of the 32 $\times$ 32 elements InSb CID from General Electric are analyzed. The results of the first infrared speckle observations with this array are presented.

034.128 **A new InSb charge amplifier for application to a spectrometer array.**
D. Hartill.
Second ESO infrared workshop, (see 012.060), p. 215 - 225 (1982).

034.129 **High sensitivity with a germanium detector in integrating mode.**
G. Olofsson.
Second ESO infrared workshop, (see 012.060), p. 227 - 231 (1982).

034.130 **Infrared array photometer.**
W. Luinge.
Second ESO infrared workshop, (see 012.060), p. 233 - 236 (1982).
An array photometer being constructed at present as a focal plane instrument for the UKIRT telescope on Hawaii is described.

034.131 **A cryogenic Fabry-Pérot for far infrared astronomy.**
J. J. Wijnbergen, G. Ploeger, G. de Groot.
Second ESO infrared workshop, (see 012.060), p. 359 - 370 (1982).
A small Fabry-Pérot module has been built which accommodates reflectors with an effective aperture of 22 mm, driven by an electromagnetic actuator. The complete optical system includes this module with a predisperser and two separate wide-band photometers. It accepts an f/6 beam and is mounted on a small 13-cm diameter base plate. Equipped with Ge:Ga photon detectors the F.P. operates at 4 K around a preselected wavelength in the 30–120 μm infrared range with a resolution in the order of 2 - 4000. The measured optical NEP is below 1×10^{-13} W $Hz^{-1/2}$ at room temperature backgrounds.

034.132 **Development and work of a new receiving equipment of the telescope of the National Astronomical Observatory of the Bulgarian Academy of Sciences.**
B. Kovachev.
Spisanie Blg. AN, Vol. 28, No. 1, p. 67 - 73 (1982). In Bulgarian. – Abstr. in Ref. zh., 51. Astron., 10.51.820 (1982).

034.133 **Photomultipliers FEU-79 and FEU-106 in the earth's magnetic field.**
V. Bobinas.
Bull. Vilnius Astron. Obs., No. 61, p. 46 - 51 (1982). In Russian.
The influence of the horizontal component of the earth's magnetic field on photomultipliers FEU-79 and FEU-106 operating in a pulse-counting mode is investigated.

034.134 **An automated Seya-Namioka spectrophotometer.**
Ü. Ibrus, H. Ibrus.
Tartu Astrofüüs. Obs. Publ., Köide 49, 178 - 181 (1982).
An automated spectrum scanner for use at the Cassegrain focus of the *f*/16 telescope is described. The scanner is designed to measure energy distribution in the spectra of stars.

034.135 **IR photometer for the spectral region λλ 0.9 - 2.5 microns.**
S. I. Bekhtev, Yu. N. Ivanov, V. V. Kir'yan, M. A. Pogodin, M. I. Rybakov, Yu. L. Shakhbazyan.
Izv. Glav. Astron. Obs. Pulkovo, Astrometr. Astrofiz., No. 199, p. 125 - 128 (1982). In Russian.
A one-channel IR photometer for the λλ 0.9 - 2.5 microns region has been designed. Basic peculiarities of the construction, optics and electronics of the photometer are given.

034.136 **Differential astronomical photometer usable in nonphotometric sky conditions.**
E. F. Milone, R. M. Robb, C. H. Hansen, F. M. Babott.
J. Opt. Soc. America, Vol. 72, 1104 (1982). – Abstract.

034.137 **Astronomical observations with Schottky staring system.**
E. L. Dereniak, R. W. Capps, W. Ewing.
J. Opt. Soc. America, Vol. 72, 1116 (1982). – Abstract.

034.138 **Prototype for a long-baseline Michelson stellar interferometer.**
J. Davis.
J. Opt. Soc. America, Vol. 72, 1759 (1982). – Abstract.

034.139 **Infrared charge-coupled-device test and evaluation console.**
M. L. Gronberg, E. L. Dereniak, R. A. Florence.
J. Opt. Soc. America, Vol. 72, 1823 (1982). – Abstract.

034.140 **Portable, high-speed photoelectric photometer.**
D. M. Varney.
J. Opt. Soc. America, Vol. 72, 1827 (1982). – Abstract.

034.141 **Astronomical spectroscopy with an echelle-interferometer Cassegrain spectrograph.**
B. Bates, R. E. L. Bankhead, W. Brown-Kerr, F. N. Byrne, D. L. Giaretta, J. A. McQuoid.
Appl. Opt., Vol. 21, 2794 - 2798 (1982).
A novel Cassegrain spectrograph has been constructed which may be used to produce either echelle spectra or high-spectral resolution Fabry-Perot interferograms. A description of the spectrograph design is given, and results from first observational tests are used to discuss the merits and limitations of the technique for studies in high-resolution astronomical spectroscopy.

034.142 **Blackbody heterodyne receiver for NEP measurements and wideband photodetector characterization.**
E. R. Brown.
Appl. Opt., Vol. 21, 3602 - 3606 (1982).
A blackbody radiometer is designed to measure the noise equivalent power of 10.6 μm heterodyne-receiver front ends as a function of intermediate frequency from 100 to 6000 MHz. The system includes noise diode injection and IF response normalization, making it particularly useful for characterizing high-speed photodetectors coupled to wideband preamplifiers. Experimental results for some 77-K HgCdTe photodiodes indicate amplifier-limited NEP values $< 2 \times 10^{-19}$ W/Hz at 4.0 GHz. The NEP data agree well with those obtained by the standard blackbody heterodyne technique.

034.143 **Universal spectrophotometer for determining the efficiency of optical components and systems.**
D. Enard, H. Visser.
Appl. Opt., Vol. 21, 4459 - 4464 (1982).
An instrument has been developed to measure the absolute reflectance or transmittance of optical elements of systems. Contrary to existing spectrophotometers, it is not limited by the size or by the optical power of the element to be analyzed.

034.144 **Non-polarizing far-infrared filters.**
G. Dall'Oglio, B. Melchiorri, P. de Bernardis, S. Masi.
Infrared Phys., Vol. 22, 307 - 308 (1982).
Polyethylene embedded with glass microspheres has shown the same optical properties as Fluorogold™ in the far-infrared. The polarizing properties of Fluorogold (laminated in thin foils) are avoided by pressing the polyethylene/glass powder between hot plates. The new filters are suitable for long-wave pass filters with intrinsic polarization less than 10^{-4}.

034.145 **New microphotometer.**
A. Pliska.
Říše hvězd, Vol. 63, 144 - 148 (1982). In Czech.

034.146 **Electron bombarded charge-coupled device (CCD) detectors for the vacuum ultraviolet.**
J. L. Lowrance, G. R. Carruthers.
Proc. SPIE Int. Soc. Opt. Eng., Vol. 279, (see 012.072), 123 - 128 (1981). – Abstr. in Phys. Abstr., Vol. 85, Abstr. 114470 (1982).

034.147 **Use of charge-coupled device (CCD) detectors for imaging in the vacuum ultraviolet.**
D. Michels, P. Burstein.
Proc. SPIE Int. Soc. Opt. Eng., Vol. 279, (see 012.072), 140 - 145 (1981). – Abstr. in Phys. Abstr., Vol. 85, Abstr. 114471 (1982).

034.148 **Infrared Fourier spectrometer for use at the Cassegrain focus.**
H. L. Buijs, R. P. Lowe, J. M. Moorhead, W. H. Wehlau.
Proc. SPIE Int. Soc. Opt. Eng., Vol. 289, (see 012.073), 62 - 65 (1981). – Abstr. in Phys. Abstr., Vol. 85, Abstr. 117594 (1982).

034.149 **Peripheral system of a multichannel polarimeter.**
S. Kikuchi.
Astron. Her., Vol. 75, 110 - 112 (1982). In Japanese. – Abstr. in Phys. Abstr., Vol. 85, Abstr. 117601 (1982).

034.150 **Development and evaluation of integrated infrared arrays for astronomical applications.**
C. R. McCreight.
Proc. SPIE Int. Soc. Opt. Eng., Vol. 304, (see 012.074), 187 - 195 (1981). – Abstr. in Phys. Abstr., Vol. 85, Abstr. 117603 (1982).

034.151 **Near infrared photopolarimeter for astronomical observations.**
T. Sajo.
Exp. Tech. Phys., Vol. 30, 315 - 322 (1982). – Abstr. in Phys. Abstr., Vol. 85, Abstr. 117604 (1982).

034.152 **Digital recording system for cosmic ray ion chambers.**
M. Wada, T. Imai, K. Imai, S. Kawasaki. M. Kusunose, L. S. Chuang.
Nucl. Instrum. Methods Phys. Res., Vol. 200, 457 - 461 (1982). – Abstr. in Phys. Abstr., Vol. 85, Abstr. 117606 (1982).

034.153 **Further developments of electrographic image detectors.**
G. R. Carruthers.
Proc. SPIE Int. Soc. Opt. Eng., Vol. 279, (see 012.072), 112 - 222 (1981). – Abstr. in Phys. Abstr., Vol. 85, Abstr. 117607 (1982).

034.154 **A pure single particle detecting counter telescope to search for fractionally charged particles.**
I. Yamamoto, T. Tomiyama, S. Furukawa, S. Endou, Y. Iga, T. Komatsubara, T. Wada, Y. Yamashita.
Nucl. Instrum. Methods Phys. Res., Vol. 201, 457 - 464 (1982). – Abstr. in Phys. Abstr., Vol. 86, Abstr. 5118 (1983).

034.155 **Photoelastic-modulator polarimeters in astronomy.**
J. C. Kemp.
Proc. SPIE Int. Soc. Opt. Eng., Vol. 307, (see 012.080), 83 - 88 (1981). – Abstr. in Phys. Abstr., Vol. 86, Abstr. 10223 (1983).

034.156 **Advances in detectors for astronomical spectroscopy.**
A. Boksenberg.
Philos. Trans. R. Soc. London, Ser. A, Vol. 307, 531 - 548 (1982). – Abstr. in Phys. Abstr., Vol. 86, Abstr. 15847 (1983).

034.157 **A more efficient faint-object spectrograph.**
C. G. Wynne.
Opt. Acta, Vol. 29, 1557 - 1571 (1982). – Abstr. in Phys. Abstr., Vol. 86, Abstr. 20247 (1983).

034.158 **A photoelectric radial-velocity spectrometer on the 1.2-m telescope of the Dominion Astrophysical Observatory.**
J. M. Fletcher, H. C. Harris, R. D. McClure, C. D. Scarfe.
Publ. Astron. Soc. Pacific, Vol. 94, 1017 - 1028 (1982).

A photoelectric radial-velocity spectrometer has been constructed for the coudé spectrograph of the 1.2-m telescope at the Dominion Astrophysical Observatory. The design of this instrument is discussed, including an innovation which eliminates the problem of "mismatch" for observations of high-velocity stars. Performance of the instrument is discussed, and results are given of a comparison of the spectrometer velocity system with the systems of IAU, LICK, Griffin, and Beavers and Eitter.

Radiation detectors in terrestrial astronomy.
See Abstr. 003.005.

Real time control with the TRS-80.
See Abstr. 003.040.

Software for photometric astronomy.
See Abstr. 003.041.

Astronomical photometry.
See Abstr. 003.056.

Diffraction gratings.
See Abstr. 003.064.

Infrared instrumentation on Calar Alto.
See Abstr. 009.025.

Solid state imagers for astronomy.
See Abstr. 012.098.

Infrared and submillimetre astronomical spectroscopy at Berkeley.
See Abstr. 013.062.

High resolution imagery and slitless spectroscopy at the prime focus of the CFH telescope.
See Abstr. 031.026.

The catadioptric camera as used for astronomical spectroscopy with image tubes and similar detectors: a limited review.
See Abstr. 031.027.

Blaze characteristics of echelle gratings.
See Abstr. 031.061.

Spectroscopic measurements of Mars and Venus with heterodyne techniques.
See Abstr. 031.519.

Computer controlled high speed photometry presented by the lunar occultation of 57 Orionis. I.
See Abstr. 031.603.

Determination of accurate radial velocities of early-type stars by cross-correlating spectra scanned on the ESO-S-3000 measuring machine.
See Abstr. 031.631.

Speckle interferometry with the BTA telescope.
See Abstr. 031.639.

Experiments in differential speckle interferometry.
See Abstr. 031.640.

Measurements of photon statistics with nanosecond resolution.
See Abstr. 031.642.

Interferometric alignment of optical surfaces and rotational axes.
See Abstr. 031.674.

Rapid alternate detection system of the Rothney Astrophysical Observatory.
See Abstr. 031.675.

The equivalent-width calibration of the KPNO echelle spectrograph.
See Abstr. 031.685.

Information efficiencies of telescopes and their instrumentation.
See Abstr. 032.047.

The C.F.H. (*Canada-France-Hawaii*) telescope as an infrared telescope.
See Abstr. 032.048.

Rocket-borne instrument with a high-resolution microchannel plate detector for planetary UV spectroscopy.
See Abstr. 032.580.

Intensified charged coupled devices for use as a spaceborne spectrographic image-plane detector system.
See Abstr. 032.582.

Stable and rugged étalon for the Dynamics Explorer Fabry-Pérot interferometer. 1: Design and construction.
See Abstr. 032.584.

Stable and rugged étalon for the Dynamics Explorer Fabry-Pérot interferometer. 2: Performance.
See Abstr. 032.585.

Photon-counting array detectors for space and ground-based studies at ultraviolet and vacuum ultraviolet (VUV) wavelengths.
See Abstr. 032.598.

Observations with a low-temperature, resonant mass, gravitational radiation detector.
See Abstr. 066.020.

Magnetic field in solar prominences measured with a new spectrally scanning magnetograph.
See Abstr. 073.060.

Fabry-Perot/CCD observations of [S III] and [S II] emissions from the Jupiter plasma torus.
See Abstr. 099.026.

Infrared spectroscopy in Australia.
See Abstr. 114.018.

Interacting binary degenerate stars.
See Abstr. 117.058.

Recent submillimetre wavelength studies of molecules in star formation regions.
See Abstr. 131.187.

Airborne far-infrared and submillimeter spectroscopy.
See Abstr. 131.220.

A search for diffuse band profile variations in the Rho Ophiuchi cloud.
See Abstr. 131.252.

The diffuse interstellar medium and the CES (*Coudé Echelle Spectrograph*).
See Abstr. 131.278.

Deep photography of radio sources with the Cambridge CCD system.
See Abstr. 141.022.

Continuum observations of extragalactic sources.
See Abstr. 141.127.

Observational results on Seyfert and radio galaxies with the Lick Observatory image tube-image dissector scanner.
See Abstr. 158.174.

Erratum

034.901 **Erratum: "Informational criterion for quality estimate of paralysable photon counters"** [Bull. Vilnius Astron. Obs., No. 53, p. 3 - 12 (1980). In Russian].
D. Ralys, R. Kalytis.
Bull. Vilnius Astron. Obs., No.61, p. 56 (1982). – See Abstr. 28.034.077.

035 Clocks and Frequency Standards

035.001 **Cylindrical sundials.**
D. K. Vedernikov.
Zemlya Vselennaya, 1982, No. 4, p. 71 - 72. In Russian.

035.002 **The developments of atomic time and frequency standards at Shanghai Observatory and some applications.**
Z.-C. Zhai, C.-H. Chuang.
J. Inst. Electron. Telecommun. Eng., Vol. 27, 500 - 503 (1981). – Abstr. in Phys. Abstr., Vol. 85, Abstr. 79525 (1982).

035.003 **Improved precision solar and sidereal clock.**
N. U. Shankar, M. Selvamani.
J. Inst. Electron. Telecommun. Eng., Vol. 27, 514 - 516 (1981). – Abstr. in Phys. Abstr., Vol. 85, Abstr. 79536 (1982).

035.004 **Time synchronization system in a two-station search for primordial black holes.**
C. L. Bhat, I. K. Kaul, R. C. Yadav.
J. Inst. Electron. Telecommun. Eng., Vol. 27, 556 - 564 (1981). – Abstr. in Phys. Abstr., Vol. 85, Abstr. 79541 (1982).

035.005 **Time synchronization for a 3-site interplanetary scintillation (IPS) experiment using ATA time signal transmission.**
A. D. Bobra, M. Selvaraj, S. K. Alurkar, R. V. Bhonsle.
J. Inst. Electron. Telecommun. Eng., Vol. 27, 569 - 572 (1981). – Abstr. in Phys. Abstr., Vol. 85, Abstr. 89155 (1982).

035.006 **Application aspects of time and frequency to astronomy.**
J. C. Bhattacharyya.
J. Inst. Electron. Telecommun. Eng., Vol. 27, 593 - 595 (1981). – Abstr. in Phys. Abstr., Vol. 85, Abstr. 89156 (1982).

035.007 **Mono-ion oscillator as potential ultimate laser frequency standard.**
H. G. Dehmelt.
IEEE Trans. Instrum. Meas., Vol. IM-31, 83 - 87 (1982). Abstr. in Phys. Abstr., Vol. 85, Abstr. 89656 (1982).

035.008 **Shuttle experiment to demonstrate high-accuracy global time and frequency transfer.**
D. W. Allan, C. O. Alley, Jr., R. Decher, R. F. C. Vessot, G. Winkler.
IEEE Trans. Geosci. Remote Sensing, Vol. GE-20, 321 - 325 (1982). – Abstr. in Phys. Abstr., Vol. 85, Abstr. 101387 (1982).

035.009 **Recent activities on time and frequency at the Radio Research Laboratories.**
Y. Yasuda, Y. Saburi.
J. Inst. Electron. Telecommun. Eng., Vol. 27, No. 10, Part 1, (see 012.044), p. 345 - 350 (1981). – Abstr. in Phys. Abstr., Vol. 85, Abstr. 89657 (1982).

035.010 **An overview of the frequency activities at National Physical Laboratory, New Delhi, India.**
B. S. Mathur.
J. Inst. Electron. Telecommun. Eng., Vol. 27, No. 10, Part 1, (see 012.044), p. 350 - 354 (1981). – Abstr. in Phys. Abstr., Vol. 85, Abstr. 89658 (1982).

035.011 **Present status of time and frequency in Indonesia.**
R. W. Matindas.
J. Inst. Electron. Telecommun. Eng., Vol. 27, No. 10, Part 1, (see 012.044), p. 355 - 357 (1981). – Abstr. in Phys. Abstr., Vol. 85, Abstr. 89659 (1982).

035.012 **Time and frequency in New Zealand.**
B. Lohrey.
J. Inst. Electron. Telecommun. Eng., Vol. 27, No. 10, Part 1, (see 012.044), p. 357 - 361 (1981). – Abstr. in Phys. Abstr., Vol. 85, Abstr. 89660 (1982).

035.013 **Reliability, serviceability and performance of frequency standards.**
H. Hellwig.
J. Inst. Electron. Telecommun. Eng., Vol. 27, No. 10, Part 1, (see 012.044), p. 362 - 365 (1981). – Abstr. in Phys. Abstr., Vol. 85, Abstr. 89661 (1982).

035.014 **The establishment of time and frequency services in Kenya.**
G. N. Ochodo.
J. Inst. Electron. Telecommun. Eng., Vol. 27, No. 10, Part 1, (see 012.044), p. 365 - 367 (1981). – Abstr. in Phys. Abstr., Vol. 85, Abstr. 89662 (1982).

035.015 **Standard time and frequency in Korea and dissemination by AM/FM broadcasting network.**
Nak Sam Chung, Suk Tai Kang, Jin Ok Kim.
J. Inst. Electron. Telecommun. Eng., Vol. 27, No. 10, Part 1, (see 012.044), p. 367 - 372 (1981). – Abstr. in Phys. Abstr., Vol. 85, Abstr. 89663 (1982).

035.016 **Time and frequency in Japan.**
S. Iijima.
J. Inst. Electron. Telecommun. Eng., Vol. 27, No. 10, Part 1, (see 012.044), p. 372 - 378 (1981). – Abstr. in Phys. Abstr., Vol. 85, Abstr. 89664 (1982).

035.017 **Some methods of maintaining and/or generating time and frequency at arbitrary points on the surface of the Earth.**
D. W. Allan.
J. Inst. Electron. Telecommun. Eng., Vol. 27, No. 10, Part 1, (see 012.044), p. 383 - 388 (1981). – Abstr. in Phys. Abstr., Vol. 85, Abstr. 94614 (1982).

035.018 **Time/frequency services of the US National Bureau of Standards and some alternatives for future improvement.**
R. E. Beehler.
J. Inst. Electron. Telecommun. Eng., Vol. 27, No. 10, Part 1, (see 012.044), p. 389 - 402 (1981). – Abstr. in Phys. Abstr., Vol. 85, Abstr. 89666 (1982).

035.019 **Precise T & F intercomparison via VLF phase measurements.**
A. S. Gupta, G. K. Goel, B. S. Mathur.
J. Inst. Electron. Telecommun. Eng., Vol. 27, No. 10, Part 1, (see 012.044), p. 403 - 407 (1981). – Abstr. in Phys. Abstr., Vol. 85, Abstr. 89667 (1982).

035.020 **VLF studies at NPL, New Delhi – experimental set-up and morphology of the data.**
A. S. Gupta, G. K. Goel, B. S. Mathur.
J. Inst. Electron. Telecommun. Eng., Vol. 27, No. 10, Part 1, (see 012.044), p. 414 - 421 (1981). – Abstr. in Phys. Abstr., Vol. 85, Abstr. 89669 (1982).

035.021 **The LASSO experiment and its operational organization.**
B. E. H. Serene.
J. Inst. Electron. Telecommun. Eng., Vol. 27, No. 10, Part 1, (see 012.044), p. 422 - 432 (1981). – Abstr. in Phys. Abstr., Vol. 85, Abstr. 89670 (1982).

035.022 **Satellite laser ranging precise time transfer techniques: some comments on the possible Indian contribution.**
P. S. Dixit, B. S. Mathur.
J. Inst. Electron. Telecommun. Eng., Vol. 27, No. 10, Part 1, (see 012.044), p. 433 - 438 (1981). – Abstr. in Phys. Abstr., Vol. 85, Abstr. 89671 (1982).

035.023 **A review of primary caesium clock development and satellite time comparison techniques at NRC.**
C. C. Costain, A. G. Mungall, H. Daams, J. S. Boulanger.
J. Inst. Electron. Telecommun. Eng., Vol. 27, No. 10, Part 1, (see 012.044), p. 438 - 442 (1981). – Abstr. in Phys. Abstr., Vol. 85, Abstr. 89672 (1982).

035.024 **Time and frequency dissemination system of moderate accuracy, sophistication and cost.**
J. Bonanomi.
J. Inst. Electron. Telecommun. Eng., Vol. 27, No. 10, Part 1, (see 012.044), p. 443 - 449 (1981). – Abstr. in Phys. Abstr., Vol. 85, Abstr. 89673 (1982).

035.025 **Precise international T/F comparisons via ground and satellite TV systems.**
B. Kovacevic, B. Dimitrijevic, M. Arsic, L. Golubobic.
J. Inst. Electron. Telecommun. Eng., Vol. 27, No. 10, Part 1, (see 012.044), p. 450 - 456 (1981). – Abstr. in Phys. Abstr., Vol. 85, Abstr. 89674 (1982).

035.026 **Time and frequency transfer using SABCTV.**
R. Lake.
J. Inst. Electron. Telecommun. Eng., Vol. 27, No. 10, Part 1, (see 012.044), p. 456 - 459 (1981). – Abstr. in Phys. Abstr., Vol. 85, Abstr. 89675 (1982).

035.027 **Time dissemination capabilities via navigation satellites.**
K. Putkovich.
J. Inst. Electron. Telecommun. Eng., Vol. 27, No. 10, Part 1, (see 012.044), p. 460 - 466 (1981). – Abstr. in Phys. Abstr., Vol. 85, Abstr. 89676 (1982).

035.028 **Precise T & F intercomparison between NPL, India and PTB, Federal Republic of Germany via Satellite Symphonie-1.**
B. S. Mathur, P. Banerjee, P. C. Sood, M. Saxena, N. Kumar, A. K. Suri.
J. Inst. Electron. Telecommun. Eng., Vol. 27, No. 10, Part 1, (see 012.044), p. 466 - 469 (1981). – Abstr. in Phys. Abstr., Vol. 85, Abstr. 89677 (1982).

035.029 **A direct reception system for time synchronization via INSAT-1.**
C. L. Jain, K. Kumar.
J. Inst. Electron. Telecommun. Eng., Vol. 27, No. 10, Part 1, (see 012.044), p. 470 - 472 (1981). – Abstr. in Phys. Abstr., Vol. 85, Abstr. 89678 (1982).

035.030 **An NNSS satellite timing receiver.**
C. L. Jain, K. Kumar, H. I. Andharia, M. Singh, V. D'Souza, V. K. Goel, A. K. Sisodia.
J. Inst. Electron. Telecommun. Eng., Vol. 27, No. 10, Part 1, (see 012.044), p. 473 - 477 (1981). – Abstr. in Phys. Abstr., Vol. 85, Abstr. 89679 (1982).

035.031 **A time-frequency signal discriminator.**
S.-h. Lu.
Publ. Shaanxi Astron. Obs., No. 1, p. 56 - 61 (1982). In Chinese.

035.032 **Anniversaries of Czechoslovak time signals.**
V. Ptáček.
Říše hvězd, Vol. 63, 232 - 234 (1982). In Czech.

Die Sonnenuhr des Augustus.
See Abstr. 003.029.

Role of International Radio Consultative Committee in co-ordination of time and frequency activities.
See Abstr. 013.071.

The transformation of geodetic coordinates and the calculation of long wave propagation range.
See Abstr. 044.025.

On methods for clock synchronization.
See Abstr. 066.016.

Gravitation and relativity experiments using atomic clocks.
See Abstr. 066.036.

Clocks and relativity experiments.
See Abstr. 066.318.

Statistical analysis of the disturbances of the lower ionosphere (D-layer).
See Abstr. 083.032.

036 Photographic Materials and Techniques

036.001 **Komeetfotografie met eenvoudige apparatuur.**
E. P. Bus.
Zenit, 9. Jaarg., 378 - 380 (1982).

036.002 **La diapositiva nella fotografia astronomica.**
E. Moltisanti.
Orione, Vol. 3, 116 - 123 (1982).

036.003 **Astrophotographie mit dem Kodak Ektachrome 400.**
P. Riepe, W. Celnik.
Sterne Weltraum, 21. Jahrg., 317 - 320 (1982).

036.004 **Die Tiefkühlkamera (Cold Camera).**
H. Haffter.
Orion, 40. Jahrg., 161 - 162 (1982).

036.005 **A new twist in planetary photography – unsharp masking.**
J. D. Beish, D. C. Parker.
Strolling Astron., Vol. 29, 145 - 147 (1982).

036.006 **Emulsion grain noise and the accuracy of stellar image positions.**
J. F. Lee, W. F. van Altena.
Bull. American Astron. Soc., Vol. 14, 645 (1982). – Abstract.

036.007 **Astronomical photography at 1.1 micrometers.**
A. G. Smith.
J. Appl. Photogr. Eng., Vol. 7, 167 - 170 (1981). – Abstr. in Phys. Abstr., Vol. 85, Abstr. 68628 (1982).

036.008 **Photographic facilities offered by the Schmidt telescope at Calern Observatory.**
J. L. Heudier.
The need for coordinated ground-based observations of Halley's comet, (see 012.025), p. 293 - 295 (1982).

036.009 **Astrofotografie.**
L. Vanhoeck.
Zenit, 9. Jaarg., 463 - 466 (1982).

036.010 **Field effects in UK Schmidt telescope plates.**
A. W. Campbell.
Observatory, Vol. 102, 195 - 199 (1982).

The existence of a field effect, in the form of a radial background density gradient on filtered, hypersensitized plates, is confirmed. Its character, magnitude and possible causes are investigated. It is concluded that it results from the hypersensitization/filter/plate-holder combination in use at the observatory. It is expected that a large proportion of IIIa-*J* and IIIa-*F* plates exposed in the telescope are affected in this way.

036.011 **Mondphotographie mit dem Celestron 8.**
D. Haesloop.
Sterne Weltraum, 21. Jahrg., 484 - 485 (1982).

036.012 **Gashypersensibilisierung von Filmmaterial.**
P. Stättmayer.
Sterne Weltraum, 21. Jahrg., 532 - 536 (1982).

036.013 **Comparison of Kodak 2415 and IIIa-F emulsions.**
R. M. West, O. M. Kurtanidze, L. A. Geonjian (*Geondzhyan*), G. N. Kimeridze.
AAS Photo-Bull., No. 28, p. 3 - 7 (1981).

For longer (e.g. stellar) exposures, nitrogen baked 2415 emulsion is at best about as fast as untreated IIIa-F and does not appear to offer any obvious advantage at medium-to-large, wide-field telescopes (scale: 1″ = 10 to 15 μm) except under very good seeing conditions. However, 2415 emulsion may be of considerable interest to users of smaller, wide-field telescopes because of its superior resolution.

036.014 **Optimum development times for Pandol developer.**
K. Hamajima, T. Soyano.
AAS Photo-Bull., No. 28, p. 8 - 12 (1981).

Optimum development times for Fuji Pandol developer have been examined for Kodak spectroscopic plates, types IIa-O, IIIa-J, IIa-D, 103a-E, 103a-F, IIa-F, IIIa-F, and I-N.

036.015 **Some unusual films for solar photography.**
L. B. Gilliam.
AAS Photo-Bull., No. 29, 9 - 10 (1982).

The purpose of this paper is to discuss three Kodak films found useful that are not commonly used in astronomy. The films are Kodak AHU microfilm 2460, Kodak RAR film 2498, and Eastman fine grain sound recording film 5375.

036.016 **High contrast copying with Kodak technical pan film 2415.**
C. R. Martys.
AAS Photo-Bull., No. 29, p. 11 - 13 (1982).

Prints made from copies on 2415 film of original photographs show higher contrast and record faint detail that cannot be reproduced by a conventional, straight print from the original. These threshold features are not readily visualized on inspection of the original negative by transillumination with diffuse light.

036.017 **Kodak IIIa-J plates – astrometry at Goethe Link Observatory.**
H. J. Wood.
AAS Photo-Bull., No. 30, p. 3 - 5 (1982).

Inexpensive methods for hypersensitization of large Kodak spectroscopic plates, type IIIa-J, are described. On nights with high humidity the telescope tube is filled with dry nitrogen gas during exposure. Gains of more than a factor of 8 have been obtained over untreated type 103a-O plates.

036.018 **Comparison of the absolute sensitivity of Kodak Technical Pan Film 2415 with standard astronomical emulsions.**
A. G. Smith.
AAS Photo-Bull., No. 30, p. 6 - 8 (1982).

The absolute sensitivities of hypersensitized and unhypersensitized samples of Kodak Technical Pan Film 2415 have been measured for exposures of astronomical duration. The results are compared with hypersensitized and unhypersensitized samples of the common Kodak astronomical emulsions IIIa-F, IIIa-J, 103a-O, and 098.

036.019 **Color negative films for astrophotography.**
E. Everhart.
AAS Photo-Bull., No. 30, p. 9 - 11 (1982).

Tested are several Kodak color negative films: Kodacolor 400, Vericolor S and L, and Kodacolor II. Of these, only Kodacolor 400 film responds well to hypersensitization procedures. It can be made as fast as most black-and-white films and plates commonly used in celestial photography.

036.020 **Are microfilms useful in astrophotography?**
E. Everhart.
AAS Photo-Bull., No. 30, p 12 - 15 (1982).

Four emulsions of ultrahigh resolution were hypersensitized and evaluated for use at the telescope: Recordak AHU Microfilm 1460 (Estar Base), Fuji Microfilm Negative HRII,

Agfa-Gevaert Copex Pan AHU TRI 13, and Eastman High Contrast Panchromatic Film 5369. Fastest of these is HRII, which becomes over one-third as fast as treated Kodak IIIa-F plates or Kodak 2415 films. It is doubtful whether any of these microfilms will find extensive use in low-level astrophotography.

036.021 **Astronomical colour printing at ESO.**
C. Madsen, M. Tarenghi.
Messenger, No. 30, p. 15 - 17 (1982).

036.022 **Le pellicole bianco e nero per la fotografia a lunga posa.**
M. Milani.
Orione, Vol. 3, 162 - 167 (1982).

036.023 **Hypersensitization by forming gas baking (II).**
T. Aoki.
Tokyo Astron. Obs. Rep., Vol. 19, 590 - 595 (1981). In Japanese.

036.024 **Hypersensitization of ZP1, ZP3 and I 850 plates by silver nitrate.**
I. Eglitis.
Astron. Tsirk., No. 1197, p. 4 - 6 (1981). In Russian.

036.025 **Class F emulsions for solar system astrometry under adverse conditions.**
J. Gibson.
Publ. Astron. Soc. Pacific, Vol. 94, 753 (1982). – Abstract.

036.026 **Direct photographic image enhancement in astronomy.**
D. F. Malin.
J. Photogr. Sci., Vol. 29, 199 - 205 (1981). – Abstr. in Phys. Abstr., Vol. 86, Abstr. 10224 (1983).

Photometric calibration of astronomical photographic plates. II. The Anglo-Australian Telescope prime focus spot sensitometer.
See Abstr. 034.002.

The SRC near-infrared survey of the galactic plane.
See Abstr. 156.003.

Positional Astronomy, Celestial Mechanics

041 Astrometry

041.001 **The FK4 orientation from the Washington observations of the sun and planets.**
M. L. Sveshnikov.
Pis'ma Astron. Zh., Tom 8, 381 - 384 (1982). In Russian. English translation in Soviet Astron. Lett., Vol. 8.

Corrections to the system of right ascensions and declinations of stars from the FK4 catalogue are obtained. One of possible causes of the equinox motion is pointed out.

041.002 **Reduction of observations of declinations of bright stars and cepheids.**
A. G. Oborneva.
Tr. Gos. Astron. Inst. Shternberga, Tom 51, 216 - 233 (1982). In Russian.

Observations of bright stars in the declination zone +70 to +90° and cepheids (−10 to +70°) were obtained. The methods of observations and reductions are described. The catalogues were tied to the FK4 system.

041.003 **Right ascensions of the sun, Mercury and Venus observed with the transit instrument at the Nikolaev Observatory in 1980.**
G. M. Petrov, R. T. Fedorova, P. N. Fedorov.
Glav. astron. obs. AN SSSR. Leningrad, 1982, 9 pp. In Russian. – Abstr. in Ref. zh., 51. Astron., 7.51.132 (1982).

041.004 **Declinations of the sun and Venus corrected for anomalous refraction.**
V. P. Sibilev.
Glav. astron. obs. AN SSSR. Leningrad, 1982, 21 pp. In Russian. – Abstr. in Ref. zh., 51. Astron., 7.51.133 (1982).

041.005 **Astrometric results obtained with the Crimea-Pushchino radio interferometer.**
V. E. Zharov, L. R. Kogan, E. N. Fedoseev.
Inst. kosm. issled. AN SSSR. Prepr., 1982, No. 687, 22 pp. In Russian. – Abstr. in Ref. zh., 51. Astron., 7.51.139 (1982).

041.006 **Precise optical positions of radio sources in the FK 4-system. II. Results from 28 sources on the northern hemisphere and a preliminary comparison of the optical-radio reference frame.**
C. de Vegt, U. K. Gehlich.
Astron. Astrophys., Vol. 113, 213 - 218 (1982).

Precise optical positions of 28 optical counterparts of extragalactic radio sources in the northern hemisphere have been derived using different long focus telescopes. The source positions are referred to the FK 4-system by the AGK 3 RN reference star catalogue as a primary reference frame; a system of secondary reference stars of intermediate brightness $m_v = 12 - 14$ has been obtained from the 23-cm astrograph of the Hamburg Observatory. A preliminary comparison of the optical and quasi absolute radio reference frame has been performed using a provisional averaged catalogue of radio positions. Significant systematic differences with local amplitudes up to 0.″2 have been found, mainly in declination.

041.007 **Modern problems of astrometry.**
V. V. Podobed.
Zemlya Vselennaya, 1982, No. 4, p. 27 - 32. In Russian.

041.008 **Block adjustment in photographic astrometry.**
J. Stock.
Rev. Mexicana Astron. Astrofis., Vol. 6, (see 012.017), 115 - 118 (1981).

A new method for block adjustment, based on a conversion to three dimensional cartesian coordinates and subsequent rotation around three axes has been tested. The measured material consists of 22 plate pairs taken with objective prism. As reference systems, the SAO Catalogue, the Yale Catalogue, the Perth 70 Meridian Circle Catalogue, and the FK 4 have been used.

041.009 **Un análisis de la transformación de coordenadas en placas Schmidt.**
M. A. Peñaloza M.
Rev. Mexicana Astron. Astrofis., Vol. 6, (see 012.017), 129 - 133 (1981).

The author analyzes the transformations which relate the celestial equatorial coordinates to the rectangular coordinates measured on a plate taken with a Schmidt telescope without objective prism. For this, he applied the concentric projection theory and made an adjustment by means of third-degree polynomials. The results demonstrate the presence of distortions which are likely due to random defects of the photographic emulsion, to systematic defects in the quality of the catalog used as reference and/or to defects introduced by anomalous atmospheric refraction and by the kind of filter used. The author finds that, within a circular plate zone of approximately one degree in radius, the mean deviation detected in the actual position of any stellar image reaches a value of 0.64 arcsec.

041.010 **Problem of determination of the orientation of coordinate systems and of periodic errors of star catalogues from optical observations.**
D. P. Duma.
Astron. Zh., Tom 59, 1012 - 1019 (1982). In Russian. English translation in Soviet Astron., Vol. 26, No. 5.

The obtained determination accuracy of the equinox correction was illustrated and its values for the FK 4 were evaluated. According to these data the correction to the right ascensions of FK 4 doesn't exceed +0.ˢ02.

041.011 **Precise positions in the FK4 system for 120 radio source reference stars.**
S. J. Dick, E. R. Holdenried.
Astron. J., Vol. 87, 1374 - 1378 (1982).

Positions are given in the FK4 system for 120 primary reference stars grouped around the optical counterparts of 11 extragalactic radio sources. The observations were made in 1979 and 1980 with the 6-in. transit circle of the U.S. Naval Observatory. The results reported here represent the first step in a three-tiered process to determine astrometric positions of the optical counterparts referred to the system of the FK4. One application of such positions is the determination of a nearly inertial optical reference frame, which eventually may be compared to a reference frame at radio wavelengths.

041.012 **Results of observation of the Sun, Mercury and Venus from 1974 to 1979.**
S. Sadžakov, M. Dačić, D. Šaletić.
Publ. Astron. Obs. Sarajevo, Vol. 1, No. 1, (see 012.043), p. 69 - 78 (1981).

Description is given of the equipment and the procedure of observation, along with the mean (O–C) values of the right ascensions and declinations, according to years, and computed corrections to the orbital elements.

041.013 **On the observations of the bright polar stars with the large transit instrument of the Belgrade Observatory.**
I. Pakvor.
Publ. Astron. Obs. Sarajevo, Vol. 1, No. 1, (see 012.043), p. 79 - 84 (1981).

The first regular observation program which is being carried out at the LTI of the Belgrade Observatory contains the determination of absolute right ascension of stars in the northern declination zone 70°–90°. The application of the vacuum meridian marks gives special interest to this program.

041.014 **Determination of declinations with a uniformly rotating vertical or meridian circle.**
D. Bozhichkovich.
Publ. Astron. Obs. Sarajevo, Vol. 1, No. 1, (see 012.043), p. 103 - 110 (1981).

In the present paper a possibility is considered of improving the accuracy of the declination determination alike with the classical meridian instruments and vertical circles, and with the horizontal meridian circles.

041.015 **Comparison of the micrometer screw corrections of the Belgrade Zenith Telescope in the different systems.**
M. Djokić.
Publ. Astron. Obs. Sarajevo, Vol. 1, No. 1, (see 012.043), p. 163 - 166 (1981).

041.016 **Characteristics of the preliminary results obtained in elaborating the catalogue of NPZT stars.**
S. Sadžakov, D. Šaletić, M. Dačić, V. Trajkovska.
Publ. Astron. Obs. Sarajevo, Vol. 1, No. 1, (see 012.043), p. 167 - 172 (1981).

041.017 **A pool of faint stars applied to star catalogue formation.**
R. Hering, H. G. Walter.
Astron. Astrophys., Vol. 115, 197 - 202 (1982).

The plans for extending the current fundamental reference system of positions and proper motions (FK4) to fainter stars have prompted the establishment of a data bank containing about 190,000 meridian circle observations of nearly 50,000 stars, and the implementation of a retrieval system. The authors examine the observational history of these stars as well as their distribution in right ascension and declination, in magnitude and spectral type. This allows to sort out a subset satisfying to a first approximation the criteria for selection of supplementary stars. The subset is indicative of a short list of about 12,000 stars forming an upper limit of potential candidates for system extension.

041.018 **The introduction of the improved stellar catalogue into the Ondřejov PZT results.**
J. Vondrák.
Geodesy and physics of the earth, (see 012.053), p. 175 - 180 (1981).

Observations obtained with a Zeiss PZT at Ondřejov Observatory in the years 1973 - 1978 were used to improve mean places and proper motions of 304 stars. In order to obtain the proper motions with higher accuracy, the PZT observations were combined with the positions of stars in the AGK2 and AGK3 catalogues. All the PZT observations from the years 1973 - 1979 were uniformly re-reduced in the system of the new catalogue. It is shown that the results are in better agreement with the BIH 79 system than with the old BIH 68 system.

041.019 **On the importance of photographic catalogues.**
W. Fricke.
Publ. Belgrade Obs., No. 29, 8 pp. (1982) = Astron. Rechen-Inst. Heidelberg, Mitt. Ser. A, Nr. 143.

The differences in the aims of photographic and meridian catalogues are outlined. Furthermore, the blinking of pairs of plates with much fainter stars has allowed one to compile catalogues of stars with large proper motions which are candidates for parallax programmes. Finally, high accuracy photographic astrometry and radio-interferometry provide the link between the extragalactic reference system of positions and the fundamental system which rests upon the dynamics of the planetary system.

041.020 **Comparison of the Belgrade General Catalogue of Latitude Stars (IKSZ) with the General Catalogue of the AOE [Engelhardt Astronomical Observatory] and Poltava Latitude Program stars.**
I. A. Urasina.
Astron. Tsirk., No. 1173, p. 6 - 7 (1981). In Russian.

041.021 **Propagation of positional errors in the corrections to star places.**
S. Takagi.
Publ. Astron. Soc. Japan, Vol. 34, 553 - 559 (1982).

Propagation of positional errors is investigated for the results of the determination of corrections to star places obtained by two typical methods using the results of astrolabe observations. It is suggested that the double passage stars should be used to get exact corrections to star places.

041.022 **A determination of absolute right ascensions of stars from observations at the earth's equator.**
G. M. Petrov.
Izv. Glav. Astron. Obs. Pulkovo, Astrometr. Astrofiz., No. 199, p. 28 - 29 (1982). In Russian.

It is shown that highly precise absolute right ascensions of stars of a narrow equatorial zone can be obtained using observations from the earth's equator.

041.023 **On star observations in areas with galaxies using meridian instruments.**
G. D. Baturina.
Izv. Glav. Astron. Obs. Pulkovo, Astrometr. Astrofiz., No. 199, p. 30 - 35 (1982). In Russian.

Some additional information is given on the compilation of the general catalogue of supplementary KSZ stars. In the compiled catalogue the right ascensions are based on 1178 star positions from 8 catalogues, declinations on 700 positions from 6 catalogues with an error $\pm 0\overset{s}{.}012 \sec\delta$ and $\pm 0\overset{''}{.}18$ respectively. It is shown that there is a shortage of observations in the areas with galaxies.

041.024 **A comparison of the AGK3 star catalogue declinations with the results of latitude observations with zenith telescopes and photographic zenith tubes.**
S. Sadzhakova, V. A. Fomin.
Izv. Glav. Astron. Obs. Pulkovo, Astrometr. Astrofiz., No. 199, p. 36 - 40 (1982). In Russian.

A comparison of the AGK3 catalogue declinations with the declinations obtained from 27 northern sky latitude programs observations during the last 15 - 20 years is made. The results of the comparison show that the systematic errors $\Delta\delta_\alpha$

may be present both in the declination system of the AGK3 catalogue and in the declinations deduced from latitude observations.

041.025 **BD +37°442 or BD +37°443?**
E. Rebeirot.
Astrophys. Space Sci., Vol. 88, 253 - 254 (1982).
The author corrects the correction brought by Bartolini et al. (1982) to the identification of the star BD +37°442.

Catalogue of declinations of the zenith zone obtained on the Moscow zenith telescope ZTL-180.
See Abstr. 002.038.

Catalogue of right ascensions of 586 FKSZ stars in the declination zone from +90° to −20°.
See Abstr. 002.040.

Catalogue of declinations of Pulkovo latitude stars.
See Abstr. 002.041.

A compilation catalog in the direction of the galactic center.
See Abstr. 002.058.

The fourth meridian catalogue of Besancon Observatory.
See Abstr. 002.078.

Documentation for the machine-readable version of the Third Santiago-Pulkovo Fundamental Stars Catalogue (SPF 3).
See Abstr. 002.083.

On the coverage in the Cordoba zones of the Astrographic Catalogue. Comments on a note by Herald.
See Abstr. 002.104.

A computer catalog of the Kiso Schmidt plates.
See Abstr. 002.106.

The SAOC and the AGK3.
See Abstr. 002.120.

Catalogue of absolute right ascensions of 1960 bright (FK4) and faint (FKSZ) stars of the southern sky.
See Abstr. 002.121.

Catalogue of absolute right ascensions of 2256 stars from the Backlund-Hough list from +32° to −90° declination observed at the Melbourne Observatory during 1928 - 1941, $Me_{50}I$ (α).
See Abstr. 002.122.

Catalogue of absolute right ascensions of 144 latitude stars from the Kimura List observed at the Melbourne Observatory during 1933 - 1935, $Me_{50}II$ (α).
See Abstr. 002.123.

Catalogue of right ascensions of 35 stars from +32° to −90° declination from the A. Kopff List observed at the Melbourne Observatory during 1928 - 1940, $Me_{50}III$ (α).
See Abstr. 002.124.

A preliminary RA catalogue of PZT stars on the International Parallel.
See Abstr. 002.125.

Deuxième complément à la bibliographie des publications ayant rapport avec l'astrolabe Danjon ou avec d'autres astrolabes de types nouveaux.
See Abstr. 002.128

Perth 75. A catalogue of positions of 2589 FK4 and FK4S stars.
See Abstr. 002.135.

General astrometry.
See Abstr. 003.098.

Quince años de actividad del Astrolabio Danjon de Santiago.
See Abstr. 013.052.

On an approach to representation of systematic differences of catalogues.
See Abstr. 021.019.

A critical comparison of the accuracy of proposed methods for small-angle and wide-angle astrometry.
See Abstr. 031.588.

Beobachtungsmethoden der photographischen Astrometrie für Zwecke des fundamentalen Referenzsystems – Einfluß des Kassettentyps auf die Genauigkeit der Resultate.
See Abstr. 031.611.

Étude préliminaire à l'établissement d'un catalogue fondamental par observation diurne des étoiles brillantes.
See Abstr. 031.629.

On joint and separate reduction of equations in astrometry.
See Abstr. 031.670.

Conversion of positions and proper motions from B1950.0 to the IAU system at J2000.0.
See Abstr. 043.003.

The connection of a catalogue of stars with an extragalactic reference frame.
See Abstr. 043.004.

Effect of different sources of variation of latitude data on meridian circle catalogues.
See Abstr. 045.001.

Hipparcos: space-age astrometry.
See Abstr. 051.001.

Astrometric observations with the Space Telescope.
See Abstr. 051.015.

Photographic positions of major planets from observations at the Nikolaev Observatory in 1976 - 1979.
See Abstr. 091.007.

Right ascensions of Mercury determined with the RATAN-600 radio telescope.
See Abstr. 092.005.

Results of photographic position observations of Venus in 1975, 1977 at Nikolaev.
See Abstr. 093.035.

Positions of Mars using photographic observations with the normal astrograph at Pulkovo in the opposition of 1975 - 1976 and 1977 - 1978.
See Abstr. 097.062.

Observations of selected minor planets at the Nikolaev Observatory in 1974 - 1975.
See Abstr. 098.007.

Observations of 1 Ceres and 2 Pallas at centimeter wavelengths.
See Abstr. 098.072.

Photographic observations of Jupiter and Saturn satellites at the Nikolaev Observatory in 1976 - 1979.
See Abstr. 099.007.

Positional photographic observations of Jupiter and the Galilean satellites.
See Abstr. 099.151.

Photographic observations of Saturn satellites at the Engelhardt Astronomical Observatory.
See Abstr. 100.012.

On the detection of other planetary systems by astrometric techniques.
See Abstr. 118.039.

Apparent proper motions of the galactic center compact radio source and PSR 1929+10.
See Abstr. 141.053.

Precise optical positions of southern radio sources.
See Abstr. 141.093.

Optical identification of radio sources in a $8\overset{\circ}{.}5 \times 8\overset{\circ}{.}5$ zone near the South Galactic Pole.
See Abstr. 141.094.

Accurate optical positions for Markarian galaxies 1303–1399.
See Abstr. 158.225.

042 Celestial Mechanics, Figures of Celestial Bodies

042.001 **Rotating post-Newtonian configurations of a homogeneous magnetized liquid similar to ellipsoids. II.**
A. N. Tsirulev, V. P. Tsvetkov.
Astron. Zh., Tom 59, 666 - 675 (1982). In Russian. English translation in Soviet Astron., Vol. 25, No. 4.
On the basis of the method developed in part I, parameters of configurations are obtained with account for post-Newtonian effects and a magnetic field. The solutions of the equations for the coefficients of the equation of the surface and corrections to the value $\omega^2/2\pi G\rho$ have been investigated. The numerical results are tabulated. The application of the results obtained to the theory of neutron stars is discussed.

042.002 **Dynamics of formation of the resonance rotations of natural celestial bodies.**
A. A. Khentov.
Astron. Zh., Tom 59, 769 - 778 (1982). In Russian. English translation in Soviet Astron., Vol. 25, No. 4.
The regularity of formation in the course of evolution of the observed type 1 : 1 and 3 : 2 commensurabilities between the rotation and the orbital motion of some bodies in the solar system is grounded. An interpretation of the Venus rotation is suggested.

042.003 **Development of the perturbation function in the problem of two fixed centres.**
V. A. Tamarov.
Astron. Zh., Tom 59, 779 - 787 (1982). In Russian. English translation in Soviet Astron., Vol. 25, No. 4.
The development in Eulerian elements of the perturbation function of the third-body attraction and the primary planet nonsphericity was obtained with the accuracy to mixed terms of the order of the product of the earth's oblateness by the perturbation factor. The formulae are trigonometric series using Kaula's inclination functions and Hansen's coefficients.

042.004 **Development of the coordinates of the elliptic motion in powers of the parameter of Levi-Civita.**
B. I. Kaminskij.
Astron. Zh., Tom 59, 788 - 793 (1982). In Russian. English translation in Soviet Astron., Vol. 25, No. 4.
General expressions are given for the coefficients of the development of the coordinates of the elliptic motion in powers of the parameter of Levi-Civita instead of eccentricity. The series converge for any eccentricity in the domain $0 \leqslant e < 1$. As an example one gives the general expressions of the development of the radius-vector of the elliptic motion.

042.005 **On the stability of liquid gravitational bodies containing small solid cores.**
M. A. Grinfel'd, A. A. Movchan.
Astron. Zh., Tom 59, 794 - 801 (1982). In Russian. English translation in Soviet Astron., Vol. 25, No. 4.
Well-known Ramsey–Lighthill results concerning the instability of gravitational two-phase liquid bodies with small cores are generalized for the case of a solid elastic core.

042.006 **Libration points of central configurations.**
E. M. Nezhinskij.
Astron. Zh., Tom 59, 802 - 809 (1982). In Russian. English translation in Soviet Astron., Vol. 25, No. 4.
The paper is concerned with the proof of the existence of central configurations for an arbitrary number of bodies with standard libration point positions.

042.007 **Method for constructing periodic orbits.**
C. Edelman.
Astron. Astrophys., Vol. 111, 220 - 223 (1982). In French.
The properties of quasi linear differential systems are used to construct periodic orbits in a fixed frame for the elliptical spatial restricted three-body problem. This system is studied in the vicinity of the orbits of non periodic comets. Several numerical investigations are made, the sun and Jupiter being the two primaries. Some periodic solutions are found which correspond to very highly eccentric orbits orthogonal to the orbital plane of the disturbing body. Others are in resonance with Jupiter.

042.008 **Radiation pressure and dust particle dynamics.**
F. Mignard.
Icarus, Vol. 49, 347 - 366 (1982).
The dynamics of small dust grains orbiting a planet is investigated when solar radiation pressure forces are added to the planet's gravitational central field.

042.009 **A manifold of periodic orbits in the planar general three-body problem with equal masses.**
E. Davoust, R. Broucke.
Astron. Astrophys., Vol. 112, 305 - 320 (1982).
The article describes about two dozen families of periodic orbits in the planar general problem of three bodies with three equal masses. These orbits have been obtained with a numerical integration of the regularized equations of motion. The regularization is of the Levi-Civita type and it handles only the three binary collisions. The stability of the periodic solutions is determined on the basis of the two coefficients (a_1, a_2) of the characteristic equation of the monodromy matrix. A detailed study of the stability coefficients reveals the existence of a large number of critical periodic orbits and bifurcations between different families.

042.010 **Stokes' theorem for many isolated bodies.**
M. A. Aleksidze.
Ivz. AN SSSR. Fiz. Zemli, 1982, No. 2, p. 40 - 51. In Russian. – Abstr. in Ref. zh., 51. Astron., 7.51.82 (1982).

042.011 **Gravitating bodies of Ramsay type.**
M. A. Grinfel'd.
Dokl. AN SSSR, Tom 262, 1339 - 1344 (1982). In Russian. Abstr. in Ref. zh., 51. Astron., 7.51.217 (1982).

042.012 **Hemelmechanica.** Recente ontwikkelingen in een dynamische wetenschap.
A. Ollongren.
Zenit, 9. Jaarg., 362 - 366 (1982).

042.013 **Existence of quasi-periodic orbits about an oblate planet.**
S. Kasperczuk, Z. Kasperczuk.
Artif. Satell., Vol. 16, No. 3, p. 21 - 29 (1981).
In this paper the authors consider the existence of quasi-periodic solutions about an oblate planet. It is shown that the typical orbits lie on invariant tori.

042.014 **Equilibrium rotation of a free elastic system.**
S. A. Alekseev, B. A. Ivchenko.
Mekh. kosm. poleta. Tr. 15 Chtenij, posvyashch. razrab. nauchn. naslediya i razvitiyu idej K. Eh. Tsiolkovskogo, Kaluga, 12 - 15 sent., 1980, Moskva, 1981, p. 64 - 67. In Russian. – Abstr. in Ref. zh., 62. Issled. kosm. prostranstva, 8.62.124 (1982).

042.015 **On the stability of periodic motions of a satellite relative to the proper mass center.**

Yu. V. Barkin, A. A. Kozlov, S. N. Lelyavin.
Kosm. Issled., Tom 20, 631 - 634 (1982). In Russian.

042.016 **The singular invariant manifold of the restricted three-body problem.**
S. Kasperczuk.
Acta Astron., Vol. 31, 485 - 491 (1981).

042.017 **Periodic solutions of the problem of the motion of the heavy rigid body around the fixed point in Kovalevskaya's case and their stability.**
F. M. El-Sabaa.
Celestial Mech., Vol. 27, 215 - 223 (1982).
The equations of motion of the heavy rigid body around the fixed point in Kovalevskaya's case were reduced to equations of plane motion of the fictive point under the action of some potential force. The periodic solutions of the new equations were found and their stability discussed.

042.018 **On the small divisors. I. The convergence of the modified Zeipel's series and the resonant quasi-periodic solutions.**
G. A. Merman.
Celestial Mech., Vol. 27, 225 - 265 (1982). In French.
The convergence of a modification of the Zeipel's series of fixed frequences taking into account the short period perturbations of the right hand terms of the differential equations of celestial mechanics has been shown. By means of these series the existence of quasi-periodic solutions depending on the degenerate spectrum of frequencies without any complementary parameteres has been established.

042.019 **On regular solutions in the problem of two rigid bodies.**
G. N. Doubochine (*Duboshin*).
Celestial Mech., Vol. 27, 267 - 284 (1982). In French.
In the present paper the problem of translatory-rotatory motion of two rigid bodies is discussed. The author has shown that this problem admits particular solutions, when each body possesses axial symmetry. In these solutions the centre of mass of one body describes a circular orbit around the other body and each body keeps the invariable orientation about this orbit.

042.020 **Virial oscillations of celestial bodies. II. A general approach to the solution of the perturbed oscillation problem and electrodynamic effects.**
V. I. Ferronsky, S. A. Denisik, S. V. Ferronsky.
Celestial Mech., Vol. 27, 285 - 304 (1982).
A general approach to the solution of the perturbed oscillation problem for celestial bodies is considered. The solution sought describes unperturbed virial oscillations (zero approximation) affected by external perturbing effects. In the general case, these perturbations can be expressed by an arbitrary given function of time, Jacobi's function and its first derivative. It is shown that while studying the evolution of a celestial body as a dissipative system in the framework of perturbed virial oscillations, the analytical expression for perturbing function can be derived, assuming the celestial body to be an oscillating electrical dipole. The general covariant form of Jacobi's equation is derived. It is shown that the scalar form of Jacobi's equation appears to be more universal than Newton's laws of motion.

042.021 **On the relation between resonance and instability in planetary systems.**
J. D. Hadjidemetriou.
Celestial Mech., Vol. 27, 305 - 322 (1982).
The factors which affect the linear stability of a periodic planetary orbit in the plane are studied. It is proved that planetary systems with two planets describing nearly circular orbits in the same direction are linearly stable and no perturbation exists which destroys the stability, unless a resonance of the form $^1/_3$, $^3/_5$, $^5/_7$, . . . among the orbits of the planets occurs. This latter resonant case is always unstable. Retrograde motion is always linearly stable. Planetary systems with three or more planets in nearly circular orbits in the same direction are proved to be unstable. It is also proved that planetary systems with nearly elliptic orbits of the planets are unstable.

042.022 **On eccentricity functions for eccentric orbits.**
A. Szeto, K. Lambeck.
Celestial Mech., Vol. 27, 325 - 337 (1982).
Situations arise in celestial mechanics where orbital eccentricities are large and yet it is desirable to maintain the Darwin–Kaula Fourier decomposition of the perturbing function. Evaluation of the appropriate eccentricity functions $G_{lpq}(e)$ requires a double summation which, for practical purposes, must be truncated. In this note criteria have been established for truncation of the expansion for eccentricities $\leqslant 0.75$.

042.023 **Elimination of nodes in the Newtonian four-body problem.**
F. Boigey.
Celestial Mech., Vol. 27, 399 - 414 (1982). In French.
The author applies the method of canonical transformations with imposed variables to the reduction of the Newtonian four-body problem. After the elimination of the center of gravity, the problem is reduced to that of three fictitious bodies. Then it is proceeded to the actual reduction using the integrals of angular momentum, in Hamiltonian formulation, and considering the geometrical aspects of the elimination of the nodes advocated by Jacobi.

042.024 **Resultados numéricos en el problema de tres cuerpos.**
F. López-García.
Rev. Mexicana Astron. Astrofis., Vol. 6, (see 012.017), 107 - 109 (1981).
Final evolution of a three-body system with finite masses is studied numerically. Elliptic-hyperbolic and elliptic-parabolic cases are considered. The formation of a binary system is also considered. The numerical integration is made using analytic prolongation methods. Possible binary collisions are considered. The variation of elliptic elements after the formation of a binary is also studied.

042.025 **On the regularization of the Kepler problem.**
M. Kummer,
Commun. Math. Phys., Vol. 84, 133 - 152 (1982). – Abstr. in Phys. Abstr., Vol. 85, Abstr. 68601 (1982).

042.026 **A simple criterion for determining the dynamical stability of three-body systems.**
D. C. Black.
Astron. J., Vol. 87, 1333 - 1337 (1982).
Coplanar, prograde three-body systems are discussed emphasizing the specification of general criteria for determining whether such systems are dynamically stable. Comparison of the Graziani-Black criteria with data for eight well studied triple-star systems indicates that the observed lower limit ($\cong 5.0$ - 5.5) for the ratio of periastron distance of the tertiary orbit to the semimajor axis of the binary orbit is due to dynamical instability rather than to cosmogonic processes.

042.027 **Expansion of the mutual distance between two planets raised to any negative power.**
O. M. Kamel.
Moon Planets, Vol. 26, 239 - 277 (1982).
An expression for Δ^{-s} in terms of the Poincaré variables L, λ, H, K, P, Q has been evaluated. The inclinations of the two planets are referred to a common fixed plane. The author

neglects in the final formula powers higher than the third of the Poincaré variables.

042.028 **Trojan-type orbits in the system of two gravitational centers with variable masses and separation.**
B. B. Djakov *(D'yakov)*, B. I. Reznikov.
Moon Planets, Vol. 26, 371 - 382 (1982).
Trojan type orbits in the system of two gravitational centers with variable masses and separation are studied within the framework of the restricted problem of three bodies. The asymmetry of the libration amplitudes about the triangular points as well as the values of periods and amplitudes within the limits for real Trojan asteroids are obtained. Trojans could be supposedly formed inside the Proto-Jupiter and escape during its intensive mass loss.

042.029 **Improvement of the theories of Jupiter and Saturn by harmonic analysis.**
J. L. Simon, G. Francou.
Astron. Astrophys., Vol. 114, 125 - 130 (1982). In French.
The authors present a method for computing the important terms which are still missing, in the third order theory of Jupiter and Saturn. They compute the perturbations by harmonic analysis using an iterative process.

042.030 **New equilibrium sequences bifurcating from Maclaurin sequence.**
Y. Eriguchi, I. Hachisu.
Prog. Theor. Phys., Vol. 67, 844 - 851 (1982). – Abstr. in Phys. Abstr., Vol. 85, Abstr. 79448 (1982).

042.031 **Floquet reference solutions for the lunar theory and the Jovian moons.**
W. E. Wiesel.
J. Guid. Control, Vol. 4, 586 - 590 (1981). – Abstr. in Phys. Abstr., Vol. 85, Abstr. 84068 (1982).

042.032 **Perturbation methods based upon varying action integrals.**
M. Rajan, J. L. Junkins.
Astrodynamics 1981, (see 012.027), p. 134 (1982). – Abstract.

042.033 **The ideal resonance problem at first order.**
A. Deprit.
Astrodynamics 1981, (see 012.027), p. 521 - 526 (1982).
Perturbations of the first order are removed jointly by a canonical transformation representing the hunting effect, and by a change of the time to synchronize the perturbed pendulum. Both operations are expressed in elementary functions. The reduced system is a simple pendulum integrable by elliptic functions.

042.034 **Modal control of an unstable periodic orbit.**
W. E. Wiesel, W. Shelton.
Astrodynamics 1981, (see 012.027), p. 527 (1982). – Abstract.

042.035 **Three-dimensional, periodic, "halo" orbits.**
K. C. Howell.
Astrodynamics 1981, (see 012.027), p. 528 (1982). – Abstract.

042.036 **Collisional lifetimes of Trojan asteroids.**
D. R. Davis, S. J. Weidenschilling.
Astrodynamics 1981, (see 012.027), p. 530 (1982). – Abstract.

042.037 **The stability of a rotating liquid cylinder as an equilibrium figure.**
A. S. Baranov.
Astron. Zh., Tom 59, 870 - 875 (1982). In Russian. English translation in Soviet Astron., Vol. 26, No. 5.
The dynamics of a homogeneous liquid cylinder with finite section has been investigated, the Lagrangian coordinates of particles depending linearly on the initial ones. It is shown that the semi-axes of the section depend periodically on time. In particular, this means that the system does not tend to disintegrate. The angular velocity of the turn of the contour of the section has been found. The dependence of the period of the lateral oscillations on the amplitude has been determined in the case of small oscillations.

042.038 **Theory for the motion of all the planets. The VSOP82 solution.**
P. Bretagnon.
Astron. Astrophys., Vol. 114, 278 - 288 (1982). In French.
The VSOP82 solution is made of the perturbations developed up to the third order of the masses for all the planets. Perturbations up to the sixth order obtained by an iterative method complete the theory of the four outer planets. It also contains the perturbations of the Moon onto the Earth–Moon barycenter and the relativistic perturbations expressed in isotropic and standard coordinates. The integration constants are determined by adjustment to the numerical integration DE200.

042.039 **The resurrection of Laplace's method of initial orbit determination.**
L. G. Taff.
Bull. American Astron. Soc., Vol. 14, 727 - 728 (1982). Abstract.

042.040 **Comparison of starting values for iterative solutions to a universal Kepler's equation.**
M. J. Bergam.
J. Astronaut. Sci., Vol. 30, 75 - 84 (1982). – Abstr. in Phys. Abstr., Vol. 85, Abstr. 94585 (1982).

042.041 **Dumb-bell-shape equilibria and mass-shedding pear-shape of a self-gravitating incompressible fluid.**
Y. Eriguchi, I. Hachisu, D. Sugimoto.
Prog. Theor. Phys., Vol. 67, 1068 - 1075 (1982). – Abstr. in Phys. Abstr., Vol. 85, Abstr. 95181 (1982).

042.042 **Regularized integration of gravity-perturbed trajectories – a numerical efficiency study.**
L. G. Kraige, J. L. Junkins, L. D. Ziems.
J. Spacecr. Rockets, Vol. 19, 291 - 293 (1982). – Abstr. in Phys. Abstr., Vol. 85, Abstr. 112308 (1982).

042.043 **Regularization of the Kepler motion by spinor calculus.**
B. D. Jovanović.
Publ. Astron. Obs. Sarajevo, Vol. 1, No. 1, (see 012.043), p. 251 - 267 (1981).

042.044 **La plej simpla formo de perturboj de la ekiraj vektoraj elementoj de planedetorbito.**
B. J. Popović.
Publ. Astron. Obs. Sarajevo, Vol. 1, No. 1, (see 012.043), p. 279 - 285 (1981).

042.045 **Bounded motion for a modified three-body problem.**
G. Bozis, M. Michalodimitrakis.
Astrophys. Space Sci., Vol. 86, 377 - 396 (1982).
The energy and the angular momentum integral of motion for the planar three point problem cannot assure bounded motion. In this paper it is shown that if one of the points is replaced by a homogeneous sphere then bounded motion can be found. The zero velocity surfaces for this modified problem are found and their evolution is described.

042.046 **Central configurations of four bodies with one inferior mass.**
R. F. Arenstorf.
Celestial Mech., Vol. 28, (see 012.049), 9 - 15 (1982).

The number of equivalence classes of central configurations (abbr. c. c.) in the planar 4-body problem with three arbitrary and a fourth small mass is investigated. These c. c. are derived according to their generic origin in the 3-body problem. It is shown that each 3-body collinear c. c. generates exactly 2 non-collinear c. c. (besides 4 collinear ones) of 4 bodies with small $m_4 \geqslant 0$; and that any 3-body equilateral triangle c. c. generates exactly 8 or 9 or 10 (depending on m_1, m_2, m_3) planar 4-body c. c. with $m_4 = 0$. Further, every one of these c. c. can be continued uniquely to sufficiently small $m_4 > 0$ except when there are just 9; then exactly one of them is degenerate, and it is conjectured that it is not continuable to $m_4 > 0$.

042.047 **Collinear relative equilibria of the planar n-body problem.**
J. I. Palmore.
Celestial Mech., Vol. 28, (see 012.049), 17 - 24 (1982).
The following theorem is proved. *Theorem.* For any $n \geqslant 2$, the set of collinear relative equilibria classes of the n-body problem generates by analytical continuation a total of $n!\,(n+3)/2$ relative equilibria classes of the n + 1 body problem. Together with Arenstorf's results the author states a general theorem for the 4 body problem with 3 arbitrary masses and 1 inferior mass.

042.048 **Motion near total collapse in the planar isosceles three-body problem.**
R. L. Devaney.
Celestial Mech., Vol. 28, (see 012.049), 25 - 36 (1982).
The author describes in detail the qualitative behavior of solutions of the planar isosceles problem which come close to triple collision. Using a method of McGehee, one may describe a neighborhood of triple collision for any mass ratios. For sufficiently small mass ratios, the author exhibits infinitely many distinct periodic and collision/ejection solutions of this problem.

042.049 **Boundary manifolds for energy surfaces in celestial mechanics.**
E. A. Lacomba, C. Simó.
Celestial Mech., Vol. 28, (see 012.049), 37 - 48 (1982).
The authors complete McGehee's picture of introducing a boundary (total collision) manifold to each energy surface. This is done by constructing the missing components of its boundary as other submanifolds, representing now the asymptotic behavior at infinity. It is necessary to treat each case $h = 0$, $h > 0$ or $h < 0$ separately. In the first case, the authors repeat the known result that the behavior at total escape is the same as in total collision. In particular, they explain why the situation is radically different in the $h > 0$ case compared with the zero energy case. In the case $h < 0$ they have many infinity manifold components, and the general situation is not quite well understood. Finally, the results for $h \geqslant 0$ are shown to be valid for general homogeneous potentials.

042.050 **Analysis of some degenerate quadruple collisions.**
C. Simó, E. Lacomba.
Celestial Mech., Vol. 28, (see 012.049), 49 - 62 (1982).
The authors consider the trapezoidal problem of four bodies. This is a special problem where only three degrees of freedom are involved. The blow up method of McGehee can be used to deal with the quadruple collision. Two degenerate cases are studied in this paper: the rectangular and the collinear problems. They have only two degrees of freedom and the analysis of total collapse can be done in a way similar to the one used for the collinear and isosceles problems of three bodies. For the collinear case, from which arises a one parameter family of equations, the analysis for extreme values of the parameter is done and numerical computations fill up the gap for the intermediate values. Dynamical consequences for possible motions near total collision as well as for regularization are obtained.

042.051 **On a theorem of stability for a homogeneous potential.**
L. Losco.
Celestial Mech., Vol. 28, (see 012.049), 63 - 68 (1982).
In French.
The McGehee's study of the triple collision of the 3-body problem is applied for the stability of an equilibrium. The author considers the homogeneous Lagrangian:
$L = (\dot{x}^2 + \dot{y}^2)/2 + U(x, y)$, where U is polynomial, with degree k, and establishes a necessary and sufficient condition on U for the stability of ω ($x = y = \dot{x} = \dot{y} = 0$).

042.052 **Symmetric and regularized coordinates on the plane triple collision manifold.**
J. Waldvogel.
Celestial Mech., Vol. 28, (see 012.049), 69 - 82 (1982).
The purpose of this paper is to introduce and discuss a set of variables well adapted to the planar problem of three bodies, in particular to triple collision. Transformations due to Murnaghan (1936), Lemaître (1954) and McGehee (1974) are applied to the plane three-body problem, as suggested earlier by Waldvogel (1979, 1982). Further the plane triple collision manifold is introduced, and some of its properties are discussed. In the final part an overview of results regarding equilibrium points is given.

042.053 **On the restricted three-body problem when the mass parameter is small.**
J. Llibre.
Celestial Mech., Vol. 28, (see 012.049), 83 - 105 (1982).
The author studies some aspects of the restricted three-body problem when the mass parameter μ is sufficiently small. First, he describes the global flow of the two-body rotating problem, $\mu = 0$, and uses it for the analysis of the collision and parabolic orbits when $\mu \gtrsim 0$. Also he shows that for any fixed value of the Jacobian constant and for any $\epsilon > 0$, there exists a $\mu_0 > 0$ such that if the mass parameter $\mu \in [0, \mu_0]$, then the set of bounded orbits which are not contained in the closure of the set of symmetric periodic orbits has Lebesgue measure less than ϵ.

042.054 **On the existence of periodic solutions of Poincaré's third sort in the general problem of three bodies in three dimensions.**
P. J. Message.
Celestial Mech., Vol. 28, (see 012.049), 107 - 118 (1982).
A proof is offered of the existence of periodic solutions of the general problem of three bodies, of the third sort envisaged by Poincaré, that is, arising by analytic continuation from unperturbed Keplerian motion of each of two bodies about a primary, in which the two orbits are of commensurable periods, of zero eccentricity, but lying in different planes, provided that the inclination of the two planes is sufficiently small (but not zero).

042.055 **Quasi-periodic solutions of the plane three-body problem near Euler's orbits.**
J. Scheurle.
Celestial Mech., Vol. 28, (see 012.049), 141 - 151 (1982).
In (Scheurle, 1981) two of the masses are regarded as being given, whereas the third one, say m_3, is used as a parameter. In this way it turns out that there are quasi-periodic solutions for almost all values of m_3. More precisely, over the m_3-axis these solutions form a variety of branches which emanate at Euler's solution. Now the author shows that, provided a certain non-degeneracy assumption is fulfilled, a specific value of m_3 is reached by a large number of solution branches. The author also analyzes the geometric behaviour of the solution branches.

042.056 **Global sensitivity to velocity errors at the libration points.**
V. Szebehely, R. Premkumar.
Celestial Mech., Vol. 28, (see 012.049), 195 - 200 (1982).
The global effects on non-linear stability are investigated at and around the triangular libration points. The model of the circular restricted problem of three bodies is used, considering the Earth and the Moon as the primaries. Areas of the initial conditions in the configuration space leading to stability with zero initial velocity around the equilibrium points are compared to stable areas in the phase space with variable initial velocity and zero initial deviations at the equilibrium points. The bahavior of the system concerning errors due to initial conditions in the configuration space is studied for 6 years and the effects of initial velocity errors are considered for 7.73 years.

042.057 **A generalization of Szebehely's equation for three dimensions.**
B. Érdi.
Celestial Mech., Vol. 28, (see 012.049), 209 - 218 (1982).
Szebehely's equation for the potential generating a prescribed family of orbits in two dimensions is generalized for three-dimensional orbits. A simultaneous system of first-order linear partial differential equations is derived for the determination of the potential in the three-dimensional case. Solutions of this system are found in several cases including Kepler's problem, too.

042.058 **Geometrical properties of the Euler-Poisson equations of a heavy rigid body about a fixed point.**
M. Langlois.
Celestial Mech., Vol. 28, (see 012.049), 219 - 232 (1982).
In this paper, the author studies the geometrical properties of the Euler-Poisson equations of the motion of a heavy rigid body about a fixed point. Two cases are studied. In the first one the rigid body has a gyrostatic moment, constant with respect to itself. Secondly the rigid body is considered without its gyrostatic moment.

042.059 **A new derivation of the Kustaanheimo-Stiefel variables.**
H. Yoshida.
Celestial Mech., Vol. 28, (see 012.049), 239 - 242 (1982).
First introducing the parabolic coordinates in R^3, the author derives the Kustaanheimo-Stiefel (KS) variables quite simply and naturally. Through this derivation it becomes clearer where and how the additional dimension is introduced and what the bilinear relation means.

042.060 **On resonance in celestial mechanics.**
B. Garfinkel.
Celestial Mech., Vol. 28, 275 - 290 (1982).
This brief survey of the author's contribution to the theory of resonance in celestial mechanics begins with the genesis of the Small Divisor. The fundamental distinction between the shallow and deep resonance is illustrated by the 5:2 Jupiter–Saturn and the 3–2 Neptune–Pluto resonances in the planetary system. The search for a global solution through a removal of the small divisor is put into a historical perspective through the work of Laplace, Bohlin, and Poincaré. The author's own contribution to the methodology is the formulation and the solution of the Ideal Resonance Problem. If the resonance is simple, all the singularities in the solution are removed by means of a regularizing function. On the other hand, if the resonance is double, the second critical divisor seems irremovable, and a global solution may be precluded.

042.061 **A method of construction of families of spatial periodic orbits in Hill's problem.**
M. L. Lidov.
Kosm. Issled., Tom 20, 787 - 807 (1982). In Russian.

042.062 **Lösung von Parameterbestimmungsproblemen in Himmelsmechanik und Satellitengeodäsie mit modernen Hilfsmitteln.**
G. Beutler.
Astron.-Geod. Arb. Schweiz, 34. Band, 7 + 257 pp. (1982).

042.063 **Recent progress in the theory of Trojan asteroids.**
B. Garfinkel.
Dynamical astronomy, (see 012.056), p. 13 - 15 (1982).
Abstract.

042.064 **On the variation of the Jacobi constant of Trojan asteroids in the elliptic restricted problem of three bodies.**
B. Érdi.
Dynamical astronomy, (see 012.056), p. 17 - 36 (1982).
A relation corresponding to the Jacobian integral of the circular restricted problem of three bodies is derived in cylindrical coordinates for the elliptic restricted three-body problem. The unknown integral appearing in this relation is evaluated for the case of Trojan asteroids using an asymptotic solution for their motion.

042.065 **Twice averaged restricted three-body problem and its application to study the evolution of orbits of asteroids.**
F. Veres.
Dynamical astronomy, (see 012.056), p. 101 - 105 (1982).
The paper deals with the twice averaged restriced three body problem. First the author shows a general analytical solution for the planar case and then he gives a particular solution for the spatial case. The results of the analytical solution are compared with results obtained by numerical integration.

042.066 **Nonsingular oscillator elements.**
W. T. Kyner.
Dynamical astronomy, (see 012.056), p. 107 - 117 (1982).

042.067 **Orbital eccentricity in a logarithmic potential.**
K. A. Innanen, L. Tähtinen, M. J. Valtonen.
Astron. J., Vol. 87, 1606 - 1607 (1982).
It is shown both numerically as well as analytically that the orbital pseudo-eccentricity of a particle moving in a logarithmic potential is close to one-half of the analogous Keplerian eccentricity.

042.068 **Construction of periodic orbits of satellites in a moving system of axes. I.**
I. Stellmacher.
Celestial Mech., Vol. 28, 351 - 365 (1982). In French.
The author gives an algorithm for the construction of periodic orbits in a rotating frame for the cases of satellites moving around an oblate planet. The orbits are near to the circular case; the asymptotic developments of the periodic solutions are completely calculated for the terms J_2 and J_4 of the potential. The solutions for small inclinations are given up to J_2^2. These solutions can be applied to the motion of the Saturnian satellites. From the observed longitudes of the ascending nodes of Mimas and Tethys, the author estimates the values J_2 and J_4 of the Saturnian potential, the value of J_2 very strongly depends on the adopted value of the planet's equatorial diameter.

042.069 **Compatibility conditions for a non-quadratic integral of motion.**
G. Bozis.
Celestial Mech., Vol. 28, 367 - 380 (1982).
Conditions are found which are satisfied by the coef-

ficients of the expression $\Phi = A\dot{x}^2 + 2B\dot{x}\dot{y} + \Gamma\dot{y}^2 + \Delta\dot{x}^2\dot{y}^2 + E$, being a second integral of the motion of an autonomous dynamical system with two degrees of freedom. The coefficients A, B, Γ, Δ, E are differentiable functions of the cartesian position coordinates x, y. The velocity components are denoted by $\dot{x}$, $\dot{y}$. It is shown that Δ must be constant and B must be of the form $B = f(x + y) + g(x - y)$ where f, g are arbitrary. Given Δ and B one can always find the remaining coefficients A, Γ, E and also the corresponding potential and the second integral. Depending on the specific case at hand a certain number of arbitrary constants (or arbitrary functions) enter into the potential and the second integral. To each potential there corresponds one integral of the above form.

042.070 **An analytic solution to the classical two-body problem with drag.**
D. Mittleman, D. Jezewski.
Celestial Mech., Vol. 28, 401 - 413 (1982).
An analytic solution to the two-body problem with a specific drag model is obtained. The model treats drag as a force proportional to the vector velocity and inversely proportional to the square of the distance to the center of attraction. The solution is expressed in terms of known functions and is of a simple and compact form. The time-of-flight is expressed as a quadrature in the "true anomaly".

042.071 **The Fourier-Chebychev approximation for time series with a great many terms.**
J. Chapront.
Celestial Mech., Vol. 28, 415 - 430 (1982).
A new type of approximation is proposed to evaluate Fourier series with a great many terms. It combines Fourier series and Chebychev polynomials to represent time series over a finite time interval. Application is made in case of the lunar theory ELP which contains about 30 000 trigonometric terms.

042.072 **Some remarks on the stability analysis in Robe's three body problem.**
R. Meire.
Int. J. Math. Math.Sci., Vol. 5, 195 - 202 (1982) = Rijksuniv. Gent, Sterrenkundig Obs., Meded. Nr. 48.
An improved technique is presented for the stability analysis of Robe's 3-body problem which gives more accurate results for the transition curves in the parameter plane than does Robe's paper. A novel property of the system of differential equations describing the motion is used, which reduces the computer time by more than 50%.

042.073 **Gravitationswechselwirkung ausgedehnter Körper.**
H. Gleixner.
Deutsche Geod. Komm. Bayerischen Akad. Wiss., Reihe A, Theor. Geod., Heft Nr. 95 = Mitt. Inst. Astron. Phys. Geod. Tech. Hochsch. München, Nr. 156, 48 pp. (1982).

042.074 **L'évaporation de systèmes triples à l'origine de binaires peu serrées.**
D. Benest, M. Fulconis.
Mitt. Astron. Ges., Nr. 57, p. 265 - 273 (1982).

042.075 **Characteristics of resonant periodic orbits.**
N. Caranicolas.
Astrophys. Space Sci., Vol. 87, 237 - 245 (1982).
The author studies the families of periodic orbits in a time-independent two-dimensional potential field symmetric with respect to both axes. By numerical calculations he finds characteristic curves of several families of periodic orbits when the ratio of the unperturbed frequencies is 2/1. There are two groups of characteristic curves: (1) The basic characteristic and the characteristics which bifurcate from it. (2) The characteristics which start from the boundary line and the axis $x = 0$.

042.076 **The averaged hyperbolic restricted three-body problem.**
A. B. Sorokovich.
Astron. Zh., Tom 59, 1200 - 1211 (1982). In Russian. English translation in Soviet Astron., Vol. 26, No. 6.
The flat single-averaged hyperbolic restricted three-body problem in the case of small eccentricities of the perturbed body is considered. The author obtained the general solution for the linearized system of differential equations in the form of small-parameter power series for the case when the first term of the perturbing function is taken into account.

042.077 **On some particular solutions of the restricted problem of three rigid bodies.**
E. N. Eremenko.
Astron. Zh., Tom 59, 1218 - 1224 (1982). In Russian. English translation in Soviet Astron., Vol. 26, No. 6.
The problem of the translational-rotational motion of a triaxial rigid body in the restricted problem of three rigid bodies is formulated. The motion of the body is described by canonical osculating elements. On the basis of Gauss' averaging scheme the existence of first-approximation particular regular solutions for the body is established. The problem of the stability of the solutions found after Lyapunov in the first approximation is studied.

042.078 **On the critical stability case of the motion around the equilateral libration points of the Earth-Moon system. II.**
M. Popescu.
Rev. Roumaine Sci. Tech. Ser. Mec. Appl., Vol. 27, No. 1, p. 3 - 14 (1982). – Abstr. in Phys. Abstr., Vol. 85, Abstr. 117571 (1982).

042.079 **Rapidly rotating polytropes and concave hamburger equilibrium.**
I. Hachisu, Y. Eriguchi, D. Sugimoto.
Prog. Theor. Phys., Vol. 68, 191 - 205 (1982). – Abstr. in Phys. Abstr., Vol. 85, Abstr. 117664 (1982).

042.080 **Bifurcation and fission of three dimensional, rigidly rotating and self-gravitating polytropes.**
I. Hachisu, Y. Eriguchi.
Prog. Theor. Phys., Vol. 68, 206 - 221 (1982). – Abstr. in Phys. Abstr., Vol. 85, Abstr. 117665 (1982).

042.081 **Equilibrium of a self-gravitating and rotating fluid.**
Y. Eriguchi.
Astron. Her., Vol. 75, 86 - 90 (1982). In Japanese. – Abstr. in Phys. Abstr., Vol. 86, Abstr. 5103 (1983).

042.082 **Shape of a polytropic primary.**
M. J. Miketinac, J. Miketinac.
Comput. Phys. Commun., Vol. 27, 285 - 297 (1982). – Abstr. in Phys. Abstr., Vol. 86, Abstr. 10396 (1983).

042.083 **Hill's stability in the elliptic restricted three-body problem in which the shape of a body is considered.**
C. Huang.
Acta Astron. Sinica, Vol. 23, 390 - 397 (1982). In Chinese.
The paper studies the elliptical restricted three-body problem in which the shape of a body is considered. Hill's stability problem of a small body which revolves around a primary body, the shape of which must be considered, is discussed by using the invariable relative formula of the mechanical system and following the qualitative method of the circular problem in connection with the perturbation theory. Particularly, the difference between Hill's stability in the mechanical system and that in the circular problem is explored.

042.084 **On the stability of the inner planets, satellites and close binaries.**
X. Zheng,
Acta Astrophys. Sinica, Vol. 2, 215 - 223 (1982). In Chinese.
On the basis of the circular restricted three-body problem in which the bigger primary is a oblate spheroid, the effect of oblateness on the stable conditions of the third particle in the Liapunov-Hill sense is discussed. The results are applied to the solar system, the earth–moon system, and the close binary system.

Selected research projects in celestial mechanics at the University of Texas at Austin.
See Abstr. 013.082.

The concept of integrability on Cantor sets for Hamiltonian systems.
See Abstr. 021.034.

Melnikov's method and averaging.
See Abstr. 021.035.

The Lyapunov transformation of Hill's equation and its dynamic interpretation.
See Abstr. 021.037.

Ein FORTRAN-Programm zur Berechnung translatorischer Mondephemeriden nach der Theorie von E. W. Brown.
See Abstr. 021.047.

Calculation of the Laplace coefficients and their derivatives in the case of $0.5 \leqslant \alpha \leqslant 0.9$.
See Abstr. 021.048.

A note to the computation of perturbations with tesseral harmonics.
See Abstr. 021.050.

Several notes on the transformation methods in non-linear systems.
See Abstr. 021.057.

Time-dependent quadratic constants of motion, symmetries, and orbit equations for classical particle dynamical systems with time-dependent Kepler potentials.
See Abstr. 022.058.

Periodic travelling waves in a non-integrable one-dimensional lattice.
See Abstr. 022.132.

The effects of diffuse solar radiation on the motion of an artificial satellite.
See Abstr. 052.018.

Long-term motion of resonant satellites with arbitrary eccentricity and inclination.
See Abstr. 052.019.

Short period elimination for the tesseral harmonics.
See Abstr. 052.027.

Perturbations in the nodal period of artificial satellites caused by the terrestrial infrared radiation pressure.
See Abstr. 052.038.

Construction of a consistent semianalytic theory of a planetary or moon orbiter perturbed by a third body.
See Abstr. 052.040.

Comments on Aksnes' intermediary.
See Abstr. 052.041.

Containment of a diffuse ionized mass orbiting around a magnetized central body.
See Abstr. 062.103.

Rupture d'étoiles par les forces de marée.
See Abstr. 065.011.

Physical distinctions between normalized solutions of the two-body problem of general relativity.
See Abstr. 066.096.

Equivalence principle for massive bodies in a generalized theory of gravitation. II.
See Abstr. 066.335.

Loi de Titius-Bode et formalisme ondulatoire.
See Abstr. 091.020.

Perturbations of collisional particles discs.
See Abstr. 091.032.

On the stability of certain planetary ringlets.
See Abstr. 091.033.

Quantum formalism in gravitation – quantitative application to the Titius-Bode law.
See Abstr. 091.039.

The dynamical evolution of the solar system.
See Abstr. 091.053.

Poynting-Robertson drag and orbital resonance.
See Abstr. 091.055.

Relativistic perturbations of the Moon in ELP 2000.
See Abstr. 094.028.

The determination of the derivatives in Brown's lunar theory.
See Abstr. 094.029.

Newtonsche Bewegungsgleichungen des Mondes.
See Abstr. 094.032.

Recent progress in the theory of Trojan asteroids.
See Abstr. 098.053.

Numerical determination of proper inclinations of Hilda-type asteroids.
See Abstr. 098.054.

A qualitative analysis of motion of Hecuba-type asteroids.
See Abstr. 098.081.

Saturn's rings: resonance about an oblate planet.
See Abstr. 100.030.

Periodic orbits of satellites. II. Libration of the Mimas-Tethys type.
See Abstr. 100.087.

The theory of motion of Hyperion.
See Abstr. 100.104.

Nonlinear dynamical astronomy. Applications and extrapolations.
See Abstr. 151.021.

Numerical study of periodic orbit properties in a dynamical system with three degrees of freedom.
See Abstr. 151.064.

Invariant surfaces and orbital behaviour in dynamical systems of 3 degrees of freedom. II.
See Abstr. 151.087.

043 Astronomical Constants, Reference Systems

043.001 **New light on Brouwer's project of observations of minor planets and on Pierce's determination of a dynamical reference frame.**
W. Fricke.
Astron. J., Vol. 87, 1338 - 1340 (1982).

Within the work on the FK 5, the equator and equinox have been determined such that the celestial coordinate system represented by the data of the FK 5 corresponds to the dynamical reference frame as closely as possible. After the completion of this determination, it was discovered that the results expressed in corrections to the equator and equinox of the FK 4 are in excellent agreement with Pierce's systematic corrections to catalogs determined from observations of 15 minor planets at a time when it was impossible to recognize their significance.

043.002 **Inertial frames determined from the rotational laws observed in spiral galaxies.**
Z. Horák.
Acta Astronaut., Vol. 9, 85 - 90 (1982). – Abstr. in Phys. Abstr., Vol. 85, Abstr. 94582 (1982).

043.003 **Conversion of positions and proper motions from B1950.0 to the IAU system at J2000.0.**
E. M. Standish, Jr.
Astron. Astrophys., Vol. 115, 20 - 22 (1982).

The complete transformation of positions and proper motions from B1950.0 to J2000.0 on the basis of IAU recommendations including the transition from FK4 to FK5 equinox may be rigorously accomplished with a simple vector equation. The equation is first presented in this paper and then subsequently derived. The transformation of the FK4 to FK5 system described by functions $f(\alpha,\delta,m)$ is in preparation in Heidelberg.

043.004 **The connection of a catalogue of stars with an extragalactic reference frame.**
M. Froeschlé, J. Kovalevsky.
Astron. Astrophys., Vol. 116, 89 - 94 (1982).

The discussion of the existing and future quasi-inertial reference systems leads to the conclusion that the best system available in the near future will be based upon the positions of quasars as observed by VLBI. In parallel, consistent accurate positions and proper motions of 100,000 stars will be obtained by HIPPARCOS, but they will not be a priori in a fixed system. Three methods of linking the HIPPARCOS system to VLBI are described and simulated in this paper. It appears that the use of radio-stars is the most powerful method.

043.005 **On basic reference coordinate systems for earth dynamics.**
Ya. S. Yatskiv.
Geodesy and physics of the earth, (see 012.053), p. 181 - 195 (1981). In Russian.

043.006 **Corrections of fundamental constants from photoelectric observations of lunar occultations.**
G. Rosselló.
Mon. Not. R. Astron. Soc., Vol. 201, 1009 - 1013 (1982).

Observations of photoelectric timings of occultations of stars by the moon in the period 1955 to 1978 are analysed in order to determine corrections to the FK4 stellar reference frame and the constants of the lunar theory.

043.007 **On the terrestrial coordinate system.**
G.-c. Shi.
Publ. Shaanxi Astron. Obs., No. 1, p. 10 - 13 (1982). In Chinese.

The history of the astronomical system of terrestrial coordinates is reviewed. The definition and the future practical realization of this system are discussed.

043.008 **Fundamental celestial reference coordinate system.**
S.-x. Wu.
Publ. Shaanxi Astron. Obs., No. 1, p. 14 - 22 (1982). In Chinese.

The progress of FK5 and the prospective tendency on the celestial reference system is given.

043.009 **New values for Oort's constant B of galactic rotation from proper motions in galactic longitude.**
W. Dieckvoss.
Mitt. Astron. Ges., Nr. 57, p. 277 - 278 (1982). – Abstract.

043.010 **On the selection of the axis for the Ephemeris Nutation of the earth.**
V. S. Gubanov.
Pis'ma Astron. Zh., Tom 8, 743 - 746 (1982). In Russian. English translation in Soviet Astron. Lett., Vol. 8.

It is shown that the Ephemeris Nutation of the earth should be referred to the axis of the angular momentum of the earth as usual but not to its dynamical figure as it was proposed by Atkinson (1973, 1975) and recommended by the XVIIth General Assembly of the IAU in 1979.

043.011 **Luni-solar precession constant and the accuracy of the Earth's principal moments of inertia.**
M. Burša.
Stud. Geophys. Geod., Vol. 26, 107 - 114 (1982).

The Stokes constants in the Earth's gravitational field together with the precession-nutation constant make possible to determine the principal moments of inertia. The accuracy of their recent values has been discussed; the relative low accuracy of the Earth's mass is here a limiting factor. The moments of inertia of the Earth's core, mantle and crust have been computed.

The Eros parallax, 1930 - 31.
See Abstr. 004.001.

Problem of determination of the orientation of coordinate systems and of periodic errors of star catalogues from optical observations.
See Abstr. 041.010.

A pool of faint stars applied to star catalogue formation.
See Abstr. 041.017.

On the definition of Universal Time.
See Abstr. 044.037.

Orientation of the JPL ephemerides, DE 200/LE 200, to the dynamical equinox of J 2000.
See Abstr. 047.016.

Why a variable G?
See Abstr. 066.118.

On the determination of the gravitational constant and the search for gravitational radiation in a cosmic laboratory.
See Abstr. 066.178.

Variation of gravity with depth in the Earth.
See Abstr. 081.023.

044 Time, Rotation of the Earth

044.001 **Time and latitude results of observations made at Merate Observatory with the astrolabe for the year 1981.**
L. Buffoni, F. Carta, F. Chlistovsky, A. Manara, F. Mazzoleni.
Astron. Astrophys., Suppl. Ser., Vol. 49, 509 - 510 (1982).
Results of observations made with the Astrolabe Danjon OPL no. 32 are given. The results are in the FK4 system.

044.002 **Irregular changes of gravity and the rotation of the earth.**
N. N. Parijskij.
Pis'ma Astron. Zh., Tom 8, 378 - 380 (1982). In Russian. English translation in Soviet Astron. Lett., Vol. 8.
A correlation trend is found between irregular changes of the earths' rotation and the changes of gravity at the stations: Sèvre (1966–1973) and Potsdam, Ledovo (near Moscow) and Novosibirsk (1975–1981).

044.003 **Comparison between two trigonometric models for the long-period variations in the Wolf numbers and in the length of day.**
G. Picchio.
Astron. Astrophys., Vol. 111, 326 - 332 (1982).
A trigonometric model by Luo Shi-fang et al. (1977) reproducing the decade variations in the l.o.d. from 1821 to 1970 is re-examined and criticized with the aid of new data from 1971 to 1980. The possibility of a correspondence between periods, amplitudes, and phase angles from such model and from a similar one reproducing the Wolf numbers from 1749 to 1977 is investigated, in order to support the hypothesis of an influence of the solar activity on the Earth's rotation. Works from previous authors on the subject are reviewed to add further arguments to the comparison.

044.004 **Zur Länge des Urtages.**
K. Bretterbauer.
Gerlands Beitr. Geophys., Band 91, 200 - 204 (1982).
Assuming invariability of the gravitational constant the derivation of a lower boundary for the length of the primeval day is attempted. Using well-known results of the theory of equilibrium figures and an empirical relation between mean density and flattening of the planets, the minimum length of the primeval day is found to be 11 hours. The corresponding mean density of the primeval earth could have been 1 g cm^{-3}, its flattening $^1/_{13}$.

044.005 **The earth's nutational motion determined from radio interferometry.**
G. H. Kaplan.
Bull. American Astron. Soc., Vol. 14, 658 (1982). – Abstract.

044.006 **The role of the core in irregular fluctuations of the Earth's rotation and the excitation of the Chandler wobble.**
S. K. Runcorn.
Philos. Trans. R. Soc. London, Ser. A, Vol. 306, (see 012.014), 261 - 270 (1982).
The nature of the coupling between the core and the mantle of the Earth is controversial: if it is electromagnetic, the quantitative difficulty has not been resolved. The author has shown that if the torques resulting from electric currents induced in the lower semiconducting mantle are of an impulse character, those parallel to the axis (required to change the length of the day) are of the same order of magnitude as those in the equatorial plane, which may be the long-sought origin of the excitation of the Chandler wobble. Recent observational evidence of the geomagnetic secular variation, and from the Earth's rotation, supports this view: the short timescale changes in the geomagnetic field at the core surface have been greatly underestimated.

044.007 **Secular variations of the earth's rotation.**
Zh. S. Erzhanov.
Sovrem. probl. mekh. aviatsii. Moskva, 1982, p. 130 - 137. In Russian. – Abstr. in Ref. zh., 51. Astron., 9.51.86 (1982).

044.008 **The steady march of atomic time.**
H. Smith.
New Scientist, Vol. 93, 382 - 384 (1982). – Abstr. in Phys. Abstr., Vol. 85, Abstr. 69126 (1982).

044.009 **The tendency of the recent variations in the rate of the Earth's rotation – on the leap second.**
S. Okazaki.
Astron. Her., Vol. 75, 50 - 52 (1982). In Japanese. – Abstr. in Phys. Abstr., Vol. 85, Abstr. 78519 (1982).

044.010 **Status of astronomical time (generation and analysis) in Survey of India.**
K. L. Khosla, M. G. Arur, P. S. Bains.
J. Inst. Electron. Telecommun. Eng., Vol. 27, 527 - 529 (1981). – Abstr. in Phys. Abstr., Vol. 85, Abstr. 84064 (1982).

044.011 **The longest darkness.**
M. Hoffmann.
Observatory, Vol. 102, 208 - 210 (1982).

044.012 **A comparative spectral analysis of the earth's rotation and the solar activity.**
F. Carta, F. Chlistovsky, A. Manara, F. Mazzoleni.
Astron. Astrophys., Vol. 114, 388 - 393 (1982).

A spectral analysis of an homogeneous rotational data set $(UT\ 1-TA)_{BIH}$, for the period 1962.0–1978.9, performed with Vaníček's method, reveals several periodicities. A similar analysis of solar emission at 2800 MHz data, emphasizes possible significant analogies between rotation and solar activity. Owing to the various peaks recorded in the range 0.7 yr–2.2 yr in the rotational data spectrum, the authors look for a possible effect of seasonal terms amplitude modulation.

044.013 **Résultats des observations faites à Paris avec l'astrolabe. Temps et latitude 1981.**
F. Chollet, S. Débarbat, J.-C. Hascoët, S. K. Lam, P. Texier, M. Tomas.
Astron. Astrophys., Suppl. Ser., Vol. 50, 195 - 197 (1982).
Results are given for the observations made with the astrolabe APP during the year 1981. No change has been made in the star catalogue since 1956. The observing technique and computation method have also been unchanged since 1972. The results are in the FK4 system.

044.014 **Length of day and obliquity of the ecliptic 850 Ma ago: preliminary results of a stromatolite growth model.**
J. P. Vanyo, S. M. Awramik.
Geophys. Res. Lett., Vol. 9, 1125 - 1128 (1982).
A sinusoidal growth pattern in a columnar stromatolite from the approximately 850 million year-old Bitter Springs Formation suggests a method for deducing length of day and other geophysical parameters. Phototropic microorganisms following the seasonal inclination of the Sun are proposed as responsible for the estimated 410 or more laminae (daily) per sine wave (annual). A preliminary paleomagnetic analysis yields results consistent with the proposed model.

044.015 **The evolution of TAI and UTC.**
H. M. Smith.
J. Inst. Electron. Telecommun. Eng., Vol. 27, No. 10, Part 1, (see 012.044), p. 407 - 414 (1981). – Abstr. in Phys. Abstr., Vol. 85, Abstr. 89668 (1982).

044.016 **Earth rotation and nutations in regard to liquid core-mantle and ocean-lithosphere tidal interactions.**
P. Melchior.
Geodesy and physics of the earth, (see 012.053), p. 4 - 36 (1981). – Review paper.

044.017 **Longitude variations of Borowiec system.**
M. Lehmann.
Geodesy and physics of the earth, (see 012.053), p. 128 - 140 (1981).

044.018 **Fluctuations in the seasonal variations of the length of the day and the earth's wobble.**
A. Poma, E. Proverbio.
Geodesy and physics of the earth, (see 012.053), p. 148 - 160 (1981).

044.019 **On some errors of the astronomical determination of time and longitude.**
J. Melicher, J. Hefty.
Geodesy and physics of the earth, (see 012.053), p. 141 - 146 (1981). In Russian.

044.020 **The earth's rotation and atmospheric circulation: 1958–1980.**
K. Lambeck, P. Hopgood.
Geophys. J. R. Astron. Soc., Vol. 71, 581 - 587 (1982).
The excitation of the earth's rotation by the zonal wind circulation has been evaluated for a 22 yr period and the astronomical observations of the length-of-day from 1958 to 1980 have been corrected for this contribution as well as for tidal effects.

044.021 **I moti della terra nella misura del tempo.**
R. Pannunzio.
Orione, Vol. 3, 139 - 153 (1982).

044.022 **Uniformity of coordinated time scales.**
V. Ptáček.
Bull. Astron. Inst. Czechoslovakia, Vol. 33, 371 - 379 (1982).
The long-term stability of the rate of a group of 17 local coordinated time scales was analysed using the method of two-sample variance over an interval of 1440 d, from the beginning of 1977 to the end of 1980, and the data published in the Annual Reports of the B.I.H. Paris, 1977 - 1980. A marked improvement of the stability was observed for $\tau \geqq 120$ d as a result of the regulation process from the noise polygons for sampling intervals $\tau = 10$ d to 320 d with most time scales. The residual drift of the rate is evaluated and a simple test is used to verify the stationarity of the processes affecting the stability of the rate of coordinated time scales.

044.023 **Greenwich Time Report.** 1981 October - 1982 March.
A. Boksenberg.
R. Greenwich Obs., Time Latitude Serv., p. 289 - 316 (1982).

044.024 **Time and Frequency Services Bulletin,** 1982 April - September.
Shaanxi Astronomical Observatory, Academia Sinica (1982).

044.025 **The transformation of geodetic coordinates and the calculation of long wave propagation range.**
J.-t. Zhang.
Publ. Shaanxi Astron. Obs., No. 1, p. 35 - 38 (1982). In Chinese.

044.026 **Ephemeris Time obtained from photographic measurements of positions of lunar craters relative to stars.**
N. A. Vasilenko, V. S. Kislyuk, R. L. Semerenko.
Astron. Tsirk., No. 1174, p. 6 - 8 (1981). In Russian.

044.027 **Zeit- und Breitenbestimmungen, Zeitsysteme, Präzisionszeitvergleiche.** 1981 May - August.
Akad. Wiss. DDR, Zentralinst. Phys. Erde, Geod.-Astron. Obs., Potsdam, Jahrg. 1981, Nos. 3 - 4 (1982).

044.028 **UT2–UT1 for 1983.**
B. I. H., Paris, Circ. A (1982).

044.029 **Bureau International de l'Heure (B. I. H.) Circular D.**
B. I. H., Paris, Circ. D188 - D194 (1982/83).
Contents: UTC, TAI, UT and coordinates of the pole, rotation of the earth. Data for 1982 May - November.

044.030 **Time and Latitude Bulletins.** 1982 January - June.
Tokyo Astron. Obs., Time Latitude Bull., Vol. 56, 1 - 37 (1982).

044.031 **On the variations of the earth's rotation rate on short time scale during 1973 - 1976.**
N. Sekiguchi, T. Hara.
Ann. Tokyo Astron. Obs., Second Ser., Vol. 19, 86 - 120 (1982).

044.032 **Daily time differences and relative phase values.** 1982 July - December.
U. S. Naval Obs., Washington, D.C. Time Serv. Publ., Ser. 4, Nos. 804 - 830 (1982).

044.033 **Time and earth orientation information.** 1982 July - December.
U. S. Naval Obs., Washington, D.C. Time Serv. Publ., Ser. 7, Nos. 757 - 783 (1982).

044.034 **Time Service report.**
U. S. Naval Obs., Washington, D.C. Time Serv. Publ., Ser. 11, No. 226 (1982).

044.035 **Time Service announcement.**
G. M. R. Winkler.
U. S. Naval Obs., Washington, D.C. Time Serv. Publ., Ser. 14, Nos. 33 - 34 (1982).

044.036 **Earth rotation parameters deduced from Starlette laser ranging.**
M. He, B. D. Tapley, R. J. Eanes.
Sci. Sinica, Ser. A, Vol. 25, 1090 - 1098 (1982).

044.037 **On the definition of Universal Time.**
B. Xu, J. Ren, Z. Sun.
Sci. Sinica, Ser. A, Vol. 25, 1197 - 1204 (1982).
The authors discuss the definition of universal time by Newcomb and the given comments and proposals on the new definition after 1984. It is shown that the conceptional differences between two kinds of precessions in RA and those of nonrotating origins should be distinguished, and the ambiguity of the time arguments is the chief problem of the current definition.

044.038 **Mathematical models of the processes of irregularity of the earth's rotation.**
G. P. Pil'nik, V. Ya. Galkin, E. L. Zhukovskij, I. G. Zhurbenko, S. Yu. Pliskin, Yu. S. Romanchuk.
Sb. rabot vychisl. tsentra MGU, 1982, No. 37, p. 121 - 134. In Russian. – Abstr. in Ref. zh., 51. Astron., 10.51.88 (1982).

044.039 **Time service of the International Latitude Station at Kitab.**
V. A. Vytnov, E. A. Litvinenko, Z. M. Malkin, V. A. Naumov.
Izv. Glav. Astron. Obs. Pulkovo, Astrofiz. Astrometr., No. 200, p. 105 - 106 (1982). In Russian.
In 1975 at the International Latitude Station at Kitab a Time Service was organized. A short description of the instruments and methods of determination of Universal Time from observations with the PZT in the stellar time scale is given.

044.040 **Nonuniformity in the earth's rotation during a day.**
G. P. Pil'nik.
Astron. Zh., Tom 59, 1184 - 1189 (1982). In Russian. English translation in Soviet Astron., Vol. 26, No. 6.
The possibility to use the astronomical observations of the International Time Service for the study of the nonuniformity in the earth's rotation during a day is discussed. The spectral analysis of the data for 1968 - 1978 shows that these observations contain waves with periods near one day and half a day.

044.041 **On the theory of the rotational motion of a deformable Earth.**
M. Burša.
Travaux Inst. Géophys. Acad. Tchécosl. Sci., No. 517. Travaux Géophysiques, Tom 27, 9 - 25 (1979). Academia, Praha 1981.
The solution of the equations describing the rotation dynamics of a deformable Earth has been discussed; the Lagrange's method of variation of constants has been used. It has been stated that the spectral analysis of the prime integrals and of integration constants being functions of time needs to introduce two more unknowns – the coordinates of the pole of the Earth's ellipsoid of inertia in the conventional system CIO.

044.042 **Variations in the Earth's rotation and crustal deformations.**
B. Burša.
Stud. Geophys. Geod., Vol. 26, 215 - 230 (1982).
The deformations of the Earth's equipotential surfaces as well as of the Earth's crust due to variations in the vector of Earth's rotation have been expressed theoretically in detail. Their zonal, tesseral and sectorial parts have been treated separately. The total effect may reach of about 10% from the luni-solar tidal deformations.

044.043 **Determination of the earth rotation between 1971.6 and 1979.0 by lunar laser ranging data observed at the McDonald Observatory.**
W.-j. Jin.
Acta Astron. Sinica, Vol. 23, 198 - 202 (1982). In Chinese.
With lunar laser ranging observations including 1658 normal points of the reflector at the McDonald Observatory in 1971 - 1979, the curve of the earth rotation is calculated and the offset of UT_1 - UTC is given.

044.044 **The effect of some systematic errors on the determination of time and latitude, and group corrections.**
J.-y. Xu.
Acta Astron. Sinica, Vol. 23, 274 - 279 (1982). In Chinese.
The effects of some systematic errors on the determination of time and latitude as well as on group corrections are discussed. In addition to catalog errors, the effects of the errors of magnitudes for photoelectric transit instruments as well as the effects of the errors of spectral types for photoelectric astrolabes are calculated. These effects are all significant.

Data on time and polar motion: immediate accessibility.
See Abstr. 013.011.

Le 25e anniversaire de la troisième opération des longitudes mondiales. (1 juillet 1957 - 31 décembre 1959).
See Abstr. 013.072.

Time service in China during the last three years.
See Abstr. 013.084.

Über Anwendung von Ergebnissen der Laserentfernungsmessungen zum Monde bei Auflösung einiger Probleme der Astronomie und Geodynamik.
See Abstr. 031.613.

The developments of atomic time and frequency standards at Shanghai Observatory and some applications.
See Abstr. 035.002.

Recent activities on time and frequency at the Radio Research Laboratories.
See Abstr. 035.009.

Time/frequency services of the US National Bureau of Standards and some alternatives for future improvement.
See Abstr. 035.018.

Precise T & F intercomparison via VLF phase measurements.
See Abstr. 035.019.

Joint use of cosmic and astrogeodetic methods for geodynamical investigations.
See Abstr. 045.012.

Astronomical determinations of latitude and longitude in 1976–1980.
See Abstr. 045.019.

PZT reductions based on the method of least-squares.
See Abstr. 045.021.

The dynamical effects of a torsion field with gravitation.
See Abstr. 066.274.

On electromagnetic core-mantle coupling.
See Abstr. 084.102.

045 Latitude Determination, Polar Motion

045.001 **Effect of different sources of variation of latitude data on meridian circle catalogues.**
T. J. Rafferty.
Astron. Astrophys., Suppl. Ser., Vol. 50, 27 - 47 (1982).
Variation of latitude data, which must be applied to fundamental meridian circle observations, have been taken from various sources by different observatories. This study shows that the differences in the polar motion data applied can have a significant effect on the individual observations, as well as the overall catalogue resulting from the meridian circle program.

045.002 **The effects of the atmosphere and oceans on the earth's wobble – I. Theory.**
J. M. Wahr.
Geophys. J. R. Astron. Soc., Vol. 70, 349 - 372 (1982).
The theory of wobble excitation for a non-rigid earth is extended to include the effects of the earth's fluid core and of the rotationally induced pole tide in the ocean. The response of the solid earth and oceans to atmospheric loading is also considered.

045.003 **On the ellipticity of the Chandler wobble of the earth's pole.**
R. M. Rasulov.
Dokl. AN USSR, 1982, A, No. 2, p. 61 - 65. In Russian.
Abstr. in Ref. zh., 51. Astron., 9.51.84 (1982).

045.004 **Refinements of the polar motion frequency modulation hypothesis.**
W. E. Carter.
J. Geophys. Res., Vol. 87, 7025 - 7028 (1982).
In a previous paper (Carter, 1981) the author suggested that the Chandler component of polar motion may be frequency modulated, as a function of the instantaneous polar motion magnitude. A composite of observational records compiled by S. C. Chandler and the International Latitude Service series suggests a variation in the magnitude at a frequency of approximately 0.0118 cycles per year (cpy). The spectrum produced by the resulting frequency modulation would depend on the modulation index. Comparison of spectra synthesized with various degrees of frequency modulation and those derived from observational data, suggests a frequency increase of 0.15 to 0.30 cpy/arc sec decrease in the polar motion magnitude. This degree of frequency modulation would cause 'splitting' of the Chandler component, producing dual spectral peaks at periods of 427 and 439 days, similar to that reported by several investigators beginning with Chandler (1901).

045.005 **The Chandlerian nutation from 1900 to 1980.**
B. Guinot.
Geophys. J. R. Astron. Soc., Vol. 71, 295 - 301 (1982).
Estimates of the amplitude, of the phase and of the ellipticity of the Chandlerian nutation are given. The ellipticity is larger than that expected from the equilibrium pole tide. On the other hand, in 1980 the amplitude of the Chandlerian nutation again changed when the total motion of the pole was at a minimum.

045.006 **A search for Chandler and nearly diurnal free wobbles using Liouville equations.**
J. Hinderer, H. Legros, M. Amalvict.
Geophys. J. R. Astron. Soc., Vol. 71, 303 - 332 (1982).
The basic equations describing the dynamical effects of the Earth's fluid core (Liouville, Navier-Stokes and elasticity equations) are derived for an ellipsoidal earth model without axial symmetry but with an homogeneous and deformable fluid core and elastic mantle. These equations are restricted in order to obtain the usual Chandler and nearly diurnal eigenfrequencies. The authors propose a method for calculating the perturbations of these eigenfrequencies.

045.007 **On the influence of the external factors on the accuracy of determination of latitude variations.**
V. Milovanović, G. Teleki, R. Grujić.
Publ. Astron. Obs. Sarajevo, Vol. 1, No. 1, (see 012.043), p. 131 - 141 (1981).

045.008 **Systematic errors in residuals in case of determination of the geographic longitude and latitude by means of astronomical observations.**
M. Jovanović, L. Durović.
Publ. Astron. Obs. Sarajevo, Vol. 1, No. 1, (see 012.043), p. 149 - 152 (1981). In Croatian.

045.009 **Variations of the geographical latitude at Józefosław - Observatory and their comparison with gravimetrical determinations.**
M. Barlik, R. Galas, J. B. Rogowski.
Geodesy and physics of the earth, (see 012.053), p. 36 - 44 (1981).

045.010 **Spectral analysis of modeled latitude variations.**
A. Brzeziński, B. Kołaczek.
Geodesy and physics of the earth, (see 012.053), p. 46 - 52 (1981).

045.011 **A new determination of the tidal coefficient Λ from latitude observations with the Paris astrolabe.**
N. Capitaine, F. Chollet, S. Debarbat.
Geodesy and physics of the earth, (see 012.053), p. 53 - 54 (1981).

045.012 **Joint use of cosmic and astrogeodetic methods for geodynamical investigations.**
N. Georgiev, C. Gergov.
Geodesy and physics of the earth, (see 012.053), p. 94 - 103 (1981).

045.013 **Variation of the Chandler period derived from latitudes observed at Potsdam Observatory.**
J. Höpfner.
Geodesy and physics of the earth, (see 012.053), p. 114 - 117 (1981).

045.014 **On variations of the period of Chandler-wobble.**
H. Jochmann.
Geodesy and physics of the earth, (see 012.053), p. 118 - 121 (1981).

045.015 **Relative translocations of initial systems applied to the calculation of the coordinates of the earth's pole.**
A. A. Korsun', A. I. Emets.
Geodesy and physics of the earth, (see 012.053), p. 122 - 127 (1981). In Russian.

045.016 **Is the Chandler period variable?**
S. Okubo.
Geophys. J. R. Astron. Soc., Vol. 71, 629 - 646 (1982).

The paper investigates whether the Chandler period is variable in time. The first step is to develop an Instantaneous Frequency Analysis to trace the time-varying spectrum for a non-stationary, complex-valued time series. The method is then applied to the observed polar motion to find no significant fluctuation of the Chandler period. Next, the observational grounds of the variable period hypothesis are reexamined. The authors cannot find any positive evidence for the hypothesis.

045.017 **Theoretical and observed *Q* of the Chandler wobble-Love number approach.**
S. Okubo.
Geophys. J. R. Astron. Soc., Vol. 71, 647 - 657 (1982).

The paper treats two principal problems on the Chandler wobble $Q(Q_w)$; (1) the most probable estimate for Q_w (present estimates range from 30 to 600), and (2) the relation between the mantle $Q(Q_m)$ and Q_w (present estimates vary over an order of magnitude).

045.018 **Breitenbestimmungen.** 1982 April - September.
Tech. Univ. Dresden, Lohrmann-Obs., Zirk. Nr. 99 - 100 (1982).

045.019 **Astronomical determinations of latitude and longitude in 1976–1980.**
M. Ollikainen.
Publ. Finnish Geod. Inst., No. 95, 68 pp. (1982).

045.020 **Monthly Notes of the International Polar Motion Service.**
Mon. Notes Int. Polar Motion Serv., Nos. 5 - 10, p. 41 - 108 (1982).

Announcement of values of latitudes observed at the collaborating stations during May - September 1982.

045.021 **PZT reductions based on the method of least-squares.**
S. Manabe, H. Kitago.
Publ. Int. Latitude Obs. Mizusawa, Vol. 15, 1 - 22 (1981).

A new method for the reduction of PZT observations is proposed. In this method, the UT0-UTC and latitude for individual stars, and the plate parameters are simultaneously estimated by solving a non-linear least-squares problem.The present method is applied to the data of the Mizusawa PZT. The scale value and the orientation angle are little affected by errors in the apparent places of stars. For stars with large zenith distances at the meridian transit, differences in estimated times and latitudes amount to 4 ms and 0.″03, respectively, between the proposed reduction method and the one that is currently in use at Mizusawa.

045.022 **Die Analyse der Polbewegung mit Hilfe meteorologischer Erregerfunktionen.**
H. Jochmann.
Veröff. Zentralinst. Phys. Erde, Nr. 67, 91 pp. (1981).

045.023 **Results of latitude determinations in Józefosław.** 1981 July - 1982 June.
L. Pieczyński.
Warsaw Tech. Univ., Latitude Circ., Nos. 80 - 83 (1982).

045.024 **Fortnightly nutation from ILS data (1955.0 - 1966.9).**
A. I. Emets, A. A. Korsun', V. K. Taradij, G. T. Yanovitskaya.
Astron. Tsirk., No. 1201, p. 6 - 8 (1981). In Russian.

045.025 **Nonpolar variations of latitudes obtained from observations with two zenith telescopes in Pulkovo.**
L. D. Kostina, N. R. Persiyaninova, E. Ya. Prudnikova.
Izv. Glav. Astron. Obs. Pulkovo, Astrofiz. Astrometr., No. 200, p. 89 - 90 (1982). In Russian.

A comparison of nonpolar latitude variations derived from observations with the zenith telescopes ZTL-180 and ZTF-135 during the period 1968.1 - 1974.4 is made. Notwithstanding different programes and types of instruments these variations are similar.

045.026 **On the *z*-term of the Pulkovo ZTL-180 zenith-telescope.**
E. Ya. Prudnikova.
Izv. Glav. Astron. Obs. Pulkovo, Astrofiz. Astrometr., No. 200, p. 91 - 95 (1982). In Russian.

A spectral analysis of the *z*-term of the ZTL-180 is made. Parameters of harmonics with an annual and semi-annual periods were determined by the method of coordinate removals of spectral lines. The annual component is rather stable and is detected reliably. The variation of the phase of the semi-annual component permits to refer it to the type of the quasi-periodic oscillations.

045.027 **A comparison of the coordinate systems of the earth's pole calculated at the Main Astronomical Observatory of the Ukrainian Academy of Sciences and the BIH (France).**
V. I. Sakharov.
Izv. Glav. Astron. Obs. Pulkovo, Astrofiz. Astrometr., No. 200, p. 96 - 100 (1982). In Russian.

A comparison is made of the coordinate systems of the earth's pole, calculated at the Main Astronomical Observatory of the Ukrainian Academy of Sciences and the Bureau International de l'Heure. Both the systems overlap on the interval of 12.5 years from the mid of 1956 to 1968. The comparison showed the absence of the systematic difference between them.

The Kitab international latitude station "Ulugh Begh".
See Abstr. 009.020.

Data on time and polar motion: immediate accessibility.
See Abstr. 013.011.

A new method of determination of the pole motion in a uniform system.
See Abstr. 031.503.

A method of simultaneous determination of latitude and azimuth by stars' observations near elongation.
See Abstr. 031.600.

Über Anwendung von Ergebnissen der Laserentfernungsmessungen zum Monde bei Auflösung einiger Probleme

der Astronomie und Geodynamik.
See Abstr. 031.613.

On the investigation of geodynamic parameters by means of laser measurements.
See Abstr. 031.615.

Comparison of the Belgrade General Catalogue of Latitude Stars (IKSZ) with the General Catalogue of the AOE [Engelhardt Astronomical Observatory] and Poltava Latitude Program stars.
See Abstr. 041.020.

Time and latitude results of observations made at Merate Observatory with the astrolabe for the year 1981.
See Abstr. 044.001.

The role of the core in irregular fluctuations of the Earth's rotation and the excitation of the Chandler wobble.
See Abstr 044.006.

Résultats des observations faites à Paris avec l'astrolabe. Temps et latitude 1981.
See Abstr. 044.013.

Greenwich Time Report.
See Abstr. 044.023.

Zeit- und Breitenbestimmungen, Zeitsysteme, Präzisionszeitvergleiche.
See Abstr. 044.027.

Time and Latitude Bulletins.
See Abstr. 044.030.

The effect of some systematic errors on the determination of time and latitude, and group corrections.
See Abstr. 044.044.

046 Astronomical Geodesy, Satellite Geodesy, Navigation

046.001 **Regional satellite networks – aspects and problems.**
R. Dietrich.
Geod. Kartogr., Tom 30, 83 - 94 (1981).
Different variants of relative point determination are described and analysed. It could be seen, that a decrease of simultaneity and station numbers leads to an increase of needed orbit information. Station configurations and satellite geometry must be analysed for an optimum network determination. Nonmobility of the devices (Laser) and conform orbit geometry of the satellites (Doppler) limit the optimum design of a network.

046.002 **A theoretical basis for simultaneous adjustment of Doppler and terrestrial network using a satellite reference ellipsoid.**
I. Gajderowicz.
Artif. Satell., Vol. 16, No. 1, p. 27 - 55 (1981).

046.003 **Applications of time and frequency in geodesy.**
W. Schlüter, K. Nottarp.
J. Inst. Electron. Telecommun. Eng., Vol. 27, 587 - 593 (1981). – Abstr. in Phys. Abstr., Vol. 85, Abstr. 88724 (1982).

046.004 **Statistical tests of ARIES data.**
S. Musman.
J. Geophys. Res., Vol. 87, 5553 - 5562 (1982).
Statistical tests are performed on Project ARIES preliminary baseline measurements in the Southern California triangle formed by the Jet Propulsion Laboratory, the Owens Valley Radio Observatory, and the Goldstone tracking complex during 1976–1980.

046.005 **Gyroscop theodolite in geodetic astronomy.**
B. Kilar.
Publ. Astron. Obs. Sarajevo, Vol. 1, No. 1, (see 012.043), p. 143 - 148 (1981).
By means of gyroscop theodolite, the direction towards the geographical north can be determined independently, relatively quick and in a simple manner. The determination does not depend upon natural and meteorological conditions, which is especially important in geodesy at solving different geodetic problems.

046.006 **Geodynamic effects in physical geodesy.**
H. Moritz.
Geodesy and physics of the earth, (see 012.053), p. 146 - 147 (1981). – Abstract.

046.007 **Research in satellite geodesy in the frame of the Intercosmos program.**
A. G. Massevitch *(Masevich)*, I. Almár, N. Georgiev, K. Hamal, H. Montag, S. Tatevian *(S. K. Tatevyan)*, Y. Zielinsky *(J. B. Zieliński)*.
Geodesy and physics of the earth, (see 012.053), p. 401 - 415 (1981).

046.008 **Requirements for observations of geostationary satellites.**
K.-G. Steinert.
Geodesy and physics of the earth, (see 012.053), p. 476 - 480 (1981) = Mitt. Lohrmann-Obs. Tech. Univ. Dresden Nr. 44.

046.009 **Sonderforschungsbereich 78 Satellitengeodäsie der Technischen Universität München. Abschlußberichte zu den Teilprojekten A1: Richtungsmessungen, A3: Dopplermessungen, B1: Modelle und Methoden der Nutzung von Meßwerten in der Satellitengeodäsie, B3: Figur- und Feldparameterbestimmung.**
M. Schneider.
Deutsche Geod. Komm. Bayer. Akad. Wiss., Reihe B: Angew. Geod., Heft Nr. 261, 198 pp. (1982). ISBN 3-7696-8556-3.

046.010 **Die Arbeiten des Sonderforschungsbereiches 78 Satellitengeodäsie der Technischen Universität München im Jahre 1981.**
M. Schneider.
Veröff. Bayer. Komm. Int. Erdmessung, Bayer. Akad. Wiss., Astron.-Geod. Arb., Heft Nr. 42, 213 pp. (1982). ISBN 3-7696-9785-5.
Contents: Entfernungsmessungen, Mobiles Laserentfernungsmeßsystem, Aufbau einer Empfangsanlage für Radio-

interferometrie, Geopotential, Dynamik des Erde-Mond-Systems, Kinematik geodätischer Punktfelder.

046.011 **Studies on the Metsähovi satellite laser ranging system.**
M. Paunonen.
Rep. Finnish Geod. Inst., No. 82:2, 47 pp. (1982). ISBN 951-711-079-0.

046.012 **Deflections of the vertical on the territory of Europe from satellite data.**
M. Burša, Z. Šíma.
Stud. Geophys. Geod., Vol. 26, 308 - 312 (1982).
On the basis of harmonics in the Earth's gravitational field obtained from satellite orbit dynamics (Smithsonian Standard Earth II), the deflections of the vertical from the normals to the best-fitting tri-axial Earth's ellipsoid have been computed. The coefficients for computing the components of deflections in meridian and prime vertical have been tabulated; the deflections have been represented graphically.

046.013 **On the concepts of combination of Doppler satellite and terrestrial geodetic networks.**
J. Adam, F. Halmos, M. Varga.
Acta Geod. Geophys. Montan., Vol. 17, 147 - 170 (1982).
Abstr. in Phys. Abstr., Vol. 86, Abstr. 4744 (1983).

046.014 **Laser ranging observations at the Satellite Geodetic Observatory (SGO).**
I. Kardos, T. Borza, T. Vass.
Acta Geod. Geophys. Montan., Vol. 17, 237 - 253 (1982).
Abstr. in Phys. Abstr., Vol. 86, Abstr. 4748 (1983).

046.015 **Study of a satellite laser ranging system for point precise measurement.**
H. Drewes, C. Reigber.
Z. Vermessungswes., Vol. 107, 396 - 405 (1982). In German.
Abstr. in Phys. Abstr., Vol. 86, Abstr. 4755 (1983).

046.016 **Determination of the longitude orientation of the Doppler (laser) network.**
L.-d. Wu, Z.-h. Li.
Acta Astron. Sinica, Vol. 23, 147 - 155 (1982). In Chinese.
The paper refers to the determination of the longitude orientation if only a few optical observations are provided. Both dynamical and geometrical methods are proposed. The optimum conditions for the observations are also discussed.

046.017 **The determination of the relative coordinates between stations by the means of the satellite Doppler translocation technique.**
W.-y. Zhu, H.-g. Xu, X. Chen, G.-q. Tang.
Acta Astron. Sinica, Vol. 23, 156 - 162 (1982). In Chinese.
The main error sources of satellite Doppler positioning, ephemeris error and refraction error, are correlated between stations simultaneously tracking a satellite pass. The technique taking advantage of this correlation to improve the positioning accuracy is called "translocation" technique. The authors studied the correlation of each error source using simultaneous observation data with the JSZ-4 Doppler receiver made in China and present the observational and computational programs of the three dimensional, multipass and multistation translocation. The authors have performed a test of the translocation between the Shanghai and the Beijing Station.

Astrogeodetic work on the geoid in Austria.
See Abstr. 013.067.

Project of improved drive of carriage with plate in SBG cameras.
See Abstr. 031.529.

Laser satellite ranging using the automated SBG Camera.
See Abstr. 031.614

On the investigation of geodynamic parameters by means of laser measurements.
See Abstr. 031.615.

Zur Leistungsfähigkeit von transportablen Zenitkameras bei der Lotabweichungsbestimmung.
See Abstr. 032.034.

Lösung von Parameterbestimmungsproblemen in Himmelsmechanik und Satellitengeodäsie mit modernen Hilfsmitteln.
See Abstr. 042.062.

Gravitationswechselwirkung ausgedehnter Körper.
See Abstr. 042.073.

Contributions of satellite geodesic measurements towards a study of the physics of the Earth.
See Abstr. 081.021.

047 Ephemerides, Almanacs, Calendars, Chronology

047.001 **1983 Abridged Nautical Almanac.** Pub. No. 683.
Published by Hydrographic Office of Japan, Maritime Safety Agency, Tokyo, Japan. 3 + 242 + 7 pp. (1982).

047.002 **Almanac for Geodetic Engineers 1982.**
Prepared by the Astronomical Observation Division, National Geophysical and Astronomical Office, PAGASA, under the supervision of R. L. Kintanar.
For sale by the PAGASA (Weather Bureau), Quezon City, Republic of the Philippines. 10 + 27 pp. Price P 7.00 (1981). ISSN 0569-0838.

047.003 **The Astronomical Almanac for the year 1983.**
Data for astronomy, space sciences, geodesy, surveying, navigation and other applications.
Issued by the Nautical Almanac Office, United States Naval Observatory, Washington; and Her Majesty's Nautical Almanac Office, Royal Greenwich Observatory, London. For sale by the Superintendent of Documents, U. S. Government Printing Office, Washington, D. C., 20402 and Her Majesty's Stationery Office, London. 9 + 519 pp. (1982). ISBN 0-11-886914-0.

047.004 **The Handbook of the British Astronomical Association 1983.**
Prepared by the Computing Section of the Association under the supervision of G. E. Taylor.
Office of the Association; Burlington House, Piccadilly, London, W1V 0NL. 106 pp. Price £ 5.00 (1982).

047.005 **Das Himmelsjahr 1983.** Sonne, Mond und Sterne im Jahreslauf.
Herausgegeben von Hans-Ulrich Keller unter Mitarbeit von E. Karkoschka.
Astro-Kosmos, Kosmos, Gesellschaft der Naturfreunde, Franckh'sche Verlagshandlung, Stuttgart. 142 pp. (1982). ISBN 3-440-05092-0.

047.006 **Efemérides Astronômicas 1982.**
Published by Observatório Nacional, Rio de Janeiro, Brazil. Ano 98, 18 + 519 pp.

047.007 **Observer's Handbook 1983.**
R. L. Bishop (Editor).
Published by the Royal Astronomical Society of Canada, 124 Merton Street, Toronto, Ontario M4S 2Z2. 176 pp. (1982). ISSN 0080-4193. – Review in J. R. Astron. Soc. Canada, Vol. 76, 392 - 395; 1982 (*J. R. Percy*).

047.008 **Nautisches Jahrbuch 1983.**
Published by Seehydrographischer Dienst der Deutschen Demokratischen Republik, Rostock. 33. Jahrg. 45 + 365 pp. (1982). ISSN 0433-681X.

047.009 **Himmelskalender 1983.** Ein kleines astronomisches Jahrbuch für Österreich.
H. Mucke.
Astronomisches Büro, Hasenwartgasse 32, A-1238 Wien, Österreich. 113 pp. (1982).

047.010 **Efemérides Astronómicas 1983.**
Published by Instituto y Observatorio de Marina, San Fernando (Cádiz), Spain. Vol. 192, 15 + 398 pp. (1982). ISBN 84-7469-016, ISSN 0080-5971.

047.011 **Almanaque Náutico, 1983. Con suplemento para la navegación aerea.**
Published by Instituto y Observatorio de Marina, San Fernando (Cádiz), Spain. 418 + 30 + 5 pp. (1982). ISBN 84-7469-015-3, ISSN 0210-735X.

047.012 **Approximation formulas for computer evaluation of acyclic types of calendars.**
M. Gossler.
Astron. Nachr., Band 303, 213 - 214 (1982). In German.
Formulas for calculating the French Republican Calendar are described.

047.013 **How accurate is the Gregorian calendar?**
C. Kluepfel.
Sky Telesc., Vol. 64, 417 - 418 (1982).

047.014 **Annuaire de l'Observatoire Royal de Belgique, 1983.** (Jaarboek van de Koninklijke Sterrenwacht van België).
Imprimerie Hayez, Rue Fin 4, 1080 Bruxelles. 150e année, 241 pp. (1982).

047.015 **1983 Nautical Almanac.** Pub. No. 681.
Published by Hydrographic Office of Japan, Maritime Safety Agency, Tokyo, Japan. 6 + 466 + 7 pp. (1982).

047.016 **Orientation of the JPL ephemerides, DE 200/LE 200, to the dynamical equinox of J 2000.**
E. M. Standish, Jr.
Astron. Astrophys., Vol. 114, 297 - 302 (1982).
The lunar and planetary ephemerides, DE 200/LE 200 have been recently produced at JPL. They will form the basis of the ephemerides in the "Astronomical Almanac" starting in the year 1984. The origin has been referenced to the J 2000 dynamical equinox of the ephemerides themselves. The procedure for this orientation is described in detail. Analyses of the ephemerides also provide comparisons with the equinox of the FK 4 and with presently adopted values for the Earth's obliquity.

047.017 **Anuarul Astronomic 1983.**
G. Stănilă, M. Stavinschi, S. Dinulescu (Editors).
Centrul de Astronomie şi ştiinţe spaţiale, Bucureşti.
Editura Academiei Republicii Socialiste România, R 79717, Bucuresti, Calea Victoriei 125. 312 pp. Price Lei 14.50 (1982).

047.018 **Éphémérides Nautiques pour l'an 1983.** Ouvrage publié par le Bureau des Longitudes spécialement a l'usage des marins.
Gauthier-Villars, Paris, France. 485 pp. (1982). ISBN 2-04-015492-2.

047.019 **Efemérides Astronómicas para o ano de 1982.**
Published by Observatório Astronómico da Universidade de Coimbra, Coimbra, Portugal. 16 + 260 pp. (1981).

047.020 **1983 Polaris Almanac for Azimuth Determination.**
Pub. No. 685.
Published by Hydrographic Office of Japan, Maritime Safety Agency, Tokyo, Japan. 15 pp. (1982).

047.021 **Le calendrier Grégorien.**
H. Naudot.
Astronomie, Vol. 96, 515 - 521 (1982).

047.022 **Kalender für Sternfreunde 1983.** Kleines astronomisches Jahrbuch.
P. Ahnert.
Johann Ambrosius Barth, Leipzig, German Democratic Republic. 168 pp. Price M 5.70 (1982). ISSN 0075-4706.

047.023 **Astronomische Grundlagen für den Kalender 1984.**
Compiled by T. Lederle, edited by Astronomisches Rechen-Institut, Heidelberg.
Verlag G. Braun, Karlsruhe. 88 pp. Price DM 56.00 (1982).

047.024 **Der Sternenhimmel 1983.** Astronomisches Jahrbuch für Sternfreunde.
W. Burgat (Editor).
43. Jahrg. Verlag Sauerländer, Aarau, Switzerland–Frankfurt am Main, F. R. Germany. 192 pp. (1982). ISBN 3-7941-2359-X.

047.025 **Gedanken zur vierhundertsten Wiederkehr der Einführung des Gregorianischen Kalenders.**
M. Gossler.
Sternenbote, 25. Jahrg., 170 - 180 (1982).

047.026 **Almanacco Astronomico della Rivista Coelum per l'anno 1983.**
P. Battistini.
Coelum, Suppl. al fasc. 11-12, 43 + 38 pp. Price L 8,000 (1982).

047.027 **Hvězdářská ročenka 1983.**
Compiled by P. Příhoda, J. Vondrák, B. Onderlička, Z. Pokorný.
Ročník 59. Academia, nakladatelství Československé akademie věd, Praha, Czechoslovakia. 165 pp. Price 25.00 Kčs. (1982).

047.028 **Dados Astronómicos para os Almanaques de 1983 para Portugal.**
Published by Observatório Astronómico de Lisboa. 51 pp. (1982).

047.029 **The Air Almanac 1983, January - June.**
Air Publication 1602, issued by Her Majesty's Nautical Almanac Office, London and Nautical Almanac Office, United States Naval Observatory, Washington.
Her Majesty's Stationery Office, London. For sale by the Superintendent of Documents, US Government Printing Office, Washington, D. C. 20402. p. 1 - 364, A104 + F4 pp. (1982). ISBN 0-11-772268-5.

047.030 **Almanac for Computers 1983.**
Nautical Almanac Office, U. S. Naval Observatory, Washington, D. C. 20390. A14 + B17 + C23 + D33 + E10 pp. (1982).

047.031 **Application of Fourier-Chebychev approximation to calculation of ephemerides.**
H. Nakai, H. Kinoshita.
Tokyo Astron. Obs. Rep., Vol. 19, 525 - 531 (1981). In Japanese.

047.032 **Astronomical handbook for 1982.**
R. J. Dodd.
Carter Obs., Astron. Bull. No. 97, 32 pp. (1981).

047.033 **Chinese Astronomical Ephemeris 1983.**
Published by Purple Mountain Observatory, Academia Sinica, Nanking, China. 539 pp. (1982). In Chinese.

047.034 **Les 400 ans de la réforme grégorienne.**
J. Denoyelle.
Ciel Terre, Vol. 98, 271 - 282 (1982).

047.035 **Datoene som forsvant 5. - 14. oktober 1582.**
G. Marthinsen.
Astron. Tidsskr., Årg. 15, 145 - 151 (1982).

047.036 **Zur Frage der inneren Reform des Kalenderjahres.**
M. Gossler.
Sterne, 58. Band, 298 - 300 (1982).

047.037 **Tables of sunrise, sunset, twilight, moonrise and moonset 1983.**
Prepared by the Astronomical Observation Division, National Geophysical and Astronomical Office under the supervision of R. L. Kintanar.
Published by Philippine Atmospheric, Geophysical and Astronomical Services Administration, Quezon City, Republic of the Philippines. 14 + 57 pp. (1982). ISSN 0115-3307.

047.038 **Ephémérides astronomiques et calendrier des événements célestes pour 1983.**
P. Cugnon, A. Koeckelenbergh.
Ciel Terre, Vol. 98, No. 5, 88 pp. (1982).

047.039 **Annuaire du Bureau des Longitudes. Ephémérides 1983.** Calendriers–soleil–lune–planètes–satellites–étoiles–marées–déclinaison magnétique.
Gauthier-Villars, Paris, France. 15 + 266 pp. (1982).
ISBN 2-04-015496-5 = Supplément à l'Astronomie de Janvier 1983.

047.040 **Rocznik Astronomiczny Obserwatorium Krakowskiego 1983.** International Supplement Nr. 54.
K. Rudnicki, P. Flin (Editors).
Państwowe Wydawnictwo Naukowe, Warszawa – Kraków. 5 + 137 pp. Price zł 220.00 (1982). ISBN 83-01-04568-X. ISSN 0075-7047. See abstracts 047.041, 119.108, 119.109, 122.167.

047.041 **Geocentric ephemerides for the year 1983 of the libration points L_4 and L_5 in the Earth-Moon system and in the Sun-Venus system.**
J. Kordylewski, R. Szafraniec.
Rocznik Astronomiczny Obserwatorium Krakowskiego 1983, (see 047.040), p. 131 - 136 (1982).

047.042 **Ephémérides des huit premiers satellites de Saturne pour 1983.**
Supplément à la Connaissance des Temps, Bureau des Longitudes, Paris, France. 46 pp. (1982).

047.043 **Satellites galiléens de Jupiter.** Phénomènes et configurations pour 1983, suivis d'une méthode permettant de calculer les phénomènes pour 1984.
B. Morando (directeur du Service des Calculs), J. E. Arlot, Y. Jannot, W. Thuillot, D. T. Vu (Rédaction et calculs).
Supplément à la Connaissance des Temps. Bureau des Longitudes, Paris, France. 71 pp. (1982).

047.044 **Astronomical Calendar of the Sofia Observatory for the year 1983.**
B. Kovachev, D. Rajkova, Z. Krajcheva, V. Ivanova, A. Antov; edited by A. Bonov.
Izdatelstvo na Blgarskata Akademiya na Naukite, Sofiya. 111 pp. Price 1.25 Lv. (1982). In Bulgarian.

047.045 **400 years of the Gregorian calendar.**
J. Bouška.
Říše hvězd, Vol. 63, 206 - 207 (1982). In Czech.

047.046 **Muster-Kalender für das Jahr 1984.**
Compiled by H. Haupt.
Published by Verlag Georg Fromme & Co., Wien, Austria. 36 pp. (1982).

047.047 **Earliest global visibility of the new moon 1981–85.**
M. Ilyas.
J. R. Astron. Soc. Canada, Vol. 76, 371 - 381 (1982).
Tabulated data on geographical longitudes meeting the minimum astronomical criterion for visibility of the new moon, and corresponding local dates are given at 10° latitude intervals. Conjunction data for the period under study are also included. A brief description of the system of calculation is given together with instructions for the use of the data for construction of calendrical global Lunar Date Lines (LDL).

047.048 **Almanacco 1983.**
Astronomia, N. 4, p. 1 - 124 (1982).

047.049 **1984 Japanese Ephemeris.** Pub. No. 684.
Compiled by A. Yamazaki, Y. Kubo, Y. Tano, A. Senda, H. Saito, T. Takemura, Y. Ganeko, Y. Harada, K. Inoue, K. Nagamori, T. Uchiyama, H. Nishimura, T. Kanazawa.
Published by Hydrographic Department, Maritime Safety Agency, Tokyo, Japan. 6 + 466 + 21 pp. (1982).

047.050 **Astronomisk Årsbok 1982.**
M. Hedin, A. Larsson, J. Schildt.
Bokförlaget INOVA, Stockholm. 96 pp. Price Sv. kr. 20.00 (1981). – Review in Astron. Tidsskr., Årg. 15, 90; 1982 (*S. Söderhjelm*).

047.051 **L'agenda del cielo 1982.**
A. Vari.
Ed. Menabò, Como. 447 pp. Price L 17,000 (1982). – Review in Orione, Vol. 3, 178 - 179; 1982 (*W. Ferreri*).

047.052 **Sky calendar.**
Abrams Planetarium, Michigan State University, E. Lansing, Mich. 48824. Price $ 5.00. ISSN 0733-6314. Review in Sky Telesc., Vol. 64, 559 (1982).

047.053 **Weltkalender 1983.**
Aldebaran Verlags GmbH, Hamburg. 200 pp. Price DM 14.80 (1982). – Review in Sterne Weltraum, 21. Jahrg., 490; 1982 (*M. Sarcander*).

Kalender und Chronologie.
See Abstr. 003.128.

Luigi Lilio and the Gregorian reform of the calendar.
See Abstr. 005.014.

Programas para el cálculo de anuarios astronómicos.
See Abstr. 021.025.

The TAGFER routine set for computing the ephemerides of the Sun, Mercury, Venus and Mars for the RATAN-600 radiotelescope.
See Abstr. 021.040.

Space Research

051 Extraterrestrial Research, Spaceflight Related to Astronomy and Astrophysics

051.001 **Hipparcos: space-age astrometry.**
A. M. Cruise.
Nature, Vol. 298, 324 - 325 (1982).

051.002 **The Far Ultraviolet Spectroscopic Explorer.**
A. Boggess.
Advances in ultraviolet astronomy, (see 012.006), p. 345 - 346 (1982).
NASA is beginning to define in some detail the scientific objectives and performance characteristics of a new astronomy mission referred to as the Far Ultraviolet Spectroscopic Explorer, or FUSE. The principical objective of FUSE is to obtain astronomical spectra at wavelengths shorter than is possible with the Space Telescope.

051.003 **Magellan: high resolution spectroscopy at FUV and EUV wavelengths.**
M. Grewing, S. di Serego Alighieri, W. Burton, C. I. Coleman, R. Hoekstra, C. Jamar, A. Labeque, C. Laurent, A. Vidal-Madjar, P. Rafanelli.
Advances in ultraviolet astronomy, (see 012.006), p. 347 - 354 (1982).
The aim of the Magellan mission is to provide high resolution spectra of celestial sources down to sixteenth magnitude over the extreme ultraviolet wavelength range (between 50 and 140 nm). This region of the electromagnetic spectrum is of fundamental importance in the study of a variety of astrophysical objects. The range extends from studies of interstellar matter in the disc and halo of our and other galaxies, to stellar envelopes, hot and evolved stars, clusters, intergalactic matter, nuclei of galaxies, quasars, and, finally planets and satellites.

051.004 **Le satellite astrométrique Hipparcos et les miras.**
J. Guibert, M.-O. Mennessier.
Bull. AFOEV, No. 21, p. 105 - 106 (1982).

051.005 **Das Tagebuch der Columbia.**
J. von Puttkamer.
Umschau, 82. Jahrg., 488 - 490, 492 (1982).

051.006 **On an investigation of the sun with automatic probes.**
V. P. Kaznevskij, V. M. Zavadskij, Yu. M. Fatkin.
K. E. Ziolkovskij and scientific-technical progress, (see 003.004), p. 19 - 25 (1982). In Russian. – Abstr. in Ref. zh., 62. Issled. kosm. prostranstva, 8.62.90 (1982).

051.007 **Some results of the program of scientific investigations during the long spaceflight of Salyut 4.**
V. I. Sevast'yanov.
K. E. Ziolkovskij and scientific-technical progress, (see 003.004), p. 10 - 14 (1982). In Russian. – Abstr. in Ref. zh., 62. Issled. kosm. prostranstva, 8.62.94 (1982).

051.008 **Encounter with comet Halley.**
L. S. Marochnik, G. A. Skuridin.
Priroda, 1982, No. 8, p. 2 - 15. In Russian.

051.009 **Space flight of the Soviet-French international crew.**
S. A. Nikitin.
Priroda, 1982, No. 9, p. 6 - 9. In Russian.

051.010 **Deutsche Beiträge zur Raumfahrt-Technologie.**
D. E. Koelle.
Umschau, 82. Jahrg., 598 (1982).

051.011 **Ein Vierteljahrhundert Raumfahrt.**
J. von Puttkamer.
Umschau, 82. Jahrg., 610 - 614, 616 (1982).

051.012 **Die europäische Raumfahrt am Scheideweg.**
P. Langereux.
Umschau, 82. Jahrg., 617 - 622 (1982).

051.013 **Airborne astronomy: a video introduction.**
A. W. Meyer.
Bull. American Astron. Soc., Vol. 14, 609 (1982). – Abstract.

051.014 **An astrometric earth orbiter for detection of other planetary systems.**
S. M. Rathjen, D. C. Black.
Bull. American Astron. Soc., Vol. 14, 656 - 657 (1982). Abstract.

051.015 **Astrometric observations with the Space Telescope.**
R. L. Duncombe, G. F. Benedict, P. D. Hemenway, W. H. Jefferys, P. J. Shelus.
The Space Telescope Observatory, (see 012.013), p. 114 - 120 (1982).
Astrometry with the Space Telescope (ST) is performed using one of the Fine Guidance Sensors (FGS). The FGS, which is based on a pair of Koester's prism interferometers, one for each axis, is capable of measuring the position of one object relative to another with an accuracy of 0.002 arcseconds. Astrometric Data Reduction Software, which will be available to the astrometric user of ST, is described. Finally, the authors discuss the kinds of problems the Space Telescope Astrometry Team plans to investigate using ST.

051.016 **Hinotori (ASTRO-A).**
K. Tanaka.
Astron. Her., Vol. 74, 307 - 315 (1981). In Japanese. – Abstr. in Phys. Abstr., Vol. 85, Abstr. 68587 (1982).

051.017 **The Giotto mission to Halley's comet.**
R. Reinhard.
The need for coordinated ground-based observations of Halley's comet, (see 012.025), p. 21 - 55 (1982).

051.018 **An artificial comet experiment.**
G. Haerendel, B. Häusler, H. Föppl, G. Paschmann, E. Rieger, A. Valenzuela.
The need for coordinated ground-based observations of Halley's comet, (see 012.025), p. 181 - 183 (1982).

051.019 **25 years of the cosmic era.**
V. P. Glushko.
Zemlya Vselennaya, 1982, No. 5, p. 4 - 10. In Russian.

051.020 **Piloted space flights in the USSR.**
V. A. Shatalov.
Zemlya Vselennaya, 1982, No. 5, p. 11 - 18. In Russian.

051.021 **PLANET-A project.**
H. Matsuo.
J. Soc. Instrum. Control Eng., Vol. 21, 218 - 222 (1982). In Japanese. – Abstr. in Phys. Abstr., Vol. 85, Abstr. 89109 (1982).

051.022 **Mission concepts for Venus surface investigation.**
S. J. Hoffman, H. Feingold, A. Friedlander.
Astrodynamics 1981, (see 012.027), p. 707 - 728 (1982).

051.023 **Near-earth asteroids: a survey of ballistic rendezvous and sample return missions.**
M. L. Stancati, J. K. Soldner.
Astrodynamics 1981, (see 012.027), p. 729 - 744 (1982).
Earth-approaching asteroids are a ready source of interesting exploration targets. An exhaustive search is made for all good ballistic trajectories in the period 1990–2000. Mass performance is assessed for both rendezvous and sample return missions. Trajectory search techniques are described, and tables of low-energy outbound and return paths are presented. Analysis of the rendezvous mission shows that there exist several opportunities each calender year which can easily deliver a nominal seven hundred kg payload.

051.024 **An astronomer's guide to EXOSAT.**
J. L. Culhane.
Observatory, Vol. 102, 108 - 109 (1982). – Abstract.

051.025 **Current NASA studies for a Far-Ultraviolet Spectrographic Explorer (FUSE).**
J. Linsky, A. Boggess, S. Bowyer, J. Caldwell, W. Cash, J. Cohen, A. Dupree, R. Green, E. Jenkins, M. Jura, D. Leckrone, H. W. Moos, B. Savage, M. Shull, T. Snow, J. G. Timothy, E. Weiler, D. York.
Third European IUE Conference, (see 012.029), p. 473 - 485 (1982).
This report summarizes the current status of planning by a NASA science working group for the proposed Far-Ultraviolet Spectrographic Explorer (FUSE). It summarizes the important new scientific problems that can be studied by FUSE, but cannot be addressed by IUE or ST, which are sensitive only to wavelengths longward of 1200 Å. Also described are two new optical designs – a grazing incidence echelle and a hybrid echelle – to accomplish these scientific goals with high throughput, large simultaneous spectral range, and low background photon-counting statistics.

051.026 **Magellan – far and extreme ultraviolet spectrographic observatory.**
S. di Serego Alighieri.
Third European IUE Conference, (see 012.029), p. 491 - 495 (1982).
Magellan is one of the future space projects being studied by the European Space Agency. The aim is to provide high resolution spectra in the far and extreme UV (between 500 and 1550 Å) of faint galactic and extragalactic objects ($V \leqslant 16^m$). The instrument consists of a mechanical collimator, a concave holographic grating and a bidimensional photon-counting detector. A low resolution mode will provide spectra of objects as faint as $18^m\!.5$.

051.027 **Exotic stellar objects: research directions for the proposed satellite Magellan.**
M. Hack.
Third European IUE Conference, (see 012.029), p. 497 - 501 (1982).
The research programs which can be tackled with the Magellan mission are described. The importance of observing peculiar stellar objects, like young contracting stars or far-evolved objects is discussed.

051.028 **The interest of observing main sequence objects with Magellan.**
F. Praderie.
Third European IUE Conference, (see 012.029), p. 503 - 505 (1982).
A survey of some of the problems to be attacked in the spectral range of Magellan, and at high spectral resolution for main sequence stars, is presented. The main sequence stars are considered in their successive layers: photospheres, chromospheres, coronae, expanding envelopes and shells.

051.029 **Observations in the solar system with the proposed earth-orbiting observatory Magellan.**
R. Prange.
Third European IUE Conference, (see 012.029), p. 507 - 512 (1982).
Spectacular discoveries have been made on the environments of the solar system planets through the planetary missions. New improvements in our understanding of the underlying basic physical phenomena now require a more global view of the system and a long-term survey of its dynamic behaviour. This makes earth-orbiting, wide field of view, high spectral resolution observatories a necessary complementary approach, and a very effective tool for spectroscopic plasma diagnostic in the UV range.

051.030 **Submillimetre astronomy from space platforms.**
T. B. H. Kuiper, D. J. Hollenbach.
The scientific importance of submillimetre observations, (see 012.030), p. 91 - 96 (1982).
Nearly six years of study into the scientific need and the technical feasibility of a 10 m class far-infrared telescope in space have culminated in the recommendation of the Large Deployable Reflector (LDR) by the Astronomy Survey Committee of the U. S. National Academy of Sciences. The scientific considerations for such a project are described and the general characteristics of LDR are derived. A possible scientific programme culminating in LDR is suggested.

051.031 **A far UV orbiting observatory of solar system objects: the SOLSTIS mission proposal.**
P. Bruston.
Bull. American Astron. Soc., Vol. 14, 711 (1982). – Abstract.

051.032 **Saturn ring rendezvous mission.**
K. T. Nock.
Bull. American Astron. Soc., Vol. 14, 712 - 713 (1982). Abstract.

051.033 **Mariner Mark II.**
R. A. Wallace.
Bull. American Astron. Soc., Vol. 14, 713 (1982). – Abstract.

051.034 **Saturn and Titan probes in the 1990's.**
D. A. Steffy.
Bull. American Astron. Soc., Vol. 14, 766 (1982). – Abstract.

051.035 **Advanced automation for space missions.**
R. A. Freitas, Jr., T. J. Healy, J. E. Long.
J. Astronaut. Sci., Vol. 30, 1 - 11 (1982). – Abstr. in Phys. Abstr., Vol. 85, Abstr. 94579 (1982).

051.036 **Disco – a solar-seismology and heliospheric-structure observatory.**
V. Domingo.

ESA Bull., No. 32, p. 16 - 18 (1982).

Investigation of the Sun's interior and the three-dimensional structure of the solar wind and the heliosphere are the main scientific objectives of ESA's Disco mission. It is proposed to study the Sun's internal structure by observing oscillations of the solar surface and variations, both short- and long-term, in solar radiation. The three-dimensional structure of the heliosphere is to be studied by simultaneous measurement of particles and fields in the ecliptic plane by Disco, and at high latitude by the International Solar-Polar Mission.

051.037 **The Infrared Space Observatory (ISO) – a study for a cooled telescope in space for infrared astronomy.**
R. Emery.
ESA Bull., No. 32, p. 19 - 21 (1982).

This is an ESA study for an infrared observatory satellite, consisting of a cooled 60 cm diameter telescope with a range of photometric and spectroscopic instruments for making measurements at wavelengths from 2 μm out to beyond 100 μm. ISO is a three-axis-stabilised free-flyer, designed for an Ariane launch, and to have an operational lifetime of at least 1.5 years.

051.038 **Kepler – a mission to the planet Mars.**
R. Grard.
ESA Bull., No. 32, p. 22 - 24 (1982).

A scientific spacecraft in orbit around a planet other than the Earth is a goal that ESA has been contemplating for a number of years. If this mission is approved, the Kepler spacecraft will be launched by Ariane in July 1988, to begin orbiting Mars in January 1989.

051.039 **Magellan – a far- and extreme-ultraviolet spectrographic observatory.**
S. di Serego Alighieri.
ESA Bull., No. 32, p. 25 - 27 (1982).

Magellan is intended to provide high-resolution spectra in the far and extreme UV of faint galactic and extragalactic objects. A low-resolution mode will provide spectra of objects as faint as $18^{m}.5$. Magellan is planned as an observatory, operated in real time, and allowing interaction with the observer.

051.040 **X-80 – a spectroscopy, transient and timing mission for X-ray astrophysics.**
B. G. Taylor.
ESA Bull., No. 32, p. 28 - 30 (1982).

X-80, a mission to perform in-depth studies of the spectral features and temporal behaviour of cosmic X-ray sources, ranging from nearby active stars to the cosmologically distant quasars, is a candidate for selection as one of the next satellites in the ESA scientific programme. It is seen as the logical successor to Exosat.

051.041 **The IRAS project.**
Ingenieur, Vol. 94, No. 1, p. 38 (1982). In Dutch. Abstr. in Phys. Abstr., Vol. 85, Abstr. 106474 (1982).

051.042 **Description of the dedicated gravitational satellite mission (GRAV SAT).**
V. L. Pisacane, J. C. Ray, J. L. MacArthur, S. E. Bergeson-Willis.
IEEE Trans. Geosci. Remote Sensing, Vol. GE-20, 315 - 321 (1982). – Abstr. in Phys. Abstr., Vol. 85, Abstr. 106475 (1982).

051.043 **The Extreme Ultraviolet Explorer.**
R. F. Malina, S. Bowyer, M. Lampton, D. Finley, F. Paresce, G. Penegor, H. Heetderks.
Opt. Eng., Vol. 21, 764 - 768 (1982). – Abstr. in Phys. Abstr., Vol. 85, Abstr. 106477 (1982).

051.044 **The Cosmic Background Explorer (COBE).**
J. C. Mather.
Opt. Eng., Vol. 21, 769 - 774 (1982). – Abstr. in Phys. Abstr., Vol. 85, Abstr. 106478 (1982).

051.045 **Spectral mapping of Jupiter and the Galilean satellites in the near infrared.**
R. W. Carlson.
Proc. SPIE Int. Soc. Opt. Eng., Vol. 268, (see 012.040), 29 - 34 (1981). – Abstr. in Phys. Abstr., Vol. 85, Abstr. 112300 (1982).

051.046 **Production of secondary particles by neutral and ionised cometary gas and dust impacting on the shield of the GIOTTO spacecraft.**
N. Pailer, E. Grün.
The Giotto spacecraft gas and plasma environments, (see 012.050), p. 1 - 14 (1982).

A plasma cloud is created around the spacecraft by impacting cometary dust and neutrals. The total ion and electron production rates by both effects have been estimated. The interaction processes result in the release of secondary particles. The way in which these particles may be extrapolated to account for the plasma cloud and for the different chemical composition of cometary wind are stated.

051.047 **Electrostatic charging of the GIOTTO spacecraft due to neutral gas impact.**
H. Maaßberg, U. Isensee.
The Giotto spacecraft gas and plasma environments, (see 012.050), p. 21 - 24 (1982).

The high energy impact of neutral gas results in strong secondary ion/electron emission from the front shield of GIOTTO spacecraft. The influence of this plasma emission on spacecraft charging and potential structure has been described by Maaßberg and Isensee (1981) in detail for the case of aluminium at the front shield. In this short paper, the case of a gold coating at the outer surface of the front shield is discussed. All calculations are carried out using the spherically symmetric model (Maaßberg et al., 1980; Maaßberg and Isensee, 1981).

051.048 **Differential charging model of solar panels.**
H. Maaßberg, U. Isensee.
The Giotto spacecraft gas and plasma environments, (see 012.050), p. 25 - 33 (1982).

The solar panels of the Giotto spacecraft are differentially charged. In the near vicinity of the nucleus, the differential charging of solar panels will be within the range of 10 Volts. Because the plasma analyser is located in the conducting part at the top of the Giotto spacecraft, the influence of this differential charging on measuring the electron distribution function will not be important.

051.049 **Estimates of a spacecraft induced gas cloud and its effects on measurement of ambient particles during a fast flyby of comet Halley.**
D. T. Young.
The Giotto spacecraft gas and plasma environments, (see 012.050), p. 35 - 46 (1982).

A gas cloud is produced by the Giotto dust bumper shield due to sputtering, dust impact vaporisation and outgasssing as it encounters the neutral coma of the comet at the high flyby velocity of 69 km/s. It is compressed between the coma gas and the onrushing satellite. Cometary neutral gas can collide inelastically with the Giotto cloud, producing spurious ionization and dissociation of cometary material. Under the assumptions used here this will amount to ~10% of the total ambient ion flux at $r \approx 1000$ km.

051.050 **Considerations on surface potentials, wake effects and related experimental investigations associated**

with a cometary fly-by.
R. J. L. Grard.
The Giotto spacecraft gas and plasma environments, (see 012.050), p. 47 - 57 (1982).

The surface of the GIOTTO spacecraft can be divided into a number of sections which interact in different ways with the cometary environment. A qualitative approach shows that differential charging may occur if the surface is not entirely conductive but that the potential difference between any two points will probably not exceed 40V. This study may be supported by experimental investigations performed on a planetary orbiter with a low periapsis, or in the laboratory with a vacuum chamber. The ultimate test will be achieved by measuring the electric characteristic of a selected surface element of GIOTTO in the comet Halley environment.

051.051 **HELIOS-mission: mission objectives, mission verification, selected results.**
H. Porsche.
The solar system and its exploration, (see 012.054), p. 43 - 50 (1981).

The mission HELIOS, HE-1 launched 10 Dec 1974 and HE-2 launched 15 Jan 1976, is the first endeavour to investigate the innermost parts of interplanetary space by means of in situ experiments. The mission objectives are described and illustrated by examples or selected results. Mission characteristics are given and some technical aspects are discussed.

051.052 **The International Solar Polar Mission (ISPM).**
R. G. Marsden, K.-P. Wenzel.
The solar system and its exploration, (see 012.054), p. 51 - 59 (1981).

The International Solar Polar Mission (ISPM) will for the first time allow exploration of the heliosphere within a few astronomical units of the sun over the full range of heliographic latitudes. The paper presents an overview of the scientific objectives of the instrumentation chosen jointly by ESA and NASA to carry out the exploratory investigations unique to ISPM. A brief description of the mission time-lines and trajectory is also given.

051.053 **The scientific objectives of the GEOS mission.**
R. Grard, K. Knott.
The solar system and its exploration, (see 012.054), p. 85 - 95 (1981).

The scientific objective of the two GEOS satellites is to contribute to our understanding of the magnetosphere response to the highly variable flow of solar plasma which approaches the Earth magnetosphere with an average speed of 400 km/s. The two payloads are identical and consist of a comprehensive set of sophisticated instruments to study in-situ waves, fields and particles over wide frequency and energy ranges.

051.054 **Space missions to Venus.**
J. L. Bertaux.
The solar system and its exploration, (see 012.054), p. 151 - 156 (1981).

Space missions to planet Venus have been exclusively conducted by two countries: the United States of America and the U.S.S.R. A list of the space missions from both countries is presented, with the main results in form of a table of basic data about Venus. Soviet Venera missions and the US Pioneer Venus mission are briefly described.

051.055 **The Venus Orbital Imaging Radar (VOIR) mission.**
F. W. Leberl.
The solar system and its exploration, (see 012.054), p. 189 - 197 (1981).

The Venus Orbital Radar (VOIR) mission is currently being prepared by NASA to map the surface of Venus with an orbiting side-looking radar imaging device from an altitude of about 250 km. This will provide near global and detailed data on the surface comparable or even superior to that available on other terrestrial planets.

051.056 **The Giotto project – a fast fly-by of Halley's comet.**
R. Reinhard.
The solar system and its exploration, (see 012.054), p. 221 - 227 (1981).

051.057 **Magnetic field measurements onboard Venera 13, Venera 14 and Venera-Halley.**
R. Schmidt, K. Schwingenschuh.
The solar system and its exploration, (see 012.054), p. 245 - 247 (1981).

Magnetometers for the spacecraft Spacelab 1, Venera 13, Venera 14 and Venera-Halley were developed and installed into the spacecraft. The project MISCHA (Magnetic Fields in the Interplanetary Space during Comet Halley's Approach) for the exploration of the magnetic field structure of the comet Halley (Venera-Halley) is in its definition phase. A brief description of the technical realization of each of them as well as the envisaged scientific objectives are given.

051.058 **Outlook on future missions.**
G. P. Haskell.
The solar system and its exploration, (see 012.054), p. 251 - 255 (1981).

The paper summarises European capabilities and limitations with regard to planetary space missions.

051.059 **New achievements in investigation of the cosmos.**
This paper was prepared by the editorial staff of Zemlya i Vselennaya on the basis of a report given by V. A. Kotel'nikov, vice-president of the USSR Academy of Sciences.
Zemlya Vselennaya, 1982, No. 6, p. 4 - 10. In Russian.

051.060 **The Giotto project – a fast flyby of Halley's comet.**
R. Reinhard.
ESA J., Vol. 5, 273 - 285 (1981).

The Giotto mission to Halley's comet is a fast flyby in March 1986, some four weeks after the comet's perihelion passage. Halley has been selected for ESA's first cometary mission because of its enormous activity, which is comparable to that of 'new' comets, its well-known orbital characteristics, the moderate launch-energy requirements, and its outstanding role in history. In-situ measurements in the cometary atmosphere are needed to unravel the complex physical and chemical processes that occur in it, to provide the elemental and isotopic compositions of the volatile components and dust, to investigate the solar-wind/comet interaction, and to image the comet nucleus, which is too small to be seen from Earth. These scientific objectives are addressed by the 10 experiments to be flown on the spacecraft.

051.061 **The European X-ray Observatory Satellite – Exosat.**
B. G. Taylor, R. D. Andresen, A. Peacock, R. Zobl.
ESA J., Vol. 6, 1 - 19 (1982).

The scientific mission of Exosat is outlined and the principal characteristics of the spacecraft and the scientific instruments are described. The performance limits for X-ray source detection, spectroscopy, and variability are presented and arrangements for observatory operations and observation planning are outlined.

051.062 **DISCO.** Result of the phase A study.
European Space Agency, SCI(82)2, 79 pp. (1982).

The prime objective of DISCO (Dual Spectral Irradiance and Solar Constant Orbiter) is to perform space-based solar seismology, observing at the same time fluctuations in both the global velocity field and solar luminosity.

051.063 **X-80: spectroscopy/timing-transient mission in X-ray astronomy**. Report on the phase A study.
European Space Agency, SCI(82)3, 93 pp. (1982).
The X-80 mission is planned as a successor of EXOSAT and ROSAT. It will be a high throughput mission with emphasis on narrow- and broad-band spectroscopy over the energy range 0.1–10 keV and with extended broad band spectroscopy up to 200 keV, on X-ray source variability and timing.

051.064 **MAGELLAN**. Report on the phase A study.
European Space Agency, SCI(82)4, 98 pp. (1982).
The MAGELLAN mission aims to provide high resolution spectra of galactic and extragalactic objects down to magnitude $V = 16$ in the far and extreme ultraviolet range (500 - 1400 Å). A complementary low resolution mode will provide spectra of fainter objects ($V \leqslant 18.5$) over an enlarged wavelength band (300 - 2000 Å).

051.065 **KEPLER: an interdisciplinary Mars orbiter mission**. Report on the phase A study.
European Space Agency, SCI(82)5, 111 pp. (1982).
The KEPLER spacecraft will be placed into a polar orbit about Mars where it will take scientific data for one Martian year. It will explore the Martian upper atmosphere and ionosphere, the Martian magnetic field, its magnetosphere and the interactions of the solar wind with the planet.

051.066 **Infrared Space Observatory**. Report on the phase A study.
European Space Agency, SCI(82)6, 124 pp. (1982).
ISO will consist of a cooled 0.6 m diameter telescope in a 3-axis stabilised spacecraft. The telescope and scientific instruments are cooled by a dual cryogen system involving liquid helium and liquid hydrogen. The payload consists of a near infrared camera, two Michelson interferometers with spectral resolving powers up to 10^5 and a multi-band photometer.

051.067 **Interstellar propulsion systems.**
F. Dyson.
Extraterrestrials: where are they? (see 012.057), p. 41 - 45 (1982).

051.068 **Expectations of Giotto.**
M. K. Wallis.
Proceedings of the workshop on interstellar comets, (see 012.058), p. 56 - 60 (1982).
Plasma and atmospheric phenomena, and nucleus outgassing concepts are surveyed in relation to measurements planned for the Giotto mission to comet Halley. How far fluid models of the plasma and gas are appropriate, what mechanisms produce the impressive and variable structures, and what are the primary constituents less volatile than H_2O, are key questions to resolve.

051.069 **A review of celescope observations from outside the atmosphere.**
W.-z. Su.
Publ. Purple Mountain Obs., Vol. 1, 73 - 80 (1982). In Chinese.

051.070 **The German Infrared Laboratory GIRL.**
D. Lemke.
Second ESO infrared workshop, (see 012.060), p. 373 - 382 (1982).

051.071 **IRAS and its mission.**
P. E. Clegg.
Second ESO infrared workshop, (see 012.060), p. 383 - 390 (1982).

051.072 **IRAS follow-up: problems and prospects.**
R. D. Joseph, N. A. Robertson.
Second ESO infrared workshop, (see 012.060), p. 391 - 399 (1982).

051.073 **IR observations of quiescent comets.**
J. Degewij.
Second ESO infrared workshop, (see 012.060), p. 401 (1982). – Abstract.

051.074 **SIRTF – the Shuttle Infrared Telescope Facility.**
M. W. Werner, F. C. Witteborn.
Second ESO infrared workshop, (see 012.060), p. 405 - 419 (1982).

051.075 **A large deployable reflector for infrared and submillimeter astronomy from space.**
M. W. Werner, T. B. H. Kuiper.
Second ESO infrared workshop, (see 012.060), p. 421 - 423 (1982).

051.076 **Infrared Space Observatory (ISO) – a European Space Agency study for a cryogenically cooled telescope in space.**
R. Emery.
Second ESO infrared workshop, (see 012.060), p. 425 - 435 (1982).

051.077 **Development of the lunar and interplanetary programs in the design bureau of S. P. Korolev.**
M. Yu. Glebov.
Iz istor. aviatsii i kosmonavt. Moskva, 1980, No. 42, p. 67 - 75. In Russian. – Abstr. in Ref. zh., 62. Issled. kosm. prostranstva, 10.62.13 (1982).

051.078 **Comments on the 'TYCHO' photometry proposal.**
A. N. Argue.
Bull. Inf. Cent. Données Stellaires, No. 23, p. 59 - 62 (1982).
The proposal to make photometric observations in three colours with the 'TYCHO' photometers on the HIPPARCOS Astrometric Satellite offers an opportunity to acquire a substantial body of photometric data. Briefly, the proposal envisages the use of the star mappers on the satellite to obtain positional and UBV data for at least 400,000 stars brighter than $B = 11^m$ with accuracies at $B = 10^m$ of 0.03 arc sec, $0^m.03$ in V and $0^m.05$ in B–V and U–B.

051.079 **STARLINK.**
M. J. Disney, P. T. Wallace.
Q. J. R. Astron. Soc., Vol. 23, 485 - 504 (1982).

051.080 **Stellar X-ray astronomy with the Advanced X-ray Astrophysics Facility.**
M. V. Zombeck.
Smithsonian Astrophys. Obs., Spec. Rep. 392, Vol. II, (see 012.065), p. 237 - 245 (1982).
The Advanced X-ray Astrophysics Facility, although conceived as an observatory principally for extragalactic X-ray astronomy, will have unique capabilities for the observation of X-ray emission from stars. It will be shuttle-launched, serviceable in-orbit, and capable of retrieval for ground refurbishment.

051.081 **High throughput X-ray studies of stellar coronae with a LAMAR facility.**
P. Gorenstein.
Smithsonian Astrophys. Obs., Spec. Rep. 392, Vol. II, (see 012.065), p. 247 - 252 (1982).
A practical means for the realization of a high throughput, moderate cost, cosmic X-ray facility was first described several years ago (Gorenstein 1973) and is known as the Large Area Modular Array of Reflectors, or LAMAR. One of

the principal areas of investigation for a LAMAR facility is the behavior of stellar coronae as observed in the X-ray band. In this application, the high throughput of the LAMAR permits a systematic study of the photometric and spectroscopic properties of stellar X-ray sources and makes a large number of objects accessible.

051.082 **STCOEX: The stellar X-ray coronal Explorer.**
G. S. Vaiana, L. Golub, C. W. Maxson, R. Rosner, M. V. Zombeck.
Smithsonian Astrophys. Obs., Spec. Rep. 392, Vol. II, (see 012.065), p. 253 - 260 (1982).

The authors have developed a list of projects which they believe may not be completed by further analysis of available Einstein data, and which also appear to be of particular importance for the study of stellar surface "activity"; a natural outgrowth of this analysis was the development of design criteria for a follow-up mission to Einstein, one whose primary emphasis would be on stellar observations, and whose general capabilities lay within the scope of the NASA Explorer program. The authors briefly review the criteria which lead them to the desired instrumental capabilities; and provide an overview of the proposed instrument and some of its capabilities, as indicated by their simulations.

051.083 **Perspectives of solar studies with a high resolution in the UV region from the stratosphere.**
R. Sh. Gafarova, L. Z. Dul'kin, V. N. Karpinskij, B. I. Kondrat'ev, V. A. Krat.
Izv. Glav. Astron. Obs. Pulkovo, Astrofiz. Astrometr., No. 200, p. 132 - 134 (1982). In Russian.

Systematic solar observations with high angular resolution (up to 0".16) in a broad range of the spectrum from 1800 Å up to the EIR-region can be made from the stratosphere (altitude 30 - 32 km) using a balloon-borne astronomical station. A soft landing after its mission will return the data obtained during the flight and instruments and will give an opportunity to use the station as a shuttle with different auxiliary equipment.

051.084 **Observational capabilities of EXOSAT and ROSAT.**
B. Aschenbach.
Accreting neutron stars, (see 012.069), p. 309 - 313 (1982).

The EXOSAT and ROSAT missions will continue the X-ray astronomical research into the 1980's. EXOSAT is almost ready for launch. The launch of ROSAT is scheduled for midth 1987. The scientific objectives and instrumentation of EXOSAT and ROSAT are described in this paper.

051.085 **Japanese X-ray astronomy program.**
S. Hayakawa.
Accreting neutron stars, (see 012.069), p. 314 - 315 (1982).

A brief description of the instrumentation of the future Japanese X-ray satellites ASTRO-B and ASTRO-C is given.

051.086 **Signal analysis and error analysis studies for a Geopotential Research Mission (GRM).**
W. D. Kahn, T. L. Felsentreger.
NASA Tech. Memo., NASA TM-83970, 17 pp. (1982).

The Geopotential Research Mission has been proposed for the mapping of the Earth's gravity field. The objectives of this paper are to discuss the signal characteristics and the geopotential parameter recovery capability of the SST Doppler sensor flown on GRM.

051.087 **Astronautics in 1981.**
M. Grün, P. Koubský.
Říše hvězd, Vol. 63, 177 - 181 (1982). In Czech.

051.088 **Astronautics in Japan.**
I. Hudec.
Kozmos, Vol. 13, 131 (1982). In Czech.

051.089 **25 years of space research.**
A. Hajduk.
Kozmos, Vol. 13, 112 - 118 (1982). In Slovak.

051.090 **Satellite astronomy and exploration of the distant universe.**
P. Koubský.
Kozmos, Vol. 13, 191 - 199 (1982). In Czech.

051.091 **The Extreme Ultraviolet Explorer.**
S. Bowyer, R. Malina, M. Lampton, D. Finley, F. Paresce, G. Penegor.
Proc. SPIE Int. Soc. Opt. Eng., Vol. 279, (see 012.072), 176 - 182 (1981). – Abstr. in Phys. Abstr., Vol. 85, Abstr. 117592 (1982).

051.092 **Future directions in X-ray/gamma-ray observations.**
D. A. Kniffen.
The galactic center, (see 012.076), p. 208 - 216 (1982).

This paper reviews the capabilities which will be available for X- and γ-ray observations of the galactic center in the 1980's.

051.093 **Gamma-ray astronomy beyond the COS-B mission.**
G. F. Bignami.
Europhys. News, Vol. 13, No. 7, p. 1 - 5 (1982). – Abstr. in Phys. Abstr., Vol. 86, Abstr. 5086 (1983).

051.094 **Space Report.**
Spaceflight, Vol. 24, 348 - 357, 402 - 412, 450 - 456 (1982).

Data and data retrieval in space astronomy.
See Abstr. 002.024.

NSSDC Data Listing.
See Abstr. 002.087.

Planetary exploration.
See Abstr. 003.132.

Space in the 1980's and beyond.
See Abstr. 012.090.

Perspectives on space astronomy.
See Abstr. 013.028.

The international space arena.
See Abstr. 013.030.

The new ocean of space.
See Abstr. 013.031.

Science operations with Space Telescope.
See Abstr. 013.046.

Reflections on the Space Telescope.
See Abstr. 013.048.

Report presented by the European Space Agency to the 24th COSPAR meeting, Ottawa, Canada, May 1982.
See Abstr. 013.058.

Update of the impact ionization problem.
See Abstr. 022.133.

The Space Telescope Observatory.
See Abstr. 032.512.

High resolution imaging systems for spin-stabilized Probe spacecraft.
See Abstr. 032.562.

The solar system: a cluttered laboratory for gravity research.
See Abstr. 066.299.

Predictions of geomagnetic activity by an interplanetary spacecraft.
See Abstr. 084.041.

In-situ and retrieved cometary dust sampling within the solar system.
See Abstr. 102.089.

Comets, their evolution and origin.
See Abstr. 102.094.

052 Astrodynamics, Navigation of Space Vehicles

052.001 **On a method for calculating the flight trajectories around the moon.**
L. I. Gusev, A. M. Nikulin.
K. Eh. Tsiolkovskij i nauchn.-tekh. progress. Moskva, 1982, p. 76 - 82. In Russian. – Abstr. in Ref. zh., 62. Issled. kosm. prostranstva, 7.62.152 (1982).

052.002 **Sperical motions of a solid body according to nonholonomic programmes.**
V. V. Dobronravov.
K. Eh. Tsiolkovskij i nauchn.-tekh. progress. Moskva, 1982, p. 67 - 71. In Russian. – Abstr. in Ref. zh., 62. Issled. kosm. prostranstva, 7.62.153 (1982).

052.003 **Periodic motions of a dynamically extended satellite in the regime of a uniaxial gravitational orientation.**
V. V. Sazonov, M. E. Sidoryuk.
Inst. prikl. mat. AN SSSR. Prepr., 1981, No. 163, 28 pp. In Russian. – Abstr. in Ref. zh., 62. Issled. kosm. prostranstva, 7.62.157 (1982).

052.004 **Lifetime prediction for satellites in low-inclination transfer orbits.**
D. G. King-Hele.
J. British Interplanet. Soc., Vol. 35, 339 - 344 (1982).

052.005 **The geometry of satellite clusters.**
J. G. Walker.
J. British Interplanet. Soc., Vol. 35, 345 - 354 (1982).

052.006 **The value of photographic observations in improving the accuracy of satellite orbits.**
D. G. King-Hele.
J. British Interplanet. Soc., Vol. 35, 355 - 362 (1982).

052.007 **Analysis of the orbit of 1966 - 63A near 15th-order resonance.**
D. M. C. Walker, A. N. Winterbottom.
J. British Interplanet. Soc., Vol. 35, 363 - 367 (1982).

052.008 **Analysis of the orbital inclination of HEOS 2 second-stage rocket 1972-05B.**
P. Moore, D. Holland.
J. British Interplanet. Soc., Vol. 35, 368 - 379 (1982).

052.009 **Theory of orientation and stabilization of an axial-symmetrical satellite in its orbiting reference system.**
A. Brzeziński.
Artif. Satell., Vol. 16, No. 1, p. 5 - 25 (1981).
In the paper the problem of stability of a symmetrical satellite in an orbiting reference frame is examined. Uniaxial stabilities are investigated, in which a symmetry axis has a stationary position in an orbiting frame and besides a uniform rotation around this axis takes place. All equilibrium solutions are established, and conditions for their stability are determined.

052.010 **On two variants of the planar problem of four fixed centres.**
I. S. Kozlov, V. S. Naniev.
Mekh. kosm. poleta. Tr. 15 Chtenij, posvyashch. razrab. nauchn. naslediya i razvitiyu idej K. Eh. Tsiolkovskogo, Kaluga, 12 - 15 sent., 1980, Moskva, 1981, p. 107 - 113. In Russian. – Abstr. in Ref. zh., 62. Issled. kosm. prostranstva, 8.62.107 (1982).

052.011 **Forecasting some configurations of a flight to comets.**
A. V. Labunskij, A. V. Pankratov, O. K. Margorin.
Mekh. kosm. poleta. Tr. 15 Chtenij, posvyashch. razrab. nauchn. naslediya i razvitiyu idej K. Eh. Tsiolkovskogo, Kaluga, 12 - 15 sent., 1980, Moskva, 1981, p. 42 - 45. In Russian. – Abstr. in Ref. zh., 62. Issled. kosm. prostranstva, 8.62.115 (1982).

052.012 **Construction of equivalent systems of equations of motion of a dynamically compressed satellite relative to the mass center.**
I. S. Astakhova.
Vestn. LGU, 1982, No. 7, p. 100 - 104. In Russian. – Abstr. in Ref. zh., 62. Issled. kosm. prostranstva, 8.62.120 (1982).

052.013 **Motion of a satellite relative to the mass center under the action of Lorentz forces.**
G. V. Lyakhovka.
Mekh. kosm. poleta. Tr. 15 Chtenij, posvyashch. razrab. nauchn. naslediya i razvitiyu idej K. Eh. Tsiolkovskogo, Kaluga, 12 - 15 sent., 1980, Moskva, 1981, p. 87 - 93. In Russian. – Abstr. in Ref. zh., 62. Issled. kosm. prostranstva, 8.62.121 (1982).

052.014 **On the stability of rotational motion of a space vehicle with generalized precession of the angular momentum.**
Yu. A. Gartung, Yu. M. Loenko.
Mekh. kosm. poleta. Tr. 15 Chtenij, posvyashch. razrab. nauchn. naslediya i razvitiyu idej K. Eh. Tsiolkovskogo, Kaluga, 12 - 15 sent., 1980, Moskva, 1981, p. 76 - 80. In Russian. – Abstr. in Ref. zh., 62. Issled. kosm. prostranstva, 8.62.125 (1982).

052.015 **Determination of the orientation of the stations of the Prognoz series from data of telemetric measurements.**
N. A. Ehjsmont.
Obrab. inf. poluch. po programme Interkosmos. Moskva, 1982, p. 156 - 159. In Russian. – Abstr. in Ref. zh., 62. Issled. kosm. prostranstva, 8.62.127 (1982).

052.016 **Deep-space navigation – a new flight-dynamics discipline for ESA.**

F. Hechler, M. Rosengren.
ESA Bull., No. 31, p. 92 - 99 (1982).

052.017 **Perturbations of the orbit of Explorer 19 due to solar radiation.**
C. J. Brookes, F. C. E. Ryland.
Celestial Mech., Vol. 27, 339 - 352 (1982).
The orbit of the balloon satellite, Explorer 19, is analysed to determine the effects of direct solar radiation pressure over one revolution of the satellite (111 min) for MJD 42822 and MJD 42966. At the earlier date, the satellite entered the Earth's shadow, presenting an opportunity to examine the effectiveness of two different shadow models. The reflectivity of the surface of the satellite was estimated from analysis of the variations in orbital eccentricity over a period of 236 days. The paper concludes with an examination of the perturbations in orbital eccentricity and inclination due to incident, specularly reflected, and diffusely reflected radiation.

052.018 **The effects of diffuse solar radiation on the motion of an artificial satellite.**
C. J. Brookes, F. C. E. Ryland.
Celestial Mech., Vol. 27, 353 - 373 (1982).
Previous analyses of the orbits of spherical balloon satellites have attempted to satisfy residuals in observed perturbations in Keplerian elements, assumed to be caused by diffuse radiation pressure. It is difficult to distinguish, however, between those perturbations caused by diffuse radiation and those caused by reflected radiation. Following the derivation by Lucas of exact expressions for both incident and reflected radiation forces on a prolate spheroidal satellite, and the subsequent work of Aksnes pertaining to spherical satellites, the theory is extended to include the effects of diffuse radiation whilst at the same time qualifying the assumption that the radiation force acting along the Sun-satellite line can be taken as parallel to the Sun-Earth line.

052.019 **Long-term motion of resonant satellites with arbitrary eccentricity and inclination.**
P. E. Nacozy, R. E. Diehl.
Celestial Mech., Vol. 27, 375 - 397 (1982).
A first-order, semi-analytical method for the long-term motion of resonant satellites is introduced. The method provides long-term solutions, valid for nearly all eccentricities and inclinations, and for all commensurability ratios. The method allows the inclusion of all zonal and tesseral harmonics of a nonspherical planet. The authors present here an application of the method to a synchronous satellite including only J_2 and J_{22} harmonics. Global, long-term solutions for this problem are given for arbitrary values of eccentricity, argument of perigee and inclination.

052.020 **Periodic oscillations of the symmetry axis of a dynamically symmetric satellite in a slightly elliptical orbit.**
M. E. Sidoryuk.
Inst. prikl. mat. AN SSSR. Prepr., 1981, No. 165, 28 pp. In Russian. – Abstr. in Ref. zh., 62. Issled. kosm. prostranstva, 9.62.113 (1982).

052.021 **On the stability of resonance regimes of the motion of space vehicles during the landing phase in the atmosphere of a planet.**
V. N. Penya.
Tr. 6 Chtenij, posvyashch. razrab. nauchn. naslediya i razvitiyu idej F. A. Tsandera. Sekts. Astrodin., Moskva, 1980, p. 60 - 67. In Russian. – Abstr. in Ref. zh., 62. Issled. kosm. prostranstva, 9.62.114 (1982).

052.022 **Calculation of the trajectory of not-controlled motion of an axisymmetric body taking into account the motion around the mass center.**
V. M. Belokonov, Yu. M. Zabolotnov.
Tr. 6 Chtenij, posvyashch. razrab. nauchn. naslediya i razvitiyu idej F. A. Tsandera. Sekts. Astrodin., Moskva, 1980, p. 68 - 74. In Russian. – Abstr. in Ref. zh., 62. Issled. kosm. prostranstva, 9.62.115 (1982).

052.023 **On a method for calculation of the spatial motion of a vehicle with arbitrary axisymmetric form.**
V. S. Aslanov, B. I. Mottsulov, I. A. Timbaj.
Tr. 6 Chtenij, posvyashch. razrab. nauchn. naslediya i razvitiyu idej F. A. Tsandera. Sekts. Astrodin., Moskva, 1980, p. 75 - 81. In Russian. – Abstr. in Ref. zh., 62. Issled. kosm. prostranstva, 9.62.116 (1982).

052.024 **The artificial Earth satellite motion in an oblate, rotating atmosphere with symmetrical diurnal effect.**
L. Burs, Á. Pál.
Astron. Nachr., Band 303, 269 - 275 (1982).
Equations describing the disturbed motion of an artificial Earth satellite in the atmosphere are integrated using a new variable instead of the true anomaly. The atmospheric flattening and rotation, the linear variation of the density scale height with the altitude, the symmetrical diurnal effect and the variation of its amplitude with the height are taken into account. Approximate formulae for the perturbations in all the orbital elements over a revolution of the satellite are given.

052.025 **The use of ground based observations of Halley's comet in the Giotto navigation.**
M. Hechler, F. Hechler.
The need for coordinated ground-based observations of Halley's comet, (see 012.025), p. 57 - 66 (1982).

052.026 **Obtaining geostationary orbit from an off-nominal low transfer orbit.**
M. Rosengren.
J. Spacecr. Rockets, Vol. 19, 80 - 83 (1982). – Abstr. in Phys. Abstr., Vol. 85, Abstr. 89112 (1982).

052.027 **Short period elimination for the tesseral harmonics.**
S. Coffey, K. T. Alfriend.
Astrodynamics 1981, (see 012.027), p. 87 - 101 (1982).
An analytic theory for the motion of an artificial satellite is developed which includes the short period perturbations due to the tesseral harmonics in the gravitational potential of the earth. The J_2 zonal harmonic is considered to be a first order perturbation of the Keplerian problem and the tesseral harmonics to be second order perturbations. The elimination of the short period terms from the Hamiltonian for this problem is accomplished by constructing two canonical transformations of the Lie type.

052.028 **An expression for the nodal period correction introduced by earth oblateness (J_2).**
R. C. Reynolds, S. J. McKenna.
Astrodynamics 1981, (see 012.027), p. 103 - 110 (1982).

052.029 **The first-order short-periodic motion of an artificial satellite due to third-body perturbations.**
M. S. Slutsky, W. D. McClain.
Astrodynamics 1981, (see 012.027), p. 131 (1982). – Abstract.

052.030 **Long-term prediction for high-altitude orbits.**
S. K. Collins, P. J. Cefola.
Astrodynamics 1981, (see 012.027), p. 132 (1982). – Abstract.

052.031 **Interplanetary navigation in the 1980's and 1990's.**
L. J. Wood, J. F. Jordan.
Astrodynamics 1981, (see 012.027), p. 137 - 161 (1982).

Interplanetary space missions of the 1980's and 1990's will require more accurate and more automated navigation than the missions of the 1970's have required. Starting with the Voyager navigation system as a baseline, anticipated navigation requirements and performance for some of these candidate future missions are discussed, along with navigation system designs and strategies. Various developments in navigation technology, required or desirable for these missions are treated.

052.032 **Application of unconnected phase narrowband ΔVLBI to deep space navigation.**
J. Ellis, B. Moultrie, C. B. Smith.
Astrodynamics 1981, (see 012.027), p. 163 - 179 (1982).

052.033 **Interplanetary trajectory optimization.**
L. A. D'Amario, D. V. Byrnes, R. H. Stanford.
Astrodynamics 1981, (see 012.027), p. 203 - 221 (1982).
A procedure for minimizing total impulsive ΔV for constrained multiple-flyby trajectories, which was originally developed for application to satellite tours, has been modified for application to interplanetary trajectories.

052.034 **Some comments on the use of Newton's method for the solution of transcendental equations in astrodynamics.**
F. R. Hoots, R. F. Morris.
Astrodynamics 1981, (see 012.027), p. 531 (1982). – Abstract.

052.035 **Touring the satellites of Saturn.**
R. E. Diehl.
Astrodynamics 1981, (see 012.027), p. 791 - 810 (1982).
A set of possible options is being considered for the next mission to Saturn after the Voyager flybys. Many of these options will include a Saturn orbiter which will provide detailed investigations of the atmosphere, magnetosphere, ring system and satellites of Saturn. The purpose of this paper is to demonstrate the feasibility of designing Saturnian satellite tours which satisfy the science objectives and are consistent with mission requirements and constraints.

052.036 **Time variation of orbital parameters of an equatorial satellite.**
S. K. Gurtu.
Def. Sci. J., Vol. 31, 277 - 284 (1981). – Abstr. in Phys. Abstr., Vol. 85, Abstr. 106473 (1982).

052.037 **Influence of the atmospheric resistence on the uniaxial gravitational orientation of an artificial satellite.**
V. A. Sarychev, V. V. Sazonov.
Kosm. Issled., Tom 20, 659 - 673 (1982). In Russian.

052.038 **Perturbations in the nodal period of artificial satellites caused by the terrestrial infrared radiation pressure.**
V. Mioc, E. Radu.
Astron. Nachr., Band 303, 313 - 315 (1982).
Considering the terrestrial infrared radiation pressure as a disturbing factor, its influence on the nodal period of an artificial satellite moving in a quasi-circular orbit is estimated.

052.039 **A systematic method of generating Galilean satellite-to-satellite transfers for orbiter/lander missions.**
J. K. Soldner, H. Feingold.
Astrodynamics 1981, (see 012.027), p. 223 - 236 (1982).
A Galilean satellite tour design strategy is presented which minimizes the approach velocities at the target satellites. A technique is developed such that once a Hohmann transfer is established between any two adjacent Galilean satellites, transfer trajectories to the remaining Galilean satellites can be derived in a systematic manner.

052.040 **Construction of a consistent semianalytic theory of a planetary or moon orbiter perturbed by a third body.**
E. A. Roth.
Celestial Mech., Vol. 28, (see 012.049), 155 - 169 (1982).
The third body perturbation of an orbiter of a planet or moon is considered. A very convenient form of the Lagrange equations is given allowing an easy derivation of the various terms of the expansion of the perturbed elements. A careful analysis of the order of magnitude of these terms indicates which ones are required for a consistent theory. It follows that in many practical cases the main term of the disturbing function has to be carried to the second order of the perturbation theory.

052.041 **Comments on Aksnes' intermediary.**
A. Deprit, D. L. Richardson.
Celestial Mech., Vol. 28, 253 - 273 (1982).
The Hamiltonian offered by Aksnes as an intermediary to the main problem in the theory of an artificial satellite is proved to be separable in spherical coordinates. But it is not of Staeckel type, and its latitude equation cannot be integrated by a quadrature. To circumvent the difficulty, a new concept of intermediary is proposed. From this standpoint, Aksnes' theory is revised (1) to simplify a possible automation of Aksnes' literal developments and (2) to bring forth the connection with the radial intermediary based on a preparatory elimination of the parallax from the main problem.

052.042 **Mission techniques and orbital mechanics for interplanetary spacecraft.**
A. Perret.
The solar system and its exploration, (see 012.054), p. 13 - 22 (1981).
A few important problems concerning planetary missions are discussed. After presenting general equations of spacecraft motion, the author treats the interplanetary rendez-vouz problem in a simplified force model. The principle of the use of gravitational perturbations during a fly-by of a celestial body, the gravity assist, is exposed and some examples with Venus, the Earth, Mars and Jupiter are proposed. The principle of orbit determination by the use of range and range-rate measurements is described. Finally a few elements concerning propulsion means and attitude control are given.

052.043 **Review of Ariane capabilities for interplanetary missions.**
J.-B. Boissard, W. Flury, G. Janin, S. Vaghi.
ESA J., Vol. 5, 1 - 13 (1981).

052.044 **Orbit determination methods used in the Satellite Geodetic Observatory.**
I. Almár, S. Mihály, T. Borza, J. Ádám.
Dynamical astronomy, (see 012.056), p. 135 - 136 (1982).
Abstract.

052.045 **Effect of upper atmospheric variations on satellite lifetimes.**
E. M. Both.
Dynamical astronomy, (see 012.056), p. 137 - 139 (1982).
Abstract.

052.046 **On the problem of optimization in two gravitational fields.**
N. A. Pronyushkina.
Probl. mekh. upravlyaem. dvizheniya. Nelinejn. dinam. sistemy. Perm', 1982, p. 151 - 156. In Russian. – Abstr. in Ref. zh., 62. Issled. kosm. prostranstva, 11.62.131 (1982).

052.047 **On the motion of a solar orbital electric station in a geosynchronous orbit.**
N. D. Dzhumanaliev, Sh. Zh. Zhumadinov, M. I. Kiselev.
Probl. mekh. upravlyaem. dvizheniya. Nelinejn. dinam sistemy. Perm', 1982, p. 68 - 71. In Russian. – Abstr. in Ref. zh., 62. Issled. kosm. prostranstva, 11.62.132 (1982).

052.048 **On a non-traditional method of determination of the elements of orbits of unknown cosmic objects based on review photograph observations processed by a computer.**
V. I. Kuryshev, N. I. Perov.
Astron. Zh., Tom 59, 1212 - 1217 (1982). In Russian. English translation in Soviet Astron., Vol. 26, No. 6.

A method of computing Kepler elements of orbits of artificial satellites from five positions is under consideration. A distinguished feature of this method is the absence of time power series in its formulae.

052.049 **Application of an extended Kalman filter to the precise prediction of satellite orbits.**
M.-f. He.
Acta Astron. Sinica, Vol. 23, 163 - 174 (1982). In Chinese.

052.050 **Combined perturbations of the orbit of a close earth satellite by considering the variance of the drag coefficient.**
C.-y. Xia, Y.-y. Wu, L. Liu.
Acta Astron. Sinica, Vol. 23, 175 - 184 (1982). In Chinese.

052.051 **The air drag perturbation of a satellite in an orbit with small eccentricity.**
B.-k. Lu, C.-a. Sun.
Acta Astron. Sinica, Vol. 23, 264 - 273 (1982). In Chinese.

Using the analytical model of air density proposed by the authors of this paper, new formulae for computing the air drag perturbation of an artificial earth satellite have been deduced. They are suitable for orbits with small eccentricity as $0 \leqslant e \leqslant 0.3$.

052.052 **Unsteady aerodynamics and motions of the Pioneer Venus probes.**
A. Seiff, C. E. DeRose, V. U. Muirhead.
J. Spacecr. Rockets, Vol. 19, 404 - 411 (1982). – Abstr. in Phys. Abstr., Vol. 86, Abstr. 15813 (1983).

052.053 **Comparison of trajectories of the mother and daughter satellites Interkosmos 18 and Magion.**
P. Lala.
Acta Astronaut., Vol. 9, (see 012.085), 425 - 430 (1982). Abstr. in Phys. Abstr., Vol. 86, Abstr. 20198 (1983).

Review of the scientific usage of INTERCOSMOS satellite observations at the Astronomical Institute of the Czechoslovak Academy of Sciences.
See Abstr. 013.049.

Preliminary results of the determination of orientation of Intercosmos 17.
See Abstr. 054.001.

A readjustment of lumped coefficients from inclination of 1974 - 70A.
See Abstr. 081.018.

Determination of satellite drag coefficient from the orbital analysis of the ANS satellite (1974 - 70A).
See Abstr. 082.035.

053 Lunar and Planetary Probes and Satellites

053.001 **Venera 13 and Venera 14.**
R. Z. Sagdeev, V. I. Moroz.
Pis'ma Astron. Zh., Tom 8, 387 - 390 (1982). In Russian. English translation in Soviet Astron. Lett., Vol. 8.
Some general data concerning lander missions of Venera 13 and Venera 14 and the list of scientific experiments are presented.

053.002 **Russische sondes op Venus.**
N. Giesinger.
Zenit, 9. Jaarg., 404 - 408 (1982).

053.003 **Voyager 1 Saturn targeting strategy.**
R. J. Cesarone.
J. Spacecr. Rockets, Vol. 19, 72 - 79 (1982). – Abstr. in Phys. Abstr., Vol. 85, Abstr. 89111 (1982).

053.004 **Voyager-2: a grand tour of the giant planets.**
A. B. Sergeyevsky.
Astrodynamics 1981, (see 012.027), p. 769 - 790 (1982).
The design of the current Voyager-2 multiple gravity assist mission, which encounters Jupiter, Saturn, Uranus and Neptune (JSUN) is described, presenting the historical background of the mission, a review of available Uranus and Neptune encounter options, their characteristics and constraints, and a summary of the trajectory selection process.

053.005 **Venusian probes subsonic drag determination from flight data.**
R. C. Blanchard, W. P. Phillips, G. M. Kelly, J. T. Findlay.
J. Spacecr. Rockets, Vol. 19, 139 - 144 (1982). – Abstr. in Phys. Abstr., Vol. 85, Abstr. 94577 (1982).

Planetary vehicles.
See Abstr. 003.067.

054 Artificial Earth Satellites

054.001 **Preliminary results of the determination of orientation of Intercosmos 17.**
M. L. Pivovarov.
Obrab. inf., poluch. po programme Interkosmos, Moskva, 1982, p. 139 - 145. In Russian. – Abstr. in Ref. zh., 62. Issled. kosm. prostranstva, 7.62.158 (1982).

054.002 **S. P. Korolev and the first artificial satellite of the earth.**
G. A. Skuridin.
Zemlya Vselennaya, 1982, No. 5, p. 57 - 61. In Russian.

054.003 **The International Ultraviolet Explorer.**
A. Boggess.
Proc. SPIE Int. Soc. Opt. Eng., Vol. 279, (see 012.072), 168 - 175 (1981). – Abstr. in Phys. Abstr., Vol. 85, Abstr. 117591 (1982).

054.004 **Satellite digest – 157 - 159.**
Compiled by R. D. Christy.
Spaceflight, Vol. 24, 370 - 372, 417 - 418, 469 - 470 (1982).

Providing a data base with results of primary reduction of information from the Interkosmos 13 satellite.
See Abstr. 002.045.

Visual observations of geostationary satellites.
See Abstr. 031.521.

Theoretical Astrophysics

061 General Aspects (Nucleosynthesis, Neutrino Astronomy, etc.)

061.001 **Gravitational instability: the second approach of the linear theory of long wavelengths.**
Ya. B. Zel'dovich.
Astron. Zh., Tom 59, 636 - 638 (1982). In Russian. English translation in Soviet Astron., Vol. 25, No. 4.
An analytic expression is constructed to describe an increasing mode of the linear theory of gravitational instability. It takes into account a pressure gradient occurring in a perturbed medium and being a small correction at wavelengths much longer than the Jeans wavelength.

061.002 **Production of ^{9}Be induced by neutrinos generated in gravitational collapse of stars.**
G. V. Domogatskij, S. V. Imshennik.
Pis'ma Astron. Zh., Tom 8, 351 - 357 (1982). In Russian. English translation in Soviet Astron. Lett., Vol. 8.
The process of ^{9}Be production in a stellar envelope induced by neutrinos generated in gravitational collapse of its central core is studied. The physical conditions are found which lead to concentrations of ^{9}Be in an expelled stellar envelope sufficient to account for the observed mean abundance of the isotope in the Galaxy.

061.003 **Cosmological neutrino background as a possible source of energy in celestial bodies.**
A. Sapar.
Publ. Tartu Astrofiz. Obs., Tom 48, 271 - 279 (1981). In Russian.
Based on the model of a hot universe and taking into account the fact that there exists a neutrino background to which, according to Lyubimov and his associates, the rest mass of 5×10^{-32} g is ascribed, it has been found that there exists a background of almost degenerated neutrinos in the universe as well as in galaxies. This background is centred around celestial bodies and their systems, filling nearly all the energy states up to the Fermi level. If there exists such a background of neutrinos with the above non-zero rest mass then the background contributes an essential input into the energy balance of white dwarfs, planets, quasars, pulsars and galactic nuclei.

061.004 **Is there a local source of magnetic monopoles?**
S. Dimopoulos, S. L. Glashow, E. M. Purcell, F. Wilczek.
Nature, Vol. 298, 824 - 825 (1982).
The authors conclude that the Cabrera flux of magnetic monopoles can be understood in terms of an orbiting cloud of monopoles in the solar system, which is replenished by a small "monopole wind" from the sun or from the Galaxy. Monopoles may have an important role in solar energetics.

061.005 **Neutrons on the earth and in the universe.**
N. A. Vlasov.
Atom. ehnergiya, Tom 52, 87 - 91 (1982). In Russian. – Abstr. in Ref. zh., 51. Astron., 7.51.208 (1982).

061.006 **Die Entstehung der schweren chemischen Elemente.**
W. Hillebrandt.
Phys. Bl., Vol. 38, 189 - 192 (1982).

061.007 **Neutrinos of mean energies in the universe.**
G. S. Bisnovatyj-Kogan, Z. F. Seidov.
Inst. kosm. issled. AN SSSR. Prepr., 1981, No. 675. 26 pp. In Russian. – Abstr. in Ref. zh., 51. Astron., 8.51.847 (1982).

061.008 **Ca-Ti-Cr anomalies in an Allende inclusion and the nβ-process.**
D. G. Sandler, S. E. Koonin, W. A. Fowler.
Astrophys. J., Vol. 259, 908 - 919 (1982).
A neutron-capture/beta-decay (nβ) process has been studied. The seed nuclei consisted of all isotopes of the elements from Si to Cr with normal solar system abundances. When this process operates at neutron densities of $\approx 10^{-7}$ mole cm^{-3}, temperatures of ≈ 30 keV, and exposure times of $\approx 10^3$ s, small admixtures ($\lesssim 10^{-4}$) of the exotic material produced are sufficient to account for most of the Ca and Ti isotopic anomalies found in the Allende meteorite inclusion EK-1-4-1.

061.009 **Neutrinos in the universe.**
G. S. Bisnovatyj-Kogan.
Priroda, 1981, No. 10, p. 26 - 31. In Russian.

061.010 **Low-energy neutrinos in astrophysics.**
W. Hampel.
Neutrino physics and astrophysics, (see 012.010), p. 61 - 79 (1982).
This review discusses the current status of solar neutrino research and the prospects for the detection of collapsing star neutrinos.

061.011 **Supernovae induced by axion-like particles.**
D. N. Schramm, J. R. Wilson.
Astrophys. J., Vol. 260, 868 - 874 (1982).
It is shown that a new type of particle which may have been seen in a recent accelerator experiment may, if truly present, provide a mechanism whereby gravitationally collapsing massive stars may eject their outer mantles and envelopes in supernova explosions of $\sim 10^{51}$ ergs while leaving the cores to form neutron star remnants. These particles are "axion-like", which means they interact semi-weakly, decay to two photons with lifetimes $\sim 10^{-3}$ s, and have masses $0.15 \leqslant M_a \lesssim 1$ MeV.

061.012 **Massive neutrino halos.**
J. L. Zhang, W. Y. Chau, K. Lake.
Bull. American Astron. Soc., Vol. 14, 631 (1982). – Abstract.

061.013 **M/L constraints on the mass of the neutrino.**
B. M. Lewis.
Bull. American Astron. Soc., Vol. 14, 631 (1982). – Abstract.

061.014 **Massive neutrinos?**
A. De Rujula.
Nucl. Phys. A, Vol. A374, p. 619c - 33 (1982). – Abstr. in Phys. Abstr., Vol. 85, Abstr. 59077 (1982).

061.015 **Counting neutrinos in p$\bar{p}$ collider experiments.**
I. H. Dunbar.

Nucl. Phys., B, Vol. B197, 189 - 200 (1982). – Abstr. in Phys. Abstr., Abstr. 59193 (1982).

061.016 **Neutrinos in the Kerr and Robertson-Walker geometries.**
S. V. Dhurandhar, C. V. Vishveshwara, J. M. Cohen.
J. Phys. A, Vol. 15, 1643 - 1652 (1982). – Abstr. in Phys. Abstr., Vol. 85, Abstr. 63872 (1982).

061.017 **$\nu\bar{\nu}\rightarrow\gamma\gamma$ with massive neutrinos.**
R. J. Crewther, J. Finjord, P. Minkowski.
Nukleonika, Vol. 26, 311 - 316 (1981). – Abstr. in Phys. Abstr., Vol. 85, Abstr. 64315 (1982).

061.018 **Is parity violation a cosmological evolution effect? .**
M. Novello, A. Santoro, H. Heintzmann.
Phys. Lett. A, Vol. 89A, 266 - 268 (1982). – Abstr. in Phys. Abstr., Vol. 85, Abstr. 73835 (1982).

061.019 **Massless neutrinos and the stellar stopping power via the V-A and Weinberg interactions.**
P. T. Leung, M. L. Rustgi.
Phys. Rev. A, Vol. 25, 2588 - 2591 (1982). – Abstr. in Phys. Abstr., Vol. 85, Abstr. 73836 (1982).

061.020 **Decaying neutrinos as a photoionization source in galactic halos.**
D. W. Sciama, A. L. Melott.
Phys. Rev. D, Vol. 25, 2214 - 2217 (1982). – Abstr. in Phys. Abstr., Vol. 85, Abstr. 74066 (1982).

061.021 **Neutrino astronomy.**
D. N. Schramm.
Ninth Texas Symposium on Relativistic Astrophysics, (see 012.020), p. 380 - 388 (1980).

061.022 **Possible time variations in ^{37}Cl solar-neutrino data and neutrino oscillations.**
R. Ehrlich.
Phys. Rev. D, Vol. 25, 2282 - 2286 (1982). – Abstr. in Phys. Abstr., Vol. 85, Abstr. 74679 (1982).

061.023 **Massive neutrinos.**
P. H. Frampton, P. Vogel.
Phys. Rep., Vol. 82, 339 - 357 (1982). – Abstr. in Phys. Abstr., Vol. 85, Abstr. 74680 (1982).

061.024 **Massive neutrinos. II. Searches for neutrino masses and oscillations.**
P. Vogel.
Phys. Rep., Vol. 82, 358 - 388 (1982). – Abstr. in Phys. Abstr., Vol. 85, Abstr. 74681 (1982).

061.025 **Neutrino masses, intermediate mass scales, and cosmology.**
A. Masiero, M. Roncadelli.
Phys. Rev. D, Vol. 25, 2612 - 2619 (1982). – Abstr. in Phys. Abstr., Vol. 85, Abstr. 84825 (1982).

061.026 **Implications of photons from massive neutrinos.**
D. Sciama.
Observatory, Vol. 102, 175 - 176 (1982). – Abstract.

061.027 **Neutrino bursts from collapsing stars – results from the Homestake burst search.**
M. L. Cherry, M. Deakyne, T. Daily, K. Lande, C. K. Lee, R. I. Steinberg, E. J. Fenyves.
J. Phys. G, Vol. 8, 879 - 885 (1982). – Abstr. in Phys. Abstr., Vol. 85, Abstr. 89099 (1982).

061.028 **Cosmic turbulence and the angular momenta of astronomical systems.**
R. C. Fleck, Jr.
Astrophys. J., Vol. 261, 631 - 635 (1982).

It is suggested that gravitationally bound astronomical systems ranging from asteroids to galaxy superclusters may derive their rotation from a hierarchy of cosmic turbulence, thus explaining the empirical specific angular momentum-mass relationship $j \sim M^{3/4}$ exhibited by these systems. It is shown that many of the properties of these systems can be accounted for if astronomical objects form in a turbulent environment.

061.029 **Neutrino emissivities and mean free paths of degenerate quark matter.**
N. Iwamoto.
Ann. Physics, Vol. 141, 1 - 49 (1982). – Abstr. in Phys. Abstr., Vol. 85, Abstr. 89932 (1982).

061.030 **^{176}Lu: cosmic clock *and* stellar thermometer.**
H. Beer.
Astrophys. J., Vol. 262, 739 - 741 (1982).

The s-process nucleosynthesis of the isobar pair ^{176}Lu/^{176}Hf, a long-lived cosmic clock, is examined. Expressions are derived which allow to calculate the s-process age from measured quantities alone. A formula is specified which does not contain the population probability of the 3.68 hr isomeric state in ^{176}Lu. This allows to decide whether ^{176}Lu is, in addition to a cosmic clock, also a stellar thermometer.

061.031 **Massive neutrinos and ultra-violet astronomy.**
D. W. Sciama.
Progress in cosmology, (see 012.047), p. 75 - 85 (1982).

Two astronomical ionisation processes are tentatively attributed to photons emitted by massive neutrinos. In the first, the intergalactic medium is assumed to be ionised by photons emitted by a cosmological distribution of neutrinos. The observations then require a neutrino mass lying between 50 and 110 eV and a radiative lifetime of order 10^{27} seconds. These values are compatible with cosmological constraints and with particle physics expectations. The second ionisation process involves the production of C IV and Si IV high up in the galactic halo. This is attributed to photons emitted by neutrinos dominating the halo. It would require a neutrino mass exceeding 96 eV and a lifetime of order 10^{27} seconds. These values are compatible with those required by the intergalactic hypothesis.

061.032 **Nuclear forces and the properties of matter at high temperature and density.**
M. Rayet, M. Arnould, F. Tondeur, G. Paulus.
Astron. Astrophys., Vol. 116, 183 - 187 (1982).

The authors present two Skyrme-type forces which are particularly well suited for the description of presupernova core or matter in a nascent neutron star. They are compared to other forces currently used in this field, with regard to finite nuclei and infinite matter properties, and to the coexistence of nuclei in a hot and dense nucleon gas.

061.033 **Elemental and nuclidic abundances in the solar system.**
A. G. W. Cameron.
Essays in nuclear astrophysics, (see 003.013), p. 23 - 43 (1982).

A new table of abundances of the elements and their isotopes is presented, which is intended to represent abundances of material in the solar system, specifically in the primitive solar nebula. The primary source of abundance information is the C1 carbonaceous meteorites. The Sun has been employed for the volatile gases.

061.034 **A comparison of stellar observations and the theory of nucleosynthesis.**
J. L. Greenstein.
Essays in nuclear astrophysics, (see 003.013), p. 45 - 63

(1982).

A review of the present state of the theory of nucleosynthesis in stars is given. Predictions of the theory are compared with element abundances in stellar atmospheres. Isotope ratios determined from molecular bands, isotopes resulting from the CNO cycle and oxygen anomalies are discussed in some detail.

061.035 **Solar power.**
R. W. Kavanagh.
Essays in nuclear astrophysics, (see 003.013), p. 159 - 170 (1982).

The present state of knowledge of the most significant nuclear data involved in the chains of hydrogen burning in the Sun is critically assessed.

061.036 **Hydrogen burning in massive stars.**
W. S. Rodney, C. Rolfs.
Essays in nuclear astrophysics, (see 003.013), p. 171 - 191 (1982).

A review is given of recent work on the hydrogen-burning cycles of energy-producing nuclear reactions in massive stars.

061.037 **Nuclear reactions in highly evolved stars.**
C. A. Barnes.
Essays in nuclear astrophysics, (see 003.013), p. 193 - 232 (1982).

This paper reviews the laboratory determination of nuclear reaction rates important in helium, carbon, neon, oxygen, and silicon burning.

061.038 **The rate of the $^{12}C(\alpha, \gamma)^{16}O$ reaction in stellar helium burning.**
T. A. Tombrello, S. E. Koonin, B. A. Flanders.
Essays in nuclear astrophysics, (see 003.013), p. 233 - 241 (1982).

The problems involved in obtaining the $^{12}C(\alpha, \gamma)^{16}O$ cross section in the temperature range appropriate for stellar helium burning are reviewed.

061.039 **^3He, red dwarfs, and our next trillion years.**
E. E. Salpeter.
Essays in nuclear astrophysics, (see 003.013), p. 287 - 299 (1982).

The author discusses the production of ^3He in red dwarfs and its connection with flare activity in these stars.

061.040 **The s-process.**
R. K. Ulrich.
Essays in nuclear astrophysics, (see 003.013), p. 301 - 323 (1982).

Nuclear data important for the s-process synthesis of heavy elements in stellar interiors are critically reviewed.

061.041 **The r-process and nucleocosmochronology.**
D. N. Schramm.
Essays in nuclear astrophysics, (see 003.013), p. 325 - 353 (1982).

The current understanding of the role of r-processes in the formation of neutron-rich heavy elements at various astrophysical sites is reviewed. Applications of the process in long- and short-lived nucleochronometric dating are discussed.

061.042 **The origin of the light elements.**
J. Audouze, H. Reeves.
Essays in nuclear astrophysics, (see 003.013), p. 355 - 375 (1982).

The authors review the situation concerning the observed abundances of the light elements, Li, Be and B, the physics of the spallation reactions, and the nucleosynthesis induced by the observed galactic cosmic ray fluxes. They then concentrate on the nucleosynthesis of ^7Li, which leads to interesting aspects of different questions including the effect of the low-energy galactic cosmic ray fluxes that are strongly modulated by the solar wind and magnetic activity, red giant nucleosynthesis, novae outbursts, and especially the big bang.

061.043 **Cosmic radioactivity: a gamma-ray search for the origins of atomic nuclei.**
D. D. Clayton.
Essays in nuclear astrophysics, (see 003.013), p. 401 - 426 (1982).

The present state of the physics of explosive nucleosynthesis during supernova outbursts is reviewed. The formation of radioactive heavy nuclei and the possibility to detect gamma-ray line radiation from their decay is analysed in detail.

061.044 **The thermal pair annihilation spectrum: a detailed balance approach.**
R. Svensson.
ESO Sci. Prepr., No. 216, 17 pp. (1982). – Submitted to Astrophys. J.

061.045 **Light element nucleosynthesis.**
J. Audouze.
Pré-publ. Inst. Astrophys. Paris, No. 13, 6 pp. (1982). Invited contribution presented at the colloquium "The neutron and its application", Cambridge, 13 - 17 September 1982.

061.046 **Nucleosynthesis and chemical evolution of galaxies.**
J. Audouze.
Pré-publ. Inst. Astrophys. Paris, No. 23, 46 pp. (1982). Paper presented at Cargèse Summer School 1982.

061.047 **Warm CNO nucleosynthesis as a possible enrichment mechanism for oxygen and fluorine isotopes.**
M. Wiescher, K.-U. Kettner.
Astrophys. J., Vol. 263, 891 - 901 (1982).

New rates for proton capture reactions on the radioactive nuclei ^{17}F and ^{18}F were calculated and compared with previous ones. The influence of these new rates as well as new experimentally determined rates for proton capture on ^{18}O on isotopic abundances in mass range $A = 17–19$ are discussed in terms of reaction-network calculations.

061.048 **Periodic system and cosmogeochemical element distribution.**
Yu. G. Shcherbakov.
Geol. i geofiz., 1982, No. 1, p. 77 - 84. In Russian. – Abstr. in Ref. zh., 51. Astron., 10.51.174 (1982).

061.049 **Lepton structure and astrophysical experiment.**
V. F. Dushenko.
Izv. AN MSSR. Ser. fiz.-tekh. i mat. nauk, 1982, No. 1, p. 62 - 64. In Russian. – Abstr. in Ref. zh., 51. Astron., 11.51.140 (1982).

061.050 **Nuclear cosmochronology or the age of the elements.**
J. F. Minster.
Formation of planetary systems, (see 012.064), p. 365 - 383 (1982).

Nuclear cosmochronology consists in calculating the rate and duration of the nucleosynthesis which contributed the protosolar nebula from the abundance of the radionuclides present in the early solar system. The various methods and assumptions and the data are presented, which give an age for the Galaxy of 10 to 22 Ga. The "interval of time" between the last contamination of the protosolar nebula by freshly nucleosynthetized elements and the formation of the solar system is of the order of 100 Ma, although some fractions of the solar nebula must have been added less than 10 Ma before this formation.

061.051 **Astrophysical consequences of neutron-antineutron oscillations.**
V. Krishan, C. Sivaram.
Astrophys. Space Sci., Vol. 87, 205 - 210 (1982).
Grand unified theories predict baryon number violating interactions and one of the implications of this is the possible existence of neutron-antineutron oscillations. The neutron-antineutron oscillations have been considered in the neutron rich astrophysical sources such as solar flares, supernovae explosions, neutron stars and the nucleosynthetic phase of the early universe in order to partly account for the antiproton flux of the cosmic rays at low energies and the γ-ray emission, at GeV energies. Low magnetic fields and high neutron concentrations provide the right environment for the production of antineutrons and hence antiprotons and GeV γ rays.

061.052 **Comment on "Does the standard hot-big-bang model explain the primordial abundances of helium and deuterium?"**
P. Langacker, G. Segrè.
Phys. Rev. Lett., Vol. 49, 1363 (1982). – For reference see Abstr. 31.061.055.

061.053 **Magnetic monopoles and the survival of galactic magnetic fields.**
M. S. Turner, E. N. Parker, T. J. Bogdan.
Phys. Rev. D, Vol. 26, 1296 - 1305 (1982). – Abstr. in Phys. Abstr., Vol. 85, Abstr. 117584 (1982).

061.054 **Two- and three-body contributions of cosmological monopole annihilation.**
D. A. Dicus, D. N. Page, V. L. Teplitz.
Phys. Rev. D, Vol. 26, 1306 - 1316 (1982). – Abstr. in Phys. Abstr., Vol. 85, Abstr. 117585 (1982).

061.055 **Nuclei in neutrino-degenerate dense matter. I. Cold case.**
R. Ogasawara, K. Sato.
Prog. Theor. Phys., Vol. 68, 222 - 235 (1982). – Abstr. in Phys. Abstr., Vol. 85, Abstr. 117687 (1982).

061.056 **On the neutrino mass, the abundance of the right-handed neutrino and the closure of the Universe.**
C.-r. Ching, Y.-s. Wu, T.-h. Ho, C.-h. Chang, Z.-l. Zou.
Commun. Theor. Phys., Vol. 1, 221 - 227 (1982). – Abstr. in Phys. Abstr., Vol. 85, Abstr. 117819 (1982).

061.057 **Transport of neutrinos, radiation and energetic particles in accretion flows.**
R. Cowsik, M. A. Lee.
Proc. R. Soc. London, Ser. A, Vol. 383, 409 - 437 (1982). Abstr. in Phys. Abstr., Vol. 86, Abstr. 141 (1983).

061.058 **Constraints on light axions.**
A. Barroso, G. C. Branco.
Phys. Lett. B, Vol. 116B, 247 - 250 (1982). – Abstr. in Phys. Abstr., Vol. 86, Abstr. 582 (1983).

061.059 **Stellar nucleosynthesis and (n, γ) reactions.**
J. Audouze.
Neutron-capture gamma-ray spectroscopy and related topics 1981, (see 012.077), p. 551 - 557 (1982). – Abstr. in Phys. Abstr., Vol. 86, Abstr. 792 (1983).

061.060 **Population of isomeric states by keV neutron capture: clues to stellar s-process nucleosynthesis.**
H. Beer, F. Kappeler.
Neutron-capture gamma-ray spectroscopy and related topics 1981, (see 012.077), p. 558 - 572 (1982). – Abstr. in Phys. Abstr., Vol. 86, Abstr. 793 (1983).

061.061 **^{178,179,180}Hf and ^{180}Ta(n, γ) cross sections and their contribution to stellar nucleosynthesis.**
H. Beer, R. L. Macklin.
Phys. Rev. C, Vol. 26, 1404 - 1416 (1982). – Abstr. in Phys. Abstr., Vol. 86, Abstr. 6234 (1983).

061.062 **Astrophysical constraints on a new light axion and other weakly interacting particles.**
M. Fukugita, S. Watamura, M. Yoshimura.
Phys. Rev. D, Vol. 26, 1840 - 1853 (1982). – Abstr. in Phys. Abstr., Vol. 86, Abstr. 10204 (1983).

061.063 **Statistical calculation of the ^{8}B solar neutrino capture cross sections for ^{97}Mo and ^{98}Mo.**
N. Itoh, Y. Kohyama.
Prog. Theor. Phys., Vol. 68, 677 - 679 (1982). – Abstr. in Phys. Abstr., Vol. 86, Abstr. 15793 (1983).

061.064 **The annihilation process $\nu\bar{\nu} \to \gamma\gamma$ with massive neutrinos in cosmology.**
R. J. Crewther, J. Finjord, P. Minkowski.
Nucl. Phys. B. Part Phys., Vol. B207, 269 - 287 (1982). Abstr. in Phys. Abstr., Vol. 86, Abstr. 16630 (1983).

061.065 **New limits on the density of superheavy magnetic monopoles in the interstellar space.**
A. K. Drukier.
Monopoles in quantum field theory, (see 012.088), p. 407 - 409 (1982). - Abstr. in Phys. Abstr., Vol. 86, Abstr. 20238 (1983).

Nucleosynthesis in galaxies.
See Abstr. 004.065.

17th International Cosmic Ray Conference. HE session: high energy interactions; MN session: muons and neutrinos.
See Abstr. 012.070.

Effective charges of ions and the stopping power of dense media.
See Abstr. 022.062.

Cross sections and thermonuclear reaction rates for ^{42}Ca(p, γ)^{43}Sc, ^{44}Ca(p, γ)^{45}Sc, ^{44}Ca(p, n)^{44}Sc and ^{45}Sc(p, n)^{45}Ti.
See Abstr. 022.073.

Nucleosynthesis in two 25 $M_\odot$ stars of different population.
See Abstr. 065.091.

Neutrinos in gravitational collapse: the Dirac formalism.
See Abstr. 066.050.

An account of the development of the solar neutrino problem.
See Abstr. 080.042.

Some short-lived nuclides in the early solar system – a connection with the placental ISM.
See Abstr. 105.077.

Meteorite actinide chemistry and cosmochronology.
See Abstr. 105.078.

Mean-field calculations of the equation of state of supernova matter II.
See Abstr. 125.012.

Primordial synthesis of ^{6}Li and the rate of $\alpha(d, \gamma)^6$Li.
See Abstr. 162.030.

Pairing in the cosmic neutrino background.
See Abstr. 162.063.

A model for neutrino galaxies.
See Abstr. 162.078.

Cosmological impact of the neutrino rest mass.
See Abstr. 162.090.

The early universe and clustering of the relic neutrinos.
See Abstr. 162.091.

Particle physics in the very early universe.
See Abstr. 162.092.

Grand unification and possible matter-antimatter domain structure in the universe.
See Abstr. 162.093.

Massive neutrinos and primordial nucleosynthesis.
See Abstr. 162.109.

Cosmology and neutrino physics.
See Abstr. 162.114.

Massive neutrino halos in an expanding universe.
See Abstr. 162.133.

Galactic neutrino models.
See Abstr. 162.135.

The rest mass of neutrinos and clustering in the early Universe.
See Abstr. 162.143.

Neutrino halos.
See Abstr. 162.155.

Cosmology and particle physics.
See Abstr. 162.184.

Abundances of elements of cosmological interest.
See Abstr. 162.194.

Free and bound quarks – accelerators and the early universe.
See Abstr. 162.206.

Particle physics and cosmology.
See Abstr. 162.207.

Nucleosynthesis in bouncing cosmologies.
See Abstr. 162.245.

Particle physics meets cosmology – the search for decaying neutrinos.
See Abstr. 162.295.

Neutrinos in perfect fluid spacetimes with local rotational symmetry.
See Abstr. 162.320.

Primordial nucleosynthesis including radiative, Coulomb, and finite-temperature corrections to weak rates.
See Abstr. 162.323.

062 Hydrodynamics, Magnetohydrodynamics, Plasma

062.001 **Self-similar solutions for the interaction of stellar ejecta with an external medium.**
R. A. Chevalier.
Astrophys. J., Vol. 258, 790 - 797 (1982).
The interaction of a uniformly expanding gas with a stationary ambient medium is studied. In the case where the density profile of the ejected matter is a steep power law with radius and that of the ambient medium is a shallow power law with radius, the structure of the interaction region tends toward a self-similar solution. It is the purpose of this paper to describe the properties of the self-similar solutions for a variety of power law exponents. Results are applied to two models describing the evolution of type I and type II supernova remnants.

062.002 **Transient and transport coefficients for radiative fluids.**
M. A. Schweizer.
Astrophys. J., Vol. 258, 798 - 811 (1982).
The generalized hyperbolic hydrodynamical equations of motion, proposed for general fluids by Israel and Stewart, are derived for radiative fluids. This provides the transient coefficients (relaxation terms) and the general form of the transport coefficients. This approach is based on a Grad 9-moment approximation of the Boltzmann equation.

062.003 **Invariant structures of a magnetic flux rope.**
A. A. Solov'ev.
Astrofizika, Tom 18, 300 - 310 (1982). In Russian. English translation in Astrophysics, Vol. 18, No. 2.
The basic properties of a screened magnetic flux rope with finite radius of the curvature are discussed. An analysis of the equilibrium conditions for the rope has been given in terms of the average values. The variation of the magnetic field structure in the rope caused by the change of its geometrical sizes has been studied. The invariant structure was found so that the transition of the rope from one equilibrium state to another is followed by variation of the length and pitch of the helix in proportion with the radius of the cross-section.

062.004 **Blowing up of two-dimensional magnetohydrostatic equilibria by an increase of electric current or pressure.**
J. Heyvaerts, J. M. Lasry, M. Schatzman, P. Witomsky.
Astron. Astrophys., Vol. 111, 104 - 112 (1982).
General properties of series of two-dimensional magnetohydrostatic configurations, described by the equation $-\Delta A = \lambda f(A)$ are deduced when $f(A)$ is positive, bounded, and has compact support. An existence theorem for the Dirichlet boundary conditions on the frontier of a semi-infinite domain is established, and the magnetic topology of the always existing solution is shown to be of the open type. On the other hand the existence of a closed topology solution is proved for small currents, but it is also shown that this solution disappears at a turning point for some finite value of λ. The relevance of these results for eruptive solar flares and other solar active phenomena is discussed.

062.005 **Transport of energetic electrons in a fully ionized hydrogen plasma.**
T. Bai.
Astrophys. J., Vol. 259, 341 - 349 (1982).
In order to study the behavior of energetic electrons in astrophysical plasmas, the author derives relationships among the Coulomb energy loss, travel distance, and pitch angle deflection due to Coulomb collisions, which hold when the Coulomb energy loss is only a small fraction of the initial energy. He develops a Monte Carlo method of calculating how the pitch angle and spatial distributions of the energetic electrons change in a uniformly magnetized plasma as these electrons lose energy by Coulomb collisions. This method is applied to a beam of monoenergetic electrons released along the magnetic field lines.

062.006 **Propagation of waves in an atmosphere in the presence of a magnetic field. V. The theory of magneto-acoustic-gravity oscillations.**
B. Leroy, S. J. Schwartz.
Astron. Astrophys., Vol. 112, 84 - 92 (1982).
The authors present the theoretical details of oscillations in an exponential atmosphere permeated by a uniform, vertical magnetic field in the non-WKBJ limit. In addition to the Alfvén mode, they find the inhomogeneous analogs of the fast and slow MHD waves which propagate in a homogeneous medium. The authors discuss the boundary conditions and energy flux calculations which are appropriate for these modes, as well as some general considerations of the series solutions themselves.

062.007 **Propagation of waves in an atmosphere in the presence of a magnetic field. VI. Application of magneto-acoustic-gravity mode theory to the solar atmosphere.**
S. J. Schwartz, B. Leroy.
Astron. Astrophys., Vol. 112, 93 - 103 (1982).
The theory of magneto-acoustic-gravity oscillations in an exponential atmosphere permeated by a uniform vertical magnetic field, presented by the authors in a companion paper, is applied to the solar atmosphere. Numerical parameters typical of both sunspots and coronal hole regions are used in the investigations over a wide frequency range of the energy flux carried by such oscillations from the photosphere to the corona. The authors conclude that the combined effects of stratification, partially transverse propagation and the transition region make it unlikely that such modes dominate coronal energetics.

062.008 **Magnetic Prandtl number and astrophysical convection.**
J. O. Murphy, J. M. Lopez.
Proc. Astron. Soc. Australia, Vol. 4, 208 - 209 (1981).

062.009 **A cyclotron theory for the beaming pattern of Jupiter's decametric radio emission.**
R. G. Hewitt, D. B. Melrose, K. G. Rönnmark.
Proc. Astron. Soc. Australia, Vol. 4, 221 - 226 (1981).
The authors are developing a general theory for cyclotron instabilities, i.e. one which includes all plasma effects and all relativistic effects. In this paper they discuss some aspects of the theory from a qualitative viewpoint. Their main purpose is to show that when applied to Jupiter's decametric radio emission (DAM) the theory leads naturally to a prediction of emission concentrated on a thin hollow cone. This makes the theory a particularly attractive candidate for the interpretation of DAM.

062.010 **Nonlinear shear instabilities in an infinite slab.**
M. Nepveu.
Astron. Astrophys., Vol. 112, 223 - 228 (1982).
The dynamical evolution of an infinite slab moving in denser and noisy (turbulent) surroundings is investgated with a $2D$ hydrodynamic code. The applicability of the results to astrophysical jets is discussed.

062.011 **Jets from a geometrically thick disk.**
J. Fukue.
Publ. Astron. Soc. Japan, Vol. 34, 163 - 171 (1982).

A steady isentropic flow in nozzles along the rotation axis of a thick disk is investigated in the framework of general relativity. There exists a transonic flow like a stellar wind. The flow has a terminal velocity of the order of the light speed if the transonic point of the flow is located near the central object of the disk. This type of steady flow will be related to jet phenomena in active galactic nuclei and in SS 433.

062.012 **Meridional circulation in a rotating star revisited.**
Y. Osaki.
Publ. Astron. Soc. Japan, Vol. 34, 257 - 268 (1982).

The author defends the Eddington-Sweet formulation of meridional circulations by arguing that it is still valid if it is properly interpreted as an initial-value problem of time-dependent circulation and rotation and that the quasi-steady slow meridional circulations do exist in a rotating star. It is shown that Busse's formulation of steady rotation and the conventional Eddington–Sweet formulation describe two different aspects of one and the same problem of the time-dependent hydrodynamics of a rotating star in opposite limits.

062.013 **Relativistic thermal plasmas: pair processes and equilibria.**
A. P. Lightman.
Noyaux actifs des galaxies, (see 012.004), 19 pp.

The author considers processes that create and destroy electron-positron pairs, and determines the equilibrium particle and photon distributions in a thermal plasma as a function of its temperature, size, and proton number density.

062.014 **Shock-wave structure in collisionless plasmas from results of laboratory experiments.**
V. G. Eselevich.
Space Sci. Rev., Vol. 32, (see 012.005), 65 - 81 (1982).

Results of laboratory experiments on the study of collisionless shock wave structure in plasmas with and without a magnetic field are summarized, and comparisons with theoretical inferences are made. The relationship between laboratory experiments and the Earth bow shock measurements is also examined.

062.015 **Physical mechanisms for turbulent dissipation in collisionless shock waves.**
C. S. Wu.
Space Sci. Rev., Vol. 32, (see 012.005), 83 - 97 (1982).

Dissipation processes in collisionless shock waves are discussed in terms of simple physical models. Two cases are considered: one is concerned with quasi-parallel shocks and the other deals with quasi-perpendicular shocks. A brief comparison of the theory with recent observations is given.

062.016 **Shock acceleration of energetic particles in the heliosphere.**
M. A. Lee, L. A. Fisk.
Space Sci. Rev., Vol. 32, (see 012.005), 205 - 228 (1982).

The theory of shock acceleration of energetic particles is briefly discussed and reviewed with an emphasis on clarifying the apparent distinction between the '$\mathbf{V} \times \mathbf{B}$' and 'Fermi' mechanisms. Attention is restricted to those situations in which the energetic particles do not themselves influence the given shock structure. In particular, application of the theory to the acceleration of energetic particles in corotating interaction regions in the solar wind is presented. Calculations and resulting energy spectra are also presented for shock acceleration of energetic particles in large solar flare events.

062.017 **Effect of a twisted magnetic field on the equilibrium structure of polytropic and degenerate fluid cylinders.**
A. M. Karnik, S. P. Talwar.
Astrophys. Space Sci., Vol. 84, 363 - 371 (1982).

The distribution of density and mass within an infinite gravitating cylinder is studied for cylinders made up of both polytropic and degenerate material. The equilibrium structure of the cylinder is found from a study of these properties. The effects of rotation of the cylinder are also considered.

062.018 **Effects of mass transfer and free-convection currents on the flow past an infinite porous plate in a rotating fluid.**
A. A. Raptis, C. P. Perdikis.
Astrophys. Space Sci., Vol. 84, 457 - 461 (1982).

An exact analysis of mass transfer and free-convection currents on the flow past a vertical infinite porous plate in a rotating fluid is presented. The examined fluid is assumed to rotate in a steady state with a constant angular velocity. An exact solution for the velocity field is derived and the effects of E (Ekman number) and Gm (Modified Grashof number) on the velocity field are discussed.

062.019 **A generalised theory of stability of superposed fluids in hydromagnetics.**
B. K. Shivamoggi.
Astrophys. Space Sci., Vol. 84, 477 - 484 (1982).

This paper gives a generalised theory of hydromagnetic stability of the interface between two infinitely-conducting, compressible plasmas with the latter accelerated perpendicular to the interface and streaming parallel to the interface and subjected to a constant magnetic field parallel to the streaming direction.

062.020 **Nonlinear Landau damping of Alfvén waves in a high β plasma.**
H. J. Völk, C. J. Cesarsky.
Z. Naturforsch., Band 37a, 809 - 815 (1982).

In this paper the authors consider nonlinear Landau damping in the presence of many waves and calculate the trapping in the combined electrostatic and magnetic mirror fields of the beat waves. This results in a modified growth rate taking into account the finite wave dispersion.

062.021 **Plasma-magnetospheric interaction in X-ray sources: an analysis of the linear Kelvin-Helmholtz instability.**
Y.-M. Wang, G. L. Welter.
Astron. Astrophys., Vol. 113, 113 - 117 (1982).

The Kelvin-Helmholtz instability is analyzed for the case of a magnetized plasma streaming over a vacuum magnetic field with arbitrary orientation in the plane parallel to the interface. It is shown that the presence of even a very weak magnetic field within the plasma may suppress the instability, suggesting that turbulent mixing of plasma into the magnetosphere around an accreting X-ray source is probably not initiated solely through the growth of Kelvin-Helmholtz modes.

062.022 **Stability of differential rotation in stars.**
E. Knobloch, H. C. Spruit.
Astron. Astrophys., Vol. 113, 261 - 268 (1982).

The authors review the linear theory of the Goldreich-Schubert-Fricke (GSF) instability, using the elegant and powerful formalism introduced by McIntyre (1970). The condition for stability, for arbitrary values of the Prandtl number, can be derived explicitly. It shows that the stability depends on a buoyancy number $R\sigma$. The marginally stable state has rotation constant on cylinders only if $R\sigma \ll 1$. The value of $R\sigma$ is of the order unity in slow rotators like the sun, so that significant deviations from constant rotation on cylinders can occur without causing the GSF instability. The authors discuss how this affects the angular momentum distribution in the final stages of the spin-down of a star.

062.023 **Structure and dynamics of supersonic jets.**
M. L. Norman, L. Smarr, K.-H. A. Winkler,

M. D. Smith.
Astron. Astrophys., Vol. 113, 285 - 302 (1982).

The Blandford and Rees (BR) beam model for extragalactic radio sources is critically examined within the context of axisymmetric, nonrelativistic ideal gas dynamics. Thus the authors assume that radio jets contain thermal matter that governs the structure and dynamics of the jet. They conduct numerical experiments that study the structure and stability of the 3 basic jet components conceived by BR-beam, working surface, and cocoon – for various beam Mach numbers and temperature ratios with respect to a constant ambient medium (the intergalactic medium).

062.024 **Frequency correlation of strong intensity fluctuations in a turbulent medium.**
V. U. Zavorotnyj.
Izv. vuzov. Radiofiz., Tom 24, 601 - 608 (1981). In Russian. Abstr. in Ref. zh., 51. Astron., 8.51.122 (1982).

062.025 **On a theory of strong shock wave propagation in inhomogeneous astrophysical systems.**
S. A. Silich, P. I. Fomin.
Inst. Teor. fiz. AN USSR. Prepr., 1982, No. 27, 11 pp. In Russian. – Abstr. in Ref. zh., 51. Astron., 8.51.123 (1982).

062.026 **Rädler effect in a collisionless plasma and field-aligned currents in the magnetotail.**
T. Namikawa, H. Hamabata, S. Matsushita.
Planet. Space Sci., Vol. 30, 875 - 878 (1982).

062.027 **Magnetohydrodynamic equilibrium. III. Helically symmetric fields.**
K. C. Tsinganos.
Astrophys. J., Vol. 259, 820 - 831 (1982).

The time-independent equations of hydromagnetics appropriate for an inviscid fluid of high electrical conductivity embedded in magnetic and gravity fields with helical symmetry are considered. Simple solutions of the equation are presented to illustrate the applications. The result allows a systematic approach to the treatment of the dynamics of helically symmetric, isentropic and nonisentropic, hydromagnetic flows.

062.028 **Magnetohydrodynamic equilibrium. IV. Nonequilibrium of nonsymmetric hydrodynamic topologies.**
K. C. Tsinganos.
Astrophys. J., Vol. 259, 832 - 843 (1982).

It is shown that when the magnetic field and the fluid motions in an ideal magnetofluid have a translational symmetry then, apart from isolated topologies, invariance along this principal direction of the fields is a necessary condition for magnetohydrodynamic equilibrium. This theorem is analogous to the well-established Taylor-Proudman theorem in fluid dynamics and runs parallel to Parker's theorem for the necessary conditions for equilibrium in magnetostatics.

062.029 **Irreversible thermodynamics in a radiating fluid.**
A. D. Verga.
Astrophys. J., Vol. 260, 286 - 298 (1982).

The author investigates dissipative processes taking place in a radiating fluid with a boundary surface, as a consequence of radiation-matter interaction. It is assumed that the medium is in a hydrodynamic regime, while the radiation field is in a nonequilibrium state. The kinetic equations are consistently solved with the entropy balance equation. This leads to a generalization of the photohydrodynamic equations by taking into account the role of photon thermalization in the irreversible processes of energy and momentum transport.

062.030 **Bremsstrahlung spectra at relativistic temperatures.**
A. Górecki, W. Kluźniak.
Acta Astron., Vol. 31, 457 - 469 (1981).

Spectra due to electron-proton bremsstrahlung in a Maxwellian plasma have been computed for relativistic temperatures up to $T_e = 50$ MeV.

062.031 **Nonlinear effects in a gravitating plasma disk being in a poloidal magnetic field.**
A. G. Morozov.
Pis'ma Astron. Zh., Tom 8, 636 - 640 (1982). In Russian. English translation in Soviet Astron. Lett., Vol. 8.

At a nonlinear regime the magnetic field in the model of a gravitating plasma disk does not influence the outburst instability and the conditions of soliton existence at harmonics excited at the minimum of the dispersion curve. The deformation of solitons resulting from the nonlinearity of order higher than cubic is investigated. Some applications of the obtained results to gaseous subsystems of flat galaxies and to plasma disks rotating around black holes are discussed.

062.032 **Stability of radiative shocks.**
J. N. Imamura, M. T. M. Wolff, R. H. Durisen.
Bull. American Astron. Soc., Vol. 14, 620 (1982). – Abstract.

062.033 **A remarkable general property of the MHD equations and its relation to coronal heating mechanisms.**
K. C. Tsinganos.
Bull. American Astron. Soc., Vol. 14, 623 (1982). – Abstract.

062.034 **Magnetic fields in astrophysics.**
R. D. Blandford.
Bull. American Astron. Soc., Vol. 14, 646 - 647 (1982). Abstract.

062.035 **Magnetospheric phenomena of celestial bodies – introductory talk.**
Y. Uchida.
Magnetospheric phenomena of celestial bodies, (see 012.011), p. 1 - 5 (1982). In Japanese.

A brief survey of the celestial objects showing the magnetospheric phenomena is given, and the types of plasmas interacting with their magnetic field are categorized. These include the interaction of magnetic field with directed plasma flow as in earth's and planetary magnetospheres in the solar wind, the interaction of magnetic field with plasma outflow like the corona and coronal holes of the sun and perhaps of stars.

062.036 **Contraction of magnetic clouds and the origin of stellar magnetic fields.**
T. Nakano, T. Umebayashi.
Magnetospheric phenomena of celestial bodies, (see 012.011), p. 63 - 65 (1982). In Japanese.

The plasma drift in a contracting very dense cloud is investigated and it is found that the magnetic field decouples from the neutral gas at $n \approx 10^{11}$ cm^{-3}. By considering the nonhomologous nature in the final stage of cloud contraction, the magnetic field in a newborn star is estimated to be about 0.03 G. If the star of 1 $M_\odot$ with 100 $R_\odot$ and 0.03 G contracts to the main sequence conserving the magnetic flux, the magnetic field grows up to 300 G, the field strength at the surface of a typical magnetic star.

062.037 **On the magnetic accretion disk consisting of a perfectly conducting non-viscous plasma.**
R. Horiuchi.
Magnetospheric phenomena of celestial bodies, (see 012.011), p. 107 - 108 (1982). In Japanese.

The vertical equation of a magnetic accretion disk is analytically solved on the approximation that the disk is geometrically thin, and its whole structure is investigated. It is found that the matter flow in the accretion disk should always have the super-Alfvénic velocity everywhere. The corotation radius must be much larger than the Alfvén one in order to obtain stable accretion disks in the super-Alfvénic region.

062.038 **Magnetohydrodynamic phenomena in quasar plasmas.**
F. Takahara.
Magnetospheric phenomena of celestial bodies, (see 012.011), p. 130 - 139 (1982). In Japanese.

The role and behavior of the magnetic fields in the accretion disk around a supermassive black hole are reviewed in connection with active phenomena in quasars and active galactic nuclei. The magnetic fields play an important role in the angular momentum transfer in the disk, non-thermal activities and the emission of cyclotron-synchrotron radiation. Elementary processes such as flux cell dynamics, magnetic buoyancy and turbulent dynamo mechanism are discussed.

062.039 **Formation of a collimated beam in radio galaxies.**
M. Yokosawa.
Magnetospheric phenomena of celestial bodies, (see 012.011), p. 140 - 144 (1982). In Japanese.

The collimated flows which are non-thermally driven by high energy particles and magneto-hydrodynamic (MHD) waves are presented. The galactic nucleus surrounded by a cool gas is investigated. The cool gas accretes onto the nucleus and the accretion matter can confine the wave zone around the nucleus in which the high energy particles are completely locked to the MHD waves. When a quasi-radial magnetic field is embedded in the accretion flow, the MHD wave packets are collimated into the direction of the symmetry axis of the galactic nuclear disc. The fluid around the nucleus is considered to be accelerated and heated by the MHD waves, and ejected along the axis.

062.040 **Magnetic reconnection in toroidal plasmas.**
M. Wakatani.
Magnetospheric phenomena of celestial bodies, (see 012.011), p. 153 - 160 (1982). In Japanese.

Tearing instabilities and the related non-linear phenomena in Tokamaks and reverse field pinches are discussed. Similar processes would be seen in solar plasma physics.

062.041 **On the role of rotation in the generation of magnetic fields by fluid motions.**
R. Hide.
Philos. Trans. R. Soc. London, Ser. A, Vol. 306, (see 012.014), 223 - 234 (1982).

It is generally accepted that the magnetic fields of planets and stars are produced by the self-exciting dynamo process (first proposed by Larmor) and that observed near-alignments of magnetic dipole axes with rotation axes are due to the influence of Coriolis forces on underlying fluid motions. The detailed role of rotation in the generation of cosmical magnetic fields has yet to be elucidated but useful insight can be obtained from general considerations of the governing magnetohydrodynamic equations.

062.042 **Progress in dynamo theory: kinematic and dynamic models.**
F. Krause.
Philos. Trans. R. Soc. London, Ser. A, Vol. 306, (see 012.014), 255 - 260 (1982).

The results of dynamo theories that have been deduced during the last 10 years have often been used for deriving statements relating properties of the magnetic field with those of the core of the Earth. These results are critically analysed, and the reliability of such statements is discussed.

062.043 **Magnetohydrodynamic oscillations in a rotating spherical shell.**
S. D. London.
Geophys. Astrophys. Fluid Dyn., Vol. 18, 227 - 242 (1981). Abstr. in Phys. Abstr., Vol. 85, Abstr. 62885 (1982).

062.044 **Astrophysical plasmas.**
R. J. Tayler.
Plasma physics and nuclear fusion research, (see 003.007), p. 619 - 641 (1981). – Abstr. in Phys. Abstr., Vol. 85, Abstr. 63243 (1982).

062.045 **Residual fields from extinct dynamos.**
E. N. Parker.
Geophys. Astrophys. Fluid Dyn., Vol. 18, 175 - 195 (1981). Abstr. in Phys. Abstr., Vol. 85, Abstr. 63449 (1982).

062.046 **On the chain of transfer equations for a collisionless turbulent plasma.**
K. G. Ivanov, A. F. Kharshiladze.
Solnech. veter i magnitosfer. issled. Moskva, 1980, p. 166 - 172. In Russian. – Abstr. in Ref. zh., 51. Astron., 9.51.284 (1982).

062.047 **On the relation between Kolmogoroff's theory and the direct interaction approximation in two-dimensional plasma turbulence.**
R. M. Kulsrud, R. N. Sudan.
Comments Plasma Phys. Controlled Fusion, Vol. 7, No. 2, p. 47 - 52 (1982). – Abstr. in Phys. Abstr., Vol. 85, Abstr. 65726 (1982).

062.048 **Some nonlinear problems in astrophysics.**
I. Lerche, B. C. Low.
Physica D, Vol. 4D, 293 - 318 (1982). – Abstr. in Phys. Abstr., Vol. 85, Abstr. 68613 (1982).

062.049 **Non-steady state of a hydromagnetic $\alpha\omega$-dynamo and its application to the geomagnetic reversals.**
H. Watanabe.
J. Geomagn. Geoelectr., Vol. 33, 531 - 543 (1981). – Abstr. in Phys. Abstr., Vol. 85, Abstr. 78543 (1982).

062.050 **Interaction of small-amplitude fluctuations with a strong magnetohydrodynamic shock.**
H.-C. Zhuang, C. T. Russell.
Phys. Fluids, Vol. 25, 748 - 758 (1982). – Abstr. in Phys. Abstr., Vol. 85, Abstr. 78927 (1982).

062.051 **The evolution of a test particle distribution in a strongly magnetized plasma.**
J. G. Kirk, D. J. Galloway.
Plasma Phys., Vol. 24, 339 - 359 (1982). – Abstr. in Phys. Abstr., Vol. 85, Abstr. 84087 (1982).

062.052 **Transport equations for multicomponent anisotropic space plasmas: a review.**
A. R. Barakat, R. W. Schunk.
Plasma Phys., Vol. 24, 389 - 418 (1982). – Abstr. in Phys. Abstr., Vol. 85, Abstr. 84088 (1982).

062.053 **The mean electromotive force generated by random Alfvén waves in a collisionless plasma.**
T. Namikawa, H. Hamabata.
J. Plasma Phys., Vol. 27, Part 3, p. 415 - 425 (1982). – Abstr. in Phys. Abstr., Vol. 85, Abstr. 86404 (1982).

062.054 **Selfgravitating magnetofluid in cylindrically symmetric spacetime.**
T. H. Date.
Bull. Astron. Soc. India, Vol. 10, 44 (1982). – Abstract.

062.055 **Cerenkov radiation in a magneto-active plasma.**
D. Patro.
Bull. Astron. Soc. India, Vol. 10, 45 (1982). – Abstract.

062.056 **Thermal convection instability of a two-component fluid layer in hydromagnetics.**
K. M. Srivastava, Sudhanshu, V. K. Katiyar.
Astrophys. Space Sci., Vol. 85, 121 - 135 (1982).
The thermal convection instability of a two component fluid layer subjected to a temperature gradient has been studied in the presence of an applied magnetic field. The associated thermal diffusion separation has a predominant effect even when the separations are small. Solutions for the non-oscillatory marginal states have been obtained. It is shown that the concentration gradient has a stabilizing or destabilizing effect according as $\Delta T <$ or > 0. Approximate solutions for the oscillatory solutions have been obtained by the method of variational principle and the dispersion relation has been solved numerically.

062.057 **The flux ejection dynamo with small diffusivity. I. Basic properties.**
E. N. Parker.
Astrophys. Space Sci., Vol. 85, 167 - 182 (1982).
The flux ejection dynamo in an infinitely broad convective layer of finite depth is treated in the limit of large electrical conductivity. The upper surface of the convective layer is open to empty space while the lower surface is closed to the passage of fluid and field. An idealized convective overturning is employed. The author discusses the boundary conditions to be applied to an astronomical body with the flux ejection dynamo operating at its surface. It is pointed out that if there were a flux ejection dynamo operating in the Sun, the implications of the magnetic bipolar regions emerging through the surface would be the opposite of the usual.

062.058 **The flux ejection dynamo with small diffusivity. II. Illustrative examples.**
E. N. Parker.
Astrophys. Space Sci., Vol. 85, 183 - 201 (1982).
A number of examples are worked out to illustrate the consequences of reverse flux ejection from the surface of a convective layer of conducting fluid. Generally the reverse flux ejection has the opposite effect of magnetic buoyancy, tending to bury the fields rather than bringing them through the surface. Even a weak flux ejection effect prevents the escape of magnetic field through the surface. Reverse flux ejection at the surface of an $\alpha\omega$-dynamo profoundly alters the character of the solutions of the dynamo equations. Altogether, flux ejection serves to obscure the interpretation of magnetic observations.

062.059 **Gravitational instability of a Hall plasma in the presence of suspended particles.**
K. C. Sharma.
Astrophys. Space Sci., Vol. 85, 263 - 269 (1982).
The gravitational instability of an infinite homogeneous and infinitely conducting self-gravitating gas particle medium in the presence of suspended particles of a Hall plasma is considered. The particular cases of the effects of Hall currents and suspended particles on the waves propagated along and perpendicular to magnetic field have been discussed. Jeans's criterion determines the gravitational instability.

062.060 **Magnetohydrodynamic free convection effects on the Stokes problem for an incompressible viscous fluid past an infinite vertical limiting surface.**
N. G. Kafousias, G. A. Georgantopoulos.
Astrophys. Space Sci., Vol. 85, 297 - 307 (1982).
An analysis of the transverse magnetic field effects on the free convective flow of an incompressible, electrically conducting viscous fluid past an infinite non-conducting and non-magnetic, vertical limiting surface (e. g., of a star), has been carried out. The limiting surface is assumed to move after receiving an initial impulse. Exact solutions of equations governing the flow are derived with the help of the Laplace transform technique. The velocity, the induced magnetic field, the skin-friction and the electric current density are shown graphically. The effects of the Grashof number G, the Prandtl number P, and the magnetic parameter M are discussed.

062.061 **Equilibrium of a static charged perfect fluid sphere.**
D. D. Dionysiou.
Astrophys. Space Sci., Vol. 85, 331 - 343 (1982).
A study is developed for treating Einstein-Maxwell's equations, applied to a static perfect fluid sphere with charge, in such a manner to generalize the Oppenheimer and Volkoff hydrostatic equilibrium equation. The author suggests reasons which could make the generalization of this equation inapplicable. It is shown that Einstein-Maxwell's equations predict such special solutions.

062.062 **Propagation of magnetogasdynamic spherical shock waves in an exponential decreasing medium.**
B. G. Verma, J. P. Vishwakarma, V. Sharan.
Astrophys. Space Sci., Vol. 85, 345 - 350 (1982).
By use of the approximate method of Whitham (1958) the effect of magnetic field is investigated on a point explosion in a medium exhibiting exponential decrease of density and temperature. It has been found that the shock velocity and shock Mach number first decrease, but after a certain distance they start increasing.

062.063 **Influence of fluctuating field on the configuration of force-free field. I. Conventional turbulence.**
W.- R. Hu.
Astrophys. Space Sci., Vol. 85, 351 - 360 (1982).
The consistent equations for random force-free magnetic fields are obtained from the fluctuation and average momentum and induction equations based on the statistical approach. Some features of average force-free fields are discussed. This idea may be applied to explain many configurations of magnetic field in the astronomical environment.

062.064 **Free convection effects on the oscillatory flow in the Stokes problem past an infinite porous vertical limiting surface with heat sources.**
V. C. Ifantis, G. A. Georgantopoulos.
Astrophys. Space Sci., Vol. 85, 361 - 368 (1982).
The authors study the free convection flow past an impulsively started infinite, vertical porous limiting surface (e. g. of a star) which is subjected to a constant suction velocity, with heat "sources" when the free stream velocity oscillates in time about a constant mean. A perturbation solution is sought in order to obtain the expressions for mean velocity and mean temperature. The results are compared with the corresponding ones without heat sources.

062.065 **Self-similar flows behind shock wave with increasing energy in magnetogasdynamics. I.**
J. B. Singh, R. N. Lal Srivastava.
Astrophys. Space Sci., Vol. 85, 427 - 435 (1982).
Self-similar solutions for adiabatic and isothermal flows driven out by a propelling contact surface, moving into a quiet solar wind region, are investigated in the presence of magnetic field. The total energy of the flow between the shock and the contact surface is taken to be time-dependent obeying a power-law. The shock is assumed to be strong and propagating into a perfect gas at rest with non-uniform density and magnetic field.

062.066 **Self-similar flows behind a spherical radiation-driven shock wave. II.**
J. B. Singh, R. N. Lal Srivastava.
Astrophys. Space Sci., Vol. 85, 437 - 441 (1982).
Self-similar flows, behind a radiation-driven shock wave, have been investigated in non-uniform atmosphere. The total

energy content of the flow behind the shock increases due to the absorption of radiation and it is assumed to be time-dependent.

062.067 **Propagation of blast waves in Roche model.**
B. G. Verma, R. C. Srivastava, A. H. Khan.
Astrophys. Space Sci., Vol. 85, 465 - 476 (1982).

A non-similarity solution for propagation of magnetogasdynamic blast waves in Roche model is obtained. The density outside the central core varies according to some inverse power of distance from the centre of explosion. The strength of the shock surface heading the disturbance as well as the total energy of the wave vary with time.

062.068 **Free-convection effects on MHD flow past a porous plate with step function change in suction velocity.**
S. T. Revankar.
Astrophys. Space Sci., Vol. 86, 13 - 23 (1982).

Free convection effects on MHD flow past a semi infinite porous flat plate is studied when the time dependent suction velocity changes in step function form. The solution of the problem is obtained in closed form for the fluid with unit Prandtl number. It is observed that for both cooling and heating of the plate the suction velocity enhances the velocity field. The heat transfer is higher with increase in suction velocity.

062.069 **Unsteady MHD flow past a porous plate with step function change in suction at slip flow regime.**
S. T. Revankar, V. M. Korwar.
Astrophys. Space Sci., Vol. 86, 25 - 31 (1982).

Unsteady two-dimensional hydromagnetic flow of an electrically conducting viscous incompressible fluid past a semi-infinite porous flat plate with step function change in suction velocity is studied allowing a first order velocity slip at the boundary condition. The solution of the problem is obtained in closed form and the results are discussed with the aid of graphs for various parameters entering in the problem.

062.070 **Free convection in MHD with vertical and horizontal temperature gradients.**
R. Singhal, K. M. Srivastava.
Astrophys. Space Sci., Vol. 86, 33 - 42 (1982).

The occurrence of oscillatory mode for magnetized fluid in the presence of vertical and horizontal temperature gradients has been studied. In both the cases (highly viscous and low-viscosity fluids) the occurrence of oscillatory solutions to the Bénard problem has been investigated as a function of Prandtl number, magnetic number, resistivity number and the ratio of applied temperature gradients.

062.071 **Free convection and mass transfer effects on the oscillatory flow past an infinite moving vertical isothermal plate with constant suction and heat sources.**
A. A. Raptis.
Astrophys. Space Sci., Vol. 86, 43 - 53 (1982).

An analysis of the two-dimensional flow of an incompressible, viscous fluid past an infinite, porous, vertical plate is presented under the following conditions: (1) the suction velocity is constant; (2) the free stream oscillates in time about a constant mean; (3) the plate moves in the upward direction in its own plane; (4) the temperature of the plate is constant; (5) there are heat generation (absorption) in the fluid. Approximate solutions for the coupled non-linear equations are obtained for the velocity, temperature and their related quantities and the influence of the various parameters entering into the problem is discussed.

062.072 **A magnetohydrodynamical theorem for steady axisymmetric systems with implications to ring galaxies.**
E. Battaner, M. L. Sánchez-Saavedra.
Astrophys. Space Sci., Vol. 86, 55 - 62 (1982).

A magnetohydrodynamical theorem can be deduced for steady axisymmetric magnetized systems in situations which are usually accepted as representative in interstellar plasmas. It shows that for particular conditions detailed below, magnetic fields have either the direction of the symmetry axis of the system or have field lines encircling this axis of symmetry. Ring galaxies are probably of the second type. Our galaxy itself would be of this second type slightly separated from steady state; the magnetic field radial distribution compatible with the gas content is estimated.

062.073 **Non-linear analysis of the stability of the Screw Pinch.**
K. M. Srivastava, D. N. Vyas.
Astrophys. Space Sci., Vol. 86, 71 - 89 (1982).

Non-linear effects on the stability of the Screw Pinch in cylindrical system surrounded by a uniform pitch magnetic field in an incompressible pressureless plasma have been studied. A linear dispersion relation governing the stability of interface has been obtained with the help of normal mode analysis. Perturbation technique has been used to obtain the non-linear Schrödinger equation for the time variation of amplitude. Non-linear analyses have also been discussed.

062.074 **Finite amplitude Alfvén waves in a multi-ion plasma: propagation, acceleration, and heating.**
P. A. Isenberg, J. V. Hollweg.
J. Geophys. Res., Vol. 87, 5023 - 5029 (1982).

The authors derive an expression for the wave action flux of finite amplitude Alfvén waves in a multi-ion plasma. The expression is valid in the presence of dissipative forces and allows an arbitrary angle between the average magnetic field and the wave vector. Applying the conservation of wave action and the first law of thermodynamics yields the following results for a multi-ion plasma: (1) an expression for the spatial evolution of Alfvén wave amplitude in the absence of dissipation, (2) the relationship between the wave amplitude and the dissipative heating, and (3) an expression for the acceleration of an ion species by finite amplitude Alfvén waves.

062.075 **The influence of buoyancy on the stability of jets.**
A. Achterberg.
Astron. Astrophys., Vol. 114, 233 - 237 (1982).

The stability of jets in a stratified medium is considered in the long wavelength limit. It is shown that buoyancy effects will be of some importance for light jets confined by a heavy medium near the nucleus of the parent galaxy.

062.076 **Local coupling of surface MHD waves with kinetic Alfvén waves in jets.**
G. Bodo, A. Ferrari.
Astron. Astrophys., Vol. 114, 394 - 399 (1982).

In the framework of the study of morphology and energetics of jets in extragalactic radiosources, the authors discuss the coupling of MHD surface modes generated by Kelvin-Helmholtz instabilities with body waves in the jet's plasma. In particular they show that a resonant mode conversion into kinetic Alfvén waves is possible whenever an inhomogeneous boundary layer is present. These kinetic Alfvén waves propagate transversally to the magnetic field toward the jet's interior, where they can provide relativistic electron acceleration and plasma heating. The authors calculate that the energy input is adequate for supporting radio emission from typical jets (M 87, Cen A, etc.).

062.077 **Linear analysis of an oscillatory instability of radiative shock waves.**
R. A. Chevalier, J. N. Imamura.
Astrophys. J., Vol. 261, 543 - 549 (1982).

The authors present a linear stability analysis of a planar radiative shock structure. They investigate an oscillatory instability similar to that found in numerical calculations of accretion onto degenerate dwarfs.

062.078 **Beam-plasma instability in the presence of low-frequency turbulence.**
M. V. Goldman, D. F. DuBois.
Phys. Fluids, Vol. 25, 1062 - 1072 (1982). – Abstr. in Phys. Abstr., Vol. 85, Abstr. 91631 (1982).

062.079 **On perturbations of magnetic field configurations.**
R. Rosner, E. Knobloch.
Astrophys. J., Vol. 262, 349 - 357 (1982).
The authors explore the behavior of stationary equilibrium solutions to the MHD equations possessing a well-defined symmetry to perturbations lacking that symmetry. They consider two distinct situations of astrophysical interest: solutions of the magnetostatic equations and of the magnetoconvection equations. The results show that in these cases changes in solution topology are not accessible to small-parameter expansions, so that such expansions do not describe the full range of behavior.

062.080 **Magnetohydrodynamics of atmospheric transients. IV. Nonplane two-dimensional analyses of energy conversion and magnetic field evolution.**
S. T. Wu, Y. Nakagawa, S. M. Han, M. Dryer.
Astrophys. J., Vol. 262, 369 - 376 (1982).
The evolution of the magnetic field and the manner of conversion of thermal energy into different forms in the corona following a solar flare are examined by a nonplane magnetohydrodynamic analysis. In the analysis all three components of magnetic field and velocity are treated in a physically self-consistent manner. The difference due to the initial magnetic field, either twisted (force-free) or nontwisted (potential), is demonstrated.

062.081 **The theory of particle acceleration in astrophysical objects containing shock waves and turbulent plasma motions.**
A. M. Bykov, I. N. Toptyghin.
J. Geophys., Vol. 50, 221 - 226 (1982). – Abstr. in Phys. Abstr., Vol., 85, Abstr. 100672 (1982).

062.082 **Physics of dense plasmas and the enhancement of thermonuclear reaction rates due to strong screening.**
N. Itoh.
Prog. Theor. Phys. Suppl., No. 70, p. 132 - 141 (1981). Abstr. in Phys. Abstr., Vol. 85, Abstr. 100725 (1982).

062.083 **On the magnetospheric structure of pulsars.**
W.-r. Hu.
Chinese Phys., Vol. 2, 677 - 687 (1982). – Abstr. in Phys. Abstr., Vol. 85, Abstr. 100968 (1982). – See Abstr. 30.062.092.

062.084 **A boundary-coupled generalization of the Newcomb stability criterion.**
A. Ray, G. Van Hoven.
Phys. Fluids, Vol. 25, 1355 - 1357 (1982). – Abstr. in Phys. Abstr., Vol. 85, Abstr. 109494 (1982).

062.085 **The local dispersion relation for magneto-atmospheric waves.**
J. H. Thomas.
Astrophys. J., Vol. 262, 760 - 767 (1982).
It is shown that the local dispersion relation for magneto-atmospheric waves has no unique form and has frequently been applied outside of its range of validity with incorrect results. The particular problem of the effect of a uniform horizontal magnetic field on the cutoff frequency is treated in detail. The local dispersion relation predicts that the cutoff frequency *decreases* with increasing magnetic field strength; the WKB approximation predicts that the cutoff frequency *increases* with increasing magnetic field strength; and the exact solution of the wave equation shows that in fact the cutoff frequency is *unaffected* by the magnetic field. A new standard form of the local dispersion relation is proposed which is both simple and unambiguous.

062.086 **On turbulent heat transport in rotating convective zones.**
G. Rüdiger.
Astron. Nachr., Band 303, 293 - 303 (1982).
This paper analyses the tensorial aspect of heat conduction which becomes important for gravitationally stratified and rotating turbulent media. The structure of the eddy conductivity tensor is indeed a non-trivial one for turbulences with correlation times not too short in comparison with the rotation period. Since this situation may be realized in the solar convection zone, a more complex formulation of solar heat-flux problems is required rather than that with a scalar heat conductivity. The author neglects all other energy transporters and treats a solar heat-flux problem with a heuristic mean-field ansatz of the eddy conductivity tensor.

062.087 **Nonlinear waves in superposed fluids.**
R. Kant, S. K. Malik.
Astrophys. Space Sci., Vol. 86, 345 - 360 (1982).
A weakly nonlinear theory of wave propagation in superposed fluids in the presence of magnetic fields is presented in this paper. The authors derive the equations governing the evolution of the amplitude of the progressive as well as the standing waves. It is demonstrated that the waves can be unstable against modulation in the presence of the magnetic fields. The authors also obtain the nonlinear cut off wave number which separates the region of stability from instability.

062.088 **Hydromagnetic gravitational instability of an infinite cylinder of a compressible fluid.**
B. K. Shivamoggi.
Astrophys. Space Sci., Vol. 86, 371 - 376 (1982).
The gravitational instability of an infinite cylinder of a self-gravitating, perfectly-conducting, compressible fluid subjected to a uniform axial magnetic field is investigated. A generalised dispersion relation describing the stability characteristics of the configuration is given. The gravitational instability of the cylinder is found to be worsened in the presence of the compressibility effects in the fluid.

062.089 **Study of toroidal magnetic configurations in a spherically symmetric gravitational field with applications to coronal loops and transients.**
V. A. Osherovich.
Astrophys. Space Sci., Vol. 86, 453 - 469 (1982).
One procedure for solving MHD equations is to search for a solution in an area that is restricted by boundary surfaces. For astrophysically relevant plasma, an approach has been formulated by the author. The proposed approach leads to an eigenvalue problem that is treated analytically. The complete set of exact MHD solutions with multi-toroidal structure is obtained. These solutions are applied to coronal loops and transients, using the similarity assumption for time-dependent solutions.

062.090 **Cyclotron emission in strongly magnetized plasmas.**
H. Herold, H. Ruder, G. Wunner.
Astron. Astrophys., Vol. 115, 90 - 96 (1982).
A fully relativistic, quantum-electrodynamical derivation of the cyclotron emission rates in arbitrary magnetic fields is presented. Special attention is paid to the determination of the correct polarization states of the excited initial electron. Employing different electron distribution functions for strongly magnetized plasmas, photon emission spectra are explicitly obtained in the range around the cyclotron reso-

nance which exhibit remarkable angle-dependent structures originating from relativistic kinematics.

062.091 **Compressibility effects on hydromagnetic surface waves.**
K. Somasundaram, C. Uberoi.
Sol. Phys., Vol. 81, 19 - 24 (1982).
The dispersion equation for hydromagnetic surface waves along a plasma-plasma interface has been solved as a function of the compressibility factor c_1/v_{A1}, where c_1 and v_{A1} are the acoustic and Alfvén wave speed in one of the medium, for general wave propagation direction. Both slow and fast magnetosonic surface waves can exist. For low-β plasmas only fast mode exists. The slow mode does not propagate below a critical value of c_1. When $c_1 \to \infty$ the phase velocity of the slow wave tends to the Alfvén surface wave velocity in the incompressible media and for large θ the phase velocity of the fast wave approaches this value. The phase velocity of the slow wave increases whereas for the fast wave it decreases with increase in the angle θ.

062.092 **Numerical simulation of the weak turbulence excited by a beam of electrons in the interplanetary plasma.**
R. J.-M. Grognard.
Sol. Phys., Vol. 81, 173 - 180 (1982).
The electron distribution functions measured at 1 AU in an electron stream passing the ISEE-3 spacecraft (Lin et al., 1981) are used as input data to a programme which simulates in a one-dimensional model the interaction between fast electrons and plasma waves ('quasi-linear' relaxation) together with the plasma wave scattering off the background ions. While the computed spectral energy density of the plasma waves excited in resonance with the streaming of electrons is below Zakharov's threshold of strong turbulence, it is sufficiently high to undergo a fast induced scattering off the background ions. The resulting spectrum is concentrated around the wave vector $k = 0$. Some simple analytic considerations show that this stage leads necessarily to the crossing of Zakharov's threshold and therefore this indirect excitation of strong turbulence seems to play an essential role in understanding the Langmuir turbulence generated by the motion of fast electron beams in the interplanetary plasma.

062.093 **On MHD wave propagation in inhomogeneous plasmas and the mechanism of resonant absorption.**
I. C. Rae, B. Roberts.
Mon. Not. R. Astron. Soc., Vol. 201, 1171 - 1182 (1982).
The propagation of MHD waves in inhomogeneous plasmas is discussed using both the eikonal and differential equation approaches. The cusp resonance, which arises in a compressible plasma, and has been largely ignored in the literature, is discussed in some detail. Resonant heating may take place at those locations where the phase-speed of an incident wave matches the local Alfvén or cusp speeds. The Alfvén resonance can only be excited by wave tunnelling, whereas the cusp resonance is excited by propagating slow waves. Cusp resonance by slow waves may therefore be important as a means of heating the solar corona.

062.094 **Zur Berechnung von Astropausenflächen.**
H. J. Fahr, W. Neutsch, G. Lay.
Mitt. Astron. Ges., Nr. 57, p. 294 - 301 (1982).

062.095 **Magnetic reconnection: a problem of general physical and astrophysical interest, with special implications in solar physics.**
J. Heyvaerts.
Pré-publ. Inst. Astrophys. Paris, No. 11, 13 pp. (1982).
Submitted to COSPAR Proceedings 1982.

062.096 **Full implicit continuous Eulerian scheme in the spherical coordinates and its applications to solar phenomena.**
S. Wang, Y. Hu, S. T. Wu.
Sci. Sinica, Ser. A, Vol. 25, 1305 - 1318 (1982).
The full implicit continuous Eulerian scheme is extended to the spherical coordinates so as to study global problems in solar atmosphere. As a numerical example, the authors discuss the physical process in the solar atmosphere due to a radial mass ejection in the neighbourhood of the equator on the solar surface, with the initial magnetic field being taken to be a dipole field. The results may elucidate the ejecting prominence and the accompanying coronal transient.

062.097 **Numerical study of line-tied magnetic reconnection.**
T. G. Forbes, E. R. Priest.
Sol. Phys., Vol. 81, 303 - 324 (1982).
A two-dimensional configuration, analogous to that at the start of the main phase in two-ribbon flares, is modelled numerically by self-consistently solving the time-dependent MHD equations.

062.098 **Structure and expansion law of a hypersonic beam.**
M. Yokosawa, S. Ikeuchi, S. Sakashita.
Publ. Astron. Soc. Japan, Vol. 34, 461 - 481 (1982).
The dynamical interaction of a hypersonic beam with a homogeneous ambient medium is studied by means of a time-dependent, two-dimensional, hydrodynamic code. The beam velocity ranges from a non-relativistic to a relativistic one.

062.099 **Radio jets and Kelvin-Helmholtz instabilities.**
T. P. Ray.
Irish Astron. J., Vol. 15, (see 012.061), 123 - 124 (1981).

062.100 **Effects of nuclear forces on ion thermalization in high-temperature plasmas.**
R. J. Gould.
Astrophys. J., Vol. 263, 879 - 890 (1982).
The effects of nuclear forces in the thermalization of ion in high-temperature plasmas are analyzed. The energy loss rate is computed for a fast nonrelativistic ion traversing a plasma, with special emphasis on the basic problem of proton-proton (p-p) interactions. A brief discussion of the effects of energy loss due to interactions with the plasma electron gas is given. The phenomenon of "nuclear breakdown" of a high-temperature plasma is considered briefly.

062.101 **Waves in magnetic structures.**
B. Roberts.
The physics of sunspots, (see 012.063), p. 369 - 383 (1981).
The theoretical nature of wave propagation in a magnetically structured and gravitationally stratified atmosphere is discussed. It is emphasized that a magnetic structure such as a sunspot or photospheric intense flux tube supports the propagation of both surface and body waves.

062.102 **Processus physiques dans les sphères célestes et le milieu interstellaire.**
J.-P. J. Lafon.
Mécanismes de production d'énergie dans le milieu interstellaire, (see 012.068), p. 95 - 139 (1981).
L'auteur expose les éléments de dynamique des fluides nécessaires à la compréhension des phénomènes. Puis il analyse, à la fois de façon théorique et sur la base d'exemples et d'applications astrophysiques, le comportement de milieux fluides soumis à diverses contraintes, ainsi que la propagation d'ondes de compression ou de raréfaction, les phénomènes de déferlement, la création et la propagation d'ondes de choc.

062.103 **Containment of a diffuse ionized mass orbiting around a magnetized central body.**
V. Banfi.

Moon Planets, Vol. 27, 165 - 177 (1982).

The containment of a diffused and ionized mass orbiting around a magnetized central body is studied and the condition equation is established. Some qualitative and quantitative applications to the planetary cosmogony problems are developed.

062.104 **Carbon-oxygen plasma: parameter domain for a strong screening effect.**
A. E. M. Khairozzaman.
Astrophys. Space Sci., Vol. 87, 3 - 11 (1982).

A critical study relating to the enhancement factor of nuclear reaction rates in the carbon-oxygen plasma is made. Plots are given of the various dimensionless parameters in the temperature-density plane after the recent analysis of Itoh et al. (1977). The enhancement factor of thermonuclear reaction rates critically depends on $3\Gamma/\tau$. The enhancement factors also have been calculated for the carbon-oxygen plasma studied by Iben (1978).

062.105 **Dynamo action in cosmic bodies.**
J. H. Piddington.
Astrophys. Space Sci., Vol. 87, 89 - 104 (1982).

This paper is largely a reply to Cowling's review of the present status of cosmic dynamo theory and its alternatives of primordial or fossil field models. Central is the question of turbulent diffusion. Turbulence does not shred or divide fields into small-scale elements as claimed; instead it creates these elements in addition to the large-scale field which remains after Ohmic diffusion has destroyed the small fields. The significance of the existence of a terrestrial dynamo is stressed and various objections to the existence of a solar-type dynamo are discussed, including the steady divergence of theory and observational evidence over a quarter century. Cowling's criticisms of the primordial field theory are discussed; these include turbulent diffusion, the timing of the solar magnetic cycle, and the importance attached to observations in active and quiet magnetic regions.

062.106 **Dynamo action in cosmic bodies.**
J. H. Piddington.
Astrophys. Space Sci., Vol. 87, 477 - 478 (1982).

A copy of the paper (Piddington, 1982) was sent to Professor T. G. Cowling whose reply is of particular interest. Cowling and the author are in agreement to the extent that convective motions destroy cosmic magnetic fields much less speedily than assumed in formal dynamo theory. On the main point: the actual time of decay of a primordial stellar field or of a contemporary dynamo field there is disagreement. A primordial field will be shredded, or stretched and tangled, during the Hayashi phase and Cowling concludes that it then has ample time to decay by Ohmic diffusion and vanish entirely, while the author maintains that only the small-scale loop and filament field created by convection will be destroyed leaving a global primordial field.

062.107 **The propagation of strong-plane magnetogasdynamic shock waves in an optically-thin atmosphere.**
B. G. Verma, R. C. Srivastava, A. H. Khan.
Astrophys. Space Sci., Vol. 87, 297 - 304 (1982).

The propagation of plane magnetogasdynamic shock waves in an optically-thin grey atmosphere of non-uniform density has been discussed by the use of the similarity method, by use of Planck's diffusion approximation. The distributions of pressure, density, magnetic field, velocity, temperature, and radiation flux have been illustrated through graphs. The numerical integration has been done on a DEC-1090 computer under a RKGS programme.

062.108 **A self-similar flow of self-gravitating gas behind a shock wave with increasing energy. I.**
J. B. Singh.
Astrophys. Space Sci., Vol. 87, 347 - 355 (1982).

Self-similar flows of self-gravitating gas behind a spherical shock wave which are driven out by a propelling contact surface, propagating in a nonuniform atmosphere at rest, are investigated. The total energy content of the flow between a shock front and contact surface is taken to be time-dependent. In brief the self-similar homothermal flows of self-gravitating gas behind a shock wave and Roche's model case are also studied in this paper.

062.109 **Nonlinear astrophysical dynamo: three-mode interaction.**
S. Hinata.
Astrophys. Space Sci., Vol. 87, 403 - 406 (1982).

The effect of mode-coupling on the temporal variation of the linearly unstable magnetic mode is investigated for weakly unstable dynamos. It is shown that the nonlinear coupling of the unstable mode to itself is the dominant effect which stabilizes the mode at a certain amplitude rather than the coupling to other damped modes.

062.110 **Vorticity and astrophysical magnetism.**
P. F. Browne.
Astrophys. Space Sci., Vol. 87, 407 - 423 (1982).

Viscous resistance to differential rotation causes a current whose magnetic field is proportional to the vorticity of the medium. The magnetic fields of stars and galaxies could arise in this manner, provided that the time scale for development of the field is reasonable. The latter condition (assuming Ohmic rather than synchrotron dissipation) requires that the scale length for a galactic field be less than 3×10^{13} cm. It is suggested that there may be continual generation of field within the core of a vortex of this dimension in the galactic nucleus, the field lines then being carried outwards by expanding plasma. The main observational evidence in connection with solar, stellar and galactic magnetic fields is appraised in the context of the above theory.

062.111 **MHD free-convection flow in the Stokes problem for a porous vertical plate.**
A. K. Singh.
Astrophys. Space Sci., Vol. 87, 455 - 461 (1982).

The effect of a uniform transverse magnetic field on the free-convection flow of an electrically-conducting fluid past an infinite, vertical, porous plate for both classes of impulsive as well as uniformly-accelerated motion of the plate is discussed. The magnetic lines of force are assumed to be fixed relative to the plate. Expressions for the velocity field and skin friction for both cases are obtained by the Laplace transform technique. The influence of the various parameters, entering into the problem, on the velocity field and skin-friction is extensively discussed with the help of graphs and tables.

062.112 **Electron-photon interaction in hot strongly magnetized plasmas.**
H. Herold.
Accreting neutron stars, (see 012.069), p. 296 - 298 (1982).

The author presents results of calculations of electron-photon interaction processes in hot strongly magnetized plasmas, with special emphasis on the relativistic effects which turn out to be of considerable importance.

062.113 **Solution of gravitational polytrope equations by the quasi-invariance group.**
S. Bouquet, M. Feix.
C. R. Acad. Sci. Paris, Tome 295, Sér. II, 993 - 996 (1982).
In French.

The transformation of quasi-invariance gives some peculiar solutions for the non linear equations of the dynamic evolution of a gaseous mass (protostar, novae, supernovae). Contractions are unstable, expansions are stable. Furthermore the allowed expansion rates are correlated with the spatial structure of the configuration.

062.114 **Nonlocal stability analysis of the MHD Kelvin-Helmholtz instability in a compressible plasma.**
A. Miura, P. L. Pritchett.
J. Geophys. Res., Vol. 87, 7431 - 7444 (1982).
A general stability analysis is performed for the Kelvin-Helmholtz instability in sheared magnetohydrodynamic flow of finite thickness in a compressible plasma.

062.115 **Generation of atmospheric turbulence by solar radiation.**
K. L. Vithal, R. P. Vats.
Astrophys. Space Sci., Vol. 88, 205 - 217 (1982).
A statistical framework of weak turbulence is applied to investigate the maintenance of atmospheric turbulence during a long period, even after the external energy supply from solar radiations has stopped. Thus, the problem of hydrodynamical turbulence without any mean motion is dealt with. Main attention is drawn to one-dimensional Burgers equation, together with a discussion devoted mainly to Millionshtchikov's hypothesis, which may be applied as a consequence of the assumed weak turbulence. Remarkably, this leads to an explicit proof of the existence of the 'cascade process' in turbulence. The results are illustrated by numerical calculations.

062.116 **Dynamic stabilisation of Rayleigh-Taylor instability.**
B. K. Shivamoggi.
Astrophys. Space Sci., Vol. 88, 255 - 259 (1982).
Residual instabilities (associated with disturbances of small wavelengths in the vertical direction) present in a vertically-oscillating system comprised of a stratified fluid with density increasing exponentially in a vertically upward direction and subjected to a gravitational field acting vertically downward are shown to be removable by a magnetic field applied along the vertical direction.

062.117 **A self-similar flow in generalized Roche model with increasing energy.**
J. B. Singh.
Astrophys. Space Sci., Vol. 88, 269 - 275 (1982).
In the present paper the author investigated the self-similar flows in generalized Roche model with increasing energy of the wave. A typical solar-wind compression in the interplanetary space is produced by the propulsion of an enhanced fresh solar plasma following the onset of a flare and the total energy of the driven wave increases as it propagates from the Sun. Numerical integration has been made for adiabatic case only.

062.118 **Propagation of spherical shock waves in an exponential medium with radiation heat flux.**
J. B. Singh, S. K. Srivastava.
Astrophys. Space Sci., Vol. 88, 277 - 282 (1982).
In this paper propagation of spherical shock waves with radiation heat flux is considered in an exponentially increasing medium. The shock wave moves with variable velocity and the total energy of the wave is variable. For different values of radiation parameter, the numerical solution has been made and the nature of the field variables are illustrated by the tables.

062.119 **Comment on "Note on the stability of parallel magnetic fields".**
C. Uberoi.
Astrophys. Space Sci., Vol. 88, 511 - 512 (1982).
It is shown that the conclusions arrived at regarding the instability of an incompressible fluid cylinder in the presence of the magnetic field and the streaming velocity in a recent communication easily follow from the study of propagation characteristics of Alfvén surface waves along cylindrical plasma columns made earlier.

062.120 **Infinitely conducting rotating sphere in a magnetic field.**
A. Georgiou.
Nuovo Cimento B, Vol. 70B, Ser. 11, 221 - 230 (1982).
Abstr. in Phys. Abstr., Vol. 85, Abstr. 112701 (1982).

062.121 **Propagation of shock waves in hot stars.**
S. N. Ojha, S. N. Singh.
Nuovo Cimento B, Vol. 70B, Ser. 11, 212 - 220 (1982).
Abstr. in Phys. Abstr., Vol. 85, Abstr. 117582 (1982).

062.122 **The moments of the non-collisional parallel flux.**
M. Fridman.
Canadian J. Phys., Vol. 60, 1471 - 1476 (1982). In French.
Abstr. in Phys. Abstr., Vol. 86, Abstr. 2204 (1983).

062.123 **The mean electromotive force generated by random Alfvén waves in a collisionless plasma under a non-uniform mean magnetic field.**
T. Namikawa, H. Hamabata.
J. Plasma Phys., Vol. 28, Part 2, p. 293 - 298 (1982). – Abstr. in Phys. Abstr., Vol. 86, Abstr. 7719 (1983).

062.124 **The mean electromotive force generated by random hydromagnetic waves in a collisionless plasma.**
T. Namikawa, H. Hamabata, Y. Hosoya.
J. Plasma Phys., Vol. 28, Part 2, p. 299 - 307 (1982). – Abstr. in Phys. Abstr., Vol. 86, Abstr. 7720 (1983).

062.125 **The mean electromotive force generated by random Alfvén waves in a collisionless non-uniform plasma.**
H. Hamabata, T. Namikawa, Y. Hosoya.
J. Plasma Phys., Vol. 28, Part 2, p. 309 - 315 (1982). – Abstr. in Phys. Abstr., Vol. 86, Abstr. 7721 (1983).

062.126 **Fast magnetic field-line reconnexion in a compressible fluid. I. Coplanar field lines.**
A. M. Soward, E. R. Priest.
J. Plasma Phys., Vol. 28, Part 2, p. 335 - 367 (1982). – Abstr. in Phys. Abstr., Vol. 86, Abstr. 7738 (1983).

062.127 **Dielectrophoretic, thermal instability in a spherical shell of fluid.**
W. J. Markiewicz, K. D. Aldridge.
Geophys. Astrophys. Fluid Dyn., Vol. 21, 225 - 236 (1982).
Abstr. in Phys. Abstr., Vol. 86, Abstr. 10238 (1983).

062.128 **The structure of untwisted magnetic flux tubes.**
P. K. Browning, E. R. Priest.
Geophys. Astrophys. Fluid Dyn., Vol. 21, 237 - 263 (1982).
Abstr. in Phys. Abstr., Vol. 86, Abstr. 10289 (1983).

062.129 **A numerical study of α^2-dynamos with anisotropic α-effect.**
E. Weisshaar.
Geophys. Astrophys. Fluid Dyn., Vol. 21, 285 - 301 (1982).
Abstr. in Phys. Abstr., Vol. 86, Abstr. 10290 (1983).

062.130 **The role of the helicity spectrum function in turbulent dynamo theory.**
H. K. Moffatt, M. R. E. Proctor.
Geophys. Astrophys. Fluid Dyn., Vol. 21, 265 - 283 (1982).
Abstr. in Phys. Abstr., Vol. 86, Abstr. 10314 (1983).

062.131 **Computation and analysis for relativistic shock waves and hot spots in extragalactic double radio sources.**
X.-y. Kong, Y.-h. Xu, Z.-x. Zhong.
Acta Astrophys. Sinica, Vol. 2, 81 - 90 (1982). In Chinese.

A model is applied to two kinds of media colliding with relativistic relative velocity, and according to the theory of shock waves for relativistic fluids early main parameters are calculated including the separation velocities of the components and the velocities of the shock fronts when double radio components are formed. The evolutional processes of hot spots from intragalaxy to intergalaxy are qualitatively discussed. Then the parameters for the quasisteady state of the radio components are approximately computed by means of expressions pertaining to a strong relativistic shock. The results obtained are in accord with observational data mainly.

062.132 **The influence of a flow field on the stability of a force free magnetic field.**
L. Wu, F. Wu.
Acta Astrophys. Sinica, Vol. 2, 152 - 158 (1982). In Chinese.

062.133 **Numerical simulations of magnetohydrodynamic chemically reactive flows in three dimensions on a vector computer.**
K. Hain, G. Hain.
Comput. Phys. Commun., Vol. 27, 325 - 333 (1982). – Abstr. in Phys. Abstr., Vol. 86, Abstr. 17983 (1983).

The Alfvén wave.
See Abstr. 003.049.

Solar magneto-hydrodynamics.
See Abstr. 003.100.

Plasma astrophysics at Santa Barbara.
See Abstr. 011.006.

Observations of structuring in the downstream region of a large spherical model in a laboratory simulated solar wind plasma.
See Abstr. 022.080.

On the solution of the time-dependent inertial-frame equation of radiative transfer in moving media to O(v/c).
See Abstr. 063.050.

Merging of hydrogen spectral lines near the Balmer limit.
See Abstr. 063.053.

Radiation from a strongly-magnetized plasma: the case of predominant scattering.
See Abstr. 063.054.

The motion and radiation of electrons in a neutral sheet.
See Abstr. 063.066.

Physics of stellar magnetospheres.
See Abstr. 064.036.

Numerical modeling of magnetospheres of rotating stars.
See Abstr. 064.037.

On the compatibility of thermal and hydrostatic equilibrium in thin radiative accretion disks.
See Abstr. 064.042.

New results on the mechanism of magnetic flux pumping by three-dimensional convection.
See Abstr. 064.059.

The Biermann mechanism and spontaneous magnetic field generation in stars.
See Abstr. 065.061.

A contribution to the study of the stability and oscillations of stars with a magnetic field.
See Abstr. 065.066.

Evolution of radiating anisotropic spheres in general relativity.
See Abstr. 066.097.

On the physics of relativistic double layers.
See Abstr. 066.225.

Coronal loop transients in streamer configurations.
See Abstr. 074.067.

Electromagnetic velocity gradient instabilities in the magnetosphere.
See Abstr. 084.050.

Solutions to the equations for corotating magnetospheric convection.
See Abstr. 099.098.

The plasma phase of the evolution of the solar system.
See Abstr. 107.031.

In situ acceleration in extragalactic radio jets.
See Abstr. 141.138.

The structure of cosmic ray shocks.
See Abstr. 143.002.

Non-linear theory of cosmic ray shocks including self-generated Alfvén waves.
See Abstr. 143.058.

Periodic orbits and gas flow in barred spirals.
See Abstr. 151.035.

Vlasov equation?
See Abstr. 151.038.

Turbulence, plasma containment, and galaxies.
See Abstr. 151.040.

Gravothermal and gravogyro catastrophes of rotating and self-gravitating gaseous disks.
See Abstr. 151.047.

Numerical gas dynamics in the galactic scales.
See Abstr. 151.051.

063 Radiative Transfer, Scattering

063.001 **Broadband linear polarization from magnetized stellar atmospheres. Numerical tables for the magnetic intensification mechanism.**
E. Landi Degl'Innocenti, G. Calamai.
Astron. Astrophys., Suppl. Ser., Vol. 49, 677 - 685 (1982).
Some numerical tables are presented concerning the broadband linear polarization to be expected from a magnetized stellar atmosphere through the mechanism of magnetic intensification.

063.002 **Approximate methods for solution of radiative transfer problems in a strongly magnetized cool plasma.**
A. D. Kaminker, G. G. Pavlov, N. A. Silant'ev, Yu. A. Shibanov.
Astrofizika, Tom 18, 283 - 299 (1982). In Russian. English translation in Astrophysics, Vol. 18, No. 2.
Two approximate methods are developed for the solution of the equations of radiative transfer in an optically thick plasma for any orientation of a uniform magnetic field. Using the approaches developed simple formulae are derived convenient for calculations as well as for qualitative analysis of spectral, angular and polarization characteristics of the outgoing radiation.

063.003 **A model for the negative polarization of light scattered by cosmic bodies deprived of atmospheres.**
Yu. G. Shkuratov.
Astron. Zh., Tom 59, 817 - 822 (1982). In Russian. English translation in Soviet Astron., Vol. 25, No. 4.
Formulae are obtained describing the polarization of light scattered by planets deprived of atmospheres for small phase angles (up to 30°). It is suggested that the negative polarization is due to a combination of the shadow effect with single double non-coplanar Fresnel reflections from microfacets of planetary soil particles.

063.004 **Velocity fields and spectral line asymmetries: a linearized analytical approach to the theory of the line bisector in a Milne-Eddington atmosphere.**
B. Buonaura, B. Caccin.
Astron. Astrophys., Vol. 111, 113 - 116 (1982).
A method using the bisector of a spectral line profile for the analysis of macroscopic velocity fields in the solar atmosphere, based on the linear approximation derived by Canfield (1976) for the case of velocities substantially smaller than the thermal velocity, is applied to get a fully analytical solution for the case of an exponentially stratified Milne-Eddington atmosphere. Ultimate limits to the information on the depth dependence of the velocity field that might be retrieved in this way from data obtained with infinite spatial and temporal resolution are derived from the intrinsic smoothing due to radiative transfer.

063.005 **Simplified expressions for the gyrosynchrotron radiation from mildly relativistic, nonthermal and thermal electrons.**
G. A. Dulk, K. A. Marsh.
Astrophys. J., Vol. 259, 350 - 358 (1982).
The authors present simplified expressions for the gyrosynchrotron radiation from mildly relativistic electrons, both for nonthermal (power-law) and for thermal (Maxwellian) distributions. They give expressions for the emissivity, absorptivity, effective temperature, frequency of peak brightness, and degree of circular polarization. Of particular interest are the ~10 keV to ~1 MeV electrons that are probably important in solar and stellar flares.

063.006 **The source function in a finite inhomogeneous atmosphere with internal sources.**
T. Viik.
Publ. Tartu Astrofiz. Obs., Tom 48, 10 - 24 (1981).
The source function in a finite inhomogeneous atmosphere with isotropic internal sources and with a reflecting boundary is obtained via the invariant inbedding equations. Specific numerical results for an atmosphere with given parameters are presented.

063.007 **Asymptotic behaviour of the resolvent function in radiative transfer.**
R. Rõõm, T. Viik, K. Kärner.
Publ. Tartu Astrofiz. Obs., Tom 48, 25 - 40 (1981).
In Russian.
The asymptotic behaviour of the resolvent function in a slab of large optical thickness is studied.

063.008 **Radiation fields for isotropically scattering finite atmospheres by reflecting surfaces.**
T. Viik, A. Heinlo.
Publ. Tartu Astrofiz. Obs., Tom 48, 63 - 80 (1981).
The aim of this paper is to study the influence of a reflecting surface on the radiation field in a homogeneous atmosphere in which both absorption and isotropic scattering take place.

063.009 **Radiative opacities and transfer in magnetic X-ray pulsars.**
J. Ventura.
Galactic X-ray sources, (see 003.001), p. 241 - 252 (1982).
Contents: Scattering and bremsstrahlung opacities in a plasma. Propagation effects of the medium. Magnetic vacuum effects. Spectra emitted from a selfradiating cylinder.

063.010 **Super-critical X-ray luminosities: the structure and stability of a radiation-supported plasma layer.**
Y.-M. Wang.
Astron. Astrophys., Vol. 112, 24 - 34 (1982).
The structure and stability of a plane-parallel Thomson-opaque plasma layer, supported against gravity by radiation and thermal pressures, are examined. The effects of Compton heating by the X-rays incident from below and of thermal bremsstrahlung by the layer itself are taken into account, and an approximate model for the radiative energy transfer is developed. The layer is shown to be dynamically unstable, suggesting that additional stabilizing factors must be present in X-ray sources with steady luminosities substantially exceeding the Eddington limit.

063.011 **Line profile fluctuations in a turbulent atmosphere.**
M. L. Loucif, C. Magnan.
Astron. Astrophys., Vol. 112, 287 - 291 (1982).
In a previous work the effective cell approximation was introduced in order to treat the line transfer problem in the presence of velocity fields with a view of taking full account of the spatial correlation of the field. Results relative to the mean emergent intensity were obtained. Here are presented calculations pertaining to the fluctuations of the intensity. The profiles are studied as a function of the velocity of the last emitting layer. It is shown that the individual profiles strongly deviate from the mean value as soon as the optical depth over one correlation length reaches values about unity.

063.012 **Induced Compton scattering in radiative transfer.**
D. B. Wilson.
Mon. Not. R. Astron. Soc., Vol. 200, 881 - 906 (1982).
The effect of induced Compton scattering – a non-

linear radiation scattering mechanism relevant when the brightness temperature $T_b > 10^{10}$ K – on the observed properties of sources is studied. The basic equations, including polarization are derived. Radiative transfer, particularly in a spherical geometry, is studied. The application to astronomical sources, and compact extragalactic sources in particular, is discussed.

063.013 **Fluctuations of radiation intensity and the hypothesis of "freezing".**
Eh. M. Krikunova, F. A. Markus.
Izv. vuzov. Radiofiz., Tom 24, 1038 - 1039 (1981). In Russian. Abstr. in Ref. zh., 51. Astron., 7.51.158 (1982).

063.014 **On the statistical theory of back scattering in randomly inhomogeneous media.**
B. M. Shevtsov.
Izv. vuzov. Radiofiz., Tom 24, 1351 - 1355 (1981). In Russian. Abstr. in Ref. zh., 51. Astron., 7.51.160 (1982).

063.015 **Radiative transfer and remote sensing.**
B. J. Conrath.
Vibrational-rotational spectroscopy for planetary atmospheres, (see 012.007), p. 47 - 62 (1982).

Radiative transfer is the basic theoretical tool for the quantitative interpretation of planetary infrared spectra. It links the remotely sensed data to the properties of the atmosphere. This paper gives a brief overview of the remote sensing problem as it pertains to the interpretation of planetary spectra.

063.016 **Rayleigh, Raman and particulate scattering.**
W. D. Cochran.
Vibrational-rotational spectroscopy for planetary atmospheres, (see 012.007), p. 63 - 85 (1982).

To understand the information contained in the absorption spectrum of a planetary atmosphere one must first understand the various scattering processes these photons undergo within the atmosphere. There are three different physical processes: Rayleigh scattering and Raman scattering by the gas molecules, and scattering by any cloud or dust aerosol particles in the atmosphere. The author examines the physics of each of these processes.

063.017 **On spectral line profiles for a one-dimensional expanding medium. I. Orthogonal and Doppler absorption coefficients.**
S. I. Grachev.
Vestn. LGU, 1982, No. 1, p. 77 - 86. In Russian. – Abstr. in Ref. zh., 51. Astron., 8.51.109 (1982).

063.018 **Monte-Carlo calculations of formation of resonance lines in plane layers.**
Yu. P. Vlasov, L. G. Titarchuk.
Pis'ma Astron. Zh., Tom 8, 566 - 572 (1982). In Russian. English translation in Soviet Astron. Lett., Vol. 8.

The output radiation spectra resulting from scattering of the input parallel flux of photons in a plane layer is calculated in the cases of complete and non-complete redistribution. The spectra in these two cases differ when the frequency of input photons coincides with the line center. Radiative source functions at different optical depths are obtained as function of the angle of incident photon flux. The transparence functions and number of collisions as functions of optical depth are calculated.

063.019 **Reflection and transmission by a vertically inhomogeneous planetary atmosphere.**
K. Stamnes.
Planet. Space Sci., Vol. 30, 727 - 732 (1982).

In a recent paper (Stamnes and Swanson, 1981) simple expressions were derived for the albedo and transmissivity of a homogeneous plane parallel atmosphere. It was also pointed out that these expressions – obtained by appealing to the reciprocity principle – lead to substantial savings in computing time when only integrated quantities such as albedo and transmissivity are required. The purpose of the present paper is to derive and verify similar expressions for an inhomogeneous atmosphere both for the standard problem with no ground reflection and the planetary problem in which the lower boundary is partially reflecting.

063.020 **Numerical evaluation of optical single-scattering properties using multiple-scattering inverse transport methods.**
R. Sanchez, N. J. McCormick.
J. Quant. Spectrosc. Radiat. Transfer, Vol. 28, 169 - 184 (1982).

Five inverse methods for determining the single-scattering properties from monochromatic optical experiments in a homogeneous multiple-scattering slab geometry are presented and compared using analytical calculations. The methods require only measurements of the specific intensities entering and leaving the target.

063.021 **Regularities and similarities in plasma broadened spectral line widths (Stark widths).**
W. L. Wiese, N. Konjevic.
J. Quant. Spectrosc. Radiat. Transfer, Vol. 28, 185 - 198 (1982).

Regularities and similarities in plasma broadened line widths have been studied by a comprehensive analysis of existing experimental data.

063.022 **Time-dependent radiative transfer with automatic flux limiting.**
D. Mihalas, R. Weaver.
J. Quant. Spectrosc. Radiat. Transfer, Vol. 28, 213 - 222 (1982).

The authors discuss a simple method for solving the time-dependent transfer problem. This scheme is automatically flux-limited and affords physical insight into how flux limitation occurs. The authors develop a second-order, time-dependent radiation energy equation that is similar in form to the diffusion limit radiation energy equation. This time-dependent energy equation approaches physically reasonable equations in optically thick and thin regions. Computational aspects of solving this energy equation are discussed.

063.023 **Two-dimensional linearly anisotropic scattering in a semi-infinite cylindrical medium exposed to a laser beam.**
A. L. Crosbie, R. L. Dougherty.
J. Quant. Spectrosc. Radiat. Transfer, Vol. 28, 233 - 263 (1982).

A modification of Ambarzumian's method is used to develop the integro-differential equations for the source function, flux, and intensity at the boundary of a two-dimensional, semi-infinite cylindrical medium which scatters linearly. The incident radiation is collimated, normal to the top surface of the medium, and is dependent only on the radial coordinate. The radial variation is assumed to be a Bessel function or a Gaussian distribution. The Gaussian boundary condition is used to simulate a laser beam. Numerical results are presented in graphical and tabular forms for both boundary conditions. Results for forward and backward scattering phase functions are compared with those for isotropic scattering. A method is presented for extending these results to the problem of a strongly anisotropic phase function which is made up of a spike in the forward direction superimposed on a linear phase function.

063.024 **The internal radiation of a planetary atmosphere.**
A. H. Karp.

J. Quant. Spectrosc. Radiat. Transfer, Vol. 28, 265 - 270 (1982).

The author shows how to find the spherical harmonics moments of the radiation at any point within a homogeneous layer embedded in an inhomogeneous planetary atmosphere given only the moments at the boundaries of the layer. The result is used to determine the heating rate within a cloud in a realistic model atmosphere. It is also used to estimate the visibility of the ground from various points in the cloud as a function of view angle. The strength of the method is that the multiple scattering problem for the full atmosphere need be solved only once, independently of the proposed applications.

063.025 **Generalization of the spherical harmonic method to radiative transfer in multi-dimensional space.**
S.-C. S. Ou, K.-N. Liou.
J. Quant. Spectrosc. Radiat. Transfer, Vol. 28, 271 - 288 (1982).

063.026 **Information content of a planetary spectrum.**
J. C. Buriez, Y. Fouquart.
J. Quant. Spectrosc. Radiat. Transfer, Vol. 28, 327 - 342 (1982).

The information content of equivalent widths and line profiles is investigated when all of the pressure-broadened lines belong to a unique spectrum. The parameters which can be derived without a priori knowledge of the planetary atmosphere are emphasized. In general, the analysis of equivalent widths alone does not allow one to distinguish between different atmospheric models. A scattering indicator, which is a dimensionless number derived from the shape of line wings, is found to determine the influence of scattering in the absorption process.

063.027 **Spectral polarization properties of optically thin synchrotron radiation.**
C.-I. Björnsson, G. R. Blumenthal.
Astrophys. J., Vol. 259, 805 - 819 (1982) = Lick Obs. Bull., No. 915.

This paper considers optically thin synchrotron emission from an arbitrary energy distribution of relativistic electrons in an arbitrary magnetic field distribution. It is demonstrated that for optically thin synchrotron sources, information about the underlying magnetic field distribution can be obtained using accurate measurements of the polarization and spectral index as functions of frequency.

063.028 **Electron-cyclotron masers as the source of certain solar and stellar radio bursts.**
D. B. Melrose, G. A. Dulk.
Astrophys. J., Vol. 259, 844 - 858 (1982).

High brightness temperatures ($\gtrsim 10^{10}$ K) and high degrees of circular polarization in solar microwave spike bursts and in radio bursts from some stars can be explained in terms of an electron-cyclotron maser. A theory is developed for maser emission due to energetic electrons with a relatively mild loss cone anisotropy.

063.029 **On redistribution and the equations for radiative transfer.**
J. Cooper, R. J. Ballagh, K. Burnett, D. G. Hummer.
Astrophys. J., Vol. 260, 299 - 316 (1982).

The authors outline the derivation of the equations of statistical equilibrium, starting from the quantum density-matrix equations, drawing particular attention to the approximations and assumptions used in the development of tractable expressions. Using the quantum-fluctuation-regression theorem, they obtain emission and absorption coefficients for multilevel atomic systems which are nondegenerate except for m-substates (thus excluding hydrogen).

063.030 **The influence of dielectronic-like recombination at low temperatures on the interpretation of interstellar, radio recombination lines of carbon.**
C. M. Walmsley, W. D. Watson.
Astrophys. J., Vol. 260, 317 - 325 (1982).

Calculations which include dielectronic-like captures due to fine structure excitations are presented at temperatures less than 100 K for the departure coefficients associated with the radio recombination lines of carbon. Results are presented for a wide range of principal quantum numbers ($n \approx 75-900$) and are especially applicable in the context of two astronomical topics related to the interstellar gas: (a) the observation of low frequency absorption lines ($n > 500$), and (b) the relative strengths of carbon recombination lines at high and low frequencies in the direction of H II regions. The authors show that in diffuse interstellar clouds, the absorption line strength for α transitions near $n \approx 600$ can be enhanced by factors of 50–100 due to dielectronic recombination.

063.031 **Concise and accurate solutions for Chandrasekhar's X and Y functions.**
W. L. Dunn, R. D. M. Garcia, C. E. Siewert.
Astrophys. J., Vol. 260, 849 - 854 (1982).

The F_N method is used in a new way to develop, for the case of isotropic scattering, an efficient and accurate procedure for computing the X and Y functions introduced by Chandrasekhar to solve the standard problem in radiative transfer.

063.032 **Polarization properties of a source in relativistic motion.**
C.-I. Björnsson.
Astrophys. J., Vol. 260, 855 - 867 (1982) = Lick Obs. Bull., No. 917.

The polarization properties of synchrotron radiation from a source in relativistic motion are considered using a three-dimensional magnetic field distribution. It is assumed that the variation of the polarization is primarily due to change in the velocity of the source. The Stokes parameters are evaluated in the rest frame of the moving source and then transformed to the observer's frame. Relativistic effects are best distinguished by correlating the flux with the polarization properties. It is suggested that the polarization properties of the knots in M87 might indicate relativistic motion along a jet axis close to the line of sight.

063.033 **The transport of polarized synchrotron radiation in a turbulent medium.**
S. R. Spangler.
Astrophys. J., Vol. 261, 310 - 320 (1982).

The author has developed a model to describe the transport of polarized synchrotron radiation in an optically thin source possessing small amplitude fluctuations in thermal electron density, energetic electron density, and vector magnetic field. Expressions are derived relating the (observable) autocorrelations and cross-correlations of the Stokes parameters Q and U to the correlation functions of the aforementioned physical quantities.

063.034 **Cerenkov line radiation and the emission-line spectra of QSOs.**
J.-h. You, F.-h. Cheng.
Chinese Phys., Vol. 2, 17 - 24, 16 (1982). – Abstr. in Phys. Abstr., Vol. 85, Abstr. 68927 (1982) – See Abstr. 28.063.050.

063.035 **Comptonization in a radiation-dominated shock wave and spectra of X-ray pulsars.**
Yu. Eh. Lyubarskij, R. A. Syunyaev.
Pis'ma Astron. Zh., Tom 8, 612 - 622 (1982). In Russian. English translation in Soviet Astron. Lett., Vol. 8.

A bulk of accretion energy under high accretion rate is released by a radiation-dominated shock. Radiation stops plasma flow and takes energy away. Multiple scattering of photons on electrons in the braking flow form a characteristic spectrum of radiation and lead to the heating of electrons. This mechanism results in the formation of a hard tail in X-ray pulsars spectra.

063.036 **Escape probabilities, mean number of scatterings and net radiative bracket for resonance lines.**
H. Frisch.
Astron. Astrophys., Vol. 114, 119 - 124 (1982).
Multiple scattering of resonance line photons in non-conservative media, where photons have a small probability ϵ to be destroyed at each scattering, is examined. An approximation for the mean number of scatterings, which takes into account both destruction of photons and escape through the boundaries, is obtained from scaling arguments. It is used to show that the fraction of the emitted energy which escapes from the medium, known as the mean escape probability, is almost independent of the value of ϵ. Analytical solutions valid deep in the interior and accurate numerical solutions of the transfer problem are used to check the validity of this result and to examine the errors made when scaling laws for the mean escape probability are transformed into scaling laws for the depth variation of the net radiative bracket.

063.037 **On the interpretation of optically thin synchrotron spectra.**
S. Pineault.
Astron. Astrophys., Vol. 114, 177 - 181 (1982).
A general formalism is presented which allows for the description of the optically thin spectrum of unresolved inhomogeneous synchrotron sources, including the effect of a radially varying maximum energy cut-off in the particle energies. The role of the source geometry, nature of the particle energy distribution (mono-energetic versus power-law type), and frequency range affected are discussed. It is shown that the formalism can be applied to a subclass of radio sources exhibiting convex (C^{-}) spectra.

063.038 **A new look at the discrete ordinate method for radiative transfer calculations in anisotropically scattering atmospheres. II. Intensity computations.**
K. Stamnes, H. Dale.
J. Atmos. Sci., Vol. 38, 2696 - 2706 (1981). – Abstr. in Phys. Abstr., Vol. 85, Abstr. 78947 (1982).

063.039 **Corrigendum on the numerical evaluation of the *H*-functions of transfer problems in multiplying media.** (Astrophysics and Space Science 72, 71, 1980).
S. R. Das Gupta, Z. Islam, B. Majee, K. S. Kölbig.
Astrophys. Space Sci., Vol. 86, 131 - 138 (1982). – See Abstr. 28.063.038.

063.040 **A generalized principle of invariance and radiation field in multilayer atmospheres.**
T. Viik.
Astrophys. Space Sci., Vol. 86, 169 - 178 (1982).
A generalized principle of invariance is derived for a plane-parallel atmosphere. On the basis of this principle a method for determining the radiation field in a multilayer atmosphere is proposed. This method, the first part of which is the well-known adding method, permits the application to problems involving optically finite as well as semi-infinite atmospheres. The reflecting boundaries may be incorporated, though in that case it is not possible to use the adding method. Some numerical results are given for the standard and Milne problems and for the problem with internal sources.

063.041 **Theory for thermal microwave emission from a homogeneous layer with rough surfaces containing spherical scatterers.**
R. T. Shin, J. A. Kong.
J. Geophys. Res., Vol. 87, 5566 - 5576 (1982).

063.042 **Pumping of H II/OH masers: IR line overlaps and collisional excitation by H_2.**
D. R. Flower, S. Guilloteau.
Astron. Astrophys., Vol. 114, 238 - 244 (1982).
The authors present the first quantitative model of H II/OH maser sources in which both radiative and collisional processes are taken into account in a coherent manner. The model incorporates recently computed rates of rotational excitation of OH by para-H_2 and the effects of "thermal" overlap of the far infrared lines of OH. The authors illustrate first the characteristics of collisional pumping in the absence of line overlap, then the global effects when overlap is included. When both mechanisms operate, the observed properties of ground and excited state maser emission of H II/OH sources can be reproduced, whereas this is not the case when only one of the mechanisms is included. It is shown that thermal far infrared radiation from local dust effectively quenches the masers if the temperature of the dust is comparable to or greater than the kinetic temperature of the gas.

063.043 **The Lumme-Bowell photometric parameters: reality or fantasy?**
B. Hapke.
Bull. American Astron. Soc., Vol. 14, 726 (1982). – Abstract.

063.044 **A reply to Hapke's criticism of the Lumme-Bowell photometric theory.**
K. Lumme, E. Bowell.
Bull. American Astron. Soc., Vol. 14, 726 (1982). – Abstract.

063.045 **A method of Cauchy integral equation for non-coherent transfer in half-space.**
H. Frisch, U. Frisch.
J. Quant. Spectrosc. Radiat. Transfer, Vol. 28, 361 - 375 (1982).
A technique is presented which allows easy construction of solutions for various half-space problems arising in non-coherent radiative transfer with complete redistribution. By use of an inverse Laplace transform method, Wiener-Hopf integral equations are reduced to Cauchy-type singular integral equations. The factorization technique used by Case and Zweifel for coherent scattering can then be carried over to non-coherent transfer. The method is applied to the inhomogeneous integral equation for the source function of a two-level atom, previously solved by Ivanov. It is also applied to the conservative, homogeneous case and to singular Wiener-Hopf equations arising from asymptotic expansions in the limit of vanishing probability of collisional destruction ϵ.

063.046 **Non-LTE transfer with complete redistribution. Scaling laws for a slab.**
H. Frisch.
J. Quant. Spectrosc. Radiat. Transfer, Vol. 28, 377 - 381 (1982).
Scaling laws for conservative scattering in a finite slab are extracted from an asymptotic analysis of the integral equation for the source function. The solution is separated into an interior and a surface boundary layer part. The matching between the two parts provides a scaling law for the surface value of the source function. When expressed in terms of a mean number of scatterings, this scaling law is generalizable to non-conservative scattering and the empirical formula of Jones and Skumanich is recovered.

063.047 **Photon path length distributions for an isotropically scattering planar medium.**
R. D. Skocypec, R. O. Buckius.
J. Quant. Spectrosc. Radiat. Transfer, Vol. 28, 425 - 439 (1982).

063.048 **Luminosity limits for funnels in thick accretion discs.**

R. Nityananda, R. Narayan.
Mon. Not. R. Astron. Soc., Vol. 201, 697 - 705 (1982).

A narrow, highly luminous funnel in a thick accretion disc is a common feature of many models for active galactic nuclei. The authors examine the constraints on the luminosity imposed by the effects of radiation forces on the funnel. Their treatment allows properly for the irradiation of any part of the funnel by the rest, an important effect for this problem. The authors show that the resulting large unbalanced forces at the funnel surface cause a significant outflow of matter which should be incorporated in the model for consistency. The results do not depend on the detailed angular momentum distribution over the disc surface but only on the funnel geometry.

063.049 **Inference of nebular density and luminosity structure from polarization maps.**
J. C. Brown.
Mon. Not. R. Astron. Soc., Vol. 201, 735 - 743 (1982).

The problem of interpreting polarimetric maps of nebulae is discussed in terms of deconvolving the nebular structure from the observed distribution of Stokes' parameters on the sky. It is shown that for an optically thin self-luminous nebula in which light from a single star is scattered from spherical particles, the problem can be expressed by a pair of integral equations of Volterra type, to be solved for the effective (two-dimensional equivalent) spatial distribution of scattering particles $n_0(\mathrm{cm}^{-3})$ and of nebular emission $l_0(\mathrm{erg\ cm}^{-3}\mathrm{s}^{-1})$ from the observed distribution of total and polarized intensities I, I_p.

063.050 **On the solution of the time-dependent inertial-frame equation of radiative transfer in moving media to O(v/c).**
D. Mihalas, R. I. Klein.
J. Comput. Phys., Vol. 46, 97 - 137 (1982). – Abstr. in Phys. Abstr., Vol. 85, Abstr. 97355 (1982).

063.051 **An iterative solution to Chandrasekhar X and Y equations for isotropic scattering.**
J. Caldwell.
J. Appl. Phys., Vol. 53, 5479 - 5483 (1982). – Abstr. in Phys. Abstr., Vol. 85, Abstr. 103006 (1982).

063.052 **Compton heating and cooling and astrophysical jets.**
M. Calvani, L. Nobili.
Lett. Nuovo Cimento, Vol. 35, Ser. 2, 7 - 13 (1982). – Abstr. in Phys. Abstr., Vol. 85, Abstr. 112318 (1982).

063.053 **Merging of hydrogen spectral lines near the Balmer limit.**
V. Ruždjak, V. Vujnović.
Publ. Astron. Obs. Sarajevo, Vol. 1, No. 1, (see 012.043), p. 209 - 214 (1981).

063.054 **Radiation from a strongly-magnetized plasma: the case of predominant scattering.**
A. D. Kaminker, G. G. Pavlov, Yu. A. Shibanov.
Astrophys. Space Sci., Vol. 86, 249 - 297 (1982).

On the basis of diffusion approach for normal modes, solutions of the radiative transfer problem are obtained and analysed for an optically thick tenuous plasma with a strong magnetic field. The case is considered when the scattering processes without change of photon frequency are dominant. The effects of electron-positron vacuum polarization are taken into account and shown to be very significant. The results obtained are necessary for analysing X-ray and gamma-ray radiation from strongly magnetized neutron stars.

063.055 **Non-LTE resonance line polarization with partial redistribution effects.**
D. E. Rees, G. J. Saliba.
Astron. Astrophys., Vol. 115, 1 - 7 (1982).

The physics of formation of non-LTE resonance line polarization is studied for a two-level atom. Partial redistribution effects are included via an approximate treatment of the case of coherent scattering in an atomic rest frame in which the upper atomic level is naturally broadened. The transfer equations are solved in finite and semi-infinite model atmospheres and the relevance of the solutions to solar observations is discussed. In particular, qualitative agreement is obtained with the linear polarization profiles of Ca I 4227 Å, NaD_2, and Ca II K observed by Stenflo et al. (1980). These profiles have maxima in the line core and the line wings, contrary to Wiehr's (1981) NaD_2 data in which the core maximum is not observed.

063.056 **X-ray nebular models.**
T. R. Kallman, R. McCray.
Astrophys. J., Suppl. Ser., Vol. 50, 263 - 317 (1982).

Theoretical models are presented for the temperature and ionization structure of spherically symmetric, constant density, gaseous nebulae surrounding compact X-ray sources and for the optical, UV, and X-ray spectra emerging from the nebulae. Model results are presented for a wide variety of gas densities and X-ray source spectra, scaling laws which allow these results to be generalized to a wide variety of astrophysical solutions are discussed, and column densities of multiply charged species are tabulated.

063.057 **On the integral form of the equation of transfer for a homogeneous sphere.**
G. C. Pomraning, C. E. Siewert,
J. Quant. Spectrosc. Radiat. Transfer, Vol. 28, 503 - 506 (1982).

Basic considerations are used to derive the integral form of the equation of transfer relevant to a homogeneous sphere with internal sources as well as external illumination and a specularly and diffusely reflecting surface.

063.058 **Exact expressions for radiative transfer in a three-dimensional rectangular geometry.**
A. L. Crosbie, R. G. Schrenker.
J. Quant. Spectrosc. Radiat. Transfer, Vol. 28, 507 - 526 (1982).

The equations for the source function, flux, and scattered intensity normal to the surface are formulated in cartesian coordinates for a 3-D rectangular absorbing, emitting, isotropically scattering medium exposed to both diffuse and collimated radiation. Simplifications of these equations for certain important geometries and uniform loading are presented. Also, superposition of these equations and radiative equilibrium are discussed. For pure scattering, the source function at the center of the square and cubic geometries is analytically determined for the diffuse boundary condition. The generalized 3-D equations are shown to reduce to the familiar 1-D results.

063.059 **An analytical approximation of the resolvent.**
T. Viik.
Tartu Astrofüüs. Obs. Teated, No. 66, p. 38 - 50 (1982).

The resolvent of the integral equation of radiative transfer for a plane-parallel homogeneous isotropically scattering atmosphere is approximated by an analytical formula. To obtain the resolvents for the case of semi-infinite and finite atmospheres the FORTRAN codes are listed in the Appendix.

063.060 **Second-order escape probability approximations in radiative transfer.**
D. G. Hummer, G. B. Rybicki.
Astrophys. J., Vol. 263, 925 - 934 (1982).

Second-order escape probability approximations make some allowance for the transfer of radiation between the point where a photon is created and that where it escapes or is absorbed. An approximation of this kind has recently been formulated by Puetter et al. for planar atmospheres of finite thickness, in the form of a first-order differential equation relating the integrated mean intensity to the source function. The authors give two alternative normalizations to the one proposed by these authors; the first of these enforces global conservation of photons in each transition, and the second gives reasonably accurate results for much less computational effort than the first.

063.061 **The emergent spectra of sources in scattering clouds.**
J. G. Peters, S. M. Lea.
Publ. Astron. Soc. Pacific, Vol. 94, 756 (1982). – Abstract.

063.062 **On spectral line profiles forming in a one-dimensional expanding medium. II. Attenuation range coefficient.**
S. I. Grachev.
Vestn. LGU, 1982, No. 7, p. 85 - 92. In Russian. – Abstr. in Ref. zh., 51. Astron., 10.51.114 (1982).

063.063 **On an approximation formula in radiative transfer.**
T. Viik.
Tartu Astrofüüs. Obs. Publ., Köide 49, 28 - 39 (1982).
A non-linear approximation formula for radiative transfer in a finite slab is derived.

063.064 **Bound-free and free-free transitions in strong magnetic fields.**
G. Wunner.
Accreting neutron stars, (see 012.069), p. 299 - 301 (1982).
The processes of photoionization, electron impact ionization, photorecombination, three-body recombination and bremsstrahlung are calculated for the presence of a strong magnetic field ($\sim 10^{10}$ gauss). The results should be applicable to plasma conditions near the surfaces of magnetized neutron stars.

063.065 **Second-order scattering approximation to X-ray photon transport at balloon altitudes.**
G. Auriemma, M. Badiali, D. Cardini.
Astrophys. Space Sci., Vol. 88, 15 - 20 (1982).
A numerical solution of the transport equation, based on the iterative method of expansion in order of scattering, is proposed to evaluate the effect of multiple Compton scattering on the cosmic X-ray spectrum from balloon experiments. Computation up to the second order of scattering is performed in the hard X-ray range (10–200 keV). It is shown that this approximation is sufficient to evaluate the X-ray flux at typical balloon atmospherical depths.

063.066 **The motion and radiation of electrons in a neutral sheet.**
S.-Y. An, L.-F. Luo.
Astrophys. Space Sci., Vol. 88, 159 - 171 (1982).
The motion and radiation of an electron in the neighbourhood of a neutral sheet are discussed. Formulae for the synchrotron and cyclotron radiation of an electron in an inhomogeneous field are obtained. The results are compared with radiation in a homogeneous field. It is predicted that the momentum of charged particles flying from a neutral sheet may take discrete eigenvalues.

063.067 **On approximating the solutions of the Chandrasekhar H-equation.**
S. R. Vatsya.
J. Math. Phys., Vol. 23, 1728 - 1731 (1982). – Abstr. in Phys. Abstr., Vol. 85, Abstr. 117581 (1982).

063.068 **Turbulence in planetary occultations: a strong scattering formulation including an inhomogeneous background.**
B. S. Haugstad.
Radio Sci., Vol. 17, 565 - 573 (1982). – Abstr. in Phys. Abstr., Vol. 86, Abstr. 10205 (1983).

063.069 **Correlations in synchrotron-self Compton spectra.**
J.-h. You, G.-z. Xie, M.-x. Bao, K.-h. Li.
Acta Astron. Sinica, Vol. 23, 353 - 362 (1982). In Chinese.
In recent years, the UV-X-ray radiation has been observed in some synchrotron radio sources. Some authors suggest that the radio and X-ray radiation may be originated from the synchrotron-self Compton effect. A series of definite relations between the synchrotron and Compton spectrum are given in the paper. An analysis of these correlations gives some semi-quantitative criteria for the observational tests of the correctness of the synchrotron-self Compton effect.

063.070 **Motion and radiation of electrons in an inhomogeneous magnetic field.**
S.-y. An, L.-f. Luo.
Acta Astron. Sinica, Vol. 23, 363 - 375 (1982). In Chinese.
The motion and radiation of electrons in the neighbourhood of a neutral sheet are discussed. Formulae for synchrotron and cyclotron radiation of electrons in an inhomogeneous field are obtained. The results are compared with the radiation in a homogeneous field. It is predicted that the momenta of charged particles flying from a neutral sheet may take discrete eigenvalues.

063.071 **Inverse Compton scattering in an intense magnetic field.**
Z. Li.
Acta Astrophys. Sinica, Vol. 2, 321 - 326 (1982). In Chinese.
Herold (1979) has discussed the Compton scattering in an intense magnetic field for a general case. In this paper, the author has generalized his results to cases of the inverse Compton scattering, in which the initial electrons have very high velocities and the final electrons could be situated in any Landau excited states. The author gets the energy distribution of scattered photons for the most general cases. The standard techniques of quantum electrodynamics are adopted. From the solutions of the Dirac equation in a homogeneous magnetic field, the author obtains scattering matrix elements, transition probabilities, and general formulae for the differential and total cross sections. The results could be applied to the mechanism of pulsar radiation.

A Monte Carlo approach to scattering off surfaces of triaxial ellipsoids.
See Abstr. 021.002.

Laboratory tests of proposed scattering laws for particulate surfaces.
See Abstr. 022.087.

Transient and transport coefficients for radiative fluids.
See Abstr. 062.002.

The propagation of strong-plane magnetogasdynamic shock waves in an optically-thin atmosphere.
See Abstr. 062.107.

Monte Carlo study of highly polarized cool stars.
See Abstr. 064.002.

Numerical simulation of radiative transfer in circumstellar dust shells. I. Spherical shells.
See Abstr. 064.048.

Line formation in winds with enhanced equatorial mass-loss rates and its application to the Wolf-Rayet star

HD 50896.
See Abstr. 064.056.

On the theory of gamma-ray amplification through stimulated annihilation radiation (GRASAR).
See Abstr. 066.128.

Radiative transfer in the surfaces of atmosphereless bodies. III. Interpretation of lunar photometry.
See Abstr. 094.001.

UV resonance line formation in Zeta Aur/VV Cep systems.
See Abstr. 119.038.

The polarization of supernova light: a measure of deviation from spherical symmetry.
See Abstr. 125.105.

Study of CO line profiles in molecular clouds: relaxing the assumption of complete frequency redistribution.
See Abstr. 131.083.

Radiative stability of interstellar masers – a variational technique.
See Abstr. 131.258.

Radiative transfer calculations for the C IV λ155 nm resonance lines in planetary nebulae.
See Abstr. 135.028.

Spectral formation in galactic X-ray sources.
See Abstr. 142.013.

The influence of X-rays on surrounding matter.
See Abstr. 142.159.

Transport and propagation of cosmic rays in galaxies. II. The effect of a galactic wind on the mean lifetime and age distribution of non-decaying cosmic rays.
See Abstr. 143.055.

Non-linear theory of cosmic ray shocks including self-generated Alfvén waves.
See Abstr. 143.058.

The intensity and spectrum of galactic center β^+ annihilation photons after Compton scattering.
See Abstr. 157.020.

Thermal interaction between matter and radiation in the early universe.
See Abstr. 162.011.

064 Stellar Atmospheres, Stellar Envelopes, Mass Loss, Accretion

064.001 **Stellar model chromospheres. XIII. M dwarf stars.**
M. S. Giampapa, S. P. Worden, J. L. Linsky.
Astrophys. J., Vol. 258, 740 - 760 (1982).
The authors construct single-component, homogeneous model chromospheres that are consistent with high-resolution profiles of the Ca II K line calibrated in surface flux units for three dMe and 2 dM stars observed at quiescent times. The models indicate a large amount of nonradiative heating present in the upper photospheres of M dwarf stars. The authors also compute Hα and the Mg II h and k lines, and estimate relative active region filling factors on the basis of the observed fluxes.

064.002 **Monte Carlo study of highly polarized cool stars.**
J.-Y. Daniel.
Astron. Astrophys., Vol. 111, 58 - 70 (1982).
The author presents a theoretical analysis of the polarization of highly polarized cool stars. A Monte Carlo treatment turns out to be the most convenient way to treat the problem since multiple scattering is very necessary to explain high intrinsic polarization. The cool stars studied are: VY CMa, NML Cyg, R Aqr, L_2Pup, V CVn, RU Her, IRC 30021, IRC 20052, IRC 10216. To these stars the author has added CRL 2688, which is the most polarized object we know (up to 50%). With his quantitative method the author obtains satisfactory explanations for the real behaviour of the polarization of these objects. From the polarimetric results he derives information on the circumstellar dust extinction, thus on the luminosity and even the nature of these objects.

064.003 **Stellar winds in binary X-ray systems.**
K. B. MacGregor, P. A. J. Vitello.
Astrophys. J., Vol. 259, 267 - 281 (1982).
Accretion from a strong stellar wind by a compact object may be responsible for the X-ray emission from binary systems containing a massive early-type primary. To investigate the effect of X-ray heating and ionization on the mass transfer process in systems of this type, the authors have constructed an idealized model for the flow of a radiation-driven wind in the presence of an X-ray source of specified luminosity L_x. The dependence of the wind parameters on the value of L_x is investigated.

064.004 **The theory of radiatively driven stellar winds. I. The line acceleration.**
D. C. Abbott.
Astrophys. J., Vol. 259, 282 - 301 (1982).
Numerical values of the radiation pressure on spectral lines are presented for the envelopes of stars with spectral type O–G, and with any luminosity, metallicity, or velocity structure. The line acceleration was calculated using a tabulation of atomic lines that is complete for the elements H–Zn. The line acceleration is remarkably constant over the temperature range $50{,}000 \geqslant T_{\rm eff} \geqslant 10{,}000$ K, but drops off sharply for cooler stars. The anomalous ionization observed in OB stars decreases the acceleration by up to a factor of 2 from radiative equilibrium values. Predicted and observed mass loss rates are compared for OB stars, Wolf-Rayet stars and F and G supergiants.

064.005 **Neutrino emission from a hot, dense, plane-parallel atmosphere in hydrostatic equilibrium. I.**
P. J. Schinder, S. L. Shapiro.
Astrophys. J., Vol. 259, 311 - 329 (1982).
The authors numerically solve the plane-parallel, semi-infinite, static neutrino atmosphere problem. They assume that the atmosphere consists of free nucleons, electrons, positrons, and photons with electron neutrinos and anti-neutrinos flowing through this gas. Neutrinos and anti-neutrinos are allowed to interact with the matter via beta emission-absorption processes, neutrino-nucleon and neutrino-electron scattering, and pair production-absorption by electron-positron pairs and plasmons.

064.006 **Twisted accretion disks.**
J. A. Petterson.
Galactic X-rays sources, (see 003.001), p. 217 - 223 (1982).
Contents: The disk in Hercules X-1. Numerical simulation of the Hercules disk. Results and conclusions.

064.007 **Stellar winds and the evidence for coronae around early type stars.**
J. P. Cassinelli.
Galactic X-ray sources, (see 003.001), p. 257 - 268 (1982).
Contents: Introduction to the theory of stellar winds. Stellar winds of early type stars. The oxygen VI problem and stellar coronae.

064.008 **Meridional circulation in optically thick accretion disks.**
W. Cabot, M. P. Savedoff.
Astron. Astrophys., Vol. 112, L1 - L2 (1982).
Thermal imbalances in stars due to rotation are known to drive mass motions in the meridional plane. The authors have made a preliminary analytic investigation of a similar effect in optically thick accretion disks using conventional thin-disk approximations.

064.009 **Shock fronts in wide binary systems.**
R. Q. Huang, A. Weigert.
Astron. Astrophys., Vol. 112, 281 - 286 (1982).
The authors consider the interaction of the two stellar winds in a binary system containing a late supergiant and an early main sequence star. The properties of the resultant shock front are studied assuming that the shock thickness and the stellar orbital motion can be neglected. A system of differential equations is derived. The results are applied to the system α Sco, where the calculated shock front can easily explain certain observed features.

064.010 **On the observational test of the temperature distribution in the atmospheres of hot stars.**
N. A. Sakhibullin.
Pis'ma Astron. Zh., Tom 8, 485 - 487 (1982). In Russian.
English translation in Soviet Astron. Lett., Vol. 8.
Non-LTE calculations of C IV lines show that the profile of the resonance doublet near 1550 Å may be used as an observational test of the temperature rise in upper layers of stars predicted theoretically by the current stellar atmosphere models.

064.011 **Penetration and circulation currents in irradiated spherically symmetric stellar atmospheres.**
H. Kirbiyik.
Mon. Not. R. Astron. Soc., Vol. 200, 907 - 923 (1982).
The author studies the motions driven in a stellar atmosphere by heating from a companion in a close binary system. Earlier work is extended by considering a spherical atmosphere, so as to be able to investigate the penetration of the motions into the unilluminated part of the star. The amount of energy carried by circulation into the dark side is estimated and the result supports the view that enough energy can be carried to explain the light curve of HZ Herculis.

064.012 **Mass loss and mass gain of stars immersed in a supermassive star.**

M. Kato.
Publ. Astron. Soc. Japan, Vol. 34, 173 - 182 (1982).

In central parts of quasars, the temperatures and densities of gaseous matter are expected to be very high. Stars moving in these regions are heated by the surrounding gaseous matter, and the stellar matter is expected to evaporate. The author has calculated numerically the structure and the mass-loss rate of the mass-losing stars which are immersed in a supermassive star. He finds that the star evaporates when it lies in the inner part of the supermassive star, while it accretes the surrounding gaseous matter when it lies in the outer part.

064.013 **The photospheric velocity field of Procyon.**
C. de Jager, L. Neven.
Astrophys. Space Sci., Vol. 84, 297 - 306 (1982).

BUSS observations of the profiles of two well observed spectral lines in the ultraviolet spectrum of αCMi (Procyon; F5 IV - V) are analysed with a Fourier transform method in order to determine values of various parameters of the velocity field of the upper photosphere.

064.014 **"Discrepant asymmetry" stars: the role of unsteady magnetic flux loops in the atmospheres of late-type giant stars.**
D. J. Mullan, R. E. Stencel.
Advances in ultraviolet astronomy, (see 012.006), p. 235 - 238 (1982).

The authors draw attention to a number of spectroscopic peculiarities of K giants and other stars which lie in a "wedge" in the HR diagram. These peculiarities can be understood in terms of unsteady magnetic flux loops emerging into the stellar atmosphere from beneath the surface.

064.015 **On the correlation between chromospheric and coronal emission.**
R. Hammer, J. L. Linsky, F. Endler.
Advances in ultraviolet astronomy, (see 012.006), p. 268 - 272 (1982).

The authors have shown that the observed correlations between chromospheric and coronal emission and the pressure of the transition region may be explained by the fact that existing models of chromosphere and corona fulfill the relation $d \log F_M / d \log p < d \log F_{Loss} / d \log p_{TR}$; and that this relation is a necessary requirement for stability.

064.016 **An alternative model for the mantles of hot stars.**
A. B. Underhill.
Advances in ultraviolet astronomy, (see 012.006), p. 619 (1982). – Abstract.

064.017 **The outer atmosphere structure of three late type stars.**
E. de Castro, M. J. Fernández-Figueroa, M. Rego.
Astron. Astrophys., Vol. 113, 94 - 98 (1982).

Far-ultraviolet emission lines of three late type stars, two dwarfs (χ Her and β Com) and one subgiant (ι Vir), have been analysed to determine the structure of their outer atmospheres. Emission line fluxes are used to find the emission measure distributions. Using three boundary values of the electron pressure, calculated in a previous paper, a sample of transition region models has been derived.

064.018 **A model atmosphere analysis of Procyon (α CMi, F5 IV-V).**
K. Kato, K. Sadakane.
Astron. Astrophys., Vol. 113, 135 - 141 (1982).

The spectrum of Procyon (F5 IV-V) in the visual region is analysed relative to the sun with a line-blanketed convective model atmosphere. Chemical compositions determined for 30 elements, Na to Os, characterize Procyon as a normal Population I star. Only Ba is slightly overabundant with respect to the solar abundance.

064.019 **White dwarf stars with chemically stratified atmospheres.**
D. Muchmore.
Astrophys. J., Vol. 259, 749 - 758 (1982).

Recent observations and theory suggest that some white dwarfs may have chemically stratified atmospheres – thin layers of hydrogen lying above helium-rich envelopes. Models of such atmospheres show that a discontinuous temperature inversion can occur at the boundary between the layers. On the basis of their continuous extreme ultraviolet spectra, it is possible to distinguish observationally between uniform and layered atmospheres for hot white dwarfs.

064.020 **Energy balance of stellar coronae. I. Methods and examples.**
R. Hammer.
Astrophys. J., Vol. 259, 767 - 778 (1982).

Simplified theoretical models of open regions in hot stellar coronae are considered. The energy balance is described by thermal conduction, optically thin radiation, stellar wind, and mechanical heating. Such models should fulfill appropriate boundary conditions at the base of the atmosphere, at the critical point, and at infinity. They are determined by four parameters: by the stellar mass and radius and by the amount and location of coronal heating. Several examples of solutions are presented, and the energy balance is discussed as a function of height.

064.021 **Energy balance of stellar coronae. II. Effect of coronal heating.**
R. Hammer.
Astrophys. J., Vol. 259, 779 - 791 (1982).

Models of magnetically open regions in hot stellar coronae are computed; and their dependence on the total amount and the location of coronal heating is investigated in terms of the following characteristic quantities: the pressure at the base of the transition region; the maximum coronal temperature and the temperature at the critical point; the heights of the temperature maximum and of the critical point; the coronal energy losses due to radiation, wind, and thermal conduction; the mass loss; and the type of the asymptotic behavior of the solution. The relations between these quantities are discussed.

064.022 **Red giant model atmospheres. III. Carbon-enriched models.**
H. R. Johnson.
Astrophys. J., Vol. 260, 254 - 271 (1982).

The author has calculated a series of model stellar atmospheres, for both solar-like and carbon-enriched compositions, based on the standard assumptions of LTE, flux constancy, and hydrostatic equilibrium in a plane-parallel geometry. Besides the usual hydrogenic opacities the lines of 33 of the more abundant molecules and of 17 atomic species are included. Three series of models are computed, each with $\log g = 0.0$, having $T_{eff} = 2500$, 3000, and 3500 K. In each series the carbon abundance is varied to give C/O ratios ranging from 0.60 to 2.00. Twenty theoretical models are tabulated.

064.023 **A study of convective model stellar atmospheres using a modified mixing-length theory.**
J. B. Lester, M. C. Lane, R. L. Kurucz.
Astrophys. J., Vol. 260, 272 - 285 (1982).

The authors have modified the model atmosphere program ATLAS to incorporate two recently suggested changes to the standard mixing-length theory of convection. These are (1) the use of horizontally averaged opacity in place of the opacity at the average horizontal temperature and (2) the use of a variable mixing length. The resulting changes in the structure of the atmosphere are studied by systematically varying effective temperature, surface gravity, and composition.

064.024 **Chromospheres of F, G, K type stars. VII. Upper chromosphere and transition region.**
Z. Musielak, J. Sikorski.
Acta Astron., Vol. 31, 493 - 503 (1981).

Basing on detailed energy balance between radiative losses and amount of mechanical energy (with different flux spectra) the authors made an attempt to reconstruct conditions for the formation of transition region chromosphere-corona. The authors confirmed that the shape of mechanical energy spectrum may be different for stars with different effective temperatures and they found that the gradient of microturbulence velocity in the chromosphere has a significant influence on formation of the transition region.

064.025 **X-radiation limited accretion flow. I. Steady flow regime.**
R. F. Stellingwerf, J. Buff.
Astrophys. J., Vol. 260, 755 - 767 (1982).

The accretion flow likely to develop onto a compact gravitating object remains a most promising model for many cases of observed X-ray sources. The authors treat the case of spherically symmetric, optically thin accretion by constructing detailed numerical flow models that include X-ray heating and cooling processes.

064.026 **X-radiation limited accretion flow. II. Stability and time dependence.**
R. F. Stellingwerf.
Astrophys. J., Vol. 260, 768 - 779 (1982).

The author presents local and global stability analyses of spherically symmetric, X-ray heated, optically thin accretion flow. It is shown that solutions exist that are globally stable, but become subject to a finite amplitude drift instability due to bremsstrahlung cooling for high X-ray luminosities.

064.027 **Accretion of grains and element abundances in cool, helium-rich white dwarfs.**
F. Wesemael, J. W. Truran.
Astrophys. J., Vol. 260, 807 - 814 (1982).

The authors consider the problem posed by the occasional presence of metal features in the spectra of cool, helium-rich white dwarfs. They explore a model where metals from the interstellar medium are accreted in the form of grains by a slowly rotating, weakly magnetic white dwarf, and where the propeller mechanism of Illarionov and Sunyaev provides the screening mechanism required to prevent the accretion of interstellar hydrogen.

064.028 **Convective accretion disks and the onset of dwarf nova outbursts.**
J. K. Cannizzo, P. Ghosh, J. C. Wheeler.
Astrophys. J., Lett., Vol. 260, L83 - L86 (1982).

The authors have constructed vertically integrated, steady state convective models of accretion disks to explore the mechanism of instability in dwarf novae. The models and observations of dwarf novae suggest a picture in which transferred matter piles up in an optically thin torus. The torus eventually becomes optically thick and the resulting convective structure is thermally unstable. Matter then flows inward, and the sudden conversion of gravitational potential energy to radiation is identified as the dwarf nova outburst.

064.029 **Wave-driven winds from cool stars. II. Models for T Tauri stars.**
L. Hartmann, S. Edwards, E. Avrett.
Astrophys. J., Vol. 261, 279 - 292 (1982).

The Alfvén wave-driven wind theory of Hartmann and MacGregor is applied to T Tauri variables, including modifications which permit the calculation of wind temperatures. It is shown that large wave fluxes generate low-temperature winds, which can radiate strongly in Balmer and other optical emission lines. If wave fluxes are restricted to be less than the stellar luminosity, mass loss rates are restricted to $\dot{M} \leqslant 10^{-8} M_{\odot}\ yr^{-1}$. It is shown that the wind models produce optical and ultraviolet emission roughly consistent with observations.

064.030 **Radiation-driven winds in X-ray binaries.**
D. B. Friend, J. I. Castor.
Astrophys. J., Vol. 261, 293 - 300 (1982).

The authors discuss the properties of a radiation-driven stellar wind in an X-ray binary system. The Castor, Abbott, and Klein line-driven wind model is used, but the effects of the compact companion and the centrifugal force due to orbital motion are included. These forces destroy the spherical symmetry of the wind and can make the mass loss and accretion strong functions of the size of the primary relative to its critical potential lobe. The authors estimate the accretion rate and X-ray luminosity of the known massive X-ray binaries and find that in most systems the wind alone could power the X-ray emission.

064.031 **Self-similar magnetohydrodynamics. II. The expansion of a stellar envelope into a surrounding vacuum.**
B. C. Low.
Astrophys. J., Vol. 261, 351 - 369 (1982).

The axisymmetric self-similar expansion of a stellar envelope into a surrounding vacuum is treated. Two exact solutions are presented to describe the time-dependent magnetohydrodynamics of the envelope. A basic assumption is that the self-similar expansion is inertial, with balance among the Lorentz force due to a dipole-like stellar magnetic field, plasma pressure, and gravity. Simple applications to the problems of novae and supernovae are given.

064.032 **The magnetohydrostatic atmosphere around a gravitating body.**
V. A. Osherovich.
Astrophys. J., Vol. 261, 370 - 374 (1982).

The axisymmetric magnetic structure around a gravitating body is considered. Using the similarity assumption, the author reduces the problem from a nonlinear elliptic partial differential equation to an ordinary nonlinear differential equation. The particular solution with a topology including a nondipole field structure is discussed. Gas pressure, density, and temperature, corresponding to this magnetic atmosphere, are obtained.

064.033 **Neutrino emission from a hot, dense, plane-parallel atmosphere in hydrostatic equilibrium. II. Numerical methods and interaction functions.**
P. J. Schinder, S. L. Shapiro.
Astrophys. J., Suppl. Ser., Vol. 50, 23 - 38 (1982).

The authors solved the plane-parallel, semi-infinite, static neutrino atmosphere problem in an earlier paper. In this paper, they discuss in detail the numerical technique used. They assemble the functions which describe neutrino and antineutrino emission, absorption, and scattering with nucleons, scattering and pair production-annihilation with electrons and positrons, and pair production-annihilation with plasmons.

064.034 **Angular momentum loss from accretion disks due to Compton-heated winds.**
C. F. McKee, M. Begelman.
Bull. American Astron. Soc., Vol. 14, 618 (1982). – Abstract.

064.035 **Neutrino emission from a hot, dense, plane-parallel atmosphere in hydrostatic equilibrium.**
P. J. Schinder, S. L. Shapiro.
Bull. American Astron. Soc., Vol. 14, 619 (1982). – Abstract.

064.036 **Physics of stellar magnetospheres.**
O. Kaburaki.

Magnetospheric phenomena of celestial bodies, (see 012.011), p. 81 - 90 (1982). In Japanese.

Stellar magnetospheric theories and stellar wind theories are reviewed from a unified stand point where a few physical parameters specify the characteristics of individual cases. They are divided into two categories. One is called the 'equipotential' magnetospheric theory because the electric field in the rotating frame vanishes ($E' = 0$ or Φ = const.). The other is called the 'non-equipotential' magnetospheric theory in which $E' \neq 0$ (or $\Phi \neq$ const.). Though almost all theories except several recent works belong to the former, it is pointed out that there are some inconsistencies in the treatment especially as for the particle inertia.

064.037 **Numerical modeling of magnetospheres of rotating stars.**
S. Shibata.
Magnetospheric phenomena of celestial bodies, (see 012.011), p. 91 - 93 (1982). In Japanese.

Magnetospheres of rotating stars have a model sequence characterized by the plasma density. An axisymmetric magnetosphere is studied in a low density situation.

064.038 **The evolution of viscous discs – III. Giant discs in symbiotic stars.**
G. T. Bath, J. E. Pringle.
Mon. Not. R. Astron. Soc., Vol. 201, 345 - 355 (1982).

The structure of time-dependent accretion discs in giant binaries with separation $\sim 10^{13}$ cm is examined. Radiative α–viscosity discs with α of order unity accreting on to main-sequence stars at accretion rates which generate luminosities greater than a giant companion decay on time-scales of the same order as the binary period, unlike those in dwarf nova binaries which decay on time-scales 100 times longer than the binary period. This results from the lower gravitational potential and consequent larger disc thickness (relative to the radius) of luminous 'giant' discs accreting at high accretion rates. The eruptions of the symbiotic binary CI Cygni are modelled.

064.039 **Transition region models for Be stars.**
J. M. Fontenla, M. Rovira, A. E. Ringuelet.
Rev. Mexicana Astron. Astrofis., Vol. 6, (see 012.017), 209 - 214 (1981).

A preliminary model that reproduces the general characteristics of equivalent widths and profiles of ultraviolet spectral lines corresponding to ions such as C IV and Si IV, observed in a number of stars, particularly in early type objects undergoing mass loss, has been worked out. The authors make simplifying assumptions and consider radiative energy loss, radiative acceleration in an approximate form, and rotation. Agreement, at least as to the order of magnitude, between the model and the observations is found if one postulates the existence of some kind of braking mechanism.

064.040 **Summary of workshop on accretion in astrophysics.**
P. Mészáros.
Ninth Texas Symposium on Relativistic Astrophysics, (see 012.020), p. 590 - 597 (1980).

064.041 **Accretion disks in stellar X-ray sources. A review of the basic theory of accretion disks and its problems.**
F. Verbunt.
Space Sci. Rev., Vol. 32, 379 - 404 (1982).

This review presents an introduction to the theory of accretion disks. After explaining the importance of accretion disks in stellar X-ray sources, it considers observational and theoretical indications of the formation of accretion disks in close binary systems. The simple Shakura and Sunyaev model and its problems are discussed. A survey of other models that try to improve this model is given and critically discussed.

064.042 **On the compatibility of thermal and hydrostatic equilibrium in thin radiative accretion disks.**
R. Kippenhahn, H.-C. Thomas.
Astron. Astrophys., Vol. 114, 77 - 84 (1982).

It is shown that in thin Keplerian disks similar to rotating stars both equilibria cannot be maintained simultaneously. For small values $\alpha < H/s$ of the viscosity parameter, α (H half-thickness of the disk, s distance from the axis of rotation) the disk is near hydrostatic equilibrium and thermally driven circulation has to fulfil energy conservation. For values of α with $H/s < \alpha < 1$ the disk is near thermal equilibrium but hydrostatic equilibrium is not fulfilled in horizontal direction. Therefore dynamically driven circulation occurs. In both cases the circulation is strongly diminished after it has produced a slight baroclinicity. Some of the consequences are discussed.

064.043 **Element segregation in stellar outer layers.**
S. Vauclair, G. Vauclair.
Annu. Rev. Astron. Astrophys., Vol. 20, (see 003.009), 37 - 60 (1982).

The diffusion theory. Observed and predicted abundance anomalies in stellar atmospheres.

064.044 **Stochastic mass loss rate fluctuations and evolution of massive stars.**
E. Nasi.
Astrophys. Space Sci., Vol. 85, 17 - 23 (1982).

Stochastic fluctuations were superimposed on the rate of mass loss as determined according to the radiation-pressure stellar-wind theory to evaluate the effects on the evolution of a 60-solar masses star. If the variations in the rate are of the order of those observed, there are no significant modifications of the stellar parameters, and the agreement between theory and observations cannot be improved taking into account stochastic fluctuations.

064.045 **Models of pulsating low-massive yellow supergiants.**
Yu. A. Fadeyev *(Fadeev)*.
Astrophys. Space Sci., Vol. 86, 143 - 155 (1982).

The nonlinear self-excited oscillations of the envelopes of low-massive highly luminous stars are described. The parameters for these models were $M = 0.8\,M_{\odot}$, $M_{bol} = -5.5$, -5.84 mag, $T_{eff} = 4500$, 5000, 5500 K. The oscillations have been found to consist of the standing wave pulsation near the envelope bottom and running waves in outer layers. The frequency of oscillations near the photosphere is found to be in close agreement with the critical frequency for running waves. Mass loss from these stars is caused by shocks. It has been shown that agreement between FG Sge's period change observed during the last decade and the period-luminosity relation for double shell stars takes place when FG Sge's luminosity is $M_{bol} = -5.96$ mag.

064.046 **Stellar winds.**
C. de Loore.
Third European IUE Conference, (see 012.029), p. 611 - 614 (1982).

In this paper our actual knowledge on mass loss rates, the dependence of the mass loss rates on the stellar parameters L, R, M, and velocity laws are summarized. Mass loss rates derived from ultraviolet data, the variability of the mass loss, episodic mass loss, two component models, are discussed.

064.047 **The influence of divergent geometries on stellar winds.**
N. P. M. Kuin, A. G. Hearn.
Astron. Astrophys., Vol. 114, 303 - 308 (1982).

The effect of increasing the divergence of the geometry of a corona is studied using the minimum flux corona theory. Unlike the studies made with pure conduction models or using the polytrope approximation the energy losses include the radiation from the corona and energy losses to the transition region as well as through the stellar wind. Further in the present work the effect of increasing the divergence of the geometry is studied keeping the total mechanical heating constant.

064.048 **Numerical simulation of radiative transfer in circumstellar dust shells. I. Spherical shells.**
J. Lefèvre, J. Bergeat, J.-Y. Daniel.
Astron. Astrophys., Vol. 114, 341 - 346 (1982).

A computer program is described which solves the radiative transfer problem in spherical shells of circumstellar dust, using a method of numerical simulation. Multiple scattering and anisotropic phase functions are fully considered for spherical dust grains. The computation gives the radial distribution of dust temperature throughout the shell, the spectral distribution of the outward net flux and the radial profile or image of the source at different wavelengths. A few results are discussed in connection with observational data and compared to analytical calculations of previous investigators.

064.049 **On the instability of thick accretion disks.**
S. Hacyan.
Astrophys. J., Vol. 262, 322 - 329 (1982).

A general method, based on a variational principle, is developed for the study of adiabatic oscillations in a gas cloud with negligible self-gravity, rotating in an external gravitational field. Second order virial equations are obtained. As an example, the formalism is applied to disk models with a barotropic equation of state; such disks turn out to be unstable.

064.050 **Theory of steady accretion disks.**
R. Hoshi.
Prog. Theor. Phys. Suppl., No. 70, p. 181 - 201 (1981).
Abstr. in Phys. Abstr., Vol. 85, Abstr. 100913 (1982).

064.051 **Corrections for hydrostatic atmospheric models: radii and effective temperatures of Wolf-Rayet stars.**
C. de Loore, P. Hellings, H. J. G. L. M. Lamers.
Wolf-Rayet stars: observations, physics, evolution, (see 012.042), p. 53 - 56 (1982).

With the assumption of planeparallel hydrostatic atmospheres, used generally for the computation of evolutionary models, the radii of WR stars are seriously underestimated. The true atmospheres may be very extended, due to the effect of the stellar wind. Instead of these hydrostatic atmospheres the authors consider dynamical atmospheres adopting a velocity law. The equation of the optical depth is integrated outwards using the equation of continuity. The "hydrostatic" radii are to be multiplied with a factor 2 to 8, and the effective temperatures with a factor 0.8 to 0.35 when Wolf-Rayet characteristics for the wind are considered, and WR mass loss rates are used. With these corrections the effective temperatures of the theoretical models, which are helium burning Roche lobe overflow remnants, range between 30 000 K and 50 000 K.

064.052 **An explanation of the radio flux mystery of HD 192163 and empirical models for WN stars.**
T. Nugis.
Wolf-Rayet stars: observations, physics, evolution, (see 012.042), p. 127 - 130 (1982).

The radio flux value of the star HD 192163 (WN6) measured by Dickel et al. (1980) imposes strong restrictions on the possible mass outflow region models of this star.

064.053 **Theories for the winds from Wolf-Rayet stars.**
J. P. Cassinelli.
Wolf-Rayet stars: observations, physics, evolution, (see 012.042), p. 173 - 183 (1982).

The massive and fast winds of Wolf-Rayet stars present a serious momentum problem for the line-driven wind theories that are commonly used to explain the fast winds of early type stars. It is perhaps possible for the winds to be driven by lines, if multiple scattering occurs and if there is a sufficient number of lines in the spectrum so that a large fraction of the continuum is blocked by line opacity in the winds. Several other mechanisms are discussed, in particular two that rely on strong magnetic fields: (1) Alfvén wave driven wind models like those recently developed by Hartmann and MacGregor for late type supergiants and (2) the "Fast Magnetic Rotator" model that was developed by Belcher and MacGregor for the winds from pre-main sequence stars.

064.054 **The theory of radiation driven stellar winds and the Wolf-Rayet phenomenon.**
D. C. Abbott.
Wolf-Rayet stars: observations, physics, evolution, (see 012.042), p. 185 - 196 (1982).

The topics the author considers are: (1) The calculated radiation pressure in OB stars, and its dependence on temperature, density, and chemical composition. (2) A comparison between predicted and observed mass loss rates and terminal velocities for OB stars. (3) The applicability of the standard radiation driven wind models to Wolf-Rayet stars. (4) Speculations on how Wolf-Rayet stars achieve their enormous mass loss rates within the context of the radiation pressure mechanism.

064.055 **A hot corona model for O-stars and WR stars.**
W. Robbrecht, C. de Loore, G. Olson.
Wolf-Rayet stars: observations, physics, evolution, (see 012.042), p. 209 - 213 (1982).

Atmospheric models are presented for the outer layers of hot stars, O, Of and Wolf-Rayet stars. The model is a two component hybrid model, consisting of a rapidly expanding component and a slower component. The corona as well as the shocks generated by the interaction of the two components can explain the observed X-rays.

064.056 **Line formation in winds with enhanced equatorial mass-loss rates and its application to the Wolf-Rayet star HD 50896.**
W. M. Rumpl.
Wolf-Rayet stars: observations, physics, evolution, (see 012.042), p. 243 - 247 (1982).

064.057 **Mass loss in Alpha Cygni: synthetic H-alpha profiles.**
P. B. Kunasz, N. D. Morrison.
Astrophys. J., Vol. 263, 226 - 238 (1982).

The Hα profile in α Cygni (A2 Ia) is characterized by shallow, shortward-shifted absorption, with an intensity longward of the rest wavelength that is at or above the continuum level. The authors interpret these characteristics in terms of a model stellar envelope that is spherically symmetric, expands with monotonically increasing velocity, and obeys mass flux continuity. The principal free variables are the mass loss rate, the wind velocity law, and the Doppler broadening velocity law. The synthetic profiles are matched to the observed profiles. Depending on model assumptions, a mass loss rate in the range 1.7 to $4.0 \times 10^{-7} M_\odot$/yr is derived.

064.058 **Mass loss by Wolf-Rayet stars.**
Kh. F. Khaliullin.
Ejection and accretion of matter in binary systems, (see 012.051), p. 99 - 108 (1982).

The problem of the mass loss from the Wolf-Rayet stars is solved. The radial-symmetric mass outflow is described by a general formula which contains three independent variables: the density, the velocity of the outflow and the distance from the centre of the star. The main problem consists in obtaining the observational data for the density and for the radial velocity in a definite distance from the centre of the star. In the case of binary systems the study of the period changes can be a favourable method for the determination of the mass transfer and mass outflow, too. A critical review of the observational data and that of the theoretical interpretation is presented. It has been shown that the radiation pressure is responsible for the mass loss from the WR stars.

064.059 **New results on the mechanism of magnetic flux pumping by three-dimensional convection.**
W. Arter, M. R. E. Proctor, D. J. Galloway.
Mon. Not. R. Astron. Soc., Vol. 201, 57P - 61P (1982).

New calculations are presented to show that the mechanism proposed by Drobyshevski and Yuferev for pumping magnetic flux at the base of a stellar convection zone is radically altered in the case of high magnetic Reynolds number. In consequence the process as they stated it is unlikely to be operating in the sun, though the revised mechanism may actually be more effective in maintaining magnetic fields at the bottom of the convecting layer.

064.060 **Models of extended atmospheres of late-type stars.**
T. Kipper.
Tartu Astrofüüs. Obs. Teated, No. 66, p. 3 - 15 (1982).

Static model atmospheres for late-type supergiants are calculated assuming spherical symmetry. The extension of the atmosphere leads to cooling of the surface layers if compared with the corresponding plane-parallel models. The variations of the emergent flux and the enhancement of the formation of some molecules are discussed.

064.061 **Models for stellar coronae: the effects of coronal heating with long dissipation scale lengths.**
A. G. Hearn.
Astron. Astrophys., Vol. 116, 296 - 302 (1982).

The calculations made by Hearn and Vardavas (1981) of models of coronae for OB supergiants used a dissipation mechanism which heated the corona substantially at its base. These calculations have been extended to heating mechanisms with very long dissipation scale lengths. The calculations show that heating a corona by a heating mechanism with a longer dissipation scale length gives substantial increases in the maximum coronal temperature and mass loss, but the pressure of the transition region is reduced. When the dissipation scale length becomes very long, the maximum coronal temperature and mass loss decline. Heating a corona with a mechanism with a longer dissipation scale length increases the energy taken up by the mass loss at the expense of the energy lost by radiation. With a suitable mechanical heating distribution, energy lost through the stellar wind can become the dominant energy loss of the corona. Suitably long dissipation scale lengths for coronal heating can probably explain the observations of coronal holes.

064.062 **The effect on empirical atmospheric modeling of the mass-flux as an independent parameter.**
R. N. Thomas.
Astrophys. J., Vol. 263, 870 - 878 (1982).

Observational data on atmospheric structure and mass fluxes from the sun and Be stars are applied to test the adequacy of the original Parker "hot corona" approach to predicting atmospheric structure and the size of the mass flux from only the radiative and nonradiative energy fluxes, and from gravity, and imposing the condition that thermal and escape points must coincide. Observations do not support this latter condition. It is concluded that the Parker approach is an asymptotic approximation to the very low mass flux limit in a nonvariable stellar atmosphere.

064.063 **Unsteady phenomena in an accretion disk.**
J. Fukue.
Publ. Astron. Soc. Japan, Vol. 34, 483 - 494 (1982).

Shock propagations caused by a strong explosion at the center of geometrically thin disks are investigated. The shock propagates almost self-similarly if the thin-disk approximation holds after the shock has passed. The inner region of the disk is swept out temporarily by such unsteady phenomena. A transient destruction of the disk by the shock may often take place around compact stars with accretion disks.

064.064 **Chromospheres of F, G, K type stars. VIII. Energy balance in transition region.**
M. Promiński, Z. Musielak, J. Sikorski.
Acta Astron., Vol. 32, 111 - 120 (1982).

Radiative losses, thermal conductivity from corona, convective energy flux, and dissipation, refraction and reflection of magnetohydrodynamic (MHD) waves are calculated for known empirical and theoretical models of transition region for quiet and active (flare) Sun and ϵ Eri. It appeared that only slow MHD modes may effectively heat the transition regions because the acoustic and the fast waves are strongly refracted. It seems the corona may be reached only by the Alfvén waves.

064.065 **Turbulent compressible convection in a deep atmosphere. I. Preliminary two-dimensional results.**
K. L. Chan, S. Sofia, C. L. Wolff.
Astrophys. J., Vol. 263, 935 - 943 (1982).

Two-dimensional numerical computations are used to simulate turbulent convection of a compressible fluid in a deep atmosphere. The authors find that "cells" with sizes ranging from the total depth of the convection zone to the smallest scale height at the top coexist. While the biggest cells traverse the whole zone, the smaller cells cluster toward the top.

064.066 **Magnetic field-related heating instabilities in the surface layers of the sun and stars.**
A. Ferrari, R. Rosner, G. S. Vaiana.
Astrophys. J., Vol. 263, 944 - 951 (1982).

The authors investigate the stability of a magnetized low-density plasma to current-driven filamentation instabilities and apply the results to the surface layers of stars. The analysis shows that the surface layers of main-sequence stars (including the Sun) which are "threaded" by magnetic fields are unstable. These results suggest that relatively modest surface motions, in conjunction with the presence of magnetic fields, suffice to account for the presence of inhomogeneous chromospheric and coronal plasma overlying a star's surface.

064.067 **Magnetostatic atmospheres with variations in three dimensions.**
B. C. Low.
Astrophys. J., Vol. 263, 952 - 969 (1982).

The paper treats the static equilibrium of a fully ionized atmosphere with an embedded magnetic field in the presence of a uniform gravity. The magnetic field lines are assumed to lie in parallel vertical planes. Except for this assumption, the system is allowed to vary in all three dimensions. Selected explicit solutions are presented to illustrate various structural properties of prominence-like density enhancements, coronal magnetic arcades, and discrete bipolar plasma loops. The paper presents for the first time, explicit equilibrium solutions for plasma loops with three-dimensional extensions.

064.068 **Effect of spots on a star's radius and luminosity.**
H. C. Spruit.
The physics of sunspots, (see 012.063), p. 480 - 487 (1981).

First the author discusses how the temperature and heat flux distribution in an effective convective envelope evolve in time after a sudden thermal disturbance. This is done by calculating the thermal decay modes for a polytropic model. On the basis of these findings, the author then calculates the effect of starspots in a model star in which the spots are small, numerous, and evenly distributed over the surface. Finally, he calculates the effect of a single spot by a 3-D heat flow calculation.

064.069 **Rotating magnetic winds from early-type stars.**
D. B. Friend.
Publ. Astron. Soc. Pacific, Vol. 94, 752 - 753 (1982). Abstract.

064.070 **Compton heated winds from accretion disks in QSOs.**
G. A. Shields, M. Begelman, C. F. McKee.
Publ. Astron. Soc. Pacific, Vol. 94, 758 (1982). – Abstract.

064.071 **Vertical structure of an accretion disc when gas pressure exceeds radiation pressure.**
L. G. Filipov.
Dokl. Bolg. AN, Vol. 35, 277 - 280 (1982). – Abstr. in Ref. zh., 51. Astron., 11.51.127 (1982).

064.072 **The differential emission measure of dynamic coronal loops.**
S. K. Antiochos.
Smithsonian Astrophys. Obs., Spec. Rep. 392, Vol. II, (see 012.065), p. 115 - 123 (1982).
Although most of the observations of soft X-ray and UV emission from stellar sources can be adequately explained in terms of static models of coronal loops, there are now several observations in soft X-rays of a solar-like flare from a stellar flare source (e.g. Johnson 1981, Kahler 1981; Stern et al. 1981). The author discusses the effects of time dependent phenomena, such as flare energization and decay, the temperature and density structure of the transition region and, in particular, the form of the differential emission measure. He finds that unlike the case of the static models, the form of the differential emission measure can be used to determine the important physical mechanisms in the dynamic models.

064.073 **Détermination des taux de perte de masse dans les étoiles chaudes.**
S. Vauclair.
Mécanismes de production d'énergie dans le milieu interstellaire, (see 012.068), p. 81 - 93 (1981).

064.074 **Two temperature spherical accretion onto a black hole: the X- and gamma-ray spectra.**
M. Colpi, L. Maraschi, A. Treves.
Accreting neutron stars, (see 012.069), p. 143 - 145 (1982).
The authors treat the coupled equations for the energy balance of both protons and electrons in a spherically accreting plasma around a black hole.

064.075 **Non-adiabatic self-consistent spherical accretion.**
A. Wandel.
Accreting neutron stars, (see 012.069), p. 146 - 149 (1982).
New self-consistent solutions for spherical accretion onto a black hole are presented. These solutions involve high accretion rates, yielding high luminosities up to a few percent of the Eddington limit.

064.076 **On the compatibility of thermal and hydrostatic equilibrium in thin radiative accretion disks.**
R. Kippenhahn, H.-C. Thomas.
Accreting neutron stars, (see 012.069), p. 159 - 160 (1982).
It is shown that in thin disks with an angular velocity distribution according to Kepler's law both equilibria cannot be fulfilled simultaneously. Depending on the amount of viscosity in the disk either thermally or dynamically driven circulation will occur. The velocity of this circulation is kept low because of a slightly baroclinic structure of the disk, which is produced by the circulation currents.

064.077 **Limit cycle instabilities in accretion disks.**
F. Meyer.
Accreting neutron stars, (see 012.069), p. 333 - 337 (1982).
Recent progress in the understanding of accretion disk instabilities which are related to a change in the vertical structure of an accretion disk is described.

064.078 **The temperature distribution in the atmosphere of the Wolf-Rayet star of the system V444 Cygni derived from infrared measurements.**
M. B. Bogdanov.
Pis'ma Astron. Zh., Tom 8, 725 - 728 (1982). In Russian. English translation in Soviet Astron. Lett., Vol. 8.
Source functions and temperature distribution in the atmosphere of the Wolf-Rayet star of the system V444 Cyg are derived from the analysis of IR-brightness distributions across its disk. The results are compared with data of model calculations of extended atmospheres.

064.079 **Determination of mass-loss rates from the first order moment W_1 of unsaturated P Cygni line profiles.**
J. Surdej.
Astrophys. Space Sci., Vol. 88, 31 - 54 (1982).
The author investigates the relation between W_1 and $\dot{M}$ under various physical and geometrical conditions. He first establishes for the case of a point-like source the general expression of the nth order moment W_n in terms of the usual parameters inherent to Sobolev-type theories. He shows how these first results are altered when considering separately the possible effects caused by collisions and rotation by the presence of an underlying photospheric absorption line, by a doublet line transition, by the finite size of the central source and by the limb-darkening of the stellar core. A discussion and general conclusions are included.

064.080 **Vertical structure of a magnetic accretion disk.**
R. Horiuchi.
Prog. Theor. Phys., Vol. 68, 541 - 553 (1982). – Abstr. in Phys. Abstr., Vol. 86, Abstr. 20351 (1983).

064.081 **Are thick accretion disks the 'central engine' for astrophysical jets?**
L. Nobili, R. Turolla, M. Calvani.
Lett. Nuovo Cimento, Vol. 35, Ser. 2, 335 - 340 (1982). Abstr. in Phys. Abstr., Vol. 86, Abstr. 20416 (1983).

Early emission line stars.
See Abstr. 003.069.

Improving the reliability of experimental atomic transition probabilities.
See Abstr. 022.145.

Self-similar flows behind a spherical radiation-driven shock wave. II.
See Abstr. 062.066.

The propagation of strong-plane magnetogasdynamic shock waves in an optically-thin atmosphere.
See Abstr. 062.107.

Broadband linear polarization from magnetized stellar atmospheres. Numerical tables for the magnetic intensification mechanism.
See Abstr. 063.001.

Velocity fields and spectral line asymmetries: a linearized analytical approach to the theory of the line bisector in a Milne-Eddington atmosphere.
See Abstr. 063.004.

Line profile fluctuations in a turbulent atmosphere.
See Abstr. 063.011.

Non-LTE resonance line polarization with partial redistribution effects.
See Abstr. 063.055.

Magnetic star models with a consistent $\alpha\omega$-effect core-dynamo and radiative envelope.
See Abstr. 065.033.

A singular perturbation approach to the effect of a weak magnetic field on stellar oscillations.
See Abstr. 065.060.

The combined effect of mass loss and overshooting. I. The evolution of 35 $M_\odot$ to 100 $M_\odot$ stars during core hydrogen burning.
See Abstr. 065.092.

The temperature profile of an accretion disk around a Kerr black hole.
See Abstr. 066.273.

Solar winds with multiple critical points and shock transitions.
See Abstr. 074.031.

Mass loss from cool stars – facts, fads, and fallacies.
See Abstr. 112.008.

Empirical atmospheric velocity patterns from combined IUE and visual observations: the Be-similar stars.
See Abstr. 112.013.

Mass loss from extreme helium stars. Detailed UV-line fits for HD 160641, BD −9°4395 and BD +10°2179.
See Abstr. 112.047.

Atmospheric parameters and chemical compositions of eighteen halo OB stars.
See Abstr. 114.010.

The structure and energy balance of cool star atmospheres.
See Abstr. 114.022.

Chromospheric, transition layer and coronal emission of metal deficient stars.
See Abstr. 114.025.

Progress report of an IUE survey of the Hyades star cluster.
See Abstr. 114.026.

Fine analysis of the intermediate helium-star CPD−46°3093.
See Abstr. 114.041.

Ultraviolet continuum of a sample of Be stars.
See Abstr. 114.076.

Observed and computed UV spectral distribution of A and F stars. Determination of T_e and log g.
See Abstr. 114.084.

The spectrum of Canopus – II. Analysis and composition.
See Abstr. 114.109.

Hot mantles, moderate photospheres for Wolf-Rayet stars.
See Abstr. 114.118.

Synthetic spectra for WN stars.
See Abstr. 114.126.

He I lines in B stars: comparison of non-local thermodynamic equilibrium models with observations.
See Abstr. 114.141.

Carbon, nitrogen, and oxygen abundances in G8–K3 giant stars.
See Abstr. 114.143.

Studies on the spectra of K-giants – I. Physical parameters and Fe and Ti abundances for 26 K-giants.
See Abstr. 114.167.

Stellar activity.
See Abstr. 114.211.

Accretion and inner excretion disks in close binaries.
See Abstr. 117.049.

Hydrodynamical calculations of accretion disks in close binary systems. I. Method.
See Abstr. 117.078.

Hydrodynamical calculations of accretion disks in close binary systems. II. Models.
See Abstr. 117.079.

Rapidly oscillating Ap stars.
See Abstr. 122.022.

Accretion disk model of dwarf novae.
See Abstr. 122.055.

Starspots: a review of observations and theory.
See Abstr. 122.148.

On the wind model of novae.
See Abstr. 124.004.

Comments on radial white dwarf accretion.
See Abstr. 126.015.

Atmospheric parameters and carbon abundance of white dwarfs of spectral types C_2 and DC.
See Abstr. 126.029.

Dynamique des sphères célestes.
See Abstr. 131.313.

Vents stellaires et dynamiques des régions H II extragalactiques.
See Abstr. 132.050.

X-ray spectra of accreting degenerate stars.
See Abstr. 142.077.

065 Stellar Structure and Evolution

065.001 **The influence of rotation and duplicity on the frequency of radial oscillations of a homogeneous star.**
K. A. Sidorov.
Astrofizika, Tom 18, 151 - 160 (1982). In Russian. English translation in Astrophysics, Vol. 18, No. 1.
The equation of small adiabatic radial oscillations of a rotating rigid body has been derived. This equation is then applied to a homogeneous star. All radial modes are considered. The frequencies of radial oscillations of a rotating star are found to be less than those of a nonrotating star of the same mass. The ratios of overtones only slightly perturb. The frequencies of the radial pulsations are not affected by tidal forces.

065.002 **Hot accreting white dwarfs in the quasi-static approximation.**
I. Iben, Jr.
Astrophys. J., Vol. 259, 244 - 266 (1982).
Properties of white dwarfs which are accreting hydrogen-rich matter at rates in the range 1.5×10^{-9} to 2.5×10^{-7} $M_\odot$ yr^{-1} are investigated in several approximations. Steady-burning models, in which matter is processed through nuclear-burning shells as rapidly as it is accreted, provide a framework for understanding the properties of models in which thermal pulses induced by hydrogen burning and helium burning are allowed to occur. Expressions are obtained for estimating, as functions of accretion rate and white dwarf mass, the thermal pulse recurrence period and the duration of hydrogen-burning phases. Suggestions for detecting observational counterparts of the theoretical models and suggestions for further theoretical investigations are offered.

065.003 **Postshock neutrino transport and electron capture in stellar collapse.**
A. Burrows, T. J. Mazurek.
Astrophys. J., Vol. 259, 330 - 340 (1982).
The authors show that the large reservoir of thermal energy inside the neutrino-sphere left behind by the shock formed in stellar collapse cannot be tapped on shock time scales by either mu, tau, or electron-type neutrino diffusion. Hence, the shock cannot be strengthened or energized by neutrino transport. It is further shown that after the shock breaks out of the neutrino-sphere, there is a significant and unavoidable decrease in electron fraction due to rapid electron capture in this quasi-transparent region. The dependence of the time history and profile of the electron fraction on shock speed, the magnitude and decay time of the neutrino flux, and the matter density profile are investigated.

065.004 **The energetics of adiabatic shocks in stellar collapse.**
T. J. Mazurek.
Astrophys. J., Lett., Vol. 259, L13 - L17 (1982).
The energetics of shocks from stellar collapse are examined analytically. A general minimal condition is derived for outward shock propagation. The dominant factor behind the failure of the shock is nuclear dissociation. It is shown that dissociation must occur over essentially the whole outer mantle if a respectable shock propagates. This indicates that collapsing iron cores more massive than ~ 1.3 $M_\odot$ do not explode.

065.005 **Meridional circulation in rotating stars. I. A boundary layer analysis of mean steady motions in early-type stars.**
J.-L. Tassoul, M. Tassoul.
Astrophys. J., Suppl. Ser., Vol. 49, 317 - 350 (1982).
The problem of rotationally driven meridional currents in a nonmagnetic star is reconsidered, taking (radiative or eddy) viscosity into account. Uniformly valid expressions are found for the circulation velocities in the radiative zone of a chemically homogeneous, early-type star, when departure from spherical symmetry is not too large. The transport of angular momentum by these currents is made to balance exactly the viscous friction arising from differential rotation, so that the mean state of motion is a steady one.

065.006 **The evolution of massive stars. II. The influence of initial composition and mass loss.**
W. M. Brunish, J. W. Truran.
Astrophys. J., Suppl. Ser., Vol. 49, 447 - 468 (1982).
The authors have studied the evolution of 15, 30, and 40 $M_\odot$ stars with compositions Y = 0.28 and Z = 0.01, 0.001, and 0.0002, and of 50 $M_\odot$ stars with Y = 0.28, Z = 0.0002. These models have been evolved from the zero-age main sequence through core helium burning and up to the point of carbon ignition. All models were evolved with and without the inclusion of mass loss. A moderate mass loss rate has been used with different efficiency factors for main-sequence and blue supergiant stars. The results indicate that all massive stars ignite helium as blue supergiants. Mass loss causes all models to evolve redward more rapidly during core helium burning.

065.007 **The influence of CN abundances on the evolution of main sequence of low-mass stars.**
A. Bazzano, F. Caputo, M. Sestili, V. Castellani.
Astron. Astrophys., Vol. 111, 312 - 316 (1982).
Theoretical isochrones have been derived for mild metal-poor ([Fe/H] $\cong -0.7$) stars under various assumptions on the relative amount of CNO elements. Cluster ages in the range 8–18 billion yr have been explored, showing that the turn-off luminosity decreases with increasing the CNO content by $\Delta \log L^{TO} = -0.06 \{[C/Fe] + [N/Fe]\}$. Consequences on the determination of cluster ages are discussed.

065.008 **Non-linear stellar oscillations. Two-mode interactions.**
J. Perdang, S. Blacher.
Astron. Astrophys., Vol. 112, 35 - 48 (1982).
Numerical experiments reported in the paper demonstrate that the adiabatic non-linear coupling between pairs of adiabatic radial linear modes can lead to chaotic stellar oscillations. The results thus corroborate a prediction based on KAM theory (Perdang, 1978). It is observed that the new class of chaotic motions continue to display an approximate periodicity, the amplitude undergoing however long time scale random fluctuations.

065.009 **Stellar evolution, mass loss and nucleosynthesis on the asymptotic giant branch.**
P. R. Wood.
Proc. Astron. Soc. Australia, Vol. 4, 145 - 148 (1981).
The author concentrates on problems related to the evolution of stars on the asymptotic giant branch (AGB). The need for models of galactic nucleosynthesis has stimulated the search for a quantitative understanding of evolution, mass loss and nucleosynthesis in AGB stars. Some of the results of theoretical studies of AGB stars are described.

065.010 **Physical characteristics of a polytrope index 5 with finite radius.**
J. O. Murphy.
Proc. Astron. Soc. Australia, Vol. 4, 205 - 208 (1981).

065.011 **Rupture d'étoiles par les forces de marée.**
J.-P. Luminet.
Noyaux actifs des galaxies, (see 012.004), 22 pp.

065.012 **An observational approach to the early stages of stellar evolution.**
V. A. Ambartsumian *(Ambartsumyan)*, L. V. Mirzoyan.
Astrophys. Space Sci., Vol. 84, 317 - 330, 331 - 345 (1982). In English and Russian.

During the last three decades an observational approach has been applied at the Byurakan Astrophysical Observatory to the problems of the evolution of astronomical bodies and systems. In the past, the observational approach has led to the prediction of an expansion of some stellar associations confirmed later by the analysis of observations. The new approach has also led to the concept of the activity of galactic nuclei. The observational approach considers as a phenomenon of primary importance the formation of nebulae as a consequence of the activity of dense bodies. The new approach in application to the early stages of stellar evolution is discussed. The T Tauri-stage is considered as a phase following the more dense protostellar state. The flare stars are regarded as the next phase of evolution. The phenomena of fuors (FU Ori-type brightenings) can be considered as an expression of the same tendency.

065.013 **Multimode stellar pulsations. II: Anharmonic oscillations.**
M. Barranco, J. R. Buchler, O. Regev.
Astrophys. Space Sci., Vol. 84, 463 - 475 (1982).

The authors apply the two time method to weakly anharmonic (nonlinear) stellar pulsations. The problem contains two small parameters: ϵ – the ratio of the dynamical and the thermal timescale and μ – the measure of the anharmonicity in the oscillator equation. The $0(\epsilon^0)$ system is solved up to $0(\mu)$ by the use of an asymptotic perturbation method. This solution is then used within the framework of the two time method to obtain the slow evolution of the amplitudes and entropy. Both monomode and double-mode pulsations are treated in a non-resonance case.

065.014 **Evolution of low mass stars through mass loss: transition from the main sequence to the degenerate phase.**
F. D'Antona, I. Mazzitelli.
Astron. Astrophys., Vol. 113, 303 - 310 (1982).

The evolution of low mass stars through mass loss at constant rate is followed, starting from fully convective structures of $M \cong 0.2\, M_\odot$ down to the masses for which nuclear energy sources are no longer supporting the structure.

065.015 **Fundamental mode of nonradial nonadiabatic pulsations of a star.**
Yu. D. Zhugzhda.
Pis'ma Astron. Zh., Tom 8, 562 - 565 (1982). In Russian. English translation in Soviet Astron. Lett., Vol. 8.

High-l nonradial nonadiabatic fundamental pulsations of a star are considered. It is found that the f-mode 5 minute oscillations may be excited only by turbulent convection.

065.016 **On the problem of stellar rotation.**
F. H. Busse.
Astrophys. J., Vol. 259, 759 - 766 (1982).

It is shown that initially steady Eddington-Sweet circulations in rotating stars change into a time dependent state within a period of rotation. Constraints on the redistribution of angular momentum by axisymmetric motions are derived, and their importance in determining the asymptotic steady solution is emphasized. A simple model of rotating stars is presented which exhibits the necessary degree of freedom for the accomodation of general initial distributions of angular momentum.

065.017 **The role of semiconvection in bringing carbon to the surface of asymptotic giant branch stars of small core mass.**
I. Iben, Jr., A. Renzini.
Astrophys. J., Lett., Vol. 259, L79 - L83 (1982).

In a low-mass asymptotic giant branch star which has a small carbon-oxygen core, a layer of carbon-rich material is produced just below the hydrogen-helium discontinuity during the peak of a thermal pulse. It is demonstrated in this paper that a semiconvection region will bring fresh carbon to the star surface for core masses smaller than previously calculated. The long-standing "carbon star mystery" may therefore be close to a solution.

065.018 **On the secular evolution of accreting white dwarfs and Type I supernovae.**
M. Y. Fujimoto, R. E. Taam.
Astrophys. J., Vol. 260, 249 - 253 (1982).

The growth of a carbon-oxygen white dwarf is investigated within the framework of the general shell-flash theory of accreting white dwarfs. Account is taken of the envelope restructuring in response to the shell flash and to possible mass loss from the white dwarf. Estimates are given for the amount of helium accumulated between successive hydrogen shell flashes. The results are discussed in the context of exploding white-dwarf models of Type I supernovae.

065.019 **CNO abundances in Cepheids and supergiants: theoretical implications.**
S. A. Becker, A. N. Cox.
Astrophys. J., Vol. 260, 707 - 715 (1982).

Models of a 9 $M_\odot$ star have been evolved to see if the observed nitrogen enhancement and the carbon and oxygen depletions in yellow supergiants and Cepheids can be explained. In the standard (first dredge-up phase) evolutionary picture, the normal or near normal CNO abundances for the envelopes of stars on the main sequence are altered when such stars become red giants with deeply mixed convective envelopes prior to core He ignition. It is found that in order to explain the observed O depletion, there must be some form of mixing between the convective hydrogen-burning core and the inner part of the radiative envelope.

065.020 **Secular evolution of cataclysmic binaries: evolution of the secondary and the period gap.**
F. D'Antona, I. Mazzitelli.
Astrophys. J., Vol. 260, 722 - 734 (1982).

The evolution of low-mass main-sequence stars subject to mass loss at constant rates is followed, starting from several initial degrees of hydrogen depletion in the central regions and imposing several different rates of mass loss. During this evolution, stars of progressively smaller mass become fully convective. The information derived from these computations is applied to the interpretation of the secular evolution of cataclysmic binaries. Main-sequence stars which are secondary components of these systems become fully convective when the orbital period is close to 3 hours, near the upper edge of the gap in the period distribution of cataclysmic binaries.

065.021 **Low mass asymptotic giant branch evolution. I.**
I. Iben, Jr.
Astrophys. J., Vol. 260, 821 - 837 (1982).

It is shown that convective dredge-up of carbon to the surface of asymptotic giant branch stars does not occur in standard models of small mass until core mass approaches $M_c \sim 0.7\, M_\odot$, large compared to a value $M_c \lesssim 0.64\, M_\odot$ demanded by the observations. However, following pulse peak, the contribution of carbon to the opacity becomes important at the outer edge of the carbon-rich region for models with $M_c \gtrsim 0.6\, M_\odot$. It is conjectured that inclusion of the physics of recombination and of the opacity appropriate to a mixture of helium and partially ionized carbon may lead to the formation of a new convective shell which carries carbon-rich matter further toward the surface than during pulse peak, perhaps reaching into hydrogen-rich layers.

065.022 **A plea for reexamining heavy element opacities in stars.**
N. R. Simon.
Astrophys. J., Lett., Vol. 260, L87 - L90 (1982).
It is shown that increasing the opacity due to heavy elements by a factor of 2 - 3 leads to classical Cepheid models which reproduce observed period ratios at evolutionary masses and luminosities. Thus the mass anomalies are removed in both the double-mode and bump Cepheid regimes. The proposed increases may also serve to energize β Cephei variables, thus solving yet another important problem in the theory of pulsating stars.

065.023 **Meridional circulation in rotating stars. II. Mean steady motions in rotationally and tidally distorted stars.**
J.-L. Tassoul, M. Tassoul.
Astrophys. J., Vol. 261, 265 - 272 (1982).
This paper deals with the problem of circulatory currents in the radiative envelope of a non-magnetic star, which is a synchronously rotating, detached component of a double star. Uniformly valid solutions are found when departure from spherical symmetry is not too large.

065.024 **Meridional circulation in rotating stars. III. The reflection effect in close binaries.**
J.-L. Tassoul, M. Tassoul.
Astrophys. J., Vol. 261, 273 - 278 (1982).
An analysis is given of the large-scale photospheric currents in an early-type star, which is a detached component of a close binary, and whose surface is nonuniformly heated by the radiation from its companion. Uniformly valid solutions are constructed, assuming that the ratio of the incident to outgoing fluxes at the surface remains less than unity.

065.025 **On the work integral in stellar pulsation theory.**
J. R. Buchler, O. Regev.
Astrophys. J., Vol. 261, 301 - 309 (1982).
The two-time formalism which serves to define rigorous time-averages for a pulsating star, undergoing slow thermal evolution, is used to derive the expression for the work integral. A continuum description is introduced instead of a previous discrete representation. In the case of infinitesimal oscillations the usual derivation of the work integral is shown to be incomplete.

065.026 **Finite propagation time in multidimensional thermonuclear runaways.**
B. A. Fryxell, S. E. Woosley.
Astrophys. J., Vol. 261, 332 - 336 (1982) = Lick Obs. Bull., No. 922.
The importance of multidimensional effects for thermonuclear runaways that occur in thin stellar shells and surface layers is discussed. The runaway will initiate at a point instead of simultaneously over the entire layer and will propagate around the star as a burning front that requires a finite time to ignite all of the fuel in the layer. The burning velocity and the time scale for the front to propagate around the star are calculated for several different types of events. For novae, X-ray bursts, and γ-ray bursts, the finite propagation time can be important in determining the rise time of duration of the event.

065.027 **Gravitational collapse of massive stars and supernovae.**
S. W. Bruenn.
Bull. American Astron. Soc., Vol. 14, 620 (1982). – Abstract.

065.028 **The structure and evolution of low mass stars.**
G. D. Neece.
Bull. American Astron. Soc., Vol. 14, 665 (1982). – Abstract.

065.029 **Is the "mixing length" physically meaningful?**
K. L. Chan, S. Sofia, C. L. Wolff.
Bull. American Astron. Soc., Vol. 14, 665 (1982). – Abstract.

065.030 **Convection and the thermonuclear runaway in the core helium flash.**
P. W. Cole, R. G. Deupree, P. Demarque.
Bull. American Astron. Soc., Vol. 14, 665 (1982). – Abstract.

065.031 **An eigenvalue approach to the solution of non-equilibrium ionization and applications to models of supernova remnants.**
J. P. Hughes.
Bull. American Astron. Soc., Vol. 14, 666 (1982). – Abstract.

065.032 **Zero age main sequence models with variable *G*.**
A. D. Pinotsis, P. G. Laskarides.
Bull. Astron. Inst. Czechoslovakia, Vol. 33, 229 - 233 (1982).
The physical characteristics and the method of obtaining initial models that were on the ZAMS one, three or five Byr ago are presented. These models are to be used for theoretical evolution with variable gravitational constant G. The three ZAM Sequences are also compared to the ZAMS obtained with constant G.

065.033 **Magnetic star models with a consistent $\alpha\omega$-effect core-dynamo and radiative envelope.**
D. Moss.
Mon. Not. R. Astron. Soc., Vol. 201, 385 - 392 (1982).
Models of middle main-sequence magnetic stars are constructed which have a convective core in which the field is generated by a dynamically consistent '$\alpha\omega$-effect' dynamo. The envelope is at first assumed to be in strict radiative equilibrium with no meridional circulation. Later more general models are constructed, which include a thermally driven meridional circulation in the envelope consistent with the MHD equation. The magnetic field is restricted to have dipolar angular form, with its axis aligned with the rotation axis. These models are direct analogues of earlier models in which the field was implicitly assumed to be of fossil origin.

065.034 **Convective mixing length and the galactic carbon to oxygen ratio.**
A. Serrano, M. Peimbert.
Rev. Mexicana Astron. Astrofis., Vol. 6, (see 012.017), 41 - 44 (1981).
The authors have studied chemical evolution models, assuming instantaneous recycling, and considering: (1) the effects of mass loss both in massive stars and in intermediate mass stars, and (2) the initial mass function of the solar neighborhood. From these models the authors have derived the yields of carbon and oxygen. It is concluded that the condition C/O $\sim$ 0.58 in the solar neighborhood can only be satisfied if, during advanced stages of stellar evolution of intermediate mass stars, the ratio of the convective mixing length to the pressure scale height is $\gtrsim 2$.

065.035 **Screened $_1H^1\,(P\beta^+\nu)\,_1H^2$ reaction rates in stellar interiors.**
A. E. M. Khairozzaman.
Natl. Acad. Sci., Lett., Vol. 4, 87 - 92 (1981). – Abstr. in Phys. Abstr., Vol. 85, Abstr. 68612 (1982).

065.036 **Evolution and explosion of massive stars.**
T. A. Weaver, S. E. Woosley.
Ninth Texas Symposium on Relativistic Astrophysics, (see 012.020), p. 335 - 357 (1980).
This paper is concerned with the attempt to build self-

consistent evolutionary models of complete massive stars ($M > 10\ M_\odot$), starting from their observable zero-age main sequence configurations, and evolving through their various hydrostatic nuclear burning stages, iron core collapse, bounce, outward-going shock formation, and finally, explosive nucleosynthesis and supernova light curve formation.

065.037 **Neutrino flow and stellar core collapse.**
J. R. Wilson.
Ninth Texas Symposium on Relativistic Astrophysics, (see 012.020), p. 358 - 365 (1980).

This paper presents the results of a detailed numerical calculation of the collapse of the iron core of a fifteen solar mass star at the end of its normal evolution.

065.038 **The physics of gravitational collapse: an overview.**
W. D. Arnett.
Ninth Texas Symposium on Relativistic Astrophysics, (see 012.020), p. 366 - 379 (1980).

065.039 **Summary of workshop on stellar collapse and supernovae.**
W. Hillebrandt.
Ninth Texas Symposium on Relativistic Astrophysics, (see 012.020), p. 395 - 401 (1980).

065.040 **The fate of massive stars: collapse, bounce and shock formation.**
G. E. Brown.
Supernovae: a survey of current research, (see 012.022), p. 13 - 33 (1982).

Supernovae of type II are believed to occur at the end of the evolution of stars with masses in the range of 8 to perhaps 100 $M_\odot$. In this paper the basic physics of collapse, core "bounce" and formation of a hydrodynamic shock in massive stars is analyzed. The general patterns of extensive computer calculations in the framework of stellar collapse models are outlined.

065.041 **Evolution and nucleosynthesis in population III stars.**
W. W. Ober, M. F. El Eid, K. J. Fricke.
Supernovae: a survey of current research, (see 012.022), p. 293 - 301 (1982).

The authors report on evolution and nucleosynthesis calculations for massive primordial stars having no metals initially in the mass range $80 < M/M_\odot < 500$. The aim of this work is to determine the mass range in which the evolution terminates explosively and to assess the kind of enrichment caused by the stars in this mass range.

065.042 **The boundary between explosion and collapse in very massive objects.**
J. R. Bond, W. D. Arnett, B. J. Carr.
Supernovae: a survey of current research, (see 012.022), p. 303 - 311 (1982).

A simple model of the evolution of very massive objects in the oxygen core phase is developed. Calculations of the effects of the pair instability, oxygen and silicon burning, and alpha quenching on a global effective potential enable the authors to predict the critical oxygen core mass for black hole formation.

065.043 **The effect of non-adiabatic layers on the vibrational behaviour of stars.**
J. R. Buchler, O. Regev.
Astron. Astrophys., Vol. 114, 188 - 191 (1982).

The authors extend the two-time formalism for radial stellar pulsations to include the treatment of the highly non-adiabatic outer layers. This enables them to overcome a major difficulty in the method's application to realistic multizone stellar models (e.g. Cepheids). In particular, the authors derive the linear vibrational stability coefficient and they show that the non-adiabatic layers not only can influence it, but can, in principle, actually change its sign.

065.044 **External magnetic field of a static spherically symmetric star in Rosen's bimetric theory of gravitation.**
A. R. Prasanna.
Phys. Rev. D, Vol. 25, 2701 - 2703 (1982). – Abstr. in Phys. Abstr., Vol. 85, Abstr. 89143 (1982).

065.045 **Magnetic stars.**
E. F. Borra, J. D. Landstreet, L. Mestel.
Annu. Rev. Astron. Astrophys., Vol. 20, (see 003.009), 191 - 220 (1982).

Observations of magnetic fields in nondegenerate stars. Theory of magnetic stars.

065.046 **Polytropic spheres with different constitution in the core.**
M. G. Durgapal, K. Pandey, A. K. Pande.
Bull. Astron. Soc. India, Vol. 10, 26 (1982). – Abstract.

065.047 **Effect of central condensation on the structure of rotating polytropes.**
M. Singh, G. R. Singh.
Bull. Astron. Soc. India, Vol. 10, 26 (1982). – Abstract.

065.048 **Small adiabatic radial oscillations of Sterne and Prasad's models.**
V. K. Gurtu, D. A. Umate.
Bull. Astron. Soc. India, Vol. 10, 26 (1982). – Abstract.

065.049 **Effect of rotational and tidal distortions on the periods of small adiabatic oscillations of a stellar model.**
C. Mohan, V. P. Singh.
Bull. Astron. Soc. India, Vol. 10, 27 (1982). – Abstract.

065.050 **The problem of nonradial oscillations of highly centrally condensed stellar models.**
C. Mohan, K. Singh.
Bull. Astron. Soc. India, Vol. 10, 27 (1982). – Abstract.

065.051 **Effect of rotational and tidal distortions on the periods of small adiabatic oscillations of stellar models.**
C. Mohan, V. P. Singh.
Astrophys. Space Sci., Vol. 85, 83 - 98 (1982).

Kopal's method of representing the inner structure of a rotationally and tidally-distorted star by a rotationally-and tidally-distorted Roche model has been used in conjunction with an averaging concept introduced by Kippenhahn and Thomas of representing the equipotential surfaces of a rotationally and tidally distorted stellar model by the equipotential surfaces of an equivalent spherical model to determine the combined effects of rotation and tidal distortion on the periods of small adiabatic oscillations of a stellar model.

065.052 **Abundance distribution of heavy nuclides up to a subnuclear density and the equation of state for gravitationally collapsing stars.**
T. Ohnishi.
Astrophys. Space Sci., Vol. 85, 243 - 262 (1982).

The abundance distribution of neutron-excess nuclides which appear up to some subnuclear density along the collapsing trajectory of a stellar core is calculated with an assumption of the β-equilibrium condition. The quantities affecting the hydrodynamics of the core, such as the average mass number and the mass fraction of heavy nuclides, the mass fraction of free neutrons, the number of electrons per baryon, the average excitation energy per nucleus, and the entropy, the pressure and the adiabatic index of the system are then calculated.

065.053 **Stellar evolution after IUE: 1. Massive stars during core hydrogen burning and massive close binaries.**
C. de Loore, A. Maeder.
Third European IUE Conference, (see 012.029), p. 15 - 20 (1982).
Recently observations with IUE allowed accurate mass loss determinations for a large number of stars. The influence of these results on stellar evolution theory is outlined. New computations for massive close binaries taking into account the effects of stellar wind, mass transfer and mass loss during Roche lobe overflow and accretion are described.

065.054 **Stellar evolution after IUE: 2. A renewal for massive stars and WR stars.**
A. Maeder, C. de Loore.
Third European IUE Conference, (see 012.029), p. 21 - 29 (1982).
The authors examine the consequences of mass loss on the evolution of massive stars ($M \geqslant 9 M_\odot$) in the MS-phase and especially in the He-burning phase. Abundances determined from IUE observations are compared with theoretically expected values. The importance of nitrogen (and neon) abundance determinations in WC stars is emphasized, in relation with possible scenarios for forming Wolf-Rayet stars.

065.055 **The ultimate fate of Wolf-Rayet stars as supernovae.**
A. Maeder, J. Lequeux.
Astron. Astrophys., Vol. 114, 409 - 413 (1982).
The evolution of massive stars ($M \geqslant 9 M_\odot$) in the $\log T_c$ vs $\log \rho_c$ diagram is quite similar for models with mass loss at the observed rates and for models of constant mass. This indicates that the final outcome of the centres of OB stars is not substantially modified by stellar winds. Thus, most mass-losing OB stars, that may become Wolf-Rayet (WR) stars after various stages, are very likely to explode as supernovae. However a fraction, probably quite small, of the WR stars may end their life as small-mass planetary nuclei. Three different determinations of the fraction of supernovae originating from WR progenitors are made.

065.056 **The critical mass of a charged star.**
J. L. Zhang, W. Y. Chau, T. Y. Deng.
Bull. American Astron. Soc., Vol. 14, 778 (1982). – Abstract.

065.057 **Localized thermonuclear runaways and volcanoes on degenerate dwarf stars.**
M. M. Shara.
Astrophys. J., Vol. 261, 649 - 660 (1982).
Practically all studies to date of thermonuclear runaways on degenerate dwarf stars in binary systems have considered only spherically symmetric eruptions. The author shows that even slightly non-spherically symmetric accretion leads to transverse temperature gradients in the dwarfs' accreted envelopes. Thus localized thermonuclear runaways rather than spherically symmetric global eruptions are likely to occur on many degenerate dwarfs. Localized thermonuclear runaways on neutron stars give rise to deflagrations or detonations which incinerate the envelope in less than 100 s. Localized runaways on white dwarfs lead to the volcanic-like outpouring of matter onto the dwarf's surface in ~ 1 day. Observed properties of X-ray transients and dwarf novae may be explainable by this model.

065.058 **Homologous collapse and deleptonization of an evolved stellar core.**
S. A. Bludman, I. Lichtenstadt, G. Hayden.
Astrophys. J., Vol. 261, 661 - 676 (1982).
One-zone calculations of deleptonization during stellar collapse are reported using the Lamb, Lattimer, Pethick, and Ravenhall equation of state and electron capture rates incorporating neutron shell blockage and neutrino trapping. The authors find a final lepton fraction and predict formation of a large homologous core and strong shock. Other recent hydrodynamic calculations are compared critically with the presented one and reasons suggested why they failed to produce a supernova explosion.

065.059 **Numerical studies of nonspherical carbon combustion models.**
E. Müller, W. D. Arnett.
Astrophys. J., Lett., Vol. 261, L109 - L115 (1982).
The authors have performed a set of axisymmetric hydrodynamic calculations of the core carbon flash, which incorporate the treatment of two-dimensional convective flow. The computations show that, in the beginning, the nuclear burning slowly propagates outward from the center of the star in the form of a very nonspherical "salt finger" shaped combustion front. Calculations done with different spatial resolutions indicate that the convectively unstable burning region cascades to smaller and smaller scales, limited only by the resolution of the computational grid. The nuclear burning at this stage is likely to be turbulent and to propagate as a roughly spherical but inhomogeneous combustion front.

065.060 **A singular perturbation approach to the effect of a weak magnetic field on stellar oscillations.**
D. Biront, M. Goossens, A. Cousens, L. Mestel.
Mon. Not. R. Astron. Soc., Vol. 201, 619 - 633 (1982).
The authors study the effect of a weak, axisymmetric magnetic field on linear, adiabatic oscillations of a star. Over the bulk of the star a regular perturbation treatment is adequate, but this breaks down near the surface if the field does not vanish there but extends into the low density exterior. Singular perturbation methods are applied to the case of a star with a polytropic atmosphere and a curl-free, dipolar field in the surface region, oscillating in a mode that is radial in the absence of the field.

065.061 **The Biermann mechanism and spontaneous magnetic field generation in stars.**
J. C. Kemp.
Publ. Astron. Soc. Pacific, Vol. 94, 627 - 633 (1982).
How magnetic fields can arise spontaneously in stars and nebular clouds, without the initial presence of primordial or seed fields, is a topic which has been somewhat bypassed in the solar and astrophysical literature. It is known that at least one efficient spontaneous process exists, the one proposed by L. Biermann in 1950. In this paper the author attempts to furnish particularly to nonspecialists a clear, physical explanation of this interesting mechanism. Certain other mechanisms will also be mentioned.

065.062 **Astrophysical constraints on broken supersymmetry.**
M. Fukugita, N. Sakai.
Phys. Lett. B, Vol. 114B, 23 - 26 (1982). – Abstr. in Phys. Abstr., Vol. 85, Abstr. 94723 (1982).

065.063 **The vanishing act of carbon giants.**
M. Cohen.
Astronomy, Vol. 10, No. 6, p. 66 - 71 (1982). – Abstr. in Phys. Abstr., Vol. 85, Abstr. 94735 (1982).

065.064 **Convection in pulsating stars. I. Nonlinear hydrodynamics.**
R. F. Stellingwerf.
Astrophys. J., Vol. 262, 330 - 338 (1982).
A nonlinear, nonlocal, time-dependent treatment of convection suitable for use in models of cool giant stars is presented. Local conservation equations plus a diffusive transport equation are used to derive the convective hydrodynamic equations for the case in which turbulent pressure,

energy, and viscosity cannot be ignored. The effects of convective overshooting, superadiabatic gradients, convection/pulsation interaction, and time dependence enter this treatment in a natural way.

065.065 **Consistency of the mixing length theory.**
D. Narasimha, H. M. Antia.
Astrophys. J., Vol. 262, 358 - 368 (1982).

The structure of the solar convection zone calculated according to the mixing length theory is found to be consistent with the transport of convective flux by a linear superposition of statistically independent unstable convective modes, provided the effects of turbulent conductivity and viscosity are taken into account. The resultant vertical velocity is in reasonable agreement with observed granular velocity in the atmosphere.

065.066 **A contribution to the study of the stability and oscillations of stars with a magnetic field.**
M. Goossens.
Meded. K. Acad. Wet. Lett. Schone Kunsten Belgie, Vol. 44, No. 1, p. 23 - 45 (1982). – Abstr. in Phys. Abstr., Vol. 85, Abstr. 100908 (1982).

065.067 **The evolution of single stars.**
R. J. Tayler.
Prog. Theor. Phys. Suppl., No. 70, p. 77 - 100 (1981). – Abstr. in Phys. Abstr., Vol. 85, Abstr. 100910 (1982).

065.068 **Development of the stellar convection theory.**
W. Unno.
Prog. Theor. Phys., Suppl. No. 70, p. 101 - 114 (1981) = Contrib. Dep. Astron., Univ. Tokyo, No. 305.

Theories of convection are reviewed in regard mostly to the applicability to the stellar convection zone. Various attempts to study the stellar convection without the use of the convectional mixing-length theory are critically reviewed. A method of constructing stellar convection zones with the help of the turbulent diffusivities is suggested.

065.069 **Stable numerical method in computation of stellar evolution.**
D. Sugimoto, K. Nomoto, Y. Eriguchi.
Prog. Theor. Phys. Suppl., No. 70, p. 115 - 131 (1981). Abstr. in Phys. Abstr., Vol. 85, Abstr. 100912 (1982).

065.070 **Low-entropy adiabats for stellar collapse.**
S. Marcos, M. Barranco, J. R. Buchler.
Nucl. Phys. A, Vol. A381, 507 - 518 (1982). – Abstr. in Phys. Abstr., Vol. 85, Abstr. 100961 (1982).

065.071 **On the chemical abundances, evolutionary stage and bolometric magnitude of WR stars.**
D. Vanbeveren.
Wolf-Rayet stars: observations, physics, evolution, (see 012.042), p. 117 - 120 (1982).

The author comments on 3 problems concerning WR stars: (1) Are WR stars core hydrogen/core helium burning? (2) Are WR stars H-poor or do they have normal abundances? (3) Can we use B.C. for WR stars which are similar as for O stars?

065.072 **The evolution of massive stars: the production of "single" WR stars.**
C. Chiosi.
Wolf-Rayet stars: observations, physics, evolution, (see 012.042), p. 323 - 341 (1982).

The author summarizes the basic observational properties of WR stars, and the current status of the evolution of massive stars, giving particular emphasis to those physical mechanisms that may lead O type stars to the WR stage. The theoretical results are compared with the observations, pointing out several still unsettled aspects of the problem.

065.073 **Properties of the scenario for the formation of WR stars as post-red supergiants.**
A. Maeder.
Wolf-Rayet stars: observations, physics, evolution, (see 012.042), p. 371 - 375 (1982).

065.074 **Origin and evolution of Wolf-Rayet stars.**
A. Tutukov, L. Yungelson *(Yungel'son)*.
Wolf-Rayet stars: observations, physics, evolution, (see 012.042), p. 377 - 381 (1982).

The larger part of close binary components with initial mass exceeding $\sim 20\ M_\odot$ becomes WR stars in the core helium burning stage. Some of the most massive WR stars may be products of evolution of single massive stars with initial masses exceeding $\sim 50\ M_\odot$ if the mass loss in the infrared supergiant stage is effective enough. The Ledoux criterion of convective stability seems more promising to explain the observed properties of WR stars.

065.075 **The various scenarios leading to WR stars: their relative importance and the role of mixing.**
A. Maeder.
Wolf-Rayet stars: observations, physics, evolution, (see 012.042), p. 405 - 409 (1982).

065.076 **Low mass Wolf-Rayet stars: theory.**
A. Renzini.
Wolf-Rayet stars: observations, physics, evolution, (see 012.042), p. 413 - 422 (1982).

065.077 **Observations and the Wolf-Rayet evolutionary scenario.**
C. Firmani.
Wolf-Rayet stars: observations, physics, evolution, (see 012.042), p. 499 - 513 (1982).

Results of a statistical analysis of massive stars and of WR stars show general agreement with the scenario that identifies the majority of the WR stars with the He-burning evolutionary phases. The duplicity, the morphological properties and the galactic distribution of the WR stars are valuable pieces of evidence that suggest correlations between the different WR types and the "channels" defined in the scenario. A rough analysis of the "traffic through each channel" is attempted.

065.078 **A numerical model for stellar core collapse calculations.**
R. L. Bowers, J. R. Wilson.
Astrophys. J., Suppl. Ser., Vol. 50, 115 - 159 (1982).

A method for calculating stellar core collapse is discussed which emphasizes physical approximations and numerical methods. Collapse is modeled in terms of hydrodynamics coupled to radiation transport of all neutrino types. The latter are not assumed to be in thermal equilibrium. Analytic approximations used for the equations of state are discussed. The overall approach includes all of the physical processes which we believe are important for stellar core collapse.

065.079 **Generalized polytropes: stellar structure with a Boltzmann factor.**
D. K. Callebaut, N. van den Bergh, M. Verbruggen.
Astrophys. Space Sci., Vol. 86, 397 - 417 (1982).

In order to study the structure of a chemically homogeneous star in equilibrium, a density profile of the form $\rho \sim T^N \exp(-\mu m(\phi - \phi_*)/kT)$ is suggested. It is shown that one reobtains for $N = 3$ Eddington's standard model, whereas for $N < 3$ nearly polytropic models result. When $N > 3$, the effective polytropic index varies appreciably over the star. From a comparison of the degree of precision of a polytropic approximation with that of the newly proposed model it follows that the new approximation is definitely better than the polytropic one. It is suggested that the model may be useful to study the structure of stellar clouds, clusters and (spherical) galaxies.

065.080 **Linear and nonlinear oscillations of simple stellar models.**
E. Antonello.
Astrophys. Space Sci., Vol. 86, 485 - 491 (1982).

A general property for the stability of the one zone model is derived in the case of linear nonadiabatic radial oscillations. The problem of the nonlinear adiabatic radial oscillations of the homogeneous model with a slow rotational velocity is discussed.

065.081 **Evolution of low mass zero metal giants up to the helium flash.**
F. D'Antona.
Astron. Astrophys., Vol. 115, L1 - L3 (1982).

If low mass stars having no metals were formed in the first stellar generation, their giant branch evolution is very different from the evolution of pop. II stars. As a consequence of the lack of CNO nuclei, the hydrogen shell burning is sustained by the proton-proton chain up to the helium flash, which occurs at luminosities about a factor 10 lower than in pop. II. However this behaviour should not produce important consequences for what concerns the observability of zero metal giants, as the reduction of the space volume in which these stars should be observable in a statistical survey is compensated by the fact that the evolutionary lifetime of the entire hydrogen shell burning phase is much longer.

065.082 **On the difference between the initial mass function of single stars and of primaries of binaries.**
D. Vanbeveren.
Astron. Astrophys., Vol. 115, 65 - 68 (1982).

A statistical study of the influence of the formation of close binaries on the initial mass function is presented. Conditions are given in order to have similar slopes for the initial mass functions of single stars and of primaries of close binaries. These conditions appear to be so restrictive that it seems very hard to believe that the complicated physics of star and binary formation will provide them. A simple model of binary formation is then introduced in order to get an idea of the expected differences between the initial mass function of single stars and primaries of close binaries.

065.083 **Evolution of stars in the post-main sequence stage.**
R. Kippenhahn.
Binary and multiple stars as tracers of stellar evolution, (see 012.046), p. 3 - 18 (1982).

In his report the author confines himself to the evolution of stars of intermediate mass (1.4 - 10 $M_\odot$) and of low mass ($M < 1.4 M_\odot$). A review appeared in vol. 12 of Annual Review of A & A by Icko Iben seven years ago (Iben, 1974) is recommended.

065.084 **A comparison of observed and theoretical luminosity functions of carbon stars and late M giants.**
G. E. Miller, J. M. Scalo.
Astrophys. J., Vol. 263, 259 - 268 (1982).

Simulated populations of double-shell phase stellar models undergoing mixing due to helium shell flashes are compared with observations of carbon stars and M giants in the Galaxy, the Large Magellanic Cloud, the Small Magellanic Cloud, and the Fornax dwarf galaxy. The comparisons are based on the means and variances of the luminosity functions. The main parameters are the metal abundance, the fraction of intershell carbon mixed up per flash, the minimum core mass M_c^{min} at which mixing begins, the core mass M_c^{max} above which mixing does not occur, and the duration of star formation. The observed mean luminosity of carbon stars and its constancy among the four regions with available data can be reproduced by the simulations only if low-mass stars begin mixing essentially when the first helium shell flash occurs, $M_c^{min} \approx 0.55\ M_\odot$. The observed near-equality of M and C star mean luminosities can be matched only if $M_c^{min} \approx 0.5\ M_\odot$ and $M_c^{max} \lesssim 1.0\ M_\odot$, and is predicted to be independent of the metal abundance.

065.085 **Oscillations of an extended ionization region in a star.**
J. R. Buchler, O. Regev.
Astrophys. J., Vol. 263, 312 - 319 (1982).

The authors describe a mechanism by which an extended ionization region, although vibrationally stable, can undergo oscillations of an unexpectedly complex nature. The oscillations can be quasi-steady periodic oscillations in either of two possible states; they can be of the relaxation type, periodic or aperiodic, or of a limit cycle type with possible bumps, depending sensitively on the model parameters. The values and range of the one-zone model parameters suggest an identification with the long-period variables.

065.086 **The nonadiabatic analysis of nonradial modes of stellar oscillation in the presence of slow rotation.**
B. W. Carroll, C. J. Hansen.
Astrophys. J., Vol. 263, 352 - 365 (1982).

The linearized equations governing nonadiabatic, nonradial pulsations of a slowly but uniformly rotating star have been derived. A large number of numerical examples are presented for near main sequence evolution of a 12 $M_\odot$ star. All modes with $l = 2$ and 3 and with periods between two and 15 hours were examined for this sequence. Other masses and values of l were studied, but to a lesser extent. For all modes and sequences it was found that slow, uniform rotation increases the pulsation frequency of prograde modes, and conversely for retrograde modes. Possible implications for β Cephei and line profile variables are discussed.

065.087 **Collapse of iron stellar cores.**
R. Bowers, J. R. Wilson.
Astrophys. J., Vol. 263, 366 - 376 (1982).

The collapse of the iron core of evolved 10, 15, and 20 $M_\odot$ stars has been calculated using a spherically symmetric numerical model. No bounce occurs and no matter is ejected. Sensitivity to some less well-established details, and to zoning, has been examined and the overall qualitative results are unchanged. A black hole is the presumed outcome of the model.

065.088 **On the formation of carbon star characteristics and the production of neutron-rich isotopes in asymptotic giant branch stars of small core mass.**
I. Iben, Jr., A. Renzini.
Astrophys. J., Lett., Vol. 263, L23 - L27 (1982).

Further calculations on semiconvective diffusion of asymptotic giant branch (AGB) stars are reported. It is demonstrated that a large region rich in ^{13}C and poor in ^{14}N forms at the base of the semiconvective zone. During a subsequent thermal pulse this region acts via the reaction $^{13}C(\alpha, n)^{16}O$ as a potent source of neutrons for the production of neutron-rich heavy isotopes in low-mass AGB stars.

065.089 **The effect of rapid rotation on stellar pulsation modes.**
M. J. Clement.
J. R. Astron. Soc. Canada, Vol. 76, 317 (1982). – Abstract.

065.090 **Discovery and implications of the dramatic rotation brake at G5III.**
D. F. Gray.
J. R. Astron. Soc. Canada, Vol. 76, 319 - 320 (1982). Abstract.

065.091 **Nucleosynthesis in two 25 $M_\odot$ stars of different population.**
S. E. Woosley, T. A. Weaver.
Essays in nuclear astrophysics, (see 003.013), p. 377 - 399 (1982).
The authors discuss the physics and evolution of two 25 $M_\odot$ star models, up to the point of iron core collapse and supernova explosion, with particular attention to the composition and distribution of "neutron excess" in these stars at the ends of their lives.

065.092 **The combined effect of mass loss and overshooting. I. The evolution of 35 $M_\odot$ to 100 $M_\odot$ stars during core hydrogen burning.**
C. Doom.
Astron. Astrophys., Vol. 116, 303 - 307 (1982).
The observed absence of late type supergiants with $M_{bol} < -9.8$ is interpreted in terms of a combined effect of mass loss by stellar wind and overshooting from convective cores. A simple parameter model, describing overshooting is developed. With this model, the evolution of massive stars is modeled in such a way that no late type supergiants are formed with $M_{bol} < -9.8$. This leads to convective cores that are 15–50% larger than when the Schwarzschild criterion is used. It is shown that the derived constraints on the amount of overshooting imply that stars more massive than 33 $M_\odot$ evolve into transition WN stars at the end of core hydrogen burning.

065.093 **The combined effect of mass loss and overshooting. II. The evolution of 10 $M_\odot$ to 30 $M_\odot$ stars during core hydrogen burning.**
C. Doom.
Astron. Astrophys., Vol. 116, 308 - 311 (1982).
The observed extent of the main sequence band around $\log L/L_\odot = 5.1$ is used to put constraints on the amount of overshooting in a 20 $M_\odot$ star. It follows that the amount of overshooting in 10–30 $M_\odot$ stars can be determined by a linear relation between the total stellar mass and the mass of the convective core on the ZAMS. It is shown that stars with hydrogen deficient atmospheres can be present in the whole upper main sequence band, corresponding to the location of most of the OBN and Of stars. The influence of overshooting on the evolution of massive close binaries is discussed.

065.094 **The evolutionary status of Be stars.**
A. B. Underhill.
Be Star Newsl., No. 6, p. 21 - 26 (1982).

065.095 **Models of convective interiors surrounded by thin radiative envelopes.**
R. Sienkiewicz.
Acta Astron., Vol. 32, 37 - 48 (1982).
A description of the method used to obtain such models is given. Their parameters are tabulated as a function of central density and temperature for given chemical composition. These results have already been used to follow the evolution of short period cataclysmic binaries.

065.096 **Excitation of nonspherical thermal instability in nonhomogeneous contracting stars.**
Yu. V. Vandakurov, M. Kozłowski.
Acta Astron., Vol. 32, 49 - 61 (1982).
An approximate equation is derived which permits to study excitation of both convection and nonradial thermal instability in a chemically nonhomogeneous stellar model. This equation is applied to investigate stability of contracting near-main sequence stars, having a nonhomogeneous core, with decreasing outwards molecular weight.

065.097 **Neutrino escape, nuclear dissociation, and core collapse and/or explosion.**
W. D. Arnett.
Astrophys. J., Lett., Vol. 263, L55 - L57 (1982).
An extensive set of numerical calculations of spherically symmetric stellar collapse has been completed in which a careful treatment of neutrino transport, equation of state, weak interaction reaction rates, and relativistic hydrodynamics has been attempted. With currently accepted physics of the collapse, the outcome is continued collapse for initial models with large iron cores and a marginal explosion for those with iron cores near 1.4 $M_\odot$.

065.098 **An isothermal shock wave in a perfectly-conducting and non-homogeneous stellar interior.**
B. G. Verma, J. P. Vishwakarma, V. Sharan.
Astrophys. Space Sci., Vol. 88, 125 - 134 (1982).
In the present paper, the authors have obtained some exact analytic self-similar solutions for a zero-temperature gradient behind a magnetogasdynamic shock wave produced by stellar explosions. The initial density of the medium is taken to vary as some power of the distance from the point of explosion. The solutions are obtained for the cases when the energy of the shocked gas is constant, the energy is varying, and the shock velocity is constant. General solutions are also obtained. The authors have also analytically obtained the position of the singular surface behind the shock wave.

065.099 **The theory of Savić about the origin of rotation of celestial bodies and the Savić–Kašanin theory of the behavior of materials under high pressures.**
V. Čelebonović.
Annales Universitatis Scientiarum Budapestinensis de Rolando Eötvös nominatae. Separatum Sectio Geologica, Tomus XXIII, 31 - 43 (1981).
The aim of this paper is to describe the theory proposed by Savić and Kašanin as a possible explanation of the origin of rotation, magnetic fields and layered internal structures of stars and planets. The theory has been applied to the sun and the planets up to Jupiter. A review of work by other authors concerning the problems of the origin of rotation, magnetic fields and internal structures of celestial bodies is presented.

065.100 **Stellar rotation and gravitational collapse: the a/m issue.**
F. de Felice, Y. Yu.
J. Phys. A, Vol. 15, 3341 - 3349 (1982). – Abstr. in Phys. Abstr., Vol. 86, Abstr. 5278 (1983).

065.101 **Nonradial oscillations of a gaseous sphere.**
C. Mohan, K. Singh.
Indian J. Theor. Phys., Vol. 29, 129 - 140 (1981). – Abstr. in Phys. Abstr., Vol. 86, Abstr. 15908 (1983).

065.102 **The structure and stability of relativistic polytropes with a compact core.**
L.-z. Fang, S.-p. Xiang.
Acta Phys. Sinica, Vol. 31, 1223 - 1234 (1982). In Chinese. Abstr. in Phys. Abstr., Vol. 86, Abstr. 20229 (1983).

065.103 **General relativistic gravitational collapse of a rotating star with magnetic fields: the formalism and the initial value equations.**
K. Maeda, K. Oohara.
Prog. Theor. Phys., Vol. 68, 567 - 579 (1982). – Abstr. in Phys. Abstr., Vol. 86, Abstr. 20350 (1983).

Physics of stellar evolution and cosmology.
See Abstr. 003.044.

Physik der Sterne.
See Abstr. 003.065.

Condensed matter – genesis of a concept.
See Abstr. 004.038.

Computation of the lowest eigenvalue arising from the problem of stellar stability by the method of Schwartz iteration.
See Abstr. 021.014.

The $^{15}N(p, \alpha_0)^{12}C$ reaction at stellar energies.
See Abstr. 022.117.

The $^{4}He(^{12}C, \gamma)^{16}O$ reaction at stellar energies.
See Abstr. 022.184.

Rapidly rotating polytropes and concave hamburger equilibrium.
See Abstr. 042.079.

Bifurcation and fission of three dimensional, rigidly rotating and self-gravitating polytropes.
See Abstr. 042.080.

A comparison of stellar observations and the theory of nucleosynthesis.
See Abstr. 061.034.

Solar power.
See Abstr. 061.035.

Hydrogen burning in massive stars.
See Abstr. 061.036.

Nuclear reactions in highly evolved stars.
See Abstr. 061.037.

The rate of the $^{12}C(\alpha, \gamma)^{16}O$ reaction in stellar helium burning.
See Abstr. 061.038.

^{3}He, red dwarfs, and our next trillion years.
See Abstr. 061.039.

The s-process.
See Abstr. 061.040.

The r-process and nucleocosmochronology.
See Abstr. 061.041.

Transport of neutrinos, radiation and energetic particles in accretion flows.
See Abstr. 061.057.

Meridional circulation in a rotating star revisited.
See Abstr. 062.012.

Stability of differential rotation in stars.
See Abstr. 062.022.

Propagation of blast waves in Roche model.
See Abstr. 062.067.

Solution of gravitational polytrope equations by the quasi-invariance group.
See Abstr. 062.113.

The role of the helicity spectrum function in turbulent dynamo theory.
See Abstr. 062.130.

Neutrino emission from a hot, dense, plane-parallel atmosphere in hydrostatic equilibrium. I.
See Abstr. 064.005.

Neutrino emission from a hot, dense, plane-parallel atmosphere in hydrostatic equilibrium. II. Numerical methods and interaction functions.
See Abstr. 064.033.

Stochastic mass loss rate fluctuations and evolution of massive stars.
See Abstr. 064.044.

Gravitational radiation from collapsing rotating stellar cores.
See Abstr. 066.065.

On the collapse dynamics of cold stars with a phase transition in the interior.
See Abstr. 066.067.

Gravitational radiation from collapsing rotating stellar cores.
See Abstr. 066.069.

Space-time continuum of a relativistic linear stellar model.
See Abstr. 066.091.

Gravothermal aspects in evolution of the stars and the Universe.
See Abstr. 066.144.

Numerical relativity.
See Abstr. 066.145.

Properties of relativistic linear stellar models.
See Abstr. 066.180.

Relativistic stars in close binary systems and masses of neutron stars.
See Abstr. 066.188.

Initial neutrino loss in neutron star formation.
See Abstr. 066.520.

An account of the development of the solar neutrino problem.
See Abstr. 080.042.

Ultraviolet observations of yellow giant stars.
See Abstr. 114.032.

Subgiant CH stars. I. s-process abundances, C/O ratios, and evolutionary status.
See Abstr. 114.049.

The Wolf-Rayet stars.
See Abstr. 114.075.

Wolf-Rayet phenomena.
See Abstr. 114.115.

Subluminous Wolf-Rayet stars: observations.
See Abstr. 114.133.

Forced oscillations in binary systems. Toroidal modes.
See Abstr. 117.007.

On the mass of the exploding stars in massive close binary systems.
See Abstr. 117.009.

Theoretical aspects of X-ray binaries.
See Abstr. 117.055.

The evolution of massive stars – the production of binary WR stars.
See Abstr. 117.092.

On the structure and evolution of the OB-companions in Wolf-Rayet binaries.
See Abstr. 117.094.

The binary system Sirius in the context of stellar evolution.
See Abstr. 118.020.

On the internal structure of main-sequence stars.
See Abstr. 119.054.

A hypothesis on the hydromagnetic activity of T Tau-type stars and related objects.
See Abstr. 121.020.

The chemical composition of R Coronae Borealis and XX Camelopardalis.
See Abstr. 122.071.

Convection in pulsating stars. II. RR Lyrae convection and stability.
See Abstr. 122.079.

Magnetorotational model of supernovae explosions.
See Abstr. 125.036.

Supernova models.
See Abstr. 125.038.

Supernova shocks and neutrino diffusion.
See Abstr. 125.040.

Supernovae for pedestrians.
See Abstr. 125.041.

Shock stagnation and neutrino losses in stellar collapse.
See Abstr. 125.042.

Theoretical models for supernovae.
See Abstr. 125.043.

Computer simulations of stellar collapse and supernovae explosions: non-rotating and rotating models.
See Abstr. 125.044.

Theoretical models of Type I supernovae.
See Abstr. 125.045.

The origin of the Crab Nebula and electron capture supernova of 8–10 $M_\odot$ stars.
See Abstr. 125.046.

Carbon-oxygen white dwarf presupernova models.
See Abstr. 125.047.

On the neutrino luminosity from a type II supernova.
See Abstr. 125.080.

Hot dense matter and stellar collapse.
See Abstr. 125.081.

Supernova theory.
See Abstr. 125.100.

Effects of neutrino trapping on supernova explosions.
See Abstr. 125.109

Nitrogen overabundances in Population II dwarfs.
See Abstr. 126.028.

Structure and evolution of dynamical protostellar envelopes in the process of massive star formation.
See Abstr. 131.303.

Luminosities and masses for three central stars of planetary nebulae in the Magellanic Clouds from ultraviolet spectroscopy with the IUE.
See Abstr. 135.035.

Gamma-ray bursts and the collapse of a white dwarf.
See Abstr. 142.522.

The globular cluster metallicity scale: evidence from stellar models.
See Abstr. 154.007.

Nitrogen variations on the main sequence of 47 Tucanae: implications from stellar structure theory.
See Abstr. 154.008.

A statistical method for determining ages of globular clusters by fitting isochrones.
See Abstr. 154.027.

Primary nucleosynthesis in the galactic disk.
See Abstr. 155.033.

G- und K-Sterne als Indikatoren der galaktischen Evolution.
See Abstr. 155.052.

The occurrence of carbon stars and Cepheid variables in common fields of the Magellanic Clouds.
See Abstr. 159.008.

Pregalactic very massive objects and their cosmological consequences.
See Abstr. 162.099.

Study of the evolutions of the Universe, stars and the solar system: personal retrospect and prospect.
See Abstr. 162.147.

Cosmic cooking, evolution of stars and our Metagalaxy.
See Abstr. 162.255.

Errata

065.901 **Erratum: "Stability of toroidal flux tubes in stars"** [Astron. Astrophys., Vol. 106, 58 - 66 (1982)].
H. C. Spruit, A. A. van Ballegooijen.
Astron. Astrophys., Vol. 113, 350 (1982). – See Abstr. 31.065.013.

065.902 **Erratum: 'Multimode stellar pulsations. III. Resonances'** [Astrophys. J., Vol. 257, 715 - 723 (1982)].
O. Regev, J. R. Buchler, M. Barranco.
Astrophys. J., Vol. 263, 483 (1982). – See Abstr. 31.065.040.

066 Relativistic Astrophysics, Gravitation Theory, Background Radiation, Black Holes, Neutron Stars

066.001 **Inhomogeneity of the early universe and formation of primordial black holes.**
N. A. Zabotin, L. S. Marochnik, P. D. Nasel'skij.
Astrofizika, Tom 18, 161 - 175 (1982). In Russian. English translation in Astrophysics, Vol. 18, No. 1.
The formation of black holes in the universe due to the collapse of primeval adiabatic inhomogeneities is considered. As necessary condition of creation of a separate black hole the Carr-Hawking criterion is used. Three possible modifications of this criterion are suggested based on the studies of eruptions with large amplitude. The relation between the parameters of the primordial black hole mass spectrum and the amplitude and spectral exponent of initial perturbations is obtained.

066.002 **Structure of the field of inhomogeneities and formation of primordial black holes.**
N. A. Zabotin, P. D. Nasel'skij.
Astron. Zh., Tom 59, 647 - 656 (1982). In Russian. English translation in Soviet Astron., Vol. 25, No. 4.
This paper treats the most probable profile of the large-amplitude disturbance as a function of parameters of the spatial spectrum of the stochastic field of inhomogeneities. The conditions, when the effects of the accretion of matter and the decay of the strong disturbance have essential influence on the formation of the mass spectrum of primordial black holes, are derived.

066.003 **An example of induced centrifugal force in general relativity.**
J. M. Cohen, W. J. Sarill, C. V. Vishveshwara.
Nature, Vol. 298, 829 (1982).
The authors demonstrate that there are solutions to Einstein's equations which exhibit induced centrifugal force.

066.004 **Bursts of gravitational radiation from active galactic nuclei and globular clusters.**
I. G. Dymnikova, A. S. Zentsova, A. K. Popov.
Fiz.-tekh. inst. AN SSSR. Prepr., 1981, No. 742, 19 pp. In Russian. – Abstr. in Ref. zh., 51. Astron., 7.51.191 (1982).

066.005 **General relativistic equations of rotational motion of a body with variable mass.**
O. P. Arkhipkin.
The problem of motion in Einstein's gravitation theory, (see 003.002), 1981, p. 69 - 77. In Russian. – Abstr. in Ref. zh., 51. Astron., 7.51.202 (1982).

066.006 **On the relativistic equations of translational motion of a system of bodies with arbitrary shape taking into account their internal structure and rotation.**
A. A. Yunusov.
The problem of motion in Einstein's gravitation theory, (see 003.002), 1981, p. 142 - 148. In Russian. – Abstr. in Ref. zh., 51. Astron., 7.51.203 (1982).

066.007 **General relativistic equations of motion of a body with variable mass with charge and magnetic moment.**
D. Z. Taipov.
The problem of motion in Einstein's gravitation theory, (see 003.002), 1981, p. 83 - 100. In Russian. – Abstr. in Ref. zh., 51. Astron., 7.51.204 (1982).

066.008 **The variational principle and PPN-theories.**
N. Kojshibaev.
The problem of motion in Einstein's gravitation theory, (see 003.002), 1981, p. 118 - 128. In Russian. – Abstr. in Ref. zh., 51. Astron., 7.51.205 (1982).

066.009 **On the relativistic effect in the motion of many bodies having variable masses in a post-Newtonian gravitational field.**
N. G. Magnaradze.
Tr. Tbilis. univ., Tom 218, 183 - 216 (1981). In Russian. Abstr. in Ref. zh., 51. Astron., 7.51.206 (1982).

066.010 **Evolution of the orbits of stars passing through a massive accretion disc.**
Eh. Ya. Vil'koviskij, N. M. Bekbasarov.
The problem of motion in Einstein's gravitation theory, (see 003.002), p. 100 - 108 (1981). In Russian. – Abstr. in Ref. zh., 51. Astron., 8.51.134 (1982).

066.011 **Analysis of some mechanical problems of general relativity.**
M. M. Abdil'din.
The problem of motion in Einstein's gravitation theory, (see 003.002), p. 3 - 41 (1981). In Russian. – Abstr. in Ref. zh., 51. Astron., 8.51.141 (1982).

066.012 **On the difference in relativistic equations of rotational motion deduced by the first and second methods of Fock.**
M. M. Abdil'din, M. S. Omarov.
The problem of motion in Einstein's gravitation theory, (see 003.002), p. 42 - 69 (1981). In Russian. – Abstr. in Ref. zh., 51. Astron., 8.51.142 (1982).

066.013 **Non-existence of high-order integrals of motion relative to velocities.**
I. L. Genkin, L. M. Genkina.
The problem of motion in Einstein's gravitation theory, (see 003.002), p. 77 - 83 (1981). In Russian. – Abstr. in Ref. zh., 51. Astron., 8.51.143 (1982).

066.014 **On the equations of motion of a system of spin bodies in general relativity.**
L. M. Chechin.
The problem of motion in Einstein's gravitation theory, (see 003.002), p. 132 - 141 (1981). In Russian. – Abstr. in Ref. zh., 51. Astron., 8.51.144 (1982).

066.015 **On some kinematic relations in the special theory of relativity.**
V. G. Ganaga.
Vestn. Kiev. Univ., Astron. Vyp. 24, p. 90 - 105 (1982). In Russian.

066.016 **On methods for clock synchronization.**
V. G. Ganaga.
Vestn. Kiev. Univ., Astron. Vyp. 24, p. 105 - 112 (1982). In Russian.

066.017 **Relativistic astrophysics and the theory of gravitation.**
Ya. B. Zel'dovich.
Zemlya Vselennaya, 1982, No. 4, p. 16 - 20. In Russian.

066.018 **On the quadrupole anisotropy of the microwave background in the far-infrared.**
C. Ceccarelli, G. Dall'Oglio, B. Melchiorri, F. Melchiorri, L. Pietranera.

Astrophys. J., Vol. 260, 484 - 487 (1982).

Infrared data on large-scale anisotropy have been analyzed with spherical harmonics formalism. The authors have found Q_2 and Q_5 terms that are above the noise. A substantial fraction, if not all, of the Q_5 term is due to an extended infrared source at $l'' = 285$ and $b'' = 75$ (Virgo cluster). The possibility of having detected the Sunyaev-Zel'dovich effect in the Virgo cluster is discussed.

066.019 **Gravitational radiation from nonspherical infall into black holes.**
S. L. Shapiro, I. Wasserman.
Astrophys. J., Vol. 260, 838 - 848 (1982).

The authors perform strong-field perturbation calculations of the gravitational radiation emitted when spheroidal dust shells of mass m and arbitrary eccentricity fall radially into Schwarzschild black holes of mass $M \gg m$. They demonstrate that the total energy emitted in gravitational radiation by a nonspherical dust cloud falling into a black hole is always less than the energy radiated by a point particle of the same mass falling into the hole. The suppression results from interference between waves emitted from different regions of the extended, infalling mass.

066.020 **Observations with a low-temperature, resonant mass, gravitational radiation detector.**
S. P. Boughn, W. M. Fairbank, R. P. Giffard, J. N. Hollenhorst, E. R. Mapoles, M. S. McAshan, P. F. Michelson, H. J. Paik, R. C. Taber.
Astrophys. J., Lett., Vol. 261, L19 - L22 (1982).

The noise level at the output of a highly sensitive antenna has been used to establish a new observational upper limit on the flux of 1 kHz gravity wave bursts.

066.021 **3 mm measurement of the anisotropy in the background radiation.**
P. M. Lubin, G. L. Epstein, G. F. Smoot.
Bull. American Astron. Soc., Vol. 14, 631 (1982). – Abstract.

066.022 **Search for small-scale anisotropy in the cosmic microwave background.**
J. M. Uson, D. T. Wilkinson.
Bull. American Astron. Soc., Vol. 14, 631 (1982). – Abstract.

066.023 **Thick supercritical accretion disks and active galactic nuclei.**
P. J. Wiita.
Comments Astrophys., Vol. 9, 251 - 260 (1982).

066.024 **The spatial anisotropies of the microwave background.**
B. Melchiorri, F. Melchiorri.
Acta Cosmologica, Zesz. 11, (see 012.012), 27 - 54 (1982).

066.025 **Distortions in the Rayleigh-Jeans region of the cosmic background radiation spectrum.**
G. De Zotti.
Acta Cosmologica, Zesz. 11, (see 012.012), 55 - 64 (1982).

The theory of the origin and evolution of distortions in the Rayleigh-Jeans region of the cosmic background radiation spectrum is reviewed. Some proposed experiments, designed to substantially improve our knowledge of that portion of the spectrum, are briefly described.

066.026 **Angular anisotropy of the cosmic microwave background radiation.**
J. Silk.
Acta Cosmologica, Zesz. 11, (see 012.012), 75 - 87 (1982).

The theory of fluctuations in the cosmic microwave background radiation is reviewed. Anisotropy on large-scale (dipole and quadrupole) and on small scales is discussed. The smoothing effects of secondary ionization (fractional ionization x) are found to be unimportant over an angular scale greater than $\sim 5\ (\Omega x)^{1/3}$ degrees.

066.027 **Pregalactic stars and the microwave background.**
B. J. Carr.
Acta Cosmologica, Zesz. 11, (see 012.012), 131 - 142 (1982).

066.028 **Off-equatorial circular orbits in the Kerr-Newman space-time.**
M. Calvani, F. de Felice, R. Fabbri, R. Turolla.
Nuovo Cimento B, Vol. 67B, Ser. 11, 1 - 10 (1982). – Abstr. in Phys. Abstr., Vol. 85, Abstr. 58666 (1982).

066.029 **Pulsars' timing noise and external disturbances: gravitational waves? .**
M. Cerdonio, S. Vitale.
Nuovo Cimento B, Vol. 67B, Ser. 11, 185 - 201 (1982).
Abstr. in Phys. Abstr., Vol. 85, Abstr. 58671 (1982).

066.030 **Massive spin-half particle in the de Sitter Universe.**
U. Khanal, N. Panchapakesan.
Ann. Physics, Vol. 138, 260 - 274 (1982). – Abstr. in Phys. Abstr., Vol. 85, Abstr. 58672 (1982).

066.031 **Weyl's geometry and physics.**
N. Rosen.
Found. Phys., Vol. 12, 213 - 248 (1982). – Abstr. in Phys. Abstr., Vol. 85, Abstr. 58673 (1982).

066.032 **Solitons in the theory of gravity.**
V. A. Belinskij.
Usp. fiz. nauk, Tom 136, 541 - 542 (1982). In Russian.
Abstr. in Ref. zh., 51. Astron., 9.51.132 (1982).

066.033 **Some astrophysical consequences of the Jordan-Brans-Dicke theory.**
I. S. Nurgaliev.
Izv. vuzov. Fiz., Tom 25, No. 2, p. 24 - 27 (1982). In Russian.
Abstr. in Ref. zh., 51. Astron., 9.51.133 (1982).

066.034 **The spectrum of the extragalactic background light.**
G. Bruzual A.
Rev. Mexicana Astron. Astrofis., Vol. 6, (see 012.017), 19 - 21 (1981).

The observed spectrum of the extragalactic background light in the range from ultraviolet to optical wavelengths is compared with a model prediction. The model uses the locally observed luminosity function of galaxies as well as evolutionary models for galaxy spectral energy distributions. The prediction is too faint by a factor of about 10.

066.035 **Real-time analysis system for a gravitational wave antenna.**
B. L. Brown, A. P. Mills, Jr., J. A. Tyson.
Rev. Sci. Instrum., Vol. 53, 479 - 484 (1982). – Abstr. in Phys. Abstr., Vol. 85, Abstr. 63871 (1982).

066.036 **Gravitation and relativity experiments using atomic clocks.**
R. F. C. Vessot.
J. Phys. Colloq., Vol. 42, No. C-8, p. 359 - 371 (1981).
Abstr. in Phys. Abstr., Vol. 85, Abstr. 63884 (1982).

066.037 **Macroscopic rotors and gravitational effects.**
R. C. Ritter.
J. Phys. Colloq., Vol. 42, No. C-8, p. 429 - 440 (1981).
Abstr. in Phys. Abstr., Vol. 85, Abstr. 63885 (1982).

066.038 **Gravitational experiments with stable electromagnetic devices.**
P. Tourrenc.

J. Phys. Colloq., Vol. 42, No. C-8, p. 441 - 449 (1981). Abstr. in Phys. Abstr., Vol. 85, Abstr. 63886 (1982).

066.039 **Demands on laser purity in an interferometric gravitational wave detector.**
A. Rüdiger, R. Schilling, L. Schnupp, W. Winkler, H. Billing, K. Maischberger.
J. Phys. Colloq., Vol. 42, No. C-8, p. 451 - 459 (1981). Abstr. in Phys. Abstr., Vol. 85, Abstr. 63887 (1982).

066.040 **High precision laser interferometry for detection of gravitational radiation.**
K. Maischberger, A. Rüdiger, R. Schilling, L. Schnupp, W. Winkler, H. Billing.
Laser spectroscopy V, (see 012.016), p. 25 - 32 (1981). Abstr. in Phys. Abstr., Vol. 85, Abstr. 63888 (1982).

066.041 **Optical cavity laser interferometers for gravitational wave detection.**
R. W. P. Drever, J. Hough, A. J. Munley, S. A. Lee, R. Spero, S. E. Whitcomb, H. Ward, G. M. Ford, M. Hereld, N. A. Robertson, I. Kerr, J. R. Pugh, G. P. Newton, B. Meers, E. D. Brooks III, Y. Gursel.
Laser spectroscopy V, (see 012.016), p. 33 - 40 (1981). Abstr. in Phys. Abstr., Vol. 85, Abstr. 63889 (1982).

066.042 **Distorted black holes.**
R. Geroch, J. B. Hartle.
J. Math. Phys., Vol. 23, 680 - 692 (1982). – Abstr. in Phys. Abstr., Vol. 85, Abstr. 68809 (1982).

066.043 **An interior solution for a charged sphere in general relativity.**
A. L. Mehra.
Phys. Lett. A, Vol. 88A, 159 - 161 (1982). – Abstr. in Phys. Abstr., Vol. 85, Abstr. 69059 (1982).

066.044 **Nonanalytic terms in the slow-motion expansion of a radiating scalar field on a Schwarzschild background.**
R. E. Kates, L. S. Kegeles.
Phys. Rev. D, Vol. 25, 2030 - 2037 (1982). – Abstr. in Phys. Abstr., Vol. 85, Abstr. 69064 (1982).

066.045 **Gauge-invariant definition of gravitational radiation in Robertson-Walker cosmologies.**
J. M. Niedra.
Phys. Rev. D, Vol. 25, 2049 - 2052 (1982). – Abstr. in Phys. Abstr., Vol. 85, Abstr. 69065 (1982).

066.046 **On scattering of scalar waves in static space-times, particularly Schwarzschild.**
R. Beig.
Acta Phys. Austriaca, Vol. 54, 129 - 141 (1982). – Abstr. in Phys. Abstr., Vol. 85, Abstr. 69066 (1982).

066.047 **Spin-1/2 particles in the presence of a Heavisidian field.**
A. Singh.
Lett. Nuovo Cimento, Vol. 33, Ser. 2, 505 - 508 (1982). Abstr. in Phys. Abstr., Vol. 85, Abstr. 69069 (1982).

066.048 **Relativistic detonation waves and bubble growth in false vacuum decay.**
P. J. Steinhardt.
Phys. Rev. D, Vol. 25, 2074 - 2085 (1982). – Abstr. in Phys. Abstr., Vol. 85, Abstr. 69073 (1982).

066.049 **A radiating metric for evaporating black holes.**
R. Balbinot, R. Bergamini.
Nuovo Cimento B, Vol. 68B, Ser. 11, 104 - 110 (1982). Abstr. in Phys. Abstr., Vol. 85, Abstr. 74009 (1982).

066.050 **Neutrinos in gravitational collapse: the Dirac formalism.**
B. R. Iyer, S. V. Dhurandhar, C. V. Vishveshwara.
Phys. Rev. D, Vol. 25, 2053 - 2064 (1982). – Abstr. in Phys. Abstr., Vol. 85, Abstr. 74010 (1982).

066.051 **Cosmic microwave background – present status and future prospects.**
R. A. Muller.
Ninth Texas Symposium on Relativistic Astrophysics, (see 012.020), p. 116 - 124 (1980).
The author discusses the status of present measurements of the spectrum and the large-angular-scale isotropy of the cosmic microwave background, and finally indicates what the prospects are for improved measurements in the next decade.

066.052 **Isolated systems in general relativity.**
J. Ehlers.
Ninth Texas Symposium on Relativistic Astrophysics, (see 012.020), p. 279 - 294 (1980).

066.053 **Supergravity.**
J. Wess.
Ninth Texas Symposium on Relativistic Astrophysics, (see 012.020), p. 295 - 298 (1980).

066.054 **Summary of workshop on gauge theories of gravitation.**
F. W. Hehl.
Ninth Texas Symposium on Relativistic Astrophysics, (see 012.020), p. 299 - 300 (1980).

066.055 **Summary of workshops on quantum gravity.**
C. J. Isham.
Ninth Texas Symposium on Relativistic Astrophysics, (see 012.020), p. 301 - 306 (1980).

066.056 **The confrontation between general relativity and experiment.**
C. M. Will.
Ninth Texas Symposium on Relativistic Astrophysics, (see 012.020), p. 307 - 321 (1980).

066.057 **Summary of workshop on gravitational radiation.**
E. Amaldi.
Ninth Texas Symposium on Relativistic Astrophysics, (see 012.020), p. 322 - 333 (1980).

066.058 **Problems and solutions in numerical relativity.**
T. Piran.
Tenth Texas Symposium on Relativistic Astrophysics, (see 012.021), p. 1 - 14 (1981).
In general relativity present analytical methods fail when strong fields and dynamical configurations are involved, as in the formation of black holes and generation of gravitational radiation. Numerical relativity aims at providing answers to such questions.

066.059 **Progress in supergravity.**
Y. Ne'eman.
Tenth Texas Symposium on Relativistic Astrophysics, (see 012.021), p. 15 - 31 (1981).
Supergravity is an embedding of Einstein's theory in a broader framework, with two main aims: renormalization of quantum gravity and unification with other interactions. Recent developments in this field are reviewed.

066.060 **The cosmic microwave background radiation.**
J. Silk.
Tenth Texas Symposium on Relativistic Astrophysics, (see 012.021), p. 188 - 209 (1981).
This review assesses recent observations of the microwave

background. The evidence for spectral distortions is reviewed, and the theoretical implications are described. The evidence for anisotropy is then discussed, and, finally, the theoretical predictions of radiation anisotropy are summarized and compared with the data.

066.061 **Gravitational lenses in astrophysics.**
J. E. Gunn.
Tenth Texas Symposium on Relativistic Astrophysics (see 012.021), p. 287 - 296 (1981).
The observational properties and relativistic models for the gravitational lenses Q 0957 + 561 and Q 1115+080 are reviewed.

066.062 **Summary of the workshop on gravitational radiation detectors.**
J. Weber, H. Hirakawa.
Tenth Texas Symposium on Relativistic Astrophysics, (see 012.021), p. 446 - 448 (1981).

066.063 **Summary of the workshop on general relativity in astrophysics.**
R. A. Matzner.
Tenth Texas Symposium on Relativistic Astrophysics, (see 012.021), p. 449 - 458 (1981).

066.064 **Summary of the workshop on radiation processes in strong gravitational fields.**
J. M. Cohen.
Tenth Texas Symposium on Relativistic Astrophysics, (see 012.021), p. 459 - 464 (1981).

066.065 **Gravitational radiation from collapsing rotating stellar cores.**
E. Müller.
Supernovae: a survey of current research, (see 012.022), p. 157 - 166 (1982).
The author calculates the gravitational radiation produced by the axisymmetric, Newtonian stellar collapse models of Müller and Hillebrandt (1981) within a post-Newtonian multipole formalism developed by Wagoner.

066.066 **The expansion of scalar spherical waves to the second order approximation in a gravitational field with quadrupol structure.**
R. Mankin.
Astron. Nachr., Band 303, 201 - 210 (1982). In German.
By means of the perturbation expansion scheme with respect to the gravitational constant the author gives a retarded solution of the covariant wave equation for the potential of a free massless scalar field in an axially symmetric gravitational background field of a point mass with a quadrupole moment. The results show that far from the source the spherical symmetric second order wave tail is effectively generated by the loss of mass, caused by gravitational radiation, and that all the other second order tail terms are compensable by first order multipole moments corresponding to a change of the boundary conditions.

066.067 **On the collapse dynamics of cold stars with a phase transition in the interior.**
B. Kämpfer.
Astron. Nachr., Band 303, 231 - 236 (1982).
Cold stars can become unstable due to a phase transition in the centre. By small perturbations the collapse is triggered toward a new stable configuration. The features of the collapse are studied by means of a parameter-independent model with piece by piece uniform density. Possible consequences of such a collapse for neutron stars are: extreme speed-up, heating and a supernova-like release of binding energy. The dynamical stability properties of the model are considered in the framework of General Relativity. In particular the very beginning of the collapse can be treated by analytical means. In a certain range relativistic effects can prevent the configuration to collapse.

066.068 **Evolutionary aspects of the cosmic black body radiation.**
D. W. Sciama.
Evolution in the Universe, (see 012.024), p. 105 - 122 (1982).
The canonical hot big-bang picture of the evolution of the Universe is briefly outlined, and some recent indications of non-canonical behaviour, both observational and theoretical, are described. The speculative, but important, possibility that neutrinos have a non-zero rest-mass is then considered, and some cosmological and astrophysical consequences of this possibility, when combined with the hot big-bang hypothesis, are explored.

066.069 **Gravitational radiation from collapsing rotating stellar cores.**
E. Müller.
Astron. Astrophys., Vol. 114, 53 - 59 (1982).
The gravitational radiation produced by the axisymmetric, Newtonian collapse models of rotating stellar cores examined by Müller and Hillebrandt (1981) is considered within a post-Newtonian multipole formalism. Waveforms and energy losses of the two lowest order multipoles together with the energy spectrum of the quadrupole radiation are calculated.

066.070 **Do black holes exist at the centres of globular clusters?**
V. G. Gurzadyan.
Astron. Astrophys., Vol. 114, 71 - 73 (1982).
The role of stochastic processes in the dynamics of a massive black hole situated at the centre of a star cluster is discussed. It is shown that under certain conditions the black hole is able to escape from the mass centre of the system and has to execute a Brownian motion influenced by fluctuating stellar perturbations.

066.071 **Radar astronomy and the special theory of relativity.**
Z. Bay, J. A. White.
Acta Phys. Acad. Sci. Hungaricae, Vol. 51, 273 - 297 (1981). Abstr. in Phys. Abstr., Vol. 85, Abstr. 74206 (1982).

066.072 **Spontaneous symmetry breaking in noninertial frames and curved space-time.**
T. Padmanabhan.
Phys. Lett. A, Vol. 89A, 131 - 132 (1982). – Abstr. in Phys. Abstr., Vol. 85, Abstr. 74279 (1982).

066.073 **Space-times admitting Killing-Yano tensors. II.**
W. Dietz, R. Rüdiger.
Proc. R. Soc. London, Ser., A, Vol. 381, 315 - 322 (1982). Abstr. in Phys. Abstr., Vol. 85, Abstr. 74281 (1982).

066.074 **Some properties of the symmetry axis in a superposition of two Kerr solutions.**
M. Kihara, A. Tomimatsu.
Prog. Theor. Phys., Vol. 67, 349 - 352 (1982). – Abstr. in Phys. Abstr., Vol. 85, Abstr. 74284 (1982).

066.075 **Thermal stress tensors in static Einstein spaces.**
D. N. Page.
Phys. Rev. D, Vol. 25, 1499 - 1509 (1982). – Abstr. in Phys. Abstr., Vol. 85, Abstr. 74292 (1982).

066.076 **Fine-structure constant: is it really a constant?**
J. D. Bekenstein.
Phys. Rev. D, Vol. 25, 1527 - 1539 (1982). – Abstr. in Phys. Abstr., Vol. 85, Abstr. 74293 (1982).

066.077 **On the coupling of the gravitational field in the quantum field theory.**
E. Kreisel.
Ann. Physik, Band 39, 63 - 68 (1982). In German. – Abstr. in Phys. Abstr., Vol. 85, Abstr. 74298 (1982).

066.078 **Exact conformally flat solution of Einstein equations for neutral superfluids.**
G. Papini, M. Weiss.
Phys. Lett. A, Vol. 89A, 329 - 331 (1982). – Abstr. in Phys. Abstr., Vol. 85, Abstr. 76775 (1982).

066.079 **A variational approach to dense relativistic matter using functional techniques.**
P. Hoodbhoy.
Ann. Physics, Vol. 139, 68 - 92 (1982). – Abstr. in Phys. Abstr., Vol. 85, Abstr. 79134 (1982).

066.080 **On the Hawking evaporation of Dirac particles in Kerr-Newman space-time.**
Z. Zhao, Y.-x. Guei, L. Liu.
Chinese Phys., Vol. 2, 386 - 391 (1982). – Abstr. in Phys. Abstr., Vol. 85, Abstr. 79140 (1982). – See Abstr. 29.066.032.

066.081 **Reversible transformations of shapes and sizes of uncharged ergospheres.**
N. Spyrou.
J. Phys. A, Vol. 15, 1979 - 1986 (1982). – Abstr. in Phys. Abstr., Vol. 85, Abstr. 79141 (1982).

066.082 **Renormalization of the stress-energy tensor in black hole spacetimes with boundaries.**
T. Elster.
Phys. Lett. A, Vol. 89A, 125 - 128 (1982). – Abstr. in Phys. Abstr., Vol. 85, Abstr. 79142 (1982).

066.083 **Gravitational radiation induced by a particle falling along the Z-axis into a Kerr black hole.**
T. Nakamura, M. Sasaki.
Phys. Lett. A, Vol. 89A, 185 - 189 (1982). – Abstr. in Phys. Abstr., Vol. 85, Abstr. 79143 (1982).

066.084 **Quasi-local mass and angular momentum in general relativity.**
R. Penrose.
Proc. R. Soc. London, Ser. A, Vol. 381, 53 - 63 (1982). Abstr. in Phys. Abstr., Vol. 85, Abstr. 79438 (1982).

066.085 **Classical self-force on an electron near a charged, rotating black hole.**
D. Lohiya.
J. Phys. A, Vol. 15, 1815 - 1819 (1982). – Abstr. in Phys. Abstr., Vol. 85, Abstr. 79444 (1982).

066.086 **An exact solution of rotating Julia-Zee Dyon black-hole.**
M. Kasuya.
Prog. Theor. Phys., Vol. 67, 499 - 507 (1982). – Abstr. in Phys. Abstr., Vol. 85, Abstr. 79447 (1982).

066.087 **On high frequency background quantization of gravity.**
H.-H. von Borzeszkowski.
Found. Phys., Vol. 12, 633 - 643 (1982). – Abstr. in Phys. Abstr., Vol. 85, Abstr. 79455 (1982).

066.088 **Exact solution of quantum gravity in 1 + 1 dimensions.**
S. G. Rajeev.
Phys. Lett. B, Vol. 113B, 146 - 150 (1982). – Abstr. in Phys. Abstr., Vol. 85, Abstr. 79457 (1982).

066.089 **Quality factors of a three-mode 50-kg gravitational radiation antenna at liquid helium temperature.**
C. Cosmelli, J.-P. Richard.
Rev. Sci. Instrum., Vol. 53, 674 - 677 (1982). – Abstr. in Phys. Abstr., Vol. 85, Abstr. 79461 (1982).

066.090 **A mechanism for supersymmetry breaking.**
B. A. Ovrut, J. Wess.
Phys. Lett. B, Vol. 112B, 347 - 350 (1982). – Abstr. in Phys. Abstr., Vol. 85, Abstr. 79760 (1982).

066.091 **Space-time continuum of a relativistic linear stellar model.**
V. Ureche.
Rev. Roumaine Phys., Vol. 27, No. 1, p. 93 - 99 (1982). Abstr. in Phys. Abstr., Vol. 85, Abstr. 84089 (1982).

066.092 **Heuristic extension of the Schwarzschild metric.**
J. M. Espinosa.
American J. Phys., Vol. 50, 508 - 513 (1982). – Abstr. in Phys. Abstr., Vol. 85, Abstr. 84409 (1982).

066.093 **Point-source idealization in classical field theories. I. Electromagnetic radiation damping of a system of two perturbed Reissner-Nordström singularities in the slow-motion limit.**
R. E. Kates.
Phys. Rev. D, Vol. 25, 2487 - 2493 (1982). – Abstr. in Phys. Abstr., Vol. 85, Abstr. 84488 (1982).

066.094 **Point-source idealization in classical field theories. II. Mechanical energy losses from electromagnetic radiation reaction.**
R. E. Kates, A. Rosenblum.
Phys. Rev. D, Vol. 25, 2494 - 2498 (1982). – Abstr. in Phys. Abstr., Vol. 85, Abstr. 84489 (1982).

066.095 **Fluid with heat flux in a conformally flat space-time.**
S. R. Maiti.
Phys. Rev. D, Vol. 25, 2518 - 2520 (1982). – Abstr. in Phys. Abstr., Vol. 85, Abstr. 84491 (1982).

066.096 **Physical distinctions between normalized solutions of the two-body problem of general relativity.**
M. Cantor, A. D. Kulkarni.
Phys. Rev. D, Vol. 25, 2521 - 2526 (1982). – Abstr. in Phys. Abstr., Vol. 85, Abstr. 84492 (1982).

066.097 **Evolution of radiating anisotropic spheres in general relativity.**
M. Cosenza, L. Herrera, M. Esculpi, L. Witten.
Phys. Rev. D, Vol. 25, 2527 - 2535 (1982). – Abstr. in Phys. Abstr., Vol. 85, Abstr. 84493 (1982).

066.098 **Gravitational radiation reaction from small-angle scattering of slowly moving compact bodies, calculated by matched asymptotic expansions.**
R. E. Kates, R. G. Madonna.
Phys. Rev. D, Vol. 25, 2499 - 2508 (1982). – Abstr. in Phys. Abstr., Vol. 85, Abstr. 84496 (1982).

066.099 **Mass spectrum in a conformally invariant Brans-Dicke theory.**
R. T. Hammond.
Phys. Rev. D, Vol. 25, 2699 - 2700 (1982). – Abstr. in Phys. Abstr., Vol. 85, Abstr. 84500 (1982).

066.100 **Broken symmetry theories of gravity and implications for cosmology.**
R. Critchley, P. C. W. Davies, G. Kennedy.

Phys. Lett. B, Vol. 112B, 331 - 335 (1982). – Abstr. in Phys. Abstr., Vol. 85, Abstr. 84505 (1982).

066.101 **Supergravity in superspace.**
P. Howe.
Nucl. Phys. B, Vol. B199, 309 - 364 (1982). – Abstr. in Phys. Abstr., Vol. 85, Abstr. 84507 (1982).

066.102 **Supergravity, R invariance and spontaneous supersymmetry breaking.**
R. Barbieri, S. Ferrara, D. V. Nanopoulos, K. S. Stelle.
Phys. Lett. B, Vol. 113B, 219 - 222 (1982). – Abstr. in Phys. Abstr., Vol. 85, Abstr. 84510 (1982).

066.103 **Orbit selection for the Stanford Relativity Gyroscope experiment.**
R. Vassar, J. V. Breakwell, C. W. F. Everitt, R. A. VanPatten.
J. Spacecr. Rockets, Vol. 19, 66 - 71 (1982). – Abstr. in Phys. Abstr., Vol. 85, Abstr. 84511 (1982).

066.104 **Search for correlations between the University of Maryland and the University of Rome gravitational radiation antennas.**
V. Ferrari, G. Pizzella, M. Lee, J. Weber.
Phys. Rev. D, Vol. 25, 2471 - 2486 (1982). – Abstr. in Phys. Abstr., Vol. 85, Abstr. 84512 (1982).

066.105 **Gravitational energy of condensed astrophysical objects.**
S. M. Chitre.
Proc. Indian Natl. Sci. Acad., Part A, Vol. 47, Suppl. 2, 112 - 123 (1981). – Abstr. in Phys. Abstr., Vol. 85, Abstr. 89144 (1982).

066.106 **Limits of the charge of a collapsed object.**
R. S. Hanni.
Phys. Rev. D, Vol. 25, 2509 - 2514 (1982). – Abstr. in Phys. Abstr., Vol. 85, Abstr. 89294 (1982).

066.107 **Black hole physics in globally hyperbolic spacetimes.**
P. S. Joshi, J. V. Narlikar.
Pramāna, Vol. 18, 385 - 396 (1982). – Abstr. in Phys. Abstr., Vol. 85, Abstr. 89295 (1982).

066.108 **Black-hole entropy fountains.**
D. Lynden-Bell.
Observatory, Vol. 102, 131 (1982). – Abstract.

066.109 **Relativistic jets from black holes.**
M. A. Abramowicz.
Observatory, Vol. 102, 132 (1982). – Abstract.

066.110 **Propagation of fast and slow jets.**
E. S. Phinney.
Observatory, Vol. 102, 132 - 133 (1982). – Abstract.

066.111 **On the thermodynamics of a collapsed object.**
N. Chaubey, V. de Sabbata.
Bull. Astron. Soc. India, Vol. 10, 42 - 43 (1982). – Abstract.

066.112 **Scattering of scalar tardyons and tachyons from a Schwarzschild black hole.**
S. Banerji, D. R. Mandal.
Bull. Astron. Soc. India, Vol. 10, 43 (1982). – Abstract.

066.113 **New exact solutions in general relativity.**
M. C. Durgapal.
Bull. Astron. Soc. India, Vol. 10, 43 (1982). – Abstract.

066.114 **Perfect fluid spheres in general relativity.**
D. C. Srivastava, S. S. Prasad.
Bull. Astron. Soc. India, Vol. 10, 44 (1982). – Abstract.

066.115 **Motion of ultrarelativistic particles.**
A. K. Pande, M. C. Durgapal.
Bull. Astron. Soc. India, Vol. 10, 44 (1982). – Abstract.

066.116 **Magnetic fields and accretion disks around black holes.**
N. Dadhich, P. J. Wiita.
Bull. Astron. Soc. India, Vol. 10, 45 (1982). – Abstract.

066.117 **Bursts of gravitational radiation from active galactic nuclei and globular clusters.**
I. G. Dymnikova, A. K. Popov, A. S. Zentsova.
Astrophys. Space Sci., Vol. 85, 231 - 241 (1982).

The characteristics of gravitational bursts from active galactic nuclei and globular clusters are obtained for three astrophysical situations: (1) scattering of stars by massive black holes resided at the centers of galaxies and globular clusters; (2) the close encounters of stars in the nuclear regions of these objects; (3) scattering of stars by black holes of stellar mass situated in the stellar population of galactic nuclei and clusters. The most effective source of gravitational bursts appears to be the scattering of stars by the massive central black holes.

066.118 **Why a variable G?**
V. M. Canuto, I. Goldman.
Astrophys. Space Sci., Vol. 86, 225 - 227 (1982).

Since a possible time variability of G has received renewed attention (Wesson and Goodson, 1981), the authors think it is important to stress a conceptual aspect so far not sufficiently appreciated and which puts the variability of G in a much wider context. Variable G is a popular but incomplete representation of a much deeper problem: Is the Strong Equivalence Principle valid?

066.119 **A unified Einstein-Maxwell theory and some of its immediate consequences.**
N. A. Batakis.
Bull. American Astron. Soc., Vol. 14, 776 (1982). – Abstract.

066.120 **Experimental tests of the linear equations for the gravitational field.**
A. Singh.
Lett. Nuovo Cimento, Vol. 34, Ser. 2, 193 - 196 (1982).
Abstr. in Phys. Abstr., Vol. 85, Abstr. 89597 (1982).

066.121 **Electrically resonant transducer for a gravitational wave antenna.**
L. Narici.
J. Appl. Phys., Vol. 53, 3941 - 3946 (1982). – Abstr. in Phys. Abstr., Vol. 85, Abstr. 89605 (1982).

066.122 **The fourth test of general relativity.**
K. Nordtvedt.
Rep. Prog. Phys., Vol. 45, 631 - 651 (1982). – Abstr. in Phys. Abstr., Vol. 85, Abstr. 89607 (1982).

066.123 **The gravitational field inside a long hollow cylinder of finite length.**
Y. T. Chen.
Proc. R. Soc. London, Ser. A, Vol. 382, 75 - 82 (1982).
Abstr. in Phys. Abstr., Vol. 85, Abstr. 89608 (1982).

066.124 **On a non-eclipse optical determination of the gravitational deflection of light by the Sun.**
F. A. Handler, R. A. Matzner.
Acta Astronaut., Vol. 9, 57 - 62 (1982). – Abstr. in Phys. Abstr., Vol. 85, Abstr. 94591 (1982).

066.125 **New results from the Viking relativity experiment.**
R. D. Reasenberg, I. I. Shapiro, R. B. Goldstein, P. E. MacNeil.
Acta Astronaut., Vol. 9, 91 - 93 (1982). – Abstr. in Phys. Abstr., Vol. 85, Abstr. 94592 (1982).

066.126 **Rate of black hole formation in a thermal box.**
T. Piran, R. M. Wald.
Phys. Lett. A, Vol. 90A, 20 - 22 (1982). – Abstr. in Phys. Abstr., Vol. 85, Abstr. 94768 (1982).

066.127 **Relativistic plasmas.**
R. J. Gould.
AIP Conf. Proc., No. 77, (see 012.034), p. 169 - 177 (1982).
Recent work on the kinetic theory of a highly-relativistic electron gas and the associated photon opacity is summarized. Results are given for the opacity from Compton scattering, pair production in the fields of electrons and ions, inverse bremsstrahlung, and synchrotron self absorption.

066.128 **On the theory of gamma-ray amplification through stimulated annihilation radiation (GRASAR).**
R. Ramaty, J. M. McKinley, F. C. Jones.
AIP Conf. Proc., No. 77, (see 012.034), p. 231 - 240 (1982).
The theory of photon emission, absorption and scattering in a relativistic plasma of positrons, electrons and photons is studied. Expressions for the emissivities and absorption coefficients of pair annihilation, pair production and Compton scattering are given and evaluated numerically. The conditions for negative absorption are investigated.

066.129 **Light rays in gravitating, refractive media.**
T. W. Noonan.
Astrophys. J., Vol. 262, 344 - 348 (1982).
The field-to-particle method of H. P. Robertson is applied to the general-relativistic Maxwell equations in order to obtain the general-relativistic equation of motion for a photon in a refractive medium. For the special case of an uncharged, refractive, spherically symmetric mass, the exact first-order differential equation for the light-ray path is given.

066.130 **Conformally flat, anisotropic spheres in general relativity.**
B. W. Stewart.
J. Phys. A, Vol. 15, 2419 - 2427 (1982). – Abstr. in Phys. Abstr., Vol. 85, Abstr. 95153 (1982).

066.131 **Spinor sources in cosmology.**
R. T. Jantzen.
J. Math. Phys., Vol. 23, 1137 - 1146 (1982). – Abstr. in Phys. Abstr., Vol. 85, Abstr. 95156 (1982).

066.132 **The investigation of some self-similar solutions of Einstein's equations.**
O. I. Bogoyavlensky *(Bogoyavlenskij)*, G. Moschetti.
J. Math. Phys., Vol. 23, 1353 - 1358 (1982). – Abstr. in Phys. Abstr., Vol. 85, Abstr. 95161 (1982).

066.133 **New exact static solutions to Einstein's equations for spherically symmetric perfect fluid distributions.**
J. Ibanez, J. L. Sanz.
J. Math. Phys., Vol. 23, 1364 - 1365 (1982). – Abstr. in Phys. Abstr., Vol. 85, Abstr. 95163 (1982).

066.134 **Mach's principle and the structure of dynamical theories.**
J. B. Barbour, B. Bertotti.
Proc. R. Soc. London, Ser. A, Vol. 382, 295 - 306 (1982). Abstr. in Phys. Abstr., Vol. 85, Abstr. 95164 (1982).

066.135 **Gravitational perturbation of the hydrogen spectrum.**
L. Parker, L. O. Pimentel.
Phys. Rev. D, Vol. 25, 3180 - 3190 (1982). – Abstr. in Phys. Abstr., Vol. 85, Abstr. 95167 (1982).

066.136 **Axially symmetric gravitational two-body problem of Cooperstock, Lim, and Hobill.**
M. Walker, C. M. Will
Phys. Rev. D, Vol. 25, 3433 - 3437 (1982). – Abstr. in Phys. Abstr., Vol. 85, Abstr. 95170 (1982).

066.137 **Reply to the comments of Walker and Will regarding the axially symmetric gravitational two-body problem.**
F. I. Cooperstock.
Phys. Rev. D, Vol. 25, 3438 - 3440 (1982). – Abstr. in Phys. Abstr., Vol. 85, Abstr. 95171 (1982).

066.138 **Axially symmetric two-body problem in general relativity. IV. Boundary conditions, time scales, and the quadrupole formula.**
F. I. Cooperstock.
Phys. Rev. D, Vol. 25, 3126 - 3136 (1982). – Abstr. in Phys. Abstr., Vol. 85, Abstr. 95178 (1982).

066.139 **Horizon-free Universe.**
A. Frenkel, K. Brecher.
Phys. Rev. D, Vol. 26, 368 - 372 (1982). – Abstr. in Phys. Abstr., Vol. 85, Abstr. 95182 (1982).

066.140 **Gravitational scattering of spinning particles: linear approximation.**
J. Ibanez, J. Martin.
Phys. Rev. D, Vol. 26, 384 - 389 (1982). – Abstr. in Phys. Abstr., Vol. 85, Abstr. 95183 (1982).

066.141 **Solution of the Planck mass problem in Kaluza-Klein theories.**
M. Pavsic.
J. Phys. G, Vol. 8, L89 - L92 (1982). – Abstr. in Phys. Abstr., Vol. 85, Abstr. 95188 (1982).

066.142 **Quantum mechanics of the gravitational field.**
C. Teitelboim.
Phys. Rev. D, Vol. 25, 3159 - 3179 (1982). – Abstr. in Phys. Abstr., Vol. 85, Abstr. 95200 (1982).

066.143 **Response of a Doppler canceling system to plane gravitational waves.**
A. Caporali.
Phys. Rev. D, Vol. 26, 345 - 355 (1982). – Abstr. in Phys. Abstr., Vol. 85, Abstr. 95212 (1982).

066.144 **Gravothermal aspects in evolution of the stars and the Universe.**
D. Sugimoto, Y. Eriguchi, I. Hachisu.
Prog. Theor. Phys. Suppl., No. 70, p. 154 - 180 (1981). Abstr. in Phys. Abstr., Vol. 85, Abstr. 100726 (1982).

066.145 **Numerical relativity.**
T. Nakamura.
Prog. Theor. Phys. Suppl., No. 70, p. 202 - 214 (1981). Abstr. in Phys. Abstr., Vol. 85, Abstr. 100727 (1982).

066.146 **Multi-soliton solutions of the Einstein equation and the Tomimatsu-Sato metric.**
A. Tomimatsu, H. Sato.
Prog. Theor. Phys. Suppl., No. 70, p. 215 - 237 (1981). Abstr. in Phys. Abstr., Vol. 85, Abstr. 100728 (1982).

066.147 **Partially self-similar spacetimes and fluid motions.**
K. Tomita.
Prog. Theor. Phys. Suppl., No. 70, p. 286 - 300 (1981).
Abstr. in Phys. Abstr., Vol. 85, Abstr. 100730 (1982).

066.148 **Collapse of radiating imperfect fluid spheres.**
K. Lake.
Phys. Rev. D, Vol. 26, 518 - 519 (1982). – Abstr. in Phys. Abstr., Vol. 85, Abstr. 100731 (1982).

066.149 **Feynman's path-integral method and Hawking evaporation.**
L. Liu.
Acta Phys. Sinica, Vol. 31, 519 - 524 (1982). In Chinese.
Abstr. in Phys. Abstr., Vol. 85, Abstr. 100971 (1982).

066.150 **Extreme Reissner-Nordström black holes can support a 'superhair'.**
R. Guven.
Phys. Rev. D, Vol. 25, 3117 - 3125 (1982). – Abstr. in Phys. Abstr., Vol. 85, Abstr. 100972 (1982).

066.151 **A class of new exact solutions in general relativity.**
M. C. Durgapal.
J. Phys. A, Vol. 15, 2637 - 2644 (1982). – Abstr. in Phys. Abstr., Vol. 85, Abstr. 101261 (1982).

066.152 **Conditions for regularity on the symmetry axis in a superposition of two Kerr-NUT solutions.**
A. Tomimatsu, M. Kihara.
Prog. Theor. Phys., Vol. 67, 1406 - 1414 (1982). – Abstr. in Phys. Abstr., Vol. 85, Abstr. 101267 (1982).

066.153 **Linearized quadrupole waves in general relativity and the motion of test particles.**
S. A. Teukolsky.
Phys. Rev. D, Vol. 26, 745 - 750 (1982). – Abstr. in Phys. Abstr., Vol. 85, Abstr. 101268 (1982).

066.154 **Amplification of gravitational waves interacting with a binary system.**
D. V. Gal'tsov, V. I. Petukhov.
Phys. Lett. A, Vol. 90A, 351 - 353 (1982). – Abstr. in Phys. Abstr., Vol. 85, Abstr. 101277 (1982).

066.155 **Experimental test of the law of gravitation.**
Y. Ogawa, K. Tsubono, H. Hirakawa.
Phys. Rev. D, Vol. 26, 729 - 734 (1982). – Abstr. in Phys. Abstr., Vol. 85, Abstr. 101296 (1982).

066.156 **Flattening spacetime near the Earth.**
R. L. Forward.
Phys. Rev. D, Vol. 26, 735 - 744 (1982). – Abstr. in Phys. Abstr., Vol. 85, Abstr. 101297 (1982).

066.157 **The Hawking evaporation of Dirac particles in general Kerr-Newman black hole background.**
C.-M. Xu, Y.-G. Shen.
Acta Phys. Sinica, Vol. 31, 207 - 212 (1982). In Chinese.
Abstr. in Phys. Abstr., Vol. 85, Abstr. 106769 (1982).

066.158 **Event horizon area in a slowly changing Weyl geometry.**
B. E. Brumbaugh, R. H. Price.
Canadian J. Phys., Vol. 60, 1155 - 1162 (1982). – Abstr. in Phys. Abstr., Vol. 85, Abstr. 106770 (1982).

066.159 **Magnetic fields and accretion discs around static black holes.**
N. Dadhich, P. J. Wiita.
J. Phys. A, Vol. 15, 2645 - 2653 (1982). – Abstr. in Phys. Abstr., Vol. 85, Abstr. 106772 (1982).

066.160 **General relativistic collapse of non-rotating, axisymmetric stars.**
T. Nakamura, H. Sato.
Prog. Theor. Phys., Vol. 67, 1396 - 1405 (1982). – Abstr. in Phys. Abstr., Vol. 85, Abstr. 106773 (1982).

066.161 **Dyon black hole and thermodynamics in the Tomimatsu-Sato-Yamazaki space-time.**
M. Kasuya.
Phys. Lett. B, Vol. 115B, 104 - 106 (1982). – Abstr. in Phys. Abstr., Vol. 85, Abstr. 106774 (1982).

066.162 **Ether formulations of relativity.**
M. C. Duffy.
Indian J. Theor. Phys., Vol. 28, 295 - 318 (1980). – Abstr. in Phys. Abstr., Vol. 85, Abstr. 107190 (1982).

066.163 **Gravitational radiation from a Kerr black hole. I. Formulation and a method for numerical analysis.**
M. Sasaki, T. Nakamura.
Prog. Theor. Phys., Vol. 67, 1788 - 1809 (1982). – Abstr. in Phys. Abstr., Vol. 85, Abstr. 107201 (1982).

066.164 **The latitudinal motion of test particles in the Kerr-Newman dyon space-time.**
M. Calvani, Z. Stuchlik.
Nuovo Cimento B, Vol. 70B, Ser. 11, 128 - 140 (1982).
Abstr. in Phys. Abstr., Vol. 85, Abstr. 107205 (1982).

066.165 **On Treder-type tetrad theories of gravitation. I. Equivalence principle and invariance properties.**
U. Bleyer.
Ann. Physik, Band 39, 203 - 208 (1982). – Abstr. in Phys. Abstr., Vol. 85, Abstr. 107206 (1982).

066.166 **On Treder-type tetrad theories of gravitation. II. Treder's tetrad theories in linear approximation.**
U. Bleyer.
Ann. Physik, Band 39, 209 - 215 (1982). – Abstr. in Phys. Abstr., Vol. 85, Abstr. 107207 (1982).

066.167 **Induced gravity and cosmology.**
J. M. Cervero, P. G. Estevez.
Ann. Physics, Vol. 142, 64 - 79 (1982). – Abstr. in Phys. Abstr., Vol. 85, Abstr. 107208 (1982).

066.168 **Dynamically broken supergravity and the hierarchy problem.**
H. P. Nilles.
Phys. Lett. B, Vol. 115B, 193 - 196 (1982). – Abstr. in Phys. Abstr., Vol. 85, Abstr. 107223 (1982).

066.169 **Positive energy in anti-de Sitter backgrounds and gauged extended supergravity.**
P. Breitenlohner, D. Z. Freedman.
Phys. Lett. B, Vol. 115B, 197 - 201 (1982). – Abstr. in Phys. Abstr., Vol. 85, Abstr. 107224 (1982).

066.170 **Quantization of Newton's constant in certain supergravity theories.**
E. Witten, J. Bagger.
Phys. Lett. B, Vol. 115B, 202 - 206 (1982). – Abstr. in Phys. Abstr., Vol. 85, Abstr. 107225 (1982).

066.171 **Variable G and the strong equivalence principle.**
V. M. Canuto.
Int. J. Theor. Phys., Vol. 21, 633 - 663 (1982). – Abstr. in Phys. Abstr., Vol. 85, Abstr. 107227 (1982).

066.172 **Line profile of relativistic Keplerian rings.**
L.-Z. Fang, T.-R. Deng.

Acta Phys. Sinica, Vol. 31, 109 - 114 (1982). In Chinese. Abstr. in Phys. Abstr., Vol. 85, Abstr. 112430 (1982).

066.173 **The boson bound states in Kerr metric.**
S.-W. Zhang, R.-K. Su.
Acta Phys. Sinica, Vol. 31, 311 - 317 (1982). In Chinese. Abstr. in Phys. Abstr., Vol. 85, Abstr. 112481 (1982).

066.174 **Vacuum polarization near the event horizon of a charged rotating black hole.**
V. P. Frolov.
Phys. Rev. D, Vol. 26, 954 - 955 (1982). – Abstr. in Phys. Abstr., Vol. 85, Abstr. 112482 (1982).

066.175 **The gravitational 'scintillation' of the microwave background due to density inhomogeneities.**
L. Z. Fang, H. Sato.
Prog. Theor. Phys., Vol. 67, 1994 - 1996 (1982). – Abstr. in Phys. Abstr., Vol. 85, Abstr. 112611 (1982).

066.176 **Meaning of a varying gravitational constant.**
S.-H. Hsieh.
Int. J. Theor. Phys., Vol. 21, 673 - 683 (1982). – Abstr. in Phys. Abstr., Vol. 85, Abstr. 112622 (1982).

066.177 **Structure and stability of rotating fluid disks around massive objects. II. General relativistic formulation.**
D. K. Chakraborty, A. R. Prasanna.
J. Astrophys. Astron., Vol. 3, 193 - 206 (1982).

The authors consider the structure of a thick perfect fluid disk of constant density rotating around a Schwarzschild black hole and its stability under axisymmetric pertubation.

066.178 **On the determination of the gravitational constant and the search for gravitational radiation in a cosmic laboratory.**
M. Steenbeck, H.-J. Treder.
Astron. Nachr., Band 303, 277 - 282 (1982). In German.

The authors describe systems of satellites giving possibilities for very exact measurements of the gravitational constant and for searches for gravitational radiation. Their systems are working by gravitational and inertial forces, only.

066.179 **A general upper limit on the mass and entropy production of a cluster of supermassive objects.**
C. Sivaram.
Astrophys. Space Sci., Vol. 86, 501 - 504 (1982).

The recent discovery of a distortion in the 3 K spectrum of the black-body background cosmic radiation has led to suggestions that a part or even all of the radiation was generated by pregalactic supermassive stars. A general upper limit on the mass of a cluster of these objects and the entropy of the radiation produced by them is obtained.

066.180 **Properties of relativistic linear stellar models.**
V. Ureche.
Binary and multiple stars as tracers of stellar evolution, (see 012.046), p. 73 - 78 (1982).

In the study of the newtonian stars, Stein (1966) has used a linear stellar model, showing that this model is useful for the determination of some representative values of the stellar characteristics. The author of the paper abstracted here has studied the properties of the linear stellar model in the frame of GRT (Ureche, 1980). Using the numerical methods, the functions describing the geometry of the space-time continuum are determined. These functions are given in tables and graphs (Ureche, 1981). For the relativistic linear stellar model an absolute minimum radius was determined.

066.181 **Population III and the microwave background.**
M. Rowan-Robinson, P. Tarbet.
Progress in cosmology, (see 012.047), p. 101 - 117 (1982).

A natural consequence of the isothermal density fluctuation picture is the possibility of a pregalactic generation of stars, Population III, where "stars" includes objects of up to 10^5 or 10^6 solar masses. A role proposed for them is the explanation of the whole microwave background. A further consequence would be the distortion of the spectrum of the microwave background at wavelengths shortward of 2 mm, in a manner similar to that observed by the Berkeley group. The third application of Population III is the production of the first heavy elements in galaxies. These three possibilities have been explored in a quantitative way. Population III light output must be dominated by accreting black holes.

066.182 **Observational aspects of the microwave cosmic background spectrum.**
D. H. Martin.
Progress in cosmology, (see 012.047), p. 119 - 143 (1982).

The author points out in this article the limitations in our present knowledge of the microwave background, identifies the observational difficulties in improving that knowledge and reports on some of the plans for future measurements. The recent critical review of background measurements by R. J. Weiss (1980) and the papers presented at a 1979 Copenhagen Symposium (Kalchar et. al. 1979) should be consulted for further detail.

066.183 **The cosmic microwave background radiation as a probe of the large-scale structure of the Universe.**
J. Silk.
Progress in cosmology, (see 012.047), p. 145 - 159 (1982).

Over scales in excess of a few percent of our present horizon, the search for anisotropy of the cosmic microwave background radiation promises to provide unique information on the existence of inhomogeneities in the matter distribution. Anisotropy at some level seems inevitable in any plausible scenario for the origin and evolution of structure in the early Universe. The author's goal here is to review the implications of galaxy formation theory for the presence of large-scale fluctuations. The dipole and recently discovered quadrupole anisotropy of the microwave background radiation severely constrain such theories. Interesting modifications of the predicted radiation anisotropy arise from the effects of spatial curvature and the possible presence of massive neutrinos.

066.184 **Relativistic electrons in galaxies and their nonthermal radiation from radio to gamma radiation.**
R. Schlickeiser.
Progress in cosmology, (see 012.047), p. 215 - 232 (1982).

Relativistic electrons are responsible for most of the nonthermal radiation in galaxies. Their dynamics is determined by interactions with cosmic magnetic, radiation and matter fields. The author discusses recent radio and gamma ray observations of our Galaxy and derives observational constraints on the propagation parameters for the electrons. He concentrates on the nature of relativistic electron sources and the nature of gamma ray point sources. A discussion of nonthermal radiation from selected extragalactic objects which includes its contribution to the extragalactic diffuse gamma ray background completes the review.

066.185 **The passage of a star by a massive black hole.**
R. A. Nolthenius, J. I. Katz.
Astrophys. J., Vol. 263, 377 - 385 (1982).

The authors have calculated the effects on a 1 $M_\odot$ star of passage by a $10^4 M_\odot$ point mass (black hole) in an initially parabolic orbit with a variety of pericenter distances. They use the gridless smoothed particle hydrodynamic method of Lucy, Gingold, and Monaghan. The tidal forces are found to induce rotation and radial and nonradial pulsations of the star. The loss of orbital energy to these internal motions leads to capture of the star by the black hole. For small pericenter distances, the outer layers of the star are disrupted, while at still smaller distances, the entire star is destroyed. These results may be applied to some X-ray sources, active galactic nuclei, and quasars.

066.186 **Stationary spherical accretion into black holes. The transition from the optically thin to the optically thick regime.**
M. H. Soffel.
Astron. Astrophys., Vol. 116, 111 - 116 (1982).
Models are constructed for the spherical accretion of fully ionized, unmagnetized, hydrogen gas into a Schwarzschild black hole with mass in the range $M_\odot < M < 10^3 M_\odot$. The accretion rate, the luminosity radiated by the infalling gas, and the characteristic temperature of the radiation are evaluated.

066.187 **On equations of motion for cross term modified gravitational field equations.**
V. Müller.
Astron. Nachr., Band 303, 321 - 327 (1982).
As proposed by Treder, possible consequences of a unitary field theory may be described phenomenologically by additional cross terms in Einstein's equations. The violation of the weak principle of equivalence and potential observable effects are discussed in deriving hydrodynamic Einstein-Infeld-Hoffmann equations. Conclusions on gravitational instabilities follow in the quasistatic approximation.

066.188 **Relativistic stars in close binary systems and masses of neutron stars.**
V. Ureche.
Ejection and accretion of matter in binary systems, (see 012.051), p. 71 - 92 (1982).
The equations which describe the hydrostatic equilibrium of a relativistic stellar configuration with spherical symmetric gravitational field are given in a non-dimensional form. The relativistic homogeneous, linear and polytropic stellar models are studied, obtaining the corresponding criteria of stability, and values for upper mass limit. The functions which measure the local effect of the gravitational field on the geometry of space-time continuum for a homogeneous relativistic star are given. From the classic virial theorem and the semiclassic limit for the internal specific energy it results an upper mass limit in agreement with the relativistic corresponding limit. The main results of the different known models for neutron stars are compared and the reliable observational data for the determination of the masses of these stars are analyzed.

066.189 **Measuring the large-scale anisotropy in the microwave background radiation.**
D. T. Wilkinson.
Philos. Trans. R. Soc. London, Ser. A, Vol. 307, (see 012.052), 55 - 65 (1982).
Measurements of large-scale anisotropy in the 2.7 K microwave background radiation are reaching a sensitivity of $\Delta T/T = 10^{-4}$ in the amplitudes of low-order spherical harmonics. The microwave instruments and techniques are discussed with the emphasis on the reduction and evaluation of spurious effects. Current results for the dipole and quadrupole moments are compared and discussed.

066.190 **Anisotropy measurements and cosmological implications.**
R. Fabbri, I. Guidi, V. Natale.
Philos. Trans. R. Soc. London, Ser. A, Vol. 307, (see 012.052), 67 - 75 (1982).
The authors summarize the present knowledge of the anisotropies of the cosmic background radiation at angular scales over 1° and present recent data on the dipole and quadrupole harmonics from the Florence group. They give a critical discussion of the connection of the background anisotropies with the large-scale properties of the Universe.

066.191 **Spectrum of the microwave background radiation.**
P. L. Richards.
Philos. Trans. R. Soc. London, Ser. A, Vol. 307, (see 012.052), 77 - 85 (1982).
A review is given of the present status of measurements of the spectrum of the microwave background. Factors that limit experimental accuracy are discussed with particular reference to high-frequency measurements. A selection of the available measurements yields a data set that is reasonably consistent with the black-body spectrum for a temperature of 2.9 K. A simple statistical analysis suggests either that there are errors in the data set, or that deviations from a black-body spectrum exist. Prospects for improved measurements are summarized.

066.192 **LMC X-3: a possible black hole.**
IAU Circ., No. 3751 (1982).

066.193 **Gravitational waves and activity of galactic nuclei and quasars.**
V. M. Lipunov, M. V. Sazhin.
Pis'ma Astron. Zh., Tom 8, 649 - 652 (1982). In Russian. English translation in Soviet Astron. Lett., Vol. 8.
A new astrophysical process is considered, namely, a disruption of stars by gravitational waves. A source of gravitational waves is proposed to be a non-symmetrical collapse of a supermassive object or a merging of two supermassive black holes. The expansion of a strong gravitational wave through a stellar cluster around the black hole results in a "QSO supernova".

066.194 **Gravitational lenses.**
P. V. Bliokh, A. A. Minakov.
Priroda, 1982, No. 11, p. 59 - 69. In Russian.

066.195 **Notas sobre relatividad. No. 5 – La interpretación geométrica de Minkowski (1).**
J. T. D'Alessio.
R Muscae, Vol. 6, 53 - 61 (1981).

066.196 **La constante gravitacional varia?**
J. R. García.
R Muscae, Vol. 6, 62 - 75 (1981).

066.197 **Thick accretion disks around black holes.** (Karl-Schwarzschild-Vorlesung 1981).
B. Paczynski.
Mitt. Astron. Ges., Nr. 57, p. 27 - 44 (1982).
Theory of thick low viscosity disks accreting onto black holes is reviewed. Usefulness of a simplified pseudo-Newtonian potential for studies of disk models is emphasized. It is shown how very high accretion rate leads to a large disk thickness, large pressure gradients in the radial direction, formation of a narrow open funnel around the rotation axis, and a decrease of efficiency of convection of rest mass into radiation. It is possible that the well collimated powerful radiation beam emerging from the funnel may accelerate some gas to moderately relativistic velocity in a form of a twin jet. The process is not efficient if the funnel is optically thin, but it is hoped that large optical depth of gas in the funnel may increase the fraction of total power coming out as kinetic energy of the jet. This class of models may be applied to SS 433, and possibly to other compact X-ray sources like Sco X-1 and Cyg X-1.

066.198 **On multiple-center gravitational lenses.**
I. L. Andronov.
Astron. Tsirk., No. 1189, p. 5 - 7 (1981). In Russian.

066.199 **Active galactic nuclei as clusters of accreting black holes.**
A. G. Pacholczyk, W. R. Stoeger.
Prepr. Steward Obs., No. 423, 5 pp. (1982).

066.200 **The generalized Einstein-Maxwell theory of gravitation.**
V. N. Ponomariov, Ju. (*Yu.*) Obuchov.
Gen. Relativ. Gravitation, Vol. 14, 309 - 330 (1982).

066.201 **Black holes and everyday physics.**
J. D. Bekenstein.
Gen. Relativ. Gravitation, Vol. 14, 355 - 359 (1982).

066.202 **Gravitational radiation and the motion of bodies in general relativity.**
F. I. Cooperstock, D. W. Hobill.
Gen. Relativ. Gravitation, Vol. 14, 361 - 378 (1982).
The authors present a selective review of the problem of the general relativistic prediction for the gravitational radiation emitted by gravitationally interacting systems with particular emphasis upon the two-body problem.

066.203 **The cosmological constant and the color charge at energies of the order of the Planck mass.**
J. M. Cerveró.
Gen. Relativ. Gravitation, Vol. 14, 393 - 397 (1982).

066.204 **Gas dynamics in strong gravitational fields.**
S. Hacyan.
Gen. Relativ. Gravitation, Vol. 14, 399 - 410 (1982).
The steady and axially symmetric flow of a perfect fluid is studied in the context of general relativistic gas dynamics. It is assumed that the flow occurs in the background field of a rotating black hole (or any compact object). The hydrodynamic equations are referred to a locally nonrotating frame and their characteristics are found. The equations describing oblique shock waves are also obtained.

066.205 **Gravitational radiation from a binary system.**
P. H. Lim.
Gen. Relativ. Gravitation, Vol. 14, 471 - 477 (1982).
It is shown that the gravitational radiation energy flux from a quasiperiodic binary system is given by the quadrupole formula contribution plus a nonlinear contribution which has been neglected in the past. The latter is found to depend on both the intensity of the gravitational field and the history of the radiating system. For PSR 1913 + 16, the nonlinear contribution to the energy flux may be neglected relative to that of the quadrupole formula.

066.206 **Absorption of gravitational waves by nearly Newtonian systems.**
B. Linet.
Gen. Relativ. Gravitation, Vol. 14, 479 - 492 (1982).

066.207 **Remark on a toroidal detector of a gravitational wave.**
K. Omori, A. Kuriyagawa, S. Mori.
Gen. Relativ. Gravitation, Vol. 14, 493 - 501 (1982).
The object of this paper is to review the detector of a gravitational wave that was proposed by Braginsky and Mensky (1971). The derivation of the sensitivity is based on the same assumption as they proposed. It is concluded that the phase difference is linear in time and that the sensitivity of this detector is different from the result claimed by Braginsky and Mensky. The foundation to obtain the phase difference, i.e., the sensitivity, in this paper is not the frequency as they used but rather the movement of the wave front in the detector.

066.208 **Gravitational radiation antennas using the Sagnac effect.**
J. Anandan, R. Y. Chiao.
Gen. Relativ. Gravitation, Vol. 14, 515 - 521 (1982).
A new class of gravitational antennas that utilize the general relativistic Sagnac effect is proposed. These antennas may be more efficient than the Weber bar. A specific case of such an antenna consisting of a superfluid helium Josephson interferometer is considered. A general relativistic theory of the interaction of the superfluid with the gravitational field is given. Using this theory, the phase shift due to a gravitational plane wave on one such antenna is obtained.

066.209 **Quantum stationary states in the Bianchi universes.**
T. Padmanabhan.
Gen. Relativ. Gravitation, Vol. 14, 549 - 557 (1982).
A path integral formulation is used in the superspace of geometries leading to a model of quantum gravity. It is shown that this method agrees with the standard path integral technique in the special case of Friedmann universes. A Schrödinger-like equation is set up, leading to the stationary states. Some solutions to these equations are presented.

066.210 **A proposed laboratory experiment to measure $\dot{G}/G$.**
T. Dannehold.
Gen. Relativ. Gravitation, Vol. 14, 565 - 568 (1982).
An experiment is proposed in which the length of a laser cavity depends on the force of gravity. A change in the gravitational constant will cause a change in the frequency of the laser, which may be measured relative to a reference laser. If an iodine stabilized He-Ne laser is used as the reference, $\dot{G}/G$ can be measured to an accuracy of $4 \times 10^{-12} yr^{-1}$.

066.211 **A note on a local effect at the Schwarzschild sphere.**
A. Karlhede, U. Lindström, J. E. Åman.
Gen. Relativ. Gravitation, Vol. 14, 569 - 571 (1982).
Passage of the Schwarzschild radius is shown to be locally measurable by a sign change in a certain scalar. In the Kerr solution this scalar changes sign at the stationary limit. This is an example of the use of a coordinate-invariant method, based on the curvature tensor and a finite number of its covariant derivatives, for investigating gravitational fields.

066.212 **A note on the derivation of the Kerr metric.**
A. H. Klotz.
Gen. Relativ. Gravitation, Vol. 14, 727 - 730 (1982).
An elementary derivation of the Kerr solution of Einstein's empty field equations is described.

066.213 **Coupled gravitational and electromagnetic waves in arbitrary space-times with magnetized plasmas and dust, I.**
R. A. Breuer.
Gen. Relativ. Gravitation, Vol. 14, 757 - 769 (1982).

066.214 **Definitions of black holes without use of the boundary at infinity.**
A. Królak.
Gen. Relativ. Gravitation, Vol. 14, 793 - 801 (1982).
Two definitions of a black hole are given without reference to the boundary at infinity of space-time. One definition does not require any global causality condition and the other is applicable to a closed Friedmann-like space-time. The area theorem is found to hold in both cases.

066.215 **The closed trapped region and the apparent horizon of two Schwarzschild black holes.**
N. T. Bishop.
Gen. Relativ. Gravitation, Vol. 14, 817 - 823 (1982).
The problem of two Schwarzschild black holes on a time-symmetric spacelike hypersurface is investigated. The location of the closed trapped region and the apparent horizon is calculated. It is shown that the topology of the closed trapped region is not as previously believed.

066.216 **Parametrized Post-Post-Newtonian (PP^2N) formalism for the solar system.**
A. F. Sarmiento G.
Gen. Relativ. Gravitation, Vol. 14, 893 - 905 (1982).

An extension of the parametrized post-Newtonian (PPN) formalism to third order in the expansion parameter m/r is used to derive analytical expressions accurate to the same order for the motion of test particles and photons in the presence of the gravitational field of the sun represented by a static, isotropic metric. The consequences of including higher-order terms are discussed in relation to the so-called classical gravitational tests for the case of general relativity theory. Present observational or experimental data are not accurate enough to detect variations due to the inclusion of higher-order terms but a planned solar probe experiment may provide information that would make such detection possible.

066.217 **Globally regular solutions of Einstein's equations.**
W. B. Bonnor.
Gen. Relativ. Gravitation, Vol. 14, 907 - 921 (1982).
This is a review, covering known globally regular solutions describing either vacuum or fields with physically reasonable sources. The largest class is that with static spherical symmetry, but many others are known, even with $\Lambda = 0$. If $\Lambda \neq 0$ there is a variety of regular cosmological solutions.

066.218 **A description of the gravitational red shift borrowed from the electrodynamics of continuous media.**
F. Pegoraro.
Gen. Relativ. Gravitation, Vol. 14, 931 - 934 (1982).

066.219 **Static conformally flat solutions in a general scalar-tensor theory of gravitation.**
V. U. M. Rao, D. R. K. Reddy.
Gen. Relativ. Gravitation, Vol. 14, 1017 - 1022 (1982).
Explicit field equations in the general scalar-tensor theory of gravitation proposed by Nordtvedt are obtained with the aid of a static spherically symmetric conformally flat metric. Exact static solutions of Nordtvedt–Barker field equations both in vacuum and in the presence of a source-free electromagnetic field are presented and studied.

066.220 **On the physical significance of coordinates and the solutions of the field equations in Einstein's theory of gravitation.**
P. Zhou.
Sci. Sinica, Ser. A, Vol. 25, 628 - 643 (1982).

066.221 **Transonic accretion flow in a thin disk around a black hole.**
B. Muchotrzeb, B. Paczyński.
Acta Astron., Vol. 32, 1 - 11 (1982).
The authors present "alpha disk models" accreting onto a $10\, M_\odot$ black hole at a critical, and one tenth of a critical rate, with a = 0.001. The authors find a transition from subsonic to supersonic radial flow near the inner disk radius, r_{in}. They also find that the "no torque at r_{in}" boundary condition is very good.

066.222 **Transonic disk accretion of barytropic gas onto black holes.**
Z. Loska.
Acta Astron., Vol. 32, 13 - 24 (1982).
A model of transonic flow from an accretion disk rotating around a black hole is constructed. For given equation of state, mass of the black hole, angular momentum of the inner parts of the disk, and the accretion rate, the equations of transonic flow designate explicitly the position of the sonic point, velocity of sound and a thickness of the stream at this point. Several properties of the stream are described.

066.223 **High energy γ-rays from black holes.**
F. Giovannelli, S. Karakula, W. Tkaczyk.
Acta Astron., Vol. 32, 121 - 130 (1982).
The authors determine the γ-ray production spectrum in the comoving plasma system for different temperatures. The observable γ-ray spectrum in the case of spherically symmetric accretion of matter onto a black hole is evaluated. Relativistic effects have been taken into account. The upper limit to the number of active black holes in the Galaxy is evaluated.

066.224 **Gravitational radiation damping.**
J. D. McCrea.
Irish Astron. J., Vol. 15, (see 012.061), 129 (1981).
Abstract.

066.225 **On the physics of relativistic double layers.**
P. Carlqvist.
Astrophys. Space Sci., Vol. 87, 21 - 39 (1982).
A model of a strong, time-independent, and relativistic double layer is studied. Besides double layers having the electric field parallel to the current the model also describes a certain type of oblique double layers. The 'Langmuir condition' (ratio of ion current density to electron current density) as well as an expression for the potential drop of the double layer are derived. Furthermore, the distributions of charged particles, electric field, and potential within the double layer are clarified and discussed. It is found that the properties of relativistic double layers differ substantially from the properties of corresponding non-relativistic double layers.

066.226 **Post-Newtonian approximations in scale covariant gravitation.**
P. Bouvier.
Astrophys. Space Sci., Vol. 87, 105 - 116 (1982).
The scale covariant theory of gravitation, outlined by Dirac (1973), later developed by Canuto et al. (1977), revisited by Maeder and Bouvier (1978, 1979), takes into account the possible relative changes in the system of units associated with different physical interactions; in this respect, it represents a generalization of the General Relativity Theory. In the line of the above papers, the author studies the case of a weak gravitational field, well suited to the inner motions of a star or galaxy cluster, in order to see whether the post-Newtonian approximation scheme can consistently fit the scale covariant formalism. Such a task turns out to be feasible when the gauge terms met in the field equations are handled in an appropriate way, but only if the gauge or metrical connection vector is inversely proportional to cosmic time, as it should be in consequence of the outer boundary condition imposed on the solution of the field equations describing the Newtonian and the first post-Newtonian approximation.

066.227 **Cosmological molecular hydrogen and the distortion of the relict radiation spectrum.**
V. K. Khersonskii (*Khersonskij*).
Astrophys. Space Sci., Vol. 88, 21 - 30 (1982).
The distortions of the relict radiation spectrum in the region of the wavelength $\lambda < 120\ \mu$ are considered. These distortions are due to the emission of photons under the formation of molecular hydrogen in the expanding universe in the cosmological epoch $40 \leqslant z \leqslant 200$. It is shown that the real intensity of the relict radiation in the region of the wavelength under consideration must significantly exceed Planck's intensity, with a radiation temperature amounting to 2.8 K.

066.228 **The influence of a net charge on the critical mass of a neutron star.**
J. L. Zhang, W. Y. Chau, T. Y. (*T. R.*) Deng.
Astrophys. Space Sci., Vol. 88, 81 - 88 (1982).
The authors derived a set of equations determining the structure of a spherically symmetric charged star within the framework of general relativity (modified Oppenheimer-Volkoff equations). The equations have been solved for a completely degenerate Fermi gas with a charge density assumed to be proportional to the matter density. It is shown that the presence of a net charge does not affect the existence

of a critical mass. The value, however, could be substantially altered, in some cases doubled.

066.229 **Observable blueshifts near compact objects.**
J. M. Cohen, M. F. Struble, K. R. Pechenick, B. Kuharetz.
Astrophys. Space Sci., Vol. 88, 307 - 312 (1982).
The authors show that observable blueshifted radiation can emanate from material freely falling toward compact objects. Using a fully relativistic treatment and considering possible blocking of photon trajectories by a neutron star or black hole, they demonstrate that blueshifts are observable.

066.230 **Massive configurations with constant densities.**
M. C. Durgapal, A. K. Pande, K. Pandey.
Astrophys. Space Sci., Vol. 88, 469 - 475 (1982).
Two new classes of solutions with constant observed proper and rest mass densities are described. Unlike the well-known solution of constant coordinate mass density, these solutions pertain to realistic physical situations. For these solutions, the various relevant parameters, viz. the redshifts $(dP/d\rho)_0$ and binding coefficients have been calculated.

066.231 **Static spherically-symmetric scalar-field theory in general relativity.**
D. D. Dionysiou.
Astrophys. Space Sci., Vol. 88, 493 - 499 (1982).
A spherically-symmetric static scalar field in general relativity is considered. The field equations are defined by $R_{ik} = -\mu\varphi_i\varphi_k$, $\varphi_i = \partial\varphi/\partial x^i$, $\varphi^i = g^{ik}\varphi_k$, where $\varphi = \varphi(r, t)$ is a scalar field. The same problem was considered by Bergmann and Leipnik (1957) and Buchdahl (1959) with the assumption that $\varphi = \varphi(r)$ is independent of t and recently by Wyman (1981) with the assumption $\varphi = \varphi(r, t)$. The object of this paper is to give explicit results with a different approach and under a more general assumption.

066.232 **Resonant absorption of cosmic-ray neutrinos by the relic-neutrino background.**
T. Weiler.
Phys. Rev. Lett., Vol. 49, 234 - 237 (1982).
Within the framework of standard big-bang cosmology and the standard electroweak model, it is shown that $\nu\bar{\nu}$ annihilation on the Z resonance is the only cosmic-ray process sensitive to relic neutrinos.

066.233 **Schiff's conjecture on gravitation.**
A. Coley.
Phys. Rev. Lett., Vol. 49, 853 - 855 (1982).
Considered here is a class of theories of gravity characterized by a set of equations which represent the gravitational and electromagnetic structure of the theories in a spherically symmetric and static gravitational field. If one demands that the weak equivalence principle and the principle of universality of gravitational red shift be satisfied, it is found that the theories under investigation must be metric.

066.234 **Constraints on cosmic magnetic monopoles imposed by the galactic magnetic field.**
E. E. Salpeter, S. L. Shapiro, I. Wasserman.
Phys. Rev. Lett., Vol. 49, 1114 - 1117 (1982).
Constraints on a hypothetical galactic halo of magnetic monopoles imposed by the galactic magnetic field are discussed.

066.235 **Search for small-scale anisotropy in the cosmic microwave background.**
J. M. Uson, D. T. Wilkinson.
Phys. Rev. Lett., Vol. 49, 1463 - 1465 (1982).

066.236 **Entropy evaporated by a black hole.**
W. H. Zurek.
Phys. Rev. Lett., Vol. 49, 1683 - 1686 (1982).
It is shown that the entropy of the radiation evaporated by an uncharged, nonrotating black hole into vacuum in the course of its lifetime is approximately $^4/_3$ times the initial entropy of this black hole. Also considered is a thermodynamically reversible process in which an increase of black-hole entropy is equal to the decrease of the entropy of its surroundings. Implications of these results for the generalized second law of thermodynamics and for the interpretation of black-hole entropy are pointed out.

066.237 **Null test of the gravitational inverse square law.**
H. A. Chan, M. V. Moody, H. J. Paik.
Phys. Rev. Lett., Vol. 49, 1745 - 1748 (1982).
An experiment has been performed to demonstrate a new source-independent null test of the inverse square law of gravitation. A single-axis superconducting gravity gradiometer was rotated into three orthogonal orientations to measure the Laplacian of the gravitational potential produced by a 1600-kg lead pendulum at an average distance of 2.3 m. The result is that if one assumes a potential of the form $\varphi(r) = -(GM/r)$ $[1 + \alpha\exp(-\mu r)]$, the value of α is $+ 0.024 \pm 0.036$ at $\mu^{-1} = 1$ m.

066.238 **Preliminary determination of the sun's gravitational quadrupole moment from rotational splitting of global oscillations and its relevance to tests of general relativity.**
H. A. Hill, R. J. Bos, P. R. Goode.
Phys. Rev. Lett., Vol. 49, 1794 - 1797 (1982).
The sun's internal angular velocity is estimated from observations of rotational splitting of low-order, low-degree global oscillations detected as fluctuations in the limb-darkening function. Values of 0.987 ± 0.006 and 0.991 ± 0.006 are obtained for $^1/_3$ $(2 + 2\gamma-\beta)$, a quantity equal to 1 in the general theory of relativity.

066.239 **Gravitational lenses.**
P. Hadrava.
Kozmos, Vol. 13, 183 - 186 (1982). In Czech.

066.240 **Semiclassical expanding discrete space-times.**
W. K. Cobb, L. L. Smalley.
Int. J. Theor. Phys., Vol. 21, (see 012.075), 757 - 763 (1982).
Abstr. in Phys. Abstr., Vol. 85, Abstr. 112754 (1982).

066.241 **Topological mass generation of scalar particles in the static flat Clifford-Klein-Robertson-Walker spacetimes.**
Yu. P. Goncharov.
Phys. Lett. A, Vol. 91A, 153 - 156 (1982). – Abstr. in Phys. Abstr., Vol. 85, Abstr. 112760 (1982).

066.242 **Anisotropic fluid spheres in general relativity.**
S. S. Bayin.
Phys. Rev. D, Vol. 26, 1262 - 1274 (1982). – Abstr. in Phys. Abstr., Vol. 85, Abstr. 112763 (1982).

066.243 **Late stages in the collapse of radiating fluid spheres.**
J.-L. Zhang, K. Lake.
Phys. Rev. D, Vol. 26, 1479 - 1480 (1982). – Abstr. in Phys. Abstr., Vol. 85, Abstr. 112770 (1982).

066.244 **Time-dependent embeddings for Schwarzschild-like solutions to the gravitational field equations.**
C. F. Chyba.
J. Math. Phys., Vol. 23, 1662 - 1667 (1982). – Abstr. in Phys. Abstr., Vol. 85, Abstr. 112779 (1982).

066.245 **The equations of motion in a conformally invariant theory of gravitation and electromagnetism.**
D. Gregorash, G. Papini.

Nuovo Cimento B, Vol. 70B, Ser. 11, 259 - 267 (1982). Abstr. in Phys. Abstr., Vol. 85, Abstr. 112780 (1982).

066.246 **Results of a 440-day search for gravitational radiation.**
B. L. Brown, A. P. Mills, Jr., J. A. Tyson.
Phys. Rev. D, Vol. 26, 1209 - 1218 (1982). – Abstr. in Phys. Abstr., Vol. 85, Abstr. 112797 (1982).

066.247 **On black holes and uniform magnetic fields.**
U. S. Pandey, S. S. Prasad.
Prog. Theor. Phys., Vol. 68, 314 - 316 (1982). – Abstr. in Phys. Abstr., Vol. 85, Abstr. 117701 (1982).

066.248 **Hawking evaporation and space-time structure.**
R. Balbinot, R. Bergamini, B. Giorgini.
Nuovo Cimento B, Vol. 70B, Ser. 11, 201 - 211 (1982). Abstr. in Phys. Abstr., Vol. 85, Abstr. 117705 (1982).

066.249 **Vacuum polarization by a massive scalar field in Schwarzschild spacetime.**
V. P. Perolov, A. I. Zel'nikov.
Phys. Lett. B, Vol. 115B, 372 - 374 (1982). – Abstr. in Phys. Abstr., Vol. 85, Abstr. 117706 (1982).

066.250 **Tensor spherical and pseudo-spherical harmonics in four-dimensional spaces.**
K. Tomita.
Prog. Theor. Phys., Vol. 68, 310 - 313 (1982). – Abstr. in Phys. Abstr., Vol. 85, Abstr. 117808 (1982).

066.251 **Physics of black holes.**
K. S. Thorne.
The galactic center, (see 012.076), p. 165 (1982). – Abstract. See Abstr. 31.066.103.

066.252 **Positron production near a $10^6 M_\odot$ black hole.**
R. D. Blandford.
The galactic center, (see 012.076), p. 177 - 179 (1982).
The annihilation line coming from the direction of the galactic center may be produced by positrons originating from the magnetosphere of a $\sim 10^6 M_\odot$ black hole.

066.253 **A subquantum origin of gravitational waves.**
A. Ceapa.
Phys. Lett. A, Vol. 92A, 17 - 22 (1982). – Abstr. in Phys. Abstr., Vol. 86, Abstr. 139 (1983).

066.254 **Contribution to the study of gravitational interaction in invariantive mechanics.**
M. Agop.
Rev. Roumaine Phys., Vol. 27, 219 - 234 (1982). – Abstr. in Phys. Abstr., Vol. 86, Abstr. 142 (1983).

066.255 **The screen effect of the Earth in the TETG. Theory of a screening experiment of a sample body at the equator, using the Earth as a screen.**
I. A. Adamuti.
Nuovo Cimento C, Vol. 5C, Ser. 1, 189 - 208 (1982). – Abstr. in Phys. Abstr., Vol. 86, Abstr. 147 (1983).

066.256 **Solutions of the gravitational Poisson equation in spherical coordinates.**
M. Claudius.
Comput. Phys. Commun., Vol. 27, 119 - 128 (1982). – Abstr. in Phys. Abstr., Vol. 86, Abstr. 5105 (1983).

066.257 **Gravitational aspects of a conformally coupled field theory in five space-time dimensions.**
S. D. Unwin.
Proc. R. Soc. London, Ser. A, Vol. 383, 457 - 464 (1982). Abstr. in Phys. Abstr., Vol. 86, Abstr. 5108 (1983).

066.258 **Proof of uniqueness of the Kerr-Newman black hole solution.**
P. O. Mazur.
J. Phys. A, Vol. 15, 3173 - 3180 (1982). – Abstr. in Phys. Abstr., Vol. 86, Abstr. 5326 (1983).

066.259 **Scattering of scalar tardyons and tachyons from a Schwarzschild black hole.**
S. Banerji, D. R. Mandal.
J. Phys. A, Vol. 15, 3181 - 3185 (1982). – Abstr. in Phys. Abstr., Vol. 86, Abstr. 5327 (1983).

066.260 **Particle creation by a black hole as a consequence of the Casimir effect.**
R. M. Nugayev *(Nugaev)*.
Phys. Lett. A, Vol. 91A, 216 - 218 (1982). – Abstr. in Phys. Abstr., Vol. 86, Abstr. 5328 (1983).

066.261 **Motion of the charged particles off the equatorial plane of a Kerr black hole in an electromagnetic field.**
D. K. Chakraborty, A. R. Prasanna.
Pramāna, Vol. 19, 141 - 150 (1982). – Abstr. in Phys. Abstr., Vol. 86, Abstr. 5329 (1983).

066.262 **Dispersion relaxations for zero-helicity gravitational waves in a relativistic expanding medium.**
J. Plinio Baptista, D. Gerbal.
J. Phys. A, Vol. 15, 3351 - 3365 (1982). – Abstr. in Phys. Abstr., Vol. 86, Abstr. 5459 (1983).

066.263 **Interaction of a gravitational wave with a Newtonian system in the Brans-Dicke theory.**
B. Linet.
Phys. Lett. A, Vol. 92A, 63 - 64 (1982). – Abstr. in Phys. Abstr., Vol. 86, Abstr. 5671 (1983).

066.264 **Gravitational field equations with cross-terms and static equilibrium configurations.**
V. Müller.
Ann. Physik, Band 39, 271 - 280 (1982). – Abstr. in Phys. Abstr., Vol. 86, Abstr. 10648 (1983).

066.265 **A new vacuum state in de Sitter space.**
J. D. Pfautsch.
Phys. Lett. B, Vol. 117B, 283 - 286 (1982). – Abstr. in Phys. Abstr., Vol. 86, Abstr. 10693 (1983).

066.266 **Fermi normal co-ordinate system and electromagnetic detectors of gravitational waves. I. Calculation of the metric.**
P. L. Fortini, C. Gualdi.
Nuovo Cimento B, Vol. 71B, Ser. 11, 37 - 54 (1982). – Abstr. in Phys. Abstr., Vol. 86, Abstr. 5698 (1983).

066.267 **Stability properties of gravity theories.**
S. Deser.
Unified theories of elementary particles, (see 012.084), p. 152 - 166 (1982). – Abstr. in Phys. Abstr., Vol. 86, Abstr. 10710 (1983).

066.268 **Neutrino constraints that transform black holes into grey holes.**
M. Ruderfer.
Speculations Sci. Technol., Vol. 5, 433 - 442 (1982). – Abstr. in Phys. Abstr., Vol. 86, Abstr. 15938 (1983).

066.269 **Magnetic properties of relativistic Fermi gas.**
E. M. Chudnovsky *(Chudnovskij)*.
Ann. New York Acad. Sci., Vol. 373, 208 - 210 (1981). Abstr. in Phys. Abstr., Vol. 86, Abstr. 5705 (1983).

066.270 **Radiation from a black hole: a Vaidya-metric-based computation.**
R. Balbinot, R. Bergamini, B. Giorgini.
Nuovo Cimento B, Vol. 71B, Ser. 11, 27 - 36 (1982). – Abstr. in Phys. Abstr., Vol. 86, Abstr. 10381 (1983).

066.271 **Experimental testing of Mach's principle.**
H. J. Treder.
Ann. Physik, Band 39, 310 - 312 (1982). In German. – Abstr. in Phys. Abstr., Vol. 86, Abstr. 10487 (1983).

066.272 **The entropies of collapsing stars and of black holes.**
Y.-x. Gui, L. Liu.
Acta Astrophys. Sinica, Vol. 2, 97 - 106 (1982). In Chinese.
The authors put forth three modes of black hole formation: (1) A black hole kern forms initially inside the collapsing star. (2) The different mass shells of the collapsing star fulfil the Schwarzschild condition simultaneously. (3) Only the outmost mass shell of the collapsing star fulfils the Schwarzschild condition. The authors then calculate the entropy of the collapsing star for modes (1) and (3) and find that they are only 10^{-19} times the entropy of a black hole. Mode (1) may occure during supernova explosions or galaxy explosions. Mode (3) may occur in the formation of galactic black holes.

066.273 **The temperature profile of an accretion disk around a Kerr black hole.**
J. Zhang, S. Jiang.
Acta Astrophys. Sinica, Vol. 2, 277 - 284 (1982). In Chinese.
The authors used the more rigorous general relativistic viscose hydrodynamical equation to discuss and calculate the temperature profile in the accretion disk around a Kerr black hole.

066.274 **The dynamical effects of a torsion field with gravitation.**
Y. Liu.
Acta Astrophys. Sinica, Vol. 2, 315 - 320 (1982). In Chinese.
The author discusses the dynamical effects of a torsion field with gravitation. In particular he discusses the variation of the precession of a spinning body in a static spherically symmetric metric torsion field, and he gets three constants of motion. The author proves that in a space with absolute parallelism, all the components of the spin of a top remain constant. The constant is independent of the orbit. The author calculates time variation of the rotation of the earth which moves in an ellipse orbit in a static spherically symmetric metric-torsion field, and compares his results with experiments. He also discusses the gyroscope experiments which can test his theory.

066.275 **The variation of G and the quantum theory.**
P. A. M. Dirac.
Proceedings of the Second Marcel Grossmann Meeting on General Relativity, (see 012.081), p. 1 - 6 (1982). – Abstr. in Phys. Abstr., Vol. 86, Abstr. 5686 (1983).

066.276 **Spontaneous supersymmetry breaking in supergravity.**
S. Ferrara.
Proceedings of the Second Marcel Grossmann Meeting on General Relativity, (see 012.081), p. 69 - 77 (1982). – Abstr. in Phys. Abstr., Vol. 86, Abstr. 5693 (1983).

066.277 **Gravitational radiation and the binary pulsar.**
J. H. Taylor.
Proceedings of the Second Marcel Grossmann Meeting on General Relativity, (see 012.081), p. 15 - 19 (1982). – Abstr. in Phys. Abstr., Vol. 86, Abstr. 5700 (1983).

066.278 **Recent progress in gravitational wave detectors.**
E. Amaldi.
Proceedings of the Second Marcel Grossmann Meeting on General Relativity, (see 012.081), p. 21 - 65 (1982). – Abstr. in Phys. Abstr., Vol. 86, Abstr. 5701 (1983).

066.279 **Some recent developments in asymptotic structure.**
S. Persides.
Proceedings of the Second Marcel Grossmann Meeting on General Relativity, (see 012.081), p. 245 - 251 (1982). Abstr. in Phys. Abstr., Vol. 86, Abstr. 10661 (1983).

066.280 **Spacetimes with null symmetry as models for gravitating systems.**
J. R. Ray.
Proceedings of the Second Marcel Grossmann Meeting on General Relativity, (see 012.081), p. 253 - 255 (1982). Abstr. in Phys. Abstr., Vol. 86, Abstr. 10662 (1983).

066.281 **Why is the line element of spacetime quadratic in the coordinate differentials?**
D. Wilkins.
Proceedings of the Second Marcel Grossmann Meeting on General Relativity, (see 012.081), p. 271 (1982). – Abstr. in Phys. Abstr., Vol. 86, Abstr. 10664 (1983).

066.282 **Separability of Einstein space-times: the state of the art.**
M. Francaviglia.
Proceedings of the Second Marcel Grossmann Meeting on General Relativity, (see 012.081), p. 301 - 340 (1982). Abstr. in Phys. Abstr., Vol. 86, Abstr. 10667 (1983).

066.283 **Measurement of the one-way speed of light and the earth's absolute velocity.**
S. Marinov.
Proceedings of the Second Marcel Grossmann Meeting on General Relativity, (see 012.081), p. 547 - 550 (1982). Abstr. in Phys. Abstr., Vol. 86, Abstr. 10672 (1983).

066.284 **Surfaces of constant mean curvature in Schwarzschild, Reissner-Nordström, and lattice spacetimes.**
D. R. Brill, J. M. Cavallo, J. A. Isenberg.
Proceedings of the Second Marcel Grossmann Meeting on General Relativity, (see 012.081), p. 565 - 573 (1982). Abstr. in Phys. Abstr., Vol. 86, Abstr. 10674 (1983).

066.285 **Numerical integration of axially symmetric systems.**
T. Nakamura, K. Maeda, S. Miyama, M. Sasaki.
Proceedings of the Second Marcel Grossmann Meeting on General Relativity, (see 012.081), p. 675 - 677 (1982). Abstr. in Phys. Abstr., Vol. 86, Abstr. 10676 (1983).

066.286 **Coupled modes on generic spherically symmetric space-times.**
U. H. Gerlach.
Proceedings of the Second Marcel Grossmann Meeting on General Relativity, (see 012.081), p. 713 - 716 (1982). Abstr. in Phys. Abstr., Vol. 86, Abstr. 10678 (1983).

066.287 **Gravitational synchrotron radiation in the metric of a new theory of gravitation.**
R. B. Mann, J. W. Moffat.
Proceedings of the Second Marcel Grossmann Meeting on General Relativity, (see 012.081), p. 553 (1982). – Abstr. in Phys. Abstr., Vol. 86, Abstr. 10679 (1983).

066.288 **Generation of gravitational waves in the source theory.**
Y. Kim.
Proceedings of the Second Marcel Grossmann Meeting on General Relativity, (see 012.081), p. 705 - 710 (1982). Abstr. in Phys. Abstr., Vol. 86, Abstr. 10680 (1983).

066.289 **Generalised master equations for wave equation separation in a Kerr or Kerr-Newman black hole background.**
B. Carter, R. G. McLenaghan.
Proceedings of the Second Marcel Grossmann Meeting on General Relativity, (see 012.081), p. 575 - 585 (1982).
Abstr. in Phys. Abstr., Vol. 86, Abstr. 10681 (1983).

066.290 **Surface effects in black hole physics.**
T. Damour.
Proceedings of the Second Marcel Grossmann Meeting on General Relativity, (see 012.081), p. 587 - 608 (1982).
Abstr. in Phys. Abstr., Vol. 86, Abstr. 10682 (1983).

066.291 **A 'constant of the motion' for the equation of geodetic deviation.**
P. Dolan, P. Choudhuri, J. L. Safko.
Proceedings of the Second Marcel Grossmann Meeting on General Relativity, (see 012.081), p. 609 - 611 (1982).
Abstr. in Phys. Abstr., Vol. 86, Abstr. 10683 (1983).

066.292 **Techniques for linearized pertubations of Kerr-Newman black holes.**
E. D. Fackerell.
Proceedings of the Second Marcel Grossmann Meeting on General Relativity, (see 012.081), p. 613 - 634 (1982).
Abstr. in Phys. Abstr., Vol. 86, Abstr. 10684 (1983).

066.293 **Stationary axially symmetric Brans-Dicke-Maxwell fields.**
T. Singh, L. N. Rai.
Proceedings of the Second Marcel Grossmann Meeting on General Relativity, (see 012.081), p. 259 (1982). – Abstr. in Phys. Abstr., Vol. 86, Abstr. 10688 (1983).

066.294 **Physical interpretation of a new theory of gravitation.**
J. W. Moffat.
Proceedings of the Second Marcel Grossmann Meeting on General Relativity, (see 012.081), p. 551 (1982). – Abstr. in Phys. Abstr., Vol. 86, Abstr. 10690 (1983).

066.295 **Initial value formulation in a new theory of gravity.**
J. W. Moffat.
Proceedings of the Second Marcel Grossmann Meeting on General Relativity, (see 012.081), p. 555 (1982). – Abstr. in Phys. Abstr., Vol. 86, Abstr. 10691 (1983).

066.296 **Vacuum solutions with double duality properties of a quadratic Poincaré gauge field theory.**
P. Baekler, F. W. Hehl, E. W. Mielke.
Proceedings of the Second Marcel Grossmann Meeting on General Relativity, (see 012.081), p. 413 - 453 (1982).
Abstr. in Phys. Abstr., Vol. 86, Abstr. 10698 (1983).

066.297 **Quantum gravity removes classical singularities and shortens the life of black holes.**
V. P. Frolov, G. A. Vilkovisky.
Proceedings of the Second Marcel Grossmann Meeting on General Relativity, (see 012.081), p. 455 - 469 (1982).
Abstr. in Phys. Abstr., Vol. 86, Abstr. 10699 (1983).

066.298 **Spectra asymmetry and quantum field theory in curved spacetime.**
G. W. Gibbons.
Proceedings of the Second Marcel Grossmann Meeting on General Relativity, (see 012.081), p. 471 - 473 (1982).
Abstr. in Phys. Abstr., Vol. 86, Abstr. 10700 (1983).

066.299 **The solar system: a cluttered laboratory for gravity research.**
R. D. Reasenberg.
Proceedings of the Second Marcel Grossmann Meeting on General Relativity, (see 012.081), p. 1019 - 1038 (1982).
Abstr. in Phys. Abstr., Vol. 86, Abstr. 10714 (1983).

066.300 **Experimental searches for gravitational effects outside of general relativity.**
R. C. Ritter.
Proceedings of the Second Marcel Grossmann Meeting on General Relativity, (see 012.081), p. 1039 - 1070 (1982).
Abstr. in Phys. Abstr., Vol. 86, Abstr. 10715 (1983).

066.301 **Topics in gravitational radiation antennas.**
J. Weber.
Proceedings of the Second Marcel Grossmann Meeting on General Relativity, (see 012.081), p. 1073 - 1081 (1982).
Abstr. in Phys. Abstr., Vol. 86, Abstr. 10716 (1983).

066.302 **Noise investigations in a laser interferometer for the detection of gravitational radiation.**
K. Maischberger, A. Rüdiger, R. Schilling, L. Schnupp, W. Winkler, H. Billing.
Proceedings of the Second Marcel Grossmann Meeting on General Relativity, (see 012.081), p. 1083 - 1100 (1982).
Abstr. in Phys. Abstr., Vol. 86, Abstr. 10717 (1983).

066.303 **The gravitational radiation project at Perth, Western Australia.**
D. G. Blair, M. J. Buckingham, C. Edwards, J. Ferreirinho, R. N. James, A. G. Mann, F. J. Van Kann, P. Veitch.
Proceedings of the Second Marcel Grossmann Meeting on General Relativity, (see 012.081), p. 1101 - 1124 (1982).
Abstr. in Phys. Abstr., Vol. 86, Abstr. 10718 (1983).

066.304 **Nonlinear excitation of gravitational radiation antennae.**
D. G. Blair.
Proceedings of the Second Marcel Grossmann Meeting on General Relativity, (see 012.081), p. 1125 - 1132 (1982).
Abstr. in Phys. Abstr., Vol. 86, Abstr. 10719 (1983).

066.305 **Progress report of the gravitational wave experiment in Beijing-Guangzhou.**
H. Renan, J. Nanfeng, L. Yicheng, Q. Rongxian, T. Dajun, T. Jingfa, W. Guozong, Z. Pinghua, Z. Zhishu, Z. Lilin, C. Jiayan, G. Tonren, H. Quingxiang, L. Yanfeng, Q. Zhongxing, Y. Xiaojian, Y. Po.
Proceedings of the Second Marcel Grossmann Meeting on General Relativity, (see 012.081), p. 1133 - 1144 (1982).
Abstr. in Phys. Abstr., Vol. 86, Abstr. 10720 (1983).

066.306 **Metering instrument of quality factor Q of gravitational wave antennae.**
C. Jia-yan, G. Tong-ren, Q. Rong-xian.
Proceedings of the Second Marcel Grossmann Meeting on General Relativity, (see 012.081), p. 1145 - 1148 (1982).
Abstr. in Phys. Abstr., Vol. 86, Abstr. 10721 (1983).

066.307 **Gravitational wave detector program at the University of Rochester.**
D. H. Douglass, M. W. Cromar, R. Q. Gram, W. W. Johnson, C. Lam, D. Macaluso.
Proceedings of the Second Marcel Grossmann Meeting on General Relativity, (see 012.081), p. 1149 - 1156 (1982).
Abstr. in Phys. Abstr., Vol. 86, Abstr. 10722 (1983).

066.308 **The Stanford low temperature gravitational radiation detector.**
W. M. Fairbank, M. Bassan, C. Chun, R. P. Giffard, J. N. Hollenhorst, E. Mapoles, M. S. McAshan, P. F. Michelson, R. C. Taber.
Proceedings of the Second Marcel Grossmann Meeting on

General Relativity, (see 012.081), p. 1157 - 1167 (1982). Abstr. in Phys. Abstr., Vol. 86, Abstr. 10723 (1983).

066.309 **A revised upper limit on gravitational radiation from the Crab pulsar.**
M.-K. Fujimoto.
Proceedings of the Second Marcel Grossmann Meeting on General Relativity, (see 012.081), p. 1169 - 1174 (1982). Abstr. in Phys. Abstr., Vol. 86, Abstr. 10724 (1983).

066.310 **The LSU cryogenic facility for gravitational radiation research.**
W. O. Hamilton, D. Darling, J. Kadlec, D. DeWitt.
Proceedings of the Second Marcel Grossmann Meeting on General Relativity, (see 012.081), p. 1175 - 1178 (1982). Abstr. in Phys. Abstr., Vol. 86, 10725 (1983).

066.311 **Tests of a superconducting, L band, dual cavity accelerometer displaying parametric gain.**
D. Darling, W. O. Hamilton.
Proceedings of the Second Marcel Grossmann Meeting on General Relativity, (see 012.081), p. 1179 - 1184 (1982). Abstr. in Phys. Abstr., Vol. 86, Abstr. 10726 (1983).

066.312 **Low frequency antennas of the University of Tokyo.**
H. Hirakawa.
Proceedings of the Second Marcel Grossmann Meeting on General Relativity, (see 012.081), p. 1185 - 1191 (1982). Abstr. in Phys. Abstr., Vol. 86, Abstr. 10727 (1983).

066.313 **Unified theory of gravitational radiation detectors for pulses and monochromatic signals.**
J. P. Ho.
Proceedings of the Second Marcel Grossmann Meeting on General Relativity, (see 012.081), p. 1193 - 1209 (1982). Abstr. in Phys. Abstr., Vol. 86, Abstr. 10728 (1983).

066.314 **Progress report on the gravitational wave experiment in Rome.**
E. Amaldi, P. Bonifazi, F. Bordoni, G. Castellano, C. Cosmelli, V. Ferrari, S. Frasca, M.-K. Fujimoto, F. Fuligni, U. Giovanardio, V. Iafolla, I. Modena, G. V. Pallottino, B. Pavan, G. Pizzella, P. Rapagnani, F. Ricci, S. Ugazio, G. Vannaroni.
Proceedings of the Second Marcel Grossmann Meeting on General Relativity, (see 012.081), p. 1211 - 1225 (1982). Abstr. in Phys. Abstr., Vol. 86, Abstr. 10729 (1983).

066.315 **Detecting gravitational radiation with quartz crystals.**
D. M. Strayer, G. Papini, B. Ramadan, E. Tward.
Proceedings of the Second Marcel Grossmann Meeting on General Relativity, (see 012.081), p. 1227 - 1234 (1982). Abstr. in Phys. Abstr., Vol. 86, Abstr. 10730 (1983).

066.316 **Quantum limit for a quartz crystal antenna.**
D. M. Strayer, G. Papini, E. Tward.
Proceedings of the Second Marcel Grossmann Meeting on General Relativity, (see 012.081), p. 1235 - 1238 (1982). Abstr. in Phys. Abstr., Vol. 86, Abstr. 10731 (1983).

066.317 **A multimode antenna for the detection of short pulses of gravitational radiation.**
J.-P. Richard.
Proceedings of the Second Marcel Grossmann Meeting on General Relativity, (see 012.081), p. 1239 - 1244 (1982). Abstr. in Phys. Abstr., Vol. 86, Abstr. 10732 (1983).

066.318 **Clocks and relativity experiments.**
R. F. C. Vessot.
Proceedings of the Second Marcel Grossmann Meeting on General Relativity, (see 012.081), p. 1247 - 1268 (1982). Abstr. in Phys. Abstr., Vol. 86, Abstr. 10733 (1983).

066.319 **Some recent results of relativistic continuum mechanics pertinent to the astrophysics of dense objects.**
G. A. Maugin.
Proceedings of the Second Marcel Grossmann Meeting on General Relativity, (see 012.081), p. 673 (1982). – Abstr. in Phys. Abstr., Vol. 86, Abstr. 15839 (1983).

066.320 **Gravitationally collapsed objects.**
R. Ruffini.
Proceedings of the Second Marcel Grossmann Meeting on General Relativity, (see 012.081), p. 701 - 703 (1982). Abstr. in Phys. Abstr., Vol. 86, Abstr. 15840 (1983).

066.321 **Perturbations of the Reissner-Nordström black hole.**
J. Bicak.
Proceedings of the Second Marcel Grossmann Meeting on General Relativity, (see 012.081), p. 277 - 286 (1982). Abstr. in Phys. Abstr., Vol. 86, Abstr. 15939 (1983).

066.322 **On massive black holes.**
B. Mashhoon.
Proceedings of the Second Marcel Grossmann Meeting on General Relativity, (see 012.081), p. 647 - 671 (1982). Abstr. in Phys. Abstr., Vol. 86, Abstr. 15940 (1983).

066.323 **Measurement of intermediate and large scale anisotropies in the microwave background.**
R. Fabbri, F. Melchiorri, V. Natale.
Proceedings of the Second Marcel Grossmann Meeting on General Relativity, (see 012.081), p. 889 - 918 (1982). Abstr. in Phys. Abstr., Vol. 86, Abstr. 20533 (1983).

066.324 **Polarization of the cosmic background radiation: an experimental approach.**
A. Blanco, S. Fonti, F. Melchiorri, V. Natale.
Proceedings of the Second Marcel Grossmann Meeting on General Relativity, (see 012.081), p. 919 - 925 (1982). Abstr. in Phys. Abstr., Vol. 86, Abstr. 20534 (1983).

066.325 **Origin of distortions of the cosmic microwave background spectrum.**
S. Aiello, S. Cecchini, N. Mandolesi, F. Melchiorri.
Proceedings of the Second Marcel Grossmann Meeting on General Relativity, (see 012.081), p. 927 - 935 (1982). Abstr. in Phys. Abstr., Vol. 86, Abstr. 20561 (1983).

066.326 **The effect of a non-minimally coupled scalar field upon classical thermodynamics.**
M. D. Pollock.
Nucl. Phys. B, Part. Phys., Vol. B208, 182 - 188 (1982). Abstr. in Phys. Abstr., Vol. 86, Abstr. 16190 (1983).

066.327 **Nonlinear approximation scheme for the Einstein evolution equations.**
M. Bovyn.
Phys. Rev. D, Vol. 26, 2557 - 2563 (1982). – Abstr. in Phys. Abstr., Vol. 86, Abstr. 16195 (1983).

066.328 **A slice theorem for the space of solutions of Einstein's equations.**
J. Isenberg, J. E. Marsden.
Phys. Rep., Vol. 89, 179 - 222 (1982). – Abstr. in Phys. Abstr., Vol. 86, Abstr. 16202 (1983).

066.329 **The energy-momentum pseudotensor of the gravitational field and the gravitational radiation emitted by an axisymmetric system.**
Y.-T. Wang.

Acta Phys. Sinica, Vol. 31, 1215 - 1222 (1982). In Chinese. Abstr. in Phys. Abstr., Vol. 86, Abstr. 16204 (1983).

066.330 **Origin of the gravitational constant and particle masses in a scale-invariant scalar-tensor theory.**
Y. Fujii.
Phys. Rev. D, Vol. 26, 2580 - 2588 (1982). – Abstr. in Phys. Abstr., Vol. 86, Abstr. 16208 (1983).

066.331 **The Bogomolny inequality for Einstein-Maxwell theory.**
G. W. Gibbons.
Monopoles in quantum field theory, (see 012.088), p. 137 - 138 (1982). – Abstr. in Phys. Abstr., Vol. 86, Abstr. 16210 (1983).

066.332 **Negative energy radiation and the second law of thermodynamics.**
D. Deutsch, A. C. Ottewill, D. W. Sciama.
Phys. Lett. B, Vol. 119B, 72 - 74 (1982). – Abstr. in Phys. Abstr., Vol. 86, Abstr. 16214 (1983).

066.333 **Antigraviting black hole solitons with scalar hair in $N = 4$ supergravity.**
G. W. Gibbons.
Nucl. Phys. B, Part. Phys., Vol. B207, 337 - 349 (1982). Abstr. in Phys. Abstr., Vol. 86, Abstr. 16219 (1983).

066.334 **Equivalence principle for massive bodies in a generalized theory of gravitation. I.**
J. C. McDow, J. W. Moffat.
Canadian J. Phys., Vol. 60, 1545 - 1555 (1982). – Abstr. in Phys. Abstr., Vol. 86, Abstr. 20230 (1983).

066.335 **Equivalence principle for massive bodies in a generalized theory of gravitation. II.**
J. C. McDow, J. W. Moffat.
Canadian J. Phys., Vol. 60, 1556 - 1560 (1982). – Abstr. in Phys. Abstr., Vol. 86, Abstr. 20231 (1983).

066.336 **Second-order contributions to gravitational deflection of light in the parametrised post-Newtonian formalism. II. Photon orbits and deflections in three dimensions.**
G. W. Richter, R. A. Matzner.
Phys. Rev. D, Vol. 26, 2549 - 2556 (1982). – Abstr. in Phys. Abstr., Vol. 86, Abstr. 20234 (1983).

066.337 **Sign of the induced gravitational constant.**
N. N. Khuri.
Phys. Rev. D, Vol. 26, 2664 - 2670 (1982). – Abstr. in Phys. Abstr., Vol. 86, Abstr. 20235 (1983).

066.338 **Induced gravity and Planck zeros.**
N. N. Khuri.
Phys. Rev. D, Vol. 26, 2671 - 2680 (1982). – Abstr. in Phys. Abstr., Vol. 86, Abstr. 20236 (1983).

066.339 **Interior Reissner-Nordström metric and the scalar wave equation.**
N. Zamorano.
Phys. Rev. D, Vol. 26, 2564 - 2574 (1982). – Abstr. in Phys. Abstr., Vol. 86, Abstr. 20417 (1983).

An introduction to tensor calculus, relativity and cosmology.
See Abstr. 003.074.

Schwarze Löcher im All.
See Abstr. 003.091.

Relativität – Gruppen – Teilchen. Spezielle Relativitätstheorie als Grundlage der Feld- und Teilchenphysik.
See Abstr. 003.111.

Theory and experiment in gravitational physics.
See Abstr. 003.126.

Histoire et singularités de la solution de Schwarzschild (1915 - 1923).
See Abstr. 004.034.

Relativity and gravitation.
See Abstr. 012.094.

General relativity primer.
See Abstr. 014.014.

Specific entropy and the sign of the energy.
See Abstr. 022.124.

Ein Äther für Elektrizität, Magnetismus und Gravitation - Faradays gravo-elektrische Induktion und Riemanns Gravitationsäther als Programm der großen Unitarisierung der Physik.
See Abstr. 022.172.

Observational aspects of the millimetre-wave cosmic background radiation.
See Abstr. 032.544.

Time synchronization system in a two-station search for primordial black holes.
See Abstr. 035.004.

Massive neutrino halos.
See Abstr. 061.012.

Neutrinos in the Kerr and Robertson-Walker geometries.
See Abstr. 061.016.

Jets from a geometrically thick disk.
See Abstr. 062.011.

Bremsstrahlung spectra at relativistic temperatures.
See Abstr. 062.030.

Cyclotron emission in strongly magnetized plasmas.
See Abstr. 062.090.

Infinitely conducting rotating sphere in a magnetic field.
See Abstr. 062.120.

Two temperature spherical accretion onto a black hole: the X- and gamma-ray spectra.
See Abstr. 064.074.

Non-adiabatic self-consistent spherical accretion.
See Abstr. 064.075.

Are thick accretion disks the 'central engine' for astrophysical jets?
See Abstr. 064.081.

Rupture d'étoiles par les forces de marée.
See Abstr. 065.011.

External magnetic field of a static spherically symmetric star in Rosen's bimetric theory of gravitation.
See Abstr. 065.044.

The critical mass of a charged star.
See Abstr. 065.056.

Collapse of iron stellar cores.
See Abstr. 065.087.

General relativistic gravitational collapse of a rotating star with magnetic fields: the formalism and the initial value equations.
See Abstr. 065.103.

On the possibility of generating gravitational radiation by the Sun.
See Abstr. 073.057.

Response of the earth to gravitational radiation.
See Abstr. 081.037.

Relativistic perturbations of the Moon in ELP 2000.
See Abstr. 094.028.

The radiation damping force of binary stars due to emission of gravitational waves.
See Abstr. 117.228.

A note on gravitational radiation from close binary systems.
See Abstr. 117.229.

Spheroidal gravitational lenses and the triple quasar.
See Abstr. 141.029.

Quasars as clusters of accreting black holes.
See Abstr. 141.073.

Detection of new compact radio component near Q0957 + 561B: quasar image or galactic core?
See Abstr. 141.075.

Evolutionary luminosity functions of extragalactic sources driven by gravitational power.
See Abstr. 141.104.

Theory of electron-positron showers in double radio sources.
See Abstr. 141.132.

A semiquantitative general-relativistic model of quasar Markarian 205 interpreted as a massive black hole ejected from NGC 4319.
See Abstr. 141.144.

Another gravitational lens?
See Abstr. 141.160.

Are there black holes in quasars?
See Abstr. 141.165.

On the origin of relativistic particles and gamma-rays in quasars.
See Abstr. 141.200.

On the gravitational radiation of the binary system with the pulsar PSR 1913 + 16.
See Abstr. 141.502.

Evidence for the existence of gravitational radiation from measurements of the binary pulsar PSR 1913 + 16.
See Abstr. 141.528.

Timing observations of the binary pulsar PSR 1913 + 16.
See Abstr. 141.536.

A pulsar model from an oscillating black hole.
See Abstr. 141.538.

Trapped line radiation in X-ray sources of Population II.
See Abstr. 142.088.

X-rays from quasars and active galaxies.
See Abstr. 142.148.

Modèles de trous noirs massifs dans les noyaux actifs de galaxies.
See Abstr. 151.011.

The distribution of stars around a black hole: numerical solution of the kinetic equation with collisions.
See Abstr. 151.016.

The compact source at the galactic center.
See Abstr. 155.072.

Activity in the centers of galaxies.
See Abstr. 158.294.

Dust-like stages in the early universe and limits on the primordial black hole spectrum.
See Abstr. 162.006.

"Stiff" universe and the mass spectrum of primordial black holes.
See Abstr. 162.007.

Some cosmological implications of Dirac's large number hypothesis.
See Abstr. 162.016.

Anisotropy of the microwave background due to the mass distribution in an open cosmological model.
See Abstr. 162.044.

Pregalactic star formation and spectral distortions of the cosmic blackbody radiation.
See Abstr. 162.051.

The cosmological constant and the gauge-covariant theory of gravitation.
See Abstr. 162.062.

Correspondence between the Brans-Dicke theory and general relativity in a closed cosmological model.
See Abstr. 162.079.

The early Universe in a generalized theory of gravitation.
See Abstr. 162.118.

Universe expansion according to the analytic equation of state of the statistical-bootstrap model 107 Fiore,R.
See Abstr. 162.141.

Creation of particles by shell-focusing singularities.
See Abstr. 162.154.

Spin, circulation, angular momentum and cosmological large numbers.
See Abstr. 162.156.

Inhomogeneous cosmological models with flat slices generated from the Einstein-de Sitter universe.
See Abstr. 162.179.

The precision on the measure of q_0 using the gravitational lensing effect.
See Abstr. 162.181.

A new view of baryon symmetric cosmology based on grand unified theories.
See Abstr. 162.183.

Breaking of gauge symmetries in the cooling Universe.
See Abstr. 162.185.

Large-scale background temperature and mass fluctuations due to scale-invariant primeval perturbations.
See Abstr. 162.191.

On cosmological testing of the Machian principle.
See Abstr. 162.192.

Interpretation of anisotropy in the cosmic background radiation.
See Abstr. 162.197.

The solar motion and interpretation of the 3K microwave background.
See Abstr. 162.229.

Manifestations of a cosmological density of compact objects in quasar light.
See Abstr. 162.233.

Cosmological models in the theory of gravitation with disturbed conformal invariance.
See Abstr. 162.242.

On Mach's relation in cosmology.
See Abstr. 162.244.

Towards a unification of the parameters underlying elementary particles and cosmology.
See Abstr. 162.249.

Gravitational effects upon cosmological phase transitions.
See Abstr. 162.256.

Quantum field theories and relativistic cosmology.
See Abstr. 162.290.

Homogeneous cosmologies and their stability behaviour.
See Abstr. 162.302.

Etheronics – a possible reappraisal of the ether concept.
See Abstr. 162.308.

Bubble collisions in the very early universe.
See Abstr. 162.322.

Neutron Stars

066.501 **Thermonuclear processes and accretion onto neutron star envelopes: X-ray burst and transient sources.**
S. Starrfield, S. Kenyon, W. M. Sparks, J. W. Truran.
Astrophys. J., Vol. 258, 683 - 695 (1982).

The authors have used a Lagrangian, fully implicit, one-dimensional, hydrodynamic computer code to investigate the evolution of thermonuclear runaways in the thick, accreted, hydrogen-rich envelopes of 1.0 $M_\odot$ neutron stars with radii of 10 km and 20 km. The simulations produce outbursts which range in time scale from about 2000 seconds to longer than 1 day. The authors have also studied the evolution of a gas cloud impacting the surface of a 20 km, 1 $M_\odot$ neutron star, in an attempt to simulate the magnetospheric gate model for the X-ray burst sources.

066.502 **The thermonuclear model for X-ray transients.**
R. K. Wallace, S. E. Woosley, T. A. Weaver.
Astrophys. J., Vol. 258, 696 - 715 (1982) = Lick Obs. Bull., No. 911.

The thermonuclear evolution of a 1.41 $M_\odot$ neutron star accreting both solar and metal-deficient mixtures of hydrogen, helium, and heavy elements at rates ranging from about 10^{-11} to $10^{-10}\,M_\odot$ per year is examined using a one-dimensional numerical model. For such accretion rates and metallicities, hydrogen burning, mediated by the β-limited CNO cycle, is stable and leads to the accumulation of a thick helium layer. Helium ignition occurs under extremely degenerate circumstances and is catastrophically violent. In all models, greatly super-Eddington luminosities in the outer layers of the neutron star lead to a sustained epoch of radiatively driven mass loss. Observationally, such models may correspond to rapid X-ray transients.

066.503 **The thermonuclear model for γ-ray bursts.**
S. E. Woosley, R. K. Wallace.
Astrophys. J., Vol. 258, 716 - 732 (1982) = Lick Obs. Bull., No. 912.

The production of γ-ray bursts by thermonuclear explosions on strongly magnetized neutron stars is examined. For a neutron star with a magnetic field strength of several times 10^{12} gauss, accretion at $\sim 10^{-13} M_\odot\ yr^{-1}$ is focused onto kilometer sized regions. Stable hydrogen burning leads to a critical helium mass that, depending upon model parameters, explodes either by convective deflagration or detonation, liberating 10^{38}–10^{40} ergs km^{-2}. Hard emission comes both from the magnetically confined photosphere and from relativistic electrons accelerated by magnetic field recombination. The hard γ-ray outburst of several seconds is followed by an enduring, softer emission of X rays.

066.504 **A two-dimensional model for γ-ray bursts.**
B. A. Fryxell, S. E. Woosley.
Astrophys. J., Vol. 258, 733 - 739 (1982) = Lick Obs. Bull., No. 913.

A thermonuclear model of a γ-ray burst is calculated using a two-dimensional hydrodynamic code. Accretion onto the magnetic pole of a neutron star leads to the formation of a helium lake 90 m deep and 400 m in radius. A thermonuclear runaway initiates at a point at the base of the helium layer and results in a detonation wave. This wave propagates across the lake with a velocity of 9000 km s^{-1} without dying out, igniting all of the helium and liberating 2.4×10^{39} ergs.

066.505 **Helium and combined hydrogen-helium shell flashes in the envelope of an accreting neutron star.**
R. E. Taam.
Astrophys. J., Vol. 258, 761 - 769 (1982).

Nuclear shell flashes in the envelope of an accreting neutron star of 1.41 $M_\odot$ have been studied in the hydrostatic and, where necessary, hydrodynamic approximation. Particular emphasis is placed on the variation of the X-ray burst properties as a function of the mass accretion rate at rates near the critical value which delineate a combined hydrogen-helium shell flash from a pure helium shell flash. The transition from the helium shell flash mode to the combined hydrogen-helium shell flash mode or vice versa may be relevant to those X-ray burst sources which exhibit irregular recurrence patterns.

066.506 **Effect of vacuum polarization by a magnetic field on radiative thermal conductivity of surface layers of neutron stars.**
G. G. Pavlov, D. G. Yakovlev.
Astrofizika, Tom 18, 119 - 131 (1982). In Russian. English translation in Astrophysics, Vol. 18, No. 1.

The radiative (Thomson) thermal conductivity is calculated for a non-relativistic non-degenerate plasma in a strong magnetic field with allowance for polarization of an electron-positron vacuum. The performed calculations are required for constructing models of X-ray pulsars as well as for investigating various phenomena concerned with heat transport in surface layers of neutron stars.

066.507 **Convection in degenerate envelopes of neutron stars.**
V. A. Urpin.
Astrofizika, Tom 18, 133 - 142 (1982). In Russian. English translation in Astrophysics, Vol. 18, No. 1.

A condition of convective instability in the degenerate layers of neutron stars is derived. It is shown that in degenerate layers a smaller temperature gradient is necessary for convection than in non-degenerate ones. Thanks to this fact the extensive regions may undergo convection in the envelopes of sufficiently hot neutron stars.

066.508 **Quantum oscillations of electron thermal and electrical conductivities of neutron stars in a magnetic field.**
D. G. Yakovlev.
Astron. Zh., Tom 59, 683 - 692 (1982). In Russian. English translation in Soviet Astron., Vol. 25, No. 4.

Electron thermal and electrical conductivities as well as the thermoelectric coefficient along a quantizing magnetic field are calculated for a degenerate non-relativistic electron gas in the surface layers of neutron stars for three main electron scattering mechanisms: on ions above the melting temperature of an ion crystal, and on high-temperature phonons and impurities in the presence of crystallization. Quantum oscillations of the kinetic coefficients due to magnetic field, density and temperature variations are analysed.

066.509 **Neutron star envelopes.**
E. H. Gudmundsson, C. J. Pethick, R. I. Epstein.
Astrophys. J., Lett., Vol. 259, L19 - L23 (1982).

The authors point out that the separate study of neutron star envelopes is a valuable technique for gaining insights into neutron star cooling calculations. They show that the temperature at the inner boundary of the envelope, T_b, is a function only of T_s^4/g_s, where T_s is the effective temperature and g_s is the surface gravity.

066.510 **Hydrogen-helium flashes on accreting neutron stars as a possible origin of gamma-ray bursts.**
J. M. Hameury, S. Bonazzola, J. Heyvaerts, J. Ventura.
Astron. Astrophys., Vol. 111, 242 - 251 (1982).

Accretion of interstellar matter on a galactic disc population of old neutron stars can lead to thermonuclear runaway offering a very plausible explanation for the phenomenon of gamma-ray bursts. The authors have studied numerically the possibility of having hydrogen and helium flashes in a system with low accretion rates $\dot{M} < 10^{-12} M_\odot yr^{-1}$. They find instabilities when the accretion rate exceeds $3 \times 10^{-13} M_\odot yr^{-1}$. Each burst is then triggered by electron capture on protons, when the accreted envelope reaches a mass of 3×10^{23} g. The resulting hydrogen flash will lead to helium detonation after a short heating time of several minutes. Several conclusions are presented.

066.511 **A puzzle concerning the quadrupole formula for gravitational radiation.**
E. Balbinski, B. F. Schutz.
Mon. Not. R. Astron. Soc., Vol. 200, 43P - 46P (1982).

A comparison is made between the damping of the fundamental mode of neutron star due to gravitational radiation reaction calculated by an approximate method based on the standard quadrupole formula for the energy loss, and the damping given by previous calculations using the full theory of general relativity. Numerical results for a mildly relativistic neutron star with $2\,GM/Rc^2 < 10$ per cent show the surprising result that the damping times differ by a factor of about 3 by these two methods.

066.512 **Line emission from neutron star surfaces.**
J. Trümper.
Galactic X-ray sources, (see 003.001), p. 231 - 240 (1982).

The author concentrates mainly on the pulsating X-ray sources which have proven to be a rich source of information on neutron star physics. He discusses the potential of line spectroscopy of neutron stars and reviews a few results which have been obtained so far. He concentrates on the surface emission from neutron stars.

066.513 **Neutron star collisions and the r-process.**
E. Symbalisty, D. N. Schramm.
Astrophys. Lett., Vol. 22, 143 - 145 (1982).

It is shown that a natural consequence of the binary pulsar's evolution is a neutron star collision. Such a collision is expected to eject neutron-rich matter of an r-process character. Taking reasonable estimates for the number of such events over the history of the Galaxy, it may be that they account for all of the r-process nuclei.

066.514 **Magnetic field strongly reduces critical luminosity of neutron stars and degenerate dwarfs.**
I. G. Mitrofanov, G. G. Pavlov.
Mon. Not. R. Astron. Soc., Vol. 200, 1033 - 1037 (1982).

In a magnetic field the electron cyclotron resonance in absorption of radiation leads to a strong increase of radiative force. It results in a considerable decrease of the critical value of the radiative flux for which the radiative force acting on electrons becomes as large as the gravitational attraction of protons. When taken into account this effect may change the traditional views on the objects with magnetized neutron stars and degenerate dwarfs.

066.515 **On X-ray and gamma-ray emission of accreting neutron stars.**
V. V. Zheleznyakov.
Usp. fiz. nauk., Tom 136, 345 - 346 (1982). In Russian. Abstr. in Ref. zh., 51. Astron., 7.51.567 (1982).

066.516 **Neutron stars and general equation of nuclear matter.**
M. C. Durgapal, R. Bannerji.

Astrophys. Space Sci., Vol. 84, 409 - 419 (1982).

It has been shown that the mass of neutron stars obtained from equations of state based on nuclear theories depends upon the number of baryons assembled in it but not on the type of interactions considered. The results obtained for $N = 1.75$ cover the entire range of neutron star masses obtained from the equations of state based on nuclear theories and give a maximum mass of $2.8\,M_\odot$. Depending upon various mechanisms for energy output the mass of Crab pulsar is estimated to range from $0.32\,M_\odot$ to $1.5\,M_\odot$.

066.517 **Kelvin-Helmholtz instabilities at the interface between an accretion disk and the magnetosphere of a neutron star.**
U. Anzer, G. Börner.
Z. Naturforsch., Band 37a, 723 - 727 (1982).

The material in an accretion disk moves on Kepler orbits whereas the magnetosphere of a neutron star rotates like a solid body. Therefore large velocity differences will occur at the interface between the two media. These give rise to Kelvin-Helmholtz instabilities. The authors present a stability analysis for a simplified geometry. The consequences of the instability for the mass flow from the disk into the magnetosphere are discussed. The resulting transfer of angular momentum to the neutron star is compared with the observed changes of the rotation periods of these stars.

066.518 **On the gravitational field of a neutron star.**
A. P. Efremov.
Izv. vuzov. Fiz., Tom 25, 53 - 57 (1982). In Russian. – Abstr. in Ref. zh., 51. Astron., 8.51.131 (1982).

066.519 **Nonradial pulsations of neutron stars.**
P. N. McDermott, H. M. van Horn, B. W. Carroll.
News Lett. Astron. Soc. N.Y., Vol. 2, No. 2, p. 26 (1982). Abstract.

066.520 **Initial neutrino loss in neutron star formation.**
B. T. Goodwin.
Astrophys. J., Vol. 261, 321 - 331 (1982).

The nonlinear diffusion equation which describes the initial loss of neutrinos from the inner core of a newly formed neutron star is solved numerically. The diffusion coefficient used is that appropriate to degenerate neutrinos in dense matter taking into account both absorption and scattering of neutrinos by nucleons. It is found that the loss times are too long for neutrino escape to be dynamically useful in matter ejection processes which occur on millisecond time scales.

066.521 **Nonradial g-mode oscillations of warm neutron stars.**
P. N. McDermott, H. M. van Horn, B. W. Carroll, J. F. Scholl.
Bull. American Astron. Soc., Vol. 14, 665 (1982). – Abstract.

066.522 **Magnetospheric structures and instabilities around accreting neutron stars.**
A. Tomimatsu.
Magnetospheric phenomena of celestial bodies, (see 012.011), p. 98 - 106 (1982). In Japanese.

The paper reviews theoretical models of mass accretion onto a magnetized neutron star, around which a magnetosphere is formed. Structures of the magnetosphere and internal flows of the plasma, in the case of spherical or disk accretion, are investigated. The author is mainly interested in some magnetohydrodynamic instabilities, which make accreted matter enter across the magnetopause for producing X-ray emissions, and he discusses whether the X-ray source is observed as a steady or burst-like one.

066.523 **Adiabatically stable neutron star model.**
P. S. Rawat, M. C. Durgapal.
Bull. Astron. Soc. India, Vol. 10, 43 (1982). – Abstract.

066.524 **Neutron stars and general equations for nuclear matter.**
M. C. Durgapal, R. Bannerji.
Bull. Astron. Soc. India, Vol. 10, 43 - 44 (1982). – Abstract.

066.525 **The universal diagram for magnetized neutron stars in the Galaxy.**
V. M. Lipunov.
Astrophys. Space Sci., Vol. 85, 451 - 457 (1982).

A convenient diagram for an analysis of astrophysical appearances and evolution of magnetized neutron stars has been proposed. The diagram gives a general approach to such different objects as radiopulsars, X-ray pulsars, gamma-ray bursters, the SS 433 source, and others.

066.526 **The role of cyclotron resonance processes in the spectra of accreting neutron stars.**
J. Trümper.
AIP Conf. Proc., No. 77, (see 012.034), p. 179 - 188 (1982).

The author reviews the observational situation regarding cyclotron resonance features in the spectra of pulsating sources and then turns to problems arising in the context of gamma-ray burst spectra. He discusses some recent work done on the radiative transfer in homogeneous plasmas under super-strong magnetic field conditions.

066.527 **Surface and magnetospheric physics of neutron stars and gamma-ray bursts.**
D. Q. Lamb.
AIP Conf. Proc., No. 77, (see 012.034), p. 249 - 272 (1982).

This paper reviews some aspects of the surface and magnetospheric physics of neutron stars that may be relevant to gamma-ray bursts if they originate on strongly magnetic neutron stars.

066.528 **The thermonuclear model for γ-ray bursts.**
S. E. Woosley.
AIP Conf. Proc., No. 77, (see 012.034), p. 273 - 292 (1982).

The evolution of magnetized neutron stars with field strengths of $\sim 10^{12}$ gauss that are accreting mass onto kilometersized polar regions at a rate of $\sim 10^{-13} M_\odot\,\mathrm{yr}^{-1}$ is examined. Owing to the extreme degeneracy of the electron gas providing pressure support, helium burning occurs as a violent thermonuclear runaway which may propagate either as a convective deflagration or as a detonation wave.

066.529 **Gamma-ray bursts from young neutron stars.**
K. Brecher.
AIP Conf. Proc., No. 77, (see 012.034), p. 293 - 297 (1982).

The authors argue that young, single neutron stars, not undergoing significant accretion, are plausible candidates for sporadic energetic radiative events such as gamma-ray bursts. They take the March 5, 1979 gamma-ray burst as the prototype source.

066.530 **A two dimensional model for γ-ray bursts.**
B. A. Fryxell, S. E. Woosley.
AIP Conf. Proc., No. 77, (see 012.034), p. 299 - 307 (1982).

Accretion focused onto a neutron star by a strong magnetic field results in the formation of a lake of helium at the magnetic pole. After a critical mass of helium has accumulated, a thermonuclear runaway initiates at a point at the base of the helium layer resulting in a detonation wave which propagates across the lake igniting all the helium. The interaction of the hot plasma behind the detonation front with the magnetic field produces the hard radiation observed in γ-ray bursts.

066.531 **Gamma bursts from neutron stars and stellar flares.**
S. A. Colgate.
AIP Conf. Proc., No. 77, (see 012.034), p. 309 - 310 (1982).

If gamma bursts are locally galactic, then the implied fluxes from a localized region of a neutron star surface are close to the blackbody limit. One reasonable model is the accretion of an asteroid or comet (Harwit and Salpeter 1973) onto a magnetized neutron star. Some implications of this model are briefly discussed.

066.532 **Precursors to γ-ray bursts in the asteroid impact scenario.**
D. Van Buren.
AIP Conf. Proc., No. 77, (see 012.034), p. 311 - 313 (1982).
The extremely large energy densities in the fireball of a γ-ray burst lead to copious pair production and annihilation independent of the burst mechanism. Asteroid impacts onto neutron stars can give rise to burst precursors as material from the compressed and heated asteroid is attached to magnetic field lines and free-falls onto the stellar surface while the asteroid orbit decays over a much longer timescale. The properties of these precursors are different enough from any precursors arising from endogenic mechanisms that observations can provide a good test of competing theories.

066.533 **Proton deceleration in a neutron star atmosphere.**
J. G. Kirk, D. J. Galloway.
AIP Conf. Proc., No. 77, (see 012.034), p. 315 - 318 (1982).
The energy loss rate of fast protons in a strongly magnetized electron gas has been calculated using the Fokker-Planck equation to describe the dominant effect of small-angle Coulomb scattering. The possibility of maser action at the gyrofrequency is indicated.

066.534 **Nuclear flash models for X-ray burst sources.**
P. C. Joss.
AIP Conf. Proc., No. 77, (see 012.034), p. 349 - 350 (1982).
Abstract.

066.535 **A numerical simulation of the magnetospheric gate model for the X-ray bursters.**
S. Starrfield, S. Kenyon, J. W. Truran, W. M. Sparks.
AIP Conf. Proc., No. 77, (see 012.034), p. 351 - 354 (1982).
The authors have used a Lagrangian, fully implicit, one-dimensional hydrodynamic code to investigate the evolution of a gas cloud impacting the surface of a 20 km, 1 $M_\odot$ neutron star. The infall results in a burst which lasts about 0.1 seconds and reaches a peak luminosity of $2.4 \times 10^5 L_\odot$.

066.536 **Thermonuclear runaways in thick hydrogen rich envelopes of neutron stars.**
S. Starrfield, S. Kenyon, J. W. Truran, W. M. Sparks.
AIP Conf. Proc., No. 77, (see 012.034), p. 355 - 361 (1982).
The authors have used a Lagrangian, fully implicit, one-dimensional hydrodynamic code to evolve thermonuclear runaways in the accreted hydrogen rich envelopes of 1.0 $M_\odot$ neutron stars with radii of 10 km and 20 km. The simulations produce outbursts which last from about 750 seconds to about one week.

066.537 **Metastability of dense neutron matter.**
P. Haensel, R. Schaeffer.
Nucl. Phys. A, Vol. A381, 519 - 543 (1982). – Abstr. in Phys. Abstr., Vol. 85, Abstr. 100970 (1982).

066.538 **The hyperon composition of neutron stars.**
N. K. Glendenning.
Phys. Lett. B, Vol. 114B, 392 - 396 (1982). – Abstr. in Phys. Abstr., Vol. 85, Abstr. 106768 (1982).

066.539 **Coherent ultraviolet radiation from accreting neutron stars.**
J. G. Kirk, R. J. Stoneham.
Mon. Not. R. Astron. Soc., Vol. 201, 1183 - 1198 (1982).
Amplified gyromagnetic radiation at the proton cyclotron frequency is predicted for accreting, strongly magnetized neutron stars. This mechanism is restricted to low luminosity sources ($< 10^{37}$ erg s^{-1}) with a relatively cool emission region ($\lesssim$ 5 keV). Neutron stars accreting at $10^{-13} M_\odot$ yr^{-1} (possible γ-ray burst sources) fulfil these conditions, and should be continuously observable in the UV band. X-ray pulsars are borderline cases which may or may not emit coherently. Typically a few per cent of the accretion luminosity is radiated at frequencies close to the proton gyrofrequency in a coherently emitting source.

066.540 **Irreducible representation of shell flashes on accreting neutron stars.**
T. Hanawa, M. Y. Fujimoto.
Publ. Astron. Soc. Japan, Vol. 34, 495 - 506 (1982).
Parameter dependences are studied for shell burnings induced by the accretion onto the surface of neutron stars. It is shown that their progress can be formulated with two parameters of the surface gravity and the column accretion rate, although the model parameters involve the mass and the radius of neutron stars and the mass accretion rate. By applying this formulation, the property of hydrogen shell burning and the condition to ignite helium shell flash are investigated. Implications to X-ray burst phenomena, and especially the relevant range of the column accretion rate are discussed.

066.541 **Nuclear fusion on accreting neutron star.**
M. Czerny, E. V. Ergma.
Acta Astron., Vol. 32, 25 - 35 (1982).
Several aspects of nuclear burning on a surface of an accreting neutron star are investigated in a steady-state approximation. For accretion rates in the examined range of 10^{16} - 10^{18} g/s nuclear burning is probably unstable. Overlapping of hydrogen and helium shells occurs for $\dot{M} \gtrsim 10^{16.5}$ g/s. If nuclear reactions proceed in the strong magnetic field the time scales of burning are strongly increased. Also in this case burning is probably unstable. Enlargement of CNO isotopes does not change the problem qualitatively. For high accretion rates neutron-induced reactions speed up the burning by a factor of ~2.

066.542 **Tunnel effect in molecules in strong magnetic fields of neutron stars.**
V. K. Khersonskij.
Astrophys. Space Sci., Vol. 87, 61 - 67 (1982).
It is shown that, in the strong magnetic field of the neutron star $B = 10^{12}$ - 10^{13} G, the probability of the tunnel effect in the molecules increases significantly. It is quite probable that this effect can catalyze nuclear reactions at the neutron star surface.

066.543 **Highly nonstationary accretion onto magnetized neutron stars.**
J. Trümper, G. Morfill, P. Bodenheimer, G. Tenorio-Tagle.
Accreting neutron stars, (see 012.069), p. 134 - 142 (1982).
The authors consider highly nonstationary accretion onto a magnetized neutron star from a surrounding accretion disc.

066.544 **Accretion discs around neutron stars.**
J. E. Pringle.
Accreting neutron stars, (see 012.069), p. 150 - 154 (1982).
The observational evidence for accretion discs in compact X-ray sources is reviewed and theoretical models of accretion onto neutron stars are examined.

066.545 **Accretion onto rotating magnetic neutron stars: the inner edge of the disk.**
U. Anzer, G. Börner.
Accreting neutron stars, (see 012.069), p. 155 (1982).
Abstract.

066.546 **Disk accretion onto weakly magnetized neutron stars.**

M. Camenzind.
Accreting neutron stars, (see 012.069), p. 156 - 158 (1982).
Low-mass X-ray binary systems are the most suitable objects to study the physics of the inner accretion disk. The author shows that mass-loss from the inner disk can explain the X-ray spectrum, X-ray light curve and the fluctuation time scales of the close binary systems Sco X-1, Cyg X-3, 4U 1822-37 and 4U 2129 + 47.

066.547 **Thermonuclear model for high energy transients.**
S. E. Woosley,
Accreting neutron stars, (see 012.069), p. 189 - 208 (1982).
The thermonuclear model for X-and γ-ray bursts is discussed. Different regimes of nuclear burning are reviewed, each appropriate to a given range of (steady state) accretion rate. Accretion rates in the range 10^{-14} to $10^{-8} M_\odot\, y^{-1}$ all appear capable of producing X-ray transients of various durations and intervals. Results from a numerical simulation of a combined hydrogen-helium runaway initiated at pycnonuclear density are presented for the first time. The thermonuclear model for γ-ray bursts is also reviewed and updated. Solely on the basis of nuclear instability, γ-ray bursts of various types appear possible for a very broad variety of accretion rates ($\sim 10^{-17}$ to $\sim 10^{-11} M_\odot\, y^{-1}$).

066.548 **Rise times and durations of X-ray and γ-ray bursts.**
B. A. Fryxell.
Accreting neutron stars, (see 012.069), p. 216 - 218 (1982).
If a sufficiently dense layer of matter is accreted onto the surface of a neutron star, a nuclear runaway can occur at the base of the layer. The effects of a finite propagation time of the nuclear detonation wave are discussed in this paper.

066.549 **A 1D thermonuclear model for X-ray transients.**
R. K. Wallace.
Accreting neutron stars, (see 012.069), p. 219 - 222 (1982).
The thermonuclear evolution of a 1.41 $M_\odot$ neutron star, with a radius of 14.3 km, accreting various mixtures of hydrogen, helium, and heavy elements at rates of 10^{-11} to $10^{-10} M_\odot$/yr is examined.

066.550 **Thermonuclear runaways at the surface of slowly accreting neutron stars.**
S. Bonazzola, J. M. Hameury, J. Heyvaerts, J. Ventura.
Accreting neutron stars, (see 012.069), p. 241 - 243 (1982).
The authors investigate the possibility of producing hydrogen-helium flashes in accreting neutron stars. It is shown that depletion of the C, N and O nuclei due to gravitational settling seriously affect the results.

066.551 **Continuous UV radiation from gamma-ray burst sources.**
J. G. Kirk, R. J. Stoneham.
Accreting neutron stars, (see 012.069), p. 244 - 245 (1982).
Current models for gamma-ray burst sources contain a magnetized neutron star accreting matter, between bursts, at a rather slow rate (10^{-13} - $10^{-16} M_\odot yr^{-1}$). The authors predict that such a system should be a continuous source of non-thermal UV photons.

066.552 **Radiation from the accretion column of magnetized neutron stars.**
P. Mészáros.
Accreting neutron stars, (see 012.069), p. 253 - 274 (1982).
The author reviews recent developments in radiative transfer for magnetized neutron star conditions, and their application in models of the structure and properties of self-consistent polar cap emission regions. Several of the assumptions and uncertainties involved are discussed, and present problems are indicated.

066.553 **Heating of the accreting neutron star interior by the thermonuclear burning and the steady core temperature.**
K. Nomoto, S. Tsuruta.
Accreting neutron stars, (see 012.069), p. 275 - 280 (1982).
The authors examine the thermal history of the accreting neutron star interior prior to the thermonuclear flash. In the models the accretion forms a steady hydrogen burning shell near the surface from which heat flows into the core. The results justify the isothermal approximation for the neutron star interior at the initiation of the helium flash.

066.554 **Mass loss from the neutron stars associated with X-ray bursts.**
T. Ebisuzaki, T. Hanawa, D. Sugimoto.
Accreting neutron stars, (see 012.069), p. 281 (1982).
Abstract.

066.555 **Upper limits on "propeller" spin-down at rapidly rotating neutron stars imposed by magnetic torque considerations.**
V. M. Vasyliunas.
Accreting neutron stars, (see 012.069), p. 288 (1982).
Abstract.

066.556 **On the mechanism of thermalization of a plasma infalling onto magnetized neutron stars.**
R. Z. Yahel.
Accreting neutron stars, (see 012.069), p. 289 - 292 (1982).
The purpose of this note is to present preliminary results of deceleration calculations for matter infalling onto strongly magnetized neutron stars.

066.557 **Proton cyclotron emission near accreting magnetic neutron stars.**
K. M. V. Apparao, S. M. Chitre, M. J. Rees.
Accreting neutron stars, (see 012.069), p. 293 - 295 (1982).
The process of incoherent proton cyclotron emission occurring near the polar regions of an accreting neutron star is examined with a view to account for the soft X-ray flux observed in Her X-1 and the possibility of detecting soft X-ray bursts accompanying gamma ray bursts.

066.558 **Towards model spectra of accreting magnetic neutron stars.**
W. Nagel.
Accreting neutron stars, (see 012.069), p. 302 - 305 (1982).
A short outline of the major difficulties in computing X-ray model spectra from accreting neutron stars is given.

066.559 **Radiative transport through an inhomogeneous accretion column.**
K. F. Bickert.
Accreting neutron stars, (see 012.069), p. 306 - 308 (1982).
The transport of X-radiation through the lower few kilometers of the accretion column of a highly magnetized neutron star is simulated using a Monte-Carlo photon-transport simulation scheme.

066.560 **Observable properties of neutron stars.**
F. K. Lamb.
Accreting neutron stars, (see 012.069), p. 316 - 332 (1982).
This paper reviews the information on neutron stars that has been obtained from observations of pulsars and pulsing X-ray stars, with emphasis on results from timing studies. Included are summaries of the most recent theoretical work on neutron star structure and the current evidence concerning neutron star masses and radii. Also included are discussions of large frequency jumps and small-scale timing irregularities, the description of the latter phenomenon in terms of random noise processes, and the use of both as probes of the internal structure and dynamical properties of neutron stars.

066.561 **Neutron star explosions?**
D. N. Page.
Phys. Lett. A, Vol. 91A, 201 - 202 (1982). – Abstr. in Phys. Abstr., Vol. 85, Abstr. 117700 (1982).

066.562 **The stability of quark matter.**
K. A. Olive.
Phys. Lett. B, Vol. 116B, 137 - 140 (1982). – Abstr. in Phys. Abstr., Vol. 85, Abstr. 117702 (1982).

066.563 **Abnormal neutron stars with abnormal protons. II. Inhomogeneous stellar models.**
Z.-r. Wang, Q.-y. Qu, T. Lu, Y. Chu.
Commun. Theor. Phys.,Vol. 1, 371 - 377 (1982). – Abstr. in Phys. Abstr., Vol. 86, Abstr. 5323 (1983).

066.564 **Abnormal neutron stars with abnormal protons. II. An inhomogeneous stellar model.**
Z. Wang, Q. Qu, T. Lu, Y. Chu.
Acta Astrophys. Sinica, Vol. 2, 285 - 290 (1982). In Chinese.

066.565 **Magnetic susceptibility of a neutron star crust.**
R. D. Blandford, L. Hernquist.
J. Phys. C, Vol. 15, 6233 - 6243 (1982). – Abstr. in Phys. Abstr., Vol. 86, Abstr. 20412 (1983).

Landolt-Börnstein. Group VI. Vol. 2c. Interstellar matter – Galaxy – Universe.
See Abstr. 003.055.

QCD and neutron stars.
See Abstr. 022.061.

Supranuclear densities in physics and astrophysics.
See Abstr. 022.076.

Neutrino emissivities and mean free paths of degenerate quark matter.
See Abstr. 061.029.

Nuclear forces and the properties of matter at high temperature and density.
See Abstr. 061.032.

New limits on the density of superheavy magnetic monopoles in the interstellar space.
See Abstr. 061.065.

Magnetic fields in astrophysics.
See Abstr. 062.034.

Radiation from a strongly-magnetized plasma: the case of predominant scattering.
See Abstr. 063.054.

Bound-free and free-free transitions in strong magnetic fields.
See Abstr. 063.064.

Summary of workshop on accretion in astrophysics.
See Abstr. 064.040.

Homologous collapse and deleptonization of an evolved stellar core.
See Abstr. 065.058.

Stellar rotation and gravitational collapse: the a/m issue.
See Abstr. 065.100.

Pulsars' timing noise and external disturbances: gravitational waves? .
See Abstr. 066.029.

On the collapse dynamics of cold stars with a phase transition in the interior.
See Abstr. 066.067.

A variational approach to dense relativistic matter using functional techniques.
See Abstr. 066.079.

A class of new exact solutions in general relativity.
See Abstr. 066.151.

Relativistic stars in close binary systems and masses of neutron stars.
See Abstr. 066.188.

The influence of a net charge on the critical mass of a neutron star.
See Abstr. 066.228.

Massive configurations with constant densities.
See Abstr. 066.230.

Magnetic properties of relativistic Fermi Gas.
See Abstr. 066.269.

HZ Her – possible line profiles.
See Abstr. 117.163.

Can gamma ray bursts originate from low mass binaries?
See Abstr. 117.218.

Late stages of close binary evolution – comments on alternative scenarios.
See Abstr. 117.219.

What can we learn from the Crab nebula?
See Abstr. 125.037.

Thermal and electrical conductivities of crystals in neutron stars and degenerate dwarfs.
See Abstr. 126.036.

The origin of nulls, mode changes and timing noise in pulsars.
See Abstr. 141.512.

Pulsar radiation mechanisms.
See Abstr. 141.527.

Summary of workshop on neutron stars and pulsars.
See Abstr. 141.529.

The magnetosphere of an X-ray star.
See Abstr. 142.070.

Thermonuclear-flash models for X-ray burst sources.
See Abstr. 142.075.

A survey of the bright galactic bulge X-ray sources.
See Abstr. 142.099.

Fast coherent oscillations in variable X-ray sources and bursters.
See Abstr. 142.130.

A model for A 0538–66, the fast flaring pulsar.
See Abstr. 142.156.

X-ray bursts: observation versus theory.
See Abstr. 142.158.

Why are most X-ray pulsars slow?
See Abstr. 142.161.

Search for instabilities in the accretion column of Her X-1 and in the accretion disk of Cyg X-1.
See Abstr. 142.170.

The low energy spectra of gamma-ray bursts.
See Abstr. 142.539.

Theoretical problems raised by gamma-ray bursts and related phenomena.
See Abstr. 142.551.

Neutron stars as sources of gamma-ray bursts: constraints from X-ray observations of source locations.
See Abstr. 142.555.

Errata

066.901 **Erratum: "Synchrotron neutrino-pair radiation in neutron stars"** [Astron. Nachr., Band 302, 167 - 176 (1981)].
D. G. Yakovlev, R. Tschaepe.
Astron. Nachr., Band 303, 218 (1982). – See Abstr. 30.066.506.

066.902 **Erratum: 'Near-millimeter spectrum of the microwave background'** [Astrophys. J., Vol. 248, 18 - 37 (1981)].
D. P. Woody, P. L. Richards.
Astrophys. J., Vol. 261, 747 (1982). – See Abstr. 30.066.005.

066.903 **Erratum: "The Einstein-Maxwell field equations, II"** [Astrophys. Space Sci., Vol. 77, 383 - 389 (1981), Vol. 82, 255 (1982)]
D. D. Dionysiou.
Astrophys. Space Sci., Vol. 86, 513 (1982). – See abstracts 29.066.053, 31.066.902.

Sun

071 Photosphere, Spectrum

071.001 **On the magnitude and the height dependence of the granular vertical flow velocity.**
M. Bässgen, F.-L. Deubner.
Astron. Astrophys., Vol. 111, L1 - L3 (1982).
Granular velocities determined from excellent quality solar spectra in two spectral lines are corrected for instrumental and atmospheric image degrading effects by application of a modulation transfer function derived from the same material by comparing the observed brightness fluctuation spectra with various "true" spectra from the literature. k–ω spectra are used to separate granular and oscillatory fluctuations. The results favour a rather flat gradient with photospheric (CI 5380 line) rms granular velocities of at least 1.0 km s^{-1}.

071.002 **Large-scale brightness fluctuations in the solar photosphere. III.**
K. S. Tavastsherna, I. V. Yudina.
Soln. Dannye 1982 Byull., No. 3, p. 111 - 116 (1982). In Russian.
The photospheric non-uniformity in the solar photosphere has been studied in the interval 7000 - 60000 km. The m. r. s. contrast of the features of 7000 - 14000 km, 14000 - 33000 km and 33000 - 62000 km in relation to the mean photospheric background has been determined.

071.003 **The CN red system in the solar spectrum.**
C. Sneden, D. L. Lambert.
Astrophys. J., Vol. 259, 381 - 391 (1982).
Solar CN lines from leading bands of the $\Delta v = -2, -1, 0, +1, +2$, and +3 sequences of the $A\,^2\Pi - X\,^2\Sigma^+$ red system are analyzed to provide band absorption oscillator strengths, radiative lifetimes of the $A\,^2\Pi$ state and the electronic transition moment. The results are compared with recent experimental results and an ab initio calculation.

071.004 **The influence of temperature inhomogeneities in the solar atmosphere on abundance determinations.**
W. Hermsen.
Astron. Astrophys., Vol. 111, 233 - 238 (1982).
Since the number densities of the elements and their line intensities are nonlinear functions of temperature, a neglect of the temperature inhomogeneities in the solar atmosphere causes an error in abundance determinations. To estimate the magnitude of these errors due to the granular temperature inhomogeneities, the author constructed several two-stream models of the solar photosphere, based on the Holweger-Müller atmosphere as the mean model. Solar f-values were derived from the two-stream models and the mean model for a set of absorption lines of Fe I, Fe II, and O I. The spread of the f-values for the different models gives the error of the abundance determination originating from the use of homogeneous models.

071.005 **Vertical structure of the solar photosphere. II. The small-scale velocity field.**
C. J. Durrant, A. Nesis.
Astron. Astrophys., Vol. 111, 272 - 278 (1982) = Mitt. Kiepenheuer Inst. Nr. 206.
The paper is a preliminary report of detailed analyses of the fine structure of the photospheric brightness and velocity field, utilizing the material from the 1975 flight of the balloon-borne Spectrostratoskop. The authors look at the velocity fields at the centre-of-the-disk. They look at two characteristics of the field; the first being the probability density distributions in physical space, the second being the power and coherence distributions in Fourier space. The authors thence derive an empirical interpretation of the granulation as a convective phenomenon, and remark on consequences in the upper atmosphere.

071.006 **Analysis of Fe I lines (0.00 eV < χ < 2.6 eV) in the solar spectrum using improved damping constants and accurate oscillator strengths: test of a solar model atmosphere.**
G. J. Simmons, D. E. Blackwell.
Astron. Astrophys., Vol. 112, 209 - 214 (1982).
An analysis is made of the equivalent widths of 26 well defined Fe I lines in the solar spectrum measured at five disk positions and in the radiation from the integrated disk. Damping constants based on the measurement of strong solar lines, and on laboratory experiments, together with oscillator strengths measured at Oxford, are used in the analysis. Iron abundances and microturbulent velocities are determined using an LTE analysis with the Holweger and Müller solar model atmospheres.

071.007 **Large-scale brightness inhomogeneities in the solar atmosphere.**
W. E. Mitchell, Jr.
Sol. Phys., Vol. 80, 3 - 14 (1982).
Limb-darkening measurements made in the visual and ultraviolet show large-scale brightness inhomogeneities. This paper examines (1) the residuals obtained by differencing the observed and computed limb-darkening and (2) the center-to-limb behavior of the inhomogeneities.

071.008 **Solar rotation and variability observed in the Ca II K line.**
S. L. Keil, S. P. Worden.
Bull. American Astron. Soc., Vol. 14, 623 - 624 (1982). Abstract.

071.009 **Photospheric spectrum line asymmetries and wavelength shifts.**
D. Dravins.
Annu. Rev. Astron. Astrophys., Vol. 20, (see 003.009), 61 - 89 (1982).
The discrepancy between solar and laboratory wavelengths. Solar line profile asymmetries. Origin of line asymmetries and shifts. Models of inhomogeneous atmospheres. Stellar line asymmetries and shifts. Outlook.

071.010 **The Compton effect in the chromosphere. III.**
M. Missana.
Astrophys. Space Sci., Vol. 85, 137 - 157 (1982).
It is shown that in the solar spectrum at the centre of the disc there is evidence of a red shift due to generalized Compton

effect and Thomson scattering in the chromosphere. It is also shown that there is no proof of the Einstein effect. The analysis of the emission lines of the limb spectrum confirms this point of view and is in accord with the previous results obtained from the study of the ultraviolet limb spectrum. At last a study of the spectra of β Orionis suggests the possibility that a quite large Compton redshift is present in the spectrum of the star.

071.011 **Interpretation of observations of weak molecular lines on the sun.**
A. P. Sarychev.
Astron. Zh., Tom 59, 985 - 990 (1982). In Russian. English translation in Soviet Astron., Vol. 26, No. 5.

A method of profile interpretation of weak molecular lines on the sun is developed which makes it possible to find on the basis of the depths measured and the line half-widths the mean parameters of the molecular layer: the temperature, the column density, the radial and tangential components of turbulent velocity.

071.012 **An integral supergranular contribution to the global scale photospheric line-of-sight velocity field.**
A. Kubičela, M. Karabin.
Publ. Astron. Obs. Sarajevo, Vol. 1, No. 1, (see 012.043), p. 185 - 189 (1981).

Taking the contribution of an unresolved supergranular line-of-sight velocity of a given sign as proportional to the mean velocity and to the relative area under the same given velocity, and taking into account curvature of the photosphere within a supergranule, a new integral effect of higher contribution of approaching velocities in supergranules has been found. It is not negligible near the limb and must be of certain importance in photospheric global scale velocity and limb effect observations.

071.013 **Ca II K_{2V} spectral features and their relation to small-scale photospheric magnetic fields.**
K. R. Sivaraman, W. C. Livingston.
Sol. Phys., Vol. 80, 227 - 231 (1982).

Although the Ca II K_{232} network is known to be cospatial with magnetic elements there has been doubt as to the magnetic origin of the fainter K_{2V} points. The authors demonstrate that weak magnetic elements also lie at the roots of the K_{2V} points, and because the latter are numerous they may contribute sensibly to the integrated light profile of Ca II K.

071.014 **Empirical NLTE analyses of solar spectral lines. III. Iron lines versus LTE models of the photosphere.**
R. J. Rutten, R. I. Kostik (*Kostyk*).
Astron. Astrophys., Vol. 115, 104 - 114 (1982).

The authors compare observational indications of departures from LTE in solar Fe I lines with published NLTE computations, in the context of discrepancies between empirical LTE and NLTE models of the solar atmosphere. They demonstrate with numerical experiments that the peculiarities of the LTE models are artifacts due to the neglect of NLTE departures. However, they show also that the NLTE formation of most optical metal lines is fortuitously well-mimicked by LTE computation when using LTE models.

071.015 **The solar abundance of tungsten.**
H. Holweger, K. Werner.
Sol. Phys., Vol. 81, 3 - 8 (1982).

Recent measurements of W I oscillator strengths (Obbarius and Kock, 1982) lead to a solar photospheric abundance of tungsten, $\log \epsilon_W = 1.06 \pm 0.15$ on the scale $\log \epsilon_H = 12$. The solar W/Si abundance ratio, 0.32 W atoms/10^6 Si, coincides with that found in carbonaceous chondrites. Implications for solar-system *r*-process abundances are pointed out.

071.016 **Proposed Mg V, Fe XIV and forbidden Fe XVII line identifications in the EUV solar spectrum.**
S. O. Kastner.
Sol. Phys., Vol. 81, 59 - 61 (1982).

Some line identifications in solar disk and limb spectra are proposed on the basis of recent laboratory and theoretical results reported in the literature, including allowed lines of Mg I and Fe XIV in the EUV spectrum and an expected forbidden line of Fe XVII near 1190 Å.

071.017 **The asymmetry of photospheric absorption lines. I. An analysis of mean solar line profiles.**
M. Kaisig, C. J. Durrant.
Astron. Astrophys., Vol. 116, 332 - 340 (1982) = Mitt. Kiepenheuer-Inst. Nr. 220.

The authors investigate the information content of the shifts and asymmetries of mean line profiles and analyse in detail, with the aid of a perturbation analysis and a two-stream model the solar spectrum observed at the centre of the disk. The degree of asymmetry (and shift) is not only a function of the one-point correlation between temperature and velocity – which measures the convective flux – but depends also strongly on two-point correlations. The latter can produce an asymmetry in the absence of heat transport along the line of sight – at the limb, for instance. The authors demonstrate the usefulness of measurements of mean line profiles as a test of convective models by examining the general characteristics of the asymmetry of the Fe I λ557.6 nm line in both the Sun and examples of other main-sequence stars.

071.018 **The design of a photographic system for fine structure observations of the solar photosphere and first observations.**
Y.-x. Fan, Z.-q. Chen, Z.-t. Xu, Q.-d. Wu.
Publ. Purple Mountain Obs., Vol. 1, 31 - 38 (1982). In Chinese.

071.019 **The profiles of solar lines for 1978.**
V. A. Krat, E. K. Kokhan, K. A. Kandaurova.
Izv. Glav. Astron. Obs. Pulkovo, Astrometr. Astrofiz., No. 199, p. 71 - 77 (1982). In Russian.

True profiles and central intensities of lines in the solar Fraunhofer spectrum obtained according to the program of observations of an 11-year cycle of solar activity are given for 1978.

071.020 **Long-period variations of central intensity, half-width and equivalent width of selected Fraunhofer lines.**
E. K. Kokhan, V. A. Krat.
Izv. Glav. Astron. Obs. Pulkovo, Astrofiz. Astrometr., No. 200, p. 3 - 14 (1982). In Russian.

At the Pulkovo Observatory the unique data on Fraunhofer lines have been obtained with the double-pass spectrophotometer during a solar activity cycle. The material has been obtained in non-excited regions of the solar center and at 1'.6, 0'.6 distances from the west limb. Using the observed profile,the central intensity, half-width and equivalent width are evaluated. Tables and graphs are given, characterizing the variation of these parameters from the solar center to limb in dependence on the time of observation.

Atlas of South Pole IR solar spectra.
See Abstr. 002.127.

New lifetime measurements for Nb I and Rh I and the solar photospheric abundances of Nb and Rh.
See Abstr. 022.014.

Absolute oscillator strengths for lines of neutral cobalt between 2276 Å and 9357 Å and a redetermination of

the solar cobalt abundance.
See Abstr. 022.042.

Oscillator strengths for Y I and Y II and the solar abundance of yttrium.
See Abstr. 022.106.

Active regions from the photosphere to the chromosphere.
See Abstr. 072.031.

The Hanle effect and the diagnostics of turbulent magnetic fields in the solar atmosphere.
See Abstr. 075.017.

Terrestrial O_2 lines used as wavelength references: comparison of measurements and model computations.
See Abstr. 082.054.

Erratum

071.901 **Erratum: "Solar luminosity variation. IV. The photospheric lines, 1976–1980"** [Astrophys. J., Vol. 252, 375 - 385 (1982)].
W. Livingston, H. Holweger.
Astrophys. J., Vol. 258, 904 (1982). – See Abstr. 31.071.001.

072 Sunspots, Faculae, Activity Cycles, Solar Patrol

072.001 **On the velocity dispersions of single sunspots.**
R. E. White, III.
Astron. J., Vol. 87, 1073 - 1075 (1982).
Analysis of Greenwich sunspot data (from the years 1915 - 1964) indicates that the velocity dispersions of single sunspots are much greater for spot areas less than $\sim 2 \times 10^8$ km^2 than for larger areas. The velocity dispersions are asymmetric in that the maximum longitudinal velocity dispersion of 1 deg/day is twice as large as the maximum latitudinal velocity dispersion. No latitude dependence is observed for these results.

072.002 **Correlation between active solar regions according to Hα observations on May 25, 28 and 29, 1980.**
M. B. Ogir'.
Soln. Dannye 1982 Byull., No. 3, p. 91 - 96 (1982). In Russian.
The correlation between flare activity in the regions 16862 - 16864 with the activity in other solar regions has been considered. The Hα solar disc movies were obtained at the Crimean Astrophysical Observatory. It was found that the flares appeared either simultaneously in different solar regions or in the maximum phase of the preceding flare in the same sunspot group or elsewhere.

072.003 **On the problem of prediction of the maxima of solar activity using polar fields on the sun.**
E. E. Benevolenskaya.
Soln. Dannye 1982 Byull., No. 3, p. 108 - 110 (1982). In Russian.
On the basis of an investigation of the relationships between polar magnetic fields on the sun during the minimum of solar activity and intensity of the cycles it is shown that the surface polar field of the sun during the minimum of solar activity is not a physically meaningful parameter for prediction the maximum of the following cycles.

072.004 **On the relation between the dynamics of filament bands and Hale's and Spoerer's laws.**
V. I. Makarov.
Soln. Dannye 1982 Byull., No. 4, p. 89 - 94 (1982). In Russian.
From an analysis of synoptic maps of the large-scale magnetic field of the sun it is shown that the formation of active regions is associated with the background field, which resulted from the evolution of active regions of the previous solar cycle.

072.005 **Some peculiarities of the development of active regions associated with proton events.**
N. K. Pereyaslova, M. N. Nazarova, I. E. Petrenko, L. A. Kolesnikova.
Soln. Dannye 1982 Byull., No. 4, p. 94 - 97 (1982). In Russian.
Evolution characteristics of active regions in association with parameters of proton events are investigated. A relationship is found between the values of the total and maximum flux and the proton spectrum in solar cosmic ray events. A correlation between the intensity of maximum proton flux and the variation of the active region magnetic field strength is established.

072.006 **Forecast of solar cycle No. 22.**
V. F. Chistyakov.
Soln. Dannye 1982 Byull., No. 4, p. 97 - 101 (1982). In Russian.

072.007 **Sunspot numbers and solar cycles.**
S. G. Wallenhorst.
Sky Telesc., Vol. 64, 234 - 236 (1982).

072.008 **Quasi-stationary primary magnetic field of the sun and intensity variations of the solar cycle.**
M. I. Pudovkin, E. E. Benevolenskaya.
Pis'ma Astron. Zh., Tom 8, 506 - 509 (1982). In Russian. English translation in Soviet Astron. Lett., Vol. 8.
The behaviour of solar activity cycles is studied in the framework of Leighton's model assuming the presence of a primary constant non-decaying poloidal magnetic field. It is shown that the primary poloidal field significantly effects the behaviour of solar activity cycles, leading to an alternation of more or less intense cycles.

072.009 **Smoothing of the Zürich series of monthly numbers by Whittaker's formula.**
L. V. Zhukov, Yu. S. Muzalevskij.
Researches on stellar physics, 1981, (see 31.003.010), p. 60 - 80 (1981). In Russian. – Abstr. in Ref. zh., 51. Astron., 7.51.410 (1982).

072.010 **Sunspot geometry and pressure balance.**
R. G. Giovanelli.
Sol. Phys., Vol. 80, 21 - 31 (1982).
Observations and analysis show that, beyond the penumbra, the field from a sunspot forms an almost horizontal magnetic canopy, the average height of the canopy base being about 180 km at the outer penumral edge. An estimate of the penumbral thickness, 250 km, puts the upper penumbral surface about 430 km above $\tau^{\odot} = 1$ $(z = 0)$ in the photosphere. Because of the curvature of the penumbral surface, the Wilson depression may be obtained only from limb-side measurements. Chistyakov's (1962) measurements lead to a value of 1040 km for the overall average Wilson depression for regular sunspots, the corresponding umbral level $\tau^{*} = 1$ being $z = -610$ km. The average shape of the penumbra in the sunspots is shown. Full pressure balance between internal and external gas pressures and the magnetic pressure on the sunspot axis is achieved within the uncertainties of measurement. Hα superpenumbral fibrils are inclined at angles of typically 13° to the horizontal near the outer penumbral edge (Beckers, 1962). This is fully compatible with an almost horizontal magnetic canopy base if the height of the Hα fibrils there is ~ 1400 km. The geometry of the resulting sunspot model is summarized.

072.011 **Multiple wavelength observations of a solar active region.**
F. Chiuderi-Drago, R. Bandiera, R. Falciani, E. Antonucci, K. R. Lang, R. F. Willson, K. Shibasaki, C. Slottje.
Sol. Phys., Vol. 80, 71 - 85 (1982).
The Solar Maximum Mission Satellite, the Sacramento Peak Vacuum Tower Telescope, the Very Large Array and the Westerbork Synthesis Radio Telescope have been used to observe active region AR 2490 on two consecutive days at soft X-ray, ultraviolet, optical and radio wavelengths (2, 6, and 20 cm), with comparable angular resolution (2″ to 15″) and field of view (4′ × 4′). The authors interpret the observations in terms of a two component model in which one component is associated with bright Hα plage, UV, and X-ray enhanced emission and the other component is associated only with sunspots.

072.012 **The sun in 1980.**
B. E. Stonehouse.
South. Stars, Vol. 29, 143 - 144 (1982).

072.013 **Sur quelques observations du plus récent maximum d'activité solaire.**
D. Yvergneaux.
Ciel Terre, Vol. 98, 229 - 235 (1982).
The author gives some comments about observations obtained around the recent maximum solar cycle and he shows some photographs of sunspots.

072.014 **Five-minute oscillations as a subsurface probe of sunspot structure.**
J. H. Thomas, L. E. Cram, A. H. Nye.
News Lett. Astron. Soc. N.Y., Vol. 2, No. 2, p. 25 (1982). Abstract.

072.015 **The Fraunhofer spectrum of the McMath 15526 active region in the stage of flare activity.**
N. N. Kondrashova, P. N. Polupan, K. V. Alikaeva, T. I. Redyuk, S. I. Gandzha.
Vestn. Kiev. Univ., Astron. Vyp. 24, p. 18 - 29 (1982). In Russian.

072.016 **On the variation of spot number in a group with time.**
U. M. Lejko.
Vestn. Kiev. Univ., Astron. Vyp. 24, p. 56 - 60 (1982). In Russian.

072.017 **Measurements of the limb darkening of faculae near the solar limb.**
G. A. Chapman, D. P. Klabunde.
Astrophys. J., Vol. 261, 387 - 395 (1982).
Measurements of the contrast at 525 nm of solar faculae are presented. These measurements were obtained with an extreme limb photometer at the San Fernando Observatory. The average value of the mean contrast of facular regions is 2.1%. This mean contrast, when corrected for a filling factor of 10–15% implies a contrast for individual faculae of 30–50%. Implications of these results for optical measurements of the solar oblateness are discussed.

072.018 **Observations of sunspot bright rings and comparisons with a convective blocking model.**
L. A. Fowler, P. Foukal.
Bull. American Astron. Soc., Vol. 14, 624 (1982). – Abstract.

072.019 **The sun as a star: correlations between Ca II K line shape parameters and other measures of solar activity.**
R. D. Holt, H. H. Stroke, P. Lonergan.
Bull. American Astron. Soc., Vol. 14, 639 - 640 (1982). Abstract.

072.020 **Death of a sunspot.**
K. P. Topka, S. G. Wallenhorst.
Bull. American Astron. Soc., Vol. 14, 640 (1982). – Abstract.

072.021 **Theory of the solar cycle. An example of a celestial dynamo process.**
H. Yoshimura.
Magnetospheric phenomena of celestial bodies, (see 012.011), p. 147 - 152 (1982).

072.022 **The occurrence of sunspot groups and the velocity of solar rotation at heliographic latitudes ⩾ 40°.**
M. Kopecký.
Bull. Astron. Inst. Czechoslovakia, Vol. 33, 285 - 293 (1982).
Based on the Greenwich Photoheliographic Results a catalogue has been compiled of sunspot groups which had an average heliographic latitude $B \geqslant 40°$ in the years 1874 - 1976. This catalogue is used to study some patterns of occurrence of high-latitude sunspot groups and the velocity of the solar rotation at these high heliographic latitudes.

072.023 **A magnetogasdynamical analysis of the shock transition model of the solar cycle.**
M. H. Gokhale, P. Venkatakrishnan.
Bull. Astron. Soc. India, Vol. 10, 35 (1982). – Abstract.

072.024 **Behaviour of some neutral molecular absorption lines in facula models.**
B. M. Tripathi, K. Sinha, M. C. Pande.
Bull. Astron. Soc. India, Vol. 10, 35 (1982). – Abstract.

072.025 **Carbon monoxide and facula models.**
B. M. Tripathi, M. C. Pande, K. Sinha.
Bull. Astron. Soc. India, Vol. 10, 35 - 36 (1982). – Abstract.

072.026 **Submillimetre-wave brightness temperatures of solar active zones and sunspots measured from a balloon-borne platform.**
F. Cartier, F. K. Kneubühl, D. Huguenin, E. A. Müller.
The scientific importance of submillimetre observations, (see 012.030), p. 175 - 179 (1982).
Submillimeter-wave brightness temperatures of the quiet sun, sunspot regions and solar active zones were measured with the aid of balloon-borne 60 cm-diameter Cassegrain telescope from a height of 38 km on May 8, 1981.

072.027 **Diode laser heterodyne observations of silicon monoxide in sunspots.**
D. A. Glenar, D. Deming, D. E. Jennings, T. Kostiuk, M. J. Mumma.
Bull. American Astron. Soc., Vol. 14, 776 (1982). – Abstract.

072.028 **Fluid motions in the solar chromosphere-corona transition region. II. Active region flows in C IV from narrow slit Dopplergrams.**
R. G. Athay, J. B. Gurman, W. Henze, R. A. Shine.
Astrophys. J., Vol. 261, 684 - 699 (1982).
From a study of Dopplergrams made with two narrow slits in the opposite wings of the C IV line, λ1548, the authors find widespread evidence for steady flow patterns associated with large sunspots and, on a larger scale, with active region magnetic field patterns. The characteristic sunspot flow is in the reverse Evershed sense with a substantial vertical component.

072.029 **On the magnetic and thermodynamic consequences of the return-flux sunspot model.**
T. Flå, V. A. Osherovich, A. Skumanich.
Astrophys. J., Vol. 261, 700 - 709 (1982).
The authors have applied the return-flux model to a sunspot with typical observational data for pressure, maximum field strength, and size. The basic nonlinear force-balance equation was solved to obtain self-consistent magnetic field, pressure, and temperature distributions. The resulting distributions appear to yield improved representations of sunspots, including a natural representation of umbra-penumbra and penumbra-quiet Sun boundaries.

072.030 **The overall structure and evolution of active regions.**
N. R. Sheeley, Jr.
Solar active regions, (see 003.011), p. 17 - 42 (1981).
Contents: Introduction. Background. Surface fields. Sunspots and faculae. Chromospheric tracers. The transition region and lower corona. Epilogue.

072.031 **Active regions from the photosphere to the chromosphere.**
G. A. Chapman.
Solar active regions, (see 003.011), p. 43 - 82 (1981).
In the paper there are discussed solar plage models, faculae, filigree and flux tubes multidimensional radiative transfer, radiative cooling of the quiet sun and plage chromospheres,upper chromosphere and the transition zone.

072.032 **The dynamics of active regions.**
G. E. Brueckner.
Solar active regions, (see 003.011), p. 113 - 127 (1981).
The paper summarizes observations of the dynamics of active regions obtained by the Naval Research Laboratory's

High Resolution Telescope and Spectrograph during two rocket flights in 1975 and 1978. The general appearance of the spectra is discussed as well as chromospheric and transition-zone flows in sunspots and plages. Coronal flow patterns above an active region are given. Possible consequences of the observed flow pattern on the formation of spectral lines are discussed. The behavior of the He II 1640 Å line is described. Finally the author tries to evaluate qualitatively the mass and energy balance of an active region.

072.033 **On rapid variations of the contrast of bright features in a sunspot penumbra.**
L. D. Parfinenko.
Soln. Dannye 1982 Byull., No. 5, 86 - 92 (1982). In Russian.
It has been found from observations of fine structure in sunspot penumbrae that small bright long-lived features (size < 1″) can be often detected in a region with a strong gradient of the magnetic field. These features have a rapidly alternating impuls character of brightness variation which can last several hours. The contrast of bright features can increase to 30% in several minutes, decreasing very quickly afterwards. It is suggested that these bright features can be potential origins of solar flares.

072.034 **On the problem of measurement of the maximum intensities of magnetic fields in sunspots during the investigation of their rapid variations.**
G. F. Vyal'shin, I. V. Yudina.
Soln. Dannye 1982 Byull., No. 5, p. 97 - 101 (1982). In Russian.
The results of a comparison of the variations of maximum intensities of magnetic fields in five sunspots with variations in quality and oscillation of the solar image are given. It is shown that there are rapid intensity variations in the studied sunspots, the detected variations in the sunspot magnetic field are real.

072.035 **Active region magnetic fields inferred from simultaneous VLA microwave maps, X-ray spectroheliograms, and magnetograms.**
E. J. Schmahl, M. R. Kundu, K. T. Strong, R. D. Bentley, J. B. Smith, Jr., K. R. Krall.
Sol. Phys., Vol. 80, 233 - 249 (1982).
The authors compare for the first time simultaneous 6 cm maps from the Very Large Array and X-ray spectroheliograms from the X-ray Polychromator on the Solar Maximum Mission. In combination with photospheric magnetograms and magnetic field calculations, these microwave and X-ray observations provide the best evidence yet obtained for the identification of the generating mechanism of radio emission from active regions at short centimeter wavelengths.

072.036 **The structure of sunspots. IV. Magnetic field strengths in small sunspots and pores.**
J. J. Brants, C. Zwaan.
Sol. Phys., Vol. 80, 251 - 258 (1982).
Magnetic field strengths in small umbrae and pores are measured using the line Ti I λ6064.6 Å, which is formed purely in umbrae. The authors find field strengths between 1900 and 2600 G in the darkest parts of small umbrae and of well established pores. The field strength in diffuse transient 'protopores' amounts to 1500 ± 250 G. They demonstrate that usage of the well-known magnetic line λ6173.3 Å and other Fe I lines yield systematically smaller magnetic field strengths than Ti I λ6064.6 Å.

072.037 **On long-term periodicities in the sunspot cycle.**
S. G. Wallenhorst.
Sol. Phys., Vol. 80, 379 - 384 (1982).
The author investigates the possible existence of periodicities other than that of 11 years in the mean annual sunspot number. The power spectrum for the sunspot numbers for the years 1711–1966 is compared with power spectra for the same period in which the solar cycles have been permuted randomly. Peaks in the power spectra at periodicities other than 11 years are found to appear and disappear in a random way. He concludes that the 11 year periodicity is the only statistically significant one present in the sunspot data.

072.038 **Angular velocity of sunspots along the butterfly diagram.**
G. Godoli, F. Mazzucconi.
Astron. Astrophys., Vol. 116, 188 - 189 (1982).
Sunspots in the left part of the butterfly diagram (i.e. to the left of the line representing the average latitudes of sunspots) tend to have greater angular velocities than those in the right part. This fact seems to indicate the possibility that sunspots respond to the torsional oscillations of the unmagnetized plasma found by Howard and LaBonte (1980).

072.039 **Size variations in regular sunspots.**
R. D. Robinson, D. C. Boice.
Sol. Phys., Vol. 81, 25 - 31 (1982).
A study was made of the time development for 31 simple sunspots. The growth and decay rates varied between spots, but were approximately constant during the lifetime of individual, relatively short-lived spots. Long-lived spots showed oscillations in umbral size, with periods of three to five days. Occasionally, rapid changes in the spot size were observed.

072.040 **On the disappearance of a small sunspot group.**
S. G. Wallenhorst, K. P. Topka.
Sol. Phys., Vol. 81, 33 - 46 (1982).
The small sunspot group associated with Hale active region 17694 was observed jointly by the Mount Wilson and the Big Bear Solar Observatories, through the time of sunspot disappearance. The magnetic flux from the region was seen to decrease by about 10% per day during the observing interval. This was accompanied by fragmentation of the dominant spot as a supergranule network formed. No evidence for spreading or diffusion of the active region field was found.

072.041 **An emission measure analysis of two sunspots observed by the UVSP instrument on the SMM spacecraft.**
A. E. Kingston, J. G. Doyle, P. L. Dufton, J. B. Gurman.
Sol. Phys., Vol. 81, 47 - 58 (1982).
The EUV observations from the SMM satellite of two sunspots are presented. These observations show the sunspots (a) to be regions of lower intensity than the surrounding plage, contrary to that found by previous authors, and (b) to have line intensities which vary little over a period of several hours. An upper limit to mass flows of 2 km s^{-1} is derived. Electron densities derived from N IV to C IV line ratios imply electron pressures ($\log N_e T_e$) of 15.0 to 15.3.

072.042 **Analysis of the high resolution Mg XI X-ray spectra. III. Non-thermal interpretation of some spectra.**
M. Siarkowski, J. Sylwester, G. Bromboszcs, V. V. Korneev, S. L. Mandelshtam (*Mandel'shtam*), S. N. Oparin, A. M. Urnov, I. A. Zhitnik.
Sol. Phys., Vol. 81, 63 - 68 (1982).
This paper contains the analysis of the INTERCOSMOS 16 ADP spectra. It is shown that by assuming the existence of a small admixture (1%) of non-thermal electrons in the active-region plasma it is possible to improve the agreement between measured and calculated fluxes for some spectra. The analysis follows the suggestion contained in the paper by Karev et al. (1980).

072.043 **Provisional sunspot numbers for June - November 1982.**
Yamamoto Circ., Nos. 1978, 1981, 1982, 1984, 1985 (1982).

072.044 **Two Cr II multiplets around 1430 Å appearing in absorption in the spectrum of a solar active region.**
S. Johansson.
Mon. Not. R. Astron. Soc., Vol. 201, 63P - 67P (1982).

In a current analysis of Cr II two new multiplets have been identified around 1430 Å. The identification of 27 lines establishes the new terms $y\,^6F^0$ and $y\,^6D^0$ of the $3d^3 4s 4p$ configuration by their combinations with $3d^4 4s\, a^6D$. Most of the lines appear in absorption in the spectrum of a solar active region.

072.045 **Carbon monoxide and facula models.**
B. M. Tripathi, M. C. Pande, K. Sinha.
Bull. Astron. Soc. India, Vol. 10, 150 - 153 (1982).

Equivalent widths for a few lines of the first overtone band of carbon monoxide have been calculated in the solar infrared spectrum for five facular and two photospheric models. The calculations show a decrease in equivalent widths in the facular models of Stellmacher and Wiehr (1973), Stenflo (1975) and Chapman (1979) and an increase in the models of Schmahl (1967) and Shine and Linsky (1974). The models are also discussed in the light of Polonskij's (1966) and Tsuji's (1977) observations.

072.046 **Phase difference between the two components of the solar cycle.**
M. H. Gokhale, K. R. Sivaraman.
Bull. Astron. Soc. India, Vol. 10, 154 - 158 (1982).

It is confirmed that near each solar minimum the latitude zone over which the small sunspot groups are distributed expands a few years before the expansion of the latitude zone of the large spot groups takes place. It is found that this expansion of the latitude zone of the small spot groups also preceeds the increase in the mean latitude of the small spot groups. These observations indicate that the thinner flux tubes of each fresh solar cycle start emerging earlier than the thicker flux tubes of the same cycle.

072.047 **Catalogue of solar activity for the year 1977.**
R. S. Gnevysheva.
Akademiya Nauk SSSR, Glavnaya Astronomicheskaya Observatoriya. Nauka, Leningradskoe Otdelenie, Leningrad. 88 pp. Price 60 Kop. (1981). In Russian.

072.048 **Catalogue of solar activity for the year 1978.**
R. S. Gnevysheva.
Akademiya Nauk SSSR, Glavnaya Astronomicheskaya Observatoriya. Nauka, Leningradskoe Otdelenie, Leningrad. 148 pp. Price 1 Rbl. (1982). In Russian.

072.049 **Die Sonnenaktivität im zweiten Quartal 1982.**
R. Beck.
Sterne Weltraum, 21. Jahrg., 474 - 475 (1982).

072.050 **Solar activity during 1978.**
K. J. Medway.
J. British Astron. Assoc., Vol. 93, 29 - 31 (1982).

072.051 **Magnetic field development in an isolated active region (McMath No. 13 736).**
V. Bumba, M. Klvaňa, P. Tomášek.
Bull. Astron. Inst. Czechoslovakia, Vol. 33, 321 - 331 (1982).

Magnetic field development in an isolated active region is investigated in its relation to the dynamics of the background magnetic field. Formation of a secondary active region during the postmaximum phase of the development of the main active region is considered to be a natural phase of the evolution of the main active region's magnetic field. The studied active region represents the final product of a secondary active longitude.

072.052 **Irregular rotation of the main sunspot in active region Hale 17 570 of 5–13 April 1981.**
A. Kučera.
Bull. Astron. Inst. Czechoslovakia, Vol. 33, 345 - 349, plates 3, 4 (1982).

The irregular rotation of the main sunspot in the active region Hale 17 570 was investigated on 5–13 April 1981. The characteristics of this motion indicate that damped oscillatory motion was involved. No correlation was found between the characteristics of the irregular rotation and flare activity.

072.053 **Observations solaires. Rotations 1703 - 1716, 17 décembre 1980 - 2 janvier 1982.**
E. Ţifrea, V. Dinulescu, A. Dimitriu, S. Dinulescu, G. Mariş, A. Óncica.
Cent. Astron. Sci. Spat., Acad. Repub. Socialiste România, Bucarest. 96 pp. Price Lei 7.50 (1982).

072.054 **Solare Beobachtungsergebnisse. Solar Data.**
Solar Radio Emission.1982 January - April.
HHI Sol. Data, Vol. 33, 1 - 59 (1982).

072.055 **Magnetic polarities and maximum field strengths of one selected sunspot group with time distances of about thirty minutes during the period 1979 August 13 - August 16, 1980 September 2 - September 4.**
H. Künzel.
HHI Suppl. Ser. Sol. Data, Vol. 5, 27 - 71 (1982).

072.056 **Solar observations made at Catania Astrophysical Observatory during 1980.**
C. D'Arrigo, R. A. Zappalà, F. Zuccarello, E. Catinoto, G. Domina, S. Sciuto, G. Celeani, G. Sapienza, A. Del Bufalo.
Oss. Astrofis. Catania, Pubbl. N. 166, 117 pp. (1981).

Contents: Sunspots, Hα and K faculae, Hα flares, Hα quiescent prominences, Hα active prominences on disc and at limb, Hα disc and limb patrol hours.

072.057 **Solar activity in the years 1969 - 1973.**
A. Özgüç.
Kandilli Obs., Heliophys. Serv., Istanbul. Publ., Second Ser., No. 1, 36 pp. (1976).

072.058 **Solar activity in the years 1974 - 1977.**
T. Tekman, L. Altaş.
Kandilli Obs., Heliophys. Serv., Istanbul. Publ., Second Ser., No. 4, 29 pp. (1982).

072.059 **Solar activity in the years 1978 - 1979.**
T. Ataç, S. Vatandaslar.
Kandilli Obs., Heliophys. Serv., Istanbul. Publ., Second Ser., No. 5, 21 pp. (1982).

072.060 **Solar maps and activity.** 1982 March - September.
F. J. Heyden, V. L. Badillo.
Manila Obs., Sol. Div.

072.061 **Actividad solar en 1980.** I. Números relativos de Wolf; II. Estadística de manchas y superficie de las mismas.
M. Lopez Arroyo.
Bol. Astron. Obs. Madrid, Vol. 10, No. 5, p. 1 - 74 (1980).

072.062 **On the origin of multipolar sunspots.**
E. F. Shaposhnikova.
Astron. Tsirk., No. 1173, p. 2 - 4 (1981). In Russian.

072.063 **Cartes synoptiques de la chromosphère solaire et catalogues des filaments et des centres d'activité.**
M.-J. Martres, G. Zlicaric.
Obs. Paris, Sect. Astrophys. Meudon, Vol. 6, Fasc. 4,

Années 1980 - 1981 (1982).
Rotations Nos. 1690 à 1716 du 27 décembre 1979 au 2 janvier 1982.

072.064 **Observations of sunspots, faculae and prominences.**
K. Fujimori.
Tokyo Astron. Obs. Rep., Vol. 19, 578 - 589 (1981). In Japanese.

072.065 **Mt. Purple No. 496 sunspot group and rare-founded Hα flare loop.**
Q.-r. Su, F.-m. Hu, T. Cao.
Publ. Purple Mountain Obs., Vol. 1, 17 - 24 (1982). In Chinese.

072.066 **Potential models of the unipolar sunspot magnetic field.**
V. V. Denisenko, V. A. Kotov, V. A. Romanov, V. S. Sokolov.
Sol. Phys., Vol. 81, 217 - 229 (1982).
The potential models of the unipolar sunspot magnetic field are calculated on the basis of magnetographic measurements of the magnetic field made in the three spectral lines of different intensities, Hα, Ca I λ6103 and Fe I λ4808. The computed distributions of the magnetic field vector are compared with actual distributions observed at these three levels.

072.067 **Observations of the longitudinal magnetic field in the transition region and photosphere of a sunspot.**
W. Henze, Jr., E. Tandberg-Hanssen, M. J. Hagyard, B. E. Woodgate, R. A. Shine, J. M. Beckers, M. Bruner, J. B. Gurman, C. L. Hyder, E. A. West.
Sol. Phys., Vol. 81, 231 - 244 (1982).
The Ultraviolet Spectrometer and Polarimeter on the Solar Maximum Mission spacecraft has observed for the first time the longitudinal component of the magnetic field by means of the Zeeman effect in the transition region above a sunspot. The data were obtained on three days in one sunspot, have spatial resolutions of 10 arc sec and 3 arc sec, and yield maximum field strengths greater than 1000 G above the umbrae in the spot. The method of analysis is described in some detail in an appendix. The transition region data for one day are compared with photospheric magnetograms from the Marshall Space Flight Center. Vertical gradients of the magnetic field are computed from the two sets of data.

072.068 **On the dynamics of the chromosphere above sunspots.**
Yu. D. Zhugzhda, V. I. Makarov.
Sol. Phys., Vol. 81, 245 - 251 (1982).
The brightness oscillations of a sunspot umbra in the Hα and Ca^+ K lines are studied. The observational results are explained in terms of the resonance theory of slow-mode magnetohydrodynamic waves in the sunspot chromosphere. The thickness of the chromosphere above a sunspot varies quasi-periodically from 420 km to 1000 km.

072.069 **High resolution EUV structure of the chromosphere-corona transition region above a sunspot.**
K. R. Nicolas, O. Kjeldseth-Moe, J.-D. F. Bartoe, G. E. Brueckner.
Sol. Phys., Vol. 81, 253 - 280 (1982).
High resolution simultaneous observations (1" and 0.05 Å) of ion emission line intensities and wavelength shifts between 1170 and 1700 Å have illuminated the flow and temperature structure of the transition region above a sunspot.

072.070 **On the sunspot distribution with respect to coronal holes.**
N. N. Petrova.
Soln. Dannye 1982 Byull., No. 7, p. 96 - 100 (1982). In Russian.
Using the superposed epoch method the distribution of sunspot areas in the central zone of the solar disc with respect to the reference date – the central meridian passage by a coronal hole (CMP) was obtained. The distribution turned out to be statistically nonhomogeneous and has the minimum a day after and the maximum 4 days before the hole CMP.

072.071 **On weekly variations in solar activity.**
R. M. Arslanova, E. S. Vernova, N. G. Ptitsyna, M. I. Tyasto.
Soln. Dannye 1982 Byull., No. 7, p. 108 - 110 (1982). In Russian.
The periodic structure of the solar activity from 2 to 12 days during 1972-1978 is considered. The results testify a stable wave of changes in total sunspot areas with period (T) of 7-8 days during long-term intervals on the descending branch of solar cycle N 20 (1972 - 1973) and on the ascending branch of solar cycle N 21 (1977 - 1978).

072.072 **On the lifetime of sunspot groups during the maximum epoch of an 11-year solar cycle.**
B. O. Berlyand.
Soln. Dannye 1982 Byull., No. 8, p. 96 - 99 (1982). In Russian.
The empirical distribution of sunspot group lifetimes T in the maximum of the 20th solar cycle (1967 - 1969) is obtained. It's a nonmonotonous distribution with a statistically significant minimum at T = 5 days.

072.073 **Daily maps of the sun and magnetic fields of sunspots.**
Soln. Dannye 1982 Byull., No. 3, p. 1 - 90; No. 4, p. 1 - 76; No. 5, p. 1 - 71; No. 6, p. 1 - 111; No. 7, p. 1 - 95; No. 8, p. 1 - 79 (1982). In Russian.

072.074 **The birth and evolution of sunspots: observations.**
P. S. McIntosh.
The physics of sunspots, (see 012.063), p. 7 - 54 (1981).
The structure, dimensions and dynamics of sunspots and sunspot groups are reviewed as they are observed in white light.

072.075 **On the dissolution of sunspot groups.**
S. G. Wallenhorst, R. Howard.
The physics of sunspots, (see 012.063), p. 55 - 58 (1981).
The behavior of magnetic fluxes from active regions is investigated for times near sunspot disappearance. It is found that the magnetic fluxes decrease on or near the date the spot vanishes. The authors investigate this effect and conclude that it is due to changes in the field, rather than through dissipation of the active region fields. This is important in considerations of the large-scale behavior of solar magnetic fields.

072.076 **An emerging active region: some preliminary results.**
J. J. Brants, L. E. Cram, C. Zwaan.
The physics of sunspots, (see 012.063), p. 60 - 63 (1981).

072.077 **Analysis of the vector magnetic fields of complex sunspots.**
S. R. Patty.
The physics of sunspots, (see 012.063), p. 64 - 74 (1981).
The vector magnetic field structures of the complex sunspot regions 2469 and 2470, May 1980, were analyzed. Both regions exhibited areas of magnetic classification δ during their passage across the solar disk. An investigation of the vector magnetic field characteristics of each complex region in the area of the δ showed a relationship between the longitudinal gradient, the amount of shear in and strength of the transverse component, and the activity of the region. Signatures for δ regions are reported.

072.078 **Magnetic features in active regions.**
J. K. Lawrence, S. C. Thorman.

The physics of sunspots, (see 012.063), 75 - 79 (1981).

A program is described for studying the evolution of sunspot groups by observing the motions of small magnetic features in evolving active regions. The magnetic features are identified with small (≲1.5") bright points in magnesium filtergrams. The filtergrams are made at intervals of 15s to 2m for several hours on successive days during the evolution of sunspot groups and combined with polarity information supplied by daily magnetograms. Results are presented for a brief observing period on 22 June 1980. These indicate life-time of the bright features of 1 - 1.5 hr and transverse velocities up to ~0.5 km/s.

072.079 **Theories of sunspot structure and evolution**
P. R. Wilson.
The physics of sunspots, (see 012.063), p. 83 - 97 (1981).

072.080 **A cluster model for sunspots.**
H. C. Spruit.
The physics of sunspots, (see 012.063), p. 98 - 103 (1981).

The aim of the work reported here is to construct a sunspot model consisting of a bundle of thin flux tubes, which at the surface are merged into a homogeneous field, but at a greater depth are separated by field-free fluid. The author demands from the model that it explains the variation of B across the spot, and the variation of the umbral (central) field strength with the size of the spot.

072.081 **On the magnetic field topology in the return-flux sunspot model.**
A. Skumanich, V. A. Osherovich.
The physics of sunspots, (see 012.063), p. 104 - 114 (1981).

The return-flux sunspot model of Osherovich (1981) is discussed and applied to the determination of the magnetic structure for a long-lived sunspot from an observationally derived pressure deficit. The analysis yields the identification of the magnetic surface which separates the set of 'open' field lines from the set that return to the surface with the umbral-penumbral boundary. The basic non-linear differential equation, which represents a special case of the exact force-balance equation, for the return-flux model is solved. The choice of the depth of the $\tau_5 = 1$ level relative to the quiet sun photosphere, i.e. the Wilson depression, was determined from a self-consistency requirement.

072.082 **Empirical sunspot models: statement of the problem.**
C. Zwaan.
The physics of sunspots, (see 012.063), p. 123 - 126 (1981).

072.083 **Constraints on umbral core models as derived from broad-band intensity observations.**
F. Albregtsen, P. Maltby.
The physics of sunspots, (see 012.063), p. 127 - 139 (1981).

The umbra/photosphere intensity ratio for sunspots with umbral radii larger than 5 sec of arc is discussed. The intensity ratio is derived for a "typical", large sunspot in the wavelength region 0.387 – 3.8 μm. The data are used in constructing a new, one-component model for the umbral core.

072.084 **Energy transport in deep umbral layers.**
A. A. van Ballegooijen.
The physics of sunspots, (see 012.063), p. 140 - 141 (1981).
Abstract.

072.085 **Model of the chromosphere over a dark umbral feature.**
H. S. Yun, H. A. Beebe, W. E. Baggett.
The physics of sunspots, (see 012.063), p. 142 - 151 (1981).

Spectra of several strong lines of sunspots have been obtained. With a variety of model atmospheres, CaII H, K and λ8498 line profiles were calculated using a non-LTE line formation procedure. The authors examined only a specific part of the sunspot umbra which is thought to be coolest over the spot. An optimum model representing such a region is presented and its physical properties are discussed.

072.086 **An empirical view of the chromospheric temperature structure above a sunspot umbra.**
B. W. Lites, A. Skumanich.
The physics of sunspots, (see 012.063), p. 152 - 166 (1981).

The authors outline the reasoning that led them to choose a model of a sunspot umbra with a shallow temperature gradient, rather than the steeper temperature gradient that has been suggested by previous models. The observed line ratios lead to this conclusion. By adopting this type of chromosphere, the authors compromise the agreement of the computed and observed line profile shapes. The observed Lα and Lβ intensities constrain the pressure in the upper chromosphere and transition region to values somewhat lower than the average quiet sun coronal pressure.

072.087 **The sunspot chromosphere-corona transition region.**
K. R. Nicolas, O. Kjeldseth-Moe, J.-D. F. Bartoe, G. E. Brueckner.
The physics of sunspots, (see 012.063), p. 167 - 190 (1981).

Rocket observations are reported for a sunspot at high spatial (1") and spectral (0.06 Å) resolution in the extreme ultraviolet (EUV) wavelength region between 1170 and 1715 Å. They provide simultaneous coverage over this spectral range. Simultaneous observation of all spectral lines from the chromosphere to the corona is imperative, both to determine the true dynamics of the plasma and to compare with detailed models.

072.088 **Comments on the thermodynamic structure and dynamics of the cool solar corona over sunspots.**
P. Foukal.
The physics of sunspots, (see 012.063), p. 191 - 209 (1981).

The author summarizes the physically most interesting results on the structure of the sunspot atmosphere obtained from continuing analysis of EUV sunspot observations made with the SO55 spectrophotometer on ATM during 1973 and 1974.

072.089 **Magnetic field strengths in umbrae and pores.**
C. Zwaan, J. J. Brants.
The physics of sunspots, (see 012.063), p. 210 - 212 (1981).

072.090 **The photospheric vector magnetic field of a sunspot and its vertical gradient.**
M. J. Hagyard, E. A. West, E. Tandberg-Hanssen, J. E. Smith, W. Henze, Jr., J. M. Beckers, E. C. Bruner, C. L. Hyder, J. B. Gurman, R. A. Shine, B. E. Woodgate.
The physics of sunspots, (see 012.063), p. 213 - 234 (1981).

On October 23, 1980, observations were made at the Marshall Space Flight Center Observatory to obtain quantitative measurements of the vector magnetic field of a sunspot at the photospheric level. A coordinated observing program to simultaneously measure the line-of-sight component of this sunspot field in the transition region was carried out with the Ultraviolet Spectrometer and Polarimeter instrument on the Solar Maximum Mission satellite. The authors describe the results of these coordinated observations which include: the derived vertical gradients of the sunspot's line-of-sight field and their distribution across the sunspot; the photospheric vector field distribution throughout the sunspot; the inclination of the photospheric field to the line-of-sight and its variations across the sunspot; and magnitudes and gradients of the photospheric field across penumbral-photospheric boundaries.

072.091 **Reference model atmosphere calculation: the Sunspot sunspot model.**
E. H. Avrett.
The physics of sunspots, (see 012.063), p. 235 - 255 (1981).
The author presents computational details of a sunspot umbral model extending from the deepest observable photospheric layers through the overlying chromosphere-corona transition region. The model is determined from a composite of photospheric, chromospheric, and transition region temperature distributions proposed by Albregtsen and Maltby, Lites and Skumanich, and Nicolas. The author calculates the continuous spectrum, the profiles of various lines, and internal properties including the chromospheric net radiative cooling rates and the radiative flux as a function of depth in the photosphere. For this model he finds that H^- provides net cooling in the temperature minimum region and that the integrated radiative flux is essentially constant with depth in the photosphere.

072.092 **Dynamic phenomena in sunspots.**
R. L. Moore.
The physics of sunspots, (see 012.063), p. 259 - 311 (1981).
The empirical properties of the various fine-scale, time-dependent structures observed in sunspots are reviewed and interrelated with emphasis on recent observational results.

072.093 **Umbral oscillations at the photospheric level.**
W. Livingston, C. Mahaffey.
The physics of sunspots, (see 012.063), p. 312 (1981).
Abstract.

072.0 94 **Observations of dynamical phenomena in sunspots.**
A. H. Nye, L. E. Cram, J. H. Thomas,
J. M. Beckers.
The physics of sunspots, (see 012.063), p. 313 - 317 (1981).
The authors present a preliminary description of an observing program designed to carry out spectroscopic studies of the fine structure of sunspots. They describe some of the time-dependent phenomena seen in the velocity and magnetic fields in the umbral photosphere.

072.095 **The chromosphere above sunspot umbrae. III. Spatial and temporal variations of chromospheric lines.**
F. Kneer, W. Mattig, M. v. Uexküll.
The physics of sunspots, (see 012.063), p. 318 (1981).
Abstract.

072.096 **UVSP/SMM observations of transition region oscillations in sunspots.**
J. B. Gurman, J. W. Leibacher, R. A. Shine, B. E. Woodgate, W. Henze.
The physics of sunspots, (see 012.063), p. 319 - 335 (1981).
Time series observations of the profile of the C IV resonance line λ 1548.19 have been obtained in 8 sunspots with the Ultraviolet Spectrometer and Polarimeter (UVSP) on the Solar Maximum Mission. All of the sunspots display significant oscillations in line-of-sight velocity with frequencies in the range 5.8 mHz-7.8 mHz. Significant intensity oscillations were observed at the same periods in four of the time series.

072.097 **The chromospheric Evershed flow observed in the EUV spectrum.**
K. R. Nicolas.
The physics of sunspots, (see 012.063), p. 336 - 339 (1981).

072.098 **The fine structure of sunspot penumbras.**
R. Muller.
The physics of sunspots, (see 012.063), p. 340 - 342 (1981).

072.099 **Theories of dynamical phenomena in sunspots.**
J. H. Thomas.
The physics of sunspots, (see 012.063), p. 345 - 358 (1981).
The review covers attempts to understand and explain observed dynamical phenomena in sunspots within the framework of magnetohydrodynamic theory.

072.100 **Small scale phenomena in umbras and penumbras. The role of convective processes.**
H. C. Spruit.
The physics of sunspots, (see 012.063), p. 359 - 368 (1981).
In the review the author discusses small scale phenomena in sunspots, in as far as they seem to be related to convection.

072.101 **Conjectures regarding the structure of a sunspot penumbra.**
L. E. Cram, A. H. Nye, J. H. Thomas.
The physics of sunspots, (see 012.063), p. 384 - 388 (1981).
Observations suggest that some of the dark filaments located on the outer edge of penumbras may lie several hundred kilometers above the base of the quiet photosphere. By studying simple models for radiative transfer in such structures, The authors have concluded that elevated dark filaments probably havehigher densities and temperatures than the surrounding atmosphere. They discuss a possible connection between the dark filaments, the photospheric Evershed flow, and umbral dots.

072.102 **Observations of the wavelength dependence of the average contrast of sunspots.**
G. A. Chapman, A. D. Meyer.
The physics of sunspots, (see 012.063), p. 446 - 452 (1981).

072.103 **Sunspot distribution analysis based on longitudinal hemispheres.**
A. Sanchez.
Publ. Astron. Soc. Pacific, Vol. 94, 757 (1982). – Abstract.

072.104 **C_2 lines in sunspots.**
H. S. Yun, H. A. Beebe, W. E. Baggett.
Publ. Astron. Soc. Pacific, Vol. 94, 761 - 762 (1982).
Abstract.

072.105 **Choice of the most probable run of the H and K Ca II Doppler width in the chromosphere above a sunspot penumbra.**
S. A. Grigor'eva, R. B. Teplitskaya.
Issled. po geomagn. aehron. i fiz. Solntsa, Moskva, 1982, No. 60, p. 100 - 143. In Russian. – Abstr. in Ref. zh., 51. Astron., 11.51.309 (1982).

072.106 **Technique and preliminary results of investigation of element distribution of a magnetic field in active regions.**
N. V. Klochek.
Issled. po geomagn., aehron. i fiz. Solntsa, Moskva, 1982, No. 60, p. 140 - 144. In Russian. – Abstr. in Ref. zh., 51. Astron., 11.51.319 (1982).

072.107 **The longitudinal distribution of sunspots and the RS CVn starspot model.**
T. Özkan, E. Derman.
Astrophys. Space Sci., Vol. 87, 211 - 216 (1982).
The longitudinal distribution of the sunspots has been investigated by using the observations made at the University Observatory of Istanbul in 1979. In order to take into account the activity as a whole, the authors developed a new weighting system which depends on the area of spots and used it in their calculations. Differential rotation appeared to effect the considered distribution. It was determined that this longitudinal distribution of spots is not uniform but forms

two maxima for two different longitudes. The obtained result verifies one of the main assumptions proposed in the starspot model in which accounts for the distortion wave appeared in the light curves of RS CVn-type binaries.

072.108 **An investigation of proper motions, magnetic field and flare activity in a sunspot group. II.**
R. N. Ikhsanov, G. P. Shchegoleva.
Izv. Glav. Astron. Obs. Pulkovo, Astrofiz. Astrometr., No. 200, p. 22 - 29 (1982). In Russian.

Complex sunspot group N 251 is studied. The pattern of the magnetic field and its variations with time, proper motions of sunspots and flare activity in the sunspot group are explained in terms of a consecutive rise of the magnetic field, showing up with the occurrence of several new complexes of sunspots.

072.109 **On the statistical stability of solar cyclicity.**
K. S. Vojchishin, M. I. Stodilka.
Astron. Zh., Tom 59, 1171 - 1183 (1982). In Russian.
English translation in Soviet Astron., Vol. 26, No. 6.

The form function of 11-year solar activity cycles with an account for their asymmetry is introduced on the interval equal to the duration of the cycles mentioned. A mathematical model to describe the time structure of the solar cycling is proposed. A method and an algorithm of optimal estimation of the parameters are described. A statistical interconnection between the parameters of 11-year cycles is discussed. Stability of their fine structure is illustrated. The conclusion that the statistical stability of the 11-year cycles of solar activity is real within the asymmetry changes is made.

072.110 **The most interesting sunspot groups published in the Greenwich Photoheliographic Results.**
M. Kopecký.
Říše hvězd, Vol. 63, 181 - 182 (1982). In Czech.

072.111 **Visual observation of the Sun in Czechoslovakia in 1981.**
L. Schmied.
Říše hvězd, Vol. 63, 184 - 185 (1982). In Czech.

072.112 **Sunspot observation for the year 1979.**
N. Doğan.
Commun. Fac. Sci. Univ. Ankara, Sér. A_3, Tome 29, 1 - 38 (1981).

Sunspot groups seen between the solar rotation number 1677 and 1689, inclusive, and their development and the distribution of these sunspot groups in the northern and southern latitudes are given. In addition, the sum of the number of the spot groups in each rotation is also given for each type at the end of their development.

072.113 **The axial distribution of the rotational velocity of a sunspot filament.**
B.-s. Li, X.-m. Gu, Y.-j. Ding.
Acta Astron. Sinica, Vol. 23, 219 - 225, plate 1 (1982). In Chinese.

The distribution of the rotational angular velocities, the integral translational velocities and the radii along the filamentary axes of six different sunspot filaments are given in the paper.

072.114 **The ancient solar activity maxima epochs of various periods.**
Y.-j. Ding, B.-r. Luo, Y.-m. Feng.
Acta Astron. Sinica, Vol. 23, 287 - 298 (1982). In Chinese.

The authors have examined the 11 year, 60 year and about 250 year periods of solar activity and the ancient solar activity maxima epochs of various periods are determined.

072.115 **Spectral character of the solar radio SVC in a solar proton active region.**
J.-x. Yao.
Acta Astron. Sinica, Vol. 23, 309 - 316 (1982). In Chinese.

Japanese scientists have worked out a statistical analysis of the relation between a solar proton event and the spectrum of the slowly varying component (SVC). They point out: if the following two conditions are fulfilled, active regions are likely to produce solar proton events. These two conditions are: (1) the flux of the SVC at λ3 cm, $S_3 > 25$ sfu (solar flux units), (2) the ratio of λ3 cm to λ8 cm of the SVC flux, $S_3/S_8 \gtrsim 1$. In this paper, this kind of the SVC spectrum has been analyzed and calculated. From this the author obtained properties of the proton active region.

072.116 **The rapidly emerging of the sunspot group in the SESC 2372 region in April 1980.**
Q.-f. Hong, Y.-j. Ding, D. Luan.
Acta Astron. Sinica, Vol. 23, 327 - 332, plates 1 + 2 (1982). In Chinese.

072.117 **A forecast of the trend of solar activity in the second half of cycle 21.**
B. Luo.
Acta Astrophys. Sinica, Vol. 2, 245 - 246 (1982). In Chinese.

Based on a regression correlation between the peak values and the descent times the final epoch of cycle 21 is given, and the smoothed monthly mean relative numbers of descent epoch are given.

072.118 **Sonnenfleckenrelativzahlen des S. I. D. C. (Sunspot Index Data Center).**
Sterne Weltraum, 21. Jahrg., 338, 390, 442, 494, 546 (1982).

072.119 **Sunspot numbers.**
Sky Telesc., Vol. 64, 197, 295, 396, 502, 617 (1982).

072.120 **L'activité solaire.**
M.-J. Martres, G. Zlicaric.
Astronomie, Vol. 96, 367 - 369, 426 - 427, 478 - 479, 530 - 531, 583 (1982).

Solar magneto-hydrodynamics.
See Abstr. 003.100.

Future observations of active regions.
See Abstr. 013.066.

Franck-Condon factors for some abundant umbral molecules.
See Abstr. 022.137.

Bestimmung der Längenausdehnung von Sonnenfleckengruppen.
See Abstr. 031.534.

Speckle image reconstruction of solar features.
See Abstr. 031.539.

Comparison between two trigonometric models for the long-period variations in the Wolf numbers and in the length of day.
See Abstr. 044.003.

A comparative spectral analysis of the earth's rotation and the solar activity.
See Abstr. 044.012.

Waves in magnetic structures.
See Abstr. 062.101.

A model of a sunspot chromosphere based on OSO 8 observations.
See Abstr. 073.010.

June - July 1974 proton-flare region. III. The geomagnetic activity during the declining phase of the complex process.
See Abstr. 073.038.

On the problem of the relationship between brightness fields and radial velocity fields in undisturbed and active solar regions.
See Abstr. 073.102.

Solar transition zone pressures from EUV observations of O IV and N IV.
See Abstr. 074.006.

Complex time analysis of recurrent activity of coronal index 1971 - 1976.
See Abstr. 074.039.

Observations of coronal structure during sunspot maximum.
See Abstr. 074.087.

Some properties of the solar wind plasma at extreme solar activities.
See Abstr. 074.096.

Open magnetic fields and the solar cycle. I: Photospheric sources of open magnetic flux.
See Abstr. 075.001.

The three-dimensional structure of atmospheric magnetic fields in two active regions.
See Abstr. 075.003.

A note on the relation of flare activity with rapid variations of the magnetic field.
See Abstr. 075.004.

On the techniques for measurements of sunspot magnetic fields.
See Abstr. 075.010.

The magnetic field of active regions with longitudinal current.
See Abstr. 075.011.

The relations between the toroidal magnetic field in the solar convection zone and the bipolar magnetic field of sunspots.
See Abstr. 075.015.

A model for slender flux tubes and its application to sunspots.
See Abstr. 075.020.

Configurations of the magnetic field in sunspot groups and solar flares. I.
See Abstr. 075.021.

Fluctuations of active region microwave radiation in the solar activity minimum.
See Abstr. 077.026.

Differential rotation and meridional motions of sunspots from 1874 to 1902.
See Abstr. 080.004.

Solar irradiance modulation by active regions during 1980.
See Abstr. 080.010.

Motion of the sun around the barycenter of the solar system and solar activity.
See Abstr. 080.020.

Facular influences on the apparent solar shape.
See Abstr. 080.021.

Solar irradiance variations and solar activity.
See Abstr. 080.028.

On the coefficient of continuous absorption in the solar atmosphere and active regions.
See Abstr. 080.046.

Measurement of the Mount Wilson solar white-light plate collection.
See Abstr. 080.054.

Sunspots and changes in the global output of the sun.
See Abstr. 080.055.

Sunspots and solar variability.
See Abstr. 080.057.

Variations in solar luminosity.
See Abstr. 080.059.

On the influence of solar faculae on solar flux variations.
See Abstr. 080.062.

The solar differential rotation from the eighteenth solar cycle.
See Abstr. 080.063.

Model of the ionospheric F_1 layer for high solar activity.
See Abstr. 083.033.

Periodic oscillations of geomagnetic and solar activity.
See Abstr. 084.015.

Variations in ion and neutral composition at Venus: evidence of solar control of the formation of the predawn bulges in H^+ and He.
See Abstr. 093.108.

Geometrical relationship of flare-generated solar wind structures to the magnetic axes of bipolar sunspot groups adjacent to their originating solar flares.
See Abstr. 106.001.

Stellar activity measured at Ca II H and K.
See Abstr. 114.212.

073 Chromosphere, Flares, Prominences

073.001 **Evidence for a poleward meridional flow on the sun.**
K. Topka, R. Moore, B. J. LaBonte, R. Howard.
Sol. Phys., Vol. 79, 231 - 245 (1982).
The authors define for observational study two subsets of all polar zone filaments, which they call polemost filaments and polar filament bands. The behavior of the mean latitude of both the polemost filaments and the polar filament bands is examined and compared with the evolution of the polar magnetic field over an activity cycle as recently distilled by Howard and LaBonte (1981) from the past 13 years of Mt. Wilson full-disk magnetograms.

073.002 **Scaling-law equilibria for calcium in canopy-type models of the solar chromosphere.**
H. P. Jones.
Sol. Phys., Vol. 79, 279 - 290 (1982).
Scaling laws for resonance line formation are used to obtain approximate excitation and ionization equilibria for a three-level model of singly ionized calcium. The method has been developed for and is applied to the study of magnetograph response in the 8542 Å infrared triplet line to magnetostatic canopies which schematically model diffuse, nearly horizontal fields in the low solar chromosphere. For this application, the method is shown to be efficient and semi-quantitative, and the results indicate the type and range of effects on calcium-line radiation which result from reduced gas pressure inside the magnetic regions.

073.003 **The determination of vector magnetic fields in prominences from the observations of the Stokes profiles in the D_3 line of helium.**
E. Landi Degl' Innocenti.
Sol. Phys., Vol. 79, 291 - 322 (1982).
A method is presented to measure the magnetic field vector in optically thin prominences by means of the polarimetric observations in the D_3 line of He obtained with the High Altitude Observatory Stokes polarimeter. The theory of the Hanle effect is generalized to take into account the detailed wavelength separation of the various Zeeman components across the D_3 line profile. The inclusion of this effect appears to be particularly important for the interpretation of the complex circular polarization profile observed in the D_3 line. The circular polarization observations can thus be added to the linear polarization ones to give additional information about the magnetic field vector in solar prominences.

073.004 **Study of the brightness distribution across spicules from observations of the spicule occultation by the Moon at the partial solar eclipse.**
G. K. Ajmanova, A. K. Ajmanov, R. A. Gulyaev.
Sol. Phys., Vol. 79, 323 - 326 (1982).
Cinematographic observations of the spicule occultation by the Moon have been carried out at the partial solar eclipse of October 2, 1978. The relative brightness distribution across a spicule in the light of $H\alpha + 0.8$ Å is first obtained. The diameter values for two spicules have been found to be of 280 ± 40 km and 1200 km.

073.005 **A triggering of a solar flare by magnetosonic waves in a neutral sheet plasma.**
J.-I. Sakai, H. Washimi.
Astrophys. J., Vol. 258, 823 - 834 (1982).
A theoretical model of the triggering of a solar flare by magnetosonic waves in a neutral sheet plasma is discussed. It is shown that the ponderomotive force due to the magnetosonic waves strongly excites the plasma convection flow in the magnetic neutral sheet which in turn enhances the tearing instability.

073.006 **The extreme-ultraviolet spectrum of a solar impulsive burst.**
K. G. Widing.
Astrophys. J., Vol. 258, 835 - 845 (1982).
The spectrum of the 1973 December 22 flare in the 170–345 Å range is presented as an example of a class of impulsive EUV events observed with the NRL spectroheliograph on Skylab. The spectrum shows strongly enhanced emission of He II and transition-zone ions. A number of new solar identification of N IV, O IV, and O V are given. The EUV emission has a fine-scale structure of the order of 2 km in O IV and O V and 60 km in Fe IX and Si X.

073.007 **Mass motions in the solar chromosphere and transition zone.**
P. Mein, G. Simon, J. C. Vial, R. A. Shine.
Astron. Astrophys., Vol. 111, 136 - 139 (1982).
A comparison between $H\alpha$ and C IV observations of Active Region 2717 (October 9, 1980) shows that: (1) Upward velocities are present above sunspots in the chromosphere-corona transition zone (20 km s^{-1}). (2) Downward velocities are well correlated in both lines. (3) Dopplershift ratios between C IV and $H\alpha$ levels ($\sim$10) are much smaller than they could be expected from density ratio estimates. This result indicates that flow lines are probably far from vertical in the transition zone but relies on model densities which may not be correct. A simple method is presented for comparison of matter flows.

073.008 **A determination of the electron temperature and electron concentration in the solar chromosphere using H lines observations.**
V. M. Sobolev.
Soln. Dannye 1982 Byull., No. 4, p. 77 - 84 (1982). In Russian.
Data on the intensities of the H_ϵ, H_8 and H_9 lines obtained from total solar eclipse observations are given. It is shown that the electron temperature does not practically vary at heights from 450 km to 4200 km and is equal to 6700 - 8100°; the electron concentration at the heights 500 km, 2000 and 4000 km equals (7.5 - 9.0) × 10^{11}, (2.9 - 3.7) × 10^{11} and (6.5 - 9.0) × 10^{10}, respectively.

073.009 **Gamma-ray and optical observations of the 1979 November 8 solar flare.**
G. R. Riegler, J. C. Ling, W. A. Mahoney, T. A. Prince, W. A. Wheaton, J. B. Willett, H. Zirin, A. S. Jacobson.
Astrophys. J., Vol. 259, 392 - 396, plate 11 (1982).
The solar flare on 1979 November 8 was observed by the Tel Aviv telescope of the Big Bear Solar Observatory and the High Resolution Gamma-Ray Spectrometer on HEAO 3. Photographs in $H\alpha$ show the development of the flare and a subsequent Moreton wave. A statistically significant signal was detected above 80 keV, from 420 to 585 keV, and above 3.8 MeV. The temporal structure of microwave, optical, X-ray, and gamma-ray emission is consistent to within $\sim$1 s with a simultaneous flare response at all energies.

073.010 **A model of a sunspot chromosphere based on OSO 8 observations.**
B. W. Lites, A. Skumanich.
Astrophys. J., Suppl. Ser., Vol. 49, 293 - 316, plate 4 (1982).
OSO 8 observations of the profiles of the resonance lines of H I, Mg II, Ca II, and of C IV for a large quiet sunspot (1975 November 16 - 17) are analyzed along with near-simultaneous ground-based Stokes measurements to yield an umbral chromosphere and transition region model. The similarity of the upper chromosphere of the sunspot umbra to the quiet-Sun chromosphere suggests that the intense magnetic field

plays only a passive role for the chromospheric heating mechanism.

073.011 **On solar hydrogen lines in the far-infrared and submillimeter spectrum.**
D. Hoang-Binh.
Astron. Astrophys., Vol. 112, L3 - L5 (1982).
Strengths of Hnα lines, $5 \leqslant n \leqslant 20$, have been evaluated, using a recent model of the solar chromosphere. It is found that they are much stronger than coronal ion lines, formed by the process of dielectronic recombination; and their wavelength dependence is strongly affected by pressure broadening. Observation of these lines may provide useful diagnostics of the physical conditions in the chromosphere, as well as valuable information on the behaviour of atoms in high Rydberg states.

073.012 **Do "magnetic transients" exist in solar flares?**
N. I. Lozitskaya, V. G. Lozitskij.
Pis'ma Astron. Zh., Tom 8, 500 - 505 (1982). In Russian. English translation in Soviet Astron. Lett., Vol. 8.
On the basis of analysis of the solar flare spectrum on July 26, 1981 it is concluded that the "magnetic transients" in flares recently found with the Big Bear observatory videomagnetograph are rather of brightness then of magnetic nature. According to the direct measurements of Zeeman splitting of the emissions of the lines Fe I λλ 5233, 5324, 5447, and 5587 the magnetic field strength has amounted 3000 Gauss.

073.013 **Strong plasma turbulence and its applications to solar flares.**
V. M. Tomozov, V. N. Tsytovich.
Sib. inst. zem. magnetizm, ionos. i rasprostr. radiovoln SO AN SSSR. Prepr., 1981, No. 3, 12 pp. – Abstr. in Ref. zh., 51. Astron., 7.51.372 (1982).

073.014 **On possible manifestations of a turbulent Stark effect in different flare models.**
V. P. Maksimov, V. M. Tomozov.
Sib. inst. zem. magnetizm, ionos. i rasprostr. radiovoln SO AN SSSR. Prepr., 1981, No. 8, 9 pp. – Abstr. in Ref. zh., 51. Astron., 7.51.373 (1982).

073.015 **Hydrodynamic shock wave formation in the solar chromosphere and corona during flares.**
B. V. Somov, A. R. Spektor.
Space Sci. Rev., Vol. 32, (see 012.005), 27 - 41 (1982).
Basic mechanisms of the hydrodynamic shock wave formation in the solar atmosphere during flares are considered. Hydrodynamic plasma flows during flares arise due to fast energy release which is accumulated in the magnetic field of currents in the solar atmosphere. Shock waves arise as a result of rapid heating of the chromospheric upper layers from accelerated particles or heat fluxes. Powerful hydrodynamic phenomena can also arise due to explosive current sheet disruption in the region of strong magnetic field reconnection. Fundamental questions of shock wave formation and propagation in a non-homogeneous emitting solar atmosphere are discussed.

073.016 **A sequence of five proton producing flares in McMath Plage 15266 28 April - 7 May 1978 and their interplanetary consequences.**
S. M. P. McKenna-Lawlor.
Space Sci. Rev., Vol. 32, (see 012.005), 235 - 250 (1982).
The paper recounts solar circumstances leading up to the formation of 'parent plage' 15266. The magnetic affiliations and optical characteristics of the flares are reviewed, together with their associated particle phenomena. The possible general effect of the alignement of the flare axis, relative to the solar meridian, on the coronal propagation of solar cosmic rays is discussed.

073.017 **Vector magnetic fields in prominences. I. Preliminary discussion of polarimeter observations in He I D_3.**
L. L. House, R. N. Smartt.
Sol. Phys., Vol. 80, 53 - 69 (1982).
The authors present a sample of measurements of Stokes parameter spectral profiles in a quiescent prominence. These observations, made in the He I D_3 multiplet, are used to infer vector magnetic fields using the Hanle effect.

073.018 **Temperature minimum heating in solar flares by resistive dissipation of Alfvén waves.**
A. G. Emslie, P. A. Sturrock.
Sol. Phys., Vol. 80, 99 - 112 (1982).
The authors examine the possibility that the strong heating produced at temperature-minimum levels during solar flares is due to resistive dissipation of Alfvén waves generated by the primary energy release process in the corona. It is shown how, for suitable parameters, these waves can carry their energy essentially undamped into the temperature-minimum layers and can then produce a degree of heating consistent with observations.

073.019 **A continuous spectrum of a white-light flare.**
E. Hiei.
Sol. Phys., Vol. 80, 113 - 127 (1982).
White-light continuum was observed in a 2B flare of 10 September 1974 in the spectral region between 3600 Å and 4000 Å. The continuum shows a Balmer free-bound component, but the main contribution to the continuum between 3646 Å and 4000 Å is H^-emission. The white-light continuum, therefore, is thought to be of photospheric origin. The energy loss in the continuum is 10^{27} erg s^{-1}.

073.020 **Dissipation and stability of return currents in solar flares.**
J. C. Brown, J. Hayward.
Sol. Phys., Vol. 80, 129 - 141 (1982).
The dynamics of an electron beam, under the effects of Coulomb collisions and classical Ohmic dissipation of the return current, is analysed for a background plasma with a temperature which is time dependent due to the heating effect of beam dissipation offset by thermal conductive cooling. It is shown that the plasma is heated toward a steady state, in time scales short compared to typical flare beam switch on times, and that in this steady state only two regimes of beam dynamics arise.

073.021 **Theoretical mechanisms for generating solar spicules.**
M. N. England.
South. Stars, Vol. 29, 130 - 135 (1982).

073.022 **Modification of the ionization balance of the upper chromosphere due to XUV irradiation in flares.**
G. Chambe.
Astron. Astrophys., Vol. 113, 31 - 38 (1982).
The author discusses the effects on the ionization balance of the upper chromosphere of the irradiation from an overlying XUV flare source. If one excludes absorption of soft X-rays through K-shell ionization, which has little effect on the ionization balance in general, one finds that radiation up to 1100 Å is absorbed through L-shell ionization of H, He, and C I, and that this process induces large modifications of the ionization balance of these species. Calculations are made for He I, II, III, C I to C IV and O I to O IV, for an a-priori specified hydrogen model, and in various irradiation conditions. An approximate analytical method is developed, which is useful in the case where the main absorber is the ionizing ion itself.

073.023 **Explosive destruction of a current layer and flares on the sun.**

A. G. Frank.
Priroda, 1982, No. 9, p. 78 - 79. In Russian.

073.024 **Propagation of low-energy electrons depending on the position of a flare relative to the interplanetary current layer.**
M. A. Zel'dovich, O. M. Kovrizhnykh, M. A. Livshits.
Kosm. Issled., Tom 20, 560 - 565 (1982). In Russian.

073.025 **Velocity of the shock generated by a large east limb flare on August 18, 1979.**
H. V. Cane, R. G. Stone, R. Woo.
Geophys. Res. Lett., Vol. 9, 897 - 900 (1982).
The large flare of August 18, 1979 resulted in a shock which produced intense type II radio emission detected at kilometer wavelengths using the radio astronomy experiment on the ISEE-3 spacecraft. The drift rate of the type II burst provides estimates of the velocity of the shock as a function of coronal height.

073.026 **Formation of the O I resonance triplet and intercombination doublet in the solar chromosphere.**
D. L. Skelton, R. A. Shine.
Astrophys. J., Vol. 259, 869 - 879 (1982).
Spectrum synthesis calculations are presented for the O I resonance triplet at 1304 Å and the intercombination doublet at 1358 Å for the solar atmosphere of Vernazza, Avrett, and Loeser and several variants, allowing for triplet fluorescence by the Lyβ emission of H I. Profiles, synthesized from a seven-level plus continuum O I atom are compared to observations taken with the high-resolution spectrometer on OSO 8. It is found that the O I triplet emission is dominated by the Lyβ fluorescence.

073.027 **The appearance of arched systems of chromospheric flares in spectral lines.**
V. A. Ostapenko, P. Palush *(Pal'uš)*, A. N. Yakobchuk.
Vestn. Kiev. Univ., Astron. Vyp. 24, p. 29 - 41 (1982). In Russian.

073.028 **Forecast of solar flares with a term from one to seven days at the Astronomical Observatory of the Kiev University.**
P. R. Romanchuk, I. Yu. Izotova, V. N. Krivodubskij, A. S. Adamenko, V. P. Babij.
Vestn. Kiev. Univ., Astron. Vyp. 24, p. 50 - 56 (1982). In Russian.

073.029 **The motion of chromospheric spicules.**
T. G. Gadzhiev, G. M. Nikol'skij.
Pis'ma Astron. Zh., Tom 8, 633 - 635 (1982). In Russian. English translation in Soviet Astron. Lett., Vol. 8.
A time sequence of Hα spectra of chromospheric spicules has been studied. Radial, tangential and total velocities of 15 spicules at the height of 4000 km have been determined. Spicule trajectories are obtained as projected on the photosphere. The spicules move along cyclic trajectories similar to ellipses in the ideal case. Thus the oscillation of a spicule as a whole is "elliptically polarized".

073.030 **Microwaves and hard X rays from solar flares: multithermal and nonthermal interpretations.**
G. A. Dulk, B. R. Dennis.
Astrophys. J., Vol. 260, 875 - 884 (1982).
The authors consider impulsive solar flares whose X-ray spectra are pure power law and conclude that, in general, they cannot result from single, homogeneous sources with power-law electron energy distributions. They consider a multithermal flare model in which the average electron energy and magnetic field strength decrease with distance from a hot core and develop expressions relating the temperature distribution, density, and field strength to the microwave and X-ray fluxes and spectral indices. They apply these expressions to the flare of 1980 June 29, 0230 UT.

073.031 **High-resolution X-ray spectra of solar flares. VI. Impulsive soft X-ray flares.**
U. Feldman, G. A. Doschek, R. W. Kreplin.
Astrophys. J., Vol. 260, 885 - 893 (1982).
The authors discuss soft X-ray spectra of a peculiar class of flares which are characterized by rapid rise and decay times in soft X rays. The spectra were recorded by Bragg crystal spectrometers flown on the P78-1 spacecraft. The lines discussed are produced by 1s–2p type transitions in iron and calcium near 1.86 and 3.2 Å, respectively. It is concluded that the 20×10^6 K plasma cools by in situ radiation, rather than by conduction into the chromosphere and subsequent radiative cooling at chromospheric temperatures.

073.032 **Correlations amongst flux-vs.-time profiles at various wavelengths during solar flares and their relation to flare heating models.**
A. G. Emslie.
Bull. American Astron. Soc., Vol. 14, 607 (1982). – Abstract.

073.033 **Bimodel energy release in solar flares.**
P. A. Sturrock, D. F. Smith.
Bull. American Astron. Soc., Vol. 14, 608 (1982). – Abstract.

073.034 **Acceleration mechanism for prompt protons and relativistic electrons from solar flares.**
T. Bai, B. R. Dennis, A. L. Kiplinger, L. E. Orwig.
Bull. American Astron. Soc., Vol. 14, 640 (1982). – Abstract.

073.035 **Physical conditions in spicules and prominences.**
D. A. Landman.
Bull. American Astron. Soc., Vol. 14, 640 (1982). – Abstract.

073.036 **Still more observations of prominence spectra.**
A. P. Bernat, D. A. Landman.
Bull. American Astron. Soc., Vol. 14, 640 (1982). – Abstract.

073.037 **Collisional tearing and twisting modes in a current sheet with a normal magnetic field and its application to the sudden disappearance of eruptive prominences.**
K.-I. Nishikawa, J.-i. Sakai, Y. Terashima.
Magnetospheric phenomena of celestial bodies, (see 012.011), p. 161 - 163 (1982).
A set of coupled eigenmode equations in a current sheet with a normal magnetic field is solved numerically by the shooting method for collisional tearing and twisting modes. As an application, a model of 'disparitions brusques' (sudden disappearance of eruptive prominences) is discussed based on Kippenhahn and Schlüter's current sheet.

073.038 **June - July 1974 proton-flare region. III. The geomagnetic activity during the declining phase of the complex process.**
V. Bumba.
Bull. Astron. Inst. Czechoslovakia, Vol. 33, 215 - 219 (1982).
In this third part of a series of papers describing the regularities in the magnetic field as well as solar activity development in a large portion of the solar atmosphere in which the processes related to the June - July 1974 proton-flare region formation take place, the author studies some characteristics of the solar wind which emanates from this portion of the solar atmosphere. It is shown that during the time when the fast disintegration and disappearance of the large-scale characteristic magnetic field patterns occurs and a sudden ceasing of sunspot-, flare- and coronal-activity is observed, the daily geomagnetic character figures C 9 reach their maximum. This solar wind enhancement is taken as one of the last observable manifestations of the whole complex process.

073.039 **Eruptive prominence of August 18, 1980.**
V. Rušin, M. Rybanský.
Bull. Astron. Inst. Czechoslovakia, Vol. 33, 219 - 229, plates 1 - 4 (1982).

An analysis of the disparition brusque phase of the prominence on August 18, 1980 is presented. Some parts of the prominence material moved out through the solar corona with a great velocity ($v \approx 400$ - 500 km s^{-1} at $2R_\odot$) and were ejected into interplanetary space. The decrease of the intensity of the 530.3 nm emission coronal line, observed after the event at the same location, was interpreted as a fast coronal transient with a mass of 3×10^{12} kg.

073.040 **June - July 1974 proton-flare region. IV. The dynamics of the local magnetic field evolution during the proton-flare events rotation.**
V. Bumba.
Bull. Astron. Inst. Czechoslovakia, Vol. 33, 281 - 285, plates 2 - 4 (1982).

The day by day evolution of the June - July 1974 proton-flare region's local magnetic field, composed originally of at least two individual local fields, is investigated and the functioning of its "magnetic centres of activity" is studied. Two such main and at least one "secondary" centres have been found to which the development of new magnetic fluxes could be related. The tendency of the complex magnetic field to form one simple bipolar field, properly inclined toward the equator, has been estimated as one of the main reasons for the changes of the observed field topology.

073.041 **On a possibility of recording chromospheric flares from intensity change of the regular background noise of ELF-VLF radiation.**
N. N. Murzaeva.
Byul. NTI. Yakut. fil. SO AN SSSR, 1981, iyul', probl. kosmofiz. i aehron., p. 22 - 23. In Russian. – From Ref. zh., 51. Astron., 9.51.294 (1982).

073.042 **Coronal transport of solar flare particles.**
J. Martinell, J. Pérez-Peraza.
Rev. Mexicana Astron. Astrofis., Vol. 6, (see 012.017), 351 - 356 (1981).

The authors develop a quantitative description of the azimuthal propagation of solar flare particles. The transport equations are solved within the frame of a two-step model. A scenario for the global phenomenon is proposed.

073.043 **Dynamics of the eruptive prominence of 6 May 1980 and its relationship to the coronal transient.**
N. Mein, B. Schmieder, G. Simon, E. Tandberg-Hanssen, S. T. Wu.
Astron. Astrophys., Vol. 114, 192 - 199 (1982).

The active prominence of the 6 May 1980 has been observed between 5.23 and 10.22 UT with the Ultraviolet Spectrometer and Polarimeter (UVSP) on board the Solar Maximum Mission (SMM) satellite. Intensities of 1548 Å line of C IV and dopplershifts have been derived. The authors note a motion of the magnetic tube maintaining the prominence material, this motion is followed by a coronal transient observed with the Coronagraph and Polarimeter (C/P) between 11 and 13 UT. The authors suggest that the event is related to a MHD wave induced by a flare occurring behind the solar disk and they propose a MHD modeling of the perturbation.

073.044 **X-ray classification of flares in large and small magnetic structures.**
C. G. Rapley.
Observatory, Vol. 102, 117 (1982). – Abstract.

073.045 **Present theories for simple-loop and two-ribbon flares.**
E. Priest.
Observatory, Vol. 102, 118 (1982). – Abstract.

073.046 **Optical observations of simple and complex flares.**
D. M. Rust.
Observatory, Vol. 102, 118 - 119 (1982). – Abstract.

073.047 **Impulsive acceleration and heating in flares.**
P. Hoyng.
Observatory, Vol. 102, 119 (1982). – Abstract.

073.048 **Hard X-ray and radio discriminators of flare classification.**
R. Dennis.
Observatory, Vol. 102, 120 (1982). – Abstract.

073.049 **Observations of flares in loops.**
R. Pallavicini.
Observatory, Vol. 102, 120 - 121 (1982). – Abstract.

073.050 **Soft X-ray discriminators of flare classification.**
E. Antonucci.
Observatory, Vol. 102, 121 - 122 (1982). – Abstract.

073.051 **Ultraviolet observations of solar flares.**
A. I. Poland.
Observatory, Vol. 102, 123 (1982). – Abstract.

073.052 **Flare classification – fact or fancy. A review of the day's discussion.**
L. W. Acton.
Observatory, Vol. 102, 123 - 124 (1982). – Abstract.

073.053 **Mass ejections associated with flare behind the solar limb on 1978 December 20.**
A. Bhatnagar, R. M. Jain, R. N. Shelke.
Bull. Astron. Soc. India, Vol. 10, 36 (1982). – Abstract.

073.054 **Study of optical, radio and X-ray emissions associated with the solar flare of 1980 March 28.**
A. Bhatnagar, R. M. Jain, R. N. Shelke.
Bull. Astron. Soc. India, Vol. 10, 36 - 37 (1982). – Abstract.

073.055 **The peculiar double solar flare of 1981 May 13.**
K. R. Bondal, P. Pant, M. C. Pande, G. M. Ballabh.
Bull. Astron. Soc. India, Vol. 10, 37 (1982). – Abstract.

073.056 **Electron and thermal conductivity of the solar flare plasma due to short wavelength drift tearing modes.**
V. Krishan.
Bull. Astron. Soc. India, Vol. 10, 39 (1982). – Abstract.

073.057 **On the possibility of generating gravitational radiation by the Sun.**
L. Křivský, T. Zeithamer.
Astrophys. Space Sci., Vol. 85, 309 - 313 (1982).

In the present paper the gravitational radiation emitted from large solar energetic flares and from the Sun as a whole in the course of the generation shock waves and quadrupole oscillations of the solar body are studied. The expected densities of the energy fluxes of the gravitational radiation at the distance of the Earth are calculated.

073.058 **A study of solar flare soft X rays and their relation to particle events observed with IMP 8.**
J. H. Nonnast, T. P. Armstrong, J. W. Kohl.
J. Geophys. Res., Vol. 87, 4327 - 4337 (1982).

Data from the Johns Hopkins University Applied Physics Laboratory charged particle measurement experiment aboard IMP 8 was used to study solar flare events and find correlations between soft X rays (4–17 Å) and particle events. The results show that a greater probability exists of observing a proton event if the observed soft X-ray flare has a relatively large decay time. It is also found that the Hα brightening area is a good indicator of the strength of the soft X-ray burst. Results also show that the strength of the soft X-ray burst can be used to estimate a probability that a proton event will occur, along with the strength of that particle event.

073.059 **Calculation of the Hα profile in bright points in flares.**
O. G. Badalyan, M. A. Livshits.
Astron. Zh., Tom 59, 975 - 984 (1982). In Russian. English translation in Soviet Astron., Vol. 26, No. 5.

The Hα profile has been computed on the basis of a model obtained from gas-dynamic calculations. Modeling of the events of different power is considered. The anomalous width of the Hα and Hβ lines in the strongest events is connected with a hypothetical rising jet of dense cool gas.

073.060 **Magnetic field in solar prominences measured with a new spectrally scanning magnetograph.**
I. S. Kim, S. Koutchmy, G. M. Nikolsky (*Nikol'skij*), G. Stellmacher.
Astron. Astrophys., Vol. 114, 347 - 350 (1982).

Observations of the longitudinal field strength in quiescent prominences obtained with the new redesigned spectrally scanning magnetograph installed in the Coudé focus of the 53 cm coronograph of the Kislovodsk station are presented. With the instrument one can obtain magnetographic data as well as central intensities, equivalent widths, line widths, and shifts of the analysed prominence lines. A first discussion, correlating magnetographic and spectroscopic data, is given.

073.061 **Temporal evolution of soft X-ray emitting plasma in solar flares.**
R. Pallavicini, G. Peres.
Bull. American Astron. Soc., Vol. 14, 776 - 777 (1982). Abstract.

073.062 **Measurements on a shock wave generated by a solar flare.**
A. Maxwell, M. Dryer.
Nature, Vol. 300, 237 - 239 (1982).

The authors review data on a high-velocity shock generated by a flare on 18 August 1979, 1400 UT, comment on some previously deduced velocities for the shock, and then present a model, based on current computer programs to account for the overall characteristics of the shock as it propagated through the corona and the interplanetary plasma.

073.063 **Detection of solar flares when the sun is observed as a star.**
W. C. Livingston, B. Ye.
Publ. Astron. Soc. Pacific, Vol. 94, 713 - 714 (1982).

Solar flares are detectable in unfocused sunlight from minute, transient, enhancements in the center of Fraunhofer Ca II K spectrum line. A survey of flare-frequency in other solar-type stars could establish the normalcy of the sun's present activity cycle.

073.064 **Observations of mass motions in active regions.**
R. G. Athay.
Solar active regions, (see 003.011), p. 83 - 112 (1981).

There is a high probability that the wavelength shifts and lateral translations are due to mass motions. In the discussion presented, the author classifies flows in five categories: (1) steady, (2) quasisteady, (3) transient, (4) oscillatory, and (5) nonthermal line widths. Steady and quasisteady flows are distinguished by the lifetimes of the flow. Thus, quasisteady flows are those that are short- or intermediate-lived, but occur often, and steady flows are those that are long-lived.
This review covers only those aspects of active region motions that are not exclusively associated with flares or energetic particles.

073.065 **Active region structures in the transition region and corona.**
D. F. Webb.
Solar active regions, (see 003.011), p. 165 - 198 (1981).

The author describes the basic structure of an active region in terms of morphological and thermal classifications of active region loops, including umbral structures and discusses what is known observationally about the thermal structure of loops. The temporal evolution of active region loop structures is reviewed with an emphasis on ephemeral regions and the emergence of active regions. A brief discussion of scheduled spaceborne observations of the transition region and corona of active regions is given.

073.066 **Second-phase acceleration versus second-step acceleration in solar flares.**
T. Bai.
AIP Conf. Proc., No. 77, (see 012.034), p. 409 - 417 (1982).

Observations and theoretical interpretations leading to two phases of acceleration and two steps of acceleration are reviewed. The distinction between the second-phase acceleration and the second-step acceleration is outlined.

073.067 **Solar flare energetics.**
R. P. Lin.
AIP Conf. Proc., No. 77, (see 012.034), p. 419 - 431 (1982).

The author reviews current knowledge of the energetics of solar flares. Recent observations by the Solar Maximum Mission and by balloon-borne instrumentation indicate that the flare hard X-ray emission arises from non-thermal bremsstrahlung. Under this interpretation, most of the energy released for many flares is initially contained in energetic electrons. In large flares a shock wave may result from explosive heating of the solar atmosphere by these electrons. This shock wave can accelerate nuclei to relativistic energies.

073.068 **The impulsive flux transfer solar flare model.**
P. J. Baum, A. Bratenahl.
AIP Conf. Proc., No. 77, (see 012.034), p. 433 - 442 (1982).

Recent modifications of the impulsive flux transfer model for solar flares are described. It involves nonsteady reconnection between independent magnetic field systems of bipolar sunspot groups and the formation of electric double layers.

073.069 **Distribution of solar flares around the Sun. I.**
Badruddin, R. S. Yadav.
Indian J. Phys. Part B, Vol. 56B, 68 - 70 (1982). – Abstr. in Phys. Abstr., Vol. 85, Abstr. 106659 (1982).

073.070 **Some internal mass motions in prominences indicated by special spectral features.**
V. Ruždjak, B. Vršnak.
Publ. Astron. Obs. Sarajevo, Vol. 1, No. 1, (see 012.043), p. 191 - 197 (1981).

073.071 **Inclined spectral features in flare spectra taken at Observatory Ondrejov.**
H. Božić, V. Ruždjak.
Publ. Astron. Obs. Sarajevo, Vol. 1, No. 1, (see 012.043), p. 199 - 203 (1981).

073.072 **The prominence of February 8th, 1979.**
V. Ruždjak, V. Vujnović.
Publ. Astron. Obs. Sarajevo, Vol. 1, No. 1, (see 012.043), p. 205 - 208 (1981).

073.073 **Note on a method of determination of true half-widths of chromospheric lines.**

V. M. Sobolev, G. F. Vyal'shin, Yu. A. Nagovitsyn.
Soln. Dannye 1982 Byull., No. 5, p. 72 - 75 (1982). In Russian.
The method developped 1980 is elaborated for the case when lines of different absolute intensities are used simultaneously. A comparison of the results obtained with the proposed method and from extra-eclipsing observations with a slit spectrograph is made.

073.074 **On physical conditions in solar faculae.**
M. M. Musaev.
Soln. Dannye 1982 Byull., No. 5, p. 76 - 80 (1982). In Russian.
On the basis of a comparison of model calculations with observational data the conclusion is made that the observed variation in the contrast of photospheric faculae may be due to a combined effect of the temperature increase and the absorption factor.

073.075 **A determination of the electron temperature and electron concentration in the chromosphere from results of He line observations.**
V. M. Sobolev.
Soln. Dannye 1982 Byull., No. 6, p. 112 - 119 (1982). In Russian.
A comparison of the observed intensities of the lines 3889 Å and 4713 Å of He I and λ4686 Å of He II in the chromosphere was made with the theoretically calculated values. It was found that the electron temperature is greater than 16000° in regions emitting He, and greater than 30000° in regions where emission is formed at λ4686 Å.

073.076 **Neutral line motion due to reconnection in two-ribbon solar flares and magnetospheric substorms.**
T. G. Forbes, E. R. Priest.
Planet. Space Sci., Vol. 30, 1183 - 1197 (1982).
Two kinematic models of line-tied reconnection are considered which describe the motion of a magnetic neutral line during the main phase of a two-ribbon solar flare and during the recovery phase of a magnetospheric substorm in the geomagnetic tail.

073.077 **Na I D brightening phenomena in quiescent prominences.**
Y. Nikaidou.
Sol. Phys., Vol. 80, 259 - 277 (1982).
The Na I D emission lines are found to brighten temporarily in restricted regions of quiescent prominences. The comparison of observed intensities of Na I D lines with model calculations shows that the phenomena is attributed to a kind of local activation of quiescent prominences accompanied by mass motions.

073.078 **Impulsive brightenings and velocity transients in prominences. I. Large events.**
J. M. Malville, G. D. Toot.
Sol. Phys., Vol. 80, 279 - 294 (1982).
Observations of quiescent prominences with the Zeiss Universal Birefringent Filter at Sacramento Peak show short-lived brightenings and velocity transients in Hα and D_3. The larger events range in area from 25 to 170 square arc sec, have lifetimes of approximately 30 min, velocities of 30 km s^{-1}, and total energy excesses up to 7×10^{27} ergs. These events do not disrupt the stable structure of the prominence, and are interpreted as either condensation events or low energy flares.

073.079 **Excitation of high-*m* tearing modes at the solar flare site.**
V. Krishan.
Sol. Phys., Vol. 80, 313 - 316 (1982).
It is shown that high-*m* drift tearing modes can be excited under the conditions prevalent at the solar flare sites. Since the growth rate of the high-*m* tearing modes is larger than that for low-*m* macroscopic tearing modes and smaller than that of microscopic ion-acoustic instability, these modes warrant accommodation in the scheme of instabilities possibly operating in the hybrid model of solar flares suggested by Spicer.

073.080 **Chromospheric evaporation in a well-observed compact flare.**
L. W. Acton, R. C. Canfield, T. A. Gunkler, H. S. Hudson, A. L. Kiplinger, J. W. Leibacher.
Astrophys. J., Vol. 263, 409 - 422, plate 9 (1982).
The authors have observed the flare of 1980 May 7 1456 UT with several Solar Maximum Mission instruments, in coordination with the Sacramento Peak Observatory vacuum tower telescope. From the X-ray data they determine the total emission measure of all material at $T > 2 \times 10^6$ K, and show that chromospheric evaporation has taken place concurrently with the appearance of the X-ray plasma. Possible driving-mechanisms for chromospheric evaporation are discussed.

073.081 **Electron acceleration in impulsive solar flares.**
S. R. Kane, A. O. Benz, R. A. Treumann.
Astrophys. J., Vol. 263, 423 - 432 (1982).
Simultaneous observations of the hard X-ray, microwave, and type III and decimetric radio bursts associated with the 1978 December 4 solar flare have been used to study the physical parameters relevant to the acceleration and propagation of energetic electrons during the impulsive phase of a solar flare. The observations are consistent with an interpretation where the impulsive-phase decimetric emission is a signature of the primary energy release mechanism and where the systematic increase in the high frequency cutoff during the impulsive phase is caused by a downward shift of the electron acceleration/injection region.

073.082 **Two types of jets and origin of macrospicules.**
K. Shibata.
Sol. Phys., Vol. 81, 9 - 17 (1982).
The purpose of this paper is to elucidate the origin of macrospicules, especially to clarify whether the two types of macrospicules, Hα macrospicules and EUV macrospicules unseen in Hα have the same physical origin. This investigation is made on the basis of numerical jet models.

073.083 **Microwave emission from hot X-ray kernels in solar flares.**
V. V. Zheleznyakov, Yu. V. Tikhomirov.
Sol. Phys., Vol. 81, 121 - 135 (1982).
Spectral and polarization characteristics of radio emission from hot X-ray kernels are considered. It is shown that the frequency spectrum of thermal cyclotron radio emission and bremsstrahlung from these regions may contain a set of lines at cyclotron harmonics and the maximum at plasma frequency. This makes possible the diagnostics of the X-ray kernel plasma and magnetic fields according to spectral and polarization observations of microwave radio emission in solar flares.

073.084 **X-ray imaging of three flares during the impulsive phase.**
A. Duijveman, P. Hoyng, M. E. Machado.
Sol. Phys., Vol. 81, 137 - 157 (1982).
The impulsive phases of three flares that occurred on April 10, May 21, and November 5, 1980 are discussed. Observations were obtained with the Hard X-ray Imaging Spectrometer and other instruments aboard SMM. All three flares have in common that they show hard X-ray brightenings at distinctly separate locations during the impulsive phase. The authors compare the hard X-ray observations with magnetograms, Hα and soft X-ray observations. From this they draw

conclusions about the nature of the bright hard X-ray locations.

073.085 **The peculiar solar flare of 1981 May 13.**
K. R. Bondal, P. Pant, M. C. Pande, G. M. Ballabh.
Bull. Astron. Soc. India, Vol. 10, 137 - 142 (1982).

Time-lapse photographic observations with 0.7 Å passband H-alpha filter and the morphological-cum-temporal behaviour of a large double solar flare which occurred on 1981 May 13 are described. The two strands, constituting the flare, which initially appeared to cross each other, with a bright region at the location of crossing, later separated out, ultimately become roughly parallel. The mean separation between the two strands at various intervals of time have been measured. It is concluded that apart from the oscillating behaviour of the strands, there is shearing in them probably due to rotation of the footpoints.

073.086 **The new parameter A in the classification of the solar Ca plages.**
A. Bakry, S. Yousef.
J. Astron. Soc. Egypt, Vol. 3, 1 - 12 (1981).

073.087 **On a relationship between the chromospheric flares and type I noise storms.**
M. Karlický, P. Kotrč, H. Urbarz.
Bull. Astron. Inst. Czechoslovakia, Vol. 33, 331 - 338, plate 1 (1982).

A relationship of chromospheric flares to type I noise storms is studied. Both the radio data and H-alpha pictures from Sept. 4, 1980 partly show coincidence with the changes of type I noise storm activity and the occurence of chromospheric flares. A better coincidence is attained by the positional selection of the flares. This suggests a whole range of relationships of chromospheric flares and type I noise storms which will depend on the positional and power characteristics of both the events compared. The results are interpreted in terms of the most advanced model of type I noise storms.

073.088 **Spectral manifestations of the spatial structure and dynamic processes of chromospheric flares on the Sun.**
V. A. Ostapenko. P. Pal'uš.
Bull. Astron. Inst. Czechoslovakia, Vol. 33, 338 - 345, plate 2 (1982).

The nature of spectrographic manifestations of ascending elongated and point arc patterns with descending plasma flows is discussed. A conclusion is drawn regarding the relation between the rate of ascent of the arc and the velocities of the directional motion of the radiating plasma in the same arc. Qualitative agreement of the calculated profiles of spectral lines of the point arc patterns with the line profiles of Severnyj's "whiskers" was obtained. The elongated arc patterns appear in the spectrum in the form of characteristic emission "peaks", analogous to those observed in flares in the H and K CaII lines. The obtained results elucidate the nature of the origin of the "flare nodes", of the "emission peaks" in the H and K CaII lines and also indicate the necessity for photometric investigations of flares.

073.089 **Observations of solar flares at Kandilli Observatory during the year 1966.**
E. Soytürk, A. Özgüç.
Kandilli Obs., Heliophys. Serv., Istanbul. Publ., Second Ser., No. 2, 43 pp. (1979).

073.090 **On a solar flare model explaining observed peculiarities of radio emission.**
I. E. Pogodin.
Astron. Tsirk., No. 1172, p. 4 - 6 (1981). In Russian.

073.091 **Structure and physics of solar faculae. III. The densities in the chromosphere-corona transition zone.**
S. Dumont, Z. Mouradian, J.-C. Pecker, J.-C. Vial, E. Chipman.
Pré-publ. Inst. Astrophys. Paris, No. 7, 43 pp. (1982).
Accepted in Sol. Phys.

073.092 **Magnetic field in solar prominence measured with a new spectrally scanning magnetograph.**
I. S. Kim, S. Koutchmy, G. M. Nikolsky (*Nikol'skij*), G. Stellmacher.
Pré-publ. Inst. Astrophys. Paris, No. 9, 12 pp. (1982).
Accepted by Astron. Astrophys.

073.093 **On the electron density of faint prominences observed during the solar eclipse of July 31, 81.**
S. Koutchmy, C. Lebecq, G. Stellmacher.
Pré-publ. Inst. Astrophys. Paris, No. 22, 12 pp. (1982).
Accepted by Astron. Astrophys.

073.094 **The energy source of the large flare (class 3B) of 13 May, 1981.**
Q.-s. Guo, X.-z. Chen, A.-d. Zhao.
Publ. Purple Mountain Obs., Vol. 1, 1 - 9 (1982). In Chinese.

073.095 **The large flare (class 3B) of May 13, 1981.**
Z.-y. Zhu, Y.-w. Jiang, Y. Liang.
Publ. Purple Mountain Obs., Vol. 1, 11 - 15 (1982). In Chinese.

073.096 **A rare event of flare mass ejection on July 19, 1981.**
G.-f. Xie.
Publ. Purple Mountain Obs., Vol. 1, 25 - 29 (1982). In Chinese.

073.097 **Hydrodynamic response of the solar chromosphere to an elementary flare burst. II. Thermal model.**
B. V. Somov, B. J. Sermulina (*B. Ya. Sermulinya*), A. R. Spektor.
Sol. Phys., Vol. 81, 281 - 292 (1982).

Impulsive heating of the upper chromosphere by a very powerful thermal flux is studied as the cause of hard X-rays during a solar flare.

073.098 **A measurement of the magnetic field direction at the site of major flares.**
H. Lundstedt.
Sol. Phys., Vol. 81, 293 - 301 (1982).

Lundstedt et al. (1981) showed that the direction of the photospheric magnetic field at the site of a flare is a good predictor of the solar wind velocity observed at earth four days later. The author describes how the field direction was obtained, and discusses possible errors involved in the determination of the angle. The discussion also includes a characterization of the solar active regions.

073.099 **Helium radiation diffusion in prominences.**
N. A. Yakovkin, M. Yu. Zeldina (*Zel'dina*), Ch. Lhagvazhav (*Lkhagvazhav*).
Sol. Phys., Vol. 81, 339 - 354 (1982).

The paper is dedicated to the solution of the integral equations describing the diffusion of multilevel helium atom radiation, taking into consideration the absorption of radiation by hydrogen, and to elucidating the mechanism of the excitation of helium radiation in prominences.

073.100 **Operative prediction of solar flare radiative consequences on the basis of their radio emission.**
S. I. Avdyushin, Yu. F. Barabanshchikov, A. P. Molchanov, I. E. Pogodin, A. G. Stupishin, S. G. Slizkov.
Soln. Dannye 1982 Byull., No. 7, p. 100 - 107 (1982). In Russian.

Current state and possibilities of solar flare cosmic ray diagnostical predictions in real time are considered. On the basis of statistical investigations some requirements to the methods and parameters of the diagnostical system are determined.

073.101 **On diameters of thin chromospheric spicules.**
G. K. Ajmanova.
Soln. Dannye 1982 Byull., No. 8, p. 80 - 83 (1982). In Russian.

Pioneering work has been done on the brightness distribution deduction across a very thin spicule using motion pictures of the spicule occultations by the moon. The spicule diameter is equal to 200 km.

073.102 **On the problem of the relationship between brightness fields and radial velocity fields in undisturbed and active solar regions.**
G. I. Kornienko.
Soln. Dannye 1982 Byull., No. 8, p. 88 - 92 (1982). In Russian.

Radial velocities in large-scale details of the chromosphere have been studied. It has been found that there is a positive correlation between brightness and radial velocity in the undisturbed chromosphere and in active regions.

073.103 **Relative intensities of the H_8 and He 3889 lines in the solar chromosphere according to data of the total solar eclipse on June 30, 1973.**
V. M. Sobolev, G. F. Vyal'shin, Yu. A. Nagovitsyn.
Soln. Dannye 1982 Byull., No. 8, p. 93 - 96 (1982). In Russian.

On the basis of the materials of the total solar eclipse in Mauritania on 30 June 1973 a high correlation between total intensities of the He 3889 line and H_8 line for spicules has been found with a high coefficient of correlation.

073.104 **X-ray spectroscopy of solar flares.**
J. L. Culhane.
Irish Astron. J., Vol. 15, 69 - 86 (1981).

The X-ray output of solar flares is discussed in some detail with particular reference to emission line excitation mechanisms and to the information that can be obtained about the flare plasma from studies of the intensities, profiles and wavelengths of these lines. The X-ray crystal spectrometers on the NASA Solar Maximum Mission spacecraft are described and the interpretation of the spectra obtained with these instruments is discussed.

073.105 **Responses of transition region models to magnetic field geometry and downflow velocities.**
R. G. Athay.
Astrophys. J., Vol. 263, 982 - 986 (1982).

Models of the chromosphere-corona transition region in which energy is supplied by thermal conduction and downflow of hot material and lost by radiation are investigated for different magnetic field geometries and for different dependences of downflow velocity on the local temperature.

073.106 **A comparative analysis of the Hα and K Ca II chromospheric networks.**
R. T. Sotnikova.
Issled. po geomagn., aehron. i fiz. Solntsa, Moskva, 1982, No. 60, p. 76 - 79. In Russian. – Abstr. in Ref. zh., 51. Astron., 11.51.298 (1982).

073.107 **Spectral analysis of inhomogeneity sizes and brightness dependence on the magnetic field strength in a plage from Ba II 4554 Å filtergrams.**
V. E. Merkulenko, M. N. Mishina, V. I. Polyakov.
Issled. po geomagn., aehron. i fiz. Solntsa, Moskva, 1982, No. 60, p. 146 - 154. In Russian. – Abstr. in Ref. zh., 51. Astron., 11.51.313 (1982).

073.108 **Plasma processes and energy release in flares.**
B. V. Somov, V. E. Stepanov, N. N. Stepanyan, V. M. Tomozov.
Issled. po geomagn., aehron. i fiz. Solntsa, Moskva, 1982, No. 60, p. 112 - 131. In Russian. – Abstr. in Ref. zh., 51. Astron., 11.51.337 (1982).

073.109 **Some features in the August 4, 1972 flare development.**
V. G. Banin, V. P. Maksimov, V. D. Trifonov.
Issled. po geomagn., aehron. i fiz. Solntsa, Moskva, 1982, No. 60, p. 3 - 18. In Russian. – Abstr. in Ref. zh., 51. Astron., 11.51.338 (1982).

073.110 **Flow of material at the chromosphere-corona transition zone of the Sun.**
W. M. Glencross.
Space Sci. Rev., Vol. 33, (see 012.067), 151 - 160 (1982).

Early models of the chromosphere-corona transition zone of the Sun considered it to be a static plane-parallel region. From these it became clear that the layer was extremely thin and had an important role in the conduction of energy from the corona. More recent observations show mass motions of order 10 km s^{-1}, which means that heating and cooling of the moving plasma has an important effect on the energy balance, while transient effects producing far higher velocities are also common. Studies of plasma motion through the zone are clearly relevant to the initial heating of material which enters the solar wind.

073.111 **Plasma stability of quiescent prominences.**
S. Migliuolo.
J. Geophys. Res., Vol. 87, 8057 - 8063 (1982).

The stability of the Kippenhahn-Schlüter model prominence to ideal MHD perturbations is analyzed. The analysis is done by self-consistently taking into account the spatial variation of equilibrium quantities (density, magnetic field) and looking for bounded eigenmodes. Some consequences (of the derived stability of perturbations) are drawn with regards to observed prominence phenomena.

073.112 **The flares of spotless regions.**
B.-r. Luo.
Acta Astron. Sinica, Vol. 23, 95 - 101, plates 1 + 2 (1982). In Chinese.

In the present article the 20 flares of sunspotless regions observed at Yunnan Observatory in Cycle 21 are analysed.

073.113 **The Rayleigh-Taylor instability in filaments and "spotless" two ribbon flares.**
T.-j. Cao, A.-a. Xu, C.-c. Cheng.
Acta Astron. Sinica, Vol. 23, 203 - 210 (1982). In Chinese.

The energy storage for a spotless two ribbon flare is discussed from the morphology of chromospheric fibrils in the surrounding of a filament prior to the flare. Also, on the basis of the Kippenhahn-Schlüter model of filaments, the authors discuss the instability of the magnetic structure in these filaments.

073.114 **The determination of the α factor of a force-free magnetic field and the forecast of solar proton flares.**
G.-x. Ai, F.-x. Kong.
Acta Astron. Sinica, Vol. 23, 211 - 218 (1982). In Chinese.

073.115 **An idea of a flare dynamo mechanism.**
D.-q. Zhou, J.-k. Qian.
Acta Astron. Sinica, Vol. 23, 226 - 231 (1982). In Chinese.

073.116 **Particle acceleration during the explosive phase of proton flares.**
C.-y. Zhang, C.-s. Li.
Acta Astron. Sinica, Vol. 23, 317 - 326 (1982). In Chinese.
Starting with the quasilinear equation of the distribution function of particles with a regular electric field, the combined diffusion coefficient in the momentum space, which contains the coupling effects of a regular electric field and turbulent fluctuation fields, has been obtained, and the combined mechanism of acceleration by a regular electric field and turbulence in the neutral sheet of solar proton flares has been proposed.

073.117 **Statistical properties of logarithmic flare indices: distribution and evolution.**
S. Dong.
Acta Astrophys. Sinica, Vol. 2, 209 - 214 (1982). In Chinese.

073.118 **The great eruptive loop-prominence on the solar western limb on April 27, 1981.**
F. Hu, T. Cao.
Acta Astrophys. Sinica, Vol. 2, 291 - 298 (1982). In Chinese.

073.119 **From the spectra of prominences to a Planck function of open systems.**
D. Zhao, M. Jiang, Z. Fan.
Acta Astrophys. Sinica, Vol. 2, 334 - 339 (1982). In Chinese.

Solar magneto-hydrodynamics.
See Abstr. 003.100.

Hvar Observatory report on the FBS/SERF (*Flare Build-up Study/Study of Energy Release in Flares*) trial period of the solar maximum year May 1979.
See Abstr. 013.069.

Dielectronic satellite spectra for highly-charged helium-like ions – VII. Calcium spectra – theory and comparison with SMM observations.
See Abstr. 022.135.

Measurements of solar transition zone velocities and line broadening using the Ultraviolet Spectrometer and Polarimeter on the Solar Maximum Mission.
See Abstr. 031.607.

Measurements of the magnetic field in solar prominences with a spectrally scanning magnetograph.
See Abstr. 034.065.

Hinotori (ASTRO-A).
See Abstr. 051.016.

Astrophysical consequences of neutron-antineutron oscillations.
See Abstr. 061.051.

Blowing up of two-dimensional magnetohydrostatic equilibria by an increase of electric current or pressure.
See Abstr. 062.004.

Transport of energetic electrons in a fully ionized hydrogen plasma.
See Abstr. 062.005.

Shock acceleration of energetic particles in the heliosphere.
See Abstr. 062.016.

Magnetohydrodynamics of atmospheric transients. IV. Nonplane two-dimensional analyses of energy conversion and magnetic field evolution.
See Abstr. 062.080.

Numerical study of line-tied magnetic reconnection.
See Abstr. 062.097.

Simplified expressions for the gyrosynchrotron radiation from mildly relativistic, nonthermal and thermal electrons.
See Abstr. 063.005.

The Compton effect in the chromosphere. III.
See Abstr. 071.010.

The overall structure and evolution of active regions.
See Abstr. 072.030.

Active regions from the photosphere to the chromosphere.
See Abstr. 072.031.

The dynamics of active regions.
See Abstr. 072.032.

Cartes synoptiques de la chromosphère solaire et catalogues des filaments et des centres d'activité.
See Abstr. 072.063.

Observations of sunspots, faculae and prominences.
See Abstr. 072.064.

Mt. Purple No. 496 sunspot group and rare-founded Hα flare loop.
See Abstr. 072.065.

On the dynamics of the chromosphere above sunspots.
See Abstr. 072.068.

Model of the chromosphere over a dark umbral feature.
See Abstr. 072.085.

An empirical view of the chromospheric temperature structure above a sunspot umbra.
See Abstr. 072.086.

The chromosphere above sunspot umbrae. III. Spatial and temporal variations of chromospheric lines.
See Abstr. 072.095.

The chromospheric Evershed flow observed in the EUV spectrum.
See Abstr. 072.097.

Choice of the most probable run of the H and K Ca II Doppler width in the chromosphere above a sunspot penumbra.
See Abstr. 072.105.

The axial distribution of the rotational velocity of a sunspot filament.
See Abstr. 072.113.

Characteristics of shocks in the solar corona, as inferred from radio, optical, and theoretical investigations.
See Abstr. 074.015.

Unusual coronal activity following the flare of 6 November 1980.
See Abstr. 074.017.

Transient behavior of flare-associated solar wind: II. In a non-radial open-field region.
See Abstr. 074.030.

Coronal emission associated with a quiescent prominence.
See Abstr. 074.035.

Flare-related coronal transients.
See Abstr. 074.044.

Physics of static loop structures and scaling laws.
See Abstr. 074.058.

The growth of filaments by the condensation of coronal arches.
See Abstr. 074.077.

Ballooning instability driven by fast magnetosonic waves and its application to coronal loop transients associated with a flare.
See Abstr. 074.079.

Processes affecting abundances in the solar wind.
See Abstr. 074.086.

Coronal transient phenomena.
See Abstr. 074.088.

A note on the relation of flare activity with rapid variations of the magnetic field.
See Abstr. 075.004.

Development of solar energetic particle sources within the background magnetic fields.
See Abstr. 075.006.

Dynamics of post-flare ejections and magnetic loop geometry.
See Abstr. 075.008.

Relaxation of a twisted magnetic loop as mechanism of a solar flare. I.
See Abstr. 075.016.

Configurations of the magnetic field in sunspot groups and solar flares. I.
See Abstr. 075.021.

Wavelength dependence of the weakening of the solar EUV line emission.
See Abstr. 076.009.

Gamma-ray lines from solar flares and cosmic transients.
See Abstr. 076.010.

Solar energetic photon transients (50 keV - 100 MeV).
See Abstr. 076.011.

Solar hard X-ray images observed by Astro-A.
See Abstr. 076.012.

Evidence for delayed second phase acceleration in solar flares.
See Abstr. 076.013.

EUV arcades: signatures of filament instability.
See Abstr. 076.014.

Hard X-rays from the sun.
See Abstr. 076.015.

Radio wave heating of the corona and electron precipitation during flares.
See Abstr. 077.006.

Radio burst characteristics of solar proton flares.
See Abstr. 077.018.

Solar microwave bursts with outward motion observed at the limb.
See Abstr. 077.019.

Comparison of helium and heavy ion spectra in ^{3}He-rich solar flares with model calculations based on stochastic Fermi acceleration in Alfvén turbulence.
See Abstr. 078.002.

Injection onsets of ~2 GeV protons, ~1 MeV electrons, and ~100 keV electrons in solar cosmic ray flares.
See Abstr. 078.004.

Study of particle generation in low energy events.
See Abstr. 078.005.

A model of propagation of solar flare particles and its application to cosmic ray ground level events.
See Abstr. 078.007.

The propagation of solar flare proton and its diffusion coefficient in the interplanetary medium.
See Abstr. 078.008.

A direct observation of solar neutrons following the 0118 UT flare on 1980 June 21.
See Abstr. 078.009.

Geometrical relationship of flare-generated solar wind structures to the magnetic axes of bipolar sunspot groups adjacent to their originating solar flares.
See Abstr. 106.001.

A study of some interplanetary shock wave tendencies.
See Abstr. 106.005.

Structure and evolution of flare-generated shock waves in interplanetary space.
See Abstr. 106.006.

Observation of flare-generated shock waves by Helios-2 near the Sun.
See Abstr. 106.007.

Experimental study of flare-generated collisionless interplanetary shock wave propagation.
See Abstr. 106.008.

Measurements of low energy electrons and ions during long-lived solar particle events.
See Abstr. 106.009.

Plasma shocks and energetic particles in the outer solar system: trapping and asymmetry observations from Pioneer 10 and Pioneer 11.
See Abstr. 106.033.

High-energy observations of stellar flares: comparison with the sun.
See Abstr. 122.076.

A study of the modulating effect of solar flares on the cosmic ray intensity using time series analysis.
See Abstr. 143.051.

074 Corona, Solar Wind

074.001 **Enthalpy flux cooling of the solar corona. II. Applications to closed coronal structures.**
S. G. Wallenhorst.
Sol. Phys., Vol. 79, 333 - 341 (1982).
Models of the solar corona which include the effects of hot downflowing material are considered. Temperature-height profiles of the quiet and flaring corona are derived, under the assumptions of hydrostatic equilibrium and that the dominant cause of transition region heating is due to the enthalpy of the downflowing matter. In addition, scaling laws for the lengths of coronal loops are derived.

074.002 **A low-β coronal loop model. II: Kink instabilities in loops surrounded by plasma.**
A. Kattenberg, R. Sillen.
Sol. Phys., Vol. 79, 343 - 351 (1982).
The low-β coronal loop model of Sillen and Kattenberg (1980) is extended to include a surrounding current-free plasma. The authors calculated the dispersion curves of kink modes by solving the linearized MHD-equations of motion. They found a strong stabilizing influence on the growth rates of kink instabilities due to the surrounding plasma.

074.003 **Hydromagnetic stability of coronal arcade structures: the effects of photospheric line tying.**
A. Ray, G. Van Hoven.
Sol. Phys., Vol. 79, 353 - 364 (1982).
A model is given of the magnetic-field equilibrium and possible dynamic excitations of a solar coronal arcade. Such structures are well observed in the spectral range from Hα to X-rays and often give rise to two-ribbon flares. However, the preflare state must be stable to ideal magnetohydrodynamic disturbances, and the authors treat this problem with particular attention to the necessary foot-point boundary conditions. Having devised a reasonably general perturbation set, they use an energy-principle analysis to show the strong stabilizing influence of inertial field-line tying at the photosphere.

074.004 **Electron acceleration and radiation signatures in loop coronal transients.**
L. Vlahos, T. E. Gergely, K. Papadopoulos.
Astrophys. J., Vol. 258, 812 - 822 (1982).
A model for electron acceleration in loop coronal transients is suggested. It is proposed that in these transients an erupting loop moves away from the solar surface, with a velocity greater than the local Alfvén speed, pushing against the overlying magnetic fields and driving a shock in the front of the moving part of the loop. The authors discuss some of the conditions under which moving or stationary type IV bursts are expected to be associated with loop coronal transients.

074.005 **Active region coronal evolution.**
L. Golub, G. Noci, G. Poletto, G. S. Vaiana.
Astrophys. J., Vol. 259, 359 - 365, plates 3 - 9 (1982).
Scaling relations between coronal base pressure and longitudinal photospheric magnetic field strength are further tested for the case of a single active region which was observed for five solar rotations from Skylab. The authors trace the evolution of measurable quantities such as coronal thermal energy content, total longitudinal photospheric magnetic flux, region scale size, and peak energy density, throughout the five rotations observed. They conclude that this study provides additional justification for the belief that magnetic field-related heating is the operative mechanism in the solar corona.

074.006 **Solar transition zone pressures from EUV observations of O IV and N IV.**
K. P. Dere, J.-D. F. Bartoe, G. E. Brueckner.
Astrophys. J., Vol. 259, 366 - 371, plate 10 (1982).
NRL High Resolution Telescope and Spectrograph observations of the O IV and N IV EUV lines emitted from the Sun are examined. The instrument provides spectra with a spatial resolution of 1″ and a spectral resolution of 0.05 Å in the range 1170–1710 Å. Intensity ratios of the intersystem lines of O IV near 1400 Å are used as the primary diagnostic to measure electron pressures in a sunspot and plage.

074.007 **X-ray lines of helium-like oxygen and neon in the solar corona.**
D. L. McKenzie, P. B. Landecker.
Astrophys. J., Vol. 259, 372 - 380 (1982).
Line ratios from the helium-like ions O VII and Ne IX are useful as plasma diagnostics for both flaring and nonflaring solar coronal active regions. The authors have analyzed 29 X-ray spectra obtained by the SOLEX B spectrometer on the USAF P78-1 satellite. Line ratios of O VII and Ne IX are used for estimating electron densities and column densities of the ions in the corona above the active regions.

074.008 **On the nature of the solar corona near the maximum of cycle 21.**
R. R. Fisher.
Astrophys. J., Vol. 259, 431 - 436 (1982).
K-coronameter observations of the white light corona, made in 1980–1981, show that a persistent global pattern in the distribution of electron density existed during this phase of the sunspot cycle. The amplitude of the density distribution is time dependent, and there is a pronounced difference of persistence observed between the equatorial structure and high-latitude coronal features. Both the high-latitude and equatorial structure of the corona rotated with a synodic period of 28ᵈ0.

074.009 **An association between coronal structures and type III burst sources.**
G. Trottet, M. Pick, L. House, R. Illing, C. Sawyer, W. Wagner.
Astron. Astrophys., Vol. 111, 306 - 311 (1982).
Observations of the coronal structures overlying type III radio burst producing regions, obtained with the HAO Coronograph/Polarimeter on board the Solar Maximum Mission, have been combined with observations of the type III bursts from the Mark III Nançay Radioheliograph at 169 MHz. A first qualitative comparison of the data, performed for two periods, indicates that the corona overlying type III producing sites is composed of small, discrete, over-dense structures, which are sometimes diverging and short-lived. The type III burst sources are complex, exhibiting elementary components. It is suggested that the type III burst structure follows that of the corona and that discrete structures may play a fundamental role for type III production.

074.010 **Diagnostic of coronal magnetic fields from microwave polarization reversal.**
R. Bandiera.
Astron. Astrophys., Vol. 112, 52 - 60 (1982).
A method is proposed to determine constraints on magnetic fields (and densities) in coronal layers above solar active regions. The method is based on the interpretation of the radio emission as due to the interaction between wave modes (coupling). The first part of the paper is devoted to the study of mode coupling effects on the wave polarization. Then a theory is presented, that allows to deduce intrinsic physical quan-

tities directly from radio maps during an observation of a polarization inversion.

074.011 **Solar activity in the corona.**
D. J. McLean.
Proc. Astron. Soc. Australia, Vol. 4, 132 - 138 (1981).
The solar corona is an extensive hot plasma permeated by and structured by fairly complex magnetic fields. For the various types of radio bursts which the author has selected the observations provide a clear key to the nature of the exciting disturbance which causes them.

074.012 **Two colour photometry and polarimetry of the solar corona of 16 February 1980.**
J. Dürst.
Astron. Astrophys., Vol. 112, 241 - 250 (1982).
Eclipse photographs obtained with a new type of eclipse camera during the solar eclipse of February 16, 1980 at Yellapur (India) are analysed, applying special care to relative and absolute calibration and to the coronal aureole. Due to good and stable atmospheric conditions at the observing site, reliable values of coronal intensities, up to $r = 10r_\odot$ are obtained. Polarization is measured up to $r = 3.3r_\odot$ and used to separate K and F corona.

074.013 **Observed polarization of the Fe XIV 5303 coronal emission line.**
J. Arnaud.
Astron. Astrophys., Vol. 112, 350 - 354 (1982).
The polarization of the Fe XIV λ 5303 Å line has been observed between 1.1 and 2.0 $R_\odot$ from disk center at Pic-du-Midi during 1977–1980. The polarization direction is found to be mainly radial. The average polarization degree is 1.6%. This low polarization may be due to a significant contribution of collisional excitation to the radiation in the denser coronal structures, as well as to magnetic depolarization when the magnetic field is not in the radial direction. The average values of polarization and intensity agree well with a strongly (but realistically) inhomogeneous density model of the corona.

074.014 **On cool coronal loops.**
P. C. H. Martens, N. P. M. Kuin.
Astron. Astrophys., Vol. 112, 366 - 368 (1982).
The authors consider a relatively "cool" equilibrium state of the plasma in coronal loops and show that this state exists for loops whose heights do not exceed a few times the pressure scale height at the temperature for which the radiation losses are maximum. This maximum height does not depend on the absolute magnitude of the heating and only very weakly on its distribution.

074.015 **Characteristics of shocks in the solar corona, as inferred from radio, optical, and theoretical investigations.**
A. Maxwell, M. Dryer.
Space Sci. Rev., Vol. 32, (see 012.005), 11 - 25 (1982).
Solar radio bursts of spectral type II provide one of the chief diagnostics for the propagation of shocks through the solar corona. Radio data on the shocks are compared with computer models for propagation of fast-mode MHD shocks through the solar corona. Data on coronal shocks and high-velocity ejecta from solar flares are then discussed in terms of a general model consisting of three main velocity regimes.

074.016 **Numerical simulation of magnetohydrodynamic shock propagation in the corona.**
S. T. Wu.
Space Sci. Rev., Vol. 32, (see 012.005), 115 - 129 (1982).
Recent developments in the field of numerical simulation models for the study of shock wave propagation in the corona are presented. The characteristics and physical interpretations of the results derived from these models are discussed in some detail. The most significant physical results obtained to date are provided by the two-dimensional non-planar, time-dependent, MHD numerical simulation model. In this model, the non-linear interaction among the three essential MHD waves, i.e., fast-, slow-, and Alfvén waves are demonstrated. The physical relevance of these numerical simulation models in relation to observed solar activity is presented.

074.017 **Unusual coronal activity following the flare of 6 November 1980.**
Z. Švestka, B. R. Dennis, M. Pick, A. Raoult, C. G. Rapley, R. T. Stewart, B. E. Woodgate.
Sol. Phys., Vol. 80, 143 - 159 (1982).
For almost 30 hr after the major (gamma-ray) two-ribbon flare on 6 November 1980, 03 : 30 UT, the Hard X-Ray Imaging Spectrometer aboard the SMM satellite imaged in > 3.5 keV X-rays a gigantic arch extending above the active region over the limb. 6.5 hr after the flare a coronal region below the arch started quasi-periodic pulsations in X-ray brightness. These brightness variations correlated with similar variations in brightness at 169 MHz.

074.018 **Plasma flow around coronal loops.**
S. Hinata.
Sol. Phys., Vol. 80, 173 - 183 (1982).
Current dissipation models of coronal loop heating are studied. Turbulent current dissipation is shown to lead to a time dependent process because of an enormous mass motion induced in the current layer. A stationary heating process involves only ohmic heating, which requires a large current layer. To insure MHD stability, the loop must be composed of many elements with the oppositely directed currents. A stationary current dissipation process induces the plasma motion across the magnetic field into the loop and down the loop with the speeds $v_\perp \sim 10^4$ cm s^{-1} and $v_\parallel \sim 2 \times 10^4$ cm s^{-1}, respectively.

074.019 **Coronal rotation during solar cycle 20.**
G. D. Parker, R. T. Hansen, S. F. Hansen.
Sol. Phys., Vol. 80, 185 - 198 (1982).
Using *K*-coronameter observations during 1964 - 1976, the authors determine the apparent recurrence period of white-light solar coronal features as a function of latitude, height, and time. A technique based on maximum entropy spectral analysis is used to produce rotational period estimates. The analysis reaffirms the existence of differential rotation in the corona and describes both its average behavior and its large year-to-year variations.

074.020 **The termal stability of solar coronal loops in hydrostatic equilibrium.**
M. A. Wragg, E. R. Priest.
Astron. Astrophys., Vol. 113, 269 - 276 (1982).
Many static models of solar coronal loops have been set up, and more recently there have been attempts to look at their stability. In the present paper the effects of gravity are included by considering a basic state in hydrostatic equilibrium. It is found that the inclusion of gravity is a stabilising influence because it lowers the density (and hence the radiation) in the upper parts of a loop. Thermally isolated loops are shown to be thermally unstable when they are short but stable when they are long enough. Thus one would expect small coronal loops to exhibit dynamic fluctuations and longer ones to possess a steadier emission.

074.021 **Resonant electrodynamic heating and the thermal stability of coronal loops.**
P. C. H. Martens, M. Kuperus.
Astron. Astrophys., Vol. 113, 324 - 327 (1982).
A theory for the thermal stability of hot coronal loops is presented, which is based on the resonant electrodynamic heating theory of Ionson (1982) and the evaporation/condensation scenario of Krall and Antiochos (1980). The theory

predicts that gradual changes in the length of a loop or in its magnetic field strength can trigger catastrophic changes in the X-ray visibility of the loop, without the need for a change in the magnetic field topology. A natural explanation is thereby given for the observations of X-ray brightenings in loops and loop evacuations with coronal rain.

074.022 **The solar coronal magnetic field and need for its consideration in the problem of coronal plasma dissipation.**
A. R. Bakhalbashyan, G. M. Nedyalkova.
Researches on stellar physics, 1981, (see 31.003.010), p. 89 - 96 (1981). In Russian. – Abstr. in Ref. zh., 51. Astron., 8.51.277 (1982).

074.023 **Ejection of slow neutrons as a probable mechanism of solar wind and corona formation.**
I. M. Matora.
Obedin. inst. yader. issled. Dubna, Soobshch., 1982, No. R4-82-41, 3 pp. In Russian. – Abstr. in Ref. zh., 51. Astron., 8.51.328 (1982).

074.024 **On the effective and true distribution of velocities in the solar wind.**
V. A. Alimov.
Izv. vuzov. Radiofiz., Tom 24, 643 - 645 (1981). In Russian. Abstr. in Ref. zh., 62. Issled. kosm. prostranstva, 8.62.257 (1982).

074.025 **Observations of mass addition to the shocked solar wind of the Venusian ionosheath.**
D. S. Intriligator.
Geophys. Res. Lett., Vol. 9, 727 - 730 (1982).
The author has continued an earlier study of the heavy ion component in the plasma observations downstream from Venus by the plasma analyzer experiment on the Pioneer Venus Orbiter. He presents information on the heavy ions including intensities, energy distributions, speed, flow directions, and the relation of the heavy ion observations to the simultaneous proton observations.

074.026 **Unsteady, thermally conductive coronal flow.**
S. T. Suess.
Astrophys. J., Vol. 259, 880 - 899 (1982).
A numerical algorithm using implicit time-differencing is applied to the solar wind equations allowing, for the first time, solutions including thermal conduction to be found for time-dependent flow traversing the subsonic to supersonic velocity transition region. Sample solutions are shown that demonstrate the distinctive differences introduced by including thermal conduction, in comparison to the commonly available solutions assuming polytropic flow.

074.027 **Measurements of outflow from the base of solar coronal holes.**
G. J. Rottman, F. Q. Orrall, J. A. Klimchuk.
Astrophys. J., Vol. 260, 326 - 337, plate 17 (1982).
New evidence is presented that EUV emission lines formed at the levels of the base of the corona and the transition region are systematically shifted to shorter wavelengths within coronal holes relative to the rest of the solar disk, and that moreover this shift increases with height in the atmosphere. This evidence is contained in measurements made with a rocket-borne EUV spectrometer. The maximum measured shift corresponded to a velocity of 12 km s^{-1} in λ625 Mg X and 7 km s^{-1} in λ629 O V.

074.028 **On the source of energy of solar wind turbulence.**
I. V. Chashej, V. I. Shishov.
Geomagn. Aehron., Tom 22, 529 - 535 (1982). In Russian.

074.029 **The dynamics of accelerating coronal bullets.**
J. T. Karpen, E. S. Oran, J. T. Mariska, J. P. Boris, G. E. Brueckner.
Astrophys. J., Vol. 261, 375 - 386 (1982).
EUV spectroscopic observations of the Sun recently have revealed jets or bullets of cool, dense plasma ($T_e \sim 10^5$ K, $n_e \sim 10^{10}$ cm^{-3}), with lifetimes of order 60 s, which accelerate through the corona at velocities of 50 to 400 km s^{-1}. The present work is focused on these coronal bullets in an attempt to understand their dynamical progress and the underlying energy source. The evolution of discrete masses of cool, dense plasma injected into the corona has been studied using the Naval Research Laboratory Dynamic Flux Tube Model. Results indicate that the bullets may comprise a significant fraction of the coronal mass flux but constitute a negligible component of the coronal energy budget.

074.030 **Transient behavior of flare-associated solar wind: II. In a non-radial open-field region.**
F. Nagai.
Bull. American Astron. Soc., Vol. 14, 606 - 607 (1982). Abstract.

074.031 **Stellar winds with multiple critical points and shock transitions.**
S. R. Habbal, K. Tsinganos.
Bull. American Astron. Soc., Vol. 14, 608 (1982). – Abstract.

074.032 **Detailed structure and energetics of accelerating coronal bullets.**
J. T. Karpen, E. S. Oran, J. P. Boris, J. T. Mariska.
Bull. American Astron. Soc., Vol. 14, 622 (1982). – Abstract.

074.033 **Self-propulsion and broadening of coronal loop transients.**
T. C. Mouschovias, R. A. Fiedler.
Bull. American Astron. Soc., Vol. 14, 623 (1982). – Abstract.

074.034 **Are coronal loops stable?**
S. K. Antiochos.
Bull. American Astron. Soc., Vol. 14, 623 (1982). – Abstract.

074.035 **Coronal emission associated with a quiescent prominence.**
R. N. Smartt, Z. Zhang.
Bull. American Astron. Soc., Vol. 14, 623 (1982). – Abstract.

074.036 **Coronal differential rotation.**
G. D. Parker.
Bull. American Astron. Soc., Vol. 14, 640 (1982). – Abstract.

074.037 **Computer simulation of the solar wind.**
H. Washimi, T. Ogino.
Magnetospheric phenomena of celestial bodies, (see 012.011), p. 36 - 38 (1982). In Japanese.
Using a 2-dimensional MHD code, the authors have obtained a stationary solution of the solar wind near the sun which shows the formation of the neutral sheet along the equatorial plane and the fast stream at high latitudes.

074.038 **Solar corona: the magnetosphere of the sun.**
T. Sakurai.
Magnetospheric phenomena of celestial bodies, (see 012.011), p. 43 - 58 (1982). In Japanese.
The magnetic field plays a vital role in various processes that take place in the solar corona. The corona is an aggregate of coronal loops, whose million degree temperature is to be explained in terms of magnetic-field-related heating mechanisms. Streamers, coronal transients, and eruptive prominences are examples of magnetohydrodynamical phenomena in the corona. Solar flares, the most energetic events in the corona, are not yet fully understood but several models have

been proposed to account for the fundamental process of magnetic energy liberation.

074.039 **Complex time analysis of recurrent activity of coronal index 1971 - 1976.**
V. Letfus, E. M. Apostolov.
Bull. Astron. Inst. Czechoslovakia, Vol. 33, 294 - 302 (1982).
The aim of this paper was not only the time analysis of the coronal index data, but also an attempt to develop a more detailed and complex analytical method for a better understanding of the results of the analysis. The following analytical methods have been used: autocorrelation analysis, complex demodulation and sequentional ranging of data in Bartels rotations. For each of the methods the observed variation of the recurrence period, connected with the Sun's rotation, has been discussed. The application of these methods gives the possibility to estimate the recurrence period statistically. The detected shifts of activity in longitude may be explained by the generation of new activity associated with the character of the convection of subphotospheric matter and with the generation of subphotospheric magnetic fields.

074.040 **Quiet-time solar-wind ionic composition.**
V. Formisano, S. Orsini.
Nuovo Cimento C, Vol. 4C, Ser. 1, 682 - 697 (1981). – Abstr. in Phys. Abstr.,Vol. 85, Abstr. 68586 (1982).

074.041 **Configuration of the magnetic field and electric conductivity of the plasma in the solar wind.**
A. D. Chertkov.
Magnitosfer. issled. No. 1, Moskva, 1982, p. 5 - 16. In Russian. Abstr. in Ref. zh., 62. Issled. kosm. prostranstva, 9.62.228 (1982).

074.042 **Structure and origin of the magnetic field of recurrent high-velocity streams in the solar wind.**
E. M. Fomina, A. D. Chertkov.
Magnitosfer. issled. No. 1, Moskva, 1982, p. 36 - 44. In Russian. – Abstr. in Ref. zh., 62. Issled. kosm. prostranstva, 9.62.271 (1982).

074.043 **Simultaneous radio scattering and white light observations of a coronal transient.**
R. Woo, J. W. Armstrong, N. R. Sheeley Jr., R. A. Howard, D. J. Michels, M. J. Koomen.
Nature, Vol. 300, 157 - 159 (1982).
The authors report that a coronal transient observed by the Solwind coronograph off the west limb of the sun on 24 October 1979 was also seen in the spectral broadening observations of the Helios-2 2.3 GHz radio signal. A comparison of these simultaneous data suggests that in this case the shock front does not coincide with the white-light front but in fact precedes it.

074.044 **Flare-related coronal transients.**
R. M. E. Illing.
Observatory, Vol. 102, 122 - 123 (1982). – Abstract.

074.045 **Multi-station observations of interplanetary scintillations.**
S. Ananthakrishnan, J. J. Kaufman.
Bull. Astron. Soc. India, Vol. 10, 1 - 15 (1982).
Solar wind velocity estimation using the method of multi-station interplanetary scintillation observations is reviewed. It is shown that the inferred model for the electron density fluctuations in the solar wind differs significantly from that for the proton density. The high latitude solar wind has a strong correlation with the presence of coronal hole regions.

074.046 **On the estimation of the amount of heating for solar coronal loops.**
S. Chandra, U. Narain.
Bull. Astron. Soc. India, Vol. 10, 18 - 25 (1982).

074.047 **On the estimation of the amount of heating for solar coronal loops.**
S. Chandra.
Bull. Astron. Soc. India, Vol. 10, 38 (1982). – Abstract.

074.048 **Two colour polarization studies of the solar corona.**
K. A. Raju, K. D. Abhyankar.
Bull. Astron. Soc. India, Vol. 10, 38 (1982). – Abstract.

074.049 **Turbulent velocities in the solar corona – an estimate from Fabry-Perot interferometric observations of the green coronal line during the eclipse of 1980 February 16.**
J. N. Desai, T. Chandrasekhar.
Bull. Astron. Soc. India, Vol. 10, 38 (1982). – Abstract.

074.050 **An indication for an intrinsically different coronal temperature.**
A. Geranios.
Astrophys. Space Sci., Vol. 85, 443 - 449 (1982).
As is well known, the interplanetary magnetic field is an extension of the solar magnetic field, and the solar wind is an extension of the solar corona (Biermann, 1951; Parker, 1958). Consequently, knowledge of these two interplanetary structures reflects a knowledge of the solar corona itself. Calculations of radial proton temperature gradients within magnetic sector boundaries, observed by Helios-1 and Helios-2 heliocentric satellites between 0.3 and 1.0 AU, revealed lower temperatures in the solar corona in the region where these boundaries merge.

074.051 **On the screening of the interstellar electric and magnetic fields by the solar wind.**
I. S. Veselovsky *(Veselovskij)*.
Astrophys. Space Sci., Vol. 86, 209 - 213 (1982).
Fast-streaming solar-wind plasma with high conductivity screens the heliosphere from the penetration of the interstellar electric and magnetic fields. The simplest model with the constant solar wind conductivity and radial velocity is considered and the boundary electrodynamic problem is solved for static external fields. The results show that screening of the external fields takes place in the heliosphere according to the exponential law.

074.052 **A three-dimensional model of corotating streams in the solar wind. 3. Magnetohydrodynamic streams.**
V. J. Pizzo.
J. Geophys. Res., Vol. 87, 4374 - 4394 (1982).
The focus of the paper is two-fold: (1) to examine how the presence of the spiral magnetic field affects the evolution of interplanetary corotating solar wind streams, and (2) to ascertain the nature of secondary large-scale phenomena likely to be associated with streams having a pronounced three-dimensional (3-D) structure. The dynamics are presumed to be governed by the nonlinear, polytropic, single-fluid, 3-D MHD equations. Solutions are obtained with an explicit, Eulerian, finite differences technique that makes use of a simple form of artificial diffusion for handling shocks.

074.053 **Wave heating and acceleration of solar wind ions by cyclotron resonance.**
E. Marsch, C. K. Goertz, K. Richter.
J. Geophys. Res., Vol. 87, 5030 - 5044 (1982).
In order to explain the ion temperature anisotropies and differential speeds, as observed in solar wind high-speed streams, a fluid type model is presented that takes into account ion heating and acceleration by resonant wave-particle interactions. Ion-cyclotron and fast magnetosonic waves propagating away from the sun parallel to the inter-

planetary magnetic field are considered. The radial evolution of the spectral wave energy density and of the bi-Maxwellian model ion distributions is calculated self-consistently for a spherically symmetric solar wind geometry. Numerical results are given for the dependence on heliocentric radial distance for the ion parallel and perpendicular temperatures and ion speeds relative to their center of mass frame. The respective ion flow speeds in the inertial frame are also calculated, based on momentum equations that include the self-consistent temperature gradients.

074.054 **On the complex state of the interplanetary medium of July 28 - 29, 1977.**
J. H. King, R. P. Lepping, J. D. Sullivan.
J. Geophys. Res., Vol. 87, 5881 - 5887 (1982).
Plasma and magnetic field variations observed on July 28 - 30, 1977, in the near-earth solar wind are presented and discussed. Both a corotating stream and a driven shock are present.

074.055 **Measurement of the rugged invariants of magnetohydrodynamic turbulence in the solar wind.**
W. H. Matthaeus, M. L. Goldstein.
J. Geophys. Res., Vol. 87, 6011 - 6028 (1982).
Measurements of the total energy, cross helicity, and magnetic helicity of the solar wind at 1, 2.8, and 5 AU are presented. These quantities are the three rugged invariants of three-dimensional ideal incompressible MHD turbulence theory. The theoretical technique for measuring the magnetic helicity from the matrix of two-point correlations is shown.

074.056 **Whistler mode turbulence in the disturbed solar wind.**
F. V. Coroniti, C. F. Kennel, F. L. Scarf, E. J. Smith.
J. Geophys. Res., Vol. 87, 6029 - 6044 (1982).

074.057 **Geomagnetic and solar wind cycles, 1900 - 1975.**
J. Feynman.
J. Geophys. Res., Vol. 87, 6153 - 6162 (1982).
Recent research has established the reality of the 80 to 100-year cycle (long cycle) in solar-terrestrial phenomena that for many years has been suspected to occur. This paper investigates some of the systematic changes that took place during the most recent long cycle. The variations of one hundred years of data on the aa index of geomagnetic activity are reinvestigated. Since the solar wind drives geomagnetic activity the results of the study are interpreted in terms of the solar wind.

074.058 **Physics of static loop structures and scaling laws.**
G. L. Withbroe.
Solar active regions, (see 003.011), p. 199 - 212 (1981).
This paper discusses the theory of models of solar coronal loops, some applications to observations, and some of the insights that they provide concerning coronal physics.

074.059 **Theory of loop flows and instability.**
E. R. Priest.
Solar active regions, (see 003.011), p. 213 - 275 (1981).
Contents: Introduction. Siphon flow. Evaporation and draining. Thermal stability of static loops. Downflow and upflow in intense flux tubes. Magnetohydrodynamic instability of loops. Conclusion.

074.060 **Mechanisms of energy supply.**
J. V. Hollweg.
Solar active regions, (see 003.011), p. 277 - 318 (1981).
The following physical processes are examined: (1) the heating of coronal active region loops by anomalous dissipation of parallel currents in thin sheaths; (2) the heating of coronal active region loops by minireconnections; (3) the role of surface waves in active region loops; (4) spatially dependent heating in the corona and the formation of coronal active region loops as influenced by refraction and Landau/transit-time damping of fast-mode MHD waves; (5) the feasibility of propagating Alfvén waves on magnetic flux tubes as a means of carrying energy from the convection zone into the chromosphere and corona on open and closed magnetic flux tubes.

074.061 **Penetration of the magnetic field of the solar wind into cometary ionospheres.**
L. S. Marochnik.
Kosm. Issled., Tom 20, 779 - 783 (1982). In Russian.

074.062 **Slow shocks around the sun.**
Y. C. Whang.
Geophys. Res. Lett., Vol. 9, 1081 - 1084 (1982).
It is inferred from this study that magnetohydrodynamic slow shocks can exist in the vicinity of the sun. The study uses a two-hole corona model, the sub-Alfvenic streams originating from the edge of the polar open-field regions are forced to turn towards equator in coronal space following the curved boundary of the closed field region. When the streamlines from the opposite poles merge at a neutral point, their directions become parallel to the neutral sheet. An oblique slow shock can develop near or at the neutral point, the shock extends polewards to form a surface of discontinuity around the sun.

074.063 **Lower hybrid waves in the solar wind.**
E. Marsch, T. Chang.
Geophys. Res. Lett., Vol. 9, 1155 - 1158 (1982).
The authors demonstrate that the frequently observed broad band low frequency electrostatic noise in the solar wind generally has a dominant lower hybrid component.

074.064 **Properties of coronal arches.**
J. M. Davis, A. S. Krieger.
Sol. Phys., Vol. 80, 295 - 307 (1982).
The properties of coronal arches located on the peripheries of active regions, observed during a sounding rocket flight on March 8, 1973, are discussed. The arches are found to overlie filament channels and their footprints are traced to locations on the perimeters of supergranulation cells. Comparison of the size of the chromospheric footprint with the arch width indicates that arches do not always expand as they ascend into the corona. The electron temperatures and densities of the plasma contained in the arches were measured and the pressure calculated. The variation of these parameters with position along the length of the arch indicates that the arches are not in hydrostatic equilibrium.

074.065 **Thermally isolated coronal loops in hydrostatic equilibrium.**
M. A. Wragg, E. R. Priest.
Sol. Phys., Vol. 80, 309 - 312 (1982).
Numerical solutions are presented for the summit temperature and heating in a thermally isolated coronal loop that is in hydrostatic equilibrium. The extent to which gravity modifies the usual uniform-pressure scaling law is shown, and plots of the differential emission measure are also given.

074.066 **Transient response of the solar wind to changes in flow geometry. Flows in coronal holes.**
S. S. Hasan, P. Venkatakrishnan.
Sol. Phys., Vol. 80, 385 - 403 (1982).
The transient response of the solar wind to changes in geometry is examined. An initial stationary flow in a configuration that diverges as r^2 is assumed. This state corresponds to the usual solar wind solution. The effect on the flow through a tube whose area $A(r, t)$ diverges faster than r^2, with the degree of divergence increasing in time, is considered. The asymptotic form of $A(r, t)$ is chosen to mimic the form

inferred in coronal holes. A detailed parameter study relating the form of $A(r, t)$ to the pattern of flow in the tube is presented.

074.067 **Coronal loop transients in streamer configurations.**
R. S. Steinolfson.
Astron. Astrophys., Vol. 115, 39 - 49 (1982).
Numerical simulations of the formation and propagation of mass-ejection coronal loop transients in coronal streamer configurations are presented. The procedure discussed by Steinolfson et al. (1982) is used to calculate the streamer configurations. An increase in thermal pressure at the base of the closed-field region in the coronal streamer simulates the energy release in a solar event (i.e., eruptive prominence, flare, etc.). The coronal transient resulting from the energy release is then simulated in the meridional plane with numerical solutions of the single-fluid, time-dependent magnetohydrodynamic equations of motion.

074.068 **Coronal response to a solar event in a corona evacuated by a prior transient.**
R. S. Steinolfson.
Astron. Astrophys., Vol. 115, 50 - 53 (1982).
The pre-event corona is well recognized as playing an important role in determining the geometry and thermodynamics of the coronal response to a solar event. Additionally, the disturbed corona following a solar event may be substantially different than the pre-event corona for several hours (or days). In this note the numerically-simulated coronal response to a solar event two hours after an identical solar event at the same location is compared to (1) the coronal response to just the first event and (2) the coronal response that would have existed if the second event had not occurred.

074.069 **The physics of the solar wind.**
F. M. Neubauer.
The solar system and its exploration, (see 012.054), p. 41 (1981). – Abstract.

074.070 **Solar wind observations with high time resolution.**
G. N. Zastenker, Yu. I. Ermolaev, S. Pinter, Z. Němeček, I. Šafránková, A. B. Belikova, A. V. Lejbov, V. I. Prokhorenko, A. E. Stefanovich, A. G. Bedrikov, B. T. Karimov.
Kosm. Issled., Tom 20, 900 - 906 (1982). In Russian.

074.071 **Visible light observations of a dense plasmoid associated with a moving Type IV solar radio burst.**
R. T. Stewart, G. A. Dulk, K. V. Sheridan, L. L. House, W. J. Wagner, C. Sawyer, R. Illing.
Astron. Astrophys., Vol. 116, 217 - 223 (1982).
On 1980 April 27 a coronal transient rising above the eastern limb of the Sun was observed simultaneously by both the Coronagraph/Polarimeter aboard the Solar Maximum Mission spacecraft and by the Culgoora radioheliograph. The Coronagraph/Polarimeter observed an outward-moving loop transient followed by a plasmoid of dense, ionized material, while the radioheliograph observed several types of radio bursts, including a moving Type IV burst at 80 MHz whose position coincided with the bright plasmoid. From the coronagraph observations the authors estimate what they consider to be a reasonable lower limit of the electron density in the moving Type IV source. This estimate indicates that the moving Type IV burst could have resulted from emission at either the fundamental or the second harmonic of the plasma frequency. The derived densities in the plasmoid require a confining magnetic field $B > 0.6$ G at $R = 2.5\ R_{\odot}$.

074.072 **The analysis of Fe XIV 5303 coronal emission-line polarization measurements.**
J. Arnaud.
Astron. Astrophys., Vol. 116, 248 - 254 (1982).
The author discusses polarization measurements obtained in September 1978 in the green coronal line with the Pic-du-Midi coronameter. The minimum value of iron abundance deduced is 7×10^{-5}; it appears to be larger than recent photospheric abundances. The interest of simultaneous measurements in the 10747 Å line of Fe XIII is emphasized. The problem of inferring the magnetic field direction is discussed.

074.073 **Coronal heating by phase-mixed shear Alfvén waves.**
J. Heyvaerts, E. R. Priest.
Pré-publ. Inst. Astrophys. Paris, No. 14, 48 pp. (1982). Accepted by Astron. Astrophys.

074.074 **Empirical model of the large scale structures of the solar corona using solar-cycle related parameters.**
S. Koutchmy.
Pré-publ. Inst. Astrophys. Paris, No. 18, 12 pp. (1982). Contributed paper presented at the fifth International Symposium on solar-terrestrial physics.

074.075 **Short period coronal oscillations: observations and interpretation.**
S. Koutchmy, Y. D. Žugžda (*Yu. D. Zhugzhda*), V. Ločans.
Pré-publ. Inst. Astrophys. Paris, No. 20, 33 pp. (1982). Accepted by Astron. Astrophys.

074.076 **On TV observations of polarization of the green coronal line 5303 Å at the time of the eclipse of 1981.07.31.**
O. S. Popov, V. N. Ivchenko, V. P. Lapchuk, G. P. Milinevskij, K. I. Osminkina, S. Koutchmy, J. Begot, C. Lebecq, G. Stellmacher.
Astron. Tsirk., No. 1202, p. 1 - 3 (1981). In Russian.

074.077 **The growth of filaments by the condensation of coronal arches.**
J. M. Davis, A. S. Krieger.
Sol. Phys., Vol. 81, 325 - 338 (1982).
A model of filament formation based on the condensation of coronal arches is described. The condensation results from initiating the radiative instability within an arch by superimposing a transient energy supply upon the steady state heating mechanism. The transient energy supply increases the density within the arch so that when it is removed the radiative losses are sufficient to lead to cooling below the minimum in the power loss curve. Times from the initial formation of the condensation to its temperature stabilization as a cool filament have been calculated for various initial conditions. Using experimentally derived parameters, filament densities of 10^{11} to 10^{12} cm^{-3} can be obtained.

074.078 **Investigation of periodic (20 - 40 min) scintillations in the magnetic field of the solar wind.**
V. N. Obridko, Ya. I. Fel'dshtejn, B. D. Shel'ting, N. G. Varfolomeeva.
Probl. kosm. ehlektrodinam. Moskva, 1981, p. 12 - 20. In Russian. – Abstr. in Ref. zh., 62. Issled. kosm. prostranstva, 10.62.392 (1982).

074.079 **Ballooning instability driven by fast magnetosonic waves and its application to coronal loop transients associated with a flare.**
J.-I. Sakai.
Astrophys. J., Vol. 263, 970 - 981 (1982).
A theoretical model for the triggering of coronal loop transients associated with a flare or eruptive prominence is presented. It is suggested that coronal loop transients are due to a ballooning instability in a coronal current loop, which can be driven by fast magnetosonic waves originating from a solar flare – or eruptive prominence – induced disturbance.

074.080 **The observation of a coronal transient directed at Earth.**
R. A. Howard, D. J. Michels, N. R. Sheeley, Jr., M. J. Koomen.
Astrophys. J., Lett., Vol. 263, L101 - L104, plates L5, L6 (1982).

The authors report the observation of a large coronal transient that must be interpreted as a three-dimensional structure. Its form is one which has not been observed before – a gradually expanding, Sun-centered disk of excess brightness, whose projected radius increased from 4 $R_\odot$ to 8 $R_\odot$ during 0832–0958 UT on 1979 November 27. This Earth-directed transient originated with the sudden disappearance of a large filament and a relatively minor solar flare. It was the source of an interplanetary shock that reached the earth at 0738 UT, November 30.

074.081 **Connection of the velocity of solar wind streams with coronal hole dimensions.**
V. A. Kovalenko, S. I. Molodykh.
Issled. po geomagn., aehron. i fiz. Solntsa, Moskva, 1982, No. 60, p. 135 - 140. In Russian. – Abstr. in Ref. zh., 51. Astron., 11.51.346 (1982).

074.082 **Optical observations of the solar corona.**
R. R. Fisher.
Space Sci. Rev. Vol. 33, (see 012.067), 9 - 16 (1982).

The history of optical observations of the solar corona is traced through the development of four topics: the observation of naturally occurring total solar eclipses, the discovery of forbidden emission lines in the solar corona, the development of narrow spectral bandpass observing techniques, and the invention of the coronagraph. These events occurred over a span of time from the middle of the last century until the present.

074.083 **Probing the solar wind acceleration region using spectroscopic techniques.**
G. L. Withbroe, J. L. Kohl, H. Weiser, R. H. Munro.
Space Sci. Rev., Vol. 33, (see 012.067), 17 - 52 (1982).

The authors discuss how measurements of spectral line radiation in conjunction with measurements of the white light *K*-corona can provide information on electron, proton and ion temperatures and velocity distribution functions; densities; chemical abundances and mass flow velocities. They provide a comprehensive review of the formation of coronal resonance line radiation, with particular emphasis on the H I Lα line, and discuss observational considerations such as requirements for rejection of stray light and effects of emission from the geocorona and interplanetary dust. Finally they summarize some results of coronal H I Lα and white light observations acquired on sounding rocket flights.

074.084 **X- and XUV-radiation of the solar corona.**
G. Elwert.
Space Sci. Rev., Vol. 33, (see 012.067), 53 - 82 (1982).

A survey of the ionization states and emission lines of the ions existing in the corona is given. Instruments for taking pictures of the Sun in the X- and XUV-region as well as for measuring spectra emitted in interesting locations on the Sun are presented. Methods of plasma diagnostics, in particular for the determination of the mean temperature and the differential emission measure are described. A review of observations, types of coronal structures, especially coronal holes, active regions and large scale structures are described. Their relations to the photospheric magnetic fields are dealt with; methods to calculate coronal magnetic fields are briefly discussed. As for temporal variations results of the analysis of expanding X-ray arches and of structures becoming visible in the outer corona in white light are mentioned. Finally, plasma diagnostics by means of high-resolution spectra are dealt with.

074.085 **Acceleration of the solar wind.**
E. Leer, T. E. Holzer, T. Flå.
Space Sci. Rev., Vol. 33, (see 012.067), 161 - 200 (1982).

The authors discuss recent research on the acceleration of the solar wind, giving emphasis to high-speed solar wind streams emanating from solar coronal holes. They explain why thermally driven wind models constrained by solar and interplanetary observations encounter substantial difficulties in explaining high speed streams. Then, through a general discussion of energy addition to the solar wind above the coronal base, they indicate a possible resolution of these difficulties. Finally, they consider the question of what role MHD waves might play in transporting energy through the solar atmosphere and depositing it in the solar wind, and they conclude by examining the specific mechanism of solar wind acceleration by Alfvén waves and the related problem of accelerating massive stellar winds with Alfvén waves.

074.086 **Processes affecting abundances in the solar wind.**
J. Geiss.
Space Sci. Rev., Vol. 33, (see 012.067), 201 - 217 (1982).

Data on composition in the solar wind are summarized and compared with best estimates of abundances in the outer convective zone of the Sun. Several mechanisms of element and isotope fractionation are discussed in relation to observed abundances and their variations. The evidence available so far indicates that in addition to ion fractionation in the corona there is a separation mechanism operating at low solar altitude that affects solar wind composition. It is suggested that the systematic depletion of helium observed in the solar wind is in part caused by ion-neutral separation in the chromosphere-transition zone. The ion-neutral separation may be responsible for the general depletion that is observed in the slow interstream solar wind as well as in the fast streams coming out of coronal holes. However, the occurrences of very low He/H ratios are probably caused in the corona.

074.087 **Observations of coronal structure during sunspot maximum.**
N. R. Sheeley, Jr., R. A. Howard, M. J. Koomen, D. J. Michels, K. L. Harvey, J. W. Harvey.
Space Sci. Rev., Vol. 33, (see 012.067), 219 - 231 (1982).

This paper presents some of the results that have been obtained from the Kitt Peak observations of coronal holes and the NRL observations of coronal transients during the recent years near sunspot maximum (1979 - 1981). On the average, low-latitude coronal holes of comparable size contained 3 times more flux near sunspot maximum than near the previous minimum. In the outer corona, transients occurred at the observed rate of at least 2 per day, and quiet conditions persisted during less than 15% of the observed days. The authors describe a sample of the more than 800 events that they have observed so far, including the observations of a comet apparently colliding with the Sun.

074.088 **Coronal transient phenomena.**
M. Dryer.
Space Sci. Rev., Vol. 33, (see 012.067), 233 - 275 (1982).

A series of coronal transient observations is discussed in the spirit of a representative overview following some introductory remarks on the background solar wind. One of these observations is chosen because its interplanetary signature – the shock wave – was detected by two spacecraft at different heliocentric radii. Other cases are chosen because of the extended observations of embedded eruptive prominences. Progress is also being made in the interdisciplinary areas of optical imagery complemented with radio astronomical techniques. Finally, several recent theoretical models and MHD computer simulation studies are summarized. It is suggested that further comparison of specific events with such models promises a rich harvest of physical understanding of the origin, structure and interplanetary progeny of coronal transients.

074.089 **Quantitative tests of a steady state theory of solar wind electrons.**
W. C. Feldman, J. R. Asbridge, S. J. Bame, J. T. Gosling.
J. Geophys. Res., Vol. 87, 7355 - 7362 (1982).

IMP electron data are compared to several predictions of a steady state theory of solar wind electrons which assumes that transport is controlled by the macroscopic interplanetary electric and magnetic fields and elastic Coulomb collisions with the solar wind protons and thermal electrons.

074.090 **Helium, hydrogen, and oxygen velocities observed on ISEE 3.**
K. W. Ogilvie, M. A. Coplan, R. D. Zwickl.
J. Geophys. Res., Vol. 87, 7363 - 7369 (1982).

The velocities of helium, oxygen, and hydrogen ions have been recorded over a full range of solar wind conditions between August 1978 and December 1979. Interspecie velocity differences were observed frequently. For solar wind velocities between 300 and 400 km s^{-1} the helium velocity exceeded the hydrogen velocity by 5 km s^{-1} on the average. For solar wind velocities between 400 and 500 km s^{-1} the average difference was 14 km s^{-1}; however, no evidence was found for a systematic nonzero average velocity difference between helium and oxygen ions even at the higher velocities. Velocity differences were examined in a number of streams and across a number of interplanetary shocks.

074.091 **Helium abundance enhancements in the solar wind.**
G. Borrini, J. T. Gosling, S. J. Bame, W. C. Feldman.
J. Geophys. Res., Vol. 87, 7370 - 7378 (1982).

Helium abundance (A(He)) enhancements observed with Los Alamos instruments on IMP 6, 7, and 8 during the 1972–1978 interval have been investigated. Statistical analysis of 73 large events with A(He) $\geqslant$ 10% (HAEs) provides evidence for a close link between helium enhancements at 1 A.U. and transient coronal mass ejections. HAE events are sporadic, sometimes clustered in time, and their frequency of occurrence is approximately in phase with the solar cycle. Nearly 50% of HAEs are associated with interplanetary shocks and/or geomagnetic activity sudden commencements, but the plasma pattern associated with A(He) enhancements is independent of shock occurrence. Evidence of a significant association between helium enhancements at 1 A.U. and type II and IV radio bursts in the corona is presented. The authors interpret their results to mean that most HAE events originate in transient coronal disturbances in which the magnetic field strength is enhanced.

074.092 **He^+ and other unusual ions in the solar wind: a systematic search covering 1972–1980.**
R. D. Zwickl, J. R. Asbridge, S. J. Bame, W. C. Feldman, J. T. Gosling.
J. Geophys. Res., Vol. 87, 7379 - 7388 (1982).

Following the recent discovery of large fluxes of He^+ in the solar wind, a systematic search of solar wind energy/charge (E/Q) spectra was initiated, using plasma data from IMP 7, IMP 8, ISEE 1, and ISEE 3 that covered the time period from October 1972 to February 1980. A technique of vertically stacking successive E/Q spectra made it possible to detect systematic trends in the spectra that were indicative of ion species beyond the He^{++} peak down to the one count level. A search of nearly 8 years of data yielded only three distinct events with identifiable levels of He^+. In addition to the unusual appearances of large fluxes of He^+ ions, iron ions ranging from Fe^{+10} to Fe^{+5} were observed. Evidence was found for the appearance of C^{+4} ions, and in one event the data can be interpreted as containing O^{+2} and O^{+3} at detectable levels. One of the phenomenological results following from these observations is that certain unusual transient events eject both hot and unusually cold coronal plasma mixed and coexisting on the same magnetic field lines.

074.093 **Surface waves on solar wind tangential discontinuities.**
J. V. Hollweg.
J. Geophys. Res., Vol. 87, 8065 - 8076 (1982).

The author demonstrates that (tangential) discontinuities in the magnetic field direction can support MHD surface waves. The surface waves resemble the usual Alfvén wave, but there are some important differences which are discussed in detail.

074.094 **Generation of electron plasma waves in the upstream solar wind.**
S. F. Fung, K. Papadopoulos, C. S. Wu.
J. Geophys. Res., Vol. 87, 8077 - 8080 (1982).

Electron plasma waves with amplitudes as high as 10 mV/m have been measured in the electron foreshock region of the earth's bow shock. It is shown that a population of stable suprathermal electrons of the type observed can produce the measured level of enhanced fluctuations without becoming unstable. The steady state wave level is given by the balancing between emission and absorption. The apparent association of fluctuation enhancements with sharp changes in electron fluxes can be understood in terms of modification of the electron distribution due to acceleration processes. The expected $2\omega_{pe}$ radiation from the model is consistent with the observations.

074.095 **Stationarity of magnetohydrodynamic fluctuations in the solar wind.**
W. H. Matthaeus, M. L. Goldstein.
J. Geophys. Res., Vol. 87, 10347 - 10354 (1982).

Solar wind research and studies of charged particle propagation often assume that the interplanetary magnetic field represents a stationary random process. In this paper the authors investigate the extent to which ensemble averages of the solar wind magnetic fields follow the asymptotic behavior predicted by the ergodic theorem. Several time periods, including a span of nearly two years, are analyzed. The results are illustrated by using magnetometer data from the ISEE 3, Voyager and IMP spacecraft.

074.096 **Some properties of the solar wind plasma at extreme solar activities.**
A. Geranios.
Astrophys. Space Sci., Vol. 88, 99 - 114 (1982).

Hourly interplanetary proton plasma data, measured by Helios-1 and Helios-2 satellites over the period extending the interval between the sunspot minimum and maximum of the 21st solar cycle are analysed. This analysis emphasizes the presence of the third type solar wind at 450 km s^{-1}, appearing at solar minimum, during which large coronal holes are dominating in the Sun. This type of solar wind is hardly to be observed during the solar maximum period. The data give an average speed of the slow solar wind of 350 km s^{-1} for the period between these two extremes of solar activities. After correlation of the plasma temperature with its speed in different heliocentric distances, it comes out the stronger heating which takes place in distances shorter than 0.6 AU than in distances between 0.6 and 1.0 AU. A different behaviour of the radial proton temperature gradient in different solar activities appears after the calculation of the gradients as a function of solar wind speed and radial distance.

074.097 **Physics of the solar corona.**
V. Rušin.
Kozmos, Vol. 13, 154 - 155 (1982). In Slovak.

074.098 **What heats the solar corona?**
L. Golub.
Astronomy, Vol. 10, No. 9, p. 74 - 79 (1982). – Abstr. in Phys. Abstr., Vol. 85, Abstr. 117645 (1982).

074.099 **Charge exchange in solar wind cometary interactions.**
T. I. Gombosi, M. Horanyi, K. Kecskemety, T. E. Cravens, A. F. Nagy.
Report KFKI-1982-68, Hungarian Acad. Sci., Budapest, Hungary. 21 pp. (1982). – Abstr. in Phys. Abstr., Vol. 86, Abstr. 10180 (1983).

074.100 **The acceleration role of whistler waves from coronal holes for the solar wind.**
Z.-d. Zhang, Y.-r. Huang, X.-q. Li.
Acta Astron. Sinica, Vol. 23, 102 - 109 (1982). In Chinese.
To investigate the possibility of an additional acceleration of the high speed solar wind due to the presence of whistler waves propagating outward from a coronal hole, the authors study a stationary, spherically symmetric model assuming a radial wind flow and also a radial magnetic field.

074.101 **The transient accelerated motions of a solar coronal loop.**
X. Liu.
Acta Astrophys. Sinica, Vol. 2, 240 - 244 (1982). In Chinese.
The mass ejecting events of the corona in the form of loop transients were discovered by Skylab in 1973. The white light coronagraph showed that the outward motions of the coronal transients are accelerated or remain approximately at a constant velocity. Recently, some researches have explained the driving mechanism of the transient processes. Anzer discussed a ring current model, in which the influence of Lorentz force produced by the axial current and the azimuthal magnetic field is discussed. In this paper, Anzer's model is restudied.

074.102 **Solar wind and its interaction with planetary ionospheres.**
D. C. Agarwal, D. C. Pande.
Indian J. Theor. Phys., Vol. 29, 151 - 174 (1981). – Abstr. in Phys. Abstr., Vol. 86, Abstr. 15800 (1983).

Solar wind and magnetospheric investigations.
See Abstr. 003.075.

Solar magneto-hydrodynamics.
See Abstr. 003.100.

Observations of structuring in the downstream region of a large spherical model in a laboratory simulated solar wind plasma.
See Abstr. 022.080.

Coronal line intensities for ions with fine-structured ground states: Si X.
See Abstr. 022.146.

Spectroscopic diagnostics of the active region: transition zone and corona.
See Abstr. 031.590.

Coronal investigations with occulted spacecraft signals.
See Abstr. 031.665.

Measurements of the properties of solar wind plasma relevant to studies of its coronal sources.
See Abstr. 031.666.

Design and performance of a new emission-line coronagraph.
See Abstr. 032.020.

Coronal observations from the Solar Maximum Mission satellite.
See Abstr. 032.565.

Narrow-band GaAsSb heterojunction detectors for coronal polarimetry.
See Abstr. 034.068.

Shock acceleration of energetic particles in the heliosphere.
See Abstr. 062.016.

A remarkable general property of the MHD equations and its relation to coronal heating mechanisms.
See Abstr. 062.033.

Transport equations for multicomponent anisotropic space plasmas: a review.
See Abstr. 062.052.

Self-similar flows behind shock wave with increasing energy in magnetogasdynamics. I.
See Abstr. 062.065.

Beam-plasma instability in the presence of low-frequency turbulence.
See Abstr. 062.078.

Magnetohydrodynamics of atmospheric transients. IV. Nonplane two-dimensional analyses of energy conversion and magnetic field evolution.
See Abstr. 062.080.

Study of toroidal magnetic configurations in a spherically symmetric gravitational field with applications to coronal loops and transients.
See Abstr. 062.089.

On MHD wave propagation in inhomogeneous plasmas and the mechanism of resonant absorption.
See Abstr. 062.093.

A self-similar flow of self-gravitating gas behind a shock wave with increasing energy. I.
See Abstr. 062.108.

A self-similar flow in generalized Roche model with increasing energy.
See Abstr. 062.117.

The moments of the non-collisional parallel flux.
See Abstr. 062.122.

The influence of divergent geometries on stellar winds.
See Abstr. 064.047.

Models for stellar coronae: the effects of coronal heating with long dissipation scale lengths.
See Abstr. 064.061.

The effect on empirical atmospheric modeling of the mass-flux as an independent parameter.
See Abstr. 064.062.

The overall structure and evolution of active regions.
See Abstr. 072.030.

On the sunspot distribution with respect to coronal holes.
See Abstr. 072.070.

Comments on the thermodynamic structure and dynamics of the cool solar corona over sunspots.
See Abstr. 072.088.

Mass motions in the solar chromosphere and transition zone.
See Abstr. 073.007.

Hydrodynamic shock wave formation in the solar chromosphere and corona during flares.
See Abstr. 073.015.

A sequence of five proton producing flares in McMath Plage 15266 28 April - 7 May 1978 and their interplanetary consequences.
See Abstr. 073.016.

June - July 1974 proton-flare region. III. The geomagnetic activity during the declining phase of the complex process.
See Abstr. 073.038.

Eruptive prominence of August 18, 1980.
See Abstr. 073.039.

Coronal transport of solar flare particles.
See Abstr. 073.042.

Dynamics of the eruptive prominence of 6 May 1980 and its relationship to the coronal transient.
See Abstr. 073.043.

Observations of mass motions in active regions.
See Abstr. 073.064.

Active region structures in the transition region and corona.
See Abstr. 073.065.

Responses of transition region models to magnetic field geometry and downflow velocities.
See Abstr. 073.105.

Flow of material at the chromosphere-corona transition zone of the Sun.
See Abstr. 073.110.

The great eruptive loop-prominence on the solar western limb on April 27, 1981.
See Abstr. 073.118.

EUV arcades: signatures of filament instability.
See Abstr. 076.014.

Determination and analysis of coronal hole radio spectra.
See Abstr. 077.001.

Possible evidence for a rigidity-dependent release of relativistic protons from the solar corona.
See Abstr. 078.003.

Surface waves and the heating of the corona.
See Abstr. 080.023.

The solar atmosphere – from photosphere to corona.
See Abstr. 080.037.

On the dependence of the western electrojet on the solar wind parameters.
See Abstr. 084.003.

Atmospheric interactions of planetary bodies with the solar wind.
See Abstr. 091.004.

Viscous dissipation at the Venus ionopause.
See Abstr. 093.100.

The dynamics of the Venus ionosphere. 1. A simulation of the solar wind compression of the upper dayside ionosphere.
See Abstr. 093.101.

The distant interplanetary wake of Venus: plasma observations from Pioneer Venus.
See Abstr. 093.103.

Comparison of gas dynamic model with steady solar wind flow around Venus.
See Abstr. 093.107.

The solar wind interaction with Mars as seen by the Viking retarding potential analyzers.
See Abstr. 097.023.

The solar wind interaction with Mars revisited.
See Abstr. 097.104.

Decametric emission of Jupiter and solar activity.
See Abstr. 099.037.

Evidence for solar wind control of Saturn radio emission.
See Abstr. 100.035.

Solar wind interaction with comets.
See Abstr. 102.007.

Solar wind interaction with comets: lessons from Venus.
See Abstr. 102.030.

Solar wind magnetic field penetration into cometary ionospheres.
See Abstr. 102.043.

The cometary ion tail and its relation to solar wind phenomena.
See Abstr. 102.050.

Interaction of turbulent solar wind with cometary plasma tails.
See Abstr. 102.054.

Plasma dynamics in ion tails – a discussion of the interaction of solar wind plasma with cometary plasma.
See Abstr. 102.090.

Helical wave and K-H instability in type I comet tails. I. Waves of infinitesimal amplitude in incompressible plasma.
See Abstr. 102.091.

Solvindens inverkan på kometerna.
See Abstr. 102.093.

Comet Bradfield 1979 X event on 1980 February 6: correlation with an interplanetary solar wind disturbance.
See Abstr. 103.201.

Interaction of non-perpendicular/parallel solar wind shock waves with the earth's magnetosphere.
See Abstr. 106.004.

The acceleration of charged particles in interplanetary shock waves.
See Abstr. 106.010.

The structure of the interplanetary medium during the first STIP interval (September - October 1975) and prerelease of energetic solar particles.
See Abstr. 106.011.

Magnetic clouds: Voyager observations between 2 and 4 AU.
See Abstr. 106.041.

Frictional mechanism for particle acceleration under conditions of the interplanetary space.
See Abstr. 106.047.

Energetic protons accelerated at corotating shocks: Pioneer 10 and 11 observations from 1 to 6 AU.
See Abstr. 106.048.

Caution! High winds beyond this point.
See Abstr. 112.027.

Cosmic ray intensity variations and two types of high speed solar streams.
See Abstr. 143.059.

Erratum

074.901 **Erratum: "A low β coronal loop model. I. Kink instabilities in the $\beta = 0$ limit'** [Sol. Phys., Vol. 67, 47 - 56 (1980)].
R. M. J. Sillen, A. Kattenberg.
Sol. Phys., Vol. 79, 401 (1982). – See Abstr. 28.074.003.

075 Magnetic Fields

075.001 **Open magnetic fields and the solar cycle. I: Photospheric sources of open magnetic flux.**
R. H. Levine.
Sol. Phys., Vol. 79, 203 - 230 (1982).

The purpose of this paper is to explore the morphology and evolution of open magnetic fields on the Sun at various times in the solar cycle, and to suggest possible physical explanations for the similarities and differences found. Because of limitations on the observations of magnetic fields on the Sun the existence of open magnetic field lines in the solar atmosphere must be inferred using either spectroheliograms or theoretical extrapolations of magnetograph observations. The present work relies on both of these methods, with an emphasis on the latter.

075.002 **Magnetograph response to canopy-type fields.**
H. P. Jones, R. G. Giovanelli.
Sol. Phys., Vol. 79, 247 - 266 (1982).

The response of longitudinal-field magnetographs to magnetic fields which are semi-infinite or confined to a horizontal layer is discussed with respect to the interpretation of solar diffuse fields, observed towards the limb, in terms of magnetic canopy models. Numerical results are presented for several reference solar models and typical 'calibration' curves are shown for the C I 9111 Å, Fe I 8688 Å, and Ca II 8542 Å lines in magnetostatic atmospheres derived from a mean model. A procedure is developed for determining the base heights of magnetic canopies from observations with an uncertainty not exceeding the order of a pressure scale height.

075.003 **The three-dimensional structure of atmospheric magnetic fields in two active regions.**
R. G. Giovanelli, H. P. Jones.
Sol. Phys., Vol. 79, 267 - 278 (1982).

The magnetic field above two unrelated active regions on 11 and 12 September, 1974 has been studied using magnetograms obtained in C I 9111, Fe I 8688, Ca II 8542, and Hα. In C I 9111, originating low in the photosphere, the fields are strong and sharply defined. In Ca II 8542 and Hα they are very diffuse, with significant diffuseness also in Fe I 8688, due to the spreading of the field with height to form almost horizontal magnetic canopies over regions free of field at lower levels.

075.004 **A note on the relation of flare activity with rapid variations of the magnetic field.**
L. D. Parfinenko.
Soln. Dannye 1982 Byull., No. 3, p. 103 - 107 (1982). In Russian.

Video maps of the longitudinal component of the magnetic field have been obtained for a sunspot group. In the region of the leader of the group rapid variations of the magnetic field structure took place. This fact made the structure of the field more complex, however, it did not lead to an increase of the flare activity.

075.005 **Magnetic and velocity fields of emerging flux regions on the sun.**
M. -J. Martres, J. Rayrole, M. Semel, I. Soru-Escaut, K. Tanaka, M. Makita, F. Moriyama, W. Unno.
Publ. Astron. Soc. Japan, Vol. 34, 299 - 310 (1982).

Emerging flux regions (EFR's) on the Sun observed in the Meudon-Mitaka Collaborative Solar Observation Program are discussed.

075.006 **Development of solar energetic particle sources within the background magnetic fields.**
V. Bumba.
Space Sci. Rev., Vol. 32, (see 012.005), 229 - 233 (1982).

The greatest coronal and interplanetary disturbances are considered consequences of complex processes accompanying development of large-flare regions. Such regions not only possess a specific magnetic field configuration, but their magnetic topology develops following certain rules within the frame of slowly changing large-scale distribution of the background magnetic field patterns as a result of the mutual influences of new magnetic flux appearance as well as old field weakening and dissipation. In this paper the author tries to demonstrate the individual phases of the large-scale long lasting magnetic field pattern formation with their morphological characteristics and magnetic field configurations.

075.007 **The magnetic flux in the quiet Sun network.**
B. J. LaBonte, R. Howard.
Sol. Phys., Vol. 80, 15 - 19 (1982).

The Ca II K line emission from the quiet Sun network does not vary with the 11-year cycle (White and Livinston, 1981). The authors confirm this result from direct magnetic measurements. This effect is not simply explained by present empirical models of the evolution of surface magnetic fields.

075.008 **Dynamics of post-flare ejections and magnetic loop geometry.**
P. Mein, N. Mein.
Sol. Phys., Vol. 80, 161 - 171 (1982).

Flare-associated mass ejections have been observed at the solar limb on June 29, 1980 in the Hα line. Radial velocities were measured as a function of time in a two dimensional field, and kinematics investigated in one selected fine structure. A simple model of locally dipole-type magnetic field increasing with time can be fitted to the data. It can be checked from extrapolation that the model is consistent with an ejection starting roughly from the same point at the same time.

075.009 **The overshoot layer at the base of the solar convective zone and the problem of magnetic flux storage.**
A. A. van Ballegooijen.
Astron. Astrophys., Vol. 113, 99 - 112 (1982).

The strict polarity rules of solar active regions imply that the sub-surface magnetic field is highly organized and apparently not subjected to strong deformation by convective motions. This leads the author to study possibilities for storing horizontal magnetic field in the overshoot layer at the base of the solar convective zone, or in the radiative zone below. A one-dimensional model of the overshoot layer is described and the equilibrium and stability of horizontal flux tubes stored in the overshoot layer is discussed. It is suggested, that flux tubes with field strength $B \gtrsim 10^4$ Gauss may be stable in the overshoot layer for solar cycle periods.

075.010 **On the techniques for measurement of sunspot magnetic fields.**
D. I. Barausov.
Researches on stellar physics, 1981, (see 31.003.010), p. 96 - 101 (1981). In Russian. – Abstr. in Ref. zh., 51. Astron., 8.51.292 (1982).

075.011 **The magnetic field of active regions with longitudinal current.**
E. A. Kornitskaya, Den O Gym, M. M. Molodenskij.
Inst. zem. magn., ionos. i rasprostr. radiovoln AN SSSR. Prepr., 1982, No. 5, 16 pp. In Russian. – Abstr. in Ref. zh., 51. Astron., 8.51.301 (1982).

075.012 **On the finished change of the sign of the polar magnetic field of the sun in the 21st solar cycle.**
V. I. Makarov, M. P. Fat'yanov.
Pis'ma Astron. Zh., Tom 8, 631 - 632 (1982). In Russian. English translation in Soviet Astron. Lett., Vol. 8.

The change of sign of the polar magnetic field from plus to minus in the 21st solar cycle took place at the north pole in December 1980. At the south pole the sign changed from minus to plus in September 1981. The sign inversion was accompanied by formation of arch prominences at the poles.

075.013 **High resolution observations of a moving magnetic feature.**
A. H. Nye, J. H. Thomas, L. E. Cram.
Bull. American Astron. Soc., Vol. 14, 624 - 625 (1982). Abstract.

075.014 **Relation between the change of magnetic field and the boundary surface.**
Y. Nakagawa.
Magnetospheric phenomena of celestial bodies, (see 012.011), p. 59 - 61 (1982). In Japanese.

The fundamental problem of magnetic energy build-up in the solar atmosphere is examined as a MHD initial-boundary value problem with the use of the method of projected characteristics within the approximation of two dimensional non-plane analysis (i.e., with all three components of magnetic and velocity fields). Numerical results show that with the lateral movements of magnetic foot-points (simulating a relative movement of sunspots), magnetic energy is rapidly built up in the atmosphere and the evolutional sequence differs considerably from the sequence of force-free fields.

075.015 **The relations between the toroidal magnetic field in the solar convection zone and the bipolar magnetic field of sunspots.**
D.-q. Zhou.
Chinese Phys., Vol. 2, 668 - 676 (1982). – Abstr. in Phys. Abstr., Vol. 85, Abstr. 100857 (1982). – See Abstr. 30.075.016.

075.016 **Relaxation of a twisted magnetic loop as mechanism of a solar flare. I.**
A. A. Solov'ev.
Soln. Dannye 1982 Byull., No. 5, p. 80 - 86 (1982). In Russian.

On the basis of the variational solution it is shown that a homogeneous magnetic rope (if field fluxes and outer pressure are known) has the energy minimum not only inside some definite configuration but also with respect to any cylindrical symmetric distribution of force-free fields.

075.017 **The Hanle effect and the diagnostics of turbulent magnetic fields in the solar atmosphere.**
J. O. Stenflo.
Sol. Phys., Vol. 80, 209 - 226 (1982).

Although the main aim of the paper is to elucidate the physics of the Hanle effect as applied to the Sun, the author's results indicate the existence of hidden or turbulent magnetic flux near the temperature minimum of the solar atmosphere, with a field strength between 10 and 100 G. This field is hidden in the sense that it is not seen in measurements of the longitudinal Zeeman effect (solar magnetograms). It carries more total magnetic flux than the kG network fields.

075.018 **A comparison of different solar magnetic field extrapolation procedures.**
N. Seehafer.
Sol. Phys., Vol. 81, 69 - 80 (1982).

In order to study the influence of the different treatment of the boundary conditions in different extrapolation schemes, the procedures of Schmidt (1964), Nakagawa and Raadu (1972), and Seehafer (1978) have been applied to the same (line-of-sight) magnetogram. The main field structure is similar for all three procedures, whereas in details there are clear differences.

075.019 **Approaching ranges of magnetic fields of N and S polarity on the sun as sources of increased mm-emission.**
M. I. Ryabov.
Astron. Tsirk., No. 1197, p. 7 - 8 (1981). In Russian.

075.020 **A model for slender flux tubes and its application to sunspots.**
A. A. van Ballegooijen.
The physics of sunspots, (see 012.063), p. 115 - 119 (1981).

A model of slender flux tubes below the solar photosphere is briefly described. The tubes are rooted in the stably stratified region below the base of the convective zone. Assuming hydrostatic equilibrium and an adiabatic stratification inside the tubes, the field strength B(z) as function of depth z below the solar surface is obtained. The magnetic tension in the horizontal part of the tubes causes a horizontal motion of the emerged flux tubes through the convective zone. This horizontal drift is of potential importance for theories of sunspots.

075.021 **Configurations of the magnetic field in sunspot groups and solar flares. I.**
R. N. Ikhsanov.
Izv. Glav. Astron. Obs. Pulkovo, Astrofiz. Astrometr., No. 200, p. 15 - 21 (1982). In Russian.

It is shown that the configuration of the magnetic field which leads to a flare forms as a result of the rise of a new magnetic complex inside or near the existing sunspot group. A classification is made of the types of magnetic configurations causing flares. Large flares, particularly proton ones, show an evident concentration to types I and IIa. The weaker the flare is the oftener it occurs in configurations of type III - V.

075.022 **Structure of the heliospheric current sheet in the early portion of sunspot cycle 21.**
J. T. Hoeksema, J. M. Wilcox, P. H. Scherrer.
J. Geophys. Res., Vol. 87, 10331 - 10338 (1982).

The structure of the heliospheric current sheet on a spherical source surface of radius 2.35 R_S has been computed via the use of a potential field model during the first year and a half after the last sunspot minimum. The solar polar magnetic field that is not fully observed in conventional magnetograph scans was included in the computation. The computed heliospheric current sheet had a quasi-stationary structure consisting of two northward and two southward maxima in latitude per solar rotation. The magnetic field polarity (away from the sun or toward the sun) at the sub-terrestrial latitude on the source surface agreed with the interplanetary magnetic field polarity observed or inferred at the earth on 82% of the days. The interplanetary field structure observed at the earth at this time is finely tuned to the structure of low-latitude fields on the source surface.

075.023 **Quadrupole distortions of the heliospheric current sheet in 1976 and 1977.**
R. Bruno, L. F. Burlaga, A. J. Hundhausen.
J. Geophys. Res., Vol. 87, 10339 - 10346 (1982).

Latitudinal variations in the position of the heliospheric current sheet were investigated using interplanetary measurements from Helios A, Helios B, IMP 8, and white light corona measurements in the period May 1976 to May 1977. The latitude of the heliospheric current sheet between 0.3 AU and 1 AU is consistent with that of the maximum brightness curve of white light polarization brightness at 1.75 solar radii within approximately ±7°. The combined data sets show that the four-sector magnetic sector pattern observed in the ecliptic plane for most of this epoch could be attributed to a warped current sheet whose shape near the sun resembles a saddle surface that is nearly symmetrical about the equatorial plane. These results can be interpreted as a consequence of the solar magnetic field, the dominant components of the field being due to a dipole nearly aligned with the sun's spin axis and a quadrupole component whose strength is ~17% of the dipole.

075.024 **Solar eclipse observations of a magnetic loop structure at wavelengths 2, 3.2, 8.2 and 21 cm.**
X. Luo, D. Yiao, Y. Yan.
Acta Astrophys. Sinica, Vol. 2, 184 - 190 (1982). In Chinese.

The total solar eclipse of February 16, 1980, was observed at Yunnan Observatory. During the lunar occultation of the active region 75 composed of a bipolar sunspot group, core sites of microwave radio sources along the occultation path were determined from the peak points of the eclipse slope curves at 2, 3.2, 8.2 and 21 cm wavelengths. These sites were shifted separately to the corresponding heights of the effective radiation layers, and connected through several sets of closed arclines, then the magnetic loop structure of the solar radio source was obtained. The corresponding correlations between the sites, angular diameters, fluxes and space distributions of the radio sources and the intensities, extensions, solid angles and shapes of the magnetic fields for the bipolar sunspot were analysed.

Solar magneto-hydrodynamics.
See Abstr. 003.100.

Magnetic reconnection: a problem of general physical and astrophysical interest, with special implications in solar physics.
See Abstr. 062.095.

Vorticity and astrophysical magnetism.
See Abstr. 062.110.

A numerical study of α^2-dynamos with anisotropic α-effect.
See Abstr. 062.129.

Ca II K_{2V} spectral features and their relation to small-scale photospheric magnetic fields.
See Abstr. 071.013.

On the problem of prediction of the maxima of solar activity using polar fields on the sun.
See Abstr. 072.003.

On the relation between the dynamics of filament bands and Hale's and Spoerer's laws.
See Abstr. 072.004.

The overall structure and evolution of active regions.
See Abstr. 072.030.

Active regions from the photosphere to the chromosphere.
See Abstr. 072.031.

On the problem of measurement of the maximum intensities of magnetic fields in sunspots during the investigation of their rapid variations.
See Abstr. 072.034.

Active region magnetic fields inferred from simultaneous VLA microwave maps, X-ray spectroheliograms, and magnetograms.
See Abstr. 072.035.

Magnetic field development in an isolated active region (McMath No. 13 736).
See Abstr. 072.051.

Magnetic polarities and maximum field strengths of one selected sunspot group with time distances of about thirty minutes during the period 1979 August 13 - August 16, 1980 September 2 - September 4.
See Abstr. 072.055.

Potential models of the unipolar sunspot magnetic field.
See Abstr. 072.066.

Observations of the longitudinal magnetic field in the transition region and photosphere of a sunspot.
See Abstr. 072.067.

On the dissolution of sunspot groups.
See Abstr. 072.075.

Analysis of the vector magnetic fields of complex sunspots.
See Abstr. 072.077.

Magnetic features in active regions.
See Abstr. 072.078.

On the magnetic field topology in the return-flux sunspot model.
See Abstr. 072.081.

Magnetic field strengths in umbrae and pores.
See Abstr. 072.089.

The photospheric vector magnetic field of a sunspot and its vertical gradient.
See Abstr. 072.090.

Technique and preliminary results of investigation of element distribution of a magnetic field in active regions.
See Abstr. 072.106.

Evidence for a poleward meridional flow on the sun.
See Abstr. 073.001.

Scaling-law equilibria for calcium in canopy-type models of the solar chromosphere.
See Abstr. 073.002.

The determination of vector magnetic fields in prominences from the observations of the Stokes profiles in the D_3 line of helium.
See Abstr. 073.003.

June - July 1974 proton-flare region. III. The geomagnetic activity during the declining phase of the complex process.
See Abstr. 073.038.

June - July 1974 proton-flare region. IV. The dynamics of the local magnetic field evolution during the proton-flare events rotation.
See Abstr. 073.040.

A measurement of the magnetic field direction at the site of major flares.
See Abstr. 073.098.

Spectral analysis of inhomogeneity sizes and brightness dependence on the magnetic field strength in a plage from Ba II 4554 Å filtergrams.
See Abstr. 073.107.

Diagnostic of coronal magnetic fields from microwave polarization reversal.
See Abstr. 074.010.

Theory of loop flows and instability.
See Abstr. 074.059.

Mechanisms of energy supply.
See Abstr. 074.060.

Evidence for an oblique magnetic solar rotator.
See Abstr. 080.009.

The effect of a magnetic field on solar luminosity.
See Abstr. 080.017.

The energy of electric current sheets. I. Models with moving magnetic dipoles.
See Abstr. 080.035.

The structure and energy balance of cool star atmospheres.
See Abstr. 114.022.

076 UV, X, Gamma Radiation

076.001 **Rocket photographs of fine structure and wave patterns in the solar temperature minimum.**
R. M. Bonnet, M. Bruner, L. W. Acton, W. A. Brown, M. Decaudin, B. Foing.
Astron. Astrophys., Vol. 111, 125 - 129 (1982).

A new series of high resolution pictures of the sun has been obtained during the second flight of the Transition Region Camera which occurred on 23 September 1980. The qualitative analysis of the results indicates that a substantial portion of the solar surface at the temperature minimum radiates in non-magnetic regions and from features below one arcsec in size. Wave patterns are observed on the 160 nm temperature minimum pictures. They are absent on the Lα pictures. Their physical characteristics are compatible with those of gravitational and acoustic waves generated by exploding granules.

076.002 **On impulsive X-ray flares on the sun.**
V. Sh. Dolidze, I. V. Ehstulin, G. Vedrenne, M. Niel, A. Talon, G. Chambon.
Pis'ma Astron. Zh., Tom 8, 488 - 494 (1982). In Russian. English translation in Soviet Astron. Lett., Vol. 8.

Data on impulsive solar flares on December 3 and 4, 1978 and April 27, 1979 with energy $E_x > 100$ keV observed by the Prognoz 7 and Venera 11 and 12 stations in the Soviet-French experiment SIGNE-2M are given.

076.003 **An upper limit for the continuous 2.223 MeV solar gamma-ray line.**
J. H. Beall, R. L. Kinzer, G. H. Share, D. J. Forrest, E. L. Chupp, J. M. Ryan, G. Kanbach, C. Reppin, E. Rieger.
Bull. American Astron. Soc., Vol. 14, 608 (1982). – Abstract.

076.004 **Gamma-ray lines, interplanetary neutrons, and the shock acceleration of ions in solar flares.**
M. E. Pesses, R. B. Decker.
Bull. American Astron. Soc., Vol. 14, 608 (1982). – Abstract.

076.005 **Extreme ultraviolet observational data on the solar spectrum.**
G. M. Nikol'skij.
Handbuch der Physik. Geophysik III/6, (see 003.006), p. 309 - 377 (1982).

Contents: Spectral observations. XUV solar images. Observation of X-ray solar radiation over wide spectral ranges. Identification and prediction of extreme ultraviolet lines. Chemical composition of the solar atmosphere derived from EUV data.

076.006 **Recent studies in the diagnostic aspects of solar EUV line emission.**
P. K. Raju.
Proc. Indian Natl. Sci. Acad., Part A, Vol. 47, Suppl. 2, 77 - 84 (1981). – Abstr. in Phys. Abstr., Vol. 85, Abstr. 89225 (1982).

076.007 **A correlation between observed VLF phase deviation and solar X-ray flux during solar flares.**
P. Pant, H. S. Mahra, M. C. Pande.
Bull. Astron. Soc. India, Vol. 10, 37 (1982). – Abstract.

076.008 **Comparison of nonflare solar soft X-ray flux with 10.7-cm radio flux.**
R. F. Donnelly.
J. Geophys. Res., Vol. 87, 6331 - 6334 (1982).
The similarities and differences of the nonflare solar 1- to 8-Å X-ray flux and the daily 10.7-cm Ottawa solar radio flux are examined. The radio flux is shown to be much less sensitive than the soft X-ray flux on the average to the coronal emission of active regions located near or beyond the solar chromospheric limb relative to regions near the center of the solar disk. This is caused by the solar soft X-ray emission's being optically thin while much of the 10.7-cm active region emission is from optical depths of $\tau \sim 1$. The radio flux includes a large quiet sun flux whereas the solar soft X-ray flux has a very small quiet sun component.

076.009 **Wavelength dependence of the weakening of the solar EUV line emission.**
M. Kanno, Y. Suematsu.
Publ. Astron. Soc. Japan, Vol. 34, 449 - 459 (1982).
The solar EUV emission from transition-region lines with wavelengths less than 912 Å is weakened by Lyman continuum absorption. Absorbers causing the weakening have not been detected in spatially resolved EUV and Hα observations. Two suggestions have been made on the origin of the weakening: the cloud model (Schmahl and Orrall, 1979) and the overlapping model (Kanno, 1979). In this paper the empirical wavelength dependence of the weakening is studied for several types of regions of the solar atmosphere in order to know which of the models is consistent with observations.

076.010 **Gamma-ray lines from solar flares and cosmic transients.**
R. Ramaty, R. E. Lingenfelter, B. Kozlovsky.
AIP Conf. Proc., No. 77, (see 012.034), p. 211 - 229 (1982).
Gamma-ray line emission processes in solar flares and cosmic transients are reviewed and implications of recent line observations are discussed. The gamma-ray line emission from solar flares results from nuclear interactions of accelerated particles with the solar atmosphere. The gamma-ray emission lines observed from gamma-ray transients appear to result from both the annihilation of positrons produced in photon-photon interactions and from deexcitation of nuclear levels and capture of neutrons produced in nuclear interactions.

076.011 **Solar energetic photon transients (50 keV - 100 MeV).**
E. L. Chupp.
AIP Conf. Proc., No. 77, (see 012.034), p. 363 - 381 (1982).
This paper reviews observations of the impulsive emissions of energetic photons associated with solar flares. A major observational conclusion from the latest solar flare measurements is that both ions and electrons are accelerated (or released) and interact with matter simultaneously or within seconds of one another.

076.012 **Solar hard X-ray images observed by Astro-A.**
K. Ohki, S. Tsuneta, N. Nitta, T. Takakura, K. Makishima, T. Murakami, Y. Ogawara, M. Oda.
AIP Conf. Proc., No. 77, (see 012.034), p. 395 - 399 (1982).
The solar X-ray Telescope (SXT) on board the Astro-A has observed many events since its launch on February 21, 1981. In this paper, two typical large events are presented. A disk event on April 2, 1981, shows a single source with a very small diameter, while a April 27, 1981, limb event shows a double source structure with unbalanced intensities.

076.013 **Evidence for delayed second phase acceleration in solar flares.**
J. B. Willett, J. C. Ling, W. A. Mahoney, G. R. Riegler, A. S. Jacobson.
AIP Conf. Proc., No. 77, (see 012.034), p. 401 - 408 (1982).
Twenty-two solar flares with emission at or above 0.5 MeV have been observed by the HEAO-3 gamma-ray spectrometer during the period 1 October 1979 through 1 July 1980. In two of these flares, the peak emission in a band around 0.5 MeV follows the peak of the impulsive hard X-ray emission at energies greater than 0.1 MeV by as much as 30 seconds. The remaining 20 flares gave no indication of a delay in either peak emission or onset time between the high energy and low energy components.

076.014 **EUV arcades: signatures of filament instability.**
E. J. Schmahl, Z. Mouradian, M. J. Martres, I. Soru-Escaut.
Sol. Phys., Vol. 81, 91 - 105 (1982).
The authors have undertaken a two-fold study of the EUV corona around active region filaments. The first goal is to define the geometry and temperature distribution around those filaments which appear to be 'stable' (i.e. lifetimes ⩾2 days). The second goal is to study time changes in structures associated with other filaments which are 'unstable' (i.e. disappear or have Hα velocities). It is important to note that the filaments and associated EUV loops reported were not flare-related.

076.015 **Hard X-rays from the sun.**
E. Haug.
Space Sci. Rev., Vol. 33, (see 012.067), 83 - 97 (1982).
Temporal and spectral characteristics of solar hard X-ray bursts are briefly reviewed. The merits of non-thermal and thermal flare models are discussed. The validity of these models may be checked by future measurements of X-ray polarization. Finally, some important results of recent satellite experiments are described providing information on the spatial distribution of hard X-ray sources: the multi-spacecraft of X-ray bursts and the imaging of X-ray sources by means of the HXIS instrument.

076.016 **A three-component model of the variability of the solar ultraviolet flux: 145–200 nm.**
J. L. Lean, O. R. White, W. C. Livingston, D. F. Heath, R. F. Donnelly, A. Skumanich.
J. Geophys. Res., Vol. 87, 10307 - 10317 (1982).
A three-component model has been developed to examine the variation with solar activity of the far ultraviolet irradiance between 145 and 200 nm. This model is based on spatially resolved observations of the Ca II K chromosphere and includes the contributions to the full disk flux from both plage and active network emission.

076.017 **Active-region evolution and solar rotation variations in solar UV irradiance, total solar irradiance, and soft X rays.**
R. F. Donnelly, D. F. Heath, J. L. Lean.
J. Geophys. Res., Vol. 87, 10318 - 10324 (1982).
Variations in the solar UV spectral irradiance, the total solar irradiance, and solar soft X-ray emission over days and weeks are analyzed by using concurrent measurements from the NIMBUS 7 and GEOS satellites. The UV variations at wavelengths (160–400 nm) of interest to ozone photochemistry and the physics of the middle atmosphere are emphasized. The nonflare variations studied are caused by two types of

processes: (1) the birth, growth, peak, and decay of solar active regions and (2) solar rotation, for example, when portions of active regions are occulted near the solar limb. The observations are interpreted by using simple empirical models that relate ground-based observations of the size and location of sunspots and plages to the full-disk temporal variations.

The effect of Fe XXIII on Fe XXIV satellite line ratios.
See Abstr. 022.051.

Spectroscopic diagnostics of the active region: transition zone and corona.
See Abstr. 031.590.

Gas scintillation proportional counters for Japanese astronomical satellites.
See Abstr. 032.548.

Performance of the hard X-ray imaging spectrometer.
See Abstr. 032.560.

A solar flare X-ray polarimeter for the Space Shuttle.
See Abstr. 032.567.

The Pinhole/Occulter Facility. Executive summary.
See Abstr. 032.589.

Proposed Mg V, Fe XIV and forbidden Fe XVII line identifications in the EUV solar spectrum.
See Abstr. 071.016.

Multiple wavelength observations of a solar active region.
See Abstr. 072.011.

The dynamics of active regions.
See Abstr. 072.032.

Active region magnetic fields inferred from simultaneous VLA microwave maps, X-ray spectroheliograms, and magnetograms.
See Abstr. 072.035.

An emission measure analysis of two sunspots observed by the UVSP instrument on the SMM spacecraft.
See Abstr. 072.041.

Analysis of the high resolution Mg XI X-ray spectra. III. Non-thermal interpretation of some spectra.
See Abstr. 072.042.

High resolution EUV structure of the chromosphere-corona transition region above a sunspot.
See Abstr. 072.069.

Gamma-ray and optical observations of the 1979 November 8 solar flare.
See Abstr. 073.009.

Microwaves and hard X rays from solar flares: multithermal and nonthermal interpretations.
See Abstr. 073.030.

High-resolution X-ray spectra of solar flares. VI. Impulsive soft X-ray flares.
See Abstr. 073.031.

Correlations amongst flux-vs.-time profiles at various wavelengths during solar flares and their relation to flare heating models.
See Abstr. 073.032.

Bimodel energy release in solar flares.
See Abstr. 073.033.

Acceleration mechanism for prompt protons and relativistic electrons from solar flares.
See Abstr. 073.034.

X-ray classification of flares in large and small magnetic structures.
See Abstr. 073.044.

Hard X-ray and radio discriminators of flare classification.
See Abstr. 073.048.

Soft X-ray discriminators of flare classification.
See Abstr. 073.050.

A study of solar flare soft X rays and their relation to particle events observed with IMP 8.
See Abstr. 073.058.

Temporal evolution of soft X-ray emitting plasma in solar flares.
See Abstr. 073.061.

Observations of mass motions in active regions.
See Abstr. 073.064.

Active region structures in the transition region and corona.
See Abstr. 073.065.

Second-phase acceleration versus second-step acceleration in solar flares.
See Abstr. 073.066.

Solar flare energetics.
See Abstr. 073.067.

Chromospheric evaporation in a well-observed compact flare.
See Abstr. 073.080.

Electron acceleration in impulsive solar flares.
See Abstr. 073.081.

Two types of jets and origin of macrospicules.
See Abstr. 073.082.

Microwave emission from hot X-ray kernels in solar flares.
See Abstr. 073.083.

X-ray imaging of three flares during the impulsive phase.
See Abstr. 073.084.

X-ray spectroscopy of solar flares.
See Abstr. 073.104.

Solar transition zone pressures from EUV observations of O IV and N IV.
See Abstr. 074.006.

X-ray lines of helium-like oxygen and neon in the solar corona.
See Abstr. 074.007.

Unusual coronal activity following the flare of 6 November 1980.
See Abstr. 074.017.

X- and XUV-radiation of the solar corona.
See Abstr. 074.084.

077 Radio, Infrared Radiation

077.001 **Determination and analysis of coronal hole radio spectra.**
M. D. Papagiannis, K. B. Baker.
Sol. Phys., Vol. 79, 365 - 374 (1982).

Solar radio maps over a wide wavelength range (millimeter to meter) and over a considerable time span (1973 - 1978) have allowed the authors to compute the radio spectrum of an average coronal hole, i.e., the brightness temperature inside a coronal hole normalized by the brightness temperature of the quiet Sun outside the coronal hole measured at several different radio wavelengths. The analysis indicates that inside an average coronal hole the following changes occur: the upper chromosphere expands by about 20% and its electron density and temperature decrease by about 10%. The transition zone experiences the largest change, expanding by a factor of about 6, its electron density decreases by a similar factor, and its temperature decreases by about 50%. Finally in the corona the electron density decreases by about 20% and the temperature by about 15%.

077.002 **Is there a common explanation for scattering of type III radio bursts and solar radar?**
D. G. Wentzel.
Sol. Phys., Vol. 79, 375 - 377 (1982).

The apparently common source heights of type III fundamental and harmonic components and the source height of the solar 38 MHz radar echoes may all refer to scattering at a coronal level where (radio wavenumber) X (electron gyroradius) $\simeq$ 1, that is, where radio frequency = (2 to 5) X plasma frequency.

077.003 **On the theory of type IV solar radio emission.**
V. G. Ledenev.
Astron. Zh., Tom 59, 742 - 749 (1982). In Russian. English translation in Soviet Astron., Vol. 25, No. 4.

Plasma wave excitement in a magnetic trap (arch) in the presence of a continuously acting source of energetic particles is considered. Estimates of plasma wave energy and energetic particles density in the trap are derived. Plasma wave coupling yields the observed flux of radio emission.

077.004 **A diffuse component of solar electron streams as a possible source of decametric and hectometric continuum.**
B. N. Levin.
Astron. Astrophys., Vol. 111, 71 - 75 (1982).

During decametric and hectometric noise storms a frequent injection of electron streams leads to the formation of a quasi-stationary diffuse distribution of hot particles close to the source region. An increased density of hot particles and a velocity anisotropy are typical of this distribution due to the breaking of the one-dimensional regime of the stream propagation. The continuum emission is interpreted as the amplification of the coronal plasma spontaneous radiation at a frequency close to the local plasma frequency because of the presence of such a diffuse electron stream component. The basic properties of the continuum are explained in terms of the proposed generation mechanism.

077.005 **On peculiarities of polarized solar radio emission bursts with respect to the frequency and solar cycle phase.**
I. E. Pogodin.
Soln. Dannye 1982 Byull., No. 4, p. 102 - 108 (1982). In Russian.

Some results of an investigation of circularly polarized radio emission burst distributions over the solar disc with respect to the radio frequency and cycle phase are interpreted.

077.006 **Radio wave heating of the corona and electron precipitation during flares.**
D. B. Melrose, G. A. Dulk.
Astrophys. J., Lett., Vol. 259, L41 - L45 (1982).

Electron-cyclotron masers, excited while energy release is occurring in a flaring magnetic loop, are likely to generate extremely intense radiation at decimeter wavelengths. The energy in the radiation can be comparable with that in the electrons associated with hard X-ray bursts, i.e., a significant fraction of the total energy in the flare.

077.007 **The theory of solar radio bursts: Can it be understood by the non-expert?**
D. B. Melrose.
Proc. Astron. Soc. Australia, Vol. 4, 139 - 144 (1981).

The author outlines the plasma physical concepts which are important in the theory of solar radio bursts, and discusses the current status of this theory. He emphasizes the qualitative interpretations.

077.008 **Relative positions of type I and type III solar metre-wave bursts.**
R. A. Duncan.
Proc. Astron. Soc. Australia, Vol. 4, 230 - 235 (1981).

The author concludes that those type I emissions that are accompanied by type III emissions (and perhaps all type I emissions) arise on open or quasi-open magnetic field lines.

077.009 **The importance of plasma effects on electron-cyclotron maser-emission from flaring loops.**
R. R. Sharma, L. Vlahos, K. Papadopoulos.
Astron. Astrophys., Vol. 112, 377 - 385 (1982).

Electron cyclotron maser instability has been suggested as the cause of the observed short (10-20 msec), intense ($\sim$ brightness temperature $\approx 10^{15}$K) and up to 100% polarized microwave solar emission. It is shown that plasma effects and thermal cyclotron damping, ignored in previous theories, play an important role in controlling the frequency range of the emission.

077.010 **On the stratification of electron beams generating solar type III radio bursts.**
A. B. Eremin, V. V. Zajtsev.
Pis'ma Astron. Zh., Tom 8, 495 - 499 (1982). In Russian. English translation in Soviet Astron. Lett., Vol. 8.

A possibility is considered for stratification of fast electron beams generating solar type III radio bursts. The stratification occurs as a result of development of electromagnetic instability in the direction perpendicular to the beam velocity and leads to the oscillating space distribution of the plasma wave energy density. Based on the analysis carried out an explanation is given for results of direct observations of plasma waves in sources of solar type III radio bursts.

077.011 **On the nature of damping of type III solar radio bursts.**
B. N. Levin.
Pis'ma Astron. Zh., Tom 8, 573 - 576 (1982). In Russian. English translation in Soviet Astron. Lett., Vol. 8.
It is shown that damping of type III solar radio bursts at meter and decameter wavelengths may be explained by stabilization of fast electrons relative to Langmuir waves due to excitation of the cyclotron branch of plasma waves.

077.012 **Peak-flux-density spectra of large solar radio bursts.**
E. W. Cliver, L. C. Gentile, J. M. Fink.
Bull. American Astron. Soc., Vol. 14, 607 (1982). – Abstract.

077.013 **Radio emission very near the plasma frequency.**
D. G. Wentzel.
Bull. American Astron. Soc., Vol. 14, 607 (1982). – Abstract.

077.014 **Kilometric type III burst directivity from simultaneous spacecraft observations.**
R. J. MacDowall, R. G. Stone, M. L. Kaiser.
Bull. American Astron. Soc., Vol. 14, 607 (1982). – Abstract.

077.015 **Observation of solar radio bursts of type II and III at kilometer wavelengths from Prognoz-8 during STIP Interval XII.**
S. Pinter, V. P. Grigorieva, K. Kecskemety, K. Kudela.
Report KFKI-1982-20, Hungarian Acad. Sci., Budapest. 13 pp. (1982). – Abstr. in Phys. Abstr., Vol. 85, Abstr. 68721 (1982).

077.016 **Solar microwave bursts – a review.**
M. R. Kundu, L. Vlahos.
Space Sci. Rev., Vol. 32, 405 - 462 (1982).
The authors review the observational and theoretical results on the physics of microwave bursts that occur in the solar atmosphere. They particularly emphasize the advances made in burst physics over the last few years with the great improvement in spatial and time resolution especially with instruments like the NRAO three element interferometer, Westerbork Synthesis Radio Telescope and more recently the Very Large Array.

077.017 **High spatial resolution solar microwave observations.**
K. A. Marsh, G. J. Hurford.
Annu. Rev. Astron. Astrophys., Vol. 20, (see 003.009), 497 - 516 (1982).
Techniques. Emission and propagation of radio waves. Quiet sun. Solar limb. Filaments. Active regions. Flares.

077.018 **Radio burst characteristics of solar proton flares.**
S. W. Kahler.
Astrophys. J., Vol. 261, 710 - 719 (1982).
To isolate the properties of solar radio bursts indicative of energetic proton acceleration, the author compares several groups of solar bursts. Intense centimetric bursts and metric type II bursts associated with solar flares at magnetically well-connected longitudes, but not resulting in detected prompt proton events, are compared with the radio bursts of proton flares. It is concluded that interplanetary energetic protons are not significantly produced in most type II burst shocks. The occurrence of a metric or decimetric stationary type IV burst appears to be a requirement for most proton flares.

077.019 **Solar microwave bursts with outward motion observed at the limb.**
H. Nakajima.
Publ. Astron. Soc. Japan, Vol. 34, 351 - 363 (1982).
Four radio bursts with conspicuous outward motion were observed at the solar limb with a 17-GHz solar radio interferometer. Each of the four radio bursts consisted of polarized "impulsive" components and a weakly polarized gradual (post-burst increase) component. The gradual components show outward motion. Their growth velocities and heights are similar to those observed for Hα post-flare loop-prominence systems or soft X-ray loops. A comparison of the height of a radio burst with that of the associated Hα loop-prominence system (Hα LPS) reveals that the relatively cool Hα LPS is located just below the hotter radio loops. It is further inferred that energy must be continuously supplied to the radio loops in the decay phase of the long-duration events.

077.020 **Time structure of solar decametre Type III radio bursts.**
G. Thejappa, C. V. Sastry.
J. Astrophys. Astron., Vol. 3, 151 - 159 (1982).
The time structure of solar radio decametre Type III bursts occurring during the periods of enhanced emission is investigated. It is shown that the profiles are probably not manifestations of fundamental-harmonic pairs.

077.021 **Time delays in solar bursts measured in the mm-cm range of wavelengths.**
P. Kaufmann, J. E. R. Costa, F. M. Strauss.
Sol. Phys., Vol. 81, 159 - 172 (1982).
Various solar bursts have been analysed with high sensitivity (0.03 sfu, rms) and high-time resolution (1 ms) at two frequencies in the millimeter wave range (22 GHz and 44 GHz), and with moderate time resolution (100 ms) by a patrol telescope at a frequency in the microwave range (7 GHz). It was found that, in most cases, burst maximum emission is not coincident in time at those frequencies. Time delays ranged from about 3 s to near coincidence, defined within 10 ms.

077.022 **Results of investigations of the phenomenon of growth of long-period pulsations of solar centimeter radio radiation before intense flares.**
M. M. Kobrin, S. V. Semenova, V. V. Pakhomov, O. A. Pakhomova, V. M. Fridman.
Astron. Tsirk., No. 1201, p. 1 - 3 (1981). In Russian.

077.023 **The quiet sun brightness temperature at 127 MHz.**
K. M. Borkowski.
Sol. Phys., Vol. 81, 207 - 215 (1982).
A correction and analysis of routine total flux measurements made in Toruń at 127 MHz allowed for the determination of the peak brightness temperature of the sun (above 9×10^5 K) in three periods grouped around the last minimum, and for the estimation of the brightness temperature of the coronal holes: $(7.3 \pm 1.4) \times 10^5$ K.

077.024 **On the spectra of type-III solar radio bursts observed at low frequencies.**
H. Alvarez.
Sol. Phys., Vol. 81, 355 - 362 (1982).
The spectra of strong bursts observed at low frequencies by OGO-5 during 1968–1970 are presented.

077.025 **The 810 MHz solar radio emission in the years 1968 - 1970.**
St. Zięba, M. Jaszczewska, A. Michalec.
Acta Astron., Vol. 32, 93 - 110 (1982).
From the obtained records the daily values of solar flux density and the solar radio bursts were compiled. For each of 572 observed bursts the following characteristics were evaluated: beginning and maximum intensity of burst, duration, profile

intensity, type of burst and correlation with chromospheric flares.

077.026 **Fluctuations of active region microwave radiation in the solar activity minimum.**
V. G. Zandanov.
Issled. po geomagn., aehron. i fiz. Solntsa, Moskva, 1982, No. 60, p. 104 - 109. In Russian. – Abstr. in Ref. zh., 51. Astron., 11.51.317 (1982).

077.027 **Mechanism of gyro-resonance radiation for the slowly varying component of the solar radio radiation.**
R.-y. Zhao, S.-j. Qian.
Acta Astrophys. Sinica, Vol. 2, 114 - 122 (1982). In Chinese.
The slowly varying component of the solar radio emission originates from the combined mechanism of bremsstrahlung and gyro-resonance radiation. In this paper the authors have only studied the mechanism of gyro-resonance for the active regions of a unipolar sunspot and a bipolar sunspot group, because this radiation mechanism plays an overwhelmingly dominant part in the combined mechanism.

077.028 **The nonlinear wave-wave interacting mechanism of the spectrum of the type IV solar decimetric radio burst.**
J. Yao.
Acta Astrophys. Sinica, Vol. 2, 299 - 307 (1982). In Chinese.
The author has shown in an earlier paper that the spectral character of solar type IV_{dm} radio bursts is explained by the mechanism of Langmuir's nonlinear scattering by thermal ions. In this paper the author tries to explain this character by a wave-wave interacting nonlinear process. His ideas are as follows. The nonthermal electrons accelerated in an explosive phase are trapped in a coronal mirror, it forms the source of the type IV_{dm} radio burst. These nonthermal electrons excite simultaneously a low-frequency whistler wave and a high-frequency Langmuir wave. These two waves' nonlinear coalescence process produces a transverse wave of a type IV_{dm} solar radio burst.

Erforschung der Sonne mit Mitteln der Radioastronomie.
See Abstr. 013.026.

Radioastronomische Beobachtungen für den Amateur. Teil 1: Die Radiosonne.
See Abstr. 031.525.

Solar radio observation at millimeter wavelength – description of instruments and preliminary results of observation.
See Abstr. 033.036.

Absolute calibration of the solar radio flux density at 35 GHz.
See Abstr. 033.057.

Absolute calibration of the solar emission in the centimeter band and the daily calibration.
See Abstr. 033.058.

Electron-cyclotron masers as the source of certain solar and stellar radio bursts.
See Abstr. 063.028.

Multiple wavelength observations of a solar active region.
See Abstr. 072.011.

Active region magnetic fields inferred from simultaneous VLA microwave maps, X-ray spectroheliograms, and magnetograms.
See Abstr. 072.035.

Solare Beobachtungsergebnisse. Solar Data.
See Abstr. 072.054.

Spectral character of the solar radio SVC in a solar proton active region.
See Abstr. 072.115.

On solar hydrogen lines in the far-infrared and submillimeter spectrum.
See Abstr. 073.011.

Propagation of low-energy electrons depending on the position of a flare relative to the interplanetary current layer.
See Abstr. 073.024.

Velocity of the shock generated by a large east limb flare on August 18, 1979.
See Abstr. 073.025.

Microwaves and hard X rays from solar flares: multithermal and nonthermal interpretations.
See Abstr. 073.030.

Study of optical, radio and X-ray emissions associated with the solar flare of 1980 March 28.
See Abstr. 073.054.

Electron acceleration in impulsive solar flares.
See Abstr. 073.081.

Microwave emission from hot X-ray kernels in solar flares.
See Abstr. 073.083.

On a relationship between the chromospheric flares and type I noise storms.
See Abstr. 073.087.

Operative prediction of solar flare radiative consequences on the basis of their radio emission.
See Abstr. 073.100.

Electron acceleration and radiation signatures in loop coronal transients.
See Abstr. 074.004.

An association between coronal structures and type III burst sources.
See Abstr. 074.009.

Solar activity in the corona.
See Abstr. 074.011.

Characteristics of shocks in the solar corona, as inferred from radio, optical, and theoretical investigations.
See Abstr. 074.015.

Visible light observations of a dense plasmoid associated with a moving Type IV solar radio burst.
See Abstr. 074.071.

Solar eclipse observations of a magnetic loop structure at wavelengths 2, 3.2, 8.2 and 21 cm.
See Abstr. 075.024.

Comparison of nonflare solar soft X-ray flux with 10.7-cm radio flux.
See Abstr. 076.008.

078 Cosmic Radiation

078.001 **Correlation between the maximum intensity of solar proton events and the time of proton propagation to the earth.**
O. V. Epifanova, V. F. Mel'nikov.
Soln. Dannye 1982 Byull., No. 4, p. 85 - 89 (1982). In Russian.
It is found that the intensity of solar proton events decreases with the increase of Sun-Earth propagation time of a bulk of protons. It is noted that the majority of events are accompanied by type IV radio bursts and have a higher intensity than other events for the same time of propagation.

078.002 **Comparison of helium and heavy ion spectra in ^{3}He-rich solar flares with model calculations based on stochastic Fermi acceleration in Alfvén turbulence.**
E. Möbius, M. Scholer, D. Hovestadt, B. Klecker, G. Gloeckler.
Astrophys. J., Vol. 259, 397 - 410 (1982).
The energy spectra of the He isotopes and O and Fe have been investigated for six ^{3}He-rich solar particle events from 1977 through 1979. The slope of the ^{3}He spectra is generally smaller than that of ^{4}He spectra at the same energy per nucleon and the O spectra exhibit a smaller slope than the Fe spectra. At energies greater than 4 MeV per nucleon the occurrence of the anomalous cosmic ray component obscures the original ^{4}He spectrum from observation. The harder spectra of ^{3}He compared with ^{4}He can be explained quantitatively in terms of a model based on stochastic Fermi acceleration in a region with Alfvén turbulence.

078.003 **Possible evidence for a rigidity-dependent release of relativistic protons from the solar corona.**
M. A. Shea, D. F. Smart.
Space Sci. Rev., Vol. 32, (see 012.005), 251 - 271 (1982).
A model of the time evolving relativistic solar proton spectra for the 7 May 1978 ground level solar cosmic ray event is presented. The authors have concluded that observational data give evidence for a rigidity dependent release of relativistic solar protons from the solar corona during the very early stages of this event.

078.004 **Injection onsets of ~2 GeV protons, ~1 MeV electrons, and ~100 keV electrons in solar cosmic ray flares.**
E. W. Cliver, S. W. Kahler, M. A. Shea, D. F. Smart.
Astrophys. J., Vol. 260, 362 - 370 (1982).
The authors review the data for all 32 ground-level cosmic-ray events (GLEs) observed from 1942 through 1978 and infer injection onset times for the ~2 GeV protons, ~1 MeV electrons, and ~100 keV electrons. They find no compelling evidence for a systematic delay in GLE onset times. The data indicate a systematic phase relationship among the injection onsets of the three particle species considered. A picture is suggested in which the earliest observed particles are injected when an outward moving acceleration region at a shock front intersects the open field lines connecting to Earth.

078.005 **Study of particle generation in low energy events.**
J. Pérez-Peraza.
Rev. Mexicana Astron. Astrofis., Vol. 6, (see 012.017), 357 (1981). – Abstract.

078.006 **Exploration of solar cosmic ray sources by means of proton energy spectra.**
J. Pérez-Peraza.
Rev. Mexicana Astron. Astrofis., Vol. 6, (see 012.017), 358 (1981). – Abstract.

078.007 **A model of propagation of solar flare particles and its application to cosmic ray ground level events.**
J. A. Lockwood, H. Debrunner, E. Flückiger, H. Neuenschwander, M. Schubnell.
J. Geophys. Res., Vol. 87, 4338 - 4344 (1982).
A Monte Carlo simulation of the propagation of relativistic solar flare protons in the interplanetary magnetic field was used to understand the intensity-time profiles and anisotropies for the GLEs on November 22, 1977 and May 7, 1978.

078.008 **The propagation of solar flare proton and its diffusion coefficient in the interplanetary medium.**
F.-s. Kuo.
Bull. Geophys., No. 22, p. 1 - 10 (1981). – Abstr. in Phys. Abstr., Vol. 85, Abstr. 100675 (1982).

078.009 **A direct observation of solar neutrons following the 0118 UT flare on 1980 June 21.**
E. L. Chupp, D. J. Forrest, J. M. Ryan, J. Heslin, C. Reppin, K. Pinkau, G. Kanbach, E. Rieger, G. H. Share.
Astrophys. J., Lett., Vol. 263, L95 - L99 (1982).
Energetic solar neutrons have been detected at the earth following a solar flare which occurred on the west limb on 1980 June 21 at 01:18:20 UT. Impulsive photon emission from 10 keV to greater than 65 MeV lasting over a period of ~66 s was followed by a transient flux of ~50–600 MeV neutrons incident over a ~17 minute period.

078.010 **Prediction of radiation characteristics of solar cosmic rays using the Bayes theorem.**
V. V. Bengin, L. I. Miroshnichenko, V. M. Petrov.
Inst. zem. magn., ionos. i rasprostr. radiovoln AN SSSR. Prepr., 1982, No. 14, 22 pp. In Russian. – Abstr. in Ref. zh., 51. Astron., 10.51.357 (1982).

078.011 **Propagation effects of solar cosmic rays (III). Influence on the ratio of hydrogen to helium.**
G. Zhang.
Acta Astrophys. Sinica, Vol. 2, 201 - 208 (1982). In Chinese.
The influences of the propagation on the ratio of hydrogen to helium (p/a) are discussed in terms of the dimensional solution of the propagation equation for solar cosmic rays in the interplanetary space with the equivalent diffusion coefficients derived from the solar proton data studied by Hamilton.

17th International Cosmic Ray Conference. SH sessions: Sun and heliosphere.
See Abstr. 012.070.

Solar particles, Comet Halley and EPONA.
See Abstr. 032.576.

Cellulose triacetate track detectors for the measurement of abundances of solar cosmic ray heavy ions of helium to iron.
See Abstr. 034.062.

Some peculiarities of the development of active regions associated with proton events.
See Abstr. 072.005.

A sequence of five proton producing flares in McMath Plage 15266 28 April - 7 May 1978 and their interplanetary consequences.
See Abstr. 073.016.

Radio burst characteristics of solar proton flares.
See Abstr. 077.018.

Cosmic rays, solar activity and neutrino flux from the sun.
See Abstr. 080.016.

Measurements of low energy electrons and ions during long-lived solar particle events.
See Abstr. 106.009.

The structure of the interplanetary medium during the first STIP interval (September - October 1975) and pre-release of energetic solar particles.
See Abstr. 106.011.

Investigation of electron streams with $E_e = 40$ - 500 keV in quiet periods of solar activity with methods of the correlation and spectral analysis.
See Abstr. 106.038.

079 Solar Eclipses

079.001 **Calculation of solar eclipses at the Main Astronomical Observatory of the USSR Academy of Sciences in Pulkovo.**
D. D. Polozhentsev.
Glav. astron. obs. AN SSSR. Leningrad, 1982, 18 pp. In Russian. – Abstr. in Ref. zh., 51. Astron., 7.51.115 (1982).

079.002 **Investigations of solar scintillations and eclipse shadow bands.**
E. J. Seykora, L. A. Marschall.
Bull. American Astron. Soc., Vol. 14, 623 (1982). – Abstract.

079.003 **Past and future total solar eclipses over some Canadian cities.**
A. Lowe.
J. R. Astron. Soc. Canda, Vol. 76, 309 - 313 (1982).

079.004 **Solar eclipse expeditions of the Skalnaté Pleso Observatory.**
V. Rušin.
Kozmos, Vol. 13, 157 - 164 (1982). In Slovak.

Early days of astronomy at Toronto – Part IV. The total solar eclipse of August 7, 1869.
See Abstr. 004.025.

Historical eclipses.
See Abstr. 004.032.

Effects of gravity waves on eclipse time absorption of radio waves.
See Abstr. 082.041.

The system of orbital constants of the Saturn satellites.
See Abstr. 100.016.

Solar eclipse 1980 February 16

079.101 **Impression of total solar eclipse in TIROS-N AVHRR imagery.**
K. R. Rao, R. K. Gupta.
Remote Sensing Environ., Vol. 11, 349 - 357 (1981). – Abstr. in Phys. Abstr., Vol. 85, Abstr. 68489 (1982).

079.102 **Observation of the total eclipse of the sun on February 16, 1980.**
M. Waldmeier.
Vierteljahresschr. Naturforsch. Ges. Zürich, Vol. 127/1, 1 - 18 (1982). In German.

Two colour photometry and polarimetry of the solar corona of 16 February 1980.
See Abstr. 074.012.

Turbulent velocities in the solar corona – an estimate from Fabry-Perot interferometric observations of the green coronal line during the eclipse of 1980 February 16.
See Abstr. 074.049.

Solar eclipse observations of a magnetic loop structure at wavelengths 2, 3.2, 8.2 and 21 cm.
See Abstr. 075.024.

The influence of a solar eclipse on the daily variation of the temperature.
See Abstr. 082.095.

Effects of 16 Feb. 1980 solar eclipse on ionosphere as observed at Gauhati.
See Abstr. 083.011.

Electron density changes during solar eclipse of 16 Feb. 1980 at Gauhati.
See Abstr. 083.035.

Solar eclipse 1981 July 31

079.201 **Die Beobachtung der totalen Sonnenfinsternis in Bratsk.**
K. Kirsch, W. Weise.
Jenaer Rundsch., 27. Jahrg., 135 - 137 (1982).

Expedition zur Sonnenfinsternis vom 31. Juli 1981 an den Amur.
See Abstr. 011.003.

Solar eclipse 1982 December 15

079.301 **Globale Betrachtung partieller Sonnenfinsternisse.**
M. Steffen.
Sterne Weltraum, 21. Jahrg., 478, 480 (1982).

079.302 **L'eclisse di sole del 15 dicembre 1982.**
A. Cavallo.
Orione, Vol. 3, 169 - 172 (1982).

Solar eclipse 1983 June 11

079.401 **Total solar eclipse of 11 June 1983.**
A. D. Fiala, M. R. Lukac.
United States Naval Obs. Circ., No. 165, 47 pp. (1982).

080 Atmosphere, Figure, Internal Constitution, Neutrinos, Rotation, etc.

080.001 **Influence of the acoustic radiation pressure on the atmosphere of the Sun.**
O. Bschorr.
Sol. Phys., Vol. 79, 327 - 331 (1982).
The objective of this paper is to calculate the radiation pressure of sound waves in the atmosphere of the Sun and to estimate its influence on the hydrostatic pressure balance. This effect is used for the explanation of the nonstationary processes, especially for the spicule phenomena.

080.002 **Vertical flows of supergranular and mesogranular scale observed on the sun with OSO 8.**
L. J. November, J. Toomre, K. B. Gebbie, G. W. Simon.
Astrophys. J., Vol. 258, 846 - 859 (1982).
Steady flows have been observed at disk center on the quiet Sun using the Un. of Colorado Ultraviolet Spectrometer on OSO 8 and the diode-array instrument at Sacramento Peak Observatory. Simultaneous observations in Fe I λ5576, Mg I λ5173, and Si II λ1817 allow comparison of time-averaged Doppler velocities over a height range of 1400 km from the photosphere to the middle chromosphere. It is shown that patterns of steady vertical velocity with the largest spatial scales of supergranulation are present in the middle chromosphere, where they correlate well with those seen lower in the atmosphere.

080.003 **On long-period oscillations of the solar rotation.**
V. F. Chistyakov.
Soln. Dannye 1982 Byull., No. 3, p. 97 - 102 (1982). In Russian.
An analysis of Fay's formulae for sunspots in 1861 - 1979 has been made.

080.004 **Differential rotation and meridional motions of sunspots from 1874 to 1902.**
M. J. Arévalo, R. Gomez, M. Vázquez, H. Balthasar, H. Wöhl.
Astron. Astrophys., Vol. 111, 266 - 271 (1982).
Position data of sunspot groups from the period 1874 - 1902 are used to investigate the differential rotation of the sun and the meridional motions of the sunspots. The equatorial velocity is higher than in the period 1940 - 1968 investigated by Balthasar and Wöhl (1980), while differences between the northern and southern hemispheres are less. For cycle 12 the authors find meridional motions toward north while in cycle 13 the motions are directed towards south like in the period 1940 - 1968. The other results are comparable to those of Balthasar and Wöhl (1980) for the period 1940 - 1968.

080.005 **Transformation of magnetogravitational waves in the solar atmosphere.**
Y. D. *(Yu. D.)* Zhugzhda, N. S. Dzhalilov.
Astron. Astrophys., Vol. 112, 16 - 23 (1982).
A theory of mutual transformations is developed for all types of magnetogravitational waves in an isothermal atmosphere. The transformation of such waves into fast magnetosonic waves is examined considering the tunneling effect in the case of propagation from regions with weak magnetic fields to regions with strong magnetic fields. The transformation ratios are used to calculate the spectra for slow and fast waves in the case of the given spectrum of the sound noise in the convective zone. It has been found that the elements with a relatively fine structure in the solar atmosphere may be heated effectively by the fast waves, while the large-scale structures are heated by the slow magnetosonic waves.

080.006 **The solar neutrino problem revisited.**
J. Lattanzio.
Proc. Astron. Soc. Australia, Vol. 4, 227 - 230 (1981).
It is the aim of this study to verify and extend the preliminary results given by Prentice (1976) using a modified version of the Mt. Stromlo Stellar Structure Program.

080.007 **100 to 200 year solar periodicities.**
A. Long.
Nature, Vol. 298, 223 - 224 (1982).

080.008 **Internal rotation and gravitational quadrupole moment of the sun.**
D. O. Gough.
Nature, Vol. 298, 334 - 339 (1982).
The internal angular velocity of the sun is estimated from observations of apparent rotational splitting of low-order low-degree global oscillations detected in fluctuations in the limb-darkening function. A sidereal rotation of 3 μHz is inferred for the deep interior; this is more than six times greater than the equatorial rotation frequency of the photosphere. The inferred angular velocity distribution does not vary strongly with latitude, and yields a gravitational quadrupole moment $J_2 \cong 3.6 \times 10^{-6}$.

080.009 **Evidence for an oblique magnetic solar rotator.**
D. O. Gough.
Nature, Vol. 298, 350 - 354 (1982).
Isaak (1982) concluded that the Birmingham data provide the first clear empirical evidence for an intense internal solar magnetic field. Here the author examines this evidence in more detail, and shows that the conclusion is premature. If the magnetic core does exist, the 5-min oscillations provide no clear evidence that it is rotating rapidly. Furthermore, unless one accepts a contrived magnetic-field configuration, an explanation for the $2l + 1$ components of the multiplets is still lacking.

080.010 **Solar irradiance modulation by active regions during 1980.**
S. Sofia, L. Oster, K. Schatten.
Sol. Phys., Vol. 80, 87 - 98 (1982).
The authors estimate the uncertainties caused by ground-based data used as input in the modeling effort, and by our currently incomplete knowledge of the proper parameters that describe the angular variation of sunspot and facular contrasts. They conclude that the most significant uncertainties are due to errors in area measurements and, possibly, varying spot and facular brightness. A "standard model" is derived by a best-fit technique of the currently available ACRIM irradiance data and the predictions of the models with appropriately varied parameters. The expected irradiance for the entire year of 1980 is computed.

080.011 **The turbulent sun.**
G. Simnett.
J. British Interplanet. Soc., Vol. 35, 434 - 440 (1982).

080.012 **New features of the oscillation spectrum of the sun.**
F. Kneer, G. Newkirk, Jr., M. von Uexküll.
Astron. Astrophys., Vol. 113, 129 - 134 (1982) = Mitt. Kiepenheuer-Inst. No. 216.
The authors analyze oscillations observed in the intensity of the CN band at 3883 Å determined from filtergrams. In addition to features familiar in the spectrum of oscillations of the velocity of the solar surface, previously undetected resonances at high frequency and wave number appear. The authors conclude that these represent surface gravity waves and the "3 min" trapped chromospheric oscillation.

080.013 **Wie rotieren Sterne?**
H. Wöhl.
Sterne Weltraum, 21. Jahrg., 299 - 303 (1982).

080.014 **The solar structure and the low *l* five-minute oscillation. II.**
R. Scuflaire, M. Gabriel, A. Noels.
Astron. Astrophys., Vol. 113, 219 - 222 (1982).
Models of solar envelope have been computed with different values of Y, Z, and l/H (respectively abundances of helium and heavy elements and ratio of the mixing-length to the pressure scale height in the convective zone). These parameters are fitted so that the model gives the frequencies of solar oscillations observed by Grec et al. (1980) at the geographic south pole. The authors find that the observations are compatible with the usually accepted solar abundances but they require a convective zone somewhat deeper than the one found in standard solar models.

080.015 **A possible mechanism of origin of inhomogeneous velocity structures in the solar atmosphere.**
Ya. Ya. Valdmanis.
Magn. gidrodinam., 1982, No. 1, p. 130 - 143. In Russian.
Abstr. in Ref. zh., 51. Astron., 8.51.266 (1982).

080.016 **Cosmic rays, solar activity and neutrino flux from the sun.**
G. A. Bazilevskaya, Yu. I. Stozhkov, T. N. Charakhch'yan.
Pis'ma v ZhEhTF, 1982, No. 6, p. 273 - 275. In Russian.
Abstr. in Ref. zh., 51. Astron., 8.51.349 (1982).

080.017 **The effect of a magnetic field on solar luminosity.**
J. N. Tandon, M. K. Das.
Astrophys. J., Vol. 260, 338 - 341 (1982).
A simple model of the sun in the presence of different interior magnetic field configurations (purely poloidal field, a general magnetic field having both poloidal and toroidal components, and a toroidal field) has been computed. The effects of these three types of magnetic field on the luminosity, radius, and effective temperature have been calculated.

080.018 **The effect of perturbation of convective energy transport on the luminosity and radius of the Sun.**
A. S. Endal, L. W. Twigg.
Astrophys. J., Vol. 260, 342 - 352 (1982).
The response of solar models to perturbations of the efficiency of convective energy transport is studied for a number of cases. Independent of the details of the perturbation scheme, the resulting change in the solar radius is always very small compared to the change in luminosity. Changes of the solar radius have been inferred by Dunham et al. from historical observations of solar eclipses in 1715 and 1925. It is concluded that these changes must be indicative of changes in the solar structure at substantial depths below the superadiabatic layer of the convective envelope.

080.019 **Solar core rotation.**
A. Claverie, G. R. Isaak, C. P. McLeod,
H. B. van der Raay, P. L. Pallé, T. Roca Cortes.
Nature, Vol. 299, 704 - 707 (1982).
In the continuing study of solar oscillation by means of optical resonant scattering, two sets of data were obtained between June and August 1981. The new data on solar line of sight velocity measurements reported here, establish the existence of a 13.1 ±0.2 day (synodic) signal of amplitude 6.5 m s^{-1}. This is variously interpreted in terms of a rapidly rotating solar core.

080.020 **Motion of the sun around the barycenter of the solar system and solar activity.**
A. I. Khlystov.
Vestn. Kiev. Univ., Astron. Vyp. 24, p. 61 - 65 (1982). In Russian.

080.021 **Facular influences on the apparent solar shape.**
K. H. Schatten, S. Sofia.
Bull. American Astron. Soc., Vol. 14, 624 (1982). – Abstract.

080.022 **Solar variability.**
S. Sofia, A. S. Endal.
Comments Astrophys., Vol. 9, 277 - 300 (1982).
The authors review the lively and often controversial subject of solar variability, dealing with changes in the total radiation and global structure of the sun on time scales ranging from days to centuries.

080.023 **Surface waves and the heating of the corona.**
I. C. Rae, B. Roberts.
Geophys. Astrophys. Fluid Dyn., Vol. 18, 197 - 226 (1981).
Abstr. in Phys. Abstr., Vol. 85, Abstr. 63350 (1982).

080.024 **Solar variability.**
S. Sofia.
Rev. Mexicana Astron. Astrofis., Vol. 6, (see 012.017), 359 - 362 (1981).
Direct measurements carried out by space-borne detectors have shown variations of the solar constant at the 0.2 percent level, having timescales of days to tens of days. It is argued that these changes do not necessarily reflect variations of the solar luminosity, and that in general, direct measurements have not yet been able to establish (or exclude) longer timescale solar luminosity changes. However, indirect techniques, radius measurements in particular, indicate that solar luminosity variations of up to ~0.7 percent have occurred within tens to hundred of years.

080.025 **The sun – an ordinary star?**
V. G. Surdin.
Zemlya Vselennaya, 1982, No. 5, p. 25 - 26. In Russian.

080.026 **Spectral line transfer effects in Lambdameter measurements of solar short-period oscillations.**
F.-L. Deubner, C. J. Durrant, J. Kaltenbacher.
Astron. Astrophys., Vol. 114, 85 - 94 (1982).
Power spectra of short-period oscillations in the solar atmosphere observed in the Na D_1 line and a medium strong Fe I line are obtained from various sets of Lambdameter records with different exit-slit settings. The structure in the power spectra is remarkably stable with regard to these changes of slit geometry which, therefore, seems to have but little bearing on the generation of the spectral pattern. With recourse to numerical modelling of short-period wave trains the authors are able to demonstrate that such a structure in the power spectra of photospheric lines is absent for rms wave amplitudes less than about 0.1 km s^{-1}, but emerges with higher amplitudes as a consequence of non-linearities in the line transfer process. Some characteristic features of the observed power spectra can also be produced with the synthetic data.

080.027 **Latest rocket measurements of the solar constant.**
C. H. Duncan, R. C. Willson, J. M. Kendall,
R. G. Harrison, J. R. Hickey.
Sol. Energy, Vol. 28, 385 - 387 (1982). – Abstr. in Phys. Abstr., Vol. 85, Abstr. 79072 (1982).

080.028 **Solar irradiance variations and solar activity.**
R. C. Willson.
J. Geophys. Res., Vol. 87, 4319 - 4326 (1982).
A mean value for the 1 AU total solar irradiance of 1368.2 W/m^2 and a downward trend of 0.05% per year were derived from measurements by the Active Cavity Radiometer Irradiance Monitor (ACRIM) experiment on the Solar

Maximum Mission during 1980. Distinct temporary solar irradiance decreases associated with solar activity maxima were observed with a series of nine dips from April to October recurring at fairly regular intervals averaging 24 days. The decreases correlated inversely with sunspot area, 2800-MHz flux, and Zürich sunspot number. Dominant periods common to the irradiance and sunspot area power spectra link the irradiance decreases to sunspot flux deficit in solar active regions. Evidence of significant total irradiance modulation by facular flux excess is cited. A persistent radiative cycle of active regions consistent with the ACRIM irradiance results and the morphology of solar active regions was found. The pattern of regularly recurrent active region maxima between April and October suggests an asymmetry in solar activity generation during this period.

080.029 **A dispersion law for solar oscillations.**
T. L. Duvall, Jr.
Nature, Vol. 300, 242 - 243 (1982).

The pressure or p-modes are acoustic vibrations trapped in a resonant cavity below the solar surface. The waves are most easily observed as primarily vertical velocities at the solar surface by Doppler shift techniques. The vertical velocities vary harmonically in time and in space across the solar surface. Here the authors compare a simple model of the oscillations discussed by Leibacher and Stein (1981) with observations.

080.030 **Standard solar models and the uncertainties in predicted capture rates of solar neutrinos.**
J. N. Bahcall, W. F. Huebner, S. H. Lubow, P. D. Parker, R. G. Ulrich.
Rev. Mod. Phys., Vol. 54, 767 - 799 (1982). – Abstr. in Phys. Abstr., Vol. 85, Abstr. 100865 (1982).

080.031 **Statistical hydromechanics of the solar differential rotation.**
A. S. Monin, L. M. Simuni.
Proc. Natl. Acad. Sci. USA, Vol. 79, 3903 - 3907 (1982). Abstr. in Phys. Abstr., Vol. 85, Abstr. 106661 (1982).

080.032 **Solar rotation measurements at Mount Wilson. III. Meridional flow and limbshift.**
B. J. LaBonte, R. Howard.
Sol. Phys., Vol. 80, 361 - 372 (1982).

The 'ears' velocity pattern described in Paper I (Howard et al., 1980) had no physical explanation. A reanalysis shows that the large scale solar velocity patterns are better described by a nonmonotonic limbshift and a meridional flow. The results of the new analysis imply that the study of solar velocity pattern at the level of a few ms^{-1} required that magnetic regions be treated separately from nonmagnetic regions.

080.033 **Are the high-latitude torsional oscillations of the sun real?**
B. J. LaBonte, R. Howard.
Sol. Phys., Vol. 80, 373 - 378 (1982).

A numerical test is made to determine if the high-latitude torsional wave is generated from the low-latitude torsional pattern as a result of the authors' reduction procedures. The results indicate that the high-latitude motions are not an artifact of the analysis, but are true solar features. It is demonstrated that the one-wave-per-hemisphere torsional oscillation does not result from the reduction procedure. These results place the observations in conflict with the predictions of $\alpha-\omega$ models of the solar cycle.

080.034 **Internal gravity waves in the solar atmosphere. II. Effects of radiative damping.**
B. W. Mihalas, J. Toomre.
Astrophys. J., Vol. 263, 386 - 408 (1982).

Radiative damping of internal gravity waves in the solar atmosphere is considered in a linearized approximation with a height-dependent Newtonian cooling time. A linear fit is made to Stix damping times for the photosphere, and this linear relation is simply extended into the chromosphere. Initial energy fluxes of 10^8, 10^7, and 10^6 ergs $cm^{-2} s^{-1}$ are assumed for monochromatic gravity waves with a wide range of frequencies and horizontal wavenumbers. These waves may play an important role in the broadening and strengthening of photospheric spectral lines.

080.035 **The energy of electric current sheets. I. Models with moving magnetic dipoles.**
Y. Q. Hu, B. C. Low.
Sol. Phys., Vol. 81, 107 - 119 (1982).

This paper treats two problems on the formation of electric current sheets in the highly electrically conducting solar atmosphere. The first problem concerns a vertical current sheet formed by decreasing the distance between a pair of parallel magnetic line-dipoles lying on the photosphere. The solution to this problem was given previously by Priest and Raadu. They derived a formula for the energy stored through the presence of the current sheet. It is shown that this formula is incorrect. The second problem of the paper concerns the current sheet formed by increasing the distance between the pair of line-dipoles. A different field configuration results, with a curved current sheet lying transverse to the vertical. An analysis of the energy properties is given, to compare with the properties of the Priest-Raadu model.

080.036 **The sun as a star.**
W. Deinzer.
The solar system and its exploration, (see 012.054), p. 25 - 28 (1981).

The properties discussed are basic for the great variety of phenomena observed in the solar atmosphere: mass, radius, luminosity; rotation; magnetic field. Since these properties are general enough to be known for many other stars too, they might be called stellar.

080.037 **The solar atmosphere – from photosphere to corona.**
C. Jordan.
The solar system and its exploration, (see 012.054), p. 29 - 39 (1981).

A review is given of recent developments in solar physics, with emphasis on areas where observations from space have made substantial contributions. Progress in establishing the temperature and density structure of the outer solar layers has been made through measurements of emission line fluxes in the extreme ultraviolet and X-ray parts of the spectrum, combined with spatial information from imaging instruments. The processes by which the corona and other hot plasma is heated remain elusive. The further observations required to limit the range of possible mechanisms are discussed.

080.038 **Global pulsations of the sun and magnetosphere.**
V. M. Mozhzherin.
Zemlya Vselennaya, 1982, No. 6, p. 16. In Russian.

080.039 **Beobachtungen zur differentiellen Rotation der Sonne.**
G. Heyer, H. Eylers, H. Lenhardt, U. Ritschel.
Sterne Weltraum, 21. Jahrg., 524 - 525 (1982).

080.040 **A golden age for solar physics.**
A. B. C. Walker, Jr.
Phys. Today, Vol. 35, No. 11, p. 60 - 67 (1982).

New space probes – and a possible journey to the sun – will maintain the fast pace of major advances in our understanding of the sun's interior, coronae, winds, magnetic explosions and cycles of activity.

080.041 **La "constante solar" es constante? (1).**
A. Papetti.
R Muscae, Vol. 6, 78 - 81 (1981).

080.042 **An account of the development of the solar neutrino problem.**
J. N. Bahcall, R. Davis, Jr.
Essays in nuclear astrophysics, (see 003.013), p. 243 - 285 (1982).
This paper is a summary of some of the ideas and events that have led to what has come to be known as the solar neutrino problem.

080.043 **The 5-min oscillations of the sun are incompatible with a rapidly-rotating core.**
R. H. Dicke.
Nature, Vol. 300, 693 - 697 (1982).
Hill's suggestion that Einstein's general relativity may be invalid requires a rapidly spinning solar core. The triplet structure observed in the solar 5-minute power spectrum seems incompatible with this rotating core. But the triplet structure can be accounted for as the effect of a magnetically distorted core rotating with a ~12.5-day period.

080.044 **On the nature of the 160-min solar oscillations.**
Yu. V. Vandakurov.
Astron. Tsirk., No. 1173, p. 1 - 2 (1981). In Russian.

080.045 **Variation of the solar radius and the motion of the sun about the barycentre.**
V. S. Prokudina.
Astron. Tsirk., No. 1179, p. 6 - 8 (1981). In Russian.

080.046 **On the coefficient of continuous absorption in the solar atmosphere and active regions.**
G. F. Sitnik.
Astron. Tsirk., No. 1182, p. 3 - 4 (1981). In Russian.

080.047 **Solar neutrino and solar particles.**
D. Basu.
Sol. Phys., Vol. 81, 363 - 365 (1982).
Examination of the most recent data on solar neutrino shows that there is a statistically significant relationship between solar neutrino and solar particles.

080.048 **On turbulent conductivity and magnetic permeability of the solar plasma.**
V. N. Krivodubskij.
Soln. Dannye 1982 Byull., No. 8, p. 99 - 109 (1982). In Russian.
This paper deals with the problem of turbulent plasma conductivity and magnetic permeability in the photosphere, sunspots and convective zone of the sun, turbulence inhomogeneity and anisotropy being taken into account.

080.049 **Solar neutrinos.**
I. W. Roxburgh.
Irish Astron. J., Vol. 15, (see 012.061), 106 - 109 (1981).
A brief review of the present state of the solar neutrino problem and the proposed solutions for the discrepancy between theory and experiment is given.

080.050 **Inferences from solar oscillations.**
D. Gough.
Irish Astron. J., Vol. 15, (see 012.061), 116 - 119 (1981).
A brief review of the principal dynamical modes of solar oscillation observed so far is given. In particular, observational evidence for degeneracy splitting by rotation and its implication for a possible quadrupole moment of the solar gravitational field is analysed.

080.051 **Global oscillations of the sun.**
G. R. Isaak.
Irish Astron. J., Vol. 15, (see 012.061), 119 - 120 (1981). Abstract.

080.052 **Rotational splitting of global solar oscillations.**
Preliminary implications for the sun's gravitational quadrupole moment and the general theory of relativity.
H. Hill, R. Bos, P. Goode.
Irish Astron. J., Vol. 15, (see 012.061), 120 - 121 (1981). Abstract.

080.053 **The solar "constant".**
D. J. Mullan.
Irish Astron. J., Vol. 15, 143 - 147 (1981).
The current state of precision measurements of the solar irradiation from spacecraft is reviewed and their implications for solar models are outlined.

080.054 **Measurement of the Mount Wilson solar white-light plate collection.**
R. Howard.
The physics of sunspots, (see 012.063), p. 59 (1981). Abstract.

080.055 **Sunspots and changes in the global output of the sun.**
P. Foukal.
The physics of sunspots, (see 012.063), p. 391 - 423 (1981).
The paper reviews recent observational developments in radiometry of the solar constant, and discusses the theoretical interpretation of relatively large dips in the irradiance caused by sunspots.

080.056 **On the variability of the solar diameter.**
A. D. Wittmann, J. A. Bonet Navarro, H. Wöhl.
The physics of sunspots, (see 012.063), p. 424 - 433 (1981).
Visual and photoelectric measurements of the solar semidiameter R have been made at Locarno since 1972. In May and June 1981 a total of 2091 visual drift times (and 339 photoelectric scans) were obtained at Izaña and Locarno. The long-term variability of the solar semidiameter has been investigated not only from the authors own material, but also by compiling and partly re-analyzing a large fraction of the observations published since 1640.

080.057 **Sunspots and solar variability.**
H. S. Hudson, R. C. Willson.
The physics of sunspots, (see 012.063), 434 - 445 (1981).
The data obtained from the total solar irradiance monitor on the Solar Maximum Mission show variations with amplitudes of as much as few tenths of a percent. The larger variations have a clear association with sunspots. The authors have extended the analysis of the effects of solar active regions on the total solar irradiance, using the synoptic sunspot data as a basis for comparison.

080.058 **Transformation of magneto-gravitational waves in the solar atmosphere.**
Yu. D. Zhugzhda, N. S. Dzhalilov.
Probl. kosm. ehlektrodin., Moskva, 1981, p. 6 - 11. In Russian. Abstr. in Ref. zh., 51. Astron., 11.51.299 (1982).

080.059 **Variations in solar luminosity.**
P. V. Foukal.
Smithsonian Astrophys., Obs., Spec. Rep. 392, Vol. II, (see 012.065), p. 17 - 29 (1982).
Important advances have been made in the past few years in confirming short-term variability of the sun's total luminosity, and in identifying the physical mechanisms responsible for this variation. In this paper the author reviews the evidence on such relatively short-term variations, and

uses it to speculate briefly on the properties of longer term, as yet undetectable changes over the 11-year cycle and beyond. Observations have confirmed the existence of transient depressions of the solar constant, lasting 5 - 30 days. The author discusses the implications for possible changes of $L_\odot$ caused specifically by surface magnetic activity (spots and faculae) over the 11-year cycle and its modulation over centuries.

080.060 **On some regularities of long-term variation of the differential rotation of the sun.**
R. N. Ikhsanov, Yu. I. Vitinskij.
Izv. Glav. Astron. Obs. Pulkovo, Astrometr. Astrofiz., No. 199, p. 78 - 85 (1982). In Russian.

Long-term variations of the solar differential rotation (1875 - 1954) have been studied separately for pre-minimum and maximum phases of 11-year cycles of solar activity. For the pre-minimum phases it is found that the velocity of the rotation increases with a decrease of the cycle height for mean and small Wolf numbers in the equatorial zone and the gradient of the differential solar rotation decreases with the increasing height of the following 11-year cycle. For the maximum of the cycle the gradient of the solar differential rotation increases with the height of the cycle. It is shown that the found regularities of the differential rotation operated during the Maunder minimum.

080.061 **The sizes of turbulent eddies in the solar atmosphere.**
R. I. Kostyk.
Astron. Zh., Tom 59, 1167 - 1170 (1982). In Russian.
English translation in Soviet Astron., Vol. 26, No. 6.

The mean sizes of turbulent eddies in the solar atmosphere decrease with height from 670 km ($h = 160$ km) to 440 km ($h = 250$ km). Their mean square turbulent velocity is depth-independent and is equal to 1.7 km s^{-1}.

080.062 **On the influence of solar faculae on solar flux variations.**
Eh. V. Kononovich, I. V. Mironova.
Pis'ma Astron. Zh., Tom 8, 740 - 742 (1982). In Russian.
English translation in Soviet Astron. Lett., Vol. 8.

Recent measurements by Willson et al. (1981) of solar constant variation are analysed. Their strong correlation (0.77) with sunspots is confirmed. Incorporation of faculae data increases the coefficient of correlation up to 0.80 - 0.83. Some considerations concerning faculae models are given.

080.063 **The solar differential rotation from the eighteenth solar cycle.**
P. Rovithis.
Astrophys. Space Sci., Vol. 88, 5 - 13 (1982).

The sidereal daily rotation of the Sun, $\xi(\varphi)$, depends on the data used. From an appropriate selection of the data – sunspots with regular motion – it is found that $\xi(\varphi) = 14.31 - 2.70 \sin^2 \varphi$, where φ denotes the heliographic latitude. Moreover, it seems that there is a variation, of the order of 3%, with the solar activity.

080.064 **Decimeter radio wave propagation in the turbulent plasma near the Sun, using Venera 10 spacecraft.**
M. A. Kolosov, O. I. Yakovlev, A. I. Efimov, V. I. Rogal'sky *(Rogal'skij)*, V. M. Razmanov, V. K. Shtrykov.
Radio Sci., Vol. 17, 664 - 674 (1982). – Abstr. in Phys. Abstr., Vol. 86, Abstr. 10295 (1983).

Solar magneto-hydrodynamics.
See Abstr. 003.100.

Laboratory exercises in astronomy – the rotation of the sun.
See Abstr. 014.015.

Ground state mass of ^{81}Kr and the solar neutrino problem.
See Abstr. 022.054.

The ^{3}He(α, γ)^{7}Be reaction and the solar neutrino problem.
See Abstr. 022.055.

Absolute cross section for ^{7}Li(d, p)^{8}Li and solar neutrino capture rates.
See Abstr. 022.078.

Seeing-independent definitions of the solar limb position.
See Abstr. 031.626.

Solar Diameter Monitor: an instrument to measure long-term changes.
See Abstr. 032.069.

DISCO.
See Abstr. 051.062.

Low-energy neutrinos in astrophysics.
See Abstr. 061.010.

Massless neutrinos and the stellar stopping power via the V-A and Weinberg interactions.
See Abstr. 061.019.

Possible time variations in ^{37}Cl solar-neutrino data and neutrino oscillations.
See Abstr. 061.022.

Solar power.
See Abstr. 061.035.

Statistical calculation of the ^{8}B solar neutrino capture cross sections for ^{97}Mo and ^{98}Mo.
See Abstr. 061.063.

Propagation of waves in an atmosphere in the presence of a magnetic field. VI. Application of magneto-acoustic-gravity mode theory to the solar atmosphere.
See Abstr. 062.007.

Magnetic reconnection in toroidal plasmas.
See Abstr. 062.040.

The local dispersion relation for magneto-atmospheric waves.
See Abstr. 062.085.

On turbulent heat transport in rotating convective zones.
See Abstr. 062.086.

Compressibility effects on hydromagnetic surface waves.
See Abstr. 062.091.

Full implicit continuous Eulerian scheme in the spherical coordinates and its applications to solar phenomena.
See Abstr. 062.096.

Non-LTE resonance line polarization with partial redistribution effects.
See Abstr. 063.055.

New results on the mechanism of magnetic flux pumping by three-dimensional convection.
See Abstr. 064.059.

Magnetic field-related heating instabilities in the surface layers of the sun and stars.
See Abstr. 064.066.

Magnetostatic atmospheres with variations in three dimensions.
See Abstr. 064.067.

Consistency of the mixing length theory.
See Abstr. 065.065.

Preliminary determination of the sun's gravitational quadrupole moment from rotational splitting of global oscillations and its relevance to tests of general relativity.
See Abstr. 066.238.

Solar rotation and variability observed in the Ca II K line.
See Abstr. 071.008.

The asymmetry of photospheric absorption lines. I. An analysis of mean solar line profiles.
See Abstr. 071.017.

Measurements of the limb darkening of faculae near the solar limb.
See Abstr. 072.017.

The occurrence of sunspot groups and the velocity of solar rotation at heliographic latitudes ⩾ 40°.
See Abstr. 072.022.

The sunspot chromosphere-corona transition region.
See Abstr. 072.087.

Reference model atmosphere calculation: the Sunspot sunspot model.
See Abstr. 072.091.

On the possibility of generating gravitational radiation by the Sun.
See Abstr. 073.057.

Observations of mass motions in active regions.
See Abstr. 073.064.

The overshoot layer at the base of the solar convective zone and the problem of magnetic flux storage.
See Abstr. 075.009.

Extreme ultraviolet observational data on the solar spectrum.
See Abstr. 076.005.

Mass loss from cool stars – facts, fads, and fallacies.
See Abstr. 112.008.

Solar calibration of stellar rotation tracers.
See Abstr. 116.005.

Earth

081 Structure, Figure, Gravity, Orbit, etc.

081.001 **Aspherical Earth structure from fundamental spheroidal-mode data.**
G. Masters, T. H. Jordan, P. G. Silver, F. Gilbert.
Nature, Vol. 298, 609 - 613 (1982).
Free-oscillation data reveal heterogeneity in the Earth's mantle whose geographical pattern is dominated by spherical harmonics of angular degree two and correlates well with the hydrostatically referenced geoid. The heterogeneity can be modelled as localized in the transition zone (420–670 km depth) and may be related to a large-scale component of mantle convection.

081.002 **The geoid and geodynamics.**
B. H. Hager.
Nature, Vol. 299, 104 - 105 (1982).

081.003 **Oxygen in the earth's metallic core?**
R. Jeanloz.
Nature, Vol. 299, 108 - 109 (1982).

081.004 **Astronomical theory of the Pleistocene ice ages: a brief historical review.**
J. Imbrie.
Icarus, Vol. 50, (see 012.003), 408 - 422 (1982).
This paper traces the evolution of the astronomical theory of the Pleistocene ice ages from its origin in the work of Adhémar and Croll, through its quantitative transformation in the hands of Milankovitch, to its current renaissance. Emphasis is given to the interplay between climatic theory and geologic observation.

081.005 **Pre-Pleistocene glaciation on Earth: implications for climatic history of Mars.**
N. Christie-Blick.
Icarus, Vol. 50, (see 012.003), 423 - 443 (1982).
The history of ice ages on Earth, extending back more then 2Ga (10^9 years), has been established by the recognition in strata of many ages of the assemblage of erosional features and deposits generated by glacial activity. The most useful indicators are widespread diamictite, striated and faceted clasts, polished and striated pavements, and laminites containing "dropstones" inferred to be ice-rafted.

081.006 **Comments on the long-term stability of the Earth's obliquity.**
W. R. Ward.
Icarus, Vol. 50, (see 012.003), 444 - 448 (1982).
Continued tidal evolution of the Earth-Moon system will lengthen the day and expand the lunar orbit. Both of these changes increase the present 26,000-year equinoctial precession period. In less than 2 billion years, this period will become comparable to the ~49,000- and ~69,000-year periods of important terms describing the precession of the Earth's orbit plane due to planetary perturbations.

081.007 **The use of satellite-to-satellite tracking data for global estimation of the gravity field.**
J. Kryński.
Artif. Satell., Vol. 16, No. 3, p. 53 - 63 (1981).

081.008 **Evolution of the earth at volumetrical differentiation of its interior.**
A. S. Monin, O. G. Sorokhtin.
Dokl. An SSSR, Tom 263, 572 - 575 (1982). In Russian.
Abstr. in Ref. zh., 51. Astron., 8.51.176 (1982).

081.009 **Comparative planetology and the early history of the earth.**
V. L. Barsukov.
Vestn. AN SSSR, 1982, No. 4, p. 42 - 54. In Russian. – From Ref. zh., 51. Astron., 8.51.178 (1982).

081.010 **A new derivation of the ratio Q (wobble)/Q_μ (mantle).**
A. Ben-Menahem.
Geophys. J. R. Astron. Soc., Vol. 70, 535 - 537 (1982).

081.011 **Modelling the effect of atmospheric pressure variations on gravity.**
R. S. Spratt.
Geophys. J. R. Astron. Soc., Vol. 71, 173 - 186 (1982).
A model is developed to describe the effects of local and global atmospheric pressure fluctuations on local gravity.

081.012 **Planetare Dynamik der Erde.**
H. Kautzleben.
Sitzungsber. Akad. Wiss. DDR, No. 3N, 24 pp. (1982).

081.013 **The constitution of the core: seismological evidence.**
B. A. Bolt.
Philos. Trans. R. Soc. London, Ser. A, Vol. 306, (see 012.014), 11 - 20 (1982).

081.014 **The Earth's core and the phase diagram of iron.**
O. L. Anderson.
Philos. Trans. R. Soc. London, Ser. A, Vol. 306, (see 012.014), 21 - 35 (1982).

081.015 **Constraints on core composition from shock-wave data.**
T. J. Ahrens.
Philos. Trans. R. Soc. London, Ser. A, Vol. 306, (see 012.014), 37 - 47 (1982).

081,016 **Chemical aspects of the formation of the core.**
C. J. Allègre, B. Dupré, O. Brévart.
Philos. Trans. R. Soc. London, Ser. A, Vol. 306, (see 012.014), 49 - 59 (1982).

081.017 **Finding core motions from magnetic observations.**
D. Gubbins.
Philos. Trans. R. Soc. London, Ser. A, Vol. 306, (see 012.014), 247 - 254 (1982).

081.018 **A readjustment of lumped coefficients from inclination of 1974 - 70A.**
J. Klokočník.
Bull. Astron. Inst. Czechoslovakia, Vol. 33, 259 - 262 (1982).
The determination of the lumped coefficients from in-

clination variations of 1974 - 70A near the 15th-order orbital resonance is revised, since the inclusion of the normal lift effect (Sehnal 1981) in the already corrected inclinations (Wakker et al, 1980) led to changes in their values greater than their accuracy.

081.019 **On the gravimetric inversion problem.**
D. O Riska.
Acta Polytech. Scandinavica, Appl. Phys. Ser., No. Ph 129, p. 1 - 30 (1981). – Abstr. in Phys. Abstr., Vol. 85, Abstr. 68230 (1982).

081.020 **Present day status of gravimetry and significance for the Earth and surveying.**
W. Torge.
Z. Vermessungswes., Vol. 106, 636 - 638 (1981). In German. Abstr. in Phys. Abstr., Vol. 85, Abstr. 68236 (1982).

081.021 **Contributions of satellite geodesic measurements towards a study of the physics of the Earth.**
C. Reigber.
Z. Vermessungswes., Vol. 106, 628 - 631 (1981). In German. Abstr. in Phys. Abstr., Vol. 85, Abstr. 68237 (1982).

081.022 **A precise definition and implementation of the geoid and related problems.**
E. Groten.
Z. Vermessungswes., Vol. 107, 26 - 32 (1982). – Abstr. in Phys. Abstr., Vol. 85, Abstr. 68239 (1982).

081.023 **Variation of gravity with depth in the Earth.**
F. A. Dahlen.
Phys. Rev. D, Vol. 25, 1735 - 1736 (1982). – Abstr. in Phys. Abstr., Vol. 85, Abstr. 73422 (1982).

081.024 **Heat flow and reversals of the Earth's magnetic field.**
J. A. Jacobs.
J. Geomagn. Geoelectr., Vol. 33, 527 - 529 (1981). – Abstr. in Phys. Abstr., Vol. 85, Abstr. 78604 (1982).

081.025 **Analytical expressions for gravity due to homogeneous revolutional compartments in the Gaussian divergence approach.**
M. Okabe.
Geophys. Prospect., Vol. 30, 166 - 187 (1982). – Abstr. in Phys. Abstr., Vol. 85, Abstr. 83736 (1982).

081.026 **Modes of mantle convection and the removal of heat from the earth's interior.**
T. Spohn, G. Schubert.
J. Geophys. Res., Vol. 87, 4682 - 4696 (1982).
Earth thermal histories based on two-layer and whole-mantle parameterized convection models have been explored to determine the influence of the mode of mantle convection on the content and distribution of mantle radioactive heat sources and on the temperatures and heat transfer properties of the mantle.

081.027 **Mean earth ellipsoid determined from SEASAT 1 altimetric observations.**
G. B. West.
J. Geophys. Res., Vol. 87, 5538 - 5540 (1982).
The radius and the origin of a mean earth ellipsoid have been determined using geoid heights derived from SEASAT-1 radar altimetric observations of the ocean surface and potential coefficients. While assuming that the altimetric geoid is the best available, adjustments are made to the geoid derived from potential coefficients so that the geoid differences are minimized. This adjustment is then applied to the ellipsoid which best fits the geoid derived from potential coefficients. A description of the geoid height data used in the solutions for the ellipsoid parameters, results of the solutions, and a summary are presented.

081.028 **Information theory lateral density distribution for earth inferred from global gravity field.**
D. P. Rubincam.
J. Geophys. Res., Vol. 87, 5541 - 5552 (1982).

081.029 **Multispectral observations of a terrestrial basaltic assemblage: Uwekahuna laccolith, Kilauea caldera.**
R. B. Singer, P. G. Lucey, P. L. Blake, T. B. McCord.
Bull. American Astron. Soc., Vol. 14, 757 (1982). – Abstract.

081.030 **Marées terrestres.**
P. Melchior (Editor).
Bull. Inf. (Obs. R. Belgique, Bruxelles), No. 88, p. 5626 - 5734 (1982).

081.031 **Investigations on the effect of the resonance of the liquid outer core of the earth in gravimetric tidal variations.**
H.-J. Dittfeld.
Geodesy and physics of the earth, (see 012.053), p. 72 - 81 (1981).

081.032 **The need for a high accuracy geoid on land and sea.**
P. V. Angus-Leppan, C. Rizos.
Geodesy and physics of the earth, (see 012.053), p. 211 - 225 (1981).

081.033 **A method for the construction of spheroidal mass distributions consistent with the harmonic part of the earth's gravity potential.**
C. C. Tscherning, H. Sünkel.
Geodesy and physics of the earth, (see 012.053), p. 481 - 500 (1981).

081.034 **Determination of the earth's figure by solving an inverse astro-gravi-magnetic-geodesic problem.**
D. Zidarov.
Geodesy and physics of the earth, (see 012.053), p. 504 - 514 (1981).

081.035 **Project DIDEX *(Differential Doppler Experiment).***
J. B. Zieliński.
Geodesy and physics of the earth, (see 012.053), p. 515 - 527 (1981).
For improving the Earth gravitational potential the low-low satellite tracking experiment has been proposed. The paper presents the general idea of the experiment, scientific objectives, accuracy analysis, outlines of technical realization. With an assumed measurement accuracy ±0.05 mm/sec the significant improvement of the geopotential model is expected.

081.036 **Planetary albedo estimates from Meteosat data.**
M. Gube.
ESA J., Vol. 6, 53 - 69 (1982).
Planetary albedo values have been derived from Meteosat data by the use of radiative-transfer calculations. Some preliminary results are presented showing the albedo modulation caused by various surface and cloud types and a number of patterns of diurnal albedo change are described.

081.037 **Response of the earth to gravitational radiation.**
K. C. Heaton.
J. R. Astron. Soc. Canada, Vol. 76, 320 (1982).

081.038 **The next development in satellite determination of the Earth's gravity field.**
W. M. Kaula.
Dynamical astronomy, (see 012.056), p. 127 - 134 (1982).

081.039 **A refined gravity model from Lageos (GEM-L2).**
F. J. Lerch, S. M. Klosko, G. B. Patel.
Geophys. Res. Lett., Vol. 9, 1263 - 1266 (1982).

081.040 **Disputes about the earth's interior.**
Nature, Vol. 300, 681 (1982).

081.041 **The initial stage of the earth's evolution. Geochemistry.**
O. L. Kuskov, N. I. Khitarov.
Izv. AN SSSR. Fiz. Zemli, 1982, No. 6, p. 69 - 82. In Russian. Abstr. in Ref. zh., 51. Astron., 10.51.185 (1982).

081.042 **Influence of astronomical factors on the change of the relief and climate of the earth.**
A. I. Petrov.
Izv. Vses. geogr. o-va, 1982, 114, No. 4, p. 336 - 340. In Russian. – Abstr. in Ref. zh., 51. Astron., 11.51.104 (1982).

081.043 **Morphology of volcanoes.**
H. J. Schuver.
Advances in planetary geology, (see 003.014), p. 565 - 567 (1982). – Abstract.

081.044 **A kinematic thermal history of the earth's mantle.**
R. T. Langan, N. H. Sleep.
J. Geophys. Res., Vol. 87, 9225 - 9235 (1982).

081.045 **Geodynamical and geodetic consistency tests of Earth density models.**
M. A. Khan.
Phys. Earth Planet. Inter., Vol. 28, 291 - 301 (1982) = Hawaii Inst. Geophys., Publ. No. 1259 (1982).

Theory of a stratified heterogeneous spheroid is applied to compute the gravity and potential parameters of selected Earth density models, and compare them with the corresponding quantities of a modern geodetic reference system based on observational data. These tests show considerable intermodel and model-reference discrepancies, the latter being attributable largely to the deviation of the Earth from hydrostatic equilibrium.

081.046 **Hydrostatic theory of the Earth and its mechanical implications.**
S. M. Nakiboglu.
Phys. Earth Planet. Inter., Vol. 28, 302 - 311 (1982).

A second-order hydrostatic theory is developed on the assumption that the trace of the Earth's inertia tensor, its mass and mean radius are invariant under any process causing deviations from the hydrostatic state. The hydrostatic flattening and the zonal coefficients of the hydrostatic gravitational field are obtained. The small discrepancy between the hydrostatic flattening and the value deduced from PREM suggests that the density is underestimated at large depths and/or it is slightly overestimated in shallow regions of the Earth.

081.047 **Whole mantle convection and the thermal evolution of the Earth.**
W. R. Peltier, G. T. Jarvis.
Phys. Earth Planet. Inter., Vol. 29, 281 - 304 (1982).

081.048 **Cosmical geodynamics in 1981.**
J. Klokočník.
Říše hvězd, Vol. 63, 186 - 189 (1982). In Czech.

081.049 **Geocentric gravitational constant.**
M. Burša.
Říše hvězd, Vol. 63, 182 - 184 (1982). In Czech.

081.050 **The Kilmichael structure, Mississippi: evidence for a meteorite impact origin.**
P. B. Robertson, M. D. Butler.
J. Geol., Vol. 90, 589 - 601 (1982). – Abstr. in Phys. Abstr., Vol. 86, Abstr. 19999 (1983).

081.051 **Astronomic geology: its subject and general problems.**
G. N. Katterfeld *(Katterfel'd)*, I. V. Galibina.
Mod. Geol., Vol. 8, 127 - 147 (1982). – Abstr. in Phys. Abstr., Vol. 86, Abstr. 20000 (1983).

Interior of the earth: its structure, constitution and evolution.
See Abstr. 003.026.

The Cambridge encyclopedia of Earth sciences.
See Abstr. 003.115.

Astrogeodetic work on the geoid in Austria.
See Abstr. 013.067.

Gravitationswechselwirkung ausgedehnter Körper.
See Abstr. 042.073.

Luni-solar precession constant and the accuracy of the Earth's principal moments of inertia.
See Abstr. 043.011.

Irregular changes of gravity and the rotation of the earth.
See Abstr. 044.002.

The role of the core in irregular fluctuations of the Earth's rotation and the excitation of the Chandler wobble.
See Abstr. 044.006.

Signal analysis and error analysis studies for a Geopotential Research Mission (GRM).
See Abstr. 051.086.

The recent secular variation and the motions at the core surface.
See Abstr. 084.033.

The source of the Earth's magnetic field.
See Abstr. 084.034.

A comparative study of the Earth's magnetic and gravity fields.
See Abstr. 084.092.

Lunar and terrestrial crusts: a contrast in origin and evolution.
See Abstr. 094.041.

Chemistry and accretion history of the earth and the inner planets.
See Abstr. 107.036.

082 Atmosphere (Refraction, Scintillation, Extinction, Airglow, Site Testing)

082.001 **Blue sunlight extinction and scattering by dust in the 60-km altitude atmospheric region.**
M. Ackerman, C. Lippens, C. Muller, P. Vrignault.
Nature, Vol. 299, 17 - 20 (1982).
Twilight data obtained photographically from a stratospheric balloon platform in the autumns of 1980 and 1981 and in the spring of 1982 are presented for blue and red light. They indicate the presence of a light absorbing layer and of a scattering layer in the mesosphere at altitudes near 60 ± 10 km with a low scattering albedo (0.1) at 0.44 μm.

082.002 **Climate models and the astronomical theory of the ice ages.**
I. M. Held.
Icarus, Vol. 50, (see 012.003), 449 - 461 (1982).
Attempts at determining the climatic response to perturbations in the Earth's orbital parameters are reviewed.

082.003 **A model study of the role of variable ice albedo in the climate response of the Earth to orbital variations.**
G. E. Birchfield, J. Weertman.
Icarus, Vol. 50, (see 012.003), 462 - 472 (1982).
Using the atmosphere-hydrosphere component of a paleoclimate model, the model sensitivity on a short time scale to changes in the parameterization of the surface albedo is examined in the context of response to orbitally produced insolation anomalies.

082.004 **The relevance of the IUE results on young stars for earth's paleoatmosphere.**
V. M. Canuto, J. S. Levine, T. R. Augustsson, C. L. Imhoff, M. S. Giampapa.
Advances in ultraviolet astronomy, (see 012.006), p. 293 - 296 (1982).
The authors have computed the levels of oxygen and ozone in the earth's paleoatmosphere as functions of the early sun's UV flux as indicated by IUE observations and of the CO_2 level determined by paleoclimatological considerations. The values the authors have calculated are high enough to allow the observed state of oxidation and reduction in early rocks but small enough to allow chemical evolution of amino acids.

082.005 **Concentration of oxygen and hydrogen atoms at heights of 80 - 100 km from hydroxyl emission observations.**
M. V. Shagaev.
Polyar. siyaniya i svechenie noch. neba, Moskva, 1981, No. 29, p. 26 - 37. In Russian. – Abstr. in Ref. zh., 62. Issled. kosm. prostranstva, 7.62.465 (1982).

082.006 **Continuous radiation and its connection with the components of the upper atmosphere.**
V. G. Sobolev.
Polyar. siyaniya i svechenie noch. neba, Moskva, 1981, No. 29, p. 84 - 92. In Russian. – Abstr. in Ref. zh., 62. Issled. kosm. prostranstva, 7.62.467 (1982).

082.007 **Results of investigations of internal gravitational waves with an optical method.**
V. I. Krasovskij.
Polyar. siyaniya i svechenie noch. neba, Moskva, 1981, No. 29, p. 5 - 8 . In Russian. – Abstr. in Ref. zh., 51. Astron., 7.51.468 (1982).

082.008 **Internal gravitational waves near the mesopause. III. Method for determining the height of the radiating layer.**
B. P. Potapov, N. N. Shefov.
Polyar. siyaniya i svechenie noch. neba, Moskva, 1981, No. 29, p. 38 - 41. In Russian. – Abstr. in Ref. zh., 62. Issled. kosm. prostranstva, 7.62.469 (1982).

082.0Q9 **Internal gravitational waves near the mesopause. IV. Measurement of the height of the hydroxyl emission layer.**
B. P. Potapov, N. N. Novikov.
Polyar. siyaniya i svechenie noch. neba, Moskva, 1981, No. 29, p. 42 - 43. In Russian. – Abstr. in Ref. zh., 62. Issled. kosm. prostranstva, 7.62.470 (1982).

082.010 **Internal gravitational waves near the mesopause. V. Application of the theory of scintillations of the earth's atmosphere.**
E. P. Chunchuzov.
Polyar. siyaniya i svechenie noch. neba, Moskva, 1981, No. 29, p. 44 - 58. In Russian. – Abstr. in Ref. zh., 62. Issled. kosm. prostranstva, 7.62.471 (1982).

082.011 **Internal gravitational waves near the mesopause. VI. Polar region.**
N. N. Novikov.
Polyar. siyaniya i svechenie noch. neba, Moskva, 1981, No. 29, p. 59 - 67. In Russian. – Abstr. in Ref. zh., 62. Issled. kosm. prostranstva, 7.62.472 (1982).

082.012 **Internal gravitational waves near the mesopause. VII. Numerical modelling of processes of excitation of hydroxyl radiation.**
M. V. Shagaev.
Polyar. siyaniya i svechenie noch. neba, Moskva, 1981, No. 29, p. 68 - 75. In Russian. – Abstr. in Ref. zh., 62. Issled. kosm. prostranstva, 7.62.466 (1982).

082.013 **Internal gravitational waves near the mesopause. VIII. On tropospheric sources.**
V. A. Sukhodoev.
Polyar. siyaniya i svechenie noch. neba, Moskva, 1981, No. 29, p. 76 - 79. In Russian. – Abstr. in Ref. zh., 62. Issled. kosm. prostranstva, 7.62.473 (1982).

082.014 **The atmospheric transmission at La Silla at 230 GHz.**
J. Brand.
Messenger, No. 29, p. 20 - 23 (1982).

082.015 **Astrophysical investigations on Mt. Majdanak.**
V. S. Shevchenko.
Astrofiz. i kosm. fiz. Moskva, 1982, p. 293 - 313. In Russian. Abstr. in Ref. zh., 51. Astron., 8.51.862 (1982).

082.016 **Radiative-photochemical models of the primitive terrestrial atmosphere.**
G. Visconti.
Planet. Space Sci., Vol. 30, 785 - 793 (1982).
The vertical thermal structure of a primitive terrestrial atmosphere is investigated with a radiative-convective-photochemical model. The radiative code includes the short wave contribution from water vapor and ozone, and long wave contribution from methane, carbon dioxide, water vapor and ozone.

082.017 **The regulation of carbon dioxide and climate: Gaia or geochemistry.**
J. E. Lovelock, A. J. Watson.

Planet. Space Sci., Vol. 30, 795 - 802 (1982).

The authors review the Gaia hypothesis which postulates a condition of planetary homeostasis affecting chemical composition and climate. Some criticisms are answered and a new model is introduced for the long term regulation of the mean surface temperature through the biological control of CO_2 partial pressure.

082.018 **Temporal variations of polar cap OI 5577-Å airglow.**
S. Ismail, L. L. Cogger.
Planet. Space Sci., Vol. 30, 865 - 873 (1982).

It is the purpose of the paper to extend the study of polar cap 5577-Å airglow variations by including photometer results obtained from the ISIS-2 satellite. Data from the satellite complement that from the ground by providing spatially averaged intensities over the polar cap.

082.019 **Radiation field in the troposphere and stratosphere from 240 - 1000 nm – I. General analysis.**
R. R. Meier, D. E. Anderson, Jr., M. Nicolet.
Planet. Space Sci., Vol. 30, 923 - 933 (1982).

A method for rapid calculation of the multiply-scattered solar induced radiation field in the troposphere and stratosphere is present. The method is described in sufficient detail so that the mathematical model can be incorporated in a straightforward manner into photochemical models of the troposphere and stratosphere. Results are described which show that multiple scattering and ground albedo can result in large amplifications of the radiation field in the photochemically active region 240 - 1000 nm.

082.020 **Radiation field in the troposphere and stratosphere II. Numerical analysis.**
M. Nicolet, R. R. Meier, D. E. Anderson, Jr.
Planet. Space Sci., Vol. 30, 935 - 983 (1982).

The results of calculations of the multiply-scattered solar-induced radiation field in the troposphere and stratosphere are presented for direct application to photochemical models. The enhancement factors due to multiple scattering are given for the heights, solar zenith angles and wavelengths (between 800 and 300 nm) which play a role in the photodissociation of various atmospheric constituents.

082.021 **Altitude of peak emission of (O I) 5577 Å in the lower thermosphere: Chapman versus Barth mechanisms.**
M. N. M. Rao, G.S. N. Murty, V. C. Jain.
J. Atmos. Terr. Phys., Vol. 44, 559 - 566 (1982).

Heights of peak (O I) 5577 Å emission rates Z_C and Z_B due to Chapman and Barth mechanisms respectively have been computed for different model atomic oxygen profiles in the lower thermosphere, generated by assigning different values to temperature (T), mixing factor (S) and peak atomic oxygen density $[O]_m$.

082.022 **Theoretical description of the main neutral constituents in the earth's upper atmosphere.**
M. N. Vlasov, V. E. Davydov.
J. Atmos. Terr. Phys., Vol. 44, 641 - 647 (1982).

The solution of time-dependent continuity and motion equations for the main neutral constituents O, O_2 and N_2 has been obtained and compared with different approximations. Numerical calculations have been carried out for different solar fluxes and eddy diffusion coefficients in the altitude range 80 - 300 km.

082.023 **Direct investigations of heavy positive ions (1 - 2000 a. e. m.) in the upper mesosphere.**
Yu. A. Bragin, B. F. Shamakhov.
Kosm. Issled., Tom 20, 644 - 645 (1982). In Russian.

082.024 **Broad band airborne water vapor radiometry.**
P. M. Kuhn.
Geophys. Res. Lett., Vol. 9, 621 - 624 (1982).

An infrared radiometer with a pass band of 280 to 520 cm^{-1} (35.7 to 19.2 μm) is employed on the NASA Ames Research Center U-2 and C-141A aircraft in the measurement of water vapor burden in the upper troposphere and stratosphere.

082.025 **Vertical column abundance of atmospheric OH at solar maximum from Fritz Peak, Colorado.**
C. R. Burnett, E. B. Burnett.
Geophys. Res. Lett., Vol. 9, 708 - 711 (1982).

New measurements of the vertical column abundance of atmospheric hydroxyl have been taken at Fritz Peak Observatory, Colorado, through the period of maximum solar activity of Solar Cycle 21.

082.026 **Corrections to "Preliminary results on 27-day solar rotation variation in stratospheric zonal winds" by G. D. Nastrom and A. D. Belmont.**
D. E. Venne, G. D. Nastrom, A. D. Belmont.
Geophys. Res. Lett., Vol. 9, 759 - 760 (1982). – See Abstr. 22.082.151.

082.027 **Observation of NO_2 diurnal variation in the stratosphere.**
J. P. Pommereau.
Geophys. Res. Lett., Vol. 9, 850 - 853 (1982).

082.028 **Radiation and chemistry in the stratosphere: sensitivity to O_2 absorption cross sections in the Herzberg continuum.**
L. Froidevaux, Y. L. Yung.
Geophys. Res. Lett., Vol. 9, 854 - 857 (1982).

082.029 **The 2145 Å dayglow and the high altitude nitric oxide density.**
G. A. Victor, A. Dalgarno.
Geophys. Res. Lett., Vol. 9, 866 - 868 (1982).

082.030 **Empirical model of the electron temperature for lower and middle latitudes in the region of heights between 100 - 200 km.**
Yu. K. Chasovitin, N. M. Klyueva, L. S. Mironova.
Geomagn. Aehron., Tom 22, 582 - 585 (1982). In Russian.

082.031 **Recording of OH (10, 4) emission at 6686 Å in the radiation of the night atmosphere.**
O. T. Yurchenko.
Geomagn. Aehron., Tom 22, 687 - 688 (1982). In Russian.

082.032 **On the effectivity of fluorescence of H_2O^+ ions in the upper atmosphere.**
N. N. Shefov.
Geomagn. Aehron., Tom 22, 688 - 689 (1982). In Russian.

082.033 **The neutral and ion chemistry of the upper atmosphere.**
L. Thomas.
Handbuch der Physik. Geophysik III/6, (see 003.006), p. 7 - 127 (1982).

Contents: Neutral and ionized atmospheric structure and composition. Interactions of solar radiations with atmospheric gases. Collision processes and chemical reactions. Laboratory measurements of relevant rate coefficients. Experimental methods for measurement of neutral and ion composition. Aeronomical models of neutral composition. Aeronomical models of ion composition: characteristics of ionospheric regions, the ion chemistry of the D region, the ion chemistry of the E region, the ion chemistry of the F region. The production of excited species and their roles in atmospheric and ionospheric processes.

082.034 **Diffusion in the high atmosphere.**
T. Yonezawa.
Handbuch der Physik. Geophysik III/6, (see 003.006), p. 129 - 245 (1982).
Contents: Introduction and the general theory of diffusion. Diffusive equilibrium, mixing, and turbulence in the high atmosphere. Vertical distributions of molecular and atomic oxygen. Ambipolar diffusion including the formation of the F2-layer. Vertical distributions of minor constituents in the high atmosphere.

082.035 **Determination of satellite drag coefficient from the orbital analysis of the ANS satellite (1974 - 70A).**
L. Sehnal.
Bull. Astron. Inst. Czechoslovakia, Vol. 33, 244 - 249 (1982).
The drag coefficient C_D is derived from the analysis of the changes of the mean daily motion of the satellite 1974 - 70A. The detailed analysis, together with previously made investigations (Sehnal, 1982), shows a changing thermal accomodation coefficient α during the first part of the satellite lifetime. The drag coefficient is then computed from formulas based on the values of α.

082.036 **The surface albedo of the Earth in the near ultraviolet (330 - 340 nm).**
J. E. Frederick, R. B. Abrams.
Remote Sensing Environ., Vol. 11, 337 - 347 (1981). – Abstr. in Phys. Abstr., Vol. 85, Abstr. 68488 (1982).

082.037 **An analytical static model of temperature and composition from 20 to 2000 km altitude.**
D. Alcayde.
Ann. Géophys., Vol. 37, 515 - 528 (1981). – Abstr. in Phys. Abstr., Vol. 85, Abstr. 68543 (1982).

082.038 **Intensity covariations of the oxygen green and red lines in the nightglow.**
K. Misawa, I. Takeuchi, Y. Kato, I. Aoyama.
Ann. Géophys., Vol. 37, 549 - 556 (1981). – Abstr. in Phys. Abstr., Vol. 85, Abstr. 68545 (1982).

082.039 **Morphological characteristics of inhomogeneous structures of the upper atmosphere.**
G. K. Solodovnikov, A. S. Bakaj, V. M. Russkin.
Fiz. polyar. ionos. Leningrad, 1982, p. 78 - 90. In Russian. Abstr. in Ref. zh., 62. Issled. kosm. prostranstva, 9.62.278 (1982).

082.040 **Meteor wind radar studies at Waltair.**
P. C. S. Devara, M. I. Ahmed, M. Srirama Rao, B. Ramachandra Rao.
Indian J. Radio Space Phys., Vol. 10, 228 - 236 (1981). Abstr. in Phys. Abstr., Vol. 85, Abstr. 78892 (1982).

082.041 **Effects of gravity waves on eclipse time absorption of radio waves.**
M. Devi, A. K. Barbara, P. Talukdar.
Indian J. Radio Space Phys., Vol. 11, 42 - 44 (1982). – Abstr. in Phys. Abstr., Vol. 85, Abstr. 78893 (1982).

082.042 **On the inertial stability of the equatorial middle atmosphere.**
T. J. Dunkerton.
J. Atmos. Sci., Vol. 38, 2354 - 2364 (1981). – Abstr. in Phys. Abstr., Vol. 85, Abstr. 78894 (1982).

082.043 **Anisotropy and polarization of hard atmospheric X-rays.**
P. N. Khosa, R. R. Rausaria, K. L. Moza.
Indian J. Radio Space Phys., Vol. 10, 241 - 245 (1981). Abstr. in Phys. Abstr., Vol. 85, Abstr. 78903 (1982).

082.044 **Optical depth of atmospheric aerosols and the valid range of Angstrom's formula.**
G.-c. Wang.
Sci. Atmos. Sinica, Vol. 6, 52 - 57 (1982). In Chinese. – Abstr. in Phys. Abstr., Vol. 85, Abstr. 89006 (1982).

082.045 **Determination of atmospheric aerosol size distribution from spectral solar irradiance measurements.**
A. Pons, A. Corrons.
Opt. Pura Apl., Vol. 14, No. 1, p. 19 - 21 (1981). – Abstr. in Phys. Abstr., Vol. 85, Abstr. 89048 (1982).

082.046 **High resolution imaging from the ground.**
N. J. Woolf.
Annu. Rev. Astron. Astrophys., Vol. 20, (see 003.009), 367 - 398 (1982).
Explanation of image profile. Observations and seeing theory. Measurements of atmospheric seeing. Local effects. Active optics. The cost of image size.

082.047 **Catalytic processes in the atmospheres of Earth and Venus.**
W. B. DeMore, Y. L. Yung.
Science, Vol. 217, 1209 - 1213 (1982).

082.048 **Natthimmelens lys – spesielt i nordlysområder.**
H. K. Myrabö.
Astron. Tidsskr., Årg. 15, 108 - 115 (1982).

082.049 **Spectroscopy of the O I 989- and 7990-Å multiplets in the dayglow and aurora.**
R. R. Meier.
J. Geophys. Res., Vol. 87, 6307 - 6316 (1982).
The spectroscopy of the $2p^4\,{}^3P$–$2p^3\,3s'\,{}^3D°$ (989 Å) transition of atomic oxygen and the intermediate branch transition $2p^3 3p\ {}^3P$–$2p^3 3s'\ {}^3D°$ (7990 Å) has been modelled for an optically thick thermosphere using Monte Carlo simulation. The model reproduces rocket observations of 989 Å in the dayglow and aurora if the $e + O$ electron impact excitation cross section is 8.4×10^{-18} cm^2 at 25 eV. The individual 989-Å multiplet strengths agree with dayglow observations which partially resolve the line structure. The 7990-Å singlet to doublet ratio observed in auroras is also consistent with the model.

082.050 **High-resolution dayglow O I (1304 Å) and O I (989 Å) rocket observations.**
A. B. Christensen, R. W. Eastes, P. D. Feldman, E. P. Gentieu.
J. Geophys. Res., Vol. 87, 6317 - 6323 (1982).
High-resolution (~1 Å) dayglow observations of the O I (1304 Å) and O I (989 Å) multiplets were made from an Astrobee-F rocket payload. Three components of the O I (1304 Å) multiplet were measured at zenith angles of approximately 50° and 140° at 1.1 Å resolution. Over the entire altitude range of observation, i.e., 100 - 260 km, the three components were found to be equal to within ±15%. The shape of the O I (989 Å) multiplet distribution observed on the flight at 1.3-Å resolution was indistinguishable from an optically thin source with the energy levels of the excited state populated according to their statistical weights in spite of the large optical depths of the O I (989 Å) multiplet at rocket altitudes. Experiments with a laboratory UV source also showed the insensitivity of the multiplet distribution to the optical depth of the source. These results are entirely consistent with a recently developed PFR radiative transport code (Meier, 1982) in the absence of selective absorption by rotational lines of N_2.

082.051 **Volume emission rate profiles of the 6300-Å tropical nightglow obtained from the AE-E satellite: latitudinal and seasonal variations.**
V. J. Abreu, G. A. Schmitt, P. B. Hays, T. P. Dachev.
J. Geophys. Res., Vol. 87, 6346 - 6352 (1982).

Surface brightnesses at 6300-Å measured by the visible airglow experiment on board the AE-E satellite have been inverted to obtain altitude profiles of volume emission rate. A morphological picture of the emission at low latitudes is developed and qualitatively interpreted in terms of the behavior of electric fields and neutral winds in the nighttime ionosphere. Average bottomside electron density altitude profiles are derived from the 6300-Å volume emission profiles.

082.052 **A rocket-borne photometric measurement of O_3 and NO_2 in the equatorial stratosphere.**
P. Schlyter, G. Witt.
J. Geophys. Res., Vol. 87, 7228 - 7234 (1982).

A determination of the NO_2 and O_3 mixing ratios has been made by measuring the attenuation of sunlight at 409 nm and 254, 288, 316 nm from an ascending rocket launched from Kourou, French Guiana (5°N) during the morning twilight of March 21, 1974.

082.053 **Investigation of processes in the thermosphere during magnetic disturbances.**
B. E. Serebryakov.
Geomagn. Aehron., Tom 22, 776 - 781 (1982). In Russian.

082.054 **Terrestrial O_2 lines used as wavelength references: comparison of measurements and model computations.**
H. Balthasar, U. Thiele, H. Wöhl.
Astron. Astrophys., Vol. 114, 357 - 359 (1982) = Mitt. Kiepenheuer-Inst. No. 215.

High-precision spectral measurements of the wavelength of terrestrial O_2 lines within solar spectra using the Kitt Peak 1 m Vacuum Fourier Transform Spectrometer are compared with model computations of the shifts of these lines due to wind up to 20 km. Although the wind data available existed only from radio sonde measurements about 65 km away and only from two launches per day, the shifts observed during two days could be explained well. Maximum shifts of terrestrial O_2 lines of ±0.3 mÅ at about 6300 Å were observed. It is shown that it may be possible to correct a general trend of the observed shifts with a precision of ±3 m s^{-1} during a day.

082.055 **NO infrared radiation in the upper atmosphere.**
G. E. Caledonia, J. P. Kennealy.
Planet. Space Sci., Vol. 30, 1043 - 1056 (1982).

A kinetic model is developed for the prediction of upper atmospheric i.r. radiation from the vibrational bands of NO. The model is appropriate to both the quiescent and aurorally-excited nighttime atmosphere and has been exercised to examine the variation in NO radiation levels which can result from both natural atmospheric variability and uncertainties in kinetic parameters. Comparisons between model predictions and i.r. radiance data are presented.

082.056 **The importance of atmospheric differential refraction in spectrophotometry.**
A. V. Filippenko.
Publ. Astron. Soc. Pacific, Vol. 94, 715 - 721 (1982).

The effects of atmospheric differential refraction on astronomical measurements are much more important than is generally assumed. In particular, it is shown that relative line and continuum intensities in spectrophotometric work may be erroneous if this phenomenon is neglected. To help observers minimize these errors, the relation between object position and optimal slit or aperture orientation is derived, and practical tables and graphs are presented for use at the telescope.

082.057 **A study of atmospheric extinction at Cerro Tololo Inter-American Observatory.**
A. Gutiérrez-Moreno, H. Moreno, G. Cortés.
Publ. Astron. Soc. Pacific, Vol. 94, 722 - 728 (1982).

The behavior of atmospheric extinction at CTIO from 1964 to 1980 is analyzed. Rayleigh scattering, aerosol extinction and ozone absorption parameters are determined. It is concluded that no increase of the effects of atmospheric pollution at the observing site are noticeable. The errors produced in the stellar monochromatic magnitudes when using average extinction values are discussed.

082.058 **Detection of minor constituents of the stratosphere by far infrared emission spectroscopy.**
D. J. W. Kendall, T. A. Clark.
Int. J. Infrared Millim. Waves, Vol. 2, 783 - 808 (1981).
Abstr. in Phys. Abstr., Vol. 85, Abstr. 100553 (1982).

082.059 **Ozone air broadening coefficients in the 9 μm region.**
J. M. Hoell, C. N. Harward, C. H. Bair, B. S. Williams.
Proc. SPIE Int. Soc. Opt. Eng., Vol. 288, (see 012.035), 376 - 382 (1981). – Abstr. in Phys. Abstr., Vol. 85, Abstr. 106309 (1982).

082.060 **Anomalous water-vapor absorption: implications for radiative-convective models.**
J. R. Hummel.
J. Atmos. Sci., Vol. 39, 879 - 885 (1982). – Abstr. in Phys. Abstr. Vol. 85, Abstr. 106313 (1982).

082.061 **Atmospheric optical propagation – an integrated approach.**
W. S. Ross, W. P. Jaeger, J. Nakai, T. T. Nguyen, J. H. Shapiro.
Opt. Eng., Vol. 21, 775 - 785 (1982). – Abstr. in Phys. Abstr., Vol. 85, Abstr. 106316 (1982).

082.062 **Observations of the Green Flash and slant range 'seeing'.**
R. Gerharz.
Pure Appl. Geophys., Vol. 119, 998 - 1002 (1981). – Abstr. in Phys. Abstr., Vol. 85, Abstr. 106318 (1982).

082.063 **Evolution of the earth's atmosphere and the astronomical refraction.**
G. Teleki.
Publ. Astron. Obs. Sarajevo, Vol. 1, No. 1, (see 012.043), p. 111 - 114 (1981).

082.064 **Emission of the atmosphere below 40 km in the region of the vacuum ultraviolet.**
L. I. Abramova, L. S. Sorokin, A. I. Filatov.
Kosm. Issled., Tom 20, 765 - 767 (1982). In Russian.

082.065 **On one of the possible causes of the upper atmosphere's hydrogen emission.**
N. M. Martsvaladze, L. M. Fishkova.
Kosm. Issled., Tom 20, 773 - 775 (1982). In Russian.

082.066 **The altitude dependence of the $O_2(A^3\Sigma_u^+)$ vibrational distribution in the terrestrial nightglow.**
I. C. McDade, E. J. Llewellyn, R. G. H. Greer, D. P. Murtagh.
Planet. Space Sci., Vol. 30, 1133 - 1136 (1982).

A simple vibrational relaxation model which reproduces the observed altitude integrated vibrational distribution of the Herzberg I bands in the nightglow is used to derive the altitude profiles of the individual vibrational levels at 1 km intervals in the 85–115 km height range. The possible errors associated with using rocket-borne photometer measurements of a limited number of bands in the $O_2(A^3\Sigma_u^+ - X^3\Sigma_g^-)$ system to infer the total Herzberg I emission profile are assessed.

082.067 **Noctilucent clouds: simulation studies of their genesis, properties and global influences.**
R. P. Turco, O. B. Toon, R. C. Whitten, R. G. Keesee, D. Hollenbach.
Planet. Space Sci., Vol. 30, 1147 - 1181 (1982).

The formation, evolution and properties of noctilucent clouds are studied using a time-dependent one-dimensional model of ice particles at mesospheric altitudes. The model treats ice crystals, meteoric dust, water vapor and air ionization as fully interactive cloud elements. For ice particles, the microphysical processes of nucleation, condensation, coagulation and sedimentation are included: the crystal habits of ice are also accounted for. Meteoric dust is analyzed in the manner of Hunten et al. (1980). The simulated particle sizes range from 10 Å to 2.6 μm. The chemistry of water vapor and the charge balance of the mesosphere are also analyzed in detail. Model simulations are compared with a wide variety of noctilucent cloud data.

082.068 **Halo's in onze directe omgeving.**
P. P. Hattinga Verschure.
Zenit, 9. Jaarg., 510 - 512 (1982).

082.069 **High-resolution observations of atmospheric emission spectra in the 20-micron window from Mauna Kea: comparison with theoretical spectra.**
D. A. Naylor, R. T. Boreiko, T. A. Clark, R. Emery, B. Fitton.
J. R. Astron. Soc. Canada, Vol. 76, 323 (1982). – Abstract.

082.070 **The physical basis of ground-based radiometric sensing of temperature profiles at 60 GHz.**
G. Skoog.
Res. Lab. Electron. Onsala Space Obs., Res. Rep. No. 144, 44 pp. (1982).

082.071 **On the increase of sodium twilight glow during meteor showers.**
G. G. Mateshvili, Yu. D. Mateshvili, T. G. Megrelishvili.
Astron. Tsirk., No. 1172, p. 6 - 8 (1981). In Russian.

082.072 **Twilight glow of metallic emissions after the passage of the Lyrid meteor shower.**
G. G. Mateshvili.
Astron. Tsirk., No. 1173, p. 4 - 5 (1981). In Russian.

082.073 **Spectral atmospheric extinction in Pulkovo.**
G. A. Alekseeva, V. V. Novikov.
Astron. Tsirk., No. 1175, p. 4 - 5 (1981). In Russian.

082.074 **PSM measured seeing comparison for several mountain summits.**
P. V. Shcheglov.
Astron. Tsirk., No. 1176, p. 1 - 2 (1981). In Russian.

082.075 **Interferometric r_0 measurements on the Crimea and their weather dependence.**
A. S. Kutyrev, A. A. Tokovinin, P. V. Shcheglov.
Astron. Tsirk., No. 1176, p. 2 - 5 (1981). In Russian.

082.076 **Atmospheric parameter r_0 and 60-cm reflector image diameter comparison on the Crimea.**
A. S. Kutyrev, V. G. Metlov, P. V. Shcheglov.
Astron. Tsirk., No. 1176, p. 5 - 6 (1981). In Russian.

082.077 **On astronomical seeing effects caused by the atmospheric ground layer and the whole atmosphere at Mt. Dushak-Erekdagh.**
A. Eh. Gur'yanov, Yu. V. Khan.
Astron. Tsirk., No. 1176, p. 6 - 8 (1981). In Russian.

082.078 **On simultaneous seeing and scintillation measurements of α UMi at Mt. Dushak-Erekdagh in 1979.**
Yu. V. Khan.
Astron. Tsirk., No. 1177, p. 2 - 4 (1981). In Russian.

082.079 **On the visibility of bright stars at Mt. Majdanak during daytime.**
G. I. Popov, N. F. Minyajlo, A. S. Kharin.
Astron. Tsirk., No. 1177, p. 4 - 5 (1981). In Russian.

082.080 **On the relation of impulsive bursts of 3-cm radio emission with meteor phenomena in the earth's atmosphere.**
O. V. Epifanova.
Astron. Tsirk., No. 1182, p. 5 - 7 (1981). In Russian.

082.081 **H_2O^+ emission in winter 1980/1981.**
O. A. Matveeva.
Astron. Tsirk., No. 1197, p. 4 (1981). In Russian.

082.082 **Determination of the atmospheric water vapour content above La Silla and the prospects for FIR observations.**
W. A. Sherwood, A. Greve,
Second ESO infrared workshop, (see 012.060), p. 173 - 177 (1982).

082.083 **Frequency spectra of scintillation of star images.**
N. Tugzhsuren, E. Donev, St. Panchev.
Blg. geofiz. spisanie, Vol. 8, No. 1, p. 42 - 50 (1982). In Bulgarian. – Abstr. in Ref. zh., 51. Astron., 11.51.823 (1982).

082.084 **The effects of atmospheric turbulence on telescopic observations.**
F. K. Brunner.
Bull. Géod., Vol. 56, 341 - 355 (1982).

Telescopic observations to terrestrial and extraterrestrial objects are affected by atmospheric turbulence. The structure of atmospheric turbulence is reviewed in terms of the refractive index structure parameter. The mean and fluctuating refraction effects of the turbulent atmosphere on the propagation of electromagnetic waves are addressed. The known formulae for the variance and the spectrum of the angle-of-arrival fluctuations are presented and experimental results are summarised.

082.085 **Noctilucent clouds.**
M. Gadsden.
Space Sci. Rev., Vol. 33, 279 - 334 (1982).

Noctilucent clouds appear during the summertime at high latitudes near the top of the mesosphere. In this review, the observational facts about them, obtained from ground level, by rocket sounding and from orbiting spacecraft, are reviewed. The clouds seem to be ice particles nucleated at very low pressures and temperatures by either meteoric smoke or by atmospheric ions. Wavepatterns in the clouds may well result from quite close relations between the troposphere and the mesosphere. The very existence of the clouds leads to difficulties in explaining why there is so much water vapour at this great height in the atmosphere. To try to predict the microscopic behaviour of the cloud particles leads one into assessment of the relative importance of radiometer effects, radiation balance, Brownian movement, electric polarization and the influence of Coulomb attraction on the growth of large clustered ions.

082.086 **A determination of the coefficient of atmospheric dispersion and correction for its influence.**
G. A. Plyugin.
Izv. Glav. Astron. Obs. Pulkovo, Astrometr. Astrofiz., No. 199, p. 48 - 52 (1982). In Russian.

The difficulties and errors in correction for atmospheric dispersion are shown. It is noted that the computed coefficient of atmospheric dispersion is not free from large systematic errors. The coefficient is more reliable if determined from observations.

082.087 **The long-period scintillation effect on the accuracy of spectrophotometric observations.**
G. A. Alekseeva, L. A. Kamionko.
Izv. Glav. Astron. Obs. Pulkovo, Astrofiz. Astrometr., No. 200, p. 63 - 67 (1982). In Russian.
Scintillations of stellar images with periods 1 min - 2 s were measured at three different stations. Mean scintillation amplitudes and Fourier spectra were obtained. A comparison shows that the character of scintillations is different for different stations and can be considered an important characteristic of the local astronomical climate.

082.088 **Spatial-frequency- and wavelength-dependent effects of aerosols on the atmospheric modulation transfer function.**
N. S. Kopeika.
J. Opt. Soc. America, Vol. 72, 1092 - 1094 (1982).

082.089 **Ray tracing in a refracting spherically symmetric atmosphere.**
D. A. Thompson, T. J. Pepin, F. W. Simon.
J. Opt. Soc. America, Vol. 72, 1498 - 1501 (1982).
Satellite- and balloon-borne instruments may be used to measure the total transmission of electromagnetic radiation along a slant-path trajectory traversing the earth's atmosphere. Inversions of this type of measurement data to determine the altitude- and wavelength-dependent atmospheric extinction require accurate ray-tracing techniques. The ray-tracing equations are derived. These equations describe the refraction angle and generate the path-length matrix for such radiation traversing a refracting, spherically symmetric atmosphere. Furthermore, they are presented in forms suitable for numerical integration.

082.090 **Stratospheric infrared emission observed by balloonborne cryogenic interferometer spectrometer.**
H. Sakai, G. A. Vanasse.
J. Opt. Soc. America, Vol. 72, 1724 (1982). – Abstract.

082.091 **Atmospheric ozone determination by solar occultation using the UV spectrometer on the Solar Maximum Mission.**
A. C. Aikin, B. Woodgate, H. J. P. Smith.
Appl. Opt., Vol. 21, 2421 - 2424 (1982).

082.092 **Radiative cooling computed for model atmospheres.**
T. S. Eriksson, C. G. Granqvist.
Appl. Opt., Vol. 21, 4381 - 4388 (1982).

082.093 **Airglow characteristics of equatorial plasma depletions.**
M. Mendillo, J. Baumgardner.
J. Geophys. Res., Vol. 87, 7641 - 7652 (1982).
A new low-light-level, all-sky, 6300 Å airglow imaging system has been developed for ground-based studies of the optical signatures of equatorial depletions. The system was designed for narrow-band interference filters (6 Å full width at half power) with interchangeable all-sky (180°) and narrow-field (60°) lenses and records images photographically by using a standard 35 mm camera. A field test of the apparatus was conducted on Ascension Island (8.0°S, 14.4°W) from January 24 to February 10, 1981. The photographs were analyzed for broad morphological characteristics by using simple overlay techniques augmented by digital image processing methods to extract more quantitative detail. The initial results coming from these observations are presented.

082.094 **Mg^+ morphology from Visual Airglow Experiment observations.**
C. G. Fesen, P. B. Hays.
J. Geophys. Res., Vol. 87, 9217 - 9223 (1982).
Observations of low-latitude Mg^+ by the Visual Airglow Experiment on Atmosphere Explorer E were analyzed by a two-dimensional inversion technique.

082.095 **The influence of a solar eclipse on the daily variation of the temperature.**
M. Waldmeier.
Meteorol. Rundsch., 33. Jahrg., 157 - 158 (1980). In German.

082.096 **Near-Earth ultraviolet environment.**
C. S. Weller.
Proc. SPIE Int. Soc. Opt. Eng., Vol. 279, (see 012.072), 216 - 222 (1981). – Abstr. in Phys. Abstr., Vol. 85, Abstr. 117513 (1982).

082.097 **Slant path atmospheric extinction measurements.**
W. M. Gutman, D. L. Walters, S. T. Hanley.
Proc. SPIE Int. Soc. Opt. Eng., Vol. 289, (see 012.073), 206 - 208 (1981). – Abstr. in Phys. Abstr., Vol. 86, Abstr. 10054 (1983).

082.098 **High resolution spectroscopy of the Earth's upper atmosphere from space.**
C. B. Farmer.
Proc. SPIE Int. Soc. Opt. Eng., Vol. 289, (see 012.073), 299 (1981). – Abstr. in Phys. Abstr., Vol. 86, Abstr. 10055 (1983).

082.099 **Note on the scattering of radiation by moderately nonspherical particles.**
P. Chylek, V. Ramaswamy, W. J. Wiscombe.
J. Atmos. Sci., Vol. 39, 1886 - 1888 (1982). – Abstr. in Phys. Abstr., Vol. 86, Abstr. 10061 (1983).

082.100 **Stochastic parameterisation of gravity wave stresses.**
T. J. Dunkerton.
J. Atmos. Sci., Vol. 39, 1711 - 1725 (1982). – Abstr. in Phys. Abstr., Vol. 86, Abstr. 10105 (1983).

082.101 **Spectral and diurnal variations in clear sky planetary albedo.**
B. Briegleb, V. Ramanathan.
J. Appl. Meteorol., Vol. 21, 1160 - 1171 (1982). – Abstr. in Phys. Abstr., Vol. 86, Abstr. 15619 (1983).

082.102 **Airglow emission as a temperature probe in Earth's atmosphere.**
P. V. Kulkarni.
Curr. Sci., Vol. 51, 637 - 639 (1982). – Abstr. in Phys. Abstr., Vol. 86, Abstr. 15622 (1983).

082.103 **Effective photodissociation cross sections for molecular oxygen and nitric oxide in the Schumann-Runge bands.**
M. Allen, J. E. Frederick.
J. Atmos. Sci., Vol. 39, 2066 - 2075 (1982). – Abstr. in Phys. Abstr., Vol. 86, Abstr. 15676 (1983).

Atmospheric transmission.
See Abstr. 012.096.

Calculation of the absorption coefficient of the 15-μm ν_2 band of CO_2 using the theory of overlapping spectral lines.
See Abstr. 022.079.

Absorptions of the $\rho\sigma\tau$, 0.8 μm and a bands of the water vapor.
See Abstr. 022.138.

Magnetic dipole infrared atmospheric oxygen bands.
See Abstr. 022.160.

The Schumann-Runge continuum of oxygen at wavelengths greater than 175 nm.
See Abstr. 022.164.

O_2 absorption cross sections (187–225 nm) from stratospheric solar flux measurements.
See Abstr. 022.165.

Methods of analysis of stratospheric ozone data.
See Abstr. 031.595.

The struggle against sky noise in ground based and air-borne observations.
See Abstr. 031.644.

Two-dimensional inversion technique for satellite airglow data.
See Abstr. 031.678.

Fourier transform spectroscopy in the submillimeter region.
See Abstr. 031.680.

On the domeless solar telescope in Hida Observatory.
See Abstr. 032.072.

Effect of upper atmospheric variations on satellite lifetimes.
See Abstr. 052.045.

Generation of atmospheric turbulence by solar radiation.
See Abstr. 062.115.

Second-order scattering approximation to X-ray photon transport at balloon altitudes.
See Abstr. 063.065.

Noble gases in planetary atmospheres: implications for the origin and evolution of atmospheres.
See Abstr. 093.070.

The evolution of an impact-generated atmosphere.
See Abstr. 107.009.

Photographic surface photometry of the Milky Way. II. Surface photometry in the region of the dark cloud "Coalsack" in U, B, V, R.
See Abstr. 155.026.

083 Ionosphere

083.001 **Alfvén's critical velocity effect tested in space.**
G. Haerendel.
Z. Naturforsch., Band 37a, 728 - 735 (1982).
The implications of a critical velocity experiment in the upper ionosphere are discussed.

083.002 **A theoretical study of the distribution of ionization in the high-latitude ionosphere and the plasmasphere: first results on the mid-latitude trough and the light-ion trough.**
S. Quegan, G. J. Bailey, R. J. Moffett, R. A. Heelis, T. J. Fuller-Rowell, D. Rees, R. W. Spiro.
J. Atmos. Terr. Phys., Vol. 44, 619 - 640 (1982).
A model of the O^+ and H^+ distributions in the plasmasphere and high-latitude ionosphere is described and first results are presented.

083.003 **Investigations of the sporadic radio radiation of the sun and of the parameters of the earth's ionosphere aboard Intercosmos-Copernicus 500. 6. Morphological and spectral characteristics of inhomogeneities of the ionosphere's electron concentration.**
V. I. Aksenov, G. M. Artem'eva, G. P. Komrakov, L. A. Skrebkova.
Kosm. Issled., Tom 20, 542 - 551 (1982). In Russian.

083.004 **[He^+] height profiles in the outer ionosphere.**
A. P. Yaichnikov.
Kosm. Issled., Tom 20, 641 - 643 (1982). In Russian.

083.005 **Seasonal variations of aeronomical parameters in the D-region.**
A. D. Danilov, A. G. Simonov.
Geomagn. Aehron., Tom 22, 560 - 564 (1982). In Russian.

083.006 **Self-consistent calculation of the oscillation temperature of nitrogen in the ionospheric F-region.**
A. G. Kolesnik.
Geomagn. Aehron., Tom 22, 601 - 607 (1982). In Russian.

083.007 **Interaction of neutral and plasma motions in the ionosphere.**
P. Stubbe.
Handbuch der Physik. Geophysik III/6, (see 003.006), p. 247 - 308 (1982).
Contents: Introduction. Fundamental equations: equation of motion for the neutral gas, auxiliary transport equations for wind calculations, applicability of the transport equations in the upper atmosphere. Calculation of wind velocities: methods of calculating wind velocities, characteristics of calculated wind velocities. Observed wind velocities. Effects of winds on the ionosphere: direct effects, indirect effects. Effects of plasma motions on the neutral wind. Conclusion.

083.008 **Rocket observations of electron density profiles in the polar ionosphere.**
S. Miyazaki, T. Ogawa, H. Mori.
Rev. Radio Res. Lab., Vol. 26, 855 - 861 (1980). In Japanese. Abstr. in Phys. Abstr., Vol. 85, Abstr. 73779 (1982).

083.009 **Dynamics of ionospheric plasma at latitudes of the subauroral zone.**
V. M. Filippov, D. D. Reshetnikov, V. I. Nikitin.
Neodnorodnosti v ionos. Yakutsk, 1981, p. 79 - 87. In Russian.

Abstr. in Ref. zh., 62. Issled. kosm. prostranstva, 9.62.261 (1982).

083.010 **Ionospheric effects of internal gravitational waves generated by an auroral electrojet.**
V. M. Smertin, A. A. Namgaladze.
Ionos. prognozir. Moskva, 1982, p. 107 - 113. In Russian. Abstr. in Ref. zh., 62. Issled. kosm. prostranstva, 9.62.307 (1982).

083.011 **Effects of 16 Feb. 1980 solar eclipse on ionosphere as observed at Gauhati.**
P. Talukdar, M. Devi, A. K. Barbara.
Indian J. Radio Space Phys., Vol. 11, 20 - 22 (1982). – Abstr. in Phys. Abstr., Vol. 85, Abstr. 78907 (1982).

083.012 **Daytime E-region ion and nitric oxide densities.**
M. C. Azpiazu, S. Duhau.
J. Phys. D, Vol. 15, 933 - 940 (1982). – Abstr. in Phys. Abstr., Vol. 85, Abstr. 78910 (1982).

083.013 **Thermal plasma dynamical interaction between the plasmasphere and the ionosphere at mid-latitudes. A review.**
J. Grochulska.
Acta Geophys. Polonica, Vol. 29, 251 - 273 (1981). – Abstr. in Phys. Abstr., Vol. 85, Abstr. 83987 (1982).

083.014 **Basic concepts in the electrodynamics of the auroral ionosphere.**
A. Hruška.
Astrophys. Space Sci., Vol. 85, 389 - 404 (1982).
Equations governing the large-scale electrodynamic processes in the auroral ionosphere are systematically discussed and the limits to drawing conclusions from incomplete sets of equations are evaluated. The global structure of the auroral ionosphere is determined by the state of equilibrium between the pressure gradients, the inertial forces and the $j \times B$-force associated with the auroral electrojets flowing along the auroral oval.

083.015 **On the heating mechanism of the origin of ionospheric F-layer irregularities.**
L. M. Erukhimov, L. M. Kagan, E. N. Myasnikov.
Geomagn. Aehron., Tom 22, 721 - 724 (1982). In Russian.

083.016 **A three-dimensional model of the high-latitude F-region with account for non-coincidence of the geographic and geomagnetic coordinates.**
A. G. Kolesnik, I. A. Golikov.
Geomagn. Aehron., Tom 22, 725 - 731 (1982). In Russian.

083.017 **A dynamical model of the main types of convection, longitudinal currents and volumetric structure of the polar ionosphere.**
N. M. Maksimova, N. K. Osipov.
Geomagn. Aehron., Tom 22, 732 - 737 (1982). In Russian.

083.018 **On a connection of the parameters of the F2 and F1 ionospheric layers in day-to-day changes.**
A. V. Vinitskij, G. I. Sukhomazova, L. A. Shchepkin.
Geomagn. Aehron., Tom 22, 862 - 864 (1982). In Russian.

083.019 **Characterization of the auroral ionosphere using optical remote-sensing techniques.**
R. D. Sears.
AIAA J., Vol. 20, 390 - 394 (1982). – Abstr. in Phys. Abstr., Vol. 85, Abstr. 94549 (1982).

083.020 **Solar zenith angle dependence of ionospheric absorption during different epochs of the solar cycle.**
B. L. Acharya, R. K. Rai.
Indian J. Radio Space Phys., Vol. 11, 76 - 80 (1982). – Abstr. in Phys. Abstr., Vol. 85, Abstr. 100641 (1982).

083.021 **Mid-latitude F-region ionosphere plasma stability in the presence of photoelectrons.**
V. B. Ivanov, A. A. Trukhan, G. V. Khazanov.
Ann. Géophys., Vol. 38, 33 - 36 (1982). – Abstr. in Phys. Abstr., Vol. 85, Abstr. 106384 (1982).

083.022 **Studies on ionospheric electron content at the magnetic equator.**
C. Raghava Reddi, S. Vaidyanathan, B. V. Krishna Murthy.
Ann. Géophys., Vol. 38, 37 - 49 (1982). – Abstr. in Phys. Abstr., Vol. 85, Abstr. 106385 (1982).

083.023 **Lunar tidal variation in ionospheric absorption at a low latitude during solar minimum.**
G. Datta, K. M. Kotadia.
Ann. Géophys., Vol. 38, 55 - 59 (1982). – Abstr. in Phys. Abstr., Vol. 85, Abstr. 106387 (1982).

083.024 **Using the intensity of the 1356 Å emission for determination of the parameters of the F-region.**
K. B. Serafimov.
Kosm. Issled., Tom 20, 736 - 741 (1982). In Russian.

083.025 **Rocket investigations of the optical emission of the equatorial ionosphere during moderate and high geomagnetic activity.**
M. M. Gogoshev, St. Sargojchev, Kr. Kanev.
Kosm. Issled., Tom 20, 776 - 779 (1982). In Russian.

083.026 **A global view of F-region electron density and temperature at solar maximum.**
L. H. Brace, R. F. Theis, W. R. Hoegy.
Geophys. Res. Lett., Vol. 9, 989 - 992 (1982).
The purpose of the paper is to present selected early results from the DE-2 Langmuir probe instrument and to make an initial evaluation of how the thermal structure of the ionosphere at solar maximum differs from that observed at solar minimum.

083.027 **The S 3–4 ionospheric irregularities satellite experiment: probe detection of multi-ion component plasmas and associated effects on instability processes.**
E. P. Szuszczewicz, J. C. Holmes, M. Singh.
Astrophys. Space Sci., Vol. 86, 235 - 248 (1982).
The pulsed plasma probe technique has been expanded to include simultaneous determinations of absolute electron density, density fluctuations, electron temperature, and mean-ion-mass with resolution limited only by probe geometry, sheath size, and telemetry. The technique has been designed to test for coupling of electron density variations and ion composition irregularities in multi-component plasmas by the comparison of electron density fluctuation power spectra and a newly-developed diagnostic parameter, the mean-ion-mass fluctuation spectra. In addition, the experiment extends satellite-borne irregularity spectral analyses down to the 5 - 20 m range while attempting to identify F-region plasma instability processes.

083.028 **Combined mass spectrometric composition measurements of positive and negative ions in the lower ionosphere–I. Positive ions.**
F. Arnold, A. A. Viggiano.
Planet. Space Sci., Vol. 30, 1295 - 1305 (1982).

083.029 **Combined mass spectrometric composition measurements of positive and negative ions in the lower ionosphere–II. Negative ions.**

F. Arnold, A. A. Viggiano, E. E. Ferguson.
Planet. Space Sci., Vol. 30, 1307 - 1314 (1982).

083.030 **The ionosphere of the lower and equatorial latitudes at 500 km height during magnetospherically-ionospheric perturbations in September-December 1977 (according to Cosmos 900 data).**
G. L. Gdalevich, I. S. Vsekhsvyatskaya, V. D. Ozerov, T. N. Soboleva.
Kosm. Issled., Tom 20, 872 - 880 (1982). In Russian.

083.031 **Ion kinetics, small neutral and disturbed components in the D-region with increased ionization level.**
S. I. Kozlov, V. A. Vlaskov, N. V. Smirnova.
Kosm. Issled., Tom 20, 881 - 891 (1982). In Russian.

083.032 **Statistical analysis of the disturbances of the lower ionosphere (D-layer).**
D.-m. Li, S.-q. Wang.
Publ. Shaanxi Astron. Obs., No. 1, p. 39 - 47 (1982). In Chinese.

083.033 **Model of the ionospheric F_1 layer for high solar activity.**
A. V. Vinitskij, G. M. Kuznetsova, G. P. Kushnarenko, L. A. Shchepkin.
Issled. po geomagn., aehron. i fiz. Solntsa. Moskva, 1982, No. 59, p. 159 - 166. In Russian. – Abstr. in Ref. zh., 62. Issled. kosm. prostranstva, 10.62.463 (1982).

083.034 **Electromagnetic resonances in the equatorial ionosphere.**
A. B. Shvartsburg.
Space Sci. Rev., Vol. 33, 361 - 413 (1982).

The diagnostic and communication possibilities, based on resonant properties of ionospheric magnetoplasma near the equator, are discussed for a wide spectral range of electromagnetic waves. The utilization of geometrical resonances of both natural and artificial origin and plasma eigenmodes is considered for a model of anisotropic heterogeneous collisional magnetoplasma at low latitudes. The possibilities of linear and non-linear wave transformation and scattering, together with the orthogonality of ionospheric gradients and geomagnetic fields, are illustrated. The threshold and self-oscillation phenomena, stimulated by heating, produced by an equatorial electrojet, photo-electrons and powerful transmitters, are examined. Non-stationary electromagnetic processes, accompanied by the transversal transport phenomena in a magnetoplasma near the equatorial plane, are discussed and the tendencies of resonant effects employment for active diagnostic and remote communication at low latitudes are also considered.

083.035 **Electron density changes during solar eclipse of 16 Feb. 1980 at Gauhati.**
G. Bharathi, P. E. Raj, B. Narasimha Murphy, C. Jogulu.
Indian J. Radio Space Phys., Vol. 11, No. 3, p. 96 - 99 (1982). Abstr. in Phys. Abstr., Vol. 85, Abstr. 117521 (1982).

083.036 **Ionospheric absorption of the extraordinary wave.**
E. O. Olatunji.
Radio Sci., Vol. 17, 1335 - 1342 (1982). – Abstr. in Phys. Abstr., Vol. 86, Abstr. 15742 (1983).

083.037 **Electron transport in inhomogeneous anisotropically scattering medium; application to the terrestrial atmosphere.**
A. F. Torabi, M. Kanal.
Transp. Theory Stat. Phys., Vol. 11, No. 1, p. 53 - 74 (1982). Abstr. in Phys. Abstr., Vol. 86, Abstr. 15744 (1983).

Physics of the polar ionosphere.
See Abstr. 003.101.

Inhomogeneities in the ionosphere.
See Abstr. 003.130.

A method for calculation of electric and magnetic fields in the high-latitude ionosphere.
See Abstr. 021.026.

Measurement of the line-of-sight velocity of high-altitude barium clouds: a technique.
See Abstr. 031.677.

Volume emission rate profiles of the 6300-Å tropical nightglow obtained from the AE-E satellite: latitudinal and seasonal variations.
See Abstr. 082.051.

Near-Earth ultraviolet environment.
See Abstr. 082.096.

On midlatitude oscillations of the geomagnetic field and their connection with dynamics of processes in the ionosphere.
See Abstr. 084.060.

Main results of the joint French-Soviet space project ARCAD-1 and ARCAD-2 for magnetospheric, auroral and ionospheric physics.
See Abstr. 084.101.

Upper atmosphere dynamics. The mission of the Dynamics Explorer satellites.
See Abstr. 084.103.

084 Aurorae, Geomagnetic Field, Magnetosphere

084.001 **New light on the equatorial source of pulsating aurora.**
M. P. Gough, A. Korth.
Nature, Vol. 298, 253 - 255 (1982).
The authors present the first evidence showing that the ELF hiss at the equator of the auroral geomagnetic field is directly related to pitch angle scattering of electrons >22 keV into the loss cone with subsequent precipitation into the auroral atmosphere.

084.002 **Polar magnetic disturbances and field-aligned currents.**
O. A. Troshichev.
Space Sci. Rev., Vol. 32, 275 - 360 (1982).
The results of research of the morphology and physics of polar magnetic disturbances and their connection with three-dimensional magnetospheric currents are reviewed. Magnetic disturbance current systems are examined, also their relation to solar wind parameters and magnetic activity level and their seasonal dependence. On the basis of numerical model calculations it is shown that magnetospheric field-aligned currents observed by the TRIAD and ISIS-2 satellites are the main generation mechanism of high-latitude magnetic disturbances. Plasma pressure gradients are examined as a source of energy for driving field-aligned currents in the closed magnetosphere.

084.003 **On the dependence of the western electrojet on the solar wind parameters.**
A. A. Danilov, G. A. Makarov, G. P. Sergeeva.
Byul. NTI Yakut. fil. SO AN SSSR, 1981, iyul', probl. kosmofiz. i aehron., p. 5 - 9. In Russian. – Abstr. in Ref. zh., 62. Issled. kosm. prostranstva, 7.62.413 (1982).

084.004 **Polarlicht.**
G. Gustafsson.
Phys. Bl., Vol. 38, 287 - 292 (1982).

084.005 **Magnetic field and field-aligned current variations in the polar cusp controlled by interplanetary medium parameters.**
Y. I. Feldstein (*Ya. I. Fel'dshtejn*), R. G. Afonina, B. A. Belov, A. E. Levitin.
Planet. Space Sci., Vol. 30, 635 - 640 (1982).

084.006 **A study of geomagnetic storms at Beijing in terms of solar wind.**
N. Lin, K.-K. Tschu.
Planet. Space Sci., Vol. 30, 669 - 675 (1982).

084.007 **The detection and interpretation of the orthohelium emission at 5876 Å in aurorae.**
K. Henriksen, P. Ya. Sukhoivanenko.
Planet. Space Sci., Vol. 30, 695 - 699 (1982).
A review of the published observations of the He I emission at 5876 Å in aurorae is presented. The emission is considered to originate from energetic precipitating helium ions, neutralized and excited several times before the atoms become thermalized in the upper atmosphere.

084.008 **Geomagnetic lunar analysis – the estimation of errors.**
R. Sellek, S. R. C. Malin.
Geophys. J. R. Astron. Soc., Vol. 70, 793 - 796 (1982).
It has recently been proposed that a reliable estimate of the errors of tidal harmonics present in geomagnetic data can be obtained by the following method: (1) randomly assign each observation to one of 10 subsets, (2) determine the tidal harmonics separately for each subset, and (3) calculate the standard deviation of the mean from the scatter among the 10 determinations. This method is valid if the subsets are statistically independent, but will lead to an underestimate of the errors if they are not. Here the authors show that, for real geomagnetic data, the assumption of statistical independence is valid.

084.009 **Palaeosecular variation and the current loop model of the geomagnetic field.**
M. Roy., J.-J. Wagner.
Geophys. J. R. Astron. Soc., Vol. 71, 269 - 273 (1982).
The geomagnetic palaeosecular variation has been studied in terms of two current loops which change their positions and orientations with time. The results broadly agree with the observed data. Comparison with the existing models is made.

084.010 **Intensity and anisotropy variations of streams of particle precipitation with energies above 30 keV.**
V. I. Altyntseva, A. V. Dronov, A. S. Kovtyukh, M. I. Panasyuk, N. M. Polekh, S. Ya. Rejzman, Eh. N. Sosnovets.
Kosm. Issled., Tom 20, 552 - 559 (1982). In Russian.

084.011 **Energetic electrons in the region of the plasmapause.**
V. E. Tsirs.
Kosm. Issled., Tom 20, 634 - 636 (1982). In Russian.

084.012 **Optical signatures of some magnetospheric processes on the dayside.**
G. G. Sivjee, G. J. Romick, C. S. Deehr.
Geophys. Res. Lett., Vol. 9, 676 - 679 (1982).
Ground-based daytime photometric observations of auroral optical emissions from the cusp region over Longyearbyen ($\Lambda = 75°$N) Svalbard are used to illustrate spatial variations in the average energy of particles precipitating in this region.

084.013 **Dynamics of evening-morning asymmetry of the ring current during the magnetic storm of December 1 - 2, 1977 from data of the artificial earth satellite Cosmos 900.**
T. I. Morozova, M. I. Panasyuk, Eh. N. Sosnovets, L. V. Tverskaya, O. V. Khorosheva.
Geomagn. Aehron., Tom 22, 608 - 611 (1982). In Russian.

084.014 **Numerical modelling of the magnetospheric convection in the region of closed force lines of the magnetic field.**
L. M. Alekseeva, A. V. Getling, V. V. Savel'ev.
Geomagn. Aehron., Tom 22, 612 - 616 (1982). In Russian.

084.015 **Periodic oscillations of geomagnetic and solar activity.**
V. I. Afanas'eva, A. I. Laptukhov.
Geomagn. Aehron., Tom 22, 617 - 622 (1982). In Russian.

084.016 **On space-time distribution of 60-year geomagnetic variations.**
N. P. Ben'kova, V. P. Golovkov, N. V. Kulanin, T. N. Cherevko.
Geomagn. Aehron., Tom 22, 638 - 642 (1982). In Russian.

084.017 **Some characteristics of the secular variations of the geomagnetic moment from archeomagnetic data.**
Yu. D. Kalinin, T. S. Rozanova.
Geomagn. Aehron., Tom 22, 643 - 645 (1982). In Russian.

084.018 **The nature of long-period variations of the geomagnetic field.**
G. F. Zagnij.
Geomagn. Aehron., Tom 22, 646 - 651 (1982). In Russian.

084.019 **On the equation defining the axes of geomagnetic multi-poles.**
Yu. P. Stepanovskij.
Geomagn. Aehron., Tom 22, 652 - 655 (1982). In Russian.

084.020 **High-energy electrons in the inner radiation belt of the earth.**
R. N. Basilova, A. A. Gusev, G. I. Pugacheva, A. F. Titenkov.
Geomagn. Aehron., Tom 22, 671 - 672 (1982). In Russian.

084.021 **The *aa* geomagnetic activity index and parameters of the interplanetary medium.**
T. E. Val'chuk, Ya. I. Fel'dshtejn.
Geomagn. Aehron., Tom 22, 691 - 693 (1982). In Russian.

084.022 **On the applicability of one-liquid magnetohydrodynamics to the description of a magnetospheric plasma.**
M. V. Samokhin.
Geomagn. Aehron., Tom 22, 693 - 695 (1982). In Russian.

084.023 **Formation and structure of the terrestrial magnetosphere.**
K. Tamao.
Magnetospheric phenomena of celestial bodies, (see 012.011), p. 7 - 16 (1982). In Japanese.

084.024 **Field-aligned currents in terrestrial magnetosphere.**
T. Iijima.
Magnetospheric phenomena of celestial bodies, (see 012.011), p. 17 - 21 (1982). In Japanese.

084.025 **A comparison of geomagnetic secular variation as recorded by historical, archaeomagnetic and palaeomagnetic measurements.**
R. Thompson.
Philos. Trans. R. Soc. London, Ser. A, Vol. 306, (see 012.014), 103 - 112 (1982).

Palaeomagnetic methods can extend the documentary record of changes in the Earth's magnetic field far into the past. Tolerable agreement is found between various methods, demonstrating the geophysical value of palaeomagnetic experiments. A mathematical approach to formulating an empirical description of global geomagnetic field behaviour is proposed and applied to palaeomagnetic data spanning the last 10 ka.

084.026 **A revised magnetic polarity timescale for the Cretaceous and Cainozoic.**
W. Lowrie.
Philos. Trans. R. Soc. London, Ser. A, Vol. 306 (see 012.014), 129 - 136 (1982) = Contrib. No. 364, Inst. Geophys., ETH-Zürich.

084.027 **Geomagnetic polarity in the early Cretaceous and Jurassic.**
J. E. T. Channell, J. G. Ogg, W. Lowrie.
Philos. Trans. R. Soc. London, Ser. A, Vol. 306 (see 012.014), 137 - 146 (1982).

084.028 **The testing of geomagnetic reversal models: recent developments.**
K. A. Hoffman.
Philos. Trans. R. Soc. London, Ser. A, Vol. 306, (see 012.014), 147 - 159 (1982).

084.029 **Modelling the core magnetic field of the Earth.**
C. G. A. Harrison, H. M. Carle.
Philos. Trans. R. Soc. London, Ser. A, Vol. 306, (see 012.014), 179 - 191 (1982).

The magnetic field generated in the core of the Earth is often represented by spherical harmonics of the magnetic potential. It has been found from looking at the equations of spherical harmonics, and from studying the values of the spherical harmonic coefficients derived from data from Magsat, that this is an unsatisfactory way of representing the core field. The authors suggest that a better way is to use radial off-centre dipoles located within the core of the Earth. They discuss some of the radial dipole models that have been proposed for the non-dipole portion of the Earth's field to arrive at a model that agrees with observations of secular variation and excursions.

084.030 **Normal vector models of the geomagnetic secular variation during late Tertiary time.**
R. E. Dodson.
Philos. Trans. R. Soc. London, Ser. A, Vol. 306, (see 012.014), 193 - 201 (1982).

084.031 **Spectral analysis of palaeomagnetic time series and the geomagnetic spectrum.**
C. E. Barton.
Philos. Trans. R. Soc. London, Ser. A, Vol. 306, (see 012.014), 203 - 209 (1982).

084.032 **Geomagnetic secular variation and fluid motion at the core surface.**
K. A. Whaler.
Philos. Trans. R. Soc. London, Ser. A, Vol. 306, (see 012.014), 235 - 246 (1982).

084.033 **The recent secular variation and the motions at the core surface.**
T. Madden, J.-L. Le Mouël.
Philos. Trans. R. Soc. London, Ser. A, Vol. 306, (see 012.014), 271 - 280 (1982).

084.034 **The source of the Earth's magnetic field.**
C. R. Carrigan, D. Gubbins.
Ceskoslovensky Cas. Fyz. Sekce A, Vol. 31, 433 - 448 (1981). In Czech. – Abstr. in Phys. Abstr., Vol. 85, Abstr. 62883 (1982).

084.035 **The geomagnetic field during Palaeozoic time.**
A. N. Khramov, V. P. Rodionov.
J. Geomagn. Geoelectr., Vol. 32, Suppl. 3, p. 99 - 115 (1980). Abstr. in Phys. Abstr., Vol. 85, Abstr. 68257 (1982).

084.036 **Reversals and excursions of the geomagnetic field as defined by palaeomagnetic data from upper Palaeozoic-Lower Mesozoic sediments and igneous rocks from Argentina.**
D. A. Valencio.
J. Geomagn. Geoelectr., Vol. 32, Suppl. 3, p. 137 - 142 (1980). Abstr. in Phys. Abstr., Vol. 85, Abstr. 68259 (1982).

084.037 **Seasonal dependence of lunar daily geomagnetic variations in different regions of the world.**
M. Shiraki.
J. Geomagn. Geoelectr., Vol. 33, 467 - 501 (1981). – Abstr. in Phys. Abstr., Vol. 85, Abstr. 68260 (1982).

084.038 **Spherical harmonic analysis in the presence of high harmonics.**
L. R. Alldredge, K. Kawasaki.
J. Geomagn. Geoelectr., Vol. 33, 503 - 515 (1981). – Abstr. in Phys. Abstr., Vol. 85, Abstr. 68261 (1982).

084.039 **The instantaneous relationship between polar cap and oval auroras at times of northward interplanetary magnetic field.**
J. S. Murphree, C. D. Anger, L. L. Cogger.
Canadian J. Phys., Vol. 60, 349 - 356 (1982). – Abstr. in Phys. Abstr., Vol. 85, Abstr. 68546 (1982).

084.040 **Cusp region and auroral zone field aligned currents.**
F. Primdahl, F. Spangslev.
Ann. Géophys., Vol. 37, 529 - 538 (1981). – Abstr. in Phys. Abstr., Vol. 85, Abstr. 68561 (1982).

084.041 **Predictions of geomagnetic activity by an interplanetary spacecraft.**
S. Watanabe.
Rev. Radio Res. Lab., Vol. 26, 929 - 939 (1980). In Japanese. Abstr. in Phys. Abstr., Vol. 85, Abstr. 73444 (1982).

084.042 **Distribution of polar aurorae and large-scale currents in the magnetosphere at different orientations of the interplanetary magnetic field.**
Ya. I. Fel'dshtejn, N. G. Vorfolomeeva.
Soln. veter i magnitosfer. issled. Moskva, 1980, p. 110 - 115. In Russian. – Abstr. in Ref. zh., 62. Issled. kosm. prostranstva, 9.62.256 (1982).

084.043 **On the nature of recurrent geomagnetic activity.**
M. I. Pudovkin, D. I. Ponyavin, A. D. Chertkov.
Magnitosfer. issled. No. 1. Moskva, 1982, p. 24 - 28. In Russian. – Abstr. in Ref. zh., 62. Issled. kosm. prostranstva, 9.62.268 (1982).

084.044 **On the nonlinear theory of instability of the auroral electrojet.**
A. V. Kustov, A. V. Volosevich, V. A. Liperovskij.
Neodnorodnosti v ionos., Yakutsk, 1981, p. 16 - 22. In Russian Abstr. in Ref. zh., 62. Issled. kosm. prostranstva, 9.62.308 (1982).

084.045 **Observations of reverse polarity flux transfer events at the earth's dayside magnetopause.**
R. P. Rijnbeek, S. W. H. Cowley, D. J. Southwood, C. T. Russell.
Nature, Vol. 300, 23 - 26 (1982).

084.046 **About the connection between longer period geomagnetic variation and the interplanetary magnetic field.**
L. Hollo.
Act. Geod. Geophys. Montan., Vol. 16, 397 - 403 (1981). Abstr. in Phys. Abstr., Vol. 85, Abstr. 78915 (1982).

084.047 **Geomagnetic storm variations and the average characteristics of the interplanetary magnetic field.**
P. Ochabova.
Contrib. Geophys. Inst. Slovak Acad. Sci., Vol. 12, 9 - 23 (1982). – Abstr. in Phys. Abstr., Vol. 85, Abstr. 84018 (1982).

084.048 **Large scale and small scale electric field distribution in the magnetosphere.**
J. Lemaire.
Proc. Indian Natl. Sci. Acad., Part A, Vol. 47, Suppl. 2, 49 - 63 (1981). – Abstr. in Phys. Abstr., Vol. 85, Abstr. 89083 (1982).

084.049 **Interaction between a magnetized plasma flow and a strongly magnetized celestial body with an ionized atmosphere: energetics of the magnetosphere.**
S.-I. Akasofu.
Annu. Rev. Astron. Astrophys., Vol. 20, (see 003.009), 117 - 138 (1982).

Solar wind-magnetosphere dynamo. Electrodynamics of the heliosphere. Global energetics of the magnetosphere. Magnetospheric plasma processes.

084.050 **Electromagnetic velocity gradient instabilities in the magnetosphere.**
G. S. Lakhina.
Astrophys. Space Sci., Vol. 85, 369 - 380 (1982).

The velocity gradients of the contrastreaming electron beams observed in the Earth's magnetosphere can excite three types of ordinary mode instabilities. Near the equatorial plane of the bouncing electron beams region the ∇B-resonance electron instability can generate electromagnetic waves having frequencies $f \sim (0.045 - 0.2)$ Hz and wavelengths $\lambda_\perp \sim (0.5 - 10)$ km. It can also occur in the plasma sheet region during the earthwards and tailwards plasma flow events and can generate waves in the frequency range $f \sim (0.007 - 0.02)$ Hz with $\lambda_\perp \sim (10 - 100)$ km near $R = -35\ R_E$. The ion cyclotron instability can generate waves up to 5th harmonic or so of ion cyclotron frequency. The unmagnetized ion instability can generate electromagnetic waves having frequencies from 5 to 50 Hz and typical wavelengths $\lambda_\perp \sim (0.5 - 6)$ km.

084.051 **Polar cusp ion drift studies through high-resolution interferometry of O^+ 7320-Å emission.**
R. W. Smith, G. G. Sivjee, R. D. Stewart, F. G. McCormac, C. S. Deehr.
J. Geophys. Res., Vol. 87, 4455 - 4460 (1982).

084.052 **Generation of the auroral kilometric radiation.**
C. S. Wu, H. K. Wong, D. J. Gorney, L. C. Lee.
J. Geophys. Res., Vol. 87, 4476 - 4487 (1982).

084.053 **Auroral studies with a chain of meridian-scanning photometers. 1. Observations of proton and electron aurora in magnetospheric substorms.**
A. Vallance Jones, F. Creutzberg, R. L. Gattinger, F. R. Harris.
J. Geophys. Res., Vol. 87, 4489 - 4503 (1982).

084.054 **Overview of the IMS (*international magnetospheric study*) July 29, 1977, magnetic storm analysis.**
R. H. Manka, T. A. Fritz, R. G. Johnson, R. A. Wolf, M. J. Teague, J. I. Vette.
J. Geophys. Res., Vol. 87, 5871 - 5880 (1982).

084.055 **A dynamic model of the magnetospheric magnetic and electric fields for July 29, 1977.**
W. P. Olson, K. A. Pfitzer.
J. Geophys. Res., Vol. 87, 5943 - 5948 (1982).

A quantitative vector representation of the magnetospheric field was developed specifically for the complex event of July 29, 1977. The magnetic field associated with each of the three major magnetospheric current systems was 'controlled' by solar wind and interplanetary field parameters and a magnetic index. This dynamic field model was compared with observed field data and found to be in good agreement.

084.056 **A time dependent three-dimensional simulation of the earth's magnetosphere: reconnection events.**
S. H. Brecht, J. G. Lyon, J. A. Fedder, K. Hain.
J. Geophys. Res., Vol. 87, 6098 - 6108 (1982).

084.057 **Estimate of velocity of a return streaming when the geomagnetic dipole is oriented along the non-disturbed solar wind velocity.**
M. V. Samokhin.
Geomagn. Aehron., Tom 22, 716 - 720 (1982). In Russian.

084.058 **Characteristics of auroral electrons as observed at the artificial earth satellite Meteor.**
Yu. B. Kazanskaya, V. I. Lazarev, M. V. Tel'tsov, V. A. Lipovetskij, V. F. Tulinov, V. M. Fejgin.
Geomagn. Aehron., Tom 22, 765 - 769 (1982). In Russian.

084.059 **Dependence of the magnetic activity on the orientation of the geomagnetic dipole relative to the interplanetary magnetic field.**
A. A. Danilov, G. A. Makarov.
Geomagn. Aehron., Tom 22, 791 - 796 (1982). In Russian.

084.060 **On midlatitude oscillations of the geomagnetic field and their connection with dynamics of processes in the ionosphere.**
L. S. Al'perovich, V. I. Drobzhev, V. M. Sorokin, V. A. Troitskaya, G. V. Fedorovich.
Geomagn. Aehron., Tom 22, 797 - 802 (1982). In Russian.

084.061 **Space-time analysis of 60-year variations of the geomagnetic field from world network observations.**
N. M. Rotanova, N. E. Papitashvili, A. N. Pushkov.
Geomagn. Aehron., Tom 22, 847 - 851 (1982). In Russian.

084.062 **Analytical description of variations of the geomagnetic field for the 20th century.**
S. I. Braginskij.
Geomagn. Aehron., Tom 22, 852 - 855 (1982). In Russian.

084.063 **On a possibility of excitation of geomagnetic pulsations with radial direction of the interplanetary magnetic field.**
V. B. Lyatskij, Yu. P. Mal'tsev.
Geomagn. Aehron., Tom 22, 875 - 877 (1982). In Russian.

084.064 **Heavy O^+ ions in the radiation belt of the earth as a possible source of geomagnetic Pc 1–2 pulsations.**
A. L. Kalisher, Yu. P. Kurchashov, V. A. Troitskaya, F. Z. Fejgin.
Geomagn. Aehron., Tom 22, 879 - 881 (1982). In Russian.

084.065 **11-year variation of the west drift of the magnetic centre of the earth.**
Yu. D. Kalinin, T. S. Rozanova.
Geomagn. Aehron., Tom 22, 883 - 884 (1982). In Russian.

084.066 **On the interpretation of the 60-year variations of the geomagnetic field by a quasi-harmonic process.**
N. M. Rotanova, N. E. Papitashvili, A. N. Pushkov.
Geomagn. Aehron., Tom 22, 885 - 887 (1982). In Russian.

084.067 **Comparison between the hot spot and geomagnetic field reference frames.**
C. G. A. Harrison, T. Lindh.
Nature, Vol. 300, 251 - 252 (1982).

Hot spots are located on the earth's surface and are caused by hot plumes of upwelling mantle material. It has been suggested that hot spots are relatively stationary with respect to each other, and also that they represent a frame of reference with which to measure the absolute movement of the lithospheric plates over the surface of the earth. Another frame of reference can be partially defined by the position of the mean magnetic dipole field of the earth. The authors now report a comparison of the two frames of reference and show that over the past 200 Myr there has been relative movement between the two frames of reference of about 20°, and that this motion has not been constant with time.

084.068 **Determination of the magnetospheric current system parameters and development of experimental geomagnetic field models based on data from IMP and HEOS satellites.**
N. A. Tsyganenko, A. V. Usmanov.
Planet. Space Sci., Vol. 30, 985 - 998 (1982).

084.069 **Distances from auroral zones to the magnetic and geographic equators.**
K. D. Cole, P. F. B. Williams.
Planet. Space Sci., Vol. 30, 1073 - 1076 (1982).

084.070 **The annual wave in geomagnetic activity as a magnetospheric modulation effect.**
D. Damaske.
J. Geomagn. Geoelectr., Vol. 33, 623 - 633 (1981). – Abstr. in Phys. Abstr., Vol. 85, Abstr. 94564 (1982).

084.071 **Currents on the surface of the Earth's bow shock.**
T.-f. Chen.
Bull. Geophys., No. 22, p. 11 - 24 (1981). – Abstr. in Phys. Abstr., Vol. 85, Abstr. 100 652 (1982).

084.072 **Altitude characteristics of radar aurora as seen by a 90-MHz double-altitude radar system operated at Karmaselga, Karelia.**
E. Timofeev, Y. *(Yu.)* Miroshnikov.
J. Geophys., Vol. 51, 44 - 54 (1982). – Abstr. in Phys. Abstr., Vol. 85, Abstr. 106381 (1982).

084.073 **Variations in the geomagnetic dipole. I. The past 50000 years.**
M. W. McElhinny, W. E. Senanayake.
J. Geomagn. Geoelectr., Vol. 34, 39 - 51 (1982). – Abstr. in Phys. Abstr., Vol. 85, Abstr. 111858 (1982).

084.074 **Electrons with energies of hundreds of MeV in the circumterrestrial cosmic space.**
N. L. Grigorov.
Kosm. Issled., Tom 20, 705 - 713 (1982). In Russian.

084.075 **Dynamics of energetic electrons from observations aboard the artificial earth satellite Intercosmos 19 in April 1979.**
Yu. V. Mineev, I. N. Senchuro, E. S. Spir'kova, P. I. Shavrin, G. A. Glukhov, Yu. P. Kratenko.
Kosm. Issled., Tom 20, 767 - 771 (1982). In Russian.

084.076 **Polarization measurements of auroral kilometric radiation by Dynamics Explorer-1.**
S. D. Shawhan, D. A. Gurnett.
Geophys. Res. Lett., Vol. 9, 913 - 916 (1982).

084.077 **Global observations of a SAR arc.**
J. D. Craven, L. A. Frank, K. L. Ackerson.
Geophys. Res. Lett., Vol. 9, 961 - 964 (1982).

A SAR arc has been observed at $\lambda = 630$ nm with global auroral imaging instrumentation on board the spacecraft Dynamics Explorer 1. The arc is first detected at about 1908 UT on October 20, 1981 and is then observed in four subsequent, consecutive orbits spanning 28 hours.

084.078 **Observation of a heated electron population associated with the 6300Å SAR arc emission.**
C. Gurgiolo, D. W. Slater, J. D. Winningham, J. L. Burch.
Geophys. Res. Lett., Vol. 9, 965 - 968 (1982).

084.079 **Thermospheric response to the 23 October 1981 SAR-arc and aurora as observed from Fritz Peak, Colorado and Calgary, Alberta during the Dynamics Explorer (DE-2) and NOAA-6 satellite overflights.**
G. Hernandez, J. W. Meriwether, C. A. Tepley, P. B. Hays, L. L. Cogger, D. W. Slater, R. G. Roble, B. A. Emery, D. S. Evans.
Geophys. Res. Lett., Vol. 9, 969 - 972 (1982).

084.080 **Characteristics of a stable auroral red arc event.**
J. U. Kozyra, T. E. Cravens, A. F. Nagy, M. O. Chandler, L. H. Brace, N. C. Maynard, D. W. Slater, B. A. Emery, S. D. Shawhan.
Geophys. Res. Lett., Vol. 9, 973 - 976, with a correction p. 1298 (1982).

084.081 **Polar views of the earth's aurora with Dynamics Explorer.**
L. A. Frank, J. D. Craven, J. L. Burch, J. D. Winningham.
Geophys. Res. Lett., Vol. 9, 1001 - 1004 (1982).

Preliminary examination of the first global auroral images gained with the vacuum-ultraviolet imaging photometer on board DE 1 reveals a remarkable spatial configuration of auroral luminosities.

084.082 **The two-dimensional structure of diffuse ions associated with the Earth's bow shock.**
G. Skadron, M. A. Lee.
Astrophys. J., Vol. 263, 468 - 476 (1982).
The authors investigate the lateral distribution of upstream, diffuse energetic ions resulting from shock acceleration at an idealized model of the Earth's bow shock. The model consists of a uniform, planar shock along which steady, source ion injection over a finite region of the shock takes place. The ions are subsequently assumed to move diffusively in an upstream solar wind and experience adiabatic compression at the shock. The analysis is based upon a two-dimensional Gleeson-Axford equation which includes the effects of ion convection and cross-field diffusion.

084.083 **Boundaries between solar wind and planetary magnetospheres.**
G. Paschmann.
The solar system and its exploration, (see 012.054), p. 75 (1981). – Abstract.

084.084 **Magnetospheric substorm morphology.**
G. Kremser.
The solar system and its exploration, (see 012.054), p. 77 - 84 (1981).
Magnetospheric substorms are the most pronounced manifestation of unsteady processes in the magnetosphere. A recent definition of substorms is explained and some substorm signatures are illustrated that can be observed from the ground. Substorm phenomena in space are summarized with the help of models that were constructed in order to interpret observations. Recently obtained results from a balloon-satellite study are presented as an example of measurements at the onset of an individual substorm.

084.085 **Measurement of plasma drifts in planetary magnetospheres.**
R. Grard, G. Haerendel, F. Melzner.
The solar system and its exploration, (see 012.054), p. 97 - 104 (1981).
One of the scientific objectives of GEOS, the magnetospheric geostationary satellite of the European Space Agency, is to measure the motion of the ionized gas which surrounds the Earth. The dynamics of this plasma is directly associated with the interactions which take place between the solar wind and the Earth magnetic field.

084.086 **Theory of the magnetic field of the earth.**
S. I. Braginskij.
Zemlya Vselennaya, 1982, No. 6, p. 40 - 44. In Russian.

084.087 **Boundaries for capture and loss of particles of the outer radiation belt due to the magnetospheric magnetic belt.**
V. A. Sergeev, N. A. Tsyganenko.
Kosm. Issled., Tom 20, 866 - 871 (1982). In Russian.

084.088 **Determination of the electron abundance of the plasmasphere with the help of coherent signals of the ATS-6 satellite recorded in Neustrelitz.**
N. Jakowski, H.-G. Kugland.
Kosm. Issled., Tom 20, 892 - 899 (1982). In Russian.

084.089 **On the positions of sources of sporadic radio radiation observed with the Electron 2 and Electron 4 satellites in the 0.7 - 2.3 MHz frequency region.**
G. M. Artem'eva, N. A. Mityakov, V. V. Pisareva, A. F. Tarasov.
Kosm. Issled., Tom 20, 907 - 912 (1982). In Russian.

084.090 **The aurora: new light on an old subject.**
S.-I. Akasofu.
Sky Telesc., Vol. 64, 534 - 537 (1982).

084.091 **Über den Ursprung des Nordlichts.**
G. Haerendel.
Sterne Weltraum, 21. Jahrg., 508 - 513 (1982).

084.092 **A comparative study of the Earth's magnetic and gravity fields.**
G. Barta.
Dynamical astronomy, (see 012.056), p. 119 - 126 (1982).

084.093 **Three-dimensional magnetosphere current system during substorm.**
R. Xu, S. Gu.
Sci. Sinica, Ser. A, Vol. 25, 644 - 653 (1982).

084.094 **Observations of inverted-V electron precipitation.**
C. S. Lin, R. A. Hoffman.
Space Sci. Rev., Vol. 33, 415 - 457 (1982).
The characteristics of inverted-V electron precipitation fluxes deduced predominantly from observations by the Atmosphere Explorer satellites are reviewed. The energy and pitch angle distributions are presented and shown to be generally in agreement with acceleration by a parallel electrostatic potential. Characteristics of secondary electrons are examined, and effects of beam plasma instabilities on these electrons are discussed. The properties of the monoenergetic component are compared with theoretical models of creating parallel DC electric fields, and found to favor the anomalous resistivity model. The article also discusses relations of inverted-V events with other auroral phenomena including auroras, electrostatic shocks, convective electric field reversals, field-aligned currents and wave emissions.

084.095 **Momentum balance and flux conservation in model magnetospheric magnetic fields.**
R. J. Walker, D. J. Southwood.
J. Geophys. Res., Vol. 87, 7460 - 7466 (1982).
The authors have tested four recent quantitative magnetospheric magnetic field models to determine whether the magnetic field configuration is consistent with a balance of forces between the $\mathbf{J} \times \mathbf{B}$ force and the gradient of the pressure.

084.096 **The narrow-channel approximation for magnetospheric flows.**
G. Atkinson.
J. Geophys. Res., Vol. 87, 7489 - 7503 (1982).

084.097 **Recurrent fast motions of pulsating auroral patches. 1. A case study on optical and quantitative characteristics during a slightly active period.**
T. Yamamoto, T. Oguti.
J. Geophys. Res., Vol. 87, 7603 - 7614 (1982).

084.098 **A quantitative model of geomagnetic activity.**
R. E. Holzer, J. A. Slavin.
J. Geophys. Res., Vol. 87, 9054 - 9058 (1982).
A quantitative model of geomagnetic activity is developed and utilized to investigate the causes of the diurnal, seasonal, and IMF sector variations in the *AL* index records.

084.099 **The worldwide character of the 1969–1970 impulse of the secular acceleration rate.**
J.-L. Le Mouël, J. Ducruix, C. Ha Duyen.
Phys. Earth Planet. Inter., Vol. 28, 337 - 350 (1982) = Contrib. Inst. Phys. Globe No. 574 (1982).

This paper continues earlier studies to ascertain whether the distribution of the 1969–1970 jump of the secular variation rate has a worldwide character. The geometry of the jump presents some similarities with the geometry of the secular variation itself. The authors give a new estimation of the westward drift rate and information about spatial and temporal variations of the secular variation field.

084.100 **A note on the secular variation of the geomagnetic field.**
M. G. McLeod.
Phys. Earth Planet. Inter., Vol. 29, 119 - 134 (1982).
Geomagnetic variations with (apparent) periods from 13 to 30 y have been reported by Alldredge (1977) who has argued that the origin of these signals must be found in the core of the Earth rather than outside the Earth. It is shown in this paper that a portion of these geomagnetic variations (perhaps most of the variations) might well be due to geomagnetic signals of much longer period, originating in the core, that appear to have periods of 13–30 y because of an artifact of the data processing.

084.101 **Main results of the joint French-Soviet space project ARCAD-1 and ARCAD-2 for magnetospheric, auroral and ionospheric physics.**
F. Cambou, Yu. I. Galperin *(Gal'perin)*.
Ann. Geophys., Vol. 38, 87 - 110 (1982). – Abstr. in Phys. Abstr., Vol. 86, Abstr. 5033 (1983).

084.102 **On electromagnetic core-mantle coupling.**
M. Stix.
Geophys. Astrophys. Fluid Dyn., Vol. 21, 303 - 313 (1982). Abstr. in Phys. Abstr., Vol. 86, Abstr. 9833 (1983).

084.103 **Upper atmosphere dynamics. The mission of the Dynamics Explorer satellites.**
D. Rees.
Phys. Bull., Vol. 33, 404 - 406 (1982). – Abstr. in Phys. Abstr., Vol. 86, Abstr. 15737 (1983).

084.104 **Geomagnetic and solar data.**
H. E. Coffey (Editor).
J. Geophys. Res., Vol. 87, 4571, 5310, 6357, 7671, 8343, 9224, 10549 (1982).

Solar wind and magnetospheric investigations.
See Abstr. 003.075.

Wandel in der Kenntnis des Polarlichtes.
See Abstr. 004.028.

Das Polarlicht im Wandel der Zeiten.
See Abstr. 004.033.

Sesquicentenary of Gauss's first measurement of the absolute value of magnetic intensity.
See Abstr. 004.036.

Historical observations of the geomagnetic field.
See Abstr. 004.037.

Annual report for magnetic observatories – 1979.
See Abstr. 009.029.

17th International Cosmic Ray Conference. SH sessions: Sun and heliosphere.
See Abstr. 012.070.

Nuclear spins in the Earth's magnetic field.
See Abstr. 014.107.

Synthetic spectra for auroral studies: the N_2 Vegard–Kaplan band system.
See Abstr. 022.166.

A 'jamjar' magnetometer.
See Abstr. 034.067.

The scientific objectives of the GEOS mission.
See Abstr. 051.053.

Shock wave structure in collisionless plasmas from results of laboratory experiments.
See Abstr. 062.014.

Rädler effect in a collisionless plasma and field-aligned currents in the magnetotail.
See Abstr. 062.026.

Magnetospheric phenomena of celestial bodies – introductory talk.
See Abstr. 062.035.

On the role of rotation in the generation of magnetic fields by fluid motions.
See Abstr. 062.041.

Progress in dynamo theory: kinematic and dynamic models.
See Abstr. 062.042.

Non-steady state of a hydromagnetic $\alpha\omega$-dynamo and its application to the geomagnetic reversals.
See Abstr. 062.049.

Interaction of small-amplitude fluctuations with a strong magnetohydrodynamic shock.
See Abstr. 062.050.

The mean electromotive force generated by random Alfvén waves in a collisionless plasma under a non-uniform mean magnetic field.
See Abstr. 062.123.

The mean electromotive force generated by random hydromagnetic waves in a collisionless plasma.
See Abstr. 062.124.

The mean electromotive force generated by random Alfvén waves in a collisionless non-uniform plasma.
See Abstr. 062.125.

Neutral line motion due to reconnection in two-ribbon solar flares and magnetospheric substorms.
See Abstr. 073.076.

Geomagnetic and solar wind cycles, 1900 - 1975.
See Abstr. 074.057.

Generation of electron plasma waves in the upstream solar wind.
See Abstr. 074.094.

Global pulsations of the sun and magnetosphere.
See Abstr. 080.038.

Finding core motions from magnetic observations.
See Abstr. 081.017.

Heat flow and reversals of the Earth's magnetic field.
See Abstr. 081.024.

Natthimmelens lys – spesielt i nordlysområder.
See Abstr. 082.048.

Spectroscopy of the O I 989- and 7990-Å multiplets in the dayglow and aurora.
See Abstr. 082.049.

Ionospheric effects of internal gravitational waves generated by an auroral electrojet.
See Abstr. 083.010.

Basic concepts in the electrodynamics of the auroral ionosphere.
See Abstr. 083.014.

Planetary and lunar magnetism.
See Abstr. 091.041.

Interaction of non-perpendicular/parallel solar wind shock waves with the earth's magnetosphere.
See Abstr. 106.004.

Analysis of a connection of the components of the vector of the interplanetary magnetic field with geomagnetic field variations at high latitudes (equinox, 1968).
See. Abstr. 106.022.

Parameters of the interplanetary magnetic field and short-term forecast of substorms.
See Abstr. 106.023.

Spectra of variations of the B_z-component of the interplanetary magnetic field in high-velocity solar wind streams and intensitites of polar magnetic disturbances.
See Abstr. 106.024.

On the geoefficiency of the B_z-component of the interplanetary magnetic field.
See Abstr. 106.025.

Energetic interplanetary nucleon flux anisotropies: the effect of earth's bow shock and magnetosheath on sunward flow.
See Abstr. 106.035.

Effect of the northward interplanetary magnetic field on the geomagnetic field inside the region surrounded with the auroral oval.
See Abstr. 106.036.

085 Solar-terrestrial Relations

085.001 **On the changes of the atmospheric circulation after the entry of the earth into a corpuscular stream.**
Eh. R. Mustel', N. B. Mulyukova, V. E. Chertoprud.
Astron. Zh., Tom 59, 750 - 762 (1982). In Russian. English translation in Soviet Astron., Vol. 25, No. 4.

On the basis of an analysis of daily maps which give the distribution of the surface atmospheric pressure over the northern hemisphere of the earth for the period of 90 years the authors analyse the changes of variances of daily pressure differences as well as the changes of the pressure itself after the entering of the earth into solar corpuscular streams.

085.002 **Flare induced geomagnetic activity and orientation of the photospheric magnetic field.**
C. S. Wright, L. F. McNamara.
Nature, Vol. 299, 42 - 44 (1982).

The authors have investigated the relationships between geomagnetic disturbances, flare energies and flare-site magnetic field directions for all large flares that occurred between 1968 and 1979. They find no significant differences between the average energies of flares with northward, southward or east-west magnetic fields. Furthermore they find that the relationship between geoefficiency and field direction is unclear.

085.003 **Solar-terrestrial influences on weather and climate.**
J. Gregory.
Nature, Vol. 299, 401 - 402 (1982).

085.004 **Il sole: I, II.**
P. Stroppa.
Coelum, Vol. 50, 205 - 211, 255 - 261 (1982).

085.005 **Disappearing solar filaments and geomagnetic activity.**
L. F. McNamara, C. S. Wright.
Nature, Vol. 299, 537 - 538 (1982).

Early attempts to link solar filaments and their disappearances with geomagnetic activity were inconclusive. Therefore the authors have investigated this problem using a large data sample obtained between 1974 and 1980 inclusive. They found that the average level of geomagnetic activity increases 3–6 days after the disappearance of large solar filaments. The authors report that both the magnitude of the disturbance and the delay in its occurrence seem to depend on the filament size. These results support the inclusion of disappearing filaments within techniques that are used for predicting geomagnetic activity.

085.006 **A mechanism for the influence of solar variations on climate.**
J. R. Bates.
Irish Astron. J., Vol. 15, (see 012.008), 50 - 51 (1981).

085.007 **Ring current energy injection rate and solar wind-magnetosphere energy coupling.**
L. C. Lee, J. R. Kan, S.-I. Akasofu.
Planet. Space Sci., Vol. 30, 627 - 634 (1982).

085.008 **On a connection of long-period variations of electrons in the outer radiation belt of the earth with the parameters of the solar wind.**
I. P. Bezrodnykh, Yu. G. Shafer.
Kosm. Issled., Tom 20, 639 - 641 (1982). In Russian.

085.009 **Generation of wave disturbances in the lower ionosphere by solar flares.**
G. I. Gerasimov, B. T. Zhumabaev, I. D. Kozin,
B. M. Rubinshtejn, V. M. Somsikov, B. A. Turkeeva.
Geomagn. Aehron., Tom 22, 554 - 559 (1982). In Russian.

085.010 **Computer simulation of the interaction of the solar wind with the earth's magnetosphere.**
T. Ogino.
Magnetospheric phenomena of celestial bodies, (see 012.011), p. 22 - 25 (1982). In Japanese.

Global formation of the earth's magnetosphere interacting with the solar wind was simulated in the meridian plane by a two-dimensional MHD model when the interplanetary magnetic field was imposed. The magnetosphere corresponding to Dungey's reconnection model was formed in the quasi-steady state for the southward magnetic field, however, the plasma sheet was frequently brought a change.

085.011 **Zonal circulation of the atmosphere and solar activity.**
A. A. Dmitriev, V. A. Malinnikov.
Vopr. klimata i pogody Nizh. Povolzh'ya, Saratov, 1982, No. 7/14, p. 3 - 13. In Russian. – Abstr. in Ref. zh., 51. Astron., 9.51.297 (1982).

085.012 **An estimate of some parameters for calculation of the scheme of effect of solar UV radiation on condensation processes in the atmosphere.**
A. A. Dmitriev, V. S. Tyurin, Vl. S. Tyurin.
Vopr. klimata i pogody Nizh. Povolzh'ya, Saratov, 1982, No. 7/14, p. 26 - 39. In Russian. – Abstr. in Ref. zh., 51. Astron., 9.51.300 (1982).

085.013 **Longitudinal currents controlled by the parameters of solar wind plasma.**
B. A. Belov, V. Yu. Gajdukov, N. I. Gershenzon.
Soln. veter i magnitosfer. issled. Moskva, 1980, p. 96 - 101. In Russian. – Abstr. in Ref. zh., 62. Issled. kosm. prostranstva, 9.62.255 (1982).

085.014 **On the geoefficiency of various regions of a high-velocity stream in the solar wind.**
B. V. Zhiromskij.
Magnitosfer. issled. No. 1, Moskva, 1982, p. 32 - 35. In Russian. – Abstr. in Ref. zh., 62. Issled. kosm. prostranstva, 9.62.270 (1982).

085.015 **Correlation between the energy of the solar wind and the indices of magnetic activity obtained with the canonical correlation method.**
N. I. Dvinskikh, N. M. Rudneva.
Issled. po geomagn. aehron. i fiz. Solntsa. Moskva, 1982, No. 59, p. 105 - 109. In Russian. – Abstr. in Ref. zh., 62. Issled. kosm. prostranstva, 9.62. 275 (1982).

085.016 **Solar – tropospheric relations and the cold film of the world ocean.**
V. E. Chertoprud.
Astron. Zh., Tom 59, 1030 - 1032 (1982). In Russian. English translation in Soviet Astron., Vol. 26, No. 5.

A possible mechanism of growth of instability of the troposphere after the arrival of solar corpuscular streams to the earth is discussed. The cold film of oceans plays an essential role in this mechanism.

085.017 **Dependence of an equinoctial analytical model of the electron concentration of the ionosphere on solar activity.**
A. G. Izraitel', E. E. Tsedilina.
Geomagn. Aehron., Tom 22, 738 - 742 (1982). In Russian.

085.018 **On the nature of oscillations of the geomagnetic field generated by solar flares.**
E. V. Metelkin, V. M. Sorokin, G. V. Fedorovich.
Geomagn. Aehron., Tom 22, 803 - 808 (1982). In Russian.

085.019 **Geoeffectivity of the meridional component of a large-scale magnetic field of the sun.**
D. I. Ponyavin, M. I. Pudovkin.
Geomagn. Aehron., Tom 22, 856 - 858 (1982). In Russian.

085.020 **On non-stationarity of a connection between the parameters of the solar wind with AE and a_p indices of magnetic activity.**
N. I. Dvinskikh, V. Yu. Pisarskij, N. M. Rudneva.
Geomagn. Aehron., Tom 22, 868 - 870 (1982). In Russian.

085.021 **Solar, volcanic, and CO_2 forcing of recent climatic changes.**
R. L. Gilliland.
Clim. Change, Vol. 4, 111 - 131 (1982). – Abstr. in Phys. Abstr., Vol. 85, Abstr. 100537 (1982).

085.022 **The solar constant and surface temperature.**
J. A. Eddy.
AIP Conf. Proc., No. 82, (see 012.037), p. 247 - 262 (1981). Abstr. in Phys. Abstr., Vol. 85, Abstr. 112108 (1982).

085.023 **11 - year solar and quasi-biannual cycles and long-term weather forecasting.**
M. A. Nuzhdina.
Soln. Dannye 1982 Byull., No. 5, p. 101 - 109 (1982). In Russian.

085.024 **Solar interaction regions, geomagnetic indices and the troposphere.**
V. P. Bhatnagar, T. Jakobsson, R. L. Rosenberg
Planet. Space Sci., Vol. 30, 1079 - 1082 (1982).

085.025 **Influence of solar wind variability on the recurrence of droughts.**
P. Venkatakrishnan.
Sol. Phys., Vol. 81, 193 - 195 (1982).

The observed effects of solar flares and interplanetary sector crossings seem to indicate that particle precipitation in the Earth's upper atmosphere decreases cyclonic activity in the troposphere. As an extrapolation to longer term effects, it is suggested that the recurrence of prolonged periods of enhanced solar wind particle precipitation in the upper atmosphere during alternate solar minima could cause the recurrence of extreme droughts.

085.026 **A relationship in manifestation of solar activity and some geophysical phenomena for long periods.**
R. V. Smirnov.
Astron. Tsirk., No. 1202, p. 3 - 4 (1981). In Russian.

085.027 **Manifestation of a relationship between solar-geomagnetic activity and the circulation forms repeating themselves in the atmosphere.**
R. V. Smirnov.
Astron. Tsirk., No. 1202, p. 6 - 7 (1981). In Russian.

085.028 **Dispersion characteristics of solar activity and geophysical phenomena in the 80-year cycle.**
R. V. Smirnov.
Astron. Tsirk., No. 1203, p. 3 - 4 (1982). In Russian.

085.029 **Correlations of magnetospheric ion composition with geomagnetic and solar activity.**
D. T. Young, H. Balsiger, J. Geiss.
J. Geophys. Res., Vol. 87, 9077 - 9096 (1982).

The beginnings of solar-terrestrial studies.
See Abstr. 004.015.

Earth radiation science seminars.
See Abstr. 012.092.

Planetary System

091 Physics of the Planetary System (Dynamics, Figure, Rotation, Interiors, Atmospheres, Magnetic Fields, etc.)

091.001 **Constraints on the formation of ring impact structures, based on terrestrial data.**
R. A. F. Grieve, P. B. Robertson, M. R. Dence.
Multi-ring basins, (see 012.001), p. 37 - 57 (1981) = Contrib. Earth Phys. Branch No. 920.
This contribution considers the evidence of the terrestrial impact record as it pertains to the formation of large complex impact structures, of which multi-ring basins are the high energy variants. The thesis is that various types of data from well-documented terrestrial structures favor proportional growth for the depth of the original cavity. This cavity was subsequently modified by massive uplift in the center and by exterior collapse. The rings produced during the modification stage are related to the same process that forms central peaks in smaller structures with a complex form.

091.002 **The modification stage of basin formation: conditions of ring formation.**
S. K. Croft.
Multi-ring basins, (see 012.001), p. 227 - 257 (1981).
The sequence of crater and basin morphologies has been re-examined to establish a classification scheme of basin ring structural types. Rings were classified on the basis of morphology, morphometry, transitional basin types, characteristic geophysical signatures, theoretical scaling and collapse models, and structural analyses of terrestrial impacts. A scenario of basin formation is presented utilizing observed particle flow in impact and explosion craters and numerical simulations as bases for the high-pressure and excavation stages. Events and energy sources in the modification stage are constrained by the initial boundary conditions set by the nature of the transient crater and the final boundary conditions of ring structure.

091.003 **Variation of albedo with solar incidence angle on planetary surfaces.**
S. W. Squyres, J. Veverka.
Icarus, Vol. 50, 115 - 122 (1982).
While it has been known for a long time that in general the albedo of a surface depends on incidence angle, this fact is commonly neglected in many calculations of planetary surface temperatures. The authors show that the effect is especially pronounced for bright surfaces. For objects such as Ganymede and Io, the effect produces substantially cooler temperatures near the poles and terminators than would be calculated under the assumption of a constant albedo–a factor which may be important in determining the stability of frosts on such surfaces.

091.004 **Atmospheric interactions of planetary bodies with the solar wind.**
M. K. Wallis, W.-H. Ip.
Nature, Vol. 298, 229 - 234 (1982).
Orbiting spacecraft about Venus have revealed a partially penetrable boundary sheath with induced fields limiting solar wind effects on the atmosphere structure and evolution. Mars differs structurally, the explanation awaiting an orbiter mission such as the European Space Agency is considering. In weak comets gyro-radii exceed other scale lengths, while in active comets the gas and solar plasma couple over the largest scale, boundary sheaths being unimportant – as occurred during early rapid-outgassing phases of Venus and Mars.

091.005 **Sharp edges of planetary rings.**
N. Borderies, P. Goldreich, S. Tremaine.
Nature, Vol. 299, 209 - 211 (1982).
The ring systems of Saturn and Uranus exhibit several sharp edges across which the optical depth drops from order unity to essentially zero. At least two and perhaps all of these features are associated with the location of orbital resonances between a satellite and the ring particles. It is remarkable that the optical depth varies on a distance scale which is much finer than that over which angular momentum can be transferred between a satellite and the ring material.

091.006 **Control of the homopause level.**
C. B. Leovy.
Icarus, Vol. 50, (see 012.003), 311 - 321 (1982).
This paper reviews the hypothesis that internal gravity wave breaking is the factor responsible for upper atmosphere turbulence generation and control of the number density at the homopause level.

091.007 **Photographic positions of major planets from observations at the Nikolaev Observatory in 1976 - 1979.**
V. I. Voronenko, G. K. Gorel'.
Glav. astron. obs. AN SSSR. Leningrad, 1982, 17 pp. In Russian. – Abstr. in Ref. zh., 51. Astron., 7.51.134 (1982).

091.008 **IUE observations of solar system objects and processes.**
R. M. Nelson.
Advances in ultraviolet astronomy, (see 012.006), p. 33 - 45 (1982).
IUE has provided significant and unique information on every solar system object it has observed. New atmospheric absorption features have been identified in the atmospheres of Venus, Mars, Jupiter and Saturn. Emission processes have also been studied and provided new insights into the physics of magnetosphere-atmosphere processes on Jupiter and Saturn. The comet studies have provided the basis for the development of coherent volatile evolution models. The possibility now exists for new classes of asteroids based on their ultraviolet albedos. New surface constituents have been identified on planetary satellites.

091.009 **Atmosphärische Superrotation.**
W. Kundt.
Phys. Bl., Vol. 38, 267 - 268 (1982).

091.010 **Reference systems of heights for planets and satellites.**
Yu. S. Tyuflin.

Geod. i kartogr., 1982, No. 4, p. 49 - 51. In Russian. – Abstr. in Ref. zh., 52. Geod. Aehrosemka, 8.52.185 (1982).

091.011 **Origin and evolution of planetary atmospheres: an introduction to the problem.**
R. G. Prinn.
Planet. Space Sci., Vol. 30, 741 - 753 (1982).
The author draws an integrated picture of the planetary atmospheric evolutionary process being careful to define the outstanding problems, to differentiate theory from fact, and to emphasize the strengths and weaknesses of a priori and a posteriori approaches to these problems.

091.012 **Thermal and nonthermal escape mechanisms for terrestrial bodies.**
D. M. Hunten.
Planet. Space Sci., Vol. 30, 773 - 783 (1982).
A survey is given of the classical thermal escape (Jeans or blowoff) and some eight nonthermal processes involving ions. Diffusional, energetic, and other limitations on the escape flux are discussed. Isotopic fractionation, due to differences in thermal escape rates or diffusive separation, is considered. The dominant processes acting at the five terrestrial planets and six major satellites are summarized. In general, nonthermal processes dominate over the classic thermal ones.

091.013 **A review of Blagg's formula in the light of recently discovered planetary moons and rings.**
G. G. Lobban, A. E. Roy, J. C. Brown.
J. British Astron. Assoc., Vol. 92, 260 - 263 (1982).
Blagg's formula is a generalized form of Bode's law which. unlike Bode's law, can be equally well applied to the satellite systems. This paper is a review of Blagg's formula in the light of the newly discovered satellites and rings of Jupiter, Saturn and Uranus. In the case of Uranus, new constants for Blagg's formula are suggested.

091.014 **Planetary rings.**
J. A. Burns.
Bull. American Astron. Soc., Vol. 14, 606 (1982). – Abstract.

091.015 **Far infrared photometry of Jupiter, Saturn, Uranus, and Callisto.**
R. F. Loewenstein, D. A. Harper, R. H. Hildebrand, J. Keene, S. Whitcomb.
Bull. American Astron. Soc., Vol. 14, 612 (1982). – Abstract.

091.016 **Planetary magnetism.**
Sh. Sh. Dolginov.
Itogi nauk. i tekh. VINITI. Issled. kosm. prostranstva, Tom 18, 130 pp. (1982). In Russian. – From Ref. zh., 51. Astron., 9.51.150 (1982).

091.017 **Planetary atmospheres and the search for life.**
T. Owen.
Phys. Teach., Vol. 20, 90 - 96 (1982). – Abstr. in Phys. Abstr., Vol. 85, Abstr. 68991 (1982).

091.018 **Equation of state experiments and theory relevant to modeling the major planets.**
M. Ross, W. J. Nellis.
AIP Conf. Proc., No. 78, (see 012.019), p. 226 - 230 (1981). Abstr. in Phys. Abstr., Vol. 85, Abstr. 73878 (1982).

091.019 **A supposed new law for planetary distances.**
E. Badolati.
Moon Planets, Vol. 26, 339 - 341 (1982).
In this paper it is remarked that the so called Basano-Hughes law of planetary distances coincides with a formula suggested by G. Armellini in 1921. Moreover some additional remarks are made on the laws of planetary distances.

091.020 **Loi de Titius-Bode et formalisme ondulatoire.**
R. Louise.
Moon Planets, Vol. 26, 389 - 398 (1982).
As far as the Titius-Bode law is concerned the author has the feeling that orbits of planets around the Sun and of satellites around their primaries do not depend on starting conditions. If it is correct, then one may expect to formulate similarity between the T-B law and the Bohr law in the early quantum theory. Such a similarity is found (Louise, 1982). This similarity suggests the existence of an unknown wave process in the solar system. The aim of the present paper is to investigate the possibility of such a process. The first approach is to study a steady wave encountered in a special membrane, showing node rings similar to the Prentice's rings (1976) which obey the T-B law. In the second part, the author tries to apply the Lindblad-Lin density wave theory of spiral galaxies to the solar nebula case.

091.021 **Topography of the terrestrial planets.**
M. Kobrick.
Astronomy, Vol. 10, No. 5, p. 18 - 22 (1982). – Abstr. in Phys. Abstr., Vol. 85, Abstr. 89174 (1982).

091.022 **Magnetic field generation in planetary interiors.**
R. Narain Singh.
Proc. Indian Natl. Sci. Acad., Part A, Vol. 47, Suppl. 2, 156 - 164 (1981). – Abstr. in Phys. Abstr., Vol. 85, Abstr. 89176 (1982).

091.023 **Experimental results on the upper atmospheres and ionospheres of Venus, Mars and Jupiter.**
K. K. Mahajan.
Proc. Indian Natl. Sci. Acad., Part A, Vol. 47, Suppl. 2, 124 - 145 (1981). – Abstr. in Phys. Abstr., Vol. 85, Abstr. 89178 (1982).

091.024 **Planetary magnetospheres.**
D. P. Stern, N. F. Ness.
Annu. Rev. Astron. Astrophys., Vol. 20, (see 003.009), 139 - 161 (1982).
Magnetospheric morphology. Magnetospheric processes.

091.025 **The dynamics of planetary rings.**
P. Goldreich, S. Tremaine.
Annu. Rev. Astron. Astrophys., Vol. 20, (see 003.009), 249 - 283 (1982).
Contents: The rings of Saturn – early history. The rings of Saturn – properties. The rings of Uranus. The ring of Jupiter. Dynamical effects. Lifetimes.

091.026 **Cometary and meteorite swarm impact on planetary surfaces.**
J. D. O'Keefe, T. J. Ahrens.
J. Geophys. Res., Vol. 87, 6668 - 6680 (1982).
The authors examine the impact-induced deformation of planetary surfaces from hypothetical cometary objects having initial densities in the 0.01 to 1 g/cm^3 and heats of vaporization of ~2 kJ/g (corresponding to water) to $\sim 10^7$ J/g range for impacts in the 5 to 45 km/s range. The direct effect of an atmosphere is neglected, although the atmosphere may in fact cause a cometary object to break up into a shower or equivalent very porous impactor. In addition to examining the partitioning of impact energy into internal energy of the impacted planet and impacting cometary material the authors have also calculated the relative efficiency of shock-induced melting and vaporization by comets on planetary surface materials and the mass loss from a given planet for various escape velocities.

091.027 **IUE observations of giant planets, satellites and asteroids.**
T. Encrenaz.

Third European IUE Conference, (see 012.029), p. 47 - 57 (1982).

UV observations of the giant planets, satellites and asteroids are reviewed since the beginning of UV experiments. The contribution of IUE is presented and the results are analysed versus other recent UV observations, expecially the Voyager data.

091.028 **Observation of giant planets and comets in the far infrared and submillimetric range.**
T. Encrenaz, M. Combes, J. Crovisier.
The scientific importance of submillimetre observations, (see 012.030), p. 167 - 173 (1982).

A brief review of the existing observations is presented for the four Giant Planets and Titan. The authors analyse some possible experiments – spaceborne, airborne or ground-based – to be performed in the future on Giant Planets, on bright comets and on Titan.

091.029 **Earth orbital planetary astronomy: "tricks of the trade" that have produced the scientific yield from IUE.**
A. L. Lane, R. M. Nelson.
Bull. American Astron. Soc., Vol. 14, 711 (1982). – Abstract.

091.030 **Nature of ice on satellites and rings.**
R. Smoluchowski.
Bull. American Astron. Soc., Vol. 14, 745 (1982). – Abstract.

091.031 **The viscosity of rock-ice mixtures: applications to the evolution of icy satellites.**
J. A. Friedson, D. J. Stevenson.
Bull. American Astron. Soc., Vol. 14, 745 (1982). – Abstract.

091.032 **Perturbations of collisional particles discs.**
A. Brahic, B. Sicardy, S. Clairemidi.
Bull. American Astron. Soc., Vol. 14, 750 (1982). – Abstract.

091.033 **On the stability of certain planetary ringlets.**
V. R. Eshleman, J. V. Breakwell, G. L. Tyler, E. A. Marouf.
Bull. American Astron. Soc., Vol. 14, 750 (1982). – Abstract.

091.034 **On the electric charge carried by planetary dust grains.**
J.-P. Lafon.
Bull. American Astron. Soc., Vol. 14, 766 (1982). – Abstract.

091.035 **Planetary applications of ion induced erosion of condensed-gas frosts.**
R. E. Johnson, L. J. Lanzerotti, W. L. Brown.
Nucl. Instrum. Methods Phys. Res., Vol. 198, 147 - 157 (1982). – Abstr. in Phys. Abstr., Vol. 85, Abstr. 100781 (1982).

091.036 **General considerations of the physics of planetary interiors.**
G. H. A. Cole.
Geophys. Surv., Vol. 5, 89 - 139 (1982). – Abstr. in Phys. Abstr., Vol. 85, Abstr. 106571 (1982).

091.037 **Earth-based radar studies of planetary surfaces and atmospheres.**
R. F. Jurgens.
IEEE Trans. Geosci. Remote Sensing, Vol. GE-20, 293 - 305 (1982). – Abstr. in Phys. Abstr., Vol. 85, Abstr. 106574 (1982).

091.038 **Earth-based remote sensing of planetary surfaces and atmospheres at radio wavelengths.**
J. R. Dickel.
IEEE Trans. Geosci. Remote Sensing, Vol. GE-20, 306 - 314 (1982). – Abstr. in Phys. Abstr., Vol. 85, Abstr. 106575 (1982).

091.039 **Quantum formalism in gravitation – quantitative application to the Titius-Bode law.**
R. Louise.
Moon Planets, Vol. 27, 59 - 62 (1982).

Quantum conception of virtual energy exchange between masses leads to the Newton's law. A guide wave similar to the De Broglie's one (1923) is able to account quantitatively for the Titius-Bode law occurring in the planetary system as well as the satellite systems.

091.040 **Planeetkraters waargenomen vanaf de aarde.**
L. Aerts.
Zenit, 9. Jaarg., 520 - 522 (1982).

091.041 **Planetary and lunar magnetism.**
J. Pohl.
The solar system and its exploration, (see 012.054), p. 115 - 120 (1981).

Knowledge on intrinsic planetary magnetic fields is summarized. For the earth, both the spatial configuration and the present and past time variations of the internally generated field are known with much detail. Important features are secular changes of non-dipole components, of the dipole orientation and strength and frequent polarity reversals. Today the moon has no detectable dipole moment. However, magnetic field measurements show widespread magnetization of the lunar crust and lunar rocks have strong and stable remanent magnetizations, which indicates the presence of magnetic fields on the moon between 3 and 4 Gy ago.

091.042 **The physics of impact cratering.**
J. Pohl.
The solar system and its exploration, (see 012.054), p. 127 - 128 (1981). – Abstract.

091.043 **Surface history of the terrestrial-type planets.**
G. Neukum.
The solar system and its exploration, (see 012.054), p. 129 - 137 (1981).

Data on the cratering records and surface histories – mainly on the basis of cratering studies – of the terrestrial-type planets Mercury, earth's moon, Mars and the moons of Jupiter and Saturn are discussed. In addition to those data from the study of the surface structure of the terrestrial-type planets, the author gives a short discussion on multispectral data of Mars and the earth's moon.

091.044 **Physics of planetary atmospheres and ionospheres.**
S. J. Bauer.
The solar system and its exploration, (see 012.054), p. 157 - 160 (1981).

The lecture deals with the physical processes controlling the structure, composition, and dynamics of planetary atmospheres. It centers on the traditional atmospheric regions, the distinction between homosphere and heterosphere, and the changing atmospheric composition, the validity of the barometric law for the major part of a planetary atmosphere and its breakdown in the exosphere due to escape of atmospheric particles. In addition, the formation and maintenance of ionospheric layers, both photochemical and diffusion-controlled ones, and their applicability for the observed ionospheres of the planets and the spatial extent of ionospheres for magnetic and nonmagnetic planets is discussed. A brief survey of the thermal and nonthermal processes responsible for the mass loss of planetary atmospheres is also given.

091.045 **Optical and infrared measurements on planetary atmospheres.**
H.-J. Bolle.

The solar system and its exploration, (see 012.054), p. 165 - 172 (1981).

Imaging, photometric and radiometric techniques are discussed, which are used in planetary exploration. Examples are given for the capability of these methods to determine atmospheric structure, composition and dynamical features.

091.046 **Triple conjunctions: twins and triplets.**
G. P. Können, J. Meeus.
J. British Astron. Assoc., Vol. 93, 20 - 24 (1982).

Triple conjunctions between the bright outer planets are presented for the years AD 0 to 3000. It is shown that for Mars-Jupiter and Mars-Saturn, two subsequent triple conjunctions are possible with a time separation of one synodic period.

091.047 **Solar-system abundances of the elements.**
E. Anders, M. Ebihara.
Geochim. Cosmochim. Acta, Vol. 46, 2363 - 2380 (1982).

A new abundance table has been compiled, based on a critical review of all C1 chondrite analyses up to mid-1982. Where C1 data were inaccurate or lacking, data for other meteorite classes were used, but with allowance for fractionation among classes. In a number of cases, interelement ratios from meteorites or lunar and terrestrial rocks as well as solar wind were used to check and constrain abundances. A few elements were interpolated (Ar, Kr, Xe, Hg) or estimated from astronomical data (H, C, N, O, He, Ne). It appears that the new set of abundances is accurate to at least 10%, as irregularities of 5 - 10% are readily detectable. Accordingly, C1 chondrites seem to match primordial solar-system matter to $\leq 10\%$, with only four exceptions. Br and I are definitely and B is possibly fractionated by hydrothermal alteration, whereas Eu seems to be enriched by nebular condensation or regolith contamination.

091.048 **Problems concerning the outer planets.**
P. K. Seidelmann, R. L. Duncombe.
Dynamical astronomy, (see 012.056), p. 37 - 78 (1982).

091.049 **Some main regularities in the composition and geologic evolution of terrestrial planets in connection with their location in the solar system.**
E. E. Milanovskij, A. M. Nikishin.
Dokl. AN SSSR, Tom 265, 420 - 424 (1982). In Russian. Abstr. in Ref. zh., 62. Issled. kosm. prostranstva, 11.62.272 (1982).

091.050 **Einschlagkrater auf Planeten und Monden.**
W. Sandner.
Sterne, 58. Band, 296 - 297 (1982).

091.051 **The origin of planetary satellites.**
E. L. Ruskol.
Izv. AN SSSR. Fiz. Zemli, 1982, No. 6, p. 40 - 51. In Russian. Abstr. in Ref. zh., 51. Astron., 10.51.182 (1982).

091.052 **On determination of orientation of the main axes of inertia of planets (Earth, Moon, Mars, Venus, Jupiter).**
A. N. Marchenko, O. A. Abrikosov, A. L. Tserklevich.
L'vov. politekh. inst. L'vov, 1982. 48 pp. In Russian. – Abstr. in Ref. zh., 51. Astron., 11.51.74 (1982).

091.053 **The dynamical evolution of the solar system.**
J. A. Burns.
Formation of planetary systems, (see 012.064), p. 403 - 501 (1982).

Contents: Orbital evolution. Radiation forces. Tides. The past solar system. The three body problem. Rotational dynamics in the solar system.

091.054 **A brief survey of the solar system.**
T. Owen.
Formation of planetary systems, (see 012.064), p. 585 - 650 (1982).

Contents: Introduction. The organization of matter in the solar system. Debris. Planetary atmospheres. Mars. Venus. Goldilocks and the three planets. The outer planets. Visitors to Earth? The search for extraterrestrial intelligence.

091.055 **Poynting-Robertson drag and orbital resonance.**
R. Gonczi, Ch. Froeschlé, Cl. Froeschlé.
Icarus, Vol. 51, 633 - 654 (1982).

Both the Poynting-Robertson drag and resonant orbits appear to be very important for the motion of small grains in the early solar system. The authors study numerically the competition between the Poynting-Robertson drag and the gravitational interaction of grains with Jupiter near the 2/1 resonance. Computations are based on the plane-restricted problem.

091.056 **The accretional heating of the terrestrial planets: a review.**
G. A. Ransford.
Phys. Earth Planet. Inter., Vol. 29, 209 - 217 (1982).

Accretional heating in forming planets results from the transfer of kinetic energy of objects striking the proto-planet surface. By accounting for all energy transfer for every cratering event it is theoretically possible to determine the thermal state of newly-formed planets. Various models of the thermodynamics of planetary formation are presented.

091.057 **Convection experiments in fluids with highly temperature-dependent viscosity and the thermal evolution of the planets.**
H. C. Nataf, F. M. Richter.
Phys. Earth Planet. Inter., Vol. 29, 320 - 329 (1982).

091.058 **The permeability-controlled accumulation of primary magma in a planetary mantle.**
S. Maaløe, Å. Scheie.
Phys. Earth Planet. Inter., Vol. 29, 344 - 353 (1982).

091.059 **The giant planets.**
R. Hide.
Phys. Bull., Vol. 33, 357 - 361 (1982). – Abstr. in Phys. Abstr., Vol. 86, Abstr. 5208 (1983).

Bibliography of the moon and the planets.
See Abstr. 002.074.

Organizing a data base for solving planetodesic problems on the ES computer.
See Abstr. 002.099.

A bibliography of planetary geology principal investigators and their associates, 1981 - 1982.
See Abstr. 002.131.

The Cambridge photographic atlas of the planets.
See Abstr. 002.144.

Atlas of the solar system.
See Abstr. 002.148.

Essays of comparative planetology.
See Abstr. 003.024.

Planets of rock and ice: from Mercury to the moons of Saturn.
See Abstr. 003.031.

Introduction to planetary geology.
See Abstr. 003.043.

Planets and moons.
See Abstr. 003.066.

Basaltic volcanism on the terrestrial planets.
See Abstr. 003.083.

In the world of aggregates of moons (satellites of planets).
See Abstr. 003.114.

Planetary rings explained and unexplained.
See Abstr. 011.011.

Mysteries of the ringed planets.
See Abstr. 011.020.

A round-table discussion on prospects for observations of planets in the next decade (1980-1990).
See Abstr. 013.022.

Planetary science in Europe.
See Abstr. 013.042.

Contemporary planetary science.
See Abstr. 013.078.

On the possibility of using Laplace series for the gravitational potential on the surface of a planet. II.
See Abstr. 021.001.

A Monte Carlo approach to scattering off surfaces of triaxial ellipsoids.
See Abstr. 021.002.

Some possibilities of a machine analysis of space photos of planets.
See Abstr. 021.051.

Laboratory band strengths of methane and their application to the atmospheres of Jupiter, Saturn, Uranus, Neptune, and Titan. II. The red region 6000–7600 Å.
See Abstr. 022.002.

Multi-ring basin formation: possible clues from impact cratering calculations.
See Abstr. 022.008.

Z-model analysis of impact cratering: an overview.
See Abstr. 022.009.

The excavation stage of basin formation: a qualitative model.
See Abstr. 022.010.

New theoretical and experimental methods for pressure-broadened linewidths and their interpretation.
See Abstr. 022.019.

Laboratory molecular spectroscopy.
See Abstr. 022.020.

Infrared absorption of sulfur-bearing compounds.
See Abstr. 022.023.

Collision-induced sharp features in the infrared spectrum of HD.
See Abstr. 022.025.

Far infrared spectra of H_2 and mixtures of H_2-CH_4 and H_2-He.
See Abstr. 022.026.

A review of acetylene, ethylene and ethane molecular spectroscopy for planetary applications.
See Abstr. 022.027.

Current studies of CH_4 from 2.5 to 7 μm.
See Abstr. 022.028.

Line intensities from band and diode laser measurements: ν_4 of CH_4.
See Abstr. 022.029.

Random walks with the 6420- and 6825-Å features of methane: lost in the wilderness.
See Abstr. 022.030.

Current band model studies of CH_4 at wavelengths less than 2.5 microns.
See Abstr. 022.031.

Long-pathlength absorption spectroscopy of CH_4.
See Abstr. 022.032.

Current studies of CH_3D: ye olde drive.
See Abstr. 022.033.

Current studies of PH_3.
See Abstr. 022.035.

Organic chemistry in Titan's atmosphere.
See Abstr. 022.036.

Shock wave apparatus for studying minerals at high pressure and impact phenomena on planetary surfaces.
See Abstr. 022.070.

Laboratory spectral measurements at low temperatures for better abundance determinations of ethane, ethylene, and propane in outer solar system atmospheres.
See Abstr. 022.088.

High-resolution spectroscopy in the 6825 Å region.
See Abstr. 022.090.

Improvements in the line parameters of molecules involved in the radiative transfer of the giant planets: line parameters of CH_4 and NH_3.
See Abstr. 022.091.

Implications of the new H_2 pressure broadening data for models of the outer planet atmospheres.
See Abstr. 022.094.

Direct computational of atmospheric normal modes and instabilities.
See Abstr. 022.100.

Spectral properties of mixtures of ice and particulates: implications for remote sensing.
See Abstr. 022.101.

Ice poor regolith development/destruction on small icy rocky bodies.
See Abstr. 022.102.

The loss-cone driven electron-cyclotron maser.
See Abstr. 022.142.

Possible applications of time domain reflectometry in planetary exploration missions.
See Abstr. 022.161.

Two methods for determining volumes for explosion and impact craters.
See Abstr. 022.163.

Toward an atomic model of the planetary system.
See Abstr. 022.173.

Molecular spectroscopy and planetary exploration from space.
See Abstr. 031.518.

On the possibility of application of a photoelectric zenith tube placed on the moon or planets for investigating their nutations.
See Abstr. 031.530.

CCD observations of faint planetary satellites.
See Abstr. 031.546.

Taking into account the conditions of orthogonality in planetodesic reduction of photographs.
See Abstr. 031.656.

Numbering of photos and condensation points of a block in the construction of reference nets on the moon and planets.
See Abstr. 031.657.

Main dependences of photogrammetry in the reduction of space photos of planets.
See Abstr. 031.658.

A complex of algorithms and programs for the preparation of foundations for photographic maps of planetary surfaces.
See Abstr. 031.659.

Fourier spectroscopy on planetary missions including Voyager.
See Abstr. 031.681.

Le télescope de 2 mètres du Pic-du-Midi: premières observations des planètes.
See Abstr. 032.033.

High-temperature electronics applications in space exploration.
See Abstr. 032.533.

Method for constructing periodic orbits.
See Abstr. 042.007.

On the stability of the inner planets, satellites and close binaries.
See Abstr. 042.084.

Observations in the solar system with the proposed earth-orbiting observatory Magellan.
See Abstr. 051.029.

A far UV orbiting observatory of solar system objects: the SOLSTIS mission proposal.
See Abstr. 051.031.

Review of Ariane capabilities for interplanetary missions.
See Abstr. 052.043.

Elemental and nuclidic abundances in the solar system.
See Abstr. 061.033.

Dielectrophoretic, thermal instability in a spherical shell of fluid.
See Abstr. 062.127.

A model for the negative polarization of light scattered by cosmic bodies deprived of atmospheres.
See Abstr. 063.003.

Radiative transfer and remote sensing.
See Abstr. 063.015.

Rayleigh, Raman and particulate scattering.
See Abstr. 063.016.

Reflection and transmission by a vertically inhomogeneous planetary atmosphere.
See Abstr. 063.019.

The internal radiation of a planetary atmosphere.
See Abstr. 063.024.

Information content of a planetary spectrum.
See Abstr. 063.026.

The Lumme-Bowell photometric parameters: reality or fantasy?
See Abstr. 063.043.

A reply to Hapke's criticism of the Lumme-Bowell photometric theory.
See Abstr. 063.044.

The theory of Savić about the origin of rotation of celestial bodies and the Savić–Kašanin theory of the behavior of materials under high pressures.
See Abstr. 065.099.

The solar system: a cluttered laboratory for gravity research.
See Abstr. 066.299.

Solar wind and its interaction with planetary ionospheres.
See Abstr. 074.102.

Morphology of volcanoes.
See Abstr. 081.043.

Astronomic geology: its subject and general problems.
See Abstr. 081.051.

Application of ring tectonic theory to Mercury and other solar system bodies.
See Abstr. 092.001.

Discovery of multi-ring basins: gestalt perception in planetary science.
See Abstr. 094.501.

Natural radioactivity of lunar and planetary rocks.
See Abstr. 094.520.

A comparison of Saturn's electrostatic discharges with lightning on various planets.
See Abstr. 100.062.

Solar wind magnetic field penetration into cometary ionospheres.
See Abstr. 102.043.

Influence of the gravitational field of planets on the structural characteristics of meteor showers.
See Abstr. 104.012.

The evolution of an impact-generated atmosphere.
See Abstr. 107.009.

Numerical experiment applicable to the latest stage of planet growth.
See Abstr. 107.010.

Chemistry and accretion history of the earth and the inner planets.
See Abstr. 107.036.

Blow-off of planetary protoatmospheres and of the protoplanetary nebula.
See Abstr. 107.054.

The chemistry of interstellar molecules.
See Abstr. 131.329.

092 Mercury

092.001 **Application of ring tectonic theory to Mercury and other solar system bodies.**
W. B. McKinnon.
Multi-ring basins, (see 012.001), p. 259 - 273 (1981).

Multiringed structures, by their presence or absence, provide a powerful tool for deciphering the thermal histories of the solid planets. The theory of ring tectonics establishes the framework of that undertaking. The purpose of this paper is to apply this conceptualization in detail to the multiringed basins on Mercury, and to briefly review our present understanding of ring tectonics on the moon, Mars, Earth, Ganymede, and Callisto. The small, icy satellites of Saturn are also discussed.

092.002 **Main features of the structure and development of the surface of Mercury.**
G. N. Katterfel'd.
Geomorfologiya, 1982, No. 1, p. 13 - 21. In Russian. – Abstr. in Ref. zh., 62. Issled. kosm. prostranstva, 7.62.284 (1982).

092.003 **Possible correlation of mineralogical variations with surface geology on Mercury.**
F. Vilas, M. A. Leake, W. W. Mendell.
Bull. American Astron. Soc., Vol. 14, 756 (1982). – Abstract.

092.004 **Map of the surface of Mercury.**
Priroda, 1982, No. 11, p. 107. In Russian.

092.005 **Right ascensions of Mercury determined with the RATAN-600 radio telescope.**
P. M. Afanas'eva, V. A. Fomin.
Astron. Tsirk., No. 1189, p. 8 (1981). In Russian.

092.006 **A comment on the insolation at Mercury.**
R. Landau.
Icarus, Vol. 52, 202 - 204 (1982).

E. Van Hemelrijk and J. Vercheval presented calculations of the insolation at Mercury and Venus which neglect the finite angular size of the Sun. To determine the temperature structure in the subsurface a more accurate calculation is needed, especially at longitudes ± 90° on Mercury, where the Sun takes 18 days to rise or set. These calculations are presented here.

092.007 **The intercrater plains of Mercury and the Moon: their nature, origin, and role in terrestrial planet evolution.**
M. A. Leake.
Advances in planetary geology, (see 003.014), p. 3 - 535 (1982).

Contents: Geologic mapping of Mercury and the Moon. Cratering histories of the intercrater plains. Geologic map analyses: correlation of geologic and cratering histories. Remote sensing and physical data of Mercury and the Moon. Thermal histories of Mercury and the Moon. Discussion of the nature, origin, and role of the intercrater plains of Mercury and the Moon.

Mercury's perihelion from Le Verrier to Einstein.
See Abstr. 003.105.

Study of peculiarities of crater distribution on Mars, Mercury and the moon from maps of crater density constructed by a computer.
See Abstr. 097.004.

Interpreting the cratering record: Mercury to Ganymede and Callisto.
See Abstr. 099.133.

093 Venus

093.001 **Results of radar observations of Venus at 39 cm wavelength in 1980.**
V. A. Kotel'nikov, Yu. N. Aleksandrov, R. A. Andreev, A. S. Vyshlov, V. M. Dubrovin, V. A. Grishmanovskij, A. L. Zajtsev, V. I. Kaevitser, A. A. Krymov, G. M. Petrov, O. N. Rzhiga, A. F. Khasyanov, A. M. Shakhovskoj.
Astron. Zh., Tom 59, 625 - 630 (1982). In Russian. English translation in Soviet Astron., Vol. 25, No. 4.

Over the period from February 15 to April 15, 1980 radar observations of Venus were carried out. Distances from the planet, its motion velocities, surface altitude profiles in the longitude range of 70°–230° were measured. The deviation of the measured distance from the predicted one, calculated on the basis of a new numerical theory of motion, did not exceed 6 km for rms values of measurement errors of 0.3–1.5 km. Two mountain areas divided by an extensive hilly plain were observed.

093.002 **Preliminary results of mass-spectrometric measurements on board the Venera 13 and Venera 14 probes.**
V. G. Istomin, K. V. Grechnev, V. A. Kochnev.
Pis'ma Astron. Zh., Tom 8, 391 - 398 (1982). In Russian. English translation in Soviet Astron. Lett., Vol. 8.

Mass-spectrometric measurements of chemical and isotopic composition of the Venusian atmosphere's minor constituents made on board the Venera 13 and Venera 14 landers are presented.

093.003 **Gas-chromatographical analysis of the chemical composition of the atmosphere of Venus by Venera 13 and Venera 14 probes.**
L. M. Mukhin, B. G. Gel'man, N. I. Lamonov, V. V. Mel'nikov, D. F. Nenarokov, B. P. Okhotnikov, V. A. Rotin, V. N. Khokhlov.
Pis'ma Astron. Zh., Tom 8, 399 - 403 (1982). In Russian. English translation in Soviet Astron. Lett., Vol. 8.

A description of the gas-chromatographical experiment on board the Venera 13 and Venera 14 probes is given. 9 analyses of the atmosphere of Venus were made beginning from the height of 58 km. Volume concentrations of some components of the Venus atmosphere were obtained.

093.004 **Spectrophotometric experiment on Venera 13 and Venera 14.**
V. I. Moroz, B. E. Moshkin, A. P. Ehkonomov, Yu. M. Golovin, V. I. Gnedykh, A. V. Grigor'ev.
Pis'ma Astron. Zh., Tom 8, 404 - 410 (1982). In Russian. English translation in Soviet Astron. Lett., Vol. 8.

Spectra of solar radiation in the region $0.45 < \lambda < 1.2\ \mu m$ were registered by the spectrophotometers of Venera 13 and Venera 14 at 6 different angles on all heights from 62 km down to the surface. Evaluation of the water vapor abundance was obtained from the intensity of H_2O bands between 0.7 and 1 μm.

093.005 **Measurement of water vapor concentration in the Venus atmosphere by Venera 13 and Venera 14.**
Yu. A. Surkov, V. F. Ivanova, A. N. Pudov, V. A. Pavlenko, N. A. Davydov, D. M. Shejnin.
Pis'ma Astron. Zh., Tom 8, 411 - 413 (1982). In Russian. English translation in Soviet Astron. Lett., Vol. 8.

Results of determination of water vapor concentration in the Venus atmosphere by Venera 13 and Venera 14 are described.

093.006 **Estimates of wind velocity from Doppler measurements on Venera 13 and Venera 14 spaceprobes.**
V. V. Kerzhanovich, Yu. F. Makarov, E. P. Molotov, V. P. Sorokin, K. G. Sukhanov, N. M. Antsibor, V. D. Kustodiev, I. A. Matsygorin, V. F. Tikhonov.
Pis'ma Astron. Zh., Tom 8, 414 - 418 (1982). In Russian. English translation in Soviet Astron. Lett., Vol. 8.

Doppler measurements of the relative velocity of the fly-by and descent modules are accomplished at the spaceprobes Venera 13 and Venera 14. Preliminary estimates of wind speed and turbulence in the Venus atmosphere are made. They reveal a qualitative agreement with previous Veneras data.

093.007 **Wind velocity at the Venera 13 and Venera 14 landing sites from acoustic measurements.**
L. V. Ksanfomaliti, N. V. Goroshkova, M. K. Naraeva, A. P. Suvorov, V. K. Khondyrev, L. V. Yabrova.
Pis'ma Astron. Zh., Tom 8, 419 - 423 (1982). In Russian. English translation in Soviet Astron. Lett., Vol. 8.

Estimates of wind velocity on the surface of Venus obtained for the landing sites of Venera 13 and Venera 14 are within the range of 0.3 to 0.6 m/s.

093.008 **Low-frequency electromagnetic field in the atmosphere of Venus from Venera 13 and Venera 14 measurements.**
L. V. Ksanfomaliti, N. M. Vasil'chikov, N. V. Goroshkova, E. V. Petrova, A. P. Suvorov, V. K. Khondyrev.
Pis'ma Astron. Zh., Tom 8, 424 - 428 (1982). In Russian. English translation in Soviet Astron. Lett., Vol. 8.

Data on the low-frequency electromagnetic field in the Venusian atmosphere and the height dependence of the field strength obtained by means of Venera 13 and Venera 14 as well as Venera 11 and Venera 12 are analyzed. The refraction calculations for electromagnetic waves in the atmosphere of Venus are presented.

093.009 **Analysis of panoramas of the Venera 13 and Venera 14 landing sites.**
K. P. Florenskij, A. T. Bazilevskij, V. P. Kryuchkov, R. O. Kuz'min, O. V. Nikolaeva, A. A. Pronin, I. M. Chernaya, A. S. Selivanov, M. K. Naraeva, Yu. S. Tyuflin.
Pis'ma Astron. Zh., Tom 8, 429 - 432 (1982). In Russian. English translation in Soviet Astron. Lett., Vol. 8.

Geologic-morphologic analysis shows that at the Venera 13 and 14 landing sites rocks with fine subhorizontal bedding are outcropped. Characteristic morphologic features of these rocks are close to that of the Venera 9 and Venera 10 landing sites. These features indicate a probable sedimentary nature of these formations.

093.010 **Evolution of the images received from Venera 13.**
A. S. Selivanov, Yu. M. Gektin, M. K. Naraeva, A. S. Panfilov, A. B. Fokin.
Pis'ma Astron. Zh., Tom 8, 433 - 436 (1982). In Russian. English translation in Soviet Astron. Lett., Vol. 8.

Preliminary results of a comparable analysis of a sequence of the panoramas obtained during the operation of Venera 13 are presented. Changes of image contrast and positions of some elements are noted.

093.011 **Preliminary results of the determination of element composition of rocks on Venus by Venera 13 and Venera 14.**
Yu. A. Surkov, L. P. Moskaleva, O. P. Shcheglov, V. P. Kharyukova, O. S. Manvelyan, G. G. Smirnov.
Pis'ma Astron. Zh., Tom 8, 437 - 443 (1982). In Russian. English translation in Soviet Astron. Lett., Vol. 8.

An elemental composition of the rocks on Venus was determined for the first time. The determination of rock composition was carried out at the landing sites of Venera 13 and Venera 14. The rock composition at the Venera 13 landing site proved to be close to potassium alkaline basalt, that at the Venera 14 landing site close to tholeitic basalt of the earth's crust.

093.012 **Micro-seismic activity at the landing sites of Venera 13 and Venera 14.**
L. V. Ksanfomaliti, V. M. Zubkova, N. A. Morozov, E. V. Petrova.
Pis'ma Astron. Zh., Tom 8, 444 - 447 (1982). In Russian. English translation in Soviet Astron. Lett., Vol. 8.

A description of the device registering small vertical removals (of about 1 μm) of soil, which was mounted on Venera 13 and Venera 14, is given. The data are compared with the working process of the apparatus systems and with the wind velocity. The obtained results indicate several microseismos.

093.013 **Thermal periodicities in the Venus atmosphere.**
J. Apt, J. Leung.
Icarus, Vol. 49, 427 - 437 (1982).

A search was made for periodic fluctuations in the thermal brightness temperatures recorded by the Pioneer Venus orbiter's infrared radiometer. Data were averaged in 10 X 10° latitude–longitude bins for each of the 72 days the instrument was in operation. This time series of thermal brightness temperatures was then analyzed to determine the amplitude of fluctuations at periods from 2 to 64 days at four levels in the atmosphere (at the cloud tops and at approximately 70, 80, and 90 km). The amplitude of such fluctuations is small at equatorial latitudes and increases to a maximum at 60-70° latitude at most altitudes.

093.014 **Orientation of planetary O^+ fluxes and magnetic field lines in the Venus wake.**
H. Pérez-de-Tejada, D. S. Intriligator, C. T. Russell.
Nature, Vol. 299, 325 - 326 (1982).

The authors show that in the outer regions of the venusian far wake the displacement of planetary O^+ particles, characteristic of the Venus upper ionosphere, does not occur necessarily along the magnetic field lines but approximately in the direction of the shocked solar wind.

093.015 **A first order approximation model of CO_2 infrared bands in the Venusian lower thermosphere.**
E. Battaner, R. Rodrigo, M. López-Puertas.
Astron. Astrophys., Vol. 112, 229 - 234 (1982).

The transfer equation of 4.3 and 15 μm photons corresponding to carbon dioxide vibrational transitions has been solved in the Venusian upper atmosphere. The numerical method was the "diffusion approximation" which was adequate and precise enough for altitudes below about 140 km. The continuity equations of the excited levels were simultaneously solved. Calculations about intensities, anisotropy, and cooling produced by the divergence of the vertical fluxes were carried out for a model atmosphere based on Pioneer Venus measurements.

093.016 **Thermal balance of Venus.**
M. Tomasko.
Vibrational-rotational spectroscopy for planetary atmospheres, (see 012.007), p. 211 - 227 (1982).

093.017 **Neutral species in the atmosphere of Venus.**
A. Young.
Vibrational-rotational spectroscopy for planetary atmospheres, (see 012.007), p. 229 - 242 (1982).

About 97% of the Venusian atmosphere is CO_2, with the balance comprised of various trace constituents. The author first discusses species other than CO_2 that have either been found in Venus' atmosphere or have been sought unsuccessfully, and then returns to problems connected with the spectrum of CO_2 itself.

093.018 **Gaseous infrared absorption in the lower atmosphere of Venus.**
C. Chackerian, Jr., R. W. Boese.
Vibrational-rotational spectroscopy for planetary atmospheres, (see 012.007), p. 255 - 269 (1982).

It is the object of this paper to identify the possibly important sources of infrared opacity in the Venusian atmosphere. CO_2 is the dominant infrared opacity source.

093.019 **The spectroscopy of Venus.**
R. Beer.
Vibrational-rotational spectroscopy for planetary atmospheres, (see 012.007), p. 271 - 276 (1982).

093.020 **On possible condensates of the main cloud layer of planet Venus.**
V. P. Volkov, Yu. I. Sidorov, I. L. Khodakovskij, V. L. Barsukov.
Geokhimiya, 1982, No. 1, p. 3 - 22. In Russian. – Abstr. in Ref. zh., 62. Issled. kosm. prostranstva, 7.62.287 (1982).

093.021 **On rock types on Venus.**
K. P. Florenskij, O. V. Nikolaeva.
Dokl. AN SSSR, Tom 262, 1245 - 1249 (1982). In Russian. Abstr. in Ref. zh., 51. Astron., 8.51.163 (1982).

093.022 **Radiation field in the Venus atmosphere for a two-layer model of its structure.**
Zh. M. Dlugach, Eh. G. Yanovitskij.
Physics of planetary atmospheres, (see 003.003), p. 3 - 27 (1981). In Russian. – Abstr. in Ref. zh., 51. Astron., 8.51.164 (1982).

093.023 **Atmosphere-surface interactions on Venus and implications for atmospheric evolution.**
I. L. Khodakovsky *(Khodakovskij)*.
Planet. Space Sci., Vol. 30, 803 - 817 (1982).

The physico-chemical processes controlling the Venusian tropospheric chemical composition surface rock mineral assemblages and volatile element distribution in the atmosphere and planetary crust is considered.

093.024 **The Venus atmosphere according to optical measurements aboard the Venera 11 and 12 stations.**
Yu. M. Golovin, V. I. Moroz, B. E. Moshkin, N. F. San'ko, A. P. Ehkonomov.
Kosm. Issled., Tom 20, 572 - 594 (1982). In Russian.

093.025 **Formation of the plasma mantle in the Venus magnetosphere.**
L. M. Zelenyj, O. L. Vajsberg.
Kosm. Issled., Tom 20, 604 - 619 (1982). In Russian.

093.026 **On the equipotential of the Venus ionopause.**
L. M. Zelenyj, O. L. Vajsberg.
Kosm. Issled., Tom 20, 645 - 647 (1982). In Russian.

093.027 **Composition of the Venus atmosphere at 110 - 200 km.**
V. A. Krasnopol'skij.
Kosm. Issled., Tom 20, 647 - 650 (1982). In Russian.

093.028 **Venus: comparison of gravity and topography in the vicinity of Beta Regio.**
R. D. Reasenberg, Z. M. Goldberg, I. I. Shapiro.
Geophys. Res. Lett., Vol. 9, 637 - 640 (1982).

The Doppler tracking data obtained from the Pioneer Venus Orbiter when it passed near Beta Regio yielded a peak vertical anomaly of 150 mGal when analyzed by the authors' two stage procedure. A comparison of maps of the gravity and topography at comparable resolution shows a striking correlation. A scatter plot shows that the observed gravity anomaly is approximately 0.4 of that expected from uncompensated topography of half the mean density of Venus. However, the spectral admittance shows that the gravity anomalies can not be explained either by Airy compensation at fixed depth or by a model comprising an elastic plate atop an inviscid fluid. The gravity and topography variations may signify deep compensation or dynamic support for Beta Regio and more shallow compensation for other features.

093.029 **Loss of oxygen from Venus.**
M. B. McElroy, M. J. Prather, J. M. Rodriguez.
Geophys. Res. Lett., Vol. 9, 649 - 651 (1982).

Ionization of thermal and non-thermal oxygen atoms above the plasmapause on Venus supplies an escape flux for O averaging 6×10^6 atoms $cm^{-2} sec^{-1}$. Hydrogen and oxygen atoms escape with stoichiometry characteristic of water. It is argued that escape of H is controlled by the oxidation state of the atmosphere, regulated by escape of O.

093.030 **A new concept of plasma motion and planetary magnetic field for Venus.**
W. C. Knudsen, P. M. Banks, K. L. Miller.
Geophys. Res. Lett., Vol. 9, 765 - 768 (1982).

The authors show that the magnetohydrodynamic conditions of the Venus ionosphere near the terminator favor convection of magnetic field rather than diffusion. Consequently, any planetary magnetic field which Venus may possess will be strongly affected by the global antisunward flow of the ionosphere which has been revealed by the Pioneer-Venus retarding potential analyzer. The magnetic flux from an internal magnetic field will accumulate in the night hemisphere. Details of the structure and dynamics of such accumulations depend on particular details of the magnetic field source and the time-dependent plasma flow pattern, but a simple interpretation of observational data yields a magnetic dipole moment of 7×10^{20} G cm^3 directed along the planet spin vector.

093.031 **Production of nitric oxide by lightning on Venus.**
J. S. Levine, G. L. Gregory, G. A. Harvey,
W. E. Howell, W. J. Borucki, R. E. Orville.
Geophys. Res. Lett., Vol. 9, 893 - 896 (1982).

The first measurements of the production of nitric oxide by a laboratory discharge in a simulated Venus atmosphere are presented.

093.032 **Television on Venus.**
A. S. Selivanov, M. K. Naraeva.
Zemlya Vselennaya, 1982, No. 4, p. 4 - 6. In Russian.

093.033 **Far infrared spectrophotometry of Venus,**
E. F. Erickson, M. R. Haas, D. Goorvitch,
D. D. McKibbin.
Bull. American Astron. Soc., Vol. 14, 612 (1982). – Abstract.

093.034 **Topography of Venus and Earth: a test for the presence of plate tectonics.**
J. W. Head, S. E. Yuter, S. C. Solomon.
American Sci., Vol. 69, 614 - 623 (1981). – Abstr. in Phys. Abstr., Vol. 85, Abstr. 63289 (1982).

093.035 **Results of photographic position observations of Venus in 1975, 1977 at Nikolaev.**
G. K. Gorel'.
Nikolaev. otd. GAO AN SSSR. Leningrad, 1982. 10 pp. In Russian. – Abstr. in Ref. zh., 51. Astron., 9.51.79 (1982).

093.036 **Optical properties of the Venusian atmosphere and radiative heat exchange.**
K. Ya. Kondratyev *(Kondrat'ev)*, N. I. Moskalenko.
Nuovo Cimento C, Vol. 4C, Ser. 1, 698 - 735 (1981). – Abstr. in Phys. Abstr., Vol. 85, Abstr. 68645 (1982).

093.037 **Rare gases on Venus.**
V. G. Istomin.
Zemlya Vselennaya, 1982, No. 5, p. 19 - 25. In Russian.

093.038 **Ionosphere and upper atmosphere of Venus.**
K. K. Mahajan.
Indian J. Radio Space Phys., Vol. 10, 211 - 222 (1981). Abstr. in Phys. Abstr., Vol. 85, Abstr. 78997 (1982).

093.039 **Optical emission on the nightside ionosphere of Venus.**
R. P. Singhal, S. A. Haider.
Indian J. Radio Space Phys., Vol. 11, 15 - 19 (1982). – Abstr. in Phys. Abstr., Vol. 85, Abstr. 78998 (1982).

093.040 **On Venus impact basins: viscous relaxation of topographic relief.**
S. C. Solomon, S. K. Stephens, J. W. Head.
J. Geophys. Res., Vol. 87, 7763 - 7771 (1982).

The authors begin with a brief summary of the evidence cited in the literature for craters and basins of possible impact origin on the Venus surface. They then outline a simple analytical model of viscous relaxation that includes the important effects of a decrease in viscosity with depth and partial to complete isostatic compensation of initial topographic relief. Applying the model to Venus under a set of conservative assumptions, the authors show that ancient basins on Venus should currently have negligible topographic relief if presently high surface temperatures date from the time of basin formation. This result implies that large, quasi-circular, topographic depressions on Venus are much younger than 3.8 b.y. in age and must presumably be of tectonic rather than impact origin.

093.041 **Venus: surface composition from radar reflectivity.**
G. H. Pettengill, P. G. Ford, S. Nozette.
Bull. American Astron. Soc., Vol. 14, 739 - 740 (1982). Abstract.

093.042 **Venus gravity harmonics: tectonic and dynamical implications.**
B. G. Bills, M. Kobrick, W. L. Sjogren.
Bull. American Astron. Soc., Vol. 14, 740 (1982). – Abstract.

093.043 **The Venus boundary layer.**
A. R. Dobrovolskis.
Bull. American Astron. Soc., Vol. 14, 740 (1982). – Abstract.

093.044 **Power spectrum analysis of Pioneer Venus extended mission UV images.**
A. D. Del Genio, W. B. Rossow.
Bull. American Astron. Soc., Vol. 14, 740 (1982). – Abstract.

093.045 **Variations of winds on Venus.**
W. B. Rossow, E. Kinsella.
Bull. American Astron. Soc., Vol. 14, 740 (1982). – Abstract.

093.046 **Polarization features on Venus.**
S. S. Limaye.
Bull. American Astron. Soc., Vol. 14, 740 - 741 (1982). Abstract.

093.047 **Venus cloud features compared with models of planetary-scale waves.**
C. Covey, G. Schubert.
Bull. American Astron. Soc., Vol. 14, 741 (1982). – Abstract.

093.048 **Wave instability in the polar region of Venus.**
L. S. Elson.
Bull. American Astron. Soc., Vol. 14, 741 (1982). – Abstract.

093.049 **Local variability in the stratospheric thermal structure of Venus.**
J. F. Appleby, D. J. McCleese, R. D. Haskins, L. S. Elson.
Bull. American Astron. Soc., Vol. 14, 741 (1982). – Abstract.

093.050 **Thermal tides in the Venus atmosphere: a model in agreement with Pioneer Venus OIR data.**
J. B. Pechmann, A. P. Ingersoll.
Bull. American Astron. Soc., Vol. 14, 741 - 742 (1982). Abstract.

093.051 **High resolution spectroscopy and mapping photometry of Venus near 5 μm.**
R. Beer, M. A. Janssen, J. Apt.
Bull. American Astron. Soc., Vol. 14, 742 (1982). – Abstract.

093.052 **Interpretation of Venera 11 data: the water vapor band at 1.1 microns.**
L. D. G. Young.
Bull. American Astron. Soc., Vol. 14, 742 (1982). – Abstract.

093.053 **A spectroscopic study of water vapor on Venus.**
N. M. Schneider, D. M. Hunten, R. A. Brown.
Bull. American Astron. Soc., Vol. 14, 742 - 743 (1982). Abstract.

093.054 **The chemistry of sulfur on Venus.**
G. T. Sill.
Bull. American Astron. Soc., Vol. 14, 743 (1982). – Abstract.

093.055 **Physics and chemistry of the Venus clouds.**
O. B. Toon, J. B. Pollack, R. C. Whitten, R. P. Turco.
Bull. American Astron. Soc., Vol. 14, 743 (1982). – Abstract.

093.056 **The latitudinal and temporal variability of the atmosphere of Venus between 40 and 90 km altitudes.**
A. J. Kliore.
Bull. American Astron. Soc., Vol. 14, 743 (1982). – Abstract.

093.057 **A model atmosphere of Venus for the day and night sides.**
S. T. Massie, D. M. Hunten.
Bull. American Astron. Soc., Vol. 14, 743 (1982). – Abstract.

093.058 **Mg II airglow in the atmosphere of Venus: observations and theory.**
M. A. Lecompte, L. J. Paxton, A. I. F. Stewart, J.-C. Gerard.
Bull. American Astron. Soc., Vol. 14, 743 - 744 (1982). Abstract.

093.059 **Atomic carbon in the atmosphere of Venus: observations and theory.**
L. J. Paxton, A. I. F. Stewart, D. E. Anderson.
Bull. American Astron. Soc., Vol. 14, 744 (1982). – Abstract.

093.060 **Temporal monitoring and spatial mapping of millimeter CO lines on Venus during the 1982 conjunction.**
F. P. Schloerb, J. C. Good.
Bull. American Astron. Soc., Vol. 14, 744 (1982). – Abstract.

093.061 **Distribution of hot oxygen atoms in the exosphere of Venus.**
A. I. F. Stewart, L. J. Paxton.
Bull. American Astron. Soc., Vol. 14, 746 (1982). – Abstract.

093.062 **Nonthermal escape of hydrogen and deuterium from Venus and implications for loss of water.**
S. Kumar, D. M. Hunten, J. B. Pollack.
Bull. American Astron. Soc., Vol. 14, 746 (1982). – Abstract.

093.063 **Investigation of the D/H ratio on Venus using ground-based high resolution spectroscopy.**
M. W. Buie, U. Fink, P. Connes.
Bull. American Astron. Soc., Vol. 14, 746 (1982). – Abstract.

093.064 **Loss of a massive early ocean from Venus.**
J. F. Kasting, J. B. Pollack.
Bull. American Astron. Soc., Vol. 14, 746 - 747 (1982). Abstract.

093.065 **Time history of outgassing on Venus, Earth, and Mars.**
J. B. Pollack, D. C. Black.
Bull. American Astron. Soc., Vol. 14, 747 (1982). – Abstract.

093.066 **On the 5577 Å night emission and electron streams in the Venus night atmosphere.**
V. A. Krasnopol'skij.
Kosm. Issled., Tom 20, 742 - 747 (1982). In Russian.

093.067 **Observations of energetic ions near the Venus ionopause.**
W. T. Kasprzak, H. A. Taylor, L. H. Brace, H. B. Niemann, F. L. Scarf.
Planet. Space Sci., Vol. 30, 1107 - 1115 (1982).

Ions (primarily O^+) with spacecraft rest frame energies > 40 eV have been observed by the Pioneer Venus Neutral Mass Spectrometer. The energetic ions coincide in location with superthermal ions observed by the Ion Mass Spectrometer and more rarely occur in some of the plasma clouds observed by the Electron Temperature Probe. These observations in conjunction with measurements by the Plasma Wave Instrument near the ionopause suggest that the ions are accelerated out of ionospheric plasma by the shocked solar wind through plasma wave-particle interactions.

093.068 **Could plate tectonics on Venus be concealed by volcanic deposits?**
W. M. Kaula, L. M. Muradian.
Geophys. Res. Lett., Vol. 9, 1021 - 1024 (1982).

The low upper limit on heat removal by plate tectonics inferrred by Kaula and Phillips (1981) has been questioned on the grounds that volcanic deposits may conceal the characteristic shape of topography arising from plate tectonic spreading. But a quantitative analysis indicates that volcanism sufficient for concealment must bring a major part of the heat to the surface, thus making plate tectonics the secondary process for heat removal. This inference is reinforced by consideration of the topographic structure and thickness of lava flows necessary. It thus appears that the principal process bringing heat to the surface of Venus is either lithospheric conduction or lava flows.

093.069 **On geological processes on Venus: analysis of the relationship between altitude and degree of surface roughness.**
A. T. Bazilevski (*Bazilevskij*), N. N. Bobina, V. P. Shashkina, Yu. G. Shkuratov, Yu. V. Kornienko, A. Ya. Usikov, D. G. Stankevich.
Moon Planets, Vol. 27, 63 - 89 (1982).

In order to study the relationship between Venus surface heights and surface roughness, the Pioneer Venus surface altitude map and map of r.m.s. slope in m-dkm scale have been analysed for the Beta and Ishtar regions using a system of digital image processing. To integrate the data obtained, the results of geomorphological analysis of Venera 9 and 10 TV panoramas as well as gamma-spectrometric and photometric measurements were used. The analysis proves that Venera 9 and 10 landing sites represent geologic-morphologic situations typical of Venus, thus enabling the results of observations made at landing sites to be extended to large provinces. Apparently this conclusion is also applicable to the Venera 8 landing site.

093.070 **Noble gases in planetary atmospheres: implications for the origin and evolution of atmospheres.**
J. B. Pollack, D. C. Black.
Icarus, Vol. 51, (see 012.048), p. 169 - 198 (1982).

In this paper, the authors analyze the abundance of radiogenic noble gases in the atmospheres of Venus and Mars to obtain bounds on the interval during which gases were emplaced into these atmospheres and on their retention of light gases. They next turn their attention to the primordial noble gases of the terrestrial planets. They begin by summarizing the available data and their ambiguity and then discuss data from other solar system objects. The authors present and assess alternative hypotheses for the origin of planetary atmospheres, with particular emphasis on the "grain-accretion" hypothesis that they previously advanced (Pollack and Black, 1979).

093.071 **Photochemistry of the stratosphere of Venus: implications for atmospheric evolution.**
Y. L. Yung, W. B. Demore.
Icarus, Vol. 51, (see 012.048), 199 - 247 (1982) = Contrib. No. 3692 Div. Geol. Planet. Sci. Calif. Inst. Technol.

The photochemistry of the stratosphere of Venus was modeled using an updated and expanded chemical scheme, combined with the results of recent observations and laboratory studies. The authors examined three models, with H_2 mixing ratio equal to 2×10^{-5}, 5×10^{-7}, and 1×10^{-13}, respectively. All models satisfactorily account for the observations of CO, O_2, $O_2(^1\Delta)$, and SO_2 in the stratosphere, but only the last one may be able to account for the diurnal behavior of mesospheric CO and the UV albedo. The modeling reveals a number of intriguing similarities between the chemistry of the stratosphere of Venus and that of the Earth.

093.072 **The chemistry of metastable species in the Venusian ionosphere.**
J. L. Fox.
Icarus, Vol. 51, (see 012.048), 248 - 260 (1982).

Reactions of metastable species are important in determining the densities of minor ions in the Venusian ionosphere. Calculations are carried out in which the coupled continuity and momentum equations are solved for twelve ions and four neutral species in the dayside ionosphere, including $O^+(^2D)$, $O^+(^2P)$, $N(^2D)$, and $N(^2P)$. Altitude profiles of these metastable species are presented. Their reactions are shown to be a significant source of several minor ions, especially N_2^+, CO^+, and N^+. The discrepancies which existed between model and measured densities of these ions are resolved.

093.073 **The Venus ionosphere at grazing incidence of solar radiation: transport of plasma to the night ionosphere.**
R. C. Whitten, B. Baldwin, W. C. Knudsen, K. L. Miller, K. Spenner.
Icarus, Vol. 51, (see 012.048), 261 - 270 (1982).

Using a quasi-two-dimensional model of the Venus ionosphere, the authors calculated the ion number densities and horizontal ion bulk velocities expected for a range of solar zenith angles near the terminator (80 to 100°), and compared them with data obtained from the Pioneer Venus Orbiter retarding potential analyzer. The calculated ion bulk velocity arises entirely from the solar EUV-induced plasma pressure gradient and has a magnitude consistent with observations. The authors find that while photoionization is the dominant source of ionospheric plasma for solar zenith angles less than 92°, plasma transport from the dayside is the dominant plasma source for solar zenith angles greater than 95°.

093.074 **Disappearing ionospheres on the nightside of Venus.**
T. E. Cravens, L. H. Brace, H. A. Taylor, Jr., C. T. Russell, W. L. Knudsen, K. L. Miller, A. Barnes, J. D. Mihalov, F. L. Scarf, S. J. Quenon, A. F. Nagy.
Icarus, Vol. 51, (see 012.048), 271 - 282 (1982).

Instruments on the Pioneer Venus Orbiter have detected a substantial ionosphere on the nightside of Venus during most orbits. However, during some orbits the nightside ionosphere seems to have almost disappeared. The solar wind dynamic pressure on these occasions is greater than average. The authors have correlated data from several instruments for a number of orbits during which the nightside ionosphere had disappeared and suggest mechanisms which might explain the reasons for the disappearance of the ionosphere when the solar wind dynamic pressure is large.

093.075 **Observed composition of the ionosphere of Venus: implications for the ionization peak and the maintenance of the nightside ionosphere.**
H. A. Taylor, Jr., R. E. Hartle, H. B. Niemann, L. H. Brace, R. E. Daniell, Jr., S. J. Bauer, A. J. Kliore.
Icarus, Vol. 51, (see 012.048), 283 - 295 (1982).

Across the nightside of Venus, daily measurements from the PV Orbiter Ion Mass Spectrometer often indicate an ionosphere of relatively abundant concentration, with a composition characteristic of the dayside ionosphere. Such conditions are interspersed by other days on which the ionosphere appears to largely " disappear". These characteristics, coupled with observations of strong day to night flows of O^+ in the upper ionosphere, support arguments that iron transport from the dayside is important for the maintenance of the nightside ionosphere. In the range 140–160 km, strong concentrations of O_2^+ and NO^+ indicate that the ionization peak is at times composed of at least two prominent ion species. Nightside concentrations of O_2^+ and NO^+ as large as 10^5 and $10^4/cm^3$, respectively, appear to require sources in addition to that provided by transport.

093.076 **The dynamics of the Venus ionosphere. II. The effects of the time scale of the solar wind dynamic pressure variations.**
R. F. Stein, R. S. Wolff.
Icarus, Vol. 51, (see 012.048), 296 - 301 (1982).

The effects on the upper dayside Venus ionosphere of a slow increase in solar wind dynamic pressure are simulated numerically with a 1-dimensional (spherically symmetric) Lagrangian hydrodynamical code. It is found that, for rise times greater than about 10 min, the compression of the ionopause is nearly adiabatic, with the ionopause moving downward at velocities of ~1-2 km/sec until it reaches a maximally compressed state, at which time the motion reverses. For short rise times the compression produces a shock wave similar to that occurring in the case of a sudden increase in pressure. The global implications of these processes are discussed within the context of Pioneer Venus observations and future theoretical work on this problem is outlined.

093.077 **Predicted electrical conductivity between 0 and 80 km in the Venusian atmosphere.**
W. J. Borucki, Z. Levin, R. C. Whitten, R. G. Keesee, L. A. Capone, O. B. Toon, J. Dubach.
Icarus, Vol. 51, (see 012.048), 302 - 321 (1982).

Calculations of the space charge, ion density, and conductivity in the Venus atmosphere were made. The presence of the cloud particles on Venus causes a profound reduction in the calculated values of the ion density and conductivity compared to the values that are obtained without consideration of the cloud particles. When the cloud particles are included in the calculations, the results for the ion density and conductivity are approximately the same as those of the terrestrial atmosphere at the same pressure-altitude. Because the particles span such a large range of sizes and are abundant over a substantial range of pressure, the space charge varies strongly with altitude and particle size. Differential settling of the particles is expected to produce weak electric fields in the clouds.

093.078 **Sulfuric acid vapor and other cloud-related gases in the Venus atmosphere: abundances inferred from observed radio opacity.**
P. G. Steffes, V. R. Eshleman.
Icarus, Vol. 51, (see 012.048), 322 - 333 (1982).

Laboratory measurements of the microwave opacity of gaseous sulfuric acid under Venus atmospheric conditions indicate that it is an exceptionally strong absorber. The authors conclude that sulfuric acid vapor is likely to be the principal microwave absorber in the 30- to 50-km-altitude range of the middle atmosphere of Venus. Carbon dioxide, the predominant atmospheric gas, is the main absorber below about 30 km, while sulfur dioxide is an important but secondary absorber in both regions. Water vapor and cloud particulates appear to be only minor contributors to the total opacity. While gaseous sulfuric acid has not been directly measured in any of the in situ probe experiments, its presence at an abundance of the deduced order of magnitude is implied by these and other observations.

093.079 **The composition and vertical structure of the lower cloud deck on Venus.**
J. Cimino.
Icarus, Vol. 51, (see 012.048), 334 - 357 (1982) = Contrib. No. 3731 Div. Geol. Planet. Sci. Calif. Inst. Technol.

The opportunity to determine the planetwide temperature and cloud structure arose with Pioneer Venus. This orbiting spacecraft contains a dual-frequency radio system which has been used to collect radio occultation data for more than two years. The primary goal of this paper is to use the absorption coefficient profiles derived from the radio occultation experiment to determine the mass content and composition of the Venusian lower cloud deck.

093.080 **The ultraviolet absorber on Venus: amorphous sulfur.**
O. B. Toon, R. P. Turco, J. B. Pollack.
Icarus, Vol. 51, (see 012.048), 358 - 373 (1982).

The authors discuss a new physical-chemical model of the Venus clouds which shows that sulfur should form there. They next describe physical characteristics of the cloud particles that, in comparison with model predictions, indicate the presence of sulfur. Then optical data suggesting that elemental sulfur occurs in the clouds is discussed. Following that, the authors describe cloud morphological evidence for sulfur. Finally, they summarize their conclusions and make recommendations for laboratory studies which are needed to demonstrate conclusively that sulfur is the UV absorber.

093.081 **Polarization studies of the Venus UV contrasts: cloud height and haze variability.**
L. W. Esposito, L. D. Travis.
Icarus, Vol. 51, (see 012.048), 374 - 390 (1982).

Polarimetry is able to show direct evidence for compositional differences in the Venus clouds. The authors present observations of the polarization in four colors of the bright and dark ultraviolet features. Exact calculations of the polarization, including multiple scattering and vertical inhomogeneity near the Venus cloud tops, are able to match the observations. The results give a straightforward interpretation of the polarization differences in terms of known constitutents of the Venus atmosphere. The authors propose two models for the UV contrasts consistent with their results.

093.082 **Temporal variability of ultraviolet cloud features in the Venus stratosphere.**
A. D. Del Genio, W. B. Rossow.
Icarus, Vol. 51, (see 012.048), 391 - 415 (1982).

Statistics on the temporal variability of UV cloud features on Venus during 66 days of nominal mission imaging by the Pioneer Venus Orbiter Cloud Photopolarimeter reveal at least five types of systematic variability on large scales: (1) a low-latitude global-scale wave of period 3.94 ± 0.1 days corresponding to longitudinal motion of the dark equatorial band; (2) a midlatitude wave of period 5.20 ± 0.2 days corresponding to wavenumber 1 oscillations of the latitude of the bright polar bands; (3) ~2- to 3-week fluctuations in the slope of longitudinal cloud brightness power spectra at intermediate wavenumbers manifested by variations in the intensity of large bow-shaped features; (4) ~2-month variations in polar region brightness; (5) a monotonic decrease in the disk-integrated brightness of Venus during the nominal mission. Small-scale features appear to correlate with large-scale albedo patterns.

093.083 **Zonal mean circulation at the cloud level on Venus: spring and fall 1979 OCPP observations.**
S. S. Limaye, C. J. Grund, S. P. Burre.
Icarus, Vol. 51, (see 012.048), 416 - 439 (1982).

Cloud motions were obtained from a number of images acquired in reflected solar ultraviolet light during spring and fall of 1979 from the Pioneer Venus Orbiter Cloud Photopolarimeter (OCPP) to determine the zonal mean circulation of the atmosphere of Venus at the cloud top level.

093.084 **The influence of thermospheric winds on exospheric hydrogen on Venus.**
R. R. Hodges, Jr., B. A. Tinsley.
Icarus, Vol. 51, (see 012.048), 440 - 447 (1982).

Monte Carlo models of the distribution of atomic hydrogen in the exosphere of Venus were computed which simulate the effects of thermospheric winds and the production of a "hot" hydrogen component by charge exchange of H^+ and H and O in the exosphere, as well as classic exospheric processes. A thermosphere wind system that is approximated by a retrograde rotating component with equatorial speed of 100 m/sec superimposed on a diurnal solar tide with cross-terminator day-to-night winds of 200 m/sec is shown to be compatible with the thermospheric hydrogen distribution deduced from Pioneer Venus orbiter measurements.

093.085 **Venus gravity: analysis of Beta Regio.**
P. B. Esposito, W. L. Sjogren, N. A. Mottinger, B. G. Bills, E. Abbott.
Icarus, Vol. 51, (see 012.048), 448 - 459 (1982).

Radio tracking data acquired over Beta Regio were analyzed to obtain a surface mass distribution from which a detailed vertical gravity field was derived. In addition, a corresponding vertical gravity field was evaluated solely from the topography of the Beta region. A comparison of these two maps confirms the strong correlation between gravity and topography which was previously seen in line-of-sight gravity maps. It also demonstrates that the observed gravity is a significant fraction of that predicted from the topography alone. The effective depth of complete isostatic compensation for the Beta region is estimated to be 330 km, which is somewhat deeper than that found for other areas of Venus.

093.086 **New data on aerosol in the cloud layer of Venus (Preliminary results of a study carried out with Venera 14).**
Yu. A. Surkov, F. F. Kirnozov, V. N. Glazov, A. G. Dunchenko, V. V. Atrashkevich.

Pis'ma Astron. Zh., Tom 8, 700 - 704 (1982). In Russian. English translation in Soviet Astron. Lett., Vol. 8.

Apparatus, technique and operation of the experiment on the study of aerosol in the cloud layer of Venus carried out on board the Venera 14 station are described. Preliminary results of the determination of aerosol elemental composition as well as of distribution of aerosol as a function of altitude are given.

093.087 **The Venus stratosphere from accelerometric measurement data of the Venera 11 and Venera 12 automatic interplanetary stations.**
V. S. Avduevskij, V. V. Semenchenko, G. R. Uspenskij, Z. P. Cheremukhina.
Kosm. Issled., Tom 20, 913 - 920 (1982). In Russian.

093.088 **Investigation of Venus is continued.**
V. L. Barsukov.
Priroda, 1982, No. 12, p. 10 - 17. In Russian.

093.089 **On the 834 Å line emission in the upper atmosphere of Venus.**
V. S. Sholokhov.
Astron. Tsirk., No. 1174, p. 1 - 3 (1981). In Russian.

093.090 **He^+ in the upper atmosphere of Venus.**
V. S. Sholokhov.
Astron. Tsirk., No. 1174, p. 3 - 4 (1981). In Russian.

093.091 **O I 1304 Å glowing in the upper atmosphere of Venus.**
M. S. Burgin, V. S. Sholokhov.
Astron. Tsirk., No. 1203, p. 1 - 2 (1982). In Russian.

093.092 **Large-scale photos of the Venus surface from Venera 13 and Venera 14 stations.**
A. S. Selivanov, Yu. S. Tyuflin, M. K. Naraeva, G. M. Garanin, S. A. Kadnichanskij.
Geod. kartogr., 1982, No. 6, p. 13 - 19. In Russian. – Abstr. in Ref. zh., 52. Geod. Aehrosemka, 11.52.229 (1982).

093.093 **Images of the relief of Venus from orbital radar data.**
Yu. V. Kornienko, D. G. Stankevich, A. T. Bazilevskij, Yu. G. Shkuratov.
Dokl. AN USSR, 1982, A, No. 5, p. 87 - 88. In Russian. Abstr. in Ref. zh., 51. Astron., 10.51.187 (1982).

093.094 **Numerical modelling of the high-altitude profile of radiative heat transfer in the undercloud atmosphere of Venus.**
V. P. Shari.
Inst. prikl. mat. AN SSSR. Prepr., 1982, No. 31, 18 pp. In Russian. – Abstr. in Ref. zh., 51. Astron., 10.51.192 (1982).

093.095 **Analysis of the relation between the surface height and roughness on Venus from Pioneer Venus radar measurements.**
A. Ya. Usikov, Yu. V. Kornienko, Yu. G. Shkuratov, D. G. Stankevich, S. P. Red'kin, A. V. Il'inskij, V. I. Bychinskij, A. T. Bazilevskij, N. N. Bobina.
Dokl. AN SSSR, Tom 264, 591 - 595 (1982). In Russian. Abstr. in Ref. zh., 51. Astron., 11.51.171; 62. Issled.kosm. prostranstva, 11.62.274 (1982).

093.096 **A description of the atmosphere of Venus.**
J. Blamont.
Formation of planetary systems, (see 012.064), p. 741 - 782 (1982).

Contents: Composition of the lower atmosphere. Structure. The clouds. Radiative properties. Dynamics. Evolution. Problems. Strategy of exploration.

093.097 **Dynamical implications of the observed thermal contrasts in Venus' upper atmosphere.**
A. Seiff.
Icarus, Vol. 51, 574 - 592 (1982).

This paper gives a detailed analysis of the vigorous dynamics in Venus' mesosphere and thermosphere, which is caused by the extremely large day–night pressure contrasts observed by Pioneer instruments.

093.098 **Transmission of CO_2 in the atmosphere of Venus for the spectral region near 7 microns.**
L. G. Young.
Icarus, Vol. 51, 606 - 609 (1982).

The transmission in the 7-μm "window" of Venus is calculated for a 7-layer model atmosphere.

093.099 **The transmission to space of the light produced by lightning in the clouds of Venus.**
M. A. Williams, L. W. Thomason, D. M. Hunten.
Icarus, Vol. 52, 166 - 170 (1982).

A Monte Carlo program by Thomason and Krider has been adapted to the Venus cloud geometry and optical properties. If lightning flashes occur within or just below the clouds, the fraction of photons of visible light escaping to space is 0.1–0.4, depending on the location of the flash. Failure of the Pioneer Venus Orbiter to unambiguously detect optical lightning signals from space is not due to attenuation by the atmosphere and clouds.

093.100 **Viscous dissipation at the Venus ionopause.**
H. Pérez-de-Tejada.
J. Geophys. Res., Vol. 87, 7405 - 7412 (1982).

A study of the effects of the transfer of momentum of the shocked solar wind to the Venus ionosphere near the terminator region is presented.

093.101 **The dynamics of the Venus ionosphere. 1. A simulation of the solar wind compression of the upper dayside ionosphere.**
R. S. Wolff, R. F. Stein, H. A. Taylor, Jr.
J. Geophys. Res., Vol. 87, 8118 - 8130 (1982).

The effects of the solar wind compression of the dayside Venus ionosphere are simulated numerically. The initial ionosphere is assumed to be in pressure equilibrium with the (shocked) solar wind at the upper boundary of the ionosphere. Composition, densities, and temperatures of ions and electrons in the ionosphere are chosen in accordance with Pioneer Venus data. A spherically symmetric Lagrangian hydrodynamic code using a two-fluid model of the ionosphere consisting of O^+ and electrons is employed to simulate the effects on the ionosphere of rapid changes in solar wind pressure. Comparison of density profiles from the authors' simulation with select Pioneer Venus ion mass spectrometer data suggest possible agreement between shock produced ionospheric ledges resulting from rapid solar wind compression and observed ionospheric ledges.

093.102 **Compression of the Venusian ionosphere on May 10, 1979, by the interplanetary shock generated by the solar eruption of May 8, 1979.**
M. Dryer, H. Pérez-de-Tejada, H. A. Taylor, Jr., D. S. Intriligator, J. D. Mihalov, B. Rompolt.
J. Geophys. Res., Vol. 87, 9035 - 9044 (1982).

An interplanetary shock wave that was produced by a solar eruption and its associated coronal transient on May 8, 1979, has been 'tracked' through interplanetary space to a rendezvous 2 days later with Venus. The interaction of the shock wave with the ionospheric obstacle at Venus produced a significant compression of the dayside ionosphere. It is believed that the tracking, as it were, was accomplished for the first time via the diagnostic observations provided by Hα and white light imagery near the sun and the plasma and field measurements of two, nearly radially aligned, spacecraft.

093.103 **The distant interplanetary wake of Venus: plasma observations from Pioneer Venus.**
J. D. Mihalov, A. Barnes.
J. Geophys. Res., Vol. 87, 9045 - 9053 (1982).

The authors report the detailed observations of the Pioneer Venus orbiter plasma analyzer during the first wake passage in June 1979, coinciding with the period of magnetotail observations. They find frequent episodic disappearance of the plasma, almost always inside the magnetotail, suggesting a tattered, filamentary extension of the solar wind cavity. On some days, abundant O^+ was found both inside and outside the magnetotail, with or without superposed magnetosheath; on other days, almost no O^+ was detected. The authors discuss the detailed morphology of these phenomena and attempt to assess the intensity and thermal spread of the streaming O^+. They briefly summarize their findings and discuss the implications for loss of Venus atmosphere to the solar wind.

093.104 **Pioneer Venus observations of plasma and field structure in the near wake of Venus.**
J. G. Luhmann, C. T. Russell, L. H. Brace, H. A. Taylor, W. C. Knudsen, F. L. Scarf, D. S. Colburn, A. Barnes.
J. Geophys. Res., Vol. 87, 9205 - 9210 (1982).

Ionospheric plasma density depletions or 'holes' are observed by the Pioneer Venus orbiter in association with radial magnetic fields in the near wake of Venus. This report presents examples of the collected observations of these unexpected features of the Venus nightside ionosphere obtained by the Langmuir probe, magnetometer, ion mass spectrometer, retarding potential analyzer, plasma analyzer, and electric field experiments. The connection between plasma density depletions and temperature changes, changes in ion composition, plasma wave emissions, and magnetic fields with a substantial radial component is illustrated. Mechanisms that may be responsible for the formation and maintenance of holes are suggested.

093.105 **Atomic carbon in the atmosphere of Venus.**
J. L. Fox.
J. Geophys. Res., Vol. 87, 9211 - 9216 (1982).

The densities of atomic carbon in the Venusian thermosphere are computed for a model which includes both chemistry and transport. The maximum density of C is 2.8×10^7 cm^{-3} near 150 km for an assumed O_2 mixing ratio of 1×10^{-4}. Photoionization of atomic carbon is found to be the major source of C^+ above 200 km, and resonance scattering of sunlight by atomic carbon may be the major source of the C I emissions at 1561 Å, 1657 Å, and 1931 Å. The computed C^+ densities are found to be in substantial agreement with those measured by Pioneer Venus.

093.106 **Mechanisms for lithospheric heat transport on Venus: implications for tectonic style and volcanism.**
S. C. Solomon, J. W. Head.
J. Geophys. Res., Vol. 87, 9236 - 9246 (1982).

The tectonic and volcanic characteristics of the surface of Venus are poorly known, but these characteristics must be closely related to the mechanism by which Venus rids itself of internal heat. On the other solid planets and satellites of the solar system, lithospheric heat transport is dominated by one of three mechanisms: (1) plate recycling, (2) lithospheric conduction, and (3) hot spot volcanism. The authors evaluate each mechanism as a candidate for the dominant mode of lithospheric heat transfer on Venus, and they explore the implications of each mechanism for the interpretation of Venus surface features.

093.107 **Comparison of gas dynamic model with steady solar wind flow around Venus.**
J. D. Mihalov, J. R. Spreiter, S. S. Stahara.
J. Geophys. Res., Vol. 87, 10363 - 10371 (1982).

A gas dynamic model for solar wind flow around Venus has been compared with Pioneer Venus orbiter plasma analyzer measurements from times when the solar wind flow seemed steadiest. The comparisons were made near the terminator. When the observed and model bow shock locations are matched, the model agrees fairly well with the observed parameters: the components of the flow velocity and magnetic field, and the proton number density and isotropic temperature. However, the Mach numbers required to fit the observed bow shock locations are less than 2/3 those estimated from the measured parameters of the free-stream flow, because the bow shock locations (near the terminator) are farther from Venus than expected. Sometimes the measured flow speeds appear to be retarded near the ionopause.

093.108 **Variations in ion and neutral composition at Venus: evidence of solar control of the formation of the predawn bulges in H^+ and He.**
H. A. Taylor, Jr., H. Mayr, H. Brinton, H. Niemann, R. Hartle, R. E. Daniell, Jr.
Icarus, Vol. 52, (see 012.048), 211 - 220 (1982).

A comparison of ion and neutral composition measurements at Venus for periods of greatly different solar activity provides qualitative evidence of solar control of the day-to-night transport of light ion and neutral species. Concentrations of H^+ and He in the predawn bulge near solar maximum in November, 1979, exhibit a depletion signature correlated with a pronounced modulation in the solar $F_{10.7}$ and EUV fluxes. This perturbation is interpreted as resulting from pronounced changes in solar heating and photoionization on the dayside, which in turn modulate the transport of ions and neutrals into the bulge region.

093.109 **Altitude profile of H in the atmosphere of Venus from Lyman α observations of Venera 11 and Venera 12 and origin of the hot exospheric component.**
J. L. Bertaux, V. M. Lepine, V. G. Kurt, A. S. Smirnov.
Icarus, Vol. 52, (see 012.048), 221 - 244 (1982).

Two extreme ultraviolet (EUV) spectrophotometers flown in December 1978 on Venera 11 and Venera 12 measured the hydrogen Lyman α emission resonantly scattered in the atmosphere of Venus. Measurements were obtained across the dayside of the disk, and in the exosphere up to 50,000 km. They were analyzed with spherically symmetric models for which the radiative transfer equation was solved. The H content of the Venus atmosphere varies from optically thin to moderately thick regions. A shape fit at the bright limb allowed to determine the exospheric temperature T_c and the number density n_c.

093.110 **Net global thermal emission from the Venusian atmosphere.**
J. T. Schofield, F. W. Taylor.
Icarus, Vol. 52, (see 012.048), 245 - 262 (1982).

Improved calculations of net emission from the northern hemisphere of Venus are presented. These are based on temperature profiles, water vapor mixing ratio profiles, and cloud models retrieved in 120 solar-fixed latitude-longitude bins from infrared measurements in six spectral channels made over a period of 72 days by the orbiter infrared radiometer instrument of the Pioneer Venus mission. The sensitivity of the calculations to extreme alternative cloud models, measurement errors, and calibration errors is also discussed. Mean net emission from the northern hemisphere corresponds to an equivalent temperature of 229.4 ± 2.5°K. If this figure is characteristic of the whole planet and if thermal balance is assumed, the bolometric albedo of Venus is 0.762 ± 0.011.

093.111 **The global distribution of water vapor in the middle atmosphere of Venus.**
J. T. Schofield, F. W. Taylor, D. J. McCleese.
Icarus, Vol. 52, (see 012.048), 263 - 278 (1982).

The mean, solar-fixed horizontal and vertical distribution of water vapor in and above the Venusian cloud layer is presented. This is derived from measurements made by the Orbiter Infrared Radiometer instrument of the Pioneer Venus mission. The sensitivity of the results to measurement errors and cloud microphysical properties is also considered. Mean water vapor column abundances above cloud unit optical depth at 11.5 μm are found to be greatest at equatorial latitudes in the early afternoon, reaching 50 ± 20 precipitable microns (100 ppm), and fall to less than 3 ± 2 precipitable microns (6 ppm) on the nightside of the planet. On the nightside mixing ratios fall monotonically with altitude, whereas dayside mixing ratios frequently increase with altitude near cloud unit optical depth.

093.112 **Reassessment of net radiation measurements in the atmosphere of Venus.**
H. E. Revercomb, L. A. Sromovsky, V. E. Suomi.
Icarus, Vol. 52, (see 012.048), 279 - 300 (1982).

Net radiative flux measurements by instruments on the Pioneer Venus Day, North, and Night probes are too large below 30 km to be consistent with present estimates of atmospheric opacity. The authors evaluate the only known mechanisms which could potentially have caused significant errors in the deep atmosphere, namely, (1) radiation field perturbations behind each probe due to its thermal wake (2) cloud particle deposition on the sensor windows, and (3) thermal perturbations within the radiation sensor produced by gas flow through the sensor window retainers. They find that the third mechanism is the most likely source of the large net flux measurements.

093.113 **Comparison of ground-based and spacecraft observations of the infrared emission from Venus: the nature of thermal contrasts.**
D. J. Diner, J. Apt, L. S. Elson.
Icarus, Vol. 52, (see 012.048), 301 - 319 (1982).

Comparisons are made between observations of spatial variations in the thermal emission from Venus obtained with ground-based telescopes and those from spacecraft. In particular, the authors concentrate on measurements of solar-related structure at low and mid-latitudes, limb-darkening, and on the high-contrast polar structure.

093.114 **Thermal structure of the atmosphere of Venus from Pioneer Venus radio occultations.**
A. J. Kliore, I. R. Patel.
Icarus, Vol. 52, (see 012.048), 320 - 334 (1982).

Eighty-seven measurements of the thermal structure in the atmosphere of Venus between the altitudes of about 40 and 85 km were derived from Pioneer Venus Orbiter radio occultation data taken during four occultation seasons from December 1978 to October 1981. These measurements cover latitudes from −68 to 88° and solar zenith angles of 8 to 166°. The results indicate that the characteristics of the thermal structure in both the troposphere and stratosphere regions are dependent predominantly on the latitude and only weakly on solar illumination conditions.

093.115 **Small-scale turbulence in the atmosphere of Venus.**
R. Woo, J. W. Armstrong, A. J. Kliore.
Icarus, Vol. 52, (see 012.048), 335 - 345 (1982).

The purpose of this paper is to present the results of further analysis of the Pioneer Venus radio scintillation observations. First, the morphology of the upper region of scintillations is studied and it is seen that the poleward increase in scintillations represents a poleward increase in the intensity of small-scale temperature fluctuations. Second, the temperature fluctuations in the lower region of scintillations near 45 km are examined and found to be considerably weaker than those at 60 km. It appears that the temperature fluctuations are produced by wind-shear-generated turbulence. Estimates of turbulence and vertical transport via small-scale eddy diffusion are obtained.

093.116 **A one-dimensional model of the atmospheric electric field near the Venusian surface.**
I. Tzur, Z. Levin.
Icarus, Vol. 52, (see 012.048), 346 - 353 (1982).

A numerical model is used to simulate the buildup of an electric field from below the Venusian cloud layer to the surface. The steady-state profiles of the ion concentration, net space charge, diffusion and conduction current, and electric field are calculated. Two electric field sources are considered. The first is that produced by the higher diffusivity of positive ions relative to negative ions. The second is assumed to be charge separation inside Venusian clouds.

093.117 **Comparison of Venusian lightning observations.**
W. J. Borucki.
Icaurs, Vol. 52, (see 012.048), 354 - 364 (1982).

A review of spacecraft observations of lightning on Venus is presented. These observations imply that lightning is occurring on both the day- and nightsides of Venus.

093.118 **Magma vesiculation and pyroclastic volcanism on Venus.**
J. B. Garvin, J. W. Head , L. Wilson.
Icarus, Vol. 52, (see 012.048), 365 - 372 (1982).

Theoretical consideration of the magma vesiculation process under observed and inferred venusian surface conditions suggests that vesicles should form in basaltic melts, especially if CO_2 is the primary magmatic volatile. However, the high surface atmospheric pressure (~90 bars) and density on Venus retard bubble coalescence and disruption sufficiently to make explosive volcanism unlikely. The products of explosive volcanism should be rare on Venus, and effusive eruptions should dominate. The volume fraction of vesicles in basaltic rocks on Venus are predicted to be less than in chemically similar rocks on Earth. Detection of pyroclastic landforms or eruption products on Venus would indicate either abnormally high volatile contents of Venus magmas or different environmental conditions in previous geologic history.

093.119 **X-ray fluorescence spectrometry on the surface of Venus.**
Yu. A. Surkov, O. P. Shcheglov, L. P. Moskalyeva *(Moskaleva)*, V. S. Kirichenko, A. D. Dudin, V. L. Gimadov, S. S. Kurochkin, V. N. Rasputny *(Rasputnyj)*.
Anal. Chem., Vol. 54, 957A - 966 (1982). – Abstr. in Phys. Abstr., Vol. 85, Abstr. 117619 (1982).

093.120 **Venus.**
B. M. Cordell.
Astronomy, Vol. 10, No. 9, p. 6 - 15, 18 - 22 (1982). – Abstr. in Phys. Abstr., Vol. 85, Abstr. 117620 (1982).

093.121 **In-situ measurements in the ionosphere of Venus.**
K. Rawer, K. Spenner, W. C. Knudsen.
Z. Flugwiss. Weltraumforsch., Vol. 6, 147 - 151 (1982). In German. – Abstr. in Phys. Abstr., Vol. 86, Abstr. 5184 (1983).

Venus, near neighbor of the sun.
See Abstr. 003.019.

The planet Venus.
See Abstr. 003.063.

Infrared absorption by OH-containing compounds.
See Abstr. 022.022.

Direct computational of atmospheric normal modes and instabilities.
See Abstr. 022.100.

An interpretation of the near-ultraviolet absorption spectrum of SO_2: implications for Venus, Io, and laboratory measurements.
See Abstr. 022.154.

Spectroscopic measurements of Mars and Venus with heterodyne techniques.
See Abstr. 031.519.

Mission concepts for Venus surface investigation.
See Abstr. 051.022.

Space missions to Venus.
See Abstr. 051.054.

The Venus Orbital Imaging Radar (VOIR) mission.
See Abstr. 051.055.

Venera 13 and Venera 14.
See Abstr. 053.001.

Russische sondes op Venus.
See Abstr. 053.002.

Observations of mass addition to the shocked solar wind of the Venusian ionosheath.
See Abstr. 074.025.

Catalytic processes in the atmospheres of Earth and Venus.
See Abstr. 082.047.

Results of scientific-research works of the TsNIIGAiK in the field of mapping the moon and planets.
See Abstr. 094.017.

Transient species on Mars and Venus.
See Abstr. 097.015.

Atomic carbon in the atmospheres of Mars and Venus.
See Abstr. 097.018.

Solar excitation of the ν_3 state of CO_2 in the mesospheres of Mars and Venus: an NLTE model.
See Abstr. 097.046.

Solar wind interaction with comets: lessons from Venus.
See Abstr. 102.030.

094 Moon (Dynamics, General Aspects, Local Properties)

Moon (Dynamics, General Aspects)

094.001 **Radiative transfer in the surfaces of atmosphereless bodies. III. Interpretation of lunar photometry.**
K. Lumme, W. M. Irvine.
Astron. J., Vol. 87, 1076 - 1082 (1982) = Contrib. No. 473, Five College Astron. Dep.

Narrowband and *UBV* photoelectric phase curves of the entire lunar disk and surface photometry of some craters have been interpreted using a newly developed generalized radiative transfer theory for planetary regoliths (Lumme and Bowell, 1981). The data are well fitted by the theory, yielding information on both macroscopic and microscopic lunar properties. Derived values for the integrated disk geometric albedo, $0.097 \leqslant p \leqslant 0.27$ for $0.36 \leqslant \lambda \leqslant 1.06$ μm, are considerably higher than quoted previously, because of the present inclusion of an accurately determined opposition effect. The mean surface roughness – defined as the ratio of the height to the radius of a typical irregularity – is found to be 0.9 ± 0.1.

094.002 **Nature and origin of basin-forming projectiles.**
G. W. Wetherill.
Multi-ring basins, (see 012.001), p. 1 - 18 (1981).

The formation of the observed lunar multi-ring basins is discussed in the context of current theories of terrestrial planet formation, particularly those in which these planets formed by the accumulation of large planetesimals. The observed number, size, and timing of lunar basin-forming impacts is in the range expected for such theories.

094.003 **On the origin of the planetesimals that produced the multi-ring basins.**
R. B. Baldwin.
Multi-ring basins, (see 012.001), p. 19 - 28 (1981).

It is concluded that the multi-ring basins of the moon came from impacts of the largest bodies in a family of bodies of all sizes in eccentric solar orbits. Such bodies abounded in the early solar system out to at least Saturn.

094.004 **On the tsunami theory of the origin of multi-ring basins.**
R. B. Baldwin.
Multi-ring basins, (see 012.001), p. 275 - 288 (1981).

The theory that best seems to fit the observations of the multi-ring basins on the moon is that these structures were formed by impacts of giant planetesimals and that the energies released fluidized the area in some fashion so that tsunami-like waves could radiate from ground zero and after the energy density from the impact had decayed sufficiently with distance from the impact point so that the slow moving tsunami could no longer distort the surface, the wave froze into position as a ripple in the moon's crust. It is concluded that the tsunami model of the origin of ring structures at the largest pre-mare basins on the moon has much to recommend it.

094.005 **On the use of the moon for the calibration of space-borne radiometers.**
R. Kolk.
Publ. Tartu Astrofiz. Obs., Tom 48, 262 - 270 (1981).
In Russian.

Calibration of space-borne radiometers by means of the moon has been discussed. Formulae for the calculation of the spectral radiance of the moon at various phase angles are presented. Ground-based measurements of the moon's spectral radiance have been carried out with a radiometer at wavelengths of 1.35, 1.9 and 2.2 μm.

094.006 **Development of a selenocentric system of coordinates according to Zond 8 photographs.**

V. I. Chikmachev.
Tr. Gos. Astron. Inst. Shternberga, Tom 51, 208 - 215 (1982). In Russian.

This paper presents method and results of the selenodetic reduction of original Zond 8 photographs for a phototriangulation network of the lunar far side in the Mare Orientale region. 105 positions of craters with diameter from 1 to 5 km and photomaps of the area Mare Orientale are given. Analogy with other lunar areas is made.

094.007 **Nickel for your thoughts: Urey and the origin of the moon.**
S. G. Brush.
Science, Vol. 217, 891 - 898 (1982).

The theories of Harold C. Urey (1893 - 1981) on the origin of the moon are discussed in relation to earlier ideas, especially Georg Howard Darwin's fission hypothesis. Urey's espousal of the idea that the moon had been captured by the earth and has preserved information about the earliest history of the solar system led him to advocate a manned lunar landing. Results from the Apollo missions, in particular the deficiency of siderophile elements in the lunar crust, led him to abandon the capture selenogony and tentatively adopt the fission hypothesis.

094.008 **Planetological consequences of accretional lunar surface melting.**
D. Möhlmann.
Gerlands Beitr. Geophys., Band 91, 269 - 270 (1982).

094.009 **A new approach to lunar origin.**
D. Möhlmann.
Gerlands Beitr. Geophys., Vol. 91, 361 - 362 (1982).

094.010 **Calculation of the second derivative of the lunar elongation and statistical regularities in distribution of some astronomical data.**
A. T. Fomenko.
Issled. operatsii i ASU, Kiev, 1982, No. 20, p. 98 - 113. In Russian. – Abstr. in Ref. zh., 51. Astron., 8.51.83 (1982).

094.011 **Die Geologie des Mondes.**
W. v. Engelhardt.
Naturwissenschaften, 69. Jahrg., 405 - 419 (1982).

094.012 **Deflections of the vertical on the farside of the Moon.**
M. Burša, Z. Šíma.
Bull. Astron. Inst. Czechoslovakia, Vol. 33, 233 - 243 (1982).

The deflections of the vertical on the nearside and farside of the Moon has been computed and the mass interpretation of fifty anomalous formations carried out. The average depth of the anomalous mass centres is 362 km on the nearside and 379 km on the farside of the Moon. There is nonsignificant difference in the anomalous structure of the gravity field on the farside and nearside of the Moon if Ferrari's set of Stokes constants (1977) is used.

094.013 **Construction of thematic lunar maps of various scales from Zond 6 photographs.**
V. D. Bol'shakov, N. P. Lavrova, B. V. Krasnopevtseva, N. I. Konstantinova, V. V. Usova.
Sovr. napravleniya razvitiya geod. i kartogr. Mater. Vses. nauchn.-tekh. konf. i torzhestv. sobraniya, posvyashch. 60-letiyu Lenin. dekreta "Ob uchrezhdenii Vysshego geod. upr., 14 - 15 marta, 1979." Moskva, 1982, p. 115 - 117. In Russian. – Abstr. in Ref. zh., 51. Astron., 9.51.189; 52. Geod. Aehrosemka, 9.52.164 (1982).

094.014 **Colour differences on the moon and chemical element content in lunar soils.**
Yu. G. Shkuratov.
Astron. vestn., Tom 16, 69 - 76 (1982). In Russian. – Abstr. in Ref. zh., 51. Astron., 9.51.190 (1982).

094.015 **Colorimetric mapping of the lunar surface in the 0.62 - 0.95 μm spectral region.**
N. N. Evsyukov, D. I. Shestopalov.
Astron. vestn., Tom 16, 77 - 86 (1982). In Russian. – Abstr. in Ref. zh., 51. Astron., 9.51.191 (1982).

094.016 **Covariance analysis of the correlation of the relief and acceleration of gravity of the moon.**
A. N. Sanovich, Kh. G. Tadzhidinov.
Astron. vestn., Tom 16, 87 - 92 (1982). In Russian. – Abstr. in Ref. zh., 51. Astron., 9.51.192 (1982).

094.017 **Results of scientific-research works of the TsNIIGAiK in the field of mapping the moon and planets.**
Yu. S. Tyuflin.
Sovremen. napravleniya razvitiya geod. i kartogr. Mater. Vses. nauchn.-tekh. konf. i torzhestv. sobraniya, posvyashch. 60-letiyu Lenin. dekreta Ob uchrezhdenii Vyssh. geod. upr., 14 - 15 marta, 1979. Moskva, 1982, p. 112 - 115. In Russian. Abstr. in Ref. zh., 52. Geod. Aehrosemka, 9.52.165 (1982).

094.018 **Avstandsbestemmelse til månen ved hjelp av laser.**
B. R. Pettersen.
Astron. Tidsskr., Årg. 15, 97 - 99 (1982).

094.019 **Seismic *Q* of the lunar upper mantle.**
Y. Nakamura, J. Koyama.
J. Geophys. Res., Vol. 87, 4855 - 4861 (1982).

The authors have determined the frequency dependence of anelastic attenuation for both *P* and *S* waves in the upper mantle of the moon in the frequency range of roughly 3 to 8 Hz.

094.020 **The deep lunar electrical conductivity profile: structural and thermal inferences.**
L. L. Hood, F. Herbert, C. P. Sonett.
J. Geophys. Res., Vol. 87, 5311 - 5326 (1982).

Simultaneous Apollo 12 surface and Explorer 35 orbital magnetometer records have been reexamined for intervals suitable for measuring the lunar inductive response in the solar wind and terrestrial magnetosheath.

094.021 **Comment on 'Lunar seismology: the internal structure of the moon' by N. R. Goins, A. M. Dainty, and M. N. Toksöz.**
J. R. Cleary, with a reply by N. R. Goins, A. M. Dainty, M. N. Toksöz.
J. Geophys. Res., Vol. 87, 5495 - 5499 (1982). – See Abstr. 30.094.001.

094.022 **Transport processes on the lunar surface: comparison of model calculations with radionuclides data.**
Y. Langevin, J. R. Arnold, K. Nishiizumi.
J. Geophys. Res., Vol. 87, 6681 - 6691 (1982).

The most useful radionuclides for deciphering the recent history of the lunar surface are ^{53}Mn and ^{26}Al. On the time scales of ~0.1 to 10 m.y. on which these species give information, the lunar surface seems to be disturbed to depths of ~1 to 50 g/cm^2. The most likely agent of such disturbances is meteoritic mixing. Since both large-scale and small-scale events are important, the use of Monte Carlo methods was indicated. Experimental evidence is now available for seven lunar cores, representing a significant statistical sample. The authors consider the evidence for disturbance recorded in these cores on the time scales of ^{26}Al and ^{53}Mn and match it against the predictions of a meteoritic mixing model. They show that if secondary processes of meteoritic impacts such as the downslope movement of ejecta are taken into account, such a model is consistent with the experimental results, which, however, exhibit a tendency for deeper depth disturbances.

094.023 **Calculation of the electric field intensity above the hydro-generated lunar surface.**
E. K. Kolesnikov, A. S. Manujlov.
Astron. Zh., Tom 59, 996 - 998 (1982). In Russian. English translation in Soviet Astron., Vol. 26, No. 5.
The velocity distribution function of lunar photoelectrons is evaluated on the basis of the measured data on the short-wave solar radiation for the period of mean activity. The electrostatic field as a function of the height above the lunar surface is obtained analytically.

094.024 **A critical analysis of the results of treatment of heliometric observations of the moon.**
I. V. Gavrilov, A. S. Duma.
Astron. Zh., Tom 59, 999 - 1005 (1982). In Russian. English translation in Soviet Astron., Vol. 26, No. 5.
The disagreement in the results obtained from various heliometric observations of the moon is explained by correlations between the unknown parameters and differences of reductions. The new solution of the normal equations leads to the noncorrelated selenographic positions of the crater Mösting A $\lambda = -5°11'05'' \pm 1''$, $\beta = -3°11'52'' \pm 4''$ and the inclination of lunar equator to the ecliptic $I = 1°32'28'' \pm 5''$.

094.025 **High resolution radar maps of the moon at 70 cm wavelength.**
T. W. Thompson.
Bull. American Astron. Soc., Vol. 14, 754 (1982). – Abstract.

094.026 **Radar scattering laws for the lunar surface.**
R. A. Simpson, G. L. Tyler.
IEEE Trans. Antennas Propag., Vol. AP-30, 438 - 449 (1982). Abstr. in Phys. Abstr., Vol. 85, Abstr. 94631 (1982).

094.027 **Factors influencing the morphology of volcanic landforms: an Earth-Moon comparison.**
J. L. Whitford-Stark.
Earth-Sci. Rev., Vol. 18, 109 - 168 (1982). – Abstr. in Phys. Abstr., Vol. 85, Abstr. 100778 (1982).

094.028 **Relativistic perturbations of the Moon in ELP 2000.**
J. F. Lestrade, M. Chapront-Touzé.
Astron. Astrophys., Vol. 116, 75 - 79 (1982).
The differential acceleration between the Earth and the Moon in General Relavity is derived from Estabrook's formula applied to the case of the Sun, Earth and Moon considered as point-masses. The method to integrate the equations of motion is then described, the complete problem being considered as a perturbation to the Newtonian main problem. The resulting relativistic perturbations in the geocentric motion of the Moon are given as semi-analytical series with an accuracy of 10^{-6} arc sec (or 1 mm). They are compared to the results obtained by Brumberg and Ivanova (1981) with a very different method of integration. The comparison shows a good agreement.

094.029 **The determination of the derivatives in Brown's lunar theory.**
K. R. Meyer, D. S. Schmidt.
Celestial Mech., Vol. 28, (see 012.049), 201 - 207 (1982).
The fundamental matrix solution T for the variational equations of a Hamiltonian system is symplectic. The authors use this fact to complete T when it is only partially known. They discuss three cases. The last one gives an easy proof for the method invented by Brown in his lunar theory.

094.030 **The moon: its structure, origin, and development.**
A. B. Binder.
The solar system and its exploration, (see 012.054), p. 175 - 178 (1981).
Analysis of the large volume of data and the large number of surface samples provided mainly by the Apollo program has resulted in a rather complete understanding of the general characteristics and history of the moon.

094.031 **Could the earth's core and moon have formed at the same time?**
D. J. Andrews.
Geophys. Res. Lett., Vol. 9, 1259 - 1262 (1982).
During accretion of the earth, a layer of liquid iron may have formed above a cold undifferentiated nucleus. Taylor instability of this configuration leads to a deformation with spherical-harmonic order $l = 1$, a lower symmetry than assumed heretofore in fission analyses. When this deformation exceeds a few tens of kilometers, strength of the materials becomes insignificant, and the deformation may grow catastrophically, the cold nucleus being displaced laterally at a velocity limited only by inertia. The asymmetric distribution of angular velocity leads to large dynamic pressure in regions of velocity convergence, so that the excess angular momentum of the iron layer may be transferred to a jet of mantle material. Sufficient energy and angular momentum might have been released by core formation to eject material to form the moon.

094.032 **Newtonsche Bewegungsgleichungen des Mondes.**
K.-P. Stoffels.
Mitt. Geod. Inst. Rheinischen Friedrich-Wilhelms-Univ. Bonn, Nr. 62, p. 1 - 25 (1980).
In accordance with Brown's Lunar Theory equations of motion are stated in consideration of gravitational effects due to the earth, the sun and the planets. Different from Brown's theory a method is applied concerning Newton's equations of motion with use of the geocentric (cartesian) position vector of the moons centre of mass. This derivation variant is more direct and more clear in opposition to the introduction of Lagrange's equations of motion.

094.033 **The threefold synchronization of the rotation of the moon.**
K. S. Shakirov.
Astron. Tsirk., No. 1173, p. 7 - 8 (1981). In Russian.

094.034 **Ephemerides of the triangular libration points of the earth-moon system for the winter 1981 - 1982.**
A. V. Zakharov, L. G. Luk'yanov.
Astron. Tsirk., No. 1179, p. 4 - 6 (1981). In Russian.

094.035 **On the antipodes on the moon.**
Zh. F. Rodionova, V. G. Surdin.
Astron. Tsirk., No. 1180, p. 4 - 6 (1981). In Russian.

094.036 **Stationary oscillations of the viscous moon.**
A. S. Dubrovskij, Yu. A. Chikanov.
Nelinejn. kolebaniya i teor. upr., Izhevsk, 1981, No. 3, p. 154 - 160. In Russian. – Abstr. in Ref. zh., 52. Geod. Aehrosemka, 11.52.212 (1982).

094.037 **Application of the method of collocation when constructing selenodetic reference nets.**
S. G. Valeev, N. A. Tkachenko, M. G. Shamarin.
Mol. uchenye i spets. Kuzbassa v 10 pyatiletke. Kemerovo, 1981, p. 20 - 23. In Russian. – Abstr. in Ref. zh., 52. Geod. Aehrosemka, 11.52.220 (1982).

094.038 **On the history of the lunar orbit.**
B. A. Conway.
Icarus, Vol. 51, 610 - 622 (1982).
A frequency-dependent model of tidal friction is used in the determintion of the time rate of change of the lunar orbital elements and the angular velocity of the earth. The

variational equations consider eccentricity, the solar tide on the earth, earth oblateness, and higher-order terms in the earth's tidal potential. Results for the history of the earth–moon system over geological time are presented and are identified as consistent with origin of the moon by capture.

094.039 **Thermoluminescence of the lunar surface.**
M.-L. Chanin, V. Lepine, J. E. Blamont.
Moon Planets, Vol. 27, 143 - 163 (1982).
Observations of the lunar luminescence are reported for a dozen of specific Moon features using the 'line-depth' method with a high resolution spectroscopic technique. The data indicate a variation of the Moon proper emission as a function of the phase angle which is interpreted as a proof of the thermoluminescent origin of this emission.

094.040 **Primeval displacements of the lunar pole.**
S. K. Runcorn.
Phys. Earth Planet. Inter., Vol. 29, 135 - 147 (1982).
The hypothesis that the magnetic field which magnetized the lunar crust was generated by the dynamo process in a small fluid iron core can now be tested. The palaeoequators corresponding to ages 4 Ga and 3.85 Ga show close relationships with the circular maria or mascons on the near side and multi-ring basins of corresponding age on the far side. The polar displacements indicated from lunar palaeomagnetism can be explained by the changes in the moment of inertia tensor consequent on the excavation and later flooding of these circular mare. Small moons in the primeval Earth–Moon system are inferred to be the impacting bodies.

094.041 **Lunar and terrestrial crusts: a contrast in origin and evolution.**
S. R. Taylor.
Phys. Earth Planet. Inter., Vol. 29, 233 - 241 (1982).

094.042 **Transient lunar phenomena and triboelectricity.**
D. K. Davies, S. P. Kendall.
J. Electrostat., Vol. 13, No. 1, p. 81 - 84, (1982). – Abstr. in Phys. Abstr., Vol. 85, Abstr. 117617 (1982).

094.043 **The Moon after Apollo.**
H. Masursky.
Endeavour New Ser., Vol. 6, No. 2, p. 48 - 58 (1982). Abstr. in Phys. Abstr., Vol. 86, Abstr. 10233 (1983).

094.044 **The Moon's deceptive tranquillity.**
K. Runcorn.
New Scientist, Vol. 96, 174 - 180 (1982). – Abstr. in Phys. Abstr., Vol. 86, Abstr. 10235 (1983).

094.045 **The stability of eccentric polar orbits around the Moon.**
G. Janin.
Acta Astronaut., Vol. 9, (see 012.085), 421 - 423 (1982). Abstr. in Phys. Abstr., Vol. 86, Abstr. 20265 (1983).

094.046 **A review of lava flow processes related to the formation of lunar sinuous rilles.**
G. Hulme.
Geophys. Surv., Vol. 5, 245 - 279 (1982). – Abstr. in Phys. Abstr., Vol. 86, Abstr. 20266 (1983).

On the preparation of a systematic index of scientific and technical information concerning the exploration of the moon.
See Abstr. 002.069.

Bibliography of the moon and the planets.
See Abstr. 002.074.

The basic problems of the Babylonian lunar theory.
See Abstr. 004.079.

A proposed photometric program for A. L. P. O. lunar observers.
See Abstr. 013.044.

Ein FORTRAN-Programm zur Berechnung translatorischer Mondephemeriden nach der Theorie von E. W. Brown.
See Abstr. 021.047.

On the relationships between impact crater diameters and projectile kinetic energy.
See Abstr. 022.065.

On the possibility of application of a photoelectric zenith tube placed on the moon or planets for investigating their nutations.
See Abstr. 031.530.

Experience of construction of a lunar colour map by digital image processing methods.
See Abstr. 031.569.

Über Anwendung von Ergebnissen der Laserentfernungsmessungen zum Monde bei Auflösung einiger Probleme der Astronomie und Geodynamik.
See Abstr. 031.613.

Floquet reference solutions for the lunar theory and the Jovian moons.
See Abstr. 042.031.

The Fourier-Chebychev approximation for time series with a great many terms.
See Abstr. 042.071.

Die Arbeiten des Sonderforschungsbereiches 78 Satellitengeodäsie der Technischen Universität München im Jahre 1981.
See Abstr. 046.010.

A model for the negative polarization of light scattered by cosmic bodies deprived of atmospheres.
See Abstr. 063.003.

Reference systems of heights for planets and satellites.
See Abstr. 091.010.

Planetary and lunar magnetism.
See Abstr. 091.041.

The intercrater plains of Mercury and the Moon: their nature, origin, and role in terrestrial planet evolution.
See Abstr. 092.007.

Lunar anorthosite 60025, the petrogenesis of lunar anorthosites, and the composition of the Moon.
See Abstr. 094.508.

Study of peculiarities of crater distribution on Mars, Mercury and the moon from maps of crater density constructed by a computer.
See Abstr. 097.004.

Lunarlike thermal emission and the standard asteroid radiometric model.
See Abstr. 098.033.

Heat transfer and the development of planetary structure.
See Abstr. 107.035.

The origin of the Moon: a reappraisal.
See Abstr. 107.051.

Moon (Local Properties)

094.501 **Discovery of multi-ring basins: gestalt perception in planetary science.**
W. K. Hartmann.
Multi-ring basins, (see 012.001), p. 79 - 90 (1981).
Early selenographers resolved individual structural components of multi-ring basin systems but missed the underlying large-scale multi-ring basin patterns. The recognition of multi-ring basins as a general class of planetary features can be divided into five steps. This sequence is an example of gestalt recognition whose implications for scientific research are discussed.

094.502 **The origin of lunar mascons: analysis of the Bouguer gravity associated with Grimaldi.**
R. J. Phillips, J. Dvorak.
Multi-ring basins, (see 012.001), p. 91 - 104 (1981).
The authors present a brief description of the geologic setting of Grimaldi, a summary of the available lunar gravity and topographic data for Grimaldi, and a comparison of the resultant Bouguer gravity profiles with several model calculations proposed for Grimaldi which are based on inferences made about the possible structure of this lunar basin from the surface geology. They discuss these results in view of previous analyses of the Bouguer gravity associated with large lunar craters.

094.503 **The lunar Procellarum basin.**
E. A. Whitaker.
Multi-ring basins, (see 012.001), p. 105 - 111 (1981).
The existence of the Procellarum basin can be inferred from the presence of portions of three distinct, concentric topographic rings visible as mare shorelines, mare ridges and isolated scarps. Further evidence of its reality is provided by the correspondence between its topographical outline and both the surface radioactivity and the potassium content. The basin's weak topographic expression suggests that it was formed at a time when the lunar lithosphere was fairly thin compared with the depth of the transient crater permitting isostatic adjustment through asthenospheric movements. The unique area of magnetic fields and other anomalies northwest of Van de Graaff may be an indirect result of the Procellarum impact.

094.504 **Modification of multi-ring basins – the Imbrium model.**
J. L. Whitford-Stark.
Multi-ring basins, (see 012.001), p. 113 - 124 (1981).
The paper describes the main features of the Imbrium multi-ring basin and attempts to show that the irregularities in the ring structure have arisen, in part, as a result of a variable lithosphere thickness.

094.505 **Chemical rings of lunar basins from orbital X-ray data.**
C. G. Andre.
Multi-ring basins, (see 012.001), p. 125 - 132 (1981).

094.506 **Apollo 17 impact melts and their relation to the Serenitatis basin.**
P. D. Spudis, G. Ryder.
Multi-ring basins, (see 012.001), p. 133 - 148 (1981).
Geologic, chemical, and petrographic data suggest that the Apollo 17 highlands have undergone a complex, multiple impact history. In contrast to prevailing opinion, the authors suggest that the Apollo 17 highland melt breccias are not all derived from the Serenitatis basin impact.

094.507 **Petrochemical constraints on lateral transport during lunar basin formation.**
P. H. Warren, G. J. Taylor.
Multi-ring basins, (see 012.001), p. 149 - 154 (1981).

094.508 **Lunar anorthosite 60025, the petrogenesis of lunar anorthosites, and the composition of the Moon.**
G. Ryder.
Geochim. Cosmochim. Acta, Vol. 46, 1591 - 1601 (1982).
The author made a new, detailed study of sample 60025, in particular its mineralogy and mineral chemistry, and established that there are systematic variations and relationships among its minerals. Literature data on petrography and chemistry demonstrates that the same features exist for anorthosites as a whole. The data strongly suggest that the anorthosites are samples of a suite of related near-perfect adcumulate rocks, i.e., all crystals in a given unmixed sample crystallized from a liquid of constant composition and temperature and are unzoned; little or no liquid was trapped. The data allow strong inferences to be made about the parental liquid and its evolution, and about the bulk composition of the Moon and its evolution.

094.509 **Methods and results of photometric investigations in the transition zone "mare-continent" from Lunokhod 2.**
V. V. Shevchenko.
Tr. Gos. Astron. Inst. Shternberga, Tom 51, 195 - 207 (1982). In Russian.
The principles of measurements of lunar features brightness with the help of Lunokhod 2 board telephotometers are stated. The results of the complex treatment of the measurements and the obtained characteristics of the transitional zone mare-continent are given.

094.510 **Records of ancient cosmic radiation in extraterrestrial rocks.**
N. Bhandari.
Proc. Indian Acad. Sci. Earth Planet Sci., Vol. 90, 359 - 382 (1981).
Recent results on cosmic ray interactions in lunar samples and meteorites resulting in production of stable and radionuclides, particle tracks and thermoluminescence are reviewed.

094.511 **New concepts for fossil track studies in lunar silicates.**
J. P. Duraud, Y. Langevin, M. Maurette.
Nucl. Tracks Methods Instrum. Appl., Vol. 5, (see 012.015), 341 - 342 (1981). – Summary. Abstr. in Phys. Abstr., Vol. 85, Abstr. 63283 (1982).

094.512 **A preliminary analysis of lunar extra-mare basalts: distribution, compositions, ages, volumes, and eruption styles.**
J. L. Whitford-Stark.

Moon Planets, Vol. 26, 323 - 338 (1982).

Extra-mare basalts occupy 8.5% of the lunar basalt area and comprise 1% of the total mare basalt volume. They are preferentially located where the crust is thin and topographically low. In terms of age, eruption style, and composition they are as variable as the mare basalts. In some instances extrusion in extra-mare craters was preceded by floor-fracturing whereas in other cases it apparently was not. The volume of lava erupted may have been controlled more by the volume of magma produced than by hydrostatic effects. A minimum of nearly 1300 separate basalt eruptions is indicated; the true value could be nearer 30000 separate eruptions.

094.513 **Spectral studies of lunar dark-halo craters.**
J. F. Bell, B. R. Hawke.
Bull. American Astron. Soc., Vol. 14, 754 (1982). – Abstract.

094.514 **An unusual lunar composition at the Aristarchus Plateau.**
P. G. Lucey, B. R. Hawke, T. B. McCord, C. M. Pieters, J. W. Head.
Bull. American Astron. Soc., Vol. 14, 756 - 757 (1982). Abstract.

094.515 **Analysis of crater distributions in mare units on the lunar far side.**
A. S. Walker, F. El-Baz.
Moon Planets, Vol. 27, 91 - 106 (1982).

Mare material is asymmetrically distributed on the Moon. The Earth-facing hemisphere, where the crust is believed to be 26 km thinner than on the farside, contains substantially more basaltic mare material. Using Lunar Topographic Orthophoto Maps, the authors calculated the thickness of the mare material in three farside craters, Aitken (0.59 km), Isaev (1.0 km), and Tsiolkovskiy (1.75 km). They also studied crater frequency distribution in five farside mare units (Aitken, Isaev, Lacus Solitudinis, Langemak, and Tsiolkovskiy) and one light plains unit (in Mendeleev). Nearly 10000 farside craters were counted. The results suggest that mare basalt emplacement on the far side ceased before it did on the near side.

094.516 **Morphological characteristics of lunar craters with moderate depth/diameter ratio. II. ($0.12<d/D<0.15$).**
M. Moutsoulas, P. Preka.
Moon Planets, Vol. 27, 111 - 130 (1982).

The morphological characteristics of craters, the depth/diameter ratio of which is between 0.12 and 0.15, are discussed. Many small secondary craters belong to that class – results of low velocity impacts – as well as young craters created by low angle impacts. Revised values for the craters' selenographic coordinates are also presented.

094.517 **A list of spatial coordinates of points on the lunar far side in the region of crater Apollo.**
V. I. Chikmachev.
Astron. Tsirk., No. 1189, p. 1 - 4 (1981). In Russian.

094.518 **Selenodetic coordinates of 117 objects on the near side of the moon.**
S. G. Valeev.
Astron. Tsirk., No. 1203, p. 5 - 7 (1982). In Russian.

094.519 **On chromium and titanium abundance in lunar mare basalts.**
G. G. Kochemasov.
Petrol. i rudonosn. prirod. assotsiatsij gorn. porod. Moskva, 1981, p. 41 - 47. In Russian. – Abstr. in Ref. zh., 62. Issled. kosm. prostranstva, 10.62.375 (1982).

094.520 **Natural radioactivity of lunar and planetary rocks.**
Yu. A. Surkov.
Geokhimiya, 1982, No. 6, p. 755 - 766. In Russian. – Abstr. in Ref. zh., 51. Astron., 10.51.243; 62. Issled. kosm. prostranstva, 11.52.273 (1982).

094.521 **Die Gesteine des Mondes.**
W. v. Engelhardt.
Sterne, 58. Band, 339 - 351 (1982).

094.522 **Reevaluation of lunar and Martian spectra in the mid-IR region.**
J. N. Plendl, H. S. Plendl.
Appl. Opt., Vol. 21, 4404 - 4410 (1982).

A reference point method has been developed to correct infrared spectra from the moon and other celestial objects for selective absorption in the earth's atmosphere. The method is applied to lunar spectra that were obtained 2.3 km above sea level within the two atmospheric IR windows. The results indicate that SiO_2 and Al_2O_3 are major mineral constituents in the four large surface areas analyzed in agreement with the localized probings at spacecraft landing sites. In addition, IR spectra from Martian dust clouds that were observed from the Mariner 9 spacecraft are examined. The principal sources of radiation in this case appear to be Al_2O_3 and sulfur.

094.523 **Ejecta distribution around a small lunar crater.**
S. J. Wetzel.
Advances in planetary geology, (see 003.014), p. 574 - 578 (1982). – Abstract.

094.524 **Ferromagnetic resonance g-factor measurement on LUNA 20 soil.**
Z. Frait.
J. Geophys., Vol. 51, 113 - 118 (1982). – Abstr. in Phys. Abstr., Vol. 86, Abstr. 15860 (1983).

NASA Catalogue of Lunar Nomenclature.
See Abstr. 002.130.

Laboratory simulation of volatilisation from melts induced by micrometeoroid impacts.
See Abstr. 022.064.

Preparation of lunar sample and meteorite mock-ups: or how to make rocks.
See Abstr. 022.122.

Nuclear reactions in Rb, Sr, Y, and Zr targets.
See Abstr. 022.178.

High-precision instrumental neutron activation analysis of meteoritic, lunar, and terrestrial samples.
See Abstr. 022.188.

Colour differences on the moon and chemical element content in lunar soils.
See Abstr. 094.014.

Isotopic composition of rare gases of the Tsarev meteorite and some lunar soil probes.
See Abstr. 105.017.

The isotopic composition of noble gases in the Tsarev meteorite and some probes of lunar soil.
See Abstr. 105.041.

Isotopic studies of noble gases from the Tsarev meteorite and from samples of lunar soil.
See Abstr. 105.081.

X-ray emission from young stars and implications for the early solar system.
See Abstr. 107.011.

095 Lunar Eclipses

095.001 **Observations of a lunar eclipse.** 1982, January 9. Zemlya Vselennaya, 1982, No. 4, p. 74 - 75. In Russian.

095.002 **Photoelectric photometry of the July 6, 1982, total lunar eclipse.**
J. E. Westfall.
Strolling Astron., Vol. 29, 168 - 171 (1982).

095.003 **Weather and occultations during the lunar eclipse of 1982 July 6.**
D. W. Dunham.
Occultation Newsl., Vol. 3, 35 - 36 (1982).

095.004 **Observation of the lunar eclipse of 9 January 1982.**
J. Bouška.
Říše hvězd, Vol. 63, 119 - 124, 141 - 143 (1982). In Czech.

Measurement of a stellar angular diameter by a high-time-resolving photometric observation of a lunar occultation.
See Abstr. 031.552.

Occultations during the lunar eclipse of 1982 December 30.
See Abstr. 096.016.

The system of orbital constants of the Saturn satellites.
See Abstr. 100.016.

096 Lunar and Planetary Occultations

096.001 **Photoelectric lunar occultations. II. Data reduction and analysis.**
G. L. Blow.
South. Stars, Vol. 29, 115 - 129 (1982).
A discussion of the means by which photoelectric data obtained at lunar occultations may be reduced to yield occultation times, and where appropriate, information concerning double stars, lunar limb slopes and stellar angular diameters, is presented.

096.002 **Occultations rasantes en France en 1983.**
J. Meeus.
Astronomie, Vol. 96, 339 - 342 (1982).

096.003 **Observations of stellar occultations by the moon at the Astronomical Observatory of the Kiev University in 1980.**
A. K. Osipov, V. I. Mazur.
Vestn. Kiev. Univ., Astron. Vyp. 24, p. 84 - 90 (1982). In Russian.

096.004 **Planetary and asteroidal occultations in 1982.**
M. Soma.
Astron. Her., Vol. 75, 9 - 12 (1982). In Japanese. – Abstr. in Phys. Abstr., Vol. 85, Abstr. 73823 (1982).

096.005 **Photometric observations of the occultation of BD – 19°4222 by Uranus.**
H. S. Mahra, U. C. Joshi, A. K. Pande, V. Mohan.
Bull. Astron. Soc. India, Vol. 10, 32 (1982). – Abstract.

096.006 **A photographic search for stars to be occulted by the larger asteroids in 1983.**
L. H. Wasserman, R. L. Millis, O. G. Franz, E. Bowell.
Bull. American Astron. Soc., Vol. 14, 729 (1982). – Abstract.

096.007 **Voyager stellar occultation by Ganymede: relevance to surface ice temperature and volatile transport.**
J. R. Spencer.
Bull. American Astron. Soc., Vol. 14, 739 (1982). – Abstract.

096.008 **Photoelectric lunar occultations. III. Some real results.**
G. L. Blow.
South. Stars, Vol. 29, 160 - 168 (1982).

096.009 **Les occultations de Jupiter par la Lune en 1983.**
J. Meeus.
Astronomie, Vol. 96, 553 - 560 (1982).

096.010 **Recent asteroidal occultation attempts, and updates for upcoming events.**
D. W. Dunham.
Occultation Newsl., Vol. 3, 4 - 6 (1982).

096.011 **Lunar occultations of planets.**
Occultation Newsl., Vol. 3, 6 - 8 (1982).

096.012 **Planetary occultations during 1983.**
D. W. Dunham.
Occultation Newsl., Vol. 3, 9 - 22 (1982).

096.013 **Selection of asteroids for occultation searches.**
D. W. Dunham.
Occultation Newsl., Vol. 3, 23 - 25 (1982).

096.014 **Asteroidal occultations of uncatalogued stars during 1983.**
R. L. Millis, O. G. Franz, L. H. Wasserman, E. Bowell, D. W. Dunham.
Occultation Newsl., Vol. 3, 25 - 27 (1982).

096.015 **1978 total occultation tally.**
J. E. Carroll.
Occultation Newsl., Vol. 3, 27 - 29 (1982).

096.016 **Occultations during the lunar eclipse of 1982 December 30.**
D. W. Dunham.
Occultation Newsl., Vol. 3, 29 - 33 (1982).

096.017 **Planetary occultations during 1983.**
D. W. Dunham.
Occultation Newsl., Vol. 3, 37 - 44 (1982).

096.018 **Photoelectric observations of lunar occultations. XIII.**
G. L. Blow, P. C. Chen, D. A. Edwards, D. S. Evans, M. Frueh.
Astron. J., Vol. 87, 1571 - 1584 (1982).

Occultation observations of 319 events occurring between 15 January 1981 and 7 February 1982 made with the 0.76 m (30 in.) reflector at McDonald Observatory are reported. Angular diameters are given for three stars.

096.019 **Cloudcroft occultation summary. II. April 1980-December 1981.**
R. R. Radick, J. L. Africano, M. R. Flores, D. A. Klimke, E. T. Tyson.
Astron. J., Vol. 87, 1874 - 1880 (1982).

The authors present further results from the Cloudcroft occultation observation program. The 75 occultations summarized include 24 events involving stars brighter than seventh magnitude, 48 reappearances, and 6 observations of binary stars or stars which may be double.

096.020 **Possible occultations by satellites of Uranus and Neptune: 1983–1985.**
D. J. Mink, A. Klemola.
Astron. J., Vol. 87, 1881 - 1883 (1982) = Lick Obs. Bull., No. 936.

Predictions are presented for 15 possible occultations by the satellites of Uranus and Neptune from 1983 through 1985. Umbriel, the third satellite of Uranus, might occult a 10.4-mag star (Hyd $-20° 51699$) on 25 March 1983 which will be occulted by Uranus 14 hr earlier. Uncertainties in star positions and ephemerides of planets and satellites are quite large in comparison to the size of these bodies, and these predictions are to be taken as possibilities only.

Photoelectric lunar occultations. I. History, uses and limitations.
See Abstr. 004.027.

Historical eclipses.
See Abstr. 004.032.

Early planning for the occultation of 1 Vulpeculae by Pallas on 1983 May 29.
See Abstr. 013.081.

Computer controlled high speed photometry presented by the lunar occultation of 57 Orionis. I.
See Abstr. 031.603.

Corrections to Watts' datum from photoelectric occultations.
See Abstr. 031.604.

Computer controlled high speed photometry presented by the lunar occultation of 57 Orionis. II.
See Abstr. 031.608.

Methods of photoelectric observations of lunar occultations and their application for the study of close binaries.
See Abstr. 031.609.

Some hints for timing occultations.
See Abstr. 031.620.

A portable system for high-speed photometry of occultations.
See Abstr. 032.017.

Corrections to fundamental constants from photoelectric observations of lunar occultations.
See Abstr. 043.006.

Turbulence in planetary occultations: a strong scattering formulation including an inhomogeneous background.
See Abstr. 063.068.

Observations of the occultation of HD 197999 by the minor planet 105 Artemis.
See Abstr. 098.006.

The occultation diameter of 88 Thisbe.
See Abstr. 098.028.

The occultation of star KME 14 by Uranus rings (22 April 1982).
See Abstr. 101.015.

The upper atmosphere of Uranus: results of the 22 April 1982 occultation.
See Abstr. 101.023.

Angular diameters by the lunar occultation technique. VI. Limb darkening of α Tauri.
See Abstr. 115.001.

097 Mars, Mars Satellites

097.001 **Schiaparelli basin, Mars: morphology, tectonics and infilling history.**
P. J. Mouginis-Mark, V. L. Sharpton, B. R. Hawke.
Multi-ring basins, (see 012.001), p. 155 - 172 (1981).
In an attempt to address some of the large-scale characteristics of martian basins, this paper consequently documents the interior and exterior deposits associated with the basin, and interprets Schiaparelli's tectonic features as a basis for future analyses of large impact structures on Mars.

097.002 **Tidal rigidity of Phobos.**
C. F. Yoder.
Icarus, Vol. 49, 327 - 346 (1982).
The Martian satellites Phobos and Deimos move along nearly circular coplanar, stable orbits and have cratered surfaces older than $\sim 10^9$ years. The accretion hypothesis suggests that their primordial orbits were also very regular. However, tides raised on Mars and Phobos can substantially alter the semimajor axis a and eccentricity e of Phobos' orbit over time. A scenario is proposed in this paper in which the observed e of Phobos results from several gravitational resonance excitations within the last 10^9 years, assuming tidal friction in Phobos has had only a small effect on its orbit.

097.003 **Martian surface properties indicated by the opposition effect.**
T. E. Thorpe.
Icarus, Vol. 49, 398 - 415 (1982).
Opposition measurements made by the Viking Orbiter television cameras in the Arabia, Syrtis Major, and Elysium Planitia regions have been combined with observations previously reported to provide a photometric comparison of these areas and several generic features. Radiative transfer expressions were used to derive average surface particle single-scattering albedos, phase functions and porosities.

097.004 **Study of peculiarities of crater distribution on Mars, Mercury and the moon from maps of crater density constructed by a computer.**
D. A. Kazimirov, B. D. Sitnikov, Zh. F. Rodionova, T. P. Skobeleva, K. I. Dekhtyareva, G. A. Poroshkova, V. A. Shuvaeva.
Tr. Gos. Astron. Inst. Shternberga, Tom 51, 178 - 194 (1982). In Russian.
A method of the investigation of peculiarities of the crater distribution on Mars, Mercury and the Moon by means of a data bank is described. Maps of density of crater distribution are given.

097.005 **Periodic climate change on Mars: review of evidence and effects on distribution of volatiles.**
M. H. Carr.
Icarus, Vol. 50, (see 012.003), 129 - 139 (1982).
This paper reviews briefly the current knowledge of the geology of the polar regions and examines how periodic climate changes might have affected geologic processes there. Different theories for the formation of the layered terrain and its peculiar topography are examined.

097.006 **Topography and stratigraphy of Martian polar layered deposits.**
K. R. Blasius, J. A. Cutts, A. D. Howard.
Icarus, Vol. 50, (see 012.003), 140 - 160 (1982).
This paper reports initial results of detailed studies of topography and stratigraphy of north polar perennial ice and layered deposits from high-resolution images returned by Viking Orbiter 2.

097.007 **Stratigraphic relationships within Martian polar cap deposits.**
A. D. Howard, J. A. Cutts, K. R. Blasius.
Icarus, Vol. 50, (see 012.003), 161 - 215 (1982).
The authors describe mapping units for polar cap features, their relationships to topography, and their probable origin. They introduce newly recognized features (banded terrain, steep scarps, grooved and unbedded terrain, and scarp-paralleling bands and ridges). The second section of the paper focuses upon examples of diagnostic stratigraphy. This section includes evidence for recent deposition of layers beneath smooth and banded terrain and for erosion of the scarps exposing layered terrain. The study sites reveal relationships between topographic position and processes of layer deposition and erosion, and evidence for both polewide and local alternation between erosion and deposition.

097.008 **Models of climate cycles recorded in Martian polar layered deposits.**
J. A. Cutts, B. H. Lewis.
Icarus, Vol. 50, (see 012.003), 216 - 244 (1982).
Models for the accumulation of polar deposits were used to predict the sequences of layer thicknesses. Continuous deposition and climate-modulated deposition models were presented and compared. Two physical models of the climate modulated type were developed in detail. The sensitivity of the sequence of layer thicknesses to model type and threshold were examined.

097.009 **Quasi-periodic climate changes on Mars: a review.**
J. B. Pollack, O. B. Toon.
Icarus, Vol. 50, (see 012.003), 259 - 287 (1982).
The authors critically review our present understanding of the impact of quasi-periodic variations in axial obliquity and direction and orbital eccentricity on the climate of Mars and the relationship between these climatic changes and the occurrence of polar layered terrain.

097.010 **Martian great dust storms: an update.**
R. W. Zurek.
Icarus, Vol. 50, (see 012.003), 288 - 310 (1982).
Observations by the Viking Orbiters and Landers have made substantial contributions to our understanding of the episodic, planetary-scale dust storms on Mars. These and other observations pertinent to the great dust storms are reviewed in this paper; most of the emphasis is on the atmospheric/climatic aspects of these great storms. Specifically, observations concerning the optical properties of the airborne dust, the frequency of occurrence of great dust storms, and the kinematics of their evolution are summarized.

097.011 **Some effects of global dust storms on the atmospheric circulation of Mars.**
R. M. Haberle, C. B. Leovy, J. B. Pollack.
Icarus, Vol. 50, (see 012.003), 322 - 367 (1982).
The response of the Martian atmosphere to various dust loads and the ability of its meridional circulation to transport dust globally are numerically simulated using a zonally symmetric primitive equation model on a sphere.

097.012 **Martian south polar cap boundary: 1971 and 1973 data.**
P. B. James, K. Lumme.
Icarus, Vol. 50, (see 012.003), 368 - 380 (1982).
International Planetary Patrol photographs of Mars obtained during the 1971 and 1973 apparitions have been used to produce regression curves for the Martian south polar cap for those years. There is remarkable consistency between the 1971

IPP and 1977 Viking data points, and the 1973 regression does not appear to have differed significantly from those of 1971 or 1977.

097.013 **Mars: the regolith-atmosphere-cap system and climate change.**
F. P. Fanale, J. R. Salvail, W. B. Banerdt, R. S. Saunders.
Icarus, Vol. 50, (see 012.003), 381 - 407 (1982).
A new model for predicting the behavior of the Martian regolith-atmosphere-cap CO_2 regime and describing its role in climate change is derived. The model describes the time-temperature histories of 90 regolith "chunks" on a latitude-depth grid, accounting for CO_2 exchange by means of laboratory-derived expressions relating temperature, CO_2 pressure (P_{CO_2}), and adsorbed CO_2 mass.

097.014 **On some types of continental deposites of Mars.**
N. V. Makarova, Ya. G. Kats.
Izv. vuzov. Geol. i razvedka, 1982, No. 3, p. 68 - 72. In Russian. – Abstr. in Ref. zh., 51. Astron., 7.51.229 (1982).

097.015 **Transient species on Mars and Venus.**
Y. L. Yung.
Vibrational-rotational spectroscopy for planetary atmospheres, (see 012.007), p. 243 - 251 (1982).
The author discusses transient species on Mars and Venus and compares them to the earth as well as to each other. The chemistry of the earth's stratosphere has striking similarities to that of the atmosphere of Mars and the stratosphere of Venus.

097.016 **Marte.**
R. Salvatori.
Coelum, Vol. 50, 181 - 191 (1982).

097.017 **Regolith-atmosphere exchange of water and carbon dioxide on Mars: effects on atmospheric history and climate change.**
F. P. Fanale, B. M. Jakosky.
Planet. Space Sci., Vol. 30, 819 - 831 (1982).
The paper summarizes recent studies of the role of regolith-atmosphere volatile exchange in seasonal and long term climate variation on Mars.

097.018 **Atomic carbon in the atmospheres of Mars and Venus.**
V. A. Krasnopol'skij.
Kosm. Issled., Tom 20, 595 - 603, with a correction p. 950 (1982). In Russian.

097.019 **Mars: gravity data analysis of the crater Antoniadi.**
W. L. Sjogren, S. J. Ritke.
Geophys. Res. Lett., Vol. 9, 739 - 742 (1982).
Topography and gravity information for this 370-km crater are analyzed to determine a depth of compensation using an Airy isostatic model. A least squares fit to the gravity profile produces an estimate of 115 km for the depth of compensation.

097.020 **Observing Mars X – reporting Mars observations.**
C. F. Capen.
Strolling Astron., Vol. 29, 133 - 138 (1982).

097.021 **A comparison of observed lee-waves on Earth and Mars.**
A. O. Pickersgill, G. E. Hunt.
Weather, Vol. 37, No. 4, p. 98 - 108 (1982). – Abstr. in Phys. Abstr., Vol. 85, Abstr. 84137 (1982).

097.022 **The amazing Olympus Mons.**
J. K. Beatty.
Sky Telesc., Vol. 64, 420 - 422 (1982).

097.023 **The solar wind interaction with Mars as seen by the Viking retarding potential analyzers.**
B. L. Cragin, W. B. Hanson, S. Sanatani.
J. Geophys. Res., Vol. 87, 4395 - 4404 (1982).
In addition to the published retarding potential analyzer (RPA) data that were taken in the Martian ionosphere, there exist many Viking RPA measurements of low-energy (<75 eV) electron fluxes out to 16,000 km above the Mars surface. A brief description of the RPA instrumentation and its operational modes is presented first. The nature of the raw data returned is described, followed by summary plots of different measured parameters versus time (altitude). Among these data are the energy spectra of the low-energy electrons, as well as derived characteristics of the variability of the low-energy electron fluxes. It is argued that these variations are related to whistler mode oscillations, and it is concluded that there may be a highly turbulent shock structure that could mask a clear signature of the bow shock in the time-averaged data. Finally, a summary of the observations and their interpretation is given.

097.024 **Comment on 'Aureole deposits of the Martian volcano Olympus Mons' by E. C. Morris.**
C. A. Hodges, H. J. Moore, with a reply by E. C. Morris.
J. Geophys. Res., Vol. 87, 6692 - 6693 (1982). – See Abstr. 31.097.087.

097.025 **Mars water vapor, near-surface.**
J. A. Ryan, R. D. Sharman, R. D. Lucich.
J. Geophys. Res., Vol. 87, 7279 - 7284 (1982).
In a previous paper the authors concluded that the temperature sensors aboard the Viking landers (VL-1 and VL-2) were detecting the water vapor frost point. In the present paper the authors extend the analysis to cover one Mars year. They also compare the data with measurements of total water vapor column abundance at the VL-1 and VL-2 sites and atmospheric opacity for the year.

097.026 **On the polarimetric properties of the Martian surface and atmosphere.**
V. P. Tishkovets, Yu. G. Shkuratov.
Astron. Zh., Tom 59, 991 - 995 (1982). In Russian. English translation in Soviet Astron., Vol. 26, No. 5.
Some physical characteristics of the Martian atmosphere aerosol particles are analyzed using ground-based polarimetric measurements. The average size of aerosol particles is about 0.5 μm which is close to the Viking data.

097.027 **Normal reflectance, colors, and UV signatures of Phobos and Deimos: implications on surface petrology.**
K. D. Pang, J. M. Ajello, J. K. Wagner, B. W. Hapke, E. N. Wells.
Bull. American Astron. Soc., Vol. 14, 718 - 719 (1982). Abstract.

097.028 **New high-resolution spectrophotometry of the Martian bright region Arabia, 0.35 to 1.0 μm.**
R. B. Singer, P. L. Blake.
Bull. American Astron. Soc., Vol. 14, 755 (1982). – Abstract.

097.029 **Spatially resolved 2.7 - 4.3 micron spectra of Mars.**
J. Apt, R. B. Singer, F. P. Fanale.
Bull. American Astron. Soc., Vol. 14, 755 (1982). – Abstract.

097.030 **Spectral evidence for aluminous hematite on Mars.**
R. V. Morris, H. V. Lauer, Jr., W. W. Mendell.
Bull. American Astron. Soc., Vol. 14, 755 (1982). – Abstract.

097.031 **Characterization of the Mars surface texture by reflectance photopolarimetry.**

A. Dollfus, M. Deschamps, L. V. Ksanfomaliti.
Bull. American Astron. Soc., Vol. 14, 755 (1982). – Abstract.

097.032 **Surface roughness estimates from Viking bistatic radar data in Mars equatorial region.**
R. A. Simpson, G. L. Tyler, G. G. Schaber.
Bull. American Astron. Soc., Vol. 14, 756 (1982). – Abstract.

097.033 **Radar altimetry and polarization measurements of the northern hemisphere of Mars.**
J. K. Harmon, D. B. Campbell, S. J. Ostro.
Bull. American Astron. Soc., Vol. 14, 756 (1982). – Abstract.

097.034 **Speculative evidence for a third year dust storm on Mars.**
S. D. Wall.
Bull. American Astron. Soc., Vol. 14, 756 (1982). – Abstract.

097.035 **A search for local Martian dust storms.**
A. R. Peterfreund.
Bull. American Astron. Soc., Vol. 14, 757 (1982). – Abstract.

097.036 **Mars: influence of obstacle type and size on wind streak production.**
S. Lee, P. Thomas, J. Veverka.
Bull. American Astron. Soc., Vol. 14, 757 (1982). – Abstract.

097.037 **An analytic Martian Hadley circulation model: comparison with eolian features.**
J. Magalhaes.
Bull. American Astron. Soc., Vol. 14, 757 - 758 (1982). Abstract.

097.038 **Eolian sediments as possible sources and sinks of atmospheric water vapor on Mars.**
S. M. Clifford.
Bull. American Astron. Soc., Vol. 14, 758 (1982). – Abstract.

097.039 **Radiative properties of the subliming seasonal polar caps of Mars.**
D. A. Paige.
Bull. American Astron. Soc., Vol. 14, 758 (1982). – Abstract.

097.040 **Martian north polar water-ice clouds.**
P. R. Christensen, R. W. Zurek.
Bull. American Astron. Soc., Vol. 14, 758 (1982). – Abstract.

097.041 **The transport of water off the north polar cap.**
R. M. Haberle.
Bull. American Astron. Soc., Vol. 14, 758 - 759 (1982). Abstract.

097.042 **The seasonal behavior of Mars atmospheric water vapor: the role of the polar caps and regolith.**
B. M. Jakosky.
Bull. American Astron. Soc., Vol. 14, 759 (1982). – Abstract.

097.043 **Variability of carbon monoxide in the Martian atmosphere.**
R. T. Clancy, D. O. Muhleman, B. M. Jakosky.
Bull. American Astron. Soc., Vol. 14, 759 (1982). – Abstract.

097.044 **Mars: seasonal variation of global atmospheric opacity.**
T. Z. Martin.
Bull. American Astron. Soc., Vol. 14, 759 (1982). – Abstract.

097.045 **The influence of atmospheric tides on the zonal-mean Martian circulation.**
R. W. Zurek, R. M. Haberle.
Bull. American Astron. Soc., Vol. 14, 764 (1982). – Abstract.

097.046 **Solar excitation of the ν_3 state of CO_2 in the mesospheres of Mars and Venus: an NLTE model.**
D. Deming. F. Espenak, D. Jennings, T. Kostiuk, M. Mumma.
Bull. American Astron. Soc., Vol. 14, 764 (1982). – Abstract.

097.047 **The effect of solar tides on the general circulation of the Martian atmosphere.**
K. Hamilton.
J. Atmos. Sci., Vol. 39, 481 - 485 (1982). – Abstr. in Phys. Abstr., Vol. 85, Abstr. 100800 (1982).

097.048 **Spectral investigations of Mars from Viking.**
N. Evans.
Proc. SPIE Int. Soc. Opt. Eng., Vol. 268, (see 012.040), 35 - 41 (1981). – Abstr. in Phys. Abstr., Vol. 85, Abstr. 112365 (1982).

097.049 **The relief of the Memnonia Fossae – Margaritifer Sinus region according to CO_2-altimetry measurements aboard Mars 5.**
L. V. Zasova, V. M. Zubkova, V. S. Zhegulev, L. V. Ksanfomaliti, V. I. Moroz, E. V. Petrova.
Kosm. Issled., Tom 20, 921 - 927 (1982). In Russian.

097.050 **Surface marking variations of selected areas on Mars.**
G. de Mottoni y Palacios, A. Dollfus.
Astron. Astrophys., Vol. 116, 323 - 331 (1982).

The authors' long-term record of shape and contrast changes in the features observed at the surface of Mars with telescopes is based upon photographs taken during 70 yr, which are summarized by 32 maps published by de Mottoni (1975). This analysis suggests that darkening has some relationship with the seasonal increase of solar radiation flux and surface heating. High solar heatings may favour high velocity winds which in turn can lift in the air the soil dust and move the grains at the surface. Thus, telescopic observations are compatible with the model of wind transportation and deposition of dust proposed on the basis of MARINER 9 images to explain the surface feature variations on Mars. Local, environmental, and topographic influences are noted.

097.051 **Photographic observations of Mars during the 1979 - 1980 opposition.**
T. Akabane, M. Matsui, K. Ishiura, K. Iwasaki, Y. Saito, T. Kitahara.
Contrib. Kwasan Hida Obs., No. 250, 40 pp. (1982).

097.052 **The motion of Mars 1751–1969.**
R. E. Laubscher.
Astron. Pap., Vol. 22, Part IV, 363 - 490 (1981).

097.053 **Dark halos of the Phobos craters and shock reproduction of carbon chondrites.**
A. A. Pronin, O. V. Nikolaeva.
Dokl. AN SSSR, Tom 265, 429 - 432 (1982). In Russian. Abstr. in Ref. zh., 62. Issled. kosm. prostranstva, 10.62.387 (1982).

097.054 **Zusammensetzung und Entstehung des Bodens des Planeten Mars.**
C. C. Allen, K. Keil.
Sterne, 58. Band, 326 - 338 (1982).

097.055 **Mars: crust constitution, volcanic processes, geomorphology.**
Ya. G. Kats, V. V. Kozlov, N. V. Makarova, E. D. Sulidi-Kondrat'ev.
Byull. Mosk. o-va ispyt. prirody. Otd. geol., Tom 57, No. 4, p. 134 (1982). In Russian. – Abstr. in Ref. zh., 51. Astron., 10.51.197 (1982).

097.056 **Gravitational field and topography of Mars.**
L. L. Polyakovskaya.
L'vov. politekh. inst. L'vov, 1982. 44 pp. In Russian. – Abstr. in Ref. zh., 51. Astron., 11.51.177 (1982).

097.057 **The geologic evolution of the planet Mars.**
P. Masson.
Formation of planetary systems, (see 012.064), p. 797 - 816 (1982).

The author presents a brief summary of the present knowledge on the Martian geology based on the results published by the members of Mariner 9 and Viking Orbiter Imaging Teams, the NASA Planetary Geology Principal Investigators and the scientists involved in the Mars Data Analysis Program. A special emphasis is given to the geologic evolution (volcanism and tectonism) related to the present knowledge on the internal structure of the planet.

097.058 **Phobos and Deimos.**
A. Dobrovolskis.
Formation of planetary systems, (see 012.064), p. 817 - 823 (1982).

Contents: Discovery. Appearance. Ejecta from Phobos and Deimos.

097.059 **The Martian satellites.**
F. Mignard.
Formation of planetary systems, (see 012.064), p. 825 - 832 (1982).

The orbital history of Phobos and Deimos is studied by means of an easy-to-describe model of tidal interaction. This allows to put some constraints on their initial orbits.

097.060 **Internal stresses in Phobos and other triaxial bodies.**
A. R. Dobrovolskis.
Icarus, Vol. 52, 136 - 148 (1982).

In this paper the Martian satellites are modeled as homogeneous, elastic triaxial ellipsoids subject to tidal, rotational, and self-gravitational stresses. A novel semianalytical treatment then gives the stress and strain fields throughout their interiors. The results indicate that Phobos and Deimos have always been stable with respect to tidal fracture or disruption, but that Phobos will probably break up before colliding with Mars.

097.061 **Dual-polarization radar observations of Mars: Tharsis and environs.**
J. K. Harmon, D. B. Campbell, S. J. Ostro.
Icarus, Vol. 52, 171 - 187 (1982).

Observations of the Tharsis region of Mars with the 12.6-cm radar at Arecibo Observatory have yielded radar echoes which are highly depolarized. The strongest depolarization appears to come from the Tharsis Ridge. The peak of the depolarization enhancement coincides with the longitudes of the major Tharsis volcanoes. While it is clear that the depolarization cannot be confined solely to these volcanic shields, it appears likely that most of the enhanced depolarization is due to extreme decimeter-scale roughness of volcanic origin. The existence of extensive lava flows over the Tharsis plains tends to support this view.

097.062 **Positions of Mars using photographic observations with the normal astrograph at Pulkovo in the opposition of 1975 - 1976 and 1977 - 1978.**
L. V. Zhukov.
Izv. Glav. Astron. Obs. Pulkovo, Astrofiz. Astrometr., No. 200, p. 109 - 113 (1982). In Russian.

097.063 **Carbon monoxide in the atmosphere of Mars.**
C. J. Macris, B. C. Petropoulos.
C. R. Acad. Sci. Paris, Tome 295, Sér. II, 989 - 992 (1982). In French.

The intensities of the transitions CO ($A^1\Pi - X^1\Sigma$) ($v''=0$, $v'=1$) have been computed as a function of altitude in the atmosphere of Mars.

097.064 **Assessing the volcanic probability of Martian landforms.**
M. M. O'Toole.
Advances in planetary geology, (see 003.014), p. 556 - 558 (1982). – Abstract.

097.065 **Thick shell tectonics on one-plate planets: applications to Mars.**
W. B. Banerdt, R. J. Phillips, N. H. Sleep, R. S. Saunders.
J. Geophys. Res., Vol. 87, (see 012.071), 9723 - 9733 (1982).

A theory for stress distributions in thick lithospheric shells on one-plate planets is developed based on the zero frequency equations of a self-gravitating elastic spherical shell overlying a strengthless fluid. Stress distributions in lithospheres are reviewed for both the compensated and flexural modes. Computational analyses are carried out for Mars, and it is found that isostatically compensated models correctly predict the graben structure in the immediate Tharsis region and a flexural loading model is satisfactory in explaining the graben in the regions surrounding Tharsis. A three-stage model is hypothesized for the evolution of Tharsis: isostasy with north-south graben formation on Tharsis, followed by flexural loading and radial graben formation on the perimeter of Tharsis, followed by a last stage of loading with little or no regional deformation. This model is consistent with the Martian lithosphere monotonically thickening over geologic time.

097.066 **Gravity field model of Mars in spherical harmonics up to degree and order eighteen.**
G. Balmino, B. Moynot, N. Valès.
J. Geophys. Res., Vol. 87, (see 012.071), 9735 - 9746 (1982).

A new model of Mars gravity field in spherical harmonics up to eighteenth degree and order has been computed from all available Doppler tracking data of Mariner 9 and Viking 1 and 2 orbiters. The model has been extensively tested from the point of view of orbit representation over periods of time ranging from a few hours to 12 days, and it is currently used for geophysical studies of the Martian lithosphere.

097.067 **New radar-derived topography for the northern hemisphere of Mars.**
G. S. Downs, P. J. Mouginis-Mark, S. H. Zisk, T. W. Thompson.
J. Geophys. Res., Vol. 87, (see 012.071), 9747 - 9754 (1982).

Earth-based radar altimetry data for the northern equatorial belt of Mars (6°S - 23°N) have recently been reduced to a common basis corresponding to the 6.1-mbar reference surface. A first look at these data indicates that the elevations of Tharsis, Elysium, and Lunae Planum are lower (by 2–5 km) than has been suggested by previous estimates. These differences show that the required amount of tectonic uplift (or constructional volcanism) for each area is less than has been previously envisioned. Atmospheric or surficial conditions are suggested which may explain the discrepancies between the radar topography and elevations measured by other techniques. The topographies of Chryse Planitia, Syrtis Major, and Valles Marineris are also described.

097.068 **Evolution of the Tharsis province of Mars: the importance of heterogeneous lithospheric thickness and volcanic construction.**
S. C. Solomon, J. W. Head.
J. Geophys. Res., Vol. 87, (see 012.071), 9755 - 9774 (1982).

The authors propose a new model for the origin and evolution of the Tharsis province which emphasizes that the topography of Tharsis may be produced largely by construction, rather than uplift.

097.069 **Tectonic history of the Tharsis region, Mars.**
J. B. Plescia, R. S. Saunders.
J. Geophys. Res., Vol. 87, (see 012.071), 9775 - 9791 (1982).
The faults of the Tharsis region of Mars have been mapped and delineated on the basis of age. The orientation of each group of faults was analyzed to determine if they were radial to a point. The results indicate that there are at least four discrete centers of faulting within the Tharsis area which are of significantly different age. Each episode of tectonism was separated from the next by a period of major basaltic volcanism. The tectonic features of the Tharsis area are thus viewed as being the cumulative result of the superposition of several discrete episodes of faulting, rather than the result of one episode about a single point.

097.070 **The role of lithospheric stress in the support of the Tharsis rise.**
R. J. Willemann, D. L. Turcotte.
J. Geophys. Res., Vol. 87, (see 012.071), 9793 - 9801 (1982).
The authors show that the topography, geoid height, and fracture patterns associated with the Tharsis rise are consistent with the hypothesis that the Tharsis rise is a partially supported load on a spherical lithosphere of uniform effective elastic thickness.

097.071 **The structure and evolution of ancient impact basins on Mars.**
P. H. Schultz, R. A. Schultz, J. Rogers.
J. Geophys. Res., Vol. 87, (see 012.071), 9803 - 9820 (1982).
The authors consider five examples of modified impact basins and delineate regions around each that have undergone similar processes such as fracturing, collapse, and channeling. They compare these processes among the different basins and correlate similar zones of modification with concentric basin rings. The authors consider the implications of these observations for current models of basin formation and the role of impact basins in controlling regional tectonics. The results suggest that large multiring impact scars leave a major but sometimes subtle imprint on the geologic structure of stable crustal regions on Mars.

097.072 **Martian volcanic materials: preliminary thickness estimates in the eastern Tharsis region.**
R. A. De Hon.
J. Geophys. Res., Vol. 87, (see 012.071), 9821 - 9828 (1982).
The paper explores the adaptation of methods used to estimate the thickness of volcanic plains-forming materials. Techniques that were developed for thickness studies of lunar mare materials are evaluated and adapted for Martian application. Martian volcanic materials are evaluated in terms of their susceptibility to thickness characterization. Finally, the materials of the eastern portion of the Tharsis dome are examined as representative samples of Martian volcanics.

097.073 **Tharsis volcanoes: separation distances, relative ages, sizes, morphologies, and depths of burial.**
J. L. Whitford-Stark.
J. Geophys. Res., Vol. 87, (see 012.071), 9829 - 9838 (1982).
Correlations between volcano size and separation distance, volcano style and separation distance, and separation distance and tectosphere thickness for the Tharsis volcanoes imply a genetic link among all three parameters. It is suggested that the greater the tectosphere thickness, the larger will be the volcano and the greater the distance between it and its nearest neighbors. These findings are likely to be applicable to other areas where adjacent volcanoes may be tapping the same magma source.

097.074 **Volcanoes and volcanic provinces: Martian western hemisphere.**
D. H. Scott.
J. Geophys. Res., Vol. 87, (see 012.071), 9839 - 9851 (1982).
Regional geologic mapping of the western hemisphere of Mars from Viking images has revealed many more probable volcanoes and volcanotectonic features than were recognized on Mariner 9 pictures. These abundant volcanoes have been assigned to several distinct provinces on the basis of their areal distribution. Volcanoes in any one province vary in morphology, size, and age, but volcanoes in each province tend to have common characteristics that distinguish that particular group.

097.075 **Syrtis Major: a low-relief volcanic shield.**
G. G. Schaber.
J. Geophys. Res., Vol. 87, (see 012.071), 9852 - 9866 (1982).
This report complements a recent study by Simpson et al. (1982) that describes radar measurements of small-scale surface roughness at Syrtis Major and correlates the roughness and surface slope data with geomorphic and eolian features as well as with thermal and bolometric albedo observations.

097.076 **Subkilometer Martian volcanoes: properties and possible terrestrial analogs.**
H. Frey, M. Jarosewich.
J. Geophys. Res., Vol. 87, (see 012.071), 9867 - 9879 (1982).
Kilometer to subkilometer cones with large central craters are the most numerous volcanic structures in the northern plains of Mars. The authors undertook a systematic survey of all available high-resolution Viking orbiter imagery in order to locate and measure the properties of the subkilometer cones. The paper describes the results of this survey, presents regional and intraregional comparisons, and examines possible terrestrial analogs.

097.077 **Absence of silicic volcanism on Mars: implications for crustal composition and volatile abundance.**
P. W. Francis, C. A. Wood.
J. Geophys. Res., Vol. 87, (see 012.071), 9881 - 9889 (1982).
The authors summarize the characteristics of terrestrial explosive volcanism, the resulting landforms, and the tectonic and petrologic conditions leading to their formation. With this background, they then discuss photogeologic evidence for some postulated examples of martian explosive and/or silicic volcanism, examine the relevant petrological and tectonic conditions prevailing on Mars, and speculate on the significance of the apparent absence of silicic volcanism on the planet.

097.078 **Explosive volcanism on Hecates Tholus, Mars: investigation of eruption conditions.**
P. J. Mouginis-Mark, L. Wilson, J. W. Head III.
J. Geophys. Res., Vol. 87, (see 012.071), 9890 - 9904 (1982).
The authors have reexamined the medium and high-resolution Viking images of Amazonian and Hesperian age volcanic centers on Mars and believe that an excellent example of well-preserved explosive activity does indeed exist close to the summit of Hecates Tholus, the northernmost of the three large Elysium Planitia constructs. A mantled region to the west of the summit caldera appears to be an example of a geologically very recent plinian air fall ash deposit. The authors present their morphological evidence for describing this as explosively generated material and also employ numerical models of magma ascent and eruption in the Martian environment to estimate the physical characteristics for this event. They also discuss the implications of this eruptive style for the composition of the erupted magma.

097.079 **Mars: stratigraphy and gravimetry of Olympus Mons and its aureole.**
K. H. Hiller, P. Janle, G. P. O. Neukum, J. E. Guest, R. M. C. Lopes.
J. Geophys. Res., Vol. 87, (see 012.071), 9905 - 9915 (1982).
The paper deals with the relative ages of the major geologic units on and around Olympus Mons and with an interpretation of the gravity anomaly found for this area.

097.080 **Further evidence for a mass movement origin of the Olympus Mons aureole.**
R. Lopes, J. E. Guest, K. Hiller, G. Neukum.
J. Geophys. Res., Vol. 87, (see 012.071), 9917 - 9928 (1982).
Analysis of the ridge, graben, and fracture patterns on the aureole material surrounding Olympus Mons lends further support to an origin by gravity sliding of the outer flanks of the shield.

097.081 **Relict drainages, conical hills, and the eolian veneer in southwest Egypt – applications to Mars.**
C. S. Breed, J. F. McCauley, M. J. Grolier.
J. Geophys. Res., Vol. 87, 9929 - 9950 (1982).
Relations between the fretted channels in the upland plateau, the plains created by backwearing of the plateau scarp, and the relative abundance of dune sand in the northern hemisphere on Mars may be at least partly explained in terms of an interplay between fluvial processes, wind, and mass wasting, such as described for the Gilf Kebir region of southwest Egypt.

097.082 **Ice sculpture in the Martian outflow channels.**
B. K. Lucchitta.
J. Geophys. Res., Vol. 87, (see 012.071), 9951 - 9973 (1982).
Many landforms in Martian outflow channels have characteristics that suggest sculpture by glaciers, ice streams, or ice sheets. Viking Orbiter and terrestrial satellite images were examined at similar resolution to compare features of the Martian outflow channels to features produced by the movement of ice on earth. Many resemblances were found. They include the anastomoses, sinuosities, and U-shaped cross profiles of valleys; hanging valleys; and linear scour marks on valley walls, grooves and ridges on valley floors, and streamlining of bedrock highs. The question of whether ice could have moved in the Martian environment is investigated.

097.083 **A model of Martian slope winds: implications for eolian transport.**
J. Magalhães, P. Gierasch.
J. Geophys. Res., Vol. 87, (see 012.071), 9975 - 9984 (1982).
Observations of dark Martian wind streaks not associated with visible topographic points of origin ('coalesced dark streaks') show these features to favor regions with slopes of 1°-10°. To investigate the circulation over such slopes, the authors adopt a Boussinesq fluid and assume a constant eddy diffusivity parameterization for small-scale mixing. The equations of motion are scaled to determine the range of slopes over which a strong local circulation due to slope winds can exist. For nighttime conditions, the estimated range compares well with the observed range of slopes favored by coalesced dark streaks. For daytime conditions, the slope wind solution is not valid. A simple one-dimensional analytic model is solved to determine friction velocities and velocity and temperature profiles.

097.084 **Martian dust mantling and surface composition: interpretation of thermophysical properties.**
P. R. Christensen.
J. Geophys. Res., Vol. 87, (see 012.071), 9985 - 9998 (1982).
The surface block abundance, fine component thermal inertia, thermal emissivity, and relative atmospheric dust content have been determined for selected regions on Mars, using broadband infrared spectral properties measured at 7, 9, 11, and 20 μm by the Viking infrared thermal mapper. Each of the surface and atmospheric properties produce characteristic spectral signatures which have been successfully modeled.

097.085 **Present wind activity on Mars: relation to large latitudinally zoned sediment deposits.**
P. Thomas.
J. Geophys. Res., Vol. 87, (see 012.071), 9999 - 10008 (1982).
The relation of present Martian winds to large latitudinally zoned sediment deposits has been investigated using global wind streak data and mapping of large sand and dust deposits.

097.086 **Rate of wind abrasion on Mars.**
R. Greeley, R. N. Leach, S. H. Williams, B. R. White, J. B. Pollack, D. H. Krinsley, J. R. Marshall.
J. Geophys. Res., Vol. 87, (see 012.071), 10009 - 10024 (1982).
The authors briefly describe the experiments performed to obtain data on windblown particles and abrasion of rocks in a simulated martian environment. Preliminary results are presented and combined with Viking meteorological data to estimate rates of wind abrasion at the Viking Lander 1 site on Mars. Finally, the authors discuss the implications that the results have for martian surface history.

097.087 **Wind streaks in Tharsis and Elysium: implications for sediment transport by slope winds.**
S. W. Lee, P. C. Thomas, J. Veverka.
J. Geophys. Res., Vol. 87, (see 012.071), 10025 - 10041 (1982).
Detailed maps of wind streaks in Tharsis and Elysium have been compiled from Viking Orbiter observations spanning one complete Martian year. The streak pattern is controlled by slope winds on the central volcanoes and on the flanks of the Tharsis bulge, while the global circulation dominates in Elysium. Dust erosion by downslope winds occurs over much of Tharsis and in the vicinity of Elysium Mons. Erosional streaks are largely absent in Elysium Planitia. Surface properties such as slope, thermal inertia, and roughness may influence the efficiency of slope wind production sufficiently to account for the pronounced differences in streak types and patterns present in these two regions.

097.088 **A summary of Viking sample-trench analyses for angles of internal friction and cohesions.**
H. J. Moore, G. D. Clow, R. E. Hutton.
J. Geophys. Res., Vol. 87, (see 012.071), 10043 - 10050 (1982).
Analyses of sample trenches excavated on Mars, using a theory for plowing by narrow blades, provide estimates of the angles of internal friction and the cohesions of the Martian surface materials.

097.089 **Two Mars years of surface changes seen at the Viking Landing sites.**
E. A. Guinness, C. E. Leff, R. E. Arvidson.
J. Geophys. Res., Vol. 87, (see 012.071), 10051 - 10058 (1982).
Color and brightness changes have occurred at both Viking Landing sites during the first two Mars years of the Lander mission. The contrast between adjacent soil units decreased, and the surface generally became brighter and redder. These changes are attributed to the deposition of a layer of bright, red dust that is most likely only several micrometers in thickness. Analysis of the spectrophotometric properties of the new dust deposited at the Viking Lander 1 site suggests that the new deposits are the same as the bright, red material that only partially covered the site at the beginning of the mission. Certain areas of the Viking 1 site also exhibited an increase in contrast during the second year. The contrast increase is interpreted to be due to the partial removal of the recently deposited dust cover by high velocity winds.

097.090 **Chemical composition of Martian fines.**
B. C. Clark, A. K. Baird, R. J. Weldon, D. M. Tsusaki, L. Schnabel, M. P. Candelaria.
J. Geophys. Res., Vol. 87, (see 012.071), 10059 - 10067 (1982).
The authors present their findings for the major and

minor element concentrations in all samples taken at the two Viking landing sites. These are based upon extensive data correction procedures and upon laboratory simulations using a flight-qualified instrument identical to those on Mars, operated under Martian conditions of temperature, pressure, and atmospheric composition.

097.091 **Chemical weathering and the Viking biology experiments on Mars.**
R. L. Huguenin.
J. Geophys. Res., Vol. 87, (see 012.071), 10069 - 10082 (1982).

The activity observed by the three Viking lander life detection experiments is reviewed, and the active agents are characterized. It is proposed that the activity in all three experiments may have been caused by a single active agent in the Martian soil, resulting from contemporary chemical weathering of mafic silicates by low-temperature frost and adsorbed H_2O. Some of the possible consequences and implications of contemporary weathering for the present state and evolution of the regolith and atmosphere are briefly discussed.

097.092 **Hydrothermally altered impact melt rock and breccia: contributions to the soil of Mars.**
C. C. Allen, J. L. Gooding, K. Keil.
J. Geophys. Res., Vol. 87, (see 012.071), 10083 - 10101 (1982).

097.093 **Shock-induced color changes in nontronite: implications for the Martian fines.**
R. J. Weldon, W. M. Thomas, M. B. Boslough, T. J. Ahrens.
J. Geophys. Res., Vol. 87, (see 012.071), 10102 - 10114 (1982).

Riverside nontronite, a candidate for the major mineral in the Martian fines, becomes both redder and darker upon shock loading between 180 and 300 kbar. The change from olive-yellow to strong brown in the 300-kbar sample brackets the range of color observed at the Viking lander sites. Optical microscopy, X-ray diffraction, optical, infrared, and ^{57}Fe Mössbauer spectroscopy were applied to understand the physical basis of the color change.

097.094 **Magnetic changes accompanying the thermal decomposition of nontronite (in air) and its relevance to Martian mineralogy.**
B. M. Moskowitz, R. B. Hargraves.
J. Geophys. Res., Vol. 87, (see 012.071), 10115 - 10128 (1982).

097.095 **Mars: definition and characterization of global surface units with emphasis on composition.**
T. B. McCord, R. B. Singer, B. R. Hawke, J. B. Adams, D. L. Evans, J. W. Head, P. J. Mouginis-Mark, C. M. Pieters, R. L. Huguenin, S. H. Zisk.
J. Geophys. Res., Vol. 87, (see 012.071), 10129 - 10148 (1982).

The article presents the major results of a study to determine the nature and regional extent of compositional units on a global scale for the Martian surface. A secondary purpose was to develop and use interactive computer image processing techniques on planetary spacecraft data sets to map lithologic units. The procedures and techniques used are discussed in some detail because they represent a new approach in some cases.

097.096 **Classification of surface units in the equatorial region of Mars based on Viking Orbiter color, albedo, and thermal data.**
R. E. Arvidson, E. A. Guinness, A. P. Zent.
J. Geophys. Res., Vol. 87, (see 012.071), 10149 - 10157 (1982).

Eight statistically significant surface units, classified on the basis of Mars Consortium color, albedo, and thermal inertia data, have been identified for the region located between $+15°$ and $-15°$ of the equator and in longitude from 280° to 360°W and 0° to 50°W. The data were decorrelated and subtle clusters were enhanced by utilizing principal components techniques to stretch the original data swarm to a five-dimensional spherical shape. When plotted in a triangular diagram with the decorrelated red/violet ratio, albedo, and thermal inertia as apices, the dominant clusters define a trend from dark, grayish areas with high thermal inertia to bright, reddish areas with low thermal inertia. Mixing apparently takes place between the units, indicating that boundaries are gradational. The brightest, reddest unit, with the lowest thermal inertia, is correlated with the presence of a thick (tens of meters) debris mantle seen in high-resolution Viking Orbiter images. Such a finding implies a correlation between color, albedo, and thermal inertia, which are probes of near-surface materials, and the presence of debris mantles, which extend to depths of meters.

097.097 **Surface properties of ancient cratered terrain in the northern hemisphere of Mars.**
J. R. Zimbelman, R. Greeley.
J. Geophys. Res., Vol. 87, (see 012.071), 10181 - 10189 (1982).

The paper presents an analysis of the surface properties of ancient cratered terrain in the northern hemisphere of Mars obtained from infrared data with more than an order of magnitude improvement in spatial resolution from previously published results.

097.098 **Band-pass filtering of one year of daily mean pressures on Mars.**
D. S. Niver, S. L. Hess.
J. Geophys. Res., Vol. 87, (see 012.071), 10191 - 10196 (1982).

The sol-mean atmospheric pressures measured at the two Viking landers exhibit fluctuations ranging in period from seasonal to a few days. The authors take advantage of the rather precise annual periodicity of these data to design sharp numerical filters and use them to split the series at each lander into six time series, each containing oscillations associated with a descrete frequency band. This analysis reveals 2- to 4-sol waves during the cold seasons at both landers, and 8-sol waves during the spring at Lander 2, confirming previous results from spectral analysis. As previously predicted, high-frequency waves appear suppressed during the second dust storm. Certain low-frequency signals, probably associated with the global dust storms, are detected and demonstrated to be significant for the first time.

097.099 **Daily temperature variations on Mars.**
R. Ditteon.
J. Geophys. Res., Vol. 87, (see 012.071), 10197 - 10214 (1982).

The purpose of the paper is to discuss the relationship between the Viking thermal observations of Mars and the Viking thermal model and to modify the model to understand the observations better.

097.100 **Remote sensing evidence for regolith water vapor sources on Mars.**
R. L. Huguenin, S. M. Clifford.
J. Geophys. Res., Vol. 87, (see 012.071), 10227 - 10251 (1982).

A survey of ground-based and spacecraft observations at visible wavelengths of condensate activity in the Martian atmosphere is presented. The observations reveal that Solis Lacus has been a center of condensate activity throughout the observational record, along with sites in Noachis-Hellespontus and in border regions of Syrtis Major. The condensate activity did not appear to be controlled by the distribution of aerosol dust, surface albedo, thermal inertia, or topography. Rather, the patterns of cloud and fog occurrence, together with observed and inferred surface winds, point to these sites as primary H_2O vapor source regions for condensate development.

097.101 **Water vapor in Mars' arctic: seasonal and spatial variability.**
D. W. Davies.
J. Geophys. Res., Vol. 87, (see 012.071), 10253 - 10263 (1982).

The seasonal and spatial behavior of water vapor in Mars' arctic is described. These data were obtained by the Mars atmosphere water detector (MAWD) instruments on the Viking Orbiters. Maps of the spatial distribution of water vapor throughout the spring and summer, high resolution one-dimensional scans, and seasonal histories of selected areas are presented.

097.102 **Martian north polar cap 1979–1980.**
K. Iwasaki, Y. Saito, T. Akabane.
J. Geophys. Res., Vol. 87, (see 012.071), 10265 - 10269 (1982).

A regression curve for the north polar cap of Mars is extracted from the photographs obtained at Kwasan Observatory and at Hida Observatory during the 1979–1980 apparition.

097.103 **The seasonal CO_2 cycle on Mars: an application of an energy balance climate model.**
P. B. James, G. R. North.
J. Geophys. Res., Vol. 87, (see 012.071), 10271 - 10283 (1982).

Energy balance climate models of the Budyko-Sellers variety are applied to the carbon-dioxide cycle on Mars. Recent data available from the Viking mission, in particular the seasonal pressure variations measured by Viking landers, are used to constrain the models. No set of parameters was found for which a one-dimensional model parameterized in terms of ground temperature gave an adequate fit to the observed pressure variations. A modified, two-dimensional model including the effects of dust storms and the polar hood reasonably reproduces the pressure curve, however. The implications of these results for martian climate changes are discussed.

097.104 **The solar wind interaction with Mars revisited.**
J. A. Slavin, R. E. Holzer.
J. Geophys. Res., Vol. 87, (see 012.071), 10285 - 10296 (1982).

Owing to a paucity of observational data, no clear consensus has been reached concerning the general nature of the solar wind interaction with Mars. In particular, the previous analyses are still at odds regarding the existence of a small intrinsic field magnetosphere at Mars as opposed to a Venus-type ionospheric interaction. This study contributes to the resolution of this question in three ways. First, an improved determination of effective obstacle altitude and shape is obtained from the Mars 2, 3, and 5 bow shock encounters through the use of a recently published catalog of gasdynamic flow solutions. Second, building upon the Pioneer Venus findings at a field-free planet it is shown that the Martian ionosphere cannot support a Venus-type ionopause at the obstacle altitudes inferred through our modeling of the bow wave observations even when maximal induced ionospheric magnetic fields and solar maximum EUV levels are assumed. Third, a search of published Mars and Mariner radio occultation measurements produced no evidence for the existence of an ionopause at Mars. After a review of the various arguments in the literature, as supplemented by the results of this study, the authors conclude that Mars most probably possesses a small intrinsic field magnetosphere.

097.105 **Infrared observations of Phobos and Deimos from Viking.**
J. I. Lunine, G. Neugebauer, B. M. Jakosky.
J. Geophys. Res., Vol. 87, (see 012.071), 10297 - 10305 (1982).

The surface thermal properties of Phobos and Deimos have been determined from observations made with the Viking Orbiter Infrared Thermal Mapper (IRTM), at wavelengths ranging from 6 to 20 μm. The data, composed of both global and high-resolution infrared photometry of the satellite surfaces as well as eclipse observations, indicate surface material of low thermal conductivity comparable to that of the earth's moon. It is concluded that both satellites are covered with a vertically uniform layer of finely divided material at least several centimeters thick. Observed differences between brightness temperatures at different wavelengths on Phobos are due mainly to topographic slopes and to the presence of ~5% by area high inertia or blocky material.

Viking Lander Atlas of Mars.
See Abstr. 002.129.

Relief, tectonics and volcanism of Mars.
See Abstr. 003.073.

Infrared absorption by OH-containing compounds.
See Abstr. 022.022.

Simulation of Martian chaotic terrain and outflow channels.
See Abstr. 022.052.

Release of volatiles from possible Martian analogs.
See Abstr. 022.153.

The effects of varying surface roughness on aerodynamic and aeolian processes.
See Abstr. 022.162.

Spectral evidence for the mineralogy of high-albedo soils and dust on Mars.
See Abstr. 022.167.

Spectral characteristics of the iron oxides with application to the Martian bright region mineralogy.
See Abstr. 022.168.

Seasonal carbon dioxide exchange between the regolith and atmosphere of Mars: experimental and theoretical studies.
See Abstr. 022.169.

Photoclinometric determination of the topography of the Martian north polar cap.
See Abstr. 031.512.

Spectroscopic measurements of Mars and Venus with heterodyne techniques.
See Abstr. 031.519.

Infrared heterodyne spectroscopy *(Mars atmosphere)*
See Abstr. 031.558.

Photogrammetry of the Viking Lander imagery.
See Abstr. 031.592.

Kepler – a mission to the planet Mars.
See Abstr. 051.038.

KEPLER: an interdisciplinary Mars orbiter mission.
See Abstr. 051.065.

Pre-pleistocene glaciation on earth: implications for climatic history of Mars.
See Abstr. 081.005.

Time history of outgassing on Venus, Earth and Mars.
See Abstr. 093.065.

Noble gases in planetary atmospheres: implications for the origin and evolution of atmospheres.
See Abstr. 093.070.

Reevaluation of lunar and Martian spectra in the mid-IR region.
See Abstr. 094.522.

098 Minor Planets

098.001 **Catalogue of minor planet identities. I. Identities with planets (1)-(2297).**
L. D. Schmadel.
Astron. Astrophys., Suppl. Ser., Vol. 49, 691 - 707 (1982).
The orbits of 2297 numbered minor planets from the 1982 Leningrad ephemeris volume are compared with some 43000 positions of unnumbered objects from a tape file of the Minor Planet Center in order to establish identities by a uniform procedure. The catalogue lists 1884 identifications concerning 1100 planets. The methods of computation as well as some statistical results are presented.

098.002 **Minor planets discoveries at the GPO, ESO-La Silla. Dependences of stars for catalogue improvement and future perturbation studies.**
H. Debehogne.
Astron. Astrophys., Suppl. Ser., Vol. 49, 775 - 776, microfiche (1982).
This paper contains 50 discoveries and observations of numbered asteroids, some of them being observed for orbits improvements and theoretical work. For the others, the authors are in advance on future requirements. It is also a contribution to the IRAS Project. They give the dependences of the reference stars for future studies (stellar paths perturbations, sub-stellar and planetary detectability, unseen astrometric companions, ...) and for future improvements (asteroid position and orbits, star places and proper motions, ...).

098.003 **Compositional structure of the asteroid belt.**
J. Gradie, E. Tedesco.
Science, Vol. 216, 1405 - 1407 (1982).
The distribution of compositional types among the asteroids is found to vary systematically with heliocentric distance. Seven distinct peaks in the relative proportion of the compositional types E, R, S, M, F, C, P, and D are found from 1.8 to 5.2 astronomical units. The inferred composition of the asteroids in each semimajor axis region is consistent with the theory that the asteroids accreted from the solar nebula at or near their present locations.

098.004 **A systematic exploration of three-dimensional asteroidal motion at the 2/1 resonance.**
C. Froeschlé, H. Scholl.
Astron. Astrophys., Vol. 111, 346 - 356 (1982).
Using Schubart's three-dimensional averaging model, the authors calculated 96 orbits at the 2/1 Kirkwood gap over 17000 years. A classification of the resulting orbits with respect to the behaviour of ω and $\tilde{\omega}$ yields 5 classes out of which 3 classes have real representatives among the known asteroids. All the orbits are stable in Hill's sense. Three major mechanisms determine orbital stability: σ-libration, e-ω phase coupling and ω-libration. The behaviour of the orbits is interpreted on the basis of results obtained in the planar case.

098.005 **Photoelectric observations of 44 Nysa during 1981 opposition.**
J. O. Piironen.
Astron. Astrophys., Vol. 112, 172 - 173 (1982).
Photoelectric observations of 44 Nysa were made in March 1981. A $0\overset{m}{.}33$ amplitude was observed between the secondary maximum and minimum. The colour indices $(B-V)$ and $(U-B)$ were $0\overset{m}{.}70$ and $0\overset{m}{.}24$ respectively.

098.006 **Observations of the occultation of HD 197999 by the minor planet 105 Artemis.**
P. B. Byrne, I. M. Coulson, J. G. Doyle, G. E. Taylor.
Mon. Not. R. Astron. Soc., Vol. 200, 65P - 68P (1982).
A minimum value of 110 km for the diameter of Artemis is derived from observations of its occultation of a star.

098.007 **Observations of selected minor planets at the Nikolaev Observatory in 1974 - 1975.**
V. I. Voronenko, G. K. Gorel', L. A. Gudkova, F. F. Kalikhevich.
Glav. astron. obs. AN SSSR. Leningrad, 1982, 53 pp. In Russian. – Abstr. in Ref. zh., 51. Astron., 7.51.135 (1982).

098.008 **Physical studies of asteroids VII: the unusual rotation on M and CMEU asteroids.**
V. Zappalà, H. Debehogne, C.-I. Lagerkvist, H. Rickman.
Astron. Astrophys., Suppl. Ser., Vol. 50, 23 - 26 (1982).
This paper presents observational results on the CMEU asteroid 125 Liberatrix. It shows a very short rotational period, $P_{syn} = 3\overset{h}{.}969 \pm 0\overset{h}{.}001$, and a rather large light curve amplitude, $A = 0.32 \pm 0.02$ mag.

098.009 **Positions of asteroids obtained during August and September 1981 with the GPO telescope at ESO, Chile.**
H. Debehogne, G. Hahn, C.-I. Lagerkvist.
Astron. Astrophys., Suppl. Ser., Vol. 50, 73 - 75 (1982).
The authors present positions of 22 numbered and 3 unnumbered asteroids.

098.010 **Positions of selected minor planets (1980 - 1981).**
G. De Sanctis, W. Ferreri, V. Zappalà.
Astron. Astrophys., Suppl. Ser., Vol. 50, 87 - 91 (1982).
365 precise positions are presented for 29 asteroids observed during July 1980 - October 1981 at the Observatory of Torino. The reductions were made by the method of dependences, using 6 reference stars.

098.011 **The origin of Apollo-Amor objects.**
Y. Yamada.
J. British Interplanet. Soc., Vol. 35, 459 - 461 (1982).
Although Apollo-Amor asteroids have been regarded as plausible candidates for extinct comets, their aphelia are well inside Jupiter's orbit, unlike those of short-period comets. This aphelion reduction is difficult to explain by Whipple's jet effect. Hence the author hypothesizes that encounters with the Earth or Venus have reduced the aphelion distances. It is found that there is a remarkable agreement between the Jacobi constants of A-A objects and those of hypothetical comets that dip inside or graze the orbits of the Earth or Venus.

098.012 **Asteroid rotation – hunting for a record: 1689 Floris-Jan.**
H. J. Schober, J. Surdej.
Messenger, No. 29, p. 19 - 20 (1982).

098.013 **Two new minor planets – Moore and Taylor.**
J. British Astron. Assoc., Vol. 92, 259 (1982).

098.014 **Gli asteroidi.**
M. A. Barucci, E. Perozzi.
Coelum, Vol. 50, 237 - 246 (1982).

098.015 **The diameter-albedo relationship among Themis family members: variations in regolith thickness?**
J. Gradie, E. F. Tedesco.
Bull. American Astron. Soc., Vol. 14, 621 (1982). – Abstract.

098.016 **Variation of the mean and median inclinations in the numbered minor planet sample.**

Z. Knežević.
Bull. Astron. Inst. Czechoslovakia, Vol. 33, 267 - 281 (1982).

Variations of the mean and median minor planet inclinations with the increase of the numbered minor planet sample size are investigated. An analysis is presented of the role of real features of the minor planet system and of the selectional effects connected with discoveries and inclusion in the list of numbered objects. Available data are discussed and prediction of the future behaviour of variations attempted.

098.017 **An interesting difference in distributions according to the averaged motions of the largest S- and C-asteroids.**
D. I. Shestopalov.
Astron. vestn., Tom 16, 93 - 100 (1982). In Russian. – Abstr. in Ref. zh., 51. Astron., 9.51.138 (1982).

098.018 ***UBV* photometry of M-type asteroids 16 Psyche and 22 Kalliope.**
D. F. Lupishko, I. N. Bel'skaya, F. A. Tupieva, G. P. Chernova.
Astron. vestn., Tom 16, 101 - 108 (1982). In Russian. – Abstr. in Ref. zh., 51. Astron., 9.51.170 (1982).

098.019 **Three characteristic parameters of orbits of Hilda-type asteroids.**
J. Schubart.
Astron. Astrophys., Vol. 114, 200 - 204 (1982).

The 3/2-resonance case is studied by results of numerical integration of 34 real Hilda-type orbits and of some fictitious examples. In analogy to the regular case of asteroid motion, it is possible to characterize these orbits by three numerical parameters. The ways of deriving a proper inclination and a proper eccentricity are described. The orbits show libration of a special angular argument, and the amplitude caused by the typical period of libration in this argument corresponds to the third characteristic parameter. Values of the three parameters appear in a table. Rigorous computations of the effects by Jupiter and Saturn cover several 10^4 yr and indicate that the three characteristic parameters are nearly independent of time during long intervals.

098.020 **Preliminary narrowband reflectance spectra of outer solar system asteroids.**
F. Vilas, B. A. Smith, U. Fink.
Bull. American Astron. Soc., Vol. 14, 719 (1982). – Abstract.

098.021 **Outer solar system asteroids and comets: new VJHK data and interpretations.**
W. K. Hartmann.
Bull. American Astron. Soc., Vol. 14, 719 (1982). – Abstract.

098.022 **Asteroid taxonomy from cluster analysis of spectrophotometry: a progress report.**
D. J. Tholen.
Bull. American Astron. Soc., Vol. 14, 719 (1982). – Abstract.

098.023 **Infrared (JHK) photometry of asteroids.**
G. J. Veeder, D. L. Matson, C. Kowal.
Bull. American Astron. Soc., Vol. 14, 719 (1982). – Abstract.

098.024 **The 3 micron spectrum of asteroid 2 Pallas.**
M. A. Feierberg, H. P. Larson, L. A. Lebofsky.
Bull. American Astron. Soc., Vol. 14, 719 - 720 (1982). Abstract.

098.025 **Hirayama families: chips off the old block or collections of rubble piles?**
D. R. Davis, C. R. Chapman, R. Greenberg, S. J. Weidenschilling.
Bull. American Astron. Soc., Vol. 14, 720 (1982). – Abstract.

098.026 **The Nysa family: remnants of a strongly differentiated asteroid?**
E. F. Tedesco, J. Gradie, D. J. Tholen, B. Zellner.
Bull. American Astron. Soc., Vol. 14, 720 (1982). – Abstract.

098.027 **Asteroids with very long rotation periods.**
A. W. Harris, J. W. Young.
Bull. American Astron. Soc., Vol. 14, 724 (1982). – Abstract.

098.028 **The occultation diameter of 88 Thisbe.**
R. L. Millis, L. H. Wasserman, O. G. Franz, E. Bowell.
Bull. American Astron. Soc., Vol. 14, 725 (1982). – Abstract.

098.029 **The size, shape, and rotational pole of 433 Eros from speckle interferometry.**
J. D. Drummond, W. J. Cocke, E. K. Hege, P. A. Strittmatter.
Bull. American Astron. Soc., Vol. 14, 725 (1982). – Abstract.

098.030 **Radar detection of 2 Pallas, 8 Flora, and 2100 Ra-Shalom.**
S. J. Ostro, D. B. Campbell, I. I. Shapiro, M. R. Showalter.
Bull. American Astron. Soc., Vol. 14, 725 (1982). – Abstract.

098.031 **Upper limit on the radar cross section of a Pallas satellite.**
M. R. Showalter, S. J. Ostro, I. I. Shapiro, D. B. Campbell.
Bull. American Astron. Soc., Vol. 14, 725 (1982). – Abstract.

098.032 **The nature of the Kirkwood gaps in the asteroid belt.**
S. F. Dermott, C. D. Murray.
Bull. American Astron. Soc., Vol. 14, 725 - 726 (1982). Abstract.

098.033 **Lunarlike thermal emission and the standard asteroid radiometric model.**
W. W. Mendell, L. A. Lebofsky.
Bull. American Astron. Soc., Vol. 14, 726 (1982). – Abstract.

098.034 **Modelling speckle interferometric observations of asteroids.**
K. Lumme, E. Bowell, L. H. Wasserman, B. L. Morgan, H. A. Vine.
Bull. American Astron. Soc., Vol. 14, 726 (1982). – Abstract.

098.035 **A photoelectric light curve survey of small main belt asteroids.**
R. P. Binzel.
Bull. American Astron. Soc., Vol. 14, 728 (1982); – Abstract.

098.036 **Photometric geodesy of asteroids: a progress report.**
C. R. Chapman, D. R. Davis, R. Greenberg, S. J. Weidenschilling.
Bull. American Astron. Soc., Vol. 14, 728 - 729 (1982). Abstract.

098.037 **Physical studies of asteroids. VIII. Photoelectric photometry of the asteroids 42, 48, 93, 105, 145 and 245.**
H. Debehogne, C.-I. Lagerkvist, V. Zappalà.
Astron. Astrophys., Suppl. Ser., Vol. 50, 277 - 281 (1982).

The 50 cm photometric telescope at ESO, La Silla, was used for photoelectric photometry of the asteroids 42, 48, 93, 105, 145 and 245 during 6 nights in January 1980 and 8 nights during September 1981. The observed asteroids were of taxonomic types S, C and U. The rotational properties agree well with those expected for objects having such sizes; moreover, one possible LASPA (93 Minerva) is suggested by the results. The *B-V* and *U-B* colour indices obtained from the observations agree well with those adopted by Bowell et al. (1979).

098.038 **Photometry of the asteroid (5) Astraea.**
K. Pavlovski, Z. Knežević, M. Muminović.
Publ. Astron. Obs. Sarajevo, Vol. 1, No. 1, (see 012.043), p. 225 - 230 (1981).

098.039 **Dynamics of minor planets and comets: a review.**
L. Kresák.
Publ. Astron. Obs. Sarajevo, Vol. 1, No. 1, (see 012.043), p. 247 - 250 (1981).

098.040 **The dependence of asteroid light curves on the orientation parameters and the shapes of asteroids.**
M. A. Barucci, M. Fulchignoni.
Moon Planets, Vol. 27, 47 - 57 (1982).

The dependence of asteroidal light curves has been derived from the obliquity, aspect and phase angles. The effects of an angle variation are discussed taking into account the possible geometry of an asteroidal body. The amplitude-aspect relation is discussed for different asteroidal shapes. On the basis of this relation a graphical attempt to determine the value of the aspect from the light curves' amplitude is described.

098.041 **Physical studies of asteroids. IX. The light curve of the M asteroid 77 Frigga.**
C.-I. Lagerkvist, H. Rickman.
Moon Planets, Vol. 27, 107 - 110 (1982).

The results of photoelectric *UBV* observations of asteroid 77 Frigga during the 1982 opposition are presented. From eight nights of observations at phase angles smaller than 2°.8 a synodic period of $0^d.3755 \pm 0^d.0006$ is derived. The light curve appears very symmetric with two maxima per period and an amplitude of $0^m.19$. The primary maximum corresponds to $V(0°) = 8^m.58$, and the colour indices are: $B-V = 0^m.738 \pm 0^m.003$ and $U-B = 0^m.200 \pm 0^m.002$.

098.042 **The six-day rotation period of 1689 Floris-Jan: a new record among slowly rotating asteroids.**
H. J. Schober, J. Surdej, A. W. Harris, J. W. Young.
Astron. Astrophys., Vol. 115, 257 - 262 (1982).

The small sized asteroid 1689 Floris-Jan was observed photoelectrically in *UBV* at ESO and CTIO, Chile, and at Mt. Table Mountain JPL, California, during its opposition in 1980, between Oct. 7 and Nov. 6, 1980. A unique synodic rotation period $P = 145^h.0 \pm 0^h.5$ corresponding to $6^d.042 \pm 0^d.021$ could be derived from a lightcurve observed during 0.6 of the rotational phase. The lightcurve should show the usual double wave characteristic with an amplitude of 0.4 mag or slightly more. Absolute magnitudes were computed with a linear extrapolation yielding $\bar{V}(1,0) = 12.08$ and $V_0(1,0) = 11.88$; colors were derived as $B-V = 0.70 \pm 0.04$ and $U-B = 0.25 \pm 0.05$. A histogram is given for 300 published asteroid rotation rates in order to show the exceptional position of 1689 Floris-Jan among other asteroids.

098.043 **On asteroid classifications in families.**
A. Carusi, G. B. Valsecchi.
Astron. Astrophys., Vol. 115, 327 - 335 (1982).

The idea of asteroid families was born in 1918 with Hirayama. Since then seven classifications of asteroids into families have been given by Hirayama himself, by Brouwer, Arnold, Lindblad and Southworth, Carusi and Massaro, Kozai, Williams. These classifications differ from each other in the original data samples, in the adopted computational methods and in their results. Owing to the recent availability of physical data on asteroids contained in the TRIAD file, that allows not only a dynamical but also a physical approach to this very important problem, it is necessary to re-examine the history of these classifications, looking for similarities and differences, in order to develop a synoptic view of the situation.

098.044 **Observations of minor planets.**
Minor Planet Circ., (M. P. C.), Nos. 6992 - 6999, 7005 - 7010, 7059 - 7135, 7171 - 7224, 7282 - 7356, 7396 - 7444 (1982).

Observations made at the following stations are published: Belgrade, Caussols, Centro Astron. Yebes, Cerro el Roble, Cerro Tololo Interamerican Obs., Chamberlin Obs., Crimea (49th - 51st report), El Leoncito, European Southern Obs., Fabra Obs., Falkensee, Figl Obs., Flagstaff (U. S. Naval Obs.), Geisei, Gissar, Goethe Link Obs., Göttingen, Haute Provence, Hemingford Abbots, Kitt Peak, Kleť, Le Creusot, Lick Obs., Lincoln Lab., Lowell Obs., Lowell Obs. Anderson Mesa Stn., Meschede, Mt. John Obs., Mt. Palomar, Oak Ridge Obs., Ordubad, Peking, Perth, Pino Torinese, Piszkestetö (Hungary), Purple Mountain Obs., Reintal, S. Vittore (Bologna), Siding Spring, Sonneberg, Special Astrophys. Obs., Steward Obs., Tartu, Tautenburg, Turku, Uccle, Zimmerwald.

098.045 **Identifications and identification changes of minor planets.**
Minor Planet Circ., (M. P. C.), Nos. 6990 - 6991, 7055, 7167, 7280, 7395 - 7396 (1982).

098.046 **Ephemerides of minor planets and comets.**
Minor Planet Circ., (M. P. C.), Nos. 7028 - 7054, 7159 - 7166, 7241 - 7278, 7375 - 7394, 7475 - 7514 (1982).

098.047 **New names of minor planets.**
Minor Planet Circ., (M. P. C.), Nos. 7156 - 7159, 7471 - 7474 (1982).

098.048 **Orbital elements of one-opposition minor planets.**
Minor Planet Circ., (M. P. C.), Nos. 7011, 7136 - 7139, 7224 - 7225, 7356 - 7360, 7444 - 7447 (1982).

098.049 **Index to orbital elements of comets and minor planets.**
Minor Planet Circ., (M. P. C.), Nos. 7000 - 7005 (1982).

Index concerning all elements published in M. P. C., Nos. 6000 - 6999.

098.050 **Critical List of minor planets.**
Minor Planet Circ., (M. P. C.), No. 7279 (1982).

098.051 **Orbital elements of numbered minor planets.**
Minor Planet Circ., (M. P. C.), Nos. 6989 - 7514 (1982).

The minor planets are listed according to their definitive number. Newly numbered objects are indicated by an asterisk. The names of the computers are given behind the respective M. P. C. numbers. (15) 7447 L. D. Schmadel; (944) 7448 W. Landgraf; (1009) 7225 - 7226 L. K. Kristensen; (1284), (1313) 7448 L. D. Schmadel; (2060) 7449 W. Landgraf; (2677)* - (2688)* 7011 - 7015 C. M. Bardwell; (2689)* - (2696)* 7017 - 7019 B. G. Marsden; (2697)* - (2711)* 7022 - 7027, 7140 T. Urata; (2712)* - (2731)* 7141 - 7148 C. M. Bardwell; (2732)* - (2736)* 7150 - 7151 B. G. Marsden; (2737)* 7226 L. D. Schmadel; (2738)* 7226 - 7227 S. Nakano; (2739)* - (2748)* 7228 - 7232 T. Urata; (2749)* - (2763)* 7234 - 7239 C. M. Bardwell; (2764)* - (2766)* 7361 - 7362 T. Urata; (2767)* - (2773)* 7363 - 7365 B. G. Marsden; (2774)* - (2782)* 7369 - 7371 C. M. Bardwell; (2783)* - (2788)* 7450 - 7452 T. Urata; (2789)* - (2800)* 7455 - 7459 B. G. Marsden; (2801)* - (2818)* 7462 - 7467 C. M. Bardwell.

098.052 **Orbital elements of unnumbered minor planets.**
Minor Planet Circ., (M. P. C.), Nos. 6989 - 7514 (1982).

The unnumbered minor planets are listed according to

their preliminary designation. Objects from the Palomar-Leiden Survey are sorted by number. The names of the orbit computers are given behind the respective M. P. C. numbers. [1930 YV] 7467 C. M. Bardwell; [1931 TK] 7233 T. Urata; [1931 TY_1] 7468, [1933 QU] 7239, [1935 QF] 7468 C. M. Bardwell; [1937 GG] 7019 B. G. Marsden; [1938 DN] 7468, [1938 SL] 7148, [1938 SD_1] 7148, [1940 WG] 7371 - 7372 C. M. Bardwell; [1940 YF] 7449 S. Nakano; [1941 UP] 7468, [1942 EB] 7239, [1953 NB] 7015, [1955 QP_1] 7015 - 7016, [1957 UK_1] 7239 C. M. Bardwell; [1959 VF] 7452 T. Urata; [1961 TA], [1964 TH_1] 7372 C. M. Bardwell; [1964 VE] 7459 B. G. Marsden; [1964 XA] 7233 T. Urata; [1965 SB] 7365 - 7366 B. G. Marsden; [1966 AA] 7233 T. Urata; [1969 TD_2] 7240 C. M. Bardwell; [1969 TE_2] 7227 S. Nakano; [1970 PA] 7366 B. G. Marsden; [1971 MG] 7240 C. M. Bardwell; [1971 UP] 7366 B. G. Marsden; [1972 NW], [1972 QM] 7372 C. M. Bardwell; [1972 RX_3] 7151 B. G. Marsden; [1973 DP] 7468 - 7469 C. M. Bardwell; [1973 SM_2] 7453 T. Urata; [1973 SD_3] 7373 C. M. Bardwell; [1973 SX_3] 7366, [1974 OA_1] 7152 B. G. Marsden; [1974 SP] 7227 S. Nakano; [1974 SO_2] 7362 T. Urata; [1974 VQ_2] 7240 C. M. Bardwell; [1975 PA] 7155 - 7156 G. R. Kastel'; [1975 SF] 7362 T. Urata; [1975 VD] 7459 B. G. Marsden; [1975 VS_5] 7140 T. Urata; [1975 VB_9] 7227 S. Nakano; [1976 GN_3] 7453 T. Urata; [1976 GN_8] 7152 B. G. Marsden; [1976 UF_1] 7469 C. M. Bardwell; [1977 DT_4], [1977 EO_1] 7228 S. Nakano; [1977 SS_1] 7459 - 7460, [1978 QJ] 7152, [1978 QB_3] 7367 B. G. Marsden; [1978 RO] 7469 C. M. Bardwell; [1978 RY] 7152 - 7153, [1978 RW_1] 7153, [1978 SE_1] 7367 B. G. Marsden; [1978 SA_3], [1978 SB_5] 7453 T. Urata; [1978 SP_7] 7153 B. G. Marsden; [1978 TB_7] 7228 S. Nakano; [1978 UC] 7367 B. G. Marsden; [1978 VB_5] 7140 T. Urata; [1979 OB] 7020 B. G. Marsden; [1979 QU_9] 7140 - 7141 T. Urata; [1979 SF_2] 7240 C. M. Bardwell; [1979 SD_7] 7141 T. Urata; [1979 SG_9] 7156 G. R. Kastel'; [1979 SJ_9] 7153 B. G. Marsden; [1979 SX_9] 7148 - 7149, [1979 SY_9] 7149, [1979 SS_{11}], [1979 SV_{11}], [1979 SW_{11}], [1979 SA_{12}] 7373 C. M. Bardwell; [1979 UC] 7154 B. G. Marsden; [1979 WO] 7374 C. M. Bardwell; [1980 DA] 7233 T. Urata; [1980 DC] 7367, [1980 LA] 7154, [1980 MD] 7367 - 7368 B. G. Marsden; [1980 RT_2] 7149 C. M. Bardwell; [1980 SH] 7020 B. G. Marsden; [1980 TB_4] 7141, [1980 TX_5] 7454 T. Urata; [1980 UA] 7020 B. G. Marsden; [1980 VN_1] 7016 C. M. Bardwell; [1980 XE] 7368 B. G. Marsden; [1980 YL] 7449 S. Nakano; [1981 AA] 7154 B. G. Marsden; [1981 ET_3] 7234 T. Urata; [1981 GX] 7460 B. G. Marsden; [1981 JM] 7016 C. M. Bardwell; [1981 JR] 7154, [1981 JS] 7155 B. G. Marsden; [1981 JA_3] 7016 C. M. Bardwell; [1981 KE] 7460 B. G. Marsden; [1981 QJ] 7360 - 7361 L. D. Schmadel; [1981 QJ_1] 7460 B. G. Marsden; [1981 QU_2] 7027 T. Urata; [1981 RF] 7368 B. G. Marsden; [1981 SM_1] 7362 T. Urata; [1981 UV_9] 7469 C. M. Bardwell; [1981 WE] 7449 - 7450, [1981 WD_4], [1981 XF_2], [1981 XK_2] 7450 S. Nakano; [1981 YB] 7469 - 7470, [1981 YO] 7241 C. M. Bardwell; [1982 BH] 7020 B. G. Marsden; [1982 BG_1] 7016 C. M. Bardwell; [1982 FK] 7028 T. Urata; [1982 FR] 7149 C. M. Bardwell; [1982 FT] 7234 T. Urata, 7461 B. G. Marsden; [1982 FE_1] 7141, [1982 HL] 7363 T. Urata; [1982 HS] 7461 B. G. Marsden; [1982 HV] 7374, [1982 HE_1] 7017, [1982 HN_1] 7374 C. M. Bardwell; [1982 KM] 7363 T. Urata; [1982 KD_1] 7374 C. M. Bardwell; [1982 RA] 7234 T. Urata, 7368, 7461 B. G. Marsden; [1982 RB] 7368, 7461, B. G. Marsden; [1982 SK], [1982 SL] 7470 C. M. Bardwell; [1982 TA] 7369, 7461 B. G. Marsden; [1982 TH] 7470, [1982 UH] 7470 - 7471 C. M. Bardwell; [2017 P-L] 7461, [3042 P-L], [3071 P-L] 7155, [4260 P-L] 7020 - 7021, [4585 P-L] 7021 B. G. Marsden; [7633 P-L] 7374 - 7375 C. M. Bardwell.

098.053 **Recent progress in the theory of Trojan asteroids.**
B. Garfinkel.
Celestial Mech., Vol. 28, (see 012.049), 153 - 154 (1982). Abstract.

098.054 **Numerical determination of proper inclinations of Hilda-type asteroids.**
J. Schubart.
Celestial Mech., Vol. 28, (see 012.049), 189 - 194 (1982).

Proper inclination and proper eccentricity are constants of integration in the theory of secular perturbations of an asteroid. This theory is based on simplified differential equations, and therefore it is not applicable to Hilda-type motion, a case of 3/2 resonance with respect to Jupiter. Numerical integration is necessary for an analogous study of the eccentric Hilda-type orbits. Values of proper eccentricity of real Hildas had resulted from such a study. Here the conditions of nature are approximated stepwise by proceeding from a simplified model to the rigorous problem Sun–Jupiter–Saturn–asteroid. This procedure leads to a way of determining proper inclinations of Hildas. Long-period integrations do not show variations of these characteristic values.

098.055 **(1009) Sirene.**
IAU Circ., No. 3714 (1982).

098.056 **1982 RA.**
IAU Circ., Nos. 3725, 3726, 3728, 3744 (1982).

098.057 **1982 RB.**
IAU Circ., Nos. 3725, 3727, 3732, 3742 (1982).

098.058 **1982 TA.**
IAU Circ., Nos. 3735, 3738 (1982).

098.059 **(9) Metis.**
IAU Circ., Nos. 3739, 3753 (1982).

098.060 **Occultation of BD +29° 579 by (93) Minerva.**
IAU Circ., No. 3746 (1982).

098.061 **Occultation of BD +24° 522 by (690) Wratislavia.**
IAU Circ., No. 3747 (1982).

098.062 **Occultation of BD +38° 542 by (375) Ursula.**
IAU Circ., No. 3751 (1982).

098.063 **1982 XB.**
IAU Circ., Nos. 3754, 3755, 3758 (1982).

098.064 **1982 YA.**
IAU Circ., No. 3758 (1982).

098.065 **1982 RA**
Yamamoto Circ., No. 1982 (1982).

098.066 **1982 RB**
Yamamoto Circ., No. 1982 (1982).

098.067 **1982 TA**
Yamamoto Circ., Nos. 1983, 1984 (1982).

098.068 **Orbital elements of some unnumbered asteroids.**
J.-x. Zhang, J.-x. Yang, S.-l. Wei, Q. Wang, Y.-l. Ge, C.-l. Yuan.
Astron. Circ., No. 14, p. 1 - 22 (1982). In Chinese and English.

098.069 **Lightcurves of variable asteroids (V).**
X.-h. Zhou, X.-y. Yang, Z.-x. Wu.
Acta Astron. Sinica, Vol. 23, 349 - 352 (1982). In Chinese.

The results of photoelectric observations of the lightcurves of some asteroids are presented and discussed.

098.070 **Independent discovery of the fast-moving asteroid 1982 BB at the Crimean Astrophysical Observatory.**
N. S. Chernykh.
Komet. Tsirk., Kiev, No. 285 (1982). In Russian.

098.071 **Observations of the asteroid (40) Harmonia.**
H. K. Raudsaar.
Komet. Tsirk., Kiev, No. 288 (1982). In Russian.

098.072 **Observations of 1 Ceres and 2 Pallas at centimeter wavelengths.**
K. J. Johnston, P. K. Seidelmann, C. M. Wade.
Astron. J., Vol. 87, 1593 - 1599 (1982)

The minor planets 1 Ceres and 2 Pallas were observed with the VLA during May and December 1981. The brightness temperature of 1 Ceres appears to be significantly lower than expected for a rapidly rotating blackbody, while that of 2 Pallas is near the expected value. The observed temperature indicates that the diameter of 1 Ceres is 818 ± 82 km. The position of 1 Ceres was found to be $-0\farcs3 \pm 0\farcs1$ from the predicted optical position, while 2 Pallas was found to be $0\farcs8 \pm 0\farcs1$ from its predicted optical position, disregarding the yet undetermined detailed relationship of the FK4 system to the quasar reference frame.

098.073 **Photographic positions of minor planets 1980.**
J.-x. Yang, S.-l. Wei, Q. Wang, Y.-l. Ge, C.-l. Yuan.
Publ. Purple Mountain Obs., Vol. 1, 55 - 72 (1982). In Chinese.

098.074 **Positions of main belt asteroids.**
H. Debehogne, G. De Sanctis, V. Zappalà.
Acta Astron., Vol. 32, 137 - 139 (1982).

114 positions of 23 main belt asteroids are given.

098.075 **Minor planets positions in April 1979.**
H. Debehogne, L. Houziaux.
Acta Astron., Vol. 32, 141 - 142 (1982).

Positions of asteroids 426, 639, 997, 1321, 1322, 1491 and 1979 HC are given.

098.076 **The distribution of orbital eccentricities of minor planets.**
R. A. Beck.
Irish Astron. J., Vol. 15, 87 - 92 (1981).

Plummer's expressions, describing some of the main features of the eccentricity distribution of minor planet orbits, are discussed. This distribution appears to be independent of the physical size of the minor planets in a sample. Other classes of solar system objects such as short-period comets and planetary satellites seem to show orbital eccentricity distributions quite different from that for the minor planets; thus the latter is unique according to our present knowledge.

098.077 **Asteroids.**
H. Scholl.
Formation of planetary systems, (see 012.064), p. 833 - 855 (1982).

Contents: Introduction. Mass, diameter, density. The taxonomic system. Surface mineralogy. Rotation, shape, asteroidal satellites. Dynamical classification. Hirayama families and groups. Resonances. Interrelation with meteorites and comets. Origin of asteroids. Spaceflights to asteroids.

098.078 **Calibration of the radiometric asteroid scale using occultation diameters.**
R. H. Brown, D. Morrison, C. M. Telesco, W. E. Brunk.
Icarus, Vol. 52, 188 - 195 (1982).

The authors report a new approach to the calibration of the radiometric asteroid scale using the recent accurate occultation measurements of the diameters of 2 Pallas and 3 Juno, and the Voyager diameter of J4 Callisto, and new infrared photometry of these objects obtained with the NASA 3-m Infrared Telescope Facility. This calibration is internally consistent to better than 5% and probably has an absolute accuracy of ± 5%.

098.079 **Photoelectric photometry of asteroids 33 Polyhymnia and 386 Siegena.**
V. Zappalá, F. Scaltriti, C.-I. Lagerkvist, H. Rickman, A. W. Harris.
Icarus, Vol. 52, 196 - 201 (1982).

Photometric parameters of the asteroids 33 Polyhymnia and 386 Siegena were obtained during an international campaign performed at three observatories: Table Mountain Observatory, European Southern Observatory, and Catania Astrophysical Observatory. The observations were carried out in the period August 15 - September 14, 1980. Color-indices, rotational periods and amplitudes, and derived albedos are presented.

098.080 **Rotation of asteroids and planetary axial rotation theory.**
N. A. Barricelli.
Moon Planets, Vol. 27, 179 - 185 (1982).

A theory (Barricelli, 1972) developed for the interpretation of planetary axial rotation is applied to an interpretation of the axial rotations of major asteroids. The interpretation is based on the assumption that also asteroids can have satellite-systems, which may influence the axial rotation of the respective primaries. The reason why smaller asteroids tend to have slower axial rotation than the major ones is discussed. Predictions of the theory can be tested by space-craft exploration of asteroids.

098.081 **A qualitative analysis of motion of Hecuba-type asteroids.**
I. A. Gerasimov.
Astron. Zh., Tom 59, 1190 - 1199 (1982). In Russian.
English translation in Soviet Astron., Vol. 26, No. 6.

The Delaunay - Hill plane-version averaging method is applied to the circular restricted three-body problem when the mean motions are close to the 2 : 1 commensurability. A qualitative analysis is carried out when the first four terms in the perturbation function are taken into account. Five types of phase trajectories are discovered. Stability of stationary solutions is considered.

098.082 **Planet-crossing asteroid survey.**
C. A. Atallah.
Advances in planetary geology, (see 003.014), p. 541 - 544 (1982). – Abstract.

A new version of TRIAD (*Tucson Revised Index of Asteroid Data*).
See Abstr. 002.076.

Asteroid data files for mission analysis.
See Abstr. 002.077.

Storage of the TRIAD-file (Tucson Revised Index of Asteroid Data) at the Centre de Données Stellaires (CDS)?
See Abstr. 002.094.

Early planning for the occultation of 1 Vulpeculae by Pallas on 1983 May 29.
See Abstr. 013.081.

A Monte Carlo approach to scattering off surfaces of triaxial ellipsoids.
See Abstr. 021.002.

Close-approach predictions for comets and asteroids.
See Abstr. 021.042.

The eight-color asteroid survey: standard stars.
See Abstr. 031.625.

Digital speckle interferometry of Juno, Amphitrite and Pluto's moon Charon.
See Abstr. 031.679.

A portable system for high-speed photometry of occultations.
See Abstr. 032.017.

Asteroid occultation astrometry with a lensless Schmidt.
See Abstr. 034.018.

Trojan-type orbits in the system of two gravitational centers with variable masses and separation.
See Abstr. 042.028.

Collisional lifetimes of Trojan asteroids.
See Abstr. 042.036.

The resurrection of Laplace's method of initial orbit determination.
See Abstr. 042.039.

Recent progress in the theory of Trojan asteroids.
See Abstr. 042.063.

On the variation of the Jacobi constant of Trojan asteroids in the elliptic restricted problem of three bodies.
See Abstr. 042.064.

Twice averaged restricted three-body problem and its application to study the evolution of orbits of asteroids.
See Abstr. 042.065.

New light on Brouwer's project of observations of minor planets and on Pierce's determination of a dynamical reference frame.
See Abstr. 043.001.

Near-earth asteroids: a survey of ballistic rendezvous and sample return missions.
See Abstr. 051.023.

Planetary and asteroidal occultations in 1982.
See Abstr. 096.004.

A photographic search for stars to be occulted by the larger asteroids in 1983.
See Abstr. 096.006.

Recent asteroidal occultation attempts, and updates for upcoming events.
See Abstr. 096.010.

Selection of asteroids for occultation searches.
See Abstr. 096.013.

Asteroidal occultations of uncatalogued stars during 1983.
See Abstr. 096.014.

Planetary occultations during 1983.
See Abstr. 096.017.

Astrometric observations of outer planets and minor planets: 1980 - 1982.
See Abstr. 101.001.

The rotation of comet nuclei.
See Abstr. 102.017.

Do comets evolve into asteroids? Evidence from physical studies.
See Abstr. 102.033.

Comets, asteroids and their interrelations.
See Abstr. 102.083.

The early evolution of planets.
See Abstr. 107.033.

Accretion of the jet stream and the formation of asteroids.
See Abstr. 107.055.

099 Jupiter, Jupiter Satellites

099.001 **Results of the PHEMU79 observation campaign of mutual phenomena of the Galilean satellites of Jupiter in 1979.**
J.-E. Arlot, A. Bernard, P. Bouchet, J. Daguillon, G. Dourneau, A. Figer, G. Helmer, J. Lecacheux, P. Merlin, C. Meyer, P. Mianes, B. Morando, D. Naves, J. Rousseau, G. Soulié, A. Terzan, W. Thuillot, L. Vapillon, G. Wlérick.
Astron. Astrophys., Vol. 111, 151 - 170 (1982). In French.
In 1979 and 1980, mutual phenomena of the Galilean satellites of Jupiter occurred, the jovian declinations of the earth and the sun being equal to zero during this period. Nine phenomena were observed. In 1979: January 14 (J2 occults J3); January 28 (J2 eclipses J3 and J2 occults J3); June 16 (J3 eclipses J1); October 1(J1 eclipses J2 and J1 occults J2); October 6 (J2 eclipses J3); October 8 (J1 eclipses J2) and in 1980: April 18 (J2 occults J1). Except for the occultation on January 28, 1979 an analysis was possible. A discussion is made for each observation and several methods of reduction are used.

099.002 **Jupiter "without a visible satellite".**
R. W. Sinnott.
Sky Telesc., Vol. 64, 214 - 215 (1982).

099.003 **A narrow-band splitting at the Jovian decametric cutoff frequency.**
Y. Leblanc, M. Rubio.
Astron. Astrophys., Vol. 111, 284 - 294 (1982).
The authors study the characteristics of the narrow-band splitting in the Jovian decametric emission as it can be observed on the dynamic spectra in a wide range of frequencies (from 20 to 40 MHz), and the implications on the theories. The authors give a brief description of the instrumentation and the conditions of observations. Then they develop the different kinds of splitting, either associated with *S*-bursts or with continuum emission and their occurrence over two years of observations. Finally the authors consider the radiophysics of Jupiter related to the splitting and the arc structure models.

099.004 **Possible terrestrial analogs of Valhalla and other ripple-ring basins.**
C. A. Wood.
Multi-ring basins, (see 012.001), p. 173 - 180 (1981).
The most remarkable feature on Callisto is Valhalla, a 3000 km wide structure comprised of dozens of concentric ridges and scarps surrounding a central smooth zone. The close morphologic resemblance of Valhalla to an unusual type of collapse structure in Icelandic glaciers is documented to suggest that processes other than impact may have contributed to the formation of multiple-ringed basins on the Galilean satellites.

099.005 **An estimate of the PH_3, CH_3D, and GeH_4 abundances on Jupiter from the Voyager IRIS data at 4.5 μm.**
P. Drossart, T. Encrenaz, V. Kunde, R. Hanel, M. Combes.
Icarus, Vol. 49, 416 - 426 (1982).
The abundances of PH_3, CH_3D, and GeH_4 are derived from the 2100- to 2250-cm^{-1} region of the Voyager 1 IRIS spectra. No evidence is seen for large-scale variations of the phosphine abundance over Jovian latitudes between -30 and $+30°$. In the atmospheric regions corresponding to 170–200°K, the derived PH_3/H_2 value is $(4.5 \pm 1.5) \times 10^{-7}$ or 0.75 ± 0.25 times the solar value. This result, compared with other PH_3 determinations at 10 μm, suggests that the PH_3/H_2 ratio on Jupiter decreases with atmospheric pressure.

099.006 **Io: geochemistry of sulfur.**
J. S. Lewis.
Icarus, Vol. 50, 103 - 114 (1982).
The evidence from Voyager imaging, Earth-based spectral reflectivity studies, and thermal emission measurements combine to suggest an extremely fresh, volcanically recycled sulfur-rich crust for Io, with very shallow large-scale melting. Two present styles of volcanism are possible, depending on the thickness of local deposits of sulfur: shallow liquid sulfur magma generation with quiescent flooding, and high-temperature volcanism with violent eruption of a sulfur-iron magma driven by SO_2.

099.007 **Photographic observations of Jupiter and Saturn satellites at the Nikolaev Observatory in 1976 - 1979.**
V. I. Voronenko, G. K. Gorel'.
Glav. astron. obs. AN SSSR. Leningrad, 1982, 16 pp. In Russian. – Abstr. in Ref. zh., 51. Astron., 7.51.138 (1982).

099.008 **IUE observations of the Jovian H I Lyman α emission (1979–1982).**
T. E. Skinner, S. T. Durrance, P. D. Feldman, H. W. Moos.
Advances in ultraviolet astronomy, (see 012.006), p. 297 - 299 (1982).
Observations of the Jovian H I Lyman-α emission made with the IUE observatory beginning in December 1978 just before the time of the Voyager encounters and extending through January 1982 are presented. The results argue strongly that the hydrogen bulge has been a permanent, though variable, feature of the Jovian atmosphere for the 3 year period covered. Moreover, the equatorial region of the planet away from the bulge has shown a remarkable consistency in Lα brightness over this same period. No significant change in the atomic hydrogen abundance in the Jovian upper atmosphere has occurred between the time of the Voyager encounters and now.

099.009 **Investigations of the ultraviolet albedo of Jupiter and Saturn.**
P. Winkelstein, J. Caldwell, T. Owen, M. Combes, T. Encrenaz, G. Hunt, V. Moore.
Advances in ultraviolet astronomy, (see 012.006), p. 300 (1982). – Abstract.

099.010 **Ultraviolet spectroscopy of the Jovian and Saturnian aurorae.**
S. T. Durrance, P. D. Feldman, H. W. Moos.
Advances in ultraviolet astronomy, (see 012.006), p. 301 - 305 (1982).
For Jupiter the authors present the results of a series of IUE observations of the north polar aurora. From these data a spectrum of the aurora with high signal-to-noise ratio, and a resolution of about 8 Å is obtained, making possible the identification of many H_2 Lyman- and Werner-bands. For Saturn the authors present the results of several IUE observations of the full disk. The H_2 Lyman- and Werner-bands were observed near the north pole in two of them.

099.011 **High dispersion IUE observations of Jupiter.**
R. Wagener, T. Owen, J. Caldwell, P. Winkelstein.
Advances in ultraviolet astronomy, (see 012.006), p. 311 (1982). – Abstract.

099.012 **Thermal structure of Jupiter and Saturn.**
G. Orton.
Vibrational-rotational spectroscopy for planetary atmospheres, (see 012.007), p. 333 - 362 (1982).
The author discusses the means by which one derives

the vertical temperature profiles for Jupiter and Saturn, the molecular spectroscopic data which form the basis for this type of analysis, and the problems involved with these models for molecular absorption and with the recovery technique in general. He gives some results for Jupiter and Saturn from the Pioneer and Voyager infrared experiments.

099.013 **Composition of Jupiter.**
R. Prinn.
Vibrational-rotational spectroscopy for planetary atmospheres, (see 012.007), p. 363 - 385 (1982).
The author reviews our understanding of the chemistry of Jupiter's atmosphere. He indicates which remote observations of the planet are most likely to yield new knowledge.

099.014 **Voyager IRIS measurements of the Jovian atmosphere.**
V. G. Kunde.
Vibrational-rotational spectroscopy for planetary atmospheres, (see 012.007), p. 387 - 405 (1982).
The author reviews some of the Jovian atmospheric observations from the Voyager infrared investigation (IRIS). He emphasizes several spectroscopic problem areas presently limiting scientific interpretation of the data.

099.015 **Jupiter and Saturn from 2 to 6 μm.**
H. P. Larson.
Vibrational-rotational spectroscopy for planetary atmospheres, (see 012.007), p. 407 - 427 (1982).
This paper summarizes some recent earth-based spectroscopic observations that support current compositional and structural analyses of planetary atmospheres.

099.016 **Gaseous SO_2 on Io.**
J. Pearl.
Vibrational-rotational spectroscopy for planetary atmospheres, (see 012.007), p. 657 - 672 (1982).
The author discusses the evidence for SO_2 on Io, beginning with the thermal infrared, then the near infrared, some recent and ongoing work in the ultraviolet, and finally some charged particle measurements which support the presence of SO_2.

099.017 **Observations of the Jupiter S-bursts between 3.2 and 32 MHz.**
G. R. A. Ellis.
Australian J. Phys., Vol. 35, 165 - 175 (1982).
Observations of the Jupiter S-bursts in the frequency range 3.2 - 32 MHz are described. Their general properties including bandwidth, time duration, frequency-time behaviour, and rate of occurrence within this frequency range are analysed and discussed.

099.018 **Jupiter: présentation 1981.**
F. Jetzer.
Orion, 40. Jahrg., 128 - 129 (1982).

099.019 **Results of studies of the reflection power of Jovian disc features in the period 1977 - 1979.**
V. M. Klimenko.
Physics of planetary atmospheres, (see 003.003), p. 63 - 91 (1981). In Russian. – Abstr. in Ref. zh., 51. Astron., 8.51.189 (1982).

099.020 **Structure of the upper layers of the Jovian atmosphere. I. Correctness of interpretation of spectroscopic observations.**
L. A. Bugaenko, L. M. Kislyuk, A. V. Morozhenko.
Physics of planetary atmospheres, (see 003.003), p. 27 - 44 (1981). In Russian. – Abstr. in Ref. zh., 51. Astron., 8.51.190 (1982).

099.021 **Structure of the upper layers of the Jovian atmosphere. II. Content of methane and ammonia, rotational temperature and pressure.**
L. A. Bugaenko, L. M. Kislyuk, A. V. Morozhenko.
Physics of planetary atmospheres, (see 003.003), p. 44 - 62 (1981). In Russian. – Abstr. in Ref. zh., 51. Astron., 8.51.191 (1982).

099.022 **Temperature regime of the surface of the Galilean satellites and some properties of their atmospheres.**
A. F. Steklov.
Physics of planetary atmospheres, (see 003.003), p. 91 - 108 (1981). In Russian. – Abstr. in Ref. zh., 51. Astron., 8.51.200 (1982).

099.023 **On the structure of the Io torus.**
C. K. Goertz, W.-H. Ip.
Planet. Space Sci., Vol. 30, 855 - 864 (1982).
The authors propose that the configuration of the Io torus is largely regulated by a field-aligned current system established in its vicinity as a result of mass loading of the magnetosphere. The principal characteristics and possible physical consequences of this current system are discussed.

099.024 **The spectrum of the Jovian aurora 1150–1700 Å.**
S. T. Durrance, P. D. Feldman, H. W. Moos.
Geophys. Res. Lett., Vol. 9, 652 - 655 (1982).
The authors present the results of a series of IUE observations of the north polar aurora of Jupiter obtained during a substantial fraction of one complete rotation of the planet. From these data a spectrum of the aurora with high signal-to-noise ratio and a resolution of about 8 Å is obtained, making possible the identification of many H_2 Lyman- and Werner-bands. The spectrum is of sufficient quality to provide reliable quantitative data for a comparison with the model atmosphere calculations. The lack of an observable absorption signature makes it possible to set an upper limit on the column density of CH_4 and C_2H_6 above the auroral emissions and hence an upper limit on the primary particle energies. A comparison of this spectrum with a laboratory spectrum of discharge-excited H_2 shows a remarkable similarity.

099.025 **Evidence for a distant ($> 8{,}700\ R_J$) Jovian magnetotail: Voyager 2 observations.**
R. P. Lepping, L. F. Burlaga, M. D. Desch, L. W. Klein.
Geophys. Res. Lett., Vol. 9, 885 - 888 (1982).
A correlative survey of Magnetometer and Planetary Radio Astronomy 1.2 kHz continuum radiation measurements from Voyager 2 provide evidence for at least eight distant Jovian magnetotail sightings occurring about once a month over the first 2/3 of 1981 at distances of $\sim 5{,}000$ to $\sim 9{,}000\ R_J$.

099.026 **Fabry-Perot/CCD observations of [S III] and [S II] emissions from the Jupiter plasma torus.**
F. L. Roesler, R. J. Oliversen, F. Scherb, J. Lattis, T. B. Williams, D. G. York, E. B. Jenkins, J. L. Lowrance, P. Zucchino, D. Long.
Astrophys. J., Vol. 259, 900 - 907, plates 16, 17 (1982).
The authors have observed the emissions [S II]λ6716 and [S III]λ9531 from the Jupiter plasma torus with an imaging spectrometer consisting of a narrow-band, tunable Fabry-Perot interferometer coupled to a CCD camera. The extension of the [S III] and [S II] images perpendicular to the symmetry plane of the torus was used to estimate the temperatures of the sulfur ions. A [S III] image obtained on 1980 March 13 gave a S^{2+} temperature between 2.2×10^5 K and 6.5×10^5 K. A [S II] image obtained on 1980 March 14 gave a S^+ temperature of $(5.8 \pm 0.8) \times 10^4$ K.

099.027 **Laboratory studies of charged particle erosion of SO_2 ice and applications to the frosts of Io.**

L. J. Lanzerotti, W. L. Brown, W. M. Augustyniak, R. E. Johnson, T. P. Armstrong.
Astrophys. J., Vol. 259, 920 - 929 (1982).

The removal and/or redistribution of SO_2 frosts on the surface of the first Galilean satellite, Io, can occur through the erosion of these frosts by the magnetosphere particle environment of the satellite. The authors have studied in laboratory experiments the energy, species, and temperature dependence of the erosion rates of SO_2 ice films by charged particles. Rutherford backscattering and thin film techniques are used in the experiments. The laboratory results are used together with measured magnetosphere particle fluxes in the vicinity of Io to estimate the erosion rates of SO_2 ice films from the satellite and implications therefrom on an SO_2 atmosphere on Io.

099.028 **Rheology of ices: a key to the tectonics of the ice moons of Jupiter and Saturn.**
J. P. Poirier.
Nature, Vol. 299, 683 - 688 (1982).

The polymorphic varieties of water ice are important rock-forming minerals in the outer solar system. The Voyager 1 and 2 missions have shown that most of the ice moons of Jupiter and Saturn have had some form of tectonic or internal activity. Understanding of the tectonics and building reasonable models of the interior of these planets depend on our knowledge of the rheological properties and viscosity of high-pressure and low-temperature forms of ice.

099.029 **Voyager photometry of Europa and the satellites of Saturn.**
B. Buratti, J. Veverka.
News Lett. Astron. Soc. N.Y., Vol. 2, No. 2, p. 18 - 19 (1982). Abstract.

099.030 **Possible long term changes in the equatorial zone of Jupiter.**
R. Tatum.
Strolling Astron., Vol. 29, 141 - 143 (1982).

099.031 **Some current events on Jupiter.**
Strolling Astron., Vol. 29, 166 - 168 (1982).

099.032 **Radio emission from Io.**
I. de Pater, W. J. Jaffe, R. A. Brown, G. L. Berge.
Astrophys. J., Vol. 261, 396 - 401 (1982).

From radio observations of Io taken with the VLA at 2, 6, and 21 cm disk brightness temperatures of T_b (2 cm) = = 98 ± 17 K, T_b (6 cm) = 85 ± 16 K, and T_b (21 cm) < 400 K are derived. These radio brightness temperatures are consistent with thermal emission expected from Io's surface on the basis of its infrared brightness temperature and its radio emissivity derived from radar studies. No evidence for non-thermal radio emission from Io has been found.

099.033 **The meteorology of Jupiter and Saturn.**
P. Gierasch.
Bull. American Astron. Soc., Vol. 14, 606 (1982). – Abstract.

099.034 **Jovian decameter emissions.**
C. St. Cyr.
Bull. American Astron. Soc., Vol. 14, 621 (1982). – Abstract.

099.035 **Studies of Io's sodium cloud.**
B. A. Goldberg, R. W. Carlson, T. V. Johnson, S. K. Lavoie, D. L. Matson, J. W. Young.
Bull. American Astron. Soc., Vol. 14, 621 (1982). – Abstract.

099.036 **Jovian magnetosphere.**
A. Nishida.
Magnetospheric phenomena of celestial bodies, (see 012.011), p. 26 - 35 (1982). In Japanese.

The author summarizes unique characteristics of the Jovian magnetosphere, reviews suggested mechanisms of particle acceleration in the magnetosphere, and introduces his recent work on the energy conversion process associated with compressions or expansions of the magnetosphere.

099.037 **Decametric emission of Jupiter and solar activity.**
Z. Pokorný.
Bull. Astron. Inst. Czechoslovakia, Vol. 33, 193 - 201 (1982).

On the basis of the new catalogue of the comprehensive index of the Jovian decametric activity (covering the time interval 1958 - 1974) it has been proved that some solar activity phenomena exert influence upon the occurrence probability and intensity of the non-source decametric emission. This is particularly true of the solar wind plasma, travelling with a velocity of the order of 10^2 km s^{-1}. The modulation of the non-source emission is more conspicuous in the time after the opposition. The results presented in this paper are of a statistical nature.

099.038 **Comparison of the thermal and nonthermal radiation characteristics of Jupiter at 6, 11, and 21 cm with model calculations.**
I. de Pater, S. Kenderdine, J. R. Dickel.
Icarus, Vol. 51, 25 - 38 (1982).

Four different data sets on Jupiter, one at 6, one at 11, and two at 21 cm, are compared to each other and with the synchrotron radiation model of the magnetosphere developed by I. de Pater (1981). The model was used to derive and interpret the characteristics of the thermal radiation component at all three wavelengths. The disk temperatures are 233 ± 17, 280 ± 20, and 340 ± 26°K at 6, 11, and 21 cm, respectively. A comparison of the data with atmospheric model calculations strongly suggests that the disk is uniform at λ6 and λ11 cm near the center of the disk, where $\mu > 0.6–0.7$. This may indicate a nonuniform distribution of ammonia at layers at and above the visible cloud layers.

099.039 **The oblateness effect on the solar radiation incident at the top of the atmospheres of the outer planets.**
E. Van Hemelrijck.
Icarus, Vol. 51, 39 - 50 (1982).

Calculations of the daily solar radiation incident at the top of the atmospheres of Jupiter, Saturn, Uranus, and Neptune, with and without the effect of the oblateness, are presented in a series of figures illustrating the seasonal and latitudinal variation of the ratio of both insolations. In addition, the author also numerically studied the mean daily solar radiation.

099.040 **Numerical model of long-lived Jovian vortices.**
A. P. Ingersoll, P. G. Cuong.
J. Atmos. Sci., Vol. 38, 2067 - 2076 (1981). – Abstr. in Phys. Abstr., Vol. 85, Abstr. 73895 (1982).

099.041 **Electromagnetic radiation from the Jovian magnetosphere.**
R. N. Singh.
Proc. Indian Natl. Sci. Acad., Part A, Vol. 47, Suppl. 2, 146 - 155 (1981). – Abstr. in Phys. Abstr., Vol. 85, Abstr. 89190 (1982).

099.042 **The satellites of Jupiter and Saturn.**
D. Morrison.
Annu. Rev. Astron. Astrophys., Vol. 20, (see 003.009), 469 - 495 (1982).

099.043 **Jupiters indre struktur og utvikling.**
B. Haugstad.
Astron. Tidsskr., Årg. 15, 103 - 107 (1982).

099.044 **A model of energy deposition of energetic electrons and EUV emission in the Jovian and Saturnian atmospheres and implications.**
J.-C. Gérard, V. Singh.
J. Geophys. Res., Vol. 87, 4525 - 4532 (1982).
A model of the interaction between incident electron precipitation and H_2 atmospheres is described. The local degraded primary and secondary electron energy distributions are calculated by using the continuous slowing down approximation. The altitude distribution of the ionization rate and various H and EUV H_2 emissions are calculated for four different incident electron spectra. Comparison with the Voyager Jupiter observations indicates that an incident energy flux of about 8 ergs $cm^{-2} s^{-1}$ was present at the time of the encounter if the emission is located in an H_2-dominated region. The local thermospheric heating rate was about 4 ergs $cm^{-2}s^{-1}$ for Jupiter and of the order of 0.1 erg $cm^{-2} s^{-1}$ for Saturn. A globally averaged atomic hydrogen production rate of $\sim 1 \times 10^{10}$ atoms/$cm^2 s^{-1}$ is induced by the Jovian auroral electron precipitation, largely exceeding the solar EUV dissociation rate.

099.045 **Spatial irregularities in Jupiter's upper ionosphere observed by Voyager radio occultations.**
D. P. Hinson, G. L. Tyler.
J. Geophys. Res., Vol. 87, 5275 - 5289 (1982).
Dual frequency radio occultation experiments carried out with Voyagers 1 and 2 provided data on the spatial irregularities in Jupiter's ionosphere at four different locations. Sample spectra of weak fluctuations in amplitude and phase of the 3.6-cm and 13-cm wavelength radio signals can be interpreted by using the theory for scattering from an anisotropic power law phase screen. Least squares solutions for ionospheric parameters derived from the observed fluctuation spectra yielded estimates of (1) the axial ratio, (2) angular orientation of the anisotropic irregularities, (3) the power law exponent of the spatial spectrum of irregularities, and (4) the magnitude of the spatial variations in electron density. The authors conclude that the measured angular orientation of the anisotropic irregularities indicates magnetic field direction and may provide a basis for refining Jovian magnetic field models.

099.046 **SO_2 ionization and dissociation in the Io plasma torus.**
A. F. Cheng.
J. Geophys. Res., Vol. 87, 5301 - 5304 (1982).
Escape of SO_2 from Io may be driven by corotating ion impact and energy deposition by the Io flux tube current. The yield of O I per SO_2 is estimated in steady state ionization and dissociation of SO_2. An average SO_2 injection rate of 10^{28} s^{-1} is found to be consistent with O I observations in the hot plasma torus. The average densities of S I and SO_2 are estimated, but the expected oxygen to sulfur ion ratios from SO_2 ionization and dissociation are uncertain. There is substantial loss of neutrals from charge exchange and electron impact on molecular and negative ions.

099.047 **On Jupiter's whistler emission.**
M. P. Leubner.
J. Geophys. Res., Vol. 87, 6335 - 6338 (1982).
A theory explaining Jupiter's banded whistler emission in particular low-frequency hiss and chorus is proposed. It is shown that superthermal electrons described by a kappa distribution function cause whistler instability below one-third of the electron gyrofrequency, whereas a two temperature Maxwellian leads to a band of unstable growing modes just below one-half of the electron gyrofrequency. A superposition of both yields almost exactly the structure of whistler mode hiss and chorus detected by the Voyager 1 and 2 plasma wave instrument.

099.048 **Ammonia in the Jovian stratosphere: a comparison of photochemical calculations with IUE observations.**
K. H. Fricke, U. von Zahn, M. Combes, T. Encrenaz.
Third European IUE Conference, (see 012.029), p. 455 - 461 (1982).
Several bands of the 2168 Å electronic transition of ammonia were identified in long-duration, high-resolution exposures of Jupiter using the long wavelength spectrograph of IUE. Using one-dimensional, steady-state photochemical models the authors generated for two temperature profiles and a range of eddy coefficients vertical distributions of ammonia, and from these reflectivities.

099.049 **Gravity parameters of the Jovian system from Voyager and Pioneer data.**
J. K. Campbell, S. P. Synnott.
Bull. American Astron. Soc., Vol. 14, 720 - 721 (1982). Abstract.

099.050 **Automatic computation of Jupiter interior models.**
W. B. Hubbard. G. P. Horedt.
Bull. American Astron. Soc., Vol. 14, 721 (1982). – Abstract.

099.051 **Theory of motions in Jupiter's and Saturn's atmospheres: evidence for deep internal circulations.**
A. P. Ingersoll.
Bull. American Astron. Soc., Vol. 14, 721 (1982). – Abstract.

099.052 **A convective model for driving global circulations on Jupiter and Saturn.**
D. H. Hathaway.
Bull. American Astron. Soc., Vol. 14, 721 - 722 (1982). Abstract.

099.053 **Jovian atmospheric dynamics: global-scale motion and shear instability for a thin, nearly adiabatic upper weather layer.**
M. D. Allison, P. J. Gierasch.
Bull. American Astron. Soc., Vol. 14, 722 (1982). – Abstract.

099.054 **The meridional energy balance of Jupiter and Saturn.**
J. Pirraglia, B. Conrath.
Bull. American Astron. Soc., Vol. 14, 722 (1982). – Abstract.

099.055 **Characteristics of morphologically-distinct regions of the Jovian atmosphere from Voyager visible and infrared observations.**
D. J. Diner, J. V. Martonchik, G. S. Orton.
Bull. American Astron. Soc., Vol. 14, 722 (1982). – Abstract.

099.056 **Kinetics of the reaction of NH_2 with phosphine and implications for Jovian atmospheric modeling.**
S. R. Bosco, W. D. Brobst, D. F. Nava, L. J. Stief.
Bull. American Astron. Soc., Vol. 14, 722 (1982). – Abstract.

099.057 **A summary of the winds of Jupiter and Saturn.**
R. F. Beebe, A. P. Ingersoll, G. W. Garneau.
Bull. American Astron. Soc., Vol. 14, 722 - 723 (1982). Abstract.

099.058 **Analysis of motions of the Red Spot and White Ovals using ground-based and Voyager observations.**
H. Guitar, R. Beebe.
Bull. American Astron. Soc., Vol. 14, 723 (1982). – Abstract.

099.059 **Analysis of the long-lived feature in Jupiter's equatorial zone from Voyager and ground-based observations.**
J. Christou, R. Beebe.
Bull. American Astron. Soc., Vol. 14, 723 (1982). – Abstract.

099.060 **Measurements of ethane abundance in the Jovian stratosphere.**
T. Kostiuk, M. J. Mumma, F. Espenak, D. Deming, D. Jennings, W. Maguire, D. Zipoy.
Bull. American Astron. Soc., Vol. 14, 723 (1982). – Abstract.

099.061 **Colored sulfur ions in the atmosphere of Jupiter.**
W. T. Huntress, Jr., L. Thorne.
Bull. American Astron. Soc., Vol. 14, 723 - 724 (1982). Abstract.

099.062 **Analysis of the Pioneer observations of Jupiter's polarization.**
P. H. Smith, M. G. Tomasko.
Bull. American Astron. Soc., Vol. 14, 724 (1982). – Abstract.

099.063 **Longitudinal variation of high altitude cloud distributions in the Jovian atmosphere.**
C. R. Stoker, R. A. West, C. W. Hord.
Bull. American Astron. Soc., Vol. 14, 724 (1982). – Abstract.

099.064 **Spatially-resolved ultraviolet spectra of selected Jovian features obtained with the International Ultraviolet Explorer Satellite.**
M. Allen, J. Trauger, R. Gladstone, J. Pingree.
Bull. American Astron. Soc., Vol. 14, 729 (1982). – Abstract

099.065 **Observations of Jupiter, Uranus and Neptune from the International Ultraviolet Explorer.**
P. Winkelstein, J. Caldwell.
Bull. American Astron. Soc., Vol. 14, 729 (1982). – Abstract.

099.066 **The Jovian spectrum in the 2.8 μ window.**
P. Drossart, T. Encrenaz.
Bull. American Astron. Soc., Vol. 14, 729 (1982). – Abstract.

099.067 **The relative contribution of reflected solar and thermal emission to the 5 μm spectrum of Jupiter and Saturn.**
G. L. Bjoraker, H. P. Larson, U. Fink, V. G. Kunde.
Bull. American Astron. Soc., Vol. 14, 730 (1982). – Abstract.

099.068 **Relative abundance of isotopic carbon in the atmospheres of the outer planets.**
K. Fox.
Bull. American Astron. Soc., Vol. 14, 730 (1982). – Abstract.

099.069 **The Jovian $^{12}C/^{13}C$ ratio from the infrared experiment on board Voyager.**
R. Courtin, D. Gautier, A. Marten, V. G. Kunde.
Bull. American Astron. Soc., Vol. 14, 730 (1982). – Abstract.

099.070 **Ortho and para molecular hydrogen in Jupiter's belts and zones.**
C. C. Cunningham, D. M. Hunten, R. A. Brown.
Bull. American Astron. Soc., Vol. 14, 730 (1982). – Abstract.

099.071 **The hydrogen distribution in the Jovian atmosphere.**
R. M. Killen.
Bull. American Astron. Soc., Vol. 14, 730 (1982). – Abstract.

099.072 **Temporal variation of the Jovian H I Lyman-alpha emission (1979 - 1982).**
T. E. Skinner, S. T. Durrance, P. D. Feldman, H. W. Moos.
Bull. American Astron. Soc., Vol. 14, 730 - 731 (1982). Abstract.

099.073 **Jovian infrared aurora I. Spatial characteristics.**
J. Caldwell, A. Tokunaga, G. Orton.
Bull. American Astron. Soc., Vol. 14, 732 (1982). – Abstract.

099.074 **Jovian infrared aurora II. Spectral characteristics.**
S. J. Kim, J. Caldwell, A. Rivolo.
Bull. American Astron. Soc., Vol. 14, 732 (1982). – Abstract

099.075 **The Io 4.8 μm flicker power spectrum from 6 μHz to 17 mHz.**
W. Sinton.
Bull. American Astron. Soc., Vol. 14, 734 (1982). – Abstract.

099.076 **Colors of lava flows on Io.**
D. C. Pieri, S. M. Baloga.
Bull. American Astron. Soc., Vol. 14, 734 - 735 (1982). Abstract.

099.077 **The auras of sulfur flows on Io.**
S. M. Baloga, D. L. Matson, D. C. Pieri.
Bull. American Astron. Soc., Vol. 14, 735 (1982). – Abstract.

099.078 **Io's 4 μm band and the role of adsorbed SO_2.**
D. B. Nash.
Bull. American Astron. Soc., Vol. 14, 736 (1982). – Abstract.

099.079 **Photochemical modeling of the atmosphere of Io.**
M. E. Summers, Y. L. Yung.
Bull. American Astron. Soc., Vol. 14, 736 (1982). – Abstract.

099.080 **Spectral evidence for magnetospheric contaminents on the surfaces of the Galilean satellites.**
T. B. McCord, M. L. Nelson, R. N. Clark, T. V. Johnson, D. Matson, J. A. Mosher, L. Soderblom.
Bull. American Astron. Soc., Vol. 14, 736 (1982). – Abstract.

099.081 **Volcanism in ice on Europa.**
A. F. Cook, E. M. Shoemaker, L. A. Soderblom, K. F. Mullins, R. Fiedler.
Bull. American Astron. Soc., Vol. 14, 736 - 737 (1982). Abstract.

099.082 **Spectral unit map of Europa.**
T. B. McCord, M. L. Nelson, R. N. Clark, A. Meloy, W. Harrison, T. V. Johnson, D. Matson, J. A. Mosher, L. Soderblom.
Bull. American Astron. Soc., Vol. 14, 737 (1982). – Abstract.

099.083 **Liquid water and active resurfacing on Europa.**
S. W. Squyres, R. T. Reynolds, P. M. Cassen, S. J. Peale.
Bull. American Astron. Soc., Vol. 14, 737 (1982). – Abstract.

099.084 **Ultraviolet spectral geometric albedos of selected Jovian and Saturnian satellites: IUE results.**
R. M. Nelson, A. L. Lane, D. L. Matson, G. J. Veeder, M. D. Morrison.
Bull. American Astron. Soc., Vol. 14, 738 (1982). – Abstract.

099.085 **Eight-color photometry of planetary satellites.**
B. Zellner, D. J. Tholen.
Bull. American Astron. Soc., Vol. 14, 738 (1982). – Abstract.

099.086 **Charged particle erosion of the icy satellites of Jupiter and Saturn.**
R. E. Johnson, J. W. Boring, J. Garrett, C. Reimann, L. J. Lanzerotti, W. L. Brown.
Bull. American Astron. Soc., Vol. 14, 745 (1982). – Abstract.

099.087 **Effects of ice creep relations on thermal histories and viscous relaxation for icy satellites.**
S. K. Croft.
Bull. American Astron. Soc., Vol. 14, 745 (1982). – Abstract.

099.088 **Structure in the Jovian ring?**
V. R. Haemmerle, G. E. Danielson, D. G. Currie,

M. Ravine, A. Hatzes.
Bull. American Astron. Soc., Vol. 14, 747 (1982). – Abstract.

099.089 **Structure of the Io plasma torus.**
C. B. Pilcher, J. S. Morgan, J. H. Fertel.
Bull. American Astron. Soc., Vol. 14, 761 - 762 (1982). Abstract.

099.090 **Models of the electrodynamic interaction between the Jovian magnetosphere and a non-global ionosphere at Io.**
F. Herbert.
Bull. American Astron. Soc., Vol. 14, 762 (1982). – Abstract.

099.091 **The escape and ionization of atomic oxygen and sulfur from Io.**
W. H. Smyth, D. E. Shemansky.
Bull. American Astron. Soc., Vol. 14, 762 (1982). – Abstract.

099.092 **Dynamics of the Io sodium cloud.**
B. A. Goldberg, R. W. Carlson, G. W. Garneau, T. V. Johnson, S. K. LaVoie, J. J. Lorre, D. L. Matson.
Bull. American Astron. Soc., Vol. 14, 762 (1982). – Abstract.

099.093 **The Io plasma torus does not corotate with Jupiter.**
R. A. Brown.
Bull. American Astron. Soc., Vol. 14, 762 - 763 (1982). Abstract.

099.094 **On the nature of S II emission from Jupiter's hot plasma torus.**
R. A. Brown, D. E. Shemansky.
Bull. American Astron. Soc., Vol. 14, 763 (1982). – Abstract.

099.095 **The Jovian [S II] - [S III] nebula.**
J. T. Trauger.
Bull. American Astron. Soc., Vol. 14, 763 (1982). – Abstract.

099.096 **A deficiency of O III in the Io plasma torus.**
D. E. Shemansky, R. A. Brown, R. E. Johnson.
Bull. American Astron. Soc., Vol. 14, 763 (1982). – Abstract.

099.097 **Rocket detection of ultraviolet emissions from neutral oxygen and sulfur in the Io torus.**
S. T. Durrance, P. D. Feldman, H. A. Weaver.
Bull. American Astron. Soc., Vol. 14, 763 - 764 (1982). Abstract.

099.098 **Solutions to the equations for corotating magnetospheric convection.**
D. Summers, G. L. Siscoe.
Astrophys. J., Vol. 261, 677 - 683 (1982).

The equations of centrifugally driven convection appropriate to a treatment of the transport of Io plasma in Jupiter's magnetosphere have been solved for two special but instructive examples of the source distribution. In the corotating frame, the mass flows outward from the source, forming spiral-like streamlines. The pitch of the spiral depends on the amount of resident mass and the ionospheric conductivity. The outflow speed depends on the size of the source region, the resident mass, and the source strength of new ions.

099.099 **Tidal heating of Io and orbital evolution of the Jovian satellites.**
A. Gailitis (*Gajlitis*).
Mon. Not. R. Astron. Soc., Vol. 201, 415 - 420 (1982).

The energy-momentum balance relates tidal heating to orbital expansion as in Yoder's secular equation. The limited energy of radial motion has preserved Io from a runaway type of melting.

099.100 **Limit on possible narrow rings around Jupiter.**
E. Dunham, J. L. Elliot, D. Mink, A. R. Klemola.
Astron. J., Vol. 87, 1423 - 1427 (1982).

The authors report an upper limit to the optical depth of the Jovian ring at high spatial resolution, determined from stellar occultation data. The spatial resolution of the observation is limited to about 13 km in Jupiter's equatorial plane by the projection of the Fresnel zone on the equatorial plane in the radial direction. At this resolution, the normal optical depth limit is about 8×10^{-3}. This limit applies to a strip in the Jovian equatorial plane that crosses the orbits of Amalthea, 1979J1, 1979J3, and the ring. An upper limit on the number density of kilometer-size boulders has been set at one per 11 000 km^2 in the equatorial plane.

099.101 **Copernicus measurement of the Jovian Lyman-alpha emission and its aeronomical significance.**
S. K. Atreya, M. C. Festou, T. M. Donahue, R. B. Kerr, E. S. Barker, W. D. Cochran, J. L. Bertaux, W. L. Upson II.
Astrophys. J., Vol. 262, 377 - 387 (1982).

Observations of Jupiter made with the high-resolution ultraviolet spectrometer of Copernicus in 1980 April and May yield the intensity of the Jovian Lyman-alpha emission to be 7 ± 2.5 kR. These measurements indicate that the Lyman-alpha intensity has decreased by about a factor of 2 from the time of the Voyager ultraviolet spectrometer measurements, nearly a year earlier. It is suggested that at the time of the Voyager Jupiter encounter, unusually high auroral activity existed, and it was perhaps linked to the high concentration of the Io plasma torus.

099.102 **The abundances of CH_4, CH_3D, NH_3, and PH_3 in the troposphere of Jupiter derived from high-resolution 1100–1200 cm^{-1} spectra.**
R. F. Knacke, S. J. Kim, S. T. Ridgway, A. T. Tokunaga.
Astrophys. J., Vol. 262, 388 - 395 (1982).

High-resolution spectra of the 1100–1200 cm^{-1} region of the central part of Jupiter were obtained in 1980 March and 1981 April. The authors evaluate the NH_3 and PH_3 vertical distributions and the CH_4 and CH_3D mixing ratios. A D/H ratio of 3.0×10^{-5} is derived. There is no evidence for NH_3 ice particles.

099.103 **Jovian winds from Voyager 2. I. Zonal mean circulation.**
S. S. Limaye, H. E. Revercomb, L. A. Sromovsky, R. J. Krauss, D. A. Santek, V. E. Suomi, S. A. Collins, C. C. Avis.
J. Atmos. Sci., Vol. 39, 1413 - 1432 (1982). – Abstr. in Phys. Abstr., Vol. 85, Abstr. 112378 (1982).

099.104 **Jovian winds from Voyager 2. II. Analysis of eddy transports.**
L. A. Sromovsky, H. E. Revercomb, V. E. Suomi, S. S. Limaye, R. J. Krauss.
J. Atmos. Sci., Vol. 39, 1433 - 1445 (1982). – Abstr. in Phys. Abstr., Vol. 85, Abstr. 112379 (1982).

099.105 **On charge exchange and knock-on processes in the exosphere of Io.**
W.-H. Ip.
Astrophys. J., Vol. 262, 780 - 785 (1982).

One direct consequence of magnetospheric interaction of Io is the strong dynamical coupling of its neutral atmosphere with the corotating plasma. Exospheric interactions like atom-ion knock-on collision and charge exchange recombination could be a significant source of the neutral clouds in the Jovian system. Using a current electrodynamic model of Io, both the absorption rate of the corotating thermal plasma and the production rates of new exospheric ions and the fast neutrals are considered.

099.106 **On the nature of S II emission from Jupiter's hot plasma torus.**
R. A. Brown, D. E. Shemansky.
Astrophys. J., Vol. 263, 433 - 442 (1982).

Hot-torus emission rates at the Voyager 1 encounter in the 6731, 1256, 911, and the reclassified 765 Å transitions of S II indicate an effective electron temperature $T_e \approx 80{,}000$ K. This temperature is consistent with independent Voyager findings in the same region. The 6716, 6731 Å spectra obtained at 5.9 R_J show large intensity variability but display a doublet ratio which is invariant within the uncertainties. The S II emission variability must be caused by changes in S II mixing ratio and therefore in the [S II]/[S III] ratio. The authors suggest this decoupling of the S II and S III ion populations may be caused by as yet unidentified ion-ion or ion-atom charge exchange reactions.

099.107 **The tropospheric gas composition of Jupiter's North Equatorial Belt (NH_3, PH_3, CH_3D, GeH_4, H_2O) and the Jovian D/H isotopic ratio.**
V. Kunde, R. Hanel, W. Maguire, D. Gautier, J. P. Baluteau, A. Marten, A. Chedin, N. Husson, N. Scott.
Astrophys. J., Vol. 263, 443 - 467 (1982).

The gas composition of the troposphere of Jupiter in the clearest regions of the North Equatorial Belt (NEB) was derived from the Voyager 1 IRIS data. The infrared spectrum for this homogeneous "cloud-free" region was modeled to infer altitude profiles for NH_3, PH_3, CH_3D, GeH_4, and H_2O. Using IRIS inferred values for the mole fractions of CH_3D and CH_4 a value of $D/H = 3.6^{+1.0}_{-1.4} \times 10^{-5}$ is derived. Assuming this Jovian D/H ratio is representative of the protosolar nebula, and correcting for chemical galactic evolution, yields a value of $5.5-9.0 \times 10^{-5}$ for the primordial D/H ratio and an upper limit of $1.8-2.4 \times 10^{-31}$ g cm^{-3} for the present-day baryon density.

099.108 **Jupiter VI (Himalia)**
IAU Circ., No. 3717 (1982).

099.109 **The atmospheres of Jupiter, Saturn, and Titan.**
S. J. Bauer.
The solar system and its exploration, (see 012.054), p. 161 - 163 (1981).

A comparison of the detailed composition, thermal structure, and dynamics of the atmospheres of Jupiter and Saturn, based on the recent Voyager observations is made, particularly of their zonal circulations and smaller-scale "weather systems." The massive atmosphere of Saturn's moon Titan, its composition and structure as well as indications of its dynamics, are also discussed based on the most recent Voyager data.

099.110 **Jupiter's moons.**
W. Kiesl.
The solar system and its exploration, (see 012.054), p. 179 - 185 (1981).

Studies of the color photographs of the Jovian satellites and recent physical studies have demonstrated that the Galilean satellites are four different worlds which resemble in size the four terrestrial planets and the moon. Of these objects Io is the most interesting one, since it is the only object in the solar system – except the earth – showing still volcanic activity. This phenomenon is extensively discussed with respect to the chemical composition of Io. Models of the structure and chemical composition are presented also for Europa, Ganymede, Callisto, Amalthea as well as for the outer, irregular, satellites. It is argued finally that all the regular satellites of Jupiter as well as Jupiter itself have been formed 4.5×10^9 a ago. A description of the presumable chemical processes during formation is also given in this paper.

099.111 **Astrometric positions of Jupiter and of the Galilean satellites in April 1979.**
H. Debehogne, L. Houziaux.
Acta Astron., Vol. 32, 143 - 144 (1982).

099.112 **Giove.**
E. Perozzi.
Coelum, Vol. 50, 293 - 303 (1982).

099.113 **Stability of latitudinal winds in the Jovian atmosphere.**
N. N. Morozov, O. P. Pogutse.
Dokl. AN SSSR, Tom 263, 1354 - 1356 (1982). In Russian.
Abstr. in Ref. zh., 51. Astron., 10.51.211 (1982).

099.114 **The hydrogen to helium ratio on Jupiter and Saturn.**
D. Gautier.
Formation of planetary systems, (see 012.064), p. 725 - 739 (1982).

Contents: Why to measure the hydrogen to helium ratio on Jupiter and Saturn. Does the H_2/H_e mixing ratio in the outer atmospheres of Jupiter and Saturn reflect the composition of the primitive nebula? The hidden assumptions. Methods to determine the hydrogen to helium ratio in the giant planets. The hydrogen to helium ratio on Jupiter and Saturn from Voyager. Conclusions.

099.115 **Thermal- and plasma-induced molecular redistribution on the icy satellites.**
E. M. Sieveka, R. E. Johnson.
Icarus, Vol. 51, 528 - 548 (1982).

The molecular transport of condensed gas species across the surfaces of the icy satellites of Jupiter and Saturn is examined. Molecular redistribution induced by thermal sublimation and magnetospheric plasma-ion impact on satellites with negligible atmospheres is calculated by assuming that the molecules follow ballistic trajectories and by statistically selecting initial molecular velocities and points of origin. The results are applied to polar frost on Ganymede, the observation of embedded SO_2 on Europa and the deposition of polar SO_2 frost on Io. The icy Saturnian satellites are also briefly discussed.

099.116 **Pedestal craters on Ganymede.**
V. M. Horner, R. Greeley.
Icarus, Vol. 51, 549 - 562 (1982).

The high-resolution Voyager images of Ganymede show a class of fresh craters 6–89 km in diameter which is distinguished by an ejecta blanket similar to those seen for some types of Martian craters. One hundred and eighty-five were identified and studied for trends with respect to latitude, longitude, and terrain type. It is concluded that the ejecta morphology probably results from impact into an icy target.

099.117 **Io: a volcanic flow model for the hot spot emission spectrum and a thermostatic mechanism.**
W. M. Sinton.
Icarus, Vol. 51, 563 - 573 (1982).

The hot spots of Io are modeled as a steady state of active areas at 600°K, continuing creation of new lava flows and calderas, cooling off of recent flows and calderas, and the cessation of radiation of old flows and calderas from the accumulation of insulation added by resurfacing.

099.118 **Formation of the Galilean satellites in a gaseous nebula.**
J. I. Lunine, D. J. Stevenson.
Icarus, Vol. 52, 14 - 39 (1982) = Div. Geol. Planet. Sci., Calif. Inst. Technol., Contrib. 3684.

A model for Galilean satellite formation is analyzed in which the satellites accrete in the presence of a dense, gaseous disk-shaped nebula and rapidly form optically thick, gravitationally bound primordial atmospheres. Upper-bound temperatures expected during accretion lead to partially differentiated structures for both Ganymede and Callisto. The model is chosen to be consistent with the observed densities of the Galilean satellites and our current understanding of Jupiter formation.

099.119 **Mass-radius relationships in icy satellites after Voyager.**
M. J. Lupo.
Icarus, Vol. 52, 40 - 53 (1982).
Using improved data for the masses and radii of the satellites of Jupiter and Saturn, models accounting for self-compression effects are presented for the interiors of Europa, Ganymede, Callisto, Rhea, and Titan.

099.120 **On the C/H and D/H ratios in the atmospheres of Jupiter and Saturn.**
T. Encrenaz, M. Combes.
Icarus, Vol. 52, 54 - 61 (1982).
The authors reestimate the C/H ratio in the atmospheres of Jupiter and Saturn by the measurements of the weak visible CH_4 bands, the $CH_4 3\nu_3$ band, and the (3–0) and (4–0) quadrupole bands of H_2.

099.121 **Motion in the interiors and atmospheres of Jupiter and Saturn: scale analysis, anelastic equations, barotropic stability criterion.**
A. P. Ingersoll, D. Pollard.
Icarus, Vol. 52, 62 - 80 (1982).
If Jupiter's and Saturn's fluid interiors were inviscid and adiabatic, any steady zonal motion would take the form of differentially rotating cylinders concentric about the planetary axis of rotation. The authors first develop a self-consistent scaling analysis of deep convection in the presence of differential rotation. Using its results they derive the anelastic equations that describe motions in the planets' interiors. The simplified problem of an adiabatic fluid with a steady zonal wind varying as a function of cylindrical radius is then solved in detail. The main conclusion of the paper is that winds in the interiors of the planets are of the same magnitude as those in the visible layers. Models that place a solid surface at cloud base are incomplete because they require a density structure that is incompatible with the existence of an internal heat source and fluid interior.

099.122 **Airborne spectroscopy of Jupiter in the 100- to 300-cm^{-1} region: global properties of ammonia gas and ice haze.**
G. S. Orton, H. H. Aumann, J. V. Martonchik, J. F. Appleby.
Icarus, Vol. 52, 81 - 93 (1982).
The authors present Fourier transform spectrometric observations of the whole disk of Jupiter in the 100- to 300-cm^{-1} (300–100 μm) spectral range made from the Kuiper Airborne Observatory. The observations have a spectral resolution corresponding to 1.65 cm^{-1}. The opacity of the Jovian atmosphere in this spectral region is dominated by NH_3 gas and the influence of a haze of NH_3 ice particles, with some additional opacity provided by the translational component of the H_2 collision-induced dipole.

099.123 **The effect of ammonia ice on the outgoing thermal radiance from the atmosphere of Jupiter.**
G. S. Orton, J. F. Appleby, J. V. Martonchik.
Icarus, Vol. 52, 94 - 116 (1982).
The authors examine the effects of NH_3 ice particle clouds in the atmosphere of Jupiter on outgoing thermal radiances. The cloud models are characterized by a number density at the cloud base, by the ratio of the scale height of the vertical distribution of particles to the gas scale height and by an effective particle radius. A large range of possible models is examined in order to understand their influence on various spectral regions.

099.124 **Color photometry of surface features on Ganymede and Callisto.**
S. W. Squyres, J. Veverka.
Icarus, Vol. 52, 117 - 125 (1982).
Voyager imaging data demonstrate that the scattering properties ("phase curves") of all major terrain types on Ganymede and Callisto are not significantly wavelength dependent between 0.4 and 0.6 μm. This can be easily understood in terms of a naive model which considers the surface to be made up of two photometric elements: (1) clean water frost, and (2) a dark, possibly carbonaceous material.

099.125 **Introduction to the satellites of Jupiter.**
D. Morrison.
Satellites of Jupiter, (see 012.066), p. 3 - 43 (1982).
The sixteen known satellites of Jupiter are a diverse and fascinating set of planetary bodies, categorized into four groups of four objects each: (1) the Galilean satellites, ranging in size from sublunar to larger than Mercury; (2) the smaller inner satellites, three of which have recently been discovered by Voyager; (3) the small prograde outer satellites; and (4) the small retrograde outer satellites. This introduction sketches the history of satellite studies, including the Pioneer and Voyager missions of the late 1970s.

099.126 **The rings of Jupiter.**
D. C. Jewitt.
Satellites of Jupiter, (see 012.066), p. 44 - 64 (1982).
The properties of the Jovian ring system are reviewed. The system has three morphologically distinct components. The bright ring is planar and circular. It merges into the faint sheet component at its inner edge. A broad halo of ring particles envelops the bright ring and the faint sheet, extending some 10^4 km above and below the plane of the bright ring, and apparently extending down to the atmosphere of the planet. The system is bounded on the outside by the small satellite Adrastea (1979 J1).

099.127 **Orbital evolution of the Galilean satellites.**
R. Greenberg.
Satellites of Jupiter, (see 012.066), p. 65 - 92 (1982).
The orbital motions of the Galilean satellites exert dramatic control over their physical properties (most notably Io's) through tidal heating. In turn, tidal dissipation in the satellites, as well as in Jupiter, has governed the evolution of the orbits and, in particular, of the Laplace resonance.

099.128 **Structure and thermal evolution of the Galilean satellites.**
P. M. Cassen, S. J. Peale, R. T. Reynolds.
Satellites of Jupiter, (see 012.066), p. 93 - 128 (1982).
The authors first discuss Io, for which intense internal heating, apparently derived from the dissipation of tidal energy, dominates all discussion of its interior. They then skip to Ganymede and Callisto, whose similar gross properties invite combined treatment. Europa is dealt with last because its evolution has been influenced by factors relevant to both Io and the outer pair. In each case, the authors' approach is to describe models based on the simplest assumptions and then identify the most important complications that are likely to exist and explore their consequences.

099.129 **The outer satellites of Jupiter.**
D. P. Cruikshank, J. Degewij, B. H. Zellner.
Satellites of Jupiter, (see 012.066), p. 129 - 146 (1982).
The eight known irregular, outer satellites of Jupiter can be divided into two dynamically different groups. The physical properties of the satellites themselves are known only superficially because of their faintness. There is marginal evidence that the two groups are of separate compositional classes, corresponding to two compositional classes of Trojan asteroids and other asteroids in the main belt. The

few data presently available do not yet answer such fundamental questions as the origin of the outer Jovian satellites and their mineralogical compositions.

099.130 **Amalthea.**
P. Thomas, J. Veverka.
Satellites of Jupiter, (see 012.066), p. 147 - 173 (1982).

Voyager images have revealed Amalthea to be an irregular object 270 × 165 × 150 km in size. The spin period is probably synchronous with the orbital period of 11.9 hr, with the long axis pointing toward Jupiter. The satellite's surface is heavily scarred by impact craters. Amalthea is very dark (reflectance ~ 5 - 6%) and very red, but isolated bright spots (reflectance up to 20%) occur. It is likely that the surface of Amalthea has been severely altered by its environment and by contamination from Io. It may, therefore, be very difficult to obtain definitive information on the composition of the intrinsic Amalthea material from remote sensing measurements.

099.131 **Composition of the surfaces of the Galilean satellites.**
G. T. Sill, R. N. Clark.
Satellites of Jupiter, (see 012.066), p. 174 - 212 (1982).

Most of what we know of the compositions of Io, Europa, Ganymede, and Callisto has come through measurements of the spectral and photometric properties of these surfaces. By comparing the spectral and photometric behavior of laboratory samples with that of the Galilean satellites, one may draw certain conclusions, some more firm than others, as to the chemical compositions of the satellites.

099.132 **Radar properties of Europa, Ganymede, and Callisto.**
S. J. Ostro.
Satellites of Jupiter, (see 012.066), p. 213 - 236 (1982).

This paper summarizes the radar properties of Europa, Ganymede, and Callisto and documents our comprehension of these properties. The author describes the radar techniques, reviews the observational results, and discusses salient aspects of the radar data set. Then he focuses on theoretical interpretation of the satellites' anomalous radar properties. He discusses the physics of the circular polarization inversion and suggests several refinements of the Goldstein-Green scattering model, arriving at a tentative description of the small-scale geomorphology of the icy Galilean satellites.

099.133 **Interpreting the cratering record: Mercury to Ganymede and Callisto.**
A. Woronow, R. G. Strom, M. Gurnis.
Satellites of Jupiter, (see 012.066), p. 237 - 276 (1982).

099.134 **Cratering time scales for the Galilean satellites.**
E. M. Shoemaker, R. F. Wolfe.
Satellites of Jupiter, (see 012.066), p. 277 - 339 (1982).

This paper is an inquiry into the available observations and theory that can be brought to bear to assess the cratering history of the Galilean satellites. The larger solid bodies that collide with the satellites at the present time are primarily active and extinct nuclei of comets. Relatively abundant high-quality information is available on the orbits of active comets, and we can calculate the probability of their collision with the Galilean satellites with fairly good accuracy. On the other hand, very little reliable information is available on the physical properties of the comet nuclei that we need to know in order to calculate cratering rates. This difficulty can be partly overcome with bounds that may be set by the cratering record of the Earth. When these bounds are introduced, it is found that roughly half the present production of large impact craters on the Galilean satellites may be due to collision with objects of asteroidal appearance on Jupiter-crossing orbits.

099.135 **Experimental simulation of impact cratering on icy satellites.**
R. Greeley, J. H. Fink, D. E. Gault, J. E. Guest.
Satellites of Jupiter, (see 012.066), p. 340 - 378 (1982).

Voyager images of Ganymede and Callisto reveal impact craters morphologically distinct from those on the terrestrial planets. To simulate cratering processes that might have occurred on icy satellites, the authors performed a series of 102 laboratory impact experiments involving a wide range of target materials. The experiments suggest that the physical properties of the target media (strength, viscosity) relative to the gravitationally induced stresses (lithostatic, hydrostatic) determine the final crater morphology.

099.136 **Craters and basins on Ganymede and Callisto: morphological indicators of crustal evolution.**
Q. R. Passey, E. M. Shoemaker.
Satellites of Jupiter, (see 012.066), p. 379 - 434 (1982).

Craterform and related features on Ganymede and Callisto include bowl-shaped craters, craters with nearly flat floors, craters with central peaks, craters with central pits, basins, crater palimpsests and penepalimpsests, and giant multiring systems of ridges and furrows. The large majority of all craters > 20 km diameter have a central pit. The pits are interpreted as formed by prompt collapse of transient central peaks. Most craters, in all size ranges, are highly flattened as a consequence of topographic relaxation by slow viscous or plastic flow.

099.137 **The geology of Ganymede.**
E. M. Shoemaker, B. K. Lucchitta, D. E. Wilhelms, J. B. Plescia, S. W. Squyres.
Satellites of Jupiter, (see 012.066), p. 435 - 520 (1982).

The authors report here the broad outlines of the geologic history of Ganymede, obtained from a first attempt to map the geology at a global scale and to interpret the characteristics of the observed geologic units.

099.138 **The geology of Europa.**
B. K. Lucchitta, L. A. Soderblom.
Satellites of Jupiter, (see 012.066), p. 521 - 555 (1982).

The surface of Europa consists of two basic units, plains, and mottled terrain. The plains units are generally bright, of uniform color, and smooth. The mottled terrain units have a dense population of spots and patches of dark material. Both plains and mottled terrain units are transected by numerous varied bands, usually brown or dark gray, that appear to have formed as dark in-filling fractures, grabens, or gaps in the crust. The light central stripes found in some dark bands may result from dike-like filling of fracture by fluids that expanded on freezing. Numerous narrow ridges visible near the terminator may have similarly formed as dikes.

099.139 **The geology of Io.**
G. G. Schaber.
Satellites of Jupiter, (see 012.066), p. 556 - 597 (1982).

Io's surface displays a wide variety of colors resulting from the spectral reflectance of various allotropes of sulfur, as well as volcanic vent craters, fissures, and other morphologic forms attributed to volcanic processes that continue to shape its surface. Complete burial of craters attributed to impact origin is associated with Io's volcanism over billions of years. A preliminary geologic map of 34.8% of the surface of Io has been compiled using best-resolution (0.5 to 5 km/line pair) Voyager 1 images. Nine volcanic units are identified, including materials of mountains, plains, flows, cones, and crater vents, in addition to six types of structural features.

099.140 **Volcanic eruption plumes on Io.**
R. G. Strom, N. M. Schneider.
Satellites of Jupiter, (see 012.066), p. 598 - 633 (1982).

The detection of active volcanism on Io by the Voyager

spacecraft established Io as the most volcanically active planet-sized object so far detected in the solar system. Voyager 1 and 2 images have revealed nine active eruptions evidenced by large volcanic plumes viewed at the limb regions of the satellite. In general the plumes appear larger in the ultraviolet and in forward-scattered rather than back-scattered illumination, suggesting the presence of a shell of very fine-grained particles at high altitudes. The dimensions of one plume (Loki) changed significantly during a relatively short time period, indicating that short-term fluctuations in the intensity of the eruptions may be common.

099.141 **Volcanic eruptions on Io: implications for surface evolution and mass loss.**
T. V. Johnson, L. A. Soderblom.
Satellites of Jupiter, (see 012.066), p. 634 - 646 (1982).
Active volcanism on Io results in a continual resurfacing of the satellite. Analysis of required burial rates to erase impact craters, the mass production in the observed plumes, and the energy requirements for the volcanic activity suggest resurfacing rates of 10^{-3} to 10 cm yr^{-1} in recent geologic time. If this rate is typical of the last 4.5 Gyr, then extensive recycling of the upper crust and mantle must have occurred.

099.142 **Dynamics and thermodynamics of volcanic eruptions: implications for the plumes on Io.**
S. W. Kieffer.
Satellites of Jupiter, (see 012.066), p. 647 - 723 (1982).
The plumes on Io are assumed volcanic in origin. Observable characteristics, such as temperature, shape and velocity, depend in a complex way on the subsurface features of the volcanic systems. A generalized model is developed in which a volcanic system is considered to be conceptually separable into five regions: supply region, reservoir, conduit, crater and plume. The fluid dynamic and thermodynamic behavior of fluids moving through these regions is examined.

099.143 **Hot spots of Io.**
J. C. Pearl, W. M. Sinton.
Satellites of Jupiter, (see 012.066), p. 724 - 755 (1982).
High-temperature regions related to volcanism on the surface of Io have not only been observed by the Voyager 1 IRIS spectrometer, but have been observed for the last decade in groundbased measurements. Detailed comparison of the IRIS observations with the Voyager imaging data reveals two kinds of associations: (1) features at < 400 K, which generally correlate with dark calderas and flows; and (2) a small 650 K area at the source of the Pele plume.

099.144 **Io's surface: its phase composition and influence on Io's atmosphere and Jupiter's magnetosphere.**
F. P. Fanale, W. B. Banerdt, L. S. Elson, T. V. Johnson, R. W. Zurek.
Satellites of Jupiter, (see 012.066), p. 756 - 781 (1982).
This paper deals with evidence and interpretations pertaining to the surface phase composition of Io and the mechanism(s) by which Io's surface influences its atmosphere. The authors also briefly discuss the mechanism(s) by which Io's surface and/or atmosphere supplies neutral and ionic species to the region around Io and ultimately to the Jovian magnetosphere.

099.145 **The atmospheres of Io and other satellites.**
S. Kumar, D. M. Hunten.
Satellites of Jupiter, (see 012.066), p. 782 - 806 (1982).

099.146 **Emissions from neutrals and ions in the Jovian magnetosphere.**
C. B. Pilcher, D. F. Strobel.
Satellites of Jupiter, (see 012.066), p. 807 - 845 (1982).
Jupiter is surrounded by a system of neutrals and plasma that originate at Io. Emission from these particles has been observed in the extreme ultraviolet. The collisionally excited emissions carry information about the electrons that are the immediate source of the excitation energy. Study of these spectral features thus yields information on the bulk of the plasma in the emitting region. The neutrals, by virtue of their insensitivity to the motion of the rotating Jovian magnetic field, more clearly illustrate the direct effects of the processes that remove material from Io. The study of these emissions in toto thus yields information on a wide range of characteristics and processes concerning Io, the Jovian magnetosphere, and their interaction.

099.147 **In situ observations of Io torus plasma.**
J. D. Sullivan, G. L. Siscoe.
Satellites of Jupiter, (see 012.066), p. 846 - 871 (1982).
The authors describe in some detail the physical properties of the Io plasma formation deduced from in situ observations. This information together with that on remote sensing observation of the Io torus should establish the basis for determining the effects that the Io plasma formation has on the satellite-ring system.

099.148 **Origin and evolution of the Jupiter satellite system.**
J. B. Pollack, F. Fanale.
Satellites of Jupiter, (see 012.066), p. 872 - 910 (1982).
According to almost all theoretical models of the origin of the Jupiter system, the gaseous envelope of the protoplanet was initially several orders of magnitude larger than the current dimensions of Jupiter. The most likely, but still far from proven, mechanism for capturing the outer, irregular satellites is by means of gas drag experienced by planetesimals passing through the outer envelope of the protoplanet just prior to the onset of the hydrodynamic collapse.

099.149 **Cartography and nomenclature for the Galilean satellites.**
M. E. Davies.
Satellites of Jupiter, (see 012.066), p. 911 - 933 (1982).

099.150 **Determination of the mass of Jupiter on the basis of orbital perturbations of P/Wolf in its sphere of action in 1922.**
E. I. Kazimirchak-Polonskaya.
Tr. Inst. Teor. Astron., Vyp. 18, p. 78 - 93 (1982). In Russian.
A new method for determining Jupiter's mass is proposed with utilization of the large perturbations of the orbital elements of a short-period comet in Jupiter's sphere of action as well as the observations of the comet after its close approach with Jupiter. For the preliminary solution of the problem comet P/Wolf has been chosen with its transit through Jupiter's sphere of action in 1922 and its observations in 1925. The limits within which lies the most probable value of the mass of Jupiter $(1:1047.355) \leq m_J \leq (1:1047.345)$ have been obtained on the basis of all researches, the upper limit being preferable.

099.151 **Positional photographic observations of Jupiter and the Galilean satellites.**
N. M. Bronnikova, T. P. Kiseleva.
Izv. Glav. Astron. Obs. Pulkovo, Astrometr. Astrofiz., No. 199, p. 19 - 27 (1982). In Russian.
The results of reduction of positional photographic observations of Jupiter and the Galilean satellites with the Normal Astrograph at Pulkovo in 1976 - 1978 are given. The observations affirm high accuracy of Sampson's theory.

099.152 **Ultraviolet photometry and modeling of the Jovian atmosphere.**
S. J. Martinek, M. G. Tomasko.
J. Opt. Soc. America, Vol. 72, 1724 (1982). – Abstract.

099.153 **Charged particle motions in the distended magnetospheres of Jupiter and Saturn.**
T. J. Birmingham.
J. Geophys. Res., Vol. 87, 7421 - 7430 (1982).
Charged particle motion in the guiding center approximation is analyzed for models of the Jovian and Saturnian magnetospheric magnetic fields based on Voyager magnetometer observations.

099.154 **First evidence for a Europa plasma torus.**
D. S. Intriligator, W. D. Miller.
J. Geophys. Res., Vol. 87, 8081 - 8090 (1982).
The paper summarizes the evidence from the Pioneer 10 plasma analyzer that plasma derived from Europa was present in the Jovian magnetosphere in December 1973. Plasma detected between 1900 UT and 2100 UT on December 3, 1973, shows a number of significant phenomena near the expected position of Europa's L shell. Mass addition to the magnetospheric plasma is suggested by a local increase in density apparently superimposed on the density gradient of Iogenic plasma. This increase in plasma density is unlike any phenomenon observed when the spacecraft is not close to a lunar L shell. The density shows fluctuations which allow an estimate of the net outflow speed of magnetospheric ions per Jovian rotation.

099.155 **Injection and loss mechanisms for energetic ions in the inner Jovian magnetosphere.**
R. M. Thorne.
J. Geophys. Res., Vol. 87, 8105 - 8110 (1982).
The pronounced depletion of energetic ions observed by Voyager outside the orbit of Io can be understood as a steady state balance between particle loss and inward diffusion.

099.156 **Field line distortion and Joule heating by the corotation enforcement current in the Jovian magneto-ionosphere.**
Y. Watanabe, A. Nishida.
J. Geophys. Res., Vol. 87, 8111 - 8117 (1982).
When radial flow is generated in the rapidly rotating magnetosphere of Jupiter, angular velocity of the magnetospheric plasma departs from the corotational value on account of the conservation of the angular momentum. This causes twisting of magnetic field lines in the magnetosphere and Joule heating in the ionosphere. The authors have estimated the energy storage and dissipation rates by using a model in which finite transit time of the signals between the equatorial plasma disc and the ionosphere as well as the reflection of the signal at the ionosphere are taken into account.

099.157 **Observations of Jupiter's distant magnetotail and wake.**
W. S. Kurth, J. D. Sullivan, D. A. Gurnett, F. L. Scarf, H. S. Bridge, E. C. Sittler, Jr.
J. Geophys. Res., Vol. 87, 10373 - 10383 (1982).
In the period from mid-1980 to August 1981 the Voyager 2 plasma wave instrument occasionally detected clear signatures of Jovian nonthermal continuum radiation while the spacecraft was in the general downstream direction from Jupiter at distances of $5000 < R < 9000\, R_J$ (up to ~4.5 AU). The authors conclude that this series of continuum radiation events is best interpreted as the passage of Voyager 2 through the extended magnetotail and wake of Jupiter. Although the analysis of the events does not lead to a unique and unambiguous understanding of the structure of the tail, the authors can speculate on the various allowed tail configurations consistent with the observations.

099.158 **Charge exchange in the Io torus and exosphere.**
R. E. Johnson, D. F. Strobel.
J. Geophys. Res., Vol. 87, 10385 - 10393 (1982).
Cross sections for charge-exchange reactions of potential importance in the Io torus and the interaction of torus plasma with Io's atmosphere are estimated with an accuracy of about a factor of 2. The size of the cross sections for a given collision velocity are shown to depend entirely on the spacing of the energy levels between the initial and final states of the reactions. Reaction rates for charge-exchange cross sections, calculated at the corotation velocity in Io's reference frame, are compared to reaction rates for electron impact ionization and recombination at a number of radii from $5.1\, R_J$ to $7.1\, R_J$ for a model torus. Expressions are given to obtain the velocity dependence of the charge-exchange cross sections and to calculate cross sections for reactions not considered in this paper.

099.159 **The proton concentration in the vicinity of the Io plasma torus.**
R. L. Tokar, D. A. Gurnett, F. Bagenal.
J. Geophys. Res., Vol. 87, 10395 - 10400 (1982).
The paper uses an improved model for the plasma distribution in Jupiter's magnetosphere to determine the light ion concentration in the vicinity of the Io plasma torus from whistler dispersion measurements. The model used assumes that the plasma is in diffusive equilibrium under the action of centrifugal, gravitational, and ambipolar electric field forces. The study provides an estimate of the plasma concentration at intermediate and high latitudes along field lines through the Io plasma torus. The method employed is to combine the Voyager I plasma wave instrument whistler observations with the Voyager 1 plasma instrument heavy ion charge concentrations throughout the Io torus to determine the light ion charge concentration along the whistler propagation paths.

099.160 **Formation of Jovian decametric *S* bursts by modulated electron streams.**
D. H. Staelin, P. W. Rosenkranz.
J. Geophys. Res., Vol. 87, 10401 - 10406 (1982).
The authors assume that electrons accelerated near Jupiter's ionosphere move outward along the magnetic field lines nearly adiabatically and with only modest conversion of longitudinal energy to transverse energy due to various plasma interactions. A modest fraction of that transverse energy is then converted to electromagnetic radiation. The authors further assume that this acceleration is concentrated in a zone near the ionosphere and consists of a steady term plus components at ~1–100 Hz. These simple assumptions lead to predictions of complex behavior for *S* bursts, much of which has already been observed. They further lead to more precise statements about the spatial distribution of radiating electrons.

099.161 **On a convective model of Jupiter.**
I. M. Yavorskaya, Yu. N. Belyaev.
Acta Astronaut., Vol. 9, (see 012.085), 481 - 486 (1982).
Abstr. in Phys. Abstr., Vol. 86, Abstr. 20288 (1983).

Barnard's satellite of Jupiter.
See Abstr. 004.012.

Barnard and his observations of Io.
See Abstr. 004.041.

Io: laboratory studies of possible surface materials.
See Abstr. 022.096.

Sulfur dioxide on Io: physical state and distribution.
See Abstr. 022.097.

Jupiter's satellite Io: a surface with sulfur trioxide?
See Abstr. 022.098.

Low temperature studies of proton irradiated SO_2: implications for Io.
See Abstr. 022.099.

Observational evidence of errors in the currently accepted transition probabilities for [S II] λλ 6716, 6731.
See Abstr. 022.103.

Erosion of frozen sulfur dioxide by ion bombardment: applications to Io.
See Abstr. 022.127.

An interpretation of the near-ultraviolet absorption spectrum of SO_2: implications for Venus, Io, and laboratory measurements.
See Abstr. 022.154.

High resolution near-infrared spectroscopy from the Kuiper Airborne Observatory.
See Abstr. 032.509.

Rocket-borne instrument with a high-resolution microchannel plate detector for planetary UV spectroscopy.
See Abstr. 032.580.

Very Long Baseline Interferometry (VLBI) system for observation of Jupiter's decametric emissions.
See Abstr. 033.054.

Floquet reference solutions for the lunar theory and the Jovian moons.
See Abstr. 042.031.

Satellites galiléens de Jupiter.
See Abstr. 047.043.

Spectral mapping of Jupiter and the Galilean satellites in the near infrared.
See Abstr. 051.045.

A cyclotron theory for the beaming pattern of Jupiter's decametric radio emission.
See Abstr. 062.009.

Voyager stellar occultation by Ganymede: relevance to surface ice temperature and volatile transport.
See Abstr. 096.007.

Detection of the 5520 Å NH_3 band and the 8620 Å S(0) 3 - 0 H_2 dipole band on Saturn.
See Abstr. 100.050.

Planetary rings.
See Abstr. 100.096.

Hyperbolic orbits of meteors and Jupiter.
See Abstr. 104.003.

Jovian modulation of interplanetary electrons as observed with Voyagers 1 and 2.
See Abstr. 106.046.

Did Jupiter once have a giant primordial moon?
See Abstr. 107.024.

Scaling from Jupiter to pulsars and mass spectrum of pulsars.
See Abstr. 141.517.

Errata

099.901 **Correction to 'Visible aurora in Jupiter's atmosphere?'** [J. Geophys. Res., Vol. 86, 8793 - 8796 (1981)].
A. F. Cook, A. Vallance Jones, D. E. Shemansky.
J. Geophys. Res., Vol. 87, 6356 (1982). – See Abstr. 30.099.133.

099.902 **Erratum: "The atmosphere and ionosphere of Jupiter"** [Vistas Astron., Vol. 25, 315 - 335 (1982)].
S. K. Atreya, T. M. Donahue.
Vistas Astron., Vol. 26, Part 1, III - IV (1982). – See Abstr. 31.099.007.

100 Saturn, Saturn Satellites

100.001 **Resonances in Saturn's rings.**
J. J. Lissauer, J. N. Cuzzi.
Astron. J., Vol. 87, 1051 - 1058 (1982).

The authors calculate the locations and strengths of the major resonances of Saturn's known moons with particles orbiting within Saturn's rings. The forcing at an $l:(m-1)$ resonance depends on the moon's eccentricity to the $|l-m|$ power. As most of Saturn's inner moons are in very nearly circular orbits, only their strongest resonances, with $l=m$ and $l=m+1$, are calculated. For Mimas, which has a somewhat larger e, resonances with $l=m+2$ are also computed.

100.002 **Microwave edge diffraction by features in Saturn's rings: observations with Voyager 1.**
E. A. Marouf, G. L. Tyler.
Science, Vol. 217, 243 - 245 (1982).

Classical edge diffraction patterns are formed at centimeter wavelengths by several features of Saturn's rings. These patterns were discovered in 3.6- and 13-centimeter radio signals from Voyager 1 during occultation by the rings. The observed shapes are in agreement with theoretical patterns computed for screens of perfectly abrupt edges having large but finite opacity.

100.003 **Delay-Doppler radar observations of Saturn's rings.**
S. J. Ostro, G. H. Pettengill, D. B. Campbell, R. M. Goldstein.
Icarus, Vol. 49, 367 - 381 (1982).

The relative radar reflectivities of Saturn's classical ring sections were estimated from delay-Doppler observations made at 12.6-cm wavelength. The A and B rings are responsible for most, if not all, of the radar echo. The average radar reflectivity per unit projected area of the A ring is nearly (~90%) as large as that of the B ring. The outer half of the B ring contains the most reflective part of the ring system. There is no firm evidence for detection of radar backscattering from particles interior to the B ring, exterior to the A ring, or from the planet itself. The radar reflectivity of the C ring is certainly no more than one-half that of the B ring.

100.004 **Infrared observations of the dark side of Iapetus.**
L. A. Lebofsky, M. A. Feierberg, A. T. Tokunaga.
Icarus, Vol. 49, 382 - 386 (1982).

Broadband (J, H, K, L, and L') and narrowband (2.8–3.8 μm) circularly variable filter photometric observations were made of the dark side of Iapetus (Eastern Elongation). The data show a 3-μm absorption feature which is too deep to be a residual water frost feature from the bright material around the poles. Laboratory simulations show that the dark material must have a strong 3-μm absorption and could be a hydrated silicate mixed with a small amount of water frost.

100.005 **Surface composition and radius of Hyperion.**
D. P. Cruikshank, R. H. Brown.
Icarus, Vol. 50, 82 - 87 (1982).

New spectrophotometric data for Hyperion in the region 1.5–2.6 μm obtained in 1981 confirm the presence of water ice bands. The bands are now shown with sufficient clarity to permit improved comparisons with other ice-bearing satellites of Jupiter and Saturn and with laboratory samples. Comparisons with Ganymede and Rhea are shown, and Hyperion is found to differ from both satellites. The effective radius of Hyperion (considered circular in cross section) derived from a 20-μm flux measurement and a revised value of V(1,0) = 4.62 is $r = 140 \pm 19$ km.

100.006 **VLA observations of Saturn at 1.3, 2, and 6 cm.**
I. de Pater, J. R. Dickel.
Icarus, Vol. 50, 88 - 102 (1982).

Radio maps with a resolution of 1".5 were made of Saturn at 1.3, 2, and 6 cm. A fraction of 0.49 ± 0.08 of Saturn's emitted light is transmitted through the rings at $\lambda 2$ and $\lambda 6$ cm. This value converts to an effective head-on or normal optical depth of 0.07 ± 0.02. The transparency at this small inclination angle must be provided entirely by the regions of very low optical depth, e.g., the Cassini and Encke Divisions, and to achieve our number this requires either a relatively large fraction of gaps or a strong forward scattering by the ring particles.

100.007 **Observation of 2 mutual events involving the satellites of Saturn in April 1980.**
G. Dourneau.
Astron. Astrophys., Vol. 112, 73 - 75 (1982).

2 mutual events were observed: 3O4P.20 (satellite No. 3 occults No. 4 partially to 20%) on April 5, 1980, and 5E3P.49 (satellite No. 5 eclipses No. 3 partially to 49%) on April 20, 1980. The analysis of the results and a comparison with theory apparently show that these 2 events occurred respectively about 1 min (for 3O4P) and 3 min (for 5E3P) in advance of predictions (Aksnes and Franklin, 1978). As Tethys is involved in these 2 events, a small difference between the true longitude and the theoretical longitude of this satellite can exist.

100.008 **Voyager and the origin of the Saturnian system.**
A. J. R. Prentice.
Proc. Astron. Soc. Australia, Vol. 4, 164 - 171 (1981).

100.009 **A possible resurfacing mechanism for icy satellites.**
J. Klinger.
Nature, Vol. 299, 41 (1982).

Photographs from Voyager Missions reveal a great variety of surface features for the icy satellites of Saturn. Thus some of Saturn's icy satellites seem to have undergone important surface modifications during their history while others did not. The author suggests that the irreversible phase transition between amorphous and cubic ice and similar phase transitions may provide a low energy resurfacing mechanism for these icy satellites.

100.010 **The source of Saturn electrostatic discharges.**
D. R. Evans, J. H. Romig, C. W. Hord, K. E. Simmons, J. W. Warwick, A. L. Lane.
Nature, Vol. 299, 236 - 237 (1982).

During both Voyager encounters with the saturnian system, the Planetary Radio Astronomy experiment detected strong discrete episodic bursts of radio emission, termed Saturn electrostatic discharges (SED). An examination of Voyager 2 photopolarimeter data now reveals a narrow feature (possibly a gap) in Saturn's B ring. A single, unique object appears to be responsible for both the SED and this feature.

100.011 **New interpretation of Saturn rings.**
V. D. Davydov.
Inst. kosm. issled. AN SSSR. Prepr., 1981, No. 645, 52 pp. In Russian. – From Ref. zh., 51. Astron., 7.51.274 (1982).

100.012 **Photographic observations of Saturn satellites at the Engelhardt Astronomical Observatory.**
V. N. Kitkin, I. G. Chugunov.
Mordov. univ. Saransk 1982. 6 pp. In Russian. – Abstr. in Ref. zh., 51. Astron., 7.51.277 (1982).

100.013 **Titan on the eve of Voyager encounter.**
J. Caldwell.
Vibrational-rotational spectroscopy for planetary atmospheres, (see 012.007), p. 673 - 677 (1982).
A decade of intense scientific study of Titan is reviewed. The atmosphere is not well understood at the time of this writing, but it is confidently expected that great progress will be made by the Voyager spacecraft.

100.014 **Evidence of production of H_2CN^+-N_2. Application to the atmosphere of Titan.**
C. V. Speller, M. Fitaire, A.-M. Pointu.
C. R. Acad. Sci. Paris, Tome 294, Sér. II, p. 1225 - 1228 (1982). In French.

100.015 **Monografías sobre planetas: Saturno.**
A. J. Camponovo.
Rev. Astron., Tomo 54, No. 221, p. 14 - 21 (1982).

100.016 **The system of orbital constants of the Saturn satellites.**
I. G. Chugunov.
Mordov. univ. Saransk, 1982. 41 pp. In Russian. – Abstr. in Ref. zh., 51. Astron., 8.51.77 (1982).

100.017 **Absolute electrophotometry of features of Saturn's disc.**
A. P. Vid'machenko.
Physics of planetary atmospheres, (see 003.003), p. 113 - 132 (1981). In Russian. – Abstr. in Ref. zh., 51. Astron., 8.51.205 (1982).

100.018 **Oriented particles in the upper layers of Saturn's atmosphere.**
O. I. Bugaenko, A. V. Morozhenko.
Physics of planetary atmospheres, (see 003.003), p. 108 - 113 (1981). In Russian. – Abstr. in Ref. zh., 51. Astron., 8.51.206 (1982).

100.019 **Absolute electrophotometry of the B ring of Saturn.**
A. P. Vid'machenko.
Physics of planetary atmospheres, (see 003.003), p. 132 - 138 (1981). In Russian. – Abstr. in Ref. zh., 51. Astron., 8.51.207 (1982).

100.020 **The composition and origin of Titan's atmosphere.**
T. Owen.
Planet. Space Sci., Vol. 30, 833 - 838 (1982).
The atmosphere of Titan is now known to be composed primarily of molecular nitrogen. But the derived mean molecular weight suggests the presence of a significant amount of some heavier gas, most probably argon. It is shown that this argon must be primordial, and that one can understand the evolution of Titan's atmosphere in terms of degassing of a mixed hydrate dominated by CH_4, N_2 and ^{36}Ar. This model satisfactorily explains the absence of neon and makes no special requirements on the satellite's surface temperature.

100.021 **Chemistry and evolution of Titan's atmosphere.**
D. F. Strobel.
Planet. Space Sci., Vol. 30, 839 - 848 (1982).
The chemistry and evolution of Titan's atmosphere is reviewed in the light of the scientific findings from Voyager mission.

100.022 **Eddy mixing coefficient on Saturn.**
S. K. Atreya.
Planet. Space Sci., Vol. 30, 849 - 854 (1982).
Data on the composition and thermal structure, and the Lyman-alpha dayglow of Saturn when analyzed in conjunction with photochemical models of the hydrocarbons and the atomic hydrogen production yield the homopause value of the eddy diffusion coefficient to be approximately 10^8 cm^2 s^{-1}. The equatorial value of the eddy diffusion coefficient at the homopause of Saturn is thus found to be approximately 100 times greater than on Jupiter. The mesosphere (and presumably, troposphere) of Saturn appears to be considerably more turbulent than the upper atmosphere of Jupiter.

100.023 **Detection of nonthermal continuum radiation in Saturn's magnetosphere.**
W. S. Kurth, F. L. Scarf, J. D. Sullivan, D. A. Gurnett.
Geophys. Res. Lett., Vol. 9, 889 - 892 (1982).
A detailed analysis of high resolution wideband data from the Voyager 1 and 2 plasma wave receivers has revealed the presence of heretofore undiscovered nonthermal continuum radiation trapped within the Saturnian magnetosphere. The discovery of Saturnian trapped continuum radiation fills a disturbing void in the Saturnian radio spectrum.

100.024 **The ambidextrous satellites: shepherding and sculpting in the F-ring of Saturn.**
M. A. Showalter, J. A. Burns.
News Lett. Astron. Soc. N.Y., Vol. 2, No. 2, p. 17 (1982). Abstract.

100.025 **Puzzling radio source on Saturn.**
L. V. Ksanfomaliti.
Zemlya Vselennaya, 1982, No. 4, p. 13. In Russian.

100.026 **Saturn as we know him now.**
M. S. Bobrov.
Zemlya Vselennaya, 1982, No. 4, p. 39 - 44. In Russian.

100.027 **Some European visual observations of Saturn in 1981.**
G. Adamoli.
Strolling Astron., Vol. 29, 143 - 145 (1982).

100.028 **Spatially resolved methane band photometry of Saturn. I. Absolute reflectivity and center-to-limb variations in the 6190-, 7250-, and 8900-Å bands.**
R. A. West, M. G. Tomasko, B. A. Smith, M. P. Wijesinghe, L. R. Doose, H. J. Reitsema, S. M. Larson.
Icarus, Vol. 51, 51 - 64 (1982).
The authors present spatially resolved measurements of Saturn's absolute reflectivity in methane bands at 6190, 7250, and 8900 Å and in nearby continuum regions. Images were obtained through narrow-band interference filters. Several data processing techniques enhanced the quality of the observations. The observations indicate fewer high-altitude aerosols in the south, a possible result of atmospheric dynamics or seasonal sublimation of NH_3 crystals. The data are tabulated and presented in a form suitable for quantitative scattering model analyses.

100.029 **Photometry and polarimetry of Titan: Pioneer 11 observations and their implications for aerosol properties.**
M. G. Tomasko, P. H. Smith.
Icarus, Vol. 51, 65 - 95 (1982).
This paper is based primarily on the Pioneer 11 observations of Titan, expanding upon the preliminary reduction and analysis of Tomasko (1980). In the time since this preliminary analysis the authors completed the data reduction process including all the known factors. They derived photometric fluxes for the integrated disk in the red and blue over the range of phase angles (22–96°) of the observations. At the closest approach to Titan, maps of the brightness of the disk were obtained in red and blue light with a resolution of 180 km. The authors refine the scattering geometries of these data, examine them for latitudinal asymmetry, and derive an equatorial limb-darkening profile. Layer doubling calculations for vertically homogeneous semi-infinite models are

used to convert the multiple-scattering observations to constraints on the single-scattering intensity and polarization produced by the aerosols. Vertically inhomogeneous models are compared to the data, too.

100.030 **Saturn's rings: resonance about an oblate planet.**
W. Wiesel.
Icarus, Vol. 51, 149 - 154 (1982).
Classical resonance theory is extended to include corrections due to Saturn's oblateness. A single classical resonance splits into a band structure, with individual resonances almost evenly spaced in radius from the planet. When applied to Saturn's rings this theory predicts, in detail, the structure of Cassini's division.

100.031 **The photometric function for Saturn's rings.**
H. Jantunen.
Moon Planets, Vol. 26, 383 - 387 (1982).
An inhomogeneous vertical distribution of matter in Saturn's rings is capable of producing the tilt effect which is observed in ring B. In this model the effective thickness of the rings is 8 to 9 times the radius of the particles.

100.032 **The Saturnian satellites.**
R. Burnham.
Astronomy, Vol. 9, No. 12, p. 6 - 24 (1981). – Abstr. in Phys. Abstr., Vol. 85, Abstr. 79020 (1982).

100.033 **De 'botsende' Saturnusmaantjes.**
I. Naudts.
Zenit, 9. Jaarg., 474 - 475, 477 (1982).

100.034 **Far-infrared brightness temperature of Saturn's disk and rings.**
R. R. Daniel, S. K. Ghosh, K. V. K. Iyengar, T. N. Rengarajan, S. N. Tandon, R. P. Verma.
Bull. Astron. Soc. India, Vol. 10, 30 - 31 (1982). – Abstract.

100.035 **Evidence for solar wind control of Saturn radio emission.**
M. D. Desch.
J. Geophys. Res., Vol. 87, 4549 - 4554 (1982).
Using data collected by the Voyager 1 and 2 spacecraft in 1980 and 1981, strong evidence is presented for a direct correlation between variations in the solar wind at Saturn and the level of activity of Saturn's nonthermal radio emission. During both 160-day analysis intervals studied, the radio emission exhibits a long-term periodicity of 25 days, identical to the periodicity seen in the solar wind at this time and consistent with the solar rotation period. The energy coupling efficiency between the solar wind and the Saturn radio emission is estimated and compared with that for the earth.

100.036 **Saturnian kilometric radiation: source locations.**
M. L. Kaiser, M. D. Desch.
J. Geophys. Res., Vol. 87, 4555 - 4559 (1982).
Using Voyager 1 and 2 Planetary Radio Astronomy data and reasonable assumptions about radiation beam geometry, the authors deduce the source locations of both polarization components of the Saturn kilometer-wavelength radiation. The resulting radio source footprints are at high latitudes north and south, and near the noon meridian. These footprints are compared with the surface locations of Saturn's UV aurorae, polar cap boundary, and polar cusp.

100.037 **Charged particle sputtering of ice surfaces in Saturn's magnetosphere.**
A. F. Cheng, L. J. Lanzerotti, V. Pirronello.
J. Geophys. Res., Vol. 87, 4567 - 4570 (1982).
Following recent investigations of the Saturn system in situ by Pioneer 11 and Voyager 1 and 2 and remotely by the International Ultraviolet Explorer, the authors have reconsidered the sputtering of water ice surfaces by magnetospheric ions as a possible source of hydrogen atoms and heavy ions. They use results from laboratory measurements of water ice sputtering and dissociation rates as a function of ice temperature. The authors also use Voyager charged particle data from the low energy charged particle experiment to obtain more accurate assessments of energetic ion fluxes and loss rates.

100.038 **An analysis of the structure of Saturn's magnetic field using charged particle absorption signatures.**
D. L. Chenette, L. Davis, Jr.
J. Geophys. Res., Vol. 87, 5267 - 5274 (1982).
A new technique is derived for determining the structure of Saturn's magnetic field. This technique uses the observed positions of charged particle absorption signatures due to the satellites and rings of Saturn to determine the parameters of an axially symmetric, spherical harmonic model of the magnetic field using the method of least squares. Absorption signatures observed along the Pioneer 11, Voyager 1, and Voyager 2 spacecraft trajectories are used to derive values for the orientation of the magnetic symmetry axis relative to Saturn's axis of rotation, the axial displacement of the center of the magnetic dipole from the center of Saturn, and the magnitude of the external field component. Comparing these results with the magnetic field model parameters deduced from analyses of magnetometer data leads the authors to prefer models that incorporate a northward offset of the dipole center by about 0.05 R_s.

100.039 **Stability of negatively charged dust grains in Saturn's ring plane.**
T. G. Northrop, J. R. Hill.
J. Geophys. Res., Vol. 87, 6045 - 6051 (1982).
Will a charged dust grain orbiting in Saturn's ring plane remain there? If given a small deflection normal to the ring plane, does it merely oscillate back and forth through the ring plane, or is it unstable and does it therefore leave the vicinity of the rings entirely? In this paper the authors study this question and show that either behavior may occur, depending on charge to mass ratio and distance from Saturn. The radius at which the instability sets in for large negative charge to mass ratio is located at one of the prominent features of the B ring.

100.040 **CO_2 abundance and upper limits for oxygen-containing molecules in Titan's atmosphere.**
V. G. Kunde, A. C. Aikin, R. A. Hanel, D. E. Jennings, W. C. Maguire, R. E. Samuelson, Y. L. Yung.
Bull. American Astron. Soc., Vol. 14, 713 (1982). – Abstract.

100.041 **Post-accretional evolution of Titan's surface and atmosphere.**
J. I. Lunine, D. J. Stevenson.
Bull. American Astron. Soc., Vol. 14, 713 (1982). – Abstract.

100.042 **Oceans on Titan?**
F. M. Flasar.
Bull. American Astron. Soc., Vol. 14, 713 (1982). – Abstract.

100.043 **The tide in the seas of Titan.**
C. Sagan, S. F. Dermott.
Bull. American Astron. Soc., Vol. 14, 713 - 714 (1982).
Abstract.

100.044 **Production and condensation of organic gases in the atmosphere of Titan.**
C. Sagan, W. R. Thompson.
Bull. American Astron. Soc., Vol. 14, 714 (1982). – Abstract.

100.045 **Aerosol extinction profiles in Titan's atmosphere.**
K. A. Rages, J. B. Pollack.
Bull. American Astron. Soc., Vol. 14, 714 (1982). – Abstract.

100.046 **Some ion-molecule reactions in the atmosphere of Titan.**
C. V. Speller, M. Fitaire, A. M. Pointu.
Bull. American Astron. Soc., Vol. 14, 714 - 715 (1982). Abstract.

100.047 **Photochemical and thermal modeling of the upper atmosphere of Titan.**
Y. L. Yung, M. Allen, J. A. Friedson.
Bull. American Astron. Soc., Vol. 14, 715 (1982). – Abstract.

100.048 **Theoretical intensities of optically thick N_2 emissions in Titan's dayglow.**
R. R. Conway.
Bull. American Astron. Soc., Vol. 14, 715 (1982). – Abstract.

100.049 **Some dynamical properties of vortex streets in Saturn's atmosphere from analyses of Voyager images.**
G. E. Hunt, D. Godfrey, V. E. Suomi.
Bull. American Astron. Soc., Vol. 14, 723 (1982). – Abstract.

100.050 **Detection of the 5520 Å NH_3 band and the 8620 Å S(0) 3 - 0 H_2 dipole band on Saturn.**
R. G. Danehy.
Bull. American Astron. Soc., Vol. 14, 731 (1982). – Abstract.

100.051 **Saturn's pole vector from combined ring occultation data.**
R. A. Simpson, G. L. Tyler, J. B. Holberg.
Bull. American Astron. Soc., Vol. 14, 731 (1982). – Abstract.

100.052 **The scattering properties of Saturn's atmosphere at 2640 Å.**
R. A. West, H. Hart, C. W. Hord, K. E. Simmons, L. W. Esposito, A. L. Lane, R. B. Pomphrey, M. Sato, D. L. Coffeen.
Bull. American Astron. Soc., Vol. 14, 732 (1982). – Abstract.

100.053 **The distribution of aerosols in Saturn's atmosphere.**
M. G. Tomasko, L. R. Doose.
Bull. American Astron. Soc., Vol. 14, 732 (1982). – Abstract.

100.054 **The NH_3 spectrum in Saturn's 5 μm window.**
U. Fink, H. P. Larson, G. L. Bjoraker, J. R. Johnson.
Bull. American Astron. Soc., Vol. 14, 732 (1982). – Abstract.

100.055 **Thermal structure of Saturn from Voyager infrared measurements.**
B. J. Conrath, J. A. Pirraglia.
Bull. American Astron. Soc., Vol. 14, 732 - 733 (1982). Abstract.

100.056 **Spatial irregularities in Saturn's ionosphere observed during Voyager 2 radio occultation.**
D. P. Hinson, G. L. Tyler.
Bull. American Astron. Soc., Vol. 14, 733 (1982). – Abstract.

100.057 **The control networks of Mimas and Enceladus.**
M. E. Davies.
Bull. American Astron. Soc., Vol. 14, 737 (1982). – Abstract.

100.058 **Enceladus: implications of its unusual photometric properties.**
B. Buratti, J. Veverka, P. Thomas.
Bull. American Astron. Soc., Vol. 14, 737 - 738 (1982). Abstract.

100.059 **Phoebe: Voyager 2 observations.**
P. Thomas, J. Veverka, D. Morrison, M. Davies.
Bull. American Astron. Soc., Vol. 14, 738 (1982). – Abstract.

100.060 **Saturn's satellites : infrared spectrophotometry (0.65 - 2.5 μm) of the leading and trailing sides and compositional implications.**
R. N. Clark, P. D. Owensby.
Bull. American Astron. Soc., Vol. 14, 738 - 739 (1982). Abstract.

100.061 **The albedo asymmetry of Iapetus.**
S. W. Squyres, C. Sagan.
Bull. American Astron. Soc., Vol. 14, 739 (1982). – Abstract.

100.062 **A comparison of Saturn's electrostatic discharges with lightning on various planets.**
J. A. Burns, R. H. Durisen, W. J. Borucki, M. R. Showalter, J. N. Cuzzi.
Bull. American Astron. Soc., Vol. 14, 747 (1982). – Abstract.

100.063 **Dynamics of spokes in Saturn's B ring.**
R. E. Eplee, Jr., B. A. Smith.
Bull. American Astron. Soc., Vol. 14, 747 (1982). – Abstract.

100.064 **Dynamical effects of non-uniform particle size in Saturn's rings.**
G. R. Stewart.
Bull. American Astron. Soc., Vol. 14, 748 (1982). – Abstract.

100.065 **Saturn's C-ring: morphology and structure from the Voyager 2 photopolarimeter.**
A. L. Lane, C. W. Hord, K. E. Simmons, A. L. Graps, L. W. Esposito, R. B. Pomphrey, R. A. West.
Bull. American Astron. Soc., Vol. 14, 748 (1982). – Abstract.

100.066 **The structure of Saturn's rings: implications from the Voyager stellar occultation.**
L. W. Esposito, M. O'Callaghan, R. A. West, C. W. Hord, K. E. Simmons, A. L. Lane, R. B. Pomphrey, D. L. Coffeen, M. Sato.
Bull. American Astron. Soc., Vol. 14, 748 (1982). – Abstract.

100.067 **Solution of forward phase-functions of Saturn ring features in the presence of multiple scatter.**
H. A. Zebker, G. L. Tyler, E. A. Marouf.
Bull. American Astron. Soc., Vol. 14, 748 (1982). – Abstract.

100.068 **Observations of eccentric features in Saturn's rings.**
R. J. Terrile, A. L. Lane, J. B. Holberg.
Bull. American Astron. Soc., Vol. 14, 748 - 749 (1982). Abstract.

100.069 **Voyager observations of eccentric ring features.**
C. Porco.
Bull. American Astron. Soc., Vol. 14, 749 (1982). – Abstract.

100.070 **Implications of the three-dimensional distribution of material in Saturn's E ring.**
W. A. Baum, T. J. Kreidl.
Bull. American Astron. Soc., Vol. 14, 749 (1982). – Abstract.

100.071 **Transmitted light through low-density areas in Saturn's rings as a model to explain the 1980 edge-on observations.**
K. Lumme, W. A. Baum, T. J. Kreidl.
Bull. American Astron. Soc., Vol. 14, 749 (1982). – Abstract.

100.072 **Dawn–dusk asymmetry of the tail region of the magnetosphere of Saturn and the interplanetary magnetic field.**
S.-I. Akasofu, M. Roederer, S. M. Krimigis.

Planet. Space Sci., Vol. 30, 1061 - 1063 (1982).

It is suggested that as a result of the dawn–dusk asymmetry of the magnetospheric structure of Saturn, caused by the interplanetary magnetic field, the Voyager 1 spacecraft remained in the closed region (the extended plasma sheet) during its entire traversal through the tail region, so that it did not have an opportunity to penetrate into the high latitude lobe (the open region).

100.073 **Identification of 1980S27 and 1980S26 resonances in Saturn's A ring.**
J. B. Holberg.
Astron. J., Vol. 87, 1416 - 1422 (1982).

The Voyager 2 ultraviolet spectrometer observation of a stellar occultation by Saturn's rings contains evidence for features associated with at least 31 resonances due to the F ring shepherd satellites 1980S27 and 1980S26. A limited analysis of density wave patterns associated with eight of these resonances yields surface mass densities for the A ring consistent with previous estimates.

100.074 **Scheme of the origin of the Saturn rings and of the Cassini division.**
V. D. Davydov.
Kosm. Issled., Tom 20, 748 - 759 (1982). In Russian.

100.075 **On the dust ring current of Saturn's F-ring.**
J. R. Hill, D. A. Mendis.
Geophys. Res. Lett., Vol. 9, 1069 - 1071 (1982).

As a consequence of the relative motion between the fine charged dust in Saturn's F-ring and the surrounding plasma, it is estimated that this ring carries a current of over 10^5 A. It is shown that such a current can alter the magnetospheric field morphology in the vicinity of the ring to an extent which could be detected by a spacecraft at a height of over 100 km above the ring plane. This in principle provides a method of estimating the dust ring current and consequently the grain charge. Some consequences of this novel current in the stability of the dust ring are briefly discussed.

100.076 **Charged particle periodicity in the Saturnian magnetosphere.**
J. F. Carbary, S. M. Krimigis.
Geophys. Res. Lett., Vol. 9, 1073 - 1076 (1982).

The authors report the first definitive evidence for charged particle modulations near the magnetic rotation period at Saturn. This periodicity is apparent in the ratios (and spectra) of low energy charged particles in the Saturnian magnetosphere.

100.077 **Eddy diffusion at Saturn's homopause.**
B. R. Sandel, J. C. McConnell, D. F. Strobel.
Geophys. Res. Lett., Vol. 9, 1077 - 1080 (1982).

Measurements of Saturn's He 584 Å dayglow and the CH_4 density profile deduced from stellar occultation data near the homopause have been combined to infer an eddy diffusion coefficient of $8 \pm 4 \times 10^7\,cm^2\,s^{-1}$ and a temperature of 125 (+40/−25) K near the homopause at Voyager 2 encounter. It appears the eddy diffusion coefficient may have increased between the Voyager encounters. Saturn's H Lyα dayglow is qualitatively compatible with this increase and the interpretation of the He 584 Å dayglow and CH_4 absorption measurement.

100.078 **The spokes in Saturn's rings: a critical evaluation of possible electrical processes.**
A. J. Weinheimer, A. A. Few, Jr.
Geophys. Res. Lett., Vol. 9, 1139 - 1142 (1982).

Among the mechanisms proposed to explain the novel spoke phenomenon occurring in Saturn's B-ring, the alignment of particles by an electric field has been suggested. In this and related theories, the particles are assumed to be adequately conducting; or at least it is the case that electrical conductivity generally acts to enhance the feasibility of the proposed processes. These ideas are appraised in light of the fact that the particles are most likely composed of ice. The authors conclude that the electrical torque required to align the particles is many orders of magnitude larger than that which is likely to exist. Also, the conductivity of ice at ring temperatures is probably so small as to inhibit some of the proposed processes.

100.079 **Two-component proton spectra in the inner Saturnian magnetosphere.**
S. M. Krimigis, T. P. Armstrong.
Geophys. Res. Lett., Vol. 9, 1143 - 1146 (1982).

100.080 **Voyager ultraviolet stellar occultation measurements of the composition and thermal profiles of the Saturnian upper atmosphere.**
M. C. Festou, S. K. Atreya.
Geophys. Res. Lett., Vol. 9, 1147 - 1150 (1982).

Occultation of the star δ Scorpii by Saturn as recorded by the Voyager Ultraviolet Spectrometer yields the value of the exospheric temperature in the equatorial region to be 800 (+150, −120) K at an altitude of 1540 km above the 1-bar atmospheric pressure level; the H_2-density at 1540 km is determined to be $5\ (+3.6, -1.8) \times 10^9\,cm^{-3}$. Temperature gradient in the thermosphere is found to be 1.25 (+0.05, −0.07) K km^{-1}. Methane volume mixing ratio at 965 km above the 1-bar pressure level is determined to be 1.5×10^{-4}.

100.081 **Self-gravitation in Saturn's rings.**
H. Salo. J. Lukkari.
Moon Planets, Vol. 27, 5 - 12 (1982).

In a ring-shaped collisional system self-gravitation reduces the equilibrium values of the geometric and optical thickness. In Saturn's rings both effects are appreciable. The previously found discrepancy between the calculated profile and the observed profile of the rings (Hämeen-Anttila, 1982) is chiefly caused by the omission of self-gravitation.

100.082 **Saturn VIII (Iapetus).**
IAU Circ., No. 3719 (1982).

100.083 **Comparison with observations and ephemeris of Phoebe.**
A. Bec-Borsenberger, P. Rocher.
Astron. Astrophys., Suppl. Ser., Vol. 50, 423 - 449 (1982). In French.

After having collected the topocentric observations of Phoebe (from 1904 to 1981), a numerical integration taking into account the planetary perturbations and those by Titan has been computed and the initial conditions have been adjusted with these observations. The final integration obtained with initial conditions adjusted in this way has been extrapolated to 1990. An ephemeris for the ten years 1981–1990 has been computed.

100.084 **The satellites of Saturn.**
M. Fulchignoni.
The solar system and its exploration, (see 012.054), p. 187 - 188 (1981). – Abstract.

100.085 **A post-Voyager view of Saturn's environment.**
S. M. Krimigis.
Johns Hopkins APL Tech. Dig., Vol. 3, 180 - 188 (1982).

Some of the important results obtained in the magnetosphere of the planet Saturn by the two Voyager spacecraft and by Pioneer 11 are discussed. These include the absence of a significant tilt between the rotational and magnetic axes of the planet, and the discovery of a torus of very hot (about 6×10^8 K) dilute plasma associated with the orbits of the satellites Rhea and Dione.

100.086 **Saturn – its rings and satellites, a celestial mechanics laboratory.**
P. K. Seidelmann.
Dynamical astronomy, (see 012.056), p. 79 - 100 (1982).

100.087 **Periodic orbits of satellites. II. Libration of the Mimas-Tethys type.**
I. Stellmacher.
Celestial Mech., Vol. 28, 381 - 400 (1982). In French.
In a previous paper (Stellmacher, 1981) the author has given an algorithm for the construction of periodic orbits in a rotating frame, for satellites around an oblate planet. In the present paper, he applies this theory to the Mimas-Tethys case.

100.088 **$H_2CN^+ \times nN_2$ clustering formation and the atmosphere of Titan.**
C. V. Speller, M. Fitaire, A. M. Pointu.
Nature, Vol. 300, 507 - 509 (1982).
The authors report the results of laboratory measurements of equilibrium constants for the clustering of N_2 with H_2CN^+ leading to $H_2CN^+ \times nN_2$ (n = 1–6) formation, which should help in understanding the ion–molecule chemistry in the atmosphere of Titan.

100.089 **The tide in the seas of Titan.**
C. Sagan, S. F. Dermott.
Nature, Vol. 300, 731 - 733 (1982).
If Titan has oceans of liquid CH_4, they are probably $>$ 100 m deep. The authors argue that, if this ocean exists, the present high eccentricity of Titan requires its depth to be $>$400 m, and that such an ocean can be detected by radar techniques.

100.090 **Saturn's excess luminosity as a consequence of a recent explosion of Titan's ices.**
Eh. M. Drobyshevskij.
Astron. Tsirk., No. 1177, p. 1 - 2 (1981). In Russian.

100.091 **Saturn's geometric albedo in 1980.**
V. G. Tejfel', G. A. Kharitonova.
Astron. Tsirk., No. 1178, p. 7 - 8 (1981). In Russian.

100.092 **Observations of the mutual phenomena of Saturnian satellites in 1980.**
M. Sôma, T. Nakamura.
Tokyo Astron. Bull., Second Ser., No. 267, p. 3039 - 3051 (1982).

100.093 **Brightness coefficients of Saturn at λλ 0.5 - 1.1 μm.**
K. S. Kuratov.
Astron. Tsirk., No. 1197, p. 1 - 2 (1981). In Russian.

100.094 **On the structure of Saturn's rings.**
A. M. Fridman, V. L. Polyachenko.
Astron. Tsirk., No. 1204, p. 1 - 6 (1981). In Russian.

100.095 **The formation and evolution of the "radial spokes" in Saturn's B-ring.**
J. R. Hill, D. A. Mendis.
Publ. Astron. Soc. Pacific, Vol. 94, 754 (1982). – Abstract.

100.096 **Planetary rings.**
A. Brahic.
Formation of planetary systems, (see 012.064), p. 651 - 724 (1982).
This paper is an abstract of two lectures given respectively by P. Goldreich and the author. Some of the observations of Voyager spacecraft have been included. A short summary of the present knowledge on rings is given and the main observational results and the dynamical problems which are of interest for cosmogonical questions are emphasized.

100.097 **The spectrum of Titan in the far-infrared and microwave regions.**
R. Courtin.
Icarus, Vol. 51, 466 - 475 (1982).
The spectrum of Titan from the microwave to the far-infrared region (0.1–600 cm^{-1}) is modeled using new opacity data and a temperature profile of Titan's atmosphere derived from the Voyager 1 radio occultation experiment. The model atmosphere is composed of N_2 and CH_4, their relative proportions being determined by the vapor pressure law of CH_4. The effective temperature of Titan is estimated at $T_e = 83.2 \pm 1.4$°K.

100.098 **Far-infrared spectrophotometry of Saturn and its rings.**
M. R. Haas, E. F. Erickson, D. D. McKibbin, D. Goorvitch, L. J. Caroff.
Icarus, Vol. 51, 476 - 490 (1982).
The spectrum of Saturn was measured from 80 to 350 cm^{-1} (29 to 125 μm) with ~6 cm^{-1} resolution using a Michelson interferometer aboard NASA's KAO. These observations are of the full disk, with little contribution from the rings. The effective temperature is 96.8 ± 2.5°K, implying that Saturn emits 3.0 ± 0.5 times as much energy as it receives from the Sun. By combining the present data with those of an earlier investigation the authors obtain the spectrum of the rings.

100.099 **Findings on rings and inner satellites of Saturn by Pioneer 11.**
J. A. Van Allen.
Icarus, Vol. 51, 509 - 527 (1982).
This paper gives a short account of the theory of absorption by planetary rings and satellites of energetic charged particles that are trapped in a planet's magnetic field. The previously published observations of absorption features during Pioneer 11's passage through Saturn's inner radiation belt on 1 September 1979 are critically reanalyzed and related to other evidence on rings and satellites inside 2.9 R_s, especially that from the imaging system on Voyagers 1 and 2.

100.100 **The dynamical evolution of the Saturnian ring spokes.**
J. R. Hill, D. A. Mendis.
J. Geophys. Res., Vol. 87, 7413 - 7420 (1982).
It is proposed that the so-called radial spokes observed in rotation across Saturn's B ring are composed of fine charged dust leviated off the surfaces of larger bodies by electrostatic forces when the latter are sporadically charged to high electrostatic potentials. From the observed wedge-shaped morphology of these spokes it is deduced that the dust grains are negatively charged. The dynamical structure of a spoke feature consisting of grains of different sizes is considered.

100.101 **The plumes of Titan.**
A. Eviatar, G. L. Siscoe, J. D. Scudder, E. C. Sittler, Jr., J. D. Sullivan.
J. Geophys. Res., Vol. 87, 8091 - 8103 (1982).
The authors interpret the plasma density enhancements observed during the Voyager 1 Titan encounter in terms of a single plume of plasma wrapped around Saturn by corotation. They consider the radial motions of the plume driven by fluctuations in solar wind pressure and the dispersal of the plasma by the centrifugal interchange instability, by heating and by azimuthal acceleration. The authors show that Saturn cannot readily impose corotation directly on the plume and incidentally, that the total dissipation associated with the Titan-magnetosphere interaction is insufficient to supply the power to produce the observed Titan ultraviolet airglow.

100.102 **Satellites of Saturn.**
J. Bouška.
Vesmír, Vol. 61, 376 (1982). In Czech.

100.103 **Observation of the Saturnian system from Voyager.**
K. Iwasaki.
Astron. Her., Vol. 75, 128 - 133 (1982). In Japanese. – Abstr. in Phys. Abstr., Vol. 85, Abstr. 117629 (1982).

100.104 **The theory of motion of Hyperion.**
S.-p. Zhang, L. Liu.
Acta Astron. Sinica, Vol. 23, 185 - 197 (1982). In Chinese.
The paper first surveys the state of motion of the natural satellites in the solar system. Then, its main part deals with the founding of the theory of motion of Hyperion. The difficulties about $a \approx a_1$ and the commensurability ($n/n_1 \approx 3/4$) are solved in proper ways and the first-order solution is given.

Keeler's gap in Saturn's A ring.
See Abstr. 004.011.

J. E. Keeler's discovery of a gap in Saturn's A-ring.
See Abstr. 004.043.

Die koorbitalen Saturn-Monde 1980 S1 und 1980 S3.
See Abstr. 014.004.

Die kleinen Monde des Saturn.
See Abstr. 014.023.

Organic chemistry in Titan's atmosphere.
See Abstr. 022.036.

Molecular analysis of tholins produced under simulated Titan conditions.
See Abstr. 022.081.

Three dimensional Thomas-Fermi-Dirac calculations for dense binary mixtures.
See Abstr. 022.082.

Thermal infrared absorption measurements for methane and acetylene at planetary atmospheric temperatures.
See Abstr. 022.093.

Evidence of production of H_2CN^+-N_2. Application to the atmosphere of Titan.
See Abstr. 022.123.

Image processing applied to CCD images of Saturn's rings.
See Abstr. 031.587.

Construction of periodic orbits of satellites in a moving system of axes. I.
See Abstr. 042.068.

Ephémérides des huit premiers satellites de Saturne pour 1983.
See Abstr. 047.042.

Saturn ring rendezvous mission.
See Abstr. 051.032.

Saturn and Titan probes in the 1990's.
See Abstr. 051.034.

Touring the satellites of Saturn.
See Abstr. 052.035.

Voyager 1 Saturn targeting strategy.
See Abstr. 053.003.

Photographic observations of Jupiter and Saturn satellites at the Nikolaev Observatory in 1976 - 1979.
See Abstr. 099.007.

Investigations of the ultraviolet albedo of Jupiter and Saturn.
See Abstr. 099.009.

Ultraviolet spectroscopy of the Jovian and Saturnian aurorae.
See Abstr. 099.010.

Thermal structure of Jupiter and Saturn.
See Abstr. 099.012.

Jupiter and Saturn from 2 to 6 μm.
See Abstr. 099.015.

Rheology of ices: a key to the tectonics of the ice moons of Jupiter and Saturn.
See Abstr. 099.028.

Voyager photometry of Europa and the satellites of Saturn.
See Abstr. 099.029.

The meteorology of Jupiter and Saturn.
See Abstr. 099.033.

The oblateness effect on the solar radiation incident at the top of the atmospheres of the outer planets.
See Abstr. 099.039.

The satellites of Jupiter and Saturn.
See Abstr. 099.042.

A model of energy deposition of energetic electrons and EUV emission in the Jovian and Saturnian atmospheres and implications.
See Abstr. 099.044.

Theory of motions in Jupiter's and Saturn's atmospheres: evidence for deep internal circulations.
See Abstr. 099.051.

A convective model for driving global circulations on Jupiter and Saturn.
See Abstr. 099.052.

The meridional energy balance of Jupiter and Saturn.
See Abstr. 099.054.

A summary of the winds of Jupiter and Saturn.
See Abstr. 099.057.

The relative contribution of reflected solar and thermal emission to the 5 μm spectrum of Jupiter and Saturn.
See Abstr. 099.067.

Relative abundance of isotopic carbon in the atmospheres of the outer planets.
See Abstr. 099.068.

Ultraviolet spectral geometric albedos of selected Jovian and Saturnian satellites: IUE results.
See Abstr. 099.084.

Eight-color photometry of planetary satellites.
See Abstr. 099.085.

Charged particle erosion of the icy satellites of Jupiter and Saturn.
See Abstr. 099.086.

The atmospheres of Jupiter, Saturn, and Titan.
See Abstr. 099.109.

The hydrogen to helium ratio on Jupiter and Saturn.
See Abstr. 099.114.

Thermal- and plasma-induced molecular redistribution on the icy satellites.
See Abstr. 099.115.

Mass-radius relationships in icy satellites after Voyager.
See Abstr. 099.119.

On the C/H and D/H ratios in the atmospheres of Jupiter and Saturn.
See Abstr. 099.120.

Motion in the interiors and atmospheres of Jupiter and Saturn: scale analysis, anelastic equations, barotropic stability criterion.
See Abstr. 099.121.

Charged particle motions in the distended magnetospheres of Jupiter and Saturn.
See Abstr. 099.153.

101 Uranus, Neptune, Pluto, Transplutonian Planets

101.001 **Astrometric observations of outer planets and minor planets: 1980 - 1982.**
A. R. Klemola, E. A. Harlan.
Astron. J., Vol. 87, 1242 - 1243 (1982) = Lick Obs. Bull., No. 925.

Equatorial coordinates for Uranus, Neptune, Pluto, and selected minor planets have been measured on direct photographs taken with the 0.51-m Carnegie double astrograph of the Lick Observatory during the period 1980 - 1982.

101.002 **Speckle interferometric observations of Pluto and Charon.**
E. K. Hege, E. N. Hubbard, J. D. Drummond, P. A. Strittmatter, S. P. Worden, T. Lauer.
Icarus, Vol. 50, 72 - 81 (1982).

The authors report speckle interferometric observations of Pluto and its moon (1978 P1) Charon obtained on 5 June 1980 with a single 1.8-m mirror of the Multiple Mirror Telescope.

101.003 **Inclinations of the uranian rings.**
R. G. French, J. L. Elliot, D. A. Allen.
Nature, Vol. 298, 827 - 829 (1982).

The authors report a feature that they have found for the uranian ring orbits, not predicted by current theoretical models: at least seven of the nine rings have orbits inclined to the equatorial plane of Uranus.

101.004 **Intense Lyα emission from Uranus.**
S. T. Durrance, H. W. Moos.
Nature, Vol. 299, 428 - 429 (1982).

The existence of intense atomic hydrogen Lyα emission from Uranus is demonstrated here by utilizing the monochromatic imaging capabilities of the International Ultraviolet Explorer (IUE) spectrograph. The observations show increased emission in the vicinity of Uranus superimposed on the geocoronal/interplanetary background. Precipitation of trapped charged particles – aurora – could explain the emissions. This would imply a planetary magnetic field.

101.005 **Orbital elements of Nereid from new observations.**
C. Veillet.
Astron. Astrophys., Vol. 112, 277 - 280 (1982).

An observing run on the Danish-ESO 1.5-m reflector in April 1981 provided the author with good plates of Nereid, faint satellite of Neptune. Orbital elements of a Keplerian orbit as well as of a disturbed one from Mignard's theory are determined. The inferred mass of Neptune is found as $m^{-1} = 19383 \pm 110$, a value in good agreement with the mass determinations made with other techniques.

101.006 **Observations of Uranus and Neptune with the IUE.**
J. Caldwell, P. Winkelstein, T. Owen, M. Combes, T. Encrenaz, G. Hunt, V. Moore.
Advances in ultraviolet astronomy, (see 012.006), p. 306 (1982). – Abstract.

101.007 **Spectroscopy of Triton and Pluto: current status and prospects.**
D. P. Cruikshank.
Vibrational-rotational spectroscopy for planetary atmospheres, (see 012.007), p. 699 - 708 (1982).

Near-infrared spectrophotometry of Triton and Pluto at low spectral resolution and signal precision reveals methane absorption on both bodies. The absorption on Triton is probably gaseous CH_4, while that on Pluto is a combination of gas and ice of CH_4. Using present detectors and telescopes, spectra of Triton and Pluto can be obtained which are 5–10 times better than those published, but such data will not be sufficient to distinguish between gaseous and solid methane on these bodies.

101.008 **Comments on Pluto's atmosphere.**
L. Trafton.
Vibrational-rotational spectroscopy for planetary atmospheres, (see 012.007), p. 709 - 715 (1982).

A pure CH_4 atmosphere would rapidly escape from Pluto. For such an atmosphere, even CH_4 frosts on Pluto's surface would completely sublimate on a time scale short compared to Pluto's life. Observations of CH_4 on Pluto therefore imply that its atmosphere must also contain another gas in significant quantity.

101.009 **On the atmospheres of Triton and Pluto.**
G. S. Golitsyn, A. F. Steklov.
Physics of planetary atmospheres, (see 003.003), p. 139 - 147 (1981). In Russian. – Abstr. in Ref. zh., 51. Astron., 8.51.218 (1982).

101.010 **Reflection properties of Pluto's surface.**
A. N. Abramenko, V. V. Avramchuk,
V. A. Kucherov, L. R. Lisina, V. V. Prokof'eva.
Physics of planetary atmospheres, (see 003.003), p. 148 - 158 (1981). In Russian. – Abstr. in Ref. zh., 51. Astron., 8.51.219 (1982).

101.011 **Evidence for a ring system of Neptune.**
E. F. Guinan, C. C. Harris, F. P. Maloney.
Bull. American Astron. Soc., Vol. 14, 658 (1982). – Abstract.

101.012 **Photometry of Pluto during the 1982 opposition.**
J. D. Mulholland, R. P. Binzel.
Bull. American Astron. Soc., Vol. 14, 658 (1982). – Abstract.

101.013 **Some problems with planetary rotations.**
H. J. Smith.
Recent advances in observational astronomy, (see 012.028), p. 63 - 69 (1981).

Rotation periods of Uranus and Neptune are in wide dispute. These planets appear to show longitudinal asymmetry in the intensity of strong CH_4 bands. Several years of observations using a differential photometric technique, comparing the integrated planetary brightness in the 6100 Å band with that of the flanking continuum, were used to establish most-probable periods of 23ʰ92 for Uranus and 18ʰ43 for Neptune.

101.014 **Observation of the Lyman-alpha albedo of Uranus.**
K. H. Fricke, J. Darius.
Third European IUE Conference, (see 012.029), p. 463 - 469 (1982).

The Lyman-alpha emission from Uranus was measured for the first time using IUE. The effective geometric albedo sets a lower limit of 0.5 ppm for the hydrocarbon mixing ratio in the high upper stratosphere, as well as an upper limit on the atomic hydrogen column density of $10^{16}\,cm^{-2}$ above the methane layer.

101.015 **The occultation of star KME 14 by Uranus rings (22 April 1982).**
B. Sicardy, A. Brahic, J. Lecacheux, P. Bouchet, C. Perrier, P. Laques, J. Legait.
Bull. American Astron. Soc., Vol. 14, 749 (1982). – Abstract.

101.016 **Inclinations of the Uranian rings.**
R. G. French, J. L. Elliot.
Bull. American Astron. Soc., Vol. 14, 749 - 750 (1982). Abstract.

101.017 **Rings around Neptune?**
J. L. Elliot.
Bull. American Astron. Soc., Vol. 14, 750 (1982). – Abstract.

101.018 **Bond albedo and energy budget for Uranus: 1961 - 1981.**
B. L. Lutz, G. W. Lockwood, D. T. Thompson, A. Warnock III.
Bull. American Astron. Soc., Vol. 14, 759 - 760 (1982). Abstract.

101.019 **The visible phase curves of Uranus and Neptune from Voyager.**
D. Wenkert, G. E. Danielson.
Bull. American Astron. Soc., Vol. 14, 760 (1982). – Abstract.

101.020 **Interpretation of the 6818.9 Å CH_4 feature observed on Uranus and Neptune.**
K. H. Baines.
Bull. American Astron. Soc., Vol. 14, 760 (1982). – Abstract.

101.021 **Study of weak visible methane bands on Uranus and Neptune, 1973 to 1981.**
K. S. Noll, R. G. Danehy, T. Owen.
Bull. American Astron. Soc., Vol. 14, 760 (1982). – Abstract.

101.022 **Thermal constraints on atmospheric structure of Uranus and Neptune from observations in the 10 μm spectral region.**
G. S. Orton, A. T. Tokunaga, J. Caldwell.
Bull. American Astron. Soc., Vol. 14, 760 (1982). – Abstract.

101.023 **The upper atmosphere of Uranus: results of the 22 April 1982 occultation.**
M. Combes, B. Sicardy, J. Lecacheux, L. Vapillon.
Bull. American Astron. Soc., Vol. 14, 760 - 761 (1982). Abstract.

101.024 **IUE observations of auroral H Ly α emission from Uranus.**
J. T. Clarke.
Bull. American Astron. Soc., Vol. 14, 761 (1982). – Abstract.

101.025 **Intense Lyman-alpha emission from Uranus.**
H. W. Moos, S. T. Durrance.
Bull. American Astron. Soc., Vol. 14, 761 (1982). – Abstract.

101.026 **Aurorae on Uranus.**
A. J. Dessler.
Bull. American Astron. Soc., Vol. 14, 761 (1982). – Abstract.

101.027 **CCD images of Uranus and Neptune.**
C. B. Pilcher, R. R. Howell, J. N. Heasley.
Bull. American Astron. Soc., Vol. 14, 761 (1982). – Abstract.

101.028 **The Uranian satellites: a new look.**
R. H. Brown.
Bull. American Astron. Soc., Vol. 14, 765 (1982). – Abstract.

101.029 **Methane on Triton: physical state and distribution.**
D. P. Cruikshank, J. Apt, R. H. Brown, C. B. Pilcher.
Bull. American Astron. Soc., Vol. 14, 765 (1982). – Abstract.

101.030 **On the origin of Triton.**
W. B. McKinnon.
Bull. American Astron. Soc., Vol. 14, 765 (1982). – Abstract.

101.031 **Diameters of Triton and Pluto.**
D. Morrison, D. P. Cruikshank, R. H. Brown.
Bull. American Astron. Soc., Vol. 14, 765 - 766 (1982). Abstract.

101.032 **The radii and albedos of Triton and Pluto.**
L. A. Lebofsky, G. H. Rieke, M. J. Lebofsky.
Bull. American Astron. Soc., Vol. 14, 766 (1982). – Abstract.

101.033 **The global distribution of Pluto's atmosphere.**
L. Trafton, S. A. Stern.
Bull. American Astron. Soc., Vol. 14, 766 (1982). – Abstract.

101.034 **Distant planetary satellites more accurately measured.**
I. Gatley.
Nature, Vol. 300, 406 (1982).

101.035 **Diameters and albedos of satellites of Uranus.**
R. H. Brown, D. P. Cruikshank, D. Morrison.
Nature, Vol. 300, 423 - 425 (1982).

New albedo measurements show that Umbriel, Titania and Oberon are similar in albedo to J4 Callisto while Ariel is very similar in albedo to S7 Hyperion; the diameters of all four satellites are similar to those of the large, icy saturnian satellites S5 Rhea and S8 Iapetus.

101.036 **Diameters of Triton and Pluto.**
D. Morrison, D. P. Cruikshank, R. H. Brown.
Nature, Vol. 300, 425 - 427 (1982).

The authors report new upper limits to the thermal IR emission from these objects that permit to set significant upper limits on their diameters and to demonstrate that both are high albedo objects. These results clearly exclude the possibility that Triton is the largest planetary satellite, and they are consistent with the small size of Pluto deduced from other data.

101.037 **Detection of auroral hydrogen Lyman-alpha emission from Uranus.**
J. T. Clarke.
Astrophys. J., Lett., Vol. 263, L105 - L109 (1982).

A series of observations of Uranus obtained with the short-wavelength spectrograph of the IUE Observatory in 1982 April and June have revealed unexpectedly strong H Lyα emission which varied between 430 and 850 rayleighs in observed disk-averaged brightness over the course of these observations. The variability of the emission alone indicates that much of the emission must be produced by charged particle excitation of H in Uranus' upper atmosphere. Solar EUV photodissociation of H_2 is insufficient to produce the necessary high column abundance of H. These data offer the first strong evidence for the presence of aurorae and therefore a magnetic field on Uranus.

101.038 **Neptune: projected evolution of Triton's orbit.**
J. Boynton.
Publ. Astron. Soc. Pacific, Vol. 94, 749 - 750 (1982).
Abstract.

101.039 **The upper atmosphere of Uranus: a critical test of isotropic turbulence models.**
R. G. French, J. L. Elliot, B. Sicardy, P. Nicholson, K. Matthews.
Icarus, Vol. 51, 491 - 508 (1982).

Observations of the 15 August 1980 Uranus occultation of KM 12, obtained from Cerro Tololo Inter American Observatory, European Southern Observatory, and Cerro Las Campanas Observatory, are used to compare the atmospheric structure at points separated by $\sim$140 km along the planetary limb. The results reveal striking, but by no means perfect, correlation of the light curves, ruling out isotropic turbulence as the cause of the light curve spikes. The atmosphere is strongly layered.

101.040 **Stability of Pluto's atmosphere.**
D. M. Hunten, A. J. Watson.
Icarus, Vol. 51, 665 - 667 (1982).

L. Trafton has pointed out that a substantial methane atmosphere, observed on Pluto by U. Fink et al. (1980, Icarus, Vol. 44, 62 - 71), appears to be unstable against blow-off. The difficulty is shown to disappear if the actual heat balance and thermal structure are considered, instead of the classic assumption that the upper atmosphere is isothermal.

101.041 **Maps of the rings of Uranus at a wavelength of 2.2 microns.**
K. Matthews, G. Neugebauer, P. D. Nicholson.
Icarus, Vol. 52, 126 - 135 (1982).

Maps of the ring system of Uranus, as seen in reflected sunlight at a wavelength of 2.2 μm, are presented for May 1978 and July 1979. Large azimuthal brightness variations revealed by these maps are consistent with the variable width established for the ϵ ring from occultation studies and support the existing elliptical model for this ring. In addition, multi-aperture broadband infrared photometry of Uranus is reported.

101.042 **The planet Pluto.**
Z. Pokorný.
Říše hvězd, Vol. 63, 161 - 163 (1982). In Czech.

Geburtstag einer Entdeckung – zum 200. Jahrestag der Auffindung des Planeten Uranus.
See Abstr. 004.031.

Die Entdeckung des Planeten Pluto.
See Abstr. 004.061.

Pluto and a transplutonian planet as predicted by Venkatesha Ketakara.
See Abstr. 004.076.

Digital speckle interferometry of Juno, Amphitrite and Pluto's moon Charon.
See Abstr. 031.679.

Planetary and asteroidal occultations in 1982.
See Abstr. 096.004.

Possible occultations by satellites of Uranus and Neptune: 1983–1985.
See Abstr. 096.020.

The oblateness effect on the solar radiation incident at the top of the atmospheres of the outer planets.
See Abstr. 099.039.

Observations of Jupiter, Uranus and Neptune from the International Ultraviolet Explorer.
See Abstr. 099.065.

Planetary Rings.
See Abstr. 100.096.

102 Comets (Origin, Structure, Atmospheres, Dynamics)

102.001 **OH fluorescence in comets: fluorescence efficiency of the ultraviolet bands.**
D. G. Schleicher, M. F. A'Hearn.
Astrophys. J., Vol. 258, 864 - 877 (1982).
The authors have carried out calculations of the fluorescence equilibrium of OH in the solar radiation field as a function of heliocentric radial velocity. Comparison of detailed band profiles with observed, high-resolution spectra of Comet Bradfield (1979 X) and Comet Seargent (1978 XV) shows very good agreement.

102.002 **The stellar perturbations of orbital elements of long-period comets.**
S. Yabushita, I. Hasegawa, K. Kobayashi.
Mon. Not. R. Astron. Soc., Vol. 200, 661 - 671 (1982).
Perturbations by nearby stars on orbital elements of comets with large heliocentric distances are calculated by the impulse approximation and by accurate numerical integration using Cowell's method. It is shown that the agreement is sufficiently close for elliptic orbits but rather poor for parabolic orbits. For random encounters, statistical parameters which specify the distributions of orbital elements are calculated by the Monte Carlo method. The results may have an implication for cometary origin.

102.003 **Vaporization in comets; outbursts from comet Schwassmann-Wachmann 1.**
J. J. Cowan, M. F. A'Hearn.
Icarus, Vol. 50, 53 - 62 (1982).
The authors propose that the outbursts frequently observed from Comet P/Schwassmann-Wachmann 1 do not require storage of energy as suggested by many authors. The authors present revised estimates to show that the total mass and kinetic energy in a typical outburst are lower than previously estimated and they show that this mass is comparable to the mass of gas vaporized as inferred from recent observations of CO^+ in this comet. The authors propose simple equilibrium vaporization of CO_2 or CO which is suddenly exposed on a nucleus which is otherwise composed primarily of H_2O.

102.004 **The effect of star passages on cometary orbits in the Oort cloud.**
H. Scholl, A. Cazenave, A. Brahic.
Astron. Astrophys., Vol. 112, 157 - 166 (1982).
The authors investigate systematically by numerical integrations close encounters between stars and comets in the Oort cloud without restrictions with respect to orbital elements. The dependency of changes in cometary orbits after close stellar encounters on the parameters of the problem, namely stellar mass, relative velocity, missdistance, and orbital configuration, is derived quantitatively. A comparison with the impulse formula for changes in cometary velocities after close stellar encounters gives the limits for the validity of this formula.

102.005 **Ion-molecular clusters in cometary nuclei.**
L. M. Shul'man.
Inst. teor. fiz. AN USSR. Prepr., 1981 (1982), No. 141, 21 pp. In Russian. – Abstr. in Ref. zh., 51. Astron., 7.51.299 (1982).

102.006 **On the phenomenon of disconnection in cometary plasma tails.**
O. V. Dobrovol'skij.
Dokl. AN TadzhSSR, Tom 24, 347 - 351 (1981). In Russian. Abstr. in Ref. zh., 51. Astron., 7.51.305 (1982).

102.007 **Solar wind interaction with comets.**
T. K. Breus.
Space Sci. Rev., Vol. 32, 361 - 376 (1982).
The planned missions to Comet Halley, which will arrive at the nearest space of the Sun in 1986, have recently revived interest in studying solar wind interaction with comets. Several unsolved problems exist and the most urgent of them are as follows: (1) The character of the solar wind interaction with comets: bow shocks and contact surface formation near comets; similarities and differences of solar- wind interaction with comets and with Venus. (2) The anomalous ionization in cometary heads. (3) The problem of the anomalously high accelerations of ions in the plasma tails of comets. (4) The variability of plasma structures observed in cometary tails.

102.008 **Speculations on the infrared molecular spectra of comets.**
M. J. Mumma.
Vibrational-rotational spectroscopy for planetary atmospheres, (see 012.007), p. 717 - 744 (1982).
The author considers certain aspects of cometary physics and non-LTE spectroscopy, and attempts to predict some aspects of the infrared signatures of cometary molecules.

102.009 **On negative ions in comets.**
O. V. Dobrovol'skij, N. M. Khashimov, Kh. I. Ibadinov, V. L. Tarakanov.
Dokl. AN TadzhSSR, Tom 24, 543 - 547 (1981). In Russian. From Ref. zh., 51. Astron., 8.51.222 (1982).

102.010 **The photochemical heating of the cometary atmosphere.**
M. L. Marconi, D. A. Mendis.
Astrophys. J., Vol. 260, 386 - 394 (1982).
A self-consistent solution of the thermodynamic structure of an H_2O dominated, multispecies cometary atmosphere at 1 AU has been obtained by solving the simultaneous set of differential equations representing conservation of number density, momentum, energy, and the flux of solar UV radiation in three streams responsible for the major photolytic processes.

102.011 **The need for cometary photometry.**
J. N. Marcus.
I.A.P.P.P. Commun., No. 8, p. 26 - 30 (1982).

102.012 **Overview of comet observations.**
S. Wyckoff.
Comets, (see 012.009), p. 3 - 55 (1982).
Observations presently indicate that the nuclei of comets are ice-dust conglomerates with masses $\sim 10^{13}$ to 10^{19} g, radii $\sim$ few km, average rotation periods $\sim$ 15 hr and tensile strengths $\sim 10^5$ dyne cm^{-2}. The latter indicates that cometary nuclei are very fragile entities. All observations support the basic concept of a comet nucleus based on Whipple's icy conglomerate model of H_2O ice plus an admixture of other ices and dust.

102.013 **Comet discoveries, statistics, and observational selection.**
Ľ. Kresák.
Comets, (see 012.009), p. 56 - 82 (1982).
This paper is an overview of the distribution of cometary orbits in their individual parameters, to distinguish between real and apparent deviations from randomness and to specify those features that are relevant to the origin, evolution, and constitution of the system of comets.

102.014 **Chemical composition of cometary nuclei.**
A. H. Delsemme.
Comets, (see 012.009), p. 85 - 130 (1982).

Cometary stuff appears to be made up of an icy conglomerate, with a volatile and refractory fraction. Circumstantial arguments will be shown to be in favor of the undifferentiated character of this icy conglomerate, at least before it starts vaporizing in the solar heat. The first step towards a quantitative understanding of this conglomerate is the assessment of its refractory-to-volatile mass ratio; this is also the dust-to-gas mass ratio of the dusty gas that sublimes from the nucleus, if the latter is undifferentiated throughout, and only when a steady state is reached.

102.015 **What are comets made of? A model based on interstellar dust.**
J. M. Greenberg.
Comets, (see 012.009), p. 131 - 163 (1982).

A model for pristine comet composition is derived on the basis of aggregated interstellar dust. The chemical and physical constituents of the precomet dust are derived by observing the evolution of clouds of dust and molecules into the molecular cloud phase and extrapolating theoretically to the ultimate composition of the dust. The observational characteristics of the dust are interpreted by a combination of theory with the results of laboratory simulation of the photochemical evolution of interstellar dust materials.

102.016 **Structure and origin of cometary nuclei.**
B. Donn, J. Rahe.
Comets, (see 012.009), p. 203 - 226 (1982).

There is strong evidence that a comet nucleus consists of a single object whose basic structure is Whipple's icy conglomerate. In this review the authors consider only such models. Derived radii fall in the range from 0.3 to 16 km. With an adopted density of 1 g cm^{-3}, masses are between 10^{13} and 10^{19}g. A number of cometary phenomena indicate that the nucleus is a low-density, fragile object with a large degree of radial uniformity in structure and composition. Details of the ice-dust pattern are more uncertain. A working model is proposed based on theories of accumulation of larger objects from grains.

102.017 **The rotation of comet nuclei.**
F. L. Whipple.
Comets, (see 012.009), p. 227 - 250 (1982).

The miniscule history of spin-vector research on comet nuclei is reviewed. Major emphasis is placed on actual determinations of rotation period and of spin-axis orientation. The latter is indicated by asymmetrical comae and by the directions of near-nucleus jets. The spin axis is known for seven periodic comets, including P/Encke for which a rapid precession of the spin axis has been established. Rotation periods can be measured by photometry (as for asteroids), by near-nucleus jets, and by halo diameters. The halo method, developed by the author, has been by far the most prolific but has been little tested. The spin periods P for 47 comets, almost all new determinations, are presented and compared with those of 41 small asteroids of diameter $\leqslant$ 40 km. The median P for the comets is 15.0 hr versus 6.8 hr for these asteroids.

102.018 **The problem of split comets in review.**
Z. Sekanina.
Comets, (see 012.009), p. 251 - 287 (1982).

The progress in the investigation of cometary splitting is reviewed from the dynamical and physical standpoints. It is shown that the rate of recession of cometary fragments is determined by the momentum from outgassing, the net differential force thence being of the same nature as the nongravitational perturbations detected in the motions of other comets. The differential radial deceleration γ and the time of splitting are the only parameters of this simple model that succesfully represents the positional observations of almost all of the 21 known split comets.

102.019 **Relationships between comets, large meteors, and meteorites.**
G. W. Wetherill, D. O. ReVelle.
Comets, (see 012.009), p. 297 - 319 (1982).

Available data relevant to the use of bright meteors (fireballs) to obtain information regarding the nature of cometary nuclei are reviewed.

102.020 **Optical and infrared observations of bright comets in the range 0.5 μm to 20 μm.**
E. P. Ney.
Comets, (see 012.009), p. 323 - 340 (1982).

Infrared observations have been obtained on seven bright comets and P/Encke since 1965. Five comets were dusty with pronounced Type II tails. All these dusty comets showed a silicate signature at 10 and 18 μm indicating the presence of small refractory grains. Measurement of coma colors from visual to infrared wavelengths allows determination of the grain temperature and of the albedo and its dependence on scattering angle.

102.021 **Interpreting the thermal properties of cometary dust.**
H. Campins, M. S. Hanner.
Comets, (see 012.009), p. 341 - 356 (1982).

The characteristics of the thermal emission from cometary dust are discussed and the observations compared with models for absorbing and silicate grains.

102.022 **Dusty gas-dynamics in real comets.**
M. K. Wallis.
Comets, (see 012.009), p. 357 - 369 (1982).

The mathematical models of dust grains as dragged out by radially expanding cometary gases are reviewed.

102.023 **Cometary dust in the solar system.**
H. Fechtig.
Comets, (see 012.019), p. 370 - 382 (1982).

By two basic methods, studies of microcraters on lunar samples and in situ dust measurements, information has been obtained on cometary dust in the solar system. This paper reviews recent observations of interplanetary dust in space and estimates the amount of cometary dust.

102.024 **Comet head photometry: past, present, and future.**
D. D. Meisel, C. S. Morris.
Comets, (see 012.009), p. 413 - 432 (1982).

Comet photometry has reached the point of convergence of new and old methodologies. The present difficulties with scale-length determination do not appear insurmountable so we predict that both modern observational and theoretical techniques will eventually provide a consistent, physically-based photometric model as a basis for studies of brightness fluctuations.

102.025 **Spectrophotometry of comets at optical wavelengths.**
M. F. A'Hearn.
Comets, (see 012.009), p. 433 - 460 (1982).

A review of recent technological advances indicates that we are on the verge of a breakthrough in cometary spectrophotometry. Homogeneous spectrophotometry can now be obtained for many comets, even faint ones. Our understanding of fluorescent emission is improving, but the extent to which other processes such as collisions must be taken into account is still unclear.

102.026 **Ultraviolet spectroscopy of comae.**
P. D. Feldman.
Comets, (see 012.009), p. 461 - 479 (1982).

Vacuum ultraviolet observations from sounding rockets and satellite observatories of the gaseous comae of several recent comets are reviewed.

102.027 **Photochemical processes in the inner coma.**
W. F. Huebner, P. T. Giguere, W. L. Slattery.
Comets, (see 012.009), p. 496 - 515 (1982).

Very little evidence exists on the physical structure and chemical composition of comet nuclei. Thus, one purpose of modeling the coma is to characterize the chemical and physical properties of the nucleus by fitting model calculations to observations. Models for inner coma chemistry are reviewed.

102.028 **Observations and dynamics of plasma tails.**
J. C. Brandt.
Comets, (see 012.009), p. 519 - 537 (1982).

A tentative morphology of plasma tail evolution has been developed; it can be interpreted as resulting from interaction of a comet with sector boundaries in the solar wind, utilizing magnetic reconnection.

102.029 **Plasma flow and magnetic fields in comets.**
H. U. Schmidt, R. Wegmann.
Comets, (see 012.009), p. 538 - 560 (1982).

The plasma flow and field amplification resulting from interaction of the solar wind with a comet are studied by model calculations. Two surfaces of discontinuity, the shock front and the contact surface, are discussed. The influence of the embedded magnetic field is investigated first in an axisymmetric model since the transverse field in the solar wind changes its direction frequently. A three-dimensional MHD-calculation then shows how the magnetic field induces an asymmetry in the plasma flow.

102.030 **Solar wind interaction with comets: lessons from Venus.**
C. T. Russell, J. G. Luhmann, R. C. Elphic, M. Neugebauer.
Comets, (see 012.009), p. 561 - 587 (1982).

The authors examine the data that have been obtained on the solar wind interaction with the planet Venus for the purpose of comparison with some of the similar processes that may be occurring in comets.

102.031 **Theories of physical processes in the cometary comae and ion tails.**
W.-H. Ip, W. I. Axford.
Comets, (see 012.009), p. 588 - 634 (1982).

A number of physical processes are involved in the comet-solar wind interaction. In the outer region of the cometary comae, there is the pickup of the new cometary ions by the solar wind. Space probe observations must explore the questions concerning the manner in which the cometary ions are being assimilated into the inward streaming plasma flow and what their resulting velocity distributions are, even though certain limits can be set by theoretical arguments. Another basic issue in cometary plasmas concerns the nature of the ionospheric outflow from the central source. Two possible flow patterns (supersonic wind versus subsonic flow) are discussed here.

102.032 **Evolution of long- and short-period orbits.**
E. Everhart.
Comets, (see 012.009), p. 659 - 664 (1982).

The orbits of short-period comets belong to the class of chaotic orbits. This type, always of inclination less than 35°, includes near-parabolic orbits with perihelia near Jupiter's orbit or beyond, near-circular orbits at Jupiter or Saturn's distance, some unstable Trojan and horseshoe orbits, temporary satellite captures by Jupiter and Saturn, and the orbits of visible short-period comets. There are frequent changes from one of these forms to another. The long-period orbits, meaning both long-period and intermediate-period orbits of any inclination with small enough perihelia for the comets to be visible, do not interact with the chaotic orbits.

102.033 **Do comets evolve into asteroids? Evidence from physical studies.**
J. Degewij, E. F. Tedesco.
Comets, (see 012.009), p. 665 - 695 (1982).

Physical observations of low-activity periodic comets and asteroids in distant and peculiar orbits are reviewed. In particular the authors discuss magnitude studies and reflection spectra at visible and infrared wavelengths.

102.034 **Comets and origin of life.**
C. Ponnamperuma, E. Ochiai.
Comets, (see 012.009), p. 696 - 703 (1982).

The observation of carbon compounds in comets is suggestive of chemical evolution. However, it appears unlikely that prebiological processes would have gone much further than the stage of monomer formation. The proposition that comets may be a vehicle for panspermia is not warranted by currently available data.

102.035 **Photochemistry and energy balance in comet comae.**
J. J. Keady, W. F. Huebner.
Bull. American Astron. Soc., Vol. 14, 621 (1982). – Abstract.

102.036 **A review of radio observations of comets.**
L. E. Snyder.
Icarus, Vol. 51, 1 - 24 (1982).

The current status of cometary radio observations is reviewed. Radio continuum observations made at different wavelengths can be used to model the properties of cometary particles. Radar observations are important for determining the size, angular momentum, direction of motion, and surface properties of the cometary nucleus. In order to clarify the sometimes confusing observations which have been reported, cometary radio spectroscopy is reviewed in chronological order, comet by comet. The results of current OH ultraviolet pumping models are briefly discussed and several formalisms for computing molecular production rates are presented.

102.037 **Could life have evolved in cometary nuclei? .**
A. Bar-Nun, A. Lazcano-Araujo, J. Oro.
Origins of Life, Vol. 11, 387 - 394 (1981). – Abstr. in Phys. Abstr., Vol. 85, Abstr. 63319 (1982).

102.038 **Geometry of the distribution of plasma tail rays in a cometary type 1 tail.**
H. Pérez de Tejada.
Rev. Mexicana Astron. Astrofis., Vol. 6, (see 012.017), 327 - 331 (1981).

A study of the orientation of plasma tail rays in type 1 cometary tails is carried out by considering a solar wind interaction similar to that observed in the planet Venus. Application of the most recent experimental and theoretical results of the flow configuration in the Venus wake lead to the prediction that the cometary tail rays must be preferably formed and evolve in the vicinity of the midnight plane of the comet. The suggested coplanarity of the system of tail rays makes its observation difficult for comets located at high ecliptic latitudes, and thus provides a selfconsistent explanation of the observed precipitous decrease of the detection frequency of such structures at high latitudes.

102.039 **Light curves of some periodic comets.**
I. Ferrín, E. Guzmán.
Rev. Mexicana Astron. Astrofis., Vol. 6, (see 012.017), 339 - 345 (1981).

The authors have derived some reliable light curves for several periodic comets, combining observations made at different apparitions. The authors present their analysis of these light curves for comets Encke and Tempel 2.

102.040 **Comets.**
D. W. Hughes.
Contemp. Phys., Vol. 23, 257 - 283 (1982). – Abstr. in Phys. Abstr., Vol. 85, Abstr. 73909 (1982).

102.041 **Dynamical aspects of the origin of comets.**
J. A. Fernández.
Astron. J., Vol. 87, 1318 - 1332 (1982).
Several possible birthplaces of comets are investigated numerically from the viewpoint of their dynamical feasibility as sources of the current flux of new comets. Cometary origins both in the planetary region and in situ (from where "new" comets seem to come) are considered. The orbital evolution of hypothetical comets is followed up to the present taking into account the perturbations of the four giant planets and of random passing stars.

102.042 **Non-radial flow near the cometary surface.**
M. L. Marconi, D. A. Mendis.
Moon Planets, Vol. 26, 345 - 351 (1982).
It is shown that, as a consequence of the non-uniform temperature distribution of the cometary nucleus, large lateral pressure gradients are set up, which in turn drive strong lateral flows. However, at small heliocentric distances the onset of turbulence within a thin boundary layer destroys these steady lateral flows and the eventual outflow of gas from within the outer boundary of this layer is expected to be more or less radial. On the other hand, at large heliocentric distances, turbulence is unlikely to set in, and the lateral flows that are set up, may persist. Consequently, it is expected that the gas flow out of the cometary nucleus at these large distances to be highly non-radial.

102.043 **Solar wind magnetic field penetration into cometary ionospheres.**
L. S. Marochnik.
Moon Planets, Vol. 26, 353 - 370 (1982).
It is shown that when the solar wind interacts with comets its magnetic field may penetrate through the discontinuity surface into the cometary ionosphere. In case of Venus-like planets the effect of regular motion of magnetic field lines through that surface is not present, and the presence of magnetic fields in the ionosphere of non-magnetic planets may be due to ionopause boundary instabilities.

102.044 **The nuclear region of comets.**
D. Malaise.
The need for coordinated ground-based observations of Halley's comet, (see 012.025), p. 95 - 113 (1982).

102.045 **Observational needs for thermal modeling of comet nuclei.**
J. Klinger.
The need for coordinated ground-based observations of Halley's comet, (see 012.025), p. 115 - 121 (1982).

102.046 **The coma of comets.**
M. C. Festou.
The need for coordinated ground-based observations of Halley's comet, (see 012.025), p. 123 - 130 (1982).

102.047 **Fluorescence of water molecules in comets.**
J. Crovisier.
The need for coordinated ground-based observations of Halley's comet, (see 012.025), p. 149 - 154 (1982).

102.048 **Isotopic abundances in comets.**
A. C. Danks.
The need for coordinated ground-based observations of Halley's comet, (see 012.025), p. 155 - 170 (1982).

102.049 **Comet ion formation and dynamics.**
M. K. Wallis.
The need for coordinated ground-based observations of Halley's comet, (see 012.025), p. 171 - 179 (1982).
Solar photon and proton ionization may well supply average ion fluxes, but closer observation correlated with solar wind conditions is needed to distinguish creation of ion structures by intrinsic mechanisms (such as auroral discharges or ionizing flow instability) from the variable solar wind 'snowplough'.

102.050 **The cometary ion tail and its relation to solar wind phenomena.**
K. Jockers.
The need for coordinated ground-based observations of Halley's comet, (see 012.025), p. 193 - 216 (1982).
Most cometary ions observed in the visible range are the ions of presumed mother molecules like H_2O, CO, CO_2, N_2. After reviewing the emissions of these ions an overview of our present understanding of the comet-solar wind interaction is given. Modulation effects on the different ions introduced by a change in solar wind flux are studied. Further topics include the general structure of the ion tail, tail rays and large-scale tail disturbances (tail disconnections). Recommendations for future ground-based observations are given.

102.051 **The significance of photopolarimetric measurements for the exploration of cometary dust.**
R. H. Zerull, R. H. Giese.
The need for coordinated ground-based observations of Halley's comet, (see 012.025), p. 257 - 260 (1982).

102.052 **On the brightness law for comets with $r > 5$ A. U.**
R. M. West.
The need for coordinated ground-based observations of Halley's comet, (see 012.025), p. 263 - 271 (1982).
An investigation of 34 visually estimated magnitudes of comets 1974 XII, 1975 II and 1977 IX at large heliocentric distances (5.6 - 9.7 A. U.) shows that the brightness variations conform with $n = 7.1 + 1.0$. This is supported by upper limits for the brightness of 1975 II at 13 A. U., established with the ESO 3.6 m telescope.

102.053 **Effect of turbulence on helical waves in cometary plasma tails.**
T. V. Subbaraman, M. Mohan, B. Buti.
Bull. Astron. Soc. India, Vol. 10, 31 - 32 (1982). – Abstract.

102.054 **Interaction of turbulent solar wind with cometary plasma tails.**
B. Buti.
Astrophys. Space Sci., Vol. 85, 35 - 41 (1982).
The longitudinal electric field associated with the observed electrostatic turbulence in the solar wind is shown to modify the dispersive characteristics of the hydromagnetic waves propagating along the interface between the solar wind and the cometary plasma. Extremely weak turbulence has a tendency to stabilize these surface waves, whereas turbulence of moderate level can be stabilizing or destabilizing depending on the strength of the cometary magnetic field B_{OC} relative to the interplanetary magnetic field B_{OS}. For $B_{OC} \gtrsim B_{OS}$, destabilization is not possible.

102.055 **The spectrum of comets as derived from IUE observations.**
M. C. Festou, P. D. Feldman, H. A. Weaver, H. U. Keller.
Third European IUE Conference, (see 012.029), p. 445 - 449 (1982).
Between October 1978 and March 1981, eight comets were observed with the IUE at various heliocentric and geocentric distances. The UV spectra of these eight comets are remarkably similar despite the large differences in the dust to

gas ratios. Since all the dominant atomic species (but N), radicals and ions of the coma are detected in this spectral region, it is now possible to estimate the total gaseous output of the nucleus.

102.056 **Are comets made of H_2O-ice?**
M. K. Wallis.
Third European IUE Conference, (see 012.029), p. 451 - 453 (1982).

The heliocentric dependence of OH-band intensities in comet Bradfield (1979 X) in the near UV was significantly steeper than expected for H_2O-ice over 0.7–1.4 AU. More realistic models for sublimating icy comets are examined. The results imply the presence of less volatile outgassing material than H_2O, primarily of organic composition.

102.057 **The relative abundance of silicate/absorbing grains in comets.**
M. S. Hanner.
Bull. American Astron. Soc., Vol. 14, 750 (1982). – Abstract.

102.058 **Dark comet nuclei.**
H. Campins.
Bull. American Astron. Soc., Vol. 14, 750 - 751 (1982). Abstract.

102.059 **Cometary impacts with the sun: physical and dynamical considerations.**
P. R. Weissman.
Bull. American Astron. Soc., Vol. 14, 751 - 752 (1982). Abstract.

102.060 **Coronagraphic observations of two new sungrazing comets.**
N. R. Sheeley, Jr., R. A. Howard, M. J. Koomen, D. J. Michels.
Nature, Vol. 300, 239 - 242 (1982).

The authors report observations of two sungrazers that encountered the sun on 27 January 1981 and 20 July 1981, respectively. These two new comets seem to have been members of the Kreutz group of sungrazers, and like 1979 XI the new comets did not reappear after their encounters with the sun.

102.061 **On the flaring of cometary plasma tails.**
A. I. Ershkovich, M. B. Niedner, Jr., J. C. Brandt.
Astrophys. J., Vol. 262, 396 - 406, plate 8 (1982).

A model of cometary plasma tail flaring is presented which explains the observed variability in flaring on the basis of the gas pressure of tail ions and the strength of the magnetic field at the flanks of the cometary ionopause. Both properties depend on the location and shape of the cometary ionopause. The flaring angles generated by the hypersonic pressure balance adopted are in good agreement with observations of tail flaring in comets Kohoutek and Bradfield.

102.062 **The photochemistry and dynamics of a dusty cometary atmosphere.**
M. L. Marconi, D. A. Mendis.
Moon Planets, Vol. 27, 27 - 46 (1982).

A self-consistent solution of the dynamical and thermal structure of an H_2O-dominated, two-phase, dusty-gas cometary atmosphere has been obtained by solving the simultaneous set of differential equations representing conservation of number density, momentum and energy together with the transfer of solar radiation in the streams responsible for the major photolytic processes and the heating of the nucleus. The validity of the model is restricted to the collision-dominated region where all the gas species are assumed to attain a common velocity and common temperature. Two models are considered for the transfer of solar radiation through the circum-nuclear dust halo.

102.063 **Introduction to cometary physics.**
H. U. Keller.
The solar system and its exploration, (see 012.054), p. 211 (1981). – Abstract.

102.064 **Aeronomical processes in cometary comae.**
M. Festou.
The solar system and its exploration, (see 012.054), p. 213 - 219 (1981).

Cometary observations reveal the real nature of a comet: what we see is a huge atmosphere, which has only a temporary existence, while the real comet is a small body which is the source of the observed material. In order to determine what is the composition of the "nucleus", it is necessary to investigate a large variety of physical phenomena which belong to aeronomy.

102.065 **How Černis and Petrauskas detected a comet.**
K. I. Churyumov.
Zemlya Vselennaya, 1982, No. 6, p. 65 - 68. In Russian.

102.066 **On the distribution of the longitudes of aphelia of comets of the Jupiter family.**
A. S. Guliev.
Komet. Tsirk., Kiev, No. 285 (1982). In Russian.

102.067 **On cometary dust of matrix origin.**
I. S. Lizunkova, V. A. Doanevich.
Komet. Tsirk., Kiev, No. 286 (1982). In Russian.

102.068 **Peculiarities of the brightness decrease of short-period comets of the Jupiter family.**
A. S. Guliev, A. A. Demenko, I. M. Demenko.
Komet. Tsirk., Kiev, No. 287 (1982). In Russian.

102.069 **Radii of nuclei of periodic comets and their orbital inclination.**
Yu. N. Ivashchenko, K. I. Churyumov.
Komet. Tsirk., Kiev, No. 287 (1982). In Russian.

102.070 **Statistics of long-period comets.**
S. K. Vsekhsvyatskij, A. A. Demenko.
Komet. Tsirk., Kiev, No. 288 (1982). In Russian.

102.071 **On the size of cometary nuclei.**
E. A. Kajmakov, I. S. Lizunkova.
Komet. Tsirk., Kiev, No. 289 (1982). In Russian.

102.072 **Analysis of photometric parameters of comets.**
M. Z. Markovich, R. S. Osherov.
Komet. Tsirk., Kiev, No. 289 (1982). In Russian.

102.073 **Why do cometary nuclei not content methane, ammonia, carbon dioxide and other matter?**
L. M. Shul'man.
Komet. Tsirk., Kiev, No. 290 (1982). In Russian.

102.074 **Possible mechanism of interaction between solar corpuscular streams and icy particles of cometary atmospheres.**
E. A. Kajmakov, I. S. Lizunkova.
Komet. Tsirk., Kiev, No. 290 (1982). In Russian.

102.075 **Clustering of comets on the diagram photometric index – heliocentric distance.**
G. K. Nazarchuk, L. M. Shul'man.
Komet. Tsirk., Kiev, No. 290 (1982). In Russian.

102.076 **Why do cometary nuclei not content clathrate hydrates of methane, ammonia, carbon dioxide and similar matter?**

L. M. Shul'man.
Komet. Tsirk., Kiev, No. 290 (1982). In Russian.

102.077 **A semi-empirical photometric theory of cometary gas and dust production: application to P/Halley's gas production rates.**
R. L. Newburn.
The comet Halley dust and gas environment, (see 012.055), p. 3 - 18 (1981).

The semi-empirical photometric theory of cometary gas and dust production has been completely recalibrated using ultraviolet observations from 14 comets and new, uniform visual photometry from eight comets. The calibration has not changed significantly but is now much more secure. A detailed presentation is made of the complete theory with all approximations evaluated and explained. The theory is applied to P/Halley, using a new light curve without the artifact caused by the close approach to Earth in 1910. Gas production rates are predicted for the 1985 - 86 apparition.

102.078 **Definition of model parameters and numerical flyby simulations.**
R. Hellmich, H. U. Keller.
The comet Halley dust and gas environment, (see 012.055), p. 31 - 39 (1981).

An energy balance at the surface of the cometary nucleus leads to initial conditions for a hydrodynamic calculation from which the authors obtain the structure of the inner dust coma. The feedback of the insolation on the sublimation process is taken into account. In the outer dust coma the spatial structure of the dust density is influenced by solar radiation pressure. The GIOTTO flyby is numerically simulated. So the authors get informations about dust flux and fluence. Numerical results illustrating the structure of the dust halo are presented.

102.079 **Coma opacity and surface contrast.**
R. Hellmich, H. U. Keller.
The comet Halley dust and gas environment, (see 012.055), p. 41 - 46 (1981).

Parameters are derived which describe the visibility of the cometary nucleus within the dust coma.

102.080 **A simple radiation model of cometary dust for P/Halley.**
N. Divine.
The comet Halley dust and gas environment, (see 012.055), p. 47 - 53 (1981).

Algebra, parameter values, and sample results for a simple model of the radiation from the dust environment of Comet P/Halley are presented. The model includes absorption and single scattering of sunlight, and infrared emission, for the dust and for the nucleus. It employs dust particle photometric properties which are independent of wavelength, particle size, and position within the coma, separately in the visible and infrared regions of the spectrum.

102.081 **Effect of the size distribution on the thermal emission from cometary dust.**
M. S. Hanner.
The comet Halley dust and gas environment, (see 012.055), p. 67 - 75 (1981).

The thermal emission from cometary dust is analyzed in terms of two size distribution functions, "Sekanina A" and a modified Sekanina-Miller distribution. The resulting thermal emission spectra are illustrated for heliocentric distances 0.15 AU - 1.6 AU. The Sekanina A distribution does not predict the observed heliocentric variation in the flux ratio I(4.8 μm)/I(3.5 μm) for recent bright comets. A size distribution intermediate between the two presented here is probably a better representation of the dust in recently observed comets.

102.082 **Interstellar comets: some possibilities.**
S. V. M. Clube.
Proceedings of the workshop on interstellar comets, (see 012.058), p. 6 - 8 (1982).

102.083 **Comets, asteroids and their interrelations.**
D. W. Hughes.
Proceedings of the workshop on interstellar comets, (see 012.058), p. 9 - 14 (1982).

102.084 **Massive nebulae and the Oort cloud of comets.**
W. M. Napier.
Proceedings of the workshop on interstellar comets, (see 012.058), p. 15 - 18 (1982).

Tidal forces exerted by molecular clouds during close encounters eject large numbers of comets from the Solar System. The Oort comet cloud could not have survived intact for 4.5 Byr with currently adopted masses and structure of the Galactic molecular cloud system. Most probably, the Oort cloud is episodically replenished by the capture of interstellar comets when the Solar System penetrates a molecular cloud.

102.085 **What are comets made of? Very dirty snowballs.**
J. M. Greenberg.
Proceedings of the workshop on interstellar comets, (see 012.058), p. 24 - 31 (1982).

102.086 **Photodesorptive jet thrusting – a mechanism for efficient formation of kilometre sized bodies in molecular clouds?**
C. M. Humphries.
Proceedings of the workshop on interstellar comets, (see 012.058), p. 32 - 36 (1982).

The renewed suggestion that comets and possibly other kilometre sized objects in the solar system have an interstellar origin from molecular clouds carries the implication that such objects can be formed in a time-scale much shorter than the lifetime of a molecular cloud, and hence that the condensation mechanism must be both abundant and efficient (more efficient, for example, than either radiation pressure or the self-gravitation of the parent cloud). One possible mechanism, the jet thrust effect (JTE), is considered exclusively in this paper.

102.087 **Testing an interstellar origin for comets by stable isotope measurements (studies on primitive meteorites).**
C. T. Pillinger.
Proceedings of the workshop on interstellar comets, (see 012.058, p. 44 - 49 (1982).

102.088 **Prospects for millimetre-wave observations of comets.**
A. Webster.
Proceedings of the workshop on interstellar comets, (see 012.058), p. 50 - 51 (1982).

102.089 **In-situ and retrieved cometary dust sampling within the solar system.**
J. A. M. McDonnell.
Proceedings of the workshop on interstellar comets, (see 012.058), p. 52 - 55 (1982).

102.090 **Plasma dynamics in ion tails – a discussion of the interaction of solar wind plasma with cometary plasma.**
D. Chen, L. Liu, C. Stoll.
Sci. Sinica, Ser. A, Vol. 25, 654 - 659 (1982).

On the assumption that the distribution function of ions in a comet tail is nearly Maxwellian and the distribution function of electrons and protons of solar mind is Gaussian, the interaction between the solar wind plasma and the cometary plasma and its stability are discussed. The results show that the mechanism discussed cannot give rise to the instability. Neither might the electron-ion transient instability initially occur. This indicates that the high acceleration and instability occurring within the comet tail should be produced by other mechanisms.

102.091 **Helical wave and K-H instability in type I comet tails. I. Waves of infinitesimal amplitude in incompressible plasma.**
D. Chen, L. Liu.
Sci. Sinica, Ser. A, Vol. 25, 971 - 982 (1982).

The helical wave in the type I comet tails is discussed using the MHD approach. The ion tail is still considered as a plasma cylinder immersed in the solar wind. But instead of a tangential discontinuity surface used in Ershkovich's theory, a boundary layer with finite thickness is introduced between the solar wind plasma and the comet tail plasma.

102.092 **The dynamics and thermodynamics of a dusty cometary atmosphere.**
M. L. Marconi, D. A. Mendis.
Publ. Astron. Soc. Pacific, Vol. 94, 755 (1982). – Abstract.

102.093 **Solvindens inverkan på kometerna.**
H. Lundstedt.
Astron. Tidsskr., Årg. 15, 164 - 175 (1982).

102.094 **Comets, their evolution and origin.**
R. Lüst.
Formation of planetary systems, (see 012.064), p. 857 - 879 (1982).

Contents: Introduction. Evolution of cometary orbits. Structure of a cometary nucleus. Chemistry of a comet. Model for a cometary nucleus. Atomic abundances in comets. Origin of comets. Space probes to a comet.

102.095 **Where do comets come from?**
T. Van Flandern.
Mercury, Vol. 11, 189 - 193 (1982).

102.096 **Chemistry in comets.**
R. Lüst.
Cosmo- and geochemistry, (see 003.015), p. 73 - 98 (1981).

Contents: Introduction. Oort's cloud. Biermann's solar wind concept. Whipple's icy conglomerate model. Molecules of the nucleus. The spectrum of the coma. Model calculations of chemical abundances. Dust in comets and their relation to meteorites. Discussion and outlook.

102.097 **Probing the presently tenuous link between comets and the origin of life.**
R. W. Hobbs, J. M. Hollis.
Origins of Life, Vol. 12, 125 - 132 (1982). – Abstr. in Phys. Abstr., Vol. 86, Abstr. 15877 (1983).

102.098 **Observations of the red auroral oxygen lines in nine comets.**
H. Spinrad.
Publ. Astron. Soc. Pacific, Vol. 94, 1008 - 1016 (1982).

The $\lambda\lambda$6300, 6363 auroral and red doublet of [O I] has been measured on digital sky-subtracted spectra of nine cometary nuclei observed with the Lick Shane 3-m telescope and IDS system. The total oxygen production rate is computed with a Haser model. The derived oxygen production rates have been compared to H_2O production rates derived from UV observations of the comets with the *IUE* satellite. For seven of the comets, the average $Q(H_2O)/Q(O\ I)$ ratio is about 20. This is in excellent accord with theory, assuming H_2O to be the sole oxygen parent. The comets have been placed on an absolute oxygen production-rate system, at a standard common distance.

Close-approach predictions for comets and asteroids.
See Abstr. 021.042.

NH^+ – a candidate for comets and interstellar space.
See Abstr. 022.039.

The infrared spectral properties of frozen volatiles.
See Abstr. 022.044.

Laboratory studies of photochemical and spectroscopic phenomena related to comets.
See Abstr. 022.045.

Radar detectability of comets.
See Abstr. 031.537.

The radio and radar observations of comets: a review.
See Abstr. 031.574.

Large-scale photographic observations, photometry, colorimetry and polarimetry of cometary tails.
See Abstr. 031.576.

Review of magnitude sources for visual cometary photometry. II.
See Abstr. 031.661.

IR observations of quiescent comets.
See Abstr. 051.073.

Penetration of the magnetic field of the solar wind into cometary ionospheres.
See Abstr. 074.061.

Observation of giant planets and comets in the far infrared and submillimetric range.
See Abstr. 091.028.

Outer solar system asteroids and comets: new VJHK data and interpretations.
See Abstr. 098.021.

Dynamics of minor planets and comets: a review.
See Abstr. 098.039.

Dynamical history of the Oort Cloud.
See Abstr. 107.008.

Formation of comets among the outer planets.
See Abstr. 107.022.

Giant molecular clouds and the solar system comets.
See Abstr. 107.038.

The chemistry of interstellar molecules.
See Abstr. 131.329.

Erratum

102.901 **Erratum: "Thermal modeling of cometary nuclei" [Icarus, Vol. 47, 302 - 311 (1981)].**
P. R. Weissman, H. H. Kieffer.
Icarus, Vol. 52, 206 (1982). – See Abstr. 30.102.051.

103 Comets (Individual Objects)

103.001 **Les comètes de l'année 1980.**
C. Bertaud.
Astronomie, Vol. 96, 409 - 422 (1982).

103.002 **Observaciones de cometas en el infrarrojo.**
H. Campins, G. H. Rieke.
Rev. Mexicana Astron. Astrofis., Vol. 6, (see 012.017), 333 (1981). – Abstract.

103.003 **Infrared observations of faint comets: a search for ice.**
A. T. Tokunaga, M. F. A'Hearn, E. Dwek.
Bull. American Astron. Soc., Vol. 14, 751 (1982). – Abstract.

103.004 **Variations of CN column densities in recent faint comets.**
E. S. Barker, A. L. Cochran.
Bull. American Astron. Soc., Vol. 14, 751 (1982). – Abstract.

103.005 **Observations of recent comets with IUE.**
P. D. Feldman, M. F. A'Hearn, M. C. Festou.
Bull. American Astron. Soc., Vol. 14, 752 (1982). – Abstract.

103.006 **Probable sungrazing comet.**
IAU Circ., No. 3718 (1982).

103.007 **Another probable sungrazing comet.**
IAU Circ., No. 3719 (1982).

103.008 **Occultations by comets.**
IAU Circ., No. 3751 (1982).

103.009 **Sungrazing comets.**
Yamamoto Circ., No. 1980 (1982).

103.010 **Observations of comets.**
Minor Planet Circ., (M. P. C.), Nos. 6991 - 6992, 7056 - 7058, 7167 - 7170, 7280 - 7281, 7396 - 7397 (1982).

Observations made at the following stations are published: Alma-Ata, Athens, Boyden Obs., Cerro Tololo Inter-american Obs., Chamberlin Obs., Climenhaga Obs., European Southern Obs., Geisei, Hemingford Abbots, Kambah (near Camberra), Kazan (Engelhardt Obs.), Kiev, Kleť, Lowell Obs. Anderson Mesa Stn., Mt. John Obs., Mt. Palomar, Nice, Oak Ridge Obs., Odessa, Perth, Siding Spring, Stakenbridge, Tautenburg, Turku, Woolston Obs., Zvenigorod. Concerning observations of the following comets: 1945 I Väisälä, 1945 II du Toit-Hartley, 1975 VII Smirnova-Chernykh, 1976 III Gunn, 1979 I Shajn-Schaldach, 1979 X Bradfield, 1979k Schwassmann-Wachmann 2, 1980 X Stephan-Oterma, 1980 XI Encke, 1980 XII Meier, 1980 XIII Tuttle, 1980b Bowell, 1980u Panther, 1981a Longmore, 1981c Elias, 1981g González, 1981k Howell, 1982a Grigg-Skjellerup, 1982e d'Arrest, 1982f Churyumov-Gerasimenko, 1982g Austin, 1982h Peters-Hartley.

103.011 **Ephemerides of comets.**
Minor Planet Circ., (M. P. C.), Nos. 7028 - 7032, 7159, 7375 - 7377, 7475 - 7478 (1982).

Concerning ephemerides of the following comets: 1966 VI Neujmin, 1970 XIII du Toit-Neujmin-Delporte, 1974 II Schwassmann-Wachmann 1, 1976 XIV Pons-Winnecke, 1977 I Johnson, 1977 IV Faye, 1980b Bowell, 1981c Elias, 1981h Kearns-Kwee, 1981i Slaughter-Burnham, 1981*l* Väisälä 1, 1982d Tempel 2, 1982g Austin, 1982h Peters-Hartley, 1982j Tempel 1, 1982k Kopff.

103.012 **Orbital elements of comets.**
Minor Planet Circ., (M. P. C.), Nos. 6989 - 7514 (1982).

The comets are listed according to their Roman numeral designation or preliminary designation. The names of the authors are given behind the respective M. P. C. numbers. 1956 VI Crommelin 7454, 1966 VI Neujmin 1 7455, 1970 XIII du Toit-Neujmin-Delporte 7454, 1975 VII Smirnova-Chernykh 7454 B. G. Marsden; 1976 XIV Pons-Winnecke 7021, 1977 I Johnson 7022 S. Nakano; 1980 XI Encke 7455 B. G. Marsden; 1982g Austin 7021 D. W. E. Green; 1982g Austin 7150, 7454, 1982h Peters-Hartley 7149 B. G. Marsden; 1982j Tempel 1 7022, 1982k Kopff 7022 S. Nakano.

103.013 **Photometric parameters of comets: 1948 - 1954.**
M. J. Hendrie, C. S. Morris.
J. British Astron. Assoc., Vol. 93, 1 - 6 (1982).

Observations of comets passing perihelion between 1948 and 1954 have been examined to extract reliable visual total-magnitude estimates, and these have been used to determine photometric parameters. A short summary of the behaviour of each comet is also presented.

103.014 **Comet and meteor notes 1980/81.**
J. C. Bennett.
Mon. Notes Astron. Soc. South. Africa, Vol. 41, 47 - 49 (1982).

103.015 **Roman designations of the comets of 1980.**
Komet. Tsirk., Kiev, No. 288 (1982). In Russian.

103.016 **Pre-discovery encounters between short-period comets and Jupiter estimated by a Keplerian approximation.**
J. Karm, H. Rickman.
Bull. Astron. Inst. Czechoslovakia, Vol. 33, 359 - 370 (1982).

A comprehensive and coherent list of pre-discovery encounters with Jupiter for all short-period comets is presented. Keplerian orbits have been used for the calculation of encounter dates and distances. The accuracy of this approximation is acceptable over time spans up to about 50 years, as shown by a comparison with published data from orbital integrations. Encounterdistances less than 0.5 AU are found for 60% of the Jupiter family comets and about 20% of the Jupiter family comets have encountered Jupiter to within 0.5 AU during the first orbital revolution preceding discovery. The percentage of Jupiter family comets discovered immediately after an encounter with Jupiter is found to have increased recently, indicating an increased efficiency of discovering newly captured comets.

103.017 **Tabulation of comet observations.**
Int. Comet Q., Vol. 4, 44 - 52, 64 - 76, 105 - 110 (1982).

Concerning comets: 1954 VII Pons-Brooks, 1974 II Schwassmann-Wachmann 1, 1980 XII Meier, 1980 XV Bradfield, 1980 b Bowell, 1980 u Panther, 1981 j Swift-Gehrels, 1982 a Grigg-Skjellerup, 1982 e d'Arrest, 1982 f Churyumov-Gerasimenko, 1982 g Austin.

103.018 **Observations of comets, 1960 - 1977.**
J. E. Bortle.
Int. Comet Q., Vol. 4, 79 - 101 (1982).

Concerning comets: 1960 II Burnham, 1962 VIII Humason, 1964 VI Tomita-Gerber-Honda, 1964 VIII Ikeya, 1964 IX Everhart, 1965 VIII Ikeya-Seki, 1966 IV Ikeya-Everhart, 1966 V Kilston, 1967 II Rudnicki, 1968 I Ikeya-Seki, 1968 IV Tago-Honda-Yamamoto, 1968 V Whitaker-

Thomas, 1968 VI Honda, 1968 VII Bally-Clayton, 1969 I Thomas, 1969 V P/Honda-Mrkos-Pajdušáková, 1969 VI P/Faye, 1969 VII Fujikawa, 1969 VIII P/Comas-Solá, 1969 IX Tago-Sato-Kosaka, 1970 II Bennett, 1970 III Kohoutek, 1970 X Suzuki-Sato-Seki, 1970 XV Abe, 1971 V Toba, 1972 V P/Tempel 1, 1972 VI P/Giacobini-Zinner, 1972 VIII Heck-Sause, 1972 IX Sandage, 1972 XI P/Kearns-Kwee, 1973 II Kojima, 1973 VI P/Tuttle-Giacobini-Kresák, 1973 VII Kohoutek, 1974 III Bradfield, 1974 V P/Encke, 1974 IX P/Forbes, 1975 I P/Boethin, 1975 IV P/West-Kohoutek-Ikemura, 1975 IX Kobayashi-Berger-Milon, 1975 X Suzuki-Saigusa-Mori, 1975 XI Bradfield, 1975 XII Mori-Sato-Fujikawa, 1976 IV Bradfield, 1976 VI West, 1976 X P/Klemola, 1976 XI P/d'Arrest.

103.019 **Infrared photometry of periodic comets Encke, Chernykh, Kearns–Kwee, Stephan–Oterma, and Tuttle.**
H. Campins, G. H. Rieke, M. J. Lebofsky.
Icarus, Vol. 51, 461 - 465 (1982).
Nearly simultaneous photometry of the reflected and thermal infrared spectra of periodic comets Encke, Chernykh, Kearns-Kwee, Stephan–Oterma, and Tuttle are presented. The 10-μm silicate emission feature has been detected for the first time in periodic comets and was observed in three of these objects. The albedo of the dust particles in the comae of these comets is calculated and compared to that of Comet Kohoutek.

103.020 **Comets in 1981.**
J. Bouška.
Vesmír, Vol. 61, 315 (1982). In Czech.

103.021 **Comet digest.**
J. E. Bortle.
Sky Telesc., Vol. 64, 198, 294 - 295, 397, 504 - 507, 615 (1982).

Index to orbital elements of comets and minor planets.
See Abstr. 098.049.

A review of radio observations of comets.
See Abstr. 102.036.

Erratum

103.099 **Erratum: "Observations of the OH radical in comets at 18 cm wavelength"** [Icarus, Vol. 47, 464 - 469 (1981)].
D. Bockelée-Morvan, J. Crovisier, E. Gerard, I. Kazès.
Icarus, Vol. 50, 125 (1982). – See Abstr. 30.103.016.

Comet 1979 XI Howard-Koomen-Michels

103.101 **The path and surviving tail of a comet that fell into the sun.**
Z. Sekanina.
Astron. J., Vol. 87, 1059 - 1072 (1982).
A satisfactory orbital solution for Comet Howard-Koomen-Michels 1979 XI, detected on several exposures taken with a satellite-borne coronagraph, is found on the assumption that the comet's line of apsides coincided with that of the Kreutz sungrazing comet group. The derived perihelion distance then shows that this is the first known case of a comet falling into the Sun. It is suggested that shortly before it was to enter the photosphere, the comet's nucleus began to fracture precipitously, owing in part to the pressure exerted by intensely subliming ices. A dust tail that survived the comet is studied as a particle flow phenomenon controlled by no force other than solar gravity and solar radiation pressure. The outline of the observed tail is then interpreted in terms of an onset of dust production (tail's sharp boundary to the west-southwest of the Sun), a peak repulsive force on the particles (its extent in the northwestern quadrant), and a circumsolar dust-free zone due to particle sublimation (its extent in the northeastern quadrant).

103.102 **Comet Howard-Koomen-Michels (1979 XI).**
Komet. Tsirk., Kiev, No. 286 (1982). In Russian.

Observations of coronal structure during sunspot maximum.
See Abstr. 074.087.

Periodic comet Grigg-Skjellerup

103.121 **Linkage of the last seven apparitions of the periodic comet Grigg-Skjellerup.**
G. Sitarski.
Acta Astron., Vol. 31, 479 - 484 (1981).
Orbit of the comet was improved using 76 observations made in the period 1947 - 1977 during its last seven apparitions; a nongravitational acceleration in the comet's motion as a daily change of the semi-major axis of the orbit was taken into account.

103.122 **Nuclear rotation of comet Grigg-Skjellerup.**
S. M. Larson.
Bull. American Astron. Soc., Vol. 14, 753 (1982). – Abstract.

103.123 **Comet Grigg-Skjellerup: radar detection of nucleus.**
P. Kamoun, G. Pettengill, I. Shapiro, D. Campbell.
Bull. American Astron. Soc., Vol. 14, 753 (1982). – Abstract.

103.124 **Continuum and [O I] (6300 Å) brightness profiles of comet P/Grigg-Skjellerup.**
E. W. Dunham.
Bull. American Astron. Soc., Vol. 14, 754 (1982). – Abstract.

103.125 **Short-period comet Grigg-Skjellerup (1982a).**
Komet. Tsirk., Kiev, Nos. 285, 288 (1982). In Russian.

Periodic comet Tuttle

103.141 **Positions of comet P/Tuttle (1980h).**
M. Grossmann, H. W. Duerbeck.
Acta Astron., Vol. 31, 515 - 516 (1981).
12 positions of comet P/Tuttle (1980h) are derived from photographic observations made with the GPO telescope of ESO between 1980 December 26 and 1981 January 9.

Comet 1937 II Wilk

103.161 **Über die Bahn des Kometen 1937 II Wilk.**
W. Landgraf.
Acta Astron., Vol. 31, 517 - 528 (1981).

Comet 1979 X Bradfield

103.201 **Comet Bradfield 1979 X event on 1980 February 6: correlation with an interplanetary solar wind disturbance.**
J.-F. Le Borgne.
The need for coordinated ground-based observations of Halley's comet, (see 012.025), p. 217 - 225 (1982).
The tail event on comet Bradfield 1979 X on 1980 February 6 is shown to be due to an interplanetary solar wind disturbance detected by instruments aboard Helios 2 and at Earth. The spacecraft was at 0.15 AU from the comet. The tail follows the bulk orientation of solar wind. Acceleration of cometary ions seems to result from small scale hydrodynamic instabilities. Cometary ion density and temperature are deduced.

103.202 **Observations of comet Bradfield (1979 X) at the Main Astronomical Observatory of the Ukrainian Academy of Sciences.**
G. V. Moroz, G. R. Kastel'.
Komet. Tsirk., Kiev, No. 287 (1982). In Russian.

OH fluorescence in comets: fluorescence efficiency of the ultraviolet bands.
See Abstr. 102.001.

Are comets made of H_2O-ice?
See Abstr. 102.056.

Comet 1974 III Bradfield

On the flaring of cometary plasma tails.
See Abstr. 102.061.

Comet 1977 XIV Kohler

103.221 **The light curve of comet Kohler 1977 XIV.**
D. W. E. Green.
Strolling Astron., Vol. 29, 150 - 155 (1982).

103.222 **Parabolic orbit of the comet 1977m Kohler.**
V. Protić-Benišek.
Publ. Astron. Obs. Sarajevo, Vol. 1, No. 1, (see 012.043), p. 287 - 291 (1981).

Comet 1976 VI West

103.241 **Comet West 1976 VI: observations of the great comet of 1976 (II).**
D. Wallentinsen.
Strolling Astron., Vol. 29, 155 - 163 (1982).

103.242 **The 0.9 - 2.5 μ spectrum of comet West (1976 VI).**
J. R. Johnson, U. Fink, H. P. Larson.
Bull. American Astron. Soc., Vol. 14, 753 (1982). – Abstract.

103.243 **Changes in the emission intensity in the spectrum of comet West (1976 VI).**
V. P. Tarashchuk, N. I. Il'chishina.
Komet Tsirk., Kiev, No. 287 (1982). In Russian.

103.244 **A CO_2-rich coma model applied to the neutral coma of Comet West.**
G. F. Mitchell, W. T. Huntress, M. B. Swift.
Astron. J., Vol. 87, 1600 - 1605 (1982).
Models of the cometary coma in which the dominant volatile is CO_2 have been constructed for a range of heliocentric distances. Model coma abundances of C_2, C_3, and CN are compared with the abundances observed in Comet West and are found to be in good agreement. Furthermore, the variation with heliocentric distance of C_2, C_3, and CN model abundances agree well with the observed variation in Comet West. The present work lends detailed support to a previous suggestion that a substance more volatile than water, such as CO_2, controls the evaporation of the nucleus of Comet West. The implications for cometary formation are briefly discussed.

103.245 **High resolution spectra of C_2 Swan bands from Comet West 1976 VI.**
D. L. Lambert, A. C. Danks.
ESO Sci. Prepr., No. 218, 54 pp. (1982). – Submitted to Astrophys. J.

Comet 1973 XII Kohoutek

103.261 **On the CN(0,0) spectrum of comet Kohoutek 1973 XII.**
I. Ferrín, O. Escalona.
Rev. Mexicana Astron. Astrofis., Vol. 6, (see 012.017), 335 - 337 (1981).
An analysis of the CN(0,0) band of comet Kohoutek 1973 XII, solving the equations of statistical equilibrium including collisions, has given the following results: (1) The rotational temperature of this comet at a solar distance of $r = 0.83$ AU is $T_{rot} = 445 \pm 30°K$. (2) Collisions are important in populating the lower vibrational level. (3) The constant R_1 in this and other comets is close to 1.0×10^{-4}.

On the flaring of cometary plasma tails.
See Abstr. 102.061.

Periodic comet Tempel 2

103.281 **Rediscovery of short-period comet Tempel 2 (1982d).**
Komet. Tsirk., Kiev, No. 288 (1982). In Russian.

Light curves of some periodic comets.
See Abstr. 102.039.

Comet 1978 XV Seargent

OH fluorescence in comets: fluorescence efficiency of the ultraviolet bands.
See Abstr. 102.001.

Periodic comet Stephan-Oterma

103.321 **Narrowband photometry of Comet P/Stephan-Oterma and the backscattering properties of cometary grains.**

R. L. Millis, M. F. A'Hearn, D. T. Thompson.
Astron. J., Vol. 87, 1310 - 1317 (1982).

The results of narrowband filter photometry of Comet P/Stephan-Oterma are presented. For heliocentric distances between 1.95 and 1.58 AU prior to perihelion, OH, CN, and C_2 production rates varied approximately as r^{-5}. Following perihelion, an even steeper dependence of production rate on heliocentric distance was observed for these species. A pronounced brightening of the comet's continuum near opposition is interpreted in terms of preferential backscattering by the cometary dust.

103.322 **Chemical evolution of the coma of comet P/Stephan-Oterma.**
A. L. Cochran.
Bull. American Astron. Soc., Vol. 14, 751 (1982). – Abstract.

103.323 **Position observations in Alma-Ata concerning short-period comet Stephan-Oterma 1980g and comet Panther 1980u.**
A. A. Semenikin, N. S. Gorodetskaya, D. I. Gorodetskij.
Komet. Tsirk., Kiev, No. 285 (1982). In Russian.

103.324 **Visual and infrared observations of the distant comets P/Stephan-Oterma (1980g), Panther (1980u), and Bowell (1980b).**
D. C. Jewitt, B. T. Soifer, G. Neugebauer, K. Matthews, G. E. Danielson.
Astron. J., Vol. 87, 1854 - 1866 (1982) = Div. Geol. Planet. Sci., Calif. Inst. Technol., Contrib. No. 3694.

Broadband observations of comets P/Stephan-Oterma (1980g), Bowell (1980b), and Panther (1980u) in the visual [$0.5 \lesssim \lambda$ (μm) $\lesssim 0.9$] and infrared [$1.2 \lesssim \lambda(\mu m) \lesssim 20$] wavelength regions are reported together with measurements in the 1.5–2.4-μm wavelength range having 5% spectral resolution. The visual data indicate the existence of solid grains in extended halos around the nuclei of the three comets. Broadband near-infrared and thermal infrared measurements of comet Panther suggest the presence of 2–4-μm radius particles in the coma. The near-infrared spectrum of P/Stephan-Oterma is a featureless solar-reflection continuum. The near-infrared spectra of Bowell and Panther exhibit features which are similar in the two comets. The spectrum of solid ammonia provides the best match to the near infrared; it is nevertheless significantly different from the comet spectra. The synthesis of the visual data with the infrared data is attempted in terms of a model involving a mantle of volatile material on the nuclei of Bowell and Panther, but not on P/Stephan-Oterma.

Comet 1980b Bowell

103.341 **The orbital evolution of Comet Bowell (1980b).**
L. Buffoni, M. Scardia, A. Manara.
Moon Planets, Vol. 26, 311 - 315 (1982).

Planetary perturbations and orbital evolution of the elements of the comet Bowell (1980b) are calculated. The sudden change of all the elements of the orbit on February 1981 is caused by Jupiter's perturbation.

103.342 **UV to IR reflectivity of grains in comet Bowell.**
M. F. A'Hearn, E. Dwek, P. D. Feldman, R. L. Millis, D. G. Schleicher, A. T. Tokunaga.
Bull. American Astron. Soc., Vol. 14, 752 - 753 (1982). Abstract.

103.343 **Gaseous species in comet Bowell (1980b).**
D. G. Schleicher, M. F. A'Hearn, R. L. Millis, D. T. Thompson, P. D. Feldman.
Bull. American Astron. Soc., Vol. 14, 753 (1982). – Abstract.

103.344 **Comet Bowell (1980b).**
D. Jewitt.
Bull. American Astron. Soc., Vol. 14, 753 (1982). – Abstract.

103.345 **Comet Bowell (1980b).**
IAU Circ., Nos. 3710, 3733, 3745 (1982).

103.346 **Comet Bowell (1980b).**
Yamamoto Circ., Nos. 1978, 1984 (1982).

103.347 **Asymmetry of the coma of comet Bowell (1980b).**
V. K. Rozenbush.
Astron. Tsirk., No. 1180, p. 7 - 8 (1981). In Russian.

103.348 **A search for frosts in comet Bowell (1980b).**
H. Campins, G. H. Rieke, M. J. Lebofsky, L. A. Lebofsky.
Prepr. Steward Obs., No. 405, 21 pp. (1982). – Submitted to Astron. J.

103.349 **Ice in comet Bowell.**
H. Campins, G. H. Rieke, M. J. Lebofsky.
Prepr. Steward Obs., No. 411, 7 pp. (1982).

103.350 **A search for frosts in Comet Bowell (1980b).**
H. Campins, G. H. Rieke, M. J. Lebofsky, L. A. Lebofsky.
Astron. J., Vol. 87, 1867 - 1873 (1982).

Infrared observations of Comet Bowell represent the first search for frost signatures in a comet beyond 2 AU from the sun. The authors have obtained broad- and narrowband photometry as well as CVF spectrophotometry of this comet and there is no evidence for absorption features in the spectral area between 1.25 and 2.3 μm. The authors have generated models of the coma which constrain the volatile content of the grains and are in agreement with the observed albedo. The darkness of the coma particles at large heliocentric distances indicates a low albedo nucleus as well. Brightness variations during the observing period seem to indicate an active nucleus at 4.5 AU from the sun.

Visual and infrared observations of the distant comets P/Stephan-Oterma (1980g), Panther (1980u), and Bowell (1980b).
See Abstr. 103.324.

Periodic comet Ashbrook-Jackson

103.361 **Orbit determination of comet Ashbrook-Jackson from observational data during the last two apparitions.**
A. Manara, L. Buffoni, M. Scardia.
Astron. Nachr., Band 303, 211 - 212 (1982).

New orbital elements of the comet Ashbrook-Jackson have been calculated. The number of observations (196) – obtained during the last two apparitions in 1971 and 1978 – enabled to take into account planetary perturbations. The authors also have prepared an ephemeris with planetary perturbations which is available on request.

103.362 **Investigation of the orbit of periodic comet Ashbrook-Jackson.**
L. Buffoni, F. Chlistovsky, A. Manara, M. Scardia.
Astron. Nachr., Band 303, 309 - 310 (1982).

The mechanical energy and planetary perturbations of the periodic comet Ashbrook-Jackson have been calculated for the time interval from 24th April 1970 to 11th May 1982 which encloses the last two passages through the perihelion.

103.363 **Linking of five P/Ashbrook-Jackson (1948 IX) apparitions over the time interval of 1948 - 1979 and prediction of its apparition in 1985 - 1986.**
E. I. Kazimirchak-Polonskaya.
Tr. Inst. Teor. Astron., Vyp. 18, p. 94 - 106 (1982). In Russian.

The paper presents a short review of the discovery of comet P/Ashbrook-Jackson, the story of its observations as well as the conditions under which it had been observed over 1948 - 1979. Investigations by M. A. Merzlyakova concerning the linking of 4 incomplete appearances during 1948 - 1970 on the basis of 161 observations taking into account perturbations of 6 planets (Venus - Uranus) with a mean error $\sigma = \pm 1''\!.74$ have been outlined. 246 observations from 1970 to 1972 and from 1977 to 1979 have been additionaly collected. Utilizing the author's method and programs the linking of 5 apparitions of the comet for the time period from 1948 to 1979 was carried out considering the perturbations from nine planets (Mercury - Pluto) with $\sigma = \pm 1''\!.17$ which fairly well represents 407 observations for these years. A precise and detailed study of the cometary orbit transformation which occurred in 1945 in Jupiter's sphere of action has been performed. The paper gives a brief summary of the investigation by N. A. Belyaev and Merzlyakova concerned with the evolution of the comet's orbit over 400 yrs time span (1660 - 2060). The capture of the comet from the Saturn's family into that of Jupiter as well as cometary orbit fluctuations on secular scale have been supplemented and illustrated by tables and a figure. This has been compared with similar captures and regularities which took place in the evolution of other cometary orbits.

Periodic comet Schwassmann-Wachmann 1

103.401 **Observations of the periodic comet Schwassmann-Wachmann 1.**
Komet. Tsirk., Kiev, No. 286 (1982). In Russian.

103.402 **Recent outbursts of P/Comet Schwassmann-Wachmann 1.**
M. Huruhata.
Int. Comet Q., Vol. 4, 59 - 60 (1982).

Vaporization in comets; outbursts from comet Schwassmann-Wachmann 1.
See Abstr. 102.003.

Comet 1982g Austin

103.421 **Comet Austin (1982g).**
IAU Circ., Nos. 3708, 3709, 3711, 3716, 3717, 3721 - 3724, 3727, 3730, 3740, 3748 (1982).

103.422 **Comet Austin (1982g).**
Yamamoto Circ., Nos. 1978 - 1985 (1982).

103.423 **New comet Austin (1982g).**
Komet. Tsirk., Kiev, Nos. 289, 290 (1982). In Russian.

103.424 **Comet Austin 1982g.**
J. Bouška.
Říše hvězd, Vol. 63, 229 - 231 (1982). In Czech.

Orbital elements of comets.
See Abstr. 103.012.

Periodic comet Peters-Hartley

103.441 **Periodic comet Peters-Hartley (1846 VI = 1982 h)**
IAU Circ., Nos. 3710, 3713, 3715 (1982).

103.442 **Comet Peters-Hartley (1982h).**
Yamamoto Circ., Nos. 1978 - 1980 (1982).

Orbital elements of comets.
See Abstr. 103.012.

Periodic comet Neujmin 3

103.461 **Periodic comet Neujmin 3.**
IAU Circ., No. 3724 (1982).

103.462 **Short-period comet Neujmin 3.**
Komet. Tsirk., Kiev, No. 290 (1982). In Russian.

Periodic comet Churyumov-Gerasimenko

103.481 **Periodic comet Churyumov-Gerasimenko (1982 f).**
IAU Circ., Nos. 3731, 3741, 3743, 3748, 3753 (1982).

103.482 **P/Churyumov-Gerasimenko (1982f).**
Yamamoto Circ., Nos. 1983 - 1985 (1982).

103.483 **Short-period comet Churyumov-Gerasimenko (1982f)**
Komet. Tsirk., Kiev, Nos. 285, 288 (1982). In Russian.

Comet 1957 III Arend-Roland

103.491 **The structure of cometary dust tails. II. Tail brightness profiles and dust characteristics of comet Arend-Roland 1957 III.**
C.-p. Liu, H. Kimura.
Acta Astron. Sinica, Vol. 23, 232 - 242 (1982). In Chinese.

The modified gamma-distribution function of dust particles, $f(\gamma)$, which can be determined from tail brightness profiles on the basis of mechanical theory, is discussed with special regard to its reliability and accuracy. Physical significance of $f(\gamma)$ is also discussed in terms of dust model parameters, and it is shown that $f(\gamma)$, if treated carefully, will serve as an effective tool in studying cometary dust. Four isophotes of comet Arend-Roland 1957 III in the orange-red light ($\lambda\lambda$ 0.53 - 0.68 micron) obtained by Ceplecha (1958), are analysed by the numerical method described in Paper I (Kimura and Liu 1975) with some improvements and higher approximations.

Periodic comet Halley

103.501 **Limits on the nucleus of Halley's comet.**
M. J. S. Belton, H. Butcher.
Nature, Vol. 298, 249 - 251 (1982).

The authors report a limiting magnitude to the brightness of the nucleus of Halley's comet which, together with published interpretations of the comet's visual light curve in 1910 and estimates of the magnitude of non-gravitational forces that have modified the comet's orbital motion throughout the past five apparitions, places some interesting limits on the physical properties of the nucleus.

103.502 **Halley's comet: AD 1986 to 2647 BC.**
J. L. Brady.
J. British Astron. Assoc., Vol. 92, 209 - 215 (1982).

103.503 **Deep sounding with electronographic camera at the prime focus of the CFHT: upper limit to the visual brightness of comet P/Halley during 1981/1982 opposition.**
P. Felenbok, J. P. Picat, A. Chevillot, J. Guérin, M. Combes, E. Gerard, J. Lecacheux, G. Lelièvre.
Astron. Astrophys., Vol. 113, L1 - L2 (1982).

The authors report an attempt to recover periodic comet Halley during the 1981/1982 opposition using an electronographic camera at the prime focus of the CFHT 3.6 m telescope. The upper limit to the visual magnitude of P/Halley is 25.2 (5 σ) or 25.8 (3 σ) and places severe constraints on the radius and geometric albedo of the nucleus.

103.504 **Der Halley'sche Komet wird erwartet.**
S. Marx.
Jenaer Rundsch., 27. Jahrg., 143 - 145 (1982).

103.505 **Halley's comet in 1986.**
J. B. Tatum.
Mercury, Vol. 11, 126 - 131 (1982).

103.506 **Infrared observations of comet Halley.**
M. Combes.
The need for coordinated ground-based observations of Halley's comet, (see 012.025), p. 83 - 91 (1982).

103.507 **An attempt at detecting comet P/Halley at the 3.60 m C. F. H. Telescope in December 1981.**
J. P. Picat, J. Guérin, M. Combes, E. Gérard, J. Lecacheux, G. Lelièvre.
The need for coordinated ground-based observations of Halley's comet, (see 012.025), p. 273 - 275 (1982).

103.508 **Comet Halley – ordinary and unique object of the solar system.**
B. Yu. Levin.
Zemlya Vselennaya, 1982, No. 5, p. 38 - 46. In Russian.

103.509 **Magnetic field distribution in the tail of comet Halley obtained from the kinematic behaviour of a plasma condensation.**
Eh. M. Dubinin, P. L. Izrajlevich, I. M. Podgornyj.
Astron. Zh., Tom 59, 1006 - 1011 (1982). In Russian. English translation in Soviet Astron., Vol. 26, No. 5.

The distribution of directions of magnetic field vectors is obtained from the values of accelerations of a plasma condensation in comet Halley's tail.

103.510 **Limits on the nucleus of Halley's comet.**
M. J. S. Belton.
Bull. American Astron. Soc., Vol. 14, 752 (1982). – Abstract.

103.511 **Infrared fluorescence of molecules in Halley's comet.**
J. Crovisier, J. Le Bourlot, T. Encrenaz, M. Combes.
Bull. American Astron. Soc., Vol. 14, 752 (1982). – Abstract.

103.512 **The recovery of Halley's comet.**
D. W. Hughes.
Nature, Vol. 300, 318 (1982).

103.513 **Estimates of the magnetic field in Halley's comet.**
I. M. Podgorny (*Podgornyj*), E. M. (*Eh. M.*) Dubinin, P. L. Israelevich (*Izrajlevich*).
Moon Planets, Vol. 27, 135 - 138 (1982).

The estimates of the magnetic field in the Halley's comet tail based on the observed acceleration of the matter and the pressure balance are discussed.

103.514 **Periodic comet Halley (1982i).**
IAU Circ., Nos. 3737, 3742, 3753, 3758 (1982).

103.515 **Periodic comet Halley (1982i).**
Yamamoto Circ., Nos. 1983, 1985 (1982).

103.516 **Comet Halley found.**
D. di Cicco.
Sky Telesc., Vol. 64, 551 (1982).

103.517 **On the history of comet Halley.**
S. K. Vsekhsvyatskij.
Komet. Tsirk., Kiev, No. 286 (1982). In Russian.

103.518 **The derivation of Halley parameters from observations.**
R. L. Newburn, R. Reinhard.
The comet Halley dust and gas environment, (see 012.055), p. 19 - 24 (1981).

A set of nominal value parameters for the nucleus, the dust and gas for P/Halley at 0.9 AU post-perihelion is derived from the light curve and spectra of Halley and by modelling the effect of the nongravitational forces. In those cases where Halley observations are not sufficient, the average value derived from a larger set of other comets has been used, or data from comet Bennett, Halley's best analogue, have been taken.

103.519 **Numerical models for Halley dust environments.**
N. Divine.
The comet Halley dust and gas environment, (see 012.055), p. 25 - 30 (1981).

The algebra, parameter values, and sample results of a Reference Model for the dust environment of Comet P/Halley are presented. The model employs simplified treatments of the icy conglomerate nucleus and of the hydrodynamic outflow of gas and dust.

103.520 **Properties of dust particles in comet Halley from observations made in 1910 during its encounter with the earth.**
Z. Sekanina.
The comet Halley dust and gas environment, (see 012.055), p. 55 - 65 (1981).

Analysis of observations of the dust tail of comet Halley about the time of the earth's transit across the orbit plane of the comet on May 19, 1910 yields information on particle properties which has so far been inaccessible by other means.

103.521 **Optical and physical properties and dynamics of dust grains released by comet Halley.**
G. H. Schwehm, B. Kneissel.
The comet Halley dust and gas environment, (see 012.055), p. 77 - 84 (1981).

For the comet Halley flyby the possible existence and the extent of an icy halo in the coma of the comet is studied. Deriving the temperature distribution and sublimation rate of ice grains two models with pure ice and 'dirty' - ice covered - core-mantle grains are presented.

103.522 **A theoretical study of Comet Halley's spectrum in the infrared range.**
T. Encrenaz, J. Crovisier, M. Combes, J. F. Crifo.
Icarus, Vol. 51, 660 - 664 (1982).

Synthetic spectra of Comet Halley between 2.5 and 15 μm are calculated on the basis of current cometary models. This study shows that molecules which are most likely detectable in the infrared range include H_2O, CO, CO_2, CH_4, NH_3, N_2H_4, and H_2CO.

103.523 **International Halley Watch.**
Int. Halley Watch Newsl., No. 1, 3 + 24 pp. (1982).
Contents: History of the International Halley Watch

(*R. Newburn*). Organization and staff of the IHW (*R. Newburn, J. Rahe*). Data rights and the Halley archive (*R. Newburn, J. Rahe*). Astrometry (*D. K. Yeomans, R. M. West*). Infrared spectroscopy and radiometry (*R. Knacke, T. Encrenaz*). Large scale phenomena (*J. Brandt, J. Rahe, M. B. Niedner*). Near-nucleus studies (*J. Rahe, Z. Sekanina, S. Larson*). Photometry and polarimetry (*M. F. A'Hearn, V. Vanysek*). Radio science (*W. M. Irvine, F. P. Schloerb, E. Gerard*). Spectroscopy and spectrophotometry (*S. Wyckoff, P. Wehinger, M. C. Festou*). The Amateur Observation Net (*S. J. Edberg*).

Landolt-Börnstein. Group VI. Vol. 2c. Interstellar matter – Galaxy – Universe.
See Abstr. 003.055.

La comète de Halley et Mira Ceti dans le Talmud?
See Abstr. 004.029.

La comète de Halley a son retour de 1835.
See Abstr. 004.039.

Why should we observe Halley's comet?
See Abstr. 013.054.

A worldwide photographic network for wide-field observations of Halley's comet in 1985 - 1986.
See Abstr. 013.055.

The contribution of French astronomers to the orbit determination of Halley's comet.
See Abstr. 013.056.

The International Halley Watch.
See Abstr. 013.057.

Status of the International Halley Watch.
See Abstr. 013.065.

Proposed radio observations of inverted OH in Halley's comet.
See Abstr. 031.575.

Astrometric observations and the motion of comet Halley.
See Abstr. 031.577.

Infrared sounding of comet Halley from the flyby Venera Halley probes.
See Abstr. 032.570.

Observations of Halley's comet with intensified photodiode arrays.
See Abstr. 034.041.

Encounter with comet Halley.
See Abstr. 051.008.

The Giotto mission to Halley's comet.
See Abstr. 051.017.

PLANET-A project.
See Abstr. 051.021.

Production of secondary particles by neutral and ionised cometary gas and dust impacting on the shield of the GIOTTO spacecraft.
See Abstr. 051.046.

Estimates of a spacecraft induced gas cloud and its effects on measurement of ambient particles during a fast flyby of comet Halley.
See Abstr. 051.049.

The Giotto project – a fast fly-by of Halley's comet.
See Abstr. 051.056.

Magnetic field measurements onboard Venera 13, Venera 14 and Venera-Halley.
See Abstr. 051.057.

The Giotto project – a fast flyby of Halley's comet.
See Abstr. 051.060.

Expectations of Giotto.
See Abstr. 051.068.

The use of ground based observations of Halley's comet in the Giotto navigation.
See Abstr. 052.025.

Non-radial flow near the cometary surface.
See Abstr. 102.042.

A semi-empirical photometric theory of cometary gas and dust production: application to P/Halley's gas production rates.
See Abstr. 102.077.

A simple radiation model of cometary dust for P/Halley.
See Abstr. 102.080.

Periodic comet d'Arrest

103.541 **Periodic comet d'Arrest (1982 e).**
IAU Circ., Nos. 3732, 3738, 3749 (1982).

103.542 **P/d'Arrest (1982e).**
Yamamoto Circ., Nos. 1982 - 1985 (1982).

103.543 **Short-period comet d'Arrest (1982e).**
Komet. Tsirk., Kiev, Nos. 285, 287 (1982). In Russian.

Periodic comet Tempel 1

103.561 **Periodic comet Tempel 1 (1982j).**
IAU Circ., No. 3757 (1982).

Orbital elements of comets.
See Abstr. 103.012.

Periodic comet Kopff

103.581 **Periodic comet Kopff (1982k).**
IAU Circ., No. 3757 (1982).

Orbital elements of comets.
See Abstr. 103.012.

Periodic comet Encke

103.601 **The ultraviolet spectrum of periodic comet Encke.**
P. D. Feldman, H. A. Weaver, M. C. Festou, H. U. Keller.
Advances in ultraviolet astronomy, (see 012.006), p. 307 - 310 (1982).

A comparison of the ultraviolet spectrum of periodic comet Encke, recorded by the IUE between 1980 October 24 and November 5, with similar spectra of short and long-period comets shows the gaseous composition of P/Encke to be virtually identical to that of the other comets. This similarity implies a homogeneous structure for the cometary ice nucleus.

103.602 **The disappearance of OH from comet Encke.**
D. T. Thompson, R. L. Millis, M. F. A'Hearn.
Bull. American Astron. Soc., Vol. 14, 754 (1982). – Abstract.

103.603 **Photometric parameters of comet Encke in the apparition of 1980.**
M. Z. Markovich.
Komet. Tsirk., Kiev, No. 288 (1982). In Russian.

Light curves of some periodic comets.
See Abstr. 102.039.

Orbital elements of comets.
See Abstr. 103.012.

Periodic comet Kearns-Kwee

103.611 **Periodic comet Kearns-Kwee (1981h).**
Komet. Tsirk., Kiev, No. 286 (1982). In Russian.

Comet 1980 XV Bradfield

103.621 **On the identification of CO in the visible spectrum of comet Bradfield 1980t.**
C. B. Cosmovici, L. Biermann, C. Arpigny.
The need for coordinated ground-based observations of Halley's comet, (see 012.025), p. 131 - 147 (1982).

Spectra taken with the 122 and 182 cm Asiago telescopes permitted the first identification of CO emission lines between 4000 and 9000 Å. Absolute fluxes of 15 lines could be determined using a Reticon detector. Preliminary theoretical calculations on possible parent molecules and production rates are presented.

103.622 **On the spectrum of Comet Bradfield 1980t.**
C. B. Cosmovici, C. Barbieri, C. Bonoli, F. Bortoletto, E. Hamzaoglu.
Astron. Astrophys., Vol. 114, 373 - 387 (1982).

Twenty spectra of Comet Bradfield 1980 t have been obtained in the period January 8–16, 1981, using the 122 and 182 cm Asiago telescopes. Three photographs have also been taken on January 8 with the 90 cm Schmidt telescope. 192 neutral and 41 plasma lines (CO^+, CO_2^+, NH^+, CH^+, and H_2O^+) have been identified in the coma and in the tail of the comet. A new cometary ion, H_2S^+, could be present. The wavelengths of 26 unidentified lines in the coma coincide with transitions of neutral CO. The first certain identification of the CO-Triplet system ($d^3\Delta \rightarrow a^3\Pi$) is reported. Another important result achieved is the determination of absolute fluxes of 86 lines in the range 5850–9100 Å, using a Reticon 512 detector.

Comet 1970 II Bennett

103.641 **Dynamic coma models for comet Bennett 1970 II.**
A. Cucchiaro, D. Malaise.
Astron. Astrophys., Vol. 114, 102 - 108 (1982).

The authors have introduced the effect of time and space variations of the source function of the gas at the nuclei of comets. These variations are normally observed as outbursts and jets in many comets and can be measured on photometric profiles obtained by the authors for comet Bennett. Several models are given to compare with the observations. The bursts decay exponentially with various time constants and are directed toward the sun.

103.642 **Determination of velocities of cyan molecule production in the head of comet Bennett 1970 II.**
I. N. Matveev.
Komet. Tsirk., Kiev, No. 286 (1982). In Russian.

103.643 **Spectrum of comet Bennett (1970 II).**
I. R. Bejtrishvili.
Komet. Tsirk., Kiev, No. 288 (1982). In Russian.

Periodic comet Schorr

103.661 **Improved orbital elements for periodic comet Schorr (1918 III).**
C. de Vegt, L. Kohoutek, B. G. Marsden.
Astron. Astrophys., Vol. 114, 147 - 150 (1982).

The observed positions of periodic comet Schorr (1918 III) have been improved by remeasuring all available photographic plates and using newly derived reference star positions in the FK4-system. A new orbit has been calculated supplying predictions for the comet's return in the second half of 1982; the brightness of the comet is not expected to be higher than 19 mag.

Periodic comet Swift-Tuttle

103.701 **La comète Swift-Tuttle.** Observations anciennes et recherche contemporaine.
A.- C. Levasseur-Regourd.
Astronomie, Vol. 96, 405 - 407 (1982).

Comet 1957 V Mrkos

103.721 **Study of a burst of ray generation – comet 1957 V.**
F. D. Miller.
Bull. American Astron. Soc., Vol. 14, 752 (1982). – Abstract.

103.722 **Two dust populations of particle fragments in the striated tail of Comet Mrkos 1957 V.**
Z. Sekanina, J. A. Farrell.
Astron. J., Vol. 87, 1836 - 1853 (1982).

A total of 26 striae in the dust tail of Comet Mrkos 1957 V is investigated on five small-scale photographs taken 14–17 August 1957. The dynamical parameters of the striae are determined on the premise that these formations are products of the ejection of dust particles that subsequently break up in the tail. The authors establish the existence of two kinds of striae, consisting of submicron-sized particle fragments made of strongly absorbing and essentially dielectric materials. The authors point out differences between the dust tails of Comets West 1976 VI and Mrkos and compare their particle-fragmentation model with a hypothesis that advocates a strong coupling between charged dust and comet plasma.

Periodic comet Crommelin

Orbital elements of comets.
See Abstr. 103.012.

Periodic comet Neujmin 1

Orbital elements of comets.
See Abstr. 103.012.

Periodic comet Wolf-Harrington

103.801 **Linkage of the last five apparitions of the periodic comet Wolf-Harrington (1952 II).**
G. Sitarski.
Acta Astron., Vol. 31, 471 - 478 (1981).
Orbital elements of the comet were improved using 97 observations made during its last five apparitions. To link all the observations by one system of elements a non-gravitational anomaly in the comet's motion was assumed.

Periodic comet du Toit-Neujmin-Delporte

Orbital elements of comets.
See Abstr. 103.012.

Periodic comet Lexell

103.831 **Perturbations by Jupiter of the particles ejected from Comet Lexell.**
A. Carusi, M. Kresáková, G. B. Valsecchi.
Astron. Astrophys., Vol. 116, 201 - 209 (1982).
The motion of particles ejected at different velocities from the nucleus of Comet Lexell at its 1770 perihelion passage – the first approach to the Sun after a considerable reduction of the perihelion distance – is followed through the next close encounter with Jupiter in 1779. Perturbed jovicentric and heliocentric orbits are computed for 45 model objects after a time lapse of 12 yr including the strong perturbation, and the subsequent evolution of the particle stream is discussed. A complete dispersion of the part of the stream passing closest to Jupiter, similar to an explosion, is accompanied by a very simple and continuous sequence of jovicentric trajectories.

Periodic comet Smirnova-Chernykh

Orbital elements of comets.
See Abstr. 103.012.

Comet 1980 XII Meier

103.851 **Comet Meier 1980q.**
Komet. Tsirk., Kiev, No. 287 (1982). In Russian.

Periodic comet Pons-Winnecke

Orbital elements of comets.
See Abstr. 103.012.

Periodic comet Johnson

Orbital elements of comets.
See Abstr. 103.012.

Periodic comet Gunn

103.891 **Short-period comet Gunn.**
Komet. Tsirk., Kiev, No. 288 (1982). In Russian.

Periodic comet Wolf

103.901 **Numerical theory of the motion of P/Wolf (1884 III) over the time span 1884 - 1984.**
E. I. Kazimirchak-Polonskaya.
Tr. Inst. Teor. Astron., Vyp. 18, p. 3 - 77 (1982). In Russian.
A review of the studies of the motion of P/Wolf made by A. Thraen, M. Kamienski and the author and published in 67 papers over the time interval from 1885 to 1974 is given. The author has developed the continuous numerical theory of the motion of P/Wolf with account of perturbations from nine planets (Mercury-Pluto) as well as the nongravitational effects over a century (1884 - 1984). This theory represented fairly well 1482 observations from 1884 to 1975 and enabled to predict its return to perihelion in 1984. The perturbing influence of 1735 minor planets has been investigated. The approach of the comet with asteroid Hidalgo in 1922 and 14 close approaches with the Galilean satellites of Jupiter – Ganymed and Callisto – have been established. A great transformation of the cometary orbit in 1875 previous to its discovery as well as the evolution of the orbit over 400 y (1660 - 2060) have been investigated.

Determination of the mass of Jupiter on the basis of orbital perturbations of P/Wolf in its sphere of action in 1922.
See Abstr. 099.150.

Periodic comet Tempel-Swift

103.921 **Short-period comet Tempel-Swift.**
Komet. Tsirk., Kiev, No. 285, (1982). In Russian.

Comet 1981 II Panther

103.931 **Observations of comet Panther (1980u).**
Komet. Tsirk., Kiev, Nos. 285, 287 (1982). In Russian.

Visual and infrared observations of the distant comets P/Stephan-Oterma (1980g), Panther (1980u), and Bowell (1980b).
See Abstr. 103.324.

Periodic comet du Toit-Hartley

103.941 **Short-period comet du Toit-Hartley (1945 II - 1982b, 1982c).**
Komet. Tsirk., Kiev, No. 287 (1982). In Russian.

103.942 **New comet Hartley (1982b).**
Komet. Tsirk., Kiev, Nos. 286, 290 (1982). In Russian.

103.943 **New comet Hartley (1982c).**
Komet. Tsirk., Kiev, No. 286 (1982). In Russian.

104 Meteors, Meteor Streams

104.001 **On the reflection of radio waves from a meteor train containing two kinds of ions.**
G. G. Novikov, S. F. Tsygankov.
Izv. vuzov. Radiofiz., Tom 24, 1459 - 1466 (1981). In Russian. Abstr. in Ref. zh., 51. Astron., 7.51.313 (1982).

104.002 **The Geminid meteor stream in 1980.**
G. H. Spalding.
J. British Astron. Assoc., Vol. 92, 227 - 233 (1982).

104.003 **Hyperbolic orbits of meteors and Jupiter.**
A. M. Kazantsev, L. M. Sherbaum.
Vestn. Kiev. Univ., Astron. Vyp. 24, p. 72 - 74 (1982). In Russian.

104.004 **On some parameters characterizing the disruption of meteor bodies.**
V. V. Benyukh.
Vestn. Kiev. Univ., Astron. Vyp. 24, p. 74 - 80 (1982). In Russian.

104.005 **Activity of the Perseids remains high.**
V. V. Martynenko, A. S. Levina.
Zemlya Vselennaya, 1982, No. 4, p. 73 - 74. In Russian.

104.006 **Occurrence of minor particles in summer meteor streams of the northern hemisphere.**
V. Znojil.
Bull. Astron. Inst. Czechoslovakia, Vol. 33, 201 - 210 (1982).
In a set of 10224 telescopic and visual meteors the author identified 23 meteor streams and associations. For most of them he determined the position of the radiant and the values of the mass coefficient s, in some cases even in a number of brightness intervals. In the conclusion the author has attempted to derive some more general rules for the relation between s and the character of the stream trajectory and its age from his own and published data.

104.007 **Analysis methods and the results of radio meteor magnitude-height surveys.**
W. J. Baggaley.
Bull. Astron. Inst. Czechoslovakia, Vol. 33, 210 - 215 (1982).
A description is given of an experimental radio-meteor survey carried out to determine the relationship between meteor ionization height and meteor magnitude. Comparisons are made with previous non-shower surveys and limitations in analysis techniques are indicated. It is also shown that reported diurnal variations in observed meteoroid mass distribution index can be explained as a consequence of the characteristics of the height-magnitude distribution.

104.008 **Southern and northern hemisphere observations of the Eta Aquarid meteor shower in 1969 - 1978.**
A. Hajduk, M. Buhagiar.
Bull. Astron. Inst. Czechoslovakia, Vol. 33, 262 - 266 (1982).
The shower activity along the solar longitude is derived from data of 2 100 Aquarid meteors observed visually in Australia and from 87 000 radar echo records obtained in Czechoslovakia during the shower periods in years 1969 - 1978. A double peak in the shower activity was confirmed, from both sets of data with a central dip at $\lambda_\odot \approx 45.5$. The width of the central zone, corresponding to the core of the stream is proportional to the size distribution of the observed particles.

104.009 **Estimate of observed radio meteors with the method of ordinal statistics.**
Yu. I. Voloshchuk, B. L. Kashcheev, A. A. Tkachuk.
Astron. vestn., Tom 16, 109 - 115 (1982). In Russian. – Abstr. in Ref. zh., 51. Astron., 9.51.202 (1982).

104.010 **Physical parameters of meteoroids undergoing continuous surface-crushing in the atmosphere. I. Methods for determination of the parameters.**
V. V. Kalenichenko.
Astron. vestn., Tom 16, 116 - 123 (1982). In Russian. – Abstr. in Ref. zh., 51. Astron., 9.51.203 (1982).

104.011 **Determination of the basic parameters of radio meteors from amplitude characteristics with weakened diffraction.**
K. K. Kostylev.
Astron. vestn., Tom 16, 124 - 127 (1982). In Russian. – Abstr. in Ref. zh., 51. Astron., 9.51.204 (1982).

104.012 **Influence of the gravitational field of planets on the structural characteristics of meteor showers.**
G. V. Andreev, P. B. Babadzhanov.
Dokl. AN TadzhSSR, Tom 24, 661 - 667 (1981). In Russian. Abstr. in Ref. zh., 51. Astron., 9.51.205 (1982).

104.013 **Two-station television observations of Perseid meteors.**
M. Hapgood, P. Rothwell, O. Royrvik.
Mon. Not. R. Astron. Soc., Vol. 201, 569 - 577 (1982).
The predictions of the dustball meteor ablation theory of Hawkes & Jones have been tested using data from two-station television observations of meteors made during the 1977 and 1978 displays of the Perseid meteor shower.

104.014 **Investigation of the disruption of meteor particles in the motion in the atmospheres of the earth, Mars and Venus.**
Eh. Z. Apshtejn, N. N. Pilyugin, N. V. Vartanyan.
Kosm. Issled., Tom 20, 730 - 735 (1982). In Russian.

104.015 **Der Quadrantiden-Schauer 1982.**
W. Schippke.
Sterne Weltraum, 21. Jahrg., 526 - 527 (1982).

104.016 **Influence of reduction of the mass of the sun in modelling condensations in a meteor stream.**

E. A. Reznikov.
Komet. Tsirk., Kiev, No. 289 (1982). In Russian.

104.017 **Observations of the π-Coronid meteor shower.**
Yu. V. Nesterov.
Komet. Tsirk., Kiev, No. 290 (1982). In Russian.

104.018 **Geminid meteor shower as observed on the long base.**
M. Šimek, R. P. Chebotarev, Sh. O. Isamutdinov, P. Pecina, V. Znojil.
Bull. Astron. Inst. Czechoslovakia, Vol. 33, 349 - 358 (1982).
The paper deals with the radar observations of the Geminid meteor shower at Dushanbe, Ondřejov and Ottawa, offering a long baseline for extending the observational time. The method of collecting the particular data into one scheme as well as the radar systems used are described. It is shown that the presented method gives good shower activity maxima positions but poor amplitudes. Influence of the Poynting-Robertson effect is illustrated.

104.019 **High activity of the Perseid meteor shower in 1980.**
A. P. Savrukhin.
Astron. Tsirk., No. 1181, p. 6 - 7 (1981). In Russian.

104.020 **Atmospheric multiple bolide of February 11, 1976.**
M. Jõeveer, T. Tõnnisson.
Tartu Astrofüüs. Obs. Publ., Köide 49, 155 - 160 (1982). In Russian.

104.021 **A note on the Delta Aurigid meteor stream.**
J. D. Drummond.
Icarus, Vol. 51, 655 - 659 (1982).
Seven doubly photographed meteors reported in the literature are shown to be members of the Delta Aurigids, which now appear to extend from September 29 to October 18. The mean orbital elements of the stream point to an unknown short period (115 years) retrograde comet as a parent.

104.022 **Interplanetary bodies.**
Z. Ceplecha.
Vesmír, Vol. 61, 368 - 371 (1982). In Czech.

104.023 **Sporadic meteors.**
J. Štohl.
Kozmos, Vol. 13, 187 - 188 (1982). In Slovak.

BMS meteor observer's handbook. Vol. 1. Naked eye work.
See Abstr. 003.079.

Photographing of bright meteors-fireballs.
See Abstr. 013.068.

Ergebnisse aus der Meteorforschung in der Tschechoslowakei.
See Abstr. 013.076.

The instrumental error in radar detection of meteor heights.
See Abstr. 031.580.

How to automate a 35 mm all-sky camera for an automatic fireball network.
See Abstr. 032.028.

The Meteor Wind Radar System at Waltair.
See Abstr. 033.051.

Device for automatic recording of a radar echo from meteors.
See Abstr. 034.008.

Das Zweikoordinatenmeßgerät ASCORECORD 3 DP im Dienste des Europäischen Feuerkugel-Programms.
See Abstr. 034.011.

Meteor wind radar studies at Waltair.
See Abstr. 082.040.

On the increase of sodium twilight glow during meteor showers.
See Abstr. 082.071.

Twilight glow of metallic emissions after the passage of the Lyrid meteor shower.
See Abstr. 082.072.

On the relation of impulsive bursts of 3-cm radio emission with meteor phenomena in the earth's atmosphere.
See Abstr. 082.080.

Relationships between comets, large meteors, and meteorites.
See Abstr. 102.019.

Comet and meteor notes 1980/81.
See Abstr. 103.014.

105 Meteorites, Meteorite Craters

105.001 **Atmospheric breakup of terrestrial impactors.**
H. J. Melosh.
Multi-ring basins, (see 012.001), p. 29 - 35 (1981).
This paper shows that kilometer-size meteoroids entering the earth's atmosphere may be fragmented and significantly dispersed before they strike the ground.

105.002 **Breccia dikes: geological constraints on the formation of complex craters.**
P. Lambert.
Multi-ring basins, (see 012.001), p. 59 - 78 (1981).
A classification of breccia dikes in crystalline rocks is proposed on the basis of texture, structure, and composition with respect to the country rock. Two major types – A and B – can be distinguished. This study shows: 1. Breccia dikes affect large volumes of target material below the crater floor and are not restricted to a central zone. 2. Several successive generations of fractures and brecciation are produced in the target during a single impact. 3. Modification of the initial cavity takes place after the A dikes, shatter cones and other fractures at all scales are already developed within the target.

105.003 **The Lappajärvi meteorite crater, Finland: petrography, Rb-Sr, major and trace element geochemistry of the impact melt and basement rocks.**
W. U. Reimold.
Geochim. Cosmochim. Acta, Vol. 46, 1203 - 1225 (1982).
Impact melt lithologies of the 77 m.y. old Finnish meteorite crater Lappajärvi as well as the Precambrian target rocks have been studied in detail, to identify and characterize different impact melt types (clast-poor, clast-rich, suevitic melt) and to study their chemical (major and trace elements) and isotopic (Rb-Sr) compositions in comparison to the composition of the target rocks.

105.004 **Origin and evolution of the Nakhla meteorite inferred from the Sm-Nd and U-Pb systematics and REE, Ba, Sr, Rb and K abundances.**
N. Nakamura, D. M. Unruh, M. Tatsumoto, R. Hutchison.
Geochim. Cosmochim. Acta, Vol. 46, 1555 - 1573 (1982).
The agreement in the age determined by three independent radiometric methods and the high initial ϵ_{Nd} value strongly suggest that the 1.3 b.y. age dates one thorough igneous event in the parent body which did not only reset these isotopic systems but also established the chemical and petrologic characteristics observed for the Nakhla meteorite. The significantly younger age of Nakhla than the youngest lunar rock; the young differentiation age inferred from the U-Th-Pb data, and the estimated LIL element abundances (including those of K, U and Th) in the source suggest that the Nakhla meteorite may have been derived from a relatively large, well-differentiated planetary body such as Mars.

105.005 **Chemical composition of HAL, an isotopically-unusual Allende inclusion.**
A. M. Davis, T. Tanaka, L. Grossman, T. Lee, G. J. Wasserburg.
Geochim. Cosmochim. Acta, Vol. 46, 1627 - 1651 (1982).
The authors report the results of a detailed chemical study of HAL (Hibonite Allende), which shows that the unusual history of this inclusion indicated by its textures, minerals and isotopic composition is reflected in a number of exceptional chemical peculiarities. They explore models for the formation of HAL that are consistent with all available data.

105.006 **Trace element concentrations in the isotopically unique Allende inclusion, EK 1-4-1.**
H. Nagasawa, D. P. Blanchard, H. Shimizu, A. Masuda.
Geochim. Cosmochim. Acta, Vol. 46, 1669 - 1673 (1982).
The main purpose of this paper is to report the results of instrumental neutron activation analysis and mass-spectrometric isotope dilution analysis of REE and other trace elements for EK 1-4-1 and its mineral separates, and to examine the distribution of trace elements among co-existing mineral phases in which the mass-fractionated O isotope ratios were observed.

105.007 **An analysis of the physical, chemical, optical, and historical impacts of the 1908 Tunguska meteor fall.**
R. P. Turco, O. B. Toon, C. Park, R. C. Whitten, J. B. Pollack, P. Noerdlinger.
Icarus, Vol. 50, 1 - 52 (1982).
A detailed analysis of the physical nature and photochemical aftereffects of the explosive cometary meteor Tunguska is presented.

105.008 **A search for evidence of superheavy-element fission in chondritic metal.**
R. K. Bull, P. Mold.
Nature, Vol. 298, 634 - 635 (1982).
The authors describe a search for superheavy elements (SHE) in metal fragments within ordinary chondrites. The absence of excesses of tracks along the boundaries between olivines and metal grains allows upper limits to the SHE content of the metal of $\sim 2.5 \times 10^{-13}$ to $\sim 2.5 \times 10^{-15}$ kg per kg to be set at a time $\sim 4 \times 10^9$ yr ago, determined from the ^{244}Pu fission track densities in chondritic phosphates.

105.009 **The first meteorite stream.**
D. W. Hughes.
Nature, Vol. 299, 14 - 15 (1982).

105.010 **Intergrown mica and montmorillonite in the Allende carbonaceous chondrite.**
K. Tomeoka, P. R. Buseck.
Nature, Vol. 299, 326 - 327 (1982).
Calcium-, aluminium-rich inclusions (CAIs) in carbonaceous chondrites such as Allende are of considerable interest because of their primitive character. The authors present high-resolution transmission electron microscopy observations of a complex mixture of mica and montmorillonite that occurs within a fine-grained CAI in the Allende C3 (V) meteorite.

105.011 **An unusual layered mineral in chondrules and aggregates of the Allende carbonaceous chondrite.**
K. Tomeoka, P. R. Buseck.
Nature, Vol. 299, 327 - 329 (1982).
The authors have found that an unusual Fe-, Ni- and O-rich layered mineral, related to serpentine, occurs within the opaque assemblage; it is formed by alteration of olivine. This is the first report of a serpentine-related mineral in the Allende meteorite.

105.012 **The cracking history of Serrania de Varas: an alternative view.**
H. J. Axon, E. Smith, F. Knowles.
Meteoritics, Vol. 17, 49 - 53 (1982).
Buchwald's (1975) theory of thermally induced cracking in Serrania de Varas is critically examined from the standpoint of elementary stress analysis. New metallographic observations are presented. It is suggested that thermal cracking is unlikely to have been operating in this material.

105.013 **Calcium-aluminium-rich chondrules in the unequilibrated ordinary chondrites.**

H. Nagahara, I. Kushiro.
Meteoritics, Vol. 17, 55 - 63 (1982).

105.014 **A new LL3 chondrite, Allan Hills A79003, and observations on matrices in ordinary chondrites.**
E. R. D. Scott, G. J. Taylor, P. Maggiore.
Meteoritics, Vol. 17, 65 - 75 (1982).

105.015 **Glanggang and Selakopi: two new paired (?) Indonesian chondrites.**
K. Fredriksson, G. S. Peretsman.
Meteoritics, Vol. 17, 77 - 86 (1982).

105.016 **The Meteoritical Bulletin, No. 60.**
Sponsored by The Meteoritical Society.
A. L. Graham (Editor).
Meteoritics, Vol. 17, 93 - 97 (1982).
Place of fall/find, date of fall/find, class and type, number of individual specimens, total weight and circumstances of fall/find of some meteorites are given.

105.017 **Isotopic composition of rare gases of the Tsarev meteorite and some lunar soil probes.**
Dang Vu Min, V. I. Simonovskij, M. M. Fugzan, N. I. Zaslavskaya, Yu. A. Shukolyukov.
Dokl. AN SSSR, Tom 264, 204 - 208 (1982). In Russian.
Abstr. in Ref. zh., 62. Issled. kosm. prostranstva, 8.62.230 (1982).

105.018 **Problems and methods of investigation of cosmic matter on the earth.**
E. L. Krinov.
Meteoritika, vyp. 40, p. 3 - 5 (1982). In Russian.

105.019 **New meteorites of the USSR: Tsarev, Kutais, Bakhardok.**
R. L. Khotinok.
Meteoritika, vyp. 40, 6 - 9, 159, plates 1 - 2 (1982). In Russian.

105.020 **The Raguli chondrite.**
G. V. Baryshnikova, V. Ya. Kharitonova, L. D. Barsukova.
Meteoritika, vyp. 40, p. 10 - 15, 159, plates 3 - 4 (1982). In Russian.

105.021 **Petrography, mineralogy and chemistry of the Andreevka chondrite.**
E. G. Osadchij, G. V. Baryshnikova, V. Ya. Kharitonova, L. D. Barsukova.
Meteoritika, vyp. 40, p. 16 - 27, 159, 160, plates 5 - 6 (1982). In Russian.

105.022 **Meteorite Pilistvere.**
V. P. Semenenko, L. G. Samojlovich.
Meteoritika, vyp. 40, p. 28 - 33, 160, plate 7 (1982). In Russian.

105.023 **Chemical-mineralogical investigation of the Iudoma, Repeev Khutor, Norin-Shibir iron meteorites.**
N. I. Zaslavskaya, G. M. Kolesov, E. G. Osadchij.
Meteoritika, vyp. 40, p. 34 - 40, 160, plates 8 - 12 (1982). In Russian.

105.024 **The chemical composition of the meteorites Aliskerovo, Aprel'skij, Vetluga and Kuleshovka.**
V. Ya. Kharitonova, L. D. Barsukova.
Meteoritika, vyp. 40, p. 41 - 44, 160, plate 13 (1982). In Russian.

105.025 **Determination of the microelement content in iron meteorites by the neutron-activation method using pneumatic tube conveyers.**
G. M. Kolesov, N. I. Zaslavskaya, A. Mashtalka, Ya. Frana.
Meteoritika, vyp. 40, p. 45 - 48 (1982). In Russian.

105.026 **On the composition of silicate minerals in iron meteorites.**
A. A. Yavnel'.
Meteoritika, vyp. 40, p. 49 - 56 (1982). In Russian.

105.027 **Can the carbonaceous chondrites be the source of rare gases for the earth and internal planets?**
L. K. Levskij.
Meteoritika, vyp. 40, p. 57 - 60 (1982). In Russian.

105.028 **Peculiarities of magnetic properties of ureilites and shocked chondrites.**
E. G. Gus'kova.
Meteoritika, vyp. 40, p. 61 - 66 (1982). In Russian.

105.029 **On the problem of central uplift formation in impact craters.**
B. A. Ivanov, A. T. Bazilevskij, L. V. Sazonova.
Meteoritika, vyp. 40, p. 67 - 81 (1982). In Russian.

105.030 **Meteoritic crater evolution from the point of view of a random walk process.**
I. T. Zotkin, A. I. Dabizha.
Meteoritika, vyp. 40, p. 82 - 90 (1982). In Russian.

105.031 **Geophysical characteristic of some astroblemes of the USSR.**
A. I. Dabizha, V. I. Fel'dman.
Meteoritika, vyp. 40, p. 91 - 101 (1982). In Russian.

105.032 **On a characteristic of the internal texture of allogenic breccias and suevites of large astroblemes.**
A. N. Danilin.
Meteoritika, vyp. 40, p. 102 - 106 (1982). In Russian.

105.033 **Inclusions in impactites.**
V. L. Masajtis.
Meteoritika, vyp. 40, p. 107 - 112 (1982). In Russian.

105.034 **Petrochemical peculiarities of impactites and the role of vaporization and condensation processes at formation of impact craters.**
O. I. Yakovlev, O. V. Parfenova.
Meteoritika, vyp. 40, p. 113 - 121 (1982). In Russian.

105.035 **Impact glasses in suevites of Kara and Ust-Kara astroblemes.**
T. V. Selivanovskaya, M. S. Mashchak, V. A. Ezerskij, N. B. Reshetnyak.
Meteoritika, vyp. 40, p. 122 - 130, 160 - 161, plate 14 (1982). In Russian.

105.036 **Daubréelite and its associations in the iron meteorites Sikhote-Alin and Boguslavka.**
G. P. Kudryavtseva, L. T. Soshkina, N. I. Zaslavskaya.
Meteoritika, vyp. 40, p. 131 - 133, 161, plates 15 - 16 (1982). In Russian.

105.037 **Irradiation history and atmospheric ablation of meteorites: results from cosmic ray track studies.**
J. N. Goswami.
Proc. Indian Acad. Sci. Earth Planet Sci., Vol. 90, 345 - 358 (1981).
Studies of cosmic ray tracks and other cosmogenic effects in meteorites give us information on the irradiation history of these meteorites and enable us to estimate the extent of ablation during their atmospheric transit, and hence their pre-atmospheric masses.

105.038 **Fission tracks and cooling rates of meteorites.**
G. Crozaz.
Proc. Indian Acad. Sci. Earth Planet Sci., Vol. 90, 383 - 388 (1981).

Recent work on fission track studies of meteorite samples to obtain cooling rates of meteorite parent bodies is reviewed.

105.039 **Siderophiles in the Brachina meteorite: impact melting?**
G. Ryder.
Nature, Vol. 299, 805 - 807 (1982).

The published high abundances and unusual ratios of siderophiles in the chassignite Brachina in comparison with its apparent relatives Chassigny, shergottites, and nakhlites, suggest that Brachina might be an impact melt. The author shows that all these meteorites could have formed on Mars, with Brachina ejected as a large tektite-like melt blob, along with shocked and unshocked target rocks.

105.040 **Terrestrial-type xenon in meteoritic troilite.**
G. Hwaung, O. K. Manuel.
Nature, Vol. 299, 807 - 810 (1982).

Recently it has been realized that the isotopic composition of Xe in different phases of chondrites is not uniform and that AVCC Xe is a mixture of the different nucleogenetic components trapped in these phases. The authors show here a similar abundance pattern for the nonradiogenic xenon isotopes in air and in meteoritic troilite (FeS), which suggests that the isotopic composition of atmospheric Xe was not produced by unique events in the history of terrestrial material but represents a particular mix of the different nucleogenetic components of Xe that was dominant in a central Fe- and S-rich region of the protoplanetary nebula.

105.041 **The isotopic composition of noble gases in the Tsarev meteorite and some probes of lunar soil.**
Dang Vu Min, V. I. Simonovskij, M. M. Rugzan, N. I. Zaslavskaya, Yu. A. Shukolyukov.
Dokl. AN SSSR, Tom 264, 204 - 208 (1982). In Russian. Abstr. in Ref. zh., 51. Astron., 9.51.216 (1982).

105.042 **Thermodynamic properties and origin of daubréelite $FeCr_2S_4$ in meteorites.**
M. I. Petaev, T. A. Stolyarova, E. G. Osadchij, Yu. V. Semenov, Yu. I. Sidorov, V. M. Zhdanov, V. A. Turdakin, I. L. Khodakovskij.
Geokhimiya, 1982, No. 5, p. 690 - 704. In Russian. – Abstr. in Ref. zh., 51. Astron., 9.51.227 (1982).

105.043 **Comparison of the mass spectrum of slag-ashes of volcanic jets with the mass spectrum of meteorite particles.**
B. A. Trubnikov, Yu. V. Adamchuk.
Volkanol. i sejsmol., 1982, No. 2, p. 96 - 99. In Russian. Abstr. in Ref. zh., 51. Astron., 9.51.230 (1982).

105.044 **On a reference list of meteorite craters on the earth.**
H. Eelsalu.
Astron. vestn., Tom 16, 128 (1982). In Russian. – Abstr. in Ref. zh., 51. Astron., 9.51.232 (1982).

105.045 **Caswellsilverite, $NaCrS_2$: a new mineral in the Norton County enstatite achondrite.**
A. Okada, K. Keil.
American Mineral., Vol. 67, 132 (1982). – Abstr. in Phys. Abstr., Vol. 85, Abstr. 68681 (1982).

105.046 **Plutonium-244 concentrations in chondritic phosphates and their significance in fission-track dating of meteorites.**
P. Mold, R. K. Bull, S. A. Durrani.
Nucl. Tracks Methods Instrum. Appl., Vol. 5, (see 012.015), 401 (1981). – Summary. Abstr. in Phys. Abstr., Vol. 85, Abstr. 68685 (1982).

105.047 **Fission tracks in chondritic meteorites and the search for siderophile superheavy elements.**
R. K. Bull, P. Mold.
Nucl. Tracks Methods Instrum. Appl., Vol. 5, (see 012.015), 402 (1981). – Summary. Abstr. in Phys. Abstr., Vol. 85, Abstr. 68686 (1982).

105.048 **The oldest meteorite of our country *(USSR).***
V. P. Semenenko.
Zemlya Vselennaya, 1982, No. 5, p. 65 - 66. In Russian.

105.049 **Mössbauer spectroscopy of chondrules and matrix from a stony meteorite (ALHA 77278).**
H. Roy-Poulsen, M.-L. Andersen, L. Larsen, N. O. Roy-Poulsen, L. Vistisen, J. M. Knudsen.
Phys. Scr., Vol. 26, 248 - 256 (1982).

Mössbauer investigations of the unequilibrated chondrite ALHA 77278 are reported. The chondrules studied are found to contain the silicates olivine and pyroxene, and one of the chondrules furthermore contains kamacite. The results are discussed within the framework of the theories about the condensation sequence of minerals in the early solar system.

105.050 **A stochastic study of the cosmic ray irradiation record of the meteorites.**
J. N. Goswami, R. Jha, D. Lal.
Bull. Astron. Soc. India, Vol. 10, 31 (1982). – Abstract.

105.051 **Fission track studies of xenolithic chondrites: implications regarding brecciation and metamorphism.**
B. K. Kothari, R. S. Rajan.
Geochim. Cosmochim. Acta, Vol. 46, 1747 - 1754 (1982).

The authors have studied fission tracks in phosphates from one gas-poor chondrite and three gas-rich chondrites to determine their thermal history and brecciation time scales. More than 70 percent of the tracks in whitlockites in these meteorites are due to the decay of extinct Pu^{244}. The authors conclude that their observed fission track ages date the end of metamorphic cooling in the meteorite parent bodies and support the planetesimal model for the formation of xenolithic chondrites.

105.052 **Microchondrule-bearing clast in the Piancaldoli LL3 meteorite: a new kind of type 3 chondrite and its relevance to the history of chondrules.**
A. E. Rubin, E. R. D. Scott, K. Keil.
Geochim. Cosmochim. Acta, Vol. 46, 1763 - 1776 (1982).

In the Piancaldoli LL3 chondrite, the authors found a mm-sized clast containing ~100 chondrules 0.2–64 μm in apparent diameter (much smaller than any previously reported) that are all of the same textural type (radial pyroxene; Fs_{1-17}). Microchondrules (those chondrules $\leqslant$100 μm in diameter) are also abundant in another new kind of type 3 chondrite clast in the Rio Negro L chondrite regolith breccia. Microchondrules probably formed by the same process that formed normal-sized droplet chondrules: melting of pre-existing dustballs.

105.053 **On neutron-induced and other noble gases in Allende inclusions.**
R. Göbel, F. Begemann, U. Ott.
Geochim. Cosmochim. Acta, Vol. 46, 1777 - 1792 (1982).

Noble gases have been measured in one bulk sample, six light inclusions, and one black xenolith from four specimens of the carbonaceous chondrite Allende which cover a range in their ^{60}Co content of about a factor of ten. The release pattern obtained upon heating the samples at different temperature steps is discussed.

105.054 **The geochemical coherence of Pu and Nd and the $^{244}Pu/^{238}U$ ratio of the early solar system.**
J. H. Jones.
Geochim. Cosmochim. Acta, Vol. 46, 1793 - 1804 (1982).
Two very different sets of $^{244}Pu/^{238}U$ ratios have been reported for early solar system materials. One group of samples yields high (0.015–0.016) ratios (Podosek, 1970a, 1972; Drozd et al., 1977) and calculations based on another group of analyses yield low ratios (~0.004) (Marti et al., 1977). Recently measured partition coefficients for Pu and Sm are used to evaluate the data of Marti et al. and $^{244}Pu/^{238}U$ ratios from other sources are also considered. A low $^{244}Pu/^{238}U$ ratio (~0.005) is favored, and some implications of this low ratio to galactic nucleosynthesis and meteorite age dating are briefly discussed.

105.055 **Are Cl chondrites chemically fractionated? A trace element study.**
M. Ebihara, R. Wolf, E. Anders.
Geochim. Cosmochim. Acta, Vol. 46, 1849 - 1861 (1982).
Six Cl chondrite samples and a C2 xenolith from the Plainview H5 chondrite were analyzed by radiochemical neutron activation for the elements Ag, Au, Bi, Br, Cd, Ce, Cs, Eu, Ge, In, Ir, Lu, Nd, Ni, Os, Pd, Pt, Rb, Re, Sb, Se, Sn, Tb, Te, Tl, Yb, and Zn. The data were combined with 9 earlier analyses from this laboratory and examined for evidence of chemical fractionation in Cl chondrites. The authors present revised mean Cl abundances for 35 elements, based on the new data and a critical selection of literature data.

105.056 **Petrology and shock metamorphism of Pampa del Infierno chondrite.**
N. Z. Boctor, P. M. Bell, H. K. Mao, G. Kullerud.
Geochim. Cosmochim. Acta, Vol. 46, 1903 - 1911 (1982).
Pampa del Infierno, an L6 chondrite, displays strong evidence of impact metamorphism. Rare chondrules and two types of dark-colored clasts occur in a light-colored matrix. Granular clasts are similar in mineralogy and chemistry to the host meteorite, but display shock metamorphic features. The light-colored matrix mainly displays evidence of shock metamorphism under subsolidus conditions. Pertinent shock-wave data used to interpret the metamorphic history of the Pampa del Infierno chondrite suggest metamorphism by impact at a minimum peak pressure greater than 300 kbar.

105.057 **Systematic compositional variations in the Cape York iron meteorite.**
K. H. Esbensen, V. F. Buchwald, D. J. Malvin, J. T. Wasson.
Geochim. Cosmochim. Acta, Vol. 46, 1913 - 1920 (1982).
Concentrations of Re, Ir and Au are nearly constant within individual masses of the Cape York IIIAB iron meteorite, but differences between masses can be as large as a factor of 2. The S concentration shows a still larger range from 13 mg/g in Agpalilik to 1.4 mg/g in Savik. Concentrations of Ir vary by ~10% and Au by ~3% between the ends of an 85-cm section from the Agpalilik mass of Cape York, but other sections through Agpalilik show smaller variations. Large gradients in Re and Ir and small gradients in Ni and Au were also observed in samples within 2 cm of a large (100 cm^3) troilite nodule. These gradients may reflect rapidly changing solid/liquid distribution coefficients during the final crystallization of S-rich liquid.

105.058 **A direct measurement of the distribution in depth of ^{26}Al in the Estacado meteorite.**
J. C. Barton, A. H. Watson, A. G. Wright.
Geochim. Cosmochim. Acta, Vol. 46, 1963 - 1967 (1982).
The two-dimensional distribution of ^{26}Al across a slice of the Estacado meteorite was measured using non-destructive γ-γ coincidence counting. The maximum activity was found to be 60 ± 2 dpm kg^{-1} and the variation across the slice rather small. The intensity and distribution are in moderate agreement with existing theoretical models if the pre-atmospheric shape of the meteorite had been a sphere of radius 0.35 ± 0.05 m, but it was more likely to have been slightly non-spherical.

105.059 **Comment on "Thermoluminescence of meteorites and their terrestrial ages".**
S. W. S. McKeever, P. D. Townsend.
Geochim. Cosmochim. Acta, Vol. 46, 1997 - 1998 (1982).
Melcher (1981) recently proposed a method to determine the terrestrial age of chondritic meteorites using thermoluminescence. This paper suggests that Melcher's analysis of the second-order thermoluminescence signals is inappropriate and hence his conclusions regarding the age estimates are wrong.

105.060 **Thermoluminescence of meteorites and their terrestrial ages: reply to comments by S. W. S. McKeever and P. D. Townsend.**
C. L. Melcher.
Geochim. Cosmochim. Acta, Vol. 46, 1999 - 2000 (1982).
The usefulness of an "effective" trap depth to approximate the group of closely spaced trap depths responsible for the low temperature thermoluminescence in meteorites is evaluated on the basis of its success when used with the TL equivalent dose method to determine terrestrial ages of Antarctic meteorites. The ages determined in this way agree in every instance with terrestrial age determinations based on ^{14}C, ^{26}Al, ^{36}Cl, and ^{53}Mn. The application of an "effective" trap depth in this TL dating technique in no way implies that the TL signal actually arises from a single trap depth.

105.061 **^{40}Ar-^{39}Ar ages of rocks and glasses from the Nördlinger Ries Crater and the temperature history of impact breccias.**
T. Staudacher, E. K. Jessberger, B. Dominik, T. Kirsten, O. A. Schaeffer.
J. Geophys., Vol. 51, 1 - 11 (1982). – Abstr. in Phys. Abstr., Vol. 85, Abstr. 106121 (1982).

105.062 **Distribution of Ni, Co, Ga, and Cu in iron meteorites.**
J. Okano, H. Nishimura,
Secondary ion mass spectrometry, (see 012.036), p. 426 - 430 (1982). – Abstr. in Phys. Abstr., Vol. 85, Abstr. 106637 (1982).

105.063 **Absolute age of formation of chondrites studied by the ^{87}Rb–^{87}Sr method.**
J.-F. Minster, J.-L. Birck, C. J. Allègre.
Nature, Vol. 300, 414 - 419 (1982).
The ^{87}Rb–^{87}Sr isochrons for H-, E- and LL-, chondrites are undistinguishable from each other. The joint isochron yields an age of 4,498 ± 15 Myr if the rubidium decay constant $\lambda\, ^{87}Rb = 1.42 \times 10^{-11}\, yr^{-1}$, and a strontium initial ratio $^{87}Rb/^{87}Sr = 0.69885 \pm 0.00010$. For the age to agree with U–Th–Pb data, $\lambda\, ^{87}Rb$ would have to be changed to $1.402 \pm 0.008 \times 10^{-11}\, yr^{-1}$.

105.064 **Aminosäuren in Meteoriten der Antarktis.**
A. Liebold.
Sterne Weltraum, 21. Jahrg., 454 - 455 (1982).

105.065 **Geology and terrestrial age of the Derrick Peak meteorite occurrence, Antarctica.**
P. J. J. Kamp, D. J. Lowe.
Meteoritics, Vol. 17, 119 - 127 (1982).

105.066 **The Derrick Peak, Antarctica, iron meteorites.**
R. S. Clarke, Jr.
Meteoritics, Vol. 17, 129 - 134 (1982).

105.067 **Two new iron meteorites from Western Australia.**
J. R. De Laeter.
Meteoritics, Vol. 17, 135 - 140 (1982).
Two new iron meteorites from Western Australia are described: Cosmo Newberry – a 2.156 kg meteorite of unusual spiky shape, and Gnowangerup – a 33.6 kg pear-shaped meteorite. X-ray fluorescence spectrometry shows that Cosmo Newberry can be classified in Group IIA, whilst Gnowangerup is a member of Group IIIAB.

105.068 **The Manegaon diogenite.**
K. Fredriksson.
Meteoritics, Vol. 17, 141 - 144 (1982).
A new analysis of the Manegaon hypersthene achondrite is presented and compared with the other eight known falls of this class.

105.069 **Distribution of Georgia tektites.**
H. Povenmire.
Meteoritics, Vol. 17, 145 - 148 (1982).

105.070 **Petrology of EETA79006 and implications for the formation of polymict eucrites.**
S. B. Simon, J. J. Papike, M. Fuhrman.
Meteoritics, Vol. 17, 149 - 162 (1982).

105.071 **Canadian Arctic meteorite search: 1981.**
P. B. Robertson, L. C. Coleman, E. J. Olsen, S. A. Kissin.
Meteoritics, Vol. 17, 163 - 170 (1982).
The recovery of several thousand meteorite fragments from Antarctica has led to speculation that accumulations may occur on the Earth's other major ice caps. A search on foot and by helicopter in July, 1981, failed to recover any specimens.

105.072 **Petrography, mineral chemistry and origin of type I enstatite chondrites.**
C. A. Leitch, J. V. Smith.
Geochim. Cosmochim. Acta, Vol. 46, 2083 - 2097 (1982).

105.073 **Crystallization sequences of Ca-Al-rich inclusions from Allende: an experimental study.**
E. Stolper.
Geochim. Cosmochim. Acta, Vol. 46, 2159 - 2180 (1982).

105.074 **The compositional classification of chondrites. III. Ungrouped carbonaceous chondrites.**
G. W. Kallemeyn, J. T. Wasson.
Geochim. Cosmochim. Acta, Vol. 46, 2217 - 2228 (1982).
Seven carbonaceous chondrites with uncertain classifications were analyzed by instrumental radiochemical neutron activation analysis for 29 elements. Five of these chondrites (A77307, Adelaide, Al Rais, Karoonda and Renazzo) are unique 'grouplets', not closely related to other groups or to each other. Only Coolidge (CV4) and Grosnaja (CV3-an) are members of previously established groups.

105.075 **Chronology and petrogenesis of young achondrites, Shergotty, Zagami, and ALHA77005: late magmatism on a geologically active planet.**
C.-Y. Shih, L. E. Nyquist, D. D. Bogard, G. A. McKay, J. L. Wooden, B. M. Bansal, H. Wiesmann.
Geochim. Cosmochim. Acta, Vol. 46, 2323 - 2344 (1982).
A study was undertaken to determine the chronology, petrogenesis, and relationships among the shergottites, Shergotty and Zagamy, and the unique achondrite ALHA77005. These meteorites are the product of a variety of complex processes.

105.076 **On the siting of noble gases in E-chondrites.**
J. Crabb, E. Anders.
Geochim. Cosmochim. Acta, Vol. 46, 2351 - 2361 (1982).
The authors have investigated the siting of noble gases in 6 E-chondrites, by analyzing fractions separated by density, grain size, and chemical resistance from Qingzhen (E3), Indarch (E4), Abee and Saint Sauveur (E4–5), and Yilmia and North West Forrest (E6).

105.077 **Some short-lived nuclides in the early solar system – a connection with the placental ISM.**
G. J. Wasserburg, D. A. Papanastassiou.
Essays in nuclear astrophysics, (see 003.013), p. 77 - 140 (1982).
By investigating observational evidence for short-lived radioactive nuclides in the early solar system a temporal connection can be established between stellar nucleosynthesis and the debris that made up the solar system. A short review of the status of the field prior to the work on the Allende meteorite is given, followed by a summary of the observations and ideas about Allende. The steps are reviewed, that led to the discovery of extinct ^{26}Al. A detailed summary of the existing facts and the conceptual basis of the relationships between Mg (24, 25, 26) and Al (26, 27) is presented.

105.078 **Meteorite actinide chemistry and cosmochronology.**
D. S. Burnett, M. I. Stapanian, J. H. Jones.
Essays in nuclear astrophysics, (see 003.013), p. 141 - 158 (1982).
This paper summarizes the present knowledge of solar system abundances of U, Th, and ^{244}Pu.

105.079 **Meteorites: their mineralogical, chemical and isotopic affinities.**
R. Hutchison.
Proceedings of the workshop on interstellar comets, (see 012.058), p. 19 - 23 (1982).

105.080 **Isotope ratios from non-solar system projectiles.**
A. E. Fallick.
Proceedings of the workshop on interstellar comets, (see 012.058), p. 37 - 43 (1982).

105.081 **Isotopic studies of noble gases from the Tsarev meteorite and from samples of lunar soil.**
Yu. A. Shukolyukov, Dang vu min, V. I. Simonovskij, M. M. Fugzan, L. F. Migdisova, N. I. Zaslavskaya.
Geokhimiya, 1982, No. 8, p. 1078 - 1093. In Russian. – Abstr. in Ref. zh., 51. Astron., 11.51.241; 62. Issled. kosm. prostranstva, 11.62.262 (1982).

105.082 **Meteorite – Erscheinung, Beschaffenheit, Entwicklung.**
G. Hoppe.
Sterne, 58. Band, 352 - 364 (1982).

105.083 **Determination of the direction of the Kaali meteorite shower with geophysical methods.**
A. Aaloé, V. Andra, H. Andra.
Izv. AN EhSSR. Geol., Tom 31, No. 2, p. 56 - 61 (1982). In Russian. – Abstr. in Ref. zh., 51. Astron., 10.51.255 (1982).

105.084 **The role of S in the evolution of the parental cores of the iron meteorites.**
A. Kracher, J. T. Wasson.
Geochim. Cosmochim. Acta, Vol. 46, 2419 - 2426 (1982).
The authors first estimate the S contents of iron meteorite parent bodies on the basis of cosmochemical relationships to undifferentiated meteorites. They then compare these to observations on the major magmatic iron meteorite groups and present evidence that S/Ni ratios in their parent melts were much lower than those inferred for the parent body. The authors offer several alternative models to account for the discrepancy. Finally, they discuss the fate of

the S-rich meteoroidal material and why it seems to be missing from their meteorite collections.

105.085 **Chemical systematics among the moldavite tektites.**
J. W. Delano, D. H. Lindsley.
Geochim. Cosmochim. Acta, Vol. 46, 2447 - 2452 (1982).

105.086 **Actinide microdistributions in the enstatite meteorites.**
M. T. Murrell, D. S. Burnett.
Geochim. Cosmochim. Acta, Vol. 46, 2453 - 2460 (1982).
The authors report their results on the mineralogical distribution of Th, U and ^{244}Pu in the enstatite meteorites. This information will be used to: (1) define the behavior of Th and U under highly reducing conditions; (2) evaluate the reported variations of Th/U in enstatite chondrites and (3) evaluate the suitability of enstatite meteorites for $^{244}Pu/^{238}U$ chronology.

105.087 **Analytical electron microscope study of eight ataxites.**
P. M. Novotny, J. I. Goldstein, D. B. Williams.
Geochim. Cosmochim. Acta, Vol. 46, 2461 - 2469 (1982).
The purpose of the investigation is to use the analytical electron microscope as well as other electron optical techniques to study the microstructure of a representative number of ataxites. The thermal histories of these meteorites are discussed in terms of the micro-structural evidence obtained.

105.088 **Chemical and physical studies of type 3 chondrites-I: Metamorphism related studies of Antarctic and other type 3 ordinary chondrites.**
D. W. Sears, J. N. Grossman, C. L. Melcher.
Geochim. Cosmochim. Acta, Vol. 46, 2471 - 2481 (1982).
Thermoluminescence sensitivity measurements have been made on 18 unequilibrated ordinary chondrites; 12 finds from Antarctica, 5 non-Antarctic finds and 1 fall. The authors make petrologic type assignments on the basis of TL sensitivity, and on silicate heterogeneity and discuss the correlation that exists between TL sensitivity and glow curve shape.

105.089 **The metal content of the eucrite parent body: constraints from the partitioning behavior of tungsten.**
H. E. Newsom, M. J. Drake.
Geochim. Cosmochim. Acta, Vol. 46, 2483 - 2489 (1982).
The paper reexamines the question of the abundance and distribution of metal in the eucrite parent body and addresses the question of the early igneous history of that object. The authors report solid metal/silicate melt partition coefficients for W determined experimentally under conditions appropriate to eucrite petrogenesis. They utilize these data to examine metal-silicate fractionation in the EPB.

105.090 **Chemical studies of L chondrites – II. Shock-induced trace element mobilization.**
T. M. Walsh, M. E. Lipschutz.
Geochim. Cosmochim. Acta, Vol. 46, 2491 - 2500 (1982).
Previous studies of chondrites heated in the laboratory for extended periods under conditions approximating those in shock-heated collisional debris indicate that Au, Co, Se, Ga, Rb, Cs, Te, Bi, In, Ag, Zn, Tl and Cd progress in mobility. The authors report data for these 13 trace elements in 14 L4-6 chondrites of established shock history and discuss these and 13 additional chondrites studied earlier.

105.091 **Extinct I^{129} in C3 chondrites.**
J. Crabb, R. S. Lewis, E. Anders.
Geochim. Cosmochim. Acta, Vol. 46, 2511 - 2525 (1982).
Eight C3 chondrites were examined by the I^{129}-Xe^{129} dating method, to see whether their I-Xe "ages" correlate with any other properties.

105.092 **The nature and origin of type B1 and B2 Ca-Al-rich inclusions in the Allende meteorite.**
D. A. Wark, J. F. Lovering.
Geochim. Cosmochim. Acta, Vol. 46, 2581 - 2594 (1982).

105.093 **Evolution of Ca-Al-rich bodies in the earliest solar system: growth by incorporation.**
D. A. Wark, J. F. Lovering.
Geochim. Cosmochim. Acta, Vol. 46, 2595 - 2607 (1982).
Some of the strangest features of the white, centimetre-sized Ca-Al-rich inclusions (CAI's) in carbonaceous chondrite meteorites are long chains, rings and clumps of spinel crystals that have no known counterparts in other meteorites or in lunar or terrestrial rocks. In this work the authors describe these unusual features and argue that some of them are the by-products of CAI formation processes in which earlier bodies were captured or incorporated into the present CAI's.

105.094 **Non-cosmogenic lithium-6 in iron meteorites.**
S. V. S. Murty, P. N. Shukla, P. S. Goel.
Earth Planet. Sci. Lett., Vol. 60, 1 - 7 (1982).

105.095 **Variation of the phase composition of C2 chondrites on heating.**
T. V. Malysheva, K. I. Tobelko, E. Ya. Shcherbovsky, D. A. Khramov, A. I. Malyshev.
Earth Planet. Sci. Lett., Vol. 60, 8 - 16 (1982).

105.096 **An ultra-refractory inclusion from the Ornans carbonaceous chondrite.**
H. Palme, F. Wlotzka, K. Nagel, A. El Goresy.
Earth Planet. Sci. Lett., Vol. 61, 1 - 12 (1982).
A small refractory inclusion was found in the Ornans carbonaceous chondrite. The inclusion is extremely enriched in the most refractory REE (Lu, Er, Ho), the more volatile light REE having more than two orders of magnitude lower concentrations. The inclusion may either represent a first condensate from a cooling gas of approximately solar composition, or the very last residue from an evaporation process.

105.097 **A new calcium-aluminate from a refractory inclusion in the Leoville carbonaceous chondrite.**
M. Christophe-Michel-Lévy, G. Kurat, F. Brandstätter.
Earth Planet. Sci. Lett., Vol. 61, 13 - 22 (1982).

105.098 **Allan Hills 77081 – an unusual stony meteorite.**
L. Schultz, H. Palme, B. Spettel, H. W. Weber, H. Wänke, M. Christophe-Michel-Lévy, J. C. Lorin.
Earth Planet. Sci. Lett., Vol. 61, 23 - 31 (1982).
Allan Hills (ALHA) 77081 is achondritic in texture while the mineral composition and the chemistry are chondritic with the exception of a few elements. An assignment to one specific group of ordinary chondrites is therefore difficult. Gas retention ages are 3.5 ± 0.5 AE for U, Th-He and 4.50 ± 0.15 AE for K-Ar.

105.099 **Investigations of cosmic-ray produced nuclides in iron meteorites, 4. Identification of noble gas abundance anomalies.**
H. Voshage.
Earth Planet. Sci. Lett., Vol. 61, 32 - 40 (1982).

105.100 **Meteorites – evidence for the interrelationships of materials in the solar system of 4.55 Ga ago.**
R. Hutchison.
Phys. Earth Planet. Inter., Vol. 29, 199 - 208 (1982).
It is concluded that the period of formation of ordinary chondrites continued until after differentiation had occurred within some planetary bodies. Such early melting may have

been caused by the decay of ^{26}Al. Some of the differentiated materials were probably incorporated as plastic rocks, liquid chondrules and solid clasts into the accreting parent bodies of the ordinary chondrites.

105.101 **Planet(oid) core crystallisation and fractionation– evidence from the Agpalilik mass of the Cape York iron meteorite shower.**
K. H. Esbensen, V. F. Buchwald.
Phys. Earth Planet. Inter., Vol. 29, 218 - 232 (1982).

105.102 **Noble gases in meteorites.**
N. Takaoka.
Astron. Her., Vol. 75, 199 - 204 (1982). In Japanese. – Abstr. in Phys. Abstr., Vol. 85, Abstr. 117635 (1982).

105.103 **Meteoritic bismuth and ^{208}Pb microdistributions.**
D. S. Woolum, D. S. Burnett.
Trans. American Nucl. Soc., Vol. 41, 191 - 192 (1982). Abstr. in Phys. Abstr., Vol. 86, Abstr. 10277 (1983).

105.104 **Organic compounds in meteorites and their origins.**
R. Hayatsu, E. Anders.
Cosmo- and geochemistry, (see 003.015), p. 1 - 37 (1981).
Contents: Introduction. Formation conditions of carbonaceous chondrites. Behavior of carbon in a solar gas. Organic compounds in meteorites. Isotopic data. Feasibility of the FTT synthesis in the solar nebula. Interstellar molecules. Origin of volatiles on inner planets.

Meteoritical Bulletin, Index 1957 - 1982, Issues 1 - 60.
See Abstr. 002.049.

Catalogue of the meteorite collection of the Leningrad mining museum on January 1, 1979.
See Abstr. 002.053.

Iron shower.
See Abstr. 003.071.

TOODY-WONDY calculations of penetration events *(meteorite craters)*.
See Abstr. 022.069.

Preparation of lunar sample and meteorite mock-ups: or how to make rocks.
See Abstr. 022.122.

Neutron-capture resonances for ^{82}Se.
See Abstr. 022.177.

Metal-silicate extraction technique for the analysis of geological and meteoritic samples.
See Abstr. 022.180.

Transmission electron microscopy on meteoritic troilite.
See Abstr. 022.182.

The application of INAA to microsamples of meteoritic metal.
See Abstr. 022.187.

High-precision instrumental neutron activation analysis of meteoritic, lunar, and terrestrial samples.
See Abstr. 022.188.

Ca-Ti-Cr anomalies in an Allende inclusion and the nβ-process.
See Abstr. 061.008.

The Kilmichael structure, Mississippi: evidence for a meteorite impact origin.
See Abstr. 081.050.

Solar-system abundances of the elements.
See Abstr. 091.047.

Noble gases in planetary atmospheres: implications for the origin and evolution of atmospheres.
See Abstr. 093.070.

Records of ancient cosmic radiation in extraterrestrial rocks.
See Abstr. 094.510.

Dark halos of the Phobos craters and shock reproduction of carbon chondrites.
See Abstr. 097.053.

Relationships between comets, large meteors, and meteorites.
See Abstr. 102.019.

X-ray emission from young stars and implications for the early solar system.
See Abstr. 107.011.

Supernovae and the origin of the solar system.
See Abstr. 107.014.

Cosmochemistry and primitive evolution of planets.
See Abstr. 107.046.

Presolar matter in meteorites.
See Abstr. 107.047.

Condensation of grains.
See Abstr. 131.305.

Cosmogenic nuclides in recently fallen meteorites: evidence for galactic cosmic ray variations during the period 1967 - 1978.
See Abstr. 143.035.

Traces of heavy and superheavy cosmic nuclei in olivines of extraterrestrial origin.
See Abstr. 143.079.

Nuclear tracks, meteorites, and the early solar system.
See Abstr. 143.081.

Erratum

105.901 **Erratum: 'Spallogenic rare gases in iron meteorites with anomalous silver'** [Earth Planet. Sci. Lett., Vol. 56, 9 - 18 (1981)].
I. M. Villa, J. C. Huneke, G. J. Wasserburg.
Earth Planet. Sci. Lett., Vol. 61, 209 (1982). – See Abstr. 30.105.113.

106 Interplanetary Matter, Interplanetary Magnetic Field, Zodiacal Light

106.001 **Geometrical relationship of flare-generated solar wind structures to the magnetic axes of bipolar sunspot groups adjacent to their originating solar flares.**
K. G. Ivanov, L. V. Evdokimova, N. V. Mikerina.
Sol. Phys., Vol. 79, 379 - 384 (1982).

Occurences of interplanetary shock waves near the Earth after the powerful isolated flares of 1957 - 1978 are investigated. The close connection between the occurrences of shock waves and the positions of magnetic axes of bipolar groups of sunspots is suggested on the basis of a statistical study.

106.002 **On interplanetary electric and magnetic fields.**
I. I. Alexeev *(Alekseev)*, A. P. Kropotkin, I. S. Veselovsky *(Veselovskij)*.
Sol. Phys., Vol. 79, 385 - 397 (1982).

A kinematic model of the stationary electromagnetic fields in interplanetary space with finite conductivity is considered. The electrodynamic problem is solved for a medium with uniform conductivity and radial plasma outflow from a spherical source. Simple analytical formulae are obtained for electric and magnetic fields, currents and charges in the case of a uniformly-magnetized rotating sphere.

106.003 **Influence of the large-scale inhomogeneities of the interplanetary plasma on the form of temporal spectra of scintillations.**
T. D. Shishova.
Astron. Zh., Tom 59, 732 - 741 (1982). In Russian. English translation in Soviet Astron., Vol. 25, No. 4.

The variations of the form of experimental temporal spectra of the interplanetary scintillations can be explained by large-scale inhomogeneities of the interplanetary plasma with a single power-law spectrum of turbulence over all the space.

106.004 **Interaction of non-perpendicular/parallel solar wind shock waves with the earth's magnetosphere.**
S. A. Grib.
Space Sci. Rev., Vol. 32, (see 012.005), 43 - 48 (1982).

The interaction of travelling interplanetary shock waves with the bow shock-magnetosphere system is considered. The author considers the general case when the interplanetary magnetic field is oblique to the Sun-planetary axis, thus, the interplanetary shock is neither parallel nor perpendicular. He finds that an ensemble of shocks are produced after the interaction for a representative range of shock Mach numbers.

106.005 **A study of some interplanetary shock wave tendencies.**
K. G. Ivanov.
Space Sci. Rev., Vol. 32, (see 012.005), 49 - 63 (1982).

The configuration, and the structure of shock waves from isolated solar flares and from a series of flares are discussed and classified.

106.006 **Structure and evolution of flare-generated shock waves in interplanetary space.**
C. d'Uston.
Space Sci. Rev., Vol. 32, (see 012.005), 99 - 114 (1982).

Observations of flare associated shock waves have led to a better understanding of the general features occurring during the interaction between the fast ejected gas and the slow ambient plasma. A description of the different regions observed in shock waves is briefly presented as well as the characteristics of the discontinuities between them. The configuration of these waves as they expand through the interplanetary medium is discussed from multi-spacecraft data as well as from IPS observations, and the results are compared with the shape and behavior predicted by theoretical models.

106.007 **Observation of flare-generated shock waves by Helios-2 near the Sun.**
P. M. Volkmer, F. M. Neubauer, R. Schwenn.
Space Sci. Rev., Vol. 32, (see 012.005), 131 - 144 (1982).

The authors present the observation of three flare generated shock waves which were measured by Helios-2 at a radial distance from the Sun of 0.476 AU, 0.444 AU, and 0.297 AU. These results show that two of the shocks fulfill the classification scheme of 'F-events' stated in the past while the third shock profile is that of a 'R-event' being interpreted as a driven wave. Because of the close distance to the Sun a flare-association was possible. Only in one case there is some evidence for a piston in the downstream region of the shock. One of the shocks is characterized by a distinct hole in the frequency distribution of directional discontinuities in the downstream region.

106.008 **Experimental study of flare-generated collisionless interplanetary shock wave propagation.**
S. Pintér.
Space Sci. Rev., Vol. 32, (see 012.005), 145 - 168 (1982).

This paper presents a review of the general properties of flare-generated collisionless interplanetary shock wave propagation, determined from multiple spacecraft plasma and magnetic field observations and by means of interplanetary scintillation of radio sources.

106.009 **Measurements of low energy electrons and ions during long-lived solar particle events.**
K. A. Anderson, R. P. Lin, D. W. Potter.
Space Sci. Rev., Vol. 32, (see 012.005), 169 - 184 (1982).

Following a solar flare in April 1979, a stream of ions and electrons appeared in interplanetary space for about 8 days. The ions follow a classic ESP pattern. Large fluxes of low energy (2–11 keV) electrons are also present throughout the event. Several distinct populations of these electrons can be identified in association with filaments of interplanetary magnetic field. The pitch angle distribution of the low energy electrons are complex and undergo many changes. Weak pitch angle scattering and adiabatic effects play a role in shaping these distributions. The low energy electron fluxes increase following the strong interplanetary shock on 5 April 1979.

106.010 **The acceleration of charged particles in interplanetary shock waves.**
M. E. Pesses, R. B. Decker, T. P. Armstrong.
Space Sci. Rev., Vol. 32, (see 012.005), 185 - 204 (1982).

The theory and observations of energetic ion acceleration in interplanetary shock waves is reviewed. The shock acceleration of the solar wind plasma and particle transport effects are discussed.

106.011 **The structure of the interplanetary medium during the first STIP interval (September - October 1975) and prerelease of energetic solar particles.**
P. Kakaletris, X. Moussas, R. Pérez-Enriquez, D. Psarros.
Astrophys. Space Sci., Vol. 84, 485 - 493 (1982).

An analysis of the interplanetary medium structure is made during STIP 1. (September - October 1975). Using a simple extrapolation method a reconstruction of the stream lines is made which shows that the interplanetary space during this time period was very quiet. The evolution of two fast solar wind streams, which dominated the interplanetary medium for very long time periods, is studied. A peculiar solar proton event is explained according to Elliot mechanism.

106.012 **Estimate of the parameters of the electron component of the interplanetary and circumplanetary plasma ac-**

cording to data of a broad-band analyzer of electrons.
M. I. Verigin, A. V. D'yachkov.
Obrab. inf., poluch. po programme Interkosmos, Moskva, 1982, p. 71 - 78. In Russian. – Abstr. in Ref. zh., 62. Issled. kosm. prostranstva, 7.62.100 (1982).

106.013 **MHD-structure of the interplanetary stream according to measurements of the plasma and the magnetic field aboard Prognoz 7 (VI STIP interval, 25 April 1979).**
G. N. Zastenker, A. N. Omel'chenko, E. G. Eroshenko, K. G. Ivanov, V. A. Styazhkin.
Kosm. Issled., Tom 20, 566 - 571 (1982). In Russian.

106.014 **Plasma wave levels and IMF orientations preceding observations of interplanetary shocks by ISEE-3.**
E. W. Greenstadt, F. L. Scarf, C. F. Kennel, E. J. Smith, R. W. Fredricks.
Geophys. Res. Lett., Vol. 9, 668 - 671 (1982).

106.015 **Correlation analysis of variations of interplanetary scintillation indices.**
V. I. Vlasov.
Geomagn. Aehron., Tom 22, 536 - 542 (1982). In Russian.

106.016 **On the discovery of boundary layers in the interplanetary magnetoplasma.**
K. G. Ivanov, Ya. I. Fel'dshtejn.
Geomagn. Aehron., Tom 22, 672 - 675 (1982). In Russian.

106.017 **Laboratory studies of interplanetary dust.**
P. Fraundorf, D. E. Brownlee, R. M. Walker.
Comets, (see 012.009), p. 383 - 409 (1982).
Samples of interplanetary dust have been successfully collected for laboratory study, from the stratosphere in the form of micrometeorites. Elemental and isotopic abundances of He, Ne, and Ar in the stratospheric particles suggest a space exposure as small particles in the inner solar system for at least tens of years. Most of the extraterrestrial particles in the stratospheric collections are fine-grained, black aggregates with major and trace element abundances similar to those of chondritic meteorites.

106.018 **Structure of the polarity-reversal-surface of the IMF.**
O. Kaburaki, T. Imai.
Magnetospheric phenomena of celestial bodies, (see 012.011), p. 39 - 42 (1982). In Japanese.
The structure of the polarity-reversal-surface of the interplanetary magnetic field (IMF) is visualized based on a phenomena-logical model, the 'wave model', proposed by Kaburaki and Yoshii (1979). It is shown that the combination of the solar centered-dipole moment and an equatorial quadrupole moment can explain the transition from a four-sectored structure near the ecliptic plane to a two-sectored one according to the inclination of the dipole axis from the rotational axis.

106.019 **The 3 January 1978 interplanetary shock event as observed by energetic particle, plasma and magnetic field devices on board of HELIOS-1, HELIOS-2 and PROGNOZ-6.**
A. K. Richter, M. I. Verigin, V. G. Kurt, V. G. Stolpovsky *(Stolpovskij),* K. I. Gringauz, E. Keppler, H. Rosenbauer, F. M. Neubauer, T. Gombosi, A. Somogyi.
J. Geophys., Vol. 50, 101 - 109 (1981). – Abstr. in Phys. Abstr., Vol. 85, Abstr. 63203 (1982).

106.020 **The perpendicular component to the solar equatorial plane of the interplanetary magnetic field and the high-latitude field of the sun.**
T. E. Val'chuk, M. A. Livshits.
Solnech. veter i magnitosfer. issled. Moskva, 1980, p. 5 - 18. In Russian. – Abstr. in Ref. zh., 51. Astron., 9.51.289 (1982).

106.021 **Ice particles in circular orbits around the sun.**
A. Parravano, I. Ferrín.
Rev. Mexicana Astron. Astrofis., Vol. 6, (see 012.017), 347 - 350 (1981).
The authors have obtained theoretical curves for the evolution of the radius of spherical particles composed of water ice, following circular or elliptical orbits around the sun. These curves were integrated numerically. The absorption efficiency was obtained using the exact Mie (1908) theory, while the temperature was calculated using the equation of energy balance. The computations are valid for spherical particles of radius from 0.2 to 100 μ, and at distances to the sun between 0.25 and 10 AU.

106.022 **Analysis of a connection of the components of the vector of the interplanetary magnetic field with geomagnetic field variations at high latitudes (equinox, 1968).**
R. G. Afonina, Yu. Z. Demidova, M. Yu. Markova, A. E. Levitin.
Soln. veter i magnitosfer. issled. Moskva, 1980, p. 54 - 71. In Russian. – Abstr. in Ref. zh., 62. Issled. kosm. prostranstva, 9.62.248 (1982).

106.023 **Parameters of the interplanetary magnetic field and short-term forecast of substorms.**
L. T. Metlyaev, N. A. Dmitrieva, O. A. Troshichev.
Magnitosfer. issled. No. 1. Moskva, 1982, p. 70 - 76. In Russian. – Abstr. in Ref. zh., 62. Issled. kosm. prostranstva, 9.62.267 (1982).

106.024 **Spectra of variations of the B_z-component of the interplanetary magnetic field in high-velocity solar wind streams and intensitites of polar magnetic disturbances.**
B. V. Zhiromskij.
Magnitosfer. issled. No. 1, Moskva, 1982, p. 29 - 31. In Russian. – Abstr. in Ref. zh., 62. Issled. kosm. prostranstva, 9.62.269 (1982).

106.025 **On the geoefficiency of the B_z-component of the interplanetary magnetic field.**
O. U. Ivanov, D. I. Ponyavin, M. I. Pudovkin, M. A. Shukhtina.
Magnitosfer. issled. No. 1, Moskva, 1982, p. 45 - 49. In Russian. – Abstr. in Ref. zh., 62. Issled. kosm. prostranstva, 9.62.272 (1982).

106.026 **Factors affecting the motion of interplanetary dust particles.**
L. Grimshaw.
Moon Planets, Vol. 26, 305 - 310 (1982).
In this paper the dynamics of individual dust particles and the effects on their motion caused by insolation and consequent evaporation are considered. Evaporation rates and the radii of dust-free zones have been computed using thermodynamic data from various sources. Some doubt is thrown on the validity of the process of matching observed thermal emission peaks with theoretical evaporation zone radii.

106.027 **Diagnostics of the interplanetary magnetic field N-S component orientation.**
A. Prigancova.
Contrib. Geophys. Inst. Slovak Acad. Sci., Vol. 12, 35 - 44 (1982). – Abstr. in Phys. Abstr., Vol. 85, Abstr. 84057 (1982).

106.028 **Interplanetary scintillation and its astrophysical significance.**
S. K. Alurkar.
Proc. Indian Natl. Sci. Acad., Part A, Vol. 47, Suppl. 2, 85 -

91 (1981). – Abstr. in Phys. Abstr., Vol. 85, Abstr. 89103 (1982).

106.029 **Interplanetary acceleration of energetic particles at 1 and 5 AU.**
X. Moussas, J. J. Quenby, J. F. Valdes Galicia.
Astrophys. Space Sci., Vol. 85, 99 - 120 (1982).
Energetic particle (0.1 to 100 MeV protons) acceleration is studied by using high resolution interplanetary magnetic field and plasma measurements at 1 AU (HEOS-2) and at ~5 AU (Pioneer 10). Energy changes of a particle population are followed by computing test particle trajectories and the energy changes through the particle interaction with the time varying magnetic field. The results show that considerable particle acceleration takes place throughout the interplanetary medium, both in the corotating interaction regions (CIR) (5 AU), and in quiet regions (1 AU), Although shocks may contribute to acceleration the authors suggest statistical acceleration within the CIRs is sufficient to explain most energetic particle observations.

106.030 **Study of parallel diffusion of energetic particles in the interplanetary medium using spacecraft data at 1 and 5 AU.**
X. Moussas, J. J. Quenby, J. F. Valdes Galicia.
Astrophys. Space Sci., Vol. 86, 185 - 195 (1982).
Energetic particle (1 - 100 MeV) pitch angle scattering in the Interplanetary Magnetic Field (IMF) is studied using spacecraft magnetometer data at 1 AU (IMP 7 and HEOS 2) and at 5 AU (Pioneer 10). Particle trajectories are followed by a computer simulation of their movement in a realistic model of the IMF. Results at 5 AU lead to a radial mean free path which is between 2 to 6 times smaller than at 1 AU, probably indicating a greater importance for perpendicular diffusion at large heliodistances.

106.031 **Drift motion and perpendicular diffusion of energetic particles in interplanetary space based on spacecraft data.**
X. Moussas, J. J. Quenby, J. F. Valdes-Galicia.
Astrophys. Space Sci., Vol. 86, 197 - 207 (1982).
A realistic model of the interplanetary magnetic field (IMF) is constructed based on measurements taken by Pioneer 10 magnetometer at 5 AU. Energetic particle (0.1 - 100 MeV) propagation in this field is studied by a computer simulation of its motion in order to calculate $K_\perp$, the perpendicular diffusion coeffecient, and the average drift velocity of an ensemble of particles. Determinations of $K_\perp$ lie in the range 3×10^{19} - 8×10^{20} cm^2 s^{-1} for the energies considered. Drift velocity calculations are very close to predictions of guiding centre theory (within 30%).

106.032 **Large-scale variations of the interplanetary magnetic field: Voyager 1 and 2 observations between 1-5 AU.**
L. F. Burlaga, R. P. Lepping, K. W. Behannon, L. W. Klein, F. M. Neubauer.
J. Geophys. Res., Vol. 87, 4345 - 4353 (1982).
Observations by the Voyager 1 and 2 spacecraft of the interplanetary magnetic field between 1 and 5 AU have been used to investigate the large-scale structure of the IMF in the years 1977 to 1979, a period of increasing solar activity. This complements the Pioneer 10, 11 investigation between 1 and 8.5 AU during 1972–1976 when the sun was less active. In contrast to the good agreement of the Pioneer observations with the ideal field configuration of the Parker spiral model during near solar minimum conditions, the Voyager spacecraft found notable deviations from that configuration. The authors attribute these deviations both to temporal variations associated with increasing solar activity, and to the effects of fluctuations of the field in the radial direction.

106.033 **Plasma shocks and energetic particles in the outer solar system: trapping and asymmetry observations from Pioneer 10 and Pioneer 11.**
D. S. Intriligator, W. D. Miller.
J. Geophys. Res., Vol. 87, 4354 - 4364 (1982).
Energetic protons (0.5–20 MeV) appear to be trapped between a pair of shocks associated with the major solar flares of April 15 and April 28, 1978. Prolonged trapping (for a period of weeks) is implied by the large count rate enhancement and the large range of radial distances and longitudinal angles over which the trapping is observed. Shocks associated with both flares are detectable in the Pioneer 10 and Pioneer 11 plasma analyzer data. These shock/flare associations are different from those previously published by others studying the interplanetary events. The combination of the observations provides evidence for unexpectedly complex interactions in the outer solar system between energetic particles and solar wind plasma.

106.034 **An analysis of shock wave disturbances observed at 1 AU from 1971 through 1978.**
G. Borrini, J. T. Gosling, S. J. Bame, W. C. Feldman.
J. Geophys. Res., Vol. 87, 4365 - 4373 (1982).
An analysis has been made of all the interplanetary shock wave disturbances detected with Los Alamos plasma instruments aboard IMP 6, 7, and 8 from 1971 through 1978.

106.035 **Energetic interplanetary nucleon flux anisotropies: the effect of earth's bow shock and magnetosheath on sunward flow.**
S. P. Christon.
J. Geophys. Res., Vol. 87, 5045 - 5062 (1982).

106.036 **Effect of the northward interplanetary magnetic field on the geomagnetic field inside the region surrounded with the auroral oval.**
N. Iwasaki.
J. Geomagn. Geoelectr., Vol. 34, 9 - 25 (1982). – Abstr. in Phys. Abstr., Vol. 85, Abstr. 112293 (1982).

106.037 **Active experiments in space, laboratory experiments and numerical simulation.**
I. M. Podgorny *(Podgornyj)*,
J. Geomagn. Geoelectr., Vol. 34, 53 - 92 (1982). – Abstr. in Phys. Abstr., Vol. 85, Abstr. 112294 (1982).

106.038 **Investigation of electron streams with E_e = 40 - 500 keV in quiet periods of solar activity with methods of the correlation and spectral analysis.**
M. A. Zel'dovich, G. A. Trebukhovskaya.
Kosm. Issled., Tom 20, 700 - 704 (1982). In Russian.

106.039 **Diffusion of Keplerian motions by a stochastic force. II. Lorentz scattering of interplanetary dusts.**
P. Barge, R. Pellat, J. Millet.
Astron. Astrophys., Vol. 115, 8 - 19 (1982).
Charged dust grains orbiting around the Sun at heliocentric latitudes less than 15° are submitted to a random Lorentz force due to recurrent sectors of the interplanetary magnetic field and high-speed-streams of the solar wind. Using a formalism previously developed (Barge et al., 1982) the problem is reinvestigated and leads to a detailed description of the scattering process. The changes in the orbital elements are estimated for a time scale of 10 yr; they are found to be less important than believed until now and to vary rapidly with eccentricity and semi-major-axis. Only the very small dielectric dust grains, with size less than 0.5 μm, may be scattered by the Lorentz force fluctuations; the main diffusion occurs in inclination and near the Sun (20–60 $R_\odot$).

106.040 **A scattering model for the zodiacal light particles.**
R. Schiffer, K. O. Thielheim.

Astron. Astrophys., Vol. 116, 1 - 9 (1982).

Light scattering by a cloud of large irregular randomly oriented particles is considered. The authors calculate the intensity and the polarization of the scattered light in the case of large scale surface roughness, i.e. the roughness structure is assumed to vary slowly on the scale of the incident wavelength. The resulting curves are compared with zodiacal light measurements carried out recently, and it is shown that large ($>10\,\mu$) slightly absorbing particles with a very rough surface can satisfactorily reproduce the empirical scattering functions of the interplanetary dust particles.

106.041 **Magnetic clouds: Voyager observations between 2 and 4 AU.**
L. F. Burlaga, K. W. Behannon.
Sol. Phys., Vol. 81, 181 - 192 (1982).

Magnetic clouds were observed in the solar wind between 2–4 AU by Voyagers 1 and 2, indicating that they are stable enough to persist without major changes out to such distances. The magnetic field configuration cannot be determined uniquely, but it is highly ordered and consistent with the passage of some kind of loop.

106.042 **Interplanetary dust: sources, sinks and dynamics.**
H. Fechtig.
The solar system and its exploration, (see 012.054), p. 203 - 210 (1981).

Two major sources, lunar microcraters and in situ measurements, provide information on the dynamics of the interplanetary dust in the solar system.

106.043 **Exploring the night sky by photography from the Saliout orbital station.**
S. Koutchmy, G. M. Nikolski (*Nikol'skij*).
Pré-publ. Inst. Astrophys. Paris, No. 15, 17 pp. (1982).
Accepted by Sky Telesc.

106.044 **The radial evolution of the IMF fluctuations: a comparison with theoretical models.**
U. Villante, M. Vellante.
Sol. Phys., Vol. 81, 367 - 374 (1982).

The authors conducted an analysis of the radial evolution of the wave phenomena observed in the inner solar system and associated with the high velocity solar wind stream pattern.

106.045 **Ultraviolet spectroscopy of the zodiacal light.**
R. P. Cebula, P. D. Feldman.
Astrophys. J., Vol. 263, 987 - 992 (1982).

The zodiacal light was observed by a rocket-borne ultraviolet spectrometer at 25 Å resolution in the spectral range 1700–3150 Å on 1979 September 24. Between 1900 and 2500 Å airglow emission due to NO and O II, identified spectroscopically and by its altitude variation, is significant. The authors report a zodiacal light color that is slightly reddened with respect to a solar-type spectrum for the wavelength range 2600–3150 Å. At 1800 Å an upper limit on the zodiacal light brightness of 2.2×10^{-2} R Å^{-1} is reported. Although this value is approximately 4 times higher than that expected from the solar flux at this wavelength, it cannot be considered a true indication of a brightening of the zodiacal light in the far-ultraviolet.

106.046 **Jovian modulation of interplanetary electrons as observed with Voyagers 1 and 2.**
A. W. Schardt, F. B. McDonald, J. H. Trainor.
NASA Tech. Memo., NASA TM-83977, 50 pp. (1982).

The release of magnetospheric electrons from Jupiter into interplanetary space is modulated by the Jovian rotation period. The Voyager 1 and 2 observations showed that the modulation period agrees on the average with the synodic period of Jupiter, but over intervals of weeks it can differ from the synodic period by several minutes. The lack of exact synchronization is attributed to changes of the plasma population in the Jovian magnetosphere. The Jovian modulation appears to be a persistent feature of the interaction between the solar wind and the magnetosphere and the disappearance of the modulation away from Jupiter is attributed to interplanetary propagation conditions.

106.047 **Frictional mechanism for particle acceleration under conditions of the interplanetary space.**
E. G. Berezhko.
Pis'ma Astron. Zh., Tom 8, 747 - 750 (1982). In Russian. English translation in Soviet Astron. Lett., Vol. 8.

An acceleration of charged particles in the region of solar wind high-speed streams is investigated. Spatial distribution and energy spectrum of accelerated particles are obtained and their intensity is estimated. The obtained results are in a good agreement with the available experimental data.

106.048 **Energetic protons accelerated at corotating shocks: Pioneer 10 and 11 observations from 1 to 6 AU.**
B. T. Tsurutani, E. J. Smith, K. R. Pyle, J. A. Simpson.
J. Geophys. Res., Vol. 87, 7389 - 7404 (1982).

The Pioneer 10 and 11 Jet Propulsion Laboratory vector helium magnetometer and the University of Chicago 0.5- to 1.8-MeV proton telescope data are used to examine the relationship between low-energy proton increases and corotating interaction regions (CIR's) in the heliosphere between 1 and 6.3 AU. The 0.5- to 1.8-MeV proton flux enhancements are correlated with the forward and reverse shocks bounding the CIR's. A schematic figure, which incorporates many of the features deduced in the report, is given to illustrate the relationship between the energetic protons, the forward and reverse shocks, and the interplanetary magnetic field structure. The predictions of various theories and mechanisms for interplanetary nucleon acceleration are discussed in light of the experimental results presented.

106.049 **Can solar wind induce the luminiscence emission in zodiacal grains?**
J. Svatoš.
Říše hvězd, Vol. 63, 157 - 158 (1982). In Czech.

17th International Cosmic Ray Conference. SH sessions: Sun and heliosphere.
See Abstr. 012.070.

Fast ion bombardment of ices and its astrophysical implications.
See Abstr. 022.141.

Study of the solar wind with the interplanetary scintillation at UHF.
See Abstr. 033.055.

Time synchronization for a 3-site interplanetary scintillation (IPS) experiment using ATA time signal transmission.
See Abstr. 035.005.

Magnetic field measurements onboard Venera 13, Venera 14 and Venera-Halley.
See Abstr. 051.057.

Shock acceleration of energetic particles in the heliosphere.
See Abstr. 062.016.

Numerical simulation of the weak turbulence excited by a beam of electrons in the interplanetary plasma.
See Abstr. 062.092.

A self-similar flow in generalized Roche model with increasing energy.
See Abstr. 062.117.

Propagation of low-energy electrons depending on the position of a flare relative to the interplanetary current layer.
See Abstr. 073.024.

Multi-station observations of interplanetary scintillations.
See Abstr. 074.045.

An indication for an intrinsically different coronal temperature.
See Abstr. 074.050.

On the screening of the interstellar electric and magnetic fields by the solar wind.
See Abstr. 074.051.

A three-dimensional model of corotating streams in the solar wind. 3. Magnetohydrodynamic streams.
See Abstr. 074.052.

On the complex state of the interplanetary medium of July 28 - 29, 1977.
See Abstr. 074.054.

Whistler mode turbulence in the disturbed solar wind.
See Abstr. 074.056.

Stationarity of magnetohydrodynamic fluctuations in the solar wind.
See Abstr. 074.095.

Some properties of the solar wind plasma at extreme solar activities.
See Abstr. 074.096.

Structure of the heliospheric current sheet in the early portion of sunspot cycle 21.
See Abstr. 075.022.

Quadrupole distortions of the heliospheric current sheet in 1976 and 1977.
See Abstr. 075.023.

Gamma-ray lines, interplanetary neutrons, and the shock acceleration of ions in solar flares.
See Abstr. 076.004.

A model of propagation of solar flare particles and its application to cosmic ray ground level events.
See Abstr. 078.007.

Decimeter radio wave propagation in the turbulent plasma near the Sun, using Venera 10 spacecraft.
See Abstr. 080.064.

Polar magnetic disturbances and field-aligned currents.
See Abstr. 084.002.

Magnetic field and field-aligned current variations in the polar cusp controlled by interplanetary medium parameters.
See Abstr. 084.005.

The *aa* geomagnetic activity index and parameters of the interplanetary medium.
See Abstr. 084.021.

Distribution of polar aurorae and large-scale currents in the magnetosphere at different orientations of the interplanetary magnetic field.
See Abstr. 084.042.

About the connection between longer period geomagnetic variation and the interplanetary magnetic field.
See Abstr. 084.046.

Geomagnetic storm variations and the average characteristics of the interplanetary magnetic field.
See Abstr. 084.047.

Dependence of the magnetic activity on the orientation of the geomagnetic dipole relative to the interplanetary magnetic field.
See Abstr. 084.059.

Computer simulation of the interaction of the solar wind with the earth's magnetosphere.
See Abstr. 085.010.

Poynting-Robertson drag and orbital resonance.
See Abstr. 091.055.

Dawn–dusk asymmetry of the tail region of the magnetosphere of Saturn and the interplanetary magnetic field.
See Abstr. 100.072.

Modifikation der interstellaren Gasparameter im Randgebiet des Sonnensystems.
See Abstr. 131.285.

Interplanetary scintillation (IPS) observations of 23 radio sources with arcsecond structure.
See Abstr. 141.117.

The effect of corotating interaction regions on the propagation of relativistic cosmic rays in the heliosphere.
See Abstr. 143.029.

Unusual characteristics of the cosmic ray intensity increase of September 17 - 18, 1979.
See Abstr. 143.068.

Cosmic ray north-south anisotropy: the role of the interplanetary magnetic field.
See Abstr. 143.069.

Cosmic-ray 13 day oscillation and two-sector interplanetary magnetic field.
See Abstr. 143.082.

107 Cosmogony

107.001 **Solid particles at an early stage of evolution of a protoplanetary cloud.**
V. S. Kessel'man.
Astron. Zh., Tom 59, 810 - 816 (1982). In Russian. English translation in Soviet Astron., Vol. 25, No. 4.

A calculation of the sizes of solid particles grown up during the coagulation process in a protoplanetary nebula is carried out. The influence of the change of the particles' mean sizes on the process of formation of massive solid bodies in a protoplanetary nebula is considered.

107.002 **Evolution of giant gaseous protoplanets embedded in the primitive solar nebula.**
A. G. W. Cameron, W. M. DeCampli, P. Bodenheimer.
Icarus, Vol. 49, 298 - 312 (1982).

Evolutionary calculations are presented, in a spherically symmetric approximation, for a protoplanet of 1 Jovian mass with homogeneous solar composition during the early phase of quasi-static contraction prior to the dissociation of molecular hydrogen.

107.003 **Orbital resonances and planetary formation sites.**
M. Torbett, R. Greenberg, R. Smoluchowski.
Icarus, Vol. 49, 313 - 326 (1982).

Orbital resonances may have played an important role in determining the locations where the planetesimal swarm eventually accreted into full-size planets. The mass loss and redistribution that occurred in the early solar system evolution can substantially alter the positions of planets and planetary embryos within the swarm. A cascaded resonance structure is hypothesized where planetesimal growth was accelerated at 2 : 1 interior and 1 : 2 exterior resonances with an early-formed Jupiter producing runaway growth of planetary embryos. These embryos produce their own resonances which, in turn, lead to additional embryos in a process that successively propagates inward and outward to generate a resonant configuration of embryos. In this manner, the early presence of Jupiter imposed a harmonic structure on the accumulating planetesimal swarm.

107.004 **Growth of planetesimals in a gaseous ring.**
K. Hourigan.
Proc. Astron. Soc. Australia, Vol. 4, 226 - 227 (1981).

107.005 **Where has that Moon been?**
K. Lambeck.
Nature, Vol. 298, 704 - 705 (1982).

107.006 **Formation of the giant planets.**
D. J. Stevenson.
Planet. Space Sci., Vol. 30, 755 - 764 (1982).

Observational constraints on interior models of the giant planets indicate that these planets were all much hotter when they formed and they all have rock and/or ice cores of ten to thirty earth masses. These cores are probably soluble in the envelopes above, especially in Jupiter and Saturn, and are therefore likely to be primordial. They persist despite the continual upward mixing by thermally driven convection throughout the age of the solar system, because of the inefficiency of double-diffusive convection. Thus, these planets most probably formed by the hydrodynamic collapse of a gaseous envelope onto a core rather than by direct instability of the gaseous solar nebula. Problems remain however, and no current model is entirely satisfactory in explaining the properties of the giant planets and simultaneously satisfying the terrestrial planet constraints. Satellite systematics and protoplanetary disk nebulae are also discussed and related to formation conditions.

107.007 **Gas capture and rare gas retention by accreting planets in the solar nebula.**
H. Mizuno, K. Nakazawa, C. Hayashi.
Planet. Space Sci., Vol. 30, 765 - 771 (1982).

The physico-chemical effects of the nebula gas on the planets are reviewed from a standpoint of planetary formation in the solar nebula.

107.008 **Dynamical history of the Oort Cloud.**
P. R. Weissman.
Comets, (see 012.009), p. 637 - 658 (1982).

The hypothesis of a cloud of comets surrounding the solar system and extending out to interstellar distances as suggested by Oort has been successful in explaining the observed distribution of orbits of the long-period comets. Numerical studies employing Monte Carlo techniques have expanded the understanding of the dynamics of comets in the Oort Cloud. The population of the cloud has been depleted over the history of the solar system. The numerical simulations are used to find the distributions of perihelia, aphelia, and energy for comets in the present Oort Cloud, and to comment on the population of interstellar comets.

107.009 **The evolution of an impact-generated atmosphere.**
M. A. Lange, T. J. Ahrens.
Icarus, Vol. 51, 96 - 120 (1982).

Shock wave and thermodynamic data for rock-forming and volatile-bearing minerals are used to determine minimum impact velocities and minimum impact pressures required to form a primary H_2O atmosphere during planetary accretion from chondritelike planetesimals. The escape of initially released water from an accreting planet is controlled by the dehydration efficiency. Attenuation of impact-induced shock pressure is modeled. The two primary processes considered are the impact release of water bound in hydrous minerals and the subsequent reincorporation of free water by hydration of forsterite and enstatite. These processes are described in terms of model calculations for the accretion of the earth. The concept of dehydration efficiency is of dominant importance in determining the degree to which an accreting planet acquires an atmosphere during its formation.

107.010 **Numerical experiment applicable to the latest stage of planet growth.**
A. Cazenave, B. Lago, K. Dominh.
Icarus, Vol. 51, 133 - 148 (1982).

A three-dimensional numerical model was developed with the goal of studying limited dynamical problems relevant to the latest stage of planet growth in the accretion theory. A small number of large protoplanets (~Moon size) of different masses, moving around the Sun, is considered. The dynamical evolution and growth of the population is studied under mutual gravitational perturbations, accretion, and collisional fragmentation processes. Gravitational encounters are treated exactly by numerical integration of the N-body problem.

107.011 **X-ray emission from young stars and implications for the early solar system.**
E. D. Feigelson.
Icarus, Vol. 51, 155 - 163 (1982).

Recent observations of soft X-ray emission from solar-type stars obtained with the Einstein X-Ray Observatory indicate that X-ray luminosity is inversely correlated with stellar age. If this result is applied to the Sun and if X-ray emission is a valid indicator of other manifestations of solar activity, then past solar wind and flare levels can be inferred. It can qualitatively explain the excess xenon and nitrogen found in the lunar regolith. The proton flux from solar flares during the 10^6 to 10^7-year pre-main-sequence phase would be sufficient to account for the ^{26}Al anomaly in meteorites.

107.012 **Physico-chemical approach to the problem of origin of the solar system.**
K. G. Miesserov.
Inst. neftekhim. sinteza AN SSSR. Moskva, 1982. 26 pp. In Russian. – Abstr. in Ref. zh., 51. Astron., 9.51.135 (1982).

107.013 **Heterogeneous growth of meteorites and planets, especially the Earth and Moon.**
J. V. Smith.
J. Geol., Vol. 90, 1 - 48 (1982). – Abstr. in Phys. Abstr., Vol. 85, Abstr. 73865 (1982).

107.014 **Supernovae and the origin of the solar system.**
D. D. Clayton.
Supernovae: a survey of current research, (see 012.022), p. 535 - 564 (1982).

The author reviews observational evidence for the hypothesis that nearby supernovae might have been linked with the formation of the solar system. The possible role of supernovae in explaining the origin of isotopic anomalies of ^{26}Al, ^{26}Mg and titanium found in carbonaceous meteorites is discussed in detail.

107.015 **The origin of the solar system.**
H. Alfvén.
Evolution in the Universe, (see 012.024), p. 31 - 49 (1982).

The first systematic attempt to base a cosmogonic theory on electromagnetic and hydromagnetic effects was done 1942 (Alfvén). It led to the discovery of two important observational regularities in the solar system: (1) the band structure, and (2) the cosmogonic shadow effect (the two-thirds fall-down effect). Both have been discussed in detail in Alfvén and Arrhenius (1975 and 1976). Further, the band structure has more recently been discussed by Alfvén (1981). The second observational regularity, the cosmogonic shadow effect,is the subject of this paper.

107.016 **Den unga jorden.**
K. Rynefors.
Astron. Tidsskr., Årg. 15, 100 - 102 (1982).

107.017 **Chemistry of the early solar system.**
W. V. Boynton.
Bull. American Astron. Soc., Vol. 14, 716 (1982). – Abstract.

107.018 **The chemical composition of a possible proto-solar nebula.**
W. M. Irvine.
Bull. American Astron. Soc., Vol. 14, 716 (1982). – Abstract.

107.019 **Tidal barriers and the viscous evolution of the solar nebula.**
K. Hourigan, W. R. Ward.
Bull. American Astron. Soc., Vol. 14, 716 - 717 (1982). Abstract.

107.020 **Nonisothermal, axisymmetric models of the presolar nebula.**
A. P. Boss.
Bull. American Astron. Soc., Vol. 14, 717 (1982). – Abstract.

107.021 **Initial sizes of planetesimals.**
S. J. Weidenschilling, R. Greenberg.
Bull. American Astron. Soc., Vol. 14, 717 (1982). – Abstract.

107.022 **Formation of comets among the outer planets.**
R. Greenberg, S. J. Weidenschilling, C. R. Chapman, D. R. Davis, W. K. Hartmann.
Bull. American Astron. Soc., Vol. 14, 717 (1982). – Abstract.

107.023 **The explanation of Bode's Law and its applicability.**
J. B. Davies.
Bull. American Astron. Soc., Vol. 14, 717 - 718 (1982). Abstract.

107.024 **Did Jupiter once have a giant primordial moon?**
J. Boynton.
Bull. American Astron. Soc., Vol. 14, 718 (1982). – Abstract.

107.025 **Models of the formation of the solar nebula.**
P. M. Cassen, A. L. Summers.
Bull. American Astron. Soc., Vol. 14, 718 (1982). – Abstract.

107.026 **The gaseous flow around a protoplanet in the primordial solar nebula.**
S. Miki.
Prog. Theor. Phys., Vol. 67, 1053 - 1067 (1982). – Abstr. in Phys. Abstr., Vol. 85, Abstr. 100774 (1982).

107.027 **Origin of the solar system: planetary growth in the gaseous nebula.**
K. Nakazawa, Y. Nakagawa.
Prog. Theor. Phys. Suppl., No. 70, p. 11 - 34 (1981). – Abstr. in Phys. Abstr., Vol. 85, Abstr. 100775 (1982).

107.028 **Structure of the solar nebula, growth and decay of magnetic fields and effects of magnetic and turbulent viscosities on the nebula.**
C. Hayashi.
Prog. Theor. Phys. Suppl., No. 70, p. 35 - 53 (1981). – Abstr. in Phys. Abstr., Vol. 85, Abstr. 100776 (1982).

107.029 **On the evolution of convective accretion disk models of the primordial solar nebula.**
D. N. C. Lin, P. Bodenheimer.
Astrophys. J., Vol. 262, 768 - 779 (1982) = Lick Obs. Bull., No. 926.

The evolution of the primordial solar nebula under the action of convectively driven turbulent viscous stress is studied numerically under various physical conditions. The mass of the nebula is assumed to be small compared with that of the central Sun, and the nebula is assumed to be in pressure equilibrium in the vertical direction. It is found that the central regions of the nebula always evolve into a quasi-stationary state such that the mass influx is independent of radius.

107.030 **Formation of solar system as a result of evolution of protostar of second or subsequent generation.**
K. G. Miesserov.
Moon Planets, Vol. 27, 13 - 25 (1982).

The formation of the planets of the solar system is represented in the present paper as a regular process of the evolution of matter. Since the main role in this process is ascribed to the surface tension of heavy metals (predominantly of iron), and the latter are formed in the course of thermonuclear reactions proceeding in the interior of stars, the formation of planetary systems is principally possible only as a result of evolution of second generation protostars and of protostars of subsequent generations, i.e., of such protostars which have been formed by way of condensation and subsequent contraction of interstellar matter, enriched with the products of evolution of stars of the first and next generations.

107.031 **The plasma phase of the evolution of the solar system.**
H. Alfvén.
The solar system and its exploration, (see 012.054), p. 3 - 10 (1981).

A new unified plasma theory, valid both for laboratory and space plasma is emerging. The new paradigm to the evolution of the solar system shows that plasma dominated processes were essential for the early phases, viz. the evolution of interstellar clouds and later the evolution of the circumsolar cloud up to the transition from the plasma to the planetesimal phase. Later phases are not affected by plasma phenomena.

107.032 **Nuclear origin of the solar system.**
M. Cassé.
The solar system and its exploration, (see 012.054), p. 11 - 12 (1981). – Abstract.

107.033 **The early evolution of planets.**
H. Alfvén.
The solar system and its exploration, (see 012.054), p. 65 - 73 (1981).

For a study of the evolution of the solar system small planets are in many respects just as important as large ones. Hence the more than 2000 known minor planets deserve more interest than the 9 "ordinary" planets. If analysed in an appropriate way the asteroidal data shows that the belt has a very regular structure, which reveals the same signature of the plasma – planetesimal transition as the Saturnian rings.

107.034 **Formation of planets.**
E. L. Ruskol.
The solar system and its exploration, (see 012.054), p. 107 - 113 (1981).

The evolution of a protoplanetary gas-dust cloud of mass 0.05 - 0.1 M$_\odot$, laminarly rotating around the sun in the boundaries of the present planetary system, results in the formation of solid planetesimals grouped in concentric zones of growing embryo-planets. The growth of terrestrial planets takes about 10^8 yr, several times longer than the period of dissipation of gas from the inner parts of the protoplanetary cloud. The formation of giant planets is a two-stage process: the formation of nuclei from the solid material with the mass of 2 - 3 Earth's masses in 10^7 - 10^8 yr and then the gaseous accretion on nuclei in 10^5 - 10^6 yr. The origin of satellites is a process accompanying the growth of planets on the most active stage, via circumplanetary swarms for terrestrial planets and accretional discs for giants.

107.035 **Heat transfer and the development of planetary structure.**
D. C. Tozer.
The solar system and its exploration, (see 012.054), p. 121 - 126 (1981).

The planetary accretion process is mainly reexamined as a possible explanation of early lunar crust and highland formation. From both observational evidence and theoretical argument it is concluded that the Benfield type models of the process widely quoted in the lunar science literature are unsatisfactory, the fault being traceable to an assumption that impact energy is immediately converted to heat at a growing planet's surface. A model is suggested which has a higher accretional energy trapping efficiency, and which is therefore able to avoid any appeal to short ($\leq 10^4$ years) accretion times and/or short period radiogenic heating.

107.036 **Chemistry and accretion history of the Earth and the inner planets.**
H. Wänke.
The solar system and its exploration, (see 012.054), p. 141 - 150 (1981).

The Earth, the Moon and the eucrite parent asteroid are depleted in moderately volatile and volatile elements compared to C1 chondrites. The Earth's mantle and the Moon are slightly enriched in refractory lithophile elements and in magnesium. It might be that this enrichment is fictitious and only due to the normalization to Si and that in fact, both Earth's mantle and Moon are depleted in Si which partly entered the Earth's core in metallic form. The depletion of V, Cr, and Mn in the Earth's mantle is explained by their partitial removal into the core. The similar depletion of these elements on the Moon indicates the strong genetic relationship of Earth and Moon.

107.037 **L'origine du système solaire.**
A. H. Delsemme.
Le Ciel, Vol. 44, 216 - 222 (1982).

107.038 **Giant molecular clouds and the solar system comets.**
S. van den Bergh.
J. R. Astron. Soc. Canada, Vol. 76, 303 - 308 (1982).

Tidal forces by passing giant molecular clouds will decimate the Oort comet cloud. This suggests that the Oort cloud has been regenerated by gravitational perturbations that have ejected comets from a reservoir of objects with semi-major axes having $a < 2 \times 10^4$ AU. It is shown that the Oort cloud cannot be replenished by capture of interstellar cometesimals.

107.039 **Ultimate strength of solids and formation of planetary cores.**
G. F. Davies.
Geophys. Res. Lett., Vol. 9, 1267 - 1270 (1982).

107.040 **Diskussionsforum: Ursprung des Sonnensystems.**
H. J. Völk (Editor), with contributions by
F. Begemann, H. J. Fahr, G. Morfill, T. V. Ruzmaikina *(Ruzmajkina)*, V. S. Safronov, H. J. Schober, W. M. Tscharnuter.
Mitt. Astron. Ges., Nr. 57, p. 45 - 63 (1982).

107.041 **Torus as a foregoing state of the solar system.**
I. L. Genkin.
Astron. Tsirk., No. 1203, p. 4 - 5 (1982). In Russian.

107.042 **Present state of the theory of origin of the earth.**
V. S. Safronov.
Izv. AN SSSR. Fiz. Zemli, 1982, No. 6, p. 5 - 24. In Russian. Abstr. in Ref. zh., 51. Astron., 10.51.183 (1982).

107.043 **Fractionation of matter during the formation and evolution of the earth.**
A. V. Vityazev.
Izv. AN SSSR. Fiz. Zemli, 1982, No. 6, p. 52 - 68. In Russian. Abstr. in Ref. zh., 51. Astron., 10.51.184 (1982).

107.044 **Problem of the evolution of the hydrosphere and expansion of the earth.**
V. F. Blinov.
Byull. Mosk. o-va ispyt. prirody. Otd. geol., Tom 57, No. 4, p. 17 - 29 (1982). In Russian. – Abstr. in Ref. zh., 51. Astron., 10.51.186 (1982).

107.045 **A few relevant facts on the problem of the origin of the solar system.**
H. Reeves.
Formation of planetary systems, (see 012.064), p. 175 - 187 (1982).

Very few things are known with a reasonable degree of certainty about the origin of the solar system. It is not because of the lack of measured quantities but rather because of the difficulty in extracting the appropriate – and useful – information. The author discusses a few facts concerning formation of the solar system which are in relation with the atomic composition and physicochemistry of the solar system.

107.046 **Cosmochemistry and primitive evolution of planets.**
C. J. Allegre.

Formation of planetary systems, (see 012.064), p. 283 - 364 (1982).

Contents: Chronologies of the solar system. What is a meteorite? Why are they important objects for studying the formation of the solar system? An elementary introduction. Chemical theories concerning the formation of the solar system: an elementary review of present day ideas.

107.047 **Presolar matter in meteorites.**
J. P. Bibring, Y. Langevin.
Formation of planetary systems, (see 012.064), p. 385 - 399 (1982).

In this paper the author briefly presents and discusses the major experimental steps which have drastically modified the models accounting for the early history of the solar system. Tentative conclusions concerning the existence of unaltered presolar grains in meteorites are proposed.

107.048 **Formation of planetesimals from dust: simple considerations.**
A. Brahic.
Formation of planetary systems, (see 012.064), p. 503 - 514 (1982).

In this paper the author shows that simple arguments can give some hints on the initial conditions in the solar nebula. From the mass of the actual planets as the only data, he tries to get some ideas on time scales and processes. Computations are in order of magnitude only, numerical multipliers of order 1 are generally omitted. Many important effects are omitted. The author wants to get some ideas on the properties of the protoplanetary disc and to find out what could be a good starting point.

107.049 **Planetesimals to planets.**
R. Greenberg.
Formation of planetary systems, (see 012.064), p. 515 - 569 (1982).

Contents: Introduction and background. Collisional processes. Analytical models of planet growth. A numerical simulation of planet growth: nominal model. Numerical results: implications, modifications. Resonances and planet growth. Regoliths on planetesimals.

107.050 **Late stage of planet growth in accretion theory.**
A. Cazenave, B. Lago.
Formation of planetary systems, (see 012.064), p. 571 - 582 (1982).

The authors summarize the main aspects of two characteristic models of planet growth and report their computer program modelling the very final stage of the planet growth.

107.051 **The origin of the Moon: a reappraisal.**
Y. Langevin, J. P. Bibring.
Formation of planetary systems, (see 012.064), p. 783 - 796 (1982).

In this paper the authors restrict themselves to a description of the results most relevant to the problems of the formation and early evolution of the Moon: internal structure, bulk chemical composition and chronology of the major events. They present the major models proposed for the origin of the Earth-Moon system: fission, capture, and accretion in orbit. These are confronted with the new available experimental data, so as to assess the present status of the problem. Finally, the authors outline studies further needed to discriminate unambiguously between the various models.

107.052 **Hydrodynamical models of presolar nebula formation.**
A. P. Boss.
Icarus, Vol. 51, 623 - 632 (1982).

Numerical three-dimensional calculations are presented which demonstrate the possibility of formation of a co-orbital triple protostellar system, which is unstable to decay to a binary and an ejected single star. The calculations are used to construct a plausible scenario for presolar nebula formation based on a hierarchy of collapse and fragmentation.

107.053 **Nucleation and condensation in the primitive solar nebula.**
A. G. W. Cameron, M. B. Fegley.
Icarus, Vol. 52, 1 - 13 (1982).

Using the accretion disk physics of Lynden-Bell and Pringle, the authors construct a section of a model of the primitive solar nebula, subject to the assumption that the optically thick part of the disk is convective and isentropic and that there is a steady mass flow toward the axis of the disk at the rate of 10^{-5} solar masses per year. The portion of the disk is studied, which includes the interface between the region where high-temperature solid condensates can exist and where they will be totally evaporated.

107.054 **Blow-off of planetary protoatmospheres and of the protoplanetary nebula.**
G. P. Horedt.
Phys. Earth Planet. Inter., Vol. 29, 252 - 260 (1982).

The mass loss from planetary protoatmospheres is in the order of 10^{-2}–10^{-4} parts of the planetary mass for the terrestrial planets and about 10^{-7}–10^{-8} parts for the giant planets. The minimum time intervals needed for the disappearance of the protoplanetary nebula are in the order of 10^6 y in the case of heating by a T-Tauri-like solar wind, 10^8 y for heating by far UV radiation from the Sun and 10^9 y for heating by extreme UV radiation.

107.055 **Accretion of the jet stream and the formation of asteroids.**
Z. Hu, Y. Tong.
Acta Astrophys. Sinica, Vol. 2, 308 - 314 (1982).
In Chinese.

The cosmic serpent. A catastrophist view of earth history.
See Abstr. 003.032.

Introduction to planetary geology.
See Abstr. 003.043.

The cosmogony of the solar system.
See Abstr. 003.062.

Theories of the origin of the solar system: some historical remarks.
See Abstr. 004.073.

Nuclear cosmochronology or the age of the elements.
See Abstr. 061.050.

Containment of a diffuse ionized mass orbiting around a magnetized central body.
See Abstr. 062.103.

The initial stage of the earth's evolution. Geochemistry.
See Abstr. 081.041.

Loi de Titius-Bode et formalisme ondulatoire.
See Abstr. 091.020.

The viscosity of rock-ice mixtures: applications to the evolution of icy satellites.
See Abstr. 091.031.

The origin of planetary satellites.
See Abstr. 091.051.

The dynamical evolution of the solar system.
See Abstr. 091.053.

A brief survey of the solar system.
See Abstr. 091.054.

Nature and origin of basin-forming projectiles.
See Abstr. 094.002.

On the origin of the planetesimals that produced the multi-ring basins.
See Abstr. 094.003.

Could the earth's core and moon have formed at the same time?
See Abstr. 094.031.

Formation of the Galilean satellites in a gaseous nebula.
See Abstr. 099.118.

Voyager and the origin of the Saturnian system.
See Abstr. 100.008.

On the origin of Triton.
See Abstr. 101.030.

The geochemical coherence of Pu and Nd and the $^{244}Pu/^{238}U$ ratio of the early solar system.
See Abstr. 105.054.

Some short-lived nuclides in the early solar system – a connection with the placental ISM.
See Abstr. 105.077.

The nature and origin of type B1 and B2 Ca-Al-rich inclusions in the Allende meteorite.
See Abstr. 105.092.

Evolution of Ca-Al-rich bodies in the earliest solar system: growth by incorporation.
See Abstr. 105.093.

Meteorites–evidence for the interrelationships of materials in the solar system of 4.55 Ga ago.
See Abstr. 105.100.

Planet(oid) core crystallisation and fractionation–evidence from the Agpalilik mass of the Cape York iron meteorite shower.
See Abstr. 105.101.

Organic compounds in meteorites and their origins.
See Abstr. 105.104.

A search for fragmentation debris near Ursa Major Stream stars.
See Abstr. 118.006.

Gravitational instabilities in protostellar nebulae.
See Abstr. 131.205.

From clouds to stars.
See Abstr. 131.304.

Condensation of grains.
See Abstr. 131.305.

The cosmic ray track record of events in the early solar system.
See Abstr. 143.025.

Steady wave model of spiral galaxies and its application in cosmogony.
See Abstr. 151.059.

Cosmic cooking, evolution of stars and our Metagalaxy.
See Abstr. 162.255.

Erratum

107.901 **Erratum: "Another derivation of the Titius-Bode law"** [Astron. Nachr., Band 302, 93 - 98 (1981)].
A. Dobó.
Astron. Nachr., Band 303, 218 (1982). – See Abstr. 29.107.023.

Stars

111 Parallaxes, Proper Motions, Radial Velocities, Space Motions, Distances

111.001 **Radial velocities of 617 stars belonging to four stellar fields of 4° × 4°.**
C. Fehrenbach, R. Burnage, C. Agniel, J. Figuière, L. Rolland, G. Traversa.
Astron. Astrophys., Suppl. Ser., Vol. 49, 483 - 496 (1982). In French.

The authors give a radial velocity catalogue of 617 stars including exact positions, spectral types and approximate magnitudes. These stars belong to four fields of 4° × 4°. The radial velocities are in the standard system and the authors discuss the accuracy of the results. They give a list of 9 stars with high radial velocities and a list of 17 probable variable stars with their radial velocities plate by plate. For each field they publish two separate histograms for B to F4 stars and F5 to M stars. The authors have studied the dispersion of radial velocities for these fields.

111.002 **An odd behavior of nearby stars velocity components in the direction $l = 330°, b = 0°$.**
S. M. de Freitas.
Astron. Astrophys., Suppl. Ser., Vol. 49, 687 - 690 (1982).

The author has taken a population of 631 nearby stars and computed their velocity components in the galactic plane, both in longitude directions 60° and 330° (positive senses), using data given by the Catalogue of Nearby Stars (Gliese, 1969). Average and standard deviations of such components were calculated for the whole population and for samples obtained by stratifying the positions of the stars in parallel strips. Dispersions in components along 330°- 150° are found to present inhomogeneities, no matter how wide the strips are, when the stratification is orthogonal to this direction. The same does not happen when the strips are parallel to 330°- 150°. For components along 240°- 60° the authors only find inhomogeneities in dispersions using narrow strips, due to random fluctuations. The odd behavior shown by the 330° velocity components is probably caused by a greater frequency of high but variegated values in a lane which is parallel to the 240°- 60° direction.

111.003 **A search for radial-velocity variations in ε Leonis (HD 84441).**
N. R. Evans.
Publ. Astron. Soc. Pacific, Vol. 94, 565 - 566 (1982).

A search was made for radial-velocity variations on the G0 II I. A. U. radial-velocity standard ε Leo. No variations larger than ± 1 km sec^{-1} were found.

111.004 **The Hβ distance scale for B stars: the Orion association.**
B. J. Anthony-Twarog.
Astron. J., Vol. 87, 1213 - 1222 (1982).

The large body of Strömgren photometry collected by Warren and Hesser (1977) for Orion B stars has been analyzed following Crawford's (1978) revised calibration for B stars. No evidence is found for any bias due to rotation in the determination of absolute magnitudes and distance moduli. Comparison to previous studies in Orion and in the Sco-Cen association illustrates the need for caution in the treatment of reddening and suggests that the revision of the standard ZAMS due to the increase in the accepted Hyades distance modulus ought not to be extended to very young clusters.

111.005 **The distance of the hot white dwarf HD 149499B.**
P. A. Ianna, J. R. Rohde, E. B. Newell.
Astrophys. J., Lett., Vol. 259, L71 - L73 (1982).

A first parallax is reported for HD 149499 supporting the main-sequence nature of the primary and the degenerate nature of HD 149499B. From PDS measurements of plates obtained with the 66 cm refractor at the Mount Stromlo Observatory, the parallax is found to be $\pi_{abs} = 0''.0297 \pm 0''.0059$ placing the system at 34 pc.

111.006 **The astrometric perturbations of G208-44, L850-62, W922, and Barnard's Star.**
R. S. Harrington, V. V. Kallarakal.
Bull. American Astron. Soc., Vol. 14, 612 (1982). – Abstract.

111.007 **Radial velocity variations in cool supergiants.**
H. C. Harris.
Bull. American Astron. Soc., Vol. 14, 613 (1982). – Abstract.

111.008 **Orbital and trigonometric parallax comparison.**
M. J. Halliwell
Bull. American Astron. Soc., Vol. 14, 645 (1982). – Abstract.

111.009 **Radial velocities of distant OB stars and galactic rotation to 26 kiloparsecs.**
F. R. Chromey.
Bull. American Astron. Soc., Vol. 14, 655 (1982). – Abstract.

111.010 **Radial velocity observations of yellow supergiants.**
L. A. Balona.
Mon. Not. R. Astron. Soc., Vol. 201, 105 - 110 (1982).

Photoelectric radial velocities are presented for seven F–G supergiants suspected of Cepheid-like variations. Four of these stars have variable velocities with ranges from 5 to 11 km s^{-1}. No definite periodicity can be assigned to any member of the group.

111.011 **Combinación de material fotográfico obtenido con diferentes telescopios Schmidt para la determinación de movimientos propios.**
F. Della Prugna.
Rev. Mexicana Astron. Astrofis., Vol. 6, (see 012.017), 119 - 122 (1981).

Six plates centered on $\alpha = 20^h00^m$, $\delta = -32°30'$ were measured with the purpose of determining proper motions for all objects common to the six exposures. The results obtained using the plates taken with objective-prism as first epoch, show that different Schmidt telescopes and observational techniques can successfully be employed to determine proper motions.

111.012 **A proper motion survey in the area of the galactic cluster in Coma Berenices.**
J. Cova S.
Rev. Mexicana Astron. Astrofis., Vol. 6, (see 012.017), 123 - 128 (1981).

New proper motions were derived for stars brighter than

m_{pg} = 14.0 in the area of the open star cluster in Coma Berenices.

111.013 **On the variable radial velocity of ϕ Phoenicis.**
M. M. Dworetsky, D. J. Stickland, G. W. Preston, A. H. Vaughan.
Observatory, Vol. 102, 145 - 146 (1982).

111.014 **Proper motions of ultraviolet-excess objects.**
T. Noguchi, M. Yutani, H. Maehara.
Publ. Astron. Soc. Japan, Vol. 34, 407 - 415 (1982).
Proper motions of the ultraviolet-excess objects detected with the Kiso Schmidt telescope are determined in two pilot fields of the Kiso plates. Glass copies of the Palomar Sky Survey are utilized as a counterpart in which the same fields were taken about 30 yr ago. The accuracy of the position determination on a single plate is about 0."43 and 0."33 for Kiso plates and PSS copies, respectively, and the total proper motions are obtained to an accuracy of 0."018 yr^{-1}. Since there exist a considerable number of objects whose proper motions fully exceed the quoted error, nearby stars are dominant among the objects.

111.015 **Absolute proper motions of blue objects near the galactic north pole.**
E. Schilbach.
Astron. Nachr., Band 303, 335 - 340 (1982). In German.
The absolute proper motions of 74 blue objects (from $16^m.5$ up to $19^m.5$) and 28 additional faint stars ($17^m.5$) near the galactic north pole are presented. They were derived with reference to 45 galaxies as the mean value from 5 plates taken in the B-system with the Tautenburg Schmidt telescope. The mean errors of the proper motion of a star are equal to $\bar{m}_x$ = 0.″0058 and $\bar{m}_y$ = 0.″0069. The errors of a blue object with a B-magnitude between $17^m.0$ and $18^m.0$ are $\bar{m}_x$ = 0.″0059 and $\bar{m}_y$ = 0.″0051 and increase with decreasing brightness. For 16 of the investigated blue objects it was possible to show that proper motions exist with a probability of 0.99.

111.016 **Recherche d'étoiles à grande vitesse dans la direction du Pôle Galactique Nord à l'aide du prisme-objectif de Fehrenbach.**
A. Florsch, J. Marcout.
Mitt. Astron. Ges., Nr. 57, p. 279 - 281 (1982).

111.017 **Equations relating mean stellar proper motions to galaxies.**
H. Eelsalu.
Tartu Astrofüüs. Obs. Publ., Köide 49, 133 - 139 (1982).
Rival programmes for relating individual proper motions of stars to galaxies were initiated by the Pulkovo Observatory and the Lick Observatory. The author tries to outline a theoretical appraoch to the reduction equations consistent with general theory of stellar statistics.

111.018 **A determination of trigonometric parallaxes of stars from observations in hour angles.**
A. A. Kiselev.
Izv. Glav. Astron. Obs. Pulkovo, Astrometr. Astrofiz., No. 199, p. 3 - 12 (1982). In Russian.
The problem of determination of trigonometric parallaxes of stars is considered on the basis of observations in hour angles. The method of reduction of observations enables a determination of colour and brightness of reference stars relative to the object for taking into account atmospheric dispersion. The results of determination of parallaxes of 5 red dwarfs from observations with the Pulkovo 26-inch refractor are given. The parallaxes were determined using 30 - 40 plates.

111.019 **Proper motion survey with the 48-inch Schmidt telescope. LX. Proper motions for 7120 stars in the Hyades.**
W. J. Luyten, S. Morris, G. Hill.
Sep. print University of Minnesota, Minneapolis, Minn., USA. 120 pp. (1981).

111.020 **Proper motion survey with the forty-eight inch Schmidt telescope. LXI. The galactic distribution of proper motion stars.**
W. J. Luyten.
Sep. print University of Minnesota, Minneapolis, Minn., USA. 6 pp. (1982).

111.021 **Proper motion survey with the 48-inch Schmidt telescope. LXII. Proper motions for 374 faint stars. The alleged cluster near the North Galactic Pole.**
W. J. Luyten, H. S. Hughes.
Sep. print University of Minnesota, Minneapolis, Minn., USA. 9 pp. (1982).

111.022 **The kinematic characteristics of Trapezium-type systems.**
A. I. Yatsenko.
Pis'ma Astron. Zh., Tom 8, 714 - 716 (1982). In Russian.
English translation in Soviet Astron. Lett., Vol. 8.
Galactic coordinates of the solar apex $L_\odot = 43° \pm 18°$, $B_\odot = +28° \pm 13°$ and velocity of the sun $V_\odot = 13 \pm 4$ km/s were determined on the basis of analysis of the proper motions of 34 Trapezium-type systems. The velocity dispersion along the galactic coordinates of the stars under study is $\sigma_l = \pm 11$ km/s, $\sigma_b = \pm 7$ km/s, the dispersion of internal space velocities of the system stars is about ± 5 km/s.

Data for the compilation of the "Third Catalogue of Nearby Stars",
See Abstr. 002.032.

Fundamental data for FK4/FK4Sup stars.
See Abstr. 002.033.

Catalogue bibliographique de vitesses radiales stellaires.
See Abstr. 002.035.

Documentation for the machine-readable version of the Lowell Proper Motion Survey, Northern Hemisphere, The G Numbered Stars.
See Abstr. 002.085.

Documentation for the machine-readable version of the General Catalogue of Trigonometric Stellar Parallaxes and Supplement.
See Abstr. 002.086.

Documentation for the machine-readable version of the Catalogue of Nearby Stars, Edition 1969.
See Abstr. 002.088.

Space velocity vectors of 2595 B and A stars.
See Abstr. 002.110.

Perth 75. A catalogue of positions of 2589 FK4 and FK4S stars.
See Abstr. 002.135.

Landolt-Börnstein. Group VI. Vol. 2c. Interstellar matter – Galaxy – Universe.
See Abstr. 003.055.

Application of mathematical modelling to the estimation of errors of photographic determinations of proper motions of stars.
See Abstr. 021.013.

A comparison of reductions obtained from galaxies at Pulkovo.
See Abstr. 031.671.

A discussion on the measuring method of radial velocities with a direct vision mosaic objective prism.
See Abstr. 031.682.

A photoelectric radial-velocity spectrometer on the 1.2-m telescope of the Dominion Astrophysical Observatory.
See Abstr. 034.158.

Conversion of positions and proper motions from B1950.0 to the IAU system at J2000.0.
See Abstr. 043.003.

Radial velocity information content in solar-type spectra.
See Abstr. 114.061.

A determination of the radial velocity of the magnetic star HD 153 882.
See Abstr. 114.220.

Absolute magnitudes of F-, G-, K-, and M-type stars from maximum-likelihood statistical parallaxes.
See Abstr. 115.020.

On the detection of other planetary systems by astrometric techniques.
See Abstr. 118.039.

A determination of the dynamic parallax and an estimate of the orbital inclination of a double star from its short-arc observation.
See Abstr. 118.042.

Proper motions of Herbig-Haro objects. II. The relationship of HH-39 to R Monocerotis and NGC 2261.
See Abstr. 121.001.

Variable stars in the General Catalogue of Trigonometric Parallaxes.
See Abstr. 123.030.

Motion of the sun in the interstellar medium.
See Abstr. 131.060.

112 Circumstellar Matter (Shells, Dust, Masers, Stellar Winds, etc.)

112.001 **On the structure of the outer layers of cool carbon stars.**
F. Querci, M. Querci, R. F. Wing, A. Cassatella, A. Heck.
Astron. Astrophys., Vol. 111, 120 - 124 (1982).

The authors report the first, unsuccessful, attempts to record the ultraviolet spectra of cool carbon stars with the IUE satellite. They give the details of the observations, and discuss the implications that can be derived from the negative results.

112.002 **High resolution observations of the Hα profile from η Car.**
J. Melnick, M. T. Ruiz, J. Maza.
Astron. Astrophys., Vol. 111, 375 - 376 (1982).

Comparison of a high resolution Hα-line profile of η Carinae with published UV spectra of the star allows the Hα profile to be decomposed in three P-Cygni like components having widths of about 1 Å, 4.5 Å, and 14.4 Å corresponding to 45 km s^{-1}, 200 km s^{-1}, and 650 km s^{-1} respectively. These components can be interpreted in terms of massive shells of matter ejected by the star in the past.

112.003 **Models of protoplanetary disks around stars of the F–G type.**
A. V. Vityazev, G. V. Pechernikova.
Pis'ma Astron. Zh., Tom 8, 371 - 377 (1982). In Russian. English translation in Soviet Astron. Lett., Vol. 8.

Some questions concerning formation and early evolution of protoplanetary disks about single solar-type stars are discussed. A new class of models of circumstellar protoplanetary disks is proposed, the parameters of the model being the disk's mass and angular momentum and the central star's mass and luminosity.

112.004 **Observations of the 30 micron feature in IRC + 10216.**
T. Herter, D. A. Briotta, Jr., G. E. Gull, J. R. Houck.
Astrophys. J., Lett., Vol. 259, L25 - L27 (1982).

Infrared observations from 30 to 37 μm of IRC +10216 extend the wavelength of the 30 μm feature found in four carbon stars and two planetary nebulae by Forrest, Houck, and McCarthy. The present data confirm the existence of this newly observed feature and determine its shape beyond 30 μm. A combined spectrum of the excess is presented and shows the feature to extend to at least 37 μm.

112.005 **Hα and Hβ line profiles of Be stars.**
G. Fontaine, B. Villeneuve, J. D. Landstreet, R. H. Taylor.
Astrophys. J., Suppl. Ser., Vol. 49, 259 - 266 (1982).

Hα line profiles for 25 emission-line and shell stars are presented. A total of 39 Hα scans have been obtained, including multiple observations of 11 of the program stars. In addition, Hβ measurements for 10 of those objects are given. These data were obtained with the Universtiy of Western Ontario intensifier dissector scanner.

112.006 **On the Balmer emission lines of the Herbig Be star HD 200775.**
J. Köppen, U. Finkenzeller, R. Mundt, M. Beltrametti.
Astron. Astrophys., Vol. 112, 174 - 177 (1982).

Recent observational material on the Balmer emission lines of the Herbig Be star HD 200775 is reviewed. The authors find no time variations of the profile shape or the strength of Hα. Numerical calculations of the level populations of hydrogen have been done to explain the observed line ratios Hα/Hβ, Hγ/Hβ and Hδ/Hβ. The authors conclude that the Balmer emission lines are produced in a compact (characteristic size $10 R_\odot$) and fairly dense (10^9 cm^{-3}) envelope.

112.007 **MERLIN spectral line observations of circumstellar shells around two OH/IR stars.**
R. P. Norris, P. J. Diamond, R. S. Booth.
Nature, Vol. 299, 131 - 134 (1982).

The authors present MERLIN observations of two OH/IR stars: OH127.8–0.0 and OH104.9+2.4. The spatial structure of OH127.8–0.0 revealed by Booth et al. (1981) was consistent with a simple expanding shell model. The authors' maps of OH127.8–0.0 confirm this interpretation, whereas the circumstellar shell that they find around OH104.9+2.4 exhibits a more complex motion, which they interpret in terms of rotation.

112.008 **Mass loss from cool stars – facts, fads, and fallacies.**
A. K. Dupree.
Advances in ultraviolet astronomy, (see 012.006), p. 3 - 16 (1982).

Sufficient observational material has accumulated to enable the authors to discern the presence and character of mass loss across the cool half of the H-R diagram and to establish constraints on theoretical models. Two determinants of atmospheric wind structure – temperature and gravity – may suffice in a most superficial way to define the wind and atmospheric structure in a star, however it is apparent that there is still a missing parameter which may stem from magnetic activity and its particular configuration. Successive observations of an active binary (λ And – G8III-IV) and a supergiant star, α Aqr (G2 Ib) have revealed that magnetic activity and perhaps mass loss occur on restricted regions of a stellar surface and that long-term structures are present in the wind. These phenomena are present in the solar atmosphere and wind and may be considered a general characteristic of stellar winds.

112.009 **An ultraviolet view of Be stars.**
T. P. Snow.
Advances in ultraviolet astronomy, (see 012.006), p. 61 - 67 (1982).

Several types of ultraviolet observations of Be stars are briefly summarized, with a view towards identifying the characteristics, if any, that distinguish Be stars from other early-type objects.

112.010 **Ultraviolet and optical spectra of the outer shell of Eta Carinae.**
K. Davidson, N. R. Walborn, T. R. Gull.
Advances in ultraviolet astronomy, (see 012.006), p. 433 (1982). – Abstract.

112.011 **Mass loss at discrete velocities in Be stars: evidence for a non-radiatively driven wind.**
G. J. Peters.
Advances in ultraviolet astronomy, (see 012.006), p. 575 - 578 (1982).

The temporal behavior of the high velocity relatively narrow absorption components in C IV, Si III, IV observed in the "pole-on" Be stars ω Ori, 66 Oph, and HR 4009 and what these observations contribute toward our understanding of the Be phenomenon are discussed. Ground-based and X-ray data on the above mentioned stars are compared as well as their UV differences with similarly active Be stars with larger projected rotational velocities.

112.012 **Ultraviolet spectra of some bright later-type Be stars and A–F shell stars.**
A. Slettebak.
Advances in ultraviolet astronomy, (see 012.006), p. 579 - 583 (1982).

This investigation concentrates on the ultraviolet spectra of later-type Be stars and A–F shell stars, with emphasis on mass-loss effects and anomalous ionization.

112.013 **Empirical atmospheric velocity patterns from combined IUE and visual observations: the Be-similar stars.**
V. Doazan, R. Stalio, R. N. Thomas.
Advances in ultraviolet astronomy, (see 012.006), p. 584 - 587 (1982).

The authors present a summary of the velocity pattern which pan-spectral observations of the Be stars suggest, and an outline of its extension via similar pan-spectral observations of the Be-similar stars. Emphasis is placed on the time-dependent interaction between the stellar mass flux and the local environment produced jointly by the star itself and its original environment.

112.014 **Mass loss from the central star of the planetary nebula IC 3568.**
J. P. Harrington.
Advances in ultraviolet astronomy, (see 012.006), p. 610 - 614 (1982).

Low resolution IUE spectra of IC 3568 have shown several stellar P Cygni features. A high resolution spectrum has been obtained which shows a saturated N V λ1240 profile from which the terminal velocity of the stellar wind is found. The author also observes P Cygni features due to C IV λ1549, O V λ1371 and N IV λ1719. The unsaturated O V λ1371 line seems best for determination of the mass loss rate. The level of ionization in the stellar wind appears to increase outwards. The author discusses the possible effects of such stellar winds from central stars on the surrounding nebulae.

112.015 **On the properties of the circumstellar matter around the bright young variable shell star HR 5999.**
J. Andersen, G. F. Gahm, J. Krelowski.
Astron. Astrophys., Vol. 113, 176 - 177 (1982).

The composition of the interstellar matter around the bright young variable shell star HR 5999 has been studied by comparing spectrograms taken near minimum light, when the increase in circumstellar dust extinction amounted to about 1.5 mag in the V band, with similar spectra taken at maximum. The authors conclude that the dust associated with HR 5999 is different from the normal interstellar dust and apparently does not carry the constituent(s) producing the interstellar diffuse absorption bands.

112.016 **Interpretation of the VLBI maps of maser spots.**
S. Deguchi.
Astrophys. J., Vol. 259, 634 - 646 (1982).

A spatial distribution of OH/H_2O maser emission in very long baseline interferometry (VLBI) maps is discussed by a term of velocity fluctuations in an expanding cloud. A detailed model is constructed for the OH 1612 MHz emission in OH/IR sources. Intensity distributions of maser emission are calculated for a model of an expanding circumstellar envelope with three-dimensional velocity fluctuations. An application of the theory to the OH/H_2O sources associated with H II regions is also discussed.

112.017 **The morphology of dust shells around extreme carbon stars.**
M. Cohen, G. D. Schmidt.
Astrophys. J., Vol. 259, 693 - 701 (1982).

Optical spectropolarimetry of three dust-enshrouded carbon stars reveals very large degrees of polarization, requiring highly organized circumstellar dust shells around each object. The extremely peculiar spectral behavior of polarization for GL 1403 is interpreted in terms of a cool dust torus and bipolar scattering lobes. Such a structure, together with a 635 day photometric period derived from infrared observations, supports evolution of high-mass carbon variables into bipolar nebulae.

112.018 **IUE observations of 89 Herculis: variability and mass loss.**
S. A. Lamb, C. L. Neese.
Bull. American Astron. Soc., Vol. 14, 614 (1982). – Abstract.

112.019 **The peculiar OH/IR star IRC + 10420.**
P. F. Bowers, K. J. Johnston, J. H. Spencer.
Bull. American Astron. Soc., Vol. 14, 637 (1982). – Abstract.

112.020 **Corona-warm wind model for the X-ray emission from Of stars and OB supergiants.**
W. L. Waldron.
Bull. American Astron. Soc., Vol. 14, 651 (1982). – Abstract.

112.021 **Long-term monitoring of SiO maser time variations in four transitions.**
A. P. Lane.
Bull. American Astron. Soc., Vol. 14, 653 (1982). – Abstract.

112.022 **Submillimeter observations of evolved stars.**
R. J. Sopka, B. Zuckerman, D. Jaffe, R. Hildebrand, T. Roellig, M. Werner, I. Gatley.
Bull. American Astron. Soc., Vol. 14, 653 (1982). – Abstract.

112.023 **Double-shell circumstellar dust shells.**
M. Rowan-Robinson.
Mon. Not. R. Astron. Soc., Vol. 201, 281 - 288 (1982).
Models are presented for high optical depth Type II OH-IR sources with deep silicate absorption. The models consist of a cool star at the centre of a spherically symmetric $n(r) \propto r^{-2}$ dust shell together with a second shell of cold absorbing dust. The origin of the second shell is discussed and the only plausible explanation is that it represents a previous phase of ejection by the star.

112.024 **Las envolturas circunestelares de las supergigantes del tipo F y G en la Nube Mayor de Magallanes.**
W. Hagen, R. M. Humphreys, R. E. Stencel.
Rev. Mexicana Astron. Astrofis., Vol. 6, (see 012.017), 59 - 63 (1981).
The authors have obtained high-dispersion échellograms at 2.5 and 5.1 Å/mm of four F and G-type supergiants ($M_V \sim -9$) in the Large Magellanic Cloud for the purpose of studying their outer atmospheres as compared to their Milky Way counterparts. Line doubling at Na I D indicates extensive circumstellar envelopes and mass loss rates in excess of $10^{-5} M_\odot yr^{-1}$ with outflow velocities of 10 - 60 km s^{-1}. Deep exposures at Ca II H and K reveal new information about the chromospheres of extragalactic stars. The presence of H and K wing emission lines augments this, and also provides an independent way of estimating M_V.

112.025 **Mass loss from hot stars.**
J. A. de Freitas Pacheco.
Rev. Mexicana Astron. Astrofis., Vol. 6, (see 012.017), 181-187 (1981).
Current problems in the understanding of the mechanisms of mass-loss from early-type stars are reviewed. The special cases of OB and Be stars, as well as central stars of planetary nebulae are considered and mass-loss rates are derived for these objects.

112.026 **V923 Aquilae.**
A. E. Ringuelet, J. Sahade.
Rev. Mexicana Astron. Astrofis., Vol. 6, (see 012.017), 208 (1981). – Abstract.

112.027 **Caution! High winds beyond this point.**
D. Mullan.
Astronomy, Vol. 10, No. 1, p. 74 - 78 (1982).
The HR diagram is discussed in terms of mass loss and coronae. The processes involved are considered. The solar coronal processes and wind are discussed. – From Phys. Abstr., Vol. 85, Abstr. 68737 (1982).

112.028 **A study of line profile variability in O stars.**
R. G. Costero, M. L. Franco, E. Kontizas, M. Kontizas, R. Stalio.
Third European IUE Conference, (see 012.029), p. 283 - 285 (1982).
Five O-type stars have been monitored with IUE in order to detect profile variability of a selected sample of lines which are sensitive indicators of outer atmospheric structures. Some of the results of this investigation and a possible model of these outer structures are presented.

112.029 **Empirical atmospheric velocity patterns from combined IUE and visual observations: the Be stars.**
V. Doazan, R. N. Thomas.
Third European IUE Conference, (see 012.029), p. 287 - 290 (1982).
Four years of monthly observations of 59 Cyg, simultaneously in the far UV, IR, and visual regions, show the strongest displacements of the far UV superionized lines at epochs of minimum Hα emission. UV emission lines are compared for a variety of Be stars. Velocities in the chromosphere-corona range up to 1500 km/s; those in the Hα and Fe II emission envelope, and absorption shell, do not exceed 1–200 km/s.

112.030 **Behavior of the energy distribution of 59 Cyg in the far-ultraviolet and in the visible.**
L. Divan, J. Zorec.
Third European IUE Conference, (see 012.029), p. 291 - 293 (1982).
Using three low-dispersion IUE spectra of 59 Cyg and continuous energy distributions obtained in the visible spectral range, the authors determine the interstellar extinction and the total luminosity of 59 Cyg at the beginning of an emission phase and during the emission phase. Comparing the behavior of the visible continuum of 59 Cyg to that of other Be stars, they suggest a possible explanation for the appearance and disappearance of the circumstellar spectrum.

112.031 **Envelope structure of the cyclic V/R variable shell stars.**
A. M. Hubert-Delplace, M. Mon, V. Ungerer, R. Hirata, F. Paterson-Beeckmans, H. Hubert, D. Baade.
Third European IUE Conference, (see 012.029), p. 295 (1982). – Abstract.

112.032 **Polarization properties of the 86 GHz SiO maser emission from R Cassiopeiae.**
F. O. Clark, T. H. Troland, D. R. Johnson.
Astrophys. J., Vol. 261, 569 - 575 (1982).
The authors have measured the polarization properties of the SiO circumstellar maser emission from R Cas over a period of nearly 2 years. Velocity profiles for both the Stokes parameter I and the total linear polarization change relatively smoothly but not periodically with time.

112.033 **Observations of circumstellar shells around OH/IR stars.**
R. S. Booth, R. P. Norris, P. J. Diamond.
Submillimetre wave astronomy, (see 012.031), p. 137 - 144 (1982).
Interferometric observations of the 1612 MHz OH maser emission from late type stars have revealed circumstellar shells of gas some $10^3 - 10^4$ A.U. in diameter. Using the velocity information of the OH spectrum and an estimate of the molecular density it has been possible to derive mass loss rates from these stars which are of the order of $10^{-5} M_\odot - 10^{-4} M_\odot$ per year.

112.034 **A dust shell around the yellow supergiant CoD –61°3326.**
M. Tapia.
Publ. Astron. Soc. Pacific, Vol. 94, 669 - 671 (1982).
Broad-band photometry from 1 to 4 μm and low-resolution spectrometry, in the 2-μm atmospheric window, of the F-type supergiant CoD –61°3326 = Hz 747 are reported. The presence of an optically thin dust shell around this star is deduced and its properties are discussed with reference to the

well-known yellow supergiant 89 Her. A similar post-main-sequence evolutionary status is proposed for both stars.

112.035 **Observations of mass loss from OB and Wolf-Rayet stars.**
M. J. Barlow.
Wolf-Rayet stars: observations, physics, evolution, (see 012.042), 149 - 172 (1982).
In the review, three observationally accessible parameters of the winds of OB and Wolf-Rayet stars are discussed.
(1) terminal velocities, (2) velocity laws, (3) mass loss rates. In addition, some discussion of the ionisation structure of the winds is included.

112.036 **The influence of the ratio of total to selective extinction on the determination of the mass loss rate of Wolf-Rayet stars from infrared excess measurements.**
P. S. The, K. A. van der Hucht, M. Arens.
Wolf-Rayet stars: observations, physics, evolution, (see 012.042), p. 197 - 201 (1982).
It is shown that the mass loss rate of the WN7 star HD 93162 decreases with larger values of the ratio of total to selective extinction R. For HD 93162 the mass loss rate will change one order of magnitude, only if $\Delta R \sim 2$. Mass loss rates are derived for nine other WR stars of which visual, red and near-infrared photometric observations were obtained.

112.037 **On the properties of Wolf-Rayet stars and their mass loss.**
N. Panagia, M. Felli.
Wolf-Rayet stars: observations, physics, evolution, (see 012.042), p. 203 - 207 (1982).
From consideration of the observed properties of the envelopes produced by mass loss in the WR stars the authors find that: (1) The velocity at the optical photosphere is in the range 200–800 km s^{-1}. (2) The effective photospheric radius for the continuous radiation capable to ionize helium twice ($\lambda < 228$ Å) is typically 5 to 15 times the optical photospheric radius. (3) The radiation temperature in the Lyman continuum is around 5×10^4 K. (4) Multiple scattering of radiation in the interval 228–504 Å can provide most of the momentum needed to accelerate the wind up to the observed terminal velocities.

112.038 **Radio continuum measurements of mass loss from Wolf-Rayet stars.**
D. C. Abbott, J. H. Bieging, E. Churchwell.
Wolf-Rayet stars: observations, physics, evolution, (see 012.042), p. 215 - 220 (1982).
The purpose of the paper is to examine the relationship between the mass loss rate and the spectral appearance of O, Of, and WR stars.

112.039 **Mass loss from Wolf-Rayet stars: an analysis of radio and infrared observations of MR 111≡AS 422.**
M. Felli, N. Panagia.
Astrophys. J., Vol. 262, 650 - 656 (1982).
VLA radio observations at 1.3 cm of the Wolf-Rayet star MR 111 are reported. The collected fluxes over a frequency range from 1.65 μm to 6 cm are considered and separated into two components: a stellar blackbody and an extended envelope emission. The envelope emission is interpreted in terms of an ionized accelerated outflow due to mass loss. MR 111 is found to be a WN-type star, with a radius of 25 $R_\odot$ losing mass at a rate of $2.5 \times 10^{-5} M_\odot yr^{-1}$.

112.040 **The resolution of P Cygni's stellar wind.**
R. L. White, R. H. Becker.
Astrophys. J., Vol. 262, 657 - 662 (1982).
The authors have resolved the 6 cm radio emission from P Cygni's stellar wind using the VLA. The wind is apparently spherically symmetric, and the radial surface brightness distribution agrees with that predicted by the constant velocity, isothermal wind model. The authors determine the wind temperature to be 18,000 ± 2000 K. Their technique makes it possible to study the adiabatic cooling of the wind.

112.041 **International Ultraviolet Explorer observations of the R Aquarii jet.**
A. G. Michalitsianos, M. Kafatos.
Astrophys. J., Lett., Vol. 262, L47 - L51 (1982).
Ultraviolet spectra were obtained with the International Ultraviolet Explorer of the newly discovered optical-radio jet feature in the symbiotic variable R Aquarii. The far-UV continuum of the jet is characterized by strong continuum which rises with decreasing wavelength in the 1200–2000 Å wavelength range and is considerably different in appearance from the relatively flat continuum exhibited by ionized nebulosity in the central star. The overall excitation of the jet suggests that it is cooler than the nebulosity that engulfs the central UV object.

112.042 **Properties and nature of Be and shell stars. 7B. 88 Her – an important clue to understanding the Be phenomenon?**
V. Doazan, P. Harmanec, P. Koubsky, J. Krpata, F. Zdarsky.
Astron. Astrophys., Vol. 115, 138 - 144 (1982).
Long series of ground-based observations of 88 Her show: (1) Periodic variations of the radial velocities of the shell Balmer lines and of the *V/R* ratio of the double Hα emission line with a period of 86.7221 d, which seems to be stable since 1912. (2) Doubling of the Hα absorption core on the descending branch of the RV curve. (3) Long-term variations of the brightness and colour of the star correlated with substantial changes of the intensities of absorption and emission lines in the optical spectrum. Limitations which these data impose on various models of the Be phenomenon are discussed. Systematic simultaneous observations of the object in the visible, far UV and X-ray regions are highly needed.

112.043 **Stokes polarimetry of main-line OH emission from stellar masers.**
M. J. Claussen, J. D. Fix.
Astrophys. J., Vol. 263, 153 - 159 (1982).
Main-line OH emission has been measured in all four Stokes parameters from seven late-type variable stars and the F8 supergiant IRC + 10420. Linearly polarized features were detected in UX Cyg, U Ori, and IRC + 10420 at 1665 MHz. A polarization pattern in IRC + 10420 is probably the best example of a complete Zeeman pattern yet observed in stellar masers. This study, combined with other recent work, shows that only about 10% of the stellar OH sources display linearly polarized features. As an aid in accounting for the observed polarization properties of stellar OH masers, model mass flows were calculated using magnetic field structures similar to that of the solar wind.

112.044 **Mass loss rates for Wolf-Rayet stars from radio continuum observations.**
J. H. Bieging, D. C. Abbott, E. B. Churchwell.
Astrophys. J., Vol. 263, 207 - 214 (1982).
The authors observed 13 Wolf-Rayet (W-R) stars for radio continuum emission at 4885 MHz with the VLA. Eight stars were detected. The mass loss rates derived from the radio fluxes are nearly the same, with a mean value of $(2.0 \pm 0.6) \times 10^{-5} M_\odot yr^{-1}$. There is no evidence for a correlation between $\dot{M}$ and other stellar properties (e.g., spectral type, luminosity, binary nature). The kinetic power of the winds is 2%–10% of the bolometric luminosity, so that the mass loss represents a nonnegligible part of the stellar energy budget.

112.045 **M1-67: a wind-blown bubble carried along by the high-velocity WR star 209 BAC?**
J. Solf, U. Carsenty.
Astron. Astrophys., Vol. 116, 54 - 59 (1982).
High-resolution long-slit coudé spectra from the peculiar nebula M1-67 in the red are presented. The velocity field is explained by a nearly spherical thin shell of 90" diameter expanding at a velocity of 42 km s^{-1}. The observations can be explained either by a nearly isotropic mass ejection from the star some 2×10^4 yr ago, or by the action of a strong stellar wind producing an expanding bubble of swept-up ambient material. Due to the space motion of the nebula, additional mass from the ambient interstellar medium has been accreted on the leading edge.

112.046 ***VBLUW* measurements of the shell star Pleione.**
F. van Leeuwen, J. J. M. Meys, P. Alphenaar.
Inf. Bull. Variable Stars, No. 2173, 3 pp. (1982).

112.047 **Mass loss from extreme helium stars. Detailed UV-line fits for HD 160641, BD −9°4395 and BD +10°2179.**
W.-R. Hamann, D. Schönberner, U. Heber.
Astron. Astrophys., Vol. 116, 273 - 285 (1982).
Indications of mass loss are found in the UV spectra of three extreme helium stars obtained with IUE in high resolution. Some of the resonance lines are obviously formed or affected by the expanding envelope. These lines are fitted with theoretical profiles in order to construct empirical wind models. Final velocities of 400–600 km s^{-1} and supersonic turbulences (60–100 km s^{-1}) are estimated. The observed ionization ratios are compared with the "warm", "cool" and "corona" wind model, and mass loss rates are determined. The mass loss rates for the three extreme helium stars are found to be of similar order-of-magnitude as expected for normal stars with the same luminosity. However they show a decrease with increasing radius, while mass and luminosity of the three program stars are similar. This contradicts all theoretical predictions as well as empirical correlations established for early-type stars of normal composition.

112.048 **New infrared spectroscopy and infrared photometry of a new WC9 star.**
A. C. Danks, M. Dennefeld, W. Wamsteker, P. A. Shaver.
ESO Sci. Prepr., No. 206, 21 pp. (1982). – Submitted to Astron. Astrophys.

112.049 **The Hα emission of helium-rich stars.**
R. Nakajima.
Sci. Rep. Tôhoku Univ., Eighth Ser., Vol. 2, 130 - 142 (1981).
The circumstellar gas of helium-rich stars will be trapped in the outer region of the stellar magnetosphere where the centrifugal force dominates. Only in rapidly rotating stars, the outer region can contain the gas much enough to emit the observed Hα emission energy flux. As the circumstellar gas corotates with the star, the Hα emission profile becomes very broad and so the emission will appear in the wing of Hα absorption line which originates in the photosphere of the star.

112.050 **Ultraviolet shell formation at V1016 Cygni.**
W. A. Feibelman.
Astrophys. J., Lett., Vol. 263, L69 - L71 (1982).
A recent 330 minute high-resolution IUE spectrogram of V1016 Cygni shows splitting of the λ1550 component of the C IV resonance doublet, suggesting that a nebular shell is in the process of formation. An expansion velocity of 15 km s^{-1} is indicated by the data. IUE data of V1016 Cyg, taken approximately 1 year earlier, do not show split C IV lines. Thus, V1016 Cyg may have undergone an event which forms a planetary nebula.

112.051 **Molecular emission bands in the ultraviolet spectrum of HD 44179.**
M. L. Sitko.
Publ. Astron. Soc. Pacific, Vol. 94, 758 (1982). – Abstract.

112.052 **Observations of SiO maser emission from R Leo during its summer 1981 maximum.**
F. O. Clark, J. A. Waak, J. M. Bologna.
Astron. J., Vol. 87, 1803 - 1805 (1982).
The 7-mm $J = 1 \rightarrow 0$, $v = 1$ and $v = 2$ SiO maser emission from the Mira variable R Leo has been closely sampled in time near the summer 1981 optical maximum in order to observe possible phase-dependent effects. A significant broadening of both $v = 1$ and $v = 2$ emission, accompanied by profile changes, was observed between phases 0.95 and 0.98. If these changes are attributed to the expected passage of the shock front, the line formation region can be inferred to be very close to the photosphere of the star, and the radial extent of the line formation region may be crudely estimated.

112.053 **The photoproduction of circumstellar OH maser shells.**
P. J. Huggins, A. E. Glassgold.
Astron. J., Vol. 87, 1828 - 1835 (1982).
The authors investigate the structure of OH shells formed from the photodestruction of H_2O by ambient UV photons in the thick, expanding envelopes around cool evolved stars. The properties of the shells are governed mainly by the envelope shielding which in turn is primarily controlled by the mass-loss rate $\dot{M}$. The peak OH densities and column densities through the shells are, respectively, slowly decreasing and increasing functions of $\dot{M}$. The characteristic radii of the shells also depend on $\dot{M}$; this dependence is well matched by recent observational data, and lends support to the OH photoproduction mechanism.

112.054 **Circumstellar material around rapidly rotating B stars. I. Nonemission-line stars.**
W. R. Oegerle, R. S. Polidan.
Publ. Astron. Soc. Pacific, Vol. 94, 997 - 1000 (1982).
The authors have searched for evidence of low-velocity circumstellar material around a number of rapidly rotating nonemission-line B stars. They find no evidence for circumstellar shell lines in any of the stars observed. If circumstellar material is present, then the results place constraints on the physical properties of the envelope.

Zur Berechnung von Astropausenflächen.
See Abstr. 062.094.

Inference of nebular density and luminosity structure from polarization maps.
See Abstr. 063.049.

Stellar winds and the evidence for coronae around early type stars.
See Abstr. 064.007.

Mass loss in Alpha Cygni: synthetic H-alpha profiles.
See Abstr. 064.057.

Mass loss by Wolf-Rayet stars.
See Abstr. 064.058.

Détermination des taux de perte de masse dans les étoiles chaudes.
See Abstr. 064.073.

The vanishing act of carbon giants.
See Abstr. 065.063.

Charge exchange in solar wind cometary interactions.
See Abstr. 074.099.

Observations of cool stars at 20, 25, and 33 μm.
See Abstr. 113.062.

IUE observations of symbiotic stars.
See Abstr. 114.023.

Spectral types and rotational velocities of the brighter Be stars and A - F type shell stars.
See Abstr. 114.054.

High-resolution 3 μm spectroscopy of IRC+10216.
See Abstr. 114.071.

Different stellar winds in O-type stars.
See Abstr. 114.080.

Fe II absorption lines in the IUE high-resolution spectra of F supergiants and shell stars.
See Abstr. 114.093.

Mass loss from extreme helium stars.
See Abstr. 114.100.

SK-67/111 and R 50, two Magellanic Cloud emission line stars.
See Abstr. 114.101.

The iron curtain of WC9 stars.
See Abstr. 114.130.

Ultraviolet emission in the Mg II h and k lines in Be stars.
See Abstr. 114.142.

Near IR spectra of symbiotic stars.
See Abstr. 114.147.

UV line emission of symbiotic stars.
See Abstr. 114.151.

Z Andromedae: the prototype?
See Abstr. 114.152.

Shell and photosphere of σ Ori E: new observations and improved model.
See Abstr. 114.159.

Properties and nature of Be and shell stars. 7A. 88 Her. Observational data, their reduction and basic evaluation.
See Abstr. 114.169.

Spectrophotometric study of Be and shell stars with the TD-1 S2/68 experiment.
See Abstr. 114.179.

IUE high dispersion spectra of luminous stars in symmetric nebulae.
See Abstr. 114.202.

Evolution of chromospheres and coronae in solar mass stars: a far-ultraviolet and soft X-ray comparison of Arcturus (K2 III) and Alpha Centauri A (G2 V).
See Abstr. 114.205.

Spectroscopic analysis of Alpha Andromedae.
See Abstr. 114.224.

Magnetospheres and winds of the helium strong stars: dependence on rotation.
See Abstr. 116.003.

The structure, energy balance, and winds of cool stars.
See Abstr. 116.010.

Radio observations of symbiotic stars.
See Abstr. 116.017.

Has P Cygni generated a shock front which emits nonthermal radiation?
See Abstr. 116.019.

Radio emission from the extended chromosphere of Alpha Orionis.
See Abstr. 116.021.

Infrared studies of symbiotic stars.
See Abstr. 117.100.

Properties of symbiotic stars from studies in the optical region.
See Abstr. 117.102.

Ultraviolet properties of the symbiotic stars.
See Abstr. 117.103.

A review of the properties of the symbiotic star CH Cygni.
See Abstr. 117.104.

Models for symbiotic stars in the light of the data.
See Abstr. 117.126.

Models of the circumstellar matter in binary systems.
See Abstr. 117.162.

Observational aspects of the circumstellar matter in the binary systems.
See Abstr. 117.165.

Circumstellar matter in the close binary SZ Camelopardalis.
See Abstr. 117.166.

A spectroscopic investigation of the eclipsing variable ER Vulpeculae.
See Abstr. 119.042.

The spectroscopic orbit of Wra 977 (= X-ray source 4U 1223-62).
See Abstr. 120.006.

Broad carbon monoxide line wings near T Tauri stars.
See Abstr. 121.004.

On the sources of variability in T Tauri stars.
See Abstr. 121.012.

Alfvén wave-driven winds as a source of mass loss from T Tauri stars.
See Abstr. 121.047.

The circumstellar extinction of RY Sagittarii.
See Abstr. 122.004.

Further *VBLUW* photometry of the S Doradus type variables S Dor and HDE 269006 in the LMC and a discussion

on their temperatures.
See Abstr. 122.015.

AG Car: a galactic S Dor variable.
See Abstr. 122.016.

Possible correlations of expansion velocity with period and 1 μm intensity variation in Mira variables.
See Abstr. 122.018.

Properties and nature of Be and shell stars. 11. A notable correlation between the long-term spectral and photometric variations of V 1294 Aql (HD 184 279).
See Abstr. 122.050.

Spectroscopic analysis of the extended shells around the novae CP Puppis and T Pyxidis.
See Abstr. 124.701.

X-ray and UV-emission from supernova shock waves in stellar winds.
See Abstr. 125.002.

The absence of 1584 MHz maser emission from ^{18}OH.
See Abstr. 131.030.

Wind-cloud interactions in broad-winged CO sources.
See Abstr. 131.120.

High-velocity CO line wings associated with T Tauri stars.
See Abstr. 131.123.

Linear polarization of astronomical masers due to geometry.
See Abstr. 131.125.

High velocity gas components associated with supergiant stars and interstellar gas.
See Abstr. 131.180.

Spectral scan of Orion A and IRC + 10216 – astrophysics and chemistry in a star formation region and in the envelope of a carbon star.
See Abstr. 131.189.

The Orion SiO maser: a unique object in the Galaxy.
See Abstr. 131.239.

Shock fronts produced by stellar winds in the interstellar gas.
See Abstr. 131.282.

The return of mass and energy to the interstellar medium by winds from early-type stars.
See Abstr. 131.300.

The structure and dynamics of a 3.5 pc diameter filamentary shell on the edge of the NGC 6334 interstellar complex.
See Abstr. 132.006.

The Bubble Nebula: far-infrared and radio molecular observations of NGC 7635.
See Abstr. 134.023.

Mass loss and the H-R diagram of globular clusters.
See Abstr. 154.004.

113 Photometric Properties

113.001 **Infrared standard stars.**
J. H. Elias, J. A. Frogel, K. Matthews, G. Neugebauer.
Astron. J., Vol. 87, 1029 - 1034, with a correction p. 1893 (1982).

Lists of standard stars suitable for *JHKL*, H_2O, and CO photometry are given. These include stars faint enough for use on large telescopes. Most of the stars have estimated uncertainties in their magnitudes of ±0.01 mag or less. The stars are accessible from both hemispheres, and cover a large enough range in color to permit definition of color transformations. The magnitudes are given on the "CIT" system.

113.002 **A photoelectric *UBV* sequence in a low extinction Puppis field.**
B. C. Reed, M. P. FitzGerald.
Astron. Astrophys., Suppl. Ser., Vol. 49, 521 - 527 (1982) = Contrib. Univ. Waterloo Obs. No. 83.

UBV photoelectric photometry and spectral classifications are presented for a sequence of 116 stars in a field in Puppis. The interstellar extinction is found to be remarkably low for a field in the Galactic Plane, amounting to $E_{B-V} \simeq 0^m\!.30$ at $r \sim 6.3$ kpc.

113.003 **Four-color and Hβ photometry for O-A0 type stars in three regions near the galactic equator.**
T. N. G. Westin.
Astron. Astrophys., Suppl. Ser., Vol. 49, 561 - 570 (1982).

A catalogue of *uvby* and Hβ photometry for 355 stars within three areas selected in the northern sky is presented. The areas are located near the galactic equator, at the galactic longitudes 135° and 175°. The stars, which are of spectral types O-A0, have *V* magnitudes between 2.89 and 10.46.

113.004 **A catalog of red stars near L1454.**
R. Duerr, E. R. Craine.
Publ. Astron. Soc. Pacific, Vol. 94, 567 - 573 (1982).

Duerr and Craine (1982) have discussed the nature of the dark cloud L1454 as deduced from analysis of star counts made utilizing Near Infrared Photographic Sky Survey data. One product of that study was compilation of a list of stars in the region for which $(V-I) \gtrsim 2^m\!.5$. Since many of these stars may be potentially interesting as individual objects of study, the authors present a catalog of those stars.

113.005 **A photometric system for limiting *JF* photography. II. Secondary sequences near 14 rich clusters of galaxies.**
W. J. Couch, E. B. Newell.

Publ. Astron. Soc. Pacific, Vol. 94, 610 - 617 (1982).

The authors present observations that define 14 secondary sequences on the photoelectric *JF* system of Couch and Newell (1980). In addition to their main role of providing calibration sequences for photometry in 14 distant clusters of galaxies, these observations extend the *JF* system to fainter magnitudes and make it accessible to Northern Hemisphere observers.

113.006 **Infrared excesses of stars with proper polarization.**
Yu. K. Melik-Alaverdyan, T. A. Movsesyan.
Astrofizika, Tom 18, 275 - 282 (1982). In Russian. English translation in Astrophysics, Vol. 18, No. 2.

The results of infrared observations of 10 stars with proper optical polarization are listed. A correlation between the infrared excess and the optical polarization of the stars is revealed. This correlation has been explained by the existence of dust envelopes around the studied stars.

113.007 **Infrared energy distribution of Cyg OB2 No. 12.**
P. Persi, M. Ferrari-Toniolo.
Astron. Astrophys., Vol. 111, L7 - L8 (1982).

The authors report the results of their IR observations from 2.3 up to 20 microns of the B supergiant star No. 12 in the Cygnus OB2 (VI Cyg) association.

113.008 **A search for Ap stars in the Scorpio-Centaurus association: additional evidence for a slow metal enrichment.**
E. F. Borra, G. Joncas, P. Wizinowich.
Astron. Astrophys., Vol. 111, 117 - 119 (1982).

The authors have used the "Δa" peculiarity index that measures the strength of the 5200 Å continuum depression, characteristic of the Ap stars, to observe a large number of normal and peculiar stars in the Scorpio-Centaurus association. They find that the Ap stars in Upper Scorpius are significantly less peculiar than the older Ap stars in the solar neighborhood. Some helium-weak stars are detected by the photometry. The observations confirm that metal enrichment takes place on time scales of a few 10^7 yr.

113.009 **Reddening relations of the *VBLUW* and *UBV* systems for objects with emission line spectra.**
A. Greve, A. M. van Genderen.
Astron. Astrophys., Vol. 111, 185 - 192 (1982).

For the Walraven *VBLUW* and Johnson *UBV* system the authors investigate reddening relations for emission line objects: emission nebulae, planetary nebulae, supernova remnants. In contrast with the reddening relations derived empirically for supernova remnants, the reddening relations calculated for planetary nebulae and emission nebulae seem to be different from the known reddening relations derived for stars, i.e. continuum sources.

113.010 **Discovery of two FK Comae stars in the southern hemisphere.**
A. C. Collier.
Mon. Not. R. Astron. Soc., Vol. 200, 489 - 496 (1982).

Radial velocity measurements, high resolution Hα spectra and broad band photometry are presented for two late G—early K southern giant stars, HD 32918 and HD 36705. Both stars are known Ca II emitters; they are members of the FK Comae class of fast-rotating surface-active late-type giants. Photometry of HD 32918 shows an RS CVn-like distortion wave with an amplitude of 0.3 mag in *V*, and a period of 9.55 days. Variable Hα emission is present in both stars. The space density of FK Comae stars is discussed; the results are broadly consistent with the hypothesis that the FK Comae stars represent a stage in the post-main-sequence evolution of W Ursae Majoris systems.

113.011 **Faint red stars at the Galactic Centre.**
R. M. Catchpole.
Mon. Not. R. Astron. Soc., Vol. 200, 33P - 37P (1982).

JHK photometry is presented and discussed for a sample of the stars discovered on a 2.2 μm survey of a small region of the Sgr I Baade window. The stars, which appear redder and fainter than those identified in previous photographic infrared surveys, occur at high density and show a well-defined giant branch.

113.012 **Mean relationships between colour indices *U–B* and *B–V* and the temperature for population II stars.**
V. Malyuto, P. Traat.
Publ. Tartu Astrofiz. Obs., Tom 48, 199 - 216 (1981).
In Russian.

At temperatures below 10000°K, in which range the *UBV* colors of stars are strongly affected by their chemical composition, the *U–B* and *B–V* colours predicted from stellar atmospheres unsatisfactorily match observations. Hence in this paper the mean relationships are derived for metal-deficient stars based essentially on observational data.

113.013 **RGU-photometry of the field Vela II.**
W. Becker, A. Marsoglu.
Astron. Astrophys., Vol. 112, 133 - 140 (1982).

Down to a limiting magnitude of $G = 16^{m}.8$: a total of 2132 stars have been measured in the RGU-system in a field of 0.31 square degrees in the direction of Vela. The line of sight passes through a 4 kpc interarm section between two spiral arms. The interstellar reddening function, the density functions for different luminosity groups and the luminosity function have been determined.

113.014 **Geneva [*U, B, V*] intrinsic colours of B-type stars.**
N. Cramer.
Astron. Astrophys., Vol. 112, 330 - 336 (1982).

An empirical technique is used to calibrate, in a first stage, the (X, Y) plane in terms of intrinsic Geneva [*U–B*] and [*B–V*] indices. Analytic expressions of these relations are derived, and discussed. The mean intrinsic loci of the MK types from O to A2 of classes V and III are mapped into the [*U, B, V*] plane. A comparison is made with the synthetic colours derived from Kurucz's (1979) models.

113.015 **Faint blue stars in the region of NGC 7000 = IC 5068-70.**
K. P. Tsvetkova, M. K. Tsvetkov.
Dokl. BAN, Vol. 34, 1197 - 1200 (1981). – Abstr. in Ref. zh., 51. Astron., 7.51.735 (1982).

113.016 **Ultraviolet photometry of A-type stars at high galactic latitudes.**
W. B. Landsman, R. C. Henry, P. D. Feldman.
Advances in ultraviolet astronomy, (see 012.006), p. 597 - 601 (1982).

IUE has been used in a study of four stars located in one target ($l = 179°$, $b = 65°$) of an earlier "diffuse background" sounding-rocket experiment. The resulting stellar correction is much smaller than that previously estimated, giving a higher diffuse background at this target.

113.017 **Photoelectric *UBV* photometry in the Large Magellanic Cloud.**
J. Isserstedt.
Astron. Astrophys., Suppl. Ser., Vol. 50, 7 - 11 (1982). In German.

Photoelectric photometry in the *UBV*-system for 245 members of the Large Magellanic Cloud is presented.

113.018 ***RGU*-Photometrie eines Sternfeldes in + 23.1° galaktischer Breite in Richtung zum galaktischen Zentrum (NGC 6171).**

D. Wiedemann.
Astron. Astrophys., Suppl. Ser., Vol. 50, 93 - 100 (1982).
2375 stars on plates taken with the 48″ Palomar Schmidt telescope, were photometrically measured using the *RGU* system in an area of 0.93 square degrees not far from the globular cluster NGC 6171 with central coordinates $l = 3.3°$, $b = +23.1°$. Two-colour diagrams TCD are presented, and from these an interstellar reddening was determined with the constant value of $E(G\text{-}R) = 0.30$. Different types of stars were classified into the following categories, according to their position in the TCD: main sequence stars of population I, late type giants of population I and metal poor main sequence stars. Density functions of the different groups of stars were determined. The luminosity functions of the main sequence stars were derived for the two distance ranges 0 - 1 kpc and 1 - 2 kpc. The two functions derived in this paper are valid for the galactic latitude + 23.1°. The luminosity function of the solar neighbourhood fits better for this particular field, when corrected for this latitude.

113.019 **Two photoelectric *UBVRI* sequences in the bar of the Small Magellanic Cloud.**
J. Vigneau, M. Azzopardi.
Astron. Astrophys., Suppl. Ser., Vol. 50, 119 - 127 (1982).
Photoelectric *UBVRI* measurements have been made of 33 stars located in two crowded regions of the Small Magellanic Cloud. *V* magnitudes, $B-V$ and $U-B$ colors are given in the Johnson system while $V-R$ and $V-I$ colors are given in the Cousins system. The *V* magnitudes range from approximately $11^m_.0$ to $15^m_.8$. 1975 equatorial coordinates and finding charts are provided.

113.020 **Dependence of the parameters of the Vilnius photometric system on spectral type, luminosity, chemical composition and interstellar absorption.**
E. E. Belyaeva.
Kazan. univ. Kazan', 1982. 6 pp. In Russian. – Abstr. in Ref. zh., 51. Astron., 8.51.888 (1982).

113.021 **VRI photometry of nearby dwarf M stars.**
E. W. Weis.
Bull. American Astron. Soc., Vol. 14, 656 (1982). – Abstract.

113.022 **Search for rapid variability of 53 Cam.**
J. Zverko.
Bull. Astron. Inst. Czechoslovakia, Vol. 33, 314 - 318 (1982).
Photoelectric observations of a magnetic Ap star 53 Cam made at the Skalnaté Pleso Observatory in 1978 and 1979 are analysed from the point of view of rapid variability. Each observation run is analysed in detail and it was concluded that all observed variations are only apparent and due to the variability of atmospheric extinction above the observation site.

113.023 **Narrow-band photometry in the 1 - 4 μm region: calibration and applications.**
R. F. Wing, C. P. Rinsland.
Rev. Mexicana Astron. Astrofis., Vol. 6, (see 012.017), 145 - 158 (1981).
The technique of narrow-band photometry has the important advantage that it can provide both photometric and spectroscopic information. Longward of 1 μm, the absolute calibration of the photometry has generally been based on theoretical energy distributions, but a program now in progress at Kitt Peak National Observatory is providing direct measurements of absolute monochromatic fluxes in a set of standard stars at 13 wavelengths. Color temperatures from narrow band measurements at 1.04 and 4.00 μm are especially well suited for very cool stars and are in good agreement with the new effective temperature scale established from lunar occultation observations. Spectral features that can be measured by narrow-band photometry in the 1 - 4 μm region include bands of CO, CN, C_2, OH, H_2O, and SiO, and they offer the possibility of determining two-dimensional spectral classifications for late-type stars from infrared data alone.

113.024 **Thirteen-color photometry of Be stars.**
M. Alvarez, W. J. Schuster.
Rev. Mexicana Astron. Astrofis., Vol. 6, (see 012.017), 163 - 172 (1981).
Thirteen-color photometry for a number of spectroscopically variable Be and shell stars is presented. Several of these stars also show photometric variability in the ultraviolet and/or infrared over a time base of two to three years. The authors analyze the more interesting stars in terms of color-color diagrams, color excesses, spectral characteristics and changes in their energy distributions. Prospects for future research are discussed.

113.025 **Properties of G-type stars from *uvby* photometry.**
A. Ardeberg, H. Lindgren.
Rev. Mexicana Astron. Astrofis., Vol. 6, (see 012.017), 173 - 180 (1981).
Because of the many interesting aspects attached to the study of stars of spectral type G, a preliminary investigation has been made of the suitability of the *uvby* photometric system for such stars.

113.026 **Photometric properties of Ap stars in the Geneva system.**
B. Hauck, P. North.
Astron. Astrophys., Vol. 114, 23 - 40 (1982).
More than 600 Ap stars have now been measured in the Geneva photometric system, and their list is given. Their properties in some photometric diagrams are examined; it is confirmed that the Balmer discontinuity is smaller than for normal stars. The link between the peculiarity parameter $\Delta(V1-G)$, the rotational velocity and the effective magnetic field is confirmed. The standard deviations for visual magnitudes are examined.

113.027 **Near-infrared photometric observations of Be stars.**
N. M. Ashok, H. Bhatt, P. V. Kulkarni.
Bull. Astron. Soc. India, Vol. 10, 33 (1982). – Abstract.

113.028 **Near-infrared observations of some of the IRC objects.**
K. V. K. Iyengar, F. Strafella, D. Lorenzetti, C. B. Cosmovici.
Bull. Astron. Soc. India, Vol. 10, 47 (1982). – Abstract.

113.029 **Near-infrared photometry of unidentified EIC-1 sources.**
S. K. Ghosh, M. V. K. Apparao, R. R. Daniel, K. V. K. Iyengar, T. N. Rengarajan, S. N. Tandon, R. P. Verma.
Bull. Astron. Soc. India, Vol. 10, 50 (1982). – Abstract.

113.030 **Some evidence for variation of the spectrophotometric standard 109 Virginis.**
B. J. Taylor.
Publ. Astron. Soc. Pacific, Vol. 94, 663 - 668 (1982).
The existing spectrophotometric data for the standard star 109 Vir show evidence for intermittent variation during the years 1966 - 68, and again after 1977. During both epochs, the variation has been of a nature to increase $(b-y)$ and decrease m_1 by roughly equal amounts. Evidence for variation of this type also appears in Strömgren data obtained by Perry during the early 1960s. The amount of this latter variation is very similar to that predicted from the later-epoch scans.

113.031 **UBV–Hβ photometry of luminous stars between $l = 335°$ and $l = 6°$.**
J. Dachs, D. Kaiser, A. Nikolov, W. A. Sherwood.
Astron. Astrophys., Suppl. Ser., Vol. 50, 261 - 275 (1982).
In order to study galactic spiral structure and the distribution of interstellar dust photoelectric *UBV* and Hβ photometry

was carried out for a total of 316 OB stars (including several supergiants of later spectral types) located in the southern Milky Way between galactic longitudes 335° and 6°. *UBV* magnitudes are presented for these stars supplemented by Hβ photometry for 37 of them. For 161 stars with measured β indices or known MK spectral types, the absolute magnitudes, distances and colour excesses were determined. The results are discussed in terms of galactic spiral structure. New distances are derived to five open clusters from program stars supposed to be cluster members.

113.032 **Preliminary results of a new study of the intrinsic properties of Wolf-Rayet stars.**
D. G. Turner.
Wolf-Rayet stars: observations, physics, evolution, (see 012.042), p. 57 - 61 (1982).

Results are presented of a semi-empirical investigation of the transformation relations between colors and magnitudes in Lindsey Smith's narrow band photometric system for WR stars and their corresponding parameters in the broad band *UBV* system. A preliminary application of these relations to the study of WR stars in open clusters indicates that the intrinsic colors of WR stars are slightly bluer than the values tabulated by Smith, and that single WN stars are less luminous than previously supposed.

113.033 **Near-infrared photometry of southern galactic Wolf-Rayet stars.**
C. Sterken, C. de Loore.
Wolf-Rayet stars: observations, physics, evolution, (see 012.042), p. 67 - 72 (1982).

113.034 **Absolute spectrophotometry of Wolf-Rayet stars: are the colors the same?**
P. Massey.
Wolf-Rayet stars: observations, physics, evolution, (see 012.042), p. 121 - 125 (1982).

113.035 **The near infrared properties of selected WN stars in the Large Magellanic Cloud.**
N. Panagia, E. G. Tanzi, M. Tarenghi.
Wolf-Rayet stars: observations, physics, evolution, (see 012.042), p. 231 - 235 (1982).

The authors report on preliminary infrared photometry of six WN stars in the Large Magellanic Cloud. The results do not confirm the extreme infrared colours of magellanic WN stars reported by Hyland et al (1978) but, on the contrary, indicate their similarity to galactic WN stars in the continuous energy distribution from the visible to the infrared. From the present data the authors derive the parameters of the stellar wind. The computed mass loss rates compare well to those found for galactic WN stars.

113.036 **The Wolf-Rayet stars in 30 Doradus.**
J. Melnick.
Wolf-Rayet stars: observations, physics, evolution, (see 012.042), p. 545 - 549 (1982).

113.037 **Simulated numerical transformation between close photometric systems.**
K. Pavlovski.
Publ. Astron. Obs. Sarajevo, Vol. 1, No. 1, (see 012.043), p. 184 (1981). – Abstract.

113.038 **Supplementary Strömgren photometry of the Ap stars HD 30849 and HD 53116.**
H. Hensberge, W. W. Weiss, F. A. Catalano, H. Schneider.
Inf. Bull. Variable Stars, No. 2170, 5 pp. (1982).

113.039 **A note on two Be stars.**
A. Heck, J. Manfroid.
Inf. Bull. Variable Stars, No. 2198, 7 pp. (1982).

113.040 **HR 3562 and HR 3600, two new multi-periodic B-type stars.**
G. Burki, M. Burnet, A. S. Magalhães, P. North, F. Rufener, C. Waelkens.
Inf. Bull. Variable Stars, No. 2211, 4 pp. (1982).

113.041 **BP Octantis – a variable Am star.**
D. W. Coates, S. Dieters, J. L. Innis, T. T. Moon, K. Thompson.
Inf. Bull. Variable Stars, No. 2233, 4 pp. (1982).

113.042 **Short time scale light variations of 21 Comae.**
B. Musielok, T. Kozar.
Inf. Bull. Variable Stars, No. 2237 (1982).

113.043 **A photometric and spectroscopic study of faint OB stars in the southern Milky Way.**
L. P. Bassino, V. H. Dessaunet, J. C. Muzzio, S. Waldhausen.
Mon. Not. R. Astron. Soc., Vol. 201, 885 - 900 (1982).

In order to check the results derived previously from purely photometric investigations of faint OB stars the authors obtained image-tube spectra of 69 early-type stars in the Vela, Centaurus, Circinus and Norma sections of the Milky Way. Besides, the study of the Vela and Norma regions was extended with new *UBV* photoelectric observations. The present study lends support to the existence of groups of OB stars already suggested by the photometry and adds new members to them. The association of some of these groups, in particular a very distant one in Circinus, with WR stars is likely. Several stars seem to be affected by anomalous extinction.

113.044 **The intrinsic colours of stars in the ultraviolet.**
D. J. Carnochan.
Mon. Not. R. Astron. Soc., Vol. 201, 1139 - 1153 (1982).

Twenty-five intrinsic ultraviolet colours for normal stars, covering most spectral types and luminosity classes, between O6.5 V and K3 III have been derived from observations by the S2/68 sky survey telescope. The 25 colours are based on absolute fluxes that cover the wavelength range 1400 to 2740 Å. The fluxes have been corrected from the absolute calibration of S2/68 to the standard calibration of Bohlin et al. (1980) for the star η UMa. Complementary tables giving the number of stars, the mean error, $(B-V)_0$, $(U-B)_0$, absolute visual magnitude, and some reddening-free colours, are also given. The local interstellar extinction curve has been re-examined and the mean values of $A(\lambda)/E(B-V)$ calculated for all 25 passbands.

113.045 **Note on HD 20722, a K giant in nebulosity.**
B. S. Carter, G. Roberts, M. W. Feast.
Mon. Notes Astron. Soc. South. Africa, Vol. 41, 52 - 53 (1982).

113.046 **Stars of spectral types A and B in the southern galactic halo. I. *UBV* photometry.**
J. R. Pier.
Astron. J., Vol. 87, 1515 - 1526 (1982).

The results of photoelectric *UBV* photometry of 234 AB stars in the southern galactic halo are presented. The stars were chosen from the Curtis Schmidt objective-prism survey of Preston and Shectman and comprise a sample of halo objects selected without kinematic bias. The majority of the stars lie in the region of the $(U-B, B-V)$ plane occupied by field and globular cluster blue horizontal-branch stars.

113.047 **Evidence for starspots on single solar-like stars.**
J. D. Dorren, E. F. Guinan.
Astron. J., Vol. 87, 1546 - 1557 (1982).

Hα intermediate-band and narrowband, and λ 4530 photoelectric photometry has been carried out for five of the single, main-sequence stars shown by Wilson to have significant variations in Ca II H + K flux on a time scale of years. For three of the stars studied, HD 149661 (12 Ophiuchi),

HD 152391, and HD 201091 (61 Cygni A), low-amplitude variations are found that are attributable to the presence of starspots, at least in the cases of 12 Oph and 61 Cyg A where an anti-correlation between the light variation and Hα and Ca II emission is found. The spot temperature is about 200 K cooler than the photosphere, and the total spot area is roughly 5% of the total stellar surface area.

113.048 **Photoelectric sequences for three open clusters: C0126+630 (NGC 559), C2137+572 (Tr 37), C2152+623 (NGC 7160).**
P. Cardon de Lichtbuer.
Vatican Obs. Publ., Vol. 2, 1 - 13 (1982).
UBVRI photoelectric photometry is given for a total of 88 stars, thus extending existing sequences in the clusters C0126+630, C2137+572 and C2152+623 to fainter magnitudes and to the *R* and *I* passbands. A tentative spectral classification is made and the results of the photometry as well as of the classification are compared to other published data.

113.049 **Near-infrared photometry, Paper I: Homogenization of near-infrared data from southern bright stars.**
J. Koornneef.
ESO Sci. Prepr., No. 222, 39 pp. (1982). – Submitted to Astron. Astrophys., Suppl. Ser.

113.050 ***UBV* photometric observations of the ultraviolet-excess objects and the Kiso photometric standard stars. I.**
M. Kondo, E. Watanabe, M. Yutani, T. Noguchi.
Publ. Astron. Soc. Japan, Vol. 34, 541 - 552 (1982).
UBV observations of the KUV's (ultraviolet-excess objects found from the Kiso *UGR*-survey) and the KPS's (some stars selected as the standard stars for photographic photometry at the Kiso Observatory) are continued. Results for 47 KUV's and 36 KPS's are given.

113.051 **Photometric behavior of Be stars.**
A. Feinstein, H. G. Marraco.
Be Star Newsl., No. 1, p. 5 - 6 (1980).

113.052 **Short-period light variations in Be stars.**
J. R. Percy, S. M. Jakate, J. M. Matthews.
Be Star Newsl., No. 1, p. 7 (1980).

113.053 **X Persei.**
M. Ferrari-Toniolo, A. Giangrande, P. Persi, R. Viotti.
Be Star Newsl., No. 2, p. 2 - 3 (1980).

113.054 **Near infrared excess in Be stars.**
A. Feinstein, H. G. Marraco.
Be Star Newsl., No. 3, p. 2 (1981). – Abstract.

113.055 **Infrared photometry of southern Be stars.**
J. Dachs, W. Wamsteker.
Be Star Newsl., No. 4, p. 2 (1981). – Abstract.

113.056 **Short-period variability in Be stars.**
J. R. Percy.
Be Star Newsl., No. 6, p. 8 - 9 (1982).

113.057 **Near infrared photometric studies of Be stars.**
N. M. Ashok, H. C. Bhatt, P. V. Kulkarni, S. C. Joshi
Be Star Newsl., No. 6, p. 19 (1982).

113.058 ***UBV* photometry of stars in Kapteyn's Selected Area No. 57.**
M. Kondo, E. Watanabe, M. Yutani.
Tokyo Astron. Obs. Rep., Vol. 19, 488 - 502 (1981). In Japanese.

113.059 **Spectral classification and extinction using ANS ultraviolet photometry.**
P. R. Wesselius.
Ultraviolet stellar classification, (see 012.062), p. 15 - 23 (1982).
The ultraviolet photometer in the Dutch satellite ANS observed 4700 positions in the sky, mostly stars. The major results of the attempts to establish a stellar classification based on UV photometry and to study interstellar extinction in the UV region are reported.

113.060 **Observations of nearby M dwarf stars in *VRI* colors.**
E. W. Weis, A. R. Upgren.
Publ. Astron. Soc. Pacific, Vol. 94, 821 - 827 (1982).
Photoelectric photometry on the Kron system is presented for 85 nearby M dwarf stars, nearly all of which have well-determined trigonometric parallaxes. For 33 stars originally observed by Kron, comparison with the present photometry reveals some differences which may indicate significant variability for many of these stars, particularly those with hydrogen emission. The main sequence defined in the M_v, $R-I$ plane by this sample biased toward higher proper motion lies about $0\overset{m}{.}5$ below the lower main sequence of the Hyades cluster.

113.061 ***UBVRI* photoelectric sequences for M 83 and NGC 5128.**
D. B. McElroy, R. M. Humphreys.
Publ. Astron. Soc. Pacific, Vol. 94, 828 - 831 (1982).
UBVRI photoelectric sequences are presented for stars in regions near the two galaxies M 83 and NGC 5128 (Centaurus A). The sequences have a range in visual magnitude from about $9\overset{m}{.}7$ to $18\overset{m}{.}2$.

113.062 **Observations of cool stars at 20, 25, and 33 μm.**
W. Hagen.
Publ. Astron. Soc. Pacific, Vol. 94, 835 - 844 (1982).
Broad-band 20 - 33 μm photometry is presented for a sample of cool stars. The long-wavelength flux is interpreted in terms of dust-shell models. It is found that the amount of silicate dust necessary to produce the observed 10 μm features can also produce the long-wavelength flux. The dust observations are used to derive lower limits for mass-loss rates. Sufficient 35-μm flux exists to pump the 1612 MHz OH masers for all stars observed in this sample, with the exception of R Aql. The data do not rule out the possibility of infrared pumping of the main-line masers.

113.063 **Photographic photometry of stars in the region of the open cluster IC 4996 in the Vilnius photometric system.**
A. Pučinskas.
Bull. Vilnius Astron. Obs., No. 59, 45 pp. (1982). In Russian.
The results of photographic photometry of stars in the area of 1 square degree around the open cluster IC 4996 in the Vilnius photometric system are given. The final catalogue presents the magnitude V, color indices $U-P$, $P-X$, $X-Y$, $Y-Z$, $Z-V$, and $V-S$, photometrically determined spectral class and absolute magnitude, color excess E_{B-V}, extinction A_V and distance r for 888 stars down to $V \approx 14.5$.

113.064 **The new calibration of the Vilnius photometric system in spectral types and absolute magnitudes. II.**
V. Straižys, G. Kurilienė, E. Jodinskienė.
Bull. Vilnius Astron. Obs., No. 60, p. 3 - 15 (1982). In Russian.
The reddening-free diagrams Q, Q of the seven-color Vilnius photometric system are calibrated in MK spectral types and absolute magnitudes determined from trigonometric and statistical parallaxes and using the stars in clusters as summarized by Straižys and Kurilienė (1981).

113.065 **The calibration of the Vilnius photometric system in temperatures and gravities.**

V. Straižys, E. Jodinskienė, G. Kurilienė.
Bull. Vilnius Astron. Obs., No. 60, p. 16 - 49 (1982). In Russian.

Reddening-free diagrams Q, Q and color indices of the Vilnius photometric system are calibrated in effective temperatures T_e and surface gravities g for the solar chemical composition stars.

113.066 **Classification properties of the Vilnius-Geneva photometric system. I. Preliminary calibration.**
V. Straižys, E. Jodinskienė, B. Hauck.
Bull. Vilnius Astron. Obs., No. 60, p. 50 - 66 (1982).

The reddening-free parameters F of the seven-color medium band photometric system combined from slightly modified magnitudes of the Vilnius and the Geneva photometric systems are calibrated in spectral types and absolute magnitudes. Additionally, the numerical values of response curves of the system, intrinsic colors, and color excess ratios are given.

113.067 **Determination of temperatures, metallicities, and interstellar reddening for G–K giants and subgiants in the Vilnius photometric system.**
V. Straižys, A. Bartkevičius.
Bull. Vilnius Astron. Obs., No. 61, p. 22 - 31 (1982). In Russian.

The problem of simultaneous determination of temperatures, metallicities, and color excesses of late-type giants and subgiants purely from photometric data has not been solved until now. The authors present the solution of the problem using new observations of 130 metal-deficient giants in the Vilnius photometric system published recently by Sperauskas et al. (1981).

113.068 ***WBVR* photometry of SS 433: the "normal" star and accretion disk spectra.**
A. M. Cherepashchuk, A. A. Aslanov, V. G. Kornilov.
Astron. Zh., Tom 59, 1157 - 1166 (1982). In Russian.
English translation in Soviet Astron., Vol. 26, No. 6.

The results of *WBVR* photometry of SS 433 at different phases of the 13-day period are presented. The light curves in the *WBVR* filters are qualitatively similar, the depth of the primary minimum being $0^m\!.8$, $0^m\!.7$, $0^m\!.55$, $0^m\!.4$ in the W, B, V, R filters respectively. The interstellar reddening $A_V = 7^m\!.4 - 8^m\!.3$ for SS 433 is determined. Using infrared observations the authors have obtained non-reddened spectra of the "normal" star and of the accretion disk at the phases 0.5 and 0 of the 164-day precession period. The spectrum of the "normal" star for the value $A_V = 7^m\!.4 - 8^m\!.3$ has the colour temperature T_C = 13 000 - 43 000 K. The mass of the "normal" star is $> 10 M_\odot$. The analogy between the processes in the accretion disk in the SS 433 system and processes observed in galactic nuclei and quasars is pointed out.

113.069 **The photometric variability of solar-type stars. I. Preliminary results for the Pleiades, Hyades, and the Malmquist field.**
R. R. Radick, L. Hartmann, D. Mihalas, S. P. Worden, J. L. Africano, A. Klimke, E. T. Tyson.
Publ. Astron. Soc. Pacific, Vol. 94, 934 - 944 (1982).

The authors present results from the first year of a program of precision photometry undertaken at Cloudcroft Observatory to search for variability among solar-type stars. Young main-sequence stars in the Pleiades and Hyades clusters and old-disk stars in the Malmquist Field near the north galactic pole were observed. Significant variations are found for five of 42 Pleiades stars, five of 44 Hyades stars, and possibly one of 41 stars in the Malmquist Field.

113.070 **Photoelectric standard stars near V1343 Aquilae (SS 433).**
E. M. Leibowitz, H. Mendelson.
Publ. Astron. Soc. Pacific, Vol. 94, 977 - 978 (1982).

The V, $(B-V)$, and $(U-B)$ color indices of 26 stars in the immediate neighborhood of V1343 Aql are presented, with a finding chart. The magnitudes of the standard stars are in the range of the variability of SS 433.

ANS ultraviolet photometry, catalogue of point sources.
See Abstr. 002.001.

A catalogue of photometric sequences.
See Abstr. 002.108.

Supplement No. 3: Catalogue of Photometric Sequences.
See Abstr. 002.117.

Atlas of photometric standards of stellar fields.
See Abstr. 002.149.

Astronomical photometry.
See Abstr. 003.056.

A simple and accurate on-line data acquisition program.
See Abstr. 021.059.

Large scale photometric surveys using archival plates.
See Abstr. 031.505.

Geneva photometric boxes. IV. A refined method for direct access.
See Abstr. 031.506.

Temperature effect in zero-point drift.
See Abstr. 031.508.

Photometry among the climatically underprivileged.
See Abstr. 031.522.

Color equations for the United Kingdom Schmidt Telescope.
See Abstr. 031.589.

A new search technique for M and C stars.
See Abstr. 031.667.

What can infrared photometry tell you?
See Abstr. 034.005.

Infrared broad-band filters optimised for observations of Wolf-Rayet stars.
See Abstr. 034.110.

Comments on the 'TYCHO' photometry proposal.
See Abstr. 051.078.

A dust shell around the yellow supergiant CoD −61°3326.
See Abstr. 112.034.

Spectrophotometry of peculiar B and A stars. XII. HD 10783, 56 Tauri, HD 43819, 53 Aurigae, 49 Camelopardalis, HD 64486, HD 147550, HD 184905 and HD 192913.
See Abstr. 114.001.

Spectroscopic and photometric study of barium stars.
See Abstr. 114.065.

Carbon and nitrogen abundances in Population I G and K giant stars from DDO photometry.
See Abstr. 114.073.

Z Andromedae: the prototype?
See Abstr. 114.152.

Optical observations of ultraviolet objects – II. Classification and photometry (l = 0 to 145°).
See Abstr. 114.168.

Correlations between line profile and photometric variations in the B2 $IV_{[e]}$ star HD 45677.
See Abstr. 114.180.

UV spectrophotometry of early-type stars.
See Abstr. 114.198.

New light on faint stars – I. The luminosity function in the solar neighbourhood.
See Abstr. 115.008.

New light on faint stars – II. A photometric study of the low luminosity main sequence.
See Abstr. 115.009.

The two-colour diagram of luminous stars in the Magellanic Clouds.
See Abstr. 115.016.

Magnetic structure in cool stars. VI. Ca II H and K fluxes from evolved stars.
See Abstr. 116.004.

Photometry of the newly identified AM Herculis system: CW 1103 + 254.
See Abstr. 117.051.

Properties of Wolf-Rayet stars in eclipsing binary systems.
See Abstr. 117.099.

IR photometry of symbiotic stars at Calar Alto Observatory.
See Abstr. 117.101.

Properties and nature of Be and shell stars. 11. A notable correlation between the long-term spectral and photometric variations of V 1294 Aql (HD 184 279).
See Abstr. 122.050.

Photometric variability of single stars with activity cycles.
See Abstr. 122.152.

A photometric map of interstellar reddening within 300 parsecs.
See Abstr. 131.296.

A photometric map of interstellar reddening within 100 pc.
See Abstr. 131.309.

Objetos con exceso ultravioleta en la vecindad del polo galáctico sur.
See Abstr. 141.091.

***UBV* photometry of SMC X-1 and Cen X-3. Some regularities in the observational appearance of X-ray binary systems.**
See Abstr. 142.048.

The nature of the 1E1145.1–6141 optical counterpart.
See Abstr. 142.083.

***UBV* photometry of the X-ray binary HD 153919 = 4U1700–37.**
See Abstr. 142.090.

Photoelectric photometry of Ap stars in the Orion and upper Scorpius associations: preliminary results.
See Abstr. 152.005.

New *UBV* photometry for the open cluster Stock 14 and its cepheid-like variable V810 Centauri (= HR 4511).
See Abstr. 153.018.

Photometry of low-mass stars in Praesepe.
See Abstr. 153.019.

***UBVRI* photometry in the open cluster NGC 3532.**
See Abstr. 153.020.

A photoelectric investigation of Ap-stars in open clusters. III. NGC 2362, NGC 2546, and NGC 3228.
See Abstr. 153.024.

The F0 Ia stars in NGC 457 and IC 2581.
See Abstr. 153.043.

A kinematic and abundance survey at the galactic poles.
See Abstr. 155.062.

Photographic surface photometry of the Milky Way. IV. The northern Milky Way in the ultraviolet spectral region.
See Abstr. 157.011.

The Bok and Tifft UBV sequence in the Large Magellanic Cloud: revised and extended.
See Abstr. 159.010.

A negative search for young stars associated with the Magellanic Stream.
See Abstr. 159.015.

114 Spectra, Temperatures, Chemical Composition, etc.

114.001 **Spectrophotometry of peculiar B and A stars. XII. HD 10783, 56 Tauri, HD 43819, 53 Aurigae, 49 Camelopardalis, HD 64486, HD 147550, HD 184905 and HD 192913.**
S. J. Adelman.
Astron. Astrophys., Suppl. Ser., Vol. 49, 663 - 672 (1982).
Optical region spectrophotometry of nine magnetic Ap stars is presented. A new scan of HD 10783 better defines the λ5200 feature profile as it has more values than previous scans. For the large amplitude variables 56 Tauri and HD 184905, scans at different parts of their cycles reveal large changes in the sizes of their Balmer jumps relative to those in their Paschen continua. HD 43819 and 49 Cam have energy distributions similar to other Ap stars in their color temperature domains. 53 Aurigae, HD 64486, and HD 147550 are found to be spectrophotometric mild Ap stars. Observations of HD 192913 about its cycle of variability show that the size of the Balmer jump and the λ4200 feature vary together.

114.002 **The Mg II resonance line emission at 2800 Å in stars with different rotational velocities and different metal abundances.**
E. Böhm-Vitense.
Astrophys. J., Vol. 258, 628 - 638 (1982).
The author has studied Mg II *k* line (2795.7 Å) emission fluxes in stars of different effective temperatures, gravities, rotational velocities, and metal abundances. Stars with variable radial velocities show enhanced fluxes, while rapidly rotating F stars show reduced fluxes as compared to slowly rotating stars. A reduction of *k* emission for metal deficient stars is found, which is nearly independent of the deficiency factor, except for extremely low abundances. The implication for the heating of the chromospheres is discussed.

114.003 **An application of the method of wavelength coincidence statistics to the ultraviolet spectrum of κ Cancri.**
D. J. Bord, J. P. Davidson.
Astrophys. J., Vol. 258, 674 - 682 (1982).
The method of wavelength coincidence statistics (WCS) is successfully applied to the ultraviolet spectrum of the Hg-Mn star κ Cnc. Methodological checks involving searches for the elements Cr, Mn, and Fe yield positive results. A search for heavy elements with $74 \leqslant Z \leqslant 81$ in the region λλ1960–3230 provides weak evidence to support the identification of Os II and Hg I; no confirmation of the presence of Au or Pt is forthcoming from this study.

114.004 **New Hα emission stars in galactic dark clouds.**
M. Kun.
Astrofizika, Tom 18, 63 - 66 (1982).
110 Hα emission objects not published previously were detected in two fields containing several dark clouds. The centres of these fields are at $\alpha = 2^h 04^m$, $\delta = +75°$, and at $\alpha = 22^h 35^m$, $\delta = +75°$. Most of the emission stars appear to be located near the edges of the dark regions. Their apparent red magnitudes are between 11^m and 16^m.

114.005 **On the nature of SS433.**
V. M. Lipunov, N. I. Shakura.
Astron. Zh., Tom 59, 631 - 635 (1982). In Russian. English translation in Soviet Astron., Vol. 25, No. 4.
A model of the object SS433 is considered. The observational properties of SS433 are assumed to be due to supercritical disk accretion onto the neutron star having a strong magnetic field.

114.006 **Scanner studies of composite spectra. II. Giants and dwarfs.**
W. I. Beavers.
Astrophys. J., Suppl. Ser., Vol. 49, 273 - 292 (1982).
In this second scanner study of composite spectra, observations with the low resolution scanner at CTIO in the region 3500–4400 Å have been employed to calibrate a set of line and continuum indices as a function of spectral type over the range late O to early G for giants. The observed spectrum scans of 58 visual, spectroscopic, and eclipsing pairs have been compared with a grid of calculated composites. Giant-giant, giant-dwarf, or dwarf-dwarf pairs of composites generated by a computer program, with appropriate ranges of spectral types and assumed magnitude differences are compared with the observations in order to find the combination which fits best.

114.007 **A search for medium *Z* elements in the ultraviolet spectrum of κ Cancri.**
J. P. Davidson, D. J. Bord.
Astron. Astrophys., Vol. 111, 362 - 364 (1982).
The authors have applied the method of wavelength coindicence statistics to search for selected elements having $12 \leqslant Z \leqslant 40$ in the long wavelength (1960–3230 Å) IUE spectrum of the Hg-Mn star κ Cancri. Evidence to support the identifications of the even *Z* elements Mg, Si, Ca, Sr, and Zr, in addition to the odd *Z* element Cl, has been found. The results are compared with those of other authors in an attempt to confirm the existence of elemental sequences in which violations of the usual even-odd effect occur.

114.008 **Optical observations of ultraviolet objects – I. Spectral classification of 103 stars ($l = 200°$ to 275°).**
M. M. Dworetsky, P. A. Whitelock, D. J. Carnochan.
Mon. Not. R. Astron. Soc., Vol. 200, 445 - 454 (1982).
The authors give results of a programme of spectral classification of 103 stars originally selected as 'ultraviolet objects' from *TD-1* satellite photometry with the S2/68 experiment. Most of the objects appear to be spectroscopically normal stars; the method of selection yielded a sample of relatively unreddened B stars at distances up to ~ 2 kpc. This result is compared with recent studies of the spatial distribution of interstellar extinction in the same regions of the sky.

114.009 **Effective temperatures and radii of Large Magellanic Cloud supergiants.**
G. I. Thompson, K. Nandy, D. H. Morgan, A. J. Willis, R. Wilson, L. Houziaux.
Mon. Not. R. Astron. Soc., Vol. 200, 551 - 562 (1982).
The absolute fluxes and flux distributions of 25 Large Magellanic Cloud supergiants are compared to those of similar galactic stars. The effective temperatures and radii of the LMC stars are found by comparison with models. No significant difference between the LMC and our Galaxy is found.

114.010 **Atmospheric parameters and chemical compositions of eighteen halo OB stars.**
F. P. Keenan, P. L. Dufton, C. D. McKeith.
Mon. Not. R. Astron. Soc., Vol. 200, 673 - 685 (1982).
Stellar equivalent widths and Hє line profiles, measured from 5 and 10 Å mm^{-1} IPCS spectra obtained at the AAT, are presented for 18 halo OB stars. Effective temperatures and gravities have been estimated using Strömgren $[c_1]$ and Hβ photometry in conjunction with these data. Also derived are the abundances relative to hydrogen of helium, carbon, nitrogen, oxygen and calcium. The authors conclude that the stars are not subluminous and show no evidence of any other pecularities. Using conservative and non-conservative evolutionary tracks, masses, ages and distances are deduced for the

stars, implying that several of them have been ejected from the galactic plane with velocities in excess of 100 km s^{-1}.

114.011 **Dependence of absorption line profiles on the parameters of rotating stars.**
M. Ruusalepp.
Publ. Tartu Astrofiz. Obs., Tom 48, 81 - 99 (1981).
In Russian.

A grid of rotationally broadened line profiles for He I λ 4471 Å and Mg II λ 4481 Å has been computed using theoretical calculations made by Stoeckley and Mihalas (1973). A Roche model and solid-body rotation have been assumed, limb and gravity darkenings have been taken into account. Von Zeipel's theorem has been used to obtain the effective temperature at any latitude of the star, which can be observed at various angles of inclination *i* and which is rotating at various values of the fractional angular velocity ω/ω_c. Half-intensity widths have been measured from the computed theoretical line profiles and have been plotted against the angle of inclination *i*.

114.012 **A survey of Hα emission in Oe and Oef stars.**
U. Hänni.
Publ. Tartu Astrofiz. Obs., Tom 48, 100 - 108 (1981).

An Hα survey of a group of the Oe and Oef stars was undertaken by the author. The primary data of the observations and the observed Hα profiles are presented, and a discussion follows.

114.013 **The nature of five Wolf-Rayet stars.**
T. Nugis.
Publ. Tartu Astrofiz. Obs., Tom 48, 109 - 137 (1981).

The author lists first more or less reliable characteristics of the WR stars and next he describes a method to determine poorly known parameters such as T_{eff} ,T_e ,N_e and others. The main physical parameters of the stars studied are presented and their reliability is discussed. C and N abundance estimates are given, and the specific features of stellar winds from the WR stars studied are cleared up.

114.014 **A search for TiO isotopic bands in the spectra of late-type giants.**
L. Hänni.
Publ. Tartu Astrofiz. Obs., Tom 48, 138 - 149 (1981).

The primary purpose of the paper is to search for isotopic band heads of the γ(1, 0) band and, if possible, to determine the relative abundances of ^{46}Ti and ^{47}Ti. The spectral material does not enable to study the (0, 1) band region, where the band heads of ^{49}TiO and ^{50}TiO can be found.

114.015 **Development of a technique of quantitative spectral classification of F and G stars with determination of the metallicity parameter.**
G. N. Dzhimshelejshvili, V. D. Malyuto.
Publ. Tartu Astrofiz. Obs., Tom 48, 183 - 198 (1981).
In Russian.

The authors undertook a search for criteria of a quantitative spectral classification for a wider spectral class interval and with the use of more extensive observational material. The basic results of previous work (1977) have been confirmed, some new criteria specially for later G-type stars have been found.

114.016 **The peculiar object SS433.**
B. Margon.
Galactic X-ray sources, (see 003.001), p. 417 - 435 (1982).

114.017 **Stellar surface phenomena: the Hα profile in RS CVn stars.**
S. E. Smith, B. W. Bopp.
Astrophys. Lett., Vol. 22, 127 - 130 (1982).

The authors report high-resolution (~0.3 Å) observations of the Hα profile in RS CVn and related Ca II emission stars. They find the Hα profiles of such stars to be readily distinguishable, at this resolution, from those of MK standards. RS CVn systems, which represent possible new radio, soft X-ray, and UV emission objects, can be identified on the basis of visual examination of a single high-resolution Hα spectrogram, and are thus detectable in Hα survey programs.

114.018 **Infrared spectroscopy in Australia.**
T. J. Jones.
Proc. Astron. Soc. Australia, Vol. 4, 152 - 155 (1981).

The author briefly describes several projects involving IR spectra performed in Australia. Special emphasis is placed on those experiments that cannot be done by any means other than IR spectroscopy. Also presented is a description of Australia's next generation of spectroscopic instrumentation for the IR.

114.019 **On the non-stability of the precession period of SS433.**
I. M. Kopylov, R. N. Kumajgorodskaya, V. B. Nebelitskij, T. A. Somova, O. I. Spiridonova.
Pis'ma Astron. Zh., Tom 8, 461 - 466 (1982). In Russian.
English translation in Soviet Astron. Lett., Vol. 8.

Spectral observations of SS433 made in June - July 1980 and June 1981 have shown that a 165-day period of spectral and photometric variability identified with the precessional period of relativistic jets ejected from this system decreased noticeably during 1978 - 1981.

114.020 **On the law of variation of the precession period of SS433 in 1978 - 1981.**
I. M. Kopylov, R. N. Kumajgorodskaya, T. A. Somova.
Pis'ma Astron. Zh., Tom 8, 467 - 471 (1982). In Russian.
English translation in Soviet Astron. Lett., Vol. 8.

From spectral observations of SS433 and other available data an attempt is made to find an analytical form for the law of decrease of the precessional period *P* of SS433 in 1978 - 1981.

114.021 **HDE 283809 – a B-type star in the direction of the dark Taurus cloud.**
V. Straižys.
Pis'ma Astron. Zh., Tom 8, 481 - 484 (1982). In Russian.
English translation in Soviet Astron. Lett., Vol. 8.

Spectral and photometric observations of the heavily reddened star HDE 283809 in the dark Taurus cloud show that it belongs to the spectral class B3. At the same time it has a strong K Ca II line and a peculiar spectral energy distribution characterized by weakened radiation in the violet and ultraviolet region and enhanced intensity in the red and infrared.

114.022 **The structure and energy balance of cool star atmospheres.**
J. L. Linsky.
Advances in ultraviolet astronomy, (see 012.006), p. 17 - 32 (1982).

A broad theme emerging from IUE observations of cool stars is that magnetic fields control the structure and energy balance of the outer atmospheres of these stars. The author summarizes the phenomena associated with magnetic fields in the sun and shows that similar phenomena occur in cool stars.

114.023 **IUE observations of symbiotic stars.**
M. Hack.
Advances in ultraviolet astronomy, (see 012.006), p. 89 - 101 (1982).

The main photometric and spectroscopic characteristics in the ultraviolet and visual range of the most extensively studied symbiotic stars are reviewed.

114.024 **Synoptic studies of chromospheric variability in F - K dwarfs with the IUE.**
K. L. Hallam, C. L. Wolff, J. R. Sewall.
Advances in ultraviolet astronomy, (see 012.006), p. 227 - 230 (1982).

Time-sequential series of IUE spectra for ten F, G and K dwarfs were obtained in 1980 and 1981 to study the rotational dependence of chromospheric flux in the ultraviolet.

114.025 **Chromospheric, transition layer and coronal emission of metal deficient stars.**
E. Böhm-Vitense.
Advances in ultraviolet astronomy, (see 012.006), p. 231 - 234 (1982).

The author shows that while Mg II k line emission decreases for metal deficient stars, the Lyα emission increases. The total chromospheric energy loss is estimated to be $4 \times 10^{-5} \times F_{bol}$. The chromospheric energy input does not seem to decrease for increasing age if the age is larger than $\sim 2 \times 10^9$ years. The transition layer emission is reduced for metal deficient stars. Coronal X-ray emission was measured for 4 metal deficient stars.

114.026 **Progress report of an IUE survey of the Hyades star cluster.**
M.C. Zolcinski, L. Kay, S. Antiochos, R. Stern, A. B. C. Walker.
Advances in ultraviolet astronomy, (see 012.006), p. 239 - 242 (1982).

11 of the brightest X-ray stars (F-K dwarfs) in the Hyades have been observed with the IUE satellite. Combining the IUE results with the X-ray observations from the Hyades survey with the Einstein Observatory the authors estimate the differential emission measure function for each of the 7 stars which show evidence of emission lines. They derive constraints on stellar atmospheric parameters and discuss the implications of their results in the context of loop models for the corona and transition region of these stars.

114.027 **Density sensitive C II lines in cool giant stars.**
R. E. Stencel, K. G. Carpenter.
Advances in ultraviolet astronomy, (see 012.006), p. 243 - 246 (1982).

The density sensitivity of the emission lines within the UV 0.01 multiplet of C II near 2325 Å has been examined in additional late type giants and supergiants with deep LWR high dispersion exposures. The new data support the original contention based on these lines that noncoronal red giants possess geometrically extended chromospheres.

114.028 **Chromospheric, transition layer and X-ray emission for stars with different rotational velocities.**
E. Böhm-Vitense.
Advances in ultraviolet astronomy, (see 012.006), p. 247 - 250 (1982).

In agreement with previous findings for the Mg II k line emission in F stars the author does not find an increase of Lyα and transition layer emission with increasing $v_r \sin i$, if $v_r \sin i > 30$ km/sec. For $v_r \sin i \leqslant 30$ km/sec the measured line intensities are consistent with an increase in emission with increasing $v_r \sin i$. Such a relation between emission and rotation for single stars is also in agreement with X-ray observations. For the young F stars in the Hyades the author finds generally enhanced emission independently of rotational velocities.

114.029 **The ultraviolet spectra of R and N type stars.**
G. T. O'Brien, H. R. Johnson.
Advances in ultraviolet astronomy, (see 012.006), p. 255 - 258 (1982).

The authors have obtained low dispersion LWR spectra of three early N-type carbon stars (BL Orionis, TX Piscium and T Indi) on March 20 and October 26, 1981 with the IUE satellite. Observations of several R-type stars (HD 182040 (R0), HD 16115 and HD 137613 (R3), HD 52432 (R5) and HD 37212 (R8)) were also obtained on the same dates. A high dispersion spectrum of HD 182040 was obtained, but the other R stars were observed in the low dispersion (6 Å resolution) mode.

114.030 **High dispersion far ultraviolet spectra of cool stars.**
R. E. Stencel, J. L. Linsky, T. R. Ayres, C. Jordan, A. Brown, O. Engvold.
Advances in ultraviolet astronomy, (see 012.006), p. 259 - 262 (1982).

The authors discuss recent far UV high dispersion spectra of two cool supergiant stars, Beta Dra (G2 Ib) and Alpha Ori (M2 Iab). These stars show very different outer atmosphere structure. Beta Dra has a geometrically thin transition region with bright emission lines of 10^5 K plasma that are red-shifted, indicating downflow in magnetic flux tubes. Alpha Ori has a cool extended chromosphere and circumstellar envelope with large mass loss.

114.031 **UV emission from the M1 supergiant TV Gem.**
A. G. Michalitsianos, M. Kafatos.
Advances in ultraviolet astronomy, (see 012.006), p. 263 - 267 (1982).

The authors discuss low and high dispersion ultraviolet spectra obtained of the M1 supergiant TV Gem with IUE.

114.032 **Ultraviolet observations of yellow giant stars.**
T. Simon, J. L. Linsky.
Advances in ultraviolet astronomy, (see 012.006), p. 273 - 276 (1982).

Low-dispersion spectra of 18 yellow giant stars of spectral types G4–K0 were obtained with IUE. The authors find that the normalized C IV flux is typically 10^{-7} or smaller, indicating very feeble stellar transition regions. By combining these results with earlier data from IUE, the authors show that there is nearly a two orders of magnitude spread in the normalized C IV flux among the yellow giants. Several likely reasons for the observed range in high-temperature emission line strengths are discussed; the more likely appears to be that the majority of the yellow giant stars observed are slow rotators evolving across the Hertzsprung Gap for the second time along a blue loop.

114.033 **Solar analogs in the 2600 to 3200 Å region.**
J. Hardorp, J. Caldwell, R. Wagener.
Advances in ultraviolet astronomy, (see 012.006), p. 277 - 280 (1982).

No new criteria for the selection of solar analogs are found. At low dispersion, the spectra of 16 Cyg A, 16 Cyg B, 18 Sco and α Cen A look identical to each other and to sunlight reflected from Galilean satellites with the possible exception of Mg II λ 2800 being stronger in the stars. The use of α Cen A or 16 Cyg B as solar substitutes for planetary photometry is expected to be far superior to using published solar irradiances.

114.034 **High dispersion IUE spectra of active chromosphere G and K dwarfs.**
T. R. Ayres, J. L. Linsky, A. Brown, C. Jordan, T. Simon.
Advances in ultraviolet astronomy, (see 012.006), p. 281 - 284 (1982).

The authors analyze IUE far ultraviolet echelle spectra of three active chromosphere dwarf stars: χ^1 Orionis (G0 V), ξ Boötis A (G8 V), and ϵ Eridani (K2 V), utilizing spectra of α Cen A (G2 V) and α Cen B (K1 V) as quiet chromosphere comparisons.

114.035 **Short-term UV line profile variation in 59 Cyg.**
C. A. Grady, V. Doazan, G. J. Peters, A. Willis, T. P. Snow, D. Aitken, P. K. Barker, C. T. Bolton, H. Henrichs, C. R. Kitchen, L. V. Kuhi, J. M. Marlborough, P. Meikle, J. Mendzies, W. Oegerle, R. S. Polidan, R. Rosner, P. Selvelli, R. Stalio, R. N. Thomas, G. Vaiana, P. Whitelock, R. Wilson, C. Wu.
Advances in ultraviolet astronomy, (see 012.006), p. 425 - 428 (1982).

IUE high dispersion spectra of 59 Cyg obtained as part of the long term monitoring program have shown that noticeable variation can occur in C IV and N V on timescales 3 hours $< t <$ 24–28 hours. The January 1982 spectra show no short term variation which may be consistent with sporadic rather than continuous variation.

114.036 **The hot components of hydrogen-deficient binaries.**
J. S. Drilling, D. Schoenberner.
Advances in ultraviolet astronomy, (see 012.006), p. 546 - 549 (1982).

The authors compare low-resolution IUE observations of the hot components of three very similar but peculiar objects: LSS 4300, Upsilon Sgr, and KS Per. They discuss the possible evolutionary scenarios in view of these observations and the extremely low hydrogen contents of the visible stars.

114.037 **The effective temperatures of early O and Wolf-Rayet stars.**
A. B. Underhill.
Advances in ultraviolet astronomy, (see 012.006), p. 588 (1982). – Abstract.

114.038 **UV fluxes and effective temperatures of extreme helium stars.**
D. Schoenberner, J. S. Drilling, A. E. Lynas-Gray, U. Heber.
Advances in ultraviolet astronomy, (see 012.006), p. 593 - 596 (1982).

The authors present low-resolution IUE spectra of a complete ensemble of extreme helium stars and discuss their appearance in comparison with normal stars. They have determined effective temperatures from these observations by means of line-blanketed model atmospheres. The temperatures are in accordance with earlier results from ground-based observations.

114.039 **On the UV variability of LSI+61°303 (≡GT 0236).**
E. G. Tanzi, G. F. Bignami, P. A. Caraveo, L. Maraschi, F. Sormani, A. Treves.
Advances in ultraviolet astronomy, (see 012.006), p. 615 - 618 (1982).

114.040 **A search for UV line variability in R136a.**
D. C. Ebbets, B. D. Savage, M. R. Meade.
Advances in ultraviolet astronomy, (see 012.006), p. 620 - 624 (1982).

The authors have been observing R136a with IUE since 1978, and thought it worthwhile to search for variability in the ultraviolet lines. The detection of variability would favor the single star hypothesis. The addition of many spectra has produced a composite of high quality, from which line strengths and profiles can be accurately measured.

114.041 **Fine analysis of the intermediate helium-star CPD–46°3093.**
D. Groote, J. P. Kaufmann, A. Lange.
Astron. Astrophys., Suppl. Ser., Vol. 50, 77 - 85 (1982).

A fine analysis of the intermediate He-star CPD–46°3093 was carried out by means of LTE-model-atmosphere techniques and spectroscopy of 20 Å/mm spectrograms in the range of 3600 - 4900 Å. Derived parameters are: T_{eff} = 26000 K, $\log g = 3.95$, $n_H = 0.73$, $n_{He} = 0.27$, $v \times \sin i = 60$ km s^{-1}, $v_r = 50$ km s^{-1}. A minimum distance results from interstellar lines and E_{B-V} estimates: $d \geqslant 1400$ pc. Minimum mass and radius are $M/M_\odot \geqslant 3.7$ and $R/R_\odot \geqslant 3.4$. Apart from H and He the elements O (factor 10 underabundant) as well as Si and Mg (slightly underabundant) show abundance anomalies. Deviations from circular velocity in the galactic plane are small. The star shows variability in H- and He I-lines like Be-stars. Most probably it is a young object near the main sequence and is a rapidly rotating (though seen nearly pole-on) variable He-rich star like σ Ori E.

114.042 **Bismuth abundance anomaly in a Hg–Mn star.**
J. M. Jacobs, M. M. Dworetsky.
Nature, Vol. 299, 535 - 536 (1982).

Strong absorption lines due to Bi II have been found in the Hg–Mn star HR 7775 (HD 193452) in high-resolution spectra obtained with the IUE. The authors have adopted a fully blanketed model atmosphere with solar composition, T_{eff} = 11,000 K, log g = 4.0. The overabundance of Bi obtained from spectrum synthesis calculations is $\sim 10^6$. Close examination of the spectra of several other Hg–Mn stars shows that Bi II is not detected in any other star in this group.

114.043 **Extreme helium stars.**
A. E. Lynas-Gray.
Irish Astron. J., Vol. 15, (see 012.008), 42 - 49 (1981).

Research into the nature of extreme helium stars is important for a more general understanding of stellar evolution. This paper is a brief review of published and unpublished work carried out over the last two decades.

114.044 **R 136: WN or O spectral characteristics?**
J. M. Vreux, M. Dennefeld, Y. Andrillat.
Astron. Astrophys., Vol. 113, L10 - L12 (1982).

Near infrared spectroscopic observations of R 136 indicate the absence of striking Wolf-Rayet characteristics but, on the contrary, a striking similarity with those of the most luminous galactic O star, HD 93129.

114.045 **The absorption line effect in A - F type stars.**
E. E. Belyaeva.
Kazan. univ. Kazan', 1982. 13 pp. In Russian. – Abstr. in Ref. zh., 51. Astron., 8.51.107 (1982).

114.046 **On the N V and O VI lines in the spectra of hot stars.**
I. N. Portnova, N. K. Bekeeva.
Kazan. univ., Kazan', 1982, 23 pp. In Russian. – Abstr. in Ref. zh., 51. Astron., 8.51.423 (1982).

114.047 **Spectrophotometry of three peculiar emission-line stars.**
V. P. Arkhipova, A. P. Ipatov.
Pis'ma Astron. Zh., Tom 8, 554 - 561 (1982). In Russian. English translation in Soviet Astron. Lett., Vol. 8.

Results of spectrophotometric observations of the emission-line stars with infrared excesses HD 51585, MWC 342 and MWC 623 made in the range λλ 3300 - 7560 with a resolution of ≈ 48 Å are presented.

114.048 **Beryllium abundances in Hg-Mn stars.**
A. M. Boesgaard, W. D. Heacox, S. C. Wolff, J. Borsenberger, F. Praderie.
Astrophys. J., Vol. 259, 723 - 729 (1982).

The authors have observed the Be II resonance doublet at λλ 3130, 3131 in 43 Hg-Mn stars and 10 normal stars in the same temperature range with the coudé spectrograph of the 2.24 m University of Hawaii telescope at Mauna Kea. Measured equivalent widths of the two lines and/or the blend of the doublet have been compared with predictions from (1) LTE model atmospheres and (2) non-LTE line formation in non-LTE model atmospheres. Overabundances of factors of

$20 - 2 \times 10^4$ relative to solar have been found for 75% of the Hg-Mn stars.

114.049 **Subgiant CH stars. I. s-process abundances, C/O ratios, and evolutionary status.**
R. E. Luck, H. E. Bond.
Astrophys. J., Vol. 259, 792 - 804 (1982).
The authors report abundance analyses of seven relatively bright subgiant CH stars and the red-giant CH star HD 26, based on Palomar and Mount Wilson photographic coudé spectrograms in the blue spectral region and on Mount Wilson coudé scans in the red region. The program stars were all found to be mildly metal-deficient. All of the stars show enhanced strontium-peak abundances, but the barium-peak enhancements vary from star to star, indicating a considerable range in the s-process neutron exposures. Carbon and oxygen abundances were determined in five subgiant CH stars from the near-infrared C I and O I lines. The results are compared with current models of stellar evolution and nucleosynthesis.

114.050 **The effect of line blocking on the light curves of the Ap stars HD 119213, HD 170000 and HD 173650.**
B. Musielok.
Acta Astron., Vol. 31, 443 - 456 (1981).
The paper presents results of line blocking measurements for three Ap stars, and a discussion of their light and line blocking variations in terms of the blanketing model.

114.051 **λ10830 He I observations of 455 stars.**
H. Zirin.
Astrophys. J., Vol. 260, 655 - 669, plates 20, 21 (1982).
Measurements of λ10830 equivalent widths in 890 image-tube spectra of 455 stars are presented and compared with K line, X-ray, and other data.

114.052 **Outer atmospheres of cool stars. XII. A survey of IUE ultraviolet emission line spectra of cool dwarf stars.**
J. L. Linsky, P. L. Bornmann, K. G. Carpenter, R. F. Wing, M. S. Giampapa, S. P. Worden, E. K. Hege.
Astrophys. J., Vol. 260, 670 - 694 (1982).
The authors present and discuss IUE low dispersion spectra (1150–3200 Å) of four dM and six dMe stars, together with spectra of the first 10 Balmer lines and the Ca II H and K lines. From these calibrated spectra, they extract absolute line fluxes and surface fluxes for emission lines formed in the chromospheres ($T < 20{,}000$ K) and transition regions ($T = 20{,}000–200{,}000$ K) of these stars. These data, together with data for four G–M dwarfs previously observed by IUE and the quiet Sun and solar plage regions, are intercompared to search for systematic trends.

114.053 **Low-resolution Mg II h and k observations of solar type stars.**
D. C. Barry, S. A. Schoolman.
Astrophys. J., Vol. 261, 220 - 227 (1982).
The authors describe an attempt to measure relative chromospheric radiative loss rates in F and G dwarf stars with low-resolution IUE spectra. The ratio of the unresolved Mg II h and k central depth to the continuum is shown to be related to chromospheric activity. The possibility of predicting h and k loss rates from ground-based Ca II H and K measures is investigated. The "young" (high chromospheric activity) group of program stars includes members of the Hyades, Coma, Ursa Major, and Pleiades clusters. The "old" (low chromospheric activity) group contains the Sun and four other solar type stars. The old stars are clearly separated from the young stars by their respective h and k central depth values.

114.054 **Spectral types and rotational velocities of the brighter Be stars and A - F type shell stars.**
A. Slettebak.
Astrophys. J., Suppl. Ser., Vol. 50, 55 - 83 (1982).
Spectral types and rotational velocities have been estimated for the Oe, Be, and A - F type shell stars brighter than ~6.0 mag over the entire sky. Spectra of the 183 program stars were obtained at Lowell Observatory and CTIO on fine grain, high contrast emulsion to enhance the visibility of faint and diffuse features. The distribution of stars by spectral type shows a single maximum at B2, with a very steep decline to earlier spectral types and a gradual decline to later B types. Mean values of the rotational velocity show little variation with either spectral type or luminosity class among the Be stars. Descriptions of the spectra plus a summary of recent observations of the program stars are given.

114.055 **Spectrophotometry of magnetic peculiar A stars.**
S. J. Adelman.
Bull. American Astron. Soc., Vol. 14, 613 (1982). – Abstract.

114.056 **BF Orionis – a peculiar pre-main sequence A star.**
S. C. Barden, J. A. Nousek, L. W. Ramsey.
Bull. American Astron. Soc., Vol. 14, 614 (1982). – Abstract.

114.057 **Fluorescence in stellar chromospheres.**
K. G. Carpenter, R. F. Wing, R. E. Stencel.
Bull. American Astron. Soc., Vol. 14, 614 (1982). – Abstract.

114.058 **The faintest Wolf-Rayet stars in the LMC.**
P. Massey, P. S. Conti.
Bull. American Astron. Soc., Vol. 14, 614 - 615 (1982). Abstract.

114.059 **Carbon, nitrogen and oxygen abundances in the mercury-manganese stars κ Cancri, μ Lepus and ι CrB.**
S. W. Roby.
Bull. American Astron. Soc., Vol. 14, 615 (1982). – Abstract.

114.060 **The strength of the calcium K line in halo A stars at the SGP.**
A. G. D. Philip.
Bull. American Astron. Soc., Vol. 14, 615 (1982). – Abstract.

114.061 **Radial velocity information content in solar-type spectra.**
W. J. Merline.
Bull. American Astron. Soc., Vol. 14, 625 (1982). – Abstract.

114.062 **Abundances of very heavy elements in the extreme population II giant star HD 122563.**
C. Sneden, M. Parthasarathy.
Bull. American Astron. Soc., Vol. 14, 650 (1982). – Abstract.

114.063 **Carbon isotope ratios in two giants in M 67.**
J. Brown.
Bull. American Astron. Soc., Vol. 14, 650 (1982). – Abstract.

114.064 **A search for technetium in M giants and supergiants.**
I. R. Little-Marenin, W. Hagen, S. J. Little, P. Gaeta.
Bull. American Astron. Soc., Vol. 14, 650 (1982). – Abstract.

114.065 **Spectroscopic and photometric study of barium stars.**
P. K. Lü.
Bull. American Astron. Soc., Vol. 14, 650 - 651 (1982). Abstract.

114.066 **Far-ultraviolet flux distributions of Orion stars, revisited.**
G. R. Carruthers, H. M. Heckathorn.
Bull. American Astron. Soc., Vol. 14, 651 (1982). – Abstract.

114.067 **IUE observations of the upper main sequence helium rich stars.**
P. K. Barker, D. N. Brown, J. D. Landstreet, S. N. Shore.
Bull. American Astron. Soc., Vol. 14, 651 (1982). – Abstract.

114.068 **Gas flow in the chromosphere of α Ori.**
O. Kjeldseth Moe, O. Engvold, E. Jensen, J. L. Linsky, R. E. Stencel.
Bull. American Astron. Soc., Vol. 14, 651 (1982). – Abstract.

114.069 **A high resolution EUV spectrum of α Ori.**
O. Engvold, O. Kjeldseth Moe, E. Jensen, J. L. Linsky, R. E. Stencel.
Bull. American Astron. Soc., Vol. 14, 651 - 652 (1982). Abstract.

114.070 **Infrared spectroscopy of 104.9 + 2.4.**
D. Mozurkewich, R. D. Gehrz.
Bull. American Astron. Soc., Vol. 14, 652 (1982). – Abstract.

114.071 **High-resolution 3 μm spectroscopy of IRC + 10216.**
R. E. S. Clegg, K. H. Hinkle, D. L. Lambert.
Mon. Not. R. Astron. Soc., Vol. 201, 95 - 103 (1982).

High-resolution ($\lambda/\Delta\lambda \cong 100000$) spectra of IRC+10216 have been obtained with a Fourier transform spectrometer. A search for HCl fundamental vibration–rotation lines near 3 μm and an analysis of the $CH_4 \nu_3$ band are described. The CH_4 expansion velocity and temperature suggest that methane is formed in cool gas near the terminal velocity of the circumstellar wind.

114.072 **Anomalies in sodium and CN in 47 Tuc stars.**
T. L. Evans, J. W. Menzies, G. H. Smith.
Mon. Not. R. Astron. Soc., Vol. 201, 137 - 143 (1982).

The sodium D lines have been measured in 19 giant stars in 47 Tuc, and show a range of 60 per cent in equivalent width at a given colour. These variations show a positive correlation with variations in CN strength on the section of the giant branch with $1.0 < (V–I_K) < 1.4$, $12.1 < V < 12.6$. Both giant branch and asymptotic giant branch stars show this behaviour. This confirms and extends the correlation found by Cottrell & Da Costa and appears to support the suggestion that primordial abundance variations are responsible.

114.073 **Carbon and nitrogen abundances in Population I G and K giant stars from DDO photometry.**
J. Bonnell, R. A. Bell.
Mon. Not. R. Astron. Soc., Vol. 201, 253 - 263 (1982).

It is shown that synthetic DDO photometric indices can be used with precise DDO photometric observations of Population I G and K giants to determine carbon and nitrogen abundances and temperatures for these stars. Results for 16 stars are given. They are compared with theoretical calculations of C/N ratios in giant stars after convective mixing of CN cycle processed material with the stellar atmosphere. Good agreement is found. The predicted stellar mass dependence of the C/N ratio is also observed. The results are in good agreement with the results of spectroscopic analyses of CN abundances by Lambert & Ries for the six stars in common.

114.074 **Fourier transform spectroscopy of six stars.**
E. E. Mendoza V.
Rev. Mexicana Astron. Astrofis., Vol. 6, (see 012.017), 137 - 143 (1981).

The paper outlines results from a digital analysis of the Fourier transform spectroscopy of six stars: ε Aur, β Ori, α Lyr, ζ Aql, α Aql and α Cyg. Nearly 1200 different spectral lines have been identified in the spectra of these six stars in the wavelength interval 4800 - 10200 Å where the spectra are of high quality, less than one per cent level of noise versus signal. α Lyr and α Cyg show spectral line and profile variations.

114.075 **The Wolf-Rayet stars.**
J. Sahade.
Rev. Mexicana Astron. Astrofis., Vol. 6, (see 012.017), 189 - 199 (1981).

Aspects of the problems of the Wolf-Rayet stars related to their chemical composition, their evolutionary status and their apparent dichotomy in two spectral sequences are discussed.

114.076 **Ultraviolet continuum of a sample of Be stars.**
A. E. Ringuelet, M. Rovira, J. M. Fontenla.
Rev. Mexicana Astron. Astrofis., Vol. 6, (see 012.017), 215 - 218 (1981).

The energy distribution of α Dra, τ Ori, 53 Per, 48 Lib, π Aqr and 27 CMa is analyzed in the UV. α Dra and τ Ori show no discrepancies with the effective temperature and gravity obtained from the photographic region; the continuum of 53 Per may be variable; 48 Lib fits log g = 4 or 4.5 and T_{eff} = 18 000°K model; π Aqr and 27 CMa fit models with T_{eff} = 30 000°K and log g = 5.

114.077 **El espectro de la estrella Ap HD 3473.**
Z. López-García.
Rev. Mexicana Astron. Astrofis., Vol. 6, (see 012.017), 223 - 226 (1981).

The spectrum of the Ap star HD 3473 has been studied from microphotometric registration of two plates with a dispersion of 16.9 Å mm^{-1} taken in the blue at Kitt Peak National Observatory. Abundances are determined using model atmospheres and the WIDTH 5 program. The presence of emission fill-in in the cores of the stronger lines of Si II and Mg II is confirmed.

114.078 **The cooling trend in FG Sagittae.**
J. Smolinski, J. L. Climenhaga.
Rev. Mexicana Astron. Astrofis., Vol. 6, (see 012.017), 231 (1981). – Abstract.

114.079 **Hα emission in F-K high luminosity stars.**
J. L. Climenhaga, B. L. Harris, J. Smolinski.
Rev. Mexicana Astron. Astrofis., Vol. 6, (see 012.017), 233 - 235 (1981).

The authors investigate the presence and variability of Hα emission in very luminous F-K supergiants and determine the lower limit of M_V where Hα begins to appear in emission.

114.080 **Different stellar winds in O-type stars.**
R. Costero, R. Stalio.
Rev. Mexicana Astron. Astrofis., Vol. 6, (see 012.017), 237 - 242 (1981).

The authors present some preliminary results on the analysis of the far UV spectra of twenty-nine O-type stars. They suggest that the possible effects of CNO chemical abundance variations may explain differences of maximum wavelength displacements in the N V and C IV resonance lines in hot stars. The CNO relative abundance differences would be due to evolutionary effects.

114.081 **Two-dimensional spectral classifications for O stars in the southern Milky Way.**
N. R. Walborn.
Astron. J., Vol. 87, 1300 - 1303 (1982).

Spectral classifications in the two-dimensional system are given for 53 additional O stars in the southern hemisphere. A number of these stars are members of the interesting clusters NGC 3324, Trumpler 14/16, NGC 3603, and NGC 6334, which are discussed in particular. A 4-m Hα photograph of the remarkable nebulosities associated with NGC 6334 is reproduced.

114.082 **The fastest runaway Wolf-Rayet star of population I in the Galaxy, 209 BAC: evidence for a low**

mass companion.
A. F. J. Moffat, R. Lamontagne, W. Seggewiss.
Astron. Astrophys., Vol. 114, 135 - 146 (1982).

The narrow lined WN8 star 209 BAC has a peculiar radial velocity component of about 156 km s^{-1}, the highest known of any galactic Wolf-Rayet (WR) star. Several series of spectroscopic and photometric observations reveal the star to have a low-mass, probably compact companion. It is concluded that 209 BAC was accelerated by the recoil of a supernova explosion in a binary system in which the present WN 8 component was the original secondary, and the presumed neutron star companion left after the explosion was the original primary. The matter associated with the relatively young H II emission nebulosity M 1 – 67 surrounding 209 BAC might have been ejected during a rapid transition from an O-type progenitor of the present WR star via non-conservative mass transfer. A comparison is made with several other WR + compact candidates.

114.083 **Spectrophotometry of Wolf-Rayet star candidates in M 33.**
E. J. Wampler.
Astron. Astrophys., Vol. 114, 165 - 169 (1982).

Spectra of twenty-four Wolf-Rayet star candidates in M 33 show that one object is a supernova remnant, four are early nitrogen sequence Wolf-Rayet stars and nineteen are early carbon sequence Wolf-Rayet stars. Except for the fact that these stars are associated with spiral arms the carbon sequence stars show no pronounced affinity for associations while three of the four nitrogen sequence stars are projected against associations previously delineated by Humphreys and Sandage (1980).

114.084 **Observed and computed UV spectral distribution of A and F stars. Determination of T_e and log *g*.**
M. L. Malagnini, R. Faraggiana, C. Morossi, L. Crivellari.
Astron. Astrophys., Vol. 114, 170 - 176 (1982).

The authors implemented an automatic and fast procedure to determine T_e and log *g* from the comparison of the UV S 2/68 spectra with Kurucz' models. The method has been applied to all A and F stars with $(B-V)_0 \geqslant 0$ and luminosity classes III, IV, IV–V, and V, included in the Ultraviolet Bright-Star Spectrophotometric Catalogue. From the analysis, it appears that the models match the observed fluxes fairly well and that the effective temperatures derived from the UV data agree with those derived from Strömgren photometric indices within 200 K. Only rough indications of log *g* have been derived for each star.

114.085 **Einstein observations of late-type stars of different calcium II activity.**
R. Mewe.
Observatory, Vol. 102, 109 - 110 (1982). – Abstract.

114.086 **Absolute energy distribution in Be stars.**
P. S. Goraya, S. C. Joshi.
Bull. Astron. Soc. India, Vol. 10, 32 (1982). – Abstract.

114.087 **The spectrum of the Ap star HR 465 near the rare-earth minimum phase.**
B. N. G. Guthrie.
Astrophys. Space Sci., Vol. 85, 315 - 324 (1982).

Four spectra of the long-period Ap star HR 465 were taken in 1969 - 70 near its rare-earth minimum phase and analysed. Equivalent widths or upper limits were obtained for lines of 23 elements.

114.088 **The generalized Compton effect in the spectrum of Deneb.**
M. Missana.
Astrophys. Space Sci., Vol. 85, 419 - 425 (1982).

The radial velocity and the redshift caused by the scattering of photons in the chromosphere of Deneb and in the interstellar matter is obtained from the wavelengths and FWHM of about 30 spectral lines measured in two spectra observed in 1979.

114.089 **Energy distribution in the spectra of seven standard stars used in the process of creation of a spectrophotometric star catalogue.**
I. N. Glushneva, S. L. Ovchinnikov.
Astron. Zh., Tom 59, 908 - 917 (1982). In Russian. English translation in Soviet Astron., Vol. 26, No. 5.

New energy distribution data E_{81} in the spectra of seven stars: β Ari, γ Ori, β Tau, α Leo, η UMa, α Aql, α Peg in the region $\lambda\lambda$ 3200 - 7600 Å were obtained on the basis of spectrophotometric comparison with the main standard α Lyr, carried out from May 1973 to November 1979.

114.090 **Non-LTE analysis of carbon lines in spectra of hot stars. I. The triplet C III lines at λ 4650 Å and λ 9710 Å in the spectra of O stars.**
N. A. Sakhibullin, L. H. Auer, K. van der Hucht.
Astron. Zh., Tom 59, 933 - 943 (1982). In Russian. English translation in Soviet Astron., Vol. 26, No. 5.

An analysis of the triplet C III lines in spectra of O stars has been carried out on the basis of a solution of the coupled statistical-equilibrium and transfer equation. Special attention was paid to the mechanism responsible for the formation of the 4650 Å and 9710 Å lines. Calculations reproduce the observed emission at λ 9710 Å ($3p - 3d$) and absorption at λ 4650 Å ($3s - 3p$). The presence of lines in emission is not a sufficient argument for the existence of extended or expanding envelopes.

114.091 **Ultraviolet resonance line emissions observed with IUE.**
C. Blanco, L. Bruca, S. Catalano, E. Marilli.
Third European IUE Conference, (see 012.029), p. 117 - 121 (1982).

Observations of ultraviolet emission line spectra obtained with the IUE in the region 1200 – 2000 Å are reported. Low resolution spectra for a representative sample of main sequence stars in the spectral range A5-K5 and the supergiant α Aqr are analysed. The detection of stellar chromospheres and transition region features in stars earlier than F0 is discussed.

114.092 **Chromospheric Mg II emission in A5 to K5 main sequence stars from high resolution IUE spectra.**
C. Blanco, L. Bruca, S. Catalano, E. Marilli.
Third European IUE Conference, (see 012.029), p. 123 (1982). Abstract.

114.093 **Fe II absorption lines in the IUE high-resolution spectra of F supergiants and shell stars.**
D. P. Gilra.
Third European IUE Conference, (see 012.029), p. 139 - 141 (1982).

The UV spectra of α Car (F supergiant) show very strong absorptions due to Fe II lines from the ground state and low lying metastable levels. The spectra of B and A shell stars show some striking resemblance to that of α Car.

114.094 **Outer atmospheric properties of Beta Draconis (G2 Ib-II).**
A. Brown, C. Jordan, T. R. Ayres, J. L. Linsky, R. E. Stencel.
Third European IUE Conference, (see 012.029), p. 142 - 148 (1982).

New observations of the supergiant Beta Draconis, made with the IUE satellite, are reported. The spectra show a range of chromospheric and transition region lines. The emission measure distribution is derived and estimates of the electron pressure are made. The radiation losses, thermal conduction, dynamics and non-thermal energy transport are briefly discussed.

114.095 **A search for Mg II h and k emission core variability in six main sequence stars.**
J. E. Beckman, M. L. Franco, P. Molaro, G. Vladilo, L. Crivellari.
Third European IUE Conference, (see 012.029), p. 149 - 155 (1982).

Spectra of six main sequence stars with spectral types from F0 to G8 have been obtained with the IUE long wavelength spectrograph at high resolution. A careful quantitative analysis of the Mg II h and k emission cores from these spectra has been performed, to search for sporadic or periodic variability. No variations greater than the ±10% level set by the statistical fluctuations have been found.

114.096 **The transition region structure of "solar-like" stars.**
M. J. Fernandez-Figueroa, E. de Castro, M. Rego.
Third European IUE Conference, (see 012.029), p. 157 - 159 (1982).

Far ultraviolet emission lines of three late type stars, two dwarfs (χ Her and β Com) and one subgiant (i Vir), have been analysed to determine the structure of their outer atmospheres. Emission line fluxes are used to find the emission measure distributions.

114.097 **High resolution EUV spectroscopy of 56 Pegasi (K0 II p + wd).**
C. Jordan, A. Brown, T. R. Ayres, J. L. Linsky, R. E. Stencel.
Third European IUE Conference, (see 012.029), p. 161 - 164 (1982).

Observations of 56 Pegasi have been obtained at high spectral resolution using the IUE satellite. Schindler et al. have proposed a model that may explain the observed X-ray and EUV emission by accretion of the cool star wind onto the white dwarf companion. The present paper considers the alternative hypothesis that the observed emission originates in a conventional chromosphere, transition region and corona.

114.098 **IUE high resolution observations of hot stars emitting coronal X rays.**
L. Bianchi.
Third European IUE Conference, (see 012.029), p. 251 - 257 (1982).

The far UV resonance lines of a sample of 21 early-type stars, which were observed in the soft X-ray band with the Einstein satellite, are examined, using IUE high resolution spectra, to search for possible correlation between the X-ray coronal emission and far UV spectral properties.

114.099 **Does the WC6 star HD 76536 have a compact companion?**
G. E. Bromage, W. M. Burton, K. A. van der Hucht, F. Macchetto, C. C. Wu.
Third European IUE Conference,(see 012.029), p. 269 - 271 (1982).

The WC6 star HD 76536 was observed by IUE as part of a study of variability in WR stars. Photometric variability is confirmed. Profile changes in the ultraviolet emission lines and stellar wind absorptions are described and discussed. The available data tend to favour a binary model with a low-luminosity secondary.

114.100 **Mass loss from extreme helium stars.**
U. Heber, W.-R. Hamann, D. Schönberner.
Third European IUE Conference, (see 012.029), p. 273 - 276 (1982).

Indications of mass loss are found in the UV spectra of the three extreme helium stars HD 160641, BD −9°4395 and BD +10°2179, obtained with IUE in high resolution. By fitting the UV resonance lines with theoretical profiles the authors construct empirical wind models. The calculations are performed with the comoving-frame technique and include the formation of blended doublets.

114.101 **SK-67/111 and R 50, two Magellanic Cloud emission line stars.**
F. J. Zickgraf, B. Wolf, R. Viotti, O. Stahl, O. Ricciardi, G. Muratorio, M. Friedjung, A. Cassatella.
Third European IUE Conference, (see 012.029), p. 277 - 279 (1982).

The authors observed SK-67/111 (O5f) of the LMC and the peculiar emission line star R 50 of the SMC with the IUE and also with the ESO 1.5 m telescope. Mass loss rates for both stars were derived.

114.102 **The UV stellar classification programme.**
A.Heck, P. Benvenuti, L. Bianchi, J. C. Blades, C. Cacciari, A. Cassatella, A. Cucchiaro, D. Egret, C. Jaschek, M. Jaschek, K. Nandy, P. Patriarchi, W. Wamsteker, C. C. Wu.
Third European IUE Conference, (see 012.029), p. 327 - 330 (1982).

The UV stellar classification programme based on IUE data is described, as well as its rationale. A report on the present status of the work is also given. It will provide in the near future sequences of calibrated spectra covering all spectral types.

114.103 **The IUE spectra of ET And – their reduction and variation.**
M. Barylak.
Third European IUE Conference, (see 012.029), p. 601 - 607 (1982).

Seventeen IUE spectra of the chemically peculiar (CP) star ET And (HR8861) were studied within the TV image processing system. Using the Fourier routines of this system, filtering, deconvolution of instrumental profiles and determination of the rotational velocity could be applied to the IUE spectra under investigation.

114.104 **New molecular bands in an R type carbon star.**
J. H. Goebel, J. D. Bregman, D. Cooper, D. Goorvitch, S. Langhoff, F. C. Witteborn.
Bull. American Astron. Soc., Vol. 14, 776 (1982). – Abstract.

114.105 **New radial velocity limits for FK Comae.**
J. K. McCarthy, L. W. Ramsey.
Bull. American Astron. Soc., Vol. 14, 780 (1982). – Abstract.

114.106 **The detection of ultraviolet photospheric absorption in the spectra of two Wolf-Rayet stars.**
E. L. Fitzpatrick.
Astrophys. J., Lett., Vol. 261, L91 - L95 (1982).

Narrow, unshifted UV absorption lines, principally of Fe V, have been detected in the spectra of two Wolf-Rayet stars, HD 93162 (WN 7) and HD 193793 (WC 7). Both stars are known to show upper Balmer-line absorption but exhibit no periodic Doppler shifts that might indicate the presence of a companion. If intrinsic to the Wolf-Rayet stars, these lines presumably form near the base of the envelope in the photosphere.

114.107 **The carbon and nitrogen abundances in WN and WC Wolf-Rayet stars.**
L. J. Smith, A. J. Willis.
Mon. Not. R. Astron. Soc., Vol. 201, 451 - 472 (1982).

New ultraviolet spectroscopy obtained with the *IUE* satellite of helium, carbon and nitrogen lines in six WN and four WC stars have been analysed using non-LTE Escape Probability techniques to derive the stellar atmospheric C/N, C/He and N/He ratios. The results are in fair agreement with recent predicted values from stellar evolutionary models in which mass loss reveals nuclear processed CNO products

(WNL, WNE stars) and He-burning products (WC stars). This agreement strengthens the scenario that the WR stars are chemically evolved objects at different stages of H-burning and He-burning in massive hot stars.

114.108 **Spectral types in the direction of the Magellanic Stream.**
E. Recillas-Cruz.
Mon. Not. R. Astron. Soc., Vol. 201, 473 - 478 (1982).

Deep hypersensitized UKSTU objective-prism plates have been used in a search for faint blue spectra belonging to early-type stars in the direction of the Magellanic Stream. From apparent magnitude and spectral type estimates, the author concludes that the number of early A-type stars found per square degree is almost a factor of 10 lower than what one should expect if the Stream originated out of a tidal stripping by gravitational interaction between one of the Magellanic Clouds and our own Galaxy some 2×10^8 yr ago. Furthermore, the absence of such a stellar component is discussed in terms of the assumed distance to the Magellanic Stream.

114.109 **The spectrum of Canopus – II. Analysis and composition.**
K. Desikachary, J. B. Hearnshaw.
Mon. Not. R. Astron. Soc., Vol. 201, 707 - 721 (1982).

The atmospheric parameters for the F-type supergiant star Canopus are obtained from spectrophotometry and Balmer line profile measurements to be $T_{eff} = 7350 \pm 30$ K, and $\log g = 1.80 \pm 0.05$. An absolute model atmosphere analysis of the equivalent widths published by Hearnshaw & Desikachary (1982) is carried out. The abundances are presented for 33 elements and all have an approximately solar abundance with the possible exception of nitrogen. Some tentative evidence is presented that nitrogen in Canopus may be five times more abundant than in the Sun. Depth-dependence of microturbulence was looked for but not found.

114.110 **Data on an unusual Wolf-Rayet star in the nearby galaxy IC 1613.**
K. Davidson, T. D. Kinman.
Publ. Astron. Soc. Pacific, Vol. 94, 634 - 639 (1982).

The authors report observations of an H II region in IC 1613 which is excited by a hot Wolf-Rayet star. There are several points of interest concerning both this nebula and its hot star: (1) The spectrum of the star itself is interesting: besides the broad C IV features reported by previous authors, there appear also to be O VI features which are somewhat less broad. (2) The spectrum of the H II region can be used, Zanstra-fashion, to estimate the far-ultraviolet luminosity of the star. (3) The H II region is of high excitation and has a low oxygen abundance, while the underlying stellar continuum is very weak at visual wavelengths. (4) The Wolf-Rayet star is likely to have had the same metal-poor initial composition.

114.111 **A very faint, very red, high-galactic-latitude carbon star.**
N. Sanduleak, P. Pesch.
Publ. Astron. Soc. Pacific, Vol. 94, 690 - 691 (1982).

A carbon star with $B \sim 21$ mag and $(B-V) \sim 4$ mag, at a galactic latitude of $b = +55°$, has been discovered during the course of an objective-prism survey of the northern galactic cap.

114.112 **Equivalent widths of spectral lines in B-type stars.**
P. Didelon.
Astron. Astrophys., Suppl. Ser., Vol. 50, 199 - 207 (1982). In French.

In this paper the author collects the equivalent widths (W_λ) of spectral lines in B-type stars which are found in the literature. He compares the measures made at different reciprocal dispersions, and finds that W_λ grows systematically when their reciprocal dispersion decreases. Only lines with reciprocal dispersions less than 40 Å/mm were retained. The random scatter is of the order of 10-20% of the measured W_λ. The measures are presented for the strongest lines of the most important species. Because of the large number of measures found in the literature, average curves for W_λ are provided as a function of spectral type and luminosity for all lines. He I λ 4171 and C II λ 4267 show a large scatter which prevents drawing average curves; these two cases are discussed in the text.

114.113 **The mid-ultraviolet spectrum of ϵ Aurigae.**
F. Castelli, R. Hoekstra, Y. Kondo.
Astron. Astrophys., Suppl. Ser., Vol. 50, 233 - 245, microfiche (1982).

An LTE synthetic spectrum with $T_e = 7800$ K and $\log g = 1$ has been computed for the 2455-to-3230 Å range of ϵ Aurigae (F0 Ia). This has been compared with the spectra observed with 0.2 Å resolution with the IUE experiment and with 0.1 Å resolution with the BUSS experiment. As a consequence of this comparison, the authors have obtained identifications and abundance values. Ions definitely present include Mg I, Mg II, Al I, Al II, Si I, Ca II, Ti II, V II, Cr II, Mn II, Fe I, Fe II, Co II, Ni I, Ni II, Zr II. Ions which are possibly present include Ti III, Mn I, Fe III, Nb II, Mo I. The abundances are nearly 10 times the solar values for the iron group, with exception of Fe, which has solar abundance. The abundances of the other elements are solar. Emissions in the lines of multiplet UV 1, 62 and 63 of Fe II have been observed. Shortward shifted lines and double lines are also present.

114.114 **Observations of the first-overtone silicon monoxide bands in late-type stars.**
C. P. Rinsland, R. F. Wing.
Astrophys. J., Vol. 262, 201 - 212 (1982).

Scans of the spectral region 3.98–4.07 μm, which includes the (2,0) and (3,1) rotation-vibration bands of silicon monoxide, have been made at a resolution of 5.5 cm^{-1} in 74 late-type stars with a grating spectrometer and InSb detector. All the M giants and bright giants observed have detectable SiO absorption, and the bands strengthen rapidly with advancing type until at least M6. The SiO band strength varies considerably among M supergiants of the same spectral type and is strongly time-dependent in Mira variables.

114.115 **Wolf-Rayet phenomena.**
P. S. Conti.
Wolf-Rayet stars: observations, physics, evolution, (see 012.042), p. 3 - 22 (1982).

114.116 **The effective temperatures of early Wolf-Rayet stars.**
W. Schmutz.
Wolf-Rayet stars: observations, physics, evolution, (see 012.042), p. 23 - 26 (1982).

The author found from his observations and model calculations that the Zanstra analysis is not valid in the case of the Wolf-Rayet stars.

114.117 **Line strengths in WN stars.**
E. M. Leep.
Wolf-Rayet stars: observations, physics, evolution, (see 012.042), p. 41 - 45 (1982).

The author has analyzed high-dispersion spectra of over 100 WN stars in the Galaxy and the Large Magellanic Cloud.

114.118 **Hot mantles, moderate photospheres for Wolf-Rayet stars.**
A. B. Underhill.
Wolf-Rayet stars: observations, physics, evolution, (see 012.042), p. 47 - 51 (1982).

The amount of continuous energy from Wolf-Rayet stars

and the shape of the continuous spectrum from the ultraviolet to the near infrared correspond to effective temperatures in the range 25000 to 30000 K. The value of log g is of the order of 4.0 ± 0.5. Thus the photospheres of Wolf-Rayet stars correspond to those of moderately hot stars. The line spectra of Wolf-Rayet stars, however, indicate that electron temperatures greater than 30000 K occur in the outer atmospheres or mantles of these stars. Here outflow is important.

114.119 **Infrared (1.4 – 4.1 μm) spectra of Wolf-Rayet stars.**
P. M. Williams.
Wolf-Rayet stars: observations, physics, evolution, (see 012.042), p. 73 - 78 (1982).

The spectra of a variety of Wolf-Rayet stars have been observed with ~1% spectral resolution in the 1.4 – 4.1 μm region using UKIRT. Strong lines due to ions of helium and carbon are observed and their relative strengths discussed.

114.120 **The infrared recombination-line spectra of Wolf-Rayet stars.**
D. G. Hummer, M. J. Barlow, P. J. Storey.
Wolf-Rayet stars: observations, physics, evolution, (see 012.042), p. 79 - 83 (1982).

Effective recombination coefficients have recently been calculated for recombination lines of He I, He II and C IV for densities and temperatures appropriate for Wolf-Rayet atmospheres. These have been applied to recently obtained spectra of γ Vel in order to derive the He^+/He^{+2} and $C^{+4}/He^+ + He^{+2}$ ratios.

114.121 **The chemical composition of the Wolf-Rayet stars.**
A. J. Willis.
Wolf-Rayet stars: observations, physics, evolution, (see 012.042), p. 87 - 104 (1982).

The review summarises current knowledge of the chemical composition of Pop I WR stars, concentrating on work carried out in this area since the last IAU, No. 49, symposium devoted to this stellar class.

114.122 **Chemical composition of Wolf-Rayet stars: abundant evidence for anomalies.**
C. D. Garmany, P. S. Conti.
Wolf-Rayet stars: observations, physics, evolution, (see 012.042), p. 105 - 107 (1982).

114.123 **H/He ratios for WN stars in the LMC and the Galaxy.**
D. N. Perry, P. S. Conti.
Wolf-Rayet stars: observations, physics, evolution, (see 012.042), p. 109 - 112 (1982).

114.124 **The C/N ratio in WN and WC stars.**
L. J. Smith, A. J. Willis.
Wolf-Rayet stars: observations, physics, evolution, (see 012.042), p. 113 - 116 (1982).

The authors present new determinations of the C/He, N/He and C/N ratios in 6 WN and 4 WC stars, resulting from a Sobolev analysis of new IUE UV line measurements in these species.

114.125 **Carbon abundance in WC stars.**
T. Nugis.
Wolf-Rayet stars: observations, physics, evolution, (see 012.042), p. 131 - 134 (1982).

114.126 **Synthetic spectra for WN stars.**
R. Wehrse.
Wolf-Rayet stars: observations, physics, evolution, (see 012.042), p. 135 - 138 (1982).

A computer program is described which calculates the structure and the radiation field (including high resolution spectra) of spherically extended and moving atmospheres. Some preliminary results for models of WN stars are presented and discussed.

114.127 **Carbon abundance in the WC 11 star CPD −56°8032.**
L. Houziaux, A. Heck.
Wolf-Rayet stars: observations, physics, evolution, (see 012.042), p. 139 - 145 (1982).

114.128 **IR spectroscopy of WN stars.**
J. D. Hillier.
Wolf-Rayet stars: observations, physics, evolution, (see 012.042), p. 225 - 229 (1982).

The author obtained IR spectra from 1.1 to 4.2 μ for a sample of southern WN stars using a low resolution, circular variable filter wheel and a cooled grating spectrometer (CGS) with a maximum resolution of 2000. With the CGS it is now possible to obtain spectra at resolutions sufficiently high to split blends and reveal details of line profiles. The author presents preliminary results of his observations.

114.129 **Near infrared observations of galactic WN stars.**
J. M. Vreux, M. Dennefeld, Y. Andrillat.
Wolf-Rayet stars: observations, physics, evolution, (see 012.042), p. 237 - 241 (1982).

The authors concentrate on the Paschen region in the spectra of late WN stars and this in connection with the discussion on the hydrogen to helium ratio on the surface of WR stars.

114.130 **The iron curtain of WC9 stars.**
K. A. van der Hucht, P. S. Conti, A. J. Willis.
Wolf-Rayet stars: observations, physics, evolution, (see 012.042), p. 277 - 281 (1982).

High resolution IUE spectra have been obtained of the two WC9 stars HD 164270 and HD 136488. The former star shows P Cygni profiles indicating a stellar wind terminal velocity of $v_\infty \approx -1400$ km s^{-1}, and the latter $v_\infty \approx -1800$ km s^{-1}. A common feature in the spectra of both stars is narrow displaced absorptions due to Fe III transitions arising from a metastable lower level. These features are displaced at subterminal velocities and are believed to be formed in the deceleration region of their stellar winds. The properties of these inferred Fe III circumstellar shells derived from these data are discussed.

114.131 **The WO Wolf-Rayet stars.**
M. J. Barlow, D. G. Hummer.
Wolf-Rayet stars: observations, physics, evolution, (see 012.042), p. 387 - 392 (1982).

114.132 **The ON stars: a wider spectroscopic definition.**
G. F. Bisiacchi, J. López, C. Firmani.
Wolf-Rayet stars: observations, physics, evolution, (see 012.042), p. 393 - 396 (1982).

During the last years, many authors have recognized the existence of some OB stars that show the nitrogen absorption lines enhanced.

114.133 **Subluminous Wolf-Rayet stars: observations.**
S. R. Heap.
Wolf-Rayet stars: observations, physics, evolution, (see 012.042), p. 423 - 445 (1982).

Contents: Introduction. Spectral classification. Calibration of spectral types. Characteristics of the wind. Evolution of WR central stars. Distinctive properties of WR central stars.

114.134 **The peculiar binary system HD 45166 (SdO+B8V?).**
A. J. Willis, D. J. Stickland.
Wolf-Rayet stars: observations, physics, evolution, (see 012.042), p. 447 - 451 (1982).

114.135 **The UV spectrum of the central star of NGC 40.**
P. Benvenuti, M. Perinotto, A. J. Willis.
Wolf-Rayet stars: observations, physics, evolution, (see 012.042), p. 453 - 456 (1982).

114.136 **A reclassification of WC and "O VI" central stars of planetary nebulae, and comparison with Population I WC stars.**
R. H. Méndez, V. S. Niemelä.
Wolf-Rayet stars: observations, physics, evolution, (see 012.042), p. 457 - 461 (1982).

The authors have started a study of WC central stars of planetary nebulae, including "O VI" objects which show WC characteristics. One of the purposes was to check, all along the WC spectral sequence, if the emission lines of planetary nuclei are narrower than for Population I objects, as they are in the case of the WC9 central star BD+30°3639.

114.137 **IUE observations of the WN–C star HD 62910.**
D. J. Stickland, A. J. Willis.
Wolf-Rayet stars: observations, physics, evolution, (see 012.042), p. 491 - 495 (1982).

114.138 **Near infrared observations of Magellanic WN stars.**
Y. Andrillat, M. Dennefeld, J. M. Vreux.
Wolf-Rayet stars: observations, physics, evolution, (see 012.042), p. 527 - 530 (1982).

114.139 **Spectra of the Wolf-Rayet stars in 30 Doradus.**
P. S. Conti.
Wolf-Rayet stars: observations, physics, evolution, (see 012.042), p. 551 - 554 (1982).

The central cluster of the 30 Doradus complex in the Large Magellanic Cloud contains at least a dozen Wolf-Rayet stars of the nitrogen sequence according to Melnick (1978). The author has obtained high resolution spectra of ten of these stars and discusses the spectral types.

114.140 **Line profile variations and binarity in Wolf-Rayet stars.**
G. F. Bisiacchi, C. Firmani, E. de Lara.
Wolf-Rayet stars: observations, physics, evolution, (see 012.042), p. 583 - 587 (1982).

114.141 **He I lines in B stars: comparison of non-local thermodynamic equilibrium models with observations.**
J. N. Heasley, S. C. Wolff, J. G. Timothy.
Astrophys. J., Vol. 262, 663 - 674 (1982).

Profiles of He I λλ4026, 4387, 4471, 4713, 5876, and 6678 have been obtained in 17 stars of spectral type B0–B5. Parameters of the non-local thermodynamic equilibrium models appropriate to each star are determined from the Strömgren c_0 index and fits to Hα line profiles. These parameters yield generally good fits to the observed He I line profiles, with the best fits being found for the blue He I lines where departures from local thermodynamic equilibrium are relatively small.

114.142 **Ultraviolet emission in the Mg II h and k lines in Be stars.**
F. C. Bruhweiler, T. H. Morgan, K. A. van der Hucht.
Astrophys. J., Vol. 262, 675 - 681 (1982).

Ultraviolet observations made by the Balloon-borne Ultraviolet Stellar Spectrometer of bright early Be stars reveal emission components to the Mg II doublet at 2800 Å. These profiles show central absorption features due to interstellar plus possible circumstellar contributions. The emission components may in part be due to resonance scattering within the line profile, but there are definite contributions to the emission via a Bowen mechanism drive by Lyman β. Slight apparent shifts in the absorption components relative to the emission may indicate expansion ~5–10 km s^{-1} for the line-forming region giving rise to Mg II.

114.143 **Carbon, nitrogen, and oxygen abundances in G8–K3 giant stars.**
P. Kjaergaard, B. Gustafsson, G. A. H. Walker, L. Hultqvist.
Astron. Astrophys., Vol. 115, 145 - 163 (1982).

High-resolution scans of the [O I] λ6300 line and of lines of the red CN (2,0) band and very-narrow-band-index measurements of the C_2 Swan (0,1) band head have been obtained for about 30 G8–K3III stars. CNO abundances have then been derived from these data. The analysis was performed by means of a grid of model atmospheres and synthetic-spectrum calculations. The effective temperatures were determined from a new calibration of near-infrared colours and the surface gravities from a calibration of a Ca II K-line emission widths.

114.144 **Chromospheric Mg II emission in A 5 to K 5 main sequence stars from high resolution IUE spectra.**
C. Blanco, L. Bruca, S. Catalano, E. Marilli.
Astron. Astrophys., Vol. 115, 280 - 292 (1982).

The authors present observations of the emission cores of the Mg II resonance doublet at λ 2800 Å for 11 main sequence stars extending over a range of spectral types from A 5 to K 5 and for α Tau (K 5 III) and α Aqr (G 2 Ib). The spectra have been obtained in the high resolution mode by means of the IUE Satellite. Mg II emission and therefore chromosphere are found on main sequence stars as early as A 7, $B-V = 0^m.22$. The chromospheric radiative losses from the h and k lines are discussed in terms of the effective temperature and the rotation period.

114.145 **Abundances in metal-poor stars. II. The anomalous globular cluster ω Centauri.**
R. G. Gratton.
Astron. Astrophys., Vol. 115, 336 - 346 (1982).

An analysis of the chemical composition of eight red giants in the anomalous globular cluster ω Centauri is presented. The inhomogeneity in the cluster chemical composition is confirmed. All the observed elements show variations in abundance from star to star. The metal enrichment history for ω Centauri seems different from the general process for the galactic medium. The iron abundances from high dispersion spectra correlates within the observational errors with the $R(V-K)$ index defined by Persson et al. (1980).

114.146 **Abundance of lithium in unevolved halo stars and old disk stars: interpretation and consequences.**
F. Spite, M. Spite.
Astron. Astrophys., Vol. 115, 357 - 366 (1982).

High resolution spectra with high signal-to-noise ratios have been obtained for a few halo dwarfs with the $3^m.6$ CFH telescope. These solar type stars are very old and extremely metal poor. The Li resonance line (λ 6707) is observed. The problem of lithium seems clear for the halo stars and is very confusing in Population I stars. It is suggested that the abundance of lithium in halo dwarfs is representative of the abundance of the interstellar matter which formed the stars and also that this matter itself retains the lithium abundances of the Big Bang, hardly altered. The authors deduce a primordial value of the lithium abundance $11.2 \times 10^{-11} N_H$. The baryonic density of the Universe deduced from this value is compared to the densities deduced from the primordial abundance of deuterium and helium. All these densities are small, and this shows that the Universe cannot be closed by nucleons.

114.147 **Near IR spectra of symbiotic stars.**
Y. Andrillat.
The nature of symbiotic stars, (see 012.045), p. 47 - 56 (1982).

It is very well known that symbiotic stars are displaying

at the same time spectral characteristics of hot stars, cool stars (principally of M type), nebulae and circumstellar dust envelopes. The investigation of the near IR is particularly interesting because one finds there important features of these different objects.

114.148 **Photographic infrared spectra of symbiotic stars.**
Y. Andrillat, L. Houziaux.
The nature of symbiotic stars, (see 012.045), p. 57 - 60 (1982).

The authors have observed six objects during the period 1962–1977 with a grating spectrograph attached to the Newtonian focus of the 120-cm telescope at Observatoire de Haute Provence. The reciprocal dispersion is 230 Å mm^{-1} and the region 5800 to 8800 Å has been covered using hyper-sensitized IN plates. The minimum equivalent width for an emission line to be seen is about 0.5 Å. The main spectral characteristics are briefly reviewed.

114.149 **Spectra of individual symbiotic stars.**
N. A. Oliversen, C. M. Anderson.
The nature of symbiotic stars, (see 012.045), p. 71 - 82 (1982).

In connection with the discussion of the individual symbiotic stars, the authors present low resolution, flux calibrated spectra and high resolution Hα profiles for each of 12 objects.

114.150 **Visual symbiotic spectra obtained with the Haute Provence Multiphot detector.**
G. Muratorio, M. Friedjung.
The nature of symbiotic stars, (see 012.045), p. 83 - 84 (1982).

Spectra of symbiotic stars that have been obtained with the Haute Provence Multiphot detector system are reported.

114.151 **UV line emission of symbiotic stars.**
H. Nussbaumer.
The nature of symbiotic stars, (see 012.045), p. 85 - 102 (1982).

General characteristics of emission line spectra from symbiotic stars are outlined. Data from some special line ratios in the 1000 Å - 3000 Å range, and others connecting the visual and the far UV lines are presented, and their application to symbiotic stars is discussed. Integrated fractional abundances for ions easily observed in the far UV are given to facilitate abundance determinations for nebular conditions. It is found that the physical conditions of the regions emitting the emission line spectra differ considerably among different symbiotic stars.

114.152 **Z Andromedae: the prototype?**
R. Viotti, A. Giangrande, O. Ricciardi, A. Cassatella.
The nature of symbiotic stars, (see 012.045), p. 125 - 130 (1982).

Z And is considered as a "prototype" of the symbiotic stars. Besides its symbiotic spectrum, the star is also known for its characteristic light curve (and for the related spectral variations). Since many theoretical speculations on Z And and similar objects have been based on the luminosity and spectral variations of this star, a critical analysis of the concerned observational data is presented.

114.153 **Recent studies of the spectrum of V1016 Cygni.**
G. Muratorio, M. Friedjung.
The nature of symbiotic stars, (see 012.045), p. 161 - 163 (1982).

Two coudé spectra of V1016 Cyg taken on June 24 and 27, 1979 were reduced, using a computer programme developed in Marseille. Radial velocities and full widths at half maximum were measured for the emission lines, and are summarized in this paper.

114.154 **The ultraviolet spectrum of AG Draconis.**
A. Altamore, G. B. Baratta, A. Cassatella, A. Giangrande, D. Ponz, O. Ricciardi, R. Viotti.
The nature of symbiotic stars, (see 012.045), p. 183 - 184 (1982).

The high resolution ultraviolet spectrum of AG Dra was observed with IUE in April and August 1981. The UV spectrum of the star appears rather different from that of the other classical symbiotic stars. The 1981 observations were compared with the pre-outburst 1979 and 1980 IUE spectra.

114.155 **Spectroscopic observations of AG Dra.**
C.-C. Huang.
The nature of symbiotic stars, (see 012.045), p. 185 - 189 (1982).

AG Dra could be classified as supergiant K0 Ib, according to the relative intensities of TiII, SrII, BaII and CaI on the 1966 plate. The velocity variations of emission lines suggest a period of about 35 years.

114.156 **The optical spectrum of R136a, the central object of the 30 Doradus Nebula.**
D. C. Ebbets, P. S. Conti.
Astrophys. J., Vol. 263, 108 - 115 (1982).

The authors have obtained and analyzed a spectrum of R 136a in the optical region 3700 < λ < 6700 Å. The overall morphology of the spectrum resembles a WN4.5 star. Balmer lines from Hγ through H11 are present in absorption and appear to have the same origin as the WN spectrum. They are redshifted with respect to the velocity of the 30 Doradus nebula. The gravitational redshift of a single ~ 2000 $M_\odot$ star would be approximately the same magnitude. Weak absorption lines of He I and He II are also present. They are blueshifted relative to the WN spectrum and the nebular velocity. The authors favor the single massive object interpretation of the spectrum, although the possibility of a dense cluster of several dozen very luminous O type supergiants and WN stars cannot be excluded by the data.

114.157 **Supergiants and the galactic metallicity gradient I. 27 late-type supergiants in the inner-arm regions.**
R. E. Luck.
Astrophys. J., Vol. 263, 215 - 225 (1982).

From an analysis of high-dispersion Kitt Peak echelle data, atmospheric parameters and chemical abundances have been derived for 27 supergiants of spectral type F through M. The stars studied lie within about 1 kpc of the Sun. The derived [Fe/H] ratios range from −0.6 to + 0.1 dex, show a mean value of −0.15 dex, and have an internal uncertainty of ± 0.2 dex.

114.158 **Ultraviolet and X-ray detection of the 56 Pegasi system (K0 IIp + WD): evidence for accretion of a cool stellar wind onto a white dwarf.**
M. Schindler, R. E. Stencel, J. L. Linsky, G. S. Basri, D. J. Helfand.
Astrophys. J., Vol. 263, 269 - 276 (1982).

IUE spectra of the slowly rotating mild barium star 56 Peg (HD 218356; K0 IIp) show excess continuum emission from 1300 to 2000 Å, a broad Lyα absorption feature, and emission lines usually associated with a $\sim 10^5$ K plasma. Einstein IPC observations of this system yield $L_x \approx 3 \times 10^{31}$ ergs s^{-1}, which is as bright as the RS CVn binary systems. It is proposed that the observed UV excess is due to the presence of a white dwarf and that the X-ray luminosity is caused by accretion of the stellar wind from the primary onto the white dwarf.

114.159 **Shell and photosphere of σ Ori E: new observations and improved model.**
D. Groote, K. Hunger.
Astron. Astrophys., Vol. 116, 64 - 74 (1982).
For σ Ori E, phase dependent infrared fluxes in the bands J, H, K, L, and M have been obtained plus high and low resolution IUE spectra in both wavelength ranges (single phase).From the fluxes, effective temperature, colour excess and angular diameter are determined. From optical spectra, helium abundances and gravities are derived as a function of phase, by making use of a grid of model atmospheres. With the dynamical parallax of σ Ori AB (d=400 pc), the mass of σ Ori E is 3 $M_{\odot}$ ± 1 $M_{\odot}$. From the good agreement with observations it is concluded that σ Ori E is a single oblique rotator. From the IR excess plus radial velocities a model of the shell is derived.

114.160 **A new probable Be star Alpha Leo.**
M. Singh.
Inf. Bull. Variable Stars, No. 2188, 2 pp. (1982).

114.161 **Multiple components in the Hα profile of the luminous supergiant HD 217476.**
J. Smolinski, J. L. Climenhaga, H. Funakawa, J. M. Fletcher.
Inf. Bull. Variable Stars, No. 2229, 4 pp. (1982).

114.162 **2 Vulpeculae and HR 7739.**
IAU Circ., No. 3710 (1982).

114.163 **V4024 Sagittarii = MWC 311.**
IAU Circ., No. 3712 (1982).

114.164 **2 Ceti and λ Eridani.**
IAU Circ., No. 3727 (1982).

114.165 **CPD −48° 1577.**
IAU Circ., Nos. 3730, 3749 (1982).

114.166 **HD 208311: an unusual new Be star.**
IAU Circ., No. 3756 (1982).

114.167 **Studies on the spectra of K-giants – I. Physical parameters and Fe and Ti abundances for 26 K-giants.**
L. Gratton, S. Gaudenzi, C. Rossi, R. G. Gratton.
Mon. Not. R. Astron. Soc., Vol. 201, 807 - 832 (1982).
Blue-violet high-dispersion spectra of 26 K-giants have been analysed by a semi-automatic procedure for the determination of microturbulence velocity, ξ_{μ}, effective temperature, T_e, gravity, g and Fe and Ti abundances. The procedure consists in a comparison of observed curves of growth with those obtained by model atmospheres. Particular care was devoted to the identification of the reference continuum and to the oscillator strengths of the lines employed. Radii, R, were derived from visual surface brightness and used to obtain masses, M and bolometric absolute magnitudes, M_b. Except for the radii, all the data depend exclusively on spectroscopic observations.

114.168 **Optical observations of ultraviolet objects – II. Classification and photometry (l = 0 to 145°).**
M. M. Dworetsky, P. A. Whitelock, D. J. Carnochan.
Mon. Not. R. Astron. Soc., Vol. 201, 901 - 922 (1982).
The authors present MK spectral classification and *UBV* photometry of a sample of stars originally selected as 'ultraviolet objects' from satellite photometry with the S2/68 experiment. Most of the stars are spectroscopically normal. A number of hot subluminous stars were also found; some are reported for the first time. The authors also give radial velocities for some of the stars observed. The majority of the apparently normal early B stars at high galactic latitudes are not runaway stars. The authors propose that these stars are subluminous disc-population analogues of horizontal branch B stars of Population II.

114.169 **Properties and nature of Be and shell stars. 7A. 88 Her. Observational data, their reduction and basic evaluation.**
V. Doazan, P. Harmanec, P. Koubský, J. Krpata, F. Ždárský.
Astron. Astrophys., Suppl. Ser., Vol. 50, 481 - 490 (1982).
A detailed description of the long series of ground-based observations of 88 Her is given. High-dispersion spectrograms covering the period from 1963 to 1979 and the UBV photometry from 1968 to 1981 are mostly employed. The data show: (1) Periodic variations of the radial velocities of the shell Balmer lines and of the V/R ratio of the double Hα emission line with a period of 86.7221 days, which seems to be stable since 1912. (2) Doubling of the Hα absorption core on the descending branch of the RV curve. (3) Long-term variations of the brightness and colour of the star correlated with substantial changes of the intensities of absorption and emission lines in the optical spectrum.

114.170 **SS433: new results.**
V. G. Surdin.
Priroda, 1982, No. 10, p. 101. In Russian.

114.171 **High dispersion investigation of CP stars around the Hα line.**
M. Gerbaldi.
Messenger, No. 30, p. 18 - 20 (1982).

114.172 **Interpretation of line profiles of the symbiotic star V 1016 Cyg.**
C. Kindl, N. Marxer, H. Nussbaumer.
Astron. Astrophys., Vol. 116, 265 - 272 (1982).
The authors have extracted profiles of the strongest UV emission lines of the symbiotic star V 1016 Cyg. These profiles are compatible with the assumption that V 1016 Cyg is a very young planetary nebula. The line profiles show two components. The narrow main component originates in the nebular shell. The authors conclude that the wide feet are due to emission from a spicular stellar wind. For the ejection mechanism they suggest magnetic acceleration of stellar matter.

114.173 **The radial velocities of three Ap stars.**
Ç. Bolcal.
Kandilli Obs., Heliophys. Serv., Istanbul. Publ., Second Ser., No. 7, 6 pp. (1982).
The radial velocities of three Ap stars, HD 25823, HD 224166 and HD 26385, have been determined from spectra of 12.4 Å/mm dispersion.

114.174 **The spectroscopic analysis of the silicon stars.**
Ç. Bolcal.
Kandilli Obs., Heliophys. Serv., Istanbul. Publ., Second Ser., No. 8, 20 pp. (1982).
A detailed model atmosphere analysis of the stars HD 25823, HD 26385, HD 35298 and HD 224166, which have been classified as silicon stars due to abnormal intensity of the lines Si II 4128, Si II 4131 and Si II 4200 have been performed.

114.175 **Properties of a luminous, Wolf-Rayet type object in the core of the extragalactic H II region IC 132.**
S. D'Odorico, P. Benvenuti.
ESO Sci. Prepr., No. 209, 20 pp. (1982). – Submitted to Mon. Not. R. Astron. Soc.

114.176 **Spectral classification.**
C. Jaschek.
Mitt. Astron. Ges., Nr. 57, p. 167 - 175 (1982).

114.177 **On a possibility of spectral classification of stars based on UV observations.**
N. A. Sakhibullin, V. Ya. Solov'ev.
Astron. Tsirk., No. 1172, p. 1 - 2 (1981). In Russian.

114.178 **Statistics on the properties of Be stars.**
A. M. Hubert-Delplace, M. Jaschek, C. Jaschek, H. Hubert.
Be Star Newsl., No. 1, p. 4 (1980).

114.179 **Spectrophotometric study of Be and shell stars with the TD-1 S2/68 experiment.**
F. Beeckmans, A. M. Hubert-Delplace.
Be Star Newsl., No. 1, p. 8 (1980).

114.180 **Correlations between line profile and photometric variations in the B2 $IV_{[e]}$ star HD 45677.**
J. P. Swings, R. Barbier, M. Klutz, A. Surdej, J. Surdej.
Be Star Newsl., No. 1, p. 10 (1980).

114.181 **Photoelectric scanner measurement of Balmer emission line profiles for southern Be stars. II. A survey for variations.**
J. Dachs, W. Eichendorf, H. Schleicher, T. Schmidt-Kaler, M. Stift, H. Tüg.
Be Star Newsl., No. 3, p. 2 - 4 (1981).

114.182 **The recent peculiar behaviour of the Be star HD 200120, 59 Cyg.**
A. M. Hubert-Delplace, H. Hubert.
Be Star Newsl., No. 3, p. 4 (1981).

114.183 **Variations du profil de la raie Hα dans quelques étoiles Be.**
Y. Andrillat, C. Fehrenbach.
Be Star Newsl., No. 3, p. 5 - 6 (1981).

114.184 **Be star terminology.**
M. Jaschek, A. Slettebak, C. Jaschek.
Be Star Newsl., No. 4, p. 9 - 11 (1981).

114.185 **Spectroscopic observations of three Be stars.**
W.-S. Gao, X.-Z. Guo, Y.-L. Guo, H.-L. Cao.
Be Star Newsl., No. 5, p. 2 - 4 (1982).

114.186 **Spectral types and rotational velocities of the brighter Be stars and A-F type shell stars.**
A. Slettebak.
Be Star Newsl., No. 5, p. 5 (1982).

114.187 **Studies of continua and emission lines in Be stars.**
D. Briot, L. Divan, J. Zorec.
Be Star Newsl., No. 6, p. 10 - 12 (1982).

114.188 **Le spectre des étoiles Be dans le proche infrarouge.**
Y. Andrillat.
Be Star Newsl., No. 6, p. 12 - 16 (1982).

114.189 **The nature of the high velocity, discrete mass loss in Be stars.**
G. J. Peters.
Be Star Newsl., No. 6, p. 17 - 18 (1982).

114.190 **A new probable Be star: Alpha Leo.**
M. Singh.
Be Star Newsl., No. 6, p. 27 - 29 (1982).

114.191 **Spectra of γ Cas, χ Oph, ο And in the near infrared.**
G. I. Shanin.
Astron. Tsirk., No. 1201, p. 3 - 6 (1981). In Russian.

114.192 **The mid-UV spectrum of Procyon.**
F. Castelli, R. Faraggiana, K. A. van der Hucht, Y. Kondo.
Bull. Inf. Cent. Données Stellaires, No. 23, p. 45 - 50 (1982).

114.193 **The chromospheres and coronae of late-type main sequence stars.**
C. Jordan.
Irish Astron. J., Vol. 15, (see 012.061), 110 - 111 (1981). Abstract.

114.194 **IUE spectroscopy of dwarf M stars.**
P. B. Byrne.
Irish Astron. J., Vol. 15, (see 012.061), 113 - 115 (1981).

114.195 **Ultraviolet classification from the S2/68 experiment on-board the TD 1A satellite.**
A. Cucchiaro.
Ultraviolet stellar classification, (see 012.062), p. 1 - 7 (1982).

An "Atlas of ultraviolet stellar spectra" has been established in order to illustrate a spectral classification system defined in a morphological way for the ultraviolet λλ 1350-2740 region. It is based upon the material provided by the S2/68 experiment on board the TD 1 satellite. The UV system set up in this way generally corresponds closely to the MK system.

114.196 **On the relation between spectral classification in the optical and ultraviolet regions.**
M. Jaschek, C. Jaschek.
Ultraviolet stellar classification, (see 012.062), p. 9 - 13 (1982).

Although there is a general agreement between classification obtained in the optical and in the UV, the coincidence is not perfect; about 5-10% of the stars present major discrepancies at the dispersion of TD 1. Higher spectral resolution instruments (like IUE) are required to solve these discrepancies.

114.197 **A near-infrared atlas of southern MK standards.**
A. C. Danks, M. Dennefeld.
Ultraviolet stellar classification, (see 012.062), p. 85 - 89 (1982).

Preliminary results of a program to extend the spectral classification system of MK standards into the near-infrared domain up to 11000 Å are reported.

114.198 **UV spectrophotometry of early-type stars.**
K. Nandy.
Ultraviolet stellar classification, (see 012.062), p. 91 - 100 (1982).

The spectral classification of early type stars from spectrophotometry of low dispersion IUE spectra is discussed. It is concluded that from the observed flux distributions over the IUE range together with *V* magnitudes it is possible to group stars according to effective temperature, surface gravity and the deviation from the shape of the mean extinction curve.

114.199 **Stellar UV classification: some critical remarks and examples taken from A5–B5 stars.**
F. Praderie.
Ultraviolet stellar classification, (see 012.062), p. 101 - 114 (1982).

Questions of spectral classification in the ultraviolet region are discussed with emphasis on the problem how to discriminate "normal" stars against "anomalous" ones. A stars (extended up to B5) are chosen because they offer an example of stars where temperature and opacity effects are mixed in the UV, prohibiting therefore the safe use of pure UV indices as temperature indicators.

114.200 **Miscellaneous remarks on theoretical aspects of UV stellar classification.**

B. Baschek.
Ultraviolet stellar classification, (see 012.062), p. 115 - 120 (1982).

114.201 **Suggested UV spectral classification criteria for A stars.**
E. Böhm-Vitense.
Ultraviolet stellar classification, (see 012.062), p. 121 - 130 (1982).

It is stated that for late B and A stars UV spectra offer the possibility to give a more sensitive temperature classification than do visual spectra. It is recommended to use the long wavelength wing of Ly α for the late B and A stars. For A and F stars the ratio of the Mg (2580 Å) and Mg^+ (2800 Å) resonance lines gives a good T and n_e classification criterion which is nearly Z independent. For A and F stars the Si I discontinuities at 1700 Å provide an excellent T classification criterion. If the absolute intensities of either one of the Mg lines is added to the criteria one has three classification criteria for a three dimensional classification according to T, n_e and Z.

114.202 **IUE high dispersion spectra of luminous stars in symmetric nebulae.**
H. M. Johnson.
Astrophys. J., Suppl. Ser., Vol. 50, 551 - 572 (1982).

The objects studied are +60° 2522 in NGC 7635, σ Ori AB and σ Ori E centered in a faint elliptical nebula, and AG Car in its nebula. Definite spectra are identified and classified according to photospheric, P Cygni, or interstellar types. Estimates of equivalent width are included in the tables. Velocities are given and discussed. The extreme ultraviolet edges of P Cygni profiles imply terminal wind speeds which allow improved determinations of mass loss rates for these stars.

114.203 **M supergiants and star formation at the galactic center.**
M. J. Lebofsky, G. H. Rieke, A. T. Tokunaga.
Astrophys. J., Vol. 263, 736 - 740 (1982).

High resolution spectra of six sources at the galactic center demonstrate that these objects are M supergiants. The dramatic concentration of such stars is interpreted as evidence for a recent burst of star formation which may also correctly model the ionization and energetics of the galactic center.

114.204 **The helium 10830 Å line in early-type stars: an atlas of Fabry-Perot scans.**
D. D. Meisel, B. A. Saunders, Z. A. Frank, M. L. Packard.
Astrophys. J., Vol. 263, 759 - 776 (1982).

Representative profiles of He I 10830 Å in 65 early-type (O6–A1) stars over a wide range of luminosity are presented. The atlas scans were obtained using the Vaughan Fabry-Perot interferometer and usually cover a range of ±15 Å at 1 Å resolution. The majority of the scans show very shallow (central depths less than 20%), broad features which do not agree with plane-parallel NLTE model atmosphere calculations of the 10830 line by Auer and Mihalas.

114.205 **Evolution of chromospheres and coronae in solar mass stars: a far-ultraviolet and soft X-ray comparison of Arcturus (K2 III) and Alpha Centauri A (G2 V).**
T. R. Ayres, T. Simon, J. L. Linsky.
Astrophys. J., Vol. 263, 791 - 802 (1982).

The authors compare IUE far-ultraviolet and Einstein soft X-ray observations of the red giant Arcturus and the nearby yellow dwarf α Centauri A, which are archetypes of solar mass stars in very different stages of evolution. They find no evidence for coronal soft X-ray emission from the red giant at surface flux levels of only 0.0006 that detected previously for α Cen A. There is evidence that the outer atmosphere of Arcturus is geometrically extended, tenuous, and cool. A second important difference between Arcturus and α Cen A is the prominent cool stellar wind of the red giant. The angular momentum loss is at least four orders of magnitude larger than that of the weak coronal wind of the sun. The rapid spin-down of a post-main-sequence solar mass star, owing initially to evolutionary expansion and later to the development of a strong wind, is likely to severely inhibit the generation and amplification of magnetic fields by the dynamo mechanism.

114.206 **Abundances in beryllium-deficient F stars.**
A. M. Boesgaard.
Publ. Astron. Soc. Pacific, Vol. 94, 749 (1982). – Abstract.

114.207 **The *IUE* spectrum of BD + 39°4926.**
A. Saha, J. B. Oke.
Publ. Astron. Soc. Pacific, Vol. 94, 802 - 806 (1982).

IUE spectra of the A- or F-type supergiant BD + 39°4926 have been combined with existing optical data. Comparison with model atmospheres yields an effective temperature of 7750 K, gravity log g of 0.5 and a reddening corresponding to $E(B-V)$ of 0.15. The Mg II λ 2800 feature was used to estimate the abundance of Mg^+, and the value of $\log [\epsilon(Mg^+) / \epsilon(Ca^+)]$ was found to be ~2.7.

114.208 **Two lithium-rich supergiants.**
R. E. Luck.
Publ. Astron. Soc. Pacific, Vol. 94, 811 - 814 (1982).

Two extraordinary stars of type F and G are identified. These stars, HR 7008 and HD 174104, have strong lithium doublets. A detailed analysis shows each to have the cosmic abundance of lithium. It is argued that they are unmixed stars on the redward approach to the first giant branch.

114.209 **Spectrophotometric exploration of the star α Col.**
V. A. Mineva, B. Z. Kovachev, T. B. Radoslavova.
Dokl. Bolg. AN, Tom 34, 1625 - 1628 (1981). – Abstr. in Ref. zh., 51. Astron., 10.51.460 (1982).

114.210 **Tables of recommended reference data. Energy distribution in absolute units in the spectrum of Vega.**
L. N. Samojlov, V. N. Nalimov, V. V. Mal'tsev, Yu. S. Ryzhov.
Vses. nauchn.-issled. tsentr. Gos. sluzhby standart. spravoch. dannykh. Moskva, 1981. 14 pp. In Russian. – Abstr. in Ref. zh., 51. Astron., 11.51.427 (1982).

114.211 **Stellar activity.**
A. K. Dupree.
Smithsonian Astrophys. Obs., Spec. Rep. 392, Vol. II, (see 012.065), p. 3 - 15 (1982).

Determinants and signatures of activity in cool stars are discussed with reference to recent ultraviolet and X-ray measurements. The inadequacy of temperature and luminosity as a complete set of physical constraints on stellar activity is demonstrated. Solar atmospheric analogues are addressed; the effects of rotation and evolution upon emission level are emphasized. Emissions from hot plasma appear to decay more rapidly with age in dwarf stars than chromospheric fluxes. The emissions of several classes of objects – the giants in the Hyades cluster giants and the W UMa type interacting binary systems – deviate from the commonly accepted activity scenario and may provide direct evidence for the influence of long term activity cycles and evolutionary patterns as additional determinants of stellar activity.

114.212 **Stellar activity measured at Ca II H and K.**
S. L. Baliunas.
Smithsonian Astrophys. Obs., Spec. Rep. 392, Vol. II, (see 012.065), p. 31 - 39 (1982).

Time series analysis of fluctuations of Ca II H and K chromospheric emission has given much information concerning stellar activity. On all timescales, events which parallel solar behavior can be observed: activity cycles, on timescales of years; rotation of stars and evolution of active areas on timescales of days to weeks; flare-like phenomena on timescales as short as minutes.

114.213 **Mg II k emission lines in stars with different rotational velocities and metal abundances.**
E. Böhm-Vitense.
Smithsonian Astrophys. Obs., Spec. Rep. 392, Vol. II, (see 012.065), p. 71 - 75 (1982).

The author has studied the Mg II k line emission in main sequence F stars with different rotational velocities. Stars later than F usually only show measurable rotational velocities if they are members of close binary systems. For these stars rotational periods and binary periods are strongly correlated. For early F stars this is not the case. The author finds rapidly rotating single stars and slowly rotating members of fairly close binaries. The effects of rotation and binary nature can therefore be separated.

114.214 **Chromospheric emission, stellar rotation and X-ray coronae.**
R. Pallavicini, L. Golub, R. Rosner, G. Vaiana.
Smithsonian Astrophys. Obs., Spec. Rep. 392, Vol. II, (see 012.065), p. 77 - 85 (1982).

The authors report on a series of empirical correlations between coronal X-ray emission, chromospheric radiative losses and stellar rotation, for an ensemble of late-type stars of spectral types F7 to M5 and luminosity classes III, IV, and V. The X-ray observations are all from the Einstein Observatory; rotational data and chromospheric radiative losses are from a variety of published sources. From the correlation of X-ray emission with stellar rotation and Ca II/Mg II emission the authors derive qualitative information on the heating mechanism of stellar chromospheres and coronae.

114.215 **Ultraviolet observations of young field stars.**
A. M. Boesgaard, T. Simon.
Smithsonian Astrophys. Obs., Spec. Rep. 392, Vol. II, (see 012.065), p. 161 - 169 (1982).

In order to study the changes in chromospheric activity with stellar age, the authors have obtained IUE spectra (1100–3200 Å) of several young, solar-type field stars to compare with spectra previously obtained of T Tauri stars and the sun. They have measured the strengths of the ultraviolet chromospheric and transition region lines. The lines of the high temperature ions (He II, C II, C IV, N V, Si IV) and the low temperature ions (C I, O I, Mg II, Si II) all show a steep decrease in strength with age. The fluxes are approximately in inverse proportion to age. The authors have followed one star, χ^1 Ori, with daily observations through almost one rotational cycle. Lines due to He II, C I, C IV, N V, O I, and Si II all show the same pattern of variation in line strength. The Mg II lines show no variation at all and C II and Si IV seem to vary in a different phase.

114.216 **The ultraviolet spectrum of carbon stars.**
G. T. O'Brien, H. R. Johnson.
Smithsonian Astrophys. Obs., Spec. Rep. 392, Vol. II, (see 012.065), p. 171 - 174 (1982).

The authors report their detection of the ultraviolet spectrum of BL Ori and compare it against similar spectra of oxygen-rich M stars. Similar comparisons are made between IUE spectra of the warmer carbon stars and those of oxygen-rich stars of similar temperature.

114.217 **On the formation of technetium lines in the spectra of variable stars.**
A. S. Bakhtiyarov, A. A. Nikitin, A. A. Sapar, T. Kh. Feklistova.
Tartu Astrofüüs. Obs. Publ., Köide 49, 55 - 61 (1982). In Russian.

The paper presents a short review of the properties of stars, in the spectra of which Tc I lines are discovered. The processes of the Tc nuclear synthesis are considered. It has been found that the formation of Tc in stellar atmospheres necessitates intensive neutron flows over a considerable time interval of the half-decay of Tc.

114.218 **Abundance of lithium in the atmospheres of late-type giants.**
L. Hänni.
Tartu Astrofüüs. Obs. Publ., Köide 49, 62 - 76 (1982). In Russian.

Observations of the lithium resonance line at λ 6707.8 in the spectra of 83 K5 - M5 giants were carried out. A break in the intensities of lithium line was found among the stars of each spectral class, most probably indicating a real gap of lithium abundances in the atmospheres of these stars.

114.219 **Determination of the rotational excitation temperature in solar type stars.**
M. J. Fernández-Figueroa, M. Rego, A. Perea.
Astrophys. Space Sci., Vol. 87, 81 - 87 (1982).

The OH molecular spectrum has been identified in three solar type stars, β Com, β CVn, and κ Cet. The rotational excitation temperature from molecular line intensities of OH has been determined for these stars. The observations in high resolution were made by use of the IUE satellite.

114.220 **A determination of the radial velocity of the magnetic star HD 153 882.**
T. N. Kuznetsova.
Izv. Glav. Astron. Obs. Pulkovo, Astrometr. Astrofiz., No. 199, p. 96 - 101 (1982). In Russian.

The paper reports the methods and results of the measurements of radial velocities, residual intensities, equivalent widths and magnetic broadening of lines of the magnetic star HD 153 882 for a number of phases.

114.221 **On the outer atmospheres of late-type stars.**
M. M. Katsova.
Astron. Zh., Tom 59, 1124 - 1133 (1982). In Russian.
English translation in Soviet Astron., Vol. 26, No. 6.

An analysis of *EUV* spectra of UX Ari during a flare and during a quiescent state and of RW Aur is carried out. The reliable conclusion was obtained from the independently derived electron density and from the emission line intensity that the extension of the chromosphere – corona transition region for these stars as well as for α Aur is much larger than that for the sun. Two groups of late-type stars are distinguished: 1) rotating red dwarfs, 2) subgiant members of RS CVn-type systems and T Tau-type stars.

114.222 **Mapping of chemical elements on the surfaces of Ap stars. I. Solution of the inverse problem to obtain local line profiles.**
A. V. Goncharskij, V. V. Stepanov, V. L. Khokhlova, A. G. Yagola.
Astron. Zh., Tom 59, 1146 - 1156 (1982). In Russian.
English translation in Soviet Astron., Vol. 26, No. 6.

An algorithm is developed for the solution of the ill-conditioned inverse problem to obtain local line profiles on a chemically inhomogeneous rotating star using Tikhonov's method.

114.223 **Spectral variability of 68 Cygni, the O(f) star at the center of a ring nebula.**
V. Ya. Alduseva, A. A. Aslanov, E. A. Kolotilov, A. M. Cherepashchuk.
Pis'ma Astron. Zh., Tom 8, 717 - 721 (1982). In Russian.
English translation in Soviet Astron. Lett., Vol. 8.

Results of spectroscopic observations of the O(f) star

68 Cyg situated at the center of a ring nebula are presented. The periodic ($P \cong 5^d.1$) variability of the radial velocity ($K \cong 30$ km/s) and equivalent width (~ 30%) of the absorption lines is discovered. The spectral class of 68 Cyg is found to be O7(f)V. It is suggested that 68 Cyg is a close binary system consisting of a massive O(f)V star (~$30 M_{\odot}$) and a low-mass (~ $3 M_{\odot}$) companion.

114.224 **Spectroscopic analysis of Alpha Andromedae.**
I. E. Derman.
Astrophys. Space Sci., Vol. 88, 135 - 158 (1982).
A detailed abundance analysis of the Hg-Mn star α And is carried out with the method of fine analysis based on high-dispersion spectrograms obtained at the Haute Provence Observatory. The observed energy distribution, manganese ionization equilibrium and hydrogen-line profiles indicate that the atmosphere can be represented by a model with $T_{eff} = 13850 \pm 350$ K, $\log g = 3.85 \pm 0.15$ (in CGS) and with $\xi_t = 3.0 \pm 0.5$ km s^{-1}. The analysis shows that the abundances of C and Si are normal; Mg, S, and Fe slightly overabundant; and P, Mn, Ga, Sr, Y, Zr, Eu, and Hg are considerably overabundant relative to the solar values. A comparison of the author's abundances with those by others shows large discrepancies. A discussion of the atmospheric structure of α And suggests that the star has a circumstellar envelope.

114.225 **Unidentified molecular features in the red spectrum (5800 to 7300 Å) of Pi Gruis.**
P. S. Murty.
Astrophys. Space Sci., Vol. 88, 189 - 196 (1982).
Moderate dispersion (46 Å mm^{-1} at Hα) image-tube spectrum (5800–7300 Å) of the S star π^1 Gru is found to exhibit several unidentified molecular features, in addition to the well-known bands of the oxides of s-process elements. A list is presented of 28 measured unidentified red-degraded bands, of which six, at 5998, 6128, 6143, 6223, 6257, and 6557 Å, are prominent. Of the reported 28 bands, 22 are observed for the first time. A discussion is included on the expected molecular sources of the unidentified features. The possible presence of two new molecules, Pr O and Nd O, in π^1 Gru is suggested.

114.226 **Line identifications in the spectrum of HR 1100.**
S. Malaroda.
Publ. Astron. Soc. Pacific, Vol. 94, 979 - 986 (1982).
The spectrum of the star HR 1100 has been studied. The author identified the atomic species present on two coudé spectrograms of the star. Fe in the first three stages of ionization, Si II, Si III, P II, P III, Ti II, Ti III, Mn II, Hg II, and Sr II are found among others. The results for HR 1100 are compared with the results for other anomalous objects.

The Copernicus ultraviolet spectral atlas of Beta Orionis.
See Abstr. 002.005.

The "Catalogue of Stellar Groups": a progress report.
See Abstr. 002.025.

Results from Volume 3 of the Michigan Catalogue of two-dimensional spectral types for the HD stars (−40° to −26°) and future plans.
See Abstr. 002.057.

A catalogue, spectrum and magnitude data bank for Be, Bp and Bpe stars.
See Abstr. 002.096.

A library of digitized spectra of stars classified in the MK system.
See Abstr. 002.100.

Some comments on the Spectral Classification Catalogue by W. Buscombe.
See Abstr. 002.105.

Supplement to the Second Catalogue of Am Stars with Known Spectral Types (End 1981).
See Abstr. 002.107.

IUE ultraviolet spectral atlas.
See Abstr. 002.115.

Michigan Catalogue of Two-dimensional Spectral Types for the HD stars. Volume 3. Declinations −40°.0 to −26°.0.
See Abstr. 002.126.

S 59 spectrophotometric atlas of the mid-ultraviolet spectra of 220 early-type stars with a resolution of 1.8 Å.
See Abstr. 002.151.

Early emission line stars.
See Abstr. 003.069.

Metal-deficient stars.
See Abstr. 003.117.

Acetylene bands observed in carbon stars: a laboratory study and an illustrative example of its application to IRC + 10216.
See Abstr. 022.048.

Rotational broadening of some ultraviolet lines in spectra of B stars.
See Abstr. 022.143.

The maximum entropy principle in two-dimensional spectral analysis.
See Abstr. 031.509.

On the linearity of the SWP (*short wavelength prime*) camera of the International Ultraviolet Explorer (IUE): a correction algorithm.
See Abstr. 031.510.

IUE data reduction: wavelength determinations and line identifications using a VAX/750 computer.
See Abstr. 031.514.

Faint object studies with IUE.
See Abstr. 031.515.

Application of a computer controlled intensity registering procedure to the spectral region at 4430 Å.
See Abstr. 031.573.

Determination of accurate radial velocities of early-type stars by cross-correlating spectra scanned on the ESO-S-3000 measuring machine.
See Abstr. 031.631.

Profils de la raie Hα par télévision analogique.
See Abstr. 031.632.

The reduction of spectra. I. VELMEAS, an interactive radial velocity program.
See Abstr. 031.633.

The reduction of spectra. II. VLINE, an interactive line fitting spectrophotometric reduction program.
See Abstr. 031.634.

Automated spectral classification – a survey.
See Abstr. 031.648.

Automated spectral classification.
See Abstr. 031.649.

Two-dimensional automated spectral classification.
See Abstr. 031.650.

Classification spectrale: mise au point d'un système d'analyse rapide pour l'exploitation des "boîtes photométriques" de l'Observatoire de Genève.
See Abstr. 031.651.

Classification automatique des spectres de prisme objectif de l'Observatoire de Haute-Provence.
See Abstr. 031.652.

La classification spectrale automatique en tant que problème de reconnaissance de forme.
See Abstr. 031.653.

Le programme de classification stellaire UV.
See Abstr. 031.654.

Further experiments in automatic quantitative spectral classification at Tartu.
See Abstr. 031.655.

Very wide band stellar spectrophotometry.
See Abstr. 034.049.

A comparison of stellar observations and the theory of nucleosynthesis.
See Abstr. 061.034.

Stellar model chromospheres. XIII. M dwarf stars.
See Abstr. 064.001.

The photospheric velocity field of Procyon.
See Abstr. 064.013.

"Discrepant asymmetry" stars: the role of unsteady magnetic flux loops in the atmospheres of late-type giant stars.
See Abstr. 064.014.

The outer atmosphere structure of three late type stars.
See Abstr. 064.017.

A model atmosphere analysis of Procyon (α CMi, F5 IV-V).
See Abstr. 064.018.

The temperature distribution in the atmosphere of the Wolf-Rayet star of the system V444 Cygni derived from infrared measurements.
See Abstr. 064.078.

The evolutionary status of Be stars.
See Abstr. 065.094.

Photospheric spectrum line asymmetries and wavelength shifts.
See Abstr. 071.009.

The Compton effect in the chromosphere. III.
See Abstr. 071.010.

The relevance of the IUE results on young stars for earth's paleoatmosphere.
See Abstr. 082.004.

A study of line profile variability in O stars.
See Abstr. 112.028.

Mass loss from extreme helium stars. Detailed UV-line fits for HD 160641, BD $-9°4395$ and BD $+10°2179$.
See Abstr. 112.047.

Discovery of two FK Comae stars in the southern hemisphere.
See Abstr. 113.010.

Narrow-band photometry in the 1 - 4 μm region: calibration and applications.
See Abstr. 113.023.

Absolute spectrophotometry of Wolf-Rayet stars: are the colors the same?
See Abstr. 113.034.

A photometric and spectroscopic study of faint OB stars in the southern Milky Way.
See Abstr. 113.043.

Spectral classification and extinction using ANS ultraviolet photometry.
See Abstr. 113.059.

Determination of temperatures, metallicities, and interstellar reddening for G–K giants and subgiants in the Vilnius photometric system.
See Abstr. 113.067.

The structure, energy balance, and winds of cool stars.
See Abstr. 116.010.

The temperature dependence of rotation and turbulence in giant stars.
See Abstr. 116.015.

The growth and decay of magnetic activity in lower main sequence stars.
See Abstr. 116.022.

Ultraviolet properties of the symbiotic stars.
See Abstr. 117.103.

Recent increase of the activity of the symbiotic star CH Cygni.
See Abstr. 117.105.

Spectral evolution of HBV 475 (= V1329 Cygni) in the ultraviolet.
See Abstr. 117.115.

Introductory remarks on the symbiotic star AG Draconis.
See Abstr. 117.116.

Symbiotic star UV emission and theoretical models.
See Abstr. 117.127.

Comparative chromospheric and coronal emission from close binaries.
See Abstr. 117.210.

The ultraviolet spectrum of KQ Puppis (Boss 1985).
See Abstr. 120.003.

Determination of the term symbiotic star.
See Abstr. 122.085.

Photometric variability of single stars with activity cycles.
See Abstr. 122.152.

Molecules in celestial objects – III. Study of CO in interstellar diffuse clouds.
See Abstr. 131.022.

Sk 143: an SMC star with a galactic-type ultraviolet interstellar extinction.
See Abstr. 131.055.

Ultraviolet extinction towards far OB associations.
See Abstr. 131.318.

Wolf-Rayet stars associated to giant regions of star formation.
See Abstr. 132.033.

Two micron observations of ^{12}CO and ^{13}CO in the red giant sources IRS 7, IRS 12, and IRS 19.
See Abstr. 133.024.

Spectroscopy of variable blue stellar objects.
See Abstr. 141.171.

Are QSOs local objects? I. A new interpretation of the emission and absorption spectra of a few QSOs.
See Abstr. 141.198.

Optical photometry and spectroscopy of the X-ray pulsar 1E 1145.1–6141.
See Abstr. 142.072.

The helium content of solar type stars in the Hyades.
See Abstr. 153.039.

Abundance indicators from vidicon spectra in six globular clusters.
See Abstr. 154.005.

Abundances in metal-poor stars. I. The globular clusters NGC 2808, NGC 3201, NGC 6397, and M 22.
See Abstr. 154.026.

Carbon, nitrogen, and iron-peak abundances for giants in the remote globular clusters NGC 7006 and Pal 13.
See Abstr. 154.046.

G- und K-Sterne als Indikatoren der galaktischen Evolution.
See Abstr. 155.052.

A catalog of diffuse nebulae and emission-line stars in NGC 6822.
See Abstr. 158.009.

On the nature of the stellar population in the nucleus of the Sd galaxy NGC 7793.
See Abstr. 158.224.

The peculiar, luminous early-type emission line stars of the Magellanic Clouds: a preliminary taxonomy.
See Abstr. 159.006.

115 Luminosities, Masses, Diameters, HR and other Diagrams

115.001 **Angular diameters by the lunar occultation technique. VI. Limb darkening of α Tauri.**
S. T. Ridgway, G. H. Jacoby, R. R. Joyce, M. J. Siegel, D. C. Wells.
Astron. J., Vol. 87, 1044 - 1050 (1982).

The authors report 14 angular diameter measurements of α Tauri at six wavelengths in the range 0.4 - 3.8 μm. An examination of all available data leads to an angular diameter determination for the limb-darkening-corrected diameter of $\phi_{LD} = 20.88 \pm 0.10$ milliarcsec. The wavelength dependence of the diameter confirms theoretical predictions for the wavelength dependence of limb darkening.

115.002 **HD 105262: a possible bright field horizontal-branch star.**
J. W. Glaspey.
Astrophys. J., Lett., Vol. 258, L71 - L73, with a correction in Vol. 263, L45 (1982).

Spectra of the 7th magnitude star HD 105262 identify it as a field horizontal-branch star. The Balmer lines are narrow and the Balmer jump is large, implying a low surface gravity, yet the only other lines detectable at 43 Å mm^{-1} are Ca II λ3933 and Mg II λ4481. The space motions of this object are too low for it to have been identified kinematically as a Population II star.

115.003 **The masses and origin of carbon stars.**
M. S. Bessell, P. R. Wood, T. Lloyd Evans.
Proc. Astron. Soc. Australia, Vol. 4, 201 - 203 (1981).

115.004 **Globules in the Orion Nebula. III. Age range of the stars in the Orion Nebula.**
S. Isobe, G. Sasaki.
Publ. Astron. Soc. Japan, Vol. 34, 241 - 247 (1982).

Photographic photometry of 413 stars with magnitudes of 5.0 to 16.0 distributed in an area of 30′ around the Orion Nebula was carried out. From a plot of R magnitude versus $R–I$ color, the age of stars in the Orion Nebula region was found to range from 10^4 to 3×10^7 yr. This indicates that the low-mass stars in the Orion Nebula region have the same ages as the oldest stars in the Orion association Ia.

115.005 **Das Hertzsprung-Russell-Diagramm und seine Deutung.**
H. Scheffler.
Sterne Weltraum, 21. Jahrg., 350 - 353 (1982).

115.006 **Dwarf stars: red, brown, and black.**
W. Tucker, K. Tucker.
Mercury, Vol. 11, 81 - 82 (1982).

115.007 **Observed mass spectrum in the solar neighbourhood derived from catalogues of an astronomical data center.**
S. V. Vereshchagin.
Pis'ma Astron. Zh., Tom 8, 546 - 553 (1982). In Russian. English translation in Soviet Astron. Lett., Vol. 8.

Observed mass spectra for stars in the solar neighborhood have been constructed using data on 17000 stars from catalogues on magnetic tape. The rate of star formation slowly decreased with time. For masses of stars from 1 to 75 $M_\odot$ the initial mass function may be approximated by a power law with an exponent of about −2.3. There is a deficiency of stars with masses of 1.5 - 2.5 $M_\odot$ and probably of 5 - 6 $M_\odot$.

115.008 **New light on faint stars – I. The luminosity function in the solar neighbourhood.**
N. Reid.
Mon. Not. R. Astron. Soc., Vol. 201, 51 - 71, with a correction p. 1199 (1982).

From photoelectric photometry of red dwarf stars identified in an objective prism survey the author has defined a magnitude-limited complete sample. Applying photometric parallaxes this sample is used to determine the space densities of stars with absolute magnitudes between $M_v = +7$ and $M_v = +12$. The resultant luminosity function is consistent with the Luyten and Wielen functions for the solar neighbourhood. An analysis of the stellar kinematics shows little significant evidence of a substantial local population of low space motion M-dwarfs.

115.009 **New light on faint stars – II. A photometric study of the low luminosity main sequence.**
N. Reid, G. Gilmore.
Mon. Not. R. Astron. Soc., Vol. 201, 73 - 94 (1982).

This paper presents the first purely photometric derivation of the stellar main-sequence luminosity function to absolute magnitude $M_V = +19$, which is comparable to the minimum mass for thermonuclear burning. Absolute magnitudes and distances are derived by photometric parallax from the $M_V/V-I$ and $M_V/I-K$ relations. For $+9 \leqslant M_V \leqslant +19$, the photometrically defined luminosity function is in agreement with that derived from samples of nearby stars, and by proper motion techniques. The missing mass in the solar neighbourhood, if any, does not reside in main-sequence stars brighter than $M_V \approx +17$ mag.

115.010 **Stellar angular diameters and the radiation from cool stars.**
S. T. Ridgway.
Recent advances in observational astronomy, (see 012.028), p. 81 - 87 (1981).

The author reviews the importance of stellar angular diameters for the interpretation of stellar spectrophotometry and the characterization of the emergent stellar radiation. Recent results include the calibration of cool star effective temperatures. Some measures of relative limb darkening may be available soon for the most favorable cases.

115.011 **Das untere Ende der Hauptreihe.**
H. Scheffler.
Sterne Weltraum, 21. Jahrg., 415 - 417 (1982).

115.012 **H-R diagram for early type Magellanic Cloud members.**
K. Nandy, D. H. Morgan, G. I. Thompson, A. Willis, R. Wilson, L. Houziaux.
Third European IUE Conference, (see 012.029), p. 281 - 282 (1982).

Low resolution UV spectra of a sample of 29 LMC early type members were obtained with the IUE. The sample includes 25 supergiants and 4 giants. Complementary visible data were obtained with the ESO 3.6 m and 1.5 m telescopes and the 1.95 m telescope of the SAAO. The effective temperatures and bolometric magnitudes of these objects have been determined by comparison with models and the results are presented.

115.013 **Wolf-Rayet stars with massive companions.**
P. Massey.
Wolf-Rayet stars: observations, physics, evolution, (see 012.042), p. 251 - 262 (1982).

115.014 **Absolute magnitudes of Wolf-Rayet stars: the WN3 and WN4 sub-classes in the Large Magellanic Cloud.**
J. Breysacher, M. Azzopardi.
Wolf-Rayet stars: observations, physics, evolution, (see 012.042), p. 523 - 526 (1982).

115.015 **Comments on the significance of the positions of Population I Wolf-Rayet stars in the HR diagram.**
A. B. Underhill.
Wolf-Rayet stars: observations, physics, evolution, (see 012.042), p. 571 - 575 (1982).

The size of typical errors in $\log L/L_\odot$ and $\log T_{eff}$ is investigated for WR stars, and it is found that typically $\Delta \log L/L_\odot = \pm(0.25 \pm 0.03)$, and $\Delta \log T_{eff} = \pm(0.05 \pm 0.02)$. The major sources of error are the uncertainties in the distance modulus of the star and the integrated continuous flux from about 1400 Å to 1 μm or so. The same is true for O and early B stars. With uncertainties of these sizes in $\log L/L_\odot$ and $\log T_{eff}$, it is unreliable to estimate the mass of an early-type star from its observed position in the HR diagram.

115.016 **The two-colour diagram of luminous stars in the Magellanic Clouds.**
J. Isserstedt.
Astron. Astrophys., Vol. 115, 97 - 103 (1982). In German.

The two-colour diagrams of luminous stars in the Magellanic Clouds are presented. The available methods for the determination of the intrinsic relations $(U-B)_0$ vs. $(B-V)_0$ are shown to produce various systematic errors. With an improved method and an enlarged observational material the intrinsic relations are redetermined. Using foreground reddenings of $0^m\!.07$ and $0^m\!.02$ for the LMC and SMC and distance moduli of $18^m\!.31$ and $18^m\!.62$, respectively, the intrinsic relations turn out to be the same for both clouds. The intrinsic colours are compared with those for galactic supergiants: the differences are not understood until now.

115.017 **The angular diameter of Betelgeuse.**
Y. *(Yu. Yu.)* Balega, A. Blazit, D. Bonneau, L. Koechlin, R. Foy, A. Labeyrie.
Astron. Astrophys., Vol. 115, 253 - 256 (1982).

The authors report measurements of the diameter of Betelgeuse using Labeyrie's speckle interferometer attached to the 6-m soviet telescope and the 3.6-m ESO telescope. The TiO related diameter variations evidenced in this paper are not consistent with the effective temperature $T_{eff} = 3900$ K proposed by Tsuji from model fitting to the observed energy distribution; this discrepancy is discussed. No other variation of diameter, either with wavelength or with time is meaningful within the range of errors.

115.018 **The luminous core of 30 Doradus: supermassive star or dense star cluster?**
A. F. J. Moffat.
J. R. Astron. Soc. Canada, Vol. 76, 323 (1982). – Abstract.

115.019 **Mass-radius diagram and Hertzsprung-Russell diagram.**
K. Pilowski.
Univ. Hannover, Astron. Stn., Veröff. Nr. 14, 15 pp. (1982).

The mass-radius diagram of eclipsing binaries is treated which is determined free of hypotheses (two photometric minima, double radial velocities) using a material of 124 eclipsing binaries. The previous mass-radius diagram is based on 72 stars. 50% of these are refined. The similarity is surprising. Only little corrections are to be regarded.

115.020 **Absolute magnitudes of F-, G-, K-, and M-type stars from maximum-likelihood statistical parallaxes.**
T. Mikami, A. Heck.
Publ. Astron. Soc. Japan, Vol. 34, 529 - 539 (1982).
Absolute magnitudes and their corresponding dispersions are obtained for large samples of F- to M-type stars (giants and dwarfs) from the solar neighborhood by a statistical method based on the principle of maximum likelihood. The samples are characterized by homogeneous proper motions in the AGK3 system. The results confirm that the absolute magnitude of late-type giants should be brighter by a few tenths of magnitude than the values usually adopted.

115.021 **Interferometric program on Be stars.**
P. Granes, F. Vakili.
Be Star Newsl., No. 6, p. 20 (1982).

115.022 **Angular diameters, effective temperatures, radii, and luminosities of O3, O4, and O5 stars.**
A. B. Underhill.
Astrophys. J., Vol. 263, 741 - 758 (1982).
The effective temperatures, angular diameters, linear radii, and luminosities of 24 O3, O4, and O5 stars are determined from ultraviolet and visible photometry. The effective temperatures range from 24,800 K to 63,000 K. There is no correlation between effective temperature and spectral type or luminosity class for the early O stars. A detailed analysis of the uncertainties in deriving the various stellar parameters is given. The 24 early O stars of the program and ζ Pup are shown to occupy an area in the H-R diagram which is crossed by the evolutionary tracks of model hydrogen burning stars having masses in the range from 20 to 100 solar masses.

115.023 **One or more superstars?**
H. Nováková.
Říše hvězd, Vol. 63, 133 - 134 (1982). In Czech.

Early emission line stars.
See Abstr. 003.069.

Computer-assisted reduction of photographic two-colour diagrams. Application to field stars near IC 348.
See Abstr. 021.028.

A derivation of the stellar luminosity function in the direction of the south galactic pole by a statistical method.
See Abstr. 031.551.

Measurement of a stellar angular diameter by a high-time-resolving photometric observation of a lunar occultation.
See Abstr. 031.552.

Measurement of angular diameters of stars. Possibilities of the 6-m telescope.
See Abstr. 031.566.

Effect of spots on a star's radius and luminosity.
See Abstr. 064.068.

On the chemical abundances, evolutionary stage and bolometric magnitude of WR stars.
See Abstr. 065.071.

A comparison of observed and theoretical luminosity functions of carbon stars and late M giants.
See Abstr. 065.084.

Photoelectric observations of lunar occultations. XIII.
See Abstr. 096.018.

Preliminary results of a new study of the intrinsic properties of Wolf-Rayet stars.
See Abstr. 113.032.

Observations of nearby M dwarf stars in *VRI* colors.
See Abstr. 113.060.

Effective temperatures and radii of Large Magellanic Cloud supergiants.
See Abstr. 114.009.

Shell and photosphere of σ Ori E: new observations and improved model.
See Abstr. 114.159.

Photometric investigation of visual binaries with a component above the main sequence.
See Abstr. 118.028.

LY Aurigae, NY Cephei, and the mass-luminosity anomaly within O type binaries.
See Abstr. 119.047.

The central object of the 30 Doradus nebula, a supermassive star.
See Abstr. 132.034.

A magnitude limited stellar X-ray survey and the F star X-ray luminosity function.
See Abstr. 142.044.

Further observational evidence for a coronal boundary line in the cool star region of the H-R diagram.
See Abstr. 142.118.

X-rays from stars.
See Abstr. 142.139.

Wolf-Rayet stars in open clusters and associations.
See Abstr. 153.022.

Mass loss and the H-R diagram of globular clusters.
See Abstr. 154.004.

Luminosity of M giants in the nuclear bulge of the Galaxy.
See Abstr. 155.003.

Variations in the stellar luminosity function.
See Abstr. 155.017.

The gradient of the brightest star age and density across the spiral arm S4 in the Andromeda galaxy.
See Abstr. 158.189.

116 Magnetic Fields, Polarization, Figure, Rotation, Radio Radiation

116.001 **A search for magnetic fields in normal upper–main-sequence stars.**
J. D. Landstreet.
Astrophys. J., Vol. 258, 639 - 650 (1982).
A search has been carried out for magnetic fields of the order of 10^2 gauss in 31 normal O9.5–F6 upper–main-sequence stars and 5 Am stars by measuring the circular polarization of the wings of single spectral lines of hydrogen, helium, or iron with a Cassegrain filter polarimeter or with a coudé line-profile scanner polarimeter. The reduction formula used to infer longitudinal magnetic fields from such polarization measurements is discussed. No magnetic fields are detected in any of the stars surveyed. The smallest standard error reached is 7 gauss, and the median error is about 65 gauss. It is clear that the measurement methods employed are not subject to serious zero-point errors, and it appears that longitudinal fields of more than 150 gauss are not common in upper–main-sequence stars.

116.002 **Spectral variations of two cool Ap stars: HD 25354 and HD 152107.**
M. Floquet.
Astron. Astrophys., Vol. 112, 299 - 304 (1982).
The two cool Ap stars HD 25354 and HD 152107 (52 Her) were observed over the complete spectral variation period. The spectrum variations are interpreted in terms of the Oblique Rotator Model and are correlated with the photometric and magnetic variations. Maps of elements repartition are provided. The two stars possess an orbital motion period different from the spectrum variation period. The problem of synchronism between orbital and rotational periods is discussed.

116.003 **Magnetospheres and winds of the helium strong stars: dependence on rotation.**
P. K. Barker, D. N. Brown, C. T. Bolton, J. D. Landstreet.
Advances in ultraviolet astronomy, (see 012.006), p. 589 - 592 (1982).
The authors present a few preliminary results fom an ongoing extensive program of contemporaneous UV and optical spectroscopy and magnetic field measurements. Phase resolved observations have been (or will be) obtained for a variety of helium strong stars chosen to provide a wide-ranging sample in rotation rate and magnetic field strength, in order to investigate the interaction of magnetic, rotational, and radiative forces in the winds from these stars.

116.004 **Magnetic structure in cool stars. VI. Ca II H and K fluxes from evolved stars.**
F. Middelkoop.
Astron. Astrophys., Vol. 113, 1 - 8 (1982).
Ca II H and K line-core fluxes were measured for 335 evolved stars. The results confirm the conclusions drawn by Middelkoop and Zwaan from Wilson's eye estimates of the Ca II H and K emission features: there is a large spread in the fluxes from stars with $(B-V) < 0.95$ while the Ca II H and K flux of single stars with $(B-V) > 0.95$ correlates with color with little spread, except for some intrinsically bright stars. Short-period binaries show a relatively high Ca II H and K flux indicating that high Ca II H and K fluxes result from rapid rotation, independent of spectral type. The data are consistent with the hypothesis that Ca II H and K emission depends on dynamo action in the convective envelope, the dynamo efficiency decreasing with decreasing rotation rate. The evolution of Ca II H and K emission is discussed as a function of stellar mass.

116.005 **Solar calibration of stellar rotation tracers.**
B. J. LaBonte.
Astrophys. J., Vol. 260, 647 - 654, plate 19 (1982).
The time variability of the disk-integrated solar magnetic flux is studied as a proxy for emission-line intensity variations observed in lower main-sequence stars. The equivalent width of Ca II H + K emission is calibrated in terms of the magnetic flux. Modulation of the integrated flux by solar rotation is present in 75% of all rotations. For observing intervals more than twice the lifetime of the features causing rotational modulation, the correct rotation period is identified in >90% of all cases. It is concluded that Ca II H and K monitoring is the most productive method for measuring rotation in lower main-sequence stars.

116.006 **The rotation of cool main-sequence stars.**
D. F. Gray.
Astrophys. J., Vol. 261, 259 - 264 (1982).
Rotation rates are plotted as a function of spectral type for F, G, and K main sequence stars, and interpreted as a map of the cessation of a strong dynamo braking phase expected to occur at or near the zero-age main sequence in analogy with the dynamo brake phase seen in G giant stars.

116.007 **A comparative study of rotational properties of high-velocity and low-velocity early-type stars: the nature of the fast rotators.**
L. Carrasco, L. A. Aguilar, E. Recillas-Cruz.
Astrophys. J., Lett., Vol. 261, L47 - L50 (1982).
The authors carry out an analysis of the distribution functions of stellar rotation of O-type stars for samples selected according to kinematical characteristics. It is found that there is a large overlap between the stellar group of fast rotators and the group of stars that presents large space velocities. The possibility of explaining this overlap, in terms of the various theories of "runaway" stars is discussed. It is found that 15%–22% of all O-type stars are likely Population II objects.

116.008 **Magnetic phenomena in stars.**
Y. Osaki.
Magnetospheric phenomena of celestial bodies, (see 012.011), p. 72 - 76 (1982). In Japanese.
Observed phenomena related to the magnetic fields in stars are briefly described. Particularly, magnetic stars and RS CVn binaries are considered.

116.009 **Once more about the constant component of polarized radiation of μ Cep and RW Cep.**
T. A. Polyakova.
Vestn. LGU, 1982, No. 7, p. 92 - 99. In Russian. – Abstr. in Ref. zh., 51. Astron., 9.51.345 (1982).

116.010 **The structure, energy balance, and winds of cool stars.**
J. L. Linsky.
Third European IUE Conference, (see 012.029), p. 3 - 14 (1982).
IUE observations of cool stars provide evidence that magnetic fields control the structure and energy balance of the outer atmospheres of these stars. The author summarizes the phenomena associated with magnetic fields in the Sun and shows that similar phenomena occur in cool stars. He classifies G and K giants and supergiants into three groups depending on whether their atmospheres are dominated by closed magnetic flux tubes, open field geometries, or a predominately open geometry with a few closed flux tubes embedded.

116.011 **UBV-polarimetry of the X-ray binaries HD 77581 (4U 0900–40), HD 153919 (4U 1700–37) and of HD 152667.**
R. Östreicher, R. Schulte-Ladbeck.
Astron. Astrophys., Vol. 114, 328 - 336 (1982).

Observations of the linear polarization in the UBV passbands of the X-ray binaries HD 77581 and HD 153919 and of the close binary HD 152667 are presented. The degree of polarization of all three binary stars shows significant time variations with a total amplitude around 0.3%. For HD 153919, the power spectrum of polarization variations shows a conspicuous maximum at the expected frequency corresponding to the orbital semiperiod.

116.012 **Radio observations of HD 193793.**
D. R. Florkowski.
Wolf-Rayet stars: observations, physics, evolution, (see 012.042), p. 63 - 66 (1982).

Radio observations of the Wolf-Rayet star HD 193793 are presented. Most of the observations were made during a large outburst in the infrared. The radio observations are consistent with the suggestion that a circumstellar dust shell recently formed.

116.013 **Radio emission from WR stars.**
D. E. Hogg.
Wolf-Rayet stars: observations, physics, evolution, (see 012.042), p. 221 - 224 (1982).

116.014 **On the period of the Ap variable ET And.**
K. Pavlovski, J. Grygar, M. Muminović, N. Grubić.
Publ. Astron. Obs. Sarajevo, Vol. 1, No. 1, (see 012.043), p. 183 (1981). – Abstract.

116.015 **The temperature dependence of rotation and turbulence in giant stars.**
D. F. Gray.
Astrophys. J., Vol. 262, 682 - 699 (1982).

Rotation and turbulence velocities of 23 luminosity class III giants are obtained from their line profiles by Fourier analysis. The rotational discontinuity at G5 III is clearly delineated. It is shown that a dynamo brake is the most likely explanation, that the brake turns on and off before the star evolves more than one- or two-tenths in spectral class, and that all giants leave the braking stage with the same rotation rate of 5 km s^{-1}. An expression of the macroturbulence velocity as a function of effective temperature is derived.

116.016 **On standard polarized stars.**
J.-C. Hsu, M. Breger.
Astrophys. J., Vol. 262, 732 - 738 (1982).

Linear polarimetric measurements have been made to improve the system of standard polarized stars north of $\delta = -35°$. Among previously used standards, 55 Cyg, 9 Gem, and HD 183143 showed definite polarimetric variability on the time scale ranging from days to weeks. Short time-scale polarimetric variability seems to be more common among supergiants of earlier spectral type than those of later spectral type. Accuracies of polarization and position angle are estimated for the system of standard stars.

116.017 **Radio observations of symbiotic stars.**
S. Kwok.
The nature of symbiotic stars, (see 012.045), p. 17 - 26 (1982).

The fact that most symbiotic stars do not have detectable radio emission is consistent with the hypothesis that the mass transfer is via Roche lobe and the optical and ultraviolet emission lines originate from the mass transfer region. However, in a number of slow novae, stellar winds from either the cool or the hot components can generate an extended circumstellar envelope from which radio emission arises. It is clear from radio observations that in classical novae the hot-star wind dominates and in normal symbiotic systems wind is not an important element.

116.018 **Rotational studies of late-type stars. I. Rotational velocities of solar-type stars.**
D. R. Soderblom.
Astrophys. J., Vol. 263, 239 - 251, (1982) = Lick Obs. Bull., No. 923.

The author discusses the process of deriving $v \sin i$ for slowly rotating solar-type stars from high quality line profiles. Solar data can be fitted accurately and precisely with simple LTE profiles that are broadened by rotation, macroturbulence, and microturbulence. The technique of profile modeling is applied to high resolution spectra of 76 solar-type stars. There is good agreement with other observers for $v \sin i \gtrsim 10$ km s^{-1}, in the domain where rotation is the dominant line broadener. There is disagreement below 10 km s^{-1}, mostly because of the poor resolution of some observations, but also because of inadequate data analysis. Macroturbulent velocities are also derived from the profiles.

116.019 **Has P Cygni generated a shock front which emits nonthermal radiation?**
H. J. Wendker.
Astron. Astrophys., Vol. 116, L5 - L8 (1982).

Radio observations are presented which show that the star P Cyg has generated an arc-shaped emission complex. The radiation is most likely of non-thermal origin. In a somewhat speculative interpretation it is proposed that P Cyg has ejected material in a preferential plane. Streaming motion of these ejected gases against the surrounding interstellar medium has given rise to a bow-wave type shock front in which a sufficient number of particles can be accelerated to relativistic velocities.

116.020 **Standard stars for linear polarization.**
M. Breger, J. C. Hsu.
Bull. Inf. Cent. Données Stellaires, No. 23, p. 51 - 52 (1982).

116.021 **Radio emission from the extended chromosphere of Alpha Orionis.**
R. T. Newell, R. M. Hjellming.
Astrophys. J., Lett., Vol. 263, L85 - L87 (1982).

The red supergiant α Orionis (Betelgeuse) has been detected at 1.46, 4.89, 15.0, and 22.5 GHz with the VLA. The spectrum can be modeled by thermal emission from a spherically symmetric, partially ionized, chromospheric region (electron temperature ~10,000 K) extending from 1 to 4 stellar radii. Current wave-driven models for mass loss in late-type supergiants can account for the existence of such a region.

116.022 **The growth and decay of magnetic activity in lower main sequence stars.**
R. W. Noyes.
Smithsonian Astrophys. Obs., Spec. Rep. 392, Vol. II, (see 012.065), p. 41 - 48 (1982).

In this paper the author discusses characteristics of growth and decay of stellar active areas, and overall lifetimes of active region complexes. We tacitly assumes that the Ca II modulation in lower main sequence stars is due to the rotation of chromospheric plages overlying photospheric magnetic flux concentrations, as for the Sun.

116.023 **Rotation in G and K dwarfs.**
D. R. Soderblom.
Smithsonian Astrophys. Obs., Spec. Rep. 392, Vol. II, (see 012.065), p. 197 - 198 (1982).

Several aspects of rotation in G and K stars are discussed A relation between the rotation and magnetic field generation is suggested.

116.024 **A search for resonance polarization in stars with enhanced Ca II H and K emission.**
P. C. Chen.
Publ. Astron. Soc. Pacific, Vol. 94, 948 - 949 (1982).
A search has been made for resonance polarization in the Ca II H and K lines in the stars λ And, σ Gem, η Gem, μ Gem, and α Her. A differential technique using pairs of polaroid filters on a spectrograph slit was employed. The author did not detect any linear polarization at the 10% level.

A note on Garcia-Alegre et al.'s article, "A model for constructing artificial integrated spectral lines and their Fourier transform properties relevant to the search for differential rotation of stars."
See Abstr. 031.606.

A technique for measuring magnetic fields on solar-type stars.
See Abstr. 031.684.

A simultaneous infrared/optical polarimeter.
See Abstr. 034.004.

The connection of a catalogue of stars with an extragalactic reference frame.
See Abstr. 043.004.

Contraction of magnetic clouds and the origin of stellar magnetic fields.
See Abstr. 062.036.

Vorticity and astrophysical magnetism.
See Abstr. 062.110.

Broadband linear polarization from magnetized stellar atmospheres. Numerical tables for the magnetic intensification mechanism.
See Abstr. 063.001.

Monte Carlo study of highly polarized cool stars.
See Abstr. 064.002.

An explanation of the radio flux mystery of HD 192163 and empirical models for WN stars.
See Abstr. 064.052.

Magnetic stars.
See Abstr. 065.045.

Discovery and implications of the dramatic rotation brake at G5 III.
See Abstr. 065.090.

Wie rotieren Sterne?
See Abstr. 080.013.

Radio continuum measurements of mass loss from Wolf-Rayet stars.
See Abstr. 112.038.

Mass loss rates for Wolf-Rayet stars from radio continuum observations.
See Abstr. 112.044.

Circumstellar material around rapidly rotating B stars. I. Nonemission-line stars.
See Abstr. 112.054.

The Mg II resonance line emission at 2800 Å in stars with different rotational velocities and different metal abundances.
See Abstr. 114.002.

Chromospheric, transition layer and X-ray emission for stars with different rotational velocities.
See Abstr. 114.028.

Spectral types and rotational velocities of the brighter Be stars and A - F type shell stars.
See Abstr. 114.054.

New radial velocity limits for FK Comae.
See Abstr. 114.105.

Shell and photosphere of σ Ori E: new observations and improved model.
See Abstr. 114.159.

Stellar activity.
See Abstr. 114.211.

Stellar activity measured at Ca II H and K.
See Abstr. 114.212.

Mg II k emission lines in stars with different rotational velocities and metal abundances.
See Abstr. 114.213.

Chromospheric emission, stellar rotation and X-ray coronae.
See Abstr. 114.214.

A coordinated radio and Hα survey of southern RS CVn systems and related objects.
See Abstr. 117.020.

A search for cataclysmic binaries containing strongly magnetic white dwarfs.
See Abstr. 117.027.

The chromospheric rotation-activity relation in late type close binary systems.
See Abstr. 117.033.

RS CVn binaries – interaction of magnetospheres.
See Abstr. 117.048.

Stellar activity in the short period subgroup of the RS CVn systems.
See Abstr. 117.066.

The IUE survey of W UMa-type binaries.
See Abstr. 117.076.

The polarization of ζ Aurigae during the 1979–1980 eclipse.
See Abstr. 117.087.

A search for magnetic fields in the symbiotic and VV Cephei variables.
See Abstr. 117.088.

Polarimetry of CH Cygni.
See Abstr. 117.106.

Variable polarization in HD 162679.
See Abstr. 120.027.

Evolution of stellar activity during the pre-main sequence phase.
See Abstr. 121.044.

VLA observations of M-dwarf flare stars.
See Abstr. 122.153.

117 Close Binaries (Observations, Theory)

117.001 **Light variation in the W UMa system TZ Bootis.**
H. M. K. Al-Naimiy.
Astron. J., Vol. 87, 1041 - 1043 (1982).

Photoelectric observations of the eclipsing variable TZ Bootis were obtained in yellow and blue light during the period May - June 1980. The light curve changes show that the system TZ Boo has nearly equal minima at the time of observations, and seems to be on the way to a deeper minimum at the phase 0.0.

117.002 **Photoelectric photometry of the RS Canum Venaticorum-type close binary AR Lacertae.**
I.-S. Nha, Y.-W. Kang.
Publ. Astron. Soc. Pacific, Vol. 94, 496 - 514 (1982) = Yonsei Univ. Obs. Contrib. No. 1.

The strongly interacting binary star AR Lac was observed for two years, 1976 and 1977, during which a total of 235 observations for the check star (BD + 44°4041) and a total of 1327 observations for AR Lac were made in two colors. The yellow light curve of the check star and the yellow and blue light curves and the color curve of AR Lac are presented. The depth of the primary minimum in the yellow light shows no variation in any one year but 0.08 magnitude difference between the two years. This indicates a long-term light variation of the K0 component of AR Lac. As an explanation for the cause of the light variability, tests have been made for the starspot model and for the shell model, but neither of these satisfies either the color variation during the primary eclipse or the constant color outside eclipses.

117.003 **The infrared light curve of U Geminorum.**
R. J. Panek, J. A. Eaton.
Astrophys. J., Vol. 258, 572 - 575 (1982).

Infrared observations of the dwarf nova U Geminorum show the periodic double wave of an ellipsoidal variable with an amplitude of 0.25 mag. This wave is reproduced in calculations of the rotational light curve of the tidally distorted red dwarf star in the system. As expected, this star is required to fill its Roche lobe. The amplitude of the variation is consistent with the expected geometrical elements of the system, with convective gravity darkening, and with a moderate degree of limb darkening. The infrared colors of the system are those of a middle M dwarf star.

117.004 **EX Hydrae: physical parameters derived from simultaneous spectroscopy and photometry.**
R. L. Gilliland.
Astrophys. J., Vol. 258, 576 - 588 (1982).

High time resolution spectrophotometry has been used to determine the orbital characteristics of the eclipsing dwarf nova system EX Hydrae. Simultaneous photometry and spectrophotometry show that the 67 min photometric period clearly appears spectroscopically as a modulation of emission-line intensities. Emission-line profiles are found to vary at one-half of the 67 min period as well as the orbital period. Models for the 67 min variation are discussed. An analysis of the radial velocity variations suggests masses of $M_1 \sim 1.4\ M_\odot$ and $M_2 \sim 0.17\ M_\odot$ for the binary components.

117.005 **Evolution of dwarf binaries.**
A. V. Tutukov, A. V. Fedorova, L. R. Yungel'son.
Pis'ma Astron. Zh., Tom 8, 365 - 370 (1982). In Russian. English translation in Soviet Astron. Lett., Vol. 8.

The conditions of mass exchange in close binary systems with masses of components less or equal to one solar mass have been analysed for the case, when the system radiates gravitational waves. It has been shown that the mass exchange rate depends in a certain way on the mass ratio of components and on the mass of component that fills its inner critical lobe. The comparison of observed periods, masses of contact components, and mass exchange rates of observed cataclysmic binaries have led to the conclusion that the evolution of the close binaries WZ Sge, OY Car, Z Cha, TT Ari, 2A 0311–227, and G 61–29 may be driven by the emission of gravitational waves.

117.006 **The detection of compact companions in OB-runaway stars.**
C. H. B. Sybesma, C. de Loore.
Astron. Astrophys., Vol. 111, 229 - 232 (1982).

Theoretical light curves are computed for a binary system consisting of a massive early type main sequence star and a compact supernova remnant, in order to examine the possibility of detecting compact companions to OB-runaway stars. Computations are made in the 5 colour Walraven filter system, using the Roche equipotential model. The influence of the period, and inclination is examined and an analysis is made of the chance of finding such systems.

117.007 **Forced oscillations in binary systems. Toroidal modes.**
A. Rocca.
Astron. Astrophys., Vol. 111, 252 - 254 (1982).

The forced oscillations of a star, due to a periodic varying outer gravitational potential are examined, more precisely concerning toroidal modes. The oscillations are assumed to be adiabatic and the star uniformly and slowly rotating. The results show that for most of these modes, a resonance is possible only for determined values of the ratio of the orbital to rotation frequency.

117.008 **Infrared, optical and ultraviolet observations of TT Ari.**
R. F. Jameson, A. R. King, M. R. Sherrington.
Mon. Not. R. Astron. Soc., Vol. 200, 455 - 462 (1982).

B, V, J and *K* light curves together with the *IUE* ultraviolet spectrum of the cataclysmic variable TT Arietis are presented. The interpretation of these data leads to a system distance of 100 pc and outer disc temperature of 11 to 12.5×10^3 K.

117.009 **On the mass of the exploding stars in massive close binary systems.**
W. Sutantyo.
Galactic X-ray sources, (see 003.001), p. 27 - 39 (1982).

The author studies the effects of asymmetric instantaneous mass ejection on the orbital parameters of massive close binary systems in the first and second supernova explosions. He presents three independent arguments which show that the mass of the exploding stars in the explosions preceding the X-ray and binary pulsar stages is not more than about 7 $M_\odot$.

117.010 **IUE observations of the HZ Herculis system .**
L. Hartmann.
Galactic X-ray sources, (see 003.001), p. 149 - 156 (1982).

The UV observations of the HZ Her system are consistent with the optical models including a precessing disk, and serve to place stringent constraints on disk properties. The dominant light of the system appears to be explained adequately by the X-ray heated photosphere model. The N V and C IV emission line strengths are of the same order of magnitude as those predicted by X-ray photoionization.

117.011 **Ultraviolet observations of AM Herculis with IUE.**
J. C. Raymond.
Galactic X-ray sources, (see 003.001), p. 405 - 408 (1982).

Spectra of AM Herculis taken with IUE show a continuum which is partially eclipsed in phase with the X-ray eclipse and can be separated into two components: a black-body component which accounts for the 1/4 keV X-rays and the eclipsed UV continuum, and a component which is not eclipsed.

117.012 **The effects of gravitational radiation and net mass loss on estimating the minimum period in cataclysmic variables.**
W. Y. Chau, L. A. Nelson.
Astrophys. Lett., Vol. 22, 131 - 134 (1982).
The authors generalize the condition recently suggested by Paczynski to estimate the minimum period in cataclysmic variables to allow for the effect of net mass loss from the system. Gravitational radiation (at a rate that takes into account the uncertainty in the Landau-Lifshitz formula) is assumed to be the dominant mechanism driving the evolution. They find that depending on the specific mode of mass and angular momentum loss, the condition for minimum period (more precisely, the condition for turning off from the main sequence) can be significantly altered.

117.013 **Microcomputer based photoelectric observations of 44i Boo.**
R. M. Genet, K. E. Kissell.
Mem. Soc. Astron. Italiana, Vol. 52, 647 - 655 (1981).
A low cost, easily operated, microcomputer-based photometric system is described. The system was used to gather and analyze data on the W UMa type binary 44i Boo. While no significant recent change in the period was found, the primary minima were somewhat deeper then reported earlier.

117.014 **The peculiar variable star R Aquarii and its jet.**
M. Kafatos, A. G. Michalitsianos.
Nature, Vol. 298, 540 - 542 (1982).
The authors suggest that the jet is the result of supercritical accretion of mass transferred from the cool 387-day period Mira to the hot companion in a highly elliptical orbit. Ancient Japanese astronomical records suggest a nova outburst in AD 930 may be associated with R Aquarii which formed the outer extended nebulosity. The jet may help explain the outbursts of this object as well as the excitation of the R Aquarii nebula.

117.015 **On the origin of low mass cataclysmic binaries.**
M. Livio.
Astron. Astrophys., Vol. 112, 190 - 194 (1982).
The possibility of forming a cataclysmic binary from a star-planet system is investigated. It is shown that the fate of the system is determined essentially by the ratio of the accretion drag timescale to the mass loss timescale. Under certain simplifying assumptions, the evolution of the separation is calculated analytically. The effect of mass transfer on tidal evolution of the binary is also investigated.

117.016 **Contact binaries: angular momentum loss in and out of contact.**
S. M. Rucinski.
Astron. Astrophys., Vol. 112, 273 - 276 (1982).
The magnetic wind controlled evolution of a close late-type binary should accelerate until a contact system is formed but then should slow down because of the shrinkage of the system and the implied decrease in the mass loss which leads to a drop in the angular momentum loss efficiency. This effect might explain the rarity of late-type binaries with periods 0.5–2.0 days and might also contribute to the observed preference for marginal contact.

117.017 **First ultraviolet observations of two new cataclysmic variables 1E0643-1648 and 4U1849-31.**
J. M. Bonnet-Bidaud, M. Mouchet, C. Motch.
Astron. Astrophys., Vol. 112, 355 - 360 (1982).
The two hard X-ray sources 1E0643-1648 and 4U1849-31 have been observed in the range 1200-3200 Å with IUE. They show ultraviolet spectra typical of cataclysmic variables. The distances inferred from the ultraviolet interstellar extinction are 300 pc and 600 pc respectively for 1E0643-1648 and 4U1849-31. The line spectrum is characterized for both sources by strong resonance lines in emission. In addition 1E0643-1648 shows intercombination lines more typical of classical and recurrent novae. The energy distribution from ultraviolet to optical is predicted correctly for the two sources by standard optically thick accretion disc models, with derived disc accretion rates implying the existence of a large soft X-ray emission.

117.018 **Conservative mass transfer calculations for semi-detached binaries using response functions.**
M. Hauschildt.
Astron. Astrophys., Vol. 112, 386 - 387 (1982).
Mass transfer in semidetached binaries is calculated using an approximation for the mass response function. The outcome of conservative mass transfer depends on the minimum value and on the form of the response function.

117.019 **Phase resolved optical spectroscopy of the compact X-ray binary 2A 1822–371.**
K. O. Mason, P. G. Murdin, I. R. Tuohy, P. Seitzer, G. Branduardi-Raymont.
Mon. Not. R. Astron. Soc., Vol. 200, 793 - 805 (1982).
Optical spectrophotometry has been obtained of the unusual X-ray binary 2A 1822–371 throughout its 5.57 hr cycle with a time resolution of ~10 min and a wavelength resolution of about 1 Å. The spectrum of 2A 1822–371 varies as a function of binary phase. The strength of the diffuse interstellar features in the spectrum of 2A 1822–371 implies a colour excess $E_{B-V} = 0.13 \pm 0.06$. An improved optical photometric ephemeris is derived; T_{min} = JD 2444105.667 ± 0.003 + N × (0.232111 ± 0.000002 day).

117.020 **A coordinated radio and Hα survey of southern RS CVn systems and related objects.**
A. C. Collier, R. F. Haynes, O. B. Slee, A. E. Wright, D. J. Hillier.
Mon. Not. R. Astron. Soc., Vol. 200, 869 - 880 (1982).
The authors present the results of a search for radio emission at 5 GHz from a selection of 21 southern hemisphere RS CVn binaries and related late-type emission-line stars. Three of the program stars were detected above the survey limit of 10 mJy, including the previously known RS CVn radio binary HR 1099 and two new radio binaries, HR 4492 (= HD 101379) and HD 195040. A simultaneous program of high-resolution Hα spectroscopy revealed significant emission profile variability in both the latter stars. Radio variability on time-scales of ~1 day was observed in both HR 1099 and 4492. An increase by a factor of 3 in the observed radio flux density from HR 4492 on 1981 July 18 was accompanied by a significant enhancement of the Hα equivalent width by ~0.5 Å on that date.

117.021 **Bifurcation points in close binary evolution: medium-mass binaries.**
J. P. De Grève.
Astrophys. Space Sci., Vol. 84, 447 - 456 (1982).
It is the aim of this paper to investigate the influence of the mass ratio on the evolution of medium mass close binaries. The existence of a bifurcation point in this class of binaries has specific implications for the occurrence of phenomena resulting from accretion onto the white dwarf remnant from the mass transfer of the companion.

117.022 **IUE spectroscopic investigation of interacting binary systems.**
G. E. McCluskey, Jr.
Advances in ultraviolet astronomy, (see 012.006), p. 102 - 113 (1982).

The International Ultraviolet Explorer satellite has provided us with a wealth of data on essentially every type of close binary system known. This review limits itself to a discussion of binaries consisting of at least one early-type star, to the Algol systems and to W Serpentis systems.

117.023 **Binary stars: mass transfer and chemical composition.**
D. L. Lambert.
Advances in ultraviolet astronomy, (see 012.006), p. 114 - 124 (1982).

Mass exchange (and mass loss) within a binary system should produce observable changes in the surface chemical composition of both the mass-losing and mass-gaining stars as a stellar interior exposed to nucleosyntheses is uncovered. Three topics relating mass exchange and/or mass loss to nucleosynthesis are sketched: the chemical composition of Algol systems; the accretion disk of a cataclysmic variable fed by mass from a dwarf secondary star; and the hypothesis that classical Ba II giants result from mass transfer from a more evolved companion now present as a white dwarf.

117.024 **Cataclysmic variables: disk characteristics from UV observations.**
P. Szkody.
Advances in ultraviolet astronomy, (see 012.006), p. 474 - 477 (1982).

Low dispersion IUE spectra of 3 cataclysmic variables (V442 Oph, V794 Aql and H2215–086) are discussed in terms of current disk models. The range of continuum fluxes, line emission and disk parameters ($\dot{M}$, T_{max}) of these three novalike systems are compared with past observations of dwarf novae at outburst and quiescence. Evidence of variability on orbital timescales is presented for V442 Oph and H2215–086.

117.025 **IUE spectroscopy, visible-band polarimetry, and radiometry of V641 Mon.**
R. H. Koch, B. J. Hrivnak, D. H. Bradstreet, W. Blitzstein, R. J. Pfeiffer, P. M. Perry.
Advances in ultraviolet astronomy, (see 012.006), p. 513 - 516 (1982).

117.026 **Line-profile and continuum variations of the contact binary SV Centauri.**
J. Rahe, H. Drechsel, W. Wargau.
Advances in ultraviolet astronomy, (see 012.006), p. 521 - 525 (1982).

A total of five high and ten low dispersion UV spectra of the interacting contact binary SV Centauri has been obtained between 1979 and 1982. The UV data were complemented with optical photometric and spectroscopic observations, in order to determine the structure and absolute dimensions of the system. The profiles of prominent UV resonance and metastable lines undergo drastic changes with phase angle and time. Their overall appearance indicates relatively strong mass loss from the system, exhibiting pronounced variations of the stellar wind. The far UV continuum distribution suggests the presence of a luminous hot radiation source with maximum emission in the soft X-ray range.

117.027 **A search for cataclysmic binaries containing strongly magnetic white dwarfs.**
H. E. Bond, G. Chanmugam.
Advances in ultraviolet astronomy, (see 012.006), p. 530 - 533 (1982).

The authors made IUE observations of seven candidate cataclysmic variables selected for optical similarity to AM Her binaries. Although all seven objects were detected in the UV, none display unusually strong UV continua. The authors suggest therefore that the distribution of magnetic field strengths among single white dwarfs may be different from that among binaries.

117.028 **Observations of mass accretion in binary stars.**
R. S. Polidan, G. J. Peters.
Advances in ultraviolet astronomy, (see 012.006), p. 534 - 537 (1982).

Results from high resolution observations of eight close binary stars (TX UMa, U CrB, CX Dra, TT Hya, AU Mon, KX And, HR 2142, and ϕ Per) are presented. Variable absorption lines, indicative of mass flow, are observed in all systems except ϕ Per. Emission lines are seen in KX And and ϕ Per. Variable high ionization features (N V, Si IV, and C IV) are seen in TX UMa, U CrB, CX Dra, and AU Mon. The observations are modeled using the calculations of Lubow and Shu.

117.029 **Model fitting of the Algol binaries.**
J. J. Dobias, M. J. Plavec.
Advances in ultraviolet astronomy, (see 012.006), p. 538 - 541 (1982).

Energy distributions of the component stars of the Algol-type semi-detached binaries are being studied by combining IUE low-dispersion spectra with optical scans made with about the same resolution. In most cases, the flux distributions can be very well matched by Kurucz model atmospheres with normal solar composition. A search is being made for deviations from normal atmospheres. Eclipse observations in U Cep and V356 Sgr revealed emission lines of the W Serpentis type.

117.030 **Continuum and emission lines in Beta Lyrae.**
M. J. Plavec, J. L. Weiland, J. J. Dobias.
Advances in ultraviolet astronomy, (see 012.006), p. 550 - 553 (1982).

Integral photometric characteristics of β Lyrae have been re-determined using its brightest visual companion, the star HD 174664. An attempt was made to separate the fluxes coming from the two components. The flux distributions are badly contaminated by the radiation of the circumstellar hydrogen. The secondary star cannot be represented by an atmosphere with a unique effective temperature, and will probably be best represented by a thick disk.

117.031 **Results of an IUE program of monitoring the ultraviolet emission line fluxes of four binary systems: HR 1099, II Peg, AR Lac, and BY Dra.**
N. Marstad, J. L. Linsky, T. Simon, M. Rodono, C. Blanco, S. Catalano, E. Marilli, A. D. Andrews, C. J. Butler, P. B. Byrne.
Advances in ultraviolet astronomy, (see 012.006), p. 554 - 557 (1982).

The authors present a preliminary report on a collaborative program to obtain IUE spectra and optical photometry and spectra of three RS CVn-type binaries (HR 1099, II Peg, and AR Lac) and the prototype BY Dra system. They detected periodic variations in emission line flux from II Peg and HR 1099, indicative of rotational modulation of an active region on these stars.

117.032 **The "X-ray binary", UW CMa.**
S. R. Heap.
Advances in ultraviolet astronomy, (see 012.006), p. 562 - 565 (1982).

Einstein photometry and IUE spectra support the hypothesis of a wind-star collision in UW CMa. The collision sets up a high-temperature region, which is the source of X-rays detected with Einstein. Its apparent X-ray variability is due to its location between the two stars, where it undergoes eclipses. The high-temperature region maintains an ionization cavity in the wind, as detected with IUE. The ionization cavity is the source of depletion of absorbing ions in the wind between the two stars.

117.033 **The chromospheric rotation-activity relation in late type close binary systems.**
F. M. Walter, G. S. Basri, R. Laurent.
Advances in ultraviolet astronomy, (see 012.006), p. 566 - 569 (1982).
The authors present IUE data on 36 late-type close binary stars (F8–K0). The rotation-activity relation shows a systematic increase in dependence upon rotation with increasing temperature. This supports the contention that the lower chromosphere is only weakly influenced by stellar rotation, that the TR is more affected, and that the corona is most dominated by stellar rotation.

117.034 **Time-resolved spectroscopy of VV Puppis.**
A. P. Cowley, D. Crampton, J. B. Hutchings.
Astrophys. J., Vol. 259, 730 - 739 (1982).
Single-trailed spectra of VV Puppis with coverage of more than a full orbital cycle show complex emission-line variations similar to those found in other magnetic white-dwarf binaries. Details of these are presented and discussed. Part of the emission lines appears to arise in a column of material falling onto the active magnetic pole.

117.035 **Simultaneous radio and X-ray activity in SS 433.**
E. R. Seaquist, W. S. Gilmore, K. J. Johnston, J. E. Grindlay.
Astrophys. J., Vol. 260, 220 - 232 (1982).
Simultaneous observations at radio and X-ray wavelengths of flarelike events in SS 433 were made during 1979 October 5 - 9 and 19 - 22. The radio spectra show evidence for low-frequency absorption, which may be due to either synchrotron self-absorption or free-free absorption by a stellar wind. The X-ray emission is evidently nonthermal and closely related to the radio flares. The source of the X-ray radiation is either synchrotron or inverse Compton emission. It is shown that simultaneous variability in the two spectral regions imposes stringent conditions for current theoretical models of SS 433.

117.036 **The optical spectrum of HDE 226868 = Cygnus X-1. I. Radial velocities and orbital elements.**
D. R. Gies, C. T. Bolton.
Astrophys. J., Vol. 260, 240 - 248 (1982).
The authors have measured radial velocities for HDE 226868 from 78 high-dispersion spectrograms obtained between 1971 and 1981. The orbital elements derived from these data are essentially identical to those previously reported, except that they do not confirm the short-period apsidal motion that has been suggested by variations in the shape of the ellipsoidal light curve. There is no evidence for periods other than the orbital period.

117.037 **On the emission lines from rotating gaseous disks.**
J. Smak.
Acta Astron., Vol. 31, 395 - 408 (1981).
Theoretical profiles are calculated for a flat, Keplerian disk with the emissivity law $f(r) \sim r^{-a}$ and including the effects of finite spectral resolution. The profiles are double peaked with the peak velocity being by about 6 percent larger than the rotational velocity $V_d \sin i$ of the outer edge of the disk. The effect of finite spectral resolution can partly explain the invisibility of doubling at low orbital inclinations. The r^{-a} emissivity law appears well applicable to disks in cataclysmic variables, with $a \cong 2$.

117.038 **A survey of W UMa-type systems in the four-colour *uvby* photometry.**
S. M. Ruciński, J. Kałużny.
Acta Astron., Vol. 31, 409 - 426 (1981).
Four-colour indices averaged from 3 to 8 random observations of 36 brighter W UMa-type systems are presented.

117.039 **The enigmatic Hα line of FK Comae: last stages of a coalescing binary?**
F. M. Walter, G. S. Basri.
Astrophys. J., Vol. 260, 735 - 743 (1982).
The apparently single, rapidly rotating G giant FK Comae displays an unusual Hα emission line which is very broad, strong, and variable. The authors present extensive high-resolution studies of this line with complete phase coverage. These data, when phased with optical photometry, indicate that the star has a bright, rather than dark, region on its surface. A qualitative model is considered wherein the bright spot is caused by accretion from a low mass unseen companion onto the surface of the G giant.

117.040 **A photometric study and analysis of AW Ursae Majoris.**
B. J. Hrivnak.
Astrophys. J., Vol. 260, 744 - 754 (1982).
New green and red light curves for the extremely low mass ratio contact binary AW UMa have been observed. These display intrinsic variability around primary minimum occurring with a time scale of a few days. A limiting light curve analysis indicates a change in the temperature of the secondary component and in the size of both components. The implications of these observations are discussed with respect to theories of contact binaries, particularly with regard to the stability of the A-type systems.

117.041 **Nodding motions of accretion rings and disks: a short-term period in SS 433.**
J. I. Katz, S. F. Anderson, B. Margon, S. A. Grandi.
Astrophys. J., Vol. 260, 780 - 793 (1982).
The authors present a formalism to compute frequencies and amplitudes of short-term "nodding" motions in precessing accretion disks in close binary systems. This motion occurs when the binary orbital period is not negligible compared with the precession period, as is the case, for example, in both SS 433 and Hercules X-1. The analysis predicts a $6^{d}.3$ periodic component of the "moving" spectral lines of SS 433, at an amplitude ~10% of the 164^{d} precessional motion. Examination of spectra of SS 433 obtained on more than 300 nights during a 3 year period in fact reveals the expected period at approximatelv the correct amplitude. Some applications of the formalism to Hercules X-1 are also briefly considered.

117.042 ***UBVRI* photometry of VW Cephei.**
A. P. Linnell.
Astrophys. J., Suppl. Ser., Vol. 50, 85 - 114 (1982).
High speed *UBVRI* photometry of the W UMa eclipsing variable VW Cep is reported. Analysis of light curves and color variations reveals a number of phenomena previously unrecognized.

117.043 **Spectrophotometry of cataclysmic variables.**
G. Williams.
Bull. American Astron. Soc., Vol. 14, 613 (1982). – Abstract.

117.044 **IUE observations of E1405 - 451: a new AM Her-type cataclysmic variable.**
J. A. Nousek, S. H. Pravdo.
Bull. American Astron. Soc., Vol. 14, 633 (1982). – Abstract.

117.045 **Simultaneous spectroscopy and photometry of the AM-Her-like variable H0139 - 68.**
J. R. Thorstensen, R. Schommer, P. A. Charles.
Bull. American Astron. Soc., Vol. 14, 633 (1982). – Abstract.

117.046 **The secondary of λ Tau.**
J. Tomkin, F. C. Fekel.
Bull. American Astron. Soc., Vol. 14, 633 (1982). – Abstract.

117.047 **Evolution and analysis of the complex light curves of λ And using the starspot model.**
J. D. Dorren, E. F. Guinan, B. Paczkowski.
Bull. American Astron. Soc., Vol. 14, 634 (1982). – Abstract.

117.048 **RS CVn binaries – interaction of magnetospheres.**
Y. Uchida, T. Sakurai.
Magnetospheric phenomena of celestial bodies, (see 012.011), p. 77 - 80 (1982). In Japanese.

Coronae and flare phenomena of RS CVn binaries observed in X-rays and radio are attributed to the gradual heating of plasmas and the sudden release of magnetic energy in the system of interacting magnetospheres of both stars, respectively. Model magnetic fields of the system, with the K star having star-spots on it, are calculated, and the heating and the flaring are discussed with the energy input caused by the twisting up of the magnetic tubes by slow relative rotation which is the deviation from the synchronism of the rotations with the revolution.

117.049 **Accretion and inner excretion disks in close binaries.**
S. Kříž.
Bull. Astron. Inst. Czechoslovakia, Vol. 33, 302 - 308 (1982).

This paper investigates Keplerian disks rotating differentially around binary components which are rotating at the critical velocity. It is proved that the angular momentum can be removed very effectively from such a component by an accretion disk encircling it. When there is no mass transfer via Roche-lobe overflow and at least one component rotates at the critical velocity, an inner excretion disk is built up around it. The author demonstrates that it should be possible to distinguish both cases observationally.

117.050 **X-ray and optical measurements of the cataclysmic variable CH UMa.**
R. H. Becker, A. S. Wilson, S. H. Pravdo, G. A. Chanan.
Mon. Not. R. Astron. Soc., Vol. 201, 265 - 270 (1982).

A program to identify optical counterparts of X-ray sources discovered with the Einstein Observatory has resulted in an independent identification of CH UMa as a cataclysmic variable. Spectrophotometric observations made with the Intensified Image Dissector Scanner on the 2.1-m telescope at KPNO revealed an emission line spectrum from a $m_b \sim 15$ mag object similar to the spectra of cataclysmic variables. A subsequent search of the Harvard photographic plate collection revealed outbursts of 4 mag in amplitude.

117.051 **Photometry of the newly identified AM Herculis system: CW 1103 + 254.**
H. R. Miller.
Mon. Not. R. Astron. Soc., Vol. 201, 21P - 24P (1982).

Photoelectric photometry (*UBV*) is reported for the AM Her type system CW 1103 + 254. Rapid variability, superimposed on a 113-min periodic variation is detected.

117.052 **A numerical analysis of the variation of the orbital period and the physical mechanism of fluctuation of the light curve of the close binary VW Cep.**
X. Liu, C. Luo, S. Lu, Y. Sun.
Kexue Tongbao, Vol. 26, 1096 - 1101 (1981). – Abstr. in Phys. Abstr., Vol. 85, Abstr. 63523 (1982).

117.053 **Polarimetric observations of southern binary systems.**
H. G. Luna.
Rev. Mexicana Astron. Astrofis., Vol. 6, (see 012.017), 111 - 113 (1981).

In 1978, the author started a systematic program for the observation of binary systems in the southern hemisphere with the polarimeter available at La Plata Observatory. The present paper shows the results for 23 systems. A high percentage of the observed stars showed polarimetric variations. Using the model proposed by Brown et al. (1978) it is possible to compute the inclination of the orbital plane.

117.054 **Observaciones de dos sistemas binarios con componentes WR.**
V. S. Niemela.
Rev. Mexicana Astron. Astrofis., Vol. 6, (see 012.017), 243 (1981).

117.055 **Theoretical aspects of X-ray binaries.**
R. Kippenhahn, H.-C. Thomas.
Ninth Texas Symposium on Relativistic Astrophysics, (see 012.020), p. 579 - 589 (1980).

117.056 **Summary of the workshop on SS 433.**
B. Margon.
Tenth Texas Symposium on Relativistic Astrophysics, (see 012.021), p. 403 - 413 (1981).

117.057 **UV observations of TT Arietis and the magnetic rotator hypothesis.**
R. F. Jameson, M. R. Sherrington, A. R. King, J. Frank.
Nature, Vol. 300, 152 - 155 (1982).

Since good quality IUE spectra can be made with exposure times of about 15 min, the authors decided to observe as many IUE spectra as possible round the binary period with a view to studying the behaviour of the hot spot at UV wavelengths. They report that the variations at these and X-ray wavelengths round the orbital cycle rule out a hot spot/accretion disk model for TT Ari. Instead they propose that it has a magnetized white dwarf primary rotating with the photometric period of 0.1329 days.

117.058 **Interacting binary degenerate stars.**
R. E. Nather.
Recent advances in observational astronomy, (see 012.028), p. 55 - 62 (1981).

117.059 **The relevance of EXOSAT to X-ray observations of RS CVn systems and cataclysmic variables.**
I. McHardy.
Observatory, Vol. 102, 111 - 112 (1982). – Abstract.

117.060 **Progenitors and birth-rates of cataclysmic variables and type I supernovae.**
V. Trimble.
Observatory, Vol. 102, 133 - 135 (1982).

Cataclysmic variables and type I supernovae apparently arise from rather similar parent systems; but the latter must do so with an efficiency approaching unity, while 1 per cent efficiency suffices for the former.

117.061 **Initial results of photometry of some X-ray emitting RS CVn systems.**
P. C. Agrawal, K. M. V. Apparao, V. Margan, K. P. Singh, M. B. K. Sarma.
Bull. Astron. Soc. India, Vol. 10, 28 (1982). – Abstract.

117.062 **On the evolution of low mass binary systems.**
P. R. Amnuel *(Amnuehl')*, O. H. Guseinov *(O. Kh. Gusejnov)*.
Astrophys. Space Sci., Vol. 86, 91 - 106 (1982).

The evolution of parameters of close binary systems containing a red dwarf (with mass loss) and a condensed star is investigated. The mass loss from the system and asynchronism of the red dwarf rotation are taken into consideration. The calculations show that if the initial mass ratio $q_0 \lesssim 2$ then during $\sim 10^7$ yr the instability of mass loss process arises, and a bright X-ray source forms with luminosity close to the

Eddington one. If $q_0 > 2$ then during the mass loss phase the bright X-ray source arises twice. The models explain the existence of the "forbidden" interval of orbital periods and the absence of systems with periods less than ~80 min.

117.063 **On the constancy of the light curve of CC Comae.**
R. A. Breinhorst, M. Hoffmann.
Astrophys. Space Sci., Vol. 86, 107 - 115 (1982).
New photoelectric measurements were obtained in 1980. Variability of the light curve was found, which affects mostly the minima. No period change could be detected. An estimation of the uncertainty of the parameters of the system showed an effect on the ratio of the radii of about 10%.

117.064 **A model of AM Canum Venaticorum.**
N. F. Vojkhanskaya.
Astron. Zh., Tom 59, 925 - 932 (1982). In Russian. English translation in Soviet Astron., Vol. 26, No. 5.
The spectrograms of AM CVn (HZ 29) obtained in 1978 - 1980 have been processed. It is shown that the wavelengths and intensities of helium absorption lines vary at a timescale much longer than the photometric period (17.5 min). A new model of the system is suggested: two similar DB white dwarfs, one of which has a strong magnetic field $B > 10^6$ G. It is shown that the period of 17.5 min is the period of axial rotation of the magnetic white dwarf.

117.065 **Rotational modulation of chromospheric features in late-type stars.**
P. B. Byrne, C. J. Butler, A. D. Andrews, J. L. Linsky, T. Simon, N. Marstad, M. Rodonò, C. Blanco, S. Catalano, E. Marilli.
Third European IUE Conference, (see 012.029), p. 125 - 127 (1982).
The authors observed three RS CVn-type binaries (HR 1099, II Peg, and AR Lac) and the prototype BY Dra binary system with IUE. They report on flux variations of emission lines formed in the chromospheres and transition regions of two of these stars.

117.066 **Stellar activity in the short period subgroup of the RS CVn systems.**
E. Budding, A. Giménez.
Third European IUE Conference, (see 012.029), p. 169 - 173 (1982).
The Mg II 2800 Å feature has been studied in several spectrograms obtained with the IUE satellite for selected binaries among the RS CVn systems. The surface flux in emission has been used as a measure of activity. Some indications of correlation between such activity and rotational velocity as well as the scale of irregular photometric distortions are presented.

117.067 **The spectrum of the symbiotic star CH Cyg in December 1981.**
M. Hack, M. Persic, P. L. Selvelli.
Third European IUE Conference, (see 012.029), p. 193 - 195 (1982).
The continuum energy distribution and the emission spectrum indicate that CH Cyg is still in a high state after the outburst started in 1977. Significant variations in the intensities of the permitted and forbidden emissions have been found with respect to previous spectra.

117.068 **A review of the ultraviolet results about symbiotic stars.**
M. Friedjung, R. Viotti.
Third European IUE Conference, (see 012.029), p. 203 - 208 (1982).
The ultraviolet properties of the symbiotic stars are reviewed, and new results on AG Dra and other objects are discussed. Different single and binary models are examined: hot star surrounded by a cool envelope, cool star with a hot envelope, and binary systems.

117.069 **A comparative study of IUE spectra of cataclysmic variables.**
C. la Dous.
Third European IUE Conference, (see 012.029), p. 213 (1982). Abstract.

117.070 **Strong UV excess radiation in cataclysmic variables.**
W. Wargau, H. Drechsel, J. Rahe.
Third European IUE Conference, (see 012.029), p. 215 - 218 (1982).
The absolute UV continuum flux distribution of nine cataclysmic variables is compared with several steady-state optically thick accretion disk models. For most of the systems, the flux distribution is not well explained by these approaches. The shapes of the UV continua can alternatively be interpreted as superposition of two blackbody sources: a hot component originating from the boundary layer, and a cooler component which is radiated by the accretion disk and/or the bright spot.

117.071 **Phase-dependent UV spectra of UX UMa.**
A. R. King.
Third European IUE Conference, (see 012.029), p. 219 (1982).
IUE spectra of the cataclysmic variable UX UMa reveal a deep continuum eclipse but partial or absent eclipses in some emission lines. These results are interpreted in terms of a model in which a photoionized wind is blown off the surface of an optically thick accretion disc.

117.072 **Far UV observations of MV Lyrae in two different states.**
L. Chiappetti, L. Maraschi, E. G. Tanzi, A. Treves.
Third European IUE Conference, (see 012.029), p. 221 - 223 (1982).
MV Lyrae is an X-ray emitting cataclysmic variable. IUE observations taken in 1980 Dec., at optical minimum, revealed a hot component possibly contributed by the white dwarf surface. The ultraviolet continuum at minimum is compared with archival observations, taken when the source was in a brighter state.

117.073 **The interacting binary KX And.**
F. Paterson-Beeckmans, A. M. Hubert-Delplace, H. Hubert, D. Ballereau.
Third European IUE Conference, (see 012.029), p. 261 - 265 (1982).
A high resolution far UV spectrum of the interacting binary KX And (HD 218393) obtained near the minimum radial velocity of the visual shell lines shows strong emission lines originating in an extended atmosphere and strong, relatively narrow, absorption lines possibly originating in a cold shell rotating around the primary B2 III-IV star.

117.074 **A study of CQ Cephei – far UV to infra-red.**
D. J. Stickland, W. M. Burton, G. Bromage, A. J. Willis, I. D. Howarth, E. Budding, R. F. Jameson.
Third European IUE Conference, (see 012.029), p. 335 - 338 (1982).
A reappraisal of old optical spectrograms and photometry of the close binary WR star CQ Cephei has been combined with new optical and infra-red photometry and high and low resolution data from IUE. Some constraints have been placed upon system modelling and it appears that the companion to the WN7 star is at least partially immersed in the WR wind. However the nature of the companion has still not been positively established.

117.075 **New observations of the interacting binary SV Cen.**
H. Drechsel, W. Wargau, J. Rahe.

Third European IUE Conference, (see 012.029), p. 349 - 350 (1982).

Previous optical and IUE observations of the interacting early-type contact binary SV Cen were supplemented by new UV measurements. Optical photometric data suggest variations of the mass loss rate on a time scale of a few years, while the analysis of UV resonance line profiles shows short-term stellar wind fluctuations. The release of rotational energy of asynchronously rotating stars can accout for the observed UV excess.

117.076 **The IUE survey of W UMa-type binaries.**
S. M. Rucinski, O. Vilhu.
Third European IUE Conference, (see 012.029), p. 351 - 354 (1982).

The authors present results of homogeneous reductions of old and newly obtained observations of the brighter among the W UMa-type contact binaries which are the fastest revolving and rotating among the late-type stars. The emphasis is on low resolution spectra which are analysed in terms of the transition region line fluxes. These fluxes, together with the available X-ray data, permit a discussion of the emission strength dependence on the temperature of the emitting plasma and on the period of stellar rotation.

117.077 **Mass-loss rate from IUE observations of Zeta Aur systems.**
A. Che, K. Hempe, D. Reimers.
Third European IUE Conference, (see 012.029), p. 355 - 357 (1982).

UV multiplets of Fe II offer a method for determining the mass-loss rate of the red supergiant in a Zeta Aur system. Observed equivalent widths of the absorption lines are matched with theoretical line strengths parametrized by mass-loss rate and stellar wind velocity. Preliminary results are given for 32 Cyg, Zeta Aur and 31 Cyg.

117.078 **Hydrodynamical calculations of accretion disks in close binary systems. I. Method.**
G. Hensler.
Astron. Astrophys., Vol. 114, 309 - 318 (1982).

A numerical method for calculations of the hydrodynamics of accretion disks in close binary systems is presented, which allows for good spatial resolution of structures. The gas is divided into a number of individual particles whose motion is followed within a grid consisting of a spherical inner part and a plane part. Connected with the grid geometry is a special treatment of viscosity.

117.079 **Hydrodynamical calculations of accretion disks in close binary systems. II. Models.**
G. Hensler.
Astron. Astrophys., Vol. 114, 319 - 327 (1982).

The numerical method developed for investigations of accretion in close binaries described in Paper I (Hensler, 1982) is applied to systems of different parameters. 2D hydrodynamical calculations have been carried out to obtain steady accretion disks. The results are discussed by comparison with theoretical investigations and interpretations of observations.

117.080 **Mass transfer in a low mass semidetached binary, taking into consideration nonequilibrium effects.**
M. Hauschildt.
Astron. Astrophys., Vol. 114, 407 - 408 (1982).

Using a stepwise constant approximation for the response function, mass transfer has been calculated for a semidetached binary with the initial masses $m_1 = 0.8\,M_\odot$ and $m_2 = 0.6\,M_\odot$. Taking into consideration nonequilibrium effects on the secondary's radius as well, it is shown that "case A" mass transfer in a low mass binary does not necessarily lead to a contact system.

117.081 **Evolution of low mass, close binaries with gravitational radiation and systemic mass losses.**
W. Y. Chau, L. A. Nelson.
Bull. American Astron. Soc., Vol. 14, 778 (1982). – Abstract.

117.082 **QU Carinae: orbital parameters and spectra for a nova-like variable.**
R. L. Gilliland, M. M. Phillips.
Astrophys. J., Vol. 261, 617 - 624 (1982).

QU Car is found by high time resolution spectroscopy to be a binary with a period of about 10.9 hr. The period coupled with the spectral properties (moderate C III–N III and He II and weak, variable Balmer lines) are consistent with classification as an old (classical) nova or a nova-like variable. The lack of clearly detectable absorption features in such a long-period system suggests that the light is dominated by the accretion disk and primary.

117.083 **The ellipsoidal light curve of VV Puppis.**
D. A. Allen, A. M. Cherepashchuk.
Mon. Not. R. Astron. Soc., Vol. 201, 521 - 525 (1982).

The authors present light curves of the AM Herculis-type binary VV Puppis at 1.2 μm in both active and quiet states. In its active state the light curve resembles existing optical data, with perhaps additional evidence for accretion on to the second pole. During quiescence the cyclotron continuum fades and reveals the ellipsoidal variations of the M dwarf. By fitting the ellipsoidal light curve the authors are able to specify the location of the M dwarf in the system and place limits on the inclination of our line-of-sight, $50° \leqslant i \leqslant 70°$.

117.084 **A possible outburst on AM Canum Venaticorum.**
Y. Elsworth, L. Grimshaw, J. F. James.
Mon. Not. R. Astron. Soc., Vol. 201, 45P - 47P (1982).

A transient event was observed on AM CVn during measurements of its light-curve. There was a 30 per cent increase in intensity lasting, with fluctuations, for approximately 300 s. The possible cases are discussed.

117.085 **Ultraviolet spectroscopy of F and G supergiants with *IUE*. II. The hot companions of HR 2786 and HR 2859.**
S. B. Parsons.
Publ. Astron. Soc. Pacific, Vol. 94, 642 - 646 (1982).

Hot companion stars in two binary systems are analysed by means of *IUE* low-dispersion spectra. HR 2859B is found to be a B2 IV or V star, and analysis of ultraviolet to near-infrared photometry indicates the primary to be an early G star only 1 magnitude brighter in V, with $M_v \approx -4.0$. A third star probably associated with this system was classified B9p Hg (Mn?); the author obtains a probable visual absolute magnitude ≈ -1.6 for it. HR 2786 is now confirmed to have a hot secondary of a spectral class of B9–9.5. With an assumed luminosity for this component, the G2 I*b* primary then has $M_v \approx -3.8$.

117.086 **The early-type component in ν^1 Sagittarii.**
Y. Kondo, G. E. McCluskey, C.-C. Wu.
Publ. Astron. Soc. Pacific, Vol. 94, 647 - 649 (1982).

Four low-resolution ($\Delta\lambda \cong 6$ Å) ultraviolet spectra of ν^1 Sgr (cK2) were obtained with the *IUE* satellite to investigate the nature of the hot component in this object. The *IUE* results show that the spectral energy distribution of ν^1 Sgr in the far-ultraviolet is comparable to that of a main-sequence B9 star. The authors suggest: (1) the hot component is a rapidly rotating star; or (2) they are in fact observing an accretion disk surrounding a central object which could be a collapsed star.

117.087 **The polarization of ζ Aurigae during the 1979–1980 eclipse.**
R. H. Koch, R. J. Pfeiffer.

Astron. J., Vol. 87, 1409 - 1415 (1982).

Numerous measures of the polarization of the close binary ζ Aurigae are presented, most of which occur within eclipse totality. Most of the observed polarization is interstellar, but there is a small intrinsic and variable polarization component as well. It is shown that the intrinsic polarization cannot be due to the wake trailing the B star nor to scattering of B star light in the K star atmosphere. The most suitable explanation of the intrinsic polarization attributes it to chromospheric inhomogeneities of the supergiant. Thus the binary condition does not cause the polarization.

117.088 **A search for magnetic fields in the symbiotic and VV Cephei variables.**
M. H. Slovak.
Astrophys. J., Vol. 262, 282 - 293 (1982).

The McDonald Observatory's 2.7 m photoelectric analyzer was used to examine five symbiotic and VV Cephei variables for the presence of coherent longitudinal fields with an absolute sensitivity of ± 100–200 gauss. To this level, no new evidence is found supporting the reported kilogauss fields on the quiescent symbiotics AG Pegasi and EG Andromedae, nor for the VV Cephei stars VV Cephei and WY Geminorum. Observations of CH Cygni also yielded null results. Therefore current magnetic models for symbiotics invoking large fields do not appear viable.

117.089 **On the orbital phase dependence of the turn-on times of Hercules X-1.**
A. M. Levine, J. G. Jernigan.
Astrophys. J., Vol. 262, 294 - 300 (1982).

A ring of matter at a given radius in an accretion disk in a binary system will precess if it is tilted relative to the binary orbital plane. When the effect of the time variation of the torque due to the companion star is included in the dynamics, the ring orientation is found to undergo a significant periodic perturbation with respect to the previously known uniform precession. The authors demonstrate that this disk "wobble" can be utilized to explain why the X-ray turn-ons of Hercules X-1 preferentially occur around orbital phases 0.2 and 0.7.

117.090 **Sudden loss of mass from a binary gravitating system.**
P. D. Noerdlinger.
American J. Phys., Vol. 50, 733 - 738 (1982). – Abstr. in Phys. Abstr., Vol. 85, Abstr. 107045 (1982).

117.091 **Wolf-Rayet stars with compact companions.**
A. F. J. Moffat.
Wolf-Rayet stars: observations, physics, evolution, (see 012.042), p. 263 - 275 (1982).

Evidence is accumulating in favour of a heretofore missing link in the evolution of massive binary stars –a second WR phase in which the companion to the WR star is a neutron star of mass 1–2 M⊙, or in some cases a more massive black hole. The discovery of such objects is also important in assessing the role of duplicity in the formation of WR stars.

117.092 **The evolution of massive stars – the production of binary WR stars.**
C. de Loore.
Wolf-Rayet stars: observations, physics, evolution, (see 012.042), p. 343 - 358 (1982).

Contents: Roche lobe overflow. Mass loss in binaries before the Roche lobe overflow phase. Conservative binary evolution. Non conservative evolution of massive close binaries. Remnants of the primary after Roche lobe overflow. Simultaneous evolution of close binaries. Effects of overshooting. Advanced evolution – 2nd Wolf-Rayet phase.

117.093 **On the expected frequency of WR + compact star systems.**
D. Vanbeveren, C. de Loore.
Wolf-Rayet stars: observations, physics, evolution, (see 012.042), p. 365 - 369 (1982).

According to the classical evolutionary scenario for massive close binaries it is generally assumed that two WR phases are encountered, i.e. a WR+OB stage and a WR + compact star stage. The survival probability of a massive binary after an instantaneous supernova (SN) explosion critically depends on the assumed asymmetry of the SN explosion, i.e. on the "kick" velocity V_k (Sutantyo, 1978). The purpose of this paper is to make a statistical study on the expected WR + compact star binaries compared to the WR+OB type binaries based on non-conservative massive binary evolution, as a function of V_k.

117.094 **On the structure and evolution of the OB-companions in Wolf-Rayet binaries.**
P. Hellings, F. Vansina, W. Packet, C. Doom, J. P. De Grève, C. de Loore.
Wolf-Rayet stars: observations, physics, evolution, (see 012.042), p. 397 - 401 (1982).

In the treatment of close binary evolution only the evolution of the mass losing star was normally computed in detail; the evolution of the accreting star was estimated in a very simple way by computing the evolution in the normal way just like for single stars, but taking into account the mass increase. Investigations of Benson (1970) however revealed that as a consequence of the inflow of matter the accreting star expands, so that contact systems can be formed. The authors calculate the evolution of massive close binaries simultaneously through detached, semi-detached and contact phases.

117.095 **The occurrence of different Wolf-Rayet phases in massive close binaries.**
C. Doom, J. P. De Grève.
Wolf-Rayet stars: observations, physics, evolution, (see 012.042), p. 403 (1982). – Abstract.

117.096 **WR stars with ring nebulae.**
A. Tutukov.
Wolf-Rayet stars: observations, physics, evolution, (see 012.042), p. 485 - 489 (1982).

It is shown that most of usually apparently single nitrogen WR stars with ring emission nebulae around them (WN + Neb) are a probable product of the evolution of a massive close binary with initial masses of components exceeding ~ 20 M⊙.

117.097 **The equivalence of the short periods measured in the spectrum of SS 433.**
G. H. Newsom, G. W. Collins II.
Astrophys. J., Vol. 262, 714 - 716 (1982).

The apparently different values reported for the "6^d" period in the wavelengths of the moving lines of SS 433 are shown to differ only as a result of the method of analysis. The phase modulation of the Doppler shifts of the moving lines resulting from the 164^d precession introduces sidebands in any Fourier analysis of the residuals from the kinematic model.

117.098 **The mystery of the missing boundary layer.**
G. J. Ferland, S. H. Langer, J. MacDonald, G. H. Pepper, G. Shaviv, J. W. Truran.
Astrophys. J., Lett., Vol. 262, L53 - L58 (1982).

The question of the nature of the ultraviolet and X-ray radiation field of cataclysmic binaries is addressed. The spectrum and luminosity of this radiation are important in determining the mass transfer rate and energy budget of the system and in studies of the ejecta surrounding novae. In many systems, the soft X-ray luminosity is $\sim 10^2$–10^4 times weaker than predicted by simple accretion models. The authors discuss several possible solutions to this discrepancy. The solution to this problem is of fundamental importance to accretion disk theory.

117.099 **Properties of Wolf-Rayet stars in eclipsing binary systems.**
A. M. Cherepashchuk.
Astrophys. Space Sci., Vol. 86, 299 - 319 (1982).

The knowledge of the physical characteristics of the Wolf-Rayet stars (WR) is necessary for understanding of their nature and evolution. In this paper the author presents the results of investigation of a number of eclipsing WR binary stars. His results are in good agreement with the conclusions of the modern theory concerning the evolution of massive close binary systems (Paczynski, 1967; Tutukov and Yungelson, 1973; de Loore and de Grève, 1975; Van den Heuvel, 1976). A new method for the solving of the light curves of the eclipsing binary systems with extended atmospheres was used for the interpretation of the photometric observations of the eclipsing WR binaries (Cherepashchuk, 1971, 1973, 1974).

117.100 **Infrared studies of symbiotic stars.**
D. A. Allen.
The nature of symbiotic stars, (see 012.045), p. 27 - 42 (1982).

Infrared photometry and spectroscopy of symbiotic stars is reviewed. It is shown that at wavelengths beyond 1 μm these systems are generally dominated by the cool star's photosphere and, indeed, are indistinguishable from ordinary late-type giants. About 25% of symbiotic stars exhibit additional emission due to circumstellar dust. Spectroscopy at longer wavelengths reveals information about the dust itself.

117.101 **IR photometry of symbiotic stars at Calar Alto Observatory.**
C. Eiroa, H. Hefele, Z.-y. Qian.
The nature of symbiotic stars, (see 012.045), p. 43 - 45 (1982).

Infrared photometry of symbiotic stars was made with the 2.2 m telescope of the Calar Alto Observatory on July 1981. Three stars (Z And, CI Cyg, V1329 Cyg) were observed during a phase of minimum activity.

117.102 **Properties of symbiotic stars from studies in the optical region.**
F. Ciatti.
The nature of symbiotic stars, (see 012.045), p. 61 - 70 (1982).

Contents: Definition of symbiotic stars. Photometric and spectroscopic evolution. The three-component model for SS. Evidence for the binary nature. Spectroscopic properties and anomalies. Single star interpretations. The very slow novae. Comparison of SS with other classes.

117.103 **Ultraviolet properties of the symbiotic stars.**
M. H. Slovak, D. L. Lambert.
The nature of symbiotic stars, (see 012.045), p 103 - 113 (1982).

The IUE observations of the symbiotic stars have revealed ultraviolet properties which rival diversity of the optical features. Nonetheless, the UV data have for the first time permitted the hot component to be studied relatively uncontaminated by the giant companion, which dominates the optical regime. The UV observations provide convincing evidence that many of the symbiotics have hot stellar companions, imbedded in the enshrouding nebula or accretion "shell" formed from the wind arising off of one or possibly both of the components. Another possibly impulsive, as opposed to steady-state, contributor to the nebula could come during the eruptive periods.

117.104 **A review of the properties of the symbiotic star CH Cygni.**
M. Hack, P. L. Selvelli.
The nature of symbiotic stars, (see 012.045), p. 131 - 136 (1982).

The authors summarize the results of the observations of the 1967–70 activity phase of CH Cygni in the photographic spectral range (3300–4900), and of the activity phase which started in 1977 in the spectral range 6700–3400 Å and 3100–1175 Å, and compare the phenomena observed during the two outbursts. CH Cygni does not show any evidence of periodic radial velocity variations. Instead, the authors' observations of the 1967–70 episode supported the hypothesis that shock waves heat the envelope surrounding the M6 giant. However, photometric observations by Slovak and Africano (1978) indicate that CH Cygni presents a rapid flickering, characteristic of cataclysmic variables.

117.105 **Recent increase of the activity of the symbiotic star CH Cygni.**
D. Chochol, L. Hric.
The nature of symbiotic stars, (see 012.045), p. 137 - 138 (1982).

CH Cyg has exhibited a great deal of activity in the recent times. The authors obtained two spectra with dispersion of 9 Å/mm with the 2 m telescope of the Ondrejov Observatory on the nights of July 4/5 and August 6/7, 1981. Their spectroscopic observations support a binary model in which a cool component fills its Roche lobe from time to time, thus inducing mass transfer towards the gainer which is a hot subdwarf. An accretion disk is formed around the hot companion which falls onto the star.

117.106 **Polarimetry of CH Cygni.**
V. Piirola.
The nature of symbiotic stars, (see 012.045), p. 139 - 140 (1982).

Variable linear polarization has been detected in CH Cyg. Observations in 1977–79 (UBV) and 1979–80 (UBVRI) are presented.

117.107 **UV eclipse observations of CI Cyg.**
A. G. Michalitsianos, M. Kafatos, R. E. Stencel, A. A. Boyarchuk.
The nature of symbiotic stars, (see 012.045), p. 141 - 144 (1982).

Low spectral resolution observations (~ 6 Å) of CI Cyg were obtained with the International Ultraviolet Explorer (IUE) during its eclipse phase. The observations are consistent with a binary star model that involves mass transfer from the extended cool envelope of the primary to the compact secondary. The formation of an accretion disc is a transitory phenomena.

117.108 **IUE observations of CI Cygni during 1979–1981.**
G. B. Baratta, A. Altamore, A. Cassatella, M. Friedjung, D. Ponz, O. Ricciardi.
The nature of symbiotic stars, (see 012.045), p. 145 - 146 (1982).

High and low resolution IUE spectra of CI Cyg were obtained at VILSPA during 1979–81 allowing for an analysis of the spectral variations related to the decreasing activity of the star, and to the eclipse (June 1980). An interstellar extinction of E(B–V) = 0.40 was derived from the low resolution IUE spectra.

117.109 **The eclipse of CI Cygni in 1980 on the objective prism spectra.**
J. Mikołajewska, M. Mikołajewski.
The nature of symbiotic stars, (see 012.045), p. 147 - 148 (1982).

The observations of CI Cyg were carried out with the Torun 60/90 cm Schmidt telescope with the objective prism giving a dispersion of 250 Å/mm at H-gamma. The spectra were typical for the quiet symbiotic star. So, all changes in the spectrum were caused by geometrical effects (eclipse). A preliminary analysis of obtained spectra is given.

117.110 **Variation spectrale de CI Cyg en 1975.**
C. Fehrenbach.
The nature of symbiotic stars, (see 012.045), p. 149 - 150 (1982). – Abstract.

117.111 **Spectral variations of CI Cyg between 1980 and 1981.**
C.-C. Huang.
The nature of symbiotic stars, (see 012.045), p. 151 - 152 (1982).

In the paper (C. Fehrenbach and C. C. Huang, 1981) the authors described the details of the spectroscopic observations of CI Cyg in 1980. Continuing the spectroscopic observations of CI Cyg, during summer 1981, a number of spectra of the star were obtained at the Haute Provence Observatory, using the Marly spectrograph with a dispersion of 80 Å mm^{-1}.

117.112 **Optical observations of CI Cygni.**
N. A.Oliversen, C. M. Anderson, K. H. Nordsieck.
The nature of symbiotic stars, (see 012.045), p.153 - 156 (1982).

The authors report optical emission line variations in CI Cygni during the 1980 eclipse.

117.113 **The peculiar symbiotic object V1329 Cygni: single-star versus binary models.**
J. Grygar, D. Chochol.
The nature of symbiotic stars, (see 012.045), p. 165 - 168 (1982).

Contents: Recapitulation of photometric and spectroscopic observations. Postdiscovery studies. Binary hypothesis. Alternatives. Discussion.

117.114 **Variability of HBV 475 in the near infrared.**
Y. Andrillat.
The nature of symbiotic stars, (see 012.045), p. 173 - 174 (1982).

In the spectral range λλ5800–8750, HBV 475 shows important spectral variations between 1969 and 1974. On August 4 1974, June 6 1975 and August 9 1981, the observations were extended up to 1.1 μ. It is well known that the spectral type of the hot component is WN5. HD 229227 WN9–WN10 in the LMC was observed during November 1980. In the infrared region, the spectrum shows the characteristics of a symbiotic star. Thus, another symbiotic star having a WR star as its hot component (Andrillat, Vreux, Dennefeld 1981) was found. It is interesting to note that all those WR stars which are members of symbiotic stars belong to the nitrogen sequence.

117.115 **Spectral evolution of HBV 475 (= V1329 Cygni) in the ultraviolet.**
C. Kindl, H. Nussbaumer.
The nature of symbiotic stars, (see 012.045), p. 175 - 176 (1982).

The authors present the low resolution IUE spectra of HBV 475 during 1979–81.

117.116 **Introductory remarks on the symbiotic star AG Draconis.**
N. A. Oliversen, C. M. Anderson.
The nature of symbiotic stars, (see 012.045), p. 177 - 181 (1982).

Spectral features and light curves of AG Draconis are reviewed and corresponding models are discussed.

117.117 **Introductory report on AG Pegasi.**
M. Friedjung, R. Viotti (based on notes taken during C. D. Keyes' talk).
The nature of symbiotic stars, (see 012.045), p. 201 - 202 (1982). – Abstract.

117.118 **The peculiar star RX Puppis.**
M. Kafatos, A. G. Michalitsianos.
The nature of symbiotic stars, (see 012.045), p. 203 - 206 (1982).

The authors have obtained the first high dispersion observations of RX Puppis in the wavelength region 1200–3200 Å with the "International Ultraviolet Explorer" (IUE). The anomalies observed in lines such as He II, C III], C IV, N III], N IV], O III], and Si III], that show split line profiles, Doppler displaced component(s) suggest dynamic activity in circumstellar material. The continuum may arise in an accretion disk around the hot secondary. Moreover, the line emission requires photoionization either from a hot-subdwarf or the inner accretion disk.

117.119 **Long term trends in the 3.5 μ light curve of RX Puppis.**
P. A. Whitelock, R. M. Catchpole.
The nature of symbiotic stars, (see 012.045), p. 207 - 208 (1982).

SAAO photometry of RX Pup at 3.5 μ between 1972 and 1981 shows a very well defined period of 580 days with an amplitude of 0.35 mag, which is attributed to a Mira variable component. In addition to the 580^d period, secular variations are also evident in the 3.5 μ data.

117.120 **HM Sagittae – a most remarkable star.**
S. Kwok.
The nature of symbiotic stars, (see 012.045), p. 209 - 211 (1982).

HM Sagittae is one of the most unusual objects in the Galaxy for it displays activity in every spectral band from X-ray to radio. HM Sge is probably best described by a binary system consisting of a late M giant and a hot compact object which is similar to central stars of planetary nebulae. The complex structure of the circumstellar nebula is possibly the result of wind interactions.

117.121 **Spectral evolution of HM Sagittae in the ultraviolet.**
C. Kindl, H. Nussbaumer.
The nature of symbiotic stars, (see 012.045), p. 213 - 214 (1982).

The authors observed the ultraviolet spectrum of HM Sge with the International Ultraviolet Explorer during 1978–1980. The low resolution spectra are shown.

117.122 **RR Telescopii.**
P. A. Whitelock, R. M. Catchpole, M. W. Feast.
The nature of symbiotic stars, (see 012.045), p. 215 - 216 (1982).

Most of slow novae and symbiotic stars are now thought to be interacting binary systems, in which a hot subdwarf accretes matter from a cool giant. In some cases, including RR Tel, the cool giant is a Mira variable. The SAAO infrared (JHKL) photometry of RR Tel from 1975 to 1980 clearly shows variations with a period of about 387 days. This period, confirms the suggestion of Feast and Glass (1974) that the RR Tel system contains a Mira.

117.123 **Ultraviolet observations of RR Telescopii.**
D. Ponz, A. Cassatella, R. Viotti.
The nature of symbiotic stars, (see 012.045), p. 217 - 218 (1982).

The ultraviolet spectrum of RR Tel was extensively studied with the IUE satellite since 1978 in both the high and low resolution modes. Some of the results are reviewed in this paper.

117.124 **The Kuwano peculiar object (PU Vulpeculae).**
T. S. Belyakina, R. E. Gershberg, Yu. S. Efimov, V. I. Krasnobabtsev, E. P. Pavlenko, P. P. Petrov, K. K. Chuvaev, V. I. Shenavrin.
The nature of symbiotic stars, (see 012.045), p. 221 - 222 (1982).

U, B, V, J, H, K photometry, polarimetry and spectroscopy of this peculiar object made from August 1979 to December 1980 is reported. The authors consider PU Vul to be binary with a late M giant and a component of probably low luminosity. The photometric behaviour, duplicity, and nature of the components are similar to those of novae and symbiotic stars, but no evidence for an ejected envelope is seen.

117.125 **The symbiotics as binary stars.**
M. J. Plavec.
The nature of symbiotic stars, (see 012.045), p. 231 - 251 (1982).

Symbiotic stars have become an important testing ground of various theories of binary star evolution. Several physically different models can explain them, but in each case certain fairly restrictive conditions must be met. The authors briefly discusses the PN symbiotic, the Algol symbiotic and the novalike symbiotic models. The most promising model is one of a binary system in the second stage of mass transfer, actually at the beginning of it. In spite of considerable observational effort, the symbiotics are known so poorly that it is hard to decide between the models, or even decide if all three can actually exist.

117.126 **Models for symbiotic stars in the light of the data.**
M. Friedjung.
The nature of symbiotic stars, (see 012.045), p. 253 - 267 (1982).

Different single and binary models of symbiotic stars are examined. Single star models encounter a number of problems, and binary models are probable. There are however difficulties in the interpretation of radial velocities. Accretion disks play a role in some cases, but winds especially from the cool component must be taken into account in realistic models. There is some evidence of excess heating of the outer layers of the cool component. Outbursts may be related to sudden changes in the characteristics of the cool star wind.

117.127 **Symbiotic star UV emission and theoretical models.**
M. Kafatos.
The nature of symbiotic stars, (see 012.045), p. 269 - 272 (1982).

Observations of symbiotic stars in the far UV have provided important information on the nature of these objects. High resolution observations of lines are important because they yield information on densities, temperatures and sizes of the line emitting region(s). A number of equations have been derived by the author that can yield the accretion parameters from the observable quantities. Boundary layer temperatures $\sim 10^5$ K and accretion rates $\gtrsim 10^{-5}$ $M_\odot$/yr are required for accreting main sequence companions.

117.128 **The evolutionary status of symbiotic stars.**
B. Rudak.
The nature of symbiotic stars, (see 012.045), p. 275 - 282 (1982).

The evolutionary relations between symbiotic stars and cataclysmic variables are presented. The symbiotic stars are assumed to be long period detached binaries containing a carbon-oxygen degenerate primary and a red giant losing its mass through a spherically symmetric wind. Such systems can be obtained in Case C evolution, provided a common envelope during a rapid mass transfer phase was not formed. The same way recurrent novae containing a red giant as a secondary component may be produced. The factors influencing the differences between symbiotic stars and nova-type stars are discussed.

117.129 **On the model of symbiotic stars.**
A. V. Tutukov, L. R. Yungelson (*Yungel'son*).
The nature of symbiotic stars, (see 012.045), p. 283 - 296 (1982).

The authors discuss conditions necessary for appearance and discovery of the symbiotic stars phenomenon within the model of a binary consisting of a red (super) giant 3 $M_\odot$ not filling the Roche lobe and of an accreting hot degenerate CO-dwarf 0.8 $M_\odot$. Within this model "classical" symbiotic stars may exist only within a narrow region of mass accretion rates and separations of components:
$10^{-7} \lesssim \dot{M} \lesssim 3 \times 10^{-7} M_\odot/y$ and $3 \times 10^{13} \lesssim a \lesssim 2 \times 10^{14}$ cm.
The evolutionary status of symbiotic stars and related objects and the mechanisms of their variability are discussed.

117.130 **Disk instabilities in Hubble-Sandage variables?**
F. Meyer, E. Meyer-Hofmeister.
Binary and multiple stars as tracers of stellar evolution, (see 012.046), p. 19 - 22 (1982).

The brightest individual objects in extragalactic nebulae are the Hubble-Sandage variables. The main characteristics are high luminosity ($L/L_\odot \approx 10^5$), blue color indices, F type spectra and irregular variability. Bath (1979) has suggested that the Hubble-Sandage variables contain an accreting main-sequence star with a Roche Lobe filling companion in a wide binary system. Based on this model the authors derive theoretical color indices for disks and determine the mass in the disk for different mass accretion rates. Further they discuss an instability of the disk which could explain the change in the color index observed for Var A in M 33.

117.131 **The frequency of Cepheid binaries.**
G. Russo.
Binary and multiple stars as tracers of stellar evolution, (see 012.046), p. 23 - 26 (1982).

The author summarizes the methods now available to detect the presence of companions to Cepheids, and then he analyzes the results of the application of the recent photometric method proposed by Fernie (1980). If applied to the set of VBLUW observations of southern Cepheids carried out by Pel (1976), this method gives a frequency of about 25%, which is consistent with the more recent spectroscopic determinations (Gieren, 1981). The importance of the IUE satellite to directly observe the U flux emitted by the B-companion is emphasized.

117.132 **A hypothesis on the richness of a galaxy in binary stellar systems.**
H. Zinnecker.
Binary and multiple stars as tracers of stellar evolution, (see 012.046), p. 115 - 117 (1982).

This contribution is concerned with the origin of binary stellar systems (cf. Abt 1977, Huang 1977). The hierarchical fragmentation scheme of rotating interstellar clouds, proposed by Bodenheimer (1978), is combined with the role of magnetic fields in the early stages of star formation (Mouschovias 1978, Dorfi 1981). The possible influence of the local strength of the mean interstellar magnetic field on the local fraction of binary systems in a galaxy is stressed.

117.133 **Evolution of close binary stars: observational aspects.**
M. J. Plavec.
Binary and multiple stars as tracers of stellar evolution, (see 012.046), p. 159 - 181 (1982).

Detached close binary systems define the main sequence band satisfactorily, but very little is known about the masses of giants and supergiants. High-dispersion IUE observations promise an improvement. The interesting cases of μ Sagittarii and ϵ Aurigae are discussed in more detail. The barium star abundance anomaly appears now to be due to mass transfer in interacting systems. The symbiotic stars are another type of binary systems containing late-type giants; several possible models are discussed. Evidence of hot circumstellar plasmas has now been found in several ordinary Algols; there may exist a smooth transition between very quiescent Algols and the W Serpentis stars. β Lyrae is discussed in the light of new spectrophotometric results.

117.134 **Statistical properties of Algol-type systems.**
G. Giuricin, F. Mardirossian, M. Mezzetti.
Binary and multiple stars as tracers of stellar evolution, (see 012.046), p. 183 - 186 (1982).

The authors have inspected the observational data of 114 Algol-type binaries in order to clarify some general aspects of their evolutionary scenario. Through a comparison with the corresponding available statistical analyses of large sample of eclipsing and spectroscopic binaries, they have found a larger than normal concentration of low values of masses, binary separations and specific orbital angular momenta for Algol-type stars. Furthermore, high mass ratios predominate in massive and early-type Algol systems, whereas high total masses are generally accompanied by less prominent oversized and overluminous properties of the mass-losing components.

117.135 **Mass loss in Algol-type stars: implication on their evolutionary stage.**
F. Mardirossian, G. Giuricin.
Binary and multiple stars as tracers of stellar evolution, (see 012.046), p. 187 - 189 (1982).

The authors have examined the observational data of 102 Algols in order to clarify the implications on their evolutionary scenario of various assumptions concerning mass and angular momentum loss during mass transfer. They have found that case B mass exchange is strongly favoured for Algols of relatively low total mass ($\sim M < 7 M_\odot$), while case A predominates, though not so widely as expected in Algols of higher total mass.

117.136 **Statistical models for close binaries.**
J. L. Halbwachs.
Binary and multiple stars as tracers of stellar evolution, (see 012.046), p. 199 - 204 (1982).

Statistical models for binaries as a whole were selected; then statistical properties of close binaries – the proportions of spectroscopic and eclipsing binaries and the distributions of the K velocities and of the depths of eclipse – were computed for each model.

117.137 **Numerical calculations for accretion disks in close binary systems.**
G. Hensler.
Binary and multiple stars as tracers of stellar evolution, (see 012.046), p. 219 - 230 (1982).

A numerical method for magnetohydrodynamical investigations of accretion disks in close binary systems is presented, which allows for good spatial resolution of structures (hot spot, accretion column). The author presents one result of hydrodynamical calculations for a binary applying the following parameters which are representative for cataclysmic variables: $M_1 = 1 M_\odot, r_1 = 10^{-2} R_\odot, M_2 = 0.5 M_\odot$, p = 0.2 d, $\dot{M} = 10^{-9} M_\odot \text{ y}^{-1}$.

117.138 **S Cnc as a case study of classical Algol evolution.**
N. S. Awadalla, E. Budding.
Binary and multiple stars as tracers of stellar evolution, (see 012.046), p. 239 - 259 (1982).

Basic information on empirical determinables for the system S Cnc are reviewed. Whilst the photometric information can be clearly analysed along well known lines, the spectrographic information is far from clear. The question of some physical relationship of the variable to the open cluster Praesepe is considered, but the result is in the negative. The evolutionary status of the binary is examined. S Cnc is in good overall agreement with the low mass Case B theoretical mode of binary evolution, and even the absence of a detectable rate of period variation is shown to be not in serious conflict with this picture. Interesting close comparisons with the well known example AS Eri are possible.

117.139 **Structural models for Beta Lyrae-type disks.**
R. E. Wilson.
Binary and multiple stars as tracers of stellar evolution, (see 012.046), p. 261 - 273 (1982).

Equilibrium structural models are computed for a thick, self-gravitating disk in a binary system. Accretion onto the star is limited by the star's rapid rotation (the system is a double-contact binary). The equations are integrated and solved by the fitting point method. The energy is supplied by viscosity. Energy transfer is by convection, and is appreciably superadiabatic throughout the disk. A mass of $0.5 M_\odot$ is assumed. Representative results are: "central" temperature, 67000 K; "central" pressure, 5×10^{11} dyn/cm^2; "equal volume" radius, $17 R_\odot$; luminosity, $5 \times 10^3 L_\odot$. The model "radius" is in excellent agreement with the observational value for β Lyrae. The model luminosity indicates that a lower disk mass (perhaps $0.25 M_\odot$) should be tried.

117.140 **Evolutionary effects in contact binaries.**
H. Mauder.
Binary and multiple stars as tracers of stellar evolution, (see 012.046), p. 275 - 278 (1982).

The structure and evolution of contact binaries is an extremely difficult problem. Several attempts have been made to set up a satisfactory model. There are, roughly, two lines of sight (DSC Theory and TRO Theory). It is necessary, therefore, to look for systems which might eventually show effects that are predicted or excluded by one of the theories. In this connection the author reports an observational campaign concentrating on stars with periods below 0.5 days and minima of unequal depths.

117.141 **Evolutionary possibilities and impossibilities for solar type contact binaries in NGC 188.**
F. van't Veer.
Binary and multiple stars as tracers of stellar evolution, (see 012.046), p. 279 - 287 (1982).

The author gives a great number of arguments for the hypothesis that several epochs of star formation have taken place at different times in the old galactic cluster NGC 188. From the last burst of star formation in this cluster, not more than a few times 10^8 years ago, 4 contact binaries are still now visible as W UMa stars. With the aid of a simplified probability calculation it is argued that these 4 contact binaries are physically related and that the alignment of the orbital axes is not accidental.

117.142 **Origin and evolution of contact binaries of W UMa type.**
T. Rahunen, O. Vilhu.
Binary and multiple stars as tracers of stellar evolution, (see 012.046), p. 289 - 299 (1982).

Using angular momentum loss estimates from single star studies, it is shown that detached binaries are good candidates as progenitors of contact binaries. Three theories constructed for contact binary evolution are discussed. In all theories the end-product is a single star. The stability of unequal entropy models was studied using a formula which couples the energy transfer with the depth of contact and with the entropy difference. The models experience cyclic behaviour and the contact never breaks. The most probable contact binary theory is perhaps a suitable combination of the three discussed theories.

117.143 **Some aspects of the evolutionary status of W Ursae Majoris binaries deduced from observational data.**
W. Van Hamme.
Binary and multiple stars as tracers of stellar evolution, (see 012.046), p. 301 - 304 (1982).

The author has developed a test for the evolutionary state of W Ursae Majoris binaries by comparing the observed spectral type of 31 of these systems (14 of type W and 17 of type A) with the expected one when their primary component is an unevolved main sequence star. It appears that both the W- and A-type systems have a primary with a mass and radius too large to be compatible with the observed spectral type, so there is no indication that each type should mark a different evolutionary stage.

117.144 **A possible explanation of distortions in the light curves of VW Cep by starspots.**
A. Yamasaki.
Binary and multiple stars as tracers of stellar evolution, (see 012.046), p. 305 - 308 (1982).

An attempt is made to explain the variations in the light curve of VW Cep in terms of a starspot model. Based upon Kwee's (1966) observations, an unperturbed light curve without spots is constructed and analysed for the geometrical elements of VW Cep. Then, by assuming that starspots exist on the surface of the primary component, Kwee's individual light curves are further analysed for locations and intensities of the starspots.

117.145 **The IUE observations of W UMa.**
S. M. Rucinski, J. E. Pringle, J. A. J. Whelan.
Binary and multiple stars as tracers of stellar evolution, (see 012.046), p. 309 - 315 (1982).

The low resolution, phase averaged SWP spectra of W UMa reveal the trend of increased transition-region line intensities for higher temperatures of ion formation which was previously observed for other active systems. Two series of low resolution LWR spectra covering two full orbital periods have been used to check the photometric results from the ANS satellite and a good agreement is found for the 2200 Å band. Wavelength shifts and changes in the intensity of the blended Mg II emission feature are not inconsistent with the ANS result that the more massive component is the more active one in the W UMa system.

117.146 **Mass flow due to heating in a binary system: application to U Cephei.**
Y. Kondo, J. L. Modisette.
Binary and multiple stars as tracers of stellar evolution, (see 012.046), p. 317 - 319 (1982).

The authors have investigated the possibility of mass flow due to the heating of the cooler component in a close binary system. The 2.4-day period binary U Cep, in which gas streaming has been observed, has been chosen for model calculations. Preliminary results show that such a heating of the lower atmosphere of the cooler star could lead to mass flow at an average rate of 10^{-9} to 10^{-7} solar mass per year without the star's necessarily filling its critical Roche surface.

117.147 **On a possible linkage between W-type W UMa systems and the short period RS CVn-like binaries.**
L. Milano, G. Russo, S. Mancuso.
Binary and multiple stars as tracers of stellar evolution, (see 012.046), p. 327 - 335 (1982).

The authors analyze the general properties of a group of W UMa-type binaries, which show RS CVn-like activity. The position of these stars in the colour-density and mass-orbital momentum diagrams is studied, but no definite answer can be given on the linkage between short period RS CVn systems and W UMa systems with RS CVn-like activity.

117.148 **Recent observational evidence on W Ursae Majoris stars.**
H. Rovithis-Livaniou, P. Rovithis.
Binary and multiple stars as tracers of stellar evolution, (see 012.046), p. 337 - 342 (1982).

The light curves of 44i Boo in *U, B, V, R* (observations of 1978), the *B* and *V* light curves of VW Cep (observations of 1979), the light curve of AB And (observations of 1979), the *B* and *V* light curves of BV Dra and the *B* and *V* light curves of U Peg are presented.

117.149 **The time change of the AW UMa system.**
M. Kurpińska, K. Otmianowska.
Binary and multiple stars as tracers of stellar evolution, (see 012.046), p. 343 (1982). – Abstract.

117.150 **On the spatial density of W UMa type stars.**
E. Budding.
Binary and multiple stars as tracers of stellar evolution, (see 012.046), p. 351 - 370 (1982).

Attention is directed to the anomalous incidence of W UMa stars, which can be regarded as coming from not only a disproportionately large accumulation among close binary systems with primaries later than around mid-F spectral type, but also as a deficit at early types. Doubt is placed on the necessity of a straightforward identification of W UMa type light curves with contact binaries; and this allows some reduction in the estimated spatial incidence of contact binaries. The incidence is considered as an example of the general evolution of the distribution of binary systems in the primary spectral type – orbital period plane, subject to some known mechanisms of binary evolution.

117.151 **Binary stars and SS 433.**
R. Ruffini.
Binary and multiple stars as tracers of stellar evolution, (see 012.046), p. 373 - 388 (1982).

Some of the most unique experimental features of the source SS 433 are outlined as well as some implications of the theoretical models of this source.

117.152 **Observations of secular changes in the Kinematic Model of SS 433.**
G. W. Collins, II, G. H. Newsom.
Binary and multiple stars as tracers of stellar evolution, (see 012.046), p. 389 - 397 (1982).

In this paper the authors present evidence that several of the defining parameters of the Kinematic Model for SS 433 are not constant. Recent data confirm the existence of the previously reported decrease in the precessional period. The value for this period change, when combined with the observed change in the period of the synodic spectral variations, implies that the orbital period is not significantly changing on a time scale less than 1000 years. In addition the authors find evidence for a statistically significant secular change in the cone angle θ at a rate of about -1.5×10^{-3} deg/day. It is possible to understand some of the data in terms of the motions to be expected from an object exhibiting classical precession in response to an external torque.

117.153 **Time resolved circular polarimetry of white dwarf pulsars.**
W. Krzemiński, J. D. Landstreet, I. Thompson.
Binary and multiple stars as tracers of stellar evolution, (see 012.046), p. 399 - 401 (1982).

There are two recognized subsets of cataclysmic binaries that contain white dwarfs: the AM Her stars and the DQ Her stars. Both groups, recognized also as X-ray sources, are important as probes of the accretion process, and the radiation mechanisms in the vicinity of the white dwarf. The authors have been carrying out a limited survey of circular polarization properties of five DQ Her stars.

117.154 **The minimum orbital period of hydrogen-rich cataclysmic binaries.**
R. Sienkiewicz, B. Paczyński.
Binary and multiple stars as tracers of stellar evolution, (see 012.046), p. 413 - 414 (1982).

This communication summarizes investigations of a relationship between the minimum binary period and the evolution of very short period hydrogen-rich cataclysmic binaries.

117.155 **Remarks on the evolutionary status of cataclysmic variables.**
N. Vogt.
Binary and multiple stars as tracers of stellar evolution, (see 012.046), p. 415 (1982). – Abstract.

117.156 **Variability of soft X-ray emission of EX Hydrae observed with Einstein Observatory.**
A. Kruszewski, R. Mewe, J. Heise, T. Chlebowski, W. van Dijk, R. Bakker.
Binary and multiple stars as tracers of stellar evolution, (see 012.046), p. 457 - 466 (1982).

The cataclysmic variable star EX Hydrae has been observed with the High Resolution Imager and the Imaging Proportional Counter (IPC) onboard the Einstein Observatory. The X-ray position is coincident within 3 arcsec of the optical position as measured on Schmidt survey plates. During a 15 1/2 hour observation with IPC the authors have searched for a modulation of the X-ray flux. Strong evidence for a 67 min period (one of two known optical periods) has been found in the energy range 0.1 - 3.5 keV with the IPC. The time dependence of modulations is used to discuss a model and evolutionary status of this close binary system.

117.157 **On the possible short-period irregular light fluctuations of V1357 Cyg = Cyg X-1.**
M. I. Kumsiashvili, Z. Kraicheva (*Krajcheva*).
Binary and multiple stars as tracers of stellar evolution, (see 012.046), p. 467 - 472 (1982).

Photoelectric observations in *U, B, V,* were carried out to study the possible rapid (during a night) fluctuations of the close binary system V1357 Cyg, including the X-ray source Cyg X-1. Assuming the existence of irregular fast light fluctuations in the X-ray source Cyg X-1, the conclusion can be drawn that during these observations in 1979 the amplitude of the variations did not exceed 0.04 mag.

117.158 **PU Vulpeculae (object Honda-Kuwano 1979) – a possible short-period relative of the symbiotic stars.**
D. Chochol, J. Grygar.
Binary and multiple stars as tracers of stellar evolution, (see 012.046), p. 473 - 474 (1982).

PU Vul was discovered by Honda (1979) and Kuwano (1979) due to its conspicuous increase of brightness in the years 1977 - 79. Several authors proposed that PU Vul is binary. The 76 - 80 days period found by Chochol et al. (1981) could be well explained as an orbital period. If this is the case, the system PU Vul may be regarded as a short-period relative to the "normal" symbiotic stars.

117.159 **Does the cataclysmic binary Z Cha contain a black dwarf secondary?**
J. Faulkner, H. Ritter.
Binary and multiple stars as tracers of stellar evolution, (see 012.046), p. 483 - 488 (1982).

It is shown that the assumption of a black dwarf secondary in Z Cha leads to a number of contradictions with well established theoretical and observational facts. In particular the model predicts the radial velocity K_1 to be much smaller than is observed, the radius of the accretion disk to be smaller than is physically possible and a white dwarf mass which is inconsistent with the white dwarf's radius derived from eclipse analysis. Using the same arguments as for Z Cha it is also possible to exclude a black dwarf secondary in the similar systems OY Car and HT Cas.

117.160 **Estimated absolute dimensions and the inferred lifetime and angular momentum of W Ursae Majoris contact binaries.**
W. Van Hamme.
Astron. Astrophys., Vol. 116, 27 - 34 (1982).

Absolute dimensions of 47 W UMa contact binaries (25 of type W and 22 of type A) are estimated from their observed spectral type and an empirical spectral type-mass relation. Then a kind of lifetime of the primary components, and in particular the fractional lifetime with respect to the time required for crossing the main sequence band, is found by searching their position – given their mass and radius – in the theoretical evolutionary sequences for single stars. Also the total specific angular momentum of each system is calculated. Combining the lifetime and angular momentum considerations, a further subdivision of the classical A- and W-type groups is found, each subclass corresponding to a characteristic origin and evolutionary state.

117.161 **Mass exchange and evolution of close binary stars.**
L. R. Yungelson (*Yungel'son*).
Ejection and accretion of matter in binary systems, (see 012.051), p. 11 - 25 (1982).

The main processes connected with mass exchange between the components of the close binary stars and its evolutionary consequences are discussed. The evolutionary scenarios for the systems of different mass are constructed and the place of many observed objects in the evolutionary chain is determined.

117.162 **Models of the circumstellar matter in binary systems.**
S. Kříž.
Ejection and accretion of matter in binary systems, (see 012.051), p. 27 (1982). – Abstract.

117.163 **HZ Her – possible line profiles.**
Z. Šíma.
Ejection and accretion of matter in binary systems, (see 012.051), p. 29 (1982). – Abstract. – See also Abstr. 31.117.038.

117.164 **Interpretation of the light curves for "realistic" eclipsing close binaries.**
J. Grygar.
Ejection and accretion of matter in binary systems, (see 012.051), p. 31 - 37 (1982).

An account of the classical methods of the light solutions is given, indicating the successes and failures of this approach. After some comments on the Fourier transform method advocated chiefly by Kopal (1979) an interpretation based on more realistic physical models of close binaries is emphasized. The advent of high-speed computers has caused the emergence of many iterative methods. The most successful methods by Wood (1972) and Wilson and Devinney (1971, 1973) are discussed in detail.

117.165 **Observational aspects of the circumstellar matter in the binary systems.**
J. Tremko.
Ejection and accretion of matter in binary systems, (see 012.051), p. 39 - 56 (1982).

The most representative observational data which confirm the existence of the circumstellar matter in binary systems with special attention to semidetached systems are analyzed. Several spectroscopic manifestations of the circumstellar matter are presented. The photometric effects of the circumstellar matter have been detected at different phases of the light curve. The cool and hot spots which have their origin in mass transfer produce measurable photometric effects. The occurrence and the cause of the intrinsic variable polarization among binary systems are investigated. Recently detected radio emission and X-ray are discussed.

117.166 **Circumstellar matter in the close binary SZ Camelopardalis.**
D. Chochol.
Ejection and accretion of matter in binary systems, (see 012.051), p. 57 (1982). – Abstract.

117.167 **Mass transfer and period variations in the semi-detached binary system TW Dra.**
J. M. Kreiner, J. Tremko.
Ejection and accretion of matter in binary systems, (see 012.051), p. 69 (1982). – Abstract. – See also Abstr. 30.117.017.

117.168 **On short-period irregular light fluctuations of V 1357 Cyg (Cyg X-1).**
M. I. Kumsiashvili, Z. Kraicheva.
Ejection and accretion of matter in binary systems, (see 012.051), p. 93 - 98 (1982).

The X-ray source Cyg X-1 entering the close binary system with the star V 1357 Cyg exhibits the most diverse chaotic variations. The authors report investigations of possible rapid (during a night) irregular fluctuations of the close binary system V 1357 Cyg including the X-ray source Cyg X-1. During some nights in 1979 photoelectric observations were performed with the 48 cm telescope AZT-14 of Abastumani Astrophysical Observatory. The observations were carried out in UBV system close to the standard one.

117.169 **1982 *UBVR* photometric observations of XY UMa.**
M. Zeilik, D. Batuski, R. Elston, G. Henson, P. Schmolke, P. Smith.
Inf. Bull. Variable Stars, No. 2169, 4 pp. (1982).

117.170 **High speed photometry of V794 Aquilae.**
B. Warner.
Inf. Bull. Variable Stars, No. 2175, 2 pp. (1982).

117.171 **1981 *UBVR* photometry of II Peg.**
M. Zeilik, R. Elston, G. Henson, P. Schmolke, P. Smith.
Inf. Bull. Variable Stars, No. 2177, 4 pp. (1982).

117.172 **New times of minima and the period variations of the W UMa-type variable XY Leo.**
J. Kałużny, G. Pojmański.
Inf. Bull. Variable Stars, No. 2181, 4 pp. (1982).

117.173 **A single night light curve of 44i Boo.**
R. M. Robb, E. F. Milone.
Inf. Bull. Variable Stars, No. 2187, 3 pp. (1982) = Rothney Astrophys. Obs. Publ., Ser. B, No. 9.

117.174 **Recent photometry of HR 1099 (V711 Tau).**
S. Mohin, A. V. Raveendran, M. V. Mekkaden, K. Jayakumar, M. J. Rosario.
Inf. Bull. Variable Stars, No. 2190, 2 pp. (1982).

117.175 **CSV 6984 – a new W UMa type variable star.**
G. V. Zhukov.
Inf. Bull. Variable Stars, No. 2191, 2 pp. (1982).

117.176 **Evidence for the variation of brightness within the primary minima in W UMa (BD +56°1400).**
M. Hamzaoğlu.
Inf. Bull. Variable Stars, No. 2196, 2 pp. (1982).

117.177 **Light elements of ST Indi.**
V. J. Mugherli, M. A. Cerruti.
Inf. Bull. Variable Stars, No. 2197, 4 pp. (1982).

117.178 **Multi-color photometry of V711 Tauri (HR 1099).**
E. C. Olson.
Inf. Bull. Variable Stars, No. 2206, 6 pp. (1982).

117.179 **Photoelectric observations of MT Her.**
E. Budding, I. M. Murad.
Inf. Bull. Variable Stars, No. 2207, 3 pp. (1982).

117.180 **HD 1833: a new variable star.**
R. D. Lines, G. W. Henry, J. Sherlin, D. S. Hall.
Inf. Bull. Variable Stars, No. 2213, 2 pp. (1982).

117.181 **HD 37847: a new variable star.**
G. W. Henry, J. Sherlin, D. S. Hall.
Inf. Bull. Variable Stars, No. 2214, 2 pp. (1982).

117.182 **HD 185510: a new variable star.**
G. W. Henry, S. Murray, D. S. Hall.
Inf. Bull. Variable Stars, No. 2215, 2 pp. (1982).

117.183 **HD 190540: a new variable star.**
G. W. Henry, J. Sherlin, D. S. Hall.
Inf. Bull. Variable Stars, No. 2216, 2 pp. (1982).

117.184 **HD 205249: a new variable star.**
G. W. Henry, J. Sherlin, D. S. Hall.
Inf. Bull. Variable Stars, No. 2217, 2 pp. (1982).

117.185 **1982 photometry of ER Vul.**
M. Zeilik, B. Baca, D. Batuski, S. Burke, R. Elston, P. Smith.
Inf. Bull. Variable Stars, No. 2221, 4 pp. (1982).

117.186 **Some remarks on the close binary X-ray source AM Her.**
W. Götz.
Inf. Bull. Variable Stars, No. 2226, 3 pp. (1982).

117.187 **Photoelectric observations of the peculiar system VW Cep.**
A. S. Asaad, S. M. Nawar, M. A. Soliman, N. S. Awadalla.
Inf. Bull. Variable Stars, No. 2228, 3 pp. (1982).

117.188 **Photoelectric observations of W UMa (BD +56°1400).**
M. Hamzaoğlu, E. Hamzaoğlu.
Inf. Bull. Variable Stars, No. 2230 (1982).

117.189 **CW 1103+254.**
IAU Circ., No. 3712 (1982).

117.190 **H 0139-68.**
IAU Circ., No. 3720 (1982).

117.191 **AM Herculis.**
IAU Circ., Nos. 3720, 3739 (1982).

117.192 **E 2003+225.**
IAU Circ., No. 3733 (1982).

117.193 **V1343 Aquilae.**
IAU Circ., No. 3750 (1982).

117.194 **TV Columbae.**
IAU Circ., No. 3752 (1982).

117.195 **The peculiar ellipsoidal variable TU Horologii.**
C. Waelkens.
Messenger, No. 30, p. 14 - 15 (1982).

117.196 **Radioemisión en binarias cerradas.**
J. Sahade.
Rev. Astron., Tomo 54, No. 222, p. 9 - 16 (1982).

117.197 **Dynamic processes in binary systems.**
L. Patkós.
Dynamical astronomy, (see 012.056), p. 141 - 142 (1982). Abstract.

117.198 **Analysis of cataclysmic variable star energy distributions.**
R. A. Wade.
Astron. J., Vol. 87, 1558 - 1570 (1982).

The author analyzes spectrophotometry of cataclysmic variable stars covering the wavelength range 0.32 - 1.1 μm. The contributions due to the cool secondary star and the accretion disk are separated. The results for individual stars, which are largely independent of specific disk models, are presented as limits on the absolute visual magnitude of the disk and its fractional contribution to the light at λ 5500. These limits are given as functions of the spectral type of the (often unseen) secondary star.

117.199 **CI Cyg: the stage of Case C mass transfer.**
T. Iijima.
Astron. Astrophys., Vol. 116, 210 - 216 (1982).

Spectroscopic and spectrophotometric observations of the symbiotic star CI Cyg were carried out during 1979–1981. The following new properties of this object were found out: This object consists of an M type bright giant; $M_V = -2.0$, and a main sequence star. The M type star just fills its Roche critical lobe. A mass transfer with a rate of $\sim 3 \times 10^{-3} M_{\odot} yr^{-1}$ is anticipated. The transferred mass is accreted by the main sequence star via a disk. The luminosity of the main sequence star is close to its Eddington critical luminosity. A part of the accreted mass may be lost from the main sequence star as a radiatively driven outflowing wind. The matter in the nebular component may be supplied by this wind.

117.200 **The symbiotic star H1-36.**
D. A. Allen.
Anglo-Australian Obs., Prepr. No. 173, 14 pp. (1982). – Submitted to Mon. Not. R. Astron. Soc.

117.201 **Zum Massenübertrag in halbgetrennten Doppelsternsystemen.**
M. Hauschildt.
Mitt. Astron. Ges., Nr. 57, p. 287 (1982). – Abstract.

117.202 **On a possible mechanism of the decreasing period of AM Herculis.**
I. L. Andronov.
Astron. Tsirk., No. 1178, p. 1 - 3 (1981). In Russian.

117.203 **Comparative significance of the effects of tidal dissipation and condensation of binaries.**
V. N. Sementsov.
Astron. Tsirk., No. 1197, p. 2 - 4 (1981). In Russian.

117.204 **The formation of optical CNO emission lines in cataclysmic variables.**
R. E. Williams, D. H. Ferguson.
Prepr. Steward Obs., No. 408, 14 pp. (1982).

117.205 **Magnetism in the AM Her-type variable CW1103+254.**
G. D. Schmidt, H. S. Stockman, S. A. Grandi.
Prepr. Steward Obs., No. 424, 31 pp. (1982).

117.206 **Identification of CW1103+254 as a new magnetic variable.**
H. S. Stockman, C. B. Foltz, G. D. Schmidt, S. Tapia.
Prepr. Steward Obs., No. 430, 39 pp. (1982).

117.207 **A light curve of BX Pegasi.**
M. Hoffmann.
Acta Astron., Vol. 32, 131 - 135 (1982).

Photometric measurements in B and V are reported of the W UMa star BX Peg. The eclipses have found to be complete, the system is of W type. Preliminary elements have been derived. This has been aggravated by activity of the system which can be found both from the light curve and from its variable period.

117.208 **The hot halo subdwarf binary system HZ-22.**
A. Young, S. T. Wentworth.
Publ. Astron. Soc. Pacific, Vol. 94, 815 - 820 (1982).

A new investigation of the hot subdwarf component of the binary system known as HZ-22 has been conducted using more sophisticated detection apparatus than was available to the original investigators in 1972. The results indicate that the star is tidally distorted, that its companion is most likely to be a white dwarf, and that no significant mass transfer is currently taking place. The importance of the system now shifts to the possibility that the internal mass distribution of a highly evolved subdwarf may be determinable by observing its response to a known tidal potential.

117.209 **Starspots and W Ursae Majoris stars.**
A. P. Linnell.
Smithsonian Astrophys. Obs., Spec. Rep. 392, Vol. II, (see 012.065), p. 65 - 70 (1982).

Several lines of evidence point toward the existence of starspots on the W-type W Ursae Majoris stars. Dynamo theory requires differential rotation to produce the magnetic fields associated with starspots. A horizontal shear is difficult to support if the stellar components have a common envelope. In that case, there will be continuous collision of oppositely-directed gas streams near the L_1 point. A vertical shear implies non-synchronous rotation below the photospheric level. Non-synchronism, to be maintained, would have to be driven. Eddy viscosity provides a strong synchronizing mechanism. This paper presents some conjectures concerning these points.

117.210 **Comparative chromospheric and coronal emission from close binaries.**
G. Basri, F. Walter, R. Laurent.
Smithsonian Astrophys. Obs., Spec. Rep. 392, Vol. II, (see 012.065), p. 87 - 92 (1982).

Evidence for a simple linear relation between X-ray surface flux and angular velocity was presented by Walter and Bowyer (1981) for the RS CVn stars. In this paper, the authors report on preliminary results of an effort to extend the study of Walter and Bowyer to other close binary systems and to the many chromospheric and transition region diagnostics available to the IUE satellite. The binary systems offer the possibility of better determinations of stellar rotation and a wider spread in the magnitude of this rotational velocity than is available in a typical sample of single late-type stars.

117.211 **An X-ray survey of 9 Algol systems.**
N. E. White, F. E. Marshall.
Smithsonian Astrophys. Obs., Spec. Rep. 392, Vol. II, (see 012.065), p. 93 - 100 (1982).

The observed X-ray fluxes from the Algol systems are in reasonable agreement with those predicted for the cool lobe filling star by the rotational relations given by Pallavicini et al. (1981) and Walter and Bowyer (1981). This confirms the hypothesis that the lobe filling star is an active coronal X-ray source and that there is no X-ray emission associated with the mass transfer process. Conversely it also implies that the mass transfer has little effect on the coronal structures on the lobe filling star. While these systems appear to be generically similar to the RS CVn stars, the average X-ray luminosity is lower, and similar to that expected from a single rapidly rotating non-interacting star.

117.212 **Ultraviolet and X-ray detection of the 56 Peg system (K0 IIp + WD).**
M. Schindler, R. E. Stencel, J. L. Linsky, G. Basri, D. Helfand.
Smithsonian Astrophys. Obs., Spec. Rep. 392, Vol. II, (see 012.065), p. 125 - 133 (1982).

In this paper the authors discuss 56 Peg, representing a class of evolved stars, the barium star binaries, in which the fundamental physical processes are completely different from the general picture of cool evolved stars. The key to understanding the emission from this star has been the recent discovery of the white dwarf companion to this cool bright giant (Schindler et al. 1981). Accretion onto the white dwarf of ~0.1% of the stellar wind of the primary is sufficient to power an X-ray source of the observed luminosity. Reprocessing of the X-rays in the cool dense stellar wind explains the origin of the UV emission line spectrum, and may explain the time-varying asymmetry of the Mg II k line profile that is observed.

117.213 **Evolution of RS CVn and W UMa systems.**
P. P. Eggleton.
Smithsonian Astrophys. Obs., Spec. Rep. 392, Vol. II, (see 012.065), p. 153 - 159 (1982).

Two kinds of binary figure prominently in discussions of cool stars: RS CVn binaries and W UMa binaries. The evolutionary status of the former seems fairly clear. A small proportion of systems e.g. RT Lac, depart from the norm, and are identified here as products of late case B overflow in a system which was previously a "normal" RS CVn. The internal structure of W UMa systems is more controversial. The "contact discontinuity" model of Shu, Lubow and Anderson (1976) appears to be unphysical. Models without the unphysical assumption tend to follow "thermal relaxation oscillations", though these may be suppressed if one component is somewhat nuclearly evolved, and perhaps also if there is a net loss of angular momentum from the system on a timescale corresponding to the thermal timescale of the secondary.

117.214 **Photoelectric observations of AW Ursae Majoris.**
L. F. Istomin, L. M. Orlov, V. V. Kulagin.
Perem. Zvezdy, Tom 21, 583 - 588 (1981). In Russian.

The results of photoelectric *UBVR* photometry of the W UMa type eclipsing variable star AW UMa carried out in 1979 are presented.

117.215 **Spectroscopic observations of Delta Orionis.**
M. Singh.
Astrophys. Space Sci., Vol. 87, 269 - 277 (1982).

The Hα profile in the spectrum of δ Orionis shows phase-dependent changes, with a period of variation equal to the orbital period of the binary system. The spectra have been obtained at the Cassegrain focus of Kavalur Observatory telescopes at 17.2 Å mm^{-1} reciprocal dispersion and resolution 0.3 Å at λ6562.817 Å. Assuming that the P Cygni profile is formed by a spherically-symmetrical region, the analysis gives a shell radius of 2.18 stellar radius and an electron density in the shell equal to 6.54×10^9 cm^{-3}, with the observed expansion velocity of 50 km/s^{-1}, a mass loss of $1.3 \times 10^{-7} M_\odot$ per year. It is also found that the orbital elements change. The observations confirm the presence of a third body.

117.216 **Four years of photometry of DK Draconis = HR 4665.**
J. A. Eaton, D. S. Hall, G. W. Henry, C. A. Vaucher, C. R. Chambliss, P. K. Detterline, H. J. Landis, R. P. Wasatonic, H. Louth, T. R. Renner, A. M. Bloschak, J. S. Gillman, A. C. Krahnert, B. B. McGuinness.
Astrophys. Space Sci., Vol. 87, 305 - 312 (1982).

Differential BV Photometry of the K0 III + K0 III long-period RS CVn-type binary DK Dra= HR 4665 in 1977, 1978, 1979 and 1980 at eight different observatories is presented. A least squares fit to times of maximum and minimum light yields the ephemeris JD = 2 443 219.5 + 63^d.75 E for times of minimum. This photometric period is 1% shorter than the spectroscopically determined orbital period 64^d.44. The light curve was symmetrical and roughly sinusoidal in 1979 and 1980 but was asymmetrical in 1977 and 1978, with a faster rise to maximum. During these four years DK Dra varied between V = 6^m.07 and 6^m.35.

117.217 **Structure and evolution of large groups of massive close binary systems.**
C. Doom, J. P. de Grève.
Astrophys. Space Sci., Vol. 87, 357 - 375 (1982).

Using the definition of four important events in the evolution of massive close binary systems, the authors define five observable evolutionary phases in the life of a massive close binary system: OB + OB, WR + OB, C + OB, C + WR, and WR + WR. They define and compute a number of observable phenomena for large groups of the massive close binaries. For one burst of star formation, they compute the number of systems in different evolutionary phases, and the total mass loss as functions of the time. For continuous star formation, they determine the fraction of WR binary stars, occurring in different phases of massive close binary evolution, and a number of average quantities (mass, mass ratio) of WR binary systems. The results are compared with observations of WR binaries in the Galaxy and in open clusters.

117.218 **Can gamma ray bursts originate from low mass binaries?**
J. Ventura.
Accreting neutron stars, (see 012.069), p. 250 - 252 (1982).

The author reports on recent developments in the evolution of close low mass binaries having a neutron star primary and on their possible implication for gamm ray burst models.

117.219 **Late stages of close binary evolution – comments on alternative scenarios.**
W. Kundt.
Accreting neutron stars, (see 012.069), p. 285 - 287 (1982).

The author outlines possible evolutionary sequences for accreting neutron stars born as fast rotators in massive close binaries.

117.220 **Nova-like star TT Arietis: a binary system with cone accretion?**
N. F. Vojkhanskaya, Yu. N. Gnedin.
Pis'ma Astron. Zh., Tom 8, 729 - 733 (1982). In Russian.
English translation in Soviet Astron. Lett., Vol. 8.

Observational results are presented for radial velocities of the nova-like star TT Ari in the low-luminosity state (October 1980). A sharp jump of radial velocities is discovered between phases 0.5 and 0.6 of the orbital period. This jump is interpreted as a result of line-of-sight cutting of the cone shock wave formed due to accretion in the binary system TT Ari.

117.221 **HD 187282: a possible binary Wolf-Rayet star with low-mass companion.**
I. I. Antokhin, A. A. Aslanov, A. M. Cherepashchuk.
Pis'ma Astron. Zh., Tom 8, 734 - 739 (1982). In Russian.
English translation in Soviet Astron. Lett., Vol. 8.

B, V photometric and spectroscopic observations of the WN4 star HD 187282 connected with a ring nebula and having high altitude above the galactic plane have been carried out. A periodic variability of light (0^m.04) and radial velocities with the period 3^d.85 are discovered. This variability implies that HD 187282 may be a close binary system with a low mass ($\sim 1\ M_\odot$) companion, possibly a neutron star.

117.222 **Theory of the light curves of eclipsing systems with oscillating components. II.**
Z. Kopal.
Astrophys. Space Sci., Vol. 88, 313 - 329 (1982).
In continuation of a preceding investigation (Kopal, 1982) of photometric consequences of free oscillations – radial or nonradial – of the eclipsing component on the light changes of close binary systems, the aim of the present paper is to include the effects of similar oscillations of the component which undergoes eclipse, with arbitrary period and phase.

117.223 **Periodic perturbations in close binary systems due to rotational distortion.**
B. Zafiropoulos, Z. Kopal.
Astrophys. Space Sci., Vol. 88, 331 - 353 (1982).
This paper deals with the perturbations which rotation can produce to the orbital elements of a close binary system. The rectangular components, **R**, **S**, and **W** of the disturbing accelerations due to rotation have been substituted to the Gauss form of Langrange's planetary equations to yield the first order approximation. The results obtained are exact for any value of orbital eccentricity between the values $0 < e < 1$ and for arbitrary inclinations of the rotational axes to the orbital plane. First and second order approximations are given for the special case when equators are coplanar to the orbit.

117.224 **Periodic perturbations in close binary systems due to tidal distortion (non-lagging tides).**
B. Zafiropoulos, Z. Kopal.
Astrophys. Space Sci., Vol. 88, 355 - 400 (1982).
In a previous paper (Zafiropoulos and Kopal, 1982) the authors have studied the effects of rotational distortion on the orbital elements. The aim of the present paper is to investigate the secular and periodic perturbations of the orbital elements due to tidal distortion. For tidal distortion when tides do not lag, the Gaussian form of Lagrange's planetary equations has been employed to yield the first- and second-order approximations. The results obtained include the effects produced by the second, third and fourth harmonic distortions. The first order approximation for non-lagging tides has been expressed by means of Hansen coefficients.

117.225 **Periodic perturbations in close binary systems due to tidal lag.**
B. Zafiropoulos, F. Zafiropoulos.
Astrophys. Space Sci., Vol. 88, 401 - 427 (1982).
In two previous papers (Zafiropoulos and Kopal, 1983) the authors have investigated the effects of rotational and tidal distortion (for non-lagging tides) on the orbital elements of a close binary system. The present paper deals with secular and periodic perturbations caused by dynamical tides. The components **R**, **S**, and **W** of disturbing accelerations for tidal lag have been substituted in the Gaussian form of Lagrange's planetary equations to give the first-order approximation. The results obtained have been expressed by means of Hansen coefficients and include the effects produced by the second, third and fourth harmonic dynamical tides.

117.226 **IUE observations of certain short period RS CVn-like stars.**
E. Budding, T. H. Kadouri, A. Gimenez.
Astrophys. Space Sci., Vol. 88, 453 - 468 (1982).
Nine low and one high resolution IUE (LWR) spectrograms for the short period subgroup RS CVn-like stars have been studied, particularly in connection with the Mg II 2800 Å feature. Clear emission contributions can be easily seen for some of these stars, while all nine of them appear to have the feature filled by emission to some extent. If surface flux in emission is considered as a measure of "activity" the short period subgroup is as active as the main group. The activity is not as obvious because of the generally higher photospheric temperatures. Correlations of such activity with angular velocity and the scale of irregular photometric distortions ("starspots") are present, though the sample size is low.

117.227 **A semi-detached system: RZ Dra.**
K.-c. Leung, D. Zha, Y. Huang.
Acta Astrophys. Sinica, Vol. 2, 144 - 151 (1982). In Chinese.
The 1978 photoelectric observations of the late type close binary RZ Dra were reanalyzed with the Wilson and Devinney approach. Photometric parameters were determined. The system is found to be semi-detached. The absolute dimensions of the system are found to be $M_1 = 0.61\ M_\odot$, $M_2 = 0.41\ M_\odot$, $R_1 = 1.15\ R_\odot$, and $R_2 = 0.96\ R_\odot$. Both components appear to be overluminous and oversized for their masses and spectral types. Their evolutionary stage is also discussed.

117.228 **The radiation damping force of binary stars due to emission of gravitational waves.**
N. Hu.
Proceedings of the Second Marcel Grossmann Meeting on General Relativity, (see 012.081), p. 717 - 725 (1982).
Abstr. in Phys. Abstr., Vol. 86, Abstr. 15961 (1983).

117.229 **A note on gravitational radiation from close binary systems.**
V. Trimble.
Proceedings of the Second Marcel Grossmann Meeting on General Relativity, (see 012.081), p. 1245 (1982). – Abstr. in Phys. Abstr., Vol. 86, Abstr. 15962 (1983).

117.230 **Profiles of close binary stars.**
A. Yamasaki.
Astron. Her., Vol. 75, 244 - 248 (1982). In Japanese. – Abstr. in Phys. Abstr., Vol. 86, Abstr. 15941 (1983).

117.231 **Detection of the secondary star in the eclipsing cataclysmic variable AC Cancri.**
R. A. Downes.
Publ. Astron. Soc. Pacific, Vol. 94, 950 - 951 (1982).
The author has obtained spectrophotometric observations which confirm the cataclysmic variable nature of AC Cnc and detected for the first time the spectrum of the secondary star. Thus AC Cnc is one of the rare cataclysmic systems where the masses of both components can potentially be determined.

Early emission line stars.
See Abstr. 003.069.

The application of Bates' algorithm to binary stars.
See Abstr. 031.511.

Methods of photoelectric observations of lunar occultations and their application for the study of close binaries.
See Abstr. 031.609.

Television spectrophotometry.
See Abstr. 034.046.

Shape of a polytropic primary.
See Abstr. 042.082.

On the stability of the inner planets, satellites and close binaries.
See Abstr. 042.084.

Jets from a geometrically thick disk.
See Abstr. 062.011.

Stellar winds in binary X-ray systems.
See Abstr. 064.003.

Twisted accretion disks.
See Abstr. 064.006.

Penetration and circulation currents in irradiated spherically symmetric stellar atmospheres.
See Abstr. 064.011.

Radiation-driven winds in X-ray binaries.
See Abstr. 064.030.

Angular momentum loss from accretion disks due to Compton-heated winds.
See Abstr. 064.034.

The evolution of viscous discs – III. Giant discs in symbiotic stars.
See Abstr. 064.038.

Summary of workshop on accretion in astrophysics.
See Abstr. 064.040.

Accretion disks in stellar X-ray sources. A review of the basic theory of accretion disks and its problems.
See Abstr. 064.041.

Unsteady phenomena in an accretion disk.
See Abstr. 064.063.

Secular evolution of cataclysmic binaries: evolution of the secondary and the period gap.
See Abstr. 065.020.

Meridional circulation in rotating stars. II. Mean steady motions in rotationally and tidally distorted stars.
See Abstr. 065.023.

Meridional circulation in rotating stars. III. The reflection effect in close binaries.
See Abstr. 065.024.

Stellar evolution after IUE: 1. Massive stars during core hydrogen burning and massive close binaries.
See Abstr. 065.053.

Localized thermonuclear runaways and volcanoes on degenerate dwarf stars.
See Abstr. 065.057.

On the difference between the initial mass function of single stars and of primaries of binaries.
See Abstr. 065.082.

The combined effect of mass loss and overshooting. II. The evolution of 10 $M_\odot$ to 30 $M_\odot$ stars during core hydrogen burning.
See Abstr. 065.093.

Relativistic stars in close binary systems and masses of neutron stars.
See Abstr. 066.188.

The universal diagram for magnetized neutron stars in the Galaxy.
See Abstr. 066.525.

The longitudinal distribution of sunspots and the RS CVn starspot model.
See Abstr. 072.107.

Properties and nature of Be and shell stars. 7B. 88 Her – an important clue to understanding the Be phenomenon?
See Abstr. 112.042.

Discovery of two FK Comae stars in the southern hemisphere.
See Abstr. 113.010.

The peculiar object SS433.
See Abstr. 114.016.

IUE observations of symbiotic stars.
See Abstr. 114.023.

The fastest runaway Wolf-Rayet star of population I in the Galaxy, 209 BAC: evidence for a low mass companion.
See Abstr. 114.082.

Does the WC6 star HD 76536 have a compact companion?
See Abstr. 114.099.

Near IR spectra of symbiotic stars.
See Abstr. 114.147.

Spectra of individual symbiotic stars.
See Abstr. 114.149.

Visual symbiotic spectra obtained with the Haute Provence Multiphot detector.
See Abstr. 114.150.

UV line emission of symbiotic stars.
See Abstr. 114.151.

Z Andromedae: the prototype?
See Abstr. 114.152.

The ultraviolet spectrum of AG Draconis.
See Abstr. 114.154.

Spectroscopic observations of AG Dra.
See Abstr. 114.155.

Mg II k emission lines in stars with different rotational velocities and metal abundances.
See Abstr. 114.213.

Magnetic phenomena in stars.
See Abstr. 116.008.

UBV-polarimetry of the X-ray binaries HD 77581 (4U 0900–40), HD 153919 (4U 1700–37) and of HD 152667.
See Abstr. 116.011.

Radio observations of symbiotic stars.
See Abstr. 116.017.

The age of the triple system HD 165590.
See Abstr. 118.025.

BV Dra and BW Dra: two contact systems in one visual binary.
See Abstr. 118.043.

BR Muscae: a new early-type contact binary.
See Abstr. 119.016.

Photometric elements of the eclipsing binary AZ Vir.
See Abstr. 119.030.

A spot model for VW Cephei.
See Abstr. 119.036.

The massive eclipsing binary RY Scuti.
See Abstr. 119.046.

A new light curve analysis of the eclipsing binary SX Aur.
See Abstr. 119.049.

UV time-dependent emission in SY Muscae.
See Abstr. 119.052.

A study of the orbital periods of 22 eclipsing binaries.
See Abstr. 119.055.

The evolutionary state of Zeta Aurigae.
See Abstr. 119.056.

Structure and evolution of SV Centauri.
See Abstr. 119.057.

Peculiar light-curve change in SV Cam.
See Abstr. 119.063.

Photometry of minima and period changes of the binary system TX UMa.
See Abstr. 119.064.

Theory of the light curves of eclipsing systems with oscillating components. I.
See Abstr. 119.104.

A photometric study of the short-period eclipsing binary BW Eridani.
See Abstr. 119.110.

On the distribution of young spectroscopic binary stars over the major semiaxes of their orbits.
See Abstr. 120.037.

A time-resolved spectroscopic study and modeling of the dwarf nova BV Centauri.
See Abstr. 120.040.

Close binary systems before and after mass transfer. III. Spectroscopic binaries.
See Abstr. 120.045.

Quasi-simultaneous photometric and polarimetric observations of T Tau and RY Tau.
See Abstr. 121.031.

Infrared light curves of the dwarf nova OY Carinae.
See Abstr. 122.023.

A spectroscopic study of the short-period variable Be star EM Cephei.
See Abstr. 122.025.

Low resolution ultraviolet and optical spectrophotometry of symbiotic stars.
See Abstr. 122.031.

Observations and analysis of the R Aquarii jet.
See Abstr. 122.032.

The nova-like variable KQ Mon and the nature of the UX Ursae Majoris stars.
See Abstr. 122.033.

High velocity gas outflow from the nova-like variable LSI 55°−8.
See Abstr. 122.034.

Kataklysmische Veränderliche in Beobachtung, Statistik und Analyse.
See Abstr. 122.043.

A search for binaries among short-period southern cepheids.
See Abstr. 122.051.

The infrared properties of Mira-type variables and other cool stars as determined from *JHKL* photometry.
See Abstr. 122.072.

Photometric observations of CN Orionis.
See Abstr. 122.082.

Physical properties of V1329 Cygni.
See Abstr. 122.084.

A model for the 1979 outburst of the recurrent nova U Scorpii.
See Abstr. 124.001.

CW 1103 +254: a new cataclysmic at high galactic latitude.
See Abstr. 124.003.

A spectroscopic study of the old nova HR Delphini.
See Abstr. 124.415.

Supernovae in binary systems: production of runaway stars and pulsars.
See Abstr. 125.085.

The mass distribution of white dwarfs and central stars of planetary nebulae.
See Abstr. 126.025.

Abundance peculiarities in white dwarf atmospheres as a result of binary evolution?
See Abstr. 126.026.

The formation of massive white dwarfs in cataclysmic binaries.
See Abstr. 126.027.

Discovery of highly ionized species in the ultraviolet spectrum of Feige 24.
See Abstr. 126.035.

Why are X-ray pulsars not observed coupled with Wolf-Rayet stars?
See Abstr. 142.005.

Simultaneous spectroscopy and photometry of the non-synchronous rotator H2252−035.
See Abstr. 142.008.

Optical review: X-ray binaries.
See Abstr. 142.010.

Some aspects of mass transfer in massive X-ray binaries.
See Abstr. 142.011.

Low mass, optically identified galactic X-ray sources.
See Abstr. 142.012.

Binary X-ray pulsars.
See Abstr. 142.018.

Young pulsars in binary systems.
See Abstr. 142.019.

Ultraviolet spectroscopy of galactic X-ray sources.
See Abstr. 142.022.

Galactic soft X-ray sources: RS CVn systems, flare stars and A stars.
See Abstr. 142.024.

High resolution soft X-ray spectra with the objective grating spectrometer aboard Einstein Observatory.
See Abstr. 142.029.

Color variability and optical light curve of 2S0921–630.
See Abstr. 142.031.

2A 0526–328, an optical and X-ray study of a cataclysmic variable.
See Abstr. 142.038.

On a possible model of the object Cyg X-3.
See Abstr. 142.039.

Observations of the May 1979 outburst of Centaurus X-4.
See Abstr. 142.040.

Circinus X-1: X-ray observations with SAS 3.
See Abstr. 142.051.

On the ultrahigh-energy gamma rays from Cygnus X-3.
See Abstr. 142.052.

The anomalous X-ray absorption spectrum of Vela X-1.
See Abstr. 142.055.

X-ray observations of AM Her in its low state.
See Abstr. 142.063.

Iron lines in the spectra of X-ray binaries.
See Abstr. 142.079.

The X-ray transient 3A 1431–409: a highly active RS CVn system?
See Abstr. 142.100.

The high galactic latitude fast-transient X-ray source 3A1431–409.
See Abstr. 142.101.

Infrared photometry of the X-ray binary 2A 1822–371: a model for the ultraviolet, optical, and infrared light curve.
See Abstr. 142.105.

X-ray observations of AM Herculis in its low state.
See Abstr. 142.110.

The theoretically expected X-ray luminosity and the binary nature of Wolf-Rayet runaway stars.
See Abstr. 142.112.

X-ray observations of symbiotic stars.
See Abstr. 142.113.

X-ray detection of the symbiotic star AG Draconis.
See Abstr. 142.114.

A comparison of the X-ray properties of X Persei and γ Cassiopeiae.
See Abstr. 142.119.

Simultaneous optical and X-ray bursts from 4U/MXB 1636–53.
See Abstr. 142.121.

Optical bursts from 4U/MXB 1636–53.
See Abstr. 142.122.

A study of X-ray emission from Ap and Am stars.
See Abstr. 142.143.

Dissipative evolution of collisionless stellar systems. I. Cooling and heating of a stellar system by binary stars.
See Abstr. 151.002.

Is NGC 1851-UV5 a binary star in a globular cluster?
See Abstr. 154.044.

The theoretical expected galactic distribution of WR runaway stars.
See Abstr. 155.011.

Errata

117.901 **Erratum: 'Mass, angular momentum and energy transfer in close binary stars'** [Annu. Rev. Astron. Astrophys., Vol. 19, 277 - 293 (1981)].
F. H. Shu, S. H. Lubow.
Annu. Rev. Astron. Astrophys., Vol. 20, VI (1982). – See Abstr. 30.117.070.

117.902 **Erratum: "Tidal evolution in close binary systems for high eccentricity"** [Astron. Astrophys., Vol. 110, 37 - 42 (1982)].
P. Hut.
Astron. Astrophys., Vol. 116, 351 (1982). – See Abstr. 31.117.135.

118 Visual Binaries, Multiple Stars, Astrometric Binaries

118.001 **Revised orbital elements of visual binary stars ADS 3182 and ADS 3483.**
M. Scardia.
Astron. Astrophys., Suppl. Ser., Vol. 49, 503 - 508 (1982). In French.

Revised orbital elements of the visual binary stars ADS 3182 and ADS 3483 are given. The dynamical parallaxes and total masses of the systems have been calculated.

118.002 **Statistical studies of visual double and multiple stars. I. Incompleteness of the IDS, intrinsic fraction of visual doubles and multiples, and number of optical systems.**
A. Poveda, C. Allen, L. Parrao.
Astrophys. J., Vol. 258, 589 - 604 (1982).

In an attempt to obtain a relatively uniform and homogeneous group of double stars for statistical studies, the Index Catalogue of Visual Double Stars has been subjected to a "filtering" procedure, so as to eliminate optical systems. The filtered IDS (FIDS) contains 51,201 entries out of 69,819 in IDS II. To describe the process of discovering and registering a double or multiple star, a simple stochastic model is postulated and applied to the FIDS. A major result of this study is that, if appropriately observed, practically every field star would be found to be a visual double or multiple system, and that nearly one out of three field stars would turn out to be a visual multiple.

118.003 **Multiple systems of astrophysical interest. II. Mixed doubles.**
O. J. Eggen.
Astrophys. J., Vol. 258, 605 - 611 (1982).

Photometry of the components of 10 widely separated, common proper motion systems, about one-half of which are probably members of the Local Association, is presented and discussed with respect to the ages of the individual stars. The components in the systems that are in the Local Association show the same apparent mixture of ages that occurs in clusters of the association.

118.004 **Masses, mass ratios, and parallaxes from astrometric studies of seven visual binaries.**
S. L. Lippincott.
Astron. J., Vol. 87, 1237 - 1241 (1982).

Long-interval series of plates on visual binaries taken with the Sproul 61-cm refractor are analyzed for parallax, mass, and mass ratio. The results are all the first astrometric mass-ratio determinations to be published. All are presumed to be main-sequence stars with spectral types later than F0V, except for the ADS 8939 system whose primary is classified as A6III. Three of the visual systems have a component that is a spectroscopic binary.

118.005 **Speckle interferometric measurements of binary stars. VII.**
H. A. McAlister, E. M. Hendry.
Astrophys.J., Suppl. Ser., Vol. 49, 267 - 272 (1982) = Astron. Contrib. Georgia State Univ., No. 59.

Two hundred ninety-five measurements of 201 binary stars observed by means of speckle interferometry with the 4 m telescope at KPNO are presented.

118.006 **A search for fragmentation debris near Ursa Major stream stars.**
F. C. Witteborn, J. D. Bregman, D. F. Lester, D. M. Rank.
Icarus, Vol. 50, 63 - 71 (1982).

Accretion models of planet formation and the early cratering history of the solar system suggest that planet formation is accompanied by a cloud of debris resulting from accumulation and fragmentation. A rough estimate of the infrared luminosities of debris clouds is presented for comparison with measured 10-μm luminosities of young stars. New measurements of 13 F, G, and K main-sequence stars of the Ursa Major Stream, which is thought to be about 2.7×10^8 years old, place constraints on the amount of debris which could be present near these stars.

118.007 **Alpha Centauri.**
M. S. Bessell.
Proc. Astron. Soc. Australia, Vol. 4, 212 - 214 (1981).

This paper reports the new photometric observations of α Cen A and B and the results of a reanalysis of the weak lines of Fe and Ti published by French and Powell (1975) for α Cen A and B and the solar lines of Furenlid (1979) and Hearnshaw (1977). The position of α Cen and the Sun in the HR diagram is examined and the possibility of planetary systems being associated with α Cen is considered.

118.008 **On the kinematics of Trapezium-type multiple systems.**
G. N. Salukvadze.
Soobshch. AN GruzSSR, Tom 104, 41 - 44 (1981). In Russian.
Abstr. in Ref. zh., 51. Astron., 7.51.616 (1982).

118.009 **ADS 3230: two possible solutions in the computation of the orbital elements.**
M. Scardia.
Astron. Astrophys., Suppl. Ser., Vol. 50, 19 - 21 (1982). In French.

Orbital elements of the visual binary star ADS 3230 are given. Two solutions are possible. The first group of elements, with period of 217 years, leads to a total mass of the system of 2.23 $M_\odot$; the other, with period of 856.7 years, more uncertain than the previous one, gives a total mass of the system of 5.03 $M_\odot$ and is more reliable from an astrophysical point of view.

118.010 **Measurements of double stars made in Nice. Orbits of three binary stars.**
P. Couteau.
Astron. Astrophys., Suppl. Ser., Vol. 50, 49 - 71 (1982). In French.

The author gives 717 measurements of 241 binaries made at the 74 and 50 cm refractors. These are mostly very close binaries discovered recently. Orbits of ADS 4236, ADS 5515 and Wor 19 are calculated.

118.011 **Measures of southern double stars in 1981.**
R. H. Wilson, Jr.
Astron. Astrophys., Suppl. Ser., Vol. 50, 115 - 118 (1982).

Measures of 115 double and multiple stars done with the Repsold micrometer attached to the 26-inch Innes telescope at Johannesburg are tabulated with notes.

118.012 **Infrared detection of the low-mass companion to Zeta Aquarii B.**
D. W. McCarthy, F. J. Low, S. G. Kleinmann, D. V. Arganbright.
Astrophys. J., Lett., Vol. 259, L75 - L78 (1982).

The quadruple system, Zeta Aquarii, has been resolved in the K and L photometric bands by one-dimensional speckle interferometry. An infrared component is detected with a north-south separation from Zeta Aquarii B of 0.″17 (3.9 AU), in agreement with the predicted orbit of a low-mass (0.22 $M_\odot$) companion. This companion has an absolute K-magnitude of 4.4 ± 0.1 and a $K - L$ color temperature of 1440 ± 200 K.

118.013 **Detection of other planetary systems.**
D. C. Black.
Bull. American Astron. Soc., Vol. 14, 625 (1982). – Abstract.

118.014 **Analysis of the Allegheny and Van Vleck observations of Barnard's star.**
G. D. Gatewood, J. W. Stein.
Bull. American Astron. Soc., Vol. 14, 627 (1982). – Abstract.

118.015 **Sproul astrometric study of Barnard's star, a progress report.**
P. van de Kamp.
Bull. American Astron. Soc., Vol. 14, 627 (1982). – Abstract.

118.016 **Status of the Sproul astrometric plate series on the nearest stars in the search for planetary systems.**
S. L. Lippincott.
Bull. American Astron. Soc., Vol. 14, 627 (1982). – Abstract.

118.017 **HD 65865: una binaria visual WN4.5 + A.**
V. S. Niemela.
Rev. Mexicana Astron. Astrofis., Vol. 6, (see 012.017), 244 (1981). – Abstract.

118.018 **The search for extrasolar planets.**
D. Finley.
Astronomy, Vol. 9, No. 12, p. 90 - 95 (1981). – Abstr. in Phys. Abstr., Vol. 85, Abstr. 79156 (1982).

118.019 **Light-time effect in AK Her and ER Ori.**
K. D. Abhyankar, T. Panchatsaram.
Bull. Astron. Soc. India, Vol. 10, 27 (1982). – Abstract.

118.020 **The binary system Sirius in the context of stellar evolution.**
F. D'Antona.
Astron. Astrophys., Vol. 114, 289 - 296 (1982).

The aim of the paper is to give a scheme as simple as possible for the evolution of the binary system Sirius A plus Sirius B, in which a reasonable interpretation may be found for several peculiarities present in the system, such as the large eccentricity of the orbit, the large mass of the white dwarf component, and the "metallic" spectral type of Sirius A.

118.021 **Infrared detection of low mass companions in several nearby star systems.**
D. W. McCarthy, F. J. Low, S. G. Kleinmann.
Bull. American Astron. Soc., Vol. 14, 777 (1982). – Abstract.

118.022 **Two visual-spectroscopic binary orbits: ADS 7780 and ADS 11060.**
W. D. Heintz.
Publ. Astron. Soc. Pacific, Vol. 94, 705 - 707 (1982).

The combination of micrometer and radial-velocity observations is applied to two orbits of high eccentricities and inclinations. All results on the G1 V triple star ADS 11060 (including the subsystem studied by Batten et al.) show good consistency and precision. Substantial uncertainty remains in the component masses of ADS 7780 = β LMi (G9 III) which could be also a triple star.

118.023 **Investigations in the research of multiple stars.**
G. M. Popović.
Publ. Astron. Obs. Sarajevo, Vol. 1, No. 1, (see 012.043), p. 269 - 278 (1981). In Croatian.

118.024 **Contribution to the study of composite spectra. III. Spectrum binaries: intermediate class between visual and spectroscopic binaries.**
J. M. Carquillat, R. Nadal, N. Ginestet, A. Pedoussaut.
Astron. Astrophys., Vol. 115, 23 - 29 (1982). In French.

After revising Hynek's sample of bright stars with composite spectra (Hynek, 1938) the authors have had to reconsider the existence of spectrum binaries as a real intermediate class between spectroscopic and visual binaries. To contribute to the answer to this question they present a study of the distribution in log *a* (*a*, semimajor axis of the relative orbit) from a sample of 180 visual binaries and 95 spectroscopic binaries. The distribution obtained is quite Gaussian for visual and astrometric binaries while it is irregular for spectroscopic binaries, making two separate groups of objects apparent. The global distribution obtained with all the binaries of the sample, presents two maxima. A distribution with only one maximum is proposed and an estimation is made about the number of binaries which have not been detected because of observational selection.

118.025 **The age of the triple system HD 165590.**
G. A. Bakos, J. Tremko.
Binary and multiple stars as tracers of stellar evolution, (see 012.046), p. 67 - 71 (1982).

It has been found that the close spectroscopic pair of the system HD 165590 consists of a G type primary and a late type secondary, which most likely is a T Tauri star. As such, the system is semidetached and a mass transfer takes place between the components.

118.026 **Evolutionary trends in wide binaries.**
P. van de Kamp.
Binary and multiple stars as tracers of stellar evolution, (see 012.046), p. 81 - 103 (1982).

Contents: Introduction. "Wide" binaries, orbital motion, common proper motion. Proper motion binaries. Long focus photographic astrometry. Wide eclipsing binaries. Classical binaries with one degenerate component. Proper-motion binaries with one degenerate component: statistical studies. Some thoughts of evolution. Perturbations and unseen astrometric companions. Multiplicity among binaries. Does the sun have a distant stellar companion?

118.027 **Trapezium type multiple systems and formation of stars.**
G. N. Salukvadze.
Binary and multiple stars as tracers of stellar evolution, (see 012.046), p. 109 - 113 (1982).

By a comparison of Trapezium-type multiple star systems in the Abastumani Catalogue and the famous Catalogue of Stellar Associations and Clusters, it is confirmed that the great number of Trapezium-type multiple stars, which belong mostly to spectral classes O-B2, are found in associations and clusters. In 13 T-associations, 120 Trapezium-type multiple systems and 182 common multiple and double stars could be detected. Based on present-day observational data and the author's photographic observations, the kinematics of Trapezium-type multiple stars of spectral class O-B2 is studied.

118.028 **Photometric investigation of visual binaries with a component above the main sequence.**
O. G. Franz.
Binary and multiple stars as tracers of stellar evolution, (see 012.046), p. 123 - 128 (1982).

Visual binaries containing a component above the main sequence in combination with a main-sequence star provide an excellent opportunity for determining absolute magnitudes of evolved stars. Some aspects of this technique are discussed, and preliminary results are presented.

118.029 **Double stars: the closest and the most distant systems.**
T. J. Herczeg.
Binary and multiple stars as tracers of stellar evolution, (see 012.046), p. 145 - 151 (1982).

Review and comments are presented on two extremely

different samples of binary and multiple stars: systems within 10 parsecs of the sun and systems discovered in galaxies of the Local Group.

118.030 **Orbites nouvelles.**
Circ. Inf., Nos. 87, 88 (1982).

118.031 **Etoiles doubles nouvelles.**
P. Couteau, P. Muller.
Circ. Inf., Nos. 87, 88 (1982).

118.032 **81 Cancri (φ 347) – a visual binary which is spectroscopically double-lined.**
R. Griffin, R. Griffin.
Observatory, Vol. 102, 217 - 222 (1982).

118.033 **The triple system of π Cephei.**
C. D. Scarfe.
J. R. Astron. Soc. Canada, Vol. 76, 325 (1982). – Abstract.

118.034 **The frequency of planetary systems in the Galaxy.**
J. P. Harrington.
Extraterrestrials: where are they? (see 012.057), p. 142 - 153 (1982).

Observations have not yet provided any persuasive evidence for the existence of extrasolar planetary systems. The observed angular momenta of binary stars, coupled with current ideas of how the solar system formed, suggest that most of the 40 percent or so of stars without close stellar companions may be accompanied by planets. It is doubtful whether planets could also exist in binary star systems. Such indirect arguments, however, are weak because the origin of our own planetary system is poorly understood. The proposition still cannot be dismissed that planetary systems are extremely rare.

118.035 **Verwendung der Radialgeschwindigkeitsmessungen zur Verbesserung der Bahn visueller Doppelsterne.**
A. Valbousquet.
Mitt. Astron. Ges., Nr. 57, p. 281 - 284 (1982).

The author shows how far one can use few radial velocity measurements to select an orbit in case of uncertainty caused for example by an unknown quadrant. He also shows that it is possible to introduce the measurements of relative radial velocities into the equations of conditions of differential adjustments in order to improve the accuracy of the orbit, and he shows what weight is to be given to the equations of conditions issued from those measurements.

118.036 **A study of visual double stars with early-type primaries. III. Radial velocities.**
G. F. Gahm.
Stockholms Obs. Rep., No. 20, 14 pp. (1982).

New radial velocities are derived for 81 components in 68 double or multiple stars. Additional velocities are compiled from the literature for all stars observed in a current spectroscopic and photometric survey of multiple stars. The velocities are used to obtain indications on which systems are optical and which may be physical.

118.037 **Double stars in the χ and h Persei cluster region.**
M. Muminov.
Astron. Tsirk., No. 1182, p. 7 - 8 (1981). In Russian.

118.038 **The observation of radial velocities of visual binaries.**
E. Van Dessel, J. Dommanget.
Bull. Inf. Cent. Données Stellaires, No. 23, p. 106 (1982).

118.039 **On the detection of other planetary systems by astrometric techniques.**
D. C. Black, J. D. Scargle.
Astrophys. J., Vol. 263, 854 - 869 (1982).

The authors examine the problem of detecting periodic signals in astrometric data, with specific emphasis on searching for a perturbation in a star's motion due to planetary companions. It is shown how spectral analysis of such data by means of the periodogram provides an objective statistical basis for evaluating the reality and characteristics of suspected perturbations. A brief description of the properties of the periodogram is followed by a demonstration of the technique in an analysis of 12 years of US Naval Observatory astrometric data for three stars with low-mass stellar companions. A quantitative discussion is given of an error source generally overlooked or ignored in analysis of perturbations in astrometric data, namely absorption of the linear component of the necessarily finite sample of a perturbation into the proper motion.

118.040 **The evolved stellar system Σ 2367.**
A. H. Batten, J. M. Fletcher, W. A. Fisher, R. D. McClure, C. L. Morbey, R. F. Griffin, C. D. Scarfe.
Publ. Astron. Soc. Pacific, Vol. 94, 860 - 871 (1982).

Orbital elements are derived for the visual binary Σ 2367 from a combination of visual and spectroscopic observations. The physical association of the component C with the orbital pair is confirmed and an approximate period deduced for the system AB-C. All three stars A, B, and C are evolved from the main sequence, although none is much more massive than the sun. The stars in the system appear to be about as old as the members of the cluster M 67.

118.041 **On the problem of orbital evolution of the components of triple star systems.**
A. A. Orlov, N. A. Solovaya.
Vestn. MGU. Fiz., astron., Tom 23, No. 3, p. 19 - 24. In Russian. – Abstr. in Ref. zh., 51. Astron., 10.51.536 (1982).

118.042 **A determination of the dynamic parallax and an estimate of the orbital inclination of a double star from its short-arc observation.**
O. V. Kiyaeva.
Izv. Glav. Astron. Obs. Pulkovo, Astrometr. Astrofiz., No. 199, p. 13 - 18 (1982). In Russian.

A new method is proposed for determining the dynamic parallax of a double star from short arc observations under the assumption that the star has a circular orbit. Parameters of the relative motion in the double star including the radius of the curvature of the apparent orbit are determined. The proposed method is compared with H. N. Russell's method (1928). The results of the reduction of photographic observations of double stars ADS 48, ADS 7251 and 61 Cyg are given.

118.043 **BV Dra and BW Dra: two contact systems in one visual binary.**
S. M. Rucinski, J. Kaluzny.
Astrophys. Space Sci., Vol. 88, 433 - 452 (1982).

The W UMa-type stars BV Dra and BW Dra form the only known visual binary (ADS 9537) consisting of two contact systems. *UBV* light curves obtained in 1975 for both components of ADS 9537 are presented. The authors compare both variables and conclude that they follow the main sequence dependences. They give also a brief discussion of the existing information on other contact binaries in multiple (visual) systems: the contact binaries seem to follow the same statistical distributions for occurrence in multiples as found for detached binaries but severe selection effects limit detectability for multiple systems with "wide" orbits having semi-axes below 100 a. u.

Current progress with the double star catalogues.
See Abstr. 002.026.

Second Catalogue d'Éphémérides des vitesses radiales relatives des composantes des étoiles doubles visuelles dont l'orbite est connue.
See Abstr. 002.097.

A new catalog of ephemerides of v_r for orbital visual binaries.
See Abstr. 002.111.

Visual double star data base.
See Abstr. 002.112.

La mesure des étoiles doubles par interférométrie "speckle".
See Abstr. 031.527.

Precision radial velocities with a hydrogen fluoride absorption cell.
See Abstr. 031.540.

Criteria for positive planetary detection using the radial velocity method.
See Abstr. 031.541.

Solar observations pertaining to the search for extra-solar planetary systems.
See Abstr. 031.542.

A search for low mass unseen companions in nearby binary star systems.
See Abstr. 031.543.

Detection of extra solar planets.
See Abstr. 031.664.

The LPL extrasolar-planet-detection spectrometer: a new configuration.
See Abstr. 034.019.

On the relation between resonance and instability in planetary systems.
See Abstr. 042.021.

A simple criterion for determining the dynamical stability of three-body systems.
See Abstr. 042.026.

L'évaporation de systèmes triples à l'origine de binaires peu serrées.
See Abstr. 042.074.

Shock fronts in wide binary systems.
See Abstr. 064.009.

Scanner studies of composite spectra. II. Giants and dwarfs.
See Abstr. 114.006.

A study of the orbital periods of 22 eclipsing binaries.
See Abstr. 119.055.

Secondaries of eclipsing binaries. IV. The triple system Lambda Tauri.
See Abstr. 119.061.

AF Geminorum: an eclipsing triple star.
See Abstr. 119.117.

Photoelectric radial velocities. Paper IX. The orbits of the spectroscopic binaries 81 and χ Geminorum.
See Abstr. 120.009.

A mass-eccentricity correlation in spectroscopic and visual binary orbits.
See Abstr. 120.036.

Search for pulsating stars in multiple stellar systems.
See Abstr. 122.087.

VLA observations of M-dwarf flare stars.
See Abstr. 122.153.

A note about multiple systems of dense molecular clouds.
See Abstr. 131.268.

Einstein solid state spectrometer observation of the peculiar red dwarf Wolf 630 AB.
See Abstr. 142.045.

119 Eclipsing Binaries

119.001 **Four-color photometry of eclipsing binaries, XIVB: Lightcurves of QX Carinae.**
J. Andersen, J. V. Clausen, B. Nordström, B. Reipurth.
Astron. Astrophys., Suppl. Ser., Vol. 49, 571 - 576 (1982).

uvby lightcurves of the early B-type double lined eclipsing binary QX Carinae are presented. These show that the system has an orbit of appreciable eccentricity.

119.002 **Periodic ephemerides for 49 eclipsing binary-star systems.**
J. B. Rafert.
Publ. Astron. Soc. Pacific, Vol. 94, 485 - 495 (1982).

An ephemeris and brief discussion is presented for each of a number of eclipsing binary-star systems whose orbital period variations can most reasonably be reproduced with a singly-periodic function. Analysis of the observed period variations by differential corrections indicates an anomalously large number of systems whose photometric mass functions are less than 0.07.

119.003 **Photometric study of the eclipsing binaries RY Indi and ST Carinae.**
E. Lapasset, J. J. Clariá.
Publ. Astron. Soc. Pacific, Vol. 94, 524 - 531 (1982).

A photometric *UBV* study of the β Lyrae-type eclipsing binaries RY Ind and ST Car is presented. Light and color curves are constructed from 430 and 750 individual observations in each passband, respectively, and updated ephemerides are given. Although partial eclipses in both light curves made the analysis somewhat difficult, a first classical approach to the orbital elements is given. While either a transit or an occultation is possible for RY Ind primary minimum, only a transit in this minimum yields a consistent solution for ST Car.

119.004 **A survey for gaseous disks in short-period Algol systems during primary eclipse.**
R. H. Kaitchuck, R. K. Honeycutt.
Publ. Astron. Soc. Pacific, Vol. 94, 532 - 536 (1982).

Time-resolved spectroscopy during primary eclipse was utilized in an attempt to find emission lines from gaseous disks surrounding the primary components of Algols with periods less than four days. A total of 18 systems was surveyed during 42 eclipses. The only systems which showed emission lines were RW Tau, U Cep, and TZ Eri. This was the first detection of a disk in TZ Eri. The authors confirm the findings of previous workers, namely that the disks in short-period Algol systems are both rare and transient.

119.005 **The apsidal motion of PV Cassiopeiae.**
A. Giménez, T. E. Margrave.
Astron. J., Vol. 87, 1233 - 1236 (1982).

New photoelectric times of minimum have been obtained for the slightly eccentric eclipsing binary PV Cassiopeiae that confirm the existence of a definite periastron revolution. The authors have analyzed their observed minima together with other available minima, and an apsidal motion period of approximately 90 yr has been established.

119.006 **UBVR photometry of the system CQ Cep, period change, and mass loss by a Wolf-Rayet star.**
Eh. A. Antokhina, N. A. Lipunova, A. M. Cherepashchuk.
Astron. Zh., Tom 59, 704 - 710 (1982). In Russian. English translation in Soviet Astron., Vol. 25, No. 4.

New photoelectric UBVR observations have been obtained for the eclipsing binary system CQ Cep (WN7+O7, $p = 1^{d}.641246$). These observations, together with all published photographic and photoelectric data from 1899 to 1981, allow to find out the period change for the system: $\dot{p} = (-0.019 \pm 0.006)$ s/yr. The rate of the matter outflow from the star WN7 to the companion star O7 is greater than $(0.69 \pm 0.22) \times 10^{-5} M_{\odot}$/yr.

119.007 ***VBLUW* photometry of RZ Oph (BD +7°3832): eclipse of the accretion disk.**
J. van Paradijs, H. van der Woerd, M. van der Bij, A. Lee Van Suu.
Astron. Astrophys.,Vol. 111, 372 - 374 (1982).

The authors present the results of photometric observations of RZ Oph, made during a 2-month interval centered on primary eclipse. The brightness of RZ Oph gradually decreases before, and increases after the eclipse of the primary star, indicating the presence of an accretion disk. From the duration of the disk eclipse the authors estimate that the radius of the disk is $\gtrsim$80% of that of the primary Roche lobe. The asymmetry of the light curve at ingress and egress of the disk eclipse suggests that the disk contains a bright region trailing the orbital motion, reminiscent of the hot spots observed in cataclysmic variables.

119.008 **Photoelectric photometry of AR Pav during the primary eclipse of 1980.**
J. W. Menzies, I. M. Coulson, J. A. R. Caldwell, P. M. Corben.
Mon. Not. R. Astron. Soc., Vol. 200, 463 - 471 (1982).

The authors present photoelectric $UBV(RI)_{KC}$ photometry of AR Pav obtained during the primary eclipse in 1980. The eclipse behaviour is similar to that observed by previous investigators. From the *VRI* photometry it is found that the secondary is an M3 giant, which is probably variable. A simple interpretation of the photometric colours of the primary object suggests that it comprises a cool and a hot component which radiate like blackbodies with temperatures of about 5000 K and at least $3 \cdot 10^4$ K, respectively.

119.009 **Photographic photometry of variable stars based on the Tartu photoplate collection. II. Area of μ Persei.**
V. Harvig, L. Leis.
Publ. Tartu Astrofiz. Obs., Tom 48, 172 - 176 (1981).

The authors continue publishing the results of the investigations of variable stars from Tartu plates.

119.010 **B and V photometry of XZ Cephei.**
V. Harvig.
Publ. Tartu Astrofiz. Obs., Tom 48, 177 - 182 (1981).

119.011 **UV observations of the 1981 eclipse of 32 Cygni.**
R. E. Stencel, R. D. Chapman, Y. Kondo, R. F. Wing.
Advances in ultraviolet astronomy, (see 012.006), p. 497 - 500 (1982).

Preliminary results of an extensive set of high dispersion UV spectra of the supergiant eclipsing system 32 Cyg are detailed and contrasted with spectroscopic studies of other ζ Aur systems.

119.012 **Pre-eclipse ultraviolet spectra of Epsilon Aurigae.**
R. D. Chapman, Y. Kondo, R. E. Stencel.
Advances in ultraviolet astronomy, (see 012.006), p. 505 - 508 (1982).

119.013 **An ultraviolet investigation of the unusual eclipsing binary system FF Aqr.**
J. D. Dorren, E. F. Guinan, E. M. Sion.
Advances in ultraviolet astronomy, (see 012.006), p. 517 - 520 (1982).

119.014 **A progress report on the W Serpentis binaries, with a few words on ϵ Aurigae.**
M. J. Plavec.
Advances in ultraviolet astronomy, (see 012.006), p. 526 - 529 (1982).

The author's recent observations show that two well-known Algol systems, U Cephei and V356 Sagittarii, should be included among the W Serpentis stars, characterized by strong ultraviolet emission lines. The spectra of the W Ser stars are similar to those of the T Tauri stars. A model of W Serpentis, namely a B star embedded in a thick disk, may be relevant to other "exotic" eclipsing systems, possibly even to ϵ Aurigae. In connection with ϵ Aur, the case of BM Orionis is reviewed.

119.015 **A recent time of minimum for and atmospheric-eclipse effects in the ultraviolet spectrum of the Wolf-Rayet eclipsing binary V444 Cygni.**
J. A. Eaton, A. M. Cherepashchuk, Kh. F. Khaliullin.
Advances in ultraviolet astronomy, (see 012.006), p. 542 - 545 (1982).

119.016 **BR Muscae: a new early-type contact binary.**
J. J. Clariá, E. Lapasset.
Astron. Astrophys., Suppl. Ser., Vol. 50, 13 - 17 (1982).

The eclipsing binary system BR Muscae was observed photoelectrically in 1980 - 1982. Four primary and two secondary eclipse curves are well defined, and eighteen epochs of minimum light and an updated ephemeris are presented. Although the coverage of the light curves has not been completed, it is evident that they are of the W UMa-type with typical features of contact binary systems. Since the observed $(B-V)$ and $(U-B)$ colours are consistent with a reddened main sequence star of spectral type B3, the authors conclude that BR Mus is a new member of the so-called "early-type contact binaries".

119.017 **RZ Ophiuchi. Visuelle Beobachtungen 1978 - 1982.**
R. Germann.
Orion, 40. Jahrg., 153 - 154 (1982).

119.018 **Transient photometric disturbances in Algol-like binaries.**
E. C. Olson.
Astrophys. J., Vol. 259, 702 - 718 (1982).

Primary eclipses of 16 short-period Algols have been searched for 3 years for photometric disturbances related to episodic mass transfer. Of 108 eclipses observed with five-color intermediate-band photometry, 18% showed such disturbances. Equatorial bulges, streams, and hot spots can explain many of these observations. Abrupt period changes were found in several systems, but their relation to transient mass transfer is unclear.

119.019 **The variable He 10830 Å line of Algol.**
H. Zirin, M. A. Liggett.
Astrophys. J., Vol. 259, 719 - 722, plates 14, 15 (1982).

Spectra of several eclipses of Algol in the range 10500–11000 Å, where the line contribution of Algol B is important, are presented. Strong unshifted λ10830 (2000 mÅ) absorption peaks at primary minimum but disappears between phases 0.3 and 0.7. At minimum the line must primarily arise in Algol B, but the presence of λ10830 absorption just outside eclipse, when the contribution to the total light of Algol B is small, must be due to excitation of He in the atmosphere of the primary by X-ray irradiation from Algol B.

119.020 **The eclipsing variable BS Draconis.**
C. Cristescu, G. Oprescu, M. D. Suran, D. G. Chis, I. Todoran, V. Pop.
Acta Astron., Vol. 31, 505 - 513 (1981).

Photoelectric light-curves of the eclipsing binary BS Draconis, observed from 1972 to 1974 have been analysed by means of Horàk's and Wood's models in order to obtain geometric, physical and photometric elements.

119.021 **Multicolor photometry of the eclipsing binary system VV Orionis.**
C. R. Chambliss, K.-C. Leung.
Astrophys. J., Suppl. Ser., Vol. 49, 531 - 544 (1982).

VV Orionis is an eclipsing binary consisting of two B-type components with a period of $1^d.485378$. Over 4200 observations of this system were obtained in *UBV* and *uvby* during the 1975 - 1979 seasons. All seven light curves were analyzed simultaneously using the Wilson-Devinney method. The larger component appears to be normal for a star of spectral type B1 V and is found to lie fairly close to the ZAMS. The secondary component, however, appears to be unusually small for its mass.

119.022 ***UBVR* and H-alpha photometry of RS CVn stars.**
M. Zeilik, R. Elston, G. Henson, P. Schmolke, P. Smith.
Bull. American Astron. Soc., Vol. 14, 613 (1982). – Abstract.

119.023 **Synthetic light curves of four short period Algol's.**
J. B. Rafert.
Bull. American Astron. Soc., Vol. 14, 613 (1982). – Abstract.

119.024 **Maximum entropy mapping of accretion disks in eclipsing cataclysmic binaries.**
K. Horne, R. Stiening.
Bull. American Astron. Soc., Vol. 14, 613 - 614 (1982). Abstract.

119.025 **Absolute dimensions and masses of CW CMa.**
C. H. Lacy.
Bull. American Astron. Soc., Vol. 14, 634 (1982). – Abstract.

119.026 **On the machine computation of the orbits of eclipsing binaries.**
D. M. Popper.
Rev. Mexicana Astron. Astrofis., Vol. 6, (see 012.017), 99 - 102 (1981).

The use of computer programs in the solution of light curves of eclipsing binaries marks a great advance. Some desirable practices in the use of computer programs are discussed briefly.

119.027 **Preliminary report on the IUE spectra of μ^1 Scorpii.**
J. Sahade, L. G. García.
Rev. Mexicana Astron. Astrofis., Vol. 6, (see 012.017), 205 (1981).

This paper is a preliminary report on an investigation of IUE ultraviolet spectra of the eclipsing variable μ^1 Scorpii, where the authors report on the type of lines that are observed at conjunction. The resonance lines of Si IV display a fine structure.

119.028 **Ground-based and IUE spectral observations of AU Monocerotis.**
J. Sahade, O. E. Ferrer.
Rev. Mexicana Astron. Astrofis., Vol. 6, (see 012.017), 207 (1981). – Abstract.

119.029 **Infrared photometry of Beta Lyrae: 1977 - 1982.**
M. Zeilik, P. Heckert, G. Henson, P. Smith.
Astron. J., Vol. 87, 1304 - 1309 (1982).

The authors have observed β Lyrae from 1.2 to 4.6 microns during the period from 1977 to 1982 to acquire reasonably complete light curves with much smaller errors than previous results. The light curves clearly show how the depth of the primary eclipse decreases relative to the secondary at longer wavelengths. A comparison to Wilson's thick-disk model (fitted to visual and UV data) is close for the eclipse depths from 1.2 to 3.5 μm, but not 4.6 μm. At all wavelengths, the observed shapes of the bottom of the eclipses have a sharper turnaround than that of the model.

119.030 **Photometric elements of the eclipsing binary AZ Vir.**
G. Giuricin, F. Mardirossian, M. Mezzetti.
Astron. Nachr., Band 303, 215 - 216 (1982).

Using Wood's (1972) model the authors have analyzed Meinunger's (1977) *B* and *V* photoelectric light curves – as yet unsolved – of the W UMa-type eclipsing binary AZ Vir, for which no photometric elements have as yet been determined. The results describe AZ Vir as a contact pair in which the star eclipsed at the deeper minimum is smaller than its companion. AZ Vir appears to conform to the general properties of the W-type subclass of W UMa systems.

119.031 **AN And: a detached eclipsing binary system with an Am primary member.**
G. Giuricin, F. Mardirossian, M. Mezzetti.
Astron. Astrophys., Vol. 114, 74 - 76 (1982).

Using Wood's (1972) model the authors have analyzed Bakos and Tremko's (1981) *V* lightcurve of the single-lined spectrum eclipsing binary AN And, whose nature is as yet very poorly known. Besides, absolute elements have been tentatively estimated assuming for the primary a mass of 1.8 $M_\odot$. In contrast with earlier opinions, AN And turns out to be a distinctly detached binary; it consists of an A7m primary accompanied by a considerably fainter and smaller early *G* unevolved star.

119.032 **Photometric solution for the eclipsing binary TT Aurigae.**
K. D. Abhyankar, N. Rajasekhar Rao.
Bull. Astron. Soc. India, Vol. 10, 27 (1982). – Abstract.

119.033 **The distortion wave and its period in the short period RS CVn type eclipsing binary UV Piscium.**
P. Vivekananda Rao, M. B. K. Sarma.
Bull. Astron. Soc. India, Vol. 10, 33 (1982). – Abstract.

119.034 **HD 8152 – a new W Ursae Majoris type eclipsing binary.**
M. B. K. Sarma, K. R. Radhakrishnan.
Bull. Astron. Soc. India, Vol. 10, 34 (1982). – Abstract.

119.035 **Analysis of TT Hydrae by Kopal's frequency domain method.**
J. Koul, K. D. Abhyankar.
Bull. Astron. Soc. India, Vol. 10, 34 (1982). – Abstract.

119.036 **A spot model for VW Cephei.**
A. Yamasaki.
Astrophys. Space Sci., Vol. 85, 43 - 58 (1982).

An attempt is made to interpret the variations of the light curve of VW Cephei in terms of a spot model. A light curve which is (hopefully) free from starspot effects is constructed from Kwee's (1966a) observations, and is analysed by the light-curve synthesis method. Then, Kwee's individual light curves are further analysed by assuming that starspots exist on the surface of the primary component. Properties of starspots are briefly discussed.

119.037 **Characteristics of the Wolf-Rayet star in the eclipsing binary system CQ Cep.**
N. A. Lipunova, A. M. Cherepashchuk.
Astron. Zh., Tom 59, 944 - 954 (1982). In Russian. English translation in Soviet Astron., Vol. 26, No. 5.

From the *U, V, R* light curves analysis of the close eclipsing binary system CQ Cep the inclination angle has been obtained, $i = 65° \pm 1°$. The common envelope (non-occulted) surrounding the system CQ Cep is found to have the luminosities in the *U, V, R* bands $L_3(U) \cong 0; L_3(V) = 0.10$; $L_3(R) = 0.29$, in units of the total luminosity of the system outside the eclipses. The radius of the WN 7-core is $r_0 = (11 \pm 1)\,R_\odot$, the brightness ($T_b$) and the color ($T_c$) temperatures are: $T_b \gtrsim 50\,000$ K, $T_c \sim 50\,000$ K. The mass of the WN 7-star is $\cong 31\,M_\odot$.

119.038 **UV resonance line formation in Zeta Aur/VV Cep systems.**
K. Hempe, D. Reimers.
Third European IUE Conference, (see 012.029), p. 339 - 341 (1982).

High resolution IUE spectra of Zeta Aur stars show strong P Cygni profiles, which vary with binary phases. Theoretical line profiles calculated on the basis of the escape probability method in non-spherical geometry qualitatively explain the observed lines.

119.039 **Evolution of the subdwarf-O eclipsing binary LB 3459.**
R. P. Kudritzki, K. P. Simon, A. E. Lynas-Gray, D. Kilkenny, P. W. Hill.
Third European IUE Conference, (see 012.029), p. 343 - 348 (1982).

LB 3459 is a sdO star, lying along the line-of-sight to the LMC, discovered to be an eclipsing binary. Ground-based (AAT) and IUE spectra are used in a non-LTE analysis of the primary. LB 3459 appears to be the result of mass exchange during the first giant phase. The primary is thought to have a mass of 0.3 $M_\odot$ and consists of a hydrogen shell and degenerate helium core. The secondary mass is 0.04 $M_\odot$.

119.040 **Analysis of UV spectra of Beta Lyrae.**
H. Cugier.
Third European IUE Conference, (see 012.029), p. 363 - 367 (1982).

An analysis of IUE observations of Beta Lyrae is performed. For this purpose line profiles are calculated for spherically symmetric expanding envelopes. A comparison with observations shows that the mass loss rate is about $3 \times 10^{-6} M_\odot/y$.

119.041 **Absolute dimensions and masses of eclipsing binaries. III. CW Canis Majoris.**
C. H. Lacy.
Astrophys. J., Vol. 261, 612 - 616 (1982).

The eclipsing binary CW CMa has been found to be strongly double-lined on high signal-to-noise ratio digital spectral scans obtained with the Reticon and Digicon spectrometers of the 2.7 m telescope at McDonald Observatory. Radial velocities of high precision have been obtained for both stars. Spectroscopic orbits derived from these data have been combined with the photometric orbit of Williamon to give very accurate masses and radii. An analysis of the line profiles of both components indicates that they are synchronously rotating with the orbital period.

119.042 **A spectroscopic investigation of the eclipsing variable ER Vulpeculae.**
B. J. McLean.
Mon. Not. R. Astron. Soc., Vol. 201, 421 - 427 (1982).

Spectroscopic observations of the eclipsing variable ER Vul have been analysed using digital reduction techniques. The spectroscopic elements are derived from the radial velocity curves. A variation in the strength of the spectral lines of the primary component is noted which may be explained by the presence of circumstellar material.

119.043 **Evidence for accretion activity and obscured hot component stars in W Serpentis type binaries.**
A. Young, J. A. Snyder.
Astrophys. J., Vol. 262, 269 - 281 (1982).

Recent IUE observations of some peculiar eclipsing binaries revealed the presence of strong, high-excitation emission lines, and of excessive ultraviolet continuum. In an effort to clarify the nature of such sources, the authors have made accurate four-color (*u, v, b, y*) observations of five of those objects throughout their primary eclipses. They report significant and unusual variations of the $b-y$ and $u-b$ color indices which, together with the IUE observations, lend support to the suggestion made by Plavec and Koch that in some of these systems a hot star is obscured by an opaque accretion disk. The systems for which satisfactory data have been secured are W Ser, V367 Cyg, SX Cas, RX Cas, and RW Per.

119.044 **Is HD 164270 a long-period eclipsing binary?**
I. Lundström, B. Stenholm.
Wolf-Rayet stars: observations, physics, evolution, (see 012.042), p. 289 - 293 (1982).

119.045 **The Wolf-Rayet eclipsing binary HD 5980 in the Small Magellanic Cloud.**
J. Breysacher, A. F. J. Moffat, V. S. Niemelä.
Wolf-Rayet stars: observations, physics, evolution, (see 012.042), p. 317 - 320 (1982).

The Wolf-Rayet star HD 5980, which is probably associated with the bright H II region NGC 346 of the Small Magellanic Cloud, was found to be an eclipsing binary by Hoffmann, Stift and Moffat (1978). The behaviour of the light curve outside the eclipses shows that we are dealing with a rather complex binary system. An analysis of the spectroscopic data is presented.

119.046 **The massive eclipsing binary RY Scuti.**
G. Giuricin, F. Mardirossian.
Wolf-Rayet stars: observations, physics, evolution, (see 012.042), p. 383 - 386 (1982).

The authors have analyzed Ciatti et al's (1980) UBV lightcurves of the massive eclipsing binary RY Sct. The B0 primary, being considerably overluminous and oversized for its mass, is certainly appreciably evolved. That this star is nearly filling its Roche lobe strongly supports the idea that it is undergoing a phase of mass transfer (case B), through Roche lobe overflow, towards its more massive secondary companion. In view of its properties, RY Sct, which represents an uncommon evolutionary stage, is likely to evolve into a Wolf-Rayet binary system.

119.047 **LY Aurigae, NY Cephei, and the mass-luminosity anomaly within O type binaries.**
D. M. Popper.
Astrophys. J., Vol. 262, 641 - 649 (1982).

The luminosities of the components in a number of double-lined O type binaries appear more nearly equal than predicted from their mass ratios. Critical examination shows that this anomaly is best established in the contact binaries LY Aur, V382 Cyg, and 14 Cep, and may be a general property of contact binaries of type O. The published mass ratios of LY Aur and NY Cep require considerable revision, making the two components in each system more nearly equal in mass than given heretofore.

119.048 **Analysis of the *UBV* light curves of TT Hydrae by Kopal's frequency domain method.**
J. Koul, K. D. Abhyankar.
J. Astrophys. Astron., Vol. 3, 93 - 99 (1982).

The light curves of the totally eclipsing system TT Hya in *UBV* colours observed by Kulkarni and Abhyankar during 1973–77 have been analysed by Kopal's frequency domain method with slight modification. The undersize and the ultraviolet colour excess of the secondary component are confirmed.

119.049 **A new light curve analysis of the eclipsing binary SX Aur.**
G. Giuricin, F. Mardirossian, M. Mezzetti.
Astron. Nachr., Band 303, 311 - 312 (1982).

Using Wood's (1972) model, the authors have analysed Bondarenko's (1974) two-colour photoelectric observations – till now unsolved – of the eclipsing binary SX Aur. Their photometric solutions turn out to be in substantial agreement with those derived by Chambliss and Leung (1979); they confirm that SX Aur is a post-mass exchange (case A) semi-detached system.

119.050 **A study of ultraviolet spectra of ζ Aur/VV Cep systems. I. Resonance line formation.**
K. Hempe.
Astron. Astrophys., Vol. 115, 133 - 137 (1982).

The author develops a line transfer scheme for UV resonance lines in ζ Aur systems on the basis of the escape probability method. His results show a strong dependence of line profile shapes on binary phases. Calculated profiles qualitatively explain resonance lines observed in the IUE spectra of ζ Aur and 32 Cyg.

119.051 **A photometric study of the eclipsing binary V 889 Aql: an example of relativistic apsidal motion.**
A. Giménez, F. Scaltriti.
Astron. Astrophys., Vol. 115, 321 - 326 (1982).

In this paper the authors have analyzed the existing photoelectric observations of the long-period eclipsing binary V 889 Aql in order to obtain the corresponding photometric elements and the period of apsidal motion. The results indicate that the two components of the system are very similar and the orbital inclination is close to 90°. Absolute dimensions were also estimated and the stars have been found to be almost unevolved. The movement of the periastron seems definitely established though at a very slow rate. The period of apsidal motion was found to be close to 2×10^4 yr. The relativistic term of the observed periastron advance appears to be the main contribution and is in a perfect agreement with the theoretical value.

119.052 **UV time-dependent emission in SY Muscae.**
A. G. Michalitsianos, M. Kafatos.
The nature of symbiotic stars, (see 012.045), p. 191 - 194 (1982).

UV spectra of SY Mus = HD 10036 acquired with the IUE on 20 September 1980 and 11 June 1981 indicate a substantial enhancement of UV emission over a nine month period. An interpretation is consistent with the authors' model for an eclipsing system.

119.053 **AR Pavonis: the Rosetta stone of the symbiotics.**
M. H. Slovak.
The nature of symbiotic stars, (see 012.045), p. 195 - 200 (1982).

The binary model proposed for AR Pav by Thackeray and Hutchings is reviewed in light of recent ultraviolet, infrared, and radio observations.

119.054 **On the internal structure of main-sequence stars.**
A. Giménez, J. M. García-Pelayo.
Binary and multiple stars as tracers of stellar evolution, (see 012.046), p. 37 - 46 (1982).

The authors present the main results of a study of the observed internal structure constants, k_2, for a wide set of eclipsing binaries. From the analysis of the variations in relative positions of the eclipses and the comparison with different theoretical models, they could deduce that the discrepancy, previously reported by several authors between theory and observations, is no longer supported. Moreover, a strong correlation has been found between the evolution of the parameter k_2 and the gravity at the surface of the star.

119.055 **A study of the orbital periods of 22 eclipsing binaries.**
T. Panchatsaram, K. D. Abhyankar.
Binary and multiple stars as tracers of stellar evolution, (see 012.046), p. 47 - 60 (1982).

Investigation of the variability of the orbital periods of eclipsing binaries is important not only from the evolutionary point of view but also for detecting additional components in them. The authors describe the study of 22 systems carried out by them at Hyderabad.

119.056 **The evolutionary state of Zeta Aurigae.**
R. D. Chapman.
Binary and multiple stars as tracers of stellar evolution, (see 012.046), p. 153 - 156 (1982).

Ultraviolet studies, originally undertaken to ascertain the state of the atmosphere of the K-supergiant component of the Zeta Aurigae system, have been sidetracked by the discovery of significant accretion effects. An analysis of the phase dependence of the profiles of resonance lines in Mg II and C IV has led to a qualitative model of the wind flow from the K star. At the position of the B star the flow velocity is about 100 km/sec and the density is 3×10^{-6} cm^{-3}, leading to a mass loss rate of 2×10^{-8} solar masses per year. This wind interacts with the B star in a shock, which is described, leading to accretion on the B star at a rate of 4×10^{-10} solar masses per year.

119.057 **Structure and evolution of SV Centauri.**
H. Drechsel, J. Rahe, W. Wargau, B. Wolf.
Binary and multiple stars as tracers of stellar evolution, (see 012.046), p. 205 - 216 (1982).

Optical photometric and spectroscopic observations of the eclipsing binary SV Cen were used to derive orbital elements and absolute dimensions of the peculiar early-type contact system. UV spectroscopic observations yielded the detection of mass loss forming an expanding circumbinary envelope. SV Cen exhibits a non-uniform rate of period decrease. The period changes are explained in terms of loss of angular momentum carried away by the ejected matter. Short-term fluctuations of the stellar wind were found. A model of the system is derived, and the evolutionary stage and interaction processes are discussed.

119.058 **The very unusual ultraviolet spectra of R Arae.**
Y. Kondo, G. E. McCluskey.
Binary and multiple stars as tracers of stellar evolution, (see 012.046), p. 321 - 326 (1982).

The high resolution ultraviolet spectra of the 4.4-day period binary R Arae, observed in 1980 with the International Ultraviolet Explorer, show that its continuum flux level varied outside of eclipse by more than a factor of two in ten days, and by over 50 percent within a same orbital cycle. The resonance lines of Mg II and Si IV exhibited shortward-shifted absorption components near phase 0.4, indicating the presence of a gas stream toward the observer at a velocity of some -450 to -500 km s^{-1}. The authors' observations show it to be an X-ray source.

119.059 **The short-period non-contact binary systems UU Lyn and GR Tau.**
A. Yamasaki, A. Okazaki, M. Kitamura.
Binary and multiple stars as tracers of stellar evolution, (see 012.046), p. 345 - 349 (1982).

Two binary systems UU Lyn ($P = 0\overset{d}{.}468$) and GR Tau ($P = 0\overset{d}{.}430$) are discussed on the basis of BV photoelectric observations and spectroscopic observations which were made between 1980 and 1981 at the Tokyo Astronomical Observatory. The component stars in both systems are found to be still inside their Roche lobes but very close to them.

119.060 **The subdwarf eclipsing binary LB3459 (AA Dor).**
R. W. Hilditch, G. Hill, D. Kilkenny.
Binary and multiple stars as tracers of stellar evolution, (see 012.046), p. 455 - 456 (1982).

119.061 **Secondaries of eclipsing binaries. IV. The triple system Lambda Tauri.**
F. C. Fekel, Jr., J. Tomkin.
Astrophys. J., Vol. 263, 289 - 301 (1982).

The authors have measured low-noise Reticon spectra to determine the velocity curve of the primary of this early-type semidetached system. The velocity curve of the secondary is determined for the first time. The 33 day periodicity in the residuals, discovered in previous investigations, is confirmed and is present in the secondary velocities as well as those of the primary. This means that it can be unambiguously ascribed to orbital motion around a third body. The mass of the third body is 0.7 ± 0.2 $M_\odot$; it is most probably a K dwarf.

119.062 **HD 187 459 – a new supergiant eclipsing binary?**
D. Chochol, P. Mayer.
Ejection and accretion of matter in binary systems, (see 012.051), p. 59 (1982). – Abstract. – See also Abstr. 30.119.052.

119.063 **Peculiar light-curve change in SV Cam.**
L. Patkós.
Ejection and accretion of matter in binary systems, (see 012.051), p. 61 - 66 (1982).

Intensive photometric observations of the short-period eclipsing binary SV Cam have been carried out since 1973. In addition to RS CVn-like migration waves a sudden brightness increase – caused by gas streams – is presented.

119.064 **Photometry of minima and period changes of the binary system TX UMa.**
J. M. Kreiner, J. Tremko.
Ejection and accretion of matter in binary systems, (see 012.051), p. 67 (1982). – Abstract. – See also Abstr. 28.119.087.

119.065 **Importance of *UBV* observations of AU Mon during January - February 1983.**
L. Lorenzi.
Inf. Bull. Variable Stars, No. 2162 (1982).

119.066 ***UBV* photometry of the eclipsing binary system SV Tauri.**
C. R. Chambliss.
Inf. Bull. Variable Stars, No. 2164, 4 pp. (1982).

119.067 **The primary eclipse of RW Monocerotis in five colors.**
C. R. Chambliss.
Inf. Bull. Variable Stars, No. 2165, 4 pp. (1982).

119.068 ***uvby* photometry of the Algol variable HD 224113.**
C. Sterken.
Inf. Bull. Variable Stars, No. 2166, 2 pp., with a correction on No. 2187 (1982).

119.069 **A 2.2 μm light curve of the eclipsing binary VW Cephei.**
M. Lunel, R. Garnier, J. Bergeat, F. van't Veer.
Inf. Bull. Variable Stars, No. 2176, 3 pp. (1982).

119.070 **On the period of HD 200925.**
S. K. Gupta, T. D. Padalia.
Inf. Bull. Variable Stars, No. 2194, 5 pp. (1982).

119.071 **Photoelectric epochs of minimum light, UZ Puppis.**
B. B. Bookmyer.
Inf. Bull. Variable Stars, No. 2212 (1982).

119.072 **Photometric observations of the secondary minimum of 1 Per.**
E. Poretti.
Inf. Bull. Variable Stars, No. 2239, 3 pp. (1982).

119.073 **Secular period changes in ten eclipsing binaries.**
T. Panchatsaram, K. D. Abhyankar.
Bull. Astron. Soc. India, Vol. 10, 111 - 136 (1982).

The photoelectric times of minima, available for the eclipsing binaries KO Aql, V566 Oph, AH Vir, AG Vir, 44i Boo B, RT And, SV Cam, TV Cas, AR Lac and U Peg have been analysed to study their orbital periods. KO Aql, V566 Oph, AH Vir, AG Vir and 44i Boo B show increase in their periods while RT And, SV Cam, TV Cas, AR Lac and U Peg exhibit decrease of period. Quadratic ephemeris for the primary times of minima for all the systems have been derived. The cause for the apparent secular change is conjectured to be the light-time effect, as no other mechanism seems appropriate to interpret the observed variations without difficulty.

119.074 **RZ Ophiuchi. Visuelle Beobachtungen 1978 - 1982.**
R. Germann.
Sterne Weltraum, 21. Jahrg., 476 - 477 (1982).

119.075 **The eclipsing binary HR 9049.**
J. F. Dean, J. D. Laing.
Mon. Notes Astron. Soc. South. Africa, Vol. 40, 48 - 53 (1981).

119.076 **RZ Oph-minimum 1981.**
M. Grünanger, W. Herzner.
BAV Rundbrief, 31. Jahrg., 56 - 58 (1982).

119.077 **Beta Aurigae. Beobachtungen 1982.**
M. Fernandes.
BAV Rundbrief, 31. Jahrg., 59 - 60 (1982).

119.078 **Photoelektrische Beobachtungen 1982 an V 695 (31) Cyg und Zeta Aur.**
M. Fernandes.
BAV Rundbrief, 31. Jahrg.,75 - 79 (1982).

119.079 **Periodenänderungen bei XY Leo.**
M. Fernandes.
BAV Rundbrief, 31. Jahrg., 80 (1982).

119.080 **94th - 96th list of minima of eclipsing binaries.**
Compiled by G. Boistel, R. Diethelm, D. P. Elias, R. Germann, K. Locher, G. Mavrofridis, D. Mourikis, P. Ralincourt, H. Peter, T. Schildknecht, G. Stefanopoulos, N. Stoikidis, R. Boninsegna, M. Kohl, J.-F. Le Borgne, R. Leyman, P. Louis, M. Andrakakou.
BBSAG Bull., No. 61, p. 1 - 4; No. 62, p. 1 - 5, 7; No. 63, p. 1 - 4 (1982).

119.081 **Z Herculis: a photoelectrically observed secondary contact.**
R. Diethelm.
BBSAG Bull., No. 61, p. 4 (1982).

119.082 **Tsvetkova's star in Cygnus: detection of the period.**
K. Locher.
BBSAG Bull., No. 61, p. 5 (1982).

119.083 **Normal minima of long period eclipsing binaries.**
A. D. Mallama.
BBSAG Bull., No. 61, p. 6 (1982).

119.084 **NSV 817 Cephei: detection of the period.**
K. Locher.
BBSAG Bull., No. 63, p. 5 - 6 (1982).

119.085 **Photoelectric observations of WX Eridani.**
K. D. Abhyankar, M. B. K. Sarma, A. G. Kulkarni.
Contrib. Nizamiah Japal-Rangapur Obs., No. 5, 15 pp. (1978).

119.086 **Photoelectric observations of EU Hydrae.**
A. G. Kulkarni, C. Raghavender Rao, R. Swaminathan.
Contrib. Nizamiah Japal-Rangapur Obs., No. 6, 8 pp. (1978).

119.087 **Further photoelectric observations of EU Hydrae.**
A. G. Kulkarni.
Contrib. Nizamiah Japal-Rangapur Obs., No. 7, 16 pp. (1978).

119.088 **Photoelectric observations of TT Aurigae.**
A. G. Kulkarni, B. Lokanadham.
Contrib. Nizamiah Japal-Rangapur Obs., No. 8, 37 pp. (1978).

119.089 **Photoelectric observations of TT Hydrae.**
A. G. Kulkarni, K. D. Abhyankar.
Contrib. Nizamiah Japal-Rangapur Obs., No. 9, 49 pp. (1978).

119.090 **Photoelectric observations of the eclipsing binary RU Eridani.**
M. B. K. Sarma, N. B. Sanwal.
Contrib. Nizamiah Japal-Rangapur Obs., No. 12, 31 pp. (1979).

119.091 **Photoelectric observations of TY Pyxidis.**
P. Vivekananda Rao, M. B. K. Sarma.
Contrib. Nizamiah Japal-Rangapur Obs., No. 13, 47 pp. (1981).

119.092 **Photoelectric observations of UV Pisces.**
P. Vivekananda Rao, M. B. K. Sarma.
Contrib. Nizamiah Japal-Rangapur Obs., No. 14, 62 pp. (1981).

119.093 **On a method of construction of confidence intervals for geometrical parameters of classical eclipsing binaries.**
A. V. Goncharskij, V. V. Stepanov.
Astron. Tsirk., No. 1175, p. 6 - 8 (1981). In Russian.

119.094 **RX Cassiopeiae and SX Cassiopeiae: the case of the dubious twins.**
M. J. Plavec, J. L. Weiland, J. J. Dobias, R. H. Koch.
Be Star Newsl., No. 4, p. 7 (1981).

119.095 **SX Cassiopeiae: spectral energy distribution and a model.**
M. J. Plavec, J. J. Dobias, J. L. Weiland, R. H. Koch.
Be Star Newsl., No. 4, p. 8 (1981).

119.096 **Differential photometry of the metallic-line eclipsing binary AN Andromedae.**
M. Kitamura, T.-h. Kim, Y. Nakamura, M. Nakagiri.
Tokyo Astron. Bull., Second Ser., No. 266, p. 3021 - 3038 (1982).

119.097 **Noncircular orbit of TU Cam.**
J. Žižňovský.
Inf. Bull. Variable Stars, No. 2251, 2 pp. (1982).

119.098 **X-ray eclipses in AR Lac: preliminary results.**
D. M. Gibson, F. M. Walter.
Smithsonian Astrophys. Obs., Spec. Rep. 392, Vol. II, (see 012.065), p. 135 - 138 (1982).

Observations of the nearby eclipsing RS CVn binary

AR Lac using the Imaging Proportional Counter on the Einstein X-ray Satellite have revealed both primary and secondary minima. The following preliminary conclusions appear to be indicated: (1) The surface flux in soft X-rays from both stars is approximately the same. (2) The soft X-ray corona of the G2 IV-V primary is small. (3) The soft X-ray corona of the K0 IV secondary is large and possibly asymmetric. (4) No individual small scale structures (e.g. loops) contain more than 10% of the X-ray flux. (5) Coronal temperatures apparently increase (while densities decrease) with distance from the stellar surfaces.

119.099 **Results of photoelectric observations of the eclipsing system TY Coronae Austrinae.**
V. I. Kardopolov, V. V. Sakhanenok, G. K. Filip'ev.
Perem. Zvezdy, Tom 21, 589 - 592 (1981). In Russian.

The results of *B*, *V* photoelectric observations of TY CrA in the CrA T1 association are presented. The observations show that it is an eclipsing variable with elements JD hel Min I = 2442954.29 + $2^{d}.888777 \cdot E$. No irregular variations were found.

119.100 **Photoelectric observations of RT Andromedae in 1973.**
V. P. Murnikova, N. D. Lemeshchenko.
Perem. Zvezdy, Tom 21, 593 - 598 (1981). In Russian.

The results of 307 V and 312 B photoelectric observations of the eclipsing variable star RT And are given. 4 moments of primary minimum have been determined. The observations show instabilities of the light curve out of eclipse and variability of minimum depth.

119.101 **Photoelectric observations of CM Lacertae in 1974.**
V. P. Murnikova, L. V. Makarchikova.
Perem. Zvezdy, Tom 21, 599 - 606 (1981). In Russian.

The results of photoelectric observations of the eclipsing system CM Lac are given: 690 magnitude values in the B system and 675 of those in the V system. Two moments of the primary light minimum have been determined. The observations show instability of the system light out of eclipse.

119.102 **Photometric and spectroscopic observations of HD 211853 – a Wolf-Rayet eclipsing binary.**
K. Annuk, T. Nugis.
Tartu Astrofüüs. Obs. Publ., Köide 49, 84 - 96 (1982).

The authors publish the results of their photometric and spectroscopic observations of HD 211853 made at the Tartu Observatory in 1979. They also try to find out if there exist any substantial differences in the physical and the chemical nature between the WN6 component of HD 211853 and single WN6 stars.

119.103 **Photoelectric photometry of Z Herculis.**
S. Evren, A. Y. Ertan, Z. Tunca, C. Ibanoğlu, M. Kurutaç, O. Tümer.
Astrophys. Space Sci., Vol. 87, 51 - 59 (1982).

Two-colour light curves of Z Her were obtained observing the system from 1978 to 1981. The wave-like distortion at outside eclipses were derived. The amplitudes of the wave increase and its minimum migrates toward the decreasing orbital phase with a period of about 1.4 yr. An attempt was made to represent the wave-like distortions with a truncated Fourier series. The representation was found to be unsatisfactory since the distribution of the present observations is not regular but forms separate groups along the time axis. Therefore, further observations to cover the phases evenly in between the minima are needed.

119.104 **Theory of the light curves of eclipsing systems with oscillating components. I.**
Z. Kopal.
Astrophys. Space Sci., Vol. 87, 149 - 191 (1982).

The aim of the present investigation is to develop a theory of the light changes of the eclipsing systems in which one (or both) components oscillate(s) freely in period(s) which may (though need not) be synchronized with that of their orbit. The light curves exhibited by systems with oscillating components will, in general, be asymmetric; and these oscillations may, in fact, be the cause of asymmetries observed in many eclipsing systems with evolved components. The present paper is concerned with an investigation of photometric phenomena arising from oscillations of the eclipsing components of close binary systems – cutting off (partly or wholly) the star which undergoes eclipse.

119.105 **The light curve variations in AR Lacertae.**
A. Y. Ertan, O. Tümer, Z. Tunca, C. Ibanoğlu, M. Kurutaç, S. Evren.
Astrophys. Space Sci., Vol. 87, 255 - 268 (1982).

The eclipsing binary AR Lac has been observed in two colours, *B* and *V*, in 1980 and 1981. The new times of minima obtained during the observations indicate that the period of the system is still continuing its decrease. The wave-like distortions outside eclipses have two maxima and are very similar in shape in successive two years.

119.106 **Photoelectric observations of the eclipsing binary U Pegasi.**
P. Rovithis, H. Rovithis-Livaniou.
Astrophys. Space Sci., Vol. 87, 287 - 295 (1982).

Photoelectric observations in *B* and *V* of the eclipsing binary U Pegasi, made during 1980 with the 48 in. reflector of the National Observatory of Athens, are presented and discussed. The light curve seems to have remained constant since 1958 while the period of the system continues to decrease.

119.107 **Spectral observations of Beta Lyrae in the visual region during June 1976. I. Absorption spectrum.**
A. N. Dadaev.
Izv. Glav. Astron. Obs. Pulkovo, Astrofiz. Astrometr., No. 200, p. 44 - 62 (1982). In Russian.

The observations of Beta Lyrae discussed here were made with the 50-inch reflector of the Crimean Astrophysical Observatory in June, 1976. The spectral region λ 5280 - 6870 Å was searched. For the absorption spectrum 666 lines of various elements were identified as components of the primary, and 238 lines were attributed to the secondary spectrum. Lines of the neutral neon and fluorine are registered in a considerable number. Fluorine has been detected for the first time in the atmosphere of Beta Lyrae.

119.108 **Ephemerides of eclipsing binaries for the year 1983.**
M. Kurpińska-Winiarska, K. Otmianowska-Mazur.
Rocznik Astronomiczny Obserwatorium Krakowskiego 1983, (see 047.040), p. 1 - 106 (1982).

119.109 **Ephemerides of eclipsing binaries among cataclysmic variables for the year 1983.**
J. M. Kreiner.
Rocznik Astronomiczny Obserwatorium Krakowskiego 1983, (see 047.040), p. 107 - 111 (1982).

119.110 **A photometric study of the short-period eclipsing binary BW Eridani.**
D. Baade, H. W. Duerbeck, M. T. Karimie, A. Yamasaki.
Astrophys. Space Sci., Vol. 88, 197 - 204 (1982).

UBV observations of the short-period eclipsing binary BW Eri are reported and analyzed. The photometric parameters are determined. BW Eri is a detached or semidetached system.

119.111 **Occultation of ϵ Aurigae.**
F. Vaclík.
Říše hvězd, Vol. 63, 118 - 119 (1982). In Czech.

119.112 **ϵ Aurigae.**
D. Chochol.
Kozmos, Vol. 13, 128 (1982). In Slovak.

119.113 **RT CrB: wave-like distortion and photometric solution.**
D.-s. Zhai, R.-x. Zhang, J.-t. Zhang.
Acta Astron. Sinica, Vol. 23, 125 - 134 (1982). In Chinese.

119.114 **Photoelectric light curves in two colours and the period changes of the eclipsing binary XY Leo.**
Q. Liu, J. Zhang.
Acta Astrophys. Sinica, Vol. 2, 131 - 137 (1982). In Chinese.
A total of 481 yellow and 481 blue observations were obtained from 16 to 21 January, 1980, using the 60 cm telescope and the photometer at Beijing Observatory. The comparison star is BD + 18°2306. The observations are corrected for differential extinctions. All the measurements are transformed to the standard UBV system. The present light curves are compared with those of R. H. Koch (1956).

119.115 **The eclipsing system AZ Cassiopeiae during the eclipse of 1975.**
F. Li, S. Jiang, L. Shen.
Acta Astrophys. Sinica, Vol. 2, 138 - 143 (1982). In Chinese.

119.116 **Photoelectric observations of the Algol system AD Boo and its photometric solution.**
D. Zhai, R. Zhang, J. Zhang.
Acta Astrophys. Sinica, Vol. 2, 224 - 232 (1982). In Chinese.
Approximately 1518 photoelectric BV observations of the eclipsing binary AD Boo were obtained at Beijing Observatory from April to June 1981. The observations show that (1) the system is Algol-type, (2) its period is about 2.0688112 days, twice of that given by others, (3) the light curve with flat maximum varies about $0^m_.64$ and $0^m_.68$ at primary, $0^m_.40$ and $0^m_.37$ at secondary minimum in yellow and blue, respectively, (4) there are obviously unexplained brightness fluctuations of $0^m_.05-0^m_.10$ amplitude outside the eclipses. A consistent set of geometrical elements fits both eclipses in the two colors by using the Russell-Merrill method. The ratio of radii and the orbital inclination are fixed near 0.79 and 87°, respectively.

119.117 **AF Geminorum: an eclipsing triple star.**
C. R. Chambliss.
Publ. Astron. Soc. Pacific, Vol. 94, 926 - 933 (1982).
AF Gem is an Algol-type eclipsing binary with a period of $1^d_.2435$. The period is found to have decreased slightly over the years that this object has been under observation. Approximately 800 observations in *UBV* have been analyzed by the Wood method, and orbital elements have been obtained. AF Gem has a faint visual companion situated about 2″ from the eclipsing pair. It is most probable that this star is a member of the system.

119.118 **Spectrographic material for RS Canum Venaticorum and Algol binaries.**
D. M. Popper.
Publ. Astron. Soc. Pacific, Vol. 94, 945 - 947 (1982).
A synopsis is given of unpublished spectrographic material on 22 subgiant eclipsing binaries with Ca II emission lines and nine semidetached binaries with B- or A-type primaries.

119.119 **Notes on ten southern hemisphere eclipsing binaries.**
A. D. Mallama.
Publ. Astron. Soc. Pacific, Vol. 94, 987 - 988 (1982).
UBV photometry of ten prominent southern hemisphere eclipsing binaries has been obtained during their minimum light phases. Some revised periods are given.

119.120 **Photoelectric observations of CN Andromedae.**
M. A. Seeds, D. K. Abernethy.
Publ. Astron. Soc. Pacific, Vol. 94, 1001 - 1002 (1982).
Recent observations of the W Ursae Majoris system CN And have suggested a slight decrease in the period since the 1950s. However, new observations combined with all published timings of primary minima show no evidence for a change in period for the last three decades. New light elements are derived. The light curve suggests that the eclipses are partial and that the light variation repeats well from cycle to cycle.

A catalogue of parameters for eclipsing binaries.
See Abstr. 002.081.

Les étoiles variables.
See Abstr. 003.095.

Computer analysis of light curves of eclipsing binaries. Part 3. Description of the program complex of the direct method for the Nairi K computer.
See Abstr. 021.053.

A method for evaluating the epoch of minimum of an eclipsing variable: the sliding integrations.
See Abstr. 031.504.

La mesure des étoiles doubles par interférométrie "speckle".
See Abstr. 031.527.

Possibilities of speckle interferometry for observations of eclipsing variable systems (computer modelling results).
See Abstr. 031.568.

Mass loss by Wolf-Rayet stars.
See Abstr. 064.058.

Stellar surface phenomena: the Hα profile in RS CVn stars.
See Abstr. 114.017.

The mid-ultraviolet spectrum of ϵ Aurigae.
See Abstr. 114.113.

Mass-radius diagram and Hertzsprung-Russell diagram.
See Abstr. 115.019.

Model fitting of the Algol binaries.
See Abstr. 117.029.

Continuum and emission lines in Beta Lyrae.
See Abstr. 117.030.

On the constancy of the light curve of CC Comae.
See Abstr. 117.063.

Properties of Wolf-Rayet stars in eclipsing binary systems.
See Abstr. 117.099.

The peculiar symbiotic object V1329 Cygni: single-star versus binary models.
See Abstr. 117.113.

Evolution of close binary stars: observational aspects.
See Abstr. 117.133.

Statistical properties of Algol-type systems.
See Abstr. 117.134.

Mass loss in Algol-type stars: implication on their evolutionary stage.
See Abstr. 117.135.

Statistical models for close binaries.
See Abstr. 117.136.

S Cnc as a case study of classical Algol evolution.
See Abstr. 117.138.

Does the cataclysmic binary Z Cha contain a black dwarf secondary?
See Abstr. 117.159.

Interpretation of the light curves for "realistic" eclipsing close binaries.
See Abstr. 117.164.

Circumstellar matter in the close binary SZ Camelopardalis.
See Abstr. 117.166.

Theory of the light curves of eclipsing systems with oscillating components. II.
See Abstr. 117.222.

Evolutionary trends in wide binaries.
See Abstr. 118.026.

Double stars: the closest and the most distant systems.
See Abstr. 118.029.

Changes in the UV spectrum of HD 4174.
See Abstr. 122.026.

CI Cygni since the 1980 eclipse.
See Abstr. 122.035.

De veelzijdige veranderlijke sterren.
See Abstr. 122.086.

Investigation of the pulsational peculiarities of the hotter component in the eclipsing binary system AB Cas.
See Abstr. 122.098.

Times of minima for southern hemisphere eclipsing binaries.
See Abstr. 123.013.

Photoelectric minima of eclipsing binaries.
See Abstr. 123.014.

Photoelectric minima of eclipsing variables.
See Abstr. 123.018.

Photographic observations of five eclipsing binaries.
See Abstr. 123.026.

Identification of the bright X-ray source 3A1148 + 719 with the cataclysmic variable YY Draconis.
See Abstr. 142.058.

An X-ray flare in the Hyades binary HD 27130.
See Abstr. 142.147.

Erratum

119.901 **Erratum: "Ground-based and *IUE* spectral observations of AU Monocerotis"** [Publ. Astron. Soc. Pacific, Vol. 94, 113 - 121 (1982)].
J. Sahade, O. E. Ferrer.
Publ. Astron. Soc. Pacific, Vol. 94, 872 (1982). – See Abstr. 31.119.047.

120 Spectroscopic Binaries

120.001 **HD 185151: a new active-chromosphere binary.**
B. W. Bopp, F. C. Fekel, Jr., D. S. Hall, G. W. Henry, P. V. Noah, J. Africano, M. S. Wilkerson, W. I. Beavers.
Astron. J., Vol. 87, 1035 - 1040 (1982).

The authors report spectroscopic and photometric observations of the K2III star HD 185151 which establish it as a chromospherically active binary with an orbital (and rotational) period of 40.1 days. The light curve is of special interest: it is effectively a double sine curve, and can be modeled by two starspots separated by almost exactly 180° in longitude.

120.002 **Radial velocity studies of spectroscopic binaries with the objective prism I: Improved orbital elements for binaries in and near NGC 2516.**
F. Gieseking, M. T. Karimie.
Astron. Astrophys., Suppl. Ser., Vol. 49, 497 - 502 (1982).

Thirteen new radial velocity observations of each of the spectroscopic binaries HD 65949, CPD−60°990, −59°953, HD 66656, 67084 and 67198 are presented. All available radial velocity data were used to calculate revised or improved spectroscopic orbital elements. Periodograms and radial velocity curves are shown. Short comments on each of the binaries are given.

120.003 **The ultraviolet spectrum of KQ Puppis (Boss 1985).**
A. Altamore, A. Giangrande, R. Viotti.
Astron. Astrophys., Suppl. Ser., Vol. 49, 511 - 520 (1982).

The authors analyse the high resolution spectrum of the VV Cep star KQ Pup observed with the IUE satellite during 1979 and 1980. More than 600 lines have been identified between 1190 and 3228 A. Two main line systems are present: (1) A broad absorption line system, which includes high ionization lines from N V, C IV, Si IV,and Al III. (2) Narrow P Cygni lines of neutral and singly ionized elements, mostly of Fe II.

120.004 **Radial velocity studies of spectroscopic binaries with the objective prism II: Three new massive binaries in the Scorpius OB 1 association.**
F. Gieseking.

Astron. Astrophys., Suppl. Ser., Vol. 49, 673 - 676 (1982).

The high luminous objects HD 151910, HD 152333 and HD 152590 near the very young open cluster NGC 6231, which is related to the Scorpius OB 1 association, are detected to be spectroscopic binaries. From 23 relative radial velocities for each object, the binary periods could be found unambiguously. Orbital elements and corresponding radial velocity curves are given. The perfect agreement of the barycentric radial velocities and the radial space motion of NGC 6231 suggests a kinematical relation of the binaries with the cluster and the surrounding OB-aggregate. Mass estimates from spectral types, absolute visual magnitudes and mass functions reveal combined masses larger than 30 solar masses.

120.005 **The spectroscopic binary HD 189178.**
A. H. Batten, W. A. Fisher, J. M. Fletcher.
Publ. Astron. Soc. Pacific, Vol. 94, 515 - 520 (1982).

Orbital elements for the primary component of HD 189178(B5) are presented for the first time, although the star has been known for 70 years to be a binary. Velocities derived from all lines visible in the spectrum lead to values for the elements $P = 70\overset{d}{.}22$, T=JD2439762.1, $\omega = 302°$, $e = 0.34$, $V_0 = -20.1$ km s^{-1}, $K_1 = 41.4$ km s^{-1}, and $f(M) = 0.43$. If the hydrogen-line velocities were omitted from the means, however, e would be smaller, K_1 larger, and V_0 slightly more negative. Reports of the visibility of the secondary spectrum are assessed and variations in the strengths of the hydrogen lines are shown to exist. It is suggested that the system may be eclipsing.

120.006 **The spectroscopic orbit of Wra 977 (= X-ray source 4U 1223-62).**
J. B. Hutchings, D. Crampton, A. P. Cowley, D. J. Bord.
Publ. Astron. Soc. Pacific, Vol. 94, 541 - 547 (1982).

Spectroscopic observations of the B1.5 supergiant Wra 977 from several observing seasons are described. The star has pronounced stellar wind characteristics which make orbital determination difficult. Lines which are considered most likely to represent orbital motions show velocities antiphased with the X-ray pulsar for the suggested 35.0-day period. A forced fit to the X-ray determined orbit yields a mass ratio of ~ 20 and implies masses of $> 30 M_\odot$ and $> 1.5 M_\odot$ for the stars. Weak He II λ4686 emission is detected. No evidence is found for profound periastron effects in the $e = 0.44$ orbit. Other spectroscopic information and the evolutionary status of the system are discussed. Similar considerations apply to the $41\overset{d}{.}37$ period.

120.007 **Radial-velocity orbit of HR 4006.**
R. F. Griffin, W. I. Beavers.
Publ. Astron. Soc. Pacific, Vol. 94, 557 - 559 (1982).

HR 4006 is shown to be a spectroscopic binary with a period of 173 days, and its orbit is determined.

120.008 **The WN4 + O4–6 Wolf-Rayet binary HD 90657.**
V. S. Niemela, A. F. J. Moffat.
Astrophys. J., Vol. 259, 213 - 218 (1982).

HD 90657 is shown to be a WN4 + O4–6 binary, much like the WN4 + O9 binary HD 186943 with mass ratio M(W-R)/M(O) ≈ 0.52. The best W-R orbit is derived from the emission lines of highest excitation, which show the least phase shift relative to the absorption-line orbit due to the O component. Light curve variations restrain the masses to the range ≈11–14 $M_\odot$ for the WN4 component, and 21–28 $M_\odot$ for the O-type component.

120.009 **Photoelectric radial velocities. Paper IX. The orbits of the spectroscopic binaries 81 and χ Geminorum.**
R. F. Griffin.
Mon. Not. R. Astron. Soc., Vol. 200, 1161 - 1168 (1982).

81 and χ Gem are fifth-magnitude late-type giant stars which have been known for more than half a century to be variable in radial velocity. Systematic Cambridge observations covering well over a decade now permit reliable orbits to be determined for both stars. The periods are approximately 4.2 and 6.7 yr respectively. The high γ-velocity, +83 km s^{-1}, of 81 Gem is noteworthy, but does not seem to be accompanied by any spectroscopic abnormality. It is likely that χ Gem is a triple system, as there appears to have been a small but significant change in its γ-velocity since the 1920s.

120.010 **The infrared spectrum of γ Velorum.**
D. K. Aitken, P. F. Roche, D. A. Allen.
Mon. Not. R. Astron. Soc., Vol. 200, 69P - 75P (1982).

The authors report infrared spectral observations of the Wolf-Rayet system γ Velorum. These show an extremely rich emission line spectrum in which permitted lines of helium and carbon and forbidden lines of neon and sulphur appear. Some of these lines have not hitherto been observed in any object. The data lead to estimates of the relative abundance of He I/He II and C/He (Hummer, Barlow and Storey, 1982) and present lower limits to the Ne/He and S/He abundance ratios.

120.011 **Capella revisited.**
T. R. Ayres.
Advances in ultraviolet astronomy, (see 012.006), p. 251 - 254 (1982).

The author summarizes the highlights of two IUE programs to study the active chromosphere binary, Capella (α Aurigae A: G6 III + F9 III).

120.012 **Analysis of non-interacting binaries with luminous cool primaries.**
S. B. Parsons.
Advances in ultraviolet astronomy, (see 012.006), p. 501 - 504 (1982).

High dispersion LWR spectra are being analyzed for the differential velocity between hot and cool components of heretofore single-lined spectroscopic binaries with late-type giant or supergiant primaries. Low dispersion SWP and LWR flux spectra are being used together with ground-based photometry to disentangle the composite energy distributions of the binaries.

120.013 **X-ray observations of single-line spectroscopic binaries.**
K. P. Singh, S. Naranan.
Astron. Astrophys., Vol. 113, 167 - 169 (1982).

Observations of three single-line spectroscopic binaries with the Imaging Proportional Counter on the Einstein Observatory are presented. None of the three binaries, π^5 Ori, θ^2 Cru, and 112 Her is detected as an X-ray source, making the presence of a compact secondary in any of these binaries highly unlikely. An extended X-ray source has been found at a position 28′ away from π^5 Ori. The X-ray source position is consistent with that of a distant cluster of galaxies, A 520.

120.014 **The hot component of KS Persei (HD 30353).**
J. S. Drilling, D. Schönberner.
Astrбn. Astrophys., Vol. 113, L22 - L23 (1982) = Contrib. Louisiana State Univ. Obs. No. 171.

The authors have detected the presence of a hot companion for the single-lined spectroscopic binary KS Per (HD 30353) from a study of low-resolution IUE spectra. The system as a whole is very similar to υ Sgr. The observations are consistent with a model in which the primary is a helium supergiant of about 1 $M_\odot$ and the secondary is an overluminous early-type star with a mass of 5 - 6 $M_\odot$. This model predicts that the system is now undergoing its second mass exchange.

120.015 **A search for spectroscopic binaries among the runaway O type stars.**

R. C. Stone.
Astrophys. J., Vol. 261, 208 - 219 (1982).

Numerous radial velocity measurements of medium dispersion were made for the 10 brighter stars given in Stone's list of very probable O type runaways. A new iterative reductional analysis was used to derive plate velocities, which are estimated to be 1.6 times more accurate internally than those found by using the traditional method. Four of the stars are identified as probable velocity variables, while three are classed as possible variables. If the source of this variability is Keplerian rather than atmospheric, ζ Per could be a spectroscopic binary with a black hole companion.

120.016 **HR 7578: a super-metal-rich, double-lined, K dwarf binary.**
F. C. Fekel, W. I. Beavers.
Bull. American Astron. Soc., Vol. 14, 633 (1982). – Abstract.

120.017 **IUE observations of phase-dependent spectral variations in WR binaries.**
L. H. Auer, G. Koenigsberger.
Bull. American Astron. Soc., Vol. 14, 634 (1982). – Abstract.

120.018 **Spectroscopic binary orbits from photoelectric radial velocities. Paper 45: HD 33708.**
R. F. Griffin.
Observatory, Vol. 102, 136 - 137 (1982).

120.019 **Orbit of the spectroscopic binary cobalt star HR 5049.**
M. M. Dworetsky.
Observatory, Vol. 102, 138 - 141 (1982).

120.020 **Spectroscopic binary orbits from photoelectric radial velocities. Paper 46: HD 90524.**
R. F. Griffin.
Observatory, Vol. 102, 200 - 201 (1982).

120.021 **2 Pup B: a new double-lined spectroscopic binary**
C. Raghavender Rao.
Bull. Astron. Soc. India, Vol. 10, 33 (1982). – Abstract.

120.022 **Long-term radial velocity variations of *o* And.**
L. Pastori, E. Antonello, M. Fracassini, L. E. Pasinetti.
Astrophys. Space Sci., Vol. 86, 179 - 184 (1982).

The results of radial velocity measures of the star *o* And from high dispersion plates relative to the years 1964, 1966, 1975 and 1976 are given. The averaged values of the 1964 and 1966 plates confirm the existence of the minimum in the trend of the radial velocities and fit well the curve obtained in the hypothesis of periodic long-term radial velocity variations with the period of 23.5 years (Fracassini et al., 1977). Preliminary orbital elements in the hypothesis of a long period spectroscopic binary system have been tentatively drawn.

120.023 **CV Serpentis: still eclipsing.**
I. D. Howarth, A. J. Willis, D. Stickland.
Third European IUE Conference, (see 012.029), p. 331 - 334 (1982).

The authors have found that CV Ser (WC8+O9) shows dramatic spectral variations with orbital phase. They interpret these variations in terms of eclipses of the O star by the Wolf-Rayet wind, deriving a mass-loss rate for the WC component of $7 \times 10^{-5} M_\odot$/yr.

120.024 **System parameters and mass-loss of δ Sge.**
D. Reimers, K. P. Schröder.
Third European IUE Conference, (see 012.029), p. 359 - 362 (1982).

The authors determine improved orbital elements of the δ Sge system using new radial velocity measurements. Combined with phase dependent speckle-interferometric measurements of the angular separation of the components and with a distance of 100 pc, a mass-ratio of ~3.5 has been obtained. A mass-loss rate of $2 - 3 \times 10^{-8} M_\odot$/yr has been found for the M2 II primary.

120.025 **Photoelectric radial velocities. Paper X. The orbits of four spectroscopic binaries in the Clube Selected Areas.**
R. F. Griffin.
Mon. Not. R. Astron. Soc., Vol. 201, 487 - 494 (1982).

Orbits are presented for four ninth-magnitude stars about which the only spectroscopic datum previously known is the HD type, K0 in each case. There is circumstantial evidence that all the stars are giants and that one of them (HD 149240) has a hotter companion which should be detectable spectroscopically and photometrically.

120.026 **A search for secondary lines in the visual spectra of 55 binaries.**
A. E. Gómez, H. A. Abt.
Publ. Astron. Soc. Pacific, Vol. 94, 650 - 654 (1982).

Using a CCD detector in the Kitt Peak coudé spectrograph, the authors searched for secondary lines in the visual spectral region of single-lined binaries. They were successful in finding them in one-quarter of the 23 B2–B5 dwarfs and 6% of the 32 F3–G2 dwarfs. In the cases of failures, there are no contradictions between the earliest possible secondary spectral types and the latest types allowed by the orbital mass functions.

120.027 **Variable polarization in HD 162679.**
H. G. Luna.
Publ. Astron. Soc. Pacific, Vol. 94, 692 - 694 (1982).

The author presents a discovery of time variations in the linear polarization of HD 162679, a spectroscopic binary system belonging to the open cluster NGC 6475. The variation seems to present complex features in a short time scale and preliminary results do not show any periodic behavior.

120.028 **Polarimetric determination of the inclination of the Wolf-Rayet binary system HD 152270.**
H. G. Luna.
Publ. Astron. Soc. Pacific, Vol. 94, 695 - 699 (1982).

UBV polarimetric observations have been obtained for the star HD 152270, a bright WC7 + O5 spectroscopic binary system. The observations show a phase-locked variation of high amplitude which is explained by the model of electron scattering in a dense co-rotating envelope. The observations permit to determine a value of 35° for the inclination of the orbital plane. The masses of the components of the system are then obtained: 11 $M_\odot$ for the WC7 component and 25 $M_\odot$ for the O5 star. A brief discussion about departures from the model assumptions is also presented.

120.029 **A spectroscopic search for duplicity among a complete sample of northern galactic Wolf-Rayet stars.**
R. Lamontagne, A. F. J. Moffat.
Wolf-Rayet stars: observations, physics, evolution, (see 012.042), p. 283 - 287 (1982).

The authors have analysed new spectroscopic observations of 22 population I WR stars in a sample defined by B<12.5 mag, $18^h < \alpha < 6^h$, $\delta > -30°$. This completes the search for orbital duplicity among all of the 30 Pop I WR stars contained within these limits. The authors have found several new binary systems. Some of these are low-amplitude, single-line binaries, often with runaway velocities and/or large separations from the galactic plane. The authors consider them to be candidates for a WR + collapsar stage.

120.030 **A spectral study of HD 50896.**
V. S. Niemelä, R. H. Méndez.
Wolf-Rayet stars: observations, physics, evolution, (see 012.042), p. 295 - 298 (1982).

120.031 **Observations of new Wolf-Rayet binaries.**
V. S. Niemelä.
Wolf-Rayet stars: observations, physics, evolution, (see 012.042), p. 299 - 304 (1982).

The author reports preliminary results of spectrographic observations for three southern WR stars, whose binary nature had not been previously verified.

120.032 **Spectral variations of Theta Muscae (WC 6 + O 9.5 I) in the ultraviolet.**
F. Beeckmans, C. A. Grady, F. Macchetto, K. A. van der Hucht.
Wolf-Rayet stars: observations, physics, evolution, (see 012.042), p. 311 - 315 (1982).

High and low resolution spectra of the spectroscopic triple star θ Muscae were studied for time-variability. Within 7 min no variations were found in the low resolution data. On a time scale of 8 days and longer, variations in the P Cygni profiles of the resonance lines occur. These may be explained as variable mass loss from the system, or as aspect effects in an asymmetric wind flow.

120.033 **Wolf-Rayet stars in the Magellanic Clouds: spectroscopic binaries and masses.**
A. F. J. Moffat.
Wolf-Rayet stars: observations, physics, evolution, (see 012.042), p. 515 - 521 (1982).

Repeated spectra to V-mag 13.0 (13.5) have been obtained since 1978 for the 25 (5) brightest WR stars in the LMC (SMC). More than half of these are orbiting spectroscopic binaries, mainly SB1 for the luminous subclasses WN6-9, and SB2 for the rest. Together with galactic data, the Cloud SB2's lead to a correlation between mass ratio WR/OB and WR spectral subclass for each of the WN and WC sequences. This correlation is interpreted as an evolutionary sequence in which the peeling-off process of the strong stellar wind exposes successively hotter regions and eventually converts WN to WC at a point in time which appears to depend on the ambient metal abundance.

120.034 **Spectroscopic binaries near the north galactic pole. Paper 4: HD 120803.**
R. F. Griffin.
J. Astrophys. Astron., Vol. 3, 101 - 105 (1982).

Photoelectric radial-velocity measurements show that HD 120803 is a spectroscopic binary with a rather eccentric orbit and a period of 700 days. Early DDO photographic observations, published individually here for the first time, fit the orbit well.

120.035 **Spectroscopic binaries near the north galactic pole. Paper 5: HD 117064.**
R. F. Griffin.
J. Astrophys. Astron., Vol. 3, 107 - 109 (1982).

Photoelectric radial-velocity measurements, which in 1971 showed HD 117064 to be a spectroscopic binary, have been continued and now permit the derivation of an eccentric orbit with a period of 6.08 years.

120.036 **A mass-eccentricity correlation in spectroscopic and visual binary orbits.**
J. Dommanget.
Binary and multiple stars as tracers of stellar evolution, (see 012.046), p. 119 - 122 (1982).

Looking for a possible explanation for the debated existence of a correlation between period and eccentricity in binary orbits the author made the assumption that the orbital evolution of the binaries could be a consequence of a substantial mass-loss of their components even when these components are late type main sequence stars.

120.037 **On the distribution of young spectroscopic binary stars over the major semiaxes of their orbits.**
E. I. Popova, A. V. Tutukov, B. M. Shustov, L. R. Yungelson *(Yungel'son).*
Binary and multiple stars as tracers of stellar evolution, (see 012.046), p. 129 - 131 (1982).

To know the distribution of a correlation between the masses of close binary systems (CBS) components and semiaxes of their orbits is necessary for the investigation of the origin and evolution of CBS. For such statistical investigations, a catalogue of CBS was compiled at the Astronomical Council. The authors discuss here the distribution of CBS in the M-a plane.

120.038 **Observational selection in spectroscopic binary eccentricities.**
M. S. Staniucha.
Binary and multiple stars as tracers of stellar evolution, (see 012.046), p. 133 - 143 (1982).

The discovery-and-identification probability for different shaped orbits of spectroscopic binary stars is estimated. The eccentricity distribution observed in the sample of ~1000 binaries with known orbits and appearing as strongly peaked toward e = 0 is corrected for observational selection effects. The resulting e-distribution seems to be flat for e in the range ~0.05 - 0.6 with some excess of circular (or almost circular) orbits and a deficiency of orbits with $e \gtrsim 0.6$.

120.039 **Age determination in CW Eri = BV 1000.**
H. Mauder.
Binary and multiple stars as tracers of stellar evolution, (see 012.046), p. 217 - 218 (1982).

Hejlesen, 1980, has calculated evolutionary tracks for stars of different chemical composition. In his graphs isochronic lines are given which allow for an age determination of binary stars. The components of CW Eri are lying on the respective evolutionary tracks according to their masses and the isochronic line for 1.5×10^9 years fits both stars.

120.040 **A time-resolved spectroscopic study and modeling of the dwarf nova BV Centauri.**
R. L. Gilliland.
Astrophys. J., Vol. 263, 302 - 311 (1982).

Time-resolved observations of the double-lined, spectroscopic binary BV Centauri are presented. The spectrum shows hydrogen emission lines from an accretion disk around the white dwarf primary and absorption lines from a late-type star of spectral class G5–8 IV–V. The reduction technique is discussed and yields a derived orbital period $P = 0^d\!.611179$. Results of stellar evolution computations are presented to better define the nature of the presumably evolved secondary in this long period system. Masses of the mass-losing secondary and accreting white dwarf are : $M_{sec} = 0.9 \pm 0.1\ M_\odot$ and $M_{WD} = 0.83 \pm 0.1\ M_\odot$.

120.041 **Five new variable stars.**
R. E. McCrosky, C. A. Whitney.
Inf. Bull. Variable Stars, No. 2186, 2 pp. (1982).

120.042 **Photometric variability of HR 3119.**
M. V. Mekkaden, A. V. Raveendran, S. Mohin.
Inf. Bull. Variable Stars, No. 2218, 3 pp. (1982).

120.043 **Spectroscopic binary orbits from photoelectric radial velocities. Paper 47: HR 8580.**
R. F. Griffin.
Observatory, Vol. 102, 223 - 225 (1982).

120.044 **Speckle interferometry of the spectroscopic binary 94 Aquarii A.**
H. A. McAlister, W. I. Hartkopf.
Publ. Astron. Soc. Pacific, Vol. 94, 832 - 834 (1982) = Astron. Contrib. Georgia State Univ., No. 61.
Five years of speckle observations of the spectroscopic binary 94 Aqr A are analyzed to determine the visual orbit of the Aa, b system.

120.045 **Close binary systems before and after mass transfer. III. Spectroscopic binaries.**
F. Vansina, J. P. de Grève.
Astrophys. Space Sci., Vol. 87, 377 - 401 (1982).
A method is given to derive absolute dimensions from data of spectroscopic binary systems, with application to the 7th Catalogue of Spectroscopic Binary Systems of Batten et al. (1978). The different samples of solutions are analysed with regard to existing solutions. Absolute dimensions are given for 62 systems, not included in the tables of Popper (1980), Lacy (1979), and Giannone and Giannuzzi (1974).

120.046 **A study of physical properties of spectroscopic binary stars.**
E. I. Popova, A. V. Tutukov, L. R. Yungelson (*Yungel'son*).
Astrophys. Space Sci., Vol. 88, 55 - 80 (1982).
The main results of a study of a catalogue of physical parameters of 1041 spectroscopic binaries are presented. The distribution of spectroscopic binaries over all main parameters M_1, a, e, M_1/M_2, P, and certain dependencies between some of them have been found.

120.047 **V 505 Mon: one of the most massive binaries.**
R. Stagni, R. Margoni, A. Mammano.
Astrophys. Space Sci., Vol. 88, 115 - 123 (1982).
From 59 spectra collected at Asiago from 1969 to 1976, the authors obtain the spectroscopic orbit of the binary V 505 Mon. The rather long period $P = 53.7805$ days and semiamplitude $K_1 = 83 \pm 4$ km s^{-1} of the primary component imply a fairly large mass of the secondary ($m_2 \geqslant 3.2 \pm 0.4\, m_\odot$). Apparently also lines of the secondary have been recorded, thus indicating highly massive components ($m_1 \cong 50$, $R_1 = 26$, $m_2 \cong 27$, $R_2 = 14$, in solar units) in a detached system (separation $\cong 255\, R_\odot$). The already reported variation of 0.15 mag, with period about one day, together with Hα emission detected by the authors point toward a possible massive X-ray progenitor.

120.048 **The invisible star.**
H. Couper.
New Scientist, Vol. 95, 762 - 764 (1982). – Abstr. in Phys. Abstr., Vol. 86, Abstr. 5359 (1983).

Computer controlled high speed photometry presented by the lunar occultation of 57 Orionis. II.
See Abstr. 031.608.

The hot components of hydrogen-deficient binaries.
See Abstr. 114.036.

Line profile variations and binarity in Wolf-Rayet stars.
See Abstr. 114.140.

Chromospheric Mg II emission in A 5 to K 5 main sequence stars from high resolution IUE spectra.
See Abstr. 114.144.

Wolf-Rayet stars with massive companions.
See Abstr. 115.013.

A search for resonance polarization in stars with enhanced Ca II H and K emission.
See Abstr. 116.024.

Statistical models for close binaries.
See Abstr. 117.136.

Two visual-spectroscopic binary orbits: ADS 7780 and ADS 11060.
See Abstr. 118.022.

Contribution to the study of composite spectra. III. Spectrum binaries: intermediate class between visual and spectroscopic binaries.
See Abstr. 118.024.

The triple system of π Cephei.
See Abstr. 118.033.

A spectroscopic investigation of the eclipsing variable ER Vulpeculae.
See Abstr. 119.042.

LY Aurigae, NY Cephei, and the mass-luminosity anomaly within O type binaries.
See Abstr. 119.047.

A search for light-time effects in binary cepheids: AW Persei.
See Abstr. 122.049.

Search for pulsating stars in multiple stellar systems.
See Abstr. 122.087.

The ultraviolet variability of T CrB.
See Abstr. 124.002.

The unprecedented light variations of NGC 2346.
See Abstr. 135.047.

X-ray spectrum of Capella and its relation to coronal structure and ultraviolet emission.
See Abstr. 142.047.

Radial velocities and internal motions in Galactic stellar systems.
See Abstr. 155.051.

121 Early-stage Stars (T Tauri Stars, Herbig-Haro Objects, etc.)

121.001 **Proper motions of Herbig-Haro objects. II. The relationship of HH-39 to R Monocerotis and NGC 2261.**
B. F. Jones, G. H. Herbig.
Astron. J., Vol. 87, 1223 - 1232 (1982) = Lick. Obs. Bull., No. 924.
Recent astrometric work on Herbig-Haro Objects has revealed high tangential velocities in HH-1, -2, -28, -29. In each case the velocity was directed away from the presumed exciting source. The authors report similar results for HH-39, associated with R Monocerotis and the reflection nebulosity NGC 2261.

121.002 **Photometric properties of RY Tauri.**
G. V. Zajtseva.
Astrofizika, Tom 18, 67 - 80 (1982). In Russian. English translation in Astrophysics, Vol. 18, No. 1.
The results are presented of photoelectric *UBV* observations of RY Tauri carried out in 1965 - 80. Two components of light variations are observed: fast (days) and slow (years). During fast variations the colour indices $U-B$ and $B-V$ change independently of brightness. During the slow variations only a reverse average dependence is observed. A comparison is made of variations of the intrinsic polarization and the brightness of RY Tauri.

121.003 **Spectrophotometric studies of non-stable stars. IV. On the spectrum of V1057 Cygni.**
D. Chalonge, L. Divan, L. V. Mirzoyan.
Astrofizika, Tom 18, 263 - 274 (1982).
The results of a spectrophotometric study of V1057 Cyg based on observations made in 1971 - 1973 are presented. It is shown that after the "brightening" of V1057 Cyg an absorbing shell was present around it which brought to increase the magnitude of the Balmer jump and to decrease its wave-length.

121.004 **Broad carbon monoxide line wings near T Tauri stars.**
M. L. Kutner, C. M. Leung, D. E. Machnik, K. N. Mead.
Astrophys. J., Lett., Vol. 259, L35 - L39 (1982).
Observations of carbon monoxide (CO) toward 26 T Tauri stars and related objects are reported. In 19 of these objects, broad CO line wings were detected, implying mass outflow is a common phenomenon in T Tauri stars. The velocity width (full width) of the wings ranges from ~ 10 to 80 km s^{-1} and clusters around 25 km s^{-1}. In general, the line wings are asymmetric with respect to the line core and the sense of asymmetry varies among sources.

121.005 **Ultraviolet spectra of Herbig-Haro objects and of the environment of the Cohen-Schwartz star.**
K. H. Böhm, E. Böhm-Vitense, J. A. Cardelli.
Advances in ultraviolet astronomy, (see 012.006), p. 223 - 226 (1982).
The authors discuss the ultraviolet spectrum of H–H 2 and the environment of the Cohen-Schwartz star and give a comparison of the H–H 2 and H–H 1 spectra.

121.006 **Chromospheres and coronae in the T Tauri stars.**
C. L. Imhoff, M. S. Giampapa.
Advances in ultraviolet astronomy, (see 012.006), p. 456 - 459 (1982).
The authors have employed IUE observations of the T Tauri stars to investigate the stars' nascent chromospheres, transition regions, and (indirectly) coronae.

121.007 **Dust and young stars in Puppis.**
B. Pettersson.
Messenger, No. 29, p. 8 - 11 (1982).

121.008 **An investigation of T Tauri variability.**
J. Schmelz.
News Lett. Astron. Soc. N.Y., Vol. 2, No. 2, p. 9 (1982). Abstract.

121.009 **Star formation in the λ Orionis region. I. The distribution of young objects.**
R. Duerr, C. L. Imhoff, C. J. Lada.
Astrophys. J., Vol. 261, 135 - 150 (1982).
The authors have made a deep survey for Hα emission objects in a region about 100 square degrees in area, centered on the λ Orionis OB association. They detected 83 objects, of which 62 were previously known from earlier, less extensive surveys. The vast majority of the Hα emission objects form an elongated, barlike structure $2° \times 8°$ in extent. These dimensions are similar to those of giant molecular cloud complexes which spawn OB and T associations. The authors suggest that the spatial distribution of Hα emission objects and OB association members in this region provides a fossil record of a preexistent giant molecular cloud complex. A qualitative discussion of star formation in this huge cloud complex is given.

121.010 **A search for high-velocity molecular gas around T Tauri stars.**
S. Edwards, R. L. Snell.
Astrophys. J., Vol. 261, 151 - 160 (1982).
A search for high-velocity molecular gas in the ^{12}CO $J = 1-0$ transition around 28 T Tauri and three Herbig Ae/Be stars reveals high-velocity CO emission toward T Tau, AS 353 A, and HL Tau. Maps of the spatially extended, high-velocity gas around T Tau and AS 353 A yield mass-loss rates of greater than $1 \times 10^{-7} M_\odot yr^{-1}$ and $\geq 7 \times 10^{-8} M_\odot yr^{-1}$, respectively.

121.011 **The two-photon continuum in Herbig-Haro objects.**
M. A. Dopita, L. Binette, R. D. Schwartz.
Astrophys. J., Vol. 261, 183 - 194 (1982).
The blue continuum seen in many Herbig-Haro (HH) objects is shown to be intrinsic to the source, and is due to a collisionally excited two-photon continuum of hydrogen. A strong inverse correlation between its strength and the excitation class of the HH spectrum is found. The most likely explanation of the data, based on theoretical models is that both the lines and continuum arise in a shock of finite age running into a partially preionized medium.

121.012 **On the sources of variability in T Tauri stars.**
J. T. Schmelz, A. E. Rydgren, F. J. Vrba.
Bull. American Astron. Soc., Vol. 14, 629 (1982). – Abstract.

121.013 **Photometric variations in Herbig Ae and Be stars.**
B. E. Phelps, W. Herbst.
Bull. American Astron. Soc., Vol. 14, 636 (1982). – Abstract.

121.014 **Photometric variations in T Tauri stars.**
J. Holtzman, W. Herbst.
Bull. American Astron. Soc., Vol. 14, 636 (1982). – Abstract.

121.015 **The environments of T Tauri stars.**
R. A. Chevalier.
Bull. American Astron. Soc., Vol. 14, 636 - 637 (1982). Abstract.

121.016 **Interactions between pre-main-sequence objects and molecular clouds. I. Elias 1–12.**
R. M. Levreault.
Bull. American Astron. Soc., Vol. 14, 638 (1982). – Abstract.

121.017 **Magnetic accretion model for the activities of T Tau stars.**
Y. Uchida.
Magnetospheric phenomena of celestial bodies, (see 012.011), p. 66 - 71 (1982). In Japanese.

Characteristics of T Tau stars such as the coexistence of inflow and outflow with large velocities, or as the presence of an X-ray emitting region nearer to the surface of the star than the Hα emitting region, are interpreted by a new model with magnetic accretion.

121.018 **13-color photometry of pre-main sequence stars: preliminary report and results.**
C. Chavarría-K., E. de Lara.
Rev. Mexicana Astron. Astrofis., Vol. 6, (see 012.017), 159 - 162 (1981).

Broad *(UBVRI)* and intermediate (13-color) band photometry of 160 stars selected mainly from the Herbig-Rao catalogue are being carried on currently, mainly to complement the published data of these stars in the optical window (for example shortward of the Balmer and longward of the Paschen discontinuities). The 13-color photometric system and its applications to pre-main sequence stars are briefly discussed. First results are presented.

121.019 **Dependence of the continuum energy distribution of T Tauri stars on the location of the temperature minimum.**
N. Calvet.
Rev. Mexicana Astron. Astrofis., Vol. 6, (see 012.017), 227 - 229 (1981).

The author investigates the influence of the position of the temperature minimum on the continuum flux produced by theoretical models. The models consist of a normal photosphere and a chromospheric temperature rise and are similar to those of Calvet (1981). The author finds that a deeper temperature rise than previously assumed cannot eliminate the discrepancy between theoretical fluxes and observations found in that work. He also finds an upper limit to the location of the temperature minimum for the type of models considered of $m \sim 2$ g cm^{-2}.

121.020 **A hypothesis on the hydromagnetic activity of T Tau-type stars and related objects.**
R. E. Gershberg.
Astron. Nachr., Band 303, 251 - 261 (1982).

A hypothesis on the hydromagnetic nature of T Tau-type stellar activity (Gershberg and Petrov, 1976; Gershberg 1978) that suggests a way of solving the main energetic problem of these stars is given in detail. In this model it is supposed that the whole surface of a T Tau-star is covered with structures similar to sunspots, and that the deficit of radiation from these dark spots is carried away by hydromagnetic waves. The assumptions on which the hypothesis rests are shown to be consistent with existing astrophysical and stellar evolutionary ideas. Several consequences of the hypothesis concerning the physical and evolutionary problems of T Tau-stars, FU Ori-type stars, magnetic, antiflare and classical flare star are discussed. Possibilities of testing the hypothesis observationally are considered.

121.021 **Some recent results in early stellar evolution.**
G. H. Herbig.
Recent advances in observational astronomy, (see 012.028), p. 19 - 23 (1981).

Newly-measured proper motions (by Jones and Herbig) for about 75 emission-Hα stars and variables in the Taurus Auriga dark clouds show that these objects have a small velocity dispersion (1-2 km/s in one coordinate) around their mean. Consequently they appear to be gravitationally trapped in their parent clouds. The proper motions of ordinary stars projected upon these clouds show that very few non-emission stars are moving with the cloud members. This is most simply explained if star formation in these nebulae was initiated only about 10^7 years ago, at about the last time these clouds passed through the galactic plane.

121.022 **An analysis of the ultraviolet spectrum of T Tauri.**
C. Jordan, M. C. de M. Ferraz, A. Brown.
Third European IUE Conference, (see 012.029), p. 83 - 88 (1982).

The pre-main-sequence star T Tauri has been investigated in detail with IUE. Significant changes are seen in the Mg II line profiles at different epochs. An emission measure distribution is derived and chromospheric models are computed. There is evidence for two different density components in the chromosphere. The chromospheric radiation losses are a significant fraction of the stellar luminosity.

121.023 **T Tauri stars: not one but two transition regions.**
C. Bertout.
Third European IUE Conference, (see 012.029), p. 89 - 94 (1982).

It is shown in this paper that IUE and X-ray observations of T Tauri stars can be explained by a model which takes the cooling and condensing of a moderately dense corona into account. This model includes two transition regions between coronal gas and cool gas: the usual zone between chromosphere and corona, and the region between corona and cool circumstellar gas.

121.024 **Far UV line widths in RU Lupi.**
M. V. Penston, M. T. V. T. Lago.
Third European IUE Conference, (see 012.029), p. 95 - 98 (1982).

Short wavelength high dispersion spectra of the T Tauri star RU Lupi have been analysed. Half widths of high excitation lines C IV, Si III] and C III] lie near 170 km s^{-1} similar to those of the optical helium lines. The data are consistent with a model wind in which, after a maximum near 250 km s^{-1}, the wind velocity decreases away from the star while the gas temperature rises.

121.025 **On the Mg II and Fe II resonance lines in Herbig Ae stars: preliminary results.**
A. Talavera, C. Catala, L. Crivellari, J. Czarny, P. Felenbok, F. Praderie.
Third European IUE Conference, (see 012.029), p. 99 - 105 (1982).

Three Herbig Ae stars have been observed with the IUE satellite in the range 2000 – 3200 Å; they are AB Aur, HD 250550, BD +46°3471. The Mg II resonance lines profiles all show a P Cygni structure, with an intense emission, and a shortward wing velocity in the absorption component of 300 to 600 km/s. The mass loss rate in the AB Aur is estimated as $\dot{M} \sim 10^{-11} M_\odot$/yr.

121.026 **Contribution to the study of the dust envelope of AB Aurigae.**
C. Catala.
Third European IUE Conference, (see 012.029), p. 107 - 111 (1982).

Observations by IUE of the Herbig Ae star, AB Aur, when compared with a theoretical energy distribution for the same spectral type, show the existence of an anomalous absorption. It is believed to be due to a circumstellar dust envelope surrounding the star. A model of infrared emission by the dust envelope is built and compared with infrared observations.

121.027 **The variable shell star HR 5999. VII. The extended atmosphere of a Herbig Ae star: chromospheric features.**
H. R. E. Tjin A Djie, P. S. Thé.

Third European IUE Conference, (see 012.029), p. 113 - 116 (1982).

IUE and ground-based spectroscopic observations on the nebular variable HR 5999 show strong chromospheric emission lines in the spectra of this star. The emission measures of the high temperature lines suggest a low temperature gradient and large radiation losses for the transition region.

121.028 **The visible and ultraviolet continuum from a Herbig-Haro object in the core of M 16 (NGC 6611).**
J. Meaburn.
Astron. Astrophys., Vol. 114, 367 - 372 (1982).

A Herbig-Haro object was found by Meaburn and White (1982) in the neutral clouds within the galactic H II region M 16. Two IUE spectra have now been obtained of this in which a strong continuum is present but no emission lines found. The continuum around 6670 Å appears to be of atomic recombination origin but the evidence favours the interpretation in the ultra-violet of a large excess of the continuum over that expected from atomic processes. However, this conclusion is heavily dependent on the particular law chosen to correct for the large amount of interstellar extinction.

121.029 **The ultraviolet continuous and emission-line spectra of the Herbig-Haro objects HH 2 and HH 1.**
E. Böhm-Vitense, K. H. Böhm, J. A. Cardelli, J. M. Nemec.
Astrophys. J., Vol. 262, 224 - 233 (1982).

The authors present the results of IUE observations of the Herbig-Haro object HH 2 and compare them with data for HH 1. The UV continuum of both sources rises steeply towards shorter wavelengths at least to ~1300 Å. The ultraviolet emission lines indicate a considerably higher degree of ionization than has been determined from the optical spectra. The UV-optical spectrum cannot be explained by a single plane-parallel shock wave model.

121.030 **The ultraviolet spectrum of Herbig-Haro object 2H.**
E. W. Brugel, J. M. Shull, C. G. Seab.
Astrophys. J., Lett., Vol. 262, L35 - L39 (1982).

The authors present IUE spectra of Herbig-Haro object 2H which show a strong "excess" UV continuum and prominent emission lines of C, N, O, Si, Mg, and possibly Al. The continuum exhibits a turnover shortward of about 1450 Å, confirming for the first time the H^0 two-photon nature of the emission source. Recently computed models of steady shocks into partially ionized gas reproduce the two-photon spectral shape, but its observed intensity relative to Hβ and the Balmer continuum is anomalously high.

121.031 **Quasi-simultaneous photometric and polarimetric observations of T Tau and RY Tau.**
N. P. Red'kina, G. P. Chernova.
Binary and multiple stars as tracers of stellar evolution, (see 012.046), p. 231 - 238 (1982).

Quasi-simultaneous photometric and polarimetric observations of T Tau and RY Tau show that the proper polarization of these stars arises in the visual region in their gaseous emitting envelopes. The polarization of T Tau shows cyclic variations with a period of $5^h\!.18$. That of RY has two different modes of variation at brightnesses $V > 11^m$ and $V < 11^m$; it is possibly connected with the suspected binary nature of this star.

121.032 **Ultraviolet radiation from the environment of the Cohen-Schwartz star.**
K. H. Böhm, E. Böhm-Vitense.
Astrophys. J., Lett., Vol. 263, L35 - L38 (1982).

Continuous (or quasi-continuous) radiation from the environment of the Cohen-Schwartz (C-S) star has been detected in the wavelength range 1250 Å $\lesssim \lambda \lesssim$ 1930 Å. The radiation does not come from the star itself but from a diffuse source extending to at least 11″ from the star. The energy distribution in the spectrum is similar to that of the continua in Herbig-Haro objects but, provided its reddening is not less than that of Herbig-Haro 1, it rises even more steeply toward shorter wavelengths. It is concluded that it is very improbable that the source of the UV radiation has the same large extinction as the C-S star. Possible emission and scattering mechanisms are discussed briefly.

121.033 **Some remarks on the T Tauri star DR Tau.**
W. Götz.
Inf. Bull. Variable Stars, No. 2172, 3 pp. (1982).

121.034 **HD 36705 – a post T Tauri star.**
S. M. Rucinski.
Inf. Bull. Variable Stars, No. 2203, 4 pp. (1982).

121.035 **Historical light curves of four T Tauri stars.**
B. E. Schaefer.
Inf. Bull. Variable Stars, No. 2219, 4 pp. (1982).

121.036 **Photographic photometry of V 1515 Cygni.**
K. P. Tsvetkova.
Inf. Bull. Variable Stars, No. 2236, 3 pp. (1982).

121.037 **Near-infrared photometry of protostars.**
C. V. M. Fridlund.
Messenger, No. 30, p. 21 (1982).

121.038 **Discovery of a family of Herbig objects in M42: implications for the geometry of the high velocity molecular flow.**
D. J. Axon, K. Taylor.
Anglo-Australian Obs., Prepr. No. 176, 29 pp. (1982). – Submitted to Mon. Not. R. Astron. Soc.

121.039 **A two-micron survey of southern Herbig-Haro objects.**
B. Reipurth, W. Wamsteker.
ESO Sci. Prepr., No. 214, 21 pp. (1982). – Submitted to Astron. Astrophys.

121.040 **TW Hya: a T Tauri star far from any dark cloud.**
S. M. Rucinski, J. Krautter.
ESO Sci. Prepr., No. 220, 41 pp. (1982). – Submitted to Astron. Astrophys.

121.041 **Spectral line profiles from spherical shells.**
R. Wagenblast, C. Bertout, U. Bastian.
Pré-publ. Inst. Astrophys. Paris, No. 21, 49 pp. (1982). Submitted to Astron. Astrophys.

121.042 **Jets from pre-main sequence stars: AS 353A and its associated Herbig-Haro objects.**
R. Mundt, J. Stocke, H. S. Stockman.
Prepr. Steward Obs., No. 416, 19 pp. (1982).

121.043 **Herbig-Haro objects 46 and 47: evidence for bipolar ejection from a young star.**
M. A. Dopita, R. D. Schwartz, I. Evans.
Astrophys. J., Lett., Vol. 263, L73 - L77 (1982).

A new Herbig-Haro object (HH 47C) has been discovered in the system HH 46/47. Radial velocity measurements indicate that HH 47A is approaching at -128 km s^{-1}, whereas HH 47C is receding at 105 km s^{-1}. Both components appear to have been driven clear in opposite directions from a small dark cloud which harbors HH 46 on a direct line between HH 47A and HH 47C. HH 46 exhibits a combined HH emission spectrum and a T Tauri star reflection spectrum. A model is presented wherein the HH objects are produced by bipolar ejection from a T Tauri star embedded in an accretion disk.

121.044 **Evolution of stellar activity during the pre-main sequence phase.**
L. V. Kuhi.
Smithsonian Astrophys. Obs., Spec. Rep. 392, Vol. II, (see 012.065), p. 141 - 151 (1982).

Considerable effort in recent years has been expended in establishing the behavior of stellar activity on the main sequence and during the subsequent evolution away from the main sequence into the red giant region of the HR diagram. The most obvious indicators of stellar activity are the Ca II K chromospheric emission, variability, stellar rotation, transition region emission lines in the far ultraviolet and coronal X-ray emission. Other indicators can readily be thought of but the author confines his attention to the five above since they are much better studied than most of the others. He first briefly reviews how these indicators behave during main sequence and post main sequence evolution and then attempts to take them as far back in time as possible.

121.045 **Far-ultraviolet and X-ray evidence concerning the chromospheres and coronae of the T Tauri stars.**
C. L. Imhoff, M. S. Giampapa.
Smithsonian Astrophys. Obs., Spec. Rep. 392, Vol. II, (see 012.065), p. 175 - 179 (1982).

The T Tauri stars possess very strong far-ultraviolet emission lines. Surface fluxes for these chromospheric and transition region lines are typically $10^6 - 10^7$ erg cm^{-2} s^{-1}. However their X-ray emission is not enhanced to the same degree. The authors suggest that some T Tauri stars, exhibiting weakened C IV, N V, and He II lines and undetected by Einstein, do not have coronae. The birth and development of the corona may take place during the T Tauri stage of pre-main sequence evolution.

121.046 **Eruptive events in T Tauri stars.**
R. Mundt.
Smithsonian Astrophys. Obs., Spec. Rep. 392, Vol. II, (see 012.065), p. 181 - 190 (1982).

The NaD line profiles of about 20 T Tauri stars have been observed at high spectral resolution ($\Delta\lambda$ = 0.2 Å) with high signal-to-noise (S/N = 30–100). In seven T Tauri stars the broad NaD emission lines showed relatively narrow (FWHM = 10–40 km/s) blueshifted absorption components with velocity shifts of about −50 to −150 km/s (that are all of circumstellar origin). The observed narrow and blueshifted absorption features are discussed in terms of shells ejected during outbursts of these objects. For spherical symmetric shells the masses of the outbursts have been estimated as $10^{-6} - 10^{-5}$ $M_\odot$ and their estimated kinetic energies range from about 10^{41} to 2×10^{42} erg. Average mass loss rates range from about $10^{-9} - 3 \times 10^{-7}$ $M_\odot$ per year. This means that eruptive phenomena in T Tauri stars and their associated ejected shells can be an important component for the mass loss activity in these objects.

121.047 **Alfvén wave-driven winds as a source of mass loss from T Tauri stars.**
S. Edwards, L. Hartmann, E. Avrett.
Smithsonian Astrophys. Obs., Spec. Rep. 392, Vol. II, (see 012.065), p. 191 - 196 (1982).

The Alfvén wave-driven wind model of Hartmann and MacGregor is investigated as a source of mass loss from T Tauri stars. The authors compare predicted and observed ultraviolet and optical emission properties, including Balmer line strengths and line profiles. The predictions of the wave-driven wind models are compared with those of Cram's deep chromosphere model for T Tauri stars.

121.048 **On some spectral peculiarities of SU Aurigae.**
L. V. Timoshenko.
Perem. Zvezdy, Tom 21, 559 - 571 (1981). In Russian.

The results of spectroscopic observations of the T Tau star SU Aur based on 51 spectrograms obtained in 1977 - 1978 are given. Rapid changes of the equivalent widths and radial velocities and other spectral parameters during night, from night to night and slow changes from season to season were found. The existence of a pseudo-cycle in the variations of the spectrum with duration of several years is supposed.

121.049 **T Tauri-type stars in the Vilnius photometric system.**
E. Meištas.
Bull. Vilnius Astron. Obs., No. 61, p. 32 - 45 (1982). In Russian.

The energy distribution data in the spectra of ten T Tauri-type stars are used to investigate color excess ratios and behaviour of the stars in two-color and in reddening-free *Q, Q*-diagrams of the Vilnius photometric system.

121.050 **Optical monitoring of Orion population stars. I. Results for some T Tauri and Herbig Ae/Be stars.**
W. Herbst, J. A. Holtzman, B. E. Phelps.
Astron. J., Vol. 87, 1710 - 1729 (1982).

A program of monitoring Orion population variables is described. The authors have obtained UBVRI and Hα data for seven Herbig Ae/Be stars and five T Tauri stars. All of the T Tauri stars showed significant variations in U which, for the most part, correlated with changes in Hα flux. In addition, three of the stars (SU Aur, RW Aur, and CO Ori) exhibited substantial variations at all wavelengths. In all cases, the stars become redder when fainter. The weak emission-line stars (SU Aur and CO Ori) showed an increase in Hα equivalent width when faint, which was primarily caused by the fading of the continuum. None of the Herbig Ae/Be stars studied showed substantial variation in magnitude of Hα strength.

Wave-driven winds from cool stars. II. Models for T Tauri stars.
See Abstr. 064.029.

An observational approach to the early stages of stellar evolution.
See Abstr. 065.012.

Ultraviolet observations of young field stars.
See Abstr. 114.215.

The age of the triple system HD 165590.
See Abstr. 118.025.

Observations of high-velocity molecular gas near Herbig-Haro objects: HH 24–27 and HH 1–2.
See Abstr. 131.067.

On the nature of bipolar sources in dense molecular clouds.
See Abstr. 131.091.

High-velocity CO line wings associated with T Tauri stars.
See Abstr. 131.123.

High velocity flows and molecular jets.
See Abstr. 131.227.

The high velocity molecular flows near young stellar objects.
See Abstr. 131.292.

From clouds to stars.
See Abstr. 131.304.

VLA observations of H I near the ionization boundary of the H II region surrounding the early-type star Lk Hα 101.
See Abstr. 132.039.

122 Intrinsic Variables (Pulsating Variables, Spectrum Variables, etc.)

122.001 **Photoelectric and spectrographic observations of ρ Vir (HR 4828).**
E. Antonello, L. Mantegazza.
Astron. Astrophys., Suppl. Ser., Vol. 49, 709 - 713 (1982).

Photoelectric and spectrographic observations of the suspected variable ρ Vir are reported. Analysis of all the available photometric data has been made. The results do not show a clear regular variability of this star.

122.002 **The radial velocity of the ten-day cepheid ζ Geminorum from spectra in the Hα region.**
T. S. Jacobsen, G. Wallerstein.
Publ. Astron. Soc. Pacific, Vol. 94, 471 - 474 (1982).

A new velocity curve based upon 19 10 Å mm^{-1} spectrograms in the red region of ζ Gem has been completed. The metallic lines indicate that the mean velocity is unchanged for 80 years, the shortening of the period between 1900 - 50 is continuing at a slower rate, and the amplitude may be slightly smaller now than in previous years. The ionized lines show no depth effects when their velocities are compared with those of Fe I lines. The Hα line appears in absorption at all phases but its curve is never coincident with the metallic-line curve. The γ velocity of the Hα curve is 9 km s^{-1} more positive than the metallic lines showing that the Hα line is formed in predominantly falling material.

122.003 ***UBVRI* photometry of HR 7308.**
J. D. Fernie.
Publ. Astron. Soc. Pacific, Vol. 94, 537 - 540 (1982).

UBVRI photometry on the Johnson system is reported for the remarkable $1^{d}_{.}49$ classical cepheid HR 7308, and also for its commonly used comparison star HR 7280. An interstellar reddening of $E_{B-V} = 0.04$ is obtained for HR 7308. The $(V-R)$ photometry is used to apply the Barnes-Evans technique to HR 7308. Discussion of the results leads to the following mean values: $M_V = -2.0 \pm 0.4, R = 23 \pm 5 R_{\odot}$, $T_{\rm eff} = 5800 \pm 500$ K. The distance of the star is found to be 400 ± 50 pc. Arguments, including the agreement of the derived M_V with that predicted by period-luminosity relations, are given that HR 7308 is not an overtone pulsator.

122.004 **The circumstellar extinction of RY Sagittarii.**
A. V. Holm, C.-C. Wu, L. R. Doherty.
Publ. Astron. Soc. Pacific, Vol. 94, 548 - 552 (1982).

In 1979 - 80, ultraviolet spectrophotometry of RY Sgr was obtained by the International Ultraviolet Explorer during the star's recovery from its 1977 - 78 minimum. The wavelength dependence of the extinction of this R Coronae Borealis-type variable was determined by comparison of spectra obtained at different times but with nearly the same pulsational phase. The measured wavelength dependence is compatible with theoretical predictions of extinction by spherical graphite particles having a radius of 0.043 micron and with the extinction observed for amorphous carbon smoke.

122.005 **Cataclysmic variable candidates from the Palomar Green Survey.**
R. F. Green, D. H. Ferguson, J. Liebert, M. Schmidt.
Publ. Astron. Soc. Pacific, Vol. 94, 560 - 564 (1982).

Twenty-two emission-line cataclysmic variable candidates from the Palomar Green Survey of blue stars are reported. General spectroscopic classifications, line fluxes, and finding charts are presented. Five have spectra resembling U Geminorum stars, and several high-excitation objects with relatively strong He II were discovered. Some were found in absorption (thick disk) phases.

122.006 **Photometry of RV Tauri and yellow semiregular variables. I.**
D. W. Dawson, C. R. Patterson.
Publ. Astron. Soc. Pacific, Vol. 94, 574 - 577 (1982).

The authors present *UBV* photometry for 13 members or candidates of the RV Tauri and SRd classes of pulsating semiregular variables. Their light and color curves and times of minima are discussed with reference to previous studies.

122.007 **Theoretical growth rates, periods, and pulsation constants for long-period variables.**
M. W. Fox, P. R. Wood.
Astrophys. J., Vol. 259, 198 - 212 (1982).

Theoretical values of the growth rate, period, and pulsation constant for the first three radial pulsation modes in red giants (Population II and galactic disk) and supergiants have been derived in the linear, nonadiabatic approximation. The effects of altering the surface boundary conditions, the effective temperature (or mixing length), and the opacity in the outer layers have been explored.

122.008 **R Aquarii: an extraordinary variable star.**
J. B. Kaler.
Sky Telesc., Vol. 64, 141 - 143 (1982).

122.009 **Turbulence variations for the three cepheids SV Vul, X Cyg, and δ Cep.**
W. Benz, M. Mayor.
Astron. Astrophys., Vol. 111, 224 - 228 (1982).

The spectral line broadening can be deduced from the profile of the correlation function of the scanner CORAVEL. An investigation of the correlation profile of the cepheids (SV Vul, X Cyg, δ Cep) reveals a strong variation of the line width with the pulsation phase. These variations probably result from a strong variation of turbulence during pulsation. The variation of the kinetic energy contained in the turbulent motions is not negligible in relation with the kinetic energy of the pulsation. These turbulence variations can affect the estimations of the cepheid radii according to the Baade-Wesselink method.

122.010 **A new analysis of light variations in μ Cephei.**
L. Mantegazza.
Astron. Astrophys., Vol. 111, 295 - 298 (1982).

The power spectrum analysis of the light curve of the red supergiant μ Cep shows the presence of two periodic terms (the longer having a Cepheid-like shape), which may be interpreted as the fundamental and first overtone radial modes. Moreover there are four terms due to some non-linear coupling between the first two.

122.011 **The frequency analysis of low amplitude δ Scuti stars – VII. HD 208664.**
D. W. Kurtz.
Mon. Not. R. Astron. Soc., Vol. 200, 497 - 501 (1982).

HD 208664 (A9 III/IV) is shown to be a low amplitude δ Scuti star. A frequency analysis of more than 40 hr of observations shows that more than half the light variation in this star is contained in the frequency $f_1 = 21.199\ \text{day}^{-1}$. Similarities between this star and the δ Scuti star θ Tuc are mentioned. It is demonstrated that the two comparison stars, HD 208580 (A8 IV) and HD 208680 (F0 III), both of which lie in the instability strip, show no variations greater than 3 m.mag peak-to-peak.

122.012 **The δ Scuti variability of three stars with Ap spectral classifications, HR 151, 239 and 7461.**
D. W. Kurtz.

Mon. Not. R. Astron. Soc., Vol. 200, 503 - 507 (1982).

Observations which show HR 151 and 7461 to be δ Scuti stars and which confirm the δ Scuti variability of HR 239 are presented. All three of these stars have been classified as Ap stars by at least one spectroscopist, but those classifications are insecure and need to be re-examined. If any of these stars are really magnetic Ap stars, then their δ Scuti variability is of considerable interest.

122.013 **Photographic photometry of variable stars based on the Tartu photoplate collection. I. Area of μ Persei: AS, FR, FV Per.**
A. Maripuu, L. Kalv.
Publ. Tartu Astrofiz. Obs., Tom 48, 162 - 171 (1981).

The observational data presented are based upon 257 plates collected on the time interval 1950 - 1958.

122.014 **Pulsed dwarf novae X-ray emission.**
F. A. Cordóva, T. J. Chester.
Galactic X-ray sources, (see 003.001), p. 57 - 58 (1982). Abstract.

122.015 **Further *VBLUW* photometry of the S Doradus type variables S Dor and HDE 269006 in the LMC and a discussion on their temperatures.**
A. M. van Genderen.
Astron. Astrophys., Vol. 112, 61 - 67 (1982).

New *VBLUW* photometry of the S Dor type variables S Dor and HDE 269006 in the LMC made in 1979 and 1981 are discussed. S Dor has reached the lowest brightness in 1981 of the last 8 yr, but it is still far from the minimum. HDE 269006 was at minimum brightness. Temperatures are estimated for the central star and the envelope by fitting black body energy distributions to the *VBLUW* energy distributions. Some basic properties of the shell of HDE 269006 are estimated.

122.016 **AG Car: a galactic S Dor variable.**
B. Wolf, O. Stahl.
Astron. Astrophys., Vol. 112, 111 - 115 (1982).

High dispersion spectroscopy of the luminous variable star AG Car was carried out during maximum phase. The spectrum shows A-type star characteristics although all the lines exhibit P Cygni-type profiles with an intense emission, indicating that AG Car is surrounded by a dense expanding envelope during the maximum phase. The mean expansion velocity of the envelope is about -95 km s^{-1} and the estimated mass-loss rate is about $3 \times 10^{-5} M_{\odot} \text{ yr}^{-1}$. A comparison with S Dor shows that the maximum spectra of both stars are very much alike. It is confirmed that AG Car belongs to the class of S Dor variables.

122.017 **The cepheid period-luminosity-colour relation: a most unsuitable distance indicator.**
M. J. Stift.
Astron. Astrophys., Vol. 112, 149 - 156 (1982).

It is argued that the canonical cepheid period-luminosity-colour (PLC) relation leads to an incorrect cosmic distance scale.

122.018 **Possible correlations of expansion velocity with period and 1 μm intensity variation in Mira variables.**
N. Ukita.
Astron. Astrophys., Vol. 112, 167 - 171 (1982).

Statistical studies indicate that the expansion velocities of circumstellar envelopes of oxygen-rich Mira variables increase with the amplitude of 1 μm intensity variations, and possibly decrease with the stellar period for a fixed 1 μm amplitude. These correlations may be interpreted as a levitation of the atmosphere, due to stellar pulsation, out to the dust condensation point, followed by a dust-driven wind in which the gas density at the dust condensation radius plays an important role in determining the terminal expansion velocity.

122.019 **A radial shock model for BW Vul.**
P. A. Stamford, R. D. Watson.
Proc. Astron. Soc. Australia, Vol. 4, 210 - 212 (1981).

122.020 **A new type of pulsating star.**
J. P. Cox.
Nature, Vol. 299, 402 (1982).

122.021 **The peculiar variable star CH Cygni in 1981.**
L. Luud, T. Tomov, J. Vennik, K. Panov.
Pis'ma Astron. Zh., Tom 8, 476 - 480 (1982). In Russian. English translation in Soviet Astron. Lett., Vol. 8.

Results of spectroscopic and photometric observations of the peculiar variable star CH Cygni obtained in the course of 1981 are given. At the maximum light the star was $\sim 0\overset{m}{.}5$ brighter than during the last 20 years. Hydrogen lines consisted of two emission and one absorption components. Rapid variations of brightness with an amplitude of $\lesssim 0\overset{m}{.}25$ in U and characteristic time of ~ 10 min were observed.

122.022 **Rapidly oscillating Ap stars.**
D. W. Kurtz.
Mon. Not. R. Astron. Soc., Vol. 200, 807 - 859 (1982).

Rapid light variations with periods in the range of 6 to 12 min have been discovered in five cool magnetic Ap stars. All of the data for these stars are consistent with the hypothesis that they are pulsating in $l = 1$ or $l = 2$ non-radial p-modes of very high overtone, k, with their pulsation axes and magnetic axes aligned. The author suggests an oblique pulsator model and discusses the possibility that the stars are δ Scuti stars or, alternatively, that pulsation in many high overtone modes is common among stars. The rapidly oscillating Ap stars provide the first body of data for the theoretical study of pulsation in the presence of strong global magnetic fields.

122.023 **Infrared light curves of the dwarf nova OY Carinae.**
M. R. Sherrington, R. F. Jameson, J. Bailey, A. B. Giles.
Mon. Not. R. Astron. Soc., Vol. 200, 861 - 868 (1982).

Near infrared light curves of the dwarf nova OY Carinae are presented. The light curves differ from the optical curves of this system in showing a much shallower primary minimum and in addition, a secondary minimum which is not observed at optical wavelengths. These light curves can be understood in terms of the secondary contributing a substantial proportion of the infrared radiation. OY Carinae is the shortest period cataclysmic variable in which the secondary has been directly observed. The K magnitude of the secondary is used to derive a distance of $d = 121 \pm 20$ pc for OY Carinae.

122.024 **On the period of the pulsating R Coronae Borealis variable, RY Sgr.**
D. Kilkenny.
Mon. Not. R. Astron. Soc., Vol. 200, 1019 - 1024 (1982).

An analysis of observatons of RY Sgr made between 1926 and 1978 shows the pulsation period to have decreased steadily during that time. The occurrence of pulsational minima can be well represented by a period which decreases at a rate of 5×10^{-4} day/cycle, a value in good agreement with theoretical models for the evolution of a hydrogen-deficient carbon star.

122.025 **A spectroscopic study of the short-period variable Be star EM Cephei.**
R. W. Hilditch, B. J. McLean, I. N. Reid.
Mon. Not. R. Astron. Soc., Vol. 200, 1153 - 1159 (1982).

Some spectrophotometric and radial velocity measurements of the Be star EM Cep, a member of the open cluster NGC 7160, are presented. From this limited survey, it is established that EM Cep is not a short-period binary system as was previously supposed but may be a non-radial pulsator. The possibility that EM Cep is a long-period binary system cannot be ruled out.

122.026 **Changes in the UV spectrum of HD 4174.**
R. E. Stencel.
Advances in ultraviolet astronomy, (see 012.006), p. 219 - 222 (1982).

The data argue for eclipse phenomena in the symbiotic star HD 4174, and suggest a binarity may be the key to understanding its spectrum.

122.027 **Ultraviolet spectra of R Coronae Borealis stars.**
A. V. Holm, C.-C. Wu.
Advances in ultraviolet astronomy, (see 012.006), p. 429 - 432 (1982).

The authors have obtained IUE spectra of the R CrB-type variables R CrB, RY Sgr, XX Cam, and MV Sgr. Their analysis of these spectra suggests that: 1) it should be possible to construct useful models for the atmospheres of these hydrogen-deficient, carbon-rich stars if present standards of metallic line blanketing are used; and 2) the observed wavelength dependence of the circumstellar extinction is primarily due to circumstellar grains.

122.028 **Chromospheres of classical Cepheids.**
E. G. Schmidt, S. B. Parsons.
Advances in ultraviolet astronomy, (see 012.006), p. 439 - 441 (1982).

The authors have obtained IUE spectra of five classical Cepheids in both the short wavelength region and in the long wavelength region at low dispersion and in the long wavelength region at high resolution. Ultraviolet light curves and emission line strengths have been obtained from the low resolution spectra. The emission line fluxes behave differently in the various stars and it appears that bumps in the ultraviolet light curves are related to the behavior of the chromospheres.

122.029 **Intensity changes in the UV spectrum of Beta Cephei.**
R. P. Fahey, D. Fischel, W. Sparks.
Advances in ultraviolet astronomy, (see 012.006), p. 442 - 445 (1982).

The strong changes of 6 day period in UV features of Beta Cephei reported are not related to the four and a half hour, short-period variation. Rather, it appears to be a feature of the behavior of the outer atmosphere, and may indicate temperature of pressure inhomogeneities of non-radiative origin.

122.030 **The high velocity symbiotic star AG Draconis after its 1980 outburst.**
R. Viotti, A. Altamore, G. B. Baratta, A. Cassatella, M. Friedjung, A. Giangrande, D. Ponz, O. Ricciardi.
Advances in ultraviolet astronomy, (see 012.006), p. 446 - 447 (1982). – Abstract.

122.031 **Low resolution ultraviolet and optical spectrophotometry of symbiotic stars.**
M. H. Slovak.
Advances in ultraviolet astronomy, (see 012.006), p. 448 - 451 (1982).

Low resolution IUE spectra combined with optical spectrophotometry provide absolute flux distributions for seven symbiotic variables from 1200 to 6450 Å. For five stars (EG And, BF Cyg, CI Cyg, AG Peg, and Z And) the data are representative of the quiescent/out-of-eclipse energy distributions; for CH Cyg and AX Per, the observations were obtained following their latest outbursts in 1977 and 1978, respectively. The de-reddened distributions reveal a remarkable diversity of both line spectra and continua.

122.032 **Observations and analysis of the R Aquarii jet.**
M. Kafatos, A. G. Michalitsianos.
Advances in ultraviolet astronomy, (see 012.006), p. 452 - 455 (1982).

Ultraviolet, optical and radio observations of the symbiotic star R Aquarii are discussed in the light of the discovery of a bright radio and optical jet from this star. The authors interpret the observations of R Aqr as indicating the existence of an accretion disk around an unseen companion.

122.033 **The nova-like variable KQ Mon and the nature of the UX Ursae Majoris stars.**
E. M. Sion, E. F. Guinan.
Advances in ultraviolet astronomy, (see 012.006), p. 460 - 464 (1982).

KQ Mon was observed at low dispersion with the IUE satellite. Six spectra taken with the short wavelength prime camera are dominated by strong broad absorption lines. KQ Mon exhibits no emission lines or P Cygni-type profiles and the velocity displacements appear to be smaller, suggesting the absence of a hot, high velocity wind characterizing other UX UMa stars. The relationship of KQ Mon to other UX UMa disk stars is discussed and a model is suggested to explain their observed properties and the lack of major outbursts.

122.034 **High velocity gas outflow from the nova-like variable LSI 55° –8.**
E. F. Guinan, E. M. Sion.
Advances in ultraviolet astronomy, (see 012.006), p. 465 - 469 (1982).

122.035 **CI Cygni since the 1980 eclipse.**
R. E. Stencel, A. G. Michalitsianos, M. Kafatos.
Advances in ultraviolet astronomy, (see 012.006), p. 509 - 512 (1982).

In the course of IUE programs ZACRS (1980) and ZADRS (1981) the authors have, with the addition of archival data from 1979, extended synoptic coverage of CI Cyg over more than one full orbital cycle, including the 1980 June eclipse. In this report they wish to detail striking recent changes in the UV spectrum of CI Cyg.

122.036 **Une nouvelle lumière sur V 348 Sagittarii?**
A. Heck.
Bull. AFOEV, No. 21, p. 106 - 108 (1982).

122.037 **IUE observations of dwarf novae during active phases.**
G. Klare, J. Krautter, B. Wolf, O. Stahl, N. Vogt, W. Wargau, J. Rahe.
Astron. Astrophys., Vol. 113, 76 - 84 (1982).

The authors observed spectra of six dwarf novae during active phases with the IUE in the low resolution mode; specifically AR And and WW Cet during normal outbursts, TU Men during a superoutburst, and TZ Per, RX And, and Z Cam during standstill. The main aims of the observations were to study the different types of outburst behaviour and to investigate the physical processes during outburst.

122.038 **Observations of flare star candidates and discovery of flare activity on the dM4e star Gliese 487.**
G. Asteriadis.
Astron. Astrophys., Vol. 113, 165 - 166 (1982).

Seven star candidates for flare activity were observed during the years 1976–1981. Flare activity was discovered on the dM4e star Gliese 487.

122.039 **AG Draconis.**
J. B. Kaler.
Mercury, Vol. 11, 83 - 84 (1982).

122.040 **Radial velocities of CH Cygni during the outburst started in 1977.**
M. Hack, L. Rusconi, G. Sedmak, S. Engin, N. Yilmaz.
Astron. Astrophys., Vol. 113, 250 - 260 (1982).
The radial velocities given by the M 6 III photospheric lines and by the "non-photospheric lines" are measured from the starting of the outburst of 1977 to June 1981 (spectral range 3400 - 6700 Å; spectral resolution ~ 0.2 Å). No evidence is found either of a hypothetical hot companion or of an accretion disk; the radial velocities given by the Balmer absorption cores, the Ca II absorption cores, the permitted and forbidden emission lines of Fe II, the forbidden lines of O I, S II, Ni II, and Cu II indicate effects of stratification in the envelope.

122.041 **BL Herculis stars: theoretical models for field variables.**
R. Carson, R. Stothers.
Astrophys. J., Vol. 259, 740 - 748 (1982).
Type II Cepheids with periods between 1 and 3 days, commonly designated as BL Herculis stars, have been modeled here with the aim of interpreting the wide variety of light curves observed among the field variables. For a mass of ~ 0.60 $M_\odot$, the models are able to reproduce the observed mean luminosities, dispersion of mean luminosities, periods, light amplitudes, light asymmetries, and phases of secondary features in the light curves of known BL Her stars.

122.042 **A simultaneous photometric and radial velocity study of short-period southern Cepheids. IV. Radii and masses.**
W. Gieren.
Astrophys. J., Vol. 260, 208 - 219 (1982).
Simultaneous *UBVRI* photometric and photoelectric radial velocity observations of 15 southern Cepheids have been used to derive the radii of the stars by the maximum likelihood method of Balona (1977). A new period-radius relationship for log P < 1 is constructed from these radii and its slope is found to agree closely with the value of 0.70 derived by Cogan (1978) using theoretical arguments. Calculation and comparison of theoretical masses, pulsation masses, and Wesselink radius masses of the Cepheids show that theoretical and pulsation masses are basically in agreement now.

122.043 **Kataklysmische Veränderliche in Beobachtung, Statistik und Analyse.**
A. Bruch.
Diss. Naturwiss. Westfälische Wilhelms-Univ. Münster, 305 pp. (1982).

122.044 **High-velocity winds from a dwarf nova during outburst.**
F. A. Córdova, K. O. Mason.
Astrophys. J., Vol. 260, 716 - 721 (1982).
An ultraviolet spectrum of the dwarf nova TW Vir during an optical outburst shows shortward-shifted absorption features with edge velocities as high as 4800 km s^{-1}, about the escape velocity of a white dwarf. A comparison of this spectrum with the UV spectra of other cataclysmic variables suggests that mass loss is evident only for systems with relatively high luminosities ($\gtrsim 10\ L_\odot$) and low inclination angles with respect to the observer's line of sight.

122.045 **RZ Sagittae: a new SU Ursae Majoris type dwarf nova.**
H. E. Bond, E. Kemper, J. A. Mattei.
Astrophys. J., Lett., Vol. 260, L79 - L82 (1982) = Contrib. Louisiana State Univ. Obs., No. 168.
The dwarf nova RZ Sagittae was found to be undergoing a supermaximum in 1981 October, making it a new member of the SU Ursae Majoris class of cataclysmic variables. On four consecutive nights of high-speed photometry obtained during this supermaximum, RZ Sge exhibited recurrent 0.2 mag "superhumps" with a period that decreased from 101. 3 to 100.8 minutes.

122.046 **Ultraviolet comparisons of normal outbursts and a supermaximum in two dwarf novae.**
P. Szkody.
Astrophys. J., Vol. 261, 200 - 207 (1982).
IUE spectra of AY Lyr obtained shortly after supermaximum and normal maximum and of RX And after normal maximum are compared. All spectra are compatible with steady state disk models in the ultraviolet while the declining stages show an upturn in flux longward of 2500 Å which indicates the presence of a cool component. The lines of RX And exhibit P Cyg profiles which diminish during the decline.

122.047 **Evidence for starspots on several non-T Tauri pre-main-sequence K stars.**
A. E. Rydgren, J. T. Schmelz, F. J. Vrba.
Bull. American Astron. Soc., Vol. 14, 629 (1982). – Abstract.

122.048 **Analysis of the light curves of FK Comae with the starspot model.**
J. D. Dorren, E. F. Guinan, G. P. McCook.
Bull. American Astron. Soc., Vol. 14, 634 (1982). – Abstract.

122.049 **A search for light-time effects in binary cepheids: AW Persei.**
N. R. Evans.
Bull. American Astron. Soc., Vol. 14, 634 (1982). – Abstract.

122.050 **Properties and nature of Be and shell stars. 11. A notable correlation between the long-term spectral and photometric variations of V 1294 Aql (HD 184 279).**
J. Horn, H. Božić, P. Harmanec, P. Koubský, K. Pavlovski, F. Žďárský.
Bull. Astron. Inst. Czechoslovakia, Vol. 33, 308 - 314 (1982).
The new *UBV* measurements of V 1294 Aql obtained at the Hvar Observatory in the summers of 1979, 1980 and 1981, combined with other available sources of photometric data clearly show that a strong correlation exists between the long-term photometric variations of the star and the long-term spectral variations recently disclosed by Ballereau and Hubert-Delplace (1981). At minimum radial velocity the object is the brightest, and vice versa. This behaviour is formally similar to what is observed for cepheid variables and apparently disagrees with the idea of an elongated revolving envelope. Moreover, the first brightness decrease of the star is clearly connected with the development of the shell spectrum.

122.051 **A search for binaries among short-period southern cepheids.**
W. Gieren.
Rev. Mexicana Astron. Astrofis., Vol. 6, (see 012.017), 103 - 106 (1981).
Simultaneous photoelectric radial velocity and *UBV(RI)*$_{KC}$ photometric observations of 15 southern cepheids with periods between 3 and 7 days have been carried out. Comparison of the resulting radial velocity curves with earlier data reveals the binary nature of three new cepheids, R TrA, AP Sgr and V496 Aql. The binary nature of V350 Sgr, previously announced by Evans (1971), is confirmed.

122.052 **Variations in the ultraviolet spectrum of the symbiotic star Z Andromedae.**
J. Sahade, E. Brandi, J. M. Fontenla.

Rev. Mexicana Astron. Astrofis., Vol. 6, (see 012.017), 201 - 204 (1981).

Variations in line intensity detected on low-resolution IUE spectra of Z Andromedae, taken in January and in July, 1979, are reported. The lines, all of which appear in emission, are classified in 5 groups and a very tentative model of the envelope is proposed.

122.053 **Periods for newly discovered variables in NGC 1851.**
A. Wehlau, M. H. Liller, C. C. Clement, P. Wizinowich.
Astron. J., Vol. 87, 1295 - 1299 (1982).

Blue magnitudes have been obtained for four newly discovered variables in NGC 1851 as well as for two stars lying in the instability strip but thought to be nonvariable. Periods have been derived for four of the variables, including one of those previously thought to be nonvariable. Three of the new variables appear to be RR Lyrae stars and one is a low-amplitude red variable. The fifth variable has a shorter period than the RR Lyrae stars and may be a field W Ursae Majoris star.

122.054 **HR 2724 – a new bright variable in the δ Scuti instability strip.**
D. Baade, O. Stahl.
Astron. Astrophys., Vol. 114, 131 - 134 (1982).

$5^1/_2$ consecutive nights of uvby photometry revealed HR 2724 as a new δ Scuti star of low amplitude. There are at least two periods, $P_A = 0.222$ d, $P_B = 0.148$ d. A third period, $P_C = 0.191$ d, is also likely. Using the period ratios several possible mode identifications are discussed. The pulsation constants, $Q_B = 0.034$ and $Q_C = 0.026$, are in good agreement with theories of radial stellar pulsation and suggest that P_C corresponds to the fundamental and P_B to the first harmonic mode. Four spectrograms show that the radial velocity is just slightly variable, the chemical composition is normal. Spectral type (A 3) and mean photometric data place HR 2724 close to the high-temperature high-luminosity border of the δ Scuti instability strip.

122.055 **Accretion disk model of dwarf novae.**
R. Hoshi.
Prog. Theor. Phys., Vol. 67, 179 - 189 (1982). – Abstr. in Phys. Abstr., Vol. 85, Abstr. 79127 (1982).

122.056 **Origin of dwarf nova outbursts.**
K. M. V. Apparao.
Bull. Astron. Soc. India, Vol. 10, 29 (1982). – Abstract.

122.057 **Quiescent X-ray emission from flare stars.**
P. C. Agrawal, A. R. Rao.
Bull. Astron. Soc. India, Vol. 10, 34 (1982). – Abstract.

122.058 **Study of the variability of the Delta Scuti stars. III. Radial and nonradial pulsations in KW 207.**
M. Bossi, G. Guerrero, L. Mantegazza, M. Scardia.
Astrophys. Space Sci., Vol. 85, 25 - 29 (1982).

Photoelectric observations of the Delta Scuti star KW 207 (HD 73576) are given. In spite of the low variations, the analysis of the light curves shows at least two significant periodicities, which seem to be related to non-radial pulsations. A third frequency, corresponding to the radial fundamental mode, follows probably from the data.

122.059 **A study of the variability of the Delta Scuti stars. IV. Pulsational modes in BD –6°4932.**
G. Guerrero, L. Mantegazza.
Astrophys. Space Sci., Vol. 86, 139 - 142 (1982).

A total of 321 observations of the Delta Scuti star BD –6°4932, obtained in 1968 by Hall and Mallama (1970), are analyzed. The authors find four frequencies which represent the light curves satisfactorily. The three periods: $P_1 = 0\overset{d}{.}240$, $P_3 = 0\overset{d}{.}182$ and $P_4 = 0\overset{d}{.}114$ seem to correspond to the radial modes of pulsation with $K = 0$, 1, and 3, respectively. The last period $P_2 = 0\overset{d}{.}220$ can be related to a non-radial mode.

122.060 **A detailed study of Cepheid photometric properties in the Walraven system.**
M. Coppola, G. Russo, C. Sollazzo.
Astrophys. Space Sci., Vol. 86, 157 - 168 (1982).

The authors have used the extensive set of photometric data in the *VBLUW* system (Pel, 1976) to perform a thorough statistical analysis of the physical properties of Cepheid variables, which can be inferred directly from these data. They have dealt with the problem of determining colour excesses and mean colours and temperatures; theoretical masses, radii and luminosities have been recalculated.

122.061 **Theoretical blue edges of the Delta Scuti instability strip.**
Ts. G. Tsvetkov.
Astron. Zh., Tom 59, 955 - 962 (1982). In Russian. English translation in Soviet Astron., Vol. 26, No. 5.

Models of Delta Scuti stars have been tested against radial pulsations in the four lowest modes, using a linear, non-adiabatic approach. It is shown that the theoretical blue edge location of the instability strip in the H–R diagram depends strongly on opacity interpolation and on variations of the envelope helium abundance. Other factors have small effects.

122.062 **IUE observations of classical Cepheids.**
W. Eichendorf, B. Reipurth, B. Caccin, G. Russo, C. Sollazzo.
Third European IUE Conference, (see 012.029), p. 129 - 133 (1982).

The impact of IUE UV spectroscopy on studies of Cepheid chromospheres, blue companions of Cepheids, and on mass determinations for Cepheids is discussed. Results on the individual Cepheids R Mus, V 810 Cen, and S Mus are presented.

122.063 **Search for Ti II 5 multiplet emissions in selected Delta Scuti variables.**
F. Castelli, M. Fracassini, L. E. Pasinetti.
Third European IUE Conference, (see 012.029), p. 135 - 138 (1982).

A search for simultaneous emissions in the Ti II 5 multiplet and Mg II h and k doublet has been attempted in the Delta Scuti variables ρ Pup, β Cas and τ Cyg.

122.064 **Variability of cool stars at optical and ultraviolet wavelengths. II. The binary (?) flare star AU Mic.**
J. L. Linsky, P. L. Bornmann, M. Rodonò, V. Pazzani, A. D. Andrews, C. J. Butler, P. B. Byrne.
Third European IUE Conference, (see 012.029), p. 165 - 167 (1982).

Preliminary results of a collaborative observation programme with the IUE are presented. The flare star AU Mic was observed simultaneously with ground-based optical observations. In addition to one definite flaring event, remarkable changes of the BY Dra-type optical light curve and evidence of slow variability in the chromospheric and transition region line fluxes were found. These are probably due to the rotational modulation of photospheric starspots and plages.

122.065 **Simultaneous UV and optical observations of dwarf novae through outburst.**
B. J. M. Hassall, J. E. Pringle, R. A. Wade, J. A. J. Whelan.
Third European IUE Conference, (see 012.029), p. 179 - 186 (1982).

The authors present simultaneous spectrophotometry in the range 1200–7000 Å of three dwarf novae (VW Hydri, WX Hydri and EK TrA) at quiescence and maximum. With one exception, the spectra are poorly fitted by the standard model disc spectrum. P Cygni features are seen in two objects in superoutburst.

122.066 **Line shapes from the symbiotic star V1016 Cyg.**
H. Nussbaumer.
Third European IUE Conference, (see 012.029), p. 209 - 212 (1982).

The high resolution mode on IUE allows observation of the line shapes emitted by V1016 Cyg. Half widths of these lines at half maximum, expressed in terms of velocity, vary between 10 km/s for the lowly ionised atoms, and 40 - 50 km/s for the most highly ionised ions. Feeble wings, corresponding to $v < 600$ km/s are seen on some of the lines.

122.067 **UV observations of V348 Sgr.**
A. Heck, L. Houziaux, A. Cassatella, S. di Serego Alighieri, F. Macchetto.
Third European IUE Conference, (see 012.029), p. 225 - 228 (1982).

UV IUE observations of V348 Sgr are reported together with simultaneous ground-based spectroscopic and photometric observations. The data favor a model including a B0–B1 central star with high C, He, N, O abundance in the normal stage at maximum light. The possibility for V348 Sgr to belong to the evolutionary sequence of R CrB stars is considered.

122.068 **Ultraviolet spectroscopy and effective temperature variations in the extreme helium star BD +13°3224.**
A. E. Lynas-Gray, P. W. Hill, D. Schönberner.
Third European IUE Conference, (see 012.029), p. 301 - 307 (1982).

Observations of the pulsating extreme helium star BD +13°3224 with IUE show the luminosity to vary by 33% of the maximum in a $0\overset{d}{.}1$ period. Effective temperature and angular radius variations are obtained by interpolation between two LTE models. Temperature and pressure changes in the envelope suggest ionization to be important in the pulsation.

122.069 **PS 74: the discovery of a new SU UMa type dwarf nova with high orbital inclination.**
H. Barwig, K. Hunger, R. P. Kudritzki, N. Vogt.
Astron. Astrophys., Vol. 114, L11 - L13 (1982).

The spectrum of the southern white dwarf candidate PS 74 is dominated by broad, doubled emission lines of H and He I. It resembles those of cataclysmic variables with high orbital inclination. Photoelectric photometry in June 1982 revealed the typical behaviour of a SU UMa type dwarf nova during decline from a superoutburst, i.e. superhumps with an amplitude of $\sim 0\overset{m}{.}1$ and a period of $0\overset{d}{.}08409$. Spectroscopic observations during quiescence yielded variable continuum flux, equivalent widths and radial velocities consistent with an orbital period of about 2^h and a semi-amplitude of the radial velocity curve $K_1 \approx 120$ km s^{-1} for the white dwarf primary.

122.070 **The light curves of RR Lyrae field stars.**
N. R. Simon, T. J. Teays.
Astrophys. J., Vol. 261, 586 - 594 (1982).

Fourier decompositions have been made of the light curves of a large sample of RR Lyrae field stars. The coefficients have been tabulated. Certain combinations of the low-order coefficients are plotted against period in order to reveal any regular progression similar to that displayed by classical Cepheids.

122.071 **The chemical composition of R Coronae Borealis and XX Camelopardalis.**
P. L. Cottrell, D. L. Lambert.
Astrophys. J., Vol. 261, 595 - 611 (1982).

Three R Coronae Borealis stars (R CrB, XX Cam, and RY Sgr) have been examined using extensive, high resolution, high signal-to-noise Reticon data. Atmospheric parameters and elemental abundances for a wide range of species were derived. The evolutionary history of these stars is discussed. It is concluded that both CNO cycle and 3 α processed material must have been mixed into the observable layers.

122.072 **The infrared properties of Mira-type variables and other cool stars as determined from *JHKL* photometry.**
M. W. Feast, B. S. C. Robertson, R. M. Catchpole, T. Lloyd Evans, I. S. Glass, B. S. Carter.
Mon. Not. R. Astron. Soc., Vol. 201, 439 - 450 (1982).

Extensive infrared photometry of Me(Mira) variables and of C, S, SC and related stars is used to establish two-colour diagrams and period–colour relations. The characteristic areas in these plots occupied by the different classes of objects are delineated. The main aim is to provide a norm with which other stars can be compared. The $(J-H)$ versus period relation offers a method for determining the interstellar reddenings of Me variables. The strong concentration to the longer periods of Miras which are OH masers is probably due primarily to the fact that in the sample surveyed, these stars also have the brightest apparent bolometric magnitudes. Period–amplitude plots are shown for Me variables. The Mira components in symbiotic stars are found to have exceptionally large amplitudes.

122.073 **Observations of the rapid oscillations of EM Cygni.**
R. F. Stiening, M. Dragovan, R. H. Hildebrand.
Publ. Astron. Soc. Pacific, Vol. 94, 672 - 677 (1982).

The authors have made high-speed two-color photometric observations of the dwarf nova EM Cyg and have found rapid coherent oscillations with periods between 14.6 and 16.5 seconds. The minimum period is less than the period of a Keplerian orbit at the surface of a carbon white dwarf. The color temperature of the oscillating component is much higher than that of the constant component of the luminosity. These results suggest that the source of the oscillations is the white dwarf or the inner part of the accretion disk.

122.074 **UW Gruis: a normal ab-type RR Lyrae star.**
A. Bernard.
Publ. Astron. Soc. Pacific, Vol. 94, 700 - 704 (1982).

Photoelectric *UBV* observations in 1978, 1979, and 1981 of the RR Lyrae star UW Gru are presented. The variable clearly belongs to the Bailey ab-type on the basis of the shape of the light and color curves, with a period of 0.548210 day and a mean visual magnitude $V = 13.22$, the amplitude in V being nearly one magnitude. The origin of the heliocentric epochs of light maxima is taken at HJD 2443689.640. A metal abundance [Fe/H] = −1.6 relative to the sun is indicated from the ultraviolet excess parameter $\delta(U-B) = 0.04$. Thus, UW Gru which is located about 2.9 kpc away from the sun, 2.3 kpc below the galactic plane, definitely shows the typical features of a normal halo RR Lyrae star.

122.075 **A spectroscopic study of the c-type RR Lyraes.**
E. Kemper.
Astron. J., Vol. 87, 1395 - 1408 (1982) = Lick Obs. Bull. No. 932.

Image tube scanner observations have been made for a sample of field c-type RR Lyraes. Using these spectra, the author shows how Preston's metal abundance indicator ΔS can be applied to the RRc's to derive [Fe/H] on Butler's system. He has also determined mean temperatures for the field RRc's based on reddening free colors determined from the strength of the Hβ line. These abundances and temperatures allow the delineation of Oosterhoff groups among field stars. It is shown that, as in clusters, the Oosterhoff effect for field stars occurs star by star and not just for mean properties of the group.

122.076 **High-energy observations of stellar flares: comparison with the sun.**
H. S. Hudson.
AIP Conf. Proc., No. 77, (see 012.034), p. 383 - 393 (1982).

This paper reviews recent observations of flaring activity on flare stars of the UV Ceti type, concentrating on X-ray and radio data in comparison with possible solar analogs. Extending the analogy, the author estimates the hard X-ray and γ-ray fluxes expected from typical stellar flares.

122.077 **Ultraviolet light curves of the dwarf novae U Geminorum and VW Hydri.**
C.-C. Wu, R. J. Panek.
Astrophys. J., Vol. 262, 244 - 252 (1982).

Ultraviolet light curves have been obtained for the quiescent dwarf novae U Geminorum and VW Hydri with the five-band spectrophotometer on board the ANS satellite. The amplitude of the hump associated with the hot spot is smaller in the UV than in the visible. The flux distribution of the disk of U Gem cannot be fitted by model spectra of steady state, viscous accretion disks. The flux distribution of VW Hydri can be matched roughly by model accretion disks.

122.078 **HEAO 1 observations of quiescent X-ray emission from flare stars.**
V. Tsikoudi.
Astrophys. J., Vol. 262, 263 - 268 (1982).

The HEAO A-2 data were searched at the positions of the 70 known flare stars for evidence of X-ray emission. Thirteen of them appear to be detected at the 3 σ level. There were no detectable X-ray flare events during the time of HEAO 1 observations. This detection, therefore, refers to the quiescent state X-ray emission from the coronae of these stars.

122.079 **Convection in pulsating stars. II. RR Lyrae convection and stability.**
R. F. Stellingwerf.
Astrophys. J., Vol. 262, 339 - 343 (1982).

Convective model envelopes of two RR Lyrae stars are constructed, one centrally located in the instability strip, and one just beyond the red edge of the strip. Steady state solutions and small amplitude oscillations are studied using a nonlinear pulsation code. The author finds that convection has only a small effect on the stability of the hotter model, primarily due to turbulent pressure work in the hydrogen ionization zone. The cooler model, however, shows strong reduction in both hydrogen and helium ionization driving, which quenches the pulsation.

122.080 **The mass of the anomalous Cepheid in the globular cluster NGC 5466.**
R. Zinn, C. R. King.
Astrophys. J., Vol. 262, 700 - 708 (1982).

The effective temperature and surface gravity of the anomalous Cepheid V19 in NGC 5466 have been measured from spectrophotometric scans. The results rule out the possibilities that it is an RR Lyrae or a BL Herculis variable, which strengthens the already strong case that V19 is an unusual Cepheid variable and a member of NGC 5466. V19 has the same radial velocity as NGC 5466 and a period and luminosity that are similar to those of the anomalous Cepheids in the dwarf spheroidal galaxies. A precise mass estimate for V19 is obtained, $M = 1.56^{+0.23}_{-0.30}\ M_{\odot}$, which is consistent with the theory that the large masses of the anomalous Cepheids result from the transfer of mass in close binary systems.

122.081 **The very small amplitude cepheids HD 9250 and HD 14662.**
G. Burki, W. Benz.
Astron. Astrophys., Vol. 115, 30 - 38 (1982).

Photoelectric measurements in radial velocity (spectrophotometer CORAVEL) and photometry (Geneva system) of two recently discovered bright cepheids of very small amplitude are presented. The width of the CORAVEL dip varies with the pulsation phase. The geometrical effect resulting from the radial pulsation cannot explain this variation. Another broadening mechanism, probably the turbulence, varies with the pulsation phase.

122.082 **Photometric observations of CN Orionis.**
R. Schoembs.
Astron. Astrophys., Vol. 115, 190 - 196 (1982) = Veröff. Sternw. München, Bd. 7, Nr. 34.

Extended photometric observations of CN Ori have been carried out in December 1981 during 14 of 16 consecutive nights. Two eruptions and the minimum stage in between were covered. The lightcurves show humplike variations with two periods of $0\overset{d}{.}1631$ and $0\overset{d}{.}1595$, the longer one being the orbital period. The amplitude of the orbital hump does not vary considerably during the observations. This is shown to be an argument in favour of the Osaki model for dwarf nova outbursts. System parameters of CN Ori are estimated using the orbital period and high frequency oscillation periods.

122.083 **Frequency analyses of light and radial velocity observations of α Lupi.**
P. Lampens, M. Goossens.
Astron. Astrophys., Vol. 115, 413 - 419 (1982).

Photoelectric photometric observations by Van Hoof are used to study the light variability of α Lupi. Frequency analyses of Van Hoof's observations in Y and UV light reveal the existence of only one oscillation with an amplitude larger than the noise level of the observations. The frequency of this oscillation is determined to be f = 3.848 424 ± 0.000 040 c/d and shows no variation that can be detected with Van Hoof's observations. The authors show that both the light variability and the radial velocity variations can be represented by an oscillation with a constant frequency.

122.084 **Physical properties of V1329 Cygni.**
T. Iijima.
The nature of symbiotic stars, (see 012.045), p. 169 - 171 (1982).

Grygar et al. (1979) proposed for V1329 Cygni a long period eclipsing binary model consisting of a massive M giant and a hot component. Some problems, however, have remained on the binary model. Many observed phenomena are well known in outbursts of novae (McLaughlin 1960) and nova-like variables (Swings 1970). It may be reasonable to consider that light variations in the last 10 years arised from recurrent outbursts. White dwarfs with rapid mass accretion may be also one of the possible models for V1329 Cyg.

122.085 **Determination of the term symbiotic star.**
A. A. Boyarchuk.
The nature of symbiotic stars, (see 012.045), p. 225 - 230 (1982).

The author proposed the following criteria for the use of the term symbiotic star: The symbiotic stars must have a spectrum which simultaneously present the cool star features (TiO bands or G-band, etc.), and the emission lines of HeII and/or [OIII], and/or [NeIII], and lines which require even higher ionization level. He also proposed a classification of symbiotic stars according to different types of observations.

122.086 **De veelzijdige veranderlijke sterren.**
J. W. Pel.
Zenit, 9. Jaarg., 500 - 505 (1982).

122.087 **Search for pulsating stars in multiple stellar systems.**
E. Antonello, M. Fracassini, L. E. Pasinetti, L. Pastori.

Binary and multiple stars as tracers of stellar evolution, (see 012.046), p. 27 - 31 (1982).

The importance of the search for pulsating stars of the δ Scuti type in multiple stellar systems is emphasized, in particular for the following objects: the evolutionary status of these systems, the interconnection between multiplicity and pulsation (Frolov et al., 1980), a verification of the theory of the turbulent mixing (Vauclair, 1976). A list of possible pulsating companions in visual and spectroscopic systems with foreseeable period and amplitude is given. Some preliminary results of a spectrophotometric survey begun at the Observatories of Milano – Merate and Bologna – Loiano are also given.

122.088 **An ad hoc hypothesis on the pulsation of Delta Scuti stars.**
E. Antonello.
Binary and multiple stars as tracers of stellar evolution, (see 012.046), p. 33 - 35 (1982).

Taking into account the possible influence of external factors such as the orbital motion in binary systems and the rotational velocity on the pulsation of δ Scuti stars, the author explains some properties of these variables.

122.089 **Ultraviolet light curves of U Geminorum and VW Hydri.**
C.-C. Wu, R. J. Panek, A. V. Holm, F. H. Schiffer, III.
Binary and multiple stars as tracers of stellar evolution, (see 012.046), p. 191 - 197 (1982).

Ultraviolet light curves have been obtained for the quiescent dwarf novae U Gem and VW Hyi. The amplitude of the hump associated with the accretion hot spot is much smaller in the UV than in the visible. This implies that the bright spot temperature is roughly 12000 K if it is optically thick. A hotter spot would have to be optically thin in the near UV. The flux distribution of U Gem in quiescence cannot be fitted by model spectra of steady state, viscous accretion disks. The absolute luminosity, the flux distribution, and the far UV spectrum suggest that the primary star is visible in the far UV. The optical-UV flux distribution of VW Hyi could be matched roughly by the authors' model accretion disks.

122.090 **UV spectroscopy of the novalike variable TT Arietis.**
W. Wargau, H. Drechsel, J. Rahe, G. Klare, B. Wolf, J. Krautter, N. Vogt.
Binary and multiple stars as tracers of stellar evolution, (see 012.046), p. 453 - 454 (1982).

TT Ari was detected by Strohmeier et al. (1957) and is classified as a novalike variable. It was hitherto unclear whether TT Ari is a special type of dwarf nova (Warner, 1976) or an old nova (Cowley et al., 1975). From UV spectra Krautter et al. (1981) concluded that TT Ari is a dwarf nova with extremely extended standstills as they are typical for Z Cam stars. The absolute flux distribution of the UV continuum can be interpreted as arising from two black body radiators with very different temperatures: the accretion disc with about 20000 K, and a hot component with 150000 - 200000 K.

122.091 **Did SU UMa undergo a classical nova outburst?**
M. F. Bode, A. Evans, A. Bruch.
Binary and multiple stars as tracers of stellar evolution, (see 012.046), p. 475 - 481 (1982).

Recent observations of an apparent soft X-ray halo around the dwarf nova SU UMa have led to a speculation that this may well be an evidence of the object having undergone a classical nova-like outburst within historical times (Cordova and Mason 1980). An alternative interpretation of the halo in terms of scattering of soft X-rays from dwarf nova outbursts by interstellar grains is suggested.

122.092 ***o* And: polarization observations of 1982 April evince the recent shell episode.**
D. P. Hayes.
Inf. Bull. Variable Stars, No. 2160, 2 pp. (1982).

122.093 **Variability of Kappa Cassiopeiae.**
J. K. Badalia, H. S. Gurm.
Inf. Bull. Variable Stars, No. 2163, 3 pp. (1982).

122.094 **"Cepheid" EY Sgr nonvariable?**
J. A. R. Caldwell, I. M. Coulson.
Inf. Bull. Variable Stars, No. 2167, 2 pp. (1982). – See also 123.016.

122.095 **1981 photometry of UX Arietis.**
M. Zeilik, R. Elston, G. Henson, P. Smith.
Inf. Bull. Variable Stars, No. 2168, 4 pp. (1982).

122.096 **Photoelectric observations of the flare star BY Dra in 1979.**
L. N. Mavridis, P. Varvoglis.
Inf. Bull. Variable Stars, No. 2174, 2 pp. (1982).

122.097 **The photometric constancy of 59 Piscium.**
S. F. González-Bedolla.
Inf. Bull. Variable Stars, No. 2178, 2 pp. (1982).

122.098 **Investigation of the pulsational peculiarities of the hotter component in the eclipsing binary system AB Cas.**
M. S. Frolov, E. N. Pastukhova, A. V. Mironov.
Inf. Bull. Variable Stars, No. 2179, 4 pp. (1982).

122.099 **Photoelectric observations of the flare star EV Lac in 1976.**
G. Asteriadis, M. E. Contadakis, F. Mahmoud, L. N. Mavridis, D. Stavridis, P. Varvoglis.
Inf. Bull. Variable Stars, No. 2183, 10 pp. (1982).

122.100 **August 1981 crisis of V348 Sgr.**
A. Heck, L. Houziaux, J. Manfroid.
Inf. Bull. Variable Stars, No. 2184, 4 pp. (1982).

122.101 **Photoelectric observations of BY Dra in 1980 and 1982.**
M. A. Soliman.
Inf. Bull. Variable Stars, No. 2193, 2 pp. (1982).

122.102 **An intense optical flare from XY UMa.**
M. Zeilik, R. Elston, G. Henson.
Inf. Bull. Variable Stars, No. 2200, 3 pp. (1982).

122.103 **Spectroscopic investigation of VZ Cnc.**
G. A. Garbusov *(Garbuzov)*.
Inf. Bull. Variable Stars, No. 2205, 3 pp. (1982).

122.104 **Photoelectric observations of the flare star BY Dra in 1977.**
L. N. Mavridis, P. Varvoglis.
Inf. Bull. Variable Stars, No. 2209, 2 pp. (1982).

122.105 **Photoelectric observations of the flare star BY Dra in 1980.**
G. Asteriadis, S. Avgoloupis, L. N. Mavridis, P. Varvoglis.
Inf. Bull. Variable Stars, No. 2210, 3 pp. (1982).

122.106 **Photoelectric observations of the flare star AD Leo in 1982.**
K. P. Panov, M. Grigorova, A. Tsintsarova.
Inf. Bull. Variable Stars, No. 2220, 9 pp. (1982).

122.107 **Flare stars in the Pleiades.**
M. K. Tsvetkov, K. P. Tsvetkova, A. G. Tsvetkova,

S. A. Tsvetkov, H. S. Chavushian *(O. S. Chavushyan)*.
Inf. Bull. Variable Stars, No. 2224, 2 pp. (1982).

122.108 **Is UX Dra a carbon binary star?**
M. Vetešnik.
Inf. Bull. Variable Stars, No. 2225 (1982).

122.109 **Flare stars in Orion.**
R. Sh. Natsvlishvili.
Inf. Bull. Variable Stars, No. 2231, 4 pp. (1982).

122.110 **Four colour photometry of the bright pre-main sequence shell star V856 Sco (HR 5999).**
J. Andersen, J. V. Clausen, B. Nordström.
Inf. Bull. Variable Stars, No. 2234, 2 pp. (1982).

122.111 **HD 154973, a new short period variable star.**
M. A. Hobart, J. H. Peña, R. Peniche.
Inf. Bull. Variable Stars, No. 2235, 4 pp. (1982).

122.112 **Refinement of the frequencies of pulsation of Delta Scuti.**
D. W. Coates, T. T. Moon, K. Thompson, M. L. Winsall.
Inf. Bull. Variable Stars, No. 2238, 3 pp. (1982).

122.113 **TT Arietis.**
IAU Circ., Nos. 3716, 3748 (1982).

122.114 **R Aquarii.**
IAU Circ., Nos. 3729, 3734 (1982).

122.115 **Z Andromedae.**
IAU Circ., Nos. 3738, 3743, 3748, 3750 (1982).

122.116 **TY Piscium.**
IAU Circ., Nos. 3740, 3747 (1982).

122.117 **SU Tauri.**
IAU Circ., No. 3740 (1982).

122.118 **VZ Sculptoris.**
IAU Circ., No. 3756 (1982).

122.119 **New maxima of CY Aquarii observed in August 1980 and accurate determination of a mean maximum.**
P. Ralincourt.
GEOS Circ., RR4, 5 pp. (1982).

122.120 **The variability of RR Tel.**
A. Heck.
Messenger, No. 30, p. 6 - 9 (1982).

122.121 **A note on BM Cha.**
B. Warner.
Mon. Notes Astron. Soc. South. Africa, Vol. 41, 49 (1982).

122.122 **XZ Draconis.**
K. Wenske.
BAV Rundbrief, 31. Jahrg., 81 - 83 (1982).

122.123 **Neue Beobachtungen und instantane Elemente für X Ari.**
W. Braune.
BAV Rundbrief, 31. Jahrg., 84 - 85 (1982).

122.124 **Zwergnovae.**
BAV Rundbrief, 31. Jahrg., 87 - 96 (1982).

122.125 **Is V986 Ophiuchi a tidally driven variable?**
C. T. Bolton, A. Fullerton.
J. R. Astron. Soc. Canada, Vol. 76, 316 (1982). – Abstract.

122.126 **Pulsational mode switching in HD 161796.**
J. D. Fernie.
J. R. Astron. Soc. Canada, Vol. 76, 319 (1982). – Abstract.

122.127 **On the statistics of ZZ Ceti stars.**
P. Lacombe, G. Fontaine.
J. R. Astron. Soc. Canada, Vol. 76, 322 (1982). – Abstract.

122.128 **The cepheid P-L-C-X relation – a third parameter emerges.**
R. J. McGonegal, R. A. McLaren, B. F. Madore, C. W. McAlary.
J. R. Astron. Soc. Canada, Vol. 76, 323 (1982). – Abstract.

122.129 **On the trail of new calibrating objects for the cepheid period-luminosity relation.**
D. Turner.
J. R. Astron. Soc. Canada, Vol. 76, 325 - 326 (1982). Abstract.

122.130 **Safe with cepheids?**
D. A. Hanes.
Nature, Vol. 300, 478 - 479 (1982).

122.131 **Structure and outburst models of the dwarf novae and their evolutionary status among the cataclysmic variables.**
N. Vogt.
Mitt. Astron. Ges., Nr. 57, p. 79 - 118 (1982). In German.

The canonical model of cataclysmic binaries, the main eruption properties of dwarf novae and their currently discussed outburst models are briefly summarized. A more detailed description is given for the observed properties of SU UMa stars which are characterized by superoutbursts and superhumps.

122.132 **Neue spektroskopische und photoelektrische Beobachtungen des novaähnlichen Veränderlichen AE Aqr.**
W. Wargau, H. Drechsel, J. Rahe.
Mitt. Astron. Ges., Nr. 57, p. 302 (1982).

122.133 **Interpretation of the relation "colour-brightness velocity decrease" for R CrB-type stars.**
A. F. Pugach.
Astron. Tsirk., No. 1181, p. 1 - 3 (1981). In Russian.

122.134 **Resonance phenomenon in classical cepheids.**
M. Takeuti, T. Aikawa.
Sci. Rep. Tôhoku Univ., Eighth Ser., Vol. 2, 106 - 129 (1981).

To investigate resonance phenomenon in classical cepheids, the non-linear radial oscillation of stars is studied based on the assumption that the non-adiabatic perturbation is expressed in terms of van der Pol's type damping. Two- and three-wave resonance in this system is applied to classical cepheids to describe their bump and double-mode behavior. The phase of bump and the depression of amplitude are explained for bump cepheids. The double-periodicity is shown by the enhancement of the third overtone in three-wave resonance. Non-linear effect on resonant period is also discussed briefly.

122.135 **Hydrodynamic models of classical cepheids. I. A technique to search the limiting amplitude oscillation.**
T. Aikawa, K.-i. Uji-Iye, T. Okuda, M. Takeuti.
Sci. Rep. Tôhoku Univ., Eighth Ser., Vol. 3, 82 - 88 (1982).

The limiting amplitude of hydrodynamical models for cepheid variables is searched by checking the intrinsic growth rate of the oscillation derived by the pumping up. The procedure is scarcely affected by the initial values. The modulation of oscillation appeared at the earlier stage may be enhanced by the overtones in resonance.

122.136 **Photometric observations of CH Cygni.**
M. Nakagiri, Y. Yamashita.
Ann. Tokyo Astron. Obs., Second Ser., Vol. 19, 1 - 7 (1982).

Photoelectric *UBV* observations of CH Cygni made from 1979 through February, 1981 are presented.

122.137 **Photometric and spectroscopic observations of the peculiar M giant HD 139216 (τ^4 Serpentis).**
Y. Yamashita, M. Nakagiri, Y. Norimoto.
Ann.Tokyo Astron. Obs., Second Ser., Vol. 19, 14 - 23 (1982).

122.138 **The spectrum of CH Cygni in 1981.**
K. H. Yoo, Y. Yamashita.
Ann. Tokyo Astron. Obs., Second Ser., Vol. 19, 38 - 54 (1982).

Two high-dispersion (5.4 Å mm^{-1}) plates covering the photographic region (λ3810–λ4950) were studied. In addition to the M-type absorption lines and the emission lines of ionized metals, a number of absorption lines of ionized metals were found. From their spectral behaviors they are considered of the circumstellar origin. The majority of these circumstellar lines shows the inverse P Cygni profile.

122.139 **On ultraviolet excesses of classical cepheids.**
G. R. Ivanov, A. L. Gachev, N. Petrov.
Astron. Tsirk., No. 1198, p. 6 - 8 (1981). In Russian.

122.140 ***UBV* photometry of DH Peg and RZ Cep.**
G. A. Garbuzov.
Astron. Tsirk., No. 1200, p. 5 - 6 (1981). In Russian.

122.141 **Moments of maxima of RR Lyrae.**
N. B. Kazimirskij.
Astron. Tsirk., No. 1204, p. 7 - 8 (1981). In Russian.

122.142 **Further observations of EV Lac in 1974.**
A. D. Andrews.
Inf. Bull. Variable Stars, No. 2252, 4 pp. (1982).

122.143 **Flares of EV Lac and *UBV* photometry during the quiescent phase in 1975.**
A. D. Andrews.
Inf. Bull. Variable Stars, No. 2253, 4 pp. (1982).

122.144 **Detection of 5 - 6 day outbursts in flare activity in EV Lacertae, interpreted as a rotational effect, and a tentative report of temporal relationships between flare events.**
A. D. Andrews.
Inf. Bull. Variable Stars, No. 2254, 7 pp. (1982).

122.145 **BZ Ursae Majoris – missing link between the dwarf novae of U Geminorum and WZ Sagittae type.**
W. Wenzel.
Inf. Bull. Variable Stars, No. 2256, 2 pp. (1982).

122.146 **Phase relationships in Cepheid variables.**
P. A. Wayman.
Irish Astron. J., Vol. 15, (see 012.061), 121 (1981). Abstract.

122.147 **An unusual microwave flare with 56 second oscillations on the M dwarf L726-8 A.**
D. E. Gary, J. L. Linsky, G. A. Dulk.
Astrophys. J., Lett., Vol. 263, L79 - L83 (1982).

Using the VLA, the authors have observed an unusual flare event on L726-8 A (dM5.5e), the primary star in the M dwarf system containing the prototype flare star UV Cet. This flare had a peak flux of 8 mJy at 6 cm and a corresponding brightness temperature $\gtrsim 10^{10}$ K, was almost entirely right-hand circularly polarized, showed large flux variations on the 10 s time resolution of the VLA, and exhibited quasi-periodic oscillations with a period of about 56 ± 5 s. The authors propose that the observed radiation was due to maser action, probably an electron-cyclotron maser, and that the energy release mechanism was modulated.

122.148 **Starspots: a review of observations and theory.**
S. S. Vogt.
The physics of sunspots, (see 012.063), p. 455 - 479 (1981).

The paper presents a brief review of the current understanding of spots on BY Dra and RS CVn-type variables. Methods for determining various physical parameters of the spots, such as sizes, temperatures, magnetic fields, and migratory motions are reviewed in detail. Two new observational programs currently underway at Lick Observatory to study spot-related activity are also described.

122.149 **Spectral scans of V348 Sagittarii.**
O. Dahari.
Publ. Astron. Soc. Pacific, Vol. 94, 751 (1982). – Abstract.

122.150 **HR 6434 and the factors limiting pulsational amplitudes of δ Scuti stars.**
M. Breger.
Publ. Astron. Soc. Pacific, Vol. 94, 845 - 849 (1982).

A major problem concerning our understanding of δ Scuti stars is the wide range of observed amplitudes. Various reasons for the generally small amplitudes are examined: fast stellar rotation, gravitational settling of helium, and mode-sharing of pulsational energy. It is the purpose of the paper to examine the frequency structure of HR 6434, and to relate the observed limiting amplitudes of this and other δ Scuti stars with the various hypotheses.

122.151 **Kinematic and shock phenomena in the atmosphere of AC Herculis.**
S. R. Baird.
Publ. Astron. Soc. Pacific, Vol. 94, 850 - 859 (1982).

The present work studies the kinematics of the atmosphere of AC Her at a higher dispersion than ever before and uses this data in the construction of a consistent model of photospheric activity that can explain some of the unique and puzzling features in the velocity curve.

122.152 **Photometric variability of single stars with activity cycles.**
J. D. Dorren, E. F. Guinan.
Smithsonian Astrophys. Obs. Spec. Rep. 392, Vol. II, (see 012.065), p. 49 - 64 (1982).

In this investigation the authors selected a sample of nearby F to M main sequence stars to search for evidence of photometric variability and Hα variations which might arise from starspots and the associated active regions, respectively. Systematic cyclic variations of low amplitude (from $0^m.02$ in red to $0^m.04$ in blue) were detected for three of the stars.

122.153 **VLA observations of M-dwarf flare stars.**
P. L. Fisher, D. M. Gibson.
Smithsonian Astrophys. Obs., Spec. Rep. 392, Vol. II, (see 012.065), p. 109 - 114 (1982).

The authors have made observations of six M-dwarf flare stars using the VLA at 6 and 20 cm, some simultaneously with optical observations at Mt. Palomar and Mt. Wilson. The objects include the single stars YZ CMi, CN Leo, and BD+16°2708, and the visual binaries L726-8AB (UV Ceti), Wolf 424 AB and EQ Peg AB. The authors have obtained a number of results which suggest a change is needed in the working models for activity in these systems.

122.154 **On the classification of variable stars.**
P. N. Kholopov.
Perem. Zvezdy, Tom 21, 465 - 484 (1981). In Russian.
A classification system of variable stars is considered. It is based on the development of known classification principles and on the modern conception of stellar evolution.

122.155 **Luminous variable stars in the M 31 Andromeda and the M 33 Triangulum nebulae in 1973 - 1979.**
A. S. Sharov.
Perem. Zvezdy, Tom 21, 485 - 497 (1981). In Russian.
The results of a photometric study of the luminous irregular variables AE And, AF And = V 19, V 15, and VA-1 in M 31 and Var A, Var B, Var C, Var 2, Var 83 and Romano's variable GR 290 in M 33 in 1973 - 1979 are given. It is found that $B-V$ and especially $U-B$ colors become redder at light maximum.

122.156 **A study of period changes of RR Lyrae variables in the globular cluster M 4 (NGC 6121).**
O. G. Suyarkova, S. Yu. Shugarov.
Perem. Zvezdy, Tom 21, 505 - 557 (1981). In Russian.
The period changes of 34 RR Lyrae stars in the globular cluster M 4 were investigated. Fifteen variables show period changes. Several stars appear to have an increase and a decrease in their periods. Six stars show increasing of periods after its last variation and 9 stars show decreasing ones.

122.157 **The Hα line in the spectrum of KR Aurigae.**
Eh. A. Vitrichenko.
Perem. Zvezdy, Tom 21, 573 - 578 (1981). In Russian.
Five spectrograms of the unique variable KR Aur have been obtained. An analysis of the Hα contour suggests that there are three emitting regions in the circumstellar space. Using the model of a single black hole it is shown that one region is a nonrotating gas disc and two other regions are gas streams falling on the star with velocity of order 500 km/s.

122.158 **On the distance of KR Aurigae.**
Eh. A. Vitrichenko, M. D. Popova.
Perem. Zvezdy, Tom 21, 579 - 581 (1981). In Russian.
An analysis of the valuable data on KR Aur suggests that the distance of the star may not exceed 200 - 300 pc.

122.159 **Photoelectric observations of R Aql.**
T. K. Kiseleva.
Perem. Zvezdy, Tom 21, 607 - 616 (1981). In Russian.
Narrowband and *UBVR* observations of R Aql are given. It is shown that the methods by which the physical parameters of non-variable stars are determined can be used for variable stars. It is evaluated that the mass loss in R Aql is $4.6 \times 10^{-7} M_{\odot}$/year.

122.160 **Photoelectric observations of CH Cygni. II.**
L. Luud, J. Vennik, M. Pehk.
Tartu Astrofüüs. Obs. Publ., Köide 49, 77 - 83 (1982). In Russian.
The results of *UBVIJHK* photometry of CH Cygni during Julian days 2442360 - 2442140 and the light-curve for 1967 - 1979 are given. Rapid brightness variations exceeding $0^m.1$ have not been detected during this period of observations. It has been found that the general long time photometric behaviour of CH Cygni has not changed since 1967.

122.161 **The period shortening of RY Sagittarii.**
H. G. Marraco, G. E. Milesi.
Astron. J., Vol. 87, 1775 - 1782 (1982).
Using 10000 visual estimates taken from the literature and obtained between 1897 and 1977, the 39-day period, 0.6-mag amplitude cycle of RY Sgr is analyzed. A secular period change is found. In the time interval studied the period change is from 39.4 to 38.2 days. Superimposed on this variation is a sinusoidal phase change suggesting a periodicity of about half a century. On top of these two phenomena there are phase shifts of about 10 days that happen rather randomly. The meaning of the secular period change is discussed and alternatively explained as mass loss or rapid evolution changes. A possible explanation of the random phase shifts is given.

122.162 **Rapid optical variation of the semiregular variable R Crt.**
S. H. B. Livi, T. S. Bergmann.
Astron. J., Vol. 87, 1783 - 1790 (1982).
The light curves of the semiregular variable R Crt in the optical regions of the spectrum covered by the B, V, and DDO filters, are presented. The DDO magnitudes and colors showed a short-period variations of less than one hour. The magnitudes measured with the wider V and B filters have not shown siginificant variations, indicating that these are produced by spectral features measured by the DDO filters. A short-period radio variation in the same time scale had also been detected before in the 22-GHz H_2O line.

122.163 **Red variable stars. I. UBVRI photometry and photometric properties.**
L. Celis S.
Astron. J., Vol. 87, 1791 - 1802 (1982).
The author presents UBVRI photometry for a group of red variable stars, observed without regard to phase for a continuous period of about four months. The two-color (V–R, R–I) diagram is used to describe the results and to discuss them according to the intensity of the TiO molecular absorption and their variation from minima to maxima and vice versa. The results are also discussed considering the amplitudes relation, the angular diameter measured with and without TiO molecular absorption, the visual absolute magnitudes relation, and the true temperatures of Mira variables indicated by recent work.

122.164 **Variable stars in the globular cluster M 3.**
Z. I. Kadla, A. N. Gerashchenko.
Izv. Glav. Astron. Obs. Pulkovo, Astrometr. Astrofiz., No. 199, p. 86 - 95 (1982). In Russian.
The positions of variable stars with a distance from the center of the globular cluster M 3 $r < 2'.5$ were determined. A search for variable stars led to the discovery of 15 previously unknown RR Lyr variables with $r < 1'.08$. The variability of 32 stars of the 39 variables detected by Holopov (1963, 1977) was confirmed. According to all data there are at present 231 known RR Lyr variables in the globular cluster M 3.

122.165 **On bolometric corrections for short- and long-period cepheids in globular clusters in the Galaxy.**
O. A. Mel'nikov.
Izv. Glav. Astron. Obs. Pulkovo, Astrofiz. Astrometr., No. 200, p. 30 - 38 (1982). In Russian.
RR Lyrae type stars *a* and *b* and cepheids up to $P \leqslant 10$ days in globular clusters of the Galaxy are considered. The complex of the variable stars is considered as a whole in relation to various dependences on periods: absolute value, colour and radius. These dependences were calculated in linear approximation on lg P using the published observational data. The bolometric corrections for the considered stars were obtained on their basis. The absolute values of the obtained bolometric values proved to be larger than those adopted earlier for the same spectral types of stationary stars.

122.166 **Radial pulsation modes and possible heterogeneity of Delta Scuti stars.**
Ts. G. Tsvetkov.
Astron. Zh., Tom 59, 1103 - 1114 (1982). In Russian.
English translation in Soviet Astron., Vol. 26, No. 6.
On the basis of Breger's (1979) new and numerous observational data, a radial pulsation mode for 83 Delta Scuti

stars is estimated. These stars can pulsate in four lowest modes F, $1H$, $2H$, $3H$ and perhaps in $4H$ and even $5H$.

122.167 **Ephemerides of RR Lyrae-type variables for the year 1983.**
W. P. Zessewitsch *(V.P. Tsesevich)*, B. N. Firmaniuk *(Firmanyuk)*, J. M. Kreiner.
Rocznik Astronomiczny Obserwatorium Krakowskiego 1983, (see 047.040), p. 113 - 130 (1982).

122.168 **Wide-band photometry of CH Cygni.**
O. G. Taranova, B. F. Yudin.
Pis'ma Astron. Zh., Tom 8, 722 - 724 (1982). In Russian. English translation in Soviet Astron. Lett., Vol. 8.
Preliminary results of *UBVRJHKLMNQ* photometry of CH Cyg made in 1978 - 1982 are presented. The IR emission of the cool component (M6III) in the active state reveals semiregular variations coinciding in period and phase with those observed in the quiet state. The more precise distance of CH Cyg is estimated to be of (310 ± 50) pc. From 1978 till March 1982 the optical brightness of CH Cyg has been increased by $\sim 2^{m}\!.5$.

122.169 **A PRC relation for Delta Scuti stars.**
E. Antonello, P. Conconi.
Astrophys. Space Sci., Vol. 88, 185 - 188 (1982).
The absolute radii of 78 δ Scuti variables in the Catalogue of the Geneva Observatory have been computed. A PRC relation among the period, radius and color has been obtained.

122.170 **Disk models for cataclysmic variables.**
S. C. Vila.
Astrophys. Space Sci., Vol. 88, 293 - 297 (1982).
This paper presents disk models for cataclysmic variables in which convection in the central layers has been included. The calculation of the vertical structure at different points is presented. The models have a central mass of 1 $M_{\odot}$ and matter fluxes of 10^{-9}, 10^{-8}, and $10^{-7} M_{\odot}$ yr^{-1}. The corresponding luminosities are 1.86, 1.86 × 10 and 1.86 × 10^{2} $L_{\odot}$.

122.171 **Mira Ceti – still admiring.**
J. Svatoš, M. Šolc.
Říše hvězd, Vol. 63, 227 - 229 (1982). In Czech.

122.172 **The photographic photometry of the variables No. 24, 28 and 43 of the globular cluster M 15.**
Y.-h. Chu, C.-j. Shen, X.-h. Di, M.-x. Bao, L.-w. Den, X.-f. Wu.
Acta Astron. Sinica, Vol. 23, 118 - 124 (1982). In Chinese.
In 1979 some plates of the globular cluster M 15 were taken at the focus of the *RC* optical system of the newly installed one meter reflector of Yunnan Observatory. The RR Lyr variables No. 24, 28 and 43 were measured on the improved Sartorius iris photometer and reduced on the TQ-6 computer at Purple Mountain Observatory. The measuring and reduction procedures are described.

122.173 **The photoelectric photometry of VZ Cancri and the change of its period.**
S.-y. Jiang, Z.-z. Yang.
Acta Astron. Sinica, Vol. 23, 333 - 348 (1982). In Chinese.

122.174 **Observations of RR Lyrae with the ANS satellite.**
J. Bonnell, C.-C. Wu, R. A. Bell, J. L. Hutchinson.
Publ. Astron. Soc. Pacific, Vol. 94, 910 - 915 (1982).
Photometric observations of RR Lyr in the ultraviolet have been obtained using the ANS. The observations are compared with theoretical light curves calculated using synthetic spectra and angular diameters determined as a function of phase for RR Lyr by Manduca et al. from photometry at longer wavelengths. A good agreement is found.

122.175 **Increased evidence for overtone pulsation in galactic cepheids: SU Cassiopeiae.**
W. Gieren.
Publ. Astron. Soc. Pacific, Vol. 94, 960 - 968 (1982).
A newly determined radius value, based on a homogeneous and accurate data set and two independent methods, is used together with the currently most-reliable estimates of other physical parameters to discuss the status, and in particular the pulsation mode, of the classical cepheid SU Cas. The assumption that the cepheid pulsates in an excited (most probably the first overtone) mode is able to explain satisfactorily all observed properties of the star, whereas the assumption of fundamental-mode pulsation leads to difficulties which apparently cannot be resolved at the present time.

Early emission line stars.
See Abstr. 003.069.

Les étoiles variables.
See Abstr. 003.095.

La comète de Halley et Mira Ceti dans le Talmud?
See Abstr. 004.029.

Activity in red-dwarf stars.
See Abstr. 011.007.

Narrow band Hα observations of the chromospheric activity of stars.
See Abstr. 031.602.

Le satellite astrométrique Hipparcos et les miras.
See Abstr. 051.004.

Convective accretion disks and the onset of dwarf nova outbursts.
See Abstr. 064.028.

Non-linear stellar oscillations. Two-mode interactions.
See Abstr. 065.008.

Multimode stellar pulsations. II: Anharmonic oscillations.
See Abstr. 065.013.

CNO abundances in Cepheids and supergiants: theoretical implications.
See Abstr. 065.019.

A plea for reexamining heavy element opacities in stars.
See Abstr. 065.022.

On the work integral in stellar pulsation theory.
See Abstr. 065.025.

The effect of non-adiabatic layers on the vibrational behaviour of stars.
See Abstr. 065.043.

Convection in pulsating stars. I. Nonlinear hydrodynamics.
See Abstr. 065.064.

Oscillations of an extended ionization region in a star.
See Abstr. 065.085.

The nonadiabatic analysis of nonradial modes of stellar oscillation in the presence of slow rotation.
See Abstr. 065.086.

Detection of solar flares when the sun is observed as a star.
See Abstr. 073.063.

Radial velocity observations of yellow supergiants.
See Abstr. 111.010.

International Ultraviolet Explorer observations of the R Aquarii jet.
See Abstr. 112.041.

Properties and nature of Be and shell stars. 7B. 88 Her – an important clue to understanding the Be phenomenon?
See Abstr. 112.042.

Observations of SiO maser emission from R Leo during its summer 1981 maximum.
See Abstr. 112.052.

Evidence for starspots on single solar-like stars.
See Abstr. 113.047.

The photometric variability of solar-type stars. I. Preliminary results for the Pleiades, Hyades, and the Malmquist Field.
See Abstr. 113.069.

Extreme helium stars.
See Abstr. 114.043.

The effect of line blocking on the light curves of the Ap stars HD 119213, HD 170000 and HD 173650.
See Abstr. 114.050.

Observations of the first-overtone silicon monoxide bands in late-type stars.
See Abstr, 114.114.

Photographic infrared spectra of symbiotic stars.
See Abstr. 114.148.

Recent studies of the spectrum of V1016 Cygni.
See Abstr. 114.153.

Properties and nature of Be and shell stars. 7A. 88 Her. Observational data, their reduction and basic evaluation.
See Abstr. 114.169.

Stellar activity measured at Ca II H and K.
See Abstr. 114.212.

The growth and decay of magnetic activity in lower main sequence stars.
See Abstr. 116.022.

The relevance of EXOSAT to X-ray observations of RS CVn systems and cataclysmic variables.
See Abstr. 117.059.

A review of the ultraviolet results about symbiotic stars.
See Abstr. 117.068.

Properties of symbiotic stars from studies in the optical region.
See Abstr. 117.102.

Variation spectrale de CI Cyg en 1975.
See Abstr. 117.110.

The peculiar symbiotic object V1329 Cygni: single-star versus binary models.
See Abstr. 117.113.

The peculiar star RX Puppis.
See Abstr. 117.118.

Long term trends in the 3.5 μ light curve of RX Puppis.
See Abstr. 117.119.

HM Sagittae – a most remarkable star.
See Abstr. 117.120.

RR Telescopii.
See Abstr. 117.122.

The frequency of Cepheid binaries.
See Abstr. 117.131.

Analysis of cataclysmic variable star energy distributions.
See Abstr. 117.198.

Photographic photometry of variable stars based on the Tartu photoplate collection. II. Area of μ Persei.
See Abstr. 119.009.

On the statistics of ZZ Ceti stars.
See Abstr. 126.002.

Temperature as a unique predictor of ZZ Ceti variability.
See Abstr. 126.003.

Multicolor variations of the ZZ Ceti stars.
See Abstr. 126.004.

ZZ Ceti stars: variable white dwarfs.
See Abstr. 126.005.

Observations and analysis of the low amplitude ZZ Ceti variable, L19–2.
See Abstr. 126.006.

Photometric observations of GD 358: DB white dwarfs do pulsate.
See Abstr. 126.023.

Introductory report on V1016 Cygni.
See Abstr. 135.038.

Identification of the X-ray source E 1551 + 718 with an uncataloged dwarf nova.
See Abstr. 142.001.

Galactic soft X-ray sources: RS CVn systems, flare stars and A stars.
See Abstr. 142.024.

The ultraviolet spectra of newly discovered hard X-ray cataclysmic variables.
See Abstr. 142.092.

The X-ray transient 3A 1431–409: a highly active RS CVn system?
See Abstr. 142.100.

Possible membership for the 51-day Cepheid GY Sagittae in a new OB association.
See Abstr. 152.001.

The Vela star cloud. II. Early-type stars and long-period Cepheids near the Vela pulsar.
See Abstr. 152.003.

A search for OB associations near southern long-period Cepheids. I. WZ Carinae, YZ Carinae, KK Centauri, and OO Centauri.
See Abstr. 152.010.

New *UBV* photometry for the open cluster Stock 14 and its cepheid-like variable V810 Centauri (= HR 4511).
See Abstr. 153.018.

On the possibility of cluster membership for the cepheid V Centauri.
See Abstr. 153.044.

The metal abundance of Pal 13.
See Abstr. 154.006.

Metal abundances of RR Lyrae stars in globular clusters.
See Abstr. 154.022.

Variable stars in the globular cluster NGC 6293.
See Abstr. 154.033.

Distance and absolute magnitudes of the brightest stars in the dwarf galaxy Sextans A.
See Abstr. 158.010.

Photometry and spectroscopy of long period variables in the Magellanic Clouds.
See Abstr. 159.004.

The metal content of Magellanic Cloud cepheids.
See Abstr. 159.007.

The occurrence of carbon stars and Cepheid variables in common fields of the Magellanic Clouds.
See Abstr. 159.008.

123 Variable Stars (Surveys, Lists of Observations, Charts, etc.)

123.001 **New variable stars in the direction of the bright cloud B in Sagittarius.**
A. Terzan, A. Bijaoui, K. H. Ju, C. Ounnas.
Astron. Astrophys., Suppl. Ser., Vol. 49, 715 - 743 (1982).
621 new variable stars have been detected on plates obtained with the Schmidt telescopes of the Mount Palomar Observatory (1968) and of the European Southern Observatory (1976 - 1980). The astrometrical measurements and the photometrical ones which have followed the detection of the variable stars have led to determine the parameters α, δ, l, b, R(Max), R(Min), δm. The authors notice that: 379 variable stars have an amplitude of variation in magnitude between 0.5 and 2.0 ($\delta m/N$ histogram). The number of variable stars strongly decreases towards the great amplitudes. The surface density of variable stars by square degreee in this central galactic direction increases from 2 to 21 in the explored field of 5°.5 × 5°.5 size.

123.002 **Photographic photometry of variable stars in the α Sagittae region. Semiregular and irregular variables.**
V. Harvig, L. Leis, L. Kalv, P. Kalv.
Publ. Tartu Astrofiz. Obs., Tom 48, 150 - 161 (1981).
The authors continue publishing the results of the investigations of variable stars from Tartu plates. Individual observations are presented in the list of observations and plotted in figures.

123.003 **Optical light curves of some cataclysmic and symbiotic variables.**
J. A. Mattei.
Advances in ultraviolet astronomy, (see 012.006), p. 486 (1982). – Abstract.

123.004 **Tableaux des observations reçues à l'AFOEV en avril, mai et juin 1982.**
Bull. AFOEV, No. 21, p. 111 - 157 (1982).

123.005 **X Leonis, 1970 - 1979.**
D. R. B. Saw.
J. British Astron. Assoc., Vol. 92, 220 - 226 (1982).

123.006 **CD Cygni.**
O. Klinting.
Astron. Tidsskr., Årg. 15, 122 - 126 (1982).

123.007 **Near-infrared observations of some variable stars.**
R. P. Verma, K. M. V. Apparao, R. R. Daniel, S. K. Ghosh, K. V. K. Iyengar, T. N. Rengarajan, S. N. Tandon, N. B. Sanwal, A. Rahman.
Bull. Astron. Soc. India, Vol. 10, 28 (1982). – Abstract.

123.008 **Outbursts of some southern dwarf novae – Paper 1.**
F. M. Bateson.
Publ. Variable Star Sect., R. Astron. Soc. New Zealand, No. 9, p. 1 - 9 (1981).
Several tables list the observed outbursts of the southern dwarf novae AH Eri; TU Men; AQ Eri; CW Mon; BV Pup; BX Pup and BB Vel. The results are discussed and it is shown that TU Men is the most important of these systems.

123.009 **Observations of some southern dwarf novae – Paper 2.**
F. M. Bateson.
Publ. Variable Star Sect., R. Astron. Soc. New Zealand, No. 9, p. 10, 12 - 37 (1981).

123.010 **A note on TY Velorum.**
M. Morel.
Publ. Variable Star Sect., R. Astron. Soc. New Zealand, No. 9, p. 38 - 40 (1981).

123.011 **Observations of some variable stars.**
G. Romano, M. Perissinotto.
Inf. Bull. Variable Stars, No. 2161, 2 pp. (1982).

123.012 **Variability of BD +40°4145 a member of NGC 6910.**
J. Kałuzny, G. Pojmański.
Inf. Bull. Variable Stars, No. 2182, 2 pp. (1982).

123.013 **Times of minima for southern hemisphere eclipsing binaries.**
G. W. Wolf, J. T. Kern, T. L. Hayes, C. R. Chaffin.
Inf. Bull. Variable Stars, No. 2185, 2 pp. (1982).

123.014 **Photoelectric minima of eclipsing binaries.**
E. Pohl, S. Evren, O. Tümer, C. Sezer.
Inf. Bull. Variable Stars, No. 2189, 6 pp. (1982).

123.015 **Remarks on four variable stars.**
T. Berthold.
Inf. Bull. Variable Stars, No. 2192, 2 pp. (1982).

123.016 **Misidentification of EY Sgr.**
M. H. Liller.
Inf. Bull. Variable Stars, No. 2195 (1982). – See also 122.094.

123.017 **Is HD 90994 a "reduction variable"?**
A. Heck, J. Manfroid.
Inf. Bull. Variable Stars, No. 2199, 4 pp. (1982).

123.018 **Photoelectric minima of eclipsing variables.**
Y. Yang, Q. Liu.
Inf. Bull. Variable Stars, No. 2202, 2 pp. (1982).

123.019 **Two new variable stars in M 13.**
T. Russeva, L. Iliev, R. Russev.
Inf. Bull. Variable Stars, No. 2223, 3 pp. (1982).

123.020 **The serendipitous discovery of two new bright variable stars, in the field of HD 108.**
E. F. Guinan, G. P. McCook, A. G. Weisenberger.
Inf. Bull. Variable Stars, No. 2227, 4 pp. (1982).

123.021 **Die B–R der neuesten BAV–Beobachtungszusammenstellung.**
W. Braune.
BAV Rundbrief, 31. Jahrg., 100 - 103 (1982).

123.022 **Tableux des observations reçues à l'AFOEV en juillet, août et septembre 1982.**
Bull. AFOEV, No. 22, p. 168 - 211 (1982).

123.023 **Beobachtungsergebnisse der Berliner Arbeitsgemeinschaft für Veränderliche Sterne e.V. (BAV).**
W. Braune, E. Mundry.
BAV - Mitt., Nr. 34, 9 pp. (1982).

123.024 **LD 23 Cygni: the period – rather not an eclipsing variable.**
K. Locher.
BBSAG Bull., No. 62, p. 5 - 7 (1982).

123.025 **New variable carbon stars.**
L. Duncans.
Astron. Tsirk., No. 1181, p. 7 - 8 (1981). In Russian.

123.026 **Photographic observations of five eclipsing binaries.**
I. V. Il'in, S. A. Yurkovskij, V. K. Kann, V. I. Kirichenko.
Astron. Tsirk., No. 1199, p. 6 - 8 (1981). In Russian.

123.027 **New variable object in Taurus.**
A. L. Gyul'budagyan.
Astron. Tsirk., No. 1200, p. 7 - 8 (1981). In Russian.

123.028 **Bearbeitung von 74 Veränderlichen im Feld β Hydri.**
H. Geßner.
Veröff. Sternw. Sonneberg, Band 9, 361 - 388 (1981).
For 74 variables, discovered at Sonneberg Observatory by C. Hoffmeister, the type of variation, light elements, amplitudes and light curves are given as far as the observational material permits.

123.029 **Comparison stars which turn out to be variables.**
L. E. Pasinetti.
Inf. Bull. Variable Stars, No. 2255 (1982).

123.030 **Variable stars in the General Catalogue of Trigonometric Parallaxes.**
D. Hoffleit.
Bull. Inf. Cent. Données Stellaires, No. 23, p. 53 - 58 (1982).

123.031 **Le nombre d'étoiles variables brillantes.**
C. Jaschek, E. Schweitzer.
Bull. Inf. Cent. Données Stellaires, No. 23, p. 63 - 68 (1982).
This paper presents a first discussion of the completeness of our knowledge of the brighter variables. At present less than 3×10^3 variables are known with m(max) $< 10^m$. A discussion of the different types of variables shows that a considerable bias exists against the discovery of small amplitude variables. An estimate of the number of variable stars with m(max) $< 10^m$ is made and a number of 3×10^4 results.

123.032 **Two variable stars in the cluster NGC 7128.**
A. Alksnis, L. Začs.
Perem. Zvezdy, Tom 21, 499 - 503 (1981). In Russian.
The results of photographic photometry in the *R* (0.63), *V* and *B* bands of the two variable stars NGC 7128-5 = HBV 478 = V1481 Cyg and NGC 7128-4 during the time interval 1968 - 1976 are presented. The star V1481 Cyg seems to belong to the type of variability EB (P = $2^d.7634$), while the star NGC 7128-4 probably belongs to the type Ia.

123.033 **Distribution of the stars in the *General Catalogue of Variable Stars*.**
S. B. Howell.
Publ. Astron. Soc. Pacific, Vol. 94, 969 - 976 (1982).
The distribution of nearly 20,000 stars from the *General Catalogue of Variable Stars* is presented. The stars have been classified by variable type, apparent magnitude, and galactic latitude.

New catalogue of stars suspected of variability containing data on 14810 variable stars which have not received definite nomenclature until 1980.
See Abstr. 002.039.

Un nouveau catalogue d'étoiles variables suspectes.
See Abstr. 002.046.

New catalogue of suspected variable stars.
See Abstr. 002.090.

Photoelectric photometry of variable stars.
See Abstr. 003.048.

94th - 96th list of minima of eclipsing binaries.
See Abstr. 119.080.

124 Novae

124.001 **A model for the 1979 outburst of the recurrent nova U Scorpii.**
S. Starrfield, W. M. Sparks, R. E. Williams.
Advances in ultraviolet astronomy, (see 012.006), p. 470 - 473 (1982).
The authors' conclusions are that the rapid accretion (disk instability) theory for the outburst of a recurrent nova such as U Sco does not fit the observations. The thermonuclear runaway theory is somewhat more successful but the length of the outburst was much too long to agree with the observations.

124.002 **The ultraviolet variability of T CrB.**
A.Cassatella, P. Patriarchi, P. L. Selvelli, L. Bianchi, C. Cacciari, A. Heck, M. Perryman, W. Wamsteker.
Advances in ultraviolet astronomy, (see 012.006), p. 482 - 485 (1982).
Preliminary results of the ultraviolet variability monitoring of the recurrent nova T CrB are presented. The data suggest that the UV variability is due to changes in the physical structure of the accretion disk around the secondary.

124.003 **CW 1103 + 254: a new cataclysmic at high galactic latitude.**
S. N. Shore, C. B. Foltz, A. J. Wasilewski, P. L. Byard, R. M. Wagner.
Publ. Astron. Soc. Pacific, Vol. 94, 682 - 689 (1982).
The authors discuss the discovery and observation of a probable old nova CW 1103 + 254, which is at high galactic latitude (+68°). It is shown that the properties of this star are consistent with the classical old nova phenomenology. Some estimates on the evolutionary status of the red companion and on the various system parameters are given. The object is a morphologically similar star to GK Per although not a known variable. Some suggestions for future observations are made.

124.004 **On the wind model of novae.**
S. Kwok.
Bull. American Astron. Soc., Vol. 14, 633 (1982). – Abstract.

124.005 **The UV variability of T CrB.**
A. Cassatella, P. Patriarchi, P. L. Selvelli, L. Bianchi, C. Cacciari, A. Heck, M. Perryman, W. Wamsteker.
Third European IUE Conference, (see 012.029), p. 229 - 231 (1982).
The recurrent nova T CrB has been found to be strongly variable in the ultraviolet, both in the continuum and in the emission lines, while it remained constant in the *V* band. The continuum and the rich emission line spectrum during the activity phase are briefly discussed.

124.006 **Novae as sources of nitrogen in galaxies.**
R. E. Williams.
Astrophys. J., Lett., Vol. 261, L77 - L80 (1982).
A comparison of the abundances of nova ejecta and planetary nebulae together with their masses and rates of occurrence in the Magellanic Clouds and M 31 indicates that novae currently inject more nitrogen into the interstellar medium of these galaxies than do the planetary nebulae.

124.007 **VY Aquarii.**
R. H. McNaught.
Inf. Bull. Variable Stars, No. 2232 (1982).

124.008 **RS Ophiuchi.**
IAU Circ., Nos. 3716, 3741 (1982).

124.009 **Novae in M31.**
IAU Circ., Nos. 3725, 3729 (1982).

124.010 **WX Ceti: une nova recurrente ou un nouveau type de variable?**
M. Verdenet.
Bull. AFOEV, No. 22, p. 161 - 162 (1982).

124.011 **Nuclear theory of novae.**
J. W. Truran.
Essays in nuclear astrophysics, (see 003.013), p. 467 - 493 (1982).
The picture of a nova outburst which emerges from theoretical studies of the response to runaway is presented. The evolution of a nova following runaway and its relation to the conventional "hydrostatic" evolution of prewhite dwarfs is considered and the role of the CNO reactions in distinguishing slow and fast novae is elaborated.

124.012 **Mass loss from novae and supernovae.**
M. Friedjung.
Pré-publ. Inst. Astrophys. Paris, No. 16, 12 pp. (1982).
To appear in 'Highlights of Astronomy', Vol. 6.

Hot accreting white dwarfs in the quasi-static approximation.
See Abstr. 065.002.

Radio observations of symbiotic stars.
See Abstr. 116.017.

A comparative study of IUE spectra of cataclysmic variables.
See Abstr. 117.069.

QU Carinae: orbital parameters and spectra for a nova-like variable.
See Abstr. 117.082.

The mystery of the missing boundary layer.
See Abstr. 117.098.

Properties of symbiotic stars from studies in the optical region.
See Abstr. 117.102.

The evolutionary status of symbiotic stars.
See Abstr. 117.128.

Did SU UMa undergo a classical nova outburst?
See Abstr. 122.091.

Condensation of grains.
See Abstr. 131.305.

Bulge X-ray sources and novae in M 31.
See Abstr. 158.083.

Nova Vulpeculae 1976 = NQ Vulpeculae

124.101 **Spectral study of Nova Vul 1976.**
L. V. Gudim, V. G. Karetnikov, Yu. A. Medvedev.
Astron. Zh., Tom 59, 711 - 718 (1982). In Russian. English translation in Soviet Astron., Vol. 25, No. 4.
From 13 diffraction spectrograms of Nova Vul 1976 obtained from November 4 to December 27, 1976 the intensities of about 350 spectral lines in absorption and emission were identified and measured. For the strongest lines the velocities of the envelope expansion were determined. As a whole, Nova Vul 1976 can be considered as a typical mean-fast nova star which can be described by a model of Baade and Boyarchuk–Gershberg showing more complicated main and additional envelopes.

Nova Monocerotis 1939 = BT Monocerotis

124.111 **Observations of prenovae orbital periods.**
B. E. Schaefer, J. Patterson.
Bull. American Astron. Soc., Vol. 14, 633 (1982). – Abstract.

Nova Herculis 1934 = DQ Herculis

124.201 **Why DQ Her is not an X-ray source.**
J. A. Petterson.
Galactic X-ray sources, (see 003.001), p. 409 - 410 (1982). Abstract.

Observations of prenovae orbital periods.
See Abstr. 124.111.

Erratum

124.219 **Erratum: 'Eruptive binaries, X. DQ Herculis'**
[Acta Astron., Vol. 30, 267 - 283 (1980)].
J. Smak.
Acta Astron., Vol. 31, 529 (1981). – See Abstr. 28.124.303.

Nova Vulpeculae 1979 = PU Vulpeculae

124.301 **Spectroscopic observations of PU Vulpeculae (Nova Vul 1979).**
Y. Yamashita, H. Maehara, Y. Norimoto.
Publ. Astron. Soc. Japan, Vol. 34, 269 - 280 (1982).
Spectrograms secured for PU Vulpeculae since April 1979 at Okayama (13 - 108 Å mm^{-1}) and at Lick (35 and 11 Å mm^{-1}) were examined. For interpreting photometric and spectroscopic properties observed, a veiling model is proposed.

124.302 **Photometric observations of PU Vulpeculae (Nova Vul 1979) in faint phases.**
M. Nakagiri, Y. Yamashita.
Ann. Tokyo Astron. Obs., Second. Ser., Vol. 19, 8 - 13 (1982).
Five-colour (*UBVRI*) observations made for PU Vulpeculae from September 15, 1980 to February 7, 1981 are presented.

The Kuwano peculiar object (PU Vulpeculae).
See Abstr. 117.124.

Nova Persei 1901 = GK Persei

124.311 **The peculiar energy distribution of the old nova GK Per (1901).**
A. Bianchini, F. Sabbadin.
Third European IUE Conference, (see 012.029), p. 187 - 189 (1982).
The UV spectrum of the old nova GK Per (1901) at quiescence greatly differs from those of the other observed old novae. In spite of the presence of high excitation emission lines the UV to visual continuum energy distribution is peaked at 3600 Å. The authors have constructed a grid of models of accretion discs around a magnetic white dwarf to fit the observed flux distribution.

Nova Coronae Austrinae 1981

124.401 **IUE observations of the 1981 outburst of Nova CrA.**
W. M. Sparks, S. Starrfield, S. Wyckoff, R. E. Williams, J. W. Truran, E. P. Ney.
Advances in ultraviolet astronomy, (see 012.006), p. 478 - 481 (1982).
IUE spectra of Nova CrA 1981 have been obtained from outburst in April 1981 through mid November 1981. These spectra cover the range 1150–3200 Å, and are mostly at low dispersion. Line emission fluxes and widths have been measured in all spectra. Results of a preliminary abundance analysis is reported.

Nova Delphini 1967 = HR Delphini

124.411 **The orbital variations of the UV spectrum of the old nova HR Del.**
Y. Andrillat, M. Friedjung, P. Puget.
Third European IUE Conference, (see 012.029), p. 191 - 192 (1982).
IUE spectra indicate the presence of a power law continuum, close to the classical accretion disk law of Lynden-Bell. A very high accretion rate of $10^{-6} M_\odot$/yr is found for the disk. Orbital variations of the continuum and emission line fluxes during 1980 can be interpreted as due to occultation of central regions of the disk by material associated with the outer edge of the disk.

124.412 **The UV spectrum of the old nova HR Del at different orbital phases.**
M. Friedjung, Y. Andrillat, P. Puget.
Astron. Astrophys., Vol. 114, 351 - 356 (1982).
Observations of this old nova with IUE are used to obtain a dereddened continuum energy distribution, which is quite close to that of a classical accretion disk of Lynden-Bell. A very high accretion rate of $10^{-6} M_\odot\ yr^{-1}$ is derived from this distribution. The variations of line and continuum fluxes with orbital phase are examined. They can most easily be understood as due to occultation of central parts of the disk by a splash where the accreted gas meets the disk.

124.413 **Spektroskopische Untersuchungen an der alten Nova HR Delphini.**
A. Bruch.
Mitt. Astron. Ges., Nr. 57, p. 328 - 331 (1982).

124.414 **The UV spectrum of the old nova HR Del at different orbital phases.**
M. Friedjung, Y. Andrillat, P. Puget.
Pré-publ. Inst. Astrophys. Paris, No. 10, 16 pp. (1982).
Submitted to Astron. Astrophys.

124.415 **A spectroscopic study of the old nova HR Delphini.**
A. Bruch.
Publ. Astron. Soc. Pacific, Vol. 94, 916 - 925 (1982).
Spectroscopic observations of the old nova HR Del are presented. A radial-velocity curve is given. It is shown that the currently adopted binary period is too small. An improved orbital period is given. The measurements and the canonical model for cataclysmic variables yield masses of $M_1 = 0.9\ M_\odot$ and $M_2 = 0.58\ M_\odot$ for the binary components. The spectral type of the secondary is estimated to be near K8.

Nova Pictoris 1925 = RR Pictoris

124.421 **Short-time spectral variations of the old nova RR Pic.**
P. L. Selvelli.
Third European IUE Conference, (see 012.029), p. 197 - 199 (1982).
The UV spectrum of RR Pic has been continuously monitored during about 6 hours. Significant variations are clearly evident in the continuum energy distribution. The line spectrum shows dramatic changes in the emissions intensity.

Nova Aquilae 1982

124.501 **Ultraviolet observations of Nova Aquilae 1982.**
M. A. J. Snijders, M. J. Seaton, J. C. Blades.
Advances in ultraviolet astronomy, (see 012.006), p. 625 - 628 (1982).
The authors have obtained a number LWR and SWP spectra of Nova Aquilae 1982 with the IUE. Observations made during 24 Feb and Mar 2 show that the nova has a heavily reddened spectrum, E(B–V) = 0.55 ± 0.15, and a complex short-wavelength spectrum with both narrow emission features and broad absorption troughs, indicating terminal velocities as high as 10000 km/s. Both the continuum flux and the absorption line profiles changed between the two observations.

124.502 **Nova in Aquila.**
V. A. Bronshtehn.
Zemlya Vselennaya, 1982, No. 4, p. 13. In Russian.

124.503 **Ultraviolet observations of Nova Aquilae 1982.**
M. A. J. Snijders, M. J. Seaton, J. C. Blades.
Third European IUE Conference, (see 012.029), p. 177 - 178 (1982).
A bright nova in Aquila was announced by Honda on 27.85 Jan 1982 UT, with a visual magnitude of 6 to 7 at discovery. The authors have obtained a number of spectra of the object with the IUE. Observations made during 24 Feb. and 2. Mar. show that the nova has a heavyly reddened spectrum, and a complex short wavelength spectrum with both narrow emission features and broad absorption troughs, indicating terminal velocities as high as 10000 km/s.

124.504 **Nova Aquilae 1982.**
IAU Circ., Nos. 3711, 3717, 3724, 3734 (1982).

124.505 **Did Nova Aquilae 1982 possess a supercritical wind?**
M. Friedjung.
Pré-publ. Inst. Astrophys. Paris, No. 17, 4 pp. (1982).

Paper contributed at IAU Colloquium No. 72 "Cataclysmic variables and related objects".

Nova V 4049 Sagittarii

124.511 **V 4049 Sgr.**
M. F. McCarthy, B. M. Lasker, T. D. Kinman.
The nature of symbiotic stars, (see 012.045), p. 219 - 220 (1982). – Abstract. – See 30.124.501.

Nova Sagittarii 1982

124.521 **Nova Sagittarii 1982.**
IAU Circ., Nos. 3733, 3734, 3736, 3737, 3741, 3747 (1982).

124.522 **Nova Sgr 1982.**
Yamamoto Circ., Nos. 1982 - 1985 (1982).

Nova Aquilae 1918 = V603 Aquilae

124.601 **Physical conditions in the accretion disk of V603 Aquilae.**
G. J. Ferland, D. L. Lambert, M. L. McCall, G. A. Shields, M. H. Slovak.
Astrophys. J., Vol. 260, 794 - 806 (1982).

Ultraviolet and optical spectra of the old nova V603 Aql are discussed. The UV-optical continuum is dominated by emission from the accretion disk. Emission lines from ions of H, He, C, N, and O are identified. These lines are probably formed in a circumstellar shell. This corona is probably heated by radiation emitted by the underlying accretion disk. Photoionization calculations of the structure and emission-line spectrum of the corona are presented.

124.602 **The spectrum of the old nova V603 Aql between 2000 and 3200 Å.**
P. L. Selvelli, A. Cassatella.
Third European IUE Conference, (see 012.029), p. 201 - 202 (1982).

One high resolution and five low resolution IUE spectra of V603 Aql taken consecutively indicate the presence of high excitation emission lines showing significant variations on a short time-scale. The continuum energy distribution shows similar, although less pronounced, variations.

124.603 **Periodic variations in the polarization of V 603 Aql.**
K. Metz.
Inf. Bull. Variable Stars, No. 2201, 4 pp. (1982).

124.604 **Neue photoelektrische und spektroskopische Messungen der alten Nova V 603 Aquilae.**
H. Drechsel, J. Rahe, W. Wargau, J. Krautter.
Mitt. Astron. Ges., Nr. 57, p. 301 (1982).

Nova CP Puppis

124.701 **Spectroscopic analysis of the extended shells around the novae CP Puppis and T Pyxidis.**
R. E. Williams.
Astrophys. J., Vol. 261, 170 - 182 (1982).

Spectral scans have been obtained of the spatially resolved shells surrounding the novae CP Pup and T Pyx. The CP Pup shell is very similar to the DQ Her envelope in being moderately ionized, but quite cold ($T_e \approx 800$ K). The recurrent nova T Pyx has a shell, apparently ejected in the 1966 outburst, whose spectrum is very similar to that of a planetary nebula. The shell is hot ($T_e \gtrsim 2 \times 10^4$ K) and probably photoionized by radiation from the nova.

Nova T Pyxidis

Spectroscopic analysis of the extended shells around the novae CP Puppis and T Pyxidis.
See Abstr. 124.701.

125 Supernovae, Supernova Remnants

125.001 **Radial velocities of individual filaments in the Cygnus Loop nebula.**
V. T. Doroshenko, F. Kh. Sakhibov, T. G. Sitnik.
Astron. Zh., Tom 59, 699 - 703 (1982). In Russian. English translation in Soviet Astron., Vol. 25, No. 4.

The radial velocities of 40 individual extended filaments of the supernova remnant Cyg Loop are determined.

125.002 **X-ray and UV-emission from supernova shock waves in stellar winds.**
C. Fransson.
Astron. Astrophys., Vol. 111, 140 - 150 (1982).

The radio and UV observations of SN 1979c provide the first direct evidence of circumstellar material around the pre-supernova star. The author discusses what can be inferred about the structure of the pre-supernova star and the supernova shock wave from observations of these sorts.

125.003 **Absolute photometry of supernova remnants and emission nebulae in the Galaxy and the Magellanic Clouds.**
A. Greve, A. M. van Genderen, M. Dennefeld, I. J. Danziger.
Astron. Astrophys., Vol. 111, 171 - 184 (1982).

In continuation of an earlier investigation (Greve and van Genderen, 1977) the authors give *VBLUW* photometric data for 4 galactic SNR, 2 SNR of the SMC, 4 SNR of the LMC, and 7 emission nebulae of the LMC. The calibration of the *VBLUW* system (Lub et al., 1979) allows the derivation of absolute fluxes in the filter passbands which in combination with emission line ratios allows the derivation of surface intensities of Hβ, Hα. The Hβ intensities are interpreted with the shock-wave theory which gives the pre-shock densities of the interstellar medium of the galaxy, SMC and LMC. The densities derived for the LMC confirm the model of the optical emission originating in higher density cloudlets embedded in a tenuous medium.

125.004 **A physically grounded classification of supernovae.**
I. S. Shklovskij.
Pis'ma Astron. Zh., Tom 8, 347 - 350 (1982). In Russian. English translation in Soviet Astron. Lett., Vol. 8.

Rational classification of supernovae outbursts is suggested which is based on the properties of their nuclei, on the presence or absence of hydrogen shells, and on the multiplicity characteristics. An explanation is given why the overwhelming majority of neutron stars are single.

125.005 **The radio and X-ray emission from type II supernovae.**
R. A. Chevalier.
Astrophys. J., Vol. 259, 302 - 310 (1982).

In the present paper, a model for the interaction of the outer supernova envelope with the circumstellar matter is described. The radio emission from the supernovae SN 1980k and SN 1979c is discussed in the context of this model. The supernova SN 1980k has been observed not only at radio wavelengths, but also at X-ray wavelengths. Two possible sources for the X-ray emission are discussed and it is found that the emission can be accounted for in the present model.

125.006 **Supernova remnants and Bell's acceleration mechanism.**
G. Cavallo.
Astron. Astrophys., Vol. 111, 368 - 370 (1982).

The total relativistic particles content as a function of the diameter in Supernova Remnants is compared with the theoretical predictions according to Bell's mechanism. The satisfactory agreement suggests that the latter is an acceptable model.

125.007 **Expansion of the supernova remnant 3C 10 (Tycho) and its implications for models of young remnants.**
R. G. Strom, W. M. Goss, P. A. Shaver.
Mon. Not. R. Astron. Soc., Vol. 200, 473 - 487 (1982).

The remnant of Tycho Brahe's supernova (3C 10) was observed with the Westerbork telescope at 21 cm in 1971 and 1979, and these measurements have been used to determine the radial expansion rate. The average value obtained is 0.256 ± 0.026 arcsec yr^{-1}. The authors consider two plausible explanations: the remnant is not yet fully in the adiabatic phase; or the dynamics are being modified by the evaporation of neutral material behind the shock front. Whether either of these is the correct explanation, it is clear that swept-up material now dominates the dynamics of 3C 10.

125.008 **Ultraviolet observations of the Cygnus Loop with IUE.**
J. C. Raymond, R. S. Wolff.
Galactic X-ray sources, (see 003.001), p. 297 - 299 (1982).

125.009 **Westerbork and VLA observations of G 127.1 + 0.5.**
T. Pauls, J. H. van Gorkom, W. M. Goss, P. A. Shaver, J. M. Dickey, S. Kulkarni.
Astron. Astrophys., Vol. 112, 120 - 123 (1982).

The galactic supernova remnant (SNR) G 127.1 + 0.5 contains the open cluster NGC 559 and the radio point source 0125 + 628 within its boundaries. Continuum measurements of the SNR at 608.5 MHz with the Westerbork Synthesis Radio Telescope (WSRT), and H I absorption spectra toward 0125 + 628 with the WSRT and the Very Large Array (VLA) are reported. The observations strongly suggest that there is no physical connection between the SNR, the point source, and the open cluster.

125.010 **Neutral hydrogen associated with southern supernova remnants. II. Lupus Loop.**
F. R. Colomb, G. Dubner.
Astron. Astrophys., Vol. 112, 141 - 148 (1982).

The 21 cm line observations around the Lupus Loop supernova remnant show the existence of two concentric expanding H I shells. A break in the neutral hydrogen distribution can be seen in both shells in the direction away from the galactic plane. Radii, expansion velocities, masses and ambient densities for the shells are derived from the observations. It is concluded that the outer shell was formed by the action of the stellar wind and that the inner shell is due to the supernova explosion. Suitable models are applied for each case and ages, kinetic energies, wind luminosity and initial energy of the explosion are derived.

125.011 **A spectrophotometric study of Kepler supernova remnant.**
M. Dennefeld.
Astron. Astrophys., Vol. 112, 215 - 222 (1982).

Spectrophotometry of Kepler SN is presented in the range 3700–10500 Å. The far red region, presented here for the first time for a SNR, reveals lines of [S II], [S III], [Fe II], [C I]. Recent atomic physics calculations by Nussbaumer and Storey allow for the first time a convincing identification of the observed 7378 and 7412 lines as being [Ni II]. The reddening found from the [S II] lines is different from the one found from the Balmer decrement, which might in fact indicate large variations in absorption over a small field. The spectrum is interpreted as the result of a shock propagating in a dense medium of about 300 cm^{-3} for a shock velocity of 120 km s^{-1}.

125.012 **Mean-field calculations of the equation of state of supernova matter II.**
P. Bonche, D. Vautherin.
Astron. Astrophys., Vol. 112, 268 - 272 (1982).
Further results of Hartree-Fock calculations for the equation of state of hot dense matter are presented. The authors include the contribution of neutrinos and consider matter in beta equilibrium with a value of the lepton fraction $Y_l = 0.35$ suggested by the latest estimates of electron capture. They discuss the adiabat $S = 1$ per baryon in the region just below nuclear matter density. A detailed investigation of the phase transition from nuclei to bubbles along this adiabat is carried out.

125.013 **Supernova remnant candidates in the Small Magellanic Cloud.**
B. Y. Mills, A. G. Little, J. M. Durdin, M. J. Kesteven.
Mon. Not. R. Astron. Soc., Vol. 200, 1007 - 1017 (1982).
Eighteen SNR candidates in the SMC deriving from X-ray, optical and radio observations have been examined at a frequency of 843 MHz. Six SNRs are confirmed, all with flux densities far above the radio detection limit. Other identified radio sources from this sample include a quasar and an emission nebula. Comparison with Galactic SNRs suggest differences in the Σ–D and N–D relations, but the number of SNRs in the present sample is too small for good statistics. The birthrate of SNRs in the SMC is roughly estimated as ~1 every 1000 yr.

125.014 **Structures of four supernova remnants at 1.4 GHz.**
J. L. Caswell, R. F. Haynes, D. K. Milne, K. J. Wellington.
Mon. Not. R. Astron. Soc., Vol. 200, 1143 - 1151 (1982).
Four small-diameter supernova remnants (SNRs) have been mapped with the Fleurs synthesis radio telescope; the maps are the first to reveal structure in these SNRs, all of which lie near zero declination, where most other synthesis telescopes do not provide satisfactory north–south resolution. Three of the sources (G15.9+0.2, G27.4+0.0 and G41.1–0.3) show clear indications of a shell structure; the remaining source, G39.2–0.3, may be a rarer variety of SNR showing centrally concentrated emission similar to the Crab Nebula.

125.015 **Hydrogen two photon emission in the UV spectrum of type II supernovae.**
P. Benvenuti, M. A. Dopita, S. D'Odorico.
Advances in ultraviolet astronomy, (see 012.006), p. 434 - 437 (1982).
The authors discuss a particular aspect of the UV spectrum of the type II SNe: the continuum emission.

125.016 **An atlas of UV spectra of supernovae.**
P. Benvenuti, L. Sanz Fernandez de Cordoba, W. Wamsteker.
Advances in ultraviolet astronomy, (see 012.006), p.438 (1982). – Abstract.

125.017 **Beobachtungen von Supernova-Explosionen.**
K. Hempe.
Sterne Weltraum, 21. Jahrg., 354 - 357 (1982).

125.018 **La supernova de 1054: un mythe?**
J.-P. Parisot.
Astronomie, Vol. 96, 347 - 350 (1982).

125.019 **Spherization of the remnants of an asymmetric explosion of a supernova in a uniform medium.**
G. S. Bisnovatyj-Kogan, S. I. Blinnikov.
Inst. teor. i ehksp. fiz. Moskva, Prepr., 1981, No. 124, 28 pp. In Russian. – Abstr. in Ref. zh., 51. Astron., 8.51.124 (1982).

125.020 **Gasdynamical models of type II supernovae.**
I. Yu. Litvinova, D. K. Nadezhin.
Inst. teor. i ehksp. fiz. Moskva, Prepr., 1982, No. 23, 54 pp. In Russian. – Abstr. in Ref. zh., 51. Astron., 8.51.384 (1982).

125.021 **Supernova remnant Monoceros Loop at decameter waves.**
E. A. Abramenkov, V. V. Krymkin, M. A. Sidorchuk.
Dokl. AN USSR, 1982, A, No. 3, p. 44 - 47. In Russian.
Abstr. in Ref. zh., 51. Astron., 8.51.661 (1982).

125.022 **Distance to Crab-like supernova remnant 3C58.**
D. A. Green, S. F. Gull.
Nature, Vol. 299, 606 - 608 (1982).
The observations presented here show that 3C58 is at a distance of only 2.6 kpc.

125.023 **Are young supernova remnants interacting with circumstellar gas?**
R. A. Chevalier.
Astrophys. J., Lett., Vol. 259, L85 - L89 (1982).
The young remnants of galactic Type I supernovae (SN 1006, SN 1572, and SN 1604) appear to be interacting with moderately dense gas ($n \geqslant 0.1$ cm^{-3}). The gas could either be the ambient interstellar medium or circumstellar matter lost from the supernovae progenitors. The implications of both hypotheses for observable properties are analysed. Currently available data favor the hypothesis of expansion in a uniform ambient medium for all three Type I remnants.

125.024 **The effects of ejecta on the X-ray luminosities of supernova remnants.**
K. S. Long, M. A. Dopita, I. R. Tuohy.
Astrophys. J., Vol. 260, 202 - 207 (1982).
The authors investigate the apparent density evolution of a SNR when substantial amounts (~1 $M_\odot$) of processed material are ejected by the supernova. If this material is shock heated, it will dominate X-ray emission from the remnant long after the SNR has swept up the equivalent amount of interstellar matter (ISM). Emission from metal-rich ejecta may explain why ISM densities of young remnants far from the galactic plane, deduced from standard analysis, are higher than expected, as well as why larger SNRs appear to be lying in regions of lower ISM density.

125.025 **On the nucleosynthesis beyond the iron peak in supernovae.**
D. A. Ptitsyn, V. M. Chechetkin.
Pis'ma Astron. Zh., Tom 8, 600 - 606 (1982). In Russian.
English translation in Soviet Astron. Lett., Vol. 8.
A process of synthesis of nuclei beyond the iron peak at high temperature, density and neutron excess which are characteristic for a supernova explosion, is proposed.

125.026 **Electron scattering and absorption line formation in the atmospheres of type II supernovae.**
A. M. Sobolev, N. N. Chugaj.
Pis'ma Astron. Zh., Tom 8, 607 - 611 (1982). In Russian.
English translation in Soviet Astron. Lett., Vol. 8.
Pure scattering line profiles originating in the envelopes expanding as $v \propto r$ were calculated by the Monte-Carlo method with electron scattering taken into account. The peculiar profile of the Na I λ 5892 line in the spectrum of SN 1970g was modeled. It was inferred that the absence of an emission component in this line is possibly explained in the model taking into account electron scattering in the supernova atmosphere.

125.027 **Radio emission from supernova remnants in a cloudy interstellar medium.**
R. D. Blandford, L. L. Cowie.
Astrophys. J., Vol. 260, 625 - 634 (1982).

The authors have modified the van der Laan theory of radio emission from supernova remnants to take account of the inhomogeneity of the interstellar medium and to allow for particle acceleration in shock fronts. It is proposed that most of the radio emission in remnants of radius 10–20 pc originates from cold interstellar clouds that have been crushed by the high pressure hot gas within the expanding remnant. Warm (~8000 K) interstellar gas may also be compressed within very large remnants of radius 30–100 pc. This may account for both the giant radio loops when these remnants are seen individually and the anomalously bright galactic non-thermal radio background which may be the superposition of a number of such features.

125.028 **The peculiar X-ray morphology of the supernova remnant G 292.0 + 1.8: evidence for an asymmetric supernova explosion.**
I. R. Tuohy, D. H. Clark, W. M. Burton.
Astrophys. J., Lett., Vol. 260, L65 - L68, plates L4, L5 (1982).

A high-resolution X-ray image from the Einstein Observatory of the young supernova remnant G 292.0 + 1.8 (MSH 11 - 54), previously noted as peculiar in terms of its spectral and morphological properties at optical and radio wavelengths, also shows an unusual X-ray morphology. The remnant consists of a central barlike feature superposed on an ellipsoidal disk of approximately uniform surface brightness. The authors attribute the bar emission to a ring of oxygen-rich material ejected in the equatorial plane of a massive rotating progenitor, and the uniform disk component to emission from material with roughly cosmic composition heated by the accompanying blast wave.

125.029 **Very high resolution observations of the radio source in the supernova remnant G 127.1 + 05.**
B. J. Geldzahler, D. B. Shaffer.
Astrophys. J., Lett., Vol. 260, L69 - L73 (1982).

The results of recent very long baseline interferometry of G 127.11 + 0.54 at 10.65 GHz reveal a core dual "jet" structure. The morphology and physical properties of the compact source are similar to those of SS 433. The combined data from various studies of this object suggest that G 127.11 + 0.54 may be the stellar remnant of the supernova event which also produced the supernova remnant G 127.1 + 0.5.

125.030 **VRO 42.05.01 – one supernova remnant or two?**
T. L. Landecker, S. Pineault, D. Routledge, J. F. Vaneldik.
Astrophys. J., Lett., Vol. 261, L41 - L45, plate L2 (1982).

The authors present a 1420 MHz synthesis map of VRO 42.05.01 (G166.0+4.3), of resolution 1$'$.0 × 1$'$.4. The radio source clearly possesses two components, both having nonthermal spectra. One component has the typical appearance of a shelltype supernova remnant (SNR); the other, which partly overlaps the shell, has an irregular outline. Both components can be distinguished on photographs. The simplest model for the object involves a single SNR whose expansion has proceeded at very different rates in two regions of sharply differing density.

125.031 **X-ray line spectra of the Puppis A supernova remnant: spatial variations.**
P. F. Winkler, B. C. Bromley, T. H. Markert, R. Petre.
Bull. American Astron. Soc., Vol. 14, 636 (1982). – Abstract.

125.032 **Optical observations of the peculiar CTB 80 supernova remnant.**
W. P. Blair, R. P. Kirshner, R. A. Fesen, T. R. Gull.
Bull. American Astron. Soc., Vol. 14, 636 (1982). – Abstract.

125.033 **The Crab nebula's progenitor.**
K. Nomoto, W. M. Sparks, R. A. Fesen, T. R. Gull, S. Miyaji, D. Sugimoto.
Nature, Vol. 299, 803 - 805 (1982).

Davidson et al.(1982), quoting two of the authors (K. N. and W. M. S.), state that the Crab's progenitor had a mass slightly larger than 8 $M_\odot$. Here the authors present in detail the reasoning behind this statement and suggest the explosion mechanism.

125.034 **On the nature of type I supernovae.**
I. S. Shklovskij.
Astrofiz. i kosm. fiz. Moskva, 1982, p. 186 - 215. In Russian.
Abstr. in Ref. zh., 51. Astron., 9.51.427 (1982).

125.035 **Neutral hydrogen associated with the supernova remnant Lupus Loop.**
F. R. Colomb, G. M. Dubner.
Rev. Mexicana Astron. Astrofis., Vol. 6, (see 012.017), 267 - 268 (1981).

H I observations around the Lupus Loop show the existence of two expanding concentric shells. The authors derive their physical parameters.

125.036 **Magnetorotational model of supernovae explosions.**
G. S. Bisnovatyj-Kogan.
Ninth Texas Symposium on Relativistic Astrophysics, (see 012.020), p. 389 - 394 (1980).

125.037 **What can we learn from the Crab nebula?**
W. Kundt.
Ninth Texas Symposium on Relativistic Astrophysics, (see 012.020), p. 429 - 441 (1980).

125.038 **Supernova models.**
S. E. Woosley, T. A. Weaver.
Tenth Texas Symposium on Relativistic Astrophysics, (see 012.021), p. 357 - 380 (1981).

The authors review recent progress in understanding the observed properties of Type I supernovae as a consequence of the thermonuclear detonation of white dwarf stars and the ensuing decay of the ^{56}Ni produced therein. They also discuss a qualitatively new approach to the problem of massive star death and Type II supernovae based upon a combination of rotation and thermonuclear burning.

125.039 **An optimist's guide to supernovae.**
R. P. Kirshner.
Supernovae: a survey of current research, (see 012.022), p. 1 - 12 (1982).

There are some hints that a satisfactory agreement between observations and models is developing in several areas of supernova research. The light curves and spectra of Type I supernovae may well have their origin in the synthesis of ^{56}Ni in the explosion of a compact star. Type II supernovae are found where massive stars form, have light curves that correspond to 10^{51} ergs suddenly deposited inside a red supergiant, and may in fact be the explosions of massive stars. Some young supernova remnants, such as Cas A, show abundance patterns that bear a striking resemblance to those seen deep inside models of massive stars on the verge of destruction.

125.040 **Supernova shocks and neutrino diffusion.**
H. A. Bethe.
Supernovae: a survey of current research, (see 012.022), p. 35 - 52 (1982).

The general features of an outgoing supernova shock are discussed. The effects of neutrino production and emission from the shock, nuclear dissociation, and electron capture behind the shock are computed. An analytic model of neutrino diffusion is applied.

125.041 **Supernovae for pedestrians.**
A. Yahil, J. M. Lattimer.
Supernovae: a survey of current research, (see 012.022), p. 53 - 70 (1982).

The authors present self-similar solutions to the collapse of a stellar core whose equation of state is given by a polytrope. Shock propagation in the outer core is described by a "hydrostatic model" which shows a clear division between successful and unsuccessful shocks.

125.042 **Shock stagnation and neutrino losses in stellar collapse.**
T. J. Mazurek, J. Cooperstein, S. Kahana.
Supernovae: a survey of current research, (see 012.022), p. 71 - 77 (1982).

The authors present the results of a detailed hydrodynamic study of stellar collapse. They describe two calculations, one including neutrino losses and the other neglecting them. Shock behaviors in both calculations are similar. The energy loss in neutrinos is detrimental, but it is not in itself responsible for stopping the shock.

125.043 **Theoretical models for supernovae.**
S. E. Woosley, T. A. Weaver.
Supernovae: a survey of current research, (see 012.022), p. 79 - 122 (1982).

The results of recent numerical simulations of supernova explosions are presented. Particular emphasis is given to (1) the nucleosynthesis expected from intermediate mass Type II supernovae and detonating white dwarf models for Type I supernovae, (2) a realistic estimate of the γ-line fluxes expected from this nucleosynthesis, (3) the continued evolution of intermediate mass stars wherein iron core collapse does not lead to a strong, mass-ejecting shock wave, and (4) the evolution and explosion of very massive stars ($M > 100\ M_\odot$) of both Population I and III.

125.044 **Computer simulations of stellar collapse and supernovae explosions: non-rotating and rotating models.**
W. Hillebrandt.
Supernovae: a survey of current research, (see 012.022), p. 123 - 155 (1982).

The author describes detailed numerical calculations for the stellar collapse and supernova explosion of 4 $M_\odot$ stellar models. In addition, results of a two-dimensional adiabatic core-collapse model for a rotating star with mass 25 $M_\odot$ are presented.

125.045 **Theoretical models of Type I supernovae.**
J. C. Wheeler.
Supernovae: a survey of current research, (see 012.022), p. 167 - 203 (1982).

The author presents a critical and comprehensive review of current supernovae models, including questions of progenitor evolution, stellar collapse, nucleosynthesis of iron peak elements and light curve development, both for Type I and Type II supernovae. Theoretical predictions are confronted with observational constraints.

125.046 **The origin of the Crab Nebula and electron capture supernova of 8–10 $M_\odot$ stars.**
K. Nomoto.
Supernovae: a survey of current research, (see 012.022), p. 205 - 213 (1982).

The chemical composition of the Crab Nebula is compared with several presupernova models. The small carbon and oxygen abundances in the helium-rich nebula are only consistent with a presupernova model of a star with main-sequence mass 8–9.5 $M_\odot$. More massive stars contain too much carbon in the helium layer and smaller mass stars do not leave neutron stars. The progenitor star of the Crab Nebula exploded as an electron capture supernova: the core collapsed to form a neutron star and only the extended helium-rich envelope was ejected by the weak shock wave.

125.047 **Carbon-oxygen white dwarf presupernova models.**
R. Canal, J. Isern, J. Labay.
Supernovae: a survey of current research, (see 012.022), p. 215 - 220 (1982).

The long-term evolution of accreting C–O white dwarfs as a possible source of neutron stars and/or SN I events is considered. The effects of solidification of the star's core are included, in particular carbon–oxygen separation.

125.048 **The nature of supernovae as determined from their light curves.**
W. D. Arnett,
Supernovae: a survey of current research, (see 012.022), p. 221 - 251 (1982).

An analysis of the probable nature of supernovae is carried through systematically using analytic models which are sufficiently general to include both types I and II. Many aspects of the models are quantitatively compared with observations. Preliminary results for cosmological distance determination using supernovae are presented and expressed as a Hubble ratio. Implications for the energy, mass, composition and pre-explosion structure are summarized.

125.049 **Type II supernovae photospheres and distances.**
R. V. Wagoner.
Supernovae: a survey of current research, (see 012.022), p. 253 - 266 (1982).

The kinematical method of determining distances to supernovae is briefly reviewed. A program to compute the required continuum flux from model atmospheres of Type II supernovae is outlined. The results of a perturbed gray-body calculation are presented to indicate qualitatively some of the important features to be expected.

125.050 **Type I supernovae – observational constraints.**
D. Branch.
Supernovae: a survey of current research, (see 012.022), p. 267 - 279 (1982).

Some of the constraints on models of Type I supernovae which can be inferred from the observations are discussed. Comparison of spectra for the SN I of 1981 in NGC 4536 establishes that elements other than nickel, iron, and cobalt are represented in the maximum-light spectrum. The rate of decline of the post-peak light curve appears to be correlated with expansion velocity and absolute magnitude.

125.051 **Radio supernovae.**
K. W. Weiler, R. A. Sramek, J. M. van der Hulst, N. Panagia.
Supernovae: a survey of current research, (see 012.022), p. 281 - 291 (1982).

Three supernovae have so far been detected in the radio range shortly after their optical outburst. All are of Type II. For two of the supernovae, extensive data are presented on their "light curves" and spectra and a preliminary discussion is given of possible mechanisms.

125.052 **The historical supernovae.**
D. H. Clark, F. R. Stephenson.
Supernovae: a survey of current research, (see 012.022), p. 355 - 370 (1982).

The authors report on a search for historical supernova sightings in written records from China (with Japan and Korea), Babylon, the Arab World and Europe. The historical "data" are compared with catalogs of radio supernova remnants.

125.053 **Supernova statistics and related problems.**
G. A. Tammann.

Supernovae: a survey of current research, (see 012.022), p. 371 - 403 (1982).

The frequencies of supernova events (SNe) in external galaxies and in our Galaxy are discussed. The galactic frequencies are compared with supernova remnants, pulsars, and white dwarfs. The absolute maximum magnitudes of SNeI and SNeII are discussed, as well as the evidence of SNeI being good standard candles. Speculations on the progenitors of SNe are consistent with the assumption that SNeII come from stars with masses $>8\ M_{\odot}$, while SNeI could come from accreting white dwarfs in close binary systems.

125.054 **Young supernova remnants.**
R. A. Chevalier.
Supernovae: a survey of current research, (see 012.022), p. 419 - 432 (1982).

The interaction of a supernova with its environment leads to a rich set of phenomena, which can be observed at radio, optical, ultraviolet, and X-ray wavelengths. The purpose of this paper is to review these phenomena, with particular emphasis on the interpretation of recent observations of young supernova remnants.

125.055 **The evolution of supernova remnants and the structure of the interstellar medium.**
C. F. McKee.
Supernovae: a survey of current research, (see 012.022), p. 433 - 457 (1982).

The author summarizes the results of analytic and numerical studies of supernova remnant evolution in a three-phase interstellar medium. The importance of supernova remnants for the energetics and structure of the interstellar medium is discussed.

125.056 **Non-thermal aspects of supernova remnants.**
R. D. Blandford.
Supernovae: a survey of current research, (see 012.022), p. 459 - 474 (1982).

The author reviews current theoretical models of particle acceleration and synchrotron emission in supernova remnants expanding into the interstellar medium.

125.057 **Supernova remnants in M 31.**
W. P. Blair.
Supernovae: a survey of current research, (see 012.022), p. 475 - 482 (1982).

The author has used Hα + [N II] and [S II] interference filter photography to identify nebulosities which he believes to be supernova remnants in M 31. Spectrophotometry has been used to confirm this identification for 12 of the candidates. Using the measured diameter and an estimate of the pressure in the optical filaments, an initial energy for each remnants has been calculated. The spectra have also been used to study abundance gradients in M 31.

125.058 **Optical supernova remnants in external galaxies.**
M. A. Dopita.
Supernovae: a survey of current research, (see 012.022), p. 483 - 494 (1982).

The author reports on an optical survey of supernova remnants in external galaxies, in particular in both Magellanic Clouds. Properties of these remnants are summarized and element abundances for N and S are discussed. Evidence for galactic abundance gradients is reviewed.

125.059 **Optical spectrophotometry of the Cygnus Loop.**
R. A. Fesen, W. P. Blair, R. P. Kirshner.
Supernovae: a survey of current research, (see 012.022), p. 495 - 499 (1982).

Optical spectral properties of several filaments in the Cygnus Loop are discussed with emphasis on observation-model comparisons. It is concluded that some departure from the steady-flow condition assumed in current modelling is necessary to improve agreement with the observations.

125.060 **X-ray spectroscopy and imagery of supernova remnants with the Einstein Observatory.**
P. F. Winkler, C. R. Canizares, T. H. Markert, A. E. Szymkowiak.
Supernovae: a survey of current research, (see 012.022), p. 501 - 518 (1982).

The authors report on high-resolution X-ray spectra and images of some 18 supernova remnants obtained with the SSS and FPCS instruments onboard the Einstein Observatory. Results for Puppis A and Cas A are discussed in detail.

125.061 **Einstein observations of supernova remnants.**
F. D. Seward.
Supernovae: a survey of current research, (see 012.022), p. 519 - 528 (1982).

The Einstein Observatory has detected and mapped soft X-ray emission from ~40 galactic SNRs. The detectors covered the energy range 0.2–4.0 keV, and data were obtained with the High Resolution Imager between November 1978 and April 1981.

125.062 **An X-ray survey of supernova remnants in the Large Magellanic Cloud.**
D. J. Helfand, K. S. Long.
Supernovae: a survey of current research, (see 012.022), p. 529 - 534 (1982).

The authors report preliminary results of a comprehensive soft X-ray survey (0.1–10 keV) of the Large Magellanic Cloud performed with the Einstein Observatory. Most of the 97 discrete sources detected seem to be supernova remnants. The implications of the results for supernova statistics, evolution of SN remnants and the structure of the interstellar medium are briefly discussed.

125.063 **The next supernova.**
F. Schaaf.
Astronomy, Vol. 9, No. 12, p. 50 - 54 (1981). – Abstr. in Phys. Abstr., Vol. 85, Abstr. 79129 (1982).

125.064 **Zur Deutung der Supernova-Explosionen.**
W. Hillebrandt.
Sterne Weltraum, 21. Jahrg., 406 - 411 (1982).

125.065 **A new look at SNR birthrate and evidence for a predominantly low-density interstellar medium.**
G. Srinivasan, K. S. Dwarakanath.
Bull. Astron. Soc. India, Vol. 10, 48 - 49 (1982). – Abstract.

125.066 **The evolution of the radio spectra of supernova remnants.**
R. Cowsik, S. Sarkar.
Bull. Astron. Soc. India, Vol. 10, 49 (1982). – Abstract.

125.067 **The location of supernovae inside parent galaxies.**
C. Michell, G. Le Denmat.
Astrophys. Space Sci., Vol. 85, 75 - 81 (1982).

The present study of the space location of supernovae in their parent galaxies suggests a new method to obtain indicative values of the SN-galaxy center distances. A sample of 25 type I supernovae is treated by means of two distinct procedures: the one by Barbon, Capaccioli, and Ciatti, the one defined here. A good agreement of the results is shown, specially about the surface density of supernovae. In the last part, the authors try to test the influence of the internal obscuration of the SNs by their own galaxies, on the basis of a correction suggested by Nottale. The frequency distribution and the surface density of the objects of the sample support the consistency of such a correction.

125.068 **Ultraviolet spectra of the Cygnus Loop – depletion of the heavy elements.**
M. Contini, G. Shaviv.
Astrophys. Space Sci., Vol. 85, 203 - 219 (1982).
A comparison between the observed UV spectra and detailed consistent calculations of the Cygnus Loop are presented. The spectra can be explained by supposing that the Cygnus Loop moves into a fully ionized gas. Further results are: (1) Observed shocks of higher velocity propagation move into regions of lower density; (2) the optical and UV spectra are emitted by very close and almost overlapping shocks (3) fast shocks ($v \gtrsim 240$ km s^{-1}) propagation in the intercloud medium produce the X-ray emission. The authors find that carbon depletion relative to other heavy elements is not more than a factor ~3 and they confirm that all heavy elements, relative to their solar abundances, are depleted by a factor ~10.

125.069 **Spherization of the remnants of an asymmetric supernova explosion in a uniform medium.**
G. S. Bisnovatyj-Kogan, S. I. Blinnikov.
Astron. Zh., Tom 59, 876 - 887 (1982). In Russian. English translation in Soviet Astron., Vol. 26, No. 5.
The spherization of a shock wave is considered, which propagates in a uniform interstellar medium in the presence of initial asymmetry of shape and velocity. Three different models have been considered with preference given to the "snowplough" model. The characteristic time of the spherization is found for different values of the density of the medium and the parameters of the explosion.

125.070 **Ultraviolet properties of supernovae.**
N. Panagia.
Third European IUE Conference, (see 012.029), p. 31 - 36 (1982).
The results of IUE observation of five supernovae, three Type I and two Type II are reviewed and discussed. A comparison with the spectral properties at other wavelength ranges is also presented.

125.071 **The Monoceros Loop: a supernova remnant interacting with the Rosette Nebula?**
A. Vidal-Madjar, C. Laurent, M. Pettini, J. A. Paul, M. Oliver.
Third European IUE Conference, (see 012.029), p. 421 - 425 (1982).
The B1 Ib star HD 47240 is located behind the Monoceros Loops, the expanding remnant of an old supernova. The rich interstellar spectrum of the star shows absorption by two high velocity components, in addition to the quiecent interstellar medium in line of sight. One component is identified with the receeding portion of the expanding shell swept up by a supernova shock. The other component most likely arises in the Rosette Nebula, a bright H II region in the line of sight.

125.072 **VRO 42.05.01: one supernova remnant or two?**
S. Pineault, D. Routledge, J. F. Vaneldik, T. L. Landecker.
Bull. American Astron. Soc., Vol. 14, 779 (1982). – Abstract.

125.073 **Optical identification of Balmer-dominated supernova remnants in the Large Magellanic Cloud.**
I. R. Tuohy, M. A. Dopita, D. S. Mathewson, K. S. Long, D. J. Helfand.
Astrophys. J., Vol. 261, 473 - 484, plates 8, 9 (1982).
The authors present optical emission-line images of four new supernova remnants (SNRs) discovered recently during an X-ray survey of the Large Magellanic Cloud. The images were obtained using the Anglo-Australian Telescope and show distinct filamentary shells which are prominent in the Balmer lines of hydrogen, but which are either absent or extremely weak at [O III] and [S II]. A spectrum of one of the remnants reveals the presence of a broad Hα component having a velocity dispersion of 2800 km s^{-1}. A comparison is made between the optical, X-ray, and radio properties of the LMC remnants and comparable young galactic SNRs. It is argued that the four LMC remnants all resulted from Type I supernovae, in which case they are the first such remnants to be identified outside the Galaxy.

125.074 **An X-ray image of Tycho's supernova remnant.**
P. B. Reid, R. H. Becker, K. S. Long.
Astrophys. J., Vol. 261, 485 - 491 (1982).
An image of Tycho's SNR has been obtained using the IPC aboard the Einstein Observatory, revealing a nearly circular, 8′, open shell of emission. The X-ray data are consistent with the interpretation that the SNR was the result of a Type I explosion and that the distance to Tycho's SNR is between 2.2 and 4.3 kpc.

125.075 **A model of the Cygnus Loop.**
S. A. E. G. Falle, A. R. Garlick.
Mon. Not. R. Astron. Soc., Vol. 201, 635 - 644 (1982).
The authors present a model of the Cygnus Loop based on a supernova exploding into a region of the interstellar medium containing a density discontinuity. The breakout phenomenon is then able to explain the high velocity optical emission observed in this remnant by Kirschner & Taylor (1976). It also gives rise to two X-ray emitting components whose intensity and hardness variations over the remnant are in agreement with observations.

125.076 **The supernova remnant G116.9+0.2 (CTB 1)–21-cm continuum and H I emission.**
T. L. Landecker, R. S. Roger, P. E. Dewdney.
Astron. J., Vol. 87, 1379 - 1389 (1982).
The authors present new synthesis observations of the continuum and H I-line emission at λ 21 cm of a 2° field centered on CTB 1, with complete sampling of spatial structure to the resolution of 2′. They measure a total flux density for CTB 1 of 8.3 Jy and, from comparison with other observations, a mean spectral index of 0.51 ± 0.13. A distance to CTB 1 of ~2 kpc and a diameter ~21 pc are determined.

125.077 **Spectrophotometry of the Cygnus Loop.**
R. A. Fesen, W. P. Blair, R. P. Kirshner.
Astrophys. J., Vol. 262, 171 - 188, plates 1 - 7 (1982).
The authors present new spectrophotometry for 17 filaments in the Cygnus Loop. A wide variety of spectral properties is observed within the remnant. Comparisons with published shock models indicate significant differences between the models and observations with the range of observed line intensities not well described by models with varying shock velocity.

125.078 **X-ray emission from young supernova remnants: nonionization equilibrium abundances and emissivities.**
J. M. Shull.
Astrophys. J., Vol. 262, 308 - 314 (1982).
X-ray line emission from hot, low-density plasma in young supernova remnants is strongly enhanced by departures from ionization equilibrium. The author has calculated the X-ray emission from a Sedov blast wave, using a nonequilibrium evolutionary treatment of the ionization structure, and has fitted the resulting spectrum to HEAO 2 data for Tycho's remnant. These models yield dramatically different elemental abundances for heavy elements, compared with two-temperature component equilibrium models.

125.079 **Acoustic waves in supernova remnants.**
L. Spitzer, Jr.
Astrophys. J., Vol. 262, 315 - 321 (1982).
When the Mach number of a supernova shock wave,

propagating through the hot interstellar gas, is less than the critical value 2.76, the interaction of the shock with an interstellar cloud produces a reflected pressure pulse that propagates away from the cloud as an acoustic wave. A preliminary analysis indicates that about 4% of the energy of an exploding supernova shell is converted into such waves. These waves may provide a heat source for the warm, neutral interstellar clouds.

125.080 **On the neutrino luminosity from a type II supernova.**
H. A. Bethe, A. Yahil, G. E. Brown.
Astrophys. J., Lett., Vol. 262, L7 - L10 (1982).

The initial burst of neutrino luminosity from a supernova is expected to last ~1 ms, in which $\sim 10^{51}$ ergs are emitted from the shocked material in the outer core. A detectable fraction of this energy is expected to be in the form of electron antineutrinos. During the deleptonization of the unshocked inner core, lasting ~1 s, the neutrino energy $\sim 10^{52}$ ergs is expected to be approximately blackbody radiation, with equal luminosities in neutrinos and antineutrinos.

125.081 **Hot dense matter and stellar collapse.**
K. Sato, M. Kiguchi.
Prog. Theor. Phys. Suppl., No. 70, p. 142 - 153 (1981).
Abstr. in Phys. Abstr., Vol. 85, Abstr. 100966 (1982).

125.082 **Observations of the supernova remnants HB 9 and IC 443 at 34.5 MHz.**
K. S. Dwarakanath, R. K. Shevgaonkar, C. V. Sastry.
J. Astrophys. Astron., Vol. 3, 207 - 216 (1982).

The authors have observed the extended supernova remnants HB 9 (G 160.5 + 2.8) and IC 443 (G 189.1 + 2.9) at 34.5 MHz. Integrated flux-density spectra were derived. A map of HB 9 is presented.

125.083 **Soft X-ray filter spectroscopy of the supernova remnants Vela X and Puppis A.**
W. Burkert, H. U. Zimmermann, B. Aschenbach,H. Bräuninger, F. Williamson.
Astron. Astrophys., Vol. 115, 167 - 170 (1982).

In January 1977 the supernova remnants Vela X and Puppis A were observed in the soft X-ray region by a sounding rocket experiment consisting of 12 X-ray paraboloidal collectors. To improve the energy resolution of the proportional counter detectors, filter spectroscopy has been applied. Besides the temperature and interstellar column density, flux values for the dominant lines (O VII, O VIII, Ne IX, Fe XVII, and Fe XVIII) and corresponding abundances have been derived. Further intrinsic parameters of both SNR's could be deduced from a general shock wave model for SNR's.

125.084 **Radio observations of small diameter sources in the field of the supernova remnant S 147.**
E. Fürst, W. Reich, R. Beck, W. Hirth, P. E. Angerhofer.
Astron. Astrophys., Vol. 115, 428 - 432 (1982).

Numerous small diameter sources in the field of the SNR S 147 have been reported in literature. It is amazing that most of the sources follow the remnant shell almost on a circle, suggesting that at least part of them are associated with the remnant. New observations between 2.7 and 10.7 GHz are reported, which allow the determination of the radio spectra of these sources. Except for two sources the radio spectra do not support an association.

125.085 **Supernovae in binary systems: production of runaway stars and pulsars.**
J.-P. De Cuyper.
Binary and multiple stars as tracers of stellar evolution, (see 012.046), p. 417 - 443 (1982).

The effects of an instantaneous asymmetric supernova explosion in an eccentric binary system are analyzed, taking into account the mass loss out of the system, the influence of the impact of the supernova shell on the companion star and the extra "kick" velocity which a collapsed star might receive in an asymmetric supernova explosion. As an example, the outcome of a possible supernova explosion in the ten best known WR+OB binaries is studied and a comparison is made with the observed run-away OB stars, radio pulsars and binary X-ray pulsars.

125.086 **Spectra and light curves of three recent supernovae.**
R. Barbon, F. Ciatti, L. Rosino, S. Ortolani, P. Rafanelli.
Astron. Astrophys., Vol. 116, 43 - 53 (1982).

Systematic observations at Asiago Observatory have provided B, V light and color curves as well as spectral information about the supernovae 1978-e in MCG6-49-36, 1979-b in NGC 3913, 1979-c in NGC 4321 = M 100. The first two are classified of Type I, while SN 1979-c resulted a Type II with almost linear decline. Distances and absolute magnitudes are discussed. Although no firm conclusion on M_B for the first object can be reached, the latter two supernovae seem to be among the brightest members of their types.

125.087 **Probable supernova in anonymous galaxy.**
IAU Circ., No. 3708 (1982).

125.088 **Possible supernova in anonymous galaxy.**
IAU Circ., No. 3709 (1982).

125.089 **Supernovae in anonymous galaxies.**
IAU Circ., No. 3722 (1982).

125.090 **Supernovae in anonymous galaxies.**
IAU Circ., No. 3728 (1982).

125.091 **Supernovae.**
IAU Circ., Nos. 3741, 3752 (1982).

125.092 **Supernovae.**
IAU Circ., No. 3749 (1982).

125.093 **Cygnus Superbubble – the remnant of a peculiar supernova.**
S. I. Blinnikov, V. S. Imshennik, V. P. Utrobin.
Pis'ma Astron. Zh., Tom 8, 671 - 678 (1982). In Russian.
English translation in Soviet Astron. Lett., Vol. 8.

The analytic solution for the evolution of supernova remnants on the radiative phase is obtained. It is shown that the X-ray source Cygnus Superbubble might be formed by an explosion with the energy release of 10^{52}-10^{53} erg.

125.094 **Monochromatic photographs of the Cygnus Loop supernova remnant. Isophotes of the fragment of the nebula in [O III] and [N II] + Hα lines.**
T. G. Sitnik, M. S. Toropova.
Pis'ma Astron. Zh., Tom 8, 679 - 684 (1982). In Russian.
English translation in Soviet Astron. Lett., Vol. 8.

A system of the isophotes of the 9′ region in the west part of the Cyg Loop supernova remnant using monochromatic photographs in the [O III] and [N II] + Hα lines is obtained. A relative displacement of the regions of emission in these lines is discovered and explained by temperature reduction due to radiative losses behind the shock of the supernova explosion. The morphology difference between the [O III] and [N II] + Hα filaments is explained.

125.095 **Les supernovae et leur recherche.**
F. Ciatti.
Bull. AFOEV, No. 22, p. 159 - 160 (1982).

125.096 **CO observations of SNR G78.2 + 2.1.**
L. A. Higgs, T. L. Landecker, R. S. Roger.
J. R. Astron. Soc. Canada, Vol. 76, 320 (1982).

125.097 **X-ray structure of the SNR G78.2 + 2.1.**
T. L. Landecker, L. A. Higgs, R. S. Roger, F. D. Seward.
J. R. Astron. Soc. Canada, Vol. 76, 322 (1982). – Abstract.

125.098 **VRO 42.05.01 – one supernova remnant or two?**
D. Routledge, J. F. Vaneldik, T. L. Landecker, S. Pineault.
J. R. Astron. Soc. Canada, Vol. 76, 324 - 325 (1982). Abstract.

125.099 **On the nature of supernovae.**
W. D. Arnett.
Essays in nuclear astrophysics, (see 003.013), p. 427 - 437 (1982).

Present ideas about supernovae are reexamined. The observed characteristics of both Type I and Type II supernova explosions are analyzed. The discussion is formulated in terms of analytic approximations.

125.100 **Supernova theory.**
H. A. Bethe.
Essays in nuclear astrophysics, (see 003.013), p. 439 - 466 (1982).

The basic supernova theory described in this chapter is that the outward propulsion of the mantle and envelope of the star is due to a hydrodynamic shock that is generated by the gravitational collapse of the central core to densities greater than nuclear density, and the subsequent rebound.

125.101 **An estimate of the masses of the stars that produced SN1923a and SN1957d.**
R. L. Pennington, R. J. Talbot, Jr., R. J. Dufour.
Astron. J., Vol. 87, 1538 - 1545 (1982).

The immediate stellar neighborhoods of the sites of two supernovae, SN1923a and SN1957d, in the spiral galaxy M 83 were analyzed using surface photometry and digital image processing techniques. For the H II region containing the site of SN1923a and for a square area around SN 1957d integrated colors were determined. Through the use of model cluster evolution tracks and assuming Whitford reddening, the authors determined ages for these two regions of 1.0×10^7 yr and 1.7×10^7 yr, respectively. Published stellar evolution calculations have yielded these lifetimes for $18\,M_\odot$ and $11\,M_\odot$ stars. This provides significant evidence that a type II supernova, SN1923 a, was associated with a high-mass star.

125.102 **Optical properties of supernova remnants.**
I. J. Danziger.
ESO Sci. Prepr., No. 221, 11 pp. (1982). – To appear in the proceedings of IAU symposium 101, "supernova remnants and their X-ray emission", Venice, September 1982.

125.103 **Photometric properties of type I supernovae in galaxies of different type.**
D. Yu. Tsvetkov, M. A. Smirnov.
Astron. Tsirk., No. 1175, p. 2 - 3 (1981). In Russian.

125.104 **Sporadic observations of the Tsvetkov supernova.**
S. G. Iskudaryan, A. S. Khodzhaev.
Astron. Tsirk., No. 1200, p. 6 - 7 (1981). In Russian.

125.105 **The polarization of supernova light: a measure of deviation from spherical symmetry.**
P. R. Shapiro, P. G. Sutherland.
Astrophys. J., Vol. 263, 902 - 924 (1982).

The assumption of spherical symmetry, made in most theoretical models of supernova (SN) explosions, light curves, and spectra, and in attempts to use SN explosions as cosmological distance indicators, has never been justified observationally. An observational test of this assumption is suggested here, based upon the likelihood that SN atmospheres are scattering dominated and the fact that light from an unresolved, *asymmetric*, scattering atmosphere is linearly polarized. The authors have calculated the linear polarization of SN light which a nonspherical, scattering-dominated SN atmosphere would produce as a function of its asphericity. The effects of interstellar matter on the observed SN polarization are included. Existing, preliminary observations of supernova polarization are reviewed.

125.106 **The structure of a type-II supernova remnant G 160.5 + 2.8 (HB-9) observed at $\lambda = 12$ m.**
E. *(Eh)* P. Abranin, L. L. Bazelyan.
Astrophys. Space Sci., Vol. 88, 299 - 306 (1982).

A two-dimensional brightness distribution of the old supernova remnant G 160.5 + 2.8 (HB-9) has been measured at a frequency of 25 MHz with the T-Shaped UTR-2 radio telescope, with the resolution $29' \times 27'$. The source has a distinctly pronounced, almost spherical envelope. At the same time, individual features located near the internal boundary of the source are observable. Certain considerations are put forward to validate a suggestion that the brightest feature in the decameter range (maximum at $\alpha = 4^h55^m\!.3$ and $\delta = 46^\circ\!.37$) might be of an extragalactic origin. From structure considerations, the velocity of the pulsar PSR 0459 + 47 projected on HB-9 is estimated. The estimate does not contradict the hypothesis of Damashek et al. (1978) concerning a possible genetic relation between HB-9 and PSR 0459 + 47.

125.107 **Bigger and better bangs in the sky.**
V. Trimble.
New Scientist, Vol. 95, 431 - 434 (1982). – Abstr. in Phys. Abstr., Vol. 85, Abstr. 117692 (1982).

125.108 **Supernovae. I. The events.**
V. Trimble.
Rev. Mod. Phys., Vol. 54, 1183 - 1224 (1982). – Abstr. in Phys. Abstr., Vol. 86, Abstr. 10376 (1983).

125.109 **Effects of neutrino trapping on supernova explosions.**
M. Takahara, K. Sato.
Prog. Theor. Phys., Vol. 68, 795 - 807 (1982). – Abstr. in Phys. Abstr., Vol. 86, Abstr. 20406 (1983).

Supernovae.
See Abstr. 011.009.

Automated supernova search.
See Abstr. 031.559.

The Berkeley automated supernova search.
See Abstr. 031.560.

Automated supernova search from photographic plates.
See Abstr. 031.561.

Will the next supernova in our Galaxy be discovered by a radio telescope?
See Abstr. 031.683.

Supernovae induced by axion-like particles.
See Abstr. 061.011.

Cosmic radioactivity: a gamma-ray search for the origins of atomic nuclei.
See Abstr. 061.043.

Astrophysical consequences of neutron-antineutron oscillations.
See Abstr. 061.051.

Self-similar solutions for the interaction of stellar ejecta with an external medium.
See Abstr. 062.001.

On the secular evolution of accreting white dwarfs and Type I supernovae.
See Abstr. 065.018.

Gravitational collapse of massive stars and supernovae.
See Abstr. 065.027.

An eigenvalue approach to the solution of non-equilibrium ionization and applications to models of supernova remnants.
See Abstr. 065.031.

Evolution and explosion of massive stars.
See Abstr. 065.036.

The physics of gravitational collapse: an overview.
See Abstr. 065.038.

Summary of workshop on stellar collapse and supernovae.
See Abstr. 065.039.

The fate of massive stars: collapse, bounce and shock formation.
See Abstr. 065.040.

The ultimate fate of Wolf-Rayet stars as supernovae.
See Abstr. 065.055.

Homologous collapse and deleptonization of an evolved stellar core.
See Abstr. 065.058.

Nucleosynthesis in two 25 $M_\odot$ stars of different population.
See Abstr. 065.091.

Neutrino escape, nuclear dissociation, and core collapse and/or explosion.
See Abstr. 065.097.

Supernovae and the origin of the solar system.
See Abstr. 107.014.

Progenitors and birth-rates of cataclysmic variables and type I supernovae.
See Abstr. 117.060.

Mass loss from novae and supernovae.
See Abstr. 124.012.

Kinematics of gas in the Vela supernova remnant.
See Abstr. 131.046.

Condensation of grains.
See Abstr. 131.305.

Observations du nuage moléculaire associé a Sh-2 152.
See Abstr. 131.312.

Dynamique des sphères célestes.
See Abstr. 131.313.

The interstellar medium and its energy sources.
See Abstr. 131.314.

Influence des supernovae et des nuages H I sur les propriétés du milieu interstellaire.
See Abstr. 131.315.

Deep optical imagery of the Crab Nebula's jet.
See Abstr. 134.008.

Kinematics of ring-shaped nebulae in the LMC. II. The radial velocity field of N 185.
See Abstr. 134.029.

Absolute photometry of the Crab nebula.
See Abstr. 134.030.

Papers presented at the workshop on pulsars and supernovae.
See Abstr. 141.534.

The galactic distribution of pulsars.
See Abstr. 141.535.

Discovery of radio pulsations from the X-ray pulsar in the supernova remnant G 320.4−1.2.
See Abstr. 141.539.

On the implications of the radio and X-ray pulsar in the supernova remnant MSH 15-52.
See Abstr. 141.557.

High-resolution X-ray spectroscopy of supernova remnants, clusters of galaxies and BL Lac objects.
See Abstr. 142.085.

Cosmic ray generation by shock drift and compression mechanisms.
See Abstr. 143.016.

The galactic abundance gradient from supernova remnant observations.
See Abstr. 155.043.

Does the galactic synchrotron radio background originate in old supernova remnants?
See Abstr. 156.015.

Environmental clues to the nuclear activity in galaxies and the genesis of supernovae.
See Abstr. 158.186.

Chemical abundances in the interstellar medium of galaxies from spectrophotometry of supernova remnants.
See Abstr. 158.293.

Supernova remnants in the Large Magellanic Cloud from Einstein.
See Abstr. 159.011.

Supernovae and the formation of galaxies.
See Abstr. 162.100.

Supernova in NGC 4321 (1979c)

125.201 **A photographic observation on SN 1979c in NGC 4321.**
X. Hao, B. Mei, Z. Liu.
Acta Astrophys. Sinica, Vol. 2, 123 - 130 (1982). In Chinese.
The authors made photographic observations of

SN 1979c in the spiral galaxy M 100 (NGC 4321) with the 40/200 cm double astrograph of Beijing Observatory from April 23 to December 27, 1979. The authors describe briefly the observational conditions and analyse and discuss in comparison with other observations some obtained results.

X-ray and UV-emission from supernova shock waves in stellar winds.
See Abstr. 125.002.

The radio and X-ray emission from type II supernovae.
See Abstr. 125.005.

Supernova in NGC 5485

125.221 **Supernova in NGC 5485.**
IAU Circ., Nos. 3754, 3756 (1982).

Supernova in NGC 6946 (1980k)

125.321 **Photometric observations of the bright type II supernova 1980k in NGC 6946.**
R. J. Buta.
Publ. Astron. Soc. Pacific, Vol. 94, 578 - 585, with a correction p. 744 (1982).

Photoelectric *UBV* photometry of the recent bright supernova 1980k in NGC 6946 is presented. The *U*, *B*, and *V* light curves show a linear decay in magnitude at the rates $0^{m}.100$, $0^{m}.058$, and $0^{m}.041$ day^{-1}, respectively, over the first 57 days past maximum, consistent with the L class of type II supernovae. The distance modulus of NGC 6946 determined from ten different photometric and geometric distance indicators is $\mu_0 = 28.7 \pm 0.3$. The apparent magnitude at maximum, $B(\max) = 11.7 \pm 0.1$ and $B_0(\max) = 10.2 \pm 0.2$ corrected for the large and uncertain extinction, indicates that SN 1980k reached an absolute magnitude $M_B(\max) = -18.5 \pm 0.35$.

125.322 **Spectra of SN 1980 k in NGC 6946.**
C. Barbieri, C. Bonoli, S. Cristiani.
Astron. Astrophys., Vol. 114, 216 - 218 (1982).

The paper presents spectra of the Type II SN 1980 k in NGC 6946. One spectrum extends to 9900 Å and shows the near infrared Ca II triplet in emission.

125.323 **Two bright supernovae in NGC 6946 and NGC 4536.**
R. Barbon, F. Ciatti, L. Rosino.
Astron. Astrophys., Vol. 116, 35 - 42 (1982).

Photographic and spectroscopic observations are reported in this paper for two recent supernovae, discovered in NGC 6946 (SN 1980 k) and in NGC 4536 (SN 1981 b). The first object results to be a type II SN with almost linear decline, and very similar properties in comparison with SN 1979 c in M 100. The second one is a type I ("fast") SN, with typical photometric and spectral evolution. Absolute magnitudes are derived, and they are bright in both cases with regard to the average values derived for the two types. The two supernovae described were also observed in a wide range of frequencies (X-ray, ultraviolet, infrared, radio).

125.324 **Observations of a supernova in the galaxy NGC 6946.**
D. Yu. Tsvetkov.
Astron. Tsirk., No. 1180, p. 2 - 4 (1981). In Russian.

The radio and X-ray emission from type II supernovae.
See Abstr. 125.005.

Erratum

125.391 **Erratum: "Detection of X-rays during the outburst of supernova 1980k"** [Astrophys. J., Lett., Vol. 253, L17 - L21 (1982)].
C. R. Canizares, G. A. Kriss, E. D. Feigelson.
Astrophys. J., Lett., Vol. 258, L83 (1982).
See Abstr. 31.125.201.

Supernova in NGC 4490

125.401 **Photographic observations of the supernova in NGC 4490.**
T. Yamagata, M. Iye.
Inf. Bull. Variable Stars, No. 2204 (1982).

Supernova in NGC 5457 (1970g)

125.421 **Line profiles and a model of the envelope of the type II supernova 1970g.**
N. N. Chugaj.
Astron. Zh., Tom 59, 1134 - 1145 (1982). In Russian.
English translation in Soviet Astron., Vol. 26, No. 6.

Profiles of the Hα and Na I λ 5892 lines observed in the spectrum of SN 1970g were calculated. From a comparison of the calculated and observed profiles the expansion velocity at the photosphere level was found. From the Hα profile the value of n_e was determined independently of the distance.

Supernova in NGC 4536 (1981b)

Two bright supernovae in NGC 6946 and NGC 4536.
See Abstr. 125.323.

Supernova in NGC 7713

125.501 **Supernova in NGC 7713.**
IAU Circ., Nos. 3717, 3720 (1982).

Supernova in NGC 521

125.601 **Supernova in NGC 521.**
IAU Circ., No. 3724 (1982).

Supernova in NGC 1332

125.701 **Supernova in NGC 1332.**
IAU Circ., No. 3727 (1982).

Supernova in NGC 1187

125.801 **Supernova in NGC 1187.**
IAU Circ., Nos. 3739, 3740 (1982).

Supernova in ESO 308–G16

125.901 **Possible supernova in ESO 308–G16.**
IAU Circ., No. 3745 (1982).

126 Low-luminosity Stars, Subdwarfs, White Dwarfs, Degenerate Stars

126.001 **IUE spectra of Feige 4, a suspected source of extreme-ultraviolet radiation.**
A. V. Holm, A. Boggess.
Publ. Astron. Soc. Pacific, Vol. 94, 553 - 556 (1982).

IUE spectra are presented for Feige 4, a DB-type white dwarf. The observed energy distribution is in good agreement with that predicted by a model atmosphere with T_{eff} = 15,500 K and having extremely low hydrogen and metal abundances. These results are evidence against Feige 4 being the source of the apparent extreme-ultraviolet signal detected by Prognoz 6.

126.002 **On the statistics of ZZ Ceti stars.**
G. Fontaine, J. T. McGraw, D. S. P. Dearborn, J. Gustafson, P. Lacombe.
Astrophys. J., Vol. 258, 651 - 660 (1982).

The results of a continuing survey for additional pulsating white dwarfs are summarized. In particular, the color distributions of stars investigated for variability are discussed. In terms of the Greenstein $G - R$ color index, among the hydrogen-rich white dwarfs only variable stars have thus far been found in the range $-0.45 \leqslant (G - R) \leqslant -0.38$. This color range corresponds to an effective temperature interval of 13,000 K $\gtrsim T_e \gtrsim$ 11,000 K. The implication of this observational result is that the vast majority, and probably all, of the hydrogen-rich white dwarfs evolve to become ZZ Ceti variables during an evolutionary phase of their lifetimes.

126.003 **Temperature as a unique predictor of ZZ Ceti variability.**
J. L. Greenstein.
Astrophys. J., Vol. 258, 661 - 673 (1982).

Multichannel spectrophotometry of 14 ZZ Ceti variables provides energy distributions and colors on various baselines. Their mean temperature is 11,100 K, the edges of the instability strip for DA stars are at 11,700 and 10,700 K, the mean gravity $\log g = 8.0$. The time-averaged energy distributions of the variables are indistinguishable from those of the non-variable DA's. The fraction of all DA's that vary, within the critical instability strip, is 60% to 80%; the $G - R$ color is an excellent criterion. It is possible that all DA's in a narrow temperature range are variable.

126.004 **Multicolor variations of the ZZ Ceti stars.**
E. L. Robinson, S. O. Kepler, R. E. Nather.
Astrophys. J., Vol. 259, 219 - 231 (1982).

Although the ZZ Ceti stars, or pulsating white dwarfs, are usually thought to be pulsating in the nonradial g-modes, this conclusion is based entirely on the observed periods of the pulsations. In the first part of this paper the authors calculate the theoretical color, radial velocity, and line profile variations of a white dwarf undergoing g-mode pulsations. In the second part they report the results of multicolor photometry of the ZZ Ceti star R548. Within the observational errors, the luminosity variations of R548 are consistent with temperature variations. The authors conclude that the properties of the pulsations of R548 are consistent with the expected properties of g-mode pulsations.

126.005 **ZZ Ceti stars: variable white dwarfs.**
D. E. Winget, H. M. Van Horn.
Sky Telesc., Vol. 64, 216 - 219 (1982).

126.006 **Observations and analysis of the low amplitude ZZ Ceti variable, L19–2.**
D. E. O'Donoghue, B. Warner.
Mon. Not. R. Astron. Soc., Vol. 200, 563 - 583 (1982).

The authors report 292 hr of high speed photometry of the low amplitude ZZ Ceti star L19–2. Five coherent oscillations were found, near 350, 192, 143, 118 and 113 s periods; each of these has multiple components, the largest in each case have amplitudes of 1.0, 6.0, 0.55, 1.8 and 2.2 $\times 10^{-3}$ mag respectively. The 192 s oscillation possesses three components equally separated in frequency. When combined with earlier observations, evidence is found for a small departure from equality, by an amount in accordance with recent calculations of the second-order rotational splitting of g-mode oscillations in white dwarfs. No period changes were detected over a 4 1/2 yr baseline. This establishes L19–2 as another very stable pulsating white dwarf.

126.007 **X-ray emission from white dwarfs.**
N. D. Kylafis, D. Q. Lamb.
Galactic X-ray sources, (see 003.001), p. 229 - 230 (1982).

126.008 **The OB subdwarf Feige 66, a chemical-composition twin to HD 149382.**
B. Baschek, P. Höflich, M. Scholz.
Astron. Astrophys., Vol. 112, 76 - 83 (1982).

The ultraviolet (1160 to 3230 Å) and the blue spectrum of the OB subdwarf Feige 66 is analyzed differentially to HD 149382. The effective temperature, the surface gravity and the helium abundance of Feige 66 deviate only very little from those of HD 149382. The line spectra are almost identical, and the same strange abundance pattern as in HD 149382 is found in Feige 66. In particular, silicon is extremely deficient relative to a solar mixture. A sulfur enrichment in Feige 66 is the only significant heavy-element abundance difference between the two stars.

126.009 **Magneto-optical effects in magnetic white dwarfs. II. Continuum.**
B. Martin, D. T. Wickramasinghe.
Mon. Not. R. Astron. Soc., Vol. 200, 993 - 1005 (1982).

The impact on the spectra of magnetic white dwarfs of magneto-optical effects or magnetic birefringence arising from free electrons in the continuum is investigated using model calculations. Linear polarization is the most strongly affected: in the continuum its magnitude is reduced and its sign often reversed, and in the lines its magnitude is greatly reduced. In the outer components of lines, the maximum values of circular polarization are reduced and their positions shifted inwards. Line depths are increased, especially the central component of a Zeeman triplet.

126.010 **Discovery of Ca II absorption at 1840 Å in IUE spectra of two helium-rich white dwarfs.**

D. Koester, G. Vauclair, V. Weidemann, E. M. Zeidler-K. T.
Astron. Astrophys., Vol. 13, L13 - L14 (1982).

For the first time the strong Ca II λ 1840 Å $^2D - ^2F^o$ (3d - 4f) absorption line was observed in the IUE short wavelength region of a white dwarf spectrum. The star, L 745 - 46 A, is classified DF due to the presence of Ca II H and K lines in the visible region. The same line, of weaker strength, was found in the short wavelength IUE spectrum of L 791 - 40. This star shows to be a true intermediate case between the spectral types DB, DC, C_2 and DF of helium-rich white dwarfs.

126.011 **A possible CH subdwarf.**
L.-T. G. Chiu, R. G. Kron.
Nature, Vol. 299, 702 - 704 (1982).

The authors discuss a possible CH subdwarf. First, they establish its subdwarf nature through photometry and proper motion, then they discuss spectroscopic data. Finally, they review how CH giants, CH subgiants and the CH subdwarf constitute a sequence of increasingly challenging cases for the stellar evolution theory.

126.012 **Astrometric planetary searches, the absence of low-luminosity white dwarf stars, and the age of the Galaxy.**
H. L. Shipman.
Bull. American Astron. Soc., Vol. 14, 626 (1982). – Abstract.

126.013 **Observations of ultraviolet carbon lines in the spectra of three DC white dwarfs.**
G. A. Wegner.
Bull. American Astron. Soc., Vol. 14, 652 (1982). – Abstract.

126.014 **A systematic spectroscopic survey of Hα in cool helium rich white dwarfs.**
F. H. Yackovich, G. A. Wegner.
Bull. American Astron. Soc., Vol. 14, 652 (1982). – Abstract.

126.015 **Comments on radial white dwarf accretion.**
J. Kuijpers, J. E. Pringle.
Astron. Astrophys., Vol. 114, L4 - L6 (1982).

The authors discuss radial accretion flow onto white dwarfs for which cyclotron emission is insignificant. They propose a non-stationary model in which the inhomogeneity of the accretion flow ensures no stand off shock and that the accretion energy is liberated deep in the stellar atmosphere and radiated away predominantly at XUV wavelengths.

126.016 **Non-LTE analysis of the sdO-star Feige 110.**
U. Heber, W.-R. Hamann, K. Hunger,
R. P. Kudritzki, K. P. Simon, R. H. Méndez.
Third European IUE Conference, (see 012.029), p. 297 - 300 (1982).

Effective temperature, gravity and helium abundance of the sdO-star Feige 110 are determined by a NLTE analysis of high resolution IUE spectra and visual spectra. The abundances of carbon and nitrogen are obtained from the analysis of ultraviolet resonance lines.

126.017 **Analysis of Feige 66, another extremely silicon-deficient OB subdwarf.**
B. Baschek, P. Höflich, M. Scholz.
Third European IUE Conference, (see 012.029), p. 309 - 311 (1982).

A high-resolution IUE spectrum from 1160 to 3230 Å of the OB subdwarf Feige 66 is analyzed by model atmosphere techniques differentially to HD 149382. Both stars not only have similar effective temperatures and surface gravities, but also exhibit remarkably similar abundance patterns. In particular, Feige 66 also shows an extreme deficiency of silicon.

126.018 **The metallicity of the extreme helium star BD +10°2179.**
U. Heber.
Third European IUE Conference, (see 012.029), p. 313 - 316 (1982).

Ultraviolet as well as visual spectra of the extreme helium star BD +10°2179 are analyzed using line blanketed LTE model atmospheres. The abundances of 19 elements have been obtained. The atmosphere consists mainly of helium (98.9% by number) with a small admixture of carbon (1%). The metal abundances are considerably different from those of the sun.

126.019 **The soft X-ray spectrum of Sirius B: evidence for the photospheric hypothesis.**
C. Martin, G. Basri, M. Lampton, S. M. Kahn.
Astrophys. J., Lett., Vol. 261, L81 - L85 (1982).

The authors have used the soft X-ray spectrum of Sirius obtained by the HEAO 1 A-2 experiment to investigate the origin of the X-radiation from that system. They show that a corona model conflicts severely with EUV observations and is physically untenable. They have also fitted white dwarf atmosphere models to the spectrum and show that a photospheric temperature of 28,000 K is much more consistent with the data. The relative consistency of the soft X-ray, EUV, and UV observations of Sirius B are discussed and it is concluded that the evidence favors the white dwarf photosphere as the origin of the soft X-rays.

126.020 **Detection of the 1400 Å absorption in the ultraviolet spectrum of the DA white dwarf LB 3303.**
G. Wegner.
Astrophys. J., Lett., Vol. 261, L87 - L89 (1982).

Low-resolution ultraviolet IUE spectra of the southern white dwarf LB 3303 show the presence of the λ 1400 absorption feature reported by Greenstein in the spectrum of 40 Eri B. The equivalent widths is 5.7 Å, and the measured wavelength is 1394 Å. It is confirmed that LB 3303 has an effective temperature near 16,000 K. There are still problems with the identification of this line. The star is not hot enough to explain the presence of Si IV, and the agreement with the spectrum of the H_2 molecule is not convincing.

126.021 **An estimation of the space number density of white dwarfs.**
K. Ishida, T. Mikami, T. Noguchi, H. Maehara.
Publ. Astron. Soc. Japan, Vol. 34, 381 - 388 (1982).

A catalog of 588 blue objects surveyed on 715 square degrees was used to estimate the space number density of white dwarfs.

126.022 **Analysis of the ultraviolet spectrum of RWT 152: a subluminous O star with a main-sequence visual spectrum.**
D. C. Ebbets, B. D. Savage.
Astrophys. J., Vol. 262, 234 - 243 (1982).

The authors have obtained both low dispersion and high dispersion IUE spectra of RWT 152, a faint but lightly reddened star with a visual spectrum of type O5 V. A blanketed model atmosphere analysis leads them to conclude that RWT 152 is not a young massive star, but rather a low mass object in an advanced stage of evolution. The profiles of the resonance lines of C IV and N V indicate that the star is losing mass via a stellar wind. Estimates of the stellar properties of radius, mass, luminosity, distance, and of the mass loss rate are obtained.

126.023 **Photometric observations of GD 358: DB white dwarfs do pulsate.**
D. E. Winget, E. L. Robinson, R. E. Nather, G. Fontaine.
Astrophys. J., Lett., Vol. 262, L11 - L15 (1982).

It is shown that GD 358, a helium atmosphere (DB)

white dwarf, is a pulsating variable star. The amplitude of its variations can range up to nearly 0.30 mag, and although the light curve does not repeat itself, there is a typical interval between its pulse maxima of 600 - 700 s. There are at least 28 pulsation modes simultaneously excited in GD 358 with periods ranging from 142.3 s to 952 s. All the pulsating white dwarfs found heretofore have been hydrogen atmosphere (DA) white dwarfs. It is suggested that the pulsations of DB white dwarfs are driven by helium partial ionization.

126.024 **Extreme ultraviolet spectrophotometry of the hot DA white dwarf HZ 43: detection of He II in the stellar atmosphere.**
R. F. Malina, S. Bowyer, G. Basri.
Astrophys. J., Vol. 262, 717 - 726 (1982).

The authors report a spectrum of the hot white dwarf HZ 43 in the extreme ultraviolet band from 170–500 Å. The spectrum is marked by a prominent absorption edge at 225 Å ± 15 Å. The observed edge is attributed to absorption at the 228 Å Lyman edge of singly ionized helium. A grid of blanketed LTE model atmospheres with uniform compositions of hydrogen with trace helium were used to interpret the data. The authors have evaluated possible sources of He II in the interstellar medium and conclude that they are unlikely to contribute the entire observed column density. Hence the observations provide the first detection of helium in HZ 43.

126.025 **The mass distribution of white dwarfs and central stars of planetary nebulae.**
V. Weidemann.
Binary and multiple stars as tracers of stellar evolution, (see 012.046), p. 403 - 408 (1982).

Recent results on mass distributions for white dwarfs and planetary nebulae are presented and compared with current theoretical predictions. Whereas single star evolution leads to final masses predominantely in a narrow interval around 0.6 $M_\odot$ which can be explained by current mass loss schemes degenerate stars in binaries present a larger range of masses. The average mass of the primaries in cataclysmic binaries seems to be more around 0.7 than 1 $M_\odot$.

126.026 **Abundance peculiarities in white dwarf atmospheres as a result of binary evolution?**
I. Bues.
Binary and multiple stars as tracers of stellar evolution, (see 012.046), p. 409 - 411 (1982).

The author investigates metal abundances in helium-rich white dwarf atmospheres and check the possibility of variations due to binary evolution.

126.027 **The formation of massive white dwarfs in cataclysmic binaries.**
W.-Y. Law, H. Ritter.
Binary and multiple stars as tracers of stellar evolution, (see 012.046), p. 447 - 451 (1982).

The formation of cataclysmic binaries with white dwarfs of less than about 0.8 $M_\odot$ can be understood as the result of a binary evolution according to low mass Case B or Case C with a subsequent spiraling-in in a common envelope. On the other hand the formation of massive white dwarfs of $M \gtrsim 1\,M_\odot$ can be explained as the result of a massive Case B mass transfer yielding a helium star which subsequently undergoes a second Case B mass transfer. The ultimate product of such an evolution is either a CO-white dwarf with a mass up to the Chandrasekhar limit or a neutron star.

126.028 **Nitrogen overabundances in Population II dwarfs.**
M. S. Bessell, J. Norris.
Astrophys. J., Lett., Vol. 263, L29 - L33 (1982).

From a survey and spectrum synthesis analysis of the NH features in the spectra of some 25 Population II dwarfs, two objects have been found in which nitrogen is overabundant relative to the heavy elements by a factor ~50 - 100. The two stars are HD 74000 and HD 160617, which are well-known, extremely metal-deficient subdwarfs. The authors interpret the data as suggesting a decoupling of nitrogen and heavy element production during the first stellar generations, and indicate that, in the early universe, nitrogen was produced as a primary element.

126.029 **Atmospheric parameters and carbon abundance of white dwarfs of spectral types C_2 and DC.**
D. Koester, V. Weidemann, E.-M. Zeidler-K.T.
Astron. Astrophys., Vol. 116, 147 - 157 (1982).

Recent IUE observations and new photometric and spectroscopic observations in the visual region of white dwarfs of spectral types C_2 and DC are analyzed using model atmosphere calculations covering a large range of effective temperatures and carbon abundances. Carbon abundances (C : He) are of the order of 10^{-7} – 10^{-6} for DC and between 10^{-6} and 2×10^{-3} for C_2, with most objects near the lower value. Upper limits for other elements (H, O, N, Mg, Ca) are discussed, and some speculative ideas on the origin of the observed carbon abundances are presented. For 17 DC/C_2 with known distances the masses are derived, yielding $\langle M \rangle = 0.55\,M_\odot$, similar to the result found for DA white dwarfs.

126.030 **Ergebnisse Photometrischer Untersuchungen von schwachen blauen Sternen der Giclas-Listen.**
G. Rupprecht, I. Bues.
Mitt. Astron. Ges., Nr. 57, p. 321 - 322 (1982).

126.031 **BD +38°3231 – high-velocity subdwarf, optical component of Vega.**
A. Bartkevičius, S. Bartšiūte.
Astron. Tsirk., No. 1177, p. 7 - 8 (1981). In Russian.

126.032 **Two hot, low-field magnetic DA white dwarfs.**
J. Liebert, G. D. Schmidt, R. F. Green, H. S. Stockman, J. T. McGraw.
Prepr. Steward Obs., No. 401, 35 pp. (1982).

126.033 **The peculiar cool white dwarf LHS1126: a different interpretation.**
J. Liebert, C. C. Dahn.
Prepr. Steward Obs., No. 421, 21 pp. (1982).

126.034 **The hydrogen-rich, cool DA white dwarf Ross 627.**
J. Liebert, R. Wehrse.
Prepr. Steward Obs., No. 425, 22 pp. (1982). – Submitted to Astron. Astrophys..

126.035 **Discovery of highly ionized species in the ultraviolet spectrum of Feige 24.**
A. K. Dupree, J. C. Raymond.
Astrophys. J., Lett., Vol. 263, L63 - L67 (1982).

A high-resolution ultraviolet spectrum of the nearby (90 pc) white dwarf binary Feige 24 (DA + M1–2 V) obtained with the International Ultraviolet Explorer satellite (IUE) reveals two sets of absorption features in the species C IV, Si IV, and N V. The low-velocity component (V ~ 1 km s^{-1}) arises from the local interstellar gas apparently ionized by the white dwarf. The second component, at high velocities (V ~ +83 km s^{-1}), arises in the atmosphere of the white dwarf in the Feige 24 system. Observations around the orbital period will provide a direct measurement of the gravitational redshift and the mass of the white dwarf.

126.036 **Thermal and electrical conductivities of crystals in neutron stars and degenerate dwarfs.**
M. E. Raikh, D. G. Yakovlev.
Astrophys. Space Sci., Vol. 87, 193 - 203 (1982).

Thermal and electrical conductivities due to electron

scattering on phonons are calculated for degenerate cores of white dwarfs and envelopes of neutron stars for wide ranges of density, temperature and ion charge.

New limits on the density of superheavy magnetic monopoles in the interstellar space.
See Abstr. 061.065.

White dwarf stars with chemically stratified atmospheres.
See Abstr. 064.019.

Accretion of grains and element abundances in cool, helium-rich white dwarfs.
See Abstr. 064.027.

Hot accreting white dwarfs in the quasi-static approximation.
See Abstr. 065.002.

On the secular evolution of accreting white dwarfs and Type I supernovae.
See Abstr. 065.018.

Astrophysical constraints on broken supersymmetry.
See Abstr. 065.062.

Magnetic properties of relativistic Fermi gas.
See Abstr. 066.269.

Magnetic field strongly reduces critical luminosity of neutron stars and degenerate dwarfs.
See Abstr. 066.514.

The distance of the hot white dwarf HD 149499B.
See Abstr. 111.005.

Ultraviolet and X-ray detection of the 56 Pegasi system (K0 IIp + WD): evidence for accretion of a cool stellar wind onto a white dwarf.
See Abstr. 114.158.

New light on faint stars – II. A photometric study of the low luminosity main sequence.
See Abstr. 115.009.

A search for cataclysmic binaries containing strongly magnetic white dwarfs.
See Abstr. 117.027.

A model of AM Canum Venaticorum.
See Abstr. 117.064.

Time resolved circular polarimetry of white dwarf pulsars.
See Abstr. 117.153.

The hot halo subdwarf binary system HZ-22.
See Abstr. 117.208.

Ultraviolet and X-ray detection of the 56 Peg system (K0 IIp + WD).
See Abstr. 117.212.

Evolution of the subdwarf-O eclipsing binary LB 3459.
See Abstr. 119.039.

The subdwarf eclipsing binary LB3459 (AA Dor).
See Abstr. 119.060.

A new type of pulsating star.
See Abstr. 122.020.

Observations of the rapid oscillations of EM Cygni.
See Abstr. 122.073.

On the statistics of ZZ Ceti stars.
See Abstr. 122.127.

The UV spectra of nearby white dwarfs and the nature of the local interstellar medium.
See Abstr. 131.016.

2A 0526–328, an optical and X-ray study of a cataclysmic variable.
See Abstr. 142.038.

X-ray spectra of accreting degenerate stars.
See Abstr. 142.077.

Spectroscopic identification of white dwarfs in galactic clusters. II. NGC 2516.
See Abstr. 153.030.

Interstellar Matter, Nebulae

131 Interstellar Matter, Star Formation

131.001 **Formaldehyde absorption measurements of selected galactic molecular clouds.**
J. H. Bieging, T. L. Wilson, D. Downes.
Astron. Astrophys., Suppl. Ser., Vol. 49, 607 - 630 (1982).

Detailed maps of H_2CO 4830 MHz absorption toward 12 bright galactic H II regions and 2 dark clouds are presented. The maps have 2.6′ angular resolution and cover the full range of velocities over which H_2CO absorption has been detected. The maps delineate molecular clouds toward low-surface-brightness regions which are inaccessible to aperture synthesis methods. H_2CO column densities and cloud sizes are tabulated for each of the detected absorption features. The mapping results indicate that large-scale systematic motions are not common, and that the observed line widths are probably a result of small-scale turbulence driven by the exciting stars of the H II regions.

131.002 **CN line emission from the molecular clouds associated with S 106 and NGC 7538.**
E. Churchwell, J. H. Bieging.
Astrophys. J., Vol. 258, 515 - 532 (1982).

The line intensity distributions of the N = 1–0 transition of CN are presented for the molecular clouds associated with S 106 and NGC 7538. In S 106, CN line emission is largely proportional to the column density of CN; the temperature and density structure of S 106 are fairly complex and clearly influence the CN line intensities. The CN emission in S 106 appears to be a disk of molecular gas in which an O8 or O9 star has formed, ionized, and dispersed the central part of the disk, and left a molecular torus with an expanding double-lobed H II region approximately perpendicular to the plane of the torus. In NGC 7538 the CN column density is essentially constant over the whole mapped region, and the CN line intensities vary with the excitation temperature. At least three regions of successive star formation are apparent toward NGC 7538. The most recent star formation appears to be occurring in the region of maximum CN emission.

131.003 **The excitation of interstellar C_2.**
E. F. van Dishoeck, J. H. Black.
Astrophys. J., Vol. 258, 533 - 547 (1982).

A theory is developed to describe the excitation of the diatomic carbon molecule in interstellar clouds. The probabilities of quadrupole transitions within the ground state and of intercombination transitions between levels of the lowest triplet and singlet states are computed. The populations of excited rotational levels can be used as a diagnostic probe of local temperature, density, and strength of the radiation field. Calculated rotational populations are presented for a range of physical conditions and are used to interpret existing interstellar absorption line observations of C_2.

131.004 **Hyperfine anomalies of HCN in cold dark clouds.**
C. M. Walmsley, E. Churchwell, A. Nash, E. Fitzpatrick.
Astrophys. J., Lett., Vol. 258, L75 - L78 (1982).

Observations of the J = 1 → 0 line of HCN measured toward six positions in nearby low-temperature dark clouds are reported. The measured relative intensities of the hyperfine components of the J = 1 → 0 line are anomalous in that the F = 0 → 1 transition is stronger than would be expected if all three components (F = 2 → 1, F = 1 → 1, F = 0 → 1) had equal excitation temperatures. Differences of approximately 20% in the populations per sublevel of J = 1 could account for the observations.

131.005 **Reddenings derived from H I and galaxy counts: accuracy and maps.**
D. Burstein, C. Heiles.
Astron. J., Vol. 87, 1165 - 1189 (1982).

A quantitative assessment is made of the accuracy of the H I/galaxy counts (H I/GC) method that was developed in an earlier paper for predicting line-of-sight reddenings in the galaxy. High-quality reddening estimates from two independent techniques are used in the comparison: the intrinsic colors of stars as determined by intermediate-band photometric systems, chiefly the *uvbyβ* system; and the intrinsic relationship between absorption line strength and color for the integrated spectra of old extragalactic stellar populations. The available evidence for the amount of reddening at the galactic poles, both from photometry and interstellar polarization, is assembled and discussed. Maps of reddening in the galaxy are given that cover almost the whole galactic sphere with $|b| \geq 10°$ and that permit an accurate estimation of reddening in almost all directions. A map of the smoothed Shane-Wirtanen (1967) galaxy counts is also given. Specific applications of these maps to other astronomical problems are presented.

131.006 **Sensitive ammonia observations in the southern hemisphere.**
D. F. Dickinson, S. Gulkis, M. J. Klein, T. B. H. Kuiper, M. Batty, F. F. Gardner, D. L. Jauncey, J. B. Whiteoak.
Astron. J., Vol. 87, 1202 - 1206 (1982).

Preliminary observations of the (1, 1) transition of ammonia are reported for a selection of eight southern molecular clouds. Estimates of gas density and ammonia column density are reported for six sources.

131.007 **Excitation of the rotational levels of the interstellar HCN molecule.**
I. I. Zinchenko, V. K. Khersonskij.
Astron. Zh., Tom 59, 676 - 682 (1982). In Russian. English translation in Soviet Astron., Vol. 25, No. 4.

HCN and DCN level populations and line intensities are computed for the intervals of kinetic temperatures $T_K = 10-100$ K and gas densities $N = 10^3-10^9$ cm^{-3} in the case of optically thin lines. The hyperfine sublevel population and the dependence on N and T_K are discussed.

131.008 **On the possibility of sweeping of interstellar gas from a galaxy by radiative pressure of an active nucleus.**
V. G. Gorbatskij.
Astrofizika, Tom 18, 234 - 244 (1982). In Russian. English translation in Astrophysics, Vol. 18, No. 2.

If the luminosity of a galactic nucleus in the phase of its activity is great enough, radiation pressure may sweep off the gas from the galaxy. The gas flow induced by this process is considered on the assumption that dust particles dragged by radiation immediately transfer the momentum and energy obtained to the gas. Estimates are made of luminosity values that are necessary to maintain the stationary flow.

131.009 **The temperature dependence of the HCO^+/DCO^+ abundance ratio in dense interstellar clouds.**
E. Herbst.
Astron. Astrophys., Vol. 111, 76 - 80 (1982).

The equilibrium coefficient of the reaction system $H_3^+(HD)=H_2D^+(H_2)$ has been calculated and found to disagree somewhat with earlier calculated and extrapolated values at interstellar cloud temperatures. Non-equilibrium effects under interstellar conditions that might affect the forward and reverse reaction rates have also been investigated and found to be of minor significance. The results permit determination of dense cloud temperatures from observed $[HCO^+]/[DCO^+]$ abundance ratios. The derived temperatures for selected "warm" interstellar clouds are compared with CO brightness temperatures.

131.010 **The graphite rich Cepheus OB 3 association.**
B. Barsella, N. Panagia, M. Perinotto.
Astron. Astrophys., Vol. 111, 130 - 135 (1982).

The four optically brightest stars of the Cepheus OB 3 association have been observed with the IUE satellite in the spectral range λ1150–3200 Å. The extinction suffered by the stars has been investigated. Both the shape of the bump at λ ~ 2200 Å and the behavior of the extinction at longer wavelengths are close to those of the mean interstellar curve. The behavior shortward of λ2000 Å is peculiar and reveals the existence of a component of dust which belongs to the association. A fit with theoretical extinction curves shows that a strong excess of graphite relative to silicates in the local comnent is required to explain the observed extinction in the Cepheus OB 3 association.

131.011 **NH_3 and H_2O in the S 106 molecular cloud.**
J. Stutzki, H. Ungerechts, G. Winnewisser.
Astron. Astrophys., Vol. 111, 201 - 208 (1982).

The authors simultaneously mapped the NH_3 (J, K) = (1, 1) and (2, 2) lines with an angular resolution of 40″ in the molecular cloud surrounding the "bipolar nebula" S 106. The NH_3 cloud consists of two parts located east and west of the visible nebula. There is no indication that the cloud forms a rotating ring. The intensities of the $NH_3(1, 1)$ nuclear quadrupole HfS components show strong deviations from LTE, which are position dependent. The authors did not find the H_2O maser reported by Cesarsky et al. (1978), instead they detected a new maser 13″ west of the central IR-point source.

131.012 **Interstellar iron and manganese: UV oscillator strengths and abundances.**
P. Lugger, E. Barker, D. G. York, W. Oegerle.
Astrophys. J., Vol. 259, 67 - 76 (1982).

Observations of 16 UV resonance lines of Fe II and six of Mn II in five stars are used to derive new *f*-values for the lines of these species at $\lambda < 1300$ Å. Values of [Fe/H] and [Mn/H] are derived. These new values are used to reassess mean depletions and range of variations in depletions for several lines of sight. The mean local depletion [Fe/H] is 1.65, in interstellar gas.

131.013 **A heuristic criterion for instability to fragmentation in rotating, interstellar clouds.**
A. P. Boss.
Astrophys. J., Vol. 259, 159 - 165 (1982).

A heuristic criterion, based on linear perturbation analysis, is applied to the initial growth of density perturbations in isothermal or adiabatic gas clouds, with initially uniform density and uniform rotation. The heuristic criterion is shown to be consistent with the available results from numerical calculations of cloud collapse.

131.014 **Airborne observations of the Orion molecular hydrogen emission spectrum.**
D. S. Davis, H. P. Larson, H. A. Smith.
Astrophys. J., Vol. 259, 166 - 179 (1982).

The authors have observed the Orion near-IR H_2 emission spectrum from an altitude of 12.5 km. Three new transitions [H $n = 6 \rightarrow 4$ and H_2 $v = 1 \rightarrow 0$ O(2) and Q(7)] have been detected. Analysis of the data indicate that earlier estimates of source extinction, temperature, and luminosity are all too high. The authors find $E_{2.1-2.4\,\mu m} = 0.59 \pm 0.06$, $T(v = 1) = 1540 \pm 100$ K and $L(H_2) = 200–400\ L_\odot$.

131.015 **Protostellar mass and angular momentum loss.**
L. Hartmann, K. B. MacGregor.
Astrophys. J., Vol. 259, 180 - 192 (1982).

Recent radio observations have indicated that appreciable continuing mass loss is occurring from regions of star formation. The authors investigate the conditions under which massive protostellar objects may exhibit strong, rotationally driven winds. For plausible rotational velocities and magnetic field strengths, ejection of $\sim 10^4 M_\odot\ yr^{-1}$ at speeds $\sim 10^1–10^2$ km s^{-1} can be maintained for time scales $\sim 10^{-4}$ yr. The centrifugally driven winds proposed present an alternative to fragmentation for providing the angular momentum loss required in the phase of rapid cloud contraction.

131.016 **The UV spectra of nearby white dwarfs and the nature of the local interstellar medium.**
F. C. Bruhweiler, Y. Kondo.
Astrophys. J., Vol. 259, 232 - 243 (1982).

The authors have investigated the local interstellar medium in the directions of four white dwarfs, G191–B2B, W1346, HD 149499B, and Sirius B. All the observational data were obtained at the high-resolution mode in the spectral range from about 1150 to 3200 Å with the IUE. Interstellar absorption lines of several elements in various stages of ionization are seen against the continuum of the white dwarfs. The results show, when combined with other recent observations, that: (1) the Sun is located inside a low-density ($n_{HI} \approx 0.1$ cm^{-3}) cloud; (2) beyond 2-3 pc from the Sun, this cloud is surrounded by an extended region of hot ($T \approx 10^{5-6}$ K) thin ($n \approx 10^{-2}$ to 10^{-3} cm^{-3}) interstellar plasma with no evidence for additional clouds in the lines of sight studied; (3) The Sun is moving through this cloud at a relative velocity of about 20 km s^{-1}.

131.017 **Observations of the new source of OH maser emission G43.2–0.1.**
M. I. Pashchenko.
Pis'ma Astron. Zh., Tom 8, 362 - 364 (1982). In Russian. English translation in Soviet Astron. Lett., Vol. 8.

OH 1612 MHz emission from the new maser source G43.2–0.1 in the direction of the W49 radio complex has been detected.

131.018 **An unusual OH maser associated with V 645 Cygni.**
M. Morris, I. Kazès.
Astron. Astrophys., Vol. 111, 239 - 241 (1982).

Ground state OH line observations of the biconical nebula V 645 Cygni have revealed an unusual maser at 1665 MHz. Although the characteristics are reminiscent of some type I OH masers, no polarization has been found and the 1665 MHz emission probably underwent an intensity variation. Possible interpretations of the results are discussed.

131.019 **Anticenter high velocity H I stream (Weaver jet) and colliding H I shells.**
T. Watanabe.
Astron. Astrophys., Vol. 111, 333 - 338 (1982).

Large-scale structures of H I clouds around the Weaver jet and a high negative-velocity stream of H I gas are examined. Two giant H I shells are apparently in collision in the direction of the jet. A low-intensity H I feature observed at $V_{lsr} \sim +10$ km s^{-1} in the direction of the jet is probably a cold region formed by the collision between these H I shells. The high-velocity core of the Weaver jet and the positive-velocity jet discovered by the present investigation seem to originate in the collision.

131.020 **Ammonia observations of cold cloud cores.**
H. Ungerechts, C. M. Walmsley, G. Winnewisser.
Astron. Astrophys., Vol. 111, 339 - 345 (1982).

The NH_3(1,1) and (2,2) lines have been observed towards the cores of seven dark dust clouds, mostly in the Taurus region. From these measurements the authors estimate the gas kinetic temperature to be close to 10 K. The cloud cores have sizes ~0.05–0.3 pc and densities $\sim 10^{4.0}$–$10^{5.0}$cm^{-3}. The authors find no evidence for local heating by embedded "protostellar objects".

131.021 **A survey of high-velocity H I in the Cetus region.**
R. J. Cohen.
Mon. Not. R. Astron. Soc., Vol. 200, 391 - 405, microfiche MN 200/2 (1982).

The region $02^h 16^m < \alpha (1950) < 03^h 12^m$, $0° < \delta$ (1950) $< 14°$ surrounding the Cohen & Davies complex of high-velocity clouds has been surveyed in the 21-cm line of H I using the Jodrell Bank MK II radio telescope. The observations cover a velocity range of 2100 km s^{-1} with a resolution of 7.3 km s^{-1} and an rms noise level of 0.025 K. Discussion is centred on the very-high-velocity clouds at velocities of −360 to −190 km s^{-1}. It is concluded that they are probably debris from the tidal interaction between our Galaxy and the Magellanic Clouds.

131.022 **Molecules in celestial objects – III. Study of CO in interstellar diffuse clouds.**
S. P. Tarafdar, K. S. Krishna Swamy.
Mon. Not. R. Astron. Soc., Vol. 200, 431 - 444 (1982).

The absorption lines corresponding to the $A-X$ transition of CO have been looked for in the *IUE* spectra of 14 stars with varying values of the colour excess, $E(B-V)$ and found to be present in the spectra of nine stars with $E(B-V) \geqslant 0.28$. The column density of CO has been determined towards these nine stars and its upper limit towards the rest of the stars. The curve of growth analysis has been found to show that the contribution to CO absorption is possibly from a single interstellar cloud for stars with $E(B-V) < 0.4$ and from more than one cloud for stars with $E(B-V) > 0.4$. The observed column density of CO as a function of $E(B-V)$ has been found to be in good agreement with that expected from the theory of ion-molecular chemistry.

131.023 **Magnetite and the origin of the very broad structure in interstellar extinction.**
T. J. Millar.
Mon. Not. R. Astron. Soc., Vol. 200, 527 - 533 (1982).

This paper investigates the possibility that the very broad structure (VBS) observed in the visible region of the interstellar extinction curve arises from magnetite particles. Small magnetic particles show structure at 1.8 μm^{-1} which resembles the VBS. However, they also produce structure at around 1.4 μm^{-1}, structure which is not observed in the interstellar medium. Magnetite remains a plausible candidate for the source of the VBS but the evidence is far from compelling.

131.024 **A study of element depletions in interstellar gas.**
A. P. Phillips, P. M. Gondhalekar, M. Pettini.
Mon. Not. R. Astron. Soc., Vol. 200, 687 - 703 (1982).

High-resolution ultraviolet observations of seven lightly reddened stars at intermediate galactic latitudes have been obtained with the *IUE* satellite. The observed interstellar absorption lines have been analysed to estimate abundances in the interstellar gas in each line-of-sight. The degree of correlation between depletion and density is found to vary from element to element. The results have been compared with previous observations and discussed in the light of current models for the depletion process.

131.025 **Polarimetry of the H_2 emission from the Orion Molecular Cloud.**
R. R. Joyce, T. Simon.
Mon. Not. R. Astron. Soc., Vol. 200, 39P - 42P (1982).

The authors have measured the linear polarization of the $v = 1 \rightarrow 0\ S(1)$ emission of molecular hydrogen at three positions in the Orion Molecular Cloud. The observed spatial variations in polarization may be due to small-scale structure in the magnetic field of the cloud, or to local differences in the relative temperatures of the gas and dust.

131.026 **OH cooling of interstellar shocks and masers.**
D. R. Flower, S. Guilloteau, T. W. Hartquist.
Mon. Not. R. Astron. Soc., Vol. 200, 55P - 59P (1982).

The results of quantum mechanical calculations of cross-sections for the rotational excitation of OH by H_2 have been incorporated in a radiative transfer programme for the OH rotational lines. Calculations with this programme show that infrared emission from excited rotational levels of OH is an important cooling process behind shocks in interstellar molecular regions and in H II/OH maser regions. Allowance for this cooling process may significantly reduce the estimates of the amount of CH^+ formed in interstellar shocks.

131.027 **OH observations of NH_3 sources.**
L. T. Little, D. A. Cesarsky.
Astron. Astrophys., Vol. 112, 49 - 51 (1982).

The authors present four frequency observations of OH emission in 12 sources known to be NH_3 emitters. OH emission (or absorption) was detected in all of them. Four sources show strong type I maser emission.

131.028 **Diffuse ionized gas around NGC 6193.**
J. C. Cersosimo.
Astrophys. Lett., Vol. 22, 157 - 160 (1982).

The H166 α line has been detected in the direction of NGC 6193 (central cluster of the Ara OB1 association) with a velocity resolution of 2 km s^{-1}. The central velocity of the line is −20 km s^{-1}. The region has been also observed in the continuum, physical parameters are derived for the ionized gas. The parameters suggest that the H166 α line is formed in a diffuse gas with an electron density of 11 cm^{-3} and an electron temperature of 4600 K.

131.029 **Cook's model and linewidths of OH masers.**
S. White.
Proc. Astron. Soc. Australia, Vol. 4, 235 - 237 (1981).

131.030 **The absence of 1584 MHz maser emission from ^{18}OH.**
J. L. Caswell, R. F. Haynes.
Proc. Astron. Soc. Australia, Vol. 4, 237 - 239 (1981).

131.031 **Distribution of H_2CO absorption at 4.8 GHz towards M 17.**
F. F. Gardner, J. B. Whiteoak.
Proc. Astron. Soc. Australia, Vol. 4, 240 - 243 (1981).

131.032 **A CO survey of the southern galactic plane.**
W. H. McCutcheon, B. J. Robinson, J. B. Whiteoak.
Proc. Astron. Soc. Australia, Vol. 4, 243 - 247 (1981).

The authors report preliminary results which are part of a survey of J = 1 − 0 CO emission which will cover the longitude range $294° \leqslant l \leqslant 354°$ and latitude range $-0°.05 \leqslant b \leqslant 0°.05$ at single-beamwidth spacings. They determine a radial distribution of CO and give rough estimates of the sizes and masses of four clouds.

131.033 **Gas and dust in CrA.**
K. N. R. Taylor, R. Llewellyn, P. Payne, R. A. Vaile, S. Sakellis.
Proc. Astron. Soc. Australia, Vol. 4, 247 - 250 (1981).

131.034 **Survey of NH_3 in the southern hemisphere.**
E. Scalise, Jr., R. E. Schaal, Y. Bakor, J. W. Vilas Boas, P. C. Myers.
Proc. Astron. Soc. Australia, Vol. 4, 250 - 251 (1981).

131.035 **Interstellar scintillation and pulsar velocities.**
A. G. Lyne, F. G. Smith.
Nature, Vol. 298, 825 - 827 (1982).
The authors present new observations which show that the scintillation characteristics of pulsars are closely related to their velocities, as measured by their proper motions.

131.036 **Far infrared polarimetry of W51A and M 42.**
W. Cudlip, I. Furniss, K. J. King, R. E. Jennings.
Mon. Not. R. Astron. Soc., Vol. 200, 1169 - 1173 (1982).
A far infrared polarimeter has been designed for use with a balloon-borne telescope. It uses a rapidly rotating wire grid polarizer with the chopping secondary mirror of the telescope synchronized to the rate of rotation. Observations of M 42 using this system show a far infrared polarization of 2.2 ± 0.4 per cent, nearly orthogonal to the near infrared polarization which is usually attributed to dust absorption. Observations of W51A show a much smaller value of the polarization of 0.8 ± 0.5 per cent.

131.037 **The interstellar extinction toward the galactic central region.**
T. Ichikawa, K. Hamajima, K. Ishida, B. Hidayat, M. Raharto.
Publ. Astron. Soc. Japan, Vol. 34, 231 - 239 (1982).
Interstellar extinction within 5 kpc from the sun is investigated in the field of $351° \leq l \leq 10°$ through $l = 0°$ and $|b| \leq 2°$. About 2700 late M giants were detected in the field on Schmidt objective-prism plates in near-infrared wavelengths. The effective limiting distance of the survey and the interstellar extinction are estimated from the observed surface number density of the late M giants. The resultant extinction is presented in each area of $0\overset{\circ}{.}5 \times 0\overset{\circ}{.}5$.

131.038 **A preliminary search for 2.6-mm ^{12}CO emission from Weaver's jet region.**
T. Watanabe, K. Ishii, S. Kodaira.
Publ. Astron. Soc. Japan, Vol. 34, 249 - 255 (1982).
Radio line emission from ^{12}CO molecules ($J = 1 - 0$) has been searched for toward Weaver's jet ($l = 197°$, $b = 2°$), which is known as an anticenter anomalous-velocity H I stream. A curious low-intensity H I feature is seen at $V_{lsr} \sim +10 \text{ km s}^{-1}$ in the same direction of the jet.

131.039 **The interstellar band at λ 4430 Å and the abundance of interstellar iron, titanium, and molecular hydrogen.**
H. Zimmermann.
Astrophys. Space Sci., Vol. 84, 505 - 517 (1982).
It is the purpose of the present paper to look for correlations between the abundance of interstellar gaseous iron, titanium, and molecular hydrogen and quantities related to the λ 4430 Å band strength.

131.040 **Abundance fluctuations in the interstellar medium.**
M. Jura.
Advances in ultraviolet astronomy, (see 012.006), p. 54 - 60 (1982).
The determination of abundances within the interstellar medium is reviewed. It appears that interstellar abundances within 1 kpc of the sun are uniform to within a factor of two or three, but it is not yet possible to determine whether there are real fluctuations at this level except for deuterium. Establishing the level of local fluctuations in the abundances will be of considerable importance for understanding the history of nucleosynthesis in the solar neighborhood, the evolution of the interstellar medium and the formation of stars.

131.041 **Ultraviolet observations of interstellar dust.**
R. Bohlin.
Advances in ultraviolet astronomy, (see 012.006), p. 68 - 79 (1982).
A large fraction of astronomical observations in the UV are affected by extinction due to interstellar dust in the line of sight. The shape of the extinction curve varies markedly around the sky. The variations of the shape provide clues to the nature of the dust. Additional insight into the physical properties of dust can be obtained from reflected UV starlight as observed in NGC 7023 by IUE and in the faint arms of M 101 as photographed by a sounding rocket.

131.042 **The distribution of interstellar gas within 50 pc of the Sun.**
F. C. Bruhweiler.
Advances in ultraviolet astronomy, (see 012.006), p. 125 - 141 (1982).
Previous Copernicus determinations of H I column densities in nearby stars (d ~ 5 pc) are reviewed. These results are compared with recent ultraviolet (IUE) data and data acquired at other wavelengths. From this combined data set a coherent picture for the distribution of neutral hydrogen within 50 pc of the Sun emerges. A tentative model is proposed which suggests that the bulk of the H I resides in a cloud in a direction toward $l^{II} \sim 0°$. The data are consistent with the Sun being embedded, but near the edge of the cloud.

131.043 **Distribution of gas in the halo of the Galaxy.**
J. C. Blades, L. L. Cowie, D. C. Morton, D. G. York, C.-C. Wu.
Advances in ultraviolet astronomy, (see 012.006), p. 359 - 362 (1982).
Results are presented from a high-resolution, ultraviolet study of interstellar gas situated away from the plane of our Galaxy, using the nuclei of Seyfert galaxies Mkn 509 and F9 as background probes. In these directions the authors detect low-velocity, galactic gas as well as two extragalactic clouds. These data are combined with results from other lines of sight to show that the ultraviolet species extend about 10 kpc from the plane. Complimentary observations of the optical Ca II and Na I species suggests that these do not extend as far.

131.044 **Highly ionized gas in the galactic halo.**
M. Pettini, K. A. West.
Advances in ultraviolet astronomy, (see 012.006), p. 363 - 367 (1982).
The authors summarise the major results of a study of interstellar absorption in the halo which makes use primarily of IUE observations of a sample of 24 OB stars at z distances between 0.5 and ~3 kpc. These data are analysed together with other available results in order to investigate the distribution and physical conditions of highly ionized gas in the halo.

131.045 **Abundances in the Magellanic Stream.**
M. V. Penston, P. Pettini, A. Boksenberg, T. R. Gull, M. A. J. Snijders.
Advances in ultraviolet astronomy, (see 012.006), p. 368 (1982). – Abstract.

131.046 **Kinematics of gas in the Vela supernova remnant.**
G. Wallerstein, E. B. Jenkins, J. Silk.
Advances in ultraviolet astronomy, (see 012.006), p. 369 (1982). – Abstract.

131.047 **Small scale structure in the interstellar medium: Orion's threadbare cloak.**
S. N. Shore.

Advances in ultraviolet astronomy, (see 012.006), p. 370 - 373 (1982).

The interstellar medium in the central portion of the Orion cloak dynamical feature has been examined for evidence of fine structure. Four stars in the Ori OB I association bracket the stars in the lower belt region. Two are members of the σ Ori subcluster. A lower limit to the scale of inhomogeneity can be set using these at 0.1 pc, while on a scale of about 10 pc, the high velocity component of the second spectrum ions shows optical depth variation of more than a factor of 5.

131.048 **The violent interstellar medium associated with the Carina Nebula.**
M. Pettini, C. Laurent, J. A. Paul.
Advances in ultraviolet astronomy, (see 012.006), p. 380 - 384 (1982).

The authors summarise the main results of a detailed study of the interstellar spectrum of HD 93205, an O3V star in the cluster Tr16, located inside the most active part of the Great Carina Nebula.

131.049 **IUE observations of lines of sight with peculiar ultraviolet extinction.**
D. Massa, E. L. Fitzpatrick, B. D. Savage.
Advances in ultraviolet astronomy, (see 012.006), p. 409 - 412 (1982).

Low resolution IUE data were used to derive UV extinction curves for a group of stars known to have peculiar extinction parameters from ANS data. The new extinction curves have shapes that separate into two distinct classes; those associated with clear field extinction and those associated with extinction in dense nebular environments. The range of variation in the curves is so large, the common practice of "ironing out the bump" can produce enormous errors in the resultant UV energy distributions when $E(B-V) > 0.3$.

131.050 **The peculiar UV extinction of Herschel 36.**
J. Hecht, H. L. Helfer, J. Wolf, B. Donn, J. L. Pipher.
Advances in ultraviolet astronomy, (see 012.006), p. 413 - 415 (1982).

The differential extinction curve of Herschel 36 has been determined from IUE data. It is quite unusual, characterized by a distinct 2200 Å peak with a very low far blue end at $5 - 7\ \mu^{-1}$.

131.051 **Recent results on extinction in the Small Magellanic Cloud.**
L. Prévot, M. L. Prévot, J. Lequeux, B. Rocca-Volmerange, E. Mauricen.
Advances in ultraviolet astronomy, (see 012.006), p. 416 (1982). – Abstract.

131.052 **Energetic outflows from young stars.**
C. J. Lada.
Sci. American, Vol. 247, No. 1, p. 74 - 83 (1982).

The radiation emitted by carbon monoxide molecules in clouds of gas and dust where new stars are forming discloses masses of gas moving outward at high velocity in opposed directions.

131.053 **Formaldehyde absorption in S 128.**
A. Heske, H. J. Wendker.
Astron. Astrophys., Vol. 113, 170 - 172 (1982).

6 cm H_2CO absorption line observations in the direction of S 128 are presented. Using these and other published data the authors propose that the S 128 complex belongs to the Perseus spiral arm, but possesses a rather large peculiar radial velocity of -20 km s^{-1}. The star formation in the complex is ascribed to a local event where a local shock front happened to compress a lump of material at the fringes of the spiral arm.

131.054 **Observations of interstellar CO at mm and UV wavelengths and diverse comments on interstellar diffuse lines.**
D. McNally.
Irish Astron. J., Vol. 15, (see 012.008), 51 - 56 (1981).

131.055 **Sk 143: an SMC star with a galactic-type ultraviolet interstellar extinction.**
J. Lequeux, E. Maurice, M.-L. Prévot-Burnichon, L. Prévot, B. Rocca-Volmerange.
Astron. Astrophys., Vol. 113, L15 - L17 (1982).

The authors present the ultraviolet extinction curve (λλ 1200 - 3000 Å) deduced from IUE observations of the Small Magellanic Cloud star Sk 143. This curve is rather similar to the galactic one, with a large bump at λ 2200 and a smaller rise shortward of λ 2000 compared to the SMC extinction curve of Rocca-Volmerange et al. (1981). The $N_H/E(B-V)$ ratio is estimated. Tentative explanations of the observed properties are given, including a possible binarity of Sk 143, the presence of a large amount of molecular hydrogen on the line of sight and the existence of dust having the same properties as galactic dust.

131.056 **Extended and anisotropic high-velocity gas flows in the Orion-KL region.**
H. Olofsson, J. Elldér, Å. Hjalmarson, G. Rydbeck.
Astron. Astrophys., Vol. 113, L18 - L21 (1982).

The authors present CO and HCO^+ $J=1-0$ data which are related to both the high-velocity gas and the shocked gas of Orion-KL.

131.057 **Shock induced star formation: the effects of magnetic fields and turbulence.**
G. L. Welter, M. Nepveu.
Astron. Astrophys., Vol. 113, 277 - 284 (1982).

The authors investigate the effects of magnetic fields and preshock turbulence with respect to gravitational collapse and star formation behind shocks in interstellar clouds.

131.058 **A distinct shell structure in H I-line-emission at intermediate galactic latitudes.**
L. Velden, W. Hirth.
Astron. Astrophys., Vol. 113, 340 - 343 (1982).

Galactic H I-observations in the direction of $l \cong 130°$ and $b \cong 22.^\circ5$ revealed a well defined shell structure with a distance of $\cong 450$ pc, z-distance of $\cong 177$ pc, a radius of $\cong 24$ pc and ar expansion velocity of $\cong 12.5$ km s^{-1}; the kinetic temperature o the shell-gas was found to be $\lesseqqgtr 80$ K. The possibility is considered that the nearby pulsar PSR 0809 + 74 with known proper motion is associated with the phenomenon and originat ed in a supernova-event. The pulsar hypothesis is surprisingly well compatible with the distance determination from kinematical models of the galactic differential rotation.

131.059 **Interstellar molecules.**
D. A. Varshalovich.
Astrofiz. i kosm. fiz. Moskva, 1982, p. 135 - 185. In Russian.
Abstr. in Ref. zh., 51. Astron., 8.51.613 (1982).

131.060 **Motion of the sun in the interstellar medium.**
V. G. Kurt.
Astrofiz. i kosm. fiz. Moskva, 1982, p. 268 - 292. In Russian.
Abstr. in Ref. zh., 51. Astron., 8.51.655 (1982).

131.061 **Warm hydrogen at high galactic latitudes.**
T. W. Hartquist, J. H. Black, A. Dalgarno.
Astrophys. J., Vol. 259, 591 - 594 (1982).

The physical state of gas in the transition zones between halo clouds and the hot coronal gas is investigated. The equation of state is calculated. If H_2 forms efficiently on grain surfaces, its action as a coolant can maintain an observable amount of gas at a temperature near 1000 K for a number density of the order of 1 cm^{-3}. It is suggested that the warm H I observed at high galactic latitudes resides in these warm transition zones.

131.062 **The role of H II regions during star formation and chemical enrichment in globular clusters.**
G. H. Smith.
Astrophys. J., Vol. 259, 607 - 616 (1982).
The shapes of the H II Strömgren "spheres" produced by stars located within massive gas clouds are determined by solving the equation of ionization equilibrium. The main conclusion from this work is that globular cluster protoclouds should have confined within their cores the H II regions that would have formed around OB stars. In the outer regions, removal of gas by the outbursting of H II regions may have been effective. In the case of open cluster protoclouds a significant cloud dispersal during OB star formation should have occurred.

131.063 **Kinematics of Orion-KL: aperture synthesis maps of 86 GHz SO emission.**
R. L. Plambeck, M. C. H. Wright, W. J. Welch, J. H. Bieging, B. Baud, P. T. P. Ho, S. N. Vogel.
Astrophys. J., Vol. 259, 617 - 624 (1982).
Using the Hat Creek interferometer, the authors have mapped the $2_2 - 1_1$ line of SO toward the Kleinmann-Low Nebula with an angular resolution of $6'' \times 6''$. The "pedestal" component of the SO emission arises from a region with a size of $11'' \times 19''$, centered within a few arc seconds of the infrared source IRc2. The maps show evidence for a systematic velocity structure across KL. The velocity shifts are consistent with a model in which the SO emission arises from an expanding "doughnut" of gas centered near IRc2. The doughnut consists of clumps of ambient cloud material which have been compressed and caught up in an outflow from a central source.

131.064 **VLA observations of a highly symmetric OH maser in a bipolar nebula.**
M. Morris, P. F. Bowers, B. E. Turner.
Astrophys. J., Vol. 259, 625 - 633 (1982).
The Very Large Array was used to map 1667 MHz OH maser emission from the bipolar nebula OH 231.8 + 4.2 at 23 distinct velocities. At most velocities, the maser maps display an unmistakable symmetry about the bipolar axis defined by the optical and infrared reflection nebulae. Most of the data can be accounted for by an axisymmetric model in which the masering OH is concentrated toward the system's equatorial plane and is expanding radially away from the central star.

131.065 **Far-infrared rotational emission by carbon monoxide.**
C. F. McKee, J. W. V. Storey, D. M. Watson, S. Green.
Astrophys. J., Vol. 259, 647 - 656 (1982).
Accurate theoretical collisional excitation rates are used to determine the emissivities of CO rotational lines for $n(H_2) \geqslant 10^4 cm^{-3}$, 100 K $< T <$ 3000 K, and $J \leqslant 60$ under the assumption that the lines are optically thin. An approximate analytic expression for the emissivities which is valid in this region is obtained. Interstellar shocks observed edge-on are a potential source of weak millimeter-wave CO maser emission.

131.066 **Radio and infrared observations of the OH maser source OH 351.78–0.54.**
J. D. Fix, R. L. Mutel, R. A. Gaume, M. J. Claussen.
Astrophys. J., Vol. 259, 657 - 663 (1982) = Contrib. No. 87 Iowa Radio Astron. Group.
Observations of the region associated with the highly variable maser source OH 351.78–0.54 at several radio and infrared wavelengths are reported. Strong emission from water vapor and carbon monoxide was detected at nearly the same velocity as the hydroxyl maser emission. An H II region was detected 12" from the OH maser position, and another, weaker, radio continuum object, at the position of the OH masers. Strong, extended near-IR emission was also found. The observations suggest a region of active star formation associated with a giant molecular cloud at a distance of approximately 6.5 kpc.

131.067 **Observations of high-velocity molecular gas near Herbig-Haro objects: HH 24–27 and HH 1–2.**
R. L. Snell, S. Edwards.
Astrophys. J., Vol. 259, 668 - 676 (1982).
High-velocity CO has been detected in the vicinity of the Herbig-Haro objects HH 24–27. These observations indicate that there are two sources of high-velocity outflow; one centered on an infrared source near HH 26, and the second centered roughly 2' south of HH 24. The redshifted and blueshifted wings in both sources are spatially separated suggesting that the high-velocity gas is due to energetic bipolar outflows from young stars embedded in the molecular cloud. The association of Herbig-Haro objects with regions of high-velocity gas suggests a common origin for both in the interaction of a stellar wind with the ambient molecular cloud.

131.068 **A model for the intense molecular line emission from OMC-1.**
B. T. Draine, W. G. Roberge.
Astrophys. J., Lett., Vol. 259, L91 - L96 (1982).
The authors present a model which attributes the observed H_2 and CO line emission from OMC-1 to a magnetohydrodynamic shock propagating into magnetized molecular gas. By requiring the shock to reproduce the observed line intensities, the shock speed is determined to be $v_s \approx 38$ km s^{-1} and the preshock density and (transverse) magnetic field to be $n_H \approx 7 \times 10^5 cm^{-3}$, $B_0 \approx 1.5$ milligauss.

131.069 **Molecular shock waves in the BN-KL region of Orion.**
D. F. Chernoff, D. J. Hollenbach, C. F. McKee.
Astrophys. J., Lett., Vol. 259, L97 - L101 (1982).
The H_2, CO, and OH infrared emission-line intensities in the BN-KL region of Orion are modeled by shock waves traveling at $v_s \approx 36$ km s^{-1} with respect to ambient gas of density $n_0 \approx 2 \times 10^5 cm^{-3}$, ionization fraction $x_{i0} < 3 \times 10^{-7}$, CO abundance of approximately 3×10^{-4}, and transverse magnetic field $B_0 \sim 0.45$ milligauss. A global model is described in which the shock waves are driven by strong winds emanating from KL. The mass-loss rate has been $\gtrsim 3 \times 10^{-3} M_\odot yr^{-1}$ for a time of approximately 10^3 years.

131.070 **NH_3 in Orion-KL: a new interpretation.**
R. Genzel, D. Downes, P. T. P. Ho, J. Bieging.
Astrophys. J., Lett., Vol. 259, L103 - L107 (1982).
The authors reinterpret the observations of ammonia in Orion-KL in a way which greatly simplifies the understanding of single-dish and VLA data. In this new picture, there are only two velocity components: the "spike" component from the quiescent molecular cloud and the "hot core" component from both streaming and "stationary" gas in the 18 km s^{-1} outflow from the compact infrared source IRc2.

131.071 **Diffuse interstellar clouds associated with dark clouds.**
S. R. Federman, R. F. Willson.
Astrophys. J., Vol. 260, 124 - 127 (1982).
Measurements of CH emission at 9 cm toward the bright stars α Cam and κ Cas and from dark clouds near the stars reveal an association between diffuse clouds and dark clouds. Here the term *diffuse clouds* refers to clouds seen in optical absorption; the term *dark clouds* refers to opaque regions on the Palomar Sky Survey observed through radio techniques.

It is concluded that for most lines of sight, diffuse clouds are the outer, relatively transparent portions of dark clouds.

131.072 **Neutral interstellar gas toward ϵ Persei: H I, H_2, D I, N I, O I, and Ar I.**
A. Vidal-Madjar, R. Ferlet, C. Laurent, D. G. York.
Astrophys. J., Vol. 260, 128 - 140 (1982).
The study of the interstellar medium toward ϵ Per, a moderately reddened B0.5 star, through the analysis of H I, D I, H_2, N I, O I, and Ar I absorption features, revealed the following structure of the line of sight: (1) a main interstellar, cold ($T \sim 100$ K; $b = 2.9$ km s^{-1}) component, (2) a weak probably cold component, and (3) a weak and hot component ($T \sim 8000$ K). All three components have normal abundances. High-velocity H I gas is also detected, located either in the interstellar medium or very likely in the stellar wind for the blue-shifted components.

131.073 **Observation of interstellar ammonia ice.**
R. F. Knacke, S. McCorkle, R. C. Puetter, E. F. Erickson, W. Krätschmer.
Astrophys. J., Vol. 260, 141 - 146 (1982).
An absorption band probably due to solid ammonia on interstellar grains has been detected in the infrared spectrum at 2.97 μm of the Becklin-Neugebauer object and probably in NGC 2264-IR. An ammonia-water amorphous ice mixture can explain the structure of the new band and of the 3.07 μm interstellar absorption. Laboratory data suggest that a long wavelength wing extending to 3.5 μm in interstellar dust spectra may be absorption by $NH_3 \cdot H_2O$ complexes in the ices. Arguments are given that interstellar features at 6.0 and 6.8 μm are also ammonia-related absorptions.

131.074 **Vibrationally excited cyanoacetylene in the Orion molecular cloud.**
P. F. Goldsmith, R. L. Snell, S. Deguchi, R. Krotkov, R. A. Linke.
Astrophys. J., Vol. 260, 147 - 158 (1982).
The authors have detected three rotational transitions from the ν_7 vibrationally excited state of cyanoacetylene in the Orion molecular cloud. This state lies approximately 320 K above the ground state. The width of these lines is 15 km s^{-1}, and their central velocity is 5 km s^{-1}. It is suggested that the vibrationally excited lines are produced in the same region responsible for the prominent low velocity shoulder on the ground vibrational state HC_3N emission lines. The unresolved emission region is located very near the infrared source IRC 2. The small source size and close proximity to the strong sources of infrared radiation present in the KL region make infrared absorption an important mechanism in the excitation of the vibrational levels of cyanoacetylene.

131.075 **On OMC-1 temperatures determined from methyl cyanide observations.**
J. M. Hollis.
Astrophys. J., Vol. 260, 159 - 162 (1982).
An analysis is performed on rotational transitions of methyl cyanide detected by other investigators in the direction of OMC-1. The original interpretation of these observations argues for the presence of two distinct temperature regions or possibly a temperature gradient within the cloud. The analysis presented here demonstrates that the observations of these particular molecular transitions are consistent with a single methyl cyanide emission region with a source kinetic temperature of 121.2 ± 8.2 K and a molecular rotational temperature of 16.6 ± 1.8 K.

131.076 **The violent interstellar medium associated with the Carina Nebula. I. The line of sight toward HD 93205.**
C. Laurent, J. A. Paul, M. Pettini.
Astrophys. J., Vol. 260, 163 - 182 (1982).
The authors have analyzed the interstellar absorption lines in high-resolution spectra of the Carina star HD 93205 obtained with the IUE. They have identified four high-velocity and two low-velocity components. The chemical composition and the physical conditions of the different regions are discussed. The two main high-velocity components show different relative abundance patterns. Among the two low-velocity components, one is identified with the normal interstellar material in the disk of the Galaxy between the Sun and the Carina Nebula. The other low-velocity component is identified with the approaching part of the expanding ionized nebula around the Carina OB associations.

131.077 **Radiation-driven implosions in molecular clouds.**
M. T. Sandford II, R. W. Whitaker, R. I. Klein.
Astrophys. J., Vol. 260, 183 - 201 (1982).
The authors present two-dimensional radiation-hydrodynamic calculations of the interaction of ionization and shock fronts with a geometrical inhomogeneity in a molecular cloud. These regions consist of low density clumps of masses $\leqslant 2$ $M_\odot$, for which self-gravity is negligible. The radiation of OB stellar associations is shown to be effective in forming dense mass globules from preexisting low density inhomogeneities near the edges of H II regions. The calculations suggest that the cloud environment surrounding the cluster may influence subsequent star formation within the cloud.

131.078 **Absorption spectra of interstellar grains. II. Fine structure of the 2200 hump.**
G. Chlewicki, J. Krełowski.
Acta Astron., Vol. 31, 427 - 441 (1981).
The ultraviolet TD-1 and IUE spectra of reddened stars were analyzed in order to check the possibility of some differences in the 2200 hump profile, dependent on the spectral type or luminosity of observed stars. The profiles seem to be clearly affected by stellar parameters supporting the authors' earlier hypothesis that this prominent UV spectral feature is at least in part of circumstellar origin.

131.079 **Infrared observations of the Hourglass region of M 8.**
C. E. Woodward, J. L. Pipher, S. P. Willner.
News Lett. Astron. Soc. N.Y., Vol. 2, No. 2, p. 11 (1982).
Abstract.

131.080 **Further observations of polarization in the Taurus cloud.**
A. Moneti.
News Lett. Astron. Soc. N.Y., Vol. 2, No. 2, p. 24 (1982).
Abstract.

131.081 **Interstellar turbulence and shock waves.**
A. M. Bykov.
Pis'ma Astron. Zh., Tom 8, 596 - 599 (1982). In Russian.
English translation in Soviet Astron. Lett., Vol. 8.
It is shown that random deflections of shock wave fronts propagating through a turbulent interstellar medium account for the presence of strong fluctuations of electron density with scales $l \gtrsim 10^{13}$ cm which are observed from pulsar radio scintillations. Turbulence generation in the hot component of the interstellar gas is considered.

131.082 **A study of interstellar absorption at high galactic latitudes. I. Highly ionized gas.**
M. Pettini, K. A. West.
Astrophys. J., Vol. 260, 561 - 578 (1982).
The authors present IUE observations of interstellar absorption lines of C IV, Si IV, N V, and Fe II in a sample of 24 distant early-type stars at high galactic latitudes. These data are analyzed together with other published results in order to study the distribution and physical conditions of highly ionized gas in the halo of the Galaxy. C IV and Si IV lines are detected in front of approximately two-thirds of the stars observed. The observed variation of the column densities

of C^{3+} and Si^{3+} with distance z from the plane is not easily interpreted in terms of a plane-parallel distribution of the gas, but is consistent with a previous suggestion of the existence of a Galactic transition layer between cool gas in the disk and outer halo regions at temperatures greater than those observable with IUE. If such a transition region exists, it is likely to occur within 2–3 kpc of the plane.

131.083 **Study of CO line profiles in molecular clouds: relaxing the assumption of complete frequency redistribution.**
S. Deguchi, J. Kwan.
Astrophys. J., Vol. 260, 579 - 583 (1982).
In order to simplify the radiative transfer calculation in a multilevel system, the assumption of complete redistribution of the frequency of a photon upon scattering is usually made. The authors test the validity of that assumption in calculating line profiles from molecular clouds by replacing the assumption with an actual calculation of the radiative transfer in each part of the line. They find that the emergent line profiles in the more elaborate calculation differ only slightly from those obtained under the assumption of complete frequency redistribution.

131.084 **CO (J = 6 → 5) distribution in Orion and detection in other galactic sources.**
G. A. Koepf, D. Buhl, G. Chin, D. D. Peck, H. R. Fetterman, B. J. Clifton, P. E. Tannenwald.
Astrophys. J., Vol. 260, 584 - 589 (1982).
The profile of the CO line in Orion has been studied at 691 GHz with the NASA infrared telescope at Mauna Kea using a laser heterodyne receiver. The line from the BN/KL region is very broad and has a peak temperature of $T_A = 180$ K. Away from BN/KL, the emission line becomes narrow, and the intensity falls off rapidly. Several other galactic sources were also detected.

131.085 **Sulfur chemistry in dense interstellar clouds.**
S. S. Prasad, W. T. Huntress, Jr.
Astrophys. J., Vol. 260, 590 - 598 (1982).
A model is presented for the gas phase chemistry of molecules containing sulfur in dense interstellar clouds. The linked chemistry of sulfur and oxygen in the present model requires a severe depletion of sulfur and low fractional abundances of both O and O_2 in dense clouds. Calculations are compared with observations of the KL-nebula in Orion.

131.086 **The magnetic fields in the Orion Kleinmann-Low Nebula as derived from hydroxyl maser radiation.**
S. S. Hansen.
Astrophys. J., Vol. 260, 599 - 603 (1982).
The Orion Kleinmann-Low Nebula was observed in the four hyperfine transitions of OH. The circularly polarized radiation at 1612 MHz exhibits a Zeeman pattern centered at an LSR velocity of 8.02 km s^{-1}. The frequency separation of the $\sigma^{\pm 1}$ and $\sigma^{\pm 2}$ lines is in excellent agreement with the Zeeman splitting produced by a 4 milligauss field.

131.087 **Anisotropic mass outflow in regions of star formation.**
L. F. Rodríguez, P. Carral, P. T. P. Ho, J. M. Moran.
Astrophys. J., Vol. 260, 635 - 646 (1982).
The authors present carbon monoxide observations of the molecular clouds associated with three of the red nebulous objects discovered by Gyulbudaghian, Glushkov, and Denisyuk: GGD 4, GGD 12–15, and GGD 27–28. The cluster GGD 12–15 shows the presence of blueshifted and redshifted CO wings indicating high-velocity gas. The high-velocity components come from spatially separated zones located on opposite sides of a water maser source. These results suggest that the stellar wind of the star associated with the H_2O maser is being channeled into two antiparallel streams. The phenomenon of anisotropic mass outflow has been detected recently in several regions of star formation. The authors discuss in a systematic way eight reported cases of high-velocity CO outflows.

131.088 **The detection of interstellar C I in the immediate vicinity of the sun.**
F. C. Bruhweiler, Y. Kondo.
Astrophys. J., Lett., Vol. 260, L91 - L94 (1982).
Multiple stacked IUE spectra reveal the presence of interstellar C I λ 1657 in the trough of a corresponding photospheric feature in the nearby star, α PsA (d = 7 pc). This represents the first detection of this neutral atom in the interstellar medium within the immediate vicinity of the sun. The column density is log N (C I) = 12.78.

131.089 **VLA synthesis of galactic H I absorption toward 3C 123.**
H. S. Liszt, J. M. Dickey, E. W. Greisen.
Astrophys. J., Vol. 261, 102 - 109 (1982).
The authors have used the VLA to synthesize a map of galactic 21 cm absorption across the ~20″ extent of 3C 123 with 2″.8 resolution. Three prominent absorption lines can be traced. They indicate a high velocity cloud, an intermediate - velocity diffuse cloud, and a molecular cloud in the line of sight. All three show angular structure.

131.090 **Carbon depletion in turbulent molecular cloud cores.**
W. Boland, T. de Jong.
Astrophys. J., Vol. 261, 110 - 114 (1982).
Observations of dense molecular cores indicate that about 10% of the carbon is still in the gas phase (depletion factor ~0.1) in spite of the fact that the depletion time – the time needed for heavy elements to freeze out on dust grains – is several orders of magnitude smaller than the cloud lifetime. To resolve this problem, the authors suggest that the material in molecular cloud cores is circulated by turbulence and that every time a parcel of gas and dust reaches the outer layers of the core, dust mantles that have formed by accretion in the center are evaporated and/or photodesorbed.

131.091 **On the nature of bipolar sources in dense molecular clouds.**
A. Königl.
Astrophys. J., Vol. 261, 115 - 134 (1982).
A unified interpretation is presented of bipolar sources associated with young stellar objects which are embedded in dense molecular clouds. It is suggested that such phenomena as optical and radio "jets", high-proper-motion Herbig-Haro objects, separating broad-wing CO emission-line lobes, and cometary and bipolar nebulae, are linked through a basic outflow mechanism which is manifested on different scales. The outflow is assumed to originate in a spherically symmetric stellar wind which expands in a medium with anisotropic density and pressure distributions.

131.092 **Diffuse interstellar bands toward stars with low color excess.**
C. Krishna Kumar, S. R. Federman, P. A. Vanden Bout.
Astrophys. J., Lett., Vol. 261, L51 - L54 (1982).
The equivalent widths of the 5780 Å and the 5797 Å diffuse interstellar bands have been measured in the spectra of stars with low color excesses. The agent responsible for the bands appears to be molecular, and defects in bulk material are ruled out as the band carrier. Abundance considerations appear to preclude gas phase molecules, suggesting molecular grain mantles as the carrier of the diffuse interstellar bands.

131.093 **Energetics of molecular clouds – summary of Sharpless molecular cloud observations.**

P. A. Vanden Bout, N. J. Evans II.
Bull. American Astron. Soc., Vol. 14, 604 (1982). – Abstract.

131.094 **Molecular clouds associated with reflection nebulae. II. General properties.**
D. E. Machnik, M. L. Kutner, R. L. Dickman.
Bull. American Astron. Soc., Vol. 14, 604 (1982). – Abstract.

131.095 **Column densities of CN, HCN, HNC, and HC_3N relative to NH_3 in selected molecular clouds.**
A. G. Nash, E. Churchwell.
Bull. American Astron. Soc., Vol. 14, 604 - 605 (1982). Abstract.

131.096 **The [HCO^+] / [HOC^+] ratio in interstellar clouds.**
W. M. Irvine, R. L. Dickman, P. F. Goldsmith, G. R. Huguenin, Å. Hjalmarson, L. A. Nyman, H. Olofsson, C. S. Gudeman, R. C. Woods.
Bull. American Astron. Soc., Vol. 14, 605 (1982). – Abstract.

131.097 **Shock formation of HCO^+.**
M. Elitzur.
Bull. American Astron. Soc., Vol. 14, 605 (1982). – Abstract.

131.098 **Interstellar hydrocarbon synthesis.**
E. Herbst.
Bull. American Astron. Soc., Vol. 14, 605 (1982). – Abstract.

131.099 **Effect of small particles on interstellar extinction anomalies.**
A. D. Code.
Bull. American Astron. Soc., Vol. 14, 605 (1982). – Abstract.

131.100 **Polarization of interstellar radio-frequency lines and magnetic field direction.**
N. D. Kylafis.
Bull. American Astron. Soc., Vol. 14, 605 - 606 (1982). Abstract.

131.101 **On the effects of rotation on interstellar molecular line profiles.**
L. M. Adelson, C. M. Leung.
Bull. American Astron. Soc., Vol. 14, 606 (1982). – Abstract.

131.102 **Star formation in the outer reaches of the Galaxy: the Sharpless 156 cloud.**
M. Joy, N. J. Evans II, P. M. Harvey, B. A. Wilking.
Bull. American Astron. Soc., Vol. 14, 609 (1982). – Abstract.

131.103 **Star formation along the northern edge of W 5.**
B. A. Wilking, C. J. Lada, P. M. Harvey, M. Joy.
Bull. American Astron. Soc., Vol. 14, 609 - 610 (1982). Abstract.

131.104 **Spectroscopy of Orion-KL: new far-infrared observations of CO and OH emission lines, and the far-infrared detection of NH_3.**
D. M. Watson, R. L. Genzel, C. H. Townes, J. W. V. Storey.
Bull. American Astron. Soc., Vol. 14, 610 (1982). – Abstract.

131.105 **The distribution of O I 63 μm emission from the Orion Nebula region.**
M. W. Werner, H. L. Dinerstein, D. J. Hollenbach, D. F. Lester, R. Genzel, D. M. Watson, J. W. V. Storey.
Bull. American Astron. Soc., Vol. 14, 610 (1982). – Abstract.

131.106 **The molecular cloud complexes in Orion and Monoceros.**
R. J. Maddalena, J. Moscowitz, M. Morris, P. Thaddeus.
Bull. American Astron. Soc., Vol. 14, 615 (1982). – Abstract.

131.107 **Star formation in Cep A.**
V. A. Hughes, J. G. Wouterloot.
Bull. American Astron. Soc., Vol. 14, 615 - 616 (1982). Abstract.

131.108 **The morphology and kinematics of the CEP IV star formation region.**
G. S. Rossano, E. J. Grayzeck, P. E. Angerhofer.
Bull. American Astron. Soc., Vol. 14, 616 (1982). – Abstract.

131.109 **High resolution CO observations of the western arc of Canis Major R1.**
M. S. Margulis, K. N. Mead, M. L. Kutner, D. E. Machnik.
Bull. American Astron. Soc., Vol. 14, 616 (1982). – Abstract.

131.110 **On the use of CO as a tracer for H_2.**
M. L. Kutner, C. M. Leung.
Bull. American Astron. Soc., Vol. 14, 616 (1982). – Abstract.

131.111 **A log N-log S and a S-ΔV relation for molecular clouds.**
T. M. Dame, P. Thaddeus.
Bull. American Astron. Soc., Vol. 14, 616 (1982). – Abstract.

131.112 **Positions and masses of the largest molecular clouds in the inner Galaxy.**
B. G. Elmegreen, T. M. Dame, R. S. Cohen, P. Thaddeus.
Bull. American Astron. Soc., Vol. 14, 617 (1982). – Abstract.

131.113 **Latitude distribution of molecular clouds outside the solar circle.**
K. N. Mead, M. L. Kutner.
Bull. American Astron. Soc., Vol. 14, 617 (1982). – Abstract.

131.114 **Molecular clouds in the outer Galaxy.**
M. Fich, S. Terebey, L. Blitz.
Bull. American Astron. Soc., Vol. 14, 617 (1982). – Abstract.

131.115 **The dependence of CO emission on luminosity and the rate of star formation in Sc galaxies.**
J. S. Young, N. Z. Scoville.
Bull. American Astron. Soc., Vol. 14, 617 - 618 (1982). Abstract.

131.116 **Extinction and temperature variations in the Orion molecular hydrogen emission region.**
S. C. Beck, S. V. Beckwith.
Bull. American Astron. Soc., Vol. 14, 627 (1982). – Abstract.

131.117 **High resolution mapping of CO J = 3–2 emission in Orion: detection of bipolar flow.**
N. R. Erickson, P. F. Goldsmith, B. L. Ulich, C. J. Lada, R. L. Berson, R. L. Snell, G. R. Huguenin.
Bull. American Astron. Soc., Vol. 14, 627 - 628 (1982). Abstract.

131.118 **CO line emission from Orion Molecular Cloud 1.**
W. G. Roberge, B. T. Draine.
Bull. American Astron. Soc., Vol. 14, 628 (1982). – Abstract.

131.119 **Vibrationally excited HC_3N and CH_3CN in Orion.**
P. F. Goldsmith, R. L. Snell, R. Krotkov, S. Deguchi, R. D. Brown, P. D. Godfrey, R. Linke.
Bull. American Astron. Soc., Vol. 14, 628 (1982). – Abstract.

131.120 **Wind-cloud interactions in broad-winged CO sources.**
P. R. Schwartz, H. A. Smith, J. A. Waak.
Bull. American Astron. Soc., Vol. 14, 628 (1982). – Abstract.

131.121 **The high velocity HCO^+ flow in NGC 2071.**
A. Wootten, R. Loren, Aa. Sandqvist, P. Friberg, Å. Hjalmarson.

Bull. American Astron. Soc., Vol. 14, 628 - 629 (1982). Abstract.

131.122 **High velocity CO emission in the Cepheus A molecular cloud core.**
D. B. Sanders, E. Brackman, P. M. Solomon.
Bull. American Astron. Soc., Vol. 14, 629 (1982). – Abstract.

131.123 **High-velocity CO line wings associated with T Tauri stars.**
C. M. Leung, M. L. Kutner, K. N. Mead, D. E. Machnik.
Bull. American Astron. Soc., Vol. 14, 629 (1982). – Abstract.

131.124 **Detecting saturation in astrophysical masers.**
L. K. DeNoyer, J. G. Dodd.
Bull. American Astron. Soc., Vol. 14, 638 (1982). – Abstract.

131.125 **Linear polarization of astronomical masers due to geometry.**
L. R. Western.
Bull. American Astron. Soc., Vol. 14, 638 (1982). – Abstract.

131.126 **Studies of three molecular clouds in four transitions of CS.**
L. G. Mundy, N. J. Evans II, R. L. Snell, P. F. Goldsmith.
Bull. American Astron. Soc., Vol. 14, 639 (1982). – Abstract.

131.127 **Highly excited rotational transitions of CO in the Orion molecular cloud.**
S. Smyers, G. Stacey, N. Kurtz, M. Harwit.
Bull. American Astron. Soc., Vol. 14, 639 (1982). – Abstract.

131.128 **Role of the gas in propagating star formation.**
P. E. Seiden.
Bull. American Astron. Soc., Vol. 14, 639 (1982). – Abstract.

131.129 **The role of rotation in the formation of stars and planetary systems.**
S. Terebey, F. H. Shu, P. M. Cassen.
Bull. American Astron. Soc., Vol. 14, 639 (1982). – Abstract.

131.130 **Energetic activity associated with star formation.**
C. J. Lada.
Bull. American Astron. Soc., Vol. 14, 641 (1982). – Abstract.

131.131 **A dynamically stable Bok globule in Cassiopeia?**
D. P. Clemens, R. L. Dickman.
Bull. American Astron. Soc., Vol. 14, 641 (1982). – Abstract.

131.132 **Stochastic analysis of dark cloud structure.**
S. Kleiner, R. L. Dickman.
Bull. American Astron. Soc., Vol. 14, 642 (1982). – Abstract.

131.133 **Cold cloud cores: NH_3 observations.**
P. C. Myers, P. J. Benson.
Bull. American Astron. Soc., Vol. 14, 642 (1982). – Abstract.

131.134 **Fragmentation in the dense filament TMC 1.**
R. L. Snell, P. Schloerb, J. S. Young.
Bull. American Astron. Soc., Vol. 14, 642 (1982). – Abstract.

131.135 **Large scale structure of molecular gas in Heiles Cloud 2.**
P. Schloerb, R. L. Snell.
Bull. American Astron. Soc., Vol. 14, 642 (1982). – Abstract.

131.136 **A new position for the CO emission peak in the M 17 SW molecular cloud core.**
P. M. Solomon, D. B. Sanders.
Bull. American Astron. Soc., Vol. 14, 642 (1982). – Abstract.

131.137 **Subillimeter observations of W 3.**
R. H. Hildebrand, D. T. Jaffe, J. Keene, S. E. Whitcomb.
Bull. American Astron. Soc., Vol. 14, 643 (1982). – Abstract.

131.138 **Diffuse interstellar lines in low color excess stars.**
C. K. Kumar, S. Federman, P. A. Vanden Bout.
Bull. American Astron. Soc., Vol. 14, 652 (1982). – Abstract.

131.139 **Implications of collisionally-supported giant molecular clouds for spiral galactic structure and massive star formation.**
D. Leisawitz, F. Bash.
Bull. American Astron. Soc., Vol. 14, 662 (1982). – Abstract.

131.140 **Reddenings derived from H I and galaxy counts: accuracy and maps.**
D. Burstein, C. Heiles.
Bull. American Astron. Soc., Vol. 14, 662 (1982). – Abstract.

131.141 **Hot galactic gas and narrow line quasar absorption systems.**
T. W. Hartquist, M. A. J. Snijders.
Nature, Vol. 299, 783 - 788 (1982).

Observations of the hot interstellar gas ($T \gg 10^4$ K) are numerous and imply that the temperatures of substantial amounts of the disk gas to be 2–3 $\times 10^5$ K and of the halo gas to be 8×10^4 K, but the theoretical models to explain the properties of both the disk and the halo hot gas seem to be in need of some refinement. Observations of the quasar narrow line absorption features suggest that they arise in gas with properties very similar to those of the galactic coronal gas.

131.142 **G35.2–0.74: a molecular cloud with interesting velocity structure.**
A. T. Brown, L. T. Little, G. H. Macdonald, D. N. Matheson.
Mon. Not. R. Astron. Soc., Vol. 201, 121 - 125 (1982).

Observations of the NH_3 (1, 1) and (2, 2) inversion transitions made with a 2.3-arcmin beam reveal the molecular cloud G35.2–0.74 to possess an interesting velocity structure. To the north, near an H_2O maser, the source is observed as a single component. To the south, around a red nebulosity, the NH_3 spectra are split into two components with a separation 3 km s^{-1}. This splitting is of significance because it represents very clear evidence that, in this cloud at least, systematic velocity motion is an important factor in determining the observed velocity dispersion.

131.143 **Interferometric observations of four protostellar OH maser sources.**
R. P. Norris, R. S. Booth, P. J. Diamond, N. D. Porter.
Mon. Not. R. Astron. Soc., Vol. 201, 191 - 207 (1982).

Four OH maser sources, NGC 7538, W51, W75S and W75N, have been observed with an interferometer of 24-km baseline. The relative positions of the spectral features in the 1665-MHz spectrum have been measured in each case and for NGC 7538 the 1667-MHz features have also been mapped. In addition, new determinations of the absolute positions of the first three sources have been made. From the resulting maps, estimates of the magnetic flux density and other physical parameters are deduced.

131.144 **MERLIN spectral line observations of W3OH.**
R. P. Norris, R. S. Booth, P. J. Diamond.
Mon. Not. R. Astron. Soc., Vol. 201, 209 - 222 (1982).

The authors present the results of the first spectral line observations made with the Jodrell Bank Multi-Element Radio-Linked Interferometer Network and describe the operation of MERLIN in its spectral line mode and the procedures used to analyse the data. They have observed the OH maser source W3OH at three ground-state transition frequencies, and have extended significantly the previous results on this source. In contrast with previous interpretations, the authors find that the masers are probably not directly associated with the compact H II region.

131.145 **Dirty silicate models for hot-centred clouds.**
M. Rowan-Robinson.
Mon. Not. R. Astron. Soc., Vol. 201, 289 - 302 (1982).
The author investigates the suitability of dirty silicate grain models for hot-centred dust clouds, applying them both to the mean spectrum for hot-centred clouds (Rowan-Robinson 1979) and to a number of individual clouds.

131.146 **Westerbork observations of 6-cm H_2CO in W3.**
E. M. Arnal, W. M. Goss, H. R. Dickel, J. R. Forster.
Mon. Not. R. Astron. Soc., Vol. 201, 317 - 330 (1982).
The distribution of the 4.8-GHz formaldehyde absorption against the continuum sources in the W3 core has been mapped with an angular resolution of 8.2×9.3 arcsec^2 and a velocity resolution of 1.22 km s^{-1}. Opacity variations on a linear scale size of 0.2 pc are seen against the continuum source W3A. The opacity enhancements are interpreted as being caused by clumping in the molecular material overlying the continuum source. There is a strong correlation between the optical depth of H_2CO and that of the silicate absorption at 9.7 μm. Evidence for an interaction between the ionized gas and the surrounding molecular cloud has been found.

131.147 **Detection of the $2(1,1) \rightarrow 2(1,2)$ rotational emission line of HDO in the Orion Molecular Cloud.**
J. E. Beckman, G. D. Watt, G. J. White, J. P. Phillips, R. L. Frost, J. H. Davis.
Mon. Not. R. Astron. Soc., Vol. 201, 357 - 364 (1982).
The $2(1,1) \rightarrow 2(1,2)$ rotational transition of HDO has been detected towards the KL position at the centre of the Orion Molecular Cloud (OMC-1), at a velocity V_{LSR} of +5.5 (±1.3) km s^{-1}, corresponding to emission from the central core. The authors combine this measurement with previous estimates of the H_2O abundance close to KL to obtain an [HDO] : [H_2O] ratio of 9.5 (±1.5) $\times$ 10^{-3}. The systematic model-dependence of this ratio is examined in some detail, to show how reliably it may be used in obtaining a radial gradient of [HDO] : [H_2O], and hence of [D] : [H], within the Galaxy.

131.148 **Interstellar extinction in the Small Magellanic Cloud.**
K. Nandy, A. McLachlan, G. I. Thompson, D. H. Morgan, A. J. Willis, R. Wilson, P. M. Gondhalekar, L. Houziaux.
Mon. Not. R. Astron. Soc., Vol. 201, 1P - 6P (1982).
IUE observations of three considerably reddened stars located near the core of the Small Magellanic Cloud and of two comparison stars which are also SMC members are presented. This region contains a considerable amount of dust.

131.149 **VLBI aperture synthesis observations of 1720-MHz OH in NGC 7538 (IRS1).**
J. R. Forster, D. Graham, W. M. Goss, R. S. Booth.
Mon. Not. R. Astron. Soc., Vol. 201, 7P - 12P (1982).
The 1720-MHz OH source in NGC 7538 (IRS1) has been observed with four telescopes in the European VLBI network. Aperture synthesis mapping with a beam of 0.03 arcsec who shows six discrete maser positions concentrated in an elongated region of size 0.3×0.15 arcsec^2. The maser emission arises from two groups of unresolved (<0.02 arcsec) spots, with apparent brightness temperatures $>10^{10}$ K. The two separate spatial groups correspond to two distinct velocity groups.

131.150 **High velocity clouds. Review of observational properties.**
I. F. Mirabel.
Rev. Mexicana Astron. Astrofis., Vol. 6, (see 012.017), 245 - 254 (1981).
Several authors have reviewed the properties of high velocity clouds in the past. In this paper the author gives an overview adding the results of recent high sensitivity surveys. A preliminary report of Arecibo observations with high angular and velocity resolution of a sample of high velocity clouds with extreme negative velocities is presented.

131.151 **Fine structure in high velocity clouds.**
R. Morras, E. Bajaja.
Rev. Mexicana Astron. Astrofis., Vol. 6, (see 012.017), 255 - 257 (1981).
Two clouds, HVC 287.5 + 22.5 + 240 and HVC 333–15.5–90, have been observed in the H I 21 cm line, with an angular resolution of 35′ and a velocity resolution of 2 km s^{-1}. The maps show unresolved components embedded in broad features. The profiles show also smaller velocity dispersion for the concentrations. It is concluded that neither of the two clouds belongs to the Magellanic Stream.

131.152 **Detection of two new water vapor emission sources in the southern hemisphere.**
M. A. Braz.
Rev. Mexicana Astron. Astrofis., Vol. 6, (see 012.017), 273 - 275 (1981).
Two new H_2O maser emission sources were detected during a survey made towards southern galactic H II regions. The most intense was discovered in the direction of the brightest OH maser source now in the sky, OH 351.78–054. The other source was found near the compact H II region G 345.645 + 0.010.

131.153 **Fragmentation of prestellar clouds by molecule formation.**
M. H. Ibañez S.
Rev. Mexicana Astron. Astrofis., Vol. 6, (see 012.017), 277 - 284 (1981).
The distribution of grain temperature through dense interstellar clouds has been investigated using the equations of heat balance and radiative transfer in the Eddington-like approximation derived by Giovanelli for non-homogeneous media. The aim is to determine whether density inhomogeneities can produce spatial temperature variations which would produce fluctuations in the rate formation of hydrogen molecules such as those suggested by Reddish (1975).

131.154 **Infrared mapping of the galactic plane. III. The large-scale mid-infrared structure of W 3, W 4, and W 5.**
H. A. Thronson, Jr., S. D. Price.
Astron. J., Vol. 87, 1288 - 1294 (1982).
The W 3, W 4, and W 5 complex of giant H II regions and molecular clouds has been mapped with about 9′ resolution at 11 μm and 20 μm. The dominant feature is the strong emission from W3A. There is in addition low-surface-brightness 11-μm emission associated with the very extended radio continuum and molecular emission around W 3/W 4. The authors find that the high dust temperature that characterizes the extended emission is direct evidence for ongoing star formation throughout virtually all of the vast W 3/W 4 molecular cloud. The W 5 object, although a luminous radio continuum source, shows only modest 11-μm and 20-μm emission.

131.155 **Formaldehyde absorption towards OH sources.**
J. R. Forster, W. Boland.
Astron. Astrophys., Vol. 114, 109 - 118 (1982).
Aperture synthesis observations in the 6 cm ($1_{10} \rightarrow 1_{11}$) formaldehyde line are presented for four fields containing ultracompact H II regions associated with OH maser sources. The H_2CO optical depths are found to be about an order of magnitude higher towards the ultracompact H II/OH sources

than towards other sources in the fields. It is argued that dense neutral material surrounding the H II regions is responsible for the absorption. Gas densities are estimated using an LVG radiative transfer calculation. The H_2CO absorption line velocity is compared to the velocity of the OH and H_2O maser emission, and to the recombination line and CO velocity. The neutral shell appears to be stationary, while the OH masers may be infalling.

131.156 **3D models for self-gravitating, rotating magnetic interstellar clouds.**
E. Dorfi.
Astron. Astrophys., Vol. 114, 151 - 164 (1982).

The author presents results of the first 3D numerical calculations on the dynamics of rotating, magnetized and self-gravitating gas clouds. The magnetic field is treated as frozen in the matter and the author discusses two cases where a homogeneous field of $3\mu G$ is either perpendicular to or aligned with the axis of rotation. He starts with an interstellar cloud of $10^4 M_\odot$ and an initial density of 1.73×10^{-23} g cm^{-3}. He assumes uniform galactic angular velocity, $\Omega = 10^{-15}$ s^{-1}. This isothermal cloud ($T = 100$ K) with a radius of 6.5×10^{16} cm is embedded in a tenuous medium of 1.73×10^{-24} g cm^{-3} at rest. The calculations were performed on a 3-dimensional cartesian (newly developed) "hierarchical" grid.

131.157 **Secrets of interstellar clouds.**
L. Marschall.
Astronomy, Vol. 10, No. 3, p. 6 - 22 (1982). – Abstr. in Phys. Abstr., Vol. 85, Abstr. 79193 (1982).

131.158 **Molecules between the stars.**
D. Darling.
Astronomy, Vol. 10, No. 3, p. 82 - 87 (1982). – Abstr. in Phys. Abstr., Vol. 85, Abstr. 79194 (1982).

131.159 **Synthesis of molecules in interstellar clouds and star formation.**
K. K. Ghosh, S. N. Ghosh.
Indian J. Radio Space Phys., Vol. 10, 237 - 240 (1981). Abstr. in Phys. Abstr., Vol. 85, Abstr. 79195 (1982).

131.160 **Rate coefficients of some important interstellar ion-neutral reactions.**
S. N. Ghosh, K. K. Ghosh.
Indian J. Radio Space Phys., Vol. 11, 40 - 41 (1982). – Abstr. in Phys. Abstr., Vol. 85, Abstr. 79196 (1982).

131.161 **Dust in space – microbes or minerals?**
D. Whittet, N. Henbest.
New Scientist, Vol. 93, 240 - 243 (1982). – Abstr. in Phys. Abstr., Vol. 85, Abstr. 79198 (1982).

131.162 **Interstellar molecular hydrogen.**
J. M. Shull.
Annu. Rev. Astron. Astrophys., Vol. 20, (see 003.009), 163 - 190 (1982).

The hydrogen molecule. Chemical and energy balance of H_2. Observations. Theories of excitation. Future work.

131.163 **Molecular clouds in galaxies.**
M. Morris, L. J. Rickard.
Annu. Rev. Astron. Astrophys., Vol. 20, (see 003.009), 517 - 545 (1982).

Galaxies with abundant molecular gas. The distribution of molecular material within galaxies. Molecular clouds in the central regions of galaxies. Molecular clouds in galactic disks. Extragalactic chemistry. Extragalactic masers.

131.164 **The search for infrared protostars.**
C. G. Wynn-Williams.
Annu. Rev. Astron. Astrophys., Vol. 20, (see 003.009), 587 - 618 (1982).

Searches and sources. Clustering and location. Observations of the dust. Observations of the gas. Gas flows near infrared sources. Comparison with theory.

131.165 **Identification of active star formation regions in the galactic plane.**
A. G. Ananth, B. V. Nagaraja.
Bull. Astron. Soc. India, Vol. 10, 46 (1982). – Abstract.

131.166 **Segregation of dust grains in dark clouds.**
H. C. Bhatt, J. N. Desai.
Bull. Astron. Soc. India, Vol. 10, 49 (1982). – Abstract.

131.167 **Far UV extinction and the interstellar 4430 Å band in the Large Magellanic Cloud.**
K. Nandy, D. H. Morgan, L. Houziaux.
Astrophys. Space Sci., Vol. 85, 159 - 165 (1982).

The central depths of the interstellar 4430 Å band and far UV extinction for a sample of reddened LMC members are presented here.

131.168 **Interstellar shocks and elemental depletions.**
W. W. Duley.
Astrophys. Space Sci., Vol. 85, 221 - 230 (1982).

The variation in elemental depletions seen in diffuse clouds in the interstellar medium is shown to be consistent with the exposure of grains to weak shocks. Depletions are shown to be the result of the incorporation of elements in refractory cores formed during equilibrium condensation in high temperature sources together with accretion under low ($T \ll 1000$ K) temperature conditions. Evidence is presented for the high temperature condensation of CaAlTi grains. In addition a refractory condensate containing 20% of available Mg, 60% of Si and 90% of Fe is inferred from data on depletions. An amorphous layer with a composition (2MgO) (SiO) condenses on these cores.

131.169 **On interstellar reddening law in the Taurus dark clouds.**
V. Straižys, W. Z. Wisniewski, M. J. Lebofsky.
Astrophys. Space Sci., Vol. 85, 271 - 275 (1982).

The interstellar reddening law in the Taurus dark clouds is investigated from visual, infrared, and ground based ultraviolet photometric observations of heavily reddened stars HD 29647, HDE 283701, and HDE 283812. The star HD 29647 located in the center of a very dense dark cloud Khavtassi 278 shows a reddening law which is anomalous in the wavelengths shorter than 400 nm. When combined with observations by Snow and Seab from the IUE it can be represented by a single straight line from 1 μm to 250 nm. Other two stars situated in less dense parts of the Taurus dark clouds show normal reddening law.

131.170 **On the collapse of a gaseous isothermal sphere: an analytical solution.**
G. Ballone, M. Fofi.
Astrophys. Space Sci., Vol. 85, 293 - 296 (1982).

An analytical solution in terms of elliptic integrals is found for the problem connected with the isothermal collapse of a gaseous sphere. In this way the time-dependence of spatial density distribution is promptly obtained in each point inside the sphere itself. An application for interstellar clouds is given.

131.171 **Organo-siliceous biomolecules and the infrared spectrum of the Trapezium nebula.**
F. Hoyle, N. C. Wickramasinghe, S. Al-Mufti.
Astrophys. Space Sci., Vol. 86, 63 - 69 (1982).

A close correspondence exists between the infrared properties of a mixed culture of diatoms and the infrared spectrum of dust in the Trapezium nebula. The authors argue that this correspondence points to a cosmic microbiological system in which organo-siliceous polymers are an abundant constituent. The high content of Si relative to Mg found in the Earth's crust and in Lunar and Martian surface material is readily explained on the basis of accretion of silicon-rich microbiology.

131.172 **Interstellar gas studies with IUE.**
M. Pettini.
Third European IUE Conference, (see 012.029), p. 37 - 42 (1982).

131.173 **Analysis of extinction curves towards the Sco OB1 association.**
S. Aiello, B. Barsella, A. Bonetti.
Third European IUE Conference, (see 012.029), p. 371 - 373 (1982).

The authors have analyzed the UV extinction curves of stars in Sco OB1 using different dust models. The first model uses a continuous size distribution, in the second one single radii for each kind of particles are considered. Both models appear to be not fully adequate to match the curves in the whole wavelength range.

131.174 **Anomalous extinction in the Cepheus OB 3 association.**
B. Barsella, N. Panagia, M. Perinotto.
Third European IUE Conference, (see 012.029), p. 375 - 377 (1982).

The UV extinction suffered by the four brightest stars of the Cepheus OB 3 association is investigated. While the shape of the bump at $\lambda \sim 2200$ Å and the extinction at longer wavelengths are close to the mean interstellar curve, the behavior shortward of $\lambda 2000$ Å is peculiar. A fit with theoretical extinction curves shows that a strong excess of graphite relative to silicates in the local component is required to explain the observed extinction in the Cepheus OB 3 association.

131.175 **Dust properties in stellar groups.**
P. Patriarchi, M. Perinotto, B. Barsella.
Third European IUE Conference, (see 012.029), p. 379 - 380 (1982).

An analysis of the interstellar extinction toward the Cas OB6 association shows that the mass ratio of graphite to silicates is larger than in the mean interstellar medium with values similar to those found in Cep OB3.

131.176 **Interstellar extinction in anomalous regions.**
J. C. Blades, L. A. Jacobs, D. McNally, W. B. Somerville, D. C. B. Whittet.
Third European IUE Conference, (see 012.029), p. 381 - 382 (1982).

Low resolution IUE spectra have been obtained at long and short wavelengths, to study the ultraviolet extinction in individual lines of sight and in particular to examine the strength of the $\lambda 2200$ feature and the form of the far ultraviolet rise in extinction in directions which have anomalous optical properties, in the strengths of diffuse interstellar bands or in the polarisation.

131.177 **Ultraviolet interstellar extinction in the SMC from IUE stellar observations.**
M. L. Prévot, J. Lequeux, E. Maurice, L. Prévot, B. Rocca-Volmerange.
Third European IUE Conference, (see 012.029), p. 383 (1982). Abstract.

131.178 **Interstellar lines in high resolution IUE spectra: paper II. Highlights of preliminary astronomical results.**
D. P. Gilra, T. H. Pwa, E. M. Arnal.
Third European IUE Conference, (see 012.029), p. 391 - 397 (1982).

Preliminary astronomical results of the authors' work on interstellar lines show that it is possible to use certain line ratios for diagnostics. Many "new" lines are detected. Warning is given for analysis and identification of interstellar lines in hot supergiants. The work on IUE data of the star ζ Oph shows some serious discrepancies with the Copernicus data.

131.179 **Studies of interstellar CO.**
W. B. Somerville, L. A. Jacobs, D. McNally, D. C. B. Whittet, J. C. Blades.
Third European IUE Conference, (see 012.029), p. 399 - 401 (1982).

High-resolution short wavelength IUE spectra of reddened early-type stars have been used to determine interstellar CO column densities. These are compared with CH, CH^+ and CN, from optical observations, and in a related programme CO from microwave measurements.

131.180 **High velocity gas components associated with supergiant stars and interstellar gas.**
B. Bates, W. Brown-Kerr, D. L. Giaretta.
Third European IUE Conference, (see 012.029), p. 403 - 409 (1982).

Programmes are outlined which involve studies of high velocity gas components (HVC) observed in several UV resonance lines. New data on the variability of expanding 'shells' associated with mass loss from β Ori are presented. A second study concerns red-shifted HVC observed in the interstellar gas in the line-of-sight towards HD 212593 (4 Lac).

131.181 **Depletion of interstellar zinc.**
A. W. Harris, G. E. Bromage, J. C. Blades.
Third European IUE Conference, (see 012.029), p. 411 - 413 (1982).

The authors have used high-resolution spectra taken from the IUE data-bank to investigate the abundance of interstellar Zn along sight-lines of up to 3 kpc in length. The results show that the relative gas-phase abundance of Zn is generally constant throughout the interstellar medium. A slight depletion relative to the solar abundance is indicated.

131.182 **Element depletions in interstellar gas.**
A. P. Phillips, P. M. Gondhalekar, M. Pettini.
Third European IUE Conference, (see 012.029), p. 415 - 420 (1982).

High resolution IUE observations have been analysed to derive element depletions in low-velocity interstellar gas toward several stars in the same region of sky. The results have been combined with previous observations to determine how the level of depletion of a given element depends on present densities in the interstellar medium.

131.183 **The violent interstellar medium associated with the Carina Nebula.**
C. Laurent, J. A. Paul, M. Pettini.
Third European IUE Conference, (see 012.029), p. 427 - 429 (1982).

The authors present a detailed study of the physical conditions and chemical composition of the interstellar medium in line to HD 93205, an O3 V star embedded in the great Carina Nebula. They have identified up to six components in the ultraviolet absorption lines, spanning a velocity range of 375 km s^{-1}.

131.184 **Interstellar gas in the direction of the Large Magellanic Cloud.**
D. Maslen, A. J. Willis, R. Wilson, K. Nandy, J. C. Blades.
Third European IUE Conference, (see 012.029), p. 431 - 433 (1982).

The authors present high-resolution ultraviolet spectra of

a star in the Large Magellanic Cloud obtained by coadding a number of individual IUE images. They have used high resolution AAT spectra of optical Ca II to determine the most appropriate velocity model for the sight-line and then used this model to fit the ultraviolet profiles.

131.185 **Highly ionised gas in the galactic halo.**
K. A. West, M. Pettini.
Third European IUE Conference, (see 012.029), p. 435 - 437 (1982).

The authors present initial results of a large-scale study of C IV, Si IV and N V interstellar absorption lines in IUE spectra of distant early-type stars located in the halo of our galaxy. They have detected C IV and Si IV lines towards approximately two-thirds of the present sample of 24 stars. N V is only detected towards two stars.

131.186 **Distribution of gas in the halo of the Galaxy.**
J. C. Blades, L. L. Cowie, D. C. Morton, D. G. York, C.-C. Wu.
Third European IUE Conference, (see 012.029), p. 439 - 440 (1982).

Results are presented from a high-resolution, ultraviolet study of interstellar gas situated away from the plane of our Galaxy, using the nuclei of Seyfert galaxies MKn 509 and F9 as background probes. The authors detect low-velocity, galactic gas as well as two extragalactic clouds, one probably associated with the Magellanic Stream and the other with Mkn 509. These data are combined with results from other lines of sight to show that the ultraviolet species extend about 10 kpc from the plane.

131.187 **Recent submillimetre wavelength studies of molecules in star formation regions.**
G. J. White.
The scientific importance of submillimetre observations, (see 012.030), p. 5 - 11 (1982).

The recent availability of sensitive heterodyne detectors in the submillimetre spectral region (300–1000 GHz) and large aperture optical/infrared telescopes at high altitude observatories have led to sensitive astronomical observations of a number of molecular species in galactic star formation regions. The current state of progress in this newly developing field is examined both with regard to present-day detectors and also astrophysical applications.

131.188 **The rotational populations and abundances of interstellar C_2 and C_2H.**
J. H. Black, E. F. van Dishoeck.
The scientific importance of submillimetre observations, (see 012.030), p. 13 - 20 (1982).

The excitation of the diatomic carbon molecule in interstellar clouds is discussed with attention to the probabilities for the radiative and collisional excitation processes. An unusual radio frequency spectrum of C_2 is predicted. The abundances of C_2 and C_2H in the interstellar cloud toward ζ Oph are discussed and brightness temperatures for C_2H in this cloud are computed.

131.189 **Spectral scan of Orion A and IRC + 10216 – astrophysics and chemistry in a star formation region and in the envelope of a carbon star.**
H. Olofsson.
The scientific importance of submillimetre observations, (see 012.030), p. 21 (1982).

The author has performed a spectral scan of the 72–91 GHz region toward the Orion A molecular cloud core (Orion-KL) and the carbon star IRC + 10216. About 190 lines from 23 known interstellar molecules were observed toward Orion-KL and 40 lines from 9 different species were seen in the IRC + 10216 envelope. Of these lines ~125 are first astronomical detections.

131.190 **Interstellar hydrides.**
G. Winnewisser, M. R. Aliev, K. Yamada.
The scientific importance of submillimetre observations, (see 012.030), p. 23 - 28 (1982).

The great progress, molecular astronomy has made in the past decade, has focused attention to one special class of molecules: the hydrides. The presently known spectra of the hydrides are reviewed, both interstellar and laboratory. No interstellar metal hydrides are presently known. A large number of important hydrides exhibit spectra only in the submillimetre and far-infrared region. Further technical advances in this area are expected to yield new interstellar detections.

131.191 **Submillimetre spectral line observations and condensations in molecular clouds.**
R. Padman.
The scientific importance of submillimetre observations, (see 012.030), p. 29 - 30 (1982).

The author considers the circumstances under which it will be possible to detect individual, possibly protostellar, fragments in regions of star formation. Present-day instruments such as UKIRT are approaching sensitivities at which it will be just possible to detect clumps with masses of between 0.1 and 1 $M_\odot$. A slightly more massive clump may already have been detected in OMC 1, and is possibly associated with H_2O maser emission.

131.192 **High-excitation rotational molecular lines from dense interstellar clouds.**
G. Silvestro, A. Ferrari.
The scientific importance of submillimetre observations, (see 012.030), p. 31 - 32 (1982).

131.193 **SO 2_3–1_2 emission in Orion A.**
J. P. Phillips, G. J. White.
The scientific importance of submillimetre observations, (see 012.030), p. 33 - 36 (1982).

Observations of Ori A in the 2_3–1_2 transition of SO are reported. They were taken with the NRAO 11 m antenna at Kitt Peak in November 1981. The SO line consists of two main components: a broad "plateau" feature corresponding to the outflow source, and a very much narrower and less intense line at $V_{lsr} \sim 8$ km s^{-1}.

131.194 **High-velocity emission in the Orion molecular cloud.**
P. F. Goldsmith.
The scientific importance of submillimetre observations, (see 012.030), p. 37 - 41 (1982).

The author summarizes the millimeter, submillimeter, and far infrared observations of carbon monoxide in the core of the Orion molecular cloud. The data indicate that within a region of approximately 40″ diameter, there is a bipolar outflow producing emission at velocities exceeding 50 km s^{-1} from the centroid of the line profiles. The alignment of the axis of the bipolar flow with two H_2 emission peaks suggests that this outflow is responsible for the shock excitation of the H_2 in a surrounding shell.

131.195 **Submillimetre observations of atomic carbon.**
T. G. Phillips, J. Keene.
The scientific importance of submillimetre observations, (see 012.030), p. 45 - 49 (1982).

Emission from the ground state fine structure transition of atomic carbon at 610 μm has been observed in galactic sources. From comparison of the observations with CO emission, it can be deduced that the abundance of neutral carbon relative to CO is high. The spatial and velocity distribution of C I and CO are often very similar.

131.196 **Cloud modelling with millimetre transitions – implications for submillimetre observations.**
P. A. Vanden Bout.

The scientific importance of submillimetre observations, (see 012.030), p. 109 - 113 (1982).

The implications of spectral line observations in the frequency range 200–350 GHz for techniques used to infer physical properties of galactic molecular clouds are reviewed. Millimeter observations at high frequencies have made it possible to study several transitions of abundant molecules, improving the confidence with which temperatures and densities can be determined. The submillimeter region offers a powerful extension of this advantage.

131.197 **Predictions of line intensities from hydrostatic models of molecular clouds.**
W. Boland.
The scientific importance of submillimetre observations, (see 012.030), p. 115 - 117 (1982).

The author has constructed plane-parallel and spherical models of cold dense molecular clouds. The gas temperature and molecular abundances are obtained by solving simultaneously the equations of thermal balance and of chemical equilibrium. Brightness temperatures of several millimetre and submillimetre lines of C^+, C, CO and HCO^+ were calculated for the spherical cloud models. Comparison is made with observations in the directions of L134 and L183.

131.198 **Detailed models of molecular clouds with and without massive star formation.**
L. G. Stenholm.
The scientific importance of submillimetre observations, (see 012.030), p. 119 - 122 (1982).

Detailed observationally determined models of molecular clouds with and without massive star formation are discussed. The assumption on which the models are based is discussed as well as the question of uniqueness. The author shows that the cloud structures can be explained by gravitational, magnetic and radiative processes. Differences between clouds seem to be due to the initial values of cloud mass, cloud radius and the ratio of gravitational to magnetic energy.

131.199 **Dynamical structure of regions of intermediate-mass star formation.**
F. Falgarone, J. L. Puget, M. Pérault.
The scientific importance of submillimetre observations, (see 012.030), p. 123 - 128 (1982).

The authors present the general features of an unusual type of molecular clouds, 99% filled with a warm and diluted ionized gas, the remaining volume being mostly filled with a warm neutral gas, possibly seen in H I emission. The high degree of clumpiness of the cold molecular component, in which the mass is concentrated, apparently prevents the formation of massive stars, which is in agreement with the large infrared excess of these peculiar clouds.

131.200 **Water as a probe of star-forming regions.**
E. N. Rodriguez Kuiper, T. B. H. Kuiper.
The scientific importance of submillimetre observations, (see 012.030), p. 129 - 130 (1982).

Preliminary calculations employing statistical equilibrium techniques have been carried out to determine brightness temperatures of water transitions occurring in the submillimeter band. The interesting candidates appear to be the 325 GHz, 380 GHz, and 448 GHz transitions. The 183 GHz and 380 GHz transitions have been observed in the Orion Kleinmann-Low nebula. These data indicate that the abundance of water may be about one twentieth of CO.

131.201 **Observations of the J = 3–2 and 1–0 transitions of HCO^+ in NGC 2071, Cep MC-1 & W 3.**
A. Sandqvist, A. Wootten, R. B. Loren, P. Friberg, Å. Hjalmarson.
The scientific importance of submillimetre observations, (see 012.030), p. 131 - 134 (1982).

Three molecular clouds with regions of star formation, NGC 2071, Cep MC-1 and W 3, are studied through observations of three transitions of HCO^+ and $H^{13}CO^+$ with different telescopes, yielding comparable resolution in the respective transitions.

131.202 **A time-dependent model for the chemical composition of interstellar molecular clouds.**
G. D. Watt, D. A. Williams.
The scientific importance of submillimetre observations, (see 012.030), p. 135 - 140 (1982).

The authors have constructed a computer program to deal with the time-dependent integration of molecular species formation in a variety of interstellar clouds. The data contains over 1400 chemical reactions for the formation and destruction of 89 different molecular species. It is found that over a large range of cloud density and kinetic temperature the resultant chemical abundances reach a "steady-state" equilibrium value by 10^7 years or less. The approach to this equilibrium is a highly sensitive function of the initial elemental abundances.

131.203 **Model calculations of the molecular composition of interstellar grain mantles.**
A. G. G. M. Tielens, W. Hagen.
Astron. Astrophys., Vol. 114, 245 - 260 (1982).

The chemical composition of mantles accreting on interstellar grains has been calculated numerically with a chemical reaction scheme which comprises gas phase as well as grain surface reactions. The equilibrium abundances of the molecules in the gas are calculated using gas phase reactions except for H_2 formation. The composition of the growing grain mantle is determined on the basis of the relative accretion rates of the gas phase molecules and diffusion controlled surface reactions. The results show that in most circumstances grain mantles consist of the molecules H_2O, H_2CO, N_2, O_2, CO, CO_2, H_2O_2, and NH_3. The relative concentrations of these species depend strongly on the physical conditions in the gas. The formation of H_2 on grain surfaces is examined in detail. The infrared characteristics of the calculated grain mantles are discussed with an emphasis on the observed 3250 cm^{-1} "ice" band.

131.204 **Loss of CO^+ ions by reaction with H_2 in OMC-1.**
W. T. Huntress, Jr., S. S. Prasad, P. R. Kemper, R. D. Cates, M. T. Bowers.
Astron. Astrophys., Vol. 114, 275 - 277 (1982).

The temperature dependence of the rate constant for the reaction $CO^+ + H_2 \rightarrow HCO^+$ has been measured in the temperature range 100–390 K. The rate constant has no dependence on temperature in this range, indicating that at the temperatures prevalent in OMC-1, the rate of loss of CO^+ by reaction with H_2 is much too large to allow for the observed column abundance under steady-state conditions.

131.205 **Gravitational instabilities in protostellar nebulae.**
B. F. Smith, P. M. Cassen.
Bull. American Astron. Soc., Vol. 14, 718 (1982). – Abstract.

131.206 **Hydrodynamic models of interstellar bullets.**
R. W. Whitaker, M. T. Sandford II.
Bull. American Astron. Soc., Vol. 14, 777 (1982). – Abstract.

131.207 **A large molecular cloud toward W 50 and SS 433.**
Y. L. Huang, T. M. Dame, P. Thaddeus.
Bull. American Astron. Soc., Vol. 14, 777 - 778 (1982). Abstract.

131.208 **Physical conditions and carbon monoxide abundance in the dark cloud B 5.**
J. S. Young, P. F. Goldsmith, W. D. Langer, R. W. Wilson, E. R. Carlson.

Astrophys. J., Vol. 261, 513 - 531 (1982).

Observations of the J = 2→1 and J = 1→0 transitions of ^{12}CO, ^{13}CO, and $C^{18}O$ in the dark cloud B 5 are presented. The hydrogen density and CO fractional abundances were determined along the north-south strip through the center of the cloud. The observations provide strong support for chemical isotopic fractionation in the outer parts of B 5. Evidence for the rotation of the bulk of B 5 is presented, while the core appears to be rotating in the opposite sense. In its present state, the cloud appears to be supported against further collapse.

131.209 **Theoretical models of the mass spectrum of interstellar clouds.**
M. A. Hausman.
Astrophys. J., Vol. 261, 532 - 542 (1982).

Monte Carlo simulations of the evolution of systems of colliding clouds are presented and compared with the observed interstellar cloud mass spectrum and velocity dispersion. Compared with observed clouds, whose mass spectrum is nearly a power law of index −1.5 to −2.0 and whose velocity dispersions are apparently independent of mass, these collisional models predict that either the mass spectrum or the velocity-mass relation should fall more steeply with mass than is in fact observed. Implications and possible explanations of this result are discussed.

131.210 **Energetic activity in a star-forming molecular cloud core; a disk constrained bipolar outflow in NGC 2071.**
J. Bally.
Astrophys. J., Vol. 261, 558 - 568 (1982).

The discovery of an extended, high-velocity molecular outflow associated with the infrared sources in NGC 2071 is reported. High-velocity CO emission from this star-forming molecular cloud is bipolar; the blueshifted CO emission is displaced to the northeast, while the redshifted emission is displaced to the southwest. Observations in the 98 GHz transition of CS reveal a ridge of dense gas oriented orthogonal to the high-velocity flow axis. It is suggested that the ridge is a dense, rotating disk centered on the IR source. A model of the bipolar mass outflow is briefly outlined.

131.211 **Detection of bipolar CO outflow in Orion.**
N. R. Erickson, P. F. Goldsmith, R. L. Snell, R. L. Berson, G. R. Huguenin, B. L. Ulich, C. J. Lada.
Astrophys. J., Lett., Vol. 261, L103 - L107 (1982).

The authors have used the Multiple Mirror Telescope as a phase coherent array to map the CO J = 3 - 2 emission from the central 1′ region of the Orion molecular cloud. The centroid of the high-velocity emission is located roughly 10″ northeast of IRc2. The peaks in the emission strength of the redshifted and blueshifted high-velocity CO are offset with the redshifted emission peaking east and the blueshifted emission peaking north of IRc2. The axis of the high-velocity CO emission is aligned with the two maxima of the 2 μm H_2 emission suggesting that the CO is participating in the same bipolar outflow.

131.212 **Quasistatic contraction of magnetic clouds due to plasma drift. II. The effect of grain friction.**
T. Nakano.
Publ. Astron. Soc. Japan, Vol. 34, 337 - 350 (1982).

The author investigates the quasistatic contraction of magnetic clouds taking into account the friction of charged grains. He derives the drift velocity of electrons taking into account the friction of charged grains. The author presents the numerical results for several models and discusses some related problems.

131.213 **The formation of giant cloud complexes.**
B. G. Elmegreen.
Submillimetre wave astronomy, (see 012.031), p. 3 - 14 (1982).

Various formation mechanisms for giant cloud complexes are reviewed. The author shows as an example how the giant molecular clouds and OB associations in the Orion, Perseus and Sco-Cen regions could have formed as condensations in Lindblad's expanding ring. A new model for the ring is proposed which includes these features in addition to several other prominent cloud complexes in the solar neighborhood. The author also discusses the formation of star complexes and the triggering of cloud formation by spiral density waves. A distinction between primary and secondary cloud formation mechanisms is emphasized.

131.214 **Cosmic rays and giant molecular clouds.**
A. W. Wolfendale.
Submillimetre wave astronomy, (see 012.031), p. 15 - 24 (1982).

Gamma rays have been observed coming from the direction of known giant molecular clouds, the quanta having almost certainly been produced by the interaction of cosmic rays with the atomic nuclei in the clouds. Arguments are presented claiming that the cosmic ray intensity is known in those clouds within about 2 kpc of the sun and it is thus possible to take the γ-ray fluxes and evaluate the masses of the clouds. It is found that the recipe given by Blitz (1980) relating CO data to column densities of molecular hydrogen gives a good description of the results. Application of the conversion factor to the CO surveys indicates that ~30% of the gas in the Galaxy is in molecular form.

131.215 **Far infrared large scale map of the Orion region.**
E. Caux, G. Malinie, R. Gispert, J. L. Puget, C. Ryter, N. Coron, G. Serra.
Submillimetre wave astronomy, (see 012.031), p. 25 - 31 (1982).

A large area (7°×20°) around the Orion nebula has been surveyed in the far infrared range using a balloon borne instrument with a 0.5° field of view. The instrument has been especially designed to be sensitive to low brightness gradient of extended sources. It works simultaneously in two wavelength channels: 114–196 μm and 71–95 μm. The maps obtained in each channel present two intense sources located respectively in the direction of the Orion A/M42 molecular cloud/H II region complex and the Orion B/NGC 2024 complex.

131.216 **Cool molecular clouds.**
R. D. Davies.
Submillimetre wave astronomy, (see 012.031), p. 33 - 46 (1982).

The properties of the cool molecules in the diffuse clouds distributed throughout the interstellar medium are described. These clouds contain H I, H_2, CO, OH, H_2CO and other common molecules with temperatures of 10 to a few hundred Kelvins and with densities of 10 to 10^3 H atoms cm^{-3}. Optical absorption and UV studies indicate that dust is well mixed with the gas. Significant molecule formation only occurs in the local interstellar medium when the hydrogen column density is greater than 2×10^{19} atoms cm^{-2}. The giant molecular clouds represent a more massive but less numerous group of clouds.

131.217 **Models for hot-centred galactic clouds.**
M. Rowan-Robinson.
Submillimetre wave astronomy, (see 012.031), p. 47 - 68 (1982).

Theoretical models for the infrared emission from hot-centred clouds are reviewed. Most work to date has been on the spherically symmetric case and the validity of this assumption for the observed clouds is discussed. Approximations to the equation of radiative transfer, and their validity, are discussed. A variety of radiative transfer phenomena have come to light through accurate numerical solution of the problem of the flow of radiation through dust and these are reviewed. Work on the time-dependent case and on problems with non-spherical geometry, together with other areas of development, is also reviewed. Finally some recent results for individual hot-centred clouds, selected for their compact size and isolated nature, are presented.

131.218 **Models of externally heated clouds: an interim report.**
D. W. Walker.
Submillimetre wave astronomy, (see 012.031), p. 69 - 79 (1982).

Models of externally heated clouds are presented. The dust temperature distribution is evaluated, and used to find the model surface brightness distribution, energy spectrum and colour temperature. A qualitative comparison is made between one of the models and infrared observations of M17.

131.219 **Dust in Bok globules.**
I. P. Williams, H. Bhatt.
Submillimetre wave astronomy, (see 012.031), p. 81 - 88 (1982).

The various forces which can operate on a grain either in the vicinity or within a Bok globule are reviewed and the consequences of various of these forces being important discussed. A number of globules where these forces may be in evidence are cited.

131.220 **Airborne far-infrared and submillimeter spectroscopy.**
D. M. Watson.
Submillimetre wave astronomy, (see 012.031), p. 91 - 109 (1982).

The paper is a review of airborne spectroscopic observations in the wavelength range 18 μm - 800 μm, and the instruments on the NASA Kuiper Airborne Observatory with which they were carried out. Recent measurements of spectral lines of ions, atoms and molecules are presented and discussed.

131.221 **The excitation and distribution of CO (J = 6 → 5) emission in the Orion Nebula.**
D. Buhl, G. Chin, G. A. Koepf, D. D. Peck, H. R. Fetterman.
Submillimetre wave astronomy, (see 012.031), p. 111 - 122 (1982).

131.222 **First observations of the J = 4 → 3 transition of HCO^+.**
R. Padman, P. Scott, A. Webster.
Submillimetre wave astronomy, (see 012.031), p. 123 - 125 (1982).

The J = 4 → 3 transition of HCO^+ was observed in OMC 1 and in three other galactic sources which have self-reversed CO profiles.

131.223 **Observations of the J = 2 → 1 CO line in molecular clouds showing ammonia emission.**
L. T. Little.
Submillimetre wave astronomy, (see 012.031), p. 127 - 136 (1982).

Observations of the J = 2 → 1 transitions of ^{12}CO and ^{13}CO at 230 and 220 GHz in 18 molecular clouds have been made at the UK Infrared Telescope. The sources were chosen on the basis of their ammonia emission. A comparison between ^{12}CO and ^{13}CO spectra reveals a variety of absorption effects, ranging from slight asymmetries in the ^{12}CO profiles relative to their ^{13}CO counterparts (W43S, S88) to a deep self absorption dip (S68 and possibly G34.3+0.1).

131.224 **A random view of Cygnus X.**
S. Harris.
Submillimetre wave astronomy, (see 012.031), p. 145 - 157 (1982).

The paper describes the first stage (viz. preliminary ^{12}CO and ^{13}CO J = 1 → 0 mapping) in a programme of observations designed to probe the mysteries both of the Cygnus X region and of the statistics of star formation in general. The first section describes the philosophy behind the observations; the second describes the results.

131.225 **The M 17 (SW) molecular source: internally or externally heated?**
M. P. Chown, J. E. Beckman, N. J. Cronin, G. J. White.
Submillimetre wave astronomy, (see 012.031), p. 159 - 165 (1982).

From a measurement of the J = 3 → 2 rotational emission line of hydrogen isocyanide, and a literature value for that J = 1 → 0 line the authors have derived the excitation parameters for HNC in the direction of the molecular peak as determined by CO. Kinetic temperatures for the optically thick CO and ^{13}CO transitions are ~50 K, while that pertaining to HNC line formation is between 20 K and 5 K. The inference that the central temperature of the cloud is lower than that of its surface, supports models in which the cloud is heated not from within, but by the O–B association to the north-west.

131.226 **Observations of rotational transitions of CH_3OH and NH_2 near 1.2 mm.**
T. de Graauw, S. Lidholm, W. Boland, T. J. Lee, C. de Vries.
Submillimetre wave astronomy, (see 012.031), p. 167 - 173 (1982).

Sixteen new methanol lines have been detected at frequencies near 242 GHz towards the direction of OMC-1. They correspond to the J = 5 → 4, sk = 0 transitions in the A and E stack of CH_3OH. Also towards OMC-1 the authors observed a broad emission feature, coinciding in frequency with J= $^7/_2 \rightarrow ^5/_2$ components of the $3_{13} \rightarrow 2_{20}$ rotational transition of NH_2. The observed antenna temperatures of the methanol lines can be reproduced with a thermal excitation model using one temperature T_{exc} = 90 K and a column density of N (CH_3OH) = 5×10^{16} cm^{-2}, assuming a 30" diameter methanol source.

131.227 **High velocity flows and molecular jets.**
C. J. Lada.
Submillimetre wave astronomy, (see 012.031), p. 175 - 183 (1982).

131.228 **Molecular clouds in the galactic nucleus.**
R. J. Cohen.
Submillimetre wave astronomy, (see 012.031), p. 185 - 190 (1982).

First results are presented from a large-scale survey of OH in the galactic centre carried out at Jodrell Bank. The observations sample the region $-6.^\circ0 \leq l \leq 8.^\circ6$, $-2.^\circ0 \leq b \leq 1.^\circ6$. The concentration of massive molecular clouds in the galactic nucleus is twice as extensive as hitherto realized. New galactic centre clouds have been detected out to projected distances of 1 kpc either side of the centre. The outermost clouds lie out of the galactic plane in a tilted distribution, and reach z-distances of 200 pc above the plane at negative longitudes and below the plane at positive longitudes.

131.229 **The interaction between chemistry and astronomy.**
H. W. Kroto.
Submillimetre wave astronomy, (see 012.031), p. 203 - 217 (1982).

It is only in the last decade that the true extent of the molecular components of the interstellar medium (ISM) has become evident. Suddenly chemistry has taken on a new significance for astronomy in so far as it affects the study of the ISM at low and moderate temperatures and in particular those regions where stars are in the process of being born. This article deals with this recent synergistic combination of chemistry and astronomy.

131.230 **Problems in modelling interstellar chemistry.**
D. A. Williams.
Submillimetre wave astronomy, (see 012.031), p. 219 - 225 (1982).

The paper discusses solutions that are found by modelling interstellar chemistry and argues that – though mathematically correct – they may not always provide a satisfactory description of the astronomical objects. The problem arises in the large array of data which are necessary and yet which are poorly determined or unknown. The paper illustrates the effects of uncertainties in these data. A modelling procedure which is potentially more reliable is outlined.

131.231 **Ion-grain collisions as a source for interstellar molecules.**
T. J. Millar.
Submillimetre wave astronomy, (see 012.031), p. 227 - 232 (1982).

In recent years it has been recognized that ionic surface sites exist on a class of interstellar grains. This raises the possibility that gas phase positively charged ions can react with negatively charged surface ions to produce interstellar molecules. A brief discussion of the nature of the surface sites is followed by an application to the chemistry of diffuse and dense clouds. It is shown that ion-grain collisions can be the dominant formation mechanism for several molecular species.

131.232 **Limits on the |D|:|H| ratio in the interstellar medium from molecular observations.**
R. L. Frost, J. E. Beckman, G. D. Watt, G. J. White, J. P. Phillips.
Submillimetre wave astronomy, (see 012.031), p. 253 - 259 (1982).

The abundance ratio of |D|:|H| in the interstellar medium can be derived from millimetre-wave emission in the rotational lines of a variety of molecules. It is shown how the basic assumptions of temperature distribution and geometry within a single source can cause the derived ratio $|HDO|:|H_2O|$ to vary between 5×10^{-2} and 2.5×10^{-3}, a comparable range to that found with three different molecules |DCN|:|HCN|, |DNC|:|HNC| and $|DCO^+|:|HCO^+|$ in the same source. This result should lead to caution in the interpretation of such abundance ratio and of their gradients within the galaxy.

131.233 **Dust in dense clouds: one stage in a cycle.**
J. M. Greenberg.
Submillimetre wave astronomy, (see 012.031), p. 261 - 306 (1982).

An evolutionary path is proposed for interstellar grains which takes them from diffuse clouds through molecular clouds and back as part of a repeating cycle in their development. Laboratory analog results are applied to the ultraviolet photoprocessing of grains in dense clouds.

131.234 **An interpretation of OH maser observations in W3 (OH).**
D. Field.
Mon. Not. R. Astron. Soc., Vol. 201, 527 - 549 (1982).

The main purpose of this work is to make theoretical predictions of OH maser brightness temperatures in W3 (OH) on the basis of a collisional pumping mechanism for Λ-doublet inversion and a theory of maser amplification both described elsewhere. An important conclusion is that a fully realistic evaluation of the physical conditions of the maser region must await experimental data on rate coefficients for rotationally inelastic collisions of H_2 with $OH^2\Pi$. The dynamics of the maser region are investigated and it is suggested that a maser may consist of a turbulent eddy or group of eddies in the interstellar medium.

131.235 **Star formation in turbulent molecular clouds: the initial stellar mass function.**
R. C. Fleck, Jr.
Mon. Not. R. Astron. Soc., Vol. 201, 551 - 559 (1982).

The observed flattening or turnover of the initial mass function (IMF) of stars at a mass roughly corresponding to the transition from supersonic to subsonic turbulence ($0.2 \lesssim M/M_\odot \lesssim 13$ for $10 \lesssim T(K) \lesssim 50$), suggests that turbulence within molecular clouds may play a significant role in determining the IMF. A theoretical IMF for Jeans mass protostellar fragments is derived together with a modified Jeans equation which relates the minimum mass that can undergo gravitational collapse to the velocity and temperature fields of the medium. The agreement between the derived fragment IMF and the IMFs of field stars and open clusters is quite good and suggests that the IMF depends on the properties of the associated molecular cloud material. Some implications for star formation are discussed.

131.236 **Complex ionized structure in the θ^2 Orionis region.**
J. R. Walsh.
Mon. Not. R. Astron. Soc., Vol. 201, 561 - 567 (1982).

An [S II] electron density map over a region of the Orion nebula (M42) south of θ^2 Orionis is presented. Comparison with an Hβ image tube photograph shows a triangular feature with the known high-velocity ionized knots situated at two corners. The radial velocities and electron densities reveal a complex feature, similar to a group of Herbig-Haro knots.

131.237 **On C_3 molecules in diffuse interstellar clouds.**
R. E. S. Clegg, D. L. Lambert.
Mon. Not. R. Astron. Soc., Vol. 201, 723 - 733 (1982).

A search for absorption lines of interstellar C_3 has been made. The $\tilde{A}^1\Pi_u - \tilde{X}^1\Sigma_g^+$ band near 4050 Å was not detected. Upper limits on the C_3 column density are $f = N(C_3)/N(H+H_2) \lesssim 1 \times 10^{-9}$ for several diffuse interstellar clouds and $f < 3 \times 10^{-9}$ for the cool core of the ζ Oph cloud. Useful laboratory studies for this molecule are pinpointed.

131.238 **Temperatures and their variation within interstellar H I structures.**
C. Heiles.
Astrophys. J., Vol. 262, 135 - 141 (1982).

The author uses recent H I absorption surveys to derive H I temperatures T and their variation within a number of H I shells. In these shells, T is in the domain of cold, neutral gas, and ranges down to 35 K; densities are about 2 cm^{-3}. Outside the relatively dense shells, H I temperatures are much higher.

131.239 **The Orion SiO maser: a unique object in the Galaxy.**
M. Elitzur.
Astrophys. J., Vol. 262, 189 - 197 (1982).

Unlike late-type stars, where the SiO masers are located in the upper parts of the stellar atmospheres, the Orion SiO maser is located in an expanding shell around the source IRc2. This paper is aimed at studying the properties of such an SiO maser in an attempt to understand the Orion source and its apparent uniqueness.

131.240 **Conversion form ^{12}CO integrated intensity at 2.6 millimeter wavelength to hydrogen column density.**
H. S. Liszt.
Astrophys. J., Vol. 262, 198 - 200 (1982).

Estimates of the column density N_H of interstellar hydrogen nuclei from the integrated intensity I_{12} of the ^{12}CO transition at 2.6 mm are compared for OMC-1, for giant molecular cloud complexes, for local dark clouds, and for giant molecular clouds in the inner Galaxy. It is shown that a fairly uniform ratio $N_H/I_{12} = (1.0 \pm 0.5) \times 10^{21}$ cm^{-2} $(K\ km\ s^{-1})^{-1}$ applies to all cases.

131.241 **The interstellar absorption-line spectrum of μ Ophiuchi.**
J. Cardelli, E. Böhm-Vitense.
Astrophys. J., Vol. 262, 213 - 223 (1982).
UV interstellar lines have been measured on high-resolution IUE spectra of the B8 V star μ Oph. Column densities for the observed atoms and ions have been determined as well as turbulent velocities. Relative depletion of elements in the cloud seen in the interstellar spectrum of μ Oph follows the same pattern as seen in the interstellar spectra of ζ Oph and six other stars in the ρ Oph cloud complex.

131.242 **Detection of interstellar sodium hydroxide in self-absorption toward the galactic center.**
J. M. Hollis, P. J. Rhodes.
Astrophys. J., Lett., Vol. 262, L1 - L5 (1982).
A weak self-absorbed emission line, which is identified as the $J = 4 - 3$ transition of sodium hydroxide, has been detected in the direction of Sgr B2(OH). The correspondingly weak Sgr B2(OH) emission line U75406, previously reported as an unidentified spectral feature by other investigators, is consistent with the $J = 3-2$ transition of sodium hydroxide. This detection may represent the first evidence of a grain reaction formation mechanism for simple metal hydroxides.

131.243 **Fundamental processes in star formation.**
T. Nakano.
Prog. Theor. Phys. Suppl., No. 70, p. 54 - 76 (1981). – Abstr. in Phys. Abstr., Vol. 85, Abstr. 100909 (1982).

131.244 **Energy exchange processes in molecular cloud W40.**
L.-p. Xu, J. Xing, Y.-f. Wu, S.-d. Xie, J.-z. Zhao.
Chinese Phys., Vol. 2, 688 - 694 (1982). – Abstr. in Phys. Abstr., Vol. 85, Abstr. 101022 (1982). – See Abstr. 30.131.177.

131.245 **On the interpretation of the spectral data of HCN obtained from the Orion A cloud.**
S. Ray, B. Bhattacharyya, A. K. Barua.
Indian J. Phys. Part B, Vol. 56B, 1 - 10 (1982). – Abstr. in Phys. Abstr., Vol. 85, Abstr. 106870 (1982).

131.246 **Effect of the magnetic pressure on cloud evolution.**
M. Fofi, P. Paolicchi.
Nuovo Cimento B, Vol. 70B, Ser. 11, 85 - 101 (1982). Abstr. in Phys. Abstr., Vol. 85, Abstr. 112434 (1982).

131.247 **The effect of electron-molecule collisions on the spectral intensity and the determination of hydrogen density in molecular clouds.**
B. Bhattacharyya, A. K. Barua.
Indian J. Phys. Part B, Vol. 56B, 135 - 148 (1982). – Abstr. in Phys. Abstr., Vol. 85, Abstr. 112533 (1982).

131.248 **Determination of hydrogen and electron densities in interstellar molecular clouds.**
B. Bhattacharyya, A. K. Barua.
Indian J. Phys. Part B, Vol. 56B, 149 - 155 (1982). – Abstr. in Phys. Abstr., Vol. 85, Abstr. 112534 (1982).

131.249 **On the origin and structure of isolated dark globules.**
C. M. Leung, M. L. Kutner, K. N. Mead.
Astrophys. J., Vol. 262, 583 - 589 (1982).
High resolution carbon monoxide observations of over a dozen isolated dark globules indicate that a power-law correlation exists between the internal velocity dispersion and cloud size. The power-law relation implies the presence of supersonic turbulence that is similar to that found in other diffuse clouds and molecular cloud complexes. Most of the globules are found to be gravitationally bound and in virial equilibrium. The results are consistent with the scenario that isolated dark globules represent condensed fragments from nearby filamentary cloud complexes.

131.250 **The relationship between carbon monoxide abundance and visual extinction in interstellar clouds.**
M. A. Frerking, W. D. Langer, R. W. Wilson.
Astrophys. J., Vol. 262, 590 - 605 (1982).
Carbon monoxide column densities are compared to visual extinctions toward field stars in the ρ Oph and Taurus molecular cloud complexes. The relationship of $C^{18}O$ column density to A_v is established, and new determinations for ^{13}CO column densities are given for a range of visual extinctions extended to beyond 20 mag. A prescription for determining hydrogen column densities and masses of molecular clouds from observations of CO isotopes is presented and discussed critically.

131.251 **A search for NaH in dense molecular clouds: evidence against formation on grains.**
R. L. Plambeck, N. R. Erickson.
Astrophys. J., Vol. 262, 606 - 610 (1982).
The $J = 1 - 0$ rotational transition of NaH, at 289.864 GHz, was searched for but not detected toward five dense molecular clouds. The most stringent upper limit on the NaH fractional abundance is found for Orion A, where $[NaH]/[H_2] < 10^{-11}$. This upper limit is consistent with the predictions of ion-molecule chemistry but is several orders of magnitude smaller than would be expected if NaH were formed efficiently on grains.

131.252 **A search for diffuse band profile variations in the Rho Ophiuchi cloud.**
T. P. Snow, J. G. Timothy, S. Saar.
Astrophys. J., Vol. 262, 611 - 618 (1982).
High signal-to-noise profiles of the broad diffuse interstellar band at 4430 Å were obtained on the 2.2 m telescope at the Mauna Kea Observatory, using the newly developed pulse-counting Multi-Anode Microchannel Array detector system. The purpose of the study was to determine whether the band profile varies with mean grain size as expected if the band is produced by absorbers embedded in grain lattices. The lack of profile variability over several lines of sight where independent evidence indicates that the mean grain size varies shows that λ4430 is probably not formed by the same grains that are responsible for interstellar extinction at visible wavelengths.

131.253 **Mass outflow in star formation regions: Cepheus A.**
P. T. P. Ho, J. M. Moran, L. F. Rodríguez.
Astrophys. J., Vol. 262, 619 - 635 (1982).
The velocity extents, the spatial distributions, and the masses, deduced from CO observations, are discussed for the red and blue outflows in Cepheus A. These properties can be modeled by placing the source of outflow amidst an anisotropic density distribution. It is concluded that the appearance of an outflow is strongly affected by the environment in which it is embedded. In particular, the outflow in Cepheus A may not be very well collimated. The apparently bipolar nature of outflows in star formation regions is discussed.

131.254 **Secondary electron spectra in interstellar clouds, and the bremsstrahlung γ-ray luminosity.**
G. E. Morfill.
Astrophys. J., Vol. 262, 749 - 759 (1982).
Low-energy electron spectra in interstellar clouds are

calculated. The processes considered are the following: inside the cloud, secondary production (knock-on, meson decay), ionization losses and bremsstrahlung losses; outside the cloud, particle transport by convection as well as diffusion. The bremsstrahlung γ-ray luminosity of interstellar clouds from secondary electrons is calculated.

131.255 **Determination of the parameters of the interstellar wind from observations of the Lα-background at large distances from the sun.**
M. S. Burgin.
Kosm. Issled., Tom 20, 725 - 729 (1982). In Russian.

131.256 **A model for interstellar extinction.**
F. Hoyle, N. C. Wickramasinghe.
Astrophys. Space Sci., Vol. 86, 321 - 329 (1982).

From an analysis of the interstellar extinction the authors conclude that interstellar grains are of three main kinds: graphite spheres of radii ~0.02 μm making up ~10% of the total grain mass, small dielectric spheres of radius about 0.04 μm making up ~25% of the mass, and hollow dielectric cylinders containing metallic iron with diameters of ~2/3 μm making up ~45% of the mass. The remaining ~20% consists of other metals and metal oxides. The main dielectric component of the grains appears to be comprised of organic material.

131.257 **The infrared spectrum of interstellar dust.**
F. Hoyle, N. C. Wickramasinghe, S. Al-Mufti.
Astrophys. Space Sci., Vol. 86, 341 - 344 (1982).

The observed spectrum of GC-IRS 7 over the waveband 2.9 - 13 μm can be explained on the basis of interstellar grains which are spectroscopically similar to dried-out diatomaceous organisms.

131.258 **Radiative stability of interstellar masers – a variational technique.**
K. K. Sen.
Astrophys. Space Sci., Vol. 86, 477 - 484 (1982).

A variational technique has been developed to study the stability of the radiation field in a cylindrical homogeneous interstellar maser with steady state pumping. A dynamical potential has been constructed, the extremal of which gives the radiative transfer equations and rate equations. The dispersion equations and the stability criteria are determined from the potential for specific natures of perturbations.

131.259 **Silicon monoxide: a component of interstellar grains?**
T. J. Millar.
Astrophys. Space Sci., Vol. 86, 497 - 500 (1982).

Gürtler et al. (1981) have argued that SiO cannot be the material responsible for the interstellar 9.7 μm feature. Their arguments are discussed and shown to be questionable. Experimental data on the thermal stability of SiO under reducing conditions typical of circumstellar regions could help decide whether SiO is an important constituent of interstellar grains.

131.260 **Detection of the (8,8) and (9,9) absorption lines of ammonia: the hot molecular cloud toward Sgr B2.**
T. L. Wilson, K. Ruf, C. M. Walmsley, R. N. Martin, T. A. Pauls, W. Batrla.
Astron. Astrophys., Vol. 115, 185 - 189 (1982).

Absorption in the $(J,K) = (8,8)$ and (9,9) metastable lines of ammonia is reported for the ~63 km s^{-1} cloud toward Sgr B2. Since these lines are 680 K and 835 K above the ground state and the metastable doublet levels of NH_3 are probably collisionally populated, the molecular cloud must be hot. One possible heating mechanism might be turbulence, driven by shearing motion of the cloud. The authors also find an H_2 density of $\sim 10^4$ cm^{-3}.

131.261 **The effects of the coronal gas on the champagne phase.**
G. Tenorio-Tagle, P. J. Bedjin.
Astron. Astrophys., Vol. 115, 207 - 208 (1982).

Recent observations have shown the existence of a third phase in the interstellar medium, the coronal phase. The authors present a calculation of the expansion of an H II region located at the edge of a cloud which is bordered by coronal gas. The effects of the coronal gas on the champagne phase are discussed.

131.262 **On interstellar linear polarization and grain growth.**
P. A. Aannestad.
Astron. Astrophys., Vol. 115, 219 - 222 (1982).

The effect of interstellar grain growth through simple accretion on the shape of the normalized linear polarization curve is calculated in the birefringent sphere approximation. Comparison is made to observed deviations from Serkowski's empirical curve. It is found that the observed narrowing of the polarization curve with increasing wavelength of maximum polarization can be explained by the shifting and steepening of the grain size distribution as caused by size-independent accretion. Non-spherical particles are expected to show the same "hyperbolic" relationship between width and maximum wavelength when subject to size-independent growth.

131.263 **An Effelsberg – Green Bank galactic H I absorption line survey. II. Results and interpretation.**
U. Mebold, A. Winnberg, P. M. W. Kalberla, W. M. Goss.
Astron. Astrophys., Vol. 115, 223 - 241 (1982).

21-cm line single dish emission and interferometer absorption spectra of galactic H I gas were observed towards 69 extragalactic radio sources with the MPIfR 100-m telescope and the NRAO three-element interferometer. The observations have been published in a previous paper (Mebold et al., 1981); the results are discussed here. The stray radiation contribution to the emission lines is taken into account and for most of the sources the fluctuation of the H I emission near the sources, the RMS spectra are studied together with the emission and absorption spectra. Properties of directly observed and derived parameters of the interstellar H I gas as well as correlations between these parameters are investigated after a decomposition into gaussian components of the absorption, emission, and RMS spectra.

131.264 **Fine structure in high velocity clouds near the South Celestial Pole.**
R. Morras.
Astron. Astrophys., Vol. 115, 249 - 252 (1982).

Three high velocity clouds (HVCs) in the region between the Magellanic Clouds and the Galactic Plane have been mapped in the 21-cm line of neutral hydrogen with the 30-m IAR Radiotelescope. One of these clouds, HVC 300 – 24 + 274 seems to be connected with the H I gas around the Magellanic Clouds. It could have originated as a part of the same phenomena that created the Magellanic Stream.

131.265 **Far infrared observations of a star forming region in Serpens.**
H. L. Nordh, R. J. van Duinen, A. I. Sargent, C. V. M. Fridlund, J. W. G. Aalders, D. Beintema.
Astron. Astrophys., Vol. 115, 308 - 314 (1982).

The Serpens molecular cloud has been observed at far infrared wavelengths over an area of size 20′× 20′, centered on the Serpens reflection nebula. The deduced total dust luminosity is only $400 \pm 100\ L_\odot$, implying that the embedded objects, of which there are at least seven, are all of intermediate or low mass. It is proposed that three very compact H II-regions in the cloud are associated with low-mass stars having strong stellar winds. It is suggested that stellar winds associated with the embedded young objects may give rise to further star formation.

131.266 **The law of interstellar absorption in the wave-number interval 0.95 μ^{-1} to 3.03 μ^{-1}.**
A. Ardeberg, B. Virdefors.
Astron. Astrophys., Vol. 115, 347 - 356 (1982).

The wave-number dependence of interstellar absorption has been studied in the interval $\lambda^{-1} = 0.95\ \mu^{-1}$ to $\lambda^{-1} = 3.03\ \mu^{-1}$. The study has been based on data from narrow-band photoelectric photometry and scanning. To a first approximation the interstellar absorption law can be approximated by two linear relations, intersecting around $\lambda^{-1} = 2.30\ \mu^{-1}$. The ratio of the first-order coefficients is around 0.78. For higher accuracy five first-order polynomials have to be used. The intervals join at $\lambda^{-1} = 1.55\ \mu^{-1}$, $1.80\ \mu^{-1}$, $2.20\ \mu^{-1}$, and $2.48\ \mu^{-1}$. The data at hand seem to support universality of the interstellar-absorption law for the wave-number region studied.

131.267 **The local interstellar medium as traced by gamma rays.**
A. W. Strong, G. F. Bignami, J. B. G. M. Bloemen, R. Buccheri, P. A. Caraveo, W. Hermsen, G. Kanbach, F. Lebrun, H. A. Mayer-Hasselwander, J. A. Paul, R. D. Wills.
Astron. Astrophys., Vol. 115, 404 - 412 (1982).

Gamma-ray flux measurements (70 - 5000 MeV) made with the COS-B satellite have been used to study the distribution of total gas column density in the latitude range $19° > |b| > 11°$. A close correlation with the distribution found from galaxy counts exists. A map of the distribution of molecular-hydrogen column density is presented based on the excess γ-ray flux over that expected from atomic hydrogen. Taken together with a similar map derived from galaxy counts, the molecular-hydrogen angular distribution is established (on scales $\sim 10°$) with greater confidence than previously possible. The map derived from γ-rays includes the Southern celestial hemisphere for which galaxy-count coverage is not available.

131.268 **A note about multiple systems of dense molecular clouds.**
V. Vanýsek.
Binary and multiple stars as tracers of stellar evolution, (see 012.046), p. 105 - 108 (1982).

The space distribution of some small dense clouds with point-like IR sources, resembles the clustering of young OB stars. It can be assumed that such objects contain heavy obscured high-luminosity stars on the ZAMS. From the comparison of infrared and radio data it follows that in typical cases, only one B star is the source of the radiation of the cloud. The total mass of the cloud is of the order of one solar mass. If the individual fragments of the cloud are gravitationally unstable, then in the later stage of the evolution only low-mass stars are formed. One can therefore expect that young OB stars are most frequently accompanied by low-mass pre-Main Sequence stars.

131.269 **VLBI measurements of the relative position of the 1665 MHz and 1667 MHz OH masers toward W 49N and NGC 6334N.**
S. R. Kent, R. L. Mutel.
Astrophys. J., Vol. 263, 145 - 152 (1982) = Iowa Radio Astron. Group, Contrib. No. 88.

Very long baseline interferometric observations of the main-line hydroxyl maser emission toward NGC 6334N and W 49N were conducted to determine the relative positions of the 1665 and 1667 MHz transitions. The main-line feature of NGC 6334N are distributed over an area of 1″ × 0.″5. A spatial offset of ~ 0.″5 was found between the 1665 and 1667 MHz emission. A Zeeman pair indicative of a magnetic field strength of ~4 milligauss was found in the 1667 MHz emission of NGC 6334N. For the source W 49N, no systematic spatial offset between the main-line transitions was found. The main-line features of W 49N are distributed over an area of 3″ × 6″.

131.270 **The velocity gradient of B 361.**
F. O. Clark, D. R. Johnson.
Astrophys. J., Vol. 263, 160 - 165 (1982).

Observations of a well-ordered velocity gradient that decreases with decreasing cloud radius are reported for the galactic dark cloud B 361. The source also exhibits systematic changes in spectral line width. These observations confirm the suggestion of rotation in this source reported earlier and imply the presence of an effective mechanism for redistributing angular momentum.

131.271 **The peculiar extinction of Herschel 36.**
J. Hecht, H. L. Helfer, J. Wolf, B. Donn, J. L. Pipher.
Astrophys. J., Lett., Vol. 263, L39 - L43 (1982).

The extinction of Herschel 36 has been measured and been found to be peculiar in the same sense as that observed in Orion. This can be explained by the presence of silicate and graphite grains that are larger than those normally found in the interstellar medium. Correcting the stellar flux for foreground extinction results in a residual extinction curve for the associated dust cloud with an unusually small normalized extinction at 1500 Å.

131.272 **Physical conditions in H II/OH maser regions.**
S. Guilloteau.
Astron. Astrophys., Vol. 116, 101 - 110 (1982).

A detailed comparison between the predictions of collisional and "thermal" overlap pumping mechanism and observations of H II/OH maser regions is presented, in which the W3(OH) region is taken as an example. Special attention is given to the derivation of the maser intensities. The pumping model implies the following physical conditions in the maser clouds: $T_k = 100$ K to 200 K, $n_{H_2} = 10^7\ cm^{-3}$ to $3 \times 10^7\ cm^{-3}$, $X_{OH} > 10^{-5}$ and the dust temperature must be slightly lower than the kinetic temperature of the gas for strong excited state masers.

131.273 **CO $J = 3 \rightarrow 2$ and submillimetre continuum observations of two molecular outflow sources.**
J. P. Phillips, G. J. White, P. A. R. Ade, C. T. Cunningham, K. J. Richardson, E. I. Robson, G. D. Watt.
Astron. Astrophys., Vol. 116, 130 - 136 (1982).

The authors present CO $J = 3 \rightarrow 2$ molecular line spectra, and submillimetre photometry at wavelengths $\lambda\lambda 377\ \mu m$, $811\ \mu m$, and $1136\ \mu m$ for the sources L 1551 and IRC+10216. Detailed analysis of the L 1551 spectra indicates the presence of strong velocity gradients in the CO emission zones, implying low optical depths and relatively high densities. The central source IRS 5 displays an infrared excess which cannot be explained in terms of a single temperature continuum. The far-infrared continuum of IRC + 10216 has been synthesised by assuming a distribution of optically thin grains whose emissivity varies as $\epsilon \propto \lambda^{-1}$, and the CO $J = 3 \rightarrow 2$ spectrum for this source supports earlier $J = 2 \rightarrow 1$ observations, implying a variable mass-loss rate.

131.274 **The dynamical destruction of shocked gas clouds.**
J. Nittmann, S. A. E. G. Falle, P. H. Gaskell.
Mon. Not. R. Astron. Soc., Vol. 201, 833 - 847 (1982).

The authors have studied the impact of a strong interstellar shock with a high-density gas cloud, using two different numerical techniques to solve the model equations. The results obtained have enabled the authors to distinguish four distinct stages in the evolution of the system.

131.275 **Star formation in a cooling flow.**
A. C. Fabian, P. E. J. Nulsen, C. R. Canizares.
Mon. Not. R. Astron. Soc., Vol. 201, 933 - 938 (1982).

The central galaxies in some clusters are observed to accrete cooling intracluster gas at rates of between 4 and 300 $M_\odot$ yr^{-1}. The colours and luminosities of these galaxies show evidence for star formation at less than 10 per cent of those rates, if the stellar initial-mass-function is similar to that in the solar neighbourhood. The authors show that the high-pressure dust-free environment in the cooling flow lowers the Jeans mass and thereby favours the formation of lower-mass stars. The lack of detectable H I in most of these clusters is also explained.

131.276 **A study of the Chamaeleon dark cloud complex: survey, structure and embedded sources.**
A. R. Hyland, T. J. Jones, R. M. Mitchell.
Mon. Not. R. Astron. Soc., Vol. 201, 1095 - 1117 (1982).

A 2 μm-survey of a significant fraction of the Chamaeleon dark cloud complex is presented. IR (*JHK*) photometry and CVF spectrophotometry of the sources found provide the following significant results: (1) Discrimination between background field stars and cloud members shows an increase in the number of pre-main sequence (PMS) members (over that found optically) by more than 50 per cent, corresponding to a star formation efficiency of ~12 per cent over the whole cloud. (2) The number density of PMS objects within the cloud is close to that required by Norman and Silk to account for molecular velocity fields within dark clouds. (3) The density structure of the cloud, derived from the reddening of background field stars shows that small-scale clustering of two groups of PMS members seems to occur preferentially near steep density gradients in the cloud. (4) The derived intrinsic infrared colours of the PMS objects are successfully reproduced by dust shell models. (5) The models suggest that the observation of significant far IR radiation from some T Tauri objects results from some non-continuous form of the shell density distribution.

131.277 **Neutral hydrogen observations towards the Puppis Window of the Milky Way.**
J. G. Stacy, P. D. Jackson.
Astron. Astrophys., Suppl. Ser., Vol. 50, 377 - 422 (1982).

A 21-cm neutral hydrogen survey has been carried out in a region of the galactic disk known as the "Puppis Window". Observations were conducted on the NRAO 43-m radio telescope and extend from galactic longitudes 239° to 251° and from galactic latitudes −9° to +3°, every 0°.2 in *l* and *b*. The presence of numerous, sometimes expanding, H I shells and filaments, for which the authors report estimates of shell size, density and mass, attests to the turbulent nature of the interstellar medium, even in the more distant portions of the outer Galaxy. The possible association of one such H I shell with the pulsar PSR0740-28 is reported. Correlations of H I features with optical spiral tracers are discussed, along with the corresponding implications regarding the possible spiral structure in this quadrant of the Galaxy. Gas-to-dust ratios determined along lines-of-sight to five distant optical features agree with gas-to-dust ratios for globular clusters, indicating a uniform Galaxy-wide gas-to-dust ratio. Total H I integrated brightness levels are calculated and presented for the entire region surveyed.

131.278 **The diffuse interstellar medium and the CES (*Coudé Echelle Spectrograph*).**
R. Ferlet.
Messenger, No. 30, p. 9 - 13 (1982).

131.279 **Early stages of star formation seen in Cep A.**
V. A. Hughes, J. G. A. Wouterloot.
J. R. Astron. Soc. Canada, Vol. 76, 320 (1982). – Abstract.

131.280 **The effect of a shock on diffuse interstellar-cloud abundances.**
J. N. Scrimger, G. F. Mitchell.
J. R. Astron. Soc. Canada, Vol. 76, 325 (1982). – Abstract.

131.281 **Champs magnétiques interstellaires et intergalactiques.**
J. P. Vallée.
J. R. Astron. Soc. Canada, Vol. 76, 326 (1982). – Abstract.

131.282 **Shock fronts produced by stellar winds in the interstellar gas.**
R. Q. Huang, A. Weigert.
Astron. Astrophys., Vol. 116, 348 - 350 (1982).

A simple method is described for determining the geometrical properties of the shock front which is produced by a stellar wind interacting with an interstellar medium of large relative velocity. The differential equations for this problem are solved numerically. Except for a trivial scaling factor, the resulting geometrical form of the front is the same for all combinations of stellar winds and interstellar properties.

131.283 **Star formation in Bok globules and low-mass clouds I. The cometary globules in the Gum Nebula.**
B. Reipurth.
ESO Sci. Prepr., No. 208, 40 pp. (1982). – Submitted to Astron. Astrophys.

131.284 **Prediction of the protostellar mass spectrum in the Orion near infrared cluster.**
H. Zinnecker.
Max-Planck-Inst. Phys. Astrophys., Inst. Extraterr. Phys., MPE Prepr. 6, 26 pp. (1982).

An accretion model of the Initial Mass Function for the infrared star cluster in the Kleinmann-Low region in the Orion molecular cloud core is suggested in which protostellar cores compete for the accretion of the residual gas of their parent cloud. For large masses, the model predicts an asymptotic mass distribution $dN/d\log M \sim M^{-1}$. Such a model may apply to other molecular cloud cores as well.

131.285 **Modifikation der interstellaren Gasparameter im Randgebiet des Sonnensystems.**
H. J. Fahr, H. W. Ripken.
Mitt. Astron. Ges., Nr. 57, p. 288 - 294 (1982).

131.286 **High density molecular gas in the ρ Ophiuchus cloud.**
J. Martin-Pintado, T. L. Wilson, F. F. Gardner, C. Henkel.
Mitt. Astron. Ges., Nr. 57, p. 305 - 306 (1982).

131.287 **A survey of interstellar neutral potassium. I. Abundances and physical conditions in clouds toward 188 early-type stars.**
F. H. Chaffee, Jr., R. E. White.
Astrophys. J., Suppl. Ser., Vol. 50, 169 - 198 (1982) = Five Coll. Obs., Contrib. No. 498.

This paper reports observations of interstellar absorption in the resonance doublet λλ7664, 7698 of neutral potassium toward 188 early-type stars at a spectral resolution of 8 km s^{-1}. The λ7664 line is successfully separated from nearly coincident telluric O_2 absorption for all but a few of the 165 stars for which K I absorption is detected, allowing abundance analysis by the doublet radio method. Comparisons of the K I column densities with those of Na I, total H, and H_2, respectively, are given for more than 20 of the brighter survey stars and with that of C I for seven stars. Stars in the Sco-Oph complex stand apart from the rest in most of the comparisons.

131.288 **Spectral survey of dark nebulae regions Kh. 179-185.**
O. M. Kurtanidze, M. G. Nikolashvili.
Astron. Tsirk., No. 1172, p. 3 - 4 (1981). In Russian.

131.289 **Observations of CO clouds in the galactic center region ($0°.4 \leqslant l \leqslant 1°.8$, $|b| \leqslant 0°.4$).**
J. Inatani.
Publ. Astron. Soc. Japan, Vol. 34, 515 - 528 (1982).
The distribution of molecular clouds in the galactic center region has been observed in the CO 115-GHz line emission. Main results are as follows: (1) A very massive molecular cloud has been found at $l = 1°.3$, $b = 0°.0$.(2) A plateau of CO emission has been found in $1°.0 \leqslant l \leqslant 1°.6$, $-0°.2 \leqslant b \leqslant 0°.2$. (3) A CO feature with a radial velocity of about 150 km s^{-1} has been found at all the observed longitudes.

131.290 **Hydrogen emission profiles at positive latitudes. II. On the warm H I component of the interstellar medium.**
K. Takakubo.
Sci. Rep. Tôhoku Univ., Eighth Ser., Vol. 3, 73 - 81 (1982).
The warm interstellar gas is investigated with Gaussian components of H I profiles. The column density of the warm gas is well represented, at least at positive latitudes, by a plane parallel gas layer and, in this sense, the warm gas is pervasive also in the immediate vicinity of the sun (while the cool diffuse clouds are very few in the local hot bubble detected by soft X-ray measurements). A hypothesis is proposed that the warm gas is composed of clouds with large internal motion.

131.291 **Detection of bipolar CO outflow in Orion.**
N. R. Erickson, P. F. Goldsmith, R. L. Snell, R. L. Berson, G. R. Huguenin, B. L. Ulich, C. J. Lada.
Prepr. Steward Obs., No. 403, 15 pp. (1982).

131.292 **The high velocity molecular flows near young stellar objects.**
J. Bally, C. J. Lada.
Prepr. Steward Obs., No. 418, 63 pp. (1982).

131.293 **The discovery of new embedded sources in the centrally condensed core of the Rho Ophiuchi dark cloud.**
B. A. Wilking, C. J. Lada.
Prepr. Steward Obs., No. 419, 43 pp. (1982).

131.294 **Star formation in the ρ Ophiuchi dark cloud: the formation of a bound cluster?**
B. A. Wilking, C. J. Lada.
Prepr. Steward Obs., No. 420, 36 pp. (1982).

131.295 **The M17SW molecular cloud.**
H. A. Thronson, Jr., C. J. Lada.
Prepr. Steward Obs., No. 429, 24 pp. (1982). – Submitted to Astrophys. J.

131.296 **A photometric map of interstellar reddening within 300 parsecs.**
C. L. Perry, L. Johnston.
Astrophys. J., Suppl. Ser., Vol. 50, 451 - 515(1982) = Contrib. Louisiana State Univ. Obs., No. 169.
An observational program is described which has as its goal a high resolution, three-dimensional survey of the dust component of the interstellar medium in the solar neighborhood. A network of ~3450 bright A and F stars was established in the northern hemisphere with the intent that the stars act as probes for the detection of interstellar reddening. In conjunction with empirical calibrations by Crawford, *uvbyβ* photometries yielded precise intrinsic colors and absolute magnitudes that, in turn, allowed the determination of the color excess and the position of each star in a rectangular coordinate system centered on the Sun. The results are summarized in a series of plots where several dust concentrations are readily apparent. These regions correspond to the inner portions of the Ophiuchus, Cepheus, and Taurus cloud complexes.

131.297 **A numerical study of the effects of ambipolar diffusion on the collapse of magnetic gas clouds.**
D. C. Black, E. H. Scott.
Astrophys. J., Vol. 263, 696 - 715 (1982).
The authors have calculated numerically the gravitational collapse of isothermal, nonrotating magnetic gas clouds including the effects of ambipolar diffusion. The fractional ionization in the clouds is approximated by a power-law function of the gas density n, $f_i \sim n^{-q}$. The results indicate that the asymptotic character of collapse is determined mainly by the value of q and is largely independent of the other parameters characterizing a cloud. The magnetic field strength at the center of a collapsing cloud is strongly amplified during collapse even for values of $q \approx 1$, despite extremely low values of fractional ionization. Possible implications of these results for the problems of magnetic braking of rotating protostars are presented.

131.298 **Search for maser emission by water vapor in the ν_2 excited vibrational state.**
P. C. Myers, A. H. Barrett.
Astrophys. J., Vol. 263, 716 - 717 (1982).
A search was conducted for the $4_{2,3} \rightarrow 3_{3,0}$ transition of water vapor in the ν_2 excited vibrational state at 12,008.80 MHz. Sixteen known H_2O maser sources were searched to ΔT_A rms levels ranging from 0.014 K to 0.026 K without detecting any emission.

131.299 **On the n-dependence of the reaction rate for $C^+ + C_n \rightarrow C_{n+1}^+$ in interstellar space.**
K. F. Freed, T. Oka, H. Suzuki.
Astrophys. J., Vol. 263, 718 - 722 (1982).
The rate of the radiative association reaction $C^+ + C_n \rightarrow C_{n+1}^+ + h\nu$ is considered under the conditions in interstellar space. It is shown that this reaction proceeds efficiently with the Langevin rate for $n \geqslant 4$. Thus, the carbon chain can grow very quickly once it reaches C_4. This may explain the observed abundance of long carbon chain molecules in interstellar space.

131.300 **The return of mass and energy to the interstellar medium by winds from early-type stars.**
D. C. Abbott.
Astrophys. J., Vol. 263, 723 - 735 (1982).
The kinetic energy of the stellar winds from a complete sample of B and A supergiants, O stars, and Wolf-Rayet stars has been either measured or estimated from well-defined calibrations. Within a distance of 3 kpc from the sun, these stars are currently returning mass and energy to the interstellar medium (ISM) at an average rate of $9 \times 10^{-5} M_\odot yr^{-1} kpc^{-2}$ and 2×10^{38} ergs $s^{-1} kpc^{-2}$. The input of wind energy from Wolf-Rayet stars greatly exceeds the wind input from stars of other spectral types. Stellar radiation completely dominates the heating rate of H II regions, supernovae are the major source of energy for the hot intercloud gas and the kinetic energy of the cloud motions, and stellar radiation and supernovae are of equal importance for heating neutral clouds.

131.301 **Detection of $HC^{17}O^+$ in Sagittarius B 2.**
M. Guélin, J. Cernicharo, R. A. Linke.
Astrophys. J., Lett., Vol. 263, L89 - L93 (1982).
Line emission at 87.057 GHz attributed to the $J = 1 \rightarrow 0$ rotational transition of the formyl ion $HC^{17}O^+$ has been detected in Sgr B2 (OH). A $HC^{18}O^+/HC^{17}O^+$ abundance ratio of (3.1 ± 0.6) is derived from the observed line ratio.

131.302 **Schritte zur mineralogischen Fernerkundung des interstellaren Staubes.**
J. Dorschner, J. Gürtler.
Sterne, 58. Band, 365 - 375 (1982).

131.303 **Structure and evolution of dynamical protostellar envelopes in the process of massive star formation.**
I. G. Kolesnik, S. G. Kravchuk.
Inst. teor. fiz. AN USSR. Prepr., 1982, No. 85, 39 pp. In Russian. – Abstr. in Ref. zh., 51. Astron., 11.51.472 (1982).

131.304 **From clouds to stars.**
B. G. Elmegreen.
Formation of planetary systems, (see 012.064), p. 61 - 174 (1982).
Contents: The interstellar medium. Cloud formation. Internal properties of star-forming clouds. Magnetic fields. Collapse to star formation. Young stars and their environments. The evolution of viscous disks. Two models for the formation of the solar system.

131.305 **Condensation of grains.**
J. M. Lattimer.
Formation of planetary systems, (see 012.064), p. 189 - 282 (1982).
These lectures concentrate on the physics and chemistry of the nucleation and growth of grains. The general theory of grain nucleation is explored, and an approximate analytical treatment is developed. The latter is applied to several astrophysical environments, such as novae, supernovae, red giants, planetary nebulae and the proto-solar nebula. The calculation of chemical equilibrium and disequilibrium condensation is explored and applied to the origin of meteorites, especially with regard to the question of isotopic anomalies. The relation of these problems to the origin of the solar system is discussed.

131.306 **Interstellar extinction and space densities of stars in the direction of the open cluster IC 4996 in Cygnus.**
A. Pučinskas.
Bull. Vilnius Astron. Obs., No. 61, p. 3 - 21 (1982). In Russian.
888 stars measured in the Vilnius photometric system are used to investigate interstellar extinction and distribution of stars in the area of 0.8 square degrees in the direction of the open cluster IC 4996.

131.307 **On the column density of interstellar gas along the line of sight to ζ Oph determined from integral optical depths.**
I. Pustyl'nik.
Tartu Astrofüüs Obs. Publ., Köide 49, 97 - 132 (1982). In Russian.
The author has used published data of Morton (1975) on 164 interstellar atomic lines and has redetermined the abundances of species along the line of sight to ζ Oph, applying both the integral optical depth and the curve of growth technique.

131.308 **Polarization of early-type stars in Norma.**
S. Waldhausen, H. G. Marraco.
Astron. J., Vol. 87, 1730 - 1738 (1982).
Linear polarization measurements of a number of heavily reddened early-type stars in the Norma region are shown. The wavelength dependence of the polarization has been found to be in accord with that of the general interstellar field. The ratio of total to selective interstellar extinction has been found to have a mean value of $R = 3.30 \pm 0.11$. The ratios of polarization to extinction found in this region are concentrated towards the maximum value of 3% per mag, the maximum for the interstellar medium, thus implying a very efficient polarization mechanism.

131.309 **A photometric map of interstellar reddening within 100 pc.**
C. L. Perry, L. Johnston, D. L. Crawford.
Astron. J., Vol. 87, 1751 - 1774 (1982) = Contrib. Louisiana State Univ. Obs., No. 173.
Color excesses and distances are calculated for 300 bright, northern, late F stars using uvbyβ photometric indices. The data allow an extension of the earlier maps by Perry and Johnston of the spatial distribution of interstellar reddening into the local ($r < 100$ pc) solar neighborhood. Some definite conclusions are made regarding the distribution of interstellar dust in the northern hemisphere and within 300 pc of the sun by merging these results and the polarimetric observations by Tinbergen for 180 stars within 35 pc of the sun.

131.310 **An accurate position for the 6-cm OH masers in W3.**
L. J. Rickard, D. A. Crocker, P. F. Bowers, S. R. Spangler.
Astron. J., Vol. 87, 1806 - 1809 (1982).
The authors have used the VLA as a narrowband spectrophotometer to verify that the 6-cm emission from rotationally excited OH in W3OH is maser emission, and to measure its position. The 4765-MHz emission, like the ground-state emission, arises from gas in front of the compact H II region, apparently falling into it, and is not associated with the nearby H_2O maser complex.

131.311 **Trous géants dans la distribution spatiale de l'hydrogène atomique associé aux galaxies.**
F. Viallefond.
Mécanismes de production d'énergie dans le milieu interstellaire, (see 012.068), p. 55 - 59 (1981).

131.312 **Observations du nuage moléculaire associé a Sh-2 152.**
C. Kahane, R. Lucas.
Mécanismes de production d'énergie dans le milieu interstellaire, (see 012.068), p. 75 - 80 (1981).

131.313 **Dynamique des sphères célestes.**
B. Lazareff.
Mécanismes de production d'énergie dans le milieu interstellaire, (see 012.068), p. 141 - 173 (1981).
Une revue des modèles de structure et d'évolution des nébuleuses plus ou moins sphériques est présentée. L'exposé traite de trois types d'objets, du plus classique au plus "moderne": régions H II, restes de supernovae et bulles de vents stellaires.

131.314 **The interstellar medium and its energy sources.**
G. Tenorio-Tagle, H. W. Yorke, P. Bodenheimer.
Mécanismes de production d'énergie dans le milieu interstellaire, (see 012.068), p. 181 - 199 (1981).
The paper is a review of the current ideas on the interstellar medium. The aim is to study in detail the effects produced by its energy sources, i. e. the evolution of H II regions, supernova remnants and stellar wind bubbles and to evaluate their degree of importance.

131.315 **Influence des supernovae et des nuages H I sur les propriétés du milieu interstellaire.**
J. P. Chieze.
Mécanismes de production d'énergie dans le milieu interstellaire, (see 012.068), p. 201 - 214 (1981).

131.316 **La composante chaude du milieu interstellaire local: dimensions, températures, abondances.**
M. Arnaud.
Mécanismes de production d'énergie dans le milieu interstellaire, (see 012.068), p. 267 - 273 (1981).

131.317 **Dense cloud chemistry. III. The detailed chemical network and the effects of parameter variation.**
T. J. Millar.
Astrophys. Space Sci., Vol. 87, 435 - 453 (1982).
The author presents the detailed reaction network used in Paper I of this series and discusses the sensitivity of the

calculated abundances of sulphur-bearing species to various parameters including reaction efficiency and elemental depletions. It is shown that while ion-grain reactions can provide appreciable abundances of many S-bearing species, they do not contribute significantly to the abundance of O-bearing molecules. Observations of HDS in interstellar clouds could provide important information on the formation of H_2S.

131.318 **Ultraviolet extinction towards far OB associations.**
S. Aiello, B. Barsella, I. Guidi, U. Penco, M. Perinotto.
Astrophys. Space Sci., Vol. 87, 463 - 476 (1982).
Stars in the Aur OB-2 and Sco OB-1 associations, located in the direction of the galactic anticenter and center, respectively, have been observed for interstellar extinction. No effects due to changes of grain composition across such large distances are seen. A star exhibiting a peculiar extinction is found in each association. The properties of the responsible dust grains are discussed in terms of a two-component dust mixture: graphite and uncoated silicates. The column density both of graphite and silicate atoms are found to be higher than in stars of the same associations exhibiting a regular UV extinction.

131.319 **Polarization of starlight in R-associations. Observational data.**
N. V. Voshchinnikov, P. E. Marchenko.
Astron. Zh., Tom 59, 1115 - 1123 (1982). In Russian. English translation in Soviet Astron., Vol. 26, No. 6.
Polarimetric data are collected for 95 stars illuminating reflection nebulae. Most of the stars are members of 18 known R-associations (groups of stars illuminating reflection nebulae). Using the observed wavelength dependence of polarization and the dependence of the degree of polarization on the colour excess E_{B-V} the authors conclude that polarization in R-associations probably arises when nonpolarized starlight passes through a medium of dust grains aligned by a magnetic field.

131.320 **The H_2O maser outburst in Orion A.**
L. I. Matveenko, J. M. Moran, R. Genzel.
Pis'ma Astron. Zh., Tom 8, 711 - 713 (1982). In Russian. English translation in Soviet Astron. Lett., Vol. 8.
The outburst of H_2O maser emission in Orion A has been studied in November 16, 1979. Its position is $\alpha = 05^h 32^m 46.^s6 \pm 0.^s06$, $\delta = -05°24'28''.7 \pm 1''$ for epoch 1950.0 and the radial velocity 8 km/s. The line profile is asymmetric and its width is 28 kHz.

131.321 **A possible explanation of the variation of the scattering index parameter.**
C. E. Akujor, P. N. Okeke.
Astrophys. Space Sci., Vol. 88, 487 - 491 (1982).
The variation of the scattering index parameter (ϵ) with galactic latitude and longitude is compared with the absorption measurements as well as electron density variations in the interstellar medium (ISM). The authors thus extend an earlier work (Okeke, 1981) by showing that the existence of a three-phase structure for the ISM could be responsible for the random variation of ϵ, which possibly is a good indicator of the nature of the ISM.

131.322 **Identification of some diffuse interstellar features.**
W. W. Duley.
Astrophys. Space Sci., Vol. 88, 501 - 505 (1982).
Twenty-eight of the thirty-nine diffuse interstellar bands identified by Herbig (1975) are shown to constitute three vibronic systems with origins at 14321, 15153, and 15343 cm^{-1} (vac). Structure within these three systems arises from the excitation of vibrational modes with $\nu_1 = 275$ cm^{-1}, $\nu_2 = 445.5$ cm^{-1}, and $\nu_3 = 793$ cm^{-1}. The electronic origins at 14321 and 15343 cm^{-1} correspond to narrow lines observed in the spectrum of Cr^{3+} ions at cubic sites in MgO solids while the 15153 cm^{-1} origin arises in Mn^{4+}: MgO. Hence, many of the diffuse bands in the visible likely are due to small MgO particles containing these ions. This observation is compatible with recent experimental data showing broad bands at 160 nm and 220 nm from finely divided MgO solids that match features in the interstellar extinction curve.

131.323 **Molecular gas near the galactic center – recent contributions from radioastronomy.**
R. Güsten.
The galactic center, (see 012.076), p. 9 - 11 (1982).
Parts of the inner 15 arc min of the galactic nucleus have been mapped in the two lowest inversion lines of ammonia (NH_3(J,K), J = K = 1,2), and in the $2_{11}-2_{12}$ transition of formaldehyde. Results of a sensitive search for water masers in the center region are reported and recent detections of "high-velocity" gas are summarized.

131.324 **The relative locations of the Sgr A molecular clouds and continuum sources.**
Å. Sandqvist.
The galactic center, (see 012.076), p. 12 - 17 (1982).
Emission in the 2 mm H_2CO line has been observed towards Sgr A West. The emission comes from a molecular belt near the galactic center which has concentrations at velocities of +50 and +15 km s^{-1} and which displays a large velocity gradient in the same sense as that of galactic rotation. Evidence is presented that the continuum radio source Sgr A East lies on the near side of the molecular region, while Sgr A West lies on the far side.

131.325 **Depression of molecular emission in the line of sight of Sgr A West.**
Y. Fukui, H. Ogawa, S. Deguchi, H. Suzuki.
The galactic center, (see 012.076), p. 18 - 24 (1982).
The galactic center region has been mapped in the 4-mm emission lines of HCCCN and H_2CO with a 1.5 arc min beam. The molecular lines are found to show depression in intensity in the line of sight of Sgr A West. Comparison with other molecular data indicates that NH_3 also shows a significant depression. The authors suggest that the depression means abnormally reduced abundance in HCCCN, NH_3, and H_2CO in the line of sight of Sgr A West.

131.326 **On the triggers for star formation in the Orion region.**
S. Isobe.
Astron. Her., Vol. 75, 216 - 220 (1982). In Japanese. – Abstr. in Phys. Abstr., Vol. 86, Abstr. 5271 (1983).

131.327 **A turbulent model for molecular clouds.**
F. Ferrini, F. Marchesoni, A. Vulpiani.
Phys. Lett. A, Vol. 92A, 47 - 50 (1982). – Abstr. in Phys. Abstr., Vol. 86, Abstr. 5401 (1983).

131.328 **Physical characteristics of astronomical masers.**
M. Elitzur.
Rev. Mod. Phys., Vol. 54, 1225 - 1260 (1982). – Abstr. in Phys. Abstr., Vol. 86, Abstr. 10434 (1983).

131.329 **The chemistry of interstellar molecules.**
G. Winnewisser.
Cosmo- and geochemistry, (see 003.015), p. 39 - 71 (1981).
Contents: Introduction. Astrophysical scenario. General aspects of interstellar chemistry. Chemistry in molecular clouds. Formation of complex molecules and carbon chemistry. Molecules in comets and planetary atmospheres. Conclusion.

Catalog of CO radial velocities toward galactic H II regions.
See Abstr. 002.004.

On the nomenclature of diffuse fragmentary objects.
See Abstr. 002.018.

Ionization energies of some molecules found in interstellar clouds calculated by a Green's function method.
See Abstr. 002.140.

Landolt-Börnstein. Group VI. Vol. 2c. Interstellar matter – Galaxy – Universe.
See Abstr. 003.055.

Bau und Physik der Galaxis.
See Abstr. 003.108.

The scientific significance of submillimetre astronomy.
See Abstr. 013.063.

Experiments and calculations on the extinction of starlight by iron or magnetite grains.
See Abstr. 022.001.

The far-infrared spectrum of the OH radical.
See Abstr. 022.003.

High-energy collision-induced dissociation of H_2 by H.
See Abstr. 022.004.

Laboratory simulation of interstellar chemical evolution.
See Abstr. 022.006.

Probability of observation of cosmic OH sources with linear and circular polarization.
See Abstr. 022.016.

Forbidden emission lines of Fe VII.
See Abstr. 022.037.

NH^+ – a candidate for comets and interstellar space.
See Abstr. 022.039.

An experimental determination of the cross section for photodesorption.
See Abstr. 022.047.

The mass distribution of an accreting system of particles – an analytic approach.
See Abstr. 022.053.

Millimeter and submillimeter laboratory spectroscopy: recent results of astrophysical interest.
See Abstr. 022.112.

Reaction products from a microwave discharge in N_2 and H_2S. The microwave spectrum of NS.
See Abstr. 022.118.

Far infrared fine structure lines in the interstellar medium.
See Abstr. 022.119.

Formaldehyde formation in a H_2O/CO_2 ice mixture under irradiation by fast ions.
See Abstr. 022.126.

Some H/D exchange reactions involved in the deuteration of interstellar molecules.
See Abstr. 022.130.

Algunos aspectos sobre la interpretacion de las observaciones radioastronómicas de la molécula OH.
See Abstr. 022.149.

Unusually long bonds in CCO^+ and COC^+.
See Abstr. 022.179.

Photodissociation processes in the HCl molecule.
See Abstr. 022.185.

Mobilities of CH_n^+ (n = 0 to 3) in He and of CH_3^+ in Ar.
See Abstr. 022.194.

Microwave spectra of molecules of astrophysical interest. XXI. Ethanol (C_2H_5OH) and propionitrile (C_2H_5CN).
See Abstr. 022.195.

Application of a computer controlled intensity registering procedure to the spectral region at 4430 Å.
See Abstr. 031.573.

Interstellar lines in high resolution IUE spectra: paper I. Groningen data reduction package and technical results.
See Abstr. 031.586.

18-25 GHz low-noise line receiver for the Tidbinbilla 64-m antenna.
See Abstr. 033.006.

Submillimeter observations using the MMT as a phased array.
See Abstr. 033.012.

New large radio telescope in Japan.
See Abstr. 033.040.

Linear analysis of an oscillatory instability of radiative shock waves.
See Abstr. 062.077.

Zur Berechnung von Astropausenflächen.
See Abstr. 062.094.

Processus physiques dans les sphères célestes et le milieu interstellaire.
See Abstr. 062.102.

The influence of dielectronic-like recombination at low temperatures on the interpretation of interstellar, radio recombination lines of carbon.
See Abstr. 063.030.

Pumping of H II/OH masers: IR line overlaps and collisional excitation by H_2.
See Abstr. 063.042.

On the difference between the initial mass function of single stars and of primaries of binaries.
See Abstr. 065.082.

On the screening of the interstellar electric and magnetic fields by the solar wind.
See Abstr. 074.051.

Isotope ratios from non-solar system projectiles.
See Abstr. 105.080.

Models of the formation of the solar nebula.
See Abstr. 107.025.

On the evolution of convective accretion disk models of the primordial solar nebula.
See Abstr. 107.029.

Hydrodynamical models of presolar nebula formation.
See Abstr. 107.052.

On the properties of the circumstellar matter around the bright young variable shell star HR 5999.
See Abstr. 112.015.

Interpretation of the VLBI maps of maser spots.
See Abstr. 112.016.

Long-term monitoring of SiO maser time variations in four transitions.
See Abstr. 112.021.

M1-67: a wind-blown bubble carried along by the high-velocity WR star 209 BAC?
See Abstr. 112.045.

A photoelectric *UBV* sequence in a low extinction Puppis field.
See Abstr. 113.002.

Reddening relations of the *VBLUW* and *UBV* systems for objects with emission line spectra.
See Abstr. 113.009.

The intrinsic colours of stars in the ultraviolet.
See Abstr. 113.044.

Determination of temperatures, metallicities, and interstellar reddening for G–K giants and subgiants in the Vilnius photometric system.
See Abstr. 113.067.

New Hα emission stars in galactic dark clouds.
See Abstr. 114.004.

Optical observations of ultraviolet objects – I. Spectral classification of 103 stars (l = 200° to 275°).
See Abstr. 114.008.

M supergiants and star formation at the galactic center.
See Abstr. 114.203.

A hypothesis on the richness of a galaxy in binary stellar systems.
See Abstr. 117.132.

Structure and evolution of large groups of massive close binary systems.
See Abstr. 117.217.

Star formation in the λ Orionis region. I. The distribution of young objects.
See Abstr. 121.009.

A search for high-velocity molecular gas around T Tauri stars.
See Abstr. 121.010.

Interactions between pre-main-sequence objects and molecular clouds. I. Elias 1 – 12.
See Abstr. 121.016.

Some recent results in early stellar evolution.
See Abstr. 121.021.

Radio emission from supernova remnants in a cloudy interstellar medium.
See Abstr. 125.027.

The evolution of supernova remnants and the structure of the interstellar medium.
See Abstr. 125.055.

The Monoceros Loop: a supernova remnant interacting with the Rosette Nebula?
See Abstr. 125.071.

Acoustic waves in supernova remnants.
See Abstr. 125.079.

The structure and dynamics of a 3.5 pc diameter filamentary shell on the edge of the NGC 6334 interstellar complex.
See Abstr. 132.006.

The H II region – molecular cloud complex Sh 2–269: an optical and millimeter wavelength study.
See Abstr. 132.008.

Magnetic field measurements in two expanding H I shells.
See Abstr. 132.013.

Measurements of magnetic field strengths in the vicinity of Orion.
See Abstr. 132.014.

Radioastronomical observations of ionised gas.
See Abstr. 132.031.

New infrared objects towards southern type I OH and H_2O masers.
See Abstr. 133.001.

The remarkable 400 micron source NGC 6334/I(North).
See Abstr. 133.002.

Near-infrared sources in the NGC 6334 molecular cloud.
See Abstr. 133.003.

Near-infrared spectrophotometry of polarized compact infrared sources.
See Abstr. 133.004.

The energetic molecular outflow near AFGL 961: millimeter-wave and infrared observations.
See Abstr. 133.007.

Spectroscopy of molecular cloud sources at 6 - 7 microns.
See Abstr. 133.008.

Discovery of a compact far infrared source in the globule B335.
See Abstr. 133.010.

Far-infrared sources in Cygnus X: an extended emission complex at DR 21 and unresolved sources at S 106 and ON 2.
See Abstr. 133.015.

A new near-infrared source in the molecular cloud associated with S106.
See Abstr. 133.017.

Near infrared spectroscopy of W51 IRS-2.
See Abstr. 133.018.

New members of the infrared cluster in the Orion Molecular Cloud.
See Abstr. 133.021.

The scattering phase function of interstellar grains: the case of the reflection nebula NGC 7023.
See Abstr. 134.022.

Observations of the Orion Nebula in Hα, [N II] and Hβ lines.
See Abstr. 134.035.

A catalogue of radio sources within 30′ of Cep A.
See Abstr. 141.015.

Causal relationship between pulsar long-term intensity variations and the interstellar medium.
See Abstr. 141.516.

Dynamic spectra of pulsar scintillations at frequencies near 0.34, 0.41, 0.63, 1.4, 1.7, 3.2 and 5.0 GHz.
See Abstr. 141.544.

X-ray emission from extragalactic bursts of star formation.
See Abstr. 142.009.

Observations of Cygnus X-3 in the *I* band.
See Abstr. 142.102.

The interrelation of cosmic γ-rays and interstellar gas in the range *l*: 65–180°.
See Abstr. 142.523.

Discrete sources of cosmic gamma rays.
See Abstr. 142.543.

Sources de rayonnement γ, origine du rayonnement cosmique, et régions de formation d'étoiles.
See Abstr. 142.549.

Confinement of cosmic rays in molecular clouds.
See Abstr. 143.012.

The acceleration of cosmic rays by shock waves.
See Abstr. 143.026.

On the transport and propagation of cosmic rays in galaxies – I. Solution of the steady-state transport equation for cosmic ray nucleons, momentum spectra and heating of the interstellar medium.
See Abstr. 143.056.

Star formation in protogalactic gas cloud collisions.
See Abstr. 151.006.

The structure and evolution of galacto-detonation waves: some analytic results in sequential star formation models of spiral galaxies.
See Abstr. 151.020.

The role of anisotropic propagation of star formation on the large scale structure of galaxies.
See Abstr. 151.025.

Three dimensional computer simulations of star formation in dwarf galaxies.
See Abstr. 151.028.

The importance of tidal disruption on star formation in spiral galaxies.
See Abstr. 151.029.

The evolution of galaxies: expectations and observations.
See Abstr. 151.066.

Dust cloud structure in young R associations.
See Abstr. 152.002.

On the group formation of stars in stellar associations.
See Abstr. 152.004.

The age spread and initial mass function of NGC 3293: implications for star formation in clusters.
See Abstr. 153.011.

Berkeley 87, a heavily-obscured young cluster associated with the ON2 star-formation complex and containing the WO star Stephenson 3.
See Abstr. 153.036.

Implications of collisionally supported giant molecular clouds for spiral galactic structure and massive star formation.
See Abstr. 155.002.

On a model of local gas related to Gould's belt.
See Abstr. 155.008.

Atomic hydrogen in the outer Milky Way.
See Abstr. 155.014.

Neutral hydrogen related to Gould's Belt.
See Abstr. 155.024.

The galactic abundance gradient from supernova remnant observations.
See Abstr. 155.043.

The distribution of neutral atomic hydrogen in our Galaxy beyond the solar circle.
See Abstr. 155.045.

Radio studies of galactic structure.
See Abstr. 155.053.

The effect of possible errors in light extinction on the determination of the parameters of the spiral structure of the Galaxy.
See Abstr. 155.059.

On the distribution of interstellar gas in the galactic halo.
See Abstr. 155.060.

O I and O III in Sgr A: neutral and ionized gas at the galactic center.
See Abstr. 155.068.

Identification of active star formation regions in the galactic plane.
See Abstr. 156.005.

Heliospheric and galactic magnetic field similarities–a cautionary note.
See Abstr. 156.006.

Infrared observations of the galactic center.
See Abstr. 156.017.

Large beam observations of the galactic center at 150, 200, and 300 μm.
See Abstr. 156.019.

Far-ultraviolet studies. VI. Further limits on diffuse galactic light scattered at large angles by dust.
See Abstr. 157.001.

A search for carbon monoxide in clumpy irregular galaxies.
See Abstr. 158.008.

Extragalactic CO: gas distributions which follow the light in IC 342 and NGC 6946.
See Abstr. 158.011.

Recent star-forming activity in the clumpy irregular galaxy NGC 7673.
See Abstr. 158.043.

Neutral hydrogen in M 31. III. The warping of the H I disk.
See Abstr. 158.051.

Neutral hydrogen in M 31. IV. Comparison with optical data and the star formation rate.
See Abstr. 158.052.

Temperatures and sizescales of giant cloud complexes in the spiral galaxy IC342.
See Abstr. 158.079.

Accurate optical positions of M 82 knots.
See Abstr. 158.084.

Local properties of star-forming regions in irregular galaxies.
See Abstr. 158.097.

Interstellar lines in the spectra of the globular clusters around NGC 5128.
See Abstr. 158.106.

The molecular cloud ring in the nucleus of NGC 1068.
See Abstr. 158.126.

Star formation rates and initial mass functions in disk galaxies.
See Abstr. 158.131.

On the correlation of CO line and radio continuum emission in nearby galaxies – the star-formation efficiency.
See Abstr. 158.182.

CO observations of M 51 and M 82.
See Abstr. 158.208.

CO along the minor axis of M 31.
See Abstr. 158.209.

The distribution of carbon monoxide in spiral and irregular galaxies.
See Abstr. 158.220.

The distribution of carbon monoxide in three face-on galaxies.
See Abstr. 158.222.

Infrared emission and star formation in NGC 5253.
See Abstr. 158.245.

The giant spiral galaxy M 101. VIII. Star formation in H I - H II associations.
See Abstr. 158.249.

The distribution of infrared obscuration in NGC 7331: evidence for a massive molecular ring.
See Abstr. 158.258.

Radio continuum emission: a tracer for star formation.
See Abstr. 158.260.

On the inner ring of H II regions in NGC 3351.
See Abstr. 158.274.

Neutral hydrogen in two extremely isolated galaxies.
See Abstr. 158.286.

Chemical abundances in the interstellar medium of galaxies from spectrophotometry of supernova remnants.
See Abstr. 158.293.

Past star formation in the disks of galaxies.
See Abstr. 158.298.

Star formation in the Magellanic Clouds – II. Discovery of a protostar in the Small Magellanic Cloud.
See Abstr. 159.002.

H I in the Small Magellanic Cloud re-examined.
See Abstr. 159.003.

H I in the Small Magellanic Cloud re-examined – II.
See Abstr. 159.005.

Abundances in the Magellanic Stream.
See Abstr. 159.014.

Observations of $^{12}CO(J = 2{-}1)$ emission in the Large and Small Magellanic Clouds.
See Abstr. 159.016.

Galactic deuterium as a test for cosmological models.
See Abstr. 162.004.

Pregalactic star formation and spectral distortions of the cosmic blackbody radiation.
See Abstr. 162.051.

Cosmic far-ultraviolet background radiation.
See Abstr. 162.188.

Errata

131.901 **Erratum: 'Interstellar extinction in the dark Taurus clouds. I.'** [Acta Astron., Vol. 30, 541 - 552 (1980)].
V. Straižys, E. Meištas.
Acta Astron., Vol. 31, 529 (1981). – See Abstr. 29.131.210.

131.902 **Erratum: 'High velocity molecular emission in Orion: a case for stellar winds"** [Astrophys. J., Vol. 251, 88 - 102 (1981)].
T. B. H. Kuiper, B. Zuckerman, E. N. Rodriguez Kuiper.
Astrophys. J., Vol. 261, 402 (1982). – See Abstr. 30.131.166.

131.903 **Erratum: 'Isotope-selective photodestruction of carbon monoxide'** [Astrophys. J., Vol. 255, 143 - 148 (1982)].
J. Bally, W. D. Langer.
Astrophys. J., Vol. 261, 747 (1982). – See Abstr. 31.131.052.

132 H I, H II Regions

132.001 **Observations of [O I], [O II], and [O III] lines in M 42 and M 43.**
C. T. Hua, R. Louise.
Publ. Astron. Soc. Pacific, Vol. 94, 453 - 458 (1982).

Monochromatic photographs of the Orion nebula in [O I], [O II], and [O III] lines were obtained at the 80-cm (*f*/6) and 120-cm (*f*/1.6) telescopes of the Haute Provence Observatory, combined with narrow-bandpass interference filters. Isothermal contours, as well as ratio contours of [O III]/[O II] and [O I]/[O II] are derived. Discussions are made about the observed temperature and line ratios.

132.002 **Far-infrared mapping of the double-lobed H II region S 106.**
P. M. Harvey, I. Gatley, H. A. Thronson, Jr., M. W. Werner.
Astrophys. J., Vol. 258, 568 - 571 (1982).

The results of 50 and 100 μm photometry and mapping of S 106 with 35″ resolution and a search for a nearby H_2O maser source at 2 and 10 μm are reported. The data are consistent with a distance of 900 pc and a spectral type of O9 (ZAMS) for a single central exciting star. A strong correlation of dust temperature with optical emission and anticorrelation of dust optical depth with optical emission are seen. These results support previously suggested models of S 106 as a single exciting star surrounded by a dusty disk. A local maximum in the dust column density was found to the west of the optical nebula near the position of an H_2O maser. No local near- or far-infrared flux maximum was found in this region, however.

132.003 **Excitation and extinction in the LMC H II region N159A and discovery of a highly excited "blob" in its vicinity.**
M. Heydari-Malayeri, G. Testor.
Astron. Astrophys., Vol. 111, L11 - L14 (1982).

The authors used high-resolution electronographic photographs taken with narrow-band filters to study the Large Magellanic Cloud HII region N159. They found a peculiar high-excitation compact object of size about 2 pc in this region. The object consists of two components separated by about 0.4 pc. The authors studied also N159A. Comparatively, this H II region is less excited, although its exciting star is suspected to be earlier than O5. The excitation and extinction vary in N159A. The most excited zones are the most reddened ones.

132.004 **The gas dynamics around OB associations. I. Recombining H II regions and the formation of expanding neutral shells.**
M. Beltrametti, G. Tenorio-Tagle, H. W. Yorke.
Astron. Astrophys., Vol. 112, 1 - 10 (1982).

The authors study the evolution of an H II region as the exciting stars move off the main sequence one by one. In all cases considered a shell of expanding ($u \lesssim 10\,\mathrm{km\,s^{-1}}$) neutral gas results. For the cases of an evolving OB-association an expanding ring-like nebula lasting some 5×10^6 yr is generated.

132.005 **The gas dynamics around OB associations. II. Dependence on the initial ambient density.**
G. Tenorio-Tagle, M. Beltrametti, P. Bodenheimer, H. W. Yorke.
Astron. Astrophys., Vol. 112, 104 - 110 (1982).

The authors present one and two dimensional calculations of the gasdynamical evolution of H II regions excited by an evolving OB association. The various phases (classical expansion, recombination, re-ionization and full recombination are studied and found to depend not only on the behaviour of the stellar photon flux, but also on the initial value of the ambient density. Many of the galactic and extragalactic ring-like nebulae and expanding neutral shells, usually interpreted as the result of supernova explosions or stellar winds, can be explained with the present calculations.

132.006 **The structure and dynamics of a 3.5 pc diameter filamentary shell on the edge of the NGC 6334 interstellar complex.**
J. Meaburn, N. J. White.
Mon. Not. R. Astron. Soc., Vol. 200, 771 - 784 (1982).

An unusual circular region of very filamentary nebulosity, 3.5 pc diameter, exists on the southern edge of the interstellar complex of NGC 6334. This surrounds an O6.5 star. Narrow-band, Schmidt photographs centred on the Hα and [N II] and [S II] lines, a Fabry-Perot interferogram of the [N II] line, three intermediate dispersion (10 Å $\mathrm{mm^{-1}}$) AAT spectra and VLA measurements have now been obtained of this object. Many components over a range in radial velocity of 150 km $\mathrm{s^{-1}}$ are found over the face of the nebula. As this nebula is a thermal radio source and the line ratios imply radiative ionization, energetic stellar winds and radiation pressure from Lyman photons are a possible source of these motions.

132.007 **CNO abundances in H II regions of the Magellanic Clouds and the Galaxy with implications regarding the nucleosynthesis of the CNO element group.**
R. J. Dufour, G. A. Shields.
Advances in ultraviolet astronomy, (see 012.006), p. 385 - 388 (1982).

This report summarizes the results of IUE observations of the UV spectra of three H II regions in the Small Magellanic Cloud and four H II regions in the Large Magellanic Cloud. The authors present the final abundance results and concentrate their discussion on the implications that the results have on the nucleosynthesis of the CNO group elements and the chemical evolution of galaxies.

132.008 **The H II region – molecular cloud complex Sh 2–269: an optical and millimeter wavelength study.**
M. Heydari-Malayeri, G. Testor, A. Baudry, G. Lafon, J. de la Noë.
Astron. Astrophys., Vol. 113, 118 - 128 (1982).

The paper presents the results of a study on the galactic H II region Sh 2–269 and its associated molecular cloud which is a region of ongoing star formation. Several narrow-band photographs and spectra near Hα were obtained and analyzed quantitatively. The molecular cloud was observed in several molecular lines and mapped in ^{13}CO ($J = 1 \rightarrow 0$). A noteworthy aspect of the nebula is the presence of high velocity ionized gas associated with an H_2O/OH/IR source. This gas is ejected from the nebula with a blueshifted velocity of about 200 km $\mathrm{s^{-1}}$ with respect to the main optical lines of the H II region. Dust is concentrated mainly in a dark band dividing the nebula, as well as towards the most excited zones of the nebula. The combined observational data have been used to suggest a model for the H II region Sh 2–269.

132.009 **The kinematical structure of the bipolar nebula S106.**
J. Solf, U. Carsenty.
Astron. Astrophys., Vol. 113, 142 - 149 (1982).

The observed kinematical properties confirm the basic bipolar structure of S106 originating from a newly formed early-type star still embedded in a dusty disk.

132.010 **Observations of bipolar and compact H II regions.**
T. Neckel.
Messenger, No. 29, p. 15 - 17 (1982).

132.011 **The origin of the infrared [C I] emission: H II or H I regions?**
D. A. Cesarsky.
Astron. Astrophys., Vol. 113, L7 - L9 (1982).

The auroral and nebular lines of [C I] have been observed by Hippelein and Münch in the direction of Orion A. They find that the radial velocity of these lines is redshifted by some 8 km/sec with respect to the radial velocity of the H II region M 42. This is essentially the same shift shown by the radio observations of carbon recombination line emission. The author suggests that the infrared [C I] line emission originates in the same regions where the radio C I emission is formed. Since the temperature of this region is likely to be low, electron excitation of the infrared lines is not possible. The author discusses the feasibility of excitation by electronic recombination, and he obtains results which are in good agreement with those deduced from the analysis of radio recombination carbon emission.

132.012 **A survey of ionized helium in galactic H II regions.**
F. J. Lockman, R. L. Brown.
Astrophys. J., Vol. 259, 595 - 606 (1982).

The authors have made high-frequency (10.5 and 14.7 GHz) recombination-line observations of the helium-to-hydrogen ratio, y, in Sgr B2 and in seven other bright galactic H II regions. It is shown that the observed helium "anomalies" arise in a large part from difficulties in establishing a good instrumental baseline for the helium lines. Two effects, impact broadening and confusion by recombination lines from weak H II regions along the line of sight, can cause baselines to be systematically high, apparent helium abundances to be systematically low, and abundances to vary both with frequency and with distance from the galactic center.

132.013 **Magnetic field measurements in two expanding H I shells.**
T. H. Troland, C. Heiles.
Astrophys. J., Lett., Vol. 260, L19 - L22 (1982).

The authors report the first measurements of the magnetic field in expanding H I shells. At one position in each of two shells, the measured line-of-sight field components are both about 7 microgauss. The physical nature of one of these shells, located in Eridanus is discussed in some detail.

132.014 **Measurements of magnetic field strengths in the vicinity of Orion.**
C. Heiles, T. H. Troland.
Astrophys. J., Lett., Vol. 260, L23 - L26 (1982).

The authors have measured the line-of-sight magnetic field in two regions associated with the Orion Nebula. One region is the large H I cloud which envelops the ionized and molecular clouds; the field strength is 10 microgauss. The other is the large CO cloud just south of the Orion Nebula discovered by Kutner, Tucker, and Chin; the field strength here is also 10 microgauss. These results are discussed using the virial theorem.

132.015 **Abundances of N, S, and He, and relative stellar temperatures, in low-excitation nebulae.**
J. S. Mathis.
Astrophys. J., Vol. 261, 195 - 199 (1982).

The observed abundances of He^+, N^+, O^+, O^{++}, S^+, and S^{++}, relative to H^+, are related to the total N, S, and He abundances through theoretical models. The parameters used to determine the total abundances are observable ratios, rather than theoretical quantities. Application of the method to several regions of the Orion Nebula and to various H II regions in the nearby galaxies LMC, M 33 and M 101 are described.

132.016 **Electron densities in five H II regions from S III infrared fine-structure lines.**
T. Herter, D. A. Briotta, Jr., G. E. Gull, M. A. Shure, J. R. Houck.
Bull. American Astron. Soc., Vol. 14, 609 (1982). – Abstract.

132.017 **Far-infrared observations of NGC 1977.**
P. Makinen, P. M. Harvey, B. A. Wilking, N. J. Evans II.
Bull. American Astron. Soc., Vol. 14, 611 (1982). – Abstract.

132.018 **VLA observations of the bipolar H II region S 106.**
J. Bally, R. Predmore, R. L. Snell.
Bull. American Astron. Soc., Vol. 14, 628 (1982). – Abstract.

132.019 **An annular H I remnant.**
F. J. Lockman, B. L. Ganzel.
Bull. American Astron. Soc., Vol. 14, 637 (1982). – Abstract.

132.020 **The H 166α recombination line in the Carina Nebula.**
I. N. Azcárate, J. C. Cersósimo, F. R. Colomb.
Rev. Mexicana Astron. Astrofis., Vol. 6, (see 012.017), 269 - 272 (1981).

Observations of the Carina Nebula were made at the frequency corresponding to the H 166α recombination line. This line was observed with a velocity resolution of about 2 km s^{-1} at different positions within the Carina Region. The main characteristic of these profiles is the appearance of double structure in some of them, with velocities centered at about −10 and −30 km s^{-1}, respectively. The single profiles have central velocities between −20 and −25 km s^{-1}. Electron temperatures are derived, under the assumption of local thermodynamical equilibrium, for different points in the Nebula.

132.021 **Movimientos internos en regiones H II. La pequena nebulosa Sharpless 158.**
M. A. Moreno.
Rev. Mexicana Astron. Astrofis., Vol. 6, (see 012.017), 325 - 326 (1981).

The object S 158 (NGC 7538) is a small H II region that has a peculiar filamentary structure. Using Fabry-Pérot interferometry, radial velocities in many distinct points have been obtained. A discussion of the results is presented.

132.022 **2–4 μm spectroscopy of the compact H II region G 45.13 + 0.14 A.**
J. Krassner.
Astron. Astrophys., Vol. 114, 19 - 22 (1982).

The optically obscured compact H II region G 45.13 + 0.14 A has been observed from 2–4 μm both photometrically and spectroscopically. Spectral features observed include the Bα, Bγ, and Pfγ hydrogen recombination lines, the He I line at 2.06 μm, and possibly the unidentified 3.28 μm feature. The strength of the He I line implies that the nebula is ionization-bounded. The data confirm a previously described power law spectrum which increases toward longer wavelengths. A variety of reddening estimates indicate a visual extinction of 20 mag.

132.023 **On the statistics of galactic H I clouds.**
K. R. Anantharamaiah, V. Radhakrishnan, P. A. Shaver.
Bull. Astron. Soc. India, Vol. 10, 47 - 48 (1982). – Abstract.

132.024 **Radio recombination line observations with the Ooty radio telescope.**
K. R. Anantharamaiah.
Bull. Astron. Soc. India, Vol. 10, 50 (1982). – Abstract.

132.025 **Observations of the H II region W 51 at decameter wavelengths.**
A. Deshpande, C. V. Sastry.
Bull. Astron. Soc. India, Vol. 10, 50 - 51 (1982). – Abstract.

132.026 **Observations of bright rims in NGC 281, IC 434 and M 16.**
R. Louise.
Astrophys. Space Sci., Vol. 85, 405 - 417 (1982). In French.
Monochromatic plates are obtained in Hα, [N II] λ 6584 and [O III] λ 5007 lines for three H II regions: NGC 281, IC 434 and M 16. It is shown that bright rims appear most contrasted on [N II] and [O III] plates. Maps of ratio [N II]/Hα are derived from the method described by Louise and Sapin (1972). This ratio remains practically constant within the nebula, except for bright rims where it increases. This may be understood by means of both increasing abundance $N(N^+)/N(H^+)$ and temperature following to the model of Mallik (1975). The [O III] enhancement towards bright rims in M 16 suggests the presence of inhomogeneities and the possible role played by charge exchange mechanisms as pointed out by Aller and Keyes (1980).

132.027 **H I near the ionization boundary of the H II region surrounding LkHα101.**
P. E. Dewdney, R. S. Roger.
Bull. American Astron. Soc., Vol. 14, 777 (1982). – Abstract.

132.028 **Optical surveys of H II regions in galaxies.**
P. W. Hodge.
Astron. J., Vol. 87, 1341 - 1354 (1982).
This paper reviews the history and present status of optical surveys of H II regions in galaxies. A compilation of data on Hα surveys of 223 different galaxies in which 16 293 H II regions have been discovered is given, together with appropriate references.

132.029 **Abundances in five nearby galactic H II regions from infrared forbidden lines.**
T. Herter, H. L. Helfer, J. L. Pipher, D. A. Briotta, Jr., W. J. Forrest, J. R. Houck, R. J. Rudy, S. P. Willner.
Astrophys. J., Vol. 262, 153 - 163 (1982).
Airborne measurements of the [Ar II] (6.99 μm) and [S III] (18.71 μm) lines for five compact H II regions in the solar neighborhod are presented, as well as 2–4 μm and 8–13 μm spectroscopy where available. From these data and radio data the authors deduce lower limits to the elemental abundances of Ar, Ne, and S. Within the uncertainties, four of the objects are consistent with standard abundances. A Perseus arm object, S 156, is apparently overabundant in sulfur.

132.030 **Observations of the infrared fine-structure lines of S III at 18.71 and 33.47 microns in four H II regions.**
T. Herter, D. A. Briotta, Jr., G. E. Gull, M. A. Shure, J. R. Houck.
Astrophys. J., Vol. 262, 164 - 170 (1982).
The [S III] 18.71 μm line has been measured in four H II regions, S 158A, S 158G, G 75.84 + 0.4, and W 3 IRS 1. The [S III] 33.47 μm line has also been detected in all of these nebulae except S 158G where only an upper limit to the line flux was determined. From these data densities and S III abundances are estimated.

132.031 **Radioastronomical observations of ionised gas.**
L. F. Rodriguez.
Rev. Mexicana Fis., Vol. 28, 355 - 379 (1982). – Abstr. in Phys. Abstr., Vol. 85, Abstr. 101028 (1982).

132.032 **NGC 604 – a giant H II region dominated by many WR stars.**
M. Rosa, S. D'Odorico.
Wolf-Rayet stars: observations, physics, evolution, (see 012.042), p. 555 - 556 (1982).

132.033 **Wolf-Rayet stars associated to giant regions of star formation.**
S. D'Odorico, M. Rosa.
Wolf-Rayet stars: observations, physics, evolution, (see 012.042), p. 557 - 562 (1982).
Data on Wolf-Rayet (WR) stars in extragalactic H II regions and emission line galaxies are presented and discussed. Two important points appear to be established: (1) The WR stars are more numerous than the blue supergiants at least in same phase of the evolution of the stellar clusters which ionize the giant H II regions. (2) When the WR stars are detected, two cases are apparently observed, one in which only WN, the other in which both WN and WC are present.

132.034 **The central object of the 30 Doradus nebula, a supermassive star.**
J. P. Cassinelli, J. S. Mathis, B. D. Savage.
Wolf-Rayet stars: observations, physics, evolution, (see 012.042), p. 563 - 565 (1982). – Abstract.

132.035 **Absorption of radio-waves in the H II regions S54, S49 and S45.**
J. Milogradov-Turin.
Publ. Astron. Obs. Sarajevo, Vol. 1, No. 1, (see 012.043), p. 233 - 245 (1981).
The coincidence in the direction of the saddle-like feature near $l \approx 18°$ on the 38 MHz survey and of the complex of H II regions S54, S49 and S45 suggests that these nebulae are responsible for the appearance of this region on the 38 MHz survey of Milogradov-Turin and Smith (1973). The calculation of the reduction of antenna temperature based on data from Jones and Finlay (1974) gives a result which is consistent with this interpretation.

132.036 **The velocity structure of S 54 and S 142.**
P. G. Johnson, P. Mack, R. Songsathaporn.
Astrophys. Space Sci., Vol. 86, 331 - 340 (1982).
The velocity structures of two extended, low-density and low-temperature H II regions S 54 and S 142, have been examined. Both exhibit an ionized sheet structure.

132.037 **Aperture synthesis observations of recombination lines from compact H II regions. V. NGC 7538.**
W. M Goss, J. H. van Gorkom, J. R. Forster.
Astron. Astrophys., Vol. 115, 164 - 166 (1982).
The authors have mapped the NGC 7538 (S 158) complex in the H 110α line using the Westerbork Synthesis Radio Telescope at 4.8 GHz. The compact source IRS-2 has a velocity of -68 km s^{-1} which is distinctly different from both NGC 7538 (-61 km s^{-1}) and the nearby molecular velocities. IRS-2 may be another example of a compact H II region located within a collapsing neutral cloud.

132.038 **The H II regions of Messier 8.**
B. T. Lynds, E. J. O'Neil, Jr.
Astrophys. J., Vol. 263, 130 - 144, plates 6, 7 (1982).
Long slit image tube spectra have been used to measure ratios of nebular emission lines in M 8. Electron densities across the Hourglass region are derived. It is proposed that each of the three O stars of NGC 6530 has a surrounding H II region contributing to the observed emission. Stasinska's models are used to calculate the integrated brightnesses of emission lines and are shown to be in general agreement with the observations. Galactic positions of the M 8 components are proposed.

132.039 **VLA observations of H I near the ionization boundary of the H II region surrounding the early-type star Lk Hα 101.**
P. E. Dewdney, R. S. Roger.
J. R. Astron. Soc. Canada, Vol. 76, 318 (1982). – Abstract.

132.040 **Traitement informatise des interférogrammes de Fabry-Pérot: dynamique de la région H II S142.**
G. Joncas, J.-R. Roy.
J. R. Astron. Soc. Canada, Vol. 76, 320 - 321 (1982). Abstract.

132.041 **Dynamique des complexes H II extragalactiques: une étrange région H II géante dans NGC 4631.**
J.-R. Roy, R. Arsenault.
J. R. Astron. Soc. Canada, Vol. 76, 325 (1982). – Abstract.

132.042 **Infrared images of southern H II regions.**
J. W. V. Storey, J. Bailey.
Anglo-Australian Obs., Prepr. No. 170, 7 pp. (1982). – Submitted to Publ. Astron. Soc. Pacific.

132.043 **Giant H II complexes outside our galaxy at optical wavelengths.**
M. Rosa.
ESO Sci. Prepr., No. 213, 6 pp. (1982). – To appear in "Highlights of Astronomy", Vol. 6.

132.044 **A rediscussion of sulfur abundances in Magellanic Clouds and galactic H II regions.**
M. Dennefeld, G. Stasińska.
Pré-publ. Inst. Astrophys. Paris, No. 12, 36 pp. (1982). Accepted by Astron. Astrophys.

132.045 **Bulles interstellaires galactiques: propriétés physiques en relation avec le contenu stellaire.**
M. C. Lortet.
Mécanismes de production d'énergie dans le milieu interstellaire, (see 012.068), p. 21 - 27 (1981).

132.046 **Observations de bulles géantes du Grand Nuage de Magellan.**
A. Laval, Y. M. Georgelin.
Mécanismes de production d'énergie dans le milieu interstellaire, (see 012.068), p. 29 - 35 (1981).

132.047 **Etude optique d'une coquille de gaz ionisé supergéante dans le Grand Nuage de Magellan.**
A. Caulet, L. Deharveng, Y. M. Georgelin, Y. P. Georgelin.
Mécanismes de production d'énergie dans le milieu interstellaire, (see 012.068), p. 37 - 40 (1981).

132.048 **Etude cinématique du halo de la Nébuleuse de la Dorade.**
P. Cox, L. Deharveng.
Mécanismes de production d'énergie dans le milieu interstellaire, (see 012.068), p. 41 - 53 (1981).

132.049 **Processus d'ionisation et d'excitation dans les bulles interstellaires et interprétation des spectres d'émission.**
G. Stasinska.
Mécanismes de production d'énergie dans le milieu interstellaire, (see 012.068), p. 61 - 74 (1981).

132.050 **Vents stellaires et dynamiques des régions H II extragalactiques.**
J. Lequeux.
Mécanismes de production d'énergie dans le milieu interstellaire, (see 012.068), p. 175 - 179 (1981).

132.051 **A radial velocity survey of the H II region S 101.**
P. G. Johnson, P. Mack, R. Songsathaporn.
Astrophys. Space Sci., Vol. 87, 13 - 20 (1982).
A radial velocity survey has been conducted over the H II region S 101 using a pressure-scanned Fabry-Pérot polychromator/imager. The interesting velocity structure observed and the implications for this object are discussed.

132.052 **Surface photometry of the H II regions S254 - S257.**
S. Mizuno.
Astrophys. Space Sci., Vol. 87, 121 - 148 (1982).
The surface photometry of S254 - S257 has been carried out by means of a wide range image processing technique in the reduction system. The photographic plates in the Hα + [N II] and *V*-bands are taken with the Schmidt telescope. Three kinds of calibrated maps of the H II regions are drawn: (1) *E*-map in the (Hα + [N II] + continuum), (2) *V*-map in the continuum at *V*-band, (3) $(E-V)$-map in the (Hα + [N II]) line emission. The intensity profiles across the nebular centers were also obtained. Based on calibrated maps, the morphological structure and mass distribution of S255 and S257 are discussed. The location of observed nebulae on the $(m_{H\alpha} - m_v)$ diagram, where $m_{H\alpha}$ and m_v denote the surface brightness, expressed in the magnitude per square arcmin, is shown together with that of some other nebulae. Some arguments on the age sequence of observed H II regions are presented.

Catalog of CO radial velocities toward galactic H II regions.
See Abstr. 002.004.

On the nomenclature of diffuse fragmentary objects.
See Abstr. 002.018.

The new catalogue of optical H II regions.
See Abstr. 002.036.

Landolt-Börnstein. Group VI. Vol. 2c. Interstellar matter – Galaxy – Universe.
See Abstr. 003.055.

The influence of dielectronic-like recombination at low temperatures on the interpretation of interstellar, radio recombination lines of carbon.
See Abstr. 063.030.

The Wolf-Rayet stars in 30 Doradus.
See Abstr. 113.036.

Data on an unusual Wolf-Rayet star in the nearby galaxy IC 1613.
See Abstr. 114.110.

Spectra of the Wolf-Rayet stars in 30 Doradus.
See Abstr. 114.139.

Properties of a luminous, Wolf-Rayet type object in the core of the extragalactic H II region IC 132.
See Abstr. 114.175.

IUE high dispersion spectra of luminous stars in symmetric nebulae.
See Abstr. 114.202.

The two-photon continuum in Herbig-Haro objects.
See Abstr. 121.011.

The visible and ultraviolet continuum from a Herbig-Haro object in the core of M 16 (NGC 6611).
See Abstr. 121.028.

The Monoceros Loop: a supernova remnant interacting with the Rosette Nebula?
See Abstr. 125.071.

Formaldehyde absorption measurements of selected galactic molecular clouds.
See Abstr. 131.001.

CN line emission from the molecular clouds associated with S 106 and NGC 7538.
See Abstr. 131.002.

Distribution of H_2CO absorption at 4.8 GHz towards M 17.
See Abstr. 131.031.

The role of H II regions during star formation and chemical enrichment in globular clusters.
See Abstr. 131.062.

The violent interstellar medium associated with the Carina Nebula. I. The line of sight toward HD 93205.
See Abstr. 131.076.

Radiation-driven implosions in molecular clouds.
See Abstr. 131.077.

VLA synthesis of galactic H I absorption toward 3C 123.
See Abstr. 131.089.

Star formation along the northern edge of W 5.
See Abstr. 131.103.

MERLIN spectral line observations of W3OH.
See Abstr. 131.144.

Westerbork observations of 6-cm H_2CO in W3.
See Abstr. 131.146.

VLBI aperture synthesis observations of 1720-MHz OH in NGC 7538 (IRS1).
See Abstr. 131.149.

Detection of two new water vapor emission sources in the southern hemisphere.
See Abstr. 131.152.

Infrared mapping of the galactic plane. III. The large-scale mid-infrared structure of W 3, W 4, and W 5.
See Abstr. 131.154.

The violent interstellar medium associated with the Carina Nebula.
See Abstr. 131.183.

Submillimetre observations of atomic carbon.
See Abstr. 131.195.

Far infrared large scale map of the Orion region.
See Abstr. 131.215.

A random view of Cygnus X.
See Abstr. 131.224.

Temperatures and their variation within interstellar H I structures.
See Abstr. 131.238.

The effects of the coronal gas on the champagne phase.
See Abstr. 131.261.

An Effelsberg – Green Bank galactic H I absorption line survey. II. Results and interpretation.
See Abstr. 131.263.

VLBI measurements of the relative position of the 1665 MHz and 1667 MHz OH masers toward W 49N and NGC 6334N.
See Abstr. 131.269.

Physical conditions in H II/OH maser regions.
See Abstr. 131.272.

The M17SW molecular cloud.
See Abstr. 131.295.

Dynamique des sphères célestes.
See Abstr. 131.313.

The interstellar medium and its energy sources.
See Abstr. 131.314.

Influence des supernovae et des nuages H I sur les propriétés du milieu interstellaire.
See Abstr. 131.315.

Infrared mapping and photometry of W 40.
See Abstr. 133.012.

Far-infrared sources in Cygnus X: an extended emission complex at DR 21 and unresolved sources at S 106 and ON 2.
See Abstr. 133.015.

Bα and Ne II line spectroscopy in the vicinity of the galactic center source IRS 16.
See Abstr. 133.022.

Optical polarization in M 17.
See Abstr. 134.001.

The velocity field of S 308. The ring nebula around the WN 5 star HD 50896.
See Abstr. 134.024.

Ring nebulae associated with Of stars: statistics, classification, origin.
See Abstr. 134.033.

High resolution spectroscopy of selected galactic emission nebulae.
See Abstr. 134.034.

Sulfur abundances in gaseous nebulae.
See Abstr. 135.013.

Assoziationen und Emissionsschalen in der Großen Magellanschen Wolke.
See Abstr. 152.008.

Infrared observations of the ionized gas in the galactic center.
See Abstr. 155.066.

Spatial and spectral studies of the galactic center near 10 μm.
See Abstr. 155.067.

O I and O III in Sgr A: neutral and ionized gas at the galactic center.
See Abstr. 155.068.

WSRT observations of the H110α recombination line in the galactic centre.
See Abstr. 156.004.

A catalog of diffuse nebulae and emission-line stars in NGC 6822.
See Abstr. 158.009.

The distribution of H II regions in external galaxies. I.
See Abstr. 158.024.

The impact of IUE observations on our knowledge about galaxies and quasars.
See Abstr. 158.058.

The galaxy NGC 1566: distribution and kinematics of the ionized gas.
See Abstr. 158.169.

The giant spiral galaxy M 101. VIII. Star formation in H I - H II associations.
See Abstr. 158.249.

On the inner ring of H II regions in NGC 3351.
See Abstr. 158.274.

Seyfert galaxies, planetary nebulae, and H II regions.
See Abstr. 158.323.

Observations of $^{12}CO(J = 2-1)$ emission in the Large and Small Magellanic Clouds.
See Abstr. 159.016.

133 Infrared Sources

133.001 **New infrared objects towards southern type I OH and H_2O masers.**
M. A. Braz, N. Epchtein.
Astron. Astrophys., Vol. 111, 91 - 96 (1982).

A near infrared survey towards 26 non stellar OH and/or H_2O maser sources of the southern hemisphere provided the detection of 6 new steep spectrum objects in the vicinity of RCW 36, OH 338.88−0.08, H_2O components C and D in NGC 6334, OH/H_2O 351.78−0.54, and OH/H_2O 353.41−0.36. Among the ~ 125 maser positions now searched for infrared counterparts by various authors, about 30% have yielded in objects with $K–L \gtrsim 2.5$. Infrared sources statistically tend to be found closer to H_2O than to OH positions. Probable early type stars suffering interstellar reddening were found near other sources. Some of them are tentatively interpreted as the possible exciting star of a nearby H II region.

133.002 **The remarkable 400 micron source NGC 6334/I(North).**
D. Y. Gezari.
Astrophys. J., Lett., Vol. 259, L29 - L33 (1982).

Observations at 400 μm of the NGC 6334 complex have confirmed the discovery of an unusual 1000 $M_\odot$ object, NGC 6334/I(North), which was not detected by broad-band 100 μm survey observations. The results suggest that this cool source (T = 19 ± 5 K) is an active star formation region in a very early stage of evolution. The observed infrared luminosity of the source ($7 \times 10^3 L_\odot$) requires the presence of one or more embedded B stars or the formation of a dense protostellar core.

133.003 **Near-infrared sources in the NGC 6334 molecular cloud.**
P. Persi, M. Ferrari-Toniolo.
Astron. Astrophys., Vol. 112, 292 - 298 (1982).

The authors have searched for near-infrared emission from seven selected areas near the 6 cm radio continuum peaks and far-infrared peaks in NGC 6334. Eleven infrared sources were detected. The authors report the positions of the new detected IR sources and the relative photometry between 1 and 5 μ. They discuss each source individually, and by comparison with the observations at other wavelengths, a more complete scheme of sequential star formation in NGC 6334 is presented. Finally an hypothesis about the presence of low-mass protostars in the molecular cloud is also discussed.

133.004 **Near-infrared spectrophotometry of polarized compact infrared sources.**
R. R. Joyce, T. Simon.
Astrophys. J., Vol. 260, 604 - 611 (1982).

The authors have obtained 2–4 μm spectrophotometry of polarized compact infrared sources associated with dense molecular clouds. Deep ice absorption bands at 3.1 μm appear in the most highly polarized sources, whereas shallow absorption bands predominate in the weakly polarized objects. These observations suggest that the presence of ice grains may be a necessary condition for strong polarization in young OB stars embedded in molecular clouds. A model is discussed, where the observed infrared polarization is produced by magnetically aligned core-mantle grains having a weakly ferromagnetic metallic oxide core and an ice mantle.

133.005 **Broad helium emission in the galactic center.**
D. N. B. Hall, S. G. Kleinmann, N. Z. Scoville.
Astrophys. J., Lett., Vol. 260, L53 - L57 (1982).

An infrared (K band) spectrum taken with a 3″.8 diameter aperture centered on the galactic center source IRS 16 exhibits weak CO band absorption and emission by hydrogen Brackett-γ and a singlet helium line near 4857 cm^{-1}. The helium emission profile is much broader (~ 1500 km s^{-1}) than any feature seen previously at the galactic center. The broadening is probably due to Doppler motion of gas either flowing through an ionizing layer or orbiting about a massive object.

133.006 **Discovery of three far-red objects in CCD images of the galactic center.**
G. R. Ricker, M. W. Bautz, D. L. DePoy, S. S. Meyer.
Astrophys. J., Lett., Vol. 260, L59 - L63 (1982).

Observations of the galactic center with a CCD camera in the far-red (9030 - 9270 Å) have revealed three previously unknown objects which are designated FRS 1, FRS 2, and FRS 3. The three objects have magnitudes 18.7, 19.5, and 19.7, respectively, and lie within 10″ of the direction of Sgr A West. Two interpretations of these far-red objects are: (1) they are compact clusters of giant stars, earlier than M0 or (2) they are extremely compact H II regions observed in the forbidden emission of doubly ionized sulfur. Observational tests of these two hypotheses are suggested.

133.007 **The energetic molecular outflow near AFGL 961: millimeter-wave and infrared observations.**
C. J. Lada, T. N. Gautier III.
Astrophys. J., Vol. 261, 161 - 169 (1982).

The authors report detailed millimeter-wave and near-infrared spectroscopy of the dynamically active region around the infrared source AFGL 961, near the Rosette nebula. Millimeter-wave ^{12}CO observations are used to study the high-velocity molecular flow around AFGL 961. These observations show that the high-velocity flow has a maximum extent of at least 6′. The flow is found to be anisotropic, with redshifted high-velocity emission considerably more extended than blueshifted high-velocity emission. The observations are interpreted as evidence that an energetic bipolar outflow of molecular gas is occurring near AFGL 961. The near-infrared spectrum of AFGL 961 from 1.4–2.4 μm was obtained. Emission from the Brγ hydrogen recombination line and the S(1) quadrupole line of molecular hydrogen was detected.

133.008 **Spectroscopy of molecular cloud sources at 6 - 7 microns.**
N. J. Evans II, J. Carr, S. Beckwith, M. Strutskie, J. Wyant.
Bull. American Astron. Soc., Vol. 14, 609 (1982). – Abstract.

133.009 **A new class of compact far-infrared sources.**
D. Jaffe, D. A. Harper, R. Hildebrand, J. Keene, R. F. Loewenstein.
Bull. American Astron. Soc., Vol. 14, 611 (1982). – Abstract.

133.010 **Discovery of a compact far infrared source in the globule B335.**
J. A. Davidson, D. A. Harper, R. H. Hildebrand, J. Keene, F. J. Low.
Bull. American Astron. Soc., Vol. 14, 611 (1982). – Abstract.

133.011 **Multi-wavelength observations of two classes of broad-winged sources: GL490, a bi-polar flow, and S140, a symmetric flow.**
H. A. Smith, P. R. Schwartz, S. H. Knowles, C. J. Lada, G. Wolf, H. A. Thronson, D. A. Harper, W. Glaccum, R. F. Loewenstein, D. Jaffe.
Bull. American Astron. Soc., Vol. 14, 611 - 612 (1982). Abstract.

133.012 **Infrared mapping and photometry of W 40.**
A. F. Bentley, R. D. Gehrz, G. L. Grasdalen, J. A. Hackwell.
Bull. American Astron. Soc., Vol. 14, 643 (1982). – Abstract.

133.013 **Brackett α line profiles from compact sources.**
P. M. Ogden, F. L. Roesler, F. Scherb.
Bull. American Astron. Soc., Vol. 14, 653 (1982). – Abstract.

133.014 **On the infrared sources near the galactic centre.**
M. J. Lebofsky, M. R. Deshpande, J. C. Kemp, G. H. Rieke.
Bull. Astron. Soc. India, Vol. 10, 46 - 47 (1982). – Abstract.

133.015 **Far-infrared sources in Cygnus X: an extended emission complex at DR 21 and unresolved sources at S 106 and ON 2.**
M. F. Campbell, W. F. Hoffmann, H. A. Thronson Jr., D. Niles, R. Nawfel, M. Hawrylycz.
Astrophys. J., Vol. 261, 550 - 557 (1982).
A new map of 90 μm emission with 15′ resolution is presented of a complex of sources approximately centered on DR 21. The emission is primarily from diffuse sources. Temperatures of the emitting interstellar dust components are estimated. Unresolved, probably compact, sources were observed at S 106 and ON 2 and are briefly discussed.

133.016 **Infrared observations of OH/IR stars.**
F. Willems, T. de Jong.
Astron. Astrophys., Vol. 115, 213 - 215 (1982).
As a part of a program to study in the infrared all presently known type II OH sources in a 140 square degrees area of sky along the galactic equator between longitudes 10° and 27°.5 the authors have identified several OH sources with previously unknown infrared counterparts. These identifications could only be made because accurate radio positions were available. The authors present 1–20 μm observations obtained with the ESO 1 m telescope of the three brightest sources in their sample so far.

133.017 **A new near-infrared source in the molecular cloud associated with S106.**
R. G. Hofmann, H. P. Larson.
Astron. Astrophys., Vol. 116, 179 - 182 (1982).
A new near infrared point-like source was detected in the central part of the S 106 molecular cloud, and a medium resolution (8 cm^{-1}) spectrum was measured between 1.25 and 2.50 μm. The spectrum consists of a flat portion at wavelengths below 1.82 μm and a steep part in the range above 2.04 μm. The maximum flux is 0.9 Jy at 2.22 μm. The $\upsilon = 2-0$ absorption band is detected and the upper limit to the integrated Brγ flux is $2 \times 10^{-16} W/m^2$.

133.018 **Near infrared spectroscopy of W51 IRS-2.**
G. J. White, J. P. Phillips, P. M. Williams, G. D. Watt, K. J. Richardson.
Astron. Astrophys., Vol. 116, 293 - 295 (1982).
Near-infrared spectra at 2.95–3.5 μm and 3.99–10 μm have been obtained towards W51 IRS-2 and its surroundings, in order to investigate the spatial variations in intensity of the 3.28 μm unidentified feature and the 4.05 μm Brackett-α line. The Brα and 3.28 μ features occupy a broadly similar spatial zone, which is characterised by an unresolved core responsible for most of the emission, and an extended and considerably weaker halo. Grain properties required to excite the 3.28 μm line, the nature of the 3.28 μm emission, and its relation to the source structure are discussed.

133.019 **Infrared spectroscopy of the sources in S235 and its implication for the line excess problem.**
R. I. Thompson, H. A. Thronson, Jr., B. Campbell.
Prepr. Steward Obs., No. 407, 34 pp. (1982).

133.020 **Polarization of compact sources in the galactic center.**
M. J. Lebofsky, G. H. Rieke, M. R. Deshpande, J. C. Kemp.
Astrophys. J., Vol. 263, 672 - 676 (1982).
Near-infrared polarimetry is reported for 15 compact sources within 100″ (5 pc) of the galactic center; the results support the association of all 15 with the center. The polarization at 11.5 μm is probably due to emission by aligned grains in optically thin dust clouds surrounding luminous stars; these grains are aligned by streaming driven by radiation pressure.

133.021 **New members of the infrared cluster in the Orion Molecular Cloud.**
C. J. Lonsdale, E. E. Becklin, T. J. Lee, J. M. Stewart.
Astron. J., Vol. 87, 1819 - 1827 (1982).
Near-infrared high-resolution scans (3″.5) of the core of the Orion Molecular Cloud (# 1) have revealed 26 sources. Eleven of these are identified with faint visible stars. The remainder are thought to be highly reddened stars embedded in the molecular cloud and include at least two of the previously known infrared cluster members. Comparison of the distribution of infrared and visible stars on the plane of the sky and in an infrared color-magnitude diagram shows that the obscured infrared sources form a separate cluster of stars. The newly discovered sources appear to fall into two categories: (1) optically identified stars probably on the front surface of the cloud and associated with the Trapezium cluster; and (2) unidentified infrared stars probably associated with the molecular cloud and the embedded infrared cluster.

133.022 **Bα and Ne II line spectroscopy in the vicinity of the galactic center source IRS 16.**

T. R. Geballe, S. E. Persson, J. H. Lacy, G. Neugebauer, S. C. Beck.
The galactic center, (see 012.076), p. 60 - 66 (1982).
Bα and [Ne II] (12.81 μm) line spectra at spatial resolutions of ~3″ and spectral resolutions of 80 km s^{-1} have been obtained on a grid of positions surrounding IRS 16, which may be at the galactic center. The Bα and [Ne II] line profiles agree within the uncertainties and neither set of spectra shows evidence of ionized gas associated with IRS 16 over a velocity range of −500 to +450 km s^{-1}. The data imply that no concentration of moderate velocity gas is directly associated with IRS 16 and that the ionized gas near IRS 16 is for the most part neither very dense nor of very high velocity.

133.023 **Three compact sources with unusual 2 to 4 micron spectra.**
S. P. Willner, J. L. Pipher.
The galactic center, (see 012.076), p. 77 - 81 (1982).
1.9 to 4.1-μm spectra of the compact infrared sources IRS 3, 4, and 19 in the galactic center region are presented. IRS 3 lacks a strong ice absorption to accompany its enhanced silicate absorption. IRS 4 shows a 3.3-μm emission feature thought to arise from interfaces between molecular and ionized gas. IRS 19 shows an extremely deep 3.1-μm absorption superposed on the steam band that originates in a stellar atmosphere.

133.024 **Two micron observations of ^{12}CO and ^{13}CO in the red giant sources IRS 7, IRS 12, and IRS 19.**
G. C. Augason, H. A. Smith, E. R. Wollman, H. P. Larson, H. R. Johnson.
The galactic center, (see 012.076), p. 82 - 84 (1982).
The authors have obtained spectra from 1.2 to 2.47 microns of the galactic center sources IRS 7, 12 and 19 with resolutions of 2.4, 2.4, and 4.8 cm^{-1} respectively. The spectra of these sources show the first overtone bandheads of ^{12}CO and ^{13}CO. The spectrum of IRS 7, the brightest of the sources, also shows photospheric absorptions of Ca, Na, and other common elements.

133.025 **The position of the infrared source IRS 16 in the galactic center region relative to a visual field star.**
G. Neugebauer, K. Matthews, B. T. Soifer.
The galactic center, (see 012.076), p. 107 - 108 (1982).

An infrared look at the universe.
See Abstr. 013.050.

Double-shell circumstellar dust shells.
See Abstr. 112.023.

Near-infrared photometry of unidentified EIC-1 sources.
See Abstr. 113.029.

Infrared mapping of the galactic plane. III. The large-scale mid-infrared structure of W 3, W 4, and W 5.
See Abstr. 131.154.

A note about multiple systems of dense molecular clouds.
See Abstr. 131.268.

CO $J = 3 \rightarrow 2$ and submillimetre continuum observations of two molecular outflow sources.
See Abstr. 131.273.

A study of the Chamaeleon dark cloud complex: survey, structure and embedded sources.
See Abstr. 131.276.

Aperture synthesis observations of recombination lines from compact H II regions. V. NGC 7538.
See Abstr. 132.037.

Infrared spectra of galactic center sources.
See Abstr. 155.001.

Near-infrared source counts in the galactic plane.
See Abstr. 155.035.

Mapping and imaging of the galactic centre in the near infrared.
See Abstr. 155.069.

Two color CCD observations of the galactic center.
See Abstr. 155.070.

Discovery of three near infrared objects in CCD images of the galactic center.
See Abstr. 155.071.

Discovery of a very red nucleus in the radio elliptical IC 5063 (PKS 2048−57).
See Abstr. 158.041.

134 Emission Nebulae, Reflection Nebulae

134.001 **Optical polarization in M 17.**
J. F. Chesterman, R. F. Warren-Smith, S. M. Scarrott.
Mon. Not. R. Astron. Soc., Vol. 200, 965 - 970 (1982).
An optical linear polarization map of M 17 is presented. The region of the map to the north-east of the ionization front shows a centro-symmetric pattern characteristic of a reflection nebula. The location of the illuminating stars is indicated. To the south-west of the ionization front, in the body of the H II region, significant polarization is observed with orientation predominantly north–south. It is proposed that the H II region is seen through a thin cloud of relatively cold dust, aligned by a magnetic field.

134.002 **High dispersion observations of selected regions in the Orion Nebula.**
G. O. Boeshaar, C. A. Harvel, A. D. Mallama, P. M. Perry, R. W. Thompson, B. Turnrose.
Advances in ultraviolet astronomy, (see 012.006), p. 374 - 379 (1982).
High resolution spectral observations were made of several regions of the Orion Nebula near θ^2 Ori A using the IUE. 28 spectra were obtained of the bright bar, three Taylor-Münch cloudlets, and several surrounding locations. The data show ionization variations from region to region and are suggestive of stellar wind interactions between the cloudlets and θ^2 Ori A.

134.003 **IUE observations of reflection nebulae.**
A. N. Witt, R. C. Bohlin, T. P. Stecher.
Advances in ultraviolet astronomy, (see 012.006), p. 401 - 404 (1982).

Low-resolution IUE spectra in the SWP and LWR ranges have been obtained of several reflection nebulae in the vicinity of their respective illuminating stars. New data for NGC 7023 (HD 200775), NGC 1435 (Merope, 23 Tau), NGC 1432 (Maia, 20 Tau) and the Electra (17 Tau) nebula have been found to display significant differences in the shape of their normalized nebular spectra. These differences can be most readily explained by differences in the line-of-sight dust distribution, coupled with a wavelength-dependent phase function.

134.004 **IUE observations of dust albedo in nebulae.**
J. Wolf, H. L. Helfer, J. L. Pipher, T. L. Herter.
Advances in ultraviolet astronomy, (see 012.006), p. 405 - 408 (1982).

The wavelength dependence of the effective albedo of the dust at several positions near the Hourglass region of M 8 has been determined. Accurate estimates of the contribution of the continuous gas emission can be made using the region 1400–1600 Å. The main uncertainty results from the peculiar extinction of the exciting star, Herschel 36. The albedo for some planetaries such as NGC 6543 has also been investigated.

134.005 **Der Orionnebel in Farbe.**
D. F. Malin.
Sterne Weltraum, 21. Jahrg., 295 - 298 (1982).

134.006 **Detection of the [S III] 33.47 micron line in the Orion Nebula.**
T. Herter, D. A. Briotta, Jr., G. E. Gull, M. A. Shure, J. R. Houck.
Astrophys. J., Lett., Vol. 259, L109 - L112 (1982).

A strong detection of the 33 μm infrared forbidden line of S III from the Trapezium region of the Orion Nebula is reported. When combined with a measurement of the [S III] 18.71 μm line, an electron density of ~ 9000 cm^{-3} is deduced.

134.007 **Measurements of dust albedo in M 8 and some planetaries.**
J. Wolf, H. L. Helfer, J. L. Pipher, T. L. Herter.
News Lett. Astron. Soc. N.Y., Vol. 2, No. 2, p. 10 (1982). Abstract.

134.008 **Deep optical imagery of the Crab Nebula's jet.**
T. R. Gull, R. A. Fesen.
Astrophys. J., Lett., Vol. 260, L75 - L78, plates L6, L7 (1982).

A very deep [O III] image of the Crab Nebula's faint jet-like feature shows it in much greater detail than previously recorded and reveals it to be a wide, yet remarkably collimated structure. The jet, which is situated about 3′ north of the remnant's center, extends outward from the nebula's outermost visible border and appears as a sharply defined rectangular emission region 45″ × 75″ in size.

134.009 **Far infrared mesurements of the Crab nebula.**
W. Glaccum, D. A. Harper, R. F. Loewenstein, R. Pernic, F. J. Low.
Bull. American Astron. Soc., Vol. 14, 612 (1982). – Abstract.

134.010 **The excitation and distribution of CO (J = 6→5) emission in the Orion nebula.**
D. Buhl, G. Chin, G. A. Koepf, D. D. Peck, H. R. Fetterman.
Bull. American Astron. Soc., Vol. 14, 630 (1982). – Abstract.

134.011 **Bipolar reflection nebulae: Monte Carlo simulations.**
F. Youssefzadeh, M. Morris, R. L. White.
Bull. American Astron. Soc., Vol. 14, 637 (1982). – Abstract.

134.012 **The Crab nebula's jet.**
R. A. Fesen, T. R. Gull.
Bull. American Astron . Soc., Vol. 14, 637 (1982). – Abstract.

134.013 **Extended blue and red spectrophotometry of Crab nebula filaments.**
R. B. C. Henry, G. M. MacAlpine, R. P. Kirshner.
Bull. American Astron. Soc., Vol. 14, 638 (1982). – Abstract.

134.014 **Far-infrared fine-structure [O I] transitions at 145μ in the Orion nebula.**
G. Stacey, N. Kurtz, S. Smyers, M. O. Harwit.
Bull. American Astron. Soc., Vol. 14, 653 (1982). – Abstract.

134.015 **Optical emission from ring-shaped nebulae in the LMC.**
M. Rosado, Y. M. Georgelin, G. Monnet.
Rev. Mexicana Astron. Astrofis., Vol. 6, (see 012.017), 321 - 324 (1981).

A photographic survey of ring-shaped nebulae in the LMC, performed with narrow band interference filters in Hα and [S II], permits the authors to determine their mean [S II]/Hα line-ratios. This information, together with the velocity dispersion obtained from FP interferograms and the knowledge of their stellar content allow an overall view of these regions and of their probable origin. Some specific nebulae are discussed.

134.016 **Limits on a radio shell around the Crab Nebula.**
A. S. Wilson, K. W. Weiler.
Nature, Vol. 300, 155 - 157 (1982).

The authors report a high sensitivity search at radio wavelengths for the hypothetical outer shell. No shell has been detected and the upper limits indicate that any such shell is fainter than expected from the surface brightness – linear diameter relation for shell type supernova remnants.

134.017 **Fabry-Perot interferometry of emission nebulae.**
P. Pishmish.
Recent advances in observational astronomy, (see 012.028), p. 73 - 79 (1981).

Research on the velocity field in galactic H II regions is carried out using the 83-cm and 1-m telescopes of the University of Mexico. The majority of the emission nebulae studied so far can be divided into two groups, distinguished from one another with respect to the process of their formation. Three of the nebulae are mainly produced by non-isotropic ejection from symmetrical regions on the ionizing central star, and three others are remnants from the formation of the ionizing star. Evidence is shown by the latter group for the sequential formation of H II regions within the same primordial cloud.

134.018 **Orion's stellar nursery.**
L. J. Robinson.
Sky Telesc., Vol. 64, 430 - 432 (1982).

134.019 **Observations of shocked gas in the Carina nebula.**
C. Laurent.
Observatory, Vol. 102, 171 - 172 (1982). – Abstract.

134.020 **A drop of the radiation intensity of the Crab nebula.**
V. P. Ivanov, A. P. Barabanov, K. S. Stankevich, S. P. Stolyarov.
Astron. Zh., Tom 59, 963 - 964 (1982). In Russian. English translation in Soviet Astron., Vol. 26, No. 5.

As a result of continuous (1961 - 1981) investigations of the radiation distribution over the spectrum of the Crab nebula it is stated that the intensity of the radiation decreased on the average by 3,5% in the frequency range from 0.6 GHz up to 10 GHz. It is probable that the flux did not change uniformly and a quick drop occurred in the interval between

October 1975 and October 1977. This effect may be explained by an escape of relativistic electrons from the nebula envelope.

134.021 **Ring nebulae around Of stars: observations of S 119 and its central star 68 Cygni.**
V. F. Esipov, A. Yu. Klement'eva, A. V. Kovalenko, T. A. Lozinskaya, V. M. Lyutyj, T. G. Sitnik, V. A. Udal'tsov.
Astron. Zh., Tom 59, 965 - 974 (1982). In Russian. English translation in Soviet Astron., Vol. 26, No. 5.

Spectral (6000 - 7000 Å), interferometric (Hα and [N II]) and radio (102 MHz) observations of S 119, and photoelectric *UBV* observations of 68 Cyg have been carried out. The ring nebula around 68 Cyg is a diffuse envelope of irregular form having a radius of 17 pc. Radio emission is thermal with the spectral index $\alpha = -0.1$. The source of the ionizing radiation in the nebula is the central star. The envelope structure is due to the stellar wind of 68 Cyg. Variability of 68 Cyg with the period $3^{d}.39$ due to the duplicity of the central star is discovered. The parameters of this binary system are discussed.

134.022 **The scattering phase function of interstellar grains: the case of the reflection nebula NGC 7023.**
A. N. Witt, G. A. H. Walker, R. C. Bohlin, T. P. Stecher.
Astrophys. J., Vol. 261, 492 - 509 (1982).

IUE observations of HD 200775 and its associated reflection nebula NGC 7023 covering the spectral range 1300–3100 Å, supplemented by new ground-based measurements of the nebular brightness distribution at 3500, 4100, 4700, and 5500 Å, have been used as a basis for a determination of the scattering properties of the nebular dust grains. The average grain albedo in the UV is $\langle a \rangle_{UV} \approx 0.54$. The albedo is found to increase to a level of $a \approx 0.6$ at $\lambda = 1400$ Å, after reaching a minimum of $a \approx 0.4$ near 2200 Å. The dust properties derived for the far-UV suggest that isotropically scattering particles of high albedo provide a significant contribution to interstellar scattering in the far-UV.

134.023 **The Bubble Nebula: far-infrared and radio molecular observations of NGC 7635.**
H. A. Thronson, Jr., C. J. Lada, P. M. Harvey, M. W. Werner.
Mon. Not. R. Astron. Soc., Vol. 201, 429 - 437 (1982).

The Bubble Nebula, NGC 7635, has been partially mapped at 1 arcmin resolution in the far-infrared and completely mapped at 2.2 arcmin resolution in 115-GHz emission from CO. The infrared results are consistent with the presence of only a small amount of dust within the optical part of the nebula. The dust may be fairly uniformly spread through the H II region, although some variation in density is necessary to explain the observed temperature distribution of the dust. The CO emission is weak and the CO column density is very low. Nevertheless it is surprising that any molecules can exist in the neighbourhood of the luminous exciting star of NGC 7635 and a special geometrical configuration may be necessary.

134.024 **The velocity field of S 308. The ring nebula around the WN5 star HD 50896.**
P. Pişmiş, A. Quintero.
Wolf-Rayet stars: observations, physics, evolution, (see 012.042), p. 305 - 309 (1982).

Radial velocities are determined by Fabry-Pérot interferometry at 131 points of the ring nebula S 308. The velocities have yielded a kinematic distance of 1.5 kpc for the object, and an expansion velocity of 45–60 km s^{-1}. The ring nebula has a diameter of 13 pc and the age is estimated to be about 1.5 to 2×10^{5} years.

134.025 **A new search for nebulae surrounding Wolf-Rayet stars.**
J. N. Heckathorn, F. C. Bruhweiler, T. R. Gull.
Wolf-Rayet stars: observations, physics, evolution, (see 012.042), p. 463 - 468 (1982).

The authors have used the plate data from an emission Line Survey of the Milky Way by Parker, Gull and Kirschner (1979) to make a new search for ring nebulae around Wolf-Rayet stars. The Survey consists of narrow-band interference filter direct imagery centered on the emission lines of Hα + [N II] at λ6570, [O III] at λ5007, and [S II] at λ6730. The authors have discovered at least five new ring nebulae.

134.026 **Ring nebulae associated with Wolf-Rayet stars.**
Y.-H. Chu.
Wolf-Rayet stars: observations, physics, evolution, (see 012.042), p. 469 - 472 (1982).

134.027 **Ring nebulae around WC6 stars: NGC 6357 around HD 157 504.**
M. C. Lortet, G. Testor, V. Niemelä.
Wolf-Rayet stars: observations, physics, evolution, (see 012.042), p. 473 - 477 (1982).

134.028 **Kinematics of the ring-shaped nebula N206 in the LMC.**
M. Rosado, G. Monnet, A. Laval, Y. Georgelin.
Wolf-Rayet stars: observations, physics, evolution, (see 012.042), p. 479 - 483 (1982).

Preliminary results on the radial velocity field of the ring-shaped nebula N206 in the LMC are given. Several early-type stars are located inside this nebula, one of them being a WC5–6 star. A discussion about the kinematics of the gas around the WR star is also given.

134.029 **Kinematics of ring-shaped nebulae in the LMC. II. The radial velocity field of N 185.**
M. Rosado, Y. M. Georgelin, Y. P. Georgelin, A. Laval, G. Monnet.
Astron. Astrophys., Vol. 115, 61 - 64 (1982).

The authors have studied the LMC ring-shaped nebula N 185 by means of Hα and [S II] imagery and Fabry-Perot interferometry. From an investigation of the stellar content they show that the interior stars can photoionize the nebula and this, together with the results on the Hα imagery allows to derive an electron density of about 4 cm^{-3} for the shell. They found that the [S II] (6717)/Hα line ratio varies from 0.15 to 0.4 among different filaments. This variation can be explained by differences in ionization degree. The radial velocity field could be accounted for by an expansion of 70 km s^{-1} ± 10 km s^{-1} superimposed to a systematic motion of 262 ± 11 km s^{-1}. The obtained parameters allow to discriminate between different origins of this expanding shell, a supernova explosion being more adequate to explain the energetics of the nebula than stellar winds from the present stars.

134.030 **Absolute photometry of the Crab nebula.**
A. Greve, A. M. van Genderen.
Astron. Astrophys., Vol. 115, 79 - 83 (1982).

The authors give VBLUW photometric observations of the Crab nebula. From the data they derive $A_{VJ} = 1^{m}.4$ extinction in the direction of M 1. For positions of the nebula with contaminations by line emission below 2% in the observed fluxes they derive $\alpha = 2.32^{-0.17}_{+0.13}$ for the synchrotron radiation.

134.031 **Is Eta Carinae about to explode?**
K. Davidson.
Mercury, Vol. 11, 138 - 141 (1982).

134.032 **The spectrum of 1548C27.**
E. R. Craine, D. E. Turnshek, D. A. Turnshek, B. T. Lynds, E. J. O'Neil, Jr.
Prepr. Steward Obs., No. 413, 17 pp. (1982).

134.033 **Ring nebulae associated with Of stars: statistics, classification, origin.**

T. A. Lozinskaya.
Astrophys. Space Sci., Vol. 87, 313 - 331 (1982).

The ring nebulae associated with galactic Of stars are considered on the basis of the list of Of nebulae proposed by Lozinskaya and Lomovsky (1982). Taking into account the selection effects, about 80% of Of stars are shown to be associated with H II regions and about 30 - 50% of these regions have shell structures. Four types of nebulae associated with Of stars are resolved: amorphous H II regions, ring-like H II regions, wind-blown bubbles, and stellar ejectas. These types appear to be identical with the morphology of nebulae around WR stars proposed by Chu (1981). Observational data are presented and the nature of a number of Of ring nebulae of different types is discussed.

134.034 **High resolution spectroscopy of selected galactic emission nebulae.**
U. Carsenty.
Diss. Naturwiss.-Math. Gesamtfak. Univ. Heidelberg, 7 + 104 pp. (1982).

Seven emission nebulae of various types (the planetary nebulae NGC 6772 and PK 89-0°1, the young planetary nebulae M2-9, M1-91, and AFGL 618, the ring nebula M1-67 around the WN star 209 BAC, and the H II region SH-106) were observed at high spectral (12 km/sec) and high spatial (2″) resolution, in the red spectral range. Additional low-resolution spectra were obtained. The observations are interpreted through three-dimensional spatio-kinematical models for each of the nebulae.

134.035 **Observations of the Orion Nebula in Hα, [N II] and Hβ lines.**
C. T. Hua, R. Louise.
Astrophys. Space Sci., Vol. 88, 477 - 485 (1982).

Monochromatic photographs of the Orion Nebula taken through narrow bandpass interference filters ($\Delta\lambda$ = 10 Å) centred on Hα, Hβ and [N II] lines are presented. Ratio contours of Hα/[N II] and Hα/Hβ are derived. They enable a detailed study of the point-to-point variation in ionization structure and temperature throughout the nebula. Dust located within the ionized gas is studied from the Hα/Hβ ratio which varies from point to point over the nebula. The following suggestions are presented: (1) Dust grains are completely or partially destroyed in regions close to the exciting stars. (2) Radiation pressure and stellar wind push the remaining dust up to some equilibrium distance outwards. The consequence of this action is obviously the formation of a "dust mantle" which is seen as a ring in projection.

High resolution near-infrared spectroscopy from the Kuiper Airborne Observatory.
See Abstr. 032.509.

An observational approach to the early stages of stellar evolution.
See Abstr. 065.012.

WR stars with ring nebulae.
See Abstr. 117.096.

Discovery of a family of Herbig objects in M42: implications for the geometry of the high velocity molecular flow.
See Abstr. 121.038.

Absolute photometry of supernova remnants and emission nebulae in the Galaxy and the Magellanic Clouds.
See Abstr. 125.003.

Airborne observations of the Orion molecular hydrogen emission spectrum.
See Abstr. 131.014.

Far infrared polarimetry of W51A and M 42.
See Abstr. 131.036.

The violent interstellar medium associated with the Carina Nebula.
See Abstr. 131.048.

A model for the intense molecular line emission from OMC-1.
See Abstr. 131.068.

Molecular shock waves in the BN-KL region of Orion.
See Abstr. 131.069.

NH_3 in Orion-KL: a new interpretation.
See Abstr. 131.070.

The violent interstellar medium associated with the Carina Nebula. I. The line of sight toward HD 93205.
See Abstr. 131.076.

Molecular clouds associated with reflection nebulae. II. General properties.
See Abstr. 131.094.

The violent interstellar medium associated with the Carina Nebula.
See Abstr. 131.183.

Energetic activity in a star-forming molecular cloud core; a disk constrained bipolar outflow in NGC 2071.
See Abstr. 131.210.

Detection of bipolar CO outflow in Orion.
See Abstr. 131.211.

Observations of [O I], [O II], and [O III] lines in M 42 and M 43.
See Abstr. 132.001.

Measurements of magnetic field strengths in the vicinity of Orion.
See Abstr. 132.014.

The H II regions of Messier 8.
See Abstr. 132.038.

Radio sources in the association of the reflection nebulae Monoceros-R2.
See Abstr. 141.051.

Gamma-ray emission from the galactic anticenter at MeV energies.
See Abstr. 157.013.

135 Planetary Nebulae

135.001 **IUE observations of proto-planetary and variable planetary nebulae. I. V1016 Cygni, HM Sagittae, and HBV 475.**
W. A. Feibelman.
Astrophys. J., Vol. 258, 548 - 561 (1982).
Ultraviolet observations of the "proto-planetary nebulae" V1016 Cyg, HM Sge, and HBV 475 (= V1329 Cyg) were obtained with the IUE satellite. Emission line fluxes, line ratios, line profiles, electron densities, and distances for these objects are presented. V1016 Cyg and HM Sge are showing levels of increasing excitation and ionization as a function of time, while for HBV 475 the trend is in the opposite direction. For HM Sge, dramatic changes have occurred in the UV line profiles over the past 4 years, suggesting that a shell is forming.

135.002 **IUE observations of proto-planetary and variable planetary nebulae. II. A search for variability in IC 4997 and NGC 6905.**
W. A. Feibelman.
Astrophys. J., Vol. 258, 562 - 567 (1982).
A search for ultraviolet variability was conducted for the planetary nebulae IC 4997 and NGC 6905 which have histories of variability in the visible region. Emission line fluxes, line ratios, line profiles, and central stars' temperatures are presented for both objects, as well as stratification effects for several ions in NGC 6905.

135.003 **Near-infrared spectroscopy of possible precursors to planetary nebulae : the Cygnus Egg and the Red Rectangle.**
H. A. Thronson, Jr.
Astron. J., Vol. 87, 1207 - 1212 (1982).
Moderate-resolution near-infrared spectroscopy is presented for two possible protoplanetary nebulae; the Red Rectangle (AFGL 915; HD 44179) and the Cygnus Egg (AFGL 2688). The Rectangle shows a generally featureless but very red continuum. There may be very weak CO absorption at 2.3 μm, which would suggest the presence of a late-type stellar component. The spectrum of the Egg is dominated by H_2 rotation-vibration emission characteristic of shock excitation. Apparently without an expanding H II region and no early-type exciting star, the source of this shock is probably mass loss from a rapidly evolving central star.

135.004 **Determination of the intrinsic Q(3)/S(1) line intensity ratio of molecular hydrogen.**
T. R. Geballe, R. W. Russell, D. Nadeau.
Astrophys. J., Lett., Vol. 259, L47 - L49 (1982).
Observations of the $H_2 v = 1 \rightarrow 0$ S(1) and Q(3) lines in the planetary nebula NGC 7027 yield an intrinsic Q(3)/S(1) line intensity ratio of 0.74 ± 0.08. This value is consistent with the theoretical ratio of the Einstein A coefficients of these transitions calculated by Turner, Kirby-Docken, and Dalgarno. The H I 20–5 line intensity in NGC 7027 is also reported and discussed.

135.005 **On the radio distance scale for planetary nebulae.**
D. K. Milne.
Mon. Not. R. Astron. Soc., Vol. 200, 51P - 54P (1982).
The evolution of the radio emission from a planetary nebula is examined and a method for estimating the distance to both optically thick and optically thin nebulae is suggested.

135.006 **NGC 2440: ionization structure, extinction, and near infrared spectrum.**
A. R. Condal.
Astron. Astrophys., Vol. 112, 124 - 132 (1982).
Two-dimensional data are used to study the ionization structure and reddening variations over the projected area of the planetary nebula NGC 2440. The one-dimensional data are used to study the physical conditions and chemical abundances in NGC 2440.

135.007 **Neutral hydrogen associated with the planetary nebula NGC 6302.**
L. F. Rodríguez, J. M. Moran.
Nature, Vol. 299, 323 - 325 (1982).
Observations of the 21-cm line of neutral hydrogen in absorption towards the thermal radio source in the planetary nebula NGC 6302 show two velocity components at 6 and -40 km s^{-1}. The 6 km s^{-1} component is probably due to a line-of-sight cloud, while the -40 km s^{-1} component is most likely associated with NGC 6302. The authors interpret this latter absorption component as coming from the neutral, outer part of an expanding ($\sim$10 km s^{-1}) ring whose inner part is ionized and produces the absorbed thermal continuum radiation. This is the first detection of H I associated with a planetary nebula.

135.008 **Integrated fluxes for emission lines in the ultraviolet spectra of several planetary nebulae.**
K. G. Carpenter, S. J. Czyzak.
Astrophys. Space Sci., Vol. 84, 495 - 503 (1982).
The IUE satellite observatory has been used to obtain absolutely-calibrated emission line fluxes for diagnostic lines of multiply-ionized C, N, O, Si, Ne, and Ar which occur in the ultraviolet spectral region of planetary nebulae. These data, when combined with data from the blue, visual, and near infrared, will provide improved estimates of ionic concentrations, plasma temperatures and densities, and elemental abundances.

135.009 **The UV emission line spectrum of NGC 6572.**
M. Grewing, G. Krämer, P. R. Preussner, E. Schulz-Lüpertz.
Advances in ultraviolet astronomy, (see 012.006), p. 389 - 392 (1982).
Ultraviolet spectra of the medium excitation planetary nebula NGC 6572 have been obtained with IUE both in the low and in the high resolution mode. These spectra reveal a superposition of emission lines arising in the expanding atmosphere of the central star, evidenced e.g. by their P Cygni profiles, and features arising in the planetary nebula proper. The latter have been analysed in order to determine C, N, O abundances.

135.010 **Stratification effects and IUE spectra of high excitation planetaries.**
W. Feibelman, L. H. Aller.
Advances in ultraviolet astronomy, (see 012.006), p. 393 - 396 (1982).
Individual strips across IUE low resolution images of a number of high excitation planetaries with appreciable angular disks are analyzed to assess stratification effects. The familiar enhancement of high excitation lines toward the center is well exhibited.

135.011 **IUE observations of planetary nebulae and their central stars in the Magellanic Clouds.**
S. P. Maran, T. P. Stecher, T. R. Gull, L. H. Aller, M. P. Savedoff.
Advances in ultraviolet astronomy, (see 012.006), p. 397 - 400 (1982).
The planetary nebulae LMC P40, SMC N2, and SMC N5 and their central stars were observed with IUE. The C abundances in the nebulae indicate that convective dredgeup of locally nucleosynthesized C has occurred. The progenitors

of the nebulae were C stars at the theoretical upper luminosity threshold. The central stars of the nebulae have masses $\sim 1\ M_{\odot}$, luminosities $\sim 4 \times 10^4\ L_{\odot}$, and radii $\sim 0.7\ R_{\odot}$.

135.012 **The expansion velocity field within the planetary nebulae NGC 1501 and NGC 6905.**
F. Sabbadin, E. Hamzaoglu.
Astron. Astrophys., Suppl. Ser., Vol. 50, 1 - 6 (1982).

The [O III] (λ 4959 and λ 5007 Å) and Hα expansion velocity fields within the planetary nebulae NGC 1501 and NGC 6905 have been derived from Echelle spectrograms. The data suggest that NGC 1501 is a prolate spheroid of moderate ellipticity. NGC 6905 consists of a truncated prolate spheroid and of two caps of less dense material faster expanding than the gas contained in the central part.

135.013 **Sulfur abundances in gaseous nebulae.**
A. R. Condal.
Messenger, No. 29, p. 18 - 19 (1982).

135.014 **S III, O IV, and Ne V abundances in eight planetary nebulae.**
M. Shure.
News Lett. Astron. Soc. N.Y., Vol. 2, No. 2, p. 8 (1982). Abstract.

135.015 **Near infrared polarization in two peculiar nebulae: M 2–9 and the PV Cephei nebula.**
M. G. Lacasse.
News Lett. Astron. Soc. N.Y., Vol. 2, No. 2, p. 23 (1982). Abstract.

135.016 **A statistical survey of local planetary nebulae.**
C. T. Daub.
Astrophys. J., Vol. 260, 612 - 624 (1982).

An empirically determined relation between the ionized masses of planetary nebulae and a particular function of observational quantities is used to derive distances to 299 primarily nearby planetary nebulae. These distances are used to investigate the scale height of the galactic distribution of planetaries, the rate of formation of planetaries, and certain aspects of the evolution of the nebulae.

135.017 **Discovery of a large, high-excitation planetary nebula, G136+5.**
J. N. Heckathorn, R. A. Fesen, T. R. Gull.
Bull. American Astron. Soc., Vol. 14, 637 (1982). – Abstract.

135.018 **A photographic, near-infrared survey of planetary nebulae.**
D. J. MacConnell.
Bull. American Astron. Soc., Vol. 14, 653 (1982). – Abstract.

135.019 **Radio interferometry of planetary nebulae.**
K. C. Turner, Y. Terzian.
Bull. American Astron. Soc., Vol. 14, 653 - 654 (1982). Abstract.

135.020 **Near infrared spectrophotometry of planetary nebulae.**
P. D. LeVan, R. J. Rudy.
Bull. American Astron. Soc., Vol. 14, 654 (1982). – Abstract.

135.021 **Distribution of forbidden neutral carbon emission in planetary nebulae.**
P. N. Kupferman, D. C. Jewitt, G. E. Danielson, S. P. Maran.
Bull. American Astron. Soc., Vol. 14, 654 (1982). – Abstract.

135.022 **Study of planetary nebula K1–2 and its variable nucleus.**
L. Kohoutek, G. F. O. Schnur.
Mon. Not. R. Astron. Soc., Vol. 201, 21 - 26 (1982).

New photographic, photometric and first spectroscopic observations of K1–2 indicate this object to be an old, very high-excitation planetary nebula at the distance of 2.4 kpc. The optical variability of the central star in the range B 16.5–17.7 mag has been confirmed. The nucleus has the following parameters: $L_* = 500–1000\ L_{\odot}$, T_z(He II) = 91000–79000 K, $R_* = 0.09–0.17\ R_{\odot}$.

135.023 **Ultraviolet observations of planetary nebulae. NGC 6572, NGC 5315 and BD + 30°3639.**
S. Torres-Peimbert, M. Peña.
Rev. Mexicana Astron. Astrofis., Vol. 6, (see 012.017), 301 - 308 (1981).

The authors present IUE observations of three planetary nebulae. They have obtained the chemical composition of NGC 6572. For NGC 5315 and BD + 30°3639, the stellar WR combination in the UV is very significant, and for these nebulae the authors are only able to derive the carbon abundance.

135.024 **Ultraviolet spectrum of the planetary nebula NGC 7662. Observations and models.**
M. Peña, S. Torres-Peimbert.
Rev. Mexicana Astron. Astrofis., Vol. 6, (see 012.017), 309 - 314 (1981).

The authors present IUE observations of emission lines in the spectrum of NGC 7662 in the 1150 - 3300 Å range. Combining these observations with optical data obtained by other authors, they derive the chemical composition for this object. The authors compute models of ionization structure and thermal balance to compare with the observed data.

135.025 **Chemical enrichment in halo planetary nebulae.**
S. Torres-Peimbert, J. F. Rayo, M. Peimbert.
Rev. Mexicana Astron. Astrofis., Vol. 6, (see 012.017), 315 - 319 (1981).

Photoelectric spectrophotometry of emission lines in the 3400 - 7400 Å region is presented for the planetary nebula 108–76°1 (BB1). From these observations the relative abundances of H, He, C, N, O, and Ne are derived. The abundances of the halo PN (BB 1, H4-1 and K648) are compared to those predicted by stellar evolution theory under the assumption that the envelope has the chemical composition of the matter located between the H burning shell and the surface.

135.026 **La nebulosa planetaria M4–18.**
M. Peimbert, S. Torres-Peimbert.
Rev. Mexicana Astron. Astrofis., Vol. 6, (see 012.017), 320 (1981). – Abstract.

135.027 **Detection and study of secondary structures in some planetary nebulae.**
R. Louise.
Astron. Astrophys., Vol. 114, 205 - 207 (1982).

Long exposure plates have been made on nine selected nebulae, using a large bandwidth (Δλ = 50 Å) interference filter coupled with an ITT image tube. Some peculiar features are observed. Spectrophotometric observations have been made for one nebula, IC 418. A contour map of the [N II]/Hα ratio is derived. It is shown that this ratio increases towards the outer extended envelope of the nebula. One of the typical features of secondary structures is the enhancement of the [N II] line with respect to Hα. In addition, filamentary structures appear sharper in [N II] than in Hα. This is fairly illustrated by NGC 650–1.

135.028 **Radiative transfer calculations for the C IV λ 155 nm resonance lines in planetary nebulae.**
J. Köppen, R. Wehrse.
Third European IUE Conference, (see 012.029), p. 317 - 321 (1982).

Ionization models for NGC 6210, 7009, 3242 and

II 2003 have been constructed from optical and IUE spectral data. The C IV λ155 nm resonance line is predicted about ten times stronger than observed. Possible solutions for this discrepancy are discussed.

135.029 **The planetary nebula NGC 6572.**
M. Grewing, G. Krämer, F. Dietz, P. R. Preussner.
Third European IUE Conference, (see 012.029), p. 323 - 325 (1982).

The IUE satellite has been used to observe the ultraviolet spectrum of the medium excitation planetary nebula NGC 6572 in both low and high resolution. The overall UV continuum energy distribution can be fitted by a black-body with 44000 K. The comparison of small and large aperture spectra allows a clearcut separation of the stellar from the nebular features proper. Absolute fluxes are given for a number of emission line features.

135.030 **Discovery of a large, high-excitation planetary nebula at $l = 136°$, $b = +5°$.**
J. N. Heckathorn, R. A. Fesen, T. R. Gull.
Astron. Astrophys., Vol. 114, 414 - 418 (1982).

The discovery of a new planetary nebula from [O III] interference-filter imagery is reported. The nebula, 136 + 5°.1, is 15′ in diameter and asymmetric in appearance. Spectrophotometry of the central regions indicate it is of excitation class of 8–10. A single bright condensation within the nebula exhibits emission lines of much lower ionization with [S II] λλ6717, 6731 line intensities indicating an electron density of about 350 ± 150 cm^{-3}. High resolution [O III] imagery of this planetary nebula is also presented.

135.031 **Observation of the far infrared continua of planetary nebulae.**
H. Moseley, R. Silverberg.
Bull. American Astron. Soc., Vol. 14, 776 (1982). – Abstract.

135.032 **The absolute Hβ flux from NGC 7027.**
R. A. Shaw, J. B. Kaler.
Astrophys. J., Vol. 261, 510 - 512 (1982).

The authors find the value of log F(Hβ) for NGC 7027 to be −10.12 ± 0.01 from a combination of new interference filter measurements and older independent determinations. This new Hβ flux allows them to reexamine the fit between the observed and theoretical nebular continuum, to improve and reconcile earlier values of electron temperature and extinction, and to make some comments about the central star.

135.033 **Ultraviolet spectra of planetary nebulae – IX. High-dispersion observations of NGC 7662.**
D. R. Flower, C. J. Penn, M. J. Seaton.
Mon. Not. R. Astron. Soc., Vol. 201, 39P - 43P (1982).

High-dispersion *IUE* spectra of NGC 7662 have been obtained with exposure times optimized for measurement of the relative strengths of the components of: C IV λλ 1548, 1551; C III] λλ 1907, 1909 and [Ne IV] λλ 2422, 2424. The C IV ratio (λ 1548)/(λ 1551) is expected to be 2.00, the ratio of the statistical weights of the emitting levels, and the observed value is 1.92 ± 0.15. The C III] and [Ne IV] ratios depend upon the electron density, N_e, and from the observations it is concluded that N_e for NGC 7662 is in the range 1300 to 4000 cm^{-3}. The presence of the dielectronic recombination line C III λ 2297 is established beyond doubt. The absolute calibration of *IUE* high-dispersion spectra is also discussed.

135.034 **Radio observations at 14.7 GHz of southern planetary nebulae.**
D. K. Milne, L. H. Aller.
Astron. Astrophys., Suppl. Ser., Vol. 50, 209 - 215 (1982).

14.7 GHz radio flux densities are presented for 397 planetary nebulae south of declination +27° and comparison made with data at 5 GHz to show that these nebulae are all optically thin in their radio emission between the two frequencies. Comparison is also made with the optical classification.

135.035 **Luminosities and masses for three central stars of planetary nebulae in the Magellanic Clouds from ultraviolet spectroscopy with the IUE.**
T. P. Stecher, S. P. Maran, T. R. Gull, L. H. Aller, M. P. Savedoff.
Astrophys. J., Lett., Vol. 262, L41 - L46, plate L1 (1982).

Values of $M \sim 1\ M_\odot$, $L \sim 4.5 \times 10^4\ L_\odot$, and $T \sim 1 \times 10^5$ K are determined for central stars of planetary nebulae of known distances. The results follow from IUE observations of the continuous spectra of the central stars of three nebulae in the Magellanic Clouds. If the derived value for M approximately equals the C-O core mass in the red giant stage, the mass of the progenitor star is $\sim 4.0\ M_\odot$, and thus $\sim 3\ M_\odot$ were lost, presumably in winds. Some implications of these results for stellar evolution calculations of carbon stars are discussed.

135.036 **Spectrophotometric observations of the peculiar nitrogen-rich planetary nebula NGC 2440.**
R. Louise.
J. Astrophys. Astron., Vol. 3, 145 - 150 (1982).

The author obtained 65 red spectra covering the whole surface of the planetary nebula NGC 2440. Intensities of Hα, [N II] λλ6548–6584 and [S II] λλ6717–6731 lines are derived. The unusual high value of the [N II]/Hα in the central core is due to nitrogen overabundance occurring in that part of the nebula. Its variations from the centre to the outer regions are interpreted as a consequence of small-scale ionization structure.

135.037 **Observations of bipolar planetary nebula 19W32.**
L. Kohoutek.
Astron. Astrophys., Vol. 115, 420 - 422 (1982).

New photographic and spectroscopic observations of 19W32 = No. 455 - 139 (359 + 1°3) indicate this object to be a bipolar planetary nebula of intermediate age at a distance of about 3 kpc. The diameter 0.3 pc and the [S II] electron density 5×10^3 cm^{-3} have been estimated for the slowly expanding nebular lobes. The central star consists of two components separated by only 0″.97; the fainter subluminous component is very probably a planetary nucleus having $M_B \gtrsim -0.3$ mag.

135.038 **Introductory report on V1016 Cygni.**
A. Cassatella.
The nature of symbiotic stars, (see 012.045), p. 157 - 160 (1982).

The formation of a new planetary nebula should be a rare event, and the case of V1016 Cyg may represent such an example.

135.039 **The far-UV spectrum of the low-excitation planetary nebula HD 138403.**
J. Surdej, A. Heck.
Astron. Astrophys., Vol. 116, 80 - 88 (1982).

A description of the far-ultraviolet (1200–2000 Å) spectrum of the low-excitation planetary nebula 315–13°1 has been given in this paper. The authors have stressed the importance of the mass-loss from the central nucleus HD 138403 as revealed by the presence of numerous P Cygni profiles. The data support the hypothesis that a bipolar structure is expanding around HD 138403, via the selective radiative process of edge and/or line locking mechanism(s).

135.040 **Abell 41.**
IAU Circ., No. 3714 (1982).

135.041 **NGC 2346.**
IAU Circ., No. 3736 (1982).

135.042 **Spatial-kinematical models for planetary nebulae: NGC 2371-2.**
F. Sabbadin, A. Bianchini, E. Hamzaoglu.
Astron. Astrophys., Suppl. Ser., Vol. 50, 523 - 528 (1982).

The [O III], H I and [N II] expansion velocity fields within the planetary nebula NGC 2371-2 obtained from Echelle spectrograms have been used to derive a three dimensional model for the nebula. The data indicate that it consists of a three axes ellipsoid with a bright and dense central ring (or toroid) + two external polar caps showing evidence of deceleration due to interstellar medium.

135.043 **Planetary nebulae.**
D. C. V. Mallik.
Bull. Astron. Soc. India, Vol. 10, 73 - 110 (1982).

Recent research on planetary nebulae is reviewed, with special emphasis on ultraviolet and infrared observations. Various distance scales and statistics based on them are discussed and the empirical mass-radius relation due to Maciel and Pottasch (1980) is used to compile a new list of local planetaries. Recent determinations of the effective temperatures of central stars and their new locations on the H-R diagram are described. The implications of the chemical abundance studies of these nebulae are discussed in the light of current work on stellar evolution and nucleosynthesis.

135.044 **A new planetary nebula near the Large Magellanic Cloud.**
A. Savage, P.G. Murdin, D. H. Clark.
Observatory, Vol. 102, 229 - 230 (1982).

135.045 **Effects of mass loss on the formation of planetary nebulae.**
S. Kwok.
J. R. Astron. Soc. Canada, Vol. 76, 322 (1982). – Abstract.

135.046 **M2-9. The Butterfly nebula.**
C. Purton, B. C. Kwok, S. Kwok.
J. R. Astron. Soc. Canada, Vol. 76, 324 (1982). – Abstract.

135.047 **The unprecedented light variations of NGC 2346.**
R. H. Méndez, R. Gathier, V. S. Niemela.
Astron. Astrophys., Vol. 116, L1 - L4 (1982).

New spectrograms and photometry of the binary central star of the planetary nebula NGC 2346 have confirmed the variations reported by Kohoutek (1982). It is shown that these variations did not exist before 1981. The analysis of the new data leads the authors to ascribe this drastic change to the passage of a dust cloud in front of the binary system. A tentative model is presented.

135.048 **Planetary nebulae in Local Group galaxies.**
H. C. Ford.
Space Telesc. Sci. Inst., Prepr. Ser., No. 2, 20 pp. (1982).

135.049 **Der Zentralstern des Planetarischen Nebels NGC 2346.**
K. Wälke.
BAV Rundbrief, 31. Jahrg., 60 - 61 (1982).

135.050 **On some characteristics of emission spectra of carbon and nitrogen ions in planetary nebulae.**
A. A. Nikitin, A. A. Sapar, T. Kh. Feklistova, A. F. Kholtygin.
Tartu Astrofüüs. Obs. Publ., Köide 49, 40 - 54 (1982). In Russian.

The spectra of ions C III, N III, C IV and N IV, originated from single and two-electron excited states, of some highly excited planetary nebulae have been studied. The recombinational spectrum has been found for C III, N III and N IV. The observed carbon and nitrogen abundances have been found.

135.051 **Spatial observations of dust emission in NGC 7027.**
A. F. Bentley.
Astron. J., Vol. 87, 1810 - 1818 (1982).

NGC 7027 was mapped in each of five spectral bands between 8.7 and 23 μm with a spatial resolution of 3.5 arcsec. Two new scientific results are presented: (1) Analysis of the observations shows that the dust in NGC 7027 contains two spatially distinct components, one of which is associated with and links the unidentified emission features at 8.6 and 11.3 μm; the second component may be graphite. (2) High-quality observations indicate that thermal emission by dust in NGC 7027 is spatially coextensive with emission by ionized hydrogen over a broad spectral range in the midinfrared (~8–28 μm). The intensity distribution of the observed infrared emission was successfully modeled by emission from within the walls of a hollow cylinder, similar to the model for 5-GHz radio emission proposed by Scott (1973).

Catalogue of planetary nebulae and their nuclei.
See Abstr. 002.037.

Catalogue of planetary nebula radial velocities.
See Abstr. 002.056.

A first glance at the Catalogue of the Central Stars of Planetary Nebulae.
See Abstr. 002.109.

The excitation of the triplet lines of O^{2+} in nebulae.
See Abstr. 022.015.

TAURUS: a wide-field imaging Fabry-Perot spectrometer for astronomy.
See Abstr. 034.058.

Low mass Wolf-Rayet stars: theory.
See Abstr. 065.076.

Mass loss from the central star of the planetary nebula IC 3568.
See Abstr. 112.014.

M1-67: a wind-blown bubble carried along by the high-velocity WR star 209 BAC?
See Abstr. 112.045.

Ultraviolet shell formation at V1016 Cygni.
See Abstr. 112.050.

Subluminous Wolf-Rayet stars: observations.
See Abstr. 114.133.

The UV spectrum of the central star of NGC 40.
See Abstr. 114.135.

A reclassification of WC and "O VI" central stars of planetary nebulae, and comparison with Population I WC stars.
See Abstr. 114.136.

Interpretation of line profiles of the symbiotic star V 1016 Cyg.
See Abstr. 114.172.

The mass distribution of white dwarfs and central stars of planetary nebulae.
See Abstr. 126.025.

IUE observations of dust albedo in nebulae.
See Abstr. 134.004.

Measurements of dust albedo in M 8 and some planetaries.
See Abstr. 134.007.

The Bubble Nebula: far-infrared and radio molecular observations of NGC 7635.
See Abstr. 134.023.

High resolution spectroscopy of selected galactic emission nebulae.
See Abstr. 134.034.

A catalog of diffuse nebulae and emission-line stars in NGC 6822.
See Abstr. 158.009.

Radio Sources, X-ray Sources, Cosmic Radiation

141 Radio Sources, Quasars, Pulsars

Radio Sources, Quasars

141.001 **Multifrequency observations of extended radio galaxies V: 3C 31, 3C 33.1, 3C 35, 3C 66B, 3C 129, 3C 130, 3C 223, 3C 310, 3C 390.3 and 4C 48.29.**
W. van Breugel, W. Jägers.
Astron. Astrophys., Suppl. Ser., Vol. 49, 529 - 559 (1982).
Total intensity and polarization maps obtained at several wavelengths with the WSRT are presented for 10 extended radio galaxies. The sources were selected from the 3C survey such that their largest angular extent was $\gtrsim 200''$ and their declination $\gtrsim 25°$. 4C 48.29, a large source with peculiar morphology, was also included.

141.002 **Quasar candidates near 3C 345.**
D. Crampton, M. Rensing.
Publ. Astron. Soc. Pacific, Vol. 94, 440 - 443 (1982).
The slitless spectroscopy technique was used to search for quasars in a 0.74-square-degree field near 3C 345 at the prime focus of the CFHT. Twenty-three probable quasars were discovered, including four quasars which are serendipitous Einstein sources. The results indicate that many more quasars, particularly those of low redshift, can be detected in good seeing conditions by the particular grating/emulsion combination used than has been possible with other similar instruments.

141.003 **Do quasars rotate?**
W. G. Mathews.
Astrophys. J., Vol. 258, 425 - 433 (1982) = Lick Obs. Bull. No. 914.
The kinematics of gaseous disks in quasars and Seyfert galaxies is reexamined. Line profiles of rotating disks are calculated, including relativistic beaming, gravitational redshifts, and transverse Doppler redshifts. The asymmetries of these profiles do not resemble those observed. It is concluded that rotation is an extremely unlikely means of producing broad emission lines in quasars and Seyfert galaxies.

141.004 **A precessing relativistic jet model for 3C 449.**
A. C. Gower, J. B. Hutchings.
Astrophys. J., Lett., Vol. 258, L63 - L66 (1982).
It is shown that the radio structure of 3C 449 can be matched with a model in which the jets are precessing and have relativistic ($\beta \geqslant 0.4$) velocities. The best-fit model implies a precession period of $\sim 10^5$yr and a cone angle which increases with time. A similar model may be relevant for the radio structure of 3C 31. A brief discussion of the implications for 3C 449 is given.

141.005 **Analysis of the absorption-line spectra of eleven quasars with $z_e > 2$.**
S. A. Levshakov, D. A. Varshalovich.
Astrofizika, Tom 18, 49 - 62 (1982). In Russian. English translation in Astrophysics, Vol. 18, No. 1.
High-resolution spectra of 11 quasars investigated by Sargent et al. (1980) and Roberts et al. (1978) are reanalysed. 39 certain and probable absorption systems containing CO lines as well as atom and ion lines are found. For each CO-system line the probability of accidental coincidence is estimated. It is shown that these systems may arise due to absorption by interstellar gas of intervening far galaxies ($z > 1$); the gravitational effect of these galaxies on absorption quasar spectra is discussed.

141.006 **A complete sample of intermediate-strength radio sources selected from the GB/GB2 1400-MHz surveys. I. Radio spectra and optical identifications of sources stronger than 0.55 Jy.**
J. Machalski, J. Maslowski.
Astron. J., Vol. 87, 1132 - 1149 (1982).
The paper presents a complete sample of 237 radio sources selected at 1400 MHz and comprising all sources stronger than 0.55 Jy from the GB and GB2 Green Bank surveys. All sources from the sample have been observed at a frequency of about 5 GHz. Their spectra and the present status of their optical identifications are summarized. A bibliography of references to other flux density measurements and previous identifications, as well as to existing finding charts, structure maps, and redshifts is given. The identification content and radio spectral properties of the sample are compared with corresponding characteristics of other samples, selected at 408, 1400, and 5000 MHz, and containing radio sources of comparable or 2-mag greater strength.

141.007 **A complete sample of intermediate-strength radio sources selected from the GB/GB2 1400-MHz surveys. II. High-resolution maps and optical identifications of sources with very steep spectra.**
J. Machalski, J. Maslowski, J. J. Condon, M. A. Condon.
Astron. J., Vol. 87, 1150 - 1164 (1982).
A complete sample of GB/GB2 sources with $\alpha\,(1400, 5000) \geqslant +0.95$ and stronger than 550 mJy at 1400 MHz, were observed at about 3-arcsec resolution with the VLA. Optical identifications were made using the PSS prints; seven are new. A large fraction of the program sources have a classical double structure. Maps made at 1465 MHz show that the identification rate increases sharply from 25% for narrow doubles to 90% for triple and complex structures.

141.008 **On the apparent superluminal motion of components of the fine structure of 3C 120.**
V. N. Kuril'chik.
Astron. Zh., Tom 59, 657 - 665 (1982). In Russian. English translation in Soviet Astron., Vol. 25, No. 4.
The observed change of the fine structure of the variable radio source 3C 120 is in agreement with a model of ultra-relativistic electrons injection into a magnetic field tube expanding with the distance from the galaxy nucleus. In the framework of this model the observed peculiarities of the components' motion are discussed. The component of radio emission, situated at 5 - 7 msec of arc from the central source, shows an oscillation shift of its position and apparently is nearest to the central source knot of a large-scale structure of the 3C 120 jet.

141.009 **Optical variability and radio structure of extragalactic sources. Evidence for recurrent activity.**
V. A. Hagen-Thorn, I. I. Shevchenko.
Astrofizika, Tom 18, 245 - 254 (1982). In Russian. English translation in Astrophysics, Vol. 18, No. 2.

The amplitude of optical variability of extragalactic objects is compared with the total size of associated radio sources. The former is considered as a measure of activity and the latter as a measure of the age of the sources. This comparison reveals the existence of a characteristic two-peak pattern. The authors consider this as an observational evidence for recurrent activity of the objects.

141.010 **IUE observations of quasars 3C249.1 and 3C232.**
D. Dultzin-Hacyan, L. Salas, E. Daltabuit.
Astron. Astrophys., Vol. 111, 43 - 49 (1982),

Spectra of quasars 3C249.1 and 3C232 were obtained with the International Ultraviolet Explorer Satellite. In 3C232 the Lyman series is observed up to Lyδ and two hydrogen ionizing lines shortward of Lyδ are observed. The possibility of identifying weak Fe II features is pointed out. Emissions at λλ 1116, 1146, 1166, and 1424–28 are clearly present in both spectra and have not been identified. Neither absorption lines nor the Lyman discontinuity are observed. The absolute spectral energy distribution is fitted by a power law and a comparison is made of spectral index and absolute flux with previous optical and infrared observations. Evidence is found of the possible variation of absolute flux in 3C249.1 and of spectral index in 3C232.

141.011 **Wind instability in clouds of radio galaxies.**
A. O. Vrobel', V. M. Kontorovich.
Pis'ma Astron. Zh., Tom 8, 330 - 333 (1982). In Russian. English translation in Soviet Astron. Lett., Vol. 8.

The wind instability in the magnetospheric model of radio galaxy clouds is characterized by a lower critical velocity than in the Kelvin-Helmholtz instability and by the existence of a singled out characteristic scale of most rapidly increasing surface waves. This makes it possible to determine the ratio of plasma density in the cloud to that in the gas flow of the intergalactic medium.

141.012 **Observational tests of QSO amplification by condensed objects in galactic halos.**
W. C. Keel.
Astrophys. J., Lett., Vol. 259, L1 - L5 (1982) = Lick Obs. Bull., No. 921.

The distribution of projected separations between QSOs and bright galaxies is derived, taking into account the sizes of these separations and those of search fields. This distribution is well fitted by the $R^{-1/4}$ distribution characteristic of stellar spheroids and globular cluster systems. If this is not a result of as yet unknown selection effects in optical search procedures, it is strong evidence for a connection between bright galaxies and detectability of QSOs. Gravitational lens models involving amplification of faint QSOs by individual halo stars could explain this result. An observational test for these models is described.

141.013 **Diffusion of electrons in radio galaxies.**
E. Valtaoja.
Astron. Astrophys., Vol. 111, 213 - 219 (1982).

The author investigates the possible role of diffusion in the observed morphology and spectral index distribution of radio galaxies. If the synchrotron-emitting electrons are accelerated at a moving point source, and then diffuse freely into infinite space, structures similar to observed radio lobes are produced. Model predictions for the radio galaxies DA 240 and 3C 236 agree reasonably well with observations. The estimated diffusion coefficients are compared with those proposed for other types of objects.

141.014 **Optical identification of the radio source 0104–408.**
H. G. Walter, R. M. West.
Astron. Astrophys., Vol. 111, 357 (1982).

The radio source 0104–408 which is a representative of the extragalactic radio reference frame has been optically identified with a 19^m stellar like object and its position measured.

141.015 **A catalogue of radio sources within 30′ of Cep A.**
V. A. Hughes, M. R. Viner, J. G. A. Wouterloot.
Astron. Astrophys., Vol. 111, 358 - 361 (1982).

A $1\overset{\circ}{.}0 \times 1\overset{\circ}{.}0$ region around the Cep A condensation was mapped at λ 21 cm with a resolution of 24″× 27″ and a flux density limit at the field centre of 3.75 mJy, using the Westerbork Synthesis Radio Telescope. The positions of the 20 detected sources are shown in relation to the distribution of CO emission measured by Sargent (1977). A comparison of the λ 21 cm flux densities with previously published fluxes and with new observations made at the Algonquin Radio Observatory establishes that 4 of the sources are thermal. No stars as early as B 0 appear to have formed, implying that the area around Cep A is not one where massive stars are forming.

141.016 **Radio observations of early-type galaxies.**
C. R. Jenkins.
Mon. Not. R. Astron. Soc., Vol. 200, 705 - 731 (1982).

A complete sample of 34 nearby early-type galaxies, based on the Arecibo survey by Dressel & Condon (1978) of objects from the Uppsala General Catalogue, has now been mapped at radio frequencies. New data are presented for 23 galaxies in the sample. The majority of the sources show strong jet-like structures. A strong correlation between radio luminosity and size has been found in the range $23.0 < \log P(2.7\ \mathrm{GHz}) < 25.0$ (P in W Hz^{-1}; $H_0 = 50$ km s^{-1} Mpc^{-1}). The radio luminosity function has been determined; although for a sample selected at 2.3 GHz, it agrees well with that found at lower frequencies.

141.017 **A deep radio/optical survey near the North Galactic Pole. I. The 5C12 catalogue.**
C. R. Benn, G. Grueff, M. Vigotti, J. V. Wall.
Mon. Not. R. Astron. Soc., Vol. 200, 747 - 766, microfiche MN 200/2 (1982).

A deep survey (5C12) of a region close to the North Galactic Pole has been carried out with the Cambridge One-Mile telescope. Within a region of 4° diameter, 299 sources were detected at 408 MHz, while 65 sources were detected at 1407 MHz in a concentric region of 1° in diameter. Source counts are presented and confirm the results from previous 5C surveys. Spectral index distributions are constructed for samples of sources selected at 408 and 1407 MHz and compared with results for samples selected at higher flux densities.

141.018 **Quasars, companion galaxies and Poisson statistics.**
A. Webster.
Mon. Not. R. Astron. Soc., Vol. 200, 47P - 49P (1982).

Arp has presented a sample of quasars lying close to the companion galaxies of bright spirals, from which he estimates a value of 10^{-17} for the probability that the galaxies and quasars are sited independently on the celestial sphere. The author draws attention to an obscure part of Arp's calculation which he has been unable to repeat. In consequence, it is not clear that the sample contains any evidence damaging to the popular notion that the redshifts of quasars indicate distance through the Hubble Law.

141.019 **Distribution of identified absorption features of QSOs at different emission redshifts.**
D. Basu.
Astrophys. Lett., Vol. 22, 139 - 142 (1982).

It has been reported that absorption features in QSO spectra are more likely to be observed at higher emission redshift (z_{em}) objects. The author's analysis shows that the presence of identified absorption lines of QSOs is an exponential function of z_{em}. An upper limit has been predicted for the percentage of QSOs showing identified absorption lines at various values of z_{em}. Furthermore, results strongly indicate that the appearance of absorbing clouds with individual QSOs is closely associated with the creation of the objects themselves, the two may have a common origin.

141.020 **Some features in the spectra of extragalactic radio sources.**
O. B. Slee.
Proc. Astron. Soc. Australia, Vol. 4, 184 - 187 (1981).

This paper is concerned with a spectral property which may be indicative of radio source evolution. A dependence of spectral index on redshift or emitted power is identified for the radio galaxies, and for the resolved sources a relationship is found between spectral index and linear dimensions.

141.021 **Optical identification of the strong radio source PKS 0511–48.**
J. G. Robertson, R. M. Smith.
Proc. Astron. Soc. Australia, Vol. 4, 187 - 189 (1981).

141.022 **Deep photography of radio sources with the Cambridge CCD system.**
C. D. Mackay, D. Astill, M. J. Batty, D. L. Jauncey, A. E. Wright, R. W. Hunstead, D. C. Morton, J. G. Robertson.
Proc. Astron. Soc. Australia, Vol. 4, 195 - 197 (1981).

The authors present preliminary results from a programme designed to produce deep images of radio source fields drawn from the Parkes 2700 MHz and Molongolo 408 MHz catalogues using a charge-coupled-device (CCD) camera system. The programme is directed at a search both for faint extensions and nebulosity around radio QSOs and BL Lac objects and for faint objects in otherwise empty radio source fields; a detailed examination of the morphology of selected radio galaxies is also included.

141.023 **QSO Lyman limit absorption.**
D. Tytler.
Nature, Vol. 298, 427 - 432 (1982).

The redshift distribution of QSO absorption systems which are optically thick in the Lyman continuum matches that of a non-evolving population of absorbers in a standard Friedmann cosmological model over the redshift range 0.4–3.5. The density of absorbers per unit velocity in the QSO rest frame is roughly constant for $-0.01 \lesssim v/c \lesssim 0.2$ and shows a rapid increase with QSO emission redshift, in accord with an 'intervening' origin for the absorbers but contrary to expectation were the material intrinsic to the QSOs.

141.024 **Twinkle, twinkle quasi star ....**
M. Abramowicz.
Nature, Vol. 298, 789 - 790 (1982).

141.025 **Further investigations on possible correlations between QSOs and the Lick catalogue of galaxies.**
J.-L. Nieto, M. Seldner.
Astron. Astrophys., Vol. 112, 321 - 329 (1982).

The authors analyse the different contributing factors in a previous study by Seldner and Peebles (1979) who found statistically significant evidence for a correlation between a list of 382 QSOs at $|b| > 40°$, $\delta > -23°$ and the Lick counts of galaxies ($m < 19.0$). Taking into account these different factors and using a larger sample of QSOs whose detection is not expected to bias the statistical results, the authors conclude that: (1) There is no longer statistically significant evidence for an excess of Lick galaxies close to QSOs from our list considered as a whole. (2) The whole correlation function $w(\theta)$ shows in general a hump between 1°.5 and 2°. (3) Numerical simulations seem to indicate that the error bars determined by assuming that cell numbers are independent are slightly underestimated.

141.026 **Spectra, dimensions and luminosities of radio sources in the Culgoora-3 list.**
O. B. Slee, B. C. Siegman, P. S. Mulhall.
Proc. Astron. Soc. Australia, Vol. 4, 278 - 307 (1982).

The authors present an analysis of spectra constructed by combining Culgoora-3 flux densities with measurements made at many frequencies by a large number of radio telescopes. The flux measurements have been fitted with first-, second- and third-degree polynomials in an attempt to define the spectral index more accurately than hitherto at a number of standard frequencies between 80 and 5120 MHz; by this means the authors hope to find spectral features which are common to sources associated with the same type of optical object (or optical field in the case of unidentified sources). About 300 sources in the Culgoora-3 list were resolved by the 1'.9 arc beam at 160 MHz; 163 maps of the larger sources were published with the Culgoora-3 list by Slee (1977). In this paper the authors make a systematic analysis of the maps by fitting elliptical Gaussian brightness distributions, from which angular and linear dimensions are derived. They list the remainder of the resolved sources and estimate their maximum angular dimensions from the observed beam broadening.

141.027 **An assessment of the significance of quasar alignments.**
A. S. Trew, S. V. M. Clube, A. Savage, R. G. Clowes.
Mon. Not. R. Astron. Soc., Vol. 200, 785 - 792 (1982).

The authors have undertaken a systematic search in two fields of quasar candidates for alignments similar to those reported by Arp and Hazard (1980) in order to test the statistical significance of these associations. Comparisons with control fields generated by a Monte Carlo technique showed that when appropriate allowance has been made for clustering there is only marginal evidence for a statistically significant excess of aligned triplets over that expected by chance. It has not, however, been possible to examine the significance of the redshift patterns found by Arp and Hazard (1980) as reliable redshifts do not exist for the majority of the quasar candidates.

141.028 **High-resolution studies of M 87 at long radio wavelengths.**
M. Charlesworth, R. E. Spencer.
Mon. Not. R. Astron. Soc., Vol. 200, 933 - 940 (1982).

Contour maps of the radio emission at 408 and 1666 MHz from the jet of M 87 have been produced with the MTRLI. The structure found at long wavelengths is generally similar to that shown previously at short radio and optical wavelengths, and the map at 1666 MHz reveals that most of the knots of emission in the jet have angular sizes of around 1 arcsec. Data obtained with a single-baseline interferometer at 151 MHz have been combined with the maps to produce low-frequency spectra. These show evidence of low-frequency absorption for several of the knots, which may be produced by free–free absorption in dense, filamentary external clouds.

141.029 **Spheroidal gravitational lenses and the triple quasar.**
D. Narasimha, K. Subramanian, S. M. Chitre.
Mon. Not. R. Astron. Soc., Vol. 200, 941 - 950 (1982).

A model for the triple quasar Q1115+080 A, B, C based on a single gravitational lens is proposed. It is demonstrated that the resulting five-image configuration produced by a lens galaxy having a spheroidal mass component is in reasonable accord with the observed features of the triple quasar. The model naturally leads to the resolution of the brightest member into two closely separated components of approximately equal intensity. The mass and redshift of the lens galaxy and time delays from various images are calculated as a function of the core radius of the galaxy.

141.030 **Relativistic beaming and quasar statistics.**
M. J. L. Orr, I. W. A. Browne.
Mon. Not. R. Astron. Soc., Vol. 200, 1067 - 1080 (1982).
The statistical predictions of a unified scheme for the radio emission from quasars are explored. This scheme attributes the observed differences between flat- and steep-spectrum quasars to projection and the effects of relativistic beaming of the emission from the nuclear components. The authors use a simple quasar model consisting of a compact relativistically beamed core with spectral index zero and unbeamed lobes, spectral index −1, to predict the proportion of flat-spectrum sources in flux-limited samples selected at different frequencies, and conclude that the statistical properties of quasars are entirely consistent with the predictions.

141.031 **Selected area surveys at 5 GHz – I. Catalogue for the 22^h, – 18° region.**
J. V. Wall, A. Savage, A. E. Wright, J. G. Bolton.
Mon. Not. R. Astron. Soc., Vol. 200, 1123 - 1134 (1982).
The authors present results from a deep 5-GHz survey for extragalactic radio sources over an area of 27 deg^2. The survey was carried out with the Parkes 64-m telescope, and produced a catalogue of 75 sources complete to S(5 GHz) = 32 mJy, corresponding to a source surface density of ~8000 sr^{-1}. Source counts and spectral indices are briefly discussed.

141.032 **Selected area surveys at 5 GHz – II. Optical identifications for the 22^h, −18° region.**
A. Savage, J. G. Bolton, J. V. Wall.
Mon. Not. R. Astron. Soc., Vol. 200, 1135 - 1141 (1982).
From a deep survey at 5 GHz covering 27 deg^2, new identifications are suggested for 19 radio sources, 17 with galaxies and two with QSOs. The identifications were made principally from a deep IIIa-F plate taken on the UK 1.2-m Schmidt telescope. Only a few per cent of sources with S(5 GHz) < 100 mJy can be identified with QSOs.

141.033 **Nonlinear spectra of radio sources at rigid injection.**
A. E. Kochanov, V. M. Kontorovich.
Izv. vuzov. Radiofiz., Tom 24, 1047 - 1053 (1981). In Russian.
Abstr. in Ref. zh., 51. Astron., 7.51.164 (1982).

141.034 **Setting-up time of power-law spectra of relativistic electrons and plasma radiation in discrete extragalactic radio sources.**
V. V. Vysochanskij, V. G. Pisarenko.
Kosm. issled. na Ukraine, Kiev, 1981, No. 15, p. 47 - 57. In Russian. – Abstr. in Ref. zh., 51. Astron., 7.51.165 (1982).

141.035 **Calculation of the process of forming of power-law spectra of relativistic electrons in an astrophysical turbulent synchrotron reactor.**
I. L. Verbitskij, V. V. Vysochanskij, V. G. Pisarenko.
Nelinejn. i dinam. protsessy fiz. i mekh., Kiev, 1981, p. 28 - 37. In Russian. – Abstr. in Ref. zh., 51. Astron., 7.51.166 (1982).

141.036 **Formation des raies permises et interdites dans les quasars.**
A. M. Dumont.
Noyaux actifs des galaxies, (see 012.004), 20 pp.

141.037 **A comparison of the emission line properties between quasars and type 1 Seyfert galaxies.**
C.-C. Wu, A. Boggess, T. R. Gull.
Advances in ultraviolet astronomy, (see 012.006), p. 160 - 164 (1982).
For quasars and Seyfert galaxies, the equivalent width of C IV λ1550 increases as the continuum luminosity of an object decreases. A reasonable interpretation is that the covering factor increases as luminosity decreases. This is consistent with the fact that the Lα and C IV equivalent widths for Seyferts are a factor of 2 larger than those for high redshift quasars. The C IV/C III] ratio is about 5 for many Seyferts while it is about 2 for high redshift quasars.

141.038 **Multifrequency observations of the quasar 1156+295.**
A. E. Glassgold, J. N. Bregman, P. J. Huggins.
Advances in ultraviolet astronomy, (see 012.006), p. 201 - 204 (1982).
The recent outburst in the quasar 1156+295 was discovered in April 1981. The authors obtained short and long wavelength spectra on three occasions when the object was very bright. The UV continuum in all cases is a steeply falling power law. Closely simultaneous observations were made at radio, infrared, and optical frequencies. The continuum is less steep at optical and infrared frequencies, and the overall spectra show little change in shape with time.

141.039 **X-ray and UV spectra of two quasars: PKS 0637−75 and PKS 1004+13 (4C 13.41).**
M. Elvis, G. Fabbiano.
Advances in ultraviolet astronomy, (see 012.006), p. 205 - 208 (1982).
The authors observed the two quasars 4C 13.41 and PKS 0637−75 in December 1980 with IUE. 4C 13.41 has the lowest L_x/L_{opt} measured in quasars. PKS 0637−75 was observed quasi-simultaneously with the Einstein Observatory and a spectrum obtained with the IPC. The UV and soft X-ray spectra have been combined and are discussed together with the optical and infrared data for both objects.

141.040 **I risultati delle ultime osservazioni eseguite sul quasar 3C 273.**
G. M. De Tullio.
Coelum, Vol. 50, 193 - 203 (1982).

141.041 **Spectral index – flux density relation for extragalactic radio sources found in metre-wavelength surveys.**
Gopal-Krishna, H. Steppe.
Astron. Astrophys., Vol. 113, 150 - 154 (1982).
By considering two-point spectral indices of a total of 1330 extragalactic radio sources appearing in 8 well observed metre-wavelength samples, the dependence of median (decimetre-wave) spectral index, α(median), on (408 MHz) flux density, S_{408}, is investigated. It is found that a statistically significant correlation exists between these two basic parameters of extragalactic radio sources.

141.042 **Statistical correction of projection of radio sources on the sky and application to the apparent size-redshift and linear size-line width relations.**
L. Nottale.
Astron. Astrophys., Vol. 113, 223 - 230 (1982).
The author demonstrates that it is possible to get information on the projection angle of double radio sources on the plane of the sky by using the component size/largest angular size ratio, Φ = CS/LAS, under the hypothesis that it is a characteristic parameter with not too large a dispersion.

141.043 **On the behaviour of QSO space density beyond z = 3.5.**
G. Mathez, L. Nottale.
Astron. Astrophys., Vol. 113, 336 - 339 (1982).
The predictions of Mathez' (1978) luminosity evolution model for QSOs are compared with counts of color selected QSOs and objective prism surveys. The authors find a good agreement with the compilation and new data by Koo and Kron (1982) provided that the quasar space density is constant

up to $z = 5 - 6$ at least. The model is also in good agreement with the number of quasars found in Osmer's (1982) survey, at low z as well as in the range $3.7 < z < 4.7$ where it predicts much fewer QSOs than Osmer's estimate.

141.044 **A rapid outburst of BL Lac at 2.72 GHz.**
W. Reich, P. Steffen.
Astron. Astrophys., Vol. 113, 348 - 349 (1982).

Monitoring the flux density of BL Lac at a frequency of 2.72 GHz, the authors observed in detail a rapid outburst in August/September 1980 lasting for 20 d and reaching a maximum in flux density of 2.2 Jy. The authors discuss these observations in terms of current models for compact extragalactic sources.

141.045 **High dynamic range mapping of strong radio sources, with application to 3C84.**
J. E. Noordam, A. G. de Bruyn.
Nature, Vol. 299, 597 - 600 (1982).

A new reduction procedure is described to correct data obtained by the Westerbork Synthesis Radio Telescope. The technique allows very high dynamic range mapping of strong radio sources by also using redundant interferometers. The performance is illustrated with 6-cm and 21-cm continuum maps of 3C84, which have a dynamic range of about 10,000:1.

141.046 **Optical polarization position angle versus radio source axis in radio galaxies.**
R. R. J. Antonucci.
Nature, Vol. 299, 605 - 606 (1982).

The author has obtained optical spectropolarimetry data and VLA radio maps for a sample of radio galaxies to identify and interpret alignment effects. The results show that most radio galaxies with optical nuclear emission are polarized. There is a population showing quasar alignment effect, and unexpectedly there is also a population showing a perpendicular relationship. Emission-line polarizations exhibit various behaviours but in all cases make it very likely that the polarizations of both groups arise from scattering or dust transmission rather than the emission process itself.

141.047 **0.35–3.5 micron photometry of X-ray emitting QSOs.**
M. L. Sitko, W. A. Stein, Y.-X. Zhang, W. Z. Wiśniewski.
Astrophys. J., Vol. 259, 486 - 494 (1982).

Visual and infrared photometry of X-ray emitting QSOs is reported. The strongest emitters of X-ray relative to visual emission are those with the strongest nonthermal visual-infrared emission. This result is most easily explained if the X-rays are Compton-scattered photons of lower initial frequency rather than a simple extrapolation of the visual-infrared spectrum into the X-ray region.

141.048 **Gaseous halos in radio galaxies and jet disruption.**
M. D. Smith.
Astrophys. J., Vol. 259, 522 - 529 (1982).

Jet disruption at the edge of a gaseous halo is investigated for radio galaxies. Correlations between radio power, visual magnitude of the associated galaxies, and the positions of greatest radio emissivity are obtained and investigated for D I radio galaxies. They agree with a model where accretion through cooling from a halo fuels the jet. The halo consists of accreted intergalactic gas and requires mass outflow to be a large fraction of mass inflow. The brightness peaks correspond to the halo edges, and radio power is proportional to jet power.

141.049 **Angular diameter-redshift relations for extragalactic radio sources.**
S. E. Okoye, L. I. Onuora.
Astrophys. J., Vol. 260, 37 - 43 (1982).

The relativistic formula for the observed angular sizes of cosmological objects is shown to be expressible as a power law function of $(1 + z)$ which thus provides a more meaningful framework for interpreting θ-z diagrams for extragalactic radio sources.

141.050 **Theoretical quasar emission-line ratios. V. Balmer continuum emission.**
R. C. Puetter, P. D. LeVan.
Astrophys. J., Vol. 260, 44 - 55 (1982).

The authors present isothermal, isobaric models of quasar emission-line regions. The new calculations include an improved treatment of radiative transfer in the bound-free continua based upon a generalization of frequency integrated line transfer techniques. It is found that large Balmer continuum emission can be obtained easily without compromising the observed line ratios.

141.051 **Radio sources in the association of the reflection nebulae Monoceros-R2.**
V. K. Kononov, T. B. Pyatunina.
Pis'ma Astron. Zh., Tom 8, 593 - 595 (1982). In Russian. English translation in Soviet Astron. Lett., Vol. 8.

Observations of the region of sky occupied by the association of reflection nebulae Monoceros-R2 were made with the radiotelescope RATAN-600 at wavelength 7.6 cm. Eight new sources were found, among them emission-reflection nebulae NGC 2170 and DG 92 and a compact radio source, which is probably connected with a group of reflection nebulae around NGC 2182. The spatial density of radio sources inside the association Mon R2 is 4 times higher than the statistical mean.

141.052 **Observational limitations of the Doppler theory of quasars.**
J. V. Narlikar, K. Subramanian.
Astrophys. J., Vol. 260, 469 - 483 (1982).

This paper is mainly concerned with the hypothesis that the redshift of a quasar is entirely due to the Doppler effect arising from its high speed of ejection in a relatively nearby center of explosion. The authors examine the viability of this hypothesis in the light of the available data on the aligned triplets of quasars discovered by Arp and Hazard and by Saslaw. Taking the view that the quasars in a triplet are physically associated, the authors show how to compute the various parameters of the Doppler problem and put constraints on quasar ejection scenarios. This enables the authors to examine critically Hoyle's recent hypothesis that quasars emit the bulk of their radiation in a specified backward cone. They find that the four triplets do provide prima facie evidence for such a hypothesis and suggest further checks on the Doppler model. In particular, their model predicts a very small but nonzero fraction of blueshifted quasars.

141.053 **Apparent proper motions of the galactic center compact radio source and PSR 1929+10.**
D. C. Backer, R. A. Sramek.
Astrophys. J., Vol. 260, 512 - 519 (1982).

The techniques of differential radio astrometry with a connected-element interferometer are presented. The motion of the compact, nonthermal radio source in the galactic center is found to be consistent with the secular parallax expected for an object at rest in the galactic center. The proper motion of PSR 1929+10 has been improved, and new limits on its trigonometric parallax are determined.

141.054 **PKS 2000–330: a quasi-stellar radio source with a redshift of 3.78.**
B. A. Peterson, A. Savage, D. L. Jauncey, A. E. Wright.
Astrophys. J., Lett., Vol. 260, L27 - L29 (1982).

Observations with the Image-Tube Dissector Scanner on the Anglo-Australian Telescope of the 19th magnitude quasi-stellar object identified with the radio source PKS 2000–330 show it to have an emission redshift of 3.78.

141.055 **The redshift distribution of X-ray selected quasars: is it really anomalous?**
G. Zamorani.
Astrophys. J., Lett., Vol. 260, L31 - L35 (1982).

X-ray selected active galactic nuclei have a redshift distribution which is significantly different from that of optically selected quasars with the same optical magnitude. The author shows that no qualitative difference in the X-ray properties of X-ray selected and optically selected quasars is required in order to explain the deficiency of high-redshift quasars in the available samples of X-ray selected objects. The redshift distribution of such samples is well reproduced by using evolutionary models and the optical luminosity function derived from optically selected samples plus the correlation between X-ray and optical luminosities found in a large sample of optically selected active galactic nuclei.

141.056 **Quantitative measures of slitless spectra of QSOs.**
B. G. Vaucher, T. J. Kreidl, N. G. Thomas, A. A. Hoag.
Astrophys. J., Vol. 261, 18 - 24 (1982).

The authors have developed procedures for interactively evaluating low-resolution slitless spectra obtained at the prime focus of the Mayall and CTIO 4 m telescopes. For grating prism spectra microphotometer measures are processed to give estimates of redshifts, equivalent widths, and rest-system magnitudes. The effectiveness of these procedures is checked by repeating a $\delta = -40°$ grating prism survey that has been evaluated by P. S. Osmer.

141.057 **The kinematics of the broad-line emission gas in quasars and Seyfert nuclei.**
J. Kwan, T. J. Carroll.
Astrophys. J., Vol. 261, 25 - 29 (1982).

The authors propose a kinematic model in which the broad-line emission clouds orbit the central continuum source in nearly parabolic orbits in order to produce a symmetric Lyα profile in spite of the angular asymmetry in the Lyα emission from an individual optically-thick cloud. Theoretical profiles are generated and compared with logarithmic line shapes, and the pros and cons of this model and others proposed are discussed.

141.058 **Spectroscopy of the QSO pair Q0028+003/Q0029+003.**
P. A. Shaver, A. Boksenberg, J. G. Robertson.
Astrophys. J., Lett., Vol. 261, L7 - L12 (1982).

A prominent narrow absorption-line system has been found in the spectrum of Q0029+003 ($z_{em} = 2.222$) at redshift $z_{abs} = 1.7334$, only 190 km s^{-1} from the emission-line redshift of Q0028+003 ($z_{em} = 1.7317$). It may be due to a galaxy located in a cluster containing Q0028+003; or it may arise in a halo associated with Q0028+003.

141.059 **Color imaging of QSO-galaxy interactions.**
J. B. Hutchings, B. Campbell, D. Crampton.
Astrophys. J., Lett., Vol. 261, L23 - L29 (1982).

Deep imaging in several colors has been carried out with the Canada-France-Hawaii Telescope of low-redshift QSOs 1229+204 and 0351+026. Both objects show extended luminosity about the QSO and an interacting, extended companion whose color is bluer than the QSO nucleus. Colors, luminosities, and sizes are presented, and it is suggested that the QSO phenomenon is related to the encounter, and a burst of hot star formation has occurred in the companion objects.

141.060 **Search for X-ray emission from optically quiet, compact radio sources.**
J. E. Ledden, S. L. O'Dell.
Bull. American Astron. Soc., Vol. 14, 602 - 603 (1982). Abstract.

141.061 **Cutoff of QSOs.**
J. M. Barnothy, M. F. Barnothy.
Bull. American Astron. Soc., Vol. 14, 631 - 632 (1982). Abstract.

141.062 **QSOs visible from two opposite directions.**
M. F. Barnothy, J. M. Barnothy.
Bull. American Astron. Soc., Vol. 14, 632 (1982). – Abstract.

141.063 **On the redshift and luminosity distribution of X-ray selected QSOs.**
P. Giommi, Y. Avni, I. M. Gioia, T. Maccacaro, J. Stocke, G. Zamorani.
Bull. American Astron. Soc., Vol. 14, 632 (1982). – Abstract.

141.064 **Luminosity function of high redshift quasars.**
B. G. Vaucher, D. W. Weedman.
Bull. American Astron. Soc., Vol. 14, 632 (1982). – Abstract.

141.065 **The morphology and energetics of discrete optical events in compact extragalactic objects.**
J. T. Pollock.
Bull. American Astron. Soc., Vol. 14, 634 - 635 (1982). Abstract.

141.066 **The 86 GHz polarization of variable extragalactic sources.**
R. Barvainis, R. Predmore, W. A. Dent.
Bull. American Astron. Soc., Vol. 14, 635 (1982). – Abstract.

141.067 **The distribution of absorption line redshift of quasars and its origin.**
L. Z. Fang, Y. Q. Chu.
Bull. American Astron. Soc., Vol. 14, 635 (1982). – Abstract.

141.068 **Correlated outbursts in extragalactic variable sources at radio and optical wavelengths.**
T. J. Balonek, W. A. Dent.
Bull. American Astron. Soc., Vol. 14, 635 (1982). – Abstract.

141.069 **Milliarcsecond morphology of the twin quasar 0957+561A, B.**
N. L. Cohen, M. V. Gorenstein, I. I. Shapiro, E. E. Falco, J. M. Marcaide, A. E. Rogers, B. E. Corey.
Bull. American Astron. Soc., Vol. 14, 635 (1982). – Abstract.

141.070 **Optical variability, absolute luminosity, and the Hubble diagram for QSOs.**
A. J. Pica, J. T. Pollock, A. G. Smith, J. R. Webb, R. J. Leacock.
Bull. American Astron. Soc., Vol. 14, 635 (1982). – Abstract.

141.071 **Derivation of the quasar X-ray luminosity function.**
Y. Avni.
Bull. American Astron. Soc., Vol. 14, 644 (1982). – Abstract.

141.072 **Emission lines in the spectra of broad absorption-line QSOs.**
V. T. Junkkarinen.
Bull. American Astron. Soc., Vol. 14, 647 (1982). – Abstract.

141.073 **Quasars as clusters of accreting black holes.**
A. G. Pacholczyk, W. R. Stoeger.
Bull. American Astron. Soc., Vol. 14, 648 (1982). – Abstract.

141.074 **A close look at a close pair of QSOs: VLBI observations of 1038 + 528 A, B.**
J. M. Marcaide, I. I. Shapiro, B. E. Corey, W. D. Cotton, N. Bartel, N. L. Cohen, M. V. Gorenstein, A. E. E. Rogers,

A. R. Whitney, L. Bååth, J. Romney.
Bull. American Astron. Soc., Vol. 14, 648 (1982). – Abstract.

141.075 **Detection of new compact radio component near Q0957 + 561B: quasar image or galactic core?**
M. V. Gorenstein, I. I. Shapiro, N. L. Cohen, E. E. Falco, J. M. Marcaide, A. E. E. Rogers, A. R. Whitney, B. E. Corey, R. W. Porcas, R. A. Preston, A. Rius.
Bull. American Astron. Soc., Vol. 14, 648 (1982). – Abstract.

141.076 **X-ray observations of supernova-like quasars.**
J. N. Bregman, P. J. Huggins, A. E. Glassgold.
Bull. American Astron. Soc., Vol. 14, 648 (1982). – Abstract.

141.077 **Apparent bending of relativistic jets in quasars by geometric effects.**
A. P. Marscher.
Bull. American Astron. Soc., Vol. 14, 648 - 649 (1982). Abstract.

141.078 **The stability of a magnetically confined radio jet.**
H. Cohn.
Bull. American Astron. Soc., Vol. 14, 649 (1982). – Abstract.

141.079 **Multifrequency VLA observations of the narrow angle tail radio source NGC 1265.**
C. P. O'Dea, F. N. Owen.
Bull. American Astron. Soc., Vol. 14, 659 (1982). – Abstract.

141.080 **The curvature of radio jets and tails in the intra-cluster media of Abell 1446 and 2220.**
J. O. Burns, T. J. Balonek.
Bull. American Astron. Soc., Vol. 14, 659 (1982). – Abstract.

141.081 **Radio and X-ray observations of the halo source in A1367.**
G. Gavazzi, G. Trinchieri.
Bull. American Astron. Soc., Vol. 14, 664 (1982). – Abstract.

141.082 **Preliminary results from the Arecibo-Green Bank low-frequency variability monitoring program.**
B. Dennison, J. J. Broderick, S. L. O'Dell, H. Payne, J. J. Condon, D. A. Altschuler.
Bull. American Astron. Soc., Vol. 14, 664 (1982). – Abstract.

141.083 **Radio images of eighteen small diameter galactic radio sources.**
D. Chance, R. Becker, R. White.
Bull. American Astron. Soc., Vol. 14, 666 (1982). – Abstract.

141.084 **Moving jets in SS 433.**
T. Daishido.
Magnetospheric phenomena of celestial bodies, (see 012.011), p. 109 - 111 (1982). In Japanese.

141.085 **Large-scale structures of magnetic fields of radio galaxies and quasars.**
S. Ikeuchi.
Magnetospheric phenomena of celestial bodies, (see 012.011), p. 123 - 129 (1982). In Japanese.

Large-scale structures of magnetic fields of radio galaxies and quasars are reviewed. In the first place, radio structures and morphological sequences of radio sources are summarized in relation to the active phenomena of galaxies and quasars. Secondly, the configuration and strength of magnetic fields are estimated by several methods. Some tendencies on magnetic fields are summarized.

141.086 **MERLIN observations of superluminal radio sources.**
I. W. A. Browne, R. R. Clark, P. K. Moore, T. W. B. Muxlow, P. N. Wilkinson, M. H. Cohen, R. W. Porcas.
Nature, Vol. 299, 788 - 793 (1982).

Powerful extragalactic radio sources may all be members of a single family in which those showing superluminal behaviour are the ones in which a relativistic jet is directed towards the observer. The superluminal sources are shown to have extended radio structures the form of which is consistent with such a unified scheme. This interpretation requires that the Hubble constant H_0 should be nearer to 100 than to 50 km s^{-1} Mpc^{-1}.

141.087 **An evaluation of spherically symmetric wind models for broad absorption line QSOs.**
J. Drew, J. Giddings.
Mon. Not. R. Astron. Soc., Vol. 201, 27 - 50 (1982).

The subject of this study are those QSOs exhibiting broad absorption lines in their spectra. In the absence of any firm evidence fixing the location of the absorbing matter, it was considered appropriate to evaluate the suitability of a class of models rather than attempt any detailed analysis. Line profiles have been calculated for a wide range of parameter values. An optically thick, continuous outflow model with a non-monotonic or decelerating velocity field, able to multiply scatter line photons, looks very promising. The question of the overall spatial symmetry of broad absorption line QSOs remains open.

141.088 **Do quasar ley lines really exist?**
A. Webster.
Mon. Not. R. Astron. Soc., Vol. 201, 179 - 190 (1982).

The hypothesis that the distribution of the quasars on the celestial sphere contains an unexpectedly large number of well-aligned triples is tested by applying, to the Cerro Tololo objective-prism sample, a shape-statistic which was originally developed to investigate whether neolithic standing stones were deliberately sited on ley lines. It is found that alignment in triples is not a conspicuous feature of the quasars in this sample.

141.089 **A complete sample of extragalactic radio sources at 1 Jy at 408 MHz – II. The optical observations.**
J. R. Allington-Smith, M. A. C. Perryman, M. S. Longair, J. E. Gunn, J. A. Westphal.
Mon. Not. R. Astron. Soc., Vol. 201, 331 - 344 (1982).

This is the second paper in a series investigating the properties of a complete sample of 59 B2 radio sources with flux densities between 1 and 2 Jy at 408 MHz. The authors present the results of a deep optical survey of the fields of 47 of these sources which are unidentified or have faint identifications on the National Geographic–Palomar Sky Survey. The proportion of the sample with identifications brighter than $m_r = 23$ is estimated to be ~80 per cent. The apparent magnitude distribution of galaxies peaks between 19 and 20 mag, while the distribution of QSOs (~24 per cent of the total) reaches a maximum between 20 and 21 mag. Some of the faintest identifications are possibly close associations of galaxies in the cores of distant clusters.

141.090 **Extended radio jets in the high-redshift quasars 3C 9 and 280.1.**
G. Swarup, R. P. Sinha, D. J. Saikia.
Mon. Not. R. Astron. Soc., Vol. 201, 393 - 400 (1982).

The authors present total intensity and polarization observations made with the VLA at 5 GHz of two high-redshift quasars 3C 9 and 280.1. The quasars appear to have extended one-sided radio jets, which are the most luminous yet to be reported. The authors discuss in greater detail the jet in 3C 280.1 which was observed with rather higher resolution.

141.091 **Objetos con exceso ultravioleta en la vecindad del polo galáctico sur.**
C. Torres, L. E. Campusano.
Rev. Mexicana Astron. Astrofis., Vol. 6, (see 012.017), 23 -

25 (1981).

The results of a search of ultraviolet excess objects ($U-B \lesssim -0.35$) are presented for an area of 6°5 × 6°5 centered on ($00^h 53^m$, −28°00′). The pairs of plates employed in the search were obtained with the 70/100 cm Maksutov telescope operated by the University of Chile. It is reported that 347 ultraviolet excess objects have been found, rediscovering 7 of the 17 blue stars from the Haro and Luyten (1962) list that are located in the search area.

141.092 **Hacia una muestra más completa de cuasares en un area selecta del polo galáctico sur.**
L. E. Campusano.
Rev. Mexicana Astron. Astrofis., Vol. 6, (see 012.017), 26 (1981). – Abstract.

141.093 **Precise optical positions of southern radio sources.**
H. Wroblewski. E. Costa, C. Torres.
Rev. Mexicana Astron. Astrofis., Vol. 6, (see 012.017), 27 (1981). – Abstract.

141.094 **Optical identification of radio sources in a 8°5 × 8°5 zone near the South Galactic Pole.**
L. E. Campusano, C. Torres.
Rev. Mexicana Astron. Astrofis., Vol. 6, (see 012.017), 29 - 32 (1981).

Optical counterparts for the 23 radio sources of the Parkes 2700 MHz survey located in the vicinity of ($00^h 53^m$, − 28°00′) were searched within ±2 rms errors of their improved radio positions using deep Kodak IIIaJ plates taken with the Maksutov telescope at Cerro El Roble. Twelve new identifications are proposed, of which 7 are galaxies, while the search areas of 5 radio sources appear blank to the plate limit. Although no new optical counterparts of stellar appearance were found, precise optical positions are given for the already known ones.

141.095 **Observaciones en el ultravioleta de los cuasares 3C249.1 y 3C232.**
E. Daltabuit, D. Dultzin-Hacyan.
Rev. Mexicana Astron. Astrofis., Vol. 6, (see 012.017), 33 (1981).

141.096 **Radio galaxies and quasars.**
K. I. Kellermann.
Ninth Texas Symposium on Relativistic Astrophysics, (see 012.020), p. 1 - 11 (1980).

There is a growing consensus that the sources of energy for radio galaxies and quasars lies in condensed massive objects, possibly black holes, and that the energy is efficiently transported hundreds of kiloparsecs or megaparsecs away by well collimated relativistic jets or beams.

141.097 **Absorption-line spectra of quasi-stellar objects: physical properties of absorbing gas.**
E. M. Burbidge.
Ninth Texas Symposium on Relativistic Astrophysics, (see 012.020), p. 39 - 50 (1980).

141.098 **X-ray quasars.**
H. V. Bradt.
Ninth Texas Symposium on Relativistic Astrophysics, (see 012.020), p. 59 - 93 (1980).

Seven X-ray sources have been identified with quasars or Seyfert-quasar transition objects as classified by optical observers: IIIZw 2, Fairall-9, 0241+622, 3C 273, IC 4329A, Mk 509, and MR 2251 − 178. In this paper the observational results for these seven objects at radio, infrared, optical, ultraviolet, X-ray and gamma-ray frequencies are reviewed.

141.099 **Quasars near companion galaxies and the problem of the absolute magnitudes of quasars.**
H. Arp.
Ninth Texas Symposium on Relativistic Astrophysics, (see 012.020), p. 94 - 112 (1980).

The observational evidence for the association of quasars with low-redshift galaxies is reviewed.

141.100 **List of papers presented at the workshop on quasars and galactic nuclei.**
F. Pacini.
Ninth Texas Symposium on Relativistic Astrophysics, (see 012.020), p. 113 (1980).

141.101 **Relativistic jet production and propagation in active galaxies.**
M. J. Rees, M. C. Begelman, R. D. Blandford.
Tenth Texas Symposium on Relativistic Astrophysics, (see 012.021), p. 254 - 286 (1981).

The large-scale radio structures mapped by aperture-synthesis techniques are energized by activity in galactic nuclei. A central "core", which involves a large concentrated mass accreting material from its surroundings, is the primary origin of the power for all categories of activity associated with active galaxies and quasars. In this review the authors discuss how this core could give rise to the fast-moving plasma that emerges in directed beams and energizes the radio sources, and then discuss in some detail the physics of plasma beams.

141.102 **The structure of 4C radio galaxies in poor clusters.**
J. O. Burns, S. A. Gregory.
Astron. J., Vol. 87, 1245 - 1265 (1982).

VLA radio maps at 20 and 6 cm are presented for a statistically complete sample of 20 4C radio galaxies in poor non-Abell Zwicky clusters. Another five radio sources in the directions of distant clusters or with S (2.7 GHz) < 200 mJy are also discussed. Fifty-three to sixty-five percent of these radio sources have jets. The authors also find that the poor clusters contain significant numbers of tailed radio sources. Finally, they discuss the cluster environments of these radio galaxies and present superpositions of radio structure on Einstein observatory X-ray maps for three of the largest radio sources.

141.103 **Dual-frequency VLA observations of the extended radio galaxy 3C 166.**
S. R. Spangler, A. H. Bridle.
Astron. J., Vol. 87, 1270 - 1278 (1982).

The authors have observed the radio galaxy 3C 166 at 1.4 and 4.9 GHz with an angular resolution of 1.6″ at both frequencies. The source has a basically double structure with strikingly dissimilar lobes. The south lobe is brightened along its outer edge, with a prominent hot spot, whereas the north lobe is relaxed in appearance with more feeble interior brightness maxima. The south lobe is strongly Faraday depolarized at 1.4 and possibly also at 4.9 GHz, but there is no measurable Faraday depth in the north lobe. The results are used to estimate magnetic field strengths and thermal particle densities in the lobes.

141.104 **Evolutionary luminosity functions of extragalactic sources driven by gravitational power.**
A. Cavaliere, E. Giallongo, A. Messina, F. Vagnetti.
Astron. Astrophys., Vol. 114, L1 - L3 (1982).

The authors communicate a new approach to describe and to understand the evolution of quasars and radio galaxies. In the plane luminosity-cosmic time the authors try downstream propagation along t of the object density from simple initial conditions, with the technical help of the continuity equation, and with the guide of a single, model-independent physical assumption: that the evolving sources derive their power from conversion of the gravitational energy in a compact configuration. The authors report relevant numerical results for the source population powered by a particular class of gravitational engines in this line: massive black holes accreting collisional debris from a dense star cluster up to exhaustion of the mass stockpile.

141.105 **Spectroscopic observations of thirteen optically-selected QSOs in a large field centred around NGC 5334.**
J. Surdej, J. P. Swings, H. C. Arp, R. Barbier.
Astron. Astrophys., Vol. 114, 182 - 187 (1982).

In order to further investigate the space density, distribution, and luminosity of quasars near as well as far from bright galaxies, a large field (~25 deg²) around NGC 5334 was searched for QSO candidates with ultraviolet-excess. A first step in this long term program consists in a report on spectroscopic (IDS) data obtained with the ESO 3.6 m telescope for 13 quasars identified within a sample of 23 observed candidates. The redshift of these optically selected quasars falls in the range $Z_e \in [0.23, 2.08]$ with only one possible exception. Two, and possibly three, of the quasars of largest redshift show absorption lines. The remaining 10 faint blue objects display zero-redshift stellar spectra.

141.106 **Recent observations on the absorption spectra of QSOs.**
E. M. Burbidge.
Recent advances in observational astronomy, (see 012.028), p. 45 - 50 (1981).

Data on absorption lines in the spectra of QSOs are summarized and discussed in terms of the physical conditions in the absorbing gas and its relationship to the QSOs (associated with them, or unrelated and lying in intervening galaxies or clouds.

141.107 **Fast versus slow (*radio jets*).**
P. A. G. Scheuer.
Observatory, Vol. 102, 125 (1982). – Abstract.

141.108 **Superluminal radio sources.**
T. J. Pearson.
Observatory, Vol. 102, 125 - 126 (1982). – Abstract.

141.109 **A unified scheme of radio sources.**
I. W. A. Browne.
Observatory, Vol. 102, 126 - 127 (1982). – Abstract.

141.110 **3C 273 - a fast jet?**
R. J. Davis.
Observatory, Vol. 102, 127 - 128 (1982). – Abstract.

141.111 **Kinematics of 3C 236.**
R. Saunders.
Observatory, Vol. 102, 128 (1982). – Abstract.

141.112 **Side-to-side asymmetries in radio sources.**
C. J. Lonsdale.
Observatory, Vol. 102, 128 - 129 (1982). – Abstract.

141.113 **Bends and the speed of radio jets.**
L. S. Sparke.
Observatory, Vol. 102, 129 - 130 (1982). – Abstract.

141.114 **The environment of radio galaxies.**
L. Miller.
Observatory, Vol. 102, 130 - 131 (1982). – Abstract.

141.115 **Relativistic beaming and the observed misalignment in double radio quasars.**
D. J. Saikia, V. K. Kapahi.
Bull. Astron. Soc. India, Vol. 10, 42 (1982). – Abstract.

141.116 **Hβ luminosity of QSO 0350-073.**
R. K. Thakur, R. K. Sood.
Bull. Astron. Soc. India, Vol. 10, 46 (1982). – Abstract.

141.117 **Interplanetary scintillation (IPS) observations of 23 radio sources with arcsecond structure.**
D. G. Banhatti, S. Ananthakrishnan.
Bull. Astron. Soc. India, Vol. 10, 51 (1982). – Abstract.

141.118 **Interferometric observations of strong extragalactic radio sources at 327 MHz.**
D. L. Narayana, S. Sukumar.
Bull. Astron. Soc. India, Vol. 10, 51 (1982). – Abstract.

141.119 **Quasi-stellar objects as rotating magnetic superstars. I. Luminosity and density evolution.**
W. H. Sorrell.
Astrophys. Space Sci., Vol. 85, 3 - 15, with a correction Vol. 86, 513 (1982).

In this paper the magnetic superstar model is used to discuss QSO luminosity and density evolution. The author's main hypotheses are that (1) mass loss from old stars in massive galaxies cools and then falls into the centre to form a nuclear disc (Bailey, 1980); and (2) magnetic superstars in galactic nuclei condense out of gaseous material at the centre of a supermassive-magnetised disc (Kundt, 1979). On this generalised model it is found that the non-thermal (synchrotron) optical luminosity scales as $L_{\mathrm{opt}} \propto L^3 t^{-7/3}$, where L is the total blue luminosity of old stars in the galaxy and t is cosmic time. In addition it is shown that QSO co-moving density follows the law $D(t) \propto \exp - (t/t_{\mathrm{Evol}})^{16/15}$ with an evolution timescale $t_{\mathrm{Evol}} = 1.95 \times 10^9$ yr. The model as a whole is in good agreement with observations.

141.120 **Relative abundance of elements in quasi-stellar objects and evolution of galaxies.**
R. K. Thakur, A. Dasgupta.
Astrophys. Space Sci., Vol. 85, 277 - 292 (1982).

Relative abundance of carbon with respect to silicon has been calculated in the QSOs Q 1246–057, PKS 0528–250, Q 0002–422, and PKS 2126–158. Relative abundance of aluminium with respect to silicon has been calculated in the QSOs PKS 0528–250 and PKS 2126–158. Relative abundance of iron with respect to magnesium has been also calculated in the QSOs Q 0453–423 and Q 0002–422. The ratios N(C)/N(Si) and N(Al)/N(Si) and N(Fe)/N(Mg) in QSOs considered as a class are 3.45 ± 2.33, 0.063 ± 0.054, and 1.08 ± 0.70 respectively. The comparison of N(C)/N(Si) in QSOs and normal galaxies suggests that QSOs might belong to an early phase in the evolutionary sequence of galaxies and that they evolve chemically into normal galaxies in course of time.

141.121 **Absorption spectra of quasars and molecular spectra: an attempt of identification.**
I. E. Val'tts.
Astron. Zh., Tom 59, 844 - 854 (1982). In Russian. English translation in Soviet Astron., Vol. 26, No. 5.

A new attempt is made to identify absorption lines in QSO spectra with those of H_2 and CO molecules using real spectra of the interstellar medium in our Galaxy. No molecular lines have been discovered in the spectra of 4C 05,34, OQ 172, 0830+115. None of identifications can be regarded as statistically significant for the PHL 957 spectrum.

141.122 **A statistical analysis of scintillating radio sources.**
V. S. Artyukh, V. I. Shishov.
Astron. Zh., Tom 59, 896 - 902 (1982). In Russian. English translation in Soviet Astron., Vol. 26, No. 5.

The distribution of the dispersion of interplanetary scintillations of radio sources is derived. The observations of the background scintillating flux density are made at the frequency 102 MHz.

141.123 **Computer approximation of spectrograms of radio sources.**
G. T. Smirnov, A. P. Tsivilev.
Astron. Zh., Tom 59, 1020 - 1029 (1982). In Russian. English translation in Soviet Astron., Vol. 26, No. 5.

A technique for a least-squares approximation to radio sources spectra is described. An analytical approximation to the Voigt function is used as a universal profile.

141.124 **Multifrequency observations of the OVV (*Optically-Violent Variable*) quasar 1156 + 295.**
A. E. Glassgold, J. N. Bregman, P. J. Huggins.
Third European IUE Conference, (see 012.029), p. 577 - 580 (1982).

A large outburst in the quasar 1156 + 295 occurred in early 1981 in which *B* increased to 13.2 mag. The authors obtained short and long wavelength spectra on three occasions when the object was very bright. The UV continuum in all cases is a power law with slope close to -2.0. No spectral features are apparent in the UV.

141.125 **Cosmology from IUE observations of quasars?**
C. M. Gaskell.
Third European IUE Conference, (see 012.029), p. 593 (1982). Abstract.

141.126 **The formation of the broad emission line regions in QSOs.**
J. E. Dyson, J. J. Perry.
Third European IUE Conference, (see 012.029), p. 595 - 597 (1982).

Shock waves in hypersonic optically thin QSO winds (or accretion flows) produce post-shock regions where strong Compton cooling occurs. It is suggested that the shock waves are bow shocks produced by obstacles in the wind. Supernovae are plausible candidates for these obstacles. Since the gas behind a bow shock retains a substantial fraction of the inflow velocity, there is no necessity to invoke a cloud acceleration process. The cooled gas forms numerous small optically thick condensations as a result of thermal instability.

141.127 **Continuum observations of extragalactic sources.**
E. I. Robson.
The scientific importance of submillimetre observations, (see 012.030), p. 147 - 155 (1982).

The current status of millimetre and submillimetre continuum observations of extragalactic sources is presented. Discussions of systems, sensitivities, detection achievements and calibration techniques are given. Three categories of extragalactic programmes are selected and the scientific importance of submillimetre observations is explained for each.

141.128 **Discovery and spectrophotometry of high-redshift quasars.**
G. M. MacAlpine, F. R. Feldman.
Astrophys. J., Vol. 261, 412 - 421, plates 2, 3 (1982).

The authors report on the discovery and spectrophotometry of 30 new high-redshift quasars, which were detected using the Curtis Schmidt technique. They also discuss new follow-up spectrophotometry for 23 quasar candidates from University of Michigan Lists I–IV. The program sample contains eight quasars with $z > 3$, at least five objects exhibiting broad absorption troughs, and a pair of quasars which are 1′ apart on the sky and nearly identical in redshift, at z near 2.13. The redshift distribution for the majority of quasars in UM List IV suggests that most of the single-line quasar candidates in the UM Lists have low to moderate redshifts. The data exhibit a correlation between Lyα emission line velocity widths and redshift.

141.129 **Spectra of compact radio sources in galactic nuclei.**
D. L. Jones, R. A. Sramek, Y. Terzian.
Astrophys. J., Vol. 261, 422 - 438 (1982).

The results of a 6 cm VLBI experiment are combined with previous VLBI results at 13 and 18 cm to determine approximate spectra of compact nuclear radio sources in 12 galaxies. In 11 cases a simple self-absorbed synchrotron spectrum could be fit to the available data. Three additional sources appeared to be highly variable, and reliable spectra could not be obtained. Some of the galaxies for which spectra were determined are also available, and in these cases the individual spectra are uncertain. For the sources with well-determined critical (turnover) frequencies, magnetic field strengths and radiative lifetimes are estimated.

141.130 **The Lyman-alpha emission-line profiles in high-redshift QSOs.**
B. J. Wilkes, R. F. Carswell.
Mon. Not. R. Astron. Soc., Vol. 201, 645 - 660 (1982).

High-resolution (~1 Å) spectra of six high-redshift QSOs were utilized in a study of the profile of the Lyα λ 1215.67 Å emission line. The symmetry of the profile was studied after removing any absorption lines. With one exception, Q2204–408, the Lyα profiles are symmetric to within the uncertainties of the procedure. It also became clear during the analysis that the profiles of the emission lines are not identical, in particular C IV λ 1549 generally has weaker wings and peak than Ly$\alpha\lambda$1216. It is argued that either the motion of the broad-line region gas is not purely radial, or else most of the Lyα photons originate in optically-thin gas.

141.131 **Further morphological studies of QSOs.**
J. B. Hutchings, D. Crampton, B. Campbell, A. C. Gower, S. C. Morris.
Astrophys. J., Vol. 262, 48 - 60 (1982).

Results are presented of optical mapping of 29 QSOs, including 13 reported in an earlier paper. The authors also present results of VLA observations of 16 of the optically observed objects. These data show that QSOs to $z = 0.45$ have extended nebulosity, and that fuzz morphology does not resemble E or S galaxies, but is very similar to Seyfert I galaxies. In some cases, indications of spiral structure are seen, and a large fraction of asymmetrical morphology is noted.

141.132 **Theory of electron-positron showers in double radio sources.**
M. L. Burns, R. V. E. Lovelace.
Astrophys. J., Vol. 262, 87 - 99 (1982) = Cornell Univ. School Appl. Eng. Phys. Rep. 11. (1981).

A black hole of mass $\sim 3 \times 10^8 M_\odot$ is assumed to be present in the nucleus of an active galaxy or quasar. With an axial magnetic field $\sim 10^3$ gauss, a potential drop $\sim 10^{19}$ volts is generated by the unipolar induction of a rotating accretion disk surrounding the black hole. The authors investigate the possibility that the acceleration of electrons or positrons in the unipolar fields initiates an electromagnetic cascade shower. If the cascade-initiating particles are collimated, the electrons and positrons emerge in a collimated beam. Such beams may power extragalactic double radio sources.

141.133 **Possible existence of quasi-stellar-object clusters.**
Y.-z. Liu, S.-d. Jiang, Y.-L. Bian, X.-y. Tang, S.-l. Cao, X.-h. Xiao.
Chinese Phys., Vol. 2, 695 - 699 (1982). – Abstr. in Phys. Abstr., Vol. 85, Abstr. 101090 (1982). – See Abstr. 30.141.146.

141.134 **Variability of compact radio sources at a wavelength of 1 millimeter.**

D. J. Ennis, G. Neugebauer, M. Werner.
Astrophys. J., Vol. 262, 451 - 459 (1982).

Broad-band 1 mm continuum observations of 3C 273, 3C 279, BL Lac, 3C 84, OJ 287, and 3C 345 made between 1977 March and 1981 January are reported. With the possible exception of 3C 279, all of the objects exhibited evidence for variability in the 1 mm flux density on the time scale of a few months. Flux variations at wavelengths of 1 mm and 2 cm are well correlated for the "blazars" (BL Lac objects and optically violent variable quasars) observed. The observed relationship between "blazar" emission variability at 1 mm and 2 cm is inconsistent with the canonical expanding-source model of radio variability.

141.135 **1 millimeter continuum observations of quasars.**
D. J. Ennis, G. Neugebauer, M. Werner.
Astrophys. J., Vol. 262, 460 - 477 (1982).

Broad-band 1 mm continuum observations of 37 active extragalactic objects are presented. The sources observed included radio-loud quasars, radio-quiet quasars, and "blazars" (BL Lac objects and optically violent variable quasars). The 1 mm flux density of radio-loud quasars and "blazars" agrees well with a direct extension of the observed radio continuum. No radio-quiet quasar was detected regardless of its infrared properties. The 1 mm results are discussed within the context of several models of the radio and infrared continuum emission of compact extragalactic objects.

141.136 **Relativistic precessing jets in quasars and radio galaxies: models to fit high resolution data.**
A. C. Gower, P. C. Gregory, J. B. Hutchings, W. G. Unruh.
Astrophys. J., Vol. 262, 478 - 496 (1982).

The authors present the formulation of generalized models tracing the geometry and intensity of the synchrotron emission from precessing, twin, relativistic jets as projected on the plane of the sky. They show that neither the shape of the image nor its relative intensities are altered by including the effects of a cosmological redshift and a relative velocity between the source and observer. The models are fitted to the available data for several quasars and radio galaxies and demonstrate the plausibility of the phenomenon.

141.137 **A thermal wind model for the broad emission line region of quasars.**
R. J. Weymann, J. S. Scott, A. V. R. Schiano, W. A. Christiansen.
Astrophys. J., Vol. 262, 497 - 510 (1982).

Arguments are summarized for supposing that the clouds giving rise to the broad emission lines of QSOs are confined by the pressure of an expanding thermal gas and that a flux of relativistic particles with luminosity comparable to the photon luminosity streams through this gas. The resulting heating and momentum deposition produces a transonic thermal wind whose dynamical properties are calculated in detail. The trajectories of clouds under the combined influence of ram pressure acceleration and radiative acceleration are calculated. From these trajectories emission line profiles are also calculated, as well as the wind and cloud parameters yielding profiles in fair agreement with observed profiles.

141.138 **In situ acceleration in extragalactic radio jets.**
G. V. Bicknell, D. B. Melrose.
Astrophys. J., Vol. 262, 511 - 528 (1982).

The authors have examined the energy dissipated by large-scale turbulence in an extragalactic jet. The turbulence is driven by a shear instability which does not disrupt the jet. It is found that dissipation occurs due to Fermi acceleration and that resonant acceleration plays no role. The Alfvénic component in the turbulent spectrum is dissipated by first being converted into magnetoacoustic waves. The effect of Fermi acceleration, adiabatic losses, and radiative losses on an initial power-law distribution with an upper cutoff is studied. Radio emission extending to at least 100 GHz is shown to be possible, and no spectral index gradients are introduced by the acceleration. The northern jet in NGC 315 is studied in detail. Differences in the turbulent velocity of oppositely directed jets arise from an inhomogeneous interstellar medium and cause a large ratio of jet surface brightnesses.

141.139 **An optical and radio investigation of the radio galaxy 3C 305.**
T. M. Heckman, G. K. Miley, B. Balick, W. J. M. van Breugel, H. R. Butcher.
Astrophys. J., Vol. 262, 529 - 553 (1982).

The results of a detailed investigation of the unusual radio galaxy 3C 305 are reported. The associated radio source is much smaller (~3 kpc) than typical double radio sources but is comparable to them in luminosity. It is identified with a galaxy having some of the characteristics of a spiral and a bright, extended emission-line region. The authors have carried out an intensive multiwavelength study of 3C 305 using radio and optical imaging as well as slit spectroscopy. It is suggested that the unusual characteristics of this source may be the result of a recent merger of a luminous early-type galaxy and a late-type galaxy.

141.140 **On the cosmological evolution of the X-ray emission from quasars.**
Y. Avni, H. Tananbaum.
Astrophys. J., Lett., Vol. 262, L17 - L21 (1982).

The authors derive the average dependence of the ratio of X-ray luminosity to optical luminosity as a function of redshift and optical luminosity for quasars and find that the explicit dependence of this ratio is predominantly on optical luminosity. For a wide class of models for the cosmological evolution of quasars, the results imply that the evolution of the X-ray luminosity function is weaker than the evolution of the optical luminosity function.

141.141 **Confirmation of a conspiracy: dual-band VLBI maps of the flat-spectrum radio source 2021+614.**
J. J. Wittels, I. I. Shapiro, W. D. Cotton.
Astrophys. J., Lett., Vol. 262, L27 - L30 (1982).

VLBI observations at 3.6 and 13 cm wavelengths of the flat-spectrum radio source 2021+614 reveal two components, separated by about 6.5 ± 0.6 milli-arcseconds. The dual-band observations show that the two components have spectral indices of opposite signs, indicating that the flat spectrum of 2021+614 is a result of the superposition of spectra from individual components, each having a peaked spectrum, presumably due to incoherent electron synchrotron radiation.

141.142 **VLA observations of hot spots in high luminosity radio sources.**
R. T. Schilizzi, V. K. Kapahi, S. G. Neff.
J. Astrophys. Astron., Vol. 3, 173 - 187 (1982).

VLA observations at 6 cm have been made of 16 distant luminous 3C sources that appeared to be unresolved or slightly resolved in Cambridge maps but which were known from VLBI observations to contain significant fine structure in their lobes. Most of the lobes contain low brightness tails which are often directed from the hot spots towards the associated optical object. The hot spots are generally unresolved by the VLA observations.

141.143 **The identification of CTA 21.**
D. A. Allen, A. E. Wright, J. G. Ables.
J. Astrophys. Astron., Vol. 3, 189 - 192 (1982).

CTA 21 has long been the outstanting example of a strong, compact radio source lacking an optical identification. The authors report the discovery at infrared wavelengths of the counterpart of CTA 21, and show that its spectral shape is unusual.

141.144 A semiquantitative general-relativistic model of quasar Markarian 205 interpreted as a massive black hole ejected from NGC 4319.
Z. Horák.
Astrophys. Space Sci., Vol. 86, 471 - 476 (1982).
In 1979 the author developed a special-relativistic analysis explaining the discrepancy of observed redshifts of spiral NGC 4319 and its companion quasar Markarian 205 by considering besides the Lorentz time dilatation also the gravitational redshift due to the gravitational field of Markarian 205 interpreted in terms of accretion of mass onto a black hole ejected from NGC 4319. In the present paper a general-relativistic analysis is given. Numerical results of the special and general theories do not differ from each other significantly and admit the conclusion that the radius, r, of the radiating region of Markarian 205 is of the order of the tidal radius of black hole. Several models for various values of the ratio of r to the Schwarzschild radius, r_s, are constructed.

141.145 Asymmetric emission-line regions with out-flowing mass in QSOs and the $Z_{ab} \gtrsim Z_{em}$ systems.
I. Goldman, J. N. Bahcall.
Astron. Astrophys., Vol. 115, 242 - 248 (1982).
A heuristic asymmetrical model for a quasar emission-line region, with material streaming out, is described. The physical parameters are essentially the same as those of static models; the geometry turns out to be the dominant factor in determining the emission line profiles. Calculations are carried out for a variety of parametrized flow models and for the Blumenthal and Mathews (1975) theoretical model of a radiatively driven flow.

141.146 The quasar B2 1320+29.
L. Feretti, G. Giovannini, P. Parma.
Astron. Astrophys., Vol. 115, 423 - 427 (1982).
The radio source B2 1320 + 29, identified with a 20th magnitude quasar, has been mapped at 1.4 GHz and 5 GHz with the Westerbork Synthesis Radio Telescope. This source shows a very unusual D2 type radio structure, with two outer components on the same side of the nucleus. The spectral index steepens when moving from the core to the outermost component. The magnetic field in the source is rather irregular; its orientation is transversal in the nuclear component and parallel to the source axis in the outermost one. The nature of the asymmetry of the source is briefly discussed.

141.147 Quasars at large redshifts.
M. G. Smith.
Progress in cosmology, (see 012.047), p. 275 - 290 (1982).
Quasi-stellar objects have been discovered with redshifts up to $z = 3.53$. A critical discussion of the reality of this apparent upper limit to QSO redshifts is given. A group of four QSOs at $z = 0.37$, with size of $\sim 75 h_0^{-1}$ Mpc has been discovered, but the recently reported lines of QSOs in the sky are probably chance associations. Considerable progress has been made in the last five years towards understanding the location of the material responsible for the absorption lines in QSOs. It appears that most of it is not associated with the QSOs themselves.

141.148 Coma quasars.
A. A. Hoag, N. G. Thomas, B. G. Vaucher.
Astrophys. J., Vol. 263, 23 - 31 (1982).
Two areas of 1 deg^2 each "on" and "off" the Coma cluster of galaxies have been examined by low-resolution slitless spectroscopy obtained at the prime focus of the 4 m Mayall reflector. The authors found 21 emission-line QSO candidates in the approximate ranges $0.3 \leqslant z \leqslant 3$ and $18 \leqslant m_{1475} \leqslant 22$ in the cluster field and 19 similar candidates in the "blank" field nearby.

141.149 A spectroscopic investigation of the nebulosity around low-luminosity quasars.
T. A. Boroson, J. B. Oke, R. F. Green.
Astrophys. J., Vol. 263, 32 - 42 (1982).
Off-nuclear spectra have been obtained for a sample of 12 low-redshift optically selected QSOs. Simple techniques involving spectral differences between the nucleus and the nebulosity were used to effectively remove the scattered light. Ten of the 12 objects were found to be embedded in nebulosity dominated by starlight. Arguments involving the central concentration of galaxies and the amount of gas required to produce the observed emission lines suggest that most of the underlying galaxies are spirals.

141.150 Abrupt cutoffs in the optical-infrared spectra of nonthermal sources.
G. H. Rieke, M. J. Lebofsky, W. Z. Wiśniewski.
Astrophys. J., Vol. 263, 73 - 78 (1982).
Optical and infrared photometry is reported for a number of radio sources with very faint and red optical identifications. Three types of object are indicated for the sample: highly redshifted galaxies with approximately normal colors; sources subjected to strong Galactic reddening; and violently variable sources with very sharp intrinsic spectral cutoffs toward higher frequencies.

141.151 A redshift difference between high and low ionization emission-line regions in QSOs – evidence for radial motions.
C. M. Gaskell.
Astrophys. J., Vol. 263, 79 - 86 (1982) = Lick Obs. Bull., No. 927.
The discovery of a systematic difference between the redshifts of the emission lines arising from the different ionization zones of broad-line quasar clouds is reported. The C IV, N V, C III], [Fe X], [Fe XI], and possibly Lyman-α lines are at a lower redshift than the O I, Mg II, and Balmer lines. Comparison with the redshifts of the narrow emission lines shows that it is the high-ionization broad emission lines that are systematically blueshifted by 600 km s^{-1}.

141.152 Quasars near companion galaxies – a comment on Arp's statistics.
I. W. A. Browne.
Astrophys. J., Lett., Vol. 263, L7 - L8 (1982).
Arp has assessed the probability of finding by chance 13 quasars near 34 companions to bright galaxies. A reanalysis is presented which shows that such an occurrence is 10 orders of magnitude more probable than Arp estimates. Taken at face value, the revised estimate of 10^{-7} for the probability still implies that Arp's findings are statistically significant.

141.153 Density of quasars around companion galaxies.
H. Arp.
Astrophys. J., Lett., Vol. 263, L9 - L12 (1982).
The density of quasars within R = 8 - 17 kpc of companion galaxies is 10 - 30 times the currently accepted background density.

141.154 MERLIN observations of compact sources with very steep radio spectra.
J. Roland, P. Véron, D. Stannard, T. Muxlow.
Astron. Astrophys., Vol. 116, 60 - 63 (1982).
The authors have mapped six very steep spectrum radio sources with the Jodrell Bank Multi Element Radio Linked Interferometer Network (MERLIN). Each of the sources is double, with all of the flux coming from two compact components. Their radio properties are very similar to those of small diameter high luminosity radio galaxies such as 3 C 295, except for their very high spectral indices ($\alpha > 1.3$ compared with 0.9 for 3 C 295). It is suggested that this difference is due to a rapid fading of the activity of the central engine which

is believed to power the classical double radio galaxies. Their relative scarcity could be explained by the short lifetime of the steep spectrum phase, 10^5 yr, compared to the typical source lifetime of at least 10^6 yr.

141.155 **QSO 0957+561.**
IAU Circ., No. 3747 (1982).

141.156 **CCD observations of radio sources from the 5C 6 and 5C 7 surveys.**
M. A. C. Perryman, M. S. Longair, J. R. Allington-Smith, J. Fielden.
Mon. Not. R. Astron. Soc., Vol. 201, 957 - 973 (1982).

The authors present the results of an optical survey of the fields of 34 faint radio sources belonging to two samples complete to 15 mJy at 1407 MHz and to 55 mJy at 408 MHz. A modest proportion of new identifications has been made; the resulting success rates are approximately 30 and 35 per cent for the low- and high-frequency samples respectively. Some 50 per cent of the sources unresolved at 2 arcsec resolution now have reliable optical counterparts and at least half of these identifications appear to be galaxies.

141.157 **Very extended ionized gas in radio galaxies – I. A radio, optical and ultraviolet study of PKS 2158–380.**
R. A. E. Fosbury, A. Boksenberg, M. A. J. Snijders, I. J. Danziger, M. J. Disney, W. M. Goss, M. V. Penston, W. Wamsteker, K. J. Wellington, A. S. Wilson.
Mon. Not. R. Astron. Soc., Vol. 201, 991 - 1008 (1982).

Several radio galaxies are known which exhibit extended emission-line regions covering tens of kiloparsecs. The gas which extends to a radius of 15 kpc in PKS 2158–380 is in a high state of ionization and has a spectrum which is inconsistent with local photoionization by hot stars. The observational study of PKS 2158–380 was made using a variety of radio, optical and ultraviolet techniques. The discussion addresses the problems of the ionization mechanism and the physical and dynamical states of the gas and its evolution. The authors find that the interpretation of the ionization mechanism imposes constraints on the geometrical configuration of the gas which are consistent with its recent origin and may imply that the evolution is rapid.

141.158 **The distribution of companion galaxies to mirror-symmetric extragalactic radio sources.**
J. A. Rose.
Mon. Not. R. Astron. Soc., Vol. 201, 1015 - 1019 (1982).

Blandford and Icke have proposed that the mirror-symmetric curved structure in some low-power extragalactic radio sources (such as 3C 31) can be attributed to the gravitational acceleration of the radio galaxy by a nearby companion galaxy. This idea is subjected to a statistical test in which the distribution of companion galaxies around a well-defined sample of low-power radio sources with extended structure is examined.

141.159 **Low-frequency radio lines of excited carbon in the direction of the radio source Cassiopeia A.**
A. A. Ershov, E. E. Lekht, G. M. Rudnitskij, R. L. Sorochenko.
Pis'ma Astron. Zh., Tom 8, 694 - 699 (1982). In Russian. English translation in Soviet Astron. Lett., Vol. 8.

Search for the C400 α radio line in the direction of Cas A showed the absence of this line with an upper limit $T_L/T_C \leqslant 2 \times 10^{-3}$ in terms of antenna temperature. By comparison of the data obtained with the results of successful observations of the radio lines C630α, C631α and C640α, some parameters of the C II region created presumably by a B-type star are determined.

141.160 **Another gravitational lens?**
J.-S. Chen.
Messenger, No. 30, p. 2 (1982).

141.161 **4C 18.68: a quasar with precessing jets?**
A. C. Gower, J. B. Hutchings.
Mercury, Vol. 11, 159 - 160 (1982).

141.162 **The detection of magnetic fields out to large redshifts.**
P. P. Kronberg, J. J. Perry.
J. R. Astron. Soc. Canada, Vol. 76, 321 (1982). – Abstract.

141.163 **The effect of an inhomogeneous high-energy cutt-off on the optically thin spectrum of unresolved synchrotron sources.**
S. Pineault.
J. R. Astron. Soc. Canada, Vol. 76, 324 (1982). – Abstract.

141.164 **Complete samples of active extragalactic objects. II. A deep 1.452-GHz VLA survey centered on $\alpha = 08^h 52^m 15^s$, $\delta = +17°16'$.**
J. J. Condon, K. J. Mitchell.
Astron. J., Vol. 87, 1429 - 1437 (1982).

A single primary-beam area centered on $\alpha = 08^h 52^m 15^s$, $\delta = +17°16'$ (1950.0) was surveyed with the Very Large Array at 1.452 GHz. The resulting source list is complete to a map peak flux density Sp = 0.2 mJy within 25 arcmin of the field center. Source counts between 0.2 and 5 mJy were determined. The best-fit model for the counts suggests that the faintest sources have absolute radio luminosities $L < 10^{24}$ W Hz^{-1} and do not evolve significantly. Sources stronger than 0.2 mJy appear to account for most of the radio-source background.

141.165 **Are there black holes in quasars?**
M. A. Abramowicz, L. Nobili.
Nature, Vol. 300, 506 - 507 (1982).

There are two main theoretical models for the central engine of quasars and other active galactic nuclei: the black hole model and the spinar model. Observations of the time scales of flux variations can rule out the black hole model and are therefore extremely important. The observed time scale of variations t, and the luminosity L must obey $\log t > \log L - 43$. Variations on a time scale shorter than that would rule out the black hole model. The authors point out here that this statement is wrong in view of recent developments in the accretion disk theory: much shorter periodicities are still consistent with the black hole model.

141.166 **Spectroscopy of the QSO pair Q0028+003/Q0029+003.**
P. A. Shaver, A. Boksenberg, J. G. Robertson.
ESO Sci. Prepr., No. 203, 11 pp. (1982). – Submitted to Astrophys. J., Lett.

141.167 **MERLIN observations of compact sources with very steep radio spectra.**
J. Roland, P. Véron, D. Stannard, T. Muxlow.
ESO Sci. Prepr., No. 204, 16 pp. (1982). – Submitted to Astron. Astrophys.

141.168 **MR 2251–178: a nearby QSO embedded in a giant H II envelope.**
J. Bergeron, A. Boksenberg, M. Dennefeld, M. Tarenghi.
Pré-publ. Inst. Astrophys. Paris, No. 8, 44 pp. (1982). Accepted for publication in Mon. Not. R. Astron. Soc.

141.169 **Infrared counterparts to "empty field" steep spectrum radio sources.**
M. J. Lebofsky, G. H. Rieke, D. Walsh.
Prepr. Steward Obs., No. 410, 9 pp. (1982).

141.170 **H1340#10: a high redshift QSO with very narrow emission lines.**
C. Foltz, R. Weymann, C. Hazard, D. Turnshek.
Prepr. Steward Obs., No. 414, 18 pp. (1982).

141.171 **Spectroscopy of variable blue stellar objects.**
P. D. Usher, A. Warnock, III, R. F. Green.
Prepr. Steward Obs., No. 422, 26 pp. (1982).

141.172 **Observations of H1414+087, a QSO with extremely high velocity broad-absorption-line gas.**
C. Foltz, B. Wilkes, R. Weymann, D. Turnshek, C. Hazard.
Prepr. Steward Obs., No. 426, 20 pp. (1982).

141.173 **The magnetic jet theory of double sources in a radio galaxy.**
W. Hu.
Sci. Sinica, Ser. A, Vol. 25, 511 - 526 (1982).
The two-dimensional acceleration theory about solar wind is applied to the study of the accelerating process of jet beam in a radio galaxy. The flowing features are given with the analytic method, and the basic flow is along the direction of the jet beam. The mechanism of acceleration from subsonic to supersonic flow is discussed. At the same time, some fine structures about the double sources in the radio galaxy are explained.

141.174 **Interpretation of the regular optical variability of quasars and nuclei of galaxies in a model of disk accretion.**
V. L. Oknyanskij.
Astron. Tsirk., No. 1180, p. 1 - 2 (1981). In Russian.

141.175 **A possible luminosity indicator for QSOs with flat optical spectra.**
T. Kiang, F. H. Cheng, Y. Y. Zhou.
Irish Astron. J., Vol. 15, (see 012.061), 124 - 125 (1981). Abstract.

141.176 **On "hot spots" and the question of very narrow collimation of radio jets.**
D. Eichler.
Astrophys. J., Vol. 263, 571 - 575 (1982).
It is suggested that the extremely fine collimation of fluid that is ejected from some radio galaxies results from "overcollimation" – the bending of the flow lines toward the flow axis. The energy in transverse motion may then be dissipated on impact at the axis. A model is presented in which the ambient pressure of a surrounding cloud channels the flow toward the axis.

141.177 **Circular polarization from compact extragalactic radio sources as a result of nonuniform magnetic fields.**
P. E. Hodge.
Astrophys. J., Vol. 263, 595 - 598 (1982).
Partial coupling between normal modes in a radio synchrotron source with a nonuniform magnetic field direction allows conversion of linear polarization into elliptical polarization during propagation through the source. The amount of conversion is calculated under the assumptions of quasi-transverse propagation and a magnetic field direction which is a linear function of distance within the source. For a source with opacity of order unity a change in magnetic field direction as small as 1° results in about 0.25% circular polarization, which is comparable to observed values. No linear-to-elliptical conversion takes place in the limits of very low or very high absorption depth.

141.178 **Spectropolarimetry and polarization alignment effects in radio galaxies.**
R. R. J. Antonucci.
Publ. Astron. Soc. Pacific, Vol. 94, 748 (1982). – Abstract.

141.179 ***IUE* observations of high redshift quasars.**
J. Bechtold, R. F. Green, R. J. Weymann, T. Heckman.
Publ. Astron. Soc. Pacific, Vol. 94, 749 (1982). – Abstract.

141.180 **Spectroscopic observations of the fuzz around quasars.**
T. Boronson.
Publ. Astron. Soc. Pacific, Vol. 94, 749 (1982). – Abstract.

141.181 **A high-resolution study of selected $L\alpha$ absorption lines toward the QSO PHL 957.**
F. H. Chaffee, Jr., R. J. Weymann, D. W. Latham, P. A. Strittmatter.
Publ. Astron. Soc. Pacific, Vol. 94, 750 (1982). – Abstract.

141.182 **Reddening measurements of Seyfert 1 galaxies and quasars.**
S. A. Grandi.
Publ. Astron. Soc. Pacific, Vol. 94, 753 (1982). – Abstract.

141.183 **Color imaging of QSO-galaxy interactions.**
J. B. Hutchings.
Publ. Astron. Soc. Pacific, Vol. 94, 754 (1982). – Abstract.

141.184 ***IUE* observations of high-redshift quasars: far-ultraviolet spectra of the accreting gas.**
M. Malkan.
Publ. Astron. Soc. Pacific, Vol. 94, 755 (1982). – Abstract.

141.185 **Infrared properties of quasars serendipitously selected as X-ray sources.**
G. Neugebauer, B. T. Soifer, K. Matthews, B. Margon, G. A. Chanan.
Publ. Astron. Soc. Pacific, Vol. 94, 755 (1982). – Abstract.

141.186 **0.36 - 3.5 micrometer spectral-flux distributions of galaxies with extended radio structure.**
J. J. Puschell, A. Tubbs.
Publ. Astron. Soc. Pacific, Vol. 94, 757 (1982). – Abstract.

141.187 **The broad emission-line profiles in low-redshift QSOs.**
J. M. Shuder.
Publ. Astron. Soc. Pacific, Vol. 94, 758 (1982). – Abstract.

141.188 **Visual-infrared photometry of X-ray QSOs.**
M. L. Sitko, W. A. Stein, W. Wisniewski.
Publ. Astron. Soc. Pacific, Vol. 94, 758 (1982). – Abstract.

141.189 **Infrared/optical energy distributions of high-redshift quasars.**
B. T. Soifer, G. Neugebauer, J. B. Oke, K. Matthews, J. H. Lacy.
Publ. Astron. Soc. Pacific, Vol. 94, 758 - 759 (1982). Abstract.

141.190 **Redshifts and spectroscopy of very distant radio galaxies with strong emission lines.**
H. Spinrad.
Publ. Astron. Soc. Pacific, Vol. 94, 759 (1982). – Abstract.

141.191 **Asymmetric extended structure in QSOs.**
A. Stockton, J. W. MacKenty, R. J. Hlivak.
Publ. Astron. Soc. Pacific, Vol. 94, 759 - 760 (1982). Abstract.

141.192 **Multicolor CCD surface photometry of the nebulosity around 3C 273: gravitational lens?**

J. A. Tyson, P. C. Boeshaar.
Publ. Astron. Soc. Pacific, Vol. 94, 760 (1982). – Abstract.

141.193 **Broad emission-line profiles in QSO spectra.**
B. J. Wilkes.
Publ. Astron. Soc. Pacific, Vol. 94, 761 (1982). – Abstract.

141.194 **Emissione radio delle galassie.**
B. Ballarati, G. Giovannini.
Coelum, Vol. 50, 311 - 324 (1982).

141.195 **Correlation analysis of light curves in optical and radio ranges for the extragalactic sources 3C 273 and 3C 120.**
B. V. Komberg, V. M. Lyutyj.
Inst. kosm. issled. AN SSSR. Prepr., 1982, No. 705, 18 pp. In Russian. – Abstr. in Ref. zh., 51. Astron., 11.51.747 (1982).

141.196 **0351+026: a QSO spawned by interacting galaxies?**
G. D. Bothun, J. Mould, T. Heckman, B. Balick, R. A. Schommer, J. Kristian.
Astron. J., Vol. 87, 1621 - 1627 (1982).

The authors present the results of a detailed investigation into the nature of the H I-rich quasi-stellar object 0351+026. An improved H I profile reveals the presence of two distinct components. A high-quality CCD image clearly shows the presence of two galaxies. The brighter galaxy hosts the QSO while the less luminous companion has a prominent, apparently stellar nucleus. The data suggest that this system represents a violent encounter between two low-luminosity H I-rich galaxies. The possibility exists that this collision has triggered the QSO activity in the brighter galaxy.

141.197 **Infrared properties of serendipitous X-ray quasars.**
G. Neugebauer, B. T. Soifer, K. Matthews, B. Margon, G. A. Chanan.
Astron. J., Vol. 87, 1639 - 1647 (1982).

Near-infrared measurements have been obtained of 30 quasars originally found serendipitously as X-ray sources in fields of other objects. The observations show that the infrared characteristics of these quasars do not differ significantly from those of quasars selected by other criteria. Because this X-ray-selected sample is subject to different selection biases than previous radio and optical surveys, this conclusion is useful in validating previous inferences regarding the infrared colors of "typical" quasars.

141.198 **Are QSOs local objects? I. A new interpretation of the emission and absorption spectra of a few QSOs.**
S. Banerji, G. C. Bhar, P. K. Mukherji.
Astrophys. Space Sci., Vol. 87, 217 - 236 (1982).

The similarities of the spectra of QSOs with those of Wolf-Rayet stars are pointed out. The emission spectrum of the earliest discovered QSO, 3C 273, in the ultraviolet and visible regions is interpreted as that of an object deficient in hydrogen like Wolf-Rayet stars but having no redshift. The visible emission spectra of two other QSOs, 3C 48 and 3C 280.1, are also similarly interpreted. It is further assumed that the absorption lines of QSOs are produced in an expanding atmosphere so that they are violet shifted as in Wolf-Rayet stars. Since the QSOs are here assumed to be comparatively local objects, the problems of energy supply, superluminal velocities, etc., raised by the conventional explanation do not arise in this case.

141.199 **On the dynamics of emitting gas in the quasar PHL 5200 and related objects.**
V. P. Grinin.
Astron. Zh., Tom 59, 1070 - 1074 (1982). In Russian.
English translation in Soviet Astron., Vol. 26, No. 6.

The investigations of the origin of the unusual resonance line profiles in the spectra of PHL 5200 made in the last years have shown that in the envelope of this quasar a radially symmetric outflow with a deceleration takes place. In this paper the mechanism of the deceleration and the causes of the uniqueness of the emission spectra of this object are shortly discussed. The qualitative similarity of the velocity field in the PHL 5200 envelope and in the envelopes of other QSOs and Seyfert galaxies is emphasized.

141.200 **On the origin of relativistic particles and gamma-rays in quasars.**
R. J. Protheroe, D. Kazanas.
NASA Tech. Memo., NASA TM-83983, 15 pp. (1982). – To be published in the Astrophys. J.

141.201 **Compact radio source 41.9 + 58: a young plerion.**
I. S. Shklovskij.
Pis'ma Astron. Zh., Tom 8, 707 - 710 (1982). In Russian.
English translation in Soviet Astron. Lett., Vol. 8.

Arguments are given in favour of that the compact radio source 41.9 + 58 located near the M 82 center is a young plerion with an age of about 30 years.

141.202 **Luminosity evolution of QSOs.**
S.-m. Gong, C.-l. Xia.
Acta Astron. Sinica, Vol. 23, 243 - 254 (1982). In Chinese.

Based on the results of four articles on the Hubble diagram with standard candles of QSOs, the authors first obtain the improved deceleration factor q_0 by means of the exact formula of the luminosity distance of the Friedmann universe model. Then from the QSO Catalogue of Hewitt and Burbidge, the authors derive the statistical relation of $m(z)$-z. Finally the luminosities of QSOs as a function of z or t and their respective derivatives are calculated.

141.203 **A further investigation of the relation between redshifts and magnitudes of QSOs with interplanetary scintillation, and the estimation of the deceleration parameter q_0.**
F. Cheng.
Acta Astrophys. Sinica, Vol. 2, 91 - 96 (1982). In Chinese.

The relation between redshifts and magnitudes of quasars with interplanetary scintillation was further studied. The results show that the quasars with strong scintillation and weak optical variation have probably similar luminosity. The best value of q_0 was estimated to be 1.9 using Mattig's luminosity distance formula.

141.204 **The evolution of the component size of double radio sources of class DII.**
T. Zhou.
Acta Astrophys. Sinica, Vol. 2, 173 - 183 (1982). In Chinese.

The overall angular or linear size of double radio sources had long been taken as an evolutionary parameter. In this paper, the component size of a double radio source in evolution is examined. Sixty samples of class DII sources with central optical objects and typical double-structure lobes containing hot spots are selected for investigation. The results are discussed in detail.

141.205 **The statistical analysis of Lα absorption lines from a uniform sample of nine high redshift QSOs.**
Z. Zou, J. Chen, Y. Bian, X. Tang, Z. Cui.
Acta Astrophys. Sinica, Vol. 2, 253 - 264 (1982). In Chinese.

141.206 **The distribution of absorption line redshifts of quasars and its origin.**
L. Fang, Y. Chu, Y. Liu.
Acta Astrophys. Sinica, Vol. 2, 265 - 270 (1982). In Chinese.

Current data on the absorption line redshifts of quasars have been studied by statistical analysis. All the statistical results, including the distribution of relative velocity and the power-spectrum of the absorption system redshift distribution, are in favour to the cosmological hypothesis of the origin of quasar absorption lines. Especially, it has been pointed out that the peaks in the distribution of the relative velocity are not an undoubted evidence for the intrinsic hypothesis of the origin of absorption lines in the quasar spectrum because it might also be rather good explained by the model of density wave perturbation in the early universe.

141.207 **A discussion of the Hubble diagram of QSOs.**
S. Cao, X. Xiao, Z. Shi, Y. Liu, S. Jiang, X. Tang, Y. Bian.
Acta Astrophys. Sinica, Vol. 2, 271 - 276 (1982). In Chinese.
The relationship between the QSOs' standard candles and their mean visual magnitudes has been investigated theoretically.

141.208 **Rapid X-ray variability in the nuclei of active galaxies.**
R. Hoshi.
Prog. Theor. Phys., Vol. 68, 673 - 676 (1982). – Abstr. in Phys. Abstr., Vol. 86, Abstr. 16030 (1983).

141.209 **The great quasar odyssey.**
M. Rowan-Robinson.
New Scientist, Vol. 96, 305 - 308 (1982). – Abstr. in Phys. Abstr., Vol. 86, Abstr. 20529 (1983).

The Parkes 2700 MHz survey: a resurvey of the ± 4° declination zone.
See Abstr. 002.047.

New optical identifications from the Parkes 2700 MHz resurvey of the ± 4° declination zone.
See Abstr. 002.048.

The radio properties of a complete sample of bright galaxies.
See Abstr. 002.055.

Landolt-Börnstein. Group VI. Vol. 2c. Interstellar matter – Galaxy – Universe.
See Abstr. 003.055.

Mapping the Universe.
See Abstr. 011.010.

The scientific significance of submillimetre astronomy.
See Abstr. 013.063.

General results of the Shanghai-Effelsberg VLBI experiment.
See Abstr. 013.109.

A graphical method for studying and modeling gravitational lens systems.
See Abstr. 021.021.

Spectroscopy in radio astronomy.
See Abstr. 031.555.

A search for quasars at very faint magnitudes.
See Abstr. 031.624.

Investigation of the fine structure of radio sources.
See Abstr. 033.011.

Significance of radio astronomy in cosmological research.
See Abstr. 033.022.

Astrometric results obtained with the Crimea-Pushchino radio interferometer.
See Abstr. 041.005.

Precise optical positions of radio sources in the FK 4-system. II. Results from 28 sources on the northern hemisphere and a preliminary comparison of the optical-radio reference frame.
See Abstr. 041.006.

Precise positions in the FK4 system for 120 radio source reference stars.
See Abstr. 041.011.

The connection of a catalogue of stars with an extragalactic reference frame.
See Abstr. 043.004.

Transport of neutrinos, radiation and energetic particles in accretion flows.
See Abstr. 061.057.

Nonlinear shear instabilities in an infinite slab.
See Abstr. 062.010.

Structure and dynamics of supersonic jets.
See Abstr. 062.023.

Magnetohydrodynamic phenomena in quasar plasmas.
See Abstr. 062.038.

Formation of a collimated beam in radio galaxies.
See Abstr. 062.039.

The influence of buoyancy on the stability of jets.
See Abstr. 062.075.

Local coupling of surface MHD waves with kinetic Alfvén waves in jets.
See Abstr. 062.076.

Structure and expansion law of a hypersonic beam.
See Abstr. 062.098.

Radio jets and Kelvin-Helmholtz instabilities.
See Abstr. 062.099.

Computation and analysis for relativistic shock waves and hot spots in extragalactic double radio sources.
See Abstr. 062.131.

Cerenkov line radiation and the emission-line spectra of QSOs.
See Abstr. 063.034.

Escape probabilities, mean number of scatterings and net radiative bracket for resonance lines.
See Abstr. 063.036.

On the interpretation of optically thin synchrotron spectra.
See Abstr. 063.037.

Compton heating and cooling and astrophysical jets.
See Abstr. 063.052.

Correlations in synchrotron-self Compton spectra.
See Abstr. 063.069.

Mass loss and mass gain of stars immersed in a supermassive star.
See Abstr. 064.012.

Compton heated winds from accretion disks in QSOs.
See Abstr. 064.070.

Gravitational lenses in astrophysics.
See Abstr. 066.061.

Observational aspects of the microwave cosmic background spectrum.
See Abstr. 066.182.

Relativistic electrons in galaxies and their non-thermal radiation from radio to gamma radiation.
See Abstr. 066.184.

Gravitational waves and activity of galactic nuclei and quasars.
See Abstr. 066.193.

Infrared spectroscopy in Australia.
See Abstr. 114.018.

Westerbork and VLA observations of G 127.1 + 0.5.
See Abstr. 125.009.

Supernova remnant candidates in the Small Magellanic Cloud.
See Abstr. 125.013.

The supernova remnant G116.9+0.2 (CTB 1)–21-cm continuum and H I emission.
See Abstr. 125.076.

Radio observations of small diameter sources in the field of the supernova remnant S 147.
See Abstr. 125.084.

Hot galactic gas and narrow line quasar absorption systems.
See Abstr. 131.141.

An Effelsberg – Green Bank galactic H I absorption line survey. II. Results and interpretation.
See Abstr. 131.263.

Optical identification of serendipitous Einstein X-ray sources: luminosity and redshift relations in the X-ray and optical emission from active galaxies and quasars.
See Abstr. 142.049.

Extragalactic X-ray sources.
See Abstr. 142.076.

The origin of the X-ray background.
See Abstr. 142.117.

High-energy spectra of three active extragalactic objects.
See Abstr. 142.144.

X-rays from quasars and active galaxies.
See Abstr. 142.148.

Radio measurements in the fields of gamma-ray sources. I. CG 195 + 04.
See Abstr. 142.503.

Luminosity evolution of quasars and active galaxies: theoretical models of the evolving mass supply rate.
See Abstr. 151.010.

Precessing jets in Sagittarius A: gas dynamics in the central parsec of the Galaxy.
See Abstr. 155.037.

Observations of the galactic center in the radio band.
See Abstr. 156.016.

Redshifts and spectroscopy of very distant radio galaxies with strong emission lines.
See Abstr. 158.005.

Once more on the radio emission of the nuclei of spiral galaxies.
See Abstr. 158.014.

VLA observations of the giant cluster spiral UGC 12423.
See Abstr. 158.017.

On the physical conditions and the velocity fields in Seyfert 1 galaxies and QSOs.
See Abstr. 158.028.

Neutral hydrogen in elliptical galaxies with nuclear radio sources and optical emission lines.
See Abstr. 158.029.

An investigation of the star-burst model for radio emission from Seyfert galaxies.
See Abstr. 158.031.

Optical absorption lines in the high redshift BL Lac object 0215 + 015.
See Abstr. 158.049.

Radio and mm-observations of active nuclei.
See Abstr. 158.056.

Physical conditions in the emission regions of active galactic nuclei.
See Abstr. 158.057.

The impact of IUE observations on our knowledge about galaxies and quasars.
See Abstr. 158.058.

Coordinated observations of Markarian 180 at ultraviolet, X-ray, optical, and radio wavelengths.
See Abstr. 158.070.

Coordinated observations of Markarian 501 at ultraviolet, X-ray, optical, and radio wavelengths.
See Abstr. 158.071.

Simultaneous multifrequency observations of BL Lac objects and violently variable quasars.
See Abstr. 158.073.

On the possibility of determination of masses of central bodies in active galactic nuclei and quasars.
See Abstr. 158.089.

Radio structures of Seyfert galaxies. III. Radio and optical observations of NGC 5548.
See Abstr. 158.093.

Neutral hydrogen in bright galaxies with strong radio sources.
See Abstr. 158.096.

The relation between Seyfert and radio galaxies.
See Abstr. 158.115.

Radio emission from X-ray selected AGN.
See Abstr. 158.120.

The arcsecond radio structure of BL Lacertae objects.
See Abstr. 158.130.

Optical emission knots associated with the inner X-ray jet of Centaurus A.
See Abstr. 158.137.

Properties of BL Lac objects.
See Abstr. 158.159.

Emission-line spectra of active galactic nuclei and quasars.
See Abstr. 158.160.

Infrared observations of extragalactic sources.
See Abstr. 158.161.

Implications of the variability and polarization of BL Lac objects.
See Abstr. 158.162.

Summary of the workshop on active galaxies and quasars.
See Abstr. 158.163.

The velocity of the jet in M 87.
See Abstr. 158.180.

Active galaxies observed by IUE.
See Abstr. 158.193.

UV observations of X-ray emitting QSOs and BL Lac objects.
See Abstr. 158.206.

Simultaneous multifrequency observations of BL Lac objects and violently variable quasars.
See Abstr. 158.207.

VLBI observations of the core sources of a sample of spiral galaxies.
See Abstr. 158.210.

The jet in M 87.
See Abstr. 158.216.

VLA observations of the Seyfert galaxy NGC 1068.
See Abstr. 158.218.

A 1415 MHz survey of Seyfert and related galaxies. III.
See Abstr. 158.227.

Radio jets in NGC 4151.
See Abstr. 158.232.

The nucleus of M 81: simultaneous 2.3 and 8.3 GHz Mark III VLBI observations.
See Abstr. 158.240.

Accurate optical positions of extragalactic emission-line objects: University of Michigan lists I–IV.
See Abstr. 158.242.

Compact radio cores and the relation between the radio and optical axes of elliptical galaxies.
See Abstr. 158.243.

New high resolution radio observations of NGC 4258. III. VLA and WSRT observations of the anomalous arms.
See Abstr. 158.247.

NGC 1961: stripping of a supermassive spiral galaxy.
See Abstr. 158.248.

The most distant galaxies.
See Abstr. 158.251.

The heating of dust in the broad-line regions of active galaxies and quasars.
See Abstr. 158.253.

High frequency radio continuum observations of bright spiral galaxies.
See Abstr. 158.259.

Radio continuum emission: a tracer for star formation.
See Abstr. 158.260.

An H I survey of southern galaxies.
See Abstr. 158.273.

Das Rätsel der Jets. Beobachtungen an M 87 und 3C 273 führen zu neuen Fragen.
See Abstr. 158.277.

Activity in the centers of galaxies.
See Abstr. 158.294.

Are there correlations between radio and optical axes of radio galaxies?
See Abstr. 158.306.

Radio structures of Seyfert galaxies. IV. Jets (?) in NGC 1068 and NGC 4151.
See Abstr. 158.310.

VLBI observations of the nucleus and jet of M 87.
See Abstr. 158.312.

The radio continuum morphology of NGC 4631 at 2.7 and 8.1 GHz.
See Abstr. 158.339.

Evolutionary scheme of objects with active nuclei.
See Abstr. 158.342.

A 1400-MHz survey of 1478 Abell clusters of galaxies.
See Abstr. 160.005.

Radio and X-ray observations of the Abell 2241 galaxy clusters.
See Abstr. 160.007.

A search for radio halo emission at 430 MHz in 72 rich clusters of galaxies.
See Abstr. 160.009.

Radio and X-ray galaxies in Abell 566.
See Abstr. 160.013.

Observations of X-ray clusters of galaxies at 102.5 MHz.
See Abstr. 160.015.

Coma-type radio halos and cluster X-ray morphology.
See Abstr. 160.040.

The evolution of radio galaxies: a VLA survey of high-redshift clusters of galaxies.
See Abstr. 160.051.

Common properties of clusters of galaxies containing radio halos and implications for models of radio halo formation.
See Abstr. 160.059.

The distribution and composition of intergalactic clouds at large red-shifts.
See Abstr. 161.003.

Absorption lines, Faraday rotation, and magnetic field estimates for QSO absorption-line clouds.
See Abstr. 161.004.

Resolved quasars: a test for the expansion of the Universe.
See Abstr. 162.018.

Estimation of the deceleration parameter q_0 using quasar data.
See Abstr. 162.037.

Clustering of Lyman-alpha absorption lines in quasars: like galaxies in proto-superclusters?
See Abstr. 162.046.

Do we really know how to measure the distances of extragalactic objects?
See Abstr. 162.095.

The precision on the measure of q_0 using the gravitational lensing effect.
See Abstr. 162.181.

Cosmic density wave and its observable vestiges – maximum scale of inhomogeneities in the distribution of galaxies and features in the distribution of absorption-line redshifts of quasars.
See Abstr. 162.225.

The dependence on redshift of the steepness of quasar counts.
See Abstr. 162.226.

Infrared astronomy and cosmology.
See Abstr. 162.227.

On the existence of cosmological evolutionary effects. I. A morphological selection effect in the counts of quasars.
See Abstr. 162.248.

Cosmological models and the statistical distribution of quasars.
See Abstr. 162.287.

Pulsars

141.501 **The mode-switching phenomenon in pulsars.**
N. Bartel, D. Morris, W. Sieber, T. H. Hankins.
Astrophys. J., Vol. 258, 776 - 789 (1982).

The authors report on a multifrequency, full polarization analysis of the two classical mode-switching pulsars PSR 0329+54 and PSR 1237+25 and compare properties of all known mode-switchers. They interpret the mode-switching phenomenon as a change in the inhomogeneous chemical composition and/or structure of the pulsar surface followed by an alteration of the electrostatic conditions in the polar cap region above the surface, which leads to a different distribution of the particles in the magnetosphere.

141.502 **On the gravitational radiation of the binary system with the pulsar PSR 1913 + 16.**
Yu. V. Baryshev.
Astrofizika, Tom 18, 93 - 99 (1982). In Russian. English translation in Astrophysics, Vol. 18, No. 1.

It is shown that according to the relativistic tensor theory of the gravitational field in flat space-time the gravitational radiation power from the binary pulsar PSR 1913 + 16 exceeds the general relativistic prediction by 3.5%. This small excess is due to scalar gravitational radiation. It is noticed that in the case of spherical pulsations the scalar radiation dominates.

141.503 **On a possible mechanism of period change of pulsars.**
E. A. Pamyatnykh.
Astrofizika, Tom 18, 101 - 106 (1982). In Russian. English translation in Astrophysics, Vol. 18, No. 1.

It is shown that the oscillations of the relativistic electron gas magnetic susceptibility caused by quantisation in a magnetic field can result in instabilities of the magnetisation of the electron component of neutron stars. When the magnetic field changes or the stars cool, these instabilities may be found in period changes of jump-like pulsars.

141.504 **Two types of pulsars?**
I. F. Malov, S. A. Sulejmanova.
Astrofizika, Tom 18, 107 - 118 (1982). In Russian. English translation in Astrophysics, Vol. 18, No. 1.

Some evidence is presented that a number of observational characteristics of pulsar radio emission, for example, pulse width, total change of position angle of linear polarization through an integrated profile and some others is different for two groups of pulsars. It is supposed that two alternative models are realized: Smith's model for short-period pulsars ($P \lesssim 0.3$ s) and Ruderman-Sutherland's model for pulsars with long periods ($P > 1$ s).

141.505 **The drift behaviour of PSR 0809+74.**
M. V. Popov, T. V. Smirnova.
Astron. Zh., Tom 59, 719 - 725 (1982). In Russian. English translation in Soviet Astron., Vol. 25, No. 4.

The drift features of PSR 0809+74 are discussed with particular reference to nulling regions. An analysis of the instantaneous drift rate is conducted and its dependence on the longitude is obtained. The relation between millistructure characteristics and the longitude of drifting subpulses is considered. The results are discussed on the basis of the "hollow cone" model.

141.506 **Dynamical screening of the acceleration region in a pulsar magnetosphere.**
V. S. Beskin.
Astron. Zh., Tom 59, 726 - 731 (1982). In Russian. English translation in Soviet Astron., Vol. 25, No. 4.
The question of the existence of a gap in a pulsar magnetosphere is considered. It is shown that the current growth connected with pair creation will be stopped by means of their own radiative electrical field.

141.507 **The anomalous braking index of the Crab pulsar: a plasma inertial effect.**
H. Heintzmann, E. Schrüfer.
Astron. Astrophys., Vol. 111, L4 - L6 (1982).
It is shown that a braking index n = $\Omega\ddot{\Omega}/\dot{\Omega}^2$ as low as n = 2.5 can be explained in the canonical (geometrical) pulsar slow-down model if plasma inertia is taken into account. Older pulsars where plasma effects are less important should have n $\geqslant$ 3.

141.508 **Towards a pulsar magnetosphere model – physical principles.**
R. R. Burman.
Proc. Astron. Soc. Australia, Vol. 4, 217 - 219 (1981).
The author outlines the physical principles on which he believes pulsar magnetosphere models should be based. The principles follow from the equations of continuity and motion for each species and the two sourceless members of Maxwell's set of equations, together with the steady-rotation constraint and boundary conditions at the stellar surface.

141.509 **Secular changes in the flux density of PSR 0833–45.**
W. B. McAdam.
Proc. Astron. Soc. Australia, Vol. 4, 219 - 221 (1981).

141.510 **On the self-consistent solutions of pulsar plasma waves.**
A. C.-L. Chian.
Astron. Astrophys., Vol. 112, 391 - 393 (1982).
In the theoretical study of strong pulsar electromagnetic waves in an electron-ion plasma a dynamic parameter relating electron and ion velocities necessarily appears. It is shown that the dynamic parameter is a function of the plasma stream velocity, the wave velocity, the wave amplitude and the electron-ion mass ratio. In order to obtain self-consistency in the solutions the dynamic parameter must be chosen appropriately when other parameters have been specified.

141.511 **Polarization of the coherent radio emission from pulsars.**
H. Ardavan.
Mon. Not. R. Astron. Soc., Vol. 200, 1053 - 1066 (1982).
The polarization characteristics of the radiation from a quasi-steady pulsar magnetosphere are calculated using the amplitude-modulated-noise interpretation of the data on pulse structures. The existing observational data on pulsar polarization are also discussed and certain additional tests of the quasi-steady theory are pointed out.

141.512 **The origin of nulls, mode changes and timing noise in pulsars.**
P. B. Jones.
Mon. Not. R. Astron. Soc., Vol. 200, 1081 - 1089 (1982).
A solvable polar cap model obtained previously has normal states which may be associated with radio emission, and null states. The solutions cannot be time- independent; the neutron star surface temperature T and mean surface nuclear charge Z are both functions of time. The normal and null states and the transitions between them, form closed cycles in the T–Z plane. Normal-null transitions can occur inside a fraction of the area of the neutron star surface intersected by open magnetic flux lines. The fraction increases with pulsar period and becomes unity when the pulsar nears extinction. Frequency noise, mode changes and pulse nulls have a common explanation in the transitions.

141.513 **Geometry of pulsar beams: relative orientations of rotation axis, magnetic axis, and line of sight.**
R. Narayan, M. Vivekanand.
Astron. Astrophys., Vol. 113, L3 - L6 (1982).
The observed polarisation angle variation in pulsars is analysed on the basis of the magnetic pole model. The angles α between the rotation and magnetic axes and β between magnetic axis and line of sight are estimated using (1) the shape of the polarisation angle (θ) vs. pulse longitude (ϕ) curve, (2) the magnitude of the central gradient $d\theta/d\phi$ in the main pulse, and (3) the magnitude and sign of the gradient in the interpulse.

141.514 **Thermal X-ray emission from isolated older pulsars: a new heating mechanism.**
J.-H. Huang, R. E. Lingenfelter, Q.-H. Peng, K.-L. Huang.
Astron. Astrophys., Vol. 113, 9 - 14 (1982).
A new pulsar heating mechanism, magnetic dipole radiation of superfluid neutrons, is proposed. It is also a new mechanism for pulsar spindown, $\dot{P} \propto P^2$. The expected surface temperatures are calculated for isolated, older pulsars for different models. These temperatures are compared with measurements from the Einstein Observatory (HEAO 2). According to this mechanism, the internal magnetic field strengths are estimated for older pulsars.

141.515 **A search for pulsar haloes at 843 MHz.**
M. J. Kesteven, J. M. Durdin.
Astron. Astrophys., Vol. 113, 211 - 212 (1982).
15 pulsars were observed at 843 MHz in a high sensitivity search for haloes. No extended emission associated with the pulsars was detected.

141.516 **Causal relationship between pulsar long-term intensity variations and the interstellar medium.**
W. Sieber.
Astron. Astrophys., Vol. 113, 311 - 313 (1982).
The characteristic time scales of pulsar long-term intensity variations show a strong correlation with the dispersion measure. It is argued that this points to a modulation process which is unrelated to the pulsar, but which is connected with the properties of the interstellar medium. Models which invoke at least two scales of irregularities might be best suited to give a satisfactory explanation.

141.517 **Scaling from Jupiter to pulsars and mass spectrum of pulsars.**
C. Y. Fan, J. Wu, H. Hang.
Astrophys. J., Vol. 260, 353 - 361 (1982).
The authors have compared the energy loss of Jupiter through the acceleration of high energy particles to that of Crab pulsar. A scaling law relating angular deceleration and moment of inertia for pulsars is derived. The mass spectrum of 194 pulsars, dN/dm, is found to be $\sim 1/m^2$.

141.518 **Pulsar timing irregularities.**
G. E. Gullahorn, J. M. Rankin.
Astrophys. J., Vol. 260, 520 - 537 (1982).
The gross timing behavior of pulsars can generally be well modeled as a train of pulses with a pulse period linearly in-

creasing in time. Deviations from this simple model have great interest for physical models of pulsars. The authors present here examples of such irregular timing behavior from the Arecibo program and give a systematic discussion of the various types of timing irregularities.

141.519 **Observations and interpretations of the pulsed emission from the Crab pulsar.**
F. K. Knight.
Astrophys. J., Vol. 260, 538 - 552 (1982).
The observations made by HEAO 1 of the 18–200 keV pulsed emission from the Crab pulsar show that the systematic variation of spectral slope as a function of pulse phase, discovered at lower energies (4–60 keV), continues up to at least ~150 keV, strengthening the case for the existence of at least two pulsed emission components: the emission from the two main pulse peaks and a separate interpulse emission connecting the two peaks. Combined spectra from infrared to X-ray energies for different pulsar phases are presented and compared with theoretical emission models.

141.520 **X-ray and gamma-ray upper limits for pulsed emission from radio pulsars.**
F. K. Knight, J. L. Matteson, L. E. Peterson, R. E. Rothschild.
Astrophys. J., Vol. 260, 553 - 560 (1982).
The authors present results from a search for pulsed emission from eleven radio pulsars in the energy range 15 keV to 11 MeV using the UCSD/MIT Hard X-ray and Low-Energy γ-ray Experiment on HEAO 1.

141.521 **Phase coherence in giant pulses from the Crab pulsar.**
I. R. Linscott, P. R. Backus.
Bull. American Astron. Soc., Vol. 14, 661 (1982). – Abstract.

141.522 **Twelve-frequency, simultaneous observations of single pulses from PSR 1919 + 21.**
J. H. Seiradakis, B. J. Rickett.
Bull. American Astron. Soc., Vol. 14, 665 - 666 (1982). Abstract.

141.523 **Pulsar statistics and two new mechanisms for spindown.**
K.- L. Huang, J.- H. Huang, Q.- H. Peng.
Bull. American Astron. Soc., Vol. 14, 666 (1982). – Abstract.

141.524 **Timing observations of the binary pulsar PSR 1913 + 16.**
V. Boriakoff, M. P. Haugan, Y. Terzian, S. Teukolsky.
Bull. American Astron. Soc., Vol. 14, 666 (1982). – Abstract.

141.525 **Detection of pulsars PSR 1133 + 16 and PSR 2045–16 at low radio frequencies.**
F. Reyes, T. D. Carr, J. P. Oliver, J. May, M. Bitrân, J. Aparici.
Rev. Mexicana Astron. Astrofis., Vol. 6, (see 012.017), 219 - 222 (1981).
Two pulsars have been observed at low radio frequencies using a narrow bandwidth to avoid pulse dispersion problems. PSR 1133 + 16 was observed at 26.3 MHz and PSR 2045–16 was observed at 45 MHz. Digital averaging and period-stacking techniques were used in order to increase the signal-to-noise ratio. Both pulsars were successfully detected by averaging typically 400 to 600 periods. The integrated pulse profile, pulse width and an estimate of the pulse flux density and energy were obtained.

141.526 **On the nature of pulsars.**
V. Radhakrishnan.
Contemp. Phys., Vol. 23, 207 - 231 (1982). – Abstr. in Phys. Abstr., Vol. 85, Abstr. 74003 (1982).

141.527 **Pulsar radiation mechanisms.**
M. Ruderman.
Ninth Texas Symposium on Relativistic Astrophysics, (see 012.020), p. 409 - 428 (1980).

141.528 **Evidence for the existence of gravitational radiation from measurements of the binary pulsar PSR 1913 + 16.**
J. H. Taylor, P. M. McCulloch.
Ninth Texas Symposium on Relativistic Astrophysics, (see 012.020), p. 442 - 446 (1980).
This paper summarizes the results of four years of timing observations of PSR 1913 + 16, made with the 305-m telescope at the Arecibo Observatory, at frequencies near 430 and 1410 MHz.

141.529 **Summary of workshop on neutron stars and pulsars.**
N. J. Holloway.
Ninth Texas Symposium on Relativistic Astrophysics, (see 012.020), p. 447 - 449 (1980).

141.530 **Is the PSR 1913 + 16 companion a pulsar in the propeller phase? .**
L. Baroni, G. Callegari, M. Savrie.
Lett. Nuovo Cimento, Vol. 33, Ser. 2, 545 - 548 (1982).
Abstr. in Phys. Abstr., Vol. 85, Abstr. 84258 (1982).

141.531 **A new look at pulsar birthrate and evidence for injection.**
M. Vivekanand, R. Narayan.
Bull. Astron. Soc. India, Vol. 10, 48 (1982). – Abstract.

141.532 **The existence of a "gap" in the evolutionary sequence of pulsars.**
C. E. Akujor, P. N. Okeke.
Astrophys. Space Sci., Vol. 85, 325 - 329 (1982).
A replot of period derivative against period is done for about 300 pulsars, and the main features of the plot are discussed. The significance of a "gap" in this plot is reexamined and the existence of pulsars with nulling and subpulse drifting behaviour below this "gap" is pointed out. The implications of this for pulsar evolution are also discussed.

141.533 **Fluctuations in the pulse arrival times of four pulsars at the frequency of 100 MHz.**
T. V. Shabanova.
Astron. Zh., Tom 59, 903 - 907 (1982). In Russian. English translation in Soviet Astron., Vol. 26, No. 5.
The results of observations of fluctuations in the phase of pulses of 4 pulsars at the frequency 102.5 MHz are presented.

141.534 **Papers presented at the workshop on pulsars and supernovae.**
F. Pacini.
Tenth Texas Symposium on Relativistic Astrophysics, (see 012.021), p. 414 (1981).

141.535 **The galactic distribution of pulsars.**
A. G. Lyne.
Supernovae: a survey of current research, (see 012.022), p. 405 - 417 (1982).
Statistical studies of the properties of large numbers of pulsars and proper motion measurements provide a better knowledge of pulsar evolution and ages and show that a birthrate of one pulsar every 20 to 50 years is required to sustain the observed galactic population of 300000. This rate is comparable with most recent estimates of the galactic supernova rate, but requires nearly all supernovae to produce active pulsars.

141.536 **Timing observations of the binary pulsar PSR 1913 + 16.**
V. Boriakoff, D. C. Ferguson, M. P. Haugan, Y. Terzian,

S. A. Teukolsky.
Astrophys. J., Lett., Vol. 261, L97 - L101 (1982).
Observations of the binary pulsar PSR 1913 + 16 made at the Arecibo Observatory are reported. The analysis of these data yields orbital parameters of the binary system which are in agreement with those obtained by J. Taylor and his collaborators. In particular the data are consistent with the derived rate of change of the orbital period being attributed to gravitational radiation reaction.

141.537 **The proper motions of 26 pulsars.**
A. G. Lyne, B. Anderson, M. J. Salter.
Mon. Not. R. Astron. Soc., Vol. 201, 503 - 520 (1982).
Observations of the proper motion of 26 pulsars have been made using radio interferometry. Comparison with other measurements suggests that proper motion estimates derived from timing observations contain a source of error not accounted for in their quoted uncertainties.

141.538 **A pulsar model from an oscillating black hole.**
M. Sachs.
Found. Phys., Vol. 12, 689 - 708 (1982). – Abstr. in Phys. Abstr., Vol. 85, Abstr. 94764 (1982).

141.539 **Discovery of radio pulsations from the X-ray pulsar in the supernova remnant G 320.4–1.2.**
R. N. Manchester, I. R. Tuohy, N. D'Amico.
Astrophys. J., Lett., Vol. 262, L31 - L33 (1982).
A radio counterpart to the X-ray pulsar discovered by Seward and Harnden in the supernova remnant G320.4–12 (MSH 15–52) has been detected. The radio observations confirm the very large period derivative indicated by the X-ray data. This implies that the object is not a member of a binary system and hence is an isolated pulsar similar in some ways to the Crab pulsar. Association of the pulsar and the supernova remnant is supported by the observed pulsar dispersion measure.

141.540 **New lease of life for pulsars?**
Nature, Vol. 300, 399 (1982).

141.541 **The angular beaming model of microstructure and the subpulse drifting phenomenon.**
J. Gil.
Astron. Astrophys., Vol. 115, 270 - 274 (1982).
Within the angular beaming model, it is argued that if the subpulse drift results from the rotation of the radiation beam associated with subpulse around the magnetic axis, then micropulses should not consistently drift with drifting subpulses. The general properties of microstructure are also discussed.

141.542 **A multi-frequency fluctuation spectrum analysis of the pulse to pulse intensity variations in nine pulsars.**
L. Nowakowski, J. Usowicz, A. Wolszczan, A. Kępa.
Astron. Astrophys., Vol. 116, 158 - 163 (1982).
A fluctuation spectrum analysis of single pulse sequences from nine pulsars measured at radio frequencies ranging from 0.43 GHz to 2.7 GHz has been carried out. It is shown that a gradual growth of the quasi-steady component of the pulsar emission with increasing radio frequency is a common feature of these sources. A theoretically predicted radio frequency independence of the P_3 periods is found to hold in both the systematic and the erratic drifters, except for PSR 2016+28 and possibly PSR 2020+28. It is suggested that a slow intensity modulation generated by the quasi-steady component may explain the break frequency found in the modulation index/radio frequency relationship for some pulsars and that the phenomena observed in the authors' data may be directly related to the pulse emission constituents described as the core and the conal emissions.

141.543 **Millisecond pulsar in 4C 21.53.**
IAU Circ., Nos. 3743, 3746 (1982).

141.544 **Dynamic spectra of pulsar scintillations at frequencies near 0.34, 0.41, 0.63, 1.4, 1.7, 3.2 and 5.0 GHz.**
J. A. Roberts, J. G. Ables.
Mon. Not. R. Astron. Soc., Vol. 201, 1119 - 1138 (1982).
The authors give examples of the dynamic spectra of the interstellar scintillations of radiation received from the pulsars PSR 0628–28, 0833–45, 0834+06, 0950+08, 1451–68, 1642–03, 1749–28, 1919+21 and 1929+10. The autocovariance of the intensity is calculated as a function of separation in time and frequency to yield scattering parameters for eight of these pulsars and for PSR 1933+16. For PSR 0833–45 the details of the shapes of the auto-covariance functions, and the scaling with frequency, are compared with theory. For two pulsars, PSR 0628–28 and PSR 0834+06, systematic patterns of drifting bands were found in the dynamic spectra. The occurrence of such spectra implies that on these occasions the radiation reached the earth predominantly from a few directions only. The authors discuss the implications of this result for the distribution of irregularity scales in the interstellar medium.

141.545 **On the problem of the coming out of electromagnetic waves from the magnetosphere of pulsars.**
A. B. Mikhajlovskij, O. G. Onishchenko, G. I. Suramlishvili, S. E. Sharapov.
Pis'ma Astron. Zh., Tom 8, 685 - 688 (1982). In Russian. English translation in Soviet Astron. Lett., Vol. 8.
Cyclotron damping of electromagnetic waves in a relativistic electron-positron plasma immersed in a longitudinally inhomogeneous curved magnetic field is studied. It is shown that this damping may be essential in particular for radio waves generated inside the magnetosphere of a fast-rotating pulsar.

141.546 **A superfast new pulsar.**
D. J. Helfand.
Nature, Vol. 300, 573 (1982).

141.547 **A millisecond pulsar.**
D. C. Backer, S. R. Kulkarni, C. Heiles, M. M. Davis, W. M. Goss.
Nature, Vol. 300, 615 - 618 (1982).
The combination of properties suggested that 4C21.53 was either an undetected pulsar or a member of some new class of objects. This puzzle may be resolved by the discovery and related observations of a fast pulsar, 1937+214, with a period of 1.558 ms in the constellation Vulpecula only a few degrees from the direction to the original pulsar, 1919+21.

141.548 **Optical identification of the millisecond pulsar 1937+214.**
S. Djorgovski.
Nature, Vol. 300, 618 - 619 (1982).
The author presents an optical identification candidate for the millisecond pulsar: it is a 20th mag red object, undetectable on the Palomar Sky Survey prints. Finding charts and offsets from nearby stars are given.

141.549 **A new class of radio pulsars.**
M. A. Alpar, A. F. Cheng, M. A. Ruderman, J. Shaham.
Nature, Vol. 300, 728 - 730 (1982).
Sufficiently low magnetic field neutron stars which accrete for long times from a surrounding keplerian disk can be spun up to millisecond periods. After accretion ceases, such stars could become isolated or binary pulsars. Such pulsars constitute a new class, with short periods, long apparent ages and pulsed optical, X-ray and γ-ray fluxes significantly below those expected for canonical pulsars with similar periods. The

authors propose that the recently discovered millisecond pulsar 4C21.53 belongs to this new class, together with the binaries PSRs 1913 + 16, 0820 + 02 and 0655 + 64, as well as possibly several isolated pulsars such as PSRs 1952 + 29 and 1804 – 08.

141.550 **A new look at pulsar polarization.**
C.-I. Björnsson.
ESO Sci. Prepr., No. 225, 34 pp. (1982). – Submitted to Astrophys. J.

141.551 **On the pulsar distribution in the Galaxy.**
A. A. Suchkov.
Astron. Tsirk., No. 1200, p. 1 - 3 (1981). In Russian.

141.552 **Pulsar nulling and drifting subpulse phase memory.**
A. V. Filippenko, V. Radhakrishnan.
Astrophys. J., Vol. 263, 828 - 834 (1982).

The standard polar cap model of pulsar radio emission provides acceptable explanations for a wide variety of observed pulsar characteristics. However, the persistence of drifting subpulse phase memory observed during pulsar nulling, as well as the phenomenon of nulling itself, seem to defy a simple explanation. The authors attempt to reconcile these difficulties with some modifications of the standard model. It is proposed that avalanche discharges above the polar caps of old pulsars are localized to discrete regions ("flux tubes"). Nulling is thought to be the manifestation of steady (rather than interrupted) pair production within the flux tubes, and it is shown that the absence of both subpulse drift and coherent radio radiation is consistent with this hypothesis.

141.553 **On pulsar models.**
M. E. Gertsenshtejn.
Diskus. vopr. teor. otnositel'nosti i gravitatsii. Moskva, 1982, p. 123 - 126. In Russian. – Abstr. in Ref. zh., 51. Astron., 11.51.144 (1982).

141.554 **Line features in the X-ray spectrum of the Crab pulsar?**
G. Hasinger, W. Pietsch, C. Reppin, J. Trümper, W. Voges, E. Kendziorra, R. Staubert.
Accreting neutron stars, (see 012.069), p. 130 - 133 (1982).

Balloon observations of the Crab pulsar in the energy range 30 - 80 keV are discussed. Upper limits for possible emission lines in this spectral range are given.

141.555 **The production of electron-positron bunches in a pulsar magnetosphere and their thermal radio-emission.**
Yu. A. Rylov.
Astrophys. Space Sci., Vol. 88, 173 - 184 (1982).

It is shown that the high brightness temperature of pulsar radioemission, its sharp directivity and some general features of pulsar radiospectra can be obtained as a corollary of the general statistical regularities of motion of electron-positron bunches in a strong magnetic field. The production mechanism of the bunches is considered and their energy and sizes are estimated. In the presented pulsar model, the energy of accelerated electrons is estimated.

141.556 **About the determination of the mass of the pulsar in a binary system.**
Z. Komárek.
Říše hvězd, Vol. 63, 204 - 206 (1982). In Czech.

141.557 **On the implications of the radio and X-ray pulsar in the supernova remnant MSH 15-52.**
G. Srinivasan, K. S. Dwarakanath, V. Radhakrishnan.
Curr. Sci., Vol. 51, 896 - 899 (1982). – Abstr. in Phys. Abstr., Vol. 86, Abstr. 5394 (1983).

141.558 **The frequency dependence of component separation in pulsar integrated profiles. II. High frequencies.**
J. Gil.
Acta Phys. Polonica B, Vol. B13, 313 - 320 (1982). – Abstr. in Phys. Abstr., Vol. 86, Abstr. 10377 (1983).

141.559 **The distribution of the apparent beamwidths of pulsars and evolution of their magnetic inclinations.**
X.-j. Wu, F. Wu, G.-x. Deng, G.-j. Qiao, J.-h. Cheng.
Acta Astron. Sinica, Vol. 23, 140 - 146 (1982). In Chinese.

The distribution of the apparent beamwidths α for a sample of 195 pulsars with known ages was analysed. Under the assumption that the radiation is directly emitted by the pulsar cone and that the rotation axis is randomly orientated the formulae of the conditioned distribution density of α and the distribution density of φ were derived. The results of numerical calculations are listed. Then some informations about the magnetic inclinations of the pulsars were obtained. The authors conclude that the magnetic inclinations have a distribution with a preference for small angles.

Was sind Pulsare?
See Abstr. 003.033.

Landolt-Börnstein. Group VI. Vol. 2c. Interstellar matter – Galaxy – Universe.
See Abstr. 003.055.

Pulsar research as normal science.
See Abstr. 013.105.

Techniques for the estimation of red power spectra. I. Context and methodology.
See Abstr. 021.020.

On the theory of pair production by accelerated charges.
See Abstr. 022.060.

A pulsar experiment – our experiences in time keeping.
See Abstr. 031.579.

On the magnetospheric structure of pulsars.
See Abstr. 062.083.

Pulsars' timing noise and external disturbances: gravitational waves? .
See Abstr. 066.029.

Gravitational radiation and the binary pulsar.
See Abstr. 066.277.

A revised upper limit on gravitational radiation from the Crab pulsar.
See Abstr. 066.309.

Gravitationally collapsed objects.
See Abstr. 066.320.

Neutron star collisions and the r-process.
See Abstr. 066.513.

Neutron stars and general equation of nuclear matter.
See Abstr. 066.516.

The universal diagram for magnetized neutron stars in the Galaxy.
See Abstr. 066.525.

Time resolved circular polarimetry of white dwarf pulsars.
See Abstr. 117.153.

What can we learn from the Crab nebula?
See Abstr. 125.037.

Supernovae in binary systems: production of runaway stars and pulsars.
See Abstr. 125.085.

The structure of a type-II supernova remnant G 160.5 + 2.8 (HB-9) observed at $\lambda = 12$ m.
See Abstr. 125.106.

Interstellar scintillation and pulsar velocities.
See Abstr. 131.035.

A distinct shell structure in H I-line-emission at intermediate galactic latitudes.
See Abstr. 131.058.

Apparent proper motions of the galactic center compact radio source and PSR 1929+10.
See Abstr. 141.053.

Transient emission of ultra-high energy pulsed gamma rays from PSR 0531.
See Abstr. 142.547.

The Vela star cloud. II. Early-type stars and long-period Cepheids near the Vela pulsar.
See Abstr. 152.003.

142 UV Sources, X-ray Sources, X-ray Background, Gamma-ray Sources, Gamma-ray Background

UV Sources, X-ray Sources, X-ray Background

142.001 **Identification of the X-ray source E 1551 + 718 with an uncataloged dwarf nova.**
K. O. Mason, G. A. Reichert, S. Bowyer, J. R. Thorstensen.
Publ. Astron. Soc. Pacific, Vol. 94, 521 - 523 (1982).

The X-ray source E 1551 + 718 is identified with a previously uncataloged dwarf nova. The X-ray source was observed at a flux level of $\sim 7 \times 10^{-13}$ erg cm^{-2} s^{-1} in the energy range 0.1 - 4.0 keV. The optical star is estimated to vary between V ~ 17 and V ~ 13.

142.002 **Fast variations in the period of a pulsar in the X-ray source 1 H 1909+096.**
D. Sadeh, M. Livio.
Astrophys. J., Vol. 258, 770 - 775 (1982).

The known X-ray source 1 H 1909+096 was found at a state of high intensity on 1977 October 15, when 5 ms accumulation data was transmitted by the HEAO A-1 experiment. During that time, this source was found to pulsate with a period of ~15.25 ms. This period was found to change during the 10 s of scanning through the source. A theoretical discussion and two models are presented: (1) non-radial g-mode oscillation, and (2) torsional oscillation in the solid crust of a neutron star.

142.003 **On the polarization of X-ray sources.**
V. M. Loskutov, V. V. Sobolev.
Astrofizika, Tom 18, 81 - 91 (1982). In Russian. English translation in Astrophysics, Vol. 18, No. 1.

It is supposed that the polarization of X-ray sources is due to electron scattering. For the theoretical determination of the degree of polarization multiple scattering in a plane slab is considered when the temperature varies with the optical depth. The linear integral equations for the emergent intensities are solved. The theoretical polarization is compared with observational data for the X-ray sources Cyg X-1 and Sco X-1. Parameters of the accretion disks in these systems are evaluated.

142.004 **GX 339-4: cyclotron radiation from an accretion flow.**
A. C. Fabian, P. W. Guilbert, C. Motch, M. Ricketts, S. A. Ilovaisky, C. Chevalier.
Astron. Astrophys., Vol. 111, L9 - L10 (1982).

The authors show that the fast optical flares recently discovered in the Galactic X-ray source GX339-4 are probably due to cyclotron emission from hot gas at 10^9 K. The inner accretion flow appears to be very similar to that in Cygnus X-1.

142.005 **Why are X-ray pulsars not observed coupled with Wolf-Rayet stars?**
V. M. Lipunov.
Pis'ma Astron. Zh., Tom 8, 358 - 361 (1982). In Russian. English translation in Soviet Astron. Lett., Vol. 8.

The peculiarities of an accretion by neutron stars in binary systems with the other component being a Wolf-Rayet star are analysed. It is shown that an X-ray pulsar phenomenon in the systems considered can occur but under specific conditions.

142.006 **On the spin down episodes of Vela X-1.**
D. Molteni, M. Rapisarda, S. Re, N. R. Robba.
Astron. Astrophys., Vol. 111, 365 - 367 (1982).

The ESA COS-B satellite observed the X-ray source Vela X-1 in July/August 1976 and in October/November 1979 in the energy range 2–12 keV. Different pulse periods, 282.819 and 282.782 s, were determined for the two observations respectively. These values confirm the presence of spin up and spin down processes in this source.

142.007 **Journal of high energy X-ray observations from the Ariel V satellite.**
M. J. Coe, G. F. Carpenter, A. R. Engel, J. J. Quenby.
Mon. Not. R. Astron. Soc., Vol. 200, 385 - 390, microfiche MN 200/2 (1982).

Presented here is a summary of nearly 100 high energy X-ray observations carried out within the period 1974 October to 1979 January from the Ariel V satellite. The detector covered the photon energy range 26–1200 keV and on average detected about 35 per cent of the sources studied. The positive results are used to construct a log N–log S curve for the galactic disc sources at these energies. Only two extragalactic sources were detected (apart from SMC X-1), these being NGC 4151 and Centaurus A.

142.008 **Simultaneous spectroscopy and photometry of the non-synchronous rotator H2252–035.**
D. T. Wickramasinghe, R. S. Stobie, M. S. Bessell.
Mon. Not. R. Astron. Soc., Vol. 200, 605 - 619 (1982).

The authors present simultaneous spectroscopic and photometric observations of the non-synchronous rotator H2252–035. The photometry shows the presence of pulsations with the 859 second period with a semi-amplitude of 0.026 mag and a slower variation of the mean light level with a period of 3.59 hr with semi-amplitude 0.05 mag. The spectra show complex line profile variations with multiple components present during some phases. The velocity semi-amplitude of the primary is estimated to be $K_1 \sim 60$ km s^{-1}. The authors find the radial velocity curves caused by a component which moves with a semi-amplitude $K_{SW} \sim 350$ km s^{-1} and discuss their results in terms of two models that have been proposed for the system.

142.009 **X-ray emission from extragalactic bursts of star formation.**
G. C. Stewart, A. C. Fabian, R. J. Terlevich, C. Hazard.
Mon. Not. R. Astron. Soc., Vol. 200, 61P - 64P (1982).

An Einstein X-ray observation of NGC 5408, a small irregular galaxy undergoing a violent burst of star formation, reveals that it radiates $\sim 10^{40}$ erg s^{-1} in 0.5–3 keV X-rays. Such a high X-ray luminosity means that $L_X/L_V > 0.1$. The low metal abundance of NGC 5408 may be an important factor and suggests that most young galaxies are X-ray luminous.

142.010 **Optical review: X-ray binaries.**
J. B. Hutchings.
Galactic X-ray sources, (see 003.001), p. 3 - 26 (1982).

Contents: Overview. The supergiant primaries. Spectroscopic complications. Other system quantities. Line emission. The Be primary systems. Light curve analysis.

142.011 **Some aspects of mass transfer in massive X-ray binaries.**
G. J. Savonije.
Galactic X-ray sources, (see 003.001), p. 41 - 55 (1982).

Contents: Introduction to mass transfer processes in

binary stars. The initial phase of Roche lobe overflow. The "Roche lobe process" of overflow. Estimates for the rate of transfer of mass by the Roche lobe process. The expected life time of massive X-ray binaries when powered by Roche lobe overflow. The evolutionary status of the early type components of massive X-ray binary systems. Stabilization of Roche lobe overflow by stellar winds. Mass ejection from non-synchronous contact stars. Periodic phenomena in non-synchronous binary stars.

142.012 **Low mass, optically identified galactic X-ray sources.**
A. P. Cowley.
Galactic X-ray sources, (see 003.001), p. 59 - 70 (1982).
The author considers low mass hard X-ray sources with $\Sigma M \lesssim 5 M_\odot$, which include (1) those in which the light of the disk dominates the optical spectrum, such as V 818 Sco/Sco X-1, (2) systems in which the non-degenerate star dominates the optical spectrum such as HZ Her/Her X-1 and V 1341 Cyg/Cyg X-2, (3) the strongly magnetic cataclysmic variables, such as AM Her/1814 + 498, and (4) some of the X-ray novae or transients, such as Nova Mon 1975/0620-00. Characteristics of each group are discussed with particular emphasis on the individual objects for which there is good spectroscopic and photometric data.

142.013 **Spectral formation in galactic X-ray sources.**
R. McCray.
Galactic X-ray sources, (see 003.001), p. 71 - 88 (1982).
Theories for the formation of the spectra from galactic X-ray sources are reviewed, with emphasis on the mechanisms for the production of atomic emission lines and the interpretation of the observations. Three classes of idealised models for sources are discussed. The first is a "coronal model", in which electron collisions control the ionisation and emissivity of the gas. The second "nebular model" is for the distribution of gas around a compact source of continuum X-rays, in which photoionisation and recombination of atoms in the surrounding cloud control the form of the spectrum. The third is a "diffusion model" in which the opacity is so great that electron scattering plays a significant role.

142.014 **Rapid aperiodic X-ray variability.**
H. V. Bradt, R. L. Kelley, L. D. Petro.
Galactic X-ray sources, (see 003.001), p. 89 - 112 (1982).
Data from at least ten "persistent" galactic X-ray sources have exhibited rapid, aperiodic (non-burst) X-ray variability with time constants ranging from 20s down to at least tens of milliseconds in some cases. Random variability on time scales of 30 - 50 ms has also been detected in both Type I and Type II X-ray bursts and < 0.4 s rise times have been observed in "fast transient" precursors.

142.015 **Optical photometry of galactic X-ray sources.**
S. A. Ilovaisky.
Galactic X-ray sources, (see 003.001), p. 113 - 130 (1982).
The author deals with what can be learned from photometric studies of galactic X-ray sources and discusses in detail some particular sources as examples.

142.016 **Temporal studies of galactic X-ray sources.**
M. G. Watson.
Galactic X-ray sources, (see 003.001), p. 131 - 144 (1982).
The author concentrates on the medium to long timescale variability (days to years) observed in galactic X-ray sources.

142.017 **Ariel V observations of Hercules X-1.**
A. N. Parmar.
Galactic X-ray sources, (see 003.001), p. 145 - 148 (1982).
Detailed spectra obtained with the Ariel 5 satellite for the turn-on and off of Hercules X-1 from the extended low are presented. The author finds that, except for anomalous and pre-eclipse dips, the hydrogen column density is generally $\sim 10^{22}$ atoms cm^{-2} and the spectral normalisation increases approximately linearly with time. A light curve of part of the secondary on-state is also presented.

142.018 **Binary X-ray pulsars.**
S. Rappaport.
Galactic X-ray sources, (see 003.001), p. 159 - 196 (1982).
The author presents a brief review of the state of the empirical knowledge of binary X-ray pulsar systems. He reports recent results for three X-ray pulsars: 4U 1626−67, GX 301−2 and 4U 0900−40.

142.019 **Young pulsars in binary systems.**
L. Maraschi, A. Treves.
Galactic X-ray sources, (see 003.001), p. 197 - 204 (1982).
The authors have discussed some observational characteristics of pulsars in binary systems and specifically the cases of Cyg X-3, Cir X-1 and SS433. One consequence of dealing with three young systems should be considered. In fact if the pulsar phase (10^3 - 10^4 y) were followed by a much longer X-ray active accretion phase (10^{6-7} y) the implied number of X-ray binaries would be too high.

142.020 **The pulsating X-ray source GX 1 + 4 (4U 1728−24).**
E. Kendziorra, R. Staubert, C. Reppin, W. Pietsch, W. Voges, J. Trümper.
Galactic X-ray sources, (see 003.001), p. 205 - 216 (1982).
GX 1 + 4 (4U 1728−24) was observed at hard X-rays (>22 keV). The photon number spectrum is measured up to ~ 150 keV. The pulsed fraction of the periodically modulated source flux was (52 ± 8) percent in the 22 - 80 keV interval. The authors present evidence that the true pulsational period is (227.5 ± 0.5) s rather than half this value.

142.021 **The 1975 May transition of Cygnus X-1.**
L. Chiappetti.
Galactic X-ray sources, (see 003.001), p. 225 - 228 (1982).
The MSSL instrument on the Ariel 5 satellite observed Cyg X-1 during May 1975 for part of the H I state and the subsequent gradual recovery to the quiet state. These data, the only known spectral observations of moderate energy resolution made during a transition, are presented.

142.022 **Ultraviolet spectroscopy of galactic X-ray sources.**
A. K. Dupree.
Galactic X-ray sources, (see 003.001), p. 269 - 285 (1982).
The author first discusses the ultraviolet spectroscopy of four high-luminosity X-ray sources, 4U 0900-40 (HD 77581), Cygnus X-1 (HDE 226868), 4U 1700-37 (HD 153919) and 1653-40 (HD 152667 = V861 Sco). Additionally the study of late-type binaries such as the RS CVn and W UMa systems has revealed substantial X-ray and ultraviolet fluxes in contrast to single stars of the same type. These late-type binary systems possessing enhanced coronae are also discussed.

142.023 **Features of the soft X-ray background.**
J. A. M. Bleeker.
Galactic X-ray sources, (see 003.001), p. 287 - 296 (1982).
The soft X-ray background data show us the hottest phase of the interstellar medium. The author considers possible formation and maintenance of hot plasma regions or "coronal" gas in interstellar space. He gives a brief description of what constraints can be derived from the available observational data on the properties of the hot gas like filling factor, thermal pressure, spatial and temperature distribution, in order to discriminate between theories on origin, evolution and physical state.

142.024 **Galactic soft X-ray sources: RS CVn systems, flare stars and A stars.**
P. Charles.

Galactic X-ray sources, (see 003.001), p. 301 - 316 (1982).

The author considers galactic point sources of soft X-ray emission where the bulk of the emission is in the 0.1 – 1 keV range.

142.025 **X-ray sources in globular clusters and burst sources.**
W. H. G. Lewin, G. W. Clark.
Galactic X-ray sources, (see 003.001), p. 319 - 354 (1982).

The similarities of the bulge and cluster X-ray sources lead one to consider them together as a class of high luminosity old population (II) X-ray sources. It is generally assumed that they are compact objects (i.e. white dwarfs, neutron stars, or black holes of either stellar or greater mass) which derive their X-ray power from the process of accretion. The authors argue on the basis of the observed properties that the majority of these sources are weakly magnetised neutron stars in low-mass binary systems.

142.026 **Ultraviolet observations of X-ray globular clusters.**
L. Hartmann.
Galactic X-ray sources, (see 003.001), p. 355 - 361 (1982).

The author presents a preliminary discussion of a search for UV emission from X-ray sources in the central regions of six globular clusters using the International Ultraviolet Explorer Satellite. The data are of much higher spatial and spectral resolution than previous ultraviolet studies of globular clusters. The data have not been corrected for the calibration error recently found. However, some spectra have been corrected and no major effect has been found so far.

142.027 **X-ray observations of globular clusters.**
J. E. Grindlay.
Galactic X-ray sources, (see 003.001), p. 363 - 378 (1982).

The author presents previous observations of X-ray globular clusters. He summarizes preliminary new results from the Einstein X-ray Observatory which promise to enhance our understanding of these objects and to provide significant tests of the neutron star vs. black hole models.

142.028 **Galactic X-ray sources as observed with the Einstein Observatory.**
E. J. Schreier.
Galactic X-ray sources, (see 003.001), p. 381 - 392 (1982).

Contents: Stellar X-ray sources. Supernova remnants. The galactic X-ray sources.

142.029 **High resolution soft X-ray spectra with the objective grating spectrometer aboard Einstein Observatory.**
J. Heise, A. C. Brinkman.
Galactic X-ray sources, (see 003.001), p. 393 - 404 (1982).

The authors present some preliminary data obtained from the binary compact X-ray source Hercules X-1 and from the close binary white dwarf AM Herculis.

142.030 **Ariel V observations of a 4.8 hour periodicity in the "high-state" X-ray spectrum of Cyg X-3.**
R. J. Blissett.
Galactic X-ray sources, (see 003.001), p. 411 - 415 (1982).

142.031 **Color variability and optical light curve of 2S0921–630.**
C. Chevalier, S. A. Ilovaisky.
Astron. Astrophys., Vol. 112, 68 - 72 (1982).

Analysis of optical photometry of the 2S0921–630 X-ray source counterpart obtained in 1979, 1980, and 1981 at ESO shows that the *U–B* index is modulated with a 9-d period, this being also the recurrence period of the magnitude-deep eclipses. Most of the light in this system comes from a luminous accretion disk, overwhelming the contribution from the companion, which becomes detectable only during the eclipses. The authors interpret the eclipses as due to partial occultation of the disk by a late-type companion and by a stream. The *U–B* modulation is closely correlated with the intensity modulation of the emission lines, suggesting an origin in a Balmer continuum emission region whose visibility varies in the course of an orbital cycle. This could be due to the changing aspect of an X-ray illuminated bulge.

142.032 **Ariel V hard X-ray observations of Vela X-1.**
R. Lieu, A. R. Engel, J. J. Quenby.
Astrophys. Lett., Vol. 22, 135 - 137 (1982).

The binary X-ray pulsar Vela X-1 (3U0900-40) was observed by the Ariel V satellite from 7 February 1979 for approximately 40 days. Data analysis led to a spectral point in the low energy (2.9–7.6 keV) range, and two upper limits in the high energy (26–122 keV) range. The results are compared with the high energy balloon data of the Munich group. Fit with a modified black body spectrum at kT ~10 keV is shown to be good. At energies E ~40 keV data from both observations show a steeper cut-off behavior than that predicted by the modified black body. This appear to suggest the presence of strong magnetic fields in the neutron star polar cap region responsible for the hard X-ray emission of Vela X-1.

142.033 **High energy X-ray observations of Cyg X-3 from OSO-8: further evidence of a 34.1 day period?**
J. F. Dolan, C. J. Crannell, B. R. Dennis, K. J. Frost, L. E. Orwig.
Astrophys. Lett., Vol. 22, 147 - 152 (1982).

The X-ray source Cyg X-3 (=4U2030+40) was observed with the high energy X-ray spectrometer on OSO-8 for two weeks in 1975 and in 1976 and for one week in 1977. No change in spectral shape and intensity above 23 keV was observed from year to year. No correlation is observed between the source's intensity and the phase of the 34.1-day period discovered by Molteni et al. (1980). The pulsed fraction of the 4.8-hour light curve between 23 and 73 keV varies from week to week, however, and the magnitude of the pulsed fraction appears to be correlated with the 34.1-day phase. No immediate explanation of this behavior is apparent in terms of previously proposed models of the source.

142.034 **Two-photon absorption and the cyclotron line in Her X-1.**
D. B. Melrose, A. J. Parle.
Proc. Astron. Soc. Australia, Vol. 4, 214 - 217 (1981).

The authors conclude that the two-photon absorption process should not be ignored in Her X-1. It can provide an effective coupling between the parallel motion of the electrons and the distribution of electrons amongst their Landau levels, and hence may be the important step in the ultimate generation of the cyclotron photons.

142.035 **Discovery of 0.5-s oscillations in long flat top bursts from the Rapid Burster.**
Y. Tawara, S. Hayakawa, H. Kunieda, F. Makino, F. Nagase.
Nature, Vol. 299, 38 - 41 (1982).

The authors report a search for periodic emission of X rays from MXB1730–335 (the Rapid Burster) during the long flat top bursts. Two events from the 63 bursts showed clear peaks in the frequency power spectra at ~0.5 s in period. However, this is unlikely to be a pulsar period, because the periods of these two bursts are different.

142.036 **On the nature of two modes in the type II bursts in MXB 1730–335.**
V. M. Loznikov, N. S. Yamburenko.
Pis'ma Astron. Zh., Tom 8, 472 - 475 (1982). In Russian. English translation in Soviet Astron. Lett., Vol. 8.

It is shown that the existence of two modes in the type II bursts of the MXB 1730–335 source can be naturally explained as a result of accretion onto one or both magnetic poles of a neutron star rotating with a period of about 250 s.

142.037 **X-ray observations of Circinus X-1 in its low state.**
L. Chiappetti, S. J. Bell Burnell.
Mon. Not. R. Astron. Soc., Vol. 200, 1025 - 1031 (1982).

Ariel V experiment C repeatedly observed Cir X-1 in the low-intensity portion of its 16.6-day light curve. The spectrum of the source is found to be a power law with photon spectral index ~2.3–3.0 and hydrogen column density $N_H \sim 10^{22}$ cm^{-2}. A weak iron feature may be present. The observation of such a low column density for the low-intensity state runs counter to previous assumptions; the significance of this observation for the current model of the source, in which the modulation is explained by absorption of X-rays in the primary's stellar wind, is discussed. A spectrum of 4U1510–59, in the same field as Cir X-1, is also reported.

142.038 **2A 0526–328, an optical and X-ray study of a cataclysmic variable.**
D. J. Watts, J. G. Greenhill, P. W. Hill, R. M. Thomas.
Mon. Not. R. Astron. Soc., Vol. 200, 1039 - 1051 (1982).

Spectroscopic observations of the cataclysmic variable 2A 0526–328 confirm that the emission line radial velocity period (5.48 hr) is significantly different from the photometric period (5.19 hr). Changes in the emission line velocity and profiles suggest that the accretion process is controlled by the white dwarf's magnetic field. Einstein X-ray observations show it is a hard source with a thermal spectrum of $kT \sim 19$ keV. Quasi-periodic variations on a time-scale of ~700 s are observed in the X-ray light curve and reported in the optical emissions. A model for the object is proposed in which a rotating white dwarf revolves with a 4-day period in a retrograde orbit with a subluminous red giant.

142.039 **On a possible model of the object Cyg X-3.**
A. A. Stepanian *(Stepanyan).*
Astrophys. Space Sci., Vol. 84, 347 - 361 (1982).

According to the proposed model of Cyg X-3, it should be recognized that it is a binary system consisting of a pulsar and a usual star. The author attempts to explain the whole range of electromagnetic radiation and its main characteristics using numerical calculations.

142.040 **Observations of the May 1979 outburst of Centaurus X-4.**
W. P. Blair, J. C. Raymond, A. K. Dupree.
Advances in ultraviolet astronomy, (see 012.006), p. 558 - 561 (1982).

The authors have obtained IUE spectra of the X-ray transient/X-ray burst source Cen X-4 at three intervals during the peak and decline of the May 1979 transient event. The spectrum is characterized by a blue continuum and strong emission lines of N V λ 1240, Si IV λ 1398 and C IV λ 1550. The origin of these emission components in the context of an X-ray dwarf nova model is investigated. It appears that an accretion disk plays a prominent role in the generation of the contimuum emission and that X-ray heating of the accretion disk and the companion star may be important in the formation of the emission lines.

142.041 **The cycle-to-cycle variability of Cygnus X-3.**
M. van der Klis, J. M. Bonnet-Bidaud.
Astron. Astrophys., Suppl. Ser., Vol. 50, 129 - 140 (1982).

Results are reported of a study of the variability of the shape of the light curve of Cyg X-3 from one cycle to the next on time scales down to ~ 200 s. Variations occur mainly around the maximum of the light curve, on a time scale of about half an hour or less, and give rise to a dispersion of the individual light curves around the mean which is a considerable fraction of the amplitude of the modulation.

142.042 **On the short-term variability of HD 153919 (= 4U1700–37=V884 Sco).**
J. van Paradijs, H. van der Woerd.
Astron. Astrophys., Vol. 113, 27 - 30 (1982).

The authors present the results of photometric observations of HD 153919, obtained during six nights when the orbital phase of the system was between 0.4 and 0.6. The data show brightness variations of ≤1%, on a time scale of hours, but there is no evidence for the general presence of optical pulsations at a period near ~1.5 h previously reported by Kruszewski et al. (1979). The authors conclude that the observed optical brightness variations are probably the result of dynamic processes in the envelope of HD 153919.

142.043 **20 years X-ray astronomy.**
I. S. Shklovskij.
Priroda, 1982, No. 9, p. 10 - 20. In Russian.

142.044 **A magnitude limited stellar X-ray survey and the F star X-ray luminosity function.**
K. Topka, Y. Avni, L. Golub, P. Gorenstein, F. R. Harnden, Jr., R. Rosner, G. S. Vaiana.
Astrophys. J., Vol. 259, 677 - 692 (1982).

The authors have conducted an X-ray survey of stars brighter than visual magnitude 8.5 that have serendipitously fallen into the fields of view of the IPC of the Einstein Observatory. The survey includes 227 separate 1°×1° fields, containing 274 stars with $V \leqslant 8.5$ and covering a wide range of spectral types and luminosity classes. X-ray emission was detected from 33 stars, and upper limits have been determined for the remainder of the sample. F type stars dominate the detected sample, and most of these are shown to be dwarfs. An X-ray luminosity function for dF stars has been deduced.

142.045 **Einstein solid state spectrometer observation of the peculiar red dwarf Wolf 630 AB.**
J. H. Swank, H. M. Johnson.
Astrophys. J., Lett., Vol. 259, L67 - L70 (1982).

Wolf 630 AB is a double and perhaps triple star with a predominant dM 3.5e spectrum. It is one of the relatively strong, red dwarf X-ray sources. The 0.5–4 keV spectral data for a steady, nonflaring flux are interpreted in terms of emission from thin thermal plasma with a dominant temperature of ~ 6.5 × 10^6 K. The quiescent corona is similar to that of the low-temperature component found for RS CVn binaries.

142.046 **Spectral evolution of active galactic nuclei: a unified description of the X-ray and gamma-ray backgrounds.**
D. Leiter, E. Boldt.
Astrophys. J., Vol. 260, 1 - 19 (1982).

A model for spectral evolution is presented whereby active galactic nuclei (AGN) of the type observed individually have emerged from an earlier stage at $z \approx 4$ in which they are the thermal X-ray sources responsible for most of the cosmic X-ray background (CXB). The authors pursue the conjecture that these precursor objects are initially supermassive Schwarzschild black holes with accretion disks radiating near the Eddington luminosity limit. A superposition of sources in the precursor stage can thereby account for that major portion of the CXB remaining after the contributions of usual AGN are considered, while a superposition of AGN sources at $z < 1$ can account for the gamma-ray background.

142.047 **X-ray spectrum of Capella and its relation to coronal structure and ultraviolet emission.**
R. Mewe, E. H. B. M. Gronenschild, N. J. Westergaard, J. Heise, F. D. Seward, T. Chlebowski, N. P. M. Kuin, A. C. Brinkman, J. H. Dijkstra, H. W. Schnopper.
Astrophys. J., Vol. 260, 233 - 239 (1982).

The binary system Capella (α Aur, G6 III + F9 III) has been observed with the objective grating spectrometer on board the Einstein Observatory. The spectra show evidence for a bimodal temperature distribution of emission measure in an optically thin plasma with one component ~5 × 10^6 K and the other one $\sim 10 \times 10^6$ K. Spectral features can be identified with line emissions from O VIII, Fe XVII, Fe XVIII, Fe XXIV, and Ne X ions. The data are interpreted in terms of emission from hot static coronal loops rather similar to the magnetic arch structures found on the Sun.

142.048 ***UBV* photometry of SMC X-1 and Cen X-3. Some regularities in the observational appearance of X-ray binary systems.**
A. M. Cherepashchuk.
Pis'ma Astron. Zh., Tom 8, 623 - 630 (1982). In Russian. English translation in Soviet Astron. Lett., Vol. 8.

Photoelectric *UBV* photometry of two faint X-ray binary systems: SMC X-1 and Cen X-3 has been carried out. The data concerning a regular long-term variability of eight X-ray binaries are summarised. These data imply that the long-period variability of X-ray binaries is connected with the forced precession of the rotational axis of the optical star. This may be a consequence of a slight asymmetry of a supernova explosion in the binary system.

142.049 **Optical identification of serendipitous Einstein X-ray sources: luminosity and redshift relations in the X-ray and optical emission from active galaxies and quasars.**
G. A. Reichert, K. O. Mason, J. R. Thorstensen, S. Bowyer.
Astrophys. J., Vol. 260, 437 - 468 (1982).

The authors have optically identified 40 objects from a sample of 139 serendipitous Einstein X-ray sources. Thirteen of the sources are identified with galactic counterparts, and 27 are identified with extragalactic counterparts. The galactic counterparts include normal bright stars, emission line stars, RS CVn stars, and cataclysmic variables. All but one of the extragalactic counterparts are active galactic nuclei (AGNs). The optical spectra of the X-ray selected AGNs are indistinguishable from those of AGNs discovered by their radio or optical emission. The authors have combined their sample of X-ray selected AGNs with other samples of X-ray selected and previously cataloged AGNs in order to investigate correlations between their optical and X-ray properties.

142.050 **Constraints on the parameters of X-ray burster emission regions.**
H. L. Marshall.
Astrophys. J., Vol. 260, 815 - 820 (1982).

This paper discusses general relativistic effects in the derivation of physical parameters from the observed properties of X-ray bursters. Limiting relations are established for the ratio of the total flux to the Eddington limit, the black-body emissivity and the fraction of the emitting surface near the central neutron star or black hole.

142.051 **Circinus X-1: X-ray observations with SAS 3.**
R. G. Dower, H. V. Bradt, E. H. Morgan.
Astrophys. J., Vol. 261, 228 - 250 (1982).

Eight observations of Cir X-1 with SAS 3, each lasting 1–6 days, have yielded a variety of new phenomena: a luminous state of steady emission, rapid large-intensity dips, an extremely rapid X-ray transition, and bright flares. Thorough searches for periodic X-ray pulsations were carried out. Aperiodic variability with characteristic times of 0.4–1.0 s was observed. No millisecond bursts were detected. Models previously proposed for the Cir X-1 system do not easily provide explanations for all the complex phenomena reported.

142.052 **On the ultrahigh-energy gamma rays from Cygnus X-3.**
W. T. Vestrand, D. Eichler.
Astrophys. J., Vol. 261, 251 - 258 (1982).

The authors examine how generation of ultrahigh-energy (UHE) γ-rays can be incorporated into a model of the Cyg X-3 system as a young pulsar in a close binary system. They show that the observed UHE light curve can be naturally reproduced if the companion's atmosphere acts as target material for particles accelerated near the pulsar.

142.053 **A possible optical counterpart to the X-ray pulsar 1E 2259+586.**
G. G. Fahlman, P. Hickson, H. B. Richer, P. C. Gregory, J. Middleditch.
Astrophys. J., Lett., Vol. 261, L1 - L5 (1982).

An optical, intensified CCD observation of the field of the 3.49 s X-ray pulsar 1E 2259+586 is discussed. Two stars with $B = 22.0$ and $B = 21.3$ are found to lie 4''.9 and 6''.3 away from the nominal X-ray position determined by the HRI camera of the Einstein Observatory. The fainter star is identified as the optical counterpart to the pulsar on the basis of recently detected infrared pulsations. The absence of a bright optical counterpart rules out a massive binary companion for the pulsar. If the pulsar has a binary companion, it must be a low-mass main-sequence dwarf or a degenerate star.

142.054 **Two X-ray selected BL Lacertae candidates.**
G. A. Chanan, B. Margon, D. J. Helfand, R. A. Downes, D. Chance.
Astrophys. J., Lett., Vol. 261, L31 - L34, plate L1 (1982).

The authors have identified two serendipitous Einstein Observatory X-ray sources with faint optical counterparts having the featureless spectra characteristic of BL Lacertae objects. Observations with the VLA at 6 cm establish that these sources have radio to optical luminosity ratios which are lower than the mean for BL Lac objects by one to two orders of magnitude. It is suggested that these two objects are among the first examples of radio-quiet BL Lac objects.

142.055 **The anomalous X-ray absorption spectrum of Vela X-1.**
T. R. Kallman, N. E. White.
Astrophys. J., Lett., Vol. 261, L35 - L39 (1982).

The authors present the results of observations during one orbit of the massive X-ray binary Vela X-1 (2U 0900–40/ HD 77581) using the SSS and MPC on the Einstein Observatory. The spectrum near orbital phase 0.2 compares favorably with absorption by neutral material with a column density corresponding to plausible values for the stellar wind velocity law and total primary mass loss rate. Spectra at later orbital phases show unexpected strong absorption features near 2 and 2.5 keV.

142.056 **X-ray structure of the Galaxy.**
R. D. Bleach, K. S. Wood, D. J. Yentis, J. F. Meekins, H. W. Smathers, M. Meidav.
Bull. American Astron. Soc., Vol. 14, 618 (1982). – Abstract.

142.057 **Evidence for a periodicity of ~300 days in Cyg X-1.**
W. Priedhorsky, J. Terrell.
Bull. American Astron. Soc., Vol. 14, 618 (1982). – Abstract.

142.058 **Identification of the bright X-ray source 3A1148+719 with the cataclysmic variable YY Draconis.**
J. Patterson, D. A. Schwartz, H. Bradt, R. Remillard, I. I. McHardy, J. P. Pye, G. Williams, R. A. Fesen, P. Szkody.
Bull. American Astron. Soc., Vol. 14, 618 (1982). – Abstract.

142.059 **X-ray variability of the globular cluster source 4U1820-30 (NGC 6624).**
L. Stella, R. I. Burg, J. E. Grindlay, S. M. Kahn.
Bull. American Astron. Soc., Vol. 14, 618 (1982). – Abstract.

142.060 **X-ray transients, 1969 - 1976 (the movie).**
J. Terrell, R. D. Belian, J. P. Conner, W. D. Evans,

W. C. Priedhorsky.
Bull. American Astron. Soc., Vol. 14, 619 (1982). – Abstract.

142.061 **Photometric variability of A0620-00 at quiescence.**
J. E. McClintock, R. A. Remillard, L. D. Petro, G. R. Ricker.
Bull. American Astron. Soc., Vol. 14, 619 (1982). – Abstract.

142.062 **X-ray emission from Herbig Ae/Be stars.**
W. T. Sanders, J. P. Cassinelli, C. M. Anderson.
Bull. American Astron. Soc., Vol. 14, 629 - 630 (1982). Abstract.

142.063 **X-ray observations of AM Her in its low state.**
G. Fabbiano.
Bull. American Astron. Soc., Vol. 14, 632 (1982). – Abstract.

142.064 **Cosmic ultraviolet (1700 - 2850 Å) background radiation.**
P. D. Tennyson, R. C. Henry, G. F. Hartig, P. D. Feldman.
Bull. American Astron. Soc., Vol. 14, 644 (1982). – Abstract.

142.065 **The Einstein Observatory Medium Sensitivity Survey.**
I. M. Gioia, J. Danziger, R. E. Griffiths, T. Maccacaro, J. Stocke.
Bull. American Astron. Soc., Vol. 14, 644 (1982). – Abstract.

142.066 **A sum-rule for stellar X-ray luminosity functions.**
G. P. Garmire.
Bull. American Astron. Soc., Vol. 14, 652 (1982). – Abstract.

142.067 **The optical identification content of the medium-sensitivity X-ray survey with the Einstein Observatory.**
R. E. Griffiths, T. Maccacaro, I. Gioia, J. Stocke, J. Liebert, J. Danziger, J. Lub.
Bull. American Astron. Soc., Vol. 14, 659 (1982). – Abstract.

142.068 **HEAO-1 and HEAO-2 observations of features in the 0.5 - 1.5 keV diffuse background near the galactic center.**
P. M. Fried, J. A. Nousek, G. Garmire.
Bull. American Astron. Soc., Vol. 14, 661 - 662 (1982). Abstract.

142.069 **Why are essential parts of "bursters" located in globular clusters?**
I. S. Shklovsky *(Shklovskij).*
Comments Astrophys., Vol. 9, 261 - 264 (1982).

142.070 **The magnetosphere of an X-ray star.**
N. Shibazaki.
Magnetospheric phenomena of celestial bodies, (see 012.011), p. 94 - 97 (1982). In Japanese.

Vela X-1 is an X-ray pulsar with a 283 second pulse period. MXB1730-335 is an X-ray burster which gives rise to bursts very frequently. The Japanese X-ray Astronomy Satellite HAKUCHO observed interesting phenomena from these sources, which are considered to have close connections with the magnetosphere of a neutron star. The observational data are presented and a discussion is given in terms of magnetospheric models of neutron stars.

142.071 **Discrete source contributions to the X-ray background.**
G. De Zotti.
Acta Cosmologica, Zesz. 11, (see 012.012), 65 - 74 (1982).

The contributions of known sources to the diffuse X-ray background are discussed. It is shown that they are likely to account for the bulk of it but a new astrophysical component might be needed to saturate its full intensity.

142.072 **Optical photometry and spectroscopy of the X-ray pulsar 1E 1145.1–6141.**
R. H. Densham, P. A. Charles.
Mon. Not. R. Astron. Soc., Vol. 201, 171 - 178 (1982).

High-resolution spectra of the optical counterpart of the X-ray pulsar 1E 1145.1–6141 have been obtained. The star is found to be of spectral type B2 Iae with a broad Hα emission feature and variable C, N, O and Si lines. Multicolour photometry is also presented, showing no variations on a time-scale of days although the object appears ~ 1 mag fainter than an observation two months previously. A distance of ~8.2 kpc is obtained for the system and orbital parameters estimated.

142.073 **X-ray pulsars observed with 'HAKUCHO'. I. Vela X-1.**
F. Nagase.
Astron. Her., Vol. 74, 336 - 343 (1981). In Japanese. – Abstr. in Phys. Abstr., Vol. 85, Abstr. 68814 (1982).

142.074 **Galactic bulge sources, what are they? (X-ray sources in globular clusters and burst sources).**
W. H. G. Lewin, G. W. Clark.
Ninth Texas Symposium on Relativistic Astrophysics, (see 012.020), p. 451 - 478 (1980).

Observed properties of galactic bulge X-ray sources including Type I burst sources, X-ray transients and X-ray sources in globular clusters are reviewed. It is argued that most of these sources are neutron stars in low-mass binaries.

142.075 **Thermonuclear-flash models for X-ray burst sources.**
P. C. Joss.
Ninth Texas Symposium on Relativistic Astrophysics, (see 012.020), p. 479 - 495 (1980).

142.076 **Extragalactic X-ray sources.**
K. A. Pounds, A. C. Fabian.
Ninth Texas Symposium on Relativistic Astrophysics, (see 012.020), p. 496 - 519 (1980).

142.077 **X-ray spectra of accreting degenerate stars.**
N. D. Kylafis, D. Q. Lamb, A. R. Masters, G. J. Weast.
Ninth Texas Symposium on Relativistic Astrophysics, (see 012.020), p. 520 - 549 (1980).

142.078 **Optical counterparts of compact galactic X-ray sources.**
B. Margon.
Ninth Texas Symposium on Relativistic Astrophysics, (see 012.020), p. 550 - 565 (1980).

142.079 **Iron lines in the spectra of X-ray binaries.**
M. M. Basko.
Ninth Texas Symposium on Relativistic Astrophysics, (see 012.020), p. 566 - 578 (1980).

142.080 **The X- and gamma-ray backgrounds.**
A. C. Fabian.
Tenth Texas Symposium on Relativistic Astrophysics, (see 012.021), p. 235 - 253 (1981).

The extragalactic X-ray and γ-ray backgrounds are generally considered to originate within a redshift of about three. The background spectrum thereby provides a measurement of the integrated X- and γ-ray evolution of known sources such as clusters of galaxies, active galaxies, and quasars, and limits possible new classes of sources such as accreting black holes and a hot intergalactic medium.

142.081 **Summary of the workshop on X-ray and gamma ray burst sources.**
R. McCray.

Tenth Texas Symposium on Relativistic Astrophysics, (see 012.021), p. 391 - 402 (1981).

142.082 **Summary of the workshop on ultraviolet cosmic background radiation.**
R. C. Henry.
Tenth Texas Symposium on Relativistic Astrophysics, (see 012.021), p. 428 - 441 (1981).

142.083 **The nature of the 1E1145.1–6141 optical counterpart.**
S. A. Ilovaisky, C. Chevalier, C. Motch.
Astron. Astrophys., Vol. 114, L7 - L8 (1982).
Optical photometry of the proposed optical counterpart for the 1E1145.1–6141 X-ray pulsar shows an ellipsoidal-type light curve of 0.07 mag amplitude with a probable 5.648 day period. The infrared colors are consistent with a reddened B2 supergiant with no infrared excess at a distance of 8 kpc. This may be another massive X-ray binary system akin to Vela X-1.

142.084 **Spectra of cosmic X-ray sources.**
S. S. Holt, R. McCray.
Annu. Rev. Astron. Astrophys., Vol. 20, (see 003.009), 323 - 365 (1982).
Optically thin thermal sources. Optically thick thermal sources. Nonthermal sources. Effects of strong magnetic fields.

142.085 **High-resolution X-ray spectroscopy of supernova remnants, clusters of galaxies and BL Lac objects.**
C. R. Canizares.
Observatory, Vol. 102, 110 (1982). – Abstract.

142.086 **X-ray, radio and infrared observations of the rapid burster MXB 1730 - 335 during 1979 and 1980.**
V. R. Venugopal.
Bull. Astron. Soc. India, Vol. 10, 28 - 29 (1982). – Abstract.

142.087 **The study of infrared emission from X-ray sources.**
A. G. Ananth, S. Seeta, B. V. Nagaraja.
Bull. Astron. Soc. India, Vol. 10, 29 (1982). – Abstract.

142.088 **Trapped line radiation in X-ray sources of Population II.**
A. Finzi.
Astrophys. Space Sci., Vol. 85, 381 - 387 (1982).
The author suggests that an X-ray source of Population II is produced by gas accreting onto a collapsed star from an extended quasi-transparent envelope formed in the distant past in a collision with a normal star. Most of the envelope is only weakly ionized; energy balance and hydrostatic equilibrium are ensured mainly by secondary line radiation, which is trapped there for very long times before it escapes or is re-absorbed. On the basis of the mechanism that controls the escape of line photons, the author distinguishes between lines of the "first" and "second" kind. The collapsed star is probably a moderately massive ($\lesssim 10\ M_{\odot}$) black hole.

142.089 **Magnetic fields of X-ray pulsars.**
V. M. Lipunov.
Astron. Zh., Tom 59, 888 - 895 (1982). In Russian. English translation in Soviet Astron., Vol. 26, No. 5.
An analytical model of magnetic torques applied to an accreting neutron star is considered. Within the framework of this model the values of magnetic dipole momenta have been obtained for X-ray pulsars. A new type of close binary systems with neutron stars is pointed out.

142.090 ***UBV* photometry of the X-ray binary HD 153919 = 4U1700–37.**
A. M. Cherepashchuk.
Astron. Zh., Tom 59, 918 - 924 (1982). In Russian. English translation in Soviet Astron., Vol. 26, No. 5.
The results of photoelectric *UBV* observations of the X-ray binary HD 153919 obtained March - June 1980 are presented.

142.091 **Ultraviolet, optical, and infrared observations of HDE 245770/A 0535+26.**
F. Giovannelli, C. de Loore, C. Bartolini, M. Burger, M. Ferrari Toniolo, A. Giangrande, A. Guarnieri, P. Persi, A. Piccioni, E. L. van Dessel.
Third European IUE Conference, (see 012.029), p. 233 - 239 (1982).
Simultaneous observations of the Be X-ray source A 0535+26 obtained during the last two years in the ultraviolet and optical, and coordinated infrared and radio ranges are discussed.

142.092 **The ultraviolet spectra of newly discovered hard X-ray cataclysmic variables.**
J. M. Bonnet-Bidaud, M. Mouchet, C. Motch.
Third European IUE Conference, (see 012.029), p. 241 - 244 (1982).
The two hard X-ray sources 1E 0643-1648 and 4U 1849-31 were observed with IUE at low spectral resolution. The two sources show an UV spectrum typical of cataclysmic variables, in particular their line spectrum is characterized by strong resonance lines in emission. The UV continuum is compared to the predicted energy distribution from blackbody, power-law and accretion disc models.

142.093 **Ultraviolet observations of V1341 Cyg = Cyg X-2.**
L. Chiappetti, L. Maraschi, E. G. Tanzi, A. Treves.
Third European IUE Conference, (see 012.029), p. 245 (1982). Abstract.

142.094 **Ultraviolet observations of X-ray bursters.**
G. Hammerschlag-Hensberge, J. van Paradijs, J. E. McClintock.
Third European IUE Conference, (see 012.029), p. 247 - 248 (1982).
The authors have observed the ultraviolet energy distribution of the X-ray burst source MXB 1735-44. The similarity of the optical energy distribution of this source to that of Sco X-1 extends into the far-ultraviolet.

142.095 **On the UV variability of LSI +61°.303 (≡ GT 0236).**
E. G. Tanzi, G. F. Bignami, P. A. Caraveo, L. Maraschi, F. Sormani, A. Treves.
Third European IUE Conference, (see 012.029), p. 249 (1982). Abstract.

142.096 **UV photometric and line-width variability of LMC X-4.**
M. van der Klis, G. Hammerschlag-Hensberge, J. M. Bonnet-Bidaud, S. A. Ilovaisky, M. Mouchet, W. M. Glencross, A. J. Willis, J. A. van Paradijs, E. J. Zuiderwijk, C. Chevalier.
Third European IUE Conference, (see 012.029), p. 259 - 260 (1982).
Low-resolution IUE spectra of the massive X-ray binary LMC X-4 have been obtained at several epochs in August 1979 and April 1980. Comparison of these observations with theoretical light curves for an X-ray heated tidally distorted star indicate the existence of a disk around the compact star.

142.097 **H 1409–45: a recurrent soft X-ray transient.**
K. A. Jensen, J. A. Nousek, J. J. Nugent.
Astrophys. J., Vol. 261, 625 - 630 (1982).
Observations of the soft X-ray variable H 1409–45 in recent years with different instruments are summarized. H 1409–45 appears to have several types of X-ray variability, ranging from rapid flickering on time scales of seconds to

changes between states of different intrinsic luminosity on time scales of months to years. Its X-ray spectrum appears to be softer in its high-luminosity state. These observations indicate that H 1409−45 may be a white dwarf accreting mass from a close binary companion.

142.098 **GX 1+4: pulse period measurement and detection of phase-variable iron line emission.**
M. J. Ricketts, R. Hall, C. G. Page, C. H. Whitford, K. A. Pounds.
Mon. Not. R. Astron. Soc., Vol. 201, 759 - 768 (1982).
Observations of the X-ray source GX 1+4 (1728−247) made by Ariel 6 in 1979 July and 1980 April show that the pulse period decreased at a faster rate during 1978−80 than over the period 1970−78. Compared with that measured in 1976, the pulse profile had become more sinusoidal, with an apparent phase lag ~0.1 between the 4−8 keV pulse and that seen at energies above 25 keV. The 1−4 keV range does not show pulsations and is stronger than hitherto. There is also an iron emission feature whose equivalent width varies with pulse phase.

142.099 **A survey of the bright galactic bulge X-ray sources.**
T. Ponman.
Mon. Not. R. Astron. Soc., Vol. 201, 769 - 799 (1982).
Fifteen bright galactic bulge sources were observed by the RMC experiment on Ariel V intermittently over the period 1975−78. The properties of each source are expressed as a series of parameter values, allowing quantitative intercomparison of the sources which shows that GX339−4 is quite different to the others. Also, significant correlation is found to exist between some of the parameters. Finally, a model for these sources is developed on the basis of a low mass neutron star binary.

142.100 **The X-ray transient 3A 1431−409: a highly active RS CVn system?**
L. Booth, P. A. Charles.
Mon. Not. R. Astron. Soc., Vol. 201, 25P - 30P (1982).
The authors present optical spectroscopy and photometry of two stars in the errorbox of the X-ray transient source 3A 1431−409. One of these stars has the spectrum of a very active star, with unusually strong emission at Hα, and narrow emission in the Ca II *H* and *K* cores, as seen in RS CVn systems. The other star has the spectrum of a late-type giant, with no emission features that would indicate the presence of X-ray activity in the system. The authors identify the giant with V532 Cen, and it is probably a Cepheid variable. They propose that the emission-line star, although of rather high inferred X-ray luminosity ($\gtrsim 10^{33}$ erg s^{-1}), even for an RS CVn type, is the likely optical counterpart of the X-ray source 3A 1431−409.

142.101 **The high galactic latitude fast-transient X-ray source 3A 1431−409.**
I. M. McHardy, J. P. Pye, A. P. Fairall, J. A. R. Caldwell, J. H. Spencer-Jones.
Mon. Not. R. Astron. Soc., Vol. 201, 31P - 37P (1982).
The authors present X-ray observations of the fast-transient source 3A 1431−409 and optical observations of the RS CVn-like system 'V532 Cen star 1' which they propose as a probable identification for the X-ray source. They discuss the physical parameters of the system and show that, assuming the identification is correct, then this is the most energetic X-ray event so far detected from an RS CVn-like system.

142.102 **Observations of Cygnus X-3 in the *I* band.**
T. C. Weekes, J. C. Geary.
Publ. Astron. Soc. Pacific, Vol. 94, 708 - 712 (1982).
Cygnus X-3 has been detected in the *I* band in four CCD images taken on the Multiple Mirror Telescope. The magnitude is 17.0 ± 0.2 and there is evidence for variability over the two months of observation. The column density of intervening hydrogen is in agreement with that deduced from the X-ray spectrum.

142.103 **Some topics on the X-ray pulsars and the X-ray bursts observed by the satellite Hakucho.**
M. Oda.
AIP Conf. Proc., No. 77, (see 012.034), p. 319 - 338 (1982).
Several subjects are selected from observational results on X-ray pulsars and X-ray burst sources obtained by the Japanese X-ray astronomy satellite "Hakucho". Results on Vela X-1 and A 0535 + 262 which exhibited transient or time variable pulsar behavior are discussed. Current models for these sources are critically examined.

142.104 **Association of recurrent soft X-ray transients with X-ray burst sources.**
T. Murakami.
AIP Conf. Proc., No. 77, (see 012.034), p. 339 - 348 (1982).
Hakucho observations have revealed that the soft transients, 2S 1608−522, 2S 1715−321, Cen X-4 and Aql X-1, all generate X-ray bursts. This result makes it convincing that these sources involve neutron stars. Furthermore, the optical counterpart of Aql X-1 is identified to be a K0 dwarf, thus providing evidence that Aql X-1 is a binary of a late dwarf and a neutron star.

142.105 **Infrared photometry of the X-ray binary 2A 1822−371: a model for the ultraviolet, optical, and infrared light curve.**
K. O. Mason, F. A. Córdova.
Astrophys. J., Vol. 262, 253 - 262 (1982).
The light curve of the 5.57 hr binary X-ray source 2A 1822−371 has been measured at 1.25 μm. It is found to be similar in shape and depth of modulation to the optical light curve. Measurements at 1.25 μm, 1.65 μm, and 2.2 μm show that the spectrum of the star in the near-infrared is consistent with the Rayleigh-Jeans tail of the optical spectrum. A model for the ultraviolet, optical, and infrared light curves is constructed, based on the azimuthal structure in the thickness of the outer accretion disk that was deduced previously from fits to the X-ray light curve.

142.106 **Einstein Observatory pulse-phase spectroscopy of Hercules X-1.**
R. A. McCray, J. M. Shull, P. E. Boynton, J. E. Deeter, S. S. Holt, N. E. White.
Astrophys. J., Vol. 262, 301 - 307 (1982).
The authors have obtained spectra of the Her X-1 pulsations from 0.15 to 4.5 keV with the Objective-Grating and Solid State Spectrometers on the Einstein Observatory. The spectra show no distinct atomic emission or absorption features, except for broad emission near 1 keV possibly due to unresolved Fe L shell emission lines. Spectrally resolved pulse waveforms show a dramatic 240° phase shift for $E < 0.85$ keV. The soft X-rays are interpreted as the result of the reprocessing of the hard X-ray beam by the inner disk at the magnetospheric boundary of the neutron star.

142.107 **A neutron star observed with the Hakucho satellite.**
M. Oda.
J. Soc. Instrum. Control Eng., Vol. 21, 575 - 576 (1982). In Japanese. − Abstr. in Phys. Abstr., Vol. 85, Abstr. 112501 (1982).

142.108 **Time-dependent X-ray observations of Wolf-Rayet binaries with O-type and with suspected compact companions.**
A. F. J. Moffat, C. Firmani, I. S. McLean, W. Seggewiss.
Wolf-Rayet stars: observations, physics, evolution, (see 012.042), p. 577 - 581 (1982).
Each of four WR binaries has been observed in the

0.1 - 5 keV range with the IPC on the Einstein Observatory. Neither HD 197406 (WN7) nor HD 96548 (WN8) was detected; hence L_X (0.5 - 3.0 keV) $\lesssim 10^{32}$ erg s^{-1}, which is on the low side, even for O-stars. EZ CMa (WN5) and V444 Cyg are moderately strong, variable X-ray sources with L_X (0.5 - 3.0 keV) $\simeq 10^{33}$ erg s^{-1}, kT $\simeq$ 0.5 keV and $N_H \simeq 10^{22} cm^{-2}$. These values suggest that the X-ray flux arises in the outer part of the wind. The variable component may be produced by collision of two winds. There is no evidence of harder X-rays coming from accretion onto a compact star.

142.109 **X-rays from Wolf-Rayet stars observed by the Einstein Observatory.**
W. T. Sanders, J. P. Cassinelli, K. A. van der Hucht.
Wolf-Rayet stars: observations, physics, evolution, (see 012.042), p. 589 - 595 (1982).

Preliminary results of three X-ray surveys are presented. Out of a sample of 20 stars, X-rays were detected from four Wolf-Rayet stars and two O8f⁺ stars. The detected stars have about the same mean value as O stars for the X-ray to total luminosity ratio, $L_x/L = 10^{-7}$, but exhibit a much larger variation about the mean. The spectral energy distributions are also found to be like that of O stars in that they do not exhibit large attenuation of X-rays softer than 1 keV. This indicates that for both the O stars and WR stars much of the X-ray emission is coming from hot wisps or shocks in the outer regions of the winds and not from a thin source at the base of the wind. The general spectral shape and flux level place severe restrictrions on models that attribute the lack of hydrogen emission lines to extremely high temperatures of the gas in the wind.

142.110 **X-ray observations of AM Herculis in its low state.**
G. Fabbiano.
Astrophys. J., Vol. 262, 709 - 713 (1982).

AM Her was observed with the Einstein IPC during the low state of 1980 August. The light curve shows an eclipse at $\phi_m = 0.0$ similar to the one observed in the customary high state. The spectral distribution of the X-ray photons does not show any soft component. The IPC flux suggests that the intensity of the hard component in the low state is about one-tenth of the intensity of the hard component during the high state, similar to the variation in the optical magnitude.

142.111 **Observations of 12–200 keV X-rays from GX 339–4.**
P. L. Nolan, D. E. Gruber, F. K. Knight, J. L. Matteson, L. E. Peterson, A. M. Levine, W. H. G. Lewin, F. A. Primini.
Astrophys. J., Vol. 262, 727 - 731 (1982).

The authors present X-ray spectra of GX 339–4 measured on three occasions in 1977 and 1978. These are the first reported measurements above 10 keV. The spectra can be described as the superposition of a soft component, which is dominant below ~20 keV, and a hard component at higher energy. The behavior of these two components is similar to that of the spectrum of Cygnus X–1; this reinforces the previously noted resemblance in rapid X-ray variability.

142.112 **The theoretically expected X-ray luminosity and the binary nature of Wolf-Rayet runaway stars.**
D. Vanbeveren, W. Van Rensbergen, C. de Loore.
Astron. Astrophys., Vol. 115, 69 - 78 (1982).

The X-ray luminosity expected in WR + compact star binaries is calculated by combining the production of hard X-rays by accretion and absorption by a stellar wind, showing that the wind is not sufficiently opaque to extinguish the X-rays beyond observational limits. Stating that there are as many WR + compact star binaries as WR + OB systems depends on the survival probability of the binary during the supernova explosion. This probability is critically investigated and a statistical study is presented based on massive close binary evolution in order to estimate the number of WR + compact star binaries. It is concluded that the numb frequency of WR + compact star binaries compared to WR + OB systems may be as low as 20%.

142.113 **X-ray observations of symbiotic stars.**
D. A. Allen.
The nature of symbiotic stars, (see 012.045), p. 115 - 116 (1982). – Abstract. The full paper is published in Mon. No R. Astron. Soc. (1981), Vol. 197, 739 - 743. – See Abstr. 30.142.060.

142.114 **X-ray detection of the symbiotic star AG Dracor**
C. M. Anderson, J. P. Cassinelli, N. A. Oliversen, R. V. Myers, W. T. Sanders.
The nature of symbiotic stars, (see 012.045), p. 117 - 121 (1982).

The symbiotic star AG Draconis was observed by the HEAO-2 Imaging Proportional Counter (IPC) and found to an unusually intense source of very soft X-rays.

142.115 **An optical burst in Sco X-1.**
H. Mauder.
Binary and multiple stars as tracers of stellar evolution, (see 012.046), p. 445 (1982).

On March 13th, 1979, Sco X-1 was in the normal acti state. However, at $7^h 29^m 49^s$UT a burst event was observec This optical burst is very similar to the optical bursts of X-1 bursters, lasting for only about 10 seconds with an amplitu of almost 0.5 mag. Thus it seems possible that Sco X-1 sho at least occasionally the characteristics of normal X-ray bursters.

142.116 **Prospects for X-ray observations of cosmological significance.**
K. A. Pounds.
Progress in cosmology, (see 012.047), p. 189 - 202 (1982).

The cosmic X-ray background shares with the microw background the property of originating at a high redshift. Thus, studies of the structure, spectrum and origin of the X background are potentially important cosmologically. Existing measurements of the background isotropy and deductions are reviewed and seen to provide interesting lin on the matter distribution on scales larger than that of supe clusters. Source counts from the Einstein Observatory and earlier sky survey experiments show a significant componer of the X-ray background to arise from a strongly evolving population of high redshift QSO's.

142.117 **The origin of the X-ray background.**
G. Zamorani.
Progress in cosmology, (see 012.047), p. 203 - 214 (1982).

The problem of the origin of the X-ray background (XR is not yet completely resolved. Two quite different and independent approaches have been followed in order to estin the contribution of discrete sources to the XRB. The first (Einstein Deep and Medium Surveys) has extended the observed log N - log S down to fluxes of $\sim 3 \times 10^{-14}$ erg $cm^{-2} s^{-}$ providing a firm lower limit of (26 ± 11)% for the contribut of discrete sources to the XRB. The second (observations o several hundreds of previously known extragalactic objects has clearly shown that quasars, as a class, are the most signi cant contributors to the XRB. An estimate of the allowed range of their contribution is presented.

142.118 **Further observational evidence for a coronal bou ary line in the cool star region of the H-R diagrar**
B. M. Haisch, T. Simon.
Astrophys. J., Vol. 263, 252 - 258 (1982).

The authors report soft (0.2 - 4.0 keV) X-ray observa tions of eight G-K giant stars obtained with the IPC of the Einstein Observatory. X-ray emission was detected from

γ Hya (G8 III) and 37 Com (G9 III). These results further strengthen earlier findings that $T \geqslant 10^6$ K stellar coronae are very weak, or possibly even absent, among the red giants and supergiants, and that a large range in coronal emission exists among the yellow giant and supergiant stars.

142.119 **A comparison of the X-ray properties of X Persei and γ Cassiopeiae.**
N. E. White, J. H. Swank, S. S. Holt, A. N. Parmar.
Astrophys. J., Vol. 263, 277 - 288 (1982).

The authors present the first detailed X-ray observations of γ Cas and compare them with similar observations of X Per. It is concluded that γ Cas must be a binary system containing a neutron star with an orbital period greater than 40 days. The UV-derived mass loss rates for X Per and γ Cas are compared with those required to power the X-ray sources.

142.120 **A 0535 + 26: a hard X-ray observation of the 1977 December flare-up with the Prognoz 6 Signe II experiment.**
F. Violes, M. Niel, A. Bui-Van, G. Vedrenne, G. Chambon, I. V. Estulin *(Ehstulin)*, Ch. Yu. *(Sh. Yu.)* Rakhamimov.
Astrophys. J., Vol. 263, 320 - 324 (1982).

The lateral antisolar detector of the Signe II MP 6 experiment observed the 1977 December flare-up of the transient source A 0535 + 26. The time history of the source in the 28–40 keV energy range is presented, as well as the spectral evolution in the range 28–146 keV, for the period 1977 December 20 - 1978 January 2.

142.121 **Simultaneous optical and X-ray bursts from 4U/MXB 1636–53.**
H. Pedersen, J. Lub, H. Inoue, K. Koyama, K. Makishima, M. Matsuoka, K. Mitsuda, T. Murakami, M. Oda, Y. Ogawara, T. Ohashi, N. Shibazaki, Y. Tanaka, S. Hayakawa, H. Kunieda, F. Makino, K. Masai, F. Nagase, Y. Tawara, S. Miyamoto, H. Tsunemi, K. Yamashita, I. Kondo, J. G. Jernigan, J. van Paradijs, A. Beardsley, L. Cominsky, J. Doty, W. H. G. Lewin.
Astrophys. J., Vol. 263, 325 - 339 (1982).

The authors report the detection of five coincident optical/X-ray bursts from 4U/MXB 1636–53. The optical observations were made with the Danish 1.5 m telescope at E.S.O. and the X-ray observations were made with the Hakucho X-ray observatory. These bursts are analyzed in terms of a simple general mathematical model, in which the optical emission (both during and between the bursts) is that of a blackbody and the result of thermally reprocessed X-rays. An accretion disk around a compact object is one plausible model for the reprocessing region. The derived restrictions on the parameters of the reprocessing region also constrain the parameters of a consistent low mass close binary system containing an $\sim 1.4\ M_\odot$ neutron star. The mass of the Roche lobe filling companion star is $< 2.0\ M_\odot$ corresponding to a binary period between 1 hour and 10 hours.

142.122 **Optical bursts from 4U/MXB 1636–53.**
H. Pedersen, J. van Paradijs, C. Motch, L. Cominsky, A. Lawrence, W. H. G. Lewin, M. Oda, T. Ohashi, M. Matsuoka.
Astrophys. J., Vol. 263, 340 - 351 (1982).

During the periods 1979 June - July and 1980 June - July, a total of 15 and 26 optical bursts, respectively, were detected from the X-ray burst source 4U/MXB 1636–53. From a comparison with X-ray bursts simultaneously detected with Hakucho, it appears that the integrated optical and X-ray burst fluxes are approximately linearly related, which is not expected on the basis of this simple X-ray reprocessing picture. The integrated optical burst fluxes are correlated with the waiting time since the previous burst: there is an indication that large optical bursts come after a long waiting time.

142.123 **Absence of optical pulses in HD 153919 (= 4U 1700–37 = V884 Sco).**
S. van Amerongen, J. van Paradijs.
Inf. Bull. Variable Stars, No. 2180, 4 pp. (1982).

142.124 **Possible identification of X-ray burster with Serpens X-2.**
IAU Circ., No. 3723 (1982).

142.125 **Cygnus X-3.**
IAU Circ., No. 3731 (1982).

142.126 **1E 2259+586.**
IAU Circ., Nos. 3739, 3745 (1982).

142.127 **A0538–66.**
IAU Circ., No. 3757 (1982).

142.128 **Variability of infrared radiation of the X-ray source A 0535 +262.**
Yu. N. Gnedin, G. V. Khozov, V. M. Larionov.
Pis'ma Astron. Zh., Tom 8, 689 - 693 (1982). In Russian. English translation in Soviet Astron. Lett., Vol. 8.

IR observations of the X-ray binary A 0535 +262 during 1981 - 1982 revealed strong IR variability of this source with the characteristic time scales from tens of minutes to days and also possibly regular variations of the IR flux with the period 104 s corresponding to the intrinsic rotation period of the neutron star in this system.

142.129 **The X-ray flux variations of Cygnus X-2.**
J. M. Bonnet-Bidaud, M. van der Klis.
Astron. Astrophys., Vol. 116, 232 - 236 (1982).

Results of a 47 d observation of Cygnus X-2 by the COS-B X-ray experiment are presented. The observed variability of the X-ray intensity is investigated for the presence of modulation at either the $9^d.843$ optical period or the $11^d.23$ X-ray period claimed for this source. Upper limits are given to pulse amplitudes for periods between 1 d and 200 s. Clear dips were observed during the time of highest source intensity, with transitions of up to 40% in the source flux in less than $1-2 \times 10^3$ s. It is suggested that the dips may be connected to the reported UV flaring activity.

142.130 **Fast coherent oscillations in variable X-ray sources and bursters.**
M. Livio, G. T. Bath.
Astron. Astrophys., Vol. 116, 286 - 292 (1982).

Coherent oscillations with periods of the order of tens of milliseconds have been reported from five variable X-ray sources or bursters. Possible mechanisms for the generation of the oscillations are examined. It is found that rotator models encounter difficulties in explaining period changes and short duration of periodicities. Oscillations of the neutron star's envelope, especially *g*-modes are found to give the most consistent general explanation. It is shown that it is unlikely that the pulsations originate in an accretion disk. Some observational tests are proposed and the resemblance of the observed oscillations with those in cataclysmic variables is pointed out.

142.131 **Persistent X-ray emission from a γ-ray burst source.**
J. E. Grindlay, T. Cline, U. D. Desai, B. J. Teegarden, G. Pizzichini, W. D. Evans, R. W. Klebesadel, J. G. Laros, K. C. Hurley, M. Niel, G. Vedrenne.
Nature, Vol. 300, 730 - 731 (1982).

A quiescent X-ray source detected with the Einstein X-ray Observatory in a location consistent with that of an intense γ-ray burst is shown to be also consistent with the location of the 1928 optical transient, the likely optical counterpart of the γ-ray burst source GBS0117-29.

142.132 **Optical bursts from 4U/MXB1636-53.**
H. Pedersen, J. van Paradijs, C. Motch, L. Cominsky, A. Lawrence, W. H. G. Lewin, M. Oda, T. Ohashi, M. Matsuoka.
ESO Sci. Prepr., No. 201, 47 pp. (1982). – Submitted to Astrophys. J.

142.133 **Simultaneous optical and X-ray bursts from 4U/MXB1636-53.**
H. Pedersen, J. Lub, H. Inoue, K. Koyama, K. Makishima, M. Matsuoka, K. Mitsuda, T. Murakami, M. Oda, Y. Ogawara, T. Ohashi, N. Shibazaki, Y. Tanaka, S. Hayakawa, H. Kunieda, F. Makino, K. Masai, F. Nagase, Y. Tawara. S. Miyamoto, H. Tsunemi, K. Yamashita, I. Kondo, J. G. Jernigan, J. van Paradijs, A. Beardsley, L. Cominsky, J. Doty, W. H. G. Lewin.
ESO Sci. Prepr., No. 202, 49 pp. (1982). – Submitted to Astrophys. J.

142.134 **Simultaneous X-ray/optical observations of GX 339-4 during the May 1981 optically bright state.**
C. Motch, M. J. Ricketts, C. G. Page, S. A. Ilovaisky, C. Chevalier.
ESO Sci. Prepr., No. 217, 18 pp. (1982). – Submitted to Astron. Astrophys.

142.135 **On the spectrum of the transient X-ray source 4U 0115 + 63.**
O. Eh. Aab, L. V. Bychkova, I. M. Kopylov.
Astron. Tsirk., No. 1198, p. 1 - 3 (1981). In Russian.

142.136 **Variable emission line Hβ in the spectrum of the X-ray source 2A 1052 + 606.**
O. Eh. Aab, I. M. Kopylov, R. N. Kumajgorodskaya.
Astron. Tsirk., No. 1198, p. 3 - 6 (1981). In Russian.

142.137 **The Einstein Observatory Medium Sensitivity Survey: optical identifications for a complete sample of X-ray sources.**
J. T. Stocke, J. Liebert, I. M. Gioia, R. E. Griffiths, T. Maccacaro, I. J. Danziger, D. Kunth, J. Lub.
Prepr. Steward Obs., No. 427, 70 pp. (1982). – Submitted to Astrophys. J.

142.138 **Cygnus X-3: current status of observations at $> 10^{12}$ eV photon energies.**
D. Fegan.
Irish Astron. J., Vol. 15, (see 012.061), 104 (1981). Abstract.

142.139 **X-rays from stars.**
D. J. Mullan.
Irish Astron. J., Vol. 15, 147 - 151 (1981).

Data from X-ray detectors on board the Einstein Observatory have revolutionized the concepts of hot coronae around stars. Current theories of the heating mechanisms for stellar coronae are reviewed briefly in this note.

142.140 **Cyclotron lines in the hard X-ray spectrum of Hercules X-1.**
W. Voges, W. Pietsch, C. Reppin, J. Trümper, E. Kendziorra, R. Staubert.
Astrophys. J., Vol. 263, 803 - 813 (1982).

The hard X-ray spectrum of Hercules X-1 in the energy range 26–120 keV was measured during a balloon observation on 1977 September 3 with a large phoswich detector system. The spectral feature in the 1.24 s pulsations found in a 1976 observation and interpreted as electron cyclotron line emission or absorption was detected again and confirmed with large significance. The line feature is located either at ~52 keV (as an emission line) or at ~38 keV (as an absorption line). Th corresponds to a magnetic field strength of $3–5 \times 10^{12}$ gau

142.141 **Observation of an outburst of the transient X-ray pulsar A 0535 + 26 in 1980.**
F. Nagase, S. Hayakawa, H. Kunieda, F. Makino, K. Masai, Y. Tawara, H. Inoue, N. Kawai, K. Koyama, K. Makishima, M. Matsuoka, T. Murakami, M. Oda, Y. Ogawara, T. Ohash N. Shibazaki, Y. Tanaka, S. Miyamoto, H. Tsunemi, K. Yamashita, I. Kondo.
Astrophys. J., Vol. 263, 814 - 822 (1982).

An outburst of the recurrent X-ray transient A 0535+ was detected on 1980 October 2, and the source was continually observed by the Hakucho satellite for 25 days. The present outburst seems to have occurred on time acco ing to the regular ~110 day recurrence of outbursts. Analy of the pulse arrival time gives a lower limit on the orbital period of 40 days, whereas the upper limit is ~120 days considering the mass function derived from optical data. A consistent picture is drawn with an eccentric orbit of a ~110 day period, whereby recurrent outbursts occur near periastron passages.

142.142 **A new X-ray pulsar with a 67 millisecond period the constellation Equuleus.**
D. Sadeh, M. Livio.
Astrophys. J., Vol. 263, 823 - 827 (1982).

By applying the epoch-folding technique to a large amount of data from the A-1 experiment onboard the HEA satellite the authors detected a new X-ray pulsar in the constellation Equuleus. Its period is 67.5492 ms and the expectation value for random occurrence is smaller than 5×10^{-9}.

142.143 **A study of X-ray emission from Ap and Am stars**
W. Cash, T. P. Snow, Jr.
Astrophys. J., Lett., Vol. 263, L59 - L62 (1982).

Twenty stars classified as either Ap or Am have been imaged by the IPC on the Einstein Observatory in a search evidence of coronal emission. The authors find that an Am star will not be a source unless it is in a spectroscopic binar with a period below about 10 days. The X-rays in the close binaries can be explained by the RS CVn effect in the late-type secondary. Among the nine Ap stars observed, four w detected. While it is possible that the X-rays are being generated by undetected late-type dwarf companions, the large fraction of detections makes the Ap stars themselves the likely source of emission.

142.144 **High-energy spectra of three active extragalactic objects.**
W. A. Baity.
Publ. Astron. Soc. Pacific, Vol. 94, 748 - 749 (1982). Abstract.

142.145 **The Einstein observatory medium sensitivity X-ra survey: the range of individual identifications.**
J. Liebert, J. Stocke, T. Maccacaro, I. Gioia, R. Griffiths, J. Danziger.
Publ. Astron. Soc. Pacific, Vol. 94, 755 (1982). – Abstract

142.146 **The Einstein observatory medium sensitivity X-ra survey: evidence for cosmological evolution of quasars, BL Lacs, and clusters of galaxies.**
J. Stocke, J. Liebert, T. Maccacaro, I. Gioia, R. Griffiths, J. Danziger.
Publ. Astron. Soc. Pacific, Vol. 94, 759 (1982). – Abstract

142.147 **An X-ray flare in the Hyades binary HD 27130.**
R. A. Stern, J. H. Underwood, S. K. Antiochos.
Smithsonian Astrophys. Obs., Spec. Rep. 392, Vol. II, (see 012.065), p. 101 - 108 (1982).

Over 40 X-ray emitting stars have been found in the central region of the Hyades cluster (Stern et al., 1981). In follow-up observations of these sources with the Einstein Observatory, the X-ray luminosity of one of them (HD 27130) was observed to undergo more than a 30-fold increase to $\sim 10^{31}$ erg s^{-1}. HD 27130 has recently been determined to be a double-lined eclipsing binary with a period of 5.6 d. The primary is a G dwarf, and the secondary is probably a K dwarf (McClure, 1981). The occurrence of such a large flare in a system of this type poses interesting questions regarding mechanisms of solar and stellar flares.

142.148 **X-rays from quasars and active galaxies.**
A. P. Lightman.
Space Sci. Rev., Vol. 33, 335 - 357 (1982).
The author reviews some of the most important theoretical ideas and observations for quasars and the nuclei of active galaxies, and suggest areas of future research. Emphasis is on the nature of the power source, the radiation processes, and the mechanism for formation and collimation of jets. Phenomena that produce X-rays are of particular concern. Particular topics discussed are the observed and expected time variability, the gas supply mechanisms and luminosity evolution, thermal and nonthermal radiation processes, observed and theoretical spectra, criteria for thermalization of electrons and ions, effects of electron-positron pairs on relativistic plasmas, hydrodynamic, electrodynamic and inertial methods for producing and confining jets.

142.149 **Optical observations of binary X-ray sources.**
P. Charles.
Accreting neutron stars, (see 012.069), p. 1 - 13 (1982).
A review of the optical properties of X-ray binaries containing a neutron star or possibly a black hole is given. Models proposed for explaining these properties are critically examined.

142.150 **Variations of pulse periods of X-ray pulsars.**
S. Hayakawa, F. Nagase.
Accreting neutron stars, (see 012.069), p. 14 - 28 (1982).
Among several X-ray pulsars thus far observed by HAKUCHO, complex time variations of pulse periods have been obtained for Cen X-3, Vela X-1, A 0535 + 26, and GX 301 − 2, which are known to be members of binary systems of finite eccentricities. The rates of long-term pulse period change for these pulsars are not constant and sometimes change their signs. A large variation of pulse period from one orbital period to another indicates that a long-term period variation results from an average of short-term period fluctuations.

142.151 **The observed properties of X-ray pulsars.**
N. E. White.
Accreting neutron stars, (see 012.069), p. 29 - 40 (1982).
This paper presents a unified description of the properties of 14 X-ray pulsars as seen by HEAO-1, OAO-8 and Einstein.

142.152 **Hard X-ray observations of fast pulsars from HEAO-1.**
D. E. Gruber, F. A. Primini.
Accreting neutron stars, (see 012.069), p. 41 - 52 (1982).
The authors review results for five fast pulsars: Her X-1, 4U 0115 + 63, LMC X-4, SMC X-1 and Cen X-3.

142.153 **GX 301−2. The 41.5 day period and its optical counterpart.**
M. W. Pakull.
Accreting neutron stars, (see 012.069), p. 53 - 56 (1982).
Photoelectric observations of Wra 977, the optical counterpart of GX 301−2 (= 4U 1223−62), are analysed to explore whether interaction effects due to the orbiting X-ray pulsar are present in the light curve of the B1 type supergiant.

142.154 **Hard X-ray recurrent transients.**
N. Shibazaki.
Accreting neutron stars, (see 012.069), p. 57 - 64 (1982).
Many hard X-ray transients exhibit recurrent nature. Some of them, such as 4U 0535 + 26, 4U 0538 − 66 and GX 301 − 2 are known to flare periodically. The author discusses the models for these recurrent X-ray sources phenomenologically.

142.155 **The optical outbursts of the periodic recurrent transient A 0538−66.**
R. H. Densham, P. A. Charles, J. W. Menzies, J. van Paradijs, M. van der Klis.
Accreting neutron stars, (see 012.069), p. 65 - 70 (1982).
Multicolour photometry is presented of the 16.6 day recurrent X-ray transient A 0538 − 66 over four consecutive outbursts in late 1981. There is evidence of considerable activity at least one day prior to maximum, and the magnitudes and colours measured between outbursts are significantly different from previous observations during quiescence.

142.156 **A model for A 0538 − 66, the fast flaring pulsar.**
L. Maraschi, R. Traversini, A. Treves.
Accreting neutron stars, (see 012.069), p. 71 - 74 (1982).
The discovery of rapid X-ray pulsation (P = 69 ms) in the recurrent (P = 16.6^d) transient A 0538 − 66 offers the first observational proof of the presence in a mass exchanging binary of a fast pulsar. The authors discuss an accretion model for this source and interpret the optical flare associated with the X-ray one, as due to reprocessing of X-rays in an envelope surrounding the pulsar.

142.157 **Compact X-ray sources in the Large Magellanic Cloud.**
M. W. Pakull.
Accreting neutron stars, (see 012.069), p. 75 (1982).
Abstract.

142.158 **X-ray bursts: observation versus theory.**
W. H. G. Lewin.
Accreting neutron stars, (see 012.069), p. 76 - 86 (1982).
The hypothesis that the common type I X-ray bursts (Hoffman, Marshall and Lewin, 1978) are caused by thermonuclear flashes in the surface layers of accreting neutron stars is reviewed and compared with observational data.

142.159 **The influence of X-rays on surrounding matter.**
A. C. Fabian, R. R. Ross.
Accreting neutron stars, (see 012.069), p. 87 - 93 (1982).
The X-rays emitted by an accreting neutron star usually have to pass through a significant amount of intervening matter. Much of this is associated with the mass transfer and accretion process itself. The authors discuss the general effects of an X-ray source in a stellar wind and on evaporated accretion-disk coronae in 4U 1822−37 and the so-called 'Galactic Bulge' sources. The Eddington limit is also briefly discussed.

142.160 **X-ray bursts from the GX 17 + 2 region.**
S. Hayakawa, T. Hirano, T. Murakami, F. Nagase, Y. Tawara.
Accreting neutron stars, (see 012.069), p. 94 - 96 (1982).
Three X-ray bursts were observed in May 1981 by HAKUCHO from the region containing GX 17 + 2. Characteristic properties of these bursts are summarized in this paper.

142.161 **Why are most X-ray pulsars slow?**
P. Ghosh, K. M. V. Apparao.
Accreting neutron stars, (see 012.069), p. 97 - 99 (1982).

The authors show the neutron stars in massive X-ray binaries can be born as slow rotators, because their progenitors can be spun down considerably by the angular momentum loss during the first stage of mass exchange.

142.162 **Her X-1: Ariel-5 and Ariel-6 observations.**
M. J. Ricketts, V. Stanger, C. G. Page.
Accreting neutron stars, (see 012.069), p. 100 - 105 (1982).
The results of long-term monitoring X-ray emission (2 - 15 keV) from Her X-1 are presented. Properties of the 35-day cycle are discussed.

142.163 **Accretion torques in Her X-1.**
H. Henrichs.
Accreting neutron stars, (see 012.069), p. 106 (1982).
Abstract.

142.164 **Pulse phase spectroscopy of 0115 + 63 and an observation of 4U 1223 – 62.**
R. Hall, M. J. Ricketts, R. Kloss.
Accreting neutron stars, (see 012.069), p. 107 - 111 (1982).

142.165 **Hard X-ray observation of the transient X-ray pulsar A 0535 + 26.**
F. Frontera, D. Dal Fiume, W. Dusi, E. Morelli, G. Spada, G. Ventura.
Accreting neutron stars, (see 012.069), p. 112 - 116 (1982).
The authors report observations of the recurrent X-ray transient A 0535 + 26 in the energy range 20 - 210 keV obtained with a balloon-borne instrument. The source was in an outburst phase during observations.

142.166 **Hard X-ray measurements of A 0535 + 26 during low state.**
V. F. Polcaro, A. Bazzano, C. La Padula, P. Ubertini, G. Vialetto, R. K. Manchanda.
Accreting neutron stars, (see 012.069), p. 117 - 120 (1982).
The authors present spectral measurements of the recurrent transient X-ray source A 0535 + 26 in the energy band 20 - 100 keV made during the low state in 1981.

142.167 **Hard X-ray observations of the Her X-1 line feature.**
R. K. Manchanda, G. Vialetto, A. Bazzano, C. La Padula, V. F. Polcaro, P. Ubertini.
Accreting neutron stars, (see 012.069), p. 121 - 124 (1982).
The authors have carried out two observations separated by a year in 1980 and 1981 during the mid-on phase of Her X-1. This paper presents the time-averaged spectral results of Her X-1 in the 15 - 150 keV energy range. The possible case of the centroid variability of the 56 keV cyclotron line seen during the 1980 experiment is discussed.

142.168 **Phase dependence of the line feature in the hard X-ray spectrum of Hercules X-1.**
W. Voges, W. Pietsch, C. Reppin, J. Trümper, E. Kendziorra, R. Staubert.
Accreting neutron stars, (see 012.069), p. 125 - 129 (1982).
The hard X-ray spectrum of Hercules X-1 in the energy range 26 - 120 keV was measured during a balloon observation on 1977 September 3. The spectral feature in the 1.24 s pulsations found in a 1976 observation was detected again and confirmed with large significance. The line feature is either located at ~52 keV (as an emission line) or at ~38 keV (as an absorption line). An analysis of phase dependent spectra shows a significant variation of the centroid of the cyclotron feature during the main pulse.

142.169 **X-ray transients as seen by Vela, 1969 - 1979.**
J. Terrell, W. C. Priedhorsky, R. D. Belian, J. P. Conner, W. D. Evans.
Accreting neutron stars, (see 012.069), p. 210 - 212 (1982).
The spacecraft Vela 5B monitored the entire X-ray sky for a period of 10 years, from May 1969 to June 1979. Ove the last several years the data produced have been re-analyz to produce a series of skymaps. The major results of this analysis for X-ray transients are presented.

142.170 **Search for instabilities in the accretion column oi Her X-1 and in the accretion disk of Cyg X-1.**
S. Bonazzola.
Accreting neutron stars, (see 012.069), p. 282 - 284 (1982)
X-ray observations of Her X-1 and Cyg X-1 have been analysed in a search for theoretically postulated plasma inst bility effects in the accretion disks around the compact con ponents of these massive binary systems. Some preliminary results are given.

ANS ultraviolet photometry, catalogue of point sources.
See Abstr. 002.001.

Landolt-Börnstein. Group VI. Vol. 2c. Interstella matter – Galaxy – Universe.
See Abstr. 003.055.

17th International Cosmic Ray Conference. XG session: X- and gamma rays (non-solar).
See Abstr. 012.070.

The National Space Science Data Center (NSSDC and the World Data Center A for Rockets and Satellites (WDC–A–R&S): their role in X-ray astronomy data.
See Abstr. 013.012.

Ballon-Röntgenmessungen an Neutronensternen und Quasaren.
See Abstr. 013.033.

The Universe in X-rays.
See Abstr. 013.059.

Source counts at and beyond the confusion limit.
See Abstr. 031.585.

Gas scintillation proportional counters for Japane astronomical satellites.
See Abstr. 032.548.

The Pinhole/Occulter Facility. Executive summar
See Abstr. 032.589.

X-80: spectroscopy/timing-transient mission in X-ray astronomy.
See Abstr. 051.063.

MAGELLAN.
See Abstr. 051.064.

Stellar X-ray astronomy with the Advanced X-ray Astrophysics Facility.
See Abstr. 051.080.

High throughput X-ray studies of stellar coronae with a LAMAR facility.
See Abstr. 051.081.

STCOEX: The stellar X-ray coronal Explorer.
See Abstr. 051.082.

Observational capabilities of EXOSAT and ROSA
See Abstr. 051.084.

Japanese X-ray astronomy program.
See Abstr. 051.085.

Future directions in X-ray/gamma-ray observations.
See Abstr. 051.092.

Plasma-magnetospheric interaction in X-ray sources: an analysis of the linear Kelvin-Helmholtz instability.
See Abstr. 062.021.

Radiative opacities and transfer in magnetic X-ray pulsars.
See Abstr. 063.009.

Super-critical X-ray luminosities: the structure and stability of a radiation-supported plasma layer.
See Abstr. 063.010.

Comptonization in a radiation-dominated shock wave and spectra of X-ray pulsars.
See Abstr. 063.035.

X-ray nebular models.
See Abstr. 063.056.

Second-order scattering approximation to X-ray photon transport at balloon altitudes.
See Abstr. 063.065.

Stellar winds in binary X-ray systems.
See Abstr. 064.003.

Twisted accretion disks.
See Abstr. 064.006.

X-radiation limited accretion flow. I. Steady flow regime.
See Abstr. 064.025.

X-radiation limited accretion flow. II. Stability and time dependence.
See Abstr. 064.026.

Radiation-driven winds in X-ray binaries.
See Abstr. 064.030.

Summary of workshop on accretion in astrophysics.
See Abstr. 064.040.

Accretion disks in stellar X-ray sources. A review of the basic theory of accretion disks and its problems.
See Abstr. 064.041.

Unsteady phenomena in an accretion disk.
See Abstr. 064.063.

Localized thermonuclear runaways and volcanoes on degenerate dwarf stars.
See Abstr. 065.057.

Gravitational energy of condensed astrophysical objects.
See Abstr. 066.105.

Relativistic stars in close binary systems and masses of neutron stars.
See Abstr. 066.188.

LMC X-3: a possible black hole.
See Abstr. 066.192.

Thermonuclear processes and accretion onto neutron star envelopes: X-ray burst and transient sources.
See Abstr. 066.501.

The thermonuclear model for X-ray transients.
See Abstr. 066.502.

Helium and combined hydrogen-helium shell flashes in the envelope of an accreting neutron star.
See Abstr. 066.505.

Line emission from neutron star surfaces.
See Abstr. 066.512.

Magnetospheric structures and instabilities around accreting neutron stars.
See Abstr. 066.522.

The universal diagram for magnetized neutron stars in the Galaxy.
See Abstr. 066.525.

Nuclear flash models for X-ray burst sources.
See Abstr. 066.534.

A numerical simulation of the magnetospheric gate model for the X-ray bursters.
See Abstr. 066.535.

Thermonuclear runaways in thick hydrogen rich envelopes of neutron stars.
See Abstr. 066.536.

Coherent ultraviolet radiation from accreting neutron stars.
See Abstr. 066.539.

Irreducible representation of shell flashes on accreting neutron stars.
See Abstr. 066.540.

Disk accretion onto weakly magnetized neutron stars.
See Abstr. 066.546.

Thermonuclear model for high energy transients.
See Abstr. 066.547.

Rise times and durations of X-ray and γ-ray bursts.
See Abstr. 066.548.

A 1D thermonuclear model for X-ray transients.
See Abstr. 066.549.

Towards model spectra of accreting magnetic neutron stars.
See Abstr. 066.558.

Corona-warm wind model for the X-ray emission from Of stars and OB supergiants.
See Abstr. 112.020.

The peculiar object SS433.
See Abstr. 114.016.

Chromospheric, transition layer and X-ray emission for stars with different rotational velocities.
See Abstr. 114.028.

Einstein observations of late-type stars of different calcium II activity.
See Abstr. 114.085.

IUE high resolution observations of hot stars emitting coronal X-rays.
See Abstr. 114.098.

Ultraviolet and X-ray detection of the 56 Pegasi system (K0 IIp + WD): evidence for accretion of a cool stellar wind onto a white dwarf.
See Abstr. 114.158.

Evolution of chromospheres and coronae in solar mass stars: a far-ultraviolet and soft X-ray comparison of Arcturus (K2 III) and Alpha Centauri A (G2 V).
See Abstr. 114.205.

Stellar activity.
See Abstr. 114.211.

Chromospheric emission, stellar rotation and X-ray coronae.
See Abstr. 114.214.

UBV-polarimetry of the X-ray binaries HD 77581 (4U 0900–40), HD 153919 (4U 1700–37) and of HD 152667.
See Abstr. 116.011.

On the mass of the exploding stars in massive close binary systems.
See Abstr. 117.009.

IUE observations of the HZ Herculis system.
See Abstr. 117.010.

Ultraviolet observations of AM Herculis with IUE.
See Abstr. 117.011.

First ultraviolet observations of two new cataclysmic variables 1E0643-1648 and 4U1849-31.
See Abstr. 117.017.

Phase resolved optical spectroscopy of the compact X-ray binary 2A 1822–371.
See Abstr. 117.019.

The "X-ray binary", UW CMa.
See Abstr. 117.032.

The optical spectrum of HDE 226868 = Cygnus X-1. I. Radial velocities and orbital elements.
See Abstr. 117.036.

X-ray and optical measurements of the cataclysmic variable CH UMa.
See Abstr. 117.050.

Theoretical aspects of X-ray binaries.
See Abstr. 117.055.

The relevance of EXOSAT to X-ray observations of RS CVn systems and cataclysmic variables.
See Abstr. 117.059.

On the evolution of low mass binary systems.
See Abstr. 117.062.

On the orbital phase dependence of the turn-on times of Hercules X-1.
See Abstr. 117.089.

Time resolved circular polarimetry of white dwarf pulsars.
See Abstr. 117.153.

Variability of soft X-ray emission of EX Hydrae observed with Einstein Observatory.
See Abstr. 117.156.

On the possible short-period irregular light fluctuations of V1357 Cyg = Cyg X-1.
See Abstr. 117.157.

On short-period irregular light fluctuations of V 1357 Cyg (Cyg X-1).
See Abstr. 117.168.

Some remarks on the close binary X-ray source AM Her.
See Abstr. 117.186.

H 0139-68.
See Abstr. 117.190.

E 2003+225.
See Abstr. 117.192.

Comparative chromospheric and coronal emissio from close binaries.
See Abstr. 117.210.

An X-ray survey of 9 Algol systems.
See Abstr. 117.211.

Ultraviolet and X-ray detection of the 56 Peg system (K0 IIp + WD).
See Abstr. 117.212.

The very unusual ultraviolet spectra of R Arae.
See Abstr. 119.058.

X-ray eclipses in AR Lac: preliminary results.
See Abstr. 119.098.

The spectroscopic orbit of Wra 977 (= X-ray sour 4U 1223-62).
See Abstr. 120.006.

X-ray observations of single-line spectroscopic binaries.
See Abstr. 120.013.

V 505 Mon: one of the most massive binaries.
See Abstr. 120.047.

HD 36705 – a post T Tauri star.
See Abstr. 121.034.

Evolution of stellar activity during the pre-main sequence phase.
See Abstr. 121.044.

Pulsed dwarf novae X-ray emission.
See Abstr. 122.014.

High-energy observations of stellar flares: compar son with the sun.
See Abstr. 122.076.

HEAO 1 observations of quiescent X-ray emissior from flare stars.
See Abstr. 122.078.

Did SU UMa undergo a classical nova outburst?
See Abstr. 122.091.

The effects of ejecta on the X-ray luminosities of supernova remnants.
See Abstr. 125.024.

The peculiar X-ray morphology of the supernova remnant G 292.0 + 1.8: evidence for an asymmetric supernova explosion.
See Abstr. 125.028.

X-ray line spectra of the Puppis A supernova remnant: spatial variations.
See Abstr. 125.031.

X-ray spectroscopy and imagery of supernova remnants with the Einstein Observatory.
See Abstr. 125.060.

Einstein observations of supernova remnants.
See Abstr. 125.061.

An X-ray survey of supernova remnants in the Large Magellanic Cloud.
See Abstr. 125.062.

Optical identification of Balmer-dominated supernova remnants in the Large Magellanic Cloud.
See Abstr. 125.073.

An X-ray image of Tycho's supernova remnant.
See Abstr. 125.074.

X-ray emission from young supernova remnants: nonionization equilibrium abundances and emissivities.
See Abstr. 125.078.

X-ray emission from white dwarfs.
See Abstr. 126.007.

The soft X-ray spectrum of Sirius B: evidence for the photospheric hypothesis.
See Abstr. 126.019.

La composante chaude du milieu interstellaire local: dimensions, températures, abondances.
See Abstr. 131.316.

X-ray and UV spectra of two quasars: PKS 0637−75 and PKS 1004+13 (4C 13.41).
See Abstr. 141.039.

The redshift distribution of X-ray selected quasars: is it really anomalous?
See Abstr. 141.055.

Search for X-ray emission from optically quiet, compact radio sources.
See Abstr. 141.060.

On the redshift and luminosity distribution of X-ray selected QSOs.
See Abstr. 141.063.

Derivation of the quasar X-ray luminosity function.
See Abstr. 141.071.

X-ray observations of supernova-like quasars.
See Abstr. 141.076.

Radio and X-ray observations of the halo source in A1367.
See Abstr. 141.081.

X-ray quasars.
See Abstr. 141.098.

On the cosmological evolution of the X-ray emission from quasars.
See Abstr. 141.140.

Visual-infrared photometry of X-ray QSOs.
See Abstr. 141.188.

Infrared properties of serendipitous X-ray quasars.
See Abstr. 141.197.

Rapid X-ray variability in the nuclei of active galaxies.
See Abstr. 141.208.

Observations and interpretations of the pulsed emission from the Crab pulsar.
See Abstr. 141.519.

X-ray and gamma-ray upper limits for pulsed emission from radio pulsars.
See Abstr. 141.520.

Discovery of radio pulsations from the X-ray pulsar in the supernova remnant G 320.4−1.2.
See Abstr. 141.539.

Line features in the X-ray spectrum of the Crab pulsar?
See Abstr. 141.554.

An X-ray survey of the Pleiades.
See Abstr. 153.009.

Optical and ultraviolet observations of the X-ray globular cluster Bo 158 in M 31.
See Abstr. 154.016.

Optical and ultraviolet observations of the X-ray globular cluster Bo 158 in M 31.
See Abstr. 154.021.

X-ray illumination of globular cluster puzzles.
See Abstr. 154.025.

Ultraviolet spectrum of the sky background at different galactic latitudes.
See Abstr. 157.012.

Observations of continuum X-ray and gamma-ray emission from the galactic center.
See Abstr. 157.014.

An 80 day X-ray light curve of 3C 371.
See Abstr. 158.027.

X rays from a peculiar nucleus galaxy NGC 2196.
See Abstr. 158.078.

Bulge X-ray sources and novae in M 31.
See Abstr. 158.083.

Ultraviolet and X-ray observations of the BL Lacertae PKS 0548−322.
See Abstr. 158.108.

High-resolution X-ray observations of M 87: nucleus, jet, and radio halo.
See Abstr. 158.111.

Optical and X-ray properties of X-ray selected active galactic nuclei.
See Abstr. 158.112.

The X-ray source in the nucleus of M 33.
See Abstr. 158.116.

X-ray images of late type galaxies.
See Abstr. 158.143.

X-ray emission from active galactic nuclei. Comptonization of self-absorbed cyclotron higher harmonics.
See Abstr. 158.172.

Variability and spectra of active galaxies.
See Abstr. 158.177.

X- and γ-ray observations of active galactic nuclei.
See Abstr. 158.178.

Einstein X-ray observations of 3C radio galaxies.
See Abstr. 158.179.

X-ray results on galactic haloes.
See Abstr. 158.184.

Einstein Observatory solid state spectrometer observations of M 87 and the Virgo cluster.
See Abstr. 158.229.

Time variations of the neutral hydrogen absorption spectrum of NGC 1275 (3C 84).
See Abstr. 158.234.

A new intermediate Seyfert galaxy: X-ray, optical, and radio properties.
See Abstr. 158.283.

X-ray observations to detect hot coronae around galaxies.
See Abstr. 158.309.

2 - 165 keV observations of active galaxies from HEAO-1.
See Abstr. 158.325.

The correlation between X-ray luminosity and velocity dispersion in clusters of galaxies.
See Abstr. 160.002.

Three rich clusters of galaxies with bimodal or clumpy X-ray morphologies.
See Abstr. 160.004.

Radio and X-ray observations of the Abell 2241 galaxy clusters.
See Abstr. 160.007.

Radio and X-ray galaxies in Abell 566.
See Abstr. 160.013.

Observations of X-ray clusters of galaxies at 102.5 MHz.
See Abstr. 160.015.

Diffuse X-rays as a probe to survey the gravitational mass distribution in the Coma cluster of galaxies.
See Abstr. 160.016.

Discovery of two active galactic nuclei in clusters of galaxies.
See Abstr. 160.023.

Discovery of a rich cluster of galaxies in Triangulum Australe identified with the bright X-ray source 4U1631-64.
See Abstr. 160.029.

Einstein observations of the size and structure of clusters of galaxies.
See Abstr. 160.032.

Implications of the cluster X-ray luminosity function for the dynamical state of clusters.
See Abstr. 160.033.

Extragalactic X-ray astronomy with the Einstein Observatory.
See Abstr. 160.039.

Coma-type radio halos and cluster X-ray morphology.
See Abstr. 160.040.

X-radiation of clusters of galaxies.
See Abstr. 160.041.

X-ray-imaging observations of clusters of galaxies.
See Abstr. 160.043.

Einstein X-ray imaging of clusters of galaxies.
See Abstr. 160.045.

Soft X-ray images of the Perseus cluster of galaxies.
See Abstr. 160.046.

Cooling flows in clusters of galaxies.
See Abstr. 160.047.

Evolution of the cluster X-ray luminosity function slope.
See Abstr. 160.049.

Medusa spectroscopy of A400, A576, A1767 and A2124.
See Abstr. 160.066.

The curvature of radio jets and tails in the intracluster media of Abell 1446 and 2220.
See Abstr. 160.069.

Gamma-ray Sources, Gamma-ray Background

142.501 **Emission mechanism and source distances of γ-ray bursts.**
E. P. Liang.
Nature, Vol. 299, 321 - 323 (1982).

The conventional view is that most γ-ray bursts are the optically thin free-free emission of a hot thermal plasma. The author shows here that: (1) independent of the source distance, there is an inherent difficulty with the free-free interpretation if the sources are indeed neutron stars with magnetic field $B \gtrsim 10^{12}$ G; (2) the spectral data can be fitted by optically thin thermal synchrotron spectra of mildly relativistic electrons; (3) the absence of observable low-energy turnover due to synchrotron self-absorption (or high energy Compton distortion) puts interesting upper limits to the source of luminosity and distance.

142.502 **Investigation of gamma bursts.**
I. V. Ehstulin.
Vestn. AN SSSR, 1982, No. 3, p. 82 - 89. In Russian. – Abstr. in Ref. zh., 51. Astron., 7.51.614 (1982).

142.503 **Radio measurements in the fields of gamma-ray sources. I. CG 195 + 04.**
W. Sieber, R. Schlickeiser.
Astron. Astrophys., Vol. 113, 314 - 323 (1982).

The γ-ray source CG 195 + 04 has been searched for radio counterparts at wavelengths between 2.8 cm and 18 cm with the 100 m telescope of the Max-Planck-Institut für Radioastronomie, Bonn. The authors have detected a number of sources and measured their spectra. Their positions may form the basis for future surveys in other frequency ranges. Different physical emission models suggest compactness of the γ-ray source.

142.504 **Investigation of cosmic gamma-bursts.**
E. P. Mazets, S. V. Golenetskij.
Astrofiz. i kosm. fiz. Moskva, 1982, p. 216 - 267. In Russian. Abstr. in Ref. zh., 51. Astron., 8.51.549 (1982).

142.505 **High-energy diffuse cosmic gamma-radiation from Cosmos 856 and Cosmos 914 measurements.**
L. F. Kalinkin, Yu. I. Nagornykh.
Pis'ma Astron. Zh., Tom 8, 542 - 545 (1982). In Russian. English translation in Soviet Astron. Lett., Vol. 8.

Results of Cosmos 856 and Cosmos 914 measurements of diffuse cosmic gamma-rays with energies above 100 MeV are discussed. Integral energy spectra of gamma rays within their energy range from 100 to 4000 MeV are presented for galactic latitudes $|b^{II}| \leqslant 30°$ and $|b^{II}| > 30°$. The form of the spectra indicates a probable contribution of the galactic component at high galactic latitudes.

142.506 **Physical processes in gamma-ray bursts.**
J. I. Katz.
Astrophys. J., Vol. 260, 371 - 385 (1982).

The author discusses physical processes occurring in gamma-ray bursts, with special attention to the event of 1979 March 5 in the Large Magellanic Cloud. The source of that event has been explained as a pair plasma confined in a strong magnetic field, as in a neutron star magnetosphere. Some consequences of this hypothesis are investigated and a simple model for gamma-gamma pair production and annihilation is calculated.

142.507 **Time-variable positron annihilation radiation from the galactic center direction.**
M. Leventhal, C. J. MacCallum, A. F. Huters, P. D. Stang.
Astrophys. J., Lett., Vol. 260, L1 - L5 (1982).

A balloon-borne germanium γ-ray telescope has been flown for a third time over Alice Springs, Australia to further study the galactic center source of 511 keV annihilation-line radiation. The source was found to be in an "off" or "low" state. The best fit experimental value for the line flux at Earth is zero, with a "1 σ" upper bound of 3.8×10^{-4} photons $cm^{-2} s^{-1}$. This null result is discussed in the context of other measurements and available theoretical models.

142.508 **Gamma-ray spectroscopy of the galactic center region: confirmation of the time variability of the positron annihilation line.**
W. S. Paciesas, T. L. Cline, B. J. Teegarden, J. Tueller, P. Durouchoux, J. M. Hameury.
Astrophys. J., Lett., Vol. 260, L7 - L10 (1982).

The NASA Goddard Space Flight Center Low-Energy Gamma-Ray Spectrometer observed the region of the galactic center during a balloon flight from Alice Springs, Australia on 1981 November 20. No significant excess over background was evident in the 511 keV annihilation line. The authors derive a 98% confidence upper limit for this line of 1.2×10^{-3} photons $cm^{-2} s^{-1}$. Continuum emission was detected above 100 keV.

142.509 **Detection of gamma-ray bursts with the hard X-ray and low energy gamma-ray experiment on HEAO-1.**
G. J. Hueter, H. S. Hudson, J. L. Matteson.
Bull. American Astron. Soc., Vol. 14, 619 (1982). – Abstract.

142.510 **Gamma burst observations in the 3 - 12 keV range from the STP 78-1 satellite.**
J. G. Laros, W. D. Evans, E. E. Fenimore, R. W. Klebesadel, S. Shulman, G. Fritz.
Bull. American Astron. Soc., Vol. 14, 619 - 620 (1982). Abstract.

142.511 **Gamma burst mechanisms.**
A. G. Petschek, S. A. Colgate.
Bull. American Astron. Soc., Vol. 14, 620 (1982). – Abstract.

142.512 **Localized sources of high-energy gamma rays.**
K. Pinkau.
Ninth Texas Symposium on Relativistic Astrophysics, (see 012.020), p. 234 - 247 (1980).

142.513 **Gamma-ray observations of Seyfert galaxies and the problem of the extragalactic gamma-ray background.**
V. Schönfelder.
Ninth Texas Symposium on Relativistic Astrophysics, (see 012.020), p. 267 - 277 (1980).

142.514 **Evidence for clustering of gamma-ray burst source positions.**
D. Van Buren.
Bull. American Astron. Soc., Vol. 14, 778 (1982). – Abstract.

142.515 **The April 19, 1980, cosmic gamma-ray burst and spurious absorption lines.**
E. E. Fenimore, R. W. Klebesadel, J. G. Laros.
Bull. American Astron. Soc., Vol. 14, 778 - 779 (1982). Abstract.

142.516 **Gamma ray bursts. A review of recent high-precision measurements.**
T. L. Cline.
Tenth Texas Symposium on Relativistic Astrophysics, (see 012.021), p. 314 - 329 (1981).

Observations of gamma ray bursts have advanced from accidental detection and circumstantial survey to detailed study with dedicated instruments. The author reports on recent measurements with high spatial,temporal and spectral resolution. The results are all consistent with neutron stars as the originating objects for gamma ray transients.

142.517 **Gamma ray astronomy with COS-B.**
B. N. Swanenburg.
Tenth Texas Symposium on Relativistic Astrophysics, (see 012.021), p. 381 - 390 (1981).
This paper gives a brief account of the observational results obtained to date with the COS-B satellite and summarizes some of the outstanding questions that have been raised.

142.518 **Cosmic gamma rays – present status and future prospects.**
H. Razdan.
Proc. Indian Natl. Sci. Acad., Part A, Vol. 47, Suppl. 2, 99 - 111 (1981). – Abstr. in Phys. Abstr., Vol. 85, Abstr. 89085 (1982).

142.519 **Possible effects of positronium negative ion on 0.511 MeV γ-ray line in astrophysical sources.**
C. Sivaram, V. Krishan.
Astrophys. Space Sci., Vol. 85, 31 - 34 (1982).
The astrophysical importance of the negative positronium ion, detected recently in the laboratory, has been pointed out. It is found that the presence of Ps^- ions will contribute additionally to the width of 0.511 γ-ray line formed by pair annihilation. The formation of Ps^- ion from an aggregate of electrons, positrons and positronium results in a variable positron population in the 0.511 MeV γ-ray line source.

142.520 **On the origin of gamma-ray bursts.**
V. M. Lipunov, E. I. Moskalenko, N. I. Shakura.
Astrophys. Space Sci., Vol. 85, 459 - 463 (1982).
A model for gamma-ray bursters which naturally explains the main observational features of GRB is proposed. The GRB turned out to be linked by evolution with well-known types of astrophysical objects.

142.521 **Study of diffuse cosmic and atmospheric gamma radiation using a spark chamber in the energy range 4 MeV–100 MeV.**
J. M. Lavigne, M. Niel, G. Vedrenne, B. Agrinier, E. Bonfand, B. Parlier, K. R. Rao.
Astrophys. J., Vol. 261, 720 - 731 (1982).
The Agathe Balloon experiment performed by the Commissariat à l'Energie Atomique of Saclay and the Centre d'Etude Spatiale des Rayonnements of Toulouse was launched in 1976 November and 1977 February in Brazil. The total atmospheric and cosmic γ-ray flux for various energy ranges was found using flux measurements during the balloon's ascent. In each case, it was possible to deduce the flux values of the diffuse and atmospheric components.

142.522 **Gamma-ray bursts and the collapse of a white dwarf.**
W. A. Baan.
Astrophys. J., Lett., Vol. 261, L71 - L75 (1982).
The temporal behavior of the γ-ray burst GB 790305 is interpreted in the context of a nonexplosive collapse of a white dwarf exceeding its mass limit. The energy of the impulsive phase of GB 790305 is due to the initial rapid cooling of the surface of the newly formed neutron star. A two-dimensional approximation of the cooling problem gives a cooling curve which is a good fit to the observed luminosity curve. The nature of three follow-up bursts is discussed in the context of neutron star glitches.

142.523 **The interrelation of cosmic γ-rays and interstellar gas in the range l: 65–180°.**
K. Arnaud, T. P. Li, P. A. Riley, A. W. Wolfendale, T. M. Dame, J. E. Brock, P. Thaddeus.
Mon. Not. R. Astron. Soc., Vol. 201, 745 - 758 (1982).
New data on the distribution of molecular hydrogen the first and second quadrants allow a detailed analysis of relationship between the column density of gas (in all maj forms) and the intensity of cosmic γ-rays in the Galaxy. It concluded that most of the features observed in the gamm ray map can be explained in terms of cosmic-ray interactic with the gas in the ISM, with molecular hydrogen playing important role. The cosmic-ray intensity in the important production regions appears to be roughly independent of galactocentric distance, R, for l: 65–100° and to fall off slowly with R for l: 100–180°.

142.524 **Gamma-ray burst systematics.**
R. W. Klebesadel, E. E. Fenimore, J. G. Laros, J. Terrell.
AIP Conf. Proc., No. 77, (see 012.034), p. 1 - 15 (1982).
The systematics of gamma-ray bursts are considered. Categorization within this class of events appears reasonab based upon both temporal and spectral features. The gamr ray bursts can be categorized as bursts like the 1979 Marcl event, bursts like the 20 minute transient observed by Jacobsen, and classical bursts. Further, the classical bursts be subdivided into bursts which are very brief, bursts with doublet or quasi-periodic time structure, or bursts which a long and irregular.

142.525 **A review of the 1979 March 5 transient.**
T. L. Cline.
AIP Conf. Proc., No. 77, (see 012.034), p. 17 - 33 (1982).
The understanding of the 1979 March 5 event remain problem of central importance for researchers in gamma ra transient astronomy. A consensus of opinion has not yet b reached regarding its possible origin in N 49 at 55 kpc dist versus in an invisible source 3 or 4 orders of magnitude clo although interpretations favoring N 49 appear to be preser gaining momentum. This presentation outlines the existing data in a review of what remains the most singular high-en astrophysical phenomenon of the space age.

142.526 **Frequency of gamma-ray bursts $> 3 \times 10^{-6}$ erg/c**
G. H. Share, K. Wood, J. Meekins, D. J. Yentis.
AIP Conf. Proc., No. 77, (see 012.034), p. 35 - 44 (1982).
The Large Area Sky Survey (LASS) instrument on HEAO - 1 has confirmed the detection of 12 of 18 γ-ray bu with intensities $> 2.5 \times 10^{-6}$ erg/cm² observed between August 1977 and January 1979. The LASS measurements all consistent with the reported source locations of the bur based on occultation by the Earth. Six new, and as yet unconfirmed, bursts have been detected between October 19 and January 1978.

142.527 **Observations of gamma-ray bursts from 10 keV t 9 MeV.**
G. H. Share, J. D. Kurfess, S. Dee, E. L. Chupp, J. M. Ryar D. J. Forrest, J. Lanigan, E. Rieger, G. Kanbach, C. Reppir
AIP Conf. Proc., No. 77, (see 012.034), p. 45 - 53 (1982).
The γ-ray spectrometer on the Solar Maximum Missio (SMM) satellite has detected at least 14 confirmed cosmic γ-ray bursts during its first 14 months in orbit. Similarity c the time profiles in different energy bands suggest that photons spanning two decades in energy are produced by t same mechanism. Spectra of four events falling within the instrument's aperture are presented.

142.528 **Gamma-ray burst observations by the HEAO - 3 High-Resolution Gamma-Ray Spectrometer.**
W. A. Wheaton, J. C. Ling, W. A. Mahoney, G. R. Riegler, A. S. Jacobson.
AIP Conf. Proc., No. 77, (see 012.034), p. 55 - 65 (1982).
Observations of cosmic gamma-ray bursts with the JP

High-Resolution Gamma-Ray Spectrometer on HEAO-3 are discussed. Two bursts seen on 1979 November 16 are of particular interest.

142.529 **Observations of a gamma-ray burst and other sources with a large-area, balloon-borne detector.**
R. B. Wilson, G. J. Fishman, C. A. Meegan.
AIP Conf. Proc., No. 77, (see 012.034), p. 67 - 78 (1982).
Observations of a weak cosmic gamma-ray burst of integrated intensity 2×10^{-6} erg/cm^{-2}, two solar flare events, and pulsed emission profiles of A 0535+26 and NP 0532 are reported for several energy intervals in the energy range from 45 to 520 keV. The measurements were made with a NaI(Tl) detector array flown on a balloon to 4 g cm^{-2} residual atmosphere.

142.530 **Search for associations of radio pulses and gamma-ray bursts.**
P. Inzani, G. Sironi, N. Mandolesi, G. Morigi.
AIP Conf. Proc., No. 77, (see 012.034), p. 79 - 84 (1982).
Continuous radio records obtained between July 1976 and May 1979 by automatic radiometers operating at 151 and 408 MHz from Medicina, Italy, have been scanned for radio pulses associated with gamma-ray bursts. Sixty five gamma events have been examined. For none of them a definite association with a particular radio pulse detected within ±10 min from the burst onset has been possible.

142.531 **Gamma-ray burst positions.**
K. Hurley.
AIP Conf. Proc., No. 77, (see 012.034), p. 85 - 99 (1982).
The international network of gamma-ray burst detectors has provided redundant localizations for six gamma-ray bursts with accuracies in the arcminute range. Catalog, radio, soft X-ray, and optical searches have been performed for some of these events. The results of these searches are reviewed. Although radio, X-ray, and optical candidates are found in the error boxes, no clear association between gamma-ray bursts and other forms of emission has emerged to date.

142.532 **Statistical study of gamma-ray burst source locations.**
G. Pizzichini.
AIP Conf. Proc., No. 77, (see 012.034), p. 101 - 105 (1982).
Some tests have been applied to a catalog of 65 gamma-ray burst source locations, in order to find if their distribution in galactic coordinates should be considered isotropic. The statistical significance of the overlapping of some error boxes is discussed.

142.533 **The observational consequences of an intrinsic burst luminosity distribution.**
M. C. Jennings.
AIP Conf. Proc., No. 77, (see 012.034), p. 107 - 113 (1982).
The galactic monoluminosity gamma burst models of Jennings and White are generalized to include intrinsic burst luminosity distributions. The log N - log S relation is substantially affected by distributions moderately peaked toward low luminosity and of dynamic range $\gtrsim 10^{-2}$. The generalized models are used to predict source angular distributions versus log S.

142.534 **Derivation of the luminosity of a gamma-ray burst.**
J. C. Higdon.
AIP Conf. Proc., No. 77, (see 012.034), p. 115 - 122 (1982).
The intrinsic luminosity in a gamma-ray burst is calculated using two assumptions: first, the power is independent of the burst duration, and, second, the spatial distribution of the bursters is similar to the spatial distribution of the pulsars.

142.535 **Gamma-ray burst spectra.**
B. J. Teegarden.
AIP Conf. Proc., No. 77, (see 012.034), p. 123 - 142 (1982).
A review of recent results in gamma-ray burst spectroscopy is given. Particular attention is paid to the recent discovery of emission and absorption features in the burst spectra. These lines represent the strongest evidence to date that gamma-ray bursts originate on or near neutron stars. Line parameters give information on the temperature, magnetic field and possibly the gravitational potential of the neutron star.

142.536 **10 June 1974 transient.**
J. C. Ling, W. A. Mahoney, J. B. Willett, A. S. Jacobson.
AIP Conf. Proc., No. 77, (see 012.034), p. 143 - 151 (1982).
The authors review results of a transient event which was observed in a balloon flight in 1974. During this twenty-minute event, four intense gamma-ray lines at energy 413 keV, 1.79, 2.2 and 5.9 MeV appeared in the background spectrum. No continuum emission was detected. Physical interpretations of this event are discussed.

142.537 **Time variations of an absorption feature in the spectrum of the gamma-ray burst on 1980 April 19.**
B. R. Dennis, K. J. Frost, A. L. Kiplinger, L. E. Orwig, U. Desai, T. L. Cline.
AIP Conf. Proc., No. 77, (see 012.034), p. 153 - 162 (1982).
A short but intense γ-ray burst was observed on 1980 April 19 with the Hard X-ray Burst Spectrometer on the Solar Maximum Mission. The photon number spectrum integrated over the impulsive part of the event fits a thermal bremstrahlung function with a temperature of (330 ± 70) keV at energies from 151 to 487 keV. At lower energies the data points lie considerably below this function suggesting a broad absorption feature extending down to $\leqslant$ 28 keV, the lowest energy measured.

142.538 **Search for time variations in 511 keV flux by ISEE-3 gamma-ray spectrometer.**
J. P. Norris, T. L. Cline, B. J. Teegarden.
AIP Conf. Proc., No. 77, (see 012.034), p. 163 - 168 (1982).
The ISEE-3 gamma-ray spectrometer has provided nearly continuous monitoring of the cosmic gamma-ray background in the energy regime 125 keV to 6.5 MeV since launch of the satellite in August 1978. The data has been analyzed for possible variations in the cosmic 511 keV line flux. Comparison is made with the HEAO-3 511 keV observations of the galactic center.

142.539 **The low energy spectra of gamma-ray bursts.**
R. W. Bussard, F. K. Lamb.
AIP Conf. Proc., No. 77, (see 012.034), p. 189 - 200 (1982).
The authors examine the implications of observed γ-ray burst spectra for the physical conditions and geometries of the sources. They use knowledge of the opacity of a hot plasma in a superstrong magnetic field and qualitative radiative transfer arguments to constrain the geometry of and physical conditions in the sources of these bursts.

142.540 **On the interpretation of gamma-ray burst continua and possible cyclotron absorption lines.**
E. E. Fenimore, J. G. Laros, R. W. Klebesadel, R. E. Stockdale, S. R. Kane.
AIP Conf. Proc., No. 77, (see 012.034), p. 201 - 209 (1982).
Most gamma-ray bursts have spectra which are fit by an optically thin thermal bremsstrahlung shape. The authors show that the small emitting volume implied by the observation of spectral features identified as cyclotron absorption lines from a 10^{12} gauss field prevent thermal bremsstrahlung from providing sufficient luminosities to account for the observed fluxes unless the optical depth to Compton scattering is large. In addition, they demonstrate that complicated detector response functions can introduce artifacts which mimic absorption features below 60 keV.

142.541 **Emission mechanisms of the March 5th 1979 gamma-ray transient.**
E. P. T. Liang.
AIP Conf. Proc., No. 77, (see 012.034), p. 241 - 248 (1982).
This paper reviews the spectral and temporal constraints on the emission mechanisms of the initial burst and subsequent pulses of the March 5, 1979 γ-transient. Interpretation of the hard tail above 300 keV as due to inverse comptonization, when combined with the synchrotron-pair annihilation model of Ramaty et al., allows one to derive the intrinsic luminosity of the source. This is in general agreement with that expected from N 49 in LMC.

142.542 **Cosmic gamma rays.**
A. W. Wolfendale.
Progress in cosmology, (see 012.047), p. 233 - 237 (1982).
The view of the Universe through the gamma ray window is without doubt still a hazy one but nevertheless there is interesting information to be gained already and as time goes on and improvements in technique are made very important advances can be confidently predicted. The present paper concerns just two topics in the field which are of contemporary interest: the nature of the γ-ray "sources" and the information contained in the apparently extragalactic γ-ray flux.

142.543 **Discrete sources of cosmic gamma rays.**
T. P. Li, A. W. Wolfendale.
Astron. Astrophys., Vol. 116, 95 - 100 (1982).
The question of the nature of the "sources" of cosmic γ-rays is an important and intriguing one. In this work the authors examine the extent to which "apparent sources" due to the irradiation of the clumpy interstellar medium by cosmic ray electrons and nuclei are able to contribute. They claim that out of nine unidentified sources in the 2 CG catalogue from the important COS B satellite experiment in a region of the Galaxy where they can make their analysis, six can be explained in this way. Of the six, four need an increase in cosmic ray intensity, compared with that locally, or an underestimate of H_2 column density by a negligible amount ($\cong$30%), the other two need a significant increase of intensity (by 3 and 10, respectively).

142.544 **Optical candidates for the 1978 November 19 γ-ray burst.**
IAU Circ., Nos. 3711, 3734, 3752 (1982).

142.545 **A new manifestation of activity of the gamma-ray burst source in Dorado.**
S. V. Golenetskij, E. P. Mazets, R. L. Aptekar', Yu. A. Gur'yan, V. N. Il'inskij, V. N. Panov, I. A. Sokolov, Z. Ya. Sokolova, T. V. Kharitonova, A. V. D'yachkov, N. G. Khavenson.
Pis'ma Astron. Zh., Tom 8, 657 - 662 (1982). In Russian. English translation in Soviet Astron. Lett., Vol. 8.
On December 1, 1981 and January 2, 1982 the KONUS instruments on board the Venera 13 and Venera 14 space probes have detected gamma-ray bursts the source of which being the same as that of the well-known sequence of events of March 5 and 6 and April 4 and 24, 1979.

142.546 **Gamma ray astronomy.**
B. P. Houston, A. W. Wolfendale.
Irish Astron. J., Vol. 15, (see 012.061), 101 - 102 (1981). Abstract.

142.547 **Transient emission of ultra-high energy pulsed gamma rays from PSR 0531.**
A. I. Gibson, A. B. Harrison, I. W. Kirkman, A. P. Lotts, J. H. Macrae, K. J. Orford, K. E. Turver, M. Walmsley.
Irish Astron. J., Vol. 15, (see 012.061), 103 (1981). Abstract.

142.548 **Discrete sources of gamma rays.**
T. P. Li.
Irish Astron. J., Vol. 15, (see 012.061), 105 (1981). Abstract.

142.549 **Sources de rayonnement γ, origine du rayonner cosmique, et régions de formation d'étoiles.**
T. Montmerle.
Mécanismes de production d'énergie dans le milieu interstellaire, (see 012.068), p. 215 - 266 (1981).

142.550 **Observations of gamma ray bursters.**
K. Hurley.
Accreting neutron stars, (see 012.069), p. 161 - 178 (198
The observational status of gamma ray burst astrono is reviewed with emphasis on evidence for a neutron star origin. Features in burst time histories are shown to be co patible with several models involving neutron stars. Burst energy spectra remain difficult to interpret in any self-cor tent manner. Recent searches for counterparts in the X-ra radio, and optical ranges have revealed objects of various types, but no coherent picture of burst sources has yet emerged.

142.551 **Theoretical problems raised by gamma-ray burs and related phenomena.**
M. J. Rees.
Accreting neutron stars, (see 012.069), p. 179 - 188 (198
The author outlines the general constraints on rapidl varying cosmic sources and then discusses specifically the extreme problems posed by the "March 5th 1979" event, this indeed involves a source as remote as the Large Magellanic Cloud (55 kpc distant). Finally, he mentions briefly a different class of object which may disgorge a luminosity far exceeding the conventional "Eddington lin

142.552 **An absorption feature in the spectrum of a gamma ray burst on March 25, 1978.**
G. J. Hueter, D. E. Gruber.
Accreting neutron stars, (see 012.069), p. 213 - 215 (198
The High Energy X-ray and Low Energy Gamma Ray Experiment on HEAO-1 detected a gamma ray burst on March 25, 1978. The main peak of the event lasted 10 seconds, while less intense emission was observed for appr mately 100 seconds. The burst spectrum shows an apparer absorption feature at 55 ± 5 keV with an equivalent width 13 ± 3 keV.

142.553 **Observation of a very weak gamma ray burst.**
P. Ubertini, A. Bazzano, R. K. Manchanda, C. La Padula, V. F. Polcaro, G. Vialetto.
Accreting neutron stars, (see 012.069), p. 223 - 228 (1982
The authors report the detection of a very faint burst in the hard X-ray range on July 21, 1981. The burst, havin a peak intensity of $\sim 7 \times 10^{-9}$ erg/cm^{2}s in the 20 - 120 keV range has been detected by the HXR-81 balloon borne har X-ray telescope (POKER) during a transmediterranean flig

142.554 **Observation of cosmic gamma ray bursts with tl gamma ray experiment on SMM.**
E. Rieger, C. Reppin, G. Kanbach, D. J. Forrest, E. L. Chu G. H. Share.
Accreting neutron stars, (see 012.069), p. 229 - 236 (1982
A brief account of the properties of all the gamma-ra bursts detected up to March 4, 1981, with the gamma-ray spectrometer on board the Solar Maximum Mission satellit is given.

142.555 **Neutron stars as sources of gamma-ray bursts: c straints from X-ray observations of source locati**
G. Pizzichini, T. L. Cline, U. D. Desai, B. J. Teegarden, K. Hurley, M. Niel, G. Vedrenne, W. D. Evans, E. E. Fenim

R. W. Klebesadel, J. G. Laros.
Accreting neutron stars, (see 012.069), p. 237 - 240 (1982).
An analysis of available data on gamma-ray bursts is given. The authors conclude that, if the sources of gamma-ray bursts are accreting neutron stars at a distance of a few hundred parsecs, either their conversion factor from accretion rate to X-ray luminosity is unusually low or the accretion rate is likely to be variable.

142.556 **Absorption of γ-rays and the location of the 5 March 1979 γ-ray burst.**
A. A. Zdziarski.
Accreting neutron stars, (see 012.069), p. 246 - 249 (1982).
The coordinates of the unusual 5 March, 1979 γ-ray burst source coincide with those of the supernova remnant N 49 in the Large Magellanic Cloud (LMC). An analysis of the self-absorption of γ-rays by pair production in a compact source is given. The results suggest that either the emitted γ-photons were strongly collimated or the 5 March burst source is much nearer than the LMC.

142.557 **Comments on the gamma-ray burst log $N(>S)$ – log S point of Beurle et al.**
M. C. Jennings.
Astrophys. Space Sci., Vol. 88, 429 - 432 (1982).
The isotropic cumulative burst rate determined by Beurle et al. from their observation of two gamma-ray bursts is shown to be statistically improbable. The difficulty arises from their assumption that the power law cumulative distribution function index γ equals one. Their observations are rediscussed and an upper limit of $N(> 8.47 \times 10^{-9}\ \mathrm{erg\ cm^{-2}}) < 5400\ \mathrm{yr^{-1}}$ is proposed.

A catalog of gamma-ray bursts with earth crossing times.
See Abstr. 002.054.

Comments on the gamma-ray burst catalog of Mazets et al. (1981a).
See Abstr. 002.132.

A reply to 'Comments on the gamma-ray burst catalog' of Laros et al.
See Abstr. 002.133.

Very high-energy gamma-ray astronomy.
See Abstr. 011.019.

17th International Cosmic Ray Conference. XG session: X- and gamma rays (non-solar); HE session: high-energy interactions.
See Abstr. 012.070.

Future directions in X-ray/gamma-ray observations.
See Abstr. 051.092.

Gamma-ray astronomy beyond the COS-B mission.
See Abstr. 051.093.

Cosmic radioactivity: a gamma-ray search for the origins of atomic nuclei.
See Abstr. 061.043.

The thermal pair annihilation spectrum: a detailed balance approach.
See Abstr. 061.044.

Relativistic electrons in galaxies and their non-thermal radiation from radio to gamma radiation.
See Abstr. 066.184.

High energy γ-rays from black holes.
See Abstr. 066.223.

The thermonuclear model for γ-ray bursts.
See Abstr. 066.503.

A two-dimensional model for γ-ray bursts.
See Abstr. 066.504.

Hydrogen-helium flashes on accreting neutron stars as a possible origin of gamma-ray bursts.
See Abstr. 066.510.

The universal diagram for magnetized neutron stars in the Galaxy.
See Abstr. 066.525.

Surface and magnetospheric physics of neutron stars and gamma-ray bursts.
See Abstr. 066.527.

The thermonuclear model for γ-ray bursts.
See Abstr. 066.528.

Gamma-ray bursts from young neutron stars.
See Abstr. 066.529.

A two dimensional model for γ-ray bursts.
See Abstr. 066.530.

Gamma bursts from neutron stars and stellar flares.
See Abstr. 066.531.

Precursors to γ-ray bursts in the asteroid impact scenario.
See Abstr. 066.532.

Thermonuclear model for high energy transients.
See Abstr. 066.547.

Rise times and durations of X-ray and γ-ray bursts.
See Abstr. 066.548.

Continuous UV radiation from gamma-ray burst sources.
See Abstr. 066.551.

Gamma-ray lines from solar flares and cosmic transients.
See Abstr. 076.010.

Can gamma ray bursts originate from low mass binaries?
See Abstr. 117.218.

Secondary electron spectra in interstellar clouds, and the bremsstrahlung γ-ray luminosity.
See Abstr. 131.254.

The local interstellar medium as traced by gamma rays.
See Abstr. 131.267.

On the origin of relativistic particles and gamma-rays in quasars.
See Abstr. 141.200.

Observations and interpretations of the pulsed emission from the Crab pulsar.
See Abstr. 141.519.

X-ray and gamma-ray upper limits for pulsed emission from radio pulsars.
See Abstr. 141.520.

Spectral evolution of active galactic nuclei: a unified description of the X-ray and gamma-ray backgrounds.
See Abstr. 142.046.

On the ultrahigh-energy gamma rays from Cygnus X-3.
See Abstr. 142.052.

The X- and gamma-ray backgrounds.
See Abstr. 142.080.

Summary of the workshop on X-ray and gamma ray burst sources.
See Abstr. 142.081.

On the UV variability of LSI +61°.303 (≡ GT 0236).
See Abstr. 142.095.

Persistent X-ray emission from a γ-ray burst source.
See Abstr. 142.131.

High-energy spectra of three active extragalactic objects.
See Abstr. 142.144.

Confinement of cosmic rays in molecular clouds.
See Abstr. 143.012.

COS-B observation of the Milky Way in high-energy gamma rays.
See Abstr. 157.003.

Gamma rays, cosmic rays, and the dynamics of the Galaxy.
See Abstr. 157.004.

Positron annihilation radiation from the galactic center region.
See Abstr. 157.006.

Gamma ray lines from the galactic center and gamma ray transients.
See Abstr. 157.007.

Galactic gamma-ray astronomy.
See Abstr. 157.009.

Observations of continuum X-ray and gamma-ray emission from the galactic center.
See Abstr. 157.014.

Observations of gamma-ray line emission from the galactic center region.
See Abstr. 157.015.

Time variable positron annihilation radiation from the galactic center direction.
See Abstr. 157.016.

Observations of the galactic center with the GSFC Low-Energy Gamma-Ray spectrometer: preliminary results.
See Abstr. 157.017.

Emission in the 0.3 to 1.0 MeV range from the galactic center region.
See Abstr. 157.018.

On the origin of the positron annihilation radiation from the direction of the galactic center.
See Abstr. 157.019.

The intensity and spectrum of galactic center β^+ annihilation photons after Compton scattering.
See Abstr. 157.020.

High resolution observations of Cen A between 70 and 500 keV.
See Abstr. 158.117.

Erratum

142.901 **Erratum: 'X-ray observations of symbiotic stars'** [Mon. Not. R. Astron. Soc., Vol. 197, 739 - 743 (1981)].
D. A. Allen.
Mon. Not. R. Astron. Soc., Vol. 201, 1199 (1982). – See Abstr. 30.142.060.

143 Cosmic Radiation

143.001 **On the stellar origin of the ^{22}Ne excess in cosmic rays.**
M. Cassé, J. A. Paul.
Astrophys. J., Vol. 258, 860 - 863 (1982).
The ^{22}Ne excess at the cosmic-ray source is discussed in terms of a ^{22}Ne-rich component injected and accelerated by carbon-rich Wolf-Rayet stars. The overabundance of ^{22}Ne relative to ^{20}Ne predicted at the surface of these stars is estimated to a factor ~120 with respect to solar system abundances. In order to give rise to a ^{22}Ne excess of about 3 at the cosmic-ray source as inferred from observations, the carbon-rich Wolf-Rayet contribution to the primary cosmic-ray flux is to be at maximum 1/60.

143.002 **The structure of cosmic ray shocks.**
W. I. Axford, E. Leer, J. F. McKenzie.
Astron. Astrophys., Vol. 111, 317 - 325 (1982).
The acceleration of cosmic rays by steady shock waves has been discussed in brief reports by Leer et al. (1976) and Axford et al. (1977). In the present paper the authors present a more extended version of this work. The energy transfer and the structure of the shock wave is discussed in detail, and it is shown that even for moderately strong shock waves most of the upstream energy flux in the background gas is transferred to the cosmic rays. This holds also when the upstream cosmic ray pressure is very small. For an intermediate Mach-number regime the overall shock structure is shown to consist of a smooth transition followed by a gas shock.

143.003 **Neutron oscillation as a source of excess sub-GeV antiprotons in galactic cosmic rays.**
C. Sivaram, V. Krishan.
Nature, Vol. 427 - 428 (1982).
The recent detection of an antiproton/proton ($\bar{p}$/p) ratio at energies much less than 1 GeV is several orders of magnitude above that predicted for $\bar{p}$ production from primary cosmic ray collisions. It is well recognized that there is no mechanism to explain this sub-GeV (130–320 MeV) $\bar{p}$ excess although many suggestions have been made. The authors postulate here that the phenomenon of neutron-antineutron ($n\bar{n}$) oscillations as a signature of grand unified theories may operate in neutron-rich astrophysical sources such as supernovae to produce the required low-energy galactic antiproton background.

143.004 **Solutions of the spherically-symmetric cosmic-ray transport equation in interplanetary space.**
H. Moraal, M. S. Potgieter.
Astrophys. Space Sci., Vol. 84, 519 - 533 (1982).
Numerical solutions of the well-known spherically-symmetric transport equation for interstellar cosmic rays in the interplanetary medium are presented. It is shown: (1) why nuclei and electron modulations differ, (2) that the only significant parameter that determines the level of modulation is the product of the solar wind speed and the number of diffusion mean free paths between observer and boundary, and (3) that observations do exist which cannot be explained in terms of this simple model.

143.005 **Energetic spectrum of super-high energy cosmic rays.**
G. B. Khristiansen.
Usp. fiz. nauk, Tom 136, 351 - 353 (1982). In Russian.
Abstr. in Ref. zh., 62. Issled. kosm. prostranstva, 7.62.341 (1982).

143.006 **Super-high energy cosmic rays: the astrophysical aspect.**
V. S. Berezinskij.
Usp. fiz. nauk, Tom 136, 349 (1982). In Russian. – Abstr. in Ref. zh., 62. Issled. kosm. prostranstva, 7.62.343 (1982).

143.007 **Absolute flux and nuclear composition of high-energy cosmic rays.**
S. I. Nikol'skij.
Usp. fiz. nauk, Tom 136, 349 - 351 (1982). In Russian. – Abstr. in Ref. zh., 62. Issled. kosm. prostranstva, 7.62.344 (1982).

143.008 **On the energetic spectrum of super-high energy cosmic rays.**
G. V. Kulikov, Yu. A. Fomin, G. B. Khristiansen.
Byul. NTI Yakut. fil. SO AN SSSR, 1981, iyul', probl. kosmofiz. i aehron., p. 26 - 29. In Russian. – Abstr. in Ref. zh., 62. Issled. kosm. prostranstva, 7.62.345 (1982).

143.009 **Cosmic rays and astronomy.**
A. W. Wolfendale.
Q. J. R. Astron. Soc., Vol. 23, 325 - 343 (1982).

143.010 **Modulation of galactic cosmic ray anisotropy in heliomagnetosphere: average sidereal daily variation.**
K. Nagashima, I. Morishita, S. Yasue.
Planet. Space Sci., Vol. 30, 879 - 896 (1982).
The authors formulate the modulation of galactic anisotropy of cosmic rays caused by their orbital deflection in the heliomagnetosphere.

143.011 **Cosmic rays near the heliospheric current sheet.**
J. Kota, J. R. Jokipii.
Geophys. Res. Lett., Vol. 9, 656 - 659 (1982).
The authors present a discussion of full, three-dimensional model simulations of the solar modulation of galactic cosmic rays, including drift. The authors' particular emphasis is on the variation of the cosmic-ray intensity near the heliospheric current sheet. They show that the model quite naturally produces a negative gradient away from a wavy current sheet as seen by an observer on the Earth. The authors conclude that recent observations of Newkirk and Lockwood are not in conflict with these model results, and hence are consistent with modulation models in which drift plays an important role.

143.012 **Confinement of cosmic rays in molecular clouds.**
E. G. Zweibel, J. M. Shull.
Astrophys. J., Vol. 259, 859 - 868 (1982).
The authors explore the trapping of cosmic rays produced by a supernova in a dense molecular cloud and the implications for COS B γ-ray sources. They conclude that unless the magnetic field in the cloud is as large as 10^{-3} gauss, the cosmic rays are unable to escape from the cloud before losing their energy through nuclear collisions. Buried supernovae could therefore be detectable sources of 100 MeV π^0-decay γ-rays.

143.013 **Can we detect antimatter from other galaxies?**
S. P. Ahlen, P. B. Price, M. H. Salamon, G. Tarlé.
Astrophys. J., Vol. 260, 20 - 32 (1982).
Recent developments in particle detection technology now make it possible to use well-established principles of high-order quantum electrodynamics to search for antimatter in the cosmic rays with unprecedented sensitivity. The technique is described and is shown to be superior in both collecting power and resolution to other, more conventional, techniques used in the past.

143.014 **On the origin of the low energy cosmic rays from stellar sources.**
S. Biswas, N. Durgaprasad, S. S. Trivedi.

Proc. Indian Acad. Sci. Earth Planet Sci., Vol. 90, 337 - 344 (1981).

The model of stellar origin of the anomalous component in the low energy cosmic rays for He to Fe ions observed in space vehicles is studied in the light of recent results.

143.015 **The energy spectrum of the 11-year modulation of galactic cosmic rays.**
L. I. Dorman, A. G. Zusmanovich, L. F. Churunova.
Geomagn. Aehron., Tom 22, 543 - 547 (1982). In Russian.

143.016 **Cosmic ray generation by the shock drift and compression mechanisms.**
J. S. Scott, M. E. Pesses, R. B. Decker.
Bull. American Astron. Soc., Vol. 14, 620 (1982). – Abstract.

143.017 **Where do cosmic rays come from?**
K. Sakurai.
Magnetospheric phenomena of celestial bodies, (see 012.011), p. 145 - 146 (1982). In Japanese.

In order to explain the observed chemical composition of galactic cosmic rays at their sources, it is suggested that this composition is resulted from some acceleration mechanism preferably related to the charge states of particles being ambient in the source regions. Since it is, therefore, likely that the most of these particles are partially ionized in these regions, the ambient temperature is so low as to be less than 10^6 K before the start of acceleration.

143.018 **Gigantic gamma-ray family events with $\Sigma E_\gamma \geqslant 1000$ TeV and hadronic interactions.**
M. Akashi, M. Amenomori, E. Konishi, H. Nanjo, Z. Watanabe, K. Mizutani.
Nuovo Cimento A, Vol. 67A, Ser. 11, 221 - 243 (1982).
Abstr. in Phys. Abstr., Vol. 85, Abstr. 63198 (1982).

143.019 **Seasonal variation of the tridiurnal wave in cosmic ray intensity.**
S. A. Wahab, A. M. Salem.
Fizika, Vol. 14, No. 1, p. 45 - 51 (1982). – Abstr. in Phys. Abstr., Vol. 85, Abstr. 68576 (1982).

143.020 **β on geomagnetic disturbed days.**
S. Kumar, R. S. Yadava, Badruddin.
Natl. Acad. Sci. Lett., Vol. 4, 141 - 142 (1981). – Abstr. in Phys. Abstr., Vol. 85, Abstr. 68577 (1982).

143.021 **Variation of spectral exponent of diurnal anisotropy of the cosmic ray intensity on quiet days.**
S. Kumar, R. S. Yadava, Badruddin.
Natl. Acad. Sci. Lett., Vol. 4, 175 - 177 (1981). – Abstr. in Phys. Abstr., Vol. 85, Abstr. 68578 (1982).

143.022 **The dependence of the cosmic-ray shower age on shower size for vertical sizes in the range $6 \times 10^5 < N_e < 6 \times 10^6$.**
R. W. Clay, P. R. Gerhardy, D. F. Liebing, G. J. Thornton, J. R. Patterson.
Nuovo Cimento C, Vol. 4C, Ser. 1, 668 - 672 (1981). – Abstr. in Phys. Abstr., Vol. 85, Abstr.68579 (1982).

143.023 **The abundance of zinc in the cosmic radiation.**
D. O'Sullivan, A. Thompson. P. H. Fowler, D. L. Henshaw.
Nucl. Tracks Methods Instrum. Appl., Vol. 5, (see 012.015), 403 (1981). – Summary. Abstr. in Phys. Abstr., Vol. 85, Abstr. 68580 (1982).

143.024 **Correction for efficiency of cosmic ray charge spectrum obtained with a SSNTD stack.**
J. Sequeiros, J. Medina.
Nucl. Tracks Methods Instrum. Appl., Vol. 5, (see 012.015), 403 - 404 (1981). – Summary. Abstr. in Phys. Abstr., Vol. 85, Abstr. 68581 (1982).

143.025 **The cosmic ray track record of events in the early solar system.**
D. Lal.
Nucl. Tracks Methods Instrum. Appl., Vol. 5, (see 012.015), 401 (1981). – Summary. Abstr. in Phys. Abstr., Vol. 85, Abstr. 68583 (1982).

143.026 **The acceleration of cosmic rays by shock waves.**
W. I. Axford.
Tenth Texas Symposium on Relativistic Astrophysics, (see 012.021), p. 297 - 313 (1981).

Significant progress has been made in the area of cosmic ray acceleration with the recognition of the fact that energy can be transferred directly to cosmic rays from supersonic motions of the background medium via shock waves. Recent developments in the theory of an efficient first-order Fermi mechanism for acceleration are reviewed.

143.027 **Summary of the workshop on cosmic ray astrophysics.**
M. M. Shapiro.
Tenth Texas Symposium on Relativistic Astrophysics, (see 012.021), p. 442 - 445 (1981).

143.028 **Anisotropy of super-high energy cosmic rays.**
V. I. Chesnokov.
Zemlya Vselennaya, 1982, No. 5, p. 10. In Russian.

143.029 **The effect of corotating interaction regions on the propagation of relativistic cosmic rays in the heliosphere.**
B. T. Thomas, R. Gall.
J. Geophys. Res., Vol. 87, 4542 - 4548 (1982).

The paper questions the validity of modulation theories that ignore the presence of large regions of enhanced magnetic field in the heliosphere. The authors simulate the scatter-free and diffusive motion of relativistic cosmic rays in models of the heliospheric magnetic field which include compressed field structures representing the corotating interaction regions which are believed to be present in the interplanetary medium throughout most of the solar cycle. The authors find that the strong field gradients associated with these regions greatly perturb the drift pattern anticipated for simple Parkerian fields, and thus the expected streaming of cosmic rays from over the poles or along the current sheet, during consecutive solar cycles, no longer holds.

143.030 **Primary composition of cosmic rays near 10^{17} eV.**
M. P. Chantler, M. A. B. Craig, T. J. L. McComb, K. J. Orford, K. E. Turver, G. M. Walley.
J. Phys. G, Vol. 8, L51 - L55 (1982). – Abstr. in Phys. Abstr., Vol. 85, Abstr. 84050 (1982).

143.031 **An experimental search for cosmic monopoles.**
R. Bonarelli, P. Capiluppi, I.D'Antone, G. Giacomelli, G. Mandrioli, C. Merli, A. M. Rossi.
Phys. Lett. B, Vol. 112B, 100 - 102 (1982). – Abstr. in Phys. Abstr., Vol. 85, Abstr. 84052 (1982).

143.032 **Cosmic ray penumbral characteristics at Cairo's latitude.**
S. A. Wahab, A. M. Salem.
Fizika, Vol. 14, 119 - 125 (1982). – Abstr. in Phys. Abstr., Vol. 85, Abstr. 89090 (1982).

143.033 **Cosmic rays in interplanetary space.**
S. Ramadurai.
Proc. Indian Natl. Sci. Acad., Part A, Vol. 47, Suppl. 2, 92 -

98 (1981). – Abstr. in Phys. Abstr., Vol. 85, Abstr. 89091 (1982).

143.034 **The elongation rate theorem for cosmic-ray showers and its application.**
T. J. L. McComb, K. E. Turver.
J. Phys. G, Vol. 8, 871 - 877 (1982). – Abstr. in Phys. Abstr., Vol. 85, Abstr. 89096 (1982).

143.035 **Cosmogenic nuclides in recently fallen meteorites: evidence for galactic cosmic ray variations during the period 1967 - 1978.**
J. C. Evans, J. H. Reeves, L. A. Rancitelli, D. D. Bogard.
J. Geophys. Res., Vol. 87, 5577 - 5591 (1982).

Cosmogenic radionuclides were measured on 48 fragments of 24 meteorites which fell between 1967 and 1978. The cosmogenic radioisotope and rare gas data are interpreted in terms of a time dependent modulation of galactic cosmic rays spanning one full 11 year sun spot cycle. The data demonstrate that the production rates of cosmic ray-produced nuclides in meteorites vary considerably according to modulation by the 11-year solar cycle and support the concept that variations of solar-modulated, cosmic ray flux of similar magnitude have occurred over much longer time periods.

143.036 **Modulation of the monoenergetic spectrum of galactic cosmic rays.**
L. I. Dorman, M. E. Kats, S. F. Nosov, Yu. I. Fedorov, B. A. Shakhov.
Geomagn. Aehron., Tom 22, 705 - 710 (1982). In Russian.

143.037 **Absorption of galactic cosmic ray streams in the atmosphere in the period 1961 to 1980.**
N. K. Pereyaslova, P. E. Pokrevskij, Yu. I. Stozhkov.
Geomagn. Aehron., Tom 22, 711 - 715 (1982). In Russian.

143.038 **The abundance of the actinides in the cosmic radiation as measured on HEAO 3.**
W. R. Binns, R. K. Fickle, T. L. Garrard, M. H. Israel, J. Klarmann, E. C. Stone, C. J. Waddington.
Astrophys. J., Lett., Vol. 261, L117 - L120 (1982).

The HEAO 3 detector of heavy cosmic-ray nuclei has observed one possible actinide nucleus and some 100 nuclei of the platinum-lead group of elements. The resulting upper limit of 3% for the abundance ratio of actinides to platinum-lead nuclides is significantly lower than previous results from other observations.

143.039 **Nuclear interactions and primary cosmic-ray component around 10^{15} eV viewed through the cluster analysis of γ-ray families.**
M. Amenomori, E. Konishi, H. Nanjo, K. Mizutani, K. Kasahara, S. Torii, T. Yuda, T. Shirai, N. Tateyama, T. Taira, I. Mito, M. Shibata, H. Sugimoto, K. Taira, N. Hotta.
Phys. Rev. D, Vol. 25, 2807 - 2819 (1982). – Abstr. in Phys. Abstr., Vol. 85, Abstr. 100687 (1982).

143.040 **Cosmic-ray-neutrino tests for heavier weak bosons and cosmic antimatter.**
R. W. Brown, F. W. Stecker.
Phys. Rev. D, Vol. 26, 373 - 383 (1982). – Abstr. in Phys. Abstr., Vol. 85, Abstr. 100688 (1982).

143.041 **The vanishing of order-G mechanical effects of cosmic massive neutrinos on bulk matter.**
N. Cabibbo, L. Maiani.
Phys. Lett. B, Vol. 114B, 115 - 117 (1982). – Abstr. in Phys. Abstr., Vol. 85, Abstr. 106462 (1982).

143.042 **Preliminary Fly's Eye results.**
G. L. Cassiday, R. Cady, J. Elbert, E. Loh, M. Salamon. P. Sokolsky, D. Steck, M. Ye.
AIP Conf. Proc., No. 85, (see 012.038), p. 528 - 539 (1981). Abstr. in Phys. Abstr., Vol. 85, Abstr. 112280 (1982).

143.043 **The radial electron distribution in extensive air showers.**
J. N. Capdevielle, J. Gawin.
J. Phys. G, Vol. 8, 1317 - 1335 (1982). – Abstr. in Phys. Abstr., Vol. 85, Abstr. 112281 (1982).

143.044 **Fine analysis of two double-cored extensive air showers and sub-shower's large p_T.**
M. Sakata, Y. Yamamoto, S. Dake.
Nuovo Cimento A, Vol. 70A, Ser. 11, 297 - 327 (1982). Abstr. in Phys. Abstr., Vol. 85, Abstr. 112282 (1982).

143.045 **Chirons.**
J. A. Chinellato, C. Dobrighkeit, J. Bellandi Filho, C. M. G. Lattes, A. Marques, M. J. Menon, C. E. Navia O, M. Ballester, C. Santos, K. Sawayanagi, E. Silva, E. H. Shibuya A. Turtelli Jr., N. M. Amato, F. M. Oliveira Castro, R. H. C. Maldonado, H. Aoki, Y. Fujimoto, Y. Funayama, S. Hasegawa, H. Kumano, H. Semba, T. Tabuki, M. Tamada, K. Tanaka, S. Yamashita, N. Arata, T. Shibata, K. Yokoi, A. Ohsawa.
AIP Conf. Proc., No. 85, (see 012.038), p. 500 - 527 (1981). Abstr. in Phys. Abstr., Vol. 85, Abstr. 112283 (1982).

143.046 **Ultra-high energy cross section from study of longitudinal development of air showers.**
R. W. Ellsworth, T. K. Gaisser, T. Stanev, G. B. Yodh.
AIP Conf. Proc., No. 85, (see 012.038), p. 540 - 551 (1981). Abstr. in Phys. Abstr., Vol. 85, Abstr. 112284 (1982).

143.047 **New event types in a balloon-borne cosmic ray experiment.**
T. H. Burnett, S. Dake, M. Fuki, J. C. Gregory, T. Hayashi, R. Holynski, J. Iwai, W. V. Jones, A. Jurak, J. J. Lord, O. Miyamura, T. Ogata, T. A. Parnell, T. Saito, T. Tabuki, Y. Takahashi, T. Tominaga, R. J. Wilkes, W. Wolter, B. Wosiek
AIP Conf. Proc. No. 85, (see 012.038), p. 552 - 561 (1981). Abstr. in Phys. Abstr., Vol. 85, Abstr. 112285 (1982).

143.048 **Baryon production at very high energies ~10^6 GeV.**
S. C. Tonwar.
AIP Conf. Proc., No. 85, (see 012.038), p. 562 - 570 (1981). Abstr. in Phys. Abstr., Vol. 85, Abstr. 112286 (1982).

143.049 **Review of low P_T physics from cosmic rays.**
T. K. Gaisser.
AIP Conf. Proc., No. 85, (see 012.038), p. 571 - 577 (1981). Abstr. in Phys. Abstr., Vol. 85, Abstr. 112287 (1982).

143.050 **Properties of high energy interactions deduced from cosmic rays and their comparison with results from recent accelerator experiments.**
J. Wdowczyk, A. W. Wolfendale.
AIP Conf. Proc., No. 85, (see 012.038). p. 578 - 584 (1981). Abstr. in Phys. Abstr., Vol. 85. Abstr. 112288 (1982).

143.051 **A study of the modulating effect of solar flares on the cosmic ray intensity using time series analysis.**
G. A. Bowe, C. J. Hatton.
Sol. Phys., Vol. 80, 351 - 359 (1982).

The cosmic ray 11-year variation for solar cycle 20 is attributed to the modulating effect of solar flare-induced shocks propagating through the interplanetary medium to the boundary of the heliosphere. The relative influence of these disturbances upon the cosmic ray intensity as a function of their travel time from the Sun is determined by a deconvolution of a linear system with the number of solar flares (importance $\geqslant 1$) and the observed cosmic ray intensity as the input and output respectively of this system. The 'impulse response function' so determined indicates that the solar flare-induced disturbances significantly modulate cosmic rays out to a distance of 70–90 AU where the modulating effect of the disturbances abruptly ends. This is interpreted as the boundary of the heliosphere.

143.052 **Cosmic ray origin above 10^{18} eV: galactic or extragalactic?**
A. K. Dasgupta.
Astrophys. Space Sci., Vol. 86, 361 - 369 (1982).
An analysis is made of the implications of assuming that suprathermal dust grains ($a \cong 3 \times 10^8$ cm) of intergalactic origin may acquire cosmic ray energies as high as 10^{20} eV. These dust grains may attain "suprathermal energy" ($v_g \cong 3 \times 10^8$ cm s^{-1}) by the Fermi process. Initially the dust grains are accelerated by the radiation pressure against the drag of the ambient gas of the medium, but once these dust grains attain a terminal velocity ($U \cong 10^5$ cm s^{-1}), then they may be expelled out of the galactic region into the intergalactic medium and finally acquire high energy.

143.053 **Antimatter in cosmic rays.**
P. Király.
Progress in cosmology, (see 012.047), p. 89 - 99 (1982).
The author reviews the problem of antimatter in cosmic rays. He concludes that there is no completely satisfactory explanation of the recently observed high cosmic antiproton flux values. Further checks and new, independent observations are needed in order to reach an understanding of this very interesting new area of cosmic ray research.

143.054 **The cosmic ray origin problem.**
J. Wdowczyk.
Progress in cosmology, (see 012.047), p. 161 - 175 (1982).
A survey is made of the measurements of the various components of the cosmic radiation: protons, nuclei and electrons. The form of the energy spectra and anisotropies are considered. The problem of the location of cosmic ray sources is examined and it is concluded that there is still no agreement; what does appear to be confirmed is that the bulk of the particles below 10^{14} eV is generated in the Galaxy and those above 10^{18} eV are extragalactic. The energy at which the transition occurs is very uncertain. A variety of features of cosmological interest is discussed.

143.055 **Transport and propagation of cosmic rays in galaxies. II. The effect of a galactic wind on the mean lifetime and age distribution of non-decaying cosmic rays.**
I. Lerche, R. Schlickeiser.
Astron. Astrophys., Vol. 116, 10 - 26 (1982).
Exact analytical solutions for the mean age and age distribution of non-decaying charged particles in a dynamical halo are presented. The particles are subject to diffusion, convection, adiabatic deceleration, momentum loss and/or acceleration processes after their injection from sources. The galactic wind velocity is assumed to be zero in the galactic plane and to increase with galactic height. The solutions allow for any source distribution with respect to position and momentum and any momentum dependence of the diffusion coefficient and the momentum loss or acceleration term. Resulting age distributions are always Gaussian and in most cases decrease at small ages. The basic physical effects are illustrated in the case of mono-momentum injection at a point source of cosmic ray protons and electrons.

143.056 **On the transport and propagation of cosmic rays in galaxies – I. Solution of the steady-state transport equation for cosmic ray nucleons, momentum spectra and heating of the interstellar medium.**
I. Lerche, R. Schlickeiser.
Mon. Not. R. Astron. Soc., Vol. 201, 1041 - 1072 (1982).
The authors present exact analytical solutions to the equation describing the steady-state transport, in a direction perpendicular to the galactic plane, of cosmic ray nucleons which are being convected, diffused and may be either gaining or losing momentum. The authors calculate momentum spectra for the special case of momentum power-law injection spectra $S \propto p^{-\epsilon}$ and power-law momentum dependence of the diffusion coefficient $D(p) = D_0 p^a$ and discuss these solutions in detail. The heating of interstellar gas in the galactic disc and corona is discussed. The cosmic ray pressure is calculated as a function of galactic height.

143.057 **Samples of the Milky Way.**
R. A. Mewaldt, E. C. Stone, M. E. Wiedenbeck.
Sci. American, Vol. 247, No. 6, p. 100 - 109 (1982).
Cosmic rays are nuclei from other regions of our galaxy. Satellite observations of such nuclei indicate that the isotopic composition of those regions differs from the composition of the solar system.

143.058 **Non-linear theory of cosmic ray shocks including self-generated Alfvén waves.**
J. F. McKenzie, H. J. Völk.
Astron. Astrophys., Vol. 116, 191 - 200 (1982).
Diffusive shock acceleration of cosmic rays is considered, including the scattering wave field in the dynamics; the non-linear treatment also takes the back-reaction of the accelerated particles on the plasma flow into account. The governing equations are derived in a "hydrodynamic" approximation, treating the cosmic rays also as a fluid. The scattering hydromagnetic waves are excited by the accelerated particles and adiabatically amplified in the broadened shock transition. The general steady state solutions are derived. It is shown that under many circumstances when the energy density of accelerated particles is large the fluctuation magnetic fields become comparable with or even larger than the average field in the shock transition. Unless there exists strong wave damping, this should naturally lead to preferential acceleration of high rigidity particles at the expense of low rigidity particles.

143.059 **Cosmic ray intensity variations and two types of high speed solar streams.**
D. Venkatesan, A. K. Shukla, S. P. Agrawal.
Sol. Phys., Vol. 81, 375 - 381 (1982).
This study deals with the short-term variations of cosmic ray intensity during the interval 1973–78. Daily means of high latitude neutron and meson monitors from the same station and those of a low latitude neutron monitor have been analysed using the Chree method of superposed epochs. The investigation reveals the existence of two types of cosmic ray intensity variations with distinctly different spectral characteristics. During the interval 1973–76, relative changes in the neutron and meson monitor rates are nearly equal indicating an almost flat rigidity spectrum of variation. During 1977–78, however, the spectrum acquires a negative spectral character similar to that observed for Forbush decreases. The authors suggest that events of the interval 1973–76 are essentially due to high speed streams associated with solar coronal holes and that events of the interval 1977–78 are due to fast streams from solar active regions with flare activity.

143.060 **On the measurement of the proportion of heavy nuclei in the primary cosmic ray flux.**
T. S. Stanev.
Dokl. Bolg. AN, Vol. 34, 1655 - 1658 (1981). – Abstr. in Ref. zh., 62. Issled. kosm. prostranstva, 11.62.311 (1982).

143.061 **The elements and isotopes of the cosmic radiation.**
D. O'Sullivan.
Irish Astron. J., Vol. 15, (see 012.061), 97 - 99 (1981).
Abstract.

143.062 **Bristol University studies of ultra-heavy cosmic rays.**
P. H. Fowler, M. R. W. Masheder, R. N. F. Walker, A. Worley.
Irish Astron. J., Vol. 15, (see 012.061), 99 (1981).
Abstract.

143.063 **Ultra-high energy cosmic rays – do they exist?**
A. A. Watson.
Irish Astron. J., Vol. 15, (see 012.061), 100 (1981).
Abstract.

143.064 **Ways of puzzling out the cosmic ray mystery.**
G. F. Krymskij.
Byull. NTI. Nauchn. issled. Yakut. fil. SO AN SSSR. Yakutsk, 1982, p. 7 - 9. In Russian. – Abstr. in Ref. zh., 51. Astron., 10.51.349 (1982).

143.065 **On modulation effects in cosmic rays during a solar activity cycle.**
S. K. Akopyan, G. A. Asatryan, Kh. P. Babayan.
Erevan. fiz. inst. Prepr., 1981, No. 509/52, 12 pp. In Russian. Abstr. in Ref. zh., 51. Astron., 11.51.349 (1982).

143.066 **Accélération des rayons cosmiques galactiques dans les ondes de choc.**
P. O. Lagage.
Mécanismes de production d'énergie dans le milieu interstellaire, (see 012.068), p. 275 - 278 (1981).

143.067 **Cosmic ray pulsations before solar proton flares according to ground-based observational data.**
S. A. Starodubtsev, A. T. Filippov, N. P. Chirkov.
Astron. Zh., Tom 59, 1229 - 1230 (1982). In Russian.
English translation in Soviet Astron., Vol. 26, No. 6.

An investigation of cosmic ray neutron component fluctuations during 11 solar flares for 1967 - 1981 on the basis of 5-minute data of ground stations was carried out. It was found that cosmic ray pulsations with periods of 30 - 40, 60 - 90, and rarely 120 - 160 minutes are observed 24 hours before the increase of the relativistic particle flux from the solar flares.

143.068 **Unusual characteristics of the cosmic ray intensity increase of September 17 - 18, 1979.**
S. P. Agrawal, D. Venkatesan.
J. Geophys. Res., Vol. 87, 9201 - 9204 (1982).

The data from a number of neutron and meson monitors are examined for the cosmic ray event of September 17 - 18, 1979. The profiles of the observed cosmic ray intensity increase at various stations exhibit an apparent solar flare origin, if one were to look at a limited amount of data. It is clearly shown that the 'flare-type' increase is unusual and is of nonsolar origin. A suggestion is made for explaining this in terms of interplanetary structures for which supportive evidence is provided by reference to in situ observations.

143.069 **Cosmic ray north-south anisotropy: the role of the interplanetary magnetic field.**
M. A. Pomerantz, S. P. Duggal, A. J. Owens, M. F. Tolba, C. H. Tsao.
J. Geophys. Res., Vol. 87, 10325 - 10330 (1982).

Nucleonic intensity data from the north and south polar stations Thule and McMurdo are analyzed to determine the nature of periodicities in the north-south asymmetry of relativistic galactic cosmic rays.

143.070 **On the nature of optical images of distant cosmic-ray air showers.**
R. J. Protheroe.
J. Phys. G, Vol. 8, L165 - L171(1982). – Abstr. in Phys. Abstr., Vol. 85, Abstr. 117545 (1982).

143.071 **The sensitivity of Cerenkov radiation pulses to the longitudinal development of cosmic-ray showers.**
A. M. Hillas.
J. Phys. G, Vol. 8, 1475 - 1492 (1982). – Abstr. in Phys. Abstr., Vol. 85, Abstr. 117546 (1982).

143.072 **Delayed hadrons in extensive air showers: evidence for the iron-group nuclei in primary cosmic-ray flux at energies $\sim 10^{13}$ - 10^{15} e.V.**
J. A. Goodman, R. W. Ellsworth, A. S. Ito, J. R. MacFall, F. Siohan, R. E. Streitmatter, S. C. Tonar, P. R. Viswanath, G. B. Yodh.
Phys. Rev. D, Vol. 26, 1043 - 1060 (1982). – Abstr. in Phys. Abstr., Vol. 85, Abstr. 117547 (1982).

143.073 **Cosmic rays and particle physics.**
T. K. Gaisser.
Comments Nucl. Part. Phys., Vol. 11, 25 - 39 (1982). – Abstr. in Phys. Abstr., Vol. 85, Abstr. 117550 (1982).

143.074 **Fragmentation parameters of cosmic-ray nuclei in nuclear photoemulsion.**
D. Kazimir, L. Just.
Acta Phys. Slovaca, Vol. 32, 275 - 281 (1982). – Abstr. in Phys. Abstr., Vol. 86, Abstr. 5063 (1983).

143.075 **Experimental limits on quarks, tachyons, and massive particles in cosmic rays.**
A. Marini, I. Peruzzi, M. Piccolo, F. Ronga, D. M. Chew, R. P. Ely, T. P. Pun, V. Vuillemin, R. Fries, B. Gobbi, W. Guryn, D. H. Miller, M. C. Ross, D. Besset, S. J. Freedman, A. M. Litke, J. Napolitano, T. C. Wang, F. A. Harris, I. Karliner, S. Parker, D. E. Yount.
Phys. Rev. D, Vol. 26, 1777 - 1780 (1982). – Abstr. in Phys. Abstr., Vol. 86, Abstr. 5064 (1983).

143.076 **Integral energy spectrum of primary cosmic rays at high energies.**
R. Hasan, S. C. Arora, D. Hans, M. S. Swami.
Pramāna, Vol. 19, 133 - 140 (1982). – Abstr. in Phys. Abstr., Vol. 86, Abstr. 5072 (1983).

143.077 **The primary cosmic electron spectrum and its astrophysical interpretation. I. Astrophysical aspect of the study of primary cosmic electron spectrum.**
A. A. Gusev, D. Hasegan, A. A. Marin, G. I. Pugachova.
Rev. Roumaine Phys., Vol.27, 235 - 242 (1982). – Abstr. in Phys. Abstr., Vol. 86, Abstr. 5073 (1983).

143.078 **The primary cosmic electron spectrum and its astrophysical interpretation. II. Experimental data on primary electron spectrum.**
A. A. Gusev, D. Hasegan, A. A. Marin, G. I. Pugachova.
Rev. Roumaine Phys., Vol. 27, 243 - 247 (1982). – Abstr. in Phys. Abstr., Vol. 86, Abstr. 5074 (1983).

143.079 **Traces of heavy and superheavy cosmic nuclei in olivines of extraterrestrial origin.**
R. Ignatova, D. Lkhagvasuren *(Lkhagvasurehn)*, O. Otgonsuren *(Otgonsurehn)*, P. Pelas *(Pellas)*, V. Pereligin *(Perelygin)*, K. Pero *(Perro)*, S. Stetsenko, T. Taneva, D. Yadav.
Bulgarian J. Phys., Vol. 9, 246 - 259 (1982). In Russian. Abstr. in Phys. Abstr., Vol. 86, Abstr. 10165 (1983).

143.080 **Detection of high energy neutrino sources.**
T. Kitamura.
J. Inst. Electron. Commun. Eng. Japan, Vol. 65, 534 - 537 (1982). In Japanese. – Abstr. in Phys. Abstr., Vol. 86, Abstr. 10168 (1983).

143.081 **Nuclear tracks, meteorites, and the early solar system.**
R. S. Rajan.
Trans. American Nucl. Soc., Vol. 41, 188 (1982). – Abstr. in Phys. Abstr., Vol. 86, Abstr. 10170 (1983).

143.082 **Cosmic-ray 13 day oscillation and two-sector interplanetary magnetic field.**
O. Filisetti, G. Lovera, V. Mussino.
Lett. Nuovo Cimento, Vol. 35, Ser. 2, 327 - 330 (1982).
Abstr. in Phys. Abstr., Vol. 86, Abstr. 20192 (1983).

Landolt-Börnstein. Group VI. Vol. 2c. Interstellar matter – Galaxy – Universe.
See Abstr. 003.055.

The origin of cosmic rays.
See Abstr. 011.004.

17th International Cosmic Ray Conference.
See Abstr. 012.070.

Fragmentation of ^{56}Fe nuclei on CH_2, carbon, and hydrogen targets: individual charge changing and total cross sections.
See Abstr. 022.046.

Nucleus-nucleus collisions and interpretation of cosmic-ray cascades above 100 TeV.
See Abstr. 022.068.

Nuclear reactions in Rb, Sr, Y, and Zr targets.
See Abstr. 022.178.

A measurement of cosmic-ray LET-spectra using a microprocessor supported microscope.
See Abstr. 032.522.

Isotope-measurement of cosmic ray nuclei from Ca to Fe by a plastic-electronic-experiment.
See Abstr. 032.523.

Characteristics of CR-39 for detection of cosmic ray particles.
See Abstr. 034.027.

An improved CR-39 track detector for cosmic ray applications.
See Abstr. 034.028.

A precision measuring microscope employing laser interferometry.
See Abstr. 034.031.

Cosmic ray trigger circuit for large solenoidal particle detectors.
See Abstr. 034.032.

Design optimization of liquid scintillation cosmic ray veto detector with BBQ shifter.
See Abstr. 034.033.

A large area cosmic ray isotope experiment using Cherenkov range detector techniques.
See Abstr. 034.034.

A method for measuring the isotopic abundances cosmic rays.
See Abstr. 034.035.

A pure single particle detecting counter telescope to search for fractionally charged particles.
See Abstr. 034.154.

The origin of the light elements.
See Abstr. 061.042.

Transport of neutrinos, radiation and energetic particles in accretion flows.
See Abstr. 061.057.

The theory of particle acceleration in astrophysic objects containing shock waves and turbulent plasma motio
See Abstr. 062.081.

Relativistic electrons in galaxies and their non-thermal radiation from radio to gamma radiation.
See Abstr. 066.184.

Cosmic rays and giant molecular clouds.
See Abstr. 131.214.

Secondary electron spectra in interstellar clouds, and the bremsstrahlung γ-ray luminosity.
See Abstr. 131.254.

The interrelation of cosmic γ-rays and interstellar gas in the range l: 65–180°.
See Abstr. 142.523.

Discrete sources of cosmic gamma rays.
See Abstr. 142.543.

Sources de rayonnement γ, origine du rayonneme cosmique, et régions de formation d'étoiles.
See Abstr. 142.549.

Cosmic rays, the metallicity gradient and the mas of molecular gas in the Galaxy.
See Abstr. 155.039.

Coma-type radio halos and cluster X-ray morphology.
See Abstr. 160.040.

Stellar Systems, Galaxy, Extragalactic Objects, Cosmology

151 Stellar Systems (Kinematics, Dynamics, Evolution)

151.001 **Retrograde closed orbits in a rotating triaxial potential.**
J. Heisler, D. Merritt, M. Schwarzschild.
Astrophys. J., Vol. 258, 490 - 498 (1982).

For a specific triaxial rotating potential, four sequences of closed periodic orbits have been numerically determined and their stability has been investigated by the standard Floquet method. The four sequences are: (1) The anomalous orbits which circle the long axis but are tipped to it so that they also circle the short rotation axis. (2) The unstable anomalous orbits which circle the intermediate axis but otherwise behave much like the anomalous orbits except that they are unstable. (3) The normal retrograde orbits which lie in the equatorial plane and are stable except in Binney's instability strip in which they are unstable against perpendicular perturbations. (4) The Z-axis orbits which lie on the rotation axis and are stable in their inner section, but farther out become unstable against perturbations parallel to the intermediate axis. The composite sequence may be of relevance to the dynamics of gas masses captured by rotating triaxial galaxies.

151.002 **Dissipative evolution of collisionless stellar systems. I. Cooling and heating of a stellar system by binary stars.**
L. M. Ozernoy *(Ozernoj)*, V. I. Dokuchaev.
Astron. Astrophys., Vol. 111, 1 - 15 (1982).

The aim of the paper is a detailed investigation of the formation rates for soft and hard binaries in large N-body systems as well as of the related dissipative processes, including cooling and heating of stellar systems by binaries.

151.003 **Dissipative evolution of collisionless stellar systems. II. Influence of binaries on the evolution of globular clusters and galactic nuclei.**
V. I. Dokuchaev, L. M. Ozernoy *(Ozernoj)*.
Astron. Astrophys., Vol. 111, 16 - 27 (1982).

Stellar system evolution equations taking into account evaporation of stars together with dissipation and heating effects are solved in different limiting cases, when heating of the system by binaries dominates over its cooling, or vice versa. The results are applied to consider evolution of globular clusters and galactic nuclei at stages when star collisions do not yet have a substantial effect on the evolution.

151.004 **Simulations of merging disk galaxies.**
R. T. Farouki, S. L. Shapiro.
Astrophys. J., Vol. 259, 103 - 115 (1982).

The authors attempt with N-body simulations a systematic study of the conditions under which two identical disk-halo galaxies can be expected to merge, and of the observational properties of the resulting merger remnants. A reliable criterion for the merging of these systems is formulated for the particular case when both spins are aligned with the orbital angular momentum. The hypothesis that most elliptical galaxies result from merging spiral systems is assessed.

151.005 **Gas cloud collisions in protogalaxies. I. Numerical simulations.**
C. Struck-Marcell.
Astrophys. J., Vol. 259, 116 - 126 (1982).

A series of time-dependent, numerical hydrodynamical simulations of high-velocity atomic gas cloud collisions in protogalaxies is presented. The protogalactic gas is assumed to be essentially metal-free. The processes of heating, cooling, ionization, recombination, molecule formation and destruction are included in the simulations.

151.006 **Star formation in protogalactic gas cloud collisions.**
C. Struck-Marcell.
Astrophys. J., Vol. 259, 127 - 132 (1982).

The conditions under which gas cloud collisions can induce star formation in protogalaxies or young galaxies are discussed. The specific case of fast collisions between metal-free, warm, atomic gas clouds has an enhanced chance of inducing star formation as a result of H_2 molecule formation and molecular cooling.

151.007 **Gravitational potential, space density and phase density of star clusters.**
Ü.-I. K. Veltmann.
Publ. Tartu Astrofiz. Obs., Tom 48, 232 - 261 (1981).

Models of star clusters are constructed in phase space on the basis of the gravitational potential Φ (r) or of the space density ρ (r) as functions known from observational data, r being the distance to the centre of a spherical stellar cluster.

151.008 **On the disk-thickness of spiral galaxies.**
K. Rohlfs, H.-J. Wiemer.
Astron. Astrophys., Vol. 112, 116 - 119 (1982).

The light distributions of the two edge-on galaxies NGC 4244 and NGC 5907, which seem to be pure disk systems without any noticeable trace of a bulge, have been fitted by three-dimensional models using a self-consistent disk model.

151.009 **The phase space structure of $r^{1/4}$ galaxies: Are these galaxies 'isothermal' after all?**
J. Binney.
Mon. Not. R. Astron. Soc., Vol. 200, 951 - 964 (1982).

The distribution function $f(E)$ that self-consistently generates a galaxy whose surface density profile obeys de Vaucouleurs' $r^{1/4}$ law is calculated. This distribution function is such that the number $N(E)dE$ of stars having binding energies near E is well approximated by the Boltzmann formula $N(E) = \exp(-\beta E)$. The surface density and velocity dispersion properties are presented of a galaxy whose $N(E)$ curve is exactly given by the Boltzmann formula. This new model fits the photometry of the standard elliptical NGC 3379 even better than the $r^{1/4}$ model. An interpretation is offered of the Boltzmann form of $N(E)$ in terms of the form of the density fluctuation in the early Universe which may have given rise to the modern galaxy.

151.010 **Luminosity evolution of quasars and active galaxies: theoretical models of the evolving mass supply rate.**
A. P. Lightman.
Noyaux actifs des galaxies, (see 012.004), 13 pp.

151.011 **Modèles de trous noirs massifs dans les noyaux actifs de galaxies.**
J. P. Luminet.
Noyaux actifs des galaxies, (see 012.004), 20 pp.

151.012 **A two-parameter scheme for the evolution of symmetrical galaxies.**
R. Caimmi, N. Dallaporta.
Astrophys. Space Sci., Vol. 84, 271 - 296 (1982).
A general evolutionary scheme for axisymmetrical rotationally supported equilibrium models for galaxies is considered. Its main phases are: an expansion phase of the initial protogalaxy, assumed to consist of a homogeneous gas sphere structured into clouds, from recombination to maximum expansion, during which it is surmised that angular momentum is acquired by tidal interactions by the expanding configuration; then a violent relaxation collapse phase, following maximum expansion and ending into a virialized deformed polytropic configuration.

151.013 **Chemical evolution of two-component galaxies. IV: A generalized model allowing for gas inflow.**
R. Caimmi.
Astrophys. Space Sci., Vol. 84, 373 - 400 (1982).
The author tries to insert into a single evolutionary scheme – in dealing with chemical evolution of galaxies – two different viewpoints that have been treated separately: namely, the *S* models, allowing mass conservation; and *I* models, allowing initial zero masses and no mass conservation due to gas inflow. The true evolution of a real proto-galaxy (after reaching the state of maximum expansion) is simulated. According to this scheme, the basic equations of chemical evolution are derived and models which simulate the history of solar neighborhood, other regions and galactic spheroid component are built up, in the whole range between the *S*-limit (mass conservation) and the *I*-limit (zero initial mass and subsequent accretion due to inflowing gas).

151.014 **The relaxation time of a one-dimensional self-gravitating system.**
H. L. Wright, B. N. Miller, W. E. Stein.
Astrophys. Space Sci., Vol. 84, 421 - 429 (1982).
A computer experiment on a one-dimensional self-gravitating system is described. The authors attempt to observe the dynamic time required for this system to thermalize by comparing the position and velocity densities with those predicted by the microcanonical ensemble.

151.015 **A two-component model of a spherical stellar system.**
B. P. Kondrat'ev, L. M. Ozernoy *(Ozernoj).*
Astrophys. Space Sci., Vol. 84, 431 - 446 (1982).
As a generalization of King's widely-used model for a spherical stellar system containing stars of identical masses, a two-component model of an equilibrium stellar system consisting of stars of two different masses is built proceeding from a proper generalization of the distribution function. In such a system equipartition of mean kinetic energy between light and heavy stars is absent. The radial distribution of partial density and velocity dispersion for each of the components and of the total density and velocity dispersion is obtained.

151.016 **The distribution of stars around a black hole: numerical solution of the kinetic equation with collisions.**
G. S. Bisnovatyi-Kogan *(Bisnovatyj-Kogan),*
R. S. Churayev *(Churaev)*, B. I. Kolosov.
Astron. Astrophys., Vol. 113, 179 - 191 (1982).
The distribution of stars around a massive black hole is investigated by numerical solution of the kinetic equation with the collisional term in the form of Landau-Fokker-Planck. The choice of the independent variables of the problem $(r, v_r, v_\perp)$ corresponds to the spherical symmetry in the space. The boundary distribution function is given at the outer radius of the cluster. The stellar flux into the black hole is calculated.

151.017 **Some flattened isothermal models of galaxies.**
A. Toomre.
Astrophys. J., Vol. 259, 535 - 543 (1982).
A family of exactly self-gravitating, collisionless stellar systems with axial symmetry is presented. Each model exhibits quasi-Gaussian random motions that vary not at all with distance along any given ray from its center, and the circular speed in its equatorial plane is likewise independent of the radius. Two-component versions of these scale-free models yield some instructive disk-halo systems. Their shapes and velocity fields are described in some detail.

151.018 **A numerical experiment on the equilibrium and stability of a rotating galactic bar.**
R. H. Miller, P. O. Vandervoort, D. E. Welty, B. F. Smith.
Astrophys. J., Vol. 259, 559 - 572 (1982).
A self-consistent, three-dimensional numerical experiment is performed on an N-body system whose initial state is a realization of a certain theoretical model of a rotating triaxial galaxy. The model is a stellar-dynamical counterpart of a uniformly rotating polytrope of index equal to 0.5. The equilibrium of the system and its stability are studied.

151.019 **Magnetic braking in galactic flows.**
L. S. Sparke.
Astrophys. J., Vol. 260, 104 - 110 (1982).
The nuclear fireworks of active galaxies are believed to derive their power from the kinetic energy of gas falling onto a massive central object. But this material shares the galactic rotation, and a centrifugal barrier will present it from reaching the nucleus, if its angular momentum is not removed. This paper shows that, if the large-scale galactic magnetic field has a strong enough radial component, magnetic torques can act to spin down the infalling matter. The magnetic fields in M 87 and NGC 1275, giant elliptical galaxies which are accreting from an intracluster medium, appear to be strong enough to allow magnetic braking.

151.020 **The structure and evolution of galacto-detonation waves: some analytic results in sequential star formation models of spiral galaxies.**
L. L. Cowie, G. B. Rybicki.
Astrophys. J., Vol. 260, 504 - 511 (1982).
Waves of star formation in a uniform, differentially rotating disk galaxy are treated analytically as a propagating detonation wave front. It is shown, that if single solitary waves could be excited, they would evolve asymptotically to one of two stable spiral forms, each of which rotates with a fixed pattern speed. However, the pattern of waves that develop naturally from an initially localized disturbance is more complex and dies out within a few rotation periods. These results suggest a conclusive observational test for deciding whether sequential star formation is an important determinant of spiral structure in some class of galaxies.

151.021 **Nonlinear dynamical astronomy. Applications and extrapolations.**
G. Contopoulos.
Bull. American Astron. Soc., Vol. 14, 647 (1982). – Abstract.

151.022 **The oscillations of a rotating, axisymmetric galaxy.**
P. O. Vandervoort.
Bull. American Astron. Soc., Vol. 14, 654 (1982). – Abstract.

151.023 **N-body simulation of galaxies: spiral and bar forming instabilities and disk-bulge parameters.**
R. H. Berman, J. W.-K. Mark.
Bull. American Astron. Soc., Vol. 14, 654 (1982). – Abstract.

151.024 **Three-body gravitational scattering.**
P. Hut, J. N. Bahcall.
Bull. American Astron. Soc., Vol. 14, 654 - 655 (1982). Abstract.

151.025 **The role of anisotropic propagation of star formation on the large scale structure of galaxies.**
W.-H. Chiang, B. G. Elmegreen.
Bull. American Astron. Soc., Vol. 14, 655 (1982). – Abstract.

151.026 **A dynamical basis for the classification of spiral galaxies.**
J. C. Haass, G. Bertin, C. C. Lin.
Bull. American Astron. Soc., Vol. 14, 655 (1982). – Abstract.

151.027 **Preferred planes for gas disks in the potential of a triaxial galaxy.**
T. Y. Steiman-Cameron, R. H. Durisen.
Bull. American Astron. Soc., Vol. 14, 658 - 659 (1982). Abstract.

151.028 **Three dimensional computer simulations of star formation in dwarf galaxies.**
N. Comins.
Bull. American Astron. Soc., Vol. 14, 660 (1982). – Abstract.

151.029 **The importance of tidal disruption on star formation in spiral galaxies.**
K. S. Rumstay.
Bull. American Astron. Soc., Vol. 14, 660 (1982). – Abstract.

151.030 **Dwarf elliptical galaxies.**
H. Gerola, P. Carnevali, E. E. Salpeter.
Bull. American Astron. Soc., Vol. 14, 661 (1982). – Abstract.

151.031 **Interactions between systems of dense clouds and diffuse gas in bar-driven disks.**
J. M. Huntley, H. Gerola.
Bull. American Astron. Soc., Vol. 14, 661 (1982). – Abstract.

151.032 **MHD of galaxies – arm structure and magnetic field.**
M. Tosa.
Magnetospheric phenomena of celestial bodies, (see 012.011), p. 120 - 122 (1982). In Japanese.

A MHD response of the galactic gas disc to the density wave was calculated in the presence of a bisymmetric configuration of the magnetic field. Large scale MHD shock waves are formed along the spiral arms and the magnetic field exhibits abrupt reversal in the shocked region of the spiral arm.

151.033 **Regular and irregular orbits in galactic bars.**
J. Binney.
Mon. Not. R. Astron. Soc., Vol. 201, 1 - 14 (1982).

A family of two-dimensional bar-like potentials is introduced. Potentials in this family are on the one hand related to the potential of Hénon & Heiles (1964) and to the highly regular potential studied by Binney & Spergel (1982) and by Richstone (1982). The orbital structures characteristic of different parameter values are described.

151.034 **Does Jeans' theorem apply only to spherical systems?**
J. Binney.
Mon. Not. R. Astron. Soc., Vol. 201, 15 - 19 (1982).

A general steady-state potential supports both regular and irregular orbits. The time-averaged phase-space density contributed by a particle on a regular orbit is constant on the intersection of the hypersurfaces in phase space that are picked out by the isolating integrals. From this it follows that the distribution function of a system that supports only regular orbits satisfies Jeans' theorem. Irregular orbits have complex phase-space structures. In particular the distribution functions of systems involving such orbits obey neither Jeans' theorem nor the Vlasov equation.

151.035 **Periodic orbits and gas flow in barred spirals.**
T. S. van Albada, R. H. Sanders.
Mon. Not. R. Astron. Soc., Vol. 201, 303 - 316 (1982).

A Henyey method for finding families of periodic orbits has been applied to the bar potential used by Sanders & Tubbs (1980) in two-dimensional hydrodynamical calculations. Families of periodic orbits have been found in the limits of weak and strong bars and compared qualitatively to the results of Contopoulos & Papayannopoulos (1980). Orbits in the strong bar potential are compared to the gas flow pattern determined in the earlier hydrodynamical calculations. It is shown that the pattern in which periodic orbits crowd can account for the linear off-set shocks found in the gas dynamical calculations provided that the bar is strong, that the mass distribution is centrally condensed and that the bar figure is rapidly tumbling.

151.036 **Disclinations in galactic structures.**
V. R. Parameswaran.
Speculations Sci. Technol., Vol. 4, 509 - 514 (1981). – Abstr. in Phys. Abstr., Vol. 85, Abstr. 63657 (1982).

151.037 **Percolation analysis of stochastic models of galactic evolution.**
L. S. Schulman, P. E. Seiden.
J. Stat. Phys., Vol. 27, 83 - 118 (1982). – Abstr. in Phys. Abstr., Vol. 85, Abstr. 74091 (1982).

151.038 **Vlasov equation?**
M. Hénon.
Astron. Astrophys., Vol. 114, 211 - 212 (1982).

151.039 **Evolution of the galactic shock wave in density-wave theory.**
W.-r. Hu, C. Ao.
Chinese Phys., Vol. 2, 392 - 399 (1982). – Abstr. in Phys. Abstr., Vol. 85, Abstr. 79172 (1982). – See Abstr. 29.151.045.

151.040 **Turbulence, plasma containment, and galaxies.**
C. C. Lin.
J. Res. Natl. Bur. Stand., Vol. 86, 557 - 563 (1981). – Abstr. in Phys. Abstr., Vol. 85, Abstr. 89372 (1982).

151.041 **Dynamics of elliptical galaxies and other spheroidal components.**
J. Binney.
Annu. Rev. Astron. Astrophys., Vol. 20, (see 003.009), 399 - 429 (1982).

Basic considerations. Radial structure of spheroidal components. Nonspherical systems.

151.042 **Escape velocities of interacting spherical galaxies.**
K. S. Sastry, S. M. Alladin.
Bull. Astron. Soc. India, Vol. 10, 39 (1982). – Abstract.

151.043 **Global spiral waves in disk-bulge systems.**
A. Ambastha, R. K. Varma.
Bull. Astron. Soc. India, Vol. 10, 39 - 40 (1982). – Abstract.

151.044 **Global density waves in self gravitating flat disks.**
A. Ambastha, R. K. Varma.
Bull. Astron. Soc. India, Vol. 10, 40 (1982). – Abstract.

151.045 **On the direction of rotation of the spirals in galaxies.**
I. I. Pasha, M. A. Smirnov.
Astrophys. Space Sci., Vol. 86, 215 - 224 (1982).

The detection of trailing arms in twenty spiral galaxies, completed in the 60 s, allowed to suggest the overwhelming prevalence of this sense of rotation. Pasha and Dambis (1982) have recently determined the direction of rotation in a further 54 galaxies and found two leading spirals, NGC 4490 and NGC 5395. The current sample of galaxies with an established sense of rotation contains 79 trailing and two leading systems. The two spirals found to be leading belong to M 51-type galaxies. In this paper the authors present a more detailed investigation of the system NGC 4490/85 which confirms that the arms in NGC 4490 are leading. Statistics are invoked to estimate a probable percentage X_p and an upper limit X_{up} of galaxies with leading arms. It gives $X_p \cong 3.5\%$ and $X_{up} \cong 8.2\%$.

151.046 **Periodic orbits in nearly axisymmetric stellar systems.**
N. Caranicolas, B. Barbanis.
Astron. Astrophys., Vol. 114, 360 - 366 (1982).

The authors investigate a time-independent two-dimensional dynamical system which may be considered to represent stellar motion on the plane of symmetry of a nearly axisymmetric galaxy. They find numerically the families of periodic orbits when the frequencies in the two axes are equal or nearly equal. It is shown that some features of the characteristics of periodic orbits can be explained by means of the second integral.Finally the authors compare the results of this paper with those found by Verhulst (1979) who used the theory of averaging.

151.047 **Gravothermal and gravogyro catastrophes of rotating and self-gravitating gaseous disks.**
I. Hachisu.
Publ. Astron. Soc. Japan, Vol. 34, 313 - 335 (1982).

Models of isothermal and rigidly rotating gaseous disks are calculated and their typical properties are discussed. The author formulates the linear response theory in terms of the negative specific heat and negative moment of inertia and examines the stability of these models. General properties of self-gravitating and rotating bodies are discussed on the basis of the present investigation.

151.048 **Dynamics of hot stellar systems.**
J. Binney.
Morphology and dynamics of galaxies, (see 012.032), p. 1 - 112 (1982).

Contents: The Vlasov equation. The Jeans equations. The virial equations. Jeans' theorem and its applications. Orbits in barred potentials. Non-axisymmetric stellar systems. Regularity and chaos.

151.049 **Observations of galaxy structure and dynamics.**
J. Kormendy.
Morphology and dynamics of galaxies, (see 012.032), p. 113 - 288 (1982).

Contents: Introduction. Galaxy morphology. Surface photometry of ordinary galaxies. Kinematics of ordinary galaxies. Barred galaxies. Conclusion.

151.050 **Environmental influences on galaxy structure.**
S. D. M. White.
Morphology and dynamics of galaxies, (see 012.032), p. 289 - 420 (1982).

Contents: Introduction. Techniques for studying the interactions of galaxies with their environment. The genesis of galaxies and their environment. Galaxy evolution in groups and clusters.

151.051 **Numerical gas dynamics in the galactic scales.**
S. Ikeuchi.
Prog. Theor. Phys. Suppl., No. 70, p. 261 - 285 (1981).
Abstr. in Phys. Abstr., Vol. 85, Abstr. 100729 (1982).

151.052 **The barred spiral structure of galaxies.**
Z.-c. Chen, S.-d. Weng, M. Xu.
Chinese Phys., Vol. 2, 648 - 655 (1982). – Abstr. in Phys. Abstr., Vol. 85, Abstr. 101059 (1982). – See Abstr. 29.151.044.

151.053 **The general solution of Poisson equation in thre dimensions for disk-like galaxies.**
Y. Ting, X.-t. Zheng.
Chinese Phys., Vol. 2, 656 - 663 (1982). – Abstr. in Phys. Abstr., Vol. 85, Abstr. 101060 (1982).

151.054 **Galactic shock waves and corrugation waves.**
T. Matsuda.
Prog. Theor. Phys. Suppl., No. 70, p. 249 - 260 (1981).
Abstr. in Phys. Abstr., Vol. 85, Abstr. 101071 (1982).

151.055 **On the dynamical basis of the classification of normal galaxies.**
J. Haass, G. Bertin, C. C. Lin.
Proc. Natl. Acad. Sci., USA, Vol. 79, 3908 - 3912 (1982).
Abstr. in Phys. Abstr., Vol. 85, Abstr. 106914 (1982).

151.056 **Analytic structure of the Hénon-Heiles system.**
J. Weiss.
AIP Conf. Proc., No. 88, (see 012.039), p. 243 - 277 (198
Abstr. in Phys. Abstr., Vol. 85, Abstr. 112558 (1982).

151.057 **The Jeans problem for a thin galaxy in steady st**
R. Whitley.
J. Astrophys. Astron., Vol. 3, 111 - 123 (1982).

The equation of continuity and Poisson's equation ar solved for the variation perpendicular to the Galactic plane a thin galaxy in a steady state. Simple expressions are obta for the joint density function for the mass density and velc and for the distribution function for the velocity and its moments. A formula is derived for the total mass density i the neighbourhood of the Sun.

151.058 **Global stability of disk-bulge systems: spiral stru ture of disk galaxies.**
A. Ambastha, R. K. Varma.
J. Astrophys. Astron., Vol. 3, 125 - 144 (1982).

The stability of self-gravitating annular disks surrounc the central rigid bulge component has been studied in orde to explain the transition from the tight spiral arms in Sa galaxies to rather open patterns in Sc galaxies as the centra amorphous component diminishes. Smooth spiral patterns found associated with the dominant modes of the system.

151.059 **Steady wave model of spiral galaxies and its application in cosmogony.**
R. Louise.
Astrophys. Space Sci., Vol. 86, 505 - 508 (1982).

The model is based on the assumption that logarithmi spiral arms of most spiral galaxies are due to the rotating de sity wave $\bar{\nu}(r, \theta, t)$ obeying the steady wave equation $\nabla^2 \bar{\nu}(r,\theta) + (q^2/r^2)\bar{\nu}(r, \theta) = 0$. It is shown that this equatio accounts also for the so-called Titius-Bode law in the sola system.

151.060 **Origin of stars and galaxies.**
W. H. McCrea.
Progress in cosmology, (see 012.047), p. 239 - 257 (1982).

Arguments are presented in support of the following proposition: A galaxy is produced by a collision between t portions of raw material moving under their mutual gravita tion; shocked material produces primary condensations in which short-lived exploding massive stars produce the first heavy elements; a condensation then becomes a globular cl ter of the first normal stars, the entire galaxy being thus initially a "cluster" of such clusters; tidal dissipation – and possibly other effects to be mentioned – converts most of these into a halo stellar population; subsequent evolution depends upon whether residual diffuse material falls mainly into a disk or into a nucleus.

151.061 **New developments in the theory of spiral galaxies.**
K. O. Thielheim.
Progress in cosmology, (see 012.047), p. 305 - 321 (1982).
About 30% of all galaxies exhibit spiral forms, 60% are elliptical and 10% irregular. It is the objective of galactic dynamics to explain these structural features. The author has considered a mechanism by which spiral density waves are produced in the surrounding disk as a consequence of the slow increase of the quadrupole moment of a central oval shaped equilibrium configuration immersed in the disk.

151.062 **The rotation of galaxies: clues to their formation.**
S. M. Fall.
Progress in cosmology, (see 012.047), p. 347 - 356 (1982).
The rotation of galaxies of different morphological types is closely linked with their structural features and therefore with the processes by which they formed. The author's purpose in this article is to discuss the observational results and their implications for several views of the origin and evolution of galaxies.

151.063 **Lifetime of spurs in galaxies.**
J. V. Feitzinger, H. Schwerdtfeger.
Astron. Astrophys., Vol. 116, 117 - 129 (1982).
Spurs are irregular interspiral arm structures. The authors interpret them as material features built up by stochastic star formation processes. The dynamical evolution of the spurs is calculated. They are represented by stars which move under the influence of the galactic gravitational field with a density wave potential. During their evolution the spurs merge with the preceding spiral arms; under galactic conditions they have lifetimes of about 1.8×10^8 yr in accordance with other time scales. Spurs in galaxies of different morphological types behave similarly.

151.064 **Numerical study of periodic orbit properties in a dynamical system with three degrees of freedom.**
P. Magnenat.
Celestial Mech., Vol. 28, 319 - 343 (1982).
The purpose of this paper is: (1) To make use of the knowledge as to periodic orbits gained in two-dimensional systems, in order to study the prolongation of such orbits in systems with three degrees of freedom. (2) To investigate the influence exercised on the position and stability of periodic orbits by the variation of the higher-order terms in the potential. (3) To examine the vicinity of unstable periodic orbits in the four-dimensional space of section, in order to observe the effect of instability upon the degree of stochasticity in three-dimensional systems.

151.065 **Dissipationless galaxy formation and the $r^{1/4}$ law.**
T. S. van Albada.
Mon. Not. R. Astron. Soc., Vol. 201, 939 - 955 (1982).
Results of collision-free *N*-body simulations of the collapse phase of galaxy formation are presented. A variety of irregular initial conditions produce density distributions that are in very good agreement with the $r^{1/4}$ law, provided that the collapse factor is large. This agreement holds for a region between radii containing 10 and 99 per cent of the total mass in the $r^{1/4}$ model and corresponds to a range in surface brightness of 12 mag. The central densities in the equilibrium configurations are up to a factor 1000 larger than the initial densities. The final models have a strongly anisotropic velocity distribution in their envelopes.

151.066 **The evolution of galaxies: expectations and observations.**
R. G. Kron.
Vistas Astron., Vol. 26, 37 - 65 (1982).
The light from a galaxy changes with time as new generations of stars are born and as the stellar population ages. Results from detailed studies of the solar neighborhood may be used to guess how the integrated light from other galaxies evolves. These expectations can be tested by comparison of local calibrating galaxies with distant samples seen at remote epochs.

151.067 **The preferred pitch angle of spiral galaxies: mathematical and physical implications.**
R. L. Oldershaw.
Mon. Notes Astron. Soc. South. Africa, Vol. 41, 42 - 46 (1982).

151.068 **An analytic solution for stellar space densities.**
B. C. Reed.
J. R. Astron. Soc. Canada, Vol. 76, 324 (1982). – Abstract.

151.069 **Role of random forces in stellar dynamics.**
L. G. Balázs.
Dynamical astronomy, (see 012.056), p. 143 - 153 (1982).

151.070 **Dynamics of galaxies.**
R. H. Miller.
Dynamical astronomy, (see 012.056), p. 155 - 157 (1982). Abstract.

151.071 **The attractive force between two spherical systems whose density distribution follows Schuster's law.**
J. C. Muzzio, R. E. Martinez.
Obs. Astron. Univ. Nac. La Plata, Ser. Astron., Tomo 41, 20 pp. (1982). In Spanish.
The authors find the exact formula for the attractive force between two spherical systems whose density distribution follows Schuster's law. An asymptotic expression is derived for the special case of very large distances between centers.

151.072 **Deutung der Geschwindigkeitsverteilung sonnennaher Sterne durch Diffusion von Sternbahnen.**
B. Fuchs, R. Wielen.
Mitt. Astron. Ges., Nr. 57, p. 276 (1982). – Abstract.

151.073 **Bestimmung des wahrscheinlichen Alters eines Sterns aus seiner Raumgeschwindigkeit.**
R. Wielen, B. Fuchs.
Mitt. Astron. Ges., Nr. 57, p. 276 - 277 (1982). – Abstract.

151.074 **Non-linear gross-evolution of stellar systems.**
L. P, Osipkov.
Astron. Tsirk., No. 1181, p. 4 - 6 (1981). In Russian.

151.075 **Do stellar systems with positive energy disintegrate?**
L. P. Osipkov.
Astron. Tsirk., No. 1182, p. 4 - 5 (1981). In Russian.

151.076 **Binary destruction and dynamical evolution of stellar systems.**
V. N. Sementsov.
Astron. Tsirk., No. 1199, p. 3 - 6 (1981). In Russian.

151.077 **Binary collisions in dense stellar systems.**
S. Mikkola.
Rep. Ser., Dep. Phys. Sci., Univ. Turku, FTL-R 31, 10 pp. (1982) = Turku Univ. Obs. Informo, No. 67.
Numerical integrations of random binary-binary encounters have been used to deduce experimental cross-sections for various phenomena in two-binary interactions. The strong interactions, where the outcome is two escapers and one binary, were found to be dynamically most important. Approximate expressions for the distribution of energies in

such collisions were also obtained. These results confirm the earlier estimates by other authors, that in equal mass systems binaries are not important. However, it is demonstrated that with increasing binary component mass the importance of binary reactions grows rapidly.

151.078 **Triaxial equilibrium models for elliptical galaxies with slow figure rotation.**
M. Schwarzschild.
Astrophys. J., Vol. 263, 599 - 610 (1982).

An earlier numerical construction of a triaxial model for a stellar system in dynamical equilibrium has been extended, first by adding a slow figure rotation, and second by adding rotational streaming on Z-tube orbits. The figure rotation induces weak rotational streaming, prograde on box orbits and retrograde on X-tube orbits. The resulting net streaming has a very low velocity and a complicated pattern, prograde in the inner portion and retrograde in the outer portion. The addition of prograde Z-tube orbits, circling around the short (Z) rotation axis, introduces strong rotational streaming on these orbits. The maximum net rotational streaming velocity thus permitted for the figure here used appears sufficient to model the observed rotation of most giant ellipticals.

151.079 **The warping of disk galaxies. I. Theory.**
K. A. Papp, K. A. Innanen.
Astrophys. J., Vol. 263, 639 - 653 (1982).

The authors show that if spiral galaxies formed with two inclined disks, warps in spiral galaxies are a natural result of the disks' interactions. The precession of gas clouds and their subsequent inelastic collisions determine the shape and extent of the warp. There is good agreement with the results of an analytic model and numerical experiments compared to the warps in our Galaxy and M 31. A brief argument is given showing the plausibility of galaxies forming with two inclined disks.

151.080 **The equilibrium of a galactic bar. II. Stellar-dynamical counterparts of the S-type Riemann ellipsoids.**
P. O. Vandervoort, D. E. Welty.
Astrophys. J., Vol. 263, 654 - 671 (1982).

The family of triaxial stellar systems described in Paper I of this series is generalized to include, in addition to the previous dependence on Jacobi's integral, a dependence of the distribution function on a second isolating integral of the motion of a star in the prevailing gravitational field. The second integral is approximated with the aid of a model of the stellar orbits which is valid in the absence of important resonances. The new stellar systems are stellar-dynamical counterparts of the classical S-type Riemann ellipsoids well known in the study of self-gravitating fluid systems. The study of triaxial, Riemann-like systems is related to the study of bar modes of oscillation in corresponding axisymmetric, Maclaurin-like systems. On the basis of this relationship, it is shown that an axisymmetric stellar system having the structure of a uniformly rotating polytrope of index $n = 0.5$ is dynamically unstable with respect to a bar mode if the ratio of the rotational kinetic energy to the magnitude of the gravitational potential energy exceeds 0.166.

151.081 **Stable polar gas disks in triaxial S0 galaxies.**
T. Y. Steiman-Cameron, R. H. Durisen.
Astrophys. J., Lett., Vol. 263, L51 - L54 (1982).

The authors show that the assumption of triaxiality in a nontumbling galaxy allows for the explanation of spindle galaxies as S0 galaxies with a gas disk or ring perpendicular to the stellar disk of the galaxy. The introduction of small deviations from axial symmetry allows for the availability of a significant cross section for capture of gas into a polar orbit. This cross section is available in galaxies for which the tumble period of the surface figure is much longer than the orbital period of a polar disk.

151.082 **Periodic and conditionally-periodic trajectories in conservative systems under minimal smoothness conditions. III.**
V. A. Antonov.
Vestn. LGU, 1982, No. 13, p. 86 - 96. In Russian. – Abstr. in Ref. zh., 51. Astron., 11.51.643 (1982).

151.083 **On the stability of stellar galaxy disks having differential rotation.**
S. V. Chalov.
Astron. Zh., Tom 59, 1081 - 1088 (1982). In Russian. English translation in Soviet Astron., Vol. 26, No. 6.

The problem of the stability of some models of infinitesimally thin differentially rotating stellar disks with respect to two-arm spiral disturbances is solved. Differential rotation and a spherical subsystem work for stabilizing the bar mode.

151.084 **Turbulence in a self-gravitating gaseous disk. II. Turbulent spectra.**
S. M. Churilov, I. G. Shukhman.
Astron. Zh., Tom 59, 1093 - 1102 (1982). In Russian. English translation in Soviet Astron., Vol. 26, No. 6.

The two-dimensional turbulence in a self-gravitating gaseous disk, which is a mixture of Jeans waves and vortices, is investigated. Purely Jeans turbulence spectra with constant energy flux in the inertial interval are obtained. In the general case, when both Jeans waves and vortices are excited, statistical equilibrium spectra are obtained and analyzed.

151.085 **On the nonlinear stability theory of a rotating gravitating disk.**
M. G. Abramyan.
Pis'ma Astron. Zh., Tom 8, 751 - 755 (1982). In Russian. English translation in Soviet Astron. Lett., Vol. 8.

A five degree nonlinear differential equation for the waves propagating in the plane of a rotating gaseous disk has been obtained. The criterion of the break-up instability is shown to differ from that of the cubical nonlinearity. Two-branch soliton-type solutions are obtained which differ by structure and conditions of formation as well as by their stability.

151.086 **Triaxiality and the galactic center.**
G. R. Lake, C. Norman.
The galactic center, (see 012.076), p. 189 - 193 (1982).

The authors investigate the properties of triaxial galaxies relevant to observations and inferred physical processes at the galactic center. They find: (1) velocity fields which mimic that of a supermassive object at the center, (2) the existence of stable elliptical orbits which are tilted with respect to all the symmetry planes, a possible explanation for the tilted disk observed at the galactic center, (3) the possibility of enhanced fueling due to the disappearance of all angular momentum integrals.

151.087 **Invariant surfaces and orbital behaviour in dynamical systems of 3 degrees of freedom. II.**
G. Contopoulos, P. Magnenat, L. Martinet.
Physica D, Vol. 6D, 123 - 136 (1982). – Abstr. in Phys. Abstr., Vol. 86, Abstr. 20210 (1983).

151.088 **Isolating integrals in galactic dynamics and the character of stellar orbits.**
W. I. Newman.
Nonlinear problems: present and future, (see 012.087), p. 147 - 155 (1982). – Abstr. in Phys. Abstr., Vol. 86, Abstr. 20439 (1983).

151.089 **Pair correlation in astronomical bound gravitating systems.**
K. Aizu, T. Morita.

Prog. Theor. Phys., Vol. 68, 554 - 566 (1982). – Abstr. in Phys. Abstr., Vol. 86, Abstr. 20440 (1983).

Landolt-Börnstein. Group VI. Vol. 2c. Interstellar matter – Galaxy – Universe.
See Abstr. 003.055.

Bau und Physik der Galaxis.
See Abstr. 003.108.

Geometry of the bifurcations of the normalized reduced Henon-Heiles family.
See Abstr.021.031.

Simulation of particle problems in astrophysics.
See Abstr. 021.032.

Characteristics of resonant periodic orbits.
See Abstr. 042.075.

Luminosity limits for funnels in thick accretion discs.
See Abstr. 063.048.

The passage of a star by a massive black hole.
See Abstr. 066.185.

Equations relating mean stellar proper motions to galaxies.
See Abstr. 111.017.

Globular clusters and galaxy mergers.
See Abstr. 154.002.

What has happened in the cores of globular clusters?
See Abstr. 154.029.

Dynamical evolution of globular clusters. I.
See Abstr. 154.039.

The velocity of escape from the Galaxy in the solar neighbourhood.
See Abstr. 155.036.

On the observational vulnerability of the models for the Galaxy.
See Abstr. 155.047.

IAU Commission 33: Structure and dynamics of the galactic system.
See Abstr. 155.065.

Gas motions in the central region and their interpretation.
See Abstr. 155.073.

Character of 'prolate' galaxies.
See Abstr. 158.040.

Interacting elliptical galaxies: matching models to observations.
See Abstr. 158.144.

Las masas de las galaxias espirales.
See Abstr. 158.155.

Extranuclear clues to the origin and evolution of activity in galaxies.
See Abstr. 158.173.

On the origin of the globular cluster system of M 87
See Abstr. 158.285.

Dynamics of the galaxy clusters Coma and Hydra I.
See Abstr. 160.018.

Simulations of clusters of galaxies: luminosity growth of the brightest galaxy as a function of richness.
See Abstr. 160.030.

A study of the dependence of the properties of galaxy clusters on cluster morphology.
See Abstr. 160.031.

On the mass-to-light ratio of Coma and the density of the dark matter.
See Abstr. 160.035.

Evolution of rich clusters of galaxies.
See Abstr. 160.042.

Monte-Carlo simulations of galaxy systems. IV. Static properties for galaxies in supercluster cells.
See Abstr. 160.055.

Monte-Carlo simulations of galaxy systems. V. Computer models of galaxy fields.
See Abstr. 160.056.

Formation of elliptical and spiral galaxies in rotating haloes.
See Abstr. 162.015.

Neutrino halos.
See Abstr. 162.155.

152 Stellar Associations

152.001 **Possible membership for the 51-day Cepheid GY Sagittae in a new OB association.**
D. Forbes.
Astron. J., Vol. 87, 1022 - 1028 (1982).
New *UBV* photometry and MK spectroscopy of the 51-day Cepheid GY Sagittae and 26 nearby OB stars are used to confirm the classical Cepheid nature of GY Sge and examine the possibility of membership in an anonymous OB association centered at $l = 55°$, $b = -1.5°$. Based on considerations of spatial coincidence, similar color excess, agreement in evolutionary ages, and derived absolute magnitude, it is suggested that GY Sge is a member of a loose association consisting of at least 17 luminous stars at a common distance of 3.47 ± 0.13 kpc.

152.002 **Dust cloud structure in young R associations.**
P. M. Harvey, B. A. Wilking, M. Joy.
Bull. American Astron. Soc., Vol. 14, 610 (1982). – Abstract.

152.003 **The Vela star cloud. II. Early-type stars and long-period Cepheids near the Vela pulsar.**
O. J. Eggen.
Astrophys. J., Suppl. Ser., Vol. 50, 199 - 220 (1982).
Intermediate band, Hβ and *RI* observations of some 200 stars and three Cepheids in a small region centered on the Vela pulsar have shown the presence of additional members of the Vela sheet (450 pc), as well as two more distant associations, Vel OB1 (800 pc) and Vel OB2 (1800 pc). Vel OB2 probably contains the Cepheids, SW, RZ, and SX Vel. The luminosities of the three Cepheids do not indicate a need for a major revision of the zero point of the period-luminosity relation.

152.004 **On the group formation of stars in stellar associations.**
L. V. Mirzoyan.
Binary and multiple stars as tracers of stellar evolution, (see 012.046), p. 61 - 66 (1982).
This paper deals with observational data accumulated since the discovery of stellar associations, proving the common origin of the components of stellar systems.

152.005 **Photoelectric photometry of Ap stars in the Orion and upper Scorpius associations: preliminary results.**
P. North.
Inf. Bull. Variable Stars, No. 2208, 4 pp. (1982).

152.006 **New probable members of a galactic cluster in the direction of the Large Magellanic Cloud.**
G. C. Kilambi.
Bull. Astron. Soc. India, Vol. 10, 143 - 149 (1982).
A *BV* photographic photometric analysis has established magnitudes and colours for an additional 20 stars in the vicinity of Anon. Bok group. Of these, five are new probable members of the group. An age of 500 Myr is derived for this group through the theoretical evolutionary isochrone comparison.

152.007 **Age and dimensions of star complexes.**
Yu. N. Efremov.
Pis'ma Astron. Zh., Tom 8, 663 - 670 (1982). In Russian.
English translation in Soviet Astron. Lett., Vol. 8.
Suggested is that the characteristic dimensions of OB associations are about 200 - 300 pc in our Galaxy, LMC, M 33 and M 31. The diameters of star complexes are usually twice larger than OB associations, their age dispersion as indicated by cepheids and O stars amounts up to 6×10^7 years.

152.008 **Assoziationen und Emissionsschalen in der Großen Magellanschen Wolke.**
E. Braunsfurth, J. V. Feitzinger.
Mitt. Astron. Ges., Nr. 57, p. 314 - 315 (1982).

152.009 **Association Vulpecula R1.**
N. V. Voshchinnikov.
Astron. Tsirk., No. 1200, p. 3 - 5 (1981). In Russian.

152.010 **A search for OB associations near southern long-period Cepheids. I. WZ Carinae, YZ Carinae, KK Centauri, and OO Centauri.**
S. van den Bergh, E. B. F. Brosterhus, G. Alcaino.
Astrophys. J., Suppl. Ser., Vol. 50, 529 - 549 (1982).
The authors report on a search for OB associations surrounding known Cepheids with periods longer than 11 days situated along the southern Milky Way. In the present paper they use *UBV* photometry to search for associations near WZ Car, YZ Car, KK Cen, and OO Cen. No associations were found near the first of these three Cepheids, but some possil OB stars are located in the vicinity of OO Cen.

The Hβ distance scale for B stars: the Orion association.
See Abstr. 111.004.

A search for Ap stars in the Scorpio-Centaurus association: additional evidence for a slow metal enrichment
See Abstr. 113.008.

Faint blue stars in the region of NGC 7000 = IC 5068-70.
See Abstr. 113.015.

Multiple systems of astrophysical interest. II. Mixe doubles.
See Abstr. 118.003.

Trapezium type multiple systems and formation o stars.
See Abstr. 118.027.

Radial velocity studies of spectroscopic binaries wi the objective prism II: Three new massive binaries in the Scorpius OB 1 association.
See Abstr. 120.004.

The graphite rich Cepheus OB 3 association.
See Abstr. 131.010.

Ultraviolet extinction towards far OB associations.
See Abstr. 131.318.

Polarization of starlight in R-associations. Observational data.
See Abstr. 131.319.

The gas dynamics around OB associations. I. Recombining H II regions and the formation of expanding neutral shells.
See Abstr. 132.004.

The gas dynamics around OB associations. II. Depe dence on the initial ambient density.
See Abstr. 132.005.

Radial velocities and internal motions in Galactic stellar systems.
See Abstr. 155.051.

153 Open Clusters

153.001 **Faint stars in the open cluster Trumpler 16.**
A. Feinstein.
Astron. J., Vol. 87, 1012 - 1021 (1982).
UBVRI photoelectric data on faint stars in the young open cluster Tr 16 in the magnitude range $V = 12 - 15$ are presented. Some of these stars are overluminous for their class, and others fall slightly below the reddening line for A0-type stars in the two-color diagram. The problem presented by these stars is discussed; one possibility is that they are at the stage of gravitational contraction. The new data in the $(RI)_{KC}$ system are compatible with a deviation from the normal reddening law in the region of Tr 16. Large color excesses for the member stars are found in regions where the nebula appears darker.

153.002 **Photographic *RGU* photometry of five southern open clusters in Vela II.**
L. Topaktas, R. P. Fenkart.
Astron. Astrophys., Suppl. Ser., Vol. 49, 475 - 482 (1982).
Five open clusters (Ru 77, Ru 78, Ru 83, Pi 16 and Ba 20) in the Vela II field have been investigated photometrically in the *RGU*-system. The photometric parameters (distance, reddening and earliest spectral type) are given.

153.003 **Photometry of possible members of the Hyades cluster. IV.**
E. W. Weis, A. R. Upgren.
Publ. Astron. Soc. Pacific, Vol. 94, 475 - 484 (1982).
Photoelectric photometry in *BVRI* colors has been obtained for 271 stars in the region of the Hyades cluster which are suspected to be members of that cluster on the basis of proper motion. Of these, 109 stars are judged to be probable cluster members and another 29 are considered possible members. This list and the three prior papers in this series include a total of 422 stars, of which 230 appear to be probable or possible cluster members.

153.004 **Intermediate-band photometry in the open cluster NGC 6664.**
E. G. Schmidt.
Astron. J., Vol. 87, 1197 - 1201 (1982).
Four-color and Hβ photometry is given for early-type stars in the open cluster NGC 6664. From these data the true distance modulus is found to be 10.7 and the reddening is found to range from $E(b-y) = 0.48$ to 0.63 across the cluster. There are two or perhaps three blue stragglers in the cluster and its age is estimated to be 46×10^6 yr.

153.005 **Open clusters and galactic structure.**
K. Janes, D. Adler.
Astrophys. J., Suppl. Ser., Vol. 49, 425 - 446 (1982).
The authors have compiled a catalog of open clusters with color-magnitude diagrams on the *UBV* or *RGU* systems including estimates of reddening, distance modulus, age, and number of cluster members from 610 references to 434 clusters. The catalog appears to be a representative sample of the 1200 or so known clusters in the Galaxy. The distribution of clusters in the galactic plane is dominated by the locations of dust clouds rather than by spiral structure. The youngest clusters have a patchy or clumpy distribution, suggestive of the stochastic star formation theory of spiral arms, rather than the density wave theory. The dominant pattern, however, is one of radial "spokes" centered on the Sun.

153.006 **DDO photometry of the HR 1614 moving group.**
G. H. Smith.
Proc. Astron. Soc. Australia, Vol. 4, 197 - 200 (1981).
DDO photometry is presented for 19 of the 26 stars identified by Eggen as being group members. This will provide information on their cyanogen abundances.

153.007 **The Orion Nebula Star Cluster.**
R. J. Panek.
Advances in ultraviolet astronomy, (see 012.006), p. 606 - 609 (1982).

153.008 **Integrated colors of young open clusters as a function of age.**
I. Tarrab.
Astron. Astrophys., Vol. 113, 57 - 60 (1982).
Luminosities and colors for young open clusters of known ages have been obtained. UBV colors are compared with those of theoretical model clusters. UBV RIK and 1600 Å fluxes are given as a function of age. An increasing dispersion is noticed toward long wavelengths in the color-age and in color-(B–V) diagrams.

153.009 **An X-ray survey of the Pleiades.**
J.-P. Caillault, D. J. Helfand, W. H.-M. Ku.
News Lett. Astron. Soc. N.Y., Vol. 2, No. 2, p. 21 - 22 (1982)
Abstract.

153.010 **Identification of field stars contaminating the color-magnitude diagram of the open cluster Be 21.**
C. A. Christian.
Astrophys. J., Suppl. Ser., Vol. 49, 555 - 592 (1982).
The possibility of using slitless spectroscopy to identify field stars which contaminate cluster color-magnitude diagram is explored as an alternative to astrometric and radial velocity studies. The spectra, analyzed with multivariate analysis, are combined with broad-band photometric data to determine rough spectral types and luminosity classes for individual stars. The technique is applied to the open cluster Be 21, where it is found that only the reddest stars in the magnitude-limited sample may be cluster members, presumably evolved stars.

153.011 **The age spread and initial mass function of NGC 3293: implications for star formation in clusters.**
W. Herbst.
Bull. American Astron. Soc., Vol. 14, 641 (1982). – Abstract.

153.012 **The structure and dynamics of the open cluster M 11**
R. D. Mathieu.
Bull. American Astron. Soc., Vol. 14, 646 (1982). – Abstract.

153.013 **Collinder 110: an old open cluster toward the galactic anticenter.**
D. W. Dawson, P. A. Ianna.
Bull. American Astron. Soc., Vol. 14, 646 (1982). – Abstract.

153.014 **Stellar membership and luminosity function of the Hyades cluster.**
A. R. Upgren, E. W. Weis.
Bull. American Astron. Soc., Vol. 14, 656 (1982). – Abstract.

153.015 **Simulated rotational velocity distributions in open clusters.**
J. W. Glaspey.
Bull. American Astron. Soc., Vol. 14, 663 (1982). – Abstract.

153.016 **Membership of faint galactic clusters.**
M. K. V. Bappu, G. S. D. Babu.
Bull. Astron. Soc. India, Vol. 10, 41 (1982). – Abstract.

153.017 ***UBV* analysis of the open cluster NGC 6192.**
G. Kilambi.
Bull. Astron. Soc. India, Vol. 10, 42 (1982). – Abstract.

153.018 **New *UBV* photometry for the open cluster Stock 14 and its cepheid-like variable V810 Centauri (= HR 4511).**
D. G. Turner.
Publ. Astron. Soc. Pacific, Vol. 94, 655 - 662 (1982).
New *UBV* photometry is presented for stars brighter than $V \cong 14$ lying in the central portion of the open cluster Stock 14. These data help to strengthen the cluster main-sequence fit, and result in a newly-determined cluster distance of 2.78 ± 0.12 kpc, and an inferred cluster age of ~ 6 (± 2) × 10^6 years. The derived luminosity for the member, cepheid-like, supergiant variable V810 Cen (= HR 4511 = HD 101947), which has a characteristic period of ~ 130^d, is $< M_V > = -7.9$. Its recently-discovered, hot, blue companion is comparable in luminosity to bright, giant, B-type stars lying on the evolved portion of the cluster main sequence.

153.019 **Photometry of low-mass stars in Praesepe.**
J. Stauffer.
Publ. Astron. Soc. Pacific, Vol. 94, 678 - 681 (1982).
The results of a program of *BVRI* photometry of a set of relatively faint stars in the Praesepe cluster are reported. The program stars were selected to be high-probability members of the cluster based on a new proper-motion study by Jones and Cudworth (1982). Those membership identifications appear to have been quite accurate, since all but two of the stars observed have colors and magnitudes consistent with cluster membership. The new photometry, combined with previous data from Upgren, Weis, and DeLuca (1979) and Weis (1981), is used to show that the intrinsic width of the Praesepe main sequence remains quite narrow ($\sigma_V \lesssim 0^m15$) to at least $M_v \cong 10^m5$.

153.020 ***UBVRI* photometry in the open cluster NGC 3532.**
P. Wizinowich, R. F. Garrison.
Astron. J., Vol. 87, 1390 - 1394 (1982).
New *UBVRI* photoelectric photometry is presented for 68 stars in the intermediate-age open cluster NGC 3532.

153.021 **Electronographic photometry in the galactic cluster M 37.**
A. Robin.
Astron. Astrophys., Suppl. Ser., Vol. 50, 251 - 260 (1982).
The author has measured electronographic plates of the galactic cluster M 37 in three colours of the Johnson-Morgan *UBV* system, to obtain a photometric sequence in the direction of the galactic anticenter. The limiting magnitudes of this photometry are $V = 20$, $B = 20.5$, $U = 20$, with an accuracy of 0.055 mag in V, 0.075 mag in B, and 0.10 mag in U. This sequence will be used in the future to calibrate photographic Schmidt plates of the galactic anticenter.

153.022 **Wolf-Rayet stars in open clusters and associations.**
I. Lundström, B. Stenholm.
Wolf-Rayet stars: observations, physics, evolution, (see 012.042), p. 539 - 543 (1982).
The authors have made a search for line-of-sight coincidences between WR stars and open clusters. As a measure of the separation they used the angular distance between the WR star and the cluster centre expressed with the cluster angular radius, r', as unit. By comparing the results found from the real angular distributions of WR stars and open clusters with those found if the WR stars were redistributed by changing signs in their latitudes, the authors concluded that the majority of WR stars with separations less than 3 r' are true cluster members.

153.023 **The Hyades main sequence.**
O. J. Eggen.
Astrophys. J., Suppl. Ser., Vol. 50, 221 - 239 (1982).
Intermediate band, Hβ and *RI* observations of 72 Hyad[illegible] cluster stars to $V = 11$ mag are presented and discussed. A modulus of 3.2 mag is derived from a comparison with field stars of large parallax. Observations are also presented for 98 main-sequence stars of the Hyades group that were previousl[illegible] found to be group members from kinematical considerations All of the stars discussed can be considered as members of a supercluster. A detailed examination of the evolved stars of both the nucleus and the extended cluster indicates an uppe[illegible] mass limit near 3.5 $M_\odot$ and an age spread from 2.6 to 9.7 × 10^8 yr.

153.024 **A photoelectric investigation of Ap-stars in open clusters. III. NGC 2362, NGC 2546, and NGC 322[illegible]**
H. M. Maitzen.
Astron. Astrophys., Vol. 115, 275 - 279 (1982).
The search for Ap-stars in open clusters using the photo electric approach of Maitzen (1976) is continued by an investigation of two young (NGC 2362 and NGC 2546) and o[illegible] medium age (NGC 3228) galactic clusters. No Ap-stars were detected in NGC 2362, 3 in NGC 2546 and 2 in NGC 3228. These stars were not previously known as Ap-stars. The use [illegible] $B-V$ (even of photographic photometry) is shown to be as e[illegible] fective for the determination of the Δa-index as $b-y$ in the preceding related papers. Also, systematically variable reddening has been found in NGC 2546.

153.025 **Polarization in NGC 7789 and the membership of blue stragglers.**
M. Breger.
Astrophys. J., Vol. 263, 199 - 206 (1982).
An interstellar polarization survey of the open cluster NGC 7789 has been used to determine cluster membership. The division of stars into members and nonmembers through polarization is in good agreement with proper-motion divisio[illegible] by McNamara and Solomon. Sixteen blue stragglers are foun[illegible] to be cluster members, while 18 candidates are nonmembers. The blue stragglers lie on the main sequence and are consiste[illegible] with core hydrogen burning. Blue straggler candidates to the right of the main sequence are nonmembers. A color-magnitude diagram of the cluster members of NGC 7789 is presented.

153.026 **The age spread and initial mass function of NGC 3293: implications for the formation of clusters.**
W. Herbst, D. P. Miller.
Astron. J., Vol. 87, 1478 - 1490 (1982).
UBV photographic photometry to $V \sim 16.5$ ($M_v = +3.5$) has been obtained for over 350 stars in the field of the youn[illegible] cluster NGC 3293. The main sequence of the cluster is populated with stars at least as faint as $M_v = 2.8 \pm 0.2$, implying a contraction age of at least 25 ± 10 million years. This is signif[illegible] cantly greater than its nuclear age of 6 ± 2 million years, implying that (1) there is an age spread of at least 20 ± 10 million years, and (2) the lower-mass stars formed first. The luminosity function has been derived in an approximate manner to $M_v = +5$; these data may be used to derive an initi[illegible] mass function (IMF).

153.027 **Membership of the Rosette Nebula cluster, NGC 2244.**
L. A. Marschall, W. F. van Altena, L.-T. G. Chiu.
Astron. J., Vol. 87, 1497 - 1506 (1982).
Relative proper motions for 287 stars in the region of t[illegible] young open cluster NGC 2244, located in the Rosette Nebul[illegible] have been determined from plates taken with the Yerkes 40-i[illegible] and the Allegheny 30-in. refractors. Probabilities of membership based on these proper motions are derived. Differential

extinction across the cluster, is identified as a complication in the immediate interpretation of the sample in terms of color-magnitude diagrams or luminosity functions.

153.028 **Observations of low-mass stars in the Pleiades: has a pre-main sequence been detected?**
J. R. Stauffer.
Astron. J., Vol. 87, 1507 - 1514 (1982).
New *BVRIJHK* photometry and spectroscopy of a large sample of faint, proper motion members of the Pleiades cluster are presented. Those data indicate that a significant number of those stars appear to lie on the main sequence to at least $V \cong 17^m\!.0$. Another well-populated sequence of stars is displaced above the main sequence by $\Delta V \cong 0^m\!.7$. Based on a detailed comparison of the photometric and spectroscopic properties of the stars in those two sequences, the brighter sequence appears likely to be composed of binaries rather than pre-main-sequence stars.

153.029 **Infrared photometry of Hyades dwarfs.**
B. W. Carney.
Astron. J., Vol. 87, 1527 - 1537 (1982).
JHK infrared photometry is presented for 76 main-sequence proper motion members brighter than $V = 12$ in the Hyades. Known binaries are confirmed and new candidates proposed using color-magnitude and color-color diagrams. The cluster binary frequency (~ 40%) is discussed, as well as its possible effects on the cluster distance modulus and stellar chromospheric/coronal activity.

153.030 **Spectroscopic identification of white dwarfs in galactic clusters. II. NGC 2516.**
D. Reimers, D. Koester.
Astron. Astrophys., Vol. 116, 341 - 347 (1982).
The authors have searched deep UV and red Schmidt plates of the intermediate age (turnoff mass around 5 $M_\odot$), rich southern cluster NGC 2516 for white dwarf candidates and found 9 candidates. Subsequent spectroscopic observations confirmed the white dwarf (DA) nature of 6 of these stars. A model atmosphere analysis shows that three of them are hot (T_{eff} = 30,000, 36,000 and 33,500 K) high-gravity (log g = 8.6, 8.9, and 9.0) DA white dwarfs and that their luminosities and cooling ages are consistent with cluster membership. The others are cooler (17,000, 10,000, and 9500 K) white dwarfs which are most probably foreground stars according to distance moduli and cooling ages. The authors model the expected number of evolved stars by means of the observed cluster luminosity function and evolutionary calculations and find a value of M_{WD}, the upper initial mass limit for formation of white dwarfs, of $M_{WD} = 8^{+3}_{-2}\, M_\odot$.

153.031 **Verändert sich die mittlere Größe der offenen Sternhaufen mit der galaktozentrischen Entfernung?**
B. A. Balázs.
Mitt. Astron. Ges., Nr. 57, p. 284 - 287 (1982).

153.032 **Be stars in open clusters.**
J. C. Mermilliod.
Be Star Newsl., No. 6, p. 26 (1982).

153.033 **Über die Struktur des offenen Sternhaufens NGC 6940 und die Entwicklungsphasen seiner Mitglieder.**
W. Götz.
Veröff. Sternw. Sonneberg, Band 9, 311 - 324 (1981).
Statistical investigations of the old open star cluster NGC 6940 served to complete, enlarge and test the cosmogonic and genetic relations, found in other clusters. The structure of the star cluster and the behaviour of the astrophysical parameters in the system were investigated in detail.

153.034 **Zum Verhalten der helleren Doppelsterne und der gelben Riesensterne in offenen Sternhaufen.**
W. Götz.
Veröff. Sternw. Sonneberg, Band 9, 325 - 359 (1981).
This investigation concerns the behaviour of brighter binaries and yellow giants, and groups of massive stars in open clusters. Results are given about the evolutionary phases of these stars and groups and their positions in different star clusters. It is shown that the high concentration of binaries and yellow giants toward the center of the open cluster NGC 2632 (Praesepe) is not an individual case. It was also found in the Hyades and the Coma Berenices cluster (clusters of nearly equal age) but not in younger and older objects.

153.035 **IAU numbers for some recently discovered clusters.**
G. Lyngå.
Bull. Inf. Cent. Données Stellaires, No. 23, p. 89 (1982).

153.036 **Berkeley 87, a heavily-obscured young cluster associated with the ON2 star-formation complex and containing the WO star Stephenson 3.**
D. G. Turner, D. Forbes.
Publ. Astron. Soc. Pacific, Vol. 94, 789 - 801 (1982).
UBV photometry was obtained for 105 stars (39 observed photoelectrically and 66 observed photographically) lying in the sparsely-populated open cluster Berkeley 87. The purpose of the study is to present these new data, and to discuss what they reveal about the distance and age of Berkeley 87, about the intrinsic properties of several of its individual member stars (including Stephenson 3), and about the nature of the molecular cloud complex surrounding the cluster.

153.037 **Spectral morphology in Trumpler 16.**
H. Levato, S. Malaroda.
Publ. Astron. Soc. Pacific, Vol. 94, 807 - 810 (1982).
The authors classified 32 stars in the field of Trumpler 16 and they found one composite spectrum, another two possible double-line spectroscopic binaries, one star with a weak shell, three stars with broad hydrogen lines, and one star with a peculiar spectrum (127) still uninterpreted. A binary nature for this object cannot be ruled out.

153.038 **Sequenze di magnitudini negli ammassi della Chioma di Berenice e Messier 39 (NGC 7092).**
G. Favero, I. Dalmeri.
Coelum, Vol. 50, 325 - 330 (1982).

153.039 **The helium content of solar type stars in the Hyades.**
J. Hardorp.
Smithsonian Astrophys. Obs., Spec. Rep. 392, Vol. II, (see 012.065), p. 199 - 202 (1982).
The primary mass of the eclipsing double-lined spectroscopic binary Van Bueren 22 (McClure 1982) shows that Hyades stars are too massive for their luminosities when compared with the Sun. This can be entirely accounted for by the computed evolution of the Sun and the observed metal enhancement of the Hyades if the distance modulus is 3.40. Reliable temperature determinations of Hyades G dwarfs by Branch et al. (1980) and Hardorp (1980) can then be used to put stringent limits on the helium content of the Hyades: $Y = Y_\odot - 0.01 \pm 0.01$. Distance moduli derived from matching observed masses of visual binaries to that of vB 22 are briefly discussed.

153.040 **Determination of the membership of the open cluster M 67.**
J.-l. Zhao, K.-p. Tian, Z.-h. Xu, M.-g. Yin.
Acta Astron. Sinica, Vol. 23, 135 - 139 (1982). In Chinese.
Using the relative proper motions of 1067 stars in the region of the open cluster M 67, 375 cluster member stars are found by means of the maximum-likelihood principle. The

membership based upon the discriminatory analysis principle is 412. In addition, a comparison with Sander's results has been made.

153.041 **Color-magnitude diagrams and ages of two young Magellanic Cloud clusters.**
P. J. Flower.
Publ. Astron. Soc. Pacific, Vol. 94, 894 - 900 (1982).

The principal aim of this paper is to determine the ages of two Large Magellanic Cloud star clusters, NGC 2058 and NGC 2065. Photoelectric and photographic photometry was used to obtain color-magnitude diagrams of both clusters and a nearby star field. A grid of evolutionary tracks was computed for masses bracketing those of the cluster giants. Both the red-giant distribution and the main-sequence termination were compared to the models and yielded an age of 1.2×10^8 yr for both clusters.

153.042 **The distant open cluster Tombaugh 2.**
D. S. Adler, K. A. Janes.
Publ. Astron. Soc. Pacific, Vol. 94, 905 - 909 (1982).

The authors have obtained a preliminary color-magnitude diagram of the distant open cluster Tombaugh 2 using plates from the CTIO 4-meter telescope. Interstellar reddening and the distance modulus for To 2 were derived. To 2 is about 18.5 kpc from the galactic center and -1.69 kpc from the galactic plane. It is thus one of the most distant open clusters both from the galactic plane and from the galactic center.

153.043 **The F0 Ia stars in NGC 457 and IC 2581.**
O. J. Eggen.
Publ. Astron. Soc. Pacific, Vol. 94, 952 - 959 (1982).

Intermediate-band and Hβ observations of stars in NGC 457 and IC 2581 are presented and discussed. A mean modulus of $11^m.25 \pm 0^m.20$ is derived for NGC 457. The observations of IC 2581 give a mean modulus of $12^m.20 \pm 0^m.25$. The F0 Ia star ϕ Cas in NGC 457 has $M_V = -7^m.9$ and HR 4110 (F0 Ia) in IC 2581 has $M_V = -8^m.8$.

153.044 **On the possibility of cluster membership for the cepheid V Centauri.**
D. G. Turner.
Publ. Astron. Soc. Pacific, Vol. 94, 1003 - 1007 (1982).

UBV photometry is presented for ten stars lying within 5 arc min of the bright $5^d.5$ cepheid V Cen. An analysis of these data and similar published data for members of the nearby cluster NGC 5662 leads to a newly-determined distance of 661 ± 24 pc for the cluster, and indicates that V Cen is an excellent candidate for membership in the cluster corona.

Remarks about the cataloguing of open cluster data.
See Abstr. 002.034.

A proper motion survey in the area of the galactic cluster in Coma Berenices.
See Abstr. 111.012.

Proper motion survey with the 48-inch Schmidt telescope. LX. Proper motions for 7120 stars in the Hyades.
See Abstr. 111.019.

Proper motion survey with the 48-inch Schmidt telescope. LXII. Proper motions for 374 faint stars. The alleged cluster near the North Galactic Pole.
See Abstr. 111.021.

***UBV*-Hβ photometry of luminous stars between $l = 335°$ and $l = 6°$.**
See Abstr. 113.031.

Preliminary results of a new study of the intrinsi properties of Wolf-Rayet stars.
See Abstr. 113.032.

Photoelectric sequences for three open clusters: C0126+630 (NGC 559), C2137+572 (Tr 37), C2152+62 (NGC 7160).
See Abstr. 113.048.

Photographic photometry of stars in the region o the open cluster IC 4996 in the Vilnius photometric system
See Abstr. 113.063.

The photometric variability of solar-type stars. I. Preliminary results for the Pleiades, Hyades, and the Malmquist Field.
See Abstr. 113.069.

Progress report of an IUE survey of the Hyades star cluster.
See Abstr. 114.026.

Low-resolution Mg II h and k observations of sol type stars.
See Abstr. 114.053.

Two-dimensional spectral classifications for O st in the southern Milky Way.
See Abstr. 114.081.

Structure and evolution of large groups of massiv close binary systems.
See Abstr. 117.217.

Trapezium type multiple systems and formation stars.
See Abstr. 118.027.

Radial velocity studies of spectroscopic binaries with the objective prism I: Improved orbital elements for binaries in and near NGC 2516.
See Abstr. 120.002.

Variable polarization in HD 162679.
See Abstr. 120.027.

Flare stars in the Pleiades.
See Abstr. 122.107.

Westerbork and VLA observations of G 127.1 + (
See Abstr. 125.009.

Interstellar extinction and space densities of stars the direction of the open cluster IC 4996 in Cygnus.
See Abstr. 131.306.

An X-ray flare in the Hyades binary HD 27130.
See Abstr. 142.147.

The Vela star cloud. II. Early-type stars and long-period Cepheids near the Vela pulsar.
See Abstr. 152.003.

Radial velocities and internal motions in Galactic stellar systems.
See Abstr. 155.051.

The cluster system of M 33.
See Abstr. 158.032.

Erratum

153.901 **Erratum: "Stellar coronae in the Hyades: a soft X-ray survey with the Einstein Observatory"** [Astrophys. J., Vol. 249, 647 - 661 (1981)].
R. A. Stern, M.-C. Zolcinski, S. K. Antiochos, J. H. Underwood.
Astrophys. J., Vol. 258, 904 (1982). – See Abstr. 30.153.023.

154 Globular Clusters

154.001 **The structure and stellar content of globular clusters. II. Star counts and main-sequence luminosity functions for 47 Tuc, NGC 6397, and NGC 6752.**
G. S. Da Costa.
Astron. J., Vol. 87, 990 - 1011 (1982).
Extensive star count data are presented for three nearby globular clusters: 47 Tuc, NGC 6397, and NGC 6752. At the limiting magnitude of the deepest plates, the stars counted have masses approximately half that of the cluster giants. The counts are used to determine the apparent shapes of the clusters, to compare radial density profiles of stars of different mass and to investigate main-sequence luminosity functions. Clear evidence is found for the presence of mass segregation in 47 Tuc; the less massive stars are less centrally concentrated than the cluster giants and upper main-sequence stars. The main-sequence luminosity functions for the clusters are found to be different. While part of the differences between the observed luminosity functions is undoubtedly due to mass segregation effects, it is suggested that real differences between the current mass functions of the clusters are present.

154.002 **Globular clusters and galaxy mergers.**
S. van den Bergh.
Publ. Astron. Soc. Pacific, Vol. 94, 459 - 464 (1982). – Paper presented at the M. I. T. Workshop on Interacting Galaxies held in Cambridge, Mass., 1981 November 19 - 20.
The specific cluster frequency in galaxies of type Sb is found to be an order of magnitude lower than that in ellipticals. This observation shows that typical E galaxies cannot have formed by mergers of spirals. Presently available observations suggest, but do not yet prove, that the spheroidal-bulge components of spiral galaxies have a lower specific globular-cluster frequency than do typical elliptical galaxies. Arguments are given which militate against the suggestion that spirals were formed from the merger of ancestral dwarf galaxies.

154.003 **Structural characteristics of Palomar 14.**
W. E. Harris, G. L. H. Harris, R. Canterna.
Publ. Astron. Soc. Pacific, Vol. 94, 465 - 470 (1982).
The authors describe new star counts around the sparse, distant outer halo cluster Palomar 14, which are used to derive its structural properties, i. e., core and tidal radii and central surface density. The relationship of these parameters to the galactic orbit of the cluster is discussed. Contrary to the recent study of Da Costa et al., the authors find that no useful estimates can be made for the perigalactic distance or orbital eccentricity of Pal 14.

154.004 **Mass loss and the H-R diagram of globular clusters.**
R. C. Peterson.
Astrophys. J., Vol. 258, 499 - 505 (1982).
New high-resolution observations of Hα in fainter giants in the globular clusters M5 and M71 fail to show the emission wings indicative of mass loss, which are apparently confined to the brightest giants. These and additional new data establish that the H-R domain of Population II stars which do show optical evidence for mass loss is very similar to that of Population I stars. The fact remains, however, that many stars within this domain show no such evidence.The randomness of the process recommends it as perhaps responsible for the spread in mass on the horizontal branch, and for anomalies in the ratio of blue to red horizontal branch stars in clusters of moderate metallicity.

154.005 **Abundance indicators from vidicon spectra in six globular clusters.**
R. Canterna, W. E. Harris, T. Ferrall.
Astrophys. J., Vol. 258, 612 - 627 (1982).
Spectroscopic scans with the CTIO vidicon systems are described for a total of 63 stars in six globular clusters, including 47 Tuc, NGC 288, NGC 362, M 15, M 2 and Pal 12. Several abundance-sensitive features (calcium lines, iron lines, G band, CN bands) are synthesized from the spectra and used to determine the relative compositions of the clusters. 47 Tuc is clearly the most metal rich system in the sample, followed in order of decreasing [Fe/H] by NGC 362, Pal 12, NGC 288, M 2, and M 15. NGC 288, NGC 362, and Pal 12 all have nearly similar metallicities but strongly different average ultraviolet CN strengths. The giants in both 47 Tuc and NGC 362 have similar average CN (λ 3883) band strengths and also variable star-to-star CN strengths. The more metal poor clusters show smaller CN variations.

154.006 **The metal abundance of Pal 13.**
R. Zinn, A. I. Díaz.
Astron. J., Vol. 87, 1190 - 1196 (1982).
Low-resolution spectrograms have been obtained of the three RR Lyrae variables in the distant and very sparse globular cluster Pal 13. A comparison of these spectrograms with similar ones of several RR Lyrae variables in the globular clusters M4, M5, and M22 reveals that Pal 13 is intermediate to M5 and M22 in metal abundance. A value of $[Fe/H] = -1.67 \pm 0.15$ is obtained for Pal 13. Pal 13 is another example of a distant halo object that is not extremely metal poor.

154.007 **The globular cluster metallicity scale: evidence from stellar models.**
P. Demarque, C. R. King, A. Diaz.
Astrophys. J., Vol. 259, 154 - 158 (1982).
Theoretical giant branches have been constructed to determine their relative positions for metallicities in the range $-2.3 \leqslant [Fe/H] \leqslant 0$. A relation between [Fe/H] and $(B-V)_{o,g}$ based on these models is presented which yields good agreement over the oberserved range of metallicities for galactic globular clusters and old disk clusters. The metallicity of 47 Tuc and M 71 given by this calibration is about –0.8 dex.

154.008 **Nitrogen variations on the main sequence of 47 Tucanae: implications from stellar structure theory.**
G. S. Da Costa, P. Demarque.
Astrophys. J., Vol. 259, 193 - 197 (1982).
The authors present evolutionary calculations of a 0.9 $M_{\odot}$ metal-poor star in which the outer layers are artificially mixed throughout the main-sequence lifetime. They show that

the observations of large nitrogen abundance variations among main-sequence stars and subgiants in the globular cluster 47 Tuc are unlikely to be the result of deep envelope mixing on the main sequence. It is concluded that the most plausible explanation of the observed variations is that they are primordial in origin, that is, they reflect the presence of nitrogen abundance inhomogeneities in the protocluster gas cloud.

154.009 **Carbon and nitrogen abundances in giant stars of the metal-poor globular cluster M 92.**
D. F. Carbon, G. E. Langer, D. Butler, R. P. Kraft, N. B. Suntzeff, E. Kemper, C. F. Trefzger, W. Romanishin.
Astrophys. J., Suppl. Ser., Vol. 49, 207 - 258 (1982) = Lick Obs. Bull., No. 812.
The authors report the results of a study of C and N abundances in M 92 giants, based on low resolution photoelectric spectral scans of more than 40 individual cluster members, some with absolute magnitudes as faint as $M_v \approx +1.7$. They observed both the G band (CH, $\Delta v = 0$) and the $\Delta v = 0$ bands of NH near λ 3360. Values of [C/H] and [N/H] were estimated by comparing computed and observed stellar spectra using the method of spectral synthesis based on model atmospheres. The study shows that the atmospheric C and N abundances vary from star to star in a much more complex way than earlier supposed.

154.010 **Multiaperture *JHK* photometry of the globular clusters in the Fornax dwarf spheroidal galaxy.**
T. J. Jones, A. R. Hyland.
Mon. Not. R. Astron. Soc., Vol. 200, 509 - 520 (1982).
The *JHK* colours for the Fornax globular clusters 3, 4 and 5 through 12, 18 and 30 arcsec apertures and cluster 2 through an 18 arcsec aperture are presented. The infrared colours through these apertures are consistent with published optical photometry and metal abundance determinations. Cluster 4, and to a lesser extent clusters 3 and 5, become redder in *J–H* with increasing aperture size. This is best explained by an excess of one or more late K giants to the outer portions of the clusters. A systematic difference in the *H–K* colour between the authors' photometry and the photometry of others for extended objects through large apertures is discussed.

154.011 **Photoelectric photometry of globular clusters in the Andromeda nebula. VIII.**
A. S. Sharov, V. M. Lyutyj, V. F. Esipov.
Pis'ma Astron. Zh., Tom 8, 451 - 453 (1982). In Russian. English translation in Soviet Astron. Lett., Vol. 8.
Results of photoelectric *UBV* photometry of 58 globular clusters in the Andromeda nebula obtained in 1981 are presented.

154.012 **Photometry in the central region of the globular cluster NGC 7099.**
G. Alcaino, W. Wamsteker.
Astron. Astrophys., Suppl. Ser., Vol. 50, 141 - 146 (1982).
The results of iris photometry in the *UBV* system on short exposure plates of M 30 (NGC 7099) are discussed. The photometry is given for the bright stars ($V < 15.3$) within $10'' < r < 90''$ from the cluster center. A factor-of-two deficiency of blue horizontal branch stars to red giant stars is found for the central part of the cluster, compared to similar evaluations made at $80'' < r < 300''$. These ratios render values of helium content which range from $Y \sim 0.27$ in the outer regions to $Y \sim 0.12$ in the central region.

154.013 **Search for (globular) clusters in M31. III. Structural properties: X-ray sources and comparison with Galactic globulars.**
P. Battistini, F. Bònoli, R. Buonanno, C. E. Corsi, F. Fusi Pecci.
Astron. Astrophys., Vol. 113, 39 - 45 (1982).
Exploiting the information deduced from the fit applied to the PDS scans of each cluster candidate in M31 listed by Battistini et al. (1980), and through a simulation made with the data known on the Galactic globulars, the authors have obtained two sets of homogeneous data on some basic parameters for a wide sample of clusters in the two systems including a significant number of X-ray clusters.

154.014 **The dynamics and metallicity of the M 31 globular cluster system.**
J. Huchra, J. Stauffer, L. Van Speybroeck.
Astrophys. J., Lett., Vol. 259, L57 - L61 (1982).
The authors have obtained digital, low-dispersion spectra of 61 globular clusters in M 31 with the Multiple Mirror Telescope. Preliminary analyses of these spectra show that: (1) the M 31 globular cluster system rotates with a velocity amplitude of approximately 160 ± 40 km s^{-1} peak to peak, and (2) there is a measurable but small metallicity gradient exhibited by those clusters.

154.015 **Globular clusters in the Sculptor Group galaxy NGC 55.**
G. S. Da Costa, J. A. Graham.
Astrophys. J., Vol. 261, 70 - 76 (1982).
SIT vidicon and photographic observations are presented of three globular cluster candidates in the field of the nearby edge-on late-type spiral galaxy NGC 55. All three candidates, which appear nonstellar on plates taken in excellent seeing, have radial velocities that agree with that of NGC 55. The spectrum of the brightest cluster shows strong hydrogen lines which, together with photoelectrically measured colors, suggests that this cluster is a young globular cluster analogous to those found in the LMC. The spectra of the other two clusters are consistent with the interpretation that they are old and metal deficient.

154.016 **Optical and ultraviolet observations of the X-ray globular cluster Bo 158 in M 31.**
C. Cacciari, A. Cassatella, L. Bianchi, F. Fusi Pecci, R. G. Kron.
Astrophys. J., Vol. 261, 77 - 84 (1982).
Medium-resolution optical spectra, two IUE low-resolution ultraviolet spectra, and the available optical data have been used to determine the population content of the X-ray globular cluster Bo 158 in M 31. Optical, UV, and X-ray fluxes are all consistent with those derived for the X-ray globular clusters in the Galaxy.

154.017 **Analysis of integrated spectra of red star clusters in the Magellanic Clouds.**
D. Rabin.
Astrophys. J., Vol. 261, 85 - 101 (1982).
SIT digital spectra of the integrated light of 16 "red" star clusters in the Magellanic Clouds exhibit a wide variation of hydrogen line strength from cluster to cluster. In a plot of Balmer line against Ca K line equivalent width, almost all of the clusters in the sample lie apart from an empirical sequence of Galactic globular clusters. At given metallic line strengths, the Cloud clusters have stronger Balmer lines. This is interpreted as indicating that few among the populous clusters in the Magellanic Clouds are as old as Galactic globular clusters.

154.018 **Approaching the main sequence of NGC 6712.**
D. H. Martins, D. A. Fraquelli.
Bull. American Astron. Soc., Vol. 14, 663 (1982). – Abstract.

154.019 **Optical and ultraviolet spectrophotometry of globular clusters in the Magellanic Clouds.**
C. Cacciari, F. Fusi-Pecci, C. Zavaroni, K. C. Freeman, A. Cassatella, P. Benvenuti, L. Bianchi, A. Heck, P. Patriarchi, W. Wamsteker.
Third European IUE Conference, (see 012.029), p. 519 - 522

(1982).

Preliminary results are presented on a program of ultraviolet and optical spectrophotometry of globular clusters in the Magellanic Clouds. Comparison is made with previous studies on Galactic and MC globulars, reconfirming the differences in the formation and dynamics of the parent galaxies.

154.020 **UV observations of young blue globular clusters in the LMC.**
A. Cassatella, E. H. Geyer.
Third European IUE Conference, (see 012.029), p. 523 (1982). Abstract.

154.021 **Optical and ultraviolet observations of the X-ray globular cluster Bo 158 in M 31.**
C. Cacciari, A. Cassatella, L. Bianchi, F. Fusi-Pecci, R. G. Kron.
Third European IUE Conference, (see 012.029), p. 525 - 528 (1982).

Medium resolution optical spectra, two IUE low resolution ultraviolet spectra, and the available optical data have been used to determine the population content of the X-ray globular cluster Bo 158 in M 31. Optical, UV and X-ray fluxes are all consistent with those derived for the X-ray globulars in the Galaxy. The ratio of the fluxes at $\lambda 3000$ and $\lambda 1500$ suggests for Bo 158 a morphology of the HR diagram with an extremely blue Horizontal Branch.

154.022 **Metal abundances of RR Lyrae stars in globular clusters.**
H. A. Smith, G. J. Perkins.
Astrophys. J., Vol. 261, 576 - 585 (1982).

The spectroscopic metallicity index ΔS has been determined for RR Lyrae stars in 13 globular clusters. These determinations, together with previous measures of ΔS for globular cluster stars, provide [Fe/H] values for 22 globular clusters. The [Fe/H] values argue strongly against a simple rescaling of globular cluster metal abundances to meet the very low iron abundances which have recently been found for M 71 and 47 Tuc from high dispersion spectroscopy. At the same time, published globular cluster chemical abundances which are derived from high dispersion spectroscopy are consistent with the contention that ΔS can be used to predict [Fe/H] along with [Ca/H].

154.023 **Does the Galaxy contain two populations of globular clusters?**
S. van den Bergh.
Publ. Astron. Soc. Pacific, Vol. 94, 640 - 641 (1982).

It is shown that extended globular clusters with core radii $r_c > 3$ pc have a luminosity function which differs at the 99% confidence level from that of more compact galactic globulars. It is speculated that extended globulars, which comprise ~15% of the galactic cluster population, might be distantly related to dwarf spheroidal galaxies.

154.024 **TiO band strengths in metal-rich globular clusters. V. 47 Tucanae.**
J. R. Mould, M. S. Bessell.
Astrophys. J., Vol. 262, 142 - 144 (1982).

Narrow-band photometry has been carried out on TiO band strengths of stars near the tip of the giant branch in the globular cluster 47 Tuc. The results are consistent either with a uniform metal deficiency $[M/H] = -0.5 \pm 0.1$ or, if oxygen is overabundant as implied by high-dispersion analyses, with $[Ti/H] = -0.65 \pm 0.2$.

154.025 **X-ray illumination of globular cluster puzzles.**
A. P. Lightman, J. E. Grindlay.
Astrophys. J., Vol. 262, 145 - 152 (1982).

The authors enumerate many of the puzzles presented by globular clusters, regarding their gross features, their internal dynamics, and their X-ray emission. They focus on the questions: Why do globular clusters provide a favorable environment for forming discrete X-ray sources? Are X-ray emitting globular clusters (XEGC) unusually near the galactic center, and if so, why? Why do the observed clusters never contain more than a single discrete X-ray source? A statistical analysis of the data allows plausible explanations of these puzzles.

154.026 **Abundances in metal-poor stars. I. The globular clusters NGC 2808, NGC 3201, NGC 6397, and M 22.**
R. G. Gratton.
Astron. Astrophys., Vol. 115, 171 - 184 (1982).

Abundances are derived for giant stars in the globular clusters NGC 2808, NGC 3201, NGC 6397, and M 22 by means of Echelle high dispersion spectra and a model atmosphere analysis. The data confirms the globular cluster metallicity scale found by means of previous high dispersion study. There are no obvious differences in composition between metal rich globular clusters and metal-poor galactic clusters.

154.027 **A statistical method for determining ages of globular clusters by fitting isochrones.**
B. P. Flannery, B. C. Johnson.
Astrophys. J., Vol. 263, 166 - 186 (1982).

The authors describe a statistical procedure to compare models of stellar evolution and atmospheres with color-magnitude diagrams of globular clusters. The technique allows one to determine parameters of the model, their uncertainty, and to assess goodness of fit. The method is applied to extensive observations of NGC 6752 and M 5, and to smaller data sets for the clusters M 3, M 5, M 15, and M 92. These data result in a spread in age of 5–8 Gyr (1 Gyr = 10^9 yr). The mean age is generally younger by 2–3 Gyr than previous estimates. Two uncalibratable sources of systematic error make the results suspect. These are uncertainty in the stellar temperatures induced by choice of mixing length, and known errors in the atmospheres. These effects could reduce age estimates by an additional 5 Gyr.

154.028 **Abundances of the elements in six stars in the globular cluster M 22.**
C. Pilachowski, G. Wallerstein, E. M. Leep, R. C. Peterson.
Astrophys. J., Vol. 263, 187 - 198, plate 8 (1982).

The authors have analyzed six stars near the red giant tip of M 22 using KPNO echelle spectra and Palomar 5 m coudé plates. The iron abundances range from $[Fe/H] = -1.4$ to -1.9. Much of this spread is real. All six stars show an enhancement of oxygen relative to iron. Calcium and titanium also show enhancements when compared with iron, as is common in metal-deficient giants and dwarfs.

154.029 **What has happened in the cores of globular clusters?**
A. P. Lightman.
Astrophys. J., Lett., Vol. 263, L19 - L22 (1982).

Some striking aspects of the observed data for globular cluster cores cannot be easily explained in terms of the existing theory of dynamical evolution. Observed cluster cores have an extraordinarily small range in mass for their range in central density. When evolved backward in time, these cluster cores have an unnatural initial distribution, indicating an incompleteness in the theory.

154.030 **Variable stars in the globular cluster Messier 28 (NGC 6626).**
A. Wehlau, H. Sawyer Hogg.
Inf. Bull. Variable Stars, No. 2171, 2 pp. (1982).

154.031 **Une étude photographique de NGC 376, 416 et 419 dans le Petit Nuage de Magellan.**

D. Durand, E. Hardy.
J. R. Astron. Soc. Canada, Vol. 76, 319 (1982).

154.032 **Strong CN stars in the globular cluster NGC 1851.**
J. E. Hesser, R. A. Bell, R. D. Cannon, G. L. H. Harris.
Astron. J., Vol. 87, 1470 - 1477 (1982).

Twenty-three stars near the southern globular cluster NGC 1851 (C0 512-400) have been observed with the SIT spectrograph on the CTIO 4-m telescope. Three out of eight bright red giant stars possess extremely strong λ 3883 and λ 4216 CN bands, very similar to those in ROA 253, a strong CN star in ω Cen. Their high radial velocities make these stars almost certainly cluster members. Comparison of the spectra with synthetic spectra calculated for a range of C and N abundances and a solar oxygen ratio suggest that the three CN strong stars may be carbon stars. The question of cluster membership is discussed for a group of possible asymptotic-giant-branch stars.

154.033 **Variable stars in the globular cluster NGC 6293.**
C. M. Clement, P. S. Panchhi, T. R. Wells.
Astron. J., Vol. 87, 1491 - 1496 (1982).

Three new variables have been found in NGC 6293. Of 11 variables in and around the cluster, periods have been determined for nine. Five of these are RR Lyrae variables, one is a Cepheid, and three are Mira variables. Four of the RR Lyrae stars may be members. A distance of 7.8 kpc has been determined for NGC 6293.

154.034 **IUE – UV – observations of young globular clusters of the Large Magellanic Cloud.**
E. H. Geyer, A. Cassatella.
Mitt. Astron. Ges., Nr. 57, p. 303 - 304 (1982).

154.035 **Kohlenstoff- und Stickstoffhäufigkeiten in Riesensternen der Kugelhaufen M 92, M 15, M 13, M 3 und in Halo-Feldriesen.**
R. P. Kraft, C. F. Trefzger, G. Langer, N. Suntzeff, D. Carbon.
Mitt. Astron. Ges., Nr. 57, p. 310 - 311 (1982).

154.036 **Variations of the ellipticities within some globular clusters.**
E. H. Geyer, B. Nelles, U. Hopp.
Mitt. Astron. Ges., Nr. 57, p. 311 - 312 (1982).

154.037 **On the structure of two globular clusters: NGC 1806 in the LMC and the galactic cluster NGC 3201.**
E. H. Geyer, U. Hopp.
Mitt. Astron. Ges., Nr. 57, p. 313 - 314 (1982).

154.038 **The extended giant branches of intermediate age globular clusters in the Magellanic Clouds. III.**
J. Mould, M. Aaronson.
Prepr. Steward Obs., No. 400, 39 pp. (1982).

154.039 **Dynamical evolution of globular clusters. I.**
J. S. Stodółkiewicz.
Acta Astron., Vol. 32, 63 - 91 (1982).

A method (a version of the Hénon Monte Carlo scheme) is presented which allows the study of the evolution of globular clusters during their whole life starting from the end of the violent relaxation phase far beyond the critical moment. This method describes the process of evolution and gives the rate of star escape, energies of escaping stars, mass segregation etc. The evolution of a nonisolated globular cluster (of mean relaxation time equal to 1.5×10^9 years) composed of stars with a continuous mass spectrum is computed over 20×10^9 years. Both mass loss and shock heating are included. It is shown that in the first 5×10^9 years of the cluster evolution the mass loss slows the collapse and is essential for the energy balance. The mass of the cluster at the end of the presented period of evolution is reduced to about one third of its initial value.

154.040 **Color-magnitude studies of globular clusters. I. The bright stars in NGC 362.**
W. E. Harris.
Astrophys. J., Suppl. Ser., Vol. 50, 573 - 586, plates 14,
(1982).

New photometric data in *UBV* are presented for the southern globular cluster NGC 362, including photoelect photometry of 89 sequence stars and a photographic colc magnitude study down to $V(\lim) \approx 16.6$. The new data h support the interesting "anomalous" nature of NGC 362 Various fine-structure features of the sequences in the co magnitude diagram (the red horizontal branch, giant brar and asymptotic branch) are described and compared witl results of stellar evolution models.

154.041 **The extended giant branches of intermediate a globular clusters in the Magellanic Clouds. III.**
J. Mould, M. Aaronson.
Astrophys. J., Vol. 263, 629 - 638, plates 10 - 13 (1982).

The authors report on the latest findings of a photo graphic near-infrared survey of the red globular clusters i Magellanic Clouds for upper asymptotic giant branch stai New infrared (*JHK*) photometry for some 80 cluster star: also presented. These results combined with earlier data a used to derive age estimates for a nearly complete sample Cloud clusters having $M_v < -7$. The age distribution of clusters in the Large Cloud, which shows a pronounced p at $t \sim 4$ Gyr, may be different from that in the Small Clo

154.042 ***UBV* photometry of stars in M 3.**
H. D. Ables, C. C. Dahn, A. V. Hewitt.
Publ. Astron. Soc. Pacific, Vol. 94, 748 (1982). – Abstra

154.043 **Carbon and nitrogen abundances in two outer l clusters NGC 7006 and Palomar 13.**
E. D. Friel, R. P. Kraft, N. B. Suntzeff, D. F. Carbon.
Publ. Astron. Soc. Pacific, Vol. 94, 752 (1982). – Abstra

154.044 **Is NGC 1851-UV5 a binary star in a globular cl**
G. S. Da Costa.
Publ. Astron. Soc. Pacific, Vol. 94, 769 - 773 (1982).

The radial velocity of the unusual early-type star UV the globular cluster NGC 1851 has been determined on ea seven consecutive nights to an accuracy of better than 4 k In strong contrast to previous published observations, the radial velocity of this star was not found to vary. Thus U no longer a viable candidate for a close binary star system globular cluster. Three other "UV-bright" stars in the fiel the cluster are shown to be nonmembers.

154.045 **A determination of the distance to the object NGC 5272 using variable stars.**
V. S. Popov.
Izv. Glav. Astron. Obs. Pulkovo, Astrometr. Astrofiz., No p. 102 (1982). In Russian.

Using the method developed by Melnikov (1979) th distance to the globular cluster M 3 (NGC 5272) has been determined. The distance has proven to be 13.0 kpc.

154.046 **Carbon, nitrogen, and iron-peak abundances fo giants in the remote globular clusters NGC 700 and Pal 13.**
E. Friel, R. P. Kraft, N. B. Suntzeff, D. F. Carbon.
Publ. Astron. Soc. Pacific, Vol. 94, 873 - 886 (1982) = Li Obs. Bull., No. 937.

Low-resolution image-dissector scanner spectra have been obtained for giants in the remote globular clusters NGC 7006 and Pal 13. Mean iron-peak metallicities [Fe/H are found to be −1.45 and −1.9 for NGC 7006 and Pal 1 respectively. Carbon abundances were obtained from a cc parison of observed and synthetic spectra in the region of G-band (~ λ4300), and nitrogen abundances were estimat by measuring the strength of the CN-band near λ3883. NGC 7006 and Pal 13 are often considered as "second parameter" clusters, i.e., clusters which exhibit anomalously red horizontal branches for their metallicities. The present results do not support the recent suggestion that anomalous overabundances of elements in the CNO group are actually directly responsible for the second parameter effect in NGC 7006.

The influence of CN abundances on the evolution of main sequence of low-mass stars.
See Abstr. 065.007.

Bursts of gravitational radiation from active galactic nuclei and globular clusters.
See Abstr. 066.004.

Do black holes exist at the centres of globular clusters?
See Abstr. 066.070.

Anomalies in sodium and CN in 47 Tuc stars.
See Abstr. 114.072.

Abundances in metal-poor stars. II. The anomalous globular cluster ω Centauri.
See Abstr. 114.145.

Evolutionary possibilities and impossibilities for solar type contact binaries in NGC 188.
See Abstr. 117.141.

Periods for newly discovered variables in NGC 1851.
See Abstr. 122.053.

The mass of the anomalous Cepheid in the globular cluster NGC 5466.
See Abstr. 122.080.

A study of period changes of RR Lyrae variables in the globular cluster M 4 (NGC 6121).
See Abstr. 122.156.

Variable stars in the globular cluster M 3.
See Abstr. 122.164.

On bolometric corrections for short- and long-period cepheids in globular clusters in the Galaxy.
See Abstr. 122.165.

The photographic photometry of the variables No. 24, 28 and 43 of the globular cluster M 15.
See Abstr. 122.172.

Two new variable stars in M 13.
See Abstr. 123.019.

The role of H II regions during star formation and chemical enrichment in globular clusters.
See Abstr. 131.062.

X-ray sources in globular clusters and burst sources.
See Abstr. 142.025.

Ultraviolet observations of X-ray globular clusters.
See Abstr. 142.026.

X-ray observations of globular clusters.
See Abstr. 142.027.

X-ray variability of the globular clouster source 4U1820 - 30 (NGC 6624).
See Abstr. 142.059.

Why are essential parts of "bursters" located in globular clusters?
See Abstr. 142.069.

Dissipative evolution of collisionless stellar systems. II. Influence of binaries on the evolution of globular clusters and galactic nuclei.
See Abstr. 151.003.

Origin of stars and galaxies.
See Abstr. 151.060.

IAU numbers for some recently discovered clusters.
See Abstr. 153.035.

The cluster system of M 33.
See Abstr. 158.032.

On the origin of the globular cluster system of M 87.
See Abstr. 158.285.

The oldest components of the nearest active galaxy (Cen A = NGC 5128).
See Abstr. 158.321.

Some implications for the early chemical and dynamical evolution of the Large Magellanic Cloud from observations of the oldest globular clusters.
See Abstr. 159.001.

NGC 2257 and the LMC halo.
See Abstr. 159.021.

On the symmetry of the system of companions of the Galaxy.
See Abstr. 160.073.

Erratum

154.901 **Erratum: 'Comments on the origin of the carbon and nitrogen variations within NGC 6752 and 47 Tucanae'** [Astrophys. J., Vol. 254, 594 - 605 (1982)].
G. H. Smith, J. Norris.
Astrophys. J., Vol. 261, 747 (1982). – See Abstr. 31.154.006.

155 Galaxy (Structure, Evolution)

155.001 **Infrared spectra of galactic center sources.**
E. R. Wollman, H. A. Smith, H. P. Larson.
Astrophys. J., Vol. 258, 506 - 514 (1982).

High-resolution $2\ \mu m$ spectra of five compact sources in the central parsec of the Galaxy are presented. Absorption due to CO and emission from atomic hydrogen and neutral helium provide information about the condition, distribution, and kinematics of red giant stars and ionized gas in the core. All three red giant sources observed are blueshifted. There are no detectable absorptions in the continuum of the central source IRS 16.

155.002 **Implications of collisionally supported giant molecular clouds for spiral galactic structure and massive star formation.**
D. Leisawitz, F. Bash.
Astrophys. J., Vol. 259, 133 - 153 (1982).

A study has been made of the environment experienced by ballistically moving giant molecular clouds (GMCs) as they encounter a hydrodynamically flowing interstellar medium. The hydrodynamically flowing gas is assumed to consist of an ensemble of discrete small clouds. GMCs are assumed to be born in spiral galactic shock fronts from which they are ejected at postshock velocities. Support for the GMCs occurs when the energy loss through decay of turbulence in the clouds is balanced by energy input via collisions with smaller clouds. It is assumed that an excessive energy input will destroy a GMC and that an unsupported cloud will collapse to form massive stars and H II regions. If the galactic distributions of small colliding clouds are like those observed for H I, the model predictions are in agreement with several observational characteristics of the Galaxy and M 81.

155.003 **Luminosity of M giants in the nuclear bulge of the Galaxy.**
J. A. Frogel, A. E. Whitford.
Astrophys. J., Lett., Vol. 259, L7 - L11 (1982) = Lick Obs. Bull., No. 920.

A color-luminosity diagram for 20 galactic bulge giants around NGC 6522 shows that Blanco types M7–M9 are on the upper asymptotic giant branch at $\langle M_{bol} \rangle = -4.5$, significantly fainter than the giant luminosities adopted in the Tinsley-Gunn galaxy population model but certainly more luminous than the He core flash luminosity assumed to be the upper limit in the models of Aaronson, Cohen, Mould, and Malkan. Distorted energy distributions and nonstandard colors relative to solar neighborhood M giants are attributed to a super–metal-rich composition as is observed in bulge K giants by Whitford and Rich.

155.004 **The black heart of the Milky Way.**
L. J. Robinson.
Sky Telesc., Vol. 64, 133 - 135 (1982).

155.005 **The radial velocity field of the Milky Way outside the galactic plane.**
J. V. Feitzinger, J. Kreitschmann.
Astron. Astrophys., Vol. 111, 255 - 259 (1982).

The new mass model of our Galaxy including a massive galactic halo, is used to compute radial velocities for a complete set of (l, b) values. Both a corotating and a differentially rotating corona are treated. The authors discuss cases of extreme orbits and show the upper and lower bounds of the line-of-sight velocities. Some consequences of the galactic fountain model are compared with the orbit computations. The radial velocities of interstellar ultraviolet absorption lines in the spectra of Magellanic Cloud stars are used to estimate new distances for the corona clouds.

155.006 **Kinematic moment equations as a tool of studyi the local galactic structure. I.**
H. Eelsalu.
Publ. Tartu Astrofiz. Obs., Tom 48, 217 - 225 (1981).

Equations of the centroid and dispersion equations h been derived connecting the stellar-statistical kinematic da with the corresponding descriptive variables (the compone of the centroid and the matrix of velocity dispersion).

155.007 **Kinematic moment equations as a tool of studyi the local galactic structure. II.**
H. Eelsalu.
Publ. Tartu Astrofiz. Obs., Tom 48, 226 - 231 (1981).

The special case of dispersion equations correspondin high galactic latitudes is discussed. The equations are inter ed from the viewpoint of the applicability of the least-squ method on them.

155.008 **On a model of local gas related to Gould's belt.**
C. A. Olano.
Astron. Astrophys., Vol. 112, 195 - 208 (1982).

In a model of the local gas related to Gould's belt, tha considers Gould's belt as an expanding ring of gas, the effe of the braking force due to the interaction between the expanding and the non-expanding interstellar gas is include

155.009 **Stellar contents contributing to the near-infrarec radiation of the Galaxy.**
T. Mikami, K. Ishida, K. Hamajima, K. Kawara.
Publ. Astron. Soc. Japan, Vol. 34, 223 - 229 (1982).

The stellar contents contributing to the near-infrared radiation of the Galaxy are examined on the basis of the st counts in the Scutum region, which were carried out by ne infrared photometry. The detected stars are divided into th groups with the use of objective-prism spectra and H–K co For these three groups, the observed cumulative frequency distributions in apparent K magnitude and the galactic dist butions are synthesized with models of space distributions stars.

155.010 **Die sonnennahen Sterne.**
W. Gliese.
Phys. unserer Zeit, Vol. 13, 97 - 103 (1982).

155.011 **The theoretical expected galactic distribution of WR runaway stars.**
D. Vanbeveren.
Astron. Astrophys., Vol. 113, 205 - 210 (1982).

A statistical study is presented on the theoretical expected number of WR + OB binaries, WR + compact star sy tems, WR single runaway stars and normal WR single stars. Starting from a complete sample of O type single stars and close binaries, imposing a mass ratio distribution for the binaries and an initial mass function for single stars and primaries of close binaries, evolution of massive stars is use determine the theoretical expected absolute average distan difference of the WR runaways and of the single lined WR in general, relative to the birthplace of their progenitor OB type stars.

155.012 **The galactic center – structure and kinematics fr 21-cm line measurements.**
K. Rohlfs, E. Braunsfurth.
Astron. Astrophys., Vol. 113, 237 - 249 (1982).

The major features dominating the neutral hydrogen emission in a data cube $-1^\circ\!.5 \leqslant l \leqslant +1^\circ\!.5$, $-1^\circ\!.5 \leqslant b \leqslant +1^\circ\!.5$ $-300\ \mathrm{km\ s^{-1}} \leqslant V \leqslant +300\ \mathrm{km\ s^{-1}}$ with a resolution of $9'$ in and $3.3\ \mathrm{km\ s^{-1}}$ in V are investigated with regard to their ki matics and position in space. Absorption spectra were determined for a contiguous field $-0\overset{\circ}{.}60 \leqslant l \leqslant 0\overset{\circ}{.}75$, $-0\overset{\circ}{.}15 \leqslant b \leqslant 0\overset{\circ}{.}0$. A Gaussian analysis for the region between Sgr A and Sgr B2 shows great continuity of the absorption features. The features are discussed in terms of galactic rotation expansion velocity.

155.013 **Nearby – in our Galaxy and outside its limits.**
B. Lapin.
Nauk. v SSSR, 1981, No. 6, p. 44 - 55. In Russian. – Abstr. in Ref. zh., 51. Astron., 8.51.39 (1982).

155.014 **Atomic hydrogen in the outer Milky Way.**
S. R. Kulkarni, L. Blitz, C. Heiles.
Astrophys. J., Lett., Vol. 259, L63 - L66, plates L1 - L4 (1982).

The H I in the outer Galaxy is reanalyzed using the CO rotation curve and the full latitude extent of the gas from the Weaver and Williams survey. Gas with a surface density $\gtrsim 0.1\ M_{\odot}\ pc^{-2}$ is found to a distance of 30 kpc from the center. Three distinct, well-defined spiral features of roughly constant surface density are seen, two of which extend at least 20–25 kpc along their length and can be traced to 20 kpc from the center. The scale height of the gas shows an almost linear increase from the solar vicinity to $R \approx 30$ kpc. This increase implies that the large mass in the outer Galaxy implied by the rotation curve does not reside in the disk.

155.015 **Der Aufbau unseres Milchstraßensystems.**
H. H. Voigt.
Nova Acta Leopoldina, Neue Folge, Nr. 250, Band 55, 26 pp. (1982).

155.016 **The "metal-rich" A-type stars at the south galactic pole.**
A. G. D. Philip.
News Lett. Astron. Soc. N.Y., Vol. 2, No. 2, p. 13 - 14 (1982). Abstract.

155.017 **Variations in the stellar luminosity function.**
T. E. Armandroff, A. R. Upgren.
Bull. American Astron. Soc., Vol. 14, 655 - 656 (1982). Abstract.

155.018 **A study of galactic light using Pioneer 10 observations of background starlight.**
G. N. Toller.
Bull. American Astron. Soc., Vol. 14, 656 (1982). – Abstract.

155.019 **The galactic plane in Hα + [N II], [S II] and [O III].**
T. R. Gull, R. P. Kirshner, R. A. R. Parker.
Bull. American Astron. Soc., Vol. 14, 662 (1982). – Abstract.

155.020 **Structure and luminosity function of the Galaxy from star counts.**
C. Pritchet.
Bull. American Astron. Soc., Vol. 14, 662 (1982). – Abstract.

155.021 **On the structure of nonthermal expansion of the Galaxy at 85 MHz.**
Ch. Sejitnepesov, A. Khanberdiev.
Izv. AN TurkmSSR. Ser. fiz.-tekh., khim. i geol. nauk. 1981, No. 4, p. 22 - 30. In Russian. – Abstr. in Ref. zh., 51. Astron., 9.51.575 (1982).

155.022 **The rotation curve of our Galaxy; how well do we know it?**
P. Pişmiş.
Rev. Mexicana Astron. Astrofis., Vol. 6, (see 012.017), 65 - 78 (1981).

Following an historical sketch of the relevant circumstances leading to the formulation of the rotation of the Galaxy, the differential rotation formulae are recalled. The necessity of obtaining an overall rotation curve at the advent of radioastronomy is stressed. The existence of the deviations from a smooth rotation curve are pointed out; in particular it is shown that the curve exhibits "waves". The interpretation of these waves as population effects suggested earlier by the author is emphasized once again. Recent observations of H II regions and CO clouds suggest that the sun is located close to the minimum of a wave. Another irregularity, the presumed difference in the north and south rotation curves, is also briefly discussed. Based on a plausible assumption that the spiral structure can be represented by a pair of symmetrically located logarithmic spirals, it is shown that if waves do indeed exist the rotation curve in our Galaxy and in others will be a function of direction from the galactic center. Finally it is emphasized that the rotation curve discussed usually is that of the fastest rotating system, the Population I, which contains not more than 10% of the total mass of the Galaxy. The rotation curve is therefore not unique.

155.023 **A deep near infrared objective prism survey for carbon stars toward the galactic center and anti-center.**
F. J. Fuenmayor.
Rev. Mexicana Astron. Astrofis., Vol. 6, (see 012.017), 83 - 93 (1981).

A deep near infrared objective-prism survey for carbon stars has been carried out in the direction of both the galactic center and the galactic anticenter. The limiting magnitude to which carbon stars could be identified is $I = 13.0$. A catalogue containing 283 carbon stars is presented, of which 123 have not been previously reported. A statistical analysis of the surface and space distribution of the stars in the sample has been carried out yielding the following results: (1) The number of carbon stars observed toward the galactic anticenter is larger than that observed toward the center by the factor of 3. (2) The center-anticenter asymmetry appears to be real and it seems to reach up to an approximate distance of 8 kpc. (3) The carbon stars in the sample appear to be concentrated at average distances from the sun of 5.5 kpc toward the galactic center, and 0.5 and 5.0 kpc toward the anticenter. The mean space density of carbon stars at these locations is 15,20 and 50 stars per cubic kpc, respectively. The results are combined with those obtained in other investigations to discuss the distribution of carbon stars on the galactic plane.

155.024 **Neutral hydrogen related to Gould's Belt.**
W. G. L. Pöppel, C. A. Olano, C. E. Cappa de Nicolau.
Rev. Mexicana Astron. Astrofis., Vol. 6, (see 012.017), 259 - 266 (1981).

A Gaussian analysis was made of the H I profiles of an extensive region in Ophiuchus, Scorpius and Lupus, which encloses part of the Scorpius-Centaurus Association as well as of Gould's Belt. The authors conclude that in some regions the H I related to Lindblad's feature A shows perturbations in the velocity and in the N_H distribution.

155.025 **The galactic nucleus.**
J. H. Oort.
Tenth Texas Symposium on Relativistic Astrophysics, (see 012.021), p. 347 - 356 (1981).

Infrared and radio observations of the galactic center and their implications for a consistent model of the galactic nucleus are reviewed.

155.026 **Photographic surface photometry of the Milky Way. II. Surface photometry in the region of the dark cloud "Coalsack" in U, B, V, R.**
K. J. Seidensticker, T. Schmidt-Kaler, W. Schlosser.
Astron. Astrophys., Vol. 114, 60 - 70 (1982). In German.

16 photographic plates, obtained at La Silla, Chile, by

the super-wide-angle camera with spherical mirror of the Astronomisches Institut der Ruhr-Universität Bochum were used to study the surface brightness distributions in UBVR of the Coalsack region over an area of 30°×30°, centered on $l = 303°$, $b = 0°$, with a resolution of 0°.3. Height and colors of airglow emission were also determined.

155.027 **Stellar populations in the Galaxy.**
J. R. Mould.
Annu. Rev. Astron. Astrophys., Vol. 20, (see 003.009), 91 - 115 (1982).
The halo. The disk. Toward a stellar population model of the Galaxy.

155.028 **Gas in the galactic halo.**
D. G. York.
Annu. Rev. Astron. Astrophys., Vol. 20, (see 003.009), 221 - 248 (1982).
Historical comments. Available data on our own halo. Physical conditions. Models for gas in the halo. Synthesis.

155.029 **Optical observations of halo gas.**
J. C. Blades.
Observatory, Vol. 102, 172 (1982). – Abstract.

155.030 **Hot gas at high galactic latitudes.**
M. Pettini, K. A. West.
Observatory, Vol. 102, 173 (1982). – Abstract.

155.031 **Halo gas.**
D. G. York.
Observatory, Vol. 102, 176 - 178 (1982). – Abstract.

155.032 **On space density distribution of late type stars in the solar neighbourhood.**
S. K. Ghosh.
Bull. Astron. Soc. India, Vol. 10, 40 (1982). – Abstract.

155.033 **Primary nucleosynthesis in the galactic disk.**
B. A. Twarog, J. C. Wheeler.
Astrophys. J., Vol. 261, 636 - 648 (1982).
New data on the history of production of C, O, Mg, and Fe are used to reexamine the production of heavy elements in the galactic disk. A standard infall model is used to show that a constant rate of production of heavy elements is consistent with the data and to derive net absolute amounts for several elements. Various models for nucleosynthesis by massive stars are considered.

155.034 **Density distribution of faint stars in the direction of the north galactic pole.**
Y. Yoshii.
Publ. Astron. Soc. Japan, Vol. 34, 365 - 379 (1982).
A method is proposed to separate density functions along the line of sight for both disk- and halo-population stars with three-color photometry. The observed star counts as a function of visual magnitude and color are synthesized by an integral equation, where the luminosity function is assumed to be that of van Rhijn for the disk stars and that of globular clusters M 92 and M 3 for the field halo stars, respectively. Adopting the conventional exponential density law for both disk and halo stars, the author derives the density gradient toward the north galactic pole. The halo mass density is estimated to be $(3 - 5)\times 10^{-4} M_\odot\ pc^{-3}$ in the solar neighborhood.

155.035 **Near-infrared source counts in the galactic plane.**
K. Kawara, T. Kozasa, S. Sato.
Publ. Astron. Soc. Japan, Vol. 34, 389 - 405 (1982).
Near-infrared source counts have been made by scanning 17 selected areas distributed along the galactic plane between $l = 349°$ and $l = 45°$. The total scanned area is 11.2 square degrees and 2757 sources brighter than 6.8 mag at the K-ba are analyzed. Longitude and latitude dependence of the s distribution resembles the surface brightness distribution served at 2.4 μm by balloon-borne telescopes. The source counts are effective for detection of luminous sources in distant regions of the Galaxy and the results indicate that space density of M supergiants is substantially high in the : called 5-kpc region. The source distribution peaks in the direction of the galactic center. This would suggest that luminous M giants are present in the central bulge.

155.036 **The velocity of escape from the Galaxy in the : neighbourhood.**
J. B. Alexander.
Mon. Not. R. Astron. Soc., Vol. 201, 579 - 594 (1982).
The expected properties of stars moving with extren high velocities relative to the centre of the Galaxy are dis cussed. Although the kinematic behaviour of the fastest known subdwarfs is consistent with unbound orbits, obse tional upper limits on the numbers of faint giant stars in t general field strongly suggest that these subdwarfs are bo to the Galaxy. Although all known subdwarfs are probab bound orbits, there is evidence that the mode of origin of peculiar velocities of subdwarfs with extremely large gala centric velocities is different from that of other high-velo stars.

155.037 **Precessing jets in Sagittarius A: gas dynamics ir central parsec of the Galaxy.**
R. L. Brown.
Astrophys. J., Vol. 262, 110 - 119 (1982).
It is suggested that the asymmetric velocity field an point-reflection symmetry of the radio continuum bright distribution of the inner parsec of the galaxy can be unde stood as complementary manifestations of collimated gas outflow along twin opposing beams that emanate from a central object. That object appears to be the compact no thermal radio source Sgr A*. The precessing beam descrip accounts, in a natural way, for specific observations of th gas dynamics that are unexpected from gas in circular rot tion. Specific comparisons with the radio continuum and infrared spectroscopic data are made.

155.038 **The nature of the central parsec of the Galaxy.**
J. H. Lacy, C. H. Townes, D. J. Hollenbach.
Astrophys. J., Vol. 262, 120 - 134 (1982).
The authors summarize the arguments for and agains the existence of a massive black hole at the galactic cente The discussion, based on experimental evidence and the u theoretical models of interaction between a black hole an surrounding matter, indicates that a black hole of mass greater than $10^6 M_\odot$ is probably not present. It is found th the best model to fulfill observational requirements invol cloud generation through disruption of red giants by stell collisions, ionization by a population of stars which is aff ed either by enhanced metal abundances or the death of t most massive stars, and gas disposal by star formation.

155.039 **Cosmic rays, the metallicity gradient and the m of molecular gas in the Galaxy.**
T.-p. Li, P. A. Riley, A. W. Wolfendale.
J. Phys. G, Vol. 8, 1141 - 1154 (1982). – Abstr. in Phys. Abstr., Vol. 85, Abstr. 101062 (1982).

155.040 **The galactic distribution of Wolf-Rayet stars.**
B. Hidayat, K. Supelli, K. A. van der Hucht.
Wolf-Rayet stars: observations, physics, evolution, (see 012.042), p. 27 - 40 (1982) = Contrib. Bosscha Obs. No.
On the basis of the most recent compilation of narro band photometry and absolute visual magnitudes of Wolf-Rayet stars, and adopting a normal interstellar extinction in all directions, the galactic distribution of 132 of the 15

known galactic WR stars is presented and discussed. The spiral structure is found to be more clearly pronounced than in earlier studies. Furthermore the authors find an indication of two spiral arms at r = 4 and 6 kpc. There appears to be an asymmetry of the z-distribution of single stars with respect to galactic longitude. The location of the WC8.5 and WC9 stars between 4.5 and 9 kpc from the galactic center is discussed in the context of Maeder's red supergiant to WR star scenario.

155.041 **Gradients in supergiant and WR stars across the galactic plane.**
G. Bertelli, C. Chiosi.
Wolf-Rayet stars: observations, physics, evolution, (see 012.042), p. 359 - 363 (1982).

The conclusion by Maeder et al. (1980) that the strong gradient in N_R/N_{WR} results from the effect of metal content on the rate of mass loss is reconsidered. The authors suggest that the above gradient simply reflects a similar gradient in the progenitor stars, although effects of metallicity cannot be excluded.

155.042 **Two-color CCD observations of the galactic center region.**
J. A. Biretta, K. Y. Lo, P. J. Young.
Astrophys. J., Vol. 262, 578 - 582 (1982).

The authors present intermediate band photometric and astrometric observations at 8000 Å and 9200 Å of the galactic center region. The discovery of two optical objects near the position of the compact nonthermal radio source and the infrared source IRS 16 by Grindlay and Liller is confirmed. Astrometry and two-color photometry suggests that these optical objects are foreground stars.

155.043 **The galactic abundance gradient from supernova remnant observations.**
L. Binette, M. A. Dopita, S. D'Odorico, P. Benvenuti.
Astron. Astrophys., Vol. 115, 315 - 320 (1982).

Using published data on Hα, [N II] and [S II] relative line intensities for galactic supernova remnants (SNR), the authors show that there is strong evidence for a galactic abundance gradient in nitrogen. This data has been analysed using the shock modelling sections of a new general-purpose code to yield local nitrogen and sulphur abundance gradients of -0.095 dex kpc^{-1} and -0.07 dex kpc^{-1} respectively, in a very good agreement with that obtained by different and independent techniques.

155.044 **Les étoiles à portée de la main.**
D. Benest.
Astronomie, Vol. 96, 500 - 507 (1982).

155.045 **The distribution of neutral atomic hydrogen in our Galaxy beyond the solar circle.**
A. P. Henderson, P. D. Jackson, F. J. Kerr.
Astrophys. J., Vol. 263, 116 - 122 (1982).

The authors have used the Parkes 18 m telescope and Hat Creek 26 m telescope surveys of the 21 cm line from Milky Way neutral hydrogen to determine the distribution of H I in the outer Galaxy based on a circular rotation model. Contour maps are presented for the H I surface density, the departure of the H I layer from the galactic plane, and the thickness of the H I layer, in terms of galactocentric radius and angle. The total mass of H I in the Galaxy is $4.81 \times 10^9 M_\odot$, of which more than 81% lies beyond the Sun's distance from the galactic center.

155.046 **Distribution of CO in the southern milky way.**
W. H. McCutcheon, B. J. Robinson, J. B. Whiteoak.
J. R. Astron. Soc. Canada, Vol. 76, 322 (1982). – Abstract.

155.047 **On the observational vulnerability of the models for the Galaxy.**
B. A. Balázs.
Dynamical astronomy, (see 012.056), p. 159 - 164 (1982). Abstract.

155.048 **Broad He emission in the galactic center.**
D. N. B. Hall, S. G. Kleinmann, N. Z. Scoville.
Space Telesc. Sci. Inst., Prepr. Ser., No. 1, 17 pp. (1982).

An infrared (K band) spectrum taken with a 3.8″ diameter aperture centered on the galactic center source IRS 16 exhibits weak CO band absorption and emission by hydrogen Brackett γ (Bγ) and a singlet helium line near 4857 cm^{-1}. The He emission profile is much broader (FWHM $\sim 1500 \pm 300$ km s^{-1}) than any feature seen previously at the galactic center. The broadening is probably due to Doppler motion of gas either flowing through an ionizing layer or orbiting about a massive object. The absence of H Bγ emission with a similar profile suggests the galactic center H/He abundance ratio is reduced by a factor $\gtrsim 500$ from normal cosmic values.

155.049 **The galactic abundance gradient.**
P. A. Shaver, R. X. McGee, L. M. Newton, A. C. Danks, S. R. Pottasch.
ESO Sci. Prepr., No. 210, 68 pp. (1982). – Submitted to Mon. Not. R. Astron. Soc.

155.050 **Die Bewegungen der Sterne und ihre Bedeutung in der galaktischen Astronomie.** (Karl-Schwarzschild-Vorlesung 1982).
J. Delhaye.
Mitt. Astron. Ges., Nr. 57, p. 123 - 133 (1982).

155.051 **Radial velocities and internal motions in Galactic stellar systems.**
F. Gieseking.
Mitt. Astron. Ges., Nr. 57, p. 143 - 166 (1982). In German.

The general contribution of stellar radial velocities to kinematical studies of binaries, open clusters, OB-associations and globular clusters in our galaxy is reviewed. Qualities and limits of proper motions and radial velocities are compared. Limits of conventionally determined radial velocities are discussed and illustrated by completeness estimates of recent compilation of known spectroscopic binaries. The immense potentialities of the unconventional much more effective methods of photoelectric radial velocity determinations and the objective prism method are emphasized. Some results based on the discussion of more than 12 000 single relative radial velocity measurements of nearly 900 stars obtained with the Fehrenbach-prism are given.

155.052 **G- und K-Sterne als Indikatoren der galaktischen Evolution.**
G. Cayrel de Strobel.
Mitt. Astron. Ges., Nr. 57, p. 177 - 205 (1982).

155.053 **Radio studies of galactic structure.**
D. Downes, R. Güsten.
Mitt. Astron. Ges., Nr. 57, p. 207 - 232 (1982).

The paper reviews investigations of the large-scale structure and kinematics of the Galaxy. The authors discuss the radio studies in the larger framework of galactic structure research, and with respect to the hierarchy of gas phases in the interstellar medium. They review the methods used in radio derivations of the star formation rate and the molecular and atomic gas density as functions of galactocentric radius, as well as the results on structure in the vertical (z) direction, and on element and isotope abundance gradients across the galactic disk. The authors also note recent work on the galactic rotation curve and the problem of spiral structure.

155.054 **Problems in stellar statistics.**
R. Buser.

Mitt. Astron. Ges., Nr. 57, p. 233 - 248 (1982). In German.

Three "classical" problems of stellar statistics are discussed in their modern context of galactic structure work: (1) Galactic clusters and interstellar extinction. (2) Late-type field giants and absolute magnitudes. (3) Subdwarfs and metallicity.

155.055 **Strukturen der Population I in Sonnenumgebung.**
T. Neckel, G. Klare.
Mitt. Astron. Ges., Nr. 57, p. 249 - 264 (1982).

In diesem Artikel wird die räumliche Dichteverteilung einiger wesentlicher Komponenten der extremen Population I, also der typischen "Spiralarmpopulation" betrachtet. Als "Sonnenumgebung" soll dabei ein Bereich mit 3 - 4 kpc Radius um die Sonne verstanden werden.

155.056 **Das Radialgeschwindigkeitsfeld der Milchstraße außerhalb der galaktischen Ebene.**
J. V. Feitzinger, J. Kreitschmann.
Mitt. Astron. Ges., Nr. 57, p. 278 (1982). – Abstract.

155.057 **Zwischenergebnisse im Basler Programm der Galaktischen Struktur.**
A. Spaenhauer, R. P. Fenkart, W. Becker.
Mitt. Astron. Ges., Nr. 57, p. 316 - 320 (1982).

155.058 **"Hole" formation in the center of the galactic disk.**
V. G. Surdin.
Astron. Tsirk., No. 1178, p. 6 - 7 (1981). In Russian.

155.059 **The effect of possible errors in light extinction on the determination of the parameters of the spiral structure of the Galaxy.**
E. D. Pavlovskaya, A. A. Suchkov.
Astron. Tsirk., No. 1199, p. 1 - 3 (1981). In Russian.

155.060 **On the distribution of interstellar gas in the galactic halo.**
L. M. Hobbs, W. W. Morgan, C. E. Albert, F. J. Lockman.
Astrophys. J., Vol. 263, 690 - 695, plate 14 (1982).

New and existing observations of 21-cm emission lines toward 10 distant, high-latitude OB stars are combined with existing observations of interstellar Lyα absorption lines, in order to determine the ratio, N_{21}/N_{α}, of the two different column densities of H I. This ratio, which is related to the fraction of the cool, neutral gas in the halo that lies beyond each star, decreases smoothly to about unity with increasing distance from the galactic plane. The column density of neutral gas beyond about 1 kpc can be as much as one-third of the total above the plane, but only relatively small amounts of such gas lie more than 2 kpc from the plane.

155.061 **The initial mass function for massive stars.**
C. D. Garmany, P. S. Conti, C. Chiosi.
Astrophys. J., Vol. 263, 777 - 790 (1982).

The authors have compiled a machine readable catalog of over 750 galactic O type stars with published photometry, spectral types, and luminosity classes. The catalog is probably complete to a distance of about 2.5 kpc. From this volume-limited data the authors have derived the initial mass function (IMF) for stars more massive than 20 $M_{\odot}$. Dividing the sample into stars inside and outside the solar circle, the authors find a substantial difference in the IMF of the most massive stars; there are proportionally more toward the galactic center.

155.062 **A kinematic and abundance survey at the galactic poles.**
W. I. Hartkopf, K. M. Yoss.
Astron. J., Vol. 87, 1679 - 1709 (1982).

DDO observations of 1076 G and K stars at the north and south galactic poles are combined with published photometry to give an abundance gradient perpendicular to the galactic plane to $z > 5$ kpc. Additional DDO photometry of bright field stars has been used to calibrate the DDO abundance index against [Fe/H] for metal-poor stars. The resulting gradient in [Fe/H] is presented. Radial velocities obtained for 302 polar giants, coupled with published velocities, have enabled the authors to determine the z component of the velocity dispersion to $z > 4$ kpc. In combination with the photometry, the radial velocities allow it to determine velocity dispersion as a function of abundance.

155.063 **The space density distribution of late-type giants in the solar neighbourhood.**
S. K. Ghosh, S. N. Tandon.
Astrophys. Space Sci., Vol. 87, 247 - 254 (1982).

The authors have analysed the data of 2.7 μm Equatorial Infrared Catalogue-1 (EIC-1) to study the space density distribution of K and M type giants in the solar neighbourhood by attempting to fit the EIC-1 data with the models of Mikami and Ishida (1981) for space density distribution of late-type giants.

155.064 **Galactic nucleus.**
O. Obůrka.
Říše hvězd, Vol. 63, 202 - 204 (1982). In Czech.

155.065 **IAU Commission 33: Structure and dynamics of the galactic system.** Report 1978 - 81.
J. Einasto, G. Kuzmin (*Kuz'min*) (Editors).
W. Struve Astrophysical Observatory, Tartu, Akademiya nauk Ehstonskoj SSR. 150 pp. Price 70 Kop. (1982).

This report is the extended version of the report of Commission 33 of the International Astronomical Union published in Transactions of the IAU Vol. XVIIIA (1982). See Abstr. 31.155.056.

155.066 **Infrared observations of the ionized gas in the galactic center.**
J. H. Lacy.
The galacitc center, (see 012.076), p. 53 - 59 (1982).

Observations of ionic emission lines from the galactic center are discussed. These observations provide information regarding the abundances of heavy elements, the spectrum of ionizing radiation, the distribution of ionized gas, and the gravitational potential in which the gas moves. There is evidence from the gas cloud velocities that a massive central black hole is present, but other considerations do not support this conclusion. The most plausible scenario involves a dense central star cluster in which stellar collisions disrupt stars, and recently formed OB stars provide the radiation which ionizes the gas.

155.067 **Spatial and spectral studies of the galactic center near 10 μm.**
D. K. Aitken, M. C. Allen, P. F. Roche.
The galactic center, (see 012.076), p. 67 - 71 (1982).

155.068 **O I and O III in Sgr A: neutral and ionized gas at the galactic center.**
R. Genzel, D. Watson, C. Townes, D. Lester, H. Dinerstein, M. Werner, J. Storey.
The galactic center, (see 012.076), p. 72 - 76 (1982).

The authors have mapped the fine structure line emission at 63 microns from neutral oxygen in the vicinity of the galactic center. The emission is extended over more than 4′ and is centered on Sgr A West. The bulk of the O I emission arises in a predominantly neutral region outside of the ionized central 3 pc of our galaxy. The gas temperatures in this region are > 130 K, that is, significantly higher than the dust temperature of 70 K. The authors have also detected the 88 micron fine structure line of O III in a 45″ FWHM beam centered on Sgr A West.

155.069 **Mapping and imaging of the galactic centre in the near infrared.**
J. W. V. Storey.
The galactic center, (see 012.076), p. 85 - 90 (1982).
A cooled CCD has been used to produce an image of the galactic centre at an effective wavelength of 0.9 microns. The image shows two unresolved point sources separated by 3″ along the galactic plane. These sources are very close to the compact non-thermal radio source, and the 2.2 micron source IRS 16. While it is possible that the new sources are a pair of highly reddened individual stars or compact clusters, it appears more likely that they are two compact H II regions seen via line emission.

155.070 **Two color CCD observations of the galactic center.**
J. A. Biretta, K. Y. Lo, P. J. Young.
The galactic center, (see 012.076), p. 91 - 96 (1982).
The authors present intermediate band photometric and astrometric observations at 8000 Å and 9200 Å of the galactic center region. Two unresolved objects are found near the positions of the compact non-thermal radio source in Sgr A and the infrared source IRS 16. If the objects are at the distance of the galactic center, their emission must be non-stellar in nature. The objects may also be foreground stars.

155.071 **Discovery of three near infrared objects in CCD images of the galactic center.**
G. R. Ricker, M. W. Bautz, D. L. DePoy, S. S. Meyer.
The galactic center, (see 012.076), p. 97 - 106 (1982).
The authors used a CCD detector system to detect 3 extremely red objects in the direction of the galactic center. All three objects are $<10''$ from the location of Sgr A West. If these objects do indeed lie at a distance of 10 kpc, then they are located within the innermost 1 pc of the Galaxy. One of these objects lies within the extended 2.2 μm infrared source IRS 16, which is thought to contain the dynamic center of the Galaxy. Two interpretations for these three objects are that they are either tightly knit clusters of K5–M0 giant stars, or that they are extremely compact H II regions seen in the light of forbidden, doubly-ionized sulfur.

155.072 **The compact source at the galactic center.**
M. J. Rees.
The galactic center, (see 012.076), p. 166 - 176 (1982).
Various circumstantial arguments for (and against) the presence of a massive black hole ($\lesssim 5 \times 10^6 M_\odot$) at the galactic center are reviewed. The upper limits to the hole's mass based on considerations of stellar disruption and swallowing are shown to be very uncertain. If a massive black hole indeed lurks at the galactic center a compact radio source resembling the one actually observed would be a natural consequence of low-level accretion onto it.

155.073 **Gas motions in the central region and their interpretation.**
J. H. Oort.
The galactic center, (see 012.076), p. 180 - 188 (1982).
The dynamical structure of the galactic center inferred from infrared and microwave observations of interstellar matter is reviewed. Three regimes are distinguished: (1) the central region within ~1 pc containing fast-moving compact H II regions, (2) the region within ~300 pc with its high density of molecular clouds, (3) the region between 1 and 3 kpc containing extended features with strongly non-circular motions. Recent orbital model calculations that may successfully explain at least part of the observed features are briefly discussed.

155.074 **Comparison of galactic center with other galaxies.**
G. H. Rieke, M. J. Lebofsky.
The galactic center, (see 012.076), p. 194 - 203 (1982).
It is unlikely that the optical-infrared energetics of the galactic center are dominated by a single, nonthermal source as would be the case for a Seyfert or other kind of active galaxy. The energetics of this region can, however, be explained as the result of recent formation of massive stars; in this regard, our Galaxy may be analogous to M 82, NGC 253, and similar galaxies.

Landolt-Börnstein. Group VI. Vol. 2c. Interstellar matter – Galaxy – Universe.
See Abstr. 003.055.

Bau und Physik der Galaxis.
See Abstr. 003.108.

The outer regions of galaxies.
See Abstr. 011.001.

Nucleosynthesis and galactic evolution: implications for the origin of life.
See Abstr. 015.029.

New values for Oort's constant B of galactic rotation from proper motions in galactic longitude.
See Abstr. 043.009.

Decaying neutrinos as a photoionization source in galactic halos.
See Abstr. 061.020.

Convective mixing length and the galactic carbon to oxygen ratio.
See Abstr. 065.034.

Observations and the Wolf-Rayet evolutionary scenario.
See Abstr. 065.077.

Positron production near a $10^6 M_\odot$ black hole.
See Abstr. 066.252.

Radial velocities of 617 stars belonging to four stellar fields of 4° × 4°.
See Abstr. 111.001.

An odd behavior of nearby stars velocity components in the direction $l = 330°, b = 0°$.
See Abstr. 111.002.

Radial velocities of distant OB stars and galactic rotation to 26 kiloparsecs.
See Abstr. 111.009.

Recherche d'étoiles à grande vitesse dans la direction du Pôle Galactique Nord à l'aide du prisme-objectif de Fehrenbach.
See Abstr. 111.016.

Proper motion survey with the forty-eight inch Schmidt telescope. LXI. The galactic distribution of proper motion stars.
See Abstr. 111.020.

A photoelectric *UBV* sequence in a low extinction Puppis field.
See Abstr. 113.002.

Four-color and Hβ photometry for O-A0 type stars in three regions near the galactic equator.
See Abstr. 113.003.

RGU-photometry of the field Vela II.
See Abstr. 113.013.

RGU-Photometrie eines Sternfeldes in +23.1° galaktischer Breite in Richtung zum galaktischen Zentrum (NGC 6171).
See Abstr. 113.018.

UBV–Hβ photometry of luminous stars between $l = 335°$ and $l = 6°$.
See Abstr. 113.031.

M supergiants and star formation at the galactic center.
See Abstr. 114.203.

New light on faint stars – I. The luminosity function in the solar neighbourhood.
See Abstr. 115.008.

New variable stars in the direction of the bright cloud B in Sagittarius.
See Abstr. 123.001.

Supernova statistics and related problems.
See Abstr. 125.053.

Astrometric planetary searches, the absence of low-luminosity white dwarf stars, and the age of the Galaxy.
See Abstr. 126.012.

An estimation of the space number density of white dwarfs.
See Abstr. 126.021.

Reddenings derived from H I and galaxy counts: accuracy and maps.
See Abstr. 131.005.

A survey of high-velocity H I in the Cetus region.
See Abstr. 131.021.

A CO survey of the southern galactic plane.
See Abstr. 131.032.

Abundance fluctuations in the interstellar medium.
See Abstr. 131.040.

The distribution of interstellar gas within 50 pc of the Sun.
See Abstr. 131.042.

Distribution of gas in the halo of the Galaxy.
See Abstr. 131.043.

Highly ionized gas in the galactic halo.
See Abstr. 131.044.

A distinct shell structure in H I-line-emission at intermediate galactic latitudes.
See Abstr. 131.058.

A study of interstellar absorption at high galactic latitudes. I. Highly ionized gas.
See Abstr. 131.082.

Implications of collisionally-supported giant molecular clouds for spiral galactic structure and massive star formation.
See Abstr. 131.139.

Reddenings derived from H I and galaxy counts: accuracy and maps.
See Abstr. 131.140.

Hot galactic gas and narrow line quasar absorptio systems.
See Abstr. 131.141.

High velocity clouds. Review of observational properties.
See Abstr. 131.150.

Interstellar gas studies with IUE.
See Abstr. 131.172.

Highly ionised gas in the galactic halo.
See Abstr. 131.185.

Distribution of gas in the halo of the Galaxy.
See Abstr. 131.186.

Molecular clouds in the galactic nucleus.
See Abstr. 131.228.

The local interstellar medium as traced by gamm rays.
See Abstr. 131.267.

Neutral hydrogen observations towards the Pupp Window of the Milky Way.
See Abstr. 131.277.

Observations of CO clouds in the galactic center region ($0.°4 \leq l \leq 1.°8$, $|b| \leq 0.°4$).
See Abstr. 131.289.

A photometric map of interstellar reddening witl 100 pc.
See Abstr. 131.309.

La composante chaude du milieu interstellaire lo dimensions, températures, abondances.
See Abstr. 131.316.

Molecular gas near the galactic center – recent contributions from radioastronomy.
See Abstr. 131.323.

The relative locations of the Sgr A molecular clouds and continuum sources.
See Abstr. 131.324.

Depression of molecular emission in the line of sight of Sgr A West.
See Abstr. 131.325.

Broad helium emission in the galactic center.
See Abstr. 133.005.

Discovery of three far-red objects in CCD images the galactic center.
See Abstr. 133.006.

Polarization of compact sources in the galactic center.
See Abstr. 133.020.

Bα and Ne II line spectroscopy in the vicinity of the galactic center source IRS 16.
See Abstr. 133.022.

Three compact sources with unusual 2 to 4 micr spectra.
See Abstr. 133.023.

Two micron observations of ^{12}CO and ^{13}CO in the red giant sources IRS 7, IRS 12, and IRS 19.
See Abstr. 133.024.

The position of the infrared sources IRS 16 in the galactic center region relative to a visual field star.
See Abstr. 133.025.

Twinkle, twinkle quasi star . . . .
See Abstr. 141.024.

The galactic distribution of pulsars.
See Abstr. 141.535.

X-ray structure of the Galaxy.
See Abstr. 142.056.

Time-variable positron annihilation radiation from the galactic center direction.
See Abstr. 142.507.

Gamma-ray spectroscopy of the galactic center region: confirmation of the time variability of the positron annihilation line.
See Abstr. 142.508.

On the transport and propagation of cosmic rays in galaxies – I. Solution of the steady-state transport equation for cosmic ray nucleons, momentum spectra and heating of the interstellar medium.
See Abstr. 143.056.

Samples of the Milky Way.
See Abstr. 143.057.

Chemical evolution of two-component galaxies. IV: A generalized model allowing for gas inflow.
See Abstr. 151.013.

The importance of tidal disruption on star formation in spiral galaxies.
See Abstr. 151.029.

Role of random forces in stellar dynamics.
See Abstr. 151.069.

Deutung der Geschwindigkeitsverteilung sonnennaher Sterne durch Diffusion von Sternbahnen.
See Abstr. 151.072.

Triaxiality and the galactic center.
See Abstr. 151.086.

Open clusters and galactic structure.
See Abstr. 153.005.

The distant open cluster Tombaugh 2.
See Abstr. 153.042.

Search for (globular) clusters in M31. III. Structural properties: X-ray sources and comparison with Galactic globulars.
See Abstr. 154.013.

Kohlenstoff- und Stickstoffhäufigkeiten in Riesensternen der Kugelhaufen M 92, M 15, M 13, M 3 und in Halo-Feldriesen.
See Abstr. 154.035.

Die diffuse galaktische Nahinfrarot-Emission.
See Abstr. 156.014.

Observations of the galactic center in the radio band.
See Abstr. 156.016.

Infrared observations of the galactic center.
See Abstr. 156.017.

New far infrared observations of the central 30′ of the Galaxy.
See Abstr. 156.018.

Large beam observations of the galactic center at 150, 200, and 300 μm.
See Abstr. 156.019.

Balloon observation of the central bulge of our Galaxy in near infrared radiation.
See Abstr. 156.020.

Gamma rays, cosmic rays, and the dynamics of the Galaxy.
See Abstr. 157.004.

Photographic surface photometry of the Milky Way. IV. The northern Milky Way in the ultraviolet spectral region.
See Abstr. 157.011.

On the origin of the positron annihilation radiation from the direction of the galactic center.
See Abstr. 157.019.

The mass function of blue stars, the production rate of Lyc-photons, and the rate of star formation in M 33.
See Abstr. 158.045.

The coronas of galaxies.
See Abstr. 158.075.

The Magellanic Stream and the galactic halo.
See Abstr. 159.012.

On the symmetry of the system of companions of the Galaxy.
See Abstr. 160.073.

Is there a ring around the Milky Way?
See Abstr. 160.076.

Galactic deuterium as a test for cosmological models.
See Abstr. 162.004.

Five crucial tests of the cosmic distance scale using the Galaxy as fundamental standard.
See Abstr. 162.029.

Cosmic cooking, evolution of stars and our Metagalaxy.
See Abstr. 162.255.

156 Galaxy (Magnetic Field, Radio and Infrared Radiation)

156.001 **The Maryland-Green Bank Galactic 21-cm Line Survey.**
G. Westerhout, H.-U. Wendlandt.
Astron. Astrophys., Suppl. Ser., Vol. 49, 143 - 425 (1982).
An atlas is presented of the distribution of the 21-cm line emission from neutral hydrogen in the Galaxy, covering the region from galactic longitude 11° to 235°, with galactic latitude coverage of approximately +2° to −2°. The data were obtained with the 300-foot radio telescope of the NRAO at Green Bank, W. Va. They are presented in the form of contour maps, giving cross sections perpendicular to the galactic equator at intervals of 0.°2 in galactic longitude. The spatial resolution is 0.°22, the velocity resolution 2 km/s.

156.002 **Investigation of the galactic magnetic field.**
R. R. Andreasyan.
Astrofizika, Tom 18, 255 - 262 (1982). In Russian. English translation in Astrophysics, Vol. 18, No. 2.
Rotation measures of 703 radio sources are used for the study of the structure of the galactic magnetic field. It is shown that the large-scale magnetic field of the Galaxy is directed approximately parallel to the local spiral arm and has an opposite direction in the regions $15° < b < 90°$ and $-90° < b < 15°$. The majority of the 46 unidentified radio sources having large rotation measures ($|RM| > 200$ rad/m²) and being located in the region $-15° < b < 15°$ probably are galactic H II regions. The parameters of the magnetic fields of these objects approximately coincide with the parameters of the surrounding medium.

156.003 **The SRC near-infrared survey of the galactic plane.**
M. Hartley, J. A. Dawe.
Proc. Astron. Soc. Australia, Vol. 4, 251 - 254 (1981).
The two main uses of the SRC near infrared atlas are likely to be the investigation of regions of heavy obscuration within the Galaxy and the identification of infrared sources with their optical counterparts. The first batch of 20 fields has already been issued and the atlas is scheduled to be completed by the launching of the IRAS satellite in August, 1982.

156.004 **WSRT observations of the H110α recombination line in the galactic centre.**
J. D. Bregman, U. J. Schwarz.
Astron. Astrophys., Vol. 112, L6 - L9 (1982).
H110α recombination line observations with the Westerbork Synthesis Radio Telescope of the source Sgr A-West are presented. The authors find line radiation related to the spiral pattern found by Ekers et al. (1982). At some positions there is good agreement with the velocities of the [Ne II] observations of Lacy et al. (1980), mainly in the outer spiral arms. However, some of the [Ne II] compact clouds are not detected in the recombination line.

156.005 **Identification of active star formation regions in the galactic plane.**
A. G. Ananth, B. V. Nagaraja.
Astrophys. J., Vol. 259, 664 - 667 (1982).
The galactic distribution of 2.2 μm infrared brightness derived from discrete sources exhibits higher fluxes at galactic longitudes, $l = 8°$, $l = 18°$, $l = 30°$, and $l = 42°$. At these galactic longitudes the 2.6 mm emission line of CO is strongly correlated with the 115–196 μm far-infrared brightness distribution. The dust temperatures computed for these regions indicate the presence of a local source of heating and suggest that star formation is still active at these locations in the galactic plane.

156.006 **Heliospheric and galactic magnetic field similarities–a cautionary note.**
J. P. Vallée.
Astrophys. J., Lett., Vol. 261, L55 - L58 (1982).
The theoretical configuration for the galactic magneti field proposed recently by Akasofu and Hakamada is confronted with recent observational data. This confrontation casts severe doubts on the ability of this heliospheric mode explain the galactic magnetic fields over sizes ranging from to 30 kpc. The heliospheric model for the magnetic field m be of practical use in another context, that of the recently discovered "magnetic bubbles" of sizes ranging from about 0.1 to 1 kpc, in which stars in an association build a comm shell of matter and magnetic fields via stellar ejecta.

156.007 **Large beam observations of the galactic center at 150, 200, and 300 μm.**
E. Dwek, M. T. Stier, R. F. Silverberg, M. G. Hauser, L. Cheung, T. Kelsall, D. Y. Gezari.
Bull. American Astron. Soc., Vol. 14, 656 (1982). – Abstr

156.008 **Preliminary results of a galactic background surv at 45 MHz.**
M. Bitrán, J. May, J. Aparici.
Rev. Mexicana Astron. Astrofis., Vol. 6, (see 012.017), 7 82 (1981).
Preliminary results of a survey of the galactic backgro at 45 MHz recently carried out at the Maipú Radio Astronc Observatory are shown. These first observations cover the region of the sky between declinations from −30° to −37° and they have been obtained with a multiple beam system.

156.009 **Submillimetre continuum emission of the Galaxy**
J. L. Puget.
The scientific importance of submillimetre observations, (s 012.030), p. 157 - 166 (1982).
The main properties of the thermal emission of interstellar dust are summarized. A review of the most recent submillimeter surveys is given. The main theoretical proble is the origin of the unresolved component detected from th near infrared up to 1 mm. The two models proposed so far really diffuse source associated with the low density ionize medium or the sum of extended sources associated with gia molecular clouds–H II regions complexes are discussed.

156.010 **A neutral hydrogen survey of the southern galacti plane for $|b| < 10°$.**
A. W. Strong, P. A. Riley, J. L. Osborne, J. D. Murray.
Mon. Not. R. Astron. Soc., Vol. 201, 495 - 501 (1982).
The authors describe a recent 21-cm survey of neutral hydrogen with the Parkes 64-m telescope, covering the gala tic plane over the region $12° > l > 245°$, $|b| < 10°$, with a sampling interval of 0.°5 in l and 1° in b. The survey has an rms noise of 1 K for velocity resolution 0.8 km s^{-1}. These data are now available on magnetic tape. Contours of integrated brightness temperature and hydrogen column density are given.

156.011 **Disc-to-halo nonthermal emissivity ratio accordin to a study of the radio background at 10 MHz.**
S. Ninković.
Publ. Astron. Obs. Sarajevo, Vol. 1, No. 1, (see 012.043), p. 219 - 224 (1981).

156.012 **High-resolution images of the galactic centre.**
D. A. Allen. A. R. Hyland, T. J. Jones.
Anglo-Australian Obs., Prepr. No. 171, 8 pp. (1982). – Submitted to Mon. Not. R. Astron. Soc.

156.013 **The galactic nucleus.**
J. W. V. Storey, D. A. Allen.
Anglo-Australian Obs., Prepr. No. 172, 11 pp. (1982). – Submitted to Mon. Not. R. Astron. Soc.

156.014 **Die diffuse galaktische Nahinfrarot-Emission.**
S. Drapatz, S. Hayakawa, H. Zinnecker.
Mitt. Astron. Ges., Nr. 57, p. 306 - 310 (1982).

156.015 **Does the galactic synchrotron radio background originate in old supernova remnants?**
S. Sarkar.
Mon. Not. R. Astron. Soc., Vol. 199, 97 - 108 (1982).
Observations of the galactic synchrotron radio background indicate that the emission arises in localized regions of high emissivity. Various lines of evidence suggest that these are the radiative shells of old supernova remnants in which the synchrotron emissivity is enhanced due to compression of the interstellar magnetic field along with the correlated increase in the energy density of cosmic ray electrons. The intensity of the background can then be understood without needing to evoke higher magnetic fields or cosmic ray electron fluxes. Other explanations for the origin of the background are briefly discussed.

156.016 **Observations of the galactic center in the radio band.**
K. Y. Lo.
The galactic center, (see 012.076), p. 1 - 8 (1982).
The author reviews here the observations and properties of the compact radio source at the galactic center and the distribution of the ionized gas within a parsec of the source. He also discusses the constraints on the nature of the underlying energy source.

156.017 **Infrared observations of the galactic center.**
I. Gatley.
The galactic center, (see 012.076), p. 25 - 32 (1982).
Observations of the thermal emission from dust grains in the inner Galaxy are used to determine the total bolometric luminosity of the central parsec; a value in the range of $1 - 3 \times 10^7 L_\odot$ is obtained. Much of this luminosity originates in the source which ionises the gas in the HII region Sgr A. The dust density in the central parsec is very low, providing much less than one magnitude of visual extinction across this volume. Maps of infrared color temperature show that grains are heated by a luminosity source which is centrally concentrated about the position of the galactic nucleus.

156.018 **New far infrared observations of the central 30′ of the Galaxy.**
W. A. Dent, M. W. Werner, I. Gatley, E. E. Becklin, R. H. Hildebrand, J. Keene, S. E. Whitcomb.
The galactic center, (see 012.076), p. 33 - 41 (1982).
A 45′ × 30′ region around the galactic center was mapped with 1′ resolution at 55 μm and 125 μm using the Kuiper Airborne Observatory. Peaks in temperature of the dust are correlated with centimeter wavelength thermal continuum sources. The distribution of the column density of dust shows minima at the galactic center (Sgr A) and at the position of an H II region complex 10′ to the North.

156.019 **Large beam observations of the galactic center at 150, 200, and 300 μm.**
M. T. Stier, E. Dwek, R. F. Silverberg, M. G. Hauser, L. Cheung, T. Kelsall, D. Y. Gezari.
The galactic center, (see 012.076), p. 42 - 47 (1982).
The authors present 10′ resolution observations of a 2 square degree portion of the galactic plane around the galactic center at 150, 200, and 300 μm. In the galactic center region the mean dust temperature is ~30 K. Both the temperature and the dust-to-gas mass ratio in the galactic center region are typical of a large portion of the galactic plane. The dust responsible for the far-infrared emission may only be a fraction of the dust responsible for the visual extinction toward the galactic center.

156.020 **Balloon observation of the central bulge of our Galaxy in near infrared radiation.**
T. Matsumoto, S. Hayakawa, H. Koizumi, H. Murakami, K. Uyama, T. Yamagami, J. A. Thomas.
The galactic center, (see 012.076), p. 48 - 52 (1982).
The central bulge of our Galaxy was observed by a balloon-borne infrared telescope at 2.4 and 3.4 μm with spatial resolution of 0.5° and 0.8°, respectively. The distribution of surface brightness and color are presented. The bulge is modeled by a spheroid with axial ratio of 0.75. Comparing the model with H I rotation velocity, the authors obtain the mass to luminosity ratio to be 2.0 ± 0.4.

156.021 **Radio observations of the galactic center: future directions.**
R. L. Brown.
The galactic center, (see 012.076), p. 204 - 207 (1982).
The role of a new generation of millimeter-wave radio telescopes in investigating the dynamics and the nature of the galactic center is outlined in this review.

Landolt-Börnstein. Group VI. Vol. 2c. Interstellar matter – Galaxy – Universe.
See Abstr. 003.055.

Telescope beam characteristics and temperature scale of the Maryland-Green Bank 21-cm Line Survey.
See Abstr. 033.001.

Magnetic monopoles and the survival of galactic magnetic fields.
See Abstr. 061.053.

A magnetohydrodynamical theorem for steady axisymmetric systems with implications to ring galaxies.
See Abstr. 062.072.

Vorticity and astrophysical magnetism.
See Abstr. 062.110.

Contraints on cosmic magnetic monopoles imposed by the galactic magnetic field.
See Abstr. 066.234.

Champs magnétiques interstellaires et intergalactiques.
See Abstr. 131.281.

Atomic hydrogen in the outer Milky Way.
See Abstr. 155.014.

Near-infrared source counts in the galactic plane.
See Abstr. 155.035.

The distribution of neutral atomic hydrogen in our Galaxy beyond the solar circle.
See Abstr. 155.045.

Radio studies of galactic structure.
See Abstr. 155.053.

IAU Commission 33: Structure and dynamics of the galactic system.
See Abstr. 155.065.

Galactic and intergalactic Faraday rotation.
See Abstr. 161.001.

157 Galaxy (UV, X, Gamma Radiation)

157.001 **Far-ultraviolet studies. VI. Further limits on diffuse galactic light scattered at large angles by dust.**
R. C. Anderson, R. C. Henry, W. G. Fastie.
Astrophys. J., Vol. 259, 573 - 590 (1982).
The Apollo 17 wide-field (12° X 12°) far-ultraviolet (1180–1680 Å) spectrometer was used to scan six large bands of the sky during trans-Earth coast. After accounting for internal scattered light and the light expected from stars in the field of view, no significant residual flux remains. This imposes important constraints on the scattering properties of the interstellar dust grains: either the grains are extremely strongly forward scattering, or the albedo of the grains is low.

157.002 **A large scale height galactic component of the diffuse 2–60 keV background.**
D. Iwan, F. E. Marshall, E. A. Boldt, R. F. Mushotzky, R. A. Shafer, A. Stottlemyer.
Astrophys. J., Vol. 260, 111 - 123 (1982).
The diffuse 2–60 keV X-ray background has a galactic component clearly detectable by its strong variation with both galactic latitude and longitude. This galactic component is typically 10% of the extragalactic background toward the galactic center, half that strong toward the anticenter, and extrapolated to a few percent of the extragalactic background toward the galactic poles. It is acceptably modeled by a finite radius emission disk with a scale height of several kiloparsecs. The most likely source of this emission is low luminosity stars with large scale heights such as subdwarfs.

157.003 **COS-B observation of the Milky Way in high-energy gamma rays.**
H. A. Mayer-Hasselwander, K. Bennett, G. F. Bignami, R. Buccheri, N. D'Amico, W. Hermsen, G. Kanbach, F. Lebrun, G. G. Lichti, J. L. Masnou, J. A. Paul, K. Pinkau, L. Scarsi, B. N. Swanenburg, R. D. Wills.
Ninth Texas Symposium on Relativistic Astrophysics, (see 012.020), p. 211 - 222 (1980).

157.004 **Gamma rays, cosmic rays, and the dynamics of the Galaxy.**
C. J. Cesarsky.
Ninth Texas Symposium on Relativistic Astrophysics, (see 012.020), p. 223 - 233 (1980).
The galactic gamma-ray emission has recently been observed by the SAS-2 and COS-B satellites; the author summarizes the impact of these results on our knowledge of galactic cosmic rays and the interstellar medium.

157.005 **Gamma-ray lines from the Galactic Center.**
M. Leventhal, C. J. MacCallum.
Ninth Texas Symposium on Relativistic Astrophysics, (see 012.020), p. 248 - 266 (1980).
The history of the detection and identification of the first gamma-ray lines from outside the solar system is outlined. The gamma-ray lines come from the Galactic Center direction and the principal line of interest is the 511 keV line from electron–positron annihilation.

157.006 **Positron annihilation radiation from the galactic center region.**
A. S. Jacobson.
Tenth Texas Symposium on Relativistic Astrophysics, (see 012.021), p. 330 - 337 (1981).
This is a report on measurements of 0.511 MeV radiation from the vicinity of the galactic center performed by the High Resolution Gamma-Ray Spectrometer aboard HEAO 3.

157.007 **Gamma ray lines from the galactic center and ga ray transients.**
R. Ramaty, D. Leiter, R. E. Lingenfelter.
Tenth Texas Symposium on Relativistic Astrophysics, (see 012.021), p. 338 - 346 (1981).
The authors discuss in this paper the interpretations a implications of astrophysical gamma ray line observations. Such lines have so far been seen from the galactic center a in gamma ray transients. The 0.511 MeV line from the galactic center is time variable, a result whose consequence that the positrons are produced by a discrete source, prob a massive black hole, within an annihilation region no larg than a light year at the galactic center. Gamma ray lines se in the spectra of gamma ray bursts strongly suggest that neutron stars are the sources of many of these bursts.

157.008 **Observation of galactic gamma ray lines.**
K. Okudaira.
Astron. Her., Vol. 75, 42 - 46 (1982). In Japanese. – Abst Phys. Abstr., Vol. 85, Abstr. 79245 (1982).

157.009 **Galactic gamma-ray astronomy.**
C. J. Cesarsky, J. A. Paul.
Nucl. Sci. Appl. Sect. A, Vol. 1, 191 - 225 (1981). – Abst Phys. Abstr., Vol. 85, Abstr. 106964 (1982).

157.010 **Diffuse galactic gamma-ray line emission from nucleosynthetic ^{60}Fe, ^{26}Al, and ^{22}Na: prelimina limits from HEAO 3.**
W. A. Mahoney, J. C. Ling, A. S. Jacobson, R. E. Lingenfe
Astrophys. J., Vol. 262, 742 - 748 (1982).
Data obtained during a 2 week period with the HEA(gamma-ray spectroscopy experiment have been searched f diffuse galactic plane gamma-ray line emission expected tc result from the decay of nucleosynthetic ^{60}Fe, ^{26}Al, and ^{2} With the possible exception of the 1809 keV line from ^{26}/ decay no positive detection was made. New limits have bee placed on diffuse emission in these lines from the vicinity the galactic center. These limits place new constraints on t yields of these radionuclides in explosive nucleosynthesis a on the present rate of galactic nucleosynthesis.

157.011 **Photographic surface photometry of the Milky Way. IV. The northern Milky Way in the ultraviolet spectral region.**
C. Winkler, T. Schmidt-Kaler, W. Schlosser.
Astron. Astrophys., Vol. 115, 115 - 127 (1982). In Germa
This is paper IV in a series dealing with photographic surface photometry derived from super wide-angle photographs. The data comprise the region $40° \leqslant l \leqslant 210°$ and $-41° \leqslant b \leqslant +41°$ with an angular resolution of 0.°6 × 0.°6. Since no simultaneous photoelectric measurements could t made, modified methods to determine the photographic calibration curve and the contribution of the airglow had t be developed. For the sake of clear representations, data w low intensities $I_{gal} < 60$ S 10U were smoothed over 1.°8 × 1 in the isophote map.

157.012 **Ultraviolet spectrum of the sky background at different galactic latitudes.**
A. M. Zvereva, A. B. Severny (*Severnyj*), L. V. Granitzky (*Granitskij*), C. T. Hua, P. Cruvellier, G. Courtès.
Astron. Astrophys., Vol. 116, 312 - 322 (1982).
The distributions in intensity of the sky background radiation in the range λλ 1100–1850 Å were measured in deep space (distances 70,000–200,000 km from the earth) with the photoelectric spectrometer ("Galaktika") on boar the satellite "Prognoz-6". The spectral distributions of *UV*

background radiation I_λ, after subtraction of stellar component, vary with galactic latittude in the interval $-58° < b^{II} < +27°$. The intensity $I(\lambda = 1600)$ in higher latitude range $|b^{II}| > 30°$ shows good correlation with soft X-ray brightness and neutral hydrogen (λ 21 cm) density N(H I). However there is an appreciable UV emission near the galactic poles where $N(\text{H I}) \cong 0$. The authors propose a possible explanation of the emission I_λ for higher latitudes by the hot component of interstellar gas of supernova remnants, as shown by the comparison of the observed spectral distributions with those calculated for a hot gas.

157.013 **Gamma-ray emission from the galactic anticenter at MeV energies.**
U. Graser, V. Schönfelder.
Astrophys. J., Vol. 263, 677 - 689 (1982).

An image of the anticenter region of the Galaxy between right ascension 50° and 110° and between declination +10° and +50° in the energy range 1.1–10 MeV is derived from data obtained during a balloon flight with the MPI Compton telescope. A significantly enhanced γ-ray emission is observed along the galactic plane from $l^{II} = 160°$ to $l^{II} = 197°$. Part of the emission is due to the Crab Nebula. The energy spectrum of the pulsed and total Crab emission is determined. An upper limit to the diffuse galactic γ-ray emission is determined, which leads to restrictions of the spectrum of interstellar cosmic ray electrons at MeV energies.

157.014 **Observations of continuum X-ray and gamma-ray emission from the galactic center.**
J. L. Matteson.
The galactic center, (see 012.076), p. 109 - 122 (1982).

Observations of the X-ray and gamma-ray continuum emission from the galactic center region are reviewed with emphasis on those which are most relevant to the galactic nucleus.

157.015 **Observations of gamma-ray line emission from the galactic center region.**
A. S. Jacobson.
The galactic center, (see 012.076), p. 123 - 131 (1982).

A series of observations of the galactic center 511 keV electron-positron annihilation line spanning a decade has begun to severely constrain theoretical ideas concerning the nature of the source. This paper reviews all available line-flux measurements and discusses physical limits for the size, the temperature and the mass of the line-emitting region.

157.016 **Time variable positron annihilation radiation from the galactic center direction.**
M. Leventhal, C. J. MacCallum.
The galactic center, (see 012.076), p. 132 - 138 (1982).

The authors' balloon-borne Ge gamma-ray telescope has recently observed for a third time the galactic center 511 keV positron annihilation line. The source was found to be in an "off" or "low" state. A preliminary flux estimate of $0 \pm 4 \times 10^{-4}$ photons $cm^{-2} s^{-1}$ has been made. This result is discussed in the context of other observations and theoretical models.

157.017 **Observations of the galactic center with the GSFC Low-Energy Gamma-Ray spectrometer: preliminary results.**
W. S. Paciesas, T. L. Cline, B. J. Teegarden, J. Tueller, P. Durouchoux, J. M. Hameury.
The galactic center, (see 012.076), p. 139 - 143 (1982).

The GSFC Low-Energy Gamma-Ray Spectrometer observed the galactic center on 20 November, 1981, during a balloon flight. The positron annihilation line at 511 keV showed no excess over background when the galactic center was in the field-of-view; the inferred upper limit to the intensity of such a feature is 1.2×10^{-3} photons/cm^2-s. Continuum emission was observed between 70 keV and 1 MeV.

157.018 **Emission in the 0.3 to 1.0 MeV range from the galactic center region.**
B. M. Gardner, D. J. Forrest, P. P. Dunphy, E. L. Chupp.
The galactic center, (see 012.076), p. 144 - 147 (1982).

The University of New Hampshire's Large Scientillator Gamma Ray Telescope observed the galactic center region on 21–23 November 1977, ten days after a similar observation by the Bell and Sandia Laboratories high resolution Ge gamma ray telescope. The UNH measurements are consistent with line emission near 511 keV with a flux that is at least twice as intense as the earlier measurement.

157.019 **On the origin of the positron annihilation radiation from the direction of the galactic center.**
R. E. Lingenfelter, R. Ramaty.
The galactic center, (see 012.076), p. 148 - 159 (1982).

The authors review the observations of 0.511 MeV positron annihilation radiation from the direction of the galactic center, and discuss the implications of the line intensity, shape and time dependence both on the nature of the positron annihilation region and on the source of the annihilating positrons. They find that the positrons are most likely produced by photon-photon collisions in the vicinity of a compact object at the galactic center.

157.020 **The intensity and spectrum of galactic center β^+ annihilation photons after Compton scattering.**
D. J. Forrest.
The galactic center, (see 012.076), p. 160 - 164 (1982).

The author demonstrates that the observed characteristics of the 300 - 511 keV gamma-ray spectrum from the galactic center can be explained by a model including Compton scattering of a narrow emission line at 511 keV.

17th International Cosmic Ray Conference.
OG session: origin and galactic phenomena.
See Abstr. 012.070.

Features of the soft X-ray background.
See Abstr. 142.023.

Summary of the workshop on ultraviolet cosmic background radiation.
See Abstr. 142.082.

Sources de rayonnement γ, origine du rayonnement cosmique, et régions de formation d'étoiles.
See Abstr. 142.549.

IAU Commission 33: Structure and dynamics of the galactic system.
See Abstr. 155.065.

158 Single and Multiple Galaxies, Peculiar Objects

158.001 **VLA observations at 6 and 20 cm of optical jet-like features associated with the peculiar galaxy MCG 5-29-86.**
Y. Sofue, Y. Fukui, M. Fujimoto, S. Deguchi, K. Wakamatsu.
Astron. J.,Vol. 87, 980 - 986 (1982).

The galaxy MCG 5-29-86 is associated with a peculiar pair of symmetric optical jet-like features emerging from the nuclear region perpendicular to the major axis. VLA observations at 6 and 20 cm of this peculiar object show that both the main body of the galaxy and the "jets" are not radio active. The authors suggest that the jets may be an edge-on view of a thin disk or a ring of old population stars rotating around a prolate central bulge of the galaxy.

158.002 **Do dwarf galaxies have massive halos?**
S. van den Bergh.
Astron. J., Vol. 87, 987 - 989 (1982).

It is pointed out that all presently available evidence for the existence of massive halos is derived from observations of giant galaxies. In the present paper, data from "A Revised Shapley Ames Catalog of Bright Galaxies" are used to show that faint Sc galaxies are more likely to exhibit bars than are bright Sc's. A possible interpretation of this result is that the dynamics of dwarf disk systems are less dominated by massive halos than is the case for giant spirals.

158.003 **New *UBV* parameters for 46 E-S0 galaxies in the Virgo cluster.**
R. Michard.
Astron. Astrophys., Suppl. Ser., Vol. 49, 591 - 606 (1982).

Multi-aperture *UBV* photometry of 46 Virgo cluster E-S0 galaxies has been obtained. This material has been checked and completed with other photoelectric data in the literature, to provide a hopefully improved set of the parameters defined in the Second Reference Catalogue of Bright Galaxies, i. e. asymptotic magnitude B_T, effective aperture diameters A_e, and colors within A_e. Interpolated *UBV* parameters at the standard aperture of 1′ are also given, to provide calibration of surface photometry. Various known correlations are rediscussed with this material. It appears that the color-magnitude diagram $(U-V)_e - V_T$ of E-S0 galaxies in Virgo is curved.

158.004 **High resolution H I observations of Messier 31.**
E. Bajaja, W. W. Shane.
Astron. Astrophys., Suppl. Ser., Vol. 49, 745 - 774 (1982).

High-resolution (24″ X 37″) observations of the central part of M 31 in the 21-cm line of H I are described. The results are presented as contour maps at constant velocity and as position-velocity contour maps on a rectangular grid. Published parameters are used to construct a kinematic model for comparison with the observed data. Several isolated dark clouds in the central region are also observed in H I.

158.005 **Redshifts and spectroscopy of very distant radio galaxies with strong emission lines.**
H. Spinrad.
Publ. Astron. Soc. Pacific, Vol. 94, 397 - 403 (1982).

The optical spectra of six newly observed 3CR radio galaxies, mostly with redshifts near unity, are unique with enormously strong [O II] emission. The observed equivalent widths of λ_0 3727 range from 100 Å to about 2000 Å, and the luminosity in this single line exceeds $10^{10}\ L_\odot$ in the extreme case of 3C 368 ($z = 1.13$). The coarse line ratios and linewidths of [O II], [Ne III], [Ne V], and the occasional presence of Mg II suggest a shock-excitation model for the gas in 3C 368, 3C 184, 3C 175.1, and possibly 3C 34. 3C 356 and 3C 280 have moderately strong [Ne V] and may require a UV power-law continuum. A rough inverse-correlation exists between the emission lines velocity width and relative strengths of the h ionization [Ne V] and the strong low-ionization [O II] lines The shape of the thermal continuum of 3C 280 has been m sured. The author notes that these very luminous radio gala are the ones which experience rapid evolution needed to explain the radio source counts. They are about 300 times more abundant (per unit co-moving volume) at $z \cong 1.0$ thar locally ($z \sim 0.1$).

158.006 **Photometry of southern galaxies.**
C. J. Peterson.
Publ. Astron. Soc. Pacific, Vol. 94, 404 - 408 (1982).

Multiaperture photoelectric *V* and $(B-V)$ measuremen of 74 southern galaxies are presented.

158.007 **Circular motions in the barred galaxy NGC 3504.**
C. J. Peterson.
Publ. Astron. Soc. Pacific, Vol. 94, 409 - 414 (1982).

A kinematical study of the galaxy NGC 3504 (Hubble type SABab) finds that the velocity field is dominated by circular motions, with no evidence (within the observationa error) of noncircular gas streaming motions of the type suggested by barred galaxy models. The rotation curve shows t characteristic spiral galaxy pattern of a rapid central rise to radius of 6 arc seconds beyond which point the velocity continues to rise but at a smaller rate; these new data do no confirm the form of the rotation curve found in the prior study by Burbidge, Burbidge, and Prendergast (1961).

158.008 **A search for carbon monoxide in clumpy irregula galaxies.**
M. A. Gordon, J. Heidmann, E. E. Epstein.
Publ. Astron. Soc. Pacific, Vol. 94, 415 - 420 (1982).

The failure to detect the 115 GHz line of CO in the ga axies Ma 7, Ma 8, Ma 296, Ma 297, Ma 325, Ma 432, and VV 523 leads to limits on the column density and mass of carbon monoxide. The upper limits to the masses of CO in each of these galaxies, relative to the masses of atomic hydr gen, and of the entire galaxies, tend to be comparable to or less than what is observed for our galaxy. If all spectra are averaged together to give a limit for a composite galaxy, the mass of CO relative to that of atomic hydrogen is about 8 times less than observed for our galaxy.

158.009 **A catalog of diffuse nebulae and emission-line sta in NGC 6822.**
R. M. Killen, R. J. Dufour.
Publ. Astron. Soc. Pacific, Vol. 94, 444 - 452 (1982).

A catalog of 31 diffuse nebulae and 36 stellar emission line objects in the Local Group irregular galaxy NGC 6822 i represented. Identifications of the objects were made from several wide-band and interference filter plates of NGC 682 obtained with the CTIO 4-m telescope at the prime focus under conditions of excellent seeing. The numbers and sizes of the diffuse nebulae in NGC 6822 are similar to those of the population of H II regions in the Small Magellanic Clou The 36 stellar emission-line objects found in NGC 6822 app to be emission-line stars, small H II spheres, and planetary nebulae.

158.010 **Distance and absolute magnitudes of the brightes stars in the dwarf galaxy Sextans A.**
A. Sandage, G. Carlson.
Astrophys. J., Vol. 258, 439 - 456 (1982).

The color-magnitude diagram, apparent magnitudes of brightest red and blue stars, and photometry for Cepheid variables are given for the nearby Im V dwarf galaxy, Sextans A. The mean magnitudes of the three brightest blue stars are $\langle V(3)\rangle = 17.98$ and $\langle B(3)\rangle = 17.88$. The three brightest red supergiants have $\langle V(3)\rangle = 18.09$ and $\langle B(3)\rangle = 20.14$. Periods and magnitudes measured for five Cepheids give an apparent blue distance modulus of $(m-M)_{AB} = 25.67 \pm 0.2$ via the P–L relation. With $A_B = 0.07$ for the foreground galactic absorption, the mean absolute magnitudes for the red supergiants of $M_V(3) = -7.56$ and $M_B(3) = -5.53$ provide an additional calibration for the brightest red stars as distance indicators. The potential use of these results in determining the distance of the Virgo cluster directly is outlined.

158.011 **Extragalactic CO: gas distributions which follow the light in IC 342 and NGC 6946.**
J. S. Young, N. Scoville.
Astrophys. J., Vol. 258, 467 - 489 (1982).

The CO emission in the two Scd galaxies, IC 342 and NGC 6946, has been mapped using the 14 m FCRAO telescope (HPBW = 50″). In each galaxy, the radial distribution of CO out to R ≈ 10 kpc exhibits a falloff which follows the exponential luminosity profile from the stellar disk. Neither galaxy shows a breach in the molecular gas like that found in our Galaxy between the nucleus and the molecular cloud annulus. The occurrence of this gap in our own Galaxy, but in neither of the external galaxies, suggests a possible link between the deficiency of gas and the presence of either a nuclear bulge or inner Lindblad resonance (ILR). The CO rotation curves for IC 342 and NGC 6946 agree well with previous H I observations; at the present resolution these galaxies appear to undergo nearly rigid rotation inside R = 5 kpc and therefore have no ILRs and have small nuclear bulges. A global spiral structure is not evident from the molecular data.

158.012 **Isolated triplets of galaxies. Virial mass-to-light ratios.**
V. E. Karachentseva, I. D. Karachentsev.
Astrofizika, Tom 18, 5 - 16 (1982). In Russian. English translation in Astrophysics, Vol. 18, No. 1.

For 41 triplets of galaxies with measured radial velocities of each component the velocity dispersion, linear dimension, integral luminosity, virial mass, and mass-to-luminosity ratio are presented. Similar values are determined for a sample of simulated triplets which have been constructed by application of the same triple system isolation criterion to apparent distribution of galaxies with $m_p \leqslant 15^m\!.7$. A comparison of these two samples have revealed unexpectedly a high number of spurious triplets. The problem of distinguishing physical and optical triple systems of galaxies is discussed. After elimination of spurious triple systems the rest of the suspected physical triplets have the following mean parameters: linear dimension ~ 50 kpc, peculiar velocity of a member ~70 km/sec, integral luminosity ~ $5 \times 10^{10} L_\odot$, and virial mass-to-luminosity ratio ~$10 f_\odot$.

158.013 **Morphology of compact galaxies. II.**
F. Börngen, A. T. Kalloglyan.
Astrofizika, Tom 18, 17 - 24 (1982). In Russian. English translation in Astrophysics, Vol. 18, No. 1.

The results of brightness profile investigation of 119 compact galaxies in the vicinity of the A 2255 cluster, 47 Zwicky compacts and 96 galaxies in A 2255 are given.

158.014 **Once more on the radio emission of the nuclei of spiral galaxies.**
G. M. Tovmasyan.
Astrofizika, Tom 18, 25 - 29 (1982). In Russian. English translation in Astrophysics, Vol. 18, No. 1.

The results of new radio observations of spiral galaxies prove the reality of the earlier noted correlation between the radio emission of galaxies and the optical evidence of activity of their nuclei: the centrally located compact radio sources are detected mainly in those galaxies in which split, starlike or semistellar nuclei are observed. It has also been shown that surface brightnesses of galaxies with relatively large and bright central condensations are in the mean greater, when central compact radio sources are present.

158.015 **Spectral observations of the galaxy M 82.**
A. S. Amirkhanyan, V. A. Hagen-Thorn, V. P. Reshetnikov.
Astrofizika, Tom 18, 31 - 35 (1982). In Russian. English translation in Astrophysics, Vol. 18, No. 1.

Two spectrograms of the filamentary structure of the galaxy M 82 have been obtained. The presence of Doppler ellipses in spectral lines in the southern part of the galaxy is confirmed. However, in the northern part Doppler splitting has not been found. The authors suppose that a highly energetic local explosion has occurred in the southern part of the galaxy. As a result an expanding supergiant dust shell has been formed. The optical polarization of M 82 is due to the scattering of the light of bright central regions of the galaxy by this shell.

158.016 **Galaxies of very elongated shape.**
B. I. Fesenko.
Astrofizika, Tom 18, 37 - 48 (1982). In Russian. English translation in Astrophysics, Vol. 18, No. 1.

Systems with the ratio of the smaller diameter to the larger one (D) not exceeding 0.10 and with $D \geqslant 1\!'.0$ have been called F-galaxies. The distributions of F-galaxies in the space as is evident from their angular diameter distribution are uniform without any indication on the existence of superclusters. That conclusion is true for the space distribution of ~30% of all galaxies from Nilson's catalogue (1973). Anisotropy in the distribution of the position angles of the F-galaxies is investigated and the conclusion is drawn that it is stipulated by observational errors.

158.017 **VLA observations of the giant cluster spiral UGC 12423.**
G. D. Bothun, B. Balick, E. Skillman.
Astron. J., Vol. 87, 1098 - 1105 (1982).

The authors report the results of 21-cm VLA observations of the late-type spiral galaxies UGC 12423 and NGC 7591, both located in the Pegasus I cluster of galaxies. The optical image of UGC 12423 reveals the galaxy to be highly flattened, possessing a thin optical disk that is noticeably warped. Its 21-cm observations show a similar morphology for the H I and suggests a rising rotation curve as well. These features may explain the galaxy's discrepant location in the infrared Tully-Fisher diagram. The observations of UGC 12423 reveal that the total linear extent of the H I is at least 60 kpc which exceeds the optical image of the galaxy by about 50%. Comparable conclusions apply to NGC 7591. The existence of extended H I in cluster galaxies may imply that a significant interaction with the environment has not yet occurred.

158.018 **Photometric properties of bright early-type spiral galaxies. IV. Multiaperture *UBVJHK* photometry for the inner (bulge) regions of 65 galaxies.**
D. Griersmith, A. R. Hyland, T. J. Jones.
Astron. J., Vol. 87, 1106 - 1126 (1982).

New *UBVJHK* multiaperture photometry is presented for 65 early-type spirals. The authors have derived colors and magnitudes at a small fraction of the galaxy size. They derive the following results: (1) For colors such as $U-V$, $V-K$, and $J-K$ the authors find radial color gradients in the inner (bulge) regions of normal (class A) early-type spirals that are measurable, but smaller than those measured previously in E and S0 galaxies. (2) The authors find no significant dependence of colors on either inclination or absolute magnitude. (3) The *UBVJHK* colors for the inner (bulge) regions of normal (class A) early-type spirals are constant with type T over the range $T = -1$ (late S0) to $T = 3$ (Sb). (4) The *UBVJHK* bulge colors for the majority of class A early-type spirals are very similar to the colors of ellipticals of similar luminosity. (5) In the *UVK* and *JHK* two-color diagrams, about one quarter of the class A early-type spirals have unusual colors compared to the majority of spirals or ellipticals. (6) Compared with normal (class A) galaxies without nuclear activity, emission-line (class B) and Seyfert early-type spirals generally have bluer *UBV* colors and redder *JHK* colors. (7) The color distributions for the inner (bulge) regions of all early-type spirals (classes A and B plus Seyferts) are normally distributed and similar to those of ellipticals at *UBV* wavelengths.

158.019 **Spectrophotometry and morphology of galaxies with UV excess. IV.**
M. A. Kazaryan, V. S. Tamazyan.
Astrofizika, Tom 18, 192 - 200 (1982). In Russian. English translation in Astrophysics, Vol. 18, No. 2.

The results of spectrophotometry and morphology of the galaxy No. 26 from list of Kazaryan (1979) are presented. The galaxy shows spiral structure with starlike nucleus. In one of the arms three condensations are marked. Two condensations are probably superassociations. The data of *UBV* photometry of the nucleus and condensations are given. The spectrum of the galaxy shows emission lines. Equivalent widths and relative intensities of the emission lines are obtained. The relative abundance of some elements, electron density and temperature are calculated. An investigation of the continuum spectrum is carried out.

158.020 **Physical conditions in interacting galaxies, components of isolated pairs and isolated galaxies.**
V. A. Dostal'.
Astrofizika, Tom 18, 201 - 214 (1982). In Russian. English translation in Astrophysics, Vol. 18, No. 2.

The spectra of 16 interacting systems (VV), 9 isolated pairs of galaxies and 11 isolated galaxies (IG) are investigated. The electron temperature of gas in VV and IG varies from $\sim 5000°$ to $\sim 7000°$. The electron density (n_e) of the gas in VV and in 35% of IG is about 10^2 - 10^3 cm^{-3} while in 65% of IG $n_e \lesssim 1$. A group of VV galaxies with higher L_c radiation compared to isolated galaxies is discovered.

158.021 **Observations of galaxies with ultraviolet continuum at 102 MHz.**
V. A. Artyukh, R. A. Kandalyan, M. A. Ogannisyan, V. A. Sanamyan.
Astrofizika, Tom 18, 215 - 226 (1982). In Russian. English translation in Astrophysics, Vol. 18, No. 2.

The results of the interplanetary scintillation at 102 MHz of 87 galaxies with ultraviolet continuum are presented. The angular sizes of the scintillating components for some of the galaxies are estimated. For 9 objects radio spectra are constructed for a wide range of frequencies. The radio properties of Seyfert galaxies at low frequencies are discussed.

158.022 **On the activity of nuclei of double systems.**
G. M. Tovmasyan.
Astrofizika, Tom 18, 227 - 233 (1982). In Russian. English translation in Astrophysics, Vol. 18, No. 2.

It is shown that in components of pairs of galaxies compact nuclear radio sources are observed about four times more often than in isolated galaxies. Components of double galaxies often have also split, stellar or semistellar nuclei which are in an active stage of their evolution.

158.023 **Observations of galaxies of high surface brightness at 3.66 GHz. II.**
V. G. Malumyan, G. A. Oganyan.
Astrofizika, Tom 18, 324 - 327 (1982). In Russian. English translation in Astrophysics, Vol. 18, No. 2.

The results of observations of 20 galaxies of high surface brightness with the radio telescope RATAN-600 are presented.

158.024 **The distribution of H II regions in external galaxies. I.**
S. Considère, E. Athanassoula.
Astron. Astrophys., Vol. 111, 28 - 42 (1982).

Using the method of Fourier transforms introduced b Kalnajs (1975) the authors analyse H II region distribution in terms of logarithmic spirals. This is an objective way of retrieving a lot of the information contained, and sometim well hidden, in such distributions. The authors apply it to four galaxies, M 31, M 51, NGC 2997, and M 33, of type *b* *c*, and *cd*, for which the number of H II regions observed i: large enough to permit an analysis.

158.025 **Spectrophotometry of four galaxies of high surf: brightness.**
M. A. Arakelyan, A. P. Magtesyan.
Pis'ma Astron. Zh., Tom 8, 334 - 338 (1982). In Russian. English translation in Soviet Astron. Lett., Vol. 8.

Spectrophotometry of emission lines for the nuclei of galaxies of high surface brightness Nos. 428, 449, 454 a 532 from the Arakelyan (1975) list is carried out. Clouds c ionized gas are detected at the distances of ~ 2 kpc from th nuclei of the two former galaxies. Besides there seems to b a cloud moving along the line of sight with velocity ~ 1500 km/s in the galaxy No. 449.

158.026 **Structure of the spiral arm S6 in the Andromeda galaxy.**
Yu. N. Efremov.
Pis'ma Astron. Zh., Tom 8, 339 - 346 (1982). In Russian. English translation in Soviet Astron. Lett., Vol. 8.

The distribution of stars across the spiral arm S6 in M within Baade's Field IV is investigated. It is found that the brightest stars are concentrated to the middle line of the a where the maximal star density is also observed. The simila between the structure of the arm S6 and that of the spiral arms near the sun is pointed out.

158.027 **An 80 day X-ray light curve of 3C 371.**
W. A. Snyder, K. S. Wood, D. J. Yentis, J. F. Meekins, H. W. Smathers, E. T. Byram, T. A. Chubb, H. Friedman.
Astrophys. J., Vol. 259, 38 - 47 (1982).

The Large Area Sky Survey experiment on HEAO 1 h monitored the N galaxy/BL Lacertae object 3C 371 from 1977 August 19 to November 7. Statistically significant threefold or larger intensity variations were observed in the 0.5–20 keV flux. The authors also observed an X-ray flare lasted ~ 25 days. The total 80 day X-ray light curve suggest either a superposition of X-ray flares or a variable injection rate that changes on weekly time scales. The implications o these results for radiation models based on synchrotron emission are outlined.

158.028 **On the physical conditions and the velocity field Seyfert 1 galaxies and QSOs.**
J. M. Shuder.
Astrophys. J., Vol. 259, 48 - 54 (1982) = Lick Obs. Bull., No. 916.

Measurements are presented which give the functional form of the Hβ/Hα and He I λ5876/Hα ratios with cloud velocity in Seyfert 1 galaxies. Combined with present mod of the line-emitting regions, these measurements indicate t the density or ionization parameter increases with increasir velocity. Arguments are presented which suggest that the li emitting clouds with the highest velocities are located close to the central source of ionizing radiation.

158.029 **Neutral hydrogen in elliptical galaxies with nucle radio sources and optical emission lines.**
L. L. Dressel, T. M. Bania, R. W. O'Connell.
Astrophys. J., Vol. 259, 55 - 66 (1982).

The authors have made an H I detection survey of 11 elliptical galaxies with powerful nuclear radio sources, usin the 305 m antenna of Arecibo Observatory, to test the hypothesis that large H I mass is conductive to the formation of nuclear radio sources in elliptical galaxies. They have combined their data with data from other H I surveys and spectroscopic surveys to search for correlations of H I mass with other galactic properties. Strong correlations of [O II] λ3727 emission with H I content and with nuclear radio power are found.

158.030 **The ultraviolet spectrum of the center of M 31.**
G. A. Welch.
Astrophys. J., Vol. 259, 77 - 88 (1982).

A total of twelve long-exposure low-resolution IUE images have been used to study the spectrum and surface brightness distribution of the ultraviolet radiation within the central 50 pc of M31. The presence of weak absorption features near 1300 Å demonstrates that the radiation of the well-known spectral upturn below 2000 Å does not originate in young metal-rich stars. The most likely sources are considered to be hot, highly evolved stars.

158.031 **An investigation of the star-burst model for radio emission from Seyfert galaxies.**
J. S. Ulvestad.
Astrophys. J., Vol. 259, 96 - 102 (1982).

The possibility that radio sources in Seyfert galaxies are generated by bursts of star formation has been investigated. The galactic Σ–D relation has been used to estimate the supernova rates necessary to produce the observed radio emission in Seyferts. Required rates ranging from ~1 to 100 per year have been found, with a median value of ~10 per year. Recently created radio supernovae, as distinct from older supernova remnants, cannot lower these required rates significantly. The large numbers of massive stars needed to make the high supernova rates would radiate an amount of energy in excess of the total luminosities of many Seyfert galaxies.

158.032 **The cluster system of M 33.**
C. A. Christian, R. A. Schommer.
Astrophys. J., Suppl. Ser., Vol. 49, 405 - 424 (1982).

The authors have compiled a catalog of more than 250 nonstellar objects found on high resolution direct plates of M 33. Additional classification criteria were derived from low dispersion 4 m grism plates and multicolor photographs. *BVR* photometry of 60 of these objects, believed to be star clusters, is presented, especially in comparison to similar observations of clusters in other Local Group galaxies. The M 33 clusters populate a fairly uniform region between $0.0 < B-V < 0.8$ mag and $17.0 < V < 19.5$ mag in an integrated color-magnitude diagram.

158.033 **Variability of the emission-line spectra and optical continua of Seyfert galaxies. I.**
B. M. Peterson, C. B. Foltz, P. L. Byard, R. M. Wagner.
Astrophys. J., Suppl. Ser., Vol. 49, 469 - 486 (1982).

New observations of 27 broad line emitting galaxies and six narrow line emitting galaxies have been obtained with the Ohio State University image-dissector scanner. The line strengths of Hβ and [O III] λλ4959, 5007 measured from these data are compared with earlier measurements by other investigators, and it is found that most (~80%) broad line emitting galaxies show significant changes in their optical spectra on time scales less than ~5 yr.

158.034 **On the relation between true and apparent flattenings of elliptical galaxies.**
R. Scuflaire.
Astron. Astrophys., Vol. 111, 371 (1982).

The integral equation relating the distribution of shapes of elliptical galaxies to the distribution of shapes of their images takes a simpler form when the eccentricity is used instead of the axes ratio. This equation may be reduced to Abel's equation and its solution may be expressed analytically.

158.035 **Surface photometry and mass distributions of the interacting spiral galaxies NGC 5426 and 5427.**
C. P. Blackman.
Mon. Not. R. Astron. Soc., Vol. 200, 407 - 429 (1982).

Rotation curves of the interacting spiral galaxies NGC 5426 and 5427 have been measured using long slit spectrograms and 'error free' luminosity profiles in *B, V* and *J* passbands have been derived from sky limited plates. Both galaxies have flat rotation curves outside the nucleus. Masses are $3.5 \times 10^{10} M_\odot$ ($r < 8$ kpc) and $2.9 \times 10^{10} M_\odot$ ($r < 9$ kpc) for NGC 5426 and 5427 respectively for a distance of 26 Mpc. Theoretical spiral patterns based on density wave theory and the observed rotation curves have been derived. These agree well with the observed spiral arms. The relative motion of the galaxies is such that no 'missing mass' is required for them to form a bound system.

158.036 **Optical spectra and redshifts for seventy-one galaxies.**
H. G. Corwin, Jr., D. Emerson.
Mon. Not. R. Astron. Soc., Vol. 200, 621 - 644 (1982).

Spectra of 71 galaxies have been obtained with an image tube spectrograph on the 1.9-m telescope of the South African Astronomical Observatory at a reciprocal dispersion of 215 Å mm^{-1}. Complete lists of absorption features and emission lines seen in the spectra have been compiled. The rest wavelengths of the absorption features have been revised and have been used to derive redshifts with external mean errors of the order of ± 60 km s^{-1}. Sandage's findings of systematic differences between H I radio and optical redshifts is confirmed. The redshifts include values for galaxies in 18 rich clusters in the Indus Supercluster area, and the data also include values tending to confirm suggestions of cell structure of the Coma Supercluster.

158.037 **Photometry of the cluster of galaxies A 1367.**
J. G. Godwin, J. V. Peach.
Mon. Not. R. Astron. Soc., Vol. 200, 733 - 746 (1982).

Two-band photometry is presented for 1561 galaxies in the field of the spiral-rich cluster A 1367. The faint end of the cluster luminosity function (to $M_B \approx -15.5$) is consistent with that of the Local Group when sources of incompleteness are properly taken into account. The mean mass-to-luminosity ratio for galaxies in the cluster is $M/L = 74$ solar units on the assumption that M/L is equal for them all, somewhat smaller than on the conventional assumption of equal masses. A colour-magnitude diagram is constructed.

158.038 **High resolution 2-D imaging polarimetry of the jet of M 87, in the light of *U, B* and *V*.**
N. Visvanathan, A. J. Pickles.
Proc. Astron. Soc. Australia, Vol. 4, 177 - 180 (1981).

Polarisation maps to a resolution of 1 arcsec in the jet of M 87 are presented. The results show that small-scale structure exists in the magnetic field down to at least this resolution, as evidenced by the fact that the 1 arcsec resolution percentage polarisations are ~80% higher than those obtained at a lower resolution, 2 arcsec. The position angles of the polarisation vectors in colours *U, B* and *V* agree within the errors. The jet is significantly broader in the light of *U* than in *B* or *V*.

158.039 **The neutrino content of elliptical galaxies.**
B. M. Lewis.
Proc. Astron. Soc. Australia, Vol. 4, 182 - 184 (1981).

This paper makes a preliminary estimate of the maximum neutrino content of elliptical galaxies.

158.040 **Character of 'prolate' galaxies.**
D. O. Richstone, M. D. Potter.
Nature, Vol. 298, 728 - 729 (1982).

It is still not known whether elliptical galaxies are oblate, prolate or triaxial. The prototype 'prolate' galaxy NGC 2685 has a rapid observed stellar rotation rate around the projected minor axis. The authors show here that a steady figure, exactly prolate model can have no such rotation and that a nearly prolate model fails by a factor of 7. They suggest that the galaxy is oblate and that the gas disk is the wreck of a dwarf companion.

158.041 **Discovery of a very red nucleus in the radio elliptical IC 5063 (PKS 2048–57).**
D. J. Axon, J. Bailey, J. H. Hough.
Nature, Vol. 299, 234 - 236 (1982).
The authors report the discovery of an extremely red IR source in the nucleus of the elliptical galaxy IC 5063 (PKS 2048–57). The observed (2.2–3.8) μm colour index within a 5 arc s aperture is 2.1 mag while the colours at shorter wavelengths and in the 5–10 arc s annulus around the nucleus are close to those of normal elliptical galaxies. The IR excess could arise from a BL Lac type nucleus with a steep non-thermal spectrum.

158.042 **Surface photometry of the spiral galaxy NGC 4501.**
U. Send.
Astron. Astrophys., Vol. 112, 235 - 240 (1982).
A calibrated photographic blue plate of the spiral galaxy NGC 4501 has been analyzed by numerical procedures to derive the standard geometrical and photometric parameters. A comparison with previous studies is made. The major and minor axis luminosity profiles are produced and are decomposed into disk, spheroid, and spiral arm components to give their separate parameters. With these results, a two-component rotation model is applied.

158.043 **Recent star-forming activity in the clumpy irregular galaxy NGC 7673.**
R. Duflot-Augarde, D. Alloin.
Astron. Astrophys., Vol. 112, 257 - 267 (1982).
The basic data already available for NGC 7673 make it well suited for a detailed analysis of its gaseous component and star-forming activity. In the present study the authors use higher dispersion spectrophotometric data than in previous papers. The observational data and results are presented and the intrinsic reddening and the Balmer absorption lines contamination are discussed. The physical properties of the emitting gas are shown and the interpretation of these results as well as some concluding remarks are given.

158.044 **The large scale trend of rotation curves in the spiral galaxies NGC 1068 and NGC 3310.**
G. Galletta, E. Recillas-Cruz.
Astron. Astrophys., Vol. 112, 361 - 365 (1982).
The authors present a dynamical study of the bright spiral galaxies NGC 1068 and NGC 3310. The purpose is to extend the knowledge of the rotation curves for these systems to radii as large as 10 kpc from the center in the light of the suggestion that galaxies might be surrounded by massive halos with high M/L ratios. From the authors' curves no outer decrease in the velocity is evident, but a slight increase of the velocity is found, yielding a mass around a few $10^{10} M_\odot$ for the galaxies. The inner M/L ratios within 7 kpc are quite low, with maximum values between 5 and $7 M_\odot/L_\odot$. The presence of a significant quantity of matter in the external parts of these galaxies, indicated by the observed velocity trends, cannot yet be interpreted as halos as massive as those required in dynamical models, at least within the radii explored.

158.045 **The mass function of blue stars, the production rate of Lyc-photons, and the rate of star formation in M 33.**
E. M. Berkhuijsen.
Astron. Astrophys., Vol. 112, 369 - 376 (1982).
The Present Day Mass Function and the Initial Mass Function for stars of masses $> 10 M_\odot$ in M 33 were derived by extrapolating the upper part of the Luminosity Function constructed from the catalogue of Humphreys and Sandage (1980). The rate of Lyc-photons expected from stars of 10 to $100 M_\odot$ is in good agreement with the rate needed to ionise the gas as inferred from Hα observations taking absorption effects into account. Similarly, the IMFs for the solar neighbourhood satisfactorily predict the rate of Lyc-photons obtained from radio observations of the ionised gas. Comparison of various quantities bearing on star formation in M 33 and the Galaxy is performed.

158.046 **A new ring galaxy in Canes Venatici.**
N. Brosch.
Astron. Astrophys., Vol. 112, 388 - 390 (1982).
A new object, tentatively identified as a ring galaxy, has been found 23″ away from the 9.5 mag SAO 44762 star. It resembles a face-on ring, almost fully circular, with a single bright knot on the side away from the star.

158.047 **Spectral observations of three galaxies with ultraviolet excess.**
M. A. Kazaryan, V. S. Tamazyan.
Pis'ma Astron. Zh., Tom 8, 454 - 457 (1982). In Russian.
English translation in Soviet Astron. Lett., Vol. 8.
Results are presented of spectral observations of three galaxies with UV excess: Nos. 31, 119 and 236 from lists by Kazaryan (1979). Absorption lines of H I, Na I, Mg I and Fe I are identified in the spectrum of the galaxy No. 31, as well as emission lines of H I, [N II], [O II], [O III] and [Ne III] in the spectra of the galaxies Nos. 119 and 236. Equivalent widths of all identified lines are estimated.

158.048 **Search for large-scale linear filament structures in the apparent distribution of galaxies.**
B. I. Fesenko.
Pis'ma Astron. Zh., Tom 8, 458 - 460 (1982). In Russian.
English translation in Soviet Astron. Lett., Vol. 8.
Some statistical tests for large-scale linear filament structure in the apparent distribution of galaxies are proposed. The application of these methods to the Lick Observatory counts and to the deep sample near the south galactic pole failed to reveal such structure.

158.049 **Optical absorption lines in the high redshift BL Lac object 0215+015.**
J. C. Blades, R. W. Hunstead, H. S. Murdoch, M. Pettini.
Mon. Not. R. Astron. Soc., Vol. 200, 1091 - 1111 (1982).
The authors present results from their study of 0215+015 (PKS 0215 + 015), the highest redshift and most luminous BL Lac object so far detected. Optical continuum measurements confirm an overall variability range of more than 4 mag. A review of the published radio data indicates that 0215 + 015 has a core-halo or multi-component structure, similar to that found for the BL Lac 0735 + 178. Four absorption redshift systems occur with $z_a = 1.254$, 1.345, 1.549 and 1.649. Possible origins of the absorption systems are discussed.

158.050 **The radio halo of M 87.**
K. Takarada.
Publ. Astron. Soc. Japan, Vol. 34, 183 - 187 (1982).
The author explains the characteristics of the radio halo of M 87 consistently with X-ray observations by pointing out that the radio halo is a double source whose components lie outside of the X-ray emitting gas associated with M 87 and the axis is near the line of sight. He also determines the mean line-of-sight magnetic fields of the hot gas associated with M 87 and that of the intracluster gas of the Virgo cluster.

158.051 **Neutral hydrogen in M 31. III. The warping of the H I disk.**
T. Sawa, Y. Sofue.
Publ. Astron. Soc. Japan, Vol. 34, 189 - 197 (1982).

An overall feature of the warping of the H I disk of M 31 is obtained through an analysis of H I-line data by assuming cylindrical rotation.

158.052 **Neutral hydrogen in M 31. IV. Comparison with optical data and the star formation rate.**
N. Nakai, Y. Sofue.
Publ. Astron. Soc. Japan, Vol. 34, 199 - 221, with a correction p. 561 (1982).

Various optical features of population I objects in M 31 are compared with the H I gas distribution. Most of H II regions, OB associations, and open clusters are distributed on ring-like H I main arms at galactocentric distances $R = 8 - 15$ kpc which form a "10-kpc ring". The UV emission comes mostly from the 10-kpc ring. The outermost trailing H I spiral arm at $R = 18 - 26$ kpc in the southwest is traced by OB associations and open clusters. The radial distributions of the young optical objects resemble that of the H I gas. A detailed comparison shows that younger objects are located in more inner regions than older objects.

158.053 **New method of investigation of the spiral structure of galaxies and its application to M 33.**
A. A. Smirnov, F. Kh. Sakhibov.
Dokl. AN TadzhSSR, Tom 24, 725 - 727 (1981). In Russian. Abstr. in Ref. zh., 51. Astron., 7.51.768 (1982).

158.054 **Seyfert and active galaxies.**
M. Ward.
Noyaux actifs des galaxies, (see 012.004), 27 pp.

Since the revival of interest in the study of this class of galaxy, which took place in the late 1960's, there has been a corresponding dramatic increase in the number of papers contributed to the journals on the subject of Seyfert galaxies, concerning both the observations and theoretical interpretations. In this paper the author highlights a fraction of these new data, and briefly comments on the theories advanced to explain the complex nature of their activity.

158.055 **Noyaux actifs: observations, classifications et contraintes observationnelles.**
J. Bergeron.
Noyaux actifs des galaxies, (see 012.004), 28 pp.

158.056 **Radio and mm-observations of active nuclei.**
P. Biermann.
Noyaux actifs des galaxies, (see 012.004), 32 pp.

The author reviews basic theory and recent radio and mm-observations of active nuclei in galaxies. He concentrates on radio, mm and X-ray observations in the continuum, and deals first with theoretical foundations, and then with some recent work at Bonn.

158.057 **Physical conditions in the emission regions of active galactic nuclei.**
L. Maraschi.
Noyaux actifs des galaxies, (see 012.004), 35 pp.

158.058 **The impact of IUE observations on our knowledge about galaxies and quasars.**
J. B. Oke.
Advances in ultraviolet astronomy, (see 012.006), p. 46 - 53 (1982).

In spite of the difficulty of making IUE observations of faint objects a large number of spectra have been obtained of many types of galaxies as well as quasars. In this paper many of the results obtained from studies of these spectra are summarized.

158.059 **Morphology of gaseous halos of galaxies.**
D. G. York.
Advances in ultraviolet astronomy, (see 012.006), p. 80 - 88 (1982).

Recent ultraviolet and optical observations of high latitude gas are reviewed. Hydrogen 21 cm observations of high latitude gas and extended disks of spiral galaxies are used to formulate a picture of the gaseous halo which may be relevant to the origin of QSO absorption lines.

158.060 **IUE and Einstein observations of NGC 5204.**
G. Fabbiano, N. Panagia.
Advances in ultraviolet astronomy, (see 012.006), p. 145 - 149 (1982).

158.061 **The IUE low-dispersion spectrum of the center of M 31.**
G. A. Welch.
Advances in ultraviolet astronomy, (see 012.006), p. 150 (1982). – Abstract.

158.062 **Markarian 36: a young galaxy!?**
J. Huchra, M. Geller, D. Hunter, J. Gallagher.
Advances in ultraviolet astronomy, (see 012.006), p 151 - 155 (1982).

The authors report UV, optical and IR observations of three galaxies – NGC 4214, NGC 4670 = Haro 9, and Markarian 36. Markarian 36 has few, if any, red stars. Combined with its low metal content, this lack of red stars is a very strong indication that this galaxy has only recently begun to form stars.

158.063 **Four years of IUE research on clumpy irregular galaxies.**
P. Benvenuti, C. Casini, J. Heidmann.
Advances in ultraviolet astronomy, (see 012.006), p. 156 - 159 (1982).

IUE spectra of clumpy irregular galaxies have shown that the clumps contain a very large number of early O and B type stars, with a large number of supergiants. Possibly WR stars and massive objects of the type of R 136a could also contribute to the clump luminosity. On the average, each clump radiates in the UV 100 times more than 30 Dor.

158.064 **Coordinated observations of Seyfert 1 galaxies.**
W. Wamsteker, P. Benvenuti, C. Cacciari, A.Cassatella, J. Blades, A. C. Danks, L. Bianchi, P. Patriarchi.
Advances in ultraviolet astronomy, (see 012.006), p. 165 - 168 (1982).

The authors present some preliminary results of their first observations for the Seyfert 1 galaxies ESO 113-IG45 (Fairall-9) and AKN 120.

158.065 **Ultraviolet variations of NGC 4151.**
A. Boksenberg, G. E. Bromage, J. Clavel, A. Elvius, M. V. Penston, G. C. Perola, M. Pettini, M. A. J. Snijders, E. G. Tanzi, M. Tarenghi, M. H. Ulrich.
Advances in ultraviolet astronomy, (see 012.006), p. 169 (1982). – Abstract.

158.066 **Time variations in C IV λ1550 and Lyα emission lines in NGC 5548.**
R. Ptak, R. Stoner.
Advances in ultraviolet astronomy, (see 012.006), p. 170 - 173 (1982).

The authors have obtained two short-wavelength images of NGC 5548 separated by a three month interval. Combining this data with that obtained by Wu et al.(1981) the authors have discovered time variations in the C IV λ1550 and Lyα profiles as well as in the intensity ratio. They have also obtained an excellent short-wavelength image of the very bright Seyfert galaxy Fairall 9, and they find that the C IV and Lyα lines in this object have essentially identical profiles.

158.067 **IUE observations of Markarian 3 and 6.**
M. A. Malkan, J. B. Oke.
Advances in ultraviolet astronomy, (see 012.006), p. 174 - 176 (1982). – Abstract.

158.068 **Spectroscopy of two BL Lacertae objects.**
M. Urry, S. Holt, Y. Kondo, R. Mushotzky, K. Hackney, R. Hackney.
Advances in ultraviolet astronomy, (see 012.006), p. 177 - 180 (1982).

As part of a continuing program of observations of BL Lac objects, the authors have used the IUE SWP and LWR cameras to look at PKS 0548–322 and PKS 2155–304. No lines in the ultraviolet or X-ray spectra of either object are seen. All spectra are well-described by power law models. Both objects exhibit variability, but intensity changes and spectral changes are not correlated.

158.069 **IUE observations of four BL Lacertae-type objects.**
D. M. Worrall, F. C. Bruhweiler.
Advances in ultraviolet astronomy, (see 012.006), p. 181 - 184 (1982).

The authors present spectral fits to IUE observations of four BL Lacertae-type objects. Combined with near-simultaneous measurements at longer wavelengths, positive spectral curvature within the infrared to ultraviolet energy range are found. The authors show that the data provide strong evidence for relativistic beaming in the sources. They discuss Lyman α emission and possible C IV and N V absorption features in the spectrum of 3C 371.

158.070 **Coordinated observations of Markarian 180 at ultraviolet, X-ray, optical, and radio wavelengths.**
S. L. Mufson, D. J. Hutter, K. R. Hackney, R. L. Hackney, Y. Kondo, R. F. Mushotzky, C. M. Urry, W. Z. Wisniewski, M. F. Aller, H. D. Aller, P. E. Hodge.
Advances in ultraviolet astronomy, (see 012.006), p. 185 - 188 (1982).

The authors present the results of multifrequency observations of Mrk 180 = 1133+704, a BL Lac object embedded in a giant elliptical galaxy, obtained during the period November–December 1980. The importance of Mrk 180 derives mostly from its association of a quasar-like object with a galaxy whose absorption-line redshift unambiguously determines its distance.

158.071 **Coordinated observations of Markarian 501 at ultraviolet, X-ray, optical, and radio wavelengths.**
D. J. Hutter, S. L. Mufson, K. R. Hackney, R. L. Hackney, Y. Kondo, R. F. Mushotzky, C. M. Urry, W. Z. Wisniewski, M. F. Aller, H. D. Aller, P. E. Hodge.
Advances in ultraviolet astronomy, (see 012.006), p. 189 - 192 (1982).

The authors present the results of multifrequency observations of Mrk 501 = B2 1652+39, a BL Lac object embedded in a giant elliptical galaxy, obtained during the period August 1980 - March 1981. The importance of Mrk 501 derives mostly from its association of a quasar-like object with a galaxy whose absorption-line redshift unambiguously determines its distance.

158.072 **The ultraviolet spectrum of the high redshift BL Lac object 0215+015.**
J. C. Blades, M. Pettini, R. W. Hunstead, H. S. Murdoch.
Advances in ultraviolet astronomy, (see 012.006), p. 193 - 196 (1982).

The object 0215+015 is highly variable. At optical wavelengths it shows no emission lines but has a rich absorption spectrum with at least six, separate redshift systems. The authors have combined 3 LWP low-dispersion images of the object to produce a high signal-to-noise spectrum between 2400 - 3200 Å which they use to study the hydrogen Lym lines of the various redshift systems.

158.073 **Simultaneous multifrequency observations of BL Lac objects and violently variable quasars.**
J. N. Bregman, A. E. Glassgold, P. J. Huggins.
Advances in ultraviolet astronomy, (see 012.006), p. 197 - 200 (1982).

The authors have obtained ultraviolet spectra and coordinated the acquisition of multifrequency observation of three BL Lac objects and one violently variable quasar. They present a study of the four objects: BL Lac, 0735+1' IZw-187, and 3C 446.

158.074 **21-cm line profiles of 392 spiral galaxies.**
L. Bottinelli, L. Gouguenheim, G. Paturel.
Astron. Astrophys., Suppl. Ser., Vol. 50, 101 - 113, with a correction p. 529 - 539 (1982).

The authors present observational data obtained with Nançay Radiotelescope as integrated 21-cm line profiles a individual comments on galaxies.

158.075 **The coronas of galaxies.**
K. S. de Boer, B. D. Savage.
Sci. American, Vol. 247, No. 2, p. 52 - 60 (1982).

158.076 **Are all galactic nuclear regions sodium rich?**
M. P. Véron-Cetty, P. Véron, M. Tarenghi.
Astron. Astrophys., Vol. 113, 46 - 56 (1982).

Photoelectric and spectrographic observations of the nucleus of a number of Shapley-Ames galaxies are presente They show that the reddening of the stellar continuum, th Balmer decrement and the equivalent width of the NaD lin are closely correlated. The D line equivalent width is much larger than the one observed in galactic stars having the sar reddening; one possible explanation is that the gas in the nuclear region of galaxies is sodium rich as it has been sho to be the case for the stars in the central regions of most elliptical galaxies.

158.077 **Study of spiral galaxies from 392 new measurements of 21-cm line data.**
L. Bottinelli, L. Gouguenheim, G. Paturel.
Astron. Astrophys., Vol. 113, 61 - 72 (1982).

392 spiral galaxies with type Sb or later have been detected in the 21-cm line of neutral hydrogen with the Nançay radiotelescope, and among them 172 H I fluxes are obtained for the first time. The source of any possible bias has been carefully analyzed. A mean value of each global parameter has been determined for each morphological typ and compared to previous results. The internal dispersion c individual values has been analyzed.

158.078 **X rays from a peculiar nucleus galaxy NGC 2196**
P. C. Agrawal, K. P. Singh.
Astron. Astrophys., Vol. 113, 73 - 75 (1982).

Detection of a new X-ray source with the Einstein Observatory and its identification with the early type spiral (Sa) galaxy NGC 2196 (MCG 4-15-14), a member of Sersic' list of peculiar nuclei galaxies is reported. The X-ray luminosity of the galaxy is found to be $\sim 2 \times 10^{40}$ erg s^{-1} which is an order of magnitude higher compared to that of normal spiral galaxies but comparable to that observed fror weakly-active galaxies. The optical and X-ray characteristic of the galaxy are discussed and compared with those of the other normal and weakly-active galaxies.

158.079 **Temperatures and sizescales of giant cloud complexes in the spiral galaxy IC342.**
P. T. P. Ho, R. N. Martin, K. Ruf.
Astron. Astrophys., Vol. 113, 155 - 164 (1982).

The (J, K) = (1, 1) and (2, 2) lines of NH_3 have been

detected toward the spiral galaxy IC342. Their nearly equal intensity implies the existence of warm (~50 K) gas. The mere fact that these lines were detected implies a lower limit to the sizescale of the emission complex (⩾ 25 pc). When combined with the infrared, radio continuum and CO observations, a consistent picture on star formation in the nucleus emerges.

158.080 **The hidden mass in galaxies.**
G. B. Field.
Mercury, Vol. 11, 74 - 80 (1982).

158.081 **M 31: the Andromeda galaxy.**
P. Hodge.
Mercury, Vol. 11, 118 - 119, 131 (1982).

158.082 **Multiaperture photometry of galaxies. II. Near-infrared observations of six isolated objects.**
N. Brosch, R. Isaacman.
Astron. Astrophys., Vol. 113, 231 - 236 (1982).

The authors present J, K, and L photometry of six isolated galaxies (no companions within 45′) through 7″, 14″, and 22″ apertures. The outer regions of these galaxies appear normal; the nuclear regions exhibit an excess of radiation in the ultraviolet as compared to "normal" galaxies which is not seen in the infrared. This argues in favor of having its origin in thermal emission rather than synchrotron radiation. The emission is suggested to be a consequence of intensive formation of massive stars in the nuclei.

158.083 **Bulge X-ray sources and novae in M 31.**
J. P. Vader, E. P. J. van den Heuvel, W. H. G. Lewin, R. J. Takens.
Astron. Astrophys., Vol. 113, 328 - 335 (1982).

The distribution of point X-ray sources in the bulge of M 31 is compared to that in the bulge of our galaxy, and to the distributions of novae and of the surface brightness in M 31. The bulge of M 31 seems to contain about four times as many luminous X-ray sources per unit mass as the bulge of our galaxy, independently of whether these sources are located in globular clusters or not. The field bulge X-ray sources (i. e. those apparently not located in globular clusters) and novae closely follow the light distribution in M 31, except for a central "nova hole" with radius ~2′.

158.084 **Accurate optical positions of M 82 knots.**
D. Bettoni, G. Galletta.
Astron. Astrophys., Vol. 113, 344 - 347 (1982).

The authors present accurate optical measurements of bright and dark knots visible in the peculiar galaxy M 82, with an optical error of about 0″.7. A direct comparison of the data with the VLA observations of hot spots and jet-like features is made. Some interesting coincidences between optical, radio, and infrared knots are discussed.

158.085 **The velocity field in the central region of the Mark 538 galaxy.**
V. L. Afanas'ev, S. S. Rassokhin.
Pis'ma Astron. Zh., Tom 8, 515 - 519 (1982). In Russian. English translation in Soviet Astron. Lett., Vol. 8.

Using the image tube spectrograms the velocity field in the central part of the Mark 538 galaxy has been investigated. There have been revealed noncircular motions of gas near the node line positional angle (150°). The mass of the inner 1.3 kpc region is estimated to be of $4.3 \times 10^9 M_\odot$. The central velocity gradient is 131 km/s kpc.

158.086 **Radial velocities of ring-like galaxies. II.**
I. P. Kostyuk, A. I. Kopylov.
Pis'ma Astron. Zh., Tom 8, 520 - 521 (1982). In Russian. English translation in Soviet Astron. Lett., Vol. 8.

Radial velocities of 53 ring-like galaxies from the list by Kostyuk (1975) are presented.

158.087 **Interstellar absorption in three Baade fields in the Andromeda nebula.**
A. P. Semenov.
Pis'ma Astron. Zh., Tom 8, 522 - 526 (1982). In Russian. English translation in Soviet Astron. Lett., Vol. 8.

The distribution of interstellar absorption in Baade's fields I, II and III in the Andromeda nebula is studied with the aid of cepheids. It is found that the absorbing material is concentrated towards the inner edge of the spiral arm S4.

158.088 **Explanation of some regularities of galactic clustering on the basis of a model of the genesis of a lognormal law.**
B. V. Karasev.
Pis'ma Astron. Zh., Tom 8, 527 - 534 (1982). In Russian. English translation in Soviet Astron. Lett., Vol. 8.

A possibility of applying a lognormal function to describe the apparent distribution of galaxies is discussed. A scale model is deduced suggesting a linear connection between the average number of galaxies and the inverse of logarithmic distribution dispersion. The model accounts for a number of regularities which are generally treated as effects of galactic clustering.

158.089 **On the possibility of determination of masses of central bodies in active galactic nuclei and quasars.**
A. S. Zentsova.
Pis'ma Astron. Zh., Tom 8, 535 - 537 (1982). In Russian. English translation in Soviet Astron. Lett., Vol. 8.

The absence of absorption of gamma rays with energy 1 - 10 MeV in the nuclei of the radio galaxy Cen A, Seyfert galaxy NGC 4151 and quasar 3C 273 gives a lower limit of the masses of the central bodies in the nuclei of these objects.

158.090 **Time variations in the high-redshift 21 centimeter absorption spectrum of the BL Lacertae object AO 0235 + 164.**
A. M. Wolfe, M. M. Davis, F. H. Briggs.
Astrophys. J., Vol. 259, 495 - 521 (1982).

The authors present evidence for time variations of the redshifted ($z = 0.524$) 21 cm absorption spectrum of the BL Lac object AO 0235 + 164. A 4 year monitoring program with the 305 m telescope at Arecibo shows that the line depths of the four deepest features (A → D) vary in time. The strongest changes occur in feature D, which was observed to double its line depth in 0.5 yr. Weaker variations (between 15 and 35%) with durations between 0.25 and 0.5 yr are observed in all four features. Whereas each of the features evolves independently during some epochs, they appear to vary synchronously during others. The variations are interpreted both in terms of intrinsic and extrinsic models of the absorption.

158.091 **The local velocity field and the calibration of the luminosity classification of spiral galaxies.**
R. C. Kennicutt, Jr.
Astrophys. J., Vol. 259, 530 - 534 (1982).

The presence of a large galaxian density enhancement in the Ursa Major-Virgo region renders current determinations of the galactic luminosity function highly sensitive to any peculiar solar motion in the Virgo direction. In order to illustrate the nature and possible severity of this bias, a linear Virgo-centric infall model with $W_\odot = 300$ km s^{-1} has been applied to the recent luminosity calibration for spiral galaxies by Tammann, Yahil, and Sandage. The results indicate that until the local velocity field is accurately mapped, luminosity functions and space densities based on magnitude-limited redshift catalogs should be applied with considerable caution.

158.092 **An outlying ring of neutral hydrogen around the Sc I galaxy NGC 628.**
F. H. Briggs.

Astrophys. J., Vol. 259, 544 - 558 (1982).

Observations in the 21 cm line of neutral hydrogen show that the normal spiral galaxy NGC 628 is encircled by an elliptical ring of gas. The outlying gas extends to 3 times the optical radius and lies in a plane which is inclined by about 15° with respect to the plane of the bright inner disk. Possible explanations for the origin and long-term stability of the ring are discussed in detail.

158.093 **Radio structures of Seyfert galaxies. III. Radio and optical observations of NGC 5548.**
A. S. Wilson, J. S. Ulvestad.
Astrophys. J., Vol. 260, 56 - 64 (1982).

The VLA has been used to map the nuclear radio source of the Seyfert galaxy NGC 5548 at 1465 and 4885 MHz. The source comprises an unresolved core coincident with the optical continuum nucleus and two diffuse lobes straddling it. The total source extent is ≈ 5.9 kpc. Optical spectrophotometry of the Seyfert nucleus and the inner galaxy has been obtained. The radio lobes have probably been energized from the active nucleus and represent a very low luminosity example of the double radio source phenomenon.

158.094 **The barred spiral galaxy NGC 3992: does it have a massive halo?**
S. T. Gottesman, J. H. Hunter, Jr.
Astrophys. J., Vol. 260, 65 - 69 (1982) = Dep. Astron., Univ. Florida, Contrib. No. 43.

The authors present new high-resolution H I observations of the barred spiral galaxy NGC 3992 and three nearby satellite systems. These data are employed to estimate the mass and mass distribution of NGC 3992.The disk mass is most probably $2 \times 10^{11} M_\odot$, within a radius of 19.7 kpc. It is concluded that the galaxy is unlikely to possess a massive ($\sim 10^{12} M_\odot$) halo.

158.095 **8–13 micron observations of nine emission-line galaxies.**
J. A. Frogel, J. H. Elias, M. M. Phillips.
Astrophys. J., Vol. 260, 70 - 74 (1982).

Observations designed to detect broad features in the 8–13 μm spectral region of nine southern emission-line galaxies are presented. With one possible exception, a minimum at 9.6 μm is detected in all of the galaxies including five with active nuclei. A similar result by Lebofsky and Rieke based on a survey of northern galaxies suggests that this feature is present in most galaxies having measurable 10 μm emission whether they have active nuclei or not.

158.096 **Neutral hydrogen in bright galaxies with strong radio sources.**
I. F. Mirabel.
Astrophys. J., Vol. 260, 75 - 80 (1982).

Neutral hydrogen spectra at 21 cm wavelength of 12 bright galaxies with strong nuclear radio sources are reported. The observations include 9 spirals and 3 peculiar galaxies.

158.097 **Local properties of star-forming regions in irregular galaxies.**
D. A. Hunter.
Astrophys. J., Vol. 260, 81 - 103 (1982).

In order to investigate the physical processes controlling the star formation histories and evolution of noninteracting irregular galaxies, the author presents optical data from which she derives properties of star-forming regions in a sample of 15 irregulars which are actively forming stars. Kinematics, sizes, and distributions of H II regions are obtained from Hα velocity profiles from high dispersion echelle spectra and images taken with the video camera through *b*, *y*, *I*, and Hα passbands. The young star-forming regions are found to dominate the galaxian light at all optical wavelengths, and the observed colors are redder than predicted from most theoretical population models of young systems.

158.098 **The dependence of CO emission on luminosity a the rate of star formation in Sc galaxies.**
J. S. Young, N. Scoville.
Astrophys. J., Lett., Vol. 260, L11 - L17, plates L1, L2 (1 = Five Coll. Astron. Dep. Contrib. No. 510.

A sample of eight Sc galaxies covering a large range ir total luminosity has been observed in the 2.6 mm CO line. approximately linear correlation is revealed between the C and blue luminosities measured for the same regions in bot the nuclei and the inner disks. If the blue luminosity is an indicator of the recent star formation rate, these results su that the star formation rate per nucleon in Sc galaxies is cc stant.

158.099 **Central condensations in Seyfert galaxies. I.**
V. L. Afanas'ev, A. A. Pimonov, V. Yu. Terebizl
Pis'ma Astron. Zh., Tom 8, 579 - 584 (1982). In Russian. English translation in Soviet Astron. Lett., Vol. 8.

Results of observations with a scanning photometer o 7 normal and 17 Seyfert galaxies show that in the Seyfert galaxies a circumnuclear region of 3 - 4 kpc in size exists distinguishing by high stellar density. A correlation betwee the volume luminosity of the spherical component and the luminosity of the galaxy's nucleus is found.

158.100 **An investigation of star complexes in the Andromeda galaxy.**
Yu. N. Efremov.
Pis'ma Astron. Zh., Tom 8, 585 - 592 (1982). In Russian. English translation in Soviet Astron. Lett., Vol. 8.

Preliminary V, $B-V$ diagrams for a number of van der Bergh's OB associations in M 31 are obtained. It is found tl the number ratio of blue to red supergiants increases towar the center of M 31 like it is in M 33 and our Galaxy. A nur of new star groupings either without O stars or with relati small dimensions may by distinguished on the plates with limiting magnitude of $21^m - 22^m B$.

158.101 **The noise of BL Lacertae.**
R. L. Moore, J. T. McGraw, J. R. P. Angel, R. Di M. J. Lebofsky, G. H. Rieke, W. Z. Wiśniewski, D. J. Axon, J. Bailey, J. M. Hough, I. Thompson, M. Breger, H. Schulz, G. C. Clayton, P. G. Martin, J. S. Miller, G. D. Schmidt, J. Africano, H. R. Miller.
Astrophys. J., Vol. 260, 415 - 436 (1982).

The authors present the results of an intensive optical and infrared monitoring program of the flux and polarizati of BL Lac. Observations were made over a 1 week period b 11 stations from Israel to California, allowing continuous coverage of up to 18 hours. Often the polarization and flux were being measured by more than one station at the same time, so the reality of variations could be established witho question. The power spectrum of the polarimetric noise is analysed and discussed within the context of theoretical models of BL Lac.

158.102 **The peculiar galaxy Abell 76.**
D. L. Talent, J. B. Kaler, J. S. Gallagher, D. A. Hunter.
Astrophys. J., Vol. 260, 488 - 494, plate 18 (1982).

New spectrophotometric and imaging observations are presented for the peculiar galaxy Abell 76. This is clearly a morphological member of the ring galaxy class. The presen of extensive optical line emission, high surface brightness, a a pronounced blue continuum suggest this object is current experiencing a very high star formation rate. Spectrophoto metric observations reveal a near-solar metallicity in the gas

158.103 **The UV continuum spectrum of M 81.**
G. Bruzual A., M. Peimbert, S. Torres-Peimbert.
Astrophys. J., Vol. 260, 495 - 503 (1982).

From IUE data the authors present the absolute continuum spectrum for the nucleus of M 81 in the 1200–3200 Å range. The spectrum can be fitted by an old stellar population with a small contribution from horizontal-branch stars. An upper limit for the nonthermal contribution to the UV spectrum is obtained.

158.104 **The peculiar disk galaxy UGC 7576.**
J. Mould, B. Balick, G. Bothun, M. Aaronson.
Astrophys. J., Lett., Vol. 260, L37 - L40, plate L3 (1982).

Optical photometry of UGC 7576 supports the view that this system consists of an edge-on disk galaxy with a prominent bulge aligned almost orthogonally to a low surface brightness disk. A mass of $2 \times 10^9 M_\odot$ of H I is detected in the system, which has the unusually high mass-to-light ratio of 30.

158.105 **On the origin of molecular cloud rings in spiral galaxies: CO observations of NGC 7331 and NGC 2841.**
J. S. Young, N. Scoville.
Astrophys. J., Lett., Vol. 260, L41 - L44 (1982).

CO emission has been mapped along the major axis of two Sb galaxies, NGC 7331 and NGC 2841. Molecular rings were discovered in both galaxies, with peaks at radii of ~ 4 kpc. The previously measured Hα and H I rotation curves reach their maximum velocities at the same radii as the molecular rings. In these Sb galaxies, the interstellar medium constitutes only 2% - 5% of the total mass, a factor of 2 - 3 less than the contribution seen in a sample of Sc galaxies.

158.106 **Interstellar lines in the spectra of the globular clusters around NGC 5128.**
J. G. Cohen.
Astrophys. J., Lett., Vol. 260, L45 - L47 (1982).

Three of the brightest globular clusters associated with the peculiar radio galaxy NGC 5128 have been observed spectroscopically. In addition to interstellar Ca II lines from our own Galaxy, two of the three show interstellar Ca II lines arising in NGC 5128. In one case, the gas is possibly associated with the rotating inclined disk of gas seen close to the nucleus. The interstellar clouds are between 12 and 30 kpc from the nucleus of NGC 5128.

158.107 **Broad hydroxyl emission in IC 4553.**
W. A. Baan, P. A. D. Wood, A. D. Haschick.
Astrophys. J., Lett., Vol. 260, L49 - L52 (1982).

The main line transitions of hydroxyl (OH) have been detected in emission in the peculiar galaxy IC 4553. The 1667 and 1665 MHz transitions appear in a ratio of 4.2 : 1, and both have a radial velocity of 5375 km s^{-1} and a full width at half-maximum of 108 km s^{-1}. This is the first detection of broad OH emission in an external galaxy. The features probably are due to stimulated emission from many H II/OH masers in the active nuclear region of the galaxy.

158.108 **Ultraviolet and X-ray observations of the BL Lacertae PKS 0548–322.**
C. M. Urry, R. F. Mushotzky, Y. Kondo, K. R. H. Hackney, R. L. Hackney.
Astrophys. J., Vol. 261, 12 - 17 (1982).

Spectrophotometric observations of the BL Lac object PKS 0548–322 have been made with ultraviolet and X-ray instruments on the IUE and HEAO 2 satellites. The authors present two observations in each spectral region, one of which was obtained simultaneously. A power law of energy index ~1.0 gives a good description of the data from ultraviolet through X-ray frequencies.

158.109 **C IV and Lyman-α emission lines in Seyfert 1 galaxies.**
S. Gregory, R. Ptak, R. Stoner.
Astrophys. J., Vol. 261, 30 - 34 (1982).

The authors have obtained good resolution spectra of C IV (λ 1550) and Lyα emission in the four Seyfert 1 galaxies NGC 5548, NGC 4151, NGC 3783, and Fairall 9. They report line profiles and intensity information for this emission. In the case of NGC 5548, the emission profiles, line intensities, and Lyα/C IV ratios exhibit large changes over a period of a few months.

158.110 **The time variation of broad emission-line profiles of Seyfert 1 galaxies.**
E. R. Capriotti, C. B. Foltz, B. M. Peterson.
Astrophys. J., Vol. 261, 35 - 41 (1982).

It has been established that the broad hydrogen line profiles of Seyfert 1 galaxies can vary on time scales ranging from a few months to a few years. The authors have calculated broad-line profiles for models of the emitting region in which the emitting clouds are in radial outflow. The continuum source is assumed to vary on a time scale a bit shorter than the light-travel time through the broad-line emitting region. The combination of a variable continuum source and light-travel time considerations reproduces many observed line features. As an example, variations of Hβ profiles in Arakelian 120 are compared with the calculations.

158.111 **High-resolution X-ray observations of M 87: nucleus, jet, and radio halo.**
E. J. Schreier, P. Gorenstein, E. D. Feigelson.
Astrophys. J., Vol. 261, 42 - 50 (1982).

High-resolution X-ray observations obtained with the Einstein Observatory of M 87 reveal the presence of four components: (1) broad, diffuse emission associated with hot gas trapped in the galaxy potential; (2) a diffuse, asymmetrical component between 2′ and 5′ from the nucleus associated with the outer lobes or radio halo; (3) a source centered on the nucleus that appears extended by 2″-3″, along with a possible more compact core; (4) a linear X-ray structure peaking at the brightest knot of the radio-optical jet and extended along the jet axis. An analysis of the X-ray emission mechanisms operating in the different components is given.

158.112 **Optical and X-ray properties of X-ray selected active galactic nuclei.**
G. A. Kriss, C. R. Canizares.
Astrophys. J., Vol. 261, 51 - 63 (1982).

In a systematic program of optical studies of X-ray sources discovered serendipitously with the Einstein Observatory, the authors have identified and studied 25 new X-ray selected active galactic nuclei (AGN). The X-ray selected AGN together with optically selected Seyfert 1 galaxies studied previously form a smooth distribution of objects over four orders of magnitude in absolute luminosity and share similar optical and X-ray properties. For this larger sample the authors study correlations between the X-ray continuum, the optical continuum, and various optical emission lines, and discuss their implications for the physical processes at work in AGN. Finally, the X-ray luminosity function of X-ray selected AGN and their surface density on the sky is discussed.

158.113 **Two galaxies with Wolf-Rayet features in their spectra.**
D. E. Osterbrock, R. D. Cohen.
Astrophys. J., Vol. 261, 64 - 69 (1982) = Lick Obs. Bull., No. 918.

Spectral scans are presented of two galaxies, NGC 6764 and Mrk 309, which show the Wolf-Rayet emission band λ4650 in the spectra of their nuclei. Their spectra are discussed and compared with the spectra of two previously known dwarf galaxies which show this same feature, as well as with the spectrum of the H II region NGC 604 in M 33. The numbers of Wolf-Rayet stars in the nuclei of NGC 6764 and Mrk 309 are estimated.

158.114 **Velocity fields in late-type galaxies from Hα Fabry-Perot interferometry. III. The spiral galaxy NGC 6503.**
G. de Vaucouleurs, A. Caulet.
Astrophys. J., Suppl. Ser., Vol. 49, 515 - 529, plates 5 - 8 (1982).

The velocity field in the isolated SA(s)cd spiral NGC 6503 at an estimated distance $\Delta \approx 7.0$ Mpc is derived from 836 velocities measured on nine Hα interferograms taken with the Galaxymeter attached to the 205 cm reflector at McDonald Observatory. Isovelocity maps at 5″ and 7″ resolution are presented. By fitting the observed rotation curve to a realistic three-component model a total mass $M = 1.9 \times 10^{10} M_{\odot}$ has been derived. The mass-to-luminosity ratio, corrected for extinction, is $M/L(B) = 3.0$. A combined catalog of 125 H II regions is given in an appendix.

158.115 **The relation between Seyfert and radio galaxies.**
A. S. Wilson.
Bull. American Astron. Soc., Vol. 14, 602 (1982). – Abstract.

158.116 **The X-ray source in the nucleus of M 33.**
T. Markert, A. Rallis, C. Canizares, G. Kriss, G. Ricker.
Bull. American Astron. Soc., Vol. 14, 602 (1982). – Abstract.

158.117 **High resolution observations of Cen A between 70 and 500 keV.**
N. Gehrels, T. L. Cline, W. S. Paciesas, B. J. Teegarden, J. Tueller.
Bull. American Astron. Soc., Vol. 14, 602 (1982). – Abstract.

158.118 **A strong [O IV] 26 μm emission line in NGC 1068.**
W. J. Forrest, J. R. Houck.
Bull. American Astron. Soc., Vol. 14, 603 (1982). – Abstract.

158.119 **The narrow-line region of intermediate Seyfert galaxies.**
R. D. Cohen.
Bull. American Astron. Soc., Vol. 14, 603 (1982). – Abstract.

158.120 **Radio emission from X-ray selected AGN.**
D. J. Helfand, G. A. Chanan, B. Margon, R. Downes.
Bull. American Astron. Soc., Vol. 14, 603 (1982). – Abstract.

158.121 **Low ionization active galactic nuclei: X-ray or shock heated?**
J. P. Halpern, J. E. Steiner.
Bull. American Astron. Soc., Vol. 14, 603 (1982). – Abstract.

158.122 **The space density and clustering properties of a new sample of emission-line galaxies.**
A. J. Wasilewski.
Bull. American Astron. Soc., Vol. 14, 604 (1982). – Abstract.

158.123 **Luminosity function of starburst galactic nuclei.**
V. A. Balzano.
Bull. American Astron. Soc., Vol. 14, 604 (1982). – Abstract.

158.124 **Far-infrared emission from galaxies.**
L. J. Rickard, P. M. Harvey.
Bull. American Astron. Soc., Vol. 14, 611 (1982). – Abstract.

158.125 **Detection of far-infrared [O I] and [O III] emission from the galaxy M 82.**
R. Genzel, D. Watson, C. Townes, H. Dinerstein, D. Lester, M. Werner, J. Storey.
Bull. American Astron. Soc., Vol. 14, 611 (1982). – Abstract.

158.126 **The molecular cloud ring in the nucleus of NGC 1068.**
N. Scoville, J. Young, L. Lucy.
Bull. American Astron. Soc., Vol. 14, 617 (1982). – Abstract.

158.127 **Infrared search for primeval galaxies.**
P. Saulson, S. Boughn.
Bull. American Astron. Soc., Vol. 14, 630 (1982). – Abstract.

158.128 **The small-angle galaxy correlation function as a function of magnitude.**
W. L. Sebok.
Bull. American Astron. Soc., Vol. 14, 630 (1982). – Abstract.

158.129 **Does the Shane-Wirtanen galaxy distribution have filamentary structure?**
J. R. Kuhn, J. M. Uson.
Bull. American Astron. Soc., Vol. 14, 631 (1982). – Abstract.

158.130 **The arcsecond radio structure of BL Lacertae objects.**
J. S. Ulvestad, K. J. Johnston, K. W. Weiler.
Bull. American Astron. Soc., Vol. 14, 635 - 636 (1982). Abstract.

158.131 **Star formation rates and initial mass functions in disk galaxies.**
R. C. Kennicutt.
Bull. American Astron. Soc., Vol. 14, 641 (1982). – Abstract.

158.132 **Infrared photometry of the halo of M 87.**
S. Boughn, P. Saulson.
Bull. American Astron. Soc., Vol. 14, 643 (1982). – Abstract.

158.133 **Discovery of carbon stars in the Leo dwarf spheroidals.**
E. Olszewski, M. Aaronson, P. Hodge.
Bull. American Astron. Soc., Vol. 14, 643 (1982). – Abstract.

158.134 **A0136 - 080: an S0 galaxy with a polar disk.**
B. C. Whitmore, F. Schweizer, V. C. Rubin.
Bull. American Astron. Soc., Vol. 14, 643 (1982). – Abstract.

158.135 **Flocculent and grand design spiral arm structure in field, binary, and group galaxies.**
D. M. Elmegreen, B. G. Elmegreen.
Bull. American Astron. Soc., Vol. 14, 643 - 644 (1982). Abstract.

158.136 **Optical study of the luminous knots in NGC 7385.**
S. M. Simkin, G. V. Bicknell, A. Bosma.
Bull. American Astron. Soc., Vol. 14, 644 (1982). – Abstract.

158.137 **Optical emission knots associated with the inner X-ray jet of Centaurus A.**
J. Brodie, S. Bowyer.
Bull. American Astron. Soc., Vol. 14, 649 (1982). – Abstract.

158.138 **Multifrequency observations of Mrk 180 and Mrk 501.**
S. L. Mufson, D. J. Hutter, K. R. Hackney, R. L. Hackney, Y. Kondo, R. F. Mushotzky, C. M. Urry, W. Z. Wisniewski, M. F. Aller, H. D. Aller, P. E. Hodge.
Bull. American Astron. Soc., Vol. 14, 649 (1982). – Abstract.

158.139 **Relativistic beam models of the spectra of Mrk 180, and Mrk 501.**
D. J. Hutter, S. L. Mufson.
Bull. American Astron. Soc., Vol. 14, 649 (1982). – Abstract.

158.140 **The 21 cm absorption spectrum of the BL Lac object AO 0235 + 164.**

M. M. Davis, A. M. Wolfe, F. M. Briggs.
Bull. American Astron. Soc., Vol. 14, 649 (1982). – Abstract.

158.141 **The calibration of a radio-independent search for BL Lac objects.**
C. D. Impey, P. W. J. L. Brand.
Bull. American Astron. Soc., Vol. 14, 650 (1982). – Abstract.

158.142 **On the space density of BL Lac objects.**
T. Maccacaro, J. Danziger, P. Giommi, R. E. Griffiths, J. Liebert, J. Lub, J. Stocke, H. Stockman.
Bull. American Astron. Soc., Vol. 14, 650 (1982). – Abstract.

158.143 **X-ray images of late type galaxies.**
A. MacDonald, G. Fabbiano, G. Trinchieri, L. van Speybroeck.
Bull. American Astron. Soc., Vol. 14, 659 (1982). – Abstract.

158.144 **Interacting elliptical galaxies: matching models to observations.**
K. D. Borne, J. G. Hoessel.
Bull. American Astron. Soc., Vol. 14, 660 - 661 (1982). Abstract.

158.145 **CO emission from the Sc galaxy NGC 253.**
J. Yerka, J. S. Young, N. Z. Scoville.
Bull. American Astron. Soc., Vol. 14, 661 (1982). – Abstract.

158.146 **CO observations of Sc galaxies in the Virgo cluster.**
E. Brady, J. S. Young.
Bull. American Astron. Soc., Vol. 14, 661 (1982). – Abstract.

158.147 **10 μm survey of spiral galaxies in the Virgo cluster.**
E. E. Becklin, R. W. Capps, N. Z. Scoville, J. S. Young.
Bull. American Astron. Soc., Vol. 14, 664 (1982). – Abstract.

158.148 **Magnetic fields of galaxies and their magnetospheres.**
M. Fujimoto, Y. Sofue.
Magnetospheric phenomena of celestial bodies, (see 012.011), p. 113 - 119 (1982). In Japanese.

158.149 **Infrared observations of early-type galaxies with dust-lanes.**
A. J. Longmore, R. M. Sharples.
Mon. Not. R. Astron. Soc., Vol. 201, 111 - 119 (1982).

New infrared (*JHK*) observations of 38 early-type galaxies with dust-lanes have been obtained. About one third of these have red broad-band colours when compared with a sample of 19 morphologically normal galaxies. IR excesses are found preferentially in systems exhibiting at least weak signs of a luminous disc. Eleven of the galaxies have published radio continuum fluxes or upper limits. Within this small subset there is no indication of a correlation between infrared excess and radio flux density. Two of the galaxies with infrared excesses lie within X-ray error boxes from HEAO-A2 observations.

158.150 **The ultraviolet spectra of the nuclei of spiral galaxies – I. NGC 4594, 3031, 5194 and 4258.**
R. S. Ellis, P. M. Gondhalekar, G. Efstathiou.
Mon. Not. R. Astron. Soc., Vol. 201, 223 - 251 (1982).

Absolute spectrophotometry is presented from 1250 Å to 4 μm within the same aperture in the nuclei of four spiral galaxies. The light distributions within the apertures is often irregular but no positive evidence for internal reddening has been found, except possibly for NGC 3031. The number of hot stars is adequate to photoionize the gas and explain the Balmer emission spectrum. It is argued that ultraviolet spectral features indicate these stars are young main-sequence objects. The mass functions determined from the models are not inconsistent with those seen in the solar neighbourhood.

158.151 **The Circinus galaxy: a second case of intense nuclear H_2O emission.**
F. F. Gardner, J. B. Whiteoak.
Mon. Not. R. Astron. Soc., Vol. 201, 13P - 15P (1982).

H_2O maser emission has been detected in the Circinus galaxy from within 10–15 arcsec of its nucleus. For a distance of 4 Mpc the luminosity is 37 $L_\odot$. This galaxy is the second found to have nuclear H_2O emission with a luminosity at least an order of magnitude greater than any known H_2O source in our Galaxy.

158.152 **Optical filaments around NGC 4696 in the Centaurus cluster.**
A. C. Fabian, P. D. Atherton, K. Taylor, P. E. J. Nulsen.
Mon. Not. R. Astron. Soc., Vol. 201, 17P - 19P (1982).

The authors report the discovery of optical filaments around the galaxy NGC 4696 in the Centaurus cluster. The filaments are strong in the light of [N II] λ 6583 and extend to ~6 kpc from the nucleus of the galaxy. This result, together with X-ray observations, confirms the presence of a cooling flow in the Centaurus cluster.

158.153 **Photometric observations of the nuclei of active galaxies.**
V. M. Lyutyj.
Astrofiz. i kosm. fiz. Moskva, 1982, p. 66 - 87. In Russian. Abstr. in Ref. zh., 51. Astron., 9.51.580 (1982).

158.154 **Spectrophotometric study of galaxies with high surface brightness. I. Arakelian 144.**
V. A. Mineva, G. T. Petrov, B. Z. Kovachev.
Dokl. Bolg. AN, Vol. 34, 1629 - 1632 (1981). – Abstr. in Ref. zh., 51. Astron., 9.51.601 (1982).

158.155 **Las masas de las galaxias espirales.**
L. Maupomé, P. Pişmiş, L. Aguilar.
Rev. Mexicana Astron. Astrofis., Vol. 6, (see 012.017), 45 - 51 (1981).

In an earlier paper the authors have found that the total mass of galaxies – especially of the spirals – based on values published until 1975, decreased as the Hubble type varied from Sa through Sc and Irregulars. It was also pointed out that masses determined from the hydrogen 21-cm line were higher than the optically determined masses. To investigate the cause of these tendencies the authors have estimated the masses using an analytic rotation curve of Brandt adjusted to the optical observations in order to include all the mass of a galaxy up to the last observed point.

158.156 **The galaxy NGC 55.**
J. A. Graham.
Rev. Mexicana Astron. Astrofis., Vol. 6, (see 012.017), 53 (1981). – Abstract.

158.157 **The Andromeda Galaxy. I. The H I gas.**
Y. Sofue.
Astron. Her., Vol. 74, 276 - 282 (1981). In Japanese. – Abstr. in Phys. Abstr., Vol. 85, Abstr. 68893 (1982).

158.158 **The Andromeda Galaxy. II. Kinematics.**
Y. Sofue, T. Sawa.
Astron. Her., Vol. 75, 13 - 19 (1982). In Japanese. – Abstr. in Phys. Abstr., Vol. 85, Abstr. 74088 (1982).

158.159 **Properties of BL Lac objects.**
A. M. Wolfe.
Ninth Texas Symposium on Relativistic Astrophysics, (see 012.020), p. 12 - 21 (1980).

The author examines the criteria proposed to define the

BL Lac class and shows how methods used to find BL Lacs influence properties thought to be intrinsic. He describes evidence that BL Lacs belong to a wider class of objects that include the optically violent variable QSOs, and examines the connection between BL Lacs and the galaxies in which they are embedded.

158.160 **Emission-line spectra of active galactic nuclei and quasars.**
D. E. Osterbrock.
Ninth Texas Symposium on Relativistic Astrophysics, (see 012.020), p. 22 - 38 (1980).

158.161 **Infrared observations of extragalactic sources.**
G. H. Rieke.
Ninth Texas Symposium on Relativistic Astrophysics, (see 012.020), p. 51 - 54 (1980).

158.162 **Implications of the variability and polarization of BL Lac objects.**
J. R. P. Angel, R. L. Moore.
Ninth Texas Symposium on Relativistic Astrophysics, (see 012.020), p. 55 - 58 (1980).
The authors discuss the results and interpretation of recent optical polarimetric observations obtained at Steward Observatory of BL Lac objects and violent quasar outbursts.

158.163 **Summary of the workshop on active galaxies and quasars.**
D. Weistrop.
Tenth Texas Symposium on Relativistic Astrophysics, (see 012.021), p. 415 - 418 (1981).

158.164 **Tests for the intrinsic shape of elliptical galaxies.**
D. Merritt.
Astron. J., Vol. 87, 1279 - 1287 (1982).
Two statistical tests for the intrinsic shape of elliptical galaxies are reanalyzed in light of a new set of velocity dispersion data. It is shown that such tests are capable of determining both the shapes of elliptical galaxies, as well as the dependence of surface brightness and velocity dispersion on intrinsic ellipticity. For the present sample of 61 galaxies, the surface brightness data are ambiguous, while the velocity data are consistent with the oblate and inconsistent with the prolate hypothesis at the one-sigma level. Neither hypothesis can be ruled out with confidence, however. If galaxies are oblate, the velocity data imply that nearly spherical galaxies have higher velocity dispersions than intrinsically flat ones.

158.165 **New dwarf galaxies in the M 81/M 82 group.**
F. Börngen, V. E. Karachentseva.
Astron. Nachr., Band 303, 189 - 192 (1982).
A survey for dwarf galaxies (Sculptor-type dwarfs) was carried out in the M 81/M82 group of galaxies in an area of about 195 sq. degrees by means of Tautenburg 52-inch Schmidt plates. Eight new faint dwarf galaxies were discovered. Coordinates, linear and absolute dimensions as well as finding charts are given.

158.166 **Galaxy photometry in the Virgo cluster. I.**
K. Fritze.
Astron. Nachr., Band 303, 237 - 244 (1982).
In a field of about 0.6 square degrees near the centre of the Virgo cluster of galaxies a photometry in B, V, and R has been performed by means of the microphotometer of the Zentralinstitut für Astrophysik. For 84 galaxies identified in this field magnitudes, diameters, and axis ratios have been determined.

158.167 **On the widths of dust layers in galaxies.**
G. Hacke, R. Schielicke, K.-H. Schmidt.
Astron. Nachr., Band 303, 245 - 249 (1982).= Mitt. Univ.-Sternw. Jena Nr. 156.
From prints of the Palomar Observatory Sky Survey th widths of the equatorial dust layers of a sample of 39 galaxi seen on edge are obtained to be in the range $500 \leqslant D \leqslant 1600$ p The width of a dust layer seems to be correlated with the luminosity of the galaxy. The dust lanes of field galaxies are obviously broader than those of the members of the Virgo cluster. This result may be interpreted as a dust deficiency caused by stripping of interstellar matter resulting from gala tic collisions and interaction of the galaxies with the intracluster medium.

158.168 **Étude optique de la morphologie et du champ de vitesses de la galaxie NGC 6946.**
A. Peton.
Astron. Astrophys., Vol. 114, 1 - 6 (1982).
The author presents optical kinematic results obtained from Fabry-Perot interferometry in Hα light, together with an analysis of the structure of the spiral galaxy NGC 6946.

158.169 **The galaxy NGC 1566: distribution and kinemati of the ionized gas.**
G. Comte, A. Duquennoy.
Astron. Astrophys., Vol. 114, 7 - 18 (1982).
The distribution of the ionized hydrogen in the large spiral galaxy NGC 1566 is studied from Hα narrow-band plates taken with the ESO 3.60 m telescope. The authors present a catalogue of 418 H II regions covering the main body of the galaxy, supplemented by 59 positions and estimated Hα luminosities for regions located in the pseudo-outer ring where no Hα plate is available. The luminosity fu tion, diameter distribution and spiral structure are discussed Three interferograms in Hα light, taken with the same instru mentation, allowed to determine a preliminary radial velocit field, acceptably fitted by a simple two component (bulge+ disk) dynamical model, in which the apparent disk mass-to-light ratio increases dramatically from the center to the edge

158.170 **21 cm line observations of cD galaxies.**
E. A. Valentijn, R. Giovanelli.
Astron. Astrophys., Vol. 114, 208 - 210 (1982).
The authors report on a small 21 cm line survey of five cD galaxies, all first ranked cluster members, and all, apart from N 2832, known to be centered on an extended X-ray source.

158.171 **Surveying the Universe.**
R. Ellis.
New Scientist, Vol. 92, 659 - 661 (1981). – Abstr. in Phys. Abstr., Vol. 85, Abstr. 84331 (1982).

158.172 **X-ray emission from active galactic nuclei. Compt ization of self-absorbed cyclotron higher harmoni**
F. Takahara, S. Tsuruta.
Prog. Theor. Phys., Vol. 67, 485 - 498 (1982). – Abstr. in Phys. Abstr., Vol. 85, Abstr. 84332 (1982).

158.173 **Extranuclear clues to the origin and evolution of activity in galaxies.**
B. Balick, T. M. Heckman.
Annu. Rev. Astron. Astrophys., Vol. 20 (see 003.009), 431 - 468 (1982).
The influence of the extragalactic environment. Activit and the stellar properties of galaxies. Gas and dust in active galaxies. Cosmological evolution of active nuclei.

158.174 **Observational results on Seyfert and radio galaxies with the Lick Observatory image tube-image dissector scanner.**
D. E. Osterbrock.
Recent advances in observational astronomy, (see 012.028), p. 39 - 44 (1981).

The Lick Observatory image tube-image dissector spectral scanner is described, emphasizing the ways in which it has been optimized for production of research results by a relatively small group of astronomers who use it frequently. A survey of optical spectra of radio and Seyfert galaxies made with this instrument is described, and some of the main results to date are summarized.

158.175 **Ultraviolet and visual observations of Seyfert galaxies.**
J. B. Oke.
Recent advances in observational astronomy, (see 012.028), p. 51 - 53 (1981).

Reddening by dust in the nuclei of Seyfert galaxies plays an important role in altering both the continuous spectrum and the emission lines. In the case of the type 2 Seyfert NGC 1068, correction for dust brings the emission line ratios to the radiative recombination values. The ultraviolet continuum is also reddened but the 2175 Å absorption feature is missing. In type 1 Seyferts, the presence of the 2175 Å absorption feature indicates the presence of varying amounts of dust.

158.176 **Nabije sterrenstelsels.**
P. C. van der Kruit.
Zenit, 9. Jaarg., 458 - 462 (1982).

158.177 **Variability and spectra of active galaxies.**
R. S. Warwick.
Observatory, Vol. 102, 114 - 115 (1982). – Abstract.

158.178 **X- and γ-ray observations of active galactic nuclei.**
A. J. Dean.
Observatory, Vol. 102, 115 (1982). – Abstract.

158.179 **Einstein X-ray observations of 3C radio galaxies.**
L. Miller.
Observatory, Vol. 102, 115 - 116 (1982). – Abstract.

158.180 **The velocity of the jet in M 87.**
R. E. Spencer.
Observatory, Vol. 102, 130 (1982). – Abstract.

158.181 **Photometry of faint galaxies with COSMOS.**
H. T. MacGillivray, R. J. Dodd.
Observatory, Vol. 102, 141 - 144 (1982).

Experiments indicate that measures from COSMOS can be used to obtain reliable photometric parameters, at high speed, for the faint galaxies detected on deep photographs taken with the UK Schmidt and Anglo-Australian telescopes.

158.182 **On the correlation of CO line and radio continuum emission in nearby galaxies – the star-formation efficiency.**
M. Rowan-Robinson, F. P. Israel.
Observatory, Vol. 102, 170 (1982). – Abstract.

158.183 **Dust in galaxies.**
D. P. Gilra.
Observatory, Vol. 102, 170 - 171 (1982). – Abstract.

158.184 **X-ray results on galactic haloes.**
P. Nulsen.
Observatory, Vol. 102, 174 - 175 (1982). – Abstract.

158.185 **The Fornax-Leo-Sculptor stream.**
D. Lynden-Bell.
Observatory, Vol. 102, 202 - 208 (1982).

The dwarf spheroidal galaxies Fornax, Leo I, Leo II and Sculptor lie on a great circle as seen from the Galactic centre, and Sculptor, the most tidally fragile, is elongated along that circle. All the dwarf spheroidal satellites of the Milky Way are in one of two streams of tidal débris.

158.186 **Environmental clues to the nuclear activity in galaxies and the genesis of supernovae.**
R. K. Kochhar, T. P. Prabhu.
Bull. Astron. Soc. India, Vol. 10, 41 (1982). – Abstract.

158.187 **Compact central components and the relation between optical and radio axes of elliptical galaxies.**
V. K. Kapahi, D. J. Saikia.
Bull. Astron. Soc. India, Vol. 10, 41 - 42 (1982). – Abstract.

158.188 **Optical nonthermal luminosity of the Seyfert 1 galaxy Mkn 335.**
R. K. Thakur, R. K. Sood.
Bull. Astron. Soc. India, Vol. 10, 45 - 46 (1982). – Abstract.

158.189 **The gradient of the brightest star age and density across the spiral arm S4 in the Andromeda galaxy.**
Yu. N. Efremov, G. R. Ivanov.
Astrophys. Space Sci., Vol. 86, 117 - 129 (1982).

On a plate obtained with the 2-m RC telescope at the Bulgarian National Observatory about 1400 stars in the spiral arm S4 of the Andromeda galaxy were measured. The limit of completeness is $20^{m}_{.}2$ (B magnitudes). In the central part of S4 a pronounced gradient of star luminosity and density is found. The authors have evaluated the velocity of star formation propagation across the arm, which is about 60 km s^{-1}.

158.190 **Modelling of colour indices and spectra for galaxies with normal chemical composition.**
O. K. Sil'chenko.
Astron. Zh., Tom 59, 855 - 863 (1982). In Russian. English translation in Soviet Astron., Vol. 26, No. 5.

An analysis of the evolutionary synthesis method for indices *UBV* colours of galaxies is made. The *UBV* data are shown in the general case to be insufficient to derive definite conclusions on the real parameters of galaxy evolution and star formation. *UBV* colours and continua calculated for the case of elliptical galaxies are compared with the published experimental data.

158.191 **Dissipative effects in gaseous subsystems of flat galaxies.**
A. G. Morozov.
Astron. Zh., Tom 59, 864 - 869 (1982). In Russian. English translation in Soviet Astron., Vol. 26, No. 5.

Taking into account the influence of dissipative effects on the perturbations dynamics in gravitationally stable gaseous subsystems of flat galaxies leads to damping of Jeans perturbations. Axisymmetric entropy perturbations and non-axisymmetric gradient perturbations turn out to be unstable in such models in the wavelength region of the Jeans order.

158.192 **Stellar populations in normal galaxies.**
M. Capaccioli.
Third European IUE Conference, (see 012.029), p. 59 - 68 (1982).

This paper reviews the impact of UV observations of normal galaxies in the context of our present understanding of stellar populations. UV data show that ellipticals possess a population of hot stars. The nature of this component is discussed on the basis of current models. The heterogeneous UV phenomenology of spirals is also accounted, with emphasis to the problem of galactic reddening. Consequences on stellar evolution and cosmology are briefly outlined.

158.193 **Active galaxies observed by IUE.**
M. V. Penston.
Third European IUE Conference, (see 012.029), p. 69 - 74 (1982).

IUE has extended the grasp of ultraviolet astronomy to cover active galaxies and quasars fainter than the sixteenth magnitude. Major achievements of these observations are summarized in this review.

158.194 **UV energy distribution in elliptical galaxies.**
F. Bertola, M. Capaccioli, J. B. Oke.
Third European IUE Conference, (see 012.029), p. 515 (1982).
The authors give a brief account of new observations of the UV-visual energy distributions for three nearby elliptical galaxies, namely NGC 205, the dwarf companion of the Andromeda nebula, and the two giants NGC 4125 and NGC 4374.

158.195 **The IUE spectrum of NGC 4258.**
R. Barbon, M. Capaccioli, G. Longo.
Third European IUE Conference, (see 012.029), p. 517 (1982).

158.196 **Coordinated observations of Seyfert 1 galaxies.**
W. Wamsteker, P. Benvenuti, C. Cacciari, A. Cassatella, L. Bianchi, P. Patriarchi, J. C. Blades, A. C. Danks.
Third European IUE Conference, (see 012.029), p. 529 - 531 (1982).
Some examples of instantaneous line profiles of UV and optical recombination lines are given for F-9 and AKN 120.

158.197 **Results from the NGC 4151 ultraviolet laboratory.**
G. E. Bromage, A. Boksenberg, J. Clavel, A. Elvius, M. V. Penston, G. C. Perola, M. Pettini, M. A. J. Snijders, E. G. Tanzi, M. Tarenghi, M. H. Ulrich.
Third European IUE Conference, (see 012.029), p. 533 - 536 (1982).
The nucleus of NGC 4151, the brightest Type 1 Seyfert, has shown complex variations in continuum, emission and absorption lines during three years of IUE monitoring. The authors report the main results and discuss their implications for the size and structure of the various sub-regions in the ionised nebula around the central power source.

158.198 **Monitoring the UV spectrum of the Seyfert 1 galaxy NGC 4593. II: short term variability.**
J. Clavel.
Third European IUE Conference, (see 012.029), p. 537 - 541 (1982).
The UV spectrum (1200–3200 Å) of the Seyfert 1 galaxy NGC 4593 was observed twice at an interval of 9 days. Significant variations are found both in the line and the continuum which yield an upper limit to the size of the broad line region $D < 2 \times 10^{16}$ cm. Using earlier data from 1979 and 1980, several correlations are found.

158.199 **NGC 7469, a Seyfert 1 galaxy observed with IUE.**
A. Elvius, B. A. M. Westin, J. Lind.
Third European IUE Conference, (see 012.029), p. 543 - 548 (1982).
The Seyfert 1 galaxy NGC 7469 was observed with the IUE in June 1978 and in May 1981. Comparisons are made between data taken at different epochs and also between NGC 7469 and some other galaxies.

158.200 **Line profiles and continuum radiation in Seyfert galaxies.**
C. Boisson, M. H. Ulrich.
Third European IUE Conference, (see 012.029), p. 549 - 550 (1982).
An attempt has been made to identify some general properties of the ultraviolet spectrum of Seyfert galaxies of type 1. It is found that in the Seyfert nuclei which are variable, the ultraviolet continuum is always harder when it is brighter. Regarding the line profiles, the full width at half maximum of C III] λ1909 is often found to be smaller than that of C IV λ1550.

158.201 **Secondary ultraviolet sources in the Seyfert 2 galaxy NGC 1068.**
M. A. J. Snijders, S. A. Briggs, A. Boksenberg.
Third European IUE Conference, (see 012.029), p. 551 - 55 (1982).
The authors have discovered four ultraviolet sources around the nucleus of the Seyfert 2 galaxy NGC 1068. All coincide with the inner spiral arms and three of these can be readily explained by the integrated light of large numbers of O and B type stars associated with the spiral arms. The four source has a harder spectrum then the Seyfert nucleus itself The unusual ultraviolet brightness and shallow P-Cygni type line profiles of this source resemble the starburst galaxy NGC 7714.

158.202 **IUE observations of NGC 3783.**
P. Barr.
Third European IUE Conference, (see 012.029), p. 559 - 56 (1982).
The author presents repeated IUE observations of the Seyfert galaxy NGC 3783 performed during the first three years of IUE. Variations are encountered in the line-of-sight extinction, the emission line strengths, the underlying continuum intensity and the spectral index. A highly significan correlation between spectral index and luminosity is found.

158.203 **Markarian 108, a tidally-torn off or turned on galaxy?**
P. Benvenuti, C. Casini, J. Heidmann.
Third European IUE Conference, (see 012.029), p. 565 - 56 (1982).
Mkn 108 (= IC 2458) belongs to a group of 4 galaxies and has an optical spectrum typical of H II regions. IUE observations show that Mkn 108 has undergone recently a v large burst of star formation.

158.204 **Ly α absorption at cz=8200 km s^{-1} in NGC 1275**
S. A. Briggs, M. A. J. Snijders, A. Boksenberg.
Third European IUE Conference, (see 012.029), p. 569 - 57 (1982).
The authors have obtained IUE spectra of NGC 1275 which show an absorption feature identified with Ly α at velocity of 8200 km s^{-1}. They attribute the origin of this feature to the intervening galaxy responsible for the high velocity system of filamentary emission seen around NGC 1275.

158.205 **The ultraviolet spectrum of the high redshift BL Lac object O 215+015.**
J. C. Blades, M. Pettini, R. W. Hunstead, H. S. Murdoch.
Third European IUE Conference, (see 012.029), p. 573 - 57 (1982).
The object O 215+015 is highly variable, with a range the optical band of more than 4 magnitudes. At optical wave lengths it shows no emission lines but has a rich absorption spectrum with at least six, separate redshift systems. The authors have obtained a high signal-to-noise spectrum between 2400–3200 Å which is used to study the hydrogen Lyman lines of the various redshift systems.

158.206 **UV observations of X-ray emitting QSOs and BL Lac objects.**
L. Chiappetti, L. Maraschi, E. G. Tanzi, A. Treves.
Third European IUE Conference, (see 012.029), p. 581 - 58 (1982).
The authors report on IUE observations of the BL Lac objects PKS 2155–304, 3C 66A, PKS 0521–37, PKS 0537– and of the QSOs PG 1351 + 64 and PKS 0637–75.

158.207 **Simultaneous multifrequency observations of BL Lac objects and violently variable quasars.**
J. N. Bregman, A. E. Glassgold, P. J. Huggins.

Third European IUE Conference, (see 012.029), p. 589 - 591 (1982).

Ultraviolet spectra simultaneous with radio, infrared, optical, and X-ray measurements were obtained for 0735+178, BL Lac, IZw-187, and 3C 446.

158.208 **CO observations of M 51 and M 82.**
G. Rydbeck, Å. Hjalmarson, H. Olofsson, G. Pilbratt, O. E. H. Rydbeck.
The scientific importance of submillimetre observations, (see 012.030), p. 142 (1982). – Abstract.

158.209 **CO along the minor axis of M 31.**
A. Sandqvist, T. Elfhag, P. O. Lindblad.
The scientific importance of submillimetre observations, (see 012.030), p. 143 (1982).

The authors briefly report on a series of observations of the J = 1–0 CO line along the minor axis of M 31, using the 20 m millimeter wave telescope of the Onsala Space Observatory.

158.210 **VLBI observations of the core sources of a sample of spiral galaxies.**
E. Hummel, C. Fanti, P. Parma, R. T. Schilizzi.
Astron. Astrophys., Vol. 114, 400 - 404 (1982).

VLBI observations at 6 cm are reported of a sample of spiral and irregular galaxies and compared with existing high resolution (~ 1″) observations of these galaxies. Only a small fraction of the flux density of the central sources originates in a compact core.

158.211 **Surface photometry of edge-on galaxies. IV. Global structure of disk galaxies.**
M. Hamabe.
Publ. Astron. Soc. Japan, Vol. 34, 423 - 447 (1982).

The surface brightness distributions of nine edge-on galaxies are decomposed with the help of a semi-empirical three-dimensional model, and the global structures of their two main components, the spheroid and the disk components, are investigated.

158.212 **NGC 1275: a growing elliptical galaxy?**
A. Wirth, S. J. Kenyon, D. A. Hunter.
Bull. American Astron. Soc., Vol. 14, 778 (1982). – Abstract.

158.213 **Discussion of BL Lacertae identifications.**
N. V. Zotov, S. Tapia.
Bull. American Astron. Soc., Vol. 14, 779 - 780 (1982). Abstract.

158.214 **Two multifrequency observations of the BL Lacertae object OJ 287.**
D. M. Worrall, J. J. Puschell, B. Jones, F. C. Bruhweiler, M. F. Aller, H. D. Aller, P. E. Hodge, M. L. Sitko, W. A. Stein, Y. X. Zhang, W. H.-M. Ku.
Astrophys. J., Vol. 261, 403 - 411 (1982).

The authors describe observations of OJ 287 made at frequencies between the radio and far ultraviolet on two occasions in late 1980. The ultraviolet flux decreased by a factor of ~0.7 between these occasions, with no corresponding change in centimeter wavelength emission. Combining their data with nonsimultaneous X-ray flux measurements, the authors have constrained parameters for a synchrotron self-Compton flux origin.

158.215 **Rotational properties of 23 Sb galaxies.**
V. C. Rubin, W. K. Ford, Jr., N. Thonnard, D. Burstein.
Astrophys. J., Vol. 261, 439 - 456, plates 4 - 7 (1982).

The authors have obtained major axis spectra for 23 Sb galaxies in a sample chosen to encompass a wide range of radii, masses, and luminosities. The Sb rotation curves exhibit many of the systematic trends with luminosity shown by the Sc's studied previously. Velocities in small, low-luminosity, low-mass Sb galaxies rise gently to a maximum rotational velocity in a large fraction of their isophotal radius; velocities in large, high-luminosity, high-mass Sb galaxies reach their rotational velocity in only a small fraction of their isophotal radius. Correlations between velocity extrapolated to the isophotal radius and blue absolute magnitude are analyzed.

158.216 **The jet in M 87.**
P. E. Hardee.
Astrophys. J., Vol. 261, 457 - 462 (1982).

Observations of the M 87 jet are used to investigate the nature of the jet. The jet appears to be expanding and twisted into a helix. Jet helicity is interpreted as being produced by helical instability of the flow. Jet helicity produced by instability is shown to imply a supersonic Mach 15 jet more dense than the surrounding medium. When this result is combined with the calculated jet conditions it is found that the jet flow velocity is ~30,000 km s^{-1} and energy flow is ~200 times the jet luminosity.

158.217 **The far-infrared disk of M 51.**
J. Smith.
Astrophys. J., Vol. 261, 463 - 472 (1982).

Far-infrared maps and multifrequency photometry are presented for M 51 and its companion galaxy NGC 5195. Dust reradiates about half the starlight of the M 51 + NGC 5195 system to produce the observed far-infrared (80–200 μm) luminosity of $3 \times 10^{10} L_\odot$. The maps show that roughly 70% of this luminosity is produced by the 5′ wide far-infrared disk of M 51. Dust producing the central luminosity of NGC 5195 is much less abundant and considerably warmer than the ~20 K dust producing the central luminosity of M 51. Far-infrared properties are given for several galactic scale complexes of star formation in M 51.

158.218 **VLA observations of the Seyfert galaxy NGC 1068.**
J. M. van der Hulst, E. Hummel, J. M. Dickey.
Astrophys. J., Lett., Vol. 261, L59 - L63 (1982).

VLA observations of the central region of NGC 1068 at λ 2 cm are reported. The central region of NGC 1068 resolves into an asymmetric 1″ double structure and a nuclear source. It is concluded that the lack of symmetry and the lack of a compact core source in the inner 100 pc of NGC 1068 complicate a jet model.

158.219 **The cross-correlation of the Zwicky and Shane-Wirtanen catalogs of galaxies.**
M. Seldner, J. M. Uson.
Astrophys. J., Lett., Vol. 261, L65 - L69 (1982).

The authors have cross-correlated the angular positions of the galaxies in several subsets of the Zwicky catalog of galaxies with the Shane-Wirtanen galaxy counts. They find that the cross-correlation function is given by a power law: $w(\Theta) = A \cdot \Theta^{1-\gamma}$ with index $\gamma = 1.81 \pm 0.03$. The functions break away from this power law at an angular scale of about 2°.5. It is concluded that the break is a local feature and is not intrinsic to the clustering of galaxies.

158.220 **The distribution of carbon monoxide in spiral and irregular galaxies.**
D. M. Elmegreen.
Submillimetre wave astronomy, (see 012.031), p. 191 - 195 (1982).

158.221 **Observations of far-infrared emission from late-type galaxies.**
L. J. Rickard, P. M. Harvey.
Submillimetre wave astronomy, (see 012.031), p. 197 - 198 (1982).

158.222 **The distribution of carbon monoxide in three face-on galaxies.**
L. J. Rickard, P. Palmer.
Submillimetre wave astronomy, (see 012.031), p. 199 - 200 (1982).

158.223 **Rapid variability of OJ 287 at 1.25 μm.**
R. D. Wolstencroft, G. Gilmore, P. M. Williams.
Mon. Not. R. Astron. Soc., Vol. 201, 479 - 485 (1982).

J band photometry of OJ 287 has been carried out with a time resolution of 5 s using the 3.8-m UKIRT. Significant variations in brightness of up to one magnitude have been observed. Several distinct outbursts can be identified in the photometric record and rise-times and decay-times as short as 1 min were seen. These observations provide evidence of highly anisotropic and/or relativistic effects in the central source.

158.224 **On the nature of the stellar population in the nucleus of the Sd galaxy NGC 7793.**
A. I. Díaz, B. E. J. Pagel, M. G. Edmunds, M. M. Phillips.
Mon. Not. R. Astron. Soc., Vol. 201, 49P - 55P (1982).

Profiles and equivalent widths of hydrogen, Ca^+ and some other lines imply that the blue-violet spectrum of the nucleus of this late-type spiral is dominated by A or early F stars near the main sequence and not by a mixture of globular cluster-like populations.

158.225 **Accurate optical positions for Markarian galaxies 1303–1399.**
G. Kojoian, R. Elliott, M. D. Bicay.
Astron. J., Vol. 87, 1364 - 1367 (1982) = Contrib. No. 7 Casey Obs.

Optical positions of 97 objects, Markarian 1303–1399, were measured from the Palomar Sky Survey Prints and are given with an accuracy of about 2.″0 arcsec in each coordinate.

158.226 **Observations of peculiar galaxies in clusters of galaxies.**
G. D. Bothun, R. A. Schommer.
Astron. J., Vol. 87, 1368 - 1373 (1982).

The authors present H I and optical measurements of some peculiar/interacting spiral galaxies located in clusters of galaxies that are also contained in The Atlas of Peculiar Galaxies. The color and gas content properties of these peculiar galaxies exhibit a large scatter.

158.227 **A 1415 MHz survey of Seyfert and related galaxies. III.**
A. S. Wilson, E. J. A. Meurs.
Astron. Astrophys., Suppl. Ser., Vol. 50, 217 - 231 (1982).

The authors present the third and final instalment of a high sensitivity search with the Westerbork telescope for radio continuum emission at 1415 MHz from Seyfert and related galaxies. 24 galaxies with declinations in the range $+10° > \delta > -31°$ have been observed to a flux density limit (3σ) of about 3 mJy. 19 (79%) were certainly or probably detected. The results of deep (3σ noise about 1 mJy), full 12-hour syntheses of 3 Seyfert galaxies (Mark 40, 372 and 391) which were undetected in Paper I of this series are also given. Each of these galaxies has now been detected. A summary listing of the results of the entire survey is given in right ascension order.

158.228 **Absolute ultraviolet fluxes of elliptical galaxies as observed with the Astronomical Netherlands Satellite (ANS).**
K. S. de Boer.
Astron. Astrophys., Suppl. Ser., Vol. 50, 247 - 250 (1982).

Ultraviolet photometry of 13 elliptical galaxies obtained with the ANS is presented. All galaxies but one were detected at 3300 and at 2200Å, and the average drop in intensity over this interval is 1.8 mag. Only NGC 4486, NGC 3115 and the dwarf companions of Andromeda, M32 and NGC205, were detected in the ANS 1550Å channel. There is good agreeme with the colors measured with the IUE for four ellipticals in common.

158.229 **Einstein Observatory solid state spectrometer obs vations of M 87 and the Virgo cluster.**
S. M. Lea, R. Mushotzky, S. S. Holt.
Astrophys. J., Vol. 262, 24 - 32 (1982).

The authors report X-ray observations of the galaxy M 87 and of a region in the Virgo cluster displaced 7′ from t center of M 87. They obtained X-ray spectra at these two locations with the SSS onboard the Einstein Observatory. Emission lines were observed in both locations, indicating the presence of heavy elements at solar abundances. There is evidence for an infall of gas onto M 87 with a rate of $\sim 20\ M_\odot$/yr. These results indicate the presence of an intracluster gas with density $\sim 10^{-3}\,cm^{-3}$ and temperature $\sim 3 \times 10^7$ K.

158.230 **X-ray spectroscopy of the galaxy M 87: radiative accretion of the hot plasma halo.**
C. R. Canizares, G. W. Clark, J. G. Jernigan, T. H. Markert.
Astrophys. J., Vol. 262, 33 - 43 (1982).

Using the Focal Plane Crystal Spectrometer on the Einstein Observatory, the authors have measured the strengt of seven X-ray emission lines/blends of ionized iron from th central 3′ × 30′ of the hot plasma halo around M 87. The lir indicate the presence of ionization stages ranging from Fe XVII to Fe XXIV and imply plasma temperatures from $\sim 3 \times 10^7$ K down to $\sim 2–4 \times 10^6$ K. The approximate distri bution and the quantity of emitting halo material are derive The results are in agreement with a model of radiative, pressure driven accretion onto M 87 with an infall rate of $\sim 3–4\ M_\odot$/yr.

158.231 **Double galaxy investigations. II. The redshift periodicity in optically observed pairs.**
W. G. Tifft.
Astrophys. J., Vol. 262, 44 - 47 (1982).

The new sample of optically observed double galaxies i tested for the 72 km s^{-1} periodicity. The periodicity is prese in the 200 pair high weight sample at or above the 99% con dence level. Detection varies between subsamples in accord with variation in data uncertainty as expected.

158.232 **Radio jets in NGC 4151.**
K. J. Johnston, M. Elvis, D. Kjer, B. S. P. Shen.
Astrophys. J., Vol. 262, 61 - 65 (1982).

The radio radiation associated with the nucleus of the Seyfert galaxy NGC 4151 has been found to lie on a line of extent over 10″ (950 pc). This nonthermal emission measur at λ6 and λ20 cm consists of at least six components and is similar to jets observed in other compact extragalactic radio sources. The optical line emission region appears to be coinc dent with the radio emission.

158.233 **A nuclear spectroscopic survey of disk galaxies. II Galaxies with emission lines not excited by stellar photoionization.**
J. R. Stauffer.
Astrophys. J., Vol. 262, 66 - 80 (1982).

Survey spectra have been used to identify a sample of galaxies whose nuclei exhibit emission lines with line ratios not characteristic of stellar photoionization. High S/N spect of most of those galaxies were subsequently obtained with t Lick 3 m telescope. This paper presents an analysis of those spectra to determine the excitation mechanism for the emission-line regions. A comparison of a list of AGNs relative to list of galaxies without detected nuclear emission lines show that the galaxies with active nuclei preferentially occur in groups, while isolated galaxies seldom have strong nuclear emission lines. That result supports a model where gas infall from tidal encounters provides the fuel for the observed optical activity.

158.234 **Time variations of the neutral hydrogen absorption spectrum of NGC 1275 (3C 84).**
A. D. Haschick, P. C. Crane, J. M. van der Hulst.
Astrophys. J., Vol. 262, 81 - 86 (1982).
The authors have monitored the neutral hydrogen absorption spectrum and the total continuum flux density of NGC 1275 over a 6-year period. Two features in the absorption spectrum (at 8114 and 8120 km s^{-1}) increased in flux density synchronously with the continuum source, while a third feature at 8106 km s^{-1} remained almost constant. It is concluded that the most likely source for the high velocity absorption is the disk of an intervening galaxy. This galaxy partly absorbs the small variable nuclear component and the more extended, steady core component of the radio continuum.

158.235 **Galaxies and clusters of galaxies.**
E. E. Salpeter.
Prog. Theor. Phys. Suppl., No. 70, p. 238 - 248 (1981).
Abstr. in Phys. Abstr., Vol. 85, Abstr. 101070 (1982).

158.236 **The first detection of Wolf-Rayet stars in M 31.**
M. M. Shara, A. F. J. Moffat.
Wolf-Rayet stars: observations, physics, evolution, (see 012.042), p. 531 - 537 (1982).
A search to continuum magnitude B $\sim$ 21.5 using a narrow band filter at λ4670Å and a wide B-band filter has revealed 21 Wolf-Rayet star candidates in about half the giant Sb galaxy M 31. These numbers are compatible with the total number of luminous (i.e. massive) stars in M 31. Eighteen of twenty confirmed candidate stars in M 31 lie in the direction of OB associations in the ring of prominent star formation 5 - 16 kpc from the center.

158.237 **Wolf-Rayet stars featured in emission-line galaxies.**
D. Kunth.
Wolf-Rayet stars: observations, physics, evolution, (see 012.042), p. 569 - 570 (1982).

158.238 **Lyα absorption at a high velocity in NGC 1275.**
S. A. Briggs, M. A. J. Snijders, A. Boksenberg.
Nature, Vol. 300, 336 - 337 (1982).
The authors present results of IUE spectrophotometry of NGC 1275 which show an absorption feature identified with Lyα at a velocity of 8,200 km s^{-1}. They attribute the origin of this feature to the intervening galaxy responsible for the high velocity emission filaments seen around NGC 1275.

158.239 **On gaseous disks in Seyfert 1 nuclei.**
K. S. Anderson.
Astrophys. J., Vol. 262, 554 - 555 (1982).
The author considers some implications of the suggestion that the broad-line component of a Seyfert 1 nucleus arises in a rotating disk whose plane is orthogonal to that of the parent galaxy. Observed line widths would be expected to decrease, and their dispersion about the mean relation to increase, with increasing inclination of the parent galaxy.

158.240 **The nucleus of M 81: simultaneous 2.3 and 8.3 GHz Mark III VLBI observations.**
N. Bartel, I. I. Shapiro, B. E. Corey, J. M. Marcaide, A. E. E. Rogers, A. R. Whitney, R. J. Cappallo, D. A. Graham, J. D. Romney, R. A. Preston.
Astrophys. J., Vol. 262, 556 - 563 (1982).
The authors made Mark III observations of the nucleus in the normal galaxy M 81 (NGC 3031) simultaneously at 2.3 and 8.3 GHz and found that nearly 100% of the flux density of the nuclear region originates in an elongated radio core with linear dimensions 1000–4000 AU, dependent on frequency. This galactic nucleus is the most compact observed in any extragalactic source. The results are consistent with the emission's being incoherent electron synchrotron radiation from an inhomogeneous region with an optical depth varying along the rotation axis of the galaxy.

158.241 **Broad line region clouds and the absorbing material in NGC 4151.**
G. J. Ferland, R. F. Mushotzky.
Astrophys. J., Vol. 262, 564 - 577 (1982).
The constraints imposed by X-ray, ultraviolet, and optical observations on the absorbing material in the Seyfert 1.5 galaxy NGC 4151 are discussed. These observations are compared to the predictions of photoionization models of both broad line region clouds and to models of the narrow line region gas. The calculations suggest that narrow line region clouds have a composition similar to extragalactic giant H II regions, and the hydrogen line spectrum shows that dust is mixed with narrow line region gas. It is argued that broad line region clouds are likely to be in a radiatively driven wind.

158.242 **Accurate optical positions of extragalactic emission-line objects: University of Michigan lists I–IV.**
G. Kojoian, R. Elliott, M. D. Bicay.
Astrophys. J., Suppl. Ser., Vol. 50, 161 - 167 (1982) = Casey Obs., Contrib. No. 6.
Optical positions of the extragalactic emission-line objects UM 1–420 are presented with an accuracy of ~1.″5 in each coordinate. This list includes the 1950 right ascension and declination of each object and alternate designations obtained from other compilations.

158.243 **Compact radio cores and the relation between the radio and optical axes of elliptical galaxies.**
V. K. Kapahi, D. J. Saikia.
J. Astrophys. Astron., Vol. 3, 161 - 171 (1982).
The authors have reinvestigated the reported tendency for the extended radio structures associated with bright elliptical galaxies to be oriented preferentially along the optical minor axes. It is found that such a tendency exists only for those galaxies in which the compact radio cores coincident with their nuclei are quite prominent.

158.244 **Dwarf galaxies in the M 81/82 group. New photometric data.**
F. Börngen, V. E. Karachentseva, R. Schmidt, G. M. Richter, W. Thänert.
Astron. Nachr., Band 303, 287 - 292 (1982).
The authors present a list of 39 dwarf galaxies of the M81/82 group. Three different methods are applied to determine their apparent integral magnitude: digital processing, equidensitometry, and calculation according to the formulae of Fisher and Tully. The plates used were obtained with the Tautenburg Schmidt telescope. For a distance of the group of 3.5 Mpc the absolute magnitudes of the dwarf galaxies investigated are $-15^m < M < -9^m$.

158.245 **Infrared emission and star formation in NGC 5253.**
A. F. M. Moorwood, I. S. Glass.
Astron. Astrophys., Vol. 115, 84 - 89 (1982).
NGC 5253 is a peculiar S0 galaxy containing both an extended, visible, H II region complex and a more compact source of infrared emission. New infrared observations, together with available data at longer and shorter wavelengths discussed in this paper, are consistent with the idea that massive stars are currently forming in the central core of this galaxy. It is argued that the currently observable star formation event in the central region of the galaxy is probably the first to have occurred within the last ~10^8 yr. This conclusion is of particular interest in view of the suggestion,

made by Van den Bergh (1980), that star clusters observed in the outer parts of NGC 5253 are the result of star formation triggered some $10^8 - 10^9$ yr ago following an encounter with M 83.

158.246 **The detection of extranuclear emission lines in the Seyfert galaxies Mk 10 and Mk 79.**
H. Schulz.
Astron. Astrophys., Vol. 115, 209 - 212 (1982).
Extranuclear emission line regions (ELR) of several kpc extent have been detected in the Seyfert galaxies Mk 10 and Mk 79 on long-slit photographic image tube spectrograms. A peak rotational velocity of 270 km s^{-1} has been derived for Mk 10. The ELRs of Mk 10 are unusually luminous in H_α for an Sb-galaxy and resemble the giant H II regions in Sc galaxies. They are most probably ionized by in situ hot stars rather than by the active nucleus of Mk 10. The [N II]/H_α-ratio of the ELRs in Mk 79 is very similar to the nuclear value. Although this suggests ionization of the ELRs by the nucleus, other possibilities cannot be ruled out in this case. Some new data of the nuclear spectra are also presented.

158.247 **New high resolution radio observations of NGC 4258. III. VLA and WSRT observations of the anomalous arms.**
G. D. van Albada, J. M. van der Hulst.
Astron. Astrophys., Vol. 115, 263 - 269 (1982).
Radio continuum observations of the anomalous arms of NGC 4258 obtained with the Very Large Array and the Westerbork Synthesis Radio Telescope are presented. A wealth of new details is discovered in these observations, showing that the anomalous arms are part of a large complex of filamentary structures. A number of mechanisms for the creation of these arms as proposed in the literature are discussed in the light of the new data.

158.248 **NGC 1961: stripping of a supermassive spiral galaxy.**
G. S. Shostak, E. Hummel, P. A. Shaver, J. M. van der Hulst, P. C. van der Kruit.
Astron. Astrophys., Vol. 115, 293 - 307 (9182).
The peculiar supermassive Sb spiral galaxy NGC 1961 has been mapped in the 21-cm line of neutral hydrogen with a resolution of 0.′5 using the Westerbork Synthesis Radio Telescope. An unusual "head-tail" morphology for the H I is found, consisting of a wing of gas extending 30 kpc to the NW, and a sharp edge to the SE. Radio continuum emission is observed in a ridge coincident with the SE spiral arm, and an X-ray map produced with the IPC of the Einstein satellite reveals an arc-shaped structure which is also aligned with the ridge. The authors interpret these observations as indicating that the gas in NGC 1961 is being stripped by a hot intergalactic medium of density $\geqslant 10^{-3}$ cm^{-3} and total mass $> 10^{10} M_\odot$. From the observed morphology, this gas would be located in the direction of the center of mass of a group of galaxies located SE of NGC 1961.

158.249 **The giant spiral galaxy M 101. VIII. Star formation in H I - H II associations.**
F. Viallefond, W. M. Goss, R. J. Allen.
Astron. Astrophys., Vol. 115, 373 - 387 (1982).
In paper VII of this series, the authors have discovered several correlations among the physical properties of the giant H I - H II associations in M 101. The present paper deals with the interpretation of these correlations in terms of models for the collapse of the interstellar medium into massive H I clouds and the subsequent formation of stars.

158.250 **Photographic photometry of galaxies using the INMP. I. The lenticulars NGC 404 and NGC 524.**
R. Barbon, M. Capaccioli, R. Rampazzo.
Astron. Astrophys., Vol. 115, 388 - 397 (1982).
A study of the photometric and geometrical properties of two face-on S0 galaxies, NGC 404 and NGC 524, is preser The complex pattern of the dust at the center of NGC 40 discussed in connection with similar morphologies in othe objects, as well as the influence on the results of the prese of the nearby bright star β And. In both galaxies the lumi profiles present the three distinct components characterist lenticulars, namely the bulge, the lens and the disk.

158.251 **The most distant galaxies.**
M. S. Longair.
Progress in cosmology, (see 012.047), p. 323 - 333 (1982)
The data indicate the great potential of observations the infrared waveband for the study of the evolution of th stellar component of giant elliptical galaxies over cosmolo time-scales. The present results are preliminary but suggest that the infrared properties of galaxies are less susceptible strong evolutionary effects than the optical properties. Co bining the optical and infrared properties provides additio constraints on the evolutionary effects which influence th magnitude-redshift relation.

158.252 **A spectroscopic method for determining the lu nosities of spiral galaxies and estimating their st population.**
A. P. Cowley, D. Crampton, R. D. McClure.
Astrophys. J., Vol. 263, 1 - 13 (1982).
The authors present new Reticon spectra of a sample nuclei of bright galaxies of various Hubble types. They ex ine the relationships between spectral line strengths and g luminosity and discuss the accuracy with which absolute r nitudes can be predicted from spectral characteristics of t nuclei. This calibration is applied to a study of the Virgo cluster of galaxies. In addition, the authors study the stell population in galaxy nuclei of late Hubble types to detern the relative roles of stellar age and metal abundance in aff ing these spectral characteristics.

158.253 **The heating of dust in the broad-line regions of active galaxies and quasars.**
R. J. Rudy, R. C. Puetter.
Astrophys. J., Vol. 263, 43 - 53 (1982).
The authors discuss the relevant energy sources which heat (and, in many cases, destroy) the dust grains in the en sion-line clouds which give rise to the broad-line emission s in active extragalactic objects. They compare the heating r from the external radiation field, trapped line radiation, an the diffuse bound-free continua. It is found that in hot clo Lyα dominates the dust heating rate, while in cooler cloud the external radiation field dominates. In all cases, the dust residing in the broad-line clouds is quite hot (T > 500 K). results are discussed in light of the observed infrared prope ties of active galaxies and quasars.

158.254 **Further examples of companion galaxies with discordant redshifts and their spectral peculiariti**
H. Arp.
Astrophys. J., Vol. 263, 54 - 72, plates 1 - 3 (1982).
Four more examples of companion galaxies with disc dant redshifts are presented (Δz = +4,800 to +26,200 km s On direct photographs three of these new companions app interactive with their main galaxy in ways extremely unlik to be a result of chance projection. These new discordant r shift companions follow the previously established trend f their spectra to be unusually rich in emission lines and to s early-type absorption spectra. All past cases of discordant r shifts are tabulated from the literature together with the ne ones. The list now contains 24 main galaxies and 38 discor redshift companions. The emission and absorption charact tics of the spectra of this whole group of discordant redshi companions is markedly different from what is known abo the general field galaxies.

158.255 **VLA observations of H I absorption in the nuclei of Seyfert and active galaxies.**
J. M. Dickey.
Astrophys. J., Vol. 263, 87 - 93 (1982).
Spiral galaxies with strong nuclear radio-continuum emission have been surveyed for H I absorption at 21 cm wavelength. Eight galaxies show clear or tentative absorption lines. The absorption line centers are offset from the galaxy systemic velocities by up to 300 km s^{-1}; for the entire sample, there is an overall preference for infall of gas toward the nucleus rather than outflow. For the detected lines, the total absorption is much greater than would be expected from the disk of our Galaxy.

158.256 **Low surface brightness spiral galaxies. I. Neutral hydrogen content and location in the infrared Fisher-Tully diagram.**
W. Romanishin, N. Krumm, E. Salpeter, G. Knapp, K. M. Strom, S. E. Strom.
Astrophys. J., Vol. 263, 94 - 100, (1982).
A neutral hydrogen survey has been carried out for a class of spiral galaxies characterized by low surface brightness disks; on average, these systems (denoted LSBs) have projected central surface brightness values smaller by a factor of 3 than the value typical of normal spiral galaxies. Infrared aperture photometry, combined with 21 cm line widths, permit the low surface brightness spirals to be located with respect to the H magnitude, Δv Fisher-Tully relationship; the LSBs fall below the observed relationship by 0.64 mag. A careful discussion of the factors affecting their location in the H, Δv plane suggests that LSBs have mass-to-light ratios a factor of 1.8 times that typical of normal spirals.

158.257 **Structure of the M 33 nucleus.**
J. S. Gallagher, J. W. Goad, J. Mould.
Astrophys. J., Vol. 263, 101 - 107, plates 4, 5 (1982).
New optical and infrared observations of the M 33 nucleus lead to a low mass-to-light ratio of $M/L_v \approx 2{-}3$. The M 33 nucleus is therefore a very different type of object than high M/L, high mass density nuclei found in M 31 and many luminous spirals. Spectroscopic and photometric evidence for a composite-age stellar population is presented, and it is suggested that growth of the M 33 nucleus may be continuing as gas is fed in from the surrounding galactic disk.

158.258 **The distribution of infrared obscuration in NGC 7331: evidence for a massive molecular ring.**
C. M. Telesco, I. Gatley, J. M. Stewart.
Astrophys. J., Lett., Vol. 263, L13 - L17 (1982).
Spatial variations in the near-infrared colors of the Sb galaxy NGC 7331 produced by interstellar dust indicate the presence of a massive ring of molecular gas ($6 \times 10^8\ M_\odot$) surrounding the nuclear bulge. The density, mass, radius, and thickness of this ring are comparable to similar structures observed within the Galaxy and M 81.

158.259 **High frequency radio continuum observations of bright spiral galaxies.**
I. M. Gioia, L. Gregorini, U. Klein.
Astron. Astrophys., Vol. 116, 164 - 174 (1982).
The 100-m Effelsberg telescope has been used to map 56 bright spiral galaxies at 4.8 and 10.7 GHz. The survey contains all spiral galaxies which have been previously detected at 408 MHz in the declination range $20° \leqslant \delta \leqslant 60°$ using the Bologna Northern Cross. These observations in most cases allow the determination of reliable integrated spectra between 408 MHz and 10.7 GHz. The results are presented along with various physical parameters derived for the sources. The spectral indices obtained for the 56 galaxies show an extremely narrow distribution suggesting that the production and propagation of cosmic ray electrons must be very similar in these spirals. Most spiral galaxies have relative amounts of thermal emission at 10.7 GHz below ~40%.

158.260 **Radio continuum emission: a tracer for star formation.**
U. Klein.
Astron. Astrophys., Vol. 116, 175 - 178 (1982).
The 10.7 GHz radio continuum emission from a large sample of galaxies of different type and morphology has been compared with their optical luminosity and colour. It is shown that bluer galaxies tend to have an excess in radio emission (normalized by the blue luminosity). In particular, blue compact galaxies resemble large H II regions in spiral galaxies. It is suggested that the enhanced star formation rate in blue galaxies is responsible for the excess radio emission: young and massive stars provide the ionization of the interstellar gas and produce cosmic ray electrons via supernova explosions. This leads to the intense radio emission.

158.261 **Distances of galaxies from the apparent size distribution of dark clouds. II. NGC 224, NGC 2841 and NGC 7331.**
A. M. I. Osman, M. S. Ella, I. A. Issa.
Astron. Nachr., Band 303, 329 - 333 (1982).
The distances of three Sb galaxies NGC 224, NGC 2841 and NGC 7331 have been determined using the apparent size distribution of dark interstellar clouds in each of them. The distances of NGC 224, NGC 2841, NGC 7331 were found to be 727.7 ± 63 kpc, 8.07 ± 0.16 Mpc and 7.22 ± 0.15 Mpc, respectively.

158.262 **Double galaxy investigations. I. Observations.**
W. G. Tifft.
Astrophys. J., Suppl. Ser., Vol. 50, 319 - 389 (1982).
Redshift information from 240 Å mm^{-1} spectrograms is presented for 370 double galaxy systems from Karachentsev, including all pairs in that catalog with separation less than 80″. An extensive error discussion utilizing internal and external (21 cm) comparisons provides calibration of systematic error and determines the uncertainty for a typical high weight optical redshift to be ±65 km s^{-1}. Extensive information on emission and other properties is also provided.

158.263 **NGC 4151.**
IAU Circ., No. 3723 (1982).

158.264 **NGC 3347.**
IAU Circ., No. 3750 (1982).

158.265 **IUE observations of the BL Lac object AO 0235 + 164.**
M. A. J. Snijders, A. Boksenberg, M. V. Penston, W. L. W. Sargent.
Mon. Not. R. Astron. Soc., Vol. 201, 801 - 805 (1982).
AO 0235 + 164 was observed with IUE during its 1979 February outburst. The object was detected in both the short and long wavelength cameras. Hydrogen Lyman α absorption in the $z = 0.524$ system is detected with equivalent width W = 62 ± 28 Å. From this and previous 21-cm data the authors find the spin temperature of the absorbing material is 140 (+160/−70) K.

158.266 **The calibration of a radio-independent search for BL Lac objects.**
C. D. Impey, P. W. J. L. Brand.
Mon. Not. R. Astron. Soc., Vol. 201, 849 - 866 (1982).
The first results from a radio-independent search for BL Lac objects using a linear polarizing filter are presented. IIIa-J plates taken on the 1.2-m UK Schmidt telescope were measured with the COSMOS machine and ~33 000 objects in a 10 square degree area were searched down to a level of $B \sim 19.5$.

158.267 **Further observations of the elliptical galaxy NGC 5813.**
G. Efstathiou, R. S. Ellis, D. Carter.
Mon. Not. R. Astron. Soc., Vol. 201, 975 - 990 (1982).
The authors have obtained the rotation curve and velocity dispersion profile along the major axis of the elliptical galaxy NGC 5813. The measurements extend out to almost 90 arcsec (≈ 16 kpc, taking $H_0 = 50$ km s^{-1} Mpc^{-1}) on either side of the centre of the galaxy. Surface photometry has also been obtained which reveals significant departures from the standard King (1966) or de Vaucouleurs (1959) profiles interior to 12 arcsec. The rotation curve is quite unusual. The authors compare the observations with simple dynamical models using the equations of stellar hydrodynamics (Binney, 1980).

158.268 **Flocculent and grand design spiral structure in field, binary and group galaxies.**
D. M. Elmegreen, B. G. Elmegreen.
Mon. Not. R. Astron. Soc., Vol. 201, 1021 - 1034 (1982).
A 12-division morphological system has been developed to classify nearly all spiral galaxies according to the regularity of their spiral arm structure. This system emphasizes arm continuity, length and symmetry, and it may be related to the presence or strength of a density wave. Arm classifications were tabulated for 305 barred and non-barred spiral galaxies; 79 of these are isolated, 52 are in binary systems and 174 are in groups. The fractional occurrences of irregular versus ordered spiral structures were determined for the galaxies in each sample.

158.269 **Flocculent and grand design spiral arm structure in cluster galaxies.**
D. M. Elmegreen, B. G. Elmegreen, A. Dressler.
Mon. Not. R. Astron. Soc., Vol. 201, 1035 - 1039 (1982).
A total of 829 spiral galaxies in 22 clusters having redshifts between $z = 0.02$ and 0.06 were classified according to the appearance of their spiral arm structures. The fraction of galaxies that have a grand design spiral structure was found to be higher among barred galaxies than among non-barred galaxies. Cluster galaxies and distant non-cluster galaxies have the same fraction of grand design galaxies when resolution effects are considered. The grand design fraction among cluster galaxies is also similar to the fraction observed among nearby galaxies in binary systems and in groups.

158.270 **The structure of groups of galaxies in the Ursa Major/Canes Venatici region – I. A discussion of the results of an H I survey.**
P. N. Appleton, R. D. Davies.
Mon. Not. R. Astron. Soc., Vol. 201, 1073 - 1094 (1982).
A neutral hydrogen survey has been made of galaxies in an area of ~110 degree^2 in Ursa Major and Canes Venatici, and the data yield many new accurate systemic velocities, H I masses and total masses. The spatial distribution and velocity structure of the area is presented and a detailed discussion is given of the global properties of the individual galaxies. The results of the survey show that the Ursa Major/Canes Venatici region contains possibly four associations of galaxies identifiable both in position and velocity.

158.271 **A mass estimate for the companion to the 'Cartwheel' galaxy.**
R. L. Davies, D. C. Morton.
Mon. Not. R. Astron. Soc., Vol. 201, 69P - 72P (1982).
The authors have measured the velocity and the velocity dispersion of the elliptical companion to the ring galaxy known as the 'Cartwheel'. The velocity confirms the suggestion of Fosbury and Hawarden that this companion galaxy was the interloper which created the ring structure. The mass inferred from the dispersion is only 5–10 per cent of the mass of the Cartwheel itself derived from the kinematics of the ring. This low mass is insufficient to account for the observed ring by the collisionless kinematic model of Lynds and Toomre. It seems that the underlying mechanism is an encounter of the kind suggested by Lynds and Toomre but that the ring is brightened by the star formation induced.

158.272 **Intermediate band filter spectrophotometry of bright galaxies. II. Data reductions.**
J.-E. Solheim, G. de Vaucouleurs.
Astron. Astrophys., Suppl. Ser., Vol. 50, 283 - 290 (1982).
Observations of 56 galaxies ranging through all the main morphological types, have been corrected for galactic absorption, redshift, aperture and inclination effects. The aperture effect is determined from the data, while the range in inclination and types is too small to determine the inclin› tion effect. Values transformed from the second Reference Catalogue are therefore used for this correction. The corrected spectral energy distribution is shown for each galaxy and for the mean of galaxy types.

158.273 **An H I survey of southern galaxies.**
K. Reif, U. Mebold, W. M. Goss, H. van Woerden, B. Siegman.
Astron. Astrophys., Suppl. Ser., Vol. 50, 451 - 479 (1982).
Using the Parkes 64-m telescope 359 galaxies of Hubbl type S0 to Im at declinations south of −18° have been observed in the 21-cm line. 196 Galaxies were detected. In this paper the spectra of the detected galaxies and the inte grated properties are summarized. Thirteen S0 and S0/a galaxies have been detected.

158.274 **On the inner ring of H II regions in NGC 3351.**
D. Alloin, J.-L. Nieto.
Astron. Astrophys., Suppl. Ser., Vol. 50, 491 - 504 (1982).
The authors present and discuss the results of broad and narrow band electronographic observations of the barre spiral, hot-spot nucleus galaxy NGC 3351. The star forming activity along a ring about 650 pc in diameter has been studied more particularly. The authors derive a mass of newly born stars of about $3.2 \times 10^6 M_\odot$. From a compariso between the observed colors for the central area (20″ in diameter) and the theoretically computed colors of a galaxy which generated a burst of star formation, the authors dedu that the presently observed young stars correspond to a bur which started some 5×10^7 yrs ago and lasted for about the same period of time.

158.275 **Infrared magnitude/H I velocity-width/surface brightness relation and cosmological expansion.**
M. M. Jôeveer.
Pis'ma Astron. Zh., Tom 8, 643 - 648 (1982). In Russian. English translation in Soviet Astron. Lett., Vol. 8.
The data by Aaronson et al. (1980) are used to derive the infrared absolute magnitude/H I velocity-width/surface brightness relation which predicts absolute magnitudes of spiral galaxies with the accuracy of $\pm 0^m.3$. The differences in the Hubble ratios of clusters of galaxies with new distances obtained using this relation cannot be explained within a simple Local Group infall scheme. It is concluded that eithe clusters of galaxies have peculiar velocities ≳ 500 km/s, or something is wrong with infrared magnitude scales in differe directions, or it is necessary to use some third parameter (which has different mean values in different clusters) when predicting absolute magnitudes.

158.276 **Shock waves in spiral galaxies.**
A. A. Rumyantsev.
Pis'ma Astron. Zh., Tom 8, 653 - 656 (1982). In Russian. English translation in Soviet Astron. Lett., Vol. 8.
In the framework of the conception of shock wave form tion in spiral arms it is shown that the shock front inclinatio increases during the propagation in a galactic disk and reach

a maximal value in the Mach regime. It is due to this relatively stable regime that the spiral angle proves to be roughly constant and the arms take the observed shape. The gaseous ejections from supernovae bursts support the spiral structure.

158.277 **Das Rätsel der Jets. Beobachtungen an M 87 und 3C 273 führen zu neuen Fragen.**
K. Meisenheimer.
Sterne Weltraum, 21. Jahrg., 462 - 464 (1982).

158.278 **IUE observations of variable Seyfert 1 galaxies.**
C. Boisson, M. H. Ulrich.
Messenger, No. 30, p. 4 - 6 (1982).

158.279 **A question of galaxies.**
W. Tucker, K. Tucker.
Mercury, Vol. 11, 151 (1982).

158.280 **A preliminary colour-magnitude diagram for Leo II.**
S. Demers, W. E. Harris.
J. R. Astron. Soc. Canada, Vol. 76, 318 (1982). – Abstract.

158.281 **NGC 3516 – an unusual Seyfert galaxy?**
M. DeRobertis.
J. R. Astron. Soc. Canada, Vol. 76, 318 (1982).

158.282 **The radio jet in NGC 6251.**
R. A. Perley, A. H. Bridle, A. G. Willis.
J. R. Astron. Soc. Canada, Vol. 76, 323 - 324 (1982). Abstract.

158.283 **A new intermediate Seyfert galaxy: X-ray, optical, and radio properties.**
F. D. Ghigo, S. Wyckoff, J. F. C. Wardle, N. L. Cohen.
Astron. J., Vol. 87, 1438 - 1442 (1982).
The X-ray source X 0459 + 034, discovered with the Einstein observatory, is shown to be a Seyfert galaxy of intermediate type. On PSS prints, the object appears slightly diffuse. The composite Hβ line profile characterizes X 0459 + 034 as a Type 1.5 Seyfert with broad and narrow line components of redshift $z = 0.016 \pm 0.001$. VLA observations at 20 and 6 cm show that it is a weak radio source with a steep nonthermal spectrum and an angular extent of about 3″.

158.284 **Neutral hydrogen in binary and multiple galaxies.**
G. Helou, E. E. Salpeter, Y. Terzian.
Astron. J., Vol. 87, 1443 - 1464 (1982).
The authors report on H I observations of 28 galaxies selected from the Uppsala General Catalog of Galaxies for belonging to pairs or small groups. The H I distributions are mostly resolved by the Arecibo beam and for each object the H I size, total H I flux and mass, redshift and rotation velocity, and spin orientation are given. The authors derive total indicative masses in individual galaxies and individual mass-to-light ratios.

158.285 **On the origin of the globular cluster system of M 87.**
J. C. Forte, R. E. Martínez, J. C. Muzzio.
Astron. J., Vol. 87, 1465 - 1469 (1982).
The authors show that the high specific frequency of globular clusters associated with M 87 and the observed color difference between the globulars and the halo of that galaxy are consistent with the suggestion that a significant accretion of clusters from less massive galaxies has occurred.

158.286 **Neutral hydrogen in two extremely isolated galaxies.**
N. Krumm, W. W. Shane.
Astron. Astrophys., Vol. 116, 237 - 247 (1982).
Maps made with the Westerbork Synthesis Radio Telescope of 21-cm line emission from NGC 2712 and 5301 are presented. The H I masses of these galaxies are normal or slightly high for their type (Sb). The H I diameters are normal or slightly small. The rotation curve of NGC 2712 rises rapidly and then remains flat. That for NGC 5301 rises slowly and is either flat or still rising at the limit of detectable gas. The authors discuss three tentative conclusions from this work: (1) isolated galaxies do not have more extended H I distributions than non-isolated galaxies, (2) warps and asymmetries in the gas distributions of spirals can be internally generated and sustained and (3) there are still some differences between isolated and non-isolated spirals which deserve further investigation.

158.287 **Remarkable kinematics of the ionised gas in the nucleus of NGC 1365.**
M. M. Phillips, A. J. Turtle, M. G. Edmunds, B. E. J. Pagel.
Anglo-Australian Obs., Prepr. No. 174, 16 pp. (1982). – Submitted to Mon. Not. R. Astron. Soc.

158.288 **TAURUS observations of the ring galaxy in Vela.**
K. Taylor, P. D. Atherton.
Anglo-Australian Obs., Prepr. No. 175, 8 pp. (1982). – Submitted to Mon. Not. R. Astron. Soc.

158.289 **Infrared emission and star formation in NGC 5253.**
A. F. M. Moorwood, I. S. Glass.
ESO Sci. Prepr., No. 205, 15 pp. (1982). – Submitted to Astron. Astrophys.

158.290 **The composite UV emission spectrum of Seyfert 1 galaxies.**
M.-P. Véron-Cetty, P. Véron, M. Tarenghi.
ESO Sci. Prepr., No. 207, 22 pp. (1982).
A composite spectrum is built in the wavelength range $\lambda\lambda$ 1200 - 3000 Å, from IUE spectra of 27 Seyfert 1 nuclei. Comparisons with spectra of quasars show that all emission features seen in quasars, including the Fe II UV multiplets are also observed in the much less luminous Seyfert 1 nuclei. This is taken as a confirmation that Seyfert 1 nuclei are low luminosity radio quiet or flat radio spectrum quasars.

158.291 **The UV spectrum of the BL Lac object PKS 0521-36.**
I. J. Danziger, J. Bergeron, R. A. E. Fosbury, L. Maraschi, E. G. Tanzi, A. Treves.
ESO Sci. Prepr., No. 212, 14 pp. (1982).
Ultraviolet observations (1200 - 3000 Å) with the IUE satellite of the BL Lac object PKS 0521-36 are presented. The only emission line which appears clearly in the spectrum is Lyα, which is asymmetric with a component displaced ~3000 km/s to the red. The intensity of the line and the upper limits on other lines are compared with the model calculations on QSOs and Seyfert nuclei by Kwan and Krolik (1981). The continuous energy distribution is discussed, combining non-simultaneous observations from the ultraviolet to the infrared. The spectral range of the non-thermal source from far IR to far UV can be described by a single power law of index −1.5.

158.292 **Calibrated *B*, *V* surface photometry of X-ray cD galaxies.**
E. A. Valentijn.
ESO Sci. Prepr., No. 215, 53 pp. (1982). – Submitted to Astron. Astrophys.

158.293 **Chemical abundances in the interstellar medium of galaxies from spectrophotometry of supernova remnants.**
S. D'Odorico, M. Dopita.
ESO Sci. Prepr., No. 223, 8 pp. (1982). – To appear in the

proceedings of the IAU symposium 101 "supernova remnants and their X-ray emission", Venice, September 1982.

158.294 **Activity in the centers of galaxies.**
W. Kundt.
Mitt. Astron. Ges., Nr. 57, p. 65 - 78 (1982).

The report concerns itself with the largest, most powerful phenomena in the Universe ranging through more than 11 orders of magnitude in linear extent – from 10 a. u. to 10 Mpc – and through 20 orders of magnitude in observed frequency – from radio frequencies to γ-ray and even cosmic ray energies. It addresses the 20-year-old riddle of the quasars as well as the blazars, radio-galaxies, the Seyfert-phenomenon, and the central activity in more normal galaxies. Despite a spread in radiated power by more than a factor of 10^{10} from the strongest (10^{48} erg s^{-1}) to the weakest, all these phenomena show similar morphologies, spectra, velocities, variabilities and source sizes, suggesting that the same type of engine is at work. The similarity of phenomena even extends down in scale to our Galactic center and further to the binary radio and X-ray stars Sco X-1 and SS 433 whose engines are probably magnetized neutron stars.

158.295 **Photored magnitudes of interacting and isolated galaxies.**
V. A. Dostal', V. G. Metlov.
Astron. Tsirk., No. 1178, p. 3 - 6 (1981). In Russian.

158.296 **Dependence of the local mass-luminosity ratio on the radius in the galaxies M 31 and M 81.**
J. Einasto, P. Tenjes, P. Traat.
Astron. Tsirk., No. 1179, p. 1 - 2 (1981). In Russian.

158.297 **Relation between the mass-luminosity ratio and colours for old subsystems of galaxies.**
J. Einasto, P. Tenjes.
Astron. Tsirk., No. 1179, p. 2 - 4 (1981). In Russian.

158.298 **Past star formation in the disks of galaxies.**
B. Rocca-Volmerange.
Pré-publ. Inst. Astrophys. Paris, No. 19, 10 pp. (1982). To be published in "First stellar generations", Frascati-Workshop, Vulcano, 6 - 10 September 1982.

158.299 **A list of Elite-galaxies.**
P. Tenjes, P. Kalamees, J. Einasto, M. Jõeveer, P. Brosche, F.-Th. Lentes.
Tartu Astrofüüs. Obs. Teated, No. 68, 56 pp. (1982).

158.300 **The optical emission in the counter-jet region of M 87.**
V. N. Kuril'chik.
Astron. Tsirk., No. 1182, p. 1 - 2 (1981). In Russian.

158.301 **Spectrophotometry of the nucleus of the Akn 583 galaxy.**
G. T. Petrov, V. A. Mineva, V. K. Golev, Z. I. Tsvetanov.
Astron. Tsirk., No. 1202, p. 4 - 5 (1981). In Russian.

158.302 **A search for periodicities in the light variations of the Seyfert galaxy NGC 4151.**
A. G. Pacholczyk, N. D. Lubart, W. R. Penning, D. H. Ferguson, D. Turnshek.
Prepr. Steward Obs., No. 409, 16 pp. (1982).

158.303 **Carbon stars and the seven dwarfs.**
M. Aaronson, E. W. Olszewski, P. W. Hodge.
Prepr. Steward Obs., No. 412, 33 pp. (1982).

158.304 **Accurate radial velocities for carbon stars in Draco and Ursa Minor: the first hint of a dwarf spheroidal mass-to-light ratio.**
M. Aaronson.
Prepr. Steward Obs., No. 417, 18 pp. (1982).

158.305 **The strength of Paschen alpha in the Seyfert 1.9 galaxy V Zwicky 317.**
R. J. Rudy, S. P. Willner.
Prepr. Steward Obs., No. 428, 14 pp. (1982).

158.306 **Are there correlations between radio and optica axes of radio galaxies?**
M. J. Valtonen.
Rep. Ser., Dep. Phys. Sci., Univ. Turku, FTL-R 30, 8 pp. (1982) = Turku Univ. Obs. Informo, No. 66.

No statistically significant correlation between optica major axes of the galaxy image and radio source axes is fo when all existing measurements, now exceeding 140, are combined.

158.307 **Stellar populations in the edge-on spiral galaxy NGC 4565. I. Surface brightness and color distri tions.**
E. B. Jensen, T. X. Thuan.
Astrophys. J., Suppl. Ser., Vol. 50, 421 - 450, plates 6 - 1 (1982).

New photographic and photoelectric photometry of edge-on Sb galaxy NGC 4565 is presented. The galaxy's lig is deconvolved into five components: (1) a "starlike" unresolved nucleus; (2) a "thin disk" component whose light constitutes the rounded part of the minor-axis profiles in t region z < 20″; (3) a "thick disk" component whose light dominates the minor axis profiles in the region 20″< z< 6((4) a "bulge" component which merges smoothly into the "thick disk" component; (5) a "corona" component whos light dominates the minor-axis profiles from z ≈ 60″ to at least 170″. The photometric properties of these componen and their probable stellar populations are discussed in deta The corona component (5) does not seem to be the massiv halo required to explain the flat H I rotation curve of NGC 4565.

158.308 **A nuclear spectroscopic survey of field disk gala**
J. R. Stauffer.
Astrophys. J., Suppl. Ser., Vol. 50, 517 - 527 (1982).

Flux-calibrated, digital spectra for the nuclei of 139 f disk galaxies were obtained. Absorption and emission line equivalent widths, as well as nuclear colors and magnitudes have been determined from those spectra. The emission lin data provide the basis for the estimation of median Hα equ lent widths as a function of morphological type and nuclea color. The correlation of the [N II] to Hα flux ratio with o nuclear properties of the survey galaxies is also examined.

158.309 **X-ray observations to detect hot coronae around galaxies.**
J. N. Bregman, A. E. Glassgold.
Astrophys. J., Vol. 263, 564 - 570 (1982).

X-ray observations were made of two nearby edge-on spiral galaxies with the IPC on the Einstein Observatory in effort to detect hot (10^6 K), gaseous galactic coronae. No diffuse emission from galactic coronae was detected. Uppe limits to diffuse emission (0.3–2.9 keV) are 1×10^{39} ergs s (NGC 3628) and 2×10^{38} ergs s^{-1} (NGC 4244). The measu ments indicate that less than 10^{-3} of the energy supplied b supernovae in these galaxies appears as soft X rays. Serendi tous X-ray sources from the two IPC fields are listed.

158.310 **Radio structures of Seyfert galaxies. IV. Jets (?) NGC 1068 and NGC 4151.**
A. S. Wilson, J. S. Ulvestad.
Astrophys. J., Vol. 263, 576 - 594 (1982).

This paper presents and discusses new radio maps of t Seyfert galaxies NGC 1068 (at 1.4 and 15 GHz) and

NGC 4151 (at 15 GHz) from the VLA. The radio emission of NGC 1068 comprises a central component coincident with the optical nucleus and of size 0''.7 × 0''.3, a linear structure of extent 13'' and large scale emission of size at least 2'. Linear polarization of magnitude 13% is detected at 15 GHz in the NE lobe of the linear structure. Earlier radio maps of NGC 4151 show a highly elongated source of extent ~3''.5. In the center of this structure is found a double source of separation 0''.45. The linear structures in both galaxies are interpreted in terms of oppositely directed "jets" which emanate from the nuclei. The direction of the jet changes with distance from the nucleus in both NGC 1068 and NGC 4151, giving each source an overall **S** shaped appearance. The authors consider various models for the origin of these **S** shapes, and favor bending by the ram pressure of the normally rotating interstellar gas of the galaxy. A detailed quantitative model of this picture is developed. It is argued that jet bending by rotating gas disks may also occur in radio galaxies where the interstellar medium may be either intrinsic or accreted.

158.311 **Dynamo action in a supermassive rotator and the active galactic nuclei.**
G. Belvedere, D. Molteni.
Astrophys. J., Vol. 263, 611 - 614 (1982).

It is shown that an α-ω dynamo in a supermassive non-collapsed rotator of mass $M = 10^8 M_\odot$ and radius $R = 10^{15}$ cm can give rise by a coronal emission mechanism to X-ray luminosity up to $L_x \approx 10^{45}$ ergs s^{-1} with magnetic field $B \approx 10^4$ gauss. The model might thus account for the central powerhouse of active galactic nuclei.

158.312 **VLBI observations of the nucleus and jet of M 87.**
M. J. Reid, J. H. M. M. Schmitt, F. N. Owen, R. S. Booth, P. N. Wilkinson, D. B. Shaffer, K. J. Johnston, P. E. Hardee.
Astrophys. J., Vol. 263, 615 - 623 (1982).

The authors mapped the nucleus and jet of M 87 with an eight-station VLBI array at 18 cm wavelength with high dynamic range. The nucleus of M 87 consists of a "core-jet" structure with a peak brightness temperature greater than 10^{10}K. Emission extends for more than 50 milli-arcsec along a position angle of 288°. This position angle precisely matches that of the 20'' radio/optical/X-ray jet. No "counterjet" is seen. If relativistic beaming is invoked to enhance the jet and diminish the counterjet, then the jet must be aligned within about 60° to our line of sight, and its flow velocity must exceed about 60% of the speed of light.

158.313 **The infrared emission from the elliptical galaxy NGC 1052.**
E. E. Becklin, A. T. Tokunaga, C. G. Wynn-Williams.
Astrophys. J., Vol. 263, 624 - 628 (1982).

Multi-aperture infrared photometry of the elliptical galaxy NGC 1052 shows that its infrared excess is confined to a region smaller than 2'' (300 pc) in diameter coincident with the visible nucleus. It is suggested that the emission in the 5–20 μm range arises from dust heated by the nonthermal source seen at other wavelengths.

158.314 **Luminosity classification of galaxies in the Revised Shapley-Ames Catalog.**
S. van den Bergh.
Publ. Astron. Soc. Pacific, Vol. 94, 745 - 747 (1982).

Recently Sandage and Tammann (1981) have produced A Revised Shapley-Ames Catalog of Bright Galaxies in which they give luminosity classifications, based on inspection of plate material obtained with large reflectors, for all late-type galaxies in the SA catalog. It is the purpose of the present paper to see how close the system used by Sandage and Tammann is to the original DDO System.

158.315 **Intermediate Seyfert galaxies.**
R. D. Cohen.
Publ. Astron. Soc. Pacific, Vol. 94, 750 - 751 (1982). Abstract.

158.316 **The extent of the 3.3-micron emission in the Seyfert galaxy NGC 7469.**
R. Cutri, R. Rudy, A. Tokunaga, S. Willner.
Publ. Astron. Soc. Pacific, Vol. 94, 751 (1982). – Abstract.

158.317 **NGC 3516 – an unusual Seyfert galaxy?**
M. de Robertis.
Publ. Astron. Soc. Pacific, Vol. 94, 751 (1982). – Abstract.

158.318 **Are there any shock-heated galaxies?**
G. J. Ferland, H. Netzer.
Publ. Astron. Soc. Pacific, Vol. 94, 752 (1982). – Abstract.

158.319 **Models of the time dependence of the broad-line profiles of Seyfert 1 galaxies.**
C. B. Foltz, E. R. Capriotti, B. M. Peterson.
Publ. Astron. Soc. Pacific, Vol. 94, 752 (1982). – Abstract.

158.320 **The nature of faint blue galaxies.**
D. Hamilton.
Publ. Astron. Soc. Pacific, Vol. 94, 754 (1982). – Abstract.

158.321 **The oldest components of the nearest active galaxy (Cen A = NGC 5128).**
J. E. Hesser, H. C. Harris, S. van den Bergh, G. L. H. Harris.
Publ. Astron. Soc. Pacific, Vol. 94, 754 (1982). – Abstract.

158.322 **Ionized gas in the nuclei of normal spirals: are they all active?**
W. C. Keel.
Publ. Astron. Soc. Pacific, Vol. 94, 754 - 755 (1982). Abstract.

158.323 **Seyfert galaxies, planetary nebulae and H II regions.**
D. E. Osterbrock.
Publ. Astron. Soc. Pacific, Vol. 94, 756 (1982). – Abstract.

158.324 **Variability of the emission-line spectra of Seyfert 1 galaxies.**
B. M. Peterson, E. R. Capriotti, C. B. Foltz.
Publ. Astron. Soc. Pacific, Vol. 94, 756 (1982). – Abstract.

158.325 **2 - 165 keV observations of active galaxies from HEAO-1.**
R. E. Rothschild, R. F. Mushotzky.
Publ. Astron. Soc. Pacific, Vol. 94, 757 (1982). – Abstract.

158.326 **Narrow-line profiles from active galaxies.**
M. Whittle.
Publ. Astron. Soc. Pacific, Vol. 94, 760 (1982). – Abstract.

158.327 **Photoionization and kinematic modeling of the narrow-line region.**
M. Whittle.
Publ. Astron. Soc. Pacific, Vol. 94, 760 - 761 (1982). Abstract.

158.328 **Multifrequency observations supporting relativistic beaming in BL Lac objects.**
D. M. Worrall, F. C. Bruhweiler.
Publ. Astron. Soc. Pacific, Vol. 94, 761 (1982). – Abstract.

158.329 **A comparison of galaxy images on new and old POSS prints.**
M. F. Struble, C. Ftaclas.
Publ. Astron. Soc. Pacific, Vol. 94, 763 - 764 (1982).

Both the major axes and axial ratios of galaxy images

measured on new POSS prints differ significantly from measures on older POSS prints. Diameters on new prints are systematically smaller (by ~14%) than those on old prints and images appear significantly rounder on new prints. This suggests that galaxy images on new prints are of inferior quality. For a sample of 37 E and cD galaxy images, the authors give relations between major axes and axial ratio as measured on both print sets.

158.330 **Wolf-Rayet stars and an extraordinary star-forming region in the barred spiral galaxy NGC 5430.**
W. C. Keel.
Publ. Astron. Soc. Pacific, Vol. 94, 765 - 768 (1982) = Lick Obs. Bull., No. 930.

A strong emission knot 20″ southeast of the nucleus of NGC 5430 shows spectral features attributable to Wolf-Rayet stars. If they are like those seen nearby, the knot contains 3×10^4 Wolf-Rayet stars, as well as a still larger number of O stars. Abundances of heavy elements in the ionized gas around the stars may be lower than expected for such a galaxy. A scenario is proposed in which this object results from the collision of a dwarf irregular galaxy with the disk and bar of NGC 5430 and ensuing star formation.

158.331 **The relationships between morphological type and quantitative measures in the cluster environment.**
G. D. Bothun.
Publ. Astron. Soc. Pacific, Vol. 94, 774 - 788 (1982).

The author investigates the relationships between morphological types in both the traditional Hubble-Sandage system as well as the recently developed RDDO system and quantitative measurements of the integrated properties for a sample of cluster spirals. Correlations are presented between morphological type and integrated color, integrated H I content, position in the color-gas content relation, and 20% velocity width. Similar correlations are presented as a function of visually estimated disk-to-bulge ratio.

158.332 **On a morphological characteristic of late-type spirals.**
S. G. Iskudaryan.
Dokl. AN ArmSSR, Tom 74, No. 3, p. 122 - 130 (1982). In Russian. – Abstr. in Ref. zh., 51. Astron., 10.51.680 (1982).

158.333 **Spectrophotometric study of galaxies with high surface brightness. II. Arakelyan 583.**
G. T. Petrov, V. A. Mineva, V. K. Golev.
Dokl. Bolg. AN, Vol. 35, 137 - 140 (1982). – Abstr. in Ref. zh., 51. Astron., 11.51.715 (1982).

158.334 **Analysis of the orientation of disks of galaxies to their orbital planes in groups and clusters.**
G. G. Kuz'min, M. Jõeveer, B. I. Fesenko.
Tartu Astrofüüs. Obs. Publ., Köide 49, 140 - 154 (1982). In Russian.

Theoretical expressions between radial velocities, inclinations and position angles are considered in the case of different orientation of disks of galaxies to their orbital planes in systems of galaxies.

158.335 **New bright Seyfert galaxies.**
J. P. Huchra, W. F. Wyatt, M. Davis.
Astron. J., Vol. 87, 1628 - 1633 (1982).

As part of an ongoing program of measuring galaxy redshifts at Mount Hopkins Observatory, the authors have discovered several previously unknown bright Seyfert galaxies. Most of the objects are of Seyfert type 2, but several are type 1 objects where the broad permitted line region is easily visible only for the Hα line. The existence of such bright active galactic nuclei that have been overlooked in previous color or prism searches can have a large effect on the calculated number density of active galactic nuclei.

158.336 **H I observations of high-luminosity elliptical galaxies.**
G. R. Knapp, J. E. Gunn.
Astron. J., Vol. 87, 1634 - 1638 (1982).

Sensitive 21-cm line observations were made with lar velocity coverage towards eight luminous early-type galaxi in a search for possible H I emission with a large velocity width. Only one object, the tidally interacting pair of S0's NGC 4105/4106, was detected, with M (H I)/L_B = 0.05 $M_\odot/L_\odot$. Previously published detections of H I in thr ellipticals (NGC 2974, 3904, and 3962) were not confirm Low limits on the H I content of the luminous ellipticals NGC 4472 and NGC 5322 were set.

158.337 **A study of the two-dimensional luminosity distr tion of NGC 3379.**
S. J. Ratcliff, D. Burstein.
Astron. J., Vol. 87, 1648 - 1655 (1982).

The two-dimensional surface photometry (B) of Burs (1979) is used to examine, in detail, the luminosity distrib tion of the elliptical galaxy NGC 3379 for $7'' \leqslant \alpha \leqslant 110''$ ($18 \leqslant \mu_B \leqslant 24$). The isophotes of NGC 3379 are very well f by ellipses that are concentric within the measuring error (≤0.1 arcsec). Ellipticities, orientations, and semiaxes of tl elliptical isophotes are determined and show reasonable agreement with results of previous investigations. The ellip ty parameters and the standard de Vaucouleurs $R^{1/4}$ luminosity distribution of NGC 3379. These models are compa with the data.

158.338 **The color and surface brightness of the Leo II d galaxy.**
P. W. Hodge.
Astron. J., Vol. 87, 1668 - 1670 (1982).

Photoelectric surface photometry of the central area the Leo II dwarf galaxy shows it to have a central surface brightness of $V = 23.91 \pm 0.18$ mag per arcsec^2 and a colo $B-V = 0.65 \pm 0.15$. Its color is like that of a globular clust rather than a giant elliptical galaxy. Using previously publi structural data, an integrated luminosity can be calculated from the central surface brightness, giving $V = 11.62 \pm 0.2$ The absolute magnitude is thus $M_V = -10.2$.

158.339 **The radio continuum morphology of NGC 4631 at 2.7 and 8.1 GHz.**
N. Duric, P. C. Crane, E. R. Seaquist.
Astron. J., Vol. 87, 1671 - 1678 (1982).

Radio continuum maps of NGC 4631 at 2.7 and 8.1 (are presented. The authors find the disk emission to be str and nonthermal. The radio emission in the nuclear region consists of a nonthermal triple source whose components l an average spectral index of ~−0.6. The nature of this trip structure is investigated in detail. A model which accounts for this peculiar emission is also presented.

158.340 **A determination of relative spectrophotometric gradients of a series of galaxies.**
G. D. Polyakova.
Izv. Glav. Astron. Obs. Pulkovo, Astrofiz. Astrometr., No. p. 39 - 43 (1982). In Russian.

The energy distribution is studied and relative spectrophotometric gradients ΔΦ of the continuum for 16 gala are given. Using the obtained values of ΔΦ the colour indic ΔC were determined for the wavelength range 5000 Å to 4000 Å. A comparison of ΔC with colour indices $(B-V)$ (C and $(B-V)_T$ for two reference catalogues of bright galaxie De Vaucouleurs et al. is made. On the basis of the obtained dependences the $(B-V)$ values were determined for the gal list under study.

158.341 **Spiral structure and gas motion in M 81.**
V. G. Berman, Yu. N. Mishurov.

Astron. Zh., Tom 59, 1055 - 1061 (1982). In Russian.
English translation in Soviet Astron., Vol. 26, No. 6.

In the model of galactic spirals (1972) the spiral pattern was calculated, and the parameters of a spiral density wave were determined, the angular rotational pattern velocity and the density of the effective rotating subsystem. For these parameters, a nonlinear gas flow in the gravitational field of the arms was calculated.

158.342 **Evolutionary scheme of objects with active nuclei.**
B. V. Komberg.
Astron. Zh., Tom 59, 1062 - 1069 (1982). In Russian.
English translation in Soviet Astron., Vol. 26, No. 6.

The question of typical features and necessary conditions for nuclei to be active in galaxies is considered. Arguments are given that favor the idea that quasistellar objects are the nuclei of distant galaxies being at an early most violent stage of their evolution. Attention is drawn to spectroscopic peculiarities of quasar-like objects of various types; a conclusion is made that evolutionary transitions are possible from objects with Sy 0-I-type spectra to those with Sy II-III-type spectra. A possible scheme of quasar-like objects evolution is suggested.

158.343 **Observations of galaxies of high surface brightness at 102 MHz frequency.**
V. S. Artyukh, V. G. Malumyan, M. A. Ogannisyan.
Astron. Zh., Tom 59, 1075 - 1080 (1982). In Russian.
English translation in Soviet Astron., Vol. 26, No. 6.

44 galaxies of high surface brightness from Arakelyan's lists have been observed at 102 MHz by the method of interplanetary scintillation. 10 galaxies have been detected at a level 1 - 2 J. 7 of them have scintillating components with angular sizes equal to or less than 1 second of arc. The spectra of scintillating components are given. Almost all scintillating galaxies are either elliptical and lenticular or show compactness of high degree in the photographic region of the spectra.

158.344 **On the difference between Seyfert 1- and 2-types of galaxies.**
A. S. Zentsova.
Astron. Zh., Tom 59, 1089 - 1092 (1982). In Russian.
English translation in Soviet Astron., Vol. 26, No. 6.

The hypothesis is founded that the activity of the Seyfert galactic nuclei of type 1 is connected with accretion of matter onto a massive black hole in the regime of a two-temperature accretion disk whereas the activity of nuclei of type 2 is due to accretion in the regime of an optically thin disk. The parameters of this disk are given.

158.345 **Galaxien um BL-Lacertae-Objekte.**
G. Gaida.
Diss. Naturwiss.-Math. Gesamtfak. Univ. Heidelberg, 6 + 60 pp. (1982).

Die vorliegende Arbeit berichtet über Untersuchungen zu der Hypothese, daß BL-Lacertae-Objekte in Zentren von Galaxien liegen.

158.346 **Clumpy irregular galaxies: an astronomical adventure.**
J. Heidmann.
Mercury, Vol. 11, 170 - 177 (1982).

158.347 **A list of ultraviolet excess galaxies.**
K. J. Mitchell, L. E. Brotzman, A. Warnock III, P. D. Usher.
Astrophys. Space Sci., Vol. 88, 219 - 242 (1982).

The authors present a list of 412 faint galaxies selected for their apparent ultraviolet excess (UVX) from an area of 30 square degrees centered on Kepteyn Selected Area 28. The objects have been selected from a three-colour (*U, B, V*) plate taken with the Palomar 48-inch Schmidt telescope. The same plate was used to select a companion list of predominantly starlike UVX objects (Usher and Mitchell, 1982).

158.348 **A discovery of a peculiar galaxy MCG 5-29-86.**
K. Wakamatsu.
Astron. Her., Vol. 75, 142 - 147 (1982). In Japanese. – Abstr. in Phys. Abstr., Vol. 85, Abstr. 117754 (1982).

158.349 **Galactic coronas.**
L. Marschall.
Astronomy, Vol. 10, No. 11, p. 74 - 79 (1982). – Abstr. in Phys. Abstr., Vol. 86, Abstr. 15996 (1983).

158.350 **Some properties of the centers of cD galaxies.**
R. G. Kron, C. E. Albert.
Publ. Astron. Soc. Pacific, Vol. 94, 887 - 890 (1982).

Large-scale images of the cores of supergiant galaxies have been compared with those of associated giant galaxies, in terms of apparent central surface brightness and the frequency of close companions. These observations are of interest for constraining models for the origin and evolution of cD galaxies.

158.351 **Reddening of the narrow-line regions of active galaxies and the intrinsic Balmer decrement.**
C. M. Gaskell.
Publ. Astron. Soc. Pacific, Vol. 94, 891 - 893 (1982).

The reliability of the narrow-line Balmer decrement as an indicator of reddening is examined. Estimates of the reddening and electron density of the narrow-line regions of a number of active galaxies are made by using the trans-auroral/nebular line ratios of [S II] and [O II]. It is concluded that the intrinsic narrow-line Balmer decrement is close to the Menzel-Baker case-B values and that observations of the narrow-line Balmer decrement do give a reliable indication of the reddening.

Detailed bibliography on the surface photometry of galaxies.
See Abstr. 002.003.

Metacatalogue of galaxies and large scale structures in the Universe.
See Abstr. 002.027.

Data retrieval in the Metacatalogue of galaxies.
See Abstr. 002.028.

Centralization of extragalactic data.
See Abstr. 002.030.

The compiled catalogue of galaxies in machine-readable form and its statistical investigation.
See Abstr. 002.031.

The radio properties of a complete sample of bright galaxies.
See Abstr. 002.055.

A catalog of infrared magnitudes and H I velocity widths for nearby galaxies.
See Abstr. 002.080.

A catalogue of peculiar and interacting galaxies.
See Abstr. 002.091.

Documentation for the machine-readable version of the Uppsala General Catalogue of Galaxies.
See Abstr. 002.138.

Documentation for the machine-readable version of the Morphological Catalogue of Galaxies (MCG) of

Vorontsov-Velyaminov et al. 1962–1968.
See Abstr. 002.139.

Landolt-Börnstein. Group VI. Vol. 2c. Interstellar matter – Galaxy – Universe.
See Abstr. 003.055.

Nucleosynthesis in galaxies.
See Abstr. 004.065.

The outer regions of galaxies.
See Abstr. 011.001.

Recent (1981-82) evolution of galactic haloes.
See Abstr. 011.015.

Seeing and surface photometry.
See Abstr. 031.660.

TAURUS: a wide-field imaging Fabry-Perot spectrometer for astronomy.
See Abstr. 034.058.

Nucleosynthesis and chemical evolution of galaxies.
See Abstr. 061.046.

Formation of a collimated beam in radio galaxies.
See Abstr. 062.039.

A magnetohydrodynamical theorem for steady axisymmetric systems with implications to ring galaxies.
See Abstr. 062.072.

The emergent spectra of sources in scattering clouds.
See Abstr. 063.061.

Bursts of gravitational radiation from active galactic nuclei and globular clusters.
See Abstr. 066.004.

Thick supercritical accretion disks and active galactic nuclei.
See Abstr. 066.023.

Gravitational waves and activity of galactic nuclei and quasars.
See Abstr. 066.193.

Active galactic nuclei as clusters of accreting black holes.
See Abstr. 066.199.

Absolute proper motions of blue objects near the galactic north pole.
See Abstr. 111.015.

***UBVRI* photoelectric sequences for M 83 and NGC 5128.**
See Abstr. 113.061.

Spectrophotometry of Wolf-Rayet star candidates in M 33.
See Abstr. 114.083.

Data on an unusual Wolf-Rayet star in the nearby galaxy IC 1613.
See Abstr. 114.110.

Disk instabilities in Hubble-Sandage variables?
See Abstr. 117.130.

A hypothesis on the richness of a galaxy in binar stellar systems.
See Abstr. 117.132.

Double stars: the closest and the most distant systems.
See Abstr. 118.029.

Novae as sources of nitrogen in galaxies.
See Abstr. 124.006.

Supernova statistics and related problems.
See Abstr. 125.053.

Supernova remnants in M 31.
See Abstr. 125.057.

Optical supernova remnants in external galaxies.
See Abstr. 125.058.

Photometric observations of the bright type II supernova 1980k in NGC 6946.
See Abstr. 125.321.

On the possibility of sweeping of interstellar gas from a galaxy by radiative pressure of an active nucleus.
See Abstr. 131.008.

The dependence of CO emission on luminosity a the rate of star formation in Sc galaxies.
See Abstr. 131.115.

Implications of collisionally-supported giant mol ular clouds for spiral galactic structure and massive star for tion.
See Abstr. 131.139.

Molecular clouds in galaxies.
See Abstr. 131.163.

Trous géants dans la distribution spatiale de l'hydrogène atomique associé aux galaxies.
See Abstr. 131.311.

Optical surveys of H II regions in galaxies.
See Abstr. 132.028.

Dynamique des complexes H II extragalactiques: une étrange région H II géante dans NGC 4631.
See Abstr. 132.041.

Planetary nebulae in Local Group galaxies.
See Abstr. 135.048.

Do quasars rotate?
See Abstr. 141.003.

Observational tests of QSO amplification by condensed objects in galactic halos.
See Abstr. 141.012.

Radio observations of early-type galaxies.
See Abstr. 141.016.

Twinkle, twinkle quasi star . . . .
See Abstr. 141.024.

Further investigations on possible correlations between QSOs and the Lick catalogue of galaxies.
See Abstr. 141.025.

High-resolution studies of M 87 at long radio wavelengths.
See Abstr. 141.028.

Formation des raies permises et interdites dans les quasars.
See Abstr. 141.036.

A comparison of the emission line properties between quasars and type 1 Seyfert galaxies.
See Abstr. 141.037.

A rapid outburst of BL Lac at 2.72 GHz.
See Abstr. 141.044.

The redshift distribution of X-ray selected quasars: is it really anomalous?
See Abstr. 141.055.

The kinematics of the broad-line emission gas in quasars and Seyfert nuclei.
See Abstr. 141.057.

Color imaging of QSO-galaxy interactions.
See Abstr. 141.059.

Correlated outbursts in extragalactic variable sources at radio and optical wavelengths.
See Abstr. 141.068.

X-ray quasars.
See Abstr. 141.098.

Quasars near companion galaxies and the problem of the absolute magnitudes of quasars.
See Abstr. 141.099.

List of papers presented at the workshop on quasars and galactic nuclei.
See Abstr. 141.100.

Relativistic jet production and propagation in active galaxies.
See Abstr. 141.101.

The structure of 4C radio galaxies in poor clusters.
See Abstr. 141.102.

Dual-frequency VLA observations of the extended radio galaxy 3C 166.
See Abstr. 141.103.

Continuum observations of extragalactic sources.
See Abstr. 141.127.

Spectra of compact radio sources in galactic nuclei.
See Abstr. 141.129.

Further morphological studies of QSOs.
See Abstr. 141.131.

Variability of compact radio sources at a wavelength of 1 millimeter.
See Abstr. 141.134.

1 millimeter continuum observations of quasars.
See Abstr. 141.135.

An optical and radio investigation of the radio galaxy 3C 305.
See Abstr. 141.139.

A redshift difference between high and low ionization emission-line regions in QSOs – evidence for radial motions.
See Abstr. 141.151.

Quasars near companion galaxies – a comment on Arp's statistics.
See Abstr. 141.152.

Density of quasars around companion galaxies.
See Abstr. 141.153.

The distribution of companion galaxies to mirror-symmetric extragalactic radio sources.
See Abstr. 141.158.

Interpretation of the regular optical variability of quasars and nuclei of galaxies in a model of disk accretion.
See Abstr. 141.174.

Reddening measurements of Seyfert 1 galaxies and quasars.
See Abstr. 141.182.

Color imaging of QSO-galaxy interactions.
See Abstr. 141.183.

0.36 - 3.5 micrometer spectral-flux distributions of galaxies with extended radio structure.
See Abstr. 141.186.

Redshifts and spectroscopy of very distant radio galaxies with strong emission lines.
See Abstr. 141.190.

Emissione radio delle galassie.
See Abstr. 141.194.

0351+026: a QSO spawned by interacting galaxies?
See Abstr. 141.196.

On the origin of relativistic particles and gamma-rays in quasars.
See Abstr. 141.200.

Rapid X-ray variability in the nuclei of active galaxies.
See Abstr. 141.208.

X-ray emission from extragalactic bursts of star formation.
See Abstr. 142.009.

Optical identification of serendipitous Einstein X-ray sources: luminosity and redshift relations in the X-ray and optical emission from active galaxies and quasars.
See Abstr. 142.049.

Two X-ray selected BL Lacertae candidates.
See Abstr. 142.054.

Extragalactic X-ray sources.
See Abstr. 142.076.

High-resolution X-ray spectroscopy of supernova remnants, clusters of galaxies and BL Lac objects.
See Abstr. 142.085.

X-rays from quasars and active galaxies.
See Abstr. 142.148.

Gamma-ray observations of Seyfert galaxies and the problem of the extragalactic gamma-ray background.
See Abstr. 142.513.

Transport and propagation of cosmic rays in galaxies. II. The effect of a galactic wind on the mean lifetime and age distribution of non-decaying cosmic rays.
See Abstr. 143.055.

Dissipative evolution of collisionless stellar systems. II. Influence of binaries on the evolution of globular clusters and galactic nuclei.
See Abstr. 151.003.

Simulations of merging disk galaxies.
See Abstr. 151.004.

On the disk-thickness of spiral galaxies.
See Abstr. 151.008.

Luminosity evolution of quasars and active galaxies: theoretical models of the evolving mass supply rate.
See Abstr. 151.010.

Modèles de trous noirs massifs dans les noyaux actifs de galaxies.
See Abstr. 151.011.

Magnetic braking in galactic flows.
See Abstr. 151.019.

The structure and evolution of galacto-detonation waves: some analytic results in sequential star formation models of spiral galaxies.
See Abstr. 151.020.

MHD of galaxies – arm structure and magnetic field.
See Abstr. 151.032.

On the direction of rotation of the spirals in galaxies.
See Abstr. 151.045.

Observations of galaxy structure and dynamics.
See Abstr. 151.049.

On the dynamical basis of the classification of normal galaxies.
See Abstr. 151.055.

The rotation of galaxies: clues to their formation.
See Abstr. 151.062.

Lifetime of spurs in galaxies.
See Abstr. 151.063.

The evolution of galaxies: expectations and observations.
See Abstr. 151.066.

Stable polar gas disks in triaxial S0 galaxies.
See Abstr. 151.081.

Multiaperture *JHK* photometry of the globular clusters in the Fornax dwarf spheroidal galaxy.
See Abstr. 154.010.

Search for (globular) clusters in M 31. III. Structural properties: X-ray sources and comparison with Galactic globulars.
See Abstr. 154.013.

The dynamics and metallicity of the M 31 globul cluster system.
See Abstr. 154.014.

Globular clusters in the Sculptor Group galaxy NGC 55.
See Abstr. 154.015.

Optical and ultraviolet observations of the X-ray globular cluster Bo 158 in M 31.
See Abstr. 154.016.

Optical and ultraviolet observations of the X-ray globular cluster Bo 158 in M 31.
See Abstr. 154.021.

Comparison of galactic center with other galaxie
See Abstr. 155.074.

Radio and X-ray galaxies in Abell 566.
See Abstr. 160.013.

Digital surface photometry of galaxies toward a quantitative classification. I. 20 galaxies in the Virgo cluste
See Abstr. 160.026.

The evolution of disk galaxies and the S0 probl revisited.
See Abstr. 160.027.

The surface distribution of galaxy morphological types in Coma.
See Abstr. 160.028.

Neutral hydrogen contents of Virgo galaxies.
See Abstr. 160.034.

The Lynx-Ursa Major supercluster.
See Abstr. 160.048.

An absorption feature and filamentary structure in the central galaxy of the Centaurus cluster, NGC 4696.
See Abstr. 160.062.

Spectroscopy of galaxies in distant clusters. I. Fi results for 3C 295 and 0024 + 1654.
See Abstr. 160.068.

Where are the Population III stars?
See Abstr. 162.003.

Multipole moments in the Rubin-Ford data.
See Abstr. 162.005.

Do we really know how to measure the distances extragalactic objects?
See Abstr. 162.095.

Galactic neutrino models.
See Abstr. 162.135.

Constraints on the primordial helium abundance from the optical emission spectra of dwarf galaxies.
See Abstr. 162.195.

Erratum

158.901 **Erratum: "The dark nebulae of M 31"** [Astron. J., Vol. 85, 376 - 386 (1980)].
P. W. Hodge.
Astron. J., Vol. 87, 1244 (1982). – See Abstr. 27.158.15

159 Magellanic Clouds

159.001 **Some implications for the early chemical and dynamical evolution of the Large Magellanic Cloud from observations of the oldest globular clusters.**
A. P. Cowley, F. D. A. Hartwick.
Astrophys. J., Vol. 259, 89 - 95, plates 1, 2 (1982).
Observations of giant stars in the oldest Large Magellanic Cloud globular clusters were used to deduce metal abundances and radial velocities for the clusters. From the distribution of metallicities, radial velocities, position in the sky, position in the two-color diagram, color-magnitude diagram morphology, and distribution of ellipticities, it is argued that these clusters make up a different dynamical component from the young disk.

159.002 **Star formation in the Magellanic Clouds – II. Discovery of a protostar in the Small Magellanic Cloud.**
I. Gatley, A. R. Hyland, T. J. Jones.
Mon. Not. R. Astron. Soc., Vol. 200, 521 - 526 (1982).
An infrared protostar has been discovered in the Small Magellanic Cloud in the H II region N76B. The properties of this object are similar to those of the galactic protostars typified by the young clusters in NGC 7538 and by the recently discovered protostar in the Large Magellanic Cloud. This result provides direct evidence for a current mode of star formation in the SMC, and shows that sites of star formation in the Magellanic Clouds have a morphology similar to those in the Galaxy. In particular the presence of large quantities of dust in the star forming regions is implied.

159.003 **H I in the Small Magellanic Cloud re-examined.**
R. X. McGee, L. M. Newton.
Proc. Astron. Soc. Australia, Vol. 4, 189 - 195 (1981).
The authors find that Hindman's (1967) total content H I map is a satisfactory representation of the gas in the line of sight. They find that the H I gas in the SMC exists in four distinct large masses separated from one another in radial velocity by 20 to 30 km s^{-1}. They show that there is good correlation between the radial velocities of H II regions, supergiant stars and H I.

159.004 **Photometry and spectroscopy of long period variables in the Magellanic Clouds.**
P. R. Wood, M. S. Bessell, M. W. Fox.
Proc. Astron. Soc. Australia, Vol. 4, 203 - 205 (1981).

159.005 **H I in the Small Magellanic Cloud re-examined – II.**
R. X. McGee, L. M. Newton.
Proc. Astron. Soc. Australia, Vol. 4, 308 - 318 (1982).
The reasons for re-surveying the neutral hydrogen in the Small Magellanic Cloud and the results from that survey are briefly discussed in a previous paper (McGee, Newton 1981). The authors now present the experimental details, the 489 observed lines profiles and details of the Gaussian analysis of each one. The profiles are from the 'main body' of the SMC.

159.006 **The peculiar, luminous early-type emission line stars of the Magellanic Clouds: a preliminary taxonomy.**
S. N. Shore, N. Sanduleak.
Advances in ultraviolet astronomy, (see 012.006), p. 602 - 605 (1982).
A sample of some 20 early-type emission supergiants in the Magellanic Clouds has been observed with both the SWP and LWR low resolution mode of IUE. The authors discuss some preliminary results from the more complete program now near completion on the UV spectra of this sample. They discuss in some detail the two most extreme emission line stars found in the clouds – S 134/LMC and S 18/SMC. Also discussed are some results for the 2200 Å feature in these supergiants, and evidence for shells around the most luminous stars in the clouds.

159.007 **The metal content of Magellanic Cloud cepheids.**
J. W. Pel.
Messenger, No. 29, p. 1 - 4 (1982).

159.008 **The occurrence of carbon stars and Cepheid variables in common fields of the Magellanic Clouds.**
S. A. Becker.
Astrophys. J., Vol. 260, 695 - 706 (1982).
It has been suggested that the apparent lack of carbon stars brighter than $M_{bol} = -6.5$ in the Magellanic Clouds may be due to an "age effect" in the fields that have been studied. This "age effect" implies that no recent star formation has occurred and that consequently the individual mass of any of the remaining stars is too small to produce very bright carbon stars. To test this hypothesis, these same fields are searched for Cepheid variables whose presence would on the contrary indicate that recent star formation has taken place. It is found that Cepheids occur in nearly all of the studied fields, which invalidates the "age effect" hypothesis.

159.009 **Neutral hydrogen in the Small Magellanic Cloud.**
N. Loiseau, E. Bajaja.
Rev. Mexicana Astron. Astrofis., Vol. 6, (see 012.017), 55 - 58 (1981).
The authors made a 21-cm survey with high velocity resolution of the Small Magellanic Cloud. They present a preliminary analysis of these observations.

159.010 **The Bok and Tifft UBV sequence in the Large Magellanic Cloud: revised and extended.**
G. Alcaino, W. Liller.
Astron. Astrophys., Vol. 114, 213 - 215 (1982).
The authors present a UBVRI photoelectric sequence of 15 stars in the magnitude range $9.4 < V < 15.3$, situated in the north-western section of the bar of the Large Magellanic Cloud. Most of these stars have been previously observed in UBV by Bok and Bok (1969) and in BV by Tifft and Snell (1981). This is the first published RI sequence in the LMC.

159.011 **Supernova remnants in the Large Magellanic Cloud from Einstein.**
D. H. Clark.
Observatory, Vol. 102, 111 (1982). – Abstract.

159.012 **The Magellanic Stream and the galactic halo.**
R. D. Davies, R. J. Cohen.
Observatory, Vol. 102, 173 - 174 (1982). – Abstract.

159.013 **Abundances in the Magellanic Stream.**
M. V. Penston.
Observatory, Vol. 102, 174 (1982). – Abstract.

159.014 **Abundances in the Magellanic Stream.**
M. V. Penston, M. Pettini, A. Boksenberg, T. R. Gull, M. A. J. Snijders.
Third European IUE Conference, (see 012.029), p. 441 (1982). Abstract.

159.015 **A negative search for young stars associated with the Magellanic Stream.**
K. I. Tanaka, K. Hamajima.
Publ. Astron. Soc. Japan, Vol. 34, 417 - 422 (1982).
Faint blue stars have been searched for in the Magellanic

Stream by means of the RGU photometry whose limiting magnitude is about 17. The survey area is 0.5 square degrees centered at $l = 92°$, $b = -49°$. No candidate of main-sequence stars associated with the Magellanic Stream has been found.

159.016 **Observations of $^{12}CO(J = 2-1)$ emission in the Large and Small Magellanic Clouds.**
F. P. Israel, T. de Graauw, S. Lidholm, H. van de Stadt, C. de Vries.
Astrophys. J., Vol. 262, 100 - 109 (1982).
The authors have detected $^{12}CO(J = 2-1)$ emission at three positions in the Large Magellanic Cloud, near the bright H II regions 30 Doradus, N 159, and N 160A. Four positions in the LMC and four in the SMC failed to show an unambiguous signal. It is concluded that the 30 Doradus sector of the LMC at most is marginally deficient in CO emission, as compared to the Galaxy, whereas the SMC appears to be significantly deficient in CO with respect to both the LMC and the Galaxy.

159.017 **The distance of the Magellanic Clouds.**
L. Divan.
Messenger, No. 30, p. 28 - 30 (1982).

159.018 **The flattening of clusters in the Large Magellanic Cloud.**
S. van den Bergh.
Observatory, Vol. 102, 228 (1982).

159.019 **The wavelength dependence of interstellar polarization in the LMC.**
G. C. Clayton, P. G. Martin, I. Thompson.
J. R. Astron. Soc. Canada, Vol. 76, 317 (1982). – Abstract.

159.020 **Kinematics of carbon stars in the Large Magellanic Cloud.**
G. Joslin, H. Richer.
J. R. Astron. Soc. Canada, Vol. 76, 321 (1982). – Abstract.

159.021 **NGC 2257 and the LMC halo.**
L. L. Stryker.
Publ. Astron. Soc. Pacific, Vol. 94, 760 (1982). – Abstract.

Catalogue of the Small Magellanic Cloud star members.
See Abstr. 002.089.

A comparison of observed and theoretical luminosity functions of carbon stars and late M giants.
See Abstr. 065.084.

Las envolturas circunestelares de las supergigantes del tipo F y G en la Nube Mayor de Magallanes.
See Abstr. 112.024.

Photoelectric *UBV* photometry in the Large Magellanic Cloud.
See Abstr. 113.017.

Two photoelectric *UBVRI* sequences in the bar of the Small Magellanic Cloud.
See Abstr. 113.019.

The near infrared properties of selected WN stars in the Large Magellanic Cloud.
See Abstr. 113.035.

Effective temperatures and radii of Large Magellanic Cloud supergiants.
See Abstr. 114.009.

A search for UV line variability in R136a.
See Abstr. 114.040.

The faintest Wolf-Rayet stars in the LMC.
See Abstr. 114.058.

SK-67/111 and R 50, two Magellanic Cloud emis sion line stars.
See Abstr. 114.101.

Spectral types in the direction of the Magellanic Stream.
See Abstr. 114.108.

Near infrared observations of Magellanic WN star
See Abstr. 114.138.

The optical spectrum of R136a, the central obje of the 30 Doradus Nebula.
See Abstr. 114.156.

H-R diagram for early type Magellanic Cloud members.
See Abstr. 115.012.

Absolute magnitudes of Wolf-Rayet stars: the WI and WN4 sub-classes in the Large Magellanic Cloud.
See Abstr. 115.014.

The two-colour diagram of luminous stars in the Magellanic Clouds.
See Abstr. 115.016.

The luminous core of 30 Doradus: supermassive star or dense star cluster?
See Abstr. 115.018.

Variability of HBV 475 in the near infrared.
See Abstr. 117.114.

Wolf-Rayet stars in the Magellanic Clouds: spectr scopic binaries and masses.
See Abstr. 120.033.

Further *VBLUW* photometry of the S Doradus ty variables S Dor and HDE 269006 in the LMC and a discussi on their temperatures.
See Abstr. 122.015.

Novae as sources of nitrogen in galaxies.
See Abstr. 124.006.

Absolute photometry of supernova remnants and emission nebulae in the Galaxy and the Magellanic Clouds.
See Abstr. 125.003.

Supernova remnant candidates in the Small Magellanic Cloud.
See Abstr. 125.013.

Optical supernova remnants in external galaxies.
See Abstr. 125.058.

An X-ray survey of supernova remnants in the Large Magellanic Cloud.
See Abstr. 125.062.

Optical identification of Balmer-dominated super nova remnants in the Large Magellanic Cloud.
See Abstr. 125.073.

A survey of high-velocity H I in the Cetus region.
See Abstr. 131.021.

Abundances in the Magellanic Stream.
See Abstr. 131.045.

Recent results on extinction in the Small Magellanic Cloud.
See Abstr. 131.051.

Sk 143: an SMC star with a galactic-type ultraviolet interstellar extinction.
See Abstr. 131.055.

Interstellar extinction in the Small Magellanic Cloud.
See Abstr. 131.148.

Far UV extinction and the interstellar 4430 Å band in the Large Magellanic Cloud.
See Abstr. 131.167.

Ultraviolet interstellar extinction in the SMC from IUE stellar observations.
See Abstr. 131.177.

Interstellar gas in the direction of the Large Magellanic Cloud.
See Abstr. 131.184.

Fine structure in high velocity clouds near the South Celestial Pole.
See Abstr. 131.264.

Excitation and extinction in the LMC H II region N159A and discovery of a highly excited "blob" in its vicinity.
See Abstr. 132.003.

CNO abundances in H II regions of the Magellanic Clouds and the Galaxy with implications regarding the nucleosynthesis of the CNO element group.
See Abstr. 132.007.

A rediscussion of sulfur abundances in Magellanic Clouds and galactic H II regions.
See Abstr. 132.044.

Observations de bulles géantes du Grand Nuage de Magellan.
See Abstr. 132.046.

Etude optique d'une coquille de gaz ionisé super-géante dans le Grand Nuage de Magellan.
See Abstr. 132.047.

Etude cinématique du halo de la Nébuleuse de la Dorade.
See Abstr. 132.048.

Optical emission from ring-shaped nebulae in the LMC.
See Abstr. 134.015.

Kinematics of ring-shaped nebulae in the LMC. II. The radial velocity field of N 185.
See Abstr. 134.029.

IUE observations of planetary nebulae and their central stars in the Magellanic Clouds.
See Abstr. 135.011.

Luminosities and masses for three central stars of planetary nebulae in the Magellanic Clouds from ultraviolet spectroscopy with the IUE.
See Abstr. 135.035.

A new planetary nebula near the Large Magellanic Cloud.
See Abstr. 135.044.

UV photometric and line-width variability of LMC X-4.
See Abstr. 142.096.

Compact X-ray sources in the Large Magellanic Cloud.
See Abstr. 142.157.

Assoziationen und Emissionsschalen in der Großen Magellanschen Wolke.
See Abstr. 152.008.

Color-magnitude diagrams and ages of two young Magellanic Cloud clusters.
See Abstr. 153.041.

Analysis of integrated spectra of red star clusters in the Magellanic Clouds.
See Abstr. 154.017.

Optical and ultraviolet spectrophotometry of globular clusters in the Magellanic Clouds.
See Abstr. 154.019.

UV observations of young blue globular clusters in the LMC.
See Abstr. 154.020.

Une étude photographique de NGC 376, 416 et 419 dans le Petit Nuage de Magellan.
See Abstr. 154.031.

IUE – UV – observations of young globular clusters of the Large Magellanic Cloud.
See Abstr. 154.034.

On the structure of two globular clusters: NGC 1806 in the LMC and the galactic cluster NGC 3201.
See Abstr. 154.037.

The extended giant branches of intermediate age globular clusters in the Magellanic Clouds. III.
See Abstr. 154.038.

The extended giant branches of intermediate age globular clusters in the Magellanic Clouds. III.
See Abstr. 154.041.

Erratum

159.901 **Erratum: 'Distribution of hot stars and hydrogen in the Large Magellanic Cloud'** [Astrophys. J., Vol. 248, 906 - 924 (1981)].
T. Page, G. R. Carruthers.
Astrophys. J., Vol. 260, 413 (1982). – See Abstr. 30.159.003.

159.902 **Erratum: 'Ultraviolet absorption by interstellar gas near the LMC star HD 36402 in the interstellar bubble N51D'** [Astrophys. J., Vol. 255, 447 - 457 (1982)].
K. S. de Boer, A. G. Nash.
Astrophys. J., Vol. 261, 747 (1982). – See Abstr. 31.159.009.

160 Groups of Galaxies, Clusters of Galaxies, Superclusters

160.001 **The dynamics of rich clusters of galaxies. I. The Coma cluster.**
S. M. Kent, J. E. Gunn.
Astron. J., Vol. 87, 945 - 971 (1982).
The structure and dynamics of the Coma cluster are analyzed using self-consistent equilibrium dynamical models. Observational material for Coma is culled from a variety of sources. Projected surface, density, and velocity-dispersion profiles are derived extending out to a radius of 3° from the cluster center, which are essentially free from field contamination. Segregation of galaxies by luminosity and morphology are discussed and a quantitative estimate of the latter is made. The method of constructing self-consistent dynamical models is discussed. Four different forms of the distribution function are analyzed allowing for different possible dependences of f on energy and angular momentum. Properties of typical models that might resemble actual clusters are presented, and the importance of having velocity-dispersion information is emphasized. The effect of a central massive object such as a cD galaxy on the core structure is illustrated. A comparison of these models with Coma is performed.

160.002 **The correlation between X-ray luminosity and velocity dispersion in clusters of galaxies.**
H. Quintana, J. Melnick.
Astron. J., Vol. 87, 972 - 979 (1982).
The correlation between X-ray luminosity and velocity dispersion for clusters of galaxies is reanalyzed on the basis of HEAO-1 A-2 and Einstein fluxes and published velocity dispersions for a significant number of clusters. Between these parameters a strong correlation of the form $L_x \sim \sigma^{4.0 \pm 0.7}$ for the HEAO-1 data and $L_x \sim \sigma^{3.7 \pm 0.4}$ for the Einstein data is found. The implications of these results for our understanding of the structure and origin of the X-ray-emitting gas are discussed in detail.

160.003 **Substructure within clusters of galaxies.**
M. J. Geller, T. C. Beers.
Publ. Astron. Soc. Pacific, Vol. 94, 421 - 439 (1982).
The authors present surface number density contour maps based on Dressler's studies of 65 rich clusters to demonstrate the frequent occurrence of substructure in these dense systems. They give a prescription for examining the significance of such structure and show that structure on scales $\gtrsim 0.24$ Mpc ($H_0 = 100$ km s^{-1} Mpc^{-1}) is present in approximately 40% of the clusters in this sample. Many of the remaining clusters are elongated.

160.004 **Three rich clusters of galaxies with bimodal or clumpy X-ray morphologies.**
M. P. Ulmer, R. G. Cruddace.
Astrophys. J., Vol. 258, 434 - 438 (1982).
The authors present the X-ray images of the rich clusters of galaxies A1560, A2356, and A2384. They show that the individual X-ray features correspond to hot gas with masses $\sim 10^{12}$ to $10^{13} M_\odot$. These results are consistent with the idea that clusters form by coalescence and merging of subclusters.

160.005 **A 1400-MHz survey of 1478 Abell clusters of galaxies.**
F. N. Owen, R. A. White, K. C. Hilldrup, R. J. Hanisch.
Astron. J., Vol. 87, 1083 - 1097 (1982).
Observations of 1478 Abell clusters of galaxies with the NRAO 91-m telescope at 1400 MHz are reported.

160.006 **Observations of the Coma cluster of galaxies at 102.5 MHz.**
A. G. Gubanov.
Astrofizika, Tom 18, 177 - 191 (1982). In Russian. English translation in Astrophysics, Vol. 18, No. 2.
Observations at 102.5 MHz of the Coma cluster are pre sented. The brightness distribution map and the brightness strip have been derived. The total flux and the extent of the Coma C radio source have been measured as well as the flux of the extended radio halo of the radio galaxies NGC 4869 and NGC 4874 and of the "pseudo-halo" evaluated.

160.007 **Radio and X-ray observations of the Abell 2241 galaxy clusters.**
W. Bijleveld, E. A. Valentijn.
Astron. Astrophys., Vol. 111, 50 - 57 (1982).
Radio and X-ray observations of the two independent clusters in the A 2241 region are presented. The difference i both the radio and the X-ray behaviour between the domina galaxies of the two clusters is interpreted as being due to different densities of the intra cluster medium. In the most distant cluster evidence is found for a large gaseous halo surrounding the central galaxy, which by buoyant forces als influences the shape of the 420 kpc long head-tail galaxy.

160.008 **Galaxy groups: sample-dependence of virial prope ties.**
F. Mardirossian, M. Mezzetti, G. Giuricin.
Astron. Astrophys., Vol. 111, 86 - 90 (1982).
Extending a previous study on Turner and Gott's (197 galaxy groups, the authors discuss the sample of groups sele ed by de Vaucouleurs (1975) to investigate the stability of some results concerning the virial parameters of the groups c galaxies.

160.009 **A search for radio halo emission at 430 MHz in 7 rich clusters of galaxies.**
R. J. Hanisch.
Astron. Astrophys., Vol. 111, 97 - 103 (1982).
A sample of 72 nearby (distance class $\leqslant 4$) Abell cluste of galaxies has been observed at 430 MHz with the 305-m te scope at Arecibo. No strong evidence has been found for the existence of radio halo sources similar to the one in the Com cluster in any of the clusters in the sample, and approximate upper limits to the luminosity of halo sources in these cluste have been calculated. It appears that Coma-type radio halos uncommon.

160.010 **The Hydra I cluster of galaxies. A unique case of membership definition.**
O.-G. Richter, J. Materne, W. K. Huchtmeier.
Astron. Astrophys., Vol. 111, 193 - 200 (1982).
New optical radial velocities for 71 galaxies in the Hyd cluster (Abell 1060) are presented. The recession velocity of the cluster is determined to $v_0 = 3425 \pm 34$ km s^{-1} yielding a distance of 45.7 ± 0.5 Mpc (assuming $H_0 = 75$ km s^{-1} Mpc^{-1} Results of galaxy counts could not be fitted by an isotherma model but by a two-component model. Many of the global cluster properties of Hydra I suggest a remarkable similarity with the Virgo cluster. No foreground galaxies and no background galaxies were noted in the observed area up to $v_0 \sim 8000$ km s^{-1} suggesting regions of empty space of the order of 40 - 50 Mpc extent along the line of sight.

160.011 **Intergalactic gas in clusters of galaxies: scattered radiation of a central radio galaxy and its polarization.**
R. A. Syunyaev.
Pis'ma Astron. Zh., Tom 8, 323 - 329 (1982). In Russian. English translation in Soviet Astron. Lett., Vol. 8.
In the case of a powerful compact radio source in a

cluster of galaxies the scattering of its radiation on the intergalactic gas electrons leads to a weak diffuse strongly polarized radio emission from the cluster. Parameters of such diffuse sources are computed for the Perseus and Virgo clusters of galaxies.

160.012 **The potential-estimation method of cluster mass determination.**
H. Smith, Jr.
Astrophys. J., Vol. 259, 423 - 430 (1982) = Dep. Astron., Univ. Florida, Contrib. No. 44.
A maximum likelihood method for estimating the masses of clusters with or without unseen matter is developed. It consists of fitting an approximate general potential function to the observed positions and radial velocities. The fit of the potential gives estimates of the system total mass and, less accurately, the "core" radius of the total mass distribution. Bahcall and Tremaine's projected mass method for test particles orbiting a point mass appears to be a special case of the present method.

160.013 **Radio and X-ray galaxies in Abell 566.**
D. E. Harris, J. G. Robertson, P. E. Dewdney, C. H. Costain.
Astron. Astrophys., Vol. 111, 299 - 305 (1982).
Radio and X-ray observations are presented for Abell 566. The X-ray emission indicates a relatively cool gas (~2 keV) with a fairly wide distribution (600 kpc core radius). Two enhancements in the smooth X-ray distribution are tentatively associated with 2 radio galaxies which are cluster members. One of these has a very steep radio spectrum. The pressure in this source from relativistic particles and the magnetic field is comparable to the thermal pressure of the cluster gas. Thus, the steep radio spectrum may be explained by thermal confinement and a turnoff of particle injection about 10^8 y ago. A second X-ray source 17′ from the field center, appears to be associated with a distant Seyfert. It is probably an example of the newly isolated class of Seyferts which have variable, soft X-ray emission.

160.014 **Intrinsic microwave emission from X-ray clusters, and the Sunyaev-Zeldovich effect.**
G. Cavallo, N. Mandolesi.
Astrophys. Lett., Vol. 22, 119 - 121 (1982).

160.015 **Observations of X-ray clusters of galaxies at 102.5 MHz.**
R. D. Dagkesamansky (*Dagkesamanskij*), A. G. Gubanov, A. D. Kuzmin (*Kuz'min*), O. B. Slee.
Mon. Not. R. Astron. Soc., Vol. 200, 971 - 991 (1982).
Observations with the BSA FIAN radio telescope at 102.5 MHz of 24 Ariel V X-ray sources identified with clusters of galaxies are presented. The measured flux densities have been used to determine the radio powers and spectral indices of these objects. There is weak evidence for a correlation between spectral index and X-ray luminosity, but there is no correlation between spectral index and radio power nor between X-ray luminosity and radio power. A comparison of the present results from 15 clusters common to the survey of 2A X-ray sources made by Slee and Quinn suggests that five of these clusters contain radio sources with angular diameters > 8 arcmin.

160.016 **Diffuse X-rays as a probe to survey the gravitational mass distribution in the Coma cluster of galaxies.**
K. I. Tanaka, Y. Fujishima, M. Fujimoto.
Publ. Astron. Soc. Japan, Vol. 34, 147 - 161 (1982).
Diffuse X-rays from the hot plasma in a cluster of galaxies are investigated by assuming its hydrostatic equilibrium in the gravitational potential of the cluster. By reproducing the observed X-ray brightness profile of the Coma cluster of galaxies, the authors determine its total gravitational mass as (2 - 3) × $10^{15} M_\odot$ and its core radius as about 10′, 1.5 times as large as the conventional scale for the distribution of the bright member galaxies.

160.017 **The south west extension of the Perseus supercluster.**
P. Focardi, B. Marano, G. Vettolani.
Astron. Astrophys., Vol. 113, 15 - 20 (1982).
The authors present the results of a redshift survey, almost complete to m_{ph}= 14.5, covering 270 square degrees between R. A. 23^h20^m and 01^h00^m and DEC 21°30′ and 33°30′. This area lies at the extreme south west region of the Perseus supercluster. New redshift determinations have been obtained for 44 galaxies. Adding the data available in the literature a total of 93 galaxies out of 99 with $m_{ph} \leqslant 14.5$ have a measured radial velocity. It is shown that the Perseus supercluster extends westward beyond the limits of previous surveys and that the general distribution of galaxies is suggestive of a "cell" structure.

160.018 **Dynamics of the galaxy clusters Coma and Hydra I.**
B. Fuchs, J. Materne.
Astron. Astrophys., Vol. 113, 85 - 93 (1982).
For two well observed clusters of galaxies, Coma and Hydra I, theoretical surface density distributions are proposed which follow closely the observed distributions. These are either a generalized Hubble profile or a two component model with exponentially decreasing density. The corresponding spatial density distributions and radial velocity distributions are derived. The observations constrain possible models to non-isothermal ones.

160.019 **Distant clusters of galaxies in the southern hemisphere.**
R. M. West, A. Kruszewski.
Irish Astron. J., Vol. 15, (see 012.008), 25 - 35 (1981).
The authors discuss an astronomical project which aims at a comprehensive study of distant clusters of galaxies in the southern hemisphere. Several hundred of these have so far been identified on ten ESO Schmidt plates (6300 - 7000 Å). Extensive software is being developed at the ESO VAX computer for automatic identification of clusters and objective evaluation of their properties. Deep CCD-frames have been obtained of ~25 clusters for the photometric calibration and identification of faint cluster galaxies. Digital spectra of galaxies were observed in about 30 clusters with redshifts between 0.05 and 0.37.

160.020 **Mapping the local supercluster.**
J. Silk.
Nature, Vol. 299, 577 - 578 (1982).

160.021 **How to identify and weigh virialized clusters of galaxies in a complete redshift catalog.**
W. H. Press, M. Davis.
Astrophys. J., Vol. 259, 449 - 473 (1982).
Groups or clusters of galaxies whose crossing time is small compared to the Hubble time can be "weighed" by the virial theorem. This paper develops a statistical procedure for identifying such subsystems within a magnitude-limited redshift catalog that is complete in some solid angle of sky. When the procedure is applied to the Center for Astrophysics Redshift Survey data, the most striking result is a strong approximately linear relation between the virial mass per galaxy in a cluster and the radius of that cluster. This trend is uniform from scales as small as tens of kiloparsecs to scales as large as several megaparsecs.

160.022 **Transform analysis of the high-resolution Shane-Wirtanen catalog: the power spectrum and the bispectrum.**
J. N. Fry, M. Seldner.
Astrophys. J., Vol. 259, 474 - 481 (1982).

To extract more statistical information from the high resolution (10′) Shane-Wirtanen data, and in particular to gain information about the large-scale behavior of galaxy clustering, the authors have examined the two- and three-point galaxy correlation functions in Fourier transform space for wavelengths corresponding to angular separations $1/2° \lesssim \theta \lesssim 70°$.

160.023 **Discovery of two active galactic nuclei in clusters of galaxies.**
J. E. Steiner, J. E. Grindlay, T. Maccacaro.
Astrophys. J., Vol. 259, 482 - 485, plates 12, 13 (1982).
The authors report two X-ray selected AGN found to be associated with clusters of galaxies. 1E 0412.5−0803 is associated with a group of galaxies in the outskirts of A 484 (richness class 1), while 1E 2304.9−2259 is found close to the center of cluster A 2534 (richness class 2). This cluster itself is detected as a diffuse X-ray source.

160.024 **Mass segregation in clusters of galaxies.**
R. Farouki.
News Lett. Astron. Soc. N.Y., Vol. 2, No. 2, p. 15 - 16 (1982). Abstract.

160.025 **Middle-ultraviolet photometry of Virgo cluster galaxies.**
A. M. Smith, R. H. Cornett.
Astrophys. J., Vol. 261, 1 - 11 (1982).
Photographic photometry of Virgo cluster galaxies has been performed in a wavelength band extending from λ1620 to λ3200 using sounding rocket techniques. The observational results are middle ultraviolet magnitudes, U_{2421}, or faint limits for U_{2421}, for 201 galaxies within 5°.5 of the cluster center. The measured colors are generally compatible with colors computed from previously generated composite spectra.

160.026 **Digital surface photometry of galaxies toward a quantitative classification. I. 20 galaxies in the Virgo cluster.**
M. Watanabe, K. Kodaira, S. Okamura.
Astrophys. J., Suppl. Ser., Vol. 50, 1 - 22, plates 1 - 5 (1982).
A new method of reducing and analyzing surface photometry of galaxies is developed and used to secure a large body of homogeneous digital data. Twenty galaxies in the Virgo cluster are analyzed in this new system, and their isophotes and luminosity profiles are presented. Structure parameters are derived from Fourier and moment analyses of luminosity profiles of the 20 galaxies; these turn out to show tight correlations with morphological type.

160.027 **The evolution of disk galaxies and the S0 problem, revisited.**
G. D. Bothun.
Astrophys. J., Suppl. Ser., Vol. 50, 39 - 54 (1982).
The author summarizes the properties of clusters of galaxies, which may change the course of galaxy evolution. The extensive data base of Bothun is used to reexamine the issue of the role of the environment in the evolution of disk galaxies. This data base consists of radio and optical observations of ~ 350 galaxies in the clusters Peg I, Cancer, Pisces, Coma, A 1367, Z 74-23, Hercules, A 539, and A 400. The basic conclusion is that initial conditions of formation and variations in star formation histories have been more important than environmental influences in determining the present-day character of spiral galaxies in clusters. The key parameter may well be the amount of neutral hydrogen remaining after star formation in the bulge component has ceased.

160.028 **The surface distribution of galaxy morphological types in Coma.**
S. A. Gregory, L. A. Thompson.
Bull. American Astron. Soc., Vol. 14, 659 (1982). – Abstract.

160.029 **Discovery of a rich cluster of galaxies in Triangulu Australe identified with the bright X-ray source 4U1631-64.**
D. A. Schwartz, J. Patterson, H. Bradt, R. Remillard, R. F. Mushotzky, M. D. Johnston.
Bull. American Astron. Soc., Vol. 14, 660 (1982). – Abstra

160.030 **Simulations of clusters of galaxies: luminosity growth of the brightest galaxy as a function of ricl ness.**
E. M. Malumuth, D. O. Richstone.
Bull. American Astron. Soc., Vol. 14, 663 (1982). – Abstra

160.031 **A study of the dependence of the properties of galaxy clusters on cluster morphology.**
P. M. Lugger.
Bull. American Astron. Soc., Vol. 14, 663 (1982). – Abstra

160.032 **Einstein observations of the size and structure of clusters of galaxies.**
C. Jones, W. Forman.
Bull. American Astron. Soc., Vol. 14, 664 (1982). – Abstra

160.033 **Implications of the cluster X-ray luminosity function for the dynamical state of clusters.**
W. Forman, C. Jones.
Bull. American Astron. Soc., Vol. 14, 664 (1982). – Abstra

160.034 **Neutral hydrogen contents of Virgo galaxies.**
G. Helou, C. Giovanardi, E. E. Salpeter.
Bull. American Astron. Soc., Vol. 14, 664 (1982). – Abstra

160.035 **On the mass-to-light ratio of Coma and the densit of the dark matter.**
M. E. Bailey.
Mon. Not. R. Astron. Soc., Vol. 201, 271 - 280 (1982).
Observations of Coma are used to constrain the density distribution of the dark matter. The three-dimensional galax velocity dispersion profile for the author's models is obtaine analytically, and several representative projected velocity dispersion profiles are given. The uncertainty in the cluster mass distribution allowed by observations and these models is a further significant factor affecting detailed modelling of cluster properties.

160.036 **Towards a study of southern distant clusters of galaxies.**
R. M. West.
Rev. Mexicana Astron. Astrofis., Vol. 6, (see 012.017), 35 - 38 (1981).
A brief description is given of a collaborative research project, which aims at the identification of distant clusters o galaxies on red plates (IIIa-F + RG 630) obtained with the ESO 1-m Schmidt telescope, as well as a detailed morpholog cal, photometric and spectroscopic study of selected objects The faintest clusters found this way may have red-shifts ~1, but so far spectroscopic observations have only been made o clusters with $z < 0.4$.

160.037 **Interaction in two contrasting examples of galacti groups.**
A. C. Danks, G. Alcaíno.
Rev. Mexicana Astron. Astrofis., Vol. 6, (see 012.017), 39 - 40 (1981). – Abstract.

160.038 **Observed dynamics of superclusters.**
R. J. Harms, H. C. Ford, R. Ciardullo, F. Bartko.
Tenth Texas Symposium on Relativistic Astrophysics, (see 012.021), p. 178 - 187 (1981).
Radial velocity measurements obtained in two large ricl superclusters show evidence of gravitational slowing of the Hubble flow within the superclusters. Simple models yield

mass density estimates for the superclusters from less than one to a few times the critical density. Preliminary counts of galaxies and Abell clusters indicate a density enhancement within the superclusters between 10 and 71 times the field mass density. These results strongly favor an open universe (assuming a standard Friedmann cosmology) with $\Omega_0 < 0.3$.

160.039 **Extragalactic X-ray astronomy with the Einstein Observatory.**
R. Giacconi.
Tenth Texas Symposium on Relativistic Astrophysics, (see 012.021), p. 210 - 234 (1981).

This review discusses recent results of X-ray observations of clusters of galaxies.

160.040 **Coma-type radio halos and cluster X-ray morphology.**
W. T. Vestrand.
Astron. J., Vol. 87, 1266 - 1269 (1982).

It is pointed out that there seems to be a characteristic X-ray morphology for clusters with Coma-type radio halos. This correlation may be the first evidence that the conditions conductive to formation of a radio halo influence or are influenced by the thermal gas distribution. It is suggested that heating of the intracluster gas by the cosmic rays that power the radio halo may be the source of the correlation. The anomalous gas temperature of Coma may be the result of this process. Future observations of cluster temperatures and velocity dispersions should provide a test of the cosmic-ray heating hypothesis.

160.041 **X-radiation of clusters of galaxies.**
V. G. Kurt.
Zemlya Vselennaya, 1982, No. 5, p. 26 - 31. In Russian.

160.042 **Evolution of rich clusters of galaxies.**
R. Roos, S. J. Aarseth.
Astron. Astrophys., Vol. 114, 41 - 52 (1982).

The evolution of a rich cluster of galaxies having a realistic initial distribution of masses is simulated numerically. The sizes and masses of the simulation particles are such that they represent halos presumed to be attached to galaxies initially. The ratio of the amount of dark to luminous material is assumed constant. The upper end of the mass spectrum is most strongly affected by the merging process occurring predominantly during the expansion and subsequent collapse of the cluster. The amount of mass segregation in the collapsed cluster is compared with the observed luminosity segregation in the Coma cluster. The amount of mass loss from galaxies in high velocity encounters after the collapse of the cluster is estimated.

160.043 **X-ray-imaging observations of clusters of galaxies.**
W. Forman, C. Jones.
Annu. Rev. Astron. Astrophys., Vol. 20, (see 003.009), 547 - 585 (1982).

Properties and models of the intracluster gas.
Clusters with large X-ray core radii–nXD clusters.
Clusters with dominant, central galaxies –XD clusters.
Cluster classification. The virial mass distribution. X-ray emission from distant clusters.

160.044 **Galaxienhaufen.**
R. Wiegandt.
Sterne Weltraum, 21. Jahrg., 402 - 405 (1982).

160.045 **Einstein X-ray imaging of clusters of galaxies.**
C. Jones, W. Forman.
Observatory, Vol. 102, 112 - 113 (1982). – Abstract.

160.046 **Soft X-ray images of the Perseus cluster of galaxies.**
G. Branduardi-Raymont.
Observatory, Vol. 102, 113 (1982). – Abstract.

160.047 **Cooling flows in clusters of galaxies.**
P. E. J. Nulsen.
Observatory, Vol. 102, 113 - 114 (1982). – Abstract.

160.048 **The Lynx-Ursa Major supercluster.**
R. Giovanelli, M. P. Haynes.
Astron. J., Vol. 87, 1355 - 1363 (1982).

The distribution of galaxies in the region of Lynx-Ursa Major outlines a filamentary structure. The average width of the filament is approximately 1.7 deg. The average redshift of the galaxies in the filament lies in the interval 4000–5300 km s^{-1}. Seventy-five newly determined 21-cm redshifts are presented and are used to complement the rest of the available redshift data to analyze the structure of the filament. The authors also point out the similarity of the redshift domain of the Lynx-Ursa Major filament with that of the Pisces-Perseus supercluster.

160.049 **Evolution of the cluster X-ray luminosity function slope.**
J. P. Henry, A. Soltan, U. Briel, J. E. Gunn.
Astrophys. J., Vol. 262, 1 - 8 (1982).

The authors report the results of an X-ray survey of 58 clusters of galaxies at moderate and high redshifts. Using a luminosity-limited subsample of 25 objects, they find that to a redshift of 0.5 the slope of the luminosity function of distant clusters is independent of redshift and consistent with that of nearby clusters. The time scale for change in the slope must be greater than 9 billion years.

160.050 **The halos of rich clusters of galaxies. I. An infall model for the Coma Cluster.**
S. A. Shectman.
Astrophys. J., Vol. 262, 9 - 14 (1982).

The large mass of a cluster of galaxies slows the Hubble expansion in a surrounding volume. The ratio of the matter density near the cluster to the average density at large distances grows as the cosmological density parameter, Ω, decreases. For values of $\Omega < 0.2$, the gravitational attraction of the Coma Cluster acting upon an initially homogeneous external medium is sufficient to account for a galaxy population comparable to that of the Coma Supercluster.

160.051 **The evolution of radio galaxies: a VLA survey of high-redshift clusters of galaxies.**
W. Jaffe.
Astrophys. J., Vol. 262, 15 - 23 (1982).

The author presents data from a VLA 1465 MHz survey of high-redshift galaxy clusters with redshifts of 0.25 to 0.9. Sources were detected in 13 of 55 clusters. The sample shows strong evolution of the radio luminosity function during this period even for weak cluster radio galaxies.

160.052 **A ~ 300 Mpc void of rich clusters of galaxies?**
N. A. Bahcall, R. M. Soneira.
Astrophys. J., Vol. 262, 419 - 423 (1982).

The distribution of rich Abell clusters of galaxies reveals a large void of cataloged, nearby rich clusters corresponding to the approximate redshift range $z \approx 0.03$–0.08. This apparent void is $\sim 100^\circ$ across the sky ($\sim 300 h^{-1}$ Mpc) and $\sim 150 h^{-1}$Mpc deep. Its projected area is completely devoid of nearby – but not distant – rich clusters. The void does not appear to be caused by absorption in the Galaxy.

160.053 **The neutral hydrogen deficiency of the cluster A 262.**
R. Giovanelli, M. P. Haynes, G. L. Chincarini.
Astrophys. J., Vol. 262, 442 - 450 (1982).

Seventy-two galaxies in the spiral-rich cluster A 262 have been observed in the 21 cm line at Arecibo. The H I content of individual galaxies, as compared with that of a sample of isolated galaxies, depends on the distance from the core of the cluster; while pronounced H I deficiency is observed near the central parts, normal amounts of H I are observed in the cluster periphery. With the newly available redshifts, a revised set of dynamical parameters for the cluster is given.

160.054 **On the origin of the voids in the galaxy distribution.**
Y. Hoffman, J. Shaham.
Astrophys. J., Lett., Vol. 262, L23 - L26 (1982).

The distribution of galaxies on scales larger than $\sim 10\,h^{-1}$ Mpc seems to be characterized by large voids, $(20-40)\,h^{-1}$ Mpc in diameter. It was previously argued that the mere existence of such voids poses a severe problem to all dissipationless clustering theories. The authors show that the voids may, in fact, be a natural outcome of a dissipationless clustering scenario if both adiabatic and isothermal density perturbations exist primordially. When the nonlinear evolution of spherical voids of this type is followed for adiabatic perturbations with an index $n > -1$, it is seen that they become surrounded by a shell of positive density contrast. Their structure is insensitive to Ω_0 while their dynamics is quite sensitive to it.

160.055 **Monte-Carlo simulations of galaxy systems. IV. Static properties for galaxies in supercluster cells.**
H. T. MacGillivray, R. J. Dodd.
Astrophys. Space Sci., Vol. 86, 419 - 435 (1982).

The Monte-Carlo simulation technique (MacGillivray et. al., 1982) is used to simulate static properties for galaxies in spherical supercluster "cells" for different galaxy 3-dimensional alignment schemes, and the projected galaxy properties examined. The results may be used to aid the interpretation of the observed geometric properties for galaxies in real supercluster cells. The simulation technique is also used to examine the properties of galaxies in "clusters" formed at the intersections of adjacent cells. The results may provide clues to the interpretation of the nature of observed galaxy geometric properties in rich clusters and are used in the present study for a preliminary comparison with observational data.

160.056 **Monte-Carlo simulations of galaxy systems. V. Computer models of galaxy fields.**
H. T. MacGillivray, R. J. Dodd.
Astrophys. Space Sci., Vol. 86, 437 - 452 (1982).

Computer models of galaxy fields are generated using the Monte-Carlo simulation technique described by MacGillivray et al., 1982. In the models, various scenarios for the true 3-dimensional distribution of galaxies (e. g., clustering on continuous or hierarchical scales) are incorporated and the galaxies projected onto a synthetic photographic plate. The model fields are compared, both qualitatively and statistically, with a field obtained from measures with COSMOS on a deep photograph taken with the UK Schmidt Telescope. The results indicate that the observed distribution of faint galaxies on photographic plates can only be represented by a model involving second order clustering of galaxies.

160.057 **Observations of galaxy clustering.**
T. Shanks.
Progress in cosmology, (see 012.047), p. 335 - 346 (1982).

Observations of galaxy clustering are important for cosmology because of the information they contain on the extent of inhomogeneity in the matter distribution of the Universe. The author reviews here what has been learnt about galaxy clustering through statistical studies of galaxy catalogues.

160.058 **V Zw 311: The once and future cD?**
D. P. Schneider, J. E. Gunn.
Astrophys. J., Vol. 263, 14 - 22 (1982).

The peculiar cluster V Zw 311 contains a compact group of elliptical galaxies in a luminous matrix. The authors have acquired spectrographic and photometric measurements of the system and have performed dynamical equilibrium and frictional lifetime calculations. The central object has a luminosity similar to a typical first-ranked cluster galaxy, but its structural parameters are quite extreme. Various formation mechanisms for this peculiar system are discussed and its future dynamical evolution is outlined.

160.059 **Common properties of clusters of galaxies containing radio halos and implications for models of radio halo formation.**
R. J. Hanisch.
Astron. Astrophys., Vol. 116, 137 - 146 (1982).

The author has examined the data concerning the optical, radio, and X-ray properties of the clusters of galaxies which have the best documented evidence for radio halo emission: the Coma cluster, Abell 2255, Abell 2256, and Abell 2319. Recent high dynamic range observations of 3C 84 (Noordam and de Bruyn, 1982) indicate that the Perseus cluster is also site of a radio halo similar to that in the Coma cluster. These clusters have a number of striking similarities. Various models for halo formation are discussed in light of the physical conditions implied by the cluster morphology.

160.060 **Galaxy clustering and the missing mass.**
M. Davis.
Philos. Trans. R. Soc. London, Ser. A, Vol. 307, (see 012.052), 111 - 119 (1982).

Several recently completed large-scale red-shift surveys of galaxies, have shown that the superclustering phenomenon is widespread and accompanied by large holes in space that appear to be quite deficient in galaxies. With the large Harvard-Smithsonian (CfA) red-shift survey complete to magnitude 14.5 at high galactic latitude, it is possible to determine dynamical masses of small groups as well as large clusters. The evidence for missing mass, derived from these dynamical studies, is reviewed.

160.061 **The estimation of galaxy angular correlation functions.**
P. C. Hewett.
Mon. Not. R. Astron. Soc., Vol. 201, 867 - 883 (1982).

Analysis of galaxy samples shows that significant bias in the estimation of galaxy correlation functions can occur depending on the type of estimator and edge correction applied to the sample. A general technique for calculating estimates of correlation functions that is free from systematic errors is described. A sharp break in the power-law form of the covariance function at scales of $7\,H^{-1}$ Mpc is found in two galaxy samples confirming the results from the Lick counts obtained by Groth and Peebles. Some evidence for weak anti-clustering at large scales is found.

160.062 **An absorption feature and filamentary structures in the central galaxy of the Centaurus cluster, NGC 4696.**
H. E. Jørgensen, H. U. Nørgaard-Nielsen.
Messenger, No. 30, p. 3 - 4 (1982).

160.063 **The plain man's guide to galaxy clustering.**
R. J. Dodd, H. T. MacGillivray.
South. Stars, Vol. 29, 169 - 176 (1982).

160.064 **Core radii determination for 11 southern clusters of galaxies.**
I. Semeniuk.
ESO Sci. Prepr., No. 224, 27 pp. (1982). – Submitted to Acta Astron.

160.065 **On the geometry of two superclusters Coma-A1367 and Perseus-Pisces.**
G. L. Chincarini, R. Giovanelli, M. P. Haynes.
ESO Sci. Prepr., No. 226, 18 pp. (1982). – Submitted to Astrophys. J.

160.066 **Medusa spectroscopy of A400, A576, A1767 and A2124.**
P. Hintzen, J. M. Hill, D. Lindley, J. S. Scott, J. R. P. Angel.
Prepr. Steward Obs., No. 402, 36 pp. (1982).

160.067 **Dynamical models and our Virgocentric deviation from Hubble flow.**
G. L. Hoffman, E. E. Salpeter.
Astrophys. J., Vol. 263, 485 - 500 (1982).

Dynamical models are used to calculate ΔV, the velocity deviation from pure Hubble flow of the Local Group due to the gravitational effect of the Virgo galaxy cluster and its surrounding supercluster, for various values of the cosmological density parameter Ω. The models are required to fit the detailed observations of the dispersion of systemic galaxy velocities in and around the cluster core. If reliable measurements of ΔV should turn out to give $\Delta V \lesssim 150$ km s^{-1}, and/or if the direction of ΔV deviates appreciably from that of the Virgo cluster center, models with small Ω ($\Omega < 0.1$) and "invisible mass" mainly concentrated in the cluster core would be ruled out.

160.068 **Spectroscopy of galaxies in distant clusters. I. First results for 3C 295 and 0024 + 1654.**
A. Dressler, J. E. Gunn.
Astrophys. J., Vol. 263, 533 - 545 (1982).

ISIT photometry by Butcher and Oemler has indicated an unusually large fraction of blue galaxies in the distant, concentrated clusters 3C 295 and 0024 + 1654. The authors have obtained low-resolution spectroscopy of 17 galaxies in these fields in order to ascertain cluster membership and attempt morphological classification of these galaxies by comparing their spectra with those of nearby ellipticals and spirals. The spectra of the blue galaxies that are known to be, or may be, cluster members are not the spectra of spiral galaxies. A sizeable fraction of the population of blue cluster members have spectra typical of active galactic nuclei, like type 2 Seyferts, which are rare in present-day clusters.

160.069 **The curvature of radio jets and tails in the intracluster media of Abell 1446 and 2220.**
J. O. Burns, T. J. Balonek.
Astrophys. J., Vol. 263, 546 - 556 (1982).

Einstein Observatory X-ray observations are reported for two rich clusters which contain prominent tailed radio galaxies. These X-ray data are used as probes of the cluster gas to explore mechanisms which may distort radio jets and tails. The wide-angle tailed radio structure of 1159+583 in A 1446 is explained by a combination of a buoyancy pressure gradient in the cluster gas and dynamic pressure arising from slow galaxy motion. The second cluster observed, A 2220, contains a clumped X-ray structure with peaks near two dominant galaxies. The strongly bent radio jets of 1638+538 (associated with one of the bright galaxies) are consistent with $\sim$1000 km s^{-1} motion, possibly in a binary orbit, through the X-ray emitting cluster gas. These X-ray data also suggest that the radio jets and tails are thermally confined by the hot intracluster media.

160.070 **The galactic content of groups of galaxies.**
R. E. de Souza, H. V. Capelato, L. Arakaki, C. Logullo.
Astrophys. J., Vol. 263, 557 - 563 (1982).

The distribution of morphological types of galaxies in small groups is studied. Using two independent samples, the authors show that the relation between the fraction of spirals and the local density of galaxies in groups is very nearly the same as the one proposed by Dressler for rich clusters. They show that there is a correlation between the fraction of spirals and the rate of collisions between galaxies. An estimate of the rate of destruction of spiral galaxies by collisions leads to a cross section for this process in the range of 60–80 kpc.

160.071 **Filamentary structure in the Shane-Wirtanen galaxy distribution.**
J. R. Kuhn, J. M. Uson.
Astrophys. J., Lett., Vol. 263, L47 - L50, plates L1 - L3 (1982).

The authors find statistical evidence that the observed Shane-Wirtanen galaxy distribution has significantly more structure than is described by the previously measured two-point and three-point functions. Their statistic implies that the Shane-Wirtanen large-scale galaxy distribution has significant filamentary structure throughout the north galactic polar region.

160.072 **The Local Supercluster of galaxies.**
M. Jõeveer.
Tartu Astrofüüs. Obs. Publ., Köide 49, 161 - 169 (1982).

The space distribution of nearby ($V_0 \leqslant 3000$ km s^{-1}) giant galaxies ($M \leqslant -21.0$) on the northern galactic hemisphere is studied, using the redshift as a measure of distance. The concentration of galaxies to the supergalactic plane is distinctly visible up to the distances 35 Mpc ($H_0 = 50$ km s^{-1} Mpc^{-1}) from the Virgo cluster of galaxies. Galaxies outside the supergalactic disk are concentrated to the clouds.

160.073 **On the symmetry of the system of companions of the Galaxy.**
J. Jaaniste.
Tartu Astrofüüs. Obs. Publ., Köide 49, 170 - 177 (1982). In Russian.

The spatial distribution of 27 companions of the Galaxy (10 dwarf galaxies and 17 globular clusters) is studied. It is shown that the distribution is anisotropic, the axis ratios of the scattering ellipsoid being 1 : 1.8 : 2.6. The direction of the largest axis is close to the direction toward the Virgo cluster, that of the smallest axis to M 31.

160.074 **Medusa spectroscopy of A400, A576, A1767, and A2124.**
P. Hintzen, J. M. Hill, D. Lindley, J. S. Scott, J. R. P. Angel.
Astron. J., Vol. 87, 1656 - 1667 (1982).

Galaxy velocity data taken with the Steward Observatory multiple aperture fiber optic spectrograph are presented for four Abell clusters. The root-mean-square external errors in these velocities are about 100 km/s; accuracy which compares favorably with that obtained from single-object observations. All four of the clusters observed are known X-ray sources and the data agree well with empirically derived velocity dispersion-X-ray luminosity relations for clusters of galaxies. The large microwave decrement observed in A576 was initially interpreted as due to Compton scattering of the microwave backbround by the X-ray-emitting intracluster gas. White and Silk have presented Einstein X-ray data which indicate that A576 contains too little gas to produce the observed microwave decrement by Compton scattering. The authors' velocity dispersion for 47 members of this cluster strengthens their conclusion.

160.075 **About the origin of clusters of galaxies.**
H. Nováková.
Říše hvězd, Vol. 63, 249 - 250 (1982). In Czech.

160.076 **Is there a ring around the Milky Way?**
J. J. Rawal.
Indian J. Radio Space Phys., Vol. 11, 100 - 101 (1982).
Abstr. in Phys. Abstr., Vol. 85, Abstr. 117767 (1982).

160.077 **Superclusters of galaxies.**
J. Einasto, E. Saar.
Proceedings of the Second Marcel Grossmann Meeting on General Relativity, (see 012.081), p. 875 - 882 (1982).
Abstr. in Phys. Abstr., Vol. 86, Abstr. 20513 (1983).

Metacatalogue of galaxies and large scale structures in the Universe.
See Abstr. 002.027.

Data retrieval in the Metacatalogue of galaxies.
See Abstr. 002.028.

Le Centre de Données des Amas de Galaxies.
See Abstr. 002.029.

Webb Society deep-sky observer's handbook. Vol. 5: Clusters of galaxies.
See Abstr. 003.125.

X-ray observations of single-line spectroscopic binaries.
See Abstr. 120.013.

Star formation in a cooling flow.
See Abstr. 131.275.

Spectroscopy of the QSO pair Q0028+003/ Q0029+003.
See Abstr. 141.058.

The curvature of radio jets and tails in the intracluster media of Abell 1446 and 2220.
See Abstr. 141.080.

Radio and X-ray observations of the halo source in A1367.
See Abstr. 141.081.

The structure of 4C radio galaxies in poor clusters.
See Abstr. 141.102.

Coma quasars.
See Abstr. 141.148.

Extragalactic X-ray sources.
See Abstr. 142.076.

High-resolution X-ray spectroscopy of supernova remnants, clusters of galaxies and BL Lac objects.
See Abstr. 142.085.

Environmental influences on galaxy structure.
See Abstr. 151.050.

Globular clusters in the Sculptor Group galaxy NGC 55.
See Abstr. 154.015.

New *UBV* parameters for 46 E-S0 galaxies in the Virgo cluster.
See Abstr. 158.003.

VLA observations of the giant cluster spiral UGC 12423.
See Abstr. 158.017.

Optical spectra and redshifts for seventy-one galaxies.
See Abstr. 158.036.

Photometry of the cluster of galaxies A 1367.
See Abstr. 158.037.

The local velocity field and the calibration of the luminosity classification of spiral galaxies.
See Abstr. 158.091.

The space density and clustering properties of a n sample of emission-line galaxies.
See Abstr. 158.122.

10 μm survey of spiral galaxies in the Virgo cluste
See Abstr. 158.147.

Optical filaments around NGC 4696 in the Centaurus cluster.
See Abstr. 158.152.

New dwarf galaxies in the M 81/M 82 group.
See Abstr. 158.165.

Galaxy photometry in the Virgo cluster. I.
See Abstr. 158.166.

On the widths of dust layers in galaxies.
See Abstr. 158.167.

The Fornax-Leo-Sculptor stream.
See Abstr. 158.185.

The cross-correlation of the Zwicky and Shane-Wirtanen catalogs of galaxies.
See Abstr. 158.219.

Observations of peculiar galaxies in clusters of galaxies.
See Abstr. 158.226.

Einstein Observatory solid state spectrometer obs vations of M 87 and the Virgo cluster.
See Abstr. 158.229.

Galaxies and clusters of galaxies.
See Abstr. 158.235.

Dwarf galaxies in the M 81/82 group. New photometric data.
See Abstr. 158.244.

NGC 1961: stripping of a supermassive spiral gala
See Abstr. 158.248.

Flocculent and grand design spiral arm structure i cluster galaxies.
See Abstr. 158.269.

The structure of groups of galaxies in the Ursa Major/Canes Venatici region – I. A discussion of the results of an H I survey.
See Abstr. 158.270.

The relationships between morphological type an quantitative measures in the cluster environment.
See Abstr. 158.331.

Analysis of the orientation of disks of galaxies to their orbital planes in groups and clusters.
See Abstr. 158.334.

Some properties of the centers of cD galaxies.
See Abstr. 158.350.

How well is gas mixed in clusters of galaxies?
See Abstr. 161.002.

Intergalactic gas in clusters of galaxies, microwave background and cosmology.
See Abstr. 161.005.

Improved standard cosmology: comparison with observation.
See Abstr. 162.010.

Correlations between the virial parameters of groups of galaxies.
See Abstr. 162.012.

Are we falling into the Virgo cluster?
See Abstr. 162.019.

The effect of neutrinos on the formation time of rich clusters of galaxies.
See Abstr. 162.020.

Clustering of Lyman-alpha absorption lines in quasars: like galaxies in proto-superclusters?
See Abstr. 162.046.

A re-examination of the Sandage–Tammann extragalactic distance scale.
See Abstr. 162.056.

Statistics of the distribution of galaxies.
See Abstr. 162.085.

The density of the universe from the deceleration of nearby galaxies.
See Abstr. 162.097.

Perturbation of the magnitude–redshift relation in an inhomogeneous relativistic model. II. Correction to the Hubble law behind clusters.
See Abstr. 162.134.

The gravitational evolution of structure in a scale-free universe.
See Abstr. 162.136.

On the large-scale variations of M/L.
See Abstr. 162.175.

The four-point function in the BBGKY hierarchy.
See Abstr. 162.176.

Cosmic far-ultraviolet background radiation.
See Abstr. 162.188.

Erratum

160.901 **Erratum: 'Systematic properties of compact groups of galaxies'** [Astrophys. J., Vol. 255, 382 - 391 (1982)].
P. Hickson.
Astrophys. J., Vol. 259, 930 (1982). – See Abstr. 31.160.019.

161 Intergalactic Matter

161.001 **Galactic and intergalactic Faraday rotation.**
R. C. Thomson, A. H. Nelson.
Mon. Not. R. Astron. Soc., Vol. 201, 365 - 383 (1982).
A model of the systematic component of the magnetic field of the Galaxy is fitted to a sample of 459 extragalactic rotation measures (*RM*), and the results are found to be consistent with a previous analysis of pulsar *RMs* by Thomson & Nelson. The model is then used to reduce the effect of galactic Faraday rotation on the *RMs* of 134 QSOs, and the results used to investigate the existence of a Faraday-active intergalactic medium. Three models are considered in order to explain the redshift dependence of the *RM* variance.

161.002 **How well is gas mixed in clusters of galaxies?**
M. Nepveu.
Astron. Astrophys., Vol. 114, 337 - 340 (1982).
In rich galaxy clusters the material stripped from galaxies by dynamic effects, gathered in wakes, can only mix with surroundings if the wakes are actually hit by galaxies. Thereby turbulent mixing is triggered. In these rich clusters this mixing may effectively start after 10^9 yr in the regions $r \lesssim R_{core}$; for regions $r \gtrsim R_{core}$ it would effectively start only after 10^{10} yr. The effect of IGM heating is considered; its influence is small.

161.003 **The distribution and composition of intergalactic clouds at large red-shifts.**
W. L. W. Sargent.
Philos. Trans. R. Soc. London, Ser. A, Vol. 307, (see 012.052), 87 - 96 (1982).
Most of the sharp absorption lines in the spectra of quasars of high red-shift are single Lyman-α lines. The stronger ones have associated Lyman-β or Lyman-γ lines but not lines of heavy elements. It is now thought that the single Lyman-α absorption lines are produced by primordial, intergalactic clouds. Studies now in progress are aimed at giving limits on the heavy-element composition of the clouds, as well as a value for the deuterium abundance of primordial gas.

161.004 **Absorption lines, Faraday rotation, and magnetic field estimates for QSO absorption-line clouds.**
P. P. Kronberg, J. J. Perry.
Astrophys. J., Vol. 263, 518 - 532 (1982).
The authors have estimated the extragalactic component of Faraday rotation for a sample of 37 QSOs for which there is good absorption line data. Statistical evidence is presented which suggests that a component of Faraday rotation which is occurring in the absorption clouds of some QSOs has been isolated. The authors compare the results with some derived simple theoretical models, and discuss the physical conditions, including magnetic field estimates, in some selected absorption line systems.

161.005 **Intergalactic gas in clusters of galaxies, microwave background and cosmology.**
Ya. B. Zel'dovich, R. A. Syunyaev.
Astrofiz. i kosm. fiz., Moskva, 1982, p. 9 - 65. In Russian. Abstr. in Ref. zh., 51. Astron., 10.51.769 (1982).

161.006 **The dynamics of the intergalactic medium.**
A. K. Dasgupta.
Astrophys. Space Sci., Vol. 87, 69 - 80 (1982).
It is assumed that the intergalactic medium contains fully ionized plasmas. A detailed discussion shows that it is necessary to distinguish between medium-density and low-density plasmas before it is possible to carry out any analy: The conditions are very critical for the conclusions and apr cation of any mechanism one likes to discuss for the intergalactic medium. It is shown that acceleration of high-ener particles in the intergalactic medium is possible.

161.007 **The optical activity of intergalactic space.**
J. N. Clarke, G. Karl, P. J. S. Watson.
Canadian J. Phys., Vol. 60, 1561 - 1563 (1982). – Abstr. i Phys. Abstr., Vol. 86, Abstr. 20488 (1983).

Massive neutrinos and ultra-violet astronomy.
See Abstr. 061.031.

Champs magnétiques interstellaires et intergalactiques.
See Abstr. 131.281.

Wind instability in clouds of radio galaxies.
See Abstr. 141.011.

The curvature of radio jets and tails in the intracluster media of Abell 1446 and 2220.
See Abstr. 141.080.

Einstein Observatory solid state spectrometer ob vations of M 87 and the Virgo cluster.
See Abstr. 158.229.

X-ray spectroscopy of the galaxy M 87: radiative accretion of the hot plasma halo.
See Abstr. 158.230.

Radio and X-ray observations of the Abell 2241 galaxy clusters.
See Abstr. 160.007.

Intergalactic gas in clusters of galaxies: scattered radiation of a central radio galaxy and its polarization.
See Abstr. 160.011.

The curvature of radio jets and tails in the intracluster media of Abell 1446 and 2220.
See Abstr. 160.069.

162 Universe (Structure, Evolution)

162.001 **Primeval adiabatic perturbations: effect of massive neutrinos.**
P. J. E. Peebles.
Astrophys. J., Vol. 258, 415 - 424 (1982).
The autocorrelation function of the mass distribution prior to galaxy formation is computed under the assumptions that the universe is dominated now by neutrinos with mass some tens of eV and that the primeval density fluctuations are adiabatic and linear. The goal is to discover whether the coherence length of the mass distribution prior to galaxy formation can be consistent with the observed clustering length of galaxies. The conclusion is that it can be, but that it requires rather special arrangements.

162.002 **Correction of a criticism of the phenomenological quadratic redshift-distance law.**
J. F. Nicoll, I. E. Segal.
Astrophys. J., Vol. 258, 457 - 466 (1982).
The claim of Soneira that his analysis "conclusively rules out the square (redshift-distance) law for the small redshift region" is shown to be unfounded in several respects.

162.003 **Where are the Population III stars?**
J. G. Hills.
Astrophys. J., Lett., Vol. 258, L67 - L69 (1982).
There appears to be no observational evidence for the existence of true Population III stars (zero-metal abundance, first generation stars) in our Galaxy. They may only be found in an earlier generation of galaxies which formed in the denser regions of space, such as the Virgo cluster. High-speed dust ejected from supernovae in these earlier galaxies may have contaminated our Galaxy with metals before any stars formed in it.

162.004 **Galactic deuterium as a test for cosmological models.**
B. V. Vajner, Yu. A. Shchekinov.
Astrofizika, Tom 18, 143 - 150 (1982). In Russian. English translation in Astrophysics, Vol. 18, No. 1.
The influence of galactic evolution on the deuterium abundance is considered. The authors show that observable interstellar deuterium is cosmological. This conclusion is independent of the rate of accretion of intergalactic gas by the Galaxy.

162.005 **Multipole moments in the Rubin-Ford data.**
M. Clutton-Brock.
Astron. J., Vol. 87, 1127 - 1131 (1982).
The Rubin-Ford sample of galaxies appears to be moving relative to the microwave background with a speed of perhaps 800 km s^{-1}. Clutton-Brock and Peebles have suggested that this could be produced by accelerations due to density fluctuations corresponding to the observed galaxy correlation function ξ (r); if so, then one should expect quadrupole and octupole moments as well as the dipole moment found by Rubin et al. Statistical analysis reveals that both quadrupole and octupole moments are large. The quadrupole moment is spurious, being contaminated by the effects of interstellar absorption, but the octupole moment is probably real.

162.006 **Dust-like stages in the early universe and limits on the primordial black hole spectrum.**
A. G. Polnarev, M. Yu. Khlopov.
Astron. Zh., Tom 59, 639 - 646 (1982). In Russian. English translation in Soviet Astron., Vol. 25, No. 4.
The account for possible early dust-like stages in the Universe is shown to result in the weakening of the limits on the fraction of the matter that went into primordial black holes (PBH). Restrictions on the parameters of grand unified theories are given with an account for such softening. The conclusion is made that the inhomogeneity of the early Universe on small scales may be compatible with modern observational limits on the PBH spectrum.

162.007 **"Stiff" universe and the mass spectrum of primordial black holes.**
N. A. Zabotin, P. D. Nasel'skij.
Astrofizika, Tom 18, 311 - 323 (1982). In Russian. English translation in Astrophysics, Vol. 18, No. 2.
In the paper the mass spectrum of primordial black holes (PBH) created in the universe with the stiff equation of state of matter is calculated. It is shown that the existence of small-scale inhomogeneities leads to the renormalization of the effective speed of sound in the matter affecting the dynamics of the development of the large-scale modes. The main object of investigation is the field of the initial perturbations having Gaussian distribution and the dispersion M^{-n}. It is shown that in the case of a flat spectrum of the initial distortions of the metric the distribution function of PBH in mass has a sharp maximum. The influence of matter accretion on the mass spectrum of creating objects is estimated.

162.008 **Correlation dynamics in an expanding Newtonian cosmology: the derivation of an exact equation.**
H. E. Kandrup.
Astrophys. J., Vol. 259, 1 - 8 (1982).
This paper presents a new, exact equation appropriate for the discussion of correlations and irregularities in an expanding universe. The starting point of the analysis is an N-particle Liouville equation for the evolution of an N-particle distribution function, reformulated in a suitably defined "average comoving" frame.

162.009 **Closed-form expressions for the rate of growth of adiabatic perturbations.**
D. J. Heath.
Astrophys. J., Vol. 259, 9 - 19 (1982).
A second order differential equation governing the growth of adiabatic perturbations in the radiation era is derived from the equations of relativistic hydrodynamics. Exact expressions are presented for the rate of growth of infinitely long wavelength perturbations. In the special case of an Einstein–de Sitter universe containing non-interacting radiation and matter, an arbitrary wavelength solution is presented.

162.010 **Improved standard cosmology: comparison with observation.**
P. S. Wesson.
Astrophys. J., Vol. 259, 20 - 24 (1982).
A cosmological model describing inhomogeneous clusters of galaxies embedded in a homogeneous background is compared to observation. In this model, a cluster is described as a spherically symmetric distribution of matter with an inverse-square density law and an isothermal equation of state, while the background universe is essentially the Einstein/de Sitter one of standard cosmology, but with a small pressure. The model is found to be in overall good agreement with observation, and its adjustable parameters are assigned numerical values.

162.011 **Thermal interaction between matter and radiation in the early universe.**
G. A. Reeves, P. T. Landsberg.
Astrophys. J., Vol. 259, 25 - 29 (1982).
The authors make a comparison between the classical Thomson scattering and the quantum-mechanical Klein-

Nishina approach for the scattering of electrons and photons. They compute for the two cases the rate of momentum transfer to the electrons by the radiation. The Thomson scattering is found to be in good agreement with the Klein-Nishina formula for radiation temperatures up to $\sim 10^7$ K. But at higher temperatures the Klein-Nishina scattering predicts a lower transfer rate than that predicted by the Thomson case, and at radiation temperatures $\sim 10^{12}$ K the reduction is of the order of 5×10^{-3}. Many cosmological studies begin at temperatures of this order, so the Klein-Nishina approach ought to be used in this situation.

162.012 **Correlations between the virial parameters of groups of galaxies.**
M. Mezzetti, G. Giuricin, F. Mardirossian.
Astrophys. J., Vol. 259, 30 - 37 (1982).
By using two samples of groups of galaxies, revised and selected by Rood and Dickel, the authors have investigated the significance of the correlations between the virial parameters of galaxy groups. It is shown that there is strong evidence for the existence of a significant and intrinsic correlation between the total luminosity and the velocity dispersion in groups of galaxies. Moreover, there is no serious ground for claiming that the virial radius correlates with both velocity dispersion and total luminosity.

162.013 **Entropy in an expanding universe.**
S. Frautschi.
Science, Vol. 217, 593 - 599 (1982).
The question of how the observed evolution of organized structures from initial chaos in the expanding universe can be reconciled with the laws of statistical mechanics is studied, with emphasis on effects of the expansion and gravity.

162.014 **Galaxy formation and phonons.**
G. V. Chibisov, V. F. Mukhanov.
Mon. Not. R. Astron. Soc., Vol. 200, 535 - 550 (1982).
The authors investigate the possibility that the spectrum of pregalactic inhomogeneities results from quantum density fluctuations in the early Universe. A hydrodynamical description of matter is used, which is reasonable for galactic scales. As an example the authors present a model in which the conditions needed for galaxy formation may be realized.

162.015 **Formation of elliptical and spiral galaxies in rotating haloes.**
A. Kashlinsky.
Mon. Not. R. Astron. Soc., Vol. 200, 585 - 603 (1982).
In this paper it was shown that formation of both ellipticals and spirals is possible within the frame of White & Rees' model for galaxy formation if one includes effects due to the rotation of the halo. Ellipticals according to the presented model would form in haloes with either: (1) low angular momentum. (2) high ratio of the gas-to-halo masses. (3) loosely centrally condensed haloes. In other haloes spirals with spherical bulges would form. The model seems to be in agreement with the recent measurements of the rotation of bulges (Kormendy & Illingworth 1981).

162.016 **Some cosmological implications of Dirac's large number hypothesis.**
W. Davidson.
Mon. Not. R. Astron. Soc., Vol. 200, 645 - 660 (1982).
An attempt is made to relate observational data to Dirac cosmology. Some problems are reviewed using the criteria, assuming Dirac's large numbers hypothesis, that there are two metric scales applicable in this theory: (1) the atomic or A scale in which conventional atomic and Maxwellian electromagnetic theory holds, (2) the Einstein or E scale in which conventional Einsteinian gravitational theory holds. If Dirac's hypothesis is correct several consequences follow, featuring Hubble's constant, the age of the Universe, the cosmic mas density, the deceleration parameter and the evolutionary characteristics of the cosmic medium including the formati of galaxies.

162.017 **On the time evolution of the cosmological reds**
S. Phillipps.
Astrophys. Lett., Vol. 22, 123 - 125 (1982).
It is shown that peculiar velocities will not have an important effect on attempts to determine q_0 from the e tion of cosmological redshifts, but that changes in peculia velocity will probably negate the advantages of this meth

162.018 **Resolved quasars: a test for the expansion of the Universe.**
S. Phillipps.
Astrophys. Lett., Vol. 22, 153 - 156 (1982).
A long standing proposal to use surface brightness as a test of the expansion of the Universe is reconsidered in the light of recent observations of diffuse images around distan quasars. The present data points toward the standard cosm logical models rather than "tired light" models, but is not definitive due to the effects of atmospheric seeing. Use of t Space Telescope to observe very distant quasars for such extended images should provide an unequivocal answer. Ev the present data poses severe problems for the "non-cosmo ical" hypothesis of quasar redshifts.

162.019 **Are we falling into the Virgo cluster?**
N. Visvanathan.
Proc. Astron. Soc. Australia, Vol. 4, 172 - 177 (1981).
The author presents the values of the Hubble constant derived from the data of groups and clusters distributed in t redshift range 200 to 7000 km/sec. An infall velocity of the Local Group of 500 km/sec is expected to affect the global value of the Hubble constant by a large amount in the nearl regions and by a small amount in the farther regions. The author's aim is to derive a global value of the Hubble const as well as a value for the infall motion of the Local Group towards Virgo from the Hubble constant-distance relation. The existing claims for the infall velocity vary from 0-480 km/sec and are controversial.

162.020 **The effect of neutrinos on the formation time of rich clusters of galaxies.**
B. M. Lewis.
Proc. Astron. Soc. Australia, Vol. 4, 181 - 182 (1981).

162.021 **Is the Universe rotating?**
P. Birch.
Nature, Vol. 298, 451 - 454 (1982).
From the study of the position angles and polarization of high luminosity classical-double radio sources, it appears that the difference between the position angles of elongatio and of polarization are highly organized, being generally positive in one half of the sky and negative in the other. Su a phenomenon can only have a physical explanation on a cosmic scale; an attractive theory is that it demonstrates the existence of a universal vorticity, that is, that the Universe i rotating with an angular velocity $\sim 10^{-13}$ rad yr^{-1}.

162.022 **Spontaneous generating of density perturbations the early Universe.**
M. Crawford, D. N. Schramm.
Nature, Vol. 298, 538 - 540 (1982).
The authors present a mechanism for the spontaneou creation of density perturbations during the transition in t early Universe from unconfined quark matter to hadronic matter.

162.023 **Is our vacuum metastable?**
M. S. Turner, F. Wilczek.

Nature, Vol. 298, 633 - 634 (1982).

What does determine the present vacuum state of the Universe? The authors argue that it is determined cosmologically by the dynamical evolution of the Universe from the hot, dense phase which existed shortly after the big bang to the present.

162.024 **A pregalactic Population III: primordial helium and heavy elements and the microwave background.**
P. W. Tarbet, M. Rowan-Robinson.
Nature, Vol. 298, 711 - 715 (1982).

The helium, heavy elements and dark remnants produced by a pregalactic generation of stars, Population III, are discussed quantitatively for three scenarios: (1) Population III is responsible for the distortion of the microwave background spectrum measured by Woody and Richards; (2) Population III objects generate the whole of the microwave background; (3) Population III provides the oldest metals found in our Galaxy today.

162.025 **The inflationary Universe – birth, death and transfiguration.**
J. D. Barrow, M. S. Turner.
Nature, Vol. 298, 801 - 805 (1982).

A number of exciting new experimental and theoretical developments have recently made a dramatic impact on the interdisciplinary study of cosmology and elementary particle physics. The Nuffield Foundation recently sponsored a 3-week workshop in Cambridge where theoreticians worked intensively on some of these problems. A brief review is given of the progress that was made as a result of the workshop, something of the background to the problems tackled, and the likely course of future developments.

162.026 **Phase transitions and dynamics of the Universe.**
V. Petrosian.
Nature, Vol. 298, 805 - 808 (1982).

The restoration of symmetry at grand unification in a closed contracting Robertson-Walker universe could slow down and halt the contraction, causing the universe to bounce and avoid the singular state or the big crunch. During subsequent expansion the symmetry is broken and the universe reverts back to its present (normal) state. A scenario is presented whereby the present entropy of the substance is created through a series of such bounces along with three possible initial states. The monopole, horizon and flatness difficulties faced in ordinary big bang models are also solved by the model presented.

162.027 **Galaxy formation by dissipationless particles heavier than neutrinos.**
G. R. Blumenthal, H. Pagels, J. R. Primack.
Nature, Vol. 299, 37 - 38 (1982).

The authors suggest that the gravitino dominated universe can produce galaxies by gravitational instability while avoiding several observational difficulties associated with the neutrino dominated universe.

162.028 **In search of the Hubble parameter.**
S. van den Bergh.
Nature, Vol. 299, 297 - 298 (1982).

162.029 **Five crucial tests of the cosmic distance scale using the Galaxy as fundamental standard.**
G. de Vaucouleurs.
Nature, Vol. 299, 303 - 307 (1982).

Five crucial tests of the 'long' and 'short' extragalactic distance scales based on a new approach using our Galaxy as the fundamental calibrator are presented. All confirm the short scale within 0.1–0.2 mag and the corresponding value of the Hubble constant $H_0 = 95 \pm 10$ km s^{-1} Mpc^{-1}. All reject the long scale ($H_0' = 50$ km s^{-1} Mpc) at very high significance levels.

162.030 **Primordial synthesis of ^{6}Li and the rate of $\alpha(d, \gamma)^6$Li.**
J. A. Morgan.
Mon. Not. R. Astron. Soc., Vol. 200, 925 - 932 (1982).

Network calculations of primordial nucleosynthesis show that the yield of ^{6}Li is very sensitive to the cross-section for $\alpha(d, \gamma)^6$Li in the vicinity of $E_{cm} < 240$ keV. A calculation of the cross-section for this reaction, primarily a direct radiative capture at astrophysical energies, yields a value of 10 times smaller than an earlier estimate. The dependence of the cross-section upon uncertainties in experimentally determined quantities used in its calculation is found to be insufficient to allow significant cosmological ^{6}Li production.

162.031 **The cosmic asymmetry between matter and antimatter.**
F. Wilczek.
Usp. fiz. nauk, Tom 136, 149 - 165 (1982). In Russian.
Abstr. in Ref. zh., 51. Astron., 7.51.910 (1982).

162.032 **Stochastic density fluctuations in the radiation era of the Universe.**
E. Nowotny, M. G. Corona.
Astrophys. Space Sci., Vol. 84, 401 - 407 (1982).

The authors solve the linear stochastic hydrodynamical equations of a radiation fluid in the Universe before the recombination era, including the fluctuating forces due to the viscous effects.

162.033 **The isotropy of the universe.**
J. D. Barrow.
Q. J. R. Astron. Soc., Vol. 23, 344 - 357 (1982).

The author explains and discusses the interpretation of a theorem regarding the stability of an isotropically expanding universe. It is argued that the Anthropic Principle cannot explain the present observations of high isotropy and proximity to spatial flatness in the way claimed by earlier commentators.

162.034 **Determinations of the deceleration parameter q_0.**
L. Z. Fang, T. Kiang, F. H. Cheng, F. X. Hu.
Q. J. R. Astron. Soc., Vol. 23, 363 - 387 (1982).

162.035 **Approach to scientific cosmology.**
O. Godart, M. Heller.
Astron. Q., Vol. 4, 53 - 63 (1982).

162.036 **Anfang und Ende: Evolution des Weltalls.**
J. O. Stenflo.
Orion, 40. Jahrg., 118 - 120 (1982).

162.037 **Estimation of the deceleration parameter q_0 using quasar data.**
F. H. Cheng.
Irish Astron. J., Vol. 15, (see 012.008), 36 - 41 (1981).

162.038 **The cosmic density wave and its observable vestige.**
Y.-z. Liu.
Astron. Astrophys., Vol. 113, 192 - 198 (1982).

The idea of a cosmic density wave (CD wave) based on the gravitational instability in the expanding universe was suggested by Liu and Cao (1981) in an earlier paper. According to the theory, a series of peaks in the z_{em}-distribution of the observed quasars can be interpreted as a vestige of the CD wave. In this paper the author estimates the coherence length of the CD wave-train, and discusses the maximum scale of the observable inhomogeneities in the large-scale distribution of galaxies and the other observable vestige of the CD wave which appears in the z_{ab}-distribution of the absorption redshift systems of quasars. The observational data now available are in agreement with the theoretical predictions.

162.039 **Evolution of a homogeneous universe with post-hydrodynamical energy-momentum tensor of a ultrarelativistic medium.**
Yu. S. Grishkan.
Red. zh. Izv. vuzov, ser. Fiz. Tomsk, 1982. 19 pp. In Russian. Abstr. in Ref. zh., 51. Astron., 8.51.859 (1982).

162.040 **On a class of cosmological models with rotation.**
V. G. Agakov, E. A. Grigor'ev.
Chuvash. univ. Cheboksary, 1981. 9 pp. In Russian. – Abstr. in Ref. zh., 51. Astron., 8.51.860 (1982).

162.041 **Matter and antimatter in the universe.**
A. D. Dolgov, Ya. B. Zel'dovich.
Priroda, 1982, No. 8, p. 33 - 45. In Russian.

162.042 **Cosmological models with a quasi-stable de Sitter stage.**
V. Ts. Gurovich.
Pis'ma Astron. Zh., Tom 8, 538 - 541 (1982). In Russian. English translation in Soviet Astron. Lett., Vol. 8.

The account of vacuum polarization of both nonconformal and conformal quantum fields makes the de Sitter stage quasi-stable. Transition to the scalaron stage can be triggered by inhomogeneous above-barrier perturbations. Homogeneous transition is also possible due to particle creation by curvature fluctuations.

162.043 **Evolution of perturbations in an inflationary universe.**
J. A. Frieman, C. M. Will.
Astrophys. J., Vol. 259, 437 - 441 (1982).

The authors analyze the evolution of inhomogeneous density perturbations in a model of the very early universe that is dominated for a time by a constant energy density of a false quantum-mechanical vacuum. During this period, the universe inflates exponentially and supercools exponentially, until a phase transition back to the true vacuum reheats the matter and radiation. All modes of inhomogeneous perturbations in this model either are constant or are exponentially damped during the inflationary era.

162.044 **Anisotropy of the microwave background due to the mass distribution in an open cosmological model.**
P. J. E. Peebles.
Astrophys. J., Vol. 259, 442 - 448 (1982).

The perturbation to the microwave background radiation by the gravitational field of the mass distribution is computed for an open cosmological model under the assumptions that the mass autocorrelation function differs from zero only on scales much less than the Hubble length cH_0^{-1} and that pressure may be neglected.

162.045 **Time variation of the cosmological redshift in Dicke-Brans-Jordan cosmologies.**
R. Rüdiger.
Astrophys. J., Vol. 260, 33 - 36 (1982).

In this paper the time variation $\dot{z}$ of the cosmological redshift z is discussed for Dicke-Brans-Jordan (DBJ) cosmologies. For certain combinations of cosmological parameters, which are within the present observational limitations, the DBJ terms in the function for $\dot{z}$ can dominate the general relativistic terms.

162.046 **Clustering of Lyman-alpha absorption lines in quasars: like galaxies in proto-superclusters?**
A. Dekel.
Astrophys. J., Lett., Vol. 261, L13 - L17 (1982).

The puzzle posed by the lack of measurable clustering among the Lyα absorption systems along the lines of sight quasars is removed if the structure in the universe has evolv according to a nondissipative pancake scenario. This scenar suggests that galaxies formed first from small-scale perturba tions and that superclusters have collapsed only recently, anisotropically and nondissipatively, from adiabatic perturl tions. The calculated two-point correlation of galaxies resembles the observed one and is consistent with the limit that of the Lyα clouds at $z > 1.7$.

162.047 **Self-similar solutions to Einstein's equations in tl early universe.**
R. N. Henriksen, A. G. Emslie.
Bull. American Astron. Soc., Vol. 14, 630 (1982). – Abstra

162.048 **Growth of anisotropy in cosmology.**
F. Occhionero, P. Carnevali, P. Santangelo.
Bull. American Astron. Soc., Vol. 14, 630 - 631 (1982). Abstract.

162.049 **Three-dimensional structure of the universe and regions devoid of galaxies.**
A. G. Doroshkevich, S. F. Shandarin, Ya. B. Zeldovich *(Zel'dovich).*
Comments Astrophys., Vol. 9, 265 - 273 (1982).

162.050 **Matter and radiation models in the standard universe.**
M. Heller.
Acta Cosmologica, Zesz. 11, (see 012.012), 11 - 25 (1982).

The transition from the dust era to the radiation era (backwards in time) is studied in terms of precise, formal models for the dust and radiation within the standard cosmological model.

162.051 **Pregalactic star formation and spectral distortion the cosmic blackbody radiation.**
J. Silk.
Acta Cosmologica, Zesz. 11, (see 012.012), 89 - 99 (1982).

Fragmentation of collapsing spherical gas clouds tends be inhibited on fluctuation scales where pressure support is initially important. It is shown that this conclusion is radica altered when realistic deviations from sphericity are taken into account. In particular, fragmentation is found to be an inevitable consequence if the clouds initially are even moderately flattened. The theory of opacity-limited fragmentation of primordial gas clouds is reviewed, and it is concluded that the mass spectrum of primordial stars is like to extend down to $\sim 1\ M_\odot$, with a characteristic mass of $\sim 10\ M_\odot$. The observability of dust produced by an early generation of stars is discussed, with regard to producing distortions of the cosmic microwave background radiation spectrum. Results are illustrated for several types of grain material, that produce observable distortions if a mass fraction $Z \sim 10^{-5}$ is present in grains at a redshift $z \sim 200$ and i there is a substantial input of stellar radiation to raise the grain temperature above that of the background radiation.

162.052 **Dissipational galaxy formation.**
J. Silk.
Acta Cosmologica, Zesz. 11, (see 012.012), 101 - 111 (198

Arguments are presented which suggest that the proces of galaxy formation involves both dissipational and violent relaxation processes. A protogalaxy model is developed whi consists of an aggregate of individually long-lived gas clouds and which undergoes mergers with other similar systems.

162.053 **The origin of entropy.**
B. J. Carr.
Acta Cosmologica, Zesz. 11, (see 012.012), 113 - 120 (198

162.054 **The origin of galaxies.**
B. J. Carr.
Acta Cosmologica, Zesz. 11, (see 012.012), 121 - 130 (1982).

162.055 **The anthropic principle.**
B. J. Carr.
Acta Cosmologica, Zesz. 11, (see 012.012), 143 - 151 (1982).

162.056 **A re-examination of the Sandage–Tammann extragalactic distance scale.**
D. A. Hanes.
Mon. Not. R. Astron. Soc., Vol. 201, 145 - 170 (1982).
The Sandage–Tammann chain of reasoning in the determination of the extragalactic distance scale is carefully reviewed. Various inconsistencies are identified, some of which have been previously noted by others.

162.057 **Baryon generation and unusual symmetry behaviour at high temperatures.**
V. A. Kuzmin *(Kuz'min)*, M. E. Shaposhnikov, I. I. Tkachev.
Nucl. Phys. B, Vol. B196, 29 - 44 (1982). – Abstr. in Phys. Abstr., Vol. 85, Abstr. 59072 (1982).

162.058 **On a cosmological aspect of the phase transition at the grand unification temperature.**
M. D. Pollock.
Nucl. Phys. B, Vol. B196, 461 - 474 (1982). – Abstr. in Phys. Abstr., Vol. 85, Abstr. 59074 (1982).

162.059 **Supersymmetric unification and cosmology.**
F. R. Klinkhamer.
Phys. Lett. B, Vol. 110B, 203 - 205 (1982). – Abstr. in Phys. Abstr., Vol. 85, Abstr. 59081 (1982).

162.060 **Bianchi type-II cosmological models in Brans-Dicke theory.**
A. Banerjee, N. O. Santos.
Nuovo Cimento B, Vol. 67B, Ser. 11, 31 - 40 (1982). – Abstr. in Phys. Abstr., Vol. 85, Abstr. 63725 (1982).

162.061 **Time-dependent, Bianchi II, rotating Universe.**
M. J. Reboucas.
Nuovo Cimento B, Vol. 67B, Ser. 11, 120 - 126 (1982). Abstr. in Phys. Abstr., Vol. 85, Abstr. 63726 (1982).

162.062 **The cosmological constant and the gauge-covariant theory of gravitation.**
J. Cervero, P. G. Estevez.
Nuovo Cimento B, Vol. 67B, Ser. 11, 202 - 212 (1982). Abstr. in Phys. Abstr., Vol. 85, Abstr. 63727 (1982).

162.063 **Pairing in the cosmic neutrino background.**
V. Alonso, R. Chela-Flores, R. Paredes.
Nuovo Cimento B, Vol. 67B, Ser. 11, 213 - 222 (1982). Abstr. in Phys. Abstr., Vol. 85, Abstr. 63728 (1982).

162.064 **Smoothing regularization and particle creation in curved spacetime.**
A. Kharkov *(Khar'kov)*.
Phys. Lett. A, Vol. 88A, 109 - 112 (1982). – Abstr. in Phys. Abstr., Vol. 85, Abstr. 63729 (1982).

162.065 **Is the Universe really expanding?.**
R. A. Waldron.
Speculations Sci. Technol., Vol. 4, 539 - 543 (1981). – Abstr. in Phys. Abstr., Vol. 85, Abstr. 63730 (1982).

162.066 **Phase transitions and the universe.**
A. M. Polyakov.
Usp. fiz. nauk, Tom 136, 538 - 539 (1982). In Russian. Abstr. in Ref. zh., 51. Astron., 9.51.724 (1982).

162.067 **Search for neutral hydrogen in the early universe.**
I. F. Mirabel.
Rev. Mexicana Astron. Astrofis., Vol. 6, (see 012.017), 17 (1981).
A report on the progress of searches for neutral hydrogen emission in the early universe is given.

162.068 **Spontaneous breaking of CP in SU(5), natural flavor conservation, and the baryon asymmetry.**
J. F. Nieves.
Phys. Rev. D, Vol. 25, 1417 - 1424 (1982). – Abstr. in Phys. Abstr., Vol. 85, Abstr. 64258 (1982).

162.069 **Explicit solutions of □u=0 on the Friedmann-Robertson-Walker space-times.**
S. Klainerman, P. Sarnak.
Ann. Inst. Henri Poincaré, Sect. A, Vol. 35, 253 - 257 (1981). Abstr. in Phys. Abstr., Vol. 85, Abstr. 68938 (1982).

162.070 **Some anisotropic non-static perfect fluid cosmological models in general relativity.**
S. R. Roy, O. P. Tiwari.
J. Phys. A, Vol. 15, 1747 - 1755 (1982). – Abstr. in Phys. Abstr., Vol. 85, Abstr. 68939 (1982).

162.071 **On the general solution for a 'diagonal' vacuum Bianchi type III model with a cosmological constant.**
M. A. H. MacCallum, A. Moussiaux, P. Tombal, J. Demaret.
J. Phys. A, Vol. 15, 1757 - 1758 (1982). – Abstr. in Phys. Abstr., Vol. 85, Abstr. 68940 (1982).

162.072 **Time development of the free-quark-to-baryon ratio from the very early Universe.**
Suk Koo Yun.
Lett. Nuovo Cimento, Vol. 33, Ser. 2, 529 - 534 (1982). Abstr. in Phys. Abstr., Vol. 85, Abstr. 68941 (1982).

162.073 **The baryon asymmetry of the Universe and the invisible axion.**
A. Masiero, G. Segre.
Phys. Lett. B, Vol. 109B, 349 - 352 (1982). – Abstr. in Phys. Abstr., Vol. 85, Abstr. 68942 (1982).

162.074 **Cosmological baryon generation in the minimal Wise, Georgi and Glashow model.**
A. Masiero, T. Yanagida.
Phys. Lett. B, Vol. 109B, 353 - 356 (1982). – Abstr. in Phys. Abstr., Vol. 85, Abstr. 68943 (1982).

162.075 **Composite models and the Universe's baryon asymmetry.**
A. Masiero, R. D. Peccei, T. Yanagida.
Phys. Lett. B, Vol. 109B, 357 - 360 (1982). – Abstr. in Phys. Abstr., Vol. 85, Abstr. 68944 (1982).

162.076 **Super-cosmology.**
D. V. Nanopoulos, K. Tamvakis.
Phys. Lett. B, Vol. 110B, 449 - 455 (1982). – Abstr. in Phys. Abstr., Vol. 85, Abstr. 68945 (1982).

162.077 **The contribution to baryon excess from gravitational interaction after the Planck time.**
S. A. Bonometto, S. Matarrese.
Phys. Lett. B, Vol. 110B, 456 - 460 (1982). – Abstr. in Phys. Abstr., Vol. 85, Abstr. 68946 (1982).

162.078 **A model for neutrino galaxies.**
G. Eder.
Acta Phys. Austriaca, Vol. 54, 111 - 121 (1982). – Abstr. in Phys. Abstr., Vol. 85, Abstr. 74076 (1982).

162.079 **Correspondence between the Brans-Dicke theory and general relativity in a closed cosmological model.**
A. Miyazaki.
Nuovo Cimento B, Vol. 68B, Ser. 11, 126 - 132 (1982).
Abstr. in Phys. Abstr., Vol. 85, Abstr. 74132 (1982).

162.080 **The Hartree approximation as a high-density limit.**
E. Alvarez.
Phys. Lett. B, Vol. 110B, 315 - 318 (1982). – Abstr. in Phys. Abstr., Vol. 85, Abstr. 74134 (1982).

162.081 **Origin of the Universe as a quantum tunneling event.**
D. Atkatz, H. Pagels.
Phys. Rev. D, Vol. 25, 2065 - 2073 (1982). – Abstr. in Phys. Abstr., Vol. 85, Abstr. 74135 (1982).

162.082 **Comments on the coherent-state representation of a scalar field in the early Universe.**
B. K. Berger.
Phys. Rev. D, Vol. 25, 2208 - 2210 (1982). – Abstr. in Phys. Abstr., Vol. 85, Abstr. 74136 (1982).

162.083 **The energy spectrum of the universe.**
G. Burbidge.
Ninth Texas Symposium on Relativistic Astrophysics, (see 012.020), p. 125 - 129 (1980).

162.084 **Limits to verification in cosmology.**
G. F. R. Ellis.
Ninth Texas Symposium on Relativistic Astrophysics, (see 012.020), p. 130 - 160 (1980).
The author discusses the question: How can we observationally prove the universe is well-described by a Friedmann-Robertson-Walker universe model?

162.085 **Statistics of the distribution of galaxies.**
P. J. E. Peebles.
Ninth Texas Symposium on Relativistic Astrophysics, (see 012.020), p. 161 - 171 (1980).
The author discusses statistical measures of galaxy clustering that are provided by the galaxy two- and three-point correlation functions.

162.086 **Gravitational instability theory of galaxy formation and clustering: some recent developments.**
S. M. Fall.
Ninth Texas Symposium on Relativistic Astrophysics, (see 012.020), p. 172 - 190 (1980).
The purpose of this review is to discuss some recent developments in the gravitational instability theory of galaxy formation and clustering, including a confrontation with observational data.

162.087 **Theories of galaxy formation.**
B. J. T. Jones.
Ninth Texas Symposium on Relativistic Astrophysics, (see 012.020), p. 191 - 205 (1980).
The gravitational instability theory argues that the force of gravitation has been primarily responsible for the nature of the large-scale organization of matter in the universe. Other theories, which are discussed in this review, invoke turbulence, magnetic fields, or even matter–antimatter annihilation in the early universe.

162.088 **Summary of workshop on galaxy formation.**
J. Silk.
Ninth Texas Symposium on Relativistic Astrophysics, (see 012.020), p. 206 - 210 (1980).

162.089 **Summary of workshop on Hubble diagram and expansion of the universe.**
G. A. Tammann.
Ninth Texas Symposium on Relativistic Astrophysics, (see 012.020), p. 598 - 599 (1980).

162.090 **Cosmological impact of the neutrino rest mass.**
A. G. Doroshkevich, M. Yu. Khlopov, R. A. Sunyaev (*Syunyaev*), A. S. Szalay, Ya. B. Zeldovich (*Zel'dovich*).
Tenth Texas Symposium on Relativistic Astrophysics, (see 012.021), p. 32 - 42 (1981).

162.091 **The early universe and clustering of the relic neutrinos.**
H. Sato.
Tenth Texas Symposium on Relativistic Astrophysics, (see 012.021), p. 43 - 53 (1981).
The astronomical consequences of a neutrino mass on the order of 10 eV are discussed. According to the big bang cosmology, the neutrino blackbody radiation was excited in the early universe.These neutrinos may contribute to the average density of the universe and they may also be found to be the main source of gravity in astronomical objects like galactic clusters.

162.092 **Particle physics in the very early universe.**
D. N. Schramm.
Tenth Texas Symposium on Relativistic Astrophysics, (see 012.021), p. 54 - 68 (1981).
The author reviews various events in the very early big bang universe where elementary particle physics effects may have been dominant. Two specific cases are developed in some details: (1) the generation of a net baryon number via grand unification and (2) the possible role of massive neutrinos in our understanding of various cosmological problems and the constraints on neutrino properties from cosmology.

162.093 **Grand unification and possible matter-antimatter domain structure in the universe.**
F. W. Stecker.
Tenth Texas Symposium on Relativistic Astrophysics, (see 012.021), p. 69 - 89 (1981).
This review gives some basic ideas and highlights of a cosmology in which matter and antimatter play an equal role in the universe.

162.094 **The cosmological distance scale: a comparison of two approaches to the Hubble constant.**
G. de Vaucouleurs.
Tenth Texas Symposium on Relativistic Astrophysics, (see 012.021), p. 90 - 122 (1981).
The author contends that a close approximation to the asymptotic value of the Hubble constant, derived from a multiplicity of indicators and calibration methods, is $H_0 = (100 \pm 10)$ km s^{-1} Mpc^{-1}, and that the value $H_0 = 50$, widely accepted in recent years, must be the result of an accumulation of systematic errors.

162.095 **Do we really know how to measure the distances of extragalactic objects?**
G. Burbidge.
Tenth Texas Symposium on Relativistic Astrophysics, (see 012.021), p. 123 - 156 (1981).
The author critically examines the observational base of the Hubble relation for distant extragalactic objects. Observations that may indicate noncosmological redshifts are presented. In particular, Arp's examples of galaxy-quasar pairs and aligned quasars are summarized. Two comprehensive tables of QSOs close to galaxies are included.

162.096 **The mass of the universe.**
P. J. E. Peebles.

Tenth Texas Symposium on Relativistic Astrophysics, (see 012.021), p. 157 - 168 (1981).

162.097 **The density of the universe from the deceleration of nearby galaxies.**
A. Yahil.
Tenth Texas Symposium on Relativistic Astrophysics, (see 012.021), p. 169 - 177 (1981).

A local cosmological test can determine the world model by relating the density excess in the Virgo supercluster to the peculiar velocity field, which has been generated over the age of the universe by the gravitational deceleration due to this density perturbation. It is tentatively concluded that the cosmological density factor is in the range of $0.1 \lesssim \Omega_0 \lesssim 0.2$, and the universe is open.

162.098 **Summary of the workshop on the early universe.**
S. A. Bludman.
Tenth Texas Symposium on Relativistic Astrophysics, (see 012.021), p. 419 - 427 (1981).

162.099 **Pregalactic very massive objects and their cosmological consequences.**
B. J. Carr, W. D. Arnett, J. R. Bond.
Supernovae: a survey of current research, (see 012.022), p. 313 - 318 (1982).

The authors present arguments for believing that very massive objects may have formed in the pregalactic era with important cosmological consequences for the problem of "missing mass", cosmic element abundances, the 3K microwave background and galaxy formation.

162.100 **Supernovae and the formation of galaxies.**
J. P. Ostriker.
Supernovae: a survey of current research, (see 012.022), p. 565 - 570 (1982).

The author considers the cumulative effect of many supernovae in a young galaxy, and the effect of the blastwave, formed by the combination of all individual supernovae, on the intergalactic medium. In late stages of an individual expanding galaxy shock a dense shell can form with mass many times greater than the mass of the seed galaxy. This shell may fragment due to gravitational instability and give rise to a new generation of galaxies.

162.101 **How large is the Universe?**
Nature, Vol. 300, 11 (1982).

Estimates of the Hubble constant have fluctuated frequently in recent decades, usually between 50 and 100 km s^{-1} per megaparsec. There is still much to be done to fix the value.

162.102 **Inhomogeneous cosmological models containing homogeneous inner hypersurface geometry. Changes of the Bianchi type.**
H.-J. Schmidt.
Astron. Nachr., Band 303, 227 - 230 (1982).

There are investigated such cosmological models which instead of the usual spatial homogeneity property fulfil the condition only that in a certain synchronized system of reference all spacelike sections t = const. are homogeneous manifolds. This allows time-dependent changes of the Bianchi type. Discussing differential-geometrical theorems it is shown which of them are permitted. Besides the trivial case of changing into type I there exist some possible changes between other types. However, physical reasons like energy inequalities partially exclude them.

162.103 **Particle physics in the early universe.**
L. Van Hove.
Evolution in the Universe, (see 012.024), p. 53 - 84 (1982).

The author's aim was to review in its most important aspects the role played by particle physics in the early expansion of the universe, and to point out the main problem areas. He stressed the great importance of neutrinos for the understanding of the universe, including its present state, and explained how modern particle physics has thrown a completely new light on the first millisecond of the expansion of the universe, which must have been marked by an amazing evolution in the composition of matter and its laws of interaction.

162.104 **The evolution of large-scale structures in the Universe.**
J. H. Oort.
Evolution in the Universe, (see 012.024), p. 85 - 104 (1982).

The inner parts of the rich and dense clusters of galaxies are usually regular in structure, with a smoothly decreasing density outward from the centre. Often the outermost regions extend over large distances in a layer, or sometimes a row, of galaxies, connecting with clusters lying tens of Mpc away. The superstructures thus formed have been called superclusters. It may be that at this point we are breaking away from the simple hierarchy of galaxy clusters and that a new element of constructions appears, characterized by very extended – up to 100 or 150 Mpc – structures that may be flat or elongated. These structures, and their origin and evolution, formed the main subject of the author's talk.

162.105 **The cosmology of the unitary field theory with gravo-electric induction.**
E. Kreisel.
Ann. Physik, Band 39, 112 - 116 (1982). In German. – Abstr. in Phys. Abstr., Vol. 85, Abstr. 79250 (1982).

162.106 **The invariant equations under finite representations of the de Sitter group.**
C. Gauthier.
Nuovo Cimento B, Vol. 68B, Ser. 11, 277 - 286 (1982). In French. – Abstr. in Phys. Abstr., Vol. 85, Abstr. 79257 (1982).

162.107 **Can the curvature effects be neglected in the early Universe?**
T. Padmanabhan, M. M. Vasanthi.
Phys. Lett. A, Vol. 89A, 327 - 328 (1982). – Abstr. in Phys. Abstr., Vol. 85, Abstr. 79258 (1982).

162.108 **Some considerations on the age and density of the Universe.**
J. L. Sanz.
Phys. Rev. D, Vol. 25, 1474 - 1480 (1982). – Abstr. in Phys. Abstr., Vol. 85, Abstr. 79260 (1982).

162.109 **Massive neutrinos and primordial nucleosynthesis.**
E. W. Kolb, R. J. Scherrer.
Phys. Rev. D, Vol. 25, 1481 - 1491 (1982). – Abstr. in Phys. Abstr., Vol. 85, Abstr. 79261 (1982).

162.110 **Dynamical symmetry breaking due to radiative corrections in cosmology.**
L. H. Ford, D. J. Toms.
Phys. Rev. D, Vol. 25, 1510 - 1518 (1982). – Abstr. in Phys. Abstr., Vol. 85, Abstr. 79262 (1982).

162.111 **Monopoles and the early Universe.**
E. J. Weinberg.
AIP Conf. Proc., No. 81, (see 012.026), p. 538 - 547 (1981). Abstr. in Phys. Abstr., Vol. 85, Abstr. 79790 (1982).

162.112 **Cosmological consequences of a hierarchical supersymmetric model.**
S.-Y. Pi.
Phys. Lett. B, Vol. 112B, 441 - 445 (1982). – Abstr. in Phys. Abstr., Vol. 85, Abstr. 79798 (1982).

162.113 **Viscous Universe solutions with p ~ ρ and $\eta \sim n^{1/3}$.**
B. Lukacs.
Acta Phys. Acad. Sci. Hungaricae, Vol. 51, 117 - 123 (1981). Abstr. in Phys. Abstr. Vol. 85, Abstr. 84353 (1982).

162.114 **Cosmology and neutrino physics.**
G. Steigman.
AIP Conf. Proc., No. 81, (see 012.026), p. 548 - 571 (1981). Abstr. in Phys. Abstr., Vol. 85, Abstr. 84356 (1982).

162.115 **Cosmological particle creation. Single vacuum propagators.**
C. Charach.
Prog. Theor. Phys., Vol. 67, 693 - 696 (1982). – Abstr. in Phys. Abstr., Vol. 85, Abstr. 84357 (1982).

162.116 **Particle creations by the expanding Universe and suppression of supercooling.**
M. Horibe, A. Hosoya.
Prog. Theor. Phys., Vol. 67, 816 - 827 (1982). – Abstr. in Phys. Abstr., Vol. 85, Abstr. 84358 (1982).

162.117 **On the quark-hadron transition in the early Universe.**
C. G. Kallman.
Phys. Lett. B, Vol. 112B, 213 - 214 (1982). – Abstr. in Phys. Abstr., Vol. 85, Abstr. 84362 (1982).

162.118 **The early Universe in a generalized theory of gravitation.**
J. W. Moffat, D. Vincent.
Canadian J. Phys., Vol. 60, 659 - 663 (1982). – Abstr. in Phys. Abstr., Vol. 85, Abstr. 89419 (1982).

162.119 **Hubble's law as neutrino-Compton effect.**
P. K. Biswas.
Fizika, Vol. 14, No. 2, p. 69 - 75 (1982). – Abstr. in Phys. Abstr., Vol. 85, Abstr. 89420 (1982).

162.120 **Cosmological baryon production versus intermediate mass scales?**
A. Masiero, T. Yanagida.
Phys. Lett. B, Vol. 112B, 336 - 340 (1982). – Abstr. in Phys. Abstr., Vol. 85, Abstr. 89426 (1982).

162.121 **The universe: past and present reflections.**
F. Hoyle.
Annu. Rev. Astron. Astrophys., Vol. 20, (see 003.009), 1 - 35 (1982).

162.122 **Vers une nouvelle échelle des distances extragalactiques.**
L. Gouguenheim.
Astronomie, Vol. 96, 435 - 445 (1982).

162.123 **Den ekstragalaktiske avstandskalaen og Hubbles konstant.**
T. S. Ringnes.
Astron. Tidsskr., Årg. 15, 70 - 76 (1982).

162.124 **Om inerti, gravitation og kosmologi.**
J. Teuber, P. Hjorth.
Astron. Tidsskr., Årg. 15, 116 - 121 (1982).

162.125 **The arrow of time in a bouncing universe.**
M. Clutton-Brock.
Observatory, Vol. 102, 147 - 148 (1982).

162.126 **The extragalactic distance scale and the Hubble constant.**
G. de Vaucouleurs.
Observatory, Vol. 102, 178 - 194 (1982).
The principal distance indicators and the derivation of the Hubble constant are reviewed. The recent evidence in favour of the value $H_0 = 95 \pm 10$ km/s/Mpc is discussed.

162.127 **Tilted electromagnetic Bianchi type III cosmolog solution.**
D. Lorenz.
Astrophys. Space Sci., Vol. 85, 59 - 61 (1982).
An exact cosmological solution of the Einstein-Maxwe equations for the case of Bianchi type III symmetry with "stiff" matter and an electromagnetic field is constructed. ' behaviour of the solution near the singularity and at late stages of the expansion is investigated.

162.128 **Tilted electromagnetic Bianchi type VI_0 cosmolc ical model.**
D. Lorenz.
Astrophys. Space Sci., Vol. 85, 63 - 67 (1982).
The Einstein-Maxwell equations are integrated for a tilted Bianchi type VI_0 space-time with "stiff" matter and electromagnetic field. The solution represents an anisotrop: homogeneous cosmological model which tends to isotropic expansion.

162.129 **Exact Bianchi type VI_0 cosmological models wit matter and an electromagnetic field.**
D. Lorenz.
Astrophys. Space Sci., Vol. 85, 69 - 74 (1982).
The author presents new exact solutions of the Einste: Maxwell equations for cosmological models of Bianchi type VI_0. The solutions represent anisotropic universes with sou: free electro-magnetic fields and perfect fluid matter. The be haviour near to and far from initial singularities is examinec

162.130 **Limitations of Maxwell's distribution.**
Nature, Vol. 299, 575 (1982).
A calculation of the evolution of a system of gravitatii masses (galaxies) in an expanding universe surprisingly suggests that Maxwell's velocity distribution is not an acceptable end point.

162.131 **Evolution of a homogeneous universe with post-hydrodynamical stress – energy tensor of an ultr relativistic medium.**
Yu. S. Grishkan.
Astron. Zh., Tom 59, 833 - 843 (1982). In Russian. English translation in Soviet Astron., Vol. 26, No. 5.
A covariant expression for the stress – energy tensor c an ultrarelativistic medium in post-hydrodynamical approxi tion is constructed. Exact solutions of Einstein's equations with the post-hydrodynamical stress – energy tensor are ob tained for isotropic Friedmann cosmological models and fo the anisotropic model of Bianchi I type. These equations ke asymptotically Friedmannian isotropic solutions far from tl cosmological singularity. Near the singularity the universe evolution is essentially different from the standard Friedma and Kasner models.

162.132 **Extragalactic ultraviolet astronomy and cosmolo**
M. S. Longair.
Third European IUE Conference, (see 012.029), p. 75 - 79 (1982).
A brief survey is presented of various aspects of extragalactic ultraviolet astronomy and cosmology.

162.133 **Massive neutrino halos in an expanding universe.**
R. Fabbri, R. T. Jantzen, R. Ruffini.
Astron. Astrophys., Vol. 114, 219 - 232 (1982).
The growth of massive neutrino halos around a mass condensation in an expanding universe is considered using the simplified model of an Einstein-Straus universe. This model illustrates the possibility of capturing freely moving particles in a homogeneous and isotropic cosmological gas l

the local gravitational field of a mass condensation. This capture process is particularly relevant to the neutrino background in a neutrino dominated universe and explicit formulas are given relating the maximum capture rate by a neutrino condensation to the cosmological redshift and the mass of the condensation. Analogies between self-gravitating systems of neutrinos in an asymptotically flat spacetime and the condensations which maximize the capture rate are noted and observational consequences are considered.

162.134 **Perturbation of the magnitude–redshift relation in an inhomogeneous relativistic model. II. Correction to the Hubble law behind clusters.**
L. Nottale.
Astron. Astrophys., Vol. 114, 261 - 269 (1982).
Possible perturbations to the magnitude–redshift relation by clusters of galaxies are studied in an exact solution of Einstein's equations where a cluster is represented by a section of Friedmann model of higher density than the background, from which it is separated by a Schwarzschild vacuole, the boundaries being comoving hypersurfaces. The redshift effects in this model have already been computed (Nottale, 1982). Here the magnitude effects are considered for a radial beam which, starting behind the inhomogeneity, crosses the high density region and finally reaches the observer lying in the background Friedmann model.

162.135 **Galactic neutrino models.**
Y. Rephaeli.
Astron. Astrophys., Vol. 114, 405 - 406 (1982).
It is shown that galaxies and clusters cannot be dominated by fully degenerate neutrinos. The author discusses the possibility that galactic halos are dominated by freely streaming neutrinos in elongated orbits, and calculates the effect of the implied central neutrino mass concentration on the motion of globular clusters.

162.136 **The gravitational evolution of structure in a scale-free universe.**
S. D. M. White, J. Negroponte.
Mon. Not. R. Astron. Soc., Vol. 201, 401 - 414 (1982).
The authors discuss the gravitational evolution of scale-free clustering using N-body simulations of simple hierarchical models constructed to mimic the correlation properties of the observed galaxy distribution. They find that the smooth core-halo structures formed by total or partial dissolution of a hierarchy resembling the observed galaxy distribution have density profiles which, with the limited resolution, are best represented as power laws of exponent -3. If the present large-scale distribution of matter evolved with negligible dissipation from scale-free initial conditions, the unseen dark matter must be smoothly distributed both in the halos of isolated galaxies and within the communal halos which bind compact groups and clusters of galaxies.

162.137 **Neighborhoods of Cauchy horizons in cosmological spacetimes with one Killing field.**
V. Moncrief.
Ann. Physics, Vol. 141, 83 - 103 (1982). – Abstr. in Phys. Abstr., Vol. 85, Abstr. 94870 (1982).

162.138 **Some anisotropic homogeneous models in a modified Brans-Dicke cosmology.**
T. Singh, L. N. Rai.
Acta Phys. Polonica B, Vol. B13, 3 - 16 (1982). – Abstr. in Phys. Abstr., Vol. 85, Abstr. 94871 (1982).

162.139 **A homogeneous universe with a viscous fluid.**
S. R. Maiti.
Indian J. Phys. Part B, Vol. 56B, 110 - 118 (1982). – Abstr. in Phys. Abstr., Vol. 85, Abstr. 94874 (1982).

162.140 **Comments on the cosmological-horizon problem.**
M. Calvani, S. Matarrese, M. D. Pollock.
Lett. Nuovo Cimento, Vol. 34, Ser. 2, 183 - 189 (1982). Abstr. in Phys. Abstr., Vol. 85, Abstr. 94875 (1982).

162.141 **Universe expansion according to the analytic equation of state of the statistical-bootstrap model**
107 Fiore,R.
R. Fiore, P. Fre.
Lett. Nuovo Cimento, Vol. 34, Ser. 2, 289 - 295 (1982). Abstr. in Phys. Abstr., Vol. 85, Abstr. 94876 (1982).

162.142 **On an Einstein-de Sitter universe of radiating massive photinos.**
D. W. Sciama.
Phys. Lett. B, Vol. 114B, 19 - 22 (1982). – Abstr. in Phys. Abstr., Vol. 85, Abstr. 94880 (1982).

162.143 **The rest mass of neutrinos and clustering in the early Universe.**
L. Fang, Y. Liu.
Chinese Phys., Vol. 2, 700 - 706 (1982). – Abstr. in Phys. Abstr., Vol. 85, Abstr. 101098 (1982). – See Abstr. 30.162.140.

162.144 **Was the Universe ever planar?**
N. E. Frankel.
Phys. Lett. A, Vol. 90A, 323 - 326 (1982). – Abstr. in Phys. Abstr., Vol. 85, Abstr. 101105 (1982).

162.145 **Primordial irregularities in an early Universe with vacuum phase transitions.**
K. Tomita.
Prog. Theor. Phys., Vol. 67, 1076 - 1092 (1982). – Abstr. in Phys. Abstr., Vol. 85, Abstr. 101106 (1982).

162.146 **Galaxy formation in the neutrino dominated Universe.**
K. Tomita.
Prog. Theor. Phys., Vol. 67, 1093 - 1106 (1982). – Abstr. in Phys. Abstr., Vol. 85, Abstr. 101107 (1982).

162.147 **Study of the evolutions of the Universe, stars and the solar system: personal retrospect and prospect.**
S. Hayakawa.
Prog. Theor. Phys. Suppl., No. 70, p. 1 - 10 (1981). – Abstr. in Phys. Abstr., Vol. 85, Abstr. 101108 (1982).

162.148 **Initial behavior of a quantized scalar field and the associated pair-creation in several isotropic closed and open universes.**
H. Nariai.
Prog. Theor. Phys. Suppl., No. 70, p. 301 - 321 (1981). Abstr. in Phys. Abstr., Vol. 85, Abstr. 101109 (1982).

162.149 **Calculation of cosmological baryon asymmetry in grand unified gauge models.**
J. A. Harvey, E. W. Kolb, D. B. Reiss, S. Wolfram.
Nucl. Phys. B, Part. Phys., Vol. B201, 16 - 100 (1982). Abstr. in Phys. Abstr., Vol. 85, Abstr. 101614 (1982).

162.150 **Will the Universe become supersymmetric?**
J. Ellis, C. H. Llewellyn, G. G. Ross.
Phys. Lett. B, Vol. 114B, 227 - 229 (1982). – Abstr. in Phys. Abstr., Vol. 85, Abstr. 101616 (1982).

162.151 **Further aspects of supercosmology.**
D. V. Nanopoulos, K. A. Olive, K. Tamvakis.
Phys. Lett. B, Vol. 115B, 15 - 20 (1982). – Abstr. in Phys. Abstr., Vol. 85, Abstr. 101620 (1982).

162.152 **Axion models with no domain wall problem.**
G. Lazarides, Q. Shafi.
Phys. Lett. B, Vol. 115B, 21 - 25 (1982). – Abstr. in Phys. Abstr., Vol. 85, Abstr. 101621 (1982).

162.153 **The survival of primordial color fluctuations.**
E. W. Kolb, M. S. Turner.
Phys. Lett. B, Vol. 115B, 99 - 103 (1982). – Abstr. in Phys. Abstr., Vol. 85, Abstr. 101662 (1982).

162.154 **Creation of particles by shell-focusing singularities.**
W. A. Hiscock, L. G. Williams, D. M. Eardley.
Phys. Rev. D, Vol. 26, 751 - 760 (1982). – Abstr. in Phys. Abstr., Vol. 85, Abstr. 106521 (1982).

162.155 **Neutrino halos.**
Y. Rephaeli.
Phys. Rev. D, Vol. 26, 770 - 774 (1982). – Abstr. in Phys. Abstr., Vol. 85, Abstr. 106522 (1982).

162.156 **Spin, circulation, angular momentum and cosmological large numbers.**
G. E. Dorrington.
Lett. Nuovo Cimento, Vol. 34, Ser. 2, 409 - 412 (1982). Abstr. in Phys. Abstr., Vol. 85, Abstr. 106976 (1982).

162.157 **Cosmic repulsion.**
M. Novello.
Phys. Lett. A, Vol. 90A, 347 - 348 (1982). – Abstr. in Phys. Abstr., Vol. 85, Abstr. 106980 (1982).

162.158 **Vacuum viscosity description of quantum processes in the early Universe.**
B. L. Hu.
Phys. Lett. A, Vol. 90A, 375 - 380 (1982). – Abstr. in Phys. Abstr., Vol. 85, Abstr. 106981 (1982).

162.159 **Spatial homogeneity and redshift-distance laws.**
J. F. Nicoll, I. E. Segal.
Proc. Natl. Acad. Sci. USA, Vol. 79, 3913 - 3917 (1982). Abstr. in Phys. Abstr., Vol. 85, Abstr. 106982 (1982).

162.160 **Statistical effect of interactions on particle creation in an expanding Universe.**
H. Kodama.
Prog. Theor. Phys., Vol. 67, 1415 - 1440 (1982). – Abstr. in Phys. Abstr., Vol. 85, Abstr. 106984 (1982).

162.161 **Coleman-Weinberg theory and the new inflationary universe scenario.**
A. D. Linde.
Phys. Lett. B, Vol. 114B, 431 - 435 (1982). – Abstr. in Phys. Abstr., Vol. 85, Abstr. 106985 (1982).

162.162 **Gravitational effects in Guth cosmology.**
A. Vilenkin.
Phys. Lett. B, Vol. 115B, 91 - 94 (1982). – Abstr. in Phys. Abstr., Vol. 85, Abstr. 106986 (1982).

162.163 **Thermal production of superheavy magnetic monopoles in the early Universe.**
M. S. Turner.
Phys. Lett. B, Vol. 115B, 95 - 98 (1982). – Abstr. in Phys. Abstr., Vol. 85, Abstr. 106987 (1982).

162.164 **Determination of the extragalactic distance scale.**
B. Mobasher.
Thesis Univ. Durham, England (1982). – Abstr. in Phys. Abstr., Vol. 85, Abstr. 106990 (1982).

162.165 **Supersymmetric grand unified theories and the early universe.**
M. Srednicki.
Nucl. Phys. B, Part. Phys., Vol. B202, 327 - 335 (1982). Abstr. in Phys. Abstr., Vol. 85, Abstr. 107577 (1982).

162.166 **Light singlet particles and their cosmological consequences in Witten-type supersymmetric models.**
P. Q. Hung, M. Suzuki.
Phys. Lett. B, Vol. 115B, 207 - 211 (1982). – Abstr. in Phys. Abstr., Vol. 85, Abstr. 107591 (1982).

162.167 **Static conformally flat cosmological solution of the Lyttleton-Bondi Universe.**
D. R. K. Reddy, V. U. M. Rao.
Indian J. Pure Appl. Phys., Vol. 20, 325 - 326 (1982). Abstr. in Phys. Abstr., Vol. 85, Abstr. 112619 (1982).

162.168 **Large numbers hypothesis. I. Classical formalism.**
P. J. Adams.
Int. J. Theor. Phys., Vol. 21, 607 - 632 (1982). – Abstr. in Phys. Abstr., Vol. 85, Abstr. 112620 (1982).

162.169 **Large number hypothesis and the matter-dominated Universe.**
I. Goldman.
Int. J. Theor. Phys., Vol. 21, 665 - 672 (1982). – Abstr. in Phys. Abstr., Vol. 85, Abstr. 112621 (1982).

162.170 **Scalar and neutrino fields in the Gödel Universe.**
D. A. Leahy.
Int. J. Theor. Phys., Vol. 21, 703 - 753 (1982). – Abstr. in Phys. Abstr., Vol. 85, Abstr. 112623 (1982).

162.171 **Kantowski-Sachs cosmological solution with torsion.**
D. Lorenz.
J. Phys. A, Vol. 15, 2809 - 2812 (1982). – Abstr. in Phys. Abstr., Vol. 85, Abstr. 112624 (1982).

162.172 **On the solution for a vacuum Bianchi type-III model with a cosmological constant.**
D. Lorenz.
J. Phys. A, Vol. 15, 2997 - 2999 (1982). – Abstr. in Phys. Abstr., Vol. 85, Abstr. 112625 (1982).

162.173 **Graviton creation in an inflationary universe and the grand unification scale.**
V. A. Rubakov, M. V. Sazhin, A. V. Veryaskin.
Phys. Lett. B, Vol. 115B, 189 - 192 (1982). – Abstr. in Phys. Abstr., Vol. 85, Abstr. 112628 (1982).

162.174 **The development of irregularities in a single bubble inflationary universe.**
S. W. Hawking.
Phys. Lett. B, Vol. 115, 295 - 297 (1982). – Abstr. in Phys. Abstr., Vol. 85, Abstr. 112629 (1982).

162.175 **On the large-scale variations of M/L.**
Y. Hoffman, J. Shaham, G. Shaviv.
Astrophys. J., Vol. 262, 413 - 418 (1982).

The large-scale behavior of M/L is suggested to be related to segregation between dark and luminous matter (galaxies). It is suggested that during the collapse and virialization of groups and clusters the galactic halos are tidally disrupted and a smooth background of dark matter is formed; galaxies moving through this background lose energy due to dynamical friction, which is the main process considered here.

162.176 **The four-point function in the BBGKY hierarchy.**
J. N. Fry.
Astrophys. J., Vol. 262, 424 - 431 (1982).

Higher order galaxy correlation functions are calculated in the hierarchical pattern of galaxy clustering by BBGKY (from Born, Bogoliubov, Green, Kirkwood, Yvon). The results are compared with data from the Lick and Zwicky galaxy catalogs.

162.177 **Heat death and oscillation in model universes containing interacting matter and radiation.**
P. T. Landsberg, G. A. Reeves.
Astrophys. J., Vol. 262, 432 - 441 (1982).
The general thermodynamics of oscillating and ever expanding universes is studied by means of continuum matter-radiation cosmological models and their generalizations. The consequences of various interactions between the components are traced and the connection with the irreversibility of oscillating models is discussed. In addition, a new continuum-particle model is introduced. This is suitable for a study of universes with oscillation, or indefinite expansion, and includes helium formation and the effect of the Thomson interaction between matter and radiation.

162.178 **Giant voids in the Universe.**
Ya. B. Zeldovich (*Zel'dovich*), J. Einasto, S. F. Shandarin.
Nature, Vol. 300, 407 - 413 (1982).
Recent observations indicate that most galaxies are concentrated in superclusters consisting of galaxies, and clusters of galaxies, aligned along strings. Giant volumes exist between superclusters which are almost empty of visible objects. Theories of galaxy formation predict the formation of non-spherical superclusters and giant voids. Large-scale structure changes very slowly, so the currently observed structure reflects the whole history of galaxy formation and structural evolution.

162.179 **Inhomogeneous cosmological models with flat slices generated from the Einstein-de Sitter universe.**
H.-J. Schmidt.
Astron. Nachr., Band 303, 283 - 285 (1982).
A family of cosmological models is considered which in a certain synchronized system of reference possess flat slices t = const. They are generated from the Einstein-de Sitter universe by a suitable transformation. Under physically reasonable presumptions these transformed models fulfil certain energy conditions.

162.180 **Estimation of the mean density in the Universe.**
P. Flin.
Astrophys. Space Sci., Vol. 86, 493 - 495 (1982).
The mean density of matter, as estimated from deep optical samples of galaxies, is too low to close the Universe. However, some additional considerations do not exclude such a possibility.

162.181 **The precision on the measure of q_0 using the gravitational lensing effect.**
G. Lacroix, J. Schneider.
Astron. Astrophys., Vol. 115, 54 - 60 (1982).
The authors examine whether the gravitational lensing effect could be used to measure q_0 considering the instrumental limitations and the limited precision on the relevant parameters. The problem is treated within the standard cosmological model; the mass distribution for the main deflector galaxy is chosen to follow the $r^{1/4}$ law of de Vaucouleurs; the instrumental angular resolutions adopted are characteristic of Space Telescope and *VLBI* techniques. The authors analyse the case of *Q* 0957 + 561. They show that the Space Telescope will be precious for the mapping of cluster galaxies near the main deflector.

162.182 **Null influence of possible local extragalactic perturbations on tests of redshift-distance laws.**
J. F. Nicoll, I. E. Segal.
Astron. Astrophys., Vol. 115, 398 - 403 (1982).
Non-parametric self- and cross-tests of the Hubble and Lundmark redshift-distance laws, which have raised questions about the former, are extended to include the effects of possible local dynamical perturbations. In no cases were either the absolute or relative fits of the two laws significantly affected.

162.183 **A new view of baryon symmetric cosmology based on grand unified theories.**
F. W. Stecker.
Progress in cosmology, (see 012.047), p. 1 - 20 (1982).
In grand unified theories, a scenario has been developed for the evolution of the early universe wherein the matter which eventually forms the galaxies arises as a "baryon excess" owing to baryon number non-conserving interactions at ultrahigh energies. The model adopted here leads to a domain structure in the universe with the domains evolving into separate regions of matter excess and antimatter excess. This can result in a universe in which matter galaxies are formed in some regions and antimatter galaxies are formed in others. There are advantages in this model in explaining various astrophysical data such as the cosmic γ-ray background spectrum, cosmic ray antiproton flux measurements, a low primordial He abundance and strong galaxy clustering.

162.184 **Cosmology and particle physics.**
J. D. Barrow.
Progress in cosmology, (see 012.047), p. 21 - 38 (1982).
A brief overview is given of recent work that integrates cosmology and particle physics. The observational data regarding the abundance of matter and radiation in the Universe is described. The manner in which the cosmological survival density of stable massive particles can be calculated is discussed along with the process of cosmological nucleosynthesis. Several applications of these general arguments are given. The use of nucleosynthesis to limit the number of lepton generations is described together with the implications of a small neutrino mass for the origin of galaxies and clusters.

162.185 **Breaking of gauge symmetries in the cooling Universe.**
F. R. Klinkhamer.
Progress in cosmology, (see 012.047), p. 39 - 61 (1982).
In this paper the author reviews unified interactions and their phase transitions (PTs), which may considerably change the standard picture of the earliest Universe. Then he considers the impact of PTs on (1) the creation of the observed matter-antimatter asymmetry, (2) the origin of the density perturbations required for galaxy formation, and (3) the related problem of the creation of (too many) monopoles. It appears that (1) and perhaps (3) can be dealt with, provided the PTs, especially the first order ones, preserve the homogeneity of the Universe. But (2) probably arises from the unknown epoch of quantum gravity importance.

162.186 **Particle physics aspects in cosmology.**
H. Fritzsch.
Progress in cosmology, (see 012.047), p. 63 - 74 (1982).
In this talk the author discusses those topics of particle physics which at the present time are relevant for astrophysics and cosmology.

162.187 **Time-asymmetry, cosmological uniformity and space-time singularities.**
R. Penrose.
Progress in cosmology, (see 012.047), p. 87 - 88 (1982).
The author shows that some new physical ideas are needed in order that the regions of the universe referred to as "space-time singularities" can be treated according to the methods of science. There must indeed have been very precise physical laws in operation at the big bang itself. The new physics involved is necessarily time-asymmetric.

162.188 **Cosmic far-ultraviolet background radiation.**
R. C. Henry.
Progress in cosmology, (see 012.047), p. 177 - 188 (1982).
It is demonstrated that interstellar dust grains forward-scatter far-ultraviolet radiation extremely strongly: the value of the Henyey-Greenstein scattering parameter g at 1425 Å is shown to be at least 0.75; the actual value is very likely greater than 0.9. Also, observations of the Virgo cluster of galaxies sets a limit $\tau > 2 \times 10^{25}$ sec on the lifetime of 17 - 20 eV/c^2 heavy neutrinos, if such neutrinos are responsible for the gravitational binding of the cluster.

162.189 **Galaxy formation, clustering and the "hidden mass".**
M. J. Rees, A. Kashlinsky.
Progress in cosmology, (see 012.047), p. 259 - 273 (1982).
Various topics relevant to galaxy formation are discussed, especially the nature of the "hidden mass", and the properties of "population III" pregalactic objects which are expected in models for cosmic evolution involving primordial entropy perturbations.

162.190 **The Age Problem.**
A. C. Edwards.
Progress in cosmology, (see 012.047), p. 291 - 303 (1982).
The Age Problem may be stated as follows: (1) Does the Universe have an age? (2) If so, is it possible to estimate that age? (3) How does one go about making such estimates in practice? (4) Have we any reason to believe that the various estimates made to date present us with any conflicts?

162.191 **Large-scale background temperature and mass fluctuations due to scale-invariant primeval perturbations.**
P. J. E. Peebles.
Astrophys. J., Lett., Vol. 263, L1 - L5 (1982).
The large-scale anisotropy of the microwave background and the large-scale fluctuations in the mass distribution are discussed under the assumptions that the universe is dominated by very massive, weakly interacting particles and that the primeval density fluctuations were adiabatic. The expected background temperature fluctuations are well below present observational limits.

162.192 **On cosmological testing of the Machian principle.**
H.-J. Treder.
Astron. Nachr., Band 303, 317 - 319 (1982). In German.
In a Machian universe with linear expansion the energy integral is zero and the matter density reaches the critical value in general relativistic cosmology.

162.193 **Introduction to Big Bang nucleosynthesis: open and closed models, anisotropies.**
R. J. Tayler.
Philos. Trans. R. Soc. London, Ser. A, Vol. 307, (see 012.052), 3 - 17 (1982).
There is no unique hot Big Bang theory, but the simplest version produces a primeval chemical composition that is in good qualitative agreement with the abundances deduced from observation. Whether or not any Big Bang theory will provide quantitative agreement with observations depends on a variety of factors in elementary particle physics and from observational astronomy (present mean baryon density of the Universe, the Hubble constant and deceleration parameter). The influence of these factors on the abundances is discussed, as is the effect of departures from homogeneity and isotropy in the early Universe.

162.194 **Abundances of elements of cosmological interest.**
B. E. J. Pagel.
Philos. Trans. R. Soc. London, Ser. A, Vol. 307, (see 012.052), 19 - 35 (1982).
The nuclear species of cosmological interest are ^{1}H, D, ^{3}He, ^{4}He and ^{7}Li and in this paper the author reviews the knowledge about their present abundances in the interstel medium and what estimates of their primordial abundanc can be made, when the effects of nuclear synthesis and destruction in stars are allowed for.

162.195 **Constraints on the primordial helium abundanc from the optical emission spectra of dwarf gala**
T. D. Kinman, K. Davidson.
Philos. Trans. R. Soc. London, Ser. A, Vol. 307, (see 012.(37 - 42 (1982).
At present, the best estimate of the primordial heliur abundance (Y_p) comes from helium abundances that have derived from the emission spectra of dwarf galaxies. The n Y_p deduced from the abundances in the 26 best observed dwarf galaxies is ~ 0.24.

162.196 **Constraints on the density of baryons in the Univ**
D. N. Schramm.
Philos. Trans. R. Soc. London, Ser. A, Vol. 307, (see 012.(43 - 54 (1982).
It is shown that not only does Big Bang nucleosynth provide an upper limit on the baryon density of the Unive but if one takes into account arguments concerning the production of ^{3}He in stars, one can show that the ^{3}He plu deuterium abundance can also provide a lower limit on th baryon density of the Universe. The derived constraints ar that the baryon: photon ratio, η, must be between 1.5 × 1 and 7×10^{-9} with a best fit between 3 and 6×10^{-10}. This small range for η has implications for our limits on numbe neutrino types, for Big Bang baryosynthesis, and for arguments about the nature of the dark matter in clusters of galaxies.

162.197 **Interpretation of anisotropy in the cosmic background radiation.**
C. J. Hogan, N. Kaiser, M. J. Rees.
Philos. Trans. R. Soc. London, Ser. A, Vol. 307, (see 012.0 97 - 110 (1982).
The authors review mechanisms for producing tempe ture and polarization anisotropies in the microwave background radiation, and summarize their relation to the larg scale distribution of matter and to theories of galaxy form tion. They also review possible sources of ambiguity in int preting data, in particular the unknown opacity of the pre galactic gas and the possible contribution of discrete sourc of radiation. Strategies for removing these ambiguities are discussed.

162.198 **Grand unified theories in cosmology.**
J. Ellis.
Philos. Trans. R. Soc. London, Ser. A, Vol. 307, (see 012.0 121 - 140 (1982).
This paper reviews grand unified theories and some o their possible applications to cosmology. The philosophy c grand unification to be followed is developed, some low-energy tests described, and expectations for new interactio causing baryon decay and neutrino masses are presented. A discussion is given of the possible relevance of baryon-num violating reactions in grand unified theories to understandi the problem of baryosynthesis, and a possible connection with the neutron electric dipole moment is mentioned. Pos sible interfaces between cosmology and particle physics involving neutrinos are mentioned.

162.199 **Phase transitions in the embryo Universe.**
A. H. Guth.
Philos. Trans. R. Soc. London, Ser. A, Vol. 307, (see 012.0 141 - 148 (1982).
The author discusses the phase transitions of grand u fied theories (GUTs), and the role that they may have play in the very early evolution of the Universe. In particular, h

describes a scenario that he calls the inflationary Universe. If this scenario is valid it would mean that GUT mechanisms are responsible not only for the production of all the baryons in the Universe, but for the production of all the matter, energy and entropy as well.

162.200 **Evolution of cosmological ideas.**
I. D. Novikov.
Zemlya Vselennaya, 1982, No. 6, p. 35 - 39. In Russian.

162.201 **Die Entwicklung großräumiger Strukturen im Universum.**
J. H. Oort.
Sterne Weltraum, 21. Jahrg., 456 - 461, 514 - 518 (1982).

162.202 **Energy, matter and gravitation in an unlimited, renewable Universe.**
H. Broberg.
ESA J., Vol. 6, 207 - 231 (1982).

A Universe without exterior limits, but built on the concept of a minimal energy quantum has been studied with the use of a mass-point model for the generation of energy fields. It is shown that a number of aspects of known and observed physical effects can be comprehended, on a cosmological as well as a micro-physical scale, within the framework of such a Universe.

162.203 **Relativistic cosmology I: Paradigm commitment and rationality.**
C. M. Copp.
Astron. Q., Vol. 4, 103 - 116 (1982).

162.204 **The arrow of time in a bouncing universe.**
M. Clutton-Brock.
J. R. Astron. Soc. Canada, Vol. 76, 317 (1982). – Abstract.

162.205 **On the search for large-scale structure of the universe.**
J. Einasto, J. Jaaniste.
Priroda, 1982, No. 12, p. 80 - 86. In Russian.

162.206 **Free and bound quarks – accelerators and the early universe.**
R. V. Wagoner.
Essays in nuclear astrophysics, (see 003.013), p. 495 - 518 (1982).

The implications of recent results of quantum chromodynamics for the very early stages of big bang models are reviewed. The possible production of free quarks in the early universe is considered.

162.207 **Particle physics and cosmology.**
G. Steigman.
Essays in nuclear astrophysics, (see 003.013), p. 519 - 531 (1982).

The author gives a review of the early universe as a cosmic accelerator in which all sorts of elementary particles were produced and, from which some relics may survive. The connection between the baryon asymmetry of the universe and the stability of the proton is explored and the implications of neutrino masses for a neutrino dominated universe are outlined.

162.208 **The value of H_0.**
G. A. Tammann, A. Sandage.
ESO Sci. Prepr., No. 211, 10 pp. (1982). – To be published in "Highlights of Astronomy", Vol. 6.

162.209 **A distance scale from the infrared magnitude/H I velocity-width relation. IV. The morphological type dependence and scatter in the relation; the distances to nearby groups.**
M. Aaronson, J. Mould.
Prepr. Steward Obs., No. 404, 57 pp. (1982).

162.210 **A theoretical framework for quantized redshifts and uncertainty in cosmology.**
W. J. Cocke.
Prepr. Steward Obs., No. 415, 27 pp. (1982).

162.211 **Studies on the Hubble law with emphasis on selection effects and galactic absorption.**
P. Teerikorpi.
Rep. Ser., Dep. Phys. Sci., Univ. Turku, No. D 4, 18 pp. (1982) = Turku Univ. Obs. Informo, No. 68.

162.212 **Cosmology without singularity and nonlinear gravitational Lagrangians.**
R. Kerner.
Gen. Relativ. Gravitation, Vol. 14, 453 - 469 (1982).

The author investigates the generalized Einstein equations derived from the Lagrangian which is an arbitrary function of R. The importance of the saturation phenomenon is underlined, which may replace the role of a cosmological constant. The spherically symmetric homogeneous model is analyzed in more detail, and an approximate solution without singularity is constructed using the method of matched asymptotic expansions.

162.213 **Isotropic focusing of light rays in cosmology.**
K. Rosquist.
Gen. Relativ. Gravitation, Vol. 14, 503 - 508 (1982).

162.214 **On cosmology, topology and field theory.**
S. D. Unwin.
Gen. Relativ. Gravitation, Vol. 14, 509 - 514 (1982).

It is suggested that in general the topological constraints imposed upon matter fields propagating in a cosmology need not be absolutely determined by the cosmological topology, which consequently allows for the possibility of different fields carrying different global structures. Quantum field stability is investigated for a particular model.

162.215 **Bianchi type-I solutions for modified Brans-Dicke cosmology.**
A. Banerjee, N. O. Santos.
Gen. Relativ. Gravitation, Vol. 14, 559 - 564 (1982).

Cosmological dust solutions have been obtained in Bianchi type-I homogeneous space for Brans-Dicke modified theory by the condition of the cosmological parameter.

162.216 **A point-particle model universe.**
R. P. A. C. Newman, G. C. McVittie.
Gen. Relativ. Gravitation, Vol. 14, 591 - 608 (1982).

162.217 **Low-energy behavior of a quantized scalar field in the linearly expanding universe.**
T. Azuma, A. Tomimatsu.
Gen. Relativ. Gravitation, Vol. 14, 629 - 636 (1982).

162.218 **Einstein–Cartan–Maxwell–Bianchi type-V cosmological models.**
D. Lorenz.
Gen. Relativ. Gravitation, Vol. 14, 691 - 701 (1982).

162.219 **Nonsingular cosmological models in Brans-Dicke theory.**
V. B. Johri, R. Chandra, R. C. Srivastava.
Gen. Relativ. Gravitation, Vol. 14, 771 - 780 (1982).

162.220 **A new demonstration of Bonnor's equation.**
J. Randon.
Gen. Relativ. Gravitation, Vol. 14, 825 - 828 (1982).

By perturbation of Robertson-Walker's metric the

author proves Bonnor's equation which governs the growth of density fluctuations at Newtonian approximation.

162.221 **Fluctuations produced by the cosmological constant in the empty Bianchi type-IX model.**
H. Sirousse-Zia.
Gen. Relativ. Gravitation, Vol. 14, 851 - 868 (1982).

162.222 **A static cosmological model in Einstein–Cartan theory.**
S. B. Kalyanshetti, B. B. Waghmode.
Gen. Relativ. Gravitation, Vol. 14, 923 - 930 (1982).

162.223 **The Lepton Brusselator: creation of structure in the early universe.**
R. E. Zimmermann.
Gen. Relativ. Gravitation, Vol. 14, 1051 - 1060 (1982).

162.224 **Singularities in self-similar cosmologies.**
Z. Wu.
Sci. Sinica, Ser. A, Vol. 25, 737 - 749 (1982).

162.225 **Cosmic density wave and its observable vestiges – maximum scale of inhomogeneities in the distribution of galaxies and features in the distribution of absorption-line redshifts of quasars.**
Y. Liu.
Sci. Sinica, Ser. A, Vol. 25, 856 - 865 (1982).

According to the theory of the cosmic density wave (CD wave) based on the gravitational instability in the expanding universe as was suggested by Liu and Cao, a series of maxima in the z_{em}-distribution of the observed quasars can be interpreted as a vestige of the CD wave. This paper presents some other observable vestiges of the CD wave with main results: (1) the upper limit scale of the inhomogeneities in the large-scale distribution of galaxies is of the order 200 Mpc, (2) there are a series of maxima in the z_{ab}-distribution of all the identified absorption systems of quasars, (3) the absorption-line redshifts of a quasar with rich absorption-line can be divided into a series of groups.

162.226 **The dependence on redshift of the steepness of quasar counts.**
I. E. Segal.
Publ. Astron. Soc. Japan, Vol. 34, 507 - 513 (1982).

The steepness of quasar counts as a function of redshift region and spectral index distribution is studied within the chronometric and Friedmann cosmologies. In the non-evolutionary and non-parametric chronometric cosmology, the quasar optical spectral index distribution reported by Richstone and Schmidt (1980) predicts a rapid rise in steepness for the redshift region accessible to objective prism surveys, which is generally consistent with those observed.

162.227 **Infrared astronomy and cosmology.**
M. S. Longair.
Irish Astron. J., Vol. 15, (see 012.061), 122 (1981).
Abstract.

162.228 **Relativistic cosmologies.**
M. A. H. MacCallum.
Irish Astron. J., Vol. 15, (see 012.061), 125 - 127 (1981).

The aim of this paper is to review how solutions of Einstein's equations of general relativity have been used in cosmology, and to point out some areas of current interest.

162.229 **The solar motion and interpretation of the 3K microwave background.**
R. D. Davies.
Irish Astron. J., Vol. 15, (see 012.061), 127 - 128 (1981).
Abstract.

162.230 **Determining the cosmological deceleration parameter, q_0.**
T. Shanks.
Irish Astron. J., Vol. 15, (see 012.061), 128 (1981).
Abstract.

162.231 **Cosmology and unified gauge theory.**
L. O'Raifeartaigh.
Irish Astron. J., Vol. 15, (see 012.061), 130 - 137 (1981).

162.232 **Grand unified reactions and dissipation in aniso-tropic cosmologies.**
T. Rothman, R. Matzner.
Astrophys. J., Vol. 263, 501 - 507 (1982).

The authors consider the evolution of anisotropy in universes containing particles interacting by the Grand Uni Theory cross section behavior in the interval after baryosy thesis ($T \lesssim 10^{15}$ GeV). They find that, contrary to expect tions, efficient damping of anisotropy is not a generic feat of these cosmologies, so no limit can be put on their very early anisotropy by demanding that the entropy per baryc be limited to observed values.

162.233 **Manifestations of a cosmological density of con pact objects in quasar light.**
C. R. Canizares.
Astrophys. J., Vol. 263, 508 - 517 (1982).

The gravitational lens effects of a cosmological densi compact objects with masses in the range $0.01 < M < 10^5$ l will be manifested in the properties of high-redshift quasa The manifestations result from the amplification of the qu continuum and include time variability, changes in the me ratios of line flux to optical continuum and to X-ray fluxe and an increase in the rms scatter of those ratios for high-z relative to low-z quasars. The sizes of these effects are con puted using an analytic probability distribution for amplif tion that accounts for single and, approximately, for multi lens amplification. Comparison of the expected manifesta-tions with a variety of quasar data suggests that the density compact objects in the $0.01-10^5\ M_\odot$ range is not sufficien close the universe if quasar continuum emission comes fro region $\lesssim 10^{-3}$ pc. This would exclude nuclear burning star and their remnants.

162.234 **Aufbau und Entwicklung des Weltalls. II. Das astronomische Fundament der modernen Kosmo gie.**
J. Gürtler, J. Dorschner.
Sterne, 58. Band, 283 - 295 (1982).

162.235 **Evolution of the main mode of density perturbati in a neutrino-dominated universe.**
V. I. Khlebnikov.
Zh. ehksp. i teor. fiz., Tom 82, 1361 - 1365 (1982). In Russian. – Abstr. in Ref. zh., 51. Astron., 10.51.778 (198

162.236 **Formation of the large-scale cellular structure of the universe.**
Ya. B. Zel'dovich.
Inst. prikl. mat. AN SSSR. Prepr., 1982, No. 26, 12 pp. In Russian. – Abstr. in Ref. zh., 51. Astron., 10.51.780 (198

162.237 **Cavities in the structure of the universe.**
Ya. B. Zel'dovich, S. F. Shandarin.
Nauk. v SSSR, 1982, No. 1, p. 34 - 39. In Russian. – Abst in Ref. zh., 51. Astron., 10.51.781 (1982).

162.238 **Seven numbers defining the structure of the universe.**
I. L. Rozental'.
Inst. kosm. issled. AN SSSR, Prepr., 1982, No. 704, 12 pp.

In Russian. – Abstr. in Ref. zh., 51. Astron., 10.51.784 (1982).

162.239 **Friedmann's hot universe without particle creation.**
A. D. Linde.
Pis'ma v ZhEhTF, Tom 35, 398 - 400 (1982). In Russian. From Ref. Zh., 51. Astron., 11.51.798 (1982).

162.240 **On particle creation in homogeneous and isotropic cosmological models.**
V. N. Mel'nikov.
Diskus. vopr. teor. otnositel'nosti i gravitatsii, Moskva, 1982, p. 120 - 122. In Russian. – Abstr. in Ref. zh., 51. Astron., 11.51.799 (1982).

162.241 **Noncoherent creation of fermions and self-consistent cosmological models.**
K. A. Bronnikov, V. N. Mel'nikov.
Diskus. vopr. teor. otnositel'nosti i gravitatsii, Moskva, 1982, p. 127 - 132. In Russian. – Abstr. in Ref. zh., 51. Astron., 11.51.814 (1982).

162.242 **Cosmological models in the theory of gravitation with disturbed conformal invariance.**
N. A. Zajtsev.
Diskus. vopr. teor. otnositel'nosti i gravitatsii, Moskva, 1982, p. 112 - 115. In Russian. – Abstr. in Ref. zh., 51. Astron., 11.51.815 (1982).

162.243 **Cosmological models with self-interacting conformal scalar field.**
S. V. Orlov.
Diskus. vopr. teor. otnositel'nosti i gravitatsii, Moskva, 1982, p. 116 - 119. In Russian. – Abstr. in Ref. zh., 51. Astron., 11.51.816 (1982).

162.244 **On Mach's relation in cosmology.**
V. E. Yakimov.
Diskus. vopr. teor. otnositel'nosti i gravitatsii, Moskva, 1982, p. 107 - 111. In Russian. – Abstr. in Ref. zh., 51. Astron., 11.51.819 (1982).

162.245 **Nucleosynthesis in bouncing cosmologies.**
J. V. Narlikar, K. M. V. Apparao.
Astrophys. Space Sci., Vol. 87, 333 - 346 (1982).
It is shown that certain anomalies connected with the primordial abundances of light nuclei may be resolved if it is assumed that the Universe oscillates between phases of finite densities. Since general relativity does not produce bouncing models of the Universe, such models are obtained through the introduction of a negative energy scalar field of zero rest mass. It is shown that all the relevant parameters of the dynamics of the model and the nucleosynthesis in it are determined by observations and that a self-consistent picture emerges. The model is capable of admitting more than three neutrino flavours without an embarrasingly high primordial helium content. It is also shown that the calculations could be adapted to describe production of light nuclei in compact massive bouncing objects.

162.246 **Steady-state versus viscous cosmology.**
M. Heller, M. Ostrowski, A. Wosczyna.
Astrophys. Space Sci., Vol. 87, 425 - 433 (1982).
Hoyle's steady-state cosmology and relativistic Robertson-Walker world models with a constant bulk viscosity are briefly reviewed and compared. It is shown that, although they are identical from a purely formal point of view, the interpretation of Hoyle's model in terms of viscous dissipation would be physically unrealistic.

162.247 **Dynamical role of the electromagnetic "primordial" field in spatially homogeneous cosmological models of "diagonal" Bianchi types I - IX.**
V. A. Ruban.
Astron. Zh., Tom 59, 1044 - 1054 (1982). In Russian. English translation in Soviet Astron., Vol. 26, No. 6.
The possible structure of the sourceless "primordial" electromagnetic field (allowable only for the Bianchi class "A" of "diagonal" metrics), its dynamical behaviour and role near the initial singularity are examined in anisotropic homogeneous cosmology.

162.248 **On the existence of cosmological evolutionary effects. I. A morphological selection effect in the counts of quasars.**
T. Jaakkola.
Astrophys. Space Sci., Vol. 88, 283 - 291 (1982).
The counts of quasars are examined. The recently reported counts of optically selected quasars (Green and Schmidt, 1978) probably have been biased by the use of distance-dependent morphological criteria of selection. The influence of this bias is to mimic cosmological evolution. The application of the K-correction for the counts of quasars is discussed. The K-corrected, presumably unbiased counts of objects with ultraviolet excess and broad emission lines, including quasars and type 1 Seyfert galaxies, give a uniform radial distribution. Correctness of the latter sampling of data is dependent on the solution of the redshift problem and presupposes an inverse correlation between luminosity and intrinsic redshift.

162.249 **Towards a unification of the parameters underlying elementary particles and cosmology.**
C. Sivaram.
Astrophys. Space Sci., Vol. 88, 507 - 510 (1982).
It is pointed out that the gross parameters characterizing the Universe such as the overall size and mass can be arrived at from microphysical considerations involving the fundamental interactions of elementary particle physics. Interesting relations for the Hubble radius and closure density are obtained in terms of the coupling constants underlying these interactions.

162.250 **Fluctuations in the new inflationary universe.**
A. H. Guth, S.-Y. Pi.
Phys. Rev. Lett., Vol. 49, 1110 - 1113 (1982).

162.251 **Gravitation, phase transitions, and the big bang.**
L. M. Krauss.
Phys. Rev. Lett., Vol. 49, 1459 - 1462 (1982).
Introduced is a model of the early universe based on the possibility of a first-order phase transition involving gravity, and arrived at by a consideration of instabilities in the semiclassical theory. The evolution of the system is very different from the standard Friedmann-Robertson-Walker big-bang scenario, indicating the potential importance of semiclassical finite-temperature gravitational effects. Baryosynthesis and monopole production in this scenario are also outlined.

162.252 **Cosmological effects of primeval fermion degeneracy in the grand unified theories.**
J. N. Fry, C. J. Hogan.
Phys. Rev. Lett., Vol. 49, 1873 - 1876 (1982).

162.253 **The large-scale structure of the Universe.**
Kozmos, Vol. 13, 147 - 150 (1982). In Slovak.

162.254 **Revision of the Hubble constant.**
O. Obůrka.
Říše hvězd, Vol. 63, 247 - 248 (1982). In Czech.

162.255 **Cosmic cooking, evolution of stars and our Metagalaxy.**
O. Vilhu.
Evolution – different aspects, common problems. Proceedings of the 1st Tutkijaliitto Symposium, held at Helsinki, Finland,

May 1981. Y. Haila, I. Tuominen, O. Vilhu (Editors). p. 16-35 (1982).

The author discusses the chemical composition of cosmic bodies in connection with their own evolution. What do the abundances of elements tell about the evolution of stars, galaxies and the Metagalaxy? What is the origin of the various elements and how do their relative abundances change during cosmic evolution?

162.256 **Gravitational effects upon cosmological phase transitions.**
A. Vilenkin, L. H. Ford.
Phys. Rev. D, Vol. 26, 1231 - 1241 (1982). – Abstr. in Phys. Abstr., Vol. 85, Abstr. 112769 (1982).

162.257 **Baryon asymmetry and the scale of supersymmetry breaking.**
H. E. Haber.
Phys. Rev. D, Vol. 26, 1317 - 1326 (1982). – Abstr. in Phys. Abstr., Vol. 85, Abstr. 113193 (1982).

162.258 **Enlargement of a void in the expanding universe.**
H. Sato.
Prog. Theor. Phys., Vol. 68, 236 - 245 (1982). – Abstr. in Phys. Abstr., Vol. 85, Abstr. 117770 (1982).

162.259 **Particle decays and CPT non-invariance in cosmology.**
L. H. Ford.
Nucl. Phys. B, Part. Phys., Vol. B204, 35 - 44 (1982). – Abstr. in Phys. Abstr., Vol. 85, Abstr. 117814 (1982).

162.260 **'Whimpers' in self-similar cosmologies.**
Z. C. Wu.
Commun. Theor. Phys., Vol. 1, 229 - 236 (1982). – Abstr. in Phys. Abstr., Vol. 85, Abstr. 117820 (1982).

162.261 **A finite temperature $\lambda\phi^4$ model and de Sitter-Friedmann phase transition in the early Universe.**
H.-y. Guo, Z.-l. Zou.
Commun. Theor. Phys., Vol. 1, 237 - 243 (1982). – Abstr. in Phys. Abstr., Vol. 85, Abstr. 117821 (1982).

162.262 **Dynamical importance of vorticity and shear in the Universe.**
J. L. Sanz.
J. Math. Phys., Vol. 23, 1732 - 1736 (1982). – Abstr. in Phys. Abstr., Vol. 85, Abstr. 117825 (1982).

162.263 **'Taub-NUT-like' cosmologies.**
D. A. Konkowski, T. M. Helliwell.
Phys. Lett. A, Vol. 91A, 149 - 152 (1982). – Abstr. in Phys. Abstr., Vol. 85, Abstr. 117832 (1982).

162.264 **Self-intersection of cosmic strings.**
T. W. B. Kibble, N. Turok.
Phys. Lett. B, Vol. 116B, 141 - 143 (1982). – Abstr. in Phys. Abstr., Vol. 85, Abstr. 117833 (1982).

162.265 **Quantum vacuum energy in Taub-NUT (Newman-Unti-Tamburino)-type cosmologies.**
W. A. Hiscock, D. A. Konkowski.
Phys. Rev. D, Vol. 26, 1225 - 1230 (1982). – Abstr. in Phys. Abstr., Vol. 85, Abstr. 117834 (1982).

162.266 **Comments on the detection of cosmic relic neutrinos.**
B. A. Campbell, P. J. O'Donnell.
Phys. Rev. D, Vol. 26, 1487 - 1489 (1982). – Abstr. in Phys. Abstr., Vol. 85, Abstr. 117835 (1982).

162.267 **Regularization and renormalization of a massive vector field in a Friedmann metric.**
S. Matarrese, S. A. Bonometto.
Nuovo Cimento A, Vol. 70A, Ser. 11, 355 - 370 (1982). Abstr. in Phys. Abstr., Vol. 86, Abstr. 522 (1983).

162.268 **Cosmological and astrophysical bounds on the scale of supersymmetry breaking.**
A. Bouquet, C. E. Vayonakis.
Phys. Lett. B, Vol. 116B, 219 - 222 (1982). – Abstr. in Phys. Abstr., Vol. 86, Abstr. 550 (1983).

162.269 **Cosmological bounds on a left-right symmetric model.**
R. Huerta.
Nucl. Phys. B, Part. Phys., Vol. B204, 413 - 418 (1982). Abstr. in Phys. Abstr., Vol. 86, Abstr. 555 (1983).

162.270 **B-L violating proton decay and late cosmological baryon production.**
A. Masiero, J. F. Nieves, T. Yanagida.
Phys. Lett. B, Vol. 116B, 11 - 15 (1982). – Abstr. in Phys. Abstr., Vol. 86, Abstr. 558 (1983).

162.271 **Scalar fields and dynamics of the early Universe.**
T. Dereli, M. Panahimoghaddam, R. W. Tucker.
J. Phys. A, Vol. 15, 3167 - 3172 (1982). – Abstr. in Phys. Abstr., Vol. 86, Abstr. 5458 (1983).

162.272 **The search for scale invariant cosmology.**
J. Gribbin.
New Scientist, Vol. 95, 844 - 846 (1982). – Abstr. in Phys. Abstr., Vol. 86, Abstr. 5461 (1983).

162.273 **Was the Universe ever hyperdimensional?**
N. E. Frankel.
Phys. Lett. A, Vol. 91A, 321 - 323 (1982). – Abstr. in Phys. Abstr., Vol. 86, Abstr. 5462 (1983).

162.274 **Superunification and the ultimate temperature of the Universe.**
C. E. Vayonakis.
Phys. Lett. B, Vol. 116B, 223 - 226 (1982). – Abstr. in Phys. Abstr., Vol. 86, Abstr. 5464 (1983).

162.275 **Baryon asymmetry in the inflationary universe.**
A. D. Dolgov, A. D. Linde.
Phys. Lett. B, Vol. 116B, 329 - 334 (1982). – Abstr. in Phys. Abstr., Vol. 86, Abstr. 5465 (1983).

162.276 **Scalar field fluctuations in the expanding universe and the new inflationary universe scenario.**
A. D. Linde.
Phys. Lett. B, Vol. 116B, 335 - 339 (1982). – Abstr. in Phys. Abstr., Vol. 86, Abstr. 5466 (1983).

162.277 **Massive scalar fields. I. Generation by conformal transformation.**
M. Nagaraj.
Vignana Bharathi, Vol. 6, No. 1, p. 18 - 26 (1980). – Abstr. in Phys. Abstr., Vol. 86, Abstr. 5467 (1983).

162.278 **Geodesic in a de Sitter space-time.**
M. Nagaraj, B. R. Srinivasa.
Vignana Bharathi, Vol. 6, No. 1, p. 37 - 44 (1980). – Abstr. in Phys. Abstr., Vol. 86, Abstr. 5468 (1983).

162.279 **Monopole suppression and baryon production in cosmologies based on Witten's upside down hierarchy model.**
T. Banks, V. Kaplunovsky.

Nucl. Phys. B, Part. Phys., Vol. B206, 45 - 52 (1982). – Abstr. in Phys. Abstr., Vol. 86, Abstr. 6024 (1983).

162.280 **More on cosmology for supersymmetry GUTs.**
M. Srednicki.
Nucl. Phys. B, Part. Phys., Vol. B206, 132 - 136 (1982). Abstr. in Phys. Abstr., Vol. 86, Abstr. 6025 (1983).

162.281 **Remark on an application of the Banach metric method to cosmology.**
M. Szydlowski, M. Heller.
Acta Phys. Polonica B, Vol. B13, 375 - 377 (1982). – Abstr. in Phys. Abstr., Vol. 86, Abstr. 10491 (1983).

162.282 **Creation of universes from nothing.**
A. Vilenkin.
Phys. Lett. B, Vol. 117B, 25 - 28 (1982). – Abstr. in Phys. Abstr., Vol. 86, Abstr. 10494 (1983).

162.283 **Particle production in the new inflationary cosmology.**
L. F. Abbott, E. Farhi, M. B. Wise.
Phys. Lett. B, Vol. 117B, 29 - 33 (1982). – Abstr. in Phys. Abstr., Vol. 86, Abstr. 10495 (1983).

162.284 **On the neutrino mass and the Lemaître model.**
C. Xu, L. Yang, Z. He, X. Xiao.
Acta Astrophys. Sinica, Vol. 2, 107 - 113 (1982). In Chinese.
The authors study the cosmological consequence of the neutrino mass within the framework of the Lemaître model with the cosmological constant Λ.

162.285 **A de Sitter-Friedmann phase transition model.**
H. Guo, Z. Zou.
Acta Astrophys. Sinica, Vol. 2, 165 - 172 (1982). In Chinese.
First, the authors solve the equation of field in the $T = 0$ case and get an anti-de Sitter RW metric with $k = -1$ involving the spontaneous symmetry breaking and trace anomaly. Then they consider the particle creation and temperature effects as well to show how the model could undergo a de Sitter-Friedmann phase transition. Finally, few words concern with the possibility of establishing a real model.

162.286 **Scale covariant cosmology and astrophysical applications.**
V. M. Canuto.
Proceedings of the Second Marcel Grossmann Meeting on General Relativity, (see 012.081), p. 729 - 784 (1982). Abstr. in Phys. Abstr., Vol. 86, Abstr. 20552 (1983).

162.287 **Cosmological models and the statistical distribution of quasars.**
L. Z. Fang.
Proceedings of the Second Marcel Grossmann Meeting on General Relativity, (see 012.081), p. 785 - 799 (1982). Abstr. in Phys. Abstr., Vol. 86, Abstr. 20553 (1983).

162.288 **A theory of inertial motion in space-time of constant curvature.**
G. Hanying, H. Y. Kuo.
Proceedings of the Second Marcel Grossmann Meeting on General Relativity, (see 012.081), p. 801 - 807 (1982). Abstr. in Phys. Abstr., Vol. 86, Abstr. 20554 (1983).

162.289 **Microcanonical entropy of a cosmological model.**
G. Horwitz.
Proceedings of the Second Marcel Grossmann Meeting on General Relativity, (see 012.081), p. 809 - 812 (1982). Abstr. in Phys. Abstr., Vol. 86, Abstr. 20555 (1983).

162.290 **Quantum field theories and relativistic cosmology.**
B. L. Hu.
Proceedings of the Second Marcel Grossmann Meeting on General Relativity, (see 012.081), p. 813 - 852 (1982). Abstr. in Phys. Abstr., Vol. 86, Abstr. 20556 (1983).

162.291 **Variational principles in cosmology.**
R. T. Jantzen.
Proceedings of the Second Marcel Grossmann Meeting on General Relativity, (see 012.081), p. 867 - 868 (1982). Abstr. in Phys. Abstr., Vol. 86, Abstr. 20557 (1983).

162.292 **The specific entropy of the universe in a self-consistent cosmological history.**
E. Gunzig, P. Nardone.
Proceedings of the Second Marcel Grossmann Meeting on General Relativity, (see 012.081), p. 869 (1983). – Abstr. in Phys. Abstr., Vol. 86, Abstr. 20558 (1983).

162.293 **Zero energy-momentum cosmological principle.**
D. K. Callebaut.
Proceedings of the Second Marcel Grossmann Meeting on General Relativity, (see 012.081), p. 871 (1982). – Abstr. in Phys. Abstr., Vol. 86, Abstr. 20559 (1983).

162.294 **The large-scale stability of the universe.**
R. Fabbri.
Proceedings of the Second Marcel Grossmann Meeting on General Relativity, (see 012.081), p. 883 - 888 (1982). Abstr. in Phys. Abstr., Vol. 86, Abstr. 20560 (1983).

162.295 **Particle physics meets cosmology – the search for decaying neutrinos.**
R. C. Henry.
Phys. Teach., Vol. 20, 531 - 539 (1982). – Abstr. in Phys. Abstr., Vol. 86, Abstr. 10547 (1983).

162.296 **On the inconsistency of a photon creation mechanism in an expanding universe.**
H. Ceccatto, A. Foussats, H. Giacomini, O. Zandron.
J. Math. Phys., Vol. 23, 1865 - 1872 (1982). – Abstr. in Phys. Abstr., Vol. 86, Abstr. 11004 (1983).

162.297 **Curvature collineations in certain cosmological space-times.**
S. Ram, H. S. Pandey.
Indian J. Pure Appl. Math., Vol. 13, 1200 - 1203 (1982). Abstr. in Phys. Abstr., Vol. 86, Abstr. 16042 (1983).

162.298 **Stability and instability in a dust filled universe.**
S. D. Thikine, B. B. Waghmode.
Indian J. Pure Appl. Math., Vol. 13, 1213 - 1220 (1982). Abstr. in Phys. Abstr., Vol. 86, Abstr. 16043 (1983).

162.299 **Quantum tachyon in a Friedmann universe.**
S. K. Srivastava.
J. Math. Phys., Vol. 23, 1981 - 1983 (1982). – Abstr. in Phys. Abstr., Vol. 86, Abstr. 16044 (1983).

162.300 **Initial data and wave propagation in one-dimensional inhomogeneous cosmologies.**
R. A. Matzner.
J. Math. Phys., Vol. 23, 1984 - 1991 (1982). – Abstr. in Phys. Abstr., Vol. 86, Abstr. 16045 (1983).

162.301 **Cosmological consequences of permanent quark confinement.**
M. DerSarkissian.
Lett. Nuovo Cimento, Vol. 35, Ser. 2, 285 - 291 (1982). Abstr. in Phys. Abstr., Vol. 86, Abstr. 16046 (1983).

162.302 **Homogeneous cosmologies and their stability behaviour.**
R. E. Zimmermann.
New Phys., Vol. 22, 291 - 315 (1982). – Abstr. in Phys. Abstr., Vol. 86, Abstr. 16048 (1983).

162.303 **The Big Bang nucleosynthesis and finite temperature field theory.**
A. E. I. Johansson, G. Peressutti, B.-S. Skagerstam.
Phys. Lett. B, Vol. 117B, 171 - 174 (1982). – Abstr. in Phys. Abstr., Vol. 86, Abstr. 16051 (1983).

162.304 **Dynamics of phase transition in the new inflationary universe scenario and generation of perturbations.**
A. A. Starobinsky (*Starobinskij*).
Phys. Lett. B, Vol. 117B, 175 - 178 (1982). – Abstr. in Phys. Abstr., Vol. 86, Abstr. 16052 (1983).

162.305 **Divergences and alternative vacuum states for quantized fields in cosmology.**
K. A. Bronnikov, V. G. Lapchinsky (*Lapchinskij*).
Phys. Lett. B, Vol. 117B, 387 - 391 (1982). – Abstr. in Phys. Abstr., Vol. 86, Abstr. 16053 (1983).

162.306 **Conformal fluctuations in a quantum universe with a scalar field.**
T. Padmanabhan.
Phys. Rev. D, Vol. 26, 2162 - 2165 (1982). – Abstr. in Phys. Abstr., Vol. 86, Abstr. 16054 (1983).

162.307 **Cosmology and particles.**
J. Demaret, J. Vandermeulen.
Recherche, Vol. 13, 1152 - 1162 (1982). In French. – Abstr. in Phys. Abstr., Vol. 86, Abstr. 16056 (1983).

162.308 **Etheronics – a possible reappraisal of the ether concept.**
I. Iovit-Popescu.
Stud. Cercet. Fiz., Vol. 34, 451 - 468 (1982). In Rumanian. Abstr. in Phys. Abstr., Vol. 86, Abstr. 16057 (1983).

162.309 **Cosmological baryon production and related topics.**
M. Yoshimura.
Grand unified theories and related topics, (see 012.083), p. 235 - 288 (1981). – Abstr. in Phys. Abstr., Vol. 86, Abstr. 16059 (1983).

162.310 **Cosmological consequences of first-order phase transition.**
K. Sato.
Grand unified theories and related topics, (see 012.083), p. 289 - 327 (1981). – Abstr. in Phys. Abstr., Vol. 86, Abstr. 16060 (1983).

162.311 **Why is the apparent cosmological constant zero?**
S. W. Hawking.
Unified theories of elementary particles (see 012.084), p. 167 - 175 (1982). – Abstr. in Phys. Abstr., Vol. 86, Abstr. 16061 (1983).

162.312 **Monopoles in the present and early universe.**
T. W. B. Kibble.
Monopoles in quantum field theory, (see 012.088), p. 341 - 376 (1982). – Abstr. in Phys. Abstr., Vol. 86, Abstr. 16588 (1983).

162.313 **Massive particles as a probe of the early universe.**
J. D. Barrow.
Nucl. Phys. B, Part. Phys., Vol. B208, 501 - 508 (1982). Abstr. in Phys. Abstr., Vol. 86, Abstr. 20538 (1983).

162.314 **Exact Bianchi type I solutions with a cosmolog constant.**
D. Lorenz.
Phys. Lett. A, Vol. 92A, 118 - 120 (1982). – Abstr. in Ph Abstr., Vol. 86, Abstr. 20541 (1983).

162.315 **Inflation can save the gravitino.**
J. Ellis, A. D. Linde, D. V. Nanopoulos.
Phys. Lett. B, Vol. 118B, 59 - 64 (1982). – Abstr. in Phy Abstr., Vol. 86, Abstr. 20542 (1983).

162.316 **Avoiding domain walls in SU(5) × SU(5).**
M. Claudson, P. H. Cox, A. Yildiz.
Phys. Lett. B, Vol. 118B, 65 - 67 (1982). – Abstr. in Phy Abstr., Vol. 86, Abstr. 20543 (1983).

162.317 **Hawking radiation in an inflationary universe.**
R. Brandenberger, R. Kahn.
Phys. Lett. B, Vol. 119B, 75 - 80 (1982), – Abstr. in Phy Abstr., Vol. 86, Abstr. 20544 (1983).

162.318 **The quark-hadron phase transition in the early universe.**
E. Suhonen.
Phys. Lett. B, Vol. 119B, 81 - 84 (1982). – Abstr. in Phy Abstr., Vol. 86, Abstr. 20545 (1983).

162.319 **Brans-Dicke cosmology with the cosmological constant.**
K. Uehara, C. W. Kim.
Phys. Rev. D, Vol. 26, 2575 - 2579 (1982). – Abstr. in Ph Abstr., Vol. 86, Abstr. 20546 (1983).

162.320 **Neutrinos in perfect fluid spacetimes with local rotational symmetry.**
S. V. Dhurandhar, C. V. Vishveshwara, J. M. Cohen.
Phys. Rev. D, Vol. 26, 2598 - 2610 (1982). – Abstr. in Ph Abstr., Vol. 86, Abstr. 20547 (1983).

162.321 **Study of a Bianchi type-V cosmological model with torsion.**
N. A. Batakis, D. Tsoubelis.
Phys. Rev. D, Vol. 26, 2611 - 2614 (1982). – Abstr. in Ph Abstr., Vol. 86, Abstr. 20548 (1983).

162.322 **Bubble collisions in the very early universe.**
S. W. Hawking, I. G. Moss, J. M. Stewart.
Phys. Rev. D, Vol. 26, 2681 - 2693 (1982). – Abstr. in Ph Abstr., Vol. 86, Abstr. 20549 (1983).

162.323 **Primordial nucleosynthesis including radiative, Coulomb, and finite-temperature corrections to weak rates.**
D. A. Dicus, E. W. Kolb, A. M. Gleeson, E. C. G. Sudarsha V. L. Teplitz, M. S. Turner.
Phys. Rev. D, Vol. 26, 2694 - 2706 (1982). – Abstr. in Ph Abstr., Vol. 86, Abstr. 20550 (1983).

Metacatalogue of galaxies and large scale structu in the Universe.
See Abstr. 002.027.

Glossary of terms used in cosmology.
See Abstr. 002.156.

Galakser og kosmologi.
See Abstr. 003.017.

The accidental universe.
See Abstr. 003.035.

Physics of stellar evolution and cosmology.
See Abstr. 003.044.

Landolt-Börnstein. Group VI. Vol. 2c. Interstellar matter – Galaxy – Universe.
See Abstr. 003.055.

Facts and dogmas in cosmology and elsewhere.
See Abstr. 003.060.

An introduction to tensor calculus, relativity and cosmology.
See Abstr. 003.074.

Die Sekunde der Kosmologen.
See Abstr. 003.106.

Thermo field dynamics and condensed states.
See Abstr. 003.121.

Modern cosmology.
See Abstr. 003.127.

Mapping the Universe.
See Abstr. 011.010.

Energy-loss mechanisms and the annihilation of confined monopoles.
See Abstr. 022.059.

Naturality problems *(QCD)*.
See Abstr. 022.072.

Vacuum stability and symmetry breaking in non-Minkowskian space-times.
See Abstr. 022.074.

Strings in SO(10).
See Abstr. 022.075.

The computer and the Universe.
See Abstr. 022.121.

Covariant chronogeometry and extreme distances. III. Macro-micro relations.
See Abstr. 022.174.

Roles of a 45-plet of Higgs particles in SU(5) grand unified theory.
See Abstr. 022.175.

Mass generation in finite temperature QED.
See Abstr. 022.176.

Left-right symmetric theories and vacuum domain walls and strings.
See Abstr. 022.189.

Z_2 vortex strings in grand unified theories.
See Abstr. 022.190.

On supercooling and Ginzburg fluctuations in the Coleman-Weinberg mode of the SU(5) grand unification model.
See Abstr. 022.191.

Peccei-Quinn symmetries without domains.
See Abstr. 022.192.

Tales of the GUT age.
See Abstr. 022.193.

Review of the experimental status (past and future) of monopole searches.
See Abstr. 022.196.

Broken supersymmetries in high energy physics.
See Abstr. 022.197.

Neutrinos in the universe.
See Abstr. 061.009.

M/L constraints on the mass of the neutrino.
See Abstr. 061.013.

Is parity violation a cosmological evolution effect? .
See Abstr. 061.018.

Massive neutrinos.
See Abstr. 061.023.

Neutrino masses, intermediate mass scales, and cosmology.
See Abstr. 061.025.

Cosmic turbulence and the angular momenta of astronomical systems.
See Abstr. 061.028.

Massive neutrinos and ultra-violet astronomy.
See Abstr. 061.031.

The origin of the light elements.
See Abstr. 061.042.

Astrophysical consequences of neutron-antineutron oscillations.
See Abstr. 061.051.

Comment on "Does the standard hot-big-bang model explain the primordial abundances of helium and deuterium? ".
See Abstr. 061.052.

Two- and three-body contributions of cosmological monopole annihilation.
See Abstr. 061.054.

On the neutrino mass, the abundance of the right-handed neutrino and the closure of the Universe.
See Abstr. 061.056.

The annihilation process $\nu\bar{\nu} \rightarrow \gamma\gamma$ with massive neutrinos in cosmology.
See Abstr. 061.064.

Inhomogeneity of the early universe and formation of primordial black holes.
See Abstr. 066.001.

The spatial anisotropies of the microwave background.
See Abstr. 066.024.

Distortions in the Rayleigh-Jeans region of the cosmic background radiation spectrum.
See Abstr. 066.025.

Angular anisotropy of the cosmic microwave background radiation.
See Abstr. 066.026.

Pregalactic stars and the microwave background.
See Abstr. 066.027.

Weyl's geometry and physics.
See Abstr. 066.031.

Gauge-invariant definition of gravitational radiation in Robertson-Walker cosmologies.
See Abstr. 066.045.

Relativistic detonation waves and bubble growth in false vacuum decay.
See Abstr. 066.048.

Cosmic microwave background – present status and future prospects.
See Abstr. 066.051.

The cosmic microwave background radiation.
See Abstr. 066.060.

Evolutionary aspects of the cosmic black body radiation.
See Abstr. 066.068.

Fluid with heat flux in a conformally flat space-time.
See Abstr. 066.095.

Broken symmetry theories of gravity and implications for cosmology.
See Abstr. 066.100.

Supergravity, R invariance and spontaneous supersymmetry breaking.
See Abstr. 066.102.

A unified Einstein-Maxwell theory and some of its immediate consequences.
See Abstr. 066.119.

Spinor sources in cosmology.
See Abstr. 066.131.

The investigation of some self-similar solutions of Einstein's equations.
See Abstr. 066.132.

Mach's principle and the structure of dynamical theories.
See Abstr. 066.134.

Horizon-free Universe.
See Abstr. 066.139.

Solution of the Planck mass problem in Kaluza-Klein theories.
See Abstr. 066.141.

Gravothermal aspects in evolution of the stars and the Universe.
See Abstr. 066.144.

Partially self-similar spacetimes and fluid motions.
See Abstr. 066.147.

Ether formulations of relativity.
See Abstr. 066.162.

Induced gravity and cosmology.
See Abstr. 066.167.

Positive energy in anti-de Sitter backgrounds and gauged extended supergravity.
See Abstr. 066.169.

A general upper limit on the mass and entropy production of a cluster of supermassive objects.
See Abstr. 066.179.

Population III and the microwave background .
See Abstr. 066.181.

The cosmic microwave background radiation as a probe of the large-scale structure of the Universe.
See Abstr. 066.183.

Measuring the large-scale anisotropy in the microwave background radiation.
See Abstr. 066.189.

Anisotropy measurements and cosmological implications.
See Abstr. 066.190.

The cosmological constant and the color charge at energies of the order of the Planck mass.
See Abstr. 066.203.

Quantum stationary states in the Bianchi universe
See Abstr. 066.209.

Cosmological molecular hydrogen and the distortion of the relict radiation spectrum.
See Abstr. 066.227.

Resonant absorption of cosmic-ray neutrinos by the relic-neutrino background.
See Abstr. 066.232.

Search for small-scale anisotropy in the cosmic microwave background.
See Abstr. 066.235.

Gravitational lenses.
See Abstr. 066.239.

Semiclassical expanding discrete space-times.
See Abstr. 066.240.

Topological mass generation of scalar particles in the static flat Clifford-Klein-Robertson-Walker spacetimes.
See Abstr. 066.241.

Hawking evaporation and space-time structure.
See Abstr. 066.248.

Vacuum polarization by a massive scalar field in Schwarzschild spacetime.
See Abstr. 066.249.

Tensor spherical and pseudo-spherical harmonics i four-dimensional spaces.
See Abstr. 066.250.

Dispersion relaxations for zero-helicity gravitatior waves in a relativistic expanding medium.
See Abstr. 066.262.

A new vacuum state in de Sitter space.
See Abstr. 066.265.

Stability properties of gravity theories.
See Abstr. 066.267.

Vacuum solutions with double duality properties a quadratic Poincaré gauge field theory.
See Abstr. 066.296.

Quantum gravity removes classical singularities and shortens the life of black holes.
See Abstr. 066.297.

Spectra asymmetry and quantum field theory in curved spacetime.
See Abstr. 066.298.

Experimental searches for gravitational effects outside of general relativity.
See Abstr. 066.300.

Measurement of intermediate and large scale anisotropies in the microwave background.
See Abstr. 066.323.

Origin of distortions of the cosmic microwave background spectrum.
See Abstr. 066.325.

The effect of a non-minimally coupled scalar field upon classical thermodynamics.
See Abstr. 066.326.

A slice theorem for the space of solutions of Einstein's equations.
See Abstr. 066.328.

Abundance of lithium in unevolved halo stars and old disk stars: interpretation and consequences.
See Abstr. 114.146.

QSO Lyman limit absorption.
See Abstr. 141.023.

Angular diameter-redshift relations for extragalactic radio sources.
See Abstr. 141.049.

Observational limitations of the Doppler theory of quasars.
See Abstr. 141.052.

Cutoff of QSOs.
See Abstr. 141.061.

Cosmology from IUE observations of quasars?
See Abstr. 141.125.

Possible existence of quasi-stellar-object clusters.
See Abstr. 141.133.

On the cosmological evolution of the X-ray emission from quasars.
See Abstr. 141.140.

Luminosity evolution of QSOs.
See Abstr. 141.202.

The distribution of absorption line redshifts of quasars and its origin.
See Abstr. 141.206.

A discussion of the Hubble diagram of QSOs.
See Abstr. 141.207.

Spectral evolution of active galactic nuclei: a unified description of the X-ray and gamma-ray backgrounds.
See Abstr. 142.046.

Prospects for X-ray observations of cosmological significance.
See Abstr. 142.116.

The Einstein observatory medium sensitivity X-ray survey: evidence for cosmological evolution of quasars, BL Lacs, and clusters of galaxies.
See Abstr. 142.146.

Can we detect antimatter from other galaxies?
See Abstr. 143.013.

The cosmic ray origin problem.
See Abstr. 143.054.

The phase space structure of $r^{1/4}$ galaxies: Are these galaxies 'isothermal' after all?
See Abstr. 151.009.

Origin of stars and galaxies.
See Abstr. 151.060.

The neutrino content of elliptical galaxies.
See Abstr. 158.039.

Infrared search for primeval galaxies.
See Abstr. 158.127.

Surveying the Universe.
See Abstr. 158.171.

Mapping the local supercluster.
See Abstr. 160.020.

Observed dynamics of superclusters.
See Abstr. 160.038.

The halos of rich clusters of galaxies. I. An infall model for the Coma cluster.
See Abstr. 160.050.

A ~300 Mpc void of rich clusters of galaxies?
See Abstr. 160.052.

On the origin of the voids in the galaxy distribution.
See Abstr. 160.054.

Observations of galaxy clustering.
See Abstr. 160.057.

Galaxy clustering and the missing mass.
See Abstr. 160.060.

Dynamical models and our Virgocentric deviation from Hubble flow.
See Abstr. 160.067.

Superclusters of galaxies.
See Abstr. 160.077.

Intergalactic gas in clusters of galaxies, microwave background and cosmology.
See Abstr. 161.005.

Author Index

The authors are listed in alphabetical order according to the initial letter following the first names.

Subject Index

Starting with Volume 18 of *Astronomy and Astrophysics Abstracts,* some alterations concerning formation, arrangement, and versatility of the key words have been made. In order to provide an adequate description of a paper, specific key words are used as frequently as possible. References to a whole subject category are suppressed now. The user, therefore, has to refer to the contents at the beginning of each volume.

Whenever possible, the key words are formed in such a way that there are two different supplementary terms, e.g. the pair

minor planets
  orbits.

An effort is made to choose preferably terms which can be inverted in order to increase the usefulness of this index. In the example given there are the two entries

minor planets
  orbits

and

orbits
  minor planets.

Exceptions to the rule of inversion of terms are given in all cases where the second key word is either a very specific one (e.g. Urca processes) or a general one (e.g. history). The use of substantives is preferred. In order to obtain the possibility to extend a one-term key word in a two-term one, combinations as

Mars
  atmosphere
	or	sun
		  active regions

are changed into

Mars atmosphere	and	solar active regions,

respectively.

Starting with Volume 30 of *Astronomy and Astrophysics Abstracts,* this Index is given in a four-column arrangement. The overall size further has been optimized by suppression of a repetition of the first key word in a two-term combination. Thus, the primary term is printed only once for all the following secondary key words which belong to this header.

The user is – in any case – requested to look for synonymous entries, because further references to this topic might exist elsewhere in the Index under another current astronomical term.

ASTRONOMY AND ASTROPHYSICS ABSTRACTS

A Publication of the Astronomisches Rechen-Institut Heidelberg

Member of the Abstracting Board
of the International Council of Scientific Unions

Editors: S. Böhme, W. Fricke, H. Hefele, I. Heinrich, W. Hofmann, D. Krahn, V. R. Matas, L. D. Schmadel, G. Zech

Volume 1	Literature 1969, Part 1	X+ 435 pp. (1969)
Volume 2	Literature 1969, Part 2	X+ 516 pp. (1970)
Volume 3	Literature 1970, Part 1	X+ 490 pp. (1970)
Volume 4	Literature 1970, Part 2	X+ 562 pp. (1971)
Volume 5	Literature 1971, Part 1	X+ 505 pp. (1971)
Volume 6	Literature 1971, Part 2	X+ 560 pp. (1972)
Volume 7	Literature 1972, Part 1	X+ 526 pp. (1972)
Volume 8	Literature 1972, Part 2	X+ 594 pp. (1973)
Volume 9	Literature 1973, Part 1	X+ 610 pp. (1973)
Volume 10	Literature 1973, Part 2	X+ 661 pp. (1974)
Volume 11	Literature 1974, Part 1	X+ 579 pp. (1974)
Volume 12	Literature 1974, Part 2	X+ 699 pp. (1975)
Volume 13	Literature 1975, Part 1	X+ 632 pp. (1975)
Volume 14	Literature 1975, Part 2	X+ 747 pp. (1976)
Volume 15/16	Author and Subject Indexes to Volumes 1–10	
	Literature 1969–1973	VII+ 655 pp. (1976)
Volume 17	Literature 1976, Part 1	XII+ 645 pp. (1976)
Volume 18	Literature 1976, Part 2	X+ 859 pp. (1977)
Volume 19	Literature 1977, Part 1	X+ 732 pp. (1977)
Volume 20	Literature 1977, Part 2	X+ 786 pp. (1978)
Volume 21	Literature 1978, Part 1	X+ 834 pp. (1978)
Volume 22	Literature 1978, Part 2	X+ 849 pp. (1979)
Volume 23/24	Author and Subject Indexes to Volumes 11–14 and 17–22	
	Literature 1974–1978	VIII+1127 pp. (1979)
Volume 25	Literature 1979, Part 1	X+ 872 pp. (1979)
Volume 26	Literature 1979, Part 2	X+ 794 pp. (1980)
Volume 27	Literature 1980, Part 1	X+ 939 pp. (1980)
Volume 28	Literature 1980, Part 2	X+ 841 pp. (1981)
Volume 29	Literature 1981, Part 1	X+ 853 pp. (1981)
Volume 30	Literature 1981, Part 2	X+ 792 pp. (1982)
Volume 31	Literature 1982, Part 1	X+ 776 pp. (1982)
Volume 32	Literature 1982, Part 2	X+ 848 pp. (1983)

Published for Astronomisches Rechen-Institut by
Springer-Verlag Berlin Heidelberg New York Tokyo

FSC
www.fsc.org
MIX
Papier | Fördert gute Waldnutzung
FSC® C083411

Kolibri 360 GmbH
Ferdinand-Jühlke-Straße 7
99095 Erfurt, Deutschland
produktsicherheit@kolibri360.de